Lecture Notes in Artificial Intelligence 3613

Edited by J. G. Carbonell and J. Siekmann

Subseries of Lecture Notes in Computer Science

Lipo Wang   Yaochu Jin (Eds.)

Fuzzy Systems and Knowledge Discovery

Second International Conference, FSKD 2005
Changsha, China, August 27-29, 2005
Proceedings, Part I

Springer

Series Editors

Jaime G. Carbonell, Carnegie Mellon University, Pittsburgh, PA, USA
Jörg Siekmann, University of Saarland, Saarbrücken, Germany

Volume Editors

Lipo Wang
Nanyang Technological University
School of Electrical and Electronic Engineering
Block S1, 50 Nanyang Avenue, Singapore 639798
E-mail: elpwang@ntu.edu.sg

Yaochu Jin
Honda Research Institute Europe
Carl-Legien-Str. 30, 63073 Offenbach/Main, Germany
E-mail: yaochu.jin@honda-ri.de

Library of Congress Control Number: 2005930642

CR Subject Classification (1998): I.2, F.4.1, F.1, F.2, G.2, I.2.3, I.4, I.5

ISSN   0302-9743
ISBN-10   3-540-28312-9 Springer Berlin Heidelberg New York
ISBN-13   978-3-540-28312-6 Springer Berlin Heidelberg New York

This work is subject to copyright. All rights are reserved, whether the whole or part of the material is concerned, specifically the rights of translation, reprinting, re-use of illustrations, recitation, broadcasting, reproduction on microfilms or in any other way, and storage in data banks. Duplication of this publication or parts thereof is permitted only under the provisions of the German Copyright Law of September 9, 1965, in its current version, and permission for use must always be obtained from Springer. Violations are liable to prosecution under the German Copyright Law.

Springer is a part of Springer Science+Business Media

springeronline.com

© Springer-Verlag Berlin Heidelberg 2005
Printed in Germany

Typesetting: Camera-ready by author, data conversion by Scientific Publishing Services, Chennai, India
Printed on acid-free paper   SPIN: 11539506   06/3142   5 4 3 2 1 0

Preface

This book and its sister volume, LNAI 3613 and 3614, constitute the proceedings of the Second International Conference on Fuzzy Systems and Knowledge Discovery (FSKD 2005), jointly held with the First International Conference on Natural Computation (ICNC 2005, LNCS 3610, 3611, and 3612) from August 27–29, 2005 in Changsha, Hunan, China. FSKD 2005 successfully attracted 1249 submissions from 32 countries/regions (the joint ICNC-FSKD 2005 received 3136 submissions). After rigorous reviews, 333 high-quality papers, i.e., 206 long papers and 127 short papers, were included in the FSKD 2005 proceedings, representing an acceptance rate of 26.7%.

The ICNC-FSKD 2005 conference featured the most up-to-date research results in computational algorithms inspired from nature, including biological, ecological, and physical systems. It is an exciting and emerging interdisciplinary area in which a wide range of techniques and methods are being studied for dealing with large, complex, and dynamic problems. The joint conferences also promoted cross-fertilization over these exciting and yet closely-related areas, which had a significant impact on the advancement of these important technologies. Specific areas included computation with words, fuzzy computation, granular computation, neural computation, quantum computation, evolutionary computation, DNA computation, chemical computation, information processing in cells and tissues, molecular computation, artificial life, swarm intelligence, ants colony, artificial immune systems, etc., with innovative applications to knowledge discovery, finance, operations research, and more. In addition to the large number of submitted papers, we were blessed with the presence of four renowned keynote speakers and several distinguished panelists.

On behalf of the Organizing Committee, we thank Xiangtan University for sponsorship, and the IEEE Circuits and Systems Society, the IEEE Computational Intelligence Society, and the IEEE Control Systems Society for technical co-sponsorship. We are grateful for the technical cooperation from the International Neural Network Society, the European Neural Network Society, the Chinese Association for Artificial Intelligence, the Japanese Neural Network Society, the International Fuzzy Systems Association, the Asia-Pacific Neural Network Assembly, the Fuzzy Mathematics and Systems Association of China, and the Hunan Computer Federation. We thank the members of the Organizing Committee, the Advisory Board, and the Program Committee for their hard work over the past 18 months. We wish to express our heart-felt appreciation to the keynote and panel speakers, special session organizers, session chairs, reviewers, and student helpers. Our special thanks go to the publisher, Springer, for publishing the FSKD 2005 proceedings as two volumes of the Lecture Notes in Artificial Intelligence series (and the ICNC 2005 proceedings as three volumes of the Lecture Notes in Computer Science series). Finally, we thank all the authors and participants for their great contributions that made this conference possible and all the hard work worthwhile.

August 2005

Lipo Wang
Yaochu Jin

Organization

FSKD 2005 was organized by Xiangtan University and technically co-sponsored by the IEEE Circuits and Systems Society, the IEEE Computational Intelligence Society, and the IEEE Control Systems Society, in cooperation with the International Neural Network Society, the European Neural Network Society, the Chinese Association for Artificial Intelligence, the Japanese Neural Network Society, the International Fuzzy Systems Association, the Asia-Pacific Neural Network Assembly, the Fuzzy Mathematics and Systems Association of China, and the Hunan Computer Federation.

Organizing Committee

Honorary Conference Chairs:	Shun-ichi Amari, *Japan*
	Lotfi A. Zadeh, *USA*
General Chair:	He-An Luo, *China*
General Co-chairs:	Lipo Wang , *Singapore*
	Yunqing Huang, *China*
Program Chair:	Yaochu Jin, *Germany*
Local Arrangement Chairs:	Renren Liu, *China*
	Xieping Gao, *China*
Proceedings Chair:	Fen Xiao, *China*
Publicity Chair:	Hepu Deng, *Australia*
Sponsorship/Exhibits Chairs:	Shaoping Ling, *China*
	Geok See Ng, *Singapore*
Webmasters:	Linai Kuang, *China*
	Yanyu Liu, *China*

Advisory Board

Toshio Fukuda, *Japan*
Kunihiko Fukushima, *Japan*
Tom Gedeon, *Australia*
Aike Guo, *China*
Zhenya He, *China*
Janusz Kacprzyk, *Poland*
Nikola Kasabov, *New Zealand*
John A. Keane, *UK*
Soo-Young Lee, *Korea*
Erkki Oja, *Finland*
Nikhil R. Pal, *India*

Witold Pedrycz, *Canada*
Jose C. Principe, *USA*
Harold Szu, *USA*
Shiro Usui, *Japan*
Xindong Wu, *USA*
Lei Xu, *Hong Kong*
Xin Yao, *UK*
Syozo Yasui, *Japan*
Bo Zhang, *China*
Yixin Zhong, *China*
Jacek M. Zurada, *USA*

VIII   Organization

Program Committee Members

Janos Abonyi, *Hungary*
Jorge Casillas, *Spain*
Pen-Chann Chang, *Taiwan*
Chaochang Chiu, *Taiwan*
Feng Chu, *Singapore*
Oscar Cordon, *Spain*
Honghua Dai, *Australia*
Fernando Gomide, *Brazil*
Saman Halgamuge, *Australia*
Kaoru Hirota, *Japan*
Frank Hoffmann, *Germany*
Jinglu Hu, *Japan*
Weili Hu, *China*
Chongfu Huang, *China*
Eyke Hüllermeier, *Germany*
Hisao Ishibuchi, *Japan*
Frank Klawoon, *Germany*
Naoyuki Kubota, *Japan*
Sam Kwong, *Hong Kong*
Zongmin Ma, *China*

Michael Margaliot, *Israel*
Ralf Mikut, *Germany*
Pabitra Mitra, *India*
Tadahiko Murata, *Japan*
Detlef Nauck, *UK*
Hajime Nobuhara, *Japan*
Andreas Nürnberger, *Germany*
Da Ruan, *Belgium*
Thomas Runkler, *Germany*
Rudy Setiono, *Singapore*
Takao Terano, *Japan*
Kai Ming Ting, *Australia*
Yiyu Yao, *Canada*
Gary Yen, *USA*
Xinghuo Yu, *Australia*
Jun Zhang, *China*
Shichao Zhang, *Australia*
Yanqing Zhang, *USA*
Zhi-Hua Zhou, *China*

Special Sessions Organizers

David Siu-Yeung Cho, *Singapore*
Vlad Dimitrov, *Australia*
Jinwu Gao, *China*
Zheng Guo, *China*
Bob Hodge, *Australia*
Jiman Hong, *Korea*
Jae-Woo Lee, *Korea*
Xia Li, *China*

Zongmin Ma, *China,*
Geok-See Ng, *Singapore*
Shaoqi Rao, *China*
Slobodan Ribari, *Croatia*
Sung Y. Shin, *USA*
Yasufumi Takama, *Japan*
Robert Woog, *Australia*

Reviewers

Nitin V. Afzulpurkar
Davut Akdas
Kürat Ayan
Yasar Becerikli
Dexue Bi
Rong-Fang Bie

Liu Bin
Tao Bo
Hongbin Cai
Yunze Cai
Jian Cao
Chunguang Chang

An-Long Chen
Dewang Chen
Gang Chen
Guangzhu Chen
Jian Chen
Shengyong Chen

Shi-Jay Chen	Zhonghui Hu	Ling-Zhi Liao
Xuerong Chen	Changchun Hua	Lei Lin
Yijiang Chen	Jin Huang	Caixia Liu
Zhimei Chen	Qian Huang	Fei Liu
Zushun Chen	Yanxin Huang	Guangli Liu
Hongqi Chen	Yuansheng Huang	Haowen Liu
Qimei Chen	Kohei Inoue	Honghai Liu
Wei Cheng	Mahdi Jalili-Kharaajoo	Jian-Guo Liu
Xiang Cheng	Caiyan Jia	Lanjuan Liu
Tae-Ho Cho	Ling-Ling Jiang	Peng Liu
Xun-Xue Cui	Michael Jiang	Qihe Liu
Ho Daniel	Xiaoyue Jiang	Sheng Liu
Hepu Deng	Yanping Jiang	Xiaohua Liu
Tingquan Deng	Yunliang Jiang	Xiaojian Liu
Yong Deng	Cheng Jin	Yang Liu
Zhi-Hong Deng	Hanjun Jin	Qiang Luo
Mingli Ding	Hong Jin	Yanbin Luo
Wei-Long Ding	Ningde Jin	Zhi-Jun Lv
Fangyan Dong	Xue-Bo Jin	Jian Ma
Jingxin Dong	Min-Soo Kim	Jixin Ma
Lihua Dong	Sungshin Kim	Longhua Ma
Yihong Dong	Taehan Kim	Ming Ma
Haifeng Du	Ibrahim Beklan	Yingcang Ma
Weifeng Du	Kucukdemiral	Dong Miao
Liu Fang	Rakesh Kumar Arya	Zhinong Miao
Zhilin Feng	Ho Jae Lee	Fan Min
Li Gang	Sang-Hyuk Lee	Zhang Min
Chuanhou Gao	Sang-Won Lee	Zhao Min
Yu Gao	Wol Young Lee	Daniel Neagu
Zhi Geng	Xiuren Lei	Yiu-Kai Ng
O. Nezih Gerek	Bicheng Li	Wu-Ming Pan
Rongjie Gu	Chunyan Li	Jong Sou Park
Chonghui Guo	Dequan Li	Yonghong Peng
Gongde Guo	Dingfang Li	Punpiti Piamsa-Nga
Huawei Guo	Gang Li	Heng-Nian Qi
Mengshu Guo	Hongyu Li	Gao Qiang
Zhongming Han	Qing Li	Wu Qing
Bo He	Ruqiang Li	Celia Ghedini Ralha
Pilian He	Tian-Rui Li	Wang Rong
Liu Hong	Weigang Li	Hongyuan Shen
Kongfa Hu	Yu Li	Zhenghao Shi
Qiao Hu	Zhichao Li	Jeong-Hoon Shin
Shiqiang Hu	Zhonghua Li	Sung Chul Shin
Zhikun Hu	Hongxing Li	Chonghui Song
Zhonghui Hu	Xiaobei Liang	Chunyue Song

X   Organization

Guangda Su	Jianping Wu	Chun-Hai Yu
Baolin Sun	Shunxiang Wu	Haibin Yuan
Changyin Sun	Xiaojun Wu	Jixue Yuan
Ling Sun	Yuying Wu	Weiqi Yuan
Zhengxing Sun	Changcheng Xiang	Chuanhua Zeng
Chang-Jie Tang	Jun Xiao	Wenyi Zeng
Shanhu Tang	Xiaoming Xiao	Yurong Zeng
N K Tiwari	Wei Xie	Guojun Zhang
Jiang Ping Wan	Gao Xin	Jian Ying Zhang
Chong-Jun Wang	Zongyi Xing	Junping Zhang
Danli Wang	Hua Xu	Ling Zhang
Fang Wang	Lijun Xu	Zhi-Zheng Zhang
Fei Wang	Pengfei Xu	Yongjin Zhang
Houfeng Wang	Weijun Xu	Yongkui Zhang
Hui Wang	Xiao Xu	Jun Zhao
Laisheng Wang	Xinli Xu	Quanming Zhao
Lin Wang	Yaoqun Xu	Xin Zhao
Ling Wang	De Xu	Yong Zhao
Shitong Wang	Maode Yan	Zhicheng Zhao
Shu-Bin Wang	Shaoze Yan	Dongjian Zheng
Xun Wang	Hai Dong Yang	Wenming Zheng
Yong Wang	Jihui Yang	Zhonglong Zheng
Zhe Wang	Wei-Min Yang	Weimin Zhong
Zhenlei Wang	Yong Yang	Hang Zhou
Zhongjie Wang	Zuyuan Yang	Hui-Cheng Zhou
Runsheng Wang	Li Yao	Qiang Zhou
Li Wei	Shengbao Yao	Yuanfeng Zhou
Weidong Wen	Bin Ye	Yue Zhou
Xiangjun Wen	Guo Yi	Daniel Zhu
Taegkeun Whangbo	Jianwei Yin	Hongwei Zhu
Huaiyu Wu	Xiang-Gang Yin	Xinglong Zhu
Jiangning Wu	Yilong Yin	
Jiangqin Wu	Deng Yong	

* The term after a name may represent either a country or a region.

Table of Contents – Part I

Fuzzy Theory and Models

On Fuzzy Inclusion in the Interval-Valued Sense
Jin Han Park, Jong Seo Park, Young Chel Kwun 1

Fuzzy Evaluation Based Multi-objective Reactive Power Optimization in Distribution Networks
Jiachuan Shi, Yutian Liu 11

Note on Interval-Valued Fuzzy Set
Wenyi Zeng, Yu Shi 20

Knowledge Structuring and Evaluation Based on Grey Theory
Chen Huang, Yushun Fan 26

A Propositional Calculus Formal Deductive System $\mathcal{L}^U$ of Universal Logic and Its Completeness
Minxia Luo, Huacan He 31

Entropy and Subsethood for General Interval-Valued Intuitionistic Fuzzy Sets
Xiao-dong Liu, Su-hua Zheng, Feng-lan Xiong 42

The Comparative Study of Logical Operator Set and Its Corresponding General Fuzzy Rough Approximation Operator Set
Suhua Zheng, Xiaodong Liu, Fenglan Xiong 53

Associative Classification Based on Correlation Analysis
Jian Chen, Jian Yin, Jin Huang, Ming Feng 59

Design of Interpretable and Accurate Fuzzy Models from Data
Zong-yi Xing, Yong Zhang, Li-min Jia, Wei-li Hu 69

Generating Extended Fuzzy Basis Function Networks Using Hybrid Algorithm
Bin Ye, Chengzhi Zhu, Chuangxin Guo, Yijia Cao 79

Analysis of Temporal Uncertainty of Trains Converging Based on Fuzzy Time Petri Nets
Yangdong Ye, Juan Wang, Limin Jia 89

Interval Regression Analysis Using Support Vector Machine and
Quantile Regression
   *Changha Hwang, Dug Hun Hong, Eunyoung Na, Hyejung Park,
Jooyong Shim* 100

An Approach Based on Similarity Measure to Multiple Attribute
Decision Making with Trapezoid Fuzzy Linguistic Variables
   Zeshui Xu 110

Research on Index System and Fuzzy Comprehensive Evaluation
Method for Passenger Satisfaction
   Yuanfeng Zhou, Jianping Wu, Yuanhua Jia 118

Research on Predicting Hydatidiform Mole Canceration Tendency by a
Fuzzy Integral Model
   Yecai Guo, Yi Guo, Wei Rao, Wei Ma 122

Consensus Measures and Adjusting Inconsistency of Linguistic
Preference Relations in Group Decision Making
   Zhi-Ping Fan, Xia Chen 130

Fuzzy Variation Coefficients Programming of Fuzzy Systems and Its
Application
   Xiaobei Liang, Daoli Zhu, Bingyong Tang 140

Weighted Possibilistic Variance of Fuzzy Number and Its Application
in Portfolio Theory
   Xun Wang, Weijun Xu, Weiguo Zhang, Maolin Hu 148

Another Discussion About Optimal Solution to Fuzzy Constraints
Linear Programming
   Yun-feng Tan, Bing-yuan Cao 156

Fuzzy Ultra Filters and Fuzzy G-Filters of MTL-Algebras
   Xiao-hong Zhang, Yong-quan Wang, Yong-lin Liu 160

A Study on Relationship Between Fuzzy Rough Approximation
Operators and Fuzzy Topological Spaces
   Wei-Zhi Wu 167

A Case Retrieval Model Based on Factor-Structure Connection and
λ–Similarity in Fuzzy Case-Based Reasoning
   Dan Meng, Zaiqiang Zhang, Yang Xu 175

A TSK Fuzzy Inference Algorithm for Online Identification
*Kyoungjung Kim, Eun Ju Whang, Chang-Woo Park, Euntai Kim,
Mignon Park* 179

Histogram-Based Generation Method of Membership Function for
Extracting Features of Brain Tissues on MRI Images
*Weibei Dou, Yuan Ren, Yanping Chen, Su Ruan, Daniel Bloyet,
Jean-Marc Constans* 189

Uncertainty Management in Data Mining

On Identity-Discrepancy-Contrary Connection Degree in SPA and Its
Applications
Yunliang Jiang, Yueting Zhuang, Yong Liu, Keqin Zhao 195

A Mathematic Model for Automatic Summarization
Zhiqi Wang, Yongcheng Wang, Kai Gao 199

Reliable Data Selection with Fuzzy Entropy
Sang-Hyuk Lee, Youn-Tae Kim, Seong-Pyo Cheon, Sungshin Kim ... 203

Uncertainty Management and Probabilistic Methods in Data Mining

Optimization of Concept Discovery in Approximate Information
System Based on FCA
Hanjun Jin, Changhua Wei, Xiaorong Wang, Jia Fu 213

Geometrical Probability Covering Algorithm
Junping Zhang, Stan Z. Li, Jue Wang 223

Approximate Reasoning

Extended Fuzzy ALCN and Its Tableau Algorithm
Jianjiang Lu, Baowen Xu, Yanhui Li, Dazhou Kang, Peng Wang ... 232

Type II Topological Logic $\mathbb{C}^2$ and Approximate Reasoning
Yalin Zheng, Changshui Zhang, Yinglong Xia 243

Type-I Topological Logic $\mathbb{C}^1_{\mathscr{T}}$ and Approximate Reasoning
Yalin Zheng, Changshui Zhang, Xin Yao 253

Vagueness and Extensionality
Shunsuke Yatabe, Hiroyuki Inaoka 263

Using Fuzzy Analogical Reasoning to Refine the Query Answers for Relational Databases with Imprecise Information
Z.M. Ma, Li Yan, Gui Li 267

A Linguistic Truth-Valued Uncertainty Reasoning Model Based on Lattice-Valued Logic
Shuwei Chen, Yang Xu, Jun Ma 276

Axiomatic Foundation

Fuzzy Programming Model for Lot Sizing Production Planning Problem
Weizhen Yan, Jianhua Zhao, Zhe Cao 285

Fuzzy Dominance Based on Credibility Distributions
Jin Peng, Henry M.K. Mok, Wai-Man Tse 295

Fuzzy Chance-Constrained Programming for Capital Budgeting Problem with Fuzzy Decisions
Jinwu Gao, Jianhua Zhao, Xiaoyu Ji 304

Genetic Algorithms for Dissimilar Shortest Paths Based on Optimal Fuzzy Dissimilar Measure and Applications
Yinzhen Li, Ruichun He, Linzhong Liu, Yaohuang Guo 312

Convergence Criteria and Convergence Relations for Sequences of Fuzzy Random Variables
Yan-Kui Liu, Jinwu Gao 321

Hybrid Genetic-SPSA Algorithm Based on Random Fuzzy Simulation for Chance-Constrained Programming
Yufu Ning, Wansheng Tang, Hui Wang 332

Random Fuzzy Age-Dependent Replacement Policy
Song Xu, Jiashun Zhang, Ruiqing Zhao 336

A Theorem for Fuzzy Random Alternating Renewal Processes
Ruiqing Zhao, Wansheng Tang, Guofei Li 340

Three Equilibrium Strategies for Two-Person Zero-Sum Game with Fuzzy Payoffs
Lin Xu, Ruiqing Zhao, Tingting Shu 350

Fuzzy Classifiers

An Improved Rectangular Decomposition Algorithm for Imprecise and Uncertain Knowledge Discovery
Jiyoung Song, Younghee Im, Daihee Park 355

XPEV: A Storage Model for Well-Formed XML Documents
Jie Qin, Shu-Mei Zhao, Shu-Qiang Yang, Wen-Hua Dou 360

Fuzzy-Rough Set Based Nearest Neighbor Clustering Classification Algorithm
Xiangyang Wang, Jie Yang, Xiaolong Teng, Ningsong Peng 370

An Efficient Text Categorization Algorithm Based on Category Memberships
Zhi-Hong Deng, Shi-Wei Tang, Ming Zhang 374

The Integrated Location Algorithm Based on Fuzzy Identification and Data Fusion with Signal Decomposition
Zhao Ping, Haoshan Shi 383

A Web Document Classification Approach Based on Fuzzy Association Concept
Jingsheng Lei, Yaohong Kang, Chunyan Lu, Zhang Yan 388

Optimized Fuzzy Classification Using Genetic Algorithm
Myung Won Kim, Joung Woo Ryu 392

Dynamic Test-Sensitive Decision Trees with Multiple Cost Scales
Zhenxing Qin, Chengqi Zhang, Xuehui Xie, Shichao Zhang 402

Design of T–S Fuzzy Classifier via Linear Matrix Inequality Approach
Moon Hwan Kim, Jin Bae Park, Young Hoon Joo, Ho Jae Lee 406

Design of Fuzzy Rule-Based Classifier: Pruning and Learning
Do Wan Kim, Jin Bae Park, Young Hoon Joo 416

Fuzzy Sets Theory Based Region Merging for Robust Image Segmentation
Hongwei Zhu, Otman Basir 426

A New Interactive Segmentation Scheme Based on Fuzzy Affinity and Live-Wire
Huiguang He, Jie Tian, Yao Lin, Ke Lu 436

Fuzzy Clustering

The Fuzzy Mega-cluster: Robustifying FCM by Scaling Down Memberships
Amit Banerjee, Rajesh N. Davé 444

Robust Kernel Fuzzy Clustering
Weiwei Du, Kohei Inoue, Kiichi Urahama 454

Spatial Homogeneity-Based Fuzzy c-Means Algorithm for Image Segmentation
Bo-Yeong Kang, Dae-Won Kim, Qing Li 462

A Novel Fuzzy-Connectedness-Based Incremental Clustering Algorithm for Large Databases
Yihong Dong, Xiaoying Tai, Jieyu Zhao 470

Classification of MPEG VBR Video Data Using Gradient-Based FCM with Divergence Measure
Dong-Chul Park 475

Fuzzy-C-Mean Determines the Principle Component Pairs to Estimate the Degree of Emotion from Facial Expressions
M. Ashraful Amin, Nitin V. Afzulpurkar, Matthew N. Dailey, Vatcharaporn Esichaikul, Dentcho N. Batanov 484

An Improved Clustering Algorithm for Information Granulation
Qinghua Hu, Daren Yu 494

A Novel Segmentation Method for MR Brain Images Based on Fuzzy Connectedness and FCM
Xian Fan, Jie Yang, Lishui Cheng 505

Improved-FCM-Based Readout Segmentation and PRML Detection for Photochromic Optical Disks
Jiqi Jian, Cheng Ma, Huibo Jia 514

Fuzzy Reward Modeling for Run-Time Peer Selection in Peer-to-Peer Networks
Huaxiang Zhang, Xiyu Liu, Peide Liu 523

KFCSA: A Novel Clustering Algorithm for High-Dimension Data
Kan Li, Yushu Liu 531

Fuzzy Database Mining and Information Retrieval

An Improved VSM Based Information Retrieval System and Fuzzy
Query Expansion
*Jiangning Wu, Hiroki Tanioka, Shizhu Wang, Donghua Pan,
Kenichi Yamamoto, Zhongtuo Wang* 537

The Extraction of Image's Salient Points for Image Retrieval
Wenyin Zhang, Jianguo Tang, Chao Li 547

A Sentence-Based Copy Detection Approach for Web Documents
Rajiv Yerra, Yiu-Kai Ng 557

The Research on Query Expansion for Chinese Question Answering
System
Zhengtao Yu, Xiaozhong Fan, Lirong Song, Jianyi Guo 571

Multinomial Approach and Multiple-Bernoulli Approach for
Information Retrieval Based on Language Modeling
Hua Huo, Junqiang Liu, Boqin Feng 580

Adaptive Query Refinement Based on Global and Local Analysis
Chaoyuan Cui, Hanxiong Chen, Kazutaka Furuse, Nobuo Ohbo 584

Information Push-Delivery for User-Centered and Personalized Service
Zhiyun Xin, Jizhong Zhao, Chihong Chi, Jiaguang Sun 594

Mining Association Rules Based on Seed Items and Weights
Chen Xiang, Zhang Yi, Wu Yue 603

An Algorithm of Online Goods Information Extraction with Two-Stage
Working Pattern
Wang Xun, Ling Yun, Yu-lian Fei 609

A Novel Method of Image Retrieval Based on Combination of Semantic
and Visual Features
Ming Li, Tong Wang, Bao-wei Zhang, Bi-Cheng Ye 619

Using Fuzzy Pattern Recognition to Detect Unknown Malicious
Executables Code
Boyun Zhang, Jianping Yin, Jingbo Hao 629

Method of Risk Discernment in Technological Innovation Based on
Path Graph and Variable Weight Fuzzy Synthetic Evaluation
Yuan-sheng Huang, Jian-xun Qi, Jun-hua Zhou 635

Application of Fuzzy Similarity to Prediction of Epileptic Seizures
Using EEG Signals
  Xiaoli Li, Xin Yao 645

A Fuzzy Multicriteria Analysis Approach to the Optimal Use of
Reserved Land for Agriculture
  Hepu Deng, Guifang Yang 653

Fuzzy Comprehensive Evaluation for the Optimal Management of
Responding to Oil Spill
  Xin Liu, Kai W. Wirtz, Susanne Adam 662

Information Fusion

Fuzzy Fusion for Face Recognition
  Xuerong Chen, Zhongliang Jing, Gang Xiao 672

A Group Decision Making Method for Integrating Outcome Preferences
in Hypergame Situations
  Yexin Song, Qian Wang, Zhijun Li 676

A Method Based on IA Operator for Multiple Attribute Group Decision
Making with Uncertain Linguistic Information
  Zeshui Xu 684

A New Prioritized Information Fusion Method for Handling Fuzzy
Information Retrieval Problems
  Won-Sin Hong, Shi-Jay Chen, Li-Hui Wang, Shyi-Ming Chen 694

Multi-context Fusion Based Robust Face Detection in Dynamic
Environments
  Mi Young Nam, Phill Kyu Rhee 698

Unscented Fuzzy Tracking Algorithm for Maneuvering Target
  Shi-qiang Hu, Li-wei Guo, Zhong-liang Jing 708

A Pixel-Level Multisensor Image Fusion Algorithm Based on Fuzzy Logic
  Long Zhao, Baochang Xu, Weilong Tang, Zhe Chen 717

Neuro-Fuzzy Systems

Approximation Bound for Fuzzy-Neural Networks with Bell
Membership Function
  Weimin Ma, Guoqing Chen 721

A Neuro-fuzzy Method of Forecasting the Network Traffic of Accessing
Web Server
  Ai-Min Yang, Xing-Min Sun, Chang-Yun Li, Ping Liu 728

A Fuzzy Neural Network System Based on Generalized Class Cover
Problem
  Yanxin Huang, Yan Wang, Wengang Zhou, Chunguang Zhou 735

A Self-constructing Compensatory Fuzzy Wavelet Network and Its
Applications
  Haibin Yu, Qianjin Guo, Aidong Xu 743

A New Balancing Method for Flexible Rotors Based on Neuro-fuzzy
System and Information Fusion
  Shi Liu 757

Recognition of Identifiers from Shipping Container Images Using Fuzzy
Binarization and Enhanced Fuzzy Neural Network
  Kwang-Baek Kim 761

Directed Knowledge Discovery Methodology for the Prediction of
Ozone Concentration
  Seong-Pyo Cheon, Sungshin Kim 772

Application of Fuzzy Systems in the Car-Following Behaviour Analysis
  Pengjun Zheng, Mike McDonald 782

Fuzzy Control

GA-Based Composite Sliding Mode Fuzzy Control for Double-
Pendulum-Type Overhead Crane
  Diantong Liu, Weiping Guo, Jianqiang Yi 792

A Balanced Model Reduction for T-S Fuzzy Systems with Integral
Quadratic Constraints
  Seog-Hwan Yoo, Byung-Jae Choi 802

An Integrated Navigation System of NGIMU/ GPS Using a Fuzzy
Logic Adaptive Kalman Filter
  Mingli Ding, Qi Wang 812

Method of Fuzzy-PID Control on Vehicle Longitudinal Dynamics System
  Yinong Li, Zheng Ling, Yang Liu, Yanjuan Qiao 822

Design of Fuzzy Controller and Parameter Optimizer for Non-linear System Based on Operator's Knowledge
Hyeon Bae, Sungshin Kim, Yejin Kim 833

A New Pre-processing Method for Multi-channel Echo Cancellation Based on Fuzzy Control
Xiaolu Li, Wang Jie, Shengli Xie 837

Robust Adaptive Fuzzy Control for Uncertain Nonlinear Systems
Chen Gang, Shuqing Wang, Jianming Zhang 841

Intelligent Fuzzy Systems for Aircraft Landing Control
Jih-Gau Juang, Bo-Shian Lin, Kuo-Chih Chin 851

Scheduling Design of Controllers with Fuzzy Deadline
Hong Jin, Hongan Wang, Hui Wang, Danli Wang 861

A Preference Method with Fuzzy Logic in Service Scheduling of Grid Computing
Yanxiang He, Haowen Liu, Weidong Wen, Hui Jin 865

H_∞ Robust Fuzzy Control of Ultra-High Rise / High Speed Elevators with Uncertainty
Hu Qing, Qingding Guo, Dongmei Yu, Xiying Ding 872

A Dual-Mode Fuzzy Model Predictive Control Scheme for Unknown Continuous Nonlinear System
Chonghui Song, Shucheng Yang, Hui yang, Huaguang Zhang, Tianyou Chai 876

Fuzzy Modeling Strategy for Control of Nonlinear Dynamical Systems
Bin Ye, Chengzhi Zhu, Chuangxin Guo, Yijia Cao 882

Intelligent Digital Control for Nonlinear Systems with Multirate Sampling
Do Wan Kim, Jin Bae Park, Young Hoon Joo 886

Feedback Control of Humanoid Robot Locomotion
Xusheng Lei, Jianbo Su 890

Application of Computational Intelligence (Fuzzy Logic, Neural Networks and Evolutionary Programming) to Active Networking Technology
Mehdi Galily, Farzad Habibipour Roudsari, Mohammadreza Sadri ... 900

Fuel-Efficient Maneuvers for Constellation Initialization Using Fuzzy
Logic Control
  Mengfei Yang, Honghua Zhang, Rucai Che, Zengqi Sun 910

Design of Interceptor Guidance Law Using Fuzzy Logic
  Ya-dong Lu, Ming Yang, Zi-cai Wang 922

Relaxed LMIs Observer-Based Controller Design via Improved T-S
Fuzzy Model Structure
  Wei Xie, Huaiyu Wu, Xin Zhao 930

Fuzzy Virtual Coupling Design for High Performance Haptic Display
  D. Bi, J. Zhang, G.L. Wang 942

Linguistic Model for the Controlled Object
  Zhinong Miao, Xiangyu Zhao, Yang Xu 950

Fuzzy Sliding Mode Control for Uncertain Nonlinear Systems
  Shao-Cheng Qu, Yong-Ji Wang 960

Fuzzy Control of Nonlinear Pipeline Systems with Bounds on Output
Peak
  Fei Liu, Jun Chen 969

Grading Fuzzy Sliding Mode Control in AC Servo System
  Hu Qing, Qingding Guo, Dongmei Yu, Xiying Ding 977

A Robust Single Input Adaptive Sliding Mode Fuzzy Logic Controller
for Automotive Active Suspension System
  *Ibrahim B. Kucukdemiral, Seref N. Engin, Vasfi E. Omurlu,
  Galip Cansever* 981

Construction of Fuzzy Models for Dynamic Systems Using
Multi-population Cooperative Particle Swarm Optimizer
  Ben Niu, Yunlong Zhu, Xiaoxian He 987

Human Clustering for a Partner Robot Based on Computational
Intelligence
  Indra Adji Sulistijono, Naoyuki Kubota 1001

Fuzzy Switching Controller for Multiple Model
  Baozhu Jia, Guang Ren, Zhihong Xiu 1011

Generation of Fuzzy Rules and Learning Algorithms for Cooperative
Behavior of Autonomouse Mobile Robots(AMRs)
  Jang-Hyun Kim, Jin-Bae Park, Hyun-Seok Yang, Young-Pil Park ... 1015

UML-Based Design and Fuzzy Control of Automated Vehicles
  Abdelkader El Kamel, Jean-Pierre Bourey 1025

Fuzzy Hardware

Design of an Analog Adaptive Fuzzy Logic Controller
  Zhihao Xu, Dongming Jin, Zhijian Li 1034

VLSI Implementation of a Self-tuning Fuzzy Controller Based on Variable Universe of Discourse
  Weiwei Shan, Dongming Jin, Weiwei Jin, Zhihao Xu 1044

Knowledge Visualization and Exploration

Method to Balance the Communication Among Multi-agents in Real Time Traffic Synchronization
  Li Weigang, Marcos Vinícius Pinheiro Dib,
  Alba Cristina Magalhães de Melo 1053

A Celerity Association Rules Method Based on Data Sort Search
  Zhiwei Huang, Qin Liao 1063

Using Web Services to Create the Collaborative Model for Enterprise Digital Content Portal
  Ruey-Ming Chao, Chin-Wen Yang 1067

Emotion-Based Textile Indexing Using Colors and Texture
  Eun Yi Kim, Soo-jeong Kim, Hyun-jin Koo, Karpjoo Jeong,
  Jee-in Kim 1077

Optimal Space Launcher Design Using a Refined Response Surface Method
  Jae-Woo Lee, Kwon-Su Jeon, Yung-Hwan Byun, Sang-Jin Kim 1081

MEDIC: A MDO-Enabling Distributed Computing Framework
  Shenyi Jin, Kwangsik Kim, Karpjoo Jeong, Jaewoo Lee,
  Jonghwa Kim, Hoyon Hwang, Hae-Gook Suh 1092

Time and Space Efficient Search for Small Alphabets with Suffix Arrays
  Jeong Seop Sim 1102

Optimal Supersonic Air-Launching Rocket Design Using Multidisciplinary System Optimization Approach
  Jae-Woo Lee, Young Chang Choi, Yung-Hwan Byun 1108

Numerical Visualization of Flow Instability in Microchannel
Considering Surface Wettability
   *Doyoung Byun, Budiono, Ji Hye Yang, Changjin Lee,
   Ki Won Lim* 1113

A Interactive Molecular Modeling System Based on Web Service
   Sungjun Park, Bosoon Kim, Jee-In Kim 1117

On the Filter Size of DMM for Passive Scalar in Complex Flow
   Yang Na, Dongshin Shin, Seungbae Lee 1127

Visualization Process for Design and Manufacturing of End Mills
   Sung-Lim Ko, Trung-Thanh Pham, Yong-Hyun Kim 1133

IP Address Lookup with the Visualizable Biased Segment Tree
   Inbok Lee, Jeong-Shik Mun, Sung-Ryul Kim 1137

A Surface Reconstruction Algorithm Using Weighted Alpha Shapes
   Si Hyung Park, Seoung Soo Lee, Jong Hwa Kim 1141

Sequential Data Analysis

HYBRID: From Atom-Clusters to Molecule-Clusters
   Zhou Bing, Jun-yi Shen, Qin-ke Peng 1151

A Fuzzy Adaptive Filter for State Estimation of Unknown Structural
System and Evaluation for Sound Environment
   Akira Ikuta, Hisako Masuike, Yegui Xiao, Mitsuo Ohta 1161

Preventing Meaningless Stock Time Series Pattern Discovery by
Changing Perceptually Important Point Detection
   Tak-chung Fu, Fu-lai Chung, Robert Luk, Chak-man Ng 1171

Discovering Frequent Itemsets Using Transaction Identifiers
   Duckjin Chai, Heeyoung Choi, Buhyun Hwang 1175

Incremental DFT Based Search Algorithm for Similar Sequence
   Quan Zheng, Zhikai Feng, Ming Zhu 1185

Parallel and Distributed Data Mining

Computing High Dimensional MOLAP with Parallel Shell Mini-cubes
   Kong-fa Hu, Chen Ling, Shen Jie, Gu Qi, Xiao-li Tang 1192

Sampling Ensembles for Frequent Patterns
  Caiyan Jia, Ruqian Lu 1197

Distributed Data Mining on Clusters with Bayesian Mixture Modeling
  M. Viswanathan, Y.K. Yang, T.K. Whangbo 1207

A Method of Data Classification Based on Parallel Genetic Algorithm
  Yuexiang Shi, Zuqiang Meng, Zixing Cai, B. Benhabib 1217

Rough Sets

Rough Computation Based on Similarity Matrix
  Huang Bing, Guo Ling, He Xin, Xian-zhong Zhou 1223

The Relationship Among Several Knowledge Reduction Approaches
  Keyun Qin, Zheng Pei, Weifeng Du 1232

Rough Approximation of a Preference Relation for Stochastic Multi-attribute Decision Problems
  Chaoyuan Yue, Shengbao Yao, Peng Zhang, Wanan Cui 1242

Incremental Target Recognition Algorithm Based on Improved Discernibility Matrix
  Liu Yong, Xu Congfu, Yan Zhiyong, Pan Yunhe 1246

Problems Relating to the Phonetic Encoding of Words in the Creation of a Phonetic Spelling Recognition Program
  Michael Higgins, Wang Shudong 1256

Diversity Measure for Multiple Classifier Systems
  Qinghua Hu, Daren Yu 1261

A Successive Design Method of Rough Controller Using Extra Excitation
  Geng Wang, Jun Zhao, Jixin Qian 1266

A Soft Sensor Model Based on Rough Set Theory and Its Application in Estimation of Oxygen Concentration
  Xingsheng Gu, Dazhong Sun 1271

A Divide-and-Conquer Discretization Algorithm
  Fan Min, Lijun Xie, Qihe Liu, Hongbin Cai 1277

A Hybrid Classifier Based on Rough Set Theory and Support Vector Machines
  Gexiang Zhang, Zhexin Cao, Yajun Gu 1287

A Heuristic Algorithm for Maximum Distribution Reduction
  Xiaobing Pei, YuanZhen Wang 1297

The Minimization of Axiom Sets Characterizing Generalized Fuzzy Rough Approximation Operators
  Xiao-Ping Yang 1303

The Representation and Resolution of Rough Sets Based on the Extended Concept Lattice
  Xuegang Hu, Yuhong Zhang, Xinya Wang 1309

Study of Integrate Models of Rough Sets and Grey Systems
  Wu Shunxiang, Liu Sifeng, Li Maoqing 1313

Author Index 1325

Table of Contents – Part II

Dimensionality Reduction

Dimensionality Reduction for Semi-supervised Face Recognition
Weiwei Du, Kohei Inoue, Kiichi Urahama 1

Cross-Document Transliterated Personal Name Coreference Resolution
Houfeng Wang 11

Difference-Similitude Matrix in Text Classification
Xiaochun Huang, Ming Wu, Delin Xia, Puliu Yan 21

A Study on Feature Selection for Toxicity Prediction
Gongde Guo, Daniel Neagu, Mark T.D. Cronin 31

Application of Feature Selection for Unsupervised Learning in Prosecutors' Office
Peng Liu, Jiaxian Zhu, Lanjuan Liu, Yanhong Li, Xuefeng Zhang 35

A Novel Field Learning Algorithm for Dual Imbalance Text Classification
Ling Zhuang, Honghua Dai, Xiaoshu Hang 39

Supervised Learning for Classification
Hongyu Li, Wenbin Chen, I-Fan Shen 49

Feature Selection for Hyperspectral Data Classification Using Double Parallel Feedforward Neural Networks
Mingyi He, Rui Huang 58

Robust Nonlinear Dimension Reduction: A Self-organizing Approach
Yuexian Hou, Liyue Yao, Pilian He 67

An Effective Feature Selection Scheme via Genetic Algorithm Using Mutual Information
Chunkai K. Zhang, Hong Hu 73

Pattern Recognition and Trend Analysis

Pattern Classification Using Rectified Nearest Feature Line Segment
  Hao Du, Yan Qiu Chen 81

Palmprint Identification Algorithm Using Hu Invariant Moments
  Jin Soo Noh, Kang Hyeon Rhee 91

Generalized Locally Nearest Neighbor Classifiers for Object Classification
  Wenming Zheng, Cairong Zou, Li Zhao 95

Nearest Neighbor Classification Using Cam Weighted Distance
  Chang Yin Zhou, Yan Qiu Chen 100

A PPM Prediction Model Based on Web Objects' Popularity
  Lei Shi, Zhimin Gu, Yunxia Pei, Lin Wei 110

An On-line Sketch Recognition Algorithm for Composite Shape
  Zhan Ding, Yin Zhang, Wei Peng, Xiuzi Ye, Huaqiang Hu 120

Axial Representation of Character by Using Wavelet Transform
  Xinge You, Bin Fang, Yuan Yan Tang, Luoqing Li, Dan Zhang ... 130

Representing and Recognizing Scenario Patterns
  Jixin Ma, Bin Luo 140

A Hybrid Artificial Intelligent-Based Criteria-Matching with Classification Algorithm
  Alex T.H. Sim, Vincent C.S. Lee 150

Auto-generation of Detection Rules with Tree Induction Algorithm
  Minsoo Kim, Jae-Hyun Seo, Il-Ahn Cheong, Bong-Nam Noh 160

Hand Gesture Recognition System Using Fuzzy Algorithm and RDBMS for Post PC
  Jung-Hyun Kim, Dong-Gyu Kim, Jeong-Hoon Shin, Sang-Won Lee, Kwang-Seok Hong 170

An Ontology-Based Method for Project and Domain Expert Matching
  Jiangning Wu, Guangfei Yang 176

Pattern Classification and Recognition of Movement Behavior of Medaka (*Oryzias Latipes*) Using Decision Tree
  Sengtai Lee, Jeehoon Kim, Jae-Yeon Baek, Man-Wi Han, Tae-Soo Chon 186

A New Algorithm for Computing the Minimal Enclosing Sphere in
Feature Space
   Chonghui Guo, Mingyu Lu, Jiantao Sun, Yuchang Lu 196

Y-AOI: Y-Means Based Attribute Oriented Induction Identifying Root
Cause for IDSs
   *Jungtae Kim, Gunhee Lee, Jung-taek Seo, Eung-ki Park,
Choon-sik Park, Dong-kyoo Kim* 205

New Segmentation Algorithm for Individual Offline Handwritten
Character Segmentation
   K.B.M.R. Batuwita, G.E.M.D.C. Bandara 215

A Method Based on the Continuous Spectrum Analysis for Fingerprint
Image Ridge Distance Estimation
   Xiaosi Zhan, Zhaocai Sun, Yilong Yin, Yayun Chu 230

A Method Based on the Markov Chain Monte Carlo for Fingerprint
Image Segmentation
   Xiaosi Zhan, Zhaocai Sun, Yilong Yin, Yun Chen 240

Unsupervised Speaker Adaptation for Phonetic Transcription Based
Voice Dialing
   Weon-Goo Kim, MinSeok Jang, Chin-Hui Lee 249

A Phase-Field Based Segmentation Algorithm for Jacquard Images
Using Multi-start Fuzzy Optimization Strategy
   Zhilin Feng, Jianwei Yin, Hui Zhang, Jinxiang Dong 255

Dynamic Modeling, Prediction and Analysis of Cytotoxicity on
Microelectronic Sensors
   Biao Huang, James Z. Xing 265

Generalized Fuzzy Morphological Operators
   Tingquan Deng, Yanmei Chen 275

Signature Verification Method Based on the Combination of Shape and
Dynamic Feature
   Yingna Deng, Hong Zhu, Shu Li, Tao Wang 285

Study on the Matching Similarity Measure Method for Image Target
Recognition
   Xiaogang Yang, Dong Miao, Fei Cao, Yongkang Ma 289

3-D Head Pose Estimation for Monocular Image
   Yingjie Pan, Hong Zhu, Ruirui Ji 293

The Speech Recognition Based on the Bark Wavelet Front-End Processing
Xueying Zhang, Zhiping Jiao, Zhefeng Zhao 302

An Accurate and Fast Iris Location Method Based on the Features of Human Eyes
Weiqi Yuan, Lu Xu, Zhonghua Lin 306

A Hybrid Classifier for Mass Classification with Different Kinds of Features in Mammography
Ping Zhang, Kuldeep Kumar, Brijesh Verma 316

Data Mining Methods for Anomaly Detection of HTTP Request Exploitations
Xiao-Feng Wang, Jing-Li Zhou, Sheng-Sheng Yu, Long-Zheng Cai .. 320

Exploring Content-Based and Image-Based Features for Nude Image Detection
Shi-lin Wang, Hong Hui, Sheng-hong Li, Hao Zhang, Yong-yu Shi, Wen-tao Qu 324

Collision Recognition and Direction Changes Using Fuzzy Logic for Small Scale Fish Robots by Acceleration Sensor Data
Seung Y. Na, Daejung Shin, Jin Y. Kim, Su-Il Choi 329

Fault Diagnosis Approach Based on Qualitative Model of Signed Directed Graph and Reasoning Rules
Bingshu Wang, Wenliang Cao, Liangyu Ma, Ji Zhang 339

Visual Tracking Algorithm for Laparoscopic Robot Surgery
Min-Seok Kim, Jin-Seok Heo, Jung-Ju Lee 344

Toward a Sound Analysis System for Telemedicine
Cong Phuong Nguyen, Thi Ngoc Yen Pham, Castelli Eric 352

Other Topics in FSKD Methods

Structural Learning of Graphical Models and Its Applications to Traditional Chinese Medicine
Ke Deng, Delin Liu, Shan Gao, Zhi Geng 362

Study of Ensemble Strategies in Discovering Linear Causal Models
Gang Li, Honghua Dai 368

The Entropy of Relations and a New Approach for Decision Tree
Learning
　　Dan Hu, HongXing Li 378

Effectively Extracting Rules from Trained Neural Networks Based
on the New Measurement Method of the Classification Power of
Attributes
　　Dexian Zhang, Yang Liu, Ziqiang Wang 388

EDTs: Evidential Decision Trees
　　Huawei Guo, Wenkang Shi, Feng Du 398

GSMA: A Structural Matching Algorithm for Schema Matching in
Data Warehousing
　　Wei Cheng, Yufang Sun 408

A New Algorithm to Get the Correspondences from the Image Sequences
　　Zhiquan Feng, Xiangxu Meng, Chenglei Yang 412

An Efficiently Algorithm Based on Itemsets Lattice and Bitmap Index
for Finding Frequent Itemsets
　　Fuzan Chen, Minqiang Li 420

Weighted Fuzzy Queries in Relational Databases
　　Ying-Chao Zhang, Yi-Fei Chen, Xiao-ling Ye, Jie-Liang Zheng 430

Study of Multiuser Detection: The Support Vector Machine Approach
　　Tao Yang, Bo Hu 442

Robust and Adaptive Backstepping Control for Nonlinear Systems
Using Fuzzy Logic Systems
　　Gang Chen, Shuqing Wang, Jianming Zhang 452

Online Mining Dynamic Web News Patterns Using Machine Learn
Methods
　　Jian-Wei Liu, Shou-Jian Yu, Jia-Jin Le 462

A New Fuzzy MCDM Method Based on Trapezoidal Fuzzy AHP and
Hierarchical Fuzzy Integral
　　Chao Zhang, Cun-bao Ma, Jia-dong Xu 466

Fast Granular Analysis Based on Watershed in Microscopic Mineral
Images
　　Danping Zou, Desheng Hu, Qizhen Liu 475

Cost-Sensitive Ensemble of Support Vector Machines for Effective Detection of Microcalcification in Breast Cancer Diagnosis
Yonghong Peng, Qian Huang, Ping Jiang, Jianmin Jiang 483

High-Dimensional Shared Nearest Neighbor Clustering Algorithm
Jian Yin, Xianli Fan, Yiqun Chen, Jiangtao Ren 494

A New Method for Fuzzy Group Decision Making Based on α-Level Cut and Similarity
Jibin Lan, Liping He, Zhongxing Wang 503

Modeling Nonlinear Systems: An Approach of Boosted Linguistic Models
Keun-Chang Kwak, Witold Pedrycz, Myung-Geun Chun 514

Multi-criterion Fuzzy Optimization Approach to Imaging from Incomplete Projections
Xin Gao, Shuqian Luo 524

Transductive Knowledge Based Fuzzy Inference System for Personalized Modeling
Qun Song, Tianmin Ma, Nikola Kasabov 528

A Sampling-Based Method for Mining Frequent Patterns from Databases
Yen-Liang Chen, Chin-Yuan Ho 536

Lagrange Problem in Fuzzy Reversed Posynomial Geometric Programming
Bing-yuan Cao 546

Direct Candidates Generation: A Novel Algorithm for Discovering Complete Share-Frequent Itemsets
Yu-Chiang Li, Jieh-Shan Yeh, Chin-Chen Chang 551

A Three-Step Preprocessing Algorithm for Minimizing E-Mail Document's Atypical Characteristics
Ok-Ran Jeong, Dong-Sub Cho 561

Failure Detection Method Based on Fuzzy Comprehensive Evaluation for Integrated Navigation System
Guoliang Liu, Yingchun Zhang, Wenyi Qiang, Zengqi Sun 567

Product Quality Improvement Analysis Using Data Mining: A Case Study in Ultra-Precision Manufacturing Industry
Hailiang Huang, Dianliang Wu 577

Two-Tier Based Intrusion Detection System
Byung-Joo Kim, Il Kon Kim 581

SuffixMiner: Efficiently Mining Frequent Itemsets in Data Streams by Suffix-Forest
Lifeng Jia, Chunguang Zhou, Zhe Wang, Xiujuan Xu 592

Improvement of Lee-Kim-Yoo's Remote User Authentication Scheme Using Smart Cards
Da-Zhi Sun, Zhen-Fu Cao 596

Mining of Spatial, Textual, Image and Time-Series Data

Grapheme-to-Phoneme Conversion Based on a Fast TBL Algorithm in Mandarin TTS Systems
Min Zheng, Qin Shi, Wei Zhang, Lianhong Cai 600

Clarity Ranking for Digital Images
Shutao Li, Guangsheng Chen 610

Attribute Uncertainty in GIS Data
Shuliang Wang, Wenzhong Shi, Hanning Yuan, Guoqing Chen 614

Association Classification Based on Sample Weighting
Jin Zhang, Xiaoyun Chen, Yi Chen, Yunfa Hu 624

Using Fuzzy Logic for Automatic Analysis of Astronomical Pipelines
Lior Shamir, Robert J. Nemiroff 634

On the On-line Learning Algorithms for EEG Signal Classification in Brain Computer Interfaces
Shiliang Sun, Changshui Zhang, Naijiang Lu 638

Automatic Keyphrase Extraction from Chinese News Documents
Houfeng Wang, Sujian Li, Shiwen Yu 648

A New Model of Document Structure Analysis
Zhiqi Wang, Yongcheng Wang, Kai Gao 658

Prediction for Silicon Content in Molten Iron Using a Combined Fuzzy-Associative-Rules Bank
Shi-Hua Luo, Xiang-Guan Liu, Min Zhao 667

An Investigation into the Use of Delay Coordinate Embedding
Technique with MIMO ANFIS for Nonlinear Prediction of Chaotic
Signals
  *Jun Zhang, Weiwei Dai, Muhui Fan, Henry Chung,
  Zhi Wei, D. Bi* 677

Replay Scene Based Sports Video Abstraction
  Jian-quan Ouyang, Jin-tao Li, Yong-dong Zhang 689

Mapping Web Usage Patterns to MDP Model and Mining with
Reinforcement Learning
  Yang Gao, Zongwei Luo, Ning Li 698

Study on Wavelet-Based Fuzzy Multiscale Edge Detection Method
  Wen Zhu, Beiping Hou, Zhegen Zhang, Kening Zhou 703

Sense Rank AALesk: A Semantic Solution for Word Sense
Disambiguation
  Yiqun Chen, Jian Yin 710

Automatic Video Knowledge Mining for Summary Generation Based
on Un-supervised Statistical Learning
  Jian Ling, Yiqun Lian, Yueting Zhuang 718

A Model for Classification of Topological Relationships Between Two
Spatial Objects
  Wu Yang, Ya Luo, Ping Guo, HuangFu Tao, Bo He 723

A New Feature of Uniformity of Image Texture Directions Coinciding
with the Human Eyes Perception
  Xing-Jian He, Yue Zhang, Tat-Ming Lok, Michael R. Lyu 727

Sunspot Time Series Prediction Using Parallel-Structure Fuzzy System
  Min-Soo Kim, Chan-Soo Chung 731

A Similarity Computing Algorithm for Volumetric Data Sets
  Tao Zhang, Wei Chen, Min Hu, Qunsheng Peng 742

Extraction of Representative Keywords Considering Co-occurrence in
Positive Documents
  Byeong-Man Kim, Qing Li, KwangHo Lee, Bo-Yeong Kang 752

On the Effective Similarity Measures for the Similarity-Based Pattern
Retrieval in Multidimensional Sequence Databases
  Seok-Lyong Lee, Ju-Hong Lee, Seok-Ju Chun 762

Crossing the Language Barrier Using Fuzzy Logic
  Rowena Chau, Chung-Hsing Yeh 768

New Algorithm Mining Intrusion Patterns
  Wu Liu, Jian-Ping Wu, Hai-Xin Duan, Xing Li 774

Dual Filtering Strategy for Chinese Term Extraction
  Xiaoming Chen, Xuening Li, Yi Hu, Ruzhan Lu 778

White Blood Cell Segmentation and Classification in Microscopic Bone Marrow Images
  Nipon Theera-Umpon 787

KNN Based Evolutionary Techniques for Updating Query Cost Models
  Zhining Liao, Hui Wang, David Glass, Gongde Guo 797

A SVM Method for Web Page Categorization Based on Weight Adjustment and Boosting Mechanism
  Mingyu Lu, Chonghui Guo, Jiantao Sun, Yuchang Lu 801

Fuzzy Systems in Bioinformatics and Bio-medical Engineering

Feature Selection for Specific Antibody Deficiency Syndrome by Neural Network with Weighted Fuzzy Membership Functions
  *Joon S. Lim, Tae W. Ryu, Ho J. Kim,
  Sudhir Gupta* 811

Evaluation and Fuzzy Classification of Gene Finding Programs on Human Genome Sequences
  Atulya Nagar, Sujita Purushothaman, Hissam Tawfik 821

Application of a Genetic Algorithm — Support Vector Machine Hybrid for Prediction of Clinical Phenotypes Based on Genome-Wide SNP Profiles of Sib Pairs
  *Binsheng Gong, Zheng Guo, Jing Li, Guohua Zhu, Sali Lv,
  Shaoqi Rao, Xia Li* 830

A New Method for Gene Functional Prediction Based on Homologous Expression Profile
  *Sali Lv, Qianghu Wang, Guangmei Zhang, Fengxia Wen,
  Zhenzhen Wang, Xia Li* 836

Analysis of Sib-Pair IBD Profiles and Genomic Context for Identification of the Relevant Molecular Signatures for Alcoholism
Chuanxing Li, Lei Du, Xia Li, Binsheng Gong, Jie Zhang, Shaoqi Rao 845

A Novel Ensemble Decision Tree Approach for Mining Genes Coding Ion Channels for Cardiopathy Subtype
Jie Zhang, Xia Li, Wei Jiang, Yanqiu Wang, Chuanxing Li, Qiuju Wang, Shaoqi Rao 852

A Permutation-Based Genetic Algorithm for Predicting RNA Secondary Structure – A Practicable Approach
Yongqiang Zhan, Maozu Guo 861

G Protein Binding Sites Analysis
Fan Zhang, Zhicheng Liu, Xia Li, Shaoqi Rao 865

A Novel Feature Ensemble Technology to Improve Prediction Performance of Multiple Heterogeneous Phenotypes Based on Microarray Data
Haiyun Wang, Qingpu Zhang, Yadong Wang, Xia Li, Shaoqi Rao, Zuquan Ding 869

Fuzzy Systems in Expert System and Informatics

Fuzzy Routing in QoS Networks
Runtong Zhang, Xiaomin Zhu 880

Component Content Soft-Sensor Based on Adaptive Fuzzy System in Rare-Earth Countercurrent Extraction Process
Hui Yang, Chonghui Song, Chunyan Yang, Tianyou Chai 891

The Fuzzy-Logic-Based Reasoning Mechanism for Product Development Process
Ying-Kui Gu, Hong-Zhong Huang, Wei-Dong Wu, Chun-Sheng Liu 897

Single Machine Scheduling Problem with Fuzzy Precedence Delays and Fuzzy Processing Times
Yuan Xie, Jianying Xie, Jun Liu 907

Fuzzy-Based Dynamic Bandwidth Allocation System
Fang-Yie Leu, Shi-Jie Yan, Wen-Kui Chang 911

Self-localization of a Mobile Robot by Local Map Matching Using
Fuzzy Logic
  Jinxia Yu, Zixing Cai, Xiaobing Zou, Zhuohua Duan 921

Navigation of Mobile Robots in Unstructured Environment Using Grid
Based Fuzzy Maps
  Özhan Karaman, Hakan Temelta 925

A Fuzzy Mixed Projects and Securities Portfolio Selection Model
  Yong Fang, K.K. Lai, Shou-Yang Wang 931

Contract Net Protocol Using Fuzzy Case Based Reasoning
  Wunan Wan, Xiaojing Wang, Yang Liu 941

A Fuzzy Approach for Equilibrium Programming with Simulated
Annealing Algorithm
  Jie Su, Junpeng Yuan, Qiang Han, Jin Huang 945

Image Processing Application with a TSK Fuzzy Model
  *Perfecto Mariño, Vicente Pastoriza, Miguel Santamaría,
  Emilio Martínez* 950

A Fuzzy Dead Reckoning Algorithm for Distributed Interactive
Applications
  Ling Chen, Gencai Chen 961

Intelligent Automated Negotiation Mechanism Based on Fuzzy
Method
  Hong Zhang, Yuhui Qiu 972

Congestion Control in Differentiated Services Networks by Means of
Fuzzy Logic
  Morteza Mosavi, Mehdi Galily 976

Fuzzy Systems in Pattern Recognition and Diagnostics

Fault Diagnosis System Based on Rough Set Theory and Support
Vector Machine
  Yitian Xu, Laisheng Wang 980

A Fuzzy Framework for Flashover Monitoring
  *Chang-Gun Um, Chang-Gi Jung, Byung-Gil Han, Young-Chul Song,
  Doo-Hyun Choi* 989

Feature Recognition Technique from 2D Ship Drawings Using Fuzzy
Inference System
  Deok-Eun Kim, Sung-Chul Shin, Soo-Young Kim 994

Transmission Relay Method for Balanced Energy Depletion in Wireless
Sensor Networks Using Fuzzy Logic
  Seung-Beom Baeg, Tae-Ho Cho 998

Validation and Comparison of Microscopic Car-Following Models Using
Beijing Traffic Flow Data
  Dewang Chen, Yueming Yuan, Baiheng Li, Jianping Wu 1008

Apply Fuzzy-Logic-Based Functional-Center Hierarchies as Inference
Engines for Self-learning Manufacture Process Diagnoses
  Yu-Shu Hu, Mohammad Modarres 1012

Fuzzy Spatial Location Model and Its Application in Spatial Query
  Yongjian Yang, Chunling Cao 1022

Segmentation of Multimodality Osteosarcoma MRI with Vectorial
Fuzzy-Connectedness Theory
  Jing Ma, Minglu Li, Yongqiang Zhao 1027

Knowledge Discovery in Bioinformatics and Bio-medical Engineering

A Global Optimization Algorithm for Protein Folds Prediction in 3D
Space
  Xiaoguang Liu, Gang Wang, Jing Liu 1031

Classification Analysis of SAGE Data Using Maximum Entropy Model
  Jin Xin, Rongfang Bie 1037

DNA Sequence Identification by Statistics-Based Models
  Jitimon Keinduangjun, Punpiti Piamsa-nga, Yong Poovorawan 1041

A New Method to Mine Gene Regulation Relationship Information
  De Pan, Fei Wang, Jiankui Guo, Jianhua Ding 1051

Knowledge Discovery in Expert System and Informatics

Shot Transition Detection by Compensating for Global and Local
Motions
  Seok-Woo Jang, Gye-Young Kim, Hyung-Il Choi 1061

Hybrid Methods for Stock Index Modeling
  Yuehui Chen, Ajith Abraham, Ju Yang, Bo Yang 1067

Designing an Intelligent Web Information System of Government Based
on Web Mining
  Gye Hang Hong, Jang Hee Lee 1071

Automatic Segmentation and Diagnosis of Breast Lesions Using
Morphology Method Based on Ultrasound
  In-Sung Jung, Devinder Thapa, Gi-Nam Wang 1079

Composition of Web Services Using Ontology with Monotonic
Inheritance
  Changyun Li, Beishui Liao, Aimin Yang, Lijun Liao 1089

Ontology-DTD Matching Algorithm for Efficient XML Query
  Myung Sook Kim, Yong Hae Kong 1093

An Approach to Web Service Discovery Based on the Semantics
  Jing Fan, Bo Ren, Li-Rong Xiong 1103

Non-deterministic Event Correlation Based on C-F Model
  Qiuhua Zheng, Yuntao Qian, Min Yao 1107

Flexible Goal Recognition via Graph Construction and Analysis
  Minghao Yin, Wenxiang Gu, Yinghua Lu 1118

An Implementation for Mapping SBML to BioSPI
  *Zhupeng Dong, Xiaoju Dong, Xian Xu, Yuxi Fu, Zhizhou Zhang,
  Lin He* 1128

Knowledge-Based Faults Diagnosis System for Wastewater
Treatment
  Jang-Hwan Park, Byong-Hee Jun, Myung-Geun Chun 1132

Study on Intelligent Information Integration of Knowledge Portals
  Yongjin Zhang, Hongqi Chen, Jiancang Xie 1136

The Risk Identification and Assessment in E-Business
Development
  Lin Wang, Yurong Zeng 1142

A Novel Wavelet Transform Based on Polar Coordinates for Datamining
Applications
  Seonggoo Kang, Sangjun Lee, Sukho Lee 1150

Impact on the Writing Granularity for Incremental Checkpointing
Junyoung Heo, Xuefeng Piao, Sangho Yi, Geunyoung Park, Minkyu Park, Jiman Hong, Yookun Cho 1154

Using Feedback Cycle for Developing an Adjustable Security Design Metric
Charlie Y. Shim, Jung Y. Kim, Sung Y. Shin, Jiman Hong 1158

w-LLC: Weighted Low-Energy Localized Clustering for Embedded Networked Sensors
Joongheon Kim, Wonjun Lee, Eunkyo Kim, Choonhwa Lee 1162

Energy Efficient Dynamic Cluster Based Clock Synchronization for Wireless Sensor Network
Md. Mamun-Or-Rashid, Choong Seon Hong, Jinsung Cho 1166

An Intelligent Power Management Scheme for Wireless Embedded Systems Using Channel State Feedbacks
Hyukjun Oh, Jiman Hong, Heejune Ahn 1170

Analyze and Guess Type of Piece in the Computer Game Intelligent System
Z.Y. Xia, Y.A. Hu, J. Wang, Y.C. Jiang, X.L. Qin 1174

Large-Scale Ensemble Decision Analysis of Sib-Pair IBD Profiles for Identification of the Relevant Molecular Signatures for Alcoholism
Xia Li, Shaoqi Rao, Wei Zhang, Guo Zheng, Wei Jiang, Lei Du 1184

A Novel Visualization Classifier and Its Applications
Jie Li, Xiang Long Tang, Xia Li 1190

Active Information Gathering on the Web

Automatic Creation of Links: An Approach Based on Decision Tree
Peng Li, Seiji Yamada 1200

Extraction of Structural Information from the Web
Tsuyoshi Murata 1204

Blog Search with Keyword Map-Based Relevance Feedback
Yasufumi Takama, Tomoki Kajinami, Akio Matsumura 1208

An One Class Classification Approach to Non-relevance Feedback
Document Retrieval
   Takashi Onoda, Hiroshi Murata, Seiji Yamada 1216

Automated Knowledge Extraction from Internet for a Crisis
Communication Portal
   Ong Sing Goh, Chun Che Fung 1226

Neural and Fuzzy Computation in Cognitive Computer Vision

Probabilistic Principal Surface Classifier
   Kuiyu Chang, Joydeep Ghosh 1236

Probabilistic Based Recursive Model for Face Recognition
   Siu-Yeung Cho, Jia-Jun Wong 1245

Performance Characterization in Computer Vision: The Role of Visual
Cognition Theory
   Aimin Wu, De Xu, Xu Yang, Jianhui Zheng 1255

Generic Solution for Image Object Recognition Based on Vision
Cognition Theory
   Aimin Wu, De Xu, Xu Yang, Jianhui Zheng 1265

Cognition Theory Motivated Image Semantics and Image Language
   Aimin Wu, De Xu, Xu Yang, Jianhui Zheng 1276

Neuro-Fuzzy Inference System to Learn Expert Decision: Between
Performance and Intelligibility
   Laurence Cornez, Manuel Samuelides, Jean-Denis Muller 1281

Fuzzy Patterns in Multi-level of Satisfaction for MCDM Model Using
Modified Smooth S-Curve MF
   Pandian Vasant, A. Bhattacharya, N.N. Barsoum 1294

Author Index 1305

On Fuzzy Inclusion in the Interval-Valued Sense

Jin Han Park[1], Jong Seo Park[2], and Young Chel Kwun[3]

[1] Division of Math. Sci., Pukyong National University, Pusan 608-737, South Korea
jihpark@pknu.ac.kr
[2] Department of Math. Education, Chinju National Universuty of Education,
Chinju 660-756, South Korea
parkjs@cue.ac.kr
[3] Department of Mathematics, Dong-A University, Pusan 604-714, South Korea
yckwun@dau.ac.kr

Abstract. As a generalization of fuzzy sets, the concept of interval-valued fuzzy sets was introduced by Gorzalczany [Fuzzy Sets and Systems **21** (1987) 1]. In this paper, we shall extend the concept of "fuzzy inclusion", introduced by Šostak [Supp. Rend. Circ. Mat. Palermo (Ser. II) **11** (1985) 89], to the interval-valued fuzzy setting and study its fundamental properties for some extent.

1 Introduction

After the introduction of the concept of fuzzy sets by Zadeh [10] several researchers were concerned about the generalizations of the notion of fuzzy set, e.g. fuzzy set of type n [11], intuitionistic fuzzy set [1,2] and interval-valued fuzzy set [4]. The concept of interval-valued fuzzy sets was introduced by Gorzalczany [4], and recently there has been progress in the study of such sets by Mondal and Samanta [6] and Ramakrishnan and Nayagam [7]. On the other hand, Šostak [8] defined fuzzy inclusion between two fuzzy sets A and B in order to give measure of inclusion of one in the other and applied this notion in fuzzy topological spaces defined by himself. Fuzzy inclusion in intuitionistic fuzzy sets was defined and studied by Çoker and Demirci [3]. In this paper, we define and study fuzzy inclusion in interval-valued fuzzy sets and then apply this inclusion to interval-valued fuzzy topological spaces in Šostak's sense.

2 Preliminaries

First we shall present the fundamental definitions given by Gorzalczany [4]:

Definition 1. [4] Let X be a nonempty fixed set. An interval-valued fuzzy set (**IVF** set, for short) A on X is an object having the form

$$A = \{(x, [\mu_A^L(x), \mu_A^U(x)]) : x \in X\}$$

where $\mu_A^L : X \to [0,1]$ and $\mu_A^U : X \to [0,1]$ are functions satisfying $\mu_A^L(x) \leq \mu_A^U(x)$ for each $x \in X$.

Let **D** be the set of all closed subintervals of the unit interval $[0,1]$ and consider singletons $\{a\}$ in $[0,1]$ as closed subintervals of the form $[a,a]$. An **IVF** set $A = \{(x, [\mu_A^L(x), \mu_A^U(x)]) : x \in X\}$ in X can identified to element in $\mathbf{D}^X$. Thus for each $x \in X$, $A(x)$ is a closed interval whose lower and upper end points are $\mu_A^L(x)$ and $\mu_A^U(x)$, respectively. Obviously, every fuzzy set $A = \{\mu_A(x) : x \in X\}$ on X is an **IVF** set of the form $A = \{(x, [\mu_A(x), \mu_A(x)]) : x \in X\}$. Let X_0 be a subset of X. For any interval $[a,b] \in \mathbf{D}$, the **IVF** set whose value is the interval $[a,b]$ for $x \in X_0$ and $[0,0]$ for $x \in X \setminus X_0$, is denoted by $[\widetilde{a,b}]_{X_0}$. In particular, if $a = b$ the **IVF** set $[\widetilde{a,b}]_{X_0}$ is denoted by simply $\tilde{a}_{X_0}$. The **IVF** set $[\widetilde{a,b}]_X$ (resp. $\tilde{a}_X$) is denoted by simply $[\widetilde{a,b}]$ (resp. $\tilde{a}$). For the sake of simplicity, we shall often use the symbol $A = [\mu_A^L, \mu_A^U]$ for the **IVF** set $A = \{(x, [\mu_A^L(x), \mu_A^U(x)]) : x \in X\}$.

Definition 2. [4] Let A and B be **IVF** sets on X. Then

(a) $A \subseteq B$ iff $\mu_A^L(x) \leq \mu_B^L(x)$ and $\mu_A^U(x) \leq \mu_B^U(x)$ for all $x \in X$;
(b) $A = B$ iff $A \subseteq B$ and $B \subseteq A$;
(c) The complement A^c of A is defined by $A^c(x) = [1 - \mu_A^U(x), 1 - \mu_A^L(x)]$ for all $x \in X$;
(d) If $\{A_i : i \in J\}$ is an arbitrary family of **IVF** sets on X, then

$$\cap A_i(x) = [\sup_{i \in J} \mu_{A_i}^L(x), \sup_{i \in J} \mu_{A_i}^U(x)],$$

$$\cup A_i(x) = [\inf_{i \in J} \mu_{A_i}^L(x), \inf_{i \in J} \mu_{A_i}^U(x)].$$

Definition 3. Let X and Y be two nonempty sets and $f : X \to Y$ be a function. Let A and B be **IVF** sets on X and Y respectively.

(a) The inverse image $f^{-1}(B)$ of B under f is the **IVF** set on X defined by $f^{-1}(B)(x) = [\mu_B^L(f(x)), \mu_B^U(f(x))]$ for all $x \in X$.
(b) The image $f(A)$ of A under f is the **IVF** set on Y defined by $f(A) = [\mu_{f(A)}^L, \mu_{f(A)}^U]$, where

$$\mu_{f(A)}^L(y) = \begin{cases} \sup_{x \in f^{-1}(y)} \mu_A^L(x) & \text{if } f^{-1}(y) \neq \phi \\ 0, & \text{otherwise}, \end{cases}$$

$$\mu_{f(A)}^U(y) = \begin{cases} \sup_{x \in f^{-1}(y)} \mu_A^U(x) & \text{if } f^{-1}(y) \neq \phi \\ 0, & \text{otherwise}. \end{cases}$$

for each $y \in Y$.

Now we list the properties of images and preimages, some of which we shall frequently use in Sections 3 and 4.

Theorem 1. Let A and A_i $(i \in J)$ be **IVF** sets on X and B and B_i $(i \in J)$ be **IVF** sets on Y and $f : X \to Y$ be a function. Then:

(a) If $A_1 \subseteq A_2$, then $f(A_1) \subseteq f(A_2)$.
(b) If $B_1 \subseteq B_2$, then $f^{-1}(B_1) \subseteq f^{-1}(B_2)$.

(c) $A \subseteq f^{-1}(f(A))$. If, furthermore, f is injective, then $A = f^{-1}(f(A))$.
(d) $f(f^{-1}(B)) \subseteq B$. If, furthermore, f is surjective, then $f(f^{-1}(B)) = B$.
(e) $f^{-1}(\bigcup B_i) = \bigcup f^{-1}(B_i)$, $f^{-1}(\bigcap B_i) = \bigcap f^{-1}(B_i)$.
(f) $f(\bigcup A_i) = \bigcup f(A_i)$, $f(\bigcap A_i) \subseteq \bigcap f(A_i)$ If, furthermore, f is injective, then $f(\bigcap A_i) = \bigcap f(A_i)$.
(g) $f^{-1}(\tilde{1}) = \tilde{1}$, $f^{-1}(\tilde{0}) = \tilde{0}$.
(h) $f(\tilde{0}) = \tilde{0}$ and $f(\tilde{1}) = \tilde{1}$ if f is surjective.
(i) $f^{-1}(B)^c = f^{-1}(B^c)$ and $f(A)^c \subseteq f(A^c)$ if f is surjective.

3 Fuzzy Inclusion in the Interval-Valued Sense

In this section, we shall extend the concept "fuzzy inclusion" [8,9] to the interval-valued fuzzy setting:

Definition 4. Let X be a nonempty set. Then the right fuzzy inclusion, denoted by $\sqsubseteq$, is the **IVF** set on $\mathbf{D}^X \times \mathbf{D}^X$ defined by

$$\mu_{\sqsubseteq}^L(A, B) = \inf\{((1 - \mu_A^U) \vee \mu_B^L)(x) : x \in X\},$$

$$\mu_{\sqsubseteq}^U(A, B) = \inf\{((1 - \mu_A^L) \vee \mu_B^U)(x) : x \in X\}$$

for each $A, B \in \mathbf{D}^X$. Here $\mu_{\sqsubseteq}^L(A, B)$ denote the lower limit of inclusion of A in B, while $\mu_{\sqsubseteq}^U(A, B)$ denotes the upper limit of inclusion of A in B.

Remark 1. Since $1 - \mu_A^U \leq 1 - \mu_A^L$, we may deduce the following:

$$((1 - \mu_A^U) \vee \mu_B^L)(x) \leq ((1 - \mu_A^L) \vee \mu_B^U)(x) \text{ for each } x \in X$$
$$\Rightarrow \inf_{x \in X}((1 - \mu_A^U) \vee \mu_B^L)(x) \leq \inf_{x \in X}((1 - \mu_A^L) \vee \mu_B^U)(x)$$
$$\Rightarrow \mu_{\sqsubseteq}^L(A, B) \leq \mu_{\sqsubseteq}^U(A, B).$$

Therefore, for each $A, B \in \mathbf{D}^X$, $[\mu_{\sqsubseteq}^L(A, B), \mu_{\sqsubseteq}^U(A, B)]$ is closed interval.

Definition 5. For any two **IVF** sets $A, B \in \mathbf{D}^X$, closed interval $[\mu_{\sqsubseteq}^L(A, B), \mu_{\sqsubseteq}^U(A, B)]$ will be denoted by $[A \sqsubseteq B]$, i.e.,

$$[A \sqsubseteq B] = [\mu_{\sqsubseteq}^L(A, B), \mu_{\sqsubseteq}^U(A, B)].$$

Remark 2. The closed interval $[A \sqsubseteq B]$ shows "to what extend the **IVF** set A is contained in the **IVF** set B" (cf. [3,9]). If A and B are crisp **IVF** sets on X given by $A = \tilde{1}_C$ and $B = \tilde{1}_D$ where C and D are nonempty subsets of X, then $[A \sqsubseteq B] = \tilde{1}$ iff $C \subseteq D$, and $[A \sqsubseteq B] = \tilde{0}$ otherwise.

Similar to the concept of right fuzzy inclusion, we can easily define the left fuzzy inclusion $\sqsupseteq$ as follows:

$$\mu_{\sqsupseteq}^L(A, B) = \mu_{\sqsubseteq}^L(B, A), \quad \mu_{\sqsupseteq}^U(A, B) = \mu_{\sqsubseteq}^U(B, A), \quad [A \sqsupseteq B] = [\mu_{\sqsupseteq}^L(A, B), \mu_{\sqsupseteq}^U(A, B)].$$

Definition 6. For any two **IVF** sets $A, B \in \mathbf{D}^X$, the interval-valued fuzzy equality of A to B defined as follows:

$$[A \simeq B] = [A \sqsubseteq B] \wedge [A \sqsupseteq B].$$

The interval values $\mu^L_\simeq(A, B) = \mu^L_\sqsubseteq(A, B) \wedge \mu^L_\sqsupseteq(A, B)$ and $\mu^U_\simeq(A, B) = \mu^U_\sqsubseteq(A, B) \vee \mu^U_\sqsupseteq(A, B)$, respectively, denote the lower limit of equality and the upper limit of equality of the **IVF** set A to the **IVF** set B.

Now we present some of the basic properties of interval-valued fuzzy inclusion:

Theorem 2. *For **IVF** sets A, B, C and D on X, the following properties hold:*
(a) *If $A \subseteq B$ and $D \subseteq C$, then $[A \sqsubseteq C] \geq [B \sqsubseteq D]$.*
(b) *$[A^c \sqsubseteq B^c] = [B \sqsubseteq A]$.*
(c) *$[A \cup B \sqsubseteq C \cap D] \leq [A \sqsubseteq C] \wedge [B \sqsubseteq D]$.*
(d) *$[A \sqsubseteq C] \vee [B \sqsubseteq D] \leq [A \cap B \sqsubseteq C \cup D]$.*
(e) *$[A \cap B^c \sqsubseteq \tilde{0}] = [A \sqsubseteq B] = [\tilde{1} \sqsubseteq A^c \cup B]$.*
(f) *If $\{B_i \in \mathbf{D}^X : i \in J\}$, then*

$$\wedge_{i \in J}[A \sqsubseteq B_i] = [A \sqsubseteq \cap_{i \in J} B_i] \text{ and } \wedge_{i \in J}[B_i \sqsubseteq A] = [\cup_{i \in J} B_i \sqsubseteq A].$$

Proof. (a) Let $A \subseteq B$ and $D \subseteq C$. Since $\mu^L_A \leq \mu^L_B$, $\mu^U_A \leq \mu^U_B$, $\mu^L_D \leq \mu^L_C$ and $\mu^U_D \leq \mu^U_C$, we obtain

$$[A \sqsubseteq C] = \left[\inf_{x \in X}((1 - \mu^U_A) \vee \mu^L_C)(x), \inf_{x \in X}((1 - \mu^L_A) \vee \mu^U_C)(x) \right]$$

$$\geq \left[\inf_{x \in X}((1 - \mu^U_B) \vee \mu^L_D)(x), \inf_{x \in X}((1 - \mu^L_B) \vee \mu^U_D)(x) \right] = [B \sqsubseteq D].$$

(b)

$$[B \sqsubseteq A] = \left[\inf_{x \in X}((1 - \mu^U_B) \vee \mu^L_A)(x), \inf_{x \in X}((1 - \mu^L_B) \vee \mu^U_A)(x) \right]$$

$$= \left[\inf_{x \in X}((1 - (1 - \mu^L_A)) \vee (1 - \mu^U_B))(x), \inf_{x \in X}((1 - (1 - \mu^U_A)) \vee (1 - \mu^L_B))(x) \right]$$

$$= [A^c \sqsubseteq B^c].$$

(c) By (a), $[A \cup B \sqsubseteq C \cap D] \leq [A \sqsubseteq C]$ and $[A \cup B \sqsubseteq C \cap D] \leq [B \sqsubseteq D]$ and hence $[A \cup B \sqsubseteq C \cap D] \leq [A \sqsubseteq C] \wedge [B \sqsubseteq D]$.

(d) Similar to (c).

(e)

$$[A \sqsubseteq B] = \left[\inf_{x \in X}((1 - \mu^U_A) \vee \mu^L_B)(x), \inf_{x \in X}((1 - \mu^L_A) \vee \mu^U_B)(x) \right]$$

$$= \left[\inf_{x \in X}(1 - (\mu^U_A \wedge (1 - \mu^L_B)))(x), \inf_{x \in X}(1 - (\mu^L_A \wedge (1 - \mu^U_B)))(x) \right]$$

$$= \left[\inf_{x \in X}((1 - \mu^U_{A \cap B^c}) \vee 0)(x), \inf_{x \in X}((1 - \mu^L_{A \cap B^c}) \vee 0)(x) \right]$$

$$= [A \cap B^c \sqsubseteq \tilde{0}]$$

and

$$[A \sqsubseteq B] = \left[\inf_{x \in X}((1-\mu_A^U) \vee \mu_B^L)(x), \inf_{x \in X}((1-\mu_A^L) \vee \mu_B^U))(x)\right]$$
$$= \left[\inf_{x \in X}(0 \vee \mu_{A^c \cup B}^L))(x), \inf_{x \in X}(0 \vee \mu_{A^c \cup B}^U)(x)\right] = [\tilde{1} \sqsubseteq A^c \cup B].$$

(f)

$$[A \sqsubseteq \cap_i B_i] = \left[\inf_{x \in X}(1-\mu_A^U) \vee \mu_{\cap B_i}^L)(x), \inf_{x \in X}((1-\mu_A^L) \vee \mu_{\cap B_i}^U)(x)\right]$$
$$= \left[\inf_{x \in X}((1-\mu_A^U) \vee \wedge_i \mu_{B_i}^L)(x), \inf_{x \in X}((1-\mu_A^L) \vee \wedge_i \mu_{B_i}^U)(x)\right]$$
$$= \left[\inf_{x \in X} \inf_i (1-\mu_A^U) \vee \mu_{B_i}^L)(x), \inf_{x \in X} \inf_i ((1-\mu_A^L) \vee \mu_{B_i}^U)(x)\right]$$
$$= \wedge_i \left[\inf_{x \in X}(1-\mu_A^U) \vee \mu_{B_i}^L)(x), \inf_{x \in X}((1-\mu_A^L) \vee \mu_{B_i}^U)(x)\right]$$
$$= \wedge_i [A \sqsubseteq B_i]$$

and by (b) we obtain

$$[\cup_i B_i \sqsubseteq A] = [A^c \sqsubseteq (\cup_i B_i)^c] = [A^c \sqsubseteq \cap_i (B_i)^c]$$
$$= \wedge_i [A^c \sqsubseteq (B_i)^c] = \wedge_i [B_i \sqsubseteq A].$$

Now we list the properties of interval-valued fuzzy inclusion related to images and preimages:

Theorem 3. Let A, B be **IVF** sets on X and C, D be **IVF** sets on Y. For a function $f: X \to Y$, the following properties hold:

(a) $[A \sqsubseteq B] \leq [f(A) \sqsubseteq f(B)]$. Furthermore, $[A \sqsubseteq B] = [f(A) \sqsubseteq f(B)]$ if f is injective.
(b) $[C \sqsubseteq D] \leq [f^{-1}(C) \sqsubseteq f^{-1}(D)]$. Furthermore, $[C \sqsubseteq D] = [f^{-1}(C) \sqsubseteq f^{-1}(D)]$ if f is surjective.
(c) $[A \sqsubseteq f^{-1}(f(A))] \leq [f(A) \sqsubseteq f(A)]$, $[f^{-1}(f(A)) \sqsubseteq A] \leq [A \sqsubseteq A]$, $[f(f^{-1}(C)) \sqsubseteq C] \leq [f^{-1}(C) \sqsubseteq f^{-1}(C)]$ and $[C \sqsubseteq f(f^{-1}(C))] \leq [C \sqsubseteq C]$.
(d) $[f(A) \sqsubseteq C] = [A \sqsubseteq f^{-1}(C)]$.
(e) If $\{C_i : i \in J\} \subseteq \mathbf{D}^Y$, then

$$[f(A) \sqsubseteq \cup_i C_i] = [A \sqsubseteq \cup_i f^{-1}(C_i)].$$

(f) If $\{A_i : i \in J\} \subseteq \mathbf{D}^X$, then

$$[f(\cap_i A_i) \sqsubseteq C] = [\cap_i A_i \sqsubseteq f^{-1}(C)].$$

Proof. (a) By Definition 3 (b), we obtain $\inf_{y \in Y}\left((1-\mu_{f(A)}^U) \vee \mu_{f(B)}^L\right)(y) \geq \inf_{x \in X}\left((1-\mu_A^U) \vee \mu_B^L\right)(x)$ and $\inf_{y \in Y}\left((1-\mu_{f(A)}^L) \vee \mu_{f(B)}^U\right)(y) \geq \inf_{x \in X}((1-\mu_A^L) \vee \mu_B^U)(x)$ and hence

$$[f(A) \sqsubseteq f(B)] = [\mu^L_{\sqsubseteq}(f(A), f(B)), \mu^U_{\sqsubseteq}(f(A), f(B))]$$
$$= \left[\inf_{y \in Y} \left((1 - \mu^U_{f(A)}) \vee \mu^L_{f(B)}\right)(y), \inf_{y \in Y} \left((1 - \mu^L_{f(A)}) \vee \mu^U_{f(B)}\right)(y)\right]$$
$$\geq \left[\inf_{x \in X} \left((1 - \mu^U_A) \vee \mu^L_B\right)(x), \inf_{x \in X} \left((1 - \mu^L_A) \vee \mu^U_B\right)(x)\right] = [A \sqsubseteq B].$$

Now we prove the equality in case that f is injective.

$$\inf_{y \in Y}((1 - \mu^U_{f(A)}) \vee \mu^L_{f(B)})(y)$$
$$= \inf_{y \in f(X)} \left((1 - \mu^U_{f(A)}) \vee \mu^L_{f(B)}\right)(y) \wedge \inf_{y \notin f(X)} \left((1 - \mu^U_{f(A)}) \vee \mu^L_{f(B)}\right)(y)$$
$$= \inf_{y \in f(X)} \left((1 - \mu^U_{f(A)}) \vee \mu^L_{f(B)}\right)(y) \wedge 1$$
$$= \inf_{y \in f(X)} \left((1 - \mu^U_{f(A)}) \vee \mu^L_{f(B)}\right)(y)$$
$$= \inf_{y \in f(X)} \left((1 - \sup_{x \in f^{-1}(y)} \mu^U_A)(x) \vee \sup_{x \in f^{-1}(y)} \mu^L_B)(x)\right)$$
$$= \inf_{x \in X} \left((1 - \mu^U_A) \vee \mu^L_B\right)(x).$$

Similarly, we have

$$\inf_{y \in Y}((1 - \mu^L_{f(A)}) \vee \mu^U_{f(B)})(y) = \inf_{x \in X} \left((1 - \mu^L_A) \vee \mu^U_B\right)(x).$$

Hence, from two equalities above, we obtain $[A \sqsubseteq B] = [f(A) \sqsubseteq f(B)]$.

(b) Similar to (a).
(c) From (a) and (b), the required inequalities can be easily obtained.
(d) By (b) and Theorem 1 (c), we have $[f(A) \sqsubseteq C] \leq [f^{-1}(f(A)) \sqsubseteq f^{-1}(C)] \leq A \sqsubseteq f^{-1}(C)]$. Similarly, by (a) and Theorem 1 (d), $[A \sqsubseteq f^{-1}(C)] \leq [f(A) \sqsubseteq f(f^{-1}(C))] \leq f(A) \sqsubseteq C]$. Hence $[f(A) \sqsubseteq C] = [A \sqsubseteq f^{-1}(C)]$.
(e) By (d) and Theorem 1 (e), we have $[f(A) \sqsubseteq \bigcup_i C_i] = [[A \sqsubseteq f^{-1}(\bigcup_i C_i)] = [A \sqsubseteq \bigcup_i f^{-1}(C_i)]$.
(f) Similar to (e).

4 Interval-Valued Fuzzy Families

In this section, we define the concept of interval-valued fuzzy family to obtain generalized De Morgan's laws and later interval-valued fuzzy topological spaces.

Definition 7. An **IVF** set $\mathcal{F}$ on the set $\mathbf{D}^X$ is called an interval-valued fuzzy family (**IVFF** for short) on X and denoted by the form $\mathcal{F} = [\mu^L_{\mathcal{F}}, \mu^U_{\mathcal{F}}]$.

Definition 8. Let $\mathcal{F}$ be an **IVFF** on X. Then the **IVFF** of complemented **IVF** sets on X is defined by $\mathcal{F}^* = [\mu_{\mathcal{F}^*}^L, \mu_{\mathcal{F}^*}^U]$, where $\mu_{\mathcal{F}^*}^L(A) = \mu_{\mathcal{F}}^L(A^c)$ and $\mu_{\mathcal{F}^*}^U(A) = \mu_{\mathcal{F}}^U(A^c)$ for each $A \in \mathbf{D}^X$.

Example 1. We can easily extend the fuzzy topological space, first defined by Šostak [8,9], to the case of interval-valued fuzzy sets [4]. Let τ be an **IVFF** on X. For each $A \in \mathbf{D}^X$, we can construct the closed interval $\tau(A)$ as follows:

$$\tau(A) = [\mu_\tau^L(A), \mu_\tau^U(A)].$$

In this case, an interval-valued fuzzy topology in Šostak's sense (**So-IVFT** for short) on a nonempty set X is an **IVFF** τ on X satisfying the following axioms:

(O1) $\tau(\tilde{0}) = \tilde{1}$ and $\tau(\tilde{1}) = \tilde{1}$;
(O2) $\tau(A_1 \cap A_2) \geq \tau(A_1) \wedge \tau(A_2)$ for any $A_1, A_2 \in \mathbf{D}^X$;
(O3) $\tau(\bigcup_i A_i) \geq \bigwedge_i \tau(A_i)$ for any $\{A_i : i \in J\} \subseteq \mathbf{D}^X$.

In this case the pair (X, τ) is called an interval-valued fuzzy topological space in Šostak's sense (**So-IVFTS** for short). For any $A \in \mathbf{D}^X$, the closed interval $\tau(A)$ is called the interval-valued degree of openness of A.

Of course, a **So-IVFTS** (X, τ) is fuzzy topological space in Chang's sense. So we also define a **So-IVFTS** in the sense of Lowen [5] as follows. (X, τ) is **So-IVFTS** in the sense of Lowen if (X, τ) is **IVFTS** satisfying the condition that for each **IVF** set in the form $[\tilde{a,b}]$, where $[a, b] \subseteq [0, 1]$, $\tau\left([\tilde{a,b}]\right) = \tilde{1}$ holds.

Let (X, τ) be a **So-IVFTS** on X. Then the **IVFF** τ^* of the complemented **IVF** sets on X is defined by $\tau^*(A) = \tau(A^c)$ for each $A \in \mathbf{D}^X$. The closed interval $\tau^*(A) = [\mu_{\tau^*}^L(A), \mu_{\tau^*}^U(A)]$ is called the interval-valued degree of closedness of A. Thus the **IVFF** τ^* on X satisfies the following properties:

(C1) $\tau^*(\tilde{0}) = \tilde{1}$ and $\tau^*(\tilde{1}) = \tilde{1}$;
(C2) $\tau^*(A_1 \cup A_2) \geq \tau^*(A_1) \wedge \tau^*(A_2)$ for any $A_1, A_2 \in \mathbf{D}^X$;
(C3) $\tau^*(\bigcap_i A_i) \geq \bigwedge_i \tau^*(A_i)$ for any $\{A_i : i \in J\} \subseteq \mathbf{D}^X$.

Now we extend the intersection and union of a fuzzy family [8,9] to the interval-valued fuzzy setting:

Definition 9. Let $\mathcal{F}$ be an **IVFF** on X. Then

(a) the intersection $\bigcap \mathcal{F}$ of this **IVFF** is the **IVF** set defined by $\bigcap \mathcal{F}(x) = \left[\mu_{\cap \mathcal{F}}^L(x), \mu_{\cap \mathcal{F}}^U(x)\right]$ for all $x \in X$, where

$$\mu_{\cap \mathcal{F}}^L(x) = \inf\{1 - \mu_{\mathcal{F}}^U(A) \vee \mu_A^L(x) : A \in \mathbf{D}^X\},$$
$$\mu_{\cap \mathcal{F}}^U(x) = \inf\{1 - \mu_{\mathcal{F}}^L(A) \vee \mu_A^U(x) : A \in \mathbf{D}^X\};$$

(b) the union $\bigcup \mathcal{F}$ of this **IVFF** is the **IVF** set defined by $\bigcup \mathcal{F}(x) = \left[\mu_{\cup \mathcal{F}}^L(x), \mu_{\cup \mathcal{F}}^U(x)\right]$ for all $x \in X$, where

$$\mu_{\cup \mathcal{F}}^L(x) = \sup\{\mu_{\mathcal{F}}^L(A) \wedge \mu_A^L(x) : A \in \mathbf{D}^X\},$$
$$\mu_{\cup \mathcal{F}}^U(x) = \sup\{\mu_{\mathcal{F}}^U(A) \wedge \mu_A^U(x) : A \in \mathbf{D}^X\}.$$

Theorem 4. (*Generalized De Morgan's Laws*) Let $\mathcal{F}$ be an **IVFF** on X. Then we have
(a) $(\bigcup \mathcal{F})^c = \bigcap \mathcal{F}^*$.
(b) $(\bigcap \mathcal{F})^c = \bigcup \mathcal{F}^*$.

Proof. Let $x \in X$. Then we have

$$\begin{aligned}
\mu^L_{\bigcap \mathcal{F}^*}(x) &= \inf\{1 - \mu^U_{\mathcal{F}^*}(A) \vee \mu^L_A(x) : A \in \mathbf{D}^X\} \\
&= \inf\{1 - \mu^U_{\mathcal{F}}(A^c) \vee \mu^L_A(x) : A \in \mathbf{D}^X\} \\
&= \inf\{1 - \mu^U_{\mathcal{F}}(A^c) \vee (1 - (1 - \mu^L_A))(x) : A \in \mathbf{D}^X\} \\
&= \inf\{1 - \mu^U_{\mathcal{F}}(A^c) \vee (1 - \mu^U_{A^c})(x) : A \in \mathbf{D}^X\} \\
&= 1 - \sup\{\mu^U_{\mathcal{F}}(A^c) \wedge \mu^U_{A^c}(x) : A \in \mathbf{D}^X\} \\
&= 1 - \sup\{\mu^U_{\mathcal{F}}(A) \wedge \mu^U_A(x) : A \in \mathbf{D}^X\} = 1 - \mu^U_{\bigcap \mathcal{F}}(x)
\end{aligned}$$

and

$$\begin{aligned}
\mu^U_{\bigcap \mathcal{F}^*}(x) &= \inf\{1 - \mu^L_{\mathcal{F}^*}(A) \vee \mu^U_A(x) : A \in \mathbf{D}^X\} \\
&= \inf\{1 - \mu^L_{\mathcal{F}}(A^c) \vee (1 - (1 - \mu^U_A))(x) : A \in \mathbf{D}^X\} \\
&= \inf\{1 - \mu^L_{\mathcal{F}}(A^c) \vee (1 - \mu^L_{A^c})(x) : A \in \mathbf{D}^X\} \\
&= 1 - \sup\{\mu^L_{\mathcal{F}}(A^c) \wedge \mu^L_{A^c}(x) : A \in \mathbf{D}^X\} \\
&= 1 - \sup\{\mu^L_{\mathcal{F}}(A) \wedge \mu^L_A(x) : A \in \mathbf{D}^X\} = 1 - \mu^L_{\bigcap \mathcal{F}}(x).
\end{aligned}$$

Hence $(\bigcup \mathcal{F})^c = \bigcap \mathcal{F}^*$.
(b) Similar to (a).

Finally, we shall define the image and preimage of **IVFF**'s under a function $f : X \to Y$:

Definition 10. Let $\mathcal{F}$ be an **IVFF** on X and let $f : X \to Y$ be an injective function. Then the image $\mathcal{F}^f = \left[\mu^L_{\mathcal{F}^f}, \mu^U_{\mathcal{F}^f}\right]$ of $\mathcal{F} = \left[\mu^L_{\mathcal{F}}, \mu^U_{\mathcal{F}}\right]$ under f is the **IVFF** on Y defined as follows:

$$\mu^L_{\mathcal{F}^f}(B) = \begin{cases} \mu^L_{\mathcal{F}}(f^{-1}(B)), & \text{if } B \subseteq \tilde{1}_{f(X)} \\ 0, & \text{otherwise} \end{cases}$$

$$\mu^U_{\mathcal{F}^f}(B) = \begin{cases} \mu^U_{\mathcal{F}}(f^{-1}(B)), & \text{if } B \subseteq \tilde{1}_{f(X)} \\ 0, & \text{otherwise.} \end{cases}$$

Definition 11. Let $\mathcal{F}$ be an **IVFF** on Y and let $f : X \to Y$ be a function. Then the preimage $\mathcal{F}^{f^{-1}} = \left[\mu^L_{\mathcal{F}^{f^{-1}}}, \mu^U_{\mathcal{F}^{f^{-1}}}\right]$ of $\mathcal{F} = \left[\mu^L_{\mathcal{F}}, \mu^U_{\mathcal{F}}\right]$ under f is the **IVFF** on X defined by

$$\mu^L_{\mathcal{F}^{f^{-1}}}(A) = \sup\{\mu^L_{\mathcal{F}}(B) : A = f^{-1}(B), B \in \mathbf{D}^Y\},$$
$$\mu^U_{\mathcal{F}^{f^{-1}}}(A) = \sup\{\mu^U_{\mathcal{F}}(B) : A = f^{-1}(B), B \in \mathbf{D}^Y\}.$$

Theorem 5. Let $\mathcal{F}$ be an **IVFF** on X and $f : X \to Y$ be an injective function. Then
  (a) $f(\bigcup \mathcal{F}) = \bigcup \mathcal{F}^f$.
  (b) $f(\bigcap \mathcal{F}) = \bigcap \mathcal{F}^f$.

Proof. (a) We notice that, under $B \subseteq \tilde{1}_{f(X)}$, if $B \in \mathbf{D}^Y$, then there exists a $A \in \mathbf{D}^X$ such that $A = f^{-1}(B)$ and so $f(A) = B$. Similarly, if $A \in \mathbf{D}^X$, then there exists a $B \in \mathbf{D}^Y$ such that $B \subseteq \tilde{1}_{f(X)}$ and $A = f^{-1}(B)$, and so $B = f(A)$. Let $y \in Y$. If $f^{-1}(y) \neq \phi$, then we have

$$\mu^L_{\bigcup \mathcal{F}^f}(y) = \sup\{\mu^L_{\mathcal{F}^f}(B) \wedge \mu^L_B(y) : B \in \mathbf{D}^Y\}$$
$$= \sup\{\mu^L_{\mathcal{F}}(f^{-1}(B)) \wedge \mu^L_B(y) : B \in \mathbf{D}^Y, B \subseteq \tilde{1}_{f(X)}\}$$
$$\vee \sup\{\mu^L_{\mathcal{F}}(f^{-1}(B)) \wedge \mu^L_B(y) : B \in \mathbf{D}^Y, B \not\subseteq \tilde{1}_{f(X)}\}$$
$$= \sup\{\mu^L_{\mathcal{F}}(f^{-1}(B)) \wedge \mu^L_B(y) : B \in \mathbf{D}^Y, B \subseteq \tilde{1}_{f(X)}\} \vee 0$$
$$= \{\mu^L_{\mathcal{F}}(f^{-1}(B)) \wedge \mu^L_B(y) : B \in \mathbf{D}^Y, B \subseteq \tilde{1}_{f(X)}\}$$
$$= \{\mu^L_{\mathcal{F}}(A) \wedge \mu^L_{f(A)}(y) : A \in \mathbf{D}^X\}$$
$$= \{\mu^L_{\mathcal{F}}(A) \wedge f(\mu^L_A)(y) : A \in \mathbf{D}^X\}.$$

On the other hand, whenever $f^{-1}(y) \neq \phi$, there exists a $x \in X$ such that $f(x) = y$ since f is injective. Then we have

$$\mu^L_{f(\bigcup \mathcal{F})}(y) = f(\mu^L_{\bigcup \mathcal{F}})(y) = \mu^L_{\bigcup \mathcal{F}}(x)$$
$$= \sup\{\mu^L_{\mathcal{F}}(A) \wedge \mu^L_A(x) : A \in \mathbf{D}^X\}$$
$$= \{\mu^L_{\mathcal{F}}(A) \wedge f(\mu^L_A)(y) : A \in \mathbf{D}^X\}.$$

If $f^{-1}(y) = \phi$, then we have $\mu^L_{\bigcup \mathcal{F}^f}(y) = 0$ and $\mu^L_{f(\bigcup \mathcal{F})}(y) = 0$. Therefore, we obtain $\mu^L_{\bigcup \mathcal{F}^f} = \mu^L_{f(\bigcup \mathcal{F})}$. Similarly, if $f^{-1}(y) \neq \phi$, then we have

$$\mu^U_{\bigcup \mathcal{F}^f}(y) = \sup\{\mu^U_{\mathcal{F}^f}(B) \wedge \mu^U_B(y) : B \in \mathbf{D}^Y\}$$
$$= \sup\{\mu^U_{\mathcal{F}}(f^{-1}(B)) \wedge \mu^U_B(y) : B \in \mathbf{D}^Y, B \subseteq \tilde{1}_{f(X)}\}$$
$$= \sup\{\mu^U_{\mathcal{F}}(A) \wedge f(\mu^U_A)(y) : A \in \mathbf{D}^X\}$$

and

$$\mu^U_{f(\bigcup \mathcal{F})}(y) = \mu^U_{\bigcup \mathcal{F}}(x) \quad [x \in f^{-1}(y)]$$
$$= \sup\{\mu^U_{\mathcal{F}}(A) \wedge \mu^U_A(x) : A \in \mathbf{D}^X\}$$
$$= \sup\{\mu^U_{\mathcal{F}}(A) \wedge f(\mu^U_A)(y) : A \in \mathbf{D}^X\}.$$

If $f^{-1}(y) = \phi$, since $f(\cup \mu^U_{\mathcal{F}})(y) = 0$ and $\mu^U_{\bigcup \mathcal{F}^f}(y) = 0$, we have $\mu^U_{\bigcup \mathcal{F}^f} = \mu^U_{f(\bigcup \mathcal{F})}$. Hence $f(\bigcup \mathcal{F}) = \bigcup \mathcal{F}^f$.
  (b) Similar to (a).

Theorem 6. Let $\mathcal{F}$ be an **IVFF** on Y and $f : X \to Y$ be a function. Then
(a) $f^{-1}(\bigcup \mathcal{F}) = \bigcup \mathcal{F}^{f^{-1}}$.
(b) $f^{-1}(\bigcap \mathcal{F}) = \bigcap \mathcal{F}^{f^{-1}}$.

Proof. (a) Let $\mathcal{N}(A) = \{B \in \mathbf{D}^Y : A = f^{-1}(B)\}$ for each $A \in \mathbf{D}^X$. Then $\mathbf{D}^X = \bigcup \{\mathcal{N}(A) : A \in \mathbf{D}^X\}$. Let $x \in X$. Then we have

$$\mu^L_{f(\bigcup \mathcal{F}^{f^{-1}})}(x) = \sup \{\mu^L_{\bigcup \mathcal{F}^{f^{-1}}}(A) \wedge \mu^L_A(x) : A \in \mathbf{D}^X\}$$
$$= \sup \{\sup \{\mu^L_{\mathcal{F}}(B) : A = f^{-1}(B), B \in \mathbf{D}^Y\} \wedge \mu^L_A(x) : A \in \mathbf{D}^X\}$$
$$= \sup \{\sup \{\mu^L_{\mathcal{F}}(B) \wedge \mu^L_A(x) : A = f^{-1}(B), B \in \mathbf{D}^Y\} : A \in \mathbf{D}^X\}$$
$$= \sup \{\sup \{\mu^L_{\mathcal{F}}(B) \wedge \mu^L_A(x) : B \in \mathcal{N}(A)\} : A \in \mathbf{D}^X\}$$
$$= \sup \{\sup \{\mu^L_{\mathcal{F}}(B) \wedge \mu^L_{f^{-1}(B)}(x) : B \in \mathcal{N}(A)\} : A \in \mathbf{D}^X\}$$
$$= \sup \{\mu^L_{\mathcal{F}}(B) \wedge \mu^L_B(f(x)) : B \in \mathbf{D}^Y\}$$
$$= \mu^L_{f^{-1}(\bigcup \mathcal{F})}(x).$$

Similarly, we have $\mu^U_{\bigcup \mathcal{F}^{f^{-1}}}(x) = \mu^U_{f^{-1}(\bigcup \mathcal{F})}(x)$. Hence $f^{-1}(\bigcup \mathcal{F}) = \bigcup \mathcal{F}^{f^{-1}}$.
(b) Similar to (a).

References

1. K. Atanassov, "Intuitionistic fuzzy sets", in: V. Sgurev, Ed., VII ITKR's Session, Sofia (June 1983 Central Sci. and Techn. Library, Bulg. Academy of Sciences, 1984).
2. K. Atanassov, "Intuitionistic fuzzy sets", *Fuzzy Sets and Systems*, Vol. 20, pp. 87-96, 1986.
3. D. Çoker and M. Demirci, "On fuzzy inclusion in the intuitionistic sense", *J. Fuzzy Math.*, Vol. 4, pp. 701-714, 1996.
4. M. B. Gorzalczany, "A method of inference in approximate reasoning based on interval-valued fuzzy sets", *Fuzzy Sets and Systems*, Vol. 21, pp. 1-17, 1987.
5. R. Lowen, "Fuzzy topological spaces and fuzzy compactness", *J. Math. Anal. Appl.*, Vol. 56, pp. 621-633, 1976.
6. T. K. Mondal and S.K. Samanta, "Topology of interval-valued fuzzy sets", *Indian J. pure appl. Math.*, Vol. 30, pp. 23-38, 1999.
7. P. V. Ramakrishnan and V. Lakshmana Gomathi Nayagam, "Hausdorff interval-valued fuzzy filters", *J. Korean Math. Soc.*, Vol. 39, pp. 137-148, 2002.
8. A. Šostak, "On a fuzzy topological structure", *Supp. Rend. Circ. Mat. Palermo (Ser. II)*, Vol. 11, pp. 89-103, 1985.
9. A. Šostak, "On compactness and connectedness degrees of fuzzy sets in fuzzy topological spaces", *General Topology and its Relations to Mordern Analysis and Algebra*, Helderman Verlag, Berlin, pp. 519-532, 1988.
10. L. A. Zadeh, "Fuzzy sets", *Inform. and Control*, Vol. 8, pp.338-353, 1965.
11. L. A. Zadeh, "The concept of a linguistic variable and its application to approximate reasoning -I", *Inform. Sci.*, Vol. 8, pp. 199-249, 1975.

Fuzzy Evaluation Based Multi-objective Reactive Power Optimization in Distribution Networks

Jiachuan Shi and Yutian Liu

School of Electrical Engineering,
Shangdong University, Jinan 250061, China
jc.shi@mail.sdu.edu.cn

Abstract. A fuzzy evaluation based multi-objective optimization model for reactive power optimization in power distribution networks is presented in this paper. The two objectives, reducing active power losses and improving voltage profiles, are evaluated by membership functions respectively, so that the objectives can be compared in a single scale. To facilitate the solving process, a compromised objective is formed by the weighted sum approach. The weights are decided according to the preferences and importance of the objectives. The reactive tabu search algorithm is employed to get global optimization solutions. Simulation results of a practical power distribution network, greatly improved voltage profiles and reduced power losses, demonstrated that the proposed method is effective.

1 Introduction

Reactive power optimization (RPO) aims to reduce active power losses by adjusting reactive power distribution. Generally, regulating means in distribution systems are mainly transformer tap-changers and shunt capacitors, which are both discrete variables. Therefore, it is formed as a constrained combinatorial optimization problem.

There have been several approaches, such as numerical programming, heuristic methods and AI-based methods, devised to solve the reactive power optimization problems.

Numerical programming algorithms solve the optimization problems by iterative techniques. The quadratic inter programming algorithm based on the active power losses reduction formulas are implemented in distribution networks capacitor allocation and real time control problems [1]. Numerical programming algorithms are effective in small-scale cases. Computational burden increases greatly when dealing with the practical large-scale networks.

Heuristic methods are based on the rules that are developed through intuition, experience and judgment. The heuristic methods produce near optimal solutions fast, but the results are not guaranteed to be optimal [2].

AI-based methods, including global optimization techniques [3,4] and fuzzy set theory [5-8], have been implemented in RPO. In [3], a GA-based two-stage algorithm, which combines GA and heuristic algorithm, is introduced to solve the constrained optimization problem. GA-SA-TS hybrid algorithms are introduced in [4]. Combining the advantages of individual algorithms, the local and global search strategies are joined to find better solutions within reasonable time. Fuzzy set theory is efficient in remedying for the uncertainty of data and optimization models. The applications of fuzzy set theory in distribution network shunt capacitor placement are introduced in [5]. Fuzzy set theory is combined with dynamic programming to solve distribution network voltage/var regulation problems in [6,7,8].

It should be noted that, most of the papers focus on reducing active power losses and/or energy losses, and the voltage profiles are considered in operating constraints. The single-objective optimization model often increase voltage to upper limits to reduce active power losses, which may be not acceptable in practice. Therefore, reducing active power losses may conflict with voltage profiles constraints.

A reactive power optimization method is presented in this paper. To solve the conflict between active power losses reduction and operating constraints, voltage profiles constraints are converted into objectives. That is to improve voltage profiles and reduce active power losses as much as possible. Therefore, the RPO problem is formed as a constrained combinatorial multi-objective optimization. Constraints include power flow, operating constraints, and adjusting frequencies. The membership functions are introduced to assess the objectives, so that the objectives can be compared without influenced by their original values. The weighted-sum approach is utilized to combine the objectives into one, and the constrained multi-objective optimization problem is solved by the Reactive Tabu Search (RTS) technology.

2 Problem Formulation

A fuzzy evaluation based multi-objective optimization model for the RPO problem in distribution networks is presented in this section.

The traditional RPO aims to minimize active power losses without voltage violations. The optimization process tends to minimize active power losses by increasing voltage to upper limit. Meanwhile the results are not acceptable in practice considering variations of loads and source node voltage. Furthermore, the voltage constraints are treated as "hard" constraints, that is, no voltage violation is allowed; it may be hard to find a feasible solution, especially considering two or more operating conditions.

In the new RPO model, the voltage profiles constraints are considered as an objective. That is, the objectives are to improve voltage profiles and to reduce active power losses as much as possible. The constraints include power flow and operating constraints. The objectives are estimated by fuzzy membership functions. The membership functions scale the objectives to the unit interval [0,1], so that the satisfactory of the objectives can be compared without influenced by their original values. The weighted-sum approach is introduced to combine the objectives. The weights imply the importance of objectives, that is, the preferences of objectives. After combination, the multi-objective optimization problem is transformed into a single-objective optimization problem. The optimal solution of the latter problem is a nondominated solution of the former one.

2.1 Voltage Profiles Assessment

To evaluate the voltage profiles, a trapezoid membership function shown as Fig.1 is introduced.

$$F_v(V_i) = \begin{cases} \dfrac{L_0^{upper} - V_i}{L_0^{upper} - L_1^{upper}} & (L_1^{upper} < V_i < L_0^{upper}) \\ \dfrac{V_i - L_0^{lower}}{L_1^{lower} - L_0^{lower}} & (L_0^{lower} < V_i < L_1^{lower}) \\ 1 & (L_1^{lower} \leq V_i \leq L_1^{upper}) \\ 0 & (others) \end{cases} \quad (1)$$

where V_i is the voltage of node i; L_0^{upper} and L_0^{lower} are unacceptable voltage limits; L_1^{upper} and L_1^{lower} are acceptable voltage margins.

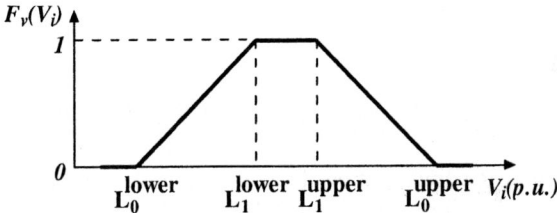

Fig. 1. Membership function for voltage profiles assessment

The membership function of the network voltage profiles is

$$F_{voltage} = \dfrac{\sum\limits_{i=1}^{N}\sum\limits_{p=1}^{3} F_v(V_{ip})}{3N} \quad (2)$$

where p represents the phase A, B and C; N is the number of secondary buses; V_{ip} is the phase voltage of node i.

To evaluate voltage eligibility of the network, three-phase voltage-eligibility ratio (TPVER) is defined as

$$TPVER = \dfrac{\sum\limits_{i=1}^{N}\sum\limits_{p=1}^{3} f(V_{ip})}{3N} \quad (3)$$

$$f(V_{ip}) = \begin{cases} 0 & (V_{ip} > V_{upper}) \text{ or } (V_{ip} < V_{lower}) \\ 1 & (V_{lower} \leq V_{ip} \leq V_{upper}) \end{cases}$$

where V_{lower} and V_{upper} are voltage limits, e.g. 0.9 and 1.07p.u., respectively.

2.2 Active Power Losses Assessment

To compare with the other objectives, the active power loss is also valued by a membership function. Different from the membership functions for voltage profiles, there is not a standard or limit for active power losses reduction.

In this paper, the membership function for active power losses is determined by two parameters, P_{l_ori} and P_{l_min}. The former one is the active power losses before optimization, and the membership value of P_{l_ori} is set 0.5. The latter one, P_{l_min}, is the minimum active power losses, and its membership value is 1.

The active power losses can be calculated by

$$P_{loss} = \sum_{i=1}^{N_b} I_i^2 R_i \tag{4}$$

where P_{loss} is the active power losses of the whole network; I_i is the current of branch i; R_i is the resistance of branch i; N_b is the number of branches.

The current can be divided into active power current I_{i_real}, and reactive power current I_{i_imag}. The reactive currents account for a portion of these losses.

$$I_i = I_{i\_real} + jI_{i\_imag}$$
$$P_{loss} = P_{loss\_real} + P_{loss\_imag} = \sum_{i=1}^{N} I_{i\_real}^2 R_i + \sum_{i=1}^{N} I_{i\_imag}^2 R_i \tag{5}$$

After reactive power compensation, branch current I_i is decreased to I_i', and the active power losses is reduced to P'_{loss}. While the losses caused by active power currents, P_{loss_real}, cannot be decreased by reactive power compensation.

$$I_i' = I_{i\_real} + j(I_{i\_imag} - I_{i\_comp})$$
$$P'_{loss} = P_{loss\_real} + P'_{loss\_imag} = P_{loss\_real} + \sum_{i=1}^{N}(I_{i\_imag} - I_{i\_comp})^2 R_i \tag{6}$$

In an ideal operating condition, enough shunt capacitors are installed and reactive power current is reduced to 0. Therefore, the P_{loss_imag} is 0 and the P_{loss_real} can be treated as the minimum active power losses, P_{l_min}.

The membership function of the active power losses shown as Fig. 2 is defined as

$$F_{loss}(P_l) = \begin{cases} 0 & P_l > (2P_{l\_ori} - P_{l\_min}) \\ 0.5 \times \dfrac{P_l - P_{l\_ori}}{P_{l\_min} - P_{l\_ori}} + 0.5 & P_{l\_min} \leq P_l \leq (2P_{l\_ori} - P_{l\_min}) \\ 1 & P_l < P_{l\_min} \end{cases} \tag{7}$$

where P_l is the active power losses.

The two parameters, P_{l_ori} and P_{l_min}, are determined by the system structure and original operating conditions. The assessment results of trial solutions under different operating conditions are comparable.

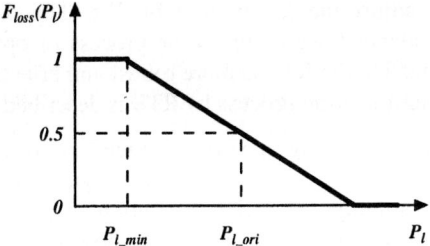

Fig. 2. Membership function for active power losses assessment

2.3 Objective Function

The multi-objective RPO can be formed as

$$\text{Max } F_{loss}$$
$$\text{Max } F_{voltage} \quad (8)$$
$$\text{s.t. } T_{k\min} \leq T_k \leq T_{k\max}$$
$$0 \leq Q_{cj} \leq Q_{cj\max}$$

where F_{loss} is the membership function for active power losses; $F_{voltage}$ is membership function for the voltage profiles; T_k is the ratio of transformer k; Q_{cj} is the capacity of capacitors at node j. The restriction of power flow is not listed here.

Normally, the solutions of a multi-objective optimization are a series of Plato solutions. A compromised objective function for operating condition m is represented as

$$F_{obj\_m} = (\alpha F_{loss} + \beta F_{voltage})_m \quad (9)$$

where F_{obj_m} is the objective function for operation condition m; α, β are weight coefficients for the two objectives.

Considering load variation, two or more different operating conditions are considered during the optimization process. The RPO problem is described as

$$\text{Max } F_{obj} = \sum_{m=1}^{M}(F_{obj\_m}) \quad (10)$$
$$\text{s.t. } T_{k\min} \leq T_k \leq T_{k\max}$$
$$0 \leq Q_{cj} \leq Q_{cj\max}$$

where M is the number of operating conditions.

3 Solution Method

Reactive Tabu Search algorithm (RTS) [9] is utilized to solve the constrained combinational optimization problem. The two searching strategies, diversification and intensification, are presented in many heuristic algorithms. In traditional tabu search, the length of tabu list is determined by experiences. The feedback mechanisms are introduced in RTS to adjust the length of tabu list. The diversification and the intensification can be balanced according to the process of optimization. Therefore, compared with traditional TS, the RTS is more robust and effective.

The reactive power optimization process by RTS is described as the following:

i) Read in initial data, including impedance of feeders, loads, regulation variables and inequality constraint conditions. Code the regulation variables.
ii) Generate initial solution. Set the regulation variables randomly without breaking the constraint conditions, including power flow restriction. Calculate the objective function $f(X)$, and set best solution vector X_{opt} as X.
iii) Generate a group of trail solutions, $X_1, X_2, \ldots, X_k$, by "move" from X. Check their feasibility and discard the unfeasible ones.
iv) Get the corresponding values by searching Hash Table, in which all the visited configurations are stored. If the trial solutions are not available from the Hash Table, calculate the objective function, $f(X_1), f(X_2), \ldots, f(X_k)$, and keep them in the Hash Table.
v) Search neighborhoods. Get the best one, X^*, from the trail solutions. Update X with X^*, if X^* is not in the Tabu list, or X^* fits aspiration criteria. Try the next solution, if the former one cannot update X.
vi) Update Tabu List. Push the record of reversed move into a FIFO (First-In-First-Out) stack, the Tabu List.
vii) Update X_{opt} with X^*, if $f(X^*)$ is better than $f(X_{opt})$.
viii) Update the length of Tabu list. If most solutions are gotten from the Hash Table, the length of Tabu list increases till an upper limit to escape from local bests. If less solution is gotten from the Hash Table, the length of Tabu list decreases till a lower limit to reduce the constraints of Tabu list.
ix) Terminate condition. Stop optimization and output results if $f(X_{opt})$ has not been improved for several iterations or number of maximum iteration is meet. If terminate conditions are not fit, iteration should continue from step iii).

4 Test Results

The effectiveness of the proposed method is verified by the application to a practical distribution system shown as Fig.3 in Jinan, China. Data are sampled by the distribution SCADA system every 15 minutes.

4.1 Operating Conditions Selection

Active power losses and voltage profiles are closely related with loads, which vary continually during different time of a day and different seasons of a year. If the voltage profiles in peak load conditions and valley load conditions are eligible, those during this period are believed to be acceptable. Consequently, the peak and valley load conditions cover the daily varying loads, and two typical days in spring and summer represent load variations in a year. Therefore, the loads variation during a long period can be covered by a few operating conditions and the optimization results fit a long-period voltage profiles variation.

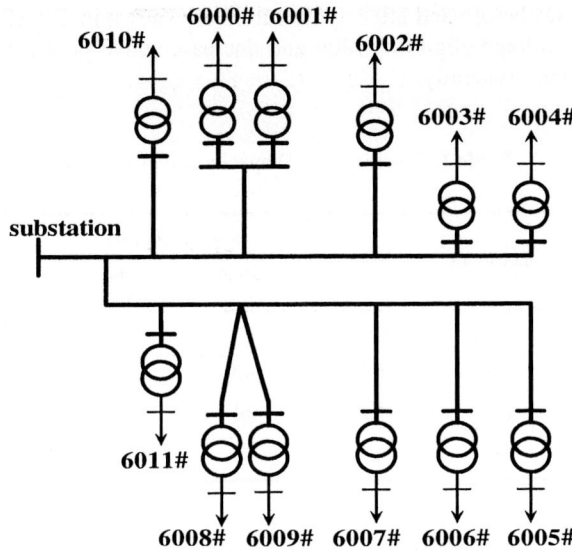

Fig. 3. Structure of case distribution network

Operating conditions considered in optimization is closely related with calculation speed. Two typical operating conditions, the maximum loads and the minimum loads in a year, are utilized to "cover" voltage and loads variations in optimization. The former one is taken from peak load in a typical day in summer, and the latter one is valley load condition in spring.

4.2 Optimization Results

Three-phase power flow is calculated by the forward-backward sweeping [10]. All 13 three-phase 10kV/0.4kV transformers in this network have no-load tap-changers (NLTC), which are 1±2.5%×2, and the original positions are 3, that is, the ratios are 1 p.u. The phase-to-neutral voltage upper and lower limits for 0.4kV distribution networks are 1+7% (235.4V) and 1-10% (198V) respectively. In Eq.1, acceptable upper and lower voltage margins are 1±2%, and unacceptable limits are 1±20%. In Eq.9, α is 3 and β is 1, which means the active power losses reduction is more emphasized than the voltage profiles improvement.

Four nodes with biggest sensitivity values are pre-selected to be compensated, shown in Table1, all of which are under heavier loads. Furthermore, 6010# is the smallest capacity one (250kVA), whose impedance is bigger than that of the others (315kVA or 400kVA). Different regulation means are compared. In scheme 1, only non-load tap-changer (NLTC) and fixed capacitors (FC) are used to regulate voltage profiles. In scheme 2, NLTC, FC and switchable capacitors (SC) are utilized. After optimization, tap changers of all the transformers are regulated from position 3 to position 1, that is, to decrease the ratios from 1 to 0.95 in p.u. The voltage profiles and active power losses before and after optimization are shown in Table2. Both schemes can improve the voltage-eligibility ratio and decrease active power losses under two operating conditions evidently.

Table 1. Sensitivity and capacity at 0.4kV sides

Locations	SC (w/kVar)	Capacity (kVar)	
		Scheme 1 (FC)	Scheme 2 (FC+SC)
6010	10.9	40	40+80
6007	8.4	40	40+80
6000	7.2	40	40+80
6008	6.1	60	60+80

Table 2. Voltage profiles and active power losses

Load Conditions	Summer Max		Spring Min	
Items	TPVER (%)	Power loss (%)	TPVER (%)	Power loss (%)
Before optimization	94.87	2.38	0	6.39
Scheme 1 (NLTC, FC)	100	2.24	100	6.36
Scheme 2 (NLTC, SC, FC)	100	2.13	100	6.36

5 Conclusions

The reactive power optimization problem is formed as a multi-objective optimization problem, which aims to improve voltage profiles and decrease active power losses. In this way, the conflict between active power losses reduction and voltage constraints can be solved compromise.

The fuzzy evaluation based weighted-sum approach is effective in solving multi-objective optimizations. The membership functions scale the objectives to the unit interval [0,1]. Therefore, the satisfaction of different objectives can be compared fairly. The fuzzy evaluation strategy releases the influence of original values.

References

1. Jin-Cheng Wang, et al. Capacitor placement and real time control in large-scale unbalanced distribution systems: Loss reduction formula, problem formulation, solution methodology and mathematical justification. IEEE Trans. on Power Delivery 1997; 12(2): 953-58.
2. H. N. Ng, M. M. A. Salama, A. Y. Chikhani. Classification of capacitor allocation techniques. IEEE Transactions on Power Delivery 2000; 15(1): 387-92.
3. KarenNan Miu, Hsiao-Dong Chiang, Gary Darling. Capacitor placement, replacement and control in large-scale distribution systems by a GA-based two-stage algorithm. IEEE Transactions on Power Systems 1997; 12(3): 1160-66.
4. Yutian Liu, Li Ma, Jianjun Zhang. Reactive power optimization by GA/SA/TS combined algorithms, Int. J. of Electric Power & Energy Systems 2002; 24(9): 765-69
5. S.F. Mekhamer, et al. Application of fuzzy logic for reactive-power compensation of radial distribution feeders; Transactions on power systems 2003; 18(1): 206-13
6. Yutian Liu, Peng Zhang, Xizhao Qiu. Optimal voltage/var control in distribution systems, Int. J. of Electric Power & Energy Systems 2002; 24(4): 271-76
7. Feng-Chang Lu, Yuan-Yih Hsu. Fuzzy dynamic programming approach to reactive power/voltage control in a distribution substation. IEEE Transactions on Power Systems 1997; 12(2): 681–88
8. Andrija T. Sarić, Milan S. Ćalović, Vladimir C. Strezoski. Fuzzy multi objective algorithm for multiple solution of distribution systems voltage control. Int. J. of Electrical Power & Energy Systems 2003; 25(2): 145-53.
9. V. J. Rayward-Smith, I.H. Osman, C. R. Reeves, G. D. Smith, Modern Heuristic search methods, John Wiley and Sons Ltd, 1996
10. Whei-Min Lin, et al, Three-phase unbalanced distribution power flow solutions with minimum data preparation. IEEE Transactions on Power Systems 1999; 14(3): 1178-83

Note on Interval-Valued Fuzzy Set

Wenyi Zeng[1,2] and Yu Shi[1]

[1] Department of Mathematics, Beijing Normal University,
Beijing, 100875, P.R. China
[2] Department of Medical Epidemiology and Biostatistics,
Karolinska Institutet, Stockholm, SE171 77, Sweden
zengwy@bnu.edu.cn
shi_yu_bnu@hotmail.com

Abstract. In this note, we introduce the concept of cut set of interval-valued fuzzy set and discuss some properties of cut set of interval-valued fuzzy set, propose three decomposition theorems of interval-valued fuzzy set and investigate some properties of cut set of interval-valued fuzzy set and mapping H in detail. These works can be used in setting up the basic theory of interval-valued fuzzy set.

1 Introduction

The theory of fuzzy set, pioneered by Zadeh[11], has achieved many successful applications in practice. As a generalization of fuzzy set, Zadeh[12, 13, 14] introduced the concept of interval-valued fuzzy set, after that, some authors investigated the topic and obtained some meaningful conclusions. For example, Biswas[2] and Li[7] in interval-valued fuzzy subgroup, Mondal[8] in interval-valued fuzzy topology, Bustince etc.[3], Chen etc.[4], Yuan etc.[10], Arnould[1] and Gorzalczany[6] in approximate reasoning of interval-valued fuzzy set, Bustince etc.[3] and Deschrijver etc.[5] in interval-valued fuzzy relations and implication, Turksen[9] in normal forms of interval-valued fuzzy set and so on. These works show the importance of interval-valued fuzzy set.

Just like that decomposition theorems of fuzzy set played an important role in the fuzzy set theory, it helped us to develop many branches such as fuzzy algebra, fuzzy measure and integral, fuzzy analysis, fuzzy decision making and so on. In this paper, our aim is to investigate decomposition theorems of interval-valued fuzzy set in order that we can do some preparation for setting up the basic theory of interval-valued fuzzy set and develop its some relative branches. In this paper, our work is organized as follows. In the section 2, we introduce the concept of cut set of interval-valued fuzzy set and discuss some properties of cut set of interval-valued fuzzy set. In section 3, we propose three decomposition theorems of interval-valued fuzzy set and give some properties of cut set of interval-valued fuzzy set and mapping H. The final is conclusion.

2 Preliminaries

In this section, we introduce the concept of cut set of interval-valued fuzzy set and give some properties of cut set of interval-valued fuzzy set. Let $I = [0,1]$ and $[I]$ be the set of all closed subintervals of the interval $[0,1]$. Then, according to Zadeh's extension principle[11], we can popularize these operations such as $\vee, \wedge$ and c to $[I]$, thus, $([I], \vee, \wedge, c)$ is a complete lattice with a minimal element $\bar{0} = [0,0]$ and a maximal element $\bar{1} = [1,1]$. Furthermore, let $\bar{a} = [a^-, a^+], \bar{b} = [b^-, b^+]$, then we have, $\bar{a} = \bar{b} \Longleftrightarrow a^- = b^-, a^+ = b^+, \bar{a} \le \bar{b} \Longleftrightarrow a^- \le b^-, a^+ \le b^+$ and $\bar{a} < \bar{b} \Longleftrightarrow \bar{a} \le \bar{b}$ and $\bar{a} \ne \bar{b}$. Considering $[I]$ is dense, therefore, $([I], \vee, \wedge, c)$ is a superior soft algebra.

Suppose X be a universal set, we call a mapping: $A : X \longrightarrow [I]$ an interval-valued fuzzy set in X. Let IVFSs stands for the set of all interval-valued fuzzy sets in X. For every $A \in$ IVFSs and $x \in X$, $A(x) = [A^-(x), A^+(x)]$ is called the degree of membership of an element x to A, then fuzzy sets $A^- : X \to [0,1]$ and $A^+ : X \to [0,1]$ are called a low fuzzy set of A and a upper fuzzy set of A, respectively. For simplicity, we denote $A = [A^-, A^+]$, and $\mathcal{F}(X)$ and $\mathcal{P}(X)$ as the set of all fuzzy sets and crisp sets in X, respectively. Therefore, some operations such as $\cup, \cap, c$ can be introduced into IVFSs, thus, (IVFSs, $\cup, \cap, c$) is a superior soft algebra.

Definition 1. Let $A \in$ IVFSs, $\lambda = [\lambda_1, \lambda_2] \in [I]$, we order:

$$A_\lambda^{(1,1)} = A_{[\lambda_1,\lambda_2]}^{(1,1)} = \{x \in X | A^-(x) \ge \lambda_1, A^+(x) \ge \lambda_2\}$$
$$A_\lambda^{(1,2)} = A_{[\lambda_1,\lambda_2]}^{(1,2)} = \{x \in X | A^-(x) \ge \lambda_1, A^+(x) > \lambda_2\}$$
$$A_\lambda^{(2,1)} = A_{[\lambda_1,\lambda_2]}^{(2,1)} = \{x \in X | A^-(x) > \lambda_1, A^+(x) \ge \lambda_2\}$$
$$A_\lambda^{(2,2)} = A_{[\lambda_1,\lambda_2]}^{(2,2)} = \{x \in X | A^-(x) > \lambda_1, A^+(x) > \lambda_2\}$$
$$A_\lambda^{(3,3)} = A_{[\lambda_1,\lambda_2]}^{(3,3)} = \{x \in X | A^-(x) \le \lambda_1 \text{ or } A^+(x) \le \lambda_2\}$$
$$A_\lambda^{(3,4)} = A_{[\lambda_1,\lambda_2]}^{(3,4)} = \{x \in X | A^-(x) \le \lambda_1 \text{ or } A^+(x) < \lambda_2\}$$
$$A_\lambda^{(4,3)} = A_{[\lambda_1,\lambda_2]}^{(4,3)} = \{x \in X | A^-(x) < \lambda_1 \text{ or } A^+(x) \le \lambda_2\}$$
$$A_\lambda^{(4,4)} = A_{[\lambda_1,\lambda_2]}^{(4,4)} = \{x \in X | A^-(x) < \lambda_1 \text{ or } A^+(x) < \lambda_2\}$$

$A_{[\lambda_1,\lambda_2]}^{(i,j)}$ is called the (i,j)th (λ_1,λ_2)-(double value) cut set of interval-valued fuzzy set A. Specially, if $\lambda = \lambda_1 = \lambda_2$, $A_\lambda^{(i,j)} = A_{[\lambda,\lambda]}^{(i,j)}$ is called the (i,j)th λ-(single value) cut set of interval-valued fuzzy set A.

For $A \in \mathcal{F}(X)$, $\lambda \in [0,1]$, we denote $A_\lambda^1 = \{x \in X | A(x) \ge \lambda\}$, $A_\lambda^2 = \{x \in X | A(x) > \lambda\}$, $A_\lambda^3 = \{x \in X | A(x) \le \lambda\}$, $A_\lambda^4 = \{x \in X | A(x) < \lambda\}$.

Therefore, we have the following properties.

Property 1.

$$A_{[\lambda_1,\lambda_2]}^{(i,j)} = (A^-)_{\lambda_1}^i \cap (A^+)_{\lambda_2}^j, \quad i,j = 1,2$$

$$A_{[\lambda_1,\lambda_2]}^{(i,j)} = (A^-)_{\lambda_1}^i \cup (A^+)_{\lambda_2}^j, \quad i,j = 3,4$$

Property 2.

$$A^{(2,2)}_{[\lambda_1,\lambda_2]} \subseteq A^{(1,2)}_{[\lambda_1,\lambda_2]} \subseteq A^{(1,1)}_{[\lambda_1,\lambda_2]}, \quad A^{(2,2)}_{[\lambda_1,\lambda_2]} \subseteq A^{(2,1)}_{[\lambda_1,\lambda_2]} \subseteq A^{(1,1)}_{[\lambda_1,\lambda_2]}$$

$$A^{(4,4)}_{[\lambda_1,\lambda_2]} \subseteq A^{(3,4)}_{[\lambda_1,\lambda_2]} \subseteq A^{(3,3)}_{[\lambda_1,\lambda_2]}, \quad A^{(4,4)}_{[\lambda_1,\lambda_2]} \subseteq A^{(4,3)}_{[\lambda_1,\lambda_2]} \subseteq A^{(3,3)}_{[\lambda_1,\lambda_2]}$$

Property 3. For $\lambda_1 = [\lambda_1^1, \lambda_1^2], \lambda_2 = [\lambda_2^1, \lambda_2^2] \in [I]$ and $\lambda_1^1 < \lambda_2^1, \lambda_1^2 < \lambda_2^2$, then $A^{(2,2)}_{\lambda_1} \supseteq A^{(1,1)}_{\lambda_2}, A^{(3,3)}_{\lambda_1} \subseteq A^{(4,4)}_{\lambda_2}$.

Property 4. For $\lambda = [\lambda_1, \lambda_2]$, then

$$\left(A^{(1,1)}_\lambda\right)^c = A^{(4,4)}_\lambda, \quad \left(A^{(1,2)}_\lambda\right)^c = A^{(4,3)}_\lambda$$
$$\left(A^{(2,1)}_\lambda\right)^c = A^{(3,4)}_\lambda, \quad \left(A^{(2,2)}_\lambda\right)^c = A^{(3,3)}_\lambda$$

Definition 2. For $[\lambda_1, \lambda_2] \in [I]$ and $A \in$ IVFSs, we order $[\lambda_1, \lambda_2] \cdot A, [\lambda_1, \lambda_2] * A \in$ IVFSs and their membership functions are defined as following.

$$([\lambda_1, \lambda_2] \cdot A)(x) \triangleq [\lambda_1 \wedge A^-(x), \lambda_2 \wedge A^+(x)]$$
$$([\lambda_1, \lambda_2] * A)(x) \triangleq [\lambda_1 \vee A^-(x), \lambda_2 \vee A^+(x)]$$

Property 5. For $A, B \in$ IVFSs and $\lambda, \lambda_1, \lambda_2 \in [I]$, then we have,

(1) $\lambda_1 \leqslant \lambda_2 \Rightarrow \lambda_1 \cdot A \subseteq \lambda_2 \cdot A, \quad \lambda_1 * A \subseteq \lambda_2 * A$
(2) $A \subseteq B \Rightarrow \lambda \cdot A \subseteq \lambda \cdot B, \quad \lambda * A \subseteq \lambda * B$

3 Decomposition Theorem

In this section, we will give three decomposition theorems of interval-valued fuzzy set and some properties of cut set of interval-valued fuzzy set and mapping H.

Theorem 1. $A = \bigcup_{[\lambda_1,\lambda_2] \in [I]} [\lambda_1, \lambda_2] \cdot A^{(1,j)}_{[\lambda_1,\lambda_2]}, \quad j = 1, 2$

Theorem 2. $A = \left(\bigcap_{\lambda \in [I]} \lambda * A^{(4,i)}_{\lambda^c}\right)^c, \quad i = 3, 4$.

Supposed H be a mapping from $[I]$ to $\mathcal{P}(X)$, $H : [I] \rightarrow \mathcal{P}(X)$, for every $\lambda = [\lambda_1, \lambda_2] \in [I]$, we have $H(\lambda) \in \mathcal{P}(X)$. Obviously, cut set of interval-valued fuzzy set, $A_\lambda \in \mathcal{P}(X)$, it means that mapping H indeed exists. Based on Theorem 1 and Theorem 2, we have the following theorem in general.

Theorem 3. For $A \in$ IVFSs, we have:

(1) If there exists $A^{(1,2)}_{[\lambda_1,\lambda_2]} \subseteq H([\lambda_1, \lambda_2]) \subseteq A^{(1,1)}_{[\lambda_1,\lambda_2]}$, then $A = \bigcup_{[\lambda_1,\lambda_2] \in [I]} [\lambda_1, \lambda_2] \cdot H([\lambda_1, \lambda_2])$.

(2) If there exists $A^{(4,4)}_{[\lambda_1,\lambda_2]} \subseteq H([\lambda_1, \lambda_2]) \subseteq A^{(4,3)}_{[\lambda_1,\lambda_2]}$, then $A = \left(\bigcap_{[\lambda_1,\lambda_2] \in [I]} [\lambda_1, \lambda_2] * H([\lambda_1, \lambda_2]^c)\right)^c$.

In the following, we investigate the relation of cut set of interval-valued fuzzy set and mapping H. For simplicity, we denote $\lambda_1 = [\lambda_1^1, \lambda_1^2], \lambda_2 = [\lambda_2^1, \lambda_2^2]$ and $\lambda_1, \lambda_2 \in [I]$, then we give some properties of cut set of interval-valued fuzzy set and mapping H.

Property 6. If there exists $A_{[\lambda_1,\lambda_2]}^{(2,2)} \subseteq H([\lambda_1,\lambda_2]) \subseteq A_{[\lambda_1,\lambda_2]}^{(1,1)}$, then we have:

(a) $\lambda_1, \lambda_2 \in [I]$ and $\lambda_1^i < \lambda_2^i, i = 1, 2 \Rightarrow H(\lambda_1) \supseteq H(\lambda_2)$,
(b) $A_{[\lambda_1,\lambda_2]}^{(1,1)} = \bigcap_{\alpha_1 < \lambda_1} \bigcap_{\alpha_2 < \lambda_2} H([\alpha_1,\alpha_2]), \quad \lambda_1 \neq 0 \text{ and } \lambda_2 \neq 0$,
(c) $A_{[\lambda_1,\lambda_2]}^{(2,2)} = \bigcup_{\alpha_1 > \lambda_1} \bigcup_{\alpha_2 > \lambda_2} H([\alpha_1,\alpha_2]), \quad \lambda_1 \neq 1 \text{ and } \lambda_2 \neq 1$.

Property 7. If there exists $A_{[\lambda_1,\lambda_2]}^{(2,2)} \subseteq H([\lambda_1,\lambda_2]) \subseteq A_{[\lambda_1,\lambda_2]}^{(2,1)}$, then we have:

(a) $\lambda_1, \lambda_2 \in [I]$ and $\lambda_1^i < \lambda_2^i, i = 1, 2 \Rightarrow H(\lambda_1) \supseteq H(\lambda_2)$,
(b) $A_{[\lambda_1,\lambda_2]}^{(2,1)} = \bigcap_{\alpha_2 < \lambda_2} H([\lambda_1,\alpha_2]), \quad \lambda_2 \neq 0$,
(c) $A_{[\lambda_1,\lambda_2]}^{(2,2)} = \bigcup_{\alpha_2 > \lambda_2} H([\lambda_1,\alpha_2]), \quad \lambda_2 \neq 1$.

Property 8. If there exists $A_{[\lambda_1,\lambda_2]}^{(2,2)} \subseteq H([\lambda_1,\lambda_2]) \subseteq A_{[\lambda_1,\lambda_2]}^{(1,2)}$, then we have:

(a) $\lambda_1, \lambda_2 \in [I]$ and $\lambda_1^i < \lambda_2^i, i = 1, 2 \Rightarrow H(\lambda_1) \supseteq H(\lambda_2)$,
(b) $A_{[\lambda_1,\lambda_2]}^{(1,2)} = \bigcap_{\alpha_1 < \lambda_1} H([\alpha_1,\lambda_2]), \quad \lambda_1 \neq 0$,
(c) $A_{[\lambda_1,\lambda_2]}^{(2,2)} = \bigcup_{\alpha_1 > \lambda_1} H([\alpha_1,\lambda_2]), \quad \lambda_1 \neq 1$.

Property 9. If there exists $A_{[\lambda_1,\lambda_2]}^{(2,1)} \subseteq H([\lambda_1,\lambda_2]) \subseteq A_{[\lambda_1,\lambda_2]}^{(1,1)}$, then we have:

(a) $\lambda_1, \lambda_2 \in [I]$ and $\lambda_1^i < \lambda_2^i, i = 1, 2 \Rightarrow H(\lambda_1) \supseteq H(\lambda_2)$,
(b) $A_{[\lambda_1,\lambda_2]}^{(1,1)} = \bigcap_{\alpha_1 < \lambda_1} H([\alpha_1,\lambda_2]), \quad \lambda_1 \neq 0$,
(c) $A_{[\lambda_1,\lambda_2]}^{(2,1)} = \bigcup_{\alpha_1 > \lambda_1} H([\alpha_1,\lambda_2]), \quad \lambda_1 \neq 1$.

Property 10. If there exists $A_{[\lambda_1,\lambda_2]}^{(1,2)} \subseteq H([\lambda_1,\lambda_2]) \subseteq A_{[\lambda_1,\lambda_2]}^{(1,1)}$, then we have:

(a) $\lambda_1, \lambda_2 \in [I]$ and $\lambda_1^i < \lambda_2^i, i = 1, 2 \Rightarrow H(\lambda_1) \supseteq H(\lambda_2)$,
(b) $A_{[\lambda_1,\lambda_2]}^{(1,1)} = \bigcap_{\alpha_2 < \lambda_2} H([\lambda_1,\alpha_2]), \quad \lambda_2 \neq 0$,
(c) $A_{[\lambda_1,\lambda_2]}^{(1,2)} = \bigcup_{\alpha_2 > \lambda_2} H([\lambda_1,\alpha_2]), \quad \lambda_2 \neq 1$.

Property 11. If there exists $A^{(4,4)}_{[\lambda_1,\lambda_2]} \subseteq H([\lambda_1,\lambda_2]) \subseteq A^{(3,3)}_{[\lambda_1,\lambda_2]}$, then we have:

(a) $\lambda_1, \lambda_2 \in [I]$ and $\lambda_1^i < \lambda_2^i, i=1,2 \Rightarrow H(\lambda_1) \subseteq H(\lambda_2)$,

(b) $A^{(4,4)}_{[\lambda_1,\lambda_2]} = \bigcup\limits_{\alpha_1 < \lambda_1} \bigcup\limits_{\alpha_2 < \lambda_2} H([\alpha_1,\alpha_2])$, $\lambda_1 \neq 0$ and $\lambda_2 \neq 0$,

(c) $A^{(3,3)}_{[\lambda_1,\lambda_2]} = \bigcap\limits_{\alpha_1 > \lambda_1} \bigcap\limits_{\alpha_2 > \lambda_2} H([\alpha_1,\alpha_2])$, $\lambda_1 \neq 1$ and $\lambda_2 \neq 1$

Property 12. If there exists $A^{(4,4)}_{[\lambda_1,\lambda_2]} \subseteq H([\lambda_1,\lambda_2]) \subseteq A^{(4,3)}_{[\lambda_1,\lambda_2]}$, then we have:

(a) $\lambda_1, \lambda_2 \in [I]$ and $\lambda_1^i < \lambda_2^i, i=1,2 \Rightarrow H(\lambda_1) \subseteq H(\lambda_2)$,

(b) $A^{(4,4)}_{[\lambda_1,\lambda_2]} = \bigcup\limits_{\alpha_2 < \lambda_2} H([\lambda_1,\alpha_2])$, $\lambda_2 \neq 0$,

(c) $A^{(4,3)}_{[\lambda_1,\lambda_2]} = \bigcap\limits_{\alpha_2 > \lambda_2} H([\lambda_1,\alpha_2])$, $\lambda_2 \neq 1$

Property 13. If there exists $A^{(4,4)}_{[\lambda_1,\lambda_2]} \subseteq H([\lambda_1,\lambda_2]) \subseteq A^{(3,4)}_{[\lambda_1,\lambda_2]}$, then we have:

(a) $\lambda_1, \lambda_2 \in [I]$ and $\lambda_1^i < \lambda_2^i, i=1,2 \Rightarrow H(\lambda_1) \subseteq H(\lambda_2)$,

(b) $A^{(4,4)}_{[\lambda_1,\lambda_2]} = \bigcup\limits_{\alpha_1 < \lambda_1} H([\alpha_1,\lambda_2])$, $\lambda_1 \neq 0$,

(c) $A^{(3,4)}_{[\lambda_1,\lambda_2]} = \bigcap\limits_{\alpha_1 > \lambda_1} H([\alpha_1,\lambda_2])$, $\lambda_1 \neq 1$

Property 14. If there exists $A^{(3,4)}_{[\lambda_1,\lambda_2]} \subseteq H([\lambda_1,\lambda_2]) \subseteq A^{(3,3)}_{[\lambda_1,\lambda_2]}$, then we have:

(a) $\lambda_1, \lambda_2 \in [I]$ and $\lambda_1^i < \lambda_2^i, i=1,2 \Rightarrow H(\lambda_1) \subseteq H(\lambda_2)$,

(b) $A^{(3,4)}_{[\lambda_1,\lambda_2]} = \bigcup\limits_{\alpha_2 < \lambda_2} H([\lambda_1,\alpha_2])$, $\lambda_2 \neq 0$,

(c) $A^{(3,3)}_{[\lambda_1,\lambda_2]} = \bigcap\limits_{\alpha_2 > \lambda_2} H([\lambda_1,\alpha_2])$, $\lambda_2 \neq 1$

Property 15. If there exists $A^{(4,3)}_{[\lambda_1,\lambda_2]} \subseteq H([\lambda_1,\lambda_2]) \subseteq A^{(3,3)}_{[\lambda_1,\lambda_2]}$, then we have:

(a) $\lambda_1, \lambda_2 \in [I]$ and $\lambda_1^i < \lambda_2^i, i=1,2 \Rightarrow H(\lambda_1) \subseteq H(\lambda_2)$,

(b) $A^{(4,3)}_{[\lambda_1,\lambda_2]} = \bigcup\limits_{\alpha_1 < \lambda_1} H([\alpha_1,\lambda_2])$, $\lambda_1 \neq 0$,

(c) $A^{(3,3)}_{[\lambda_1,\lambda_2]} = \bigcap\limits_{\alpha_1 > \lambda_1} H([\alpha_1,\lambda_2])$, $\lambda_1 \neq 1$

4 Conclusion

In this paper, we introduce the concept of cut set of interval-valued fuzzy set and discuss some properties of cut set of interval-valued fuzzy set, propose three decomposition theorems of interval-valued fuzzy set and investigate some properties of cut set of interval-valued fuzzy set and mapping H in detail. These works can be used in setting up the basic theory of interval-valued fuzzy set. The discussion of representation theorems of interval-valued fuzzy set will be studied in other papers.

References

[1] Arnould, T., Tano, S.: Interval valued fuzzy backward reasoning, IEEE Trans, Fuzzy Syst. 3(4)(1995), 425-437
[2] Biswas, R.: Rosenfeld's fuzzy subgroups with interval-valued membership functions, Fuzzy Sets and Systems 63(1994), 87-90
[3] Bustince, H., Burillo, P.: Mathematical analysis of interval-valued fuzzy relations: Application to approximate reasoning, Fuzzy Sets and Systems 113(2000), 205-219
[4] Chen, S.M., Hsiao, W.H.: Bidirectional approximate reasoning for rule-based systems using interval valued fuzzy sets, Fuzzy Sets and Systems 113(2000), 185-203
[5] Deschrijver, G., Kerre, E.E.: On the relationship between some extensions of fuzzy set theory, Fuzzy Sets and Systems 133(2003), 227-235
[6] Gorzalczany, M.B.: A method of inference in approximate reasoning based on interval-valued fuzzy sets, Fuzzy Setss and Systems 21(1987), 1-17
[7] Li, X.P., Wang, G.J.: The S_H-interval-valued fuzzy subgroup, Fuzzy Sets and Systems 112(2000), 319-325
[8] Mondal, T.K., Samanta, S.K.: Topology of interval-valued fuzzy sets, Indian J. Pure Appl. Math., 30(1999), 23-38
[9] Turksen, I.B.: Interval-valued fuzzy sets based on normal forms, Fuzzy Sets and Systems 20(1986), 191-210
[10] Yuan, B., Pan, Y., Wu, W.M.: On normal form based on interval-valued fuzzy sets and their applications to approximate reasoning, Internat. J. General Systems 23(1995), 241-254
[11] Zadeh, L.A.: Fuzzy sets, Infor. and Contr. 8(1965), 338-353
[12] Zadeh, L.A.: The concept of a linguistic variable and its application to approximate reasoning, Part 1, Infor. Sci. 8(1975), 199-249
[13] Zadeh, L.A.: The concept of a linguistic variable and its application to approximate reasoning, Part 2, Infor. Sci. 8(1975), 301-357
[14] Zadeh, L.A.: The concept of a linguistic variable and its application to approximate reasoning, Part 3, Infor. Sci. 9(1975), 43-80

Knowledge Structuring and Evaluation Based on Grey Theory

Chen Huang and Yushun Fan

CIMS ERC, Department of Automation, Tsinghua University, Beijing 100084, PR China
hc01@mails.tsinghua.edu.cn
fanyus@tsinghua.edu.cn

Abstract. It is important nowadays to provide guidance for individuals or organizations to improve their knowledge according to their objectives, especially in the case of incomplete cognition. Based on grey system theory, a knowledge architecture which consists of grey elements including knowledge fields and knowledge units is built. The method to calculate the weightiness of each knowledge unit, with regard to the user's objectives, is detailed. The knowledge possessed by the user is also evaluated with grey clustering method by whitenization weight function.

1 Introduction

Knowledge is a very important factor to knowledge workers and knowledge-intensive organizations. Guidance, which points out what the most important knowledge is according to user's objectives and gives the evaluation of current knowledge possessions, is required to improve one's knowledge more efficiently. However, the existing knowledge management technology[1, 2] can not resolve this problem since the cognition and evaluation of knowledge is indefinite and difficult.

To resolve this problem, this paper introduces grey theory[3, 4] into the knowledge management. The grey theory was founded by Prof. Julong Deng in 1982 and caused intense attention because of its original thought and broad applicability. Grey means the information is incomplete. This theory intended to use extremely limited known information to forecast unknown information. By far, it has already developed a set of technologies including system modeling, analysis, evaluation, optimization, forecasting and decision-making, and has been applied in many fields such as agriculture, environment and mechanical engineering.

To knowledge workers and knowledge-intensive organizations, the objectives are usually indefinite and changing, while the cognition of its own knowledge is incomplete. Though the cognition will be continually improved along with the accumulation of knowledge, the characteristic of grey will always exist. Therefore, it is more suitable to use grey system theory, instead of other traditional theories and methods, to model and evaluate one's knowledge.

In this paper, the knowledge architecture based on grey theory is given firstly. Then it provides a method to calculate the weightiness of each knowledge unit in the architecture, with regard to the user's objectives. Finally it presents how to evaluate the knowledge with grey clustering method.

2 Knowledge Architecture Based on Grey Theory

Since the pursuing of knowledge should be objective-driven, in this knowledge architecture, we should first define the system objective, which could change with time. The knowledge architecture mainly consists of knowledge fields (KF) and knowledge units (KU), both of which are grey elements. A knowledge field is the field that the knowledge belongs to. Each knowledge field could be divided into many sub-fields. The knowledge fields that can not be divided any more are called knowledge units. The knowledge architecture defined above is shown as Fig. 1. Here KF refers to knowledge field, while KU refers to knowledge unit.

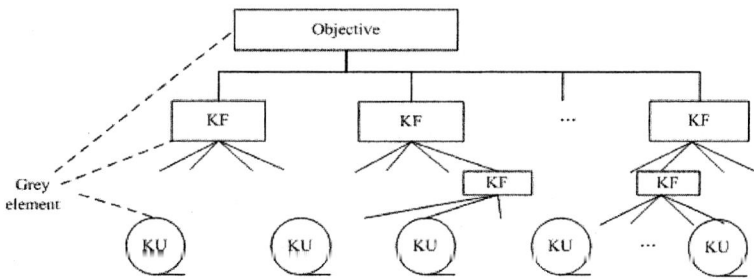

Fig. 1. Knowledge architecture

Because of the incomplete cognition, each element in this knowledge architecture is a grey element, which means the element has not been cognized completely. We use grey degree to represent the extent of grey for grey elements. Grey degree describes the grey extent of a knowledge unit. The value range is $(0,1]$. 0 represents completely known, while 1 represents totally unknown.

According to grey theory, the knowledge units will never turn 'white', in other words, the grey degree will never reach 0. But users can define that when the grey degree is less than a certain value (such as 0.1), the knowledge unit can be regarded as white approximately. What's more, since the grey degree is very hard to be measured precisely, we can define several grey clusters in $(0,1]$, thus the grey degree could be estimated by judging which grey cluster the knowledge unit belongs to.

3 The Weightiness of Knowledge Units

The weightiness of a knowledge unit is a quantitative parameter used to indicate the importance of a knowledge unit, with regard to the user's objective. The basic principle to determine the weightiness of a knowledge unit is the analysis hierarchy process. The process is detailed as following.

Firstly, construct judgement matrixes including Objective-KF/KU matrix and KF-KF/KU matrixes. Each node in the architecture can have a judgement matrix with its sub-nodes. Suppose node A has n sub-nodes. Then the judgement matrix is

$\Delta = \{\delta_{ij}\}, i, j = 1, \cdots, n.$, which means it is n-rank. Compare every two sub-nodes: sub-node i and sub-node j. If they are the same important, we have $\delta_{ij} = 1$. If i is more important than j, $\delta_{ij} = 5$. If i is extremely more important than j, $\delta_{ij} = 9$.

Secondly, we can calculate the weightiness of every mono-layer according to the judgement matrixes. In other words, for a node that has sub-nodes, calculate the weightiness of its sub-nodes. It can be given by calculating the latent root λ_{max} and eigenvector W of the judgement matrix Δ, where

$$\Delta W = \lambda_{max} W \qquad (1)$$

Finally, we calculate the overall weightiness, making use of the weightiness of every mono-layer.

Suppose the layer 1 has m elements: $A_1, \ldots, A_m$, and the weightiness of A_i to layer 0 is a_0^i, i=1,2,…,m. The layer 2 has k elements: $B_1, \ldots, B_k$, and the weightiness of B_j to A_i is b_i^j, $j = 1, 2, \ldots, k$. Here if B_j is independent of A_i, we have $b_i^j = 0$. Then the weightiness of B_j to layer 0 is

$$w_0^j = \sum_{i=1}^{m} a_0^i b_i^j, j = 1, 2, \ldots k \qquad (2)$$

Actually, since B_j only has one father node, supposing its father node is A_i, then the weightiness of B_j to layer 0 is

$$w_0^j = a_0^i b_i^j \qquad (3)$$

If there are several objectives, the weightiness of each knowledge unit to each objective can be calculated in the similar way.

4 Knowledge Evaluation with Grey Clustering Method

Since knowledge is very hard to evaluate, we use grey clustering method to classify knowledge units with whitenization weight function according to how they are mastered.

Definition: Suppose there are n objects to be clustered, m clustering criterion, s different grey clusters. According to sampling x_{ij} ($i = 1, 2, \ldots, n; j = 1, 2, \ldots, m$) of object i (i=1,2,…,n) regarding criterion j (j=1,2,…,m), classify object i into grey cluster k ($k \in \{1, 2, \ldots, s\}$). We call it grey clustering.[3]

In the knowledge architecture, each knowledge unit is an object to be clustered. Clustering criterions are observation criterions used to judge how the knowledge unit is mastered. Grey clusters refer to the grey classes defined based on the extent of knowledge mastery such as 'bad', 'medium' and 'excellent'.

In grey theory, whitenization weight function is frequently used to describe the preference extent when a grey element takes different value in its value field. Frequently used whitenization weight functions are shown in Fig. 2.

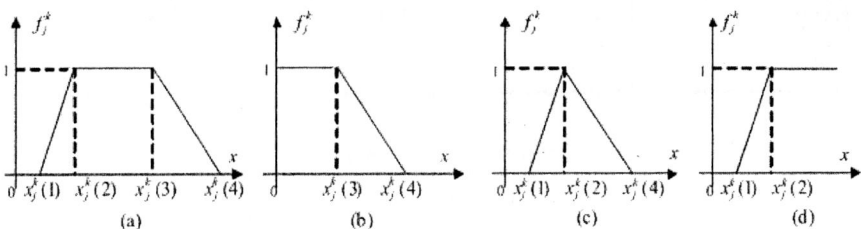

Fig. 2. Whitenization weight function

The whitenization weight function is represented as $f_j^k(\bullet)$. If $f_j^k(\bullet)$ is as shown in Fig. 2(a), Fig. 2(b), Fig. 2(c) or Fig. 2(d), then it is represented as $f_j^k[x_j^k(1), x_j^k(2), x_j^k(3), x_j^k(4)]$, $f_j^k[-,-,x_j^k(3), x_j^k(4)]$, $f_j^k[x_j^k(1), x_j^k(2),-,x_j^k(4)]$ or $f_j^k[x_j^k(1), x_j^k(2),-,-]$ separately. In the knowledge architecture, generally speaking, the whitenization weight function of 'bad' grey cluster should be like $f_j^k[-,-,x_j^k(3), x_j^k(4)]$; that of 'medium' grey cluster should be like $f_j^k[x_j^k(1), x_j^k(2),-,x_j^k(4)]$; and that of 'excellent' grey cluster should be like $f_j^k[x_j^k(1), x_j^k(2),-,-]$.

Since the significance and the dimension of each criterion are very different with each other, we adopt fixed weightiness to cluster. We call η_j the clustering weightiness of criterion j.

The steps of grey fixed-weightiness clustering are:

(1) Give the whitenization weight function of sub-cluster k of criterion j $f_j^k(\bullet)(j=1,2,...,m; k=1,2,...,s)$.

(2) Give the clustering weightiness of each criterion $\eta_j (j=1,2,...,m)$ by qualitative analysis or using the method given in Section 3.

(3) Given $f_j^k(\bullet)$, η_j and x_{ij} $(i=1,2,...,n; j=1,2,...,m)$ which are samplings of object i regarding criterion j, calculate grey fixed-weightiness clustering quotieties $\sigma_i^k = \sum_{j=1}^{m} f_j^k(x_{ij}) \cdot \eta_j$ $(i=1,2,...,n; k=1,2,...,s)$.

(4) If $\sigma_i^{k^*} = \max_{1 \leq k \leq s}\{\sigma_i^k\}$, we can conclude that object i belongs to grey cluster k^*.

For example, suppose there are 4 knowledge units a, b, c, d. Establishing criterion set as { I : number of literatures have been read; II : number of published papers; III: number of giving lectures}, we classify the four units into three clusters which are 'excellent', 'medium' and 'bad'. The samplings are shown in table 1.

Table 1. Samplings of each knowledge unit regarding criterions

	criterion I	criterion II	criterion III
KU a	5	1	0
KU b	44	4	3
KU c	20	3	1
KU d	25	5	2

Firstly, give the whitenization weight functions: $f_1^1[0,40,-,-]$, $f_1^2[0,20,-,40]$, $f_1^3[-,-,5,10]$, $f_2^1[0,10,-,-]$, $f_2^2[0,5,-,10]$, $f_2^3[-,-,4,8]$, $f_3^1[0,4,-,-]$, $f_3^2[0,2,-,4]$, $f_3^3[-,-,1,2]$. Assume the clustering weightiness of each criterion is $\eta_1 = 0.2, \eta_2 = 0.5, \eta_3 = 0.3$. Then it can be given:

$\sigma_1 = (\sigma_1^1, \sigma_1^2, \sigma_1^{3*}) = (0.075, 0.15, 1.0)$, $\sigma_2 = (\sigma_2^{1*}, \sigma_2^2, \sigma_2^3) = (0.625, 0.55, 0.5)$, $\sigma_3 = (\sigma_3^1, \sigma_3^2, \sigma_3^{3*}) = (0.385, 0.65, 0.8)$, $\sigma_4 = (\sigma_4^1, \sigma_4^{2*}, \sigma_4^3) = (0.525, 0.95, 0.375)$

The results indicate that knowledge unit b2 belongs to 'excellent' grey cluster, b4 belongs to 'medium' grey cluster, b1 and b3 belongs to 'bad' grey cluster. Combining the evaluation result with the weightiness of knowledge units, the foremost learning and developing directions can be known.

6 Conclusion

This paper gives the knowledge architecture based on grey theory. It provides the method to calculate the weightiness of knowledge and to evaluate them. This method can not only help users to establish their knowledge architecture, but also inform them which are the most important knowledge and which are the foremost learning directions. It also provides the existing knowledge management technology a new idea, which is human-oriented since the user's objective and the expansibility of cognition are well considered. Furthermore, since the method provided in this paper is just objective-oriented, the process-based knowledge modeling and evaluation will be studied in our future work.

References

1. Schreiber, G., Akkermans, H., Anjewierden, A., deHoog, R., Shadbolt, N., VandeVelde, W., Wielinga, B.: Knowledge engineering and management: the commonKADS methodology. MIT Press, Cambridge Massachusetts London England (2000)
2. Mertins, K., Heisig, P., Vorbeck, J.: Knowledge management concepts and best practices. Tsinghua University Press, Beijing (2004)
3. Liu, S., Guo, T., Dang, Y.: Theory and applications of grey systems. Science Press, Beijing (1999)
4. Deng, J.: The tutorial of grey system theory. Huazhong Science and Technology Uniuersity Press, Wuhan (1990)
5. Xia, S., Yang, J., Yang, Z.: The introduction of system engineering. Tsinghua University Press, Beijing (1995)

A Propositional Calculus Formal Deductive System $\mathcal{L}^U$ of Universal Logic and Its Completeness

Minxia Luo[1,2] and Huacan He[1]

[1] School of Computer Science, Northwestern Polytechnical,
University, Xi'an, 710072, P.R. China
minxialuo@163.com, hehuac@nwpu.edu.cn
[2] Department of Mathematics, Yuncheng University,
Yuncheng, 044000, P.R. China

Abstract. Universal logic has given 0-level universal conjunction operation, universal disjunction operation and the universal implication operation. We introduce a new kind of algebra system UBL algebra based on these operations. A general propositional calculus formal deductive system $\mathcal{L}^U$ of universal logic based on UBL algebras is built up, and its completeness is proved.

1 Introduction

In past several years, fuzzy logic has been application in fuzzy control field. However, fuzzy logic theory is not consummate. Fuzzy logic has been doubt and animadversion (see [1]) because these methods of fuzzy reasoning have not strict logical foundation.

Residuated fuzzy logic calculi are related to continuous t-norms which are used as truth functions for the conjunction connective, and their residua as truth function for the implication. Main examples are Lukasiewicz, Gödel and product logics, related to Lukasiewicz t-norm ($x*y = max(0, x + y - 1)$), Gödel t-norm ($x*y = min(x, y)$) and product t-norm ($x*y = xy$) respectively. Rose and Rosser [2] proved completeness results for Lukasiewicz logic and Dummet [3] for Gödel logic, and recently three of the authors [4] axiomatized product logic. More recently, Hájek [5] has proposed the axiomatic system BL corresponding to a generic continuous t-norm. A kind of fuzzy propositional logic was proposed by Professor Guojun Wang [6] in 1997, and its completeness has proved in 2002(see[8]). A new kind of logical system–based on strong regular residuated lattices has been proposed by Professor Daowu Pei in 2002, and its completeness has been proved(see[9]).

A new requirement for classical logic was raised with developing of computer science and modern logic. Non-classical logic and modern logic are developing rapidly. Universal logic principle was proposed by professor Huacan He in 2001(see [10]). A kind of flexible relation between fuzzy propositions has been found in the study of artificial intelligence theory. The flexible relation was described by the model of continuous universal logic operation. The main reasons to influence the flexible relation are generalized correlation and generalized autocorrelation. Some 0-level binary universal operations are defined as follows(see[10]):

$$T(x,y,h) = \text{ite}\{0|x=0 \text{ or } y=0; (\max(0, x^m + y^m - 1))^{1/m}\}$$

$$S(x,y,h) = \text{ite}\{1|x=1 \text{ or } y=1; 1-(\max(0,(1-x)^m + (1-y)^m - 1))^{1/m}\}$$

$$I(x,y,h) = \text{ite}\{1|x \leq y; 0|m \leq 0 \text{ and } y=0; (1-x^m+y^m)^{1/m}\}$$

where $m = (3-4h)/(4h(1-h))$, $h \in [0,1]$. $S = \text{ite}\{\beta|\alpha;\gamma\}$, it is a conditional expression which express that if α is true, then $S = \beta$; otherwise, $S = \gamma$.

Universal logical formal deduction system $\mathcal{B}$ in ideal condition($h \equiv 0.5$) has been established by author [11], and its completeness has been proved [12]. In this paper, we introduce a kind of algebra–UBL algebra, and its properties are discussed. A general propositional calculus formal system $\mathcal{L}^U$ based on UBL algebras is built up, and its completeness is proved. Moreover, formal deduction system $\mathcal{B}$ is a schematic extension of general universal logic system $\mathcal{L}^U$.

2 Main Properties of UBL Algebras

Some properties of the model of 0-level binary universal propositional connectives have been studied in [10]. It has been proved that the model of 0-level universal conjunction is a nilpotent t-norm for $h \in (0, 0.75)$, a strict t-norm for $h \in (0.75, 1)$ in [13]. The model of 0-level universal implication is a residuum of the model of 0-level universal conjunction, i.e., the model of 0-level universal conjunction and the model of 0-level universal implication form adjoint pair (see [13]).

In this section, we introduce UBL algebra by the model of 0-level universal conjunction, universal disjunction and universal implication. Its main properties are studied as follows.

Definition 1. *Let L be a partial order set. A UBL algebra is an algebra $(L, \otimes, \oplus, \Rightarrow, 0, 1)$ with three binary operations and two constants such that*

(1) $(L, \otimes, 1)$ *is a commutative semigroup with the unit element 1, i.e. $\otimes$ is commutative, associative, and $1 \otimes x = x$ for all x;*
(2) $(L, \oplus, 0)$ *is a commutative semigroup with the unit element 0;*
(3) $\otimes$ *and $\Rightarrow$ form an adjoint pair, i.e. $x \otimes y \leq z$ if and only if $x \leq y \Rightarrow z$;*
(4) $(x \Rightarrow y) \Rightarrow ((x \oplus z) \Rightarrow (y \oplus z)) = 1$;
(5) $x \Rightarrow y = 1$ *or $y \Rightarrow x = 1$.*

Example 1. (1) MV-algebra is an UBL algebra.
  (2) For $\forall x, y \in [0,1]$, we define the following operations:

$$x \otimes y = \text{ite}\{0|x=0 \text{ or } y=0; (\max(0, x^m + y^m - 1))^{1/m}\}$$

$$x \oplus y = \text{ite}\{1|x=1 \text{ or } y=1; 1-(\max(0,(1-x)^m + (1-y)^m - 1))^{1/m}\}$$

$$x{\Rightarrow}y = \text{ite}\{1|x\leq y;\ 0|m\leq 0 \text{ and } y=0;\ (1-x^m+y^m)^{1/m}\}$$

where $m\in R$. Then $([0,1],\otimes,\oplus,\Rightarrow,0,1))$ is UBL algebra which is called UBL unit interval.

Proposition 1. *Let L be an UBL algebra. $\forall a, b, c \in L$, the following properties are true:*

(1) $a\otimes b = b\otimes a$, $\quad a\oplus b = b\oplus a$
(2) $(a\otimes b)\otimes c = a\otimes(b\otimes c)$, $\quad (a\oplus b)\oplus c = a\oplus(b\oplus c)$
(3) $1\otimes a = a$, $\quad 0\oplus a = a$
(4) $a\otimes b \leq c$ or $a\leq b\Rightarrow c$
(5) $(a\otimes b)\Rightarrow c = a\Rightarrow(b\Rightarrow c)$
(6) $a\leq b$ or $a\Rightarrow b = 1$
(7) $1\Rightarrow a = a$
(8) $a\Rightarrow(b\Rightarrow a) = 1$
(9) $a\Rightarrow(b\Rightarrow c) = b\Rightarrow(a\Rightarrow c)$
(10) $a\leq b\Rightarrow c$ or $b\leq a\Rightarrow c$
(11) $a\Rightarrow(b\Rightarrow a\otimes b) = 1$
(12) $((a\Rightarrow b)\otimes a)\Rightarrow b = 1$
(13) $b \Rightarrow c \leq (a\Rightarrow b)\Rightarrow(a\Rightarrow c)$
(14) $a\Rightarrow b \leq (b\Rightarrow c)\Rightarrow(a\Rightarrow c)$
(15) $a\Rightarrow b \leq (a\otimes c)\Rightarrow(b\otimes c)$
(16) $a\Rightarrow b \leq (a\oplus c)\Rightarrow(b\oplus c)$
(17) $a^n \leq a^m$, $m\leq n$, where $a^{k+1} = a^k \otimes a$.

Proof omit.

3 A Propositional Calculus Formal Deductive System $\mathcal{L}^U$

The 0-level universal conjunction, universal disjunction and universal implication propositional connectives are $\wedge_h$, $\vee_h$ and $\rightarrow_h$, and written by $\wedge$, $\vee$ and $\rightarrow$ respectively.

Definition 2. *For the 0-level model of universal conjunction, universal disjunction and universal implication, a propositional calculus system $\mathcal{L}^U$ is defined as follows: The set $\mathcal{F}$ of well-formed formulas (wfs) of $\mathcal{L}^U$ is defined as usual from a countable set of propositional variables $P_1, P_2, \cdots$, three connectives $\wedge$, $\vee$, $\rightarrow$ and the truth constant $\bar{0}$ for 0. Further definable connectives are:*

$$\neg A : A \rightarrow \bar{0}$$
$$A \equiv B : (A\rightarrow B) \wedge (B\rightarrow A)$$

Definition 3. *The following formulas are axioms of universal logic system $\mathcal{L}^U$:*
(U1) $(A\rightarrow B)\rightarrow((B\rightarrow C)\rightarrow(A\rightarrow C))$
(U2) $A\wedge B\rightarrow A$
(U3) $A\wedge B\rightarrow B\wedge A$
(U4) $A\rightarrow A\vee B$

(U5) $A \vee B \to B \vee A$
(U6) $(A \wedge (A \to B)) \to (B \wedge (B \to A))$
(U7) $(A \to (B \to C)) \to ((A \wedge B) \to C)$
(U8) $((A \wedge B) \to C) \to (A \to (B \to C))$
(U9) $((A \to B) \to C) \to (((B \to A) \to C) \to C)$
(U10) $\bar{0} \to A$
(U11) $(A \vee B) \vee C \to A \vee (B \vee C)$
(U12) $(A \to B) \to (A \vee C \to B \vee C)$

The deduction rule is modus ponens(MP): i.e., from A and $A \to B$ infer B.

In a natural manner, we can introduce the concepts such as proof, theorem, deduction from a formula set Γ, Γ-consequence in the system $\mathcal{L}^U$. A theory of $\mathcal{L}^U$ is a set of wfs. $\Gamma \vdash A$ denotes that A is provable in the theory Γ. $\vdash A$ denotes that A is a theorem of system $\mathcal{L}^U$. We denote

$$\text{Thm}(\mathcal{L}^U) = \{A \in \mathcal{F} | \vdash A\} \quad \text{Ded}(\Gamma) = \{A \in \mathcal{F} | \Gamma \vdash A\}$$

Proposition 2. *The hypothetical syllogism (HS rule for short) holds, i.e. assume $\Gamma = \{A \to B, B \to C\}$, then $\Gamma \vdash A \to C$.*

Proof. Assume $\Gamma = \{A \to B, B \to C\}$, then

1^0 $A \to B$ $\qquad\qquad\qquad\qquad$ (Γ)
2^0 $(A \to B) \to ((B \to C) \to (A \to C))$ $\qquad$ $(U1)$
3^0 $(B \to C) \to (A \to C)$ $\qquad\qquad$ $(1^0, 2^0, \text{MP})$
4^0 $B \to C$ $\qquad\qquad\qquad\qquad$ (Γ)
5^0 $A \to C$ $\qquad\qquad\qquad\qquad$ $(3^0, 4^0, \text{MP})$ $\qquad\square$

Theorem 1. *The following formulae are theorems of the formal system $\mathcal{L}^U$:*

(T1) $(A \to (B \to C)) \to (B \to (A \to C))$
(T2) $(B \to C) \to ((A \to B) \to (A \to C))$
(T3) $A \to (B \to A)$
(T4) $A \to A$
(T5) $A \to (B \to (A \wedge B))$
(T6) $(A \wedge B) \to (A \to B)$
(T7) $(A \to B \wedge C) \to (A \wedge B \to C)$
(T8) $(A \vee B \to C) \to (A \to B \vee C)$
(T9) $((A \to C) \wedge (B \to C)) \to (A \wedge B \to C)$
(T10) $((A \to B) \wedge (A \to C)) \to (A \to B \vee C)$
(T11) $(A \wedge (A \to B)) \to B$
(T12) $(A \to B) \to (A \wedge C \to B \wedge C)$
(T13) $A \wedge (B \wedge C) \to (A \wedge B) \wedge C$
(T14) $(A \wedge B) \wedge C \to A \wedge (B \wedge C)$

Proof. (T1)

1⁰ $B \wedge A \to A \wedge B$ (U3)
2⁰ $(B \wedge A \to A \wedge B) \to ((A \wedge B \to C) \to (B \wedge A \to C))$ (U1)
3⁰ $(A \wedge B \to C) \to (B \wedge A \to C)$ (1⁰, 2⁰, MP)
4⁰ $(B \wedge A \to C) \to (B \to (A \to C))$ (U8)
5⁰ $(A \wedge B \to C) \to (B \to (A \to C))$ (3⁰, 4⁰, HS)
6⁰ $(A \to (B \to C)) \to (A \wedge B \to C)$ (U7)
7⁰ $(A \to (B \to C)) \to (B \to (A \to C))$ (5⁰, 6⁰, HS)

(T2)

1⁰ $(A \to B) \to ((B \to C) \to (A \to C))$ (U1)
2⁰ $((A \to B) \to ((B \to C) \to (A \to C))) \to$
    $((B \to C) \to ((A \to B) \to (A \to C)))$ (T1)
3⁰ $(B \to C) \to ((A \to B) \to (A \to C))$ (1⁰, 2⁰, MP)

(T3)

1⁰ $A \wedge B \to A$ (U2)
2⁰ $(A \wedge B \to A) \to (A \to (B \to A))$ (U8)
3⁰ $A \to (B \to A)$ (1⁰, 2⁰, MP)

(T4)

1⁰ $A \to (B \to A)$ (T3)
2⁰ $(A \to (B \to A)) \to (B \to (A \to A))$ (T1)
3⁰ $B \to (A \to A)$ (1⁰, 2⁰, MP)
4⁰ B (Let B be any axiom)
5⁰ $A \to A$ (3⁰, 4⁰, MP)

(T5)

1⁰ $B \wedge A \to A \wedge B$ (U3)
2⁰ $(B \wedge A \to A \wedge B) \to (B \to (A \to A \wedge B))$ (U8)
3⁰ $B \to (A \to A \wedge B)$ (1⁰, 2⁰, MP)
4⁰ $(B \to (A \to A \wedge B)) \to (A \to (B \to A \wedge B))$ (T1)
5⁰ $A \to (B \to A \wedge B)$ (3⁰, 4⁰, MP)

(T6)

1⁰ $A \wedge B \to B \wedge A$ (U3)
2⁰ $B \wedge A \to B$ (U2)
3⁰ $A \wedge B \to B$ (1⁰, 2⁰, HS)
4⁰ $B \to (A \to B)$ (T3)
5⁰ $A \wedge B \to (A \to B)$ (3⁰, 4⁰, HS)

(T7)

1⁰ $B \wedge C \to (B \to C)$ (T6)
2⁰ $(B \wedge C \to (B \to C)) \to ((A \to B \wedge C) \to (A \to (B \to C)))$ (T2)
3⁰ $(A \to B \wedge C) \to (A \to (B \to C))$ (1⁰, 2⁰, MP)
4⁰ $(A \to (B \to C)) \to (A \wedge B \to C)$ (U7)
5⁰ $(A \to B \wedge C) \to (A \wedge B \to C)$ (3⁰, 4⁰, HS)

(T8)

1^0 $A \to A \lor B$ (U4)
2^0 $(A \to A \lor B) \to ((A \lor B \to C) \to (A \to C))$ (U1)
3^0 $(A \lor B \to C) \to (A \to C)$ ($1^0, 2^0$, MP)
4^0 $C \to C \lor B$ (U4)
5^0 $C \lor B \to B \lor C$ (U5)
6^0 $C \to B \lor C$ ($4^0, 5^0$, HS)
7^0 $(C \to B \lor C) \to ((A \to C) \to (A \to B \lor C))$ (T2)
8^0 $(A \to C) \to (A \to B \lor C)$ ($6^0, 7^0$, MP)
9^0 $(A \lor B \to C) \to (A \to B \lor C)$ ($3^0, 8^0$, HS)

(T9)
1^0 $(A \to C) \land (B \to C) \to (A \to C)$ (U2)
2^0 $A \land B \to A$ (U2)
3^0 $(A \land B \to A) \to ((A \to C) \to (A \land B \to C))$ (U1)
4^0 $(A \to C) \to (A \land B \to C)$ ($2^0, 3^0$, MP)
5^0 $(A \to C) \land (B \to C) \to (A \land B \to C)$ ($1^0, 4^0$, HS)

(T10)
1^0 $(A \to B) \land (A \to C) \to (A \to C) \land (A \to B)$ (U3)
2^0 $(A \to C) \land (A \to B) \to (A \to C)$ (U2)
3^0 $(A \to B) \land (A \to C) \to (A \to C)$ ($1^0, 2^0$, HS)
4^0 $C \to B \lor C$ (U4)
5^0 $(C \to B \lor C) \to ((A \to C) \to (A \to B \lor C))$ (T2)
6^0 $(A \to C) \to (A \to B \lor C)$ ($4^0, 5^0$, MP)
7^0 $((A \to B) \land (A \to C)) \to (A \to B \lor C)$ ($3^0, 6^0$, HS)

(T11)
1^0 $(A \to B) \to (A \to B)$ (T4)
2^0 $((A \to B) \to (A \to B)) \to ((A \to B) \land A \to B)$ (U7)
3^0 $(A \to B) \land A \to B$ ($1^0, 2^0$, MP)
4^0 $A \land (A \to B) \to (A \to B) \land A$ (U3)
5^0 $((A \to B) \land A \to B) \to (A \land (A \to B) \to B)$ (U1)
6^0 $(A \land (A \to B)) \to B$ ($4^0, 5^0$, MP)

(T12)
1^0 $(A \land (A \to B)) \to B$ (T11)
2^0 $B \to (C \to B \land C)$ (T5)
3^0 $(A \land (A \to B)) \to (C \to B \land C)$ ($1^0, 2^0$, HS)
4^0 $((A \land (A \to B)) \to (C \to B \land C)) \to$
      $(A \to ((A \to B) \to (C \to B \land C)))$ (U8)
5^0 $A \to ((A \to B) \to (C \to B \land C))$ ($3^0, 4^0$, MP)
6^0 $((A \to B) \to (C \to B \land C)) \to$
      $(C \to ((A \to B) \to B \land C))$ (T1)
7^0 $A \to (C \to ((A \to B) \to B \land C))$ ($5^0, 6^0$, HS)
8^0 $(A \to (C \to ((A \to B) \to B \land C))) \to$
      $(A \land C \to ((A \to B) \to B \land C))$ (U7)
9^0 $(A \land C) \to ((A \to B) \to B \land C)$ ($7^0, 8^0$, MP)
10^0 $((A \land C) \to ((A \to B) \to B \land C)) \to$
       $((A \to B) \to (A \land C \to B \land C))$ (T1)
11^0 $(A \to B) \to (A \land C \to B \land C)$ ($9^0, 10^0$, MP)

(T13)
1^0 $((A\wedge B)\wedge C\to D)\to(A\wedge B\to(C\to D))$ \hfill (U8)
2^0 $(A\wedge B\to(C\to D))\to(A\to(B\to(C\to D)))$ \hfill (U8)
3^0 $((A\wedge B)\wedge C\to D)\to(A\to(B\to(C\to D)))$ \hfill $(1^0, 2^0, \text{HS})$
4^0 $(B\to(C\to D))\to(B\wedge C\to D)$ \hfill (U7)
5^0 $((B\to(C\to D))\to(B\wedge C\to D))\to$
$\quad((A\to(B\to(C\to D)))\to(A\to(B\wedge C\to D)))$ \hfill (T2)
6^0 $(A\to(B\to(C\to D)))\to(A\to(B\wedge C\to D))$ \hfill $(4^0, 5^0, \text{MP})$
7^0 $(A\to(B\wedge C\to D))\to(A\wedge(B\wedge C)\to D)$ \hfill (U7)
8^0 $(A\to(B\to(C\to D)))\to(A\wedge(B\wedge C)\to D)$ \hfill $(6^0, 7^0, \text{HS})$
9^0 $((A\wedge B)\wedge C\to D)\to(A\wedge(B\wedge C)\to D)$ \hfill $(3^0, 8^0, \text{HS})$
10^0 Let $D = (A\wedge B)\wedge C$
11^0 $A\wedge(B\wedge C)\to(A\wedge B)\wedge C$ \hfill $(9^0, 10^0, \text{MP})$
The proof of (T14) is similar to (T13). $\square$

Definition 4. *The binary relation $\sim$ on $\mathcal{F}$ is called provable equivalence relation,*

$$A\sim B \text{ if and only if } \vdash A\to B, \vdash B\to A$$

Theorem 2. *The relation $\sim$ is a congruence relation of $\mathcal{F}$. The quotient algebra $[F] = \mathcal{F}/\sim = \{[A]|\ A\in\mathcal{F}\}$ is an UBL algebra in which the partial ordering $\leq$ is defined as follows:*

$$[A]\leq[B] \text{ if and only if } \vdash A\to B$$

Proof. It is clear that the relation $\sim$ is an equivalent relation on $\mathcal{F}$. It can be proved that the relation $\sim$ is a congruence relation of $\mathcal{F}$ by $(U1), (U12), (T1)$, $(T2)$ and $(T12)$. The quotient algebra $[F] = \mathcal{F}/\sim = \{[A]|\ A\in\mathcal{F}\}$ is an UBL algebra, where

$$[A]\vee[B] = [A\vee B],\ [A]\wedge[B] = [A\wedge B],\ [A]\to[B] = [A\to B].$$

$\square$

4 The Completeness of Formal System $\mathcal{L}^U$

Now we extend the semantical concepts of the system $\mathcal{L}^U$ onto general UBL algebra.

Definition 5. *Let L be an UBL algebra. A $(\wedge, \vee, \to)$-type homomorphism, i.e.*

$$v: \mathcal{F}\to L$$

$v(A\wedge B) = v(A)\otimes v(B), v(A\vee B) = v(A)\oplus v(B), v(A\to B) = v(A)\Rightarrow v(B)$

is called an L-valuation of the system $\mathcal{L}^U$.

The set $\Omega(L)$ of all L-valuation of the system $\mathcal{L}^U$ is called the L-semantics of the system $\mathcal{L}^U$.

Definition 6. Let L be an UBL algebra, $A \in \mathcal{F}$, $\Gamma \subseteq \mathcal{F}$.

(1) A is called an L-tautology, denote by $\models_L A$, if $\forall v \in \Omega(L)$, $v(A) = 1$.
(2) A is called an L-semantics consequence of Γ, denote by $\Gamma \models_L A$, if $\forall v \in \Omega(L)$, we always have $v(A) = 1$ whenever $v(\Gamma) \subseteq \{1\}$.

A is L-tautology when $\Gamma = \emptyset$. We use $T(L)$ denote the set of all L-tautology, i.e., $T(L) = \{A \in \mathcal{F} | \models_L A\}$.

Theorem 3 (L-soundness). Let L be an UBL algebra, $\forall A \in \mathcal{F}, \forall \Gamma \subseteq \mathcal{F}$. If $\Gamma \vdash A$, then $\Gamma \models_L A$. Specially, $Thm(\mathcal{L}^U) \subseteq T(L)$.

Proof. $\forall \Gamma \subseteq \mathcal{F}$, if $v(\Gamma) \subseteq \{1\}$ and $\Gamma \vdash A$, then $A \in \text{Axm}(\mathcal{L}^U)$, or $A \in \Gamma$, or A is obtained from B and $B \to A$ by MP, where B and $B \to A$ are Γ-consequence, and $v(B) = v(B \to A) = 1$. If $A \in \text{Axm}(\mathcal{L}^U)$ or $A \in \Gamma$, then $v(A) = 1$. If A is obtained from B and $B \to A$ by MP, then $v(A) = 1$ since

$$v(A) \geq v(B) \otimes (v(B) \Rightarrow v(A)) = v(B) \otimes v(B \to A) = 1.$$

□

Definition 7. Let $L = (L, \oplus, \otimes, \Rightarrow, 0, 1)$ be an UBL algebra. A filter on L is a non-empty set $F \subseteq L$ such that for each $a, b \in L$,

$$a \in F, \text{ and } b \in F \text{ implies } a \otimes b \in F,$$

$$a \in F \text{ and } a \leq b \text{ implies } b \in F.$$

F is a prime filter iff for each $a, b \in L$,

$$(a \Rightarrow b) \in F \text{ or } (b \Rightarrow a) \in F.$$

Remark. If F is a filter on UBL algebra, $x, y \in F$, then $x \oplus y \in F$. In fact, it is true that $x \otimes y \leq x \leq x \oplus y$.

Proposition 3. Let L be an UBL algebra and F be a filter. We define the relation as follows:

$$a \sim_F b \text{ if and only if } (a \Rightarrow b) \in F \text{ and } (b \Rightarrow a) \in F.$$

then

(1) $\sim_F$ is a congruence and the quotient algebra $L/\sim_F$ is an UBL algebra.
(2) $L/\sim_F$ is linearly ordered if and only if F is a prime filter.

Proposition 4. Let L be an UBL algebra and $a \in L \setminus \{1\}$. Then there is a prime filter F on L such that $a \notin F$.

Proof. Let $\mathcal{A}$ be the set of all filters on L not containing a. Then $\mathcal{A} \neq \emptyset$ by $\{1\} \in \mathcal{A}$. There is a maximal element F in $\mathcal{A}$, because every chain $\{F_t\}_{t \in I}$ has an upper bound $\cup_{t \in I} F_t$ in partially order set $(\mathcal{A}, \subseteq)$ by Zorn's Lemma. To show that F is a prime filter on L.

If there exist $b, c \in L$, such that $b \Rightarrow c \notin F$ and $c \Rightarrow b \notin F$, let

$$F_1 = \{x \in L | \exists y \in F, \exists n \in N, y \otimes (b \Rightarrow c)^n \leq x\}$$

$$F_2 = \{u \in L | \exists v \in F, \exists m \in N, v \otimes (c \Rightarrow b)^m \leq u\}$$

We can prove that F_1 and F_2 are filters on L such that

$$b \Rightarrow c \in F_1, \quad F \subset F_1, \quad c \Rightarrow b \in F_2, \quad F \subset F_2.$$

Now we only prove that $F_1 \in \mathcal{A}$ or $F_2 \in \mathcal{A}$, i.e., $a \notin F_1$ or $a \notin F_2$.

In fact, if $a \in F_1$ and $a \in F_2$, then $\exists y, v \in F$, $\exists n, m \in N$ such that

$$a \geq y \otimes (b \Rightarrow c)^n, \quad a \geq v \otimes (c \Rightarrow b)^m.$$

Without loss of generality, we assume $n \geq m$, then $a \geq v \otimes (c \Rightarrow b)^n$. Hence

$$a \geq \max(y \otimes (b \Rightarrow c)^n, v \otimes (c \Rightarrow b)^n) \geq y \otimes v \otimes (\max((b \Rightarrow c)^n, (c \Rightarrow b)^n)) = y \otimes v$$

Thus $a \in F$ by $y \otimes v \in F$, a contradiction. Therefore, $a \notin F_1$ or $a \notin F_2$. □

Theorem 4. Let L be any an UBL algebra. Then exist UBL chain $\{L_t | t \in I\}$ such that L can be isomorphically embedded into $L^* = \Pi_{t \in I} L_t$.

Proof. Let $\mathcal{P} = \{F | F$ is a prime filter on $L\}$, then $\mathcal{P} \neq \emptyset$. We have that $L/\sim_F$ is an UBL chain for all $F \in \mathcal{P}$ by Proposition 3. We may prove that the mapping

$$i : L \to L^*, \quad x \mapsto ([x]_F)_{F \in \mathcal{P}}$$

is an embedding from L to L^*, i.e., the mapping i is monomorphism.

In fact, obviously, i is a homomorphism. $\forall a, b \in L$, if $a \neq b$, without loss of generality, we assume $a \not\leq b$, then $a \Rightarrow b \neq 1$. There exist a prime filter F on L such that $a \Rightarrow b \notin F$ by Proposition 4. Thus $[a]_F \not\leq [b]_F$ in $L/\sim_F$. Hence $[a]_F \neq [b]_F$, i.e., $i(a) \neq i(b)$. □

Proposition 5. If L_1 and L_2 are UBL algebra, $L_1 \cong L_2$, then $T(L_1) = T(L_2)$.

Proposition 6. Let L, L_1 be UBL algebra. If L_1 is a subalgebra of L, then $T(L) \subseteq T(L_1)$.

Theorem 5. Let L be an UBL chain and $A \in \mathcal{F}$. If $A \in T(L)$, then $A \in T(L_1)$, where L_1 is any UBL algebra.

Proof. We shall show that $A \in T(L^*)$ when $A \in T(L_t)(\forall t \in I)$ by Theorem 4, Proposition 5 and Proposition 6, where $L^* = \Pi_{t \in I} L_t$, $\{L_t\}_{t \in I}$ is chain.

In fact, $\forall v \in \Omega(L^*)$, then $f_t v \in \Omega(L_t)$, where $f_t : L^* \to L_t$ is a projection mapping. If $v(A) = (a_t)_{t \in I} \neq 1$, then exist $t \in I$ such that $a_t \neq 1$, i.e. $f_t v(A) = f_t(v(A)) \neq 1$, a contradiction. Thus $A \in T(L^*)$. □

Theorem 6 ($[L]$-completeness). *Let $A \in \mathcal{F}$. Then $\vdash A$ if and only if $\models_{[L]} A$.*

Proof. The necessity is obviously by Theorem 3.

$[A] \leq [B]$ if and only if $\vdash A \to B$ by Theorem 2 in $[L]$. $[A] = [B]$ if and only if $A \sim B$. If $B \in \text{Thm}(\mathcal{L}^U)$, then $\forall A \in \mathcal{F}$, $\vdash A \to B$, i.e., $[A] \leq [B]$. Thus $[B] = 1$ is the maximal element of $[L]$. Assume $[A] = 1$, then $[A] = [B]$, therefore $A \sim B$, i.e., $A \in \text{Thm}(\mathcal{L}^U)$. Hence the maximal element 1 of $[L]$ is the set of all theorems of system $\mathcal{L}^U$.

Suppose $A \in T([L])$, then $\forall v \in \Omega([L])$, $v(A) = 1$. Specially,

$$v : \mathcal{F} \to [L], \quad A \mapsto [A], \quad A \in \mathcal{F},$$

$v(A) = 1$, i.e., $[A] = 1$. Thus $A \in \text{Thm}(\mathcal{L}^U)$, i.e., $\vdash A$. □

Theorem 7 (completeness). *Let $A \in \mathcal{F}$, then the following are equivalent:*
(i) $\vdash A$;
(ii) for each linearly ordered UBL algebra L, $A \in T(L)$;
(iii) for each UBL algebra L, $A \in T(L)$.

Proof. The implications $(i) \Rightarrow (ii)$, $(ii) \Rightarrow (iii)$ and $(iii) \Rightarrow (i)$ have been proved by Theorem 3, Theorem 5 and Theorem 6 respectively. □

Theorem 8 (strong completeness). *Let $\Gamma \subseteq \mathcal{F}$, $A \in \mathcal{F}$. The following are equivalent:*
(i) $\Gamma \vdash A$;
(ii) for each linearly ordered UBL algebra L, $\Gamma \models_L A$;
(iii) for each UBL algebra L, $\Gamma \models_L A$.

Proof. The implications $(i) \Rightarrow (ii)$, $(ii) \Rightarrow (iii)$ have been proved by Theorem 3 and Theorem 5 respectively. We shall show $(iii) \Rightarrow (i)$.

We define the relation on $\mathcal{F}$ $\sim_\Gamma$:

$$A \sim_\Gamma B \text{ if and only if } \Gamma \vdash A \to B, \quad \Gamma \vdash B \to A$$

It can be proved easily that $\sim_\Gamma$ is a congruence on $\mathcal{F}$, and corresponding quotient algebra $[F]_\Gamma = \mathcal{F}/\sim_\Gamma = \{[A]_\Gamma | A \in \mathcal{F}\}$ is an UBL algebra, and $\forall v \in \Omega([F]_\Gamma)$, $v(\Gamma) \subseteq \{1\}$. We can prove that $1 = \text{Ded}(\Gamma)$ is the maximal element of $[F]_\Gamma$.

If $\Gamma \models_{[F]_\Gamma} A$, then $\forall v \in \Omega([F]_\Gamma)$, $v(A) = 1$. Specially, for the mapping

$$v : \mathcal{F} \to [F]_\Gamma, \quad A \to [A]_\Gamma, \quad A \in \mathcal{F}$$

we have $[A]_\Gamma = 1 = \text{Ded}(\Gamma)$. Hence $A \in \text{Ded}(\Gamma)$, i.e., $\Gamma \vdash A$. □

5 Conclusions

We introduce a new kind of algebra system UBL algebra based on 0-level universal conjunction operation, universal disjunction operation and universal implication operation. A general propositional calculus formal deductive system

$\mathcal{L}^U$ of universal logic based on UBL algebras is built up, and its completeness is proved. It show that syntax and semantics of formal system $\mathcal{L}^U$ is concordant. We may establish a solid logical foundation for flexible reasoning. Moreover, the formal deductive system $\mathcal{B}$ ([11]) of universal logic in ideal condition($h\equiv 0.5$) is a schematic extension of general universal logic system $\mathcal{L}^U$.

References

1. Hongxing L.: To see the success of fuzzy logic from mathematical essence of fuzzy control-on the paradoxical sussess of fuzzy logic. Fuzzy systems and mathematics **9**(1995)1–14(in chinese)
2. Rose,P.A., Rosser,J.B.: Fragments of many valued statement calculi. Trant.A.M.S. **87**(1958)1–53
3. Dummett,M.: A propositional calculus with denumerable matrix. Journal of Symbolic Logic **24**(1959)97–106
4. Hájek,P., Godo,L., Esteva,F.: A complete many-valued logic with product conjunction. Archive for Mathematical Logic **35**(1996)191–208
5. Hájek,P.: Metamathematics of fuzzy logic. Kluwer Academic Publishers(1998)
6. Guojun W.: A formal deductive system of fuzzy propositional calculus. Chinese Science Bulletin **42**(1997)10 11 10 15(in chinese)
7. Guojun W.: The full implication triple I method for fuzzy reasoning. Science in China(Series E) **29**(1999)43–53(in chinese)
8. Daowu P., Guojun W.: The completeness and applications of the formal system L* . Science in China(Series F) **45**(2002)40–50
9. Daowu P.: A logic system based on strong regular residuated lattices and its completeness. Acta Mathematica Sinica **45**(2002)745–752(in chinese)
10. Huacan He.: Principle of Universal Logic. Beijing: Chinese Scientific Press(2001)(in chinese)
11. Minxia L., Huacan H.: The formal deductive system B of universal logic in the ideal condition. Science of Computer **31**(2004)95–98(in chinese)
12. Minxia L., Huacan H.: The completeness of the formal deductive system B of universal logic in the ideal condition. Science of computer(in chinese)(in press)
13. Minxia L., Huacan H.: Some algebraic properties about the 0-level universal operation model of universal logic. Fuzzy systems and Mathematics(in chinese)(in press)

Entropy and Subsethood for General Interval-Valued Intuitionistic Fuzzy Sets

Xiao-dong Liu, Su-hua Zheng, and Feng-lan Xiong

Department of Mathematics, Ocean University of China,
Qingdao 266071, P.R. China
yingmu1229@sina.com, su-hua-2003@163.com, yingmu1229@hotmail.com

Abstract. In this paper, we mainly extend entropy and subsethood from intuitionistic fuzzy sets to general interval-valued intuitionistic fuzzy sets, propose a definition of entropy and subsethood, offer a function of entropy and construct a class of subsethood function. Then from discussing the relationship between entropy and subsethood, we know that while choosing the subsethood, we can get some kinds of function of entropy based on subsethood. Our work is also applicable to practical fields such as: neural networks, expert systems, and other.

1 Introduction

Since L.A.Zadeh introduced fuzzy sets [1] in 1965, a lot of new theories treating imprecision and uncertainty have been introduced. Some of them are extensions of fuzzy set theory. K.T.Atanassov extended this theory, proposed the definition of intuitionistic fuzzy sets (IFS, for short) ([3] [4]) and interval-valued intuitionistic fuzzy sets ($IVIFS$, for short) [5], which have been found to be highly useful to deal with vagueness out of several higher order fuzzy sets. And then, in the year 1999, Atanassov defined a Lattice-intuitionistic fuzzy set [6].

A measure of fuzziness often used and cited in the literature is entropy first mentioned in 1965 by L.A.Zadeh [2]. The name entropy was chosen due to an intrinsic similarity of equations to the ones in the Shannon entropy [7]. But they are different in types of uncertainty and Shannon entropy basically measures the average uncertainty in bits associated with the prediction of outcomes in a random experiment. This theory was extended and a non-probabilistic-type entropy measure for IFS was proposed by Eulalia Szmidt, Janusz Kacprzyk [8]. We extended it again onto general interval-valued intuitionistic fuzzy sets ($VIFS$, for short) in this paper.

We organize this paper as follows: in section 2, at first, we define the definition of $VIFS$, then discuss the relation between $VIFS$ and some other kinds of intuitionistic fuzzy sets. From the discussion, we make sure that all these sets are subset of $VIFS$ in the view of isomorphic imbedding mapping, and this ensures that our work about the theory of entropy and subsethood is a reasonable extension of FS, IFS and $IVIFS$. So according to section 2, theories in this paper are feasible on all these kinds of intuitionistic fuzzy sets. In section 3, we mainly extend entropy [7] onto $VIFS(X)$. Firstly, we give a reasonable definition of A is less fuzzy than B on $VIFS$. For there is not always a comparable relation between any two elements of L, which is a lattice. So we premise the definition of A refines B with the discussion of the incomparable condition. Then we definite entropy by giving the properties of A refines B in theorem 3.1. Finally from theorem 3.2, we get a idiographic entropy function. In section 4, we definite a class of subsethood function by three real functions. Then, from discussing the relation between entropy and subsethood, we get a class of entropy function which is useful in different conditions and eg 2 ensures that is reasonable. Thus, while choosing the subsethood, we can get some kinds of function of entropy based on subsethood.

2 Definitions and Quantities of Some Kinds of IFS

Throughout this paper, let X is a nonempty definite set, $|X| = n$.

Definition 2.1. An interval number over $[0, 1]$ is defined as an object of the form: $\underline{a} = [a^-, a^+]$ with the property: $0 \leq a^- \leq a^+ \leq 1$.

Let L denote all interval numbers over $[0, 1]$. We define the definition of relation (it is specified by " $\leq$ "):
$$\underline{a} \leq \underline{b} \Leftrightarrow a^- \leq b^- \text{ and } a^+ \leq b^+ .$$
We can easily prove that " $\leq$ " is a partially ordered relation on L. So $< L, \text{ "} \leq \text{"} >$ is a complete lattice where
$$\underline{a} \vee \underline{b} = [a^- \vee b^-, a^+ \vee b^+] , \quad \underline{a} \wedge \underline{b} = [a^- \wedge b^-, a^+ \wedge b^+] ,$$
$$\bigvee_{i \in I} \underline{a}_i = [\bigvee_{i \in I} a_i^-, \bigvee_{i \in I} a_i^+] , \quad \bigwedge_{i \in I} \underline{a}_i = [\bigwedge_{i \in I} a_i^-, \bigwedge_{i \in I} a_i^+] .$$
Let $\underline{1} = [1,1]$ denotes the maximum of $< L, \text{ "} \leq \text{"} >$ and $\underline{0} = [0,0]$ denotes the minimum of $< L, \text{ "} \leq \text{"} >$.

We define the "C" operation, where $\underline{a}^c = [1 - a^+, 1 - a^-]$.

We can easily prove that the complement operation "C" has the following properties:
1. $(\underline{a}^c)^c = \underline{a}$, 2. $(\underline{a} \vee \underline{b})^c = (\underline{a}^c \wedge \underline{b}^c), (\underline{a} \wedge \underline{b})^c = (\underline{a}^c \vee \underline{b}^c)$,
3. $(\bigvee_{i \in I} \underline{a}_i)^c = \bigwedge_{i \in I} \underline{a}_i^c, (\bigwedge_{i \in I} \underline{a}_i)^c = \bigvee_{i \in I} \underline{a}_i^c$, 4. If $\underline{a} \leq \underline{b}$, then $(\underline{a}^c \geq \underline{b}^c)$ $\forall \underline{a}, \underline{b} \in L$.
Then we will give the definition of intuitionistic fuzzy sets on L .

Definition 2.2. A general interval valued intuitionistic fuzzy sets A on X is an object of the form: $A = \{< x, \mu_A(x), \nu_A(x), x \in X >\}$,
where $\mu_A : X \to L$ and $\nu_A : X \to L$,
with the property: $\underline{0} \leq \mu_A(x) \leq \nu_A(x)^c \leq \underline{1}$ $(\forall x \in x)$.
For briefly, we denote $A = \{< x, \mu_A(x), \nu_A(x) >, x \in X\}$ by $A = < \mu_A(x), \nu_A(x) >$ and denote $\mu_A(x) = [\mu_A^-(x), \mu_A^+(x)] \in L, \nu_A(x) = [\nu_A^-(x), \nu_A^+(x)] \in L$.
Let $VIFS(X)$ denote all set of general interval valued intuitionistic fuzzy sets on L . We define the relation " $\leq$ ": $A \leq B \Leftrightarrow \mu_A(x) \leq \mu_B(x)$ and $\nu_A(x) \geq \nu_B(x)$.
We can easily prove that $\leq$ is also a partially ordered relation on $VIFS(X)$. So $< VIFS(X), \text{ "} \leq \text{"} >$ is a complete lattice where

$$A \vee B = [\mu_A(x) \vee \mu_B(x),\ \nu_A(x) \wedge \nu_B(x)],\quad \bigvee_{i \in I} A_i = [\bigvee_{i \in I} \mu_{A_i},\ \bigwedge_{i \in I} \nu_{A_i}],$$
$$A \wedge B = [\mu_A(x) \wedge \mu_B(x),\ \nu_A(x) \vee \nu_B(x)],\quad \bigwedge_{i \in I} A_i = [\bigwedge_{i \in I} \mu_{A_i},\ \bigvee_{i \in I} \nu_{A_i}].$$

We define the "C" operation, where $A^c = <\nu_A(x),\ \mu_A(x)>$. We can easily prove that the complement operation "C" has the following properties:
1. $(A^c)^c = A$, 2. $(A \vee B)^c = (A^c \wedge B^c)$, 3. $(\bigvee_{i \in I} A_i)^c = \bigwedge_{i \in I} A_i^c,\ (\bigwedge_{i \in I} A_i)^c = \bigvee_{i \in I} A_i^c$,
4. If $A \leq B$, then $(A^c \geq B^c)\ \forall A, B \in VIFS(X)$.
So we know $<VIFS(X),\ "\leq"\ >$ is a complete lattice with complement. Let I denotes the maximum of $VIFS(X)$ and θ denotes the minimum of $VIFS(X)$, Where $\mu_I(x) = \underline{1}$, $\nu_I(x) = \underline{0}$, $\mu_\theta(x) = \underline{0}$, $\nu_\theta(x) = \underline{1}$. Then we will discuss the relationship between $IVIFS(X)^{[5]}$ and $VIFS(X)$.

Definition 2.3. An interval-valued intuitionistic fuzzy set on X ($IVIFS(X)$, for short) is an object of the form $B = \{<x, M_B(x), N_B(x), x \in X>\}$.
Where $M_B : X \to L$ and $N_B : X \to L\ (\forall x \in X)$,
With the condition $0 \leq M_B^+(x) + N_B^+(x) \leq 1 \qquad (\forall x \in X)$.

Proposition 2.1. $IVIFS(X)^{[5]}$ is a subset of $VIFS(X)$.

Proof. To any B belongs to $IVIFS(X)^{[5]}$, we have
$B = \{<x, M_B(x), N_B(x), x \in X>\}$, Where $M_B : X \to L$ and $N_B : X \to L$
With the condition $0 \leq M_B^+(x) + N_B^+(x) \leq 1\ (\forall x \in X)$,
that is $0 \leq M_B^+(x) \leq 1 - N_B^+(x)\ (\forall x \in X)$. from definition 2.2 we can easily prove $0 \leq M_B(x) \leq N_B(x)^C \leq 1\ (\forall x \in X)$. That is B belongs to $VIFS(X)$. So $IVIFS(X)^{[5]}$ is a subset of $VIFS(X)$. We have an example here:

Eg1. Let $A = \{<x, [0, 0.3], [0, 0.8]>, x \in X\}$ for $[0, 0.8]^c = [0.2, 1], [0, 0.3] \leq [0.2, 1]$, we have $A \in VIFS(X)$. But from $0.3 + 0.8 \geq 1$, we get $A \notin VIFS(X)$. Then we get $IVIFS(X) \subset VIFS(X)$ and $IVIFS(X) \neq VIFS(X)$. It is shown in the Theorem 2.1 that in the view of isomorphism insertion, we consider $P(X)$ is a subset of $VIFS(X)$.

Definition 2.4. A function $P(X) \to VIFS(X)$, where
$$\mu_{\delta(A)}(x) = \begin{cases} \underline{1}, & \text{if } x \in A, \\ \underline{0}, & \text{if } x \notin A \end{cases};\quad \nu_{\delta(A)}(x) = \begin{cases} \underline{0}, & \text{if } x \in A, \\ \underline{1}, & \text{if } x \notin A \end{cases} \quad (\forall A \in P(X)).$$

Theorem 2.1. The function δ is a injection map reserving union, intersection and complement.

Proof. Let $A_i \in I$,
1. It is easy to prove that δ is a injection.
2. $\mu_{\delta(\bigvee_{i \in I} A_i)}(x) = \underline{1} \Leftrightarrow x \in \bigvee_{i \in I} A_i \Leftrightarrow \exists i_0 \in I$ satisfying $x \in A_{i_0}$,
$\bigvee_{i \in I} \mu_{\delta(A_i)}(x) = \underline{1} \Leftrightarrow \exists i_0 \in I,\ \mu_{\delta(A_{i_0})}(x) = \underline{1} \Leftrightarrow \exists i_0 \in I$ satisfying $x \in A_{i_0}$,
That is $\mu_{\delta(\bigvee_{i \in I} A_i)}(x) = \bigvee_{i \in I} \mu_{\delta(A_i)}(x)\ (\forall x \in X)$,

$\nu_{\delta(\bigvee_{i\in I} A_i)}(x) = \underline{1} \Leftrightarrow x \notin \bigvee_{i\in I} A_i \Leftrightarrow \forall i \in I$ there is $x \notin A_i$.

$\bigwedge_{i\in I} \nu_{\delta(A_i)}(x) = \underline{1} \Leftrightarrow \forall i \in I$, $\nu_{\delta(A_i)}(x) = \underline{1} \Leftrightarrow \forall i \in I$, there is $x \notin A_i$.

That is $\nu_{\delta(\bigvee_{i\in I} A_i)}(x) = \bigwedge_{i\in I} \nu_{\delta(A_i)}(x)$ ($\forall x \in X$). So we have $\delta(\bigvee_{i\in I} A_i) = \bigvee_{i\in I} \delta(A_i)$.

3. Similarly, we can prove $\delta(\bigwedge_{i\in I} A_i) = \bigwedge_{i\in I} \delta(A_i)$,

4.
$$\mu_{\delta(A^c)}(x) = \underline{1} \Leftrightarrow x \in A^c \Leftrightarrow x \notin A , \tag{1}$$
$$\nu_{\delta(A^c)} = \underline{1} \Leftrightarrow x \notin A^c \Leftrightarrow x \in A , \tag{2}$$
$$\mu_{(\delta(A))^c}(x) = \nu_{(\delta(A))}(x) = \underline{1} \Leftrightarrow x \notin A , \tag{3}$$
$$\nu_{(\delta(A))^c}(x) = \mu_{(\delta(A))}(x) = \underline{1} \Leftrightarrow x \in A . \tag{4}$$

From (1) and (3) we have

$\mu_{\delta(A^c)}(x) = \mu_{(\delta(A))^c}(x)$ ($\forall x \in X$) , $\nu_{\delta(A^c)} = \nu_{(\delta(A))^c}(x)$ ($\forall x \in X$) .

Then we get $\delta(A^c) = (\delta(A))^c$. So δ is a injection map reserving union, intersection and complement. In the view of imbedding mapping, we may consider $P(X)$ as the subset of $VIFS(X)$. Let $F(X)$ is the class of all fuzzy sets proposed by L.A.Zadeh, let $\xi: F(x) \to VIFS(X)$,$\xi(F) = A = <\mu_A(x), \nu_A(x) >$, $\mu_A(x) = [F(x), F(x)]$, $\nu_A(x) = [1 - F(x), 1 - F(x)]$.

So that the injection ξ holds union, intersection and complement.

In the same way, we may also consider $F(X)$ as the subset of $VIFS(X)$ in the view of imbedding mapping.

Definition 2.5. An intuitionistic fuzzy set on X ($IFS(X)$, for short) is an object of the form: $A = \{< x, \mu_A(x), \nu_A(x), x \in X >\}$,
where $\mu_A : X \to [0,1]$ and $\nu_A : X \to [0,1]$,
with the property: $0 \leq \nu_A(x) + \mu_A(x) \leq 1$ ($\forall x \in x$).

Let's discuss the relationship between $IFS(X)^{[4]}$ and $VIFS(X)$.
Let $\eta: IFS(X) \to VIFS(X)$ Where any C belongs to $IFS(X)^{[4]}$,
$C = \{< x, \alpha(x), \beta(x) >, x \in X\}$ ($\alpha(x) + \beta(x) \leq 1$),
$\eta(C) = A = <\mu_A(x), \nu_A(x) >, \mu_A(x) = [\alpha(x), \alpha(x)]$, $\nu_A(x) = [\beta(x), \beta(x)]$.
Similarly we have $IFS(X)$ is the subset of $VIFS(X)$ in the view of imbedding mapping. All the above ensure that what we get in this paper are feasible on $P(X), F(X), IFS(X)$ and $IVIFS(X)$.

A method of judging whether set $A \in VIFS(X)$ is a subset of $P(X)$, is given by the following theorem .

Theorem 2.2. Let $A \in VIFS(X)$, so we have $A \vee A^c = I \Leftrightarrow A \in \delta(P(X))$.

Proof. "$\Rightarrow$" Let $A = <\mu_A(x), \nu_A(x) >$, from definition 1) we have:
$A^c = <\nu_A(x), \mu_A(x) >$, $\mu_{A \vee A^c} = \mu_A(x) \vee \nu_A(x)$, $\nu_{A \vee A^c} = \mu_A(x) \wedge \nu_A(x)$.
From $\mu_A(x) \vee \nu_A(x) = \underline{1}$, we have $\mu_A^-(x) \vee \nu_A^-(x) = 1$; $\tag{1}$
From $\mu_A(x) \wedge \nu_A(x) = \underline{0}$, we have $\mu_A^+(x) \wedge \nu_A^+(x) = 0$. $\tag{2}$

There are two different cases:
1. If $\nu_A^+(x) = 0$, then we have $\nu_A^-(x) = 0$, that is $\nu_A(x) = \underline{0}$, and from (1) we have $\mu_A^-(x) = 1$. So we get $\mu_A^+(x) = 1$, that is $\mu_A(x) = \underline{1}$;
2. If $\mu_A^+(x) = 0$ then we have $\mu_A^-(x) = 0$, that is $\mu_A(x) = \underline{0}$.

And from (1) we have $\nu_A^-(x) = \underline{1}$. So we get $\nu_A^+(x) = 1$ that is $\nu_A(x) = \underline{1}$. So we know that for any x belongs to X, $\mu_A(x)$ and $\nu_A(x)$ can only be $\underline{0}$ or $\underline{1}$ with the following property: $\mu_A(x) = \underline{1} \Leftrightarrow \nu_A(x) = \underline{0}$. Let $K = \{x | x \in X, \mu_A(x) = \underline{1}\}$, from definition we have: $\mu_{\delta(K)}(x) = \underline{1} \Leftrightarrow x \in K \Leftrightarrow \mu_A(x) = \underline{1}$; $\nu_{\delta(K)}(x) = \underline{0} \Leftrightarrow x \in K \Leftrightarrow \mu_A(x) = \underline{1} \Leftrightarrow \nu_A(x) = \underline{0}$ ($\forall x \in X$). So we get $\mu_{\delta(K)}(x) = \mu_A(x), \nu_{\delta(K)}(x) = \nu_A(x)$ ($\forall x \in X$). That is $\delta(K) = A$.
" $\Leftarrow$ " It is straight-forward.

Corollary 2.1. $A \wedge A^c = \theta \Leftrightarrow A \in \delta(P(X))$.
From proposition 2.2 to proposition 2.4, we give some qualities of $VIFS$.

Proposition 2.2. Let $A \in VIFS(X)$, then we have:
1. $\mu_A(x) = \underline{1} \Rightarrow \nu_A(x) = \underline{0}$ ($\forall x \in X$), 2. $\nu_A(x) = \underline{1} \Rightarrow \mu_A(x) = \underline{0}$ ($\forall x \in X$).

Proof. 1. For $\nu_A(x) \le (\mu_A(x))^c = (\underline{1})^c = \underline{0}$, we have $\nu_A = \underline{0}$. Similarly we can prove 2.

Proposition 2.3. $A \vee A^c \ne \theta$ ($\forall A \in VIFS(X)$)

Proof. Let $A = <\mu_A(x), \nu_A(x)>$
then $A^c = <\nu_A(x), \mu_A(x)>$, $A \vee A^c = <\mu_A(x) \vee \nu_A(x), \nu_A(x) \wedge \mu_A(x)>$.
If $A \vee A^c = \theta$, then we have $\mu_A(x) \vee \nu_A(x) = \underline{0}$, $\nu_A(x) \wedge \mu_A(x) = \underline{1}$.
It is contradiction. Hence $A \vee A^c \ne \theta$ ($\forall A \in VIFS(X)$).

Proposition 2.4. Let $A \in VIFS(x)$, Then we have $\nu_{A \vee A^c}(x) = \nu_{A \wedge A^c}(x) \Leftrightarrow \mu_A(x) = \nu_A(x)$ ($\forall x \in X$).
Proof. " $\Rightarrow$ " From the definition 2.2,
$\nu_{A \vee A^c}(x) = \nu_A(x) \wedge \mu_A(x) = [\nu_A^-(x) \wedge \mu_A^-(x), \nu_A^+(x) \wedge \mu_A^+(x)]$,
$\nu_{A \wedge A^c}(x) = \nu_A(x) \vee \mu_A(x) = [\nu_A^-(x) \vee \mu_A^-(x), \nu_A^+(x) \vee \mu_A^+(x)]$.
From $\nu_{A \vee A^c}(x) = \nu_{A \wedge A^c}(x)$,
we have $\nu_A^-(x) \wedge \mu_A^-(x) = \nu_A^-(x) \vee \mu_A^-(x)$, $\nu_A^+(x) \wedge \mu_A^+(x) = \nu_A^+(x) \vee \mu_A^+(x)$.
So we get $\nu_A^-(x) = \mu_A^-(x), \nu_A^+(x) = \mu_A^+(x)$. That is $\nu_A(x) = \mu_A(x)$.
" $\Leftarrow$ " It is straight-forward.

3 Entropy on VIFS

De Luca and Termini [12] first axiomatized non-probabilistic entropy. The De Luca-Termini axioms formulated for intuitionistic fuzzy sets are intuitive and have been wildly employed in the fuzzy literature. To extend this theory, firstly we definite the function as follows:

Definition 3.1. $M: VIFS(X) \to [0,1]$, where $\forall A \in VIFS(x)$,
$$M(A) = \frac{1}{2n} \sum_{x \in X} [2 - \nu_A^-(x) - \nu_A^+(x)] \ (\forall x \in X).$$

Proposition 3.1. To any $A \in VIFS(X)$, $M(A) = 0 \Leftrightarrow \nu_A^-(x) = \nu_A^+(x) = 1 \Leftrightarrow A = \theta$ $(\forall x \in X)$.

Proof. $M(A) = 0 \Leftrightarrow 2 - \nu_A^-(x) - \nu_A^+(x) = 0$ $(\forall x \in X)$, for $\nu_A^-(x) \leq 1, \nu_A^+(x) \leq 1$, we have $M(A) = 0 \Leftrightarrow \nu_A^-(x) = \nu_A^+(x) = 1$ $(\forall x \in X)$. From proposition 2.2, we have $\mu_A(x) = \underline{0}$. So we have $\mu_A(x) = \mu_\theta(x)$, $\nu_A(x) = \nu_\theta(x)$. That is $A = \theta$, hence $M(A) = 0 \Leftrightarrow \nu_A^-(x) = \nu_A^+(x) = 1 \Leftrightarrow A = \theta$ $(\forall x \in X)$.

Proposition 3.2. Let $A, B \in VIFS(X)$, then we have $A \geq B \Rightarrow M(A) \geq M(B)$.

Proof. For $A \geq B$, then we have $\nu_A(x) \leq \nu_B(x)$ $(\forall x \in X)$. So we have
$$\nu_A^-(x) \leq \nu_B^-(x), \nu_A^+(x) \leq \nu_B^+(x).$$
Hence $\sum_{x \in X}[2 - \nu_A^-(x) - \nu_A^+(x)] \geq \sum_{x \in X}[2 - \nu_A^-(x) - \nu_A^+(x)]$. That is $M(A) \geq M(B)$.

Proposition 3.3. Let $A, B \in VIFS(X)$, then we have
$$\nu_A(x) \leq \nu_B(x) \text{ and } M(A) \geq M(B) \Leftrightarrow \nu_A(x) = \nu_B(x) \quad (\forall x \in X).$$

Proof. "$\Leftarrow$" It is straight-forward.
"$\Rightarrow$" For $\nu_A(x) \leq \nu_B(x)$, we have $\nu_A^-(x) \leq \nu_B^-(x)$ and $\nu_A^+(x) \leq \nu_B^+(x)$. So we get $2 - \nu_A^-(x) - \nu_A^+(x) \geq \nu_B^-(x) - \nu_B^+(x)$ $(\forall x \subset X)$. If there is a $x_0 \in X$, let one of the following two inequations $\nu_A^-(x) < \nu_B^-(x)$, $\nu_A^+(x) < \nu_B^+(x)$ establish, then we have
$$M(A) = \sum_{x \neq x_0, x \in X}[2 - \nu_A^-(x) - \nu_A^+(x)] + [2 - \nu_A^-(x_0) - \nu_A^+(x_0)]$$
$$> \sum_{x \neq x_0, x \in X}[2 - \nu_B^-(x) - \nu_B^+(x)] + [2 - \nu_B^-(x_0) - \nu_B^+(x_0)]$$
$$= M(B).$$
This is contrary to $M(A) = M(B)$. So for any $x \in X$, we have $\nu_A^-(x) = \nu_B^-(x)$, $\nu_A^+(x) = \nu_B^+(x)$. That is $\nu_A(x) = \nu_B(x)$ $(\forall \in X)$.

Definition 3.2. Let $A, B \in VIFS(X)$ with the following properties:
1. If $\mu_B(x) \geq \nu_B(x)$, then $\mu_A(x) \geq \mu_B(x)$ and $\nu_A(x) \leq \nu_B(x)$,
2. If $\mu_B(x) < \nu_B(x)$, then $\mu_A(x) \leq \mu_B(x)$ and $\nu_A(x) \geq \nu_B(x)$,
3. If $\mu_B(x)$ and $\nu_B(x)$ is incomparable, there are two conditions:

1). If $\mu_B^-(x) \geq \nu_B^-(x)$, then there are four inequations as follow:
$(1) \mu_A^-(x) \geq \mu_B^-(x)$, $(2) \mu_A^+(x) \leq \mu_B^+(x)$, $(3) \nu_A^-(x) \leq \nu_B^-(x)$, $(4) \nu_A^+(x) \geq \nu_B^+(x)$.
2). If $\mu_B^-(x) < \nu_B^-(x)$ then there are four inequations as follow:
$(5) \mu_A^-(x) \leq \mu_B^-(x)$, $(6) \mu_A^+(x) \geq \mu_B^+(x)$, $(7) \nu_A^-(x) \geq \nu_B^-(x)$, $(8) \nu_A^+(x) \leq \nu_B^+(x)$.
Thus we call that A refines B (that is A is less fuzzy than B).

Theorem 3.1. Let $A, B \in VIFS(X)$, and A refines B, then we have
1. $(A \wedge A^c) \leq (B \wedge B^c)$, 2. $(A \vee A^c) \geq (B \vee B^c)$.

Proof. We prove inequation 1 first:
1. If $\mu_B(x) \geq \nu_B(x)$ from definition 3.2, we have $\mu_A(x) \geq \mu_B(x)$ and $\nu_A(x) \leq \nu_B(x)$. Then we have $\mu_A(x) \geq \mu_B(x) \geq \nu_B(x) \geq \nu_A(x)$.

Hence $\mu_{A \wedge A^c}(x) = \mu_A(x) \wedge \nu_A(x) = \nu_A(x)$, $\mu_{B \wedge B^c}(x) = \mu_B(x) \wedge \nu_B(x) = \nu_B(x)$,
$\nu_{A \wedge A^c}(x) = \mu_A(x) \vee \nu_A(x) = \mu_A(x)$, $\nu_{B \wedge B^c}(x) = \mu_B(x) \vee \nu_B(x) = \mu_B(x)$.
So $\mu_{A \wedge A^c}(x) \leq \mu_{B \wedge B^c}(x)$, $\nu_{A \wedge A^c}(x) \geq \nu_{B \wedge B^c}(x)$.

2. If $\mu_B(x) < \nu_B(x)$, from definition 3.2, we have $\mu_A(x) \leq \mu_B(x)$ and $\nu_A(x) \geq \nu_B(x)$. Then we have $\mu_A(x) \leq \mu_B(x) \leq \nu_B(x) \leq \nu_A(x)$.
Hence $\mu_{A \wedge A^c}(x) = \mu_A(x) \wedge \nu_A(x) = \mu_A(x)$, $\mu_{B \wedge B^c}(x) = \mu_B(x) \wedge \nu_B(x) = \mu_B(x)$,
$\nu_{A \wedge A^c}(x) = \mu_A(x) \vee \nu_A(x) = \nu_A(x)$, $\nu_{B \wedge B^c}(x) = \mu_B(x) \vee \nu_B(x) = \nu_B(x)$.
So $\mu_{A \wedge A^c}(x) \leq \mu_{B \wedge B^c}(x)$, $\nu_{A \wedge A^c}(x) \geq \nu_{B \wedge B^c}(x)$.

3. In the case of $\mu_B(x)$ and $\nu_B(x)$ is incomparable, there are still another two different cases:

1). If $\mu_B^-(x) \geq \nu_B^-(x)$ (9), for $\mu_B(x)$ is not larger than $\nu_B(x)$, we have $\mu_B^+(x) \leq \nu_B^+(x)$ (10). From (1), (3) and (9) we get $\mu_A^-(x) \geq \mu_B^-(x) \geq \nu_B^-(x) \geq \nu_A^-(x)$. So we have $\mu_A^-(x) \geq \nu_A^-(x)$ (11) and from (2), (4) and (10) we get $\mu_A^+(x) \leq \mu_B^+(x) \leq \nu_B^+(x) \leq \nu_A^+(x)$. So we have $\mu_A^+(x) \leq \nu_A^+(x)$ (12). Then from (9) and (10), we have

$\mu_{B \wedge B^c}(x) = \mu_B(x) \wedge \nu_B(x) = [\mu_B^-(x) \wedge \nu_B^-(x), \mu_B^+(x) \wedge \nu_B^+(x)] = [\nu_B^-(x), \mu_B^+(x)]$,
$\nu_{B \wedge B^c}(x) = \mu_B(x) \vee \nu_B(x) = [\mu_B^-(x) \vee \nu_B^-(x), \mu_B^+(x) \vee \nu_B^+(x)] = [\mu_B^-(x), \nu_B^+(x)]$.
In the same way, from (11) and (12), we get $\mu_{A \wedge A^c}(x) = [\nu_A^-(x), \mu_A^+(x)]$ and $\nu_{A \wedge A^c}(x) = [\mu_A^-(x), \nu_A^+(x)]$. Then from (2) and (3), we get $\mu_{A \wedge A^c}(x) \leq \mu_{B \wedge B^c}(x)$. From (1) and (4) we get $\nu_{A \wedge A^c}(x) \geq \nu_{B \wedge B^c}(x)$.

2). If $\mu_B^-(x) < \nu_B^-(x)$ (13), for $\mu_B(x)$ is not less than $\nu_B(x)$, we have $\mu_B^+(x) \geq \nu_B^+(x)$ (14). From (5), (13) and (7), we get $\mu_A^-(x) \leq \mu_B^-(x) < \nu_B^-(x) \leq \nu_A^-(x)$. Hence $\mu_A^-(x) < \nu_A^-(x)$ (15), and from (6),(14) and (8), we get $\mu_A^+(x) \geq \mu_B^+(x) \geq \nu_B^+(x) \geq \nu_A^+(x)$. Hence $\mu_A^+(x) \geq \nu_A^+(x)$ (16). Then from (13) and (14), we have
$\mu_{B \wedge B^c}(x) = \mu_B(x) \wedge \nu_B(x) = [\mu_B^-(x) \wedge \nu_B^-(x), \mu_B^+(x) \wedge \nu_B^+(x)] = [\mu_B^-(x), \nu_B^+(x)]$,
$\nu_{B \wedge B^c}(x) = \mu_B(x) \vee \nu_B(x) = [\mu_B^-(x) \vee \nu_B^-(x), \mu_B^+(x) \vee \nu_B^+(x)] = [\nu_B^-(x), \mu_B^+(x)]$.
In the same way, from (15) and (16), we get $\mu_{A \wedge A^c}(x) = [\mu_A^-(x), \nu_A^+(x)]$ and $\nu_{A \wedge A^c}(x) = [\nu_A^-(x), \mu_A^+(x)]$. Then from (5) and (8), we get $\mu_{A \wedge A^c}(x) \leq \mu_{B \wedge B^c}(x)$. From (6) and (7), we get $\nu_{A \wedge A^c}(x) \geq \nu_{B \wedge B^c}(x)$.

Summary, we get inequation 1. $(A \wedge A^c) \leq (B \wedge B^c)$. From inequation 1., we can easily get inequation 2. $(A \vee A^c) \geq (B \vee B^c)$.

Then we can define the definition of entropy on *VIFS*.

The De Luca-Termini axioms[12] were formulated in the following way. Let E be a set-to-point mapping $E : F(X) \to [0,1]$. Hence E is a fuzzy set defined on fuzzy sets. E is an entropy measure if it satisfies the four De Luca and Termini axioms:

1. $E(A) = 0$ iff $A \in X$ (A non-fuzzy), 2. $E(A) = 1$ iff $\mu_A(x) = 0.5$ for $\forall x \in X$,
3. $E(A) \leq E(B)$ if A is less fuzzy than B, i.e., if $\mu_A \leq \mu_B$ when $\mu_B \leq 0.5$ and $\mu_A \leq \mu_B$ when $\mu_B \geq 0.5$,
4. $E(A) = E(A^c)$.

Since the De Luca and Termini axioms were formulated for fuzzy sets, we extend them for *VIFS*.

Definition 3.3. A real function $E : VIFS(X) \to [0,1]$ is an entropy measure if E has the following properties:

1. $E(A) = 0 \Leftrightarrow A \in \delta(P(X))$, 2. $E(A) = 1 \Leftrightarrow \nu_A(x) = \mu_A(x)$ $(\forall x \in X)$,
3. $E(A) \leq E(B)$ if A refines B , 4. $E(A) = E(A^c)$.

Definition 3.4. Let $\sigma: VIFS(X) \to [0,1]$, where any $A \in VIFS(X)$, we have
$$\sigma(A) = \frac{M(A \wedge A^c)}{M(A \vee A^c)} \quad (\forall x \in X).$$
From proposition 2.3, we know that $A \vee A^c \neq \theta (\forall A \in VIFS(X))$, then from proposition 3.1, we know that $M(A \vee A^c) \neq 0$, and then from $A \vee A^c \geq A \wedge A^c$ and proposition 3.2, we know $M(A \vee A^c) \geq M(A \wedge A^c)$, that is
$0 \leq \sigma(A) = \frac{M(A \wedge A^c)}{M(A \vee A^c)} \leq 1$. So the definition of is reasonable.

Theorem 3.2. σ is entropy.

Proof.
1. $\sigma(A) = 0 \Leftrightarrow M(A \wedge A^C) = 0 \Leftrightarrow A \wedge A^c = \theta \Leftrightarrow A \in \delta(P(X))$.
2. $\sigma(A) = 1 \Leftrightarrow M(A \wedge A^c) = M(A \vee A^c)$.
For $A \vee A^c \geq A \wedge A^c$, we get $\nu_{A \vee A^c}(x) \leq \nu_{A \wedge A^c}(x)$, $(\forall x \in X)$. So from proposition 3.3, we have $\nu_{A \vee A^c}(x) = \nu_{A \wedge A^c}(x) (\forall x \in X)$. And from proposition 2.4, we get $\mu_A(x) = \nu_A(x) (\forall x \in X)$.
3. If A refines B, we have $A \wedge A^c \leq B \wedge B^c$. From proposition 3.2, we have $M(A \wedge A^c) \leq M(B \wedge B^c)$. And from $A \vee A^c \geq B \vee B^c$, we have $M(A \vee A^c) \geq M(B \vee B^c)$. So we get $\frac{M(A \wedge A^c)}{M(A \vee A^c)} \leq \frac{M(B \wedge B^c)}{M(B \vee B^c)}$, that is $\sigma(A) \leq \sigma(B)$.
4. It is straight-forward.
So we constructed a class of entropy function on $VIFS$.

4 Subsethood on VIFS

Definition 4.1. A real function $Q : IFS(X) \times IFS(X) \to [0,1]$ is called subsethood, if Q has the following properties:
1. $Q(A,B) = 0 \Leftrightarrow A = I, B = \theta$, 2. If $A \leq B \Rightarrow Q(A,B) = 1$,
3. If $A \geq B$ and $Q(A,B) = 1$, then $A = B$,
4. If $A \leq B \leq C$, then $Q(C,A) \geq Q(C,B), Q(C,A) \leq (Q(B,A))$.

Definition 4.2. We define the function $f : VIFS(X) \times VIFS(X) \to [0,1]$ by
$$f(A,B) = \frac{1}{2n} \{ \sum_{x \in X} min[1, g(\varphi(\mu_A^-(x) - \mu_B^-(x) + 1), \psi(\nu_A^-(x) - \nu_B^-(x) + 1))]$$
$$+ \sum_{x \in X} min[1, g(\varphi(\mu_A^+(x) - \mu_B^+(x) + 1), \psi(\nu_A^+(x) - \nu_B^+(x) + 1))] \} .$$
where $\varphi: [0,2] \to [0,2]$ and $\psi: [0,2] \to [0,2]$ with the following properties:
1. $\alpha > \beta \Rightarrow \varphi(\alpha) > \varphi(\beta), \psi(\alpha) > \psi(\beta)$ $(\alpha, \beta \in [0,2])$,
2. $\varphi(\alpha) = 2 \Leftrightarrow \alpha = 2; \psi(\beta) = 0 \Leftrightarrow \beta = 0$,
3. $\varphi(1) = \psi(1) = 1$.
And the function $g : [0,2] \times [0,2] \to [0,2]$ with the following properties:
1. $\alpha > \beta \Rightarrow g(\alpha, \gamma) < g(\beta, \gamma)$, $g(\gamma, \alpha) > g(\gamma, \beta)$ $(\alpha, \beta, \gamma \in [0,2])$,
2. $g(\alpha, \beta) = 0 \Leftrightarrow \alpha = 2, \beta = 0$,
3. $g(1,1) = 1$.

Theorem 4.1. f is subsethood.

Proof. Let $A, B \in VIFS(X)$
1. For each $x \in X$, we have $\mu_I(x) = 1, \nu_I(x) = 0, \mu_\theta(x) = 0$ and $\nu_\theta(x) = 1$.
Then
$$g(\varphi(\mu_A^-(x) - \mu_B^-(x) + 1), \psi(\nu_A^-(x) - \nu_B^-(x) + 1)) = 0 ,$$
$$g(\varphi(\mu_A^+(x) - \mu_B^+(x) + 1), \psi(\nu_A^+(x) - \nu_B^+(x) + 1)) = 0 .$$
So we get $f(I, \theta) = 0$.

On the contrary, If $f(A, B) = 0$, Then for each $x \in X$, We have
$$g(\varphi(\mu_A^-(x) - \mu_B^-(x) + 1), \psi(\nu_A^-(x) - \nu_B^-(x) + 1)) = 0 .$$
And then $\varphi(\mu_A^-(x) - \mu_B^-(x) + 1) = 2$, $\psi(\nu_A^-(x) - \nu_B^-(x) + 1) = 0$.
So we can get $(\mu_A^-(x) - \mu_B^-(x) + 1) = 2$ (1), $(\nu_A^-(x) - \nu_B^-(x) + 1) = 0$ (2).
From (1), we have $\mu_A^-(x) = \mu_B^-(x) + 1$. for $\mu_A^-(x) \leq 1, \mu_B^-(x) \geq 0$, We have
$$\mu_A^-(x) = 1 , \quad \mu_B^-(x) = 0 .$$
From (2), we have $\nu_B^-(x) = \nu_A^-(x) + 1$. Similarly, we have $\nu_A^-(x) = 0, \nu_B^-(x) = 1$.
By the same way, we can prove $\mu_A^+(x) = 1, \nu_A^+(x) = 0, \mu_B^+(x) = 0$, and $\nu_B^+(x) = 1$. Thus we have $\mu_A(x) = 1, \nu_A(x) = 0, \mu_B(x) = 0$, and $\nu_B(x) = 1$, that is $A = I$ and $B = \theta$.

2. If $A \leq B$, then for each $x \in X$, we have $\mu_A(x) \leq \mu_B(x)$ and $\nu_A(x) \geq \nu_B(x)$.
So we have the following four inequations:
$$\mu_A^-(x) - \mu_B^-(x) + 1 \leq 1 , \quad \mu_A^+(x) - \mu_B^+(x) + 1 \leq 1 ,$$
$$\nu_A^-(x) - \nu_B^-(x) + 1 \geq 1 , \quad \nu_A^+(x) - \nu_B^+(x) + 1 \geq 1 .$$
Then we have $\varphi(\mu_A^-(x) - \mu_B^-(x) + 1) \leq 1, \psi(\mu_A^+(x) - \mu_B^+(x) + 1) \leq 1$,
$$\varphi(\nu_A^-(x) - \nu_B^-(x) + 1) \geq 1, \psi(\nu_A^+(x) - \nu_B^+(x) + 1) \geq 1 .$$
And then we have
$$g(\varphi(\mu_A^-(x) - \mu_B^-(x) + 1), \psi(\nu_A^-(x) - \nu_B^-(x) + 1)) \geq g(1, 1) = 1 ,$$
$$g(\varphi(\mu_A^+(x) - \mu_B^+(x) + 1), \psi(\nu_A^+(x) - \nu_B^+(x) + 1)) \geq g(1, 1) = 1 .$$
So we get
$f(A, B) = \frac{1}{2n} \{\sum min[1, g(\varphi(\mu_A^-(x) - \mu_B^-(x) + 1), \psi(\nu_A^-(x) - \nu_B^-(x) + 1))$
$\quad + \sum min[1, g(\varphi(\mu_A^+(x) - \mu_B^+(x) + 1), \psi(\nu_A^+(x) - \nu_B^+(x) + 1))\}$
$= 1$.

3. If $f(A, B) = 1$, then for each $x \in X$, we have
$$g(\varphi(\mu_A^-(x) - \mu_B^-(x) + 1) , \psi(\nu_A^-(x) - \nu_B^-(x) + 1)) \geq 1 ,$$
$$g(\varphi(\mu_A^+(x) - \mu_B^+(x) + 1) , \psi(\nu_A^+(x) - \nu_B^+(x) + 1)) \geq 1 .$$
And from $A \geq B$, we have $\mu_A(x) \geq \mu_B(x), \mu_A(x) \leq \mu_B(x)$ ($\forall x \in X$). So we have the following four inequations: $\mu_A^-(x) - \mu_B^-(x) + 1 \geq 1, \mu_A^+(x) - \mu_B^+(x) + 1 \geq 1, \nu_A^-(x) - \nu_B^-(x) + 1 \leq 1, \nu_A^+(x) - \nu_B^+(x) + 1 \leq 1$. If $A \neq B$, then at least one of the four inequations given above would be never equal:
1) If $\mu_A^-(x) - \mu_B^-(x) + 1 > 1$, then we have
$$g(\varphi(\mu_A^-(x) - \mu_B^-(x) + 1), \psi(\nu_A^-(x) - \nu_B^-(x) + 1))$$
$$< g(1, \psi(\nu_A^-(x) - \nu_B^-(x) + 1))$$
$$\leq g(1, 1) \leq 1. \quad \text{It is contradiction.}$$
2) If $\mu_A^+(x) - \mu_B^+(x) + 1 > 1$, then we have
$$g(\varphi(\mu_A^+(x) - \mu_B^+(x) + 1), \psi(\nu_A^+(x) - \nu_B^+(x) + 1))$$
$$< g(1, \psi(\nu_A^+(x) - \nu_B^+(x) + 1))$$
$$\leq g(1, 1) = 1. \quad \text{It is contradiction.}$$

3) If $\nu_A^-(x) - \nu_B^-(x) + 1 < 1$, then we have
$$g(\varphi(\mu_A^-(x) - \mu_B^-(x) + 1), \psi(\nu_A^-(x) - \nu_B^-(x) + 1))$$
$$< g(\varphi(\mu_A^-(x) - \mu_B^-(x) + 1), 1)$$
$$\leq g(1,1) = 1. \quad \text{It is contradiction.}$$

4) If $\nu_A^+(x) - \nu_B^+(x) + 1 < 1$, then we have
$$g(\varphi(\mu_A^+(x) - \mu_B^+(x) + 1), \psi(\nu_A^+(x) - \nu_B^+(x) + 1))$$
$$< g(\varphi(\mu_A^+(x) - \mu_B^+(x) + 1), 1)$$
$$\leq g(1,1) = 1. \quad \text{It is contradiction.}$$

So we have $A = B$.

4. Let $A \leq B \leq C$ $(A, B, C \in VIFS(X))$, then for each x belongs to X, we have $\mu_A(x) \leq \mu_B(x) \leq \mu_C(x)$ and $\nu_A(x) \geq \nu_B(x) \geq \nu_C(x)$. So we have
$$\mu_C^-(x) - \mu_A^-(x) \geq \mu_C^-(x) - \mu_B^-(x), \quad \nu_C^-(x) - \nu_A^-(x) \leq \nu_C^-(x) - \nu_B^-(x),$$
$$\mu_C^+(x) - \mu_A^+(x) \geq \mu_C^+(x) - \mu_B^+(x), \quad \nu_C^+(x) - \nu_A^+(x) \leq \nu_C^+(x) - \nu_B^+(x).$$

Then we get
$$f(C, A) = \tfrac{1}{2n}\{\sum \min[1, g(\varphi(\mu_C^-(x) - \mu_A^-(x) + 1), \psi(\nu_C^-(x) - \nu_A^-(x) + 1))]$$
$$+ \sum \min[1, g(\varphi(\mu_C^+(x) - \mu_A^+(x) + 1), \psi(\nu_C^+(x) - \nu_A^+(x) + 1))]\}$$
$$\leq \tfrac{1}{2n}\{\sum \min[1, g(\varphi(\mu_C^-(x) - \mu_B^-(x) + 1), \psi(\nu_C^-(x) - \nu_B^-(x) + 1))]$$
$$+ \sum \min[1, g(\varphi(\mu_C^+(x) - \mu_B^+(x) + 1), \psi(\nu_C^+(x) - \nu_B^+(x) + 1))]\}$$
$$= f(C, B).$$

In the same way, we can prove $f(C, A) = f(B, A)$.

Eg2. For $\varphi, \psi : [0, 2] \to [0, 2], g : [0, 2] \times [0, 2] \to [0, 2], \varphi(x) = x, \psi(y) = y$, $g(x, y) = [(2 - x) + y] \times \tfrac{1}{2}$. It is straight-forward to prove that φ and ψ has the following qualities: 1). $\alpha > \beta \Rightarrow \varphi(\alpha) > \varphi(\beta), \psi(\alpha) > \psi(\beta), (\alpha, \beta \in [0, 2])$.
2). $\varphi(\alpha) = 2 \Leftrightarrow \alpha = 2$; $\psi(\beta) = 0 \Leftrightarrow \beta = 0$. 3). $\varphi(1) = \psi(1) = 1$.
And to g, it also has qualities as follows:
1). $\alpha > \beta \Rightarrow g(\alpha, \gamma) < g(\beta, \gamma), g(\gamma, \alpha) > g(\gamma, \beta)$ $(\alpha, \beta, \gamma \in [0, 2])$,
2). $g(\alpha, \beta) = 0 \Leftrightarrow \alpha = 2, \beta = 0$, 3).$g(1,1) = 1$.
It is shown in theorem 4.2. the relationship between entropy and subsethood on $VIFS$.

Theorem 4.2. Let Q is subsethood, $\rho: VIFS(X) \to [0, 1]$, where $\rho(A) = Q(A \vee A^c, A \wedge A^c)$ $(\forall A \in VIFS(X))$, then we have ρ is entropy.

Proof. 1. $\forall A \in \delta(P(X))$ for $A \vee A^c = I, A \wedge A^c = \theta$, we have
$$\rho(A) = Q(A \vee A^c, A \wedge A^c) = Q(I, \theta) = 0.$$
On the contrary, if $\rho(A) = Q(A \vee A^c, A \wedge A^c) = 0$. Then we have $A \vee A^c = I, A \wedge A^c = \theta$, so we get $\forall A \in \delta(P(X))$.
2. If $E(A) = 1$ that is $Q(A \vee A^c, A \wedge A^c) = 1$ and $A \vee A^c \geq A \wedge A^c$. Then we have $A \vee A^c = A \wedge A^c$. So we get $\mu_A(x) = \nu_A(x)$;
On the contrary, if $\mu_A(x) = \nu_A(x)$, we have $A \vee A^c = A \wedge A^c$, So we may get
$$Q(A \vee A^c, A \wedge A^c) = E(A) = 1.$$
3. If A refines B, then we have $A \wedge A^c \leq B \wedge B^c \leq B \vee B^c \leq A \vee A^c$, so we can get $E(A) = Q(A \vee A^c, A \wedge A^c) \leq Q(B \vee B^c, A \wedge A^c) \leq Q(B \vee B^c, B \wedge B^c) \leq E(B)$.
4. $E(A) = Q(A \vee A^c, A \wedge A^c) = Q(A^c \vee A, A^c \wedge A) = E(A^c)$.

From theorem 4.1 and theorem 4.2, we can construct a class of reasonable subsethood and entropy function, which are useful in different practical conditions.

5 Conclusion

In this paper, we offered different kinds of entropy function and subsethood function and they would be practical in different experiments. would be practical in different experiments.

References

1. Zadeh, L.A.: Fuzzy sets. Inform.and Control 8 (1965) 338–353
2. Zadeh, L.A.: Fuzzy sets and Systems. In: Proc. Systems Theory, Polytechnic Institute of Brooklyn, New York (1965) 29–67
3. Atanassov, K.T.: Intuitionistic Fuzzy Sets. In: V.Sgurev(Ed.), VII ITKR's session, Sofia, June (1983), Central Sci.and Techn.Library, Bulgaria Academy of Sciences(1984)
4. Atanassov, K.T.: Intuitionistic fuzzy sets. Fuzzy Sets and Systems 20 (1986) 87–97
5. Atanassov, K.T., Gargov, G.: Interval valued intuitionistic fuzzy sets. Fuzzy Sets and Systems 31 (1989) 343–349
6. Atanassov, K.T.: Intuitionistic Fuzzy Sets. Physica-Verlag, Heidelberg, New York(1999)
7. Jaynes, E.T.: Where do We Stand on Maximum Entropy? In: Levine, Tribus (Eds.), The Maximum Entropy Formalism, MIT Press, Cambridge, MA.
8. Szmidt, E., Kacprzyk, J.: Entropy for Intuitionistic Fuzzy Sets. Fuzzy Sets and Systems 118 (2001) 467–477
9. Yu-hai Liu, Feng-lan Xiong, Subsethood on Intuitionistic Fuzzy Sets. International Conference on Machine Learning and Cybernetics V.3 (2002) 1336–1339
10. Glad Deschrijver, Etienne E.Kerre, On her Relationship between some Extensions of Fuzzy Set Theory. Fuzzy Sets and Systems 133 (2003) 277–235
11. Guo-jun Wang, Ying-ming He: Intuitionistic Sets and Fuzzy Sets. Fuzzy Sets and Systems 110 (2000) 271–274
12. Luca, A.Ed., Termini, S.: A Definition of a Non-probabilistic Entropy in the Setting of Fuzzy Sets Theory. Inform. and Control 20 (1972) 301–312

The Comparative Study of Logical Operator Set and Its Corresponding General Fuzzy Rough Approximation Operator Set

Suhua Zheng, Xiaodong Liu, and Fenglan Xiong

Department of Mathematics, Ocean University of China; Qingdao 266071,
Shandong, P.R. China
su-hua-2003@163.com
yingmu1229@sina.com

Abstract. This paper presents a general framework for the study of fuzzy rough sets in which constructive approach is used. In the approach, a pair of lower and upper general fuzzy rough approximation operator in the lattice L is defined. Furthermore, the entire property and connection between the set of logical operator and the set of its corresponding general fuzzy rough approximation operator are examined, and we prove that they are 1-1 mapping. In addition, the structural theorem of negator operator is given. At last, the decomposition and synthesize theorem of general fuzzy rough approximation operator are proved. That for how to promote general rough approximation operator to suitable general fuzzy rough approximation operator, that is, how to select logical operator , provides theory foundation.

1 Introduction

After Pawlak proposed the notion of rough set in 1982[3], many authors have investigated attribute reduction and knowledge representation based on rough set theory. However, about fuzzy rough set and fuzzy rough approximation theory, the study ([4][5][6]) is not penetrating enough. In 2002, all kinds of logical operators and their corresponding fuzzy rough sets are defined and investigated in [1]. In 2003, generalized approximation operators are studied in [10]. This paper is the development and extension, we define the upper and lower general fuzzy rough approximation operator decided by t-norm and t-conorm in the lattice L. Besides, we study the entire properties and relativity between the logical operator set and its corresponding fuzzy rough approximation operator set.

This paper is organized as follows. In section 2, we define the mapping σ from the set of t-norm operators in the lattice L to the set of general fuzzy rough approximation operators, and we prove that σ is a surjection. Furthermore, we construct 1-1 mapping from the set of t-norm equivalent classes to the set of general upper fuzzy rough approximation operators. Subsequently, in section 3, we define the mapping $\sum_N$ from t-norm operator set to the pair of lower and upper fuzzy rough approximation operator set. In addition, we give the constructional theorem of negator, so we can construct different forms of negator operators, then get the dual t-norm and t-conorm operators with respect to N and their corresponding general lower and upper fuzzy rough approximation operators. In section 4, we prove the decomposition and synthesize theorem of general fuzzy rough approximation operator.

In this paper, unless otherwise stated, we will consider L be a lattice with the largest and least element $0, 1$; X, Y and Z to be finite and nonempty set; $F_L(X)$ to be all the fuzzy sets on X whose range is L; $R \in F_L(X \times Y)$ where R is a serial fuzzy relation[10]: for every $x \in X$, there exists $y \in Y$, such that $R(x,y) = 1$.

2 The Properties and Relativity of t-norm Operator Set and General Upper Fuzzy Rough Approximate Operator Set

Definition 2.1 ([1]) A t-norm operator in L is a mapping $t : L \times L \longrightarrow L$, if for $\forall x, y, y_1, y_2 \in L, t$ satisfies the following conditions:

1) $t(x,y) = t(y,x)$; 2) $t(x,1) = x$; 3) If $y_1 \geq y_2$, then $t(x,y_1) \geq t(x,y_2)$.

From the definition 2.1, it is easy to get the follows:

4) $t(x,0) = 0$ for all $x \in L$;   5) $t(x,y) \leq x \wedge y$ for all $x, y \in L$;

Definition 2.2. ([1][3])Let t be a t-norm in L, then $\varphi_t : F_L(Y) \to F_L(X)$ is called a general upper fuzzy rough approximation operator decided by t, if for every $x \in X, A \in F_L(Y)$, it have: $\varphi_t(A)(x) = \sup_{y \in Y} t(R(x,y), A(y))$.

For briefly, we denote φ_t by φ. And, (X, Y, R) is called the general fuzzy rough approximation space.

Theorem 2.1. Let $\lambda \in L, x^i, x^j, x^k, x^e \in X$; $y^r, y^d, y^p, y^q \in Y$; then φ has the following properties:

p1) $\varphi(z_\lambda)(x) = t(R(x,z), \lambda)$.   **p2)** $\varphi(\phi) = \phi$.
p3) $\varphi(\bigvee_{i \in I} z^i_{\lambda_i}) = \bigvee_{i \in I} \varphi(z^i_{\lambda_i})$. ($z^i_{\lambda_i} \in F_L(Y), I$ is finite; and if $\bigvee_{i \in I} \lambda_i \notin \{\lambda_i | i \in I\}$,
$i, j \in I, i \neq j$, it follows $z_i \neq z_j$).
p4) If $R(x^i, y^d) = R(x^j, y^r)$, it follows that $\varphi(y^d_\lambda)(x^i) = \varphi(y^r_\lambda)(x^j)$.
p5) If $R(x^i, y^d) = 1$, it follows that $\varphi(y^d_\lambda)(x^i) = \lambda$.
p6) $\varphi(y^d_{R(x^j, y^r)})(x^i) = \varphi(y^r_{R(x^i, y^d)})(x^j)$.
p7) Let $R(x^i, y^d) \geq R(x^j, y^r)$, it follows that $\varphi(y^d_\lambda)(x^i) \geq \varphi(y^r_\lambda)(x^j)$.
p8) Let $u, v \in L$ and satisfying: $R(x^i, y^d) < u < R(x^j, y^r)$, $R(x^k, y^p) < v < R(x^e, y^q)$; it follows that: $\varphi(y^d_v)(x^i) \vee \varphi(y^p_u)(x^k) \leq \varphi(y^r_v)(x^j) \wedge \varphi(y^q_u)(x^e)$.

From now on, assume that $\forall \alpha \in L, \alpha$ and $R(x,y)(\forall x \in X, y \in Y)$ are comparative.

Theorem 2.2. Let $\overline{L} = \{R(x,y) \mid x \in X, y \in Y\}, \overline{L} \bigcup \{0\} = \{0 = \overline{L}_0 < \overline{L}_1 < \overline{L}_2 < ... < \overline{L}_s = 1\}, T = \{t | t$ is a t-norm operator in $L\}, M = \{\varphi | \varphi : F_L(Y) \to$

$F_L(X)$ and satisfies $p2) - p8)\}$; we define $\sigma : T \to M$, by $\sigma(t) = \varphi_t$. Then σ is a surjection.

Proof: Firstly, we construct $t(u,v)$ from φ:

1) If $0 \in \{u,v\}$, then we define $t(u,v) = 0$;
2) If neither of them is equal to zero and at least one of them belong to $\overline{L} - \{0\}$, we have: If $u \in \overline{L} - \{0\}$, then we can assume $u = \overline{L}_i = R(x^i, y^i)$, where $0 < i \leq s, x^i \in X, y^i \in Y$, so we define $t(u,v) = \varphi(y_v^i)(x^i)$; By theorem 2.1 p4) follows that the definition is reasonable.
   Otherwise, let $t(u,v) = t(v,u)$. We can see that: If $v = R(x^j, y^j)$, from p6) it follows that: $\varphi(y_{R(x^j,y^j)}^i)(x^i) = \varphi(y_{R(x^i,y^i)}^j)(x^j)$ i.e. $t(u,v) = t(v,u)$.
3) If $u \neq 0, v \neq 0$ and $u, v \notin \overline{L} - \{0\}$, let $R(x^i, y^i) = \overline{L}_i < u < \overline{L}_{i+1} = R(x^{i+1}, y^{i+1})$, $R(x^j, y^j) = \overline{L}_j < v < \overline{L}_{j+1} = R(x^{j+1}, y^{j+1})$. We define: $t(u,v) = \varphi(y_v^i)(x^i) \vee \varphi(y_u^j)(x^j)$. And we can know: If arbitrary $x \in X, y \in Y$, it follows that $R(x,y) \neq 0$; then we define $R(x^0, y^0) = 0 = \varphi(y_v^0)(x^0) = \varphi(y_u^0)(x^0)$.

Now, we can prove that $t(u,v)$ is a t-norm, and $\varphi_t = \varphi$. From all of the above, we obtain that σ is a surjection.

Theorem2.3. Let σ and T be defined as before $t_1, t_2 \in T$, then for every $x \in X, y \in Y, \lambda \in L$, the following conclusion holds:
$\sigma(t_1) = \sigma(t_2)$ if and only if $t_1(R(x,y), \lambda) = t_2(R(x,y), \lambda)$.

Corollary2.4. Let arbitrary $x \in X, y \in Y, \lambda \in L$, If we define the equivalent relation on T as follows: $t_1 \sim t_2 \Leftrightarrow t_1(R(x,y), \lambda) = t_1(R(x,y), \lambda)$; then $\exists \mu : T/\sim \longrightarrow M$, by $\mu([t]) = \sigma(t), \forall [t] \in T/\sim$; furthermore, μ is a 1-1 mapping.

Definition2.3. We use $\underline{t}$ to denote the t-norm that we construct only by 1),2),3) in theorem 2.2. In fact, we also can construct other t-norm by φ, we can only change in 3) as follows: Let $R(x^i, y^i) = \overline{L}_i < u < \overline{L}_i = R(x^{i+1}, y^{i+1}), R(x^j, y^j) = \overline{L}_j < v < \overline{L}_{j+1} = R(x^{j+1}, y^{j+1})$; where $0 \leq i, j \leq s - 1$. Then we can definite $t(u,v) = \varphi(y_v^{i+1})(x^{i+1}) \wedge \varphi(y_u^{j+1})(x^{j+1})$. We denote it by $\overline{t}$, we can prove that $\sigma(\overline{t}) = \sigma(\underline{t})$.

In order to show it more clearly, we give theorem 2.5.

Theorem2.5. Let arbitrary $u, v \in L, t \in T, \varphi \in M$, and $\sigma(t) = \varphi$, then we have $\underline{t}(u,v) \leq t(u,v) \leq \overline{t}(u,v)$.

3 The Structural Theorem of Negator, the t-norm Equivalent Classes and the Pair of Dual Upper and Lower General Fuzzy Rough Approximation Operator is 1-1 Corresponding

Definition 3.1. [1] A t-conorm operator in L is a mapping $g : L \times L \longrightarrow L$. If for all $x, y, y_1, y_2 \in L, g$ satisfies the following conditions:

1) $g(x,y) = g(y,x)$; 2) $g(x,0) = x$; 3) If $y_1 \geq y_2$, then $g(x,y_1) \geq g(x,y_2)$.

From the definition 3.1, it is easy to get the property:

4) $g(x,y) \geq x \vee y$ for every $x,y \in L$. We denote $G = \{g|g$ is a t-conorm in $L\}$.

Definition 3.2[1]: A function $N : L \longrightarrow L$ is called negator, if $\forall x,y \in L, N$ has the properties: 1) $N(N(x)) = x$; 2) If $x \geq y$, then $N(x) \leq N(y)$.

It is easy to know the following properties hold:

3) $N(0) = 1, N(1) = 0$; 4)[8] $\bigwedge_{i \in I} N(x_i) = N(\bigvee_{i \in I} x_i)$, where I is finite and $x_i \in L$.

In addition, we denote $\Pi = \{N|N$ is a negator in $L.\}$, and assume $\Pi \neq \Phi$

Definition 3.3[1] Let $t \in T, g \in G, N \in \Pi$, if they satisfy
$t(x.y) = N(g(N(x), N(y)))$, we call that t,g are dual with respect to N.

It is easy to know that the following property holds:
t,g is dual with respect to N if and only if $t(N(x), N(y)) = N(g(x,y))$.

Definition 3.4.[1][8]: Let arbitrary $A \in F_L(Y), x \in X; \psi_g : F_L(Y) \longrightarrow F_L(X)$ is called general lower fuzzy rough approximation operator, if
$$\psi_g(A)(x) = \inf_{y \in Y} g(N(R(x,y)), A(y)).$$
Let $x \in Y, A \in F_L(Y)$, we define $N(A) \in F_L(Y)$ by $N(A(x)) = N(A)(x)$.

Definition 3.5. Let $\varphi, \psi : F_L(Y) \to F_L(X)$ are the upper and lower general fuzzy rough approximation operator respectively, we call that they are dual with respect to N if they satisfy: for every $A \in F_L(Y)$, $\varphi(A) = N(\psi(N(A)))$.

Theorem 3.1. Let $E = \{\psi_g | g \in G\}$, and $W_N = \{(\varphi, \psi) | \varphi, \psi$ are dual with respect to $N\} \subset M \times E$, where N is given. Let arbitrary $t \in T$, we define the function $\Sigma_N : T \longrightarrow W_N$ by $\Sigma_N(t) = (\varphi_t, \psi_g)$, where g is dual to t with respect to N. Then there exists $\nu : T/\sim \to W_N$ by $\nu([t]) = \Sigma_N(t)$, and σ is a 1-1 mapping.

Theorem 3.2. Let $L = [0,1]$, $h(x)$ is a negator if and only if there is a function $f(x)$ in $[0,1]$, which is continuous, strictly monotone decreasing and satisfying $f(0) = 1$, and $\exists \xi \in (0,1)$ such that
$$h(x) = \begin{cases} f(x), & \text{if } 0 \leq x \leq \xi, \\ f^{-1}(x), & \text{if } \xi < x \leq 1. \end{cases}$$

Remark. Theorem 3.2 is different from the constructional theorem of negator in [9]. There exist examples to prove that.

4 The Decomposition and Composition Theorem of General Fuzzy Rough Approximation Operator

In this section, L is an order lattice with the largest and least element denoted by 1,0. And, it has negator operator which satisfies that $\bigvee_{\lambda < \alpha} \lambda = \alpha$, for all $\alpha \in L$.

Definition 4.1. Let $R \in P(X \times Y)$, we define the neighborhood operator $f_R : X \longrightarrow P(Y)$ by $f_R(x) = \{y | \exists y \in Y, \text{ satisfying } (x,y) \in R\}$; for all $x \in X$.

Definition 4.2. Let $R \in P(X \times Y), \forall A \in P(Y)$, then $H_R : P(Y) \longrightarrow P(X); L_R : P(Y) \longrightarrow P(X)$ are called general rough approximation operator, by

$L_R(A) = \{x | x \in X \text{ and } f_R(x) \subseteq A\}; H_R(A) = \{x | x \in X \text{ and } f_R(x) \cap A \neq \phi\}$.

And, (X, Y, R) is called general rough approximation space.

Theorem 4.1. Let arbitrary $R \in F_L(X \times Y)$; φ_t, ψ_g are decided by t, g respectively. Then for all $\lambda \in L, A \in F_L(Y), (x,y) \in (X \times Y), \xi \in L$, it follows that:

1) $(\varphi_t(A))_\lambda = H_{R_\lambda}(A_\lambda)$ if and only if $t(R(x,y), \xi) = R(x,y) \wedge \xi$;
2) $(\psi_g(A))_\lambda = L_{R_{N(\lambda)}}(A_\lambda)$ if and only if $g(N(R(x,y)), \xi) = N(R(x,y)) \vee \xi$.

Corollary 4.2 Let arbitrary $(x,y) \in X \times Y, \lambda \in L, R \in F_L(X \times Y)$, if $t(R(x,y), \xi) = R(x,y) \wedge \xi$ and $g(N(R(x,y)), \xi) = N(R(x,y)) \vee \xi$, then for $\forall A \in F_L(Y)$, it follows that: 1) $\varphi_t(A) = \bigcup_{\lambda \in L} \lambda H_{R_\lambda}(A_\lambda)$. 2) $\psi_g(A) = \bigcup_{\lambda \in L} \lambda L_{R_{N(\lambda)}}(A_\lambda)$.

More generally, we have the following conclusions.

Theorem 4.3. Let $R \in F_L(X \times Y), A \in F_L(Y), \forall \lambda \in L$; it follows that:
1) $(\varphi_t(A))_\lambda \subseteq H_{R_\lambda}(A_\lambda)$, 2) $L_{R_{N(\lambda)}}(A_\lambda) \subseteq (\psi_g(A))_\lambda$.

Theorem 4.4. Let $R \in F_L(X \times Y), S \in F_L(Y \times Z)$, we define $P = R \circ S \in F_L(X \times Z)$, where for arbitrary $(x,z) \in (X \times Z), P(x,z) = \sup_{y \in Y} t(R(x,y), S(y,z))$. t, g are dual with respect to N and t is combinative, we define, $\varphi_P, \psi_P : F_L(Z) \longrightarrow F_L(X)$, for every $A \in F_L(Z)$, $\varphi_P(A)(x) = \sup_{z \in Z} t(P(x,y), A(z)), \psi_P(A)(x) = \inf_{z \in Z} g(N(P(x,y)), A(z))$; then we have:

1) $\varphi_P(A) = \varphi_R(\varphi_S(A))$; 2) $\psi_P(A) = \psi_R(\psi_S(A))$.

5 Conclusion

This paper studies the properties and relation between logical operator and general fuzzy rough approximation operator in theory. The further research is to construct the specific logical operator and corresponding general fuzzy rough approximation operator, furthermore, to apply them to knowledge representation, attribute reduction and machine learning, and so on.

References

1. Anna, M.R.K.: A Commparative Study of Fuzzy Rough Sets. Fuzzy Sets and Systems 126 (2002) 137–155
2. Jusheng Mi, Wenxiu Zheng: Indirect Learning Based on Rough Set Theory. Computer and Science 29(6) (2002) 97–97 (in Chinese)
3. Pawlak Z.: Rough Sets. [J] International Journal of Computer and Information Science 11(5) (1982) 205–218
4. Dubios, D., Prade, H.: Rough Fuzzy Sets and Fuzzy Rough Sets. Internat.J.General System 17 (1990) 191–209
5. Dubios, D., Prade, H.: Putting Fuzzy Sets and Rough Sets Together. In: R.S lowiTnski(Ed.): Interellgent Decision Support, Kluwer Academic, Dordrecht (1992) 203–232
6. Nakamura, A.: Fuzzy Rough Sets. Note Multiple-Valuaed Logic Japan 9 (8) (1988) 1–8
7. Klir, G.J., Yuan, B., Fuzzy Logic: Theory and Applications. Prentice-Hall, Englewood CliIs, NJ (1995)
8. Guojun Wang : L-fuzzy Topological Space. Shanxi Normal University Press, xi'an (1998)
9. Trillas, E.: Sobre functions de negationen en la teoria de conjuntos difusos, Stochastics 111 (1979)47–60
10. Wenzhi Wu, Jusheng Mi, Wenxiu Zhang: Generalized Fuzzy Rough Sets. Information Science 151(2003)263–282

Associative Classification Based on Correlation Analysis*

Jian Chen, Jian Yin, Jin Huang, and Ming Feng

Department of Computer Science, Zhongshan University,
Guangzhou, China
ellachen@gmail.com

Abstract. Associative classification is a well-known technique which uses association rules to predict the class label for new data object. This model has been recently reported to achieve higher accuracy than traditional classification approaches like C4.5. In this paper, we propose a novel associative classification algorithm based on correlation analysis, ACBCA, which aims at extracting the k-best strong correlated positive and negative association rules directly from training set for classification, avoiding to appoint complex support and confidence threshold. ACBCA integrates the advantages of the previously proposed effective strategies as well as the new strategies presented in this paper. An extensive performance study reveals that the improvement of ACBCA outperform other associative classification approaches on accuracy.

1 Introduction

Association rules describe the co-occurring relationships among data item in a large transaction database. They have been extensively studied in the literature for their usefulness in many real world areas such as market baskets analysis, expert system, stocks trend prediction, even public health surveillance, etc. So association rule mining is always a major topic in data mining research community. There have been many efficient algorithms and their variants of association rules discovery.

Classification is a supervised machine learning model, which firstly explores the relationships between data attributes and its class label in a training set, and built a classifier to predict new data object for which the class label is unknown. Classification has multiple applications and has been applied in many fields such as text categorization, drug discovery and development, speech recognition, etc.

In recent years, a new classification technique, called **associative classification**, is proposed to combine the advantages of association rule mining and classification. In general, this model has three phases as follows: (1) Generates all the

---

* This work is supported by the National Natural Science Foundation of China (60205007), Natural Science Foundation of Guangdong Province (031558, 04300462), Research Foundation of National Science and Technology Plan Project (2004BA721A02), Research Foundation of Science and Technology Plan Project in Guangdong Province (2003C50118) and Research Foundation of Science and Technology Plan Project in Guangzhou City (2002Z3-E0017).

class association rules (CARs) satisfying certain user-specified *minimum support* threshold as candidate rules by association rule mining algorithm, such as Apriori or FPgrowth. These discovered rules have the form of ($attributes \Rightarrow class\_label$); (2) Evaluates the qualities of all CARs discovered in the previous phase and pruning the redundant and low effective rules. Usually the *minimum confidence* is taken as a criterion to evaluate the qualities of the rules. Just "useful" rules with higher confidence than a certain threshold are selected to form a classifier; (3) Assigns a class label for a new data object. The classifier scores all the rules consistent with new data object and select some or all suitable rules to make a prediction.

The recent studies show that this classification model achieves higher accuracy than traditional rule-based classification approaches such as C4.5 [1] and Ripper [2]. However, this model still suffers from accuracy and efficiency because of its well-known support-confidence framework. In this paper, we propose a novel associative classification algorithm based on correlation analysis, which aims at extracting the k-best strong correlated positive and negative association rules directly from training set for classification, avoiding to appoint complex support and confidence threshold. The remainder of the paper is organized as follows: Section 2 gives some related concepts and definitions on associative classification. In Section 3, we introduce our approach in detailed on literal selection, correlation analysis and how to use these correlated rules for classification. Experimental results are described in Section 4 along with the performance of our algorithm in compared with other previous rule-based classification approaches. Finally, we summarize our research work and draw conclusions in Section 5.

2 Basic Concepts and Terminology

Let $\mathcal{A} = \{A_1, A_2, \ldots, A_k\}$ is a set of k attributes. $V[\mathcal{A}] = \{v_1, v_2, \ldots, v_j\}$ is the domain of attribute $\mathcal{A}$. Each continuous and nominal attribute in it has been discretized into a categorical attribute. Let $\mathcal{C} = \{c_1, c_2, \ldots, c_m\}$ is a set of possible class label for class attributes. Let $\mathcal{T} = \{t_1, t_2, \ldots, t_n\}$ is called a dataset, where each t in $\mathcal{T}$ follows the scheme $\{v_1, v_2, \ldots, v_k\}(v_i \in V[\mathcal{A}], 1 \leq v_i \leq n)$.

Definition 1. *(Literal) A literal l is an attribute-value pair of the form $A_i = v$, $A_i \in \mathcal{A}, v \in V[\mathcal{A}]$. An example $t = \{v_1, v_2, \ldots, v_k\}$ satisfies literal l iff $v_i = v$, where v_i is the value of the ith attribute of t.*

Definition 2. *(Association Rule) An association rule r, which takes the form of $a_r \Rightarrow c_r$ with a_r of the form $l_1 \wedge l_2 \wedge \cdots \wedge l_i$ called the antecedent of rule and c_r of the form $l'_1 \wedge l'_2 \wedge \cdots \wedge l'_j$ called the consequent of rule, $a_r \cap c_r = \emptyset$. We say an example t satisfies rule's antecedent iff it satisfies every literal in a_r and it satisfies the whole rule iff it satisfies c_r as well.*

Definition 3. *(Support and Confidence) For a given rule r is of the form $a_r \Rightarrow c_r$,*

$$sup(r) = |\{t | t \in \mathcal{T}, t \ satisfies \ r\}|$$

is called the **support** of r on $\mathcal{T}$,

$$conf(r) = \frac{sup(r)}{|\{t|t \in \mathcal{T}, t\ satisfies\ a_r\}|}$$

is the **confidence** of rule r.

The main task of associative classification is to discover a set of rules from the training set with the attributes in the antecedents and the class label in the consequent, and use them to build a classifier that is used later in the classification process. So only association rules relating the rule antecedent to a certain class label are of interest. In the literature on associative classification the term class association rule has been introduced to distinguish such rules from regular association rules whose consequent may consist of an arbitrary conjunction of literals.

Definition 4. *(Class Association Rule) An class association rule r, which takes the form of $a_r \Rightarrow c_r$ with a_r of the form $l_1 \wedge l_2 \wedge \cdots \wedge l_i$ called the antecedent of rule and c_r of the form c_i called the consequent of rule, $c_i \in \mathcal{C}$, where $\mathcal{C}$ is the set of class label.*

The traditional associative classification model based on well-known support-confidence framework suffers from accuracy and efficiency due to the following facts:

- Deciding on a good value of support is not easy. Most previous work usually set the threshold value to 1%. But low support threshold may cause the generation of a very huge rules set, in part of which are useless and redundant. And it is challenging to store, retrieve, prune and sort such a large number of rules efficiently for classification. However, it is bad if the support is assigned too high, overfitting will happen and some highly predictive rules with low support but high confidence will probably be missed.
- The minimum confidence, which is used as the criterion to evaluate CARs, has been reported that may cause underfitting of rules.
- The rules pruning phase is time consuming because it must reload all of rules discovered in the previous phase and has to estimate the qualities of them one by one, which is expensive in both time and space, making the whole classification inefficient on large database.
- Typical association rules in classification are those referred to positive association rules of the form $(a_r \Rightarrow c)$. In fact, negative association like $(\overline{a_r} \Rightarrow c_r)$, $(a_r \Rightarrow \overline{c_r})$ and $(\overline{a_r} \Rightarrow \overline{c_r})$ can provide valuable information as same as positive association rules if it is not expend much more time. There are few algorithm use negative association rules to do classification.

3 Associative Classification Based on Correlation Analysis

3.1 Literal Selection

In rule generation process, most previous works on associative classification like CBA [3] or CMAR [4] usually use *minimum support* and set its value to 1% to evaluate each rule. But as mentioned above, this threshold is not easy to decide on a good value and some works already pointed out this support threshold based method may not be enough accurate in some case. Instead of generating candidate rules, we will use Foil Gain [5] to evaluate the goodness of the current rule r, and generate a small set of predictive rules directly from the dataset. This gain values are calculated by using the total weights in the positive and negative example sets instead of simply counting up the number of records in the training sets.

Definition 5. *(Positive Examples and Negative Examples)* Given a literal p and a rule $r : a_r \Rightarrow c_r$, $p \notin a_r$. The **positive examples** satisfy not only a_r but also c_r in the dataset, while the **negative examples** just satisfy a_r but does not consist with c_r in the dataset.

Definition 6. *(Foil Gain)* Given a literal p and the current rule $r : a_r \Rightarrow c_r$, $p \notin a_r$, the **foil gain** of p is defined as follows:

$$I(r) = -\log \frac{P(r)}{P(r) + N(r)} \qquad (1)$$

$$FoilGain(p) = P(r+p) \times [I(r) - I(r+p)] \qquad (2)$$

where there are $P(r)$ positive examples and $N(r)$ negative examples satisfying r. And after appending p to r, there will be $P(r+p)$ positive and $N(r+p)$ negative examples satisfying the new rule $r' : a_r \wedge p \Rightarrow c_r$.

Essentially, as mentioned in many literature, $FoilGain(p)$ represents the total number of bits saved in representing all the positive examples by adding p to r. By using this gain measure, ACBCA inherits the basic idea of CPAR [6], performing a depth-first-search rule generation process that builds rules one by one. At each step, every possible literal is evaluated and the best one is appended to the current rule. All literals in a rule must be a single form (all positive literals or all negative literals). Moreover, instead of selecting only the best one, ACBCA also keeps all close-to-the-best literals in rules generation process so that it will not miss the some important rule. After building each rule, all positive target tuples satisfying that rule are removed. ACBCA repeatedly searches for the current best rule and removes all the positive examples covered by the rule until all the positive examples in the data set are covered.

3.2 Correlation Measure

Piatetsky-Shapiro [7] argued that a rule $X \Rightarrow Y$ is *not interesting* if

$$sup(X \cup Y) \approx sup(X)sup(Y).$$

One interpretation of this proposition is that the task of association rule mining is to explore the "relationship" among attributes in rule. So a rule is called *not interesting* if its antecedent and consequent are approximately independent. Brin et. al mentioned for the first time in [8] the notion of **correlation** and studied the problem of efficiently finding strong correlated rules set of data objects from large databases. They defined the correlation on the χ^2 metric, which is used by CMAR for judging whether the current rule r is positively correlated or negative correlated. Only those with χ^2 value passing a certain threshold are used for classification.

For a given class association rule $r : a_r \Rightarrow c_r$, there are two types of correlation between antecedent a_r and class label c_r. A positive correlation is evidence of a general tendency that when the example has attribute a_r it is very likely to be classed as c_r. For example, $(wings, claws) \Rightarrow bird$. A negative correlation occurs when we discover the example does/doesn't belong to c_r while it hasn't/has the attribute a_r. For example, $(wings, bat) \Rightarrow \overline{bird}$. The mostly used correlated rules in previous classification approaches are those referred to positive association rules like $(a_r \Rightarrow c_r)$. In fact, the negative association such as $(\overline{a_r} \Rightarrow c_r)$, $(a_r \Rightarrow \overline{c_r})$ and $(\overline{a_r} \Rightarrow \overline{c_r})$ can also provide valuable information as same as positive association rules if it does not expend much more time. Unlike previous correlation mining algorithms, which mainly focus on finding strong positive correlations, our algorithm cares both positive and negative correlations in CARs.

Based on Piatetsky-Shapiro's argument, for a given class association rule $r : a_r \Rightarrow c_r$, $c_r \in \mathcal{C}$, where $\mathcal{C}$ is the set of class label, we can write the interestingness of an association between a_r and c_r in the form of their statistical dependence as equation (3) [9]:

$$Dependence(a_r, c_r) = \frac{sup(a_r \cup c_r)}{sup(a_r)sup(c_r)} = \frac{conf(r)}{sup(c_r)} \quad (3)$$

There are the following three possible cases:

1. $Dependence(a_r, c_r) = 1 \Rightarrow conf(r) = sup(c_r)$, a_r and c_r are independent.
2. $Dependence(a_r, c_r) > 1 \Rightarrow conf(r) > sup(c_r)$, a_r and c_r are positively correlated, and the following holds:

$$0 < conf(r) - sup(c_r) \leq 1 - sup(c_r)$$

In particular, we have:

$$0 < C_m = \frac{conf(r) - sup(c_r)}{1 - sup(c_r)} \leq 1$$

The nearer the ratio C_m is close to 1, the higher the positive correlation between antecedent a_r and class label c_r.

3. $Dependence(a_r, c_r) < 1 \Rightarrow conf(r) < sup(c_r)$, a_r and c_r are negatively correlated, and the following holds:

$$-sup(c_r) \leq conf(r) - sup(c_r) < 0$$

In particular, we have:

$$-1 \leq C_m = \frac{conf(r) - sup(c_r)}{sup(c_r)} < 0$$

The nearer the ratio C_m is close to -1, the higher the negative correlation between antecedent a_r and class label c_r.

To differentiate the subtlety of negative correlations, we develop a new measure, C_m(Equation (4)), as the positive and negative correlation measure:

$$C_m(a_r, c_r) = = \begin{cases} \frac{conf(r)-sup(c_r)}{1-sup(c_r)} \text{ (if } conf(r) \geq sup(c_r) \wedge sup(c_r) \neq 1) \\ \frac{conf(r)-sup(c_r)}{sup(c_r)} \text{ (if } conf(r) < sup(c_r) \wedge sup(c_r) \neq 0) \end{cases} \quad (4)$$

To evaluate and measure both positive and negative association rules, we can take C_m as the correlation of the class association rule between antecedent a_r and class label c_r.

3.3 More General Rule

To make the classification more accurate and effective, ACBCA prunes rules whose information can be expressed by other simpler but more essential rules.

Definition 7. *(More General Rule) Given two rules* $r_1 : a_{r1} \Rightarrow c_{r1}$ *and* $r_2 : a_{r2} \Rightarrow c_{r2}$, *we says* r_1 *is* ***more general than*** r_2 *iff (1) Accuracy(r_1)* $\geq$ *Accuracy(r_2); (2)* $a_{r1} \subset a_{r2}$ *and (3)* $c_{r1} = c_{r2}$.

If r_1 is more general than r_2, that means r_1 does more contribution to classification but occupies smaller memory space. ACBCA just keeps these rules have higher rank and fewer attributes in its antecedent and other more specific rules with low rank should be pruned.

3.4 Our Algorithm

In this paper, we develop a novel associative classification algorithm based on correlation analysis, ACBCA, i.e., *Associative Classification Based on Correlation Analysis*, which integrates the advantages of the previously proposed effective strategies as well as the new strategies presented in this paper. ACBCA outperforms other associative classification methods by the following advantages:

1. extracts a smaller set of rules with higher quality and lower redundancy directly from the training dataset, avoiding repeated database scan in rules generation;
2. prunes the weak correlation rules directly in the process of rules generation. The advantage is that this is not only more efficient (no post-pruning is necessary) but also more elegant in that it is a direct approach;
3. uses both positive and negative association rules of interest for classification, which take into account the *presence* as well as *absence* of attributes as a basis of rules.

Algorithm 1 gives the detailed pseudo-code for class associative rules generation algorithm on positive and negative class association rules generation process in ACBCA. For each possible class c_i in $\mathcal{C}$, CARGenerator seeks the best literal and add it to the current rule. Moreover, CARGenerator appends all close-to-the-best literals to different rules which share the same prefix in the same time. After building each rule, CARGenerator weaken the effect of all positive satisfying the current rule by a decay factor instead of removing them directly.

Algorithm 1: CARs Generator in ACBCA: CARGenerator

Input: Training set $\mathcal{T}$, Global attributes set $\mathcal{A}$
Output: Correlated Rules Set $\mathcal{R}$

```
1  begin
2      generate positive example arrays P satisfying c_i;
3      while (TotalWeight(P) > MIN_TOTAL_WEIGHT) do
4          A' ← A, T' ← T;
5          a_r ← ∅;
6          while (1) do
7              foreach literal l_i ∈ A' do CalculateFoilGain(l_i) in T';
8              l = AttributeOfBestGain();
9              if l.gain ≤ MIN_BEST_GAIN then
10                 φ = C̄_m(a_r, c_i);
11                 if φ ≥ φ_min ∧ NoMoreGeneralRule(a_r ⇒ c_i) then
12                     R ← R ∧ (a_r ⇒ c_i, φ);
13                 end
14                 if φ ≤ -φ_min ∧ NoMoreGeneralRule(ā_r ⇒ c_i) then
15                     R ← R ∧ (ā_r ⇒ c_i, φ);
16                 end
17             end
18             gainThreshold = bestGain*GAIN_SIMILARITY_RATIO;
19             foreach l' ∈ A' do
20                 if l'.gain ≥ gainThreshold then
21                     a'_r ← a_r; a'_r ← a'_r ∧ l';
22                     remove a'_r from A';
23                     remove each t ∈ T' dissatisfy a'_r;
24                     A'' ← A', T'' ← T';
25                     CARGenerator(T'', A'', c_i);
26                 end
27             end
28             a_r ← a_r ∧ l;
29             remove a_r from A';
30             remove each t ∈ T' dissatisfy a_r;
31         end
32     end
33     foreach t ∈ P satisfy a_r do reduce t.weight by a decay factor
34 end
```

Once the classifier has been established in the form of a list of rules, regardless of the methodology used to generate it, there are existing a number of strategies for using the resulting classifier to classify unseen data object as follows: (1) Choose the single best rule which matches the data object and has the highest ranks to make a prediction, (2) Collect all rules in the classifier satisfying the given unseen data and make a prediction by the "combined effect" of different class association rules, and (3) Select k-best rules for each class, evaluates the average accuracy of each class and choose the class with the highest expected accuracy as the predicted class.

In our case, ACBCA establish the classifier in the form of a list of ordered rules in the process of rules generation. Instead of taking all rules satisfying the new data object into consideration, ACBCA just selects a small set of strong correlated association rules to make prediction by the following procedure:

Step 1: For those rules satisfying the new data in antecedent, selects the first k-best rules for each class according to rules' consequents;
Step 2: Calculates the rank of each group by summing up the C_m of relevant rules. We think the *average* accuracy is not advisable because many trivial rules with low accuracy will weaken the effect of the whole group;
Step 3: The class label of new data will be assigned by that of the highest rank group.

4 Experimental Results and Performance Study

To evaluate the accuracy and efficiency of ABCBA, we have performed an extensive performance study on some datasets from UCI Machine learning Repository [10]. It would have been desirable to use the same datasets as those used by other associative classification approaches; however it was discovered that many of these datasets were no longer available in the UCI repository. All the experiments are performed on a 2.4GHz Pentium-4 PC with 512MB main memory. And a 10-fold cross validation was performed on each dataset and the results are given as average of the accuracies obtained for each fold. The parameters of ABCBA are set as the following. In the rule generation algorithm, MIN_TOTAL_WEIGHT is set to 0.05, MIN_BEST_GAIN to 0.7, $GAIN\_SIMILARITY\_RATIO$ to 0.99, ϕ_{min} to 0.2 and decay factor to 2/3. The best 3 rules are used in prediction. The experimental results for C4.5, Ripper, CBA, CMAR and CPAR are taken from [6].

The table 1 gives the average predictive accuracy for each algorithm respectively. **Bold** values denote the best accuracy for the respective dataset. **Column 2~3** present the results of traditional rule-based classification algorithms C4.5 and Ripper. **Column 4~6** show the results of several recently developed associative classification methods CBA, CMAR and CPAR. **Column 7** describes the results of ABCBA when only positive rules are considered for classification. **Column 8** shows the results of ABCBA, whichconsiders both positive rules of the form $a_r \Rightarrow c_r$ and negative rules of the form $\overline{a_r} \Rightarrow c_r$.

As can be seen, ABCBA on almost all occasions achieves best accuracy or comes very close. Moreover, the classification accuracy increases when the correlated rules are taken into consideration. This is most noticeable for the Ionosphere, the Iris Plant and the Pima Indians diabetes datasets on which the correlated rules can give better supervision to the classification.

5 Conclusions

Associative classification is an important issue in data mining and machine learning involving large, real databases. We have introduced ABCBA, a new associative classification algorithm based on correlation analysis. ABCBA differs from

Table 1. Comparison of Accuracies on UCI Different Datasets

Dataset	C4.5	Ripper	CBA	CMAR	CPAR	ABCBA_p	ABCBA_all
austral	84.7	**87.3**	84.9	86.1	86.2	85.22	85.8
breast	95	95.1	96.3	**96.4**	96	95.1	94.5
cleve	78.2	82.2	**82.8**	82.2	81.5	78.21	79.21
crx	84.9	84.9	84.7	84.9	**85.7**	85.51	85.65
diabetes	74.2	74.7	74.5	75.8	75.1	**76.67**	75.99
german	72.3	69.8	73.4	**74.9**	73.4	70.9	73.3
hepatic	80.6	76.7	**81.8**	80.5	79.4	80.83	80.83
horse	82.6	**84.8**	82.1	82.6	84.2	82.02	82.02
iono	90	91.2	92.3	91.5	92.6	92.02	**93.74**
iris	**95.3**	94	94.7	94	94.7	94.67	**95.3**
labor	79.3	84	86.3	89.7	84.7	89.17	**97.17**
led7	73.5	69.7	71.9	72.5	73.6	72.56	**73.91**
pima	75.5	73.1	72.9	75.1	73.8	74.08	**76.18**
wine	92.7	91.6	95	95	**95.5**	93.13	94.5
zoo	92.2	88.1	96.8	**97.1**	95.1	96	94
Average	83.4	83.15	84.69	85.22	84.77	84.41	**85.47**

existing algorithms for this task insofar, that it does not employ the support confidence framework and take both positive and negative rules under consideration. It uses an exhaustive and greedy algorithm based Foil Gain to extract CARs directly from the training set. During the rule building process, instead of selecting only the best literals, ABCBA inherits the basic idea of CPAR, keeping all close-to-the-best literals in rules generation process so that it will not miss the some important rule.

Since the class distribution in the dataset and the correlated relationship among attributes should be taken into account because it is conforming to the usual or ordinary course of nature in the real world, we present a new rule scoring schema named C_m to evaluate the correlation of the CARs. The experimental results of the ABCBA show that a much smaller set of positive and negative association rules can achieve best accuracy or comes very close on almost all occasions.

Actually, we just use the association rules of the type $(a_r \Rightarrow c_r)$ and $(\overline{a_r} \Rightarrow c_r)$ for classification. These two type rules have an direct association with the class label, so they can be considered together. However, if there are more than two classes in the dataset, $(a_r \Rightarrow \overline{c_r})$ and $(\overline{a_r} \Rightarrow \overline{c_r})$ just provide information that "not-belong-to", instead of giving the association between data attributes and class label directly. We are currently investigating reasonable and effective methods to use these two kinds of rules.

References

1. Quinlan, J.R.: C4.5: Programs for Machine Learning. Morgan Kaufmann (1993)
2. Cohen, W.W.: Fast effective rule induction. In Prieditis, A., Russell, S.J., eds.: ICML 1995, Tahoe City, California, USA, Morgan Kaufmann (1995) 115-123

3. Liu, B. Hsu, W. and Ma, Y.: Integrating Classification and Association Rule Mining. Proceedings KDD-98, New York, 27-31 August. AAAI. (1998) 80-86.
4. Li W., Han, J. and Pei, J.: CMAR: Accurate and Efficient Classification Based on Multiple Class-Association Rules. Proceedings of the 2001 IEEE International Conference on Data Mining, San José, California, USA, IEEE Computer Society (2001)
5. J. R. Quinlan and R. M. Cameron-Jones: FOIL: A midterm report. In Proceedings of European Conference. Machine Learning, Vienna, Austria 1993, (1993) 3-20
6. X. Yin and J. Han: CPAR: Classification based on Predictive Association Rules, Proc. 2003 SIAM Int.Conf. on Data Mining (SDM'03), San Fransisco, CA, May 2003.
7. Piatetsky-Shapiro, G.,: Discovery, Analysis, and Presentation of Strong Rules. Knowledge Discovery in Databases, G. Piatetsky-Shapiro and WJ Frawley (Eds.), AAAI/MIT Press, 1991, pp. 229-238.
8. Sergey Brin, Rajeev Motwani, and Craig Silverstein: Beyond market baskets: Generalizing association rules to correlations. SIGMOD Record (ACM Special Interest Group on Management of Data), (1997) 265-276.
9. Xindong Wu,Chengqi Zhang,Shichao Zhang: Efficient Mining of Both Positive and Negative Association Rules. ACM Transactions on Information Systems, 22(2004), 3: 381-405.
10. Blake, C.L. and Merz, C.J. (1998). UCI Repository of machine learning databases. http://www.ics.uci.edu/ mlearn/MLRepository.html, Irvine, CA: University of California, Department of Information and Computer Science.

Design of Interpretable and Accurate Fuzzy Models from Data

Zong-yi Xing [1], Yong Zhang [1], Li-min Jia [2], and Wei-li Hu [1]

[1] Nanjing University of Science and Technology, Nanjing 210094, China
[2] Beijing Jiaotong University, Beijing, 100044, China
xingzongyi@tom.com

Abstract. An approach to identify data-driven interpretable and accurate fuzzy models is presented in this paper. Firstly, Gustafson-Kessel fuzzy clustering algorithm is used to identify initial fuzzy model, and cluster validity indices are adopted to determine the number of rules. Secondly, orthogonal least square method and similarity measure of fuzzy sets are utilized to reduce the initial fuzzy model and improve its interpretability. Thirdly, constraint Levenberg-Marquardt algorithm is used to optimize the reduced fuzzy model to improve its accuracy. The proposed approach is applied to PH neutralization process, and results show its validity.

1 Introduction

During the past years, fuzzy modeling techniques have become an active research area due to its successful application to classification, data mining, pattern recognition, simulation, prediction, control, etc [1-4]. Several fuzzy modeling methods have been proposed including fuzzy clustering based algorithm [5], neuro-fuzzy systems [6-7] and genetic rules generation [8-9]. However all these technologies only focus on precision that simply fit data with highest possible accuracy, neglecting interpretability of obtained fuzzy models, which is considered as a primary merit of fuzzy systems and is the most prominent feature that distinguishes fuzzy systems from many other models [10].

In order to improve interpretability of fuzzy models, some methods have been developed. Setnes et al. [11] proposed a set-theoretic similarity measure to quantify the similarity among fuzzy sets, and to reduce the number of fuzzy sets in the model. Yen et al. [12] introduced several orthogonal transformation techniques for selecting the most important fuzzy rules from a given rule base in order to construct a compact fuzzy model. Abonyi et al. [13] proposed a combined method to create simple Takagi-Sugeno fuzzy model that can be effectively used to represent complex systems. Delgado [14] presents fuzzy modeling as a multi-objective decision making problem considering accuracy, interpretability and autonomy as goals, and all these goals are handled via single-objective $\varepsilon-$ constrained decision making problem which is solved by a hierarchical evolutionary algorithm.

This paper proposes systematic techniques to construct interpretable and accurate fuzzy models. The paper is organized as follows: section 2 describes the Takagi-Sugeno fuzzy model. The systematic fuzzy modeling approach, including fuzzy clustering algorithm, cluster validity measure, rule reduction and fuzzy sets merging and constraint Levenberg-Marquardt optimization, is stated in section 3. The proposed method is demonstrated on the PH neutralization benchmark in section 4. Section 5 concludes the paper.

2 Takagi-Sugeno Fuzzy Model

The Takagi-Sugeno (TS) Fuzzy model [1] was proposed by Takagi and Sugeno in an effort to develop a systematic approach to generating fuzzy model from a given input-output data set. A typical fuzzy rule of the model has the form:

$$R^i : \text{IF } x_1 \text{ is } A_{i,1} \text{ and } \cdots \text{ and } x_p \text{ is } A_{i,p} \tag{1}$$
$$\text{THEN } y_i = a_{i0} + a_{i1}x_1 + \cdots + a_{ip}x_p$$

where x_j are the input variables, A_{ij} are fuzzy sets defined on the universe of discourse of the input variables, y_i are outputs of rules.

The output of the TS fuzzy model is computed using the normalized fuzzy mean formula:

$$y(k) = \sum_{i=1}^{c} p_i(x)\hat{y}_i \tag{2}$$

where c is the number of rules, P_i is the normalized firing strength of the ith rule:

$$P_i(x) = \frac{\prod_{j=1}^{p} A_{ij}(x_j)}{\sum_{i=1}^{c} \prod_{j=1}^{p} A_{ij}(x_j)} \tag{3}$$

Given N input-output data pairs $\{x_k, y_k\}$, the model in (2) can be written as a linear regression problem

$$y = P\theta + e \tag{4}$$

where θ is consequents matrix of rules, and e is approximation error matrix.

In this paper, Gaussian membership functions are used to represent the fuzzy set A_{ij}

$$A_{ij}(x_j) = \exp(-\frac{1}{2}\frac{(x_j - v_{ij})^2}{\sigma_{ij}^2}) \tag{5}$$

where v_{ij} and σ_{ij} represent center and variance of Gaussian function respectively.

3 Design of Data-Driven Interpretable and Accurate Fuzzy Models

3.1 Construct Initial Fuzzy Models Using Fuzzy Clustering Algorithm

The Gustafson-Kessel [15] (GK) algorithm is employed to construct initial fuzzy models. The objective function of GK algorithm is described following:

$$J(\mathbf{Z};\mathbf{U},\mathbf{V}) = \sum_{i=1}^{C}\sum_{k=1}^{N}(\mu_{ik})^{m} D_{ik}^{2} \quad (6)$$

where $\mathbf{Z}$ is the set of data, $\mathbf{U} = [\mu_{ik}]$ is the fuzzy partition matrix, $\mathbf{V} = [\mathbf{v}_1, \mathbf{v}_2, \cdots, \mathbf{v}_c]^T$ is the set of centers of the clusters, c is the number of clusters, N is the number of data, m is the weighting exponent, μ_{ik} is the membership degree between the ith cluster and kth data, which satisfy conditions:

$$\mu_{ik} \in [0,1]; \quad \sum_{i=1}^{C}\mu_{ik} = 1; \quad 0 < \sum_{k=1}^{N}\mu_{ik} < N \quad (7)$$

The norm of distance between the ith cluster and kth data is:

$$D_{ik}^{2} = \|\mathbf{z}_k - \mathbf{v}_i\|_{A_i}^{2} = (\mathbf{z}_k - \mathbf{v}_i)^T \mathbf{A}_i (\mathbf{z}_k - \mathbf{v}_i) \quad (8)$$

where

$$\mathbf{A}_i = (\rho \det(\mathbf{F}_i))^{1/n} \mathbf{F}_i^{-1} \quad (9)$$

$$\rho = \det(\mathbf{A}_i) \quad (10)$$

F_i is the fuzzy covariance matrix of i-th cluster:

$$\mathbf{F}_i = \frac{\sum_{k=1}^{N}(\mu_{ik})^{m}(\mathbf{z}_k - \mathbf{v}_i)(\mathbf{z}_k - \mathbf{v}_i)^T}{\sum_{k=1}^{N}(\mu_{ik})^{m}} \quad (11)$$

The minimum of $(\mathbf{U}, \mathbf{V})$ is calculated as follows:

$$\mu_{ik} = \frac{1}{\sum_{j=1}^{c}(D_{ik}/D_{jk})^{2/(m-1)}} \quad (12)$$

$$\mathbf{v}_i = \frac{\sum_{k=1}^{N}(\mu_{ik})^{m}\mathbf{z}_k}{\sum_{k=1}^{N}(\mu_{ik})^{m}} \quad (13)$$

The variance of Gaussian function is calculated as:

$$\sigma_{ij}^{2} = \frac{\sum_{k=1}^{N}\mu_{ik}(x_{jk} - v_{jk})^{2}}{\sum_{k=1}^{N}\mu_{ik}} \quad (14)$$

Given the input variable X, output y and fuzzy matrix U following:

$$X = \begin{bmatrix} X_1^T \\ X_2^T \\ \vdots \\ X_N^T \end{bmatrix}, \quad y = \begin{bmatrix} y_1 \\ y_2 \\ \vdots \\ y_N \end{bmatrix}, \quad U_i = \begin{bmatrix} u_{i1} & 0 & \cdots & 0 \\ 0 & u_{i2} & \cdots & 0 \\ \vdots & \vdots & \ddots & \vdots \\ 0 & 0 & \cdots & u_{iN} \end{bmatrix} \quad (15)$$

Appending a unitary column to X gives the extended matrix X_e:

$$X_e = [X \ 1] \quad (16)$$

Then

$$\theta_i = [X_e^T U_i X_e]^{-1} X_e^T U_i y \quad (17)$$

is the consequent parameter of fuzzy model.

The procedure of constructing fuzzy model is summarized as follows:

1) Choose the number of fuzzy rules and the stop criterion $\varepsilon > 0$.
2) Generate the matrix U with the membership randomly.
3) Compute the parameters of model using (13), (14) and (17).
4) Calculate norm of distance utilizing (8.
5) Update the partition matrix U using (12);
6) Stop if $\|U^{(l)} - U^{(l-1)}\| \le \varepsilon$, else go to 3).

3.2 Cluster Validation Indices

It is essential to determine the number of rules, i.e. the number of fuzzy clusters. This problem can be solved by validation analysis using cluster validity indices.

There are two categories of fuzzy validity indices. The first category uses only the membership values of a fuzzy partition of data. On the other hand, the latter one involves both partition matrix and the data itself.

The partition coefficient (*PC*) and the partition entropy coefficient (*PE*) [16] are the typical cluster validity indices of the first category.

$$PC(c) = \frac{1}{n} \sum_{i=1}^{c} \sum_{k=1}^{N} \mu_{ik}^2 \quad (18)$$

$$PE(c) = -\frac{1}{n} \sum_{i=1}^{c} \sum_{k=1}^{N} \mu_{ik} \log_a \mu_{ik} \quad (19)$$

where $PC(c) \in [1/c, 1]$, $PE(c) \in [0, \log_a c]$. The number corresponding to significant knee is selected as the optimal number of rules.

The compactness and separation validity function proposed by Xie and Beni (*XB*) [17] is a representation of the second category. The smallest value of *XB* indicates the optimized clusters.

$$XB(c) = \frac{\sum_{i=1}^{c} \sum_{k=1}^{N} \mu_{ik}^m \|x_k - v_i\|^2}{n \cdot \min_{i,k} \|v_i - v_k\|^2} \quad (20)$$

3.3 Rule Reduction Based on Orthogonal Least Square Method

In the process of cluster validity analysis, it is possible that the noise or abnormal data are clustered to create redundant or incorrect fuzzy rules. This problem can be solved by rule reduction using orthogonal least square method (OLS) [18].

The OLS method transforms the columns of the firing strength matrix into a set of orthogonal basis vectors. Using Gram–Schmidt orthogonalization procedure, the firing strength matrix P is decomposed into

$$P = WA \tag{21}$$

where W is a matrix with orthogonal columns w_i, and A is an upper triangular matrix with unity diagonal elements.

Substituting (21) into (4) yields

$$y = WA\theta + e = Wg + e \tag{22}$$

where $g = A\theta$. Since the columns w_i of W are orthogonal, the sum of squares of $y(k)$ can be written as

$$y^T y = \sum_{i=1}^{c} g_i^2 w_i^T w_i + e^T e \tag{23}$$

Dividing N on both side of (23), it can be seen that the part of the output variance $y^T y/N$ explained by the regressors is $\sum g_i w_i^T w_i / N$, and an error reduction ratio due to an individual rule is defined as

$$[err]^i = \frac{g_i^2 w_i^T w_i}{y^T y}, \quad 1 \leq i \leq c. \tag{24}$$

The ratio offers a simple means for seeking a subset of important rules in a forward-regression manner. If it is decided that r rules are used to construct a fuzzy model, then the first r rules with the largest error reduction ratios will be selected.

If the importance measure of a fuzzy rule is far less than others, and deletion of this rule doesn't deteriorate precision performance, this fuzzy rule will be picked out to improve interpretability of fuzzy model.

3.4 Similarity Merging of Fuzzy Sets

Fuzzy models obtained above may contain redundant information in the form of similarity between fuzzy sets. This makes the fuzzy model uninterpretable, for it is difficult to assign qualitatively meaningful labels to similar fuzzy sets. In order to acquire an effective and interpretable fuzzy model, elimination of redundancy and making the fuzzy model as simple as possible are necessary.

As for two similar fuzzy sets, a similarity measure is unutilized to determine if fuzzy sets should be combined to a new fuzzy set. For fuzzy sets A and B, a set-theoretic operation based similarity measure [11] is defined as

$$S(A,B) = \frac{\sum_{k=1}^{N}[\mu_A(x_k) \wedge \mu_B(x_k)]}{\sum_{k=1}^{N}[\mu_A(x_k) \vee \mu_B(x_k)]} \qquad (25)$$

Where $X = \{x_j \mid j=1,2,\cdots,m\}$ is the discrete universe, $\wedge$ and $\vee$ are the minimum and maximum operators respectively. S is a similarity measure in [0,1]. $S=1$ means the compared fuzzy sets are equal, while $S=0$ indicates that there is no overlapping between fuzzy sets.

If similarity measure $S(A,B) > \tau$, i.e. fuzzy sets are very similar, then the two fuzzy sets A and B should be merged to create a new fuzzy set C, where τ is a predefined threshold. In a general way, $\tau = [0.4-0.7]$ is a good choice.

3.5 Optimization

After rule reduction and similar fuzzy sets merging, the precision of initial fuzzy model is improved, while its precision performance is reduced. It is essential to optimize the reduced fuzzy model to improve its precision, while preserves its interpretability.

The precision and parameters of fuzzy model are strongly nonlinear, so a robust optimization technique should be applied in order to assure a good convergence. The constraint Levenberg-Marquardt (LM) method [19] is adopted in this paper. The premise parameters are limited to change in a range of $\pm\alpha\%$ around their initial values in order to preserve the distinguishability of fuzzy sets. For the sake of maintaining the local interpretability of fuzzy model, the consequent parameters are restricted to vary $\pm\beta\%$ of the corresponding consequent parameters.

4 Example

PH neutralization [20] is a typical nonlinear system with three influent streams (acid, buffer and base) and one effluent stream. The influent buffer stream and the influent acid stream are kept constant, and the influent base stream change randomly. The output is the PH in the tank.

In order to determine the number of rules, cluster validity indices including *PC*, *PE*, *XB* and Average Partition density (*PA*) [21] are adopted. Fig. 1 diagrams the result of cluster validity analysis intuitively. Obviously, all the cluster validity indices indicate that optimal number of rules is 3.

The GK fuzzy clustering algorithm is used to constructed fuzzy model with 3 fuzzy rules. The fuzzy sets of obtained fuzzy model are illustrated in Fig 2(b). Obviously, the fuzzy sets are distinguishable, and it is easy to assign understandable linguistic term to each fuzzy set.

The OLS method is adopted to pick out unnecessary fuzzy rules. The importance measures of fuzzy rules are [0.2249 0.0250 0.7438] respectively, where the measure of the 2nd fuzzy rule is far less than the others. Without considering precision performance, the 2nd fuzzy rule can be deleted. Fig 2(a) diagrams the fuzzy sets of membership with 2 rules.

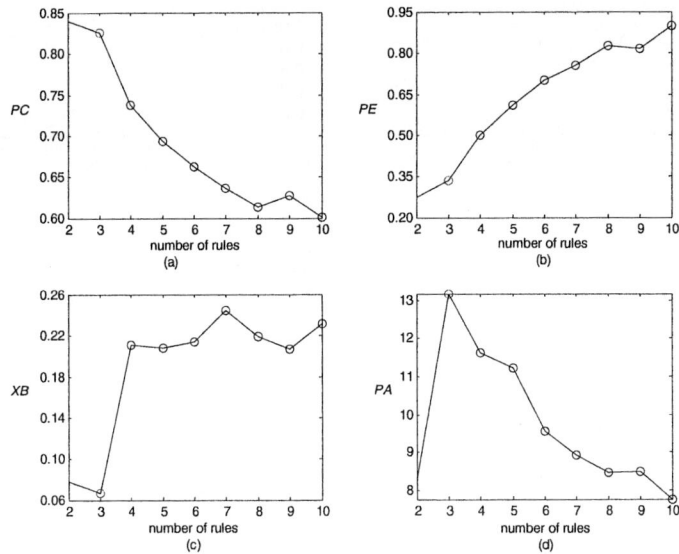

Fig. 1. Cluster validity indices

After rule reduction, the interpretability of fuzzy model is improved, while the precision is reduced. If the model with 2 rules can satisfy practical demands, this model will be employed, otherwise, the model with 3 rules is preserved. The reduced fuzzy model with 2 rules is adopted in order to demonstrate interpretability in this paper.

Fuzzy sets merging process is carried out sequentially. The similarity measure between fuzzy sets of *flow rate* is 0.3420, and the measure between fuzzy sets of *PH* is 0.0943. Without considering precision and reality, the fuzzy sets of *flow rate* can be merged to a new fuzzy set, so the *flow rate* variable can also be deleted for containing only one fuzzy set.

In practice, similarity measure between fuzzy sets of *flow rate* is small, and *flow rate* variable is independent variable, and the precision performance of model after fuzzy sets merging is deteriorative, so the fuzzy sets of *flow rate* are not merged in this paper.

In order of illustrated interpretability and precision of different fuzzy models intuitively, Fig. 2 diagrams fuzzy sets of model with 2, 3, 4,5 fuzzy rules, and Table 1 shows the corresponding errors and number of rules/fuzzy sets.

Obviously, with the increase of rules, the precision is improved, while the interpretability is reduced. When the number of rules is 4 and 5, the fuzzy sets are heavy overlapped, whereas the precision is only increased a little.

$$R^1 : \text{If } FR(k) \text{ is High and } PH(k) \text{ is High}$$
$$\text{Then } PH(k+1) = 0.0321FR(k) + 0.8614PH(k) + 0.6659$$
$$R^2 : \text{If } FR(k) \text{ is Low and } PH(k) \text{ is Low}$$
$$\text{Then } PH(k+1) = 0.1103FR(k) + 0.7521PH(k) + 0.2614$$

(26)

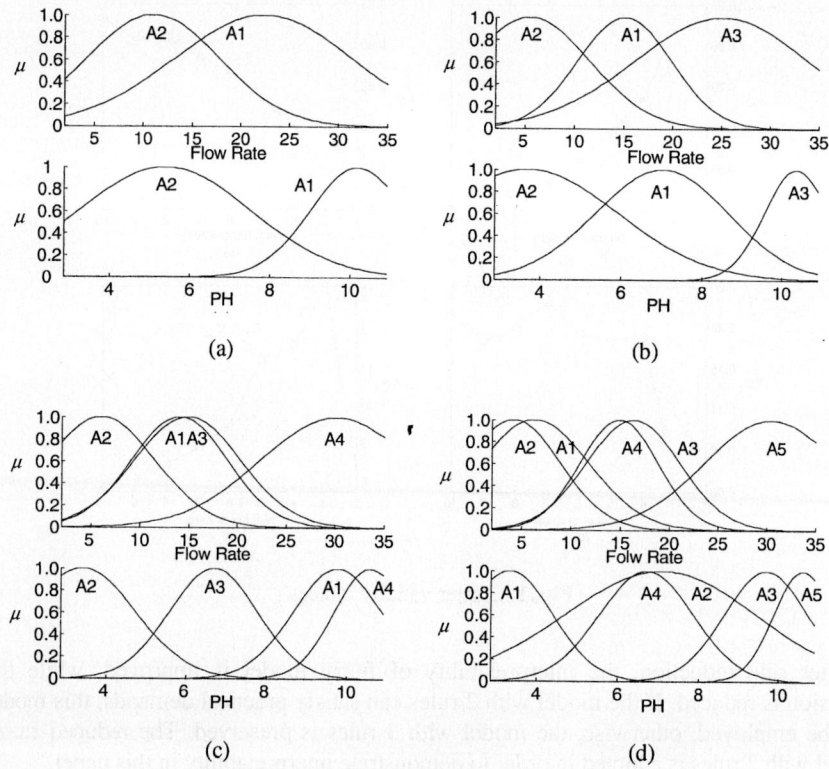

Fig. 2. Fuzzy sets of different fuzzy models

Table 1. Comparison of different fuzzy models

Number of fuzzy rules	Number of fuzzy sets	Training error	Validation error	Validation error descend
2	2	1.6290	1.7547	—
2	4	0.4020	0.3433	411.13%
3	6	0.3791	0.3160	8.64%
4	8	0.3716	0.3119	1.31%
5	10	0.3739	0.3099	0.65%

The fuzzy model is optimized by Levenberg-Marquardt method. The constraint of antecedent parameters is 5%, and the constraint of consequent parameters is 15%. The optimized fuzzy model is described as (26), where $FR(k)$ is the value of flow rate, $PH(k)$ is the value of PH. Fig. 3(a) diagrams the fuzzy sets of the model.

After optimization, the training error and validation error of the model are 0.3711 and 0.3088. Without decrease interpretability, the precision of the model is improved. Fig 3(b) illustrates the comparison of model outputs and measured outputs.

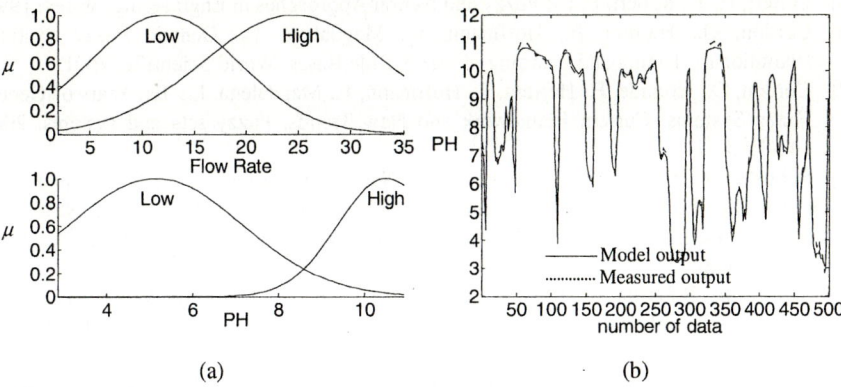

Fig. 3. Fuzzy sets of the model and comparison of model output and measured output

5 Conclusion

This paper proposes systematic techniques to construct interpretable and accurate fuzzy models. Firstly, GK fuzzy clustering algorithm is used to identify fuzzy model, and cluster validity indices are adopted to determine the number of rules. Secondly, orthogonal least square method and similarity measure of fuzzy sets are utilized to reduce the initial fuzzy model and improve its interpretability. Thirdly, constraint Levenberg-Marquardt algorithm is used to optimize the reduced fuzzy model. At last, the simulation on PH neutralization illustrates the effectiveness of proposed methods.

Acknowledgements

This paper is supported by scientific research foundation of Nanjing University of Science and Technology, and by Jiangsu Planned Projects for Postdoctoral Research Funds.

References

[1] Takagi, T., Sugeno, M.: Fuzzy identification of systems and its application to modeling and control. IEEE Trans on Systems Man and Cybernetics, 1985, 15(1): 116-132
[2] Sugeno, M., Yasukawa, T.: A fuzzy-logic- based approach to qualitative modeling. IEEE Trans on Fuzzy Systems, 1993, 1(1): 7-31
[3] Wang, L. X.: Adaptive fuzzy systems and control: design and stability analysis. Prentice Hall, 1994
[4] Min, Y. C., Linkens, D. A.: Rule-base self-generation and simplification for data-driven fuzzy models. Fuzzy sets and systems, 2004,142(2): 243-265
[5] Gomez-Skarmeta, A. F., DELGADO, M., VILA, M. A.: About the use of fuzzy clustering techniques for fuzzy model identification. Fuzzy Sets and Systems, 1999, 106(2): 179-188
[6] Jang, J. R.: ANFIS: Adaptive-network-based fuzzy inference system. IEEE Trans on Systems Man, and Cybernetics, 1993, 23:665–684

[7] Lefteri, H. T., Robert, E. U.: Fuzzy and Neural Approaches in Engineering. Wiley, 1997
[8] Cordon, O., Herrera, F., Hoffmann, F., Magdalena, L.: Genetic Fuzzy Systems: Evolutionary Tuning and Learning of Fuzzy Rule Bases. World Scientific, 2001
[9] Cordon, O., Gomide, F., Herrera, F., Hoffmann, F., Magdalena, L.: Ten Years of Genetic Fuzzy Systems: Current Framework and New Trends. Fuzzy sets and systems, 2004, 141(1): 5-31
[10] Babuska, R., Bersini, H., Linkens, D. A., Nauck, D., Tselentis, G., Wolkenhauer, O.: Future Prospects for Fuzzy Systems and Technology [EL/OB]. ERUDIT Newsletter, Aachen, Germany, 6(1), 2000. Available: http://www.erudit.de/erudit/newsletters/news 61/page5.htm
[11] Sentes, M., Babuska, R., Kaymak, U., Lemke, H. R. N.: Similarity Measures in Fuzzy Rule Base Simplification. IEEE Trans on Systems Man and Cybernetics, 1998, 28(3): 376-386.
[12] Yen, J., Wang, L.: Simplifying fuzzy modeling by both gray relational analysis and data transformation methods . IEEE Trans on Systems Man and Cybernetics, 1999, 29(1): 13-24
[13] Abonyi, J., Roubos, J. A., Oosterom, M., Szeifert, F.: Compact TS-fuzzy models through clustering and OLS plus FIS model reduction. Proc of IEEE int conf on fuzzy systems, Sydney, Australia, 2001: 1420-1423
[14] Delgado, M. R., Zuben, F. V., Gomide, F.: Multi-Objective Decision Making: Towards Improvement of Accuracy, Interpretability and Design Autonomy in Hierarchical Genetic Fuzzy Systems. Proc of IEEE int conf on fuzzy systems, Honolulu, Hawai, 2002: 1222-1227
[15] Gustafson, D., Kessel, W.: Fuzzy clustering with a fuzzy covariance matrix. Proc of IEEE conf on decision and control. San Diego, USA, 1979: 761-766
[16] Bezdek, J. C.: Pattern Recognition with fuzzy objective algorithm. New York: Plenum Press, 1981
[17] Xie, X. L., Beni, G: A Validity Measure for Fuzzy Clustering. IEEE Trans on Pattern Analysis and Machine Intelligence, 1991, 13(8): 841-847
[18] Yen, J., Wang, L.: Simplifying fuzzy modeling by both gray relational analysis and data transformation methods. IEEE Trans on Systems Man and Cybernetics, 1999, 29(1): 13-24
[19] Gaweda, A. E.: Optimal data-driven rule extraction using adaptive fuzzy neural models. University of Louisville 2002
[20] Babuska, R.: Fuzzy Modeling for Control. Boston: Kluwer Academic Publishers, 1998
[21] Gath, I., Geva, A. B.: Fuzzy clustering for the estimation of the parameters of the components of mixtures of normal distributions. Pattern Recognition Letters, 1989, 9: 77-86

Generating Extended Fuzzy Basis Function Networks Using Hybrid Algorithm

Bin Ye, Chengzhi Zhu, Chuangxin Guo, and Yijia Cao

College of Electrical Engineering, National Laboratory of Industrial Control Technology,
Zhejiang University, Hangzhou 310027, Zhejiang, China
yebin@zju.edu.cn, yijiacao@zju.edu.cn

Abstract. This paper presents a new kind of Evolutionary Fuzzy System (EFS) based on the Least Squares (LS) method and a hybrid learning algorithm: Adaptive Evolutionary-programming and Particle-swarm-optimization (AEPPSO). The structure of the Extended Fuzzy Basis Function Network (EFBFN) is firstly proposed, and the LS method is used to design it with presetting the widths of the hidden units in EFBFN. Then, to enhance the performance of the obtained EFBFN ulteriorly, a novel learning algorithm based on least squares and the hybrid of evolutionary programming and particle swarm optimization (AEPPSO) is proposed, in which we use EPPSO to tune the parameters of the premise part in EFBFN, and the LS algorithm to decide the consequent parameters in it simultaneously. In the simulation part, the proposed method is employed to predict a chaotic time series. Comparisons with some typical fuzzy modeling methods and artificial neural networks are presented and discussed.

1 Introduction

In recent years, various neural-fuzzy networks have been proposed. Among them, the fuzzy basis function networks (FBFNs) [1], which are similar in structure to radial basis function networks (RBFNs) [2], have gained much attention. In addition to their simple structure, FBFNs possess another advantage that they can readily adopt various learning algorithms already developed for RBFNs. Among the various methods for designing FBFNs, the OLS method [1] has attracted much attention due to its simple and straightforward computation with reliable performance. However, its performance largely depends on the preset input membership functions (MFs) because the parameters in the MFs remain unchanged during learning process. Furthermore, it fails to yield a meaningful fuzzy system after learning [3].

In this paper, we extend the original FBFN with singleton output fuzzy variables to be *Extended* FBFN with 1st-order fuzzy output, named EFBFN, and the least squares (LS) method is used to select the significant fuzzy rules from candidate fuzzy basis functions (FBFs [1]) based on error reduction measure which is calculated by a projection matrix instead of orthogonalization. While the LS algorithm is a computationally efficient way of constructing fuzzy system, its performance depends on the preset parameters of basis functions. Therefore, we have the necessity to tune the parameters of the obtained MFs within a small range using a tuning algorithm. Here, a tuning algorithm based on least squares and the hybrid of evolutionary programming and particle swarm optimization, named Adaptive Evolutionary-programming and Particle-swarm-optimization (AEPPSO), is proposed.

The hybrid algorithm EPPSO based on EP and PSO is performed to tune the parameters of the premise part, while the LS algorithm is used to decide the consequent parameters of the fuzzy rule base simultaneously. To demonstrate the performance of the proposed algorithm for designing EFBFN based on the hybrid of LS and AEPPSO, the prediction of a chaotic time series are performed, and the results are also compared to other methods.

2 Extended Fuzzy Basis Function Network

2.1 Structure of Extended Fuzzy Basis Function Network

The original FBFN [1] used singleton fuzzy membership functions as output fuzzy variables, while in this paper we extend the FBFN to be the following form:

$$R^i : \text{if } x_1 \text{ is } A_1^i(x_1) \text{ and } \ldots x_k \text{ is } A_k^i(x_k), \text{ then } y^i \text{ is } \xi_0^i + \xi_1^i x_1 + \ldots + \xi_k^i x_k, \quad (1)$$

where R^i represents the ith rule $(1 \le i \le r)$, $x_j (1 \le j \le k)$ is an input variable and y^i is an output variable. $A_j^i(x_j)$ are the fuzzy variables defined as in Eqn.2, and $\xi_j^i (1 \le i \le r, 0 \le j \le k)$ are the consequent parameters of the fuzzy model.

$$A_j^i(x_j) = \exp[-\frac{1}{2}*[(x_j - m_j^i)/\sigma_j^i]^2], \quad (2)$$

where m_j^i and σ_j^i are the mean value and the standard deviation of the Gaussian type MF, respectively. The fuzzy basis functions defined by Wang can also be used here:

$$p_i(\chi) = \prod_{j=1}^k A_j^i(x_j) / \sum_{i=1}^r (\prod_{j=1}^k A_j^i(x_j)), \quad i = 1, 2, \ldots r, \chi = [\chi_1, \ldots, \chi_k]. \quad (3)$$

The fuzzy basis function expansion defined as in Eqn.4 for extended FBFN is equivalent to the 1st-order T-S fuzzy model suggested by Takagi and Sugeno [4].

$$y^* = \sum_{i=1}^r p_i(\chi) C_i, \quad (4)$$

where C_i is the consequent part of the fuzzy rule, and $C_i = \xi_0^i + \xi_1^i x_1 + \ldots + \xi_k^i x_k$.

2.2 EFBFN Based on Least Squares Method

Before describing the LS method for designing EFBFN, we define the I/O data as:

$$\mathbb{R} = \{\chi_h, y_h | \chi_h = (x_{h1}, x_{h2}, \ldots x_{hk}), h = 1, 2 \ldots N\}.$$

According to the orthogonal least algorithm proposed for constructing FBFN by Wang [1], the following N candidate fuzzy basis function (FBF) nodes are generated at the beginning:

$$p_h(\chi) = \Pi_{j=1}^{k} A_j^h(x_j) / \Sigma_{h=1}^{N} (\Pi_{j=1}^{k} A_j^h(x_j)), \quad h = 1, 2, \ldots N, \chi = [\chi_1, \ldots, \chi_k]., \quad (5)$$

where in the fuzzy MFs, the widths and the candidate centers are defined as follows:

$$\sigma_j^i = \frac{\max(\mathbf{x}_j) - \min(\mathbf{x}_j)}{r}, \quad m_j^i = \chi_h, \text{ where} \quad (6)$$
$$j = 1, 2, \ldots, k, \chi_h = [x_{h1}, x_{h2}, \ldots x_{hk}], \mathbf{x}_j = [x_{1j}, \ldots, x_{Nj}].$$

Assume that r ($r < N$) EFBFN nodes have been selected by the OLS algorithm, the resultant *FBF expansion* produced by the OLS algorithm is represented by:

$$y^* = \sum_{i=1}^{r} p_{h_i}(\chi)C_i = \Sigma_{i=1}^{r} C_i(\Pi_{j=1}^{k} A_j^{h_i}(x_j)) / \Sigma_{h=1}^{N}(\Pi_{j=1}^{k} A_j^{h_i}(x_j)), \quad i = 1, 2, \ldots r. \quad (7)$$

Compare above equation with the original definition of EFBFN in Eqn. (4), i.e., the following equation:

$$y^* = \Sigma_{i=1}^{r} C_i(\Pi_{j=1}^{k} A_j^i(x_j)) / \Sigma_{i=1}^{r}(\Pi_{j=1}^{k} A_j^i(x_j)), \quad C_i = \xi_0^i + \xi_1^i x_1 + \ldots + \xi_k^i x_k, \quad (8)$$

it is clear to see that the fuzzy basis functions selected by the OLS algorithm can not be interpreted as true FIS since they retain the normalization factor in the denominator before training. While the least squares method used in this paper for constructing EFBFN can be clearly understood in terms of fuzzy rules defined in physics domains and with a computational efficiency [3].

We consider the inferred output formula, Eqn.4 as a linear regression model:

$$\mathbf{y} = \mathbf{W}\xi + \mathbf{e}, \quad (9)$$

where $\mathbf{y} = [y_1, \ldots y_l, \ldots, y_N]$ is the desired output, $\mathbf{e} = [e_1, e_2, \ldots, e_N]$ is the error signal, the matrix $\mathbf{W}$ and the vector ξ are defined as follows:

$$\mathbf{W} = [\mathbf{w}_1, \ldots \mathbf{w}_N] = [\mathbf{p}_1, \ldots \mathbf{p}_r, \mathbf{p}_1 * \mathbf{x}_1, \ldots \mathbf{p}_r * \mathbf{x}_1, \ldots \mathbf{p}_1 * \mathbf{x}_k, \ldots \mathbf{p}_r * \mathbf{x}_k]$$
$$\xi = [\xi_0^1, \ldots \xi_0^r, \xi_1^1, \ldots \xi_1^r, \ldots \xi_k^1, \ldots \xi_k^r]^T$$
$$\mathbf{p}_i = [p_i(\chi_1), \ldots, p_i(\chi_N)]^T, \quad \mathbf{x}_j = [x_{1j}, x_{2j}, \ldots x_{Nj}]^T,$$
$$\mathbf{p}_i * \mathbf{x}_j = [p_i(\chi_1)x_{1j}, \ldots, p_i(\chi_N)x_{Nj}]^T, i = 1, 2, \ldots r, j = 1, 2, \ldots, k$$

The LS algorithm selects the most significant FBFs by maximizing the following error reduction measure [err]:

$$[err] = \|\mathbf{W}\mathbf{W}^+\mathbf{d}\|, \quad (10)$$

where $\mathbf{W}^+$ denotes the pseudoinverse of $\mathbf{W}$. It can be seen that $\mathbf{W}\mathbf{W}^+$ is the orthogonal projection onto the column space of $\mathbf{W}$.

With LS algorithm applied in EFBFN, when we are selecting the first EFBFN ($r = 1$) node from the FBF sets, the N candidate FBFs are calculated using the following formula:

$$p_h(\chi_g) = \prod_{j=1}^{k} A_j^1(x_j), \quad g = 1, 2, ...N, h = 1, 2, ...N, \tag{11}$$

where the parameters in the fuzzy MFs are defined as in Eqn. (6). When s-1 EFBFN nodes have been picked from the data sets, for selecting the sth EFBFN nodes, the N-s+1 candidate FBFs for EFBFN using LS are defined as follows:

$$p_h(\chi_g) = \prod_{j=1}^{k} A_j^i(x_j) / \sum_{h=1}^{s} \prod_{j=1}^{k} A_j^i(x_j), \quad h \neq h_1, ..., h \neq h_{s-1}, g = 1, 2, ...N. \tag{12}$$

Our objective is to find r fuzzy rules from the N training data pairs. The selection procedure can be described as follows:

Step 1: for $1 \leq h \leq N$, compute

$$\mathbf{p}_1^{(h)} = [p_h(\chi_1), p_h(\chi_2), ..., p_h(\chi_N)]^T, \quad h = 1, 2, ...N,$$
$$\mathbf{W}_1^{(h)} = [\mathbf{p}_1^{(h)}, \mathbf{p}_1^{(h)}.*\mathbf{x}_1, ..., \mathbf{p}_1^{(h)}.*\mathbf{x}_k]$$
$$[err]_1^{(h)} = \left\| \mathbf{W}_1^{(h)}(\mathbf{W}_1^{(h)})^+ \mathbf{y} \right\|$$

Find $[err]_1^{(h_1)} = \max\{[err]_1^{(h)}, 1 \leq h \leq N\}$, and select $\mathbf{p}_1 = \mathbf{p}_1^{(h_1)}$.

Step s: where $s \geq 2$, for $1 \leq h \leq N, h \neq h_1, ..., h \neq h_{s-1}$, compute

$$\mathbf{p}_s^{(h)} = [p_h(\chi_1), p_h(\chi_2), ..., p_h(\chi_N)]^T, \quad h = 1, 2, ...N,$$
$$\mathbf{W}_s^{(h)} = [\mathbf{p}_1, ..., \mathbf{p}_s^{(h)}, \mathbf{p}_1.*\mathbf{x}_1, ..., \mathbf{p}_s^{(h)}.*\mathbf{x}_1, ..., \mathbf{p}_1.*\mathbf{x}_k, ..., \mathbf{p}_s^{(h)}.*\mathbf{x}_k]$$
$$[err]_s^{(h)} = \left\| \mathbf{W}_s^{(h)}(\mathbf{W}_s^{(h)})^+ \mathbf{y} \right\|$$

Find $[err]_s^{(h_s)} = \max\{[err]_s^{(h)}, 1 \leq h \leq N, h \neq h_1, ..., h \neq h_{s-1}\}$, and select $\mathbf{p}_s = \mathbf{p}_s^{(h_s)}$.

The procedure will stop when the number of the selected FBFs reaches r. Once r FBFs are found, i.e., r fuzzy rules are generated, the matrix $\mathbf{W}$ in Eqn. (9) is determined as $\mathbf{W} = [\mathbf{p}_1, ..., \mathbf{p}_r, \mathbf{p}_1.*\mathbf{x}_1, ..., \mathbf{p}_r.*\mathbf{x}_1, ..., \mathbf{p}_1.*\mathbf{x}_k, ..., \mathbf{p}_r.*\mathbf{x}_k]$, and the consequent parameters can be calculated as:

$$\xi = \mathbf{W}^+ \mathbf{d}. \tag{13}$$

The final EFBFN is:

$$y^* = \sum_{i=1}^{r} p_i(\chi) \cdot \chi' \cdot \xi_i, \tag{14}$$

where $\chi' = [1, x_1, ..., x_k]$, and $\xi_i = [\xi_0^i, \xi_1^i, ..., \xi_k^i]^T$.

3 Adaptive EPPSO for EFBFN

The EFBFN constructed by LS algorithm has the limitation that the widths of the membership functions in the hidden units are prefixed and the centers are selected only for the available train data pairs. In this section, we will use the proposed algorithm AEPPSO to tune the parameters of the MFs in the EFBFN obtained using LS algorithm, and to determine the consequent parameters simultaneously.

3.1 Population Initialization for EPPSO

(1) **Parameter representation** The fuzzy model we used in this paper has been described as in Eqn. (1), in which the parameters could be divided into the premise part and the consequent part. The parameter set of the two parts can both be expressed as a two dimensional matrix. For the premise part, the parameter set $\{m_j^i, \sigma_j^i, 1 \leq i \leq r, 1 \leq j \leq k\}$ of MFs constitute the first matrix named Q, while for the consequent part, the consequent parameter set $\{\xi_j^i, 1 \leq i \leq r, 0 \leq j \leq k\}$ constitutes the second matrix named ξ, where r is the number of the fuzzy rules and k is the number of input variables. The Gaussian type MF is used in this paper, thus the size of Q is $\mathbb{C}_1 = r \times (2k)$, and the size of ξ is $\mathbb{C}_2 = r \times (k+1)$, totally $\mathbb{C} = \mathbb{C}_1 + \mathbb{C}_2$ parameters.

(2) **Initialization** In this step, M individuals forming the population is initialized, and each consists of $\mathbb{C}_1$ parameters. The initialization of the parameter matrix Q is performed based on those MF parameters of the EFBFN obtained by LS algorithm in the above section. Assume that the obtained parameters are $\{m_j^i(0), \sigma_j^i(0)\}$, and the domains of the ith input variable in the train data set has been found to be $[\min(x_j), \max(x_j)]$, then we initialize the centers and widths for the FBFs as random values with the following domain:

$$m_j^i \in [m_j^i(0) - \delta_j, m_j^i(0) + \delta_j], \quad \sigma_j^i \in [\sigma_j^i(0) - \delta_j, \sigma_j^i(0) + \delta_j], \tag{15}$$

where δ_j is a small positive value, usually defined as $\delta_j = (\max(x_j) - \min(x_j))/10$.

These individuals could be regarded as population members in terms of EP and particles in terms of PSO, respectively.

3.2 Algorithm Description

(1) **PSO operator** In each generation, after the fitness values of each individual are evaluated, the top 50% individuals are selected as the elites and the others are discarded. Similar to the maturing phenomenon in nature, the individuals are firstly enhanced by PSO and become more suitable to the environment after acquiring the

knowledge from the society. The whole elites could be regarded as a swarm, and each elite corresponds to a particle in it. In PSO, individuals (particles) of the same generation enhance themselves based on their own private cognition and social interactions with each other. And this procedure is regarded as the maturing phenomenon in EPPSO. The selected $M/2$ elites are regarded as particles in PSO, and each elite corresponds to a particle $\mathbf{z}_\theta = (z_{\theta 1}, z_{\theta 2}, ..., z_{\theta \mathbb{C}_1})^T$.

In applying PSO, we adopt the following equation [5] to improve these individuals:

$$\mathbf{v}_\theta(t+1) = \omega \mathbf{v}_\theta(t) + c_1 r_1 (\mathbf{\Psi}_g(t) - \mathbf{z}_\theta(t)) + c_2 r_2 (\mathbf{\Psi}_g(t) - \mathbf{z}_\theta(t))$$
$$\mathbf{z}_\theta(t) = \mathbf{z}_\theta(t) + \mathbf{v}_\theta(t+1)$$
(16)

where $\mathbf{v}_\theta = (v_{\theta 1}, v_{\theta 2}, ..., v_{\theta \mathbb{C}_1})^T$ is the velocity vector of this particle, $\mathbf{\Psi}_\theta = (\psi_{\theta 1}, \psi_{\theta 2}, ..., \psi_{\theta \mathbb{C}_1})^T$ is the best previous position encountered by the ith particle, $\mathbf{\Psi}_g(t)$ is the best previous position among all the individuals of the swarm, $\theta = 1, 2, ..., M/2$, ω is a parameter called the inertia weight, t is the iteration counter, c_1 and c_2 are positive constants, referred to as cognitive and social parameters, respectively, and r_1, r_2 are random numbers, uniformly distributed within the interval [0, 1].

In Eqn. (16), $\mathbf{\Psi}_g(t)$ is the best performing individual evolved so far, either the enhanced elite obtained by PSO or the offspring produced using EP. By performing the PSO operation on the selected elites before mutation operation in EP, we may accelerate the convergence speed of the individuals and improve the search ability. The enhanced elites will be copied to the next generation and also designated as the parents of the generation for EP, copied and mutated to produce the other $M/2$ individuals.

(2) **EP operator** To produce better-performing offspring, the mutation parents are selected merely from the enhanced elites by PSO. In the EP operation, the $M/2$ particles in PSO will be $M/2$ population members with parameters $\{\mathbf{z}_\theta \mid \mathbf{z}_\theta = z_{\theta n}, 1 \leq \theta \leq M/2; 1 \leq n \leq \mathbb{C}_1\}$. The parameter mutation changes the parameters of membership functions by membership functions by adding Gaussian random numbers generated with the probability of p to them [6]:

$$z_{\theta n} = z_{\theta n} + \alpha * e^{(F_{max} - F_m)/F_{max}} * N(0,1),$$
(17)

where F_{max} is the largest fitness value of the individuals in the current generation, F_m is the fitness of the mth individual, α is a real value between 0 and 1. The combination of the offspring of EP and the elites enhanced by PSO comes to be the new generation of EPPSO, and after the fitness evaluation, the evolution will go ahead again until termination condition is satisfied.

The details of the basic procedure of EP are referred to Ref. [7].

(3) **Least Squares Estimate** The hybrid algorithm EPPSO is applied to tune the MF parameters of the EFBFN, while the LS is used to determine the $\mathbb{C}_2$ consequent

parameters ξ_j^i of the fuzzy rule base. For each individual in each generation, when the premise parameter matrix Q is determined by EPPSO, we fix these parameters and using the following procedure to obtain the consequent parameters:

i. Use the MF parameters $\{m_j^i, \sigma_j^i\}$ in Q to calculate the FBFs $p_i(\chi)$ as follows:

$$p_i(\chi) = \frac{\prod_{j=1}^{k} A_j^i(x_j)}{\sum_{i=1}^{r}(\prod_{j=1}^{k} A_j^i(x_j))}, \text{ where } A_j^i(x_j) = \exp[-\frac{1}{2}*(\frac{x_j - m_j^i}{\sigma_j^i})^2].$$

ii. Calculate matrix **W**

$$\mathbf{W} = [\mathbf{w}_1, ... \mathbf{w}_N]^T = [\mathbf{p}_1, ... \mathbf{p}_r, \mathbf{p}_1 \cdot \mathbf{x}_1, ... \mathbf{p}_r \cdot \mathbf{x}_1, ... \mathbf{p}_1 \cdot \mathbf{x}_k, ... \mathbf{p}_r \cdot \mathbf{x}_k],$$

where $\mathbf{p}_i = [p_i(\chi_1), ..., p_i(\chi_N)]^T$, $\mathbf{x}_j = [x_{1j}, x_{2j}, ... x_{Nj}]^T$,

iii. Now, we can calculate the consequent parameters of the current fuzzy rule base using Eqn. (13), and also the inferred output could be calculated using Eqn. (14).

Since the fitness evaluation is not the same for different problem, we will introduce it in the simulation part. When the fitness is evaluated, we select the top 50% individuals and till this step the program finish one generation. And the evolution will go on until the termination condition is satisfied.

4 Simulation Results

In this section, simulation results of predicting a chaotic time series using the fuzzy inference systems based on the proposed hybrid algorithm are presented.

The chaotic time series is generated from the Mackey-Glass differential delay equation [8] defined below:

$$\dot{x}(t) = 0.2x(t-\tau)/[1+x^{10}(t-\tau)] - 0.1x(t), \tag{18}$$

the problem is to use the past values of x to predict some future value of x. The same example as published in [8-10] has been adopted to allow a comparison with the published results. To obtain the time series value at each integer point, we applied the fourth Runge-Kutta method to find the numerical solution of Eqn. (18). Initial condition $x(0)=1.2$, $\tau = 17$ and the value of the signal six steps ahead $x(t+6)$ is predicted based on the values of the signal at current moment, 6, 12 and 18 steps back. The input-output data pairs are of the following format:

$$[x(t-18), x(t-12), x(t-6), x(t); x(t+6)]. \tag{19}$$

The data range of $118 \le t \le 1117$ has also been adopted, with the first 500 samples forming the training data set and the second 500 forming the validation data set. The nondimensional error index (*NDEI*) has been calculated to compare model performance, and the fitness function used in this example is defined as $F = 1/NDEI$.

During the training process of the EFBFN constructed for the chaotic time series using LS algorithm, the *NDEI* decrease of predicting the chaotic time series using EFBFN comes to be very little when the number of fuzzy rules reaches about twelve. Hence, we determine to generate a FIS with twelve fuzzy rules for predicting the chaotic time series. The final predicting *NDEI* in training and testing are 0.020 and 0.022, respectively.

We use the proposed algorithm AEPPSO to tune the MF parameters of the obtained EFBFN to achieve better performance. The twelve fuzzy rules in EFBFN result in 156 free parameters totally, in which 96 premise parameters to be tuned using EPPSO, and other 60 consequent parameters to be determined by LS method.

In applying AEPPSO, 70 individuals are initially randomly generated in a population, i.e., $M = 70$. During PSO operation, the inertia weights are set to be $w_{max} = 0.35$, $w_{min} = 0.1$, the cognitive parameter c_1 and the social parameter c_2 are both set to be 1.5; the mutation probability p and learning parameter α in EP are both set to be 0.1.

To show the superiority of AEPPSO, other two algorithms named AEP (Adaptive Evolutionary Programming) and APSO (Adaptive Particle Swarm Optimization) are also performed to the same problem.

The evolutions are processed for 100 generations and repeated for 5 runs. The training averaged best-so-far RMSE values over 5 runs for each generation are shown in Fig.1. From the figure, we can see that AEP converges with a slower speed compared to APSO and AEPPSO, while APSO converges fastest with a larger *NDEI* compared to that of AEPPSO. The phenomenon sufficiently incarnates the characteristics of each algorithm as follows:

i. AEP obtains knowledge merely from the individuals themselves without sharing the knowledge with each other, while the individuals in APSO adapt themselves according to their own knowledge and also their companions' knowledge. Therefore, the APSO is reasonably to have a faster convergence speed.
ii. APSO memorizes the previous knowledge of good solutions as flying destinations of the particles. However, the particles are restricted within the previous knowledge and thus coming to be lacking in diversity. This feature makes APSO probably converge to an unsatisfying result.
iii. AEPPSO combines the two algorithms' merits and the simulation results in Fig.1 validate its features of satisfying result and convergence speed.

Observing from the figure, the tuning performance for EFBFN based on the hybrid of LS and AEP is much worse than that of the other two, uncovering the ineffectiveness of EP in problems of pursuing high precision.

Table 1 lists different methods' generalization capabilities which are measured by using each method to predict 500 points immediately following the training set. The result of the proposed method outperforms that of any other approaches listed in the table except for that of ANFIS. Note that the generalization *NDEI*s of the hybrid algorithms are the mean values over 5 runs, and the minimum generalization *NDEI*s of the hybrid algorithms LS+AEP, LS+APSO and LS+AEPPSO are 0.015, 0.011 and 0.009, respectively.

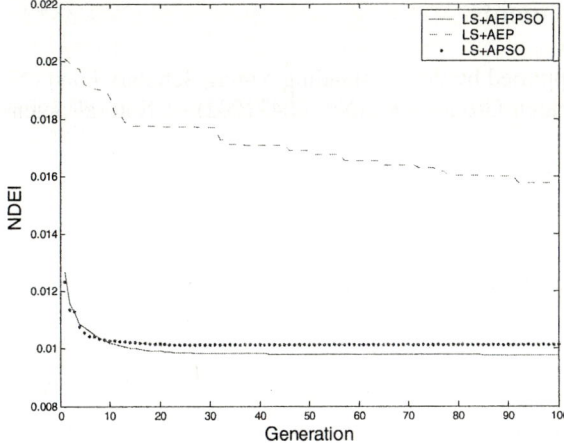

Fig. 1. Average best-so-far *NDEI* in each generation (iteration) using different methods

Table 1. Comparisons of generalization capability with published results

Method	Training Data	Error Index (NDEI)
Cascade-Correlation NN [8]	500	0.06
Back-Prop NN [8]	500	0.02
6th-order Polynomial [8]	500	0.04
ANFIS [9]	500	0.007
AEPLSE [10]	500	0.014
LS+AEP	500	0.017
LS+APSO	500	0.013
LS+AEPPSO	500	0.011

Unlike the back-propagation and LSE method of the neuro-fuzzy system ANFIS, the method proposed in this paper use the hybrid of evolution algorithm and LSE, and obtains an approximate performance. As a new evolutionary fuzzy system, the EFBFN generating method base on the hybrid of LS and AEPPSO provide another effective method of modeling and prediction using fuzzy inference system.

5 Conclusions

A novel fuzzy modeling strategy based on least squares method and the so-called Adaptive Evolutionary-programming and Particle-swarm-optimization (AEPPSO) has been presented. The LS method is firstly used to determine the fuzzy basis functions of the proposed EFBFN by presetting centers of Gaussian membership functions. In the second stage, the combined algorithm (EPPSO) is proposed to tune the obtained MF parameters in EFBFN, and the LS method is used again here to determine the consequent parameters in it simultaneously. Proposed as a new kind of Evolutionary Fuzzy System (EFS), the method has been examined on predicting a chaotic time series, and the results are compared to neuro-fuzzy system, neural networks, and other fuzzy modeling methods to demonstrate its effectiveness.

Acknowledgements

This work is supported by the Outstanding Young Scholars Fund (No. 60225006) and Innovative Research Group Fund (No. 60421002) of Natural Science Foundation of China.

References

1. Wang, L. X., Mendel, J. M.: Fuzzy Basis Functions, Universal Approximation, and Orthogonal-Least Squares Learning. IEEE Trans. Neural Networks. 3 (1992) 807-814
2. Chen, S., Cowan, C. F. N., Grant, P. M.: Orthogonal Least Squares Learning Algorithm for Radial Basis Function Networks. IEEE Trans. Neural Networks. 2 (1991) 302-309
3. Lee, C. W., Shin, Y. C.: Construction of Fuzzy Systems Using Least-Squares Method and Genetic Algorithms. Fuzzy Sets and Systems. 137 (2003) 297-323
4. Takagi, T., Sugeno, M.: Fuzzy Identification of Systems and Its Application. IEEE Trans. Systems, Man, and Cybernetics. 15 (1985) 116-132
5. Shi, Y., Eberhart, R. C.: Parameter Selection in Particle Swarm Optimization. The 7th Annual Conference on Evolutionary Programming, San Diego, USA. 7 (1998) 591-600
6. Hwang, H. S.: Automatic Design of Fuzzy Rule Bases for Modeling and Control Using Evolutionary Programming. IEE Proc. Control Theory Appl. 146 (1999) 9-16
7. Fogel, D. B.: Evolutionary Computations: Toward a New Philosophy of Machine Intelligence. New York: IEEE (1995)
8. Crowder, R. S.: Predicting The Mackey-Glass Time Series With Cascade-correlation Learning. Proc. 1990 Connectionist Models Summer School: Carngie Mellon University. (1990) 117-123
9. Jang, J. S. R., Sun, C. T., Mizutani, E.: ANFIS: Adaptive-Network-Based Fuzzy Inference System. IEEE Trans. Systems., Man and Cybernetics. 23 (1993) 665-685
10. Ye, B., Guo, C. X., Cao, Y. J.: Identification of fuzzy model using evolutionary programming and least squares estimate. Proc. 2004 IEEE Fuzzy Systems Conf. 2 (2004) 593-598

Analysis of Temporal Uncertainty of Trains Converging Based on Fuzzy Time Petri Nets*

Yangdong Ye[1], Juan Wang[1], and Limin Jia[2]

[1] Department of Computer Science, Zhengzhou University, Zhengzhou 450052, China
yeyd@zzu.edu.cn, wjuan@gs.zzu.edu.cn
[2] School of Traffic and Transportation, Beijing Jiaotong University, Beijing 100044, China
jialimin@jtys.bjtu.edu.cn

Abstract. The paper defines a fuzzy time Petri net (FTPN) which adopts four fuzzy set theoretic functions of time called fuzzy timestamp, fuzzy enabling time, fuzzy occurrence time and fuzzy delay, to deal with temporal uncertainty of train group operation and we also present different firing strategies for the net to give prominence to key events. The application instance shows that the method based on FTPN can efficiently analyze trains converging time, the possibility of converging and train terminal time in adjustment of train operation plan. Compared with time interval method, this method has some outstanding characteristics such as accurate analysis, simple computation, system simplifying and convenience for system integrating.

1 Introduction

The main issues [1][2] of train group operation modeling in the research of Railway Intelligent Transportation System (RITS) include the representation of train group concurrency, processing of multi-levels problems oriented to different control and decision, the disposal of system hybrid attributes and processing of various uncertainty factors affecting train group operation, all of which are relative to analysis of time parameter. With accelerating the speed of train group operation and the influences of objective uncertainty factors, it is increasingly important to process temporal uncertainty issues in train group operation. The analysis of temporal uncertainty in trains converging is significant for dynamic control during train operation, trains dispatching in stations, passengers changing trains, goods transferred and resources distribution of railway system.

In the process of dealing with time factors in train group operation, the usual representation of single point of time is impractical and not integrated and the representation of time interval is difficult to do quantitative analysis. Based on existed Petri net model of train group operation [2][10], the paper introduces fuzzy set theory to describe uncertainty or subjective time information, which can satisfy applications of reality. How to represent uncertainty knowledge is an attentional issue [3~7] in the research of Petri net modeling. With the existed time Petri net (TPN) [8~11], we define

---

* This research was partially supported by the National Science Foundation of China under grant number 600332020 and the Henan Science Foundation under grant number 0411012300.

a fuzzy time Petri net adopting fuzzy set theoretic functions of time to analyze trains converging issue.

The paper introduces relative concepts of fuzzy time Petri net and the fuzzy set theoretic functions of time in Section 2 and 3; then we give an example of train operation in Section 4 and make relative analysis in Section 5; finally aiming at temporal uncertainty we compare this method with reasoning algorithm of TPN in Section 6.

2 Fuzzy Time Petri Nets

In order to describe train behaviors with time constraints, the paper defines a fuzzy time Petri nets adopting four fuzzy time functions to deal with temporal uncertainty of trains converging issue. These functions appear in some publications and applications [6~8] and their definitions were given by Murata etc [3~5].

Definition 2.1. Fuzzy time Petri nets' system (FTPN's) FTPN's = $(P, T, E, \beta, A, FT, D, M_0)$, where:

(1) $P = \{p_1, p_2, \ldots p_n\}$ is a finite set of places;
(2) $T = \{t_1, t_2, \ldots t_m\}$ is a finite set of transitions, where $P \cup T \neq \emptyset$, $P \cap T = \emptyset$;
(3) $E = \{e_1, e_2, \ldots e_m\}$ is a finite set of events;
(4) $\beta: E \rightarrow T$, is a mapping function that represents an event is relevant to a transition;
(5) $A \subseteq (P \times T) \cup (T \times P)$ is a set of arcs;
(6) FT is a set of fuzzy timestamps. It is related with tokens. An unrestricted timestamp is represented by [0,0,0,0], and an empty timestamp is $\emptyset$;
(7) D is a set of fuzzy delay time that is related with outgoing arcs of transitions;
(8) $M_0: P \rightarrow FT$, is an initial marking function.

Definition 2.2. State marking of FTPN's

A marking $M_i: P \rightarrow FT$ (i=0,1,2,3...) is a description of dynamic behaviors of system. A marking of system corresponds to a vector about places. In this paper, we use sets of tokens, $\{(p, \pi(\tau)) | p \in P, \pi(\tau) \in FT\}$, to describe marking and the empty timestamp $\emptyset$ does not appear in the sets of tokens. If M_2 is directly reachable from M_1 via e_l, the sequence q is $M_1[e_l>M_2$.

Definition 2.3. Firing strategies of transitions in FTPN's

We primarily define FTPN to describe the analysis of temporal uncertainty facing train group operation. The firing of transition takes no time. Under conditions with no conflicts, transitions will fire immediately when all required tokens arrive. For conditions with conflicts, system needs to use certain firing strategies to do different analysis. The following strategies are used according to different problems in FTPN's: (1) the earliest enabling time firing strategy; (2) the strategy of multi-condition first firing; (3) the strategy of high-possibility first firing; (4) mixed strategy etc. Using these firing strategies we can give prominence to key events and simplify system analyzing.

3 Fuzzy Time Function

A fuzzy time function is a mapping function from time scale, the set of all non-negative real numbers, to the real interval [0, 1]. The value of function indicates degree of possibilities for an event on a point of time τ. Fuzzy time functions are specified by 4-tuple [3~5] and their graphs are trapezoid describing train earliest arrival or departing time, the most possible time and the latest time. The paper uses square brackets express the functions to keep consistent with time interval [10,11].

Definition 3.1. Fuzzy timestamp π(τ)

The fuzzy timestamp is the possibility distribution of a token arriving in one place on time τ.

Definition 3.2. The possibility distribution of the *latest* time *latest* is a multiple operator to calculate the possibility distribution of the latest time of fuzzy timestamps. Suppose that there are n fuzzy timestamps $\pi_i(\tau) = h_i[a_i, b_i, c_i, d_i]$, i=1,2,...,n, we calculate the possibility distribution of the *latest* time as following

$latest\{\pi_1(\tau), \pi_2(\tau)\} = min(h_1, h_2)[max(a_1, a_2), max(b_1, b_2), max(c_1, c_2), max(d_1, d_2)]$
$latest\{\pi_i(\tau), i=1,2,...,n\} = latest\{latest\{...\{latest\{\pi_1(\tau), \pi_2(\tau)\},...,\pi_i(\tau)\},...\}, \pi_n(\tau)\}$.

Definition 3.3. The possibility distribution of the *earliest* time *earliest* is an operator to calculate the possibility distribution of the earliest time among many timestamps. Suppose that there are n fuzzy timestamps $\pi_i(\tau) = h_i[a_i, b_i, c_i, d_i]$, i=1,2,...,n, the possibility distribution of the *earliest* time is following

$earliest\{\pi_1(\tau), \pi_2(\tau)\} = max(h_1, h_2)[min(a_1, a_2), min(b_1, b_2), min(c_1, c_2), min(d_1, d_2)]$
$earliest\{\pi_i(\tau), i=1,2,...,n\} = earliest\{earliest\{...\{earliest\{\pi_1(\tau), \pi_2(\tau)\},...,\pi_i(\tau)\},...\}, \pi_n(\tau)\}$.

Definition 3.4. Fuzzy enabling time e(τ)

When the occurrence of a transition or an event needs more than one token or resource, the *latest* time possibility distribution is fuzzy enabling time e(τ). The net in this paper is ordinary net (the weight of inputting arcs is 1). So if transition *t* has n input places, to fire *t* needs n tokens, i.e. the occurrence of the corresponding event needs n tokens respectively in n places. The fuzzy enabling time of *t* is

$$e(\tau) = latest\{\pi_i(\tau), i=1,2,...,n\} \qquad (1)$$

Definition 3.5. Fuzzy occurrence time o(τ)

The fuzzy occurrence time o(τ) is the possibility distribution of an event occurring time. In ordinary circumstances we adopt the principle of First Come, First Served, to give high priority to the earlier enabling event.

The fuzzy occurrence time of an event is the minimum overlapping area between fuzzy enabling time of this event and the result of *earliest*, and this operation is expressed by *min* [3][4]. In this paper the operator *Min* rounds the value of the calculated result of *min*. Suppose that there are n enabling events e_i, i=1,2,...,n, and the corresponding fuzzy enabling time is $e_i(\tau)$, i=1,2,...,n by formula (1). The fuzzy occurrence time of the event e_t with fuzzy enabling time $e_t(\tau)$ calculates as following

$$o_t(\tau) = Min\{e_t(\tau), earliest\{e_i(\tau), i=1,2,\ldots,t,\ldots,n\}\} \quad (2)$$

Definition 3.6. Fuzzy delay $d(\tau)$

The fuzzy delay is a fuzzy time function associated with outgoing arcs from transitions. It is one kind of measurement about time when describing and analyzing events and means degree of possibility of time span that system states change from one to another when an event occurs. Given fuzzy delay and fuzzy occurrence time, the fuzzy timestamp of the new token is calculated as following

$$\pi(\tau) = o(\tau) \oplus d(\tau) = [o_1,o_2,o_3,o_4] \oplus [d_1,d_2,d_3,d_4] = [o_1+d_1,o_2+d_2,o_3+d_3,o_4+d_4] \quad (3)$$

Suppose that transition t has one input place p_1 and one output place p_2, and no conflict. If the fuzzy timestamp of the token in p_1 is $\pi_1(\tau)$, then the fuzzy enabling time is $e_t(\tau) = \pi_1(\tau)$ and the fuzzy occurrence time is $o_t(\tau) = e_t(\tau) = \pi_1(\tau)$. Given fuzzy delay $d(\tau)$, the fuzzy timestamp of the new token in p_2 after the event occurs is

$$\pi_2(\tau) = o_t(\tau) \oplus d(\tau) = e_t(\tau) \oplus d(\tau) = \pi_1(\tau) \oplus d(\tau) \quad (4)$$

Now the fuzzy timestamp of new token equals the result of operator $\oplus$ between fuzzy timestamps of the token in input place and the fuzzy delay.

4 Train Group Behaviors Model

During the modeling of train group behaviors, a series of temporal uncertainty problems emerge, which are caused by objective stochastic uncertainty in the train operation departure from train operation graph.

4.1 Example of Train Group Operation

In order to explain the validity of FTPN clearly, a railway net in Fig. 1 is given, where S is station and Se is section. Train operation is denoted by following time constraints, where time (unit of time uses minute) is given by fuzzy time function.

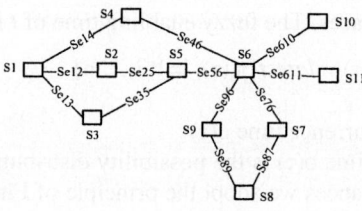

Fig. 1. Railway route net of certain area

Train Tr1 plans to start from Station S1 at 7:00am, then to arrive at S6, and after time [13,15,20,22] on Se611 to reach terminal S11. There are three paths to S6. Tr1 runs on Se12 and Se25 for time [18,20,25,27] and [13,15,20,22] to S5, or on Se13 and Se35 for [18,20,25,27] and [20,22,30,32] to S5, and then from S5 to spend [18,20, 25,27] on Se56 to S6. Another path is from S1 via Se14 and Se46 to S6 for time [26, 28,32,34] and [26,28,32,34] respectively. The problem of converging will emerge at S6; If it happens, the train will delay for [2,2,4,4]; If not, there is no delay.

Train Tr2 starts from S8 at 7:00am, via Se89, Se96 for [18,20,25,27] and [20,22,25,27] respectively to reach S6; or via Se87, Se76 for [13,15,18,20] and [23,25,30,32] to get to S6, and then spends [18,20,25,27] to S10. There is the same problem of converging at S6.

4.2 Structure of the Model

We construct FTPN model as shown in Fig. 5. The descriptions of places and transitions are listed in Table 1. FTPN's = ($P, T, E, \beta, A, FT, D, M_0$), where

(1) $P = \{ p_i | i=1,2,...,14 \}$; $T = \{ t_j | j=1,2,...,16 \}$, $E = \{ e_j | j=1,2,...,16 \}$;
(2) Fuzzy delay $D = \{$ $d_1 = [18,20,25,27]$, $d_2 = [18,20,25,27]$, $d_3 = [13,15,20,22]$, $d_4 = [20,22,30,32]$, $d_5 = [18,20,25,27]$, $d_6 = [26,28,32,34]$, $d_7 = [26,28,32,34]$, $d_8 = d_{15} = [0,0,0,0]$, $d_9 = d_{17} = [2,2,4,4]$, $d_{10} = [13,15,20,22]$, $d_{11} = [18,20,25,27]$, $d_{12} = [13,15,18,20]$, $d_{13} = [20,22,25,27]$, $d_{14} = [23,25,30,32]$, $d_{16} = [18,20,25,27]$ $\}$;
(3) M_0: $\{ (p_1, \pi_{01}(\tau)), (p_9, \pi_{02}(\tau)) \}$; $\pi_{01}(\tau) = [0,0,0,0]$, $\pi_{02}(\tau) = [0,0,0,0]$.

Table 1. Places and transitions of train operation FTPN model

$p_1, p_2, p_3, p_4, p_5, p_6$	Tr1 is in station S1, S2, S3, S4, S5 and S6.
p_7, p_{13}	Tr1 and Tr2 is ready to depart from S6.
p_8	Tr1 is in station S11.
$p_9, p_{10}, p_{11}, p_{12}$	Tr2 is in station S8, S9, S7 and S6.
p_{14}	Tr2 is in station S10.
t_1, t_2, t_3, t_4, t_5	Tr1 is running on section Se12, Se13, Se25, Se35 and Se56.
t_6, t_7	Tr1 is running on section Se14 and Se46.
t_8, t_{15}	Trains are in S6 without trains converging.
t_9	Trains stay in S6 with trains converging.
t_{10}	Tr1 is running on section Se611.
$t_{11}, t_{12}, t_{13}, t_{14}, t_{16}$	Tr2 is running on section Se89, Se87, Se96, Se76 and Se610.

5 Analysis of Temporal Uncertainty of Trains Converging

This chapter analyzes the issues in the process of adjusting train operation plan for the model in Fig. 3 including trains converging time, the possibility of converging, train terminal time and the adjustment of train operation time.

5.1 Problem Statement

The analysis of time parameters of trains converging in stations can conclude to the computation of timestamps and analysis of transition sequences from one state of FTPN's to another.

Three transition sequences, t_1, t_3, t_5; t_2, t_4, t_5; t_6, t_7, exist when one token moves from p_1 to p_6. Two sequences, t_{11}, t_{13}; t_{12}, t_{14}, exist when token moves from p_9 to p_{12}. Suppose that the possible fuzzy timestamps in p_6 are $\pi_{61}(\tau)$, $\pi_{62}(\tau)$, $\pi_{63}(\tau)$ and the fuzzy timestamps in p_{12} are $\pi_{121}(\tau)$, $\pi_{122}(\tau)$. When there are tokens in both p_6 and p_{12}, system marking is M_p:

$\{(p_6, \pi_6(\tau)), (p_{12}, \pi_{12}(\tau))\}$. M_p is the critical state whether *trains converging* happens or not. All possible fuzzy timestamps in M_p can be calculated by formula (3) (4) and the results are shown in Fig. 2.

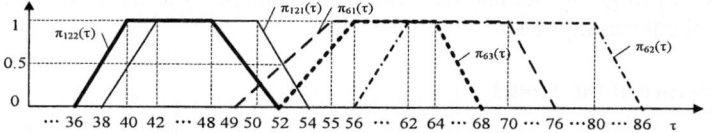

Fig. 2. All possible fuzzy timestamps in M_p

From Fig. 2 transition t_8, t_{15} and t_9 under M_p may be enabled simultaneously, so the corresponding events e_8, e_{15} and e_9 have conflict. With different firing strategies the occurrence sequence from one state to another is different. Here we adopt the principle of small subscript first to deal with the events, which have the same enabling time and are independent to each other. The behaviors of two trains after arriving in station S6 conclude as following,

q_1: $M_p[e_9>M_j[e_{10}>M_{10}[e_{16}>M_f$
q_2: $M_p[e_8>M_8[e_{15}>M_{nj}[e_{10}>M_{10'}[e_{16}>M_{nf}$

among which the markings are M_j: $\{(p_7, \pi_{j1}(\tau)), (p_{13}, \pi_{j2}(\tau))\}$, M_{nj}: $\{(p_7, \pi_{nj1}(\tau)), (p_{13}, \pi_{nj2}(\tau))\}$, M_f: $\{(p_8, \pi_8(\tau)), (p_{14}, \pi_{14}(\tau))\}$ and M_{nf}: $\{(p_8, \pi_{8'}(\tau)), (p_6, \pi_{14'}(\tau))\}$.

The trains converging issue is the computation of timestamps and analysis of transition sequences from state M_0 to M_p, then to M_j or M_{nj} and to final state M_f or M_{nf}. All the transition sequences and correlative markings can be seen in Table 2. When trains converging does not happen the values of corresponding fuzzy timestamps keep invariable, so we use the same subscripts in Table 2.

Table 2. Transition sequences and part markings of model

Occurrence sequence	Transition sequence of each case	State M_p before converging	State M_j/M_{nj} after converging
Sequence with Trains converging: q_1	q_{11}: $t_1,t_3,t_5,t_{11},t_{13},t_9,t_{10},t_{16}$	M_{p1}: $\{(p_6,\pi_{61}(\tau)),(p_{12},\pi_{121}(\tau))\}$	M_{j1}: $\{(p_7,\pi_{71}(\tau)),(p_{13},\pi_{131}(\tau))\}$
	q_{12}: $t_6,t_7,t_{11},t_{13},t_9,t_{10},t_{16}$	M_{p2}: $\{(p_6,\pi_{63}(\tau)),(p_{12},\pi_{121}(\tau))\}$	M_{j2}: $\{(p_7,\pi_{72}(\tau)),(p_{13},\pi_{132}(\tau))\}$
	q_{13}: $t_1,t_3,t_5,t_{12},t_{14},t_9,t_{10},t_{16}$	M_{p3}: $\{(p_6,\pi_{61}(\tau)),(p_{12},\pi_{122}(\tau))\}$	M_{j3}: $\{(p_7,\pi_{73}(\tau)),(p_{13},\pi_{133}(\tau))\}$
Sequence without trains converging: q_2	q_{21}: $t_2,t_4,t_5,t_{11},t_{13},t_8,t_{15},t_{10},t_{16}$	M_{p4}: $\{(p_6,\pi_{62}(\tau)),(p_{12},\pi_{121}(\tau))\}$	M_{nj1}: $\{(p_7,\pi_{62}(\tau)),(p_{13},\pi_{121}(\tau))\}$
	q_{22}: $t_2,t_4,t_5,t_{12},t_{14},t_8,t_{15},t_{10},t_{16}$	M_{p5}: $\{(p_6,\pi_{62}(\tau)),(p_{12},\pi_{122}(\tau))\}$	M_{nj2}: $\{(p_7,\pi_{62}(\tau)),(p_{13},\pi_{122}(\tau))\}$
	q_{23}: $t_6,t_7,t_{12},t_{14},t_8,t_{15},t_{10},t_{16}$	M_{p6}: $\{(p_6,\pi_{63}(\tau)),(p_{12},\pi_{122}(\tau))\}$	M_{nj3}: $\{(p_7,\pi_{63}(\tau)),(p_{13},\pi_{122}(\tau))\}$
	q_{24}: $t_1,t_3,t_5,t_{11},t_{13},t_8,t_{15},t_{10},t_{16}$	M_{p7}: $\{(p_6,\pi_{61}(\tau)),(p_{12},\pi_{121}(\tau))\}$	M_{nj4}: $\{(p_7,\pi_{61}(\tau)),(p_{13},\pi_{121}(\tau))\}$
	q_{25}: $t_6,t_7,t_{11},t_{13},t_8,t_{15},t_{10},t_{16}$	M_{p8}: $\{(p_6,\pi_{63}(\tau)),(p_{12},\pi_{121}(\tau))\}$	M_{nj5}: $\{(p_7,\pi_{63}(\tau)),(p_{13},\pi_{121}(\tau))\}$
	q_{26}: $t_1,t_3,t_5,t_{12},t_{14},t_8,t_{15},t_{10},t_{16}$	M_{p9}: $\{(p_6,\pi_{61}(\tau)),(p_{12},\pi_{122}(\tau))\}$	M_{nj6}: $\{(p_7,\pi_{61}(\tau)),(p_{13},\pi_{122}(\tau))\}$

In base of the earliest enabling time firing strategy we adopt the strategies of multi-condition first firing and high-possibility first firing to deal with temporal uncertainty of trains converging. The problems include that whether trains converging occurs, what the possibility of converging is, what the earliest or latest terminal time is and how to adjust train operation plan with analysis of trains converging.

5.2 Analysis Using the Strategy of Multi-condition First Firing

The strategy of multi-condition first firing means that comparing the number of required tokens for the enabling events the transition requiring more tokens will fire first. This strategy stands out the complex event which requires more resources.

With the strategy the occurrence sequence is q_1. The analysis of trains converging time corresponds to the computation of timestamp of e_9. The fuzzy occurrence time of e_9 under M_{p1}, M_{p2}, M_{p3} in Table 2 can be calculated by formula (1) (2) respectively.

$$o_{e91}(\tau) = 0.5[49,52,52,54], o_{e92}(\tau) = 0.25[52,53,53,54], o_{e93}(\tau) = 0.25[49,51,51,52].$$

Final fuzzy timestamps corresponding to marking M_f can be calculated by formula (4) and the results are shown with dashed line in Fig. 3(a) (b).

The analysis of trains converging can be made out from above computation.

- Trains converging time and the possibility of converging can be got by $o_{e9}(\tau)$. There is the highest occurrence possibility 0.5 at $\tau=52$ i.e. train Tr1 and Tr2 can meet in station S6 between 7: 49am and 7: 54am and the largest possibility is 0.5.
- The earliest or latest terminal time after trains converging can be got from Fig. 3(a) (b). The earlier fuzzy time functions when Tr1 arrives in S8 are $\pi_{81}(\tau) = 0.5[64,69,76,80]$ and $\pi_{83}(\tau)=0.25[64,68,73,78]$, so that the earliest terminal time of Tr1 is 8:04am and the highest possibility arriving in terminal station is 0.5.
- The paths corresponding with the earliest or latest terminal time also can be got. The fuzzy timestamp $\pi_{81}(\tau)$ must pass through the transition sequence $t_1,t_3,t_5,t_{11},t_{13},t_9,t_{10},t_{16}$, i.e. Tr1 goes only through the section Se13, Se35, Se56 and Se611 and stops for a while at station S6, and after that there is the earliest terminal time 8:04am. Similarly the time and paths of Tr2 can be analyzed in the same way.

5.3 Analysis with the Strategy of High-Possibility First Firing

The strategy of high-possibility first firing means that for the given occurrence possibility threshold δ, if there is time τ making the possibility h_i of $o_{ei}(\tau)$ greater than δ, i.e. $h_i>\delta$, then the event e_i is the key event to be analyzed. This strategy actually gives prominence to the event with higher possibility.

If $\delta=0.5$ in the strategy of high-possibility first, the trains converging event of the model will become the nonoccurrence or low-possibility event. The occurrence sequence q_2 includes all cases without trains converging.

The fuzzy occurrence times of e_8, e_{15} are $o_{e8}(\tau) = \pi_6(\tau)$ and $o_{e15}(\tau) = \pi_{12}(\tau)$ and the marking after e_8 and e_{15} is M_{nj}: $\{(p_7,\pi_{nj1}(\tau)),(p_{13},\pi_{nj2}(\tau))\}$. The corresponding transition sequences and fuzzy timestamps can be seen in Table 2. All final fuzzy timestamps calculate by formula (4) are shown in Fig 3 (a) (b) with real line.

The analysis of temporal uncertainty about trains converging can be got from above results.

- Trains converging event does not occur with this strategy. So the two trains run respectively and there is no stop in station S6.

- The earliest or latest terminal time without trains converging can be got from Fig. 3(a) (b). The earliest fuzzy time when Tr1 arrives in S8 is $\pi_{81'}(\tau)=[62,70,90,98]$, so that the earliest terminal time of Tr1 is 8:02am and the highest possibility is 1.
- Corresponding with $\pi_{81}(\tau)$ Tr1 goes only through the section Se12, Se25, Se56 and Se611 and no stop in S6, and then there is the earliest terminal time 8:02am.

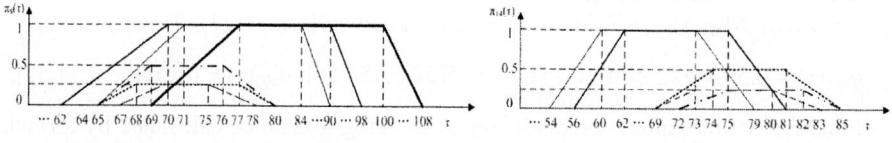

(a) All possible fuzzy timestamps of token in p_8 (b) All possible fuzzy timestamps of token in p_{14}

Fig. 3. All final fuzzy timestamps

5.4 Analysis of Train Operation Plan Adjustment Based on Trains Converging

Analysis of train operation plan adjustment based on trains converging can make out from Fig. 2. The possibility of trains not converging is higher than that of trains converging. If the system in reality needs trains converging, we can do quantitative analysis for train operation plan using the FTPN model. Among all the cases of $\pi_6(\tau)$, $\pi_{12}(\tau)$ the maximum time difference is 5. With the minimum adjustment degree when the later token arrives ahead of schedule for 5 units of time or the earlier token delays for 5 units the event e_9 has the highest possibility 1 through transition sequence t_1, t_3, t_5, t_{11}, t_{13}. That is to say that if Tr2 departs ahead of time or speed up, or Tr1 put off departing, the possibility of trains converging will be higher.

With the adjustment based on one train as benchmark and the highest possibility 1, Tr1 passes through Se12, Se25, Se56 and Tr2 goes through Se89, Se96, the time of converging is [44,50,50,54] or [49,55,55,59] which are shown in Fig. 4 (a) (b).

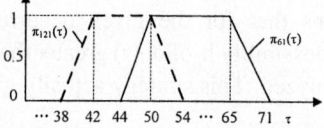 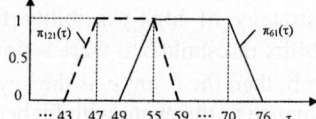

(a) Fuzzy timestamps of Tr1 ahead for 5 minutes   (b) Fuzzy timestamps of Tr2 delaying for 5 minutes

Fig. 4. Train operation adjustment for trains converging

Similarly when the later token delays for 5 units of time or the earlier token arrives ahead of time for 5 units of time e_9 has the lowest possibility 0.

6 Comparing Analysis with Reasoning Algorithm of Time Petri Net

The present research of temporal knowledge reasoning algorithm [10,11] of TPN aims at the system model of TPN, uses finite system states, creates sprouting graph of time parameters, and then with the sprouting graph analyzes time parameters.

Analysis of Temporal Uncertainty of Trains Converging Based on Fuzzy Time Petri Nets 97

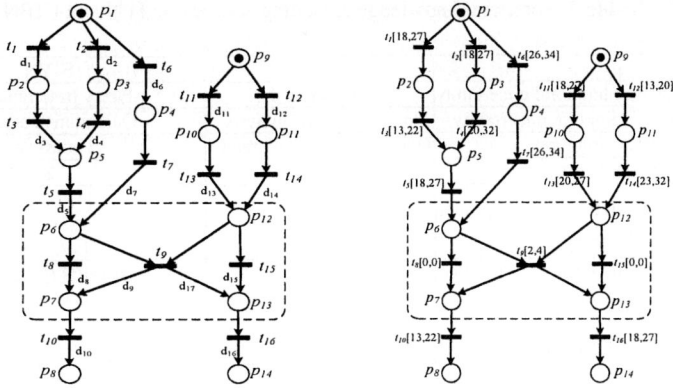

Fig. 5. FTPN model of train operation   **Fig. 6.** TPN model of train operation

6.1 Time Petri Net Model of the Example

Time Petri net (TPN) adopts time interval to deal with temporal uncertainty of train group operation [10]. The model of the example using TPN in [10] is shown in Fig. 6, where time interval parameters are marked. The corresponding sprouting graph with time parameter of the model is shown in Fig. 7.

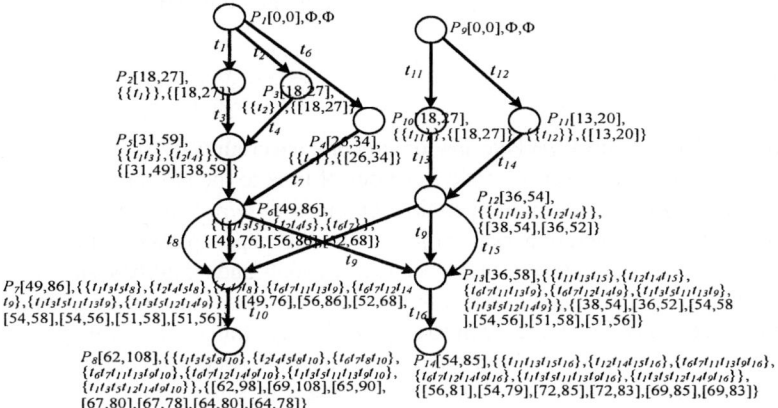

Fig. 7. Time parameter sprouting graph of train operation

6.2 Comparing Analysis

We analyze the example using the temporal knowledge reasoning algorithm of TPN and the specific contrasts with the analysis in chapter 5 can be seen in Table 3.

Table 3. Temporal knowledge reasoning contrast of TPN and FTPN

	TPN	FTPN
Representation of temporal uncertainty	Time interval	Fuzzy time function
Processing of temporal uncertainty	No Quantitative analysis	Quantitative analysis
Data structure	Dynamic sprouting graph	Timestamp
Computation procedure	4	2
Occurrence time of trains converging in q_{11}, q_{12}, q_{13}	[49,54] [52,54] [49,52]	0.5[49,52,52,54] 0.25[52,53,53,54] 0.25[49,51,51,52]
The earliest terminal time of Tr1 after converging and transition sequence	[64,78] q_{13}: $t_1, t_3, t_5, t_{12}, t_{14}, t_9, t_{10}, t_{16}$	0.25[64,68,75,78] q_{13}: $t_1, t_3, t_5, t_{12}, t_{14}, t_9, t_{10}, t_{16}$
The earliest terminal time of Tr2 after converging and transition sequence	[69,83] q_{13}: $t_1, t_3, t_5, t_{12}, t_{14}, t_9, t_{10}, t_{16}$	0.25[69,73,80,83] q_{13}: $t_1, t_3, t_5, t_{12}, t_{14}, t_9, t_{10}, t_{16}$
Adjustment of train operation plan	No	Adjustment degree is 5 units of time

The following conclusions can be deduced from Table 3: (1) The descriptions of temporal uncertainty of the two methods are consistent. (2) The fuzzy time functions of FTPN have quantitative analysis. (3) Our computation procedure is simple. The existed method has four procedures which are modeling, system states searching, creating sprouting graph and analyzing processes. But our method just needs two procedures, system states searching and time parameters calculating and analyzing when token moves. (4) Time analysis is easier in our method. FTPN adopts token structure of fuzzy timestamp which is much simpler than sprouting graph structure. (5) Our method adopts firing strategies that can give prominence to key events in different conditions. (6) The method is convenient for integration of train object models.

7 Conclusions

The paper defines a fuzzy time Petri net for temporal uncertainty of train group operation and introduces four fuzzy set theoretic functions of time to the net for representation of temporal uncertainty during train operation. Oriented to temporal uncertainty of trains converging in train operation plan adjustment, this paper presents different firing strategies for the model to give prominence to key events and simplify system modeling and analyzing. The application instance shows that the method can analyze quantitatively temporal uncertainty of trains converging including trains converging time, the possibility of converging and terminal time in train operation plan adjustment. Comparing with time interval method we conclude that our method has outstanding characteristics such as accurate analysis, simple computation, system simplifying and convenience for system integrating. So the method can be used to represent temporal knowledge and enriches the research of various railway expert systems.

References

1. Jia, L.-M., Jiang, Q.-H.: Study on Essential Characters of RITS. Proceeding of 6th International Symposium on Autonomous Decentralized Systems. IEEE Computer Society, Pisa Italy (2003) 216-221.

2. Ye, Y.-D., Zhang, L., Du, Y.-H., Jia, L.-M.: Three-dimension Train Group Operation Simulation System Based on Petri Net with Objects. Proceeding of IEEE 6th International Conference on Intelligent Transportation Systems. Vol.2. (2003) 1568-1573.
3. Murata, T.: Temporal Uncertainty and Fuzzy-timing High-Level Petri Nets. 17th International conference on Application and Theory of Petri Nets, Lecture Notes in Computer Science, vol.1091. Springer-Verlag, New York (1996) 11-28.
4. Zhou, Y., Murata, T.: Petri Net Model with Fuzzy-timing and Fuzzy-Metric. International Journal of Intelligent Systems. 14(8) (1999) 719-746.
5. Zhou, Y., Murata, T., DeFanti, T.A.: Modeling and Performance Analysis using Extended Fuzzy-timing Petri Nets for Networked Virtual Environments. IEEE Transaction on System, Man and Cybernetics-Part B: Cybernetics, 30(5) (2000) 737-756.
6. Cardoso, J., Valette, R., Dubois, D.: Possibilistic Petri Nets. IEEE Transaction on System, Man and Cybernetics-Part B: Cybernetics. 29(5) (1999) 573-582.
7. Dubois, D., Prade, H.: Processing Fuzzy Temporal Knowledge. IEEE Transaction on System, Man and Cybernetics. 19(4) (1989) 729-744.
8. Merlin P.M.: A Methodology for The Design and Implementation of Communication Protocol. IEEE Transaction on Communication. 24(6) (1976) 614-621.
9. Tsai, J.-J.P., Yang, S.-J., Chang, Y.-H.: Timing Constraint Petri Nets and Their Application to Schedulability Analysis of Real-time System Specifications. IEEE Transaction on Software Engineering. 21(1) (1995) 32-49.
10. Ye, Y.-D., Du, Y.-H., Gao, J.-W., Jia, L.-M. A Temporal Knowledge Reasoning Algorithm using Time Petri Nets and its Applications in Railway Intelligent Transportation System. Journal of the china railway society. 24(5) (2002) 5-10.
11. Jong, W.-T., Shiau, Y.-S., Horng, Y.-J., Chen, H.-H., Chen, S.-M.: Temporal Knowledge Representation and Reasoning Techniques Using Time Petri Nets. IEEE Transaction on System, Man and Cybernetics, Part-B: Cybernetics. 29(4) (1999) 541-545.

Interval Regression Analysis Using Support Vector Machine and Quantile Regression

Changha Hwang[1], Dug Hun Hong[2], Eunyoung Na[3],
Hyejung Park[3], and Jooyong Shim[4]

[1] Division of Information and Computer Sciences, Dankook University,
Yongsan Seoul, 140-714, South Korea
chwang@dankook.ac.kr
[2] Department of Mathematics, Myongji University,
Yongin Kyunggido, 449-728, South Korea
dhhong@mju.ac.kr
[3] Department of Statistical Information, Catholic University of Daegu,
Kyungbuk 712 - 702, South Korea
{ney111, hyjpark }@cu.ac.kr
[4] Corresponding Author, Department of Statistics,
Catholic University of Daegu, Kyungbuk 702-701, South Korea
jyshim@cu.ac.kr

Abstract. This paper deals with interval regression analysis using support vector machine and quantile regression method. The algorithm consists of two phases - the identification of the main trend of the data and the interval regression based on acquired main trend. Using the principle of support vector machine the linear interval regression can be extended to the nonlinear interval regression. Numerical studies are then presented which indicate the performance of this algorithm.

1 Introduction

Regression analysis has been the most frequently used technique in various areas of business, science, and engineering. But we encounter many situations where necessary assumptions for regression analysis cannot be met because they are not based on random variables. In such situations interval regression analysis, which is regarded as the simplest version of possibilistic regression analysis introduced by Tanaka et al.[9], can be a good alternative. Interval regression that can be measured its spreading extent of data in the presence of heteroscedasticity, is addressed and analyzed. In interval regression analysis, the sample involved whole interval will be estimated. However, when estimation is made only through the whole data, outlier becomes its focus; thus in this case, the analysis of tendency for the whole data would be difficult. Lee and Tanaka[6] proposed upper and lower approximation model to overcome the above problem in the linear interval regression based on the quantile regression.

The support vector machine(SVM), firstly developed by Vapnik[10][11], is being used as a new technique for regression and classification problem. SVM is gaining popularity due to many attractive features, and promising empirical performance. It has been successfully applied to a number of real world problems such as handwritten character and digit recognition, face detection, text categorization and object detection in machine vision. The aforementioned applications are related to classification problems. It is also widely applicable in regression problems. SVM was initially developed to solve classification problems but recently it has been extended to the domain of regression problems. SVM is based on the structural risk minimization(SRM) principle, which has been shown to be superior to traditional empirical risk minimization(ERM) principle. SRM minimizes an upper bound on the expected risk unlike ERM minimizing the error on the training data. By minimizing this bound, high generalization performance can be achieved. For introductions and overviews of recent developments of SVM regression can be found in Gunn[2], Smola and Schoelkopf[8], Kecman[4], and Wang[12]. SVM avoids over-fitting by using regularization technique and is robust to outlier.

Hong and Hwang[3] introduced interval regression analysis using quadratic loss SVM for crisp input and interval output. In this paper, we propose interval regression analysis for crisp input and output using the principle of SVM and quantile regression. The proposed model can be applied to both the linear and nonlinear interval regression. The remainder of this paper is organized as follows. In Section 2 we explain the estimation methods of quantile regression using SVM. In Section 3 we explain the estimation methods of interval regression using the principle of SVM and the quantile regression method in Section 2. In Section 4 we perform the numerical studies through four examples. Finally, Section 5 gives the conclusions.

2 Quantile Regression Using Support Vector Machine

The quantile regression model introduced by Koenker and Bassett[5] assumes that the conditional quantile function of the response given x_i is linearly related to the covariate vector x_i as follows

$$Q(p|x_i) = \beta(p)^t x_i \text{ for } p \in (0,1) \qquad (1)$$

where $\beta(p)$, p-th regression quantile, is defined as any solution to the optimization problem,

$$\min_{\beta} \sum_{i=1}^{n} \rho_p(y_i - \beta^t x_i) \qquad (2)$$

where $\rho_p(\cdot)$ is the check function defined as

$$\rho_p(r) = p_r I(r \geq 0) + (p-1)_r I(r < 0). \qquad (3)$$

Here we reexpress the above problem by the formulation stated in Vapnik[10][11].

$$\text{minimize } \frac{1}{2}\|w\|^2 + \gamma \sum_{i=1}^{n}(p\xi_i + (1-p)\xi_i^*) \text{ for } p \in (0,1),$$

$$\text{subject to} \quad \begin{aligned} y_i - w^t x_i &\leq \xi_i, \\ w^t x_i - y_i &\leq \xi_i^*, \\ \xi_i, \xi_i^* &\geq 0, \end{aligned} \qquad (4)$$

where the p-th regression quantile $\beta(p)$ is expressed in terms of w^t. We construct a Lagrange function as follows:

$$L = \tfrac{1}{2}\|w\|^2 + \gamma \sum_{i=1}^{n}(p\xi_i + (1-p)\xi_i^*) - \sum_{i=1}^{n}\alpha_i(\xi_i - y_i + w^t x_i)$$
$$- \sum_{i=1}^{n}\alpha_i^*(\xi_i^* + y_i - w^t x_i) - \sum_{i=1}^{n}(\eta_i \xi_i + \eta_i^* \xi_i^*). \qquad (5)$$

We notice that the positivity constraints $\alpha_i, \alpha_i^*, \eta_i, \eta_i^* \geq 0$ should be satisfied. After taking partial derivatives of the above equation with regard to the primal variables (w, ξ_i, ξ_i^*) and plugging them into the above equation, we have the optimization problem below.

$$\max_{\alpha, \alpha^*} -\frac{1}{2}\sum_{i,j=1}^{n}(\alpha_i - \alpha_i^*)(\alpha_j - \alpha_j^*)x_i^t x_j + \sum_{i=1}^{n} y_i(\alpha_i - \alpha_i^*) \qquad (6)$$

with constraints $\alpha_i \in [0, p\gamma]$ and $\alpha_i \in [0, (1-p)\gamma]$.

Solving the above equation with the constraints determines the optimal Lagrange multipliers, the α_i, α_i^*, the p-th regression quantile estimators and the conditional quantile function estimator given the covariate vector x are obtained as, respectively,

$$w = \sum_{i=1}^{n}(\alpha_i - \alpha_i^*)x_i \quad \text{and} \quad Q(p|x) = \sum_{i=1}^{n}(\alpha_i - \alpha_i^*)x_i^t x_i. \qquad (7)$$

For the nonlinear quantile regression case, the conditional quantile function requires the computations of dot products $\phi(x_k)^t \phi(x_l)$, $k, l = 1, \cdots, n$, in a potentially higher dimensional feature space. Under certain conditions(Mercer[7]), these demanding computations can be reduced significantly by introducing a kernel function K such that

$$\phi(x_k)^t \phi(x_l) = K(x_k, x_l). \qquad (8)$$

The kernels often used are given below.

$$K(x,y) = (x^t y + 1)^d, \quad K(x,y) = e^{-\frac{\|x-y\|^2}{2\sigma^2}}, \qquad (9)$$

where d and σ^2 are kernel parameters.

3 Support Vector Interval Regression

The proposed model is divided into two parts - the lower approximation model and the upper approximation model. In the lower approximation model, a main proportion of the data without extreme points is determined from the data classified into three classes. In the upper approximation model, intervals including all observations are obtained based on the already obtained lower approximation model.

3.1 Lower Approximation Model

In the lower approximation model, we identify the interval regression model from the main trend of data. We adopt the quantile regression using SVM to select the data set which we want to consider mainly. By applying the quantile regression, we can determine a main proportion of the given data without extreme points. If we want to consider $p100\%(0 < p < 1)$ center-located observations mainly, that portion of data can be obtained by the quantile regression with $p_1 = 0.5 + p/2$ and $p_2 = 0.5 - p/2$. The obtained model is insensitive to extreme data points since only $p100\%$ center-located observations are used. The lower approximation model is expressed as

$$Y_L(x_i) = B_0 + B_1 x_{i1} + \cdots + B_m x_{im}$$
$$= [y_L^-(x_i), y_L^+(x_i)], \ i \in C_2, \quad (10)$$

where the interval coefficients are denoted as $B_j = (b_j, d_j)$ $(i = 0, 1, \cdots, m)$ and $y_L^-(x_i)$ and $y_L^+(x_i)$ are bounds of $Y_L(x_i)$. Outputs in Class 2 $(i \in C_2)$ should be included in the lower approximation model $Y_L(x_i)$, which can be expressed as follows:

$$y_i \in Y_L(x_i) \iff \begin{cases} b^t x_i + d^t |x_i| \geq y_i \\ b^t x_i - d^t |x_i| \leq y_i \end{cases}, \ i \in C_2. \quad (11)$$

It is desirable that observations in C_1 are located above the upper bound of $Y_L(x_i)$, while observations in C_3 are located below the lower bound of $Y_L(x_i)$. But, for multivariate data, some observations are in C_1 or C_3 or both are included in the lower approximation model $Y_L(x_i)$. Thus, to permit some observations in C_1 or C_3 or both are included in $Y_L(x_i)$, we introduce a tolerance vector $\theta = (\theta_0, \cdots, \theta_m)^t$. Adding the tolerance vector θ to the radius vector d in the interval coefficient vector B, the following inequalities for observations in C_1 and C_3 are considered:

$$\begin{aligned} y_i \geq b^t x_i + (d^t|x_i| - \theta^t|x_i|), & \ i \in C_1, \\ y_i \leq b^t x_i - (d^t|x_i| - \theta^t|x_i|), & \ i \in C_3, \end{aligned} \quad (12)$$

where θ_j is the tolerance parameter introduced by Lee and Tanaka[6]. $\theta_j = 0$ $(j = 0, 1, \cdots, m)$ indicates that a lower approximation model $Y_L(x_i)$ classifies the data into three classes clearly, and $\theta_j \neq 0$ for some $i = \{0, 1, \cdots, n\}$ indicates that the lower approximation model $Y_L(x_i)$ includes some data points in C_1 or C_3 or both. Based on the assumptions mentioned above, the problem is to obtain the optimal interval coefficients of the lower approximation model $B_j = (b_j, d_j)$ $(i = 0, 1, \cdots, m)$ that minimize the following objective function:

$$\min \frac{1}{2}(\|b\|^2 + \|d\|^2 + \|\theta\|^2) + \gamma(\sum_{i \in C} \xi_{1i} + \sum_{i \in C} \xi_{2i} + \sum_{i \in C_2}(\xi_{3i} + \xi_{3i}^*))$$

subject to
$$\begin{cases} d^t|x_i| \leq \xi_{1i}, \ i \in C = C_1 \cup C_2 \cup C_3 \\ \theta^t|x_i| \leq \xi_{2i}, \ i \in C \\ y_i - b^t x_i \leq \xi_{3i} + \epsilon, \ b^t x_i - y_i \leq \xi_{3i}^* + \epsilon, \ i \in C_2 \\ b^t x_i - d^t|x_i| \leq y_i \leq b^t x_i + d^t|x_i|, \ i \in C_2 \\ y_i \geq b^t x_i + d^t|x_i| - \theta^t|x_i|, \ i \in C_1 \\ y_i \leq b^t x_i - d^t|x_i| + \theta^t|x_i|, \ i \in C_3 \\ d^t|x_i| - \theta^t|x_i| \geq 0, \ i \in C \\ \xi_{1i} \geq 0, \ \xi_{2i} \geq 0, \ i \in C \text{ and } \xi_{3i}^{(*)} \geq 0, \ i \in C_2. \end{cases} \quad (13)$$

Here, ξ_{1i} represent spreads of the estimated outputs, and $\xi_{2i}, \xi_{3i}, \xi_{3i}^*$ are slack variables representing upper and lower constraints on the outputs of the model. We construct a Lagrange function and differentiating it with respect to $a, c, \xi_{1i}, \xi_{2i}, \xi_{3i}, \xi_{3i}^*$, we have the corresponding dual optimization problem. we can derive the corresponding dual optimization problem.

$$\begin{aligned}
\max \ & -\tfrac{1}{2} \Big(\sum_{i,j \in C} \alpha_{1i}\alpha_{1j} |x_i|^t|x_j| + \sum_{i,j \in C} \alpha_{2i}\alpha_{2j} |x_i|^t|x_j| \\
& +2 \sum_{i,j \in C} \alpha_{7i}\alpha_{7j} |x_i|^t|x_j| + \sum_{i,j \in C_2} (\alpha_{3i} - \alpha_{3i}^*)(\alpha_{3j} - \alpha_{3j}^*) x_i^t x_j \\
& + \sum_{i,j \in C_2} (\alpha_{4i} - \alpha_{4i}^*)(\alpha_{4j} - \alpha_{4j}^*) x_i^t x_j \\
& + -2 \sum_{i \in C_1, j \in C_3} \alpha_{5i}\alpha_{6j} x_i^t x_j \\
& + \sum_{i,j \in C_2} (\alpha_{4i} + \alpha_{4i}^*)(\alpha_{4j} + \alpha_{4j}^*) |x_i|^t|x_j| \\
& +3 \sum_{i,j \in C_1} \alpha_{5i}\alpha_{5j} |x_i|^t|x_j| + 3 \sum_{i,j \in C_3} \alpha_{6i}\alpha_{6j} |x_i|^t|x_j| \\
& + \sum_{i,j \in C_2} (\alpha_{3i} - \alpha_{3i}^*)(\alpha_{4j} - \alpha_{4j}^*) x_i^t x_j \\
& -2 \sum_{i \in C_2, j \in C_1} (\alpha_{3i} - \alpha_{3i}^*)\alpha_{5j} x_i^t x_j \\
& +2 \sum_{i \in C_2, j \in C_3} (\alpha_{3i} - \alpha_{3i}^*)\alpha_{6j} x_i^t x_j \\
& -2 \sum_{i \in C_2, j \in C_1} (\alpha_{4i} - \alpha_{4i}^*)\alpha_{5j} x_i^t x_j \\
& +2 \sum_{i \in C_2, j \in C_3} (\alpha_{4i} - \alpha_{4i}^*)\alpha_{6j} x_i^t x_j \\
& -2 \sum_{i \in C, j \in C_2} (\alpha_{1i} + \alpha_{7i}^*)(\alpha_{4j} + \alpha_{4j}^*) |x_i|^t|x_j| \\
& -2 \sum_{i \in C_2, j \in C_1} (\alpha_{4i} + \alpha_{4i}^*)\alpha_{5j} |x_i|^t|x_j| \\
& -2 \sum_{i \in C_2, j \in C_3} (\alpha_{4i} + \alpha_{4i}^*)\alpha_{6j} |x_i|^t|x_j| \\
& +2 \sum_{i,j \in C} (\alpha_{1i} + \alpha_{2i}^*)\alpha_{7j} |x_i|^t|x_j| \\
& +2 \sum_{i \in C, j \in C_1} (\alpha_{1i} - \alpha_{2i}^*)\alpha_{5j} |x_i|^t|x_j| \\
& -2 \sum_{i \in C, j \in C_3} (\alpha_{1i} + \alpha_{2i}^*)\alpha_{6j} |x_i|^t|x_j| \\
& +4 \sum_{i \in C_1, j \in C_3} \alpha_{5i}\alpha_{6j} |x_i|^t|x_j| \Big) \\
& - \Big(\sum_{i \in C_2} \alpha_{3i}(\epsilon - y_i) + \sum_{i \in C_2} \alpha_{3i}^*(\epsilon + y_i) \Big) \\
& - \sum_{i \in C_2} \alpha_{4i} y_i + \sum_{i \in C_2} \alpha_{4i}^* y_i \\
& + \sum_{i \in C_1} \alpha_{5i} y_i - \sum_{i \in C_3} \alpha_{3i}^* y_i \Big)
\end{aligned} \quad (14)$$

subject to $0 \leq \alpha_{1i}, \alpha_{2i}, \alpha_{3i}^{(*)} \leq \gamma, \ \alpha_{4i}^{(*)}, \alpha_{5i}, \alpha_{6i}, \alpha_{7i} \geq 0.$

Solving the above problem, the optimal upper approximation model is obtained as follows:

$$\begin{aligned}
Y_L(x) = & \big(\sum_{i \in C_2} [(\alpha_{3i} - \alpha_{3i}^*) + (\alpha_{4i} - \alpha_{4i}^*)] x_i^t x - \sum_{i \in C_1} \alpha_{5i} x_i^t x \\
& + \sum_{i \in C_3} \alpha_{6i} x_i^t x , \ \sum_{i \in C} [-\alpha_{3i} + \alpha_{7i}] |x_i|^t |x| \\
& + \sum_{i \in C_2} (\alpha_{4i} + \alpha_{4i}^*) |x_i|^t |x| - \sum_{i \in C_1} \alpha_{5i} |x_i|^t |x| - \sum_{i \in C_3} \alpha_{6i} |x_i|^t |x| \big).
\end{aligned} \quad (15)$$

3.2 Upper Approximation Model

In the upper approximation model, we formulate the interval regression model including all observations based on the already obtained lower approximation model(main trend). The upper approximation model $Y_U(x_i)$ including all data can be expressed as

$$Y_U(x_i) = A_0 + A_1 x_{i1} + \cdots + A_m x_{im}$$
$$= [y_U^-(x_i), y_U^+(x_i)], \ i = 1, \cdots, n, \quad (16)$$

where $A_j = (a_j, c_j)$ $(j = 0, 1, \cdots, m)$, a_j and c_j are a center and a radius of the interval coefficient A_j and $y_U^-(x_i)$ and $y_U^+(x_i)$ are bounds of $Y_U(x_i)$. We have already obtained the lower approximation model $Y_L(x_i)$ representing the main trend of the given data. Since the upper approximation model $Y_U(x_i)$ should include the lower approximation model. Thus, we can formulate the upper approximation model including all observations as the following problem:

$$\begin{aligned}
\min \ & \tfrac{1}{2}(\|a\|^2 + \|c\|^2) + \gamma(\textstyle\sum_{i=1}^n \xi_{1i} + \sum_{i=1}^n \xi_{2i} + \sum_{i=1}^n \xi_{2i}^*), \\
\text{subject to} \ & c^t |x_i| \le \xi_{1i}, \\
& y_i - a^t x_i \le \xi_{2i} + \epsilon, \ a^t x_i - y_i \le \xi_{2i}^* + \epsilon, \\
& a^t x_i - c^t |x_i| \le \min(y(x_i), y_L^-(x_i)), \\
& a^t x_i + c^t |x_i| \ge \max(y(x_i), y_L^+(x_i)) \\
& \xi_{1i} \ge 0, \ \xi_{2i}^{(*)} \ge 0.
\end{aligned} \quad (17)$$

Then constructing a Lagrange function and differentiating it with respect to $a, c, \xi_{1i}, \xi_{2i}, \xi_{2i}^*$, we have the corresponding dual optimization problem.

$$\begin{aligned}
\text{maximize} \ & -\tfrac{1}{2}\textstyle\sum_{i,j=1}^n (\alpha_{2i} - \alpha_{2i}^*)(\alpha_{2j} - \alpha_{2j}^*) x_i^t x_j \\
& -\tfrac{1}{2}\sum_{i,j=1}^n (\alpha_{3i} - \alpha_{3i}^*)(\alpha_{3j} - \alpha 3j^*) x_i^t x_j \\
& -\sum_{i,j=1}^n (\alpha_{2i} - \alpha_{2i}^*)(\alpha_{3j} - \alpha_{3j}^*) x_i^t x_j \\
& -\tfrac{1}{2}\sum_{i,j=1}^n (\alpha_{3i} + \alpha_{3i}^*)(\alpha_{3j} + \alpha_{3j}^*) |x_i|^t |x_j| \\
& -\tfrac{1}{2}\sum_{i,j=1}^n (\alpha_{1i}\alpha_{1j}) |x_i|^t |x_j| \\
& +\sum_{i,j=1}^n \alpha_{1i}(\alpha_{3j} + \alpha_{3j}^*) |x_i|^t |x_j| + \sum_{i=1}^n (\alpha_{2i} - \alpha_{2i}^*) y_i \\
& +\sum_{i=1}^n \alpha_{3i} \max(y(x_i), y_L^+(x_i)) - \sum_{i=1}^n \alpha_{3i}^* \min(y(x_i), y_L^-(x_i)) \\
& -\epsilon \sum_{i=1}^n (\alpha_{2i} + \alpha_{2i}^*)
\end{aligned} \quad (18)$$

Solving the above problem, the optimal upper approximation model is obtained as follows:

$$\begin{aligned}
Y_U(x) = \ & (\textstyle\sum_{i=1}^n [(\alpha_{2i} - \alpha_{2i}^*) + (\alpha_{3i} - \alpha_{3i}^*)] x_i^t x, \\
& \sum_{i=1}^n [-\alpha_{1i} + (\alpha_{3i} + \alpha_{3i}^*)] |x_i|^t |x|).
\end{aligned} \quad (19)$$

By using kernel tricks mentioned in Section 2, we easily extend the linear interval regression into the nonlinear interval regression. We obtain the following dual optimization problem:

$$\begin{aligned}\text{maximize } &-\tfrac{1}{2}\sum_{i,j=1}^{n}(\alpha_{2i}-\alpha_{2i}^{*})(\alpha_{2j}-\alpha_{2j}^{*})K(x_{i},x_{j})\\&-\tfrac{1}{2}\sum_{i,j=1}^{n}(\alpha_{3i}-\alpha_{3i}^{*})(\alpha_{3j}-\alpha_{3j}^{*})K(x_{i},x_{j})\\&-\sum_{i,j=1}^{n}(\alpha_{2i}-\alpha_{2i}^{*})(\alpha_{3j}-\alpha_{3j}^{*})K(x_{i},x_{j})\\&-\tfrac{1}{2}\sum_{i,j=1}^{n}(\alpha_{3i}+\alpha_{3i}^{*})(\alpha_{3j}+\alpha_{3j}^{*})K(|x_{i}|,|x_{j}|)\\&-\tfrac{1}{2}\sum_{i,j=1}^{n}(\alpha_{1i}\alpha_{1j})K(|x_{i}|,|x_{j}|)\\&+\sum_{i,j=1}^{n}\alpha_{1i}(\alpha_{3j}+\alpha_{3j}^{*})K(|x_{i}|,|x_{j}|)\\&+\sum_{i=1}^{n}(\alpha_{2i}-\alpha_{2i}^{*})y_{i}\\&+\sum_{i=1}^{n}\alpha_{3i}\max(y(x_{i}),y_{L}^{+}(x_{i}))-\sum_{i=1}^{n}\alpha_{3i}^{*}\min(y(x_{i}),y_{L}^{-}(x_{i}))\\&-\epsilon\sum_{i=1}^{n}(\alpha_{2i}+\alpha_{2i}^{*})\end{aligned} \quad (20)$$

Here we should notice that the constraints $0 \le \alpha_{1i}, \alpha_{2i}, \alpha_{2i}^{*} \le \gamma$ and $\alpha_{3i}, \alpha_{3i}^{*} \ge 0$ are unchanged. Solving the above dual optimization problem determines the Lagrange multipliers, $\alpha_{1i}, \alpha_{ki}, \alpha_{ki}^{*}, k=2,3$.

For the nonlinear case, the difference from the linear case is that a and c are no longer explicitly given. However, they are uniquely defined in the weak sense by the inner products $a^t\phi(x)$ and $c^t\phi(|x|)$. Similar to the linear case, it is noted that $c^t\phi(|x|)$ is nonnegative. Therefore, interval nonlinear regression function is given as follows:

$$Y_U(x) = (\sum_{i=1}^{n}[(\alpha_{2i}-\alpha_{2i}^{*})+(\alpha_{3i}-\alpha_{3i}^{*})]K(x_i,x), \\ \sum_{i=1}^{n}[-\alpha_{1i}+(\alpha_{3i}+\alpha_{3i}^{*})]K(|x_i|,|x|)). \quad (21)$$

4 Numerical Studies

In order to illustrate the performance of the interval regression estimation using SVM, four examples are considered. Two examples are for the linear interval regression and other two examples are for the nonlinear interval regression. First, we apply the linear interval regression to the officials number data(Lee and Tanaka[6]). The officials number data are from 32 mid-size (the number of officials is between 400 and 1000) cities of Korea in 1987,

Input data :
x_1 = area (km^2)
x_2 = population (1000)
x_3 = number of district
x_4 = annual revenues (10^8 won)
x_5 = rate of water sewerage service (%)
x_6 = rate of water service (%)
x_7 = rate of housing supply (%)
Output data : y = number of officials.

We put the main trend of the number of officials in mid-size cities as 60%, so that $p_1 = 0.8$ and $p_2 = 0.2$. Our goal is to determine the lower approximation model $Y_L(x_i)$ and the upper approximation model $Y_U(x_i)$. Here, C_1, C_2, C_3 are found by quantile regression using SVM. In this data set, γ and ϵ are chosen as 30 and 0 which minimize $\sum_i d^t|x_i| + \sum_i |y_i - b^t x_i|_\epsilon$. For the illustration of the nonlinear interval regression, 51 of x are generated from $x_i = 0.04(i-1)-1$ and 51 of y are generated from $y_i = x_i + 2\exp(-16x_i^2) + 0.5e_i$, $i = 1,\cdots,51$, $e_i \sim U(-1,+1)$. C_1, C_2, C_3 are found by SVM based quantile regression and the Gaussian kernel $K(x,y) = e^{-\frac{\|x-y\|^2}{\sigma^2}}$ is used. For quantile regression, the kernel parameter σ^2 and γ are chosen as 0.1 and 500 by 10-fold cross validation, respectively. And the kernel parameter σ^2 and γ are chosen as 0.3 and 200 and $\epsilon = 0.05$ which minimize $\sum_i d^t |\phi(x_i)| + \sum_i |y_i - b^t \phi(x_i)|_\epsilon$. Figure 1 and 2 illustrate the estimation results of officials number data and the simulated nonlinear data, respectively. In figures dot line shows the main trend of data and solid line shows the interval regression including all data.

We have another real data sets, each of data set has one input variable so that they can show the difference of the linear and the nonlinear interval regression model clearly. Figure 3 illustrates the estimation result of fisheggs data (DeBlois and Leggett[1]) where the number eaten and the density of eggs are known to be linearly related. For the estimation of the fisheggs data we use $p_1 = 0.8$, $p_2 = 0.2$, $\gamma = 200$, and $\epsilon = 0.1$. Figure 4 illustrates the estimation result of the ultrasonic data available from http://www.itl.nist.gov/div898/strd/nls, where the ultrasonic response and the metal distance are known to be nonlinearly related. Here we use $p_1 = 0.8$, $p_2 = 0.2$, $\gamma = 500$, $\sigma^2 = 1$ for the quantile regression, and $\gamma = 100$, $\sigma^2 = 1$, $\epsilon = 0.05$ for the interval regression. The Gaussian kernel is used for the ultrasonic data, the values of parameters used in estimation of fisheggs data and ultrasonic data are obtained from 10-fold cross validation and minimizing $\sum_i d^t |\phi(x_i)| + \sum_i |y_i - b^t \phi(x_i)|_\epsilon$.

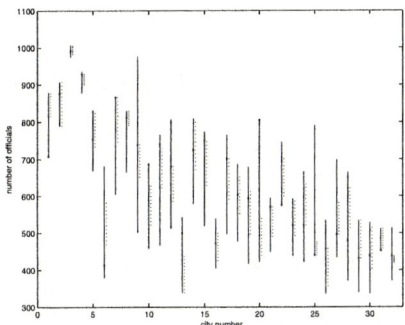
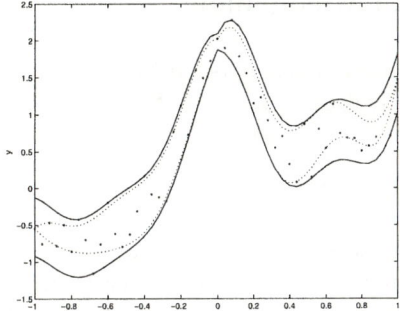

Fig. 1. Linear interval Regression 1   **Fig. 2.** Nonlinear interval Regression 1

5 Conclusions

We proposed interval regression analysis using the principle of SVM and quantile regression. We identify the main trend from the lower approximation model using the designated center-located proportion of the given data. The obtained lower approximation model can explain the relationship between the input and output

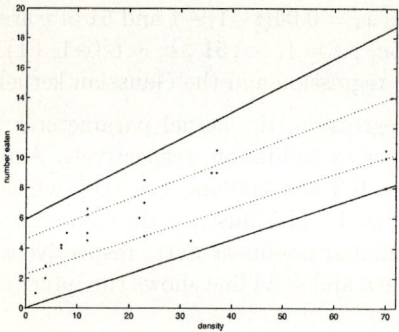 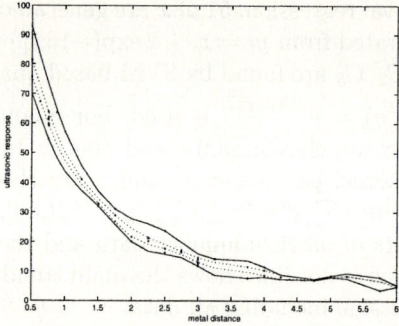

Fig. 3. Linear interval Regression 2   **Fig. 4.** Nonlinear interval Regression 2

data well, because the obtained model was not influenced by extreme points. The upper approximation model including all data was obtained on the basis of the lower approximation model. Through examples, we showed that the proposed algorithm derives the satisfying solutions and is attractive approach to interval regression. In particular, we can use this algorithm successfully when a linear model is inappropriate. By using two approximation models we could easily recognize observations whose outputs are extremely far from the main trend of the given data, which leads our proposed algorithm to be useful for the analysis of the data with possibilistic viewpoints.

Acknowledgement

This work was supported by the Korea Research Foundation Grant(KRF-2004-042-C00020).

References

1. DeBlois, E.M. and W.C. Leggett. : Functional response and potential impact of invertebrate predators on benthic fish eggs: analysis of the Calliopius laeviusculus-capelin (Mallotus villosus) predator-prey system. Marine Ecology Progress Series **69** (1991) 205–216
2. Gunn S. : Support vector machines for classification and regression. ISIS Technical Report, U. of Southampton (1998)
3. Hong D. and Hwang C. : Interval regression analysis using quadratic loss support vector machine. IEEE Transactions on Fuzzy Systems **13**(2) April (2005) 229–237
4. Kecman V. : Learning and soft computing, support vector machines, neural networks and fuzzy logic moldes. The MIT Press, Cambridge, MA (2001)
5. Koenker R. and Bassett G. : Regression quantiles. Econometrica **46** (1978) 33–50
6. Lee H. and Tanaka H. : Upper and lower approximation models in interval regression using regression quantile techniques. European Journal of Operational Research **116** (1999) 653–666

7. Mercer J. : Functions of positive and negative and their connection with the theory of integral equations. Philosphical Transactions of the Royal Society **A** (1909) 415–446
8. Smola A. and Schoelkopf B. : On a kernel-based method for pattern recognition, regression, approximation and operator inversion. Algorithmica **22** (1998) 211–231
9. Tanaka H., Koyana K. and Lee H. : Interval regression analysis based on quadratic programming. In: Proceedings of The 5th IEEE International Conference on Fuzzy Systems, New Orleans, USA (1996) 325–329
10. Vapnik V. N. : The nature of statistical learning theory. Springer, New York (1995)
11. Vapnik, V. N.: Statistical learning theory. Springer, New York (1998)
12. Wang, L.(Ed.) : Support vector machines: theory and application. Springer, Berlin Heidelberg New York (2005)

An Approach Based on Similarity Measure to Multiple Attribute Decision Making with Trapezoid Fuzzy Linguistic Variables

Zeshui Xu

College of Economics and Management, Southeast University,
Nanjing, Jiangsu 210096, China
Xu_zeshui@263.net

Abstract. In this paper, we investigate the multiple attribute decision making problems under fuzzy linguistic environment. We introduce the concept of trapezoid fuzzy linguistic variable and some operational laws of trapezoid fuzzy linguistic variables. We develop a similarity measure between two trapezoid fuzzy linguistic variables. Based on the similarity measure and the ideal point of attribute values, we develop an approach to ranking the decision alternatives in multiple attribute decision making with trapezoid fuzzy linguistic variables. We finally illustrate the developed approach with a practical example.

1 Introduction

Multiple attribute decision making under linguistic environment is an interesting research topic having received more and more attention from researchers during the last several years [1-5]. In the process of multiple attribute decision making, the linguistic decision information needs to be aggregated by means of some proper approaches so as to rank the given decision alternatives and then to select the most desirable one. Bordogna et al. [1] developed a model within fuzzy set theory by linguistic ordered weighted average (OWA) operators for group decision making in a linguistic context. Herrera and Martínez [2] established a linguistic 2-tuple computational model for dealing with linguistic information. Li and Yang [3] developed a linear programming technique for multidimensional analysis of preferences in multiple attribute group decision making under fuzzy environments, in which all the linguistic information and real numbers are transformed into triangular fuzzy numbers. Xu [4,5] proposed some methods, which compute with words directly. In this paper, we shall investigate the multiple attribute decision making problems under fuzzy linguistic environment, in which the decision maker can only provide their preferences (attribute values) in the form of trapezoid fuzzy linguistic variables. In order to do that, this paper is structured as follows. In Section 2 we define the concept of trapezoid fuzzy linguistic variable and some operational laws of trapezoid fuzzy linguistic variables, and then develop a similarity measure between two trapezoid fuzzy linguistic variables. In Section 3 we develop an approach to ranking the decision alternatives based on the similarity measure and the ideal point of attribute values. We illustrate the developed approach with a practical example in Section 4, and give concluding remarks in Section 5.

2 Trapezoid Fuzzy Linguistic Variables

In [6], Zadeh introduced the concept of linguistic variable, that is, whose values are words rather than numbers. The concept of linguistic variable has played and is continuing to play a pivotal role in decision making with linguistic information. Computing with words is a methodology in which the objects of computation are words and propositions drawn from a natural language, e.g., small, large, far, heavy, not very likely, etc., it is inspired by the remarkable human capability to perform a wide variety of physical and mental tasks without any measurements and any computations [7]. In the process of decision making with linguistic information, the decision maker generally provides his/her linguistic assessment information by using a linguistic scale. In [8], Xu defined a finite and totally ordered discrete linguistic scale as $S = \{s_i \mid i = -t, ..., t\}$, where t is a non-negative integer, s_i represents a possible value for a linguistic variable, and it requires that $s_i < s_j$ iff $i < j$. For example, a set of nine labels S could be:

$$S = \{s_{-4} = extremely\ poor,\ s_{-3} = very\ poor,\ s_{-2} = poor,$$
$$s_{-1} = slightly\ poor,\ s_0 = fair,\ s_1 = slightly\ good,$$
$$s_2 = good,\ s_3 = very\ good,\ s_4 = extremely\ good\}$$

In the process of information aggregating, some results may do not exactly match any linguistic labels in S. To preserve all the given information, Xu [8] extended the discrete label set S to a continuous label set $\overline{S} = \{s_\alpha \mid \alpha \in [-q, q]\}$, where $q (q > t)$ is a sufficiently large positive integer. If $s_\alpha \in S$, then s_α is termed an original linguistic label, otherwise, s_α is termed a virtual linguistic label. In general, the decision maker uses the original linguistic labels to evaluate alternatives, and the virtual linguistic labels can only appear in operation.

Since the decision maker is characterized by his own personal background and experience, in some situations, the decision maker may provide fuzzy linguistic information because of time pressure, lack of knowledge, and their limited expertise related with the problem domain. In the following we define the concept of trapezoid fuzzy linguistic variable

Definition 1. *Let* $\hat{s} = [s_\alpha, s_\beta, s_\gamma, s_\eta] \in \hat{S}$, *where* $s_\alpha, s_\beta, s_\gamma, s_\eta \in \overline{S}$, s_β *and* s_γ *indicate the interval in which the membership value is 1, with* s_α *and* s_η *indicating the lower and upper values of* $\hat{S}$, *respectively, then* $\hat{S}$ *is called a trapezoid fuzzy linguistic variable, which is characterized by the following member function (see Fig.1)*

$$\mu_{\hat{s}}(\theta) = \begin{cases} 0, & s_{-q} \leq s_\theta \leq s_\alpha \\ \dfrac{d(s_\theta, s_\alpha)}{d(s_\beta, s_\alpha)}, & s_\alpha \leq s_\theta \leq s_\beta \\ 1, & s_\beta \leq s_\theta \leq s_\gamma \\ \dfrac{d(s_\theta, s_\eta)}{d(s_\gamma, s_\eta)}, & s_\gamma \leq s_\theta \leq s_\eta \\ 0, & s_\eta \leq s_\theta \leq s_q \end{cases}$$

where $\hat{S}$ is the set of all trapezoid fuzzy linguistic variables. Especially, if any two of $\alpha, \beta, \gamma, \eta$ are equal, then $\hat{S}$ is reduced to a triangular fuzzy linguistic variable; if any three of $\alpha, \beta, \gamma, \eta$ are equal, then $\hat{S}$ is reduced to an uncertain linguistic variable.

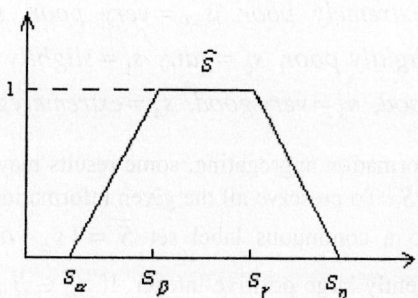

Fig. 1. A trapezoid fuzzy linguistic variable $\hat{S}$

Consider any three trapezoid fuzzy linguistic variables $\hat{s} = [s_\alpha, s_\beta, s_\gamma, s_\eta]$, $\hat{s}_1 = [s_{\alpha_1}, s_{\beta_1}, s_{\gamma_1}, s_{\eta_1}]$, $\hat{s}_2 = [s_{\alpha_2}, s_{\beta_2}, s_{\gamma_2}, s_{\eta_2}] \in \hat{S}$, and suppose that $\lambda \in [0,1]$, then we define their operational laws as follows:

1) $\hat{s}_1 \oplus \hat{s}_2 = [s_{\alpha_1}, s_{\beta_1}, s_{\gamma_1}, s_{\eta_1}] \oplus [s_{\alpha_2}, s_{\beta_2}, s_{\gamma_2}, s_{\eta_2}] = [s_{\alpha_1+\alpha_2}, s_{\beta_1+\beta_2}, s_{\gamma_1+\gamma_2}, s_{\eta_1+\eta_2}]$;
2) $\lambda \hat{s} = \lambda [s_\alpha, s_\beta, s_\gamma, s_\eta] = [s_{\lambda\alpha}, s_{\lambda\beta}, s_{\lambda\gamma}, s_{\lambda\eta}]$.

In order to measure the similarity degree between any two trapezoid fuzzy linguistic values $\hat{s}_1 = [s_{\alpha_1}, s_{\beta_1}, s_{\gamma_1}, s_{\eta_1}]$ and $\hat{s}_2 = [s_{\alpha_2}, s_{\beta_2}, s_{\gamma_2}, s_{\eta_2}] \in \hat{S}$, we introduce a similarity measure as follows:

$$s(\hat{s}_1,\hat{s}_2) = 1 - \frac{|\alpha_2-\alpha_1|+|\beta_2-\beta_1|+|\gamma_2-\gamma_1|+|\eta_2-\eta_1|}{8q} \quad (1)$$

where $s(\hat{s}_1,\hat{s}_2)$ is called the similarity degree between $\hat{s}_1$ and $\hat{s}_2$. Obviously, the greater the value of $s(\hat{s}_1,\hat{s}_2)$, the closer $\hat{s}_1$ to $\hat{s}_2$.

The properties of the similarity degree $s(\hat{s}_1,\hat{s}_2)$ are shown as follows:

1) $0 \le s(\hat{s}_1,\hat{s}_2) \le 1$;
2) $s(\hat{s}_1,\hat{s}_2) = 1$ iff $\hat{s}_1 = \hat{s}_2$, that is $\alpha_1 = \alpha_2$, $\beta_1 = \beta_2$, $\gamma_1 = \gamma_2$, $\eta_1 = \eta_2$;
3) $s(\hat{s}_1,\hat{s}_2) = s(\hat{s}_2,\hat{s}_1)$.

Below we propose an operator for aggregating trapezoid fuzzy linguistic variables.

Definition 2. Let $TFLWA: \hat{S}^n \to \hat{S}$, if

$$TFLWA_w(\hat{s}_1,\hat{s}_2,...,\hat{s}_n) = w_1\hat{s}_1 \oplus w_2\hat{s}_2 \oplus \cdots \oplus w_n\hat{s}_n \quad (2)$$

where $w = (w_1, w_2,..., w_n)$ is the weighting vector of the $\hat{s}_i$, $\hat{s}_i \in \hat{S}$, $w_i \ge 0$, $i = 1,2,...,n$, $\sum_{i=1}^{n} w_i = 1$ then TFLWA is called a trapezoid fuzzy linguistic weighted averaging (TFLWA) operator. Especially, if $w = (1/n, 1/n,...,1/n)$, then TFLWA operator is reduced to a trapezoid fuzzy linguistic averaging (TFLA) operator.

Example 1. Assume $\hat{s}_1 = [s_{-3},s_{-2},s_0,s_2]$, $\hat{s}_2 = [s_{-1},s_0,s_1,s_2]$, $\hat{s}_3 = [s_0,s_1,s_2,s_4]$, and $\hat{s}_4 = [s_{-1},s_1,s_2,s_3]$, $w = (0.3,0.1,0.2,0.4)$, then by the operational laws of trapezoid fuzzy linguistic variables, we have

$$TFLWA_w(\hat{s}_1,\hat{s}_2,\hat{s}_3,\hat{s}_4) = 0.3 \times [s_{-3},s_{-2},s_0,s_2] \oplus 0.1 \times [s_{-1},s_0,s_1,s_2]$$
$$\oplus 0.2 \times [s_{-1},s_0,s_1,s_2] \oplus 0.4 \times [s_0,s_1,s_2,s_4]$$
$$= [s_{-1.2},s_{-0.2},s_{1.1},s_{2.8}]$$

3 A Similarity Measure Based Approach

A multiple attribute decision making problem under fuzzy linguistic environment is represented as follows:

Let $X = \{x_1,x_2,...,x_n\}$ be the set of alternatives, and $U = \{u_1,u_2,...,u_m\}$ be the set of attributes. Let $w = (w_1, w_2,..., w_m)$ be the weight vector of attributes, where $w_i \geq 0$, $i = 1,2,...,m$, $\sum_{i=1}^{m} w_i = 1$. Suppose that $\hat{A} = (\hat{a}_{ij})_{m \times n}$ is the fuzzy linguistic decision matrix, where $\hat{a}_{ij} = [a_{ij}^{(\alpha)}, a_{ij}^{(\beta)}, a_{ij}^{(\gamma)}, a_{ij}^{(\eta)}] \in \hat{S}$ is the attribute value, which takes the form of trapezoid fuzzy linguistic variable, given by the decision maker, for the alternative $x_j \in X$ with respect to the attribute $u_i \in U$. Let $\hat{a}_j = (\hat{a}_{1j}, \hat{a}_{2j},...,\hat{a}_{mj})$ be the vector of the attribute values corresponding to the alternative x_j, $j = 1,2,...,n$.

Definition 3. Let $\hat{A} = (\hat{a}_{ij})_{m \times n}$ be the decision matrix with trapezoid fuzzy linguistic variables, then we call $\hat{I} = (\hat{I}_1, \hat{I}_2,..., \hat{I}_m)$ the ideal point of attribute values, where $\hat{I}_i = [I_i^{(\alpha)}, I_i^{(\beta)}, I_i^{(\gamma)}, I_i^{(\eta)}]$, $I_i^{(\alpha)} = \max_j \{a_{ij}^{(\alpha)}\}$, $I_i^{(\beta)} = \max_j \{a_{ij}^{(\beta)}\}$, $I_i^{(\gamma)} = \max_j \{a_{ij}^{(\gamma)}\}$, $I_i^{(\eta)} = \max_j \{a_{ij}^{(\eta)}\}$, $i = 1,2,...,m$.

In the following we develop an approach to ranking the decision alternatives based on the similarity measure and the ideal point of attribute values:

Step 1. Utilize the TFLWA operator

$$\hat{z}_j = TFLWA_w(\hat{a}_{1j}, \hat{a}_{2j},...,\hat{a}_{mj}) = w_1\hat{a}_{1j} \oplus w_2\hat{a}_{2j} \oplus \cdots \oplus w_m\hat{a}_{mj},$$
$$j = 1,2,..., n \quad (3)$$

to derive the overall values $\hat{z}_j (j = 1,2,...,n)$ of the alternatives $x_j (j = 1,2,...,n)$, where $w = (w_1, w_2,...,w_m)$ is the weight vector of attributes.

Step 2. Utilize the TFLWA operator

$$\hat{z} = TFLWA_w(\hat{I}_1, \hat{I}_2,..., \hat{I}_m) = w_1\hat{I}_1 \oplus w_2\hat{I}_2 \oplus \cdots \oplus w_m\hat{I}_m,$$
$$j = 1,2,..., n \quad (4)$$

to derive the overall value $\hat{z}$ of the ideal point $\hat{I} = (\hat{I}_1, \hat{I}_2,..., \hat{I}_m)$, where $w = (w_1, w_2,...,w_m)$ is the weight vector of attributes.

Step 3. By (1), we get the similarity degree $s(\hat{z}, \hat{z}_j)$ between $\hat{z}$ and $\hat{z}_j$ $(j = 1,2,...,n)$.

Step 4. Rank all the alternatives $x_j (j = 1,2,...,n)$ and select the best one in accordance with $s(\hat{z}, \hat{z}_j)$ $(j = 1,2,...,n)$.

Step 5. End.

4 Illustrative Example

In this section, a decision making problem of assessing cars for buying (adapted from [2]) is used to illustrate the developed approach.

Let us consider, a customer who intends to buy a car. Four types of cars x_j ($j=1, 2, 3, 4$) are available. The customer takes into account four attributes to decide which car to buy: 1) G_1 : economy, 2) G_2 : comfort, 3) G_3 : design, and 4) G_4 : safety. The decision maker evaluates these four types of cars x_j ($j=1, 2, 3, 4$) under the attributes $G_i (i=1,2,3,4)$ (whose weight vector is $w=(0.3, 0.2, 0.1, 0.4)$) by using the linguistic scale

$$S = \{s_{-4} = extremely\ poor,\ s_{-3} = very\ poor,\ s_{-2} = poor,$$
$$s_{-1} = slightly\ poor,\ s_0 = fair,\ s_1 = slightly\ good,$$
$$s_2 = good,\ s_3 = very\ good,\ s_4 = extremely\ good\}$$

and gives a fuzzy linguistic decision matrix as listed in Table 1.

Table 1. Fuzzy linguistic decision matrix $\hat{A}$

G_i	x_1	x_2	x_3	x_4
G_1	[s_{-3}, s_{-2}, s_0, s_1]	[s_{-2}, s_0, s_1, s_2]	[s_{-1}, s_1, s_3, s_4]	[s_0, s_1, s_2, s_4]
G_2	[s_{-1}, s_0, s_3, s_4]	[s_0, s_1, s_2, s_3]	[s_{-4}, s_{-3}, s_{-1}, s_1]	[s_{-1}, s_2, s_3, s_4]
G_3	[s_0, s_1, s_2, s_4]	[s_{-1}, s_0, s_3, s_4]	[s_1, s_2, s_3, s_4]	[s_{-2}, s_0, s_1, s_2]
G_4	[s_{-2}, s_{-1}, s_0, s_2]	[s_{-1}, s_0, s_2, s_3]	[s_{-2}, s_{-1}, s_0, s_1]	[s_1, s_2, s_3, s_4]

In the following, we utilize the approach developed in this paper to get the most desirable car:

Step 1. From Table 1, we get the vector of the attribute values corresponding to the alternative x_j ($j=1,2,3,4$), and the ideal point as follows:

1) $\hat{a}_1 = (\hat{a}_{11}, \hat{a}_{21}, \hat{a}_{31}, \hat{a}_{41})$, where $\hat{a}_{11} = [s_{-3}, s_{-2}, s_0, s_1]$,
$\hat{a}_{21} = [s_{-1}, s_0, s_3, s_4]$, $\hat{a}_{31} = [s_0, s_1, s_2, s_4]$, $\hat{a}_{41} = [s_{-2}, s_{-1}, s_0, s_2]$.

2) $\hat{a}_2 = (\hat{a}_{12}, \hat{a}_{22}, \hat{a}_{32}, \hat{a}_{42})$, where $\hat{a}_{12} = [s_{-2}, s_0, s_1, s_2]$,
$\hat{a}_{22} = [s_0, s_1, s_2, s_3]$, $\hat{a}_{32} = [s_{-1}, s_0, s_3, s_4]$, $\hat{a}_{42} = [s_{-1}, s_0, s_2, s_3]$.

3) $\hat{a}_3 = (\hat{a}_{13}, \hat{a}_{23}, \hat{a}_{33}, \hat{a}_{43})$, where $\hat{a}_{13} = [s_{-1}, s_1, s_3, s_4]$,
$\hat{a}_{23} = [s_{-4}, s_{-3}, s_{-1}, s_1]$, $\hat{a}_{33} = [s_1, s_2, s_3, s_4]$, $\hat{a}_{43} = [s_{-2}, s_{-1}, s_0, s_1]$

4) $\hat{a}_4 = (\hat{a}_{14}, \hat{a}_{24}, \hat{a}_{34}, \hat{a}_{44})$, where $\hat{a}_{14} = [s_0, s_1, s_2, s_4]$,
$\hat{a}_{24} = [s_{-1}, s_2, s_3, s_4]$, $\hat{a}_{34} = [s_{-2}, s_0, s_1, s_2]$, $\hat{a}_{44} = [s_1, s_2, s_3, s_4]$
5) $\hat{I} = (\hat{I}_1, \hat{I}_2, \hat{I}_3, \hat{I}_4)$, where $\hat{I}_1 = [s_0, s_1, s_3, s_4]$,
$\hat{I}_2 = [s_0, s_2, s_3, s_4]$, $\hat{I}_3 = [s_1, s_2, s_3, s_4]$, $\hat{I}_4 = [s_1, s_2, s_3, s_4]$.

Step 2. Utilize (3) to derive the overall values $\hat{z}_j$ ($j = 1,2,3,4$) of the alternatives x_j ($j = 1,2,3,4$):

$$\hat{z}_1 = [s_{-1.9}, s_{-0.9}, s_{0.8}, s_{2.3}], \quad \hat{z}_2 = [s_{-1.1}, s_{-0.4}, s_{1.8}, s_{2.8}]$$
$$\hat{z}_3 = [s_{-1.8}, s_{-0.5}, s_1, s_{2.2}], \quad \hat{z}_4 = [s_0, s_{1.5}, s_{2.5}, s_{3.8}]$$

Step 3. Utilize (4) to derive the overall value $\hat{z}$ of the ideal point $\hat{I}$:

$$\hat{z} = [s_{0.5}, s_{1.7}, s_3, s_4]$$

Step 4. By (1) (suppose that $q = 4$), we get the similarity degree $s(\hat{z}, \hat{z}_j)$ between $\hat{z}$ and $\hat{z}_j$ ($j = 1,2,3,4$):

$$s(\hat{z}, \hat{z}_1) = 0.72, \quad s(\hat{z}, \hat{z}_2) = 0.83, \quad s(\hat{z}, \hat{z}_3) = 0.74, \quad s(\hat{z}, \hat{z}_4) = 0.96$$

Step 5. Rank all the alternatives x_j ($j = 1,2,3,4$) in accordance with $s(\hat{z}, \hat{z}_j)$ ($j = 1,2,3,4$):

$$x_4 \succ x_2 \succ x_3 \succ x_1$$

and thus the best car is x_4.

5 Concluding Remarks

We have defined the concept of trapezoid fuzzy linguistic variable, and developed a similarity measure between two trapezoid fuzzy linguistic variables. Based on the similarity measure and the ideal point of attribute values, we have developed an approach to multiple attribute decision making with trapezoid fuzzy linguistic variables, which utilizes an aggregation operator called trapezoid fuzzy linguistic weighted averaging (TFLWA) operator to fuse all the given decision information corresponding to each alternative, and uses the similarity measure to rank the decision alternatives and then to select the most desirable one.

Acknowledgement

This work was supported by China Postdoctoral Science Foundation under Project (2003034366).

References

1. Bordogna, G., Fedrizzi, M., Pasi G.: A Linguistic Modeling of Consensus in Group Decision Making Based on OWA Operators. IEEE Transactions on Systems, Man, and Cybernetics-Part A 27 (1997) 126-132.
2. Herrera, F., Martínez, L.: An Approach for Combining Numerical and Linguistic Information Based on the 2-Tuple Fuzzy Linguistic Representation Model in Decision Making. International Journal of Uncertainty, Fuzziness and Knowledge -Based Systems 8 (2000) 539-562.
3. Li, D.F., Yang, J.B.: Fuzzy Linear Programming Technique for Multi-attribute Group Decision Making in Fuzzy Environments. Information Sciences 158 (2004) 263-275.
4. Xu, Z.S.: Uncertain Linguistic Aggregation Operators Based Approach to Multiple Attribute Group Decision Making under Uncertain Linguistic Environment. Information Sciences 168 (2004) 171-184.
5. Xu, Z.S.: Uncertain Multiple Attribute Decision Making: Methods and Applications, Tsinghua University Press, Beijing (2004).
6. Zadeh, L.A.: Outline of a New Approach to the Analysis of Complex Systems and Decision Processes. IEEE Transactions on Systems, Man, and Cybernetics 3 (1973) 28-44.
7. Zadeh, L.A.: From Computing with Numbers to Computing with Words-from Manipulation of Measurements to Manipulation of Perceptions. International Journal of Applied Mathematics and Computer Science, 12 (2002) 307-324.
8. Xu, Z.S. Deviation Measures of Linguistic Preference Relations in Group Decision Making. Omega, 33 (2005) 249-254.

Research on Index System and Fuzzy Comprehensive Evaluation Method for Passenger Satisfaction*

Yuanfeng Zhou[1], Jianping Wu[2], and Yuanhua Jia[3]

[1] School of Traffic and Transportation, Beijing Jiaotong University,
100044, Beijing, P. R. China
zyfbbb@163.com
[2] School of Civil Engineering and the Environment, University of Southampton,
SO17 1BJ, Southampton, UK
[3] School of Traffic and Transportation, Beijing Jiaotong University,
100044, Beijing, P. R. China

Abstract. Passenger satisfaction index (PSI) is one of the most important indexes in comprehensive evaluation of management performance and service quality of passenger transport corporations. Based on the investigations in China, the authors introduced the notion and method for passenger group division and the concept of index weight matrix, and made successful application for passenger satisfaction evaluation. Index weight matrix developed by applying AHP and Delphi methods gives satisfactory results. The paper ends with examples of using a fuzzy inference system for passenger satisfaction evaluation in Beijing railway station.

1 Introduction

Satisfaction is the feeling of people formed by comparing perceivable quality with that expected for services or products. At present, competition for passengers in transport markets is fierce. It is important to develop a passenger satisfaction index system (PSI), and help transport enterprises efficiently modify and improve their service quality and management performance.

The research team conducted by Claes Fornell developed a SCSB model in 1989[8]. In 1994, an ACSI model was developed in the USA and in 1999, an ECSI model was developed with the sponsorship of EOQ and EFQM. More attention was paid to the commodities quality and service in above models. In China, Zhao Ping assessed the China railway passenger satisfaction by making use of the least square method in 2001[6]. However, this model didn't consider the difference made by education and the economic background of different passenger groups, which is believed to have significant influence on the accuracy and objectivity of the evaluation results [7].

In this paper, we give an index system for passenger satisfaction evaluation. We designed a series of questionnaires firstly and investigated in stations and trains. From the statistic data, the index matrix was plotted and then passenger satisfaction was

---

* This paper is based on the China Railway User Satisfaction Evaluation Project, which was supported by the S&T Development Plan(2002X024) of the Railway Department of China.

gained by fuzzy comprehensive evaluation. Then, a real-time emulator for PS evaluation was designed and the result could reflect the true situation properly.

2 Index for Passenger Satisfaction Evaluation

Passenger satisfaction includes much element associated with a passenger starting from buying/booking a ticket to the completion of a trip. Based on the pre-research results, a passenger satisfaction evaluation index system was proposed in this study. The author proposed a two-level index system, the General Survey and the Specific Survey. A general survey will have the following characteristics: comprehensiveness, importance, generality and completeness. The indexes of a general survey includes security, economy, efficiency, convenience, comfort, service quality, staff service quality, monitoring, infrastructures, etc. The details can be found in the research of Jia and Zhou in 2003[7]. However, a specific survey is the survey, which focuses on a specific service or products of passenger transport.

3 Fuzzy Model for Evaluation of Passenger Satisfaction

As the result of the evaluation of satisfaction degree of passenger transport has involved too many factors, to use the Fuzzy theory (L.A. Zadeh, 1965) for the evaluation of the satisfaction degree is possibly a better alternative.

A fuzzy evaluation system can be represented as: $U = \{u_1, u_2, \cdots u_i\}$, where u_i is one of the indexes in the investigation. According to rundle theory, we divide passenger's subjective judgment into 5 levels from very unsatisfactory to very satisfactory, and the relevant score P_j (j=1,...5) was given for a quantitative evaluation, i.e.: $V = \{v_1, v_2, \cdots v_5\}$. This paper used the method which combines AHP with Delphi method, to decide the fuzzy set of weights of indexes: $A = \{a_1, a_2, \cdots a_i\}$. The details of calculation process of the eigenvector $W = [W_1, \cdots, W_i, \cdots, W_n]^T$, λ_{max} and C.R. can be seen in the study of Li, 1993[2].

Passenger satisfaction is the subjective judgment of passengers, which is closely related to many factors such as education background, occupancy, age, etc. The following are identified as the major elements to be considered for passenger group classification: Social attributes; Natural attributes; Geographical distribution; Nature of the travel (private or company paid travel); Travel frequency; Economic solvency and psychological enduring ability of passengers.

Different weights were given to different individuals and groups. When a passenger i of group k submitted a questionnaire, the real-time evaluation system will calculate the passenger's weight vector by AHP method according to the indexes importance ordering given by him. The synthesized index weight vector of passenger group k is:

$$A_k = \frac{1}{n}\sum_{i=1}^{n} S_i \cdot W_i^k \quad (1)$$

$$S_i = \mu^T x_i = (\mu_A,\ldots,\mu_C)x_i = \mu_A A_{ij} + \mu_E E_{ij} + \mu_I I_{ij} + \mu_O O_{ij} + \mu_F F_{ij} + \mu_C C_{ij} \quad (2)$$

Where, μ^T is the weighing factor vector. For passenger i, S_i shows his contribution to index weight distribution of passenger group k, S_i is calculated based on his individual information. A_{ij}: passenger i with age group j. Similarly, E_{ij} is for education, I_{ij} is for income, O_{ij} is for occupation, F_{ij} is for frequency of travel, C_{ij} is for the charge source.

Single Passenger Satisfaction Evaluation Model:

$$f_i = \sum_{i=1}^{m}\sum_{j=1}^{5} \sigma_{ij} P_j W_i \quad (3)$$

$$\begin{cases} \sigma_{ij} = 1 & \text{Satisfaction degree j of index } i \text{ was selected} \\ \sigma_{ij} = 0 & \text{Satisfaction degree j of index } i \text{ wasn't selected} \end{cases}$$

$$\text{s.t.} \quad f_i = \sum_{i=1}^{m}\sum_{j=1}^{5} \sigma_{ij} P_j W_i \leq 100 \quad i=1,2,\ldots,n \quad (4)$$

$$W = (w_1,\cdots,w_m)^T \geq 0 \quad (5)$$

Fuzzy Comprehensive Evaluation for Passenger Satisfaction:

Firstly, we develop the membership function of index and build the fuzzy inference matrix R_{nm}. In the matrix, r_{ij} is the percentage of passengers who give index i a judgment of j. In order to consider the different influences of factors, the weight matrix A was applied. This leads to a fuzzy inference matrix B which has the form:

$$B = A*R \quad (6)$$

$B = \{b_1, b_2, \cdots b_m\}$ is the fuzzy evaluation results. We applied the following rules: Assuming $b_k = \max b_i$, then $\bar{b} = \sum_{i=1}^{k-1} b_i$. If $\bar{b} \leq 0.5$, the passenger satisfaction is grade K, otherwise, it is grade $K-1$[3]. The passenger transport enterprises can take $\bar{b}$ as a comprehensive index of the passenger satisfaction evaluation.

4 Application Example

Taking Beijing Railway Station passenger transport service as a reference, following is the relevant parameters and evaluation result:

$\lambda_{max} = 15.387$	C.I.= 0.1067	R.I. = 1.58	C.R. = 0.0675 < 0.10
$B = A*R =$	\multicolumn{3}{l	}{$\begin{bmatrix} 0.299953 & 0.459185 & 0.089498 & 0.111913 & 0.039451 \\ 0.315847 & 0.459253 & 0.083172 & 0.103227 & 0.038501 \end{bmatrix}$}

where $\overline{b}_{12} = 0.758$, $\overline{b}_{22} = 0.774$. As far as the service quality of Beijing Station is concerned, the evaluation of the higher middle group and that of the lower group passengers are both of the similar degree of comparatively satisfied, although the lower group passengers give a slightly higher score. This result is relatively much more acceptable and proper than the evaluating result conducted by Zhao Ping(2001).

5 Conclusion and Recommendations

Based on the results above, we have the following conclusions and recommendations: Index weight reflects the differences between passengers on evaluation because of their different backgrounds. It is important to produce a credible index weight matrix. The index weight systems estimated by the method of combination of AHP and Delphi has given satisfactory and reasonable results. The fuzzy inference system has been successfully used in passenger satisfaction evaluation, and the application in Beijing Stations has shown its efficiency and credibility. The real time evaluation systems can help on management performance of transport enterprises by understanding the requests of passengers of different groups quickly.

References

1. Jianping Huang, Jin-Ming Li: Application of Consumer Satisfaction and Satisfaction Index in China. Journal of Beijing Institute of Business, Beijing (2000)
2. Guogang Li, Bao-Shan Li: Management Systems Engineering. The Publish House of the People's University of China, Beijing (1993)
3. Yongbo Lv, Yifei Xu: System Engineering. The Publish House of the Northern Jiaotong University, Beijing (2003)
4. Yong-Ling Cen: Fuzzy Comprehensive Evaluation Model for Consumer Satisfaction. Journal of Liaoning Engineering and Technology University (2001)
5. Yuanfeng Zhou, Yuan-hua Jia: Research on Fuzzy Evaluation Method for Railway Passenger Satisfaction. Journal of Northern Jiaotong University, vol.27(5), Beijing (2003)64-68.
6. Ping Zhao: Guide to China Customer Satisfaction Index. The Publish House of China Criterion, Beijing (2003)
7. Yuanhua Jia, Yuanfeng Zhou, Sufen Li: Report of China Railway Passenger Satisfaction Testing Index System and Data-collection and Real-time Evaluation system (2003)
8. Claes Fornell: A National Customer Satisfaction Barometer: The Swedish Experience. Journal of Marketing Vol56 January (1992)6~21
9. L.A. Zadeh: Fuzzy Sets. Inf. and Control. 8 (1965)38-53

Research on Predicting Hydatidiform Mole Canceration Tendency by a Fuzzy Integral Model

Yecai Guo[1], Wei Rao[1], Yi Guo[2], and Wei Ma[2]

[1] Anhui University of Science and Technology,
Huainan 232001, China
[2] Shanghai Marine University, Shanghai, China
guo-yecai@163.com

Abstract. Based on the Fuzzy mathematical principle, a fuzzy integral model on forecasting the cancerational tendency of hydatidiform mole is created. In this paper, attaching function, quantum standard, weight value of each factor, which causes disease, and the threshold value of fuzzy integral value are determined under condition that medical experts take part in. The detailed measures in this paper are taken as follows: First, each medical expert gives the score of the sub-factors of each factor based on their clinic experience and professional knowledge. Second, based on analyzing the feature of the scores given by medical experts, attaching functions are established using K power parabola larger type. Third, weight values are determined using method by the analytic hierarchy process[AHP] method. Finally, the relative information is obtained from the case histories of hydatidiform mole cases. Fuzzy integral value of each case is calculated and its threshold value is finally determined. Accurate rate of the fuzzy integral model(FIM) is greater than that of the maximum likelihood method (MLM) via diagnosing the history cases and for new cases, the diagnosis results of the FIM is in accordance with those of the medical experts.

1 Introduction

Hydatidiform mole is a benign and deutoplasmic tumo(u)r and regarded as benign hydatidiform mole. When its organization transfers to adjacent or far apparatus or invades human body of womb and becomes greater in size, volume, quantity, or scope, uterine cavity may be wore out and this will result in massive haemorrhage of abdomen. When its organization invades intraliagamentary, uterine hematoma is brought, indeed, this organization may transfer to vagina, lung, and brain, patients may die. In this case, benign hydatidiform mole has been turn into the malignant tumor and is fatal, it is called as malignant hydatidiform mole[1]. Now that malignant hydatidiform mole is the result of benign hydatidiform mole canceration, it is very necessary for us to predict that if the benign hydatidiform mole can become malignant hydatidiform mole or not. This is a question for worthful discussion.

In this paper, based on the fuzzy mathematic method, we discuss this the problem.

The organization of this paper is as follows. In section 2, we establish the model for predicting the benign hydatidiform mole canceration. Section 3 gives a example to illustrate the effective of this model..

2 Hydatidiform Mole Canceration Prediction Model

Assume that prediction set of hydatidiform mole canceration is given by

$$T=\{T_1,T_2\}. \quad (1)$$

where T_1 denotes hydatidiform mole non-canceration tendency, T_2 represents hydatidiform mole canceration tendency.

According to the correlative literatures[1][2] and the conditions of medical apparatus, assume that factor set S associated with the pathogenies of hydatidiform mole is written as

$$S = \{S_1, S_2, \cdots, S_{10}\}. \quad (2)$$

where $S_1, S_2, S_3, S_4, S_5, S_6, S_7, S_8, S_9, S_{10}$ are ages of the suffers, number of pregnancy, delitescence, enlarged rate of womb, womb size(weeks of pregnancy), hydatid mole size, highest titer value of HCG, Time Taken by turning result of pregnancy test based on crude urine into Negative Reaction(TTNR), pathological examination, and method for termination of pregnancy, respectively. The sub-factor set of the ith factor of pathogenies is denoted by

$$S_i = \{S_{i1}, S_{i2}, \cdots, S_{iL}, \cdots, S_{iK}\}. \quad (3)$$

where $i = 1, 2, \cdots, 10$; $L = 1, 2, \cdots, K$, K represents number of sub-factor of the i th factor.

2.1 Quantum Standard of Each Factor

Each factor of the pathogenies of hydatidiform mole is quantized according to the given standard in Table 1

2.2 Conformation of Attaching Function of Each Factor

Attaching function of each factor is given by following equations

$$\mu(S_{1j}) = \frac{3}{5}\log(7+S_{1j})(j=1,2,\cdots,7),$$

$$\mu(S_{2j}) = \frac{1}{3}\ln(6+S_{2j})(j=1,2,3)$$

$$\mu(S_{3j}) = \frac{1}{2}\log(4+S_{3j})(j=1,2,\cdots,6),$$

$$\mu(S_{4j}) = \frac{3}{4}\log(8+S_{4j})(j=1,2,3),$$

$$\mu(S_{5j}) = \frac{4}{5}\log(5+S_{5j})(j=1,2,3),$$

$$\mu(S_{6j}) = \frac{3}{5}\log(15+S_{6j})(j=1,2,3),$$ (4)

$$\mu(S_{ij}) = \frac{3}{5}\log(10+S_{ij})(i=7,8; j=1,2,\cdots,7),$$

$$\mu(S_{9j}) = \frac{5}{7}\log(7+S_{9j})(j=1,2,3),$$

$$\mu(S_{10,j}) = \frac{5}{9}\log(10+S_{10,j})(j=1,2,\cdots,7).$$

where S_{ij} is the j th sub-factor of the i th factor S_i. Their quantum standard is given and their attaching function values are calculated by Eq.(4) and shown by Table 1.

Table 1. Sub-factors of each factor and their scores and weights (1)

Factor	Weight	Sub-factor	Score	Attaching function
Age (S_1)	0.0345	<20 (S_{11})	5	0.6475
		20~24 (S_{12})	6	0.6684
		25~29 (S_{13})	9	0.7225
		30~34 (S_{14})	11	0.7532*
		35~39 (S_{15})	12	0.7672
		40~44 (S_{16})	13	0.7806
		>44 (S_{17})	14	0.7933
Number of Pregnancy (S_2)	0.0345	1 (S_{21})	1	0.6486
		2~3 (S_{22})	3	0.7324*
		>3 (S_{23})	7	0.8549
Delitescence (days) (S_3)	0.1034	<7 (S_{31})	7	0.5206
		8~30 (S_{32})	8	0.5396
		31~60 (S_{33})	9	0.5568
		61~<90 (S_{34})	10	0.5730*
		91~120 (S_{35})	12	0.6020
		>120 (S_{36})	11	0.5880
Enlarged rate of womb (S_4)	0.1034	Fast(>) (S_{41})	12	0.9758*
		Normal(=Month) (S_{42})	3	0.7810
		Slow(<Month) (S_{43})	0	0.6773
Womb size (S_5)	0.1034	<12 (S_{51})	7	0.8633
		12~19 (S_{52})	8	0.8911*
		>19 (S_{53})	12	0.9844

Table 1. Sub-factors of each factor and their scores and weighs (2)

Factor	Weight	Sub-factor	Score	Attaching function
Womb size (S_5)	0.1034	<12 (S_{51}) 12~19 (S_{52}) >19 (S_{53})	7 8 12	0.8633 0.8911* 0.9844
Hydatid mole Size (S_6)	0.1725	Small grape particles (S_{61}) Big grape particles (S_{62}) All grape particles (S_{63})	17 1 10	0.9030 0.7224 0.8388*
Highest titer Value(IU/L) of HCG (S_7)	0.1725	50 (S_{71}) 100 (S_{72}) 200 (S_{73}) 400 (S_{74}) 800 (S_{75}) 1600 (S_{76}) 3200 (S_{77})	2 6 7 12 14 15 18	0.6475 0.7225 0.7383 0.8055 0.8281* 0.8388 0.8682
TTNR (S_8)	0.1034	7days(S_{81}) 1~2weeks (S_{82}) 2~3weeks (S_{83}) 3~4weeks (S_{83}) 1~2months (S_{84}) 2~3months (S_{85}) >3months (S_{86})	1 4 5 7 11 12 13	0.6248 0.6877 0.7057 0.7383* 0.7933 0.8055 0.8170
Pathological examination (S_9)	0.1379	Low-grade hyperplasia (S_{91}) Middle-grade hyperplsia(S_{92}) Serve-grade hyperplasia (S_{93})	0 7 12	0.6036 0.8186* 0.9314
Method for termination of pregnancy (S_{10})}	0.0345	Nature ($S_{10,1}$) Induced labor success ($S_{10,2}$) Induced labor unsuccess($S_{10,3}$) Uterine aspiration ($S_{10,4}$) Uterine apoxesis ($S_{10,5}$) Uterine apoxesis removel ($_{10,6}$) Direct removel ($S_{10,7}$)	5 10 12 6 7 1 2	0.6534 0.7228 0.7458 0.6689 0.6836 0.5786* 0.5995

2.3 Weight of Each Factor

According to the clinic experience of medical experts and relative literatures, the weight value of each factor may be determined. Assume that weight set is expressed by

$$A = \{a_1, a_2, \cdots, a_{10}\} \quad (5)$$

where a_i is the weight value of the ith factor S_i. The weight values are given by Table 1.

2.4 Computing Fuzzy Measure $a(S_i)$

For λ-Fuzzy measure $a(S_i)$ [3],[4], we have example,

$$a(S_1) = a_1, \qquad (6a)$$
$$a(S_i) = a_1 + a(S_{i-1}) + \lambda \cdot a_i \cdot a(S_{i-1}). \qquad (6b)$$

When $\lambda = 0$, according to the Eq.(6), we also have
Using Eq.(7), fuzzy measure may be computed.

2.4 Computing Fuzzy Integral E Value

In limited ranges, fuzzy integral is denoted by the symbol "E" and given by

(8)

where $\mu(S_i)$ is a attaching function of the factor S_i and monotone. Weight value a_i is arrayed according to the depressed order of $\mu(S_i)$.

2.6 Threshold of Fuzzy Integral E

Based on the clinic experience of the medical experts, the threshold of fuzzy integral E value is determined and shown in Table 2.

Table 2. Threshold of fuzzy integral E

Hydatidiform mole canceration tendency	Canceration	Non-canceration
Threshold of fuzzy integral(E)	<0.7500	≥0.7500

For the special case, after her examined factors are given, we can compute threshold of Fuzzy integral E using above model and predict the cancerational tendency of hydatidiform mole according to the Table 2.

The fuzzy integral model(FIM) for predicting cancerational tendency of hydatidiform mole is not established until now.

3 Example Analysis

A case: her age was 32 year old, she was a two-children mother. During her pregnancy, increasing rate of her womb is greater than intermenstrual months. The value of the HCG was 800(IU/L). Trophocyte proliferation and differentiation belonged to the second level. There were small and large grape particles in hydatid mole. Weeks of pregnancy were 14 weeks. Her vagina bled and her blood was coffee color after amentia. Time Taken by turning result of pregnancy test based on crude urine into Negative Reaction(TTNR) is three weeks. This case was diagnosed as the sufferer of hydatidiform mole and its womb was cut. According to the symptoms of the sufferer, her factor set was given by

$$S = \{S_{14}, S_{22}, S_{34}, S_{41}, S_{52}, S_{63}, S_{75}, S_{82}, S_{92}, S_{10,6}\} \quad (9)$$

Prediction process is as follows:

Step 1: According to the sub-factor in Eq.(9), the score of each sub-factor can be obtained from Table 1. Using Eq.(4), the attaching function $\mu(S_i)$ of these factors can be calculated. These attaching function values, which are placed an identifying or distinctive "*", are shown in Table 1

Step 2: The attaching function $\mu(S_i)$ of the case is put into a monotone degressive order. Corresponding weight values are also arranged with attaching function $\mu(S_i)$. All results are shown in Table 3.

Step 3: Substituting the data in Table 3 into equation (8), we have

$$E = \int \mu(S_{ij}) \circ da = \bigvee_{i=1}^{10} [\mu(S_{ij}) \wedge a(S_i)]$$
$$= (0.9758 \wedge 0.1034) \vee (0.8911 \wedge 0.2068)$$
$$\vee (0.8388 \wedge 0.3793) \vee (0.8281 \wedge 0.5188) \quad (10)$$
$$\vee (0.8186 \wedge 0.6897) \vee (0.7532 \wedge 0.7237)$$
$$\vee (0.7383 \wedge 0.8271) \vee (0.7324 \wedge 0.8616)$$
$$\vee (0.586 \wedge 0.8961) \vee (0.5730 \wedge 1.0000) = 0.7383$$

Table 3. The attaching function $\mu(S_i)$ of the case according to a monotone degressive order

Sub-factor	$\mu(s_i)$	a_i	$a(s_i)$
S_{41}	0.9758	0.1034	0.1034
S_{52}	0.8911	0.1034	0.2068
S_{63}	0.8388	0.1725	0.3793
S_{75}	0.8281	0.1379	0.5188
S_{92}	0.8186	0.0345	0.6897
S_{14}	0.7532	0.1034	0.7237
S_{82}	0.7382	0.0345	0.8271
S_{22}	0.7324	0.0345	0.8616
$S_{10,6}$	0.5786	0.0345	0.8961
S_{34}	0.5730	0.1034	1.000

Based on the threshold of Fuzzy integration E in Table 2 and Eq.(10), we can come to conclusion that the hydatidiform mole of the case is not depraved and the result is in accordance with that of the clinic diagnosis.

In order to illustrate the performance of the FIM compared with the Maximum Likelihood(ML) method, we have gone on the following two-stage researches.

The first stage: The diagnosed cases, who were suffers of hydatidiform mole canceration or non-canceration, were tested by the the FIM and the ML method. 100 previous cases, who were taken ill of hydatidiform mole, were diagnosed by the the FIM and the ML method, respectively. The diagnosis results were shown in Table 4.

Table 4. Prediction results for previous cases

Method	Cases	Coincidental number	Accurate rate (%)
FIM	Canceration in 110 cases	98	89
	Non-canceration in 163 cases	142	87
ML method	Canceration in 110 cases	80	73
	Non-anceration in 163 cases	129	79

Table 5. Prediction results for 32 new cases

Method	Cases of canceration tendency	Cases of non-canceration tendency
ML method	14	18
FIM	9	23
Medical experts	10	22

Table 4 has shown that the diagnosis accurate of rate fuzzy integral model(FIM) is higher than that of the ML method.

The second stage: New cases, who were suffers of hydatidiform mole but canceration tendency of these suffers were not predicted by medical experts. We applied this the FIM to clinic predication for 32 suffers of hydatidiform mole,who come from the 10 hospitals in Huainan city, Anhui Province,China, from 2000 to 2004. The suffers were predicted by the FL method, the FIM and medical experts synchronously. The predicted results were shown in Table 5.

Table 5 told us that the predication results of the FIM is almost the same results of the medical experts.

From the above two-stage predication results, we can know that the it is effective to employ the FIM for predicting the canceration tendency of hydatidiform mole. Suddenly, the validity of fuzzy integral model(FIM) in the clinic diagnosis and the prediction depends on the following factors:

(1) The nosogeneses associated with the hydatidiform mole are reasonably determined.

(2) Attaching function, quantum standard, weight value of each factor, which cause disease, and the threshold value of fuzzy integral value are determined under condition that medical experts take part in. In this paper, we take measures as follows:

First, each medical expert gives the scores of the sub-factor of each factor based on their clinic experience and special knowledge and all scores expect for the highest score and lowest scores are averaged in order to avoid the subjective error of scores.

Second, based on analyzing the feature of the scores given by medical experts, attaching functions are established using K power parabola larger type.

Third, weight values are determined via method of comparison by twos. In other words, the importance of each factor is compared by twos by medical experts and its scores are given. The weight values are computed by the AHP method[4].

Finally, the relative information is obtained from the case history of hydatidiform mole cases. Fuzzy integral E value of each cases is calculated by Eq.(8). Threshold value of E is finally determined.

(3) The FIM is not modified again and again according to a great many of history cases until that its diagnosis accurate rate is enough high and it can be directly applied to the clinic diagnosis.

References

1. Ding Manling.: Gynecopathy Diagnosis and Differentiation. Beijing, Renmin Sanitary Book Concern. (2000) 90-100
2. Guo Yecai.: Fuzzy integral method based coronary heart disease prediction. Journal of Mathematical Medicine. 5(4), (1992) 79-81.
3. Guo Yecai.: A Model of Informational Entropy and Fuzzy iIntegral Decision of Microcomputer Medical Expert System on Radiodiagnosis of Congential Heart Disease. Chinese Journal of Medical Physics. 12(3), (1995) 158-162.
4. Li Wuwei.: Application of Fuzzy Mathematics and Anal.ytic Hierarchy Process Method. China Science and Technology Press. (1990) 125-157.
5. Xi-zhao Wang, Xiao-jun Wang. A New Methodology for Determining Fuzzy Densities in the Fusion Model Based on Fuzzy Integral. 2004 International Conference on Machine Learning and Cybernetics. Shanghai, China, Aug. 26-29. (2004) 1234-1238.

Consensus Measures and Adjusting Inconsistency of Linguistic Preference Relations in Group Decision Making

Zhi-Ping Fan and Xia Chen

School of Business Administration, Northeastern University,
Shenyang 110004, China
zpfan@mail.neu.edu.cn

Abstract. This paper presents a method for consensus measures and adjusting inconsistency of linguistic preference relations in group decision-making (GDM). Several consensus measures are defined by comparing positions of alternatives between individual ordered vectors and a collective ordered vector. In such a way, the consensus situation is evaluated in each moment in a more realistic way. Feedback mechanism would be applied if the consensus degree of all experts does not reach the required consensus level. This feedback mechanism is simple and easy computing process to help experts change their opinions in order to obtain the consensus level. It is based on the use of individual linguistic preferences, collective and individual ordered vectors of alternatives. It is also based on the use of fuzzy majority of consensus, represented by means of a linguistic quantifier.

Keywords: Group decision making (GDM); Linguistic preference relation; Consensus degree; Feedback mechanism.

1 Introduction

The research on group decision making (GDM) is getting more and more attention. It has become an important research field in decision analysis. GDM is to aggregate individual preference information to form a collective preference, and then obtain ranking of alternatives based on the derived collective preference. In GDM, the preference relation given by the expert is a most common preference format. According to element forms in preference relation, there are often two kinds of preference relations: numerical preference relation (i.e., multiplicative preference relation and fuzzy preference relation) [1-4] and linguistic preference relation [5, 6]. So far, the research on linguistic preference relation has become a very important research topic and some valuable research results have been found [5, 6, 9, 10].

However, it is often very difficult to obtain all experts' agreement of linguistic preference relations in GDM due to the difference of individual knowledge, judgment and preference. Complete agreement in all experts is not necessary in real life. It is inevitable to develop a consensus process to obtain a solution of group consensus. Research on group consensus problems based on linguistic preference relations in GDM has been found in [7] and [8]. Herrera et al. [7] propose a model of consensus model in GDM under linguistic assessments. Several linguistic consensus degrees and linguistic distances are obtained. Herrera et al. [8] present a method to solve linguistic consensus problem in GDM. But how to adjust inconsistency in GDM are not given above. If the consensus level is not acceptable, i.e., consensus degree is lower than the threshold requested, which means that there exists a great discrepancy in the experts' opinions, then the question to solve is how to substitute the actions of moderator to group discussion process in order to model automatically the whole consensus process. In this paper, we develop a new consensus model by comparing the positions of the alternatives between the individual ordered vector and collective ordered vector. There are three levels of consensus measures, i.e., consensus measures of each expert for each alternative, consensus measures of all experts on each alternative and consensus measure of all experts over the set of alternatives. Meanwhile, the feedback mechanism would be used if consensus measure of all experts over the set of alternatives had not reached the consensus level. The feedback mechanism can help the experts change their opinions on alternatives and know the direction of change in the discussion process. The consensus process is defined as a cycle process controlled automatically without using the moderator.

This paper is organized as follows. Section 2 presents the background on consensus measure. Section 3 presents consensus measures and the feedback mechanism to analyze consensus problems of linguistic preference relations in GDM. To illustrate the use of the proposed method, a numerical example is presented in Section 4. Finally, Section 5 summarizes the paper.

2 Background on Consensus Measure

2.1 Linguistic Assessment in GDM

This section describes the GDM problem with linguistic preference relations on alternatives. Let $X = \{x_1, x_2, \cdots, x_n\}$ ($n \geq 2$) be a finite set of alternatives and $E = \{e_1, e_2, \cdots, e_m\}$ ($m \geq 2$) be a finite set of experts. Let $S = \{s_i\}$, $i \in U = \{0, \cdots T\}$ be a finite and totally ordered discrete term set, where s_i denotes a possible value for a linguistic variable. For example, this is the case of the following linguistic term set: $S = \{$ $s_0 = $ I (Impossible), $s_1 = $ EU (Extremely Unlikely), $s_2 = $ VLC (Very Low Chance), $s_3 = $ SC (Small Chance), $s_4 = $ IM (It May), $s_5 = $ MC (Meaningful Chance), $s_6 = $ ML (Most Likely) $s_7 = $ EL (Extremely Likely), $s_8 = $ C (Certain). Usually, in the cases, the set has several characteristics: 1) the set is ordered: $s_i \geq s_j$ if $i \geq j$; 2) there is the negation operator: Neg $(s_i) = s_j$ such that $j = T - i$; 3) Maximization operator: Max $(s_i, s_j) = s_i$ if $s_i \geq s_j$; 4) Minimization operator: Min $(s_i, s_j) = s_i$ if $s_i \leq s_j$.

2.2 Individual Ranking of Alternatives

In this paper, the preference information of expert e_i on X is described by linguistic preference relation $P_k \subset X \times X$, $P_k = (p_{ij}^k)_{n \times n}$, in which element $p_{ij}^k \in S$ denotes the preference degree of alternative x_i over x_j: $p_{ij}^k"="s_{T/2}$ denotes indifference between x_i and x_j (i.e., $x_i \sim x_j$); $s_0"\leq"p_{ij}^k"<"s_{T/2}$ denotes that x_i preferred to x_j (i.e., $x_j \succ x$), $s_{T/2}"<"p_{ij}^k"\leq"s_T$ denotes that x_i preferred to x_j (i.e., $x_i \succ x_j$).

Definition 2.1. For linguistic preference relation $P_k = (p_{ij}^k)_{n \times n}$, if $p_{ij}^k"\leq"s_{T/2}$ and $p_{jk}^k"\leq"s_{T/2}$, $\forall i, j \in \{1,\cdots,n\}$, then $p_{ik}^k"\leq"s_{T/2}$; or if $p_{ij}^k"\geq"s_{T/2}$ and $p_{jk}^k"\geq"s_{T/2}$, $\forall i, j \in \{1,\cdots,n\}$, then $p_{ik}^k"\geq"s_{T/2}$. So P_k has the satisfactory consistency.

Individual ranking of alternatives is obtained in [9]. In the above P_k, using number m_j of alternative x_j over x_i $(i = 1,\cdots,n)$, The number m_j is larger, then alternative x_j is preferred to others. Based on number m_j of alternatives x_i $(i = 1,\cdots,n)$, the ranking or selection on alternatives is obtained.

2.3 Collective Ranking of Alternatives

According to [5], we give the definition on LOWA operator as follows.

Definition 2.2. Let $A = \{a_1, a_2, \cdots, a_m\}$ be a set of labels to be aggregated, then the LOWA operator is defined as follows:

$$\phi_L(a_1, a_2, \cdots, a_m) = W \cdot B^T = C^m\{w_k b_k, k = 1, \cdots, m\}$$
$$= w_1 \otimes b_1 \oplus (1 - w_1) \otimes C^{m-1}\{\beta_h b_h, h = 2, \cdots, m\}, \qquad (1)$$

where $W = (w_1, w_2, \cdots, w_m)$ is a weighting vector, such that $w_i \in [0,1]$ and $\sum_{i=1}^{m} w_i = 1$; $\beta_h = w_h \big/ \sum_{k=2}^{m} w_k$, $h = 2, \cdots, m$, and B is associated ordered label vector. Element $b_i \in B$ is the i th largest label in the collection $a_1, \cdots, a_n$.

Weight w_i can be obtained by means of the following expression:

$$w_i = Q(i/m) - Q((i-1)/m), \quad \forall i \in \{1, \cdots, m\}, \qquad (2)$$

where Q is a non-decreasing quantifier represented by the following membership function:

$$Q(r) = \begin{cases} 0, & r < a \\ \dfrac{r-a}{b-a}, & a \leq r \leq b, \\ 1, & r > b \end{cases} \qquad (3)$$

with $a, b, r \in [0, 1]$.

According to [6], the linguistic relative quantifier is obtained as follows:

$$Q^2 : [0,1] \to S, \tag{4a}$$

$$Q^2(r) = \begin{cases} s_0, & r < a \\ s_i, & a \le r \le b, \\ s_T, & r > b \end{cases} \tag{4b}$$

where s_0 and s_T are the minimum and maximum labels in S, respectively.

$s_i = \underset{s_q \in M}{\text{Sup}}\{S_q\}$, with $M = \{s_q \in S : \mu_{s_q}(r)\} = \underset{i \in U}{\text{Sup}}\{\mu_{s_i}(\frac{r-a}{b-a})\}$, $a, b, r \in [0,1]$.

Some examples of relative quantifiers are "most", "at least half", "as many as possible" defined by the parameters (a, b), $(0.3, 0.8)$, $(0, 0.5)$, $(0.5, 1)$, respectively.

Collective ranking of alternatives in GDM is obtained in [6]. A brief description of ranking or selection of alternatives is given below.

Firstly, the dominant preference relation $P^S = (p_{ij}^s)_{n \times n}$ is defined, where

$$p_{ij}^s = \begin{cases} s_0, & p_{ij} < p_{ji} \\ s_k, & p_{ij} \ge p_{ji} \end{cases}, \forall i, j \in \{1, \cdots, n\}, \text{ if } p_{ij} = s_l, p_{ji} = s_t, \forall i, j \in \{1, \cdots, n\}, \text{ then }$$

$k = l - t$; Then, the non-dominance degrees and dominance degrees of alternative are respectively defined as $\mu_{ND}(x_i) = \text{MIN}\{Neg(p_{ji}^s)\}$ and $LDD(x_i) = \phi_{L_{*,j}}(p_{ij})$, $\forall j \in \{1, \cdots, n\}$, where ϕ_L is LOWA operator. The alternative sets with the largest non-dominance degrees and the largest dominant degrees are respectively defined as follows:

$$x^{ND} = \{x_i \in \{1, \cdots, n\} | \mu_{ND}(x_i)\} = \underset{y \in \{1, \cdots, n\}}{\text{MAX}} [\mu_{ND}(y)],$$

$$x^{LDD} = \{x_i \in x^{ND} | LDD(x_i)\} = \underset{y \in X^{ND}}{\text{MAX}} [LDD(y)].$$

Finally, based on Q^2 operator, the strict dominance degrees of alternative is defined as $SDD(x_i) = Q^2(\frac{r_i}{n-1})$, where $r_i = \#\{p_{iq} \in S | p_{iq} > p_{qi}\}$, # is the set numbers, $q \in \{1, \cdots, n\}$. In the computing process, if $\#\{x^{ND}\} = 1$ or $\#\{x^{LLD}\} = 1$, then the process is over. The alternative with the largest non-dominance degree or the largest dominance degree is preferred to others.

3 Consensus Measures and Feedback Mechanism

3.1 Consensus Measures

Definition 3.1. Let $V^c = \{v_1^c, \cdots, v_n^c\}$ and $V^i = \{v_1^i, \cdots, v_n^i\}$ be the collective and individual (i.e., expert e_i) ordered vector of alternatives, respectively, where v_j^c and v_j^i are the position of alternative x_j in that collective ordered vector and individual are the position of alternative x_j in that collective ordered vector and individual (i.e., expert e_i) ordered vector, then consensus degree $C_i(x_j)$ and linguistic consensus degree $Q^2[C_i(x_j)]$ of each expert e_i for each alternative x_j are defined according to the following expressions:

$$C_i(x_j) = 1 - |v_j^c - v_j^i|/(n-1), \tag{5a}$$

$$Q^2[C_i(x_j)] = Q^2[1 - |v_j^c - v_j^i|/(n-1)]. \tag{5b}$$

In which, for example, if collective and individual (i.e., expert e_3) ordered vector of alternatives are $V^c = \{x_3, x_1, x_2, x_4\}$ and $V^3 = \{x_2, x_3, x_4, x_1\}$, linguistic quantifier "most" with the pair (0.3, 0.8), then $C_3(x_1)$ and $Q^2[C_3(x_1)]$ are obtained as:

$$C_3(x_1) = 1 - |2-4|/(4-1) = 1 - 2/3 = 1/3, \quad Q^2[C_3(x_1)] = s_1 = EU.$$

Definition 3.2. If $Q^2[C_i(x_j)] = s_T$, where s_T is the largest element in set S, then consensus measure $C_i(x_j)$ of expert e_i for alternative x_j reaches the consensus level. Or else consensus measure $C_i(x_j)$ of expert e_i for alternative x_j does not reach the consensus level.

Definition 3.3. Consensus degree $C_i(x_j)$ and linguistic consensus degree $Q^2[C(x_j)]$ of all experts on alternative x_j are defined according to the following expressions:

$$C(x_j) = [\sum_{i=1}^{m} C_i(x_j)]/m = [\sum_{i=1}^{m}(1-|v_j^c - v_j^i|/(n-1))]/m, \tag{6a}$$

$$Q^2[C(x_j)] = Q^2[(\sum_{i=1}^{m} C_i(x_j))/m] = Q^2[(\sum_{i=1}^{m}(1-|v_j^c - v_j^i|/(n-1))/m]. \tag{6b}$$

Definition 3.4. If $Q^2[C(x_j)] = s_T$, where s_T is the largest element in set S, then consensus measure $C(x_j)$ of all experts on alternative x_j reaches the consensus level. Or else consensus measure $C(x_j)$ of all experts on alternative x_j does not reach the consensus level.

Definition 3.5. Consensus degree C_X and linguistic consensus degree $Q^2[C_X]$ of all experts over the alternative set are defined according to the following expressions:

$$C_X = [\sum_{j=1}^{n} C(x_j)]/n = [\sum_{j=1}^{n}\sum_{i=1}^{m}(1-|v_j^c - v_j^i|/(n-1))]/nm, \tag{7a}$$

$$Q^2(C_X) = Q^2[(\sum_{j=1}^{n} C(x_j))/n] = Q^2[(\sum_{j=1}^{n}\sum_{i=1}^{m}(1-|v_j^c - v_j^i|/(n-1)))/nm]. \tag{7b}$$

Definition 3.6. If $Q^2[C_X]=s_T$, where s_T is the largest element in set S, then the consensus measure C_X of all experts over the alternative set reaches the consensus level. Or else consensus measure C_X of all experts over the alternative sst does not reach the consensus level.

Theorem 3.1. If the consensus measure of expert e_i $(i=1,\cdots,m)$ on alternative x_j reaches the consensus level, then consensus measure on alternative x_j of all experts reaches the consensus level.

Proof. First, using (5b), $Q^2[C_i(x_j)] = s_T$, $\forall i \in \{1,\cdots,m\}$. Using (4b), then $C_i(x_j) > b$, $\forall i \in \{1,\cdots,m\}$. Using (6a), we have $C(x_j) = [\sum_{i=1}^{m}(1-|v_j^c - v_j^i|/(n-1))]/m$
$= [\sum_{i=1}^{m} C_i(x_j)]/m > (\sum_{i=1}^{m} b)/m = b$, i.e., $C(x_j) > b$. Hence, $Q^2[C(x_j)] = C$. By Definition 3.4, we know that consensus measure on alternative x_j of all experts reaches the consensus.

Theorem 3.2. If consensus measure of all experts on alternative x_j $(j=1,\cdots,n)$ reaches the consensus level, then consensus measure C_X of all experts reaches the consensus level.

Proof. First, using (6b), $Q^2[C(x_j)] = s_T$, $\forall j \in \{1,\cdots,n\}$. Using (4b), then $C(x_j) > b$, $\forall j \in \{1,\cdots,n\}$. Using (7a), we have $C_X = [\sum_{j=1}^{n}\sum_{i=1}^{m}(1-|v_j^c - v_j^i|/(n-1))]/nm$
$= [\sum_{j=1}^{n}(C(x_j))]/n > (\sum_{j=1}^{n} b)/n = b$. i.e., $C_X > b$. Hence, $Q^2[C_X] = C$. By Definition 3.6, consensus measure of all experts over the alternative set reaches the consensus level.

Remark 3.1. Note that definitions 3.2, 3.4 and 3.6, as in Definitions 3.2, 3.4 and 3.6, the required consensus levels are difference if linguistic relative quantifier are difference based on the difference parameters (a, b). The required consensus levels are usually decided by a leading decision maker.

3.2 Feedback Mechanism

When consensus measure C_X has not reached the required consensus level, i.e., $C_X < C$, then the experts' opinions should be improved or modified. In this pape, the rule of feedback mechanism is developed below.

(1) If $v_j^c - v_j^i < 0$, then move forward position of alternative x_j for the ith expert, i.e., number m_j about $x_j \succ x_i$ $(i=1,\cdots,n)$ should be increased in the above P_i [9].

(2) If $v_j^c - v_j^i = 0$, do not changed position of alternative x_j for the ith expert.

(3) If $v_j^c - v_j^i > 0$, then move backward position of alternative x_j for the ith expert, i.e., numbers m_j about $x_j \succ x_i$ $(i=1,\cdots,n)$ should be decreased in the above P_i [9].

An adjusting approach is expressed in the following steps.

Step 1. Consensus measure is calculated by (5a)-(7a). If $Q^2[C_X] = s_T$, then go Step 5. Or else go next step.

Step 2. Let $\left|v_{j^*}^c - v_{j^*}^{i^*}\right| = \max_{\substack{1 \le i \le m \\ 1 \le j \le n}} \left|v_j^c - v_j^i\right|$, then the linguistic preference relation of the i^*th expert will be adjusted, which the position of alternative x_{j^*} for the i^*th expert will be changed.

Step 3. If $v_{j^*}^c - v_{j^*}^{i^*} < 0$, then the position of alternative x_{j^*} for the i^*th expert will be moved forward. The adjusting approach is expressed as follows. In the above P_{i^*}, numbers m_{j^*} about $x_{j^*} \succ x_i$ $(i=1,\cdots,n)$ will be increased. Number m_{j^*} is bigger, alternative x_{j^*} is preferred to other alternatives. If $v_{j^*}^c - v_{j^*}^{i^*} = 0$, the position of alternative x_{j^*} for the i^*th expert will be not changed. If $v_{j^*}^c - v_{j^*}^{i^*} > 0$, then the adjusting approach is same above.

Step 4. If P_{i^*} adjusted by the i^*th expert has the satisfactory consistency, then consensus measure is calculated by (5a)~(7a). If $Q^2[C_X] < C$, then go Step 2. Or else go next step.

Step 5. End.

4 Illustrative Example

To illustrate the above method for consensus measures and adjusting inconsistency of linguistic preference relations, in this paper, consider the nine linguistic label set above. There are four experts (i.e., $e_i, i=1,2,3,4$) and four alternatives (i.e., $x_j, j=1,2,3,4$). The four experts give respectively linguistic preference relations as follows:

$$P_1 = \begin{bmatrix} - & MC & MC & VLC \\ SC & - & MC & SC \\ SC & SC & - & SC \\ ML & MC & MC & - \end{bmatrix}, \quad P_2 = \begin{bmatrix} - & SC & VLC & VLC \\ MC & - & VLC & SC \\ ML & ML & - & MC \\ ML & MC & SC & - \end{bmatrix},$$

$$P_3 = \begin{bmatrix} - & ML & MC & EL \\ VLC & - & SC & MC \\ SC & MC & - & ML \\ EU & SC & VLC & - \end{bmatrix}, \quad P_4 = \begin{bmatrix} - & ML & MC & MC \\ VLC & - & SC & SC \\ SC & MC & - & VLC \\ SC & MC & ML & - \end{bmatrix}.$$

From these linguistic preferences relations, we will obtain the following collective and individuals ordered vector of alternatives, consensus measures and feedback process.

(A) Individuals alternatives vectors and collective alternatives vector

(A.1) Individuals ordered vectors of alternatives. Linguistic preference relations $P_k (k=1,2,3,4)$ verified have satisfactory consistency. Individuals ranking of alternatives are obtained as follows [9]:

$e_1 : \{x_4, x_1, x_2, x_3\}$, $e_2 : \{x_3, x_4, x_2, x_1\}$, $e_3 : \{x_1, x_3, x_2, x_4\}$, $e_4 : \{x_1, x_4, x_3, x_2\}$.

(A.2) Collective ordered vector of alternatives. Using the linguistic quantifier "at least half" with the pair (0, 0.5), LOWA operator with the weighting vector W=(1/2, 1/2, 0, 0), we obtain the following collective linguistic preference relations:

$$P = \begin{bmatrix} - & ML & MC & ML \\ IM & - & IM & IM \\ MC & MC & - & MC \\ ML & MC & MC & - \end{bmatrix}.$$

Using dominance degree of alternative, non-dominance degree of alternative and strict dominance degree of alternative [6], The collective ranking of alternatives is $\{x_1, x_4, x_3, x_2\}$.

(B) Consensus measures. Consensus degree of all experts over the alternative set calculated by (5a), (6a) and (7a) is $C_X = 0.71$. Using the linguistic quantifier "most" with the pair (0.3, 0.8), linguistic consensus degree of all experts over the alternative set is $Q^2(C_X) = ML < C$. The experts' opinions should be modified because the consensus measure C_X has not reached the required consensus level.

(C) Feedback process. The difference between collective ranking of alternatives and individual ranking of alternatives are as follows:

$v_j^c - v_j^i$	x_1	x_2	x_3	x_4
e_1	-1	1	-1	1
e_2	-3	1	2	0
e_3	0	1	1	-2
e_4	0	0	0	0

Let $\max\limits_{\substack{1\le j\le n \\ 1\le i\le m}}|v_j^c - v_j^i| = |v_1^C - v_1^2|$, then the 2th expert' opinion shall be modified. Using rule of the above adjusting approach, the position of alternative x_1 will move forward for the 2th expert. In the above P_2, numbers m_1 about $x_1 \succ x_i$ $(i=1,\cdots,4)$ will be increased, i.e., numbers m_1 should be 3. The new preference relation P_2^* and new collective preference relations are as follows:

$$P_2^* = \begin{bmatrix} - & EL & ML & MC \\ EU & - & SC & SC \\ VLC & MC & - & VLC \\ SC & MC & ML & - \end{bmatrix}, \quad P^* = \begin{bmatrix} - & ML & MC & ML \\ VLC & - & IM & IM \\ SC & MC & - & IM \\ IM & MC & ML & - \end{bmatrix}.$$

For the 2th expert, collective ranking of alternatives is $\{x_1, x_4, x_3, x_2\}$. The consensus degree of alternative $x_j (j=1,\cdots,n)$ for all experts calculated by (5a), (6a) and (6b) are as follows:

	x_1	x_2	x_3	x_4
$C(x_i)$	0.92	0.83	0.83	0.75
$Q^2[C(x_i)]$	C	C	C	EL

Consensus degree of all experts over the alternative set calculated by (7a) is $C_X = 0.83$. Using the linguistic quantifier "most" with the pair (0.3, 0.8), linguistic consensus degree of all experts over the alternative set is $Q^2(C_X) = C$.

Remark 4.1. we can obtained several conclusion: 1) The consensus degree of alternatives x_1, x_2 and x_3 for all experts have reached consensus level, but consensus degree of alternatives x_4 has not reached consensus level; 2) Consensus degree of all experts over the alternative set has reached consensus level.

5 Conclusions

In this paper, we have presented a method for consensus measures and adjusting inconsistency of linguistic preference relations in GDM. All the relevant results are expressed based on the use of individuals linguistic preferences, on the use of collective and individual ordered vectors of alternatives and on the use of fuzzy majority of consensus, represented by means of a linguistic quantifiers. There are three levels of consensus measures, which the first level is called consensus measure of each expert for each alternative, the second level is called consensus measure of all experts on each alternative, and the last level is called consensus measure of all experts over the alternative set. Meanwhile, the feedback mechanism is represented based on simple and easy rules. Adjusting inconsistency approach can be applied to help experts change their opinions in order to obtain the consensus level.

Acknowledgements

This research was partly supported by the National Natural Science Foundation of China (NSFC, Project No. 70371050), the Teaching and Research Award Program for Outstanding Young Teachers (TRAPOYT) in Higher Education Institutions and the Research Fund for the Doctoral Program of Higher Education (Project No. 20040145018), Ministry of Education, China.

References

1. Saaty T A. The analytic hierarchy process. New York: McGraw-Hill (1980).
2. Orlorski S A. Decision-making with a fuzzy preference relation. Fuzzy Sets and Systems, 3 (1978) 155-167.
3. Herrera-Viedma E, Herrera F, Chiclana F. A Consensus Model for multiperson decision making with different preference structures. IEEE Transactions on Systems, Man and Cybernetics, 3 (2002) 394-402.
4. Yager R R. On ordered weighted averaging aggregation operators in multicriteria decision making. IEEE Transactions on Systems, Man and Cybernetics, 18 (1988) 183-190.
5. Herrera F, Herrera-Viedma E, Verdegay J L. Direct approach processes in group decision making using linguistic OWA operators, Fuzzy Sets and Systems, 79 (1996) 175-190.
6. Herrera F, Herrera-Viedma E, Verdegay J L. A sequential selection process in group decision making with linguistic assessment. Information Science, 4 (1995) 223-239.
7. Herrera F, Herrera-Viedma E, Verdegay J L. A Model of Consensus in group decision making under linguistic assessments. Fuzzy Sets and Systems, 1 (1996) 73-87.
8. Herrera F, Herrera-Viedma E, Verdegay J L. Linguistic Measures based on fuzzy coincidence for reaching consensus in group decision making. Journal of Approximate Reasoning, 3 (1997) 309-334.
9. Fan Z P, Xiao S H. Consistency and Ranking method for comparison matrix with linguistic assessment. Systems Engineering-Theory & Practice, 5 (2002) 87-91.

Fuzzy Variation Coefficients Programming of Fuzzy Systems and Its Application

Xiaobei Liang[1,2], Daoli Zhu[1], and Bingyong Tang[3]

[1] School of Management, Fudan University, Shanghai, P.R. China, 200433
[2] Chain Operation School, Shanghai Business School, Shanghai, P.R. China, 200235
[3] Glorious Sun School of Business and Management, Donghua University, Shanghai, P.R. China, 200051
xbliang@mail.dhu.edu.cn

Abstract. Fuzzy set theory has grown to become a major scientific domain collectively referred to as fuzzy systems. In this paper, the fuzzy variation coefficients programming and its application are discussed. The proposed fuzzy variation coefficients programming has found application in operations management and it will find various application in other areas.

1 Introduction

Fuzzy set theory has grown to become a major scientific domain collectively referred to as fuzzy systems, which include fuzzy sets, logic, algorithms, and programming [1]. Because of many criterions of the problem, we know that the essence of the supplier's appraisal is multi-objective, but there is not so much literature about supplier's appraisal by multi-goals programming. In order to satisfy all criterions, it will often produce the conflict between the goals, can use multi-goals programming to solve the problem [2].

Tony, P. et al investigated the effect of Zarco, using a number of single-factor and multiple factor analyses of variance, the authors compared the recollection of students treated with the Zarco activity with that of controlled students [3]. Geoff, R discusses how to maintain a balance in the use of information technology in manufacturing demand chain management [4]. The research has been carried out to investigate the use of genetic algorithms as a common solution technique for solving the range of problems that arise when designing and planning manufacturing operations [5]. Ocean shipping transportation has been studied mainly from an economic and management point of view. Laqoudis, I., N. et al follows an operations management approach, adopting a Business Systems Engineering methodology with the assistance of process modeling [6].

In this paper, the fuzzy variation coefficients programming and its application are discussed. The fuzzy variation coefficients LP problem [7] combines the forecasting problem with the LP problem commendably, and can preferably solve a series of long-term planning problem.

2 Linear Programming with Altering Coefficient

The linear programming problem with altering coefficient [7] is the problem of extreme value of the condition as follows:

Evaluate the function:

$$\max F^{(k)} = \sum_{j=1}^{n} c_j^{(k)} x_j^{(k)} \qquad (1)$$

$$\text{St.} \sum_{j=1}^{n} a_{ij}^{(k)} x_j^{(k)} = b_i^{(k)} \qquad i=1,2,\cdots,m \qquad (2)$$

$$x_j^{(k)} \geq 0 \qquad j=1,2,\cdots,n \qquad (3)$$

There into: k is the flowing time, $c_j^{(k)}$ and $x_{ij}^{(k)}$ both are the coefficient of variable time, $b_i^{(k)}$ are invariable (or variable) constant, $x_j^{(k)}$ are the unknown quantity of time k.

Function (1) is instituted as the aim function, (2) and (3) is instituted as the restrictive condition. The solution satisfying the equation group is $\{x_1^{(k)}, x_2^{(k)}, \ldots, x_n^{(k)}\}$, and it can be instituted as the permissible solution if it satisfies the nonnegative condition (3). The solution which satisfies the function (1) can be instituted as the optimization solution.

Utilizing the expression of matrix and vector, if to set:

$$\vec{c}^{(k)} = (c_1^{(k)}, c_2^{(k)}, \cdots, c_n^{(k)})^T$$

$$A^{(k)} = \begin{bmatrix} a_{11}^{(k)} & \cdots & \cdots & a_{1n}^{(k)} \\ \vdots & \ddots & & \vdots \\ a_{m11}^{(k)} & \cdots & \cdots & a_{mn}^{(k)} \end{bmatrix}$$

$$\vec{b}^{(k)} = (b_1^{(k)}, b_2^{(k)}, \cdots, b_m^{(k)})^T$$

$$\vec{x}^{(k)} = (x_1^{(k)}, x_2^{(k)}, \cdots, x_n^{(k)})^T$$

$$\vec{A}_i^{(k)} = (x_1^{(k)}, x_2^{(k)}, \cdots, x_n^{(k)})^T \qquad i=1,2,\cdots,m$$

Then the question can be written as follows:
Evaluate the function:

$$\max F^{(k)} = \vec{c}^{(k)T} \vec{x}^{(k)} \qquad (4)$$

$$\text{St.} \vec{A}_i^{(k)T} \vec{x}^{(k)} = b_i^{(k)} \qquad i=1,2,\cdots;m \qquad (5)$$

$$\vec{x}^{(k)} \geq \vec{0} \qquad (6)$$

Or St. $A^{(k)T} \vec{x}^{(k)} = \vec{b}^{(k)}$ (7)

$$\vec{x}^{(k)} \geq \vec{0}$$

If the enacting aim function is $\min \vec{c}^{(k)T}\vec{x}^{(k)}$, it can be translated into $\max(-\vec{c}^{(k)T}\vec{x}^{(k)})$.

If there is the inequation in the enacting restrictive condition as follows:

$$\vec{A}_i^{(k)T}\vec{x}^{(k)} \le b_i^{(k)} \tag{8}$$

Or: $\vec{A}_i^{(k)T}\vec{x}^{(k)} \ge b_i^{(k)} \tag{9}$

Then it can be equivalently translated into:

$$\vec{A}_i^{(k)T}\vec{x}^{(k)} + x_{n+i}^{(k)} = b_i^{(k)} \qquad x_{n+i}^{(k)} \ge 0 \tag{10}$$

Or: $\vec{A}_i^{(k)T}\vec{x}^{(k)} - x_{n+1}^{(k)} = b_i^{(k)} \qquad x_{n+i}^{(k)} \ge 0 \tag{11}$

The new variable $x_{n+i}^{(k)}$ is instituted as slack variable.

Utilizing the theory of linear planning problem with altering coefficient to solve the issues in practice. Because the coefficient in it often is unbeknown and has the characteristic of variable time, the main process to solve it can be concluded as follows:

1) Summarize the practical problem discussed as a corresponding variation coefficients LP problem, and standardization it.
2) According to the historical data related to time variable coefficients and constants, and because of their different orderliness, we can use appropriate forecasting methods to do scientific forecast, and then get the prospective value of time variable coefficients and constants in different periods.
3) Put the above value in the problem, we can make the problem as LP problem of N different periods, by using the simple shape law or other methods, the optimal value of each plan year can be reached.

The variation coefficients LP problem combines the forecasting problem with the LP problem commendably, and can preferably solve a series of long-term planning problem [8].

3 Fuzzy Variation Coefficients LP Problem

In the variation coefficients LP problem discussed above, if we reconsider it under the constraints of:

$$A^{(k)}x^{(k)} = \tilde{b}^{(k)} (or \le \tilde{b}^{(k)}, or \ge \tilde{b}^{(k)}) \tag{12}$$

And $\quad x^{(k)} \ge 0 \tag{13}$

To get the maximum value of $F^{(k)} = c^{(k)T}x^{(k)}$, then the problem becomes to fuzzy variation coefficients LP problem [9].

Here, the subject function of fuzzy constraints is defined as: when each $b_i^{(k)}$ of $b^{(k)}$ increase to $b_i^{(k)} + d_i^{(k)}$, the subject degree can be added defined as: $\mu_i(d_i^{(k)}), d_i^{(k)} \geq 0$.

And let μ_i as strictly descending function.

$\varphi_C^{(k)}(x^{(k)})$, as fuzzy goals, can be discussed in the standardized region [0,1]:

$$\varphi_G^{(k)}(x^{(k)}) = C^{(k)T} x^{(k)} / v^{(k)} \tag{14}$$

Here $v^{(k)}$ defined by:

$$v^{(k)} = \max_{x^{(k)} \in R^{(k)}} C^{(k)T} x^{(k)}$$

$$R^{(k)} = \{x^{(k)} \mid A^{(k)} x^{(k)} = b^{'(k)}, x^{(k)} \geq 0\} \tag{15}$$

And
$$\begin{cases} b^{'(k)} = (b_1^{(k)} + d'_1{}^{(k)}, b_2^{(k)} + d'_2{}^{(k)}, \ldots b_m^{(k)} + d'_m{}^{(k)})^T \\ d'_a{}^{(k)} = \max\{d_i^{(k)} \mid d_a^{(k)} \in \sup p\mu_i\} \quad i = 1,2,\ldots m \end{cases}$$

The $\sup p\mu_i$ is the closure of set $\{d_i^{(k)} \mid \mu_i(d_i^{(k)}) \neq 0\}$.

When level $\alpha^{(k)}$ was given, $\alpha^{(k)}$--level set $C_\varepsilon(k)$ can be expressed as:

$$C_\varepsilon^{(k)} = \{x^{(k)} \mid A^{(k)} x^{(k)} = b^{\alpha^{(k)}}, x^{(k)} \geq 0\} \tag{16}$$

Here
$$\begin{cases} b^{\alpha^{(k)}} = (b_1^{(k)} + d^{\alpha b\alpha^{(k)}}{}_1, b_2^{(k)} + d^{\varepsilon b\alpha^{(k)}}{}_2, \ldots b_m^{(k)} + d^{\alpha b\alpha^{(k)}}{}_m)^T \\ b^{\alpha^{(k)}} = \max\{d_i^{(k)} \mid d_a^{(k)} \in N_\alpha(k)(\mu_i)\} \quad i = 1,2,\ldots m \\ N_\alpha(k)(\mu_i) = \{d_i^{(k)} \mid \mu_i(d_i^{(k)}) \geq \alpha^{(k)}\} \quad i = 1,2,\ldots m \end{cases}$$

Compare to $\alpha^{(k)}$, level set $C_\varepsilon(k)$, $\max_{x^{(k)} \in C\alpha(k)} \varphi_G^{(k)}(x^{(k)})$ can be solved by general LP method. So we can deduce the algorithm of variation coefficients LP problem as following steps:

1) Let $l = 1$, then suppose both $\alpha_1^{(k)}$, $\forall \varepsilon^{(k)} > 0$ and $0 \leq \alpha_1^{(k)} \leq 1$;

2) Calculating $\varphi_G^{(k)l} = \max_{x^{(k)} \in C_{\alpha_l^{(k)}}} \varphi^{(k)T} \cdot x^{(k)} = \max_{x^{(k)} \in C_{\alpha_l^{(k)}}} C^{(k)T} \cdot x^{(k)} / v^{(k)}$

3) Calculating $\varepsilon_l^{(k)} = \alpha_l^{(k)} - \varphi_G^{(k)}$, if $|\varepsilon_l^{(k)}| \geq \varepsilon^{(k)}$, then turn to step 4). If $|\varepsilon_l^{(k)}| \leq \varepsilon^{(k)}$, then turn to step 5).

4) Suppose $\alpha_{l+1}^{(k)} = \alpha_l^{(k)} - \gamma_l^{(k)} \cdot \varepsilon_l^{(k)}$, set $l = l+1$, then turn to the step 2).
There into, when selecting, it must satisfy $0 \leq \alpha_{l+1}^{(k)} \leq 1$;

5) Set $\overline{\alpha}^{(k)} = \alpha_l^{(k)}$, evaluate $\overline{x}^{(k)}$. If $\overline{x}^{(k)}$ achieves $\varphi_G^{(k)}(\overline{x}^{(k)}) = \max_{x^{(k)} \in C_\alpha^{(k)}} \varphi_G^{(k)}(x^{(k)})$, then $\overline{x}^{(k)}$ is the optimization solution. And the calculation ends.

By all appearances, hereinbefore the step 2) and 5) in the arithmetic is the ecumenical linear planning with altering coefficient.

4 The Applied Example About Produces Scheduling

When establishing the optimization programming about produce at tempering of one corporation, handling the fuzzy linear planning method as above to deal with the problems can be taken into account because there are some conditions, which are fuzzy. The question is: how to reasonably plan every production's output amount of each year to satisfy the conditions as above and make the total production value to be maximized?

Setting: $x_j^{(k)}$ is the output of production A_j in the k year (calculated according to standard output's amount); $a_j^{(k)}, b_j^{(k)}$ shows the output's upper and lower limit of production A_j separately in the k year; $c_j^{(k)}$ is production A_j's production value of standard output's amount in the k year; $k = 1,2,3$; $j = 1,2,3,4,5$.

Then according to the practical problem and the fuzzy linear planning method with altering coefficient, can be reduced to the fuzzy linear planning problem with altering coefficient as follows:

Evaluating the nonnegative variable of three groups:

$$x_1^{(k)}, x_2^{(k)}, x_3^{(k)}, x_4^{(k)}, x_5^{(k)} \quad (k = 1,2,3)$$

Making they to satisfy:

St. $\begin{cases} x_j^{(k)} \leq a_j^{(k)} & j = 1,2,3,4,5 \\ x_j^{(k)} \geq b_j^{(k)} & j = 1,2,3,4,5 \\ -x_1^{(k)} + 10.47\, x_2^{(k)} = 0 \\ -x_2^{(k)} + 1.375\, x_3^{(k)} = 0 \\ -x_3^{(k)} + 0.935\, x_4^{(k)} = 0 \\ -x_4^{(k)} + 0.25\, x_5^{(k)} = 0 \\ -x_5^{(k)} + 0.455\, x_1^{(k)} = 0 \end{cases}$ (17)

And maximizing the value of the aim function

$$F^{(k)} = \sum_{j=1}^{5} c_j^{(k)} x_j^{(k)} \quad (k=1,2,3). \tag{18}$$

To the mathematics model as above, the subject function of the fuzzy restraint (there are 10 in all) $\mu_i (i = 1,2,\cdots,10)$ are arranged in the order of priority. Can consider so: the multipliable quantity $d (\geq 0)$ of the right value in the fuzzy restraint condition can be defined as:

$$\mu_1^{(k)}(d) = \mu_5^{(k)}(d) = \mu_8^{(k)}(d) = \mu_9^{(k)}(d) = \begin{cases} 1-d & 0 \leq d \leq 1 \\ 0 & d \geq 1 \end{cases} \tag{19}$$

$$\mu_2^{(k)}(d) = \mu_3^{(k)}(d) = \mu_4^{(k)}(d) = \begin{cases} 1-2d & 0 \leq d \leq 0.5 \\ 0 & d \geq 0.5 \end{cases} \tag{20}$$

$$\mu_6^{(k)}(d) = \mu_7^{(k)}(d) = \mu_{10}^{(k)}(d) = \begin{cases} 1-0.2d & 0 \leq d \leq 5 \\ 0 & d \geq 5 \end{cases} \tag{21}$$

If effortful expanding the limit retrained to make the subject function's value of the function (19)-(21) is 0, then d separately is 1, 0.5 or 5. So calculating the fuzzy aim $\varphi_G^{(k)}$, making:

$$\varphi_G^{(k)} = \sum_{j=1}^{5} \beta_j^{(k)} x_j^{(k)} \quad k=1,2,3 \tag{22}$$

There into, the values of the variation coefficient $\beta_j^{(k)} (j=1,2,3,4,5)$ are as follows Table 1:

Table 1. The values of the variation coefficient $\beta_j^{(k)}$

k	$\beta_1^{(k)}$	$\beta_2^{(k)}$	$\beta_3^{(k)}$	$\beta_4^{(k)}$	$\beta_5^{(k)}$
1	0.0222	0.1260	0.0467	0.0178	0.0196
2	0.0149	0.0750	0.0629	0.0105	0.0124
3	0.0118	0.0666	0.0558	0.0073	0.0079

Set $\varepsilon^{(k)} = 0.01$, $\alpha_1^{(k)} = 0.9$, $\gamma_l^{(k)} = 0.5$, repeatedly computing by the preceding algorithm, at last we can receive the optimization solution $\overline{x}^{(k)} = (\overline{x}_1^{(k)}, \overline{x}_2^{(k)}, \overline{x}_3^{(k)}, \overline{x}_4^{(k)}, \overline{x}_5^{(k)})^T$ when $k=1,2,3$.

For example, when $k=1$, the function (22) is:

$$x_G^{(1)} = 0.0222x_1^{(1)} + 0.01260x_2^{(1)} + 0.0467x_3^{(1)} + 0.0178x_4^{(1)} + 0.0196x_5^{(1)}$$

(23)

Set: $\alpha_1^{(1)} = 0.9$, $\varepsilon^{(1)} = 0.01$, $\gamma_l^{(1)} = 0.5$, $l = 1,2,3,\cdots$. According to the algorithm, receiving the result: $x_G^{(1)} = 1.0025$, here:

$$\left|\varepsilon_1^{(1)}\right| = |0.9 - 1.0025| = |-0.1025| = 0.1025 > 0.01 = \varepsilon^{(1)}$$

(24)

So setting

$$\alpha_2^{(1)} = 0.9 - 0.5 \times (-0.1025) = 0.9 + 0.0513 = 0.9513,$$

Receiving the result again: $x_G^{(1)2} = 0.9988$, here:

$$\left|\varepsilon_2^{(1)}\right| = |0.9513 - 0.9988| = |-0.0475| = 0.0475 > 0.01 = \varepsilon^{(1)}$$

(25)

So setting again $\alpha_3^{(1)} = 0.9513 - 0.5 \times (-0.0475) = 0.9750,\cdots$, going on like this, and repeating the preceding algorithm, till receiving the result $x_G^{(1)5} = 1.0000$ when $\alpha_5^{(1)} = 0.9941$. Here there is:

$$\left|\varepsilon_5^{(1)}\right| = |0.9941 - 1.0000| = 0.0059 < 0.01 = \varepsilon^{(1)}$$

(26)

Then choosing the level $\overline{\alpha}^{(1)} = \alpha_5^{(1)} = 0.9941$ (i.e. the optimization level), then the optimum solution is reached

$$\overline{x}^{(1)} = (20.33, 2.01, 2, 1.98, 8.51)^T$$

Similarly, when $k = 2,3$, the solutions are:

$$\overline{x}^{(2)} = (28.8, 2.85, 2.83, 2.83, 12.06)^T$$
$$\overline{x}^{(3)} = (36, 3.56, 3.54, 3.51, 15.08)^T$$

5 Conclusion

The proposed fuzzy variation coefficients programming has found application in operations management and it will find various application in other areas. In other words, it can be applied to a class of time-varying systems in which conventional programming techniques have been used for many times (years, months, weeks).

References

1. Zimmerman.H.J. 1978. Fuzzy programming and linear programming with sereral objective functions. Fuzzy Sets and Systems (1), 45-55.
2. Weber.C.A, Current.J.R. 1993. A multi-objective approach to vender selection. European Journal of Operational Research (68), 173-184.
3. Tony, P., John, K., Kevin, W., 2004, Improving Operations Management Concept Recollection Via the Zarco Experiential Learning Activity. Journal of Education for Business; 79 (5): 283-287.
4. Geoff, R., 2004, Maintaining the balance. Manufacturing Engineer. 83 (3): 45-47.
5. Stockton, D., J., Quinn, L., Khalil, R., A., 2004, Use of genetic algorithms in operations management Part 1: applications; Part 2: results. Proceedings of the Institution of Mechanical Engineers -- Part B -- Engineering Manufacture; 218 (3): 315-344.
6. Laqoudis, I., N., Lalwani, C., S., Naim, M., M., 2004, A Generic Systems Model for Ocean Shipping Companies in the Bulk Sector. Transportation Journal; 43 (1): 56-77.
7. Tang.B.Y. 1985. Mathematical programming with altering coefficient and its application in farming-forest structure. Journal of Heilongjiang University (Nature Sci.) 2(2),24-30.
8. Tang.B.Y. 1990. Environmental systems engineering method. China Environmental Science Press, Beijing.
9. Tang.B.Y. 1985 Fuzzy Linear programming with altering coefficient and its application in raise-birds structure optimize programming. Fuzzy Mathematics. 5(1),87-94.

Weighted Possibilistic Variance of Fuzzy Number and Its Application in Portfolio Theory

Xun Wang[1], Weijun Xu[1], Weiguo Zhang[1,2], and Maolin Hu[3]

[1] School of Management, Xi'an Jiaotong University, Xi'an, 710049, China
{wangxun_xjtu, xuweijun75}@163.com
[2] College of Business, South China University of Technology,
Guangzhou, 510641, China
[3] Department of Mathematics, Guyuan Teachers college, Guyuan, 756000, China

Abstract. Dubois and Prade defined an interval-valued expectation of fuzzy numbers, viewing them as consonant random sets. Fullér and Majlender then proposed an weighted possibility mean value, variance and covariance of fuzzy numbers, viewing them as weighted possibility distributions. In this paper, we define a new weighted possibilistic variance and covariance of fuzzy numbers based on Fullér and Majlenders' notations. Some properties of these notations are obtained in a similar manner as in probability theory. We also consider the weighted possibilistic mean-variance model of portfolio selection and introduce the notations of the weighted possibilistic efficient portfolio and efficient frontier. Moreover, a simple example is presented to show the application of our results in security market.

1 Introduction

In 1987, Dubois and Prade defined an interval-valued expectation of fuzzy numbers as a closed interval bounded by the expectations calculated from its upper and lower distribution funtions. They also showed that this expectation remains additive in the sense of addition of fuzzy numbers [3]. In 2001, Carlsson and Fullér introduced an interval-valued mean value of fuzzy numbers, viewing them as possibility distributions [2]. In 2003, Fullér and Majlender proposed an weighted possibility mean value of fuzzy numbers, viewing them as weighted possibility distributions [6]. Moreover, Zhang considered alternative notations of the possibilistic variance and covariance of fuzzy numbers. In fact, Zhang's notations are the extension of Carlsson and Fullérs' mean values and variances [9].

In this paper, we develop the notations of the weighted possibilistic variances and covariances of fuzzy numbers based on Fullér and Majlenders' notations in Section 2, and some properties of these notations can be discussed in a similar manner as in probability theory. Section 3 uses these notations to build the weighted possibilistic model of portfolio selection. Since the future returns and risks of assets cannot be predicted accurately, the fuzzy number is a powerful tool used to describe an uncertain environment with vagueness and ambiguity. Moreover, a simple example is given to illustrate the application of our results in security market. Finally, we conclude this paper in Section 4.

2 Weighted Possibilistic Mean and Variance

A fuzzy number A is a fuzzy set of the real line $\mathcal{R}$ with a normal, fuzzy convex and continuous membership function of bounded support. The family of fuzzy numbers will be denoted by $\mathcal{F}$. Let $A \in \mathcal{F}$ be fuzzy number with $[A]^\gamma = [a_1(\gamma), a_2(\gamma)], \gamma \in [0,1]$. A γ-level set of a fuzzy number A is defined by $[A]^\gamma = \{t \in \mathcal{R} | A(t) \geq \gamma\}$ if $\gamma > 0$ and $[A]^\gamma = cl\{t \in \mathcal{R} | A(t) > 0\}$ (the closure of the support of A) if $\gamma = 0$. It is well known that if A is a fuzzy number then $[A]^\gamma$ is a compact subset of $\mathcal{R}$ for all $\gamma \in [0,1]$. Moreover, a function $f : [0,1] \to \mathcal{R}$ is said to be a weighting function if f is non-negative, monotone increasing and satisfies the normalization condition $\int_0^1 f(\gamma)d\gamma = 1$.

Fullér and Majlender defined the notations of the f-weighted lower and upper possibilistic mean values of a fuzzy number A [6]. They defined the f-weighted possibilistic mean value of A as

$$\overline{M}_f(A) = \int_0^1 f(\gamma)\frac{a_1(\gamma) + a_2(\gamma)}{2}d\gamma = \frac{M_f^L(A) + M_f^U(A)}{2},$$

where

$$M_f^L(A) = \int_0^1 f(\gamma)a_1(\gamma)d\gamma = \frac{\int_0^1 a_1(\gamma)f(Pos[A \leq a_1(\gamma)])d\gamma}{\int_0^1 f(Pos[A \leq a_1(\gamma)])d\gamma},$$

$$M_f^U(A) = \int_0^1 f(\gamma)a_2(\gamma)d\gamma = \frac{\int_0^1 a_2(\gamma)f(Pos[A \geq a_2(\gamma)])d\gamma}{\int_0^1 f(Pos[A \geq a_2(\gamma)])d\gamma},$$

where Pos denotes possibility, i.e.,

$$Pos[A \leq a_1(\gamma)] = \Pi((-\infty, a_1(\gamma)]) = \sup_{u \leq a_1(\gamma)} A(u) = \gamma,$$

$$Pos[A \geq a_2(\gamma)] = \Pi([a_2(\gamma), +\infty]) = \sup_{u \geq a_2(\gamma)} A(u) = \gamma.$$

Then the following lemma can directly be proved using the definition of f-weighted interval-valued possibilistic mean [6].

Lemma 1. *Let $A \in \mathcal{F}$, and let θ, λ be two real numbers. Then*

$$\overline{M}_f(A + \theta) = \overline{M}_f(A) + \theta, \quad \overline{M}_f(\lambda A) = \lambda \overline{M}_f(A),$$

where the addition and multiplication by a scalar of fuzzy numbers is defined by the sup-min extension principle [4].

We will now introduce the notations of the weighted possibilistic variance and covariance of fuzzy numbers.

Definition 1. *The weighted possibilistic variance of $A \in \mathcal{F}$ is defined as*

$$\overline{Var}_f(A) = \int_0^1 f(\gamma)([M_f^L(A) - a_1(\gamma)]^2 + [M_f^U(A) - a_2(\gamma)]^2)d\gamma. \tag{1}$$

Note that the equation (1) can be also rewritten as

$$\overline{Var}_f(A) = \int_0^1 f(\gamma)([M_f^L(A) - a_1(\gamma)]^2 + [M_f^U(A) - a_2(\gamma)]^2)d\gamma$$

$$= \int_0^1 f(\gamma)(a_1^2(\gamma) + a_2^2(\gamma))d\gamma - [M_f^{L^2}(A) + M_f^{U^2}(A)].$$

Definition 2. *The weighted possibilistic covariance of $A, B \in \mathcal{F}$ is defined as*

$$\overline{Cov}_f(A, B) = \int_0^1 f(\gamma)[(M_f^L(A) - a_1(\gamma))(M_f^L(B) - b_1(\gamma)) + (M_f^U(A) - a_2(\gamma))(M_f^U(B) - b_2(\gamma))]d\gamma.$$

The following theorem shows some important properties of the weighted possibilistic variance and covariance.

Theorem 1. *Let $A \in \mathcal{F}$, and let θ, λ be two real numbers. Then*

$$\overline{Var}_f(A + \theta) = \overline{Var}_f(A), \quad \overline{Var}_f(\lambda A) = \lambda^2 \overline{Var}_f(A).$$

Proof. Let $[A]^\gamma = [a_1(\gamma), a_2(\gamma)], \gamma \in [0, 1]$. From the relationship

$$[A + \theta]^\gamma = [a_1(\gamma) + \theta, a_2(\gamma) + \theta],$$

we get

$$M_f^L(A + \theta) = \int_0^1 f(\gamma)(a_1(\gamma) + \theta)d\gamma = M_f^L(A) + \theta,$$

$$M_f^U(A + \theta) = \int_0^1 f(\gamma)(a_2(\gamma) + \theta)d\gamma = M_f^U(A) + \theta.$$

Thus,
$$\overline{Var}(A + \theta) = \overline{Var}(A).$$

According to the relationship

$$[\lambda A]^\gamma = \lambda[A]^\gamma = \lambda[a_1(\gamma), a_2(\gamma)] = \begin{cases} [\lambda a_1(\gamma), \lambda a_2(\gamma)] & \text{if } \lambda \geq 0, \\ [\lambda a_2(\gamma), \lambda a_1(\gamma)] & \text{if } \lambda < 0, \end{cases}$$

we get that

$$M_f^L(\lambda A) = \begin{cases} \lambda M_f^L(A) & \text{if } \lambda \geq 0, \\ \lambda M_f^U(A) & \text{if } \lambda < 0, \end{cases}$$

$$M_f^U(\lambda A) = \begin{cases} \lambda M_f^U(A) & \text{if } \lambda \geq 0, \\ \lambda M_f^L(A) & \text{if } \lambda < 0. \end{cases}$$

Hence,
$$\overline{Var}_f(\lambda A) = \lambda^2 \overline{Var}_f(A).$$

The proof of theorem is completed.

We show that the variance of a fuzzy number is invariant to shifting. The following theorem can be proved in a similar way as Theorem 1.

Theorem 2. *Let f be a weighting function, let A, B, C be fuzzy numbers and let x and y be real numbers. Then the following properties hold*

$$\overline{Cov}_f(A,B) = \overline{Cov}_f(B,A), \ \overline{Cov}_f(A,A) = \overline{Var}_f(A),$$
$$\overline{Cov}_f(x,A) = 0, \ \overline{Var}_f(x) = 0.$$

The following theorem shows that the possibilistic variance of linear combination of fuzzy numbers can easily be computed.

Theorem 3. *Let A and B be two fuzzy numbers and let $\lambda, \mu \in \mathcal{R}$. Then*

$$\overline{Var}_f(\lambda A + \mu B) = \lambda^2 \overline{Var}_f(A) + \mu^2 \overline{Var}_f(B) + 2|\lambda\mu|\overline{Var}_f(\phi(\lambda)A, \phi(\mu)B),$$

where $\phi(x)$ is a sign function of $x \in \mathcal{R}$.

Proof. Let $[A]^\gamma = [a_1(\gamma), a_2(\gamma)]$ and $[B]^\gamma = [b_1(\gamma), b_2(\gamma)], \gamma \in [0,1]$. Suppose $\lambda < 0$ and $\mu < 0$. Then

$$[\lambda A + \mu B]^\gamma = [\lambda a_2(\gamma) + \mu b_2(\gamma), \lambda a_1(\gamma) + \mu b_1(\gamma)].$$

We obtain

$$\overline{Var}_f(\lambda A + \mu B) = \int_0^1 f(\gamma)(M_f^L(\lambda A + \mu B) - \lambda a_2(\gamma) - \mu b_2(\gamma))^2 d\gamma$$
$$+ \int_0^1 f(\gamma)(M_f^U(\lambda A + \mu B) - \lambda a_1(\gamma) - \mu b_1(\gamma))^2 d\gamma$$
$$= \int_0^1 f(\gamma)(\lambda M_f^U(A) + \mu M_f^U(B) - \lambda a_2(\gamma) - \mu b_2(\gamma))^2 d\gamma$$
$$+ \int_0^1 f(\gamma)(\lambda M_f^L(A) + \mu M_f^L(B) - \lambda a_1(\gamma) - \mu b_1(\gamma))^2 d\gamma$$
$$= \lambda^2 \overline{Var}_f(A) + \mu^2 \overline{Var}_f(B) + 2\lambda\mu \overline{Cov}_f(A,B)$$
$$= \lambda^2 \overline{Var}_f(A) + \mu^2 \overline{Var}_f(B) + 2\lambda\mu \overline{Cov}_f(-A,-B),$$

that is,

$$\overline{Var}_f(\lambda A + \mu B) = \lambda^2 \overline{Var}_f(A) + \mu^2 \overline{Var}_f(B) + 2|\lambda\mu|\overline{Cov}_f(\phi(\lambda)A, \phi(\mu)B),$$

where

$$\phi(x) = \begin{cases} 1 & \text{if } x > 0, \\ 0 & \text{if } x = 0, \\ -1 & \text{if } x < 0. \end{cases}$$

Similar reasoning holds for the case $\lambda \geq 0$ and $\mu \geq 0$.

Suppose now that $\lambda > 0$ and $\mu < 0$. Then
$$[\lambda A + \mu B]^\gamma = [\lambda a_1(\gamma) + \mu b_2(\gamma), \lambda a_2(\gamma) + \mu b_1(\gamma)].$$

We get
$$\begin{aligned}\overline{Var}_f(\lambda A + \mu B) &= \int_0^1 f(\gamma)(M_f^L(\lambda A + \mu B) - \lambda a_1(\gamma) - \mu b_2(\gamma))^2 d\gamma \\ &+ \int_0^1 f(\gamma)(M_f^U(\lambda A + \mu B) - \lambda a_2(\gamma) - \mu b_1(\gamma))^2 d\gamma \\ &= \int_0^1 f(\gamma)(\lambda M_f^L(A) + \mu M_f^U(B) - \lambda a_1(\gamma) - \mu b_2(\gamma))^2 d\gamma \\ &+ \int_0^1 f(\gamma)(\lambda M_f^U(A) + \mu M_f^L(B) - \lambda a_2(\gamma) - \mu b_1(\gamma))^2 d\gamma \\ &= \lambda^2 \overline{Var}_f(A) + \mu^2 \overline{Var}_f(B) + 2|\lambda\mu| \overline{Cov}_f(\phi(\lambda)A, \phi(\mu)B).\end{aligned}$$

Similar reasoning holds for the case $\lambda < 0$ and $\mu \geq 0$, which ends the proof.

Let A_i $(i = 1, \ldots, n)$ be fuzzy numbers. Set
$$b_{ij} = \overline{Cov}_f(A_i, A_j), \; i, j = 1, \ldots, n.$$

The matrix $\overline{Cov}_f = (b_{ij})_{n \times n}$ is called as possibilistic covariance matrix of $(A_1, A_2, \ldots, A_n)$. The following theorem shows that $\overline{Cov}_f = (b_{ij})_{n \times n}$ has the same properties as the covariance matrix in probability theory.

Theorem 4. *Let A_i $(i = 1, \ldots, n)$ be fuzzy numbers. Then $\overline{Cov}_f = (b_{ij})_{n \times n}$ is nonnegative definite matrix.*

Proof. Let $[A_i]^\gamma = [a_{i1}(\gamma), a_{i2}(\gamma)]$ $(i = 1, \ldots, n)$. Then
$$b_{ii} = \overline{Var}_f(A_i), \; b_{ij} = b_{ji}, \; i \neq j = 1, 2, \ldots, n.$$

For any $t_i \in \mathcal{R}$ $(i = 1, \ldots, n)$,
$$\begin{aligned}\sum_{i=1}^n \sum_{j=1}^n b_{ij} t_i t_j &= \sum_{i=1}^n \sum_{j=1}^n \int_0^1 f(\gamma)([M_f^L(A_i) - a_{i1}(\gamma)][M_f^L(A_j) - a_{j1}(\gamma)] \\ &+ [M_f^U(A_i) - a_{i2}(\gamma)][M_f^U(A_j) - a_{j2}(\gamma)]) t_i t_j d\gamma \\ &= \int_0^1 f(\gamma) \Big[\sum_{i=1}^n t_i(M_f^L(A_i) - a_{i1}(\gamma))\Big]^2 d\gamma \\ &+ \int_0^1 f(\gamma) \Big[\sum_{i=1}^n t_i(M_f^U(A_i) - a_{i2}(\gamma))\Big]^2 d\gamma \\ &\geq 0.\end{aligned}$$

This concludes the proof of theorem.

3 Weighted Possibilistic Portfolio Analysis

Assume that there are n risky assets available in markets. Let $r = (\overline{M}_f(r_1), \ldots, \overline{M}_f(r_n))^T$ be the vector of mean return rates where $\overline{M}_f(r_i) = \int_0^1 f(\gamma) \frac{a_{1i}(\gamma) + a_{2i}(\gamma)}{2} d\gamma$ is the weighted possibilistic mean of fuzzy number r_i that means the expected return rate of asset i. Let $x = (x_1, \ldots, x_n)^T$ be the investment weights of a portfolio to the n assets. Then the weighted possibilistic mean value of r is given by $r^T x$. $\overline{Cov}_f = (b_{ij})_{n \times n}$ is the $n \times n$ variance-covariance matrix among the return fuzzy numbers of n assets, where $b_{ij} = \overline{Cov}_f(r_i, r_j)$ is the weighted possibilistic covariance between the return fuzzy numbers of the i-th and j-th assets, and $b_{ii} = \overline{Var}_f(r_i)$ is the weighted possibilistic variance of the return fuzzy number of asset i. The weighted possibilistic variance of portfolio r is then given by $x^T \overline{Cov}_f x$.

Thus, if the weighted possibilistic variance-covariance matrix $\overline{Cov}_f$ is positive definite, the weighted possibilistic M–V portfolio model with short sales may be described by

$$\begin{aligned}\min \quad & \overline{Var}_f(x) = x^T \overline{Cov}_f x \\ \text{s.t} \quad & r^T x = \overline{r}_E \\ & e^T x = 1,\end{aligned} \quad (2)$$

where $\overline{Var}_f(x)$ is the weighted possibilistic risk (variance) of the portfolio x, $\overline{r}_E$ is the goal of expected returns for the investment, and e is the n-dimensional vector with all elements being one.

Similar to Markovitz portfolio model in [1] and [7], we can also obtain that the unique global optimal portfolio of (2) is

$$x^* = \overline{Cov}_f^{-1}(r, e) A^{-1} \begin{pmatrix} \overline{r}_E \\ 1 \end{pmatrix},$$

and is called the weighted possibilistic efficient M–V portfolio with goal $\overline{r}_E$, and

$$A = \begin{bmatrix} c & b \\ b & a \end{bmatrix}, \quad a = e^T \overline{Cov}_f^{-1} e, \quad b = e^T \overline{Cov}_f^{-1} r, \quad c = r^T \overline{Cov}_f^{-1} r.$$

The corresponding minimum weighted possibilistic risk is obtained by substituting x^* into the objective function of (2),

$$\overline{Var}_f(x^*) = \frac{a \overline{r}_E^2 - 2 b \overline{r}_E + c}{\Delta}, \quad (3)$$

where $\Delta = ac - b^2 > 0$, because of the positive definiteness of the matrix $\overline{Cov}_f$. In $\overline{Var}_f - \overline{r}_E$ plane, the equation (3) describes a hyperbola and the right-branch gives the boundary for feasible portfolios. The top half of the boundary is called the weighted possibilistic efficient frontier, and the low half is the inefficient frontier, while the extreme point G gives the portfolio that has the global minimum weighted possibilistic variance and the global minimum weighted possibilistic return rate,

$$x_G = \frac{\overline{Cov_f}^{-1} e}{a}, \quad r_G = \frac{b}{a}, \quad \overline{Var}_{fG} = \frac{1}{a}.$$

x_G is called the global minimum weighted possibilistic variance portfolio.

Now, we will give a simple example to illustrate the application of our results in the security market as follows. Four stocks in the security market are only considered with allowing short sales. Let r_j be an LR-type fuzzy number that means the return rate of security S_j ($j = 1, 2, 3, 4$). Similarly in [6], let the fuzzy number $r_j = (r_{j-}, r_{j+}, \alpha, \beta)$ be a fuzzy number of trapezoidal form with peak $[a, b]$, left width $\alpha > 0$ and right width $\beta > 0$, and let $f(\gamma) = 3\gamma^2$. A γ-level sets of r_j is computed by

$$[r_j]^\gamma = [r_{j-} - \alpha(1-\gamma), r_{j+} + \beta(1-\gamma)], \forall \gamma \in [0, 1].$$

Then we can compute out the f-weighted possibilistic mean values, variances, and covariances as follows.

stock j	return rate $r_j\%$	$M_f^L(r_j)$	$M_f^U(r_j)$	$\overline{M}_f(r_j)$	$\overline{Var}_f(r_j)$
1	[11−2.0, 12+1.2]	10.5	12.3	11.4	0.204
2	[13−1.5, 14+2.5]	12.625	14.625	13.625	0.31875
3	[12−1.0, 13+1.0]	11.75	13.25	12.5	0.075
4	[16−0.5, 17+1.5]	15.875	17.375	16.625	0.09375

and

$\overline{Cov}_f(r_1, r_2) = 0.225, \quad \overline{Cov}_f(r_1, r_3) = 0.12, \quad \overline{Cov}_f(r_1, r_4) = 0.015,$

$\overline{Cov}_f(r_2, r_3) = 0.15, \quad \overline{Cov}_f(r_2, r_4) = 0.16875, \quad \overline{Cov}_f(r_3, r_4) = 0.075.$

Thus, we can obtain the portfolio solution $x = (-1.0762, 0.0103, -0.4969, 1.5628)$.

In the financial market with uncertain factors, real world problems are not usually so easily formulated as accuratedly mathematical forms. Moreover, the investors' or experts' judgment in data analysis is very important, while the weighting function can effectively characterize their knowledge. Based on this fact, the weighted possibility analysis methods in the portfolio selection will be suitable for such an attempt as shown in this paper.

4 Summary

In this paper, we have introduced the notations of the weighted possibilistic variance and covariance of fuzzy numbers, which are consistent with the extension principle and the well-known definitions of variance and covariance in probability theory. Moreover, we use these notations to build the weighted possibilistic model of portfolio selection. Since the returns of risky assets are in a fuzzy uncertain economic environment and vary from time to time, the fuzzy number is a powerful tool used to describe an uncertain environment with vagueness and ambiguity.

Acknowledgement

The authors are grateful to the referees for their valuable comments and suggestions. Moreover, the last author would like to acknowledge the support No. 2004070 from the Fund of Higher Education Science Research Project of Ningxia in China.

References

1. Best, M.J., Hlouskova, J.: The efficient frontier for bounded assets. Mathematics of Operations Research. **52** (2000) 195–212
2. Carisson, C., Fullér, R.: On possibilistic mean value and variance of fuzzy numbers. Fuzzy Sets and Systems. **122** (2001) 315–326
3. Dubois, D., Prade, H.: The mean value of a fuzzy number. Fuzzy Sets and Systems. **24** (1987) 279–300
4. Dubois, D., Prade, H.: What are fuzzy rules and how to use them. Fuzzy Sets and Systems. **84** (1996) 169–185
5. Dubois, D., Prade, H., Sabbadin, R.: Decision-theoretic foundations of qualitative possibility theory. European Journal of Operational Research. **128** (2001) 459–478
6. Fullér, R., Majlender, P.: On weighted possibilistic mean and variance of fuzzy numbers. Fuzzy Sets and Systems. **136** (2003) 363–374
7. Markovitz H.: Portfolio selection: Efficient diversification of investment. Wiley, New York. (1959)
8. Yager, R.: On the instantiation of possibility distributions. Fuzzy Sets and Systems. **128** (2002) 261–266
9. Zhang, W., Nie, Z.: On possibilistic variance of fuzzy numbers. In: Wang, G., Liu, Q., Yao, Y., Skowron, A. (eds.): Rough Sets, Fuzzy Sets, Data Mining, and Granular Computing. Lecture Notes in Computer Science, Vol. 2639. Springer-Verlag, Berlin Heidelberg New York (2003) 398–402

Another Discussion About Optimal Solution to Fuzzy Constraints Linear Programming

Yun-feng Tan[1] and Bing-yuan Cao[1,2]

[1] Dept of Math., Shantou University, Shantou 515063 P. R. China
[2] School of Math. and Inf. Sci., Guangzhou University, Guangzhou 510405 China

Abstract. In this paper, we focus on the fuzzy constraints linear programming. First we discuss the properties of an optimal solution vector and of an optimal value in the corresponding parametric programming, and propose a method to the critical values. Then we present a new algorithm to the fuzzy constraint linear programming by associating an object function with the optimal value of parametric programming.

1 Introduction

The normal form of a linear programming problem with fuzzy constraint is

$$(\widetilde{\text{LP}}) \quad \begin{aligned} \max\ & Z = CX \\ \text{s.t.}\ & AX \lesssim b, \\ & X \geqslant 0, \end{aligned}$$

where $A = (a_{ij})_{m \times n}$ is an $m \times n$ matrix, b is an m-dimensional vector, C and X are n-dimensional vectors. Let rank$(A)=m$. "$\lesssim$" denotes the fuzzy version of "$\leqslant$" and has the linguistic interaction "essentially smaller than or equal to"[1].

The representative method to $(\widetilde{\text{LP}})$ is to turn it into a classical linear programming [2]. We will try to explain why its fuzzy decision usually is 0.5 found by Researchers [3][4][5], but we shall propose another algorithm to $(\widetilde{\text{LP}})$.

2 Analysis of Fuzzy LP

We first discuss the properties of an optimal value Z_λ in

$$(\text{LP}_\lambda) \quad \begin{aligned} \max\ & Z = CX \\ \text{s.t.}\ & AX \leqslant b + (1-\lambda)d, \\ & X \geqslant 0, \end{aligned}$$

where λ is a parameter on the interval [0,1], and $d \geqslant 0$. In this paper, X_λ denotes the optimal solution to (LP_λ), B_λ and Z_λ the optimal basis matrix and the optimal value of (LP_λ), respectively. The right side coefficient $b + (1-\lambda)d$ of the constraint condition in (LP_λ) will vary with the changing of parameter λ. The optimal solution to the parametric linear programming (LP_λ) is $B_\lambda^{-1}(b+(1-\lambda)d)$.

If we solve (LP$_\lambda$) by using a simplex method, there is no relationship between discriminate number $\sigma = C_N - C_B B^{-1} N$ and parameter λ, so the variation of an optimal basis matrix is decided only by X_λ.

Definition 1. *Let B be one of the optimal basic matrix of (LP$_\lambda$). If an interval $[\lambda_1, \lambda_2]$ exists, satisfying that B is an optimal basic matrix of (LP$_\lambda$)($\forall \lambda \in [\lambda_1, \lambda_2]$) while B is not an optimal matrix for each $\lambda \bar{\in} [\lambda_1, \lambda_2]$, we call λ_1 and λ_2 critical values of (LP$_\lambda$) and $[\lambda_1, \lambda_2]$ characteristic interval.*

Theorem 1. *(LP$_\lambda$) has finite characteristic interval on the interval $[0, 1]$.*

Proof. Let us assume B is an optimal basis matrix of LP$_\lambda$, and there are two characteristic intervals $[\lambda_{i-1}, \lambda_i]$ and $[\lambda_{i+1}, \lambda_{i+2}]$,($\lambda_i < \lambda_{i+1}$) corresponding to B. The optimal solution to LP$_\lambda$ is $(X_B, X_N)^T$, where $\lambda \in [\lambda_{i-1}, \lambda_i] \cup [\lambda_{i+1}, \lambda_{i+2}]$, $X_B = B^{-1}(b + (1-\lambda)d) \geq 0$, $X_N = 0$, $\lambda_i < \lambda_{i+1}$. So $X_B = B^{-1}(b + (1-\lambda)d) \geq 0$ when $\lambda \in [\lambda_i, \lambda_{i+1}]$, this means an optimal matrix of LP$_\lambda$ is also B on the interval $[\lambda_i, \lambda_{i+1}]$. Therefore the characteristic interval where the optimal matrix keeps invariant is $[\lambda_{i-1}, \lambda_{i+2}]$. So the optimal matrix has only one corresponding characteristic interval. Because the coefficient matrix of LP$_\lambda$ keeps invariant on the interval $[0,1]$, and an optimal matrix is finite, the number of a characteristic intervals is finite.

Theorem 2. *Let B be the optimal basis matrix of (LP$_\lambda$) on a characteristic interval $[\lambda_1, \lambda_2]$. If $(B^{-1}d)_i \neq 0, \forall i \in \{1, \ldots, m\}$, then we can obtain*

$$\lambda_1 = \max\left[\frac{[B^{-1}(b+d)]_i}{(B^{-1}d)_i},\ 0\ |\ (B^{-1}d)_i < 0, i = 1, 2, \ldots m\right], \quad (1)$$

$$\lambda_2 = \min\left[\frac{[B^{-1}(b+d)]_i}{(B^{-1}d)_i},\ 0\ |\ (B^{-1}d)_i > 0, i = 1, 2, \ldots m\right], \quad (2)$$

where $(B^{-1}(b+d))_i$ and $(B^{-1}d)_i$ are the i-th components of $B^{-1}(b+d)$ and $B^{-1}d$, respectively.

Proof. We can use partitioned matrices to represent a simplex method to an LP. Since

$$\begin{pmatrix} B & N & b+(1-\lambda)d \\ C_B & C_N & Z \end{pmatrix} \Longrightarrow \begin{pmatrix} I & B^{-1}N & B^{-1}(b+(1-\lambda)d) \\ C_B & C_N & Z \end{pmatrix}$$

$$\Longrightarrow \begin{pmatrix} I & B^{-1}N & B^{-1}(b+(1-\lambda)d) \\ 0 & C_N - C_B B^{-1}N & Z - C_B B^{-1}(b+(1-\lambda)d) \end{pmatrix},$$

there is no relationship between variable λ and the discriminate number, where N is a non-basis matrix corresponding to B. $B^{-1}(b + (1-\lambda)d) \geq 0$ is only required in order to make the optimal matrix of (LP$_\lambda$) invariant. This means $[B^{-1}(b+d)]_i - \lambda[B^{-1}d]_i \geq 0, \forall i$. By solving this inequality, we can obtain $\lambda \in [\lambda_1, \lambda_2]$, where λ_1 and λ_2 are represented with (1) and (2). It is obvious that the optimal matrix of (LP$_\lambda$) will change when $\lambda > \lambda_2$ or $\lambda < \lambda_1$. Therefore the characteristic interval, corresponding to the optimal basis matrix B, is $[\lambda_1, \lambda_2]$. □

Now we can easily get the properties of optimal value function Z_λ as follows.

Property 1. Let B be an optimal matrix of (LP_λ) on the characteristic interval $[\lambda_i, \lambda_j]$. Then $X_\lambda = B^{-1}(b + (1-\lambda)d)(\lambda_i \leqslant \lambda \leqslant \lambda_j)$ is a linear vector function about variable λ. The optimal value function $Z_\lambda = C_B B^{-1}(b + (1-\lambda)d)$ is a linear function about variable λ and decreases with the increase of variable λ.

It is obvious that every component of the bound variable $b + (1+\lambda)d$ and Z_λ decrease continuously and monotonously on every characteristic interval $[\lambda_i, \lambda_{i+1}]$ from the knowledge of property 1. Every component of the optimal solution vector function is continuous at every point λ_i.

Property 2. The optimal value function Z_λ of (LP_λ) is continuous on the interval $[0,1]$.

Based on the above conclusion, we consider an optimal solution to $(\widetilde{LP})$.

Theorem 3. *Let C be the constraint and G the objective function on domain X. Then the optimal solution x^* to the fuzzy optimal set $D = C \wedge G$ satisfies*

$$D(x^*) = \max_{x \in X} D(x) = \max_{0 \leqslant \lambda \leqslant 1} \{\lambda \wedge \max_{x \in c_\lambda} G(x)\},$$

where $C_\lambda = \{x | x \in X, C(x) \geqslant \lambda\}$[2].

The fuzzy objective function can be defined as C_λ: $Z_\lambda = Z_1 + d_0 \lambda$, so we can use the intersection of the fuzzy objective function C_λ: $Z_\lambda = Z_1 + d_0 \lambda$ and G_λ: $Z_\lambda = C_{B\lambda} B_\lambda^{-1}(b + (1-\lambda)d)$ to find an optimal decision of $(\widetilde{LP})$, shown as in Fig. 1.

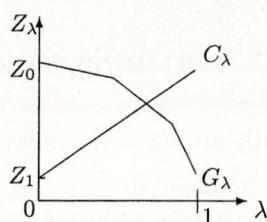

Fig. 1. The intersection of C_λ and G_λ

3 Algorithm to Fuzzy LP Problem

Let Z_1 be an optimal value of (LP_1), and Z_0 be an optimal value of (LP_0), $d_0 = Z_0 - Z_1 > 0$. Based on the above conclusions, we give a new algorithm to $(\widetilde{LP})$ problem as follows.

Step 1. Solve linear programming (LP_0) and (LP_1). Let the optimal solutions be X_0, X_1, the optimal values be Z_0, Z_1, and the optimal matrix of (LP_0) be B_0.

Step 2. Solve $[B_0^{-1}(b+(1-\lambda)d)]_i = 0$. Assume the solutions are $\lambda_1, \ldots, \lambda_{n-1}$, $(0 < \lambda_1 < \cdots < \lambda_{n-1} < 1)$, let $\lambda_0 = 0, \lambda_n = 1, \lambda = \lambda_1, k = 1$.

Step 3. Solve (LP_λ). Let the optimal value be Z_λ. If $Z_\lambda \leqslant Z_1 + d_0 \lambda$. Then turn to step 4, otherwise let $k = k+1, \lambda = \lambda_k$, turn to step 3.

Step 4. Solve the optimal decision as follows
$$\lambda^* = \frac{Z_1\lambda_k - Z_1\lambda_{k-1} - Z_{\lambda k-1}\lambda_k + Z_{\lambda k}\lambda_{k-1}}{Z_{\lambda k} - Z_{\lambda k-1} - \lambda_k d_0 + \lambda_{k-1}d_0}.$$

Step 5. Solve linear programming (LP$_{\lambda^*}$), and we can obtain the optimal solution X_{λ^*} and optimal value Z_{λ^*}.

Calculate the following example by putting the above algorithm to practice.

Example 1.
$$\max 3x_1 + 5x_2$$
$$\text{s.t. } 7x_1 + 2x_2 \lesssim 66, 5x_1 + 3x_2 \lesssim 61,$$
$$x_1 + x_2 \lesssim 16, x_1 \lesssim 8, x_2 \lesssim 5,$$
$$x_i \geq 0 \ (i=1,2), d = (0,0,0,0,7).$$

Solution. According to the above algorithm, we can obtain the optimal decision λ^* is 0.557. By calculating (LP$_{0.557}$), we obtain $X_{0.557} = (6.861, 8.899)^T$ and $Z_{0.557} = 65.077$. So the optimal solution to the example is $X^* = (6.861, 8.899)^T$ and the optimal value is $Z^* = 65.078$.

4 Conclusion

From this paper, we know the optimal decision of the fuzzy constraint linear programming does not necessarily equal 0.5. The new algorithm shows us that the following. (1) When LP$_0$ and LP$_1$ have a same optimal matrix, an optimal value function of a parametric programming corresponding to $(\widetilde{LP})$ is a segment. So the optimal decision is 0.5. (2) When LP$_0$ and LP$_1$ have not a same optimal matrix, an optimal value function corresponding to $(\widetilde{LP})$ is convex and composed with several segments. Therefore the optimal decision is more than 0.5.

Acknowledgments

Thanks to the support by National Natural Science Foundation of China (No. 70271047) and " 211" Project Foundation of Shantou University.

References

1. Cao B.Y.: The Method of Research LP Antinomy. Journal of Hunan Education Institute **9(2)** (1991) 17–22
2. Cao, B.Y.: Fuzzy Geometric Programming. Kluwer Academic Pub. Dordrecht (2002)
3. Fu G.Y.: On Optimal Solution of Fuzzy Linear Programming. Fuzzy Systems and Mathematics **4(1)** (1990) 65–72
4. Lui M.J., Cao B.Y.: The Research and Expansion on Optimal Solution of Fuzzy LP. Proc. of 1th Int. Conf. on FSKD, Vol. 2. Singapore (2002) 539–543
5. Liu T.F.: The Solution to the Problem of Parametric Linear Programming by Lumped Matrix. Journal of Electric Power **15(1)** (2000) 22–25
6. Zimmermann H.-J.: Fuzzy Programming and Linear Progamming with Several Objective Functions. Fuzzy Sets and Systems **1(1)** (1987) 49–55

Fuzzy Ultra Filters and Fuzzy G-Filters of MTL-Algebras

Xiao-hong Zhang [1,*], Yong-quan Wang [2], and Yong-lin Liu [3]

[1] Faculty of Science, Ningbo University, Ningbo 315211,
Zhejiang Province, P.R.China
zxhonghz@263.net
[2] Information Technology Center, East China University of Politics and Law,
Shanghai 200042, P.R.China
wangyquan@sina.com
[3] Department of Mathematics, Nanping Teachers College,
Nanping 353000, P.R.China
ylliun@tom.com

Abstract. The concepts of fuzzy ultra filters and fuzzy G-filters of MTL-algebras are introduced. Some examples are given and the following main results are proved: (1) a fuzzy filter of MTL-algebra is fuzzy ultra filter if and only if it is both a fuzzy prime filter and fuzzy Boolean filter; (2) a fuzzy filter of MTL-algebra is fuzzy Boolean filter if and only if it is both a fuzzy G-filter and fuzzy MV-filter.

1 Introduction

The interest in foundation of Fuzzy logic has been rapidly growing recently and several new algebras playing the role of the structures of truth values have been introduced. P.Hajek introduced the axiom system of basic logic (BL) for fuzzy propositional logic and defined the class of BL-algebras (see [1]). Ying discussed fuzzy reasoning and implication operators (see [2], [3]). Wang proposed a formal deductive system L^* for fuzzy propositional calculus, and a kind of new algebraic structures, called R_0-algebras (see [4],[5],[6]). Esteva and Godo proposed a new formal deductive system MTL, called monoidal t-norm-based logic, intended to cope with left-continuous t-norms and their residual, the algebraic semantics for MTL is based on MTL-algebras (see [7], [8]).

Zhang ([13], [14]) further studied properties of filters in MTL-algebras, and introduced the notion of Boolean filters, MV-filters and G-filters. Based on the fuzzy set theory, Kim, Zhang and Jun [9] studied the fuzzy structure of filters in MTL-algebras. Furthermore, Jun, Xu and Zhang studied characterizations of fuzzy filters in MTL-algebras, and investigated the fuzzification of Boolean filters and MV-filters. As a continuation of the paper [10], in this paper we further study fuzzy ultra filter, fuzzy prime filter and fuzzy G-filters of MTL-algebras. We give some examples of fuzzy

---

* This work was supported by the National Natural Science Foundation of P. R. China (Grant no. 60273087) and the Research Foundation of Ningbo University.

ultra filter and fuzzy G-filter, and prove the following important results: (1) a fuzzy filter of MTL-algebra is fuzzy ultra filter if and only if it is both a fuzzy prime filter and fuzzy Boolean filter; (2) a fuzzy filter of MTL-algebra is fuzzy Boolean filter if and only if it is both a fuzzy G-filter and fuzzy MV-filter.

2 Preliminaries

First of all, we recall some notions and their important properties.

Definition 2.1. (see [1]). A residuated lattice is an algebra $(L, \wedge, \vee, \odot, \rightarrow, 0, 1)$ with four binary operations and two constants such that

(i) $(L, \wedge, \vee, 0, 1)$ is a lattice with the largest element 1 and the least element 0 (with respect to the lattice ordering $\leq$).
(ii) $(L, \odot, 1)$ is a commutative semigroup with the unit element 1, i.e. $\odot$ is commutative, associative, and $1 \odot x = x$ for all $x \in L$.
(iii) $\odot$ and $\rightarrow$ form an adjoint pair, i.e. $z \leq x \rightarrow y$ iff $z \odot x \leq y$ for all $x, y, z \in L$.

Definition 2.2. (see [7]). A residuated lattice $(L, \wedge, \vee, \odot, \rightarrow, 0, 1)$ is called a MTL-algebra, if it satisfies the pre-linearity equation: $(x \rightarrow y) \vee (y \rightarrow x) = 1$, $\forall x, y \in L$.

Proposition 2.3. (see [11] and [13]). Let $(L, \wedge, \vee, \odot, \rightarrow, 0, 1)$ be a residuated lattice, then for all $x, y, z \in L$,

(1) $x \leq y$ if and only if $x \rightarrow y = 1$;
(2) $x = 1 \rightarrow x$, $x \rightarrow (y \rightarrow x) = 1$, $y \leq (y \rightarrow x) \rightarrow x$;
(3) $x \leq y \rightarrow z$ if and only if $y \leq x \rightarrow z$;
(4) $x \rightarrow (y \rightarrow z) = (x \odot y) \rightarrow z = y \rightarrow (x \rightarrow z)$;
(5) $x \leq y$ implies $z \rightarrow x \leq z \rightarrow y$, $y \rightarrow z \leq x \rightarrow z$;
(6) $z \rightarrow y \leq (x \rightarrow z) \rightarrow (x \rightarrow y)$, $z \rightarrow y \leq (y \rightarrow x) \rightarrow (z \rightarrow x)$;
(7) $(x \rightarrow y) \odot (y \rightarrow z) \leq x \rightarrow z$;
(8) $x' = x'''$, $x \leq x''$, $x' \odot x = 0$;
(9) $x' \wedge y' = (x \vee y)'$;
(10) if $x \vee x' = 1$ then $x \wedge x' = 0$;
(11) $(\vee_{i \in \Gamma} y_i) \rightarrow x = \wedge_{i \in \Gamma} (y_i \rightarrow x)$;
(12) $x \odot (\vee_{i \in \Gamma} y_i) = \vee_{i \in \Gamma} (x \odot y_i)$;
(13) $x \rightarrow (\wedge_{i \in \Gamma} y_i) = \wedge_{i \in \Gamma} (x \rightarrow y_i)$;
(14) $\vee_{i \in \Gamma} (y_i \rightarrow x) \leq (\wedge_{i \in \Gamma} y_i) \rightarrow x$;
(15) $x \leq y$ implies $x \odot z \leq y \odot z$;
(16) $y \rightarrow z \leq x \vee y \rightarrow x \vee z$.

where $x' = x \rightarrow 0$; Γ is a finite or infinite index set and assume that the corresponding infinite meets and joints exist in L.

Proposition 2.4. (see [11] and [13]) Let $(L, \wedge, \vee, \odot, \rightarrow, 0, 1)$ be a MTL-algebra, then for all $x, y, z \in L$,

(17) $(x \wedge y) \rightarrow z = (x \rightarrow z) \vee (y \rightarrow z)$;
(18) $x \vee y = ((x \rightarrow y) \rightarrow y) \wedge ((y \rightarrow x) \rightarrow x)$;
(19) $((x \rightarrow y) \rightarrow z) \rightarrow (((y \rightarrow x) \rightarrow z) \rightarrow z) = 1$;
(20) $x \rightarrow (y \vee z) = (x \rightarrow y) \vee (x \rightarrow z)$;
(21) $x \wedge (y \vee z) = (x \wedge y) \vee (x \wedge z)$, $x \vee (y \wedge z) = (x \vee y) \wedge (x \vee z)$;
(22) $x \odot y \leq x \wedge y$;
(23) $x' \vee y' = (x \wedge y)'$.

Definition 2.5 (see [7]). Let L be a MTL-algebra. A filter F is a nonempty subset of L satisfying

(F1) for any $x, y \in F$, $x \odot y \in F$;   (F2) for any $x \in F$, if $x \leq y$, then $y \in F$.

Proposition 2.6 (see [7]). Let $(L, \wedge, \vee, \odot, \rightarrow, 0, 1)$ be a MTL-algebra, $F \subseteq L$. Then F is a filter of L if and only if

(F3) $1 \in F$;   (F4) $y \in F$ whenever $x, x \rightarrow y \in F$, for any $x, y \in L$.

In what follows let L denote MTL-algebra unless otherwise specified.

Definition 2.7 ([9]). A fuzzy set μ in L is called a fuzzy filter of L if it satisfies

(i) For any $x, y \in L$, $\mu(x \odot y) \geq \min\{\mu(x), \mu(y)\}$;
(ii) μ is order-preserving, that is, For any $x, y \in L$, $x \leq y \Rightarrow \mu(x) \leq \mu(y)$.

Theorem 2.8 (see [10]). A fuzzy set μ in L is a fuzzy filter of L if and only if it satisfies

(iii) For any $x \in L$, $\mu(1) \geq \mu(x)$;   (iv) For any $x, y \in L$, $\mu(y) \geq \min\{\mu(x), \mu(x \rightarrow y)\}$.

Definition 2.9 (see [10]). A fuzzy filter μ of L is said to be Boolean if it satisfies the following equality

(v) For any $x \in L$, $\mu(x \vee x') = \mu(1)$.

Theorem 2.10 (see [10]) Let μ be a fuzzy filter of L. Then the following assertions are equivalent:

(vi) (vi) μ is Boolean;
(vii) For any $x, y, z \in L$, $\mu(x \rightarrow z) \geq \min\{\mu(x \rightarrow (z' \rightarrow y)), \mu(y \rightarrow z)\}$;
(viii) For any $x, y \in L$, $\mu(x) \geq \mu((x \rightarrow y) \rightarrow x))$.

Definition 2.11 (see [10]). A fuzzy set μ in L is called a fuzzy MV-filter of L if it is a fuzzy filter of L that satisfies the following inequality:

(ix) For any $x, y \in L$, $\mu(x \rightarrow y) \leq \mu(((y \rightarrow x) \rightarrow x) \rightarrow y)$.

3 Fuzzy Ultra Filters of MTL-Algebras

Definition 3.1 A fuzzy set μ in L is called a fuzzy ultra filter of L if it is a fuzzy filter of L that satisfies the following condition:

(x) For any $x \in L$, $\mu(x) = \mu(1)$ or $\mu(x') = \mu(1)$.

Example 3.2 Let $L = \{0, a, b, c, d, 1\}$ in which $\rightarrow, \odot$ are defined by Table 1, 2. Then $L(\wedge, \vee, \odot, \rightarrow, 0, 1)$ is a MTL-algebra.

Let μ be a fuzzy set in L given by

$$\mu(x) = \begin{cases} \alpha & \text{if } x \in \{1, a, d\} \\ \beta & \text{otherwise} \end{cases}$$

where $\alpha > \beta$ in $[0,1]$. Then μ is a fuzzy ultra filter of L.

Table 1

→	0	a	b	c	d	1
0	1	1	1	1	1	1
a	c	1	b	b	a	1
b	d	a	1	a	d	1
c	a	1	1	1	a	1
d	b	1	b	b	1	1
1	0	a	b	c	d	1

Table 2

⊙	0	a	b	c	d	1
0	0	0	0	0	0	0
a	0	d	c	0	d	a
b	0	c	b	c	0	b
c	0	0	c	0	0	c
d	0	d	0	0	d	d
1	0	a	b	c	d	1

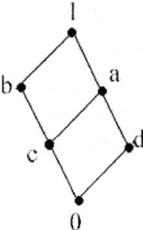

Fig. 1

Definition 3.3. A fuzzy set μ in L is called a fuzzy prime filter of L if it is a fuzzy filter of L that satisfies the following condition:

(xi) For any $x, y \in L$, $\mu(x \vee y) \leq \mu(x) \vee \mu(y)$.

In Example 3.2, μ is a fuzzy prime ultra filter of L.

Example 3.4. Let $L=[0,1]$ and define a product $\circ$ and a residuum $\to$ on L as follows

$$x \odot y = \begin{cases} x \wedge y & \text{if } x+y>1 \\ 0 & \text{otherwise} \end{cases}, \quad x \to y = \begin{cases} 1 & \text{if } x \leq y \\ (1-x) \vee y & \text{otherwise} \end{cases}$$

for all $x, y \in L$. Then L is a MTL-algebra. Let μ_1 and μ_2 be two fuzzy sets in L given by

$$\mu_1(x) = \begin{cases} \alpha & \text{if } x \in [0, 0.5] \\ \alpha + 0.5x^2 & \text{if } x \in (0.5, 1] \end{cases}, \quad \mu_2(x) = \begin{cases} \alpha & \text{if } x \in [0, 0.5] \\ \alpha + 0.3x & \text{if } x \in (0.5, 1] \end{cases}$$

where α in $[0, 0.5)$. Then μ_1 and μ_2 are fuzzy prime filter of L, but μ_1 and μ_2 are not fuzzy ultra filter of L, since $\mu(0.5) \neq \mu(1)$ and $\mu(0.5') \neq \mu(1)$.

Theorem 3.5. A fuzzy set μ in L is a fuzzy ultra filter of L if and only if it satisfies

(a) μ is a fuzzy Boolean filter of L;
(b) μ is a fuzzy prime filter of L.

Proof: Suppose that μ is a fuzzy Boolean and fuzzy prime filter of L. For any $x \in L$, by (v) and (xi) we have $\mu(x \vee x') = \mu(1) \leq \mu(x) \vee \mu(x')$. If $\mu(x) \neq \mu(1)$, since $\mu(x) \leq \mu(1)$, $\mu(x') \leq \mu(1)$, by Theorem 2.8 (iii) we have $\mu(x') = \mu(1)$. Thus, μ is a fuzzy ultra filter of L.

Conversely, let μ be a fuzzy ultra filter of L. For any $x \in L$, since $x \leq x \vee x'$, $x' \leq x \vee x'$, so $\mu(x) \leq \mu(x \vee x')$, $\mu(x') \leq \mu(x \vee x')$ by Definition 2.7 (ii). According to the Definition of fuzzy ultra filter, we have $\mu(x) = \mu(1)$ or $\mu(x') \leq \mu(1)$. Thus, $\mu(1) \leq \mu(x \vee x')$. From this and Theorem 2.8 (iii), we get $\mu(1) = \mu(x \vee x')$. This means that μ is a fuzzy Boolean filter of L.

And, by Definition 2.7 (ii) and proposition 2.4 (18) we have

$$\mu(x \vee y) = \mu(((x \to y) \to y) \wedge ((y \to x) \to x)) \leq \mu((x \to y) \to y).$$

From $0 \leq y$ and Proposition 2.3 (5) we get $(x \to y) \to y \leq x' \to y$. Thus, $\mu((x \to y) \to y) \leq \mu(x' \to y)$ by Definition 2.7 (ii). So, $\mu(x \vee y) \leq \mu(x' \to y)$.

For any $x, y \in L$, if $\mu(x)=\mu(1)$, then $\mu(x \vee y) \leq \mu(1)=\mu(x) \leq \mu(x) \vee \mu(y)$. If $\mu(x) \neq \mu(1)$, then $\mu(x')=\mu(1)$ by Definition 3.1. Thus,

$$\mu(y) \geq \min\{\mu(x'), \mu(x' \to y)\} = \min\{\mu(1), \mu(x' \to y)\} = \mu(x' \to y).$$

by Theorem 2.8 (iv) and (iii). Therefore, $\mu(x \vee y) \leq \mu(x' \to y) \leq \mu(y) \leq \mu(x) \vee \mu(y)$. This means that μ is a fuzzy prime filter of L.

4 Fuzzy G-Filters of MTL-Algebras

Definition 4.1. A fuzzy set μ in L is called a fuzzy G-filter of L if it is a fuzzy filter of L that satisfies the following inequality:
(x) For any $x, y \in L$, $\mu(x \odot x \to y) \leq \mu(x \to y)$.

Example 4.2. Let $L=\{0,a,b,c,d,1\}$ in which $\to, \odot$ are defined by Table 3,4. Then $L(\wedge, \vee, \odot, \to, 0, 1)$ is a MTL-algebra. Note that $(L, \leq)$ is not a chain.

Table 3

→	0	a	b	c	d	1
0	1	1	1	1	1	1
a	d	1	b	b	d	1
b	0	a	1	a	d	1
c	d	1	1	1	d	1
d	a	1	1	1	1	1
1	0	a	b	c	d	1

Table 4

⊙	0	a	b	c	d	1
0	0	0	0	0	0	0
a	0	a	c	c	0	a
b	0	c	b	c	d	b
c	0	c	c	c	0	c
d	0	0	d	0	0	d
1	0	a	b	c	d	1

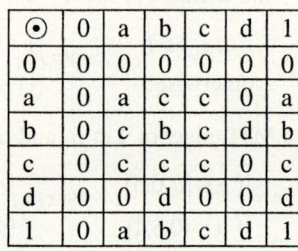

Fig. 2

Let μ be a fuzzy set in L given by

$$\mu(x) = \begin{cases} \alpha & \text{if } x \in \{1, a\} \\ \beta & \text{otherwise} \end{cases}$$

where $\alpha > \beta$ in $[0,1]$. Then it is routine to verify that μ is a fuzzy G-filter of L.

Example 4.3. Let $L=[0,1]$ and define a product $\odot$ and a residuum $\to$ on L as follows

$$x \odot y = \begin{cases} x \wedge y & \text{if } x+y>1 \\ 0 & \text{otherwise} \end{cases}, \quad x \to y = \begin{cases} 1 & \text{if } x \leq y \\ (1-x) \vee y & \text{otherwise} \end{cases}$$

for all $x, y \in L$. Then L is a MTL-algebra. Let μ_1 and μ_2 be two fuzzy sets in L given by

$$\mu_1(x) = \begin{cases} \alpha & \text{if } x \in [0, 0.5] \\ \alpha+(1-\alpha)x^2 & \text{if } x \in (0.5, 1] \end{cases}, \quad \mu_2(x) = \begin{cases} \alpha & \text{if } x \in [0, 0.5] \\ \alpha+(1-\alpha)x & \text{if } x \in (0.5, 1] \end{cases}$$

where α in $[0,1)$. Then μ_1 and μ_2 are fuzzy filter of L, but μ_1 and μ_2 are not fuzzy G-filter of L, since $\mu_1(0.3 \odot 0.3 \to 0.1) > \mu_1(0.3 \to 0.1)$, $\mu_2(0.3 \odot 0.3 \to 0.1) > \mu_2(0.3 \to 0.1)$.

Lemma 4.4. (see [10]) Every fuzzy Boolean filter μ of L satisfies the following inequality

(xi) For any $x, y, z \in L$, $\mu(x \to z) \geq \min\{\mu(x \to (y \to z)), \mu(x \to y)\}$

Theorem 4.5. A fuzzy set μ in L is a fuzzy Boolean filter of L if and only if it satisfies

(a) μ is a fuzzy G-filter of L;
(b) μ is a fuzzy MV-filter of L.

Proof: Suppose that μ is a fuzzy Boolean filter of L. By theorem 3.19 in [10], we know that μ is a fuzzy MV-filter of L. For any $x, y \in L$, by Lemma 4.4 (xi) and Theorem 2.8 (iii) we have

$$\mu(x \to y) \geq \min\{\mu(x \to (x \to y)), \mu(x \to x)\} = \min\{\mu(x \to (x \to y)), \mu(1)\}$$
$$= \mu(x \to (x \to y)) = \mu(x \odot x \to y).$$

Thus, μ is a fuzzy G-filter of L.

Conversely, let μ be a fuzzy G-filter and fuzzy MV-filter of L. For any $x, y \in L$, we have

$x \leq (x \to y) \to y$ (by Proposition 2.3 (2))
$(x \to y) \to x \leq (x \to y) \to ((x \to y) \to y)$ (by Proposition 2.3 (6))
$(x \to y) \to x \leq ((x \to y) \odot (x \to y)) \to y$ (by Proposition 2.3 (4))

So,

$$\mu((x \to y) \to x) \leq \mu((x \to y) \odot (x \to y) \to y) \leq \mu((x \to y) \to y)$$

by Definition 2.7 (ii) and Definition 4.1. And, we also have

$(x \to y) \to y \leq (x \to (x \to y)) \to (x \to y)$ (by Proposition 2.3 (6))
$((x \to (x \to y)) \to (x \to y)) \to x \leq ((x \to y) \to y) \to x$ (by Proposition 2.3 (5))

So,

$$\mu((x \to y) \to x) \leq \mu(((x \to (x \to y)) \to (x \to y)) \to x) \leq \mu(((x \to y) \to y) \to x)$$

by Definition 2.11 (ix) and Definition 2.7 (ii). Thus,

$$\mu((x \to y) \to x) \leq \min\{\mu((x \to y) \to y), \mu(((x \to y) \to y) \to x)\}.$$

Therefore, by Theorem 2.8 (iv) we get $\mu((x \to y) \to x) \leq \mu(x)$. This means that μ is a fuzzy Boolean filter of L by Theorem 2.10 (viii). The proof is end.

References

1. Hájek P.: Metamathematics of Fuzzy Logic. Kluwer Academic Publishers, Dordrecht (1998)
2. M.S.Ying: Perturbation of fuzzy reasoning. IEEE Trans. Fuzzy Sys. 7(1999) 625-629
3. M.S.Ying: Implications operators in fuzzy logic. IEEE Trans. Fuzzy Sys. 10(2002) 88-91
4. G.J.Wang: A formal deductive system for fuzzy propositional calculus. Chinese Science Bulletin 42 (1997) 1521-1526.

5. G.J.Wang: On the logic foundation of fuzzy reasoning. Information Sciences 117 (1999) 47-88
6. G.J.Wang: Non-classical Mathematical Logic and Approximate Reasoning. Science Press, Beijing (in Chinese, 2000)
7. Esteva F., Godo L.: Monoidal t-norm-based logic: towards a logic for left-continuous t-norms. Fuzzy Sets and Systems 12 (2001) (4) 271-288
8. Jenei S., Montagna F.: A proof of standard completeness for Esteva and Godo's logic MTL. Studia Logica 70 (2002) 183-192.
9. K.H.Kim, Zhang Q., Y.B.Jun: On Fuzzy filters of MTL-algebras. The Journal of Fuzzy Mathematics 10(2002) (4) 981-989
10. Y.B.Jun, Y.Xu, X.H.Zhang: Fuzzy filters of MTL-algebras. Information Sciences (in print)
11. E.Turunen: Boolean deductive systems of BL-algebras. Arch. Math. Logic 40 (2001) 467-473
12. R.S.Su, G.J.Wang: On fuzzy MP filters of R_0-algebras. Fuzzy Systems and Mathematics 18 (2) (2004) 15-23.
13. X.H.Zhang: On filters in MTL-algebras (submitted)
14. X.H.Zhang, W.H. Li: On fuzzy logic algebraic systems MTL (submitted)
15. X.H.Zhang: Fuzzy BZ-ideals and fuzzy anti-grouped ideals. The Journal of Fuzzy Mathematics 11(2003) (4) 915-932

A Study on Relationship Between Fuzzy Rough Approximation Operators and Fuzzy Topological Spaces

Wei-Zhi Wu

Information College, Zhejiang Ocean University,
Zhoushan, Zhejiang, 316004, P.R. China
wuwz@zjou.net.cn

Abstract. It is proved that a pair of dual fuzzy rough approximation operators can induce a topological space if and only if the fuzzy relation is reflexive and transitive. The sufficient and necessary condition that a fuzzy interior (closure) operator derived from a fuzzy topological space can associate with a fuzzy reflexive and transitive relation such that the induced fuzzy lower (upper) approximation operator is the fuzzy interior (closure) operator is also examined.

1 Introduction

The theory of rough set, proposed by Pawlak [1], is an extension of set theory for the study of intelligent systems characterized by insufficient and incomplete information. The notion of an approximation space consisting of a universe of discourse and an indiscernible relation imposed on it is one of the fundamental concept of rough set theory. Based on the approximation space, the primitive notion of lower and upper approximation operators can be induced. From both theoretic and practical needs, many authors have generalized the concept of approximation operators by using nonequivalent binary relations [2-6], neighborhood systems [7, 8], or by using axiomatic approaches [9-11]. More general frameworks have been obtained under fuzzy environment which involve the rough approximations of fuzzy sets (rough fuzzy sets), the fuzzy approximations of sets (fuzzy rough sets) [12-23]. Extensive research has also been carried out to compare the theory of rough sets with other theories of uncertainty, such as modal logic [18, 24, 25], conditional events [26], and Dempster-Shafer theory of evidence [5, 6, 27, 28]. On the other hand, the relationships between rough sets and topological spaces were studied by many authors [29-32]. The relationships between crisp rough sets and crisp topological spaces were studied in detail. The relationship between fuzzy rough sets and fuzzy topological spaces was investigated by Boixader et al [12], but their studies were restricted to fuzzy T-rough sets defined by fuzzy T-similarity relations which were equivalence crisp relations when they degenerated into crisp ones. In this paper, we focus mainly on the study of the relationship between fuzzy rough approximation operators and fuzzy topological spaces in general case.

2 Fuzzy Rough Approximation Operators

Let X be a nonempty set called the universe of discourse. The class of all subsets (respectively, fuzzy subsets) of X will be denoted by $\mathcal{P}(X)$ (respectively, by $\mathcal{F}(X)$). For any $A \in \mathcal{F}(X)$, the complement of A is denoted by $\sim A$.

Definition 1. *Let R be a fuzzy binary relation on U, i.e. $R \in \mathcal{F}(U \times U)$. R is referred to as a serial fuzzy relation if $\forall x \in U$, $\bigvee_{y \in U} R(x,y) = 1$; R is referred to as a reflexive fuzzy relation if $R(x,x) = 1$ for all $x \in U$; R is referred to as a transitive fuzzy relation if $R(x,z) \geq \bigvee_{y \in U} (R(x,y) \wedge R(y,z))$ for all $x, z \in U$.*

Definition 2. *If R is a fuzzy relation on U, then the pair (U, R) is referred to as a fuzzy approximation space. Let $A \in \mathcal{F}(U)$, the lower and upper approximations of A, $\underline{R}(A)$ and $\overline{R}(A)$, with respect to the fuzzy approximation space (U, R) are fuzzy sets of U whose membership functions, for each $x \in U$, are defined, respectively, by*

$$\underline{R}(A)(x) = \bigwedge_{y \in U} [(1 - R(x,y)) \vee A(y)],$$
$$\overline{R}(A)(x) = \bigvee_{y \in U} [R(x,y) \wedge A(y)].$$

The pair $(\underline{R}(A), \overline{R}(A))$ is referred to as a fuzzy rough set, and $\underline{R}, \overline{R} : \mathcal{F}(U) \to \mathcal{F}(U)$ are referred to as lower and upper fuzzy rough approximation operators, respectively.

Lemma 3 ([20]). *The lower and upper fuzzy rough approximation operators, $\underline{R}$ and $\overline{R}$, satisfy the properties: $\forall A, B \in \mathcal{F}(U)$, $\forall A_j \in \mathcal{F}(U)(\forall j \in J)$, $\forall \alpha \in I = [0, 1]$,*

(FL1) $\underline{R}(A) = \sim \overline{R}(\sim A)$,   (FU1) $\overline{R}(A) = \sim \underline{R}(\sim A)$;
(FL2) $\underline{R}(A \cup \widehat{\alpha}) = \underline{R}(A) \cup \widehat{\alpha}$,   (FU2) $\overline{R}(A \cap \widehat{\alpha}) = \overline{R}(A) \cap \widehat{\alpha}$;
(FL3) $\underline{R}(\bigcap_{j \in J} A_j) = \bigcap_{j \in J} \underline{R}(A_j)$,   (FU3) $\overline{R}(\bigcup_{j \in J} A_j) = \bigcup_{j \in J} \overline{R}(A_j)$,

where $\widehat{\alpha}$ is the constant fuzzy set, i.e. $\widehat{\alpha}(x) = a, \forall x \in U$.

Properties (FL1) and (FU1) show that the fuzzy rough approximation operators $\underline{R}$ and $\overline{R}$ are dual each other. Properties with the same number may be regarded as dual properties. It can be checked that

(FL4) $A \subseteq B \Longrightarrow \underline{R}(A) \subseteq \underline{R}(B)$,   (FU4) $A \subseteq B \Longrightarrow \overline{R}(A) \subseteq \overline{R}(B)$;
(FL5) $\underline{R}(\bigcup_{j \in J} A_j) \supseteq \bigcup_{j \in J} \underline{R}(A_j)$,   (FU5) $\overline{R}(\bigcap_{j \in J} A_j) \subseteq \bigcap_{j \in J} \overline{R}(A_j)$.

Properties (FL2) and (FU2) imply the following properties (FL2)' and (FU2)':

(FL2)' $\underline{R}(U) = U$,   (FU2)' $\overline{R}(\emptyset) = \emptyset$.

Definition 4. Let $L, H : \mathcal{F}(U) \to \mathcal{F}(U)$ be two operators. They are referred to as dual operators if for all $A \in \mathcal{F}(U)$:

$$\text{(Fl1)} \quad L(A) = \sim H(\sim A), \qquad \text{(Fu1)} \quad H(A) = \sim L(\sim A).$$

Rough set approximation operators can also be characterized by axioms. In the axiomatic approach, rough sets are axiomatized by abstract operators. For the case of fuzzy rough sets, the primitive notion is a system $(\mathcal{F}(U), \cap, \cup, \sim, L, H)$, where $L, H : \mathcal{F}(U) \to \mathcal{F}(U)$ are operators from $\mathcal{F}(U)$ to $\mathcal{F}(U)$.

Lemma 5 ([20]). Let $L, H : \mathcal{F}(U) \to \mathcal{F}(U)$ be two operators. Then there exists a fuzzy relation R on U such that for all $A \in \mathcal{F}(U)$

$$L(A) = \underline{R}(A), \quad \text{and} \quad H(A) = \overline{R}(A)$$

if and only if (iff) L and H satisfy the following axioms: $\forall A, B \in \mathcal{F}(U), \forall A_j \in \mathcal{F}(U)(\forall j \in J)$, and $\forall \alpha \in I$

(Fl1) $L(A) = \sim H(\sim A)$,   (Fu1) $H(A) = \sim L(\sim A)$;
(Fl2) $L(A \cup \widehat{\alpha}) = L(A) \cup \widehat{\alpha}$,   (Fu2) $H(A \cap \widehat{\alpha}) = H(A) \cap \widehat{\alpha}$;
(Fl3) $L(\bigcap_{j \in J} A_j) = \bigcap_{j \in J} L(A_j)$, (Fu3) $H(\bigcup_{j \in J} A_j) = \bigcup_{j \in J} H(A_j)$.

Definition 6. Let $L, H : \mathcal{F}(U) \to \mathcal{F}(U)$ be a pair of dual operators. If L satisfies axioms (Fl2) and (Fl3) or equivalently H satisfies axioms (Fu2) and (Fu3), then the system $(\mathcal{F}(U), \cap, \cup, \sim, L, H)$ is referred to as a fuzzy rough set algebra, and L and H are referred to as lower and upper fuzzy approximation operators respectively.

Lemma 7 ([20]). Assume that $L, H : \mathcal{F}(U) \to \mathcal{F}(U)$ is a pair of dual fuzzy approximation operators, i.e., L satisfies axioms (Fl1), (Fl2) and (Fl3), and H satisfies (Fu1), (Fu2) and (Fu3). Then there exists a serial fuzzy relation R on U such that $L(A) = \underline{R}(A)$ and $H(A) = \overline{R}(A)$ for all $A \in \mathcal{F}(U)$ iff L and H satisfy axioms:

(Fl0) $L(\widehat{\alpha}) = \widehat{\alpha}, \quad \forall \alpha \in I$;
(Fu0) $H(\widehat{\alpha}) = \widehat{\alpha}, \quad \forall \alpha \in I$.

Lemma 8 ([20]). Assume that $L, H : \mathcal{F}(U) \to \mathcal{F}(U)$ is a pair of dual fuzzy approximation operators. Then there exists a reflexive fuzzy relation R on U such that $L(A) = \underline{R}(A)$ and $H(A) = \overline{R}(A)$ for all $A \in \mathcal{F}(U)$ iff L and H satisfy axioms:

(Fl6) $L(A) \subseteq A, \quad \forall A \in \mathcal{F}(U)$;
(Fu6) $A \subseteq H(A), \quad \forall A \in \mathcal{F}(U)$.

Lemma 9 ([20]). Assume that $L, H : \mathcal{F}(U) \to \mathcal{F}(U)$ is a pair of dual fuzzy approximation operators. Then there exists a transitive fuzzy relation R on U such that $L(A) = \underline{R}(A)$ and $H(A) = \overline{R}(A)$ for all $A \in \mathcal{F}(U)$ iff L and H satisfy axioms:

(Fl7) $L(A) \subseteq L(L(A)), \quad \forall A \in \mathcal{F}(U)$;
(Fu7) $H(H(A)) \subseteq H(A), \forall A \in \mathcal{F}(U)$.

3 Fuzzy Topological Spaces and Fuzzy Rough Approximation Operators

Definition 10 ([33]). *A subset τ of $\mathcal{F}(U)$ is referred to as a fuzzy topology on U iff it satisfies*

(1) *If $\mathcal{A} \subseteq \tau$, then $\bigcup_{A \in \mathcal{A}} A \in \tau$,*
(2) *If $A, B \in \tau$, then $A \cap B \in \tau$,*
(3) *If $\widehat{\alpha} \in \mathcal{F}(U)$ is a constant fuzzy set, then $\widehat{\alpha} \in \tau$.*

Definition 11 ([33]). *A map $\Psi : \mathcal{F}(U) \to \mathcal{F}(U)$ is referred to as a fuzzy interior operator iff for all $A, B \in \mathcal{F}(U)$ it satisfies:*

(1) $\Psi(A) \subseteq A$,
(2) $\Psi(A \cap B) = \Psi(A) \cap \Psi(B)$,
(3) $\Psi^2(A) = \Psi(A)$,
(4) $\Psi(\widehat{\alpha}) = \widehat{\alpha}, \quad \forall \alpha \in I$.

Definition 12 ([33]). *A map $\Phi : \mathcal{F}(U) \to \mathcal{F}(U)$ is referred to as a fuzzy closure operator iff for all $A, B \in \mathcal{F}(U)$ it satisfies:*

(1) $A \subseteq \Phi(A)$,
(2) $\Phi(A \cup B) = \Phi(A) \cup \Phi(B)$,
(3) $\Phi^2(A) = \Phi(A)$,
(4) $\Phi(\widehat{\alpha}) = \widehat{\alpha}, \quad \forall \alpha \in I$.

The elements of a fuzzy topology τ are referred to as open fuzzy sets, and it is easy to show that a fuzzy interior operator Ψ defines a fuzzy topology $\tau_\Psi = \{A \in \mathcal{F}(U) : \Psi(A) = A\}$. So, the open fuzzy sets are the fixed points of Ψ.

Theorem 13. *Assume that R is a fuzzy relation on U. Then the operator $\Phi = \overline{R} : \mathcal{F}(U) \to \mathcal{F}(U)$ is a fuzzy closure operator iff R is a reflexive and transitive fuzzy relation.*

Proof. " $\Rightarrow$ " Assume that $\Phi = \overline{R} : \mathcal{F}(U) \to \mathcal{F}(U)$ is a fuzzy closure operator. Since $A \subseteq \Phi(A), \forall A \in \mathcal{F}(U)$, by Lemma 8 we know that R is a reflexive fuzzy relation. Since $\Phi^2(A) = \Phi(A)$, i.e. $\overline{R}(\overline{R}(A)) = \overline{R}(A), \forall A \in \mathcal{F}(U)$, by using the reflexivity of R we must have

$$\overline{R}(\overline{R}(A)) \subseteq \overline{R}(A), \quad \forall A \in \mathcal{F}(U).$$

Hence by Lemma 9 we conclude that R is a transitive fuzzy relation.

" $\Leftarrow$ " If R is a reflexive and transitive fuzzy relation, by Lemma 8 we have

$$A \subseteq \Phi(A), \forall A \in \mathcal{F}(U).$$

It should be noted that the reflexivity of R implies that R is serial, it follows from Lemma 7 that

$$\Phi(\widehat{\alpha}) = \widehat{\alpha}, \quad \forall \alpha \in I.$$

Since R is transitive, by Lemma 9 we have that
$$\Phi^2(A) \subseteq \Phi(A), \quad \forall A \in \mathcal{F}(U).$$
On the other hand, by using the reflexivity it is easy to see that
$$\Phi^2(A) \supseteq \Phi(A), \quad \forall A \in \mathcal{F}(U).$$
Hence
$$\Phi^2(A) = \Phi(A), \quad \forall A \in \mathcal{F}(U).$$
From Lemma 3 we have that
$$\Phi(A \cup B) = \Phi(A) \cup \Phi(B).$$
Thus we have proved that $\Phi = \overline{R} : \mathcal{F}(U) \to \mathcal{F}(U)$ is a fuzzy closure operator.

Theorem 14. *Assume that R is a fuzzy relation on U. Then the operator $\Psi = \underline{R} : \mathcal{F}(U) \to \mathcal{F}(U)$ is a fuzzy interior operator iff R is a reflexive and transitive fuzzy relation.*

Proof. It is similar to the proof of Theorem 13.

Theorem 15. *Assume that R is a reflexive and transitive fuzzy relation on U. Then there exists a fuzzy topology τ_R on U such that $\Psi = \underline{R} : \mathcal{F}(U) \to \mathcal{F}(U)$ and $\Phi = \overline{R} : \mathcal{F}(U) \to \mathcal{F}(U)$ are the fuzzy interior and closure operators respectively.*

Proof. Suppose that R is a reflexive and transitive fuzzy relation on U. Then by Theorem 13 and Theorem 14 we conclude that $\Psi = \underline{R} : \mathcal{F}(U) \to \mathcal{F}(U)$ and $\Phi = \overline{R} : \mathcal{F}(U) \to \mathcal{F}(U)$ are the fuzzy interior and closure operators respectively. Define
$$\tau_R = \{A \in \mathcal{F}(U) : \Psi(A) = \underline{R}(A) = A\}.$$
It can easily be checked that τ_R is a fuzzy topology on U. Evidently, Ψ and Φ are respectively the interior and closure operators induced by τ_R.

Theorem 16. *Let $\Phi : \mathcal{F}(U) \to \mathcal{F}(U)$ be a fuzzy closure operator, then there exists a reflexive and transitive fuzzy relation on U such that $\overline{R}(A) = \Phi(A)$ for all $A \in \mathcal{F}(U)$ iff Φ satisfies the following two conditions*

(1) $\Phi(\bigcup_{j \in J} A_j) = \bigcup_{j \in J} \Phi(A_j), \quad A_j \in \mathcal{F}(U), j \in J$,
(2) $\Phi(A \cap \widehat{\alpha}) = \Phi(A) \cap \widehat{\alpha}, \quad \forall A \in \mathcal{F}(U), \forall \alpha \in I$.

Proof. Let $\Phi : \mathcal{F}(U) \to \mathcal{F}(U)$ be a fuzzy closure operator. If their exists a reflexive and transitive fuzzy relation on U such that $\overline{R}(A) = \Phi(A)$ for all $A \in \mathcal{F}(U)$, then by Lemma 3 we know that conditions (1) and (2) hold. Conversely, if Φ satisfies the conditions (1) and (2), then, by Lemmas 5, 8, 9, and Theorem 13, there exists a reflexive and transitive fuzzy relation on U such that $\overline{R}(A) = \Phi(A)$ for all $A \in \mathcal{F}(U)$.

Theorem 17. Let $\Psi : \mathcal{F}(U) \to \mathcal{F}(U)$ be a fuzzy interior operator, then there exists a reflexive and transitive fuzzy relation on U such that $\underline{R}(A) = \Psi(A)$ for all $A \in \mathcal{F}(U)$ iff Ψ satisfies the following two conditions

(1) $\Psi(\bigcap_{j \in J} A_j) = \bigcap_{j \in J} \Psi(A_j), \quad A_j \in \mathcal{F}(U), \ j \in J$,
(2) $\Psi(A \cup \widehat{\alpha}) = \Psi(A) \cup \widehat{\alpha}, \quad \forall A \in \mathcal{F}(U), \forall \alpha \in I$.

Proof. Let $\Psi : \mathcal{F}(U) \to \mathcal{F}(U)$ be a fuzzy interior operator. If their exists a reflexive and transitive fuzzy relation on U such that $\underline{R}(A) = \Psi(A)$ for all $A \in \mathcal{F}(U)$, then by Lemma 3 we know that conditions (1) and (2) hold. Conversely, if Ψ satisfies the conditions (1) and (2), then, by Lemmas 5, 8, 9, and Theorem 14, there exists a reflexive and transitive fuzzy relation on U such that $\underline{R}(A) = \Psi(A)$ for all $A \in \mathcal{F}(U)$.

Remark. It should be pointed out that if U is a finite universe of discourse, then properties (FL3) and (FU3) in Lemma 3 can equivalently be replaced by the following properties (FL3)′ and (FU3)′ respectively [21, 22]

(FL3)′ $\underline{R}(A \bigcap B) = \underline{R}(A) \bigcap \underline{R}(B), \forall A, B \in \mathcal{F}(U)$,
(FU3)′ $\overline{R}(A \bigcup B) = \overline{R}(A) \bigcup \overline{R}(B), \forall A, B \in \mathcal{F}(U)$,

thus the condition (1) in Theorems 16 and 17 can be omitted.

4 Conclusion

We have proved that a pair of dual fuzzy rough approximation operators can induce a topological space if and only if the fuzzy relation is reflexive and transitive. On the other hand, under certain conditions a fuzzy interior (closure) operator derived from a fuzzy topological space can associate with a reflexive and transitive fuzzy relation such that the induced fuzzy lower (upper) approximation operator is the fuzzy interior (closure) operator. In this paper, we only consider the fuzzy rough sets constructed by the triangle normal $T = \min$, we will generalize the research to the $(\mathcal{I}, \mathcal{T})$-fuzzy rough sets. The analysis will facilitate further research in uncertain reasoning under fuzziness.

Acknowledgement

This work was supported by a grant from the National Natural Science Foundation of China (No. 60373078)

References

1. Pawlak, Z.: Rough sets. International Journal of Computer and Information Science **11**(1982) 341–356
2. Pomykala, J.A.: Approximation operations in approximation space. Bulletin of the Polish Academy of Sciences: Mathematics **35**(1987) 653–662

3. Wybraniec-Skardowska, U.: On a generalization of approximation space. Bulletin of the Polish Academy of Sciences: Mathematics **37**(1989) 51–61
4. Yao, Y.Y.: Generalized rough set model. In: Polkowski, L., Skowron, A.(eds): Rough Sets in Knowledge Discovery 1. Methodology and Applications. Physica-Verlag, Heidelberg, 1998, pp.286–318
5. Zhang, W.-X., Leung, Y., Wu, W.-Z.: Information Systems and Knowledge Discovery. Science Press, Beijing, 2003
6. Zhang, W.-X., Wu, W.-Z., Liang, J.-Y., Li, D.-Y.: Rough Set Theory and Approach. Science Press, Beijing, 2001
7. Wu, W.-Z., Zhang, W.-X.: Neighborhood operator systems and approximations. Information Sciences **144**(2002) 201–217
8. Yao, Y.Y.: Relational interpretations of neighborhood operators and rough set approximation operators. Information Sciences **111**(1998) 239–259
9. Lin, T.Y., Liu, Q.: Rough approximate operators: axiomatic rough set theory. In: Ziarko, W.(eds.): Rough Sets, Fuzzy Sets and Knowledge Discovery. Springer, Berlin, 1994, pp.256–260
10. Thiele, H.: On axiomatic characterisations of crisp approximation operators. Information Sciences **129**(2000) 221–226
11. Yao, Y.Y.: Constructive and algebraic methods of the theory of rough sets. Journal of Information Sciences **109**(1998) 21–47
12. Boixader, D., Jacas, J., Recasens, J.: Upper and lower approximations of fuzzy sets. International Journal of General Systems **29**(2000) 555–568
13. Dubois, D., Prade, H.: Rough fuzzy sets and fuzzy rough sets. International Journal of General Systems **17**(1990) 191–208
14. Mi, J.-S., Zhang, W.-X.: An axiomatic characterization of a fuzzy generalization of rough sets. Information Sciences **160**(2004) 235–249
15. Morsi, N.N., Yakout, M.M.: Axiomatics for fuzzy rough sets. Fuzzy Sets and Systems **100**(1998) 327–342
16. Radzikowska, A.M., Kerre, E.E.: A comparative study of fuzzy rough sets. Fuzzy Sets and Systems **126**(2002) 137–155
17. Thiele, H.: On axiomatic characterisation of fuzzy approximation operators I, the fuzzy rough set based case. RSCTC 2000, Banff Park Lodge, Bariff, Canada, Oct. 19–19, 2000. Conference Proceedings, pp.239–247
18. Thiele, H.: On axiomatic characterisation of fuzzy approximation operators II, the rough fuzzy set based case. Proceedings of the 31st IEEE International Symposium on Multiple-Valued Logic, 2001, pp.330–335
19. Thiele, H.: On axiomatic characterization of fuzzy approximation operators III, the fuzzy diamond and fuzzy box cases. The 10th IEEE International Conference on Fuzzy Systems 2001, Vol. 2, pp.1148–1151
20. Wu, W.-Z., Leung, Y., Mi, J.-S.: On characterizations of (I, T)-fuzzy rough approximation operators. Fuzzy Sets and Systems (to appear)
21. Wu, W.-Z., Mi, J.-S., Zhang, W.-X.: Generalized fuzzy rough sets. Information Sciences **151**(2003) 263–282
22. Wu., W.-Z., Zhang, W.-X.: Constructive and axiomatic approaches of fuzzy approximation operators. Information Sciences **159**(2004) 233–254
23. Yao, Y.Y.: Combination of rough and fuzzy sets based on α-level sets. In: Lin, T. Y., Cercone, N.(eds.): Rough Sets and Data Mining: Analysis for Imprecise Data. Kluwer Academic Publishers, Boston, 1997, pp.301–321
24. Nakamura, A., Gao, J. M.: On a KTB-modal fuzzy logic. Fuzzy Sets and Systems **45**(1992) 327–334

25. Yao, Y.Y., Lin, T.Y.: Generalization of rough sets using modal logic. Intelligent Automation and Soft Computing: An International Journal **2** (1996) 103–120
26. Wasilewska, A.: Conditional knowledge representation systems—model for an implementation. Bulletin of the Polish Academy of Sciences: Mathematics **37**(1989) 63–69
27. Wu, W.-Z., Leung, Y., Zhang, W.-X.: Connections between rough set theory and Dempster-Shafer theory of evidence. International Journal of General Systems **31**(2002) 405–430
28. Yao, Y. Y., Lingras, P.J.: Interpretations of belief functions in the theory of rough sets. Information Sciences **104**(1998) 81–106
29. Chuchro, M.: On rough sets in topological Boolean algebras. In: Ziarko, W.(ed.): Rough Sets, Fuzzy Sets and Knowledge Discovery, Springer-Verlag, New York, 1994, pp.157–160
30. Chuchro, M.: A certain conception of rough sets in topological Boolean algebras. Bulletin of the Section of Logic **22**(1)(1993) 9–12
31. Kortelainen, J.: On relationship between modified sets, topological space and rough sets. Fuzzy Sets and Systems **61**(1994) 91–95
32. Wiweger, R.: On topological rough sets. Bulletin of Polish Academy of Sciences: Mathematics **37**(1989) 89–93
33. Lowen, R.: Fuzzy topological spaces and fuzzy compactness. Journal of Mathematical Analysis and Applications **56**(1976) 621–633

A Case Retrieval Model Based on Factor-Structure Connection and $\lambda-$Similarity in Fuzzy Case-Based Reasoning

Dan Meng[1], Zaiqiang Zhang[2], and Yang Xu[3]

[1] School of Economics Information engineering,
Southwest University of Finance and Economics,
Chengdu, Sichuan, China, 610074
mengd_t@swufe.edu.cn
[2] School of Economics and Management, Southwest Jiaotong University,
Chengdu, Sichuan, China, 610031
nicelark@sina.com
[3] Intelligent Control Development Center, Southwest Jiaotong University,
Chengdu, Sichuan, China, 610031
xuyang@home.swjtu.edu.cn

Abstract. One of the fundamental goals of artificial intelligence (AI) is to build artificially computer-based systems which make computer simulate, extend and expand human's intelligence and empower computers to perform tasks which are routinely performed by human beings. Without effective reasoning mechanism, it is impossible to make computer think and reason. Research on reasoning is an interesting and meaningful problem. Case-based reasoning is a branch of reasoning research. There are a lot of research results and successful application on Case-based reasoning. How to index and retrieve similar cases is one of key issue in Case-based reasoning research hotspots in CBR research works. In addition, there is a lot of uncertainty information in daily work and everyday life. So how to deal with uncertainty information in Case-based Reasoning effectively becomes more and more important. In this paper, a new case indexing and retrieval model which can deal with both fuzzy information and accurate information is presented.

1 Introduction

Case-based reasoning (CBR) is an effective reasoning technique which has been widely used in many real-world applications such as disease-diagnose, decision support system, classification, planning, configuration, and design and so on [2,3,4,6,7,11,12].

There are a lot of research work on CBR[1,4,5,7,11,10,12]. However, there are still some problems. For example, some methods only are fit for qualitative attributes, but not for quantitative attribute cases efficiently, especially for the fuzzy quantitative attributes. In addition, many previous methods didn't refer to the inner logic relations which exist in each case assuredly or can't deal with inner logic relation reasonably. So a new indexing model is discussed in this paper. It will have the following characteristics. First, the presentation of cases is based on all kinds of logic relations, which make the indexing and retrieval process more clearly and simply. Second, the model can deal with case which has qualitative attributes and quantitative attribute.Third, the case is stored in case base in the form of factor and its corresponding weight, which make the storage of case base be less than the previous methods. Fourth, The determination of threshold λ and α can be dynamically.

2 The Presentation of Cases

In this paper, the presentation will be based on the factor space[9]. And we use a new unity model – representation based on universal logic relation to represent cases. This method eliminates the logic relation formally, but the virtual relation is reserved.

Definition 1. *Presentation of universal logic relation*

1. *if the relation of premises is conjunction or fuzzy conjunction or fuzzy-weight conjunction, then the case is expressed in 'IF $A_1, \cdots, A_n$, THEN B'($i = 1, 2, \cdots, n$);*
2. *if the relation of premises is disjunction or fuzzy disjunction or fuzzy-weight disjunction, then decompose 'IF $A_1 \vee A_2 \vee \cdots \vee A_n$, THEN B' into 'IF A_i THEN B'($i = 1, 2, \cdots, n$);*
3. *if the relation of premises is both conjunction or fuzzy conjunction or fuzzy-weight conjunction and disjunction or fuzzy disjunction or fuzzy-weight disjunction or several of them, then turn it into a disjunction normal form[13], then decompose it in 1,2 respectively;*
4. *if the relation of premises is feeble logic relation, then express the case into 'IF $A_1, \cdots, A_n$, THEN B'. By the above method, every case can be expressed into 'IF $A_1, \cdots, A_n$, THEN B'.*

Definition 2. *A fuzzy set consisting of atom factors and their corresponding weights is called an atom factor fuzzy set, which has the form of $\{\omega_1/f_1, \cdots, \omega_n/f_n\}$.*

In the following , we express the case (whether for query or for reasoning) in a atom factor fuzzy set $\{\omega_1/g_1, \cdots, \omega_m/g_m\}$ and its atom factor state set $\{A_1, \cdots, A_m\}$,wherein g_m is factor, ω_i is corresponding weight, A_i is fuzzy set (including classic set).

3 Indexing and Retrieval Model Based on Factor-Structure Connection and λ-Similarity

3.1 Factor-Structure Connection Among Cases

Definition 3. $SR(a_i, b_j)$ *is called semantic connection between atom factors a_i and b_j:*

$$SR(a_i, b_j) = \begin{cases} 1, & a_i = b_j \\ 0, & a_i \neq b_j. \end{cases} \qquad (1)$$

Definition 4. *Let atom factor fuzzy set of case A, B be $A = \{p_1/a_1, \cdots, p_n/a_n\}$, $B = \{q_1/b_1, \cdots, q_n/b_n\}$ respectively, such that*

1. $\sum_{i=1}^{n} p_i \leq 1, \sum_{j=1}^{n} q_i \leq 1$;
2. $SR(a_i, b_i) = 1$, *where* $i = 1, 2, \cdots, n$,

then $SR(A, B)$ is called factor-structure connection of case A,B,

$$SR(A, B) = \sum_{i=1}^{k}(1 - |p_i - q_i|) \times (p_i \otimes q_i), \qquad (2)$$

where $\otimes$ is average or min, $\times$ is multiplication.

Definition 5. *Let atom factor fuzzy set of A, B be $A = \{p_1/a_1, \cdots, p_n/a_n\}$, $B = \{q_1/b_1, \cdots, q_m/b_m\}$ respectively, and factor set of A, B be $F_A = \{a_1, \cdots, a_n\}$, $F_B = \{b_1, \cdots, b_m\}$ respectively, if $F_A \cap F_B = k, k \leq min(m, n)$, arrange the same atom factor and its corresponding weight of A, B and delete the different factors and weights. Note the worked reminder fuzzy set is $A^* = \{p_1^*/a_1, \cdots, p_k^*/a_k\}$, $B^* = \{q_1^*/a_1, \cdots, q_k^*/a_k\}$. Without loss of universality, give the definition of factor-structure connection of A, B as following*

$$SR(A, B) = SR(A^*, B^*) = \sum_{i=1}^{k}(1 - |p_i^* - q_i^*|) \times (p_i^* \otimes q_i^*) \qquad (3)$$

We can compute the factor-structure connection of existed case and problem case A_j, A_0 $SR(A_j A_0)$ by definition 4 and definition 5, The next step is to judge if their states are similar.

3.2 λ-Similarity of Cases

Definition 6. *Define adjacency field $D_\lambda(A, B)$ of fuzzy set A, B as following: $D_\lambda : F(U) \times F(U) \to P(U)$:*

$$\forall (A, B) \in F(U) \times F(U), D_\lambda(A, B) = \{u | u \in U \text{ and } d(A(u), B(u)) < \lambda\}, \qquad (4)$$

$\lambda \in [0, 1]$, $d(a, b)$ *is some distance on $[0, 1]$.*

Definition 7. *Let U be a finite discrete domain, define λ-similarity S_λ of fuzzy sets A, B as follows $S_\lambda : F(U) \times F(U) \to [0, 1]$,*

$$\forall (A, B) \in F(U) \times F(U), S_\lambda(A, B) = \frac{|D_\lambda|}{|U|} \qquad (5)$$

wherein $\lambda \in [0, 1]$, $D_\lambda(A, B)$ is adjacency field of fuzzy set A, B.

Let the remainder state value set of problem A_0 be $\{A_{01}, \cdots, A_{0k}\}$, the remainder state value set of existent cases A_j be $\{A_{j1}, \cdots, A_{jk}\}$, then we can get $S_\lambda(A_{ji}, A_{0I})$, $i = 1, 2, \cdots$ and by the following formula

$$S_\lambda(A_j, A_0) = \sum_{i=1}^{k} (\omega_{ji} \otimes w_{0i}) \circ S_\lambda(A_{ji}, A_{0i}) \qquad (6)$$

where ω_{ji}, w_{0i} is weight of the same factor of existent case and problem respectively, $\otimes$ is average or min, $\circ$ is min or multiplication.

4 Conclusion

In this paper, an indexing model in fuzzy Case-based reasoning is given. It is more suitable for the cases that have qualitative or quantitative attribute cases and for the precise or fuzzy attribute value cases theoretically. However, how to define the weights of each factor reasonably and how to define the membership of fuzzy set appropriately still should be further discussed.[1]

References

1. He Xingui: fuzzy data-base systems (in Chinese). Tsinghua university press,Beijing, (1994), 61-63
2. He Xingui: The theory and technology of fuzzy knowledge processing.Beijing: National Defence Industry Press, (1994), 23-32,178-182,
3. He Xingui: Weighted fuzzy logic and its application in different fields. Chinese Journal of Computers(in Chinese),(1989), 458-464, 12(6),
4. Li and D.Xu: Case-based Reasoning. IEEE Potentials, (1994), 10-14
5. Nguyen Hoang Phuong, Nguyen Ba Tu et al: Case Based Reasoning Using Fuzzy Set Theory and the Importance of Features in Medicine. IEEE, (2001), 872-876
6. Shu Chang etal.: The feeble logic of Compound fuzzy proposition and it's algorithm. Chinese Journal of Computers (in Chinese),(2000),272-277, 23(3)
7. Tsatoulis C and Lasbyap RL: Case-Based reasoning and learning In manufacturing with the TOLTEC planne. IEEE Transaction on Systems, Man, and cybernetics, (1993), 1010-1023, 23(4)
8. Wei-Chou Chen et al.: A Framework of Features for the Case-based Reasoning. IEEE, (2000), 1-5
9. Wang Peizhuang and Li Hongxing: Fuzzy system and fuzzy computer(in Chinese). Science press, (1996), 29-59
10. Wang Mingyang et al.: ICIMX:An expert systems with case-based reasoning. Chinese Journal of Computers (in Chinese),(1997), 105-110, 20(2)
11. Xu Ming and Hu Shouren: Research on indexing model for Case-based reasoning. Computer Science (in Chinese), (1993), 32-35, 20(4)
12. X. Z. Wang,and D. S. Yeung: Using Fuzzy Integral to Modeling Case Based Reasoning with Feature Interaction. IEEE, (2000), 3660-3665

---

[1] This work was supported by the Scientific Research Fund of Southwestern University of Finance and Economics, the National Natrual Science Fund of P.R.China (Grant No. 60474022) and Innovation Fund of SWJTU for Ph.D.

A TSK Fuzzy Inference Algorithm for Online Identification

Kyoungjung Kim[1], Eun Ju Whang[1], Chang-Woo Park[2], Euntai Kim[1], and Mignon Park[1]

[1] Department of electrical and electronic engineering, Yonsei University,
134, Shinchon-dong, Sudaemoon-gu, Seoul, Korea
{kjkim01, garung, etkim, mignpark}@yonsei.ac.kr
http://yeics.yonsei.ac.kr
[2] Korea electronics technology institute,
401-402 B/D 192, Yakdae-Dong, Wonmi-Gu, Buchon-Si, Kyunggi-Do, Korea
{drcwpark}@keti.re.kr

Abstract. This paper proposes an online self-organizing identification algorithm for TSK fuzzy model. The structure of TSK fuzzy model is identified using distance. Parameters of the piecewise linear function consisting consequent part are obtained using recursive version of combined learning method of global and local learning. Both input and output spaces are considered in the proposed algorithm to identify the structure of the TSK fuzzy model. By processing clustering both in input and output space, outliers are excluded in clustering effectively. The proposed algorithm is non-sensitive to noise not by using data itself as cluster centers. The proposed algorithm can obtain a TSK fuzzy model through one pass. By using the proposed combined learning method, the estimated function can have high accuracy.

1 Introduction

TSK fuzzy has been used in various applications. In recent years, many studies on fuzzy model based identification of nonlinear dynamic systems have been conducted. System identification can be classified into two groups roughly, that is, 1) off line identification [12], [21], [22], [23], 2) online identification [1-5], [7], [9-11], [13], [15], [20], [28-30]. Off line identification supposes that all the data is available at the start of the process of training. In online identification, model structure and parameter identification is required. Clustering method is used for structure identification mainly. In clustering, to update fuzzy rules, distance from centers of fuzzy rules, potentials of new data sample, or error from previous step have been used. Different mechanisms are used in constructing structure. Self-organizing fuzzy neural network (SOFNN) [1], self-constructing fuzzy neural network SCFNN [5], dynamic fuzzy neural network (D-FNN) [9], [10] and general dynamic fuzzy neural network (GD-FNN) [8] use error to update fuzzy rules. Dynamic evolving neural-fuzzy inference system (DENFIS) [3] and self-constructing neural fuzzy inference network (SONFIN) [4] use distance to update fuzzy rules. Evolving Takagi-Sugeno model (ETS) [2] uses potential to update fuzzy rules. There are other approaches; that is, evolving fuzzy neural networks (EFuNNs) [15] using fuzzy difference between two membership vector, [20], neural fuzzy control network (NFCN) [7] using kohonen's feature maps, neuro-fuzzy ART-based structure and parameter learning TSK model (NeuroFAST) [11] using ART concept and generalized adaptive neuro-fuzzy inference systems (GANFIS) [13] using modified mountain clustering. To learn parameters of consequent part, recursive least square (RLS) [2], [4], [9,10], weighted recursive least square (wRLS) [2] are used mainly, and also singular value decomposition (SVD) [6], δ rule [11] is used.

A new online structure identification algorithm is proposed in this paper. Generally, clustering is considered in input space only. Because not only input but also output include important information of the system, we considered clustering in both input and output space. Also, by not take data points as centers of clusters, the proposed algorithm shows non-sensitive characteristics to noise. We use maximum distance from centers to update rules. To learn parameters of consequent part, we propose a method to learn globally and locally. Parameters of consequence are calculated recursively using recursive version of the proposed learning method.

The rest of the paper is organized as follows. In section 2, we describe TSK fuzzy briefly, and a present new structure identification algorithm. New Parameter learning algorithm is proposed also in section 2. In section 3, Procedure for online structure and parameter identification algorithm is presented. In section 4, test results are given. Conclusions are presented in section 5.

2 TSK Fuzzy Modeling Algorithm

TSK fuzzy model can present static or dynamic nonlinear system. This paper deals with dynamic nonlinear system. TSK fuzzy model has premise parts of membership function and consequent parts consisting of piecewise linear functions. TSK fuzzy model has the form as

$$R^i : \quad \text{If } x_1 \text{ is } A_1^i(\vec{\theta}_1^i) \text{ and } x_2 \text{ is } A_2^i(\vec{\theta}_2^i), \ldots, x_n \text{ is } A_n^i(\vec{\theta}_n^i) \quad (1)$$

$$\text{then } h^i = b_0^i + b_1^i x_1 + b_2^i x_2 + \cdots + b_n^i x_n$$

where $i = 1, 2, \cdots, r$, r is the number of rules, $A_j^i(\vec{\theta}_j^i)$ is the fuzzy set of the j th rule for x_j with the adjustable parameter set $\vec{\theta}_j^i$, and $\vec{b}^i = (b_0^i, \cdots, b_n^i)$ is the parameter set in the consequent part.

The predicted output of the fuzzy model can be inferred from (1) as

$$\hat{y} = \frac{\sum_{i=1}^{C} h^i w^i}{\sum_{i=1}^{C} w^i} \quad (2)$$

where h^i is the output of the i th rule, w^i is the i th rule's firing strength, which is obtained as the minimum of the fuzzy membership functions

$$A_j^i(C_{ij}, \sigma_{ij}) = \exp\left\{-\frac{(x_j - C_{ij})^2}{2\sigma_{ij}^2}\right\} \quad (3)$$

where C_{ij} is cluster center, and σ_{ij} is width of membership function.

We describe the method to find centers and widths of membership functions of premise parts and parameters of consequences fast and effectively.

2.1 Structure Identification

Structure identification is conducted by clustering data in input and output space and deciding centers. We suppose fuzzy membership function has gaussian type. Clustering process is presented in 2-dimensional space.

We calculate distances between incoming data and all existing cluster centers in input and output space as

$$d_{ij}^{in} = \|x_i - C_j^x\|, \ i = 1, \cdots, n, \ j = 1, \cdots, m \quad (4)$$

$$d_{ij}^{out} = \|y_i - C_j^y\|, \ i = 1, \cdots, n, \ j = 1, \cdots, m \quad (5)$$

where x_i is input data, y_i is output data, d_{ij}^{in} is the distance between i th incoming data and j th cluster center in input space, d_{ij}^{out} is the distance between i th incoming data and the center of cluster j in output space and n, m is number of data and number of clusters respectively. C_j^x is center in input space and C_j^y is center in output space.

Let $\overline{d}^{in}$, $\overline{d}^{out}$ predefined upper bound in input space and output space respectively. $MAXD_i$ represents maximum distance between the center cluster j and data points included in cluster j.

1) Create a new cluster and take the first incoming data as a position of the cluster center. Initialize width of membership function with small random number.

$$C_1 = [x_1, y_1], \ \sigma_1 = \sigma_0$$

And, initialize maximum distance from cluster center to 0.

When k th data sample enters, calculate distance between incoming data and all existing cluster center in input space and output space, and find minimum distance.

With minimum distance obtained from the above, create new cluster and update center position and width as the following.

- $\min(d_{ij}^{in}) > \overline{d}^{in}$ and $d_{ij}^{out} > \overline{d}^{out}$: When minimum distance is greater than upper bound in input space and distance between data points and corresponding cluster is greater than upper bound in output space, incoming data is considered to outlier. In this case, the center and width of cluster and the number of cluster is not changed.

$$C_j = C_j, \sigma_j = \sigma_j, m(t+1) = m(t) \tag{6}$$

- $\min(d_{ij}^{in}) > \overline{d}^{in}$ and $d_{ij}^{out} \leq \overline{d}^{out}$: When minimum distance is greater than upper bound in input space and distance between incoming data and corresponding cluster is smaller than or equal to upper bound in output space, create a new cluster and increase the number of cluster. Initialize cluster center and width as

$$C_j = [x_k, y_k], \sigma_j = \sigma_0, m(t+1) = m(t)+1 \tag{7}$$

- $\min(d_{ij}^{in}) \leq \overline{d}^{in}$ and $d_{ij}^{out} > \overline{d}^{out}$: When minimum distance is smaller than or equal to upper bound in input space and distance between incoming data and corresponding cluster is greater than upper bound in output space, incoming data is considered to outlier and the number, center and width of cluster is maintained as (6).

- $\min(d_{ij}^{in}) \leq \overline{d}^{in}$ and $d_{ij}^{out} \leq \overline{d}^{out}$: When minimum distance is smaller than or equal to upper bound in input space and distance between incoming data and corresponding cluster center is smaller than or equal to upper bound in output space, incoming data is considered to the data points included in cluster j, and center position and width of cluster is updated. In this case, the number of cluster is maintained.

In case of $MAXD_i \geq d_{ij}^{in}$, update center position as follows

$$C_i(t+1) = C_i(t) + \alpha(D_{i\max} - d_{ij}) \tag{8}$$

$$D_{i\max} = d_{ij}^{in} \tag{9}$$

where α is update ratio.

In case of $MAXD_i < d_{ij}^{in}$, maintain center position and maximum distance between cluster j and any data points included in cluster j.

Next, we determine width using distance between cluster center and each data points included in this cluster.

$$S_j(t+1) = S_j(t) + d_{ij}^2 \tag{10}$$

$$\sigma_j^2 = S_j / g \tag{11}$$

where g represents the number of data included cluster j.

We can construct fuzzy structure from the above process. After the structure is obtained, we can learn parameters of consequence.

2.2 Parameter Identification

In this paper, we use a combined cost function of global and local approximation [6] to estimate the parameters of the consequent part.

Each piecewise linear function of consequence can be represented as

$$\hat{y} = b_0 + b_1 x_1 + b_2 x_2 + \cdots + b_q x_q \tag{12}$$

The cost function for global learning is as follows

$$J_G = \sum_{k=1}^{N} [y(k) - \hat{y}(k)]^2 \tag{13}$$

where $y(k)$ is the output of the real system, $\hat{y}(k)$ is the output of the identified model, and N is the number of training data.

The cost function for local learning is as follows

$$J_L = \sum_{i=1}^{L} \sum_{k=1}^{N} [y(k) - w_i Xb]^2 \tag{14}$$

The cost function of combined learning is as follows

$$J = \gamma_1 (\mathbf{y} - \mathbf{Xb})^T (\mathbf{y} - \mathbf{Xb}) + \gamma_2 (\mathbf{y} - \mathbf{WXb})^T (\mathbf{y} - \mathbf{WXb}) \tag{15}$$

where γ_1 and γ_2 are two positive constants satisfying (16)

$$\gamma_1 + \gamma_2 = 1 \tag{16}$$

From (15) b can be obtained as

$$\mathbf{b} = (\mathbf{X}^T (\gamma_1 \mathbf{I} + \gamma_2 \mathbf{W}^T \mathbf{W}) \mathbf{X})^{-1} (\gamma_1 \mathbf{I} + \gamma_2 \mathbf{W}) \mathbf{X}^T \mathbf{y} \tag{17}$$

We can rearrange (17) as

$$\mathbf{b} = \mathbf{P}(\gamma_1 \mathbf{I} + \gamma_2 \mathbf{W}) \mathbf{X}^T \mathbf{y} \tag{18}$$

$$\mathbf{P} = (\mathbf{X}^T (\gamma_1 \mathbf{I} + \gamma_2 \mathbf{W}^T \mathbf{W}) \mathbf{X})^{-1} \tag{19}$$

Now, we can represent (18) and (19) to a recursive way with forgetting factor as follows

$$\mathbf{b}_{k+1} = \mathbf{b}_k + (\gamma_1 + \gamma_2 w_{k+1}) \mathbf{P}_{k+1} \mathbf{x}_{k+1} (y_{k+1} - \mathbf{x}_{k+1}^T \mathbf{b}_k) \tag{20}$$

$$P_{k+1} = \frac{1}{\lambda}\left(P_k - \frac{(\gamma_1 + \gamma_2 w_{k+1}^T w_{k+1}) P_k x_{k+1} x_{k+1}^T P_k}{\lambda + x_{k+1}^T P_k x_{k+1}}\right) \quad (21)$$

where λ is a forgetting factor which typical value is between 0.8-1.

Initial value of **P** and **b** can be calculated from (20) and (21) applying to the center and the first data points included in the center.

3 Procedure for Online Identification

We proposed the structure and parameter identification algorithm in section 2. Now, we describe the procedure for the online learning.

The online learning procedure of the proposed algorithm is described as follows

Step 1) Set upper bound $\overline{d}^{in}$, $\overline{d}^{out}$ in input and output space, center update ratio α, forgetting factor λ and two positive constant γ_1 and γ_2.

Step 2) Create the first cluster and take the first data sample as the center position, and set width σ with small random number.

Step 3) When k th data point enters, calculate distances between entered data and all existing cluster centers, and update center position according to (6)-(9).

Step 4) Calculate the widths of membership functions according to (10) and (11)

Step 5) Consequence parameters can be calculated recursively from (20) and (21).

Step 6) Stop the procedure if there is no more data to learn.

By conducting above procedure, we can obtain fuzzy model that is non-sensitive to outliers and noise. Because it dose not memorize many data and conduct procedure through one pass, the proposed algorithm shows fast response.

4 Experimental Results

In this section, we test the proposed algorithm. We use Mackey-Glass (MG) data that used as a standard example in area of fuzzy system and neural network in [2], [3], [15].

Time series data to use in test are generated using MG time delay differential equation as

$$\frac{dx(t)}{dt} = \frac{0.2 x(t-\tau)}{1 + x^{10}(t-\tau)} - 0.1 x(t) \quad (22)$$

In this tests data obtained when $x(0) = 1.2$, $\tau = 17$, and $x(t) = 0$ for $t < 0$ are used. The I/O data for the proposed algorithm from the Mackey-Glass time series data is in the following format: input vector $[x(t), x(t-6), x(t-12), x(t-18)]$ and output vector $[x(t+6)]$. The task is to predict future values x(t+6) from four points spaced at six time intervals in the past. We test using 500 data samples from $t = 5001$ to $t = 5500$.

To evaluate the performance of the proposed algorithm, non-dimensional error index (NDEI) in [2], [3], [15] is used. NDEI is the value of RMSE divided by the standard deviation of the target series.

In this test, we set update ratio of cluster center, α, in the proposed algorithm to 0.5 and forgetting factor λ to 0.8. Figure 1 shows identification results of the proposed algorithm for the data samples obtained from (22). We set upper bound in input space to 0.2 and upper bound in output space to 0.8. The number of fuzzy rules is 100 in figure 1. Solid line represents real function and dotted line represents estimated function. We can see that the proposed algorithm estimate real function well from Figure 1. In Figure 1, weight γ_1, γ_2 is 0.8 and 0.2 respectively.

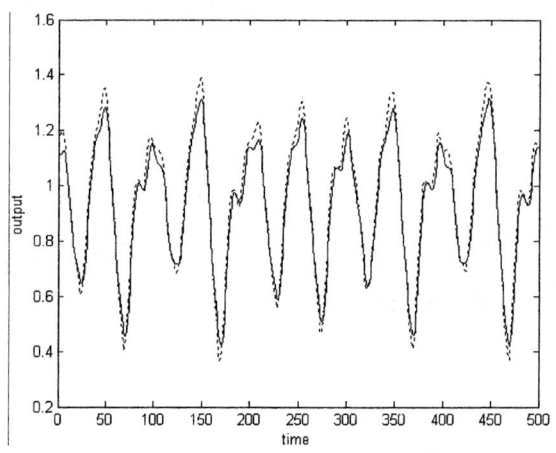

Fig. 1. Estimation results (—: real function, •••: estimated function)

Table 1 presents experimental results compared with other algorithms. We set upper bound in input space to 7, upper bound in output space to 0.8, γ_1 and γ_2 to 0.8 and 0.2 respectively. In this case, TSK fuzzy model has 72 rules. We use eTS model [2], DENFIS [3] and EFuNN [15] to compare with the proposed algorithm.

Table 1. Results compared with other algorithms

Methods	Rules	NDEI
DENFIS	58	0.276
eTS Model	113	0.0954
EfuNN	1125	0.094
The proposed algorithm	72	0.0172

From Table 1, we can know that the proposed algorithm is superior to other algorithms when it applied to data samples without noise.

We test whether the proposed algorithm can estimate with noisy data well. We obtain noisy data adding noise to data sample from (24).

$$F = (1-\varepsilon)G + \varepsilon H \qquad (24)$$

where F is distribution of noise and G and H are probability distribution of probability of $1-\varepsilon$ and ε respectively.

Figure 2 presents real function and noisy data. Here, we use $G \sim N(0,0.05)$, $H \sim N(0,0.1)$ and $\varepsilon = 0.05$.

From the above experimental results, we can know that the proposed algorithm shows superior performance to other algorithms. It has smaller number of fuzzy rules than other algorithms. And, it can approximate well with noisy data. It shows robust

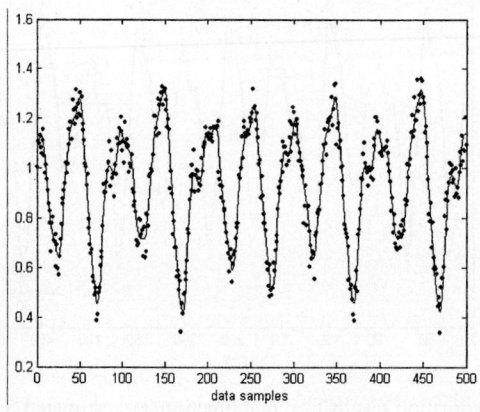

Fig. 2. Data samples with noise (—: real function, •: noisy data)

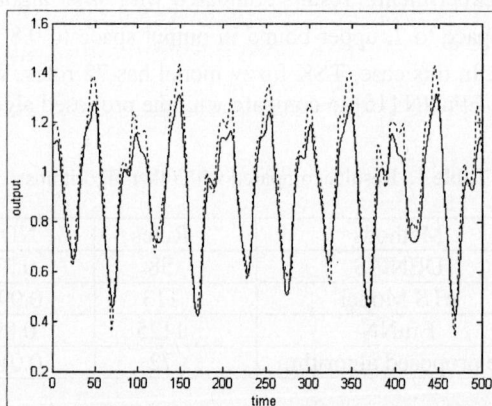

Fig. 3. Estimation results with noisy data (—: real function, •••: estimated function)

performance against noise. Because it has smaller number of fuzzy rules than other algorithms, it can reduce computational efforts.

5 Conclusions

This paper proposes self-organizing online TSK fuzzy identification algorithm. Distance is used to update fuzzy rule and recursive version of combined learning method of global and local learning is used to calculate parameters of the consequence in the proposed algorithm. The proposed algorithm has a robust performance not by using data points itself as cluster centers. Outliers are excluded from data estimation by considering both input and output space in clustering. We calculate widths of membership functions recursively with squared distance between cluster center and temporal data. The proposed algorithm shows robust performance against noise, and it is effective in identifying the system that includes noise.

References

1. Leng, G., Prasad, G. and McGinnity T. M.: An on-line algorithm for creating self-organizing fuzzy neural networks. Neural Networks, Vol. 17 (2004) 1477-1493
2. Angelov, P. P. and Filev, D. P.: An Approach to Online Identification of Takagi-Sugeno Fuzzy Models. IEEE Trans. System, Man, And Cybernetics-part B, Vol. 34 (2004) 484-498
3. Kasabov, N. and Song, Q.: DENFIS: Dynamic Evolving Neural-Fuzzy Inference System and Its Application for Time-Series Prediction. IEEE Trans. Fuzzy Systems, Vol. 10 (2004) 144-154
4. Juang, C.-F. and Lin, C.-T.: An On-line Self-Constructing Neural Fuzzy Inference Network and Its Applications. IEEE Trans. Fuzzy Systems, Vol. 6 (1998) 12-32
5. Lin, F.-J., Lin, C.-H. and Shen, P-H.: Self-Constructing Fuzzy Neural Network Speed Controller for Permanent-Magnet Synchronous Motor Drive. IEEE Trans. Fuzzy Systems, Vol. 9 (2001) 751-759
6. Yen, J., Wang, L. and Gillespie, C. W.: Improving the Interpretability of TSK Fuzzy Models by Combining Global Learning and Local learning. IEEE Trans. Fuzzy Systems, Vol. 6 (1998) 530-537
7. Lin, C.-T.: A neural fuzzy control system with structure and parameter learning. Fuzzy Sets and Systems, Vol. 70 (1995) 183-212
8. Wu, S., Er, M. J. and Gao, Y.: A Fast Approach for Automatic Generation of Fuzzy Fules by Generalized Dynamic Fuzzy Neural Networks. IEEE Trans. Fuzzy Systems, Vol. 9 (2001) 578-594
9. Er, M. J. and Wu, S.: A fast learning algorithm for parsimonious fuzzy neural systems. Fuzzy Sets and Systems, Vol. 126 (2002) 337-351
10. Wu, S. and Er, M. J.: Dynamic Fuzzy Neural Networks-A Novel Approach to Function Approximation. IEEE Trans. Systems, Man, and Cybernetics-Part B, Vol. 30 (2000) 358-364
11. Tzafestas, S. G. and Zikidis, K. C.: NeuroFAST: On-line Neuro-Fuzzy ART-Based Structure and Parameter Learning TSK Model. IEEE Trans. Systems, Man, and Cybernetics-Part B, Vol. 31 (2001) 797-802

12. Kukolj, D. and Levi, E.: Identification of Complex Systems Based on Neural and Takagi-Sugeno Fuzzy Model. IEEE Trans. Systems, Man, and Cybernetics-PART B, Vol. 34 (2004) 272-282
13. Azeem, M. F., Hanmandlu, M. and Ahmad, N.: Structure Identification of Generalized Adaptive Neuro-Fuzzy Inference Systems. IEEE Trans. Fuzzy Systems, Vol. 11 (2003) 666-681
14. Liu, P. X. and Meng, M. Q.-H.: Online Data-Driven Fuzzy Clustering With Applications to Real-Time Robotic Tracking. IEEE Trans. Fuzzy Systems, Vol. 12 (2004) 516-523
15. Kasabov, N.: Evolving Fuzzy Neural Networks for Supervised/Unsupervised Online Knowledge-Based Learning. IEEE Trans. Systems, Man, and Cybernetics-PART B, Vol. 31 (2001) 902-918
16. Angelov, P. P., Hanby, V. I., Buswell, R. A. and Wright, J. A.: Automatic generation of fuzzy rule-based models from data by genetic algorithms. in Advances in Soft Computing, R. John and R. Birkinhead, Eds. Heidelberg, Germany: Springer-Verlag (2001) 31-40
17. Angelov, P. P.: Evolving Rule-Based Models: A Tool for Design of Flexible Adaptive Systems. Heidelberg, Germany: Springer-Verlag (2002)
18. Takagi, T. and Sugeno, M.: Fuzzy identification of systems and its applications to modeling and control. IEEE Trans. Systems, Man, and Cybernetics, Vol. 15 (1985) 116-132
19. Wang, L. X.: A Course in Fuzzy Systems and Control. Prentice Hall (1997)
20. Yamakawa, T. and Matsumoto, G.: Methodologies for the Conception, Design, and Applications of Soft Computing, Eds. Singapore: World Scientific (1998) 271-274
21. Kim, K., Kim, Y.-K., Kim, E. and Park, M.: A New Fuzzy Modeling Approach. Proc. of FUZZ-IEEE 2004 (2004) 773-776
22. Kim, E., Park, M., Ji, S. and Park, M.: A new approach to fuzzy modeling. IEEE Trans. Fuzzy Systems, Vol. 5 (1997) 328-337
23. Kim, K., Kyung, K. M., Park, C.-W., Kim, E. and Park, M.: Robust TSK Fuzzy Modeling Approach Using Noise Clustering Concept for Function Approximation. LNCS, Vol. 3314 (2004) 538-543
24. Isermann, R., Lachmann, K.-H. and Matko, D.: Adaptive Control Systems. Prentice Hall (1992)
25. Ljung, L.: System Identification: Theory for the user. Prentice Hall (1998)
26. Goodwin, G. C. and Sin, K. S.: Adaptive Filtering Prediction and Control. Prentice Hall (1984)
27. Haykin, S.: Adaptive Filter Theory. Prentice Hall (1996)
28. Wai, R.-J., Chen, P.-C.: Intelligent Tracking Control for Robot Manipulator Including Actuator Dynamics via TSK-Type Fuzzy Neural Network.. IEEE Trans. Fuzzy Systems, Vol. 12 (2004) 552-559
29. Kiguchi, K., Tanaka, T., Fukuda, T.: Neuro-Fuzzy Control of a Robotic Exoskeleton With EMG Signals. IEEE Trans. Fuzzy Systems, Vol. 12 (2004) 481-490
30. Wang, L., Frayman, Y.: A dynamically generated fuzzy neural network and its application to torsional vibration control of tandem cold rolling mill spindles. Engineering Appl. Artificial Intelligence, Vol. 15 (2002) 541-550

Histogram-Based Generation Method of Membership Function for Extracting Features of Brain Tissues on MRI Images

Weibei Dou[1,2], Yuan Ren[1], Yanping Chen[3],
Su Ruan[2], Daniel Bloyet[2], and Jean-Marc Constans[4]

[1] Department of Electronic Engineering,
Tsinghua University, 100084 Beijing, China
douwb@tsinghua.edu.cn
[2] GREYC-ENSICAEN CNRS UMR 6072,
Bd. Maréchal Juin, 14050 Caen, France
wdou@greyc.ensicaen.fr
[3] Imaging Diagnostic Center, Nanfang Hospital, Guangzhou, China
[4] Unité d'IRM, EA3916, CHRU, Caen, France

Abstract. We propose a generation method of membership function for extracting features of brain tissues on images of Magnetic Resonance Imaging (MRI)[1]. This method is derived from histogram analysis to create a membership function. According to *a priori* knowledge given by the neuro-radiologist, such as the features of gray level of differentiate brain tissues in MR images, we detect the peak or valley features of the histogram of MRI brain images. Then we determine a transformation of the histogram by selecting the feature values to generate a fuzzy membership function that corresponds to one type of brain tissues. A function approximations process is used to build a continuous membership function. This proposed method is validated for extracting whiter matter (WM), gray matter (GM), cerebra spino fluid (CSF). It is evaluated also using simulated MR images with two different, T1-weighted, T2-weighted MRI sequences. The higher agreement with the reference fuzzy model has been discovered by kappa statistic.

1 Introduction

The feature extraction is very important process in a fusion system especially for the features fusion. Fuzzy set theory is an *empirical science*[1]. It represents those natural phenomenon which we observe in our real lives. A fuzzy set has been defined as a collection of some objects with membership degree [2]. A membership function represents a mapping of the elements of a universe of discourse to the unit interval [0, 1] in order to determine the degree to which each object is compatible with distinctive features to collect. The meaning and measurement to the membership degree is one of the main difficulties for determining a membership

---

[1] Project 60372023 supported by National Natural Science Foundation of China.

function. [2] had introduced six classes of some experimental methods that help to determine membership functions. We can put them into three main categories: Experiment-based fuzzy statistic; Experiment-based trichotomy; Parametric estimation based on fuzzy distribution. However the first two methods depend on some necessary experiment conditions and the limitation of the third one is how to determine the fuzzy distribution and the estimation criterion. Normally, the boundaries and the shapes of membership function must strictly correspond to an interpretation for an observation. So that an objective measurement is the first choice and the most suited to determine the membership function for representing a fuzzy set.

We have found a correlation between a histogram of MRI image and the interpretation of a radiologist for the image. It is the gist of that we propose a generation method (based on histogram) of membership function for extracting features of brain tissues on MRI images. The second section of this paper explains some of the relationship between the interpretation and gray level of MRI images. The third section presents the combination of histogram measurement and property ranking to determine the membership degree of some brain tissues. The validation is given at last by simulated MR images from BrainWeb [3] with two different, T1-weighted (T1), T2-weighted (T2) MRI sequences. The error measurement has been carried out and has shown that this is an efficient method of a membership function generation by using kappa statistic.

2   Measurement of the Membership Degree

MRI images can provide much information about brain tissues from a variety of excitation sequences. Generally there are three main tissues in brain image, white matter (WM), gray matter (GM) and cerebral spinal fluid (CSF). Due to the limited resolution of the acquisition system, the partial volume effects in which one pixel may be composed of multiple tissue types.

2.1   Observation Space and Object

Observation Space and Object. The universe of discourse $\mathbf{B} = \{v\}$ is the brain image of MRI, where $v = (x, y, z)$ is the coordinate of voxel. The various sequences images of MRI are called *signal intensity spaces* of $\mathbf{B}$, noted $\mathbf{SI} \in \mathbf{B}$. We focuss on the two subsets of $\mathbf{B}$, T1 and T2, noted as $\mathbf{T1} = \{(v, GL_{T1})\}$, $\mathbf{T2} = \{(v, GL_{T2})\}$, where GL denote the gray level or signal intensity of v. Such that $\mathbf{SI} = \{\mathbf{T1}, \mathbf{T2}\}$. The pair of v and GL, (v, GL) is our observation object.

Destination Space. Our task is to answer the question of "Which tissue does a voxel belonged to?" So the different tissues are our destination space noted as $\mathbf{Tiss} = \{\mathbf{CSF}, \mathbf{GM}, \mathbf{WM}, \ldots\}$, $\mathbf{Tiss} \in \mathbf{B}$.

Definition Fuzzy Set. Due to the partial volume effect, a fuzziness result is more suitable to answer the question mentioned above. Then this question has been mapped to some fuzzy sets of $\mathbf{B}$. View from a relation pair of the *signal* *intensity space* and the destination space, (**SI**, **Tiss**), we can construct the fuzzy sets as a map like $\mu_{\text{Tiss}}^{\text{SI}} : \mathbf{B} \rightarrow [0,1]$. It means that a voxel belongs to a destination space **Tiss** with the membership degree $\mu_{\text{Tiss}}^{\text{SI}}$ in a *signal intensity space* **SI**.

2.2 Interpretation of the Histogram Features

The gray level histogram of an image represents some statistic features. If we use GL_k to present k^{th} gray level, and the total number of gray level is L (for 8-bit image, L=256), $k = 0, 1, \ldots, L-1$. The histogram $p(GL_k)$ gives us an estimation of the probability of GL_k.

$$p(GL_k) = \frac{n_k}{N} \quad (1)$$

where n_k is the total number of pixel which their gray level is equal to GL_k. N is the total number of pixels in image.

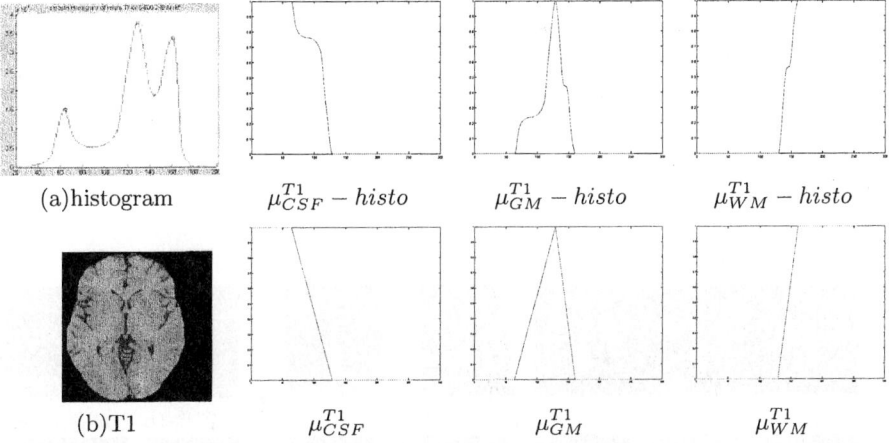

Fig. 1. Histogram and membership function. (a) is the histogram of whole sequence of the original image (b). Continuous membership function $(GL_{T1}, \mu_{CSF}^{T1})$, (GL_{T1}, μ_{GM}^{T1}), (GL_{T1}, μ_{WM}^{T1}) are shown in second line, and transformed histogram (first line) $(GL_{T1}, \mu_{CSF}^{T1} - histo)$, $(GL_{T1}, \mu_{GM}^{T1} - histo)$ and $(GL_{T1}, \mu_{WM}^{T1} - histo)$.

One example of histogram of whole sequence is shown in figure 1(a). What meaning are these peaks and valleys? By combining the gray level characteristics of the three main tissues on the different MRI sequences [4], we have found some features of the MRI image's histogram:

1. **The gray levels at the histogram peaks correspond to the feature of pure tissue types.** For example in figure 1(a), the peak maximum corresponds to the feature of pure GM, the middle-height peak corresponds to pure WM and the peak minimum corresponds to pure CSF.

2. **The gray levels at the histogram valleys correspond to the feature of mix tissues.** The valley means a transition process from one tissue type to another. The maximum valley in figure 1(a) corresponds to the mix of GM and WM, and the other one corresponds to the mix of CSF and GM.
3. **Normalized histogram provide an assessment of possibility.** The gray level histogram $p(GL_k)$ of an image provides the frequency or probability of the gray value GL_k in the image. If the histogram has been normalized with that of pure tissue corresponded, the normalized histogram can provide an assessment of possibility that the GL_k belongs to the pure tissue.

3   Generation of the Membership Function

We propose an approach for transforming the histogram into a membership function. The feature points for transformation are their peaks, noted as $(GL_{SI}^{pk}, p(GL_{SI}^{pk}))$ and their valleys, noted as $(GL_{SI}^{vl}, p(GL_{SI}^{vl}))$. Let $\mathcal{F}$ denote one of transformation operation. So the membership function is presented as equation 2, and an example of transformation from figure 1(a) is shown in the figure 2-(a).

$$\mu_{\text{Tiss}}^{\text{SI}} = \mathcal{F}\{(GL_{SI}, p(GL_{SI})), p(GL_{SI}^{pk}), p(GL_{SI}^{vl})\} \tag{2}$$

where $GL_{SI} = \{GL_{T1}, GL_{T2}\}$.

As an example, $\mu_{\text{Tiss}}^{\text{SI}}$ can be obtained with the properties ranking in **T1**,

$$\mu_{WM}^{T1} \propto GL_{T1}; \tag{3}$$

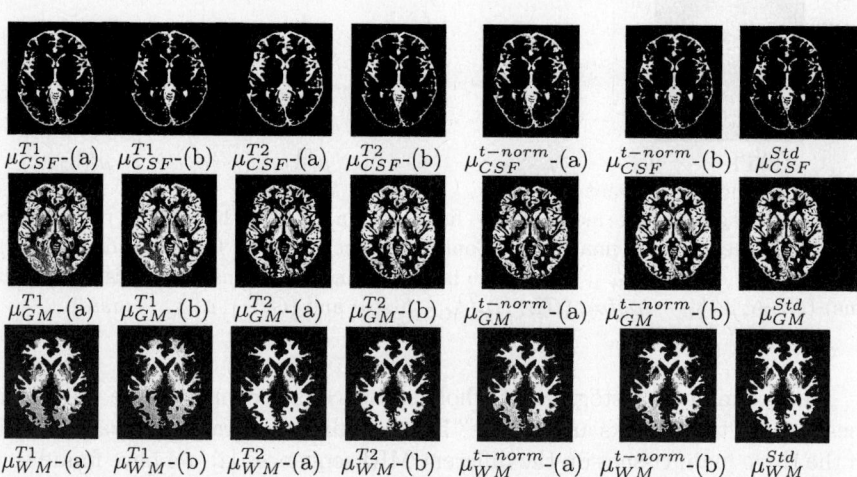

Fig. 2. Anatomical models μ_{Tiss}^{Std} and extraction results by the membership functions μ_{Tiss}^{SI}. μ_{Tiss}^{SI}-(a) are taken by transformation of histogram, μ_{Tiss}^{SI}-(b) are taken by approximating function, and μ_{Tiss}^{t-norm} are taken by operating t-norm between μ_{Tiss}^{T1} and μ_{Tiss}^{T2}

$$\mu_{CSF}^{T1} \propto \frac{1}{GL_{T1}}; \qquad (4)$$

$$\mu_{GM}^{T1} \propto \begin{cases} GL_{T1}, & if \ GL_{T1} \leq GL_{T1}^{pk} \\ \frac{1}{GL_{T1}}, & if \ GL_{T1}^{pk} < GL_{T1} \end{cases} \qquad (5)$$

For building a simplicity and continuous membership function of brain tissues on MRI images, function approximating approaches is firstly considered. According to the shape of μ_{Tiss}^{SI}, select a simpler continuous function to approximate them with a rule of minimizing error. For example, we determined a semi-trapezoid's taking lower function for approximating μ_{CSF}^{T1}, a triangle's taking middle function for μ_{GM}^{T1}, and a semi-trapezoid's taking upper function for μ_{WM}^{T1}. These approximation function are illustrated in figure 1. The parameters of these functions have been determined of course using the boundary points of μ_{Tiss}^{SI}, i.e. the feature points in histogram, such as the peaks or valleys.

4 Fuzzy Feature Extraction and Evaluation

The simulated MRI volumes, available on the BrainWeb [3], are used to evaluate our method. For extracting their features from original images, a geometric mean as t-norm operation between μ_{Tiss}^{SI} is used as a fusion operator of fuzzy information fusion system proposed in [4]. These results are shown in figure 2. Figure 2-(a) derived directly from the transformed histogram μ_{Tiss}^{SI}, and -(b) derived from the approximated functions of μ_{Tiss}^{SI} mentioned in section 3.

The fuzzy anatomic models provided by BrainWeb [3], are utilized as the standard sets μ_{Tiss}^{Std} shown in figure 2. The error measurement between μ_{Tiss}^{Std} and μ_{Tiss}^{SI} are done with two methods: detection error of binary set and a measure of agreement using kappa statistic [5]. Because of the fuzzy value $\mu = 0.5$ is a very important feature point which the maximum fuzziness value is in fuzzy set. So the measurement has done by separating into two cases: $\mu_{Tiss} > 0$ and $\mu_{Tiss} \geq 0.5$.

We define the **relative miss detection** η_{rm} in (6) and the **relative false detection** η_{rf} in (7).

$$\eta_{rm} = \frac{N_{FalseNegative}}{N_{FalseNegative} + N_{TruePositive}} \qquad (6)$$

$$\eta_{rf} = \frac{N_{FalsePositive}}{N_{FalseNegative} + N_{TruePositive}} \qquad (7)$$

where N notes the number of voxels.

Some important results are explained that: (1)The fusion operation, geometric mean proposed in [4] as t-norm give a better result. (2)Higher agreement between the μ_{Tiss}^{SI} that is derived directly from the transformed histogram, and the μ_{Tiss}^{Std}. That is because the coefficient of kappa statistic for any tissue is upon 0.67 for 10 classes separated from entire set and is upon 0.96 for two classes that are separated by 0.5-cut set. (3)Lower relative false detection η_{rf}. The maximum

η_{rf} is only 5.49% for 0.5-cut set and for WM is 0.09%. (4)Lower relative miss detection η_{rm}. The maximum η_{rm} is 5.74% for 0.5-cut set and 5.52% for entire set. For GM, η_{rm} is only 1.41% for 0.5-cut set.

5   Conclusion

The first one of measurement problem in fuzzy set theory deals with measuring the degree of membership of several subjects or objects in a single fuzzy set [1]. The proposed generation approach of membership function based on histogram gives the membership degree by statistic histogram of grey level. This process is described by three steps: calculating the histogram of whole MRI sequence; transformation the histogram into membership function by selecting some feature values such as peaks and valleys of the histogram; a function approximation process to build a continuous membership function. It has performed well the conjunction or fusion of objective measure and *a priori* knowledge or experiences of neuro-radiologist. The measuring of membership degree is objective, because it is derived from entire histogram of image volume of MRI. The transformation of histogram is derived from an integration of *a priori* knowledge, experience, objective observation for histogram and analysis of property ranking. The generated membership function can be approximated with an appropriate continuous function. There is at least geometric mean, one of appropriate t-norm operator for conjunction the different models. It gives a natural description of possibility.

The result of error measure shows a higher agreement of the constructed fuzzy model and the reference fuzzy model by resulting from the coefficient of kappa statistic. Then, this method of membership function generation is only applicable for an object or fuzzy set with larger size.

References

1. Dubois, Didier and Prade, Henri: Fundamentals of fuzzy sets. Kluwer academic publishers (2000).
2. Pedrycz, Witold and Gomide, Fernando: An introduction to fuzzy sets analysis and design. The MIT Press (1998).
3. Cocosco, A., Kollokian, V., Kwan, R. K-S., and Evans, A. C.,: Brain Web: Online interface to a 3D MRI simulated brain database. available at http://www.bic.mni.mcgill.ca/brainweb.
4. Dou, W., Ruan, S., Bloyet, D., Constans, JM., and Chen, Y.: Segmentation based on Information Fusion Applied to Brain Tissue on MRI. Proceedings of SPIE-IST Electronic Imaging Vol.5298 (2004) pp.492-503.
5. Siegel, Sidney and N. J. Castellan, Jr.,: Nonparametric Statistics for the Behavioral Sciences. Second edition. McGraw-Hill (1988).

On Identity-Discrepancy-Contrary Connection Degree in SPA and Its Applications

Yunliang Jiang [1,2], Yueting Zhuang [1], Yong Liu [1], and Keqin Zhao [3]

[1] College of Computer Science, Zhejiang University,
Hangzhou 310027, P.R. China
[2] School of Information & Engineering, Huzhou University,
Huzhou 313000, P.R. China
[3] Institute of Zhuji Connection Mathematics,
Zhuji 311811, P.R. China
{jylsy, yzhuang}@zju.edu.cn, cckaffe@yahoo.com.cn,
zjzhaokq@mail.sxptt.zj.cn

Abstract. As a kind of new uncertainty theory, the set pair analysis (SPA) researches certainties and uncertainties from the whole. The main idea of SPA is: (1) Every system is comprised of certainty knowledge and uncertainty knowledge; (2) Certainties and uncertainties are inter-related, inter-influenced, inter-restricted, and inter-transformed even under certain condition, in every system; (3) A computation formula (identity-discrepancy-contrary (IDC) connection degree formula) which can embody the above idea fully is used to depict uniformly all kinds of uncertainties such as fuzzy uncertainty, random uncertainty, indeterminate-known uncertainty, unknown and unexpected incident uncertainty, and uncertainty which is resulted from imperfective information. This paper introduces the concepts and provides the applications of IDC connection degree and IDC connection number in SPA.

1 Introduction

Set pair analysis (SPA) is a novel mathematical tool dealing with vagueness and uncertainty. It can be effectively used to analysis and process such uncertainty knowledge as imprecise knowledge, disagreement knowledge, and unintegrity knowledge, etc., find hidden knowledge, and reveal latent rule. [1]

SPA was proposed by Keqin Zhao, one of the four authors of this paper, a Chinese scholar, in 1989. From 1995 up to now, the workshop of SPA is convened annually in China. SPA has been one of the solicitant topics of the national academic conference of Chinese Artificial Intelligence Institute from the year 2001. SPA has also been one of the important topics of many academic journals. All these promote the development and application of SPA effectively. Now SPA has been a relatively new academic hot point in the field of AI (especially in the field of the uncertainty knowledge processing) in China, and has received much attention of the researchers all over China. Advances have been achieved substantially in the theory and application of SPA after more than ten years of research and development. In recent years, the IDC connection degree (number) in SPA has been successfully applied to many fields including decision making, forecasting, data fusion, etc.

2 Basic Concepts

2.1 IDC Connection Degree

Under general circumstances, it's a formula [1]

$$\mu = \frac{S}{N} + \frac{F}{N}i + \frac{P}{N}j \tag{1}$$

Here N is the total number of features, S is the number of identity features, and P is the number of contrary features, of the set pair discussed. $F = N\text{-}S\text{-}P$ is the number of features of the two sets that are neither identity nor contrary. S/N, F/N, and P/N are called **identity degree**, **discrepancy degree**, and **contrary degree** of the two sets under certain circumstances respectively. The "j" is the coefficient of the contrary degree, and is specified as -1. The "i" is the coefficient of the discrepancy degree, and is an uncertain value between -1 and 1, i.e. $i \in [-1,1]$, in terms of various circumstances. When the values of "i" and "j" do not demand to be considered, the "i" and "j" can merely be regarded as the markers of the discrepancy degree and the contrary degree respectively. In order to clarify formula (1), we set $a = S/N$, $b = F/N$, $c = P/N$, and then the formula (1) can be rewritten as

$$\mu = a + bi + cj \tag{2}$$

It is obvious that the "a", "b", and "c" satisfy the equation $a+b+c=1$.

2.2 IDC Connection Number

The IDC connection number is a kind of number coming from the IDC connection degree. Its representation form is still $\mu = a + bi + cj$, here "a", "b", and "c" are random positive numbers, $j=-1$, $i \in [-1,1]$. [1]

SPA understands the meaning of the IDC connection number from the viewpoint of layer. SPA considers the IDC connection number as a kind of number that can depict uncertain quantity, and thinks that the IDC connection number is different from constant, variable, and super uncertain quantity essentially(see Tab.1)

Table 1. Constant, variable, uncertain quantity, and super uncertain quantity

Items	Constant	Variable	Uncertain quantity	Super uncertain quantity
macro layer	certain	uncertain	certain	uncertain
micro layer	certain	certain	uncertain	uncertain
number concept	constant number	variable number	IDC connection number	chaos number
instance	circle rateπ	free dropping rate v	waiver rate bi	

3 Research Advances

The Research of SPA is currently focused on its mathematical property, its extension, the relation and mutual supplement with other uncertainty approaches, etc.

3.1 Mathematical Property and Extension of SPA

The paper [2] researches the fundamental rule and property of arithmetic operation, the equivalence relation, and the priority relation, of the IDC connection degree, and proves that the IDC connection degree is "half order".

The research of extension of SPA deals mainly with the generalized SPA model. The research of the generalized SPA model doesn't consider that the numbers of elements of two sets in a set pair should be limited as equal. Moreover, the research considers that the comparison of elements in two sets should be done in "order-pair" of elements, not in corresponding elements directly. "Whether or not has certain relation" should be adopted as the comparison style. [2]

3.2 Relation and Mutual Supplement with Other Uncertainty Approaches

The paper [3] discussed some differences of SPA and the fuzzy set theory in research object, thought method, application field, etc. The papers [4, 5] presented the fuzzy SPA way which regards certainty knowledge and uncertainty knowledge as a pair of fuzzy sets and depicts from certainty and uncertainty. The way picks up relatively certainty knowledge from the object, acknowledges and considers relatively uncertainty knowledge corresponding to a kind of depiction, and regards this knowledge as useful knowledge.

The paper [6] presented the definitions of the rough logic operator and the valuation of the upper and lower approximation sets, applied the IDC connection degree to the rough set, and put forward the concept of the rough IDC connection degree. The paper [7] proposed the concept of the IDC connection degree of the rough set of the set type, presented the calculation steps of the condition attributes reduction and the attribute redundant values reduction of the decision table using the rough set IDC connection degree, and proposed an example that illustrates that the new approach is simpler than the traditional reduction approach of the rough set theory.

4 Applications

Fei Peng et al. introduced SPA into FHW and founded the schema evaluation decision approach based on SAP and FHW DSS.

Jie Gao et al. proposed a SPA classified forecasting method which combines "the principle of selecting the closer" of IDC pattern recognition in SPA with the basic idea of clustering analysis.

Radar-ESM correlation is one of the key subjects in data fusion. Based on SPA and Multiple Hypothesis Testing (MHT), Guohong Wang et al. developed a new radar-ESM correlation algorithm suitable for the situations where both radar and ESM tracks are specified by different numbers of measurements.

More than 400 papers have been published on application of SPA till now. The application fields of SPA also include uncertainty reasoning, network planning, comprehensive evaluation, traffic, power system, and other fields. [8]

5 Conclusions

Though the development history of SPA has only more than ten years till now, many fruits of the research and application of SPA have been achieved. It has been showed that SPA is a relatively promising soft computing approach which provides a powerful analysis means for the processing of uncertainty knowledge.

Acknowledgements

This research was supported in part by the Zhejiang Provincial Natural Science Foundation of China under grant M603169, and in part by the Natural Science Foundation of China under grant 60402010, and in part by the Huzhou Natural Science Foundation of Zhejiang. We are grateful to the anonymous referees for their insightful comments and suggestions, which clarified the presentation.

References

1. Keqin Zhao. Set Pair Analysis and its Preliminary Applications. Zhejiang Science and Technology Press, Hangzhou, 2000 (In Chinese).
2. Peng Zhang, and Guangyuan Wang. New Theory of Set Pair. Journal of Harbin University of C. E. & Architecture, 33(3): 1-5, 2000 (In Chinese).
3. Bin Zhang. The Method and Thought of Set Pair Treated with Uncertainties Information. Fuzzy Systems and Mathematics, 15(2): 89-93, 2001 (In Chinese).
4. Bin Zhang, and Xiumei Zhao. The Fuzzy Set Pair Analysis Way for Processing Uncertainty Problem. Journal of University of Electronic Science and Technology of China, 26(6): 630-634, 1997 (In Chinese).
5. Bin Zhang. The Fuzzy Set Pair Analysis Way of Multi-Objective System Decision. Systems Engineering-Theory &Practice, 17(12): 108-114, 1997 (In Chinese).
6. Ping Zhang, and Decai Huang. Rough Set Based on Connection Degree. Journal of Hangzhou Institute of Electronic Engineering, 21(1): 50-54, 2001 (In Chinese).
7. Ping Zhang, and Decai Huang. Reducing Method for Knowledge Representation System Based on Rough Set Connection Degree. Journal of Zhejiang University of Technology, 30(1): 5-8, 2002 (In Chinese).
8. Yunliang Jiang, Congfu Xu, et al. Advances in Set Pair Analysis Theory and its Applications. Computer Science (In Chinese) (Accepted).

A Mathematic Model for Automatic Summarization

Zhiqi Wang, Yongcheng Wang, and Kai Gao

Department of Computer Science and Engineering, Shanghai Jiao Tong University,
200030, Shanghai, P. R. China
{shrimpwang, ycwang, gaokai}@sjtu.edu.cn

Abstract. Automatic Summarization is need of the era. Mathematics is an important tool of nonfigurative thinking. A mathematic model of automatic summarization is established and discussed in the paper. The model makes use of meta-knowledge to describe the composition of the summary and help to calculate the semantic distance between summary and source document. It is proposed that how to get meta-knowledge aggregate and their weight are the key problems in the model.

1 Introduction

With tons of text information pouring in every day, text summaries are becoming essential. It is said that a professional abstractor can edit only 55 summaries at most in one day (15 summaries at least and 27 summaries on average) [1]. This makes it necessary using computer to make summary automatically. Mathematics is an important tool of nonfigurative thinking. Research and development of almost every kind of science field must turn to mathematics as something of a tool. However, there is no any previous work concerned about mathematic description of automatic summarization, although it will be useful for both the application of automatic summarization and automatic assessment for summaries.

2 Mathematic Model for Automatic Summarization

2.1 Mathematic Description of Automatic Summarization

Suppose D represents the source document. A represents a summary of document D. L is A's length, and the error scope of length L is $\pm \delta$. Suppose Length(S) is the length measuring function of string S and TopicSet(S) is an aggregate, which contains all the information conveyed by S. So a summary is a string, which must satisfy all the conditions as follows:

Condition 1: $|\text{Length}(A_i) - L| \leq \delta$ (L<Length (D))

Condition 2: $\text{TopicSet}(A_i) \subset \text{TopicSet}(D)$

Condition 3:

$$\text{Dis}(\text{TopicSet}(D), \text{TopicSet}(A^*)) = \min_i(\text{Dis}(\text{TopicSet}(D), \text{TopicSet}(A_i))) \quad (1)$$

A* in condition 3 represents the best summary and it belongs to the summary candidate aggregate.

Then the problem of automatic summarization can be transformed into: finding the string A* which satisfies condition 1, condition 2 and condition3.

2.2 A Mathematic Model for Automatic Summarization

Given the aggregate of all meta-knowledge $\phi = \{t_1, t_2, t_3, \ldots\}$. Both TopicSet(D) and TopicSet(A_i) are the subsets of ϕ, and they must satisfy condition 2. Then, TopicSet(D) and TopicSet(A_i) can be denoted as following:

Definition 1:

$$\text{TopicSet}(D) = \{t_{d_1}, t_{d_2}, \ldots, t_{d_n}\} \qquad t_{d_j} \in \phi \quad 1 \leq j \leq n$$

$$\text{TopicSet}(A_i) = \{t_{a_1}, t_{a_2}, \ldots, t_{a_m}\} \qquad t_{a_k} \in \phi \quad 1 \leq k \leq m \leq n$$

Because TopicSet(A_i) $\subset$ TopicSet(D), definition 1 must satisfy the following condition:

Condition 4: $t_{a_k} \in \text{TopicSet}(D) \qquad \forall t_{a_k} \in \text{TopicSet}(A_i), 1 \leq k \leq m$

Suppose the weight of each meta-knowledge in D is defined as definition 2:

Definition 2:

$$\text{WeightSet}(D) = \{w_{d_1}, w_{d_2}, \ldots, w_{d_n}\} \qquad w_{d_j} \in R \quad 1 \leq j \leq n$$

$$\text{WeightSet}(A_i) = \{w_{a_1}, w_{a_2}, \ldots, w_{a_n}\} \qquad w_{a_k} \in R \quad 1 \leq k \leq m \leq n$$

The problem is transformed into calculating the distance between TopicSet(D) and TopicSet(A_i), which can be measured by the difference between the two aggregation.

$$\begin{aligned}
&\text{Dis}(\text{TopicSet}(D), \text{TopicSet}(A_i)) \\
&= \text{Dis}(\{t_{d_1}, t_{d_2}, \cdots, t_{d_n}\}, \{t_{a_1}, t_{a_2}, \cdots t_{a_m}\}) \\
&= \sum_{j=1}^{n} w_{d_j} d(t_{d_j}, \text{TopicSet}(A_i)) + \sum_{k=1}^{m} w_{a_k} d(t_{a_k}, \text{TopicSet}(D))
\end{aligned} \qquad (2)$$

$d(e, S)$ represents the distance between element e and aggregate S, which can be defined as following:

Definition 3:
$$d(e, S) = \begin{cases} 0 & \text{if } e \in S \\ 1 & \text{if } e \notin S \end{cases}$$

3 Further Discussions

3.1 Construction of Meta-knowledge Aggregate

Meta-knowledge is the basic knowledge unit which can not be segmented. In fact, such meta-knowledge is hard to get in the real applications. Concepts can be considered to substitute for meta-knowledge. However, one problem is that concepts are related with each other and they can even be cut into smaller units such as meta-concepts according to the definition of Hownet [8]. Suppose e represents a concept, and S represents a concept aggregate. The relationship between e and S can not described by "$e \in S$" or "$e \notin S$". Therefore, d (e, S) can not be defined using definition 3.

Given $S=\{c_1, c_2, ..., c_n\}$, $d(e,S)$ is defined as the degree that e does not belong to S.

Definition 4:

$$d(e, S) = 1 - \max_{i=0}^{n} \text{Concept\_Coherence}(e, c_i) \qquad (3)$$

Concept_Coherence(e,c_i) is the semantic distance between the two concepts e and c_i. It can be calculated by means of Hownet or other tools [2].

3.2 Weight of Meta-knowledge

Weight of meta-knowledge is very important to automatic summarization system. It can be evaluated by many factors. Edmundson discussed about this problem [3]. He compared the output summaries of four systems using different evaluate method. It is found that the Cue-Title-Location method has the highest mean coselection score, while the Key method in isolation is the lowest of the automatic methods (as figure 1 show).

3.3 About the Best Summary

In fact, "the best summary" as we called is not absolutely the best. Even if the summary is made by professional abstractor, it may be different when it is made by different people, at different time or under different background. But those summaries can all be seen as "the best". The best summary produced according to the mathematic model may be more than one. It brings a challenge to real applications: which one should be selected as the output of the system? It is not decided in the mathematic model mentioned above. In fact, it can be solved using different method. Some solutions are presented as following:

  Scheme 1: Choose the summary whose length is the closest to the required length as the output summary.
  Scheme 2: Choose the summary whose meta-knowledge can be the most evenly scattered in source document as output. This can ensure that the topical meta-knowledge can be included into the summary as much as possible instead of focusing on some certain topical meta-knowledge.
  Scheme 3: choose the summary which has the largest meta-knowledge weight as output.

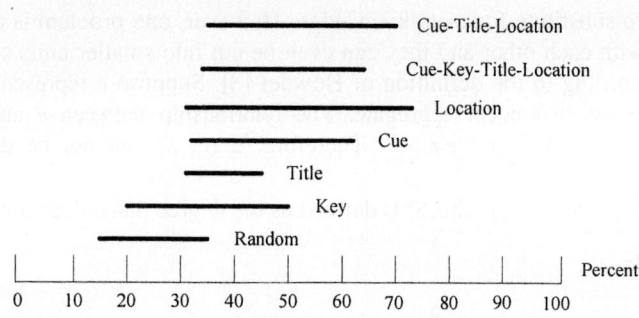

Fig. 1. Edmundson's mean coselection scores of the methods

4 Conclusion

Summary is a product of electronic document explosion, and can be seen as the condensation of the document. Mathematics is an important tool of nonfigurative thinking. There is no any mathematic description of automatic summarization till now. Against this background, a mathematic model of automatic summarization is established and discussed in this paper. In the model, meta-knowledge is used to describe the composition of summary and help to measure the distance between the summary and source document. It is pointed out that the construction of meta-knowledge aggregate and how to get weight of the meta-knowledge are the key problems in the model.

References

1. Cunningham Ann Marie, Wicks Wicks: Guide to Careers in Abstracting and Indexing. Philadelphia, PA: National Federation of Abstracting and Information Services (1992)
2. Liu Gongshen, Wang Yongcheng, Bao zhengrong, etc.: Multi-Subject Analyzing Based on Concept Coherence (CC). Journal of Information Science, Vol.21, No.1 (2002), 2-6
3. Edmundson H P: New Methods in Automatic Extracting. Journal of the association for Computing Machinery, Vol.16, No.2 (1969), 264-285

Reliable Data Selection with Fuzzy Entropy

Sang-Hyuk Lee, Youn-Tae Kim, Seong-Pyo Cheon, and Sungshin Kim

School of Electrical and Computer Engineering,
Pusan National University, Changjeon-dong, Busan 609-735, Korea
{leehyuk, dream0561, buzz74, sskim}@pusan.ac.kr

Abstract. In this paper, the selection of a data set from a universal set is carried out using a fuzzy entropy function. According to the definition of fuzzy entropy, the fuzzy entropy function is proposed and that function is proved through definitions. The proposed fuzzy entropy function calculates the certainty or uncertainty value of a data set; hence we can choose the data set that satisfies certain bounds or references. Therefore a reliable data set can be obtained using the proposed fuzzy entropy function. With a simple example we verify that the proposed fuzzy entropy function selects the reliable data set.

1 Introduction

Generally for linear systems, reliable input invokes reliable output. Hence, reliable data selection is necessary for a reliable result. Previous results concerning this field are related to pattern recognition and information theory. Pattern recognition is generally used in classifying patterns[1]. The geometrical distance between data sets is used mainly to classify patterns. Also, it is well known that entropy represents the uncertainty of the fact. Hence, entropy has been studied in the fields of information theory, thermodynamics, system theory, and others. The result that the entropy of a fuzzy set is a measure of the fuzziness of the fuzzy set has been reported by numerous researchers [2-9]. An axiomatic definition of entropy was proposed by Liu. The relation between distance measure and fuzzy entropy was investigated Kosko. Bhandari and Pal gave a fuzzy information measure for discrimination of a fuzzy set A relative to some other fuzzy set B. Pal and Pal analyzed classical Shannon information entropy. Also, Ghosh applied entropy to neural networks. Because these studies focussed on the design of entropy function and analysis of fuzzy entropy measure, distance measure, and similarity measure, we carried out the application of fuzzy entropy to the selection of a reliable data set from a universal set. In this paper, we derive fuzzy entropy with distance measure. The proposed fuzzy entropy is constructed by the Hamming distance measure, which has a simple structure compared with the previously proposed entropy [8]. With the proposed entropy, we verify the usefulness through the application of fuzziness measure to the universal data set. We also carry out calculations of the fuzziness of a sample data set. However, the entropy definition contains a complementary characteristic, which is $e(A) = e(A^C), \forall A \in F(X)$. This property invokes an unreliable result in the selection of reliable data. Hence we consider another idea to overcome this problem. We check the usefulness of fuzzy entropy and discuss the outcomes. In Chapter 2, definitions of the entropy, distance measure, and similarity measure of fuzzy sets are introduced, and the proof of the proposed entropy is discussed. In Chapter 3, a simple example is illustrated and the results are discussed. Experiments for choosing the average level 5 students among 65 students are performed, and the chosen sample data are measured by the proposed entropy measure. Finally, conclusions follow in Chapter 4.

The notations of this paper are those of Fan and Ma (1999).

2 Fuzzy Entropy

In this chapter, we introduce some preliminary results of fuzzy entropy. Measure of fuzziness is an interesting object in the fields of pattern recognition or decision theory. It is well known that measure of entropy for fuzzy sets represents the information of uncertainty. Measure of a crisp set can be determined by classical mathematical study, whereas the concepts of fuzzy measures and fuzzy integrals have been proposed by Sugeno[11]. Recently, Liu has suggested three axiomatic definitions of fuzzy entropy, distance measure, and similarity measure as Definitions 2.1-3[5]. By these definitions, we can induce an entropy that is a satisfying definition of fuzzy entropy, and compare it with the result of Fan et al.

Definition 2.1. [5] A real function, $e : F(X) \to R^+$ is called the entropy on $F(X)$, if has the following properties:

(E1) $e(D) = 0, \forall D \in P(X)$;

(E2) $e([1/2]) = max_{A \in F(X)} e(A)$;

(E3) $e(A^*) \le e(A)$, for any sharpening A^* of A; and

(E4) $e(A) = e(A^C)$, $\forall A \in F(X)$,

where $R^+ = [0, \infty)$, A^C is the complement of A, and A^* is a sharpening of A. To express entropy function explicitly, distance measure is needed. Next, we define distance measure.

Definition 2.2. [5] A real function, $e : F^2 \to R^+$ is called the distance measure on $F(X)$ if satisfies the following properties:

(D1) $d(A, B) = d(B, A), \forall A, B \in F(X)$;

(D2) $d(A, A) = 0, \forall A \in F(X)$;

(D3) $d(D, D^C) = max_{A, B \in F} d(A, B), \forall D \in P(X), \forall A, B \in F(X)$;

(D4) $\forall A, B, C \in F(X)$, if $A \subset B \subset C$; then $d(A, B) \le d(A, C)$ and $d(B, C) \le d(A, C)$.

One well known distance measure is Hamming distance. Similarity measure can be expressed as the complement of distance measure. Accordingly, the definition of similarity measure is illustrated as follows.

Definition 2.3. [5] A real function, $e : F^2 \to R^+$ is called a similarity measure if s has the following properties:

(S1) $s(A, B) = d(B, A), \forall A, B \in F(X)$;

(S2) $s(A.A^C) = 0, \forall A \in F(X)$;

(S3) $s(D, D) = max_{A,B \in F} s(A, B), \forall D \in P(X), \forall A, B \in F(X)$;

(S4) $\forall A, B, C \in F(X)$, if $A \subset B \subset C$; then $s(A, B) \geq s(A, C)$ and $s(B, C) \geq s(A, C)$.

The above definitions are axiomatic, and Liu also pointed out that there is an one-to-one relation between all distance measures and all similarity measures, $d + s = 1$. Next, some useful related definitions are listed. If we divide universal set X into two parts D and D^C in $P(X)$, then the fuzzy entropy, fuzzy distance, and similarity are obtained by the previous results. When we focus on an interesting area of a universal set, we can extend the theory of the entropy, distance measure, and similarity measure of fuzzy sets.

Definition 2.4. [7] Let e be the entropy on $F(X)$. Then, for any $A \in F(X)$,

$$e(A) = e(A \cap D) + e(A \cap D^C)$$

is the σ-entropy on $F(X)$.

Definition 2.5. [7] Let d be the distance measure on $F(X)$. Then, for any $A, B \in F(X)$, and $D \in P(X)$,

$$d(A, B) = d(A \cap D, B \cap D) + d(A \cap D^C, B \cap D^C)$$

is the σ-distance measure on $F(X)$.

Definition 2.6. [7] Let s be the similarity measure on $F(X)$. Then, for any $A, B \in F(X)$, and $D \in P(X)$,

$$s(A, B) = s(A \cap D, B \cap D^C) + s(A \cap D^C, B \cap D)$$

is the σ-similarity measure on $F(X)$.

From the properties of Definition 2.5, we can derive the following proposition.

Proposition 2.1. [7] Let d be the σ-distance measure on $F(X)$: then

(i) $d(A, A_{near}) \geq d(A^*, A_{near})$

(ii) $d(A, A_{far}) \leq d(A^*, A_{far})$.

Fan, Ma and Xie have also proposed the following theorem [8]. In their theorem, they proposed a fuzzy entropy function with distance measure. The proposed entropy contains two crisp sets A_{near} and A_{far}.

Theorem 2.1. [8] Let d be the σ-distance measure on $F(X)$; if d satisfies

(i) $d(\frac{1}{2}D,[0]) = d(\frac{1}{2}D,D), \forall D \in P(X)$ and

(ii) $d(A^C, B^C) = d(A,B), A,B \in F(X), e(A) = d(A, A_{near}) + 1 - d(A, A_{far})$

is the fuzzy entropy.

Via the defined entropy Fan and Xie derived a new entropy, which is introduces by $e^* = e/(2-e)$, where e is the entropy on $F(X)$. To discriminate between entropies, we posit another entropy using Fan's idea.

Theorem 2.2. If e is the entropy on $F(X)$, $e = e^k$ is also the entropy on $F(X)$, where real number $k \geq 1$.

Proof. It is clear that $0 \leq e(A) \leq 1$ for any $A \in F(X)$, and e satisfies Definition 2.1 as follows.

$(E1): e(D)$ is zero for $\forall D \in P(X)$, hence satisfied;

$(E2): e([\frac{1}{2}]) = max_{A \in F(X)} e(A)$ is also satisfied;

$(E3): e(A^*) \leq e(A)$ is clear;

$(E4): e(A) = e(A^C)$ is also easily proved, where $\forall A \in F(X)$.

Hence the structure of Theorem 2.2 satisfies the entropy that is induced from another entropy. It is often required that a reliable data set be selected among many data sets. In this chapter, we introduce the relation of the fuzzy membership function and fuzzy entropy. Let X be a space of objects and x be a generic element of X. A classical set A, $A \subseteq X$, is defined as a collection of elements or objects $x \in X$, such that each x can either belong or not belong to the set A, whereas a fuzzy set A in X is defined as a set of ordered pairs $A = \{(x, \mu_A(x)) \mid x \in X\}$ where $\mu_A(x)$ is called the membership function for the fuzzy set A. The membership function maps each element of X to a membership grade between 0 and 1.

According to the results of Liu, the fuzzy entropy function expressed by the following Hamming distance measure between fuzzy sets A and B,

$$d(A,B) = \frac{1}{n}\sum_{i=1}^{n}|\mu_A(x_i) - \mu_B(x_i)| \tag{1}$$

where $X = \{x_1, x_2, \cdots, x_n\}$.

Fuzzy entropy indicates the uncertainty of the fuzzy set, and represents the two times of shaded area of Fig. 1 [7],[8]. In Fig. 1, A_{near} denotes the crisp set of fuzzy set A.

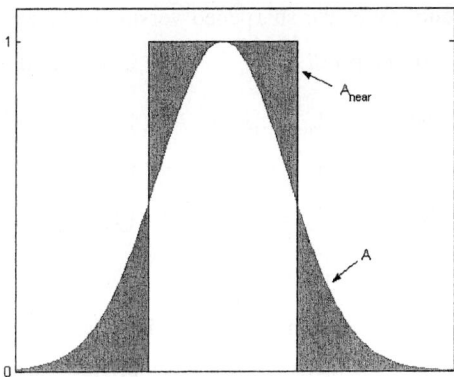

Fig. 1. Representation of entropy

The closer the fuzzy set is to the crisp set A_{near}, the more the membership function becomes certain. For the crisp set, then, there is no uncertainty. In the next theorem, we propose a fuzzy entropy function with the Hamming distance, which is different from the Theorem of Fan, Ma and Xie [8]. The proposed entropy needs only an A_{near} crisp set because of the advantage in computation of entropy.

Theorem 2.3. Let d be the σ-distance measure on $F(X)$; if d satisfies $d(A^C, B^C) = d(A,B), A,B \in F(X), A,B \in F(X),$

$$e(A) = 2d((A \cap A_{near}),[1]) + 2d((A \cup A_{near},[0]) - 2 \qquad (2)$$

is a fuzzy entropy.

Proof. For $(E1)$, $\forall D \in P(X)$, $D_{near} = D$; therefore,

$$e(D) = 2d((D \cap D_{near}),[1]) + 2d((D \cup D_{near}),[0]) - 2$$
$$= ed(D,[1]) + 2d(D,[0]) - 2 = 0.$$

$(E2)$ represents that crisp set 1/2 has the maximum entropy 1. Therefore, the entropy $e([\frac{1}{2}])$ satisfies

$$e([\tfrac{1}{2}]) = 2d(([\tfrac{1}{2}] \cap [\tfrac{1}{2}]_{near}),[1]) + 2d(([\tfrac{1}{2}] \cup [\tfrac{1}{2}]_{near}),[0]) - 2$$
$$= 2d(([\tfrac{1}{2}] \cap [1]),[1]) + 2d(([\tfrac{1}{2}] \cup [1]),[0]) - 2$$
$$= 2 \cdot \tfrac{1}{2} + 2 \cdot 1 - 2 = 1.$$

In the above equation, $[\tfrac{1}{2}]_{near} = [1]$ is satisfied.

$(E3)$ shows that the entropy of the sharpened version of fuzzy set A, $e(A^*)$, is less than or equal to $e(A)$. For the proof, $A^*_{near} = A_{near}$ is also used.

$$\begin{aligned} e(A^*) &= 2d((A^* \cap A^*_{near}),[1]) + 2d((A^* \cup A^*_{near}),[0]) - 2 \\ &= 2d((A^* \cap A_{near}),[1]) + 2d((A^* \cup A_{near}),[0]) - 2 \\ &\leq 2d((A \cap A_{near}),[1]) + 2d((A \cup A_{near}),[0]) - 2 \\ &= e(A). \end{aligned}$$

The inequality in the above equation is satisfied because $d(A, A_{near}) \geq d(A^*, A_{near})$ in Proposition 3.1 (i). Finally, $(E4)$ is proved using the assumption $d(A^C, B^C) = d(A, B)$; hence we have

$$\begin{aligned} e(A) &= 2d((A \cap A_{near}),[1]) + 2d((A \cup A_{near}),[0]) - 2 \\ &= 2d((A \cap A_{near}),[1]) + 2d((A \cup A_{near}),[0]) - 2 \\ &\leq 2d((A^C \cup A^C_{near}),[0]) + 2d((A^C \cup A^C_{near}),[1]) - 2 \\ &= e(A^C). \end{aligned}$$

Theorem 2.3 uses only the A_{near} crisp set, so we consider the complementary entropy function in the next theorem, which considers only A_{far}, and it has a more compact form than Theorem 2.3.

Theorem 2.4. Let d be the σ-distance measure on $F(X)$; if d satisfies $d(A^C, B^C) = d(A, B)$, $A, B \in F(X)$,

$$e(A) = 2d((A \cap A_{far}),[0]) + 2d((A \cup A_{far}),[1]) \qquad (3)$$

is a fuzzy entropy.

Proof. In a similar way we can prove from $(E1)$ to $(E4)$. For $(E1)$, $\forall D \in P(X)$, $D_{far} = D^C$; therefore,

$$\begin{aligned} e(D) &= 2d((D \cap D_{far}),[0]) + 2d((D \cup D_{far}),[1]) \\ &= 2d([0],[0]) + 2d([1],[1]) = 0. \end{aligned}$$

And the entropy of crisp set $[\frac{1}{2}]$ is obtained as follows; hence $(E2)$ is

$$\begin{aligned} e([\tfrac{1}{2}]) &= 2d(([\tfrac{1}{2}] \cap [\tfrac{1}{2}]_{far}),[0]) + 2d(([\tfrac{1}{2}] \cup [\tfrac{1}{2}]_{far},[1]) \\ &= 2d(([\tfrac{1}{2}] \cap [0]),[0]) + 2d(([\tfrac{1}{2}],[0]),[1]) \\ &= 0 + 2 \cdot \tfrac{1}{2} = 1. \end{aligned}$$

In this case, $[½]_{far} = [0]$ is also used. The entropy between the fuzzy set and the sharpened version is derived from the proof of $(E3)$, for the proof $A^*_{far} = A_{far}$ is also used. The property of Proposition 3.1 (ii), $d(A, A_{far}) \le d(A^*, A_{far})$, is used the inequality of the following proof.

$$\begin{aligned} e(A^*) &= 2d((A^* \cap A^*_{far}), [0]) + 2d((A^* \cup A^*_{far}), [1]) \\ &= 2d((A^* \cap A_{far}), [0]) + 2d((A^* \cup A_{far}), [1]) \\ &\le 2d((A \cap A_{near}), [0]) + 2d((A \cup A_{near}), [1]) \\ &= e(A) \end{aligned}$$

$(E4)$ is derived from the assumption $d(A^C, B^C) = d(A, B)$; then, we have

$$\begin{aligned} e(A) &= 2d((A^* \cap A_{far}), [0]) + 2d((A \cup A_{far}), [1]) \\ &= 2d((A^* \cap A_{far})^C, [0]^C) + 2d((A \cup A_{far})^C, [1]^C) \\ &= 2d((A^C \cap A^C_{near}), [1]) + 2d((A^C \cap A^C_{far}), [0]) \\ &= e(A^C). \end{aligned}$$

The proposed entropies Theorems 2.3 and 2.4 have the advantage over Fan et al.'s [8] that they need only assumption, $d(A^C, B^C) = d(A, B)$, to prove (2) and (3). Furthermore, (2) and (3) use only the crisp sets A_{near} and A_{far}, respectively. According to the computational results, (2) and (3) are the σ-entropy on $F(X)$. This σ-entropy property has an advantage in computation burden, which is shown in the literature[10].

3 Illustrative Example

We illustrate an example of reliable data set selection from a universal set. Because statistical mean and variance do not propose the "how much" of fuzziness or reliability, with the help of the definition of entropy, fuzzy measure was introduced in Chapter 2. It is assumed that one class consists of 65 students. Educational level can be classified according to the two viewpoints: one is heuristic representation and the other is grade. The mean grade of the 65 students is 53.73, and the average-level student membership function is shown in Fig. 2. As is explained in Fig. 1, the shaded area of Fig. 2 stands for the uncertainty of the average-level. The average-level students have grades of B and C, which points are between 37 and 71. In this case, the level of the chosen 5 students can be measured by the entropy function that is illustrated in (2) and (3). Furthermore, the crisp set of grades B and C is also represented in Fig. 2 by a rectangle.

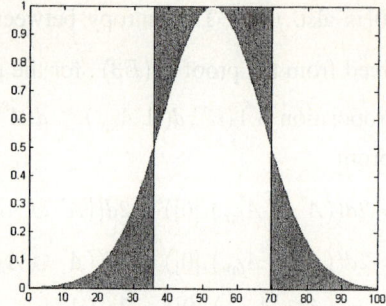

Fig. 2. Average level of student membership function, B and C Grades

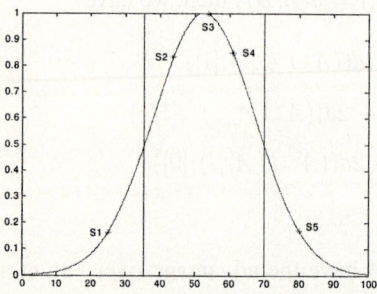

Fig. 3. Selection of 5 students(Test1)

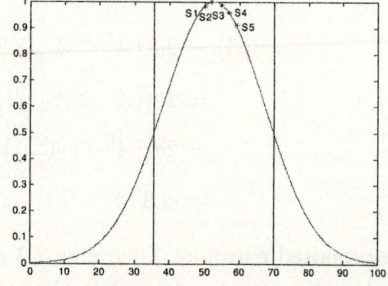

Fig. 4. Selection of 5 students(Test2)

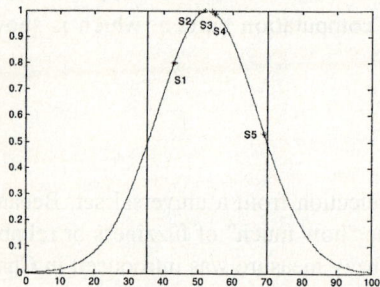

Fig. 5. Selection of 5 students(Test3)

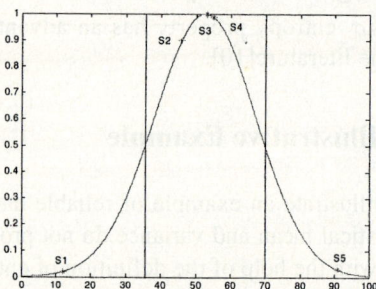

Fig. 6. Selection of 5 students(Test4)

We choose 5 students 4 times randomly. The students and their grades are illustrated in Figures 3, 4, 5, and 6. The fuzzy entropy value can be calculated by Eq. (2) or (3).

The actual values, the membership function value and the fuzzy entropy values, are shown in Table 1. As the obtained entropy value approaches zero, the student group has a higher tendency toward B and C grade. As shown in Table 1, the mean values of the trials are 52.8, 54.6, 54.6, and 51.4, respectively. The statistical results show that the sample means of each case are similar to the total average; furthermore, the 2nd and 3rd trials illustrate the same mean values. Even though the 2nd and 3rd trials have the same means, this also represents a different meaning from the heuristic viewpoint, whereas the fuzzy entropy average values of the groups are 0.260, 0.066, 0.275, and 0.066. According to the meaning of fuzzy entropy, it is clear that the 2nd and 4th trials are the most reliable. However, the 4th test result does not represent the average-level students because two students exist outside of the rectangle. This is the reason that fuzzy entropy contains E4 of Definition 2.1; the 1st and 5th students of Test 4 represent low entropy values. Hence we need another idea to solve this problem. Our object is to select 5 average-level students. Hence, even if one student is contained in the outside of the rectangle, we discard this set. As a result, the 2nd and 3rd trials result in the candidates for the average-level students among the total of 65 students. The fuzzy entropy values of 0.066 and 0.275 represent the fact that the 2nd trial represents the fittest average level.

Table 1. Sample, membership and fuzzy entropy values

		Sample	Membership value	Fuzzy entropy
		25	0.161	0.323
		44	0.835	0.329
Test 1		54	0.996	0.008
		61	0.849	0.302
		80	0.170	0.340
	Average	52.8	0.60	0.260
		50	0.983	0.034
		52	0.999	0.002
Test 2		55	0.987	0.025
		57	0.957	0.086
		59	0.090	0.180
	Average	54.6	0.80	0.066
		43	0.800	0.400
		52	0.999	0.002
Test 3		54	0.996	0.008
		55	0.987	0.025
		69	0.532	0.937
	Average	54.6	0.86	0.275
		12	0.019	0.039
		46	0.899	0.203
Test 4		53	1.000	0.000
		55	0.987	0.025
		91	0.031	0.016
	Average	51.4	0.59	0.066

4 Conclusions

In this paper, we derived fuzzy entropy with distance measure. The proposed fuzzy entropy was constructed using the Hamming distance measure, which has a simple structure compared to the previously proposed entropy. With the proposed entropy, the usefulness was verified through the application of the measured fuzziness to the sampled set from the universal data set. Through the example, we could verify that fuzzy entropy has a different meaning compared with the statistical results. Next, it would be valuable to extend this study to a multi-dimensional case or continuous data monitoring. Our results, with careful consideration, could be extended to a multi-dimensional case. For such a case, a newly defined membership function will be required.

Acknowledgement. This work has been supported by KESRI(R-2004-B-129), which is funded by MOCIE(Ministry of commerce, industry and energy).

References

1. J. C. Bezdek and S.K. Pal, *Fuzzy models for pattern recognition*, IEEE, 1992.
2. D. Bhandari, N. R. Pal, "Some new information measure of fuzzy sets", Inform. Sci. Vol. 67, 209-228, 1993.
3. A. Ghosh, "Use of fuzziness measure in layered networks for object extraction: a generalization", Fuzzy Sets and Systems Vol. 72, 331-348, 1995.
4. B. Kosko, *Neural Networks and Fuzzy Systems*, Prentice-Hall, Englewood Cliffs, NJ, 1992.
5. Liu Xuecheng, "Entropy, distance measure and similarity measure of fuzzy sets and their relations", Fuzzy Sets and Systems, Vol. 52, 305-318, 1992.
6. N. R. Pal, S. K. Pal, "Object-background segmentation using new definitions of entropy", IEEE Proc. Vol. 36, 284-295, 1989.
7. J. L. Fan, W. X. Xie, "Distance measure and induced fuzzy entropy", Fuzzy Set and Systems, Vol. 104, 305-314, 1999.
8. J. L. Fan, Y. L. Ma, and W. X. Xie, "On some properties of distance measures", Fuzzy Set and Systems, Vol. 117, 355-361, 2001.
9. S. H. Lee and S. S. Kim, "On some properties of distance measures and fuzzy entropy", Proceedings of KFIS Fall Conference, 9-12, 2002.
10. S. H Lee, K. B. Kang and S. S. Kim, "Measure of fuzziness with the fuzzy entropy", Journal of Fuzzy Logic and Intelligent Systems, Vol. 14, No. 5, 642-647, 2004.
11. M. Sugeno, *Theory of fuzzy integrals and its applications*, Ph.D. Dissertation, Tokyo Institute of Technology, 1974.
12. Z. Wang and G. T. Klir, *Fuzzy measure theory*, New York : Plenum Press, 1992.
13. Development of fault detection algorithm with the stator current, POSCON, 2003.

Optimization of Concept Discovery in Approximate Information System Based on FCA

Hanjun Jin[1,3], Changhua Wei[2], Xiaorong Wang[3], and Jia Fu[3]

[1] College of Hydropower and Information Engineering,
Huazhong University of Science and Technology, Wuhan, 430074, China
Wangp@public.wh.hb.cn
[2] Department of Information technology,
Jiangmen PolyTechnic, Jiangmen, 529000, China
chwei@mail.ccnu.edu.cn
[3] Department of Computer Science,
Central China Normal University, Wuhan, 430079, China

Abstract. This paper proposes the formal description of nondeterministic information system based on tolerance rough set theory, analyzes six cases of approximate information system, and gives the concept of strong and weak similarity. After defining tolerance rough set, combining the theories of FCA and expanding non-definable concept into non-definable attributes, non-definable objects and non-definable context, we present optimal algorithm of formal concept of approximation system. Really emulation has illustrated that the algorithm obtains a satisfied approximate concept and a shorter time complexity.

1 Introduction

Applying concept and the reasoning of the concept are the important cognitive way to know the objective world. Studying the internal and external mathematics method is very significant to information processing, knowledge express and knowledge discovering. With the development of science and technology, the study of concept has developed from traditional logic to modern logic; from determinacy concept to the seldom deterministic concept and from general concept to abstract one. Formal concept analysis that is put forward by Wille and his colleagues (Formal Concept Analysis, FCA) offers the mathematics foundation for imitating the concept and knowledge processing in the truly world [4], [5], [6]. The structure of the formal concept can be used for knowledge expressing, organizing and obtaining.

In practical application, however, on one hand we may face an incomplete, n-value one, approximate and outlier information system, that is non-deterministic problem[2]; On the other hand we may face the problem that can't be described, that is to say only A or B is known in the (A, B) concept. In such case, we must try to get optimum concept close to B or A according to the already known A or B. We divide the non-definable concept into non-definable attribute, non-definable object and non-definable context. The research of the optimum form of non-definable concept is called concept approach. It is Kent that first puts forward this concept, and its research is based on Rough Sets Theory (RST), the one that is similar to the mathematics model. J. Saquer and J. S. Deogun utilize RST and estimating approach method in their research [3].

This research is the further study of document [3]. Firstly, research object expand from deterministic information system to non-deterministic information system; secondly, research approach uses the fault-tolerant, and combines RST with FCA to discuss the concept optimizing; thirdly similar estimation is defined according to different concrete conditions of system.

2 Approximate Information System

2.1 Information System

Basic thought of Z.Pawlak is to talk about object of land can be distinguished through its indefinable relationship in order to obtain the knowledge expression and object's classification. Generally speaking, undistinguishable relation can be regarded as equivalent relation, having reflexive, getting symmetrical, but might not have transitivity.

Let $\approx$ be equivalent relation with U, as to arbitrary $x \in U$, write $[x]_\approx = \{y \in U | x \approx y\}$, among them $[x]_\approx$ is made up with the object x which can't be distinguished. To arbitrary subset $X \subseteq U$, let

$$X_\downarrow = \{x \in U \mid [x]_\approx \subseteq X\} \tag{1}$$

$$X^\uparrow = \{x \in U \mid [x]_\approx \cap X \neq \emptyset\} \tag{2}$$

where $X_\downarrow$ and $X^\uparrow$ are called lower approximation and upper approximation, and $B_n(X) = X^\uparrow - X_\downarrow$ is called boundary of X. If $B_n(X) = \emptyset$, x is certain in the system; otherwise it is rough.

Definition 2.1. X is equal to Y, if and only if $X_\downarrow = Y_\downarrow$ and $X^\uparrow = Y^\uparrow$. The equivalence is called rough set; all rough set belong to one system.

2.2 Non-deterministic Information System

Definition 2.2. A non-deterministic information system $NIS = (U, A, \{V_a\}_{a \in A})$, where U is the non-empty limited set of object, A is the non-empty limited set of attribute, $\{V_a\}_{a \in A}$ is the set of attribute value. Each attribute is a function $a : U \to \wp(V_a) - \{\emptyset\}$.

Definition 2.3. In the indefinite information system, if the attribute of the object x in $a(x)$ has only one value, that is $Card(a(x)) = 1$, but the value is indefinite.

2.3 Tolerance Rough Sets of Approximate Information System

Suppose in the approximate information system S, the jth object is x_j, and the ith attribute of x_j is $a_i(x_j)$, the ith attribute of object x_k is $a_i(x_k)$, the upper limit and the lower limit they decide their values are: $a_i^{inf}(x_j)$, $a_i^{sup}(x_j)$; $a_i^{inf}(x_k)$, $a_i^{sup}(x_k)$; $(i = 1, 2, \cdots, m;\ j, k = 1, 2, \cdots, n)$, then for approximate information system S_a has:

1) Total cover, if $(a_i^{inf}(x_j) \leq a_i^{inf}(x_k)) \wedge (a_i^{sup}(x_j) \geq a_i^{sup}(x_k))$, the object x_j covers x_k totally.
2) Left cover, if $(a_i^{inf}(x_j) = a_i^{inf}(x_k)) \wedge (a_i^{sup}(x_j) > a_i^{sup}(x_k))$, the object x_j left covers x_k.
3) Right cover, if $(a_i^{inf}(x_j) < a_i^{inf}(x_k)) \wedge (a_i^{sup}(x_j) = a_i^{sup}(x_k))$, the object x_j right covers x_k.
4) Left-partcover, if $(a_i^{inf}(x_j) > a_i^{inf}(x_k)) \wedge (a_i^{inf}(x_j) < a_i^{sup}(x_k)) \wedge (a_i^{sup}(x_j) < a_i^{sup}(x_k))$, the object x_j covers x_k by the left part.
5) Right-partcover, if $(a_i^{inf}(x_j) < a_i^{inf}(x_k)) \wedge (a_i^{sup}(x_j) > a_i^{inf}(x_k)) \wedge (a_i^{sup}(x_j) > a_i^{sup}(x_k))$, the object x_j covers x_k by the right part.
6) Separation, if $(a_i^{inf}(x_j) < a_i^{inf}(x_k)) \wedge (a_i^{sup}(x_j) \geq a_i^{inf}(x_k)) \wedge (a_i^{sup}(x_j) > a_i^{sup}(x_k))$, the object x_j separates x_k.

Definition 2.4. In the approximate information system S_a, if the same attribute value of two objects $a_i(x_j)$ and $a_i(x_k)$ is one of the above five situations, that is called $a_i(x_j)$ is similar to $a_i(x_k)$. The information system, which is formed by the similar information objects, is called approximate information system.

Definition 2.5. Let $R_a = \{R_a \subseteq V_a \mid a \in A\}$ be the set of tolerate relation under the condition of situation attribute. If and only if tolerate relation satisfy:

1) reflexive. $\forall \nu \in V_a$, so $\nu R_a \nu$.
2) symmetric. if $\nu R_a \nu'$, $\nu' R_a \nu$.

Lemma 2.1. If $x_j, x_k \in U$, the approximation of $a(x_j)$ and $a(x_k)$ is equal to $a(x_j) R_a a(x_k)$.

Definition 2.6. According to definition 2.5, tolerate relation in all domain $\tau : \forall\ x_j, x_k \in \bigcup x_j \tau(\Re_A) x_k$, when and only when $\forall\ a \in A,\ a(x_j) R_a a(x_k)$.

Definition 2.7. Let x_j and x_k are two different objects, they have the similar attributes $a \in A, \emptyset \neq B \subseteq A$ and the dual relation of object covering: x_j and x_k are strong approximation, called $(x_j, x_k) \in Sim(B)$, if and only if $a(x_j) \cap a(x_k) \neq \emptyset$ in regard to all $a \in B$; if x_j and x_k are the weak approximate, called $(x_j, x_k) \in Wim(B)$, if and only if $a(x_j) \cap a(x_k) \neq \emptyset$ for at least one $a \in B$ exists.

Definition 2.8. Let U be the set of objects, R is the tolerate relation decided by similarity, so we have lower approximation of R is $X_R = \{x \in U \mid x/R \subseteq X\}$, and the upper approximation of R is $X^R = \{x \in U \mid x/R \cap X \neq \emptyset\}$, and the border of R is $B_R(X) = X^R - X_R$, and the tolerance rough set is $Tol(X) = (X_R, X^R)$.

Table 1. A statistical table

	Paper	thesis	Outlay
Software staff	$\{12 \sim 15\}$		Pre-research
Application staff	$\{15 \sim 17\}$		$\{5.50 \sim 8.00\}^*$
Architectonic staff	$\{6 \sim 8\}^*$		$\{6.00 \sim 8.80\}$
Computer net staff	$\{3 \sim 5\}^*$		Pre-research
Foundation staff	/		$\{25.00 \sim 29.00\}$
Comm. staff	$\{12 \sim 14\}^*$		/
AI staff	$\{13 \sim 15\}$		$\{7.50 \sim 9.00\}^*$
Software Eng.	$\{22 \sim 24\}$		Pre-research
Multimedia staff	$\{16 \sim 18\}^*$		/

3 Approximate Context and Approximate Concept Lattice

The basic logic of Wille's FCA is the traditional logic, and the object is the concept in the definite information system. In this paper, we broaden FCA to approximate information system.

3.1 Approximate Context

Definition 3.1. Approximate context is a triple $C = (G, M, I)$, in which G is the limited object set, M is the limited attribute set, I is the tolerance relation of G and M, so we have $I \subseteq C^{G \times M}$.

In FCA, we define the object set $A \subseteq G$ and the attribute set $B \subseteq M$ as follows:

Derivation function $\beta(A)$ is the attribute set in the object A, written $\beta(A) = \{m \in M \mid g\text{Im}, \forall\ g \in A\}$, the derivation function $\alpha(B)$ is the object set of attribute set B, written $\alpha(B) = \{g \in G \mid g\text{Im}, \forall\ m \in B\}$.

Lemma 3.1. Let (G, M, I) be the formal context, $A_1, A_2 \subseteq G$, $B_1, B_2 \subseteq M$, so following assertions hold:

$$A_1 \subseteq A_2 \rightarrow \beta(A_1) \supseteq \beta(A_2),\ \forall\ A_1, A_2 \subseteq G;$$
$$B_1 \subseteq B_2 \rightarrow \alpha(B_1) \supseteq \alpha(B_2),\ \forall\ B_1, B_2 \subseteq M. \tag{3}$$
$$A \subseteq \alpha(\beta(A)),\ \beta(A) = \beta(\alpha(\beta(A))),\ \forall\ A \subseteq G;$$
$$B \subseteq \beta(\alpha(B)),\ \alpha(B) = \alpha(\beta(\alpha(B))),\ \forall\ B \subseteq M. \tag{4}$$
$$\beta(A) = \beta(\alpha(\beta(A))),\ \alpha(B) = \alpha(\beta(\alpha(B))). \tag{5}$$
$$A \subseteq \alpha(B) \Leftrightarrow B \subseteq \beta(A) \Leftrightarrow A \times B \subseteq I. \tag{6}$$

Definition 3.2. The closure definition of function $A \subseteq G$ is $Closure(A) = \alpha(\beta(A))$. On the contrary, when $B \subseteq M$, so $Closure(B) = \beta(\alpha(B))$.

Definition 3.3. In the context $C = (G, M, I)$, a concept is an ordered pair (A, B), so $\beta(A) = B$, $\alpha(B) = A$, here $A \subseteq G$, $B \subseteq M$, and A is called the extension of (A, B), B is called the intension of (A, B).

Table 2. The cross table

	Paper			Outlay			
	a	b	c	d	e	f	g
1	×		×				
2	×				×		×
3	×	×	×	×	×		×
4	×	×	×	×			
5						×	
6	×	×					
7	×				×		×
8	×		×				
9	×	×					

In arbitrary context $C = (G, M, I)$, no matter which subset in G or M, they each have a corresponding concept. Obviously, set A and set B have their own corresponding concept, and if and only if the set is closed.

Definition 3.4. If $B = Closure(B)$, then is the feasible intension. This indicates that B is the only intension of $(\alpha(B), B \subseteq M)$; on the contrary, if $A = Closure(A)$, so $A \subseteq G$ is the feasible extension, which shows that A is the only extension of $(A, \beta(A))$. If set X is the feasible intension of a concept, or the feasible extension of it, so it is called the feasible set of X, or it is infeasible.

Table 1 is estimation of the scientific research situation in the computer science department in the past 3 years. This is an approximate information system.

Let context $C = (G, M, I)$ be $\wp(G, M, I)$, so $\wp(G, M, I)$ has pos. In order to describe the concept systematically, we introduce subconcept and superconcept into it.

Definition 3.5. If there are two concepts (A_1, B_1) and (A_2, B_2), and $(A_1, B_1) \leq (A_2, B_2)$, so we call (A_1, B_1) is the subconcept of (A_2, B_2); if $(A_2, B_2) \geq (A_1, B_1)$, we call (A_2, B_2) is the superconcept of (A_1, B_1). The relationship between subconcept and superconcept can be described like:

$$(A_1, B_1) \leq (A_2, B_2) \Leftrightarrow A_1 \subseteq A_2 (\Leftrightarrow B_1 \supseteq B_2) \tag{7}$$

Definition 3.6. The approximate context (G, M, I) has a definable approximate concept (A, B), if and only if $A \subseteq G$, $B \subseteq M$, $\beta(A) = B$, $\alpha(B) = A$, or it can't be defined.

Definition 3.7. The approximate context (G, M, I) has an indefinable approximate concept (A, B), and $\alpha(B) \neq A$, so its object is indefinable.

Definition 3.8. The approximate context (G, M, I) has an indefinable approximate concept (A, B), and $\beta(A) \neq B$, so its attribute is indefinable.

Definition 3.9. The approximate context (G, M, I) has an indefinable approximate concept (A, B), $\alpha(B) \neq A$ and $\beta(A) \neq B$, so the context is indefinable.

For example, in table 3 if $A = \{1\}$, $B = \{b, d\}$, Because $\alpha(B) = \{1, 3, 4, 8\} \neq A$, $\beta(A) = \{b, d\} = B$, the object is indefinable; if $A = \{3, 4\}$, $B = \{c, d\}$, since $\beta(A) = \{a, b, c, d\} \neq B$, $\alpha(B) = \{3, 4\} = A$, attribute can't be defined;

Table 3. Concepts for the Table 2

Number	Concepts
1	$(\{3, 4\}, \{a, b, c, d\})$
2	$(\{1, 2, 3, 4, 6, 7, 8, 9\}, \{b\})$
3	$(\{3, 4, 6, 9\}, \{b, c\})$
4	$(\{1, 3, 4, 8\}, \{b, d\})$
5	$(\{2, 3, 7\}, \{b, e, g\})$
6	$(\{5\}, \{f\})$
7	$(\{\emptyset\}, \{a, b, c, d, e, f, g\})$
8	$(\{1, 2, 3, 4, 5, 6, 7, 8, 9\}, \{\emptyset\})$
9	$(\{3\}, \{a, b, c, d, e, g\})$

if $A = \{1, 4\}$, $B = \{b, c\}$, since $\alpha(B) = \{3, 4, 6, 8\} \neq A$, $\beta(A) = \{b, d\} \neq B$, context can't be defined. While if $A = \{3, 4, 6, 9\}$, $B = \{b, c\}$, since $\beta(A) = B$, $\alpha(B) = A$ it can be defined, that is to say $(\{3, 4, 6, 9\}, \{b, c\})$ is an approximate concept.

3.2 Approximate Concept Lattice

In the context $C = (G, M, I)$ is the set of all the concepts, that is $\mathcal{L}(G, M, I) = \{(A, B) \in G \times M \mid \beta(A) = B, \alpha(B) = A\}$, which also includes the partially ordered $(A_1, B_1) \leq (A_2, B_2) \Leftrightarrow A_1 \subseteq A_2$ (or $\Leftrightarrow B_1 \supseteq B_2$), so $\mathcal{L}(C)$ is called as the concept case of C.

Lemma 3.2. If we make $C = (G, M, I)$ be the formal context, so that

$$\mathcal{L}(C) = \{(\alpha(B), \beta(\alpha(B))) \mid B \subseteq M\} \tag{8}$$

According to the FCA theory, each concept case is a complete case, and the set of terms is closured.

Definition 3.10. The concept case of approximate context $C = (G, M, I)$ is the side-sequence of all the inferior concepts to superior concepts of concept (A, B), which $\leq$ the set. It is a complete case, noted as $\mathcal{L}(C)$, whose maximum and minimum accurate limit respectively are:

$$\bigwedge_{j\in J}(A_j, B_j) = \left(\bigcap_{j\in J} A_j, \beta\left(\alpha\left(\bigcup_{j\in J} B_j\right)\right)\right) = \left(\bigcap_{j\in J} A_j, \beta\left(\bigcap_{j\in J} A_j\right)\right) \quad (9)$$

$$\bigvee_{j\in J}(A_j, B_j) = \left(\alpha\left(\beta\left(\bigcup_{j\in J} A_j\right)\right), \bigcap_{j\in J} B_j\right) = \left(\alpha\left(\bigcap_{j\in J} B_j\right), B_j\right) \quad (10)$$

On the contrary, if $\mathcal{L}$ is a complete case, so $L \cong \mathcal{L}(C)$, only when it comes to the existence of the reflection $\gamma : G \to L$ and $\mu : M \to L$, which makes $\gamma(G)$ belong to the sets of L's supremum dense set; and $\mu(M)$ belong to the sets of L's infimum dense set. What's more $(g, m) \in L$, equals to $\gamma(g) \leq \mu(m)$, for $\forall\ g \in G$, $\forall\ m \in M$.

The corresponding approximate concept of Table 2 (the cross table) is showed in Table 3.

4 The Approximate Concept Optimization

4.1 Some Cases of Approximate Concept Optimization

Consider an arbitrary concept (A, B), $A \subseteq G$, $B \subseteq M$, our purpose is to get a best (A, B) formal concept in the concept set, whose extension is approaching A and intension is approaching B.

Definition 4.1. Given the concept (A, B), if $A \subseteq G$, $B \subseteq M$ respectively, A and B belong to the object and attribute of some concept in the concept set, that is to say $A = \alpha(\beta(A))$, $B = \beta(\alpha(B))$, therefore (A, B) can be considered as conceptually feasible.

Definition 4.2. Assuming that the concept (A, B), $A \subseteq G$, $B \subseteq M$ is given, if A is a object of some concept in the concept set, that is to say $A = \alpha(\beta(A))$, while $B \neq \beta(\alpha(B))$, therefore Object A is feasible.

Definition 4.3. Assuming that the concept (A, B), $A \subseteq G$, $B \subseteq M$ is given, if B is an attribute of some concept in the concept set, that is to say $B = \beta(\alpha(B))$, while $A \neq \alpha(\beta(A))$, therefore Attribute B is feasible.

Definition 4.4. Assuming that the concept (A, B), $A \subseteq G$, $B \subseteq M$ is given, if respectively, A and B neither belong to the object nor attribute of some concept in the concept set, that is to say, $\alpha(\beta(A)) \neq A$, $\beta(\alpha(B)) \neq B$ therefore concept (A, B) is considered as unfeasible.

Obviously, if A is feasible, $(A, \beta(A))$ is a concept; likewise, if B is feasible, $(\alpha(B), B)$ is a concept, too. Now, let's discuss the four above-mentioned cases separately.

1. (A, B) is concept feasible, there are two possibilities.

(1)$\beta(A) = B$ is already known, if $\beta(A) = B$, so $\alpha(B) = A$. Therefore, the sequential pair (A, B) is what we wanted.

Theorem 4.1. Assuming that $A \subseteq G$ and $B \subseteq M$, if A is feasible, and $\beta(A) = B$, so that B is feasible, and $\alpha(B) = A$.

Identification: Since A is feasible, $\alpha(\beta(A)) = A$; since $\beta(A) = B$ and $\alpha(B) = A$; and β operation is made on the both sides of the equality, then $\beta(\alpha(B)) = B$ is got, which means B is feasible. Approving finished.

(2) $\beta(A) \neq B$ is already known, if $\beta(A) \neq B$, which indicates that $\alpha(B) \neq A$, let $\beta(A) = A'$, $\alpha(B) = B'$, since A and B are feasible, the similar degree $f_{1C}(A,B)$ of similar concept can be defined as

$$f_{1C}(A,B) = \frac{1}{2}\left(\frac{Card(A \cap extent(C))}{Card(A \cup extent(C))} + \frac{Card(B \cap intent(C))}{Card(B \cup intent(C))}\right) \quad (11)$$

among them, C is a concept in concept set $\mathcal{L}(C)$.

2. Object A is feasible, while attribute B is unfeasible. Since object A is feasible, the similar degree of approximate concept $f_C(A)$ can be defined as

$$f_C(A) = \frac{1}{2}\left(\frac{Card(A \cap extent(C))}{Card(A \cup extent(C))} + \frac{Card(\beta(A) \cap intent(C))}{Card(\beta(A) \cup intent(C))}\right) \quad (12)$$

3. Attribute B is feasible, while object A is unfeasible. Similar to (2), since attribute B is feasible, the similar degree of approximate concept $f_C(B)$ is

$$f_C(B) = \frac{1}{2}\left(\frac{Card(B \cap intent(C))}{Card(B \cup intent(C))} + \frac{Card(\alpha(B) \cap extent(C))}{Card(\alpha(B) \cup extent(C))}\right) \quad (13)$$

4. Object A and attribute B are both unfeasible. Since they are both unfeasible, the similar degree of approximate concept $f_{2C}(A,B)$ is

$$f_{2C}(A,B) = \frac{1}{2}\left(\frac{Card(\beta(B) \cap extent(C))}{Card(\beta(B) \cup extent(C))} + \frac{Card(\alpha(A) \cap intent(C))}{Card(\alpha(A) \cup intent(C))}\right) \quad (14)$$

4.2 The Optimization Algorithm of the Approximate Concept

Now the optimization algorithm of the approximate concept is provided. Assuming that a sequential pair (A,B), $A \subseteq G$, $B \subseteq M$ in the approximate information system $\mathcal{L}(C)$ is given, optimization algorithm is trying to get a certain approximate concept (A'_i, B'_i) (and i is the serial number of concepts $\mathcal{L}(C)$ in concept lattice), whose optimum extension is approaching A, and the optimum intension is approaching B.

Then, the optimization algorithm of the approximate concept is given as followed.

Input: The context alternate form of approximate system, approximate concept lattice $\mathcal{L}(C)$ and the concept (A,B) awaiting approach.
Output: The approached (A', B') in approximate system $\mathcal{L}(C)$.

```
{
    n ← Concept diagram in concept lattice
    if (α(β(A)) == A))                      //A is feasible, accurate answer
        if (β(A) == B)
            (A', B') ← (A, B);
        else if (β(α(B)) == B)
            for (i = i, n, i++) {
                f_{1C_i}(A, B) = \frac{1}{2}\left(\frac{Card(A \cap extent(C_i))}{Card(A \cup extent(C_i))} + \frac{Card(B \cap intent(C_i))}{Card(B \cup intent(C_i))}\right);   //calculate the
            similar degree
                (A', B') ← \max_{1 \leq i \leq n}\{f'_i(A,B)\};  }    //to get optimum approximate concept
        else                                  //A is feasible, B is unfeasible
            for (i = i, n, i++) {
                f_{iC}(A) = \frac{1}{2}\left(\frac{Card(A \cap extent(C_i))}{Card(A \cup extent(C_i))} + \frac{Card(\beta(A) \cap intent(C_i))}{Card(\beta(A) \cup intent(C_i))}\right);
                (A', B') ← \max_{1 \leq i \leq n}\{f'_{iC}(A,B)\};  }    //to get optimum approximate concept
        end if;
    else if (β(α(B)) == B)                  //B is feasible, A is unfeasible
        for (i = i, n, i++) {
            f_{iC}(B) = \frac{1}{2}\left(\frac{Card(B \cap intent(C_i))}{Card(B \cup intent(C_i))} + \frac{Card(\alpha(B) \cap extent(C_i))}{Card(\alpha(B) \cup extent(C_i))}\right);
            (A', B') ← \max_{1 \leq i \leq n}\{f'_{iC}(B)\};  }    //to get optimum approximate concept
    else                                     //A is unfeasible, B is unfeasible either
        for (i = i, n, i++) {
            f_{2C_i}(A, B) = \frac{1}{2}\left(\frac{Card(\beta(B) \cap extent(C_i))}{Card(\beta(B) \cup extent(C_i))} + \frac{Card(\alpha(A) \cap intent(C_i))}{Card(\alpha(A) \cup intent(C_i))}\right);
            (A', B') ← \max_{1 \leq i \leq n}\{f'_{iC}(A,B)\};  }    //to get optimum approximate concept
    end if;
}
```

4.3 Result of Emulation Calculation

Now the result of optimizing the concept in approximate system table 1 by using the algorithm given in the paper is provided. Assuming that $A = \{3, 4, 6\}$,

Table 4. Result of emulation calculation

C_i	Extent(C_i)	Intent(C_i)	$f_{1i}(A,B)$	$f_i(A)$	$f_i(B)$	$f_{2i}(A,B)$
1	$\{3,4\}$	$\{a,b,c,d\}$	0.433	0.583	0.225	0.375
2	$\{1,2,3,4,6,7,8,9\}$	$\{b\}$	0.438	0.438	0.438	0.438
3	$\{3,4,6,9\}$	$\{b,c\}$	0.541	0.875*	0.250	0.583
4	$\{1,3,4,8\}$	$\{b,d\}$	0.367	0.367	0.250	0.250
5	$\{2,3,7\}$	$\{b,e,g\}$	0.433	0.225	0.833*	0.625
6	$\{5\}$	$\{f\}$	0	0	0	0
7	$\{\emptyset\}$	$\{a,b,c,d,e,f,g\}$	0.143	0.143	0.143	0.143
8	$\{1,2,3,4,5,6,7,8,9\}$	$\{\emptyset\}$	0.167	0.167	0.167	0.167
9	$\{3\}$	$\{a,b,c,d,e,g\}$	0.333	0.333	0.333	0.333

$B = \{b, e\}$, separately corresponding to formula (11), (12), (13) and (14), the results of $f_{1i}(A, B)$, $f_i(A)$, $f_i(B)$ and $f_{2i}(A, B)$ are as follows.

We can conclude from the result that the optimum concept of approximate concept $A = \{3, 4, 6\}$, $B = \{b, e\}$ is $C_1 = (\{3, 4, 6, 9\}, \{b, c\})$ and $C_2 = (\{2, 3, 7\}, \{b, e, g\})$, and they have the similar degree $f_3(A) = 0.875$ and $f_5(B) = 0.833$ respectively.

5 Conclusion

From the above-mentioned algorithm and result of emulation calculation, it is known that, we can deal with approximate information system well and get satisfactory optimum concept based on the combination of FCA and fault-tolerate rough sets theory. Meanwhile, we can prove the dependability of this method by inspecting the optimum concept. For example, the approximate concept in emulation calculation ($\{3, 4, 6\}, \{b, e\}$) is attribute feasible, but is object unfeasible. Since $\beta(A) = \{b, c\} \neq B$, $\alpha(\beta(A)) = \{3, 4, 6, 9\} \neq A$, while $\alpha(B) = \{2, 3, 7\} \neq A$, $\beta(\alpha(B)) = \{b, e\} = B$, we adopt (13) to calculate similar degree to optimize it and get a satisfactory result. The time complexity of this algorithm is $O(|L(C)|)$ and the worst situation is $O(|G|^2|M|^2)$.

References

1. Chen Shiquan, Chen Lichun.: Fuzzy Concept Lattice. Fuzzy System and Mathematics, Vol. 16, No. 4, (2002).
2. Jouni Jarvinen.: Knowledge Representation and Rough Sets. Turku Centre for Computer Science, TUCS Dissertations, No. 14, March (1999)53-56.
3. Jamil Saquer, Jitender S. Deogun.: Concept Approximations Based on Rough Sets and Similarity Measure. Int. j. Math. Comput. Sci., Vol. 11, No. 3, (2001)655-674.
4. Wille, R.: Restructuring lattice theory: an approach based on hierarchies of concepts. In Ordered Sets, In: Ordered Sets (ed. I. Rival). Reidel, Dordrecht-Boston (1982)445-470.
5. Ganter, B., R. Wille.: Formal Begriffsanalyse: Mathematische Grundlagen. Spring-Verlag, Berlin-Heidelberg (1996).
6. Ganter, B., R. Wille.: Formal Concept Analysis: mathematical foundations. (translated from the German by Cnenelia Franzke) Springer-Verlag, Berlin-Heidelberg (1999).
7. Kent R.: Rough Concept Analysis: A Synthesis of Rough Sets and Formal Concept Analysis. Fund. Inform., Vol. 27, No. 2, 169-181.

Geometrical Probability Covering Algorithm

Junping Zhang[1,2], Stan Z. Li[3], and Jue Wang[2]

[1] Shanghai Key Laboratory of Intelligent Information Processing,
Department of Computer Science and Engineering,
Fudan University, Shanghai 200433, China
jpzhang@fudan.edu.cn

[2] The Key Laboratory of Complex Systems and Intelligence Science,
Institute of Automation, Chinese Academy of Sciences, Beijing, 100080, China
{junping.zhang, jue.wang}@mail.ia.ac.cn

[3] National Laboratory of Pattern Recognition & Center for Biometrics and Security Research,
Institute of Automation, Chinese Academy of Sciences, Beijing, 100080, China
szli@nlpr.ia.ac.cn

Abstract. In this paper, we propose a novel classification algorithm, called geometrical probability covering (GPC) algorithm, to improve classification ability. On the basis of geometrical properties of data, the proposed algorithm first forms extended prototypes through computing means of any two prototypes in the same class. Then Gaussian kernel is employed for covering the geometrical structure of data and used as a local probability measurement. By computing the sum of the probabilities that a new sample to be classified to the set of prototypes and extended prototypes, the classified criterion based on the global probability measurement is achieved. The proposed GPC algorithm is simple but powerful, especially, when training samples are sparse and small size. Experiments on several databases show that the proposed algorithm is promising. Also, we explore other potential applications such as outlier removal with the proposed GPC algorithm.

1 Introduction

One goal of machine learning is classification, namely, identifying an unknown sample into the corresponding class. The nonparametric one, for example, the nearest neighbor (NN) algorithm [1,2,3] and the nearest feature line (NFL) [4], is to classify an unknown sample based on either nearest neighbor distance or nearest projection distance. A fundamental framework underlying these algorithms is that classification can be achieved through computing the local minimal distances from the unknown sample to labelled classes. When the number of data is small and sparse, however, the performance of these approaches may be impaired. The parametric one such as the gaussian mixture model, is to classify sample based on maximum likelihood. The potential assumption of the approach is that samples in the kth class are generated by some probability distribution, and the probability of data located in the 'middle' of distribution is higher than that of data in the boundary. Posterior probability is often used as classification criterion [5]. The main drawbacks are that the computational cost is high and convergence are hard to be guaranteed [6].

Therefore, we propose Geometrical Probability Covering (GPC) algorithm in section 2. First, the middle points (or means) of any two prototypes (training samples) in the same class are generated to build an extended prototype set with prototypes. And then for each extend prototype, gaussian kernel is introduced to compute the local probability of an unknown sample. By calculating the sum of probabilities from sample to each extended prototype set, the global probability of the unknown sample are obtained. As a result, the geometrical structure of data is approximated with the set of prototypes and extended prototypes. We also demonstrate the distinction between the proposed GPC algorithm and NFL algorithm in the next section. In the penultimate section experiments on several databases show the proposed GPC algorithm are promising, and the final section we discuss some potential problems and further researches.

2 Geometrical Probability Covering Algorithm

2.1 The Proposed Algorithm

Geometrically intuitive and computationally effective, the k-NN (nearest neighborhood) algorithm nevertheless is incapable of representing the distribution of data. Another geometrical classification approach is nearest feature line in which the classification criterion is to compute the nearest projection distance from sample to be classified to the nearest feature lines which are generated by any two prototypes. However, the NFL algorithm may form the mis-classified distance if the training data are highly twisted (The detail can be seen in section 2.2).

To overcome these disadvantages, the geometrical probability covering algorithm is proposed in this section. First, the means of any two training samples in the same class are computed to form an extended prototype set with training samples. Then the extended prototype set is used for modeling the geometrical covering of data distribution in the same class. It is not difficult to show that the total number of extended prototypes in the same class is

$$m'_k = C^2_{m_k} = \frac{m_k(m_k - 1)}{2}.$$

Where m_k is the number of training samples in the kth class, and m'_k the number of extended prototypes in the kth class. If the number of training samples in the same class is 5, for instance, then $m'_k = C^2_5 = 10$.

Intuitively, the density in the boundary of distribution are lower than that in the center of each class. If each extended prototype is given a local probability density function, we presume that the sum of probability distributions of the extended prototype set can approximate the underlying distribution of data.

Specifically, let $P(x_i|p_j, \mathcal{M}_k)$ be the conditional probability of some unknown sample x_i given the extended prototype p_j of the kth class. The corresponding local conditional probability density is formulated based on Gaussian kernel as:

$$P(x_i|p_j, \mathcal{M}_k) = \exp(\frac{-\|x_i - p_j\|^2}{2\sigma^2})$$
$$j = 1, \ldots, m'_k, k = 1, \ldots, K \qquad (1)$$

where K is the number of classes. And variance σ^2 controls the range of local probability in each extended prototype. It is not difficult to see that the range of value of each local Gaussian kernel is $[0, 1]$. From geometrical point of view, when σ^2 is rather large, for example, 1e18 (where data is assumed to be normalized), then each local probability density is equal to 1. And the probabilities that samples are classified to some extended prototypes are equal everywhere. Conversely, when σ^2 is rather small, for example, $1e-18$, then each local probability will collapse to 0 rapidly. It shows that the Gaussian kernel influences the local range of the prototype and converge to each prototype. Therefore, the suboptimal value of variance σ^2 can be found within the range of the two extreme cases.

One way to obtain the global probability from sample to be classified to some class is to sum up all the local probabilities from sample to each extended prototypes. As a result, the distribution of data can be approximated when the parameters σ^2 of Gaussian kernels are suitable. The sum of local probability density is illustrated as:

$$P(x_i|\mathcal{M}_k) = \sum_{j=1}^{m'_k} P(x_i|p_j, \mathcal{M}_k) \quad (2)$$

For satisfying the probability measure, we constrain the range of probability within the value $[0, 1]$ in this paper. So we normalize Eq. (2) as:

$$P'(x_i|\mathcal{M}_k) = \frac{P(x_i|\mathcal{M}_k)}{\max_j P(x_j|\mathcal{M}_k)} \quad j=1,\cdots,k \quad (3)$$

Once the probabilities of each unknown sample in the different classes are calculated, the classification criterion is established as follows:

$$C(x_i) = \arg\max_k P'(x_i|\mathcal{M}_k) \quad (4)$$

Actually, the proposed GPC algorithm can be used not only for classification, but also for removing the outliers or the samples in the unlabelled classes. From the probability point of view, the probability of outliers and unlabelled samples will lower than sample in the labelled classes. Hence, we propose an extension version of the GPC algorithm by introducing threshold for identifying these exceptions. The extension of the GPC algorithm can be formulated as:

$$v(x_i) = \begin{cases} 1, \forall k, P'(x_i|\mathcal{M}_k) < \delta \\ 0, \text{otherwise} \end{cases} \quad (5)$$

where δ is the predefined threshold based on the trade-off between recognition accuracy and the number of outlier removal, and the probability of each sample to be classified is:

$$P''(x_i|\mathcal{M}_k) = P'(x_i|\mathcal{M}_k) \cdot v(x_i) \quad (6)$$

The criterion for classification is therefore re-written as follows:

$$C(x_i) = \begin{cases} \arg\max_k P''(x_i|\mathcal{M}_k), & \text{if } \forall k, P''(x_i|\mathcal{M}_k) > 0 \\ l, & \text{otherwise} \end{cases} \quad (7)$$

where l means some unlabelled class or outlier. With respect to Eq. (6) and (7), we can remove outliers and unknown labelled samples effectively. It is worthy noting that the proposed algorithm is incapable of distinguish the two exceptions (outlier and unknown labelled sample) due to the absence of prior knowledge.

2.2 Comparison with NFL Algorithms

The proposed algorithm inherits the basic idea of NFL algorithms [4], that is to say, some extended prototypes are generated by employing interpolation approach. However, there exists some differences between the two algorithms. The distinction is illustrated as Figure 1.

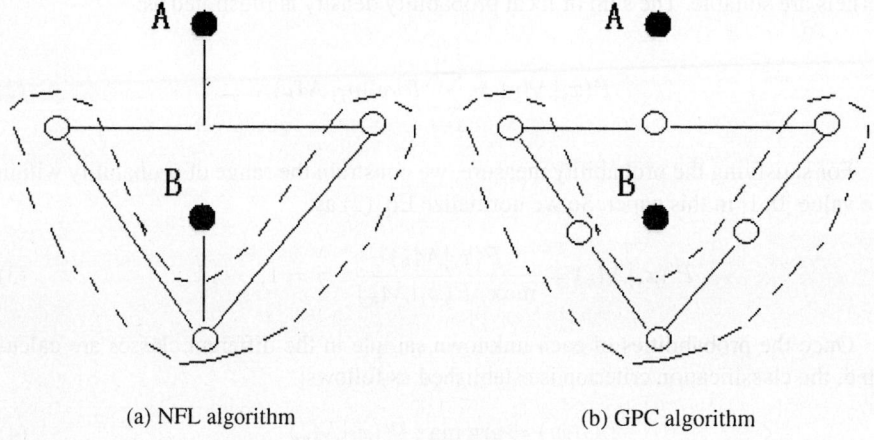

(a) NFL algorithm (b) GPC algorithm

Fig. 1. Two examples

In Figure 1, point B is equidistance to three edges of triangle and the projection distance of point A is the same projection distance to nearest line as point B, and three circular white points are the assumed prototypes. Also the dash curves are hypothesized to form the boundary of data distribution.

From Figure 1(a), it can be shown that the extended prototypes are generated through orthogonal projection from sample A to three lines which are made up of any two training samples. Meanwhile, in Figure 1(b), the extended prototypes are means of any two prototypes in the same class with the proposed GPC algorithm. Definitely, point A has the same projection distance as point B if NFL algorithm is used. And therefore, when using NFL to classify point A and B, the nearest projection distance would be equal.

From Figure 1(b) it is clear that point B is closer to the boundary of distribution. Nevertheless, point B should be more important than point A. Because of the local behavior of NFL algorithm, the property has not been embodied, especially when data are sparse and highly twisted.

Conversely, with the proposed GPC algorithm, the probability of each sample is the sum of the probabilities that sample are classified to the extended prototype set in the same class. The procedure is similar to cover the structure or distribution of data. As in Figure 1(b), the local probability densities from sample A to each extended prototypes are the same as those of sample B. But the global probability of the sample A will be smaller because the number of the covered gaussian kernel of sample in the boundary is less than that of sample in the middle of distribution when suitable variance σ^2 is selected. The reason is that the proposed GPC algorithm considers the properties of probability distribution based on the geometrical covering. More generally speaking, the data in the boundary of distribution will have lower probabilities than these samples in the "middle" of data. In particular, the GPC algorithm may obtain better classification performance than the NFL algorithm when data are highly twisted and sparse.

3 Experiments

Extensive experiments are carried out on several databases to verify the performance of the proposed GPC algorithm. Also we explore potential applications of the probability representation of the expression and outlier removal based on the extension version of the GPC algorithm.

3.1 Recognition

In the recognition experiments, several databases including three face databases and one image-segmentation database, are used for evaluating the classification ability of the proposed GPC algorithm. The face databases are ORL, UMIST and JAFFE databases, respectively [7,8,9]. Image segmentation databases are from UCI repository [10]. All the databases have been standardized to scope $[0, 1]$. It is noticeable that we employ the dimensionality reduction approaches for reducing the dimensions of the high- dimensional databases and avoiding the problem of "the curse of dimensionality". The details of experimental procedure are listed in Table 1:

Table 1. Experimental approach and the abbreviation of the corresponding algorithms: 1: PCA+NN; 2: PCA+NFL; 3: PCA+GPC; 4: LLE+NN; 5: LLE+NFL; 6: LLE+GPC; 7: PCA+LDA+NN; 8: PCA+LDA+NFL; 9: PCA+LDA+GPC; 10: ULLELDA+NN; 11: ULLELDA+NFL; 12: ULLELDA+GPC. TR: training samples (*classes); TE: test samples.

	Compared Algorithms	Experimental Average	TR	TE	classes
ORL	1-12	100 runs	5*40	5*40	40
UMIST	1-12	100 runs	10*20	375	20
JAFFE	1-12	100 runs	6*10	153	10
Image-Segmentation	1-12	Once	30*7	300*7	7

In Table 1, LDA denotes linear discriminant analysis [11], PCA principal component analysis [12], LLE locally linear embedding algorithm [13], ULLELDA is our proposed dimensionality reduction approach [14]. These dimensionality reduction approaches are used to obtain different subspaces to assure the robustness of the proposed GPC algorithm. Because the paper focuses on the proposed GPC algorithm, the detail on how to achieve dimensionality reduction can be seen in [14,15,16].

For classifiers, NN denotes 1-nearest neighbor and NFL nearest feature line. In our experiments, several combinational algorithms are thus applied for recognition. For example, PCA+NN represents the combination of principal component analysis and nearest neighbor algorithm.

As for the proposed GPC algorithm, variance σ^2 need to be predefined. Considering the geometrical explanation mentioned above, a set of parameters are generated from $1e-10$ to $1e+10$ according to logarithm basis. The final results are selected based on the lowest error rates of each mentioned algorithms. The results are reported in Table 2, where the lowest error rates have been marked with bold type.

Table 2. The Error rates (%) and Standard Deviations (%) of the classification algorithms

	ORL	JAFFE	UMIST	Image Segmentation
LLE+GPC	5.01±1.68	2.75±1.44	4.323±1.68	9.52
LLE+NFL	4.35±1.42	3.79±2.14	4.50±1.76	9.19
LLE+NN	7.01±1.83	5.78±2.49	5.96±1.97	10.91
PCA+GPC	8.39±2.51	11.21±2.86	6.55±2.06	12.62
PCA+NFL	8.27±2.41	9.65±2.70	5.44±1.74	10.61
PCA+NN	10.12±2.06	11.29±2.85	8.38±1.88	14.24
ULLELDA+GPC	**3.82±1.91**	2.08±1.39	**2.20±1.21**	**7.05**
ULLELDA+NFL	4.06±1.40	2.14±1.45	2.40±1.18	8.81
ULLELDA+NN	4.13±1.33	2.18±1.38	2.20±1.22	8.76
PCA+LDA+GPC	7.56±1.92	**0.80±0.95**	2.60±1.21	7.61
PCA+LDA+NFL	7.35±1.89	1.09±1.11	2.62±1.22	8.33
PCA+LDA+NN	7.45±1.92	0.93±1.05	2.59±1.21	9.10

From the table it is clear that the proposed GPC algorithm obtain the lowest recognition when being compared with other mentioned classification algorithms. When the dimensionality of database is large, moreover, the GPC algorithm combining with some dimensionality reduction techniques will obtain better classification ability.

We also investigate the influences of training samples for the proposed GPC algorithm which can be illustrated in Figure 2. As training samples increases, the ULLELDA+GPC algorithm reports the lowest error rates for ORL database. And when training samples of the ORL face databases are 9 each class, the error rates and standard variance of ULLELDA+GPC algorithm are 0.30% and 0.96%, respectively. However, when training samples of the JAFFE face databases are 11 each class, two algorithms (PCA+LDA+GPC and PCA+LDA+NN) achieve 100% recognition accuracy. The reason we assumed is that when the distribution of training samples tends to true one, the two algorithms make a comparable performance. We also observe that when training samples are sparse, the classification of the proposed GPC algorithm are better than those of the mentioned algorithms.

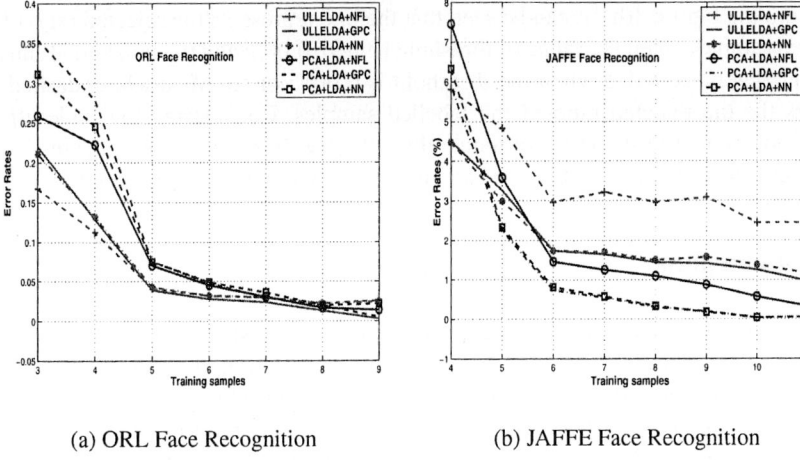

(a) ORL Face Recognition   (b) JAFFE Face Recognition

Fig. 2. The influence of training samples

3.2 Outlier Removal

Generally speaking, the categories of some samples often differ from the labelled classes when we process large-scale data. An effective face recognition system should not only recognize the face, but also exclude some unlabelled nonface images or data. Based on the extension of the GPC algorithm, the outlier or unlabelled data can be removed. In the experiment, we randomly select some images such as faces and landscapes, which are not in the JAFFE databases. The images are shown in Figure 3(a) and have the same pixels (146*111).

With respect to the proposed GPC algorithm, let variance σ^2 be 1 and the reduced dimensions be 150 using the LLE algorithm, then the relationship between the rejected rate and accepted rate with different thresholds is illustrated in Figure 3(b).

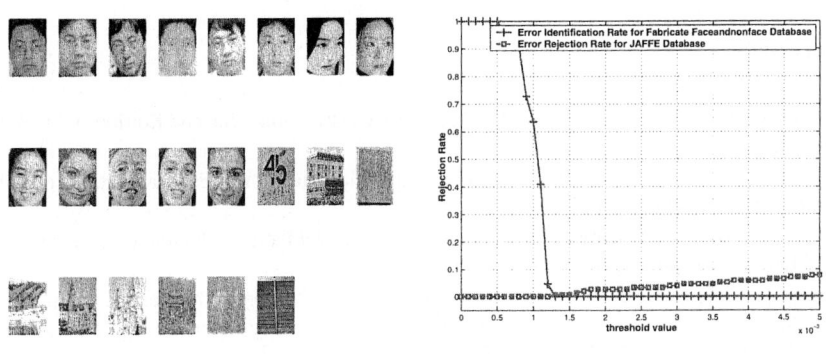

(a) Unlabelled Face and Landscapes   (b) The Rejected and Accepted Curves of the Unlabelled Samples

Fig. 3. Outlier Removal

From the Figure 3(b), it can be seen that the line represents the rejected ratio of unlabelled samples when the range of threshold is in $[0, 0.005]$ and the unlabelled samples are completely excluded when the threshold is 0.0015 or so. Meanwhile, dashed-line reflects the the rejected ratio of the labelled samples. Considering the trade-off, the threshold is equal to 0.0012. Consequently, only one of 22 unlabelled images is misclassified and all the labelled images are recognized without being removed as outliers.

4 Conclusions

In the paper, we propose the GPC algorithm to improve classification ability. With combination of geometrical covering and probability distribution, the proposed algorithm first forms extended prototypes through computing means of any two prototypes of the same class. And then the Gaussian kernel is adopted for controlling the scope of geometrical covering and generates the local probability measurement. By computing the sum of the probabilities from sample to the set of prototypes and extended prototypes, the global measurements for classification is obtained. Based on the GPC algorithm, the data in the boundary will have lower probability than in the 'middle' of distribution. Simple and geometric intuitive, the GPC algorithm is computationally effective without involving estimation of many parameters. When the training samples are sparse, the proposed algorithm obtain better performance than other classification algorithms. Experiments on several databases show the advantages of the proposed algorithm.

However, several problems remain. First of all, the computational complexity of the proposed algorithm is higher than that of NN algorithm. It is difficult for large-scale data to obtain practical application. How to decrease the complexity deserves research profoundly. Second, the choice of variances depends on human experiences. Adaptive selecting the parameter is worth studying. Finally, all local probability is represented with the same Gaussian kernel in the paper. But there may exist different kernels which lead to better classification performance. Therefore, we will research the possibility of the alternative local kernels in the future work.

References

1. Duda, R. O., Hart, P. E., and Stork, D. G.: Pattern Classification. Second Edition, John Wiley & Sons, Inc, (2001).
2. Cover, T. M., Hart, P. E.: Nearest neighbor pattern classification. IEEE Transactions on Infomation Theory. **13** (1968) 21-27
3. Dasarathy, B. V., ed.: Nearest Neighbor (NN) Norms: NN Pattern Classification Techniques. IEEE Computer Society. Washington, (1991)
4. Li, S. Z., Chan K. L. and Wang, C. L.: Performance Evaluation of the Nearest Feature Line Method in Image Classification and Retrieval. IEEE Transactions on Pattern Analysis and Machine Intelligence. **22(11)**, November, (2000), 1335-1339
5. Fraley C., Raftery, A. E.: Model-based Clustering, Discriminant Analysis, and Density Estimation. Technical Report No. **380**, University of Washington, (2000)
6. Dempster, A., Laird, N., and Rubin, D.: Maximum likelihood from incomplete data via the EM algorithm. J. Roy. Statist. Soc. B, **39** (1977) 1-38

7. Samaria, F. S.: Face Recognition Using Hidden Markov Models. PhD thesis, University of Cambridge, (1994)
8. Wechsler, H., Phillips, P. J., Bruce, V., F. Fogelman-Soulie V. and Huang T. S.(eds): em Characterizing virtual Eigensignatures for General Purpose Face Recognition. Daniel B Graham and Nigel M Allinson. In Face Recognition: From Theory to Applications. NATO ASI Series F, Computer and Systems Sciences, **163** (1998) 446-456
9. Michael, J. L., Budynek, J., and Akamatsu, S.: Automatic classification of Single Facial Images. IEEE Transactions on Pattern Analysis and Machine Intelligence. **21(12)** (1999) 1357-1362
10. Murphy, P. M., Aha, D. W., *UCI Repository of machine learning databases* [http://www.ics.uci.edu/ mlearn/MLRepository.html]. Irvine, CA: University of California, Department of Information and Computer Science, (1994)
11. Daniel, L. S., Weng, J.: Using Discriminant Eigenfeatures for Image Retrieval. IEEE Transactions on Pattern Analysis and Machine Intelligences. **18(8)** (1996) 831-836
12. Turk, M., Pentland, A.: Eigenfaces for Recognition. Journal of Cognitive Neuroscience. **3(1)** (1991) 71-86
13. Roweis, S. T., Lawrance, K. S.: Nonlinear Dimensionality reduction by locally linear embedding. Science. **290** (2000) 2323-2326.
14. Zhang, J., Shen, H., and Zhou, Z-H.: Unified Locally Linear Embedding and Linear Discriminant Analysis Algorithm (ULLELDA) for Face Recognition. Advances in Biometric Personal Authentication. Stan Z. Li, Jianhuang Lai, Tieniu Tan, Guocan Feng, Yunhong Wang (Ed.), LNCS 3338, Springer-Verlag, (2004) 209-307
15. Zhang, J., Li, S. Z., and Wang, J.: Nearest Manifold Approach for Face Recognition. The 6th IEEE International Conference on Automatic Face and Gesture Recogntion.Seoul, Korea, May, (2004)
16. Zhang, J., Li, S. Z., and Wang, J.: Manifold Learning and Applications in Recognition. in Intelligent Multimedia Processing with Soft Computing. Yap Peng Tan, Kim Hui Yap, Lipo Wang (Ed.), Springer-Verlag, Heidelberg, (2004)

Extended Fuzzy ALCN and Its Tableau Algorithm*

Jianjiang Lu[1,2], Baowen Xu[1,2,3], Yanhui Li[1], Dazhou Kang[1], and Peng Wang[1]

[1] Department of Computer Science and Engineering, Southeast University,
Nanjing 210096, China
[2] Jiangsu Institute of Software Quality, Nanjing 210096, China
[3] State Key Laboratory of Software, Wuhan University, Wuhan 430072, China
jjlu@seu.edu.cn

Abstract. Typical description logics are limited to dealing with crisp concepts. It is necessary to add fuzzy features to description logics for management of the fuzzy information. In this paper, we propose extended fuzzy ALCN to enable representation and reasoning for complex fuzzy information. We define syntax structure, semantic interpretation and reasoning problems of the extended fuzzy ALCN, and discuss the reasoning properties inexistent in typical description logics. We also design tableau algorithms of reasoning problems for extended fuzzy ALCN. The tableau algorithms are developed in the style of so-called constraint propagation method. Extended fuzzy ALCN is more expressive than the existing fuzzy description logics and present more wide fuzzy information.

1 Introduction

The Semantic Web [1] is an extension of the current Web in which web information is given well-defined semantic meaning to be machine-understandable, better enabling intelligent Web data processing. Description logics (DLs) [2] provide a logical reconstruction of object-centric and frame-based knowledge representation languages as main inferential means on the Semantic Web. Web applications often need management of fuzzy information, but typical DLs are limited to dealing with crisp concepts and crisp roles. Therefore, it is necessary to add fuzzy features to description logics. Meghini [3] proposed a preliminary fuzzy DL, which lacks reasoning algorithm, as a modeling tool for multimedia document retrieval. Straccia [4] presented Fuzzy ALC (FALC), a fuzzy extension of ALC combining fuzzy logic with typical ALC [5], and gave a constraint propagation calculus for reasoning in it. FALC just offers limited but not sufficient expressive power of complex fuzzy information. Some discussion about reducing FALC into classical ALC is given in [6], and the reduction does not extend the expressive bound of FALC. We present a more expressive fuzzy DL called Extended Fuzzy ALCN (EFALCN), which introduces the cut sets of fuzzy concepts and fuzzy roles as atomic concepts and atomic roles,

---

* This work was supported in part by the NSFC (60373066, 60425206, 90412003), National Grand Fundamental Research 973 Program of China (2002CB312000), National Research Foundation for the Doctoral Program of Higher Education of China (20020286004).

inherits the concept constructors and representation from ALCN [2], and defines a semantic interpretation for concepts and roles. EFALCN adopts a special fuzzify-method different from FALC and cover the expressive scope of FALC with more expressive power.

2 A Quick Look to FALC

Many vague concepts in the real world do not have a precise definition of membership. Such vague concepts are called fuzzy concepts. DLs are limited to dealing with crisp concepts and lack expressive power of fuzzy concepts. Straccia presented FALC, which is an extension of ALC with fuzzy features, to support fuzzy concept representation.

In the syntactic view, starting with a set of atomic concepts and a set of atomic roles, let B be an atomic concept, R be an atomic role, FALC concept descriptions C and D are inductively defined with the operation of ALC concept constructors in the following syntax rules: $C, D ::= \top \mid \bot \mid B \mid \neg C \mid C \sqcup D \mid C \sqcap D \mid \forall R.C \mid \exists R.C$. The top concept $\top$ denotes the whole domain and the bottom concept $\bot$ denotes the empty set. FALC roles can only be atomic roles. A FALC knowledge base is a pair $\Sigma(A, T)$, where A is an ABox, which is a set of fuzzy assertions, and T is a TBox, which is a set of fuzzy terminological axioms. A fuzzy assertion is in the form of $<C(a) \geq n>$ or $<R(a, b) \geq n>$. $<C(a) \geq n>$ means the membership degree of the individual a being instance of the concept C is at least n, and $<R(a, b) \geq n>$ has a similar meaning. A fuzzy terminological axiom is in the form of $<B \prec D>$ or $<B \approx D>$. $<B \prec D>$ means B is a sub concept of D, and $<B \approx D>$ means B is equivalent to D.

In the semantic view, the fuzzy interpretation for FALC is a pair $I = <\Delta^I, \cdot^I>$, where Δ^I is a nonempty domain, and $\cdot^I$ is an interpretation function satisfying:

$$(a)^I = a^I \in \Delta^I; \; B^I : \Delta^I \to [0,1]; \; R^I : \Delta^I \times \Delta^I \to [0,1]; \quad (1)$$
$$\top^I(d) = 1; \; \bot^I(d) = 0; \; (\neg C)^I(d) = 1 - C^I(d);$$
$$(C \sqcap D)^I(d) = \min\{C^I(d), D^I(d)\}; \; (C \sqcup D)^I(d) = \max\{C^I(d), D^I(d)\};$$
$$(\exists R.C)^I(d) = \sup_{d' \in \Delta^I}\{\min\{R^I(d,d'), C^I(d')\}\};$$
$$(\forall R.C)^I(d) = \inf_{d' \in \Delta^I}\{\max\{1 - R^I(d,d'), C^I(d')\}\}.$$

I satisfies a fuzzy assertion $<C(a) \geq n>$ (resp. $<R(a, b) \geq n>$) iff $C^I(a^I) \geq n$ (resp. $R^I(a^I, b^I) \geq n$). I satisfies an ABox A (written $I \mid \approx A$) iff I satisfies all fuzzy assertions in A. I satisfies a fuzzy terminological axiom $<B \prec D>$ (resp. $<B \approx D>$) iff $\forall d \in \Delta^I$, $B^I(d) \leq D^I(d)$ (resp. $B^I(d) = D^I(d)$). I satisfies a TBox T (written $I \mid \approx T$) iff I satisfies all fuzzy terminological axioms in T. I satisfies a knowledge base $\Sigma(A, T)$, iff I satisfies both A and T.

3 EFALCN

3.1 Syntax Structure of EFALCN

EFALCN introduces the cut sets of fuzzy concepts and fuzzy roles. Let ΔN_C and ΔN_R denote the sets of atomic fuzzy concepts and the set of atomic fuzzy roles. The set of atomic concepts in EFALCN is defined as $\Delta N_C^E = \{B_{[n]} \mid B \in \Delta N_C \wedge n \in (0,1]\}$, where $B_{[n]}$ is called an atomic cut concept. The set of atomic roles in EFALCN is defined as $\Delta N_R^E = \{R_{[n]} \mid R \in \Delta N_R \wedge n \in (0,1]\}$, where $R_{[n]}$ is called an atomic cut role. For any $B_{[n]}$ (resp. $R_{[n]}$), we define B (resp. R) as the prefix of $[n]$ and reversely $[n]$ as the suffix of B (resp. R).

EFALCN starts with $B_{[n]}$ and $R_{[n]}$, and inherits concept constructors of ALCN to define cut concepts in EFALCN:

1. The top and bottom concept (denoted by $\top_{[1]}$, $\bot_{[1]}$) are cut concepts;
2. For any $B_{[n]} \in \Delta N_C^E$, $B_{[n]}$ is a cut concept;
3. If $C_{[n_1,\ldots,n_k]}$ and $D_{[m_1,\ldots,m_l]}$ are cut concepts, and $R_{[n]}$ is an element of ΔN_R^E, then $\neg C_{[n_1,\ldots,n_k]}$, $C_{[n_1,\ldots,n_k]} \sqcup D_{[m_1,\ldots,m_l]}$, $C_{[n_1,\ldots,n_k]} \sqcap D_{[m_1,\ldots,m_l]}$, $\exists R_{[n]}.D_{[m_1,\ldots,m_l]}$, $\forall R_{[n]}.D_{[m_1,\ldots,m_l]}$, $\geq NR_{[n]}$, $\leq NR_{[n]}$ are cut concepts.

where $C_{[n_1,\ldots,n_k]}$ and $D_{[m_1,\ldots,m_l]}$ are abbreviation expressions for cut concepts. For example, $\exists \text{friend}_{[0.7]}.(\text{Tall}_{[0.7]} \sqcap \text{Strong}_{[0.9]})$ is a cut concept, and it can be denoted by $\exists \text{friend}.(\text{Tall} \sqcap \text{Strong})_{[0.7,0.7,0.9]}$ with collecting all the suffixes in a suffix vector, where $\exists \text{friend}.(\text{Tall} \sqcap \text{Strong})$ is called the prototype of the cut concept and [0.7, 0.7, 0.9] is called the suffix vector of the cut concept.

The cut assertions, which are in the form of $a : C_{[n_1,\ldots,n_k]}$ or $(a,b) : R_{[n]}$, build up EFALCN ABoxs, where $a : C_{[n_1,\ldots,n_k]}$ shows that the individual a belongs to the cut concept $C_{[n_1,\ldots,n_k]}$. For example, Stan: $\exists \text{friend}.(\text{Tall} \sqcap \text{Strong})_{[0.7,0.7,0.9]}$ means Stan likely has a friend who is likely tall and more likely strong. Similarly, $(a,b) : R_{[n]}$ shows that the pair of individuals (a, b) belongs to the cut role $R_{[n]}$. EFALCN TBoxes are composed of cut terminological axioms about inclusion and equivalent of cut concepts, which are in the form of $B_{[n]} \sqsubseteq C_{[f_1(n),\ldots,f_k(n)]}$ $n \in X$ and $B_{[n]} \equiv C_{[f_1(n),\ldots,f_k(n)]}$ $n \in X$, where $B_{[n]}$ is an atomic cut concept and $C_{[f_1(n),\ldots,f_k(n)]}$ is a cut concept, X is continuous and $f_i(n)$ is a linear function from domain X to $(0, 1]$, $i = 1,\ldots,k$. The suffix vectors including variables and functions are called alterable suffix vectors, and corresponding cut concepts are called alterable cut concepts in the given domain X. $R_{[f_i(n)]}$ may appears in $C_{[f_1(n),\ldots,f_k(n)]}$, such $R_{[f_i(n)]}$ is also called an alterable cur role in the given domain X. EFALCN ABox A and TBox T make up an EFALCN knowledge base $\Sigma_E(A, T)$.

3.2 Semantic Interpretation of EFALCN

The interpretation of EFALCN is defined as a pair $I = <\Delta^I, \cdot^I>$, where Δ^I is a nonempty set as the domain and $\cdot^I$ is an interpretation function. $\cdot^I$ maps every individual a into an element of the domain: $(a)^I = a^I$, and maps every atomic fuzzy concept and atomic fuzzy role into a membership degree function: $B^I : \Delta^I \to [0,1]$, $R^I : \Delta^I \times \Delta^I \to [0,1]$. Additionally, $\cdot^I$ maps cut concepts and cut roles into subsets of Δ^I and $\Delta^I \times \Delta^I$ (correspondingly called interpretations of cut concepts and cut roles).

The interpretation of the atomic cut concept $B_{[n]}$ is $(B_{[n]})^I = \{d \mid d \in \Delta^I \wedge B^I(d) \geq n\}$. In fuzzy math, $(B_{[n]})^I$ can be treated as the n-cut of fuzzy set B^I with respect to the universe Δ^I, which is the reason that we call $B_{[n]}$ a cut concept. Similarly, the interpretation of the atomic cut role $R_{[n]}$ is $(R_{[n]})^I = \{(d,d') \mid d,d' \in \Delta^I \wedge R^I(d,d') \geq n\}$.

The interpretations of cut concepts are inductively defined as:

$$(\neg C_{[n_1,\ldots,n_k]})^I = \Delta^I / (C_{[n_1,\ldots,n_k]})^I ; \tag{2}$$

$$((C \sqcap D)_{[n_1,\ldots,n_k]})^I = (C_{[n_1,\ldots,n_m]})^I \cap (D_{[n_{m+1},\ldots,n_k]})^I ; \tag{3}$$

$$((C \sqcup D)_{[n_1,\ldots,n_k]})^I = (C_{[n_1,\ldots,n_m]})^I \cup (D_{[n_{m+1},\ldots,n_k]})^I ; \tag{4}$$

$$((\exists R.C)_{[n_1,\ldots,n_k]})^I = \{d \mid d \in \Delta^I \wedge \exists d' \in \Delta^I, R^I(d,d') \geq n_1 \wedge d' \in (C_{[n_2,\ldots,n_k]})^I\} ; \tag{5}$$

$$((\forall R.C)_{[n_1,\ldots,n_k]})^I = \{d \mid d \in \Delta^I \wedge \forall d' \in \Delta^I, R^I(d,d') \geq n_1 \to d' \in (C_{[n_2,\ldots,n_k]})^I\} ; \tag{6}$$

$$(\geq NR_{[n]})^I = \{d \mid d \in \Delta^I, |\{d' \mid R(d,d') \geq n\}| \geq N\} ; \tag{7}$$

$$(\leq NR_{[n]})^I = \{d \mid d \in \Delta^I, |\{d' \mid R(d,d') \geq n\}| \leq N\} . \tag{8}$$

In equations (3-4), $[n_1,\ldots,n_m]$ and $[n_{m+1},\ldots,n_k]$ are the suffix vector of cut concept $C_{[n_1,\ldots,n_m]}$ and $D_{[n_{m+1},\ldots,n_k]}$ respectively. In equations (5-6) $[n_2,\ldots,n_k]$ is the suffix vector of $C_{[n_2,\ldots,n_k]}$. In equations (7-8), $|Z|$ denotes the number of elements in the set Z.

EFALCN also can convert cut concepts into equivalent forms in NNF. An EFALCN cut concept $C_{[n_1,\ldots,n_k]}$ is in NNF iff negation $\neg$ only occurs in front of the atomic cut concepts. Every cut concept can be converted in NNF by exhaustively applying the following rewrite rules:

$$\neg\neg C_{[n_1,\ldots,n_k]} \sim C_{[n_1,\ldots,n_k]} ; \; \neg(C_{[n_1,\ldots,n_k]} \sqcap D_{[h_1,\ldots,h_s]}) \sim \neg C_{[n_1,\ldots,n_k]} \sqcup \neg D_{[h_1,\ldots,h_s]} ;$$
$$\neg(C_{[n_1,\ldots,n_k]} \sqcup D_{[h_1,\ldots,h_s]}) \sim \neg C_{[n_1,\ldots,n_k]} \sqcap \neg D_{[h_1,\ldots,h_s]} ; \; \neg(\exists R.C)_{[n_1,\ldots,n_{k+1}]} \sim \forall R_{[n_1]}.\neg C_{[n_2,\ldots,n_{k+1}]} ; \tag{9}$$
$$\neg(\forall R.C)_{[n_1,\ldots,n_{k+1}]} \sim \exists R_{[n_1]}.\neg C_{[n_2,\ldots,n_{k+1}]} ; \; \neg(\geq NR_{[n_1]}) \sim (\leq N-1R_{[n_1]}) ;$$
$$\neg(\leq NR_{[n_1]}) \sim (\geq N-1R_{[n_1]}) .$$

For example: a cut concept $\neg((\text{Young} \sqcup \text{Tall}) \sqcap \geq 10\text{friend})_{[0.8,0.7,0.9]}$ can be converted to $(\neg \text{Young} \sqcap \neg \text{Tall} \sqcup (\leq 9\text{friend}))_{[0.8,0.7,0.9]}$, which is in NNF. For any $I = <\Delta^I, \cdot^I>$, we can prove that the above rewrite rules are semantic consistent.

Now we will talk about the relation between EFALCN interpretations I and knowledge base $\Sigma_E (A, T)$. I satisfies a cut assertion $a : C_{[n_1,\ldots,n_k]}$ (resp. $(a,b) : R_{[n]}$), iff

$a^I \in (C_{[n_1,...,n_k]})^I$ (resp. $(a^I, b^I) \in (R_{[n]})^I$). I satisfies an ABox A (written $I \approx A$), iff I satisfies all cut assertions in A; such I is called a model of A. Any interpretation I satisfies $B_{[n]} \sqsubseteq C_{[f_1(n),...,f_k(n)]}$, $n \in X$ iff for any $n_0 \in X$, $(B_{[n_0]})^I \subseteq (C_{[f_1(n_0),...,f_k(n_0)]})^I$. I satisfies $B_{[n]} \equiv C_{[f_1(n),...,f_k(n)]}$, $n \in X$ iff for any $n_0 \in X$, $(B_{[n_0]})^I = (C_{[f_1(n_0),...,f_k(n_0)]})^I$. I satisfies a TBox T (written $I \approx T$), iff I satisfies all cut terminological axioms in T; such I is called a model of T. I satisfies $\Sigma_E (A, T)$ iff I is a model of both A and T.

3.3 Reasoning Problems and Reasoning Properties of EFALCN

The main reasoning problems of EFALCN are satisfiability and consistency. In this paper, we only talk about reasoning problems without respect to TBox.

Satisfiability: a given cut concept $C_{[n_1,...,n_k]}$ is satisfiable iff there is an interpretation I such that $C^I_{[n_1,...,n_k]} \neq \emptyset$.

Consistency: a given ABox A is consistent iff there is a model I of A.

For alterable concepts allowed in EFALCN, we extend *satisfiability* of cut concepts to *satisfiability* of alterable cut concepts in a given domain X, and discuss the *sat-domain* task instead of *satisfiability* task.

Sat-domain: for an alterable cut concept $C_{[f_1(n),...,f_k(n)]}$ in a given domain $n \in X_0$, where $X_0 \subseteq [0,1]$, X_0 is continuous, and $f_i(n)$ is a linear function from domain X_0 to $(0,1]$, the s*at-domain* task computes satisfiable and unsatisfiable sub-domains of X_0. For any $n_0 \in X_0$, if $C_{[f_1(n_0),...,f_k(n_0)]}$ is satisfiable then n_0 is in the satisfiable sub-domain, otherwise n_0 is in the unsatisfiable sub-domain.

After talking about reasoning problems, we will discuss two reasoning properties dealing with suffix vectors of cut concepts and cut roles. Two properties are better enabling reasoning in EFALCN.

Property 1. For any two cut roles $R_{[n_1]}$ and $R_{[m_1]}$, if they have the same prototypes and $n_1 \leq m_1$, then for any interpretation $I = <\Delta^I, \cdot^I>$, it is true that $(R_{[n_1]})^I \supseteq (R_{[m_1]})^I$ (this implicitly means $R_{[m_1]} \sqsubseteq R_{[n_1]}$).

The constraint of suffixes: for any suffix, when its prefix is an atomic fuzzy concept, if $\neg$ occurs in front of the concept, its constraint is $\neg$; when its prefix is an atomic fuzzy role, if $\forall$, $\exists$, $\geq N$ or $\leq N$ is in front of the role, its constraint is $\forall, \exists, \geq N$ or $\leq N$ respectively.

Property 2. For any two cut concepts $C_{[n_1,...,n_k]}$ and $C_{[m_1,...,m_k]}$ with the same prototypes, for any two i^{th} suffixes n_i, m_i in the suffix vectors, they have the same prefix and constraint, if it is true that

> Condition 1: When the prefix is an atomic fuzzy concept, if the constraint is $\neg$, then $n_i \geq m_i$, otherwise $n_i \leq m_i$.
> Condition 2: When the prefix is an atomic fuzzy role, if the constrain is $\forall$ and $\leq N$, then $n_i \geq m_i$, otherwise $n_i \leq m_i$.

Condition 3: then for any interpretation $I = <\Delta^I, \cdot^I>$, it is true that $(C_{[n_1,...,n_k]})^I \supseteq (C_{[m_1,...,m_k]})^I$.

4 Tableau Algorithm for EFALCN

We design tableau algorithms of *sat-domain* and *consistency*. The algorithms are developed in the style of so-called constraint propagation method. For in the process of constrain propagation, ABoxes may contain such assertions $x: D_{[f_i(n),...,f_l(n)]}$ or $(x, y): R_{[f(n)]}$, and $n \in X$. Such ABox is called an alterable ABox, donated as a pair (A, X), where A is the set of assertional axioms and X is the domain of variable n.

In EFALCN, we define translation rules to achieve constraint propagation process with keeping consistency of alterable ABox (A, X) in its domain. By defining the translation rules, we should consider the effect of comparison between suffix functions in domain X. For example, there are two assertional axioms in current alterable ABox(A, X): $x: \forall R_{[f_i(n)]}.D_{[f_{i+1}(n),...,f_l(n)]}$ and $(x,y) \in R_{[f_j(n)]}$. The comparison between two functions $f_i(n)$ and $f_j(n)$ in domain X will affect the application of translation rules. If $f_i(n) \leq f_j(n)$ in X, then translation rules will add $y: D_{[f_{i+1}(n),...,f_l(n)]}$. In more complex case, if $f_i(n) \leq f_j(n)$ in sub-domain X^1 of X, then (A, X) will be divided into two new ABox: (A^1, X^1) with adding $y: D_{[f_2(n),...,f_k(n)]}$ in A^1 and replacing domain with X^1 and (A, X^2) with keeping A and changing domain with $X^2 = X \backslash X^1$. We define a method to achieve the comparison between two functions $f_i(n)$ and $f_j(n)$ in domain X.

Method *Comparison*
Input $(f_i(n), f_j(n), X)$
$f_i(n)$ and $f_j(n)$ are two functions, and X is the domain of variable n.
Output (X^1, X^2)
X^1 is the sub-domain where $f_i(n) \leq f_j(n)$ and X^2 is the other sub-domain of X

Fig. 1. Comparison method between functions $f_i(n)$ and $f_j(n)$ in X

Additionally, for EFALCN supports number restrictions $\geq NR_{[n]}$ and $\leq NR_{[n]}$, in the constraint propagation process, the comparison between one given function and other multi functions may be encountered. For example, current alterable ABox (A, X) may contain $x :\leq NR_{[f_0(n)]}$ and $(x, y_j) \in R_{[f_j(n)]}$ $1 \leq j \leq N+1$. If $f_0(n) \leq f_j(n)$ $1 \leq j \leq N+1$ in sub-domain X^* of X, obviously X^* is a special sub-domain which need be separated from the whole domain X. To keep the domain of alterable ABox is continuous, then X will be divided into three sub-domains: X^*, left and right adjacent domains of X^*. We also define a method to achieve the comparison between a given function $f_0(n)$ and other $N+1$ functions in domain X.

> **Method** *Multi-Comparison*
> **Input** $(f_0(n), f_1(n), \ldots, f_{N+1}(n), X)$
> $f_i(n)$ is a function, $0 \leq i \leq N+1$ and X is the domain of variable n.
> **Output** (X^1, X^2, X^3), where X^2 is the sub-domain where $f_0(n) \leq f_j(n)$, $1 \leq j \leq N+1$; X^1 and X^3 are left and right adjacent domains of X^2.

Fig. 2. Multi-Comparison method between functions $f_0(n)$ and multi functions in X

After talking about some preparation, we express our tableau algorithm. Given an alterable cut concept $C_{[f_1(n),\ldots,f_k(n)]}$ in domain X, the tableau algorithm starts with a single alterable ABox $(\{x_0: C_{[f_1(n),\ldots,f_k(n)]}\}, X)$, and applies domain-consistency preserving transform rules (fig.3.) to ABoxes, which divides the whole domain X into two sub-domains with satisfiability and unsatisfiability.

New ABoxes generated by the action of rules are called subsequences of the given ABoxes. By Applying some translation rules, the given alterable ABox (A, X) is translated into multi subsequences. Therefore we define a set of ABoxes S instead of the single ABox. The tableau algorithm starts with a single set $S_0 = \{(\{x_0: C_{[f_1(n),\ldots,f_k(n)]}\}, X_0)\}$. Translation rules are applied to S with replacing one alterable ABox by its subsequence. We will define some terms to circumscribe conditions of termination.

Definition 1. An alterable ABox (A, X) is complete iff none of translation rules could be applied to it. An alterable ABox (A, X) is not closed, iff it is not announced closed. An alterable ABox is open iff it is complete and not closed. A finite set of alterable ABoxes S is complete iff any alterable ABox (A, X) in S is complete. The satisfiable domain of S is denoted by $Sat(S)$ and defined as the union of domains of open ABox (A_i, X_i): $Sat(S) = \bigcup X_i$, $(A_i, X_i) \in S$ and (A_i, X_i) is open.

Now we refine process of tableau algorithm for *sat-domain* (fig.4.) in following steps:

(1) Let $C_{[f_1(n),\ldots,f_k(n)]}$ be a cut concept in *NNF*. The tableau algorithm starts with a set of a single alterable ABox: $S_0 = \{(\{x_0: C_{[f_1(n),\ldots,f_k(n)]}\}, X)\}$.

(2) The tableau algorithm exhaustively applies translation rules to current S_i. We denote S_{i+1} is subsequence set of ABoxes of S_i, so there is a chain of S_i by rules application: $S_0 \rightarrow S_1 \rightarrow \ldots \rightarrow S_i \rightarrow \ldots$.

(3) If the current S_i is complete, the tableau algorithm returns (X_{sat}, X_{unsat}), where $X_{sat} = Sat(S_i)$ as satisfiable sub-domain and $X_{unsat} = X \setminus X_{sat}$ as unsatisfiable subdomain.

The termination and correctness of the tableau algorithm can be guaranteed, and the complexity of *sat-domain* is *PSPACE-complete*. In this paper, we do not give the detailed proof of these results.

For any alterable ABox (A, X) which is not closed
$\sqcap$-rule
Condition: $x: D_{[f_i(n),...,f_s(n)]} \sqcap E_{[f_{s+1}(n),...,f_l(n)]} \in A$, $x: D_{[f_i(n),...,f_s(n)]}$ or $x: E_{[f_{s+1}(n),...,f_l(n)]} \notin A$.
Action: $(A^1, X) = (A \cup \{x: D_{[f_i(n),...,f_s(n)]}, x: E_{[f_{s+1}(n),...,f_l(n)]}\}, X)$.

$\sqcup$-rule
Condition: $x: D_{[f_i(n),...,f_s(n)]} \sqcup E_{[f_{s+1}(n),...,f_l(n)]} \in A$, $x: D_{[f_i(n),...,f_s(n)]}$, $E_{[f_{s+1}(n),...,f_l(n)]} \notin A$.
Action: $(A^1, X) = (A \cup \{x: D_{[f_i(n),...,f_s(n)]}\}, X)$; $(A^2, X) = (A \cup \{x: E_{[f_{s+1}(n),...,f_l(n)]}\}, X)$.

$\exists$-rule
Condition: $x: \exists R_{[f_i(n)]}.D_{[f_{i+1}(n),...,f_l(n)]} \in A$, $\forall y \in A$, $(x, y): R_{[f_i(n)]}$ or $y: D_{[f_{i+1}(n),...,f_l(n)]} \notin A$.
Action: $(A^1, X) = (A \cup \{z: D_{[f_{i+1}(n),...,f_l(n)]}, (x, z): R_{[f_i(n)]}\}, X)$, where z is new generated.

$\forall$-rule
Condition: $x: \forall R_{[f_i(n)]}.D_{[f_{i+1}(n),...,f_l(n)]}$, $(x, y): R_{[f_j(n)]} \in A$, $y: D_{[f_{i+1}(n),...,f_l(n)]} \notin A$,
$(X^1, X^2) = Comparison(f_i(n), f_j(n), X)$ and $X^1 \neq \emptyset$.
Action: $(A^1, X) = (A \cup \{y: D_{[f_{i+1}(n),...,f_l(n)]}\}, X^1)$; if $X^2 \neq \emptyset$, $(A^2, X) = (A, X^2)$.

$\neg$-rule
Condition: $x: \neg B_{[f_i(n)]}$, $x: B_{[f_j(n)]} \in A$, $(X^1, X^2) = Comparison(f_i(n), f_j(n), X)$ and $X^1 \neq \emptyset$.
Action: $(A^1, X) = (A, X^1)$, (A^1, X) is announced closed; if $X^2 \neq \emptyset$, $(A^2, X) = (A, X^2)$.

$\geq$-rule
Condition: $x: \geq NR_{[f_i(n)]} \in A$ and $|\{y_i|(x, y_i): R_{[f_i(n)]} \in A\}| < N$.
Action: $(A^1, X) = (A \cup \{(x, z_i): R_{[f_{k_0}(n)]} \mid z_i \text{ is new generated individual and } 1 \leq i \leq N\}, X)$.

$\leq$-rule
Condition: $x: \leq NR_{[f_{k_0}(n)]}$, $(x, y_i): R_{[f_{k_i}(n)]} \in A$ $1 \leq i \leq N+1$.
$(X^1, X^2, X^3) = Multi\text{-}Comparison(f_{k_0}(n), f_{k_1}(n), ..., f_{k_{N+1}}(n), X)$ and $X^2 \neq \emptyset$.
Action: if $X^1 \neq \emptyset$ or $X^3 \neq \emptyset$, $(A^2, X) = (A, X^2)$, $(A^1, X) = (A, X^1)$ or $(A^3, X) = (A, X^3)$;
if $X^2 = X$, for any $1 \leq i < j \leq N+1$, $(A^{ij}, X) = ([y_i/y_j]A, X)$, where $[y_i/y_j]A$ is generated from A by replacing each y_i in A by y_j.

$\geq$-$\leq$-rule
Condition: $x: \leq MR_{[f_i(n)]}$, $x: \geq NR_{[f_j(n)]} \in A$, $M < N$;
  $(X^1, X^2) = Comparison(f_i(n), f_j(n), X)$ and $X^1 \neq \emptyset$.
Action: $(A^1, X) = (A, X^1)$, (A^1, X) is announced closed; if $X^2 \neq \emptyset$, $(A^2, X) = (A, X^2)$.

Fig. 3. Translation rules of *sat-domain* algorithm

Consistency is a problem to tell whether An ABox A_E is consistent. It can be converted into *sat-domain*.

Algorithm *Sat-domain*

Input ($C_{[f_1(n),...,f_k(n)]}$, X_0)

$C_{[f_1(n),...,f_k(n)]}$ is the given alterable cut concept, and X_0 is the domain of variable n.

Output (X_{sat}, X_{unsat})

X_{sat} is the satisfiable sub-domain and X_{unsat} is the unsatisfiable sub-domain.

Fig. 4. Algorithm for *sat-domain*

5 Related Work

The extensions of DLs introduce probabilistic or fuzzy features (called Probabilistic or Fuzzy DLs) to increase repressive and reasoning capability of knowledge with uncertainty, enabling them to deal with probabilistic or fuzzy information.

In Probabilistic DLs area, Heinsohn [7] and Jaeger [8] presented two similar probabilistic extensions of ALC, which both allow for terminological probabilistic knowledge about concepts and roles. Giugno [9] proposed a probabilistic extension of the more expressive DL SHOQ(D), called P-SHOQ(D). P-SHOQ(D) can represent both terminological probabilistic knowledge about concepts and roles, and assertional probabilistic knowledge about concept instances and role instances.

In Fuzzy DLs area, Yen [10] extended the limited DL FL. However, it only supports concept subsumption reasoning with nonsupport of reasoning in presence of assertions. The recent work about Fuzzy DL has been discussed in introduction, and detailed comparison between Straccia's FALC and our EFALCN will be given below:

Both EFALCN and FALC are Fuzzy DLs with fuzzy features, but EFALCN has more expressive power than FALC. To explain this conclusion, we extend EFALCN with complex cut concepts $C_{[n]}$, where C is a complex fuzzy concept and $n \in [0,1]$, and call it EFALCN*. (EFALCN only allows for atomic cut concept.)

Firstly, EFALCN* can express any FALC knowledge base Σ (A, T):

(1) Each fuzzy assertion in a FALC ABox A can be equally converted to a cut assertion in EFALCN* ABox by applying the following rewrite rules: < $C(a) \geq n$ > ~ a: $C_{[n]}$ and < $R(a, b) \geq n$ > ~ (a, b): $R_{[n]}$.
(2) Each fuzzy terminological axiom in a FALC TBox T can be equally converted to a cut terminological axiom in EFALCN* TBox by applying rewrite rules: < $B \prec D$ > ~ $B_{[n]} \sqsubseteq D_{[n]}$, $n \in [0,1]$ and < $B \approx D$ > ~ $B_{[n]} \sqsubseteq D_{[n]}$ $n \in [0,1]$ and $D_{[n]} \sqsubseteq B_{[n]}$, $n \in [0,1]$.

Secondly, EFALCN and EFALCN* have the same expressive power for fuzzy information. We only have to prove that every complex cut concept in EFALCN* can be equally denoted by atom cut concepts in EFALCN. Obviously, it is true. For example, if $C_{[n]} = (D \sqcap E)_{[n]}$, $C_{[n]}$ can be denoted by $D_{[n]} \sqcap E_{[n]}$. The denotation of other forms of complex cut concept is similar.

Above all, EFALCN* can express any FALC knowledge base Σ (A, T), and EFALCN and EFALCN* have the same expressive power for fuzzy information, so all fuzzy information described by FALC can also be described by EFALCN.

EFALCN is more expressive than FALC.

Firstly, FALCN supports more powerful presentation of fuzzy concept assertion, which is not allowed in FALC. For example, $\exists R.C$ is a FALC fuzzy concept, if an interpretation I satisfies fuzzy assertion $<(\exists R.C)(a) \geq n>$, then it means $(\exists R.C)^I(a^I) \geq n$. From the definition of $(\exists R.C)^I$, we can get that $\exists b^I \in \Delta^I, R^I(a^I,b^I) \geq n$ and $C^I(b^I) \geq n$. In many application, it may encounter more complex fuzzy assertion, such as an individual a such that $\exists b^I \in \Delta^I, R^I(a^I,b^I) \geq n_1$ and $C^I(b^I) \geq n_2$, where $n_1 \neq n_2$. In this case, the membership degrees of concept and role are different and need respective presentations. FALC cannot distinguish such diverse membership degrees, but EFALCN can handle it. The above fuzzy assertion can be described as a: $\exists R_{[n_1]}.C_{[n_2]}$.

Secondly, EFALCN extends the expressive scope of fuzzy terminological information in FALC. The FALC fuzzy terminological axiom $C \prec D$ implicit means $\forall d \in \Delta^I$, $C^I(d) \geq n \rightarrow D^I(d) \geq n$, which can not express complex inclusion based on various membership degrees. For example, the fuzzy concept Young and Very-young have the following inclusion relation: if membership degree of an individual a belonging to Very-young is more than 0.6, the membership degree of this individual belonging to Young is more than 0.9, formally denoted by $\forall d \in \Delta^I$, Very-young(d) $\geq 0.6 \rightarrow$ Young(d) ≥ 0.9. Now two side of the inclusion contain various membership degree constrains, which is not allowed in FALC but can be denoted in EFALCN: Very-young$_{[0.6]} \sqsubseteq$ Young$_{[0.9]}$.

6 Conclusions

We propose the extended fuzzy description logic EFALCN, which introduces the cut sets of fuzzy concepts and fuzzy roles as atomic concepts and atomic roles. In detail, we define the syntax structure, semantic interpretation and reasoning problem of EFALCN. We also design tableau algorithms of reasoning problems for EFALCN. We compare EFALCN with other fuzzy DLs, and explain that EFALCN has more expressive power. This work can be applied as a language in current Semantic Web to enrich its representation means, and regarded as a new idea of extending DLs with fuzzy features.

References

1. Berners-Lee, T., Hendler, J., Lassila, O.: The Semantic Web. Scientific American, vol. 284, no.5 (2001) 34–43
2. Baader, F., Calvanese, D., McGuinness, D.L., Nardi, D., Patel-Schneider, P.F.(Eds.): The Description Logic Handbook: Theory, Implementation, and Applications. Cambridge University Press (2003)
3. Meghini, C., Sebastiani, F., Straccia, U.: Reasoning about the form and content for multimedia objects. In: Proceedings of AAAI 1997 Spring Symposium on Intelligent Integration and Use of Text, Image, Video and Audio, California (1997) 89–94

4. Straccia, U.: Reasoning within fuzzy description logics. Journal of Artificial Intelligence Research, no. 14 (2001) 137–166
5. Schmidt-Schau?, M., Smolka, G.: Attributive concept descriptions with complements. Artificial Intelligence, vol. 48 (1991) 1–26
6. Straccia, U.: Transforming fuzzy description logics into classical description logics. In: Proceeedings of the 9th European Conference on Logics in Artificial Intelligence, Lisbon, (2004) 385–399
7. Heinsohn, J.: Probabilistic description logics. In: Proceedings of UAI-94 (1994) 311–318
8. Jaeger, M.: Probabilistic reasoning in terminological logics. In: Proceedings of KR-94 (1994) 305–316
9. Giugno, R., Lukasiewicz, T.: A probabilistic extension of shoq(d) for probabilistic ontologies in the semantic web. In: Proceeedings of the 9th European Conference on Logics in Artificial Intelligence, Cosenza, Italy (2002) 23–26
10. 10.Yen, J.: Generalizing term subsumption languages to fuzzy logic. In: Proceedings of the 12th Int. Joint Conference on Artificial Intelligence (1991) 472–477

Type II Topological Logic $\mathbb{C}_\mathcal{T}^2$ and Approximate Reasoning*

Yalin Zheng, Changshui Zhang, and Yinglong Xia

State Key Laboratory for Intelligent Technology and Systems,
Department of Automation , Faculty of Information Science and Technology,
Tsinghua University , Beijing 100084 , The People's Republic of China
{zheng-yl, zcs}@mail.tsinghua.edu.cn
xiayl03@mails.tsinghua.edu.cn

Abstract. This paper propose a topological logic model of approximate reasoning based on the type II topological logic $\mathbb{C}_\mathcal{T}^2$ and the structure of matching function $\mathcal{C}$ and $\mathcal{C}$-match neighborhood group. The type II topological algorithm of simple approximate reasoning and multiple approximate reasoning is given in type II topological logic $\mathbb{C}_\mathcal{T}^2$ with matching function $\mathcal{C}$. We also propose the structure of type II regular topological logic $\mathbb{C}_\mathcal{T}^2$ and regular matching function $\mathcal{C}$. The type II completeness and type II perfectness of knowledge base K is investigated in type II regular topological logic $\mathbb{C}_\mathcal{T}^2$ with regular matching function $\mathcal{C}$.

1 Introduction

In *Fréchet* topology, for each knowledge $A_i \to B_i$ in knowledge base

$$K = \{A_i \to B_i | i - 1, 2, \cdots, n\},$$

the topological closure $\{A_i\}^-$ of antecedent A_i is identically itself, that is,

$$\{A_i\}^- = \{A_i\}.$$

Therefore, any input A^* that different from A_i can not active knowledge $A_i \to B_i$, which implies that *Fréchet* topology is not suitable for type I topological algorithm of approximate reasoning. Consequently, we intend to explore a reasonable topology $\mathcal{T}$ in non-*Fréchet* topologies for type I topological algorithm of approximate reasoning.

The non-*Fréchet* topology $\mathcal{T}$, which makes type I topological algorithm sensibly is valid and significant, possess the following two characteristics. On one hand, its open sets are few, which make it possible to include the antecedent A

---

* This work is supported by the project (60475001) of the National Natural Science Foundation of China.

of knowledge $A \to B$ into $\bigcap_{U \in \mathcal{U}(A^*)} U$, the intersection of all neighborhoods of A^*; On the other hand, the correspondent type I topological algorithm holds a rigid topological requirement, say, input A^* succeeds in firing knowledge $A \to B$ iff A^* is the adherent formula of $\{A\}$. Then, any adherent formula B^* of $\{B\}$ is taken as the reasoning result and it is outputted from the reasoning system.

In this paper, we significantly decrease the "topological requirement" and significantly increase the "number of open sets" so that the topological algorithm is more convenient for application.

2   The Structure of Type II Topological Logic $\mathbb{C}_\mathcal{T}^2$

Assume classical propositional logic $\mathbb{C} = \{\hat{\mathbb{C}}, \tilde{\mathbb{C}}\}$, where $\hat{\mathbb{C}}$ is the syntax of $\mathbb{C}$ and $\tilde{\mathbb{C}}$ is the semantic of $\mathbb{C}$. $F(S)$, which is the type $(2, 2, 2, 1)$ free algebra generated by non-empty countable set S, is called proposition set, or formulae set of $\mathbb{C}$. $\to, \wedge, \vee, \neg$ are implication, conjunction, disjunction and negation connectives for propositions respectively. $\emptyset^\vdash$ is the set of all theorems in $\hat{\mathbb{C}}$. $\emptyset^\vDash$ is the set of all tautologies in $\tilde{\mathbb{C}}$. According to the soundness theorem and the completeness theorem for $\mathbb{C}$, we have $\emptyset^\vdash = \emptyset^\vDash$.

Assign a topology $\mathcal{T}$ to formulae set $F(S)$ in $\mathbb{C}$, we obtain a topological space $(F(S), \mathcal{T})$. For any formula $A, B, A \to B \in F(S)$, we denote their neighborhood systems under the topology $\mathcal{T}$ with $\mathcal{U}(A), \mathcal{U}(B)$ and $\mathcal{U}(A \to B)$.

For any implication formula $A \to B \in F(S)$, we have

$$\forall W \in \mathcal{U}(A \to B), \exists U \in \mathcal{U}(A),$$
$$\exists V \in \mathcal{U}(B), s.t. U \to V \subseteq W \qquad (1)$$

where

$$U \to V = \{A^* \to B^* | A^* \in U, B^* \in U\}$$

then, the 3-tuple

$$\mathbb{C}_\mathcal{T}^2 = (\hat{\mathbb{C}}, \tilde{\mathbb{C}}, \mathcal{T})$$

is called the *type II topological propositional logic* on formulae set $F(S)$, or *type II topological logic* for short. It is also said that topology $\mathcal{T}$ is the *type II consistent topology* with respect to logic $\mathbb{C} = (\hat{\mathbb{C}}, \tilde{\mathbb{C}})$.

Constraint (1) depicts the following characteristic of type II topological logic $\mathbb{C}_\mathcal{T}^2$. For any "precision requirement" W approaching to formula $A \to B$, it always can find the "precision requirement" U approaching to formula A and the "precision requirement" V approaching to formula B which makes the approximate degree between implication formulae $A^* \to B^*$ and $A \to B$ dose not exceed the preassigned W when the approximate degree between A^* and A dose not exceed U and the approximate degree between B^* and B dose not exceed V.

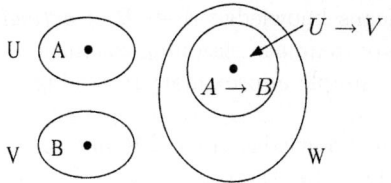

Fig. 1. The structure of type II topological logic $\mathbb{C}_T^2$

3 The Simple Approximate Reasoning Based on Type II Topological Logic $\mathbb{C}_T^2$

In type II topological logic $\mathbb{C}_T^2$, it can construct the choice function

$$\mathcal{C}: \begin{cases} \mathcal{H} \to \bigcup_{A \to B \in \mathcal{H}} (\mathcal{U}(A) \times \mathcal{U}(B) \times \mathcal{U}(A \to B)); \\ (A \to B) \mapsto (U_A, U_B, U_{A \to B}) \in \mathcal{U}(A) \times \mathcal{U}(B) \times \mathcal{U}(A \to B). \end{cases}$$

which satisfies

$$U_A \to U_B \subseteq U_{A \to B}$$

where $\mathcal{H}$ is the set of all implication formulae in form of $A \to B$, that is,

$$\mathcal{H} = \{A \to B | A \in F(S), B \in F(S)\}.$$

We call the choice function $\mathcal{C}$ as the *matching function* on formulae set $\mathcal{H}$, the 3-tuple

$$\mathcal{C}(A \to B) = (U_A, U_B, U_{A \to B})$$

as the $\mathcal{C}$−*matching neighborhood group* with respect to implication formula $A \to B$, or *matching neighborhood group* for short.

In type II topological logic $\mathbb{C}_T^2$ with matching function $\mathcal{C}$, consider following simple approximate reasoning model

$$\frac{A \to B \quad A^*}{B^*}. \tag{2}$$

where $A \to B \in \emptyset^\vdash$ is called *knowledge*; $A^* \in F(S)$ is called *input* and $B^* \in F(S)$ is called *output*, or *approximate reasoning conclusion*. Matching function $\mathcal{C}$ assigns the matching neighborhood group

$$(U_A, U_B, U_{A \to B})$$

to the knowledge $A \to B$.

If $A^* \in U_A$, it means knowledge $A \to B$ is *activated* by input A^*. Taking any $B^* \in U_B$ as the approximate reasoning conclusion, we call B^* as the *type II topological solution* of simple approximate reasoning with respect to knowledge $A \to B$ and input A^*.

If $A^* \bar{\in} U_A$, it means knowledge $A \to B$ is *not activated* by input A^*. In this case, A^* is not considered i.e. there is no type II topological solution of simple approximate reasoning with respect to knowledge $A \to B$ and input A^*.

The above algorithm is called *type II topological algorithm* of simple approximate reasoning.

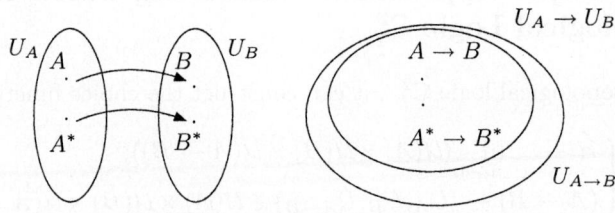

Fig. 2. The type II topological algorithm of simple approximate reasoning

The structure of type II topological logic $\mathbb{C}_T^2$ guarantees the successful construction of matching function $\mathcal{C}$; the $\mathcal{C}$-matching neighborhood group of knowledge $A \to B$ assigned by matching function $\mathcal{C}$ guarantees sufficient approaching degree between approximate knowledge $A^* \to B^*$ and the standard knowledge $A \to B$ with respect to type II topological solution B^* of simple approximate reasoning. Particularly, when the approximate degree between input A^* and the antecedent A of knowledge $A \to B$ dose not exceed U_A, the approximate degree between solution B^* obtained by type II topological algorithm of simple approximate reasoning and the descendant B of knowledge $A \to B$ will not exceed U_B. The approximate degree between approximate knowledge $A^* \to B^*$ and standard knowledge $A \to B$ will be also within $U_{A \to B}$.

It is straightforward to derive the theorem below.

Theorem 1. *In type II topological logic $\mathbb{C}_T^2$ with matching function $\mathcal{C}$, if there exists type II topological solution B^* of input A^* with respect to knowledge $A \to B$, then we have*

$$A^* \to B^* \in U_{A \to B}$$

Review a simple approximate reasoning algorithm named *Modus Ponens reappear algorithm*, or *MP reappear algorithm* for short. If

$$\frac{A \to B \quad A^*}{B^*} . \tag{3}$$

that is, when A^* is just the antecedent A of knowledge $A \to B$, the descendant B of $A \to B$ can be taken as the reasoning result.

Theorem 2. *In type II topological logic $\mathbb{C}_\mathcal{T}^2$ with matching function $\mathcal{C}$, the type II topological algorithm of simple approximate reasoning (2) is MP reappear algorithm.*

Review a simple approximate reasoning algorithm named *degenerated approximate reasoning algorithm*, or *classical reasoning algorithm* for short. If each knowledge $A \to B$ can be activated by input A^*, if and only if $A^* = A$.

Theorem 3. *In type II topological logic $\mathbb{C}_\mathcal{T}^2$ with matching function $\mathcal{C}$, type II topological algorithm of simple approximate reasoning is classical reasoning algorithm, if and only if the topology $\mathcal{T}$ in $\mathbb{C}_\mathcal{T}^2$ is a discrete topology and $\mathcal{C}$-matching neighborhood group $\mathcal{C}(A \to B)$ assigned by matching function $\mathcal{C}$ for each implication $A \to B$ satisfies*

$$(U_A, U_B, U_{A \to B}) = (\{A\}, \{B\}, \{A \to B\})$$

Two versions of above theorem is very useful.

Theorem 4. *For every knowledge $A \to B$ and input A^* which is different from A in type II topological logic $\mathbb{C}_\mathcal{T}^2$ with matching function $\mathcal{C}$, there dose not exist any type II topological solution of simple approximate reasoning for A^* with respect to $A \to B$, if and only if the topology $\mathcal{T}$ is discrete in $\mathbb{C}_\mathcal{T}^2$, and the $\mathcal{C}$-matching neighborhood group $\mathcal{C}(A \to B)$ assigned by matching function $\mathcal{C}$ for each implication formula $A \to B$ satisfies*

$$(U_A, U_B, U_{A \to B}) = (\{A\}, \{B\}, \{A \to B\})$$

Theorem 5. *In type II topological logic $\mathbb{C}_\mathcal{T}^2$ with matching function $\mathcal{C}$, there exists type II topological solution B^* of simple approximate reasoning for a knowledge $A \to B$ with respect to input A^* which is different from A, if and only if the $\mathcal{C}$-matching neighborhood group $(U_A, U_B, U_{A \to B})$ assigned by matching function $\mathcal{C}$ for $A \to B$ satisfies*

$$U_A \neq \{A\},$$

and which is independent of the fact whether the topology $\mathcal{T}$ in $\mathbb{C}_\mathcal{T}^2$ is discrete.

From above discussion, we know that the type II topological algorithm of approximate reasoning only brings little constraints on topology $\mathcal{T}$, which show that the type II topological algorithm can choose topologies from very wide scope. Specially, we will know that the Hausdorff matching with good "separation" will be favorite after the type II topological algorithm of multiple approximate reasoning is given. It illustrates the topology $\mathcal{T}$ with large numbers of open sets is more favorable for type II topological algorithm of multiple approximate reasoning because a input A^* which can activate a knowledge in knowledge base K can determine completely the final output B^* as the reasoning result.

Furthermore, an input A^* can activate a knowledge $A \to B$ if and only if it is included in the neighborhood U_A assigned for A in type II topological algorithm, that is, $A \in U_A$. Differently, a input A^* which can activate knowledge $A \to B$ must be included in the topological closure $\{A\}^-$ i.e. $A^* \in \{A\}^-$ in type I topological algorithm, that is, A should be include in all the neighborhoods of A^* i.e. $A \in \bigcap_{U \in \mathcal{U}(A^*)} U$. Therefore, type II topological algorithm is more convenient than type I topological algorithm in practice.

From the above discussion, we also known that the key phase for determining and implementing the precision requirement of type II topological algorithm of approximate reasoning is the choice of matching function $\mathcal{C}$. Tuning matching function $\mathcal{C}$ according to the requirement, we can fulfill different precision requirements of approximate reasoning.

4 The Multiple Approximate Reasoning Based on Type II Topological Logic $\mathbb{C}_\mathcal{T}^2$

In type II topological logic $\mathbb{C}_\mathcal{T}^2$ with matching function $\mathcal{C}$, consider following multiple approximate reasoning model

$$\begin{array}{c} A_1 \to B_1 \\ A_2 \to B_2 \\ \cdots \\ A_n \to B_n \\ \underline{A^*} \\ B^* \end{array} \qquad (4)$$

where $K = \{A_i \to B_i | i = 1, 2, \cdots, n\} \subseteq \emptyset^\vdash$, is called *knowledge base*, the nonempty subset of which is called *knowledge group*; $A^* \in F(S)$ is called *input*; $B^* \in F(S)$ is called *output* or *approximate reasoning conclusion*. The $\mathcal{C}$-matching neighborhood group assigned by matching function $\mathcal{C}$ for each knowledge $A_i \to B_i (i = 1, 2, \cdots, n)$ is given by

$$(U_{A_1}, U_{B_1}, \cdots, U_{A_1 \to B_1})$$
$$(U_{A_2}, U_{B_2}, \cdots, U_{A_2 \to B_2})$$
$$\cdots$$
$$(U_{A_n}, U_{B_n}, \cdots, U_{A_n \to B_n})$$

If $A^* \in U_{A_i}$, then say that knowledge $A_i \to B_i$ can be activated by input A^*. If $\exists E \subseteq \{1, 2, \cdots, n\}, E \neq \emptyset$, such that $A^* \in U_{A_i}$ for each $i \in E$, that is,

$$A^* \in \bigcap_{i \in E} U_{A_i}$$

but $A^* \bar{\in} U_{A_i}$ for each $i \in \{1, 2, \cdots, n\} - E$, that is,

$$A^* \bar{\in} \bigcup_{i \in \{1,2,\cdots,n\}-E} U_{A_i}$$

then say that knowledge group $\{A_i \to B_i | i \in E\}$ can be *activated* by input A^*. Take any

$$B_i^* \in U_{B_i}$$

for each $i \in E$ and let

$$B^* = \bigoplus_{i \in E} B_i^*$$

then

$$B^* \in \bigcup_{i \in E} U_{B_i}$$

where $\bigoplus_{i \in E} B_i^*$ is a aggregation of $\{B_i^* | i \in E\}$. B^*, as the conclusion of approximate reasoning, is called *type II topological solution* of multiple approximate reasoning for A^* with respect to knowledge base $K = \{A_i \to B_i | i = 1, 2, \cdots, n\}$.

Specially, if there exists unique $i \in \{1, 2, \cdots, n\}$ which satisfies $A^* \in U_{A_i}$ i.e. $A_i \to B_i$ can be activated by A^*, then take any

$$B^* \in U_{B_i}$$

to be the approximate reasoning conclusion, called *type II topological solution* of multiple approximate reasoning for A^* with respect to knowledge base $K = \{A_i \to B_i | i = 1, 2, \cdots, n\}$.

If $A^* \bar{\in} U_{A_i}$ for any $i \in \{i, 2, \cdots, n\}$, that is,

$$A^* \bar{\in} \bigcup_{i=1}^n U_{A_i}$$

then, knowledge base $K = \{A_i \to B_i | i = 1, 2, \cdots, n\}$ can not be activated by input A^*. Input A^* will be ignored in this case, say, there is no *type II topological solution* of multiple approximate reasoning of A^* with respect to knowledge base $K = \{A_i \to B_i | i = 1, 2, \cdots, n\}$.

This scheme is called *type II topological algorithm* of multiple approximate reasoning.

It is straightforward to shown those theorems as follows.

Theorem 6. *In type II topological logic $\mathbb{C}_T^2$ with matching function $\mathcal{C}$, there exists type II topological solution of multiple approximate reasoning for A^* with respect to knowledge base*

$$K = \{A_i \to B_i | i = 1, 2, \cdots, n\},$$

if and only if there at least exists an $i \in \{1, 2, \cdots, n\}$ which satisfies

$$A^* \in U_{A_i}$$

Theorem 7. *In type II topological logic $\mathbb{C}_T^2$ with matching function $\mathcal{C}$, there exists type II topological solution of multiple approximate reasoning for A^* with respect to knowledge base*

$$K = \{A_i \to B_i | i = 1, 2, \cdots, n\},$$

if and only if

$$A^* \in \bigcup_{i=1}^{n} U_{A_i}$$

Theorem 8. *In type II topological logic $\mathbb{C}_T^2$ with matching function $\mathcal{C}$, if there exists type II topological solution B^* of multiple approximate reasoning for A^* with respect to knowledge base*

$$K = \{A_i \to B_i | i = 1, 2, \cdots, n\},$$

then there at least exists an $i \in \{1, 2, \cdots, n\}$ which holds

$$B^* \in U_{B_i}$$

Theorem 9. *In type II topological logic $\mathbb{C}_T^2$ with matching function $\mathcal{C}$, if there exists type II topological solution B^* of multiple approximate reasoning for A^* with respect to knowledge base*

$$K = \{A_i \to B_i | i = 1, 2, \cdots, n\},$$

then

$$B^* \in \bigcup_{i=1}^{n} U_{B_i}$$

Theorem 10. *In type II topological logic $\mathbb{C}_T^2$ with matching function $\mathcal{C}$, if there exists type II topological solution B^* of multiple approximate reasoning for A^* with respect to knowledge base*

$$K = \{A_i \to B_i | i = 1, 2, \cdots, n\},$$

then there at least exists an $i \in \{1, 2, \cdots, n\}$ which holds

$$A^* \to B^* \in U_{A_i \to B_i}$$

Theorem 11. *In type II topological logic $\mathbb{C}_\mathcal{T}^2$ with matching function $\mathcal{C}$, if there exists type II topological solution B^* of multiple approximate reasoning for A^* with respect to knowledge base*

$$K = \{A_i \to B_i | i = 1, 2, \cdots, n\},$$

then there at least exists an $i \in \{1, 2, \cdots, n\}$ which holds

$$A^* \to B^* \in \bigcup_{i=1}^{n} U_{A_i \to B_i}$$

It is not ideal that a input A^* simultaneously activated more than one knowledges in knowledge base

$$K = \{A_i \to B_i | i = 1, 2, \cdots, n\},$$

that is to say, the most favorable situation is that a input A^* activates at most one knowledge and B^* is consequently regarded as the output.

Therefore, we would like to impose some constraints to matching function $\mathcal{C}$ and choose a proper topology $\mathcal{T}$. The reason is that the performance of type II topological algorithm of approximate reasoning majorally depends on the structure of matching function $\mathcal{C}$, which depends on the choice of topology $\mathcal{T}$ for type II topological logic $\mathbb{C}_\mathcal{T}^2$ to some extent.

In type II topological logic $\mathbb{C}_\mathcal{T}^2$, matching function $\mathcal{C}$ is called a *Hausdorff matching* with respect to knowledge base

$$K = \{A_i \to B_i | i = 1, 2, \cdots, n\},$$

if and only if

$$U_{A_i} \bigcap U_{A_j} = \emptyset$$

where $i, j \in \{1, 2, \cdots, n\}$ and $i \neq j$.

Theorem 12. *In type II topological logic $\mathbb{C}_\mathcal{T}^2$ with matching function $\mathcal{C}$, if $\mathcal{C}$ is a Hausdorff matching with respect to knowledge base*

$$K = \{A_i \to B_i | i = 1, 2, \cdots, n\},$$

then each input A^* can activate at most only one knowledge $A_i \to B_i$, therefore, $B^* \in U_{B_i}$, $A^* \to B^* \in U_{A_i \to B_i}$.

5 Conclusion

We explore approximate reasoning in logical framework. We propose the structure of type II topological logic $\mathbb{C}_\mathcal{T}^2$, the structure of matching function $\mathcal{C}$ and the structure of matching neighborhood group. We investigate approximate reasoning in type II topological logic $\mathbb{C}_\mathcal{T}^2$ with matching function $\mathcal{C}$, develop type II topological algorithm of simple approximate reasoning and multiple approximate reasoning, introduce the essential characteristics of those schemes.

The structure of type II regular topological logic $\mathbb{C}_\mathcal{T}^2$ and the structure of regular matching function $\mathcal{C}$ are also proposed in this paper. The type II completeness and type II perfectness of knowledge base K in type II regular topological logic $\mathbb{C}_\mathcal{T}^2$ with regular matching function $\mathcal{C}$ are also presented.

In the future, we will investigate type II approximate knowledge mass, type II knowledge universe and type II k-level knowledge base K_k. We will also build the model of type II topological expert system with the function of automatic extension. Besides, we will study the structure and properties of true k-level knowledge circle and type II (k,j)−level knowledge base $\mathbb{K}_{k,j}$, establish the model of extended type II topological expert system, research into the type II (k,j)−level perfectness of extended type II knowledge base.

We also intend to give the structure of type II strong topological logic $\mathbb{C}_\mathcal{T}^2$ and the structure of strong matching function $\mathcal{C}$. The multidimensional approximate reasoning, multiple multidimensional approximate reasoning and the correspondent approximate reasoning schemes in type II strong topological logic $\mathbb{C}_\mathcal{T}^2$ with strong matching function $\mathcal{C}$ will also be investigate.

In order to make our topological schemes more applicable, we significantly decrease the "topological requirements" and significantly increase the "number of open sets" simultaneously in our research work.

References

1. Wang, G.J.: On the logic foundation of fuzzy reasoning. Information Science, **117** (1999) 47-88.
2. Wang, G.J., Leung, Y.: Intergrated semantics and logic metric spaces. Fuzzy Sets and Systems, **136** (2003) 71-91.
3. Wang, G.J., Wang, H.: Non-fuzzy versions of fuzzy reasoning in classical logic. Information Sciences, **138** (2001) 211-236.
4. Ying, M.S.: A logic for approximate reasoning. The Journal of Symbolic Logic, **59** (1994) 830-837.
5. Ying, M.S.: Fuzzy reasoning under approximate match. Science Bulletin, **37** (1992) 1244-1245.
6. Ying, M.S.: Reasoning about probabilistic sequential programs in a probabilistic logic. Acta Informatica, **39** (2003) 315-389.
7. Zheng, Y.L.: Stratified construction of fuzzy propositional logic. Proceedings of International Conference on Fuzzy Information Processing, Tsinghua University Press, Springer Verlag, **1-2** (2003) 169-174.
8. Zheng, Y.L., Zhang, C.S., Yi, X.: Mamdaniean logic. Proceedings of IEEE International Conference on Fuzzy Systems, Budapest, Hungary, **1-3** (2004) 629-634.

Type-I Topological Logic $\mathbb{C}^1_{\mathscr{T}}$ and Approximate Reasoning*

Yalin Zheng, Changshui Zhang, and Xin Yao

State Key Laboratory of Intelligent Technology and Systems,
Department of Automation, Faculty of Information Science and Technology,
Tsinghua University, Beijing 100084, P.R. China
{zheng-yl, zcs}@mail.tsinghua.edu.cn
yaoxin99@mails.tsinghua.edu.cn

Abstract. We introduce the consistent topological structure and neighborhood structure into the logical framework for providing the logical foundation and logical normalization for the approximate reasoning. We present the concept of the *formulae mass*, the *knowledge mass* and the *approximating knowledge closure* of the knowledge library by means of topological closure. We obtain the fundamental framework of type-I topological logics. In this framework, we present the type-I topological algorithm of the simple approximate reasoning and multi-approximate reasoning. In the frameworks of type-I strong topological logics, we present the type-I topological algorithm of multidimensional approximate reasoning and multiple multidimensional approximate reasoning. We study the type-I completeness and type-I perfection of the knowledge library in the framework of topological logical frameworks. We construct the type-I *knowledge universe* and prove that the second class knowledge universe of type-I is coincident with the first class knowledge universe of type-I, therefore the type-I knowledge universe is stable. We construct a self-extensive type-I knowledge library and the type-I expert system. In this expert system, the new approximate knowledge acquired by the self-extensive type-I knowledge library K^I will not beyond the type-I approximate knowledge closure, $(K_0)^-$, of the initial knowledge library K_0. Therefore, the precision of all new acquired approximate knowledge of this automatic reasoning system will be controlled well by the type-I approximate knowledge closure $(K_0)^-$ of the initial knowledge library K_0.

1 Introduction

Many logistician have done a lot of research work for providing the logical foundation and logical normalization of the approximate reasoning. Guo-Jun Wang constructed a fuzzy propositional logical system $\mathcal{L}^*$ and gave the $\alpha-3I$ algorithm

---

* This work is supported by the project (60475001) of the National Natural Science Foundation of China.

of the approximate reasoning in [1]. In [2], Wang studied the logical metric space based on their integrate semantic theory. The approximate reasoning model in the framework of the classic propositional logic is also studied in [3]. Mingsheng Ying gave a logic model of approximate reasoning in [4] and studied approximate reasoning based on the fuzzy matching in [5]. We constructed the framework of the stratified fuzzy propositional logic in [6] and discovered an important logical property of Mandanian algorithm for fuzzy reasoning in [7]. All the above work motivate the authors to study the approximate reasoning model in the framework of topological logic.

For the approximate reasoning

$$\frac{A \to B \quad A^*}{B^*}, \tag{1}$$

the MP rule in the classic propositional logic $\mathbb{C}$ will be invalid, if $A^* \neq A$. This case occurs frequently in the theory and application of the artificial intelligence, where the matching degree is applied to deal with such case. Considering the metric $d(A^*, A)$, approximate degree $q(A^*, A)$, similarity $s(A^*, A)$, etc., the essential of all above definitions are measure of the approximation between the input A^* and the antecedent A of the knowledge $A \to B$. Based on such approximation measures, the conclusion B^* of the approximate reasoning can be obtained so as to keep close to the descendant B of the knowledge $A \to B$, according to the corresponding approximation measures. From the view of the abstract mathematics, the approximation of two objects can be described by topological structure. This thought lead up to the topics of topological logics and the topics of approximate reasoning in the topological logics. Such topological logical description of approximate reasoning may reflect the more essential relations between the approximate reasoning and the logics.

2 The Construction of Type-I Topological Logic $\mathbb{C}^1_{\mathscr{T}}$

Let $\mathbb{C} = (\widehat{\mathbb{C}}, \widetilde{\mathbb{C}})$ be the classic propositional logic, $\widehat{\mathbb{C}}$ is the syntax of $\mathbb{C}$ and $\widetilde{\mathbb{C}}$ is the semantic of $\mathbb{C}$. Let $F(S)$ be the proposition set of $\mathbb{C}$, which is also called formulae set. $\to$ is the implication connective between the propositions. $\varnothing^\vdash$ is the set of all theorems in $\widehat{\mathbb{C}}$ and $\varnothing^\vDash$ is the set of all tautologies in $\widetilde{\mathbb{C}}$. According to the soundness theorem and the completeness theorem for $\mathbb{C}$, we have $\varnothing^\vdash = \varnothing^\vDash$.

Suppose $\mathscr{T}$ is a topology on the formulae set $F(S)$, then $(F(S), \mathscr{T})$ is a topological space. For any formulae $A, B, A \to B \in F(S)$, let $\mathcal{U}(A), \mathcal{U}(B), \mathcal{U}(A \to B)$ denote their neighborhoods respectively under the topology $\mathscr{T}$. Let "$-$" denote the topological closure operator decided by $\mathscr{T}$.

If for each formula $A \to B \in F(S)$, we have

$$\{A\}^- - \{B\}^- = \{A \to B\}^- \tag{2}$$

where $\{A\}^- - \{B\}^- = \{A^* \to B^* | A^* \in \{A\}^-, B^* \in \{B\}^-\}$. Then the triple

$$\mathbb{C}^1_{\mathscr{T}} = (\widehat{\mathbb{C}}, \widetilde{\mathbb{C}}, \mathscr{T})$$

is called a *type-I topological propositional logical system* on formulae set $F(S)$ and $\mathscr{T}$ is called *type-I consistent topology* with the logic $\mathbb{C}$.

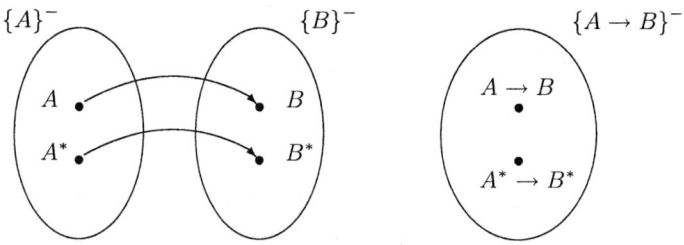

Fig. 1. The construction of the type-I topological logic $\mathbb{C}^1_{\mathscr{T}}$

The adhesive points in general topology is called adhesive formula in our context. According to the language of adhesive formulae, the condition Eq.(2) describes the following characteristic of type-I topological logic $\mathbb{C}^1_{\mathscr{T}}$: if A^* is adhesive formula of $\{A\}$ and B^* is adhesive formula of $\{B\}$, then $A^* \to B^*$ is also an adhesive formula of $A \to B$, and vise versa. In other words, if A^* and B^* approximate to A and B enough respectively, then $A^* \to B^*$ approximates to $A \to B$ enough, and vice versa.

Let us give some examples of the formula set with the topological structure.

Example 1. Given an equivalent relation R on the formula set $F(S)$ (e.g., logical equivalence relation, provable equivalence relation, and some other equivalence relations), and let $[A]_R$ denote the R−equivalence class which include formula A. The family of all the definable sets constructs a topology on the $F(S)$, $\mathscr{T}_R$. Under this topology, $\{A\}^-$, the closure of the singleton $\{A\}$, is coincident with $[A]_R$, the R−equivalence class which include A, that is

$$\{A\}^- = [A]_R \qquad (3)$$

Example 2. Given a similarity degree function of formula set $F(S)$

$$s : \begin{cases} F(S) \times F(S) \to & [0,1] \\ (A, B) & \to s(A, B) \end{cases} \qquad (4)$$

Define another mapping

$$d : \begin{cases} F(S) \times F(S) \to & [0,1] \\ (A, B) & \to d(A, B) \end{cases} \qquad (5)$$

such that $\forall (A, B) \in F(S) \times F(S)$,

$$d(A, B) = 1 - s(A, B), \tag{6}$$

then we get a pseudo-distance function d, from which a topology $\mathscr{T}_S$ can be induced naturally on the formula set $F(S)$. Under this topology, the α−closure of the singleton $\{A\}$ is

$$\begin{aligned}\{A\}^{-\alpha} &= \{A^* \in F(S) | d(A^*, A) \leq \alpha\} \\ &= \{A^* \in F(S) | s(A^*, A) \geq 1 - \alpha\},\end{aligned}$$

especially, the closure of the singleton $\{A\}$ is

$$\begin{aligned}\{A\}^- &= \{A^* \in F(S) | d(A^*, A) \leq 0\} \\ &= \{A^* \in F(S) | s(A^*, A) \geq 1\}.\end{aligned}$$

3   The Simple Approximate Reasoning Based on Type-I Topological Logic $\mathbb{C}^1_{\mathscr{T}}$

Considering the following simple approximate reasoning model in type-I topological logic $\mathbb{C}^1_{\mathscr{T}}$,

$$\frac{A \to B \quad\quad A^*}{B^*} \tag{7}$$

where $A \to B \in \varnothing^\vdash$ is called *knowledge* and $A^* \in F(S)$ is called *input*. $B^* \in F(S)$ is called *output* or *approximate reasoning conclusion*.

If $A^* \in \{A\}^-$, then it is said that A^* can *activate* knowledge $A \to B$ and take any $B^* \in \{B^-\}$ as the conclusion of approximate reasoning which is called *type-I topological solution* of the simple approximate reasoning of input A^* under the knowledge $A \to B$.

If $A^* \notin \{A\}^-$, then it is said that A^* cannot activate knowledge $A \to B$ and the input A^* will not be processed. In other words, there exists no type-I topological solution for the input A^* according to the simple approximate reasoning of knowledge $A \to B$.

The above algorithm is called *type-I topological algorithm* of simple approximate reasoning.

The construction of the type-I topological logic $\mathbb{C}^1_{\mathscr{T}}$ guarantees the sufficient approximation between the newly obtained approximate knowledge $A^* \to B^*$ and the original standard knowledge $A \to B$ based on type-I topological solution B^* of the simple approximate reasoning Eq.(7). In more exactly words, this construction guarantees that $A^* \to B^*$ is the adhesive formula of $\{A \to B\}$.

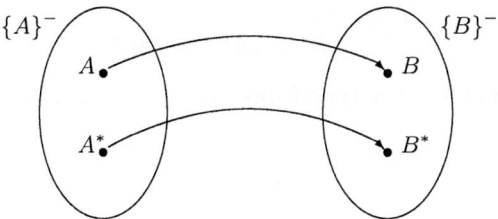

Fig. 2. Type-I topological algorithm of the simple approximate reasoning

Theorem 1. *In the type-I topological logic $\mathbb{C}^1_{\mathscr{T}}$, there exists a type-I topological solution for the simple approximate reasoning under the input A^* and the knowledge $A \to B$, if and only if,*

$$A \in \bigcap_{U \in \mathcal{U}(A^*)} U \tag{8}$$

that is, the antecedent A of knowledge $A \to B$ is contained in the intersection of all neighborhoods of input A^*, $\bigcap_{U \in \mathcal{U}(A^*)} U$.

Proof. The simple approximate reasoning Eq.(7) has type-I topological solution $\iff A^* \in \{A\}^- \iff \forall U \in \mathcal{U}(A^*), U \cap \{A\} \neq \emptyset \iff \forall U \in \mathcal{U}(A^*), A \in U \iff A \in \bigcap_{U \in \mathcal{U}(A^*)} U$.

□

Theorem 2. *In the type-I topological logic $\mathbb{C}^1_{\mathscr{T}}$, if there exists a type-I topological solution B^* for the simple approximate reasoning under the input A^* and knowledge $A \to B$, then,*

$$B \in \bigcap_{U \in \mathcal{U}(B^*)} U \tag{9}$$

that is, the descendant B of the knowledge $A \to B$ is contained in the intersection of all neighborhoods of the type-I topological solution B^*, $\bigcap_{U \in \mathcal{U}(B^*)} U$.

Theorem 3. *In the type-I topological logic $\mathbb{C}^1_{\mathscr{T}}$, if there exists a type-I topological solution B^* for the simple approximate reasoning under the input A^* and knowledge $A \to B$, then,*

$$A^* \to B^* \in \{A \to B\}^- \tag{10}$$

that is, $A^* \to B^*$ is an adhesive formula of $A \to B$.

Proof. It can be seen from the type-I topological algorithm that

$$A^* \in \{A\}^-, \quad B^* \in \{B\}^-$$

therefore,
$$A^* \to B^* \in \{A\}^- \to \{B\}^-.$$

From the construction of the type-I topological logic $\mathbb{C}^1_{\mathscr{T}}$, we have
$$\{A\}^- \to \{B\}^- = \{A \to B\}^-,$$

therefore,
$$A^* \to B^* \in \{A \to B\}^-.$$

□

The importance of the Theorem 3 lies in the fact that B^*, which is the type-I topological solution of the simple approximate reasoning Eq.(7), gives a good approximation between the newly obtained approximate knowledge $A^* \to B^*$ and the standard knowledge $A \to B$, which is guaranteed by the construction of the type-I topological logic.

Theorem 4. *In the type-I topological logic $\mathbb{C}^1_{\mathscr{T}}$, if there exists a type-I topological solution B^* for the simple approximate reasoning under the input A^* and knowledge $A \to B$, then,*

$$A \to B \in \bigcap_{U \in \mathcal{U}(A^* \to B^*)} U, \tag{11}$$

that is, the standard knowledge $A \to B$ is contained in the intersection of all neighborhoods of the newly obtained approximate knowledge $A^* \to B^*$.

A algorithm of the simple aproximate reasoning is called the *Modus Ponens possible reappearance algorithm*, for short, *MP possible recurrence algorithm*, if the input is the antecedent A of the knowledge $A \to B$, the output conclusion can be taken as the descendant of $A \to B$.

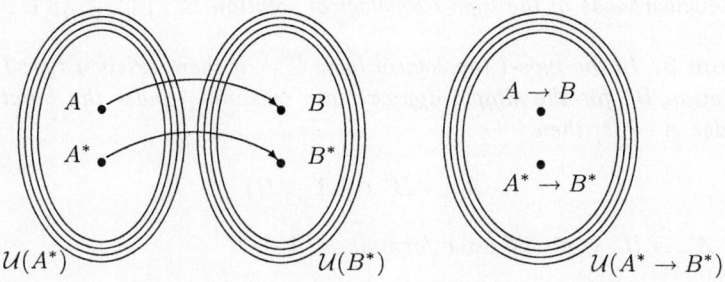

Fig. 3. The description of the neighborhoods of type-I topological algorithm

Theorem 5. *In the type-I topological logic $\mathbb{C}^1_{\mathscr{T}}$, the type-I topological algorithm of the simple approximate reasoning Eq.(7) is MP possible reappearance algorithm.*

Proof. It can be seen from $A \in \{A\}^-$ and $B \in \{B\}^-$. □

The *degenerate approximate reasoning algorithm*, which is also called *classic reasoning algorithm*, is another algorithm of the simple approximate reasoning. This algorithm can be described as follows: for each knowledge $A \to B$, it can be activated by the input A^* if and only if $A^* = A$.

Theorem 6. *In the type-I topological logic $\mathbb{C}^1_{\mathscr{T}}$, the type-I topological algorithm of the simple approximate reasoning Eq.(7) is the classic reasoning algorithm if and only if the topology $\mathscr{T}$ on $\mathbb{C}^1_{\mathscr{T}}$ is a Fréchet topology.*

Proof. The type-I topological algorithm is the classic reasoning algorithm $\iff$ $A \to B$ can be activated by $A^* \iff A^* = A \iff \{A\}^- = \{A\} \iff$ the topology $\mathscr{T}$ on $\mathbb{C}^1_{\mathscr{T}}$ is the Fréchet topology. □

Theorem 7. *In the type-I topological logic $\mathbb{C}^1_{\mathscr{T}}$, when $A^* \neq A$, there exists no type-I topological solution for the simple approximate reasoning Eq.(7), if and only if the topology $\mathscr{T}$ on $\mathbb{C}^1_{\mathscr{T}}$ is the Fréchet topology.*

Proof. There exists no type-I topological solution for simple approximate reasoning Eq.(7) if $A^* \neq A \iff A^*$ cannot activate $A \to B$ if $A^* \neq A \iff A^* \notin \{A\}^-$ if $A^* \neq A \iff \{A\}^- = \{A\}$.

Since the knowledge $a \to B$ is chosen arbitrarily from $\varnothing^\vdash$ (e.g., for each formula $A \in F(S)$, $A \to A$ is a knowledge), we have $\{A\}^- = \{A\}$ if and only if the topology $\mathscr{T}$ on $\mathbb{C}^1_{\mathscr{T}}$ is the Fréchet topology. □

The Theorem 8, which is another statement of the Theorem 7, may be more valuable.

Theorem 8. *In the type-I topological logic $\mathbb{C}^1_{\mathscr{T}}$, there exists type-I topological solution for the simple approximate reasoning Eq.(7) under some A^*, s.t., $A^* \neq A$, if and only if the topology $\mathscr{T}$ on $\mathbb{C}^1_{\mathscr{T}}$ is not the Fréchet topology.*

According to above discussion, Fréchet topology is a critical topology in type-I topological logic for the type-I topological algorithm of the simple approximate reasoning. It is only valid for the classic MP reasoning and is invalid for the approximate reasonings which allow $A^* \neq A$. The topologies with higher separability, such as Hausdorff topology, regular topology, normal topology, etc., are only valid for the classic MP reasoning and is invalid for the approximate reasonings which allow $A^* \neq A$. The topologies which are valid for the type-I topological algorithm of the simple approximate reasoning, can only be the non-Fréchet topologies. These topologies can realize not only the classic MP reasoning but also the non-classic approximate reasoning. It can activate the knowledge $A \to B$ as $A^* \neq A$.

On the one hand, the non-Fréchet topologies for the type-I topological algorithm of the simple approximate reasoning include fewer open sets such that the intersection of all neighborhoods of input A^*, $\bigcap_{U \in \mathcal{U}(A^*)} U$, may include the antecedent A of the knowledge $A \to B$. On the other hand, the corresponding type-I topological algorithms have higher requirements for the topologies, i.e., the input A^* actives the knowledge $A \to B$ successfully if and only if A^* is the adhesive formula of $\{A\}$, here taken any adhesive formula $B^* \in \{B\}$ as the conclusion of the approximate reasoning, it will guarantee that the newly obtained approximate knowledge $A^* \to B^*$ is also the adhesive formula of $A \to B$. "Fewer open sets" and "higher requirements for topologies" are the two folds of the integral system which is constituted by the type-I topological logic and type-I topological algorithm.

However, we will decrease the "topological requirements" and increase the "number of open sets" in the following study, for convenience of and accuracy.

In the following we will condsider the different influence on the approximate reasoning of two topologies, $\mathscr{T}_1$ and $\mathscr{T}_2$. Let $\mathbb{C}^1_{\mathscr{T}_1}$ and $\mathbb{C}^1_{\mathscr{T}_2}$ be two type-I topological logics. For each knowledge $A \to B$ and each input A^*, if "A^* activate $A \to B$ in $\mathbb{C}^1_{\mathscr{T}_1}$" $\Rightarrow$ "A^* activate $A \to B$ in $\mathbb{C}^1_{\mathscr{T}_2}$", then it is said that, for the simple approximate reasoning Eq.(7), the type-I topological algorithm σ in type-I topological logic $\mathbb{C}^1_{\mathscr{T}_1}$ is *finer* than the type-I topological algorithm τ in type-I topological logic $\mathbb{C}^1_{\mathscr{T}_2}$, which is denoted as

$$\sigma \prec \tau$$

If $\sigma \prec \tau$ and $\tau \prec \sigma$, then the algorithm τ is *equivalent* to σ and it is denoted by

$$\sigma \sim \tau$$

Theorem 9. *Suppose that σ and τ are the type-I topological algorithms of simple approximate reasoning Eq.(7) in type-I topologies $\mathbb{C}^1_{\mathscr{T}_1}$ and $\mathbb{C}^1_{\mathscr{T}_2}$, respectively, then σ is finer that τ, if and only if the topology $\mathscr{T}_1$ is finer than $\mathscr{T}_2$.*

Proof.

$$\sigma \prec \tau$$
$$\Longleftrightarrow \text{"}A^* \in \{A\}^{-\mathscr{T}_1} \Rightarrow A^* \in \{A\}^{-\mathscr{T}_2}\text{"}$$
$$\Longleftrightarrow \{A\}^{-\mathscr{T}_1} \subseteq \{A\}^{-\mathscr{T}_2}$$
$$\Longleftrightarrow \mathscr{T}_1 \supseteq \mathscr{T}_2$$
$$\Longleftrightarrow \mathscr{T}_1 \prec \mathscr{T}_2$$

where the "$-\mathscr{T}_1$" and "$-\mathscr{T}_2$" are the closure operators of the topologies $\mathscr{T}_1$ and $\mathscr{T}_2$, respectively. □

It is clear the the algorithms σ and τ are equivalent, $\sigma \sim \tau$, if and only if the topologies $\mathscr{T}_1$ and $\mathscr{T}_2$ are the same.

For the equivalent algorithms σ and τ, the σ−solution of A^* under $A \to B$ may not be the same as the τ−solution of A^* under $A \to B$, which is allowed by the type-I topological algorithm of simple approximate reasoning.

4 Discussion and Summation

We introduce the consistent topological structure and neighborhood structure to the framework of logic for providing the logical foundation and logical normalization for the approximate reasoning. Taking the topological closure as a tool, we present a topological logical model of the simple approximate reasoning and discuss its fundamental property in the framework of type-I topological logic $\mathbb{C}^1_{\mathscr{T}}$.

We also present the concepts of formulae mass and knowledge mass through the tool of topological closure. In the framework of topological logic $\mathbb{C}^1_{\mathscr{T}}$, we present a topological logical model for the multiple approximate reasoning. In the framework of strong topological logic $\mathbb{C}^1_{\mathscr{T}}$, we present a topological logical model for the multidimensional approximate reasoning and multiple multidimensional approximate reasoning. We give the type-I completeness and type-I perfectness of the knowledge library in the topological logic $\mathbb{C}^1_{\mathscr{T}}$. We also construct a type-I knowledge universe Ω^I and prove that type-I grade-2 knowledge universe Ω_2 will be coincident with the type-I grade-1 knowledge universe Ω_1, therefore the type-I knowledge universe Ω^I is a stable knowledge universe. We construct an automatically expansive type-I knowledge library K^I and type-I topological expert system $\mathcal{E} = (\mathbb{C}^1_{\mathscr{T}}, \Omega^I, K^I)$. In our type-I topological expert system, the knowledge which obtained by the automatically extensive type-I knowledge library K^I will not exceed the $(K_0)^-$, which is the type-I approximate knowledge closure of the initial knowledge library K_0. Therefore the precision of the approximate knowledge obtained by the automatically reasoning system is well controlled by the type-I approximate knowledge closure, $(K_0)^-$, of the initial knowledge library K_0.

Fréchet topology is a critical topology in the study of the type-I topological logics. It is only valid for classic MP reasoning and is invalid for the approximate reasoning where $A \neq A^*$. The topologies whose separability are higher than Fréchet topology, such as Hausdorff topology, regular topology and normal topology, etc., are only valid for classic MP reasoning and are invalid for approximate reasoning where $A \neq A^*$. The topologies which are valid for type-I topological algorithm of approximate reasoning are non-Fréchet topologies which can not only realized the classic MP reasoning but also make some inputs A^*, which are different with A, activate the knowledge $A \to B$.

The non-Frechet topologies $\mathscr{T}$, which make the type-I topological algorithm of the approximate reasoning valid, possess fewer open sets so that the intersection of all neighborhoods of A^*, $\cap_{U \in \mathcal{U}(A^*)} U$, may include A. At the same time, the corresponding type-I topological algorithms possess higher requirements for the topologies, i.e., the input A^* activates $A \to B$ successfully if and only if A^* is the adhesive formula of A, here taking any adhesive formula B^* of the $\{B\}$ as the conclusion of the approximate reasoning, it will guarantee not only the sufficient close between B and B^* but also the sufficient close between the knowledge $A^* \to B^*$ and $A \to B$, i.e., $A^* \to B^*$ is also the adhesive formula of $\{A^* \to B^*\}$. Here, "fewer open sets" and "higher requirements for topologies" are the two folds of the integrate system constituted by the type-I topological logics and type-I topological algorithms.

However, in our future works, we will try to decrease the requirements for topologies and increase the number of open sets which make our topological algorithms more applicable.

References

1. Wang, G.J.: On the logic foundation of fuzzy reasoning. Information Science, **117** (1999) 47-88.
2. Wang, G.J., Leung, Y.: Intergrated semantics and logic metric spaces. Fuzzy Sets and Systems, **136** (2003) 71-91.
3. Wang, G.J., Wang, H.: Non-fuzzy versions of fuzzy reasoning in classical logic. Information Sciences, **138** (2001) 211-236.
4. Ying, M.S.: A logic for approximate reasoning. The Journal of Symbolic Logic, **59** (1994) 830-837.
5. Ying, M.S.: Fuzzy reasoning under approximate match. Science Bulletin, **37** (1992) 1244-1245.
6. Zheng, Y.L.: Stratified construction of fuzzy propositional logic. Proceedings of International Conference on Fuzzy Information Processing, Tsinghua University Press, Springer Verlag, **1-2** (2003) 169-174.
7. Zheng, Y.L., Zhang, C.S., Yi, X.: Mamdaniean logic. Proceedings of IEEE International Conference on Fuzzy Systems, Budapest, Hungary, **1-3** (2004) 629-634.

Vagueness and Extensionality

Shunsuke Yatabe[1] and Hiroyuki Inaoka[2]

[1] Faculty of Engineering, Kobe University, Kobe 657-8501, Japan
yatabe@kurt.scitec.kobe-u.ac.jp
[2] The Graduate School of Humanities and Social Sciences,
Kobe University, Kobe 657-8501, Japan
hinaoka@lit.kobe-u.ac.jp

Abstract. We introduce a property of set to represent vagueness without using truth value. It has gotten less attention in fuzzy set theory. We introduce it by analyzing a well-known philosophical argument by Gearth Evans. To interpret 'a is a vague object' as 'the Axiom of Extensionality is violated for a' allows us to represent a vague object in Evans's sense, even within classical logic, and of course within fuzzy logic.

1 Introduction

In fuzzy set theory, vagueness is represented by truth value: Set x is a vague object if there is a set y such that the truth value of $y \in x$ is indeterminate, i.e. except for $0, 1$. However set has a property which can be applied to represent vagueness without using truth value; it has never been regarded as connected with vagueness in fuzzy set theory. In this paper, we focus our attention on it by analyzing a well-known philosophical argument which insists the impossibility of a vague object.

In his short paper [3], Gearth Evans defined vague objects as having vague identity statement: a is a vague object if there is some object b such that $a = b$ is of indeterminate truth value. Let us assume *there can be a vague object in his sense in the world*; we call this Evans's Vagueness Assumption (**EVA**). He proceeded his argument as follows: Let $a = b$ be indeterminate, then

(I) $\triangledown(a = b)$,   i.e. $a = b$ is indeterminate (assumption),
(II) $\lambda x[\triangledown(a = x)]_b$,   i.e. b is indeterminately equal to a (by (I)),
(III) $\neg \triangledown (a = a)$,   i.e. $a = a$ is determinate,
(IV) $\neg \lambda x[\triangledown(a = x)]_a$,   i.e. a is not indeterminately equal to a (by (III)),
(V) $a \neq b$,   i.e. a is not equal to b (from (II) and (IV)).

We note that $\triangledown \varphi$ means that the truth value of φ is indeterminate. The conclusion (V) is derived by the definition of Leibniz equality, i.e. $a = b$ if and only if $\varphi(a) \leftrightarrow \varphi(b)$ for any φ, and (V) is 'contradicting to the assumption that the identity statement "$a = b$" is of indeterminate truth value'. Hence (I) must be rejected, that is to say, any identity statement has determinate truth value. Therefore, he seems to have concluded **EVA** doesn't hold.

L. Wang and Y. Jin (Eds.): FSKD 2005, LNAI 3613, pp. 263–266, 2005.
© Springer-Verlag Berlin Heidelberg 2005

Many articles have been published for or against Evans's proof. Many philosophers defended **EVA** and denied his derivation. They implicitly supposed that classical logic is not a suitable framework to represent vagueness; to change logic (e.g. fuzzy logic) is required to represent a vague object[1].

We defend **EVA**, however we show that we need not to deny his derivation from (I) to (V) even within classical logic: To interpret '$a = b$ is indeterminate' as 'a and b are extensionally equal (they share the same elements) nevertheless $a \neq b$ holds' gives us its model. This means that the Axiom of Extensionality is violated for a and b: The Axiom insists that any set is determined precisely by its members, i.e. if x and y has the same members then $x = y$ holds.. So we can regard it as a representation of precision. In this sense, the violation of the Axiom represents a sort of vagueness.

Many non-extensional set theories, i.e. set theories without the Axiom of Extensionality, have been proposed. In particular, Hajek and Hanikova's Fuzzy Set Theory (**FST**) [5] was proposed to axiomatize our intuition of fuzzy set. It proves that Leibniz equality is crisp (i.e. its truth value is 0 or 1) but extensional equality can be a fuzzy relation, so the Axiom can't be valid. Such violation of the Axiom has been regarded as merely introduced for technical reasons, however it seems to suggest that it is a necessary feature of fuzzy set implicitly connoted by our intuition itself.

It seems that no one studying vagueness has mentioned the Axiom of Extensionality is connected with vagueness representation. We must investigate phenomenon inherently found in examining such vagueness.

2 Representation of Vagueness and Extensionality

Hereafter we employ set theory within classical logic. The famous relations in set theory are as follows:

Leibniz equality $x = y$ iff $(\forall z)[x \in z \leftrightarrow y \in z]$,
Extensional equality $x =_{\text{ext}} y$ iff $(\forall z)[z \in x \leftrightarrow z \in y]$.

Of course $x = y \to x =_{\text{ext}} y$ holds. It is necessary to consider the Axiom of Extensionality when we think about identity relations: The Axiom of Extensionality guarantees that, for any set x and y, $x =_{\text{ext}} y \to x = y$.

2.1 Vagueness as the Violation of the Axiom of Extensionality

In this subsection, we introduce our interpretation of indeterminacy. First we analyze Evans's proof. Technically speaking, his proof seems to have three implicit assumptions as follows:

---

[1] One typical approach is to analyze his proof within many-valued logic. For example, Jack Copeland tried to prove that derivation (I) from (V) isn't valid within fuzzy logic [2]. Michael Tye represented vagueness within many-valued logic [9]. However it has been objected that 'the writers who adopt this strategy rarely provide much argument for the need for a many-valued logic' [6]. Another approach is within modal logic. It requires to add new modal operators to represent indeterminacy. For example, Ken Akiba defined a vague object as a transworld object [1].

(i) For every a, $a = a$ has definite truth value ($\neg \triangledown (a = a)$),
(ii) *the Diversity of the Dissimilar* (**DD**) : if object a has a property that b lacks, then you can infer $a \neq b$,
(iii) $\vdash \varphi$ implies $\vdash \triangle \varphi$ (as the generalization law in $\mathbf{S_5}$-modal logic).

For more details, see [6]. We note that $\triangle \varphi$ means that the truth value of φ is determinate. (ii) is used to infer (V) from (II) and (IV). Now, $\triangle(a \neq b)$ is inferred from (V) and (iii). For duality ($\neg \triangle \neg \varphi \leftrightarrow \triangledown \varphi$), $\neg \triangledown (a = b)$ is inferred from $\triangle(a \neq b)$. This contradicts (I).

We don't assume (iii): (i) and (ii) are necessary to derive (V) from (I) but (iii) is nothing to do with the derivation itself. What Evans proved is merely that *vague identity statement (I) implies (V)* if we don't admit (iii). We call this, $\triangledown(a = b) \to a \neq b$ Evans Conditional (**EC**) as in [2]. We define vague equality so as to satisfy **EC**, and show the existence of a model which witnesses the consistency of **EVA** and **EC**.

We propose to define vague equality as follows:

Definition 1. *$a = b$ is indeterminate iff $a =_{ext} b$ & $a \neq b$.*

This means, the Axiom of Extensionality is violated for a and b. For more details about the justification of this consequence, see [10].

As we saw, a vague object in Evans's sense is defined by using vague identity.

Definition 2. *a is a vague object iff the Axiom of Extensionality is violated for a, i.e. $(\exists b)[a =_{ext} b \ \& \ a \neq b]$.*

Now **EC** implies contradiction only when we assume the Axiom of Extensionality. Otherwise $\triangledown(a = b)$ implies $a \neq b$ without implying $\triangle a \neq b$. So any model of set theory without the Axiom of Extensionality is a model of **EVA** and **EC**.

2.2 A Fuzzy Set Theory as an Example of Non-extensional Set Theory

We can easily generalize definition 2: The violation of the Axiom of Extensionality represents vagueness not only within classical logic but also within a greater variety of logics. So, within any logic, we insist that set theory without the Axiom of Extensionality is required to represent vague object.

Many non-extensional set theories have been proposed by now[2]. In particular, **FST** is a set theory within the framework of fuzzy logic with operator $\triangle$ which means 'determinately true', i.e. its truth value is 1 [5]. It is in the style of **ZF**, and it seems to be an attempt to axiomatize our intuition of fuzzy set.

---

[2] Traditionally, it has been studied within intuitionistic logic; the one of the most famous result is due to Harvey Friedman [4]. There are a few studies to generalize **ZF** first within intuitionistic logic, next within strengthened logic which is referred to as Gödel logic [8]. Gödel logic has truth set $[0, 1]$ as fuzzy logic, but it has a few new logical constants.

In **FST**, the Axiom of Extensionality can't be valid. It is because that Leibniz equality becomes crisp nevertheless the truth value of extensional equality can be indeterminate. We note that **FST** has the Axiom of Pairing, which guarantees that there is a *crisp* set $\{x,y\}$ for any set x,y, where X is crisp set if and only if $x \in X$ is crisp for any x. For any set x, fix the crisp set $\{x\}$ by the Axiom of Pairing: $\{x\} = \{x,x\}$. Then for any set y, the truth value of $y \in \{x\}$ is 1 or 0: if it is 1 this means that so is the truth value of $x = y$, otherwise the truth value of $x = y$ is 0. Here the Axiom of Extensionality holds for crisp set, but it might be violated for some fuzzy set.

Such violation of the Axiom of Extensionality has been regarded as merely introduced for technical reasons, but we justify it positively: We can represent two different sorts of vagueness simultaneously in **FST**. Furthermore, the violation seems to suggest that it is a necessary feature of fuzzy set implicitly connoted by our intuition of fuzziness itself. In this sense, definition 2 can be regarded as an isolation of some aspect of fuzziness so that we can represent it within a greater variety of logics.

3 Conclusion

In this paper, we gave an example of a new way of representation of vagueness in set theory by analyzing Evans's philosophical argument on a vague object. We defined a vague object in his sense as an object for which the Axiom of Extensionality doesn't hold, so we could construct a model of **EVA** and **EC** within classical logic without adding a new operator which represents indeterminacy.

Many non-extensional set theories have been proposed, and the violation of the Axiom in **FST** seems to suggest that it is a necessary feature of fuzziness. We must investigate phenomenon inherently found in examining such vagueness.

References

1. Akiba, Ken. 2000. Vagueness as a modality. *The Philosophical Quarterly* 50: 359-70.
2. Copeland, B. Jack. 1995. On Vague Objects, Fuzzy Logic and Fractal Boundaries. *Southern journal of philosophy* 33: 83-95.
3. Evans, Gearth. 1978. Can there be vague objects? *Analysis* 38: 208. Reprinted in *Vagueness: a reader*, ed. R.Keefe, P.Smith, 317. Cambridge, Mass.:MIT press.
4. Friedman, Harvey. 1973. The consistency of classical set theory relative to a set theory with intuitionistic logic. *Journal of Symbolic Logic* 38: 315-319.
5. Hajek, Petr. Hanikova, Zuzana. 2003. A development of set theory in fuzzy logic. *Theory and applications of multiple-valued logic*, 273-85, Heidelberg.: Physica
6. Keefe, Rossanna. Smith, Peter. 1997. Introduction: theories of vagueness. in *Vagueness: a reader*, ed. R.Keefe, P.Smith, 1-57. Cambridge, Mass.:MIT press.
7. Noonan, H.W. 2004. Are there vague objects? *Analysis* 64: 131-34.
8. Titani, S. Takeuti, G. Fuzzy logic and fuzzy set theory. 1992. *Arch. Math Logic* 32: 1-32.
9. Tye, Michael. 1994. Sorites paradoxes and the semantics of vagueness. Reprinted in *Vagueness: a reader*, 281-93.
10. Yatabe, S. Inaoka, H. On Evans's vague object from set theoretic viewpoint. Preprint.

Using Fuzzy Analogical Reasoning to Refine the Query Answers for Relational Databases with Imprecise Information

Z.M. Ma[1], Li Yan[1], and Gui Li[2]

[1] Northeastern University,
Shenyang, Liaoning 110004, China
[2] Liaoning Branch of China Netcom (Group) Corporation LTD.,
Shenyang, Liaoning 110044, China

Abstract. In this paper, we use the notion of equivalence degree of fuzzy data, by which we can mine the rules of fuzzy functional dependencies from the fuzzy relational databases. Following the rules of fuzzy functional dependencies, we can apply the frame of analogical reasoning to refine the imprecise query answer for the relational databases with imprecise information.

1 Introduction

Analogy is an important inference tool in human cognition and is a powerful computation tool for general inference in artificial intelligence and decision-making. Approximate reasoning refers to such an inferring process that an object S (the source) has properties P and Q, respectively, and another object T (the target) shares the property P, then we can infer that T may also possess the property Q [4]. In other words, S and T have the same property Q if S and T are matched on P to each other. This process can be expressed by the following logical rule.

IF P (S) and Q (S) and P (T) **THEN** Q (T)

Note that the properties P and Q must have an associated relationship in applying the approximate reasoning above. So there are two issues involved in approximate reasoning. One is that which property P can determine the property Q. Another is that which object S has the same property as the object T on property P such that T has the same property as S on property Q.

In relational databases, the relationship between properties P and Q in approximate reasoning is essentially the constraints of functional dependency between P and Q, where P and Q can be seen as attributes. According to semantics of functional dependency, two tuples, i.e., objects S and T, must have the same attribute values on Q if they have the same attribute values on P. Therefore, in relational databases, approximate reasoning can be emulated by functional dependency.

In real world applications, information is often vague or ambiguous. Different kinds of imperfect information have been extensively introduced into relational databases. Fuzzy information has been extensively investigated in the context of the relational model. Since the fuzzy relational databases contain fuzzy information, the query answers may be imprecise or even null [2, 3]. Applying approximate reasoning, it is clear that we can refine the imprecise query answer of tuple T on attribute Q when there exists a more precise attribute value of tuple S on attribute Q. The refinement result is that the value of T on Q is the same as the value of S on Q. However, in the fuzzy relational databases, the associated relationship between P and Q (i.e., functional dependency: $P \rightarrow Q$) may be fuzzy. In addition, the match between S and T on P is generally fuzzy.

In this paper, we use the notion of equivalence degree of fuzzy data to mine the rules of fuzzy functional dependencies from the fuzzy relational databases so that we could know which property P fuzzily determines the property Q. Also we need to determine which object S has the highest match degree with the object T on property P such that T has the same property Q as S. On the basis, we can apply the frame of analogical reasoning to refine the imprecise query answer for the relational databases with imprecise information.

The remainder of this paper is organized as follows. Section 2 of the paper gives the basic knowledge of fuzzy sets and fuzzy relational databases. Section 3 of the paper presents the method of semantic measure of fuzzy data. Section 4 investigates fuzzy functional dependency in the fuzzy relational databases. Section 5 proposes a framework to refine imprecise query answers using fuzzy analogical reasoning. Section 6 summaries this paper.

2 Fuzzy Sets and Fuzzy Relational Databases

Fuzzy data is originally described as fuzzy set by Zadeh [15]. Let U be a universe of discourse, then a fuzzy value on U is characterized by a fuzzy set F in U. A membership function $\mu_F: U \rightarrow [0, 1]$ is defined for the fuzzy set F, where $\mu_F(u)$, for each $u \in U$, denotes the degree of membership of u in the fuzzy set F. Thus the fuzzy set F is described as follows.

$$F = \{\mu(u_1)/u_1, \mu(u_2)/u_2, ..., \mu(u_n)/u_n\}$$

When the $\mu_F(u)$ above is explained to be a measure of the possibility that a variable X has the value u in this approach, where X takes values in U, a fuzzy value is described by a possibility distribution π_X [16]. Let π_X and F be the possibility distribution representation and the fuzzy set representation for a fuzzy value, respectively. It is apparent that $\pi_X = F$ is true [11].

In addition, a fuzzy data is represented by similarity relations in domain elements [1], in which the fuzziness comes from the similarity relations between two values in a universe of discourse, not from the status of an object itself. Similarity relations are thus used to describe the similarity degree of two values from the same universe of discourse. A similarity relation Sim on the universe of discourse U is a mapping: $U \times U \rightarrow [0, 1]$ such that

(a) for $\forall x \in U$, $Sim(x, x) = 1$, (reflexivity)
(b) for $\forall x, y \in U$, $Sim(x, y) = Sim_i(y, x)$, and (symmetry)
(c) for $\forall x, y, z \in U$, $Sim(x, z) \geq \max_y (\min(Sim(x, y), Sim(y, z)))$. (transitivity)

In connection to the three types of fuzzy data representations, there exist two basic extended data models for fuzzy relational databases. One of the data models is based on similarity relations [1], or proximity relation [13], or resemblance [12]. The other one is based on possibility distribution [10, 11]. The latter can further be classified into two categories, i.e. tuples associated with possibilities and attribute values represented possibility distributions [11]. The form of an n-tuple in each of the above-mentioned fuzzy relational models can be expressed, respectively, as

$$t = <p_1, p_2, ..., p_i, ..., p_n>, t = <a_1, a_2, ..., a_i, ..., a_n, d> \text{ and}$$
$$t = <\pi_{A1}, \pi_{A2}, ..., \pi_{Ai}, ..., \pi_{An}>,$$

where $p_i \subseteq D_i$ with D_i being the domain of attribute A_i, $a_i \in D_i$, $d \in (0, 1]$, π_{Ai} is the possibility distribution of attribute A_i on its domain D_i, and $\pi_{Ai}(x)$, $x \in D_i$, denotes the possibility that x is the actual value of t [A_i].

It is clear that, based on the above-mentioned basic fuzzy relational models, there should be one type of extended fuzzy relational model [12], where possibility distribution and resemblance relation arise in a relational database simultaneously. The focus of this paper is put on such fuzzy relational databases and it is assumed that the possibility of each tuple in a fuzzy relation is exactly 1.

Definition: A fuzzy relation r on a relational schema R (A1, A2, ..., An) is a subset of the Cartesian product of Dom (A1) × Dom (A2) × ... × Dom (An), where Dom (Ai) may be a fuzzy subset or even a set of fuzzy subset and there is the resemblance relation on the Dom (Ai). A resemblance relation *Res* on Dom (Ai) is a mapping: Dom (Ai) × Dom (Ai) → [0, 1] such that

(a) for all x in Dom (Ai), *Res* (x, x) = 1   (reflexivity)
(b) for all x, y in Dom (Ai), *Res* (x, y) = *Res* (y, x)   (symmetry)

3 Semantic Measure of Fuzzy Data

In [7], the notions of semantic inclusion degree and semantic equivalence degree were proposed for the measure of fuzzy data. For two fuzzy data π_A and π_B, semantic inclusion degree SID (π_A, π_B) denotes the degree that π_A semantically includes π_B and semantic equivalence degree SED (π_A, π_B) denote the degree that π_A and π_B are equivalent to each other. Based on possibility distribution and resemblance relation, the definitions of calculating the semantic inclusion degree and the semantic equivalence degree of two fuzzy data are given as follows.

Definition: Let $U = \{u_1, u_2, ..., u_n\}$ be the universe of discourse. Let π_A and π_B be two fuzzy data on U based on possibility distribution and $\pi_A(u_i)$, $u_i \in U$, denote the possibility that u_i is true. Let *Res* be a resemblance relation on domain U and α ($0 \leq \alpha \leq 1$) be a threshold corresponding to *Res*. Then

$$SID_\alpha(\pi_A, \pi_B) = \sum_{i=1}^{n} \min_{u_i, u_j \in U \text{ and } Res_U(u_i, u_j) \geq \alpha} (\pi_B(u_i), \pi_A(u_j)) / \sum_{i=1}^{n} \pi_B(u_i)$$

$$SED_\alpha(\pi_A, \pi_B) = \min(SID_\alpha(\pi_A, \pi_B), SID_\alpha(\pi_B, \pi_A))$$

Example. Let π_1 = {1.0/a, 0.95/b, 0.9/c} and π_1 = {0.95/a, 0.9/b, 1.0r/d, 0.3/e} be two fuzzy data on domain U = {a, b, c, d, e, f} and let *Res* be a resemblance relation on U given in Figure 1. Let threshold $\alpha = 0.9$. Then

SID (π_1, π_2) = {0.95 + 0.9 + 0.9}/{0.95 + 0.9 + 1.0 + 0.3} = 0.873,

SID (π_2, π_1) = {0.95 + 0.9 + 0.9}/{1.0 + 0.95 + 0.9} = 0.965,

and thus

SED (π_1, π_2) = min (SID (π_1, π_2), SID (π_2, π_1)) = min (0.873, 0.965) = 0.873.

Res	a	b	c	d	e	f
a	1.0	0.1	0.4	0.3	0.1	0.1
b		1.0	0.2	0.3	0.2	0.2
c			1.0	0.95	0.5	0.3
d				1.0	0.3	0.1
e					1.0	0.4
f						1.0

Fig. 1. A Resemblance Relation

Now Let us consider the following two particular cases, namely, between π_3 = {1.0/a, 1.0/b, 1.0/c} and π_3 = {1.0/a, 1.0/b, 1.0/d} and between π_5 = {0.9/a, 1.0/b, 0.8/c} and π_6 = {0.3/a, 0.4/b, 0.2/d}. In the similar way above, we have

SID (π_3, π_4) = {1.0 + 1.0 + 1.0}/{1.0 + 1.0 + 1.0} = 1.0,

SID (π_4, π_3) = {1.0 + 1.0 + 1.0}/{1.0 + 1.0 + 1.0} = 1.0,

SID (π_5, π_6) = {0.3 + 0.4 + 0.2}/{0.3 + 0.4 + 0.2} = 1.0,

SID (π_6, π_5) = {0.3 + 0.4 + 0.2}/{0.9 + 1.0 + 0.8} = 0.333,

and thus

SE (π_3, π_4) = min (SID (π_3, π_4), SID (π_4, π_3)) = min (1.0, 1.0) = 1.0;

SE (π_5, π_6) = min (SID (π_5, π_6), SID (π_6, π_5)) = min (1.0, 0.333) = 0.333.

It follows that π_5 semantically includes π_6 whereas π_6 does not include π_5.

The notion of the semantic inclusion (or equivalence) degree of attribute values can be extended to the semantic equivalence degree of tuples. Let t_i = <a_{i1}, a_{i2}, …, a_{in}> and t_j = <a_{j1}, a_{j2}, …, a_{jn}> be two tuples in fuzzy relational instance r over schema R (A_1, A_2, …, A_n). The semantic inclusion degree of tuples t_i and t_j is denoted

$$SID_\alpha (t_i, t_j) = \min \{SID_\alpha (t_i [A_1], t_j [A_1]), SID_\alpha (t_i [A_2], t_j [A_2]), \ldots, SID_\alpha (t_i [A_n], t_j [A_n])\}.$$

The semantic equivalence degree of tuples t_i and t_j is denoted

$$SED_\alpha (t_i, t_j) = \min \{SED_\alpha (t_i [A_1], t_j [A_1]), SED_\alpha (t_i [A_2], t_j [A_2]), \ldots, SED_\alpha (t_i [A_n], t_j [A_n])\}.$$

4 Fuzzy Functional Dependency in Fuzzy Relational Databases

Functional dependencies (*FDs*), being one of the most important data dependencies in relational databases, express the dependency relationships among attribute values in relation [5]. In classical relational databases, functional dependencies can be defined as follows [9].

Definition: For a classical relation r (R), in which R denotes the set of attributes, we say r satisfies the functional dependency $FD: X \rightarrow Y$ where $XY \subseteq R$ if

$$(\forall t \in r)(\forall s \in r)(t[X] = s[X] \Rightarrow t[Y] = s[Y]).$$

For a $FD: X \rightarrow Y$, $t[X] = s[X]$ implies $t[Y] = s[Y]$. Such knowledge exists in relational databases and can easily be mined from precise databases when we do not know it. However, how to discover such knowledge in fuzzy relational databases is difficult.

Fuzzy functional dependencies (*FFD*) can reflexively represent the dependency relationships among attribute values in fuzzy relations such as "the salary almost dependents on the job position and experience" and have been extensively discussed in the context of fuzzy relational databases [9, 6, 11]. The following gives the definition of fuzzy functional dependency according to the related work.

Definition: For a relation instance r (R), where R denotes the schema, its attribute set is denoted by U, and X, Y $\subseteq$ U, we say r satisfies the *fuzzy functional dependency* $FFD: X \hookrightarrow Y$, if

$$(\forall t \in r)(\forall s \in r)(SED_\alpha(t[X], s[X]) \leq SED_\alpha(t[Y], s[Y]))$$

Example. Consider a fuzzy relation instance r in Table 1. Assume that attribute domains Dom (X) = {a, b, c, d, e} and Dom (Y) = {f, g, h, i, j}. There are two resemblance relations Res (X) and Res (Y) on X and Y shown in Fig. 2 and Fig. 3, respectively. Let two thresholds on Res (X) and Res (Y) be $\alpha_1 = 0.90$ and $\alpha_2 = 0.95$, respectively.

Table 1. Fuzzy relation instance r

K	X	Y
1001	{0.7/a, 0.4/b, 0.5/d}	{0.9/f, 0.6/g, 1.0./h}
1002	{0.5/a, 0.4/c, 0.8/d}	{0.6/g, 0.9/h, 0.9/i}
1003	{0.3/d, 0.8/e}	{0.6/h, 0.4/i, 0.1/j}

Since SE (t [X], s [X]) = min (SID (t [X], s [X]), SID (s [X], t [X])) = min (0.824, 0.875) = 0.824 and SE (t [Y], s [Y]) = min (SID (t [Y], s [Y]), SID (s [Y], t [Y])) = min (1.0, 0.96) =0.96, so SE (t [X], s [X]) $\leq$ SE (t [Y], s [Y]) is true. Similarly, we have SE (t [X], u [X]) $\leq$ SE (t [Y], u [Y]) and SE (s [X], u [X]) $\leq$ SE (s [Y], u [Y]). Hence *FFD*: $X \hookrightarrow Y$ holds in r.

Res (X)	a	b	c	d	e
a	1.0	0.2	0.3	0.2	0.4
b		1.0	0.92	0.4	0.1
c			1.0	0.1	0.3
d				1.0	0.2
e					1.0

Fig. 2. Resemblance relation *Res* (X) on X

Res (Y)	f	g	h	i	j
f	1.0	0.3	0.2	0.96	0.2
g		1.0	0.4	0.2	0.3
h			1.0	0.3	0.1
i				1.0	0.4
j					1.0

Fig. 3. Resemblance relation *Res* (Y) on Y

Fuzzy functional dependencies have the following properties.

Proposition: A classical functional dependency *FD* satisfies the definition of *FFD*.

Proof: Let *FD*: $X \rightarrow Y$ hold in *r* (*R*). Then for $\forall\ t \in r$ and $\forall\ s \in r$, $t\ [X] = s\ [X] \Rightarrow t\ [Y] = s\ [Y]$, and *SED* ($t\ [X]$, $s\ [X]$) = *SED* ($t\ [Y]$, $s\ [Y]$) = 1.

5 Refining Imprecise Query Answers

Approximate reasoning refers to such an inferring process that an object *S* (the source) has properties *P* and *Q*, respectively, and another object *T* (the target) shares the property *P*, then we can infer that *T* may also possess the property *Q*. In other words, *S* and *T* have the same property Q if *S* and *T* are matched on *P* to each other. This process can be expressed by the following logical rule.

IF *P* (*S*) and *Q* (*S*) and *P* (*T*) **THEN** *Q* (*T*)

It is clear that there are two issues involved in approximate reasoning. One is that which property *P* can determine the property *Q*. Another is that which object *S* has the same property as the object *T* on property *P* such that *T* has the same property as *S* on property *Q*.

In order to use approximate reasoning to refine the query answers for the relational databases with imprecise information, we need to determine which property *P* imprecisely determines the property *Q* and which object *S* has the highest match degree with the object *T* on property *P* such that *T* has the same property *Q* as *S*. In relational databases, the relationship between properties *P* and *Q* of approximate reasoning is essentially the constraints of functional dependency between P and Q, where P and Q can be seen as attributes. Consider that the associated relationship between P and Q may be fuzzy and the match between S and T on P is generally fuzzy. The approximate reasoning under such conditions is called *fuzzy approximate reasoning*.

Let an fuzzy database relation with imprecise data be r (A_1, A_2, ..., A_n) which consists of tuples t_1, t_2, ..., t_m. Assume that we have an imprecise query answer t_m [A_n] and we would like to refine this query answer using approximate reasoning. For this purpose, first, we need to know which attribute A_i ($1 \leq i \leq n - 1$) can determine the attribute A_n. In other words, we have to mine such a rule of fuzzy functional dependency: $A_i \hookrightarrow A_n$ ($1 \leq i \leq n - 1$). Second, for the found attribute A_i, we need to know which tuple t_j ($1 \leq j \leq m - 1$) has the highest equivalent degree with the tuple t_m on attribute A_i. On the basis of that, we can replace t_m [A_n] with t_j [A_n].

5.1 Mining Fuzzy Functional Dependency

According to the definition of fuzzy functional dependency, the requirement SED (t_j [A_i], t_k [A_i]) $\leq SED$ (t_j [A_n], t_k [A_n]) ($1 \leq j, k \leq m - 1$) must be satisfied in a relation r (A_1, A_2, ..., A_n) with imprecise information if FFD: $A_i \hookrightarrow A_n$. In order to discover such attribute A_i from the relation r, we give the algorithm as follows.

```
For i := 1 to n – 1 do
    For j := 1 to m – 1 do
        For k := j + 1 to m – 1 do
            If SED (tj [Ai], tk [Ai]) > SED (tj [An], tk [An])) then loop next i;
        Next k;
    Next j;
    Output Ai;
Next i.
```

For any one output attribute A_i, FFD: $A_i \hookrightarrow A_n$ must hold.

5.2 Measuring Equivalence Degree Between Attribute Values

Assume that FFD: $A_i \hookrightarrow A_n$ holds. Then we should find a tuple t_j ($1 \leq j \leq m - 1$) from r such that for any tuple t_k ($1 \leq k \leq m - 1$ and $k \neq j$) in r, SED (t_j [A_i], t_m [A_i]) $\geq SED$ (t_k [A_i], t_m [A_i]). In the following we give the corresponding algorithm.

```
t := t1;
x := SED (t1 [Ai], tm [Ai]);
For j := 2 to m – 1 do
    If SED (tj [Ai], tm [Ai]) > x then
        {
        x := SED (tj [Ai], tm [Ai]);
        t := tj
        };
Next j.
```

From the algorithm above, it can be seen that tuple t has the highest equivalence degree with tuple t_m on attribute A_i.

5.3 Refining Imprecise Query Answer

Following the idea of analogical reasoning discussed above, we can infer that t_m [A_n] should be equivalent to t [A_n] because FFD: $A_i \hookrightarrow A_n$ holds and t_m [A_i] is equivalent to t [A_i] with the highest equivalence degree. If t [A_n] is more precise data than t_m [A_n], we can replace query answer t_m [A_n] with t [A_n] as an approximate answer. Viewed from the result, the query answer is refined.

In order to refine the imprecise query answer, the simplest approach is to replace t_m [A_n] with t [A_n]. However the imprecise query answer t_m [A_n] can generally be compressed to be a more precise value than t [A_n] according to the constraints of fuzzy functional dependencies [14, 9]. In the following, we give the strategies for compressing imprecise value by using fuzzy functional dependencies. Assume the imprecise values are value intervals on the continua domain of attribute. With respect to the partial values on the discrete domain of attribute, the similar method can be used. Here the value intervals and the partial values can be regarded as the special fuzzy values, where the membership degree of each value in the intervals or the partial values is exactly 1.0. In [6], fuzzy values have been transformed into intervals.

Let t [A_i] = $f1$, t_m [A_i] = $f2$, t [A_n] = $g1$, and t_m [A_n] = $g2$. Moreover, $|f1|$ = p, $|f2|$ = q, $|g1|$ = u, $|g2|$ = v, $|g1 \cap g2|$ = k, and $|f1 \cap f2|$ = k', then $|g1 \cup g2|$ = u + v − k and $|f1 \cup f2|$ = p + q − k'. Here $|\eta|$ denotes the modular of value interval, and $|\eta|$ = $b - a$ if $\eta = [a, b]$ and $b \neq a$. Assume SED ($f1, f2$) > SED ($g1, g2$). It is clear that FFD: $A_i \hookrightarrow A_n$ does not hold and we should make SED ($f1, f2$) equal SED ($g1, g2$) after compression processing.

(a) If 0 < SED ($g1, g2$) < SED ($f1, f2$) = 1, g2 can be refined to make g1 and g2 closer in semantics. At this moment, g2 is compressed into $g2'$ = $g1 \cap g2$.

(b) If 0 < SED ($g1, g2$) < SED ($f1, f2$) < 1, we can compress g2 to satisfy FFD: $A_i \hookrightarrow A_n$.

- sub ($g1$) ≥ sub ($g2$) and sup ($g1$) ≥ sup ($g2$)

  Let $g2'$ = $Rstring$ ($g2$, x) such that SED ($g1, g2'$) = SED ($f1, f2$), where $|g1|$ = u, $|g2'|$ = x, $|g1 \cap g2'|$ = k, and $|g1 \cup g2'|$ = u + x − k. Hence x can be obtained uniquely. If x ≤ k, let x = k. When sub ($g1$) ≤ sub ($g2$) and sup ($g1$) ≤ sup ($g2$), we can apply the similar method to gain the unique x.

- sub ($g1$) ≥ sub ($g2$) and sup ($g1$) ≤ sup ($g2$)

  Let $g2'$ = $Rstring$ ($Lstring$ ($g2$, x), y) such that SED ($g1, g2'$) = SED ($f1, f2$) and (n−i)/(i−j) = (sup ($g2$) − sup ($g1$))/(sub ($g1$) − sub ($g2$)), where $|g1|$ = u, $|g2'|$ = y, $|g1 \cap g2'|$ = u, and $|g1 \cup g2'|$ = y. So x and y can be obtained uniquely. If x < y ≤ k, let x = y = k. When sub (g1) ≤ sub (g2) and sup (g1) ≥ sup (g2), we cannot refine g2.

Note that there may be another attribute A_j such that $A_j \hookrightarrow A_n$ (1 ≤ j ≤ n − 1). Under the situation that multiple attributes functionally determine A_n, we can repeat the processing procedure above to obtain the more precise query answers.

6 Summary

Imprecise data may appear in databases due to data unavailability. Fuzzy sets have been used to model imprecise data in the relational databases. Functional dependencies, which are one kind of integrity constraints in relational databases, not only play a critical role in a logical database design but also can be employ to reasoning, being knowledge.

In this paper, we discuss the issues on refining imprecise query answers for the relational databases with imprecise information by using analogical reasoning. In order to do that, we use the notion of equivalence degree of fuzzy data, by which we define the fuzzy functional dependencies in the fuzzy relational databases. On the basis of that, we give the approach to mining the rules of fuzzy functional dependencies and apply the frame of analogical reasoning to refine the imprecise query answer.

References

1. Buckles, B. P., Petry, F. E., A Fuzzy Representation of Data for Relational Database. *Fuzzy Sets and Systems*, 7 (3): 213-226, 1982.
2. Codd, E. F., Extending the Database Relational Model to Capture More Meaning. *ACM Transactions on Database Systems*, 4 (4): 397-434, 1979.
3. DeMichiel, L. G., Resolving Database Incompatibility: An Approach to Performing Relational Operations over Mismatched Domains. *IEEE Transactions on Knowledge and Data Engineering*, 1 (4): 485-493, 1989.
4. Dutta, S., Approximate Reasoning by Analogy to Null Queries. *International Journal of Approximate Reasoning*, 5: 373-398, 1991.
5. Hale, J., Shenoi, S., Analyzing FD Inference in Relational Databases. *Data and Knowledge Engineering*, 18: 167-183, 1996.
6. Liao, S. Y., Wang, H. Q., Liu, W. Y., Functional Dependencies with Null Values. Fuzzy Values, and Crisp Values, *IEEE Transactions on Fuzzy Systems*, 7 (1): 97-103, 1999.
7. Ma, Z. M., Zhang, W. J., Ma, W. Y., Semantic Measure of Fuzzy Data in Extended Possibility-based Fuzzy Relational Databases. *International Journal of Intelligent Systems*, 15 (8): 705-716, 2000.
8. Ma, Z. M., Zhang, W. J., Ma, W. Y., Mili, F., Data Dependencies in Extended Possibility-Based Fuzzy Relational Databases. *International Journal of Intelligent Systems*, 17 (3): 321-332, 2002.
9. Ma, Z. M., Zhang, W. J., Mili, F., Fuzzy Data Compression Based on Data Dependencies. *International Journal of Intelligent Systems*, 17 (4): 409-426, 2002.
10. Prade, H., Testemale, C., Generalizing Database Relational Algebra for the Treatment of Incomplete or Uncertain Information and Vague Queries. *Information Sciences*, 34: 115-143, 1984.
11. Raju, K. V. S. V. N., Majumdar, A. K., Fuzzy Functional Dependencies and Lossless Join Decomposition of Fuzzy Relational Database Systems. *ACM Transactions on Database Systems*, 13 (2): 129-166, 1988.
12. Rundensteiner, E. A., Hawkes, L. W., Bandler, W., On Nearness Measures in Fuzzy Relational Data Models. *International Journal of Approximate Reasoning*, 3:267-98, 1989.
13. Shenoi, S., Melton, A., Proximity Relations in the Fuzzy Relational Databases. *Fuzzy Sets and Systems*, 31 (3): 285-296, 1989.
14. Tseng, F. S. C., Chen, A. L. P., Yang, W. P., Refining Imprecise Data by Integrity Constraints. *Data and Knowledge Engineering*, 11 (3): 299-316, 1993.
15. Zadeh, L. A., Fuzzy Sets. *Information and Control*, 8 (3): 338-353, 1965.
16. Zadeh, L. A., Fuzzy Sets as a Basis for a Theory of Possibility. *Fuzzy Sets and Systems*, 1 (1): 3-28, 1978.

A Linguistic Truth-Valued Uncertainty Reasoning Model Based on Lattice-Valued Logic

Shuwei Chen, Yang Xu, and Jun Ma

Department of Mathematics, Southwest Jiaotong University,
Chengdu 610031, Sichuan, P.R. China
chensw915@163.com

Abstract. The subject of this work is to establish a mathematical framework that provide the basis and tool for uncertainty reasoning based on linguistic information. This paper focuses on a flexible and realistic approach, i.e., the use of linguistic terms, specially, the symbolic approach acts by direct computation on linguistic terms. An algebra model with linguistic terms, which is based on a logical algebraic structure, i.e., lattice implication algebra, is applied to represent imprecise information and deals with both comparable and incomparable linguistic terms (i.e., non-ordered linguistic terms). Within this framework, some inferential rules are analyzed and extended to deal with these kinds of lattice-valued linguistic information.

1 Introduction

One of the fundamental goals of artificial intelligence (AI) is to build artificially computer- based systems which make computer simulate, extend and expand human's intelligence and empower computers to perform tasks which are routinely performed by human beings. Due to the fact that human intelligence actions are always involved with uncertainty information processing, one important task of AI is to study how to make the computer simulate human being to deal with uncertainty information. Among major ways in which human being deal with uncertainty information, the uncertainty reasoning becomes an essential mechanism in AI.

In real uncertainty reasoning problem, most information, which are always propositions with truth-values, can be very qualitative in nature, i.e., described in natural language. Usually, in a quantitative setting the information is expressed by means of numerical values. However, when we work in a qualitative setting, that is, with vague or imprecise knowledge, this cannot be estimated with an exact numerical value. Then, it may be more realistic to use linguistic truth-values instead of numerical values.

Since 1990, there have been some important conclusions on inference with linguistic terms. In 1990, Ho [3] constructed a distributive lattice-Hedge algebra, which can be used to deal with linguistic terms. He [4] gave a measure function between two linguistic terms, and obtained the fuzzy inference theory and method, which based on linguistic term Hedge algebra. In 1996, Zadeh [17] discussed the formalization of some words and proposition of natural language, and given the standard form of language propositional and production ruler, and he discussed the linguistic terms fuzzy inference based on the fuzzy sets theory and fuzzy inference method. In 1998 and 1999, Turksen [9,10] studied the formalization and inference of descriptive words, substantive words and declarative sentence. In 2003, Xu [5], and in 2004 Pei [8] proposed a kind of simple lattice implication algebra with linguistic truth-values.

Based on the symbolic approaches, a linguistic truth-valued algebra model, which is based on a logical algebraic structure, i.e., lattice implication algebra, is applied to represent imprecise information and deals with both comparable and incomparable linguistic terms (i.e., non-ordered linguistic values). Within this framework, some inferential rules are analyzed and extended to deal with these kinds of lattice-value linguistic information.

The paper is organized as follows: Section 2 analyzes the structure of lattice implication algebras with linguistic terms. Based on it, a linguistic truth-valued uncertainty reasoning model based on lattice-valued logic is proposed in Section 3 with an illustration. Section 4 comes to the conclusion.

2 Lattice Implication Algebra with Linguistic Terms

2.1 Lattice Structure and Lattice Implication Algebra

Lattice structures [1] have been successfully applied to many fields, such as reliability theory, rough theory, knowledge representation and inference etc. Among them, the introduction of L-fuzzy sets by Goguen [2] provides a general framework for Zadeh's fuzzy set theory.

Two important cases of L are of interest and often being used: when L is a finite simple ordered set; and when L is the unit interval [0, 1]. More general, L should be a lattice with suitable operations like $\wedge$, $\vee$, $\rightarrow$, $\prime$. The question of the appropriate operation and lattice structure has generated much literature. Goguen [2] established L-fuzzy logic of which truth value set is a complete lattice-ordered monoid, also called a complete residuated lattice in Pavelka and Novak's L-fuzzy logic [7,6]. Since this algebraic structure is quite general, we specify the algebraic structure to lattice implication algebras introduced by Xu [11], which was established by combining lattice and implication algebra with the attempt to model and deal with the comparable and incomparable information. There have been much work about lattice implication algebra, as well as the corresponding lattice valued logic system, lattice-valued reasoning theory and methods [11,12,13,16]. The lattice implication algebra is defined axiomatically as:

Definition 1. [11] *Let $(L, \vee, \wedge, O, I)$ be a bounded lattice with an order- reversing involution $\prime$, I and O the greatest and the smallest element of L respectively, and $\rightarrow: L \times L \rightarrow L$ be a mapping. $(L, \vee, \wedge, O, I)$ is called a lattice implication algebra (LIA) if the following conditions hold for any $x, y, z \in L$:*

(I_1) $x \to (y \to z) = y \to (x \to z)$;
(I_2) $x \to x = I$;
(I_3) $x \to y = y' \to x'$;
(I_4) $x \to y = y \to x = I$ $\quad implies \quad x = y$;
(I_5) $(x \to y) \to y = (y \to x) \to x$;
(l_1) $(x \vee y) \to z = (x \to z) \wedge (y \to z)$;
(l_2) $(x \wedge y) \to z = (x \to z) \vee (y \to z)$.

In the following, LIA structure will be used to characterize the set of linguistic terms.

2.2 Lattice Implication Algebra with Linguistic Terms

A linguistic term differs from a numerical one in that its values are not numbers, but words or sentences in a natural or artificial language. On the other hand, these words, in different natural language, seem difficult to distinguish their boundary sometime, but their meaning of common usage can be understood. A simple structure of linguistic terms is given in Fig. 1.

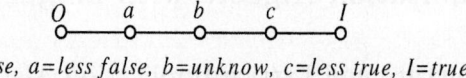

O=false, a=less false, b=unknow, c=less true, I=true

Fig. 1. Linguistic terms in a linear ordering

Note that there are some "vague overlap districts" among some words which cannot be strictly linearly ordered, a more general structure of linguistic terms [8] is given as in Fig. 2.

Note that c=less true, a=less false are incomparable. You can not collapse that structure into a linearly ordered structure, because then you would impose an ordering on a and c which was originally not present. It means the set of linguistic values may not be strictly linearly ordered. Moreover, linguistic terms can be ordered by their meanings in natural language. Naturally, it should be suitable to represent the linguistic values by a partially ordered set or lattice.

According to the feature of linguistic variables, we need to find some suitable algebras to characterize the values of linguistic variables. There is a good basic to believe that this approach would provide us with simple algorithms for reasoning. To attain this goal we characterize the set of linguistic truth-values by a LIA structure, i.e., use LIA to construct the structure of value sets of linguistic variables.

In general, the value of a linguistic variable can be a linguistic expression involving a set of linguistic terms such as "*high*," "*middle*," and "*low*," modifiers such as "*very*," "*more or less*" (called hedges [3]) and connectives (e.g., "*and*," "*or*"). Let us consider the domain of the linguistic variable "*truth*": domain (*truth*)={*true, false, very true, more or less true, possibly true, very false,* ...},

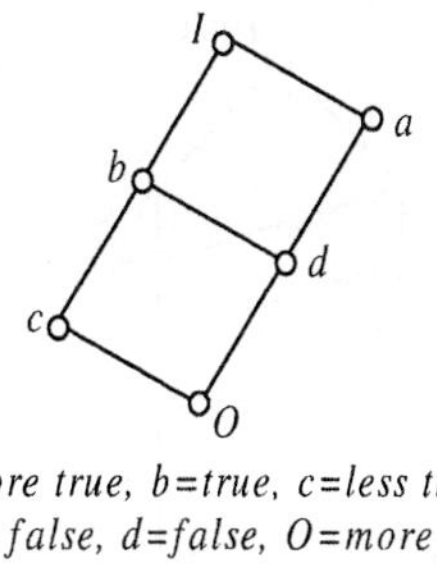

*l=more true, b=true, c=less true,
a=less false, d=false, O=more false*

Fig. 2. Linguistic terms in a nonlinear ordering

which can be regarded as a partially ordered set whose elements are ordered by their meanings and also regarded as an algebraically generated set from the generators $G=\{true, false\}$ by means of a set of linguistic modifiers $M=\{very, more\ or\ less, possibly, ...\}$. The linguistic modifiers [3] play a vital role in the representation of fuzzy sets. They are strictly related to the notion of vague concept. The generators G can be regarded as the prime terms, different prime terms correspond to the different linguistic variables.

Taking into account the above remarks, construction of an appropriate set of linguistic values for an application can be carried out step by step. Consider a set of linguistic hedges, $H=\{absolutely, highly, very, quite, exactly, almost, rather, somewhat, slightly\}$, which is selected by a careful and detailed investigation and analysis. H can be naturally ordered by the degree of strengthening or weakening the meaning of the primary terms: $true$ and $false$. We say that $a \leq b$ iff a(True)$\leq b$(True) in the natural language, where a and b are linguistic hedges.

Applying the hedges of H to the primary term "true" or "false" we obtain a partially ordered set or lattice. For example, as represented in Fig. 3, we can obtain a lattice generated from "*true*" or "*false*" by means of modifiers in H. The set of linguistic truth-values obtained by the above procedure is a lattice with the boundary, denoted by $L_{18}=\{I=absolutely\ true, a=highly\ true, b=very\ true, c=quite\ true, d=exactly\ true, e=almost\ true, f=rather\ true, g=somewhat\ true, h=slightly\ true, O=absolutely\ false, t=highly\ false, s=very\ false, r=quite\ false, q=exactly\ false, p=almost\ false, n=rather\ false, m=somewhat\ false, l=slightly\ false\}$. Moreover, one can define $\wedge, \vee$, implication $\rightarrow$ and complement operation $\prime$ on this lattice according to the LIA structure. The operations $\wedge$ and $\vee$ can be shown in the Hasse diagram Fig. 3. The complement operation is given in Table 1 according to the lattice implication algebra structure and properties respectively (the table of implication operation is not given here for its large size):

As we can show that $(L_{18}, \vee, \wedge, \prime, \rightarrow, O, I)$ constructed above is a lattice implication algebra, all the properties of lattice implication algebra hold in L. So the finite set of linguistic terms can be characterized by a finite lattice implication algebra.

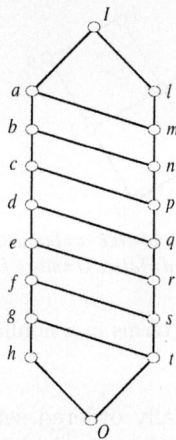

Fig. 3. The lattice-ordering structure of linguistic terms with 18 elements

Table 1. Complement operator of L_{18}

′	O	t	s	r	q	p	n	m	l	h	g	f	e	d	c	b	a	I
	I	g	f	e	d	c	b	a	h	l	t	s	r	q	p	n	m	O

3 Linguistic Truth-Valued Uncertainty Reasoning Model Based on Lattice-Valued Logic

Uncertainty reasoning can be seen as the basis of intelligent control, decision making, etc. From the viewpoint of symbolism, it is highly necessary to study and establish the logical foundation for uncertainty reasoning. In order to provide a kind of logical foundation for uncertainty reasoning and automated reasoning, Xu and his colleagues established some kinds of logical systems, especially the gradational lattice-valued propositional logic L_{vpl} [12] and the gradational lattice-valued first-order logic L_{vfl} [14,15]. Based on the lattice-valued logic, some uncertainty reasoning and automated reasoning theory and methods are also investigated [13,16].

Here we take the gradational lattice-valued propositional logic L_{vpl} as our framework for linguistic truth-valued uncertainty reasoning approach. We review some related results of L_{vpl}, detailed information can be found in [12,16].

Definition 2. [12] *Let $D_n \subseteq \mathscr{F}_p^n$, where $\mathscr{F}_p$ is the set of formulae of L_{vpl}. A mapping $r_n : D_n \longrightarrow \mathscr{F}_p$ is called an n-ary partial operation on $\mathscr{F}_p$, where D_n is the domain of r_n, also denoted by $D_n(r_n)$.*

Definition 3. [12] *A mapping $t_n : L^n \longrightarrow L$ is said to be an n-ary truth-valued operation on L, if*

(1). $\alpha \to t_n(\alpha_1, \cdots, \alpha_n) \geq t_n(\alpha \to \alpha_1, \cdots, \alpha \to \alpha_n)$ holds for any $\alpha \in L$ and $(\alpha_1, \cdots, \alpha_n) \in L^n$.
(2). t_n is isotone in each argument.

Denote

$$R_n \subseteq \{r_n \mid r_n \text{ is an n-ary partial operation on } \mathscr{F}_p\},$$
$$T_n \subseteq \{t_n \mid t_n \text{ is an n-ary truth-valued operation on } L\},$$
$$\mathscr{R}_n \subseteq R_n \times T_n, \text{ and } \mathscr{R} \subseteq \bigcup_{n=0}^{+\infty} \mathscr{R}_n.$$

If $(r,t) \in \mathscr{R}_n$, then (r,t) is called an n-ary rule of inference in L_{vpl}.

Remark 4. *It can be seen that the rules of inference in L_{vpl} are composed of two parts, one for the formal deduction of formulae, and the other for the transfer of the truth-values of the corresponding formulae. The following two definitions about α-i type closed and α-i type sound mean that the syntactical deduction and the semantic transfer should be consistent with some degree. In this sense, uncertainty reasoning can be brought into the framework of lattice-valued logic.*

Example 5. For any $p, q, g \in \mathscr{F}_p$, $\beta, \theta \in L$, define

$$r_2^0(p, p \to q) = q, \quad t_2^*(\theta, \beta) = \theta \wedge \beta,$$
$$r_2^\circ(p, q) = p \wedge q,$$
$$r_2^*(p \to g, p \to q) = p \to (g \wedge q),$$
$$r_2^\triangle(p \to q, q \to g) = p \to g,$$
$$r_2^\nabla(p, q) = (p \to q).$$

It can be check easily that (r_2^0, t_2^*), (r_2°, t_2^*), (r_2^*, t_2^*), $(r_2^\triangle, t_2^*)$ and (r_2^∇, t_2^*) are all 2-ary rule of inference in L_{vpl}.

In the following, we assume that $\mathscr{T} \subseteq \mathscr{F}_L(\mathscr{F}_p)$, $\mathscr{T}_h \subseteq \{T \mid T : \mathscr{F}_p \longrightarrow L$ is a homomorphic mapping $\} \triangleq \mathscr{T}_H$, where $\mathscr{F}_L(\mathscr{F}_p)$ is the set of all L-fuzzy subsets on $\mathscr{F}_p$.

Definition 6. [12] *Let $X \in \mathscr{F}_L(\mathscr{F}_p)$, $(r,t) \in \mathscr{R}_n$, $\alpha \in L$. If*

$$X \circ r \supseteq \alpha \otimes (t \circ \prod^n X) \tag{1}$$

holds in $D_n(r)$, then X is said to be α-I type closed w.r.t. (r,t). If

$$X \circ r \supseteq t \circ \prod^n (\alpha \otimes X) \tag{2}$$

holds in $D_n(r)$, then X is said to be α-II type closed w.r.t. (r,t).
If for any $(r,t) \in \mathscr{R}$, X is α-i type closed w.r.t. (r,t), then X is said to be α-i type closed w.r.t. $\mathscr{R}$, $i =$I, II.

Definition 7. [12] *Let $\alpha \in L$, $\mathscr{R}$ is said to be α-i type sound w.r.t. $\mathscr{T}$, if T is α-i type closed w.r.t. $\mathscr{R}$ holds for any $T \in \mathscr{T}$, $i = $ I, II.*

Proposition 8. *Let $\alpha, \theta \in L$, $i = $ I, II, $T \in \mathscr{T}_H$,*
(1). If $\alpha \leq \bigwedge_{\theta \in L}(\theta \vee \theta')$, then T is α-i type closed w.r.t. (r_2^0, t_2^*) and $(r_2^\triangle, t_2^*)$.
(2). T is α-i type closed w.r.t. $(r_2^\diamond, t_2^*)$, (r_2^*, t_2^*) and (r_2^∇, t_2^*).

The linguistic truth-valued uncertainty reasoning approach based on lattice-valued logic is described as follows:

Step 1. Transfer the real reasoning problem into formulae in L_{vpl} with linguistic truth-values, such as $(p, very\ true)$, $(p \rightarrow q, somewhat\ true)$, etc..
Step 2. Choose appropriate rules of inference in L_{vpl}.
Step 3. Use the selected rules of inference to get the reasoning consequence, which also takes the form of $(q, somewhat\ true)$.
Step 4. Transfer the obtained reasoning consequence into natural languages.

To illustrate how the proposed method works, we consider a simple example of deciding whether to buy some kind of car.

Example 9. *Suppose that we have the following rule:*

If the car is cheap, safe and comfortable, then David will highly want to buy it.

The above rule describe in natural language can be transferred into

$$(p_1, I) \wedge (p_2, I) \wedge (p_3, I) \rightarrow (q, a),$$

where $p_1 = $ "the car is cheap," $p_2 = $ "the car is safe," $p_3 = $ "the car is comfortable," $q = $ "David will buy it," $I = $ absolutely true, $a = $ highly true. By using $(r_2^\diamond, t_2^*)$ two times and (r_2^∇, t_2^*) once, we can transfer the above formula with linguistic truth-values into

$$((p_1 \wedge p_2 \wedge p_3) \rightarrow q, a).$$

Now, there is a car which is rather cheap, quite safe and very comfortable. Then what about David's decision? By taking a similar procedure, we can transfer the fact into

$$((p_1 \wedge p_2 \wedge p_3, f),$$

where $f = $ rather true. So, by applying (r_2^0, t_2^*), the reasoning consequence can be obtained as (q, f), which is interpreted in natural language as

David will rather want to buy this car.

Remark 10. *The above example can also be looked upon as a kind of evaluation or decision-making problem with uncertainties, and weights can also be added to the factors. This shows that the above method could also be applied to treat evaluation and decision-making problems.*

4 Conclusions

A linguistic truth-valued uncertainty reasoning approach based on lattice-valued logic was proposed in this paper. This method offers the advantage of not requiring the linguistic approximation step. Furthermore, it does not require the definition of the membership functions associated with the linguistic terms. Especially, we do use a finite set of linguistic terms with a rich lattice ordering algebraic structure to represent the truth-values of propositions. So this procedure has another advantage of handling incomparable linguistic terms and the logical operations to fulfill the transfer of truth-values accompanied the procedure of formal deduction. Note that lattice is a more universal structure than the set of linguistic terms, and the implication operation in LIA is much general richer, it would be reasonable and realistic to design uncertainty reasoning models based on these methodologies.

Acknowledgements

This work was partially supported by the National Natural Science Foundation of China (Grant No. 60474022).

References

1. G. Birkhoff, *Lattice Theory*, 3rd edition, American Mathematical Society, New York, 1967.
2. J. A. Goguen, "The logic of inexact concepts," *Synthese*, Vol. 19, pp. 325-373, 1968.
3. N. C. Ho, and W. Wechler, "Hedge algebras: an algebraic approach to structure of sets of linguistic truth values," *Fuzzy Sets and Systems*, Vol. 35, pp. 281-293, 1990.
4. N. C. Ho, and W. Wechler, "Extended hedge algebras and their application to fuzzy logic," *Fuzzy Sets and Systems*, Vol. 52, pp. 259-281, 1992.
5. T. T. Lee, and Y. Xu, "The consistencey of rule-bases in lattice-valued first-order logic LF(X)," *Proc. of 2003 IEEE Internatinal Conference on SMC*, pp. 4968-4973, 2003.
6. V. Novak, "First-order fuzzy logic," *Studia Logica*, Vol. 46, pp. 87-109, 1982.
7. J. Pavelka, "On fuzzy logic I: Many-valued rules of inference, II: Enriched residuated lattices and semantics of propositional calculi, III: Semantical completeness of some many-valued propositional calculi," *Zeitschr. F. Math. Logik und Grundlagend. Math.* , Vol. 25, pp. 45-52, 119-134, 447-464, 1979.
8. Z. Pei, and Y. Xu, "Lattice implication algebra model of a kind of linguistic terms and its inference," *Proc. of the 6th International FLINS Conference*, pp. 93-98, 2004.
9. I. B. Turksen, A. Kandel, and Y. Q. Zhang, "Universal truth tables and normal forms," *IEEE Trans. Fuzzy Syst.*, Vol. 6, pp. 295-303, 1998.
10. I. B. Turksen, "Computing with descriptive and verisic words," *Proc. NAFIP'99*, pp. 13-17, 1999.

11. Y. Xu, "Lattice implication algebras," *J. Southwest Jiaotong Univ.*, Vol. 89, pp. 20-27, 1993.
12. Y. Xu, K. Y. Qin, J. Liu, and Z. M. Song, "L-valued propositional logic L_{vpl}," *Inform. Sci.*, Vol. 114, pp. 205-235, 1999.
13. Y. Xu, D. Ruan, and J. Liu, "Approximate reasoning based on lattice-valued propositional logic L_{vpl}," in: D. Ruan and E. E. Kerre (Eds.) *Fuzzy Sets Theory and Applications*. Kluwer Academic Publishers, pp. 81-105, 2000.
14. Y. Xu, J. Liu, Z. M. Song, and K. Y. Qin, "On sematics of L-valued first-order logic L_{vfl}," *Int. J. Gen. Syst.*, Vol. 29, pp. 53-79, 2000.
15. Y. Xu, Z. M. Song, K. Y. Qin, and J. Liu, "Syntax of L-valued first-order logic L_{vfl}," *Int. J. Mutiple-Valued Logic*, Vol. 7, pp. 213-257, 2001.
16. Y. Xu, D. Ruan, K. Y. Qin, and J. Liu, *Lattice-Valued Logic*, Springer-Verlag, Berlin, 2003.
17. L. A. Zadeh, "Fuzzy logic = computing with words," *IEEE Trans. Fuzzy Syst.*, Vol. 4, pp. 103-111, 1996.

Fuzzy Programming Model for Lot Sizing Production Planning Problem

Weizhen Yan[1], Jianhua Zhao[2], and Zhe Cao[3]

[1] Institute of Systems Engineering, Tianjin University, Tianjin 300072, China
[2] Department of Mathematics, Shijiazhuang College, Shijiazhuang 050801, China
[3] Department of Mathematical Sciences, Tsinghua University, Beijing 100084, China

Abstract. This paper investigates lot sizing production planning problem with fuzzy unit profits, fuzzy capacities and fuzzy demands. First, the fuzzy production planning problem is formulated as a credibility measure based fuzzy programming model. Second, the crisp equivalent model is derived when the fuzzy parameters are characterized by trapezoidal fuzzy numbers. Then a fuzzy simulation-based genetic algorithm is designed for solving the proposed fuzzy programming model as well as its crisp equivalent. Finally, a numerical example is provided for illustrating the effectiveness of algorithm.

1 Introduction

As an important topic in operations research, production planning problem has been attracting attentions from both the researchers and enterprizes. In traditional production planning problems, it is assumed that the system parameters are deterministic, and the objectives and constraints are crisp. However, in reality, uncertainty is unavoidably involved in a production planning problem, which makes the objectives and constraints uncertain. In order to deal with the uncertain environment and arrange the production planning properly, an enterprize has to take into account the uncertainties involved so that it can retain a satisfactory profit level.

Recent years viewed increasing interests in the problem of production planning in fuzzy environments. In [1], Dong *et al.* presented a possibility based chance- constrained programming model for fuzzy production planning problem. In [12], Schnidis discussed the fuzzy fuzzy production planning problem by tolerance method, which was presented by Zimmermann [18]. In [13,14,15,16], fuzzy production planning problems were modelled by parametric programming, in which the possibility that the fuzzy constraints holds is described by fuzzy membership function. As far as we know, the study of fuzzy production problem is based on the fuzzy membership function and possibility that fuzzy events hold.

Recently, Liu and Liu [9] proposed the credibility measure and the credibility measure based fuzzy programming models. The purpose of this paper is to formulate the fuzzy production planning problem by credibility measure based chance-constrained programming model. Then based on the credibility theory, we derive the crisp equivalent when the fuzzy parameters are characterized by trapezoidal fuzzy numbers. Moreover, a hybrid intelligent algorithm by integrating fuzzy simulation and genetic algorithm is designed for solving the fuzzy programming model as well as its crisp equivalent model. Finally, a numerical example is provided to illustrate the effectiveness of the hybrid intelligent algorithm.

2 Preliminaries

Possibility theory [2][17] is based on a pair of dual fuzzy measures—possibility measure and necessity measure. Let Θ be a nonempty set, $\mathcal{P}(\Theta)$ the power set of Θ, and Pos a set function defined on $\mathcal{P}(\Theta)$. Pos is said to be possibility measure, if Pos satisfies the following conditions (i) $\text{Pos}(\emptyset) = 0$, $\text{Pos}(\Theta) = 1$; (ii) $\text{Pos}(\cup_k A_k) = \sup_k \text{Pos}(A_k)$ for any arbitrary collection $\{A_k\}$ in $\mathcal{P}(\Theta)$. While the necessity measure Nec is defined by

$$\text{Nec}\{A\} = 1 - \text{Pos}\{A^c\} \text{ for all } A \in \mathcal{P}(\Theta).$$

Recently, the third set function Cr, called credibility measure, was introduced by Liu and Liu [9] as follows

$$\text{Cr}(A) = \frac{1}{2}\left(\text{Pos}(A) + \text{Nec}(A)\right).$$

It is easy to verify that Cr has the following properties:

(i) $\text{Cr}(\emptyset) = 0$, $\text{Cr}(\Theta) = 1$;
(ii) $\text{Cr}(A) \leq \text{Cr}(B)$ whenever $A, B \in \mathcal{P}(\Theta)$, and $A \subset B$;
(iii) $\text{Pos}(A) \geq \text{Cr}(A) \geq \text{Nec}(A)$ for all $A \in \mathcal{P}(\Theta)$;
(iv) $\text{Cr}(A) + \text{Cr}(A^c) = 1$ for all $A \in \mathcal{P}(\Theta)$.

Remark 1. (Gao and Liu [3])A fuzzy event does not hold necessarily even though its possibility achieves 1, and may hold possibly even though its necessity is 0. However, if the credibility of a fuzzy event is 1, then it holds certainly. On the other hand, if the credibility of fuzzy event is 0, then it does not hold certainly.

3 Problem Formulation

Consider an enterprize that manufactures and supplies N types of products to customers in T production planning periods. Let ξ_i be profit per unit product i, η_i be the inventory cost per unit product i, x_{it} be the production quantity in the production period t, y_{it} be the inventory level at production period t, $i = 1, 2, \cdots, N$, $t = 1, 2, \cdots, T$. Then the total profit from the sales of all products in T periods is $\sum_{i=1}^{N}\sum_{t=1}^{T}\xi_i x_{it}$, and the total inventory cost in T periods is $\sum_{i=1}^{N}\sum_{t=1}^{T}h_i y_{it}$. Therefore, the net profit of the enterprize is

$$\sum_{i=1}^{N}\sum_{t=1}^{T}(\xi_i x_{it} - h_i y_{it}).$$

Suppose that the production consists of K working procedures. Let w_{ij} be the production capacity required by product i in working procedure j, η_{jt} the production capacity of working procedure j in period t. Then we have

$$\sum_{i=1}^{n} w_{ij} x_{it} \leq \eta_{jt}, \quad j = 1, 2, \cdots, K, \quad t = 1, 2, \cdots, T. \tag{1}$$

That is, in each period, the should not exceed the production capacity of the enterprize.

Let the initial inventory level be zero, and d_{it} the custom demand quantity of product i in period t. Then we have

$$x_{it} + y_{i(t-1)} - y_{it} \geq d_{it}, \quad y_{i0} = 0, \quad i = 1, 2, \cdots, N, \quad t = 1, 2, \cdots, T. \tag{2}$$

That is, in each period, the customer demand should be completely satisfied with out delays.

Suppose that the enterprize pursues the total profit by making decisions on the production quantities x_{it} and inventory levels y_{it} with production capacity constraints (1) and demand constraints (2) in the planning horizon. Then we have the following mathematical model:

$$\begin{cases} \max \sum_{i=1}^{N}\sum_{t=1}^{T}(\xi_i x_{it} - h_i y_{it}) \\ \text{s.t.} \sum_{i=1}^{N} w_{ij} x_{it} \leq \eta_{jt}, \quad j = 1, 2, \cdots, K, \quad t = 1, 2, \cdots, T \\ x_{it} + y_{i(t-1)} - y_{it} \geq d_{it}, \quad i = 1, 2, \cdots, N, \quad t = 1, 2, \cdots, T \\ x_{it} \geq 0, y_{it} \geq 0, \quad y_{i0} = 0, \quad i = 1, 2, \cdots, N, \quad t = 1, 2, \cdots, T, \end{cases} \tag{3}$$

In practice, demands, profit per unit production and production capacity are often subject to fluctuations and difficult to measure. Certainly, we may neglect the uncertainties and solve the problem via deterministic models. However, this simplification may be quite costly because, sometimes, the optimal solution may even be infeasible with respect to the realization of uncertain parameters. Hence, it is necessary for us to take into account explicitly the range of the possible realizations of uncertain parameters, and formulate the production planning problem under uncertainty as uncertain models. Assume that the profit of per unit production i ξ_i, customer demands d_{it} and production capacity η_{jt} are characterized by fuzzy variables. Then model (3) is meaningless. In the following, we will use fuzzy programming to model the fuzzy production planning problem.

Let

$$\boldsymbol{x} = (x_{11}, \cdots, x_{1T}, x_{21}, \cdots, x_{2T}, \cdots, x_{N1}, \cdots, x_{NT}),$$
$$\boldsymbol{y} = (y_{11}, \cdots, y_{1T}, y_{21}, \cdots, y_{2T}, \cdots, y_{N1}, \cdots, y_{NT}),$$

and
$$\boldsymbol{\xi} = (\xi_1, \cdots, \xi_N).$$

Since there fuzzy parameters, the total profit

$$f(\boldsymbol{x}, \boldsymbol{y}, \boldsymbol{\xi}) = \sum_{i=1}^{N} \sum_{t=1}^{T} (\xi_i x_{it} - h_i y_{it})$$

is fuzzy, too. In many cases, the enterprize may give a profit level, and wish to maximize the chance that the fuzzy total profit is more than the predetermined profit level. Assume that the profit level is set as T_0 by the enterprize. Then we have the objective of the enterprize as follows:

$$\max \mathrm{Cr}\{f(\boldsymbol{x}, \boldsymbol{y}, \boldsymbol{\xi}) \geq T_0\}.$$

Moreover, due to the fuzziness of the production capacity, customer demands, the constraint functions are also fuzzy, and cannot define the feasible set crisply. In practice, the enterprize often gives certain confidence levels, and regards a solution feasible when the constraint functions hold at the predetermined confidence levels. That is, the feasible set is defined by inequalities

$$\mathrm{Cr}\left\{\sum_{i=1}^{N} w_{ij} x_{it} \leq \eta_{jt}\right\} \geq \alpha, \quad j = 1, 2, \cdots, K, \quad t = 1, 2, \cdots, T$$

and

$$\mathrm{Cr}\{x_{it} + y_{i(t-1)} - y_{it} \geq d_{it}\} \geq \beta, \quad y_{i0} = 0, \quad i = 1, 2, \cdots, N, \quad t = 1, 2, \cdots, T$$

where α and β are confidence levels given by the enterprize. Then we have the following fuzzy programming model for the fuzzy production planning problem:

$$\begin{cases} \max \ \mathrm{Cr}\{f(\boldsymbol{x}, \boldsymbol{y}, \boldsymbol{\xi}) \geq T_0\} \\ \mathrm{s.t.} \\ \quad \mathrm{Cr}\left\{\sum_{i=1}^{N} w_{ij} x_{it} \leq \eta_{jt}\right\} \geq \alpha, \quad j = 1, 2, \cdots, K, \quad t = 1, 2, \cdots, T \\ \quad \mathrm{Cr}\{x_{it} + y_{i(t-1)} - y_{it} \geq d_{it}\} \geq \beta, \quad i = 1, 2, \cdots, N, \quad t = 1, 2, \cdots, T \\ \quad x_{it} \geq 0, y_{it} \geq 0, \quad y_{i0} = 0, \quad i = 1, 2, \cdots, N, \quad t = 1, 2, \cdots, T. \end{cases} \quad (4)$$

4 Crisp Equivalents

In this section, we assume that the fuzzy parameters ξ_i, η_{jt} and d_{it} are all trapezoidal fuzzy numbers. Then we derive the crisp equivalent model of (4) based on credibility theory.

Lemma 1. *Let ξ be a trapezoidal fuzzy number (r_1, r_2, r_3, r_4). Then for any given confidence level $\alpha(> 0.5)$, we have*

$$\mathrm{Cr}\{\xi \geq r\} \geq \alpha \Longleftrightarrow r \leq K_\alpha,$$

where $K_\alpha = (2\alpha - 1)r_1 + 2(1 - \alpha)r_2$.

Proof. It follows from the definition of the credibility measure, we can get the credibility distribution function of ξ is

$$\mathrm{Cr}\{\xi \geq r\} = \begin{cases} 1, & \text{if } r \leq r_1 \\ \dfrac{r - 2r_2 + r_1}{2(r_1 - r_2)}, & \text{if } r_1 \leq r \leq r_2 \\ \dfrac{1}{2}, & \text{if } r_2 \leq r \leq r_3 \\ \dfrac{r - r_4}{2(r_3 - r_4)}, & \text{if } r_3 \leq r \leq r_4 \\ 0, & \text{otherwise,} \end{cases} \quad (5)$$

It is obvious that the function $\mathrm{Cr}\{\xi \geq r\}$ is continuous in r. If $\alpha > 1/2$ and $\mathrm{Cr}\{\xi \geq r\} \geq \alpha$, then we have $\mathrm{Cr}\{\xi \geq r\} = (r - 2r_2 + r_1)/(2r_1 - 2r_2) \geq \alpha$ or $r \leq r_1$. It is easy to verify that $r \leq (2\alpha - 1)r_1 + 2(1 - \alpha)r_2$. Conversely, if $r \leq r_1$, then we have $\mathrm{Cr}\{\xi \geq r\} = 1 \geq \alpha$. If $r \leq (2\alpha - 1)r_1 + 2(1 - \alpha)r_2$, then we have $\mathrm{Cr}\{\xi \geq r\} = (r - 2r_2 + r_1)/(2r_1 - 2r_2) \geq \alpha$. That is, the crisp equivalent of $\mathrm{Cr}\{\xi \geq r\} \geq \alpha$ is $r \leq (2\alpha - 1)r_1 + 2(1 - \alpha)r_2 = K_\alpha$. The lemma is proved.

Lemma 2. *Let ξ be a trapezoidal fuzzy number (r_1, r_2, r_3, r_4). Then for any given confidence levels $beta(> 0.5)$, we have*

$$\mathrm{Cr}\{\xi \leq r\} \geq \beta \Longleftrightarrow r \geq K_\beta,$$

where $K_\beta = 2(1 - \beta)r_3 + (2\beta - 1)r_4$.

Proof. The proof is similar to that of Lemma 1, and omitted here.

Assume ξ_i, η_{jt} and d_{it} to be independent trapezoidal fuzzy variables. Now we derive the crisp equivalent model of fuzzy programming model (4).

Theorem 1. *Let $\xi_i = (r_{i1}, r_{i2}, r_{i3}, r_{i4})$, $\eta_{jt} = (c_{jt1}, c_{jt2}, c_{jt3}, c_{jt4})$ and $d_{it} = (s_{it1}, s_{it2}, s_{it3}, s_{it4})$, where $i = 1, 2, \cdots, N$, $j = 1, 2, \cdots, K$ and $t = 1, 2, \cdots, T$. Then we have the crisp equivalent of fuzzy programming model (4) as follows:*

$$\begin{cases} \max g(\boldsymbol{x}, \boldsymbol{y}) \\ \text{s.t.} \\ \quad \displaystyle\sum_{i=1}^{N} w_{ij} x_{it} \leq (2\alpha - 1)c_{jt1} + 2(1 - \alpha)c_{jt2}, \\ \qquad\qquad\qquad\qquad j = 1, 2, \cdots, K, \quad t = 1, 2, \cdots, T \\ \quad x_{it} + y_{i(t-1)} - y_{it} \geq 2(1 - \beta)s_{it3} + (2\beta - 1)s_{it4}, \\ \qquad\qquad\qquad\qquad i = 1, 2, \cdots, N, \quad t = 1, 2, \cdots, T \\ \quad x_{it} \geq 0, \; y_{it} \geq 0, \; y_{i0} = 0, \; i = 1, 2, \cdots, N, \; t = 1, 2 \cdots, T, \end{cases} \quad (6)$$

where

$$g(\boldsymbol{x},\boldsymbol{y}) = \begin{cases} 1, & \text{if } T_0 \leq R_1(\boldsymbol{x},\boldsymbol{y}) \\ \dfrac{T_0 - 2R_2(\boldsymbol{x},\boldsymbol{y}) + R_1(\boldsymbol{x},\boldsymbol{y})}{2(R_1(\boldsymbol{x},\boldsymbol{y}) - R_2(\boldsymbol{x},\boldsymbol{y}))}, & \text{if } R_1(\boldsymbol{x},\boldsymbol{y}) \leq T_0 \leq R_2(\boldsymbol{x},\boldsymbol{y}) \\ \dfrac{1}{2}, & \text{if } R_2(\boldsymbol{x},\boldsymbol{y}) \leq T_0 \leq R_3(\boldsymbol{x},\boldsymbol{y}) \\ \dfrac{T_0 - R_4(\boldsymbol{x},\boldsymbol{y})}{2(R_3(\boldsymbol{x},\boldsymbol{y}) - R_4(\boldsymbol{x},\boldsymbol{y}))}, & \text{if } R_3(\boldsymbol{x},\boldsymbol{y}) \leq T_0 \leq R_4(\boldsymbol{x},\boldsymbol{y}) \\ 0, & \text{otherwise} \end{cases} \quad (7)$$

and

$$\begin{aligned} R_1(\boldsymbol{x},\boldsymbol{y}) &= \sum_{i=1}^{N}\sum_{t=1}^{T}(r_{i1}x_{it} - h_i y_{it}), \\ R_2(\boldsymbol{x},\boldsymbol{y}) &= \sum_{i=1}^{N}\sum_{t=1}^{T}(r_{i2}x_{it} - h_i y_{it}), \\ R_3(\boldsymbol{x},\boldsymbol{y}) &= \sum_{i=1}^{N}\sum_{t=1}^{T}(r_{i3}x_{it} - h_i y_{it}), \\ R_4(\boldsymbol{x},\boldsymbol{y}) &= \sum_{i=1}^{N}\sum_{t=1}^{T}(r_{i4}x_{it} - h_i y_{it}). \end{aligned} \quad (8)$$

Proof. Firstly, it follows the linearity of $f(\boldsymbol{x},\boldsymbol{y},\boldsymbol{\xi})$ in ξ_i and Extension Principle of Zadeh that $f(\boldsymbol{x},\boldsymbol{y},\boldsymbol{\xi})$ is a trapezoidal fuzzy variable, too. Let

$$f(\boldsymbol{x},\boldsymbol{y},\boldsymbol{\xi}) = (R_1(\boldsymbol{x},\boldsymbol{y}), R_2(\boldsymbol{x},\boldsymbol{y}), R_3(\boldsymbol{x},\boldsymbol{y}), R_4(\boldsymbol{x},\boldsymbol{y})), \quad (9)$$

where $R_1(\boldsymbol{x},\boldsymbol{y}), R_2(\boldsymbol{x},\boldsymbol{y}), R_3(\boldsymbol{x},\boldsymbol{y}), R_4(\boldsymbol{x},\boldsymbol{y})$ are defined by equation (8). Then it follows from equation (5) that

$$\mathrm{Cr}\{f(\boldsymbol{x},\boldsymbol{y},\boldsymbol{\xi}) \geq T_0\} = g(\boldsymbol{x},\boldsymbol{y}).$$

Secondly, it follows from lemma 2 that

$$\mathrm{Cr}\left\{\eta_{jt} \geq \sum_{i=1}^{N} w_{ij}x_{it}\right\} \geq \alpha \iff \sum_{i=1}^{N} w_{ij}x_{it} \leq (2\alpha-1)c_{jt1} + 2(1-\alpha)c_{jt2}.$$

Thirdly, it follows from lemma 1 that

$$\mathrm{Cr}\{d_{it} \leq x_{it}+y_{i(t-1)}-y_{it}\} \geq \beta \iff x_{it}+y_{i(t-1)}-y_{it} \geq 2(1-\beta)s_{it3}+(2\beta-1)s_{it4}.$$

Lastly, we can turn the fuzzy programming model (4) to its crisp equivalent model defined by (6). The proof is completed.

5 Simulation Based Genetic Algorithm

In general cases, the fuzzy parameters are not all trapezoidal fuzzy variables, and it is difficult for us to derive their crisp equivalent models. Now, we design a simulation based genetic algorithm for solving the proposed fuzzy programming model (4) as well as its equivalent model (6).

First, we have to cope with uncertain functions (functions with fuzzy parameters)in model (4):

$$U_1(\boldsymbol{x}, \boldsymbol{y}) = \mathrm{Cr}\{f(\boldsymbol{x}, \boldsymbol{y}, \boldsymbol{\xi}) \geq T^0\},$$

$$U_2(\boldsymbol{x}) = \mathrm{Cr}\left\{\sum_{i=1}^{N} w_{ij} x_{it} \leq \eta_{jt}\right\},$$

$$U_3(\boldsymbol{x}, \boldsymbol{y}) = \mathrm{Cr}\{x_{it} + y_{i(t-1)} - y_{it} \geq d_{it}\},$$

$i = 1, 2, \cdots, N, j = 1, 2, \cdots, K, t = 1, 2, \cdots, T$. The fuzzy simulation technique was proposed and discussed by [10][11]. Now we give a fuzzy simulation procedure for computing uncertain functions.

Fuzzy Simulation for Uncertain Function U_1:

Step 1. Randomly generate u_{ik} from the ε-level set of ξ_i, $i = 1, 2, \cdots, N$, respectively,
    where $k = 1, 2, \cdots, M$ and ε is a sufficiently small positive number.
Step 2. Set $\nu_k = \min_{i,j}\{\mu_{\xi_i}(u_{ik})\}$ for $k = 1, 2, \cdots, M$.
Step 3. Return $L(T^0)$ via the following estimation formula (10)

$$L(r) = \frac{1}{2}\left(\max_{1 \leq k \leq M}\{\nu_k | f_k(\boldsymbol{x}, \boldsymbol{y}) \geq r\} + \min_{1 \leq k \leq M}\{1 - \nu_k | f_k(\boldsymbol{x}, \boldsymbol{y}) < r\}\right), \quad (10)$$

where

$$f_k(\boldsymbol{x}, \boldsymbol{y}) = \sum_{i=1}^{N}\sum_{t=1}^{T}(u_{ik} x_{it} - h_i y_{it}), \quad k = 1, 2, \cdots, M. \quad (11)$$

Uncertain functions U_2 and U_3 can be calculated similarly. For simplicity, we omitted their computing procedures here.

Now we embed fuzzy simulation into a genetic algorithm, thus producing a simulation based genetic algorithm as follows:

Fuzzy Simulation Based Genetic Algorithm

Step 1. Initialize *pop_size* chromosomes randomly.
Step 2. Update the chromosomes by crossover and mutation operations, in which
    fuzzy simulations are used to check the feasibility of chromosomes.
Step 3. Calculate the objective values for all chromosomes by fuzzy simulation.
Step 4. Compute the fitness of each chromosome according to the objective values.
Step 5. Select the chromosomes by spinning the roulette wheel.
Step 6. Repeat the second to fifth steps for a given number of cycles.
Step 7. Report the best chromosome as the optimal solution.

6 A Numerical Example

Consider an enterprize that manufactures three types of products in 6 periods. Each product needs 7 working procedures. The customer demands d_{it} and other parameters are listed in Table 1, 2 and 3.

Table 1. Customer Demands (d_{it})

i	$t=1$	$t=2$	$t=3$	$t=4$	$t=5$	$t=6$
1	(30,35,36,38)	(65,66,68,70)	(43,45,46,47)	(56,60,61,62)	(41,44,45,46)	(51,55,56,57)
2	(54,60,61,62)	(37,39,40,41)	(64,67,69,70)	(35,45,46,47)	(35,40,41,43)	(34,36,37,38)
3	(40,45,46,47)	(46,49,50,51)	(24,26,28,29)	(52,54,55,56)	(58,61,63,65)	(54,59,60,61)

Table 2. Problem Parameters w_{ij}, ξ_i and h_i

i	w_{ij}							ξ_i	h_i
	$j=1$	$j=2$	$j=3$	$j=4$	$j=5$	$j=6$	$j=7$		
1	0.45	4.43	11.91	3.18	3.53	0.42	0.85	(10,11,12,14)	1.8
2	0.41	4.54	12.87	2.90	2.80	0.34	0.87	(15,16,18,19)	4.3
3	0.44	5.00	11.7	4.20	3.51	0.43	0.81	(8,9,10,11)	1.2

Table 3. Production Capacity (η_{jt})

j	$t=1,2,\cdots,6$
$j=1$	(74,75,76,77)
$j=2$	(840,860,880,900)
$j=3$	(2200,2300,2400,2500)
$j=4$	(620,630,640,650)
$j=5$	(600,610,620,630)
$j=6$	(70,71,76,80)
$j=7$	(145,150,180,200)

Table 4. Computing Results

t	1	2	3	4	5	6
x_{1t}	45.287	72.282	51.362	67.266	51.196	61.604
x_{2t}	70.247	46.34	76.548	55.799	49.954	45.027
x_{3t}	56.837	51.078	44.128	47.814	69.721	63.673
y_{1t}	4.456	2.098	0.511	0.233	1.396	0.579
y_{2t}	0.03	0.667	0.035	0.016	0.022	0.19
y_{3t}	4.254	0.095	11.012	0.68	0.146	0.242

Since ξ_i, η_{jt} and d_{it} are all trapezoid fuzzy variables, we can derive the crisp equivalent of fuzzy programming model (4). A run of the genetic algorithm for

10000 generation with population size 30, probability of crossover 0.3, probability of mutation 0.2, $T^0 = 11500$, $\alpha = \beta = 0.9$ and number of cycles 10000, we get the optimal result in Table 4.

7 Conclusion

Lot sizing production planning problem with fuzzy unit profits, fuzzy capacities and fuzzy demands was formulated as a credibility measure based fuzzy programming model. Then its crisp equivalent model was also derived when the fuzzy parameters are characterized by trapezoidal fuzzy numbers. Moreover, a fuzzy simulation-based genetic algorithm was designed for solving the proposed fuzzy programming model as well as its crisp equivalent. Finally, a numerical example was provided for illustrating the effectiveness of algorithm.

Acknowledgments

This work was supported by National Natural Science Foundation of China (No.6042-5309), and Specialized Research Fund for the Doctoral Program of Higher Education (No.20020003009).

References

1. Dong, Y., Tang, J., Xu, B., Wang, D.: Chance-constrained programming approach to aggregate production planning. *Journal of System Engineering*, **18**(2003), 255–261 (In Chinese)
2. Dubois, D., Prade, H.: *Possibility Theory*. Plenum, New York, 1988
3. Gao, J., Liu B.: New primitive chance measures of fuzzy random event, *International Journal of Fuzzy Systems*. **3**(2001), 527–531
4. Gao, J., Lu M.: Fuzzy quadratic minimum spanning tree problem. *Applied Mathematics and Computation*, **164/3**(2005), 773-788
5. Gao, J., Lu M.: On the Randic Index of Unicyclic Graphs. *MATCH Commun. Math. Comput. Chem.*, **53**(2005), 377-384
6. Gao, J., Liu Y.: Stochastic Nash equilibrium with a numerical solution method. *Lecture Notes in Computer Science*, **3496**(2005), 811-816
7. Liu B., Iwamura, K.: Chance constrained programming with fuzzy parameters. *Fuzzy Sets and Systems*, **94**(1998), 227–237
8. Liu, B.: Dependent-chance programming in fuzzy environments. *Fuzzy Sets and Systems*, **109**(2000) 97–106
9. Liu, B., Liu, Y.K.: Expected value of fuzzy variable and fuzzy expected value models, *IEEE Transactions on Fuzzy Systems*, **10**(2002), 445–450
10. Liu, B.: *Theory and Practice of Uncertain Programming*. Physica-Verlag, Heideberg, 2002
11. Liu, B.: *Uncertainty Theory: An Introduction to its Axiomatic Foundations*. Springer-Verlag, Berlin, 2004
12. Sahinidis, N.V.: Optimization under uncertainty: state-of-the-art and opportunities. *Computers and Chemical Engineering*, **28**(2004), 971–983

13. Sakawa, M., Nishizaki, I., Uemura, Y.: Fuzzy programming and profit and cost allocation for a production and transportation problem. *European Journal of Operational Research*, **131**(2001), 1–15
14. Sakawa, M., Nishizaki, I., Uemura, Y.: Interactive fuzzy programming for two-level linear and linear fractional production and assignment problems: A case study. *European Journal of Operational Research*, **135**(2001), 142–157
15. Tang, J., Wang, D., Xu, B.: Fuzzy modelling approach to aggregate production planning with multi-product. *Journal of Management Sciences in China*, **6**(2003), 44–50 (In Chinese)
16. Wang, R.C., Fang, H.H.: Aggregate production planning with multiple objectives in a fuzzy environment. *European Journal of Operational Research*, **133**(2001), 521–536
17. Zadeh, L.A.: Fuzzy sets as a basis for a theory of possibility, *Fuzzy Sets and Systems*, **1**(1978), 3–28.
18. Zimmermann, H.J.: *Fuzzy Set and its Application*. Boston, Hingham, 1991

Fuzzy Dominance Based on Credibility Distributions

Jin Peng[1], Henry M.K. Mok[2], and Wai-Man Tse[3]

[1] Department of Mathematics, Huanggang Normal University,
Hubei 438000, China
pengjin01@tsinghua.org.cn
[2] Department of Decision Sciences and Managerial Economics,
The Chinese University of Hong Kong, Hong Kong
henry@baf.msmail.cuhk.edu.hk
[3] Department of Statistics, The Chinese University of Hong Kong, Hong Kong
wmtse@sta.msmail.cuhk.edu.hk

Abstract. Comparison of fuzzy variables is considered one of the most important and interesting topics in fuzzy theory and applications. This paper introduces the new concept of fuzzy dominance based on credibility distributions of fuzzy variables. Some basic properties of fuzzy dominance are investigated. As an illustration, the first order case of fuzzy dominance rule for typical triangular fuzzy variables is examined.

Keywords: fuzzy theory, fuzzy dominance, credibility distribution.

1 Introduction

Fuzzy set theory, initiated by Zadeh [22], has been well developed and applied in a wide variety of real problems. The term fuzzy variable was first introduced by Kaufmann [16], then it appeared in Zadeh [23] and Nahmias [18]. Meanwhile, possibility theory was proposed by Zadeh [23], and developed by Dubois & Prade [6] and other researchers [4]. Fuzzy set theory and possibility theory provide an alternative framework for modelling of real-world fuzzy decision systems mathematically.

As the average of possibility measure and necessity measure, a new type of measure—credibility measure is introduced by Liu & Liu [15]. Traditionally, possibility measure is widely used and regarded as the parallel concept of probability measure. However, as many researchers pointed out, there exist part and parcel differences between possibility measure and probability measure. As we know from Dubois and Prade [6], the necessity measure is the dual of possibility measure. But neither possibility measure nor necessity measure is self-dual. Fortunately, the credibility measure is self-dual (Liu [11]). In this aspect, the credibility measure shares some properties like the probability measure. In this way, credibility measure weights possibility and necessity measures harmoniously. In fact, it is credibility measure that plays the role of probability measure in the fuzzy world. More importantly, a new kind of expected value operator is defined by Liu & Liu [15] by means of the credibility measure. Motivated by developing an axiomatic approach based on credibility measure, credibility theory is proposed by Liu [12].

Based on the possibility measure, necessity measure, and credibility measure of fuzzy event, the optimistic and pessimistic values of fuzzy variable have been introduced by Liu [11] and investigated by Peng & Liu [20] for handling optimization problems in fuzzy environments. Traditionally, possibility measure and necessity measure, as well as the optimistic and pessimistic values of fuzzy variable with respect to possibility measure and necessity measure, are employed to model a type of fuzzy chance constrained programming [10][13][14]. In recent years, credibility measure, together with the optimistic and pessimistic values of fuzzy variable with respect to credibility measure, has been applied to modelling a new type of fuzzy chance constrained programming [11].

Uncertainty is the key ingredient in many decision problems and comparing uncertain variables is one of fundamental interests of decision theory [19]. Random variable and fuzzy variable are two basic types of uncertain variables. Stochastic dominance (see the review articles [1][9] and the references therein) is based on probability distribution and has been widely used in economics, finance, statistics, and operations research [2][3][5]. Inspired by the stochastic dominance approach, we introduce the new concept of fuzzy dominance based on credibility distribution. So far, comparison of fuzzy numbers is still considered one of the most important topics in fuzzy logic theory and numerous papers have been written about this subject [7][8][17][21]. Fuzzy dominance by means of credibility distribution, in quite a different light, provides a new approach to compare fuzzy variables in theory and applications of fuzzy logic.

The rest of this paper is organized as follows. Section 2 presents preliminaries in credibility theory. The concepts of fuzzy dominance are formally defined in Section 3. In Section 4 we investigate some basic properties of fuzzy dominance. Section 5 provides an example to show the first order case of fuzzy dominance rule for typical triangular fuzzy variables. Section 6 concludes this paper.

2 Preliminaries

In this section, we are going to review some concepts of credibility measure, credibility distribution, optimistic and pessimistic values as well as expected value operator for fuzzy variable.

Let us introduce some notation used throughout this paper. An abstract possibility space is denoted by $(\Theta, \mathcal{P}(\Theta), \text{Pos})$. The expected value operator is denoted by standard symbol E.

2.1 Fuzzy Variable

Fuzzy variable has been defined in many ways. In this paper we use the following definition of fuzzy variable in Nahmias' sense [18].

Definition 1. *A fuzzy variable is defined as a function from the possibility space* $(\Theta, \mathcal{P}(\Theta), \text{Pos})$ *to the real line* $\Re$.

Let ξ be a fuzzy variable on the possibility space $(\Theta, \mathcal{P}(\Theta), \text{Pos})$. Then its membership function is derived from the possibility measure Pos by

$$\mu(x) = \text{Pos}\{\theta \in \Theta \mid \xi(\theta) = x\}. \tag{1}$$

2.2 Possibility, Necessity, Credibility

Dubois and Prade [6] described the possibility measure and necessity measure as follows. Let r be a real number. The possibility and necessity measure of fuzzy event $\{\xi \leq r\}$ are defined as

$$\text{Pos}\{\xi \leq r\} = \sup_{x \leq r} \mu(x), \tag{2}$$

$$\text{Nec}\{\xi \leq r\} = 1 - \sup_{x > r} \mu(x), \tag{3}$$

respectively.

Liu and Liu [15] introduced a relatively new concept—credibility measure.

Definition 2. *Let r be a real number. The credibility measure of fuzzy event $\{\xi \leq r\}$ is defined as*

$$Cr\{\xi \leq r\} = \frac{1}{2}(\text{Pos}\{\xi \leq r\} + \text{Nec}\{\xi \leq r\}). \tag{4}$$

Similarly, the credibility measure of fuzzy event $\{\xi \geq r\}$ is defined as

$$Cr\{\xi \geq r\} = \frac{1}{2}(\text{Pos}\{\xi \geq r\} + \text{Nec}\{\xi \geq r\}). \tag{5}$$

2.3 Credibility Distribution, Credibility Density Function

Definition 3 (Liu [12]). *The credibility distribution $\Phi : \Re \to [0,1]$ of a fuzzy variable ξ is defined by*

$$\Phi(x) = Cr\{\theta \in \Theta \mid \xi(\theta) \leq x\}. \tag{6}$$

Definition 4 (Liu [12]). *The credibility density function $\phi \colon \Re \to [0, +\infty)$ of a fuzzy variable ξ is a function such that*

$$\Phi(x) = \int_{-\infty}^{x} \phi(y) dy \tag{7}$$

holds for all $x \in \Re$, where Φ is the credibility distribution of the fuzzy variable ξ.

Obviously, the credibility distribution of a fuzzy variable is nonnegative and bounded. However, generally speaking, it is neither left-continuous nor right-continuous. Moreover, we have

Lemma 5 (Liu [12]). *The credibility distribution Φ of a fuzzy variable is a non-decreasing function on $\Re$ with*

$$\begin{cases} \lim_{x\to-\infty} \Phi(x) \leq 0.5 \leq \lim_{x\to\infty} \Phi(x) \\ \lim_{y\downarrow x} \Phi(y) = \Phi(x) \text{ if } \lim_{y\downarrow x} \Phi(y) > 0.5 \text{ or } \Phi(x) \geq 0.5. \end{cases} \qquad (8)$$

2.4 Optimistic Value, Pessimistic Value, and Expected Value

Liu [11] introduced the following concepts of optimistic and pessimistic values.

Definition 6 (Liu [11]). *Let ξ be a fuzzy variable, and $\alpha \in (0,1]$. Then*

$$\xi_{\sup}(\alpha) = \sup \{r \mid \mathrm{Cr}\{\xi \geq r\} \geq \alpha\} \qquad (9)$$

and

$$\xi_{\inf}(\alpha) = \inf \{r \mid \mathrm{Cr}\{\xi \leq r\} \geq \alpha\} \qquad (10)$$

are called the α-optimistic value and α-pessimistic value of ξ with respect to credibility, respectively.

Remark 7. α-pessimistic value of ξ with respect to credibility is also called the generalized inverse function (g.i.f.) of credibility distribution function $\Phi(x)$, denoted by $\Phi^{-1}(\alpha)$.

Definition 8 (Liu and Liu [15]). *Let ξ be a fuzzy variable. Then the expected value of ξ is defined by*

$$E[\xi] = \int_0^{+\infty} \mathrm{Cr}\{\xi \geq r\} dr - \int_{-\infty}^0 \mathrm{Cr}\{\xi \leq r\} dr \qquad (11)$$

provided that at least one of the two integrals is finite.

3 Fuzzy Dominance

The forthcoming definitions of fuzzy dominance are based on the credibility distribution functions of fuzzy variables.

Let ξ and η be two fuzzy variables, and $\Phi(x)$ and $\Psi(x)$ denote the credibility distribution functions of ξ and η, respectively. Hereinafter, we write $\Phi^{(k)}(x) = \int_{-\infty}^x \Phi^{(k-1)}(t)dt$ and $\Psi^{(k)}(x) = \int_{-\infty}^x \Psi^{(k-1)}(t)dt$, $k = 2, 3, \cdots$ where $\Phi^{(1)}(x) = \Phi(x)$ and $\Psi^{(1)}(x) = \Psi(x)$. We focus on the set of fuzzy variables

$$L_k = \{\xi | \xi \text{ is with credibility distribution } \Phi(x) \text{ and } \Phi^{(k)}(x) < \infty\}.$$

In this way, it is guaranteed that the integral $\int_{-\infty}^x \Phi^{(k-1)}(t)dt$ sounds well defined when $\xi \in L_k (k = 2, 3, \cdots)$.

Definition 9. Let ξ and η be two fuzzy variables in L_1. We say that ξ First Order Fuzzy Dominates η, denoted by $\xi \succeq_{FFD} \eta$, if and only if $\Phi(x) \leq \Psi(x)$ for all $x \in \Re$.

Definition 10. Let ξ and η are two fuzzy variables in L_2. We say that ξ Second Order Fuzzy Dominates η, denoted by $\xi \succeq_{SFD} \eta$, if and only if $\int_{-\infty}^{x} \Phi(t)\mathrm{d}t \leq \int_{-\infty}^{x} \Psi(t)\mathrm{d}t$ for all $x \in \Re$.

Definition 11. Let ξ and η are two fuzzy variables in L_3. We say that ξ Third Order Fuzzy Dominates η, denoted $\xi \succeq_{TFD} \eta$, if and only if $\int_{-\infty}^{x}\int_{-\infty}^{t} \Phi(s)\mathrm{d}s\mathrm{d}t \leq \int_{-\infty}^{x}\int_{-\infty}^{t} \Psi(s)\mathrm{d}s\mathrm{d}t$ for all $x \in \Re$.

In a general way, we can introduce the concept of fuzzy dominance in k-order.

Definition 12. Let ξ and η are two fuzzy variables in L_k. We say that ξ k-Order Fuzzy Dominates η, denoted $\xi \succeq_{kFD} \eta$, if and only if $\Phi^{(k)}(x) \leq \Psi^{(k)}(x)$ for all $x \in \Re$.

4 Properties of Fuzzy Dominance

In this section, some basic properties related to the fuzzy dominance are presented.

Theorem 13. Let ξ be a fuzzy variable with $E[\xi] < \infty$ and $\Phi(x)$ be the credibility distribution functions of ξ. Then $\Phi^{(2)}(x) = \int_{-\infty}^{x} \Phi(t)\mathrm{d}t$ is finite for all $x \in \Re$.

Proof. We recall that $E[\xi] = \int_{0}^{+\infty} \mathrm{Cr}\{\xi \geq r\}\mathrm{d}r - \int_{-\infty}^{0} \mathrm{Cr}\{\xi \leq r\}\mathrm{d}r$. Since $E[\xi] < \infty$, the integral $\int_{-\infty}^{0} \mathrm{Cr}\{\xi \leq r\}\mathrm{d}r$ is finite. Then $\int_{-\infty}^{x} \mathrm{Cr}\{\xi \leq r\}\mathrm{d}r$ is finite for all $x \in \Re$. □

Theorem 14. Let ξ be a fuzzy variable in L_2. Then

(a) $\Phi^{(2)}(x)$ is a non-decreasing, nonnegative function of x;
(b) If ξ has credibility density $\phi(x)$, then $\Phi^{(2)}(x)$ is a convex function of x;
(c) $\lim_{x \to -\infty} \Phi^{(2)}(x) = 0$;
(d) If $\Phi(x_0) > 0$ for some $x_0 \in \Re$, then $\Phi^{(2)}(x)$ is an increasing function of $x \in [x_0, \infty)$ in strict sense.

Proof. (a) The result follows from the fact that $\Phi(x)$ is a nonnegative, monotonic integrable function of x. (b) If ξ has credibility density $\phi(x)$, then $\Phi^{(2)}(x) =$

$\int_{-\infty}^{x} \Phi(t)\mathrm{d}t = \int_{-\infty}^{x} \int_{-\infty}^{t} \phi(s)\mathrm{d}s\mathrm{d}t$. Thus we obtain that $\frac{d^2 \Phi^{(2)}}{dx^2} = \phi(x) \geq 0$, a.e. and hence $\Phi^{(2)}(x)$ is a convex function. (c) It follows immediately from Theorem 13. (d) Suppose that $x_0 \leq x_1 < x_2$. Then we have $\Phi^{(2)}(x_2) = \int_{-\infty}^{x_2} \Phi(t)\mathrm{d}t = \int_{-\infty}^{x_1} \Phi(t)\mathrm{d}t + \int_{x_1}^{x_2} \Phi(t)\mathrm{d}t \geq \int_{-\infty}^{x_1} \Phi(t)\mathrm{d}t + \Phi(x_0)(x_2 - x_1) > \Phi^{(2)}(x_1)$. Therefore $\Phi^{(2)}(x)$ is an increasing function of $x \in [x_0, \infty)$ in strict sense. □

Theorem 15. *Let ξ be a fuzzy variable in L_3. Then $\Phi^{(3)}(x)$ is a non-decreasing, convex, nonnegative function of x.*

Proof. Similar to the proofs of (a) and (b) in Theorem 14. □

Theorem 16. *Let ξ, η and ζ be fuzzy variables in L_1. Then*
(a) Reflexivity: $\xi \succeq_{FFD} \xi$;
(b) Transitivity: $\xi \succeq_{FFD} \eta$ and $\eta \succeq_{FFD} \zeta$ implies $\xi \succeq_{FFD} \zeta$.

Proof. Simple consequences from the definition of FFD. □

Remark 17. The first (second or third) order fuzzy dominance relation is a partial order but not a total order among fuzzy variables (or their distributions).

Theorem 18. *Let ξ and η be two fuzzy variables in L_3. Then $\xi \succeq_{FFD} \eta \Rightarrow \xi \succeq_{SFD} \eta \Rightarrow \xi \succeq_{TFD} \eta$.*

Proof. Let $\Phi(x)$ and $\Psi(x)$ denote the credibility distribution functions of ξ and η, respectively. Since $\xi \succeq_{FFD} \eta$, we have $\Phi(x) \leq \Psi(x)$ for all $x \in \Re$. Then $\int_{-\infty}^{x} \Phi(t)\mathrm{d}t \leq \int_{-\infty}^{x} \Psi(x)\mathrm{d}t$ for all $x \in \Re$. That is, $\xi \succeq_{SFD} \eta$. Now, we also have $\int_{-\infty}^{x} \int_{-\infty}^{t} \Phi(s)\mathrm{d}s\mathrm{d}t \leq \int_{-\infty}^{x} \int_{-\infty}^{t} \Psi(s)\mathrm{d}s\mathrm{d}t$ for all $x \in \Re$. Therefore we conclude that $\xi \succeq_{TFD} \eta$. □

A sufficient and necessary condition for FFD is given by the following theorem.

Theorem 19. *Let ξ and η be two fuzzy variables in L_1, $\alpha \in (0, 1]$. Let $\xi_{\inf}(\alpha)$ and $\eta_{\inf}(\alpha)$ denote the α-pessimistic values of ξ and η, respectively. Then $\xi \succeq_{FFD} \eta$ if and only if $\xi_{\inf}(\alpha) \geq \eta_{\inf}(\alpha)$ for all $\alpha \in (0, 1]$.*

Proof. If $\xi \succeq_{FFD} \eta$, i.e., $\Phi(x) \leq \Psi(x)$ for all $x \in \Re$, then for all $\alpha \in (0, 1]$, we have $\xi_{\inf}(\alpha) = \inf\{r \mid \Phi(r) \geq \alpha\} \geq \inf\{r \mid \Psi(r) \geq \alpha\} = \eta_{\inf}(\alpha)$.

Conversely, assume that there exists some $\alpha_0 \in (0, 1]$ such that $\Phi(x_0) > \alpha_0 > \Psi(x_0)$ for some $x_0 \in \Re$. On one side, from the definition of $\eta_{\inf}(\alpha_0)$, we have $x_0 < \eta_{\inf}(\alpha_0)$. On the other side, from the definition of $\xi_{\inf}(\alpha_0)$, we have $x_0 \geq \xi_{\inf}(\alpha_0)$. Then $\xi_{\inf}(\alpha_0) < \eta_{\inf}(\alpha_0)$ which contradicts to $\xi_{\inf}(\alpha_0) \geq \eta_{\inf}(\alpha_0)$. Therefore, $\Phi(x) \leq \Psi(x)$ for all $x \in \Re$. That is, $\xi \succeq_{FFD} \eta$. □

Theorem 20. *Let ξ and η be two continuous fuzzy variables and $\{\xi_n\}$ be a fuzzy variable sequence in L_1. If $\xi_n \succeq_{FFD} \eta$ for all $n = 1, 2, \cdots$ and $\{\xi_n\}$ converges in distribution to ξ (that is, $\Phi_n(x) \to \Phi(x)$ as $n \to \infty$ for all continuity points x of Φ), then $\xi \succeq_{FFD} \eta$.*

Proof. Suppose that $\Phi, \Psi, \Phi_1, \Phi_2, \cdots$ are the credibility distributions of fuzzy variables $\xi, \eta, \xi_1, \xi_2, \cdots$, respectively. Since $\xi_n \succeq_{FFD} \eta$, $n = 1, 2, \cdots$, we have $\Phi_n(x) \leq \Psi(x)$ for all $x \in \Re$. Because $\lim_{n \to \infty} \Phi_n(x) = \Phi(x)$ for all continuity points x of Φ, we conclude $\Phi(x) \leq \Psi(x)$ for all continuity points x of Φ. That is, $\Phi(x) \leq \Psi(x)$ holds for all $x \in \Re$. This means $\xi \succeq_{FFD} \eta$. □

Theorem 21. *Let ξ and η be two fuzzy variables in L_1, $U(x)$ is a strictly monotone increasing function defined on $\Re$. Then $\xi \succeq_{FFD} \eta$ iff $U(\xi) \succeq_{FFD} U(\eta)$.*

Proof. Let $\Phi(x)$ and $\Psi(x)$ denote the credibility distribution functions of ξ and η, respectively. Assume that $F(x)$ and $G(x)$ are the credibility distribution functions of fuzzy variables $U(\xi)$ and $U(\eta)$, respectively. Note that $F(x) = \mathrm{Cr}\{U(\xi) \leq x\} = \mathrm{Cr}\{\xi \leq U^{-1}(x)\} = \Phi(U^{-1}(x))$ and $G(x) = \mathrm{Cr}\{U(\eta) \leq x\} = \mathrm{Cr}\{\eta \leq U^{-1}(x)\} = \Psi(U^{-1}(x))$.

If $\xi \succeq_{FFD} \eta$, i.e., $\Phi(y) \leq \Psi(y)$ for all $y \in \Re$, then for all $x \in \Re$, we have $F(x) = \Phi(U^{-1}(x)) \leq \Psi(U^{-1}(x)) = G(x)$. This means $U(\xi) \succeq_{FFD} U(\eta)$.

Conversely, if $U(\xi) \succeq_{FFD} U(\eta)$, then the reverse function U^{-1} exists and is also strictly monotone. Therefore, $U^{-1}U(\xi) \succeq_{FFD} U^{-1}U(\eta)$. That is, $\xi \succeq_{FFD} \eta$. □

The above theorem shows that the first order fuzzy dominance remains invariant under monotonic transformation.

Theorem 22. *Let ξ and η be two fuzzy variables in L_1, $a, b \in \Re$ and $a > 0$. If $\xi \succeq_{FFD} \eta$, then $a\xi + b \succeq_{FFD} a\eta + b$.*

Proof. It follows immediately from Theorem 21. □

5   An Example

Generally speaking, the condition of fuzzy dominance is rather strong and difficult to verify due to the fact that it is an inequality of two credibility distribution functions. In some specific cases, the fuzzy dominance relation can be easily observed. As an illustration, the first order case of fuzzy dominance rule for typical triangular fuzzy variables are examined in the following example.

Example 23. The credibility distribution function of a triangular fuzzy variable $\xi = (r_1, r_2, r_3)$ is

$$\Phi(x) = \begin{cases} 0, & \text{if } x \leq r_1 \\ \dfrac{x - r_1}{2(r_2 - r_1)}, & \text{if } r_1 \leq x \leq r_2 \\ \dfrac{x + r_3 - 2r_2}{2(r_3 - r_2)}, & \text{if } r_2 \leq x \leq r_3 \\ 1, & \text{if } r_3 \leq x, \end{cases}$$

and for any given $\alpha(0 < \alpha \leq 1)$, we have

$$\xi_{\inf}(\alpha) = \begin{cases} r_1 + 2(r_2 - r_1)\alpha, & \text{if } 0 < \alpha \leq 0.5 \\ 2r_2 - r_3 + 2(r_3 - r_2)\alpha, & \text{if } 0.5 < \alpha \leq 1. \end{cases}$$

The credibility distribution function and optimistic value function of a triangular fuzzy variable $\eta = (s_1, s_2, s_3)$ and are expressed in similar way. A question is that under what conditions a triangular fuzzy variable fuzzy dominates another one? According to the definition of first order fuzzy dominance or Theorem 19, we can obtain the following intuitive result.

Proposition 24. *Let $\xi = (r_1, r_2, r_3)$ and $\eta = (s_1, s_2, s_3)$ be two triangular fuzzy variables. Then $\xi \geq_{FFD} \eta$ iff $r_1 \geq s_1, r_2 \geq s_2, r_3 \geq s_3$.*

6 Conclusions

This paper introduced the fuzzy dominance approach in terms of credibility distribution. Some properties of fuzzy dominance are investigated. As an illustration, the first order case of fuzzy dominance rule for typical triangular fuzzy variables is examined.

We see this short paper as a first step towards developing a route for dealing with uncertainty by means of fuzzy dominance within the axiomatic framework of credibility theory. Future research both in theoretical and applied aspects is on the way.

Acknowledgments

This work was supported by the National Natural Science Foundation of China Grant No. 60174049, the Significant Project No. Z200527001 and Scientific and Technological Innovative Group Project of Hubei Provincial Department of Education, China.

References

1. Bawa, V. S.: Stochastic dominance: A research bibliography. Management Science **28** (1982) 698–712
2. Dachraoui, K., Dionne, G.: Stochastic dominance and optimal portfolio, Economics Letters **71**(2001) 347–354
3. Dana, R. A.: Market behavior when preferences are generated by second-order stochastic dominance. Journal of Mathematical Economics **40**(2004) 619–639
4. De Cooman, G.: Possibility theory-I, II, III. International Journal of General Systems **25**(1997) 291–371.
5. Dentcheva, D., Ruszczynski, A.: Optimization with Stochastic Dominance Constraints. SIAM Journal on Optimization **14**(2003) 548–566.
6. Dubois, D., Prade, H.: Possibility Theory: An Approach to Computerized Processing of Uncertainty. Plenum Press, New York (1988)

7. Facchinetti, G., Ricci, R. G., Muzzioli, S.: Note on ranking fuzzy triangular numbers. International Journal of Intelligent Systems **13**(1998) 613–622
8. Iskander, M. G.: A suggested approach for possibility and necessity dominance indices in stochastic fuzzy linear programming, Applied Mathematics Letters **18** (2005) 395–399
9. Levy, H.: Sochastic Dominance and Expected Utility: Survey and Analysis. Management Science **38**(1992) 555–593
10. Liu, B.: Uncertain Programming. John Wiley & Sons, New York (1999)
11. Liu, B.: Theory and Practice of Uncertain Programming. Physica-Verlag, Heidelberg (2002)
12. Liu, B.: Uncertainty Theory: An Introduction to its Axiomatic Foundations. Springer-Verlag, Heidelberg (2004)
13. Liu, B., Iwamura, K.: Chance-constrained programming with fuzzy parameters. Fuzzy Sets and Systems **94**(1998) 227–237
14. Liu, B.: Dependent-chance programming in fuzzy environments. Fuzzy Sets and Systems **109** (2000) 95–104
15. Liu, B., Liu, Y.-K.: Expected value of fuzzy variable and fuzzy expected value model. IEEE Transactions on Fuzzy Systems **10**(2002) 445–450
16. Kaufmann, A.: Introduction to the Theory of Fuzzy Subsets. Academic Press, New York(1975)
17. Mitchell, H. B., Schaefer, P. A.: On ordering fuzzy numbers. International Journal of Intelligent Systems **15**(2000) 981–993
18. Nahmias, S.: Fuzzy variables. Fuzzy Sets and Systems **1**(1978) 97–100
19. Ogryczak, W., Ruszczyński, A.: From stochastic dominance to mean-risk models: Semideviations as risk measures. European Journal of Operational Research **116**(1999) 33–50
20. Peng, J., Liu, B.: Some properties of optimistic and pessimistic values of fuzzy variables. In: Proceedings of the Tenth IEEE International Conference on Fuzzy Systems **2**(2004) 292–295
21. Yager, R. R., Filev, D.: On ranking fuzzy numbers using valuations. International Journal of Intelligent Systems **14**(1999) 1249–1268
22. Zadeh, L. A.: Fuzzy sets. Information and Control **8**(1965) 338–353
23. Zadeh, L. A.: Fuzzy set as a basis for a theory of possibility. Fuzzy Sets and Systems **1**(1978) 3–28

Fuzzy Chance-Constrained Programming for Capital Budgeting Problem with Fuzzy Decisions

Jinwu Gao[1], Jianhua Zhao[2], and Xiaoyu Ji[3]

[1] School of Information, Renmin University of China, Beijing 100872, China
[2] Department of Mathematics, Shijiazhuang College, Shijiazhuang 050801, China
[3] Department of Mathematical Sciences, Tsinghua University, Beijing 100084, China

Abstract. In this paper, capital budgeting problem with fuzzy decisions is formulated as fuzzy chance-constrained programming models. Then fuzzy simulation, neural network and genetic algorithm are integrated to produce a hybrid intelligent algorithm for solving the proposed models. Finally, numerical experiments are provided to illustrate the effectiveness of the hybrid intelligent algorithm.

1 Introduction

The original capital budgeting is to choose an appropriate combination of projects under fixed assets to maximize the total net profit. Early attempts to model the capital budgeting problem was linear programming with single objective and crisp coefficients. But in practice, there often exist many uncertain elements such as future demands, the benefits of per project and the level of funds that needs to be allocated to every projects. Moreover, multiple conflicting goals make the problem more complex. For this reason, uncertain programming [12][17], the more powerful and suitable approaches are employed to deal with it.

Chance-constrained programming (CCP) [1] has long been used to model stochastic capital budgeting problems in the working capital management [8] and in the production area [9]. De *et al* [3] extended chance-constrained goal programming (CCGP) to the zero-one case and applied it to capital budgeting problems. A framework of fuzzy chance-constrained programming was presented by Liu and Iwamura [10] [11], and used to model capital budgeting problems in fuzzy environment by Iwamura and Liu [6].

Traditional mathematic programming models all produce crisp decision vectors such that some objectives achieve the optimal values. But sometimes we should provide fuzzy decisions for practical purposes. In view this reasons, Liu and Iwamura [15] provided a spectrum of CCP with fuzzy decisions. The purpose of this paper is to investigate capital budgeting problem with fuzzy decisions. Section 2 formulate capital budgeting problems with fuzzy decisions by CCP and CCGP models. Then, fuzzy simulation, neural network and genetic algorithm are integrated to produce a hybrid intelligent algorithm for solving the proposed CCP models. Lastly, two numerical examples are provided to illustrate the effectiveness of the hybrid intelligent algorithm.

2 Fuzzy Capital Budgeting Problem with Fuzzy Decisions

Consider a company which are planning to initialize a new factory. Suppose that there are several types of machines to be selected, and different types of machines can produce the same product. Certainly, the objective is to make profit by satisfying future demands, and there are resources constraints such as capital and space. The following notation will be used henceforh.

n : the number of types of machines;
A_i : denoting the i-th type of machines;
x_i : nonnegative integer representing the number of A_i selected;
b_i, b : the price of an A_i and the total capital available for buying new machines;
c_i, c : space that an A_i takes up and the total space available, $i = 1, 2, \cdots, n$;
η_j : the j-th class demand per time period, $j = 1, 2, \cdots, p$;
ξ_{ij} : the production capacity of an A_i for j-th class product, $i = 1, 2, \cdots, n, j = 1, 2, \cdots, p$;
a_i : the net profit of an A_i per time period, $i = 1, 2, \cdots, n$.

Immediately, we have the following deterministic linear integer programming model for capital budgeting problem,

$$\begin{cases} \max \sum_{i=1}^n a_i x_i \\ \text{subject to:} \\ \quad \sum_{i=1}^n b_i x_i \leq b \\ \quad \sum_{i=1}^n c_i x_i \leq c \\ \quad \sum_{i=1}^n \xi_{ij} x_i \geq \eta_j, \ j = 1, 2, \cdots, p \\ \quad x_i, \ i = 1, 2, \cdots, n, \ \text{nonnegative integers.} \end{cases} \quad (1)$$

In practice, the production capacities ξ_{ij}, future demand η_j, the net profit a_i per A_i, the level of funds that needs to be allocated to type i machine b_i, $i = 1, 2, \cdots, n, j = 1, 2, \cdots, p$, and so on are not necessarily deterministic. Here, we assume that they are fuzzy variables. In addition, management decisions sometimes cannot be precisely performed in actual cases, even if we had provided crisp values. It is thus reasonable for us to fuzzify the decision. In the fuzzy decision, each decision may be restricted in a reference collection of fuzzy sets which are determined by the property of decision systems. Then the decision maker would have more selection. In the following, we denote a fuzzy decision by a vector $\tilde{x} = (\tilde{x}_1, \tilde{x}_2, \cdots, \tilde{x}_n)$, in which each component $\tilde{x}_i$ will be assigned an element in the reference collection of fuzzy sets X_i, $i = 1, 2, \cdots, n$, respectively. Since the reference collections are all predetermined, $\tilde{x}$ must be taken from the Cartesian product

$$\mathcal{X} = X_1 \otimes X_2 \otimes \cdots \otimes X_n. \quad (2)$$

Since there exist fuzzy parameters and fuzzy decision variables, the objective and constraints in model (1) are fuzzy, too. Then then feasible set as well as the optimal objective are not defined mathematically. In the following, we will model the capital budgeting problem with fuzzy decisions by chance-constrained programming models.

3 Chance-Constrained Programming Models

Suppose that the manager wishes to maximize the total profit at a confidence level α_1. He also specified the confidence levels that the capital constraint, the space constraint, and the demand constraints should hold. Then we have the CCP with fuzzy decisions as follows

$$\begin{cases} \max \overline{f} \\ \text{subject to:} \\ \quad \text{Pos}\{\sum_{i=1}^n a_i \tilde{x}_i \geq \overline{f}\} \geq \alpha_1 \\ \quad \text{Pos}\{\sum_{i=1}^n b_i \tilde{x}_i \leq b\} \geq \alpha_2 \\ \quad \text{Pos}\{\sum_{i=1}^n c_i \tilde{x}_i \leq c\} \geq \alpha_3 \\ \quad \text{Pos}\{\sum_{i=1}^n \xi_{ij} \tilde{x}_i \geq \eta_j\} \geq \beta_j,\ j = 1, 2, \cdots, p, \end{cases} \quad (3)$$

In general, there are several elements such as the total profit, service level, capital and space available, which need to be taken into accounts by the manager. Sometimes, the manager has to take more than one of them as objectives, and the others as constraints. However, the multiple goals are often conflicting. In order to balance them, suppose that the decision maker give the following target levels and priority structure.

Priority 1: Profit goal, the total profit should achieve a given level a at confidence level α_1;

Priority 2: Budget goal, that is, the total cost spent for machines should not exceed the amount available at confidence level α_2;

Priority 3: Space goal, the total space used by the machines should not exceed the space available at confidence level α_3.

Then the fuzzy capital budgeting problem may also be modelled by the following CCGP with fuzzy decision and fuzzy coefficients.

$$\begin{cases} \text{lexmin}\{d_1^-, d_2^+, d_3^+\} \\ \text{subject to:} \\ \quad \text{Pos}\{a - \sum_{i=1}^n a_i \tilde{x}_i \leq d_1^-\} \geq \alpha_1 \\ \quad \text{Pos}\{\sum_{i=1}^n b_i \tilde{x}_i - b \leq d_2^+\} \geq \alpha_2 \\ \quad \text{Pos}\{\sum_{i=1}^n c_i \tilde{x}_i - c \leq d_3^+\} \geq \alpha_3 \\ \quad \text{Pos}\{\sum_{i=1}^n \xi_{ij} \tilde{x}_i \geq \eta_j\} \geq \beta_j,\ j = 1, 2, \cdots, p \\ \quad d_1^-, d_2^+, d_3^+ \geq 0, \end{cases} \quad (4)$$

where lexmin represents lexicographical minimization.

4 Hybrid Intelligent Algorithm

In this section, fuzzy simulation, neural network, and genetic algorithm are to be integrated to produce a hybrid intelligent algorithm for solving the chance-constrained programming models with fuzzy decisions.

4.1 Fuzzy Simulations

In our fuzzy programming models with fuzzy decisions, we summarize the following two types of uncertain functions.

$$U_1(\tilde{x}) : \tilde{x} \to \text{Pos}\{g(\tilde{x}, \xi) \leq 0\}$$
$$U_2(\tilde{x}) : \tilde{x} \to \max\left\{\overline{f} \mid \text{Pos}\{f(\tilde{x}, \xi) \geq \overline{f}\} \geq \alpha\right\} \quad (5)$$

where $\tilde{x}$ and ξ are fuzzy vectors. Due to the complexity, we will resort to fuzzy simulation for computing the uncertain functions.

For each given decision variable $\tilde{x}$, the computation procedures of the two types of uncertain functions $U_1(\tilde{x})$ and $U_2(\tilde{x})$ are given as follows.

Fuzzy Simulation for the Possibility of Fuzzy Event:

Step 1: Set $U_2(\tilde{x}) = 0$;
Step 2: Sample a crisp vector (x_0, ξ_0) from the α-level set of $(\tilde{x}, \xi)$, if $g(\tilde{x}_0, \xi_0) \leq 0$ and $U_1(\tilde{x}) < \mu(\tilde{x}_i, \xi_i)$, then let $U_1(\tilde{x}) = \mu(\tilde{x}_i, \xi_i)$, else go to step 1.
Step 3: Repeat steps 1 and 2 for N times, then return $U_1(\tilde{x})$.

Fuzzy Simulation for the Optimistic Value:

Step 1: Set $U_2(\tilde{x}) = -\infty$;
Step 2: Sample a crisp vector (x_0, ξ_0) from the α-level set of $(\tilde{x}, \xi)$, if $f(\tilde{x}_0, \xi_0) > U_2(\tilde{x})$, then let $U_2(\tilde{x}) = f(\tilde{x}_0, \xi_0)$, else go to step 1.
Step 3: Repeat steps 1 and 2 for N times, then return $U_2(\tilde{x})$.

4.2 Neural Networks

A neural network is essentially a nonlinear mapping from the input space to the output space. It is known that a neural network with an arbitrary number of hidden neurons is a universal approximator for continuous functions [2][5]. Moreover, it has high speed of operation after it is well-trained on a set of input-output data. In order to speed up the solution process, we train neural networks to approximate uncertain functions, and then use the trained neural networks to evaluate the uncertain functions in the solution process.

Training Procedure of Neural Network:

Step 1. Sample M fuzzy vectors $\tilde{x}_1, \tilde{x}_2, \cdots, \tilde{x}_M$ from the reference set $\mathcal{X} = X_1 \otimes X_2 \otimes \cdots \otimes X_n$, then calculate the value of $U_1(\tilde{x}_i)$ and $U_2(\tilde{x}_i)$, $i = 1, 2, \cdots, M$ by fuzzy simulation.
Step 2. Use these samples to train NNs by BPA to approximate these uncertain functions
Step 3. return the tained NN.

4.3 Hybrid Intelligent Agorithm

Genetic algorithm is a search method for optimization problems based on the mechanics of natural selection in the evolution of biology. The algorithm has demonstrated considerable success in providing approximate optimal solutions to many complex optimization problems. In general genetic algorithm, the feasibility and objective value of each chromosome has to be calculated by time-consuming fuzzy simulation. A natural idea is to train a neural network to approximate the uncertain functions, and then embed it into genetic algorithm, thus much improved the efficiency of the genetic algorithm. Up to now, an idea of hybrid intelligent algorithm is proposed.

The concrete procedure of the hybrid intelligent algorithm may be written as follows:

Hybrid Intelligent Algorithm:

Step 1. Generate samples of uncertain functions by fuzzy simulation;
Step 2. Train neural networks with these samples to approximate the uncertain functions;
Step 3. Initialize a population in which the feasibility of each chromosome is checked through the trained neural networks;
Step 4. Use crossover and mutation operators to generate offspring, then the trained neural networks to check the feasibility of them;
Step 5. Calculate the objective values of all chromosomes through the trained neural networks, then give each of them a fitness;
Step 6. Select the chromosomes by spinning the roulette wheel.
Step 7. Repeat Step 4 to 6 for a given number of cycles.
Step 8. Report the best chromosome as the optimal solution.

5 Numerical Examples

In this section, two numerical examples, which has been written in C language, are given to illustrate the effectiveness of the hybrid intelligent algorithm.

Firstly, we give a concrete description of a plant which is going to initiate five types of machines. The fuzzy parameters are given as follows: a_i ($i = 1, 2, \cdots, 5$) are fuzzy numbers with membership functions $exp(-|x-5|)$, $exp(-|x-6|)$, $exp(-|x-4|)$, $exp(-|x-10|)$ and $exp(-|x-7|)$, respectively; b, b_i ($i = 1, 2, \cdots, 5$) are triangular fuzzy numbers $(250, 350, 450)$, $(3, 4, 5)$, $(4, 5, 6)$, $(5, 6, 7)$, $(11, 12, 13)$ and $(6, 7, 8)$, respectively; c, $c_i (i = 1, 2, \cdots, n)$ are 700, 20, 18, 16, 12 and 15, respectively; ξ_{11}, ξ_{21}, ξ_{31}, ξ_{42} and ξ_{52} are triangular fuzzy numbers $(50, 60, 70)$, $(60, 70, 80)$, $(80, 90, 100)$, $(110, 130, 150)$, $(100, 110, 120)$, respectively. η_1, η_2 are trapezoidal fuzzy numbers $(2800, 2900, 3000, 3100)$, $(3300, 3400, 3500, 3600)$, respectively.

Suppose the decision maker wants to maximize the net profit at a predetermined confidence level subject to capital, space, and demands constraints at certain confidence level. According to the discussion in Section 3, the CCP model with fuzzy decisions for capital budgeting is

$$\begin{cases} \max \overline{f} \\ \text{subject to:} \\ \quad \text{Pos}\{\sum_{i=1}^{n} a_i \tilde{x}_i \geq \overline{f}\} \geq 0.95 \\ \quad \text{Pos}\{\sum_{i=1}^{n} b_i \tilde{x}_i \leq b\} \geq 0.95 \\ \quad \text{Pos}\{\sum_{i=1}^{n} c_i \tilde{x}_i \leq 700\} \geq 0.90 \\ \quad \text{Pos}\{\xi_{11}\tilde{x}_1 + \xi_{21}\tilde{x}_2 + \xi_{31}\tilde{x}_3 \geq \eta_1\} \geq 0.85 \\ \quad \text{Pos}\{\xi_{42}\tilde{x}_4 + \xi_{52}\tilde{x}_3 \geq \eta_2\} \geq 0.80. \end{cases} \quad (6)$$

Where fuzzy decision $\tilde{x} = (\tilde{x}_1, \tilde{x}_2, \cdots, \tilde{x}_5)$ should be taken from the Cartesian product $\mathcal{X} = X_1 \otimes X_2 \otimes \cdots \otimes X_5$, and the reference collection X_i are composed of all fuzzy numbers

$$\sum_{j=-4}^{4} (1 + 0.05j)/(t+j), \; t = 6, 7, \cdots, 15.$$

Then we train an neural network (5 input neurons, 10 hidden neurons, 5 output neurons) to approximate the uncertain functions. Finally, we integrate the trained neural network and genetic algorithm to produce a hybrid intelligent algorithm. A run of the hybrid intelligent algorithm (5000 cycles in simulation, 2000 steps in training, 500 generations in genetic algorithm) shows that the optimal solution $(\tilde{x}_1^*, \tilde{x}_2^*, \tilde{x}_3^*, \tilde{x}_4^*, \tilde{x}_5^*)$ is

$$\begin{cases} \tilde{x}_1^* = \sum_{j=-4}^{4} (1 - 0.05|j|)/(12+j) \\ \tilde{x}_2^* = \sum_{j=-4}^{4} (1 - 0.05|j|)/(6+j) \\ \tilde{x}_3^* = \sum_{j=-4}^{4} (1 - 0.05|j|)/(12+j) \\ \tilde{x}_4^* = \sum_{j=-4}^{4} (1 - 0.05|j|)/(15+j) \\ \tilde{x}_5^* = \sum_{j=-4}^{4} (1 - 0.05|j|)/(15+j), \end{cases}$$

whose objective value is $\overline{f}^* \approx 433$.

In order to balance multiple conflicting objectives, suppose that the decision maker has set the target level and priority structure as in Subsection 3. Then we can formulate the CCGP with fuzzy decisions for capital budgeting as follows:

$$\begin{cases} \text{lexmin}\{d_1^-, d_2^+, d_3^+\} \\ \text{subject to:} \\ \quad \text{Pos}\{433 - \sum_{i=1}^{n} a_i \tilde{x}_i \leq d_1^-\} \geq 0.95 \\ \quad \text{Pos}\{\sum_{i=1}^{n} b_i \tilde{x}_i - b \leq d_2^+\} \geq 0.95 \\ \quad \text{Pos}\{\sum_{i=1}^{n} c_i \tilde{x}_i - 700 \leq d_3^+\} \geq 0.90 \\ \quad \text{Pos}\{\xi_{11}\tilde{x}_1 + \xi_{21}\tilde{x}_2 + \xi_{31}\tilde{x}_3 \geq \eta_1\} \geq 0.85 \\ \quad \text{Pos}\{\xi_{42}\tilde{x}_4 + \xi_{52}\tilde{x}_5 \geq \eta_2\} \geq 0.80 \\ \quad d_1^-, d_2^+, d_3^+ \geq 0. \end{cases} \quad (7)$$

Then we train an neural network (5 input neurons, 10 hidden neurons, 3 output neurons) to approximate the uncertain function. Finally, we integrate the trained neural network and genetic algorithm to produce a hybrid intelligent algorithm. A run of the hybrid intelligent algorithm (5000 cycles in simulation, 2000 steps in training, 1500 generations in genetic algorithm) shows that the optimal fuzzy solution $(\tilde{x}_1^*, \tilde{x}_2^*, \tilde{x}_3^*, \tilde{x}_4^*, \tilde{x}_5^*)$ is

$$\begin{cases} \tilde{x}_1^* = \sum_{j=-4}^{4} (1 - 0.05|j|)/(12+j) \\ \tilde{x}_2^* = \sum_{j=-4}^{4} (1 - 0.05|j|)/(14+j) \\ \tilde{x}_3^* = \sum_{j=-4}^{4} (1 - 0.05|j|)/(6+j) \\ \tilde{x}_4^* = \sum_{j=-4}^{4} (1 - 0.05|j|)/(13+j) \\ \tilde{x}_5^* = \sum_{j=-4}^{4} (1 - 0.05|j|)/(13+j), \end{cases}$$

which can satisfy the first two goals, but the third objective is 80.

6 Conclusions

In this paper, CCP and CCGP models were used to formulate capital budgeting problem with fuzzy coefficients and decisions. A hybrid intelligent algorithm was also designed for solving the proposed fuzzy programming models. Two numerical examples showed that the hybrid intelligent algorithm was effective.

Acknowledgments

This work was supported by National Natural Science Foundation of China (No.6042-5309), and Specialized Research Fund for the Doctoral Program of Higher Education (No.20020003009).

References

1. Charnes, A., Cooper, W.W.: Chance-constrained programming. *Management Science*, **6**(1959), 73–79
2. Cybenko, G.: Approximations by superpositions of a sigmoidal function. *Mathematics of Control, Signals and Systems*, **2**(1989), 183–192
3. De, P.K., Acharya, D., Sahu, K.C.: A chance-constrained goal programming model for capital budgeting. *Journal of the Operational Research Society*, **33**(1982), 635–638
4. Gao, J., Lu, M.: Dependent-chance integer programming models for capital budgeting problems in fuzzy environments. *Proceedings of the Asia-Pacific Conference on Genetic Algorithms and Applications*, May 3-5, (2000), Global-Link Publishing Company, Hong Kong, 409–514

5. Hornik, K., Stinchcombe, M., White, H.: Multilayer feedforward networks are universal approximators. *Neural Networks*, **2**(1989), 359–366
6. Iwamura, K., Liu, B.: Chance-constrained integer programming for capital budgeting in fuzzy environments. *Journal of the Operational Research Society*, **49**(1998), 854–860
7. Iwamura, K., Liu B.: Dependent-chance integer programming applied to capital budgeting. *Journal of the Operations Research Society of Japan*, **42**(1999), 117–127
8. Keown, A.J., Martin, J.D.: A chance-constrained goal programming model for working capital management. *Engng Econ* **22**(1977) 153–174
9. Keown, A.J., Taylor, B.W.: A chance-constrained integer goal programming model for capital budgeting in the production area. *Journal of the Operational Research Society*, **31** (1980) 579–589
10. Liu, B., Iwamura, K.: A note on chance-constrained programming with fuzzy coefficients. *Fuzzy Sets and Systems*, **100**(1998), 229–233
11. Liu, B., Iwamura, K.: Chance-constrained programming with fuzzy parameters. *Fuzzy Sets and Systems*, **94**(1998), 227–237
12. Liu, B.: *Uncertain Programming*, John Wiley & Sons, New York, 1999
13. Liu, B., Iwamura, K.: Modeling stochastic decision systems using dependent-chance programming. *European Journal of Operational Research*, **101**(1997), 193–203
14. Liu, B.: Dependent-chance programming: A class of stochastic optimization. *Computers & Mathematics with Applications*, **34**(1997), 89–104
15. Liu, B.: Dependent-chance programming with fuzzy decisions. *IEEE Transactions on Fuzzy Systems*, **7**(1999), 354–360
16. Liu, B.: Dependent-chance programming in fuzzy environments. *Fuzzy Sets and Systems*, **109**(2000), 95–104
17. Liu, B.: *Theory and Practice of Uncertain Programming*, Physica-Verlag, Heidelberg, 2002

Genetic Algorithms for Dissimilar Shortest Paths Based on Optimal Fuzzy Dissimilar Measure and Applications

Yinzhen Li[1,2], Ruichun He[1], Linzhong Liu[1], and Yaohuang Guo[2]

[1] Lanzhou Jiaotong University, Lanzhou 730070, China
[2] Southwest Jiaotong University, Chengdu, 610031, China

Abstract. The derivative problems from the classical shortest path problem (SPP) are becoming more and more important in real life[1]. The dissimilar shortest paths problem is a typical derivative problem. In Vehicles Navigation System(VNS),it is necessary to provide drivers alternative paths to select. Usually, the path selected is a dissimilar path to the jammed path. In fact, "dissimilar" is fuzzy. Considering traffic and transportation networks in this paper, we put forward to the definition of dissimilar paths measure that takes into account the decision maker's preference on both the road sections and the intersections. The minimum model is formulated in which not only the length of paths but also the paths dissimilar measure is considered. And a genetic algorithm also is designed. Finally, we calculate and analyze the dissimilar paths in the traffic network of the middle and east districts of Lanzhou city in P.R. of China by the method proposed in this paper.

1 Introduction

With the development of communications, computer science, traffic and transportation systems, and so on, derivative problems from the classical SPP are becoming more and more important in real life. Many researchers have paid much attention to them[1]. In some applications, because of the restriction of some conditions, it is very difficult to make a single shortest path implementation. Thus, the second shortest path, the third shortest path,..., are needed. For example, in order to find a practicable railway freight path, it is always needed to find k-shortest paths because of the restriction of railway transport capacity. In the 0-1 assignment method for urban traffic assignment problem, it is always needed to find the k-shortest paths. There have been many works on finding k-shortest paths [2-4]. There is another kind of k-shortest paths problem called dissimilar paths in traffic and transportation systems. The dissimilar paths are a set of some shortest paths from the source node to the destination node. These shortest paths have minimum common edges or maximum difference in the geography. The maximum difference in the geography is fuzzy. For example, in Vehicles Navigation System(VNS), it is necessary to provide drivers several alternative paths to select. However, because the circumstance or traffic jam is different in different road sections, and the number of intersections or delay is different in different paths, and so on, drivers always want to choose a satisfied path synthetically according to various factors. Usually, the path chosen is a dissimilar path to the jammed path. In addition, when delivering the military supplies or special hazard goods, dispatchers always choose several dissimilar paths according to weather condition or nature disaster.

S.P.Miaou, S.M.Chin, D.R.Shier and C.K.Lee proposed that after obtaining a set of shortest paths by the traditional algorithms for finding k-shortest paths, decision makers choose a satisfied path from them according to their requirements[3,4,5]. However, only one objective is considered in the traditional algorithms, such as time, distance, cost, and so on. Thus the shortest paths in a set are very similar. In many cases, these paths share not a few of common edges or of common nodes. Generally, it is very difficult for the decision makers to choose the best dissimilar path. R.Dial, F.Glover and D.Karney putted forward to the method called GSP(Gateway Shortest Paths)[6]. The interested readers are referred to the reference [6] for a detailed survey. Obviously, this method has randomness in nominating the passing nodes and emphasizes geographical distributes of a network. In addition, K. Lombard and R.L. Church pointed out that this method would get a group of very similar paths sometimes and lose some very excellent dissimilar paths[7]. E. Erkut and N.M. Ruphail proposed the penalty iteration method[8,12]. The essence of the method consists in the penalty mechanism. Besides, there are p dispersion methods[9,10,11].

In the VNS, considering the delay caused by the traffic jam, drivers want to get a group of shortest paths that they are dissimilar in the road sections and intersections. Taking into account this practical situation, Li and He defined the paths dissimilar measure α based on road sections and intersections, and designed the node-edge penalty algorithm[14]. Further, we formulate a fuzzy dissimilar measure and a minimum α model with the limited length of the paths in this paper. And also we design a genetic algorithm. Finally, as an instance, we calculate and analyze the dissimilar paths in the real traffic network of the middle and east districts of Lanzhou city in P.R. of China.

2 Models

2.1 α Fuzzy Measure

Definition 1. *Given a directed graph $G=\{V,E,W\}$, where $V = \{v_1, v_2, \cdots, v_n\}$ is a set of nodes, $|V| = n$, $E = \{e_1, e_2, \cdots, e_m\}$ is a set of edges, $|E| = m$, and $W = \{w_{i,j} | i,j \in V\}$ is a set of weights of edges. If let $p_{1,n}$ be g^{th} path from source node 1 to destination node n, then $p_{1,n} = \{v_1, e_i, v_k, \cdots, e_j, v_n\}$ denotes a set of the nodes and edges in the path, $l^g_{1,n}$ is the length of the path, $v^g_{1,n}$ is the number of nodes in the path not including the source node and destination node. Let $P_{1,n}$ be a set of all paths from the source node 1 to the destination node n.*

Definition 2. *If $\exists p'_{1,n} \in P_{1,n}$, $\exists p''_{1,n} \in P_{1,n}$, let $l'_{1,n}$ and $l''_{1,n}$ be the length of $p'_{1,n}$ and $p''_{1,n}$ respectively, then define the following membership function*

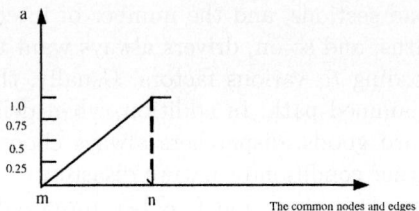

Fig. 1. The membership function where, $p'_{1,n} = p''_{1,n}$ at "m", and $p'_{1,n} \cap p''_{1,n} = \Phi$ at "n"

$$\alpha = \begin{cases} 1, & when\ p'_{1,n} = p''_{1,n}, \\ \frac{2\lambda \sum w_{i,j}}{l'_{1,n} + l''_{1,n}} + \frac{2(1-\lambda)v^*_{1,n}}{v'_{1,n} + v''_{1,n}}, & others, \\ 0, & when\ p'_{1,n} \cap p''_{1,n} = \Phi. \end{cases}$$

called paths fuzzy dissimilar measure of path $p'_{1,n}$ to $p''_{1,n}$ (or $p''_{1,n}$ to $p'_{1,n}$), simply called α fuzzy measure, where $w_{i,j}: e_k \in p'_{1,n} \cap p''_{1,n}$, and $v^*_{1,n}$ is the number of nodes(not including source node 1 and destination node n)that they present commonly in $p'_{1,n}$ and $p''_{1,n}$. λ and $1 - \lambda$ are the decision maker's preference to edges(road sections) and nodes(intersections) respectively.

2.2 Model

Assume we have gotten $m - 1$ shortest paths $p^r_{1,n}$ from node 1 to node n, $r = 1, 2, \ldots, m - 1$. The shortest path is $p^*_{1,n}$ among them, its length is $l^*_{1,n}$. To get m^{th} shortest path $p^m_{1,n}$ which has the best paths dissimilar measure to $p^r_{1,n}$, $r = 1, 2, \ldots, m - 1$, we formulate the following optimal α model with the limited length of the path.

$$Min\ \alpha = \frac{1}{C^2_m} \sum_{C^2_m} \alpha_k, \quad k = 1, 2, \ldots, C^2_m.$$

s.t. $\alpha_k = \begin{cases} 1, & when\ p^g_{1,n} = p^h_{1,n}, \\ \frac{2\lambda \sum w_{i,j}}{l^g_{1,n} + l^h_{1,n}} + \frac{2(1-\lambda)v^*_{1,n}}{v^g_{1,n} + v^h_{1,n}}, & g = 1, 2, \ldots, m;\ h = 1, 2, \ldots, m;\ g \neq h,\ , others, \\ 0, & when\ p^g_{1,n} \cap p^h_{1,n} = \Phi. \end{cases}$

$$\sum x_{i,j} - \sum x_{j,i} = \begin{cases} 1, & if\ i = 1, \\ 0, & \forall\ i,j \in V\backslash\{1,n\}, \\ -1, & if\ i = n. \end{cases}$$

$l^m_{1,n} = \sum w_{i,j} x_{i,j}$,
$l^m_{1,n} \leq (1 + \beta) l^*_{1,n}$,
$x_{i,j} = 0$ or 1.

where, the first formula minimizes paths fuzzy dissimilar measure α. The second formula calculates the paths fuzzy dissimilar measure of arbitrary two paths. The third is the constraints of the path from node 1 to node n. The 4th and 5th are the constraints of mth path length. β is a constant given.

It is obvious that the model is NP-Hard. We propose the following genetic algorithms.

3 Algorithm

The priority-based encoding genetic algorithm is a very effective method. M.Gen [13] solved the bicriterion SP problem by this method in 2004. But he only considered acyclic graph. By the use of the idea of the priority-based encoding, aiming at arbitrary networks with non-negative weights, we design the genetic algorithm for the above model.

The dissimilar paths problem is very complex. Its solution is a set of solutions, not one solution. And the set of solutions includes the shortest path $p_{1,n}^*$. Thus it is not necessary to design and calculate the fitness function of single chromosome. In other words, it is needed to formulate the fitness function of a set of chromosomes, and select several sets of chromosomes as the offspring chromosomes. Assume we want to find m optimal dissimilar paths, then each set of chromosomes includes $m-1$ chromosomes, because the shortest path $p_{1,n}^*$ is certainly included in the optimal solutions. Thus, the size of population is certainly $m-1$ multiples.

Node i	1	2	3	4	5	6	7
Priority n(i)	5	4	1	6	7	3	2

Fig. 2. The code of a chromosome

The steps of the algorithm are as follows.

Step 1. Find the shortest path $p_{1,n}^*$ from node 1 to node n by Dijkstra algorithm, and marked as m^{th} path. Let l^* be its length.

Step 2. Generate initial population.

Step 2.1. Let $n(i) = i, \forall i \in V$.

Step 2.2. Generate $\lceil \frac{n}{2} \rceil$ random integer pairs $l_1, l_2 \in [1, n]$, if $l_1 \neq l_2$, then $swap(n(l_1), n(l_2))$. Array $n()$ is a chromosome.

Step 2.3. Decode. The process of decoding from node 1 to node n is as follows.

$i = 1$

flag=false

$P = \Phi$

while not flag

    A(i)={all adjacent vertices of node i}

    repeat

        $j^* = \arg \max_j \{n(j)\}, j \in A(i)$

        if $j^* \in P$ then $A(i) - \{j^*\} \longrightarrow A(i)$

        else $p(i) = j^*; \{j^*\} + P \longrightarrow P; i = j^*$, break if $j^* = n$ then flag=true; return

if $A(i) = \Phi$ then go to Step 2.1 { array n() is a unfeasible chromosome.}

until true

loop

p() is a path from node 1 to node n.

Step 2.4. Examine the feasibility.

Calculate the length of path $p() : l = \sum w_{i,j}$, If $l \leq (1+\beta)l^*$, then n() is a feasible chromosome, otherwise, n() is not feasible, go to Step 2.1.

Step 2.5. If $size$ feasible chromosomes are obtained, go to Step3, otherwise, go to Step2.1 to continue.

Step 3. Crossover operator. The OX(Order Crossover) operator is adopted in the paper. Select chromosomes by crossover probability P_c and randomly group them by pairs. In this way, we can obtain the pairs as crossover parents. The OX is shown in Figure.3.

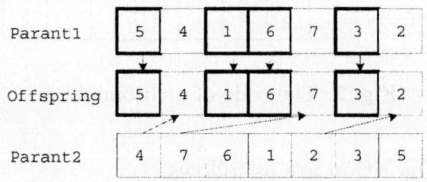

Fig. 3. Example of OX operator

Step 3.1. Random multi-position crossover is adopted here. We can get an offspring by crossing one pairs. For example, an offspring is generated by selecting preserved genes of $parent1$ at random and replace unpreserved genes of $parent1$ with corresponding genes of $parent2$.

Step 3.2. Examine the feasibility of the offspring generated in Step 3.1 by the use of Step 2.3 and Step 2.4. If it is unfeasible, repeat Step 3.1 and Step 3.2 until it is feasible or the predetermined rounds has been run. If we cannot obtain a feasible offspring by Step 3.1 and Step 3.2, the $parent1$(or $parent2$) will be reserved as an offspring.

Step 3.3. Repeat Step 3.1 and Step 3.2 until all pairs are crossed.

Step 4. Mutation operator. Select chromosomes by mutation probability P_m. The swap mutation operator is used here.

Step 4.1. Select two gene positions of a mutation parent randomly and then swap them. The process is shown in Figure.4.

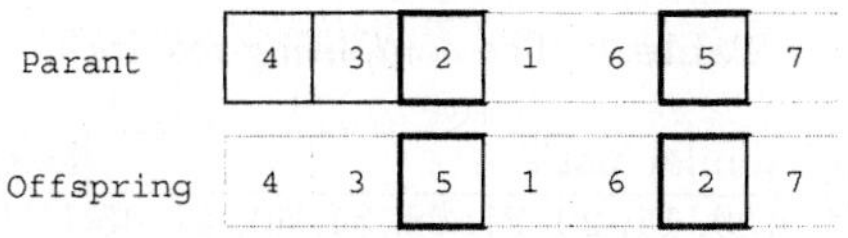

```
Parant     | 4 | 3 | 2 | 1 | 6 | 5 | 7 |
Offspring  | 4 | 3 | 5 | 1 | 6 | 2 | 7 |
```

Fig. 4. Example of mutation operator

Step 4.2. Examine the feasibility of the offspring generated in Step 4.1 by the use of Step 2.3 and Step 2.4. If it is unfeasible, repeat Step 4.1 and Step 4.2 until it is feasible or the predetermined rounds has been run. If we cannot obtain a feasible offspring by Step 4.1 and Step 4.2, the parent will be reserved as an offspring.

Step 4.3. Repeat Step 4.1 and Step 4.2 until all mutation chromosomes are processed.

Step 5. Evaluate the fitness function of chromosomes group.

Because of $0 \leq \alpha \leq 1$, we know by the definition of α that the smaller α, the better the paths dissimilar measure. Thus, the fitness function of kth chromosomes group is defined as follows:

$$eval(k) = 1 - \frac{\alpha_k}{\sum_{j=1}^{C_{size}^{m-1}} \alpha_j}, \quad k = 1, 2, \ldots, C_{size}^{m-1}.$$

where

$$\alpha_k = \frac{1}{C_m^2} \sum_{g,h} \begin{cases} 1, & \text{when } p_{1,n}^g = p_{1,n}^h, \\ \frac{2\lambda \sum w_{i,j}}{l_{1,n}^g + l_{1,n}^h} + \frac{2(1-\lambda)v_{1,n}^*}{v_{1,n}^g + v_{1,n}^h}, & g=1,2,\ldots,m; \ h=1,2,\ldots,m; \ g \neq h, \text{ others}, \\ 0, & \text{when } p_{1,n}^g \cap p_{1,n}^h = \Phi. \end{cases}$$

Step 6. We select $T(T \times (m-1) = size)$ chromosomes groups as the offspring chromosomes by wheel approach.

The probability of kth chromosomes group to be selected is

$$\frac{eval(k)}{\sum_k eval(k)}, \quad k = 1, 2, \ldots, C_{size}^{m-1}.$$

Step 7. Repeat Step 3-Step 6 until a satisfied set of solutions is obtained or predetermined rounds is run.

4 Instance

In VNS, it is very necessary for the drivers to avoid the jammed traffic routes by providing some dissimilar paths for them. Figure.5 is the simplified network of the middle and east districts of Lanzhou city in China. In terms of the real data and calculating, the route $1, 4, 9, 14, 20, 21, 28, 34, 40, 44, 45$ is always jammed among all paths from 1 to 45. Now we want to provide 4 dissimilar routes from 1 to 45 for the drivers. We designed the computer program for the algorithm and

Table 1. The computing results

no	dissimilar path	length
1	1, 4, 9, 14, 20, 21, 28, 34, 40, 44, 45	11271
2	1, 3, 8, 13, 19, 26, 33, 39, 43, 45	12932
3	1, 5, 11, 12, 45	13509
4	1, 5, 11, 22, 29, 35, 40, 44, 45	12384
5	1, 4, 9, 14, 20, 27, 33, 39, 43, 45	12398
The average dissimilar measure α=0.11145		

run it with the instance. Where $size = 40, P_c = 0.4, P_m = 0.2, \lambda = 0.5, \beta = 0.2$. The computing results are in table 1. Where no.2–no.5 are dissimilar paths that are provided for the drivers.

The decision makers may choose the required route in table 1. Figure.6 shows that the optimal solutions does not mater with the initial population. This illustrates the robustness of the algorithm proposed in the paper is good. In addition, Figure.7 shows that the convergence of the algorithm is also good.

5 Conclusion

In Vehicles Navigation System(VNS),it is necessary to provide drivers alternative paths to select. Usually, the path selected is a fuzzy dissimilar path to the jammed path. Considering traffic and transportation networks in this paper, we put forward to the definition of α fuzzy dissimilar paths measure that takes into account the decision maker's preference on both the road sections and the intersections. The minimum α model is formulated in which not only the length of paths but also the paths dissimilar measure is considered. The genetic algorithm designed has good robustness and convergence. It can find a group of dissimilar paths being of both shortest lengths and best fuzzy dissimilar measures.

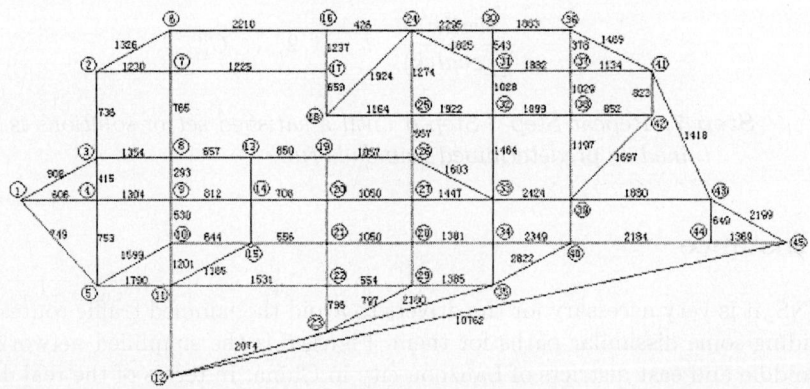

Fig. 5. The simplified traffic network of middle and east districts of Lanzhou city

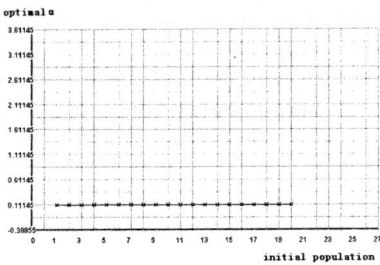

Fig. 6. The relation of optimal and initial population

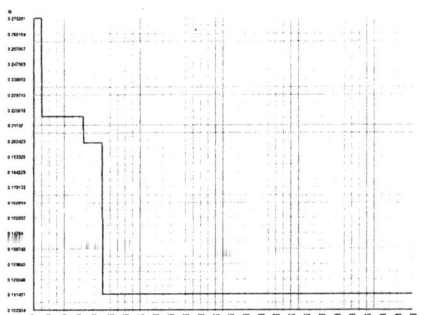

Fig. 7. The evolution process of the first population

Acknowledgments

This work is supported by National Natural Science Foundation of China(No. 70071028), and Qinglan Project of Lanzhou Jiaotong University.

References

1. B.V. Cherkassky, Andrew V. Goldberg, Tomasz Radzik. Shortest paths algorithms:Theory and experimental evaluation[J]. Mathematical programming, 1996: 129–174.
2. J.Y.Yen. Finding the k shortest loopless paths in a network[J]. Management Science, 1971: 17, 721–716.
3. S.P. Miaou, S.M. Chin. Computing k-shortest paths for nuclear spent fuel highway transportation[J]. European Journal of Operational Research, 1991: 53,64–80.
4. D.R. Shier. On algorithms for finding the k shortest paths in a network[J]. Networks,1979: 17,341–352.
5. C.K.Lee. A multiple path routing strategy for Vehicle route guidance systems[J]. Tran. Res.,1994:2c, 185–195.
6. R.Dial, F.Glover, D.Karney. A computational analysis of alternative algorithms and labeling techniques for finding shortest path trees[J]. Networks,1999:9, 214–248.

7. K. Lombard, R.L. Church. The gateway shortest path problem:Generating alternative routes for a corridor location problem[J]. Geographical systems,1993: 1,25–45.
8. Kanliang Wang,Yinfeng Xu. A model and algorithm on the path selection in dissimilar path problem [J]. TheoryMethodApplication of system engineering, 2001: 1, 8–12.
9. M.Kuby. Programming models for facility dispersion: The p-dispersion and Maximum Dispersion Problems[J]. Geographical Analysis,1987: 19, 315–329.
10. E. Erkut. The discrete p-dispersion problem[J]. European Journal of operational Research,1990: 46,48–60.
11. N.M. Ruphail, et al. A Dss for dynamic pre-trip route planning[A]. Applications of advanced technologies in transportation engineering[C]: Proc of the 4th Intl. Conference,1995: 325–329.
12. Qun Yang . Research on the path selection based on multiple rational paths[J]. Journal of managing Engineering, 2004: 4, 42–45.
13. M.Gen. Bicriterion network design problem[R]. the 3d International Information and Management Science Conference, Dunhuang, P.R.of China,2004: 5, 419–425.
14. Yinzhen Li and Ruichun He. On the models and algorithms for finding dissimilar shortest paths in a traffic network[R]. The 7th National Operation Research Conference of China, Qingdao, China, 2004: 8

Convergence Criteria and Convergence Relations for Sequences of Fuzzy Random Variables

Yan-Kui Liu[1] and Jinwu Gao[2]

[1] College of Mathematics and Computer Science, Hebei University
Baoding 071002, Hebei, China
liuyankui@tsinghua.org.cn

[2] School of Information, Renmin University of China, Beijing 100872, China
jgao@orsc.edu.cn

Abstract. Fuzzy random variable is a measurable map from a probability space to a collection of fuzzy variables. In this paper, we first present several new convergence concepts for sequences of fuzzy random variables, including convergence almost sure, uniform convergence, almost uniform convergence, convergence in mean chance, and convergence in mean chance distribution. Then, we discuss the criteria for convergence almost sure, almost uniform convergence, and convergence in mean chance. Finally, we deal with the relationship among various types of convergence.

1 Introduction

In classical measure theory [1], there are well-known convergence theorems such as Lebesgue's theorem, Riesz's theorem and Egoroff's theorem for sequences of measurable functions. Nonadditive measure theory [22] is an extension of measure theory. By using the structural characteristics of fuzzy measure, most of the convergence theorems in measure theory for both real-valued and set-valued measurable functions have been studied [10,11]. Possibility theory [2,4,23] can be regarded as a subset theory of nonadditive measure theory since it is based on two dual semicontinuous nonadditive measures—possibility and necessity measures, it is also the theoretical foundation of fuzzy programming [7].

In a practical decision-making process, we often face a hybrid uncertain environment where linguistic and frequent nature coexist. Fuzzy random variable is one of the tools to deal with this kind of twofold uncertainty. The term fuzzy random variable was introduced by Kwakernaak [6] to depict the phenomena in which fuzziness and randomness appear simultaneously. Since then, its variants as well as extensions were presented by other researchers, aiming at different purposes, e.g., Krätschmer [3], Kruse and Meyer [5], López-Diaz and Gil [17], and Puri and Ralescu [20]. In this paper, we adopt the definition of fuzzy random variable given in [12] for fuzzy random optimization such as fuzzy random expected value model [13], fuzzy random minimum-risk problem [14], and fuzzy random equilibrium chance-constrained programming [15]. For other formulation of fuzzy random programming, we may refer to Wang and Qiao [16], and

Luhandjula [18]. Our goal in this paper is to study the convergence for sequences of fuzzy random variables, which are the theoretical foundation of fuzzy random optimization.

The material is arranged into five sections. First, in Section 2, we recall some concepts in fuzzy random theory such as fuzzy variable, credibility measure, fuzzy random variable, and the mean chance of a fuzzy random event. Section 3 first presents several new convergence concepts for sequences of fuzzy random variables such as uniform convergence, convergence in mean chance, and convergence in mean chance distribution, then discusses their convergence criteria. The relationship among various types of convergence is covered in Section 4.

2   Fuzzy Random Variables

Given a universe Γ, an ample field [21] $\mathcal{A}$ on Γ is a class of subsets of Γ that is closed under arbitrary unions, intersections, and complementation in Γ. Let Pos be a set function defined on the ample field $\mathcal{A}$. The set function Pos is said to be a possibility measure if it satisfies the following conditions

(P1) $\text{Pos}(\emptyset)=0$, and $\text{Pos}(\Gamma) = 1$;

(P2) $\text{Pos}\left(\bigcup_{i \in I} A_i\right) = \sup_{i \in I} \text{Pos}(A_i)$ for any subclass $\{A_i \mid i \in I\}$ of $\mathcal{A}$, where I is an arbitrary index set.

The triplet $(\Gamma, \mathcal{A}, \text{Pos})$ will be called a *possibility space*, which was called a *pattern space* by Nahmias [19].

Based on possibility measure, a self-dual set function Cr, called *credibility measure*, was defined as follows [9]:

Definition 1. *Let $(\Gamma, \mathcal{A}, \text{Pos})$ be a possibility space. The credibility measure, denoted* Cr, *is defined as*

$$\text{Cr}(A) = \frac{1}{2}\left(1 + \text{Pos}(A) - \text{Pos}(A^c)\right), \quad A \in \mathcal{A}, \tag{1}$$

where A^c is the complement of A.

A credibility measure has the following properties:

(C1) $\text{Cr}(\emptyset)=0$, and $\text{Cr}(\Gamma) = 1$.
(C2) Monotonicity: $\text{Cr}(A) \leq \text{Cr}(B)$ for all $A, B \in \mathcal{A}$ with $A \subset B$.
(C3) Self-duality: $\text{Cr}(A) + \text{Cr}(A^c) = 1$ for all $A \in \mathcal{A}$.
(C4) Subadditivity: $\text{Cr}(A \cup B) \leq \text{Cr}(A) + \text{Cr}(B)$ for all $A, B \in \mathcal{A}$.

Using possibility space, a fuzzy variable can be defined as follows.

Definition 2. *Let $(\Gamma, \mathcal{A}, \text{Pos})$ be a possibility space. A function X from Γ to $\Re$ is said to be a fuzzy variable defined on the possibility space if*

$$\{\gamma \mid X(\gamma) \leq t\} \in \mathcal{A} \tag{2}$$

for every $t \in \Re$.

The possibility distribution of X is defined as

$$\mu_X(t) = \text{Pos}\{\gamma \in \Gamma \mid X(\gamma) = t\} \tag{3}$$

for every $t \in \Re$.

Definition 3 ([12]). *Let $(\Omega, \Sigma, \Pr)$ be a probability space. A fuzzy random variable is a map $\xi : \Omega \to \mathcal{F}_v$ such that for any Borel subset B of $\Re$, the following function*

$$\xi^*(B)(\omega) = \text{Pos}\{\gamma \mid \xi_\omega(\gamma) \in B\} \tag{4}$$

is measurable with respect to ω, where $\mathcal{F}_v$ is a collection of fuzzy variables defined on a possibility space $(\Gamma, \mathcal{A}, \text{Pos})$.

Suppose ξ is a fuzzy random variable, and f is a continuous functions defined on $\Re$. In the theory of fuzzy random optimization, a fuzzy random event is usually expressed as follows

$$f(\xi) \leq 0. \tag{5}$$

For each fixed $\omega \in \Omega$, ξ_ω is a fuzzy variable and (5) becomes a fuzzy event

$$f(\xi_\omega) \leq 0 \tag{6}$$

whose credibility is denoted as $\text{Cr}\{f(\xi_\omega) \leq 0\}$, which is a measurable function of ω [12].

Definition 4 ([14]). *Let ξ be a fuzzy random variable, and $f : \Re \to \Re$ be a continuous function. Then the mean chance, denoted Ch, of the fuzzy random event (5) is defined as*

$$\text{Ch}\{f(\xi) \leq 0\} = \int_0^1 \Pr\{\omega \in \Omega \mid \text{Cr}\{f(\xi_\omega) \leq 0\} \geq p\} \, dp. \tag{7}$$

Since $\Pr$ is an additive measure, the mean chance defined by (7) is equivalent to the following form

$$\text{Ch}\{f(\xi) \leq 0\} = \int_\Omega \text{Cr}\{f(\xi_\omega) \leq 0\} \Pr(d\omega). \tag{8}$$

3 Convergence Criteria

3.1 Modes of Convergence

The intent of this section is to present some new convergence concepts for sequences of fuzzy random variables, including convergence almost uniform, convergence almost sure, convergence in chance, and convergence in distribution. From measure-theoretic viewpoint, these concepts are important for the development of fuzzy random theory [8].

Definition 5 (Convergence almost sure). *A sequence $\{\xi_n\}$ of fuzzy random variables is said to converge almost surely to a fuzzy random variable ξ, denoted $\xi_n \xrightarrow{a.s.} \xi$, if there exist $E \in \Sigma, F \in \mathcal{A}$ with $\Pr(E) = \mathrm{Cr}(F) = 0$ such that for every $(\omega, \gamma) \in \Omega \backslash E \times \Gamma \backslash F$,*

$$\lim_{n \to \infty} \xi_{n,\omega}(\gamma) \to \xi_\omega(\gamma).$$

Definition 6 (Uniform convergence). *A sequence $\{\xi_n\}$ of fuzzy random variables is said to converge uniformly to a fuzzy random variable ξ on $\Omega \times \Gamma$, denoted $\xi_n \xrightarrow{u.} \xi$, if*

$$\lim_{n \to \infty} \sup_{(\omega,\gamma) \in \Omega \times \Gamma} |\xi_{n,\omega}(\gamma) - \xi_\omega(\gamma)| = 0.$$

Definition 7 (Almost uniform convergence). *A sequence $\{\xi_n\}$ of fuzzy random variables is said to converge almost uniformly to a fuzzy random variable ξ, denoted $\xi_n \xrightarrow{a.u.} \xi$, if there exist two nonincreasing sequences $\{E_m\} \subset \Sigma, \{F_m\} \subset \mathcal{A}$ with $\lim_m \Pr(E_m) = \lim_m \mathrm{Cr}(F_m) = 0$ such that for each $m = 1, 2, \cdots$, we have $\xi_n \xrightarrow{u.} \xi$ on $\Omega \backslash E_m \times \Gamma \backslash F_m$.*

Definition 8 (Convergence in chance). *A sequence $\{\xi_n\}$ of fuzzy random variables is said to converge in chance Ch to a fuzzy random variable ξ, denoted $\xi_n \xrightarrow{\mathrm{Ch}} \xi$, if for every $\varepsilon > 0$,*

$$\lim_{n \to \infty} \mathrm{Ch}\{|\xi_n - \xi| \geq \varepsilon\} = 0.$$

Definition 9. *Let ξ be a fuzzy random variable. The (mean) chance distribution function of ξ is defined as*

$$G_\xi(t) = \mathrm{Ch}\{\xi \geq t\}, \quad t \in \Re.$$

It is evident that G_ξ is a nonincreasing real-valued function.

Let $\{F_n\}$ and F be nonincreasing real-valued functions. The sequence $\{F_n\}$ is said to converge weakly to F, denoted $F_n \xrightarrow{w.} F$, if $F_n(t) \to F(t)$ for all continuity points t of F.

Definition 10 (Convergence in distribution). *Let G_{ξ_n} and G_ξ be chance distribution functions of fuzzy random variables ξ_n and ξ, respectively. The sequence $\{\xi_n\}$ is said to converge in distribution to ξ, denoted $\xi_n \xrightarrow{d.} \xi$, if $G_{\xi_n} \xrightarrow{w.} G_\xi$.*

3.2 Convergence Criteria

The purpose of this section is to deal with the criteria for the convergence almost sure, convergence almost uniform, and convergence in mean chance.

Proposition 1 (Convergence a.s. criterion). *Suppose $\{\xi_n\}$ and ξ are fuzzy random variables. Then $\xi_n \xrightarrow{a.s.} \xi$ if and only if for every $\varepsilon > 0$,*

$$\mathrm{Ch}\left(\bigcap_{m=1}^{\infty} \bigcup_{n=m}^{\infty} \{|\xi_n - \xi| \geq \varepsilon\}\right) = 0. \tag{9}$$

Proof. First, it is easy to check that $\xi_n \xrightarrow{a.s.} \xi$ iff the limit $\xi_{n,\omega} \xrightarrow{a.s.} \xi_\omega$ holds almost surely with respect to ω, i.e.,

$$\Pr\left\{\omega \mid \xi_{n,\omega} \xrightarrow{a.s.} \xi_\omega\right\} = 1.$$

Therefore, there exists $E \in \Sigma$ with $\Pr(E) = 0$ such that for each $\omega \in \Omega \backslash E$, $\xi_{n,\omega} \xrightarrow{a.s.} \xi_\omega$, i.e., for every $\varepsilon > 0$,

$$\mathrm{Cr}\left(\bigcap_{m=1}^{\infty} \bigcup_{n=m}^{\infty} \{\gamma \in \Gamma \mid |\xi_{n,\omega}(\gamma) - \xi_\omega(\gamma)| \geq \varepsilon\}\right) = 0, \text{ a.s.}$$

which is equivalent to

$$\mathrm{Ch}\left(\bigcap_{m=1}^{\infty} \bigcup_{n=m}^{\infty} \{|\xi_n - \xi| \geq \varepsilon\}\right) = 0.$$

The proof is complete. □

Proposition 2 (Convergence a.u. criterion). *Suppose $\{\xi_n\}$ and ξ are fuzzy random variables. If $\xi_n \xrightarrow{a.u.} \xi$, then for every $\varepsilon > 0$, the limit*

$$\lim_{m \to \infty} \mathrm{Cr}\left(\bigcup_{n=m}^{\infty} \{\gamma \in \Gamma \mid |\xi_{n,\omega}(\gamma) - \xi_\omega(\gamma)| \geq \varepsilon\}\right) = 0 \tag{10}$$

holds almost surely with respect to ω, and furthermore

$$\lim_{m \to \infty} \mathrm{Ch}\left(\bigcup_{n=m}^{\infty} \{|\xi_n - \xi| \geq \varepsilon\}\right) = 0. \tag{11}$$

Conversely, if ω is a discrete random variable assuming finite number of values, then Eq. (10) implies $\xi_n \xrightarrow{a.u.} \xi$.

Proof. If $\xi_n \xrightarrow{a.u.} \xi$, then there exist two nonincreasing sequences $\{E_m\} \subset \Sigma$, $\{F_m\} \subset \mathcal{A}$ with $\lim_m \Pr(E_m) = \lim_m \mathrm{Cr}(F_m) = 0$ such that for each $m = 1, 2, \cdots$, $\xi_n \xrightarrow{u.} \xi$ on $\Omega \backslash E_m \times \Gamma \backslash F_m$.

Let $E = \bigcap_{m=1}^{\infty} E_m$. Then $\Pr(E) = 0$, and for every $\omega \in \Omega \backslash E$, there is a positive integer m_ω such that $\omega \in \Omega \backslash E_{m_\omega}$. Since $\{E_m\}$ is nonincreasing, we have $\omega \in \Omega \backslash E_m$ whenever $m \geq m_\omega$. Therefore, there is a subsequence $\{F_m, m \geq m_\omega\}$ of $\{F_m\}$ such that for each $m_\omega, m_\omega + 1, \cdots$, the sequence $\{\xi_{n,\omega}\}$ converges to ξ_ω uniformly on F_m, i.e., $\xi_{n,\omega} \xrightarrow{a.u.} \xi_\omega$ almost surely with respect to ω.

We now show that $\xi_{n,\omega} \xrightarrow{a.u.} \xi_\omega$ implies

$$\lim_{m \to \infty} \mathrm{Ch}\left(\bigcup_{n=m}^{\infty} \{|\xi_n - \xi| \geq \varepsilon\}\right) = 0.$$

In fact, for any $\delta > 0$, there exists $F_\omega \in \mathcal{A}$ with $\mathrm{Cr}(F_\omega) < \delta$ such that $\{\xi_{n,\omega}\}$ converges to ξ_ω uniformly on $\Gamma \backslash F_\omega$. Thus, for every $\varepsilon > 0$, there exists a positive integer $m(\varepsilon, \omega)$ such that for all $\gamma \in \Gamma \backslash F_\omega$, $|\xi_{n,\omega}(\gamma) - \xi_\omega(\gamma)| < \varepsilon$ whenever $n \geq m$. Therefore

$$\Gamma \backslash F_\omega \subset \bigcap_{n=m}^{\infty} \{\gamma \in \Gamma \mid |\xi_{n,\omega}(\gamma) - \xi_\omega(\gamma)| < \varepsilon\},$$

or

$$\bigcup_{n=m}^{\infty} \{\gamma \in \Gamma \mid |\xi_{n,\omega}(\gamma) - \xi_\omega(\gamma)| \geq \varepsilon\} \subset F_\omega,$$

which implies

$$\mathrm{Cr}\left(\bigcup_{n=m}^{\infty} \{\gamma \in \Gamma \mid |\xi_{n,\omega}(\gamma) - \xi_\omega(\gamma)| \geq \varepsilon\}\right) \leq \mathrm{Cr}(F_\omega) < \delta.$$

Letting $\delta \to 0$, we have

$$\limsup_{n \to \infty} \mathrm{Cr}\left(\bigcup_{n=m}^{\infty} \{\gamma \in \Gamma \mid |\xi_{n,\omega}(\gamma) - \xi_\omega(\gamma)| \geq \varepsilon\}\right) = 0.$$

Applying dominated convergence theorem, we obtain

$$\lim_{m \to \infty} \mathrm{Ch}\left(\bigcup_{n=m}^{\infty} \{|\xi_n - \xi| \geq \varepsilon\}\right) = \int_\Omega \lim_{n \to \infty} \mathrm{Cr}\left(\bigcup_{n=m}^{\infty} \{|\xi_{n,\omega} - \xi_\omega| \geq \varepsilon\}\right) = 0.$$

Conversely, suppose Eq. (10) is valid, we prove $\xi_n \xrightarrow{a.u.} \xi$. By supposition, assume ω has the following probability distribution

$$\omega \sim \begin{pmatrix} \omega_1, \omega_2, \cdots, \omega_N \\ p_1, p_2, \cdots, p_N \end{pmatrix}$$

with $p_i > 0$ and $\sum_{i=1}^{N} p_i = 1$. Since for each $i = 1, \cdots, N$, we have

$$\lim_{m \to \infty} \mathrm{Cr}\left(\bigcup_{n=m}^{\infty} \{\gamma \in \Gamma \mid |\xi_{n,\omega_i}(\gamma) - \xi_{\omega_i}(\gamma)| \geq \varepsilon\}\right) = 0.$$

Then for every $\delta \in (0, 1)$, and each $k = 1, 2, \cdots$, there exists a positive integer m_k such that

$$\mathrm{Cr}\left(\bigcup_{n=m_k}^{\infty} \{\gamma \in \Gamma \mid |\xi_{n,\omega_i}(\gamma) - \xi_{\omega_i}(\gamma)| \geq 1/k\}\right) < \delta/2^k$$

for $i = 1, \cdots, N$. Let

$$F = \bigcup_{i=1}^{N} \bigcup_{k=1}^{\infty} \bigcup_{n=m_k}^{\infty} \{\gamma \in \Gamma \mid |\xi_{n,\omega_i}(\gamma) - \xi_{\omega_i}(\gamma)| \geq 1/k\}.$$

Then
$$\operatorname{Cr}(F) = \operatorname{Cr}\left(\bigcup_{i=1}^{N}\bigcup_{k=1}^{\infty}\bigcup_{n=m_k}^{\infty}\{\gamma \in \Gamma \mid |\xi_{n,\omega_i}(\gamma) - \xi_{\omega_i}(\gamma)| \geq 1/k\}\right)$$
$$\leq \sup_i \sup_k \operatorname{Pos}\left(\bigcup_{n=m_k}^{\infty}\{\gamma \mid |\xi_{n,\omega_i}(\gamma) - \xi_{\omega_i}(\gamma)| \geq 1/k\}\right) < \delta.$$

In addition, for each $k = 1, 2, \cdots$, one has
$$\sup_{1 \leq i \leq N} \sup_{\gamma \in \Gamma \setminus F} |\xi_{n,\omega_i}(\gamma) - \xi_{\omega_i}(\gamma)| < 1/k$$

whenever $n \geq m_k$, which implies $\xi_n \xrightarrow{a.u.} \xi$. The proof is complete. □

Proposition 3 (Convergence in chance criterion). *Suppose $\{\xi_n\}$ and ξ are fuzzy random variables. Then $\xi_n \xrightarrow{\operatorname{Ch}} \xi$ if and only if for every $\varepsilon > 0$, $\operatorname{Cr}\{\gamma \mid |\xi_{n,\omega}(\gamma) - \xi_\omega(\gamma)| \geq \varepsilon\} \xrightarrow{\operatorname{Pr}} 0$.*

Proof. Assume that $\xi_n \xrightarrow{\operatorname{Ch}} \xi$, then for every $\varepsilon > 0$ and $\eta > 0$,
$$\Pr\{\omega \mid \operatorname{Cr}\{|\xi_{n,\omega}(\gamma) - \xi_\omega(\gamma)| \geq \varepsilon\} \geq \eta\}$$
$$\leq \tfrac{1}{\eta}\int_\Omega \operatorname{Cr}\{\gamma \mid |\xi_{n,\omega}(\gamma) - \xi_\omega(\gamma)| \geq \varepsilon\}\Pr(d\omega) = \tfrac{1}{\eta}\operatorname{Ch}\{|\xi_n - \xi| \geq \varepsilon\},$$

which implies $\operatorname{Cr}\{\gamma \mid |\xi_{n,\omega}(\gamma) - \xi_\omega(\gamma)| \geq \varepsilon\} \xrightarrow{\operatorname{Pr}} 0$.

On the other hand, if $\operatorname{Cr}\{\gamma \mid |\xi_{n,\omega}(\gamma) - \xi_\omega(\gamma)| \geq \varepsilon\} \xrightarrow{\operatorname{Pr}} 0$, then for every $\alpha \in (0, 1]$,
$$\lim_{n \to \infty} \Pr\{\omega \mid \operatorname{Cr}\{|\xi_{n,\omega}(\gamma) - \xi_\omega(\gamma)| \geq \varepsilon\} \geq \alpha\} = 0.$$

Applying Lebesgue bounded convergence theorem, we obtain
$$\lim_{n \to \infty} \operatorname{Ch}\{|\xi_n - \xi| \geq \varepsilon\}$$
$$= \int_0^1 \lim_{n \to \infty} \Pr\{\omega \mid \operatorname{Cr}\{|\xi_{n,\omega}(\gamma) - \xi_\omega(\gamma)| \geq \varepsilon\} \geq \alpha\} d\alpha = 0,$$

which completes the proof. □

Corollary 1. *Suppose $\{\xi_n\}$ and ξ are fuzzy random variables. Then $\xi_n \xrightarrow{\operatorname{Ch}} \xi$ provided that the limit $\xi_{n,\omega} \xrightarrow{\operatorname{Cr}} \xi_\omega$ holds almost surely with respect to ω.*

Proof. Assume that $\xi_{n,\omega} \xrightarrow{\operatorname{Cr}} \xi_\omega$ almost surely with respect to ω. Then for every $\varepsilon > 0$, the limit
$$\lim_{n \to \infty} \operatorname{Cr}\{\gamma \mid |\xi_{n,\omega}(\gamma) - \xi_\omega(\gamma)| \geq \varepsilon\} = 0$$

holds with probability 1. Since convergence a.s. implies convergence in probability, one has
$$\operatorname{Cr}\{\gamma \mid |\xi_{n,\omega}(\gamma) - \xi_\omega(\gamma)| \geq \varepsilon\} \xrightarrow{\operatorname{Pr}} 0,$$

which, by Proposition 3, implies $\xi_n \xrightarrow{\operatorname{Ch}} \xi$. □

4 Relationship

In this section, we discuss the relationship among various types of convergence based on the convergence criteria obtained in Section 3.

The following theorem compares convergence a.u. and convergence a.s..

Theorem 1. *Suppose $\{\xi_n\}$ and ξ are fuzzy random variables. If $\xi_n \xrightarrow{a.u.} \xi$, then $\xi_n \xrightarrow{a.s.} \xi$.*

Proof. Assume $\xi_n \xrightarrow{a.u.} \xi$. By Proposition 2, we have

$$\lim_{m \to \infty} \mathrm{Ch}\left(\bigcup_{n=m}^{\infty} \{|\xi_n - \xi| \geq \varepsilon\}\right) = 0.$$

Since

$$\mathrm{Ch}\left(\bigcap_{m=1}^{\infty} \bigcup_{n=m}^{\infty} \{|\xi_n - \xi| \geq \varepsilon\}\right) \leq \bigcup_{n=m}^{\infty} \{|\xi_n - \xi| \geq \varepsilon\},$$

we obtain

$$\mathrm{Ch}\left(\bigcap_{m=1}^{\infty} \bigcup_{n=m}^{\infty} \{|\xi_n - \xi| \geq \varepsilon\}\right) = 0.$$

It follows from Proposition 1 that $\xi_n \xrightarrow{a.s.} \xi$. The proof is complete. □

The following theorem compares convergence a.u. and convergence in chance.

Theorem 2. *Suppose $\{\xi_n\}$ and ξ are fuzzy random variables. If $\xi_n \xrightarrow{a.u.} \xi$, then $\xi_n \xrightarrow{\mathrm{Ch}} \xi$.*

Conversely, if ω is a discrete random variable assuming finite number of values, then $\xi_n \xrightarrow{\mathrm{Ch}} \xi$ implies $\xi_n \xrightarrow{a.u.} \xi$.

Proof. Suppose $\xi_n \xrightarrow{a.u.} \xi$. By Proposition 2, we have

$$\lim_{m \to \infty} \mathrm{Ch}\left(\bigcup_{n=m}^{\infty} \{|\xi_n - \xi| \geq \varepsilon\}\right) = 0.$$

According to the following inequality

$$\mathrm{Ch}\{|\xi_m - \xi| \geq \varepsilon\} \leq \mathrm{Ch}\left(\bigcup_{n=m}^{\infty} \{|\xi_n - \xi| \geq \varepsilon\}\right),$$

we obtain that $\lim_{m \to \infty} \mathrm{Ch}\{|\xi_m - \xi| \geq \varepsilon\} = 0$, i.e., $\xi_n \xrightarrow{\mathrm{Ch}} \xi$.

We now assume ω is a discrete random variable whose probability distribution is

$$\omega \sim \begin{pmatrix} \omega_1, \omega_2, \cdots, \omega_N \\ p_1, p_2, \cdots, p_N \end{pmatrix}$$

with $p_i > 0$ and $\sum_{i=1}^{N} p_i = 1$. By $\xi_n \xrightarrow{\text{Ch}} \xi$, we deduce that $\xi_{n,\omega_i} \xrightarrow{\text{Cr}} \xi_{\omega_i}$ for $i = 1, 2, \cdots, N$. Then for each positive integer $k = 1, 2, \cdots$,

$$\lim_{n \to \infty} \text{Cr}\left\{\gamma \in \Gamma \mid |\xi_{n,\omega_i}(\gamma) - \xi_{\omega_i}(\gamma)| \geq 1/k\right\} = 0$$

for $i = 1, 2, \cdots, N$. Thus, for each m, there exists N_{km} such that for $i = 1, 2, \cdots, N$,

$$\text{Cr}\left\{\gamma \in \Gamma \mid |\xi_{n,\omega_i}(\gamma) - \xi_{\omega_i}(\gamma)| \geq 1/k\right\} < 1/2m$$

whenever $n \geq N_{km}$. Letting

$$E_m = \bigcup_{i=1}^{N} \bigcup_{k=1}^{\infty} \bigcup_{n \geq N_{km}} \{\gamma \in \Gamma \mid |\xi_{n,\omega_i}(\gamma) - \xi_{\omega_i}(\gamma)| \geq 1/k\},$$

then we have

$$\text{Cr}(E_m) \leq \sup_i \sup_k \sup_{n \geq N_{km}} \text{Pos}\left\{\gamma \in \Gamma \mid |\xi_{n,\omega_i}(\gamma) - \xi_{\omega_i}(\gamma)| \geq \frac{1}{k}\right\} < 1/m.$$

It is easy to show that $\{\xi_{n,\omega_i}\}$ converges to ξ_{ω_i} uniformly on each $\Gamma \setminus E_m$. Thus, $\xi_n \xrightarrow{\text{a.u.}} \xi$. The proof is complete. □

As a consequence of Theorems 1 and 2, we have the following result:

Theorem 3. *Suppose $\{\xi_n\}$ and ξ are fuzzy random variables. If ω is a discrete random variable assuming finite number of values, then $\xi_n \xrightarrow{\text{Ch}} \xi$ implies $\xi_n \xrightarrow{\text{a.s.}} \xi$.*

Finally, we compare convergence in chance and convergence in distribution.

Theorem 4. *Suppose $\{\xi_n\}$ and ξ are fuzzy random variables. If $\xi_n \xrightarrow{\text{Ch}} \xi$, then $\xi_n \xrightarrow{d.} \xi$.*

Proof. Let G_n and G be the chance distribution functions of ξ_n and ξ, respectively. Then for every $\omega \in \Omega, t \in \Re, \varepsilon > 0$ and each integer n, one has

$$\text{Cr}\{\xi_{n,\omega} \geq t\} \leq \text{Cr}\{\xi_{n,\omega} \geq t, |\xi_{n,\omega} - \xi_\omega| < \varepsilon\} + \text{Cr}\{\xi_{n,\omega} \geq t, |\xi_{n,\omega} - \xi_\omega| \geq \varepsilon\}$$
$$\leq \text{Cr}\{\xi_\omega \geq t - \varepsilon\} + \text{Cr}\{|\xi_{n,\omega} - \xi_\omega| \geq \varepsilon\}.$$

Integrating with respect to ω on the preceding inequality, we obtain

$$\text{Ch}\{\xi_n \geq t\} \leq \text{Ch}\{\xi \geq t - \varepsilon\} + \text{Ch}\{|\xi_n - \xi| \geq \varepsilon\}.$$

Letting $n \to \infty$, and then $\varepsilon \to 0$, we obtain

$$\limsup_{n \to \infty} \text{Ch}\{\xi_n \geq t\} \leq G(t^-),$$

where $G(t^-)$ is the left limit of function $G(s)$ at $s = t$.

On the other hand, according to the following inequality

$$\begin{aligned}&\mathrm{Cr}\{\xi_\omega \geq t+\varepsilon\}\\ &\leq \mathrm{Cr}\{\xi_\omega \geq t+\varepsilon, |\xi_{n,\omega}-\xi_\omega|<\varepsilon\}+\mathrm{Cr}\{\xi_\omega \geq t+\varepsilon, |\xi_{n,\omega}-\xi_\omega|\geq \varepsilon\}\\ &\leq \mathrm{Cr}\{\xi_{n,\omega}\geq t\}+\mathrm{Cr}\{|\xi_{n,\omega}-\xi_\omega|\geq \varepsilon\},\end{aligned}$$

we have
$$\mathrm{Ch}\{\xi \geq t+\varepsilon\} \leq \mathrm{Ch}\{\xi_n \geq t\} + \mathrm{Ch}\{|\xi_n - \xi| \geq \varepsilon\}.$$

Letting $n \to \infty$, and then $\varepsilon \to 0$, we deduce

$$\liminf_{n\to\infty} \mathrm{Ch}\{\xi_n \geq t\} \geq G(t^+),$$

where $G(t^+)$ is the right limit of function $G(s)$ at $s=t$.

Therefore, $G_n \xrightarrow{w.} G$, i.e., $\xi_n \xrightarrow{d.} \xi$. □

5 Concluding Remarks

The major new results of this paper can be summarized as the following three aspects:

- Several new convergence concepts such as convergence almost uniform, convergence in mean chance and convergence in mean chance distribution for sequences of fuzzy random variables were presented.
- The criteria for convergence almost sure, convergence almost uniform, and convergence in chance criterion were obtained.
- The relationship between convergence almost uniform and convergence almost sure, convergence almost uniform and convergence in chance, convergence in chance and convergence almost sure, and convergence in chance and convergence in distribution were established.

As the continuation of this work, there are several issues will be considered in our future research. For instance, we will apply the obtained convergence results to fuzzy random optimization to discuss the convergence of approximation approach to the mean chance of a fuzzy random event and the expected value of a fuzzy random variable.

Acknowledgments

This work was supported by Natural Science Foundation of Hebei Province Grant A2005000087, and National Natural Science Foundation of China (NSFC).

References

1. Cohn, D. L.: Measure Theory. Birkhäuser, Boston (1980)
2. Dubois, D., Prade, H.: Possibility Theory. Plenum Press, New York (1988)

3. Krätschmer, V.: A Unified Approach to Fuzzy Random Variables. Fuzzy Sets Syst. **123** (2001) 1–9
4. Klir, G. J.: On Fuzzy-Set Interpretation of Possibility Theory. Fuzzy Sets Syst. **108** (1999) 263–373
5. Kruse, R., Meyer, K. D.: Statistics with Vague Data. D. Reidel Publishing Company, Dordrecht (1987)
6. Kwakernaak, H.: Fuzzy Random Variables–I. Definitions and Theorems. Inform. Sciences **15** (1978) 1–29
7. Liu, B.: Uncertain Programming. John Wiley & Sons, New York (1999)
8. Liu, B.: Uncertainty Theory: An Introduction to Its Axiomatic Foundations. Springer-Verlag, Berlin Heidelberg New York (2004)
9. Liu, B., Liu, Y.-K.: Expected Value of Fuzzy Variable and Fuzzy Expected Value Models. IEEE Trans. Fuzzy Syst. **10** (2002) 445–450
10. Liu, Y.-K.: On the Convergence of Measurable Set-Valued Function Sequence on Fuzzy Measure Space. Fuzzy Sets Syst. **112** (2000) 241–249
11. Liu, Y.-K., Liu, B.: The Relationship between Structural Characteristics of Fuzzy Measure and Convergences of Sequences of Measurable Functions. Fuzzy Sets Syst. **120** (2001) 511–516
12. Liu, Y.-K., Liu, B.: Fuzzy Random Variable: A Scalar Expected Value Operator. Fuzzy Optimization and Decision Making **2** (2003) 143–160
13. Liu, Y.-K., Liu, B.: A Class of Fuzzy Random Optimization: Expected Value Models. Inform. Sciences **155** (2003) 89–102
14. Liu, Y.-K., Liu, B.: On Minimum-Risk Problems in Fuzzy Random Decision Systems. Computers & Operations Research **32** (2005) 257-283
15. Liu, Y.-K., Liu, B.: Fuzzy Random Programming with Equilibrium Chance Constraints. Inform. Sciences **170** (2005) 363-395
16. Wang, G., Qiao, Z.: Linear Programming with Fuzzy Random Variable Coefficients. Fuzzy Sets Syst. **57** (1993) 295–311
17. López-Diaz, M., Gil, M. A.: Constructive Definitions of Fuzzy Random Variables. Statistics and Probability Letters **36** (1997) 135–143
18. Luhandjula, M. K.: Fuzziness and Randomness in an Optimization Framework. Fuzzy Sets Syst. **77** (1996) 291–297
19. Nahmias, S.: Fuzzy Variables. Fuzzy Sets Syst. **1** (1978) 97–101
20. Puri, M. L., Ralescu, D. A.: Fuzzy Random Variables. J. Math. Anal. Appl. **114** (1986) 409–422
21. Wang, P.: Fuzzy Contactability and Fuzzy Variables. Fuzzy Sets Syst. **8** (1982) 81–92
22. Wang, Z., Klir, G. J.: Fuzzy Measure Theory. Plenum Press, New York (1992)
23. Zadeh, L. A.: Fuzzy Sets as a Basis for a Theory of Possibility. Fuzzy Sets Syst. **1** (1978) 3–28

Hybrid Genetic-SPSA Algorithm Based on Random Fuzzy Simulation for Chance-Constrained Programming

Yufu Ning[1,2], Wansheng Tang[1], and Hui Wang[3]

[1] Institute of Systems Engineering, Tianjin University, Tianjin 300072, China
[2] Department of Computer Science, Dezhou University, Dezhou 253023, China
[3] Department of Statistics, Henan Institute of Finance and Economics, Zhengzhou 450002, China

Abstract. In this paper, hybrid genetic-SPSA algorithm based on random fuzzy simulation is proposed for solving chance-constrained programming in random fuzzy decision-making systems by combining random fuzzy simulation, genetic algorithm (GA), and simultaneous perturbation stochastic approximation (SPSA). In the provided algorithm, random fuzzy simulation is designed to estimate the chance of a random fuzzy event and the optimistic value to a random fuzzy variable, GA is employed to search for the optimal solution in the entire space, and SPSA is used to improve the new chromosomes obtained by crossover and mutation operations at each generation in GA. At the end of this paper, an example is given to illustrate the effectiveness of the presented algorithm.

1 Introduction

In a large number of realistic decision-making systems, randomness and fuzziness usually appear simultaneously. Fuzzy random variable was introduced by Kwakernaak [2] as a function from a probability space to a collection of fuzzy variables. In contrast with fuzzy random variables, Liu [3] defined a random fuzzy variable from a possibility space to a collection of random variables. In random fuzzy environments, the chance-constrained programming models have been provided in [3,4]. The basic idea for constructing random fuzzy chance-constrained programming is making the decision to maximize the optimistic value to the random fuzzy return subject to some chance constraints.

To solve random fuzzy chance-constrained programming, hybrid intelligent algorithms (HIAs) have been designed in [3,4], integrating random fuzzy simulation, neural network (NN) and genetic algorithm (GA). In the HIAs, GA is used to search for the optimal solution in the entire space. GA performs well in global search, but it is relatively time-consuming to converge to a local optimum, especially in solving large scale problems. A local search method, on the other hand, can converge faster to a local optimum but is poor at finding global optimum. Therefore, the issue incorporating local search method into GA has attracted a lot of researchers such as [1,5,8].

SPSA is a gradient approximation method for solving optimization problems. This gradient approximation uses only two measurements of the objective function in each iteration, independent of the dimension of the problem. So the SPSA algorithm is efficient in solving large-dimensional optimization problems [6]. The details on the implementation of SPSA can be found in [7]. In essence, SPSA is a "hill-climbing" method, so it is a local search algorithm.

Inspired by the above mentioned algorithms, this paper proposes hybrid genetic-SPSA algorithm based on random fuzzy simulation for solving random fuzzy chance-constrained programming. In Section 2, random fuzzy chance-constrained programming is introduced, then an algorithm is proposed in Section 3. Finally, the presented algorithm is employed to solve an example in Section 4.

2 Random Fuzzy Chance-Constrained Programming

The details on the concept and results about random fuzzy variables can be found in [3].

It is assumed that the decision-maker wants to maximize the optimistic value to the random fuzzy return subject to some chance constraints. Then the following random fuzzy chance-constrained programming can be constructed [3]:

$$\begin{cases} \max \overline{f} \\ \text{subject to:} \\ \quad \text{Ch}\left\{f(\boldsymbol{x},\boldsymbol{\xi}) \geq \overline{f}\right\}(\gamma) \geq \delta, \\ \quad \text{Ch}\{g_j(\boldsymbol{x},\boldsymbol{\xi}) \leq 0\}(\alpha_j) \geq \beta_j, j = 1,2,\cdots,p, \end{cases} \quad (1)$$

where $\boldsymbol{x}$ is a decision vector, $\boldsymbol{\xi}$ is a random fuzzy vector, $f(\boldsymbol{x},\boldsymbol{\xi})$ is the objective function, $g_j(\boldsymbol{x},\boldsymbol{\xi})$ is the constraint functions for $j = 1,2,\cdots,p$, and α_j, β_j are specified confidence levels for $j = 1,2,\cdots,p$.

3 Hybrid Genetic-SPSA Algorithm Based on Random Fuzzy Simulation

One of the difficulties in solving model (1) is to calculate the chances of random fuzzy events, $\text{Ch}\{g_j(\boldsymbol{x},\boldsymbol{\xi}) \leq 0\}(\alpha_j), j = 1,2,\cdots,p$, and the (γ,δ)-optimistic value $\overline{f}$ to random fuzzy variable $f(\boldsymbol{x},\boldsymbol{\xi})$, for every fixed $\boldsymbol{x}$. In most practical decision-making systems, it is very difficult or impossible to obtain their exact values by analytic methods. Therefore, random fuzzy simulation is used to estimate these values (see [3]).

In this section, hybrid genetic-SPSA algorithm based on random fuzzy simulation is proposed. The procedure is summarized as follows.

Step 1) Determine the size of population *pop_size*, crossover probability P_c, mutation probability P_m in GA, and the parameters a, A, c, α, γ in SPSA.

Step 2) Generate randomly *pop_size* feasible chromosomes. To verify the feasibility of each chromosome, random fuzzy simulation is used to estimate the values of constraint functions.

Step 3) Apply SPSA to all the *pop_size* chromosomes in initial population.

Step 4) Estimate the objective values of the *pop_size* chromosomes by random fuzzy simulation, and calculate the fitness of each chromosome.

Step 5) Select the chromosomes by spinning the roulette wheel.

Step 6) Apply crossover operator to the chromosomes selected by P_c.

Step 7) Apply SPSA to the new chromosomes obtained by crossover operation.

Step 8) Apply mutation operator to the chromosomes selected by P_m.

Step 9) Apply SPSA to the new chromosomes obtained by mutation operation.

Step 10) If the termination criterion for GA is satisfied, turn to next step. Otherwise, return to Step 4.

Step 11) Apply SPSA to all the final chromosomes obtained by GA.

Step 12) Display the chromosome with maximum fitness among all the *pop_size* chromosomes improved by SPSA, then end the algorithm.

Remark 1. The choice of parameters a, A, c, α, γ in SPSA is crucial for the performance of the proposed algorithm. A detailed guideline can be found in [7].

Remark 2. To prevent SPSA from spending almost all available computation time to improve the new chromosomes generated by crossover and mutation operations in GA, a relatively small number of iterations in SPSA before GA ends are required. Besides, to further improve the final chromosomes reached by GA, a relatively large number of iterations in SPSA after GA ends are recommended.

4 An Example

Let us consider the following model:

$$\begin{cases} \max \overline{f} \\ \text{subject to:} \\ \text{Ch}\left\{\xi_1 x_1^2 - \xi_2 x_2^3 + \xi_3 x_3^3 - \xi_4 x_4^2 - \xi_2 x_1 x_2 - \xi_3 x_3 x_4 \geq \overline{f}\right\}(0.95) \geq 0.90, \\ \text{Ch}\left\{(\xi_3 x_1^2 - x_2)(\xi_2 x_3 - x_1) + \xi_4 x_2^2 + x_3 + x_4 \leq 10\right\}(0.90) \geq 0.85, \\ x_i \geq 0, \quad i = 1, 2, 3, 4, \end{cases}$$

where x_1, x_2, x_3 and x_4 are decision variables, ξ_1, ξ_2, ξ_3 and ξ_4 are random fuzzy variables defined as follows:

$$\xi_1 \sim \mathcal{U}(\rho_1 - 1, \rho_1 + 1), \quad \xi_2 \sim \mathcal{N}(\rho_2, 2), \quad \xi_3 \sim \mathcal{EXP}(\rho_3), \quad \xi_4 \sim \mathcal{N}(\rho_4, 4),$$

where ρ_1 is a fuzzy variable with membership function $\mu_{\rho_1}(x) = [1-(x-1)^2] \vee 0$, ρ_2 is a triangular fuzzy variable $(2,3,4)$, ρ_3 is a trapezoidal fuzzy variable $(1,2,3,4)$, and ρ_4 is a fuzzy variable with membership function $\mu_{\rho_4}(x) = [1-(x-4)^2] \vee 0$.

The solution obtained by the proposed algorithm (the parameters *pop_size* = 30, $P_c = 0.3, P_m = 0.2, a = 0.16, A = 100, c = 0.20, \alpha = 0.602, \gamma = 0.101$, 5000 cycles in random fuzzy simulation, 500 generations in GA, and 1500 iterations in SPSA) is

$$x_1^* = 1.0823, x_2^* = 0.0031, x_3^* = 0.2530, x_4^* = 0.2317$$

with the objective value -1.2805.

Now, we employ GA alone to search for the optimal solution, then the solution after 2000 generations is

$$x_1' = 9.5156, x_2' = 0.0700, x_3' = 1.7186, x_4' = 0.4427$$

with the objective value -59.7336.

5 Conclusion

This paper proposed hybrid genetic-SPSA algorithm based on random fuzzy simulation, which had both global search ability of GA and strong convergence property of SPSA. Finally, an example showed the validity of the provided algorithm.

Acknowledgments

This work was supported by the National Natural Science Foundation of China Grant No. 70471049 and China Postdoctoral Science Foundation No. 2004035013.

References

1. Ishibuchi, H., Murata, T.: A multi-objective genetic local search algorithm and its application to flowshop scheduling. IEEE Transactions on Systems, Man, and Cybernetics - C. **28** (1998) 392-403
2. Kwakernaak, H.: Fuzzy random variables-I: Definition and theorems. Information Sciences. **15** (1978) 1-29
3. Liu, B.: Theory and Practice of Uncertain Programming. Physica-Verlag, Heidelberg (2002)
4. Liu, Y. K., Liu, B.: Random fuzzy programming with chance measures defined by fuzzy integrals. Mathematical and Computer Modelling. **36** (2002) 509-524
5. Renders, J. M., Flasse, S.: Hybrid methods using genetic algorithm for global optimization. IEEE Transactions on Systems, Man, and Cybernetics - B. **26** (1996) 243-258
6. Spall, J. C.: Multivariate stochastic approximation using a simultaneous perturbation gradient approximation. IEEE Transactions on Automatic Control. **37** (1992) 332-341
7. Spall, J. C.: Implementation of the simultaneous perturbation algorithm for stochastic optimization. IEEE Transactions on Aerospace and Electronic Systems. **34** (1998) 817-823
8. Tsai, J. T., Liu, T. K., Chou, J. H.: Hybrid Taguchi-genetic algorithm for global numerical optimization. IEEE Transactions on Evolutionary Computation. **8** (2004) 365-377

Random Fuzzy Age-Dependent Replacement Policy

Song Xu, Jiashun Zhang, and Ruiqing Zhao

Institute of Systems Engineering, Tianjin University, Tianjin, 300072, China
dtsszhang@yahoo.com.cn, zhao@tju.edu.cn

Abstract. This paper discusses the age-dependent replacement policy, in which the interarrival lifetimes of components are characterized as random fuzzy variables. A random fuzzy expected value model is presented and shown how it can be applied to reduce the loss of system failures. To solve the proposed model, a simultaneous perturbation stochastic approximation (SPSA) algorithm based on random fuzzy simulation is developed to search the optimal solution. At the end of this paper, a numerical example is enumerated.

1 Introduction

In the past several decades, renewal processes have been extensively investigated in the literature. Most models treated in the literature restrict attention to the system lifetime which is characterized through probability measure. Based on probability theory, stochastic renewal theory considers the processes in which the interarrival maintenance times are assumed to be independent and identically distributed (iid) random variables [1]. Different from stochastic renewal theory, fuzzy renewal theory reckons that the interarrival times can not be known precisely in the real-life world. Under this theory, Huang [2] presented a genetic-evolved fuzzy system for maintenance scheduling of generating units. Huang et al [3] proposed a new approach using fuzzy dynamic programming for generator maintenance scheduling in power systems. Suresh [9] gave out a fuzzy-set model for maintenance policy of multi-state equipment.

In many practical situations, however, randomness and fuzziness are required to be considered simultaneously in one process. Zhao and Tang [10] considered a random fuzzy renewal process, in which the interarrival times were modeled as iid random fuzzy variables. Later, Zhao and Tang [11] extended the idea of Zhao and Tang [10] to random fuzzy renewal reward process and provided a theorem on the limit of the long-run expected reward per unit time.

In this paper, the lifetimes of components are treated as random fuzzy variables defined by Liu and Liu [4]. In Section 2, we recall some basic concepts and results about random fuzzy variables. In Section 3, we discuss the random fuzzy age-dependent replacement policy, and presente a mathematic model for this policy. In Section 4, we employ the random fuzzy simulation and SPSA algorithm to solve the proposed model. At last, a numerical example is enumerated.

2 Random Fuzzy Variables

Let Θ be a nonempty set, $P(\Theta)$ the power set of Θ, and Pos a possibility measure. Then $(\Theta, P(\Theta), \text{Pos})$ is called a possibility space (for details, see [7]). Based on the possibility measure, necessity and credibility of a fuzzy event are given as $\text{Nec}\{\mathcal{A}\} = 1 - \text{Pos}\{\mathcal{A}^c\}$ and $\text{Cr}\{\mathcal{A}\} = \frac{1}{2}(\text{Pos}\{\mathcal{A}\} + \text{Nec}\{\mathcal{A}\})$, respectively.

A random fuzzy variable is a function from the possibility space $(\Theta, P(\Theta), \text{Pos})$ to the set of random variables. In a practical system, the lifetimes of components should be nonnegative and iid.

Definition 1. *A random fuzzy variable ξ is nonnegative if and only if for each $\theta \in \Theta$, $\Pr(\xi(\theta) < 0) = 0$.*

Definition 2 (Liu [6]). *The random fuzzy variables $\xi_1, \xi_2, \cdots, \xi_n$ are iid if and only if*

$$(\Pr\{\xi_i(\theta \in B_1)\}, \Pr\{\xi_i(\theta \in B_2)\}, \cdots, \Pr\{\xi_i(\theta \in B_m)\}), i = 1, 2, \cdots, n$$

are iid vectors for any Borel set $B_1, B_2, \cdots, B_m$ of $\Re$ and any integer m.

Definition 3 (Liu and Liu [4]). *Let ξ be a random fuzzy variable. Then the expected value $E[\xi]$ is defined by*

$$E[\xi] = \int_0^\infty \text{Cr}\{\theta \in \Theta \mid E[\xi(\theta)] \geq r\} dr - \int_{-\infty}^0 \text{Cr}\{\theta \in \Theta \mid E[\xi(\theta)] \leq r\} dr \quad (1)$$

provided that at least one of the two integrals is finite.

3 Random Fuzzy Age-Dependent Replacement Policy

In this section, we focus on the age-dependent replacement policy in random fuzzy environment.

Let ξ_i be the interarrival time between the $(i-1)$th event and the ith event, $i = 1, 2, \cdots$, respectively. Here, we suppose that $\xi_i, i = 1, 2, \cdots$ are iid nonnegative random fuzzy variables in this section. Then the length of the ith maintenance cycle is $L_i(T) = \xi_i I_{(\xi_i \leq T)} + T I_{(\xi_i > T)}$, where $I_{(\cdot)}$ is the characteristic function of the event $(\cdot)$. And the cost of the ith cycle is $C_i(T) = c_f I_{(\xi_i \leq T)} + c_p I_{(\xi_i > T)}$, where c_f and c_p are the costs upon failures and at the predetermined age T, respectively. For the routine maintenance cost should be less than the loss of system failure, it is easy to see that $c_f > c_p$. Let $N(t)$ be the total number of the events by time t, and $C(t)$ the total reward by time t, then we have

$$C(t) = \sum_{i=1}^{N(t)} C_i(T).$$

Definition 4. *The long-run expected cost per unit time under the age-dependent replacement policy is defined as*

$$C_a(T) = \lim_{t \to \infty} \frac{E[C(t)]}{t}. \quad (2)$$

Furthermore, if $E\left[\frac{C_1(T)}{L_1(T)}\right]$ is finite, it follows from Zhao and Tang [10] that $C_a(T) = E\left[\frac{C_1(T)}{L_1(T)}\right]$. Our problem is to find an optimal value T such that $C_a(T)$ is minimized. This problem can be described as the following model.

$$\begin{cases} \min E\left[\frac{C_1(T)}{L_1(T)}\right] \\ \text{subject to:} \\ T > 0, \end{cases}$$

where $L_1(T) = \xi_1 I_{(\xi_1 \leq T)} + T I_{(\xi_1 > T)}$ is the length of the first cycle, $C_1(T) = c_f I_{(\xi_1 \leq T)} + c_p I_{(\xi_1 > T)}$ is the cost of the first cycle, and ξ_1 is the lifetime of first renewed component.

4 SPSA Algorithm Based on Random Fuzzy Simulation and Numerical Experiment

For any fixed time T, it is difficult to calculate the value of $E\left[\frac{C_1(T)}{L_1(T)}\right]$ due to the complexity of function structure. Here we employ the random fuzzy simulation technique designed by Liu and Liu [4] to estimate it. The method for estimating $E\left[\frac{C_1(T)}{L_1(T)}\right]$ is summrized as follows.

1. Set $E=0$.
2. Uniformly sample θ_i from Θ such that $\text{Pos}\{\theta_i\} \geq \xi, i = 1, 2, \cdots, N$, where ξ is a sufficiently small number.
3. Let $a = \min_{1 \leq i \leq N} E\left[\frac{C_1(T)(\theta_i)}{L_1(T)(\theta_i)}\right]$ and $b = \max_{1 \leq i \leq N} E\left[\frac{C_1(T)(\theta_i)}{L_1(T)(\theta_i)}\right]$.
4. Uniformly generate r from $[a, b]$.
5. If $r > 0$, then $E \leftarrow E + \text{Cr}\left\{\theta \in \Theta \mid E\left[\frac{C_1(T)(\theta_i)}{L_1(T)(\theta_i)}\right] \geq r\right\}$.
6. If $r < 0$, then $E \leftarrow E - \text{Cr}\left\{\theta \in \Theta \mid E\left[\frac{C_1(T)(\theta_i)}{L_1(T)(\theta_i)}\right] \leq r\right\}$.
7. Repeat the fourth to sixth steps for N times.
8. $E\left[\frac{C_1(T)}{L_1(T)}\right] = a \vee 0 + b \wedge 0 + E \cdot (b-a)/N$.

In addition, since the gradient of $C_a(T)$ may not always exist, the common gradient descent algorithms can not be used to find an optimal value of T such that $C_a(T)$ is minimized. Here, we introduce SPSA algorithm based on random fuzzy simulation to solve the problem (about SPSA algorithm, see [8]).

1. Choose initial guess of T and the nonnegative parameters a, c, A, α, and γ.
2. Generate the perturbation Δ_k, which should satisfy the conditions provided in Spall [8].
3. Compute $C_a(\hat{T}_k + c_k \Delta_k)$ and $C_a(\hat{T}_k - c_k \Delta_k)$ by random fuzzy simulation proposed above, where $c_k = c/(c+k+1)^\gamma$.

4. Generate the simultaneous perturbation approximation to the unknown gradient

$$\hat{g}_k(\hat{T}_k) = \frac{C_a(\hat{T}_k + c_k \Delta_k) - C_a(\hat{T}_k - c_k \Delta_k)}{2c_k \Delta_k}. \qquad (3)$$

5. Use $\hat{T}_{k+1} = \hat{T}_k - a_k \hat{g}_k(\hat{T}_k)$ to update $\hat{T}_k$, where $a_k = a/(A+k+1)^\alpha$.
6. Return to Step 2 with $k+1$ replacing k. Terminate the algorithm and return $\hat{T}_k$, if there is little change in several successive iterates.

Example 1. Let ξ be the lifetime of the component and defined as $\xi \sim \mathcal{EXP}(\frac{1}{\theta})$, where $\theta = (0, 4, 7, 10)$, and $c_f = 1000$, $c_p = 300$.

In order to minimize the long-run expected cost per unit time, we first initialize the value of decision variable $\hat{T}_0 = 5$ and select coefficients $A = 100$, $c = 0.2$, $\alpha = 0.602$, $\gamma = 0.101$. Then we begin the SPSA iterations presented in Section 4. After 3000 cycles iterations, we get the optimal solution is

$$T^* = 6.87385$$

with the cost $C_a(T^*) = 77.9375$.

Acknowledgments

This work was supported by National Natural Science Foundation of China Grant No. 70471049 and China Postdoctoral Science Foundation No. 2004035013.

References

1. Asmussen, S.: Applied Probability and Queues. Wiley, New York. (1987)
2. Huang, S.: A genetic-evolved fuzzy system for maintenance scheduling of generating units. Electrical Power Energy Systems. **20** (1998) 191-195
3. Huang, C., Lin, C.: Fuzzy approach for generator maintenance scheduling. Electric Power System Research. **24** (1992) 31-38
4. Liu, Y., Liu, B.: Expected value operator of random fuzzy variable and random fuzzy expected value models. International Journal of Uncertainty Fuzziness & Knowledge-Based Systems. **11** (2003) 195-215
5. Liu, B.: Uncertainty Theory: An Introduction to its Axiomatic Foundations. Springer-Verlag, Berlin (2004)
6. Liu, B.: Theory and Practice of Uncertain Programming. Phisica-Verlag, Heidelberg (2002)
7. Nahmias, S.: Fuzzy variables. Fuzzy Sets and Systems. **1** (1978) 97-110
8. Spall, J.:An overview of the simultaneous perturbation method for efficient optimization. Johns Hopkins Apl Technical Digest. **19** (1998) 482-492
9. Suresh, P.,Chaudhuri, D.: Fuzzy-set approach to select maintenance strategies for multistate equipment. IEEE Transactions on Reliability. **43** (1994) 431-456
10. Zhao, R., Tang, W.: Random fuzzy renewal process. European Journal of Operational Research. to be published
11. Zhao, R., Tang, W.: Random fuzzy renewal reward process. Information Sciences. to be published

A Theorem for Fuzzy Random Alternating Renewal Processes

Ruiqing Zhao, Wansheng Tang, and Guofei Li

Institute of Systems Engineering, Tianjin University, Tianjin 300072, China
zhao@tju.edu.cn, tang@tju.edu.cn

Abstract. In this paper, a new kind of alternating renewal processes—fuzzy random alternating renewal processes—is devoted. A theorem on the limit value of the mean chance of the fuzzy random event "system is on at time t" is presented. The two degenerate cases of the theorem, stochastic and fuzzy cases, are also analyzed. The importance of the results lies in the fact that the relation between classical alternating renewal processes and fuzzy random alternating renewal processes is established.

1 Introduction

Stochastic alternating renewal process is one of important process in stochastic renewal theory. In such a process, on and off times are usually assumed to be independent and identically distributed (iid) random variables, respectively. The results on stochastic alternating renewal processes have been discussed in detail in almost all standard text books in stochastic processes such as Asmussen [1], Birolini [2], Daley and Vere-jones [3], Ross [20] and Tijms [21].

After stochastic renewal theory, fuzzy renewal theory has been considered by several authors. In this theory, the parameters such as the interarrival times were supposed to be fuzzy variables. Dozzi et al [4] gave a limit theorem for a counting renewal process indexed by fuzzy sets. Liu [13] discussed a fuzzy-valued Markov process and gave some properties on this process. Zhao and Liu [22] discussed fuzzy renewal process and fuzzy renewal reward process in which fuzzy elementary renewal theorem and renewal reward theorem were provided.

As an extension of stochastic renewal theory and fuzzy renewal theory, fuzzy random renewal theory has been considered recently. Popova and Wu [17] considered a fuzzy random renewal reward process in which interarrival times and rewards were assumed to be fuzzy random variables and gave a theorem on the long-run average fuzzy reward per unit time. Hwang [5] investigated a renewal process in which the interarrival times were considered as iid fuzzy random variables and provided a theorem on the fuzzy rate of the fuzzy random renewal process.

As a continuation of the above-mentioned works, this paper deals with a new kind of renewal processes—fuzzy random alternating renewal processes. In Section 2, we review some basic concepts on fuzzy variables and fuzzy random variables. Finally, fuzzy random alternating renewal processes were discussed in Section 3.

2 Fuzzy Variables and Fuzzy Random Variables

Let ξ be a fuzzy variable on possibility space $(\Theta, \mathrm{P}(\Theta), \mathrm{Pos})$, where Θ is a universe, $\mathcal{P}(\Theta)$ is the power set of Θ and Pos is a possibility measure defined on $\mathcal{P}(\Theta)$. Based on the *possibility measure* Pos, *necessity* (Nec) and *credibility* (Cr) of a fuzzy event $\{\xi \geq r\}$ can be expressed by

$$\mathrm{Nec}\{\xi \geq r\} = 1 - \mathrm{Pos}\{\xi < r\},$$
$$\mathrm{Cr}\{\xi \geq r\} = \frac{1}{2}\left(\mathrm{Pos}\{\xi \geq r\} + \mathrm{Nec}\{\xi \geq r\}\right), \quad (1)$$

respectively.

Definition 1 (Liu [9]). *Let ξ be a fuzzy variable on possibility space $(\Theta, \mathcal{P}(\Theta), \mathrm{Pos})$, and $\alpha \in (0,1]$. Then*

$$\xi'_\alpha = \inf\{r \mid \mathrm{Pos}\{\xi \leq r\} \geq \alpha\} \quad \text{and} \quad \xi''_\alpha = \sup\{r \mid \mathrm{Pos}\{\xi \geq r\} \geq \alpha\} \quad (2)$$

are called α-pessimistic value and α-optimistic value of ξ, respectively.

Definition 2 (Liu and Liu [12]). *Let ξ be a fuzzy variable on the possibility space $(\Theta, \mathcal{P}(\Theta), \mathrm{Pos})$. The expected value $E[\xi]$ is defined as*

$$E[\xi] = \int_0^\infty \mathrm{Cr}\{\xi \geq r\}\mathrm{d}r - \int_{-\infty}^0 \mathrm{Cr}\{\xi \leq r\}\mathrm{d}r \quad (3)$$

provided that at least one of the two integrals is finite.

The definitions of fuzzy random variable have been discussed by several authors [6],[7], [8], [10], [11], [14], [18], [19]. In what follows, we will introduce the results in [14] that are related to our work. We assume that $(\Omega, \mathcal{A}, \mathrm{Pr})$ is a probability space. Let $\mathcal{F}$ be a collection of fuzzy variables defined on the possibility space $(\Theta, \mathrm{P}(\Theta), \mathrm{Pos})$.

Definition 3 (Liu and Liu [14]). *A fuzzy random variable is a function $\xi : \Omega \to \mathcal{F}$ such that for any Borel set B of $\Re$, $\mathrm{Pos}\{\xi(\omega) \in B\}$ is a measurable function of ω.*

Definition 4 (Liu and Liu [14]). *Let ξ be a fuzzy random variable defined on the probability space $(\Omega, \mathcal{A}, \mathrm{Pr})$. Then its expected value is defined by*

$$E[\xi] = \int_\Omega \left[\int_0^\infty \mathrm{Cr}\{\xi(\omega) \geq r\}\mathrm{d}r - \int_{-\infty}^0 \mathrm{Cr}\{\xi(\omega) \leq r\}\mathrm{d}r\right] \mathrm{Pr}(\mathrm{d}\omega)$$

provided that at least one of the two integrals is finite.

Definition 5 (Liu and Liu [16]). *Let ξ be a fuzzy random variable defined on the possibility space $(\Theta, P(\Theta), \mathrm{Pos})$. Then the mean chance, denoted by Ch, of fuzzy random event characterized by $\{\xi \leq 0\}$ is defined as*

$$\mathrm{Ch}\{\xi \leq 0\} = \int_0^1 \Pr\{\omega \in \Omega \mid \mathrm{Cr}\{\xi(\omega) \leq 0\} \geq \alpha\} \, d\alpha. \tag{4}$$

Proposition 1. *It follows that the equation (4) is equivalent to the following form*

$$\mathrm{Ch}\{\xi \leq 0\} = \frac{1}{2}\int_0^1 \left(\Pr\{\omega \in \Omega \mid \xi'_\alpha(\omega) \leq 0\} + \Pr\{\omega \in \Omega \mid \xi''_\alpha(\omega) \leq 0\}\right) d\alpha,$$

where $\xi'_\alpha(\omega)$ and $\xi''_\alpha(\omega)$ are α-pessimistic value and α-optimistic value of $\xi(\omega)$, respectively.

Proof. Note that

$$\mathrm{Ch}(\xi \leq 0) = \int_0^1 \Pr\left(\omega \in \Omega \mid \mathrm{Cr}\{\xi(\omega) \leq 0\} \geq \alpha\right) d\alpha$$

$$= \int_\Omega \mathrm{Cr}\{\xi(\omega) \leq 0\} \Pr(d\omega)$$

$$= \frac{1}{2}\int_\Omega \left[\mathrm{Pos}\left(\xi(\omega) \leq 0\right) + \mathrm{Nec}\left(\xi(\omega) \leq 0\right)\right] \Pr(d\omega)$$

$$= \frac{1}{2}\int_\Omega \mathrm{Pos}\left(\xi(\omega) \leq 0\right) \Pr(d\omega) + \frac{1}{2}\int_\Omega \mathrm{Nec}\left(\xi(\omega) \leq 0\right) \Pr(d\omega)$$

$$= \frac{1}{2}\int_\Omega \mathrm{Pos}\left(\xi(\omega) \leq 0\right) \Pr(d\omega) + \frac{1}{2}\int_\Omega \left[1 - \mathrm{Pos}\left(\xi(\omega) > 0\right)\right] \Pr(d\omega)$$

$$= \frac{1}{2}\int_0^1 \Pr\left(\omega \in \Omega \mid \mathrm{Pos}\left(\xi(\omega) \leq 0\right) \geq \alpha\right) d\alpha +$$

$$\frac{1}{2}\int_0^1 \Pr\left(\omega \in \Omega \mid \left[1 - \mathrm{Pos}\left(\xi(\omega) \geq 0\right)\right] \geq \alpha\right) d\alpha$$

$$= \frac{1}{2}\int_0^1 \Pr\left(\omega \in \Omega \mid \mathrm{Pos}\left(\xi(\omega) \leq 0\right) \geq \alpha\right) d\alpha +$$

$$\frac{1}{2}\int_0^1 \Pr\left(\omega \in \Omega \mid \mathrm{Pos}\left(\xi(\omega) > 0\right) \leq 1 - \alpha\right) d\alpha$$

$$= \frac{1}{2}\int_0^1 \Pr\left(\omega \in \Omega \mid \mathrm{Pos}\left(\xi(\omega) \leq 0\right) \geq \alpha\right) d\alpha +$$

$$\frac{1}{2}\int_1^0 \Pr\left(\omega \in \Omega \mid \mathrm{Pos}\left(\xi(\omega) > 0\right) \leq p\right) (-dp)$$

$$= \frac{1}{2} \int_0^1 \Pr\left(\omega \in \Omega \mid \text{Pos}\left(\xi(\omega) \le 0\right) \ge \alpha\right) d\alpha +$$

$$\frac{1}{2} \int_0^1 \Pr\left(\omega \in \Omega \mid \text{Pos}\left(\xi(\omega) > 0\right) \le p\right) dp$$

$$= \frac{1}{2} \int_0^1 \Pr\left(\omega \in \Omega \mid \text{Pos}\left(\xi(\omega) \le 0\right) \ge \alpha\right) d\alpha +$$

$$\frac{1}{2} \int_0^1 \Pr\left(\omega \in \Omega \mid \text{Pos}\left(\xi(\omega) > 0\right) \le \alpha\right) d\alpha$$

$$= \frac{1}{2} \int_0^1 \left[\Pr\left(\omega \in \Omega \mid \text{Pos}\left(\xi(\omega) \le 0\right) \ge \alpha\right) + \Pr\left(\omega \in \Omega \mid \text{Pos}\left(\xi(\omega) > 0\right) \le \alpha\right)\right] d\alpha$$

$$= \frac{1}{2} \int_0^1 \left[\Pr\left(\omega \in \Omega \mid \xi'_\alpha(\omega) \le 0\right) + \Pr\left(\omega \in \Omega \mid \xi''_\alpha(\omega) \ge 0\right)\right] d\alpha$$

$$= \frac{1}{2} \int_0^1 \left[\Pr\left(\xi'_\alpha(\omega) \le 0\right) + \Pr\left(\xi''_\alpha(\omega) \le 0\right)\right] d\alpha.$$

The proof is completed.

Remark 1. For convenience, the smaller one of $\Pr\{\omega \in \Omega \mid \xi'_\alpha(\omega) \le 0\}$ and $\Pr\{\omega \in \Omega \mid \xi''_\alpha(\omega) \le 0\}$ is called the α-pessimistic probability of the event $\{\xi \le 0\}$ and the other is called the α-optimistic probability of the event $\{\xi \le 0\}$.

Definition 6 (Liu and Liu [14]). *The fuzzy random variables $\xi_1, \xi_2, \cdots, \xi_n$ are indepedent and identically distributed if and only if*

$$(\text{Pos}\{\xi_i(\omega) \in B_1\}, \text{Pos}\{\xi_i(\omega) \in B_2\}, \cdots, \text{Pos}\{\xi_i(\omega) \in B_m\}), \quad i = 1, 2, \cdots, n$$

are iid random vectors for any Borel sets $B_1, B_2, \cdots, B_m$ of $\Re$ and any positive integer m.

3 Fuzzy Random Alternating Renewal Processes

Consider a system that can be in one of the two states: *on* or *off*. Initially it is on and it remains on for a period of time ξ_1; it then goes off and remains off for a period of time η_1; it then goes on for a period of time ξ_2; then off for a period of time η_2; then on, and so forth. We suppose that (ξ_i, η_i), $i = 1, 2, \cdots$ are iid fuzzy random vectors on the probability space $(\Omega, \mathcal{A}, \Pr)$, especially, ξ_i and η_i are assumed to be independent. The process characterized by $\{(\xi_i, \eta_i), i \ge 1\}$ is called a *fuzzy random alternating renewal process*.

For each $\omega \in \Omega$, $\xi_i(\omega)$ and $\eta_i(\omega)$ are clearly fuzzy variables and their α-pessimistic values and α-optimistic values are denoted by $\xi'_{i,\alpha}(\omega), \xi''_{i,\alpha}(\omega), \eta'_{i,\alpha}(\omega)$, and $\eta''_{i,\alpha}(\omega)$, respectively.

In what follows, we will pay our attention to the mean chance of the event {system is on at t}. For convenience, the α-pessimistic probability of the event {system is on at t} is denoted by $\Pr'_\alpha\{\omega \in \Omega \mid \text{system is on at } t\}$ and its α-optimistic probability is denoted by $\Pr''_\alpha\{\omega \in \Omega \mid \text{system is on at } t\}$. By Definition 5 and Proposition 1, we have

Ch{system is on at time t}

$$= \tfrac{1}{2} \int_0^1 \left(\Pr'_\alpha\{\omega \in \Omega | \text{system is on at } t\} + \Pr''_\alpha\{\omega \in \Omega | \text{system is on at } t\} \right) d\alpha. \tag{5}$$

Definition 7. *A positive random variable ζ is lattice if and only if there exists $d \geq 0$ such that $\sum_{n=0}^{\infty} \Pr\{\zeta = nd\} = 1$.*

Theorem 1. *Assume that $\{(\xi_i, \eta_i), i \geq 1\}$ be a sequence of pairs of iid positive fuzzy random variables. For any given $\omega \in \Omega$ and $\alpha \in (0,1]$, let $\xi'_{i,\alpha}(\omega)$, $\eta'_{i,\alpha}(\omega)$, $\xi''_{i,\alpha}(\omega)$, and $\eta''_{i,\alpha}(\omega)$ be α-pessimistic values and α-optimistic values of $\xi_i(\omega)$ and $\eta_i(\omega)$, $i = 1, 2 \cdots$ If the random variables $\xi'_{i,\alpha}(\omega)$, $\xi''_{i,\alpha}(\omega)$, $\eta'_{i,\alpha}(\omega)$ and $\eta''_{i,\alpha}(\omega)$ are nonlattice, $E\left[\xi'_{1,\alpha}(\omega)\right] + E\left[\eta'_{1,\alpha}(\omega)\right] < \infty$ and $E\left[\xi''_{1,\alpha}(\omega)\right] + E\left[\eta''_{1,\alpha}(\omega)\right] < \infty$, then*

$$\lim_{t \to \infty} \Pr'_\alpha\{\omega \in \Omega \mid \text{system is on at } t\} = \frac{E\left[\xi'_{1,\alpha}(\omega)\right]}{E\left[\xi'_{1,\alpha}(\omega)\right] + E\left[\eta''_{1,\alpha}(\omega)\right]}, \tag{6}$$

$$\lim_{t \to \infty} \Pr''_\alpha\{\omega \in \Omega \mid \text{system is on at } t\} = \frac{E\left[\xi''_{1,\alpha}(\omega)\right]}{E\left[\xi''_{1,\alpha}(\omega)\right] + E\left[\eta'_{1,\alpha}(\omega)\right]}. \tag{7}$$

Proof. For any given $\alpha \in (0,1]$, we define two sequences of real numbers $\{\beta_i\}$ and $\{\gamma_i\}$ such that $\alpha \leq \beta_i \leq 1$ and $\alpha \leq \gamma_i \leq 1$ for $i = 1, 2, \cdots$ Furthermore, for any given $\omega \in \Omega$, $\xi_i(\omega)$ and $\eta_i(\omega)$ are clearly fuzzy variables. Let $\xi_{\beta_i}(\omega)$ be the β_i-pessimistic value or the β_i-optimistic value of $\xi_i(\omega)$ and $\eta_{\gamma_i}(\omega)$ the γ_i-pessimistic value or the γ_i-optimistic value of $\eta_i(\omega)$. Hence, we have

$$\xi'_{i,\alpha}(\omega) \leq \xi_{\beta_i}(\omega) \leq \xi''_{i,\alpha}(\omega) \quad \text{and} \quad \eta'_{i,\alpha}(\omega) \leq \eta_{\gamma_i}(\omega) \leq \eta''_{i,\alpha}(\omega)$$

for each $\omega \in \Omega$. Moreover, $\xi_{\beta_i}(\omega)$ and $\eta_{\gamma_i}(\omega)$ are random variables when ω varies all over Ω. Thus, for any given $t > 0$, we have

$$\Pr\left\{\xi'_{i,\alpha}(\omega) > t\right\} \leq \Pr\{\xi_{\beta_i}(\omega) > t\} \leq \Pr\left\{\xi''_{i,\alpha}(\omega) > t\right\} \tag{8}$$

and

$$\Pr\left\{\eta'_{i,\alpha}(\omega) > t\right\} \leq \Pr\{\eta_{\gamma_i}(\omega) > t\} \leq \Pr\left\{\eta''_{i,\alpha}(\omega) > t\right\}. \tag{9}$$

Now let us consider the following three processes:

(1) the process A characterized by $\{(\xi'_{i,\alpha}(w), \eta''_{i,\alpha}(w)), i \geq 1\}$;
(2) the process B characterized by $\{(\xi''_{i,\alpha}(w), \eta'_{i,\alpha}(w)), i \geq 1\}$;
(3) the process C characterized by $\{(\xi_{\beta_i}(w), \eta_{\gamma_i}(w)), i \geq 1\}$.

It is easy to know that the processes A and B are two standard stochastic alternating renewal processes since $(\xi'_{i,\alpha}(w), \eta''_{i,\alpha}(w))$ and $(\xi''_{i,\alpha}(w), \eta'_{i,\alpha}(w))$, $i = 1, 2, \cdots$ are iid random vectors, respectively.

Correspondingly, let

$$P_1(t) = \Pr\{\text{process A is on at } t\}, \qquad (10)$$

$$P_2(t) = \Pr\{\text{process B is on at } t\}, \qquad (11)$$

$$P_3(t) = \Pr\{\text{process C is on at } t\}. \qquad (12)$$

Now we prove that

$$P_1(t) \leq P_3(t) \leq P_2(t). \qquad (13)$$

We just prove the left inequation of (13). To do so, we say that a renewal takes place each time the system goes on. Conditioning on the time of that last renewal prior to or at time t yields

$$P_1(t) = \Pr\{\text{process A is on at } t \mid S_A = 0\}$$
$$+ \int_0^\infty \Pr\{\text{process A is on at } t \mid S_A = y\}\,dF_{S_A}(y)$$

and

$$P_3(t) = \Pr\{\text{process C is on at } t \mid S_C = 0\}$$
$$+ \int_0^\infty \Pr\{\text{process C is on at } t \mid S_C = y\}\,dF_{S_C}(y),$$

where S_A represents the time of the last renewal prior to or at time t in the process A, S_C the time of the last renewal prior to or at time t in the process C, $F_{S_A}(y)$ the distribution function of S_A and $F_{S_C}(y)$ the distribution function of S_C. Furthermore, we have

$$\Pr\{\text{process A is on at } t \mid S_A = 0\} = \Pr\{\xi'_{1,\alpha}(w) > t \mid \xi'_{1,\alpha}(w) + \eta''_{1,\alpha}(w) > t\},$$

$$\Pr\{\text{process C is on at } t \mid S_C = 0\} = \Pr\{\xi_{\beta_1}(w) > t \mid \xi_{\beta_1}(w) + \eta_{\gamma_1}(w) > t\},$$

and for $y < t$,

$$\Pr\{\text{process A is on at } t \mid S_A = y\} = \Pr\{\xi'_{1,\alpha}(w) > t - y \mid \xi'_{1,\alpha}(w) + \eta''_{1,\alpha}(w) > t - y\},$$

$$\Pr\{\text{process C is on at } t \mid S_C = y\} = \Pr\{\xi_\beta(w) > t - y \mid \xi_\beta(w) + \eta_\gamma(w) > t - y\},$$

where $\xi_\beta(\omega)$ and $\eta_\gamma(\omega)$ are the β-pessimistic (or β-optimistic) value and γ-pessimistic (or γ-optimistic) value of the last on and off times prior to or at time t in the process C, respectively. It follows from (8), (9) and the definition of the conditional probability that

$$\Pr\left\{\xi'_{1,\alpha}(\omega) > t \mid \xi'_{1,\alpha}(\omega) + \eta''_{1,\alpha}(\omega) > t\right\} \leq \Pr\left\{\xi'_{1,\alpha}(\omega) > t \mid \xi'_{1,\alpha}(\omega) + \eta_{\gamma_1}(\omega) > t\right\}$$

$$\leq \Pr\left\{\xi_\beta(\omega) > t \mid \xi_\beta(\omega) + \eta_\gamma(\omega) > t\right\},$$

i.e.,

$$\Pr\left\{\text{process A is on at time } t \mid S_A = 0\right\} \leq \Pr\left\{\text{process C is on at } t \mid S_C = 0\right\}. \quad (14)$$

Similarly, we can prove that

$$\Pr\left\{\text{process A is on at time } t \mid S_A = y\right\} \leq \Pr\left\{\text{process C is on at } t \mid S_C = y\right\} \quad (15)$$

for any $y < t$. It follows from (14) and (15) that the left inequation of (13) is proved. The similar technique may prove the right inequation of (13).

By the arbitrariness of the process C, it follows from (13) that

$$P_1(t) = \Pr'_\alpha\{\omega \in \Omega \mid \text{system is on at } t\} \quad (16)$$

and

$$P_2(t) = \Pr''_\alpha\{\omega \in \Omega \mid \text{system is on at } t\}. \quad (17)$$

Note that the processes A and B are two standard stochastic alternating renewal processes, it follows from the random alternating renewal process that

$$\lim_{t\to\infty} P_1(t) = \frac{E\left[\xi'_{1,\alpha}(\omega)\right]}{E\left[\xi'_{1,\alpha}(\omega)\right] + E\left[\eta''_{1,\alpha}(\omega)\right]} \quad (18)$$

and

$$\lim_{t\to\infty} P_2(t) = \frac{E\left[\xi''_{1,\alpha}(\omega)\right]}{E\left[\xi''_{1,\alpha}(\omega)\right] + E\left[\eta'_{1,\alpha}(\omega)\right]} \quad (19)$$

provide that $E\left[\xi'_{1,\alpha}(\omega)\right] + E\left[\eta''_{1,\alpha}(\omega)\right] < \infty$ and $E\left[\xi''_{1,\alpha}(\omega)\right] + E\left[\eta'_{1,\alpha}(\omega)\right] < \infty$ (see Ross [20]).

Finally, it follows from (16) and (17) that the results (6) and (7) hold. The theorem is proved.

Remark 2. If $\{(\xi_i, \eta_i), i \geq 1\}$ degenerates to a sequence of iid random vectors, then the results (6) and (7) in Theorem 1 degenerate to the form

$$\lim_{t\to\infty} \Pr\{\text{system is on at } t\} = \frac{E[\xi_1]}{E[\xi_1] + E[\eta_1]},$$

which is just the conventional result in stochastic case (see Ross [20]).

Remark 3. If $\{(\xi_i, \eta_i), i \geq 1\}$ degenerates to a sequence of fuzzy vectors with the same membership function, then, for each $\alpha \in (0,1]$, α-pessimistic values and α-optimistic values of ξ_1 and η_1 degenerate to four real numbers (denoted by ξ'_α, ξ''_α, η'_α, η''_α). The results (6) and (7) in Theorem 1 degenerate to the form

$$\lim_{t \to \infty} \Pr'_\alpha \{\omega \in \Omega \mid \text{system is on at } t\} = \frac{\xi'_\alpha}{\xi'_\alpha + \eta'_\alpha}$$

and

$$\lim_{t \to \infty} \Pr''_\alpha \{\omega \in \Omega \mid \text{system is on at } t\} = \frac{\xi''_\alpha}{\xi''_\alpha + \eta''_\alpha}, \quad (20)$$

respectively.

Let ξ be one of the fuzzy variables with α-pessimistic value $E\left[\xi'_{1,\alpha}(\omega)\right]$, and α-optimistic value $E\left[\xi''_{1,\alpha}(\omega)\right]$, and η one of the fuzzy variables with α-pessimistic value $E\left[\eta'_{1,\alpha}(\omega)\right]$ and α-optimistic value $E\left[\eta''_{1,\alpha}(\omega)\right]$. Then we have the following theorem.

Theorem 2. *Assume that $\{(\xi_i, \eta_i), i \geq 1\}$ is a sequence of pairs of iid positive fuzzy random variables. For any given $\alpha \in (0,1]$, if the random variables $\xi'_{i,\alpha}(\omega)$, $\xi''_{i,\alpha}(\omega)$, $\eta'_{i,\alpha}(\omega)$ and $\eta''_{i,\alpha}(\omega)$ are nonlattice, $E[\xi + \eta] < \infty$ and $E\left[\dfrac{\xi}{\xi + \eta}\right] < \infty$, then we have*

$$\lim_{t \to \infty} \text{Ch}\{\text{system is on at } t\} = E\left[\frac{\xi}{\xi + \eta}\right]. \quad (21)$$

Proof. By Definition 5 and Remark 1, we have

Ch{system is on at t}

$$= \tfrac{1}{2} \int_0^1 \left(\Pr'_\alpha \{\omega \in \Omega | \text{system is on at } t\} + \Pr''_\alpha \{\omega \in \Omega | \text{system is on at } t\} \right) d\alpha. \quad (22)$$

It follows from Theorem 1 that

$$\lim_{t \to \infty} \Pr'_\alpha \{\omega \in \Omega \mid \text{system is on at } t\} = \frac{E\left[\xi'_{1,\alpha}(\omega)\right]}{E\left[\xi'_{1,\alpha}(\omega)\right] + E\left[\eta'_{1,\alpha}(\omega)\right]} \quad (23)$$

$$\lim_{t \to \infty} \Pr''_\alpha \{\omega \in \Omega \mid \text{system is on at } t\} = \frac{E\left[\xi''_{1,\alpha}(\omega)\right]}{E\left[\xi''_{1,\alpha}(\omega)\right] + E\left[\eta''_{1,\alpha}(\omega)\right]}. \quad (24)$$

That is,

$$\lim_{t \to \infty} \left(\Pr'_\alpha \{\omega \in \Omega \mid \text{system is on at } t\} + \Pr''_\alpha \{\omega \in \Omega \mid \text{system is on at } t\} \right)$$

$$= \frac{E\left[\xi'_{1,\alpha}(\omega)\right]}{E\left[\xi'_{1,\alpha}(\omega)\right] + E\left[\eta'_{1,\alpha}(\omega)\right]} + \frac{E\left[\xi''_{1,\alpha}(\omega)\right]}{E\left[\xi''_{1,\alpha}(\omega)\right] + E\left[\eta''_{1,\alpha}(\omega)\right]}.$$

On the one hand, we have

$$0 \leq \Pr{}'_\alpha\{\omega \in \Omega | \text{system is on at } t\} + \Pr{}''_\alpha\{\omega \in \Omega | \text{system is on at } t\} \leq 2. \quad (25)$$

It follows from the dominated convergence theorem that

$$\lim_{t\to\infty} \text{Ch}\{\text{system is on at } t\}$$

$$= \frac{1}{2}\int_0^1 \lim_{t\to\infty}\left(\Pr{}'_\alpha\{\omega \in \Omega|\text{system is on at } t\} + \Pr{}''_\alpha\{\omega \in \Omega|\text{system is on at } t\}\right) d\alpha$$

$$= \frac{1}{2}\int_0^1 \left(\frac{E\left[\xi'_{1,\alpha}(\omega)\right]}{E\left[\xi'_{1,\alpha}(\omega)\right] + E\left[\eta''_{1,\alpha}(\omega)\right]} + \frac{E\left[\xi''_{1,\alpha}(\omega)\right]}{E\left[\xi''_{1,\alpha}(\omega)\right] + E\left[\eta'_{1,\alpha}(\omega)\right]}\right) d\alpha$$

On the other hand,

$$E\left[\frac{\xi}{\xi+\eta}\right] = \int_0^\infty \text{Cr}\left\{\frac{\xi}{\xi+\eta} \geq r\right\} dr$$

$$= \frac{1}{2}\int_0^1 \left(\frac{E\left[\xi'_{1,\alpha}(\omega)\right]}{E\left[\xi'_{1,\alpha}(\omega)\right] + E\left[\eta''_{1,\alpha}(\omega)\right]} + \frac{E\left[\xi''_{1,\alpha}(\omega)\right]}{E\left[\xi''_{1,\alpha}(\omega)\right] + E\left[\eta'_{1,\alpha}(\omega)\right]}\right) d\alpha.$$

The proof is completed.

Remark 4. If $\{(\xi_i, \eta_i), i \geq 1\}$ degenerates to a sequence of iid random vectors, then the result in Theorem 2 degenerates to the form

$$\lim_{t\to\infty} \Pr\{\text{system is on at time } t\} = \frac{E[\xi_1]}{E[\xi_1] + E[\eta_1]},$$

which is just the conventional result in stochastic case (see Ross [20]). If $\{(\xi_i, \eta_i), i \geq 1\}$ degenerates to a sequence of fuzzy vectors, then, the result in Theorem 2 degenerates to the form

$$\lim_{t\to\infty} \text{Cr}\{\text{system is on at time } t\} = E\left[\frac{\xi_1}{\xi_1 + \eta_1}\right].$$

Acknowledgments

This work was supported by the National Natural Science Foundation of China Grant No. 70471049 and China Postdoctoral Science Foundation No. 2004035013.

References

1. Asmussen, S.: Applied Probability and Queues. Wiley, Now York. 1987
2. Birolini, A.: Quality and Reliability of Technical Systems. Spring, Berlin. 1994

3. Daley, D., Vere-Jones, D.: An Introduction to the Theory of Point Processes. Springer, Berlin. 1988
4. Dozzi, M., Merzbach, E., Schmidt, V.: Limit theorems for sums of random fuzzy sets. Journal of Mathematical Analysis and Applications. **259** (2001) 554-565
5. Hwang, C.: A theorem of renewal process for fuzzy random variables and its application. Fuzzy Sets and Systems. **116** (2000) 237-244
6. Kwakernaak, H.: Fuzzy random variables—I. Information Sciences. **15** (1978) 1-29
7. Kwakernaak, H.: Fuzzy random variables—II. Information Sciences. **17** (1979) 253-278
8. Kruse, R., Meyer, K.: Statistics with Vague Data. D. Reidel Publishing Company, Dordrecht. 1987
9. Liu, B.: Theory and Practice of Uncertain Programming. Physica-Verlag, Heidelberg. 2002
10. Liu, B.: Fuzzy random chance-constrained programming. IEEE Transactions on Fuzzy Systems. **9** (2001) 713-720
11. Liu, B.: Fuzzy random dependent-chance programming. IEEE Transactions on Fuzzy Systems. **9** (2001) 721-726
12. Liu, B., Liu, Y.: Expected value of fuzzy variable and fuzzy expected value models. IEEE Transactions on Fuzzy Systems. **10** (2002) 445-450
13. Liu, P.: Fuzzy-valued Markov processes and their properties. Fuzzy Sets and Systems. **91** (1997) 45-52
14. Liu, Y., Liu, B.: Fuzzy random variables: a scalar expected value. Fuzzy Optimization and Decision Making. **2** (2003) 143-160
15. Liu, Y., Liu, B.: Expected value operator of random fuzzy variable and random fuzzy expected value models. International Journal of Uncertainty, Fuzziness & Knowledge-Based Systems. **11** (2003) 195-215
16. Liu, Y., Liu, B.: On minimum-risk problems in fuzzy random decision systems. Computers & Operations Research. **32** (2005) 257-283
17. Popova, E., Wu, H.: Renewal reward processes with fuzzy rewards and their applications to T-age replacement policies. European Journal of Operational Research. **117** (1999) 606-617
18. Puri, M., Ralescu, D.: The concept of normality for fuzzy random variables. Annals of Probability. **13** (1985) 1371-1379
19. Puri, M., Ralescu, D.: Fuzzy random variables. Journal of Mathematical Analysis and Applications. **114** (1986) 409-422
20. Ross, S.: Stochastic Processes. John Wiley & Sons, Inc, New York. 1996
21. Tijms, H.: Stochastic Modelling and Analysis: A Computational Approach. Wiley, New York. 1994
22. Zhao, R., Liu, B.: Renewal process with fuzzy interarrival times and rewards. International Journal of Uncertainty, Fuzziness and Knowledge-Based Systems. **11** (2003) 573-586

Three Equilibrium Strategies for Two-Person Zero-Sum Game with Fuzzy Payoffs

Lin Xu, Ruiqing Zhao, and Tingting Shu

Institute of Systems Engineering, Tianjin University, Tianjin 300072, China
xulin93@163.com, rzhao@orsc.edu.cn, tingtingshu@hotmail.com

Abstract. In this paper, a two-person zero-sum game is considered, in which the payoffs are characterized as fuzzy variables. Based on possibility measure, credibility measure, and fuzzy expected value operator, three types of concept of minimax equilibrium strategies, r-possible minimax equilibrium strategy, r-credible minimax equilibrium strategy, and expected minimax equilibrium strategy, are defined. An iterative algorithm based on fuzzy simulation is designed to find the equilibrium strategies. Finally, a numerical example is provided to illustrate the effectiveness of the algorithm.

1 Introduction

When the game theory is applied to real world problems in the field of economics and management science, on occasions it is difficult to assess payoffs exactly because of imprecise information and fuzzy understanding of situations by experts. In such situations, it is useful to model the problems as games with fuzzy payoffs. Campos [1] studied two-person zero-sum matrix game with fuzzy payoffs, in which fuzzy linear programming models were employed to solve fuzzy matrix games. Furthermore, Nishizaki and Sakawa [10] discussed multiobjective matrix games with fuzzy payoffs. Maeda [7,8] discussed a game, in which the payoffs were characterized as fuzzy numbers, and also defined equilibrium strategies based on possibility and necessity measure. Recently, Liu and Liu [4] defined a credibility measure and gave an expected value operator of fuzzy variable. Moreover, the propositions about possibility measure, credibility measure and fuzzy expected value operator can be found in [4,5,6,9].

In this paper, we will discuss a two-person zero-sum game with fuzzy payoffs, and provide three types of minimax equilibrium strategies based on possibility measure, credibility measure and fuzzy expected value operator, respectively. An iterative algorithm based on fuzzy simulation is designed to find the equilibrium strategies. Finally we provide a numerical example to illustrate the procedure.

2 Two-Person Zero-Sum Game

In this section, we discuss a two-person zero-sum game, in which the payoffs are characterized as fuzzy variables, and then define three types of concept of minimax equilibrium strategies to the game.

Let $I \triangleq \{1, 2, \cdots, m\}$ be a set of pure strategies of player I and $J \triangleq \{1, 2, \cdots, n\}$ a set of pure strategies of player II. Mixed strategies of players I and II are represented by weights to their pure strategies, i.e.,

$$x \triangleq \{(x_1, x_2, \cdots, x_m)^T \in \Re_+^m \mid \sum_{i \in I} x_i = 1\}$$

is a mixed strategy of player I, and

$$y \triangleq \{(y_1, y_2, \cdots, y_n)^T \in \Re_+^n \mid \sum_{j \in J} y_j = 1\}$$

is a mixed strategy of player II, where $\Re_+^m = \{\mathbf{a} \in \Re^m \mid a_i \geq 0, i = 1, 2, \cdots, m\}$ and x^T is the transposition of x.

By fuzzy variable ξ_{ij}, we denote the payoff that player I receives or player II loses when player I plays the pure strategy i and player II plays the pure strategy j. Then a two-person zero-sum game is represented as fuzzy variables payoff matrix

$$\eta = \begin{bmatrix} \xi_{11} & \xi_{12} & \cdots & \xi_{1n} \\ \xi_{21} & \xi_{22} & \cdots & \xi_{2n} \\ \vdots & \vdots & \ddots & \vdots \\ \xi_{m1} & \xi_{m2} & \cdots & \xi_{mn} \end{bmatrix}. \tag{1}$$

Definition 1. *Let $\xi_{ij}(i = 1, 2, \cdots, m, j = 1, 2, \cdots, n)$ be independent fuzzy variables. Then (x^*, y^*) is called an expected minimax equilibrium strategy to the game if*

$$E\left[x^T \eta y^*\right] \leq E\left[x^{*T} \eta y^*\right] \leq E\left[x^{*T} \eta y\right]. \tag{2}$$

where η is defined by equality (1).

Definition 2. *Let $\xi_{ij}(i = 1, 2, \cdots, m, j = 1, 2, \cdots, n)$ be independent fuzzy variables and r the predetermined level of the payoffs, $r \in \Re$. Then (x^*, y^*) is called a r-possible minimax equilibrium strategy to the game if*

$$\mathrm{Pos}\left\{x^T \eta y^* \geq r\right\} \leq \mathrm{Pos}\left\{x^{*T} \eta y^* \geq r\right\} \leq \mathrm{Pos}\left\{x^{*T} \eta y \geq r\right\}. \tag{3}$$

where η is defined by equality (1).

Definition 3. *Let $\xi_{ij}(i = 1, 2, \cdots, m, j = 1, 2, \cdots, n)$ be independent fuzzy variables and r the predetermined level of the payoffs, $r \in \Re$. Then (x^*, y^*) is called a r-credible minimax equilibrium strategy to the game if*

$$\mathrm{Cr}\left\{x^T \eta y^* \geq r\right\} \leq \mathrm{Cr}\left\{x^{*T} \eta y^* \geq r\right\} \leq \mathrm{Cr}\left\{x^{*T} \eta y \geq r\right\}. \tag{4}$$

where η is defined by equality (1).

3 Iterative Algorithm Based on Fuzzy Simulation

When payoffs are fuzzy variables in a game, traditional analytical solution methods are not applicable to find a minimax equilibrium strategy. In this section, we propose an algorithm, in which fuzzy simulation is employed to estimate the possibility, credibility, and expected value, and iterative technique is used to seek the minimax equilibrium strategy. The reader interested in fuzzy simulation can consult [2,3,4].

For convenience, let $g(\xi_{ij})$ be one of the functions $E[\xi_{ij}]$, $\text{Pos}\{\xi_{ij} \geq r\}$, and $\text{Cr}\{\xi_{ij} \geq r\}$, where ξ_{ij} is the payoff that player I receives or player II loses when player I plays the pure strategy i and player II plays the pure strategy j. Similarly, let $f(\boldsymbol{x}^*, \boldsymbol{y}^*)$ be one of the functions $E[\boldsymbol{x}^{*T}\eta\boldsymbol{y}^*]$, $\text{Pos}\{\boldsymbol{x}^{*T}\eta\boldsymbol{y}^* \geq r\}$, and $\text{Cr}\{\boldsymbol{x}^{*T}\eta\boldsymbol{y}^* \geq r\}$. The procedure to find the minimax equilibrium strategy for a two-person zero-sum game with fuzzy payoffs is summarized as follows.

Step 1) Firstly, let x_{i_1} be a pure strategy selected randomly from the set of pure strategies $\{1, 2, \cdots, m\}$ by player I, and y_{j_1} a pure strategy selected randomly from the set of pure strategies $\{1, 2, \cdots, n\}$ by player II.

Step 2) After the tth game, the pure strategies selected by player I are $x_{i_1}, \cdots, x_{i_t}$ while the pure strategies selected by player II are $y_{j_1}, \cdots, y_{j_t}$. In the $(t+1)$th game, the pure strategy $x_{i_{t+1}}$ selected by player I should satisfy the condition

$$\sum_{k=1}^{t} g(\xi_{i_{t+1} j_k}) = \max_{1 \leq i \leq m} \sum_{k=1}^{t} g(\xi_{i j_k}).$$

The pure strategy $y_{j_{t+1}}$ selected by player II should satisfy the condition

$$\sum_{k=1}^{t} g(\xi_{i_k j_{t+1}}) = \min_{1 \leq j \leq n} \sum_{k=1}^{t} g(\xi_{i_k j}),$$

where $g(\xi_{ij})$ can be estimated by fuzzy simulation.

Step 3) Return to Step 2 with $t = t+1$ until the iteration number is N, where N is a sufficiently large number.

Step 4) Let $\boldsymbol{x}^* = \{x_1^*, x_2^*, \cdots, x_m^*\}$, where x_i^* is the ratio of times of the pure strategy i selected by player I in the above-mentioned steps to N, $i = 1, 2, \cdots, m$. Let $\boldsymbol{y}^* = \{y_1^*, y_2^*, \cdots, y_n^*\}$, where y_j^* is the ratio of times of the pure strategy j selected by player II in the above-mentioned steps to N, $j = 1, 2, \cdots, n$.

Step 5) Estimate $f(\boldsymbol{x}^*, \boldsymbol{y}^*)$ by fuzzy simulation, then display the minimax equilibrium strategy $(\boldsymbol{x}^*, \boldsymbol{y}^*)$ and $f(\boldsymbol{x}^*, \boldsymbol{y}^*)$.

4 Numerical Example

Let $\eta = \begin{bmatrix} \xi_{11} & \xi_{12} \\ \xi_{21} & \xi_{22} \end{bmatrix}$, where ξ_{ij} ($i = 1, 2$, $j = 1, 2$) are fuzzy variables whose membership functions are given by

$$\mu_{11}(x) = e^{-\frac{(x-1)^2}{2}}, \mu_{12}(x) = e^{-\frac{(x-3)^2}{2}}, \mu_{21}(x) = e^{-\frac{(x-4)^2}{2}}, \mu_{22}(x) = e^{-\frac{(x-2)^2}{2}},$$

respectively. By 1000 iterations (2000 sample points in fuzzy simulation), the expected minimax equilibrium strategy is

$$((x_1^*, x_2^*), (y_1^*, y_2^*)) = ((0.5230, 0.4770), (0.2530, 0.7470))$$

with $E\left[\boldsymbol{x}^{*T}\eta\boldsymbol{y}^*\right] = 2.5370$. The results of iterative algorithm are shown in Fig. 1, in which the two straight lines represent the optimal solutions of x_1 and y_1, and the curves represent the solutions obtained by different numbers of the iterative numbers.

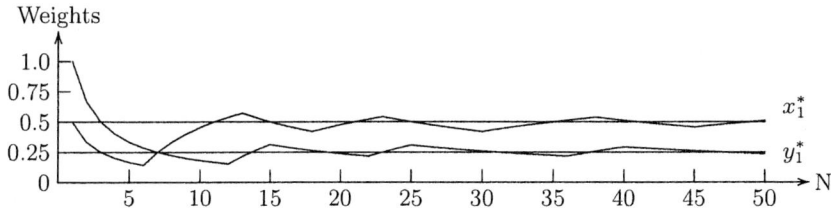

Fig. 1. An iterative process based on fuzzy simulation

5 Conclusions

In this paper, a two-person zero-sum game with fuzzy payoffs was studied. By employing fuzzy variables to describe the payoffs of the game, three types of concept of minimax equilibrium strategies were presented. An algorithm based on fuzzy simulation was designed to seek the minimax equilibrium strategies.

Acknowledgments

This work was supported by the National Natural Science Foundation of China Grant No. 70471049 and China Postdoctoral Science Foundation No. 2004035013.

References

1. Campos, L.: Fuzzy linear programming models to solve fuzzy matrix games. Fuzzy Sets and Systems. **32** (1989) 275-289
2. Liu, B., Iwamura, K.: Chance constrained programming with fuzzy parameters. Fuzzy Sets and Systems. **94** (1998) 227-237
3. Liu, B., Iwamura, K.: A note on chance constrained programming with fuzzy coefficients. Fuzzy Sets and Systems. **100** (1998) 229-233
4. Liu, B., Liu, Y.: Expected value of fuzzy variable and fuzzy expected value model. IEEE Transactions on Fuzzy Systems. **10** (2002) 445-450
5. Liu, B.: Toward fuzzy optimization without mathematical ambiguity. Fuzzy Optimization and Decision Making. **1** (2002) 43-63

6. Liu, B.: Uncertainty Theory: An Introduction to its Axiomatic Foundations. Springer-Verlag, Berlin (2004)
7. Maeda, T.: Characterization of equilibrium strategy of the bi-matrix game with fuzzy payoff. Journal of Mathematical Analysis and Applications. **251** (2000) 885-896
8. Maeda, T.: On the characterization of equilibrium strategy of two-person zero-sum game with fuzzy payoffs. Fuzzy Sets and Systems. **139** (2003) 283-296
9. Nahmias, S.: Fuzzy variables. Fuzzy Sets and Systems. **1** (1978) 97-110
10. Nishizaki, I., Sakawa, M.: Fuzzy and Multiobjective Games for Conflict Resolution. Physica-Verlag, Heidelberg (2001)

An Improved Rectangular Decomposition Algorithm for Imprecise and Uncertain Knowledge Discovery

Jiyoung Song, Younghee Im, and Daihee Park

Korea Univ. Dept. of Computer & Information Science
{songjy, yheeim, dhpark}@korea.ac.kr

Abstract. In this paper, we propose a novel improved algorithm for the rectangular decomposition technique for the purpose of performing fuzzy knowledge discovery from large scaled database in a dynamic environment. To demonstrate its effectiveness, we compare the proposed one which is based on the newly derived mathematical properties with those of other methods with respect to the classification rate, the number of rules, and complexity analysis.

1 Introduction

In the real world applications, the database is usually updated by an adding of new objects or by modification of old measurements. Maddouri[1] combined rectangular decomposition algorithm which is the incremental updates possible with fuzzy set theory to process imprecise and uncertain knowledge. The rectangular decomposition algorithms of Khcherif[2] and Maddouri[1] transform a binary matrix to a general graph first and find the maximum clique from the general graph. It is NP-hard problem[2]. However, our algorithm transforms a binary matrix to a bipartite graph first and finds bicliques from the bipartite graph by using node-deletion[3]. This can be solved in polynomial times[4].

In Section 2, we review rectangular decomposition techniques. In Section 3, we introduces a newly created rectangular decomposition algorithm and analyse the complexity. In Section 4, we present and analyze the result of experiments. Finally, concluding remarks are given in section 5.

2 Rectangular Decomposition Technique

Rectangular decomposition is a technique that finds an optimal coverage of binary relation R, where relational database is considered as binary relation, denoted $R(O, P)$, between a set of objects, O, and a set attribute value or properties P. The rectangle and the optimal coverage obtained in this step mean one rule by one-to-one mapping and minimum rule-base respectively [1] [2].

3 Improved Algorithm

In this paper, we propose a novel improved algorithm that notably reduces the costs for optimal coverage keeping the advantages of incremental update feature of the rectangular decomposition technique under dynamic environments where there are more frequent insertions, deletions, and updates. For this purpose following theorems are derived based on bipartite graph obtained from transformation of the binary matrix.

Theorems 1. Let the given bipartite graph be $B = (V_1 \bigcup V_2, E)$, and the complete bipartite graph, a subset of B, be $B_c = (W_1 \bigcup W_2, E_c)$. Where $B_c = W_1 \times W_2$, $W_1 \subset V_1$, $W_2 \subset V_2$, $E_c \subset E$. If the rectangle obtained by the node-deletion is B_c, then B_c is the maximal rectangle.

Proof. To prove it by contradiction, let's assume that the conclusion is false. Then the B_c is not a maximal rectangle and there exists B'_c such that $B_c = W_1 \times W_2 \subset W'_1 \times W'_2 = B'_c$. Hence, the only following three cases are valid when the B_c becomes the proper subset of B'_c.

i) $W'_1 = W_1 + x$ and $W'_2 = W_2$
ii) $W'_1 = W_1$ and $W'_2 = W_2 + y$
iii) $W'_1 = W_1 + x$ and $W'_2 = W_2 + y$

For the case of i), let's define the elements of V_1 as $u_1, u_2, \cdots, u_m$, elements of V_2 as $v_1, v_2, \cdots, v_n$, elements of W_1 as $w_1, w_2, \cdots, w_i$, and elements of W_2 as $z_1, z_2, \cdots, z_j$. Let x be a temporary element of $V_1 - W_1$ that satisfies $x \in V_1$ and $x \notin W_1$. If we rearrange the elements in the sets as $u_k = w_k, v_l = z_l (1 \leq k \leq i, 1 \leq l \leq j)$ then x is one of elements in the set, $u_i+1, u_i+2, \cdots, u_m$. And if we let $W_1 + x = W'_1$ then there exists $W'_1 \times W_2 = B'_c$. Since B'_c is a complete bipartite graph, x must have edges that connect all the elements of W_2. However, in the node deletion technique, only the nodes without the edges to $\forall z (\in W_2)$ are deleted. Accordingly, no element x satisfies $x \in V_1$ and $x \notin W_1$ exists. This is a contradiction to the initial condition i). Therefore, the rectangle B_c which is obtained by node-deletion is maximal rectangle. ii) and iii) can be proved likewise.

Corollary. Let the given bipartite graph be $B = (V_1 \bigcup V_2, E)$, and the complete bipartite graph, a subset of B, be $B_c = (W_1 \bigcup W_2, E_c)$. Here, we know that $B_c = W_1 \times W_2, W_1 \subset V_1, W_2 \subset V_2, E_c \subset E$. Now, let be B_{c_1} from B using node deletion and B_{c_2} be B that was obtained from $e_2 \notin E_{c_1}$, the edge that is not included in B_{c_1}. Using the same method let B_{c_1} be B_c that was obtained from $e_3 \notin (E_{c_1} \bigcup E_{c_2})$ and get B_{c_n} for all edges, $e_i \in E$. If the set of all B_c is $CV = \{B_{c_1}, B_{c_2}, B_{c_3}, \cdots, B_{c_n}\}$ then is a coverage.

Proof. Since all $B_{c_i} \in CV$ are rectangle by **theorem 1** and all the edges, $e_i \in E$ is contained in CV, $CV = \{B_{c_1}, B_{c_2}, B_{c_3}, \cdots, B_{c_n}\}$ is a coverage by the definition of coverage.

Theorems 2. Let define the maximal rectangle B_{c_i} obtained by a node-deletion as $Rec_i (i = 1, 2, \cdots, n)$ and the coverage obtained from the corollary as $CV_{opt} =$

$\{Rec_1, Rec_2, \cdots, Rec_n\}$. If we set the each maximal rectangle as optimal rectangle then CV_{opt} is the optimal coverage.

Proof. Since we define each rectangle as optimal rectangle, it is sufficient to show that all elements of Rec_i in CV_{opt} are not redundant rectangles. This is same as showing that we can't form a coverage with $n-1$ number of Rec_i. Let's assume e_i is an edge which is the first pair of nodes to get $Rec_i (i = 1, 2, \cdots, n)$ in node-deletion. Let's first find out if the Rec_i is remaining rectangle. By the definition of the coverage all the edges must belong to at least one of rectangles. However, e_n was not included in any member of $Rec_i (i = 1, 2, \cdots, n-1)$ when $Rec_i, Rec_2, \cdots, Rec_{n-1}$ belong to CV_{opt} and each member of $Rec_i (i = 1, 2, \cdots, n-1)$ is optimal rectangle hence the optimal rectangle, Rec_n, which contains e_n must be included in CV_{opt}. Therefore, Rec_n is not a redundant rectangle. Likewise, $Rec_i, Rec_2, \cdots, Rec_{n-1}$) must be included in the coverage. A rectangle can't be included in one or multiple rectangles(since other rectangles are optimal rectangles) because each rectangle is an optimal rectangle. Consequently, the coverage can't be formed with $n-1$ number of rectangles. Therefore CV_{opt} is an optimal coverage by definition.

From the theorems derived above we could prove that it's possible to get desired optimal coverage by finding biclique directly from bipartite graph without transforming the bipartite graph that was obtained from binary matrix during the process of rectangular decomposition to general graph. Because the searching space that includes the solution for the problem to find biclique from the bipartite graph is $m \times n$ matrix, it becomes significantly smaller than $(m+n) \times (m+n)$ matrix from the methodology of Maddouri[1]. Moreover, finding the maximum clique is a problem regarding NP-hard while the problem of finding biclique from bipartite graph can only be solved by polynomial time. Hence this proves the method proposed in this paper is more efficient than Maddouri's methodology that solves NP-hard problem using the heuristic.

```
Input : Information Table (Binary relation)      Input : List and Edge
Output : Optimal Coverage (Rule-base)            Output : Maximal Rectangle
Opt_Cov()                                        Max_Rec(List, Edge)
{                                                {
  Coverage = ∅;                                    Maximal = Candidate ;
  Remain_Edge = All_Edge;                          List = List - Candidate ;
  All_List = Dom ∪ Cod;                            ∀n(∈List)   s. t. (Candidate, n)∉Edge
  Candidate = ∅ ;                                    List = List - n;
  while ( Remain_Edge != ∅ )                       while ( List != ∅ );
  {                                                {
    Coverage = Coverage ∪ Max_Rec(All_List, All_Edge);   n = Best_node(List);
    Remain_Edge = Remain_Edge - edge(Max_Rec);      Maximal = Maximal ∪ {n};
    Candidate = Candidate_Node(Remain_Edge);        List = List - n;
  }                                                 ∀n'(∈List)   s. t. (n, n')∉Edge
  return(Coverage);                                    List = List - n';
}                                                }
                                                 return(Maximal);
                                                 }
          (a)                                              (b)
```

Fig. 1. (a) Algorithm that finds the optimal coverage from the bipartite graph. (b) Algorithm that finds the maximal rectangle

	Method	Complexity	Knowledge representation
Rule Induction Methods	PVM	$\sum_{k=1}^{c}(c^2d)^{2^{k-1}}$	Symbolic
	IPR	$(cd)^2(c+d)^2$	Symbolic
	FIPR	$(cd)^2(c+d)^2$	Linguistic
	Difunctional relation	$O(cd^2)$	Linguistic
	Proposed method	$O((c+d)d)$	Linguistic
Decision Tree Methods	CART	$c^{11} \times d$	Tree
	ID3	$c^2 \times d^2$	Tree
	SPINA	$c^2 \times d^2$	Latticial graph

Fig. 2. Comparison of machine learning methods

Fig. 1 show the proposed rectangular decomposition algorithm. The node of domain, node of co-domain, and edges represent property, object, and relation respectively.

The improved rectangular decomposition algorithm proposed in this paper modifies only the part of finding the optimal coverage. Therefore we can still keep the advantage of incremental updates from method proposed by Khcherif[2].

3.1 Complexity Analysis

To show its effectiveness, we compare the proposed method with other methods with respect to complexity and knowledge representation in Fig. 2

4 Experimental Results

To assess the proposed algorithm, experiments were performed using the IRIS data.

Table 1. Comparison of number of fuzzy rules and classification rate between conventional methods and proposed method

Number of training data	NM criterion[6]	RM criterion[6]	Jang [7]	Proposed method
21	89.8(71)	89.6(72)	88.3(32)	90.8(14)
30	93.0(83)	93.3(87)	91.6(36)	92.8(17)
60	93.9(105)	94.1(107)	93.3(46)	94.5(24)
90	94.8(150)	94.6(150)	95.0(46)	95.6(28)

In Table 1, the proposed method has less rules and higher classification rate than conventional methods. In newly proposed method, we created and classified the rules with 21 test data first and added 9, 30, and 30 data gradually to the initial 21 data for incremental updates on rule-base since the our method supports the incremental updates.

5 Conclusions

In this paper, we proposed a further improved algorithm for the rectangular decomposition technique with incremental updating for large scaled database in a dynamic environment. The conventional methods transform a binary matrix to a general graph first and find the maximum clique from the general graph. This is considered as an NP-hard problem. However the proposed method transforms a binary matrix to a bipartite graph first and finds bicliques from the bipartite graph by using node-deletion. This can be solved in polynomial times. The proposed algorithm is valid because it's based on the newly derived mathematical proofs. Also, it is not only effective but also has better results than conventional methods in comparisons of number of rules and the classification rates.

References

1. M. Maddouri, S. Elloumi, A. Jaoua: An Incremental Learning System for Imprecise and Uncertain Knowledge Discovery. Information Sciences. **109** (1998) 149-164
2. R. Khcherif, A. Jaoua: Rectangular Decomposition Heuristics for Documentary Databases. Information Sciences. **102** (1997) 187-202
3. M. Yannakakis: Node deletion problems on bipartite graphs. SIAM J. Comput. **10** (1981) 310-327
4. D. S. Hochbaum: Approximating Clique and Biclique Problems. Journal of Algorithm. **29** (1998) 174-200
5. R. Khcherif, M. M. Gammoudi, A. Jaoua: Using difunctional relations in information organization. Information Sciences. **125**(2000) 153-166
6. H. Ishibuchi, K. Nozaki, H. Tanaka: Efficient fuzzy partition of pattern space for classification problems. Fuzzy Sets and Systems. **59** (1993) 295-304
7. Jang, D-S., Choi, H-I.: Automatic Generation of Fuzzy Rules with Fuzzy Associative Memory. Proceeding of the ISCA 5th International Conference. (1996) 182-186

XPEV: A Storage Model for Well-Formed XML Documents*

Jie Qin[1], Shu-Mei Zhao[2], Shu-Qiang Yang[1], and Wen-Hua Dou[1]

[1] School of Computer, National University of Deference Technology,
Changsha Hunan, 410073 China
qinjie0160@sina.com
[2] Zhengzhou Railway Vocational & Technical College,
Zhengzhou Henan, 450052 China

Abstract. XML is an emerging standard for Internet data representation and exchange. There are more and more XML documents without associated schema on the Web. An XML document without associated schema is called well-formed XML document. It is difficult to store and query well-formed XML documents. This paper proposes a new XML documents storage model based on model-mapping, namely XPEV. The unique feature of model-mapping-based storage model is that no XML schema information is required for XML data storage. Through XPEV's three tables: Path table, Edge table and Value table, well-formed XML documents could be easily stored in relational databases. XPEV model can make full use of the index technology of relational databases, and give a better solution for querying and storing well-formed XML documents using relational databases.

1 Introduction

As the eXtensible Markup Language (XML) becoming a standard for Internet data representation and exchange, large amounts of XML documents are being generated, the problem of storing and querying XML documents has gained increasing attention [1],[2],[3]. There exist numerous different approaches to store XML document, such as file systems, object-oriented database systems, native XML databases systems, relational or object-relational database system. Comparative studies [4],[5] indicate that relational or object-relational database systems are competitive in terms of efficiency as compared to native XML systems, object-oriented database systems, or file systems. Significant benefits include scalability and support for value-based indexes. Consider easy to manage and realize, relational database management system (RDBMS) is chosen as one efficient way to store XML data.

Effective storage for XML documents must consider both the dynamic nature of XML data and the stringent requirements of XQuery[6]. When XML documents are stored in off-the-shelf database management systems, the problem of storage model

---

* This work was supported by National Science Foundation of China under Grant No.90412011, the National High-Tech Research and Development Plan of China under Grant No. 2003AA111020, and the National High-Tech Research and Development Plan of China under Grant No. 2004AA112020.

design for storing XML data becomes a database schema design problem. In [3], the authors categorize such database schemas into two categories: structure-mapping approach and model-mapping approach. In the former, the design of database schema is based on the understanding of DTDs (Document Type Declarations) or XML Schema that describes the structure of XML documents. In the latter, a fixed database schema is used to store any XML documents without assistance of XML schema. Obviously, the structure-mapping approach is inappropriate to the storage of a large number of dynamic(DTDs vary from time to time) and structurally-variant XML documents.

An XML document is called well-formed if it does not have an associated DTDs (or XML Schema). In many applications, the schema of XML documents is unknown. There are more and more well-formed XML documents on the Web. It is difficult to query these non-schema XML documents. Model-mapping approach gives solution to this problem.

A new model-mapping-based storage model, namely XPEV was present in this paper. XPEV is a multi-XML document storage model. It uses a three tables schema to store XML data to relational databases: Path table stores distinct (root-to-any) paths, Edge table keeps parent-child relationships, Value table stores all the values of elements and attributes. Through XPEV, any well-formed XML document can be converted to relational database schema, experiment shows that XPEV is an efficient storage model for query processing using relational database technology. And it is easy to realize. To the best of our knowledge, XPEV outperform known model-mapping methods.

The organization of this paper is as follows. Section 2 discusses XML data models. Section3 introduces XPEV data storage model. Section 4 introduces two existing model-mapping approaches, XRel[3] and Edge[8]. Section 5 gives out experimentation results. Section 6 is conclusion.

2 Data Model

The World Wide Web Consortium (W3C) proposed four XML data model: the Infoset model, the XPath data model, the DOM model, and the XQuery 1.0 and XPath 2.0 data model. Each of the models describes an XML document as a tree, but there are differences in the trees. Although these variations may not impact the models' ability to represent enterprise data for most applications, they do impact the ability to provide consistent and uniform management of documents across diverse applications. The particular contents of the four XML data model can be found in [1]. It is important to note that among the four models only the XQuery 1.0 and XPath 2.0 data model acknowledges that the data universe includes more than just a single document. Furthermore, it is also the only model that includes inter-document and intra-document links in a distinct node type (i.e., Reference). We adopt the XQuery 1.0 and XPath 2.0 data model Data Model to represent XML documents.

The XQuery 1.0 and XPath 2.0 data model models XML documents as an ordered graph using 7 types of nodes, namely, root, element, text, attribute, namespace, processing instruction and comment. In brief, here we emphasize on four types: root, element, text and attribute. The formal definition of the XML data graph is omitted. The full specification can be found in [6],[7].

```
<Proceedings>
    <Proceeding>
        <conference> SIGMOD  </conference>
        <Year> 2002 </Year>
        <Article Id=A1 >
            <Title> Storing and querying XML </Title>
            <Author> Igor </Author>
            <Author> Stratis </Author>
            <Reference>Id =A2 </Reference>
        </Article>
    </Proceeding>
    <Proceeding>
        <conference> SIGMOD  </conference>
        <Year> 2001 </Year>
        <Article Id=A2 >
            <Title> Updating XML </Title>
            <Author> Igor </Author>
        </Article>
    </Proceeding>
</Proceedings>
```

Fig. 1. A small XML document

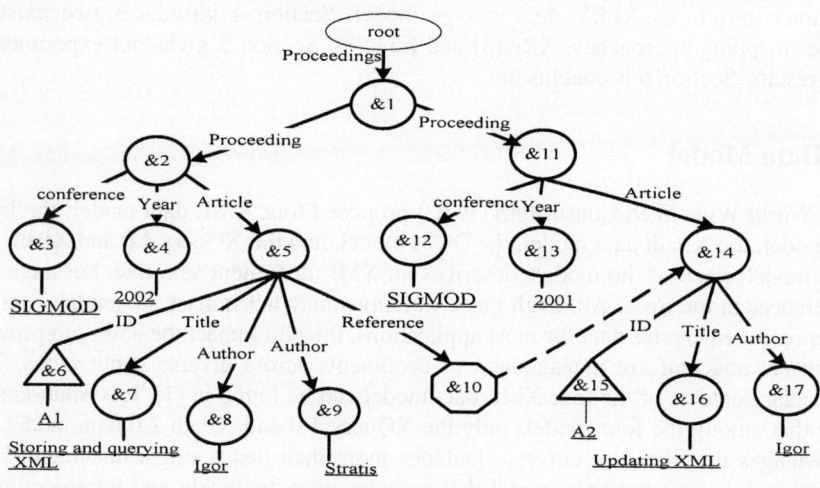

Fig. 2. A data graph for a small XML document

Fig. 1 is a simple XML document example. Fig. 2 is its data graph illustrating the data model in our research. All the round nodes are element nodes. Each node is assigned a unique preorder node traversal number. An element node has an element-type name, which is represented as an edge-label of the incoming edge pointing to the element node itself. The text nodes are marked with underline. All the triangle nodes are attribute nodes. The IDREF attribute nodes are used for intra-document references. The root node of the XML data graph has an edge, with an edge-label Proceedings, pointing to the element (&1), which is the real root element of the XML document. The IDREF attribute node (octagon) is used for intra-document (inter-document) reference. For example, the element node (&5) has an IDREF attribute (@Reference) pointing to the element node (&14) whose ID attribute (@Id) is A2. A label-path in an XML data graph is a /-separated sequence of edge-labels (element-type names), such as /Proceedings/Proceeding/Article/Author in Fig. 2.

3 XPEV Storage Model

XPEV is a new model-mapping-based storage model. XPEV storage model consists of three parts, storage schema, index methods and query evaluation algorithm.

3.1 Storage Schema

XPEV's storage schema contains three tables: Path table, Edge table and Value table. Path table stores distinct (root-to-any) path, it gives a global view on the XML documents stored in the database management system. It allows navigation through the data with regular path expressions, which can speed up querying. Edge table keeps parent-child relationships, and describes all edges of XML data graphs. Value table stores all the values of elements and attributes which are leaf nodes in XML data graph. Table 1 is Path table for the XML data graph of Fig. 2. Table 2 shows Edge table of Fig. 2. Table 3 shows Value table of Fig. 2.

Path table keeps all distinct label-paths of XML documents. For example, in Fig. 2, two data-paths, &1/&2/&3, &1/&11/&12, share the same label-path #/Proceedings/Proceeding/conference, their Path-id is 3 in Table 1. Symbol #/ denotes root node.

Edge table keeps all edges of XML data graph which describes parent-child relationships of XML data. Each edge is specified by two node identifiers, Source and Target. The Path_id indicates which path the edge belongs to. The Doc_id indicates which XML document the edge belongs to. In this example only one XML document exists whose Doc_id is X1. The Label attribute keeps the edge-label of an edge. Because elements of an XML document have order, the order of an element is the order of this element among all siblings that share the same parent. The Order attribute records the order of the Target element node among its siblings. The Flag value indicates whether the target node is an inter-object reference (ref) or points to a value (val). For example, the triple (8, "X1", &5, &9, "Author", 2, Val), describes the edge from the element node (&5) to the element node (&9). The edge has an edge-label "Author", and a value ("Stratis" in the value table), the order of element node &9 is 2 (the second author of the article "Storing and querying XML"), its path is #/Proceedings/Proceeding/Article/Author (Path_id =8), it belongs to XML document "X1".

Table 1. Path table of Fig. 2

Path_id	Pathexpress
1	#/Proceedings
2	#/Proceedings/Proceeding
3	#/Proceedings/Proceeding/Conference
4	#/Proceedings/Proceeding/Year
5	#/Proceedings/Proceeding/Article
6	#/Proceedings/Proceeding/Article/ID
7	#/Proceedings/Proceeding/Article/Title
8	#/Proceedings/Proceeding/Article/Author
9	#/Proceedings/Proceeding/Article/Reference

Table 2. Edge table of Fig. 2

Path_id	Doc_id	Source	Target	Label	Order	Flag
2	X1	&1	&2	Proceeding	1	Ref
5	X1	&2	&5	Article	1	Ref
2	X1	&1	&11	Proceeding	2	Ref
5	X1	&12	&14	Article	1	Ref
3	X1	&2	&3	Conference	1	Val
4	X1	&2	&4	Year	1	Val
6	X1	&5	&6	ID	1	Val
7	X1	&5	&7	Title	1	Val
8	X1	&5	&8	Author	1	Val
8	X1	&5	&9	Author	2	Val
9	X1	&5	&10	Reference	1	Val
3	X1	&11	&12	Conference	1	Val
4	X1	&11	&13	Year	1	Val
6	X1	&14	&15	ID	1	Val
7	X1	&14	&16	Title	1	Val
8	X1	&14	&17	Author	1	Val

Table 3. Value table of Fig. 2

Path_id	Doc_id	Source	Target	Lable	Order	Val
3	X1	&2	&3	Conference	1	SIGDOM
4	X1	&2	&4	Year	1	2002
6	X1	&5	&6	ID	1	A1
7	X1	&5	&7	Title	1	Storing and Querying XML
8	X1	&5	&8	Author	1	Igor
8	X1	&5	&9	Author	2	Stratis
9	X1	&5	&10	Reference	1	A2
3	X1	&11	&12	Conference	1	SIGDOM
4	X1	&11	&13	Year	1	2001
6	X1	&14	&15	ID	1	A2
7	X1	&14	&16	Title	1	Updating XML
8	X1	&14	&17	Author	1	Igor

Value table keeps all the leaf-nodes information of XML data graphs. For example, in table 3 the triple (8, "X1", &5, &9, "Author", 2, "Stratis"), describes the edge ("Author") from the element node (&5) to the element node (&9) has a value "Stratis", and "Stratis" is the second author of the article "Storing and querying XML".

Through the above three tables, any XML documents can be appropriate decomposed and stored into a RDBMS according to their tree structures. This model can be efficiently supported by the conventional index mechanisms.

3.2 Query Evaluation Algorithm

When retrieve XML documents from such databases, XML queries (Xquery) should be translated into SQL queries. With the help of path expressions (Pathexpress in Path table), and the Path_id in Edge table and value table, query searching scope can be efficiency reduced. When a simple XML query(query has no branch) executes, first using string match algorithm from Path table select the paths (Path_id) which satisfy the query path, then from Value table select the results which have the same Path_id. When a complex XML query(query has multi-branch) executes, both Edge table and Value table are used to find the final results. The outline of query evaluation algorithm is described below:

```
Input: An XQuery query expression
Output: result of SQL query
Algorithm:
  1.Transform XQuery query expression to SQL format;
  2.Select all the Path_ids from Path table which query
    path satisfy Pathexpress;
  3.If query is single path query
    {
      Using the selected Path_id to find all matched
      triples from Value table;
      Output all the matched triples according to SQL
      format;
      if no matched triples in Value table
        {
          Using the selected Path_id to find all matched
          triples from Edge table;
          Output all the matched triples according to SQL
          format;
        }
    }
  4.else //query is complex query which has multi-branch
    {
      Set n=the number of selected Path_id in step 2;
      Using one of selected Path_ids whose query path
      is the longest, to find all the matched triples
      from Value table or Element table;
      Store the matched triples to temp table T1;
      while n-1>0 do
```

```
        {
             Using one of the unused selected Path_id to
             find all the matched triples from Element ta-
             ble;
             Store the matched triples to temp table T2;
             for each triple of T2 ,if its Source value
             equal to any one of T1 triple's Source node's
             ancestor, and they have same Doc_id, then
             join these two triples, and store the new
             triple to T3;
             Set T2 to empty;
             Set T3 to empty;
             Set T1=T3;
             n=n-1;
        }
        Output all the triples in T1 according to SQL
        format;
    }
```

From the above describe of the query evaluation algorithm, it is easy to derive that the algorithm can be finished in time O(M+N), M is the size of query sentence, and N is the size of XML source data.

4 Related Work

Edge [8] and XRel [3] are two representatives of the model-mapping approaches. The Edge approach stores XML data graphs in a table named Edge, its Edge table is defined as follows:

Edge(Source, Target, Order, Label, Flag, value).

Because it only keeps edge-label, rather than the label paths, a large number of joins is needed to check edge-connections. It needs to concatenate the edges to form a path for processing user queries. It requires a number of join operations in proportion to the length of the path expression.

XRel is a path-based storage model. The basic XRel schema consists of the following four relational schemas:

Attribute(docID, pathID, start, end, value)
Text(docID, pathID, start, end, value)
Path(pathID, pathexp)
Element(docID, pathID, start, end, index, reindex).

Attribute table stores values of attributes of XML elements. Text table stores values of XML elements Path table stores distinct (root-to-any) paths, just like XPEV's Path table. Element table associates each path with a region (start, end). There is no edge information explicitly maintained in this schema. With a concept called region, it maintains a containment relationship. A region is a pair of start and end points of a node. A node n_i is reachable from another node n_j, if the region of n_i is included in the region of n_j. The containment relationship is not only for parent-child relationships, but rather ancestor-descendent relationships. The attributes index and reindex in the relation Element represent the occurrence order of an element node among the sibling element nodes in document order and reverse document order, respectively. The advantage of this approach is that it can easily identify whether a node is in a path from another node by using θ-joins (to test the containment- relationship using >, <). But θ-joins are more costly than equijoins[5]. In addition, off-the-self database management systems usually do not have a special index mechanism to support containments.

5 Performance Experiments

In order to check the effectiveness of XPEV method, a series of performance experiments were carried out. Here reports the outlines of the implementation and the experimental results.

All experiments were running on a PC with AMD XP1600+ CPU, 512 MB RAM and 40 GB hard disk. Operation system is Linux . The underling DBMS is IBM DB2 V7.1 with the XML Extender[9].

The Gnome XML[10] parser is used to parse XML data. A primitive query interface is provided C++. Query processing is directly implemented using the query interface.

The data sets we used are public XML databases DBLP[11] and the XML benchmark database XMARK[12]. Three subsets of DBLP, DBLP1(60MB), DBLP2(90MB), DBLP3(120MB) were used in our experiment. An XMARK dataset is a single record with a very large and complicated tree structure. In our experiments, we use an XMARK dataset generated by xmlgen[13] with scaling factor 1.0, totaling 108 MBytes of data.

In XPEV, a clustered index was built on Edge.Path_Id and Value.Path_Id to efficiently select tuples with particular Path_Ids. B+-tree indexes were created on Edge.Source and Edge.Target to speed up self-joins of the Edge table. Another B+-tree index was created on Value.Val.

For Edge, we created indexes as proposed in [8].

For XRel, a clustered composite index was built on Element(PathId,Start,End) and Text.(PathId, Start, End). A B+-tree index was created on Text.Value.

We separately use Edge, XRel, and XPEV approach to store DBLP1, DBLP2, DBLP3. Five queries selected from XML use case[14] (Query 1-2 are simple XML queries, Query 3-5 are complexity XML queries) were used to test their querying performance.

Query 1 Select papers written by Michael Stonebraker.
Query 2 Select papers written by Michael Stonebraker or Jim Gray.
Query 3 Select journal papers that have a cite entry whose label is CARE84.
Query 4 Select papers published between 1990 and 1994,with titles starting with "database".
Query 5 Select conference papers published in year 2000 on XML.

Table 4 show Query1-5(Q1-5) elapsed time. From Table 4, we can see that XPEV approach outperforms Edge and XRel, besides Query 4. This is because XRel is good at processing region query.

Table 4. Query elapsed time: Edge, XRel and XPEV using DBPL1

	Q1	Q2	Q3	Q4	Q5
Edge	0.15	0.19	14.3	19.7	16.9
XRel	0.12	0.16	10,1	8.3	11.6
XPEV	0.12	0.15	8.3	11.2	9.3

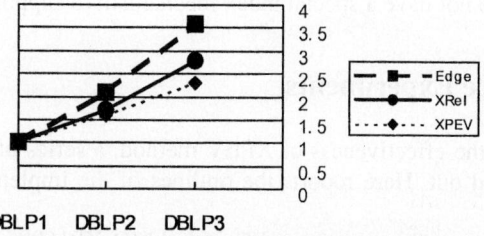

Fig. 3. The elapsed time ratios

Fig. 3 shows the average elapsed time ratios for XPEV, Edge and XRel using above five queries, respectively, where the x-axis shows the data size. The query elapsed time ratios is defined as follows: Suppose the elapsed time of a query using DBLP1 is treated as scale t, and suppose the elapsed time of the same query is T, using either DBLP2 or DBLP3.The elapsed time ratio is T/t. From Fig. 3, we can see XPEV approach has better scalability querying performance as compare with Edge and XRel approach, when data set becomes bigger.

We further tested XPEV vs XRel using XMARK1.0. Experimentation results also show that using the same query sentences, XPEV method is always better than XRel method with respect of query elapsed time.

6 Conclusions

Well-formed XML document have not associated schema. It's difficulty to query useful information from a large amount of this kind of XML documents. The aim of our work is to find a better solution to store and query well-formed XML documents using relational database technology. A new model-mapping approach called XPEV was proposed in this paper. XPEV is a three-table schema: Path table keeps all distinct label-paths of XML documents. Edge table keeps all edges of XML data graph. Value table keeps all the leaf-nodes information of XML data graphs. Experimental results confirm that the XPEV storage model outperforms known model-mapping approaches, such as Edge and XRel in most cases, and XPEV model is easier to implement than XRel approach.

References

1. Salminen, A., Wm, F.: Requirements for XML Document Database Systems. First ACM Symposium on Document Engineering, Atlanta (2001)85-94
2. Tatarinov, I., Viglas, S., Beyer, K. S., Shanmugasundaram, J., Shekita, E. J., Chun Zhang: Storing and querying ordered XML using a relational database system. SIGMOD Conference (2002)204-215

3. YoshiKawa, M., Amagasa, T., Shimura, T.: XRel: A path-based approach to storage and retrieval of XML documents using relational databases. ACM Transactions on Internet Technology, Vol. 1.(2001)110-141
4. Shanmugasundaram, J., Shekita, E. J., Kiernan, J.. A General Techniques for Querying XML Documents using a Relational Database System. SIGMOD Record (3): (2001)20-26
5. Haifeng Jiang, Hongjun Lu, Wei Wang, J. X. Yu.: Path Materialization Revisited: An Efficient Storage Model for XML Data. AICE2002
6. W3C: XQuery 1.0: An XML Query Language. W3C working draft 16 Aug. 2002. http://www.w3.org/TR/xquery
7. Berglund, A., Boag, S. Chamberlin,D.: XML Path Language (XPath) 2.0. W3C Working Draft. 16 August 2002. http://www.w3.org/TR/xpath20
8. Florescu, D., Kossman, D.: A Performance Evaluation of Alternative Mapping Schemes for Storing XML Data in a Relational Database, Rapport de Recherche No. 3680 INRIA, Rocquencourt, France, (1999)
9. IBM's DB2 extender for XML.http://www-4.ibm.com/software/data/db2/extenders/xmlext.html
10. The XML C library for Gnome libxml. http://xmlsoft.org/
11. DBLP maintained by Ley, M.: http://www.informatik.uni-trier.de/~ley/db/index.html
12. XMARK: The XML-benchmark project. http://monetdb.cwi.nl/xml
13. Schmidt, A., Waas, F., Kersten, M. L., Florescu, D., Manolescu, I., Carey, M., and Busse, R.: The XML benchmark project. Technical Report INS-R0103, Centrum voor Wiskunde en Informatica(2001)
14. W3C: XML Query use cases. W3C working draft 16 Aug. 2002. http://www.w3.org/TR/xmlquery-use-cases

Fuzzy-Rough Set Based Nearest Neighbor Clustering Classification Algorithm

Xiangyang Wang, Jie Yang, Xiaolong Teng, and Ningsong Peng

Institute of Image Processing & Pattern Recognition, Shanghai Jiaotong University,
Shanghai 200030, China
{wangxiangyang, jieyang}@sjtu.edu.cn

Abstract. We propose a new nearest neighbor clustering classification algorithm based on fuzzy-rough set theory (FRNNC). First, we make every training sample fuzzy-roughness and use edit nearest neighbor algorithm to remove training sample points in class boundary or overlapping regions, and then use Mountain Clustering method to select representative cluster center points, then Fuzzy-Rough Nearest neighbor algorithm (FRNN) is applied to classify the test data. The new algorithm is applied to hand gesture image recognition, the results show that it is more effective and performs better than other nearest neighbor methods.

1 Introduction

Nearest Neighbor (NN) algorithm is one of the most important classification methods for pattern recognition. Rough Set [2] takes the imprecise features description of the elements in universe as one of the reasons to wrongly classification. Fuzzy-Rough sets [3,4] combines rough set with fuzzy set to handle datasets with both roughness and fuzziness. In [5], a Fuzzy-Rough Nearest Neighbors (FRNN) algorithm is proposed, which introduces roughness uncertainty into Fuzzy KNN [1]. FRNN takes every training pattern as the neighbor of the test pattern with different fuzzy similarities, and it is more robust than Fuzzy KNN. However, for each test pattern it has to compute the similarity for every training pattern, which leads to more complex.

Our algorithm is proposed to reduce the complexity of FRNN and improve the speed of classification. In [9], based on the properties of rough sets, Lingras P extents the K-means and Fuzzy C-means to represent clusters as interval set. In our algorithm, Mountain Clustering method is used to select representative cluster center points from training data, which is different from Lingras P's method.

2 Fuzzy-Rough Set and Fuzzy-Rough Nearest Neighbor Algorithm

For input data set $X = \{x_1, x_2, ..., x_n\}$ with total C classes, let R an equivalence relation on X, and $[x]_R$ is the equivalence class contains $x \in X$. For any output class $c \subseteq X$, lower and upper approximation is: $\underline{R}(c) = \cup \{[x]_R \mid [x]_R \subseteq c, x \in X\}$ and

$\overline{R}(c) = \cup\{[x]_R \mid [x]_R \cap c \neq \emptyset, x \in X\}$. Rough set is the set of lower and upper approximation, i.e., $R(c) = <\underline{R}(c), \overline{R}(c)>$. When the equivalence classes in rough set are imprecise, they will be in the form of fuzzy clusters $\{F_1, F_2, ..., F_H\}$ and X is generated by fuzzy weak partition. Here, every fuzzy cluster F_j is a fuzzy set. The output class c can also be fuzzy and in the form of lower approximation $\underline{c}$ and upper approximation $\overline{c}$:

$$\mu_{\underline{c}}(F_j) = \inf_{x \in c} \max\{1 - \mu_{Fj}(x), \mu_c(x)\}, \forall x \qquad (1)$$

$$\mu_{\overline{c}}(F_j) = \sup_{x \in c} \min\{\mu_{Fj}(x), \mu_c(x)\}, \forall x \qquad (2)$$

$<\underline{c}, \overline{c}>$ is a fuzzy-rough set, $\mu_c(x) \in [0,1]$ is the fuzzy membership of x to class c. The fuzzy-rough membership function of x to class c is defined as:

$$\tau_c(x) = \frac{1}{|X|} \sum_{y \in X} \tilde{\mu}_x(y) \mu_c(y) \qquad (3)$$

where $\tilde{\mu}_x(y)$ denotes the fuzzy similarity between x and y. In our algorithm, we adopt the similarity, $\tilde{\mu}_x(y) = \exp(-\frac{\|y-x\|^2}{\beta})$ [8], $\|y-x\|$ is the distance in Euclidean norm,

and $\beta = \frac{\sum_{i=1}^{n} \|x_i - \overline{x}\|^2}{n}$ is the normalized term, where $\overline{x} = \frac{\sum_{i=1}^{n} x_i}{n}$.

Fuzzy-rough nearest neighbor algorithm (FRNN) [5] is based on fuzzy-rough set. It computes the fuzzy-rough membership value of test pattern to each class according to equation (3). The test pattern is assigned to the maximal fuzzy-rough membership value related class.

3 Fuzzy-Rough Set Based Nearest Neighbor Clustering Classification Algorithm

The steps of our new algorithm are as follows.

(1) By using FRNN, assign fuzzy-rough membership value to each training pattern.

For every pattern in the training set, use leave-one-out FRNN algorithm to compute the fuzzy-rough membership value to all classes. The value implies that to what degree others support this pattern. The higher the value is the less the fuzzy-roughness exists in the neighborhood.

(2) Use muti-edit-nearest-neighbor algorithm to filter the training set

After step (1), there may exist some samples, whose maximal fuzzy-rough membership value related class is inconsistent to their known class label. Such sample points are regarded as in the class boundary or overlapping region. We use muti-edit-nearest-neighbor algorithm [6] to edit them to improve the classifier to test set.

(3) Select cluster representative points from the edited training set.

Mountain Clustering is an approximate clustering method [7]. It firstly constructs Mountain Function from dataset, and then uses Destruction Function to destroy the Mountain step by step and acquire cluster center points. These points can represent the distribution of each class.

For class c, firstly we select the sample point x_j whose fuzzy-rough membership value to c is the highest (recorded as maxc). We take x_j as the fist cluster center point of class c. Then we use the following equation to update the fuzzy-rough membership value of all training patterns to class c:

$$\tau_c(x_l) = [1 - \tilde{\mu}_{x_j}(x_l)]\tau_c(x_l), l = 1,2,....,n \qquad (4)$$

Equation (3) can be seen as the Mountain Function, and equation (4) as Destruction Function. The new point with the highest fuzzy-rough membership value is selected as the next cluster center point. Such a process continues until the ratio of the lately selected maximal fuzzy-rough membership value to maxc is less than a given threshold.

(4) Classify the test set with selected cluster representative points with FRNN.

With the selected cluster representative points in step (3), we use FRNN algorithm to classify the test set. Since the cluster representative points are usually far less than original training set, the classification speed can be improved greatly while keep the same or even better classification accuracy than FRNN.

4 Experiments and Discussions

We apply the proposed algorithm, to hand gesture recognition. All the captured images are normalized as 36×36 pixels in gray-level. The dataset contains total 30 kinds of gestures. Each kind represents a single or double letter, as shown in Fig.1.

In the hand gesture recognition problem, since there exist much difference in the same kind of gesture images, the rough uncertainty exists in the input data. On the other hand, different kinds of gesture images may be very similar, e.g. in Fig.1, the different gestures *h*, *x* and *i* are very similar. The classifier tends to wrongly classify these similar gesture images, which is due to the fuzzy output class. We use fuzzy-rough set based FRNNC algorithm to deal with such data.

There are total 4152 images, about 150 images for each kind. 2/3 of the images (3000 samples, 100samples per class) are randomly selected as training set, and the left 1/3 as test set. The algorithms, NN, KNN, Fuzzy KNN, FRNN and FRNNC are evaluated on such dataset respectively. The experimental results are given in Table 1.

Our algorithm performs better than KNN and Fuzzy KNN. Compared with FRNN, it has less computation complexity and higher classification speed, and its classification accuracy is not bad than that of FRNN. The algorithm has two properties: (1) it considers both the fuzziness and roughness in data; (2) It is especially suitable for numerical data with good clustering character.

In addition, the similarity computation method for samples is the key, which directly affects the classification result. The improvement to it will be the future work.

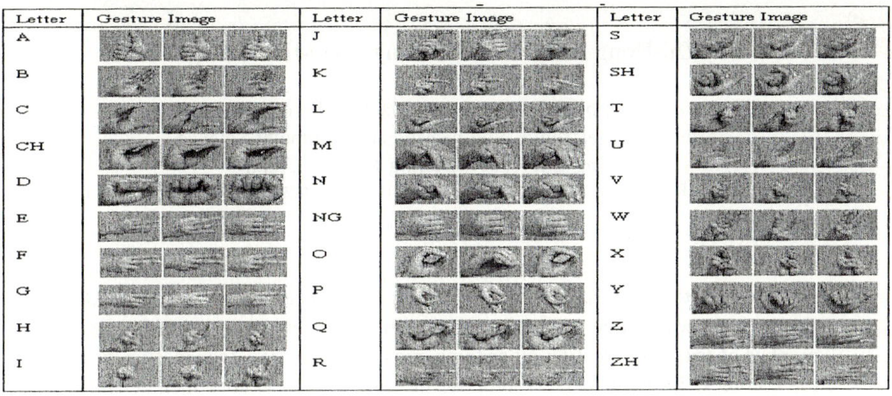

Fig. 1. The Hand Gesture Image and the Represented Letters

Table 1. Experimental results

Algorithms	NN	KNN			Fuzzy KNN			FRNN	FRNNC
Classification accuracy (%)	90.12	K=3	K=5	K=7	K=3	K=5	K=7	96.39	94.96
		90.45	89.45	88.79	90.59	89.92	89.45		

References

1. J.M., Keller, M.R., Gray: A Fuzzy K-nearest neighbor algorithm. IEEE Transactions on System, Man and Cybernetics. 15(4) (1985) 580-585
2. Z., Pawlak: Rough set approach to knowledge-based decision support. European Journal of Operational Research 99 (1997) 48-57
3. D., Dubois, H., Prade: Rough-fuzzy sets and fuzzy-rough sets. International Journal of General Systems. 17(2-3) (1990) 191-209
4. Anna, M.R, Etienne, E.K.: A comparative study of fuzzy rough sets. Fuzzy Sets and Systems. 126 (2002) 137-155
5. Sarkar, M.: Fuzzy-Rough nearest neighbors algorithm. In: Proc. Of IEEE Int. Conference on Systems, Man and Cybernetics. Nashville Tennessee USA (2000) 3556-3561
6. C.Z., Ye, J., Yang: Improving performance of decision trees with muti -edit-nearest-neighbor algorithm. Control and Decision. 18(1) (2003) 96-102
7. Yager, R.R, Filev, D.P.: Approximate clustering via the mountain method. IEEE Transactions on Systems, Man and Cybernetics. 24(8) (1994) 1279-1284
8. M.S., Yang, K.L., Wu: A Similarity Based Robust Clustering Method. IEEE Transactions on Pattern Analysis and Machine Intelligence. 26(4) (2004) 434-448
9. Lingras, P., Yan, R.: Interval clustering using fuzzy and rough set theory. In: Processing of NAFIPS '04. Banff Alberta Canada 2 (2004) 780-784

An Efficient Text Categorization Algorithm Based on Category Memberships

Zhi-Hong Deng, Shi-Wei Tang, and Ming Zhang

National Laboratory on Machine Perception,
School of Electronics Engineering and Computer Science,
Peking University, Beijing 100871, China
zhdeng@cis.pku.edu.cn, tsw@pku.edu.cn, mzhang@db.pku.edu.cn

Abstract. Text Categorization is the process of automatically assigning predefined categories to free text documents. Although there have existed a large number of text classification algorithms, most of them are either inefficient or too complex. In this paper, we propose the concept of category memberships, which stand for the degrees that words belonging to categories. Based on category memberships, a simple but efficient algorithm is presented. To evaluate our new algorithm, we have conducted experiments using Newsgroup_18828 text collection to compare it with Naive Bayes and k-NN. Experimental results show that our algorithm outperforms Naive Bayes and k-NN if a suitable category membership function is adopted.

1 Introduction

Since 1990's, we have seen an exponential growth in the volume of text documents available on the Internet. The exponential growth of these Web documents has led to a great deal of interest in developing efficient tools and software to assist users in finding relevant information. Text classification, which is the task of assigning natural language texts to predefined categories based on their context, has been proved to be useful in helping organize and search text information on the Web. A growing number of statistical classification methods and pattern recognition techniques have been applied to text categorization in recent years, including nearest neighbor classification [1], Naïve Bayes [2], decision trees [3], neural networks [4], boosting methods [5], and Support Vector Machines [6]. Most of which are either inefficient or too complex.

In this paper, we first propose the definition of category memberships. Then, a new text categorization algorithm, which is based on category memberships, is presented. The core idea of our algorithm is that it converts the problem of classifying of unlabeled documents into the problem of computing the category memberships of words.

The remainder of this paper is organized as follows. Section 2 describes the concept of category memberships and presents some function for measuring the category memberships of words. Section 3 presents a new text categorization algorithm based on category memberships in details. Section 4 describes the Newsgroup_18828 text collection and presents the experiments and results. Section 5 summarizes works in the paper and points out future research.

2 Category Memberships

Let $C = \{c_1, ..., c_m\}$ be the set of predefined categories, $F = \{f_1, ..., f_n\}$ be word set, and $TD = \cup D_i$ be the set of training documents with labeled categories, where D_i is the set of documents that are labeled category c_i. In this section, we will describe category membership and other relevant concepts.

2.1 The Definition of Category Membership Set

Definition 1. (category membership set) Given a category c ($\in C$), the category membership set S_c of c is defined by a function μ_c, which a map from word set F to $\Re^1$. μ_c is called the membership function (or grade of membership) of S_c. That is to say, μ_c stands for the degree (or grade) that word f ($\in F$) belongs to c. S_c can also be represented as the set of tuple of word f and its membership function $\mu_c(f)$, which is as follows:

$$S_C = \{< f, \mu_c(f) > | f \in F\} \tag{1}$$

In fact, Category membership set is the extension of fuzzy set, where value domain is [0, 1] instead of $\Re$. The reason that we adopt $\Re$ is that it is a more nature way and more efficient than [0, 1] in text categorization. It is obvious that a category c with a membership function specifies a category membership set distinctly.

From definition 1, we know that the most vital step for constructing category membership sets is finding suitable membership functions. Luckily, a large number of statistical methods used for feature selection have been proposed in text categorization [7][8]. These statistical methods select words useful to text categorization from large word set by measuring the relativity of words and categories. Therefore, these statistical methods are also suitable for acting as membership functions. In the following sections, we will descript these methods in detail.

2.2 MI

Mutual information (or *MI* for short) is a criterion commonly used in statistical language modeling of word associations, feature selection, and related applications [8][9][10]. Let f_i be a word in F, c_j be a category in C, and $P(A)$ be the probability that event A occurs. The mutual information criterion between f_i and c_j is defined to be

$$MI(f_i, c_j) = \log \frac{P(f_i \wedge c_j)}{P(f_i) \times P(c_j)} \tag{2}$$

For the sake of computation, we can estimate $MI(f_i, c_j)$ by using

$$MI(f_i, c_j) \approx \log \frac{X \times N}{(X + Z) \times P(X + Y)} \tag{3}$$

---

[1] $\Re$ is the set of real numbers.

where X is the number of documents that contain word f_i and belong to D_j (that is to say, they are labeled category c_j); Y is the number of documents that contain f_i but don't belong to D_j; Z is the number of documents that don't contain f_i but belongs to category c_j. $MI(f_i, c_j)$ has a natural value of zero if f_i and c_j are independent.

A characteristic of mutual information is that the score is strongly influence by the marginal probabilities of words, as is showed in following equivalent formula

$$MI(f_i, c_j) = \log P(f_i | c_j) - \log P(f_i) \tag{4}$$

If words have an equal conditional probability $P(f_i | c_j)$, these words, which occurs rarely, will have a higher score than common words.

2.3 OddsRatio

OddsRatio is commonly used in information retrieval where the problem is to rank out documents according to their relevance for the positive class with using occurrence of different words as features. It was first used as feature selection methods by Mladenic[7]. Mladenic have compare six feature scoring measures with each other on real Web documents. He found that *OddsRation* showed the best performance. This shows that *OddsRatio* is best for feature scoring and may be very suitable for acting as membership function. If one considers the two-way contingency table of a word f_i and a category c_j, where X is the number of documents that contain f_i and belong to D_j, Y is the number of documents that belong to D_j, U is the number of documents that contain f_i but don't belong to D_j, V is the number of documents that don't belong to D_j, then the *OddsRatio* between f_i and c_j is defined to be

$$OddsRatio(f_i, c_j) = \log \frac{P(f_i | c_j)(1 - P(f_i | \neg c_j))}{(1 - P(f_i | c_j)) P(f_i | \neg c_j)} \tag{5}$$

and is estimated using

$$OddsRatio(f_i, c_j) \approx \log \left(\frac{X}{Y}(1 - \frac{U}{V}) / \left((1 - \frac{X}{Y}) \frac{U}{V} \right) \right) \tag{6}$$

2.4 CHI

The *CHI* (Abbreviation for χ^2 statistic) measures the lack of independence between a word and a category and can be compared to the χ^2 distribution with one degree of freedom to judge extremeness. Given a word f_i and a category c_j, The *CHI* of f_i and c_j is given by

$$CHI(f_i, c_j) = \frac{N \times (XV - UY)^2}{(X+U) \times (Y+V) \times (X+Y) \times (U+V)} \tag{7}$$

where N is the total number of training documents; A is the number documents that contain f_i and belong to D_j; Y is the number of documents that contain f_i but don't belong to D_j; U is the number of documents that belong to D_j but don't contain f_i; V is the number of documents that neither contain f_i nor belong to D_j. The $CHI(f_i, c_j)$ has a value of zero if f_i and c_j are independent. On the other hand, the $CHI(f_i, c_j)$ has the maximal value of N if f_i and c_j either co-occur or co-absent. The more f_i and c_j are correlative the more the CHI_{ij} is high and vice versa. Yang [8] reported that CHI is one of the most effective feature selection methods. Therefore, CHI may be a good choice as a membership function. As mentioned in [11], a major weakness of CHI is that it is not reliable to low-frequency words.

3 CMB: A Category-Membership-Based Algorithm

Before describing CMB, we first present some basic concepts. These concepts include document representation and similarity of documents and categories.

3.1 Basic Concepts

For classifying documents efficiently, documents should represent by some models that are suitable to be processed by computer. In this paper, we adopt the classic models called word bag. This model considers that each document is described by a set of representative words. For differentiating the importance of each word for describing the document semantic contents, a weight is associated with each word of a document.

Definition 2. (word bag model with weights) Let d be a document. Document d is represented as the set of tuple of word f and its weight in d, which is as follows:

$$d = \{< f, freq(f,d) > | f \in F\} \qquad (8)$$

Word frequency[2] is known to provide one good measure of how well that words describes the document contents [12]. However, it is well known that words in long documents have high word frequencies than words in short documents. If we use word frequency as $freq(f, d)$ directly, it would be unfair for words in short documents. Therefore, we measure $freq(f, d)$ with the normalized frequency of word f in d. That is, $freq(f, d)$ is given by

$$freq(f,d) = \frac{wd(f,d)}{\max\{wd(f_i,d) | f_i \in F\}} \qquad (9)$$

where the maximum is computed over all words in F, $wd(f, d)$ is the word frequency of f in d. That is, $wd(f, d)$ is equal to the number of times that f occurs in d.

Given a document d represented by word bag model with weights and a category c with a membership function μ_c, the similarity of d and d is defined to be

$$sim(d,c) = \sum_{i=1}^{n} freq(f_i,d) \times \mu_c(f_i) \qquad (10)$$

3.2 Algorithm Description

Given a training document set $TD = \cup D_i$, where D_i is the set of documents that are labeled category c_i, and a membership function, such as MI, $OddsRatio$, or CHI, we

---

[2] The word frequency of a word f in a document d is the number of times that f occurs in d.

can collect all words and compute the category membership set of each category. For an unlabeled document, we get the similarity of this document and each category by formula (10). By sorting categories in similarity score descending order, we obtain a ranked list of categories. Since documents of our data set (Newsgroup_18828) have only one category, we assign the only top ranking category to the unlabeled document. Above discussions are the core ideas of CBM.

Constructing CBM classifier includes two components: one for learning classifiers and the other for classifying unlabeled documents. For the sake of description, we label the former ***Training_Phase*** and the latter ***Classifying_Phase***. The pseudo-code for CBM is shown as follows.

Training_Phase:

Input: training documents set $TD = \cup D_i$, $1 \le i \le m$, $D_i = \{$document $d \mid d$ is labeled category $c_i\}$, a membership function μ. // μ may be *MI*, *OddsRatio*, or *CHI*

Output: word set $F = \{f_1, f_2, ..., f_n\}$; $CMS = \{CMS_1, CMS_2, ..., CMS_m\}$, where CMS_i is the category membership set of c_i.

    Step 1. $F = \emptyset$, $CMS = \emptyset$.
    Step 2. Scan the training documents set *TD* once. Collect words in *TD*. Insert all these words into *F*. Let $F = \{f_1, f_2, ..., f_n\}$.
    Step 3. For $i = 1$ to m do:
        Construct CMS_i, the category membership set of c_i, according to μ;
        $CMS = CMS \cup \{CMS_i\}$;

Classifying_Phase:

Input: *F*, *CMF*, and an unlabelled document d_{new}.

Output: the category of d_{new}.

    Step 1. $Sim\_Scores = \emptyset$; Scan d_{new} once. Compute the normalized frequency of each word f_i ($\in F$) in d_{new}.
    Step 2. For $i = 1$ to m, do:
        Compute Sim_i, the Similarity of d_{new} and c_i, according to formula (10);
        $Sim\_Scores = Sim\_Scores \cup \{Sim_i\}$;
    Step 3. Sort $C = \{c_1, ..., c_m\}$ in similarity score descending order as *OC*. Let c_x be the only top category in the ranking list *OC*. c_x is outputted as the category of d_{new}.

4 Experimental Evaluation

To assess the effectiveness of our algorithm, we conduct experiments to evaluate the performance of the CBM by comparing it with *k*-NN and Naïve Bayes on text collections Newsgroup_18828.

4.1 *k*-NN

k-NN stands for k nearest neighbor classification, a well known statistical method which has been intensively studied in machine learning for over four decades [13].

k-NN has been applied to text categorization since the early stages of the research. It is one of the top-performing methods among algorithms used for text categorization [6][14][15].

The idea of k-NN algorithm is simple: given a new unlabeled document, the algorithm finds the top k nearest neighbors among the training documents, and predict the category of the unlabeled document with the categories of the k neighbors. Let d, which was represented as a vector of words with their weights, be an unlabeled document. The k-NN algorithm assigns a similarity score to each candidate category c_i using the following formula

$$s(d,c_i) = \sum_{d_i \in kNN \cap D_i} \cos(d, d_i) \qquad (11)$$

where kNN is the set of k nearest neighbors of document d, and D_i is the set of training documents labeled category c_i. By sorting the scores of all candidate categories, we obtain a ranked list of categories for document d. The only top ranking category is assigned to d.

4.2 Naïve Bayes

Naïve Bayes are also commonly used in text categorization [2] [6] [14]. The basic idea is to use the joint probabilities of words and categories to estimate the probabilities of categories given a document. The naïve part of such a model is the assumption of word independence. $C = \{c_1, ..., c_m\}$ be the set of categories, $F = \{f_1, ..., f_n\}$ be word set, and $TD = \cup D_i$ be the set of training documents with labeled categories, where D_i is the set of documents that are labeled category c_i. Given a new unlabeled document d, the Naïve Bayes algorithm assigns d to a category c^* as follows:

$$c^* = \arg\max_{c_j \in C} P(c_j) \prod_{i=1}^{n} P(f_i | c_j)^{wf(f_i, d)} \qquad (12)$$

where $P(c_j)$ is the priori probability of category c_j and $P(f_i | c_j)$ is the conditional probability of word f_i given category c_j. $wf(f_i, d)$ is the same as mentioned in section 3.1. By training documents, $P(c_j)$ and $P(f_i | c_j)$ can be estimated as follows:

$$P(c_j) \approx \frac{n_j}{N} \qquad (13)$$

$$P(f_i | c_j) \approx \frac{n_{ij} + 1}{n_j + n} \qquad (14)$$

where N is the total number of training documents in TD, n_j is the number of documents in D_j, n_{ij} is the number of documents that contain word f_i and belong to D_j, n is the total number of words in F.

4.3 Text Collection

The Newsgroup_18828[3] text collection, collected by Jason Rennie, contains 18828 documents evenly divided among 20 UseNet discussion groups. 14120 documents (about 75%) were used for training documents and the remaining 4708 documents (about 25%) for test documents. We conduct experiments by using fourfold cross-validation method. That is, Newsgroup_18828 is split into four subsets, and each subset was used once as test documents in a particular run while the remaining subsets were used as training documents for that run. The split into training and test documents for each run was the same for all algorithms. Finally, the results of experiments are averages of four runs. For each run, we used a stop word list to remove common words, and the words were stemmed using Porter's suffix-stripping algorithm [16]. Furthermore, we also skip rare frequency words that occur in less than three documents. After above text preprocessing operation, the number of words from training documents is about 26,000.

4.4 Performance Measures

In terms of the performance measures, we followed the stand recall, precision and F_1 measures. Recall (r) is defined to be the radio of correct positive predictions by the system divided by the total number of positive examples. Precision (p) is defined as the radio of correct positive predictions by the system divided by the total number of positive predictions by the system. Recall and precision reflect the different aspect of classification performance. Usually, if one of the two measures is increasing, the other will decrease. To obtain a better measure describing performance, F_1, which first introduced by van Rijsbergen [17], is adopted. It combines recall and precision in the form as follows:

$$F_1 = \frac{2rp}{r+p} \qquad (15)$$

For evaluating an average performance across categories, we used the *micro-averaging* method and *macro-averaging* method [15]. *micro-averaging* method counts the decisions for all the categories in a joint pool and computer the global recall, precision and F_1 for the global pool. *macro-averaging* method first computes recall, precision and F_1 for each category, and then averages over categories as a global measure of the average performance over all categories. In this paper, we use *micro-averaging* and *macro-averaging* F_1 for performance measures.

4.5 Primary Results

Table 1 summarizes the categorization results of our experiments. CMB-*MI* means that *MI* is selected as the membership function, and so do CMB-*CHI* and CMB-*OddsRatio*. For the parameter k in k-NN, which is the number of nearest neighbor, the values of 5, 15, 30, 60, 90 were tested. The best result select as the final performance of k-NN.

---

[3] http://www.ai.mit.edu/~jrennie/20Newsgroups/20news-18828.tar.gz.

The *micro*-level analysis suggests that

Naive Bayes > CMB-*OddsRatio* > *k*-NN > CMB-*MI* > CMB-*CHI*.

While the *macro*-level analysis suggests that

CMB-*OddsRatio* > *k*-NN > {Naive Bayes, CMB-*MI*} > CMB-*CHI*.

Combining the *micro*-level scores and *macro*-level score with each accounting for 50%, we have performance ranking as follow:

CMB-*OddsRatio* > Naive Bayes > *k*-NN > CMB-*MI* > CMB-*CHI*.

The bad performanc of CMB-*CHI* may be that its weakness mentioned in section 2.4. That is, *CHI* is not reliable to low-frequency words. As stated in [12], low-frequency words are assumed to useful for distinguishing relevant documents from non-relevant documents. This means that low-frequency words contain relatively rich information and hence play important roles in classifying text documents. The other way round, low-frequency words have higher scores than common words in *MI*. This makes *MI* to be coincident with the assumption in information retrieval. Therefore, MI performs far better than *CHI* when they associate with CMB to classify documents. *OddsRatio* measure the membership of a word f to a category c by considering the times that f occurs in positive documents (documents labeled c) and negative documents (*no*-c documents). This makes *OddsRatio* more rational. Hence, *OddsRatio* achieves the best performance. Our experimental results also show that membership functions play a vital importance in CMB for classifying documents.

Table 1. Performances summary

	k-NN	Naive Bayes	CMB-*MI*	CMB-*CHI*	CMB-*OddsRatio*
micro-averaging F_1	0.82	0.829	0.816	0.682	0.824
macro-averaging F_1	0.812	0.811	0.811	0.66	0.818

5 Conclusion

In this paper, we have presented CMB, a simple but efficient text categorization algorithm, based on category memberships. Comparison experiments on Newsgroup_18828 text collections show the effect of our method. As a future work, we need the additional research for finding better methods used to measure the category memberships. In addition, we will investigate combination of membership functions. The intuition is that different membership functions measure membership scores in qualitatively different ways. This suggests that these different functions potentially offer complementary information and the proper combination of these functions would be more effective than each one. Plentiful results from combination of classifier would provide valuable information.

Acknowledgement. This research is supported by the National Natural Science Foundation of China under grant No. 60473072. Any opinions, findings, and conclusions or recommendations expressed in this paper are the authors' and do not necessarily reflect those of the sponsor. We are also grateful to anonymous reviewers for their comments.

References

1. Yang,Y.: Expert network: Effective and efficient learning from human decisions in text categorization and retrieval. In *17th Annual International ACM SIGIR Conference on Research and Development in Information Retrieval* (1994) 13-22
2. McCallum, A., Nigam, K.: A comparison of event models for naïve bayes text classification. In *AAA-98 Workshop on Learning for Text Categorization* (1998)
3. Apte, C., Damerau, F., Weiss, S.: Text mining with decision rules and decision trees. In *proceedings of Conference on Automated Learning and Discovery, Workshop 6: Learning from Text and the Web* (1998)
4. Ng, H.T., Goh, W.B., Low, K.L.: Feature selection, perceptron learning, and a usability case study for text categorization. In *20th Annual International ACM SIGIR Conference on Research and Development in Information Retrieval* (1997) 67-73
5. Schapire, R.E., Singer, Y.: BoosTexter: A Boosting-based System for Text Categorization. *Machine Learning* 2/3 (2000) 135-168
6. Joachims, T.: Text Categorization with Support Vector Machines: Learning with Many Relevant Features. In *Proceedings of the 1998 European of conference on Machine Learning* (1998) 137-142
7. Mladenic, D., Grobelnik, M.: Feature Selection for Classification Based on Text Hierarchy. In *Working notes of Learning from Text and the Web, Conference on Automated Learning and Discovery* (1998)
8. Yang, Y., Pedersen, J.P.: A Comparative Study on Feature Selection in Text Categorization. In *Proceedings of 14th International Conference on Machine Learning* (1997) 412-420
9. Church, K.W., Hanks, P.: Word association norms, mutual information and lexicography. In *Proceedings of 27th ACL* (1989) 76-83
10. Fano, R.: Transmission of information. MIT Press, Cambridge, MA (1961)
11. Dunning, T.E.: Accurate methods for the statistics of surprise and coincidence. Computational Linguistics 1 (1993) 61-74
12. Ricardo, B.Y., Berthier, R.N.: Modern Information Retrieval. ACM Press (1999)
13. Dasarathy, B.V.: Nearest Neighbor (NN) Norms: NN Pattern Classification Techniques. MCGraw-Hill Computer Science Series. IEEE Computer Society, Las Alamitos, California (1991)
14. Yang, Y.: An evaluation of statistical approaches to text categorization. *Journal of Information Retrieval* 1/2 (1999) 67-88
15. Yang, Y., Liu, X.: A re-examination of text categorization methods. In *22nd Annual International ACM SIGIR Conference on Research and Development in Information Retrieval* (1999) 42-49
16. Porter, M.F.: An algorithm for suffix stripping. *Program* 3 (1980) 130-137
17. van Rijsbergen, C.J.: Information Retrieval. Butterworths, London (1979)

The Integrated Location Algorithm Based on Fuzzy Identification and Data Fusion with Signal Decomposition

Zhao Ping and Haoshan Shi

Institute of Electric Information, Northwestern Polytechnical University,
Xi'an, Shaanxi, 710072, P. R. China
zpnwpu@126.com, shilaoshi@nwpu.edu.cn

Abstract. In this paper, an efficient integrated location algorithm based on fuzzy identification and data fusion is presented in order to carry put precision and reliability position estimation and improve location accuracy and efficiency. In addition to the selectivity advantage gained by combining different location parameters, the use of integrated location algorithm by data fusion may increase location integrity with which a robust and anti-interference result can be obtained.

1 Introduction

Network location signals are fuzzy and unsteady in most time and vary widely due to all kind of channel noises and interferences [1]. Also due to complex circumstance in city area there are strong multipath and non line of sight (NLOS) interferences to make more uncertainty and complicacy for position estimation. In order to model a hybrid location algorithm, we propose a systematic methodology based on fuzzy identification and data fusion. The algorithm conveys three distinct features: an improved fuzzy identification model; a selected signal stratification decomposition; and an efficient algorithm with integrated location estimation based on data fusion of TOA, TDOA and AOA. So the location algorithm is no longer restricted to ordinary signal parameter processing for user position estimation.

2 The Fuzzy Identification Model

Fuzzy stratification identification for interfered signal is focused for long time. There are so many sorts of fuzzy identification algorithm with different methods, but they are almost same in their main courses except their own typical attribute and correlation character.

As an identification method for nonlinear location signals, which are non-stable and varied in wide range with low signal noise ratio (SNR) and channel interference, we use two forms of the fuzzy identification courses, directness and indirectness fuzzy identification; exploit a selected signal processing with stratification and decomposition; carry out parameter estimation based on data fusion and logic reason judgment.

2.1 Directness Fuzzy Identification

Assume that there are n fuzzy subsets in theory field:

$$\{A_1, A_2, ..., A_n\} \quad (1)$$

if for every A_i there is a subfunction $\mu_{Ai}(u_0)$, so for any one element $u_0 \in U$ we can make sure its attribute according to subjection principle as below.

If it satisfies:

$$\mu_{Ai}(u_0) = \max(\mu_{A1}(u_0), ..., \mu_{An}(u_0)) \quad (2)$$

Then we take u_0 as subjection to A_i, i.e. u_0 should be subjected to model A_i.

2.2 Indirectness Fuzzy Identification

Assume that $A_1, A_2, ..., A_n$ are n fuzzy subsets on theory field U and an object B waiting for identification is also a fuzzy subset on theory field U. The identification course can be carried out as below according to nearby choosing principle.

If there is a fuzzy set A_i, which cause:

$$(B, A_i) = \max\{(B, A_1), (B, A_2), ..., (B, A_n)\} \quad (3)$$

Then B and A_i are most close to each other, that is B may be relatively attachable to A_i. Among equation (3), (B, A_j) means adjacence degree between B and A_i.

2.3 Identification Model

Data fusion judgment can be carried out by different principles according to different signal characteristic. It can fuse different data to give out precise identification result and also reliable parameter estimation. The framework of fuzzy signal identification course based on stratification decomposition and data fusion is shown in figure 1. The modeling and identification environment carries out parameter identification through a synergistic usage of clustering techniques such as stratification decomposition, data fusion and optimization judgment etc. According to the selection and adjustment of the weighting factor, an aggregate objective function can be used to achieve a balance between approximation and generalization.

3 Signal Decomposition and Feature Extraction

A layered decomposition with signal processing for extracting features has two aims. One is to reduce original signal information set or cancel useless and unnecessary component. The other is to concentrate needed important information with known noise statistical characteristic from stratification classification. So, the needed fine feature on local signal can be extracted by stratification decomposition form original signal [2].

For the first purpose, location signal decomposition based on wavelet transform provides a set of decomposed signals in independent frequency bandwidths, which contain much independent dynamic information in different stratification due to the orthogonality of wavelet transform. For the second purpose, the identified signals are filtered by Kalman filter with known noise statistical characteristic to pick up needed components.

Figure 2 (a) shows the original signal which consists of gradual change signal piling up three different period sinusoid waves with added noise, and Figure 2 (b) shows the decomposition result by wavelet analysis with 'db1' at lever 3.

Simulation shows that a signal generated by the standard state-space stochastic model can be decomposed into sub layers at the different sampling frequencies associated to different levels of resolution. The main advantage is that these innovations are all uncorrelated with each other but they can be added together synthetically to form correlated signal.

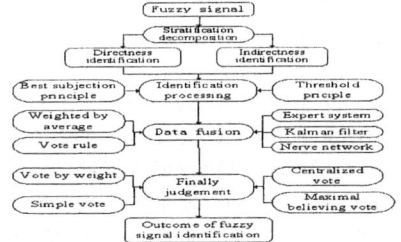
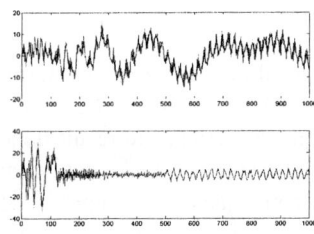

Fig. 1. Fuzzy identification framework   **Fig. 2.** Wavelet stratification decomposition

4 Integrated Location Algorithm with Data Fusion

The location algorithm makes use of statistical feature of objective characteristic to buildup different kinds of data fusion project. Here the data fusion is carried out once more as well as the parameter fuzzy identification to fulfill precise and reliable location within mobile network.

According to a typical concept model of data fusion system configuration presented in 1992, which mainly included pretreatment module, four level fusion and data management function, Kleine-Ostmann and Bell put forward a data fusion model for network location problem [3]. In this model not only three basis data fusions can be come true, which are data layer fusion, attribution layer fusion and decision-making layer fusion, but also among different fusion layer data fusing can be carried out simultaneously.

In the K-B model, the main locating data comes from independent TOA or TDOA parameters. But in some time or at certain position, location algorithm is incapable of obtaining three TOA parameters for essential calculation requirement. So we bring forward a TOA/TDOA/AOA integrated location algorithm based on data fusion and mathematical statistics. Through defining confidence function, a position-based dynamic location algorithm is obtained for multi parameters integrated location. As smart antenna or antenna array developing, angular resolution and precision of AOA is increased greatly by the frame of critical system parameters. By AOA parameter data fusion, the fuzzy problem caused by insufficiency of TOA or TDOA parameters can be avoided. The framework model of integrated location algorithm with three parameters of TOA/TDOA/AOA is shown in figure 3.

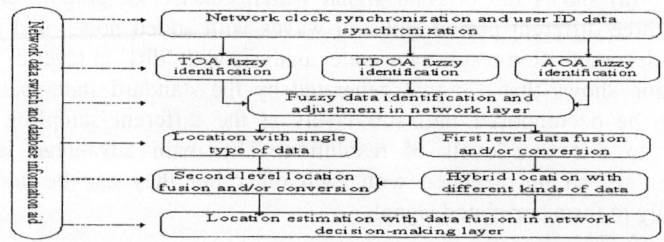

Fig. 3. The framework model of integrated location algorithm

The algorithm can carry out terminal position estimation from different layer approaches. One is to make use of single parameter among AOA, TOA and TDOA for position estimation; the result data can be as input for second level data fusion. Another approach is to make data conversion from AOA, TOA and TDOA to other form as location algorithm input; it also can transform plane equation groups into space line equation of AOA as dynamic location algorithm input. There are also several different data fusion methods to make mobile terminal position estimation such as taken the results of first level and second level data fusion as input to carry out algorithm in network decision-making layer.

5 Conclusion

From all above discussion, we can conclude that the integrated location algorithm with fuzzy identification and data fusion has remarkable abilities on location precision and accuracy, system reliability and configuration integrity, anti-interference ability and weak signal processing.

By using network stratification processing with TOA/TDOA/AOA parameter selection, many advantages can be achieved to improve location integrity and perfect location configuration as to get optimal location estimation.

References

1. Perez, Jose A. Moreno; Vega, J. Marcos Moreno; Verdegay, Jose L., Fuzzy location problems on networks, Fuzzy Sets and Systems, v 142, n 3, Mar 16, 2004, p 393-405.
2. Chiu, S., Fuzzy Model Identification Based on Cluster Estimation, Journal of Intelligent & Fuzzy Systems, Vol. 2, No. 3, Sept. 1994.
3. KLEINE-OSTMANN T, BELL A. E. A data fusion architecture for enhanced position estimation in wireless networks [J].IEEE Communications Letters, 2001, 5(8).

A Web Document Classification Approach Based on Fuzzy Association Concept

Jingsheng Lei, Yaohong Kang, Chunyan Lu, and Zhang Yan

College of Information Science and Technology, Hainan University,
Haikou 570228, P.R. China
jshlei@hainu.edu.cn

Abstract. In this paper, a method of automatically identifying topics for Web documents via a classification technique is proposed. Web documents tend to have unpredictable characteristics, i.e. differences in length, quality and authorship. Motivated by these fuzzy characteristics, we adopt the fuzzy association concept to classify the documents into some predefined categories or topics. The experimental results show that our approach yields higher classification accuracy compared to the vector space model.

1 Introduction

Due to the explosive growth of available information on the World Wide Web (WWW), users have suffered from the information overload. To alleviate the problem, many data mining techniques have been applied into the Web context. This research area is generally known as Web mining[1]. Web mining is defined as the discovery and analysis of useful information from WWW. Some examples of Web mining techniques include analysis of user access patterns[2], Web document clustering[3], classification[4], and information filtering.

In this paper, an intelligent content-based filtering that can automatically and intelligently filter Web documents based on the user preferences by utilizing topic identification is proposed. Web documents tend to have unpredictable characteristics, i.e. differences in length, quality and authorship. Motivated by these fuzzy characteristics, the fuzzy association concept in classifying Web documents into a predefined set of categories is adopted in our approach.

2 Fuzzy Association Method for Document Classification

The process of classifying Web documents in our approach is explained as follows. Given $C = \{C_1, C_2, \cdots, C_m\}$, a set of categories, where m is the number of categories.

Step 1. The first step is to collect the training sets of Web documents, $TD = \{TD_1, TD_2, \cdots, TD_m\}$, from each category in C. This step involves crawling through the hypertext links encapsulated in each document.

Step 2. To get index terms of a Chinese web document, first it needs word segment to deal with. Next, banish the stop words using stop list and decrease the terms by some algorithm. Then the document index term set is gained. The keywords from TD are extracted and put into separate keyword sets, $CK = \{CK_1, CK_2, \cdots, CK_m\}$. The document frequency-inverse category frequency (*df-icf*) strategy, adapted from the *tf-idf* concept is proposed to select and rank the keywords within each category based on the number of documents in which the keyword appears (i.e. *df*) and the inverse of the number of categories in which the keyword appears (i.e. *icf*).

$$df - icf(k, C_i) = DF(k, C_i) \times ICF(k) \tag{1}$$

where $DF(k, C_i)$ is the number of documents in which keyword k occurs at least once, $ICF(k) = \log(\frac{|C|}{CF(k)})$, $|C|$ is the total number of categories, and $CF(k)$ is the number of categories in which the keyword k occurs at least once.

Step 3. Let $A = \{k_1, k_2, \cdots, k_n\}$ be the set of all distinct keywords from CK, where n is the number of all keywords. Then, the keyword correlation matrix M is generated via Eq. (2).

$$r_{i,j} = \frac{n_{i,j}}{n_i + n_j - n_{i,j}} \tag{2}$$

where $r_{i,j}$ represents the fuzzy relation between keyword i and j, $n_{i,j}$ is the number of documents containing both *i*th and *j*th keywords, n_i is the number of documents including the *i*th keyword, and n_j is the number of documents including the *j*th keyword.

Step 4. To classify a test document d into category C_i, a set of keywords from CK_i are used to represent C_i. Then d is cleaned and its set of representative keywords is extracted from A. That is, $d = \{|k_1|, |k_2|, \cdots, |k_n|\}$, where $|k_i|$ is the frequency that k_i appeared in d. After that, the membership degree between d and C_i is calculated using the following equation

$$\mu_{d,C_i} = \sum_{\forall k_a \in d} [1 - \prod_{\forall k_b \in CK_i} (1 - r_{a,b})] \tag{3}$$

where μ_{d,C_i} is the membership degree of d belonging to C_i, and $r_{a,b}$ is the fuzzy relation between keyword $k_a \in d$ and keyword $k_b \in CK_i$.

Document d is classified into category C_i when μ_{d,C_i} is the maximum for all *i*. The keyword k_a in d is associated to category C_i if the keywords k_b in CK_i are related to k_a. Whenever there is at least one keyword in CK_i which is strongly related to $k_a \in d$ ($r_{a,b} \approx 1$), then Eq. (3) yields $\mu_{d,C_i} \approx 1$, and the keyword k_a is a good fuzzy index for the category C_i. In the case when all keywords in CK_i are either loosely related or unrelated to k_a, then k_a is not a good fuzzy index for C_i ($\mu_{d,C_i} \approx 0$).

3 Experimental Results and Discussions

3.1 Experimental Data Sets

Experiments using the predefined categories as document topics and the document sets collected from two Web portals: *Sohu*, *Sina* and *Yahoo! Chinese* are conducted. In our experiments, we only consider documents in Chinese and ignore all other non-Chinese documents, the selected categories and number of the Web documents are shown in Table 1. Based on these predefined categories, we collected 5651 documents from each of the Web directories as the training and test data sets. To avoid the problem of over-fitting the data when performing the experiments, we randomly select two-third of the document sets as the training set and one-third as the test set.

For the *Sohu*, *Sina* and *Yahoo! Chinese* training data set, 100 index terms whose *df-icf* values are the highest among all index terms are selected from each of its 11 categories. Next, we combine these index terms into the set of 1463 distinct index terms.

Table 1. Predefined category sets and the number

Category	Abbr.	The number of the Web documents
News	news	554
Sports	sport	521
finance & economics	fin	528
Entertainment	et	519
Education	edu	573
Games	gm	561
Life	life	586
Autos	auto	417
Travel	travel	482
Health	hl	478
house property	hm	432
TOTAL		5651

3.2 Experimental Results and Discussions

To compare the performance of our method (Fuzzy) to the vector space model(Vector) approach, we use the test data sets from the three Web directories. In Fig.1, the performance result based on the 11 categories of the data set is presented. As expected, our approach yields higher accuracies for most of the categories.

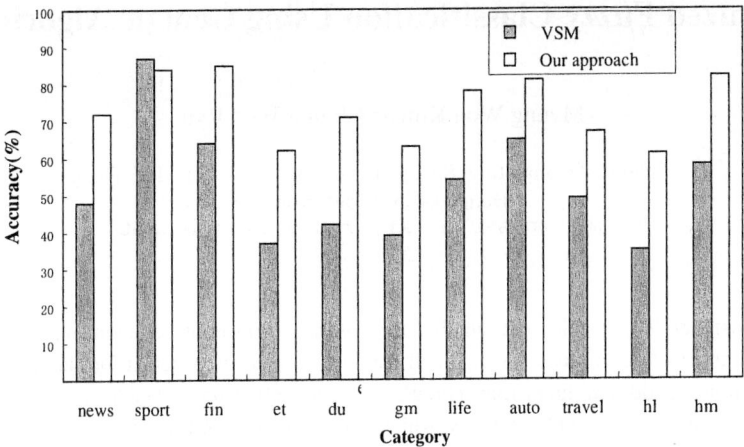

Fig. 1. Classification performance comparison by categories

4 Conclusion

In this paper, a fuzzy classification approach that automatically identifies topics for Web documents via a classification technique was proposed. Our approach adopts the fuzzy association concept as a machine learning technique to classify the documents into some predefined categories or topics. The result is that each pair of words has an associated value to distinguish itself from other pairs of words. We performed several experiments using the data sets obtained from three different Web directories: *Sohu*, *Sina* and *Yahoo! Chinese*. We compared our approach to the vector space model approach. The results show that, our approach yields higher classification accuracies compared to the vector space model when varying the number of category representation keywords.

References

1. R. Cooley, B. Mobasher and J. Srivastava, Web mining: Information and pattern discovery on the world wide web, Proc. 9th IEEE Int. Conf. Tools Artif. Intell. (ICTAI'97), Newport Beach, CA, November 1997, 558-567
2. J. Pitkow and P. Pirolli, Mining longest repeating subsequences to predict world wide web surfing, Proc. SENIX Symp. Internet Tech. Syst. (USITS'99), Boulder, CO, October 1999, 139-150.
3. A. Z. Broder, S. C. Glassman, M. S. Manasse Syntactic clustering of the web, Proc. 6th Int.World Wide Web Conf. (WWW'6), Santa Clara, CA, April 1997, 391-404.
4. S. T. Dumais and H. Chen, Hierarchical classification of web content, Proc. 23rd Int. ACM Conf. Res. Dev. Inf. Retrieval (SIGIR), Athens, Greece, August 2000, 256-263.

Optimized Fuzzy Classification Using Genetic Algorithm[1]

Myung Won Kim and Joung Woo Ryu

School of Computing, Soongsil University, 1-1, Sangdo 5-Dong,
Dongjak-Gu, Seoul, Korea
mkim@comp.ssu.ac.kr, ryu0914@orgio.net

Abstract. Fuzzy rules are suitable for describing uncertain phenomena and natural for human understanding and they are, in general, efficient for classification. In addition, fuzzy rules allow us to effectively classify data having non-axis-parallel decision boundaries, which is difficult for the conventional attribute-based methods. In this paper, we propose an optimized fuzzy rule generation method for classification both in accuracy and comprehensibility (or rule complexity). We investigate the use of genetic algorithm to determine an optimal set of membership functions for quantitative data. In our method, for a given set of membership functions a fuzzy decision tree is constructed and its accuracy and rule complexity are evaluated, which are combined into the fitness function to be optimized. We have experimented our algorithm with several benchmark data sets. The experiment results show that our method is more efficient in performance and comprehensibility of rules compared with the existing methods including C4.5 and FID3.1 (Fuzzy ID3).

1 Introduction

Data mining is a new technique which discovers useful knowledge from data [1]. In data mining important evaluation criteria are efficiency and comprehensibility of knowledge. The discovered knowledge should well describe the characteristics of the data and it should be easy to understand in order to facilitate better understanding of the data and use it effectively. Classification is one of important techniques in data mining and it is used in various applications including pattern recognition, customer relationship management, targeted marketing, and disease diagnosis.

A decision tree such as ID3 and C4.5 is one of the most widely used classification methods in data mining [2] ~ [4]. One of the difficult problems in classification is to handle quantitative data appropriately. Conventionally, a quantitative attribute domain is divided into a set of crisp regions and by doing so the whole data space is partitioned into a set of (crisp) subspaces (hyper-rectangles), each of which corresponds to a classification rule describing that a sample belonging to the subspace is classified into the representative class of the subspace. However, such a crisp partitioning is not natural to human and inefficient in performance because of the sharp boundary prob-

---

[1] This work was supported by Korea Research Foundation Grant (KRF-2004-041-D00627).

lem. Recently, fuzzy decision trees have been proposed to overcome this problem [5] ~ [11]. It is well known that the fuzzy theory not only provides natural tool for describing quantitative data but also generally produces good performance in many applications. However, one of the difficulties with fuzzy decision trees is determining an appropriate set of membership functions representing fuzzy linguistic terms. Usually membership functions are given manually, however, it is difficult for even an expert to determine an appropriate set of membership functions when the volume and dimensionality of data are large.

In this paper we investigate combining the fuzzy theory and the conventional decision tree algorithm for accurate and comprehensible classification. We propose an efficient fuzzy rule generation method using the fuzzy decision tree (FDT) algorithm for data mining, which integrates the comprehensibility of decision trees and the expressive power of fuzzy sets. We also propose the use of genetic algorithm for optimal set of fuzzy rules by determining an appropriate set of fuzzy sets for quantitative data. In our method for a given fuzzy membership function a fuzzy decision tree is constructed and it is used to evaluate classification accuracy and rule complexity. Fuzzy membership functions evolve so that they optimize the fitness function combining both classification accuracy and rule complexity.

2 Fuzzy Inference

2.1 Fuzzy Classification Rules

We use a simple form of fuzzy rules and inference for better human understanding. Each fuzzy rule is of the form *"if A then B"* where *A* and *B* are called an antecedent and a consequent, respectively. In our approach the antecedent is simple conditions conjoined while the consequent is *"Class is k."* A simple condition is of the form *"Att is Val"* where *Att* represents an attribute name and *Val* represents a value of the attribute. Each fuzzy rule is associated with a *CF* (Certainty Factor) to represent the degree of belief that the consequent is drawn from the antecedent satisfied. Rule (1) is a typical form of fuzzy classification rules used in our approach.

$$R_i : \text{if } A_{i1} \text{ is } V_{i1} \text{ and } A_{i2} \text{ is } V_{i2} ... \text{and } A_{im} \text{ is } V_{im} \text{ then 'Class' is } k \; (CF_i) \tag{1}$$

In the rule A_{ik} represents an attribute and V_{ik} represents a fuzzy linguistic term represented by a fuzzy set associated with attribute A_{ik}. Application of the rule to a sample *X* results the confidence with which *X* is classified into class *k* given that the antecedent is satisfied. In this paper among a variety of fuzzy inference methods we adopt the standard method as described in the following:

1) *min* is used to combine the degrees of satisfaction of individual simple conditions of the antecedent;
2) *product* is used to propagate the degree of satisfaction of the antecedent to the consequent;
3) *max* is applied for aggregating the results of individual rule applications.

For a given sample X, according to our method the confidence of class k is obtained as

$$Conf_k(X) = \max_{R_i \in R(k)} \left\{ \left(\min_j \mu_{V_{ij}}(x_j) \right) \cdot CF_i \right\} \quad (2)$$

where x_j is the value for attribute A_{ij} of X.

In equation (2) $\mu_V(x)$ represents the membership degree that x belongs to fuzzy set V, $R(k)$ represents the set of all rules that classify samples into class k (their consequent parts are 'Class is k'). The class of the maximum $Conf_k(X)$ is the final classification of X.

Membership functions are very important in fuzzy rules. They affect not only the performance of fuzzy rule based systems but also the comprehensibility of rules. Triangular, trapezoidal, and Gaussian membership functions are widely used, however, in this paper we adopt triangular membership functions. A triangular membership function can be represented by a triple of numbers (l, c, r), where l, c, and r represent the left, the center, and the right points of the triangular membership function, respectively.

In this paper we investigate the use of genetic algorithm to automatically generate an appropriate set of membership functions for a given set of data to classify. The membership functions are optimized in the sense that the generated rules are efficient and comprehensible and it is described in Section 3.

2.2 Fuzzy Decision Tree

A fuzzy decision tree is similar to a (crisp) decision tree. It is composed of nodes and arcs representing attributes and attribute values or value sets, respectively. The major difference is that in a fuzzy decision tree each arc is associated with a fuzzy linguistic term, which is usually represented by a fuzzy set. Also in a fuzzy decision tree a leaf node represents a class and it is associated with a certainty factor representing the confidence of the decision corresponding to the leaf node. In a fuzzy decision tree a decision is made by aggregating the conclusions of multiple rules (paths) fired as Equation (2) describes while in a crisp decision tree only a single rule is fired for a decision.

Let us assume that $A_1, A_2,..., A_d$ represent attributes in consideration for a given data set, where d represents the dimension of the data. The whole data space W can be represented as $W = U_1 \times U_2 \times ... \times U_d$, where U_i represents the domain of attribute A_i. A sample X can be represented as a point in W as $X = (x_1, x_2, , , x_d)$, where $x_i \in U_i$. In a fuzzy decision tree each arc (l, m) from node l to node m is associated with a fuzzy set $F(l, m)$ representing a fuzzy linguistic term as a value of the attribute selected for node l. Suppose we have a fuzzy decision tree and let n be a node and P_n be the path from the root node to node n in the tree. Then we can consider that node n is associated with a fuzzy subspace W_n of W defined as follows.

$$W_n = S_1 \times S_2 \times ... \times S_d$$

where
$$S_i = \begin{cases} F(l,m) & \text{if } A_i = att(l) \text{ for an arc } (l,m) \text{ in } P_n; \\ U_i & \text{otherwise}. \end{cases}$$

Here, $F(l,m)$ is a fuzzy set corresponding to arc (l,m) and $att(l)$ represents the attribute selected for node l. Let $v_n(X)$ represent the membership that X belongs to W_n, then we have the following according to the above definitions:

$$v_n(X) = \mu_{W_n}(X)$$
$$= \min_{(l,m) \text{ in } P_n} \mu_{F(l,m)}(x_i) \text{ where } A_i = att(l).$$

Our fuzzy decision tree construction algorithm is as follows.

<Step 1>
Starting with the root node, continue to grow the tree as following until the termination conditions are satisfied. If one of the following conditions is satisfied, then make node m a leaf node.

(1) $\dfrac{1}{|D|} \sum_{X \in D} v_m(X) \leq \theta_s$

(2) $\dfrac{\sum_{X \in D_{k^*}} v_m(Y)}{\sum_{X \in D} v_m(X)} \geq \theta_d$ where $k^* = \arg \max_{k \in C} (\sum_{Y \in D_k} v_m(Y))$

(3) no more attributes are available.

In condition (2) class k^* represents the representative class of the corresponding fuzzy subspace. In this case it is the leaf node whose class is k^* and the associated CF is determined by

$$CF = \dfrac{\sum_{Y \in D_{k^*}} v_m(Y)}{\sum_{X \in D} v_m(X)}$$

Otherwise,

<Step 2>

(1) Let $E(m, A_{i^*}) = \min_i (E(m, A_i))$.

   $E(m,A_i)$ represents the entropy of attribute A_i for node m and it was defined in [11].
(2) For each fuzzy membership functions of attribute A_{i^*}, make a child node of node m.
(3) Go to Step 1 and apply the algorithm to all newly generated nodes, recursively.

Node expansion ((2) in Step 2) corresponds to partitioning the fuzzy subspace corresponding to node m into fuzzy subspaces each of which corresponds to a value of the attribute selected for the node. In Step 1 the threshold parameters θ_s and θ_d determine when partitioning terminates. Condition (1) prohibits further partitioning sufficiently sparse fuzzy subspaces while condition (2) prohibits further partitioning fuzzy subspaces having sufficiently large portion of a single class samples. The parameters θ_s and θ_d are used to control overfitting by prohibiting too much detail rules to be generated.

After constructing a fuzzy decision tree, we name each fuzzy membership function with an appropriate linguistic term.

3 Membership Function Generation Using Genetic Algorithm

In fuzzy rules membership functions are important since they affect both of accuracy and comprehensibility of rules. However, membership functions are usually given manually and it is difficult even for an expert to determine an appropriate set of membership functions when the volume and dimensionality of data are large. In this paper, we propose the use of genetic algorithm to determine an optimal set of membership functions for a given classification problem.

3.1 Genetic Algorithm

Genetic algorithm is an efficient search method simulating natural evolution, which is characterized by survival of the fittest.
Our fuzzy decision tree construction using genetic algorithm is following.

(1) Generate an initial population of chromosomes of membership functions;
(2) Construct fuzzy decision trees using the membership functions;
(3) Evaluate each individual set of membership functions corresponding to a chromosome by evaluating the performance and tree complexity of its corresponding fuzzy decision tree;
(4) Test if the termination condition is satisfied;
(5) If yes, then exit;
(6) Otherwise, generate a new population of chromosomes of membership functions by applying genetic operators, and go to (2).

We use genetic algorithm to generate an appropriate set of membership functions for quantitative data. Membership functions should be appropriate in the sense that they result in a good performance and they result in as simple a decision tree as possible.

In our genetic algorithm a chromosome is of the form $<\phi_1, \phi_2, ..., \phi_d>$ where ϕ_i represents a set of membership functions associated with attribute A_i, which is, in turn, of the form $\{f_1(A_i), f_2(A_i), ,, f_l(A_i)\}$ where $f_j(A_i)$ represents a triplet of membership function for attribute A_i. The number of membership functions for an attribute may not necessarily be fixed.

3.2 Genetic Operations

We have genetic operations such as crossover, mutation, addition, and merging as described in the following.

1) **Crossover**: it generates new chromosomes by exchanging the whole sets of membership functions for a randomly selected attribute of the parent chromosomes.
2) **Mutation**: we combine random mutation and heuristic mutation. In random mutation membership functions randomly selected from a chromosome are mutated by adding Gaussian noise to the left, the center, and the right points of an individual membership function. In heuristic mutation membership functions are adjusted to classify correctly a set of randomly sampled incorrectly classified data.
3) **Addition**: for any attribute in a chromosome, the set of associated membership functions are analyzed and new appropriate membership functions are added if necessary. For example, when some attribute values are not covered or poorly covered by the current membership functions, new membership functions are added to cover those values properly.
4) **Merging**: any two close membership functions of any attribute of a chromosome are merged into one.

We apply the roulette wheel method for selecting candidate chromosomes for the crossover operation. We also adopt the elitism in which the best fit chromosome in the current population is selected for the new population.

If membership functions are determined for each attribute, we generate an fuzzy decision tree according to the algorithm described in Section 2.2. We use the performance of the generated fuzzy decision tree in fitness evaluation of a chromosome. Suppose a fuzzy decision tree $\tau(e)$ is generated from the membership functions represented by chromosome e. The fitness score of chromosome e is given by

$$Fit(c) = \alpha P(\tau(e)) - \beta C(\tau(e)) \qquad (3)$$

where $P(\tau(e))$ represents the performance of decision tree $\tau(e)$ and $C(\tau(e))$ represents the complexity of $\tau(e)$, measured in terms of the number of nodes of $\tau(e)$ in this paper.

We also can use genetic algorithm to generate fuzzy rules by evolving the form of rules (selection of attributes and their associated values) and membership functions simultaneously. However, genetic algorithm is generally time-consuming and our method trades off between classification accuracy and computational time.

4 Related Works

C4.5 is a successor to ID3 [2] and it is wildly used for applications [3], [4]. One of the important improvements made in C4.5 is that it can handle continuous valued attributes. In C4.5 continuous values are partitioned into two intervals by selecting a cut point, which maximizes the information gain indicating how well a partition classifies the data. By sorting the data according to the attribute, then identifying adjacent values of the attribute that differ in their target classification, we can generate candidate cut points. Among those cut points one that maximizes the information gain is selected. ID3 and C4.5 are a crisp decision tree in the sense that the whole data space is partitioned into a set of crisp subspaces and they suffer the sharp boundary problem.

A fuzzy decision tree is more powerful, efficient, and natural to human understanding, particularly compared with crisp decision trees. [5] proposes fuzzification of CART (Classification And Regression Tree) using sigmoidal fuzzy splits replacing Boolean tests of a crisp CART tree and applying the back-propagation algorithm to learn parameters associated with fuzzy splits to optimize the global impurity of the tree. However, the structure of fuzzy decision tree proposed in [5] is different from that we propose in this paper and it is difficult to directly compare them.

[6] proposes a method for automatically generating fuzzy membership functions for continuous valued attributes based on the principle of maximum information gain in cut point selection for each attribute.

Our method for constructing a fuzzy decision tree described in Section 2 is similar to Janikow's method proposed in [8]. A fuzzified ID3 have been proposed in [6] and [9] and fuzzy membership functions are generated automatically based on the principle of maximum information gain in selection of cut points for each continuous valued attribute. However, FID3 is still a hill-climbing method, which may get stuck to local optima. FID3 also allows the same attribute selected for different nodes to have different values associated with their outgoing arcs. It can improve the efficiency of the fuzzy decision tree generated by allowing the data space partitioned into sufficiently small subspaces if necessary. However, it can harm comprehensibility by creating many odd fuzzy linguistic terms.

[10] proposes an automatic design of fuzzy rule-based classification systems. It is approached by combining tools for feature selection, model initialization, model reduction and model tuning. However, the model initialization is derived from the data not determined by the number of clusters but clustered by the number of classes.

5 Experiments

5.1 Experiments with Manually Generated Data Sets

In this experiment we use manually generated data sets as shown in Fig.1. For the slashed pattern and the rotated generalized XOR data we notice that our algorithm can well classify data with non-axis-parallel decision boundaries. Such data are difficulty to classify efficiently using the conventional ID3 and C4.5 type decision tree based classification methods. For the generalized XOR, the histogram analysis fails to

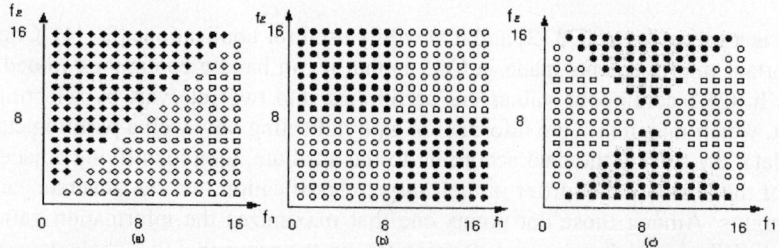

Fig. 1. (a) Slashed two-class data; (b) Generalized XOR; (c) Rotated generalized XOR

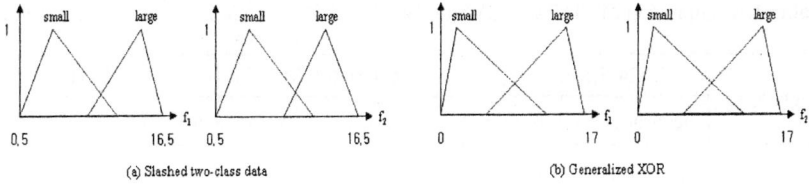

Fig. 2. Membership functions generated for the data sets (a), (b) of Fig.1

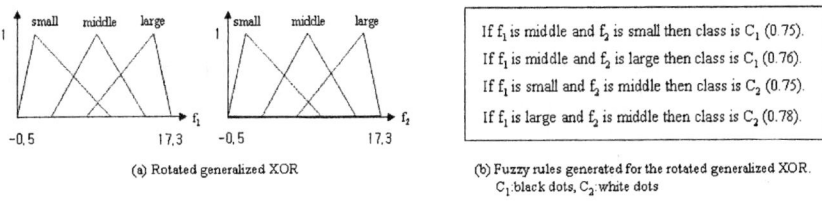

Fig. 3. Membership functions and fuzzy rules generated for data set (c) of Fig.1

generate an appropriate set of membership functions, however, the genetic algorithm well succeeds in doing it (Fig. 2(b)). For all cases our algorithm generates simple fuzzy rules which classify patterns 100% correctly. Fig. 3(b) illustrates four fuzzy rules generated for the rotated generalized XOR. Each rule corresponds to one of four pattern blocks as shown in Fig.1(c). We also notice that fuzzy rules have a kind of abstraction capability in that fuzzy rules in linguistic terms describe rough pattern classification, while all the detail classification is taken care of by membership functions. For example, for the slashed two-class data two fuzzy rules are generated and they describe upper left corner for black dots and lower right corner for white dots.

5.2 Comparison of FDT and Other Classification Algorithms

We have experimented our algorithm with several other sets of data including the credit screening data (approval of credit cards), the heart disease data (diagnosis of heart disease), and the sonar data (target object classification by analyzing the echoed sonar signals) in the UCI machine learning databases [12]. In these experiments, we use genetic algorithm to generate membership functions for the FDT algorithm. Table 1 compares the performance of FDT with that of C4.5 [3], [4] (release 7 and 8) in classification accuracy and tree size. Accuracies and tree sizes are averaged using 10-fold cross-validation for each data set. It is clearly shown that our algorithm is more efficient than C4.5 both in classification accuracy and tree size.

However, it should be noticed that for FDT the tree sizes are more sensitive to training data compared with C4.5. It can be expected by considering the nature of genetic algorithm. It is because that in genetic algorithm membership functions can evolves quite differently from others depending the training data and the initial population of chromosomes. For each data set in only one or two cases out of 10 relatively big trees are constructed.

Table 1. Comparison of FDT and C4.5(release 7 and 8) (±x indicates the standard deviation.)

Data set	C4.5(Rel. 7)		C4.5(Rel. 8)		FDT	
	accuracy(%)	tree size	accuracy(%)	tree size	accuracy(%)	tree size
Breast cancer	94.71±0.09	20.3±0.5	94.74±0.19	25.0±0.5	96.66±0.02	5.9±2.8
Credit screening	84.20±0.30	57.3±1.2	85.30±0.20	33.2±1.1	86.18±0.03	4±0.0
Heart disease	75.10±0.40	45.3±0.3	77.00±0.50	39.9±0.4	77.8±0.05	22±6.0
Iris	95.13±0.20	9.3±0.1	95.20±0.17	8.5±0.0	97.99±0.03	4.2±0.4
Sonar	71.60±0.60	33.1±0.5	74.40±0.70	28.4±0.2	77.52±0.06	14.1±3.1

Table 2. Comparison of C4.5, FID3.1 and FDT

Data set	C4.5		FID3.1		FDT	
	accuracy(%)	tree size	accuracy(%)	tree size	accuracy(%)	tree size
Iris	94.0	5.0	96.0	5.0	97.99	4.2
Bupa	67.9	34.4	70.2	28.9	70.69	12.1
Pima	74.7	45.2	71.56	23.1	76.90	4.2

In the next experiment, we compare our algorithm with C4.5 and FID3.1 using the data sets of Iris, Bupa and Pima. All experiments were performed using 10-fold cross-validation. Table 2 compares the performance in accuracy and tree size of FDT and FID3.1 [9]. It is clearly shown that FDT is consistently more efficient than both C4.5 and FID3.1.

6 Conclusions

In this paper, we propose a fuzzy rule generation algorithm based on fuzzy decision tree for data mining. Our method provides the efficiency and the comprehensibility of the generated fuzzy rules, which are important to data mining. Particularly, fuzzy rules allow us to effectively classify data of non-axis-parallel decision boundaries using membership functions properly, which is difficult to do using the conventional attribute-based methods. We also propose the use of genetic algorithm for automatic generation of an optimal set of membership functions. We have experimented our algorithm with several benchmark data sets including the iris data, the Wisconsin breast cancer data, the credit screening data, and others. The experiment results show that our method is more efficient in classification accuracy and compactness of rules compared with C4.5 and FID3.1.

We plan to investigate the use of co-evolution algorithm in place of genetic algorithm and dynamic sampling to speed up the algorithm significantly. We can incorporate FID3 into initialization of membership functions in our method for more efficient search. We are also applying our algorithm for generating fuzzy rules describing the blast furnace operation in steel making.

References

1. Fayyad, U., Mannila, H., and Piatetsky-Shapiro, G.: Data Mining and Knowledge Discovery. Kluwer Academic Publishers (1997)
2. Quinlan, J.R.: Induction of decision trees. Machine Learning, Vol.1, Issue 1 (1986) 81-106
3. Quinlan, J.R.: Improved use of continuous attributes in C4.5. Journal of Artificial Intelligence Research, Vol.4 (1996) 77-90
4. Quinlan, J.R.: C4.5: Programs for Machine Learning. Morgan Kaufmann Publishers (1993)
5. Suárez, A. and Lutsko, J.F.: Globally Optimal Fuzzy Decision Trees for Classification and Regression. IEEE Trans. on Pattern Analysis and Machine Intelligence, Vol.21, Issue 12 (1999) 1297-1311
6. Zeidler, J. and Schlosser M.: Continuous-Valued Attributes in Fuzzy Decision Trees. Proc. of the 6th Int. Conf. on Information Processing and Management of Uncertainty in Knowledge-Based Systems (1996) 395-400
7. J.Abonyi and J.A. Roubos: Structure Identification of fuzzy classifiers. 5th Online world conference on soft computing in industrial applications (2000) 4-18
8. Janikow, C.Z.: Fuzzy Decision Tree: Issues and Methods. IEEE Trans. on Systems, Man and Cybernetics - Part B, Vol.28, Issue 1 (1998) 1–14
9. Janikow, C.Z., Faifer, M.: Fuzzy Partitioning with FID3.1. Proc. of the 18th International Conference of the North American Fuzzy Information Processing Society, IEEE 1999 (1999) 467-471
10. Johannes A. Roubus, Magne Setnes, Janos Abonyi: Learning fuzzy classification rules from labeled data. Information sciences informatics and computer science: an international journal archive, Vol. 150, Issue 1-2 (2003) 77-93
11. Myung Won Kim, Joung Woo Ryu: Optimized Fuzzy Classification for Data Mining. Lecture Notes in Computer Science, Vol. 2973 (2004) 582-593
12. Blake, C.L., Merz, C.J.: UCI Repository of machine learning databases. Irvine, CA: University of California, Department of Information and Computer Science (1998)

Dynamic Test-Sensitive Decision Trees with Multiple Cost Scales

Zhenxing Qin[1], Chengqi Zhang[1], Xuehui Xie[2], and Shichao Zhang[1,2]

[1] Faculty of Information Technology, University of Technology, Sydney,
PO Box 123, Broadway, Sydney, NSW 2007, Australia
{zqin, Chengqi, zhangsc}@it.uts.edu.au
[2] Network center, Guangxi Normal University,
15 Yucai road, Guilin, Guangxi, P.R. China 541004
xxh@mailbox.gxnu.edu.cn

Abstract. Previous work considering both test and misclassification costs rely on the assumption that the test cost and the misclassification cost must be defined on the same cost scale. However, it can be difficult to define the multiple costs on the same cost scale. In our previous work, a novel yet efficient approach for involving multiple cost scales is proposed. Specifically speaking, we first introduce a new test-sensitive decision tree with two kinds of cost scales, that minimizes the one kind of cost and control the other in a given specific budget. In this paper, a dynamic test strategy with known information utilization and global resource control is proposed to keep the minimization of overall target cost. Our work will be useful in many urgent diagnostic tasks involving target cost minimization and resource consumption for obtaining missing information.

1 Introduction

Recently, researchers have begun to consider both test and misclassification costs. In [6], the cost-sensitive learning problem is cast as a Markov Decision Process (MDP), and solutions are given as searches in a state space for optimal policies. However, it may take very high computational cost to conduct the search process. Similar in the interest in constructing an optimal learner, [7] studied the theoretical aspects of active learning with test costs using a PAC learning framework. Turney [4] presented a system called ICET, which uses a genetic algorithm to build a decision tree to minimize the sum of both costs. Of these works, [1] proposed a decision tree based method that explicitly considers how to directly incorporate both types of costs in decision tree building processes and in determining the next attribute to test, should the attributes contain missing values. However, sometimes we may meet difficulty to define the multiple costs on the same cost scale. Our previous work in [9] proposed a new test cost-sensitive decision tree with two cost scales.

In this paper, we consider the dynamic tree for specific test example in the test-sensitive tree with two cost scales. The goal is to minimize one scale (called target scale) and control other one (resource scale) in a specific budget. In some situations, such as medical diagnosis, this scenario is more practical since doctors and scientists often suggest specific medical tests for specific patient.

2 Dynamic Tree Building with Multiple Cost Scales

The goal of our decision-tree learning algorithm is to minimize the sum of target cost on misclassification and test, at the same time, resource cost must less than the resource budget. We assume test and the misclassification cost contain two kinds of cost – target and resource. Both of the target and resource have been defined on two different cost scales relatively, such as dollar cost and time cost incurred in a medical diagnosis. We assume there is a maximum limit on resource, called resource budget. Table 1 shows a sample of two cost scales on "Ecoli" dataset. From table1, we can see that, there are two kinds of costs, cost1 is the target cost, and cost2 is the resource consumption. We also use the same test case (table 2) in [1] to illustrate our test strategies.

Table 1. Test and misclassification costs set for "Ecoli" dataset

	FP	FN	A1	A2	A3	A4	A5	A6
Target	800	800	50	60	60	50	50	30
Resource	150	100	10	20	10	10	10	10

Table 2. An example testing case with several unknown values. The true values are in parenthesis and can be obtained by performing the tests (with costs list in Table 1).

A1	A2	A3	A4	A5	A6	Class
? (6)	2	? (1)	2	2	? (3)	P

This strategy is exactly the idea of trade-off between target and resource. It uses the target gain ratio to choose potential splitting attributes. Assume the total target cost reduction of choosing A as a splitting attribute is G_A. G_A is actually equals to $T-T_A$ in [1], if $G_A > 0$, the attribute A is a candidate for further splitting.

In this paper, we need to consider the resource budget B. Assume the resource consumption here of A is C_A and the rest resource is B'. We first normalization the consumption as C'_A, the gain ratio of choosing A as a splitting attribute is

$$R_A = G_A / C'_A$$

This dynamic tree building strategy builds a new tree for current test example based on known values. From this tree, the testing examples can easily be classified, as values of attributes used in the tree are all known in the testing example. At the leaf node it reaches, we follow our tree building procedure to evaluate if this node should be split by the attributes with unknown values. Once an attribute is chosen to split the node, we begin to evaluate the resource consumption of all his children. The tree building is stopped as no further target cost gain or resource exhausted.

3 Experiments

We conducted experiments on five real-world datasets [1,3,5]. These datasets are chosen because they have at least some discrete attributes, binary class, and a good number of examples. The numerical attributes in datasets are discretized first using minimal entropy method [6].

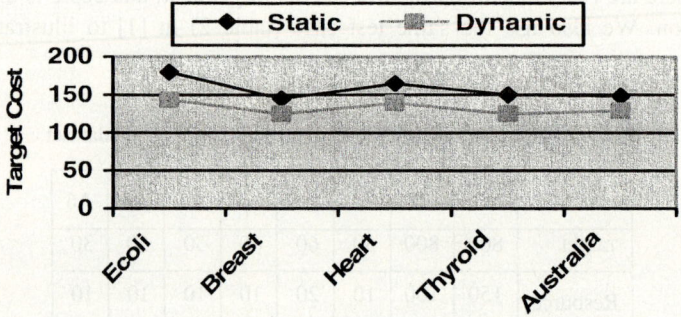

Fig. 1. Comparing of total target cost of two tree building strategies on different datasets

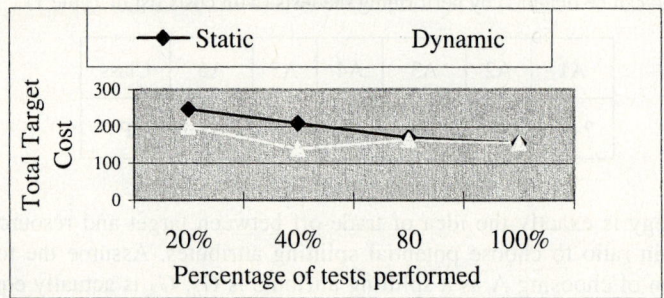

Fig. 2. Comparing of total target cost on percentage of tests performed under resource budget

First, we compare the target cost and resource consumption of the static building strategy in [1] against our dynamic strategy (we assume resource budget is 180 that only support half tests) on all five dataset in Figure1. We can see that dynamic strategy outperform the other two in target cost. It means dynamic strategy got a better overall performance with limit resource budget.

To compare the influence of resource budget on three strategies, we conducted an experiment on all the datasets with varying budget B to support a part of all needed tests from 20 to 100 percent. For the more completely usage of resource, we use OST testing strategy first, once the cost is exhaust we use M2 testing strategy to give a result. The result is shown in figure 5. From figure 5, we can see that all target cost will go down as the test examples can explore further branches, then lower total cost are obtained. The performance-first strategy also outperforms the other two in target cost with same resource consumption.

4 Conclusions and Future Work

In this paper, a new dynamic test strategy is proposed to utilize the known values and resource. Our experiments show that our new dynamic test strategy outperformed dramatically the other strategies as no enough resource for tests. In real world, resource is usually insufficient, so our new strategy is thus more robust and practical. In the future, we plan to consider how to minimize the total target cost with partial cost-resource exchanging. In some situations, such as medical diagnosis, this scenario is more practical since lot of hospitals provide VIP services. We also want to extend our test strategy to near Optimal Batch Test.

References

1. C. Ling, Q. Yang, J. Wang, and S. Zhang. Decision trees with minimal costs. In *Proceedings of 2004 International Conference on Machine Learning*, 2004.
2. Turney, P. D., Types of cost in inductive concept learning, *Workshop on Cost-Sensitive Learning at the Seventeenth International Conference on Machine Learning,* Stanford University, California.2000
3. *UCI Repository of machine learning databases.* Irvine, CA: University of California, Department of Information and Computer Science. (See [http://www.ics.uci.edu/~mlearn/MLRepository.html])
4. Turney, P. D, Cost-sensitive classication: Empirical evaluation of a hybrid genetic decision tree induction algorithm. *Journal of Articial Intelligence Research*, 2: 369-409, 1995.
5. Machine Learning. McGraw Hills
6. Fayyad, U. M., & Irani, K. B. Multi-interval discretization of continuous-valued attributes for classification learning. In Proceedings of the 13th International Joint Conference on Artificial Intelligence, pages 1022--1027. Morgan Kaufmann,
7. Greiner, R., Grove, A. J., and Roth D. (2002), Learning cost-sensitive active classiers. Articial Intelligence, 139(2): 137-174, 2002.
8. Quinlan, J. R. *C4.5: Programs for Machine Learning.* Morgan Kaufmann, San Mateo, California, 1993.
9. Z. Qin, C. Zhang and S. Zhang, Cost-sensitive Decision Trees With Multiple Cost Scales. In: Proceedings of the 17th Australian Joint Conference on Artificial Intelligence (AI 2004), Cairns, Queensland, Australia, 6-10 December, 2004.

Design of T–S Fuzzy Classifier via Linear Matrix Inequality Approach

Moon Hwan Kim[1], Jin Bae Park[1], Young Hoon Joo[3], and Ho Jae Lee[1]

[1] Department of Electrical and Electronic Engineering, Yonsei University,
Seodaemun-gu, Seoul, 120-749, Korea
{jmacs, jbpark, mylchi}@yonssei.ac.kr

[2] School of Electronic and Information Engineering, Kunsan National University,
Kunsan, Chonbuk, 573-701, Korea
yhjoo@kunsan.ac.kr

Abstract. A linear matrix inequality approach to designing accurate classifier with a compact T–S(Takagi–Sugeno) fuzzy-rule is proposed, in which all the elements of the T–S fuzzy classifier design problem have been moved in parameters of a LMI optimization problem. Two-step procedure is used to effectively design the T–S fuzzy classifier with many tuning parameters: antecedent part and consequent part design. Then two LMI optimization problems are formulated in both parts and solved efficiently by using interior-point method. Iris data is used to evaluate the performance of the proposed approach. From the simulation results, the proposed approach showed superior performance over other approaches.

1 Introduction

Patten classification plays a crucial role in a large number of applications, including printed and handwritten text recognition [1,2,3], speech recognition [4], human face recognition [5]. A great variety of conventional and computational intelligence techniques for pattern classification can be found in [6]. The pattern classification system can be implemented by using various systems: linear system, fuzzy system, neural network system, etc [9,10,11,12,13,14,15,16] . One of them, the fuzzy system, is studied by many researchers recently because it has high comprehensibility and is easy to apply to the complex classification problems. Fuzzy classification system design can be performed by supervised learning using a set of training data with fuzzy or nonfuzzy labels. When given a pattern, the fuzzy classifier computes the membership value of the pattern in each class and makes decisions based on these membership values. The fuzzy labels of a fuzzy classifier can be defuzzified and then the fuzzy classifier becomes a hard classifier but uses the idea of fuzziness in the model.

Membership function is the key point of fuzzy rule-based systems. However, the conventional method of designing membership functions for fuzzy systems relies on the manual work and experience of experts. This inevitably becomes a bottleneck in fuzzy rule-based system design. In recent years, some researchers have employed the learning ability of various optimization methods to learn the membership functions from training data for fuzzy systems. Jang developed an adaptive network-based fuzzy inference system (ANFIS) which is used as a fuzzy controller [7]. Setnes used genetic algorithm (GA) to optimize parameters in fuzzy system [17]. If flexible membership functions and fuzzy rules with both certainty grade are determined simultaneously based on accurate and compact fuzzy-rule base, the design of fuzzy classifiers can be regarded as an optimization problem with lots of system's tuning parameters. The conventional non-evolutionary optimization method is hard to applied to determine all parameters in the fuzzy classifier because these optimization problems is so complex and has nonlinear relationship between antecedent and consequent part. The performance of evolutionary optimization method, GA, also would be greatly degraded when applied to a large parameter optimization problem (LPOP) that is shown by theoretical analysis in [8]. As a result, the success of the approach to formulating the fuzzy classifier design to an LPOP mainly relies on a new powerful optimization algorithm to solve the LPOP.

In this paper, an LMI approach to designing accurate classifier with a compact T–S fuzzy-rule base is proposed. Two step design procedure is presented. All design elements in the antecedent part is converted to parameters in the convex optimization problem. The designed convex optimization problem is solved by LMI optimization method efficiently. Some theoretical base is given to design antecedent part alone without any consideration for consequent part. After all parameters in the antecedent part are determined, the consequent parameters are determined by solving LMI optimization problem converted from overdetermined problems. To show clear design procedure, two design algorithms are given.

The organization of this paper is as follows. Section 2 presents the basic approach to classifier design by LMI. Simulation results testify to the classifier's performances and the utilities of the proposed method are discussed in Section 3.

2   LMI Based Fuzzy Classifier Design

The T–S fuzzy classifier is consist of T–S type fuzzy rules described as the following form [17]:

$$R_i : \text{IF } x_1 \text{ is } A_{i1} \text{ and } \ldots \text{ and } x_n \text{ is } A_{in} \quad (1)$$
$$\text{THEN } y_i(x) = b_{i1}x_1 + \ldots + b_{in}x_n + c_i, i = 1, \ldots, l$$

where $x_i \in \mathbb{R}$ is the ith feature input, $A_{i1}, \ldots, A_{in}$ are the antecedent fuzzy sets, $y_i(x)$ is the consequent output of the ith rule, $x = [x_1, \ldots, x_n]^T \in F \subset \mathbb{R}^n$ is the input feature vector, F is the feature vector set, b_{ij} and c_i are consequent parameters, and l is the number of fuzzy rule. The output of T–S type fuzzy rule system is inferred by following equations:

$$Y(x) = \frac{\sum_{i=1}^{l} \tau_i(x) y_i(x)}{\sum_{i=1}^{l} \tau_i(x)} \quad (2)$$

$$\tau_i(x) = \prod_{j=1}^{n} \mu_{A_{ij}}(x_j) \quad (3)$$

where $\tau_i(x)$ is the firing strength of the ith rule and $\mu_{M_{ij}}(x_j) \in \mathbb{R}[0,1]$ is the membership degree of the jth feature of the ith rule. To compute the degrees of class membership for pattern x, a Gaussian membership function is adopted such that

$$\mu_{A_{ij}}(x_j) = e^{-\frac{(m_j^i - x_j)^2}{\sigma_j^i}} \qquad (4)$$

where c_j^i is the center and σ_j^i is the width of jth feature of the ith rule.

Fuzzy classifier almost always means arriving at a hard classifier because most pattern recognition systems require hard labels for objects being classified. In order to convert soft label $Y(x)$ to hard label $Y_c(x)$, we use following mapping equation,

$$Y_c(x) = \arg_g \min\{|g - Y(x)|\}, \ g \in \{1, \ldots, n\} \qquad (5)$$

where g is the index of the class and n denotes the number of classes.

Designing the T–S fuzzy classifier can be treated as parameter optimization problem with compact fuzzy rule sets. In general, the compact fuzzy rule means the minimum rule for fuzzy system with good performance. In this paper, we consider compact fuzzy rule as minimum size fuzzy rule which consist of one fuzzy rule for one class. Then the main goal of designing T–S fuzzy classifier is to determine parameter of membership functions in the antecedent and parameters in the consequent. Figure 1 presents the overall procedure of the proposed design method.

2.1 Identification of Antecedent Part

Notice that the antecedent part can not be identified without any consideration for consequent part because final output $Y_c(x)$ is calculated based on outputs

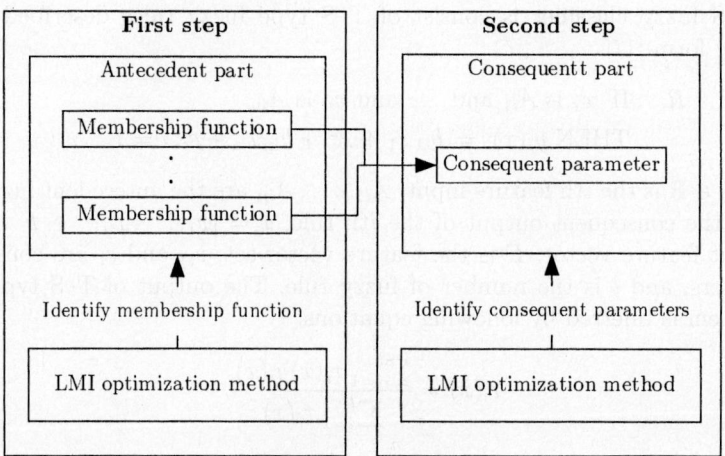

Fig. 1. Proposed Fuzzy classifier design procedure

of both part. However, the antecedent part and consequent part are hard to designed simultaneously because there are some nonlinear relationship hard to deal in mathematical method. To overcome this difficulty, the classical design goal for antecedent part is given. For computational convenience, the parameters of $\tau_i(x)$ can be reformulated as follows:

$$\tau_i(x) = e^{-\frac{(m_1^i - x_1)^2}{\sigma_1^i}} \times e^{-\frac{(m_2^i - x_2)^2}{\sigma_2^i}} \times \ldots \times e^{-\frac{(m_n^i - x_n)^2}{\sigma_n^i}}$$

$$= e^{-\sum_{j=1}^{n} \frac{(m_j^i - x_j)^2}{\sigma_j^i}}$$

$$= e^{-(x - m_i)^T \sigma_i^T \sigma_i (x - m_i)} \tag{6}$$

where $\sigma_i = \text{diag}\left(\frac{1}{\sqrt{\sigma_1^i}}, \ldots, \frac{1}{\sqrt{\sigma_n^i}}\right)$ is the diagonal matrix containing the widthes of the Gaussian membership functions in the antecedent part of the ith rule, and $m_i = [m_1^i, \ldots, m_n^i]$ represents center values of the membership function of the ith rule.

In classical fuzzy classifier design, the membership function should satisfy following conditions,

$$\tau_i(x) = 0, x \notin C_i$$
$$\tau_i(x) = 1, x \in C_i$$

where C_i means the set of data belonging to class i. Therefore, the main design objective can be defined as determining V_i and m_i satisfying,

$$(x - m_i)^T \sigma_i^T \sigma_i (x - m_i) = 0, \forall x \in C_i \tag{7}$$
$$(x - m_i)^T \sigma_i^T \sigma_i (x - m_i) = \infty, \forall x \notin C_i \tag{8}$$

The condition (7) can be formulated as condition of LMI optimization problem directly. However, it is not easy to consider condition (8) because it is non-convex form. To overcome this difficulty, we relax the condition (8) to condition which is minimizing σ_i.

Theorem 1. *If x belongs to class C_i, V_i and m_i of membership functions in the antecedent part of the rule i are determined by solving the following general eigenvalue problem (GEVP):*

$$\underset{q_i, \sigma_i}{\text{Minimize}} \quad \gamma \quad \text{subject to}$$

$$\gamma W > \sigma_i > 0 \tag{9}$$

$$\begin{bmatrix} \gamma & \star \\ \sigma_i x - q_i & \gamma \end{bmatrix} > 0, \quad \forall x \in C_i \tag{10}$$

where $q_i = \sigma_i m_i$, $W = \text{diag}\{w_1, \ldots, w_n\}$ is diagonal matrix, w_i is the variance of ith feature, and $\star$ denotes the transposed element matrix for the symmetric position.

Proof. The proof is omitted due to lack of space.

Theorem 1 gives the method to determine membership functions in the antecedent part of only one rule. we could apply this method to each fuzzy rule and obtain well-identified membership functions.

2.2 Identification of Consequent Part

After parameters of membership functions in the antecedent part is determined, the consequent parameters should be identified. For the computational convenience, $Y(x)$ can also be represented as following matrix form,

$$Y(x) = D^T(Bx + C) \tag{11}$$

where

$$D = \begin{bmatrix} d_1(x) \\ \vdots \\ d_i(x) \\ \vdots \\ d_l(x) \end{bmatrix}, \quad B = \begin{bmatrix} b_{11} & \ldots & b_{1n} \\ \vdots & & \vdots \\ b_{i1} & \ldots & b_{in} \\ \vdots & & \vdots \\ b_{l1} & \ldots & b_{ln} \end{bmatrix}, \quad C = \begin{bmatrix} c_1 \\ \vdots \\ c_i \\ \vdots \\ c_l \end{bmatrix} \tag{12}$$

$$d_i(x) = \frac{\tau_i(x)}{\sum_{j=1}^{l} \tau_j(x)}. \tag{13}$$

Assume that the parameters of antecedent is completely determined. With Given H and x we could formulated following key equation,

$$Y_d = D^T(Bx + C), \quad \forall x \in F \tag{14}$$

where Y_d is desired output of the class and is determined as one of index of class. Finally, by finding A and B satisfying (14), we could get desired output $Y(x)$. Notice that (14) can be converted LMI optimization problem directly. Theorem 3 shows the GEVP for determining A and B in the consequent part.

Theorem 2. *If x, Y_d, and H are given, A and B of the proposed T-S fuzzy classifier are determined by solving the following GEVP*

$$\underset{B,C}{\text{Minimize}} \quad \gamma \quad \text{subject to}$$

$$\begin{bmatrix} \gamma & \star \\ Y_d - D^T(Bx + C) & I \end{bmatrix} > 0, \quad \forall x \in F. \tag{15}$$

Proof. The proof is omitted due to lack of space.

3 Simulation

The iris database created by Fisher is a common benchmark in the classification and the pattern recognition studies. It has four feature variables: sepal length, sepal width, petal length, and petal width and consists of 150 feature vectors: 50 for each iris sestosa, iris versicolor, iris virinica. Figure 2 shows the design step for iris data.

Table 1 shows some results of well-known classifier systems. For example, Wang et al. [18] applied a self-adaptive neuro-fuzzy inference system (SANFIS) that is capable of self-adapting and self-organizing its internal structure to acquire a parsimonious rule-base for interpreting the embedded knowledge of a system from the given training data set. He derived three-rule SANFIS system with 97.5% correct. Abonyi et al. [19] proposed a new data-driven method to design compact fuzzy classifiers via combining a genetic algorithm, a decision-tree initialization, and a similarity-driven rule reduction technique. The final system had three fuzzy rules. The accuracy is 96.11% correct (six misclassifications). Abe et al. [20] discussed a fuzzy classifier with ellipsoidal regions. They applied clustering techniques to extract fuzzy rules, with one rule around one cluster center, and then they tuned the slopes of their membership functions to obtain a high recognition rate. Finally, they obtained a fuzzy classifier with a recognition rate of 98.7% (two misclassifications). Shi et al. [21] applied an integer-code genetic algorithm to learn a Mamdani-type fuzzy system for classifying the iris data by training on all 150 patterns. After several trials with different learning options, a four-rule fuzzy system was obtained with 98% correct recognition

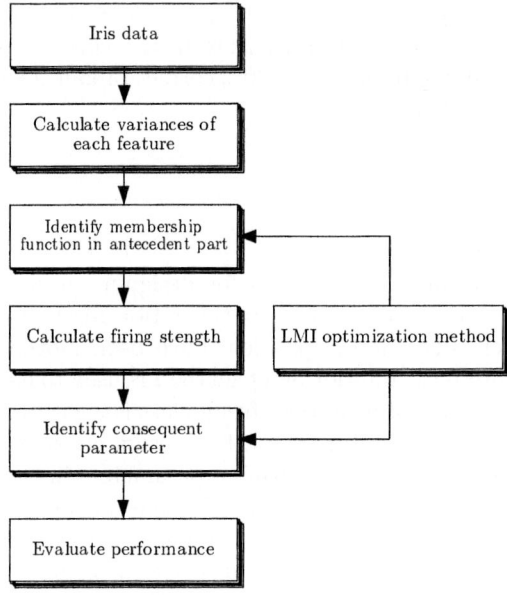

Fig. 2. Design procedure for iris classification

(three misclassifications). Russo [22] applied a hybrid GA neuro-fuzzy approach to learn a fuzzy model for the iris data. He derived a five-rule fuzzy system with 18 fuzzy sets and 0 misclassifications. Ishibuchi et al. [23] applied all 150 samples in the training process, and derived a fuzzy classifier with five rules. The resolution was 98.0% correct and three misclassifications.

For the Iris example, we use 150 patterns to design a fuzzy classifier via the proposed method. Since Iris example has three classes, the number of compact fuzzy rule is three. σ_i and m_i in antecedent part and A and B in consequent part are identified. Figure 3 shows the identified membership function. The consequent parameters is then determined as

$$B = \begin{bmatrix} -0.0000 & 0.0000 & -0.0001 \\ -0.1121 & -0.2234 & 0.0029 \\ -0.1020 & -0.0624 & 0.1276 \end{bmatrix}, \quad C = \begin{bmatrix} 0.6667 \\ 1.7547 \\ 1.8412 \end{bmatrix}.$$

Table 1 shows the comparison of results between the above fuzzy classifier system with other well-known classifier systems on number of rules and classification accuracy. The resulting system arrives at the highest degree of accuracy using the smallest number of term sets.

To estimate the performance of the proposed method on unseen data, the five-fold cross-validation experiment was performed on the iris data. In experiment, the normalized iris data were divided into five disjoint group containing 30 different patterns which consist of each ten patterns of three classes. Then we derived fuzzy classifier via the proposed method on all data outside one group and tested the resulting fuzzy classifier on the data inside that group. Finally, five fuzzy classifiers were derived. Table 2 reports the results of five-folder cross-validation. The average classification result is 97.64% correct (about 2.4 misclassifications) on the training data and 97.30% correct (about 0.8 misclassification) on the test data using 3 rules.

4 Conclusions

This paper proposed automatic method for designing accurate classifier with compact fuzzy-rule base using an LMI optimization method. All the elements of the T–S fuzzy classifier design problem have been moved in parameter of LMI optimization problem. Interior point method is used to effectively solve the design problems of various-dimensional fuzzy classifier with many tuning parameters. Two-step design procedure is used to design classifier: antecedent part and consequent part. In order to determine parameters in the antecedent part without any consideration for consequent parameter, some theoretical analysis are given. Then after antecedent part is designed, consequent parameters are determined based on firing predetermined strength of antecedent. In simulation, the high performance of the proposed method is validated by using iris data.

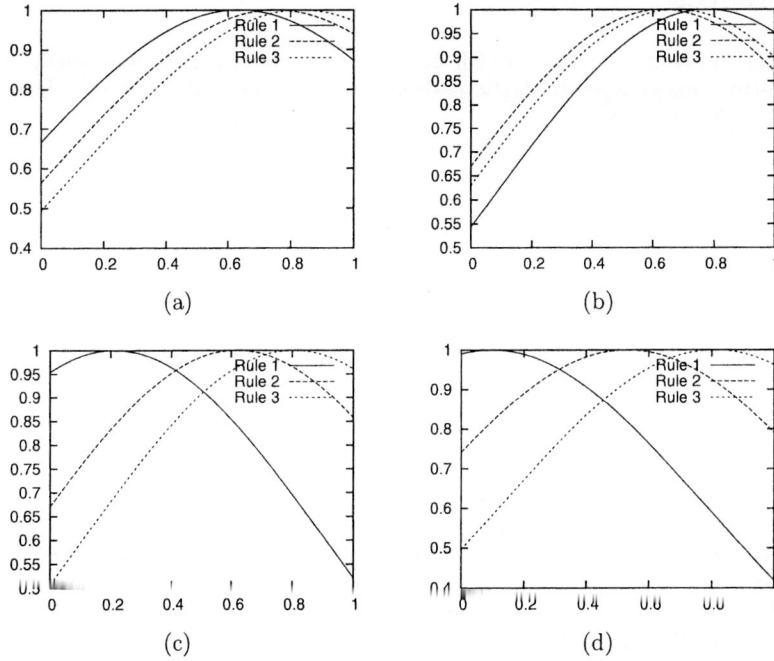

Fig. 3. Membership functions for iris data: training simulation. a) feature x_1. (b) feature x_2. (c) feature x_3. (d) feature x_4.

Table 1. Comparison of classification results for Iris data

	Rules	Classification accuracy (%)
Wang et al.	3	97.5
Abonyi et al.	3	98.3
This paper	3	98.7
Shi et al.	4	98.0
Russo	5	100
Ishibuchi et al.	5	98.0

Table 2. Five-folder cross-validation result on Iris data

	1	2	3	4	5	Average
Rules	3	3	3	3	3	3
Training patterns	120	120	120	120	120	120
Misclassifications(training)	3	1	3	2	3	2.4
Classification accuracy(training)	97.5	99.2	97.5	98.3	97.5	97.63
Testing patterns	30	30	30	30	30	30
Misclassifications(testing)	0	2	1	0	1	0.8
Classification accuracy(testing)	100	93.3	96.7	100	96.7	97.3

References

1. Chen M., Kundu A., Zhou J.: Off-line handwritten word recognition using a hidden Markov model type stochastic network. IEEE Trans. Pattern Anal. Mach. Intel. **16** (1994) 481–496.
2. Cohen E.: Computational theory for interpreting handwritten text in constrained domains. Artif. Intell. **67** (1994) 1–31.
3. Partizeau M., Plamondon R.: A fuzzy-syntactic approach to allograph modeling for cursive script recognition. IEEE Trans. Pattern Anal. Mach. Intel. **17** (1995) 702–712.
4. Bourlard H., Morgan N.: Connectionist Speech Recognition-A Hybrid Approach. Boston. MA: Kluwer Academic (1994).
5. Lam K. M., Yan H.: Locating and extracting the eye in human face images. Pattern Recog. **29** (1996) 771–779.
6. Schalkoff R.: Pattern Recognition-Statistical, Structural and Neural Approaches. New York, Wiley (1992).
7. Jang J. S. R.: Fuzzy controller design without domain experts. Proc. IEEE Int. Conf. Fuzzy Systems. San Diego, CA, Mar. (1992) 289–296.
8. Kumar K., Narayanaswamy S., Garg S.: Solving large parameter optimization problems using a genetic algorithm with stochastic coding. In Winter G., Periaux J., Galan M., Cuesta P.: Genetic Algorithms in Engineering and Computer Science. Eds. New York, Wiley (1995).
9. Ishibuchi H.,Murata T.,Turksen I. B.: Single-objective and two-objective genetic algorithms for selecting linguistic rules for pattern classification problems. Fuzzy Sets Syst. **89** (1997) 135–150.
10. Wang C. -H., Hong T. -P., Tseng S. -S.: Integrating fuzzy knowledge by genetic algorithms. Fuzzy Sets Syst. **2** (1998) 138–149.
11. Ishibuchi H., Nakashima N., Murata T.: Performance evaluation of fuzzy classifier systems for multidimensional pattern classification problems. IEEE Trans. Syst., Man, Cybern. B **29** (1999) 601–618.
12. Hall L. O., Ozyurt I. B.,Bezdek J. C.,: Clustering with genetically optimized approach. IEEE Trans. Evolut. Computing (1999) 103–112.
13. Hwang H. -S.: Control strategy for optimal compromise between trip time and energy consumption in a high-speed railway IEEE Trans. Syst., Man, Cybern. A. **28** (1998) 791–802.
14. Jagielska I., Matthews C., Whitfort T.: An investigation into the application of neural networks, fuzzy logic, genetic algorithms, and rough sets to automated knowledge acquisition for classification problems. Neurocomputing. **24** (1999) 37–54.
15. Russo M.: FuGeNeSys.A fuzzy genetic neural system for fuzzy modeling IEEE Trans. Fuzzy Syst. **6** (1998) 373–388.
16. Wang L. and Yen J.: Extracting fuzzy rules for system modeling using a hybrid of genetic algorithms and Kalman filter. Fuzzy Sets Syst. **101**, (1999) 353–362.
17. Setnes M. and Roubos H.: GA-fuzzy modeling and classification: complexity and performance. IEEE Trans. Fuzzy Syst. **8** (2000) 509–522.
18. Wang J. S., Lee G. C. S.: Self-adaptive neuro-fuzzy inference system for classification application. IEEE Trans. Fuzzy Syst. **10** (2002) 790–802.
19. Abonyi J., Roubos J. A., Szeifert F.: Data-driven generation of compact, accurate, and linguistically sound fuzzy classifiers based on a decision- tree initialization. Int. J. Approx. Reason. 32 (2003) 1–21.

20. Abe S., Thawonmas R.: A fuzzy classifier with ellipsoidal regions. IEEE Trans. Fuzzy Syst. **5** (1997) 358–368.
21. Shi Y., Eberhart R., Chen Y.: Implementation of evolutionary fuzzy system. IEEE Trans. Fuzzy Syst. **7** (1999) 109–119.
22. Russo M.: Genetic fuzzy learning. IEEE Trans. Evolut. Computat. **4** (2000) 259–273.
23. Ishibuchi H., Nakashima T., Murata T.: Three-objective geneticbased machine learning for linguistic rule extraction. Inf. Sci. **136** (2001) 109–133.

Design of Fuzzy Rule-Based Classifier: Pruning and Learning

Do Wan Kim[1], Jin Bae Park[1], and Young Hoon Joo[2]

[1] Yonsei University, Seodaemun-gu, Seoul, 120-749, Korea
{dwkim, jbpark}@yonsei.ac.kr,
[2] Kunsan National University, Kunsan, Chunbuk, 573-701, Korea
yhjoo@kunsan.ac.kr

Abstract. This paper presents new pruning and learning methods for the fuzzy rule-based classifier. For the simplicity of the model structure, the unnecessary features for each fuzzy rule are eliminated through the iterative pruning algorithm. The quality of the feature is measured by the proposed correctness method, which is defined as the ratio of the fuzzy values for a set of the feature values on the decision region to one for all feature values. For the improvement of the classification performance, the parameters of the proposed classifier are adjusted by using the gradient descent method so that the misclassified feature vectors are correctly re-categorized. Finally, the fuzzy rule-based classifier is tested on two data sets and is found to demonstrate an excellent performance.

1 Introduction

The fuzzy rule-based classifier is a popular counter part of a fuzzy control system and a fuzzy modeling [1,2], which carries out the pattern classification by using the membership grades of the feature variables. If there is a contribution of the fuzzy rule-based classifier, it would lie in guiding the steps by which one takes knowledge in a linguistic form and casts it into discriminant functions [3]. In numerous researches [9,4,10,11,5,6,12,13,14,7,8], the excellent capabilities to the pattern classification of the fuzzy rule-based classifier have been shown.

In the design of the fuzzy rule-based classifier, there are two main issues which involve the model complexity and the classification performance. If too many free parameters are used, there is a danger of overfitting; conversely, if too few parameters are used, the training set may not be learned. Thus, the design process can be divided into two strategies: the feature selection and the learning. For the simplicity of the model, one possibility is to reduce the dimensionality by selecting an appropriate subset of the existing features. The key point in the feature selection is the measure of the quality of a set of the features, which concerns some measure of the predictive power of the features. An attractive approach [4,5,6] is to measure the similarity of the overlapping degree between two fuzzy sets. For the improvement of the classification performance, the learning in the fuzzy rule-based classifier can be formulated as minimizing a cost function. In [7,8], the cost function was selected as the squared-error between the classifier output and the desired value, and then the learning rules are derived from forming the gradient. The classifier outputs for the correct class and the rest converge into the upper bound and the lower bound, respectively, as the cost function approaches zero.

Despite of the existence of the excellent previous researches [4,5,6,7,8], there are still critical issues. In [4,5,6], measuring the overlapping degree does not become the accurate criterion in the classification problem because not all data on the overlapping region can be judged as the classification error. And also, a way to measure some degree between two classes may not be efficient in the multicategory case. In [7,8], the desired values for the correct class and the rest are defined as the upper bound and the lower bound of the classifier outputs, respectively. However, these desired values are very strict conditions for the learning objective, which is that the classifier output for the correct class gets larger than one for the rest as the cost function approaches zero. These strict conditions make the learning inefficient. In addition, this approach cannot be applied in the classifier with the unbounded outputs.

This paper aims at developing the fuzzy rule-based classifier for the model complexity and the classification capability. The main contributions of this paper are to measure the qualities of the given feature variables by using a new analysis technique of the fuzzy sets, *correctness method* and to derive the learning rules for the classifier without the upper bound. In the proposed pruning method, an appropriate subset of the existing features for each fuzzy rule is selected through the iterative pruning algorithm. To measure the quality of the features, the correctness degree is defined as the ratio of the fuzzy values for a set of the feature values on the decision region to one for all feature values. In the proposed learning method, the related parameters of the proposed classifier are tuned by using the gradient descent method so that the misclassified feature vectors are correctly re-categorized.

2 Fuzzy Rule-Based Classifier

2.1 Initiallization

For a given feature vector $\mathbf{x}$, the proposed fuzzy rule-based classifier is formulated in the following form.

$$R_i : \text{IF } x_1 \text{ is } \Gamma_{i1} \text{ and } \ldots \text{ and } x_n \text{ is } \Gamma_{in},$$
$$\text{THEN } y_i = d_i^c(\mathbf{x}) \tag{1}$$

where y_i is the output vector of R_i, $\Gamma_{ih}(x_h) = \exp\left(-\frac{1}{2}\left(\frac{x_h - m_{ih}^p}{\sigma_{ih}^p}\right)^2\right)$, and $d_i^c(\mathbf{x}) = \frac{1}{(2\pi)^{\frac{n}{2}} \prod_{h=1}^{n} \sigma_{ih}^c} \exp\left(-\frac{1}{2} \sum_{h=1}^{n} \left(\frac{x_h - m_{ih}^c}{\sigma_{ih}^c}\right)^2\right)$. The unknown parameters m_{ih}^p, m_{ih}^c, σ_{ih}^p, and σ_{ih}^c are initially identified by using the arithmetic average and the standard deviation for the training data.

The ith final output of (1) is inferred as follows:

$$\widehat{y}_i(\mathbf{x}) = d_i^p(\mathbf{x}) d_i^c(\mathbf{x}) \qquad (2)$$

where $d_i^p(\mathbf{x}) = \max_{h \in \mathcal{I}_n} \Gamma_{ih}(x_h)$ by using the maximum inference engine and the singleton fuzzifier. From (2), the proposed classifier is said to assign a feature vector $\mathbf{x}$ to the class i if

$$\widehat{y}_i(\mathbf{x}) > \widehat{y}_j(\mathbf{x}), \quad \forall j \neq i \qquad (3)$$

The effect of $\widehat{y}_i(\mathbf{x})$ is to separate the feature space into m decision regions $\mathcal{R}_1, \mathcal{R}_2, \ldots, \mathcal{R}_m$.

2.2 Pruning

We suggest a pruning algorithm based on the analysis of the fuzzy sets for eliminating the unnecessary features for each fuzzy rule. The difficult point of selecting the unnecessary feature is that pruning the feature has direct impact on the classification performance. The main reason is that the decision regions are changed by eliminating the labelled feature. Thus, in the analysis of the fuzzy set, the class label and the decision region are considerable matters. In addition, for the general application, the analysis technique is efficiently usable in the multicategory case. Dealing such issues are formulated as follows:

*Problem 1 (**The analysis of the fuzzy set for pruning the feature**). If the analysis technique of the fuzzy set is used for checking to whether the feature is unnecessary for the supervised learning, it is sufficiently satisfied with following conditions:*

(i) The analysis tool must consider the class label and the decision region, which are important concepts in the supervised learning.
(ii) The analysis technique of the fuzzy set must be simply applicable in the multicategory case.

To resolve Problem 1, the correctness degree of the fuzzy set Γ_{ih} for the hth feature x_h labelled as w_i is measured by using the cardinality, and it is defined as follows.

Definition 1. *The correctness degree of Γ_{ih} is defined as*

$$C(\Gamma_{ih}) = \frac{|\Gamma_{ih}|_{(x_h \in w_i) \in \mathcal{R}_{ih}}}{|\Gamma_{ih}|_{x_h \in w_i}} \qquad (4)$$

where $|\cdot|$ denotes the cardinality of a set, and $\mathcal{R}_{ih}$ is one-dimensional fuzzy region according to $d_i^p(x_h) > d_j^p(x_h)$ for all $j \neq i$.

By applying the definition of the cardinality, (4) becomes

$$C(\Gamma_{ih}) = \frac{\sum_{(x_h \in w_i) \in \mathcal{R}_{ih}} \Gamma_{ih}(x_h)}{\sum_{x_h \in w_i} \Gamma_{ih}(x_h)} \qquad (5)$$

Specifically, if Γ_{ih} is not overlapped with the others and/or all $x_h \in w_i$ fall on $\mathcal{R}_{ih}$, then $C(\Gamma_{ih}) = 1$. Conversely, if Γ_{ih} is completely overlapped with the others and/or no $x_h \in w_i$ falls on $\mathcal{R}_{ih}$, $C(\Gamma_{ih}) = 0$.

By using the correctness degree (4), we set the criterions of selecting the fuzzy rule and the feature, which are applied in the proposed pruning algorithm. To select the fuzzy rule, the following average correctness degree is employed.

$$\bar{C}(\Gamma_{ih}) = \frac{1}{n} \sum_{h=1}^{n} \frac{\sum_{(x_h \in w_i) \in \mathcal{R}_{ih}} \Gamma_{ih}(x_h)}{\sum_{x_h \in w_i} \Gamma_{ih}(x_h)} \quad (6)$$

The proposed pruning algorithm becomes as follows.

Step 1 Select the fuzzy rule in the order of large value of (6).
Step 2 Prune any feature of the selected fuzzy rule that result in improving the recognition rate, where the feature is selected in the order of small value of (4).
Step 3 If no feature of the selected fuzzy rule is pruned, stop the algorithm; otherwise, update (4) and then repeat by going to Step 1.

Remark 1. The proposed method can be applied in the both of prepruning and postpruning.

2.3 Learning

Our goal is to develop a learning technique, which can be formulated as minimizing a cost function $\mathcal{J}$, for the parameters of (1) so that the misclassified feature vectors until the preceding step are correctly re-categorized. To this end, we formulate the following problem of adjusting the parameters for the misclassified feature vectors.

*Problem 2 (**Learning parameters for the misclassified feature vectors**).* To correctly categorize the misclassified feature vector **x** labelled as w_i, the learning method should be sufficiently satisfied with the following condition: The parameters σ_{ih}^p, σ_{ih}^c, m_{ih}^p, and m_{ih}^c of (1) should be adjusted so as to satisfy $\widehat{y}_i(\mathbf{x}) > \widehat{y}_j(\mathbf{x})$ for all $j \neq i$.

Define the cost function

$$\mathcal{J} = \frac{(\max_{j \in \mathcal{I}_m, j \neq i} \widehat{y}_j(\mathbf{x}) + \epsilon - \widehat{y}_i(\mathbf{x}))^2}{2} \quad (7)$$

It is noticed that, because of $\epsilon > 0$, $\widehat{y}_i(\mathbf{x})$ obviously gets larger than $\widehat{y}_j(\mathbf{x})$ for all $j \neq i$ as $\mathcal{J}$ approaches zero. The problem of minimizing the squared-error can be numerically solved by a gradient descent method. The following theorems suggest the gradient descent method to tune the parameters σ_{ih}^p, σ_{ih}^c, m_{ih}^p, and m_{ih}^c of (1).

Theorem 1. *Given the misclassified feature vector* $\mathbf{x}$ *labelled as* w_i, *the parameters* σ_{ih}^p, σ_{ih}^c, m_{ih}^p, *and* m_{ih}^c *of (1) can be adjusted by the following learning rules, respectively: For the class* i

$$\Delta m_{ih}^p = \alpha_1 (\max_{j \in \mathcal{I}_m, j \neq i} d_j^p(\mathbf{x}) d_j^c(\mathbf{x}) - d_i^p(\mathbf{x}) d_i^c(\mathbf{x}) + \epsilon) d_i^c(\mathbf{x}) \frac{\partial d_i^p(\mathbf{x})}{\partial m_{ih}^p} \quad (8)$$

$$\Delta \sigma_{ih}^p = \alpha_2 (\max_{j \in \mathcal{I}_m, j \neq i} d_j^p(\mathbf{x}) d_j^c(\mathbf{x}) - d_i^p(\mathbf{x}) d_i^c(\mathbf{x}) + \epsilon) d_i^c(\mathbf{x}) \frac{\partial d_i^p(\mathbf{x})}{\partial \sigma_{ih}^p} \quad (9)$$

and for the class j

$$\Delta m_{jh}^p = -\beta_1 (\max_{j \in \mathcal{I}_m, j \neq i} d_j^p(\mathbf{x}) d_j^c(\mathbf{x}) - d_i^p(\mathbf{x}) d_i^c(\mathbf{x}) + \epsilon) d_j^c(\mathbf{x}) \frac{\partial d_j^p(\mathbf{x})}{\partial m_{jh}^p} \quad (10)$$

$$\Delta \sigma_{jh}^p = -\beta_2 (\max_{j \in \mathcal{I}_m, j \neq i} d_j^p(\mathbf{x}) d_j^c(\mathbf{x}) - d_i^p(\mathbf{x}) d_i^c(\mathbf{x}) + \epsilon) d_j^c(\mathbf{x}) \frac{\partial d_j^p(\mathbf{x})}{\partial \sigma_{jh}^p} \quad (11)$$

where α_1, α_2, β_1, *and* β_2 *are the learning rates for* m_{ih}^p, σ_{ih}^p, m_{jh}^p, *and* σ_{jh}^p, *respectively.*

Proof. The proof is omitted due to lack of space.

Theorem 2. *The consequent parameters of the fuzzy model can be adjusted by the following learning rules: for the class* i

$$\Delta m_{ih}^c = \gamma_1 (\max_{j \in \mathcal{I}_m, j \neq i} d_j^p(\mathbf{x}) d_j^c(\mathbf{x}) - d_i^p(\mathbf{x}) d_i^c(\mathbf{x}) + \epsilon) d_i^p(\mathbf{x}) \frac{\partial d_i^c(\mathbf{x})}{\partial m_{ih}^c} \quad (12)$$

$$\Delta \sigma_{ih}^c = \gamma_2 (\max_{j \in \mathcal{I}_m, j \neq i} d_j^p(\mathbf{x}) d_j^c(\mathbf{x}) - d_i^p(\mathbf{x}) d_i^c(\mathbf{x}) + \epsilon) d_i^p(\mathbf{x}) \frac{\partial d_i^c(\mathbf{x})}{\partial \sigma_{ih}^c} \quad (13)$$

and for the class j

$$\Delta m_{jh}^c = -\delta_1 (\max_{j \in \mathcal{I}_m, j \neq i} d_j^p(\mathbf{x}) d_j^c(\mathbf{x}) - d_i^p(\mathbf{x}) d_i^c(\mathbf{x}) + \epsilon) d_j^p(\mathbf{x}) \frac{\partial d_j^c(\mathbf{x})}{\partial m_{jh}^c} \quad (14)$$

$$\Delta \sigma_{jh}^c = -\delta_2 (\max_{j \in \mathcal{I}_m, j \neq i} d_j^p(\mathbf{x}) d_j^c(\mathbf{x}) - d_i^p(\mathbf{x}) d_i^c(\mathbf{x}) + \epsilon) d_j^p(\mathbf{x}) \frac{\partial d_j^c(\mathbf{x})}{\partial \sigma_{jh}^c} \quad (15)$$

where γ_1, γ_2, δ_1, *and* δ_2 *are the learning rates for* m_{ih}^c, σ_{ih}^c, m_{jh}^c, *and* σ_{jh}^c, *respectively.*

Proof. The proof is omitted due to lack of space.

3 Computer Simulations: Glass Data

3.1 Iris Data

The Iris data [15] is a common benchmark in the pattern recognition studies [13,12,14]. It has four features: x_1-sepal length, x_2-sepal width, x_3-petal length,

Table 1. Parameter values of initial fuzzy rule-based classifier

i	$\sigma_{i1}^{F,B}$	$m_{i1}^{F,B}$	$\sigma_{i2}^{F,B}$	$m_{i2}^{F,B}$	$\sigma_{i3}^{F,B}$	$m_{i3}^{F,B}$	$\sigma_{i4}^{F,B}$	$m_{i4}^{F,B}$
1	0.3525	5.0060	0.3791	3.4280	0.1737	1.4620	0.1054	0.2460
2	0.5173	5.9340	0.3161	2.7640	0.4713	4.2480	0.2003	1.3220
3	0.6359	6.5880	0.3225	2.9740	0.5519	5.5520	0.2747	2.0260

and x_4-petal width and consists of 150 feature vectors: 50 for each Iris sestosa, Iris versicolor, and Iris virginica. Iris sestosa is linearly separable from the others; the latter are not linearly separable from each other. To show the effectiveness of the proposed classifier, we provides a simulation for Iris data– all 150 feature vectors are selected as training data. Now, following the design procedure, the design of the proposed fuzzy rule-based classifier is given by following steps.

Step 1 Initialization. In the premise parts of (1), m_{ih}^F and σ_{ih}^F are simply identified by the mean and the standard deviation of the Iris data, which are shown in Table 1. In the consequent parts of (1), the mean vectors, the covariance matrices, and the prior probabilities are obtained as follows:

$$\mathbf{m}_1 = \begin{bmatrix} 5.0060 \\ 3.4280 \\ 1.4620 \\ 0.2460 \end{bmatrix}, \Sigma_1 = \begin{bmatrix} 0.1242 & 0 & 0 & 0 \\ 0 & 0.1437 & 0 & 0 \\ 0 & 0 & 0.0302 & 0 \\ 0 & 0 & 0 & 0.0111 \end{bmatrix}, P(C_1) = \frac{50}{150},$$

$$\mathbf{m}_2 = \begin{bmatrix} 5.9340 \\ 2.7640 \\ 4.2480 \\ 1.3220 \end{bmatrix}, \Sigma_2 = \begin{bmatrix} 0.2676 & 0 & 0 & 0 \\ 0 & 0.0999 & 0 & 0 \\ 0 & 0 & 0.2221 & 0 \\ 0 & 0 & 0 & 0.0401 \end{bmatrix}, P(C_2) = \frac{50}{150},$$

$$\mathbf{m}_3 = \begin{bmatrix} 6.5880 \\ 2.9740 \\ 5.5520 \\ 2.0260 \end{bmatrix}, \Sigma_3 = \begin{bmatrix} 0.4043 & 0 & 0 & 0 \\ 0 & 0.1040 & 0 & 0 \\ 0 & 0 & 0.3046 & 0 \\ 0 & 0 & 0 & 0.0754 \end{bmatrix}, P(C_3) = \frac{50}{150}$$

Then, the initial classifier is given by

R_1 : IF x_1 is A_{11} and x_2 is A_{12} and x_3 is A_{13} and x_4 is A_{14},
THEN $y_1 = d_1^B(\mathbf{x})$

R_2 : IF x_1 is A_{21} and x_2 is A_{22} and x_3 is A_{23} and x_4 is A_{24},
THEN $y_2 = d_2^B(\mathbf{x})$

R_3 : IF x_1 is A_{31} and x_2 is A_{32} and x_3 is A_{33} and x_4 is A_{34},
THEN $y_3 = d_3^B(\mathbf{x})$ (16)

where $\mathbf{x} = [x_1, x_2, x_3, x_4]^T$. The recognition rate of the initial classifier (16) is 95.33%.

Table 2. Correctness degrees for initial fuzzy sets

i	A_{i1}	A_{i2}	A_{i3}	A_{i4}
1	0.8930	0.8263	1	0.9999
2	0.7759	0.5486	0.9648	0.9606
3	0.8003	0.4822	0.9730	0.9598

Table 3. Correctness degrees for fuzzy sets after pruning

i	A_{i1}	A_{i2}	A_{i3}	A_{i4}
1	∅	∅	∅	1
2	∅	∅	0.9648	1
3	1	1	0.9730	∅

Step 2 Pruning. After several iterations using the proposed pruning algorithm, x_1, x_2, x_4 in R_1, x_1, x_2 in R_2, and x_4 in R_3 are effectively eliminated. Specifically, Table 2 and 3 show the correctness degrees of the initial classifier and the pruned classifier, respectively. Thus, the complexity of (16) rapidly decreases. Nevertheless, its recognition rate rises to 96.67%.

Step 3 Learning. Until the preceding step, the fuzzy rule-based classifier has the following misclassified feature vectors: $[5.9, 3.2, 4.8, 1.8]^T$, $[6.0, 2.7, 5.1, 1.6]^T$, and $[6.7, 3.0, 5.0, 1.7]^T$ in Iris versicolor, and $[6.0, 2.2, 5.0, 1.5]^T$ and $[4.9, 2.5, 4.5, 1.7]^T$ in Iris virginica. After several iterations using the learning rules in Theorem 1 and 2, the feature vectors $[5.9, 3.2, 4.8, 1.8]^T$ and $[6.0, 2.7, 5.1, 1.6]^T$ among the misclassified feature vectors is correctly categorized. After all, the final classifier is obtained as follows:

$$R_1 : \text{IF } x_4 \text{ is } A_{14},$$
$$\text{THEN } y_1 = d_1^B(x_4)$$
$$R_2 : \text{IF } x_3 \text{ is } A_{23} \text{ and } x_4 \text{ is } A_{24},$$
$$\text{THEN } y_2 = d_2^B(x_3, x_4)$$
$$R_3 : \text{IF } x_1 \text{ is } A_{31} \text{ and } x_2 \text{ is } A_{32} \text{ and } x_3 \text{ is } A_{33},$$
$$\text{THEN } y_3 = d_3^B(x_1, x_2, x_3) \quad (17)$$

where the tuned parameter values are given in Table 4 and 5. Therefore, the recognition rate of the proposed fuzzy rule-based classifier increases from

Table 4. Premise parameters of fuzzy rule-based classifier after learning

i	σ_{i1}^F	m_{i1}^F	σ_{i2}^F	m_{i2}^F	σ_{i3}^F	m_{i3}^F	σ_{i4}^F	m_{i4}^F
1	∅	∅	∅	∅	∅	∅	0.1737	0.2460
2	∅	∅	∅	∅	0.5338	4.2857	0.2361	1.3384
3	0.6015	6.6231	0.3225	2.9740	0.5163	5.5778	∅	∅

Table 5. Consequent parameters of fuzzy rule-based classifier after learning

i	σ_{i1}^B	m_{i1}^B	σ_{i2}^B	m_{i2}^B	σ_{i3}^B	m_{i3}^B	σ_{i4}^B	m_{i4}^B
1	∅	∅	∅	∅	∅	∅	0.1054	0.2460
2	∅	∅	∅	∅	0.5175	4.2926	0.2312	1.3885
3	0.6375	6.6364	0.3616	3.0079	0.5434	5.6144	∅	∅

Table 6. Comparison of classification results on Iris data (150 training data)

Ref.	Number of rules	Number of premise fuzzy sets	Recognition rate
[12]	4	12	98%
[13]	3	6	96.67%
[14]	8	16	95.3%
Ours	3	6	98%

96.67% to 98%. Table 6 shows that the proposed classifier is superior to other fuzzy-rule-based classifiers in terms of the complexity and the recognition rate.

3.2 Glass Data

The glass data set [16] is based on the chemical analysis of glass splinters. Nine features are used to classify six types of glass: building windows float processed, building windows non float processed, vehicle windows float processed, containers, tableware, and headlamps. The features are refractive index, sodium, magnesium, aluminum, silicon, potassium, calcium, barium, and iron. The unit of measurement of all features but refractive index is weight percent in corresponding oxide.

We attempt to perform the 25 times pattern classification. One half of 214 feature vectors are randomly selected as the training data and the other half are used as the testing data. Table 7 contains the simulation results of the proposed classifier for each step of the design procedure. Although the average number of the features reduces from 9 to 5.61, the average recognition rates of the training and the testing set increase from 47.44% to 68.90% and from 39.48% to 58.28%, respectively. That definitely shows that the proposed design algorithm effectively provides the robustness for the overfitting and the decline of the dimensionality. Moreover, the classification performance of the proposed classifier is better than other classifiers as shown in Table 8.

Table 7. Classification results on glass data

Design procedure	Avg. number of features for each fuzzy rule	Avg. training recognition rate	Avg. testing recognition rate
Initial	9	47.44%	39.48%
Pruning	5.61	64.11%	53.23%
Learning	5.61	68.90%	58.28%

Table 8. Comparison of classification results on glass data

Ref.	Avg. testing recognition rate
[17]	50.95%
[7]	52.70%
Ours	58.28%

4 Conclusions

In this paper, a novel design approach to the fuzzy rule-based classifier has been proposed for the model complexity and the classification performance. Unlike other pruning methods based on the similarity analysis between two fuzzy sets, the proposed method utilizes the correctness degree, which is the major factor that improves the simplicity of the model. In addition, the problem of learning the premise parameters is formulated as minimizing the cost function, which is determined as squared-error between the classifier output for the correct class and the sum of the maximum output for the rest and a positive scalar. Finally, the computer simulations are given. The results show that the proposed fuzzy rule-based classifier has the low complexity, the very accurate classification ability, and the robustness for the overfitting in comparison with the conventional classifier. It indicates the great potential for reliable application of the pattern recognition.

References

1. Joo Y. H., Hwang H. S., Kim K. B., and Woo K. B.: Linguistic model identification for fuzzy system. Electron. Letter **31** (1995) 330-331
2. Joo Y. H., Hwang H. S., Kim K. B., and Woo K. B.: Fuzzy system modeling by fuzzy partition and GA hybrid schemes. Fuzzy Set and Syst. **86** (1997) 279-288
3. Duda R. O., Hart P. E., and Stork D. G.: Pattern classification. A wiley-interscience publishing company, inc. (2001)
4. Wu T. P. and Chen S. M.: A new method for constructing membership functions and fuzzy rules from training examples. IEEE Trans. Syst., Man, Cybern. B. **29** (1999) 25-40
5. Roubos H. and Setnes M.: Compact transparent fuzzy models and classifiers through iterative complexity reduction. IEEE Trans. Fuzzy Systems **9** (2001) 516-524
6. Setnes M. and Roubos H.: GA-fuzzy modeling and classification: complexity and performance. IEEE Trans. Fuzzy Systems **8** (2000) 509-522
7. Pal N. R. and Chakraborty S.: Fuzzy rule extraction from ID3-type decision trees for real data. IEEE Trans. Syst., Man, Cybern. B. **31** (2001) 745-754.
8. Paul S. and Kumar S.: Subsethood based adaptive linguistic networks for pattern classification. IEEE Trans. Syst., Man, Cybern. C. **33** (2003) 248-258
9. Ishibuchi H., Murata T. and Turksen I. B.: Single-objective and two-objective genetic algorithms for selecting linguistic rules for pattern classification problems. Fuzzy Sets Syst. **89**, (1997) 135-149

10. Abe S. and Thawonmas R.: A fuzzy classifier with ellipsoidal regions. IEEE Trans. Fuzzy Systems **5** (1997) 358-368
11. Thawonmas R. and Abe S.: A novel approach to feature selection based on analysis of class regions. IEEE Trans. Syst., Man, Cybern. B. **27** (1997) 196-207
12. Shi Y., Eberhart R., and Chen Y.: Implementation of evolutionary fuzzy systems. IEEE Trans. Fuzzy Systems **7** (1999) 109-119
13. Li R., Mukaidono M., and Turksen I. B.: A fuzzy neural network for pattern classification and feature selection. Fuzzy Sets Syst. **130** (2002) 101-108
14. Hong T. P. and chen J. B.: Processing individual fuzzy attributes for fuzzy rule induction. Fuzzy Sets Syst. **112** (2000) 127-140
15. Fisher R. A.: The use of multiple measurements in taxonomic problems. Ann. Eugenics. **7** (1936) 179-188
16. Merz C. J. and Murphy P. M.: UCI repository of machine learning databases. http://www.ics.uci.edu/ mlearn/MLRepository.html, Irvine, Dept. of Information and Computer Science, Univ. of California, Irvine (1996)
17. Castellano G., Fanelli A. M., and Mencar C.: An empirical risk functional to improve learning in a neuro-fuzzy Classifier. IEEE Trans. Syst., Man, Cybern. B. **34** (2004) 725-731

Fuzzy Sets Theory Based Region Merging for Robust Image Segmentation

Hongwei Zhu[1] and Otman Basir[2]

[1] Pattern Analysis and Machine Intelligence Research Group,
University of Waterloo, Waterloo, Ontario, Canada
h4zhu@pami.uwaterloo.ca

[2] Department of Electrical and Computer Engineering,
University of Waterloo, Waterloo, Ontario, Canada
obasir@uwaterloo.ca

Abstract. A fuzzy set theory based region merging approach is presented to tackle the issue of oversegmentation from the watershed algorithm, for achieving robust image segmentation. A novel hybrid similarity measure is proposed as the merging criterion, based on the region-based similarity and the edge-based similarity. Both similarities are obtained using the fuzzy set theory. To adaptively adjust the influential degree of each similarity to region merging, a simple but effective weighting scheme is employed with the weight varying as region merging proceeds. The proposed approach has been applied to various images, including gray-scale images and color images. Experimental results have demonstrated that the proposed approach produces quite robust segmentations.

1 Introduction

Many image segmentation methods have been intensively investigated, including edge-based methods, region-based methods, and hybrid methods [1,2]. Boundary-based methods operate on gradient images, and segment images using the edge information. Region-based methods segment images by grouping neighboring pixels or regions according to certain similarity (homogeneity) criteria, defined on such features as gray level, color information, or wavelet coefficients. Hybrid-based methods take in account multiple futures for image segmentation, such as edge information and region information. The watershed algorithm presented in [3] represents a typical hybrid-based method for image segmentation. It begins with creating a gradient of the image to be segmented. Applying the immersion simulation approach to the gradient image produces an initial segmentation with each local minimum of the gradient corresponding to a region. As a result, oversegmentation occurs due to the high sensitivity of the watershed algorithm to image pixel intensity variations.

As far as watershed based image segmentation approaches are concerned, their segmentation performances highly rely on the region merging criteria adopted. In [1], a region dissimilarity function is derived from an optimization point of view, based on the piecewise least-square approximation. This function takes into account the sizes and the mean pixel intensities of two regions under consideration, and the two most similar regions having the minimum dissimilarity are merged first. In [4], a dissimilarity measure is calculated for two neighboring regions as the square of the difference between the summation of the pixel intensity mean, the second-order central moment, and the third-order central moment of each region. The region merging criterions proposed in [1] and [4] take into account only region information, without considering the edge information between neighboring regions.

A segmentation algorithm that integrates region and edge information is proposed in [6] for merging neighboring regions. A merit function is defined, which takes edge length in consideration, however it ignores the important information of pixel intensity and their spatial distribution. A region merging criterion is defined in [5] using Sugeno fuzzy integral, to take into account region based features (mean, and standard deviation of pixel intensities of a region, region size) and edge based features (gradient magnitude), and the expected importance of these features. It applications sometimes suffer the difficulty in obtaining appropriate importance values for the adopted features due to the lack of domain experts' knowledge. A hybrid dissimilarity measure is proposed in [2] for color image segmentation. This measure is defined as a constant weighted combination of two components: 1) the difference of mean values between two regions, in terms of the Hue component in the HSV color space, 2) the average gradient magnitude of pixels in the shared edge between the two regions. Both region information and edge information are exploited. However, they are utilized in a very rough manner. First, the region size is not considered as region sizes are demonstrated quite important to eliminate oversegmentaions in watershed based approaches [1]. Secondly, only edge's strength is utilized while its length is overlooked. Thirdly, the use of constant weights in the hybrid dissimilarity measure is not suitable since it is not able to be adaptable to the region merging process for robust segmentation performance.

To mitigate the above drawbacks, we propose a new hybrid similarity measure based on the fuzzy feature representation of regions and edges using the fuzzy set theory. The hybrid similarity is a weighted combination of two similarities: the region based similarity and the edge based similarity, with variable weights that are adapted to the region merging process. The region based similarity takes into account the mean pixel intensity of a region and its size. The edge-based similarity accounts for edge strength and edge length. Each region is formulated as a fuzzy set. The membership assumes an normalized Gaussian function with respect to the mean pixel intensity of the region and the region size. Therefore, the region-based similarity is obtained by calculating the similarity between two fuzzy sets. It is justified that, when the region-based similarity is used as the region merging criterion, it is equivalent to the optimization derived dissimilarity measure presented in [1]. Furthermore, a second fuzzy set is formulated to represent the edge between two regions. Each pixel on the edge constitutes a member of this fuzzy set with the membership value indicating the strength of edge (degree of edgeness [12]) at this pixel. Therefore, the edge-based similarity is calculated based on the the cardinality of this fuzzy set, which not only indicates the information of the number of edge pixels, but also reflects the degree of edgeness of the edge pixels. A weight that varies with the region number adaptively adjusts the influential degrees from the two similarities during the region merging process. To examine its effectiveness, the proposed approach has been carried out to segment various images, including gray-scale images and color images. Experimental results have demonstrated that the proposed approach produces robust and meaningful segmentations.

In this paper we focus on the design of the new hybrid similarity measure without detailing the watershed based image segmentation procedure. Readers are strongly referred to [7] for the gradient image generation using the Canny edge operator, [3] for watershed transform algorithm, and [8] for region merging using the region adjacency graph (RAG) .

The remainder of this paper is organized as follows. Section 2 specifies the fuzzy set based definition of region-based similarity from a global point of view. Section 3 details how to formulate local similarity based on the edge information of a region pair. A hybrid similarity measured is defined in Section 4. Experimental results on different images are presented in Section 6, and conclusions are finally provided in Section 5.

2 Region-Based Similarity: A Global View

Let IM_0 denote a grey level image or a color component image to be segmented, with its pixel intensity denoted by $g_0(x,y)$ at point (x,y). Let IM_1 be the corresponding gradient image with pixel intensity $g_1(x,y)$ at point (x,y). Suppose that the current image region under consideration, R_0, has K neighboring regions $R_k, k = 1, \cdots K$. Let E_k denote the edge that consists of all the edge pixels between R_0 and R_k. Based on the above notations, we propose a hybrid similarity measure which takes into account both region-based features and edge-based features.

As far as the similarity of two regions is concerned, in terms of their pixel intensities, variant relevant features have been utilized in different manners [1,4,2,9]. We adapt the similarity measure originally used for image retrieval in [9] to image segmentation and employ it to calculate the region-based similarity as follows. The underlying idea is that each region is formulated as a fuzzy set, e.g., $\tilde{R}_k$ for R_k, with an associated membership function $\mu_{\tilde{R}_k}$ defined on the pixel intensity g_0. For illustration simplicity, we assume that k can take number 0, so as R_k to represent also R_0. The similarity between two regions is thus obtained using fuzzy similarity measure. For effectiveness and efficiency of region-based similarity computation, the key is to design a good membership function. Using the aforementioned notations, we define two region-based features: 1) region size of R_k: $|R_k|$, which is the number of pixels in region R_k, and 2) mean pixel intensity of R_k:

$$u_k = \frac{1}{|R_k|} \sum_{(x,y) \in R_k} g_0(x,y) \qquad (1)$$

Then we represent region R_k as a fuzzy set $\tilde{R}_k$ with the membership function assuming the normalized Gaussian function:

$$\mu_{\tilde{R}_k}(g_0) = e^{-(g_0 - u_k)^2 / d^2} \quad (2)$$

where $d = \frac{C}{\sqrt{|R_k|}}$, and it is a parameter controlling the spread of the membership function. C is an image dependent constant.

Given two regions, say R_0 and R_k, their similarity can be calculated by applying fuzzy similarity measures to the fuzzy sets $\tilde{R}_0$ and $\tilde{R}_k$. There exist many different definitions of fuzzy similarity measures [10]. Similarly as [9], the employed fuzzy similarity measure is defined as:

$$S(\tilde{R}_0, \tilde{R}_k) = \max_{g_0 \in \Re} \min\{\mu_{\tilde{R}_0}(g_0), \mu_{\tilde{R}_k}(g_0)\} \quad (3)$$

Substituting the membership function in Eq. (2) to Eq. (3) yields:

$$S(\tilde{R}_0, \tilde{R}_k) = e^{-\frac{1}{C^2} \times \frac{|R_0| \times |R_k| \times (u_0 - u_k)^2}{\left(\sqrt{|R_0|} + \sqrt{|R_k|}\right)^2}} \quad (4)$$

which indicates that the similarity of two regions depends on the difference of their mean pixel intensities and their sizes. The closer their mean pixel intensities are, the more similar they are. The smaller their sizes are, the larger their similarity is. It is interesting to see that this property is quite similar as that provided by the dissimilarity in [1]:

$$\delta(R_0, R_k) = \frac{|R_0| \times |R_k| \times (u_0 - u_k)^2}{|R_0| + |R_k|} \quad (5)$$

which is resulted from the piecewise (region-wise) least-square approximation. This is not strange if we rewrite Eq. (4) and Eq. (5) as follows for comparison:

$$S(\tilde{R}_0, \tilde{R}_k) = e^{-\delta'(R_0, R_k)/C^2} \quad \text{with}$$

$$\delta'(R_0, R_k) = \frac{(u_0 - u_k)^2}{\left(1/\sqrt{|R_0|} + 1/\sqrt{|R_k|}\right)^2} \quad (6)$$

and

$$\delta(R_0, R_k) = \frac{(u_0 - u_k)^2}{1/|R_0| + 1/|R_k|} \quad (7)$$

Obviously, both $\delta(R_0, R_k)$ and $\delta'(R_0, R_k)$ react, in a same manner, to the variation of $|R_0|$, $|R_k|$ and $(u_0 - u_k)^2$, respectively. Therefore, applying the two measures in Eq. (6) and Eq. (7) to any given set of regions produces the same order of dissimilarity between region pairs though their dissimilarity values are different. As a result, if the similarity measure in Eq. (4) individually serves as the region merging criterion, no doubt it results in the same segmentation as applying the dissimilarity measure in Eq. (5). In this sense, the region-based similarity measure is able to achieve the piecewise (region-wise) least-square approximation.

Next, we determine parameter C in Eq. (4) as:

$$C = \sigma \times \sqrt{\frac{m}{16} \times \frac{n}{16}} \qquad (8)$$

where m and n denote the IM_0's size respectively in the x and y directions, and σ is the overall standard derivation of the pixel intensities in image IM_0.

3 Edge-Based Similarity: A Local View

The above region-based similarity may be considered as a global-level measure as two regions are compared since it takes into account their mean pixel intensities and their sizes only, without considering the local information around their edge. Actually, an edge is an indicative of dissimilarity among neighboring pixels, and it provides an important feature to distinguish discrete objects, from a local point of view. Given two regions R_0 and R_k, the edge between them can be represented as a set of pixels on the edge E_k:

$$E_k = \{p_1, \cdots, p_1, \cdots, p_{|E_k|}\} \qquad (9)$$

where p_i denotes the ith pixel on edge E_k for $i = 1, \cdots, |E_k|$. $|E_k|$ denotes the number of pixels at E_k. The crisp edge representation in Eq. (9) suggests that a pixel p_i belongs to the edge with a full degree of membership. This representation may not be so informative since it overlooks the difference between edge pixels, and is not able to distinguish the degrees of membership to the edge for edge pixels that have different gradient magnitudes. In addition, due to possible influence from noise, it is not reasonable to assign the edge pixels caused by noise with a full degree of membership. In other words, edge pixels should be treated differently in certain proper manner. In view of this, it is natural for us to replace the crisp representation in Eq. (9) using a fuzzy edge representation:

$$\tilde{E}_k = \left\{ \frac{\mu_{\tilde{E}_k}(g_1(p_i))}{p_i} \;\middle|\; i = 1, \cdots, |E_k| \right\} \qquad (10)$$

where $\mu_{\tilde{E}_k}(g_1(p_i))$ denote the degree of edgenss of edge pixel p_i [12], and it further assumes a specific trapezoidal membership function, i.e.,

$$\mu_{\tilde{E}_k}(g_1(p_i)) = T(g_1(p_i); \min g_1, \max g_1) \qquad (11)$$

where $\min g_1$ and $\max g_1$ denote the minimum and the maximum of gradient gradient magnitude in IM_1. The used trapezoidal membership function is defined as:

$$T(x; a, b) = \begin{cases} 0 & \text{if } x < a \\ (x-a)/(b-a) & \text{if } a \leq x < b \\ 1 & \text{if } x \geq b \end{cases} \qquad (12)$$

Having the fuzzy edge representation in Eq. (10), the cardinality (sigma count) of fuzzy set $\tilde{E}_k$ is defined as:

$$|\tilde{E}_k| = \sum_{i=1}^{|E_k|} \mu_{\tilde{E}_k}(g_1(p_i)) \qquad (13)$$

In the context of fuzzy image processing using fuzzy geometry of images, the cardinality $|\tilde{E}_k|$ can be viewed as the length of the fuzzy set along the edge direction [11]. Therefore, $|\tilde{E}_k|$ carries the information of number of edge pixels and the degrees of edgeness of pixels on the edge. It provides an overall indicative of the edge's strength. It is also natural to map the cardinality to the range from 0 to 1, to quantify the overall strongness of the edge using a function, say $M(|\tilde{E}_k|)$. Proper $M(|\tilde{E}_k|)$ would be a monotonically increasing function of cardinality $|\tilde{E}_k|$, as will be addressed later. Therefore, the second similarity measure for region merging may be defined as the degree of non-strongness of the edge:

$$S(E_k) = 1 - M(|\tilde{E}_k|) \qquad (14)$$

$S(E_k)$ denotes the non-separability of region R_0 and R_k from the local viewpoint of their edge. The cardinality $|\tilde{E}_k|$ in Eq. (13) can be further rewritten as:

$$|\tilde{E}_k| = \sum_{i=1}^{|E_k|} \mu_{\tilde{E}_k}(g_1(p_i)) = |E_k| \times \bar{\mu}_{\tilde{E}_k}, \text{ with } \bar{\mu}_{\tilde{E}_k} = \frac{1}{|E_k|} \sum_{i=1}^{|E_k|} \mu_{\tilde{E}_k}(g_1(p_i)) \qquad (15)$$

$\bar{\mu}_{\tilde{E}_k}$ is the average degree of edgeness for edge E_k. For computational efficiency, $\min g_1$ and $\max g_1$ in Eq. (11) are respectively set as the minimum and the maximum in gradient image IM_1. Thus, the computation of $\bar{\mu}_{\tilde{E}_k}$ can be simplified as calculating the membership value of the average gradient magnitude $\bar{g}_{1k}$ over all pixels on edge E_k, i.e., $\mu_{\tilde{E}_k}(\bar{g}_{1k})$, with

$$\bar{g}_{1k} = \frac{1}{|E_k|} \sum_{i=1}^{|E_k|} g_1(p_i) \qquad (16)$$

This is because the membership function of edgeness in Eq. (11) is guaranteed linear due to the use of $\min g_1$ and $\max g_1$ respectively being the minimum and the maximum in gradient image IM_1. Therefore, using the above strategy, the cardinality $|\tilde{E}_k|$ in Eq. (13) is a simple product of the number of edge pixels and the average edge gradient magnitude:

$$|\tilde{E}_k| = |E_k| \times \mu_{\tilde{E}_k}(\bar{g}_{1k}) \qquad (17)$$

It is time to define the mapping function $M(|\tilde{E}_k|)$ used in the edge-based similarity definition of Eq. (14). Associated with the simplified representation of cardinality in Eq. (17), we define $M(|\tilde{E}_k|)$ as:

$$M(|\tilde{E}_k|) = 0.5T(|E_k|; |E|_a, |E|_b) + 0.5T(\mu_{\tilde{E}_k}(\bar{g}_{1k}); \min g_1, \max g_1) \qquad (18)$$

where $|E|_a$ and $|E|_b$ are two parameters to adjust the effect of edge length to the overall edge strongness. They are respectively set as: $|E|_a = 1$ and $|E|_b$ takes the smaller one of m, n.

4 Hybrid Similarity: A Combined View

Having obtained the region-based similarity in Eq. (4) and the edge-based similarity in Eq. (14), a hybrid similarity measure is defined as:

$$S_h(R_0, R_k) = \lambda \times S(\tilde{R}_0, \tilde{R}_k) + (1 - \lambda) \times S(E_k) \qquad (19)$$

where $\lambda \in [0, 1]$), is a weight adjusting the impact factors of the region-based similarity and the edge-based similarity to the hybrid measure. If $\lambda = 1$, the hybrid measure is fully determined by region-based similarity, and the region merging degraded to the approach used in [1] due the the previously justified equivalence between the use of Eq. (6) and that of Eq. (7) in the region merging criteria. If $\lambda = 0$, the hybrid measure considers only the local-level edge based similarity, without taking into account global-level region-based information. Surely, it is important to determine λ properly.

λ in Eq. (19) is designed to automatically vary with the region merging process, so as to dynamically adjust the influential factors of the two similarities to the hybrid similarity (i.e., the region merging criterion). The adopted strategy is that, we decrease the contribution from the region-based similarity and increase that from the edge-based similarity as the region merging process proceeds. The rationale is based on the following underlying facts.

1) There exist many small regions in the initial oversegmentation due to the watershed algorithm. They are quite homogenous (in a relative sense compared to the later generated regions). In nature, each small region may represent an area with uniform or gradually changing pixel intensities. Most of them are likely to correspond to detailed pieces of some large objects. Therefore, it is suitable to utilize a region merging criterion to merge these small regions, which is derived from the piecewise least-square approximation.
2) When serving as the region merging criterion, the fuzzy set theory based region similarity measure in Eq. (4) has been justified to have the equivalent performance as the dissimilarity measure in [1], which is an optimal region merging criterion from the viewpoint of the piecewise least-square approximation. Therefore, it is reasonable to account more on the proposed region-based similarity measure at the beginning stage of merging.
3) Most of the edges existing between regions in the initial oversegmentation are short, which are likely to be resulted from image details or image noise. These edges may not represent real borders between distinct realistic objects. In other words, the edge information in the initial segmentation is not reliable and less representative. Certainly, we can not account on it too much at the beginning stage of region merging.
4) As region merging proceeds, region sizes becomes larger and edges become longer and stronger. From an asymptotic point of view, a larger size region has a larger variance of pixel intensity. That is, very likely a large region is no more as uniform as its small component regions. In this sense, in contrast to the early merging, it is less suitable to still apply the piecewise least-square approximation derived region merging criterion.

5) As region merging proceeds, in general edges between regions become longer due to the increasing region sizes, and edges become also stronger since weak edges between similar neighboring regions disappear due to earlier region merging. Therefore, with merging proceeding the edge information becomes more and more reliable and representative to real ones. It would be reasonable to increase the influential weight of edge-based similarity, accordingly.

Based on the above analysis, we empirically determine weight λ as:

$$\lambda = 0.1 + 0.9 \times T(r; 1, 100) \tag{20}$$

where r denotes the number of surviving regions during the region merging process. When there are more than 100 regions during region merging, i.e. $r \geq 100$, λ is 1, indicating only the region-based similarity is taken as the region merging criterion. As the region number is less than 100, weight λ decreases in a proportional way to the region number, together with a base weight 0.1, so that the influential fact of region-based similarity decreases with the region number while that of the edge-based similarity increases.

5 Experimental Results

The proposed image segmentation approach has been evaluated on different type images, including gray-scale images and color images. The reported images include: Cameraman (gray), Girl (color), Lena (color), head-MRI (gray), Airplane (F-16) (color).

In case of color images, the Y component (luminance) images in the YUV system are selected as the original intensity images. All images are scaled to the size of 256 × 256. The Canny edge operator is first applied to the intensity images to generate gradient images. Gradient images are further preprocessed by setting those gradients less than 3% of the maximum gradient magnitude to 0, so as to reduce the number of regions in the initial segmentation due to the use of the watershed algorithm. Based on the gradient images, the watershed algorithm is carried to form initial segmentation, and then the proposed approach is employed for final image segmentation through region merging.

For comparisons, two region merging criteria are employed, and they are respectively the hybrid similarity measure in Eq. (19) and the region-based similarity measure in Eq. (4). In these experiments, we set the target number of regions in the final segmentations as 19. Their results are shown in Fig. 1. For each image (on a row) column (a) presents the original image to be segmented; Columns (b) and (d) present the label maps of the segmentation with each region assigned an unique number (label), respectively using the hybrid measure and the region-based measure; Columns (c) (using the hybrid measure) and (e) (using the region-based measure) highlight the boundaries (edge maps) in the final segmentations.

Both region merging criteria result in good segmentations based on our visual judgement. Important edges are preserved while meaningful image details

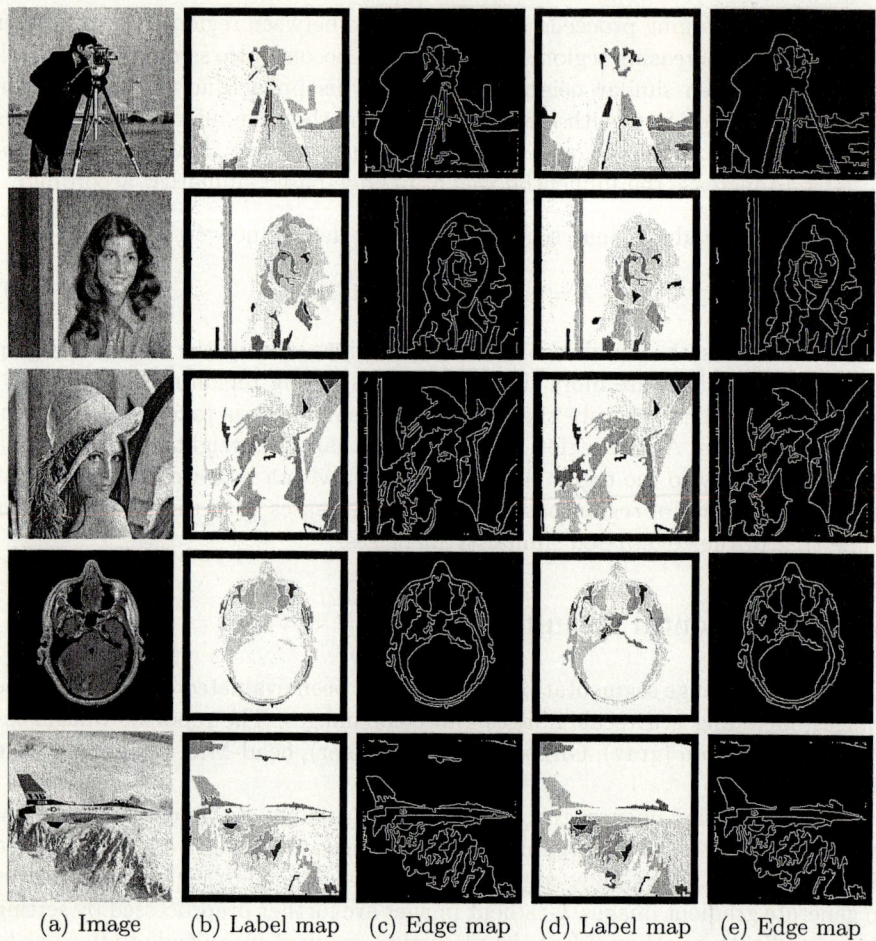

(a) Image   (b) Label map   (c) Edge map   (d) Label map   (e) Edge map

Fig. 1. Segmentation results: (b) (c) using the hybrid similarity, (d) (e) using region-based similarity

are also maintained (see facial features of Girl and Lena). However, the hybrid measure outperforms the region-based measure. For example, the far tall building with weak contrast to the sky in image Cameraman is successfully segmented using the hybrid similarity measure. For F16 image, both measures based methods are not able to segment the airplane as one individual region (mixed with the mountain in the front when using the hybrid measure, and mixed with the cloud in the upper middle part when using the region-based measure). However, based on human visual judgement, the hybrid measure produces more reasonable segmentation than the region-based measure since a more complete boundary of the plane is obtained in the column of the last row.

6 Conclusions

Based on the fuzzy set theory, we proposed an effective hybrid similarity measure for robust region merging to tackle the oversegmentation problem in the watershed algorithm based methods. Region-based features and edge-based features collaborate through the hybrid measure in an natural manner to the region merging process, for achieving good balance between the use of the global region information and the local edge information. Note that if the region-based similarity is individually used it results in the same segmentation as that using the region merging measure that is based on the the piecewise least-square approximation [1]. Experimental results on larger number of variant images have demonstrated the robustness and the effectiveness of the proposed image segmentation approach.

References

1. Haris, K., Efstratiadis, S.N., Maglaveras, N., Katsaggelo, A.K.: Hybrid image segmentation using watersheds and fast region merging. IEEE Trans. Image Processing. 7 (1998) 1684–1699
2. Navon, E., Miller, O., Averbuch, A.: Color image segmentation based on adaptive local thresholds. Image and Vision Computing. 23 (2005) 69–85
3. Vincent, L., Soille, P.: Watersheds in digital space: an efficient algorithm based on immersion simulations. IEEE Trans. PAMI. 13 (1991) 583–598
4. Kim, J.B., Kim, H.J.: Multiresolution-based watersheds for efficient image segmentation. Pattern Recognition Letters. 24 (2003) 473–488
5. Zhu, H., Basir, O:, Karray, F.: Fuzzy integral based region merging for watershed image segmentation. Proc. 10th IEEE Int. Conf. on Fuzzy Systems. 1 (2001) 27–30
6. Chu, C., Aggarwal, J.K.: The integration of image segmentation maps using region and edge information. IEEE Trans. PAMI. 15 (1993) 1241–1252
7. Canny, J.: A computational approach to edge detection. IEEE Trans. PAMI, 8 (1986) 679–698
8. Ballard, D., Brown, C.: Computer Vision. Englewood Cliffs, NJ: Prentice-Hall. (1982)
9. Chen, Y., Wang, J.Z.: A region-based fuzzy feature matching approach to content-based image retrieval. IEEE Trans. PAMI. 24 (2002) 1252–1267
10. Van der Weken, D., Nachtegael, M., Kerre, E.E.: Using similarity and homogeneity for the comparison of images. Image and Vision Computing. 22 (2004) 695–702
11. Pal, S.K., Ghosh, A.: Index of area coverage of fuzzy image subsets and object extraction. Pattern Recognition Letters. 11 (1990) 831–841
12. Tizhoosh, H.R.: Fuzzy Image Processing: Introduction in Theory and Practice. Springer-Verlag, Gemany. (1997)

A New Interactive Segmentation Scheme Based on Fuzzy Affinity and Live-Wire

Huiguang He, Jie Tian[1], Yao Lin, and Ke Lu

Medical Image Processing Group, Key Laboratory of Complex Systems and
Intelligence Science, Chinese Academy of Science, P.O.Box 2728, Beijing, 100080, China
huiguang.he@mail.ia.ac.cn
tian@doctor.com

Abstract. In this paper we report the combination of the Live-Wire method with the region growing algorithm based on fuzzy affinity. First, we employed anisotropic diffusion filter to process the images which smoothed the images while keeping the edge, and then we confined the possible boundary in applying the Live-Wire method to the over-segmentation found by the region growing algorithm. The speed and the reliability of the segmentation of the Live-Wire method are greatly improved by such combination. This method has been used for CT and MR image segmentation. The results confirmed that our method is practical and accurate in the medical image segmentation.

1 Introduction

Image segmentation is one of the most challenging problems in medical image analysis and computer vision. Many techniques are developed to fulfill satisfactory segmentation results, but there is no unified approach, yet which is suitable to all kinds of images [1]. Usually, the segmentation methods are divided into two classes: automatic method and interactive method. Automatic methods [2][3] have no user intervention, and the complete success cannot be always guaranteed. Interactive methods [4][5] range from totally manually drawing of the object boundary to the detection of object boundaries with the minimal user assistance. The automatic methods are currently used in an application-specific and tailored fashion, and they fail to different image modalities. To make it work effectively in a repeated fashion on a large number of data sets often requires considerable research and development.

Since interactive methods provide the prior information in the segmentation process and it may improve the segmentation results, it attracts more and more attention. A X. Falcao etc [6] proposed a live-wire segmentation method. In his method, the image was considered as a directed graph with the pixels as the nodes and the bound-

---

[1] Corresponding author: Jie Tian; Telephone: 8610-62532105; Fax: 8610-62527995. This paper is supported by the Project for National Science Fund for Distinguished Young Scholars of China under Grant No. 60225008, 863 hi-tech research and development program of China No. 2004AA420060, the National Natural Science Foundation of China under Grant No. 60302016,Beijing Natural Science Fund under Grant No. 4051002, 4042024.

ary of nearby pixels as the sides connecting nodes. After assigning a cost value to every pixel edge, we can trace the boundary in a desired object by graph searching. This boundary is the shortest path between two boundary points appointed by the user and can be solved via Dynamic Programming (DP). There're two shortcomings in the traditional live-wire method. Firstly, the DP module found the shortest path from the appointed node to all the other nodes without any discrimination, which made it very slow. Secondly, the segmentation result relied too much on the cost function and parameters, and training was necessary before any segmentation, which made the operation very complex and baffled its practical application. In order to overcome these shortages, we combine the live-wire method with region growing based on fuzzy affinity method. Before the live wire method has been used, we first apply anisotropic diffusion filter to smooth the images while keeping the edges; second, we employ the region growing method based on fuzzy affinity to obtain an over-segmentation of the image. During the live-wire process, when DP module is performed, it is limited on the boundary of the regions decided by the over-segmentation. For this purpose, we need only remove the sides that are inside a region from the graph before applying DP to find the shortest path.

The merit of our method lies in three aspects. First, the searching scope is limited by the specific region, which is only one quarter of which used without over-segmentation, thus, the speed is much faster. Second, the segmentation accuracy can be improved since the shortest paths are bounded as the potential boundaries of desired objects. Finally, the reliance of the segmentation result on the cost function and parameters are greatly decreased and the training has no longer needed. Figure 1 is the main steps of our method.

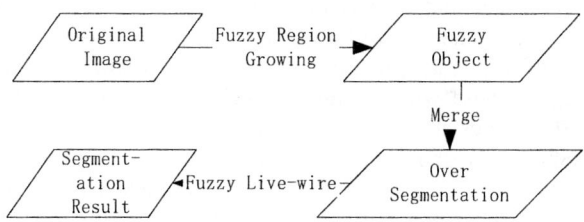

Fig. 1. The main steps of the method

This paper is organized as follows: Section 2 describes the anisotropic diffusion. Section 3 introduces the region growing method with fuzzy affinity. Section 4 demonstrates how to use improved live-wire method to extract the boundary from the over-segmentation. Finally, we present the implementation result and draw the conclusion.

2 Anistropic Diffusion

Due to the complex background and diversity of medical image, most of them have poorly defined object boundary, and nearby the desired boundary it may exist noise of strong feature. Therefore, pre-process is necessary.

Embedding the original image in a set of images derived from it, we can describe the series of images being processed as I(x,y,t). When t=0, I(x,y,0) represents the original image. Then the image filtering process may be achieved by convolving the original image with a function F(x,y,t) (see equation 1).

$$I(x,y,t) = I(x,y,0) * F(x,y,t) \tag{1}$$

Gaussian kernel $G(x,y;t)$ is a single variable function (variance t). Gaussian filter can smooth the images, but it can also distort the edge. So we need a special filter which can smooth the image while keeping the edge. Anisotropic diffusion proposed by Perona and Malik [7] can work well. The diffusion equation is

$$I_t = \frac{1}{\sqrt{1+|\nabla I|^2}}\left(\frac{1}{1+|\nabla I|^2}I_{\eta\eta} + I_{\xi\xi}\right) \tag{2}$$

Where $|\nabla I|$ is the gradient magnitude, and ξ denotes the contour direction and η stands for the gradient direction. From the equation 2, we can see that when $|\nabla I|$ is large it allows almost no smoothing in the gradient direction, while in the contour direction, it always executes maximal smoothing. In this manner, the image is smoothed without destroying the edges.

3 Region Growing with Fuzzy Affinity

The concept of fuzzy affinity and fuzzy connectedness was first presented by Rosenfeld [8] in 1979. J. K. Udupa [9] extended the theory and presented a framework of fuzzy connected object definition. The fuzzy connectedness describes the spatial and topological relationship between every pair of image elements. A local fuzzy relation as affinity is assigned to every pair(c,d) which reflects the strength of local hanging togetherness with a value in [0,1]. Fuzzy connectedness is a global fuzzy relation, which takes all paths from c to d.

Unlike gradient and threshold-based methods, which are not robust to noise within the object boundaries, our method combines both the intensity and gradient information to better estimate those boundaries. Our fuzzy affinity is defined as follows:

$$\mu_k(c,d) = \omega_1 f_1 + \omega_2 f_2 \tag{3}$$

where $f_1 = e^{-\frac{1}{2}(\frac{f(c)-f(d)-\mu_1}{\delta_1})^2}$, $f_2 = e^{-\frac{1}{2}(\frac{(f(c)+f(d))/2-\mu_2}{\delta_2})^2}$ and $\omega_1+\omega_2=1, \mu_1, \delta_1, \mu_2, \delta_2$ is the respective mean and the standard deviation of the gradient feature and intensity feature $f(c)$.

To simplify the computation, we choose 4-neighbor to describe $\mu_\alpha(c,d)$,

$$\mu_\alpha(c,d) = \begin{cases} 1, & \text{if } c=d, \text{ or } c \text{ and } d \text{ are } 4-\text{neighbor} \\ 0, & \text{otherwise} \end{cases} \tag{4}$$

The fuzzy affinity describes the probability that two points belong to the same object. From this definition, it is clear that if the density difference between c and d is smaller and the average density of c and d is closer to the mean intensity of the interest object, then the value is bigger. This definition is reasonable by intuitive understanding.

To further reduce the computation time, we use region growing method to get the result, and it is easy to prove that region growing based fuzzy affinity can get the same result as DP. The following is the pseudo code of algorithm Fuzzy-Region-Growing(FRG):

Input: Image C, fuzzy affinity threshold x
  Output: Image after being segmented by FRG
  Auxiliary Data Structure:
  A dualistic flag list L={flag, p}, flag denotes if the point p has been visited or not (1 is yes, 0 is not).
  A queue Q stores the location of the candidate point.
  Begin
  0.   Initialize L (let flag=0), Q (let Q is empty);
  Repeat 1-3 steps until the flag of each pixel in L is 1
  1.   Select a point p from L if L[p].flag=0, let L[p].flag=1;
  2.   For each neighbor c of p, and calculate (p, c)
     if (p, c)>x, then
       put c in Q, and let L[c].flag=1;
     endif
    Endfor
    While Q is not empty do:
  3.   Remove a pixel c from Q,
  4.   For each neighbor d of c, and calculate (c, d)
     if (c, d)>x, then
       put d to Q, and Let L[d].flag=1
     endif
    Endfor
  End while
  5. Output region R(o), R(o) include all the pixel which have been put in Q in the step 1-4
  End

4 Improved Live-Wire Method

We get the over-segmentation of the image after region growing based on *fuzzy affinity*. Since over-segmentation may separate one object to two or more parts, we should perform merging process for over-segmentation. The popular Fisher's test has been used to decide whether two adjacent regions can be merged. Suppose there are two adjacent regions R_1 and R_2, where $n_1, n_2, \mu_1, \mu_2, \sigma_1, \sigma_2$ are the size, sample means, and sample variances of R_1 and R_2, respectively, the squared Fisher distance is defined as follows:

$$FD^2 = \frac{(n_1+n_2)(\mu_1-\mu_2)^2}{n_1 \sigma_1^2 + n_2 \sigma_2^2} \quad (5)$$

If this statistic is smaller than a certain threshold, then the regions are merged. Then we will use improved live-wire method to extract the boundary. Our improved live-wire method is different from the original [7] in that we use fuzzy affinity as the cost function, and the searching scope is limited by the over-segmentation. First, the user selects an initial point in the boundary of the desired object, and then the other points will be specified by the interactive method. While the user defines a point in the boundary, the computer will calculate the shortest path from this node to all the other nodes in the graph, and the path will be accepted only when it is in the region computed by *FRG* algorithm. As the user moves mouse, the system will display the shortest path from previous user-defined node to the current mouse position in real time by searching graph. If this path adequately describes the boundary of the desired object, then the user can confirm that the path is a valid boundary for the desired object. Then make the current mouse position as new starting point. A complete 2D boundary is specified via a set of live-wire segments in this fashion. Figure 2 is just the example

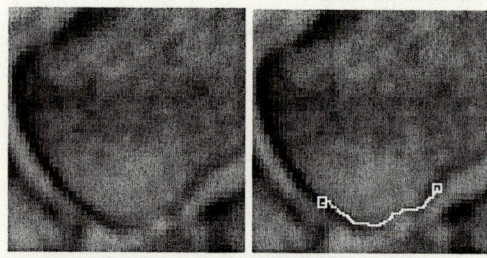

Fig. 2. The left is the CT image, and the right is the part of boundary calculated by live-wire

5 Result and Evaluation

5.1 Experimental Results

The algorithm was implemented with C++, and we run all the experiment on the 3DMED medical image processing and analyzing system developed by Medical Image Processing Group, Institute of Automation, Chinese Academy of Sciences [10]. Figure 3 shows the segmentation of the MR image. Figure 4 shows the CT images of knee joint segmentation. Figure 5 shows the tumor segmentation. These experiment results represent that our algorithm is efficient not only for CT image, but also for MR images.

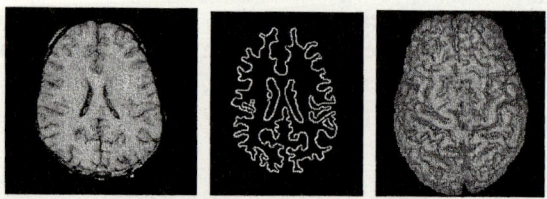

Fig. 3. The left shows the original MR image of patient's brain, and the middle is the satisfied boundary of the white matter. The right is the 3D model of the white mater reconstructed by 3Dmed.

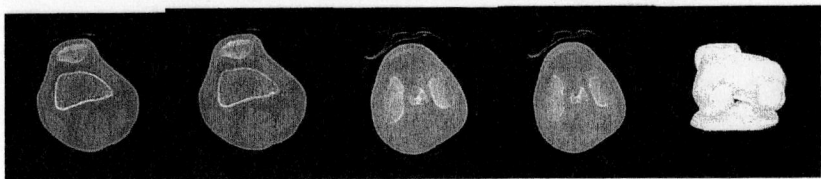

Fig. 4. From the left to the right, the first and third images are the CT images of knee joint. The second and fourth images are the boundary of the bone of the first and third images respectively. The fifth image is the 3D bone model of the knee joint.

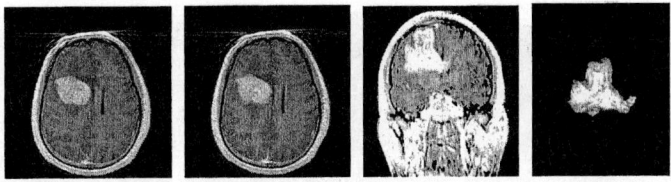

Fig. 5. From the left to the right, the first image is the original MR image of the head in the coronary direction, and we can see that there is a tumor in it. The second image is the boundary of the tumor. The third image is the image in the axial direction reconstructed by the virtual cutting technology, and 3D model of the tumor reconstructed by 3DMED is shown as the forth image.

5.2 Quantitative Evaluation

To evaluate the segmentation results quantitatively and objectively, the International Brain Segmentation Repository (IBSR) database have been used to test our algorithm. Both the real brain MR image and the corresponding ground truth segmentation can be freely downloaded from the website http://neuro-www.mgh.harvard.edu/cma/ibsr. Since the output of the edge detection is a binary image, the Baddeley's Δ_w^p [11] has been selected as the distance measure for comparing such binary "boundary pixel images" according to Thomas C.M. Leey's evaluation method [12]. Let X be a grid with N pixels, and let $A \subset X$ and $B \subset X$ be the set of all "black pixels" of a true and fitted binary image respectively.

$$\Delta_w^p(A, B) = \left[\frac{1}{N} \sum | w(d(x, A)) - w(d(x, B)) |^p \right]^{\frac{1}{p}} \quad (6)$$

Where d(x, A) is the smallest distance from x to A, w(t)=mint(t, c) is a threshold function, and p and c are parameters provided by the user. We follow Baddeley and set p=2 and c=5.

We use the dataset 788_6_m from IBSR to demonstrate the segmentation results. It contains 60 contiguous 3.0 mm slices, which scanned by 1.5 Tesla General Electric Signa. We give the segmentation result of the white matter in some slices from this

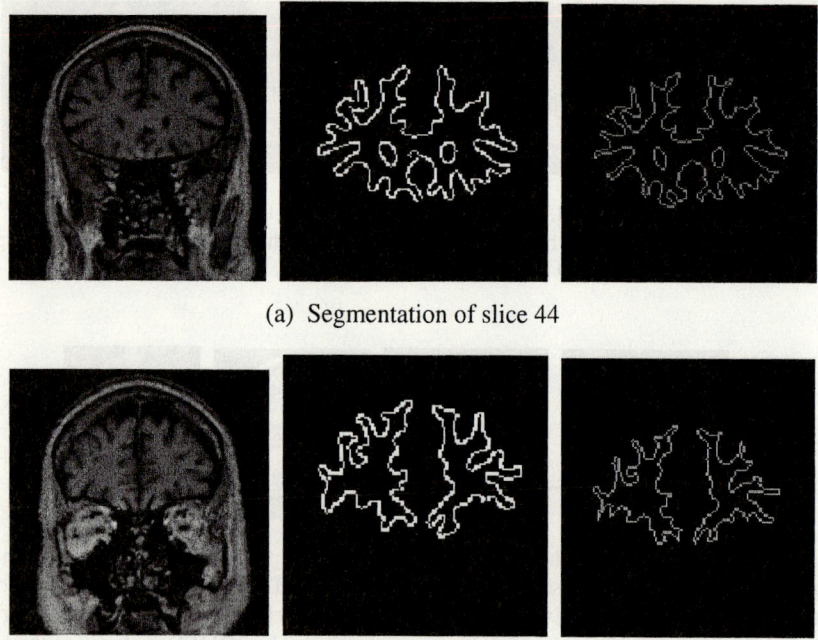

(a) Segmentation of slice 44

(b) Segmentation of slice 45

Fig. 6. Segmentation of the brain white matter in MR images from IBSR

Table 1. The Baddeley Distance Measure

	Slice 43	Slice 44	Slice 45	Slice 46	Slice 47	Slice 48	Slice 49
Distance measure	0.357	0.267	0.268	0.275	0.258	0.289	0.268

dataset (Figure 6). The left column is the original image; the middle one is the segmentation result of our method; the right one is the ground truth segmentation. The Baddeley's distance between the segmentation result of our method and the ground truth segmentation has been computed and the evaluation is enumerated in Table 1.

6 Conclusion

Interactive segmentation is one of the most important algorithms in medical image processing. The proposed method is different from other work on live-wire in three aspects; all of them are focused on time-saving with the improved accuracy of segmentation. (1) Our fuzzy affinity definition combined the intensity and the gradient of the image, which makes it robust to the noise. (2) By confining the searching scope to the FRG computed over-segmentation, we change the method to seek the shortest path from the point defined by the user to all the other points in the image, and don't allow the path to go through the inside of the object, thus, the scope of DP is only one quarter of the original. (3) Our method guarantees that the shortest path found by DP is the potential boundary. However there are areas of potential improvements, such as changing the definition of fuzzy affinity or cost function so that we can minimize the user involvement and computing time while maintaining the result of segmentation.

References

1. James S.Duncan, and Nicholas Ayache, "Medical Image Analysis: Progress over Two Decades and the Challenges Ahead", IEEE Trans. on PAMI, 2000, 22: 85-105
2. Michael Kaus, Simon K. Warfield, Arya Nabavi, Peter M. Black, Ferenc A. Jolesz, and Ron Kikinis. Automated segmentation of MRI of brain tumors. Radiology, 2001, 218:586-591.
3. Jui-Cheng Yen, Fu-Juay Chang and Shyang Chang, "A New Criterion for Automatic Multilevel Thresholding", IEEE Trans. on Image Processing, 1995, 4: 370-377
4. Falcao, A.X., Bergo, F.P.G., Interactive volume segmentation with differential image foresting transforms IEEE Transactions on Medical Imaging, 2004, 23:1100 – 1108
5. Carl-Fredrik Westin, Liana M. Lorigo, Olivier Faugeras, W.Eric L.Grimson, Steven Dawson, Alexander Norbash, and Ron Kikinis. Segmentation by adaptive geodesic active contours. MICCAI(Medical Image Computing and Computer Assisted Intervention) , 2000, 266-275.
6. P. Perona and J. Malik, Scale space and edge detection using anisotropic diffusion, IEEE Transactions on Pattern Analysis and Machine Intelligence, 1997,12:629–639.
7. X. Falcao, J. K. Udupa, S. Samarasekera and Shoba Sharma, "User-steered Image Segmentation Paradigms: Live Wire and Live Lane", Graphic models and Image Processing, 1998, 60:233-260
8. Azrier Rosenfeld, "Fuzzy Digital Topology", Information and Control, 1979, 40:76-87
9. Jayaram K. Udupa, Supun Samarasekera, "Fuzzy Connectedness and Object Definition: Theory, Algorithms, and Applications in Image Segmentation", Graphical Model and Image Processing, 1996, 58:246-261
10. Jie Tian, Huiguang He, Mingchang Zhao, Integrated 3D medical image processing and analysis system, Proc. SPIE, 2003, 4958:284-293
11. A.J. Baddeley, Errors in binary images and an Lp version of the Hausdorff metric, Nieuw Archief voor Wiskunde, 1992, 157–183.
12. T.C.M. Leey, A minimum description length based image segmentation procedure and its comparison with a crossvalidation based segmentation procedure, Department of Statistics, University of Chicago, 1997.

The Fuzzy Mega-cluster: Robustifying FCM by Scaling Down Memberships

Amit Banerjee and Rajesh N. Davé

Department of Mechanical Engineering, New Jersey Institute of Technology,
Newark, NJ 07032, USA
{ab2, dave}@njit.edu

Abstract. A new robust clustering scheme based on fuzzy c-means is proposed and the concept of a fuzzy mega-cluster is introduced in this paper. The fuzzy mega-cluster is conceptually similar to the noise cluster, designed to group outliers in a separate cluster. This proposed scheme, called the mega-clustering algorithm is shown to be robust against outliers. Another interesting property is its ability to distinguish between true outliers and non-outliers (vectors that are neither part of any particular cluster nor can be considered true noise). Robustness is achieved by scaling down the fuzzy memberships, as generated by FCM so that the *infamous* unity constraint of FCM is relaxed with the intensity of scaling differing across datum. The mega-clustering algorithm is tested on noisy data sets from literature and the results presented.

1 Introduction

Cluster analysis is a technique for grouping and finding substructure in data. The most common application of clustering methods is to partition a data set into clusters, where similar data vectors are assigned to the same cluster and dissimilar data vectors to different clusters. The immensely popular *k*-Means is a partitioning procedure that partitions data based on the minimization of a least squares type fitting functional. The fuzzy derivative of *k*-Means known as Fuzzy *c*-Means (FCM) [1], has an objective functional of the form,

$$J(X;U,v) = \sum_{i=1}^{c}\sum_{j=1}^{n} u_{ij}^{m} d^{2}(v_{i},x_{j}), \qquad (1)$$

where n is the number of data vectors, c is the number of clusters to be found, $u_{ij} \in [0,1]$ is the membership degree of the jth data vector x_j in the ith cluster, the ith cluster represented by the cluster prototype v_i, $m \in [1,\infty)$ is a weighting exponent called the fuzzifier and $d(v_i,x_j)$ is the distance of x_j from the cluster prototype v_i. The fixed-point iterative FCM algorithm (FCM-AO) guarantees a local minimum solution when $J(X;U,v)$ is minimized. In order to avoid the trivial solution, $u_{ij} = 0$, additional assumptions have to be made leading to probabilistic, possibilistic and noise clustering.

However, since FCM is based on a least squares functional, it is susceptible to outliers in the data. The performance of FCM is known to degrade drastically when the data set is noisy. This is similar to LS regression where the presence of a single outlier is enough to throw off the regression estimates. The need has therefore been to develop robust clustering algorithms within the framework of FCM (primarily because of FCM's simplistic iterative scheme and good convergence properties). The usual FCM minimization constraints are,

$$\sum_{i=1}^{c} u_{ij} = 1, \quad 0 \leq \sum_{j=1}^{n} u_{ij} \leq n, \quad i = 1,..,c \,; j = 1,..., n \qquad (2)$$

The relaxation of the equality constraint in (2) has been shown to robustify the resulting algorithm. This relaxation is usually accomplished by reformulating the objective functional in (1), in order to avoid the trivial solution $u_{ij} = 0$. The Possibilistic c-Means (PCM) algorithm [2] was developed to provide information on the relationship between vectors within a cluster. Instead of the usual probabilistic memberships as calculated by FCM, PCM provides an index that quantifies the *typicality* of a data vector as belonging to a cluster. This is also shown to impart a robust property to the procedure in the sense that noise points have less typicality in good clusters. Another effective clustering technique based on FCM is the Noise Clustering (NC) algorithm [3] which uses a conceptual class called the *noise cluster* to group together outliers in the data. All data vectors are assumed to be a constant distance, called the noise distance, away from the noise cluster. The presence of the noise cluster allows outliers to have arbitrarily small memberships in good clusters. Later modifications define a varying noise distance for every datum [4,5]. The Least Biased Fuzzy Clustering (LBFC) algorithm [6] partitions the data set by maximizing the total *fuzzy entropy* of each cluster, which in turn is a function of clustering memberships. The scaled LBFC clustering memberships are shown to be related to PCM typicalities and the resulting LBFC algorithm is robust against outliers. The Fuzzy Possibilistic c-Means (FPCM) algorithm [7] has an optimization functional which is a combination of probabilistic and possibilistic components. The algorithm uses two types of memberships, a probabilistic FCM type membership that measures the degree of sharing of a datum among the different clusters and a possibilistic component of membership that provides information on intra-cluster datum relationships. For an outlier, FPCM generates low-valued typicalities and like PCM, is a noise resistant procedure. The Credibilistic Fuzzy c-Means (CFCM) algorithm [8] uses datum *credibility* as the measure to delineate outliers from good datum in the data set. As opposed to typicality in PCM, credibility of a datum represents its typicality to the entire data set and not to any particular cluster. An outlier is shown to have a low credibility and hence is atypical to the data set. The Fuzzy Outlier Clustering (FC-O) algorithm [9] uses datum weights and a modified membership function which is inversely proportion to the datum weight; the outliers get assigned a large weight and hence have a low membership. There is another class of noise resistant clustering methods based on robust statistics, prominent among which are the Fuzzy c-Medians [10] that uses cluster median as the representative prototype and the Fuzzy Trimmed c-Prototypes [11] based on the robust least trimmed squares regression procedure. For a detailed review of robust fuzzy clustering methods and their relation to robust statistics, the reader is referred to [12].

The aforementioned techniques and algorithm have been shown to be effective in clustering noisy data but they are plagued with problems of their own. Strictly speaking PCM is not a clustering algorithm but rather a mode seeking algorithm [13], which in disguise makes it tolerant of noise. One needs to have a reasonably good estimate of cluster variances to start with, which might not be possible in all cases. The noise distance in NC is also user specified and clustering results could be sensitive to variations in the noise distance. FC-O also depends on user specified quantities like total datum weights and a weighting exponent in addition to the fuzzifier. LBFC suffers from the same anomaly as PCM; it often generates coincident clusters since the objective functional is linearly separable. The centroids generated by FPCM are often seriously affected by outliers as would be with FCM. The concept of data credibility although appealing, is fundamentally plagued with the logic of total credibility – according to the current formulation, while outliers have zero credibility, no datum can have full credibility (unity). Moreover all FCM based methods suffer from the dependency on proper initialization.

In this paper we propose an FCM based noise clustering procedure and introduce the concept of a mega-cluster. The mega-cluster is compared with Davé's noise cluster; they are similar in the sense that both are theoretical concepts and designed to cluster noise points together but are shown to be fundamentally different. The concept of memberships of data points in the mega-cluster is compared to the theory of data credibility. The FCM memberships are modified by scaling them down; the farther a datum is from a prototype, the more intense is the scaling. This scaling-down relaxes the unity constraint in (2) for good clusters. The motivation for this is derived from the Rival Checked FCM (RCFCM) [14] and Suppressed-FCM (S-FCM) [15] where selective memberships are scaled up and the rest scaled down in order to increase the convergence speed of the FCM algorithm.

2 The Concept of a Fuzzy Mega-cluster

FCM partitions the data set into overlapping clusters but in general works well with compact, well separated and spherical clusters. Outliers in the data influence the location of the prototypes; as a result, the centroids are pulled towards the outlier(s). At this point, we make a distinction between true outliers and non-conforming non-outliers (which in the course of this work would be referred to as non-outliers). While the former data vectors are noise and do not belong to any cluster in the data, non-outliers are data vectors that neither belong to any cluster in the data nor can be considered noise. In other words, non-outliers can be considered to be equally likely to belong to any cluster in the data set and because of this reason, such data vectors can not be assigned to any one particular cluster. The difference is clearly indicated in fig.1 and any clustering algorithm should have the power to treat such entities differently. Unfortunately FCM and almost all robust clustering algorithms (except noise clustering in our experience) fail to differentiate between true outliers and non-outliers; FCM would assign memberships of (0.5, 0.5) to both x' and x'' in fig. 1.

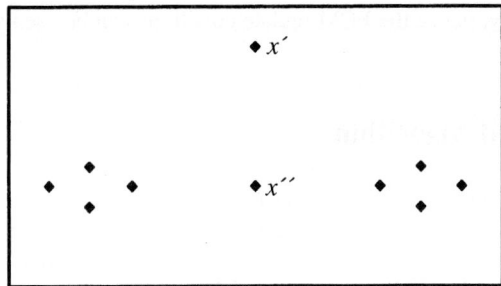

Fig. 1. True outliers and Non conforming non-outliers, x' is a true outlier, equally *unlikely* to belong to either cluster and x'' is a non-outlier, equally *likely* to lie in either cluster

We define a cluster called the mega-cluster which would view data vectors differently depending on how they belong to any good cluster in the data. Suppose in a two cluster data set, the datum x is a good representative of cluster *I*. In such a case, the membership of x in cluster *I* would be the largest, followed by its membership in the mega cluster and it would have the smallest membership in cluster *II*. On the other hand, if x' is a true outlier, its membership in the mega cluster would be largest, followed by relatively small memberships in the two good clusters *I* and *II*. This treatment is fundamentally different from the concepts of noise cluster and credibility of a data point vis-à-vis the entire data set. With the noise cluster, the membership of x would be the largest in cluster *I*, followed by its membership in cluster *II* and it would have a comparatively small membership in the noise cluster. However like the mega-cluster, x' would have a high degree of membership in the noise cluster, followed by low memberships in the two clusters. The concept of credibility as opposed to membership is defined as the degree of representativeness of a data point to the entire data set and as per definition, noise points have low data credibility and good data points have high credibility. Now if x'' is a non-outlier, its membership in the mega-cluster would be the highest followed by almost equal memberships in the two clusters; moreover if it is a symmetrically located non-outlier (as is x'' in fig. 1b) the sum of its memberships in the two clusters would equal its membership in the mega-cluster. This treatment allows for the subjective fact that such a non-outlier is equally likely to be considered part of either of the clusters but most likely considered *noise*. The mega-cluster can be thought of as a *super-group* encompassing the entire data set and views the data points differently depending on their belongingness in true clusters of the data. A further proposition would be that a mega-cluster membership is representative of both credibility and noise memberships (as well as true FCM memberships). A high mega-cluster membership would correspond to a high noise membership and low credibility and likewise a low mega-cluster membership would correspond to a low noise membership and a high credibility and thus a high true membership of the data point in one of the clusters. This cluster would not be detected by the standard FCM formulation. It is further assumed that while all data points have varying degrees of membership in the mega-cluster, they are all equally representative of the mega-cluster in the sense that distance of the data points from the mega-cluster is zero. Conceptually for the purposes of prototype calculations, the mega-cluster can be thought of as composed of n-point centers, each located exactly at the n data points. Furthermore, the memberships of a datum summed over the true clusters and the mega-cluster is unity, hence the FCM update equations can be used without any change of form.

3 The Proposed Algorithm

We seek to reduce the sensitivity of the FCM formulation towards noise by scaling down the memberships produced by FCM, in an inverse proportion to the cluster-datum distance. To speed up the convergence of FCM, two membership scaling procedures were proposed, viz. Rival Checked-FCM and the Suppressed-FCM. In every iteration of the FCM-AO scheme and for each datum, the two algorithms reward the largest membership by scaling it up by a constant factor. RCFCM then suppresses the second highest membership while SCFM suppresses all other memberships by a corresponding factor. Because of the scaling up, the two algorithms are found to be highly sensitive to noise. In fact in our experiments, RCFCM in most cases does not converge to a stable solution because it disturbs the sequence of memberships (as a result there is much oscillation between successive iterations). On the other hand, at low values of α (0.1-0.3), where SCFM behaves more like hard c-Means (HCM), we found that it generates singleton noise clusters most of the time and hence with appropriate modifications can be used as an outlier diagnostic tool.

The proposed algorithm is based on the following logic – what essentially distinguishes a good data point from an outlier is their distance (dissimilarity) from a representative prototype. This difference becomes muddled in the presence of noise in the data because of the centroid-pulling effect of the outliers. Hence for noisy data, if one could provide a mechanism which would accentuate this difference, one could conceptually reproduce results similar to FCM on a noise free data.

The proposed algorithm tries to underline this difference between good points and outliers by scaling down membership values of a data point across all clusters, in an inverse proportion to their distance from the cluster prototypes. Hence the effective membership of all points in true classes is less than one. This scaling down is more prominent for outliers which successively undergo a drastic reduction in memberships, which relates to a corresponding increase of its membership in the proposed conceptual mega-cluster.

The FCM-AO algorithm is presented below,

Initialize:
Randomly initialize centroid locations, $V^0 = v_i$. Let $k=0$; fix fuzzifier m and termination condition, $\varepsilon > 0$.
Iterate:
1. Compute prototype-data point distances, D^k as,

$$d_{ij} = \| v_i - x_j \|_A . \qquad (3)$$

2. Calculate memberships U^k using,

$$u_{ij} = \frac{1}{\sum_{k=1}^{c} \left[\frac{d_{ij}^2}{d_{kj}^2} \right]^{1/m-1}} . \qquad (4)$$

If there exists (r,j) such that $d_{rj} = 0$, then let $u_{rj} = 1$, $u_{ij} = 0$ for all $i \ne r$.

3. Calculate cluster centers, V^{k+1} using,

$$V_i = \frac{\sum_{j=1}^{n} u_{ij}^m x_j}{\sum_{j=1}^{n} u_{ij}^m}. \qquad (5)$$

4. If $\|V^{k+1} - V^k\| < \varepsilon$, then terminate. Else increment $k=k+1$ and return to step 1.

In the proposed algorithm (henceforth referred to as MC), the memberships as calculated by FCM are then modified depending on the datum-cluster center distance. For unusually large distances, the scaling is more intense and is achieved by scaling with respect to the maximum distance in the data set and is shown in (6a). This scaling repeatedly done on the outliers reduces their memberships rapidly as compared to scaling done on good datum. For reasonable datum-cluster center distances, the scaling is moderate and is done with respect to the sum of distances of the datum from all the c cluster centers as in (6b),

$$\text{intense scaling: } u_{ij} = \left[1 - \frac{\beta d_{ij}}{\max_{\substack{i=1,\ldots,c \\ j=1,\ldots,n}}(d_{ij})}\right] u_{ij}, \quad \beta \in [0,1]. \qquad (6a)$$

$$\text{moderate scaling: } u_{ij} = \left[1 - \frac{\beta d_{ij}}{\sum_{k=1}^{c} d_{kj}}\right] u_{ij}, \quad \beta \in [0,1]. \qquad (6b)$$

This modification is introduced in the FCM-AO as step 2-1, after the completion of FCM membership update in step 2. An if-else condition is used to decide whether to use a moderate or an intense scaling and the condition checks how unusually large a particular datum-cluster center distance, d_{ij}, is. The scaling is comparable in content to the credibility of a datum x_j as proposed by [8],

$$\Psi_j = 1 - \frac{(1-\theta)\alpha_j}{\max_{k=1,\ldots,n}(\alpha_k)}, \text{ where } \alpha_j = \min_{i=1,\ldots,c}(d_{ij}). \qquad (7)$$

As with credibility, $\beta=0$ reduces the formulation to FCM and at $\beta=1$ the formulation serves as a *complete noise reduction* algorithm. At levels between $\beta=1$ and $\beta=0$, the algorithm tries to balance between assigning a low membership to true outliers and assigning comparatively higher memberships to non-outliers, such as x'' in fig 1. If it is known that the data is noise free, a choice of $\beta=0$ would reproduce FCM results known to be fairly accurate in the absence of noise. In the presence of noise and general outliers, a judicious choice of β needs to be made; as inferred from the experiments presented in the next section, it is seen that any value of β in the range 1.0-0.7 generates good partitions in noisy data sets. This scaling down of memberships relaxes the unity constraint in (2); the resultant constraint is an inequality condition and the membership of a datum x_j in the mega-cluster is hence given by,

$$u_{MCj} = 1 - \sum_{i=1}^{c} u_{ij} \;.$$
(8)

4 Experiments and Results

We use the three data sets presented in [7] called X11, X^A12 and X^B14. X11 is a noise free data set consisting of 11 two-dimensional vectors while X^A12 and X^B14 are noisy versions of X11. A comparison of FCM, FPCM and CFCM on X11, X^A12 and X^B14 is presented in [8]. Here we compare performance of FCM as against the proposed algorithm on the same three data sets. The vector x_6 is a non-outlier with equal probability of belonging to the two underlying clusters in X11. The two well defined clusters lie on either side of x_6. The vectors x^A_{12} (in X^A12) and x^B_{12}, x^B_{13} and x^B_{14} (in X^B14) are true outliers. In our implementations, we have used c=2, m=2 and $\varepsilon = 0.001$ for both FCM and the proposed algorithm (for β=1).

For a prototype initialization of $v_1 = x_2\text{-}10^{-3}$ and $v_2 = x_{11}\text{-}10^{-3}$ in case of X^A12, we find that the proposed algorithm performs better than FCM, the cluster centers generated are shown in fig. 2a. The data vectors are shown by solid squares, FCM centroids are depicted by crosses and the MC centroids by small triangles. In fact the results are comparable to the ones generated by CFCM and certainly better than FPCM (see fig. 2. of [8]). FCM also fails to distinguish between x_6 and x^A_{12}. The proposed algorithm provides a higher membership for x_6 as compared to x^A_{12} in the two clusters. For the data set X^B14 shown in fig. 2b, the results provide a striking contrast; while FCM groups the three outliers in one cluster and the rest of the data set into another, the proposed algorithm finds the two real clusters. This is comparable to what FCM would generate on the noise-free data set X11. The proposed algorithm produced the same result over a wide range of distance factors for intense scaling (> 0.8 d_{max}, 0.5 d_{max} and 0.3 d_{max}) in case of X^A12 while there was a little difference in memberships for X^B14 when intense scaling was done for $d_{ij} > 0.8\ d_{max}$ as compared to the memberships obtained for $d_{ij} > 0.5\ d_{max}$. The results presented in Table I pertain to intense

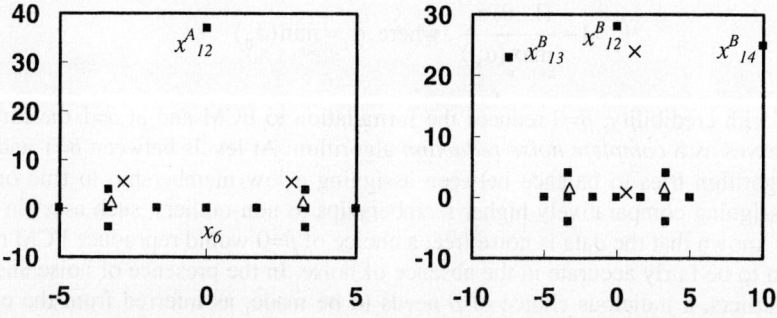

Fig. 2. Centroids generated by the proposed algorithm (v-MC) and FCM (v-FCM) on the (a)X^A12 and (b)X^B14 data set

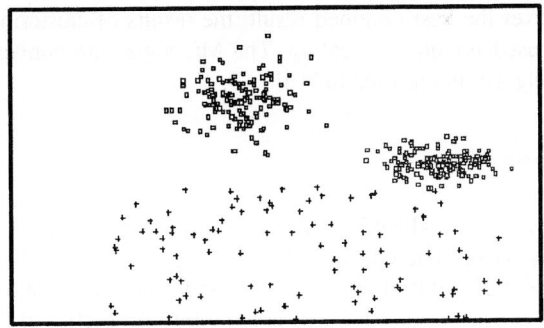

Fig. 3. The two-cluster normal data set with uniformly distributed noise

scaling done for $d_{ij} > 0.5\ d_{max}$ (the difference was however insignificant, affecting only third decimal places in the memberships). In both cases, the symmetrically located x_6 has a membership of about 0.5 in the mega-cluster and the true outliers had relatively large memberships in the mega-cluster compared to their memberships in the *good* clusters.

Table 1. Memberships for $X^B 14$ as generated by the proposed mega-clustering algorithm (compare with Table I, [8], p.1463)

Vector	Feature 1 X	Feature 2 Y	Memberships for $X^B 14$		
			u_{1j}	u_{2j}	u_{MCj}
x_1	-5.00	0.00	0.930136	0.001265	0.068599
x_2	-3.34	1.67	0.916823	0.001806	0.081371
x_3	-3.34	0.00	0.997538	0.000002	0.002460
x_4	-3.34	-1.67	0.865677	0.004842	0.129481
x_5	-1.67	0.00	0.794419	0.011815	0.193766
x_6	**0.00**	**0.00**	**0.241098**	**0.259063**	**0.499839**
x_7	1.67	0.00	0.009883	0.811054	0.179063
x_8	3.34	1.67	0.002283	0.906717	0.091000
x_9	3.34	0.00	0.000000	0.999322	0.000678
x_{10}	3.34	-1.67	0.003771	0.880852	0.115277
x_{11}	5.00	0.00	0.001369	0.927366	0.071265
x^B_{12}	**0.00**	**27.00**	**0.041163**	**0.037726**	**0.921111**
x^B_{13}	**-7.00**	**23.00**	**0.179158**	**0.094016**	**0.726826**
x^B_{14}	**10.00**	**25.00**	**0.000000**	**0.089727**	**0.910273**

The MC algorithm is also tested on a large synthetic data set with 25% noise. The data set is shown in fig. 3 and was first published in [16]. It consists of two normally distributed clusters of 150 2-D patterns each and 100 uniformly distributed noise points. For $c=2$, $\beta=1$ and intense scaling, the MC algorithm correctly identified 96 of the 100 noise points as true outliers, while FCM clustered the outliers together with one of the good cluster (the true cluster on the right). Identification of 96 out of 100 outliers is however the best obtained result; the results of clustering varying with the distance factor used for intense scaling. The MC algorithm converged in lesser time than the FLTS algorithm reported in [16].

5 Conclusions

An intuitive and easily realizable robust clustering scheme based on FCM has been presented in this paper. The concept of a fuzzy mega-cluster which is central to the proposed scheme is also introduced and discussed in detail and shown to be conceptually similar to Davé's noise cluster. While the robust properties of the proposed mega-clustering algorithm are investigated using test cases from literature, we also enunciate another interesting property of the algorithm – the power to distinguish true outliers from non-conformers. The sensitivity of FCM towards noise is reduced by scaling down the memberships and the excess membership is attributed to the mega-cluster. The scaling down of memberships in the good clusters is more intense for vectors which are perceived to have an unnaturally large distance from the prototypes and such a definition makes more intuitive sense when the data set is noisy.

Although the proposed algorithm is robust, it still suffers from typical FCM drawbacks such as dependence on fairly good initialization and the tendency to get trapped in local minima. The proposed scheme, like most robust clustering procedures, expects that the approximate amount of contamination in the data set is known beforehand. For $X^A 12$, where the contamination was less severe (~ 15%), it is found that intense scaling could be done for almost any distance factor and the results produced are identical but in the case of $X^B 14$ and the two-normal cluster set, where the contamination is almost close to 30%, the results differed when intense scaling was employed under different distance factors. This dependency needs to be further investigated with larger and more natural data sets.

References

1. Bezdek, J.C.: Pattern Recognition with Fuzzy Objective Function Algorithms, Plenum Press, New York (1981)
2. Krishnapuram, R., Keller, J.M.: A Possibilistic Approach to Clustering. IEEE Trans. Fuzzy Syst.. 1 (1993) 98-110
3. Davé, R.N.: Characterization and Detection of Noise in Clustering. Pattern Recog. Letters. 12 (1991) 657-664
4. Davé, R.N., Sen, S.: On Generalizing the Noise Clustering Algorithms. Invited paper, 7th IFSA World Congress. Prague (1997) 205-210
5. Davé, R.N., Sen, S: Noise Clustering Algorithm Revisited. In Proc. Biennial Workshop NAFIPS. Syracuse (1997) 199-204
6. Beni, G., Liu, X.: A Least Biased Fuzzy Clustering Method. IEEE Trans. Pattern Anal. Mach. Intell. 16 (1994) 954-960
7. Pal, N.R., Pal, K., Bezdek, J.C.: A Mixed c-Means Clustering Model. In Proc. 6th IEEE Conf. Fuzzy Syst. (1997) 11-21
8. Chintalapudi, K.K., Kam, M.: A Noise-Resistant Fuzzy c-Means Algorithm for Clustering. In Proc. 7th IEEE Conf. Fuzzy Syst. (1998) 1458-1463

9. Keller, A.: Fuzzy Clustering with Outliers. In Proc. 19th International Conference of NAFIPS, the North American Fuzzy Information Processing Society (2000) 143-147
10. Kersten, P.R.: Fuzzy Order Statistics and their Application to Fuzzy Clustering. IEEE Trans. Fuzzy Syst. 7 (1999) 708-712
11. Kim, J., Krishnapuram, R., Davé, R.N.: Application of the Least Trimmed Squares Technique to Prototype-based Clustering. Pattern Recog. Letters 17 (1996) 633-641
12. Davé, R.N., Krishnapuram, R.: Robust Clustering Methods: A Unified View. IEEE Trans. Fuzzy Syst. 5 (1997) 270-293
13. Krishnapuram, R., Keller, J.M.: The Possibilistic c-Means Algorithm: Insights and Recommendations. IEEE Trans. Fuzzy Syst. 4 (1996) 385-393
14. Wie, L.M., Xie, W.X.: Rival Checked Fuzzy c-Means Algorithm. Acta Electronica Sinica, 28 (2000) 63-66
15. Fan, J.L., Zhen, W.Z., Xie, W.X.: Supressed Fuzzy c-Means Clustering Algorithms. Pattern Recog. Letters 24 (2003) 1607-1612
16. Banerjee, A., Davé, R.N.: The Feasible Solution Algorithm for Fuzzy Least Trimmed Squares Clustering. In Proc. 23rd International Conference of NAFIPS, the North American Fuzzy Information Processing Society (2004) 222-227

Robust Kernel Fuzzy Clustering

Weiwei Du, Kohei Inoue, and Kiichi Urahama

Kyushu University, Fukuoka-shi, 815-8540 Japan

Abstract. We present a method for extracting arbitrarily shaped clusters buried in uniform noise data. The popular k-means algorithm is firstly fuzzified with addition of entropic terms to the objective function of data partitioning problem. This fuzzy clustering is then kernelized for adapting to the arbitrary shape of clusters. Finally, the Euclidean distance in this kernelized fuzzy clustering is modified to a robust one for avoiding the influence of noisy background data. This robust kernel fuzzy clustering method is shown to outperform every its predecessor: fuzzified k-means, robust fuzzified k-means and kernel fuzzified k-means algorithms.

1 Introduction

Practical issues in clustering are arbitrary shapes of clusters, robustness to noise or outlier data, estimation of the number of clusters, and so on. We adress in this paper to the former two issues assuming the number of clusters to be given.

The k-means algorithm is popularly used for clustering data. It is an iterative method of which solution depends on initial values. Its fuzzification such as the fuzzy c-means is known to improve the stability of the algorithm raising the chance of convergence to the global optimum solution. Hence the fuzzy clustering methods have been used for complex data such as image segmentation. However, the shape of clusters extracted with the k-means or its fuzzified version is restricted to some simple forms such as spheres or shells. Therefore the kernel techniques have been incorporated into them to enable them to adapt to arbitrary shapes of clusters[1,2]. These kernelized methods can deal with arbitrarily shaped clusters, however, the data used in the experiments there include no noise, hence their robustness is unknown. In fact, the kernel fuzzy clustering method is not robust to noise data as is shown with the experiment in this paper.

Another branch of the extension of clustering methods is their robustification. The extension of the distance measure from the popularly used Euclidean norm to a nonlinear one used in robust statistics makes the algorithm insensitive to noise data[3,4]. However, these robust clustering methods assume prescribed shapes of clusters as the same as the basic k-means or the fuzzy c-means algorithms, hence they cannot extract arbitrarily shaped clusters.

In summary, kernelized clustering algorithms can extract arbitrarily shaped clusters but are not robust to noise data, while nonlinear distance methods are robust to noise data but cannot extract arbitrarily shaped clusters.

We present, in this paper, a robust kernel fuzzy clustering algorithm which is a kernelized robust fuzzy clustering and can extract arbitrarily shaped clusters without the influence of noise data. The performance of the method is examined for 2-dimensional toy data and high dimensional face images

2 Algorithms

We consider extracting n clusters from m data d_i $(i = 1, ..., m)$. We use a kernel technique. Let the mapping of a datum d_i into a high dimensional space be $\phi_i = \phi(d_i)$. We adopt the Gaussian kernel and analyze the data in the high dimensional space without explicit computation of ϕ_i by exploiting the kernel trick: $\phi_i^T \phi_{i'} = e^{-\alpha \|d_i - d_{i'}\|^2}$ which can be computed by using only the feature vectors in the original low dimensional space. In the subsequent subsections, we start from the basic fuzzy clustering algorithm and extend it to a robust one and then kernelize it successively.

2.1 Fuzzification of k-Means with Entropic Regularization

If we add an entropic term to the objective function in the spherical clustering problem, we get the regularized objective function:

$$\sum_{i=1}^{m} \sum_{j=1}^{n} x_{ij} \|d_i - r_j\|^2 + \beta^{-1} \sum_{i=1}^{m} \sum_{j=1}^{n} x_{ij} \ln x_{ij} \quad (1)$$

The fuzzy memberships x_{ij} are the solution of the optimization problem which minimizes this objective function together with the centroid vectors r_j under the constraint $\sum_{j=1}^{n} x_{ij} = 1$ $(i = 1, ..., m)$. The x_{ij} is the membership of the datum i in the cluster j and r_j is the centroid of the cluster j. With the Lagrange multiplier method, we get from (1)

$$x_{ij} = y_{ij} (\sum_{j'=1}^{n} y_{ij'})^{-1} \quad (2)$$

$$r_j = \sum_{i=1}^{m} x_{ij} d_i / (\sum_{i=1}^{m} x_{ij})^{-1} \quad (3)$$

where $y_{ij} = e^{-\beta \|d_i - r_j\|^2}$. We set initially $x = [x_{ij}]$ $(i = 1, ..., m; j = 1, ..., n)$ randomly and compute $r = [r_1, ..., r_n]$ with eq.(3), and then we update x with eq.(2). We repeat this update of x until its convergence. The objective function (1) decreases monotonically along this iteration and stops at its local minimum. We call this the KMFE (k-means fuzzified with entropy). The entropic regularization is an alternative scheme of fuzzification to the popular fuzzy c-means where the Euclidean norm is powered instead of addition of regularization terms.

2.2 Robust KMFE

If we modify the squared Euclidean norm $\|d_i - r_j\|^2$ into $1 - e^{-\gamma\|d_i-r_j\|^2}$, then eq.(1) becomes

$$\sum_{i=1}^{m}\sum_{j=1}^{n} x_{ij}(1 - e^{-\gamma\|d_i-r_j\|^2}) + \beta^{-1}\sum_{i=1}^{m}\sum_{j=1}^{n} x_{ij}\ln x_{ij} \qquad (4)$$

This modification of the norm is popularly used in the robust statistics and has also been incorporated into the fuzzy c-means to make it robust[3]. The equations for x and r become in this case

$$x_{ij} = e^{\beta z_{ij}}(\sum_{j'=1}^{n} e^{\beta z_{ij'}})^{-1} \qquad (5)$$

$$r_j = \sum_{i=1}^{m} x_{ij}d_i z_{ij}(\sum_{i=1}^{m} x_{ij}z_{ij})^{-1} \qquad (6)$$

where $z_{ij} = e^{-\gamma\|d_i-r_j\|^2}$. We set initially x randomly and solve eq.(6) with an iterative method and update x by substituting the obtained r into eq.(5). This update is repeated until the convergence. We can prove the global convergence of this algorithm by using the Legendre transform in a similar way in [5], whose details are omitted here.

Now, before proceeding to the next subsection, we derive eq.(4) from a kernel in order to clarify the difference between this robust algorithm and the kernel algorithm in the next subsection. If we denote $\varphi_j = \phi(r_j)$, then there holds the equation $\|\phi_i - \varphi_j\|^2 = 2(1 - e^{-\alpha\|d_i-r_j\|^2})$, hence eq.(1) is transformed into eq.(4) through this kernelization. Zhang et al. called this technique the kernel method[6]. However, the centroids are unique points in the original space in their method, therefore it cannot treat arbitrarily shaped clusters. In fact, Zhang et al.[6] dealt with only ellipsoidal clusters without entangled ones.

2.3 Kernel KMFE

The KMFE is written in a high dimensional space as

$$\sum_{i=1}^{m}\sum_{j=1}^{n} x_{ij}\|\phi_i - r_j^\phi\|^2 + \beta^{-1}\sum_{i=1}^{m}\sum_{j=1}^{n} x_{ij}\ln x_{ij} \qquad (7)$$

where r_j^ϕ is the prototype of the cluster j in the high dimensional space, which is different to φ_j in the above subsection. From eq.(7), the equation for x becomes eq.(2) where $y_{ij} = e^{-\beta\|\phi_i - r_j^\phi\|^2}$ in this case, and the equation for r^ϕ becomes

$$r_j^\phi = \sum_{i=1}^{m} x_{ij}\phi_i(\sum_{i=1}^{m} x_{ij})^{-1} \qquad (8)$$

Substitution of this r_j^ϕ into $\|\phi_i - r_j^\phi\|$ leads to

$$\|\phi_i - r_j^\phi\|^2 = 1 - 2\sum_{i'=1}^m s_{ii'} x_{i'j} \left(\sum_{i'=1}^m x_{i'j}\right)^{-1} + \sum_{i'=1}^m \sum_{i''=1}^m x_{i'j} s_{i'i''} x_{i''j} \left(\sum_{i'=1}^m x_{i'j}\right)^{-2} \quad (9)$$

where $s_{ii'} = \phi_i^T \phi_{i'} = e^{-\alpha\|d_i - d_{i'}\|^2}$. Substituting eq.(9) into $y_{ij} = e^{-\beta\|\phi_i - r_j^\phi\|^2}$, we get the expression of y_{ij} as a function of x only. Thus, in this case, eq.(2) becomes the equation for x alone. We solve it with the iteration starting from random x. Contrary to the scheme in the previous subsection where the prototypes lie in the original space, they lie in the high dimensional space in this scheme. Therefore this scheme is capable to extract arbitrarily shaped clusters owing to improved separability of clusters in the high dimensional space.

2.4 Robust Kernel KMFE

Finally the same modification of the norm as that in eq.(1) into eq.(4) is applied to eq.(7), which leads to

$$\sum_{i=1}^m \sum_{j=1}^n x_{ij} (1 - e^{-\gamma\|\phi_i - r_j^\phi\|^2}) + \beta^{-1} \sum_{i=1}^m \sum_{j=1}^n x_{ij} \ln x_{ij} \quad (10)$$

In this case, if we denote $z_{ij} = e^{-\gamma\|\phi_i - r_j^\phi\|^2}$, then the equation for x becomes eq.(5) and the equation for r^ϕ becomes

$$r_j^\phi = \sum_{i=1}^m x_{ij} z_{ij} \phi_i \left(\sum_{i=1}^m x_{ij} z_{ij}\right)^{-1} \quad (11)$$

of which substitution into $\|\phi_i - r_j^\phi\|$ leads to

$$\|\phi_i - r_j^\phi\|^2 = 1 - 2\sum_{i'=1}^m s_{ii'} x_{i'j} z_{i'j} \left(\sum_{i'=1}^m x_{i'j} z_{i'j}\right)^{-1}$$
$$+ \sum_{i'=1}^m \sum_{i''=1}^m x_{i'j} z_{i'j} s_{i'i''} x_{i''j} z_{i''j} \left(\sum_{i'=1}^m x_{i'j} z_{i'j}\right)^{-2} \quad (12)$$

Substituting eq.(12) into $z_{ij} = e^{-\gamma\|\phi_i - r_j^\phi\|^2}$, we get the equation for z including x. Hence we set initially x randomly and substitute it into this equation for z and solve it by iteration, and then update x by substituting the obtained z into eq.(5). We repeat this update of x until its convergence. We can prove its convergence in a similar way to [5].

3 Experiments

We have experimented the above four schemes firstly for 2-dimensional toy data and then for high dimensional face image data.

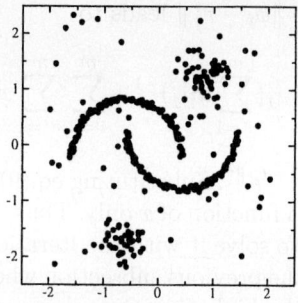

Fig. 1. Example of 2-dimensional data

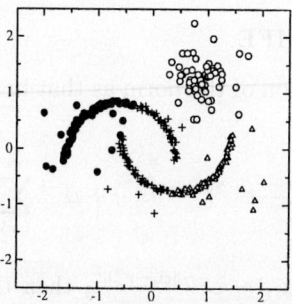

Fig. 2. Result of KMFE

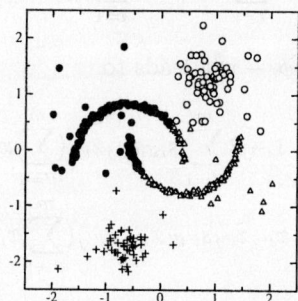

Fig. 3. Result of robust KMFE

3.1 Toy Data

The first data with which we have experimented is shown in Fig.1 which includes 312 inlier data composed of four clusters and 50 uniformly distributed noise data. The parameters were set as $\alpha = 10, \beta = 30, \gamma = 1$. A common initial value of x was used in its iterations for every scheme. A common threshold value of 0.6 was also used for defuzzifying x into crisp values 0 or 1.

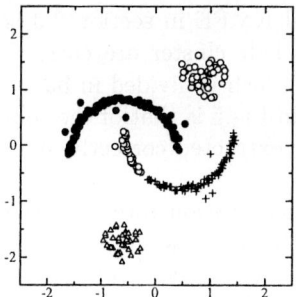

Fig. 4. Result of kernel KMFE

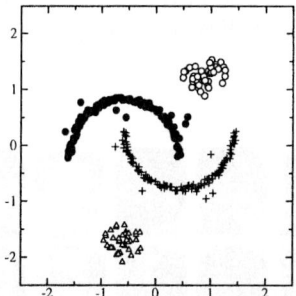

Fig. 5. Result of robust kernel KMFE

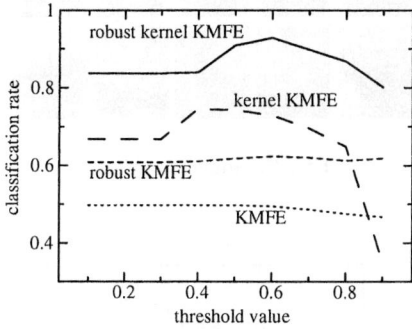

Fig. 6. Variation in classification rates with threshold value

The result of the KMFE in section 2.1 is illustrated in Fig.2. The spherical cluster at the bottom is lost and two entangled belt clusters at the center are divided into three clusters. These errors are attributed to the disturbance from noise data.

Next Fig.3 illustrates the result of the robust KMFE in section 2.2. Because the influence of noise data is weakened, no cluster is left out, but the central two belt clusters are still divided erroneously.

The result of the kernel KMFE in section 2.3 is shown in Fig.4 where two spherical clusters and one belt cluster are correctly extracted except for the remaining one belt cluster which is divided in half.

The final result shown in Fig.5 is that of the robust kernel KMFE in section 2.4. All of four clusters are extracted correctly and almost all of noise data are discarded.

The variation in the classification rate with the value of the threshold for defuzzification of the memberships is shown in Fig.6. This classification rate is defined as 1) for noise data, if every membership is below the threshold, then they are correctly classified, and 2) for inlier data, if the membership in the correct cluster is above the threshold, then their classification is correct. The dotted line in Fig.6 denotes the KMFE, finely broken line is the robust KMFE, coarsely broken line is the kernel KMFE, and the solid line represents the robust kernel KMFE which keeps the best performance in these methods.

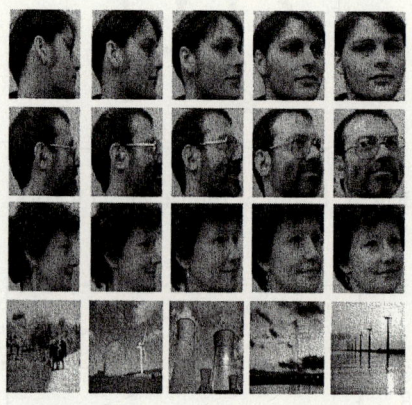

Fig. 7. Image samples

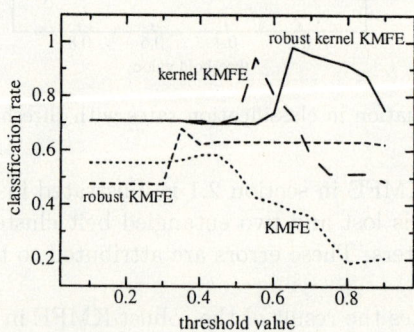

Fig. 8. Classification rates for face image data

3.2 Face Images

We have next experimented with image data sampled from the UMIST Face Database. Experimented data are face images of three persons from each of which 23, 26 and 24 images are sampled and 30 images of natural scenes are added as noise data. Some images in data are illustrated in Fig.7. Inlier data are composed of face images photographed from various viewpoints, hence they form elongated clusters. The feature vectors are arrays of gray levels of each pixel. Image size is 112×92 hence the dimensionality of feature vectors is 10304. Figure 8 illustrates the variation in the classification rate with the value of the threshold for this data. The threshold should be set to high values because noise data lie close to inlier data in this dataset. The robust kernel KMFE reveals high performance around such high threshold values, hence it is useful for these complex data.

4 Conclusion

We have presented a robust kernel fuzzy clustering method for extracting arbitrarily shaped clusters without the disturbance from noise data. Its performane has been verified with experiments for toy data and face images. Detailed theoretical analysis of the algorithm and development of the devise for determination of the number of clusters are the subjects of further study.

References

1. Girolami, M.: Mercer kernel-based clustering in feature space. IEEE Trans. Neural Netw. **13** (2002) 780–784
2. Kim, D.-W., Lee, K., Lee, D., Lee, K. H.: Evaluation of the performance of clustering algorithms in kernel-based feature space. Patt. Recog. **38** (2004) 607–611
3. Wu, K.-L., Yang, M.-S.: Alternative c-means clustering algorithms. Patt. Recog. **35** (2002) 2267–2278
4. Leski, J.: Towards a robust fuzzy clustering. Fuzzy Sets & Syst. **137** (2003) 215–233
5. Urahama, K.: Convergence of alternative c-means clustering algorithms. IEICE Trans. Inf. & Syst. **E86-D** (2003) 752–754
6. Zhang, D.-Q., Chen, S.-C.: Kernel-based fuzzy and possibilistic c-means clustering. Proc. ICANN'03 (2003) 122–125

Spatial Homogeneity-Based Fuzzy c-Means Algorithm for Image Segmentation

Bo-Yeong Kang[1], Dae-Won Kim[2], and Qing Li[1]

[1] School of Engineering, ICU, 103-6, Moonji-ro, Yuseong-gu, Daejeon, Korea
[2] Department of BioSystems, KAIST, Guseong-dong, Yuseong-gu, Daejeon, Korea
kby@icu.ac.kr

Abstract. A fuzzy c-means algorithm incorporating the notion of dominant colors and spatial homogeneity is proposed for the color clustering problem. The proposed algorithm extracts the most vivid and distinguishable colors, referred to as the dominant colors, and then used these colors as the initial centroids in the clustering calculations. This is achieved by introducing reference colors and defining a fuzzy membership model between a color point and each reference color. The objective function of the proposed algorithm incorporates the spatial homogeneity, which reflects the uniformity of a region. The homogeneity is quantified in terms of the variance and discontinuity of the spatial neighborhood around a color point. The effectiveness and reliability of the proposed method is demonstrated through various color clustering examples.

1 Introduction

The objective of color clustering is to divide a color set into c homogeneous color clusters. Color clustering is used in a variety of applications, such as color image segmentation and recognition. Color clustering is an inherently ambiguous task because color boundaries are often blurred due to the image acquisition process [1]. Fuzzy clustering models have proved a particularly promising solution to the color clustering problem. In fuzzy clustering, the uncertainty inherent in a system is preserved as long as possible before decisions are made. As a result, fuzzy clustering is less prone to falling into local optima than crisp clustering algorithms [1,2,3].

The most widely used fuzzy clustering algorithm is the FCM algorithm proposed by Bezdek [1]. This algorithm classifies a set of data points X into c homogeneous groups represented as fuzzy sets $F_1, F_2, ..., F_c$. The objective is to obtain the fuzzy c-partition $F = \{F_1, F_2, .., F_c\}$ for both an unlabeled data set $X = \{x_1, ..., x_n\}$ and the number of clusters c by minimizing the function J_m:

$$J_m(U, V : X) = \sum_{i=1}^{c} \sum_{j=1}^{n} (\mu_{F_i}(x_j))^m \|x_j - v_i\|^2 \qquad (1)$$

where $\mu_{F_i}(x_j)$ is the membership degree of data point x_j to the fuzzy cluster F_i, and is additionally an element of a $(c \times n)$ pattern matrix $U = [\mu_{F_i}(x_j)]$.

The i-th row of U, U_i, corresponds to the fuzzy cluster F_i. $V = (v_1, v_2, .., v_c)$ is a vector comprised of the centroids of the fuzzy clusters $F_1, F_2, ..., F_c$. Thus, a fuzzy partition can be denoted by the pair (U, V). $\|x_j - v_i\|$ denotes the Euclidean norm between x_j and v_i. The parameter m controls the fuzziness of membership of each datum. The goal is to iteratively improve a sequence of sets of fuzzy clusters $F(1), F(2), ..., F(t)$ (where t is the iteration step) until $J_m(U, V : X)$ shows no further improvement.

Since Bezdek first reported his fuzzy set theoretic image segmentation algorithm for aerial images [1,2,4], numerous reports have appeared detailing the superior manner in which the FCM algorithm handles uncertainties [5,6,7,8,9,10]. However, traditional FCM methods have a number of limitations. First, the initialization of cluster centroids is important because different selections of the initial cluster centroids can potentially lead to different local optima or different partitions. However, there is no general agreement regarding which initialization scheme gives the best results [1,2]. Most algorithms developed to date assign the random locations to the initial centroids because such a selection will definitely converge in the iterative process. Moreover, the traditional spatial FCM algorithms are limited in formulating the influence of the spatial neighborhood of a data point x_j on calculating a degree of membership of x_j to a given cluster F_i. As Liew pointed out [10], in the original spatial FCM algorithm put forwarded by Tolias and Panas [9], the influence of neighboring color points on a central color point is binary in that it is implemented through the addition and subtraction of a fixed constant. The adaptive spatial FCM algorithm of Liew, which uses the notion of a homogeneous regions, still does not provide an explicit mathematical formulation for handling homogeneous regions.

To solve the addressed problems, a new fuzzy c-means algorithm for clustering color data is proposed in the present study. The initial cluster centroids are selected based on the notion that dominant colors in a given color set are unlikely to belong to the same cluster (Section 2.1). Thus, we developed a scheme for identifying the most vivid and distinguishable colors in a given color data set; these dominant colors are then used to guess the initial cluster centroids for the FCM algorithm. In addition, spatial knowledge is incorporated into the clustering algorithm by employing the homogeneity of the spatial neighborhood, which is quantified in terms of the variance and discontinuity of the neighborhood of a color point. (Section 2.2). The homogeneity value of color x_j indicates the likelihood that x_j and its neighborhood belong to the same homogeneous region. This homogeneity value is used in calculating the degree of membership of x_j to fuzzy clusters in the FCM algorithm. We call the proposed method a spatial homogeneity-based fuzzy c-means (SHFCM) algorithm.

2 Spatial Homogeneity-Based Fuzzy c-Means Algorithm

2.1 Initialization of Cluster Centroids

The clustering initialization procedure aims to establish good initial centroid for each cluster. We guess the initial centroids from the dominant colors that are the most distinguishable colors in a given color data set; the number of dominant colors is set equal to the number of clusters (c). To obtain the dominant colors from a color set X, we use the notion of reference colors, assuming that these colors contain the major distinguishable colors found in natural scenes [11]. This set provides standard colors with which to compare and measure similarities between colors. The dominant colors are taken as the top c reference colors that record the higher matching score for all colors $x \in X$.

The matching scores are calculated by the degree of membership of each color point to a set of reference colors. Suppose a color point, denoted by $x = (x_L, x_a, x_b) \in X$, is a point in the CIELAB color space; the superiority of the CIELAB color space over other color spaces has been demonstrated in many color image applications [13,14]. Likewise, the i-th reference color, denoted by $r_i = (r_L^i, r_a^i, r_b^i) \in R$, is a point in the CIELAB space. The distance between x and r_i is calculated from the CIELAB color difference formula [12], denoted by $\delta(x, r_i) = \sqrt{(x_L - r_L^i)^2 + (x_a - r_a^i)^2 + (x_b - r_b^i)^2}$.

We now define a membership function $\mu_{r_i} : x \to [0, 1]$ for a given color point x that quantifies the degree of membership $\mu_{r_i}(x)$ of x to the reference color r_i.

$$\mu_{r_i}(x) = \begin{cases} 1.0 & \text{if } \delta(x, r_i) = 0 \\ 0.0 & \text{if } \delta(x, r_j) = 0 \quad (r_i \neq r_j, r_j \in R) \\ \left(\sum_{j=1}^{k} \left(\frac{\delta(x, r_i)}{\delta(x, r_j)} \right)^\lambda \right)^{-1} & \text{otherwise} \end{cases} \quad (2)$$

where λ is a positive weighting parameter for the membership of x to r_i. If a color point x exactly coincides with a reference color r_i, it has a membership degree of 1.0. Conversely, if x exactly coincides with another reference color r_j, then x has no relation to r_i and the membership degree is assigned a value of 0.0. When the color point x does not exactly coincide with any reference colors, the membership value for each reference color is determined by Eq. 2.

To determine the dominant colors, we compute the membership degrees between all color points $x_j \in X$ and reference colors $r_i \in R$. Each reference color r_i has two additional attributes, denoted μ_i and p_i. Reference color r_i is therefore defined as $r_i = ((r_L^i, r_a^i, r_b^i), \mu_i, p_i)$. Here, $\mu_i = \max \mu_{r_i}(x_j)$ indicates the highest membership degree obtained by computing $\mu_{r_i}(x_j)$ for all $x_j \in X$, and $p_i = \arg\max_{x_j} \mu_{r_i}(x_j)$ indicates the closest color point x_j to r_i. For a color point $x_j \in X$, we compute the color membership degree for each of the reference colors, and update μ_i and p_i. When the computation is completed, the reference colors are sorted by μ_i in decreasing order. The sorted list of reference colors is represented by $R^s = (r_{s1}, r_{s2}, ..., r_{sk})$ where the reference color r_{s1} has the highest value of μ_i and r_{sk} has the lowest one.

Now we can define the dominant colors as $D = \{d_i \mid d_i = r_{si}, 1 \leq i \leq c\}$. Thus the set of dominant colors consists of the first c reference colors in the sorted list, which represent the most distinguishable and vivid colors in a given color set X. Having established the dominant colors, the initial cluster centroids are assigned to the color point x_i that is closest to the dominant color d_i, i.e., $V_0 = \{v_i \mid v_i = p_i, p_i \in d_i\}$.

2.2 Spatial Homogeneity

Homogeneity is a local information that reflects the uniformity of a region [15]; when a region is homogeneous, data in the homogeneous region are likely to have similar characteristics. This concept has the potential to be very useful in color clustering because the objective of such clustering is to partition color data into several homogeneous groups. For simplicity, in the present study we consider a spatial homogeneity in two-dimensional space. Recently, Cheng and Sun proposed a homogeneity-based histogram analysis [16] in which homogeneity is formulated in terms of the gray intensity. In the present work, we extend their definition of homogeneity to the CIELAB color space. For each color point, the homogeneity value is computed in terms of two components, variance and discontinuity. In this scheme, the homogeneity value of a color increases with decreasing variance and discontinuity. In other words, the higher the homogeneity value of color $x_j \in X$, the greater the likelihood that x_j and its spatial neighborhood belong to the same homogeneous region.

The variance is a measure of the color contrasts in a predefined region. Let x_j be a color point, and let S_j denote a spatial neighborhood in two-dimensional space, specified to be a $d \times d$ window size centered at x_j. The mean values $M_j^{L^*}$, $M_j^{a^*}$, and $M_j^{b^*}$ of the color data in S_j can be computed as:

$$M_j^{L^*} = \frac{1}{d^2} \sum_{x_k \in S_j} x_{k,L}, \quad M_j^{a^*} = \frac{1}{d^2} \sum_{x_k \in S_j} x_{k,a}, \quad M_j^{b^*} = \frac{1}{d^2} \sum_{x_k \in S_j} x_{k,b} \quad (3)$$

Therefore, the variance $\sigma^2(x_j)$ of color x_j is computed as:

$$\sigma^2(x_j) = \frac{1}{d^2} \sum_{x_k \in S_j} (x_{k,L} - M_j^{L^*})^2 + (x_{k,a} - M_j^{a^*})^2 + (x_{k,b} - M_j^{b^*})^2 \quad (4)$$

The discontinuity $g(x_j)$ of color x_j is obtained through the first derivatives at spatial coordinate x_j, which is defined as the magnitude of the gradient between x_j and S_j [15,16].

$$g(x_j) = [G_{x_j,h}^2 + G_{x_j,v}^2]^{1/2} = \left[\left(\frac{\partial g}{\partial x_{j,h}}\right)^2 + \left(\frac{\partial g}{\partial x_{j,v}}\right)^2\right]^{1/2} \quad (5)$$

where $G_{x_j,h}^2$ and $G_{x_j,v}^2$ are the gradients for the horizontal and vertical directions, respectively.

The two measures, $\sigma^2(x_j)$ and $g(x_j)$, have different scales that needs to be reconciled through a normalization. Thus, the variance and discontinuity are normalized by their maximum values; specifically, $\sigma_n^2(x_j) = \sigma^2(x_j)/\max \sigma^2(x_j)$ and $g_n(x_j) = g(x_j)/\max g(x_j)$. Then the spatial homogeneity $h(x_j)$ of color x_j is defined as:

$$h(x_j) = 1 - \sigma_n^2(x_j) \times g_n(x_j) \quad (6)$$

The goal of the present study is to cluster a set of color data $X = \{x_1, x_2, ..., x_n\}$ into c homogeneous groups. To achieve this, we propose a spatial homogeneity-based fuzzy c-means (SHFCM) algorithm. The objective of SHFCM is to cluster the data X into c clusters $(F_1, F_2, .., F_c)$ by minimizing the function

$$J_m(U, V : X) = \sum_{i=1}^{c} \sum_{j=1}^{n} (\mu_{F_i}(x_j))^m \|x_j - v_i\|^2 \qquad (7)$$

where

$$\mu_{F_i}(x_j) = \frac{1}{n_j} \left(\mu_{ij} + \sum_{x_k \in S_j} \mu_{ik} h(x_k) \right) \qquad (8)$$

and

$$\mu_{ij} = \left(\sum_{z=1}^{c} \left(\frac{\|x_j - v_i\|^2}{\|x_j - v_z\|^2} \right)^{\frac{1}{m-1}} \right)^{-1} \qquad (9)$$

subject to $0 \leq \mu_{F_i}(x_j) \leq 1, \sum_{i=1}^{c} \mu_{F_i}(x_j) = 1, 0 < \sum_{j=1}^{n} \mu_{F_i}(x_j) < n$. Here, μ_{ij} is obtained by computing relative distances between clusters in a color feature space. $x_k \in S_j$ is a spatial neighborhood of x_j, defined as a $d \times d$ window. $n_j (= d^2)$ is the number of x_j and its neighbors, k $h(x_k)$ is the homogeneity of x_k.

In Eq. 8, we can see that $\mu_{F_i}(x_j)$ is influenced by the relations between $x_k \in S_j$ and F_i and, furthermore, the extent of this influence is controlled by the degree of homogeneity $h(x_k)$. Let us suppose that x_j is located in a homogeneous region. Under such circumstances, it is useful to exploit the relations between the neighborhood and the centroid because data in a homogeneous region are likely to belong to the same cluster. In contrast, when $x_k \in S_j$ lies in a non-homogeneous region such as an edge, the influence of x_k on x_j should be made as small as possible. Thus, the degree to which x_k influences x_j is determined through its homogeneity value $h(x_k)$.

3 Experimental Results

To test the effectiveness with which the proposed method clusters color data, we applied the conventional FCM algorithm and the proposed SHFCM algorithm to four image segmentation problems, and compared the performances of the algorithms. In the FCM calculations, the initial centroids were randomly selected. In the SHFCM, the initial centroids were obtained from the aforementioned dominant colors. In these experiments, the FCM and SHFCM parameters were set as follows: the weighting exponent was set to $m = 2.0$ because this value has been overwhelmingly favored in previous studies [1]; the termination criterion was set to $\epsilon = 0.001$; λ was set to be 2.0 as this value provides the best performance; and the neighborhood dimension was set to $d = 3$.

Figure 1(a) shows an original image ("pants") containing four colors ($c = 4$): white for the background, a gray for the left-side pants, a red for the right-side pants, and black for the rectangular mark on the red pants. (Please note that all figures in this section must be viewed in color to appreciate the color segmentation results.) An ideal clustering would result in four segmented objects.

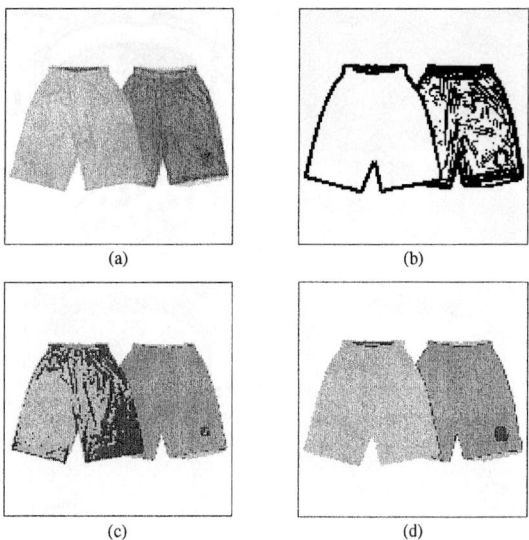

Fig. 1. Comparison of the results using FCM and SHFCM: (a) original image "pants"; (b) homogeneity (threshold = 0.9), (c) FCM clustering, (d) SHFCM clustering

Although this image appears to be simple to cluster, the clustering problem is complicated by the presence of many color points on the gray pants (e.g., those in the crease areas) that are close to those of the rectangular mark on the red pants in the color space. Figure 1(b) shows the homogeneity values extracted from the original image; in this figure, the white areas represent homogeneous regions and the black areas represent nonhomogeneous regions. These regions were determined based on a threshold homogeneity value of 0.90. Clustering based on the FCM algorithm [Fig. 1(c)] led to an over-segmentation in the left-side gray pants due to the large number of creases in those pants, and the black mark in the right-side pants is not clear. In contrast, the SHFCM algorithm [Fig. 1(d)] provided a better classification on the four objects than FCM.

Figure 2(a) shows an original image ("clown") containing various colors. Each clustering algorithm was used to segment the image into six colors ($c = 6$): white, black, red, brown, yellow, and green. The FCM algorithm failed to give correct clustering results [Fig. 2(c)]; the green hair was misclassified as a yellow color in the final image. In contrast, the SHFCM algorithm [Fig. 2(d)] successfully segmented the six colors; it clearly classified the hair as being of green color because the green color was identified as one of the dominant colors. Figure 3(a) shows an original image ("house"). Each clustering algorithm was used to segment this image into six clusters ($c = 6$): white, blue, red, black, gray, and green. The FCM algorithm failed to give a correct clustering result [Fig. 3(c)]; it over-segmented the lower part of the green grass and misclassified the gray roof of the house as being of white color. In contrast, the SHFCM algorithm successfully segmented

Fig. 2. Comparison of the results using FCM and SHFCM: (a) original image "clown"; (b) homogeneity (threshold = 0.9) (c) FCM clustering; (d) SHFCM clustering

Fig. 3. Comparison of the results using FCM and SHFCM: (a) original image "house"; (b) homogeneity (threshold = 0.9); (c) FCM clustering; (d) SHFCM clustering

the six color objects [Fig. 3(d)]; it more clearly classified the green grass and the black shade, demonstrating its superior performance.

4 Conclusion

To tackle the color clustering problems, in this paper we have proposed a new spatial homogeneity-based fuzzy c-means (SHFCM) algorithm. In this algorithm, the initial cluster centroids are selected by identifying the dominant colors in a given color set, and then placing the initial centroids at the points closest in color space to the dominant colors. In addition, the proposed method exploits spatial knowledge by inclusion of the influence of the neighborhood of a color point, as discriminated by its homogeneity value. Comparisons of the SHFCM and FCM methods showed that the SHFCM method is more effective.

References

1. Bezdek, J.C.: Fuzzy Models and Algorithms for Pattern Recognition and Image Processing. Kluwer Academic Publishers, Boston (1999)
2. Jain, A.K., Dubes, R.C.: Algorithms For Clustering. Prentice-Hall, NJ (1998)
3. Jain, A.K., Murty, M.N., Flynn, P.J.: Data Clustering: A Review. ACM Computing Surveys 31 (1999) 264–323
4. Pal, N.R., Pal, S.K.: A Review On Image Segmentation Techniques. Pattern Recognition 26 (1993) 1277–1294
5. Lim, Y.W., Lee, S.U.: On The Color Image Segmentation Algorithm Based on the Thresholding and the Fuzzy c-Means Techniques. Pattern Recognition 23 (1990) 935–952
6. Bensaid, A.M.: Partially Supervised Clustering For Image Segmentation. Pattern Recognition 29 (1996) 859–871
7. Cheng, T.W., Goldgof, D.B, Hall, L.O.: Fast Fuzzy Clustering. Fuzzy Sets and Systems 93 (1998) 49–56
8. Qzdemir, D., Akarun, L.: A Fuzzy Algorithm For Color Quantization Of Images. Pattern Recognition 35 (2002) 1785–1791
9. Tolias, Y.A., Panas, S.M.: Image Segmentation by a Fuzzy Clustering Algorithm Using Adaptive Spatially Constrained Membership Functions. IEEE Trans. Syst. Man Cybern. 28 (1998) 359–369
10. Liew, A.W.C., Leung, S.H., Lau, W.H.: Fuzzy Image Clustering Incorporating Spatial Continuity. IEE Proceedings of Vis. Image Process 147 (2000) 185–192
11. Kim, D.-W., Lee, K.H., Lee, D.: A Novel Initalization Scheme For the Fuzzy c-Means Algorithm for Color Clustering. Pattern Recognition Letters 25 (2004) 227–237
12. Wyszecki, G., Stiles, W.S.: Color Science : Concepts and Methods, Quantitative Data and Formulae. Wiley-Interscience Publication, New York (2000)
13. Paschos, G.: Perceptually Uniform Color Spaces For Color Texture and Analysis: An Empirical Evaluation. IEEE Trans. Image Processing 10 (2001) 932-937
14. Shafarenko, L., Petrou, H., Kittler, J.: Histogram-based Segmentation In a Perceptually Uniform Color Space. IEEE Trans. on Image Processing 7 (1998) 1354–1358
15. Gonzalez, R.C., Wintz, P.: Digital Image Processing. Addison-Wesley, MA (1987)
16. Cheng, H.-D., Sun, Y.: A Hierarchical Approach To Color Image Segmentation Using Homogeneity. IEEE Transactions on Image Processing 9 (2000) 2071–2082

A Novel Fuzzy-Connectedness-Based Incremental Clustering Algorithm for Large Databases

Yihong Dong[1,2], Xiaoying Tai[1], and Jieyu Zhao[1]

[1] Institute of Computer Science and Technology, Ningbo University, Ningbo 315211, China
[2] Institute of Artificial Intelligence, Zhejiang University, Hangzhou 310027, China
noel99@tom.com

Abstract. Many clustering methods have been proposed in data mining fields, but seldom were focused on the incremental databases. In this paper, we present an incremental algorithm-IFHC that is applicable in periodically incremental environment based on FHC[3]. Not only can FHC and IFHC dispose the data with numeric attributes, but with categorical attributes. Experiment shows that IFHC is faster and more efficient than FHC in update of databases.

1 Introduction

Incremental mining technique, is a technique which updates the mining result incrementally when insertions and deletions on the operational databases occur, instead of re-mining the whole changed databases.

If we can make full use of the latest mining result, we can get better efficiency. The insertion or deletion of data should be taken into account in this incremental update instead of the whole update databases. There have some reports on the research in incremental clustering algorithms. IncrementalDBSCAN[1] is the first incremental clustering algorithm whose performance evaluation has been proven to a good efficiency on a spatial database as well as on a WWW-log database. IGDCA[2] first partitions the data space into a number of units, and then deals with units instead of points. Only those units with the density no less than a given minimum density threshold are useful in extending clusters.

Based on fuzzy hierarchical clustering algorithm(FHC) proposed in literature [3], we present its incremental algorithm-IFHC based on FHC to deal with a bulk of updates.

2 Incremental Fuzzy-Connectedness-Based Clustering Algorithm

In FHC[3] method, the datasets are partitioned firstly into several sub-clusters using partitioning method, and then fuzzy graph of sub-clusters are constructed by analyzing the fuzzy-connectedness among the sub-clusters. By making λ cut graph for the fuzzy graph, we get the connected components of the fuzzy graph, which are the result of clustering we want to get. The algorithm can be performed in high-dimensional data set, clustering the arbitrary shape of clusters such as the spherical, linear, elongated or concave ones.

Definition 1: (neighborhood) Neighborhood of an object is an area with object p as center and r as radius, which is called neighborhood p, denoted by $Neig(p)$.

Definition 2: (connection) Point x is a connection between neighborhood p and neighborhood q, if and only if x is either in the neighbor of point p or in that of point q. We use the signal $connection(p,q)$ to denote this relationship. $connection(p,q) = \{x \mid x \in Neig(p), x \in Neig(q)\}$.

Definition 3: (fuzzy-connectedness) Fuzzy-connectedness is the connected intensity between neighborhood p and neighborhood q.

$$\mu(p,q) = \frac{\mid connection(p,q) \mid}{N_p + N_q - \mid connection(p,q) \mid} \qquad 0 \leq \mu(p,q) \leq 1$$

Definition 4: (directly λ-fuzzy-connection) D is a set of objects, $p \in D, q \in D$, neighborhood p and neighborhood q is directly λ-fuzzy-connection if and only if $\mu(p,q) \geq \lambda$, denoted as $p \underset{D}{\overset{\lambda}{\leftrightarrow}} q$.

Definition 5: (λ-fuzzy-connection) D is a set of objects, if a chain $p_1, p_2, ..., p_n$, $p_1 = q, p_n = p$, where where $p_i \in D(1 \leq i \leq n)$, $p_{i+1} \underset{D}{\overset{\lambda}{\leftrightarrow}} p_i$ exists, neighborhood p and neighborhood q is λ-fuzzy-connection in D, denote as $p \underset{D}{\overset{\lambda}{\cdots}} q$.

Definition 6: (affected-sub-cluster of p) Assume a set of sub-clusters in database D is $O = \{O_1, O_2, ..., O_m\}$, $p \in D$, affected-sub-cluster of p means the set of sub-clusters' centers contained in neighborhood p, which is denoted as $Affected\_subcluster(p) = \{q \mid \exists q \in D, and\ q \in Neig(p)\}$.

Definition 7: (center of p) Assume a set of sub-clusters in database D is $O = \{O_1, O_2, ..., O_m\}$, $p \in D$, center of p is the nearest sub-cluster in $Affected\_subcluster(p)$, denoted as denoted as center(p)=O_j.

Definition 8: (Affected_neig) Let D be a database of objects and p be some object(either in or not in D). We define the set of neighborhoods in D affected by the insertion or deletion of p as

$Affected\_Neig(p) = Affected\_subcluster(p) \cup \{r \mid \forall r \underset{D \cup \{p\}}{\overset{\lambda}{\cdots}} x,\ x \in Affected\_subcluster(p)\}$

Lemma: Let D be a set of objects and p be some object. Then $\forall x \in D : x \notin Affected\_Neig(p) \Rightarrow \{q \mid q \underset{D \setminus \{p\}}{\overset{\lambda}{\cdots}} x\} = \{q \mid q \underset{D \cup \{p\}}{\overset{\lambda}{\cdots}} x\}$

Proof: $\subseteq$: $\forall q \in \{q \mid q \xleftrightarrow[D\backslash\{p\}]{\lambda} x\}$, by definition of λ fuzzy-connection there exists chain $p_1, p_2, \ldots, p_n$, $p_1 = q$, $p_n = x$, where $p_i \in D\backslash\{p\}(1 \leq i \leq n)$, $p_{i+1} \xleftrightarrow{\lambda}_D p_i$. $\because D\backslash\{p\} \subseteq D \cup \{p\}$, $\therefore p_i \in D \cup \{p\}$. From the same reason by definition of λ fuzzy-connection, $q \in \{q \mid q \xleftrightarrow[D\cup\{p\}]{\lambda} x\}$, so $\{q \mid q \xleftrightarrow[D\backslash\{p\}]{\lambda} x\} \subseteq \{q \mid q \xleftrightarrow[D\cup\{p\}]{\lambda} x\}$

$\supseteq$: Assume that for each $q \in \{q \mid q \xleftrightarrow[D\cup\{p\}]{\lambda} x\}$, the formula $\{q \mid q \xleftrightarrow[D\backslash\{p\}]{\lambda} x\} \supseteq \{q \mid q \xleftrightarrow[D\cup\{p\}]{\lambda} x\}$ does not come into existence. Then a chain including neighborhood p exists at least to make neighborhood q and neighborhood x λ fuzzy-connection, i.e. $\exists p_1, p_2, \ldots, p, \ldots, p_n$, $p_1 = q$, $p_n = x$, where $p_i \in D\backslash\{p\}(1 \leq i \leq n)$, $p_{i+1} \xleftrightarrow{\lambda}_D p_i$. By the definition of the set Affected_neig, we know that $x \in Affected\_Neig(p)$ which is in contrast to the assumption of $x \notin Affected\_Neig(p)$. Thus, $\{q \mid q \xleftrightarrow[D\backslash\{p\}]{\lambda} x\} \supseteq \{q \mid q \xleftrightarrow[D\cup\{p\}]{\lambda} x\}$

By summarization, $\{q \mid q \xleftrightarrow[D\backslash\{p\}]{\lambda} x\} = \{q \mid q \xleftrightarrow[D\cup\{p\}]{\lambda} x\}$

Due to lemma, after inserting or deleting an object p, it is sufficient to reapply FHC algorithm to the set Affected_neig(p) in order to update the clustering.

Insertions. When inserting a new object p, new directly λ-fuzzy-connection may be established, but none are removed. By lemma, it is sufficient to restrict the application of the clustering procedure to the set Affected_neig(p). When inserting an object p into the database D, we can distinguish the following cases:

(1) If Affected_neig(p)=Ø, p is a noise object and nothing else is changed.
(2) If Affected_neig(p)=Center(p), and |Center(p)|<Minpts, then p is a noise object; If Affected_neig(p)=Center(p) and |Center(p)|>=Minpts, the object p is a number of cluster Center(p), and should be absorbed into cluster Center(p).
(3) $Affected\_Neig(p) = Affected\_subcluster(p) \cup \{r \mid \forall r \xleftrightarrow[D\cup\{p\}]{\lambda} x, x \in Affected\_subcluster(p)\}$

,where $Affected\_subcluster(p) = \{q \mid \exists q \in O \text{ and } q \in Neig(p)\}$, connection(p,q) should be increment by 1. $\mu(p, q)$ must be recomputed to judge whether a new directly λ-fuzzy-connection is created. If it is created, two or more clusters may be combined.

Deletions. As opposed to the insertion, when deleting an object p, directly λ-fuzzy-connection may be removed, but no new directly λ-fuzzy-connection are established. It will make a cluster split into two or more clusters. In fact, the split of a cluster is not very frequent. When deleting an object p from the database D we can distinguish the following cases:

(1) If Affected_neig(p) =Ø, then p is an outlier which is regarded as a noise object. Deletion of p does no more effect to the clustering result.
(2) If Affected_neig(p)=Center(p), object p should be deleted from the cluster Center(p).

(3) $Affected\_Neig(p) = Affected\_subcluster(p) \cup \{r \mid \forall r \overset{\lambda}{\underset{D\setminus\{p\}}{\cdots}} x, x \in Affected\_subcluster(p)\}$

connection(p,q) should be decease by 1. $\mu(p,q)$ should be recomputed to judge whether directly λ-fuzzy-connection is removed. If directly λ-fuzzy-connection is removed, the cluster will be divided into two or more clusters.

3 Comparison IFHC Versus FHC

In this section, we evaluate the efficiency of IFHC versus FHC. We present an experimental evaluation using a synthetic database of 100,000 records with 30 clusters of similar sizes and 5% noise that distributed outside of the clusters. The Euclidean distance was used as distance function and ε was set to 2 and λ was set to 0.08. All experiments have been run on PIV1.7G machine with 256M of RAM and running VC6.0.

Three experiments are designed to compare the performance of IFHC versus FHC. There are only insertions, only deletions, and both insertions and deletions as Table 1 shows. There are half insertions and half deletions in "both insertions and deletions" item. From these tables, we can see that IFHC is faster and more efficient than FHC in updates of databases.

Table 1. Run time comparison IFHC versus FHC

Number of updates		1000	2000	3000	4000	5000
Insertions	FHC	327.1	331.7	335.4	338.6	340.6
	IFHC	34.3	51.0	67.2	85.4	102.8
Deletions	FHC	318.5	316.2	313.4	311.7	307.8
	IFHC	31.5	48.5	65.9	82.3	104.4
Both insertions and deletions	FHC	325.8	322.6	326.3	324.9	320.6
	IFHC	32.2	50.8	66.3	83.7	98.5

4 Conclusion

Recently many researchers focus on clustering as a primary data mining method for knowledge discovery. In this paper, we introduce a novel fuzzy-connectedness-based incremental algorithm-IFHC to deal with a bulk of updates instead of single update. FHC and IFHC have high efficiency to discover any clusters with arbitrary shapes. The results of our experimental study in data sets show IFHC is faster and more efficient than FHC.

Acknowledgements. This work is partially supported by Scientific Research Fund of Zhejiang Provincial Education Department of China(20030485) to Yihong Dong, Natural Science Foundation of China(NSFC 60472099) to Xiaoying Tai.

References

1. Ester M., Kriegel H. –P, Sander J., Wimmer M., Xu X.: Incremental clustering for mining in a data warehousing environment. In Proc. 24th VLDB Int. Conf. New York(1998)323-333
2. Chen Ning, Chen An, Zhou Long-xiang: An incremental grid density-based clustering algorithm. Journal of Software, 2002, 13(01)1-7
3. Yihong Dong, Yueting Zhuang: Fuzzy Hierarchical Clustering Algorithm Facing Large Databases. In Proc. of 5th World Congress on Intelligent Control and Automation, Hangzhou(2004)4282-4286

Classification of MPEG VBR Video Data Using Gradient-Based FCM with Divergence Measure

Dong-Chul Park

Intelligent Computing Research Lab.,
Dept. of Information Engineering,
Myong Ji University, Korea
parkd@mju.ac.kr

Abstract. An efficient approximation of the Gaussian Probability Density Function (GPDF) is proposed in this paper. The proposed algorithm, called the Gradient-Based FCM with Divergence Measure (GBFCM (DM)), employs the divergence measurement as its distance measure and utilizes the spatial characteristics of MPEG VBR video data for MPEG data classification problems. When compared with conventional clustering and classification algorithms such as the FCM and GBFCM, the proposed GBFCM(DM) successfully finds clusters and classifies the MPEG VBR data modelled by the 12-dimensional GPDFs.

1 Introduction

Multimedia technology has applied to various areas including information and communication, education, and entertainment[1]. Recently, the research on video transmission in multimedia service has become one of the most active fields and video service will be even more active when the broadband network service becomes available [2],[3]. Video data are analyzed and classified based on its contents in many applications. However, when video data are compressed, the analysis of the video data with a compressed format becomes complicated. Furthermore, it will be even harder to classify and retrieve the video data by their contents when the digital video data are stored by various compression methods[4]. If the compressed video data can be classified without going through the decompressing procedure, the efficiency and usefulness of the video data, especially the MPEG(Moving Picture Expert Group) VBR(Variable Bit Rate) video data, can be maximized [5].

Patel and Sethi proposed a method to analyze the compressed video data directly by using a decision-tree classifier [6] and Liang and Mendel proposed a classification method by employing a fuzzy classifier that uses the Fuzzy c-Means(FCM) algorithm [5]. Patel and Sethi's approach basically considers the MPEG VBR video data as deterministic time-series data. However, according to Rose's thorough analysis on MPEG VBR data [7], it would be more realistic to consider the MPEG VBR video data as GPDF data. The probabilistic nature of the MPEG VBR video data was also accepted by Liang and Mendel. Liang and Mendel, however, only utilize FCM with mean values of the GPDF data while missing the variance information of the GPDF data. In order to utilize the entire information, mean and variance, of the GPDFs in MPEG VBR video data, this paper proposes a clustering algorithm based on the divergence measure. The Gradient-based FCM(GBFCM) is considered as a basis for the new clustering algorithm[4,5]and the resultant GBFCM with Divergence Measure, GBFCM(DM), is proposed in this paper for clustering MPEG VBR video data [8]-[13].

The MPEG video data traffic is introduced in Section 2. Section 3 summarizes several conventional algorithms for data clustering. The Gradient-based FCM with Divergence Measure is proposed in Section 4. Experimental results and performance comparisons among conventional algorithms and the GBFCM(DM) are given in Section 5. Section 6 concludes this paper.

2 Characteristics of MPEG Video Traffic

The MPEG video traffic is composed of sequences of GoP(group of picture). Each GoP includes I-frame (Infra coded frame), P-frame(Predictive coded frame), and B-frame (Bidirectional predictive coded frame). I-frames use DCT encoding only to compress a single frame without reference to any other frame in the sequence. Typically I-frames are coded with 2 bits per pixel on average. P-frames are encoded as the differences from the previous I or P frame. The new P-frame is first predicted by taking the previous I or P frame and 'predicting' the values of each new pixel. B-frames are encoded as differences from the previous or next I or P frame. B-frames use prediction as for P-frames but for each block either the previous I or P frame is used or the next I or P frame. Fig. 1 shows an example of MPEG data with I-, P-, and B-frame from the video data 'SOC1' of Table 2 [7].

3 Existing Algorithms

3.1 Fuzzy c-Means(FCM) Algorithm

The objective of clustering algorithms is to group of similar objects and separate dissimilar ones. Bezdek first generalized the *fuzzy ISODATA* by defining a family of objective functions $J_m, 1 < m < \infty$, and established a convergence theorem for that family of objective functions [11,12]. For FCM, the objective function is defined as :

$$J_m(U, v) = \sum_{k=1}^{n} \sum_{i=1}^{c} (\mu_{ki})^m (d_i(x_k))^2 \qquad (1)$$

where $d_i(x_k)$ denotes the distance from the input data x_k to v_i, the center of the cluster i, μ_{ki} is the membership value of the data x_k to the cluster i, and m is the weighting exponent, $m \in 1, \cdots, \infty$, while n and c are the number of input data and clusters, respectively. Note that the distance measure used in FCM is the Euclidean distance.

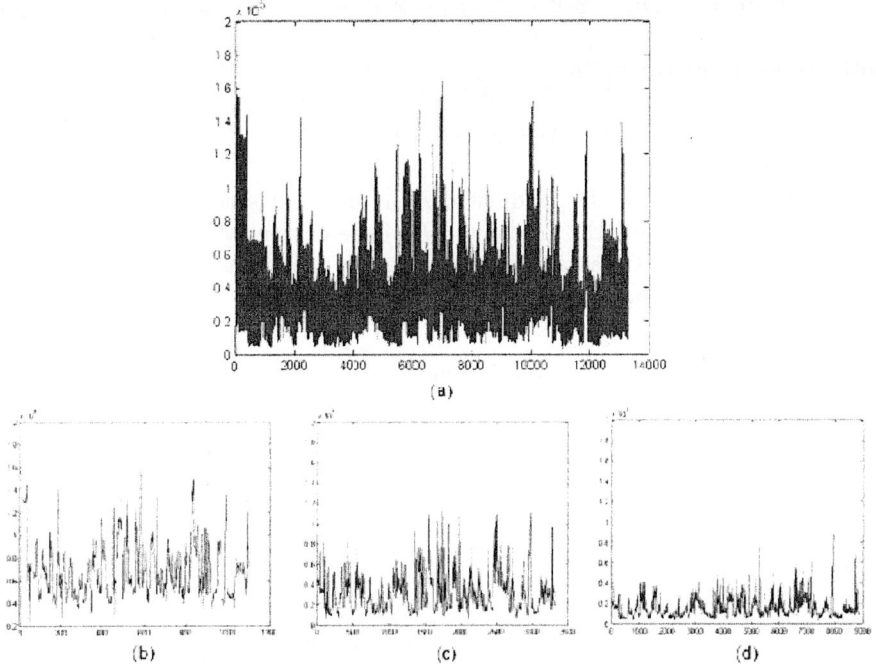

Fig. 1. Example of MPEG data: (a) whole data (b) I-frame (c) P-frame (d) B-frame

Bezdek defined a condition for minimizing the objective function with the following two equations [11,12]:

$$\mu_{ki} = \frac{1}{\sum_{j=1}^{c}\left(\frac{d_i(\boldsymbol{x}_k)}{d_j(\boldsymbol{x}_k)}\right)^{\frac{2}{m-1}}} \qquad (2)$$

$$\boldsymbol{v}_i = \frac{\sum_{k=1}^{n}(\mu_{ki})^m \boldsymbol{x}_k}{\sum_{k=1}^{n}(\mu_{ki})^m} \qquad (3)$$

The FCM finds the optimal values of group centers iteratively by applying Eq. (2) and Eq. (3) in an alternating fashion.

3.2 Gradient-Based Fuzzy c-Means(GBFCM) Algorithm

One attempt to improve the FCM algorithm was made by minimizing the objective function using one input data at a time instead of the entire input data. That is, the FCM in Eq. (2) and Eq. (3) uses all data to update the center value of the cluster, but the GBFCM that is used in this paper was developed to update the center value of the cluster with a given individual data sequentially [8,9]. Given one data $\boldsymbol{x}_i$ and c clusters with centers at $\boldsymbol{v}_j, (j = 1, 2, \cdots, c)$, the objective function to be minimized is:

$$J_i = \mu_{1i}^2(v_1 - x_i)^2 + \mu_{2i}^2(v_2 - x_i)^2 + \cdots + \mu_{ci}^2(v_c - x_i)^2 \tag{4}$$

with the following constraint:

$$\mu_{1i} + \mu_{2i} + \cdots + \mu_{ci} = 1 \tag{5}$$

The basic procedure of the gradient descent method is that starting from an initial center vector, v_0, the gradient ΔJ_i of the current objective function can be computed. The next value of v is obtained by moving to the direction of the negative gradient along the error surface such that:

$$v_{k+1} = v_k - \eta \frac{\partial J_i}{\partial v_k}$$

where k is the iteration index and

$$\frac{\partial J_i}{\partial v_k} = 2\mu_{ki}^2(v_k - x_i)$$

Equivalently,

$$v_{k+1} = v_k - 2\eta \mu_{ki}^2(v_k - x_i) \tag{6}$$

where η is a learning constant.

A necessary condition for optimal positions of the centers for the groups can be found by the following:

$$\frac{\partial J_i}{\partial \mu} = 0 \tag{7}$$

After applying the condition of Eq. (7), the membership grades can be found as:

$$\mu_{ki} = \frac{1}{\sum_{j=1}^{c}(\frac{d_i(x_k)}{d_j(x_k)})^2} \tag{8}$$

Both the FCM and GBFCM have an objective function that related the distance between each center and data with a membership grade reflecting the degree of their similarities with respect to other centers. On the other hand, they differ in the way they try to minimize it:

- As can be seen from Eq. (2) and Eq. (3), all the data should be present in the objective function in the FCM and the gradients are set to zero in order to obtain the equations necessary for minimization [14]- [16].
- As can be seen from Eq. (6) and Eq. (8), however, only one datum is present for updating the centers and corresponding membership values at a time in the GBFCM.

More detailed explanation about the GBFCM can be found in [8,9].

4 GBFCM with Divergence Measure

The MPEG VBR video data to be used in this paper are originally in a time-series format. However, previous researches confirm that the GPDF representation of MPEG VBR video data has advantages over one-dimensional time-series representation as far as data classification accuracy is concerned [7,10,13].

In distribution clustering, selecting a proper distance measure between two data vectors is very important since the performance of the algorithm largely depends on the choice of the distance measure[11,17]. After evaluating various distance measures, the Divergence distance (*Kullback-Leibler Divergence*) between two GPDFs, $\boldsymbol{x} = (x_i^\mu, x_i^{\sigma^2})$ and $\boldsymbol{v} = (v_i^\mu, v_i^{\sigma^2})$, $i = 1, \cdots, d$, is chosen as the distance measure in our algorithm[11,18]:

$$D(\boldsymbol{x}, \boldsymbol{v}) = \sum_{i=1}^{d} \left(\frac{x_i^{\sigma^2} + (x_i^\mu - v_i^\mu)^2}{v_i^{\sigma^2}} + \frac{v_i^{\sigma^2} + (x_i^\mu - v_i^\mu)^2}{x_i^{\sigma^2}} \right) \quad (9)$$

where x_i^μ and $x_i^{\sigma^2}$ denote μ and σ^2 values of the ith component of $\boldsymbol{x}$, respectively, while v_i^μ and $v_i^{\sigma^2}$ denote μ and σ^2 values of the ith component of $\boldsymbol{v}$, respectively.

The GBFCM to be used in this paper is based on the FCM algorithm. However, instead of calculating the center parameters of the clusters after applying all the data vectors in FCM, the GBFCM updates their center parameters at every presentation of data vectors. By doing so, the GBFCM can converge faster than the FCM [8,9]. To deal with probabilistic data such as the GPDF, the proposed GBFCM(DM) updates the center parameters, mean and variance, according to the distance measure shown in Eq. (9). That is, the membership grade for each data vector $\boldsymbol{x}$ to the cluster i is calculated by the following:

$$\mu_i(\boldsymbol{x}) = \frac{1}{\sum_{j=1}^{c} \left(\frac{D(\boldsymbol{x}, \boldsymbol{v}_i)}{D(\boldsymbol{x}, \boldsymbol{v}_j)} \right)^2} \quad (10)$$

After finding the proper membership grade from an input data vector $\boldsymbol{x}$ to each cluster i, the GBFCM-DM updates the mean and variance of each center as follows:

$$\boldsymbol{v}_i^\mu(n+1) = \boldsymbol{v}_i^\mu(n) - \eta \mu_i^2(\boldsymbol{x})(\boldsymbol{v}_i^\mu(n) - \boldsymbol{x}^\mu) \quad (11)$$

$$\boldsymbol{v}_i^{\sigma^2}(n+1) = \frac{\sum_{k=1}^{N_i} (\boldsymbol{x}_{k,i}^{\sigma^2}(n) + (\boldsymbol{x}_{k,i}^\mu(n) - \boldsymbol{v}_i^\mu(n))^2)}{N_i} \quad (12)$$

where

- $\boldsymbol{v}_i^\mu(n)$ or $\boldsymbol{v}_i^{\sigma^2}(n)$: the mean or variance of the cluster i at the time of iteration n
- $\boldsymbol{x}_{k,i}^\mu(n)$ or $\boldsymbol{x}_{k,i}^{\sigma^2}(n)$: the mean or variance of the kth data in the cluster i at the time of iteration n
- η and N_i : the learning gain and the number of data in the cluster i

Table 1 is a pseudocode of the GBFCM(DM).

Table 1. The GBFCM Algorithm

```
Algorithm GBFCM(DM)
  Procedure main()
    Read c, ε, m
    [c: initialize cluster, ε: is small value,
     m is a weighting exponent (m ∈ 1, ... ∞)]
    error := 0
    While (error > ε)
      While (input file is not empty)
        Read one datum x
        [Update GBFCM(DM) center Mean]
```
$$v^\mu(n+1) = v^\mu(n) - \eta\mu^2(v^\mu(n) - x^\mu)$$
[Update GBFCM(DM) membership grade]
$$\mu_i(x) = \frac{1}{\sum_{j=1}^{c}(\frac{D(x,v_i)}{D(x,v_j)})^2}$$

$$e := v^\mu(n+1) - v^\mu(n)$$
```
      End while
```
$$v_i^{\sigma^2}(n+1) = \frac{\sum_{k=1}^{N_i}(x_{k,i}^{\sigma^2}(n)+(x_{k,i}^\mu(n)-v_i^\mu(n))^2)}{N_i}$$
```
      error := e
    End while
    Output μ_i, v^μ and v^{σ²}
  End main()
End
```

5 Experiments and Results

The MPEG video traffic data sets considered in this paper are from the following internet site:

http://www3.informatik.uni-wuerzburg.de/MPEG/

The data sets are prepared by O. Rose [7] and are compressed with MPEG-1, where the size of GoP is 12 and the sequence is IBBPBBPBBPBB. 10 video streams are used for experiments. Each video stream has 40,000 frames at 25 frames/sec. Table 2 shows the subjects of the 10 video streams.

The video data prepared by Rose[7] have been analyzed and approximated with the statistical features of frames and GoP size by using Gamma or Lognormal by Manzoni et. al [10]. Later, Krunz et. al have found that Lognormal representation is the best matching method for I/P/B frames[13]. In this paper, the Lognormal representation of MPEG VBR Video data is also employed for our experiments.

For applying the divergence measure, the mean and variance of each 12 frame data in the GoP are obtained and each GoP is expressed as 12-dimensional data

Table 2. MPEG VBR Video used for experiments

MOVIE	SPORTS
"Jurassic Park"(Dino)	ATP tennis final (Atp)
"The silence of the lambs"(Lambs)	Formula 1:GP Hockenheim 1994 (Race)
"Star Wars"(Star)	Super Bowl 1995: Chargers-49ers (Sbowl)
"Terminator 2"(Term)	Two 1993 World Cup matches (Soc1)
"a 1994 movie preview"(Movie2)	Two 1993 World Cup matches (Soc2)

Table 3. Classification results of Experiment 1 in False Alarm Rate (%)

# of Clusters	FCM	GBFCM	GBFCM(DM)
3	31.1	28.6	13.2
4	28.6	26.7	12.9
5	26.8	26.8	12.7
6	30.1	24.8	14.2
7	27.8	25.6	12.8

Table 4. Classification results of Experiment 2 in False Alarm Rate (%)

# of Clusters	FCM	GBFCM	GBFCM(DM)
3	17.4	15.8	10.4
4	21.2	16.5	10.1
5	20.5	15.2	9.9
6	18.8	22.1	9.9
7	18.4	17.6	9.9

where each dimension consists of the mean and the variance values of each frame in the GoP. In our experiments, parts of available MPEG VBR video data are used for the training of algorithms and the rest of the video data are used for performance evaluation of the trained algorithms. Experiments are performed in two ways by which the training data are chosen:

- Experiment 1: Dino from movie data and ATP from sports data
- Experiment 2: randomly chosen 10 % of data from each class of data

Since Dino and ATP data have been thoroughly analyzed and successfully modeled with the GPDF, each data stream is selected for a training data set representing each class. Note that each MPEG VBR data set consists of 3,333 GPDF data. The test data sets used are the rest of the available MPEG VBR data for both cases of experiments. Note that the training data and test data sets in the experiment 2 are selected for 20 different cases for obtaining unbiased results. The False Alarm Rate (FAR) is used for the performance measure [5]:

$$FAR = \frac{\text{\# of misclassification cases}}{\text{total \# of cases}}$$

For performance evaluation, the FCM and GBFCM are compared with the proposed GBFCM(DM). In the cases of the FCM and GBFCM, however, they use a only the mean of each frame data while the GBFCM(DM) uses both the mean and the variance. In experiments, each class of movie and sports is assigned with several clusters because one cluster for each class is found insufficient. The number of clusters for each class has been increased up to 7. Classification results for the experiments are given in Table 3 and Table 4. As can be seen from Table 3 and Table 4, the proposed GBFCM(DM) outperforms both the FCM and GBFCM. This is a somewhat obvious result because the FCM and GBFCM do not utilize the variance information of the frame data while the GBFCM(DM) does. The divergence measure use in GBFCM(DM) plays a very important role for modeling and classification of MPEG VBR data streams. This result implies that the divergence measure makes it possible to utilize the research results obtained by researchers including Rose and Manzoni *et. al* [7,10,13].

6 Conclusions

An efficient clustering algorithm, called the Gradient-Based Fuzzy C-Means algorithm with Divergence Measurement (GBFCM(DM)) for the GPDF is proposed in this paper. The proposed GBFCM(DM) employs the divergence measure as its distance measurement and utilizes the spatial characteristics of MPEG VBR video data for MPEG data classification problems. When compared with conventional clustering and classification algorithms such as the FCM and GBFCM, the proposed GBFCM(DM) successfully finds clusters and classifies the MPEG VBR data modelled by the 12-dimensional GPDFs. The results show that the GBFCM(DM) gives 5-15% improvement in False Alarm Rate over the FCM and GBFCM whose FARs are in the range between 15.2% and 31.1% according to the number of clusters used. This result implies that the divergence measure used in the proposed GBFCM(DM) plays a very important role for modeling and classification of MPEG VBR video data streams. Furthermore, the GBFCM(DM) can be used as a useful tool for clustering GPDF data.

Acknowledgement

This research was supported by the Korea Research Foundation (Grant # R05-2003-000-10992-0(2004)).

References

1. Pacifici,G.,Karlsson,G.,Garrett,M.,Ohta,N.: Guest editorial real-time video services in multimedia networks, IEEE J. Select. Areas Commun. **15** (1997) 961-964
2. Tsang,D.,Bensaou,B.,Lam,S.: Fuzzy-based rate control for real-time MPEG video, IEEE Trans. Fuzzy Syst. **6** (1998) 504-516

3. Tan,Y.P.,Yap,K.H.,Wang,L.(eds.): Intelligent Multimedia Processing with Soft Computing, Series: Studies in Fuzziness and Soft Computing, Vol. 168. Springer, Berlin Heidelberg New York (2005)
4. Dimitrova,N.,Golshani,F.: Motion recovery for video content classification, ACM Trans. Inform. Sust. **13** (1995) 408-439
5. Liang,Q.,Mendel,J.M.: MPEG VBR Video Traffic Modeling and Classification Using Fuzzy Technique, IEEE Trans. Fuzzy Systems **9** (2001) 183-193
6. Patel,N.,Sethi,I.K.: Video shot detection and characterization for video databases. Pattern Recog. **30** (1977) 583-592
7. Rose,O.: Satistical properties of MPEG video traffic and their impact on traffic modeling in ATM systems, Univ. Wurzburg,Inst. Comput.Sci. Rep. **101** (1995)
8. Park,D.C.,Dagher,I.: Gradient Based Fuzzy c-means (GBFCM) Algorithm, IEEE Int. Conf. on Neural Networks, ICNN-94 **3** (1994) 1626-1631
9. Looney,C.: Pattern Recognition Using Neural Networks, New York, Oxford University press (1997) 252-254
10. Manzoni,P.,Cremonesi,P.,Serazzi,G.: Workload models of VBR video traffic and their use in resource allocation policies, IEEE Trans. Networking **7** (1999) 387-397
11. Bezdek,J.C.: Pattern recognition with fuzzy objective function algorithms, New York : Plenum, (1981)
12. Bezdek,J.C.: A convergence theorem fo the fuzzy ISODATA clustering algorithms, IEEE trans. pattern Anal. Mach. int. **2** (1980) 1-8, **24** (1975) 835-838.
13. Krunz,M.,Sass,R.,Hughes,H.: Statistical characteristics and multiplexing of MPEG streams, Proc. IEEE Int. Conf. Comput. Commun., INFOCOM'95, Boston, MA **2** (1995) 445-462
14. Kohonen,T.: Learning Vector Quantization, Helsinki University of Technology, Laboratory of Computer and Information Science, Report TKK-F-A-601 (1986)
15. Dunn.,J.C.: A fuzzy relative of the ISODATA process and its use in detecting compact well separated clusters. J. Cybern. **3** (1973):32-75
16. Windham,M.P.: Cluster Validity for the Fuzzy cneans clustering algorithm, IEEE trans. pattern Anal. Mach. int. **4** (1982) 357-363
17. Gokcay,E.,Principe,J.C.: Information Theoretic Clustering, IEEE Trans. Pattern Ana. Mach Int. **24** (2002) 158-171
18. Fukunaga,K.: Introduction to Statistical Pattern Recognition, Academic Press Inc. 2nd edition (1990)

Fuzzy-C-Mean Determines the Principle Component Pairs to Estimate the Degree of Emotion from Facial Expressions

M. Ashraful Amin[1], Nitin V. Afzulpurkar[2], Matthew N. Dailey[3],
Vatcharaporn Esichaikul[1], and Dentcho N. Batanov[1]

[1] Department of CS & IM
[2] Department of MT & ME,
Asian Institute of Technology, Thailand
amin021us@yahoo.com, nitin@ait.ac.th
{vatchara, batanov}@cs.ait.ac.th
[3] Sirindhorn International Institute of Technology,
Thammasat University, Thailand
mdailey@siit.tu.ac.th

Abstract. Although many systems exist for automatic classification of faces according to their emotional expression, these systems do not explicitly estimate the strength of given expressions. This paper describes and empirically evaluates an algorithm capable of estimating the degree to which a face expresses a given emotion. The system first aligns and normalizes an input face image, then applies a filter bank of Gabor wavelets and reduces the data's dimensionality via principal components analysis. Finally, an unsupervised Fuzzy-C-Mean clustering algorithm is employed recursively on the same set of data to find the best pair of principle components from the amount of alignment of the cluster centers on a straight line. The cluster memberships are then mapped to degrees of a facial expression (i.e. less Happy, moderately happy, and very happy). In a test on 54 previously unseen happy faces., we find an orderly mapping of faces to clusters as the subject's face moves from a neutral to very happy emotional display. Similar results are observed on 78 previously unseen surprised faces.

1 Introduction

A significant amount of research work on facial expression recognition has been performed by researchers from multiple disciplines [1], [2], [3]. In this research, we build on existing systems by applying a fuzzy clustering technique to not only determine the category of a facial expression, but to estimate its strength or degree. The clustering is also used to choose the best description of faces in a reduced dimension.

Very few researchers have considered the problem of estimating the degree or intensity of facial expressions. Kimura and Yachida [4] used the concept of a potential network on normalized facial images to recognize and estimate facial expression and its degree respectively. Pantic and Rothkrantz [5] used the famous Ekman [1] defined FACS (Facial Action Coding System) to determine facial expression and its intensity.

2 The Facial Expression Degree Estimation System

Our implementation of the facial expression recognition & degree estimation system involves four major steps (Fig-1).

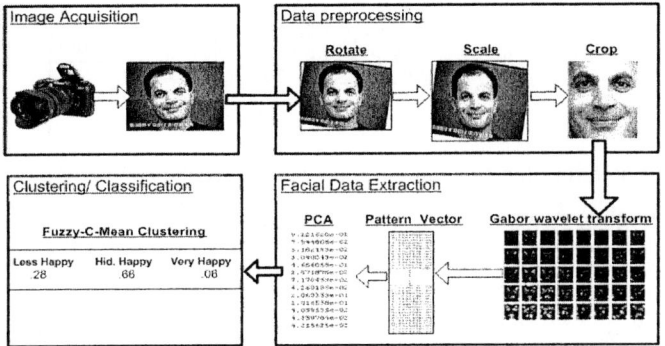

Fig. 1. The facial expression degree estimation system

2.1 Facial Data Acquisition

The training and testing data for our experimental facial expression recognition & degree estimation system is collected from the Cohn-Kanade AU-Coded Facial Expression Database [6].

2.2 Data Preprocessing

Two main issues in image processing will affect the recognition results: the brightness distribution of the facial images and facial geometric correspondence to keep face size constant across subjects. To ensure the above-mentioned criterions in facial expression images, an affine transformation (rotation, scaling and translation) is used to normalize the face geometric position and maintain face magnification invariance and also to ensure that gray values of each face have close geometric correspondence [7].

Fig. 2. Normalization of Face image using the affine-transformation

2.3 Facial Data Extraction

The Gabor wavelets, whose kernels are similar to the 2D receptive field profiles of the mammalian cortical simple cells, exhibit desirable characteristics of spatial locality and orientation selectivity and are optimally localized in the space and frequency domains [8]. The Gabor wavelet (kernel, filter) can be defined as follows:

$$\psi_{\mu,\nu}(z) = \frac{\|k_{\mu,\nu}\|^2}{\sigma^2} e^{\frac{\|k_{\mu,\nu}\|^2 \|z\|^2}{2\sigma^2}} \left[e^{ik_{\mu,\nu}z} - e^{-\frac{\sigma^2}{2}} \right] \quad (1)$$

where μ and ν define the orientation and scale of the Gabor kernel, $z = (x, y)$, $\|\cdot\|$ denotes the Euclidean norm operator, and wave vector $k_{\mu,\nu}$ is defined as follows:

$$k_{\mu,\nu} = k_\nu e^{i\phi_\mu} \quad (2)$$

where $k_\nu = k_{max}/f^\nu$ (here $\nu = \{0,1,...,4\}$) and $\phi_\mu = \pi\mu/8$ (here $\mu = \{0,1,2...,7\}$), here k_{max} is the maximum frequency, and f is the spacing factor between kernels in frequency domain [9]. We employ a lattice of phase-invariant filters at five scales, ranging between 16 and 96 pixels in width, and eight orientations, 0 to $7\pi/8$.

Fig. 3. The magnitude of Gabor kernels at five different scales

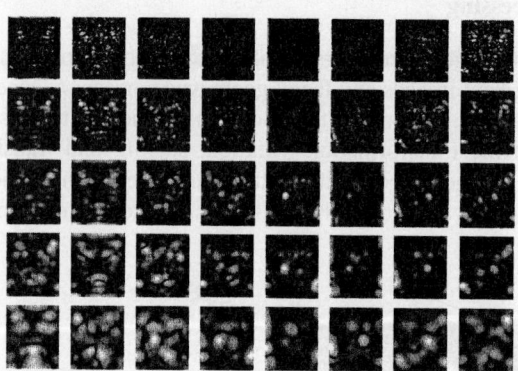

Fig. 4. Gabor magnitude representation of the face from Figure-2

$$\sigma = 2\pi, \ k_{max} = \pi/2, \text{and } f = \sqrt{2}$$

Principle Component Analysis (PCA): Even with sub-sampling, the dimensionality of our Gabor feature vector is much larger to classify. Principal components analysis is a simple statistical method to reduce dimensionality while minimizing mean squared reconstruction error [11].

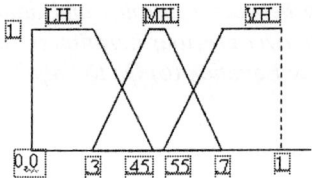

Fig. 5. An ideal membership function for the degree of an emotion

2.4 Clustering and Classification

Fuzzy C-Means (FCM) [10] is one of the most commonly used fuzzy clustering techniques for different degree estimation problems. Its strength over the famous k-Means algorithm [11] is that, given an input point, it yields the point's membership value in each of the classes. In one dimension, we would expect the technique to yield a membership function as shown in Fig. 5.

The aim of FCM is to find cluster centers (centroids) that minimize a dissimilarity function. The membership matrix U is randomly initialized as:

$$\sum_{i=1}^{c} u_{ij} = 1, \text{ for } \forall \ j=1,...n \tag{3}$$

The dissimilarity function that is used in FCM is given as:

$$J(U, c_1, c_2, ... c_c) = \sum_{i=1}^{c} J_i = \sum_{i=1}^{c} \sum_{j=1}^{n} u_{ij}^m d_{ij}^2 \tag{4}$$

Here, $u_{ij} \in [0,1]$, c_i is the centroid of ith cluster, d_{ij} is the Euclidian distance between i^{th} centroid and j^{th} data point and $m \in [1, \infty]$ is a weighting exponent.

To reach a minimum of dissimilarity function there are two conditions. These are given in Equation (5) and (6).

$$c_i = \sum_{j=1}^{n} u_{ij}^m x_j \Big/ \sum_{j=1}^{n} u_{ij}^m \tag{5}$$

$$u_{ij} = 1 \Big/ \sum_{k=1}^{c} \left(\frac{d_{ij}}{d_{kj}} \right)^{2/(m-1)} \tag{6}$$

Pseudo code for Fuzzy-C-Means follows:

I. Randomly initialize the membership matrix (U) that has constraints in Equation (3).
II. Calculate centroids (c_i) by using Equation (5).

III. Compute dissimilarity between centroids and data points using Equation (4). Stop if its improvement over previous iteration is below a threshold.
IV. Compute a new U using Equation (6) Go to Step 2.

3 Results and Observations

3.1 Neutral-Happy Faces

We applied the above-mentioned steps on 945 facial images which are classified as neutral-happy sequences portrayed by 50 actors. Initially we divided the data into two random groups to use in the PCA stage: 200 faces are used to compute the covariance matrix, and then the remaining 745 faces are projected onto this covariance matrix's principal components.

Out of these 745 faces we held out 4 randomly-selected subjects (54 faces) to test our clustering approach. We used FCM to cluster the remaining 691 faces into 3 fuzzy clusters. From this clustering we obtained the clusters. In Fig. 6., we present the scatter plot of the first three clustering, where the cluster centers are the black dots.

In the clustering process we used different combinations of principal components. We considered the principle components in groups of two in sequence: (1st, 2nd), (2nd, 3rd), (3rd, 4th), (4th, 5th) and so on in different combinations until we get a satisfactory result. Later in this chapter we describe what satisfactory result means.

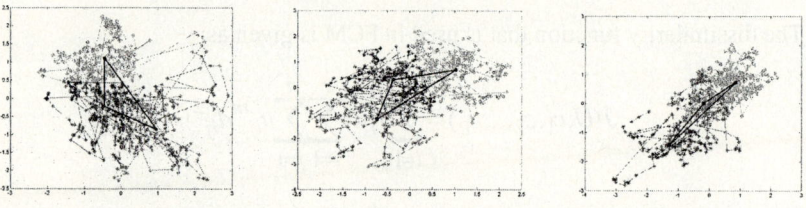

Fig. 6. Plot for principle component pairs; $(1^{st}, 2^{nd}), (2^{nd}, 3^{rd}), (3^{rd}, 4^{th})$ of Neutral-Happy faces

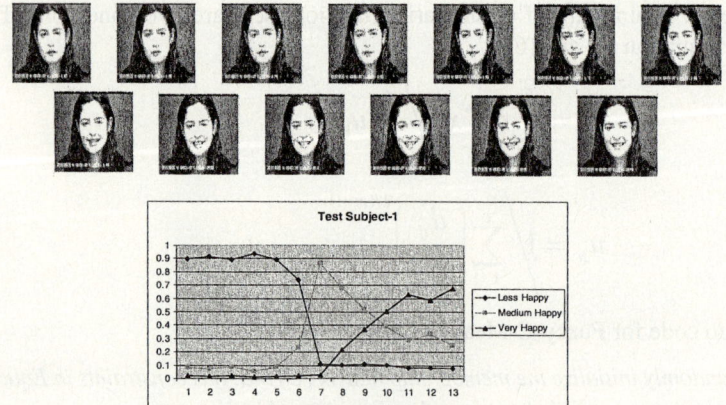

Fig. 7. Neutral-Happy photo sequence of Test Example-1; Plot of membership values

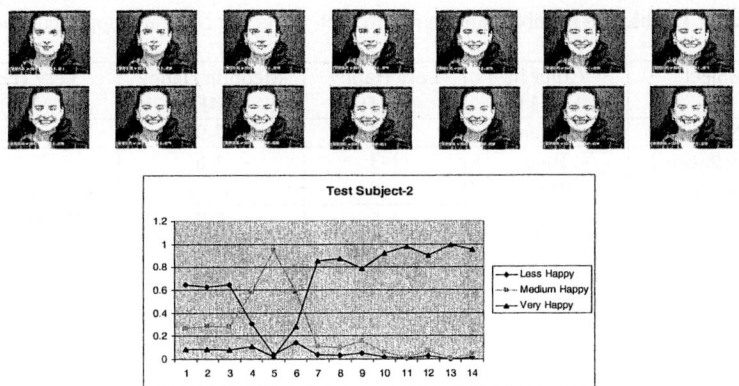

Fig. 8. Neutral-Happy photo sequence of Test Example-2; Plot of membership values

Two individual subjects Neutral-Happy facial sequences are presented in Fig.7. and Fig.8., along with the membership values for each facial image that are projected using the (3rd, 4th) principle component pairs. Few interesting points from these two figures need to be noticed, one is that the membership function is similar to the ideal trapezoidal fuzzy membership function given in Fig. 5.. The other observation is that when we use the winner-take-all criteria to assign the absolute membership; the first person goes slowly to the maximum intensity on the other hand the second individual remains longer in the maximum intensity (Table 1). Similar results are presented on Fig.9. (Due to space constrains the image sequence is skipped).

Winner-take-all strategy is applied on all the fuzzy membership values for all 4 subjects for (3rd, 4th) principle component and the result is provided in Table 1. If any face is classified as less where it is appearing after the sequence medium then it is consider as an uneven assignment of class. It absolutely follows the desired sequencing for a fuzzy clustering in the fuzzy membership representation and also reflected in the absolute class assignment.

Test subjects are viewed in trajectory plotted using the 3rd and 4th principle component cluster center. Here notice that, the individuals faces are moving from near of one cluster center to the other cluster centers in Fig.10.. This again proves that the system is able to correctly capture the fuzzy characteristic that is embedded in the degree estimation process of facial expression.

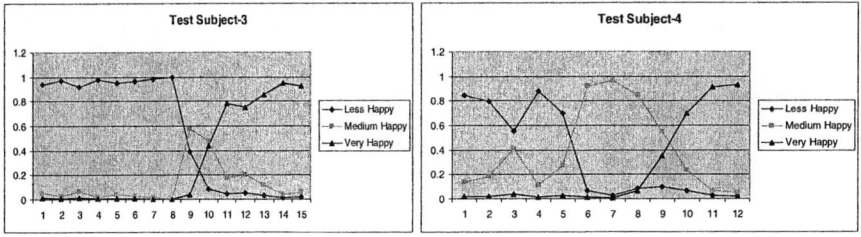

Fig. 9. Plot of Neutral-Happy sequence of Test Example-3 (left) & Example-4 (right) membership values

Table 1. Absalute membership asigned to each image for Neutral-Happy sequence

Subject	Color in Fig. 11.	Less Happy (LH)	Medium Happy (MH)	Very Happy (VH)
Test Subject-1	Green	1-6	7-9	10-13
Test Subject-2	Blue	1-3	4-6	7-14
Test Subject-3	Red	1-8	9-10	11-15
Test Subject-4	Black	1-5	6-9	10-12

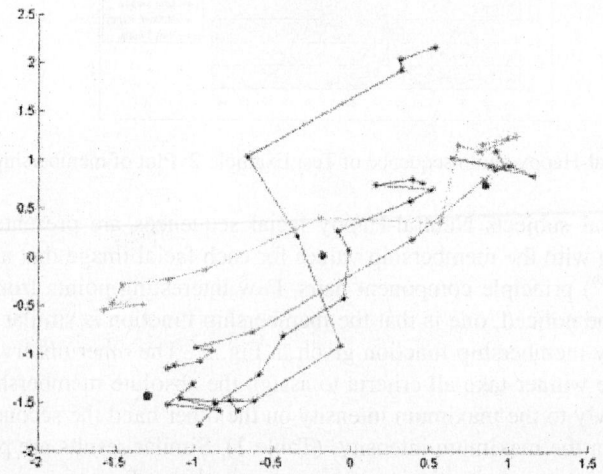

Fig. 10. Neutral-Happy (3rd, 4th principle component) trajectory of the 4 test individuals

3.2 Neutral-Surprise Faces

We applied the similar steps on 1173 facial images which are classified as Neutral-Surprise sequences as portrayed by 63 actors. This time 209 faces are kept for PCA. The remaining 964 faces are projected onto this covariance matrix's principal components. Out of these 964 faces we held out 4 randomly-selected subjects (72 faces) to test our clustering approach. Cluster center is calculated using other 892 faces (Fig. 11.). The best clustering is achieved for 2nd, 3rd principle components.

In Fig.12. and Fig.13. notice that the membership curve is rather an x-curve then a trapezoidal one. But also notice that it still follows the concept of fuzzy class assignment as only two consecutive classes are present for one individual.

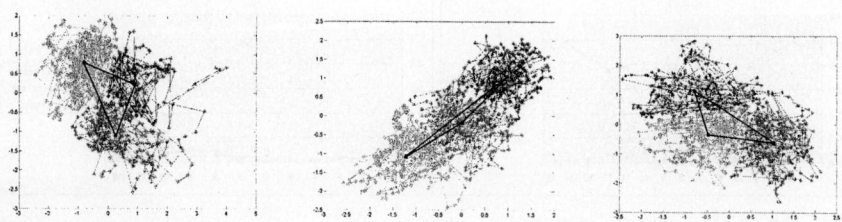

Fig. 11. Plot for (1st, 2nd), (2nd, 3rd), (3rd, 4th) principle components of Neutral-Surprise faces

Fuzzy-C-Mean Determines the Principle Component Pairs to Estimate the Degree 491

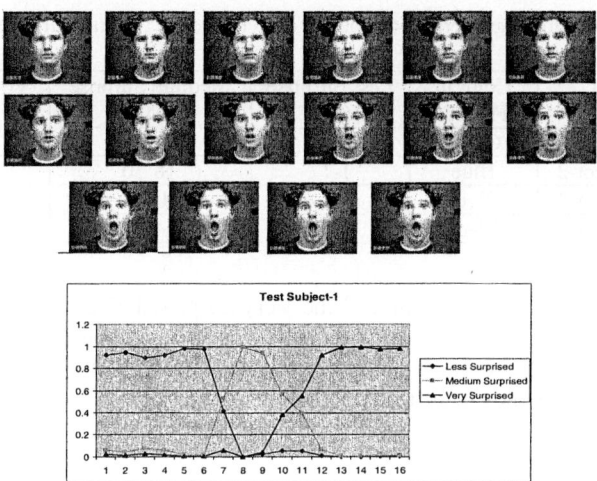

Fig. 12. Neutral-Surprise photo sequence of Test Example-1; Plot of membership values

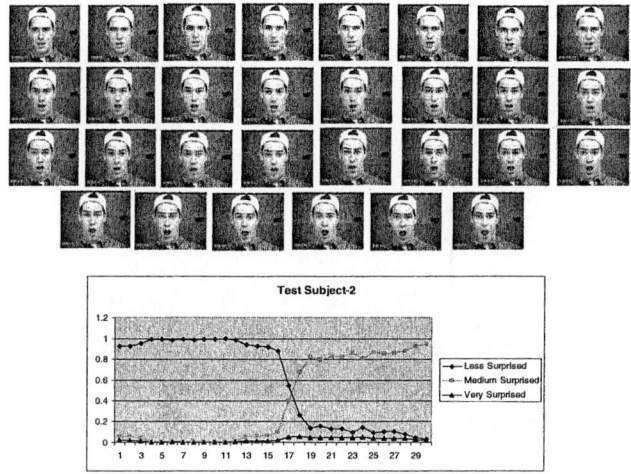

Fig. 13. Neutral-Surprised photo sequence of Test Example-2; Plot of membership values

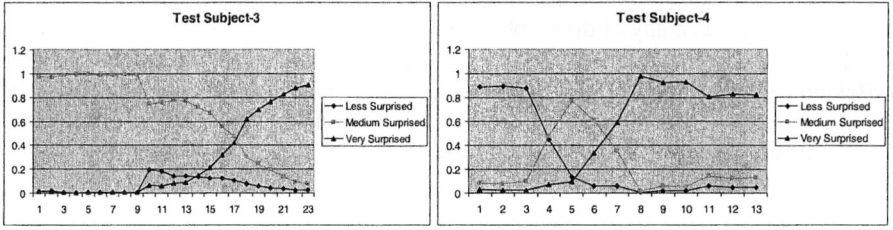

Fig. 14. Plot of Neutral-Surprise sequence of Test Example-3 (left) & Example-4 (right) membership values

Table. 2. Absalute membership asigned to each image for Neutral-Surprised sequence

Subject	Color in Fig. 17.	Less Surprise (LS)	Medium Surprise (MS)	Very Surprise (VS)
Test Subject-1	Green	1-6	7-10	11-16
Test Subject-2	Blue	1-17	18-30	-----
Test Subject-3	Red	-----	1-17	18-23
Test Subject-4	Black	1-3	4-6	7-13

Notice that in Table 2. for subject-2 the Very Surprised category is empty. And for the 3rd subject the Less Surprised is empty, which also is trivial from the membership curves. More over the result could be more suitably projected from Fig.15., here notice that two subject ends their trajectory substantially earlier.

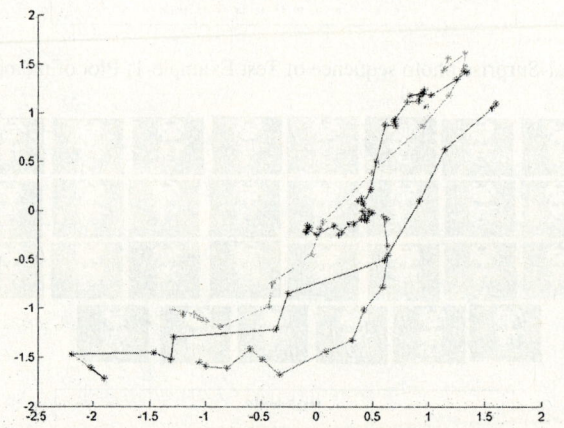

Fig. 15. Neutral-Surprise (2nd, 3rd principle component) trajectory of the 4 test individuals

3.3 Best Cluster Criterion

From the above example and evidences it is clear that our proposed system works satisfactorily. But interesting point to be noted is that how to find out the principle components that best captures the fuzzyness of data. As we could see the principle component pair for different emotion is different (Happy 3rd, 4th, Surprised 2nd, 3rd). We had to check as many of the combination possible.

The best pair of principle components is chosen depending on **minimum distance criterion (MDC)**. The criterion is, out of all combinations of principle component pairs, we cluster them using the similar initial condition and record the distance of the middle cluster center from the middle of the other two centers connecting line. This is provided in the following equation:

$$D = \sqrt{(x_2 - (x_1 + x_3)/2))^2 + (y_2 - (y_1 + y_3)/2))^2} \qquad (7)$$

Where (x_1, y_1), (x_2, y_2) and (x_3, y_3) are cluster centers and D is the distance of the middle cluster center from the midpoint of the connecting line of other two.

4 Conclusion and Future Work

Here we have shown that fuzzy clustering is a promising approach to estimating the degree of intensity of a facial expression, when a face is characterized with Gabor kernels and projected into a low-dimensional space using PCA. The best result is achieved when (3rd,4th) and (2nd, 3rd) principal components are used to describe the Neutral-Happy and Neutral-Surprise faces consecutively. The best suitable principle component is selected depending on the MDC (Minimum Distance Criteria). Satisfactory results are observed in the experimentation process.

Presently we are doing experiment on other prototypic emotions using the same approach and in future we will expand this to more sophisticated facial expressions that are already known as hard problem to characterize with computers.

Reference

1. Ekman, P. and Friesen, W.: The Facial Action Coding System. Consulting Psychologists Press, San Francisco, USA, 1978.
2. Pantic, M. and Rothkkrantz, L. J. M.: Automatic Analysis of Facial Expression: the State of Art, IEEE Trans. Pattern analysis and machine intelligence, Vol. 24, NO. 1, 2000, 1424-1445.
3. Cowic, R., Douglas, E., Tsapatsoulis, N., Votsis, G., Kollias, S., Fellenz, W. and Taylor, J. G.: EMOTION RECOGNITION in Human-Computer Interaction IEEE Signal Processing Magazine, no. 1, 2001, 32-80.
4. Kimura, S. and Yachida, M.: Facial Expression Recognition and Its Degree Estimation, Proc. Computer Vision and Pattern Recognition, 1997, 295-300.
5. Pantic, M. and Rothkrantz, L. J.M.: An Expert System for Recognition of Facial Actions and their Intensity, Proc. 12th International Conference on Innovative Applications of Artificial Intelligence, 2000, 1026-1033.
6. http://vasc.ri.cmu.edu/idb/html/face/facial_expression/, 15 May 2004.
7. Jain, A. K.: Fundamentals of Digital Image Processing, Prentice-Hall of India Private Limited, New Delhi, India, 2003.
8. Daugman, J. G.: Complete Discrete 2-D Gabor Transform by Neural Networks for Image Analysis and Compression, IEEE Trans. Acoustics, Speech and Signal Processing, vol. 36, no. 7, 1988, 1169-1179.
9. Lee, T. S.: Image Representation Using 2D Gabor Wavelets, IEEE Trans. PAMI, Vol. 18, no. 10, 1996, 959-971.
10. Höppner, F., Klawonn, F., Kruse, R. and Runkler, T.: Fuzzy Cluster Analysis, Wiley, 1999.
11. Dubes, R. C.: Cluster analysis and related issues, Handbook of Pattern Recognition & Computer Vision, Chen, C. H., Pau, L. F., and Wang, P. S. P. (Eds.): World Scientific Publishing Co., Inc., River Edge, NJ, 3–32.

An Improved Clustering Algorithm for Information Granulation

Qinghua Hu and Daren Yu

Harbin Institute of Technology, Harbin, China, 150001
huqinghua@hcms.hit.edu.cn

Abstract. C-means clustering is a popular technique to classify unlabeled data into dif-ferent categories. Hard c-means (HCM), fuzzy c-means (FCM) and rough c-means (RCM) were proposed for various applications. In this paper a fuzzy rough c-means algorithm (FRCM) is present, which integrates the advantage of fuzzy set theory and rough set theory. Each cluster is represented by a center, a crisp lower approximation and a fuzzy boundary. The Area of a lower approximation is controlled over a threshold T, which also influences the fuzziness of the final partition. The analysis shows the proposed FRCM achieves the trade-off between convergence and speed relative to HCM and FCM. FRCM will de-grade to HCM or FCM by changing the parameter T. One of the advantages of the proposed algorithm is that the membership of clustering results coincides with human's perceptions, which makes the method has a potential application in understandable fuzzy information granulation.

1 Introduction

In essence, an information granule is a cluster of physical or metal objects drawn together by indistinguishability, similarity, proximity or functionality (Zadeh, 1997). Information granulation, constraint representation and constraint propagation are main aspects of computing with words. Information granulation, the point of departure of computing with words, plays an important role in these techniques and computational theory of perceptions (Zadeh, 1999). Generally speaking, there are two kinds of information granules: crisp and fuzzy. Although crisp granules are the foundation of a variety of techniques, such as interval analysis, rough set theory, D-S theory, they fail to reflect the fact that the granules are fuzzy rather than crisp in most of perception information and human reasoning. In human cognition, fuzziness is the direct consequence of fuzziness of the perceptions of similarity. It is entailed by the finite capability of human mind and sensory organs to resolve detain and store information. However, fuzzy information granulation underlies the remarkable human ability to make rational decision in a condition of imprecision, partial knowledge, partial certainty and partial truth. Fuzzy information granulation plays an important role in fuzzy logic, computing with words and computational theory of perceptions.

Partitioning a given set of unlabeled data into granules is a most fundamental problem in pattern recognition and data mining. C means clustering algorithms are a series of popular techniques to find the structure in unlabeled sample sets. A classical clustering algorithm introduced three decades ago [1], called Hard C Means (HCM), is to assign a sample with a label according to the nearest neighbor principle and minimize the within cluster distance. Fuzzy c means (FCM), an generalization of HCM introduced by Dunn [2] and generalized by Bezdek [3] is one of the most well-known techniques in clustering analysis. The main difference between HCM and FCM is introducing the weighting index m, which is used to control the fuzziness. FCM has better global convergence and slower speed. Basically the performance of FCM clustering is dependent of some parameters, such as the fuzziness weighting exponent, number of clusters [4, 5, 6] and initialization.

Rough set theory [7] is a new paradigm to deal with uncertainty, vagueness and incompleteness and has been applied to fuzzy rule extraction [8], reasoning with uncertainty [9] and fuzzy modeling. Rough set theory is proposed for indiscernibility in classification according to some similarity, whereas fuzzy set theory characterizes the vagueness in language variables. They are complementary in some aspects. Combining fuzzy set and rough set has been an important direction in reasoning with uncertainty [10, 11].

Pawan Lingras [12] introduced a new clustering method, called rough k-means, for web users pattern mining. The algorithm describes a cluster by a center and a pair of lower and upper approximations. And the lower and upper approximations are weighted different parameters in compute the new centers in the algorithm. S. Asharaf etc. extended the technique as a Leader one [13], which doesn't require user specify the number of clusters. In this paper, a fuzzy rough c-means clustering algorithm will be presented. It combines two soft computing techniques, rough set theory and fuzzy set theory, together. Each cluster is represented by a center, a crisp lower approximation and a fuzzy boundary in this algorithm. Then a new center is a weighting average of the lower approximation and boundary. The rest of the paper is organized as follows.

2 Review on HCM and FCM

C-means clustering is defined over a real-number space, and computes its centers iteratively by minimizing the objective function. Given data set $X = \{x_1, x_2, \cdots, x_N\} \subset R^s$, $u = \{u_{ik}\}_{c \times N} \in M_{fcn}$ is a membership matrix, $v = \{v_1, v_2, \cdots, v_c\}$ are c centers of clusters. $v_i \in R^s$, $2 \leq c \leq N$. C-means partitions N data points into c clusters $C_i (i = 1, 2, \cdots, c)$. The objective function is

$$J(u,v) = \sum_{k=1}^{N} \sum_{i=1}^{c} u_{ik} \| x_k - v_i \|^2 ,$$

where $\sum_{i=1}^{c} u_{ik} = 1$ and $u_{ik} \in \{0, 1\}$ for Hard c-means clustering.

The c-means algorithm calculates cluster centers iteratively as follows:

1. Initialize the centers c^l using random sampling;
2. Decide membership of the patterns in one of the c clusters according the nearest neighbor principle from cluster center criteria;
3. Calculate new c^{l+1} centers as

$$v_i^{l+1} = \frac{\sum_{k=1}^{N} u_{ik}^{l+1} x_k}{\sum_{k=1}^{N} u_{ik}^{l+1}}$$

where $u_{ik}^{l+1} = \begin{cases} 1 & if\ i = \arg\min\{\|x_k - v_i^l\|\} \\ 0, & otherwise \end{cases}$

4. if $\max_i \|v_i^{l+1} - v_i^l\| < \varepsilon$, end, otherwise go to step 2.

Hard c-means algorithm is intuitional, easy to be implemented and has fast convergence. However the results from this method are dependent of the cluster center initialization because it will stop in a local minimum.

The most popular clustering algorithm FCM was proposed by Dunn and generalized by Bezdek in 1981. Each pattern doesn't belong to some cluster definitely in this technique, but is a member of all clusters with a membership value. Here the objective function is defined as:

$$J_m(u,v) = \sum_{k=1}^{N} \sum_{i=1}^{c} (u_{ik})^m \|x_k - v_i\|^2$$

where $\sum_{i=1}^{c} u_{ik} = 1$ and $u_{ik} \in [0, 1]$, $m \in (1, +\infty)$.

In order to minimize the objective function, the new centers and memberships are calculated as follows:

$$v_i = \frac{\sum_{k=1}^{N} (u_{ik})^m x_k}{\sum_{k=1}^{N} (u_{ik})^m}, \quad i = 1, 2, \cdots, c$$

$$u_{ik} = \frac{1}{\sum_{j=1}^{c} \left(\frac{d_{ik}}{d_{jk}}\right)^{2/m-1}}, \quad i = 1, 2, \cdots, c; k = 1, 2, \cdots, N$$

The performance of FCM depends on selection of fuzziness weighting exponent m. On one side FCM degrades to HCM when $m = 1$, on the other side, FCM will give the mass center of all data when $m \to \infty$. Bezdek suggested that the preferred value of m is over the range 1.5-2.5. FCM usually stops on a local minimum or saddle-point. The result changes with the initialization. There are a lot of variants of FCM proposed for all kinds of applications in the last decade. Different metrics [14], objective functions [15] and constraint conditions on membership and center will lead to a new clustering algorithm.

3 Rough Sets and Rough C-Means

Rough set methodology has been witnessed great success in modeling imprecise and incomplete information. The basic idea of this method hinges on classifying objects of discourse into clusters containing indiscernible objects with respect to some attributes. In this section we will review some basic definitions in rough set theory.

$<U, A, V, f>$ is called an information system (IS), where $U=\{x_1, x_2, \cdots x_n\}$ is the universe; A is a family of attributes, called knowledge on the universe; V is the value domain of A and f is an information function $f : U \times A \to V$.

Any subset B of knowledge A defines an indiscernibility relation or equivalence relation $IND(B)$ on U and generates a partition Π_B of U, where

$$IND(B) = \{(x, y) \in U \times U \mid \forall a \in B, f_a(x) = f_a(y)\}$$
$$\Pi_B = U/B = \otimes\{a \in B : U / IND(a)\}$$

where

$$A \otimes B = \{X \cap Y : \forall X \in A, \forall Y \in B, X \cap Y \neq \phi\}.$$

An indiscernibility relation on U will generate a partition of U, and a partition of U is necessarily induced by an indiscernibility relation. Indiscernibility relation and partition are corresponding one by one.

We denote the equivalence classes induced by attribute set B as

$$\Pi_B = U/B = \{[x_i]_B : x_i \in U\}$$

Knowledge B induces the elemental concepts $[x_i]_B$ of the universe, which are used to approximate arbitrary subsets in U.

We say Π_A is a refinement of Π_B if there is a partial ordering

$$\Pi_A \prec \Pi_B \Leftrightarrow \forall [x_i]_A \in \Pi_A, \exists [x_j]_B : [x_i]_A \subseteq [x_j]_B$$

Arbitrary subset X of U is characterized by a two-tuple $<\underline{B}X, \overline{B}X>$, called lower approximation and upper approximation, respectively

$$\begin{cases} \underline{B}X = \cup\{[x_i]_B \mid [x_i]_B \subseteq X\} \\ \overline{B}X = \cup\{[x_i]_B \mid [x_i]_B \cap X \neq \phi\} \end{cases}.$$

Lower approximation is the greatest union set of $[x_i]_B$ contained in X and upper approximation is the least union set of $[x_i]_B$ containing X. Lower approximation sometimes also is called positive region, denoted by $POS_B(X)$. Correspondingly, negative region is defined as $NEG_B(X) = U - \overline{B}X$.

If $\underline{B}X = \overline{B}X$, we say that set X is definable, otherwise, X is indefinable or rough set. $BN_B(X) = \overline{B}X - \underline{B}X$ is called boundary set. A set X in U is definable if it's composed by some elemental concepts, which leads X can be precisely characterized with respect to knowledge B and $BN_B(X) = \phi$.

According to the definitions of lower and upper approximations, we know that the object set $[x_i]_B$ belongs to the lower approximation means all of the objects in $[x_i]_B$ are contained by X definitely, $[x_j]_B$ belongs to upper approximation of X is to say that objects in $[x_i]_B$ probably are contained based on knowledge B. Here knowledge B classifies the universe into three cases respect to a certain object subset X: lower approximation, boundary region and negative region.

There are some elemental properties in rough set theory. Given an information system $<U, A, V, f>$, $B \subseteq A$, $U/B = \{X_1, X_2, \cdots, X_c\}$,:

- $\forall x \in U$, $x \in \underline{B}X_i \Rightarrow x \notin \underline{B}X_j$, $j = 1, 2, \cdots, c; j \neq i$
- $\forall x \in U$, $x \in \underline{B}X_i \Rightarrow x \in \overline{B}X_i$, $i = 1, 2, \cdots, c$
- $\forall x \in U$ $\forall i$, $x \notin \underline{B}X_i \Rightarrow \exists X_k, X_l : x \in \overline{B}X_k$ and $x \in \overline{B}X_l$.

Property 1 shows an object can be part of at most one lower approximation; property 2 means that objects that belong to the lower approximation necessarily are contained by the upper approximation and the third shows if an object is not part of any lower approximation, the object must belong to at least two upper approximations.

The key problem in incorporating rough set theory into c-means clustering is how to define the definitions of lower and upper approximation in real space. Assumed that the lower approximations and upper approximations have been found, the modified centroid is given by:

$$v_j = \begin{cases} \omega_{lower} \times \dfrac{\sum_{x \in \underline{A}X} x_j}{|\underline{A}X|} + \omega_{upper} \times \dfrac{\sum_{x \in \overline{A}X - \underline{A}X} x_j}{|\overline{A}X - \underline{A}X|}, & |\overline{A}X - \underline{A}X| \neq 0 \\ \omega_{lower} \times \dfrac{\sum_{x \in \underline{A}X} x_j}{|\underline{A}X|} & \text{otherwise} \end{cases}$$

where ω_{lower} and ω_{upper} correspond to the relative importance of the lower approximation and boundary. It is easy to find that the above formula is a generalization of HCM. Especially when $|\overline{A}X - \underline{A}X| = 0$, RCM will degenerates to HCM.

Let's design the criteria to determine whether an object belongs to the lower approximation, boundary or negative region.

For each object x and center point v, $D(x,v)$ is the distance from x to v. The differences between $D(x,v_i)$ and $D(x,v_j)$ are used to determine the label of x. Let $D(x,v_j) = \min_{1 \leq i \leq c} D(x,v_i)$ and $T = \{\forall i, i \neq j : |D(x,v_i) - D(x,v_j)| \leq Threshold\}$.

If $T \neq \phi \Rightarrow x \in \overline{A}(v_i)$, $x \in \overline{A}(v_j)$ and $x \notin \underline{A}(v_l)$, $l = 1, 2, \cdots, c$

If $T = \phi$, $x \in \underline{A}(v_j)$, and $x \in \overline{A}(v_j)$.

The performance of rough c-means algorithm depends on five conditions: the weighting index ω_{lower}, ω_{upper}; Threshold, number of clusters c and initialization of c centers. So it's difficult for users to manipulate the algorithm in practice. We will present a fuzzy rough c means (shortly, FRCM) to overcome this problem.

4 Fuzzy Rough Clustering

As we know, HCM assigns a label to an object definitely; the membership value is 0 or 1. While FCM maps a membership over the arrange 0 to 1; each object belongs to some or all of the clusters to some fuzzy degree. RCM classify the object space into three parts, lower approximation, boundary and negative region. Then different weighting values are taken in computing the new centers, respectively.

According RCM, all the objects in lower approximation take the same weight, and all the objects in boundary take another weighting index uniformly. In fact, the objects in boundary regions have different influence on the centers and clusters. So different weighting should be imposed on the objects.

How to d it in computing? According to the lower approximation is the object subset which belongs to a cluster without doubt. And the boundary is the object subset that maybe belongs to the cluster. So boundary is the region assigned a label with uncertainty. The fuzziness membership should be imposed on the objects in boundary.

Here we define that membership function is given by

$$u_{ik} = \begin{cases} 1, & x_k \in \underline{A}(v_i) \\ \dfrac{1}{\sum_{j=1}^{c}\left(\dfrac{d_{ik}}{d_{jk}}\right)^{2/m-1}}, & x_k \in \overline{A}(v_i) \end{cases} \quad i = 1,2,\cdots,c; k = 1,2,\cdots,N \ .$$

It is worth noting that the membership function is constructed and is not derived by the objective function. So it is not strict. However, this problem has little influence on the performance.

Then the new centers are calculated by

$$v_i = \frac{\sum_{k=1}^{N}(u_{ik})^m x_k}{\sum_{k=1}^{N}(u_{ik})^m} , \ i = 1,2,\cdots,c \ .$$

The objective function used is

$$J_m(u,v) = \sum_{k=1}^{N}\sum_{i=1}^{c}(u_{ik})^m \parallel x_k - v_i \parallel^2$$

Then the lower and upper approximations are defined the same as RCM.

For each object x and center point v, $D(x,v)$ is the distance from x to v. The differences between $D(x,v_i)$ and $D(x,v_j)$ are used to determine the label of x. Let

$$D(x,v_j) = \min_{1 \le i \le c} D(x,v_i)$$

and $T = \{\forall i, i \ne j : |\ D(x,v_i) - D(x,v_j)| \le Threshold\}$.

1. If $T \neq \phi \Rightarrow x \in \overline{A}(v_i)$, $x \in \overline{A}(v_j)$ and $x \notin \underline{A}(v_l)$, $l = 1,2,\cdots,c$
2. If $T = \phi$, $x \in \underline{A}(v_j)$, and $x \in \overline{A}(v_j)$.

It's worth noting that the definitions of lower and upper approximations are different from the classical ones. They are not defined based on any predefined indiscernible relation on the universe. What' more it is remarkable that the distance metric is not limited with Euclidean distance. So 1-norm, 2-norm, p-norm, infinite norm and some other measures can be applied.

The FRCM can be formulated as follows:

Input: unlabeled data set; number of cluster c, threshold T, exponent index m, stop criterion ε.

Output: membership matrix U.

Step 1. Let l=0, $J_m^{(0)}(u,v) = 0$; randomly make a membership $U_{c \times N}^l$;

Step 2. Compute the c centers $v_i^{(l)}, (i = 1,2,\cdots,c)$ with $U_{c \times N}^l$ and data set;

Step 3. Compute the $\underline{A}(v_i^{(l)})$, $\overline{A}(v_i^{(l)})$, $u_{ij}^{(l+1)}$ $v_i^{(l+1)}, (i = 1,2,\cdots,c)$ with $v_i^{(l)}, (i = 1,2,\cdots,c)$, and Threshold T,

Step 4. Compute $J_m^{(l+1)}(u,v)$;

Step 5. $\| J_m^{(l+1)}(u,v) - J_m^{(l)}(u,v) \| < \varepsilon$, then stop, otherwise, $l = l+1$, go to step 2.

The difference between FCM and FRCM is that the membership values of objects in lower approximation are 1, while those in boundary region are the same as FCM. In other word, FRCM first partitions the data into two classes: lower approximation and boundary. Only the objects in boundary are fuzzified.

The difference between RCM and FRCM is that, as for RCM the same weights are imposed on all of the objects in lower approximations and boundary regions, respectively. However, as to FCM, each object in boundary region is imposed a distinct weight, respectively.

5 Numeric Experiments

To illustrate the differences of these algorithms, let's look at the membership function in the one-dimensional space for a two-cluster problem.

The membership functions are showed in figure 1.

Comparing figure 1, we can find the membership generated by FRCM clustering is much similar with human conceptions. For example, we need to cluster some men into two groups according their ages. HCM and RCM will present two crisp sets of the men, which cannot reflect the fuzziness in human reasoning. Although FCM will give two fuzzy classes, but the memberships of the fuzzy sets are not understandable. As to a one year old baby, his memberships to young and old are nearly identical and close to 0.5. In fact, it's unreasonable and unacceptable according to my intuition.

Fuzzy rough c-means algorithm is implemented and applied to some artificial data. Three clusters of data are generated randomly. There are one hundred points in each cluster and the data points are satisfied a normal distribution. The data and result of HCM are showed in figure 2.

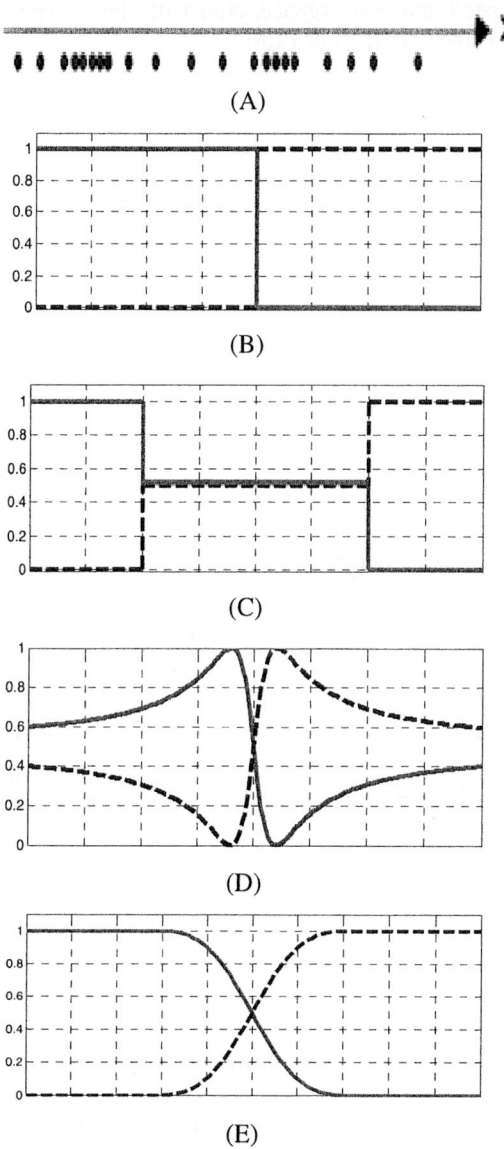

Fig. 1. (A) shows some data distributed in one-dimensional space; (B) is the membership function with HCM; (C) is the membership function with RCM; (D) is function with FCM and (E) is with FRCM. It's easy to see that HCM gets a crisp boundary line; RCM produces a boundary region; FCM and FRCM develop a soft boundary region.

Figure 3 shows the clustering results with different thresholds. And the comparison is presented in table 1. We can find the iteration number of HCM is the fastest. And global convergence, namely objective function, of FCM is the best. FRCM make a tradeoff between speed and convergence. And with the change of threshold FRCM has a similar performance as HCM or FCM.

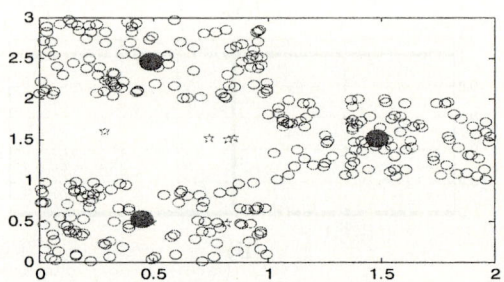

Fig. 2. Result of HCM

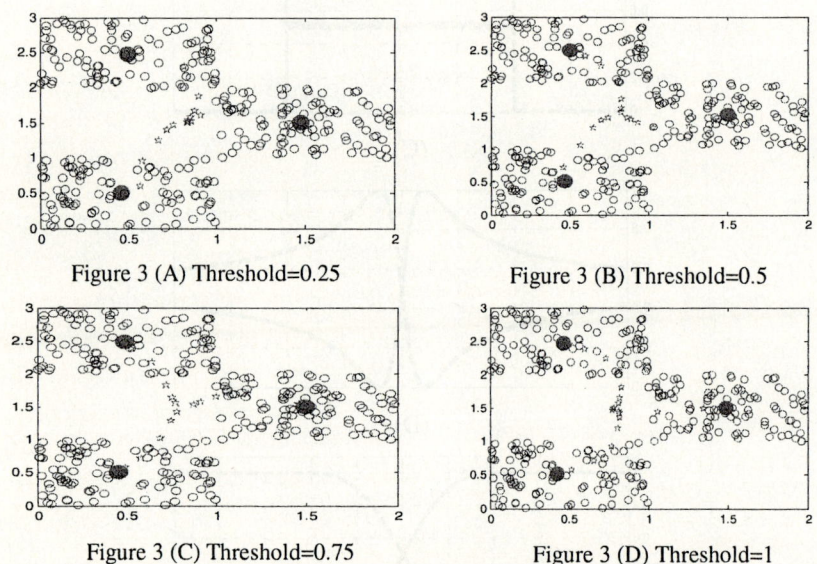

Figure 3 (A) Threshold=0.25

Figure 3 (B) Threshold=0.5

Figure 3 (C) Threshold=0.75

Figure 3 (D) Threshold=1

Fig. 3. The results with different Threshold from FRCM

Table 1. Comparison of HCM, FCM and FRCM

	HCM	FCM	FRCM	
			T=0.25	T=10
Iter. count	6	16	11	16
obj. fcn	7.0589	5.6172	6.5679	5.6172
Center1	0.4487,0.5200	0.4272,0.5153	0.4592, 0.5002	0.4272, 0.5153
Center2	0.4919,2.4553	0.4675,0.4699	0.4953, 2.4701	0.4675, 2.4699
Center3	1.4868, 1.5083	1.4873,1.5054	1.4792, 1.5097	1.4873, 1.5054

6 Conclusion

Combining two soft computing methods, a fuzzy rough c means clustering algorithm (FRCM) is proposed in this paper. FRCM characterizes each class with a positive region, a fuzzy boundary region and a negative region. According rough set methodology, the objects in positive regions and negative regions are definitely `contained or not contained by the class. Only the objects in boundary regions are fuzzy and unclassified, so a soft separating hyper-plane is required in these regions. A new membership function is constructed based on the idea that membership should be only assigned the classes which centers are near the data points. What' more, the size of fuzzy boundary regions can be controlled with a proportional index, which control the fuzziness of the clustering results. Because the clustering results are similar to human's perception, this algorithm is applicable to fuzzy information granulation and computing with words.

Reference

1. Mac Queen, J.: Some methods for classification and analysis of multivariate observations. In: Le Cam. 281-297
2. Dunn, J.C.: Some recent investigations of a new fuzzy partition algorithm and its application to pattern classification problems. J. cybernetics, 4 (1974) 1-15
3. Bezdek, J., C.: pattern recognition with fuzzy objective function algorithms. Plenum, New York. 1981
4. Nikhil R. Pal, James C. Bezdek: On cluster validity for the fuzzy c-means model. IEEE transaction on fuzzy sytems. 3 (1995) 370-379
5. Jian Yu, Qiansheng Cheng, Houkuan Huang: Analysis on weighting exponent in the FCM. IEEE Transaction on SMC, Part B—cybernetics, 31 (2004) 634-639
6. Shehroz S. Khan, Amir Ahmad: Cluster center initialization algorithms for K-means clustering. Pattern recognition letters. 25 (2004) 1293-1302
7. Pawlak Z.: rough sets—theoretical aspects of reasoning about data. Kluwer academic publishers, 1991
8. Jensen R., Shen Q.: Fuzzy-rough attribute reduction with application to web categorization. Fuzzy sets and systems. 141 (2004) 469-485
9. Wang, Yi-Fan: Mining stock price using fuzzy rough set system. Expert Systems with Applications. 24 (2003) 13-23
10. Dubois D., Prade H.: Putting fuzzy sets and rough sets together, in: R. Slowiniski (Ed.), intelligent Decision support, 1992, 203-232
11. Wu, W. Zhang, W.: Constructive and axiomatic approaches of fuzzy approximation operators. Information Sciences. 159 (2004) 233-254
12. Pawan Lingras, Chad West: Interval set clustering of web users with rough k-means. Inter. J. of intell. Inform. system. 23 (2003) 5-16
13. Asharaf S., Narasimh M. Murty: A rough fuzzy approach to web usage categorization. Fuzzy sets and systems. 148 (2004) 119-129
14. Hathaway, R. J., Bezdek J. C. C-means clustering strategies using L_p norm distance. IEEE Trans. On fuzzy systems. 8 (2000) 576-582

15. Li R.P. Mukaidon M. A maximum entropy approach to fuzzy clustering. In: proc. Of the 4th IEEE conf. on fuzzy systems, 1995. 2227-2232
16. Zadeh L. A.: Fuzzy logic equals Computing with words. IEEE Transactions on fuzzy systems 2 (1996) 103-111
17. Zadeh L. A.: Toward a theory of fuzzy information granulation and its centrality in human reasoning and fuzzy logic. Fuzzy sets and systems. 90 (1997) 111-127
18. Zadeh L A.: From computing with numbers to computing with words - From manipulation of measurements to manipulation of perceptions. IEEE Transactions on circuits and systems. 46 (1999) 105-119

A Novel Segmentation Method for MR Brain Images Based on Fuzzy Connectedness and FCM

Xian Fan, Jie Yang, and Lishui Cheng

Institute of Image Processing and Pattern Recognition,
Shanghai Jiao Tong University,
200030, Shanghai, P.R. China
Yivanne@sjtu.edu.cn

Abstract. Image segmentation is an important research topic in image processing and computer vision community. In this paper, a new unsupervised method for MR brain image segmentation is proposed based on fuzzy c-means (FCM) and fuzzy connectedness. FCM is a widely used unsupervised clustering algorithm for pattern recognition and image processing problems. However, FCM does not consider the spatial coherence of images and is sensitive to noise. On the other hand, fuzzy connectedness method has achieved good performance for medical image segmentation. However, in the computation of fuzzy connectedness, one needs to select seeds manually which is elaborative and time-consuming. Our new method used FCM as the first step to select salient seeded points and then applied fuzzy connectedness algorithm based on those seeds. Thus our method achieved unsupervised automatic segmentation for brain MR images. Experiments on simulated and real data sets proved it is effective and robust to noise.

1 Introduction

Image processing techniques are important in medical engineering, such as 3D visualization, registration and fusion, and operation planning. In image processing, object extraction or image segmentation is the most crucial step because it lays the foundation for subsequent steps such as object visualization, manipulation, and analysis. Since manual or semi-automatic segmentation takes a lot of time and energy, automatic image segmentation algorithms have received a lot of attention.

Images are by nature fuzzy [1]. This is especially true to the biomedical images. The fuzzy property of images is usually made by the limitation of scanners in the ways of spatial, parametric, and temporal resolutions. What's more, the heterogeneous material composition of human organs adds to the fuzzy property in magnetic resonance images (MRI). As the goal of image segmentation is to extract the object from the other parts, segmentation by hard means may despoil the fuzziness of images, and lead to bad results. By contrast, using fuzzy methods to segment biomedical images would respect the inherent property fully, and could retain inaccuracies and uncertainties as realistically as possible [2].

Although the object regions of biomedical images manifest themselves with heterogeneity of intensity values due to their fuzzy property, knowledgeable human observers could recognize the objects easily from background. That is, the elements in these regions seem to hang together to form the object regions in spite of their heterogeneity of intensity.

In 1973, Dunn[3] firstly developed "fuzzy c-means" (FCM) which is a fuzzy clustering method to allow one piece of data to belong to two or more clusters. In [4], Bezdek improved the algorithm so that the objective function minimizes in an iterative procedure. Clark has developed a system to provide completely automatic segmentation and labeling of MR brain images in [5]. Although FCM maintains inaccuracy of the elements' membership to every cluster, it does not take into consideration that the elements in the same object are hanging together. What's more, it is susceptible to noise and may end with a local minimum.

In [6], Rosenfeld developed fuzzy digital topological and geometric concepts. He defined fuzzy connectedness by using a min-max construct on fuzzy subsets of 2-D picture domains. Based on this, Udupa [7] introduces a fundamental concept called affinity, which combined fuzzy connectedness directly with images and utilized it for image segmentation. Thus the properties of inaccuracy and hanging togetherness are made full use of. However, the step of manual selection of seeds, although onerous and time-consuming, is unavoidable in the initialization of fuzzy connectedness.

This paper proposes a new segmentation method based on the combination of fuzzy connectedness and FCM, and applies it to segment MR brain images. The method makes full use of the advantages of the two fuzzy methods. The segmentation procedure implements automatic seed selection and at the same time guarantees the quality of segmentation. The experiments on simulated and real MR brain images prove that this new method behaves well.

The remainder of this paper is organized as follows. Section 2 introduces the fuzzy connectedness method. Section 3 describes our fuzzy connectedness and FCM based method in detail. Section 4 presents the experimental results. Conclusions are given in Section 5.

2 Background

In this section, we summarize the fuzzy connectedness method for segmentation proposed by Udupa[7].

Let X be any reference set. A *fuzzy subset* of $\mathbf{A}$ is a set of ordered pairs where $\mu_A : X \to [0,1]$. μ_A is called the *membership function* of $\mathbf{A}$ in X. A *2-ary fuzzy relation* ρ in X is a fuzzy subset of $X \times X$, $\rho = \{((x,y), \mu_\rho(x,y)) \mid x, y \in X\}$ where $\mu_\rho : X \times X \to [0,1]$.

For any $n \geq 2$, let n-dimensional Euclidean space R^n be subdivided into hypercuboids by n mutually orthogonal families of parallel hyperplanes. Assume, with no loss of generality, that the hyperplanes in each family have equal unit spacing so that the hypercuboids are unit hypercubes, and we shall choose coordinates so that the center of each hypercube has integer coordinates. The hypercubes will be called *spels* (an abbreviation for "space elements"). Z^n is the set of all spels in R^n.

A fuzzy relation α in Z^n is said to be a *fuzzy adjacency* if it is both reflexive and symmetric. It is desirable that α be such that is a non increasing function of the distance $\|c-d\|$ between c and d. We call the pair (Z^n, α), where α is a fuzzy adjacency, a *fuzzy digital space*.

A *scene* over a fuzzy digital space (Z^n, α) is a pair $C = (C, f)$ where $C = \{c \mid -b_j \leq c_j \leq b_j \text{ for some } b \in Z_+^n\}$; Z_+^n is the set of n-tuples of positive integers; f, called *scene intensity*, is a function whose domain is C. C is called the *scene domain*, whose range $[L, H]$ is a set of numbers (usually integers).

Let $C = (C, f)$ be any scene over (Z^n, α). Any fuzzy relation κ in C is said to be a *fuzzy spel affinity* (or, *affinity* for short) in C if it is reflexive and symmetric. In practice, for κ to lead to meaningful segmentation, it should be such that, for any $c, d \in C$, $\mu_\kappa(c,d)$ is a function of: 1) the fuzzy adjacency between c and d; 2) the homogeneity of the spel intensities at c and d; 3) the closeness of the spel intensities and of the intensity-based features of c and d to some expected intensity and feature values for the object. Further, $\mu_\kappa(c,d)$ may depend on the actual location of $\mu_\kappa(c,d)$ (i.e., μ_κ is shift variant).

Path strength is denoted as the strength of a certain path connecting two spels. Saha and Udupa[7] have shown that under a set of reasonable assumptions the minimum of affinities is the only valid choice for path strength. So the path strength is

$$\mu_N(p) = \min_{1 \leq i \leq l_p} [\mu_\kappa(c_{i-1}, c_i)], \qquad (1)$$

where p is the path $< c_1, c_2, ..., c_{l_p} >$ and N is denoted as the fuzzy κ-net in C. Every pair of (c_{i-1}, c_i) is a link in the path while c_{i-1} and c_i may not always be adjacent.

For any scene C over (Z^n, α), for any affinity κ and κ-net N in C, fuzzy κ-connectedness K in C is a fuzzy relation in C defined by the following membership function. For any $c, d \in C$

$$\mu_K(c,d) = \max_{p \in P_{cd}} [\mu_N(p)]. \qquad (2)$$

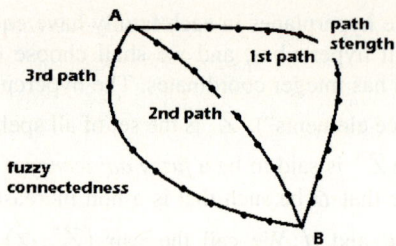

Fig. 1. Illustration of the algorithm of fuzzy connectedness

Combined with eq.(1), eq.(2) shows the min-max property of the fuzzy connectedness between each two spels, as is illustrated in Fig.1.

A physical analogy one may consider is to think that there are a lot of strings connecting spels A and B, each with its own strength (called path strength as to a certain path). Imagine that A and B are pulled apart. Under the force the strings will break one by one. As to a certain string, the affinity of the link where the string breaks is denoted as the path strength of this string (the path strength is defined as the minimum affinity of all the links in the path). When all but one string are broken, the last string behave as the strongest one and it's path strength is denoted as the fuzzy connectedness between spels A and B.

Let S be any subset of C. We refer to S as the set of seed spels and assume throughout that $S \neq \emptyset$. The fuzzy $k\theta$-object of C containing S equals

$$O_{K\theta}(S) = \left\{ c \mid c \in C \text{ and } \max_{s \in S}[\mu_K(s,c)] \geq \theta \right\} . \tag{3}$$

With eq.(3), we could extract the object we want given θ and S. This could be computed via dynamic programming [7].

3 Proposed Method

In this section, we propose a new segmentation method which is unsupervised and robust, based on the combination of fuzzy connectedness and FCM.

As Udupa has been trying to utilize the fuzzy nature of the biomedical images in both aspects of inaccuracy and hanging togetherness, users' identifying seed spels belonging to the various objects is always left to be an onerous task. Therefore, automatic selection of seeds becomes important in our research work. On the other hand, FCM, as an old fuzzy method in clustering, is likely to converge to the local minimum and thus lead to false segmentation. Our new method combines the two fuzzy methods organically, trying to implement automatic seed selection and guarantee the segmentation quality.

The outline of the new method is as follows. First, FCM is used to pre-segment the MR brain image, through which the scope of the seeds is obtained. Since the number of seeds within the scope is much more than that we need, the unqualified seeds within the scope are automatically eliminated according to their spatial information. With the left seeds as an initialization, the MR brain image is segmented by fuzzy connectedness. Here are the detailed steps of our proposed algorithm applied to MR brain images.

Step 1. Preprocess the MR brain images, including denoising, intensity correction, and inhomogeneity-correction. In intensity correction, a standardized histogram of MR brain image is acquired, and all the other images' is corrected so that their histogram would match the standardized one as best as possible [8].

Step 2. Set the number of clustering to 4, and pre-segment the image by FCM, so that the expected segmented objects will be white matter, grey matter, CSF and background.

Step 3. Compute the average intensity of each of the four clusters, which is denoted as $v_j^{(b)}, b=1,2,3,4$. Due to the biomedical knowledge that white matter's intensity is larger than other tissues in MR brain image, we find the largest $v_j^{(i)}$. The cluster $v_j^{(i)}$ belongs to is the white matter.

Step 4. Compute the deviation of the cluster obtained in step 3. The scope of the seeds is defined by the equation

$$v_j^{(i)} \pm 0.3\,\delta^{(i)} \ . \tag{4}$$

Step 5. Define N as the number of seeds in fuzzy connectedness.

Step 6. Take the spatial information of the spels within the scope into account, and class them into N clusters. Make the center spels of each cluster the seeds as initialization.

Step 7. Segment the image precisely with selected seeds by fuzzy connectedness method.

With step 1, we get the standardized histogram of each MR brain image, which guarantees that the parameter 0.3 in eq.(4) will work. Then according to the average and deviation intensity of the region of interest, the scope of the seeds could be gotten with the eq.(4). Since the spatial information of these spels is taken into account, the automatically selected seeds are effective.

4 Experimental Results

In this section, we give the experimental results with both simulated and real MR brain images.

To simulated images, as the reference result is the "ground truth", accuracy is described with three parameters: true positive volume fraction (TPVF), false positive volume fraction (FPVF), and false negative volume fraction (FNVF). These parameters are defined as follows:

$$TPVF(V,V_t) = \frac{|V \cap V_t|}{|V_t|}, \qquad (5)$$

$$FPVF(V,V_t) = \frac{|V - V_t|}{|V_t|}, \qquad (6)$$

$$FNVF(V,V_t) = \frac{|V_t - V|}{|V_t|}, \qquad (7)$$

Where V_t denotes the set of spels in the reference result, V denotes the set of spels resulting from the users' algorithms. $FPVF(V,V_t)$ denotes the cardinality of the set of spels expressed as a fraction of the cardinality of V_t that are in the segmentation result of the method but not in V_t. Analogously, $FNVF(V,V_t)$ denotes a fraction of V_t that is missing in V. We use probability of error (PE) to evaluate the overall accuracy of the simulated image segmentation. PE [9] could be described as

$$PE = FPVF(V,V_t) + FNVF(V,V_t). \qquad (8)$$

The larger PE is, the poorer the accuracy of the segmentation method is.

To the real images, as the reference result is made manually, the inaccuracy of the reference segmentation result should be taken into account. Zijdenbos's[10] Dice Similarity Coefficient (DSC), which has been adopted for voxel-by-voxel classification agreement, is proper to evaluate the segmentation result of real images. We now describe it as follows. For any type T, assuming that V_m denotes the set of pixels assigned for it by manual segmentation and V_α denotes the set of pixels assigned for it by the algorithm, DSC is defined as follows:

$$DSC = 2 \bullet \frac{|V_m \cap V_\alpha|}{|V_m| + |V_\alpha|}. \qquad (9)$$

Since manual segmentations are not "ground truth", DSC provides a reasonable way to evaluate automated segmentation methods.

4.1 Simulated Images

We applied our method on simulated MR brain images provided by McConnell Brain Image Center [11]. As the brain database has provided the gold standard, validation of the segmentation methods could be carried out. The experimental images in this paper were imaged with T1 modality, 217 * 181 (spels) resolution and 1mm slice thickness. 3% noise has been put up and the intensity-nonuniformity is 20%.

We take the 81th to 100th slices of the whole brain MRI in the database, and segment them with fuzzy connectedness (FC), FCM and our proposed method respectively. The evaluation values are listed in table 1 according to the gold standard.

Table 1. Average value of 20 simulated images using three segmentation methods (%)

Study	TPVF	FPVF	FNVF	PE
FCM	96.45	2.79	3.55	6.34
FC	98.03	1.09	1.97	3.06
Proposed Method	98.89	1.55	1.11	2.66

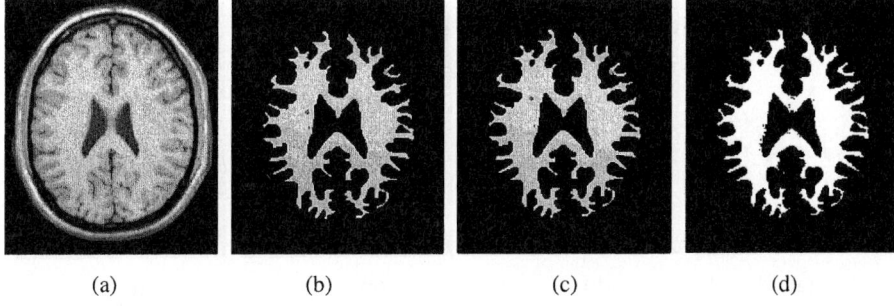

(a) (b) (c) (d)

Fig. 2. Segmentation of the white matter from simulated images (a) original image; (b) result of segmentation by fuzzy connectedness; (c) result of segmentation by our proposed method; (d) white matter provided by gold standard

It could be noticed in table 1 that our proposed method behaves best, and the second is fuzzy connectedness, the last FCM. This should be own to the high quality of the automatic seed selection. And the high accuracy of fuzzy connectedness also contributes to the good result.

Take the 97th slice as an example. When using fuzzy connectedness as the segmentation method, manual selected seeds are (79, 64), (61,131), as shown in Fig. 2(b). When our proposed method is implemented, the scope of the seeds from FCM is 0.98949 ± 0.0002878. That is, the mean value of the intensity is 0.98949, and the deviation value is 0.0009593. According to eq.(4), the scope of seeds is obtained. Considering the spatial information of the seeds from FCM, the unqualified seeds are eliminated and the two left is (85, 65) and (62,130), as is shown in Fig.2(c). It could be seen that the seeds automatically selected are located near to the manual selected ones, and thus they work as effectively as the manual ones.

4.2 Real Images

We applied our method to twelve real MRI brain data sets. They were imaged with a 1.5 T MRI system (GE, Signa) with the resolution 0:94 *1.25*1.5 mm (256* 192*

124 voxels). These data sets have been previously labeled through a labor-intensive (usually 50 hours per brain) manual method by an expert. We first applied our method on these data sets and then compared with expert results. We give the results from the 3th to the 10th of them gained by our method in Table 2 and Fig. 3.

Table 2. Average DSC of 8 real cases (%)

	1	2	3	4	5	6	7	8
FC	72	67	61	67	68	55	72	54
FCM	78	77	79	80	82	82	81	80
Proposed Method	82	83	85	84	85	86	82	81

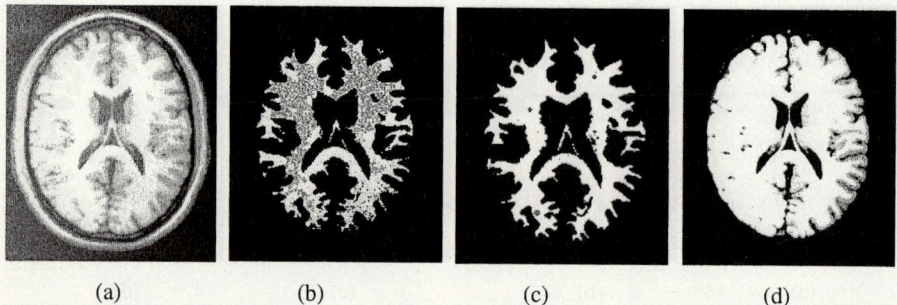

(a)   (b)   (c)   (d)

Fig. 3. Segmentation of the white matter in real images (a) original image; (b) the scope of seeds to be selected from FCM; (c) white matter segmented from proposed method; (d) white matter segmented from FCM

According to Table 2, the rank of DSC by the three methods is the same as that in simulated image experiments. It could be observed that there are several cases' DSC of FCM is rather poor. That is because the noise in real images changes widely in different cases. Since FCM is sensitive to noise, it could not segment the object correctly.

Take the 6th case as an illustration. Although the original figure is corrupted by noise severely, the scope of the seeds selected by FCM is still within the region of white matter as is marked in Fig.3 (b). Thus our proposed method behaves well with two automatically selected seeds as in Fig.3(c). Yet the result of FCM is very poor as is shown in Fig.3 (d).

5 Conclusion

Fuzzy property is the nature of images. When the images are segmented by fuzzy methods, the inaccuracies property of the elements is adopted. FCM is one of the fuzzy clustering techniques, but it ignores the hanging-togetherness property of images. Thus it is sensitive to noise and may end with a local minimum. Fuzzy connectedness is a method which takes both the inaccuracies and hanging-togetherness of the elements into consideration. Although fuzzy connectedness behaves well, the selection of seeds takes operators time and energy. In our proposed method, seed selection is made automatic by FCM. The quality of the automatically selected seeds is guaranteed by the elimination of unqualified seeds. What's more, unlike FCM, the new method is robust because of the use of fuzzy connectedness. Through simulated and real image experiments, we could see that our proposed method could not only automatically select seeds, but has a desirable accuracy.

Reference

1. Udupa, J.K., Saha, P.K.: Fuzzy Connectedness and Imaging Segmentation. Proceedings of The IEEE, Vol.91, No.10, pp: 1649-1669, 2003
2. Saha, P.K., Udupa, J.K. and Odhner, D.: Scale-based fuzzy connected image segmentation: Theory, algorithms, and validation. Computer Vision and Image Understanding, Vol.77, pp: 145-174, 2000
3. Dunn, J.C. A Fuzzy Relative of the ISODATA Process and Its Use in Detecting Compact Well-Separated Clusters. Journal of Cybernetics, Vol.3, pp: 32, 1973
4. Bezdek, J.C.: Pattern Recognition with Fuzzy Objective Function Algorithms. Plenum Press, 1981
5. Clark, M.C.: MRI segmentation using fuzzy clustering techniques. IEEE Engineering in Medicine and Biology, Vol.13, No.5, pp: 730, 1994
6. Rosenfeld, A. Fuzzy digital topology. Information and Control, Vol.40, No.1, pp: 76, 1979
7. Udupa, J.K. Samarasekera, S. Fuzzy Connectedness and Object Definition: Theory, Algorithms, and Applications in Image Segmentation. Graphical Model and Image Processing, Vol.58, No.3, pp: 246, 1995
8. Nyul, L.G., Udupa, J.K.: On standardizing the MR image intensity scale. Magn Reson Med, Vol.42, pp: 1072-1081, 1999
9. Dam, E.B.: Evaluation of diffusion schemes for multiscale watershed segmentation. MSC. Dissertation, University of Copenhagen, 2000
10. Zijdenbos, A.P., Dawant, B.M., Margolin, R.A., Palmer, A.C. : Morphometric analysis of white matter lesions in MR images: Method and validation. IEEE Transactions on Medical Imaging, Vol.13, No.4, pp: 716, 1994
11. Cocosco, C.A., Kollokian, V., Kwan, R.K.-S., Evans, A.C.: BrainWeb: Online Interface to a 3D MRI Simulated Brain Database. NeuroImage, Vol.5, No.4, pp: part 2/4, S425, 1994

Improved-FCM-Based Readout Segmentation and PRML Detection for Photochromic Optical Disks

Jiqi Jian, Cheng Ma, and Huibo Jia

Optical Memory National Engineering Research Center,
Tsinghua University, Beijing 100084, P.R. China
Jian00@mails.tsinghua.edu.cn
{macheng, jiahb}@tsinghua.edu.cn

Abstract. Algorithm of improved Fuzzy C-Means (FCM) clustering with preprocessing is analyzed and validated in the case of readout segmentation of photochromic optical disks. Characteristic of the readout and its differential coefficient and other knowledge are considered in the method, which makes it more applicable than the traditional FCM algorithm. The crest and trough segments could be divided clearly and the rising and falling edges could be located properly with the improved-FCM-based readout segmentation, which makes RLL encoding/decoding applicable to photochromic optical disks and makes the storage density increased. Further discussion proves the consistency of the segmentation method with PRML, and the improved-FCM-based detection could be regarded as an extension of PRML detection.

Keywords: Fuzzy clustering, FCM, PRML, optical storage, photochromism.

1 Introduction

With the fast development of technology and with the ever growing demand for high density optical data storage, photochromic optical disks, including multi-level and multi-wavelength photochromic optical disks, become more and more important a research area. Continuous efforts and experiments have been made on it, but most sample disks are dotted recorded, whose storage density is lower than run-length limited (RLL) encoded disks, such as CD, DVD and blu-ray disk. The nonapplication of RLL encoding to photochromic optical disks is partly because of the indistinct edges of the information characters on a photometric optical disk, different from the sharp edges of the characters of CD, DVD and blu-ray disk, which make it hard to clearly divide the readout into wave crests and troughs and locate the rising and falling edges, as RLL encoding/decoding requests. As the laboratory findings in the Optical Memory National Engineering Research Center, the information characters and readouts of a dual-wavelength photochromic sample disk are shown in fig. 1 and fig. 2.

In recent years the synthesis between clustering algorithms and fuzzy set theory has led to the development of fuzzy clustering whose aim is to model fuzzy unsupervised patterns efficiently. Fuzzy C-means (FCM) method proposed by Bezdech is the

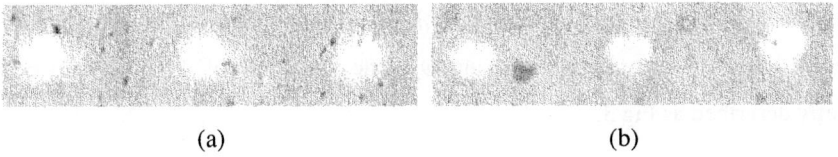

Fig. 1. Information characters on a dual-wavelength photometric sample disk (a) Information characters for the laser, 650nm; (b) Information characters for the laser, 532nm

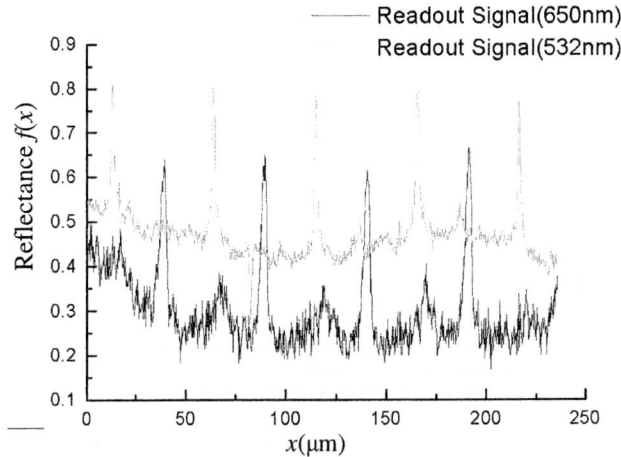

Fig. 2. Readouts of a dual-wavelength photometric sample disk

most widely used fuzzy clustering algorithm.[1-8] It may help us to divide the readout of photochromic disks into crests and troughs and locate the rising and falling edges.

2 Theoretical Background

FCM clustering is based on the theory of fuzzy sets. The most important feature of fuzzy set theory is the ability to express in numerical format the impression that stems from a grouping of elements into classes that do not have sharply defined boundaries, which is exactly fit for crest and trough classification of the readouts of photochromic disks.

For a given dataset $X=\{x_j | j=1,2,\ldots,n\}$, FCM algorithm generates a fuzzy partition providing the membership degree u_{ij} of data x_j to cluster i, ($i=1,2,\ldots,C$). The objective of the algorithm is to minimize the objective function J_m to get the optimal fuzzy partition for the given dataset, where m is a real-valued number which controls the 'fuzziness' of the resulting clusters, and $d(x_j, v_i)$ corresponds to the square value of the Euclidean distance between x_j and v_i. The procedure of the algorithm can be briefly described as Fig 3.

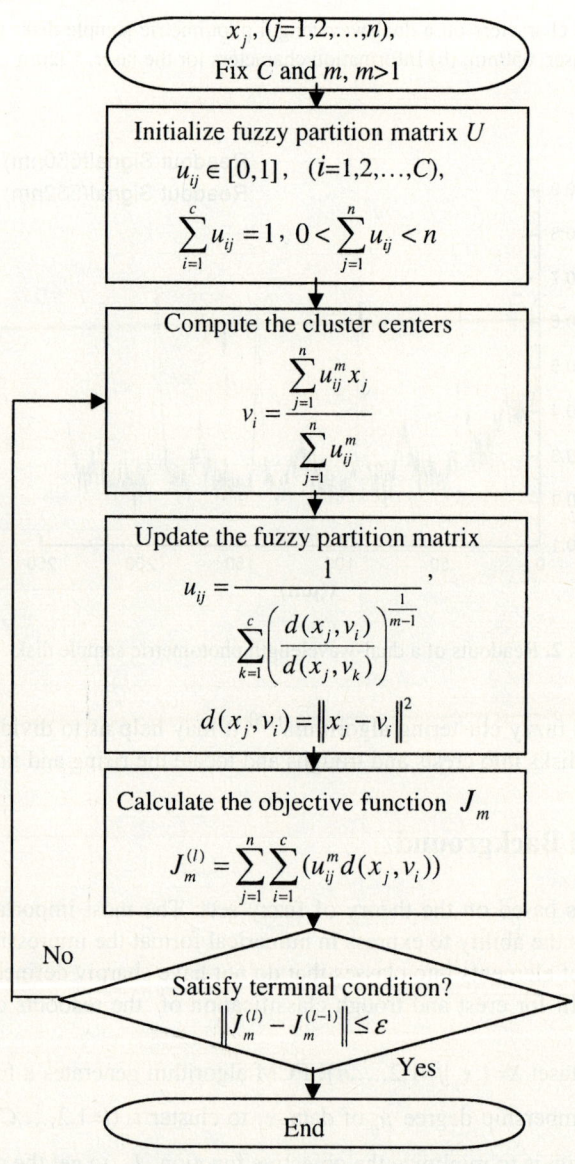

Fig. 3. Process of FCM algorithm

3 Comparison of Two Readout Segmentation Methods

The traditional FCM algorithm and improved FCM method with preprocessing are tested illustratively using the 650nm readout curve $f(x)$ of a dual-wavelength photometric sample disk shown in fig. 2. Both of $f(x)$ and its differential coefficient $f'(x)$ are shown in fig. 4.

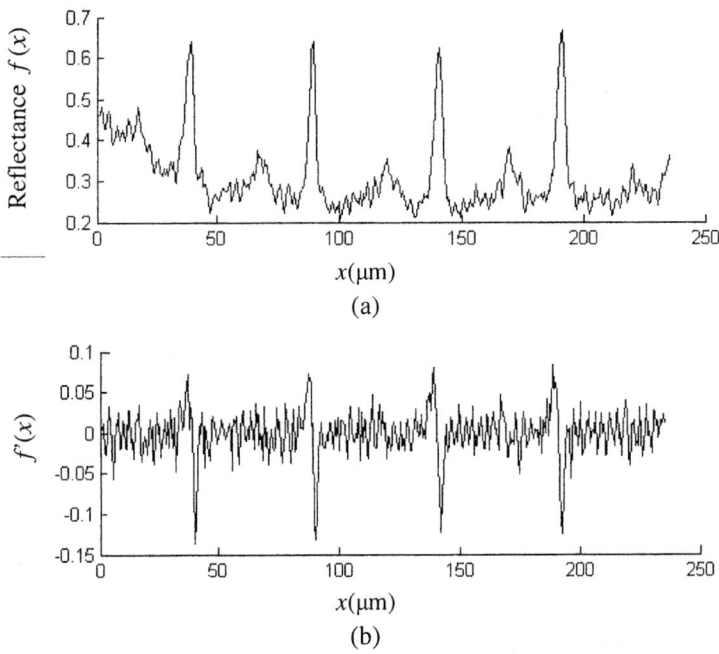

Fig. 4. 650nm readout curve and its differential coefficient

3.1 FCM-Based Readout Segmentation

Using the readout curve and its differential coefficient as dataset given, we can segment the readout with FCM method. When the exponent for the partition matrix m is 2.0[8], we can get the result shown in Fig. 5(a), where cluster centers are indicated by the large characters, and (b), where the curve segments with crosses are crests, the curve segments with dots are troughs, and between them, there are the rising and falling edges. The segmentation result of the 650nm readout is shown in Fig. 4, from which we could see that the curve is segmented unsupervisedly.

However, there are some mistakes of the result of readout segmentation based on FCM Clustering. For example, some parts of the curve are not classified properly, such as the pseudo-peaks on the left side in the figure. And some edges are located inaccurately, such as the first rising edge on the left.

The main reason of the errors is that there is no fore-known information applied in the algorithm, which limits the efficiency of the algorithm significantly, no matter how many categories the points on curve are divided into, as fig. 5 suggests.

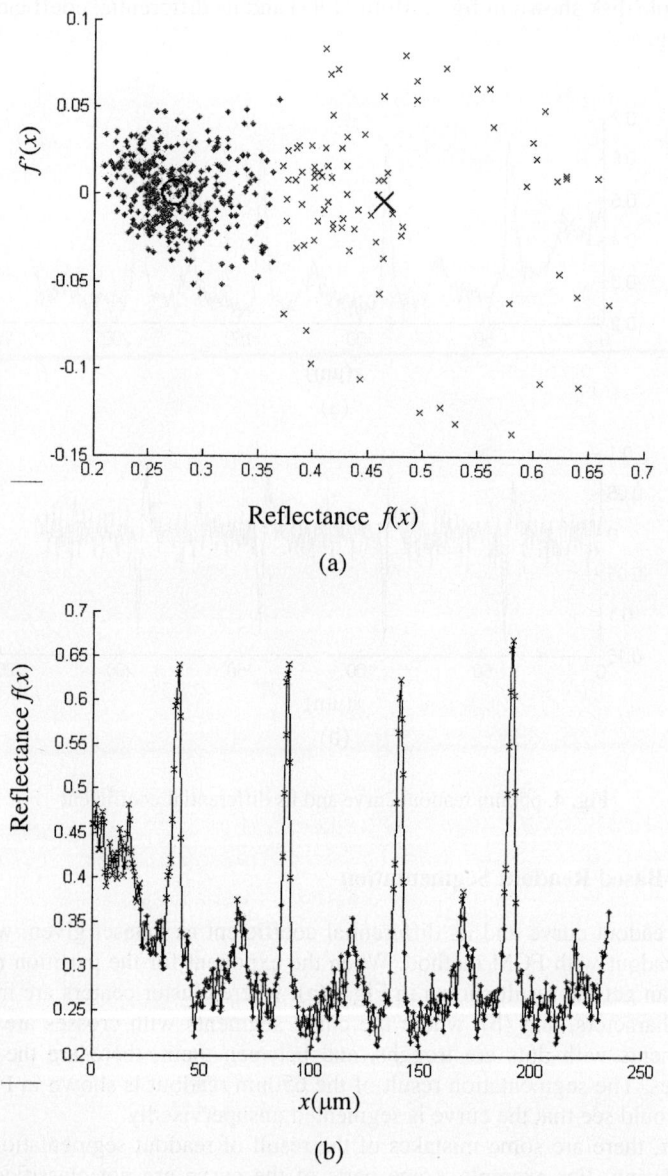

Fig. 5. Result of readout segmentation based on traditional FCM algorithm ($C=2$)

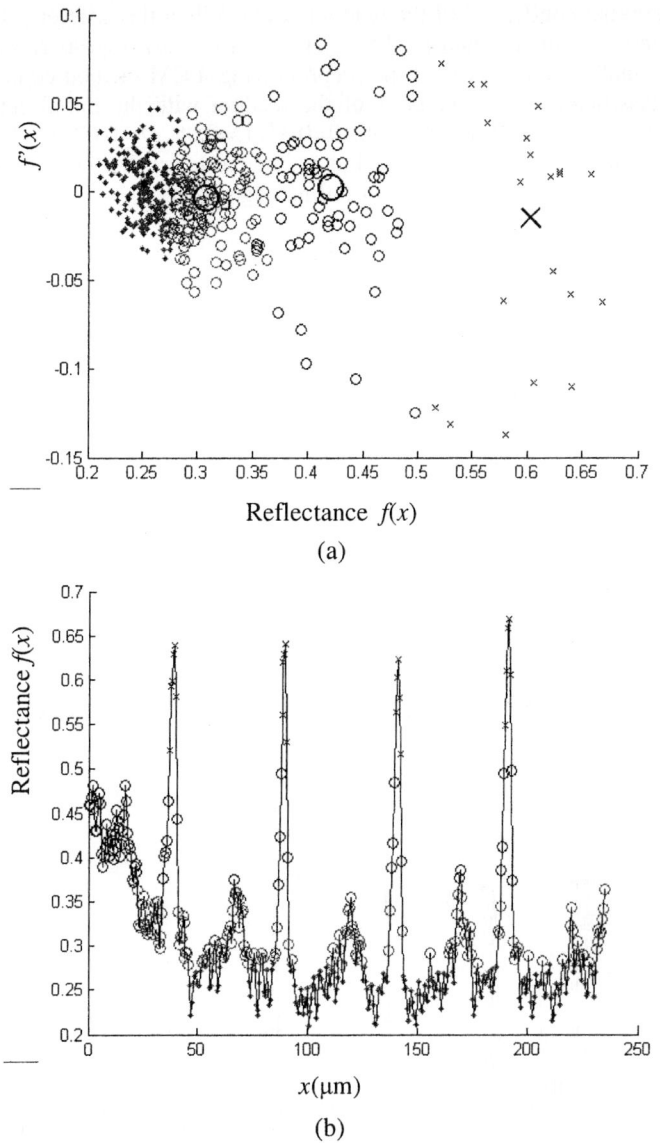

Fig. 6. Result of readout segmentation based on traditional FCM algorithm ($C=4$)

3.2 Improved-FCM-Based Readout Segmentation with Preprocessing

With the characteristic of the readout and its differential coefficient considered as well as other knowledge, we could add the following pretreatments to the segmentation based on FCM clustering. Firstly, we could give two threshold values A and B for the readout curve. A>B. When the value of a readout sample point is larger than A, the point can be classified as crest point, and when the value is smaller than B, the point can be classified as trough point. Secondly, give two threshold values C and D for the differential coefficient of the readout. C>D. When the differential coefficient is larger than C or smaller than D, the point would be classified as rising or falling edge point. Finally, on the basis of the preprocessing, FCM method could be used to judge the classification of other parts of the readout with the initial centers as the results of pre-clustering. When A=0.50, B=0.35, C=0.05, D=-0.07, m=2.0, we could get the result shown in Fig. 7, where points on the curve are divided into four categories: crest points, trough points, rising or falling edge points.

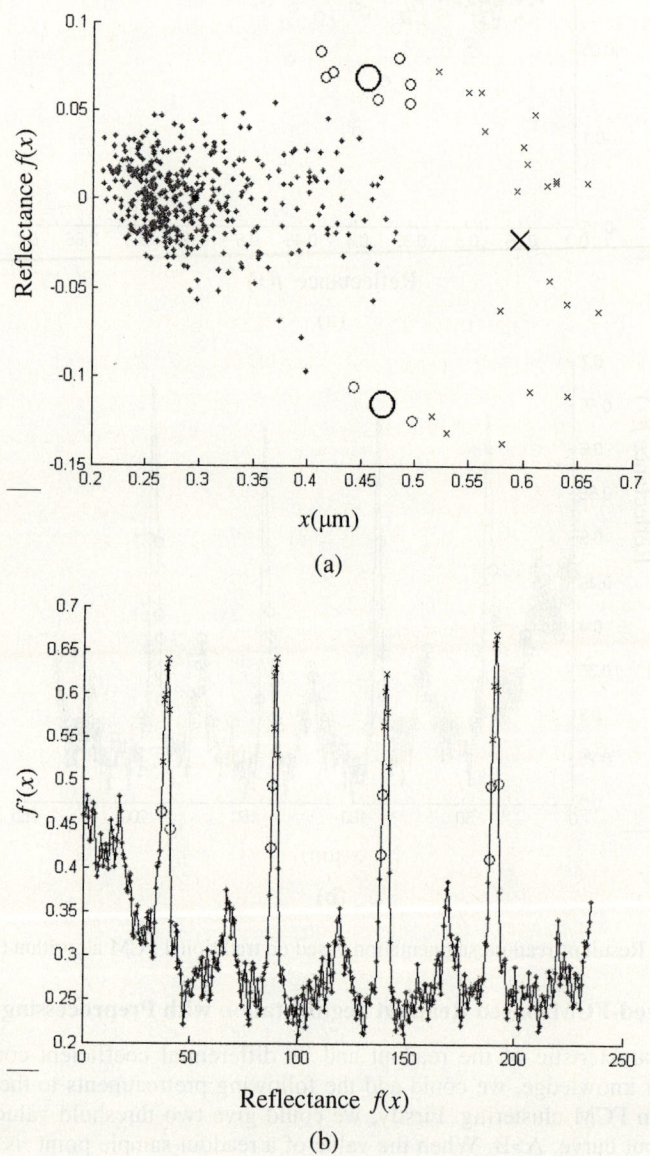

Fig.7. Result of readout segmentation based on improved FCM method

The crest and trough segments are divided clearly, and the rising and falling edges are located properly.

3.3 Consistency of the Readout Segmentation Methods with PRML Detection

As data density increases, partial response maximum likelihood (PRML) detection method has been applied more and more widely to avoid the Inter-Symbol Interference (ISI) problem.[9, 10] Instead of trying to distinguish individual peaks to find flux reversals, PRML manipulates the analog data stream coming from the disk (the "partial response" component) and then determines the most likely sequence of the bits ("maximum likelihood"). In a PRML system, the channel is truncated to a target PR mode by linear equalization, and the PR modes could be expressed as $P(D) = r_0 + r_1 D + r_2 D^2 + \cdots r_M D^M$.

Both of the two readout segmentation methods manipulate the readout $f(x)$ and its differential coefficient $f'(x)$ together, where $f'(x) = (f(x) - f(x-1))/2 = (1-D) f(x)/2$. So the detection based on the improved-FCM-based readout segmentation method could be regarded as the extension of PRML detection with the PR mode $P(D) = \alpha_0 + \alpha_1 D$, where α_0 and α_1 are variable and both have different expressions when $f(x)$ and $f'(x)$ have different relation with the threshold values A, B, C, D, etc.

4 Conclusion

Algorithm of improved FCM clustering is analyzed and validated in the case of readout segmentation of photochromic optical disks. Characteristic of the readout and its differential coefficient and other knowledge are considered in the method, which makes it more applicable than the traditional FCM algorithm. The crest and trough segments could be divided clearly and the rising and falling edges could be located properly with the improved method, which makes RLL encoding/decoding applicable to photochromic optical disks and paves the way for the advance of photochromic optical storage, as well as photochromism based multi-level and multi-wavelength optical storage. Further discussion proves the consistency of the segmentation method with PRML, and the detection method based on the improved-FCM-based readout segmentation could be regarded as the extension of the PRML detection method with the PR mode $P(D) = \alpha_0 + \alpha_1 D$ where α_0 and α_1 are variable and both have different expressions when $f(x)$ and $f'(x)$ have different relation with the threshold values.

References

1. Bezdek J. C.: Fuzzy mathematics In pattern classfication, PhD thesis, Cornell University, Ithaca (1973)
2. Bezdek J. C.: Pattern Recognition with Fuzzy Objective Function Algorithm, Plenum, New York (1981)
3. Yoo S. H., Cho S. B.: Partially Evaluated Genetic Algorithm Based on Fuzzy c-Means Algorithm. Parallel Problem Solving from Nature - PPSN VIII, Lecture Notes in Computer Science, Vol. 3242. Springer-Verlag, Berlin Heidelberg New York (2004) 440

4. Pedrycz W., Loia V., Senatore S.: P-FCM: a proximity-based fuzzy clustering. Fuzzy Sets and Systems, Vol.148. Elsevier, Hoboken (2004) 21-41
5. Tsekouras G. E., Sarimveis H.: A new approach for measuring the validity of the fuzzy c-means algorithm. Advances in Engineering Software, Vol. 35. Elsevier, Hoboken (2004) 567-575
6. Loia V., Pedrycz W., Senatore S.: P-FCM: a proximity-based fuzzy clustering for user-centered web applications. International Journal of Approximate Reasoning, Vol. 34. Elsevier, Hoboken (2003) 121-144
7. Sebzalli Y. M., Wang X. Z.: Knowledge discovery from process operational data using PCA and fuzzy clustering. Engineering Applications of Artificial Intelligence, Vol. 14, Elsevier, Hoboken (2001) 607-616
8. Paland N. R., Bezdek J. C.: On cluster validity for the fuzzy c-means model. IEEE Tans. Fuzzy Systems, Vol. 3, IEEE (1995) 370-379
9. Lee C. H., Cho Y. S.: A PRML detector for a DVDR system. IEEE Trans. Consumer Electron, Vol. 45, IEEE (1999) 278-285
10. Choi S. H., Kong J. J., Chung B. J., Kim Y. H.: Viterbi detector architecture for high speed optical storage. in Speech and Image Technologies for Computing and Telecommunications, IEEE Proceedings/TENCON, Vol. 1, IEEE (1997) 89-92

Fuzzy Reward Modeling for Run-Time Peer Selection in Peer-to-Peer Networks

Huaxiang Zhang[1], Xiyu Liu[1], and Peide Liu[2]

[1] Dept. of Computer Science, Shandong Normal Univ.
Jinan 250014, Shandong, China
Huaxzhang@hotmail.com
[2] Dept. of Computer Science, Shandong Economics Univ.
Jinan 250014, Shandong, China

Abstract. A good query plan in p2p networks is crucial to increase the query performance. Optimization of the plan requires effective and suitable remote cost estimation of the candidate peers on the basis of the information concerning the candidates' run-time cost model and online time. We propose a fuzzy reward model to evaluate a candidate peer's online reliability relative to the query host, and utilize a real-time cost model to estimate the query execution time. The optimizer is based on the run-time information to generate an effective query plan.

1 Introduction

The peer-to-peer (p2p) [1] paradigm has opened up new research areas such as networking and distributed computing. As one of the main application areas, file sharing has gained much attention. Routing the query messages to the destination peers and executing the query plan efficiently pose new challenges to the research. As each peer can provide services such as data services or computing services, a query plan can be executed with the coordination of relevant peers. The execution cost estimation of a query plan is a crucial problem, and an optimization approach should be employed to generate the plan. The cost of executing a query at remote nodes may be expensive because of the communication cost or the heavy workload at the remote sites, and different distributed computing strategies result in quite different query processing performances.

To optimize the query cost in distributed information systems is difficult. The evaluation cost models proposed can be categorized into static and dynamic cost models. The static cost model simplifies the distributed systems as static systems, and utilizes different approaches in the cost measurement metrics. As the name implies, no changes will be made in static models once they are derived. So employing the static models to estimate a query cost is not suitable to a dynamic environment such as p2p networks. In p2p networks, high-availability, fault-tolerance and scalability are main characteristics, and every peer can joint in or drop out of the systems freely. The workload of a remote peer changes as more or less tasks need to be supported, and the dynamic property of p2p requires new query cost models be proposed.

Availability of a peer is also very important since the dynamic property of p2p networks. This issue is important since many p2p systems are designed on the fundamental assumption that the availability of a peer depends on the online time of it. In p2p networks, intermittent component online time is feasible, and that a peer leaves the system and joins the system again at a later time is very common. We are also motivated to study peer availability in part to shape the evaluation of p2p systems. The primary goal of this consideration is to optimize the query cost model in p2p networks.

2 Related Work

As soft computing technologies have been adopted in communications [2], some models based on these technologies have been proposed. Bosc et al [3] propose a fuzzy data model as the basis of the design of a run-time service selection trader. It's not suitable to the p2p networks as the trader environment is static. The cost-based query optimization in distributed information sources has been studied [4], and a calibration-based approach has been proposed. Based on the sampling query running against user database systems, zhu et. Al [5] proposes a local cost model. Shahabi et. al [6] proposes a run-time cost statistics model for the remote peers by employing a probe-based strategy. Even though the approach proposed in [6] is able to reflect more run-time cost statistics of remote peers, but it lacks scalability due to the extensive probe query number. All the models mentioned above take no dynamism into consideration. Ling et. al [7] proposes a progressive "push-based" remote cost monitoring approach, and introduces a fuzzy cost evaluation metric to measure a peer's reliability. Even though their approach takes dynamism into consideration, the fuzzy reliability is not properly defined. If we say a remote peer is more reliable to a query host, it means both the remote and the query host share more common online time, instead that the remote peer has more online time. Even complex p2p query systems have been proposed in the literature recently, which employ different message routing and query location schemes, how to optimize a query plan is still a key issue needing much attention, because the query plan has great impacts on the query performance.

3 Cost Evaluation

A complex query generated by a peer can be executed in several peers having the target resources. Suppose there are candidate peers storing the target resources, the objective of the query optimization is to select n ($n \ll m$) peers from m candidates to execute the query at the minimal cost and with high execution reliability. This problem can be considered as a multi-agent coordination problem [8], and an effective coordination mechanism is essential for a query host to achieve its goal in p2p systems. The essential coordination solves the query problem in the p2p environment with distributed resources. Selection of the candidate peers should take the remote peer's reliability, the execution time and other factors into consideration. In p2p networks, a peer's reliability depends on its online time relative to other peers, and this online time can be estimated. Many factors affect the query execution time of a peer, such as the cpu processing speed, the network communication cost, the input or output, the size of information to be transmitted and the waiting time. To simplify the query cost optimization problem, we think a peer's reliability and the query execution time are the two main factors that should be utilized in a query cost model.

3.1 Relative Online Time

A peer's availability is not well modeled in p2p networks, and the reliability of one peer to another peer depends not only its online availability but also on the inter-dependence between these peers. A peer's online time determines its reliability. For example, if a peer is offline, and the host still sends message to it, then the host will get no result and have to re-transmit the message to other candidate peers. In this case, the peer is unreliable, and we set the reliability degree to 0; if a peer is online all the time and can provide other peers with software or hardware services, then it is reliable and we set its reliability to 1. The concept of reliability is relative. For example, if a peer A is online from 0 am to 2 am, and peer B is online from 2 am to 12 pm, even though peer B's online time is very long, it is not reliable to peer A as they share no common online time.

Relative reliability of each peer pair can also be characterized by using conditional probabilities. Consider both peer A and B, the conditional probability of B being available given that A is available for a given time period. The value $P(B = 1/A = 1)$ is relative reliability of B to A. In this case, these two peers are dependent.

Definition 1: $[a, b]$ is defined as an online time interval of a peer, and the length of $[a, b]$ is denoted as $\|[a, b]\|(= b - a)$. $[a, b] \wedge [c, d]$ is the common online time interval falling within both $[a, b]$ and $[c, d]$ concurrently. If $[a, b] \wedge [c, d] = \emptyset$, then $\|[a, b] \wedge [c, d]\| = 0$.

Definition 2: $[a, b] \vee [c, d]$ means a peer's two online time intervals, and $\|[a, b] \vee [c, d]\| = b - a + d - c - \|[a, b] \wedge [c, d]\|$.

Definition 3: If the jth online time interval of peer n is $[s_{nj}, e_{nj}](j = 1, \cdots, k)$ (k is the number of online time intervals), then the online time interval of peer n is $T_{no} = \vee_{j=1}^{k}[s_{nj}, e_{nj}]$, and n's online time length is $\|T_{no}\|$.

Definition 4: If T_{no} and T_{mo} are the online time intervals of peer n and m respectively, then we define $T_{no} \wedge T_{mo}$ as the online time of peer n relative to m, and denote it as T_{nmo}. $\|T_{nmo}\|$ is the relative online time length, and it's easy to prove that $\|T_{nmo}\| = \|T_{mno}\|$.

If $T_{nmo} = \oslash$, it means peer m and n share no common online time intervals. In this case, peer n is unreliable to peer m. If $T_{nmo} = T_{no}$, it means peer n's online time falls within peer m's online time. In this case, peer m is completely reliable to peer n. Else if $T_{nmo} = T_{mo}$, it means peer m's online time falls within peer n's online time, and peer n is completely reliable to peer m.

3.2 Reliability Degree

The execution time of a query is affected by several factors as we described above, and it is also affected by the dynamism of the p2p networks. A peer will take more time to finish a query with heavy workload than that it takes with light workload.

Definition 5: The relative reliability degree of peer m to n is defined as $r_{mn} = \frac{\|T_{mno}\|}{\|T_{no}\|}$.

Example 1. If m's online time intervals are $[2, 4]$ and $[6, 10]$, and n's online time intervals are $[3, 5], [7, 9], [14, 20]$. We have $T_{mno} = [3, 4] \vee [7, 9]$ and $\|T_{mno}\| = 4-3+9-7 = 3$. $T_{no} = [3, 5] \vee [7, 9] \vee [14, 20]$ and $\|T_{no}\| = 5-3+9-7+20-14 = 10$. $r_{mn} = \frac{\|T_{mno}\|}{\|T_{no}\|} = \frac{3}{10} = 0.3$. We can similarly calculate $r_{nm} = \frac{\|T_{nmo}\|}{\|T_{mo}\|} = \frac{3}{6} = 0.5$. It's clear $r_{mn} \neq r_{nm}$.

3.3 Fuzzy Reward Formulation

The query can be considered as a cooperative task requiring several peers to finish, and the query host peer has to select the peers that will engage in the coordination. Peer selection should follow a query optimization rule to select the peers from all the candidates with less query cost and more reliability. If a candidate peer is more reliable to a query host, we think it needs less reward for the query execution. Otherwise, we think it needs more reward. So the query optimization is transformed to be a problem of selecting peers to minimize the cost and rewards.

We use fuzzy set theory [9] to formulate the fuzzy reward. The reliability degree of a candidate relative to the query host falls within $[0, 1]$, and we regard it as a fuzzy set member. Then a fuzzy set $A(r)$ in the universe of discourse where it denotes all possible reliability degrees can be formulated as $A(r) = \sum_{r \in C} \mu_A(r)/r$. C is the set of all candidates' reliability degrees, and the membership function $\mu_A(r)$ in equation (1) indicates the degree that r belongs to $A(r)$.

$$\mu_A(r) = \begin{cases} 0 & r \leq \omega \\ 2(1 + (1-\omega)(r-\omega)^{-1})^{-1} & r > \omega \end{cases} \quad (1)$$

Where, ω is a threshold. A host may set a reliability degree and selects candidates with degrees no less than the set degree ω. $\mu_A(r)$ is approaching 1 if and only if r is closing to 1.

As the relation between the reliability degree and the reward is discussed above, we define the fuzzy reward function $f(r)$ in equation (2)

$$f(r) = \begin{cases} \infty & r \leq \omega \\ (1 + (1-\omega)(r-\omega)^{-1})/2 & r > \omega \end{cases} \quad (2)$$

$f(r)$ indicates the fuzzy reward needed for a candidate peer of reliability degree r to execute the query generated by a host. The objective of the host is to select suitable ones from all candidates with $r > \omega$ to minimize its cost.

The online time interval can be collected statistically by a query sent to a candidate peer.

3.4 Cost Model

Several criteria influence the cost needed for a peer to execute a remote query, such as the waiting time, the communication cost, the workload of the peer, the cpu cost and input/output cost, the size of the message to be transmitted and etc. We use the cost model proposed in [7] to estimate the cost in terms of time. It is used to calculate the time a peer required to finish a query. The model employs four evaluation criteria. Previous studies on the static query optimization assume the evaluation criteria remain as constants. This may not be true for the dynamic p2p environment.

We utilize the above cost model and describe the cost at a very high level. For each peer, we use a query execution time to indicate the cost. The time may change as the dynamic property of p2p networks. The host peer has a local cache to store the cost models of its candidates and updates the models according to their corresponding cost monitoring agents at the candidate peers. We use the cost models stored locally to estimate the query time required by a candidate peer.

How to estimate the available time of a candidate peer should be solved. We adopt a prober to periodically probe each peers to determine whether they are available or not at a particular time interval. This information is stored in the host peer and updated according newly coming information. Relative reliability of a peer is calculated based on the probed information stored in the host peer.

4 Query Optimization Algorithm

For the peers in the host candidate list, we use equation (1) and (2) to calculate their reliability degrees and fuzzy rewards. The candidates with degrees less than the threshold are ignored, and we can get the left peer number m. For a given query, we utilize the cost model to estimate the execution time t required by each peer. Then the host has m value pair $(t_i, f_i)(i = 1, \cdots, m)$. f_i is the fuzzy reward requested by the ith peer for executing the query, and t_i is the time cost. The host needs to select $n(n \ll m)$ from the m peers to finish the query. The query ending time is the largest one among all the times required by the selected n peers, and we denote the ending time as t_e. Therefore we have the following optimization problem:

$$Q = \min(\alpha t_e + \sum_{i=1}^{n} f_i) \tag{3}$$

Where, α is a coefficient, and can be gotten as $\alpha = \frac{\sum_{i=1}^{m} f_i}{\sum_{i=1}^{m} t_i}$.

(3) is a programming problem, and can be solved by a query optimization algorithm proposed in the following.

If we arrange all the time values in an ascending order, we can get an ordered tuple $(t'_1, \cdots, t'_m)$, where t'_i is the ith smallest member of the tuple. So we can conclude the maximal time t required for a host to finish the query must be a member from tuple $(t'_n, \cdots, t'_m)$. We testify each member in $(t'_n, \cdots, t'_m)$ to calculate the Q values and compare them with each other. The time that minimizes the values is the solution of equation (3). We have the optimization algorithm as shown in table 1.

Table 1. Query selection algorithm

(1) arrange all the time ts in an ascending order, and get an ordered tuple $T' = (t'_1, \cdots, t'_m)$
(2) initialize a set $S = \emptyset$, put m value pairs of (f, t) into S; arrange the pairs in an ascending list ordered by t. $S = \{(f_1, t'_1), \cdots, (f_m, t'_m)\}$
(3) arrange the values of f whose t is a member of $(t'_1, \cdots, t'_n)$ in ascendant order too. The ordered values of f are denoted as $F^0 = (f^0_1, \cdots, f^0_n)$. We calculate $U_0 = (\sum_{i=1}^{n} f^0_i) + \alpha t'_n$
(4) for $j = 1$ to $m - n$ { compute $U_j = (\sum_{i=1}^{n-1} f^{j-1}_i) + f_{n+j} + \alpha t'_{n+j}$;   Insert f_{n+j} into F^{j-1}. Finally, we get an ordered F^j, $F^j = (f^j_1, \cdots, f^j_{n+j})\}$
(5) the smallest one of U_k, where $k \in (0, \cdots, m - n)$, is the best one. The first $n - 1$ members of F^k in addition with f_{n+k} are the fuzzy rewards we should select, and t'_{n+k} is the maximal time of the selected n peers.

The analysis of computational complexity of algorithm in table one can be given as follows: the time complexities in step (1), (2), (3), (4) and (5) are

$m \log m$, m , $n \log n$, $\sum_{i=0}^{m-n} \log(n+i)$ and $m - n + 1$ separately. So the algorithm's computational complexity is the sum of step (1) to (5). We get $O(m \log m)$.

5 Conclusions

We study the problem of remote query cost optimization in p2p networks. It's important for a query host to optimize the query and generate a query strategy. As the host just needs some of the candidates to execute the query, a selection algorithm is required. The objective is to select peers that provide minimal response times and with high reliability degrees. We consider a query as a cooperative task that can be performed by multi-agent, and propose a concept of fuzzy reward to evaluate a peer's online reliability relative to the query host. We also adopt a peer cost model to estimate the dynamic cost of each peer, and use this model to calculate the query execution time. Taking the fuzzy reward and the dynamic cost model into consideration, we can utilize the run-time information to optimize the query.

We argue that existing measurements and proposed novel query cost models in p2p systems do not capture the complex time-varying nature of availability in the dynamic p2p environment, and propose the concept of relative reliability to deal with the query optimization problem. As query optimization problem is very complex in p2p environment, some extensive work should be done.

Acknowledgement

This paper is supported by the Natural Science Fund of Shandong Province of China with Grant No. Z2004G02

References

1. Risson J., Moors T.: Survey of research towards robust peer-to-peer networks search methods. In: Technical Report UNSW-EE-P2P-1-1, Univ. of New South Wales, Sydney, Australia (2004)
2. Wang, L.P.: Soft Computing in Communications. Springer, Berlin Heidelberg New York (2003)
3. Bosc P., Damiani E., Fugini M.: Fuzzy service selection in a distributed object-oriented environment. In: IEEE TRAN. on fuzzy systems 9, 5 (2001) 682–698
4. Adali S., Candan K. S., Papakonstantinou Y. and Subrahmanian V.: Query caching and optimization in distributed mediator systems. In: Proc. of the 1996 ACM SIGMOD Int. Conf. on Management of Data, Montreal, Canada (1996) 137–148
5. Zhu Q., Larson P.A.: Solving local cost estimation problem for global query optimization in multidatabase-systems. In: Distributed and Parallel Databases 6,(1998) 373–420

6. Khan L., McLeod D., Shahabi C.: An Adaptive Probe-Based Technique to Optimize Join Queries in Distributed Internet Databases. In: J. Database Manag. 12, 4 (2001) 3–14
7. Ling B. , Ng W. S. , Shu Y., Zhou A.: Fuzzy Cost Modeling for Peer-to-Peer Systems. In: LNCS, Vol 2872 (2004) 138–143
8. Bourne R. A. , Excelente-Toledo C. B., Jennings N. R.: Run-Time Selection of Coordination Mechanisms in Multi-Agent Systems. In: Proc. 14th European Conf. on Artificial Intelligence. Berlin, Germany (2000) 348–352
9. Jang J.S.R., Sun C. T., Mizutani E.: Neuro-fuzzy and soft computing. Prentice Hall (1997)

KFCSA: A Novel Clustering Algorithm for High-Dimension Data

Kan Li and Yushu Liu

Dept. of Computer Science and Engineering, Beijing Institute of Technology,
Beijing , China, 100081
alanlink@163.com
liuyushu@bit.edu.cn

Abstract. Classical fuzzy c-means and its variants cannot get better effect when the characteristic of samples is not obvious, and these algorithms run easily into locally optimal solution. According to the drawbacks, a novel mercer kernel based fuzzy clustering self-adaptive algorithm(KFCSA) is presented. Mercer kernel method is used to map implicitly the input data into the high-dimensional feature space through the nonlinear transformation. A self-adaptive algorithm is proposed to decide the number of clusters, which is not given in advance, and it can be gotten automatically by a validity measure function. In addition, attribute reduction algorithm is used to decrease the numbers of attributes before high dimensional data are clustered. Finally, experiments indicate that KFCSA may get better performance.

1 Introduction

Clustering analysis groups data points according to some distance or similarity measure in order that objects in a cluster have high similarity.The popular methods such as C-means ,fuzzy C-means and its variants[1-5]which represent clusters through centroids by optimizing the squared error function in the input space. If the separation boundaries among clusters are nonlinear, the performance of these methods will be decreased. At the same time, the outliers affect the effect of the clustering. Fuzzy c-means algorithm cannot get better clustering effect when the characteristic of samples is not obvious, and the current fuzzy clustering algorithm run easily into locally optimal solution because the number of clusters needs to be determined in advance. According to the disadvantage of fuzzy c-means, a novel kernel based fuzzy clustering self-adaptive algorithm(KFCSA) is proposed in the paper. Mercer kernel method is introduced to fuzzy c-means method. It may map implicitly the input data into the high-dimensional feature space through nonlinear mapping. Validity measure function is used to justify iteratively the number of clusters instead of number of clusters given in advance.

2 Attribute Reduction of High-Dimension Data

Rough set based attribute reduction algorithm here is used to reduce the number of attributes in the terrain database(high-dimension data) in order to improve speed of clustering. Before a decision table is reducted, the decision table should be judged whether or not it is consistent. To an inconsistent table, the table is divided into two parts: the complete consistent table and the incomplete consistent table.

2.1 Attribute Reduction of Consistent Decision Table

According to database theory, redundancy and dependency should be as few as possible in databases. Our algorithm views this rules as criteria of attribute reduction. The attribute set which average relevance is the minimum value is the last result of reduction. Conditional entropy is used to judge the relevance of attributes.

Algorithm 1. RSAR(Rough Set Based Attribute Reduction Algorithm)

Input. Decision table S={U,A,V,f}, A=C$\cup$D, condition attributes C and decision attributes D.

Output. A set of attributes REDU.

Method

Step 1. Step 1.Computes the core C_0 based on discernibility matrix M= (m_{ij}) nxn, where i,j=1,2…n. REDU= C_0;

Step 2. Step 2.The matrix element m_{ij} which does not include core builds an expression by conjunct, that is $\wedge\ m_{ij}$;

Step 3. Step 3.Converts the above expression into the extract form. Its terms S_i are the set of attribute reduction Ri={V S_i,i=1,2…n};

Step 4. Step 4. To R_i, computes the relevance of attributes in R_i based on the conditional entropy H（B|A）=- $\sum_{i=1}^{n} p(ai) \sum_{j=1}^{m} p(bj|ai) \log(p(bj|ai))$,（A,B are the elements of Ri, A（U/IND（A）={$a_1,a_2…a_n$},B（U/IND（B）= {$b_1,b_2…b_n$}）.In Ri=Ri$\cup$C$_0$, the set which value is the minimum of average of attribute relevance is REDU.

2.2 Results

Terrain data include DEM, water area, concealment, vegetation, barrier, shelter , distance and traffic capacity attributes.

The terrain data are from the city of Xiamen, which is located in the south of China. The range is from (117° 38'14'',24° 33'30'') to (117° 52'30'',24° 25'14''). After attribute reduction algorithm in high dimensional data is used, DEM, water area, concealment, vegetation and traffic capacity attributes will be used to cluster.

3 Feature Space Clustering

3.1 Algorithm 2. KFCSA (Kernel Based Fuzzy C-Means Self-adaptive Algorithm)

Clustering of data in a feature space has been previously proposed[6],where c-means was expressed by the kernel trick. It was the hard-clustering case. In the paper, we use kernel function to fuzzy c-means.

An alternative kernel function is given which is equivalent to map into a high dimensional space called feature space($x \to \varphi(x)$). The mapping is gotten by means of a replacement of the inner product.

In the feature space, cluster center can be expressed as:

$$c_j = \sum_{k=1}^{n} \gamma_{jk} \varphi(x_k) \tag{1}$$

The objective function is defined as:

$$J = \sum_{i=1}^{n}\sum_{j=1}^{c} u_{ij}^{\alpha} \| \varphi(x_i) - \sum_{k=1}^{n} \gamma_{jk}\varphi(x_k) \|^2 = \sum_{i=1}^{n}\sum_{j=1}^{c} u_{ij}^{\alpha} [k(x_i,x_i) - 2\gamma_j^T k_i + \gamma_j^T k \gamma_j] \tag{2}$$

where $\gamma_i = (\gamma_{i1}, \gamma_{i2} \cdots \gamma_{in})^T$, $k = (k_1, k_1 \cdots k_n)$, $k_i = (k_{i1}, k_{i2} \cdots k_{in})^T$

Equation u_{ij} and γ_{jk} is as follows

$$u_{ij} = \frac{1}{\sum_{g=1}^{c} (\frac{d_{ij}}{d_{ig}})^{\frac{1}{a-1}}}, \quad \gamma_j = \frac{\sum_{j=1}^{n} u_{ij}^a k^{-1} k_j}{\sum_{j=1}^{n} u_{ij}^a} \tag{3}$$

where $d_{ij} = k(x_i, x_i) - 2\gamma_j^T k_i + \gamma_j^T k \gamma_j$.

The kernel based fuzzy c-means self-adaptive algorithm is as follows

Step1. Initializes the positive parameters α, ε and iterations m=1; initialize γ_j ;fix c=2;

Step2. Computes the kernel matrix using Gaussian kernel $k(x_i, x_j) = e^{-q\|x_i - x_j\|^2}$;

Step3. Updates $u_{ij}^{(m)}$; calculate $\gamma_j^{(m)}$ again;

Step4. If $\max | u_{ij}^{(m)} - u_{ij}^{(m-1)} | \leq \varepsilon$, stop; else m=m+1,go step3;

Step5. If validity measure function s get the minimum value, the clustering is over; else c=c+1, go step3.

3.2 Clustering Validity Measure Analysis

Validity measure function is used to estimate the number of clusters. Girolami used the block diagonal structure in the kernel matrix to determine the number of clusters[7]. Girolami's method is used in c-means method. Other researchers applied the method to fuzzy c-means. But the method has its disadvantage. When distinguish of the data in the data set is not obvious, the block diagonal structure in the matrix is not also obvious. It easily arrives at locally optimal solution.

In the paper, the number of clusters is determined through the self-adaptive algorithm. The number of clusters need not be given in advanced. The initial value of number of the clusters is supposed. Then validity measure function is proposed to justify iteratively the number of clusters. Validity measure function is to estimate the correctness of the number of clusters.

Compactness of clustering is expressed as

$$comp = \sum_{i=1}^{c}\sum_{j=1}^{n} \lambda_{ij} \frac{k(x_i,x_i) - 2\gamma_j^T k_i + \gamma_j^T k\gamma_j}{nu_{ij}^2} \quad (4)$$

where λ_{ij} is used to judge the outlier.

$$\lambda_{ij} = \begin{cases} 1 &, u_{ij} > u_{lj} \ i \neq l, \\ 0 &, others. \end{cases}$$

If $\lambda_{ij} = 0$, the data point is the outlier and it will be deleted. u_{ij}^2 is introduced to compactness function to strengthen the compactness of clusters.

Separability of clustering is expressed as

$$sep = \min_{i,j}(\gamma_i^T k\gamma_i - 2\gamma_i^T k\gamma_j + \gamma_j^T k\gamma_j) \quad (5)$$

Validity measure function is defined as

$$s = \frac{comp}{sep} = \frac{\sum_{i=1}^{c}\sum_{j=1}^{n} \lambda_{ij} \frac{k(x_i,x_i) - 2\gamma_j^T k_i + \gamma_j^T k\gamma_j}{nu_{ij}^2}}{\min_{i,j}(\gamma_i^T k\gamma_i - 2\gamma_i^T k\gamma_j + \gamma_j^T k\gamma_j)} \quad (6)$$

where comp means the compactness of clustering and sep means the separability of clustering. After the number of clusters c is determined, data set may be divided into c clusters and weighted sum of squares from data in the clusters to their cluster centers arrive at the minimum value.

4 Experiments

4.1 Test in the Standard Data Set

In order to verify the validity and feasibility of the clustering algorithm, experiments are made with fuzzy c-means and our algorithm. Experimental data are iris data set(from UCI Repository). We select randomly 20 data in iris set. In order to distinguish data, cluster centers and data points are drawn with different labels in the figures. In fig.1, the number of clusters is determined(c=3) in advance. From the result of FCM algorithm, the cluster centers easily run into locally optimal solution. At the same time, the results are related to the selection of the initial value of cluster centers. In the fig.2, KFCSA algorithm is used. The number of clusters(c=3) is gotten automatically via self-adaptive method. Three clusters may be shown in the fig.2. From the figures, the clustering effectiveness with our algorithm is better than the one with fuzzy c-means algorithm. The error rate in our algorithm is lower than the one in fuzzy c-means.

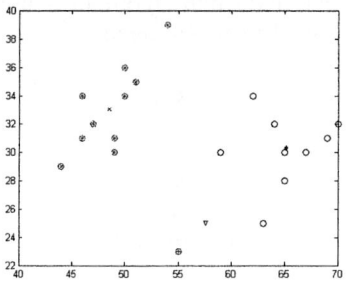 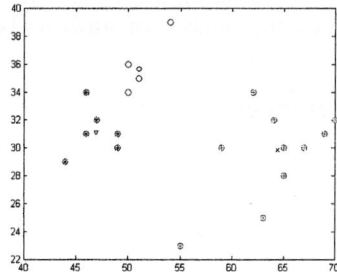

Fig. 1. FCM clustering   **Fig. 2.** KFCSA clustering

4.2 Application in the High Dimensional Data

Terrain is analyzed to determine the characteristic of terrain by the algorithm of KFCSA. DEM, water area, concealment, vegetation and traffic capacity attributes in the terrain data through RSAS algorithm are used to clustering analysis. The terrain data are also from (117° 38'14'',24° 33'30'') to (117° 52'30'',24° 25'14'') , which is located in the city of Xiamen in China. The result of terrain analysis by KFCSA algorithm is shown in fig.3. Two ellipses in fig.3 indicate two clusters which are selected by KFCSA algorithm.

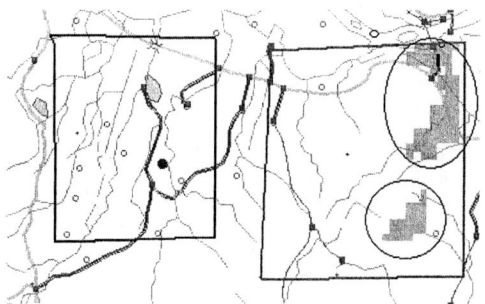

Fig. 3. KFCSA applied in high dimensional data

5 Conclusions

The validity of clustering algorithm depends on characteristic distinction along clusters. When distinction along clusters is not obvious, or overlapped, classical fuzzy c-means method cannot tackle it well. In the paper, mercer kernel based fuzzy clustering self-adaptive algorithm is proposed. Data in the input space by mercer kernel is mapped into the feature space. Characteristic of data in the feature space may be strengthened. It can realize well the clustering to the data which distinction is faint. In our algorithm, self-adaptive method is used to determine automatically the number of clusters. Experiments show that our proposed algorithm gets better performance than classical clustering algorithms. In addition, attribute reduction algorithm is used to decrease the numbers of attributes before these terrain data are clustered.

References

1. MacQueen,J.: Some methods for classification and analysis of multivariate observations. Proc. 5th Berkeley Symposium(1967) 281-297
2. Hartigan,J., Wang,M.:A K-means clustering algorithm. Applied Statistics,Vol. 28.(1979)100-108
3. Bezdek,J.C.: Pattern recognition with fuzzy objective function algorithm. Plenum Press(1981)
4. Jain,A., Dubes,R.: Algorithms for clustering data. Prentice Hall(1988)
5. Wallace,R.: Finding natural clusters through entropy minimization. Ph.D Thesis. CarnegieMellon University, CS Dept (1989)
6. Schölkopf ,B., Smola, A., Müller,K.R.: Nonlinear component analysis as a kernel eigenvalue problem. Neural Computation,Vol. 10(5) (1998)1299-1319
7. Girolami, M.:Mercer kernel based clustering in feature space.IEEE Trans Neural Network,Vol.13(3) (2002)780-784

An Improved VSM Based Information Retrieval System and Fuzzy Query Expansion

Jiangning Wu[1], Hiroki Tanioka[2], Shizhu Wang[3], Donghua Pan[1], Kenichi Yamamoto[2], and Zhongtuo Wang[1]

[1] Institute of Systems Engineering, Dalian University of Technology, Dalian, 116024, China
{jnwu, gyise, wangzt}@dlut.edu.cn
[2] Department of R&D Strategy, Justsystem Corporation, Tokushima, 771-0189, Japan
{hiroki_tanioka, kenichi_yamamoto}@justsystem.co.jp
[3] Dalian Justsystem Co.,Ltd, Dalian, 116024, China
wangshizhu@justsystem.cn

Abstract. In this paper, we propose an improved information retrieval model, where the integration of modification-words and head-words is introduced into the representation of user queries and the traditional vector space model. We show how to calculate the weights of combined terms in vectors. We also propose a new strategy to construct the thesaurus in a fuzzy way for query expansion. Through the developed information retrieval system, we can retrieve documents in a relatively narrow search space and meanwhile extend the coverage of the retrieval to the related documents that do not necessarily contain the same terms as the given query. Experiments for testing the retrieval effectiveness have been implemented by using benchmark corpora. Experimental results show that the improved information retrieval system is capable of improving the retrieval performance both in precision and recall rates.

1 Introduction

With the information explosion on the Internet, Internet users have to encounter huge amount of information junk when they retrieve documents. Therefore, there is a great need for tools and methods to filter such information junk out and meanwhile retain the documents that users really want. The information retrieval (IR) system is thought to be one of good tools for solving the problems mentioned above. So far, many models have been proposed to construct effective information retrieval systems, of which the vector space model (VSM) [1] [2] [3] is the most influential. To date, this model leads the others in terms of performance [1]. It is hereby adopted in our study to construct the improved IR system. The major problem with VSM comes from the over simplicity of its purely term-based representation of information [4]. In this model, keywords are identified, pulled out of context, and further processed to generate term vectors. Unfortunately, independent keywords cannot adequately capture the document contents, resulting in poor retrieval performance. This motivates us to find a new way to the representation of information.

It should be noted that among the identified keywords, some are nouns and verbs and some are adjectives and adverbs according to their parts of speech. From the grammatical point of view, there would be no any actual sense if an adjective or adverb appears alone. In other words, adjectives and adverbs are both constraint words rather than independent concepts. In this paper, adjectives and adverbs are named as modification-words (MWs), and nouns and verbs are named as head-words (HWs). Ignoring the constraints of MWs will result in many irrelevant documents coming from the independent meaningless modification-words. To enhance MWs in user queries, a new information representing method is proposed in this paper, which integrates MWs with HWs to form new combined terms. The combined terms bring somewhat closer to user's requests of the given queries.

Except for the above problem with VSM, there exit some other problems to be solved. In the traditional VSM, the system's relevance judgment is based on the basic assumption that a query and a document are related to each other only if there are shared words in the query and the document. However, simply by examining the terms a document and a query share, it is still difficult to determine whether the document is relevant to the user. The difficulty lies in the fact that most terms have multiple meanings (polysemy) on the one hand, and on the other hand, some concepts can be described by more than one term (synonym). Therefore, sometimes, many unrelated documents may be included in the answer set because they match some of the query terms, while some other relevant documents may not be retrieved because they do not contain any of the exact query terms [5]. To handle these problems, we propose a new method in which MWs and HWs are partially expanded based on the developed fuzzy synonym thesauruses, and then the expansion MWs and the expansion HWs are recombined in terms of the correlations between them for the search process.

Based on the proposed methods, we develop a new IR system that can retrieve the relevant documents in a relatively narrow search space and meanwhile extend the coverage of the retrieval to the related documents that do not necessarily contain the same words as the given query. The experimental results show that the retrieval results obtained from our proposed IR system have improvements in two main measures, precision and recall.

2 System Architecture

The flowchart of the proposed IR system is shown in Figure 1. This system consists mainly of three processing stages: the query expanding stage, the document representing stage, and the query-document matching stage. During the first stage, MWs and HWs are firstly identified by means of the syntactic analyzer. After that, if the given query has no MWs, the system only follows the path 2. In this case, the proposed IR system will return the same retrieval results as normal (we call this process normal search in the paper). On the other hand, if the given query contains MWs, the proposed system will follow the path 1. In this case, MWs and HWs will be integrated at first and then MWs and HWs be expanded respectively to recombine new terms according to the correlations between them as well as the fuzzy synonym thesauruses. Therefore, the system will return different retrieval results comparing with normal search (we call this process modification & expansion search in the paper). During the second stage, documents are represented by use of the modified VSM. The document vector can then be formed. During the last stage, structured search is conducted to obtain the similarities between queries and documents and the scores of the candidate documents. Finally, the retrieved documents whose scores exceed the predefined threshold return to the end user.

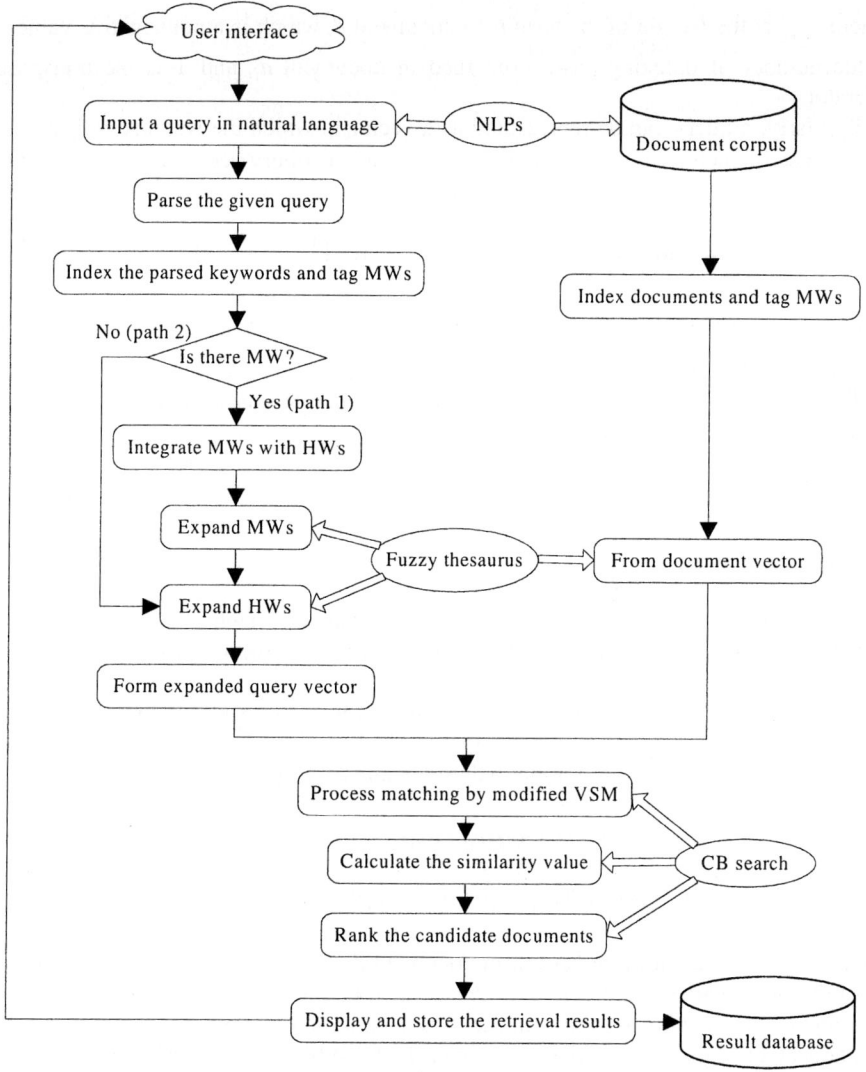

Fig. 1. The flowchart of the proposed IR system

3 Methods

3.1 General VSM

In general VSM [2], each document d_i, $i \in [1, m]$, m is the total number of documents in the corpus, can be represented as a weighted vector,

$$d_i = \{w_{1i}, w_{2i}, \ldots w_{ki}, \ldots, w_{li}\}^T , \qquad (1)$$

where w_{ki} is the weight of the term t_k to document d_i which is a nonnegative value, l is the number of indexing terms contained in document d_i, and T is the transpose operator.

Similarly, a query can be represented as a vector in which the weights indicate the importance of each term in the overall query. Thus, the query vector $q_j, j \in [1, n]$, n is the total number of queries, can be written as:

$$q_j = \{w_{1j}, w_{2j}, \ldots, w_{kj}, \ldots, w_{lj}\}^T, \qquad (2)$$

where w_{kj} is the weight of term t_k to query q_j which is a nonnegative value; l and T respectively have the same meanings as above.

The similarity $s(d_i, q_j)$ between two vectors can be measured by the cosine of the angle between d_i and q_j. That is

$$s(d_i, q_j) = \cos(d_i, q_j). \qquad (3)$$

3.2 Modified VSM

The basic idea behind the modified VSM is recalculate the weights of combined terms and the similarity values between document vectors and query vectors presented in Formulas (1) and (2).

We all know that the classic *tf_idf* is a common way to weigh the importance and uniqueness of a term in a document. In this study, we use the following formulas provided by Justsystem Corporation to define *tf* and *idf* respectively.

$$tf_{ki} = 0.5 + 0.5 \times (freq / \max\_freq) \text{ and} \qquad (4)$$

$$idf_k = 1.0 + \log(total\_doc / dist), \qquad (5)$$

where *freq* is the count of the term t_k in document d_i, max_*freq* is the maximum count of a term in the same document, *total_doc* is the total number of the document collection, and *dist* is the number of documents which contain the term t_k.

The weight of the combined term can be calculated by the following formula:

$$w_{ki} = tf_{ki} \times idf_k. \qquad (6)$$

By conducting normalization to Formula (6), we finally get

$$w_{ki} = \frac{tf_{ki} \times idf_k}{\sqrt{\sum_{k=1}^{l}(tf_{ki} \times idf_k)^2}}, \qquad (7)$$

where l is the size of the indexing term set in document d_i.

With respect to the weight of term t_k in the query q_j, denoted as w_{kj}, it is defined in a similar way as w_{kj} (that is, $tf_{kj} \times idf_k$).

Once the weights of all indexing combined terms both in the document d_i and in the query q_j are determined, the similarity between d_i and q_j can be measured by the cosine of the angle. That is

$$s(d_i, q_j) = \frac{\sum_{k=1}^{l} w_{ki} \times w_{kj}}{\sqrt{\sum_{k=1}^{l} w_{ki}^2 \times \sum_{k=1}^{l} w_{kj}^2}}. \quad (8)$$

3.3 Query Expansion and Fuzzy Thesaurus

Query expansion is a natural idea when, as is often in practice the case, the user's initial requests are very brief, regardless of whether the initial request terms are very specific or not. Thus enlarging requests allows both for more discriminating retrieval through matches on several terms and for more file coverage through getting a match at all.

Query expansion using thesauruses is proved to be an effective way for improving the performance of the IR system. Briefly, there are two types of thesauri, that is, hand-crafted thesauri and corpus-based thesauri [9]. The latter is generated based on the term co-occurrence statistics. In our study, we construct the fuzzy thesaurus by calculating the simultaneous occurrences of terms and the term-term similarities in thesauruses derived from WordNet[1].

Now let's take the determination of expansion terms and their corresponding weights by the fuzzy way into account in detail. Suppose that the given query q_j contains l terms describing l aspects of the query. Suppose that all terms in the given query can be expanded according to the fuzzy thesauruses. If all the expansion terms are directly added into the original query vector, then the new expanded query vector will be in the form of

$$q_j = \{w_{1j}, w_{1j,1}, w_{1j,2}, \ldots w_{1j,N_1}, w_{2j}, \ldots, w_{sj}, \ldots, w_{lj}, w_{lj,1}, \ldots w_{lj,N_l}\}^T, \quad (9)$$

where w_{lj,N_l} is the weight of the expansion term related to the term t_l in the query j, N_l is the number of expanded terms corresponding to the term t_l and T is the transpose operator.

To calculate the weights of the expansion terms, we must firstly obtain the values of nearness degrees between the original term and the expansion terms. In this paper, they are determined mainly based on the term co-occurrence measure.

Suppose that two terms that occur frequently together in the same document are related to the same concept. Therefore, the similarity of the original term t_k in the query and the expansion term t_e can be determined by a term-term relevance coefficient r_{ke} according to thesauruses derived from WordNet, such as Tanimoto coefficient [9]:

---

[1] WordNet http://www.cogsci.princeton.edu/~wn/

$$r_{ke} = \frac{n_{ke}}{n_k + n_e - n_{ke}}, \tag{10}$$

where n_k is the number of synonym sets in thesauruses which contain the term t_k, n_e is the number of synonym sets in thesauruses which contain the term t_e, and n_{ke} is the number of synonym sets in thesauruses which contain both terms.

Such a relevance coefficient represents the nearness degree of the term t_k to the expansion term t_e, which takes values in the interval [0, 1]. All nearness values for the term t_k and all corresponding expansion terms form a relevance matrix R_{ke} shown as following:

$$R_{ke} = \begin{matrix} t_1 \\ t_2 \\ \vdots \\ t_{N_k} \end{matrix} \begin{bmatrix} t_1 & t_2 & \cdots & t_{N_k} \\ r_{11} & r_{12} & \cdots & r_{1N_k} \\ r_{21} & r_{22} & \cdots & r_{2N_k} \\ \vdots & \vdots & r_{ke} & \vdots \\ r_{N_k 1} & r_{N_k 2} & \cdots & r_{N_k N_k} \end{bmatrix}, \tag{11}$$

where the element r_{ke} represents the nearness degree of term t_k to term t_e, $r_{ke} \in [0, 1]$, and N_k is the number of expansion terms. For $k \neq e$, $r_{ke} = r_{ek}$; for $k = e$, $r_{kk} = r_{ee} = 1$.

In real-world applications, not all expansion terms are important enough to the document collection. From this point of view, we therefore define a membership value between document d_i and expansion terms in a fuzzy way based on the term correlation matrix R_{ke}. In this research, the membership value is defined as following

$$\mu_{kd_i} = 1 - \prod_{\substack{m=1 \\ m \neq k}}^{N_k} (1 - r_{km}), \quad \text{for } r_{km} \in R_{ke} \tag{12}$$

where μ_{kd_i} ($k = 1, \ldots, N_k$) denotes the membership degree for term t_k to document d_i, which is computed as a complement of the negated algebraic product over all expansion terms involved; and r_{km} takes values from the term correlation matrix R_{ke}. Here the adoption of an algebraic sum over all expansion terms with respect to the given index term t_k (instead of the classical maximum function) allows a smooth transition for the values of μ_{kd_i} factor.

Once the relatively important expansion terms are determined, the weighting method can be applied to them. For a query j, $q_j = \{w_{1j}, w_{2j}, \ldots, w_{kj}, \ldots, w_{lj}\}^T$ and an expansion term t_e, the similarity between q_j and t_e can be defined as following [11]:

$$s(q_j, t_e) = \sum_{t_{kj} \in q_j} w_{kj} \times r_{ke} \quad \text{for } k \in [1, l] \tag{13}$$

where w_{kj} is the weight of term t_k when it is contained in the query j, r_{ke} is the nearness value for the term t_k and the expansion term t_e, and l is the total number of indexing terms in the collection.

The weight of the actual expansion term t_e, denoted as w_{ej}, with respect to the query vector q_j is defined based on $s(q_j, t_e)$ as:

$$w_{ej} = \frac{s(q_j, t_e)}{\sum_{t_{kj} \in q_j} w_{kj}} \quad . \tag{14}$$

Depending to the modification weights of all expansion terms corresponding to the original term t_k, an expansion term vector can then be formed. It is

$$q_k = \{w_{ej,1}, w_{ej,2}, \ldots, w_{ej,N_e}\}^T \quad . \tag{15}$$

where N_e is the number of expansion terms with respect to the original term t_k.

By adding the above vector into the original query vector, the expanded query vector can then be obtained.

4 Effectiveness Evaluation

We use a small size collection, LA-times, contained in the TREC[2] corpus to examine the effectiveness of the proposed IR system. The queries and a list of documents relevant to each query are randomly chosen from this collection. The statistics on them are listed in Table 1 below.

To improve the performance of the proposed IR system, we create a fuzzy synonym thesaurus that stores a number of adjectives to expand MWs occurring in the user queries and the documents. We use WordNet to determine the related terms, where only synonymy relations implied in WordNet synsets are concerned.

In our study, we use precision and recall to evaluate the effectiveness of the proposed IR system. To verify the correctness and effectiveness of the proposed IR system developed based on the methods described above, three types of experiments have been implemented at first, Normal Search (NS), Head-word WeighTing Search (HWTS) and Head-word WeighTing plus Expansion Search (HWTplusES).

In respect of NS, only CB Search tool developed by Justsystem Corp. is involved in, which is used as a baseline against the other two search approaches. The average precision and recall as benchmarks obtained by using NS for LA-times collection are presented in Tables 2 below.

In HWTS, CB search tool is also used alone. However differing from NS, for HWTS, the weights of head-words are adjusted by adding an important factor that reflects the importance of the head-words in the examined query in order to heighten the weights of head-words (mainly referring to nouns in our experiments). Therefore an increased precision is obtained against NS for LA-times collection, see Table 2.

To test our fuzzy thesaurus and query expansion method addressed previously, the other experiment named HWTplusES based on HWTS is then designed and implemented. In this experiment, the VSM modifying module and the document retrieving

---

[2] TREC http://trec.nist.gov/

module are both used. The retrieved results obtained by using HWTplusES for LA-times collection are also summarized in Table 2.

In addition, to reveal the constraints of modifiers to the associated head-words ad hoc adjectives and nouns, the experiment named Modifier Adjective Search (MAS) is accordingly conducted for LA-times collection by implementing combined search. The retrieved results as shown in Table 2 indicate that the precision of the proposed IR system increases greatly with a 5.16 percent rise against the normal search.

Figure 2 illustrates 11-point precision-recall curves using three different searches with respect to La-times collection for all 108 queries.

Except for the precision-recall curves, we also draw a bar graph for La-times collection using all 108 queries in order to make a comparison between NS, HWTS and HWTplusES in another way, see Figure 3, in which each bar is corresponding to one point in Figure 2.

Table 1. Statistics for the selected collection

Collection name	Number of documents	Number of queries	Size (Mbytes)
LA-times	3,319	108	15

Table 2. Average precision and recall obtained from four different search approaches

Name of approach	Avg. Precision (%)	Avg. Recall (%)
NS	20.01	73.73
HWTS	20.10	73.80
HWTplusES	20.08	73.71
MAS	25.17	34.84

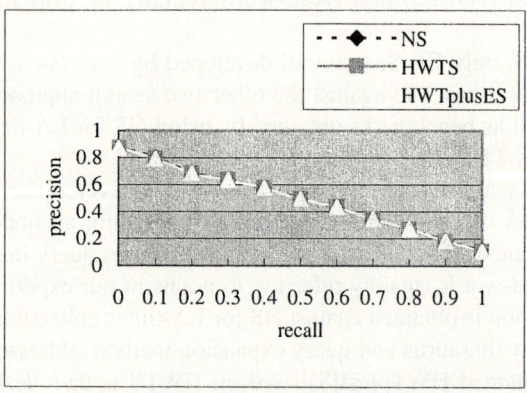

Fig. 2. Precision-recall curves using three different searches with respect to LA-times collection for all 108 queries

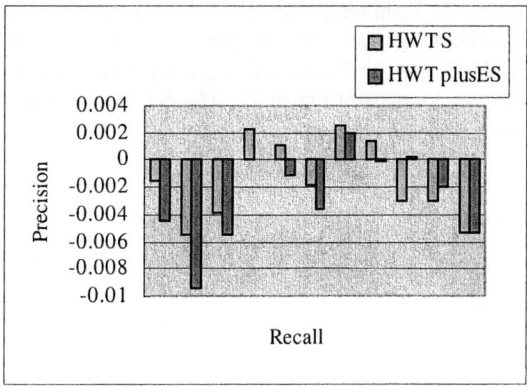

Fig. 3. A bar graph for LA-times collection using all 108 queries corresponding to 11 precision-recall points in two experiments HWTS and HWTplusES

5 Conclusion Remarks

In this paper, a new improved IR system based on VSM is proposed. The results obtained from the experiments lead us to believe that the integration of MWs and the associated HWs is indeed an effective way to improving the performance of the vector-based IR system. Our proposed system not only performs well in terms of precision and recall but also provides a fuzzy thesaurus that is very useful for query expansion. The automatic approach to constructing such a fuzzy thesaurus can save, to a great extent, the manpower and meanwhile avoid inconsistencies resulted from human mistakes. After analyzing the method and doing experiments with TREC corpus, it can be concluded that during the text retrieval, restricting HWs by the associated MWs can improve the precision of the proposed IR system compared to the normal CB Search system, and expanding MWs and HWs simultaneously can improve the recall of the proposed IR system compared to the normal CB Search system.

Acknowledgements

The work reported in this paper is subject to an international collaborative research project that is sponsored by Justsystem Corporation of Japan. The authors would like to thank master students Huinan Ma and Jun Zhang of Dalian University of technology who did much work on documentation and experimention.

References

1. Kraft D.H., Petry F.E.: Fuzzy information systems: managing uncertainty in databases and information retrieval systems. Fuzzy Sets and Systems. 90 (1997) 183-191
2. Salton G., Wong A., Yang C. S.: A vector space model for automatic indexing. Communications of the ACM. 18 (1975) 613-620

3. Salton G., Buckley C.: Term-weighting in information retrieval using the term precision model. Journal of the Association for Computing Machinery. 29 (1982) 152-170
4. Papadimitriou C.H., Raghavan P., Tamaki H., Vempala S.: Latent semantic indexing: A probabilistic analysis. Journal of Computer and System Sciences. 61 (2000) 217-235
5. Letsche T.A., Berry M.W.: Large-scale information retrieval with latent semantic indexing. Information Sciences. 100 (1997) 105-137
6. Chandren-Muniyandi R., Komputer J.S., Maklumat F.T. dan S.: Neural network: An exploration in document retrieval system. In: Proceedings of TENCON 2000. Vol. 1 (2000) 156-160
7. Ramirez C., Cooley R.: Case-based reasoning model applied to information retrieval. In: IEE Colloquium on Case Based Reasoning: Prospects for Applications. (1995) 9/1 -9/3.
8. Liu G.: The semantic vector space model (SVSM)—A text representation and searching technique. In: Proceedings of the Twenty-Seventh Hawaii International Conference on System Sciences, Vol. IV: Information Systems: Collaboration Technology Organizational Systems and Technology. Vol. 4 (1994) 928 –937
9. Mandala R., Tokunaga T., Tanaka H.: Query expansion using heterogeneous thesauri. Information Processing and Management. 36 (2000) 361-378
10. Kim M.C., Choi K.S.: A comparison of collocation-based similarity measures in query expansion. Information Processing and Management. 35 (1999) 19-30
11. Qiu Y., Frei H.: Concept based query expansion. In: Proceedings of the 16th Annual International ACM-SIGIR Conference on Research and Development in Information Retrieval. (1993) 160-169
12. Nie J.Y., Jin F.: Integrating logical operators in query expansion in vector space model. In: Workshop on Mathematical/Formal Methods in Information Retrieval, 25th ACM-SIGIR. (2002)

The Extraction of Image's Salient Points for Image Retrieval

Wenyin Zhang[1,2], Jianguo Tang[1,2], and Chao Li[1]

[1] Chengdu University of Information Technology, 610041, P.R. China
[2] Chengdu Institute of Computer Applications
Chinese Academy of Sciences, Chengdu 610041, P.R. China

Abstract. A new salient point extraction method from Discrete Cosine Transformation (DCT) compressed domain for content-based image retrieval is proposed in this paper. Using a few significant DCT coefficients, we provide a robust self-adaptive salient point extraction algorithm, and based on salient points, we extract 13 rotation-, translation- and scale-invariant moments as the image shape features for retrieval. Our system reduces the amount of data to be processed and only needs to do partial entropy decoding and partial de-qualification. Therefore, our proposed scheme can accelerate the work of image retrieval. The experimental results also demonstrate it improves performance both in retrieval efficiency and effectiveness.

Keywords: Salient Point, Image Retrieval, Discrete Cosine Transformation, DCT.

1 Introduction

Digital image databases have grown enormously in both size and number over the years [1]. In order to reduce bandwidth and storage space, most image and video data are stored and transmitted by some kind of compressed format. However, the compressed images cannot be conveniently processed for image retrieval because they need to be decompressed beforehand, and that means an increase in both complexity and search time. Therefore, it is important to develop an efficient image retrieval technique to retrieve wanted images from the compressed domain.

Nowadays, more and more attention has been paid on the compressed-domain based image retrieval techniques [2] which extract image features from the compressed data of the image. The JPEG is the image compression standard [3] using DCT and is widely used in large image databases and on the World Wide Web because of its good compression rate and image quality. However, the conventional image retrieval approaches used for JPEG compressed images need full decompression which consumes too much time and requires large amount of computation. Some new researches [4,5,6,7,8,9,10] have recently resulted in improvements in that image features can be directly extracted in the compressed domain without full decompression.

The purpose of this paper is to propose a novel compressed image retrieval method based on salient points [11] computed from DCT compressed domain. The salient points are interesting for image retrieval because they are located in visual focus points and thus they can capture the local image information and reduce the amount of data to be processed. The salient points are related to the visually most important parts of the images and lead to a more discriminant image feature than interesting points such as corners [14]. Unlike the traditional interesting points, the salient points should not be clustered in few regions. It's quite easy to understand that using a small amount of such points instead of all images reduces the amount of data to be processed. First, based on a small part of important DCT coefficients, we provide a new salient point extraction algorithm which is very robust to noise, rotation, translation and scale and most of common image processing such as lighting, darkening, blurring, compressing and so on. Then, we adaptively choose some important salient points to constitute a binary salient map of the image, which represents the shape of the objects in the image. Last, we extract 13 rotation-, translation- and scale-invariant moments [12,13] from the salient map as the shape features of the image for retrieval.

The remainder of this paper is organized as follows. In Section 2, we introduce the works related to JPEG compression image retrieval. In Section 3, we discuss in details our new scheme, followed by the experimental results and analysis. Finally, Section 5 concludes the paper.

2 Related Works

Direct manipulation of the compressed images and videos offers low-cost processing of real time multimedia applications. It is more efficient to directly extract features in the JPEG compressed domain. As a matter of fact, many JPEG compressed image retrieval methods based on DCT coefficients have been developed in recent years.

Climer and Bhatia proposed a quadtree-structure-based method [4] that organizes the DCT coefficients of an image into a quadtree structure. This way, the system can use these coefficients on the nodes of the quadtree as image features. However, although such a retrieval system can effectively extract features from DCT coefficients, its main drawback is that the computation of the distances between images will grow undesirably fast when the number of relevant images is big or the threshold value is large. Feng and Jiang proposed a statistical parameter-based method [5] that uses the mean and variance of the pixels in each block as image features. The mean and variance can be directly computed via DCT coefficients. However, this system has to calculate the mean and variance of each block in each image, including the query image and the images in the database, and the calculation of the mean value and variance value of each block is a computationally heavy load. Chang, Chuang and Hu provided a direct JPEG compressed image retrieval technique [6] based on DC difference and the AC correlation. Instead of fully decompressing the images, it only needs to do partial entropy decoding and extracts the DC difference and the AC correlation as two image features. However, although the retrieval system is faster than the method [4,5], it doesn't do well in anti-rotation.

The related techniques are not limited to the above three typical methods. Shneier [7] described a method of generating keys of JPEG images for retrieval, where a key is the average value of DCT coefficients computed over a window. Huang [8] rearranged the DCT coefficients and then got the image contour for image retrieval. B.Furht [9] and Jose A.Lay [10] made use of the energy histograms of DCT coefficients for image or video retrieval. Most image retrieval methods based on DCT compressed domain strengthened the affectivity and efficiency of image retrieval [2]. But most of these research focused on global statistical feature distributions which have limited discriminating power because they are unable to capture the local image information or shape information.

In our proposed approach, we use the image salient points computed from a small part of significant DCT coefficients to describe the image feature. The salient points give local outstanding information and on the whole provide the shape features of the image.

3 The Proposed Method

In this section, we introduce in details our retrieval methods based on salient points. The content of the section is arranged with the sequence: edge point detection→ salient point extraction→image feature extraction.

3.1 Fast Edge Detection Based on DCT Coefficients

Edges are significant local changes in the image and are important feature for analyzing image because they are relevant to estimating the structure and properties of objects in the scene. Here we provide a fast edge detection algorithm in DCT domain which directly compute the pixel gradients from DCT coefficients to get edge information. Based on it, we give the salient points extraction algorithm.

The 8×8 Inverse DCT formula is as follows:

$$f(x,y) = \frac{1}{4} \sum_{u=0}^{7} \sum_{v=0}^{7} C(x,u)C(y,v)F(u,v); \qquad (1)$$

$$where: C(x,u) = c(u) \cos \frac{(2x+1)u\pi}{16};$$

$$c(x) = \begin{cases} \frac{1}{\sqrt{2}}, & x = 0 \\ 1, & x \neq 0 \end{cases}$$

Compute derivative to formula (1), we get:

$$f'(x,y) = \frac{\partial f(x,y)}{\partial x} + \frac{\partial f(x,y)}{\partial y}$$

$$= \frac{1}{4}\sum_{u=0}^{7}\sum_{v=0}^{7} C'(x,u)C(y,v)F(u,v)$$

$$+ \frac{1}{4}\sum_{u=0}^{7}\sum_{v=0}^{7} C(x,u)C'(y,v)F(u,v) \qquad (2)$$

$$Where: \quad C'(x,u) = -\frac{u\pi}{8}c(u)\sin\frac{(2x+1)u\pi}{16}$$

From the equation (2), we can compute the pixel gradient in (x,y), its magnitude can be given by:

$$G(x,y) = |(\frac{\partial f(x,y)}{\partial x})| + |(\frac{\partial f(x,y)}{\partial y})| \qquad (3)$$

In order to simplify computation, we change the angle $\frac{(2x+1)u\pi}{16}$ to acute angle [15]. Let $(2x+1)u = 8(4k+l) + q_{x,u}$, k, l and $q_{x,u}$ are integers, in which: $q_{x,u} = (2x+1)u \bmod 8$, $k = (2x+1)u/32$, $l = (2x+1)u/8 \bmod 4$, $0 \leq q_{x,u} < 8$, $0 \leq l < 4$. Then, we can do as follows:

$$\sin(\frac{(2x+1)u\pi}{16}) = \sin(\frac{(8(4k+l)+q_{x,u})}{16})$$

$$= \begin{cases} \sin(\frac{q'_{x,u}\pi}{16}) & : q'_{x,u} = q_{x,u}, l=0 \\ \sin(\frac{q'_{x,u}\pi}{16}) & : q'_{x,u} = 8 - q_{x,u}, l=1 \\ -\sin(\frac{q'_{x,u}\pi}{16}) & : q'_{x,u} = q_{x,u}, l=2 \\ -\sin(\frac{q'_{x,u}\pi}{16}) & : q'_{x,u} = 8 - q_{x,u}, l=3 \end{cases}$$

$$= (-1)^{\lceil \frac{l-1}{2} \rceil} \sin(\frac{q'_{x,u}}{16}) \qquad (4)$$

Similarly, we can get:

$$\cos(\frac{(2x+1)u\pi}{16}) = (-1)^{\lceil \frac{l+1}{2} \rceil} \cos(\frac{q'_{x,u}}{16}) \qquad (5)$$

To the formulae (4) and (5), the sign and the $q'_{x,u}$ can be decided aforehand according to x, u. Let $ss_{x,u}$ and $cs_{x,u}$ be the signs of formulae (4) and (5). The $ss_{x,u}$ and $q'_{x,u}$ can be described as follows:

$$ss_{x,u} = \begin{Bmatrix} + + + + + + + + \\ + + + + + + - - \\ + + - - - + + + \\ + + + - - + + - \\ + + - - + + - - \\ + + - + + - + + \\ + + - + - + + - \\ + + - + - + - + \end{Bmatrix} ; \quad q'_{x,u} = \begin{Bmatrix} 0\ 1\ 2\ 3\ 4\ 5\ 6\ 7 \\ 0\ 3\ 6\ 7\ 4\ 1\ 2\ 5 \\ 0\ 5\ 6\ 1\ 4\ 7\ 2\ 3 \\ 0\ 7\ 2\ 5\ 4\ 3\ 6\ 1 \\ 0\ 7\ 2\ 5\ 4\ 3\ 6\ 1 \\ 0\ 5\ 6\ 1\ 4\ 7\ 2\ 3 \\ 0\ 3\ 6\ 7\ 4\ 1\ 2\ 5 \\ 0\ 1\ 2\ 3\ 4\ 5\ 6\ 7 \end{Bmatrix}$$

The $cs_{x,u}$ can be given the same as $ss_{x,u}$. For more time-saving, according to Taylor formula, we extend $\sin(\frac{q'_{x,u}}{16})$ and $\cos(\frac{q'_{x,u}}{16})$ at $\pi/4$:

$$\sin(\frac{q'_{x,u}}{16}) = \sum_{k=0}^{n} \sin^{(k)}(\frac{\pi}{4}) \frac{(\frac{q'_{x,u}}{16} - \frac{\pi}{4})^n}{n!} + R_n(\frac{q'_{x,u}}{16}) \qquad (6)$$

$$\cos(\frac{q'_{x,u}}{16}) = \sum_{k=0}^{n} \cos^{(k)}(\frac{\pi}{4}) \frac{(\frac{q'_{x,u}}{16} - \frac{\pi}{4})^n}{n!} + R_n(\frac{q'_{x,u}}{16}) \qquad (7)$$

$$where: |R_n| < \frac{(\frac{\pi}{4})^{n+1}|\frac{q'_{x,u}-4}{4}|^{n+1}}{(n+1)!} \leq \frac{(\frac{3\pi}{16})^{n+1}}{(n+1)!}$$

When consider to extend up to second order, the equation (6) and (7) can be approximated as follows:

$$\sin(\frac{q'_{x,u}}{16}) \approx \frac{\sqrt{2}}{2}[1 - \frac{\pi}{16}(4 - q'_{x,u}) + \frac{\pi^2}{512}(4 - q'_{x,u})^2] \qquad (8)$$

$$\cos(\frac{q'_{x,u}}{16}) \approx \frac{\sqrt{2}}{2}[1 + \frac{\pi}{16}(4 - q'_{x,u}) - \frac{\pi^2}{512}(4 - q'_{x,u})^2] \qquad (9)$$

The residue error R_2 is more less than 0.034. This suggests that the equation (8) and (9) can be approximately decided by $q'_{x,u}$ and can be calculated off-line. As such, the $C'(x,u)$ and $C(x,u)$ in the equation (2) also can be calculated approximately by the $q'_{x,u}$, $ss_{x,u}$ and $cs_{x,u}$ off-line, which means that the coefficients of the extension of equation (2) can be computed in advance and the equation (3) is only related to the DCT coefficients $F(u,v)$. So, the computation of the equation (3) is much simplified.

Further more, we need not use all the coefficients to compute the equation (2), because most of the DCT coefficients with high frequency are zero and do nothing to the values of edge points, so they can be omitted, which means we can use a small part of DCT coefficients with low frequency to compute the edge points and more computation cost is saved again. Fig.1 gives an example for edge detection using different DCT coefficients. From the Fig.1, we can see that the more the DCT coefficients used, the smoother the edge. With the decreasing of the number of DCT coefficients used, the "block effect" becomes more and more obvious. In a 8×8 block, the more gray changes, the larger every edge point value. The Fig.2 gives another example of edge detection with first 4×4 DCT coefficients.

3.2 Salient Points Computation

According to analysis in Sec.3.1 that the edge points in a block reflect the gray changes in this block, the more changes, the larger edge points value, we sum all the edge points values in one block to stand for one salient point value, which means that one $8*8$ block corresponds to one salient point. If $M \times N$ stands for

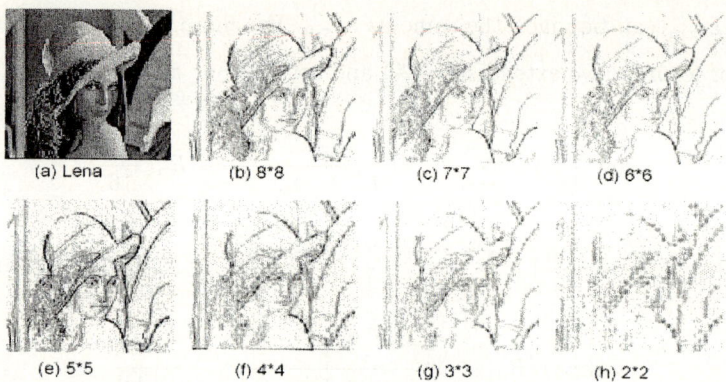

Fig. 1. An example of edge detection: (a) Lena.jpg, (b)-(h) are the edge images of Lena, which are computed respectively by first n*n DCT coefficients, $2 \leq n \leq 8$.

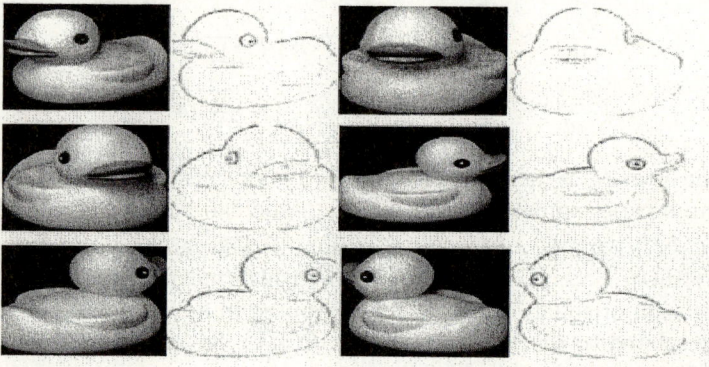

Fig. 2. Another example of edge detection, which are computed respectively by first 4*4 DCT coefficients.

the size of an image, the maximum of its salient points number is $M/8 \times N/8$. Let $Sp(x', y')$ be the salient point value in $(x', y'), 0 \leq x' < M/8, 0 \leq y' < N/8$, it can be computed as follows: (γ is a parameter, $2 \leq \gamma \leq 7$)

$$Sp(x', y') = \sum_{x=x' \times 8}^{x' \times 8 + \gamma} \sum_{y=y' \times 8}^{y' \times 8 + \gamma} |G(x, y)| \qquad (10)$$

3.3 Adaptive Selection of Salient Points

Not all salient points are important, we want to adaptively select the salient points with larger value. The number of the salient points extracted will clearly influence the retrieval results. Less salient points will not mark the image; More salient point will increase the computation cost. A given threshold T may be not suited for all images. Through experiments we have found that the gray changes in a block can be relatively reflected by the variance (denoted by σ)of AC coefficients in this block. The more changes, the larger the variance. Let M_{sp} be the mean value of $Sp(x',y')$, We adaptively select the salient points which satisfy the following condition:

$$\lambda \times Sp(x',y') > \mu \times M_{sp} \qquad (11)$$

$$Where: \lambda = \sigma/128, 0 \leq \lambda \leq 1;$$

$$\sigma \approx \frac{1}{8}\sqrt{\sum_{x=0}^{x<8}\sum_{y=0}^{y<8}F^2(x,y)}, 0 \leq \sigma \leq 128,$$

x, y are not zero simultaneously.

The condition (11) tends to choose the salient points with larger variance σ and larger $Sp(x',y')$ values. The Fig.2 shows some examples of images' salient points, from which we can see that the salient points actually describe the shape feature of the image.

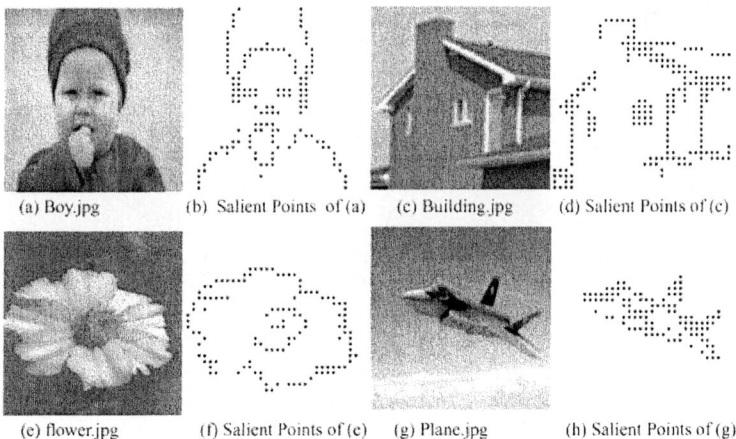

(a) Boy.jpg   (b) Salient Points of (a)   (c) Building.jpg   (d) Salient Points of (c)

(e) flower.jpg   (f) Salient Points of (e)   (g) Plane.jpg   (h) Salient Points of (g)

Fig. 3. Examples of images' salient points: (a), (c), (e) and (g) are the four JPEG images, (b), (d), (f) and (h) are their salient point images respectively. The parameter $\mu = 5.5$, and first $4*4$ coefficients in each 8×8 block are used to computed the variance σ and the edge points.

3.4 The Feature Extraction Based on Salient Points

If we assign 1 to the block with salient point and 0 to the block without salient point, then we get a binary image named saliency map with the size of $M/8*N/8$. The saliency map stands for the shape of the object in the image, so we extract the image shape feathers directly from its saliency map. Inspired by those shape description in pixel domain, invariant moments can be calculated inside the salient map to characterize the shape information. Based on the work of [12,13], we choose 13 invariant moments as the image shape features: seven Hu's moments $\varphi_1, \varphi_2, \cdots, \varphi_7$ and six contour moments $\overline{m}_1, \overline{m}_2, \overline{m}_3, \overline{\mu}_1, \overline{\mu}_2, \overline{\mu}_3$.

4 Experimental Results and Discussions

A JPEG image database is used to test the proposed approach, which comprises 1460 images from http://www.benchathlon.net. In the experiment, we compute the image features from three color channels, so we get totally 39 image features.

Firstly, we compared the results of our salient point method with those of other interest point detectors such as multi-resolution contrast based key point detector [14], the Harris' detector proposed by Schmid in [16] and the Susan's detector proposed in [17]. The Fig.3 shows the results. An outstanding characteristic of our result is that the salient points detected by our method are related not only to the corners but also to the smoothed edges while the other three methods lean to detect corners. The smoothed edge and corners are the most important visual focus parts in an image to human eyes, so they have powerful ability to character the image content for retrieval. Furthermore, our salient point extraction algorithm has more robustness than these detectors in anti-noise, rotation, translation and so on.

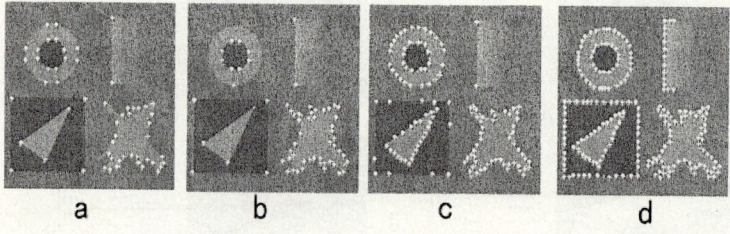

Fig. 4. An example of salient point with different methods: (a) multi-resolution contrast based key point detector [14], (b) the Harris' detector [16], (c) the Susan's detector [17],(d) our method.

In order to evaluate the performance of our retrieval system, nine kinds of images such as peppers, buildings, fires, airplane, flowers, birds, toys, animals and scenes in the image database are selected to investigate the retrieval effectiveness. We also compared our retrieval results with the three typical JPEG image retrieval methods [4,5,6] mentioned in section 2. Classical "precision and recall" scheme was used to evaluate the proposed technique and those three ones. Fig.4 shows the Recall-Precision graph for our database with the four techniques. From the graph, we can see our proposed method achieves a much better result than the methods [4,5] and slightly better than the method [6] from the whole. In Fig.5 we present an example of a retrieval results of using image $flower001.jpg$ as the query image, which shows that our system is very capable of recognizing any images as long as their shapes are not violently changed.

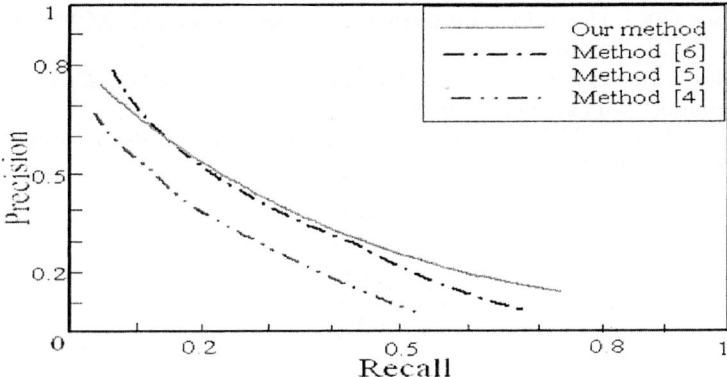

Fig. 5. Retrieval results compared with the three typical JPEG image retrieval methods [4,5,6]

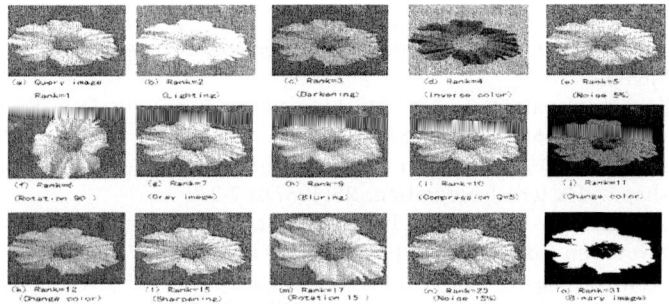

Fig. 6. An example of a retrieval results of using image $flower001.jpg$ as the query image

5 Conclusions

In this paper, we provided a scheme to extract the shape features based on salient points directly from DCT compressed domain for content based image retrieval. Because of the robustness of our salient point extraction approach in compressed domain, our retrieval system gives more improvement in efficiency and effectiveness shown by our experiments. In the future, we will go deep into our work based on our salient points for content based image retrieval.

References

1. Brunelli R., Mich O., Modena C.M.: A survey on the automatic indexing of video data. Journal of Visual Communication and Image Representation **10**(1999)78-112
2. Mandal M. K., Idris F., Panchanatha S.: A critical evaluation of image and video indexing techniques in the compressed domain. Image and Vision Computing, **17**(1999)513-529

3. Wallace G.K.: The JPEG still picture compression standard. ACM Communications. **34(4)**(1991)31-45.
4. Climer S., Bhatia S.K.: Image database indexing using JPEG coefficients. Pattern Recognition. **35(11)**(2002)2479-2488.
5. Feng G., Jiang J.: JPEG compressed image retrieval via statistical features. Pattern Recognition. **36(4)**(2002)977-985.
6. Chang C.C., Chuang J.C., Hu Y.S.: Retrieval digital images from a JPEG compressed image database. Image and Vision Computing. **22**(2004)471-484
7. Shneier M., Abdel-Mottaleb M.: Exploiting the JPEG compression scheme for image retrieval. IEEE Trans. Pattern Anal. Mach. Intell. **18**(1996)849-853.
8. Huang X.L., Song L., and Shen L.X.: Image retrieval method based on DCT domain. Journal of Electronics & Information Technology. **12**(2002)1786-1789
9. Furht B., Saksobhavivat P.: A Fast Content-Based Video and Image Retrieval Technique Over Communication Channels. Proc. of SPIE Symposium on Multimedia Storage and Archiving Systems, Boston, MA, November(1998)
10. Jose A., Ling Guan: Image Retrieval Based on Energy Histograms of the Low Frequency DCT Coefficients. IEEE International Conference on acoustics, Speech, and Signal Processing. **6**(1999) 3009-3012
11. Sebe N., Lew M.S.: Comparing salient point detectors. Pattern Recognition Letters, **24**(2003)89-96
12. Hu M.K.: Visual pattern recognition by moment invariants. IRE Transactios Information Theory. **8**(1962)179-187
13. Gupta L., Srinath M.D.: Contour sequence moments for the classification of closed planar shapes. Pattern Recognition. **23**(1987)267-272
14. Bres S., Jolion J.M.: Detection of Interest Points for Image Indexation. VISUAL¡99, Lecture Notes on Computer Science,Vol 1614(1999)427-435.
15. Jiang J., Armstrong A., Feng G.: Direct content access and extraction from JPEG compressed images. Pattern Recognition. **35**(2002)2511-2519.
16. Schmid C., Mohr R.: Local Grayvalue Invariants for Image Retrieval. IEEE Trans. on Pattern Analysis and Machine Intelligence. **19**(1997)530-535
17. Smith S.M., Brady J.M.: SUSAN - A New Approach to Low Level Image Processing. Int. Journal of Computer Vision. **23**(1997) 45-78

A Sentence-Based Copy Detection Approach for Web Documents

Rajiv Yerra and Yiu-Kai Ng

Computer Science Dept., Brigham Young University, Provo, Utah 84602, USA
ng@cs.byu.edu, rajiv_yerra@yahoo.com

Abstract. Web documents that are either partially or completely duplicated in content are easily found on the Internet these days. Not only these documents create redundant information on the Web, which take longer to filter unique information and cause additional storage space, but they also degrade the efficiency of Web information retrieval. In this paper, we present a sentence-based copy detection approach on Web documents, which determines the existence of overlapped portions of any two given Web documents and graphically displays the locations of (semantically the) same sentences detected in the documents. Two sentences are treated as either the same or different according to the degree of similarity of the sentences computed by using either the *three least-frequent 4-gram* approach or the *fuzzy-set information retrieval (IR)* approach. Experimental results show that the fuzzy-set IR approach outperforms the three least-frequent 4-gram approach in our copy detection approach, which handles wide range of documents in different subject areas and does not require static word lists.

1 Introduction

Besides piracy one of the problems on the Internet these days is redundant information, which exist due to replicated pages archived at different locations like mirror sites. As a result, the burden is on Web users to sort through retrieved Web documents to identify non-redundant data, which is a tedious and tiring process. Such documents can be found in different forms, such as documents in different versions; small documents combined with others to form a larger document; large documents co-existing with documents that are split from them. One classic example of such documents is news articles where new information is added to an original article and is republished as a new (updated) article. Since the amount of information available on the Internet increases on a daily basis, filtering redundant and similar documents becomes a more difficult task to the user. Fortunately, copy detection can play the role of a filter, which identifies and excludes documents that overlap in content with other documents.

In our copy detection approach, each document is first passed through a stopword-removal [1] and stemming [11] process, which removes all the stopwords and reduces every word to its stem. Stopwords in a sentence should first be removed since stopwords, such as articles, conjunctions, prepositions, punctuation marks, numbers, non-alphabetic characters, etc., often do not play a significant role in representing the sentence. This process reduces the size of a document for comparison and subsequently the complexity on copy detection. Since our copy detection approach is a sentence-based approach, i.e., documents are compared sentence-by-sentence, non-stop, stemmed words in each sentence in one document are compared with non-stop, stemmed words in a sentence in another document to determine whether they are the same using either the *three least-frequent 4-gram* (4-gram for short) or the *fuzzy-set information retrieval* (*IR*) approach. Same (or similar[1]) sentences in different documents are detected, and the relative locations of matched sentences in the documents are graphically depicted by a dotplot view [6], which visually reflects the degree of overlapping of two documents.

We proceed to present our results as follows. In Section 2, we discuss related work in copy detection. In Section 3, we introduce our copy detection approach. In Section 4, we include the experimental results of our copy detection approach. In Section 5, we give a concluding remark.

2   Related Work

Many efforts have been made in the past [2,4,5,7,9,12] in finding similarities among documents. Well-known copy detection methods include (i) Diff (Unix/Linux man pages), which displays the differences in two files by printing the lines in the files that are different, (ii) SCAM [12] (Stanford Copy Analysis Mechanism), which performs word-based copy detection, (iii) SIF [8], which detects similar files, (iv) COPS [2] and KOALA [9], which are designed for plagiarism detection, and (v) the copy detection system for digital documents [4].

Diff, which is designed for source code, text, and other line-oriented files, shows differences between two textual documents, even spaces. It captures the differences between two text documents one line at a time and checks lines in the same order. SCAM, which does not specify the location of overlap between documents, is geared towards small documents. Differed from SCAM, SIF finds similar files in a file system by using the fingerprinting[2] scheme to characterize documents. SIF, however, cannot gauge the extent of document overlap nor display the location of overlap, and the notion of similarity as defined in SIF is completely syntactic, e.g., files containing the same information but using different sentence structures will not be considered similar. COPS, which is developed specifically to detect plagiarism among documents, uses hash-based scheme for copy detection. It compares hash values of given documents with that in the database for copy detection. The basic scheme of SIF and COPS is similar; however, COPS generates syntactic hash units as opposed to fixed-length strings

---

[1] From now on, unless stated otherwise, whenever we use the term *similar* sentences (documents), we mean sentences (documents) that are semantically the same but different in terms of words used in the sentences (documents).

[2] Fingerprints of a document yield the set of all possible document substrings of a certain length, called *fingerprints*, and fingerprinting is the process of generating fingerprints.

adopted by SIF and does have its limitations: (i) it uses a hash function that produces large number of collisions, (ii) documents to be compared by COPS must have at least 10 sentences or more, and (iii) it has problems selecting correct sentence boundaries. KOALA, like COPS, which is specifically designed for plagiarism and is a compromise between random fingerprinting of SIF and exhaustive fingerprinting of COPS, selects substrings of a document based on their usage. This results in lower usage of memory and increase accuracy. Neither KOALA nor COPS, however, can report the location of overlap of two documents and handle documents with varying size. [7] present an approach to identify duplicated HTML documents; however, their identification approach is significantly differed from ours, since they consider HTML documents only, whereas we consider any Web documents. Furthermore, two HTML documents are treated as the same by [7] if the number of occurrences of their HTML tags are the same, which are not considered by us because tags are relatively insignificant in terms of representing document content. The copy detection approach of [4] is closer to ours than others, since it is also sentence-based; however, their copy detection is restricted to copy detection of same sentences. Overall, our copy detection approach does not require static word lists and hence is applicable to Web documents in different subject areas.

Our copy detection approach overcomes most of the limitations of existing approaches. Since our approach is not line-oriented, it overcomes the problems posed by Diff. Unlike KOALA and COPS, our approach is more general, since our approach is not particularly targeted for plagiarism, even though we can handle the problem. Compared with SIF and SCAM, which cannot measure the extent of overlap, our copy-detection approach can specify the relative positions of similar sentences in their corresponding documents graphically.

3 Our Copy Detection Approach

In our copy detection approach, a stopword-removal and stemming process is first performed on sentences in each Web document to yield what we call "refined" sentences, i.e., sentences without stopwords and words in the sentences are stemmed. Since our copy detection approach is a sentence-based approach,[3] refined sentences in one document are compared with refined sentences in another document to determine whether they are similar or different using either the three least-frequent 4-gram or the fuzzy-set information retrieval (IR) approach. Detailed discussions follow.

3.1 Eliminating Non-essential Data

Words in sentences of a Web document involved in copy detection are made to undergo stopword list removal and stemming to represent each sentence independent of its grammatical (dis)similarities so that sentences which have different

---

[3] We use the boundary disambiguation algorithm in [3], which has accuracy in the 90% range but operates in linear time, to determine sentence boundary.

structures (such as active and passive voice) but convey the same meaning are recognized as same sentences, i.e., similar sentences. During the process of stopwords removal from a Web document, each individual sentence is parsed against the list of pre-determined stopwords found in English language. (Even though our copy detection approach is restricted to English Web documents, it can easily be extended to other languages.) This step would reduce the unwanted complexity of processing insignificant/non-representative information and hence can speed up the copy detection process. Consider the following sentence S in a Web document, $Sistani_1$.html:

> *Iraq's top Shi'ite Muslim cleric, Grand Ayatollah Ali al-Sistani, travelled to Britain where he is expected to receive treatment for a heart condition, a spokesman said.*

Removing all the stopwords from S yields

> *iraqs top shiite muslim cleric grand ayatollah ali alsistani travelled britain expected receive treatment heart condition spokesman said.*

denoted S_r. Next, the stemming algorithm [11], which is a suffix removal algorithm, is applied to sentences without stopwords. During the stemming process, an explicit list of suffixes is used, and with each suffix the criterion under which it may be removed from a word to leave a valid stem based on pre-defined rules is considered to find the root of the word. Applying the stemming algorithm to S_r yields S_f as follows:

> *iraqs top shiite muslim cleric grand ayatollah ali alsistani travel britain expect receive treat heart condi spoke said.*

3.2 Different Approaches for Detecting Similar Sentences

After performing stopword removal and stemming on two Web documents, we use either (i) the three least-frequent 4-gram or (ii) the fuzzy-set IR approach to determine which sentences in the documents are similar (same) or different.

3.2.1 Sentence Representations by Three Least-Frequent 4-Grams

The 4-gram approach, which is an exhaustive fingerprinting approach, allows an intuitive, well-defined notion of similarity between documents to be defined [4]. A 4-grams of a string S is a 4-character substring of S. For example, the 4-grams of the string "novel creations" are *nove, ovel, velc, elcr, lcre, crea, reat, eati, atio, tion*, and *ions*. For comparing two sentences using substrings of sentences, two substring selection strategies are usually considered: (i) selection based on N-grams and (ii) selection based on words. We adopt the N-gram approach over word frequencies because (i) N-grams frequencies are largely independent of the type of documents from which they come from [5], (ii) N-grams handle novel words robustly and provide an elegant solution to the zero-frequency problem addressed by [13], which estimates the likelihood of the occurrence of a novel event, and (iii) N-grams do not require static-word sets and thus applicable to various domain sizes. The only disadvantage of N-grams over words is an increase in computation time. However, extracting N-grams is simple and the memory storage of selected N-grams that represent a sentence is far less than storing a whole sentence in the memory. For these reasons, we consider the 4-gram approach as an elegant choice for copy detection.

In using the N-gram approach, it is critical to choose the right value of N to provide good discrimination among sentences. Although any value of N can be considered, $N = 4$ is an ideal choice. This is because as the value of N increases, the better it becomes to distinguish words in one sentence from words in another. Since each N-gram requires N bytes of storage, the memory requirements become too large for $N = 5$. This is because for any given N, each N-gram requires N bytes of storage. For smaller values of N (i.e., $N = 1$, 2, or 3), however, it has been observed [4] that they do not provide good discrimination between sentences. To illustrate the 4-gram construction process, let's consider sentence S_f as computed in Section 3.1, which is obtained from its original sentence S after applying the stopword-removal and stemming steps. The individual 4-grams for the sentence are *iraq, raqs, aqst, qsto, stop, tops*, etc., which are shown in the 4th concatenated 4-grams in Table 1. As part of the pre-processing step, which is not included in the real time copy detection process, each document extracted from the large TREC archive data set (http://trec.nist.gov/data/t5_confusion.html), the *Gutenberg* collection, and numerous Web sites (as shown in Table 2) is registered into the database, and each sentence from the document is converted into a set of four grams and stored in the form of a tree. These 4-grams are then used to determine the three *least-frequent* 4-grams in a sentence from a document, which is the best option to represent the sentence uniquely [5]. The three least-frequent 4-grams in a sentence are concatenated to *represent* the sentence from a document to be compared with the three least-frequent 4-gram representations of sentences in another document. For example, the three least-frequent 4-grams of S in Section 3.1 are *raqs, aqst,* and *qsto*, and S is represented as *raqsaqstqsto*. Two sentences are treated the *same* if their corresponding three least-frequent 4-gram representations are the same. Converting sentences into their three least-frequent 4-gram representations speeds up the processing time of comparing sentences. All the refined sentences involved in the copy detection process are converted into their three least-frequent 4-grams for comparison.

A total number of 200 documents, which were randomly sampled from the set of 52,506 documents as shown in Table 2, were used to study the performance of our 4-gram copy detection approach. Out of the 200 sampled documents, 12.5%, 17.5%, 32.5%, and 37.5% of these documents were chosen from the archive sites TREC, Gutenberg, News articles on the Web, and Etext.org, respectively, and 33% of the 200 sampled documents are HTML documents. Each pair of sentences retrieved from different documents in the sampled set that were treated as either the same or different by using our 4-gram approach were manually verified for *false positives* (i.e., sentences that are *different* but are treated as the *same*) and *false negatives* (i.e., sentences that are the *same* but are treated as *differ-*

Table 1. List of (three least-frequent) 4-grams in sentences in $Sistani_1$.html

$Sistani_1$.html	
Concatenated 4-grams	Three Least Frequent 4-grams
(1) tops opsh pshi shii hiit iite itec tecl ecle cler leri eric ricl ... ... (4) iraq raqs aqst qsto stop tops opsh pshi shii hiit iite item	(1) opsh, pshi, hiit ... (4) raqs, aqst, qsto ...

Table 2. Documents used for constructing least-frequent 4-grams and the correlation matrix in fuzzy IR approaches

Sources	Size (GB)	Number of pages	Number of Sentences	Number of Words*
TREC (http://trec.nist.gov/data/)	1.29	4,390	4,829,653	28,983,918
Gutenberg (ftp://ftp.archive.org/pub/etext/)	1.53	5,643	8,935,215	72,681,213
News Articles on the Web	0.63	39,349	701,520	5,429,643
A Text Archive (ftp://ftp.etext.org/pub/)	1.13	3,124	6,239,154	41,624,048
Total	4.58	52,506	20,705,542	148,718,822

*Number of distinct non-stop, stemmed words.

ent), which are then plotted on the graph for increasing number of sentences. The study shows that the aggressive increase of percentage of false positives and false negatives slows down at after 5,000 sentences and becomes steady and stable thereafter at around 16% for false positives and 12% for false negatives. (See Figure 1(a) for details.) We also observe that when the total number of sentences (to be compared) is less than 500, the error percentage of false positives and false negatives is below 2%, respectively, which is relatively low and acceptable for copy detection on HTML documents to which significant number of existing Web documents on the Internet belong. In fact, the average size of an HTML document, as shown in one of our surveys, is less than 60 sentences (See Figure 1(b) for details, which is constructed by using the 39,349 HTML documents out of the 52,506 documents as shown in Table 2.), and there are significant number of Web documents that are HTML documents.[4]

The major drawback of using our three least-frequent 4-gram copy detection approach is that it cannot detect similar, but only same, sentences, i.e., similar sentences are treated as different, a deficiency that can be corrected by our fuzzy-set IR copy detection approach.

---

[4] We collected 4,389 Web documents from the Internet at one time, and almost 90% of the downloaded documents were HTML documents. This may be an isolated instance, however, since HTML is a widely used language in creating Web documents, the experimental result sounds reasonable to us.

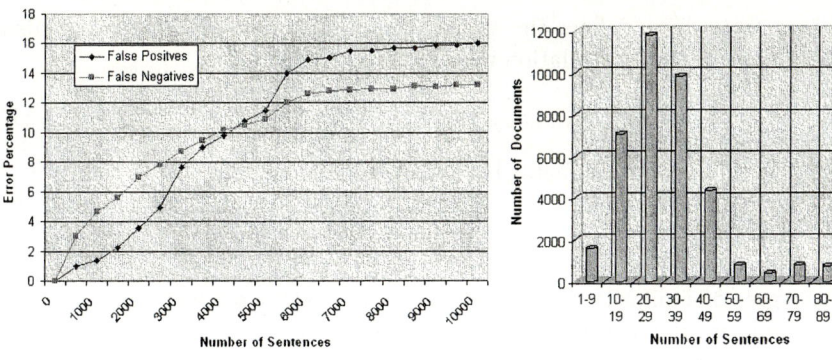

(a) Error analysis of the 4-gram approach (b) Sentence distribution in HTML documents

Fig. 1. Analysis of our 4-gram copy detection approach and sizes of HTML documents

3.2.2 Degree of Similarity in Fuzzy-Set IR Model

Apart from the three least-frequent 4-gram approach to detect same sentences, we adopt and modify the fuzzy-set IR model [10] to find similar sentences. At a high level, a sentence can be treated as a group of words arranged in a particular order. In English language, two sentences can be semantically the same but differ in structure (such as using the active versus passive voice), and matching two sentences is approximate or vague. This can be modeled by considering that each *word* in a sentence is associated with a fuzzy set that contains words with same meaning, and there is a degree of similarity (usually less than 1) between (words in) a sentence and the fuzzy set. This interpretation in fuzzy theory is the fundamental concept of various fuzzy-set models in IR. We consider the fuzzy-set IR model for copy detection, since (i) identification and measurement of similar sentences, apart from exact match, further enhances the accuracy of our copy detection approach, and (ii) the fuzzy-set model is designed and has been proved to work well for partially related semantic content, which can handle the problem of copy detection of similar sentences.

In the fuzzy-set IR model, a *term-term correlation matrix*, which is constructed in a pre-processing step of our copy detection approach, consists of words and their corresponding *correlation factors* that measure the degrees of similarity among different words, such as "automobile" and "car." Our fuzzy-set IR model obtains the degrees of similarity among sentences by computing the correlation factors between any pair of words from two different sentences in their respective documents. (It has been observed that words in sentences that are common to a number of documents discuss the same subject.) In the determination of the degree of similarity between two sentences using the fuzzy-set IR approach, a term-term correlation matrix with rows and columns associated to words[5] captures the degree of similarity among words. The following *word-word*

---

[5] From now on, unless stated otherwise, whenever we use the term *word*, we mean *non-stop, stemmed word*.

correlation factor, $c_{i,j}$, defines the extent of similarity between any two words i and j in the term-term correlation matrix:

$$c_{i,j} = n_{i,j}/(n_i + n_j - n_{i,j}), \tag{1}$$

where $c_{i,j}$ is the correlation factor between words i and j, $n_{i,j}$ is the number of documents in a collection (such as the data set as shown in Table 2) with both words i and j, n_i (n_j, respectively) is the number of documents with word i (word j, respectively) in the collection.

The degree of similarity of two sentences is the extent to which the sentences match. To obtain the degree of similarity between two sentences S_l and S_j, we first compute the *word-sentence correlation factor* $\mu_{i,j}$ of word i in S_l with all the words in S_j, which measures the degree of similarity between i and (all the words in) S_j, as

$$\mu_{i,j} = 1 - \prod_{k \in S_j}(1 - c_{i,k}), \tag{2}$$

where k is one of the words in S_j and $c_{i,k}$ is the correlation factor between words i and k as defined in Equation 1.

Based on the μ-value of each word in a sentence S_i, which is computed against sentence S_j, we define the *degree of similarity* of S_i with respect to S_j as follows:

$$Sim(S_i, S_j) = \frac{\mu_{w_1,j} + \mu_{w_2,j} + \ldots + \mu_{w_n,j}}{n}, \tag{3}$$

where w_k ($1 \leq k \leq n$) is a word in S_i, and n is the total number of words in S_i. $Sim(S_i, S_j)$ is a normalized value. Likewise, $Sim(S_j, S_i)$, which is the *degree of similarity* of S_j with respect to S_i, is defined accordingly.

Using Equation 3 as defined above, we determine whether two sentences S_i and S_j should be treated the same, i.e., equal (EQ), below.

$$EQ(S_i, S_j) = \begin{cases} 1 \text{ if } MIN(Sim(S_i, S_j), Sim(S_j, S_i)) \geq 0.825 \wedge \\ \quad\quad |Sim(S_i, S_j) - Sim(S_j, S_i)| \leq 0.15 \\ 0 \text{ otherwise,} \end{cases} \tag{4}$$

where 0.825 is called the *permission threshold value*, whereas 0.15 is called the *variation threshold value*. The permissible threshold is a value set to obtain the *minimal* similarity between any two sentences S_i and S_j in our fuzzy-set IR copy detection approach, which is used partially to determine whether S_i and S_j should be treated as equal (EQ). Along with the permissible threshold value, the variation threshold value is used to decrease the number of false positives and false negatives in determining the equality of two sentences. The variation threshold value sets the the maximal, allowable difference in sentence sizes between S_i and S_j, which is computed by calculating the *difference* between $Sim(S_i, S_j)$ and $Sim(S_j, S_i)$. The threshold values 0.825 and 0.15 were determined by testing the documents in the collection as shown in Table 2, which provide the necessary and sufficient conditions for estimating the equality of two sentences. We explain how to compute the threshold values below.

Using the same randomly sampled 200 documents for analyzing the performance of our 4-gram copy detection approach as discussed in Section 3.2.1, we determined the *permissible* and *variation threshold values*. According to each predefined threshold value V, which is in between 0.5 and 1.0 with an increment of 0.05, we categorized each pair of sentences S_1 and S_2 in the 200 documents as equal or different, depending on V and the minimum of the two degrees of similarity measures on S_1 and S_2. S_1 and S_2 are potentially *equal* if $MIN(Sim(S_1, S_2), Sim(S_2, S_1)) \geq V$; otherwise, they are *different*. Hereafter, we manually (i) examined each pair of sentences to determine the accuracy of the conclusion, i.e., *equal* or *different*, drawn on the sentences, (ii) computed the number of false positives and false negatives, and (iii) plotted the outcomes in a graph. We set the *permissible threshold value* to be 0.825, which is the "balance" point of the minimum false positives and false negatives, i.e., neither the values of false positives nor false negatives dominates the copy detection errors.

To obtain the *variation threshold value*, we examined a set PS of pairs of sentences whose minimal degrees of similarity exceed or equal to the permissible threshold value. The differences between the degrees of similarity of each pair of sentences in PS are first calculated and then sorted. The false positives and false negatives are then determined manually and plotted at regular intervals in sorted order. The intersection of the false positive and false negative curves is chosen as the variation threshold value, which is neither dominated by false positives nor false negatives.

Since the minimum permissible threshold value is set to 0.825, the difference in similarity between two documents can only range between 0 and 0.175. The greater the difference in similarity, the greater are the chances of error. For example, if the difference is 0.175, one of the similarity measures can be $sim(S_1, S_2) = 1.0$ and the other is $sim(S_2, S_1) = 0.825$, or vice versa. This is the case only when the set of words in S_1 is a subset of the words in S_2. For example, if sentence S_1 is "The boy goes to school every day on bus," whereas sentence S_2 is "The bus going to school passing by the lake stops for the boy at the corner every day." In this example, S_1 is a part of S_2, but not vice versa, and should be treated as different. According to the experimental results compiled by using the 200 sampled documents, the ideal variation threshold value between the range 0 and 0.185 is 0.15, which we adopt in our fuzzy-set copy detection approach.

3.3 Degree of Overlap

Using the EQ value as defined in Equation 4, which determines whether two sentences should be treated as the same, we derive the *degree of overlap* between documents doc_1 and doc_2, which computes the degree of overlap between doc_1 and doc_2 according to the ratios of sentences common to both documents, which is defined as follows:

$$Overlap(doc_1, doc_2) = \frac{(|doc_1| \cap |doc_2|)}{|doc_1|}, \frac{(|doc_1| \cap |doc_2|)}{|doc_2|}, \quad (5)$$

where $|doc_1| \cap |doc_2|$ is the number of common sentences in doc_1 and doc_2, and $|doc_1|$ ($|doc_2|$, respectively) is number of sentences in doc_1 (doc_2, respectively).

3.4 Complexity Analysis of Our Copy Detection Approach

The complexity of our copy detection approach is $\mathcal{O}(|A|)(|B|)$, where $|A|$ and $|B|$ are the number of sentences in two documents A and B, respectively. The time complexity for stopword removal is $\mathcal{O}(|N|)$, where N is the number of words in a document, and the stemming complexity is $\mathcal{O}(|M| + |P|)$, where M is the number of words searched and P is the number of iterations for converting each word into its corresponding stem. The last two measurements can be ignored, since $\mathcal{O}(|A|)(|B|)$ is the dominating factor. Also, the time complexity for applying the three least-frequent 4-gram or the fuzzy-set IR approach is $\mathcal{O}(w_1)(w_2)$, where w_1 (w_2, respectively) is the number of words in sentence 1 (sentence 2, respectively). Since $\mathcal{O}(w_1)(w_2) \leq \mathcal{O}(|A|)(|B|)$ holds, the time complexity of our copy detection approach is $\mathcal{O}(|A|)(|B|) \cong \mathcal{O}(|A|^2)$.

4 Experimental Results

Using a set of Web documents collected from various sources (as shown in Table 2), we evaluated the performance of our copy detection approach for Web documents. Randomly sampled pairs of sentences S_1 and S_2 from different documents were compared using our copy detection approach. The results, which include some pairs of sentences listed below, are shown in Table 3.

1) Please do not hesitate to contact us   We can reach the destination by tomorrow
2) We acknowledge receipt of your letter   We got your letter
3) I consider Bob Kim's best friend   Bob is Kim's best friend
4) John is the king   The king is intelligent
5) The student submitted papers to the company for employment   To get employed the graduate applied to the office with relevant documents
6) No one is allowed to talk to him   Please do not hesitate to contact us
7) We have pleasure enclosing our updated brochure   We have put a new brochure in this letter

As shown in Table 3, the fuzzy-set IR approach outperforms the three least-frequent 4-gram approach in detecting similar sentences, since the fuzzy-set IR approach checks for similarity between completely different words, besides same words, whereas the three least-frequent 4-gram approach detects similarity between same words only, e.g., "document" and "paper" are considered similar by the fuzzy-set IR approach, whereas they are declared different by the 4-gram.

In order to further verify the correctness of the three least-frequent 4-gram approach, as well as our fuzzy-set IR approach, we ran other test cases to compare (i) two documents that are closely related—all the sentences in one are included in the other, (ii) unrelated sentences added to the subset document in (i), (iii) two

Table 3. Experimental results of our copy detection approach for Web documemts on a set of randomly sampled pairs of sentences

Sentence Pairs	Matched	FP (4-gram)	FN (4-gram)	FP (Fuzzy)	FN (Fuzzy)
(1)	No	No	No	No	No
(2)	Yes	No	Yes	No	No
(3)	Yes	No	No	No	No
(4)	No	No	No	Yes	No
(5)	Yes	No	Yes	No	No
(6)	No	No	No	No	No
(7)	Yes	No	No	No	No
...	...	...	...	...	...
Total Count		0%	22%	5.5%	5.5%

F(alse)P(ositive): Sentences that are *different* but are treated as the *same*
F(alse)N(egative): Sentences that are the *same* but are treated as *different*

Table 4. Comparisons of successive CNN news reports on Russian hostages taken with Rus_1 reported on 09/01/2004, Rus_2 and Rus_3 on 09/02/2004, and Rus_4, Rus_5, and Rus_6 on 09/03/2004 using the 3 least-frequent 4-gram and fuzzy-set IR approaches

Case	Documents 1st	2nd	# Sentences in Documents 1st	2nd	Computed by **4-gram** Sentence Matched	Overlap % 1st	2nd	Computed by **Fuzzy-set** Sentence Matched	Overlap % 1st	2nd
1-2	Rus_1	Rus_2	99	83	42	42%	51%	49	50%	59%
2-3	Rus_2	Rus_3	83	90	36	43%	40%	37	45%	41%
3-4	Rus_3	Rus_4	90	101	39	43%	39%	43	48%	43%
4-5	Rus_4	Rus_5	101	104	77	76%	74%	77	76%	74%
5-6	Rus_5	Rus_6	104	101	96	92%	95%	96	92%	95%
1-6	Rus_1	Rus_6	99	101	42	42%	42%	46	47%	46%

documents that are highly unrelated, (iv) two documents that are (moderately) related, (v) two documents that vary in size to demonstrate the resistance of our copy detection approach to size variations, and (vi) revisions of a set of successive news reports. The degree of overslapping of any two documents are graphically depicted by using dotplot views as shown in Figure 2.

The additional test cases, along with others, confirm that both three least-frequent 4-gram and fuzzy-set IR approaches perform very well with same sentences, with only a few false positives and false negatives, and the fuzzy-set IR approach detects similar sentences significantly better than the 4-gram approach, since the latter cannot detect similar sentences. We observe that the number of sentences matched in Figure 2(b) is less than the one as shown in Figure 2(a). This is because occasional false positives are obtained when using the 4-gram approach. Since the 4-gram approach uses three least 4-grams in a sentence S to represent S, two different sentences can be represented incorrectly to be the same, if by chance the same three least 4-grams dominate the sentences.

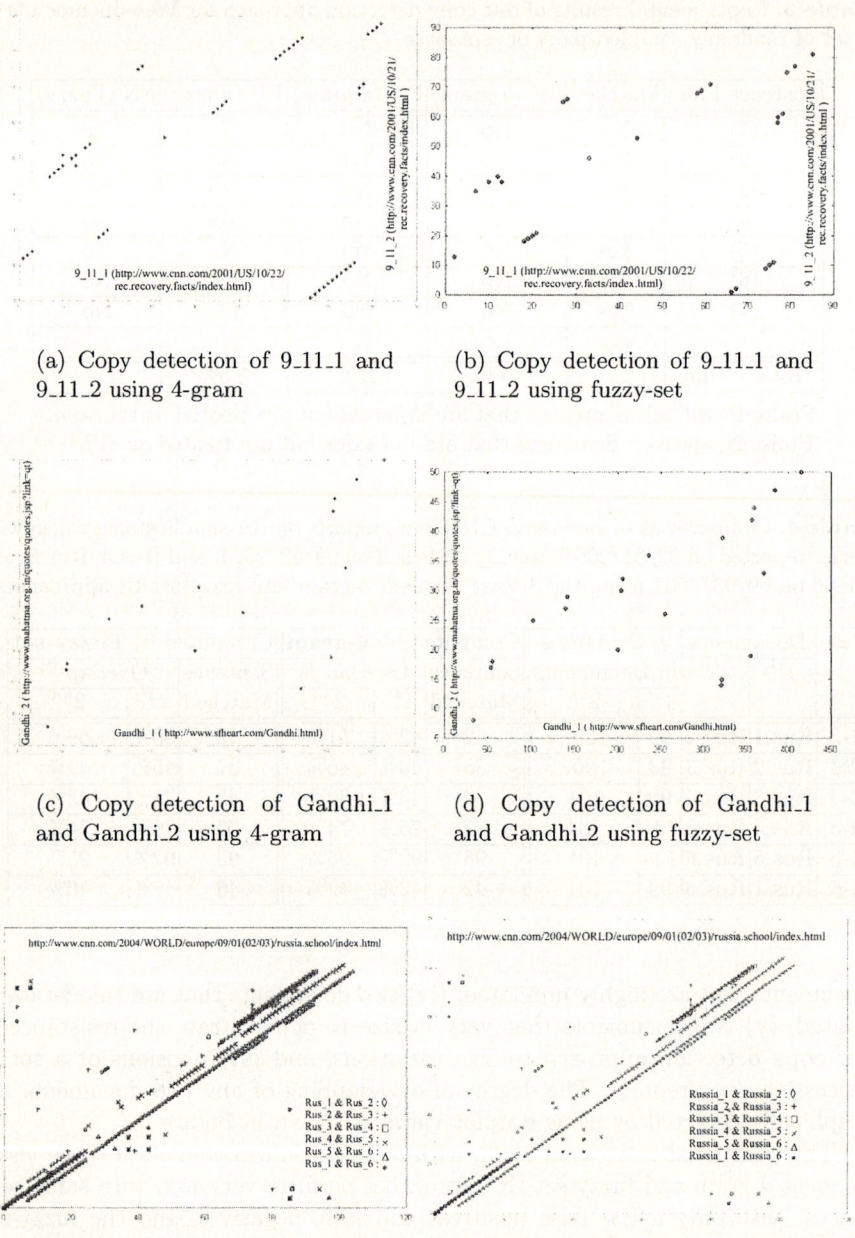

Fig. 2. Dotplot views of similar sentences in different Web documents detected by using the 3 least-frequent 4-gram and fuzzy-set IR approaches

In Figure 2(d) more sentences are detected compared to the ones as shown in Figure 2(c), since the corresponding documents contain some similar sentences apart from exact sentences. Figures 2(e) and 2(f) show almost the same results, since most of the sentences detected are exact matches and there are negligible, or very, few similar sentences. Overall, the fuzzy-set IR approach out-performs the 4-gram approach in terms of (i) detecting similar sentences, (ii) generating less *false positives* in detecting similar sentences than the 4-gram approach because unlike the 4-gram approach, the fuzzy-set IR approach does not use the least-frequent keywords to determine similarity of sentences, and (iii) producing less *false negatives* in determining similar sentences than the 4-gram approach because the 4-gram approach cannot detect similar sentences and thus treats similar sentences as different sentences.

5 Conclusions

We have presented a new approach that performs copy detection on Web documents. Our copy detection approach (i) determines similar Web documents, which can act as a filter to various Web search engines/Web crawlers to improve their search efficiency by indexing fewer documents and eliminating the ones that are redundant, (ii) detects similar sentences, apart from same sentences, by using the fuzzy-set information retrieval (IR) approach on Web documents or detects same sentences using either the three least-frequent 4-gram approach and/or the fuzzy-set IR approach, and (iii) captures similar sentences in any two Web documents graphically, which displays the location of overlapping portions of the documents. Not only our copy detection approach handles wide range of documents (such as sports, news, science, etc.), but also since it does not require static word lists, it is applicable to Web documents in different subject areas. We implemented our copy detection approach in Java on a Windows XP PC.

For future work, we would like to (i) analyze similar sentences in two Web documents using the natural-language processing approach, which could further enhance the accuracy of our copy detection approach and (ii) extend our similarity measures to handle copy detection on non-English Web documents.

References

1. Baeza-Yates, R., Ribeiro-Neto, B.: Modern Information Retrieval. Addison Wesley (1999)
2. Brin, S., Davis, J., Garcia-Molina, H.: Copy Detection Mechanisms for Digital Documents. In Proceedings of the ACM SIGMOD (1995) 398–409
3. Campbell, D.: A Sentence Boundary Recognizer for English Sentences. Unpublished Work (1997)
4. Campbell, D., Chen, W., Smith, R.: Copy Detection Systems for Digital Documents. In Proceedings of IEEE Advances in Digital Libraries (2000) 78–88
5. Damashek, M.: Gauging Similarity with N-grams: Language-Independent Categorization of Text. Science **267** (1995) 843–848

6. Helfman, J.: Dotplot Patterns: A Literal Look at Pattern Languages. Theory and Practice of Object Systems **2(1)** (1996) 31–41
7. Lucca, G.A.D., Penta, M.D., Fasolino, A.R.: An Approach to Identify Duplicated Web Pages. In Proceedings of COMPSAC (2002) 481–486
8. Manber, U.: Finding Similar Files in Large File System. In USENIX Winter Technical Conferences (1994)
9. Nevin, H.: Scalable Document Fingerprinting. In Proceedings of the 2nd USENIX Workshop on Electronic Commerce (1996) 191–200
10. Ogawa, Y., Morita, T., Kobayashi, K.: A Fuzzy Document Retrieval System Using the Keyword Connection Matrix and a Learning Method. Fuzzy Sets and Systems **39** (1991) 163–179
11. Porter, M.: An Algorithm for Suffix Stripping. Program **14(3)** (1980) 130–137
12. Shivakumar, N., Garcia-Molina, H.: SCAM: A Copy Detection Mechanism for Digital Documents. D-Lib Magazine (1995) http://www.dlib.org
13. Witten, I., Bell, T.: The Zero Frequency Problem: Estimating the Probabilities of Novel Events in Adaptive Text Compression. Technical Report, University of Calgary (1991)

The Research on Query Expansion for Chinese Question Answering System*

Zhengtao Yu[1,2], Xiaozhong Fan[2], Lirong Song[3], and Jianyi Guo[1]

[1] The School of Information Engineering and Automation,
Kunming University of Science and Technology, Kunming, P.R. China 650051
ztyu@bit.edu.cn, gjade86@hotmail.com
[2] Department of Computer Science & Engineering, Beijing Institute of Technology,
Beijing, P.R. China 100081
fxz@bit.edu.cn
[3] English Department, Beijing Institute of Technology, Beijing 100081
rongdasong@bit.edu.cn

Abstract. In document retrieval, expanding query with words that are semantically related or frequently co-occur can get good performance. In Chinese question answering system, in order to improve answer-document retrieval precision, query expansion is also necessary. Aiming at the specialty of Chinese question answering system, a method of query expansion based on related words for specific question types and synonym in HowNet is proposed. A computing method of similarity between questions and documents based on minimal matching span is presented. This method is based on vector space model, and also fully considers the position information of query words and query expansion words in the documents. Finally, the experiment results show that the effect of expanding query makes better than unexpanded one.

1 Introduction

In Chinese question answering system, when using question presented with natural language to retrieval document, query words extraction and query expansion are the key factors that affect its retrieval performance. After stop-words are removed from Chinese question segmentation, nouns, adverbs and adjectives are extracted to form the query [1]. For query already generated, in order to improve the recall rate of answer-document retrieval, it is necessary to expand query. At present, there are two kinds of query expansion methods in information retrieval [2]: first, statistically analyses the initial retrieval documents, select relevant words that frequently co-occur and add them to query, such as "Blind relevant feedback" [2]. The effect of this method deeply depends on the initial retrieval results, and it may expand more irrelevant words so that the information retrieval effect is reduced [3]. Second, make use of a certain semantic knowledge resource, and expand query words directly according to the semantic relation between query words (for example: synonymy). Query expansion based on "WordNet" and "HowNet" resources has already made good retrieval effect [3].

---

* This paper is supported by Yunnan province information technology fund (2002IT03).

In Chinese question answering system, for questions of some specific types, correct answer-documents may contain some same words. For example, for question type about "距离(distance)", the answer-document may contain some words that are related to the type of "距离(num_distance)",such as "公里,里,英尺,尺". If words which potentially occur in the answers can be expanded into query, the question query ranges can be limited, and the answer-document retrieval precision can be improve. For example, Q_1 土星距离太阳有多远？(How far away from the Saturn to the sun?) query words {土星,距离,太阳} Baidu search engine retrieves this query words and recalls 24200 WebPages in all (2004-10-30). In the first 20 WebPages, the 12nd and 13rd contain the correct answer. Similarly, if the word "公里(kilometer)" is expanded into query according to the question type "距离(num_distance)",a new query set {土星,距离,太阳,公里} is formed. Baidu (http://www.baidu.com) search engine retrieves this set and recalls 9210 WebPages. In the first 20 recall WebPages, the 1st 4th 10th 15th 18th contain the correct answer. Obviously query expansion restricts the answer range, and the precision that the first 20 recall WebPages contain the answer is higher than unexpanded query. From the above instance analysis, in Chinese question answering system, it is very necessary to do research on query expansion according to the characteristics of Chinese question types. This paper proposes a query expansion method between the above two kinds of methods. This method expands query with related words for specific question types obtained by statistical analysis and synonym in HowNet. Based on VSM, this paper proposes a computing method of similarity between question and answer-document based on minimal matching span; finally this paper does answer-document retrieval experiments on the methods proposed.

2 Query Expansion Methods for Chinese Question

Because Chinese questions are very complicated, this paper focuses on query expansion for factoid question [4]. Because different question has different features and questions can be classified, in question parsing stage, question types can be recognized according to the combinatorial rules between the interrogative and its interrelated words [1], or recognized by machine learning method [5]. The recognition of question types has already made good effects [1][5]. After the question type is recognized, answer-document can be retrieved according to the corresponding characteristics of question types. According to analysis of abundant question-answer pairs of different question types, it can be seen that for some questions of specific question types (such as "距离(num_distance)", "重量(num_weight)", "面积(num_area)", "年龄(num_age)" etc.), a certain common phenomenon may exist in correct answer-document ,i.e. correct answer-document usually contains some words related to corresponding question type. For example, the question type of "位置(location)",the words such as "位于" "地处" "坐落于" etc. often occur in the correct answer-documents. According to the statistical analysis of question-answer pairs of each question type, potential co-occurrence words of each question type are extracted to form related words for specific question type. We construct the related words for specific question types of more than 40-question types in all. Table 1 lists part of question types and their related words. There is no unified criterion for question classification. Some question types in the table can be further subdivided. Subdivided classification may have good effect on expansion. But too fine classifications will bring much larger difficulty for the recognition of question type.

Table 1. Some question types and their expansion related words

Question type	The related words for specific question types
价格(num-price)	美元，英镑，$，元，角，分
速度(num_speed)	里，公里，快，速度，时速，公里/秒，千米/秒，每
温度(num_temperature)	度，度数，华氏温度，摄氏
年龄(num_age)	岁，年，月，日，年龄
重量(num_weight)	公斤，千克，磅，吨，千吨
频率(num_frequency)	次，次数，MHz，Hz，经常，常常
距离(num_distance)	米，里，公里，英尺，码，厘米
面积(num_area)	亩，公顷，平方，平方米，平方公里
岛屿(loc-island)	位于，在，地处，坐落于，处在
年份(time_year)	于，在，年

After got the related words, the query can be expanded with these words. Due to the synonymy relation of the Chinese words, the words occurring in answer-document was not always the query words but synonymy of query words, so some relative documents can not be recalled in retrieval. So the Chinese question expansion method combining related words and HowNet is proposed. This method firstly expanded query with related words according to question type, then expanded the noun of query words with synonymy in HowNet.

An example of query expansion according to related words for specific question type and synonym in HowNet:

Q_2: 北京大学占地多少？(How much is the area of Peking University?)
Question type:"面积(num_area)"
Query words:{北京大学,占地}
Related words for specific type:{ 亩,公顷,平方,平方米,平方公里}
Synonymy of query words: {北大}
Query expansion words :{亩,公顷,平方,平方米,平方公里,北大}

Each Chinese question is expanded by related words for specific question types and Synonymy in HowNet. Query after expansion includes two parts: set of query words and set of query expansion words. The two sets have different effects on document retrieval, so the two sets must be synthetically considered when computing the similarity between questions and documents.

3 Answer-Document Retrieval Based on Query Expansion

Information retrieval system SMART is an information retrieval model based on vector space model [7]. It includes many functions, such as establishing index, retrieval and evaluation etc. So we do researches on answer-document retrieval in Chinese question answering system by using information retrieval system SMART. Because retrieval system SMART is based on vector space model and doesn't support Boolean "or" operation, when processing query expansion words of question, the former system needs to be modified so that it can adapt the answer-document retrieval based on query expansion.

3.1 Computing Method of Global Similarity Between Question and Document

In order to retrieve answer-documents that are related to question, similarity between question and document must be computed. Generally computing method of similarity adopts vector space model. This paper takes Lnu.Itc [7] weighting scheme to weigh the question and document. Its similarity is called global similarity whose computing formula is shown as follows (1) [6][8]:

$$Sim_1(q,d) = \sum_{t \in q \cap d} \frac{\frac{1+\lg(f_{t,d})}{1+\lg(a_{t' \in d} f_{t',d})} \cdot \lg(\frac{N}{n_t})}{((1-sl) \cdot pv + sl \cdot uw_d)} \qquad (1)$$

Here, $f_{t,d}$ refers to the occurrence frequency of word t in document d; $a_{t' \in d} f_{t',d}$ refers to the average number of words occurring in document d; N is the number of document in the set; n_t is the number of document containing word t in the document set; pv refers to the average length of document in all document set; sl is a weight parameter, it is an experiment empirical value and generally it is 0.2 [6], uw_d is the number of unique words in the document.

Computing method in formula (1) only considers query words, and doesn't consider the effect of query expansion words on document similarity. So, the effect or query expansion words on document similarity must be considered after query expansion. For example, for Q_2, the contributive degree of query expansion words {亩,公顷,平方,平方米,平方公里,北大} to query must be considered when computing similarity. Therefore, computing method of global similarity in formula(1) must be modified. The set of query expansion words may contain many words, but not all the words have equivalent function for answer retrieval. Therefore, the weight of each query expansion words must be computed by using formula (2).

$$w(t,d) = \frac{1+\lg(f_{t,d})}{1+\lg(a_{t' \in d} f_{t',d})} \cdot \lg(\frac{N}{n_t}) \qquad (2)$$

The weight of query expansion words depends on two factors: One is the normalization word frequency (tf) of expansion words in the document; the other is idf value of query expansion word t. According to the above formula (2), the weight of each expansion word can be computed respectively. The query expansion word with the greatest weight is selected to participate in computing the similarity between questions and documents; For example, given a set of query expansion words (t_1, t_2, t_3)

of a question and a document, assuming the document contain the query expansion words t_1, t_2. If the corresponding weights, $w(t_1, d) > w(t_2, d)$, then only t_1 is selected to participate in computing global similarity. According to the above method, query expansion word is selected. Formula (1) should be modified, and a query expansion word with greatest weight is added to compute similarity. The global similarity after expansion is computed by the formula (3):

$$Sim_2(q,d) = \sum_{t \in q \cap d} \frac{\frac{1+\lg(f_{t,d})}{1+\lg(a_{i \in d} f_{i,d})} \cdot \lg(\frac{N}{n_t})}{((1-sl).pv + sl.uw_d)} + \frac{\max_{t \in q_a \cap d} \frac{1+\lg(f_{t,d})}{1+\lg(a_{i \in d} f_{i,d})} \cdot \lg(\frac{N}{n_t})}{((1-sl).pv + sl.uw_d)} \quad (3)$$

In the above formula, the contribution of query words and a query expansion word are considered respectively in information retrieval. q' is the set of the original query words before query expansion. q_a is set of query expansion words, and the other parameters are the same as formula(1).

3.2 Computing Method of Similarity Between Question and Document Based on Minimal Matching Span

The computing method of global similarity between question and document has been discussed in the former part. Because of using VSM, word frequency statistical information in the document is only considered, while the distribution information of query words and query expansion words in the document is not considered. The distribution distance between query words and query expansion words in the document has great effect on answer-document retrieval. Therefore on the basis of global similarity, a computing method between question and document based on minimal matching span is proposed.

This method doesn't only consider number of matching words between query and document, but also consider distribution information of query in document. In order to computer distribution information of query in document, two definitions are listed in the following.

Definition 1 (matching span): Given a query q and a document d, q' is the set of query words (unexpanded), and q_a is the set of query expansion words. $q' \subseteq q$, $q_a \subseteq q$, where the function $wordpos_d(p)$ return the word occurring at position p in d. A matching span (ms) is a set of positions that contains at least one positions of each matching word from q' and one position of a matching word from q_a, i.e. $\bigcup_{p \in ms} wordpos_d(p) \in \{(q' \cup \{t\}) \cap d | t \in q_a\}$.

Definition 1 doesn't consider query expansion words, viz. q_a is empty, a matching span (ms) is a set of positions that contains at least one position of each matching word, which only embodies distribution information of query words in document. A query may contain many query expansion words, so the corresponding ms of each query expansion word must be computed respectively when computing its matching span.

Definition 2 (Minimal matching span): Given a matching span ms, let b_d be the minimal value in ms, i.e., $b_d=min(ms)$, and e_d be the maximal value in ms, i.e, $e_d=max(ms)$. A matching span ms is a minimal matching span (mms) if there is no other matching span ms' with $b_d' = min(ms')$, $e_d' = max(ms')$, such that $b_d \neq b_d'$ or $e_d \neq e_d'$, and $b_d \leq b_d' \leq e_d' \leq e_d$.

The above two definitions reflect distribution information of query words and query expansion words in document. Because similarity between question and document is mainly embodied by number of matching words and distribution information in query and document, on the basis of global similarity, a computing method of similarity based on minimal matching span is proposed. Formula (4) and formula (5) are used to compute the similarity of unexpanded query and expanded query between question and document based on minimal matching span respectively.

$$Sim_3(q,d) = \begin{cases} \lambda Sim_1(q,d) + (1-\lambda)\left(\dfrac{|q \cap d|}{1+\max(mms)-\min(mms)}\right)^\alpha \times \left(\dfrac{|q \cap d|}{|q|}\right)^\beta, & |q \cap d| > 1 \\ Sim_1(q,d), & |q \cap d| = 1 \end{cases} \quad (4)$$

In formula(4), $|q \cap d|/(1+\max(mms)-\min(mms))$ is the span size ratio, which is the number of unique matching words in the span over the total number of token in the span. It reflects distribution information; $|q \cap d|/|q|$ is word matching ratio, which is the number of unique matching words in query and document over the number of unique words in query. It reflects the numbers relationship of same words. α, β and λ are empirical values, usually $\alpha=1/8$, $\beta=1$ and $\lambda=0.4$.

$$Sim_4(q,d) = \begin{cases} \lambda Sim_2(q,d) + (1-\lambda)\left(\dfrac{|q' \cap d| + ne(q_a \cap d)}{1+\max(mms)-\min(mms)}\right)^\alpha \times \left(\dfrac{|q' \cap d| + ne(q_a \cap d)}{|q'| + ne(q_a)}\right)^\beta, & |q' \cap d| + ne(q_a \cap d) > 1 \\ Sim_2(q,d), & |q' \cap d| + ne(q_a \cap d) = 1 \end{cases} \quad (5)$$

Formula (5) is modified based on the formula (4), and it adds a query expansion words. Query words and query expansion words are considered when computing the span size ratio and word matching ratio. When counting the number of query words, the set of query expansion words (q_a) as a whole counts as one query word. The $ne(\cdots)$ function checks whether the set $q_a \cap d$ is empty. If $|q_a \cap d| > 0$ then $ne(q_a \cap d)$ returns 1, represent that query expansion word occurring in document, while 0 represents that query expansion word doesn't occurring in document.

3.3 Case Study of Computing Similarity

Give a question Q_3: 地球距离太阳有多远？(How far away from the earth to the sun?), query words are {地球,太阳,距离}, and question type is "距离(num_distance)". After query expansion, query expansion words are {米，里，公里，英尺，码，厘米，寰球，球}. The computing process of similarity between question and document based on minimal matching span is shown as follows:

Assuming that there is a document d with query words matching at following position : Pos_d(地球) = {20,35,60}, Pos_d(太阳) = {38,70}, Pos_d(距离) = {Φ}, Before query expansion, matching span (ms) is {20,35,38,60,70},{20,35,70}{20,38},{35,38}, {38,60},······,{60,70},obtained by definition 1, and minimal matching span (mms) is {35,38},obtained by definition 2, so its span size ratio is 2/(1+38-35)=0.5, and its word matching ratio is 2/3=0.67. Take $\alpha=1/8, \beta=1$, and $\lambda=0.4$, assuming the global similarity before query expansion, $Sim_1(q,d)=0.72$, is computed by formula(1), then the minimal matching span similarity of unexpanded $Sim_3(q,d)$ is $0.4 \times 0.72 + 0.6 \times (0.5)^{\frac{1}{8}} \times (0.6)^1 = 0.657$.

According to question type "距离(distance)",query is expanded. Assume that the matching positions of query expansion words in the document are as follows: Pos_d(公里) = {41}, Pos_d(里) = {Φ}, Pos_d(英尺) = {Φ}, Pos_d(码) = {Φ}, Pos_d(米) = {180}, Pos_d(厘米) = {Φ}, Pos_d(寰球) = {Φ}, Pos_d(球) = {Φ}, in which expansion words "公里" and "米" occur in the document. Therefore, distribution relationship of two words must be considered respectively, so that one is selected to compute similarity.

For expansion word "公里",its corresponding (ms) is {20,35,38,41,60,70}, {20,35,38,41},{35,38,41},{20,38,41},······,{20,,41,70},obtained by definition 1, and minimal matching span (mms) is {35,38,41},obtained by definition 2, so span size ratio is 3/(1+41-35)=0.429.

Similarity, for example word "米",its corresponding ms is {20,35,38,60,70,180}, {20,35,38,180},{35,38,180},{35,85,180},······,{60,70,180},and its minimal matching span (mms) is {60,70,180},so the span size ratio is 3/(1+180-60)=0.025.

Because the span size ratio of "公里" is bigger than that of "米", "公里" is selected as expansion word to compute similarity. The word span ratio after query expansion is 3/4=0.75, take $\alpha=1/8$, $\beta=1$, $\lambda=0.4$. Assuming the global similarity after query expansion, $Sim_2(q,d)=0.8$, computed by formula (3), then the minimal matching span similarity after query expansion $Sim_4(q,d)$ is $0.4 \times 0.8 + 0.6 \times (0.429)^{\frac{1}{8}} \times (0.75)^1 = 0.725$.

4 Experiment and Evaluation

The experiment focuses on the evaluation of performance effect that query expansion has on answer-document retrieval. At present, there is no uniform test data set of questions and answer-documents about Chinese question answering system, and the purpose of answer retrieval in question answering system is to retrieve the document containing correct answers. Therefore, data set for experiments must be collected and tagged. 10 fine-grained question types including "价格(num_price)", "速度(num_speed)","温度(num_temperature)","年龄(num_age)","重量(num_weight)", "距离(num_distance)","频率(num_frequency)","面积(num_area)","岛屿(loc_island)", and "年份(time_year)" are selected to do this experiment. We collect 20 questions for each fine-grained type respectively, obtain the query words after parsing each question, and submit them to Baidu search engine; then we extract the first 30 WebPages of each question as the candidate answer-documents, and establish 200 Chinese questions and more than 6000 corresponding document sets; finally, we process these documents, tag whether the document contains the corresponding answers, and tag the number of corresponding answer for document containing correct answers .Thus a test data set of questions and answers is established.

Generally, in information retrieval, evaluation methods are precision and recall rate. For Chinese question answering system, correct answers are needed to be further extracted from the recall documents, so evaluation emphasizes whether the recall documents contain correct answer-documents. Therefore, there is a quite simple evaluation method a@n for evaluating precision. It only checks whether the first n recall documents contain correct answer-documents. If they contain correct answer-document, then it is 1,and 0 otherwise. For all test questions, a@n refers to that the number of all questions in test set is divided by the number of documents containing correct answers in the first n recall documents. The experiment uses precision a@n to evaluate answer-document retrieval.

In order to verify the effect of the proposed method. Using vector space modole (VSM) and minimal matching span (MMS) to retrieve answer-document, we evaluate the answer-document retrieval experiment that doesn't expand query (unexpanded) and does expand query (expanded) for the 200 questions in test set respectively. We use formula (1) (VSM unexpanded), formula (3) (VSM expanded), formula (4) (MMS unexpanded) and formula (5) (MMS expanded) to compute document similarity respectively. Table 2 lists the data of answer retrieval results before query expansion and after query expansion.

From the above data, when we evaluate that the number of recall documents is 5,8,10,15 respectively, document retrieval precision shows that query expansion makes substantial improvement over an unexpanded baseline using two methods respectively. The percent in bracket is ratio improved after query expansion. The method based on MMS make better than the one based on VSM.

Table 2. Comparison of answer-document retrieval results before query expansion and after query expansion based on MMS and VSM respectively

a@n	VSM		MMS	
	unexpanded	expanded	unexpanded	expanded
a@5	0.445	0.51(+14.6%)	0.505	0.71(+40.6%)
a@8	0.595	0.665(+11.8%)	0.655	0.78(+19.1%)
a@10	0.68	0.745(+9.7%)	0.75	0.855(+14%)
a@15	0.775	0.85(+9.7%)	0.855	0.96(+12.3%)

5 Conclusions

The purpose of answer-document retrieval in Chinese question answering system is to recall the documents containing correct answer to questions. Query expansion method in information retrieval is not fully suitable for question answering system. According to the features of Chinese question answering system, this paper proposes a new query expansion method that establishes related words for specific question type through statistical method, and expands query with related words for specific question type and synonymy in HowNet. In order to verify the effect of this method, on the basis of vector space model, this paper proposes a computing method of similarity between questions and documents based on minimal matching span. This method fully considers the effects of query words and query expansion words on answer-document retrieval. The experiment results show that in Chinese question-answering system, answer-document retrieval has quite good effect using query expansion method proposed in this paper. Further research will focus on the following aspects:

1. query expansion of specific question type for extracting sentences as answer;
2. answer extraction according to question type and answer sentence model.

References

1. Zheng, S.F., Liu, T., Qin, B.: Overview of Question Answering. Journal of Chinese Information Processing, Vol.16, No.6, (2002) 46-52
2. Cui, H., Wen, J.R., Li, M.Q.: A Statistical Query Expansion Model Based on Query Logs. Journal of Software, Vol.19, No.3, (2003) 1593-1599
3. He, H.Z., He, P.L., Gao, J.F.: Query Expansion Based on the Context in Chinese Information Retrieval. Joural of Chinese Information Processing, Vol.16, No.6, (2003) 32-45
4. Voorhees, E.: Overview of the TREC 2003 Question Answering Track. In: Voorhees, E. (eds.): Proceeding of the 11th Text Retrieval Conference,NIST Special Publication, Gaithersburg, (2003)1-15
5. Li, X., Roth, D.: Learning Question Classifier. In: Tseng, S.C. (eds): Proceeding of the 19th International Conference on Computational Linguistics, Morgan Kaufmann Publishers, Taipei, (2002) 556-562
6. Singhal, A., Salton, G., Mitra, M.: Document Length Normalization. Information Processing & Management. Vol.32, No.5, (1996) 619–633
7. Buckley, C., Singhal, A., Mitra, M.: New Retrieval Approaches Using SMART. In: Harman, D. (eds.): Proceedings of the Fourth Text REtrieval Conference. NIST Special Publication, Gaithersburg, (1995) 25-48
8. Salton, G. Buckley, C.: Term Weighting Approaches in Automatic Text Retrieval. Information Processing and Management, Vol.24, No.5, (1998) 513–523

Multinomial Approach and Multiple-Bernoulli Approach for Information Retrieval Based on Language Modeling*

Hua Huo[1,2], Junqiang Liu[2], and Boqin Feng[1]

[1] Department of Computer Science, Xi'an Jiaotong University, P.R. China
hhuo@mail.xjtu.edu.cn
[2] School of Electronics and Information, Henan University of Science and Technology, P.R.China

Abstract. We present a new retrieval method based on multiple-Bernoulli model and multinomial model in this paper. We use the multiple-Bernoulli model and multinomial model to estimate the term probabilities by importing the conjugate prior and the term frequencies, and use Dirchlet method to smooth the models for solving the "zero probability" problem of the language model.

1 Introduction

Ponte and Croft's multiple-Bernoulli approach in [4] and Miller's multinomial approach in [3] are typical retrieval methods based language model. For further improving retrieval performance, we present a new multiple-Bernoulli approach and a new multinomial approach. Different from the two approaches in which the term frequencies in the query and document are not captured, we employ the term frequencies and use Dirchlet smoothing method to solve the "zero problem" in the approaches instead of the linear interpolation smoothing method.

The general idea is to build a language model M_d for each document d, and rank the documents according to how likely the query q can be generated from each of these document models, i.e. $P(q|M_d)$. In different models, the probability is calculated in different approaches [4]. In *Multinomial model*, the query is treated as a sequence of independent terms (i.e. $q = w_1, w_2, w_m$), taking into account possibly multiple occurrences of the same term. The "ordered sequence of terms assumption" behind this approach states that both queries and documents are defined by an ordered sequence of terms. A query of length k is modeled by an ordered sequence of k random variables, one for each term occurrence in the query. Based on this assumption,the query probability can be obtained by multiplying the individual term probabilities [1].

$$P(q|M_d) = \prod_i^m P(w_i|M_d) \qquad (1)$$

---

* Supported by the science research foundation program of Henan University of Science and Technology, China (2004ZY041) and the natural science foundation program of the Henan Educational Department, China (200410464004).

In *Multiple − Bernoulli model*, a query is represented as a vector of binary attributes, one for each unique term in the vocabulary, indicating the presence or absence of terms in the query. Terms are assumed to occur independently of one another in a document. So, the query likelihood $P(q|M_d)$ is thus formulated as the product of two probabilities-the probability of producing the query terms and the probability of not producing other terms [1].

$$P(q|M_d) = \prod_{w \in q} P(w_i|M_d) \prod_{w! \in q} (1 - P(w_i|M_d)) \qquad (2)$$

2 Computing the Term Probability

Both queries and documents are represented as vectors of indexed words. Given a document $d = (f_1, f_2, ..., f_V) \in [0,1]^V$, where f_i is the term frequency of the ith word w_i, and V is the size of the vocabulary. Furthermore, consider a Multiple-Bernoulli generation model or a multinomial generation model for each document, parameterized by the vector: $M_d = (M_{f_1}, M_{f_2}, ..., M_{f_V}) \in [0,1]^V$ which indicates the probabilities of emission of the different words in the vocabulary. Where $M_{f_i} = P(w_i|M_d)$ and $\sum_{i=1}^{V} M_{f_i} = 1$. Now, let us define the length of a document as the sum of their components: $n_d = \sum_i f_i$. We wish to compute the maximum likelihood by Bayesian inference and use it as the documents language model [1][5]. According the Bayesian approximate equation

$$P(M_d|d) \approx P(d|M_d)P(M_d) \qquad (3)$$

we have

$$\hat{M_d} \approx \arg\max_{M_d} P(d|M_d)P(M_d) \qquad (4)$$

where $P(d|M_d)$ is the likelihood of the document given M_d, and $P(M_d)$ is the prior of the sample distribution model.

We assume that we sample from a multinomial distribution once for each word in the document model. When M_d parameterizes a multinomial and model prior is Dirichlet, the conjugate prior for the multinomial, we get:

$$\hat{M_d} = \arg\max_{M_d}(\Gamma(n_d + \sum_{i=1}^{V} \sigma_i))/(\prod_{i=1}^{V} \Gamma(f_i + \sigma_i)) \prod_{i=1}^{V} (M_{f_i})^{f_i + \sigma_i - 1} \qquad (5)$$

where Γ is the Gamma function, $\Gamma(s) = (s-1)!$. σ_i are the parameters of the Dirichlet prior. The solution to the above Equation by using EM algorithm yields the following form of probability estimates:

$$\hat{M}_{f_i} = (f_i + \sigma_i - 1)/(n_d + \sum_{i=1}^{V} \sigma_i - V) \qquad (6)$$

The simplest method of choosing the parameters σ_i is to attribute equal probability to all words in the query. However, this leads to "zero probability" problem. So, we use the document collection model to smooth the document model and choose to set the parameters as $\sigma_i = \mu c_i/n_c + 1$. The setting of σ_i yields the popular Dirichlet smoothing [2][5]. μ is the smoothing parameter, c_i is the number of times word w_i appears in the collection, and n_c is the total number of words in the collection. So, we get the term probability

$$P(w_i|\hat{M_d}) = (f_i + \mu(c_i/n_c))/(n_d + \sum_{i=1}^{V}(\mu(c_i/n_c))) \qquad (7)$$

If we assume that we sample from a multiple-Bernoulli distribution once for each word in the document model, we get

$$\hat{M_d} = \arg\max_{M_d} \prod_{i=1}^{V} \frac{\Gamma(\alpha_i+\beta_i)}{\Gamma(\alpha_i)\Gamma(\beta_i)}(M_{f_i})^{f_i+\alpha_i-1}(1-M_{f_i})^{n_d-f_i+\beta_i-1} \qquad (8)$$

The solution to the above Equation is

$$\hat{M}_{f_i} = (f_i + \alpha_i - 1)/(n_d + \alpha_i + \beta_i - 2) \qquad (9)$$

We set $\alpha_i = \mu c_i/n_c + 1$ and $\beta_i = n_c/c_i + \mu(1 - c_i/n_c) - 1$. This setting also leads the Dirichlet smoothing, then we get the term probability

$$P(w_i|\hat{M_d}) = (f_i + \mu(c_i/n_c))/(n_d + (n_c/c_i) + \mu - 2) \qquad (10)$$

3 Experiments and Results

Two sets of experiments are performed on the four data sets from TREC. The first set of experiments investigates whether the performances of retrieval based on our new multiple-Bernoulli approach (NMB) and our new multinomial approach (NML) are sensitive to the setting of the smoothing parameter μ. The second set of experiments is to compare the performance of NMB and NML with the performance of the Ponte and Croft's traditional multiple-Bernoulli approach (TMB) and the performance of the Miller's traditional multinomial approach (TML). Results of the first set of experiments are presented in Fig. 1. It is clear that average precision is much more sensitive to μ, especially when the values of μ are small. However, the optimal values of μ seem to vary from data set to data set, though in most they are around 2000.

Results of the second set of experiments are shown in Table 1. We observe that the NMB has improvements of 11.5%, 13%, 10.5%, and 13.6% in average precision on the four data sets respectively over the TMB, and that the average improvement in average precision is 12.2%. We think that the improvements of the NMB over the TMB are attributed to the using of the term frequency and Dirchlet smoothing method in the NMB. We also find that the NML has improvements of 7.7%, 10.1%, 8.3%, and 11.6% respectively over the TML in average precision on the four data sets, and the average improvement is 9.4%.

Table 1. Average precision of TMB,NMB,TML and NML on four data sets

Data set	TMB	NMB	Chg1%	TML	NML	Chg2%
TREC5	0.2304	0.2568	11.5	0.2571	0.2768	7.7
TREC6	0.2101	0.2374	13	0.2404	0.2646	10.1
TREC7	0.2243	0.2478	10.5	0.2585	0.2799	8.3
TREC8	0.2105	0.2392	13.6	0.2386	0.2662	11.6
Avg.	0.2188	0.2453	12.2	0.2486	0.2718	9.4

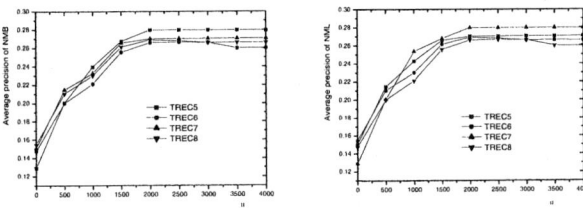

Fig. 1. Plots of average precision of NMB and NML for different μ

References

1. Lafferty,J., Zhai, C,: Document language models, query models, and risk minimization for information retrieval, In proceedings of SIGIR'01, (2001)111-119.
2. Metzler, D., Lavrenko, V., and Croft, W.B.: Formal multiple-Bernoulli models for language modeling. In proceedings of ACM SIGIR'04,(2004)231-235.
3. Miller, D.H., Leek, T. and Schwartz, R.: A hidden Markov model information retrieval system. In proceedings of ACM SIGIR'99 , (1999)214-221.
4. Ponte, J., Croft,W.B.: A language modeling approach to information retrieval. In proceedings of ACM SIGIR'98, (1998)275-281.
5. Zaragoza,H., Hiemstra,D.,et.: Bayesian extension to the language model for ad hoc information retrieval. In proceedings of ACM SIGIR'03.(2003)325-327.

Adaptive Query Refinement Based on Global and Local Analysis

Chaoyuan Cui, Hanxiong Chen, Kazutaka Furuse, and Nobuo Ohbo

University of Tsukuba, Tsukuba Ibaraki 305-8573, Japan
{cui, chx, furuse, ohbo}@dblab.is.tsukuba.ac.jp

Abstract. The goal of information retrieval (IR) is to identify documents which best satisfy users' information need. The task of formulating an effective query is difficult in the sense that it requires users to predict the keywords that will appear in the desired documents. In our study we proposed a method of query refinement by combining candidate keywords with query operators. The method uses the concept *Prime Keyword Set*, which is a subset of whole keywords and obtained by global analysis of the target database. Considering user's intension we generate rational size of candidates by local analysis based on several specified principles. The experiments are conducted to confirm the effectiveness and efficiency of our proposed method. Moreover, as an extension of our approach an online system is implemented to investigate the feasibility.

1 Introduction

The rapid growth of on-line documents increases the difficulty for users to find documents relevant to his/her information need. Unfortunately, short queries are becoming increasing common in retrieval applications, especially with the advent of the World Wide Web (WWW). According to the research by [4], queries submitted by users to the WWW search engines contains averagely only two keywords. Short query makes it difficult to distinguish relevant documents from irrelevant ones. Therefor it requires a system to provide candidates automatically for helping users to refine their initial queries. Generally, meaning-based candidates and statistics-based candidates are two main sources of related keywords.

Meaning-based approach construct thesauri manually. WordNet ([6]) provides a tool to search dictionaries conceptually, rather than merely alphabetically. It's basic object is *synset*, which is a set of synonymous words representing a certain meaning. Synsets are organized by the semantic relations defined on them. However, this kind of thesaurus is difficult to use because of ambiguity. Selecting the correct meaning for refining may be difficult, especially in case of short queries ([[10]]).

Statistics-based approach can be divided into two kinds, namely global analysis and local analysis. Automatic thesaurus construction technique grouped keywords together based on their co-occurrence in documents, keywords which often occur together in documents are assumed to be similar. These thesaurus can be used for automatic or manual query expansion. A global technique requires some corpus-wide statistics that take a considerable amount of computer resources to compute, such as co-occurrence data about all possible pairs of terms in a database.

In contrast, local strategy only processes a small number of top-ranked documents. It is based on a hypothesis that top-ranked documents is indeed relevant. and carried out by relevance feedback systems. [8] improves the quality of the initial run by re-ranking a small number of the retrieved documents by making use of the term co-occurrence information to estimate word correlation. [3] use information-theoretic term-score function within top retrieved documents to assign scores to candidate expansion terms. Query expansion has been shown to produce good retrieval performance and prevent query drift.

[11] combines the global analysis and local analysis. This approach is based on the use of noun groups instead of simple keywords, as document concepts. It works as follows. Firstly retrieve some top-ranked passages using the original query. This is accomplished by breaking up the documents retrieved by the query in fixed length passages (of say, 300 words) and ranking these passages as if they were documents. Then the similarity between each concept c in the top-ranked passages and the whole query q is computed using a variant of $tf\text{-}idf$ ranking. The query q is refined by adding some top-ranked concepts, based on the similarity between the concepts and q.

In this work, we adopted both global analysis and local analysis to generate refinement candidates based on statistics. We differ from the above work in that we analyze the relationship between documents of the database. in global analysis, which improve the precision of refinement significantly. The rest of the paper is organized as follows. Section 2 gives several principles used throughout this study, and approaches for refinement. Section 3 introduces two important concepts. Section 4 and 5 explain how global analysis and local analysis work. Section 6 outlines the test data and gives experimental results. Section 7 offers some conclusions and directions for future works.

2   Principles and Goal

The existing techniques emphasize to improve retrieval effectiveness, but there still remains a problem. Sometimes information user wants can not be reproduced by way of suggested candidates. Certainly, all keywords included in retrieval results satisfy user's intension, but large size of candidates is inconvenient for user to browse them all. Here, we offer several principles which reflect the basic requirement of user in refinement system.

- No loss of information : any of retrieved documents can be accessed through candidates or their combinations
- Screen effect : appropriate reduction of documents by refined query
- Rational size of candidates
- Response time

Based on these principles we aim at developing a refinement system as shown in Figure 1 which illustrates a running example of the prototype system. We use CRAN test collection as a database, which includes 1398 documents and 4612 keywords. In this example, suppose that a user issues a query by keyword "method". The system shows that there are 366 hits for this query, and it takes 0.01 second to retrieve. Besides the links to hit documents, refinement candidates each of which accompanied by bool operators, are provided.

For example, in the first line, "+(8) -(358) affected", "+" and "-" represent the bool operators "AND" and "NOT", respectively. Selection of this line with "+(8)" means that by adding candidate keyword "affected" to the original query, 8 documents will leave in the result. Selection of "-" means the exclusion of documents containing "affected". Here suppose that the user is interested in the field of aircraft and does not want documents that include keyword "tn", then user may choose "+ aircraft" and "- tn" which means that the query is modified to {method + aircraft - tn}. Modified query become more clear and retrieved documents decrease. User can access documents by the links, or, carry out the refinement process repeatedly until he/she is satisfied with the results.

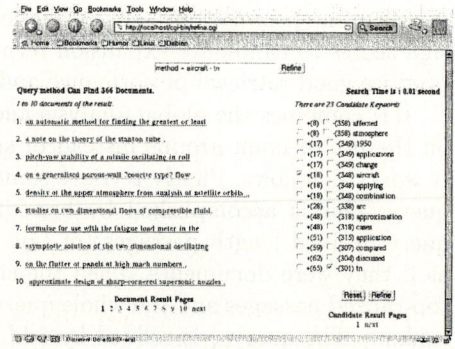

Fig. 1. An example of prototype refinement system

3 Prime Keyword Set

One of the drawback in the interactive system is that it takes lots of computation after user's query arrived. Because most of the systems search the database and have to process contents of retrieved documents. Usually the content of a document is represented approximately by keywords included in it, and content of database is thus represented by a keyword set included in it. In fact, some of these keywords are unnecessary in sense of specifying documents. On other hand, a document can be represented only by some "specific" keywords. As an extension, a database can also be represented by a "specific" keyword set. In our study, we call this kind of keyword set *prime keyword set*. If a database can be well represented by prime keyword set, the effectiveness of online computation would be improved.

For convenience we first give the notations that will be used in the following explanation. Let $\mathbb{D}$ and $\mathbb{K}$ be a set of documents and a set of keywords, respectively. ρ extracts keywords from a document $d \in \mathbb{D}$, or $\rho(d) = \{k \mid (k \in \mathbb{K}) \wedge (k \text{ is a keyword included in } d)\}$. Further, let $D \subset \mathbb{D}$. $\bigcup_{d \in D} \rho(d)$ is denoted by $\rho(D)$, and in particular $\rho(\mathbb{D}) = \mathbb{K}$. A *query* Q is simply a subset of $\mathbb{K}$. A query evaluation retrieving all the documents that contain all the given keywords in Q is defined as $\sigma(Q) = \{d \mid d \in \mathbb{D} \wedge Q \subseteq \rho(d)\}$. A refinement candidate of Q is also a subset of $\mathbb{K}$. Let X and Y be subsets of $\mathbb{K}$, the fact that "Y is a refinement candidate of X", is denoted by an association rule $X \Rightarrow Y$. The *confidence* of such a rule is defined as $|\sigma(X \cup Y)|/|\sigma(X)|$ where $|\sigma(X \cup Y)|$ is also called the *support* of the rule.

In order to ensure that system does not arbitrarily narrow the retrieved result by user's query we introduce two concepts: coverage and prime keyword set.

Coverage: Let $\mathcal{K}(\subseteq 2^{\mathbb{K}})$ be a subset of the power set of $\mathbb{K}$ and $D(\subseteq \mathbb{D})$ be a set of documents. If
$$D \subseteq \bigcup_{Q \in \mathcal{K}} \sigma(Q)$$
then we say that $\mathcal{K}$ covers D, or $\mathcal{K}$ is a coverage of D. In a similar fashion, $\mathcal{K}$ is also called a coverage of $\mathcal{K}'$ ($\subseteq 2^{\mathbb{K}}$) *iff*
$$\bigcup_{Q \in \mathcal{K}'} \sigma(Q) \subseteq \bigcup_{Q \in \mathcal{K}} \sigma(Q)$$

Naturally, coverage $\mathcal{K}$ of D is a *minimum coverage* if and only if D can not be covered by any pure subset of $\mathcal{K}$. Minimum coverage is defined in order to exclude redundant keywords from the keyword database. In general, minimum coverage is not unique for a specific database.

Prime Keyword: Let $\mathbb{D}$ and $\mathbb{K}$ be the set of all documents and all keywords, respectively. $\mathbb{K}_p$ is called the *prime keyword set* of $\mathbb{D}$ (or $\mathbb{K}$) *iff* $\{\{k\} \mid k \in \mathbb{K}_p\}$ is a minimum cover of $\mathbb{D}$ (or $2^{\mathbb{K}}$). In addition, keywords included in prime keyword set are called *prime keywords*.

By the definition of coverage, prime keywords guarantees no loss of information. In other words, any documents can be retrieved from a certain subset of prime keyword, expressed as the follows.
$$\mathbb{D} = \bigcup_{k \in \mathbb{K}_p} \sigma(\{k\}) \equiv \bigcup_{K \subset \mathbb{K}_p} \sigma(K)$$

Based on these concepts we extract prime keyword set and generate candidates according to the relationship between prime keywords.

4 Global Analysis

In global analysis, system extracts prime keyword set based on the principles mentioned in Section 2. Though no loss of information can be guaranteed by prime keywords, the practicability of system may be hurt due to the composition of database. As a refinement system, it is important to keep good screen effect. Keywords with very low support are unsuitable to be candidates because too many candidates are necessary. On the other hand, keywords with very high support can not well discriminate one document from the others. Unfortunately, some documents only contain keywords with very low or very high support. We call them outliers in sense of effectiveness of refinement. In our study, in the pre-processing phase we exclude outlier documents in which keywords are out of a specified range of support. Certainly, outlier documents can be retrieved by the system though no refinement candidates are returned.

Refinement Coefficient. Theoretically, the vector space model had the disadvantage that index terms are assumed to be mutually independent. Due to the locality of many term dependencies, their indiscriminate application to all documents in the database might in fact decline the overall performance. In order to obtain high effectiveness of refinement the relationship between keywords should be taken into consideration. In our study, we use Refinement Coefficient (RC) defined below to express the correlation of keywords.

$$RC(k,d) = \frac{tf(k,d)}{|\rho(\{d\})|} \times \frac{\sum_{k_x \in \rho(\sigma(\{k\})) - \{k\}} cnf(k_x \Rightarrow k)}{|\rho(\sigma(\{k\}))| - 1} \quad (\mid \rho(\sigma(\{k\})) \mid > 1)$$

Where $tf(k,d)$ is the term frequency of keyword k included in the document d, $\mid \rho(\{d\}) \mid$ is the number of keywords included in d, $\rho(\sigma(\{k\}))$ is a set of keywords that co-occur with k.

The formula consists in two parts, the left part shows the importance of keyword in d. In general, the higher the frequency of keyword that appears in the document is, the more important the keyword becomes. On the other hand, if the value of $\mid \rho(\{d\}) \mid$ is small, keywords included in d are considered to be important. As an extreme example, if a document contains only one keyword, then the keyword is essential. The right part means average effect of k in the refinement. k_x is a keyword included in $\rho(\sigma(\{k\})) - \{k\}$. Suppose k_x is issued as a query, the refinement degree of k is denoted as $cnf(k_x \Rightarrow k)$. So the right part reflects average effect of refinement when when each k_x is a query.

Generation of Prime Keyword Set. We use RC as a main factor to extract prime keyword set. Firstly, system select a keyword with maximal RC value from each document. Certainly, the keyword set selected is a coverage of database. Then keywords in the coverage are sorted in ascending order of RC. And keyword whose RC value is small are deleted if the rest of them can cover the whole database after this one is removed. Finally, the minimal cover is output as prime keyword set after the redundant keywords are removed. The pseudo code is as follows.

Algorithm *Generation of Prime Keyword Set*
Input : Document Set $\mathbb{D}$, Keyword Set $\mathbb{K}$, range of support $MinSpt$ and $MaxSpt$
Output : Prime Keyword Set $\mathbb{K}_p$ (initially empty)
1   **forall** $d \in \mathbb{D}$ **do**
2       $X_d = \{k \mid k \in \rho(\{d\}) \land MinSpt \leq spt(k) \leq MaxSpt\}$

```
3     select k₀ such that RC(k₀, d) = max_{k∈X_d}{RC(k, d)}
4        K_p := K_p ∪ {k₀}
5     end
6     sort K_p in ascending order of RC
7     forall k_i ∈ K_p do
8        if ⋃_{k∈K_p−{k_i}} σ({k}) = D then K_p := K_p − {k_i}
9     end
```

The algorithm extracts precise information that can make local analysis efficiency. In this algorithm, all keywords included in each document have to be processed, so the cost of computation is very high. Fortunately, this process is done during the off-line phase in advance. Therefore it does not affect the interactive processing.

5   Local Analysis

There have been a number of efforts to improve local analysis ([2], [3], [5], [7], [9]) These strategies can not work well when loss of information occurs. Almost all of existing methods ignore this problem. Our approach makes up for the drawback of existing techniques by using concept *prime keywords*. At a glance, it may take great effort and may be infeasible. In fact, the size of $\sigma(Q)$ is greatly reduced and system only aims at prime keywords included in it to carry out local analysis. Fortunately, the prime keyword set is a coverage of the whole database, and naturally prime keywords included in retrieved documents become a coverage of it. Candidates from this keyword set can guarantee no loss of information. In particular, the retrieved results contain user's intension. Keywords, especially prime keywords contained in them reflect screen effect of refinement. In addition, the size of prime keyword set is significantly reduced so time efficiency would be improved.

Local RC. Based on the prime keywords extracted from $\sigma(Q)$, system looks for refinement candidates. In local analysis, we still use RC as a criterion of evaluation to discriminate candidates from others. And we localize RC as follows in order to keep consistency.

$RC_l(k,d) = \frac{tf(k,d)}{|\rho(\{d\})|} \times cnf(Q \Rightarrow \{k\}), d \in \sigma(Q)$
RC_l reflects the effect of refinement to query Q.

Generation of Candidates. We proposed two algorithms, called *Generation of Conjunction Candidates* and *Generation of Exclusion Candidates* to generate candidates. The similarity function between document d and Query Q is defined as follows.

$$sim(d, Q) = \frac{\sum\limits_{k_x \in Q} w(k_x, d)}{\sqrt{\sum\limits_{k \in \rho(d)} w^2(k, d) \times |Q|}}$$

where $w(k, d)$ is the weight of keyword k in document d, and

$$w(k, d) = \frac{tf(k,d)}{|\rho(\{d\})|} \times log(\frac{|\mathbb{D}|}{|\sigma(k)|}).$$

Using $sim(d, Q)$, we define $\sigma_r(Q)$ as a document set which satisfies the following condition. $\sigma_r(Q) \subset \sigma(Q) \land |\sigma_r(Q)| = r \land \forall d_1 \in \sigma_r(Q), \forall d_2 \in (\sigma(Q) - \sigma_r(Q)), \quad sim(d_1, Q) \geq sim(d_2, Q)$

The pseudo code of two algorithms is as follows.

Algorithm *Generation of Conjunction Candidates*
Input: $\mathbb{K}_p$, $\sigma(Q)$, and ranking threshold r.
Output: Refinement Candidates C_a
1   sort $\sigma(Q)$ in descending order of $sim(d, Q)$
2    $C_a := \rho_p(\sigma_r(Q)) - Q$
3   **while** $\sigma(Q) - \bigcup\limits_{k \in C_a} \sigma(\{k\}) \neq \emptyset$
4   **forall** $d \in \sigma(Q) - \bigcup\limits_{k \in C_a} \sigma(\{k\})$ **do**
5   select k_0 such that $RC_l(k_0, d) = max_{k \in \rho_p(\{d\}) - Q}\{RC_l(k, d)\}$
6        $C_a := C_a \cup \{k_0\}$
7   **end**

Algorithm *Generation of Exclusion Candidates*
Input: $\mathbb{K}_p$, $\sigma(Q)$, and l low-ranked documents.
Output: Refinement Candidates C_a
1   sort $\sigma(Q)$ in ascending order of $sim(d, Q)$
2    $C_a := \rho_p(\sigma_l(Q)) - \rho_p(\sigma_{n-l}(Q)) - Q$
3   **while** $\sigma(Q) - \bigcup\limits_{k \in C_a} \sigma(\{k\}) \neq \emptyset$
4   **forall** $d \in \sigma(Q) - \bigcup\limits_{k \in C_a} \sigma(\{k\})$ **do**
5   select k_0 such that $RC_l(d, k_0) = max_{k \in \rho_p(\{d\}) - Q}\{RC_l(d, k)\}$
6        $C_a := C_a \cup \{k_0\}$
7   **end**

The former emphasize the importance of prime keywords appeared in r top-ranked documents. If prime keywords in $\sigma_r(Q)$ do not cover $\sigma(Q)$, then more keyword are chosen from uncovered documents. This process is performed repeatedly until $\sigma(Q)$ is covered. Moreover, the candidate set is a subset of $\rho_p(\sigma(Q))$, the size of it can be significantly reduced. The effect of candidates and the time of online computation will discussed in our experiments. Besides the above idea,

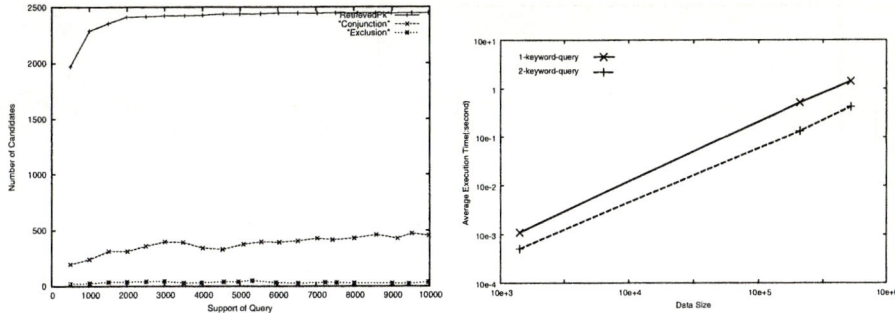

Fig. 2. Number of candidates   **Fig. 3.** Execution time vs. database size

if all irrelevant documents are excluded, the remainder are the relevant ones and become easy to be found. The latter algorithm *Generation of Exclusion Candidates* is designed based on this idea.

6 Experiments

We use C++ for our implementation. The experiments were run on a 3.0GHz Pentium IV processor with 4GB of main memory and 70GB of disk space running Debian Linux (Kernel 2.4). The proposed methods are evaluated using TREC-7,8 ad hoc dataset, which includes 527,993 documents and 72,359 keywords after the elimination of stop words and stemming, as well as 100 test questions. As mentioned in section 4, our system eliminates outliers before extracting prime keyword set. In our experiments, we set $MinSpt$ and $MaxSpt$ to [500, 10, 000] and 738 outlier documents are removed. Applying algorithm *extracting prime keyword set* to the dataset, we obtained a prime keyword set of less than $\frac{1}{30}$ of the original keyword set.

Size of Candidates and Time of Online Processing. In order to reveal time efficiency and candidate's reduction ratio of the two algorithms in local analysis, we compare them with the method which picked up all the prime keywords included in $\sigma(Q)$ as refinement candidates. In the following this method is referred to as *RetrievedPk*.

We take the following steps to carry out our experiments. First the system randomly issues 10,000 1-keyword-queries and 2-keyword-queries respectively. Then we generate candidates based on the two algorithms. For queries with same support, our system calculate the average number of candidates and the average time.

Figure 2 shows the number of candidates by different algorithms. By our local analysis, in most cases, the number of conjunction and exclusion candidates are reduced to less than 15% and 3%, respectively.

Next experiment aims at investigating the scalability of our approach. As shown in Figure 3, three datasets, ad hoc, FT(a subset of ad hoc) and CRAN are used to measure the execution time. The results show that in both heuris-

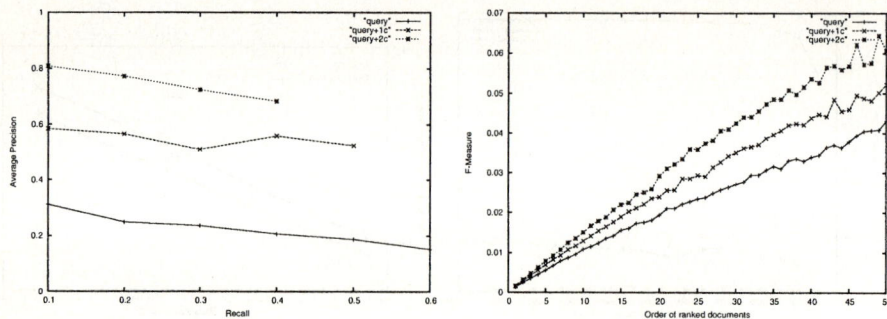

Fig. 4. Average Precision and Recall **Fig. 5.** F-Measure

tics, the processing time is almost linear to the size of database. Considering the hardware specification used in the experiments, we can conclude that the system is feasible.

Effect of Candidates. This experiment investigates the effect of refinement. Obviously, query by several keywords can narrow the intention and decrease the ambiguity. This is proved in Figure 4. Adding one candidate to the initial query improves the average precision, and an addition candidate enhances the average precision further.

A single measure which combines recall and precision might be of interest. One such measure is the harmonic mean F of recall and precision which is computed as $F(i) = 2/(r^{-1}(i) + p^{-1}(i))$ where $r(i)$ is the recall for the i-th document in the ranking, $p(i)$ is the precision for the i-th document in the ranking, Figure 5 shows the same result with Figure 4 about F-Measure.

Comparing with Rocchio. There have been a lot of approaches attempted to improve retrieval effects on query modification. Rocchio is a famous way and has been proved to be effective. Many systems use *Ide_Regular* formula to improve the performance by query expansion ([1]). $Q_{new} = \alpha Q_{orig} + \beta \sum_{\forall d_j \in D_r} d_j - \gamma \sum_{\forall d_j \in D_n} d_j$

Here, Q_{orig} and Q_{new} are the initial query vector and the expanded (refined) query vector, respectively D_r and D_n stand for the sets of relevant and irrelevant documents, respectively. Considering the fact that the information contained in the irrelevant documents is much less important, we set the parameter γ to 0. The experiments are done by the following steps.

- Look for $\sigma(Q_{orig})$.
- Sort retrieved results by $sim(d, Q_{orig})$.
- Specify relevant documents, look for expansion query by *Ide_Regular*.

Figure 6 shows that our method doubles the precisions of those of Rocchio's for all recall. As for F-measure shown in Figure 7, our method is also superior to Rocchio's remarkably.

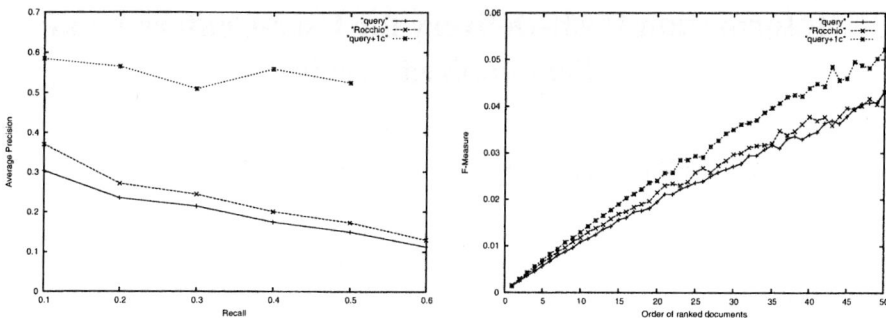

Fig. 6. Comparison with Rocchio (Average precision)

Fig. 7. Comparison with Rocchio (F-Measure)

7 Future Work

Our study use global analysis and local analysis to improve response time and quality of candidates in query refinement. It guarantees no loss of information and is proved to be effective and efficient experimentally.

In order to examine the practicability we attempt to implement a refinement system with practical dataset It is expected to applied to web search engine in the future.

References

1. Baeza-Yates, R. and Ribeiro-Neto, B.: Modern Information Retrieval. *Addison Wesley*, pp.24-34, 1999
2. Carmel, D., Farchi, E. and Petruschka, Y.: Automatic Query Refinement using Lexical Affinities with Maximal Information Gain. *ACM SIGIR*, pp.283-290, 2002
3. Carpineto, C., De Mori, R., Romano, G. and Bigi, B.: An Information-Theoretic Approach to Automatic Query Expansion. *ACM TOIS*, vol.19(1), pp.1-27, 2001.
4. Croft, W.B., Cook, R. and Wilder, D.: Providing government information on the Internet: Experience with THOMAS. In *Proceedings of the Digital Libraries Conference*, (DL'95) pp.19-24, 1995
5. Cui, C., Chen, H., Furuse, K. and Ohbo, N.: Web Query Refinement without Information Loss. *APWeb 2004*, pp.363-372, 2004
6. George, M.: Special Issue, WordNet: An On-line Lexical Database. *International Journal of Lexicography*, 3(4), 1990.
7. Kraft, R. and Zien, J.: Mining anchor text for query refinement *Proceedings of the 13th international conference on World Wide Web* pp.666-674(2004)
8. Mitra, M., Singhal, A. and Buckley, C.: Improving Automatic Query Expansion. *ACM SIGIR*, pp.206-214, 1999
9. Vélez, B., et al: Fast and Effective Query Refinement. *ACM SIGIR*, pp.6-15, 1997
10. Voorhees, E.M.: *Query Expansion Using Lexical-Semantic Relations*. In *ACM SIGIR*, pp.61-69, 1994
11. Xu, J., and Croft, W. B.: Improving the effectiveness of information retrieval with local context analysis. *ACM Transactions on Information System*, vol.18(1), pp.79-112, 2000.

Information Push-Delivery for User-Centered and Personalized Service

Zhiyun Xin[1], Jizhong Zhao[2], Chihong Chi[1], and Jiaguang Sun[1]

[1] School of Software, Tsinghua University,
100084, Beijing, China
xinzy@csrc.gov.cn

[2] Department of Computer Science and Technology, Xi'an Jiaotong University,
710049, Xi'an, China
zjz@mail.xjtu.edu.cn

Abstract. In this paper, an Adaptive and Active Computing Paradigm (AACP) for personalized information service in heterogeneous environment is proposed to provide user-centered, push-based high quality information service timely in a proper way, the motivation of which is generalized as R4 Service: the Right information at the Right time in the Right way to the Right person, upon which formalized algorithms of adaptive user profile management, incremental information retrieval, information filtering, and active delivery mechanism are discussed in details. The AACP paradigm serves users in a push-based, event-driven, interest-related, adaptive and active information service mode, which is useful and promising for long-term user to gain fresh information instead of polling from kinds of information sources. Performance evaluations based on the AACP retrieval system that we have fully implemented manifest the proposed schema is effective, stable, feasible for adaptive and active information service in distributed heterogeneous environment.

1 Introduction

During the past decades pull-based information service such as search engine and traditional full-text retrieval were studied much more[1-5], and many applications have been put to real use. However with the explosive growth of the Internet and World Wide Web, locating relevant information is time consuming and expensive, push technology[4-10] promises a proper way to relieve users from the drudgery of information searching. Some current commerce software or prototype systems such as PointCast Network, CNN Newswatch, SmartPush and ConCall serve users in a personalized way[8,10,12], while recommendation system[13-15] such as GroupLens, MovieLens, Alexa, Amazon.com, CDNow.com and Levis.com are used in many Internet commerce fields. Although kinds of personalized recommendation systems were developed, still many things left unresolved, these problems result in deficiency and low quality of information service as the systems declared. One of most important reasons of which is that single recommendation mechanism such as content-based or collaborative recommendation is difficult to serve kinds of users for their various information needs[10-15].

In this paper an Adaptive and Active Computing Paradigm (AACP) for personalized information service in wide-area distributed heterogeneous environment is proposed to provide user-centered, push-based high quality information service timely in a proper way, the motivation of which is generalized as R4 Service: the Right information at the Right time in the Right way to the Right person, Upon which formalized algorithms framework of adaptive user profile management, incremental information retrieval, information filtering, and active delivery mechanism are discussed in details. As the work of National High-Tech Research and Development Plan of China, we have fully implemented the Self-Adaptive and Active Information Retrieval System (AIRS) for scientific research use, and evaluations showed the AIRS system is effective and reliable to serves large scale users for adaptive and active information retrieval.

2 The Adaptive and Active Computing Paradigm

2.1 Abstract Model of Information Retrieval System

Usually traditional information retrieval system[1-6] is composed of Retrieval-Enabled Information Source (IS), Indexing Engine (XE), Information Model (IM), Retrieval Engine (RE), Graphic User Interface (GUI), entrance for users to retrieve for information, which includes the order option parameters.

Definition 1: *The general Information System can be defined according to its components in a system's view as:*

$$\text{IRSSysView} := \{IS, XE, IM, RE, GUI\} \tag{1}$$

Especially, in an open source information system such as search engine, the indirect IS is the WWW, while the direct IS is the abstract image of the indirect information source. In non-open information sources such as full-text retrieval system the IS is the original metadata suitable for retrieval.

Definition 2: *In a user's view, the information retrieval system is a 3-tuple framework composed of virtual or real information source (IS), user's Retrieval Input (RI) and Retrieval Output (RO), which is:*

$$\text{IRSUserView} := \{IS, RI, IO\} \tag{2}$$

Traditional information system is essentially information-centered, pull-based computing paradigm, which serves customers in a passive mode, and can't meet long-term users' demand of getting information in an active way.

2.2 Adaptive and Active Computing Paradigm

An Adaptive and Active Computing Paradigm (AACP) for personalized information service[12-15] in heterogeneous environment is user-centered, push-based high quality information service in a proper way, the motivation of which is generalized as R4 Service: the Right information at the Right time in the Right way to the Right person, that is:

$$R^4 := R \times R \times R \times R \tag{3}$$

The R4 Service serves users in an active and adaptive mode for high quality information timely and correctly, which can adjust the information content according to user's preference dynamically and adaptive to user's interest in some degree, automatically retrieve and delivery related information to users with interaction. The R4 service can be described for adaptive and active information retrieval.

2.3 The Abstract Architecture of AACP

Definition 3: *The Adaptive and Active Information System (AAIS) for personalized information service is encapsulated on the traditional information system with adaptivity and activity, and can be viewed as the following composition:*

$$AACP := IRSSysView\ AAIS \quad (4)$$

Where

$$AAIS := (VUser, VIS, C, T, P) \quad (5)$$

The semantic of each element in AAIS is described as table 1:

Table 1. Semantic of AAIS

0Symbol	1Definition
V_{USER}	Vector of User's interests or preferences
V_{IS}	Vector of Information Source
C	Condition that V_{IS} matches V_{USER}
T	Trigger for the system starts to work for U
P	Period that the system serves for U

For a long-term user in information retrieval system, what one really needs in a relative stable period is just a little part in the whole global information domain, the relationships that can be precisely described as definition 4.

Definition 4: *A Personalized Information Set (PIS) is a subset of Domain Information Set (DIS), while DIS is a subset of Global Information Set (GIS), so that:*

$$PIS \subseteq DIS \subset GIS \quad (6)$$

Considering the high-dimension of information system, usually the domains overlap each other especially in multi-subjects, the set of the above is in fact n-dimension space, and fig.4 just shows a flat model of the above relationship.

Definition 5: *Personalized Information Service over heterogeneous information source is a matching map R (where R stand for Retrieval) from VIS to VUser with the Condition C and trigger T during the period P, that is:*

$$Q = R(V_{User}, V_{IS,C}, T, P; M_{Info}) \quad (7)$$

where Minfo is the information Model for the Retrieval Map, which may be Boolean, Probable, or VSM modal. To serve users adaptively and actively, the system must know users' needs, usually user profile is the basic infrastructure, which decides the quality of adaptive and active service. Except for adaptivity and activity, the AACP paradigm also builds on the infrastructure of information indexing, information retrieval, information filtering. Moreover, automatically monitoring for retrieval and delivery is another key technology. The following section will talk about each key technology and its optimization strategy in details.

3 Infrastructure of AACP

3.1 Abstract User Profile

To serve users according to their preference and interests, user profile is needed as a image of what users require, by which then system can decide who to serve, when to serve, what to serve and how to serve. Thus a user profile can be defined in an abstract form as the following multi-tipple:

$$AUP := (ExtUID, IntUID, TrueVector, RecVectosr, \quad (8)$$
$$StartTime, EndTime, Freq)$$

where ExtUID, IntUID stand for external and internal User Identity respectively, and TrueVector stands for true and confirmed user interest vector, but RecVector is the recommended interest vector, which should be evaluated by user whether to retrieve by the vector or not, while StartTime, EndTime describe the start and end time that user want to be served, Freq is the frequency that user consumes the information that pushed to him/her. To provide high quality information it is necessary to define different interests with different weight, thus TrurVector and RecVector can be defined as following:

$$TrueVector := <(v_1,w_1),(v_2,w_2),\ldots,(v_n,w_n)> \quad (9)$$

$w_i \geq w_j, (1 \leq i, j \leq n, i<j)$, $w_i \geq w_0$ (threshold of the Weight of TrueVector)

$$RecVector := <(v'_1,w'_1),(v'_2,w'_2),\ldots,(v'_n,w'_n)> \quad (10)$$

$w'i \geq w'j, (1 \leq i, j \leq n, i<j)$, $w'0<w'i<w0$

According to the above definition, we can see that TrueVector is set of weighted keyuwords whose weight must be more than w0, and RecVector's should be not less than w'0. As for the other key problem for user profile in AACP, the profile location, the style of update user profile are important aspects to be considered. In fact, location of profile can be either at client, proxy server, or web server side, and the style of update is can be explicit or implicit, that is:

$$ProfileLoc := Client|Proxy|Server \quad (11)$$
$$UpdateStyle := Explicit|Implicit$$

3.2 Adaptive User Profile Management

Algorithm: *Adaptive Update of User Profile*

Input: IntUID,TrueVector TV=<$(v_1,w_1), (v_2, w_2), \ldots, (v_n, w_n)$>, RocVector RV=<$(v'_1,w'_1),(v'_2, w'_2), \ldots, (v'_n, w'_n)$>, w_0, w'_0

Output: TV', RV'

Step1: Select IntUID from UserDB, Extract Possible Interest Vector (PIV) from Buffer Saved in ProfileLoc, PIV=< $(v_{01},w_{01}),(v_{02},w_{02}),\ldots,(v_{0n},w_{0n})$ >, where $w'_0<w_{0i}<w_{0j}(1<i<j<n)$

Step2: for each w_{0i} check whether v_{0i} is true vector or recommendation vector

if($w_{0i}>w_0$) Insert (v_{0i},w_{0i}) into TrueVector Set

TV':=TV∪{ (v_{0i},w_{0i}) }

else RV':=RV∪{ (v_{0i}, w_{0i}) }

Step3: for the updated TV and RV, sort all the weight w_i in decrease order.

Step4: Return the TV',RV'as updated True Vector and Recommendation Vector respectively.

3.3 Incremental Information Retrieval

Algorithm: Incremental Information Retrieval

Input: IntUID,TV,RV, TS(TimeStamp),s_0(the similarity threshold)

Output: TS,R'S (Retrieval Result Set)

Step1: Select IntUID from UserDB,Extract and Normalize IV, RV to normalized form that the retrieval system can recognize.

Step2: According to TS, TV and the retrieval condition, retrieve related documents for IntUID ,and for each v_i (1<I<n), according to VSM, Calculate similarity s between IV_i and candidate document D:

$$s := \text{sim}(IV_i, D) = \frac{\sum_{i=1}^{n} v_i d_i}{\sqrt{\sum_{i=1}^{n}(v_i)^2 \sum_{i=1}^{n}(d_i)^2}} \qquad (12)$$

If s>s_0,then organize the retrieval result(s) R_i According to order style construct the true interests result set R_T : $R_T:=R_T+R_i$

Step3: For RV, Do the same Operation as Step3 to Construct Recommendation Result set R_R:

$$R_R := R_R + R_i$$

Step4: Update TS;

Step5: Return R_T and R_R

3.4 Two Phase Monitor Mechanism

Input: IntUID, P, TS

Output: bRetrieved (the Flag whether to Trigger the Retrieval Operation or not), TS

Phase 1: Change Detection of Information Source

Step1: Set Timer Period to Detect whether the Information Source Changed or not:

bSourceChanged=DetectChange(DataSource,TimePeriod)

Step2: if(bSourceChanged)

Notify the main monitor that the data source(s) has changed with bSourceChanged, and update the Time Stamp TS with the current Time.

Else Continue to monitor without set Time Stamp TS.

Phase 2: Active Retrieval and Delivery Monitor

Step3: For Ui(1<i<n), read the Ui.Period, verify whether It expired the defined Period,

bExpired=VerifyExpired(Ui.Period, CurTime)

If(bExpired) delete(Ui)

ElseIf(bSourceChanged) then goto Step4.

Step4: Retrieve related Information for Ui according to the Incremental Retrieval Algorithm described in 2.3.

Step5: Delivery the results at the specified time to Ui.

Step6: Finish the work for U_i Update Time Stamp, and deal with next User.

The Two-Phase monitor mechanism uses two independent Timer to detect whether to retrieve or not, which is valuable for reducing cost when the user group is too large.

3.5 Information Filtering

Information Filtering is the key problem to take into account in AACP which includes collaborative filtering and content-based filtering. collaborative filtering methods utilize explicit or implicit rating from many users to recommend items to a given user, which can be formally defined as:

$$RecDoc = CollFilter (U_1, U_2, \ldots, U_n, Threshold) \qquad (13)$$

While content-based filtering make recommendation by matching a user's query especially by user profile, that is:

$$RecDoc = ContFilter(Prof(u), Threshold) \qquad (14)$$

Pure collaborative filtering systems are not proper when little is known about a user, or he/she has different preferences, while content-based filtering system can't account for community endorsements. Collaborative and content-based ones together can improve the performance.

4 System Implementation

We have implemented the whole system AACP as part formal work of National High-Tech Research and Development Plan of China. The system is based on J2EE, XML Web Service, and WebSphere (as Application Server) for Self-Adaptive and Active Computing platform aim at scientific researchers to gain useful and fresh information timely and efficiently. Fig. 1 is the homepage of AACP Platform, while fig. 6 is the new and fresh retrieved results.

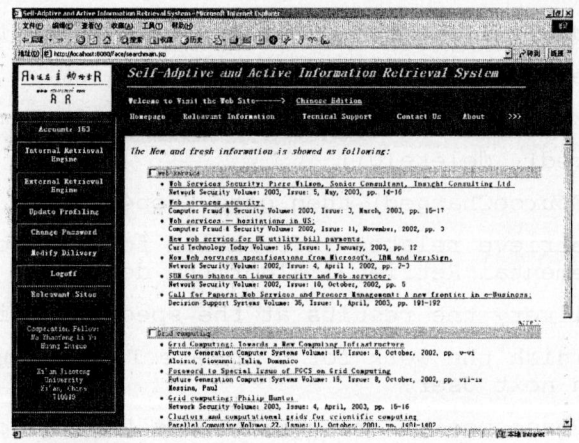

Fig. 1. Incremental and Fresh Retrieval Results

5 Performance Evaluation

Based on the AACP platform we have evaluated the performance in efficiency of system service for user(s), incremental retrieval precision in variant parameters. The AACP platform is built on the XML Database for its extensibility and reliability in heterogeneous environment, that is, when reusing the user profiles they are easy to access without any change, especially XML can work in any OS environment, another most important thing to take into account is incremental precision, which is a benchmark evaluation issue in AACP. If the system can't serve users in an incremental way for new and fresh information, then the work will lose its original motivation. As for the incremental retrieval as described, a more precise timestamp is adopted to record the start and end time of each retrieval and delivery event. Table 3 are the evaluation results from real use of the AACP platform in different run-time environment with variant parameters use.

Table 2. Performance Evaluation of Increment Retrieval (User Account=100)

Retrieval Frequency(day)	Time Consumption(ms)	NewInfo Precision(%)
1	213	100
3	328	91
7	318	79
15	389	68
30	677	54

Table 2 showed that when using XML as the data management mechanism, the efficiency is almost near to using Relational Database Management system (here ORACLE 9i is used to performance comparison with XML), but XML is extendable and reachable in the distributed heterogeneous environment. When evaluating incremental precision and time consumption, we did enough efforts to get even precise results, in which variant users are evaluated by time consumption and new and fresh information retrieved. Table 2 are the average results in the case of 10/100 user accounts, each result here are averaged from 10 groups data that are selected randomly from the datasets that are previously collected from nearly 200 doctor candidate students and more than 150 master students at school of Electronics and Information Engineering in Tsinghua and Xi'an Jiaotong University. Results of the evaluations suggest when the retrieval frequency is too much larger the incremental retrieval precision will get down, otherwise, when the interval of incremental retrieval is reasonable, the newly retrieved precision is often fairly good, which provides references for users to decide how to select optimal retrieval interval.

6 Conclusion

Comparing with pull-based information service such as DBMS, search engine, full-text retrieval system, the AACP paradigm provides a push-based, event-driven, content-sensitive, interest-related, adaptive and active information service mode, which servers users in an active style instead of passive mode, thus users needn't polling from kinds of information sources for new and fresh information. Algorithms of user profile management, incremental information retrieval, information filtering, automatic change detection are discussed in details, and system optimization is also taken into account. The adaptive and active computing paradigm for personalized information service is promising and helpful for long-term users to gain fresh information in time with high quality information.

References

1. Carpineto, Claudio; Romano, Giovanni.: Information retrieval through hybrid navigation of lattice representations. International Journal of Human-Computer Studies.(1996)553-578
2. Lalmas, Mounia: Logical models in information retrieval: introduction and overview. International Journal of Information Processing and Management.(1998)19-33
3. Sparck Jones, Karen.: Information retrieval and artificial intelligence Artificial Intelligence. (1999)257-281
4. Dominich, Sandor.: Connectionist interaction information retrieval. International Journal of Information Processing and Management. (2003)167-193
5. Olsen, Kai A.; Sochats, Kenneth M.; Williams, James G.: Full Text Searching and Information Overload. International Information & Library Review. (1998) 105-122
6. Kontogiannis, Tom; Embrey, David.: A User-Centred Design Approach for Introducing Computer-based Process information systems. Applied Ergonomics, (1997)109-119
7. Underwood, George M.; Maglio, Paul P.; Barrett, Rob.: User-centered push for timely information delivery. Computer Networks and ISDN Systems. (1998) 33-41
8. Smart, Karl L.; Whiting, Matthew E.: Designing systems that support learning and use: a customer-centered approach. Information and Management. (2001)177-190
9. Lin Xia, Chan Lois.: Personalized Knowledge Organization and Access for the Web. Library and Information Science Research. (1999)153-172
10. Cho Yoon Ho, Kim Jae Kyeong, Kim Soung Hie: A personalized recommender system based on web usage mining and decision tree induction. Expert Systems with Applications. (2002) 329-342
11. Lee Wei-Po, Liu Chih-Hung. Lu Cheng-Che.: Intelligent agent-based systems for personalized recommendations in Internet commerce. Expert Systems with Applications, (2002) 275-284
12. Yuan Soe-Tsyr,Tsao Y.W.: A recommendation mechanism for contextualized mobile advertising. Expert Systems with Applications. (2003)399-414
13. Lee Dong-Seop, Kim Gye-Young, Choi,Hyung.: A Web-based collaborative filtering system. Pattern Recognition. (2003) 519-526
14. Nick Achim, Koenemann Jürgen. Schalück Elmar. ELFI: information brokering for the domain of research funding. Computer Networks and ISDN Systems,(1998)1491-1500
15. Chen, Chen-Tung; Tai, Wei-Shen.: An information push-delivery system design for personal information service on the Internet. International Journal of Information Processing and Management. (2003) 873-888

Mining Association Rules Based on Seed Items and Weights

Chen Xiang, Zhang Yi, and Wu Yue

University of Electronic Science and Technology of China,
College of Computer Science and Engineering
`cchenx@163.com`
`Elezy@nus.du.sg`
`ywu@uestc.edu.cn`

Abstract. The traditional algorithms of mining association rules, such as *Apriori*, often suffered from the bottleneck of itemset generation because the database is too large or the threshold of minimum support is not suitable. Furthermore, the traditional methods often treated each item evenly. It resulted in some problems. In this paper, a new algorithm to solve the above problems is proposed. The approach is to replace the database with the base set based on some seed items and assign weights to each item in the base set. Experiments on performance study will prove the superiority of the new algorithm.

1 Introduction

Mining of association rules is an important technique applied in analyzing large data sets. It was first introduced in [1] and since then it has been widely studied by many authors.

Given a transaction database D, the process for mining association rules can be divided into two steps: finding large itemsets and then generating association rules. Because the second step is relatively simple, most of the researches on mining association rules are focused on finding large itemsets.

Most of the past work about mining of association rules didn't consider the importance of each item. It means that each item in the database has the same nature and importance. This results in some problems. Furthermore, an unsuitable user defined minimum support will result in too many large itemsets to be analyzed.

In this paper, we will propose a new approach to the mining of association rules based on seed items and item weights. By introducing the seed items, we will get some interested base sets smaller than the original database for the mining.

2 Base Set

The traditional algorithms of mining association rules, such as *Apriori*, often suffered from the bottleneck of itemset generation because the database is too large or the threshold of minimum support is not suitable. In this paper, we can select the base set I suitably to avoid the ad hoc problem of mining association rules.

Suppose we have a transaction database D. To start the mining of association rules from D, users often have some idea in their mind. Most of the cases may be like in this way that users are interested to find association rules related to some interested items. The next subsection will address this problem.

2.1 Finding Interesting Base Set

Give a transaction database D. Suppose that users want to find association rules associated with some interested items of D. To generate the base set, we use an iterative method for propagating counts on the elements in same tuple of D. Firstly, We seed user interested items with count equal 1. Then we propagate to other items in D that co-occur frequently with these seed items. A minimum correlation value is defined to prune those items which have little correlation with the seed items. The algorithm for finding base set is given as follows.

```
While (1){
   for each tuple r = {v,u₁,...,uᵣ} containing v{
      if (Cv = Cu₁ = ...= Cuᵣ = 0) Xₜ = 0;
      else Xₜ = 1;
      Cv = Σᵣ Xᵣ;}
   //prune the items which have little correlation with
   //the items in prior base set. corr is user defined.
   for each Ci > 0{
      // Cs is the minimum count value in prior base set
      if (Ci / Cs < corr)  Ci = 0;}
   if(C' == C) break; //the count configuration converges
   C'⇔C   //update the count configuration C}
Return;
```

2.2 Base Set with *TIDs*

In the process of finding base set we can associate each element of I with all of its *TIDs* in the transaction database D. Let us firstly employ an example to illustrate this point. Give a database as shown in table 1.

Table 1. A transaction database

TID	Items
1	ACD
2	BCE
3	ABCE
4	BE

Suppose the base set is $I = \{B, C\}$. The proposed method is to scan the database D to find out all the *TIDs* for each element of I, as shown in table 2.

Table 2. Base set I with *TIDs*

Base set	TID
B	23
C	123

Knowing all the *TIDs* for each element of the base set I will have enough information for the mining of association rules. In fact, the most important information in mining is to find out the count of each element in 2^I. The following lemma shows the count can be got if all the *TIDs* of each element in I are known.

We use 2^I to denote the set of all subsets of I. For any $X \in 2^I$, we denote the set of all *TIDs* that contain X by *tids(X)*. We use *numTids(tids(X))* to denote the number of the elements of the set *tids(X)*.

Lemma 1: For any $X=\{x_1 x_2 \ldots x_r\} \in 2^I$, where $x_i \in I (i=1,\ldots,r)$, we have $tids(X) = tids(x_1) \cap tids(x_2) \cap \ldots \cap tids(x_r)$ and $count(X) = numTids(tids(X))$.

The proof is trivial. For example, based on the above example, we have

$$\begin{aligned} count(BC) &= numTids(tids(BC)) \\ &= numTids(tids(B) \cap tids(C)) \\ &= numTids(23) \\ &= 2. \end{aligned}$$

3 Support Function

Let I be a base set. For each element of $x \in I$, we assign a real number $Wx \in [0,1]$ to x, and Wx is called the weight of x. Since the mining of association rules is to find implications between the elements in 2^I, not I, we must define weights for all elements in 2^I. This can be done in this way. For any $X \in 2^I$, suppose $X=\{x_1,x_2,\ldots,x_r\}$, where $x_i \in I (I=1,\ldots,r)$, we define the weight of x as a function of Wx_i as follows:

$$Wx = F(Wx_1, Wx_2, \ldots, Wx_r) \quad \text{where } F: R^r \Rightarrow [0,1]. \tag{1}$$

For any $X \in 2^I$, we define the support function as

$$f(X) = Wx \cdot numTids(X)/numTids(\hat{o}). \tag{2}$$

We must choose the weight function F to make f be a support function. That is for any $X \subseteq Y$, f should satisfy $f(X) \geq f(Y)$. For this aim, we should have

$$W_X / W_Y \geq numTids(Y) / numTids(X). \tag{3}$$

Since $numTids(X) \geq numTids(Y)$, if we have $W_X \geq W_Y$, the above inequation can be satisfied. We can define F to be minimum function of its elements, i.e.

$$W_X = \min(Wx_1, Wx_2, \ldots, Wx_r). \tag{4}$$

Since $X \subseteq Y$, it must follow that $W_X \geq W_Y$, so we get the desired result.

4 Algorithm of Mining Weighted Association Rules

In this section, we will give an algorithm of mining weighted association rules. To start the mining, we firstly use the method described in section 2 to find out the base set I. Then, we scan the transaction database D to find out *TIDs* for each element of I. After getting the base set I, we assign each element of I with a weight. Then we begin to mine large itemsets. The distinct significance of our algorithm is that it keeps *TIDs* for each large itemset in every step. It enables us not to scan the original transaction database after the first scan for finding *TIDs* of *I*. The algorithm scan database only once makes it different from the famous algorithm *Apriori* introduced in [2] because *Apriori* needs to scan database many times. The algorithm is described as follows.

```
L₁ = init(I);
for(k=2; Lₖ≠φ; k++)
    Lₖ = LargeItemGen(Lₖ₋₁, minsup);
L = ∪ₖLₖ;
R = AssociationRuleGen(L, minconf);
return;

//LargeItemGen(Lₖ₋₁, minsup)
for X and Y is in Lₖ₋₁{
    if first k-2 items of X and Y are same and any subset
    of X∪Y is in Lₖ₋₁{
        tids(X∪Y) = tids(X)∩tids(Y);
        f(X∪Y) = min(Wₓ,Wᵧ) · numTids(tids(X∩Y));
        if f(X∪Y)≥minsup
            insert X∪Y, tids(X∩Y) and f(X∪Y) into Lₖ;}}
return L;

//AssociationRuleGen(L, minconf)
for any X∈L and any nonempty subset Y of X
    if f(X)/f(Y)≥minconf  insert X⇒(X-Y) into R ;
return R;
```

5 Performance Study

In this section we will study the performance of the proposed algorithm of mining weighted association rules. All experiments are carried on a Pentium 4 2.66G PC with 256MB memory. The software platform is Windows XP.

Let I be a base set, for any $x \in I$, we give a method to define the weight of x as follows:

$$W_x = 1 - â \times numTids(tids(x))/\max_{z \in I}(numTids(tids(z))) \quad â \in [0,1]. \tag{5}$$

It is easy to see that for any two elements $x, y \in I$ if the frequency of x is larger than the frequency of y, that is $numTids(tids(x)) \geq numTids(tids(y))$, then $W_x \leq W_y$. This satisfies the basic requirement for using weights in the mining of association rules.

To study the performance, we use the synthetic data generated by the data generator introduced in [2]. We take the parameters shown in table 3.

For our experiments, we use two dataset summarized in table 4.

Table 3. Parameters

Arguments	T	I	P
Description	Number of transactions	Number of items	Number of patterns

Table. 4. Parameter settings

Name	T	I	P	Size in Megabytes
T10.I1.P100	10K	1K	100	3.2
T100.I10.P1000	100K	10K	1000	32.2

To show the advantage of our algorithm, we compare our algorithm called *Base* with *Apriori* algorithm. Fig. 1 shows the execution time and the number of large sets generated by two algorithms based on two different datasets.

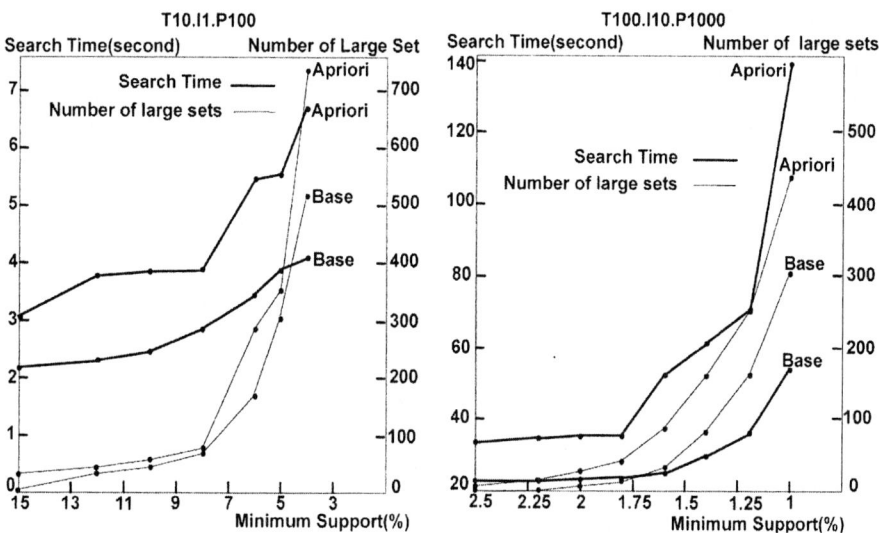

Fig. 1. Performance comparison of two algorithms on dataset T10.I1.P100 and T100.I10.P1000

To find a large base set to test our algorithm, we select the most frequent item as the seed item. From the figure we can see that our algorithm cost less time and generate less large itemsets than *Apriori* algorithm. It means that we saved time and avoided large amounts of useless results. If the correlation of the items in the data file is smaller, the advantage of our algorithm will be more obvious. This is reasonable because we can get a much smaller base set than the original database. It means that we can save more time and avoid more useless results by mining a smaller base set instead of the original database.

6 Conclusions

We proposed an algorithm for the problem of mining association rules based on seed items and weights. We carried some experiments to illustrate the performance of the proposed techniques. The proposed method can generate the user interesting results and at the same time it considers the different importance of different items. As the result, the proposed method has much superior performance in both efficiency and the generation of the useful results to *Apriori* algorithm.

References

1. Agrawal, R., Imielinski, T., Swami, A.: Mining Association Rules between Sets of Items in Large Databases. Proc. ACM SIGMOD Conf. Management of data(1993) 207-216
2. Agrawal, R., Srikant, R.: Fast Algorithms for Mining Association Rules. Proc. 20th Int'l Conf. Very Large Databases(1994) 478-499
3. Wai-chee Fu, Wang-Wai Kwong, Jian Tang: Mining N-Most Interesting Itemsets. International Syposium on Methodologies for Intelligent Systems Date(2000)
4. Wang, K., He, Y., Han, J.: Mining Frequent Itemsets Using Support Constraints. Proc. 20th Int'l Conf. Very Large Databases(2000)
5. Brin, S., Motwani, R., Ullman, J. D., Tsur, S.: Dynamic Itemset Counting and Implication Rules for Market Basket Data. Proc. ACM SIGMOD Conf. Management of data(1997)

An Algorithm of Online Goods Information Extraction with Two-Stage Working Pattern

Wang Xun, Ling Yun, and Yu-lian Fei

College of Computer and Information Engineering,
Zhejiang Gongshang University, Hangzhou 310035, china
wx@mail.zjgsu.edu.cn

Abstract. The key technology in comparison-shopping is the online goods information extraction. Based on DOM, the information extraction with two-stage working pattern and the conception of page information unit have been proposed after a large number of sample pages testing. PIU is extracted and categorized by the classifying algorithm, and information is extracted from PIU. It is implemented that the key information of online goods is extracted based on the above-mentioned information extraction algorithm. It shows that the algorithm is steady and has higher Recall and Precision rate with the sample page testing.

1 Introduction

Intelligent Comparison Shopping Systems(ICSS) based on traditional information retrieve techniques can be used to gather the product information from the Internet and compare that information according to customers' demand. The core technology is specific Web information extraction. But the tag of HTML just informs the browser how to display the information without any semantics, so the web pages written in HTML is only suitable for browsing after browser resolution, while being unsuitable for data communication processed by computers [1].

Under web environment, normally it is the wrapper that is responsible for extracting the information contained in HTML documents and transforming that information into data structure storage that can be further processed. Generating the wrapper has become the research hot spot of intelligent information processing domain [1]. In recent years, many famous international conferences such as SIGMOD, VLDB, ICDE and etc. have also published several relevant articles [2-4]. Document [5] has categorized those methods from different point of views, while most of those methods bases on different rules to generate wrapper. According to producing the difference of rules, wrapper can be roughly divided into the following two categories: (1)developing special grammars to illustrate the distribution of data in HTML pages and data extraction[6-8]; (2)applying induction technique in generating extraction rules automatically or semi-automatically [9-10].

Document [6] designs a declaration advanced language in writing templates to define extracting rules, and describes the queries wrapper would receive and the objects would be returned. Once the query matches template, the action associated with that template would be activated, transforming the query of integrated system into query of data source oriented and translating the results into the form that integrated system could discern. This method must be adapted to the designing requirements of the integrated system, so it could not be generally applied. Document [8] regards the Web as a large distributed database system composed by non-structured and semi-structured files, so constructing hypertext relation view and using query languages could fulfill queries, but the results of queries have flaws that the granules are too big and are not precise enough.

Document [9] introduces the research on incorporating induction into the wrapper, so the wrapper could analysis Web data source efficiently and has enough capability to represent real data source. But it handles mainly table-layout data source with much limitation. Document [10] uses Landmark from induction learning method to represent extraction rules, including the beginning rule and ending rule, which is especially suitable for extracting information delimited by the start tag and the end tag, but unsuitable for other complex information extraction.

After analyzing the drawbacks of various wrappers and features of comparison-shopping systems, this paper introduces the tree structure path expression of document object model to position the information in HTML pages to be extracted and proposes a two-stage information extraction algorithm, and testing results shows that the algorithm is steady and efficient.

2 Web Page Analysis

In comparison-shopping systems, the forms of product information in web vary greatly and the structures embedded in web pages differ, it is quite difficult to induce data pattern for all the web pages. In the application of ICSS, the types of information to be extracted are relatively fixed, focusing mainly on extraction of information re- lated to products while rejecting irrelevant information contained in the web pages. With the research on sampling web pages, following features have been spotted:

(1) In the web pages that contain product information, most product information is densely located under certain web page tags, such as <table>, <tr>, <td>, <th>, etc. So these four tags could be considered as key basis for extracting product information.

(2) If the web page contains information about more than one product item, then the structures of these information are usually organized in similar mode or similar classified mode.

(3) The components of product information are usually in the same hierarchy, being arranged in parallel. In the DOM tree structure, the relationship then can be drawn as sibling nodes connecting to parent node.

According to the research of above sampling web pages, the whole extraction process could be divided into two stages. At the first stage, extracting the page information unit (PIU) that contained in web pages and classifying PIU according to the different structure features when these PIU appear in the web pages. At the second stage, applying different extraction rules to extract product information from classified PIU information and knowledge domains.

Page information unit, which is separated from the web page, contains the corresponding code of the product information needs to be extracted. So PIU should possess following features:

(1) PIU should contain at least one piece of complete product information. Completeness refers that the product information elements appeared within key query words and other relevant product information are totally included. Product information elements will vary according to different product category and system requirements. In this paper, we use steels as example, whose elements always include designation, specifications, material quality, price, producer, vendor, transaction site and mode, etc.

(2) PIU contains some web page tags, whose various structural forms can classify PIU into different categories. Since the tags in PIU are closely correlated with the hierarchical layout of product information, the number of categories of corresponding DOM tree structures is relatively fewer.

(3) PIU begins from the web page start tag and finishes at the end tag, with the contents enclosed between the pair of tags. Accordingly, in DOM tree structure, tag node is considered as root node, while text node is treated as leaf node.

3 Classification of PIU

3.1 PIU Classifying Design

Information extraction rules need to be applied to PIU of various web page formats and web structures and delimit the structures and categories of product information contained. Extensive research has discovered that the usage of XML pages is not extensive enough in native commercial web sites and the sampling pages are rare, so this paper will not take it into consideration. Since other web pages such as HTML, asp, jsp, php are loaded at the client side, so there are not much differences at all. Those tags could be grouped into the same category and will not influence the extraction of information.

This paper classifies the web pages containing product information into: list type, table type, mixed type of list and table and a few irregular types. So the PIU extracted from web pages can be categorized into two types: list and table types and can be further divided according to their structural features as follows:

PIU of list type can be divided further into those with header information and those without header information, while the contents of those two PIU list types are all enclosed between tags of <table> and </table>. The text nodes of corresponding DOM trees usually contain few materials and are in uniform format, and every column represents one element of the product information, while every line represents the entire information of one product. PIU of list type with header information can analysis and distinguish the contents of every column by analyzing the header information, while PIU of list type without header information relies on the domain knowledge to distinguish the product elements of every column.

With the corresponding DOM tree for the PIU of table type, the text nodes are more complex. Since web page of table types cannot represent product information elements in uniform identification using header information, text nodes usually contain some interpretations explaining the information contained. For example, the designation: XXX, specification: XXX, etc. The text node of the PIU of table type can also include more than one product information elements, so trying to distinguish the product information of PIU of table type usually demands additional product domain knowledge.

The code fragment of web page used to extract PIU contains at least an entire product information, so the PIU of table type can be further divided into four categories according to different start web tags:

  (a) Code fragment in tags <td> or <th>;
  (b) Code fragment in tags <tr>;
  (c) Code fragment in tags <table>;
  (d) Besides the three types listed above, elements of an entire product information located at different <table> tags.

3.2 PIU Classifying Extraction Algorithm

PIU classifying extraction means to extract and classify the PIU contained in web pages using the classifying rules for further use in extracting information. The algorithm is described as follows:

```
// key words,code of source pages
Input: Keyword(key₁, key₂, ···, keyₘ), page code;
Output: classified PIU;    // classified PIU
To pretreat page;          // preprocessing of web pages
To make TableSet(table₁, table₂, ···, tableₙ);
i=i+1;
loop1:    if  key₁ ∉ tableᵢ  then goto loop3
          To make KeywordNodes(node₁, node₂,···, nodeₚ);
loop2:    if condition1 then { if condition5 then PIU∈type(a)
                              j=j+1;
                              if j<=p then goto loop2
                              else goto loop3 }
          else if condition2 then { if condition6 then PIU∈type(b)
                              j=j+1;
                              if j<=p then goto loop2
                              else goto loop3 }
          else if condition3 then { if condition4 then
                              { if condition7 then
                                {if condition8 then
```

```
                    PIU  list type with header
                 else PIU∈f list type without header} }
               else if condition7 then PIU type(c)
loop3:   { i=i+1;
           if i<=n then goto loop1 else exit }
```

The symbols used in the algorithm are explained as below:

TableSet($table_1$, $table_2$, $\cdots$, $table_n$): the set of all the <table> units in the web page, n being the number of <table> units in the web page.

Keyword(key_1, key_2, $\cdots$, key_m): the set of all the key words provided by user, while m being the number of key words, and key_1 represents the product designation.

KeywordNodes($node_1$, $node_2$, $\cdots$, $node_p$)£the DOM tree structure of $table_i$ contains the node key_1, while p being the number of nodes that contain key_1.

condition1: the columns of $node_j$ (<tr> or <th>) contain other elements of product information, and can constitute at least an entire product information, while containing explanatory text information about elements of product information.

condition2: the line of $node_j$ (<tr>) contains other elements of product information, and can constitute at least an entire product information, while containing explanatory text information about elements of product information.

condition3: the table of $node_j$ (<table>) contains other elements of product information and can constitute at least an entire product information.

condition4: $node_j$ and key_1 fully match, while nodes containing other elements of product information do not include explanatory text information about elements of product information.

condition5: the other elements of product information of the column of $node_j$ (<td> or <th>)contains key_2, $\cdots$, key_m.

condition6: the other elements of product information of the line of $node_j$ (<tr>)contains key_2, $\cdots$, key_m.

condition7: the other elements of product information of the table of $node_j$ (<table>)contains key_2, $\cdots$, key_m.

condition8: the first line would be the header information if the table of $node_j$ (<table>)and the columns of the line of $node_j$ match.

4 PIU Information Extraction Algorithm

In Section 3, the classified PIU contains several product elements, while the user is interested in comparing at least one element in order to make comparison shopping. Key elements of product information could be more than one and depend on user requirements. Various PIU information extraction algorithms are listed as follows.

4.1 List Type PIU Information Extraction Algorithm

The web page structure of list type PIU is with regular patter and the content of the text node of DOM tree is plain, so matching the key words could locate the corresponding text in the text node, while the elements of product information are at just the same line of the text (the text contained in <tr> node of DOM tree). The difference is that if it is a list with header information, then the elements of product information could be defined with the prompt of the header information; if it is a list without head information, then matching the elements of product information with related domain knowledge to define exact meanings. The information extraction algorithm for PIU with header is relatively simple, so the algorithm for PIU without header is described as follows.

```
Input: PIU, Keyword(key₁, key₂, ···, keyₘ);
Output: ProductInfo(info₁, info₂, ···, infoₜ);
To make DOM for PIU;
To make NodeSet(text₁, text₂, ···, textₙ);
i=1; j=1;
while i<=n do{
   To make Textᵢ_tr(td₁, td₂, ···, tdₜ);
   while j<=t do{
     infoⱼ = tdⱼ.toString();
     if (infoⱼ ∈ Keyword) then (To extract infoⱼ by Keyword )
     else{ r=1;
        while r<=q do{
          if (infoⱼ ∈ knowledgeᵣ) then
             {To extract infoⱼ by knowledgeᵣ; goto loop1}
          else r=r+1;}
        r=1;
        while r<=q do{
           // the info_j and knowledge_r resemble enough
           if similitude(infoⱼ, knowledgeᵣ) > ε then
             { To update knowledgeᵣ and extract infoⱼ
                by knowledgeᵣ; goto loop1}
           else r=r+1;}
     To append semantic information of infoⱼ;
     goto loop2;}
loop1:   j=j+1}
    print(ProductInfo(info₁, info₂, ···, infoₜ));
loop2:   i=i+1}
```

The symbols used in the algorithm are explained as follows:

NodeSet($text_1, text_2, \cdots, text_n$): the set of text nodes in DOM tree containing key_1, while n being the number of text nodes that contains key_1.

$Text_i\_tr(td_1, td_2, \cdots, td_t)$: lines in the list that contain $text_i$, the content between <tr> and </tr> in DOM tree that contains $text_i$, while t being the number of lines.

td_j.toString(): the string of elements of product information in the text nodes contained in td_j.

ProductInfo($info_1, info_2, !`, info_t$): the set of elements of product information extracted form PIU, while t is the number of elements and equals the number of columns in $Text_i\_tr$.

tr_k.toString(): the string of text node contained in tr_k, namely the string with text information in the kth <tr> of the <table>.

KnowledgeSet($knowledge_1, knowledge_2, \cdots, knowledge_q$): the knowledge set of a specific domain, while q being the numbers of sub items in the knowledge set.

4.2 Table Type PIU Information Extraction Algorithm

The web page structure of table type PIU is much more diversified and the content of text nodes in DOM tree is quite complicated, including not only the elements of product information, but also the explanatory text information of those elements. So when extracting information, it is necessary to distinguish between explanatory text information and elements of product information, trying to match the elements with relevant domain knowledge in order to define specific meanings.

In the web page structure of table type PIU, type (c) is the most complicated one, so the discussion below tries to give out corresponding algorithm. Since types (a) and (b) are relatively simple, they will not be discussed any more.

The elements of product information of PIU type (c) are distributed in various <tr> tags in <table>, which means that from the view of web code programming, PIU type (c) contains many pieces of product information. After analyzing sampling web pages, it can be concluded that almost all PIUs of type (c) have several pieces of product information and the text information in the same <tr> containing various products' elements listed in sequence. For example, in the nth <td> of the first <tr> would be the designation of the nth product, while the nth <td> of the second <tr> would contain the specification of the nth product. Benefiting from this feature, just counting the occurrences of key elements of product information of table type PIU (c) well determines the numbers of complete product information contained. Then through the judgment of locations of the queried key words will obtain the elements of product information. The algorithm is described as below:

```
Input: PIU, Keyword(key₁, key₂, !`, keyₘ);
Output: ProductInfo(info₁, info₂, !`, infoₜ);
To make DOM for PIU;
KeyInfo.count()=tr_KeyInfo.children().size();
To make NodeSet(text₁, text₂, ···, textₙ);
i=1;
while i<n do{
  x=1; k=1;
  To make Text_i_tr(td₁, td₂, ···, tdₜ);
  To make trList(tr₁, tr₂, !`, tr_p);
  while k<=p do{
loop:   word=tr_k.child(z+x-1).toString();
```

```
        x=x+1;
        To separate "description" from "core" in "word";
        if x <=tr_k.children().size()/KeyInfo.count() then goto loop;
        x=x+1; k=k+1;}
    i=i+1;}
print(ProductInfo(info_1, info_2, ..., info_t));
```

The symbols used in the algorithm are explained as follows:

KeyInfo: key elements of product information.

KeyInfo.count(): the number of occurrences of key elements of product information in trKeyInfo.

tr_k.children().size(): the number of kth <tr> sub-nodes in the DOM tree, namely the number of columns with kth line.

tr_k.child(j).toString(): the string of text nodes contained in the jth sub-node of kth <tr> of the DOM tree, namely the string of the jth <td> of the kth <tr> in <table>.

5 Experimental Results

The test pages are chosen randomly from web pages about two specific groups (steels and cosmetics) after casting XML pages and keeping totally 386 pages of types such as HTML, .asp, .jsp, .php. There are 216 pages about steels, mainly of list type; 170 pages about cosmetics, mainly of table type. At the same time, price is chosen to be the key element of product information.

In the first stage, test on classifying pages and extracting PIU is carried out, and the classification result is given in recall and precision rates. The precise processing rate is defined as the number of pages precisely classified divided by the number of pages processed, while the processing rate is defined as the number of pages precisely classified divided by the number of pages processed in the test. The results is listed in Table. 1.

It is shown that the classifying and extraction algorithm is efficient and its various indexes reach the criteria of comparison-shopping system.

In the second stage, comprehensive testing based on two-stage extraction algorithm is applied to sampling pages, the result is listed in Table. 2.

Table 1. The results of web pages classification and extraction

Web page type	Manual classification	Classifying algorithm		Precisely processing rate	Processing rate
		Web page processed	Web pages precisely processed		
List type	224	224	224	100	100
Table type(a)	85	85	84	98.8	98.8
Table type(b)	37	36	36	100	97.3
Table type(c)	29	26	25	96.2	86.2
Table type(d)	11	15	11	73.3	100

Table 2. The comprehensive testing results based on two-stage extraction algorithm

Web page type	Total number of page	Web page processed	Average precision rate	Average recall rate
List type web page	224	224	93.7%	98.5%
Table type web page	162	147	81.4%	92.6%
Web page of steel	216	215	95.7%	99.3%
Web page of cosmetic	170	156	78.2%	92.3%

From Table. 2, it can be concluded that the recall rate of pages containing PIU of table type is a bit lower than that of PIU of list type. Further analysis reveals the reason being that the sampling pages selected contain some pages of table type (d). The reason why the precision rate of pages of PIU of table type is much lower than that of pages of PIU of list type is that the pages of list type is more regular and the contents of text nodes of DOM tree is more plain, while parts of the outcome from information extraction algorithm applied on PIU table type needs further to be processed with manual appended semantics.

The precision rate of pages of steels is much higher than that of cosmetics; the reason is that the domain knowledge of steel is more standard than that of cosmetics besides the differences of page structures, so the domain knowledge database of cosmetics needs to be improved.

6 Conclusions

In this paper, the concept of PIU is proposed, and at the same time the PIU classifying extraction and the extraction algorithm of product elements from PIU are presented too. This algorithm can greatly reduce the workload of manual appended semantics in web page information extraction while keeping high extraction efficiency. The experiment outcome shows that the two-stage information extraction technique meets the expected results on the whole and could facilitate the implementation of a comparison shopping system.

With the further development of e-commerce, product information pages in XML will increase. The superiority of XML over HTML makes it definitely the new web information exchange standard. Further research would be centered on making the ICSS process XML and other complicated web pages, and improving the coverage of pages to be handled.

References

1. Florescu D, Levy A Y, Mendelzon A. Database techniques for the World-Wide Web: A Survey. In: ACM The SIGMOD Record, 1998.59-74.
2. Liu L, Pu C, Han W. XWRAP: An CML-enabled wrapper construction system for web information sources. In: Proc International Conference on Data Engineering(ICDE), San diego, California, 2000. 611-621.
3. Cham berlin D, Robie J, Florescu D. Quilt: an XML query language for the heterogeneous data sources. In: Proc International Workshop on the Web and Database(WebDB'2000), Dallas, Texas, 2000. 53-62.

4. Sahuguet A, Azavant F. Building light-weight wrappers for legacy web datasources using w4f. In: Proc International Conference on Very Large Database, Edinburgh, Scotland, 1999. 738-741.
5. Laender A, Ribeiro-Neto B, Silva A. A brief survey of web data extraction Tools, SIGMOD Record, 2002£31(2): 84-93.
6. Hammer J, Garcia-Molina H, Nestorov S et al. Template-based wrappers in the TSIMMIS system. In: Proc of ACM SIGMOD Conference on Management of Data, Tucson, Arizona, 1997. 523-535.
7. Gruser Jean-Robert, Raschid L, Vidal M E et al. Wrapper generation for web accessible data sources. In: Proc the Coopis, 1998. 14-23.
8. Liu M, Ling T M. A conceptual model and rule-based query language for HTML. World Wide Web, 2001, 4(1): 49-77.
9. Kushmerick N, Weld D et al. Induction for information extraction. In: Proc the 15th International Joint Conference on Artificial Intelligence, Nagoya, Japan, 1997, 2: 729-737.
10. Ashish N, Knoblock C. Wrapper generation for semi-structured internet sources. In: Proc of Workshop on Management of Semi-Structured Data, Tucson, Arizona,1997,10-17.
11. Liu L, Pu C, Han W. XWRAP: An CML-enabled wrapper construction system for web information sources. In: Proc International Conference on Data Engineering(ICDE), San diego, California, 2000. 611-621.

A Novel Method of Image Retrieval Based on Combination of Semantic and Visual Features[*]

Ming Li[1], Tong Wang[1], Bao-wei Zhang[1], and Bi-Cheng Ye[2]

[1] School of Computer and Communication,
Lanzhou University of Technology, Lanzhou, 730050, China
[2] School of Electromechanical Engineering,
Lanzhou University of Technology, Lanzhou, 730050, China
Zjxywt@163.com

Abstract. Content-based image retrieval (CBIR) and semantic-based image retrieval (SBIR) have attracted great research attentions. However, they all have disadvantages. This paper proposes a retrieval method trying to overcome them. To achieve this we first introduce an approach of rough set-based low-level features selection. We propose an approach of feedback-based semantic-level features annotation. We also introduce a corresponding computing technology of similarity. We then build a model of combination of these two kinds of methods. Experimental results show that our method is more user-adaptive, and can achieve better performance compared with another retrieval method which is only based on low-level features.

1 Introduction

With the development of computer and network technologies, there has been an explosion in the volume of multimedia database composed of digital image, video and so on. In order to make use of this vast volume of data, efficient and effective techniques to retrieve multimedia information need to be developed. Among the various media types, images are of prime importance. Thereby image retrieval (IR) has attracted great research attention.

Text-based image retrieval (TBIR) is the traditional image retrieval paradigm. Now, content-based image retrieval (CBIR) [1] and semantic-based image retrieval (SBIR) were proposed. CBIR systems use the visual content of the images, such as color, texture, and shape features, as the image index. On the other hand, if the user is searching for an object that has clear semantic meanings but cannot be sufficiently represented by combinations of available feature vectors, the content-based systems will not return many relevant results. Furthermore, the inherent complexity of the images makes it almost impossible for users to present the system with a query that fully describes their intentions.

To overcome the difficulties of this approaches, semantic-based image retrieval (SBIR) [2] is proposed, which is a kind of category search [3]. The user searches for images that belong to a prototypical category. However, it is very difficult to directly

---

[*] This paper is partially supported by Natural Science Foundation of Gansu and Foundation of Science Research of Gansu Education Office under grant 0416B-04.

extract the semantic-level features from images with the current technology in computer vision and image understanding. That is to say it is hard to bridge this gap between visual features and semantic features of images.

A way to gather information between low-level features and semantic-level features from the user interaction is called relevance feedback (RF) [4]. During the feedback step, the user is requested to specify relevant and non-relevant images [5]. The performance of a relevance feedback method can be measured in terms of the number of iterations required to retrieve a prefixed number of images relevant to the user and in terms of precision after a fixed number of iterations. Most of the previous relevance feedback research can be classified into two approached: query vector movement and [6] and updating the weight [7].

Rough set theory was proposed by Z.Pawlak in 1982[8]. The theory and its application in data mining is a new research area. It presents us a method to deal with uncertainty and fuzziness. Compared with other data mining methods, a main advantage of rough set is that it doesn't need any additional information.

In this paper, we make use of rough set theory and relevance feedback, performed on both the images' low-level features selection and the semantic contents represented by keywords. During the retrieval step, often produce much redundancy because there is too much features (visual features), and will flood the useful main classification features in a large number of useless features sometimes. Therefore, the feature selection is an essential problem in the image retrieval. Attribute reduction of rough sets is presented to get the minimum subsets of the visual features. Only using the visual features as the retrieval condition often can not get the satisfactory results, so we need extract the semantic image features. A new method using relevance feedback in image semantic index is proposed to extract semantic image features. Moreover, we provide a ranking measure that integrates both visual and semantic features-based similarities for our approaches. Finally, we find out the desired image by measuring the similarities of two feature vectors.

The rest of the paper is organized as follows: in the following section, we discuss a method of visual features selection of images based on rough set theory. A method of relevance feedback-based semantic annotation of images is presented in section 3. In sections 4, we introduce our approaches based on the combination of low level visual features and high level semantic features for image retrieval and the unified similarity ranking measure along with methods. We illustrate our approach by an example in section 5, which will also present the results. Concluding remarks and a discussion of possible future work are given in section 6.

2 Feature Selection Using Rough Set Theory

2.1 Attribute Reduction of Decision Table

Attribute reduction is an important means to deal with the information system. It can find the minimal reduction of attributes on the basis of "classificatory power" is unchangeable.

Definition 2.1. For $c \in C$, when $POS_{(C-\{c\})}(D) = POS_C(D)$, c is dispensable in C, otherwise c indispensable in C. If all $c \in C$ are indispensable, C is independent.

Definition 2.2. If R is the least attribute sets which satisfies the property $POS_R(D) = POS_C(D)$, R is the reduction of C.

Definition 2.3. CORE(C) is all the indispensable attribute sets in C, $CORE(C) = \cap RED(C)$, $RED(C)$ is the reduction of C.

2.2 Application of Attribute Reduction in Image Feature Selection

As is well known visual features of images can be described in the form of tables. Such tables can be treated as knowledge representation system (KRS). We change this table into discrete decision table of visual features firstly. It is presented as follows:

Table 1. Decision table of visual image features

I	Conditional attribute f				c
	f_1	f_2	f_3	f_4	
i_1	0	0	0	0	c_1
i_2	0	0	0	1	c_1
i_3	1	0	0	0	c_2
i_4	2	1	0	0	c_2
i_5	2	2	1	0	c_2
i_6	2	2	1	1	c_1
i_7	1	2	1	1	c_2
i_8	0	1	0	0	c_1
i_9	0	2	1	0	c_2
i_{10}	2	1	1	0	c_2
i_{11}	0	1	1	1	c_2
i_{12}	1	1	0	1	c_2
i_{13}	1	0	1	0	c_2
i_{14}	2	1	0	1	c_1

Decision table is defined as: I is composed of 14 images, which is object set $U = \{i_1, i_2, i_3, i_4, i_5, i_6, i_7, i_8, i_9, i_{10}, i_{11}, i_{12}, i_{13}, i_{14}\}$. P is visual feature set, which constitutes conditional attribute set $P = \{f_1, f_2, f_3, f_4\}$. Q is the category annotation of image i, which constitutes decision attribute set $Q = \{c_1, c_2\}$, the numbers in the table stand for specific discrete values of corresponding conditional attribute c_i, and the same numbers in different columns are of different meaning to their visual features values.

The family P and Q induce classification

$IND(P) = \{\{1\},\{2\},\{3\},\{4\},\{5\},\{6\},\{7\},\{8\},\{9\},\{10\},\{11\},\{12\},\{13\},\{14\}\}$,
$IND(Q) = \{\{1,2,6,8,14\},\{3,4,5,7,9,10,11,12,13\}\}$,

The positive region of Q with respect to P is the union of all equivalence classes of IND(P).

$$POS_P(Q) = U.$$

The decision table is consistent.

IND($P - \{f_1\}$) = {{1,3},{2},{4,8},{5,9},{6,7},{10},{11},{12,14},{13}},
IND($P - \{f_2\}$) = {{1,8},{2},{3},{4},{5,10},{6},{7},{9},{11},{12},{13},{14}},
IND($P - \{f_3\}$) = {{1},{2},{3,13},{4,10},{5},{6},{7},{8},{9},{11},{12},{14}},
IND($P - \{f_4\}$) = {{1,2},{3},{4,14},{5,6},{7},{8},{9},{10},{11},{12},{13}},

The positive region

$POS_{(P-\{f_1\})}(Q) = \{2,5,9,10,11,13\}$,

$POS_{(P-\{f_2\})}(Q) = POS_P(Q)$,

$POS_{(P-\{f_3\})}(Q) = POS_P(Q)$

$POS_{(P-\{f_4\})}(Q) = \{1,2,3,7,8,9,10,11,12,13\}$,

f_2 and f_3 are Q-dispensable in P, and f_1 and f_4 are Q-indispensable in P.

$$CORE_Q(P) = \{f_1, f_4\},$$

IND($P - \{f_2, f_3\}$) = {{1,8,9},{2,11},{3,13},{4,5,10},{6,14},{7,12}},
$POS_{(P-\{f_2,f_3\})}(Q) = \{3,4,5,6,7,10,12,13,14\}$,

f_2 and f_3 are not Q-dispensable in P at the same time.

$RED_Q(P) = \{\{f_1, f_3, f_4\}, \{f_1, f_2, f_4\}\}$,
$CORE_Q(P) = \cap RED_Q(P) = \{f_1, f_3, f_4\} \cap \{f_1, f_2, f_4\} = \{f_1, f_4\}$.

Attribute reduction only eliminates the superfluous attributes in decision table a certain extent, however, doesn't eliminate superfluous information in decision table. Value of attribute reduction is the deep reduction of decision table.

1. Elimination of the same example;
2. Elimination of some column from the decision table;
3. Elimination of superfluous values of attributes;
4. Compute the minimal reduction;
5. Get the corresponding rule induction of the decision table.

3 Semantic Feature Extraction

In this section, we analyze semantic relations among objects in images and provide our framework of image semantic annotations based on relevance feedback.

3.1 Representation of Semantic Feature of Images

We must give expression to the human's understanding of the images in a certain way. This paper provides a method of semantic image features representation based on semantic network. The semantic network in our method consists of a set of keywords having links to the images in the database, as shown pictorially in Fig. 1.

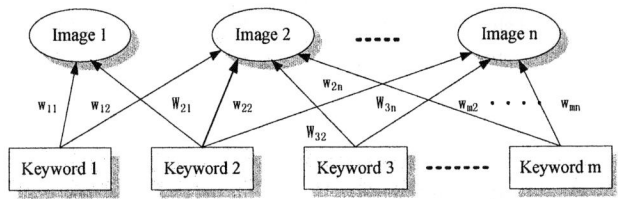

Fig. 1. Semantic network-based semantic image features representation

The links between the keywords and images provide structure for the network. The degree of relevance of the keywords to the associated images' semantic content is represented as the weight on each link. An image can be associated with multiple keywords, each of which with a different degree of relevance.

We should obtain the keywords (semantic features) after the image database is built up. There are several ways to obtain keyword associations. This paper adopts 24 category words in Corel [9] library as the keywords for the description of images.

Table 2. Image classification description words table

Agriculture	Animal	Architecture	Art
Business	China	Cloth	Education
Fashion	Holiday	Entertainment	Food
Family	Military	Nature	Plant
People	Religion	Society	Sports
Science	Travel	Vehicle	Texture

We use a method based on the semantic annotation to describe an image. An image is often described with a lot of keywords with close meaning, which has all brought the difficulty in keywords matching and management. In order to solve this problem, we adopt image hierarchical description structure, namely describe the image with keywords of two levels. Use the words in image classification description words table as the first level semantic information. On the second semantic description level, user can describe the image with various keywords.

As illustrated in the picture, though we can use different words such as Basketball, Football and Milk to describe the given image, image is classified into two sorts such as Sports and Food. Other words are extensions of these two sorts. The structure has a certain function in following keyword matching and semantic feedback.

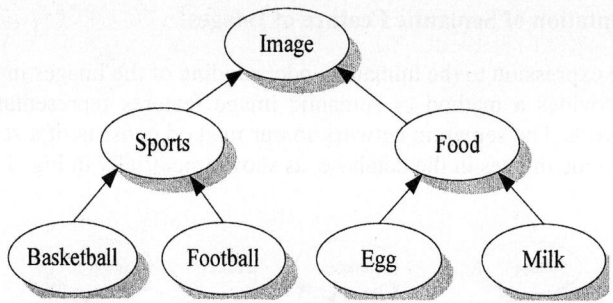

Fig. 2. A simple example of image hierarchical description structure

3.2 Feature Extraction Algorithm

Then, a relevant feedback process refines the retrieval results, updates the semantic network and feature weights in similarity measures. The algorithm can be realized by the following steps:

Algorithm 1: Update (weight)

Input: all images j, their keywords i and the user query vector q;

Output: all weights of keywords associated with each image w_{ij};

Step 1. Initialize all weights w_{ij} to 1. That is, every keyword is assigned the same importance;

Step 2. Collect the user query and the positive and negative feedback examples corresponding to the query;

Step 3. For each keyword in the input query, check to see if any of them is not in the keyword database. If so, add them into the database without creating any links;

Step 4. For each positive example, check to see if any query keyword is not linked to it. If so, create a link with weight 1 from each missing keyword to this image. For all other keywords that are already linked to this image, increment the weight by some predefined values (i.e., 1 in the implementation of this paper);

Step 5. For each negative example, check to see if any query keyword is linked with it. If so, decrease its weight in some way (i.e., we set the new weight to be one fourth of the original weight). If the weight w_{ij} on any link is less than 1, delete that link.

The weight associated with each link of a keyword represents the degree of relevance in which this keyword describes the linked image's semantic content. This effectively suggests a very simple voting scheme for updating the semantic network in which the keywords with a majority of user consensus will emerge as the dominant representation of the semantic content of their associated images.

The dimension of the keyword space is decided by the cardinal number of the keyword sets, which is defined as the union of all probably keyword in all investigated samples. Any image I_j can be represented as many dimensions keyword vector:

$$I_j = \{<k_1, w(I_j,k_1)>, <k_2, w(I_j,k_2)>, \cdots, <k_n, w(I_j,k_n)>\}$$

Coordinates $w(I_j, k_n)$ represents the weight of keywords, it is the measures of keywords in images.

4 Combination of Visual and Semantic Features for Image Retrieval Method（ConIR）

4.1 Structure of ConIR

In this section, we propose a novel method of combination visual and semantic features for image retrieval, its system structure is shown as follows:

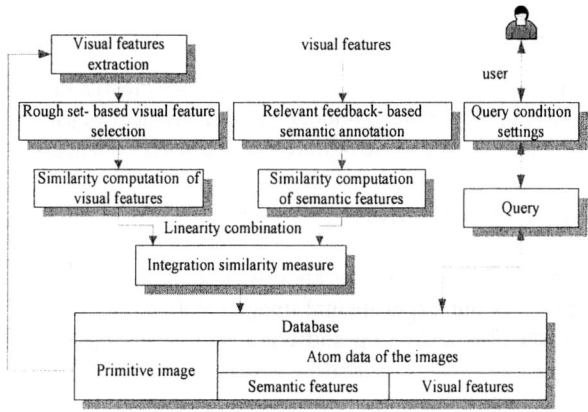

Fig. 3. System structure of ConIR

4.2 Combination of Visual and Semantic Features

Make use of rough set to lower the dimension of the feature vectors. The user query is transformed to a vector which is combined with the low-level features and semantic-level features. Then we compute the similarity between the image vectors in database and the user query vector. There are two steps to compute the similarity: First, compute the similarity of the visual feature vectors. Second, compute the similarity of semantic-level feature vectors, then, we adopt the linearity combined method that makes up to get the last similarity. The image which has the highest similarity value is the desired image.

We define a unified ranking measure function G_j to measure the relevance of any image within the image database in terms of both semantic and low-level feature content. The function G_j is defined using a modified form of the Rocchio's formula as follows:

$$G_j = SIG(j) + \beta \left\{ \frac{1}{N_R} \sum_{k \in N_R} \left[\left(1 + \frac{I_1}{A_1}\right) D_{jk} \right] \right\} - \gamma \left\{ \frac{1}{N_N} \sum_{k \in N_N} \left[\left(1 + \frac{I_2}{A_2}\right) D_{jk} \right] \right\}$$

This formula is composed with three parts. Firstly, $SIG(j)$ is the Matching measuring of the semantic features between the images j and k. N_R and N_N are the number of positive and negative feedbacks respectively; D_{jk} is simply the Euclidean distance of the low-level features between the images j and k. A_1 and A_2 are the total number of distinct keywords associated with all the positive and negative feedback images respectively; I_1 is the number of distinct keywords in common between the image and all the positive feedback images; I_2 is the number of distinct keywords in common between the image and all the negative feedback images.

If the value of I_1 and I_2 are 0, there are not semantic annotation of images, however, have the similarity of visual features. Thereby we can clearly see that more and more relevant images are being semantic annotated as the number of user feedbacks increase, and the importance of the semantic features in images is enhanced, accordingly.

5 Experiment and Results

Under Matlab 6.0 environment, we choose 500 pictures from standard Corel [9] photograph data set as our experimental image database. The visual features used are color and texture. Color histogram and color auto-correlogram [10] are used as the representations for color feature. Gabor wavelet [11] and wavelet moments are used as the texture feature representations.

The remaining experimental result is evaluated in terms of precision and recall defined as follows:

$$\text{precision} = \frac{R_r}{R_r + R_n}$$

$$\text{recall} = \frac{R_r}{R_r + N_r}$$

where R_r is the number of relevant images retrieved, R_n is the number of non-relevant images retrieved, and N_r is the number of relevant images not retrieved. In other words, for a given query, $R_r + R_n$ is the total number of retrieved images and $R_r + N_r$ is the total number of relevant images in the database.

We compare our approach of combination visual and semantic features for image retrieval (ConIR) with a method of image retrieval based on visual features (VisIR).

Table 3. Average Precision (%) of 50 queries

Method	Number of iterations				
	0	1	2	3	4
ConIR	42.60	75.34	85.27	89.61	89.87
VisIR	30.14	42.58	53.64	62.50	67.25

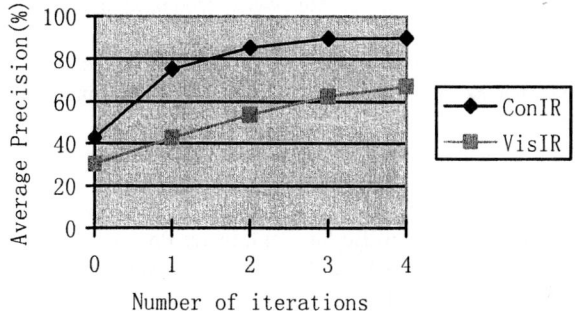

Fig. 4. Comparison of Average Precision (APR) Curve

Based on the tables, we can see that ConIR gives the best retrieval performance in comparison to VisIR. After 4 iterations, the retrieval precision obtained is 85.60% (ConIR), 81.25% (VisIR), respectively. Consistently ConIR achieves a higher APR than another method. The APR of ConIR increases quickly in the initial stage. This is a desirable property, since it provides significant improvement on the retrieval results quickly.

6 Conclusions

A method of image retrieval based on combination of visual and semantic features is proposed in this paper. Attributes reduction of rough set is used to feature selection. We use relevance feedback to semantic feature abstraction, and update the weight of semantic information in retrieval through the interaction. However, this method has some disadvantages. For example, the image annotation method plays an important role during the retrieval procedure. It's hard to guarantee all annotated images may satisfy the conditions. Because there are too many images in database and their meanings are various. So the more overall annotated image in the image database is, the higher rate of accuracy is. We should increase the number of user feedback iterations, in order to improve the rate of accuracy.

Experiment shows that the dimension of vectors space and the scale of the problem are reduced. At the same time, the retrieval accuracy is improved, which is reflected in our method. However, the least reduction of rough set is the NP –hard problem, how to adopt a high efficient algorithm need our research.

References

1. Gudivada, V.N. Raghavan, V. V; Content based image retrieval systems [J]; Computer, 1995, 28 (9):307-315.
2. B. Bradshaw. Semantic Based Image Retrieval: A Probabilistic Approach [J].Pmr. Of ACM Multimedia, 2000.23(6): 676—689.
3. Wenyin. L, Dumais .S, Sun .Y, Semi-automatic image annotation [A]. Rosa, M. Proc. International Conference on Human-Computer Interaction INTERACT'01[C].Amsterdam: IOS Press, 2001: 326-333.
4. Rui Y,Huang T S,Ortega M,et al. Relevance feedback: a power tool for interactive content-based image retrieval[J].IEEE CSVT, 1998.8(5):644-655
5. Rocchio J.J. Relevance feedback in information retrieval [A].Gerard S. The smart Retrieval System: Experiments in Automatic Document Processing[C], Amsterdam: IOS Press,1971:313—323
6. Y. Rui and T. S. Huang.: A novel relevance feedback technique in image retrieval. [A]. In Proc. ACM Multimedia[C], vol. 2, Orlando, 1990, 2 (2):67–70.
7. Y. Rui, T. S. Huang, and S. Mehrotra.: Content-based image retrieval with relevance feedback in MARS. [C]. Proc. IEEE Int. Conf. Image Processing, 1997: 815–818,
8. Pawlak Z. Rough Sets: Theoretical Aspects of Reasoning About Data[M]. Dordrecht: Kluwer Academic Publishers, 1991
9. Corel stock photo library, Corel Corp., Ontario, Canada.
10. J. Huang, S. R. Kumar, and M. Metra. Combining supervised learning with color correlograms for content-based image retrieval [J]. Proc. Of ACM Multimedia, 1997, 325-334.
11. B. S. Manjunath and W. Y. Ma. Texture features for browsing and retrieval of image data [J]. IEEE Trans. Pattern Anal. Machine Intell. 1996,18(5): 837-842

Using Fuzzy Pattern Recognition to Detect Unknown Malicious Executables Code

Boyun Zhang[1,2], Jianping Yin[1], and Jingbo Hao[1]

[1] School of Computer Science, National University of Defense Technology,
Changsha 410073, China
hnjxzby@yahoo.com.cn
[2] Department of Computer Science, Hunan Public Security College,
Changsha 410138, China

Abstract. An intelligent detect system to recognition unknown computer virus is proposed. Using the method based on fuzzy pattern recognition algorithm, a malicious executable code detection network model is designed also. This model target at Win32 binary viruses on Intel IA32 architectures. It could detect known and unknown malicious code by analyzing their behavior. We gathered 423 benign and 209 malicious executable programs that are in the Windows Portable Executable (PE) format as dataset for experiment . After extracting the most relevant API calls as feature, the fuzzy pattern recognition algorithm to detect computer virus was evaluated.

1 Introduction

Malicious code is "any code added, changed, or removed from a software system to intentionally cause harm or subvert the system's intended function"[1]. Such software has been used to compromise computer systems, to destroy their information, and to render them useless.Excellent technology exists for detecting known malicious executables. Software for virus detection has been quite successful, and programs such as McAfee Virus Scan and Norton AntiVirus are ubiquitous. These programs search executable code for known patterns. One shortcoming of this method is that we must obtain a copy of a malicious program before extracting the pattern necessary for its detection.

Our efforts to address this problem have resulted in a fielded application, built using techniques from fuzzy pattern recognition and machine learning . The Malicious Executable Classification System currently detects unknown malicious executables code without removing any obfuscation. As far as know, our experiments is the first time to established methods based on fuzzy pattern recognition applying to detect malicious executables.

In the following sections, we describe related research in the area of malicious code detection. Then we illustrate the architecture of our detect model in section 3. Section 4 details the method of extraction feature from program, and stating the detect engine work procedure. Section 5 details the implementation and experiment results. We state our plans for future work in Section 6.

2 Related Work

There have been few attempts to use machine learning and data mining for the purpose of identifying new or unknown malicious code. In an early attempt, Lo et al. [2] conducted an analysis of several programs evidently by hand and identified tell-tale signs, which they subsequently used to filter new programs. Researchers at IBM's T.J.Watson Research Center have investigated neural networks for virus detection and have incorporated a similar approach for detecting boot-sector viruses into IBM's Anti-Virus software [3].

More recently, instead of focusing on boot-sector viruses, Schultz et al. [4] used data mining methods, such as naïve Bayes, to detect malicious code.

There are other methods of guarding against malicious code, such as object reconciliation, which involves comparing current files and directories to past copies. One can also compare cryptographic hashes. One can also audit running programs and statically analyze executables using pre-defined malicious patterns. These approaches are not based on data mining.

3 Model Structure

We first describe a general framework for detecting malicious executable code. Figure 1 illustrates the proposed architecture. The framework is divided into 3 part: Application Server, Detect Server, and Virus Detect Firewall based on character code scanning.

Before a file save to the application server, it will be scanned by the virus detect firewall. If the file is infected with virus then quarantine it. Otherwise if there is no malicious information about the file, it will be replicated 2 copies. Then, one copy will be sent to the application server, another one will be sent to the detect server based on Fuzzy Pattern Recognition (FPR) detect engine.

At the following stage, the file's features is extracted in the detect server. The detect server drives detect engine based on FPR check the copy again. According to the result from detect server, if the file is infected with unknown malicious code, the application server will be remind to remove the copy from its application database.

And then quarantine it in a special database or sent it to an expert to analyze it by hand.

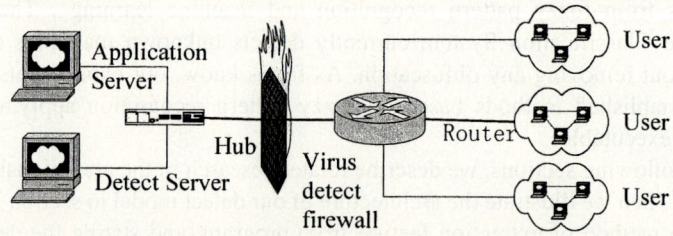

Fig. 1. Architecture

4 Malicious Code Detect Engine

4.1 Feature Extraction

Our first intuition into the problem was to extract information from the PE executables that would dictate its behavior. We choose the Windows API function calls as the main feature. Lots of API function calls by tracing the programs in the training set could be obtained. It is surely that each API calls play different role on detecting malicious code. When an API call often appears in the malicious codes but seldom in the benign codes, so it plays more important role in detection.

Here we use '*mean square deviation* 'as the main parameter to select API function calls as program's feature. The *mean square deviation* between classes computed as follow:

(1) Tracing each sample program in the training set to obtain it API calls sequence $A=\{A_1, A_2, ..., A_p\}, (1 \leq i \leq p)$, count each API function(A_i) frequency A_{ij}^V in every malicious executables V_j. And count its frequency A_{ij}^N in every benign executables N_j;

(2) Compute *average frequency* $E(A_i^V)$ and $E(A_i^N)$ of each API function(A_i) in malicious executables set and benign executables set as:

$$E(A_i^V) = \frac{1}{s}\sum_{j=1}^{s} A_{ij}^V , E(A_i^N) = \frac{1}{n}\sum_{j=1}^{n} A_{ij}^N \quad (1)$$

where s is the number of malicious executables, n is the number of benign executables.

(3) Compute total *mean frequency* $E(A_i)$ of each API function A_i as:

$$E(A_i) = \frac{E(A_i^V) + E(A_i^N)}{2} \quad (2)$$

(4) Compute *mean square deviation* $D(A_i)$ of each API function A_i as:

$$D(A_i) = \sqrt{(E(A_i) - E(A_i^V))^2 + (E(A_i) - E(A_i^N))^2} \quad (3)$$

At the last stage, we sorted the API function call sequence on $D(A_i)$, and choose the first t-th API function as the fuzzy *feature vector*. An example of feature vector shows in table 1.

Table 1. Feature Vector List Sample

№	Program's behavior	API Function Calls	DLLS reference
1	Search File	FindClose ;FindFirstFileA; FindNextFileA ;FindResourceA	KERNEL32.dll

4.2 Detection Algorithm

The result of detect a computer program is only "benign" or "malicious". We could get a set of feature from each sample file x, given C is the class set {benign, malicious}, C_1 denotes benign, C_2 denotes malicious. Our goal is to determine what class it is after the file's feature F was obtained.

In our method, a program file could be described by fuzzy set. Given $Q = \{q_1, q_2, ..., q_n\}$ is the domain of a fuzzy set

$$\tilde{M} = \{\mu_1/q_1, \mu_2/q_2, ..., \mu_n/q_n\} \tag{4}$$

Where n is the number of features, μ_i is a real number which value is between[0,1], μ_1/q is the degree of membership of the test file which has feature q_i. So the benign code and malicious code could be describe by fuzzy set on domain Q.

Given $E(A_i^V)$ is the mean frequency of a API function call in the malicious files, $E(A_i^N)$ is the mean frequency of a API function call in the benign files, The malicious file set $\tilde{V}$'s membership function create from the normal distribution of F distribution as:

$$\mu_V(E(A_i^V)) = \begin{cases} 0 & , E(A_i^V) < 0 \\ 1 - e^{-(E(A_i^V))^2/\sigma^2} & , E(A_i^V) \geq 0 \end{cases} \tag{5}$$

where $\sigma = \max\{E(A_1^V), E(A_2^V), ..., E(A_t^V)\}/3$, t is the number of features.
And the benign file set $\tilde{N}$'s membership function is:

$$\mu_N(E(A_i^N)) = \begin{cases} 0 & , E(A_i^N) < 0 \\ 1 - e^{-(E(A_i^N))^2/\sigma^2} & , E(A_i^N) \geq 0 \end{cases} \tag{6}$$

In the same way, the test file's membership function express as:

$$\mu_M(A_i) = \begin{cases} 0 & , A_i < 0 \\ 1 - e^{-(A_i)^2/\sigma^2} & , A_i \geq 0 \end{cases} \tag{7}$$

During the training step, we compute the frequency of all features over malicious code set. According to membership function $\mu_V(E(A_i^V))$, we get fuzzy set $\tilde{V}$ as:

$$\tilde{V} = \{\mu_1^V/A_1, \mu_2^V/A_2, ..., \mu_t^V/A_t\} \tag{8}$$

In the same way, we compute the frequency of all feature over benign code set. So we get fuzzy set $\tilde{N}$ as:

$$\tilde{N} = \{\mu_1^N/A_1, \mu_2^N/A_2, ..., \mu_t^N/A_t\} \tag{9}$$

For a test file M, by tracing its API function calls first, we could get the API call sequence. Then the frequency of all feature was computed too. Then $\{A_1, A_2, ..., A_t\}$ was get, where t=88 in our experiment.

According to membership function $\mu_M(A_i)$, we get fuzzy set $\tilde{M}$:

$$\tilde{M} = \{\mu_1/A_1, \mu_2/A_2, ..., \mu_t/A_t\} \tag{10}$$

On the Second step, the degree of similarity $\psi(\tilde{M},\tilde{V})$ between $\tilde{M}$ and $\tilde{V}$, $\psi(\tilde{M},\tilde{N})$ between $\tilde{M}$ and $\tilde{N}$ were computed as follow:

$$\psi(\tilde{A},\tilde{B}) = 1 - \frac{1}{\sqrt{t}}(\sum_{i=1}^{t}(\mu_i^A - \mu_i^B)^2)^{\frac{1}{2}} \tag{11}$$

Where $\psi(\tilde{A},\tilde{B})$ is Euclid degree of similarity.

At the last step, we can determine which class the test file is by Theorem 1.

Theorem 1: if $\exists i$ satisfy the follow equation: $\psi(\tilde{A}_i,\tilde{B}) = \bigvee_{1\leq j\leq n}\psi(\tilde{A}_j,\tilde{B})$, then classify $\hat{A}_i$ and $\tilde{B}$ in the same class. Where $\tilde{A}_i, \tilde{B}(i=1,2,...,n)$ is fuzzy sets, $\psi(\tilde{A},\tilde{B})$ is the Euclid degree of similarity between $\tilde{A}$ and $\tilde{B}$.

5 Experiment Results

We estimate our results over data set in table 2. The data set consisting of PE format executables was composed of 423 benign programs and 209 malicious executables. The malicious executables were downloaded from http://vx.netlux.org and http://www.cs.columbia.edu/ids/mef/ . The clean programs were gathered from a freshly installed Windows 2000 server machine. Each sample was labeled by a commercial virus scanner with the correct class label(malicious or benign) for our method. After verification of the data set the next step of our method was to extract features from the programs using API tracing tool-*APISPY.EXE* that we designed.

To evaluate our system we were interested in several quantities: (1). False Negative, the number of malicious executable examples classified as benign;(2). False Positives, the number of benign programs classified as malicious executables. We were interested in the detection rate of the classifier. In our case this was the percentage of the total malicious programs labeled malicious. We were also interested in the false positive rate. This was the percentage of benign programs which were labeled as malicious, also called false alarms. For the algorithms we plotted the detection rate using Receiver Operating Characteristic(ROC) curves. The ROC curves in Fig.2 show that our method had the lowest False Negative rate, 4.45%. Notice that the curve is down slowly when the number of samples increases. This is very fit to detect malicious code when the malicious sample obtained is difficult.

In another experiments[5], we had used a algorithm based on K Nearest Neighbor(KNN) to classify the data set in table 2. The result is shown in Fig.3.That algorithm had the lowest false positive rate, 4.8%. The Fuzzy Pattern Recognition algorithm(FPR) has better detection rates than the algorithm based on KNN. But the KNN algorithm occupies less compute resources than FPR. The trade-off between detect rate and system overhead must be think over in practical application.

Table 2. Dataset in experiment

	Sample space	Training set	Test Set
Benign Code	423	373	50
Malicious Code	209	159	50
sum	632	532	100

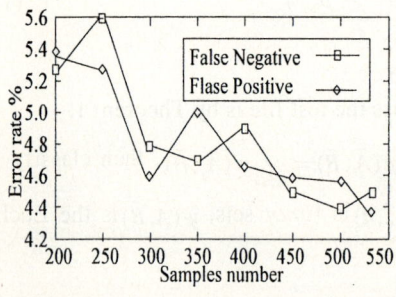

Fig. 2. Fuzzy Pattern Recognition ROC

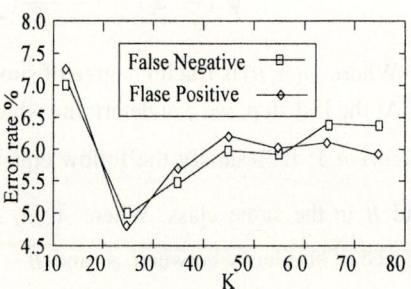

Fig. 3. K Nearest Neighbor ROC

6 Conclusion

We presented a method for detecting previously undetectable malicious executables. As our knowledge, this is the first time that using fuzzy pattern recognition algorithm to detect computer virus. However, the rate of error alert seems high in our experiment. So future work involves extending our learning algorithms to better utilize API call sequences and other feature of virus. We are planning to use Neural network to gain higher accuracy and detection rates. We also would like to implement the system on a network of computers to evaluate its performance in terms of time and space in real world environments. Finally, we are planning on testing this method over a larger set of malicious and benign executables.

References

1. McGraw,G., Morisett,G.: Attacking malicious code: A report to the Infosec Research Council. IEEE Software. 5(2000) 33-41
2. Lo,R., Levitt,K., Olsson,R.: MCF: A malicious code filter. Computers & Security.14 (1995)541-566
3. Tesauro,G., Kephart,J., Sorkin,G.: Neural networks for computer virus recognition. IEEE Expert. 11(1996)5-6
4. Schultz,M., Eskin,E., Zadok,E., Stolfo,S.: Data mining methods for detection of new malicious executables. In: Proceedings of the IEEE Symposium on Security and Privacy. IEEE Press, Los Alamitos, CA, (2001)38-49
5. ZHANG Boyun,YIN Jianping,ZHANG Dingxing,HAO Jingbo.:Unkown computer virus detection based on K-nearest neighbor algorithm. Computer Engineering and Applications. 6(2005)7-10

Method of Risk Discernment in Technological Innovation Based on Path Graph and Variable Weight Fuzzy Synthetic Evaluation

Yuan-sheng Huang, Jian-xun Qi, and Jun-hua Zhou

Department of Economy and Management,
North China Electric Power University, Baoding 071003, China
hys2656@yahoo.com.cn

Abstract. Risk in technological innovation is one of the important factors that hold enterprises from launching technological innovation. What cause the technological innovation risks is very complicated, and traditional methods of risk discernment can only draw general estimate on the risks. But enterprises need to understand the concrete links that cause technological innovation risks. For this reason, this paper puts forward a novel method of risk discernment in technological innovation, combining technological path graph with variable weight fuzzy evaluation, in order to clearerly, more accurately find the positions, in which technological innovation risks may take place, and evaluate the risks. Finally, the paper has verified the dependability of this method experimentally.

1 Introduction

Faced with intense market competition, Enterprises need to pay increasing attention to technological innovation.[1] However, technological innovation is a kind of highly risky exploration. Once the technological innovation fails, enterprises will suffer enormously. So, in consideration of the risk in technological innovation scientifically, rationally, to improve the success rate of technological innovation becomes the focus that academia and business circles pay close attention to. By far, people have proposed some methods to analyze the technological innovation risk, such as probability analysis, technological path analysis, scenario analysis etc., with the purpose of estimating overall technological innovation risk.[2][3][4]

As far as we know, technological innovation is, in essence, research activity proposed to realize technological function updating under the current technological system, with the introduction of the key elements of new technology, or seeking new combination mode of key elements. Because technological system involves many key elements and the relationship among the key elements is intricate, technological innovation is a kind of complicated exploration activity, with a lot of uncertain elements. It is these uncertain factors that cause all kinds of risks in technological innovation. So, in the final analysis, the technological innovation risk comes from the inadequacy of information such as information about factors involved in technological innovation and the interrelationship among these factors.[3]

Through the above analysis, this paper proposes to use fuzzy synthetic evaluation with variable weights, based on path graph, to analyze and discern technological innovation risk.

2 Analytic Hierarchy Process Based on Technological Path Graph

2.1 Principle of Technological Path Graph

The path graph, introduced in reference [2], was developed by the technology research institution in University of Illinois at the end of the 60s of the 20th century, and quickly used by a lot of enterprises. Technological path graph is structured according to the following information: (1) historical information of science and technology related; (2) tracing similar products and technology development among enterprises in the same line, the information of the important localization of enterprise and comparative materials for drafting technological path graph; (3) to evaluate the technological capability of enterprise correctly, and define the direction of the improving of technological capability accordingly. Hence, path graph simulated the main course of technological innovation to a certain extent.

The drawing of technological path graph can be summed up briefly as follows: (1) to determine the technology in path, which we pay close attention to; (2) to define key subsystems that contain technological problems in the complicated, multilinked technological innovation; (3) to identify important technology group that satisfying customers' demand; (4) to analyze technological capability of the enterprise, and then determine which technology make use of the enterprise's own resource, and what else need to be get from outside; (5) to draft technological path graph.

2.2 Analytic Hierarchy Process Based on Path Analysis

Vertical Analysis of Technological Path. To get the technological path graph, we need to vertically divide the R&D course of a technological innovation project into some stages at first, and so get the R&D stage drawing of the technological innovation project. It includes innovative technology derived from source technology, each R&D stage, and products promotion. Namely, intuitionisticly describe every stage that the technological innovation course must go through.

Usually, as for different kinds of technological innovations, the corresponding technological paths are also different. In order to facilitate our discussion, we set general technological innovations path based on common course of technological innovation. (In practice, we can set additional paths according to the technology project chosen.) The whole course includes putting forward the idea of innovation, survey and demonstration, research and development, middle experiment, commercialized production, and marketing. See Fig. 1.

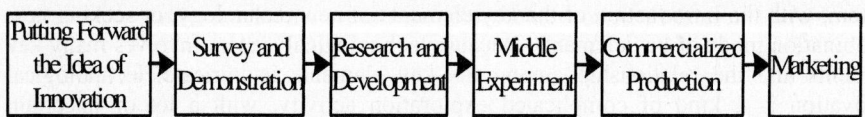

Fig. 1. General technological innovations path

Horizontal Analysis of Technological Path Graph. Vertical analysis of technological path describes all links, which technological innovation must go through, only along vertical direction. It is still not enough to show all information that technological innovation contains. So, we need decompose technological path graph horizontally for further analysis. Here, AHP is used to show and arrange the key factors that involved in technological innovation, still in order to analyze horizontally. Generally speaking, the content and the key factors involved in each stage are not the same in different technological innovation projects. To get the risk hierarchy, the following is what we do: obtain the detailed content and key factors for each stage after classifying all key elements, involved in technological innovation, according to general technological innovation. At last, two level indexes are gotten.

3 Fuzzy Synthetic Evaluation on the Risk Degree of Technological Innovation with Hierarchical Variable Weilghts

3.1 Fuzzy Analysis of the Risk Degree of Technological Innovation

Establishing the Factor Set. As discussed in the previous paragraphs, with technological path analysis and AHP, the key factors involved in technological innovation form a two-level index system in table 1. Of which, the first level of the factor set is $U = \{U_1, \cdots, U_6\}$, the second is $U_i = \{u_{i1}, \cdots, u_{ij}\}, i = 1,...,6$.

In addition, most second level indexes $\{u_{ij}\}$ in Tab. 1 are qualitative, and only a few are quantitative. It is obvious that even the quantitative indexes cannot be calculated accurately, so we transform quantitative indexes into qualitative indexes, such as setting value ranges in fuzzy ratings. There are also many other ways to deal with quantitative index in fussy analysis, such as making use of trigonometric function. As are common methods, they will be not dwelled on. Here, we regard all indexes $\{u_{ij}\}$ as qualitative indexes in this paper, and give emphasis on calculating state vectors of the qualitative indexes with fuzzy statistical method, and synthetic evaluation with variable weights.

Establishing the Comment Set. As figure 1 and table 1 show, the whole course of technological innovation is composed of a lot of links. And technological innovation risk exists in all these links. In order to reflect the risk degree of every link better, we choose the seven-grade comment set of risks. That is $V = \{v_1, v_2, \cdots, v_7\}$={very little, little, relatively little, average, relatively large, large, very large}, and its corresponding score set is $\{1,2,3,4,5,6,7\}$. The comment criteria of indexes can be seen in reference [3].

Calculating State Vectors of Level 2 Indexes. Fuzzy statistical method is adopted. The analysis is set as follows: considering the situation of this technological innovation project sufficiently, set fuzzy mapping $U_i \rightarrow F(V), i = 1,...,6$. That is, let expert panel members classify u_{ij} in the factor set U_i according to the above comment criteria, then we gather the result, count the frequency $W_{ij}^{(k)}$ that u_{ij} belongs to the rank $v_k (k = 1,2,3,4,5)$ in V, and so we can calculate:

$$r_{v_k}(u_{ij}) = W_{ij}^{(k)} / N \qquad (1)$$

Table 1. The index system of technological innovation risk

Level 1 index	U_i	Level 2 index	u_{ij}
Putting forward the idea of innovation	U_1	Enterprise's development goal	u_{11}
		Political, legal environment and their change	u_{12}
		Relevant regulations of responsible institution and departments concerned	u_{13}
		The development trend of science and technology and the impact on this enterprise	u_{14}
		The market of this trade and the development of technology	u_{15}
		The rival's technological innovation strategy	u_{16}
		Such resource states as talent, experiment and fund of enterprises, etc.	u_{17}
Survey and demonstration	U_2	Maturity of technology	u_{21}
		Advance of technology	u_{22}
		Difficulty degree of research and development and analysis of complexity	u_{23}
		Difficulty degree of middle experiment and analysis of its complexity	u_{24}
		The supply situations of raw materials and fittings	u_{25}
		The analysis of the equipment renewal	u_{26}
		The fund demand of R&D	u_{27}
		Consumer's dependence on the product and capacity of market	u_{28}
		The time analysis of the new technology being put into the market	u_{29}
		The management experience of enterprises and precautionary measures to adverse factors	u_{210}
Research and development	U_3	The knowledge and discipline involved in innovation	u_{31}
		The difficulty Degree of research on the innovation project	u_{32}
		Experimental facilities which research on the innovation project needs	u_{33}
		Talents which research on the innovation project needs	u_{34}
		The fund which research on the innovation project needs and research condition	u_{35}
		Time limit of research on the technological innovation project	u_{36}
Middle experiment	U_4	The poor condition of enterprise's technological equipment and middle experiment	u_{41}
		Insufficiency of enterprise's technology accumulated	u_{42}
		Poor ability of enterprise's technological personnel	u_{43}
		Poor technological cooperation relation of the enterprise	u_{44}
		Poor managerial ability of the enterprise	u_{45}
		The management experience of enterprises and precautionary measures to averse factors	u_{46}
Commercialized production	U_5	The adjustment that producing new products needs for the existing equipment and craft	u_{51}
		Poor correlativity between new product and the existing product	u_{52}
		The difficulty in introducing the craft, equipment and technology into the enterprise	u_{53}
		Extortionate cost of new products	u_{54}
		Low possibility of new products realizing the seriation and multi-specification	u_{55}
		The poor quality and performance of new products	u_{56}
		The demand of technological feature of raw materials or spare parts for new products	u_{57}
		The fund of upgrading the equipment and craft	u_{58}
Marketing	U_6	User's cognitive situation to new products	u_{61}
		The situation of new products utilizing enterprise's existing marketing channel	u_{62}
		Price analysis of new products	u_{63}
		Difficulty of promoting new products	u_{64}
		Poor prestige and popularity of the enterprise	u_{65}
		Poor ability of enterprise's advertisement and promoting	u_{66}
		Short life cycle of the new product	u_{67}
		The situation of the new product's rival and substitute	u_{68}
		Consumption habit's impact on new products	u_{69}
		New products' substitution and influence on the existing products of enterprise	u_{610}

Of which, $r_{v_k}(u_{ij})$ is the membership degree that u_{ij} belongs to v_k or the membership function.

Thus, the judgment vector R_{ij} of single factor u_{ij} is $\left(r_{v_1}(u_{ij}), r_{v_2}(u_{ij}), r_{v_3}(u_{ij}), r_{v_4}(u_{ij}), r_{v_5}(u_{ij})\right)$. Then the attribute value of u_{ij} is $c_{ij} = V \bullet R_{ij}^T$. Take U_1 as an example, after working out $c_{11}, \cdots, c_{17}$ with the above-mentioned method, and normalizing $C_1 = (c_{11}, \cdots, c_{17})$, we get the state vector of U_1, which is $X_1 = (x_{11}, \cdots, x_{17})$. As is the same to other factors, we can get $X_2 = (x_{21}, \cdots, x_{2,10}), \cdots, X_6 = (x_{61}, \cdots, x_{6,10})$.

3.2 Computing Constant Weights of Factor Sets and the Problem Existing in This Course

When computing constant weights of factor sets at various levels, we should try our best to answer the actual conditions. Usually, the methods are used, such as empirical data statistics, AHP, Delphi, and Sheffield. Because people carried out a great deal of empirical study on the risk of technological innovation in recent years, so a lot of statistics have been accumulated. Thus we sum up and analyze these experimental statistics (see Sect. 4.2), and then can get constant weights of the indexes at various levels.

Here, what we need to pay attention to is, in the course of actual appraisal on the risk of technological innovation projects, enterprises will present this kind of situation: When the specific key factors deviate from normal value greatly, enterprises should abandon this technological innovation project rationally. However, with the routine method of granting weight, the judgment of technological innovation can still continue arithmetically. This problem cannot be solved simply by increasing weight, so this paper proposes adopting variable weight fuzzy synthetic evaluation to overcome the problem.

3.3 Synthetic Evaluation on the Risk Degree of Technological Innovation with Hierarchical Variable Weights

Theory of Synthetic Evaluation with Hierarchical Variable Weights. Taking the two-level index system in Tab. 1 as an example, we briefly illustrate the thinking of hierarchical variable weight. The details can be found in references [6][7][8].

Known in Tab. 1, the factor set U equals $\{U_1, \cdots, U_6\}$. Select its subfactor set U_1 that equals $\{u_{11}, \cdots, u_{17}\}$, and then determine the state vector of U_1, which is $X_1 = \{x_{11}, \cdots, x_{17}\}$. Suppose its weight, $A_1 = (a_{11}, \cdots, a_{17})$, and the state variable weight vector, $S^{(1)}(x_{11}, \cdots, x_{17}) = (S_{11}(x_{11}, \cdots, x_{17}), \cdots, S_{17}(x_{11}, \cdots, x_{17}))$, have been given. Hereinto, the type of state variable weight vector can be divided into punishing type, encouraging type and mixed type, which corresponds to different variable weight model. Select the appropriate type of state variable weight vector $S^{(1)}(x_{11}, \cdots, x_{17})$, and then we get the variable weight model:

$$a'_{1j}(x_{11},\cdots,x_{17}) = \left. a_{1j}S_{1j}(x_{11},\cdots,x_{17}) \middle/ \sum_{k=1}^{7} a_{1k}S_{1k}(x_{11},\cdots,x_{17}) \right. \tag{2}$$

Accordingly, its synthetic evaluation is:

$$y_1 = \sum_{j=1}^{7} a'_{1j}(x_{11},\cdots,x_{17}) * x_{1j} \tag{3}$$

Analogously, $y_2,\cdots,y_6$ are separately worked out. Repeating to the above steps, we carry on variable weight synthesis to Y, which equals $(y_1,\cdots,y_6)$. In the end, the value of variable weight synthetic evaluation of U is gotten.

Selection of Variable Weight Model of the Risk Degree of Technological Innovation. We think that when choosing a technological innovation project, an enterprise always hopes it will be successful. So the risk expectation of the project should be as low as possible. That is to say, the enterprise is risk-averse at this moment. Thus, we should choose the model that equals to the encouraging type of variable weight in the course of synthetic evaluation on the risk degree, in order to enlarge the weights of the high-risk factors encouragingly, not to enlarge the weights of the low-risk factors punishingly. Namely, if a factor index is of high risk, its weight will be enlarged encouragingly, and finally the value of comprehensive appraisal will be reduced correspondingly. Of course, the greater the value is, the higher the innovation risk is. It reflects that, to the low-risk factor, the policymaker pays more close attention to the high-risk factor.

Variable Weight Synthetic Evaluation of the Risk Degree of Technological Innovation. Based on state vector $X_i = (x_{i1},\cdots,x_{ij})$, which is shown in Sect. 3.1, we choose the following variable weight model:

$$S_{ij}(X) = x_{ij}^{\alpha}, 0 < \alpha \leq 1 \tag{4}$$

Obviously, because x_{ij} is a converse index, when $0 < \alpha \leq 1$, $S_{ij}(X)$ equals to encouraging type of state variable weight vector. Take it into formula two, and then put the result into formula three, so the variable weight synthetic model of the risk degree of technological innovation can be gotten, which is used in this paper.

Generally, when less considering the balance problem of the factors, we can choose $\alpha > 1/2$; when relatively considering it, we choose $\alpha < 1/2$; when $\alpha = 1$, the model is constant weight type. Therefore, the smaller the coefficient α is in the model, the greater the encouraging effect on the high-risk factors is in technological innovation.

According to the above hierarchical variable weight theory, with two level variable weight synthesis, we obtain the final result of synthetic evaluation of the risk degree of technological innovation.

3.4 Reasoning Analysis Based on Risk Knowledge

Fuzzy evaluation, based on technological path graph and AHP, only analyze the risk in the course of technological innovation as a whole. But this is still not enough for the risk precaution of technological innovation. So we need use the links, illustrated in Fig.1, to carry on risk analysis. Firstly, when we decompose the technological innovation course, the links where potential risks exist must be found out; secondly, the reason that risks happen in these links must be anatomized, such as the shortage of talents, the insufficiency of the experiment condition, and the lack of fund. Moreover, we must find out the method to solve these problems that exist in the risk links, and calculate the cost of solving them. At last, the feasibility of the project has been established.

4 The Instance of Risk Discernment of Technological Innovation

A power plant's generating sets run more steadily in Shanxi, so its repair subsidiary's operation is under full capacity and the benefit decreases. In order to reverse the situation, the factory determines to develop new product. Through a great deal of analysis and discussion, they plan to use wear-resisting cast iron material with bainite structure to produce wear-resisting steel ball. First of all, the wear-resisting steel ball is consumed in a large amount in such industry as power industry and coal industry, and the market is enormous. Secondly, the composition of bainite wear-resisting material is simple, and the wear-resisting performance is very good, so the advancement and competitive advantage are obvious. Moreover, the enterprise has other advantages in such aspects as personnel, fund, equipment and raw material.

Wear-resisting performance of material with bainite structure comes from metallography of its surface. Getting this kind of metallography structure needs solving: (1) the proportions of more than ten kinds of elements such as iron, manganese, silicon and lanthanon; (2) adding order of every element, furnace temperature, and circle time; (3) the temperature of casting, quenching and tempering, and cooling rate. Thus, the key factors of the technology of this product lie in the 'value' of every craft parameter.

4.1 Determining Technological Innovation Route and Key Factors with Path Graph

The key factors, which influence technological innovation, can be deduced based on technological path division. Most factors are as follows: (1) The enterprise's technician need to master the principle of new technology quickly; (2) Grasp all parameters of bainite wear-resisting material through experiment; (3) Settle the problems generated in mass production.

4.2 Determining Constant Weights of Factors

In determining the weights of level 1 factors, we choose the data, in reference [3], that American, Japanese, Chinese scholars gain in the analysis of the unsuccessful technological innovation. The detailed steps are: (1) regard them as three groups of experts' grading data; (2) carry on normalization to the data; (3) decompose result; (4) calculate weighted average of the data; (5) finally, the distribution of the factors' weight is computed:

$$A = (0.090 \quad 0.120 \quad 0.313 \quad 0.120 \quad 0.080 \quad 0.278)$$

In determining the weights of level 2 factors, we adopt the data material in references [3] that believes technological innovation involving 58 factors. With the above similar process, the distribution of level 2 factors' weights is:

$$A_1 = (0.140 \quad 0.144 \quad 0.119 \quad 0.144 \quad 0.155 \quad 0.147 \quad 0.151)$$

Analogously, A_2, A_3, A_4, A_5, A_6 is obtained respectively. See column a_{ij} of table 3.

4.3 Determining State Vectors of Level 2 Indexes

According to the result of fuzzy judgment in column v_k of table 2, with the method in Sect. 3.3, the attribute value of level 2 indexes can be calculated. The corresponding state vectors are obtained after standardization. See column x_{ij} of table 2.

Table 2. Calculating state vectors of level 2 indexes

u_{ij}	v_1	v_2	v_3	v_4	v_5	v_6	v_7	x_{ij}	u_{ij}	v_1	v_2	v_3	v_4	v_5	v_6	v_7	x_{ij}
u_{11}	0.2	0.3	0.35	0.1	0.05	0	0	0.3571	u_{41}	0.15	0.2	0.2	0.15	0.15	0.15	0	0.4857
u_{12}	0.6	0.3	0.1	0	0	0	0	0.2143	u_{42}	0.15	0.15	0.2	0.2	0.15	0.1	0.05	0.5071
u_{13}	0.5	0.2	0.2	0.05	0.05	0	0	0.2786	u_{43}	0.5	0.3	0.1	0.1	0	0	0	0.2571
u_{14}	0	0.15	0.25	0.2	0.2	0.15	0.05	0.5857	u_{44}	0.4	0.3	0.2	0.1	0	0	0	0.2857
u_{15}	0.5	0.25	0.1	0.1	0.05	0	0	0.2786	u_{45}	0.3	0.3	0.2	0.1	0.05	0.05	0	0.3500
u_{16}	0.1	0.2	0.25	0.25	0.15	0.05	0	0.4714	u_{51}	0.1	0.2	0.3	0.15	0.1	0.1	0.05	0.4929
u_{17}	0.7	0.2	0.1	0	0	0	0	0.2000	u_{52}	0.2	0.2	0.3	0.2	0.1	0	0	0.4000
u_{21}	0.5	0.2	0.1	0.1	0.05	0.05	0	0.3071	u_{53}	0.7	0.2	0.1	0	0	0	0	0.2000
u_{22}	0.3	0.3	0.1	0.15	0.1	0.05	0	0.3714	u_{54}	0.4	0.2	0.15	0.15	0.05	0.05	0	0.3429
u_{23}	0.1	0.2	0.2	0.3	0.1	0.1	0	0.4286	u_{55}	0.6	0.3	0.1	0	0	0	0	0.2143
u_{24}	0.3	0.2	0.2	0.2	0.1	0	0	0.3714	u_{56}	0.2	0.2	0.3	0.2	0.1	0	0	0.4000
u_{25}	0.7	0.2	0.1	0	0	0	0	0.2000	u_{57}	0.6	0.3	0.1	0	0	0	0	0.2143
u_{26}	0.3	0.2	0.2	0.2	0.1	0	0	0.3714	u_{58}	0.4	0.2	0.15	0.1	0.1	0.05	0	0.3500
u_{27}	0.3	0.3	0.2	0.2	0	0	0	0.3286	u_{61}	0.4	0.3	0.2	0.1	0	0	0	0.2857
u_{28}	0.4	0.3	0.2	0.1	0	0	0	0.2857	u_{62}	0.15	0.15	0.3	0.3	0.05	0.05	0	0.4429
u_{29}	0.8	0.1	0.1	0	0	0	0	0.1857	u_{63}	0.3	0.2	0.2	0.2	0.1	0	0	0.3714
u_{210}	0.3	0.25	0.15	0.2	0.05	0.05	0	0.3714	u_{64}	0.5	0.3	0.1	0.1	0	0	0	0.2571
u_{31}	0.1	0.1	0.2	0.3	0.15	0.1	0.05	0.5429	u_{65}	0.3	0.2	0.2	0.1	0.1	0	0	0.3143
u_{32}	0.1	0.1	0.15	0.15	0.2	0.2	0.1	0.6071	u_{66}	0.1	0.2	0.3	0.2	0.1	0	0	0.3857
u_{33}	0.5	0.2	0.2	0.1	0	0	0	0.2714	u_{67}	0.7	0.25	0.05	0	0	0	0	0.1929
u_{34}	0.4	0.3	0.2	0.1	0	0	0	0.2857	u_{68}	0.1	0.2	0.3	0.25	0.1	0.05	0	0.4571
u_{35}	0.5	0.2	0.1	0.1	0.1	0	0	0.3000	u_{69}	0.2	0.2	0.3	0.2	0.1	0	0	0.4000
u_{36}	0.6	0.3	0.1	0	0	0	0	0.2143	u_{610}	0.6	0.3	0.1	0	0	0	0	0.2143

4.4 Evaluation on the Risk of Technological Innovation

Based on the state value of level 2 indexes and variable weight model in Sect. 3.3, that is $S_{ij}(X) = x_{ij}^{\alpha}$, and let $\alpha = 0.5$, we carry on appraisal with two level hierarchical variable weights. Here, the powerful function of matrix operation of Matlab6.1 is utilized to program the data. The result is shown in table 3.

In table 3, after gaining the state vector of level 1 indexes, $U^* = (U_1^*, U_2^*, \cdots, U_3^*)$, we change the weights of level 1 indexes with the same principle.

At last, after calculation, the final value of the synthetic evaluation of this enterprise's technological innovation risk is 0.3672.

According to the comment set in Sect. 3.1, we can calculate out the range of ideal risk degree is 0~0.2857, the range of preferable risk degree is 0.2857~0.4286, the range of considerable risk degree is 0.4286~0.5714, and the range of seriously dangerous risk degree is 0.5714 ~ 1. Obviously, the innovation project that the enterprise chooses is a preferable one. Considering the enterprise can remedy the factors influencing technological innovation to succeed, this project is an ideal innovation project.

Table 3. The process of hierarchical variable weight synthetic evaluation

U_i	u_{ij}	a_{ij}	x_{ij}	$S_{ij}(x)$	a_{ij}^*	U_i^*	U_i	u_{ij}	a_{ij}	x_{ij}	$S_{ij}(x)$	a_{ij}^*	U_i^*
U_1	u_{11}	0.140	0.3571	0.5976	0.1459	0.3662	U_4	u_{41}	0.199	0.4857	0.6969	0.2289	0.3882
	u_{12}	0.144	0.2143	0.4629	0.1162			u_{42}	0.190	0.5071	0.7121	0.2233	
	u_{13}	0.119	0.2786	0.5278	0.1095			u_{43}	0.219	0.2571	0.5071	0.1832	
	u_{14}	0.144	0.5857	0.7653	0.1921			u_{44}	0.192	0.2857	0.5345	0.1694	
	u_{15}	0.155	0.2786	0.5278	0.1462			u_{45}	0.200	0.3500	0.5916	0.1953	
	u_{16}	0.147	0.4714	0.6866	0.1760		U_5	u_{51}	0.114	0.4929	0.7021	0.1400	0.3478
	u_{17}	0.151	0.2000	0.4472	0.1177			u_{52}	0.132	0.4000	0.6325	0.1461	
U_2	u_{21}	0.105	0.3071	0.5542	0.1026	0.2987		u_{53}	0.110	0.2000	0.4472	0.0861	
	u_{22}	0.098	0.3714	0.6094	0.1053			u_{54}	0.141	0.3429	0.5856	0.1445	
	u_{23}	0.099	0.4286	0.6547	0.1143			u_{55}	0.109	0.2143	0.4629	0.0883	
	u_{24}	0.096	0.3714	0.6094	0.1032			u_{56}	0.151	0.4000	0.6325	0.1671	
	u_{25}	0.074	0.2000	0.4472	0.0584			u_{57}	0.109	0.2143	0.4629	0.0883	
	u_{26}	0.076	0.3714	0.6094	0.0817			u_{58}	0.135	0.3500	0.5916	0.1397	
	u_{27}	0.096	0.3286	0.5732	0.0970		U_6	u_{61}	0.108	0.2857	0.5345	0.1006	0.3468
	u_{28}	0.088	0.2857	0.5345	0.0829			u_{62}	0.107	0.4429	0.6655	0.1240	
	u_{29}	0.106	0.1857	0.4309	0.0805			u_{63}	0.092	0.3714	0.6094	0.0977	
	u_{210}	0.162	0.3714	0.6094	0.1741			u_{64}	0.100	0.2571	0.5071	0.0883	
U_3	u_{31}	0.170	0.5429	0.7368	0.2081	0.4036		u_{65}	0.110	0.3143	0.5606	0.1074	
	u_{32}	0.174	0.6071	0.7792	0.2252			u_{66}	0.100	0.3857	0.6210	0.1082	
	u_{33}	0.163	0.2714	0.5210	0.1411			u_{67}	0.094	0.1929	0.4392	0.0719	
	u_{34}	0.194	0.2857	0.5345	0.1722			u_{68}	0.103	0.4571	0.6761	0.1213	
	u_{35}	0.156	0.3000	0.5477	0.1419			u_{69}	0.101	0.4000	0.6325	0.1113	
	u_{36}	0.145	0.2143	0.4629	0.1115			u_{610}	0.086	0.2143	0.4629	0.0693	

4.5 Knowledge Deduction of the Risk Source

With the hierarchical analysis of factors in technological innovation, the company can quickly deduce the risks that may cause the technological innovation to fail. They can be generalized as follows: (1) As for the company, this project is a totally new one. So, there is a risk that whether the company can master the theories of the technology quickly and carry on correct experiments. (2) According to experimental statistics, whether the company can find out stable smelting formula and gain reasonable technological craft is another risk. (3) Whether the company can solve the various problems happening in mass production is also a risk.

To solve the above risks the company has taken the following precautionary measures: (1) Establish a special research team with a group of technological staff, and purchase essential experimental facilities; (2) Invite the inventor of the technology to explain the technological principle, and guide the experiment. After that, set down the draft scheme of the production, and carry on the experiment on the productive plan. (3) Analyze the difference between production equipment and experimental facilities, and take productive adjustment in terms of the experimental statistics; (4) Analyze the problem that continuous work causes molding sand properties changing, cooling process quickening, heat treatment prolonged, in turn alters the structure of some wear-resisting balls, and find ways to settle it; (5) Strengthen training to workers, make workers be familiar with and master the new craft better, and carry out the operational procedure of the craft correctly.

The actual conditions basically accord with the above analysis too. Though the project has run into some questions in the production link, the enterprise solves them quickly. On the whole, the innovative course is comparatively smooth. The success of researching and developing the wear-resisting steel ball with bainite structure brings good economic benefit and social benefit to the enterprise, and their products have passed ministerial level qualification, and also have gained the first prize of scientific and technological progress. Their Products enter the market with powerful competition advantage rapidly; even cause the market in short supply for a period of time. The enterprise has regained all investment less than one year.

References

1. Fu Jia-yi: Technological Innovation. Publishing House of Tsing-Hua University (2003)
2. Chen Jin, Song Jian-yuan: Unscrambling Research & Development. Publishing House of Mechanical Industry (2002)
3. Xie Ke-yuan: Risk management of technological innovation. Hebei: Publishing House of Science & Technology (2002)
4. Ma Li, Chen Xue-zhong, Yuan Xue-mei: Multifactor Hierarchical Fuzzy Synthetic Evaluation on Investment Risk of High-tech Industry. Research of Quantitative Economics & Technological Economics, Vol. 18, No 7, 2001
5. Liu Wen-qi: Balanced Function and Its Application for Variable Weight Synthesizing. Symstems Engineering-Theory & Practice, Vol. 17, No 4, 1997
6. Li De-qing, Cui Hong-mei, Li Hong-xing: Multifactor decision making based hierarchical variable weights. Journal of Systems Engineering, Vol. 19,No 3, 2004
7. Li De-qing, Li Hong-xing: The properties and construction of state variable weight vectors, Journal of Beijing Normal University, Vol. 38, No 4, 2002
8. Li Hong-xing: Factor spaces and mathematical frame of knowledge representation. Journal of Systems Engineering, Vol. 13, No 1, 1998

Application of Fuzzy Similarity to Prediction of Epileptic Seizures Using EEG Signals

Xiaoli Li and Xin Yao

The Centre of Excellence for Research in Computational Intelligence and Applications,
School of Computer Science, The University of Birmingham, Edgbaston,
Birmingham, B15 2TT, UK
{x.li, x.yao}@cs.bham.ac.uk

Abstract. The prediction of epileptic seizures is a very attractive issue for all patients suffering from epilepsy in EEG (electroencephalograph) signals. It can assist to develop an intervention system to control / prevent upcoming seizures and change the current treatment method of epilepsy. This paper describes a new method based on wavelet transform and fuzzy similarity measurement to predict the seizures by using EEG signals. One part of the method is to calculate the energy and entropy of EEG data at the different scale; another part of this method is to calculate the similarity between the features set of the reference segment and the test segment using fuzzy measure. The test results of real rats show this method detect temporal dynamic changes prior to a seizure in real time.

1 Introduction

There are millions of people with epilepsy throughout world according to an estimate of the World Health Organization [1]. Unfortunately, at least 20% of all epileptic patients still keep seizure – free due to the inefficacy of pharmacological treatment. Therefore, many researchers attempted to develop a new method to prevent the epileptic seizures, such as chronic vagal nerve stimulation (VNS) [2] and brief burst of pulse stimulation [3,4]. An alternative method is to design an intervention system to inject a drug, or have a drug released once the seizure is coming [5]. The application of these methods basically dependents on the prediction of the seizures, in particular for the intervention system of a drug. Furthermore, the prediction of seizures enables patients to take as little medicine as possible, who only take it under necessary conditions rather than constantly. A warning system of the seizure prediction also may enable patients take a drug in advance to prevent from the upcoming seizures or leave some of places where is dangerous for him/her as soon as possible such as a swimming pool, a business street. The role of seizure prediction is summarized in Fig. 1

So far, a number of characterizing measures derived from nonlinear signal processing are capable of extracting information from EEG signals to predict the seizures [6-7]. Often, this non-linear method encounters two parameters (delay time and embedding dimension) determination and computation cost problems [8-10]; so they are not always applied to real EEG data, especially when the noise contribution is significant. An intelligent system has been proposed to predict the seizures [11], which is based on a finding that wavelet – derived energy increase before electrical and clinical seizure onset. Although intelligent techniques can learn to distinguish the pre-seizure state from the normal states, unfortunately the results from the learning methods cannot be explained in more details. From an application point of view, developing a simple and reliable prediction method is still necessary.

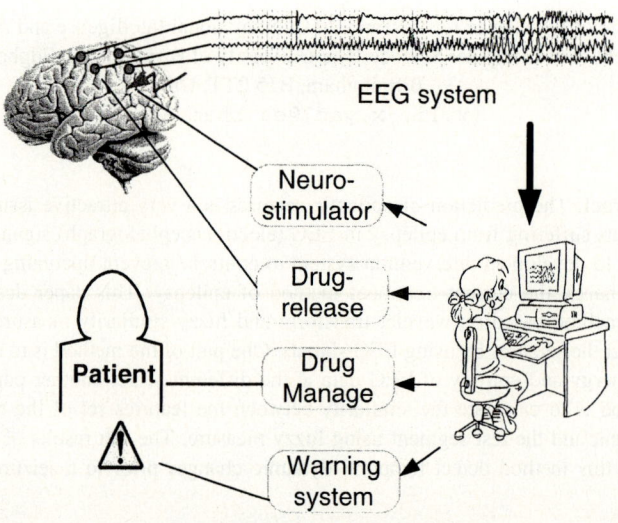

Fig.1. Block diagram of application of seizure prediction. The application of these systems is based on the seizure prediction. For a neuro-stimulator, the seizure prediction may reduce the strength of stimulation. For other three functions, they need as long prediction time as possible.

This paper proposes a new method by combining wavelet transform and fuzzy similarity measurement to predict the epileptic seizures in EEG. The first step of this method is to obtain the features from wavelet space of an EEG data, including energy and entropy of wavelet coefficients in different frequency bands. Second step is to calculate fuzzy similarity index of feature sets between the reference segment and the test segment, which integrates the energy and entropy of field potentials in different frequency bands. Finally, long-term EEG recordings of real rats are applied to test this novel prediction method. The results indicate the prediction of the epileptic seizures based on the wavelet - fuzzy similarity index could be applied in an intervention system.

2 Method

Often, there are five broad spectral bands of EEG signal from the clinical interest: delta (0.5-4 Hz), theta (4-8 Hz), alpha (8-12 Hz), beta (12-30Hz), and gamma waves (30-above Hz). Above five frequency bands can be extracted by using a discrete wavelet transform. The discrete wavelet decomposition of an EEG signal $x(t)$ is written as follows:

$$x(t) = \sum_{k=-\infty}^{+\infty} C_J(k)\phi(2^{-J}t-k) + \sum_{j=1}^{J}\sum_{k=-\infty}^{+\infty} d_j(k)\psi(2^{-j}t-k) \quad (1)$$

where $d_j(k)$ are called wavelet coefficients at the level j, which can be calculated by a fast recursive scheme [12]. In this study, we only consider the energy and entropy of wavelet coefficients at each level ($j=1,\ldots,5$) to describe the brain state:

(1) Energy features: the energy at each resolutions level $j=1,\ldots,5$ based on the wavelet coefficients for an segments with a sliding window (K is the length of wavelet coefficients at each resolution level j) with index i is written as:

$$(BE_j)^i = (\sum_{k=1}^{K} d_j^2(k))^i \quad (2)$$

(2) Entropy features: the entropy at each resolutions level $j=1,\ldots,5$ based on the wavelet coefficients for an segments with a sliding window (K is the length of wavelet coefficients at each resolution level j) with index i is given by

$$(H_j)^i = (-\sum_{k=1}^{K} p_{k,j} \ln(p_{k,j}))^i \quad (3)$$

where $p_{k,j}$ is the probability density of wavelet coefficients at each resolution level $j=1,\ldots,5$, which can be obtained by using a histogram method. Based on our experiences in [8], a biorthogonal wavelet is chosen in this study.

Then, a standard Gaussian membership function is used to describe five energy features (BE_j) and five entropy features (H_j) calculated ($j=1, 2, \ldots,5$), as shown in Fig. 2. The symmetric Gaussian function depends on two parameter σ and c as given by $f(x;\sigma,c) = \exp(-(x-c)^2/(2\sigma^2))$. The parameters σ and c are determined by mean and deviation of each feature BE_j or H_j. Six membership functions is applied to describe the fuzzy character of each feature. Applying the fuzziness process for each feature, a fuzzy sets A can be obtained from the energy and entropy feature sets.

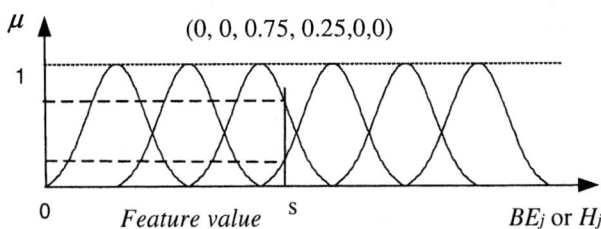

Fig. 2. Fuzzy membership functions of the features BE_j or H_j. The horizontal axis is the feature value; on the other hand the vertical axis represent the fuzzy degree that ranges from 0 to 1. The feature value s in this figure can be represented as a fuzzy set of (0,0,0.75,0.25,0,0).

Suppose two fuzzy sets A (present state) and B (reference/normal state), and each set contains N features $x_1, x_2, \ldots, x_N$, a simple and reliable method proposed by [13,14] can be used to calculate the similarity between the two fuzzy sets, A and B, the formula is

$$S(A,B) = \frac{\sum_{i=1}^{N}(1-|\mu_A(x_i)-\mu_B(x_i)|)}{N} \tag{4}$$

where, $1-|\mu_A(x_i)-\mu_B(x_i)|$ is regarded as the similarity degree of fuzzy set A and B on the features, x_i. The $S(A,B)$ is the average of the similarity degree of fuzzy set A and B, called fuzzy similarity index. The range of $S(A,B)$ is from 0 to 1, which corresponds with the different similarity degree. $S(A,B)=1$ means the two fuzzy sets is identical, otherwise there exist a difference between the two fuzzy sets.

Main purpose of this study is to design an automated algorithm to predict the seizures by means of the fuzzy similarity index S. In order to obtain a baseline for the preictal state, firstly the mean value M and standard deviation SD of S_t ($t=1,2…,M$) are calculated, which are the fuzzy similarity indexes of all interictal recordings of a subject. A local drop can be characterized by two independent parameters: its depth and duration, for any given baselines. The depth of a drop can be estimated in units of the standard deviation of the baseline epoch, whereas its duration can be quantified by the time during which the mean value of a profile drops below a certain threshold. In this study, a backward moving-average filter of width w is employed to smooth the time profiles of S_t and declared a preictal state if the smoothed profiles S_t^w dropped below the interictal mean M by more than k standard deviations SD:

$$PS = \begin{cases} 1 & \text{if } S_t^w < M-k*SD; \\ 0 & \text{otherwise} \end{cases} \tag{5}$$

where $PS=1$ stands for a preictal state. Two parameters k and w govern the mean depth of a drop over a certain time. In order to determine suitable values for the two parameters, the sensitivity and specificity rate of preictal state detection should be considered. The depth parameter k varied from 0 to 6 interictal standard deviations while the duration parameter w varied from 0 to 10 min in this study.

All of procedure of the method is summarized in Fig. 3. The procedure contains a discrete wavelet transform, feature extraction, fuzziness, reference state setting, threshold settings, fuzzy similarity index calculation and indication of the preictal state.

3 EEG Recordings

In this paper, an epilepsy model with bicuculline is applied. 14 Sprague-/Dawley rats are tested. Prior to surgery, the rats are anesthetized with an i.p. injection of Nembutal (sodium pentobarbital, 65 mg/kg body weight). One electrode is placed on the skull with dental acrylic. During the experiments, every precaution is taken to minimize suffering to the animals. Rat is initially anesthetized with pentobarbital (60 mg/kg, i.p.) while body temperature is maintained (36.5–37.5°C) with a piece of blanket. The degree of anesthesia is assessed by continuously monitoring the EEG, and additional doses of anesthetic are administered at the slightest change toward an awake pattern (i.e., an increase in the frequency and reduction in the amplitude of the EEG waves). Then, bicuculline i.p. injection is used to induce the rat epileptic seizures. EEG signals are recorded using an amplifier with band-pass filter setting 0.5-100 Hz. Seizure onset times are determined by visual identification of a clear EEG seizure discharge. The detail of the recording of EEG data and the determination of epileptic onset time can be found in [9-10].

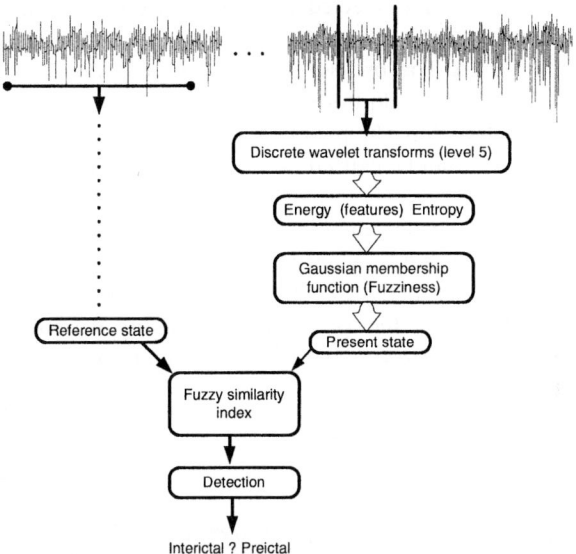

Fig. 3. Illustration of complete procedure of the prediction method for the epileptic seizures. The method is composed of two parts. (Left) one is to build the reference states from the interictal, which is so far from ictal. Another is to calculate the features of current state and the fuzzy similarity index with a reference state, then to identify the current state whether or not is a preictal state by using the detection method.

4 Results

The EEG recording of rats are analyzed using a moving – window technique, which are divided into segments of 1000 sampling points each, corresponding to a window length of 5 s at the given sampling rate (200 Hz), and the window overlapped by 50% so the distance in time between the starting points of two consecutive windows is 2.5 s. Prior to the calculation of the fuzzy similarity, two steps of data processing are carried out for each data window: (1) discrete wavelet decomposition; and (2) calculation of energy and entropy of wavelet coefficients. The prediction method is carried out for all 14 animal tests. We only show one typical example in this paper.

Fig. 4 shows the application result of the proposed method to long-term EEG recordings covering 20 minutes. At 6:41(minute:second), the bicuculline is injected to induce the epileptic seizures. The beginning of epileptic seizure occurs at

12:27(minute:second). The first three minutes EEG recordings are treated as a reference state. Hopefully the reference segment should include the most of salient features of the interictal EEG signal. Then, the fuzzy similarity index with the reference state is obtained by moving window over the entire EEG recordings, as shown in Fig. 4 (bottom) after smoothed. The three vertical lines are the injection time, the preictal start time and the seizure start time from left to right, the horizontal long line is threshold, where the k and d are selected as 3.5 and 2 according to the testing of some EEG signals. The plot reveals that the fuzzy similarity index gradually decreases during the preictal period. The preictal state is detected when the fuzzy similarity index is less than the threshold. In this test, the beginning point of preictal state is at 10:15(minute:second), therefore the preictal duration is 102 seconds. The quantification of 20 min recordings can be carried out in less than 30 s using Matlab on a P6 2 G personal computer.

Fig. 5 summarizes the seizure anticipation time of 14 animals. The results confirm that the method proposed can predict the seizures in most of the animals (13/14) several minutes (seconds) in advance (mean 109.6 s). It is seen that the various anticipation times for the 14 animals are highly variable in the duration. These results suggest that seizure emergence is a complex, non-repetitive process even for similar animals.

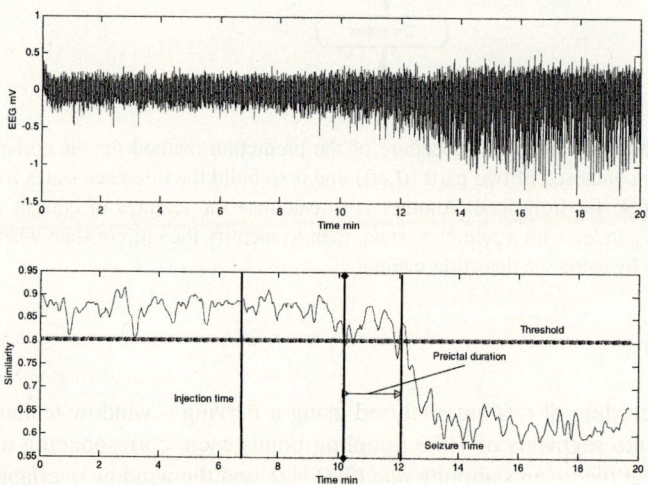

Fig. 4. Top: EEG recordings; Bottom: Preictal state detection. At 6:41, the bicuculline is injected. The epileptic seizure occurs at 12:07. The beginning point of the preictal state is at 10:15. The preictal duration is about 1:52.

5 Conclusions

A prediction method of epileptic seizure should meet following requirements [5]: (a) the method is able to run real-time robustly and on-line in a clinical setting with

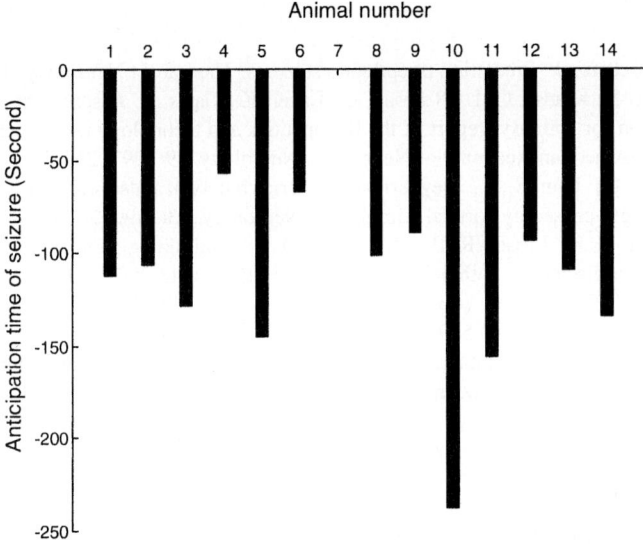

Fig. 5. The anticipation time of 14 animals by using the fuzzy similarity index. The mean time of the seizure anticipation is 109.6 s. The method gave a false prediction of seizure for animal 7.

noised EEG; (b) it possesses very good sensitivity and specificity in long-term continuous recordings across animals or patients; (c) it can provide a long-term warning before each seizure and an accurate prediction of the occurrence time of the impending seizure at the time of the warning.

The advantages of the method in this paper are summarized. (i) Features extractions with wavelet transform. Wavelet transform are very meaningful because the hidden information of EEG can be revealed; the noise effort can also be reduced because some data under some scales is omitted. (ii) Significant advantage of this method is able to check directly the fuzzy similarity index between two windows, rather than differences between quantities evaluated separately. Thus, the threshold for the detection of preictal is not very hard to determine by the fuzzy similarity index of each objective, this method is not influenced by the biological variability. (iii) The computational cost of the new method relatively low and can be used in real time for clinical application. (iv) The method does not finally require specific prior knowledge for each objective. Therefore, this novel method could be applied a practical intervention system.

Acknowledgements

We are grateful to the support of Wellcome Trust and AMD UK. We also thank John Jefferys for his helpful comments.

References

1. Litt, B., Echauz, J.: Prediction of epileptic seizures. The Lancet Neurology. 1(2002) 22-30
2. Fisher, R. S., Krauss, G. L., Ramsay, E., Laxer, K., Gates, J.: Assessment of vagus nerve stimulation for epilepsy: report of the therapeutics and technology assessment subcommittee of the American Academy of Neurology. Neurology. 49(1997) 293–297
3. Lesser, R. P., Kim, S. H., Beyderman, L.: Brief bursts of pulse stimulation terminate afterdischarges caused by cortical stimulation. Neurology. 53(1999) 2073–2081
4. Motamedi, G. K., Lesser, R. P., Miglioretti, D. L.: Optimizing parameters for terminating cortical afterdischarges with pulse stimulation. Epilepsia. 43(2002) 836–846.
5. Iasemidis, L.D.: Epileptic seizure prediction and control. IEEE Transactions on Biomedical Engineering. 50(2003) 549-558
6. Lehnertz, K.: Non-linear time series analysis of intracranial EEG recordings in patients with epilepsy - an overview. International Journal of Psychophysiology. 34(1999) 45-52
7. Winterhalder, M., Maiwald, T., Voss, H.U., Aschenbrenner-Scheibe, R., Timmer, J., Schulze-Bonhage, A.: The seizure prediction characteristic: A general framework to assess and compare seizure prediction methods. Epilepsy and Behavior. 4(2003) 318-325
8. Li, X., Guan, X., Du, R.: Using damping time for epileptic seizures detection in EEG. Modelling and Control in Biomedical Systems (Edited by David Dagan Feng and Ewart Carson), Elsevier Ltd, (2003) 255-258
9. Li, X., Kapiris, P.G., Polygiannakis, J., Eftaxias, K.A., Yao, X.: Fractal spectral analysis of pre-epileptic seizures phase: in terms of criticality. Journal of Neural Engineering, **2** (2005) 11-16
10. Li, X. Ouyang, G., Yao, X., Guan, X.: Dynamical Characteristics of Pre-epileptic Seizures in Rats with Recurrence Quantification Analysis, Physics Letters A. 333 (2004) 164–171
11. Geva, A.B., Kerem, D.H.: Forecasting generalized epileptic seizures from the EEG signal by wavelet analysis and dynamic unsupervised fuzzy clustering. IEEE Transactions on Biomedical Engineering. 45(1998) 1205–1216
12. Meyer, Y.: Wavelets, Applications and Algorithms, Siam, 1993
13. Hyung, L. K., Song, Y. S., Lee, K. M.: Similarity measure between fuzzy sets and between elements. Fuzzy Sets and Systems. 62(1994) 291-293
14. Wang, W. J.: New similarity measures on fuzzy sets and on elements. Fuzzy Sets and Systems. 85 (1997) 305-309

A Fuzzy Multicriteria Analysis Approach to the Optimal Use of Reserved Land for Agriculture

Hepu Deng[1] and Guifang Yang[2]

[1] School of Business Information Technology, RMIT University,
GPO Box 2476V, Melbourne, Victoria, 3001, Australia
hepu.deng@rmit.edu.au
[2] Department of Geography, East China Normal University,
Shanghai, 200062, China
yangguifang@126.com

Abstract. This paper presents a multicriteria analysis (MA) approach to solve the problem of the optimal use of reserved land for agriculture involving multiple criteria and subjective assessments. Linguistic terms approximated by fuzzy numbers are used to adequately model the subjectiveness and imprecision of the decision making process. The degree of similarity between fuzzy numbers is used to calculate the overall performance index for each alternative across all criteria based on the concept of fuzzy ideal solution. As a result, the unreliable and often computationally demanding process of comparing fuzzy utilities usually required in fuzzy MA is avoided, and effective decisions can be made. A case study in Shanghai, China is presented that shows the fuzzy MA approach developed is efficient in computation, simple and comprehensible in concept, and practical in solving this kind of problems.

1 Introduction

Effective decision making for the optimal use of reserved land for agriculture is of the utmost importance to the wellbeing of individuals in the society and the sustainable development of every country [2], [8], [10], [31]. This is mainly due to (a) the increasing population of the world, (b) the depletion of the valuable land resource, and (c) the decreasing quality of human habitat [18], [23].

Making effective decisions for the optimal use of reserved land involves in a systematic assessment of available lands [6]. This process often characterizes with (a) availability of alternative lands, (b) existence of numerous criteria, and (c) presence of subjective and imprecise assessments [11], [19]. To effectively solve this problem, it is desirable to have an efficient approach capable of properly handling the multi-dimensional nature of the problem and adequately modeling the subjectiveness and imprecision of decision making process.

Much research has been done for tackling this problem, leading to the development of various methodologies and their applications [24], [26], [27], [28]. For example, a classification-based approach is developed allowing the decision maker (DM) to incorporate their subjective preferences into the decision process [1], [9], [22]. This approach is however not satisfactory. An integrated index approach is proposed capable of encompassing the subjective assessments on criteria importance in a numerical scale in a given evaluation process. This approach is often criticized due to its inability to adequately tackle the uncertainty and vagueness [13], [17].

This paper presents a fuzzy multicriteria analysis (MA) approach for effectively solving the problem of the optimal use of reserved land for agriculture. The multidimensional nature of the problem is adequately handled using the MA methodology, and the subjectiveness and imprecision of the decision making process are properly modeled with fuzzy set theory. As a result, effective decisions can be made. An empirical study at Shanghai, China is presented that shows the proposed approach is effective and efficient for solving this kind of problems in real world settings.

2 The Optimal Use of Reserved Land for Agriculture in Shanghai

Shanghai is one of the largest cities in China with the population of more than 15 millions. With its temperate climate, broad and level topography, and fertile soil, Shanghai plays an important role in the modern Chinese economy [16]. In the recent several decades, Shanghai has witnessed the rapid development in impetuous urbanization, thus invariably challenged by the prominent man-land problem. In particular, the complicated natural factors and intensifying anthropogenic activities have largely aggravated the pressure on already intense agrarian resources [20], [21].

To mitigate the contradiction between limited land and rapid population expansion, local government has to regularly issue measures for better collocating agricultural land and keeping the dynamic equilibrium of available agricultural lands. In such cases, the reserved lands referring to the unexploited lands in terms of special cultivatable or reclaimable units [5], [25] have to be evaluated for their optimal use.

To effectively prioritize the alternative lands for agriculture, an overall evaluation of the available lands with respect to various criteria regarding their appropriateness for use is required. After comprehensive consultations with various stakeholders, Environment Suitability (C_1), Economic Feasibility (C_2), and Social Rationality (C_3) are identified as the three main criteria for the evaluation process. Fig. 1 shows the hierarchical structure of the evaluation problem.

The Environmental Suitability (C_1) reflects the concern of the DMs on the environmental characteristics and benefit. It consists of soil quality (C_{11}), groundwater property (C_{12}), surface water status (C_{13}), biological diversity (C_{14}), hydrodynamics condition (C_{15}), irrigation guarantee (C_{16}), water installations (C_{17}), and environmental benefit due to the exploitation (C_{18}).

The Economic Feasibility (C_2) is an indication of the extent to which the potential reserved land should be reclaimed in term of economic cost (C_{21}), and anticipative economic benefit (C_{22}). The economic cost is determined by the cost concerning land-enclosure, artificial sedimentation and reclamation, soil melioration, hydrological improvement, transportation construction, and hazard-proof works. The anticipative benefit principally refers to the economic income generated from exploitation.

The Social Rationality (C_3) shows the concerns of the DMs with the impacts of the land reclamation on the society. This is related to social disturbance resulting from land exploitation (C_{31}) and subsequent social yields (C_{32}). Usually the social disturbance reflects the land reclamation effects on the immigration, social structure. The social benefit is of significance in occupational opportunity, social stabilization.

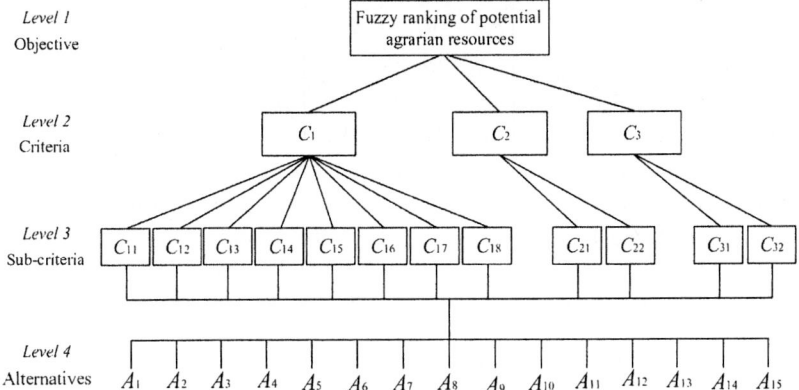

Fig. 1. The hierarchical structure of the evaluation problem

3 The Fuzzy MA Approach

Multicriteria analysis is widely used to assist DMs in prioritizing or selecting one or more alternatives from a finite set of available ones with respect to multiple, usually conflicting criteria [4], [7], [29]. With the characteristics of the problem of the optimal use of reserved land for agriculture, MA is well suited for evaluating the overall performance of the available lands on their optimal use in a specific situation with respect to the various interests of stakeholders in the decision making process.

Uncertainty in terms of subjectiveness and imprecision is often associated with human decision making [4], [12]. Methods for handling this uncertainty are developed basically along the lines of probability theory or fuzzy set theory. The former focuses on the stochastic nature of the decision making process, while the latter is concerned with the subjectiveness and imprecision of human judgments [4], [29]. It is well recognized that stochastic methods such as statistical analysis cannot adequately handle the subjectiveness and imprecision in the decision making process [4].

The application of fuzzy set theory to MA provides an effective means of formulating decision problems in a fuzzy environment, where the information available is subjective and imprecise [4], [12], [15], [29]. The DM's subjective assessments can be better handled by fuzzy numbers [4], [29]. Numerous approaches have been developed for solving problems involving subjectiveness and imprecision [15], [29]. These approaches, however, are not totally satisfactory due to the complicated process of comparing fuzzy utilities usually required [4]. To avoid these shortcomings, we present a fuzzy MA approach here to solving the problem. Several concepts, including (a) fuzzy set theory [4], (b) hierarchical analysis [7], (c) ideal solutions [15], [29], and (d) fuzzy similarity, are synthesized believing that effective decision making can be resulted from the fusion of these concepts by taking the advantages of individual concepts [4].

The general MA problem usually consists of a number of alternatives $A_i (i = 1, 2, ..., n)$ to be evaluated against a set of criteria C_j ($j = 1, 2, ..., m$). Subjective assessments are often required for determining the performance of each alternative A_i ($i = 1, 2, ..., n$) with respect to each criterion C_j ($j = 1, 2, ..., m$), denoted as x_{ij}, and the relative importance of the each criterion, represented as w_j, with respect to the overall objective of the problem. As a result, a decision matrix for the MA problem can be obtained as

$$X = \begin{bmatrix} x_{11} & x_{12} & \cdots & x_{1m} \\ x_{21} & x_{22} & \cdots & x_{2m} \\ \cdots & & \cdots & \\ x_{n1} & x_{n2} & \cdots & x_{nm} \end{bmatrix} \quad (1)$$

To facilitate the making of subjective assessments, linguistic terms defined as in Table 1 can be used. For computational simplicity, fuzzy numbers represented as (a_i, b_i, c_i) defined as in Table 1, where $1 \leq a_i \leq b_i \leq c_i \leq 9$ are used to approximate these linguistic terms [14], [29], in which b_i is the most possible value of a linguistic term, and a_i and c_i are the lower and upper bounds respectively [7], [15], [29].

Table 1. Linguistic terms used by the decision matrix

Linguistic Variables	Very Low (VL)	Low (L)	Fair (F)	High (H)	Very High (VH)
Fuzzy Numbers	(1, 1, 3)	(1, 3, 5)	(3, 5, 7)	(5, 7, 9)	(7, 9, 9)

If sub-criteria C_{jk} (k = 1, 2, ..., p_j) are existent for criterion C_j, a lower-level decision matrix can be determined for all the alternative lands, given as in (2) where y_{ik} are the DM's subjective assessments of the performance rating of alternative A_i with respect to sub-criteria C_{jk} of the criterion C_j (j = 1, 2, ..., m; k = 1, 2, ..., p_j).

$$Y_{C_j} = \begin{bmatrix} y_{11} & y_{21} & \cdots & y_{n1} \\ y_{12} & y_{22} & \cdots & y_{n2} \\ \cdots & \cdots & \cdots & \cdots \\ y_{1p_j} & y_{2p_j} & \cdots & y_{np_j} \end{bmatrix} \quad (2)$$

To reduce the cognitive demand on the DM for criteria and sub-criteria weighting, linguistic terms approximated by triangular fuzzy numbers defined as in Table 2 can be used for describing the relative importance of the criteria and sub-criteria with respect to the overall objective of the over evaluation problem.

Table 2. Linguistic terms used by the weighting vectors

Linguistic Variables	Least Important	Less Important	Important	More Important	Most Important
Fuzzy Numbers	(1, 1, 3)	(1, 3, 5)	(3, 5, 7)	(5, 7, 9)	(7, 9, 9)

As a result, the weighting vectors W and W_j (j = 1, 2, ..., m) for the criteria and their associated sub-criteria respectively can be represented as (3) and (4) where w_j and w_{jk} are the fuzzy weights of criteria C_j and sub-criteria C_{jk} (j = 1, 2, ..., m; k = 1, 2, ..., p_j).

$$W = (w_1, w_2, ..., w_j, ..., w_m) \quad (3)$$

$$W_j = (w_{j1}, w_{j2}, ..., w_{jk}, ..., w_{jp_j}) \quad (4)$$

The fuzzy MA approach starts at the generation of a weighted fuzzy performance matrix representing the overall performance of alternative with respect to every criterion. The weighted fuzzy performance matrix is the multiplication of the criteria weightings in (3) by the decision matrix in (1) based on interval arithmetic, given as follows:

$$WX = \begin{bmatrix} w_1 x_{11} & w_2 x_{12} & \cdots & w_m x_{1m} \\ w_1 x_{21} & w_2 x_{22} & \cdots & w_m x_{2m} \\ \cdots & \cdots & \cdots & \cdots \\ w_1 x_{n1} & w_2 x_{n2} & \cdots & w_m x_{nm} \end{bmatrix} \quad (5)$$

If criterion C_j consists of sub-criteria C_{jk}, the decision vector $(x_{1j}, x_{2j}, \ldots, x_{nj})$ across all the alternatives with respect to criteria C_j in (1) can be determined by

$$(x_{1j}, x_{2j}, \ldots, x_{nj}) = \frac{W_j Y_{C_j}}{\sum_{k=1}^{p_j} w_{jk}} \quad (6)$$

To rank the alternatives based on the weighted fuzzy performance matrix as above, the concept of fuzzy ideal solutions including the positive fuzzy ideal solution (alternative) and the negative fuzzy ideal solutions (alternative) [15], [29] is used. This concept is developed by extending the original concept of the ideal solution that has proved to be effective and popular to accommodate the presence of fuzziness and imprecision in decision analysis under uncertainty [7],[15], [29]. In line with the discussion above, the positive fuzzy ideal solution $A^{\square}$ and the negative fuzzy ideal solution A^- respectively representing the best and the worst fuzzy performance ratings among all the alternatives with respect to criterion C_j can be defined as follows:

$$A^+ = (M_{max}^1, M_{max}^2, \ldots, M_{max}^m) \quad (7)$$

$$A^- = (M_{min}^1, M_{min}^2, \ldots, M_{min}^m) \quad (8)$$

where M_{max}^j ($j = 1, 2, \ldots, m$) and M_{min}^j ($j = 1, 2, \ldots, m$) are respectively the best and the worst weighted fuzzy performance ratings of all alternatives with respect to criterion C_j ($j = 1, 2, \ldots, m$) [3], [30], determined by

$$M_{max}^j = (x_{min}^j, x_{max}^j, x_{max}^j), \quad j = 1, 2, \ldots, m \quad (9)$$

$$M_{min}^j = (x_{min}^j, x_{min}^j, x_{max}^j) \quad j = 1, 2, \ldots, m \quad (10)$$

where their membership functions are determined respectively by

$$\mu_{M_{max}^j}(x) = \begin{cases} \dfrac{x - x_{min}^j}{x_{max}^j - x_{min}^j}, & x_{min}^j \leq x \leq x_{max}^j, \\ 0, & \end{cases} \quad (11)$$

$$\mu_{M_{min}^j}(x) = \begin{cases} \dfrac{x_{max}^j - x}{x_{max}^j - x_{min}^j}, & x_{min}^j \leq x \leq x_{max}^j, \\ 0, & \end{cases} \quad (12)$$

where $i = 1, 2, \ldots, n; j = 1, 2, \ldots, m$.

$$x_{max}^j = \sup \bigcup_{i=1}^{n} \{x, x \in R \text{ and } 0 < \mu_{w_j x_{ij}}(x) < 1 \} \qquad (13)$$

$$x_{min}^j = \inf \bigcup_{i=1}^{n} \{x, x \in R \text{ and } 0 < \mu_{w_j x_{ij}}(x) < 1 \} \qquad (14)$$

With the introduction of the positive fuzzy ideal solution and the negative fuzzy ideal solution as above, the degree of similarity between each alternative and the positive ideal solution [4] can be calculated respectively as

$$d_i^+ = \sum_{j=1}^{m} d(w_i x_{ij}, M_{max}^j), \qquad i=1, 2, \ldots, n. \qquad (15)$$

Similarly, the degree of similarity between each alternative A_i ($i \in N = \{1, 2, \ldots, n\}$) and the negative fuzzy ideal solution [4] can be calculated as

$$d_i^- = \sum_{j=1}^{m} d(w_i x_{ij}, M_{min}^j), \qquad i=1, 2, \ldots, n. \qquad (16)$$

The most preferred alternative should not only have the highest degree of similarity with the positive fuzzy ideal solution, but also have the least degree of similarity with the negative fuzzy ideal solution [4], [29]. Within this line of thought, the overall performance index for each alternative across all criteria can be calculated as follows:

$$P_i = \frac{d_i^-}{d_i^+ + d_i^-} \qquad i = 1, 2, \ldots, n \qquad (17)$$

The larger the performance index, the more preferred the alternative.

4 An Empirical Study

An evaluation of fifteen land alternatives for their optimal use at Shanghai, China is conducted with respect to the criteria and sub-criteria shown as in Fig. 1. Subjective assessments with respect to the weights for three criteria and their associated sub-criteria and the performance of alternative lands with respect to each sub-criterion are given based on the linguistic terms defined as in Tables 1 and 2. Table 3 shows the linguistic assessment results for all the criteria and their sub-criteria. Table 4 presents the weighting vectors for the criteria and their associated sub-criteria.

With the data in Tables 1-4, the decision matrix for the criteria can be calculated. Multiplying the decision matrix with the weighting vector, the weighted fuzzy performance of all the alternatives with respect to each criterion can be obtained by (5). Accordingly, the positive fuzzy ideal solution and the negative fuzzy ideal solution can be calculated respectively based on (7) to (14) as follows

$$A^+ = \{(6.58, 123.16, 123.16), (4.67, 121.50, 121.50), (1.29, 93.33, 93.33)\}$$

$$A^- = \{(6.58, 6.58, 123.16), (4.67, 4.67, 121.50), (1.29, 1.29, 93.33)\}$$

The degree of similarity between each alternative and the positive ideal solution, together with negative ideal solution can be computed respectively based on (15) and (16). Consequently, the performance index can be easily derived from formula (17) and the ranking order can be achieved finally. Table 5 shows the results.

Table 3. Linguistic assessment results for sub-criteria

Sub-criteria	A_1	A_2	A_3	A_4	A_5	A_6	A_7	A_8	A_9	A_{10}	A_{11}	A_{12}	A_{13}	A_{14}	A_{15}
C_{11}	VH	VH	H	H	H	H	VH	H	H	H	F	H	L	L	VL
C_{12}	VH	VH	L	F	H	F	F	F	L	VL	F	VL	L	VL	VL
C_{13}	L	VL	H	H	F	VL	F	F	VL	VL	VL	VL	VL	VL	VL
C_{14}	F	H	H	VL	VL	H	F	F	F	F	F	F	L	L	L
C_{15}	H	L	L	L	L	F	L	F	L	H	VH	VL	VL	VL	VH
C_{16}	VH	VH	VH	VH	VH	VH	VH	VH	F	F	F	VH	F	VH	F
C_{17}	L	L	L	L	L	L	L	L	L	H	F	H	L	VH	H
C_{18}	VH	VH	VH	VH	VH	F	VH	VH	VH	H	H	H	F	L	L
C_{21}	F	H	H	H	H	H	H	H	H	H	F	F	F	F	VL
C_{22}	VH	VH	VH	F	F	VH	H	H	F	F	F	H	L	L	L
C_{31}	L	VL	VL	F	F	L	L	L	L	L	F	F	H	H	H
C_{32}	VH	VH	VH	VH	VH	VH	H	H	F	F	F	VH	F	F	L

Table 4. Weighting vectors for the criteria and their associated sub-criteria

Weighting vector	Fuzzy criteria weights
W	(7, 9, 9) (7, 9, 9) (1, 3, 5)
W_1	(7, 9, 9) (5, 7, 9) (5, 7, 9) (1, 3, 5) (5, 7, 9) (5, 7, 9) (3, 5, 7) (7, 9, 9)
W_2	(7, 9, 9) (5, 7, 9)
W_3	(3, 5, 7) (3, 5, 7)

Table 5. The preference index and ranking of the alternatives

Alternatives	Index (P_i)	Ranking	Integrated index system
A_1	0.411	3	Very suitable
A_2	0.424	1	Very suitable
A_3	0.418	2	Very suitable
A_4	0.371	7	Generally suitable
A_5	0.368	8	Generally suitable
A_6	0.391	5	Very suitable
A_7	0.397	4	Very suitable
A_8	0.387	6	Very suitable
A_9	0.334	11	Generally suitable
A_{10}	0.335	10	Generally suitable
A_{11}	0.320	12	Generally suitable
A_{12}	0.338	9	Generally suitable
A_{13}	0.302	15	Suitable after melioration
A_{14}	0.306	13	Suitable after melioration
A_{15}	0.305	14	Suitable after melioration

To facilitate a comparative study, Table 5 also shows the assessment results obtained using the integrated index approach. Overall a consistent ranking for all the fifteen alternative land resources are obtained shown as in Table 5 using the fuzzy MA approach and the integrated index approach.

The integrated index approach shows that the alternatives A_1, A_2, A_3, A_6, A_7, and A_8 are very suitable for being exploited into the agricultural land, while the A_{13}, A_{14} and A_{15} are most improper for land use planning. This of course is not crisp enough in some situations where a clearly identified ranking of available alternatives is desirable. In this regard, the fuzzy MA approach developed can address this issue by presenting a more rational estimation by incorporating the dominant factors into the integrated analysis, thus eliciting and producing a crisper ranking order, shown as in Table 5.

5 Conclusion

The optimal use of reserved land resources for agriculture is a complex problem that involves subjective assessments with multiple criteria. This paper has presented a fuzzy MA approach for effectively solving this problem. An empirical study in Shanghai, China has been conducted using the approach presented. It shows that the underlying concept of the approach developed is transparent and comprehensible, and the computation required is simple and efficient in solving this kind of problems.

Acknowledgement

The authors would like to thank Shanghai Institute of Geological Survey for generously contributing the detailed investigation data. Appreciation should be also given to Dr. C.H., Yao for his invaluable contribution in helping evaluation system establishment. The study is supported financially by Shanghai Municipal Bureau of Housing, Lands and Resources Administration.

References

1. Beek, K.J.: Land Evaluation for Agriculture Development: Some Exploration of Land Use System Analysis with Particular Reference to Latin America. International Institute for Land Reclamation and Improvement (ILRI), Wogeningen (1978).
2. Cannon, M., Kouvaritakis, B., Huang, G.: Modeling and Optimization for Sustainable Development Policy Assessment. Europe Journal of Operational Research. 2 (2005) 475–490
3. Chen, S.H.: Ranking Fuzzy Numbers with Maximizing Set and Minimizing Set. Fuzzy Sets and Systems. 2 (1985) 113–129.
4. Chen, S.J., Hwang, C.L.: Fuzzy Multiple Attribute Decision Making: Methods and Applications. Springer-Verlag, Berlin Heidelberg New York (1992).
5. China Agriculture Ministry.: Classification of Type Regions and Fertility of Cultivated Land in China. China Agriculture Ministry, Beijing (1996).
6. Chiou, H.K., Tzeng, G.H., Cheng, D.C.: Evaluating Sustainable Fishing Development Strategies Using Fuzzy MCDM Approach. Omega. 3 (2005) 223–234.
7. Deng, H., Yeh, C.H.: Fuzzy Ranking of Discrete Multicriteria Alternatives. In: Proceeding of the IEEE Second International Conference on Intelligent Processing Systems. Gold Coast, Australia (1998) 344–348
8. Ding, C.R.: Land Policy Reform in China: Assessment and Prospects. Land Use Policy. 20 (2003) 109–120

9. Dowall, D.E.: An Overview of the Land Management Assessment Technique. In: Gareth, J., Peper, M.W. (Eds.): Methodology for Land and Housing Market Analysis. UCL press, London (1994).
10. Garba, S., Al-Mubaiyedh, S.: An Assessment Framework for Public Urban Land Management Intervention. Land Use Policy. 4 (1999) 269–279.
11. Han, R.X., Han, T.K.: Sustainable Objectives of Reserved Land Resources Development. Territory and Natural Resources Study. 3 (1999) 17–18.
12. Herrera, F., Verdegay, J.L. Fuzzy Sets and Operations Research: Perspectives. Fuzzy Sets and Systems. 2 (1997) 207–218.
13. Jin, Z.G., Qu, X.H., Sun, P.: Investigation into the Assessment Approaches of Land Use Planning. Modern Agriculture. 4 (1997) 5–6 (in Chinese).
14. Juang, C.H., Lee, D.H.: A Fuzzy Scale for Measuring Weights of Criteria in Hierarchical Structures. IFES. (1991) 415–421.
15. Liang, G.S.: Fuzzy, MCDM Based on Ideal and Anti-ideal Concepts. European Journal of Operational Research. 112 (1999) 682–691.
16. Lin, G.C.S., Ho, S.P.S.: China's Land Resources and Land-use Change: Insights from the 1996 Land Survey. Land Use Policy. 20 (2003) 87–107.
17. Lv, X. J., Luo, L.T.: Application of Fuzzy Mathematics to the Quality Evaluation of the Added Cultivated Field through Exploitation and Upgrading. Journal of Xi'an University of Science and Technology. 1 (2004) 65–68.
18. Mohamed, A.A., Mohamed, A.S., Herman, V.K.: An Integrated Agro-economic and Agro-ecological Methodology for Land Use Planning and Policy Analysis. International Journal of Applied Earth Observation and Geoinformation. 2 (2000) 87–103
19. Ren, G.Y., Cai, Y.M.: The Characteristics and Countermeasures Related to Exploitation of Reserved Cultivated Land Resources in China. Resources Sciences. 5 (1998) 46–51.
20. Ren, W.J., Zhong, Y., Meligrana, J., et al.: Urbanization, Land Use, and Water Quality in Shanghai 1947-1996. Environment International. 29 (2003) 649–659.
21. Shanghai Municipal Bureau of Housing and Resources Administration.: Investigation of Reserved Land Resources in Shanghai, China. (2003).
22. Shao, X., Chen C.K.: Recent Process in the Study of Land Evaluation in China. Acta of Geographic Sinica. 1 (1993) 75–83.
23. Stewart, T. J., Janssen, R., Herwijnen, M.V.: A Genetic Algorithm Approach to Multiobjective Landuse Planning. Computers and Operations Research. 31 (2004) 2293–2313.
24. Stomph, T.J., Fresco, L.O., Van Keulen, H.: Land Use System Evaluation: Concepts and Methodology. Agriculture System. 44 (1994) 243–255.
25. Tang, Y.: Urban Land Use in China: Policy Issues and Options. Land Use Policy. 6 (1989) 53–63.
26. Verheye, W.H.: Land Use Planning and National Soil Policies. Agricultural Systems. 53 (1996) 161–174.
27. Wu, J.G., Zhang, T.: Evaluation Method and Application of Reserved Cultivated Land. Territory and Natural Resources Study. 9 (1999) 1–3.
28. Xiang, W.N., Gross, M., Fabos, J.G., MacDougall, E.B.: A Fuzzy-group Multicriteria Decision-Making Model and Its Application to Land-use Planning. Environment and Planning B: Planning and Design. 19 (1992) 61–84.
29. Yeh, C.H., Deng, H., Chang, Y.H.: Fuzzy Multicriteria Analysis for Performance Evaluation of Bus Companies. European Journal of Operational Research. 126 (2000) 459–473.
30. Zadeh, L.A.: Some Reflections on the Anniversary of Fuzzy Sets and Systems. Fuzzy Sets and Systems. 1-3 (1998) 5–7.
31. Zhu, Y.X., Lin, B.R.: Sustainable Housing and Urban Construction in China. Energy and Building. 12 (2004) 1287–1297.

Fuzzy Comprehensive Evaluation for the Optimal Management of Responding to Oil Spill

Xin Liu[1], Kai W. Wirtz[1,2], and Susanne Adam[1]

[1] Institute for Chemistry and Biology of the Marine Environment, Oldenburg University, Postfach 2503,
26111 Oldenburg, Germany
[2] Institute for Coastal Research GKSS Center,
21501 Geesthacht, Germany

Abstract. Studies on multi-group multi-criteria decision making problems for oil spill contingency management are in their infancy. This paper presents a second order fuzzy comprehensive evaluation (FCE) model to resolve decision-making problem in the area of contingency management after environmental disasters. To assess the performance of different oil combat strategies FCE allows the utilization of lexical information, consideration of ecological and socio-economic criteria and involvement of a variety of stakeholders. On the other hand, the approach can be validated by using internal and external checks, which refer to sensitivity tests regarding its internal setups and comparisons with other methods, respectively. Through a case study based on the Pallas oil spill occurred in German North-Sea 1998, it is demonstrated that this approach has wide application potential in the field of integrated coastal zone management.

1 Introduction

Although the North-Sea coastal region in Germany is small (shown in Fig. 1), its economic productivity (e.g. yearly gross value is over 125 billion Euros) is among the highest in Germany [1]. The main economic activities at this site are transportation, recreation, tourism, fishery and some wind energy conversion. It is also a particularly important natural ecosystem which supports breeding populations of seabirds, seals, dolphins or other marine species. Due to its ecological sensitivity, social, cultural, economic importance and scientific and educational purposes, a major part of Wadden Sea has been declared as Particularly Sensitive Sea Area (PPSA) within the framework of the International Maritime Organization (IMO). However, frequented shipping movements make this zone vulnerable to oil or chemical spills, as oil spills may lead to long lived consequences for near-shore ecosystems and economic uses. This has been demonstrated by the ecological disaster caused by the Pallas oil spill, a shipwrecked south west of the German island Amrum, 1998. Therefore, responding to emergency cases in an effective way turns out to be a critical concern in the domain of integrated coastal zone management. A golden rule of oil spill contingency management, on the one hand is to remove as much oil as possible from the sea surface in order to minimize the onshore impact, and on the other hand aims to minimize the cleanup cost. In this paper, we simulate a set of feasible combat strategies based on the Pallas case using available combat vessels as shown in Fig. 1. This creates an array of potential response measures which, in turn, can be selected after an integrated consideration of socio-economic and environmental impacts. For this, also a variety of stakeholders should be accounted for since they are directly or indirectly affected by decisions. Often their different interests cause a conflict on selecting an oil spill response strategy. Thus, we here formulate the selection of optimal combat strategies as a multi-group multi-criteria decision making problem and propose a second order fuzzy comprehensive evaluation (FCE). Hence, the specific objectives of the paper can be formulated as follows:

- To represent systematically opposing stakeholder interests within a decision support tool for oil spill contingency management.
- To re-evaluate response measures taken in a specific contingency case (Pallas, German Bight).
- To explore potentials and limitations of the FCE for future applications in the field of integrated coastal zone management.

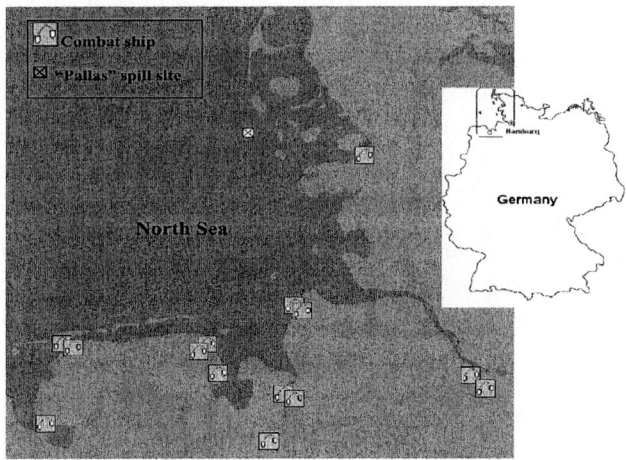

Fig. 1. German North-Sea case study area. Totally, there exist 14 oil combat vessels distributed in selected coastal administrative districts along the German North-Sea.

2 Model Inputs

Formulation of a multi-group multi-criteria decision-making problem is based on three basic components: (1) alternatives, (2) criteria and (3) stakeholders. The OSCAR (Oil Spill Contingency and Response) model system developed by SINTEF, Norway simulated a variety of combat strategies for a 60-ton crude oil

Table 1. Response strategies and combat vessels

Alternatives	#	Name of vessel
Alt.1	5	Neuwerk,Mellum,Knechtsan,Norderhever,Westensee
Alt.2	6	Neuwerk,Mellum,Knechtsan,Norderhever,Westensee,Nordsee
Alt.3	6	Neuwerk,Mellum,Knechtsan,Norderhever,Westensee,Eversand
Alt.4	6	Neuwerk,Mellum,Knechtsan,Norderhever,Westensee,Thor
Alt.5	4	Neuwerk,Mellum,Knechtsan,Norderhever

spill at the site where the accident of Pallas occurred ($54°32.5'N$; $8°17.24'E$). One major issue of the discussions in the aftermath of the accident was whether an appropriate number of response ship is in existence and, if yes, how many of these should have been used in the Pallas case. Thus, after a preliminary evaluation of these combat alternatives, five alternatives characterized by a variable number of 4-6 combat vessels are preselected. Among these five alternatives, alternative 1 can be taken as a reference as it includes all five activated combat vessels: Neuwerk, Mellum, Westensee, Knechtsand and Norderhever. Based on alternative 1, in alternatives 2, 3 and 4 one more combat vessel is assumed while in the alternative 5 only four combat vessels are considered (see Table 1). The five alternatives are evaluated with respect to a set of selected criteria, which can be regarded as representative for many coastal regions around the world with their specific economic uses and ecological values: the stranded oil, residual risk, oil collected, cleanup costs, fishery area, tourism area and duck area (details can be seen in Table 2). They reflect existing interests as well as existing background information at the German North-Sea coast, with special regard paid to oil pollution. The performances of the alternatives in terms of these criteria contribute to one major input matrix for the model of fuzzy comprehensive evaluation (FCE). The other important input matrix required by FCE are the stakeholders' preferences regarding each criterion. These weighting values can be revealed in either a quantitative or a qualitative way [8]. In many cases, it

Table 2. Selected criteria and their descriptions

	Criteria	Descriptions
SO	stranded oil	the stranded oil of tons in the coastal areas
RR	residual risk	summed amount of oil (tons) in the open sea
OC	oil collected	oil collected in tons by combat vessels
CC	cleanup costs	the costs in Euros by using the combat vessels and their equipments
F	fishery	summed amount of oil (tons) in the principal fishery areas
T	tourism	summed amount of oil (tons) in main recreation areas along the German North-Sea coastline
D	duck	summed amount of oil (tons) in important areas supporting breeding of eider ducks

is not realistic to ask participants who are from non-technical background to assign a numeric scale for the importance of criteria, although this kind of numeric scale response is quite straightforward for a further evaluation. Thus, we here use three different importance levels only. If participants are asked to select one importance level, their preferences will be directly integrated in the FCE. The FCE consists of three principal steps [6]: (a) A first order evaluation of performances of alternatives with respect to various criteria. (b) A second order evaluation with an involvement of weighting schemes assigned to the selected criteria by groups with different interests. (c) Making a rule based consensus which represents a majority view of interested groups.

3 Decision-Making Based on Fuzzy Comprehensive Evaluation

Using the performance matrix and the weighting schemes as inputs, the five different combat strategies can be assessed with the method of fuzzy comprehensive evaluation (FCE). A detailed procedures of applying FCE into the Pallas case are outlined in the following paragraphs.

3.1 Fuzzy Grades

To each criterion five lexically fuzzy grades are pre-defined: very low impact (*VLI*), low (*LI*), middle (*MI*), fairly high (*FHI*) and high impact (*HI*). Decision makers could also set another set of fuzzy grades, according to the resolution required for a specific problem. Thus, we get a fuzzy set containing a series of fuzzy grades for each criterion,

$$u^i = \{VLI^i, LI^i, MI^i, FHI^i, HI^i\} \tag{1}$$

where u^i denotes the set of lexical grades for the ith criterion.

3.2 Establishing Membership Degrees

We link values in the performance matrix to the lexical grades by using a fuzzy membership function,

$$\mu_{ij}^n = max(min(\frac{x_i^n - e_{ij}^1}{e_{ij}^2 - e_{ij}^1}, 1, \frac{e_{ij}^4 - x_i^n}{e_{ij}^4 - e_{ij}^3}), 0) \tag{2}$$

where x_i^n is the performance value of the alternative n in terms of the criterion i; μ_{ij}^n indicates the membership degree of x_i^n regarding to the jth grade of the ith criterion and $e_{ij}^{1,\cdots,4}$ are four scalar parameters for the jth fuzzy grade of the ith criterion. Therefore, a degree vector (A_i^n) is constructed,

$$\begin{cases} A_i^n = \{a_{i1}^n, a_{i2}^n, a_{i3}^n, a_{i4}^n, a_{i5}^n\} \\ a_{ij}^n = \mu_{ij}^n / \sum_{j=1}^{5} \mu_{ij}^n \end{cases} \tag{3}$$

An intuitive example for the criterion SO can be seen in Fig. 2. Its fuzzy set is defined as $u^1 = \{VLI^1, LI^1, MI^1, FHI^1, HI^1\}$. Supposed that there are 28.5 tons of spilled oil stranded, then the fuzzy degree reads $A_1 = (0, 0, 0.5, 0.5, 0)$. Namely, 28.5 tons oil pollution falls into the category of MI and FHI with the fuzzy membership of 0.5, respectively.

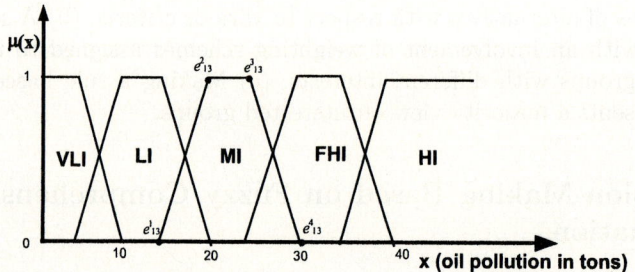

Fig. 2. The fuzzy membership function for criterion SO (stranded oil)

3.3 Defining Damage Levels

To some extent, coastal environments could be damaged due to the spilled oil. To aid decision makers to assess the performance of different combat strategies in a quantitative way. For simplicity, equally spaced oil spill damage levels (ς) ranging from 0 to 1 are used as numerical scales.

$$\varsigma = \{\varsigma_1, \varsigma_2, \cdots, \varsigma_{11}\} = \{0, 0.1, \cdots, 0.9, 1.0\} \tag{4}$$

where 0 represents no damage, while 1.0 denotes a complete damage in concerned coastal areas. In other words, an efficient strategy leads to lower damage level.

3.4 First-Order Fuzzy Evaluation

A first-order fuzzy degree assignment matrix represents fuzzy degrees of lexical grades associated with those eleven damage levels. Details can be seen in [3,6]. Though roughly representing existing expert knowledge and rules, the matrix coefficients remain empirical and can be modified for a specific application. Through combining the pre-defined first order fuzzy degree assignment matrix (R_i) and the fuzzy degree vector (A_i^n), the first-order FCE set (B_i^n) for alternative n in terms of criterion i can be obtained,

$$B^n = A_i^n \cdot R_i \tag{5}$$

3.5 Second-Order Fuzzy Evaluation

It is evident that the criteria (i.e. SO, RR, OC, CC, F, T, D) may not be equally important from the perspective of different stakeholders who are involved in using and managing coastal resources. Hence, a parameter W^s is used to denote

Table 3. Weighting schemes (⊕:highly important; ⊙:moderate; ⊖:non-important)

	Criteria						
	SO	RR	OC	CC	F	T	D
Group 1: Policy makers	⊙	⊙	⊙	⊙	⊙	⊙	⊙
Group 2: Combat organizations	⊙	⊙	⊕	⊕	⊖	⊖	⊖
Group 3: Environmentalists	⊕	⊖	⊖	⊖	⊙	⊕	⊕
Numerical value	⊕ = 0.95		⊙ = 0.5		⊖ = 0.05		

the weights for criteria according to the opinion of stakeholder s. For simplicity, three different importance levels are designed for each criterion: highly, moderate and non-important. In this paper, we supposed three different groups participating in the decision making process. Their weighting schemes are shown in Table 3 where policy makers tend to treat these criteria equally important, while group 2 and 3 put more emphasis on efficiency of the combat strategy and environmental damages, respectively. By multiplying W^s by B^n, one gets the second-order FCE set ($K^{s,n}$) for alternative n according to stakeholder s,

$$K^{s,n} = W^s \cdot B^n = k_1^{s,n}, k_2^{s,n}, \cdots, k_{11}^{s,n} \qquad (6)$$

3.6 Calculating the Overall Impact

The overall impact (OI) for a specific alternative n according to opinion of stakeholder s is determined as follows,

$$OI^{s,n} = \sum_{p=1}^{11} k_p^{s,n} \varsigma_p / \sum_{p=1}^{11} k_p^{s,n} \qquad (7)$$

A smaller value of the overall impact is preferred, since it indicates a less damage.

3.7 A Wide Consensus

With the overall impact of each alternative, the latter can be ranked. For stakeholder s, alternative e outranks alternative f, if $OI^{s,e} < OI^{s,f}$. Obviously, various rankings may be presented due to the different opinions of stakeholders. In order to make a consensus which represents a majority view of stakeholders, a mean rank for each alternative is taken into account, which represents the average of ranks according to all interested groups. As it has been argued to be a widely accepted method to reach a consensus through comparisons with other group consensus methods in a voting system [3].

4 Results and Discussions

If we compare the mean rank of five alternatives in Fig. 3, it appears that Alt.2 is the best, followed by Alt.5, 3, 1 and 4. On the other hand, the standard deviations of ranking for each alternative indicate that Alt.4 is less controversial, reflecting

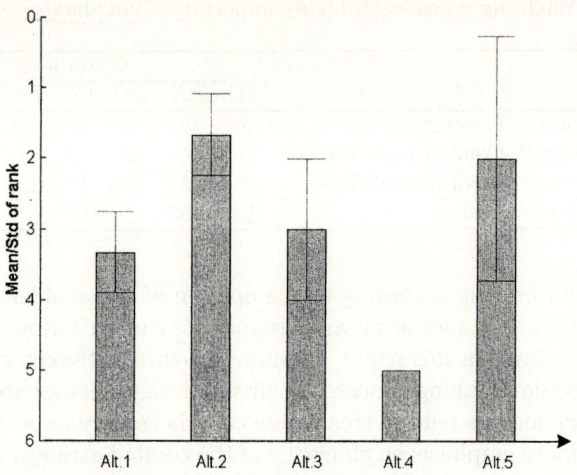

Fig. 3. Rankings of combat alternatives based on FCE. The interval plotted in solid lines indicates the standard deviation value of ranks.

that all three groups take it as the worst case. While both of Alt.1 and 2 are sort of controversial and Alt.3 and 5 are the most controversial according to different stakeholders' interests. Although the Alt.2 outranks Alt.5 with respect to the mean rank value, the overlap among them suggests they are very similar. Considering the mean rank and the interval of ranks comprehensively, Alt.2 and 5 can be grouped into the most preferred class and the Alt.4 appears to be the least preferred. In order to guarantee that the evaluation using FCE is reliable, two types of examination have to be done. One is the internal check (e.g. sensitivity test) and the other is the external check. In the internal check effects of internal setups on the result are examined. Two critical setups in FCE are the membership function and the damage level mentioned in section 3.2 and 3.3, respectively (see Fig. 4). The ranking result presented in Fig. 3 depends on the assumption that the number of damage level is 11 and the membership function is trapezoidal. In order to examine whether other possible setups could lead to different results or not, two tests are conducted separately. All criteria are considered to carry the same weight (e.g. the view of policy makers) in both test cases. According to the maximal extent at which the ranking of a specific alternative varies with the change of setups, the alternatives can be grouped as (i) not sensitive alternatives, (ii) relatively sensitive alternatives or (iii) highly sensitive alternatives,

$$alt.j \ is \begin{cases} \text{not sensitive} & if \ max(\Delta R(alt.j)) \in [0,1] \\ \text{relatively sensitive} & if \ max(\Delta R(alt.j)) = 2 \\ \text{highly sensitive} & if \ max(\Delta R(alt.j)) \in [3,4] \end{cases} \quad (8)$$

where $\Delta R(alt.j)$ indicates the difference between ranks associated with the alternative j by changing the internal setups. Firstly, three different membership

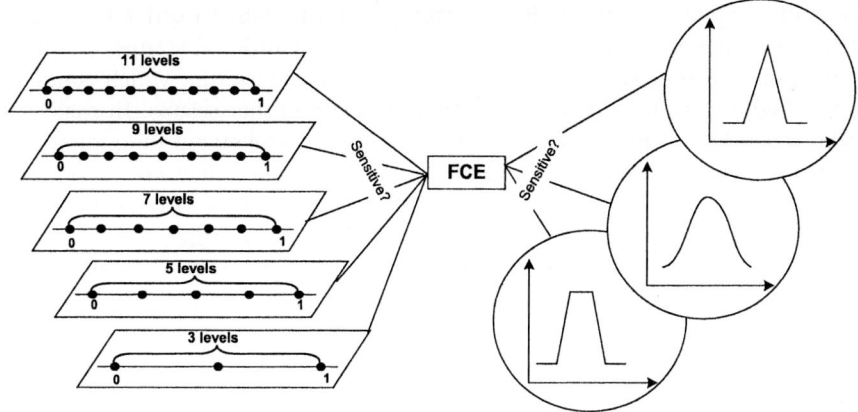

Fig. 4. The internal check of FCE. Different setups regarding the number of damage levels and various membership functions are examined.

functions are compared: the trapezoidal shape, the triangular shape and the Gaussian curve (see Fig. 4). As evident from comparing results shown in Table 4, most cases are not sensitive to the change of membership functions. The rankings of Alt.2 vary significantly when the membership function is Gaussian like. A possible reason is that the Gaussian curve has continuous tails, while the triangular and trapezoidal shaped functions are truncated at both sides. Secondly, the number of damage levels is assigned as 3, 5, 7, 9 and 11 respectively (see Fig. 4). Their effects on the rank-ordering of alternatives are presented in Table 4. Similar with the first test, all alternatives are not sensitive with the changing of such a setup according to the Eq.(8). Generally, different internal setups could

Table 4. Sensitivity tests regarding the internal setups of FCE. The ranks of alternatives are indicated by numerical numbers

Tests	Alternatives				
1.Membership function	Alt. 1	Alt. 2	Alt. 3	Alt. 4	Alt. 5
Trapezoidal shape	4	2	3	5	1
Triangular shape	4	2	3	5	1
Gaussian curve	3	5	2	4	1
Sensitive?	not	highly	not	not	not
2. Damage levels	Alt. 1	Alt. 2	Alt. 3	Alt. 4	Alt. 5
#3	4	1	3	5	2
#5	4	2	3	5	1
#7	3	2	4	5	1
#9	4	2	3	5	1
#11	4	2	3	5	1
Sensitive?	not	not	not	not	not

result in a minor change of the ordering of alternatives. In order to minimize such effects or uncertainties introduced by different internal setups, traditional correlation analyses are useful to determine a suitable setup, which could produce a highly correlated result with those based on other setups. In case of the Pallas study in this paper, the trapezoidal shaped membership function and the number of damage levels over 7 are recommended, since they may produce results which are relatively highly correlated with those derived from other setups. Additionally, the external check is also an useful way to validate FCE, since it allows people compare the result derived from FCE with those from other methods. For the Pallas case study, other methods such the multi-criteria analysis and the monetary evaluation method are also applied in the decision-making problem of combat options. Compared with FCE, they produce consistent results that the best case is Alt.2 and the worst case oscillates between Alt.4 and Alt.5 [5].

Summarizing we experienced both benefits and limitations of FCE such as it turns out, (i) as a method to deal with lexical data; (ii) to be capable of aggregating ecological and socio-economic criteria which are measured in different metrics; (iii) to provide a clear and traceable structure to integrate a variety of stakeholders into the decision-making process; (iv) to robustly differentiate the optimal and worst alternative groups; (v) that one potential difficulty of the FCE is that it requires knowledge and attentions for designing the internal setups. Thus, its setups should be examined carefully when FCE is applied to other case studies. On the other hand, decision makers will in general not completely rely on the computer-based results and constrain the final decision to the order of ranking for alternatives [2]. Since decision makers are responsible for the consequences of the decision, they must maintain the freedom to deviate from a modeled solution and may inspire suggestions for new alternatives from the results and analyses [2,7]. The attractive alternatives found by FCE are not yet the compromise alternatives, although they collect a wide consensus among the majority stakeholders. However, the utilization of FCE could be seen as a basis for a further agreement finding through a negotiation over the combat alternatives and money payments, through which stakeholders being forced to make disadvantageous agreements are compensated [4].

5 Conclusions

As a computer-aided decision-making tool, the fuzzy comprehensive evaluation (FCE) is without doubt useful for searching for efficient combat measure for the oil spill contingency management. The generic nature of this approach is capable of dealing with lexical data, considering ecological and socio-economical criteria and integrating a variety of stakeholders simultaneously in the decision-making process. These benefits are demonstrated by the Pallas case study presented in this paper, as well as applications in other field done previously. Additionally, in order to improve its validation, both of the internal and external checks are highly recommended.

Acknowledgments

We thank SINTEF Marine Environmental Technology for the permission to use the OSCAR software. This work was supported by the German Federal Ministry of Education and Research (BMBF) and the Land Niedersachsen.

References

1. Hagner, C.: The Economic Productivity at the German Coast. Tech. Rep. GKSS 2003/14, Institute for Coastal Research GKSS Center, Geesthacht,Germany (2003)
2. Lahdelma, R., Salminen, P., and Hokkanen, J.: Using Multicriteria Methods in Environmental Planning and Management. Environmental Management **26** (2000) 595-605
3. Liu, X. and Wirtz, K.W.: Multi-Group Multi-Criteria Decision Making of Oil Spill Contingency Options with Fuzzy Comprehensive Evaluation. Submitted to Environmental Modeling and Assessment (2004)
4. Liu, X. and Wirtz, K.W.: Sequential Negotiation in Multiagent Systems for Oil Spill Response Decision-Making. Marine Pollution Bulletin **50** (2005) 469-474
5. Liu, X., Wirtz, K.W. and Adam, S.: Decision Making of Oil Spill Combat Options Based on Benefit Cost Analysis. Submitted to TIES (2005)
6. Ji, S., Li, X. and Du, R.: Tolerance Synthesis Using Second-Order Fuzzy Comprehensive Evaluation and Genetic Algorithm. Int. J. Prod. Res. **38** (2000) 3471-3483
7. Ozernoy, V.: Generating Alternatives in Multiple Criteria Decision Making Problems: A Survey. In:Haimes, Y.,Chankong, V. (eds.): Decision making with multiple objectives. Springer-Verlag, New York (1984)
8. Simonovic, S.P.: Sustainable Floodplain Management-Participatory Planning in the Red River Basin, Canada. In Proceedings of IFAC, Venice (2004)

Fuzzy Fusion for Face Recognition

Xuerong Chen, Zhongliang Jing, and Gang Xiao

Institute of Aerospace Information and Control, School of Electronic,
Information and Electrical Engineering, Shanghai Jiaotong University,
Shanghai, China, 200030
{xrchen, zljing, xiaogang}@sjtu.edu.cn

Abstract. Face recognition based only on the visual spectrum is not accurate or robust enough to be used in uncontrolled environments. This paper describes a fusion of visible and infrared (IR) imagery for face recognition. In this paper, a scheme based on membership function and fuzzy integral is proposed to fuse information from the two modalities. Recognition rate is used to evaluate the fusion scheme. Experimental results show the scheme improves recognition performance substantially.

1 Introduction

In recent years face recognition has received substantial attention [1], but still remains very challenging in real applications. Recently, a number of studies have shown that infrared (IR) imagery of human faces offers a promising alternative to visible imagery due to its relative insensitive to illumination changes. However, IR has its own limitations [2][3]. One limitation is that IR is opaque to glass (Fig. 1), which makes a large portion of the face wearing eyeglasses hid. So, developing face recognition algorithms that fuse information from the two sensors is a future line of research [4]. In this paper, we propose a fuzzy fusion scheme. Our scheme is decision-based, and experimental results show recognition performance is improved substantially.

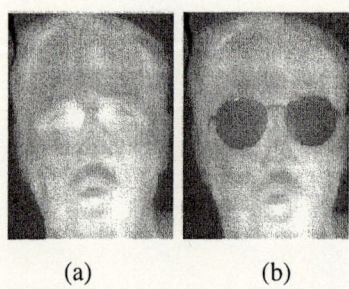

(a)   (b)

Fig. 1. The same IR face. (a) Without eyeglasses and (b) With eyeglasses

2 Fusion Scheme

Our fusion scheme is described below. We assume that each face is represented by a pair of images taken simultaneously and co-registered, one in the IR spectrum and one in the visible. The goal of fusion is to combine important information from each modality. Fig. 2 illustrates the main steps of our fusion scheme.

Step 1. We compute two eigenspaces [5], one using the IR face images and one using the visible face images. Then, each face is represented by two sets of eigenface components.

Step 2. Each set of eigenface components is treated as a vector, and normalized prior to fusion. The normalized vector v of an original vector Ω is defined as

$$v = \frac{\Omega}{\sqrt{(\Omega^T \cdot \Omega)}} . \tag{1}$$

Step 3. The purpose of fuzzification is to map input vector v from each modality to values from 0 to 1, representing evidence that the object satisfies the class hypothesis C_k. The generation of membership function is very important [6][7]. In this paper, we propose a modified histogram-based method. Let x be the distance of input object and its class, and $h(x)$ be the histogram of x, which provides information regarding the distribution of distance. Membership function $u(x)$ can be constructed as follow

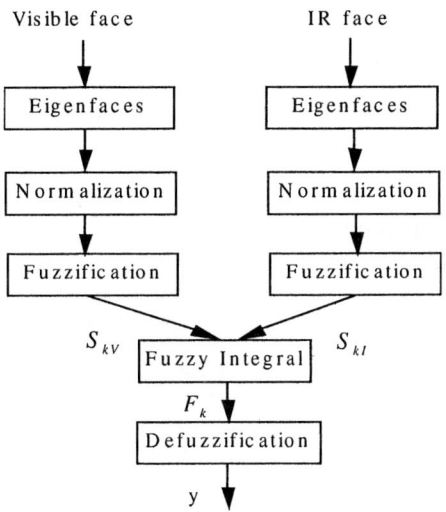

Fig. 2. Flowchart of the fusion scheme

$$u(x) = \int_x^{+\infty} h(x)dx \ . \tag{2}$$

By Eq. 2, we construct membership function $u(x)$ for each sensor. Let $\varepsilon_k = \|v - v_k\|$, where v_k is the vector describing the k th class. The fuzzification result S_k is computed as

$$S_k = u(\varepsilon_k) \ . \tag{3}$$

Step 4. Fuzzy integral considers the objective evidence supplied by each source (called the h-function) and the expected worth of each source (via a fuzzy measure) [8]. Let x_1 represent the visible source, and x_2 represent the IR source. The fuzzy density value $g^i = g\{x_i\}$ is determined via statistical measurements on recognition rate of the single source x_i. Thus the output of fuzzy integral F_k can be expressed as

$$F_k = \begin{cases} \max(\min(S_{kV}, g^1), S_{kI}) & S_{kV} \geq S_{kI} \\ \max(\min(S_{kI}, g^2), S_{kV}) & else \end{cases} \ . \tag{4}$$

where S_{kV} is the fuzzification result of visible source, and S_{kI} is the fuzzification result of IR source.

Step 5. We classify the input face into a specific class if that class had the maximum output of fuzzy integral

$$y = \arg\max_k F_k \ . \tag{5}$$

3 Experiments

In our experiments, we used the face database collected by Equinox Corporation. The format consists of image pairs, co-registered to within 1/3 pixel. We created a training set and 5 test sets for each modality. There were 30 subjects with a total of 120 images in the training set, and 270 images in each test set. Our criteria were as follow:

Training set: frontal illumination, no eyeglasses, no expression
Test_no set: frontal illumination, no eyeglasses, no expression
Test_illumination set: lateral illumination, no eyeglasses, no expression
Test_eyeglasses set: frontal illumination, with eyeglasses, no expression
Test_expression set: frontal illumination, no eyeglasses, with expression
Test_all set: lateral illumination, with eyeglasses, with expression

In experiments, we first constructed membership function for each modality from real data. The number of eigenvectors for egienface was set 60.

Table 1. Recognition performance of visible imagery, infrared imagery and fusion (%)

	Visible	Infrared	Fusion
Test_no	95.96	98.99	100
Test_illumination	30.30	93.94	91.92
Test_eyeglasses	94.95	74.77	95.96
Test_expression	95.96	97.98	100
Test_all	19.24	63.64	86.87

We set $g^1 = 0.8$, and $g^2 = 0.9$ in Eq. 4 after statistically analyzing recognition rate of the two modalities. Recognition performance of each modality (using eigenfaces and Top-match) and fusion (using the proposed scheme) are shown in Table 1. From Table 1, we can see that considerable improvements in recognition performance were achieved by fusing visible and infrared imagery.

4 Conclusions

We presented a new fusion scheme to combine visible and infrared images for the purpose of face recognition. Our scheme is decision-based, using fuzzy integral to fuse the objective evidence supplied by each modality. Experimental results show the scheme is easily realized, and improves recognition performance substantially. In the paper, we also designed a modified histogram-based method to generate membership function. Future work includes designing more effective schemes to deal with kinds of challenging situations.

References

1. Zhao, W., Chellappa, R., Rosenfeld, A., Phillips, P. J.: Face Recognition: A Literature Survey. ACM Computing Survey 35 (2003) 399–458
2. Prokoski, F.: History, Current Status, and Future of Infrared Identification. IEEE Workshop on Computer Vision Beyond the Visible Spectrum, Hilton Head (2000) 5–14
3. Wilder, J., Phillips, P. J., Jiang, C. H., Wiener, S.: Comparison of Visible and Infra-red Imagery for Face Recognition. In: Proc. 2nd Intl. Conf. on Automatic Face & Gesture Recognition, Killington (1996) 182–187
4. Singh, S., Gyaourova, A., Bebis, G., Pavlidis, I.: Infrared and Visible Image Fusion for Face Recognition. SPIE Defense and Security Symposium (Biometric Technology for Human Identification), Orlando (2004) 585–596
5. Turk, M., Pentland, A.: Face Recognition Using Eigenface. IEEE Conf. on Computer Vision and Pattern Recognition (1991) 586–591
6. Zadeh, L. A.: Fuzzy sets. Information and Control 8 (1965) 338–353
7. Medasani, S., Kim, J., Krishnapuram, R.: An overview of Membership Function Generation Techniques for Patter Recognition. International Journal of Approximate Reasoning 19 (1998) 391–417
8. Keller, J. M., Osborn, J.: Training the Fuzzy Integral. International Journal of Approximate Reasoning 15 (1996) 1–24

A Group Decision Making Method for Integrating Outcome Preferences in Hypergame Situations*

Yexin Song[1], Qian Wang[1], and Zhijun Li[2]

[1] College of Sciences, Naval University of Engineering, Wuhan, Hubei 430033, China
yxsong@21cn.com, wangmeau999@hotmail.com
[2] Wuhan University of Technology, Wuhan, Hubei 430063, China
junwenlee@sohu.com

Abstract. This paper presents a novel group decision making method for integrating outcome preferences in the first-level hypergame models where each player correctly perceives the strategy set, but perceives possibly different outcome preferences of the opponent players. To get more correct preferences information in hypergame models, each player is often consisted of a group of decision makers who can give their perception about opponent players' preferences respectively. In the face of opponent players' different linguistic preferences relations over outcome space perceived by different decision makers, a group fuzzy preferences relation is first accurately computed using standard fuzzy arithmetic operations. Concept of consensus winner is then introduced to decide the crisp outcome preference vectors. A numerical example is provided at the end to illustrate the method.

1 Introduction

A hypergame is a model to describe a conflict situation where the players involved have different understandings about the conflict [1,2]. The players involved in a hypergame may have incorrect beliefs about the other players' options, strategies, or preferences, or even be unaware of some of the players in the game. Hypergames can have many levels of perception because one player may have misperceptions about the other players' interpretations of the game. Instead of using a single traditional game to depict such a situation, a set of individual games is built up in a hypergame to reflect the players' diverse perspectives. Because the players are allowed to see a dispute differently, hypergames are more flexible and realistic to deal with real-world conflicts [3,4].

In this paper, we consider a first-level hypergame model where each player correctly perceives the strategy set, but perceives possibly different outcome preferences of the opponent players. To get more correct preferences information, each player is often consisted of a group of decision makers who can give their perception about opponent players' preferences respectively. Because of the ambiguity existing in available information as well as the essential fuzziness in human judgement, every decision maker in one player can give his (her) perception about the opponent players'

---

* This work is supported by National Natural Science Foundation of China Grant #70471031.

preferences relations over outcome space using the linguistic values or labels [5], subjectively represented by the trapezoidal fuzzy numbers. Therefore, the main objective of this paper is to integrate all the decision makers' perception information and obtain the crisp outcome preference vectors. At first, in the face of opponent players' different linguistic preferences relations perceived by different decision makers, a group fuzzy preferences relation over outcome space is accurately computed using standard fuzzy arithmetic operations. Then, based on the group fuzzy preferences relation and concept of consensus winner, a method is presented for determining the crisp outcome preference vectors in the first-level hypergame models.

2 The First-Level Hypergame Models

When a conflict is modeled as a game, the game model consists of
(1) a set of players, N;
(2) a set of strategies for each of the players, S_i, $\forall i \in N$;
(3) a set of payoff functions, v_i, $\forall i \in N$, which reflect the players' preferences over the outcome space $O = S_1 \times S_2 \times \cdots \times S_n$.

Hence, a game can be defined as

$$G = \{S_1, S_2, \cdots, S_n, v_1, v_2 \cdots, v_n\}.$$

If all the outcomes are ranked in order according to a player's payoffs by writing the most preferred outcome on the left and least preferred on the right, then a preference vector (PV) is formed for the player and denoted by V_i. Consequently, a game can be represented by a set of PVs: $G = \{V_1, V_2, \cdots, V_i, \cdots, V_n\}$. In a simple game (or a game with complete information), there are no misperceptions, each player is represented by only one PV, and all the players see the same set of PVs.

In a hypergame, the players' interpretations of the set of payoff functions can be different from one another because of misperceptions. Subscripts are used to describe such an interpretation: v_{ij} (V_{ij}) expresses player i's payoff function (PV) as interpreted by player j; the first subscript i tells to whom the payoff function (PV) belongs, while the second one j stands for the player who perceives the payoff function (PV). When there is misperception, for instance, player j's interpretation of player i's PV will be different from V_i, so that the game played by player j will be different from the one played by player i. A formal definition of an Lth-level hypergame model is given below [1]:

$$H^L = \{H_1^{L-1}, H_2^{L-1}, \cdots, H_n^{L-1}\}, \quad L = 1,2,3,\cdots, \quad \exists i, j \in N, H_i^{L-1} \neq H_j^{L-1}.$$

By using the above general formula, hypergame model on a given level can be written explicitly. In a first-level hypergame, at least one of the players is playing a different game from the others. Each player's game is formed by a set of perceived PVs, except the one for the player himself. The first-level hypergame model is formulated as

$$H^1 = \{H_1^0, H_2^0, \cdots, H_i^0, \cdots, H_n^0\} \quad (\exists i, j \in N : H_i^0 \neq H_j^0)$$

$$= \left\{ \begin{bmatrix} V_1 \\ V_{21} \\ \vdots \\ V_{i1} \\ \vdots \\ V_{n1} \end{bmatrix} \begin{bmatrix} V_{12} \\ V_2 \\ \vdots \\ V_{i2} \\ \vdots \\ V_{n2} \end{bmatrix} \cdots \begin{bmatrix} V_{1i} \\ V_{2i} \\ \vdots \\ V_i \\ \vdots \\ V_{ni} \end{bmatrix} \cdots \begin{bmatrix} V_{1n} \\ V_{2n} \\ \vdots \\ V_{in} \\ \vdots \\ V_n \end{bmatrix} \right\} \quad (1)$$

3 Group Decision Making Method for Integrating Outcome Preferences

It is important for a player to correctly determine preference vector V_{ij} in model (1). In this section, we suppose that player j is consisted of M experts $D_j = \{D_{j1}, D_{j2}, \cdots, D_{jM}\}$. Each expert $D_{jm} \in D_j$ provides his/her perception about player i's preferences over outcome space $O = \{O_1, O_2, \cdots, O_K\}$ as a fuzzy preference relation, $R^{ijm} \subset O \times O$, with membership function $\mu_{R^{ijm}} : O \times O \to [0,1]$, where $\mu_{R^{ijm}}(O_p, O_q)$ denotes the preference degree of the outcome O_p over O_q. Player j assigns a weight $\mu_{D_j}(m)$ to expert D_{jm}, defining on the expert set, D_j, a fuzzy set with membership function $\mu_{D_j} : D_j \to [0,1]$. Given an expert D_{jm}, his weight $\mu_{D_j}(m)$ is interpreted as the power or importance degree of his opinion.

We use the linguistic values or labels to express every expert' weight and opinion about player i's outcome preferences. Each linguistic value is subjectively given by a trapezoidal fuzzy number represented by the 4-tuple $(\alpha, \beta, \gamma, \delta)$, where α and δ indicate the lower and upper bounds of the distribution; β and γ indicate the interval in which the membership value is 1.0. Assume the linguistic weight of expert D_{jm} is

$$\mu_{D_j}(m) = \tilde{w}_m = (\varepsilon_m, \zeta_m, \eta_m, \theta_m), \; m = 1, 2, \cdots, M,$$

where, $0 \leq \varepsilon_m \leq \zeta_m \leq \eta_m \leq \theta_m \leq 1$, and expert D_{jm} provides his/her perception about player i's linguistic preference relation over O expressed in the matrix form as

$$R^{ijm} = [\tilde{a}_{pq}^{ijm}]_{K \times K},$$

where, $\tilde{a}_{pq}^{ijm} = (\alpha_{pq}^{ijm}, \beta_{pq}^{ijm}, \gamma_{pq}^{ijm}, \delta_{pq}^{ijm})$, $0 \leq \alpha_{pq}^{ijm} \leq \beta_{pq}^{ijm} \leq \gamma_{pq}^{ijm} \leq \delta_{pq}^{ijm} \leq 1$, $m = 1, 2, \cdots, M$; $p, q = 1, 2, \cdots, K$, denote player i's linguistic preference of the outcome O_p over O_q perceived by expert D_{jm}.

In this section, we will focus our attention on how to integrate all experts' linguistic preferences information and obtain a crisp preference vector V_{ij} in model (1).

3.1 Group Fuzzy Preferences Relation over Outcome Space

Given the data R^{ijm} and $\tilde{w}_m$, player i's linguistic preference of the outcome O_p over O_q perceived by all experts (player j), $\tilde{a}_{pq}^{ij}$, can be computed as follows

$$\tilde{a}_{pq}^{ij} = (1/M) \otimes \{(\tilde{w}_1 \otimes \tilde{a}_{pq}^{ij1}) \oplus (\tilde{w}_2 \otimes \tilde{a}_{pq}^{ij2}) \oplus \cdots \oplus (\tilde{w}_M \otimes \tilde{a}_{pq}^{ijM})\} \qquad (2)$$

where, $\oplus$ and $\otimes$ are the fuzzy addition and fuzzy multiplication operations, respectively.

Theorem 1. The graph of the membership function of $\tilde{a}_{pq}^{ij}$ is: zero to the left of α_{pq}^{ij}; $L_{pq1}^{ij} y^2 + L_{pq2}^{ij} y + \alpha_{pq}^{ij} = x$ on $[\alpha_{pq}^{ij}, \beta_{pq}^{ij}]$; $y = 1$ between $[\beta_{pq}^{ij}, \gamma_{pq}^{ij}]$; $U_{pq1}^{ij} y^2 + U_{pq2}^{ij} y + \delta_{pq}^{ij} = x$ on $[\gamma_{pq}^{ij}, \delta_{pq}^{ij}]$ and zero to the right of δ_{pq}^{ij}. $\tilde{a}_{pq}^{ij}$ can be expressed as

$$\tilde{a}_{pq}^{ij} = (\alpha_{pq}^{ij} [L_{pq1}^{ij}, L_{pq2}^{ij}] / \beta_{pq}^{ij}, \gamma_{pq}^{ij} / \delta_{pq}^{ij} [U_{pq1}^{ij}, U_{pq2}^{ij}]) \qquad (3)$$

where

$$\alpha_{pq}^{ij} = (\sum_{m=1}^{M} \alpha_{pq}^{ijm} \varepsilon_m)/M, \quad \beta_{pq}^{ij} = (\sum_{m=1}^{M} \beta_{pq}^{ijm} \zeta_m)/M,$$

$$\gamma_{pq}^{ij} = (\sum_{m=1}^{M} \gamma_{pq}^{ijm} \eta_m)/M, \quad \delta_{pq}^{ij} = (\sum_{m=1}^{M} \delta_{pq}^{ijm} \theta_m)/M,$$

$$L_{pq1}^{ij} = \{\sum_{m=1}^{M} (\beta_{pq}^{ijm} - \alpha_{pq}^{ijm})(\zeta_m - \varepsilon_m)\}/M,$$

$$L_{pq2}^{ij} = [\sum_{m=1}^{M} \{\alpha_{pq}^{ijm}(\zeta_m - \varepsilon_m) + \varepsilon_m(\beta_{pq}^{ijm} - \alpha_{pq}^{ijm})\}]/M,$$

$$U_{pq1}^{ij} = \{\sum_{m=1}^{M} (\delta_{pq}^{ijm} - \gamma_{pq}^{ijm})(\theta_m - \eta_m)\}/M,$$

$$U_{pq2}^{ij} = -[\sum_{m=1}^{M} \{\delta_{pq}^{ijm}(\theta_m - \eta_m) + \theta_m(\delta_{pq}^{ijm} - \gamma_{pq}^{ijm})\}]/M.$$

The proof is well described in the works of Dubois and Prade [6] and Buckley [7]. According to theorem 1, the membership function of $\tilde{a}_{pq}^{ij}$ is

$$\mu_{\tilde{a}_{pq}^{ij}}(x) = \begin{cases} -L_{pq2}^{ij}/2L_{pq1}^{ij} + \{(L_{pq2}^{ij}/2L_{pq1}^{ij})^2 + (x - \alpha_{pq}^{ij})/L_{pq1}^{ij}\}^{1/2}, & \alpha_{pq}^{ij} < x < \beta_{pq}^{ij}, \\ 1, & \beta_{pq}^{ij} < x < \gamma_{pq}^{ij}, \\ -U_{pq2}^{ij}/2U_{pq1}^{ij} + \{(U_{pq2}^{ij}/2U_{pq1}^{ij})^2 + (x - \delta_{pq}^{ij})/U_{pq1}^{ij}\}^{1/2}, & \gamma_{pq}^{ij} < x < \delta_{pq}^{ij}, \\ 0, & \text{otherwise.} \end{cases} \qquad (4)$$

So, from the individual fuzzy preference relations we accurately determine a group fuzzy preference relation, i.e. player i's linguistic preferences over O perceived by player j, which is expressed in the matrix form as $R^{ij} = [\tilde{a}_{pq}^{ij}]_{K \times K}$.

3.2 Crisp Preference Vector in First-Level Hypergame Model (1)

We now discuss how to determine a crisp preference vector in model (1) from the group fuzzy preference relation. Denote

$$\mu_{\tilde{a}_{pq}^{ij}}^L(x) = -L_{pq2}^{ij}/2L_{pq1}^{ij} + \{(L_{pq2}^{ij}/2L_{pq1}^{ij})^2 + (x - \alpha_{pq}^{ij})/L_{pq1}^{ij}\}^{1/2},$$

$$\mu_{\tilde{a}_{pq}^{ij}}^R(x) = -U_{pq2}^{ij}/2U_{pq1}^{ij} + \{(U_{pq2}^{ij}/2U_{pq1}^{ij})^2 + (x - \delta_{pq}^{ij})/U_{pq1}^{ij}\}^{1/2},$$

$$g_{\tilde{a}_{pq}^{ij}}^L(y) = L_{pq1}^{ij} y^2 + L_{pq2}^{ij} y + \alpha_{pq}^{ij},$$

$$g_{\tilde{a}_{pq}^{ij}}^R(y) = U_{pq1}^{ij} y^2 + U_{pq2}^{ij} y + \delta_{pq}^{ij}.$$

(5)

It is obvious that $\mu_{\tilde{a}_{pq}^{ij}}^L(x)$, which is the left membership function of the fuzzy number $\tilde{a}_{pq}^{ij}$, is continuous and strictly increasing on $[\alpha_{pq}^{ij}, \beta_{pq}^{ij}]$, and $\mu_{\tilde{a}_{pq}^{ij}}^R(x)$, which is the right membership function of the fuzzy number $\tilde{a}_{pq}^{ij}$, is continuous and strictly decreasing on $[\gamma_{pq}^{ij}, \delta_{pq}^{ij}]$. $g_{\tilde{a}_{pq}^{ij}}^L(y)$ and $g_{\tilde{a}_{pq}^{ij}}^R(y)$ are inverse functions of $\mu_{\tilde{a}_{pq}^{ij}}^L(x)$ and $\mu_{\tilde{a}_{pq}^{ij}}^R(x)$, respectively, they are all continuous on $[0,1]$, thus they are also integrable on $[0,1]$. Define

$$I_L(\tilde{a}_{pq}^{ij}) = \int_0^1 g_{\tilde{a}_{pq}^{ij}}^L(y) dy,$$

$$I_R(\tilde{a}_{pq}^{ij}) = \int_0^1 g_{\tilde{a}_{pq}^{ij}}^R(y) dy,$$

(6)

$$I_T(\tilde{a}_{pq}^{ij}) = [I_L(\tilde{a}_{pq}^{ij}) + I_R(\tilde{a}_{pq}^{ij})]/2.$$

Definition 1. $I_T(\tilde{a}_{pq}^{ij})$ is called the total integral expected value of $\tilde{a}_{pq}^{ij}$.

According to definition 1,

$$I_T(\tilde{a}_{pq}^{ij}) = [I_L(\tilde{a}_{pq}^{ij}) + I_R(\tilde{a}_{pq}^{ij})]/2$$

$$= [\int_0^1 g_{\tilde{a}_{pq}^{ij}}^L(y) dy + \int_0^1 g_{\tilde{a}_{pq}^{ij}}^R(y) dy]/2$$

$$= \frac{1}{2} [\int_0^1 (L_{pq1}^{ij} y^2 + L_{pq2}^{ij} y + \alpha_{pq}^{ij}) dy + \int_0^1 (U_{pq1}^{ij} y^2 + U_{pq2}^{ij} y + \delta_{pq}^{ij}) dy]$$

$$= \frac{L_{pq1}^{ij} + U_{pq1}^{ij}}{6} + \frac{L_{pq2}^{ij} + U_{pq2}^{ij}}{4} + \frac{\alpha_{pq}^{ij} + \delta_{pq}^{ij}}{2}.$$

(7)

Definition 2. Denote $e_{pq}^{ij} = I_T(\tilde{a}_{pq}^{ij})$, $E^{ij} = [e_{pq}^{ij}]_{K \times K}$ is called group fuzzy preference expected value relation matrix.

Define

$$g_{pq}^{ij} = \begin{cases} 1, & \text{if } e_{pq}^{ij} > 0.5, \\ 0, & \text{otherwise.} \end{cases} \quad (8)$$

which expresses whether outcome O_p defeats O_q or not, and then

$$g_p^{ij} = \frac{1}{K-1} \sum_{q=1, q \neq p}^{K} g_{pq}^{ij}, \quad (9)$$

which is the mean degree to which outcome O_p is preferred to all the other outcomes. Assume that a fuzzy linguistic quantifier $Q =$'most' to be a fuzzy set defined in [0,1] and given as

$$\mu_Q(x) = \begin{cases} 1, & \text{for } x \geq 0.8, \\ 2x - 0.6, & \text{for } 0.3 < x < 0.8, \\ 0, & \text{for } x \leq 0.3. \end{cases} \quad (10)$$

Then $z_{pQ}^{ij} = \mu_Q(g_p^{ij})$ is the extent to which O_p is preferred to Q other outcomes. Finally, the fuzzy Q-consensus winner [8] is defined as

$$W_Q = \frac{z_{1Q}^{ij}}{O_1} + \frac{z_{2Q}^{ij}}{O_2} + \cdots + \frac{z_{KQ}^{ij}}{O_K}, \quad (11)$$

i.e. as a fuzzy set of outcomes that are preferred to Q other outcomes. So, the crisp preference vector V_{ij} in model (1) can be determined according to the fuzzy Q-consensus winner W_Q.

4 Numerical Example

Considering a two-person hypergame situation where all the players have 2 strategies. Player 1 is consisted of 3 experts (D_{11}, D_{12}, D_{13}), whose respective linguistic weights are: $\tilde{w}_1 = H$, $\tilde{w}_2 = B$, $\tilde{w}_3 = AF$, and whose perception about player 2's preferences over $O=\{O_1, O_2, O_3, O_4\}$ using linguistic preference relations as follows: (here, use the linguistic term set $L=\{B, VH, H, AF, F, BF, L, VL, W\}$, where $B =$ Best= (0.8,1,1,1), $VH =$ Very High = (0.6,0.8, 1,1), $H =$ High= (0.6,0.8,0.8,1), AF=Above Fair= (0.3,0.5,0.8,1), $F =$ Fair = (0.3,0.5,0.5,0.7), BF =Below Fair =(0,0.2,0.5,0.7), L=Low= (0,0.2,0.2,0.4), $VL =$ Very Low=(0,0,0.2,0.4), $W =$ Worst= (0,0,0,0.2))

$$R^{211} = [\tilde{a}_{pq}^{211}]_{4\times 4} = \begin{bmatrix} - & AF & B & W \\ BF & - & VL & VL \\ W & VH & - & L \\ B & VH & H & - \end{bmatrix},$$

$$R^{212} = [\tilde{a}_{pq}^{212}]_{4\times 4} = \begin{bmatrix} - & AF & H & VL \\ BF & - & VL & W \\ L & VH & - & L \\ VH & B & H & - \end{bmatrix},$$

$$R^{213} = [\tilde{a}_{pq}^{213}]_{4\times 4} = \begin{bmatrix} - & F & VH & W \\ F & - & W & BF \\ VL & B & - & L \\ B & AF & H & - \end{bmatrix}.$$

At first, using Eqs. (2), (3) and (7), we can obtain the group fuzzy preference expected value relation matrix as follows:

$$E^{21} = \begin{bmatrix} 0 & 0.511 & 0.708 & 0.081 \\ 0.336 & 0 & 0.111 & 0.160 \\ 0.157 & 0.715 & 0 & 0.176 \\ 0.738 & 0.697 & 0.656 & 0 \end{bmatrix}.$$

Then, applying Eqs. (9), (10) and (11), the fuzzy *Q-consensus winner* can be computed as follows:

$$W_Q = \frac{0.733}{O_1} + \frac{0}{O_2} + \frac{0.066}{O_3} + \frac{1}{O_4}.$$

According to W_Q, we can determine the crisp preference vector V_{21} in model (1) as follows:

$$V_{21} = (4,1,3,2).$$

5 Conclusion

A novel group decision making method has been proposed in this paper for integrating outcome preferences in the first-level hypergame models where each player correctly perceives the strategy set, but perceives possibly different outcome preferences of the opponent players. In the face of opponent players' different linguistic preferences relations over outcome space perceived by different decision makers in a player, a group fuzzy preferences relation is accurately computed using standard fuzzy arithmetic operations. The concept of consensus winner is utilized to decide the crisp outcome preference vectors. An illustrative example verifies the feasibility and effectiveness of the proposed method.

References

1. Wang M., Hipel K.W. and Frase N. M.: Solution concepts in hypergames. Applied Mathematics and Computation. 34 (1989) 147-171
2. Putro U.S., Kijima K., and Takahashi S.: Adaptive learning of hypergame situations using a genetic algorithm. IEEE Transactions on Systems, Man and Cybernetics-Part A: Systems and Humans. 5 (2000) 562-572
3. Hipel K.W., Wang M., and Frase N. M.: Hypergame analysis of the Falkland Island crisis. Internat. Stud. Quart. 32 (1988) 335-358
4. Hipel K.W., Dagnino A., and Frase N. M.: A hypergame algorithm for modeling misperceptions in bargaining. J. Environmental Management. 12 (1988) 131-152
5. Herrera F., Herrera-Viedma E., Verdegay J.L., A rational consensus model in group decision making using linguistic assessments. Fuzzy Sets and Systems. 88 (1997) 31-49.
6. Dubois D. and Prade H.: Ranking of fuzzy numbers in the setting of possibility theory. Inform. Sci. 30(1983) 183-224
7. Buckley J.J.: Ranking alternatives using fuzzy numbers. Fuzzy Sets and Systems, 15(1985) 21-31
8. Kacprzyk J., Fedrizzi M., Nurmi H.: Group decision making and consensus under fuzzy preferences and fuzzy majority. Fuzzy Sets and Systems, 49(1992) 21-31

A Method Based on IA Operator for Multiple Attribute Group Decision Making with Uncertain Linguistic Information

Zeshui Xu

College of Economics and Management, Southeast University,
Nanjing, Jiangsu 210096, China
Xu_zeshui@263.net

Abstract. In this paper, we study the multiple attribute group decision making (MAGDM) problems, in which the information about the attribute weights and the expert weights are interval numbers, and the attribute values take the form of uncertain linguistic information. We introduce some operational laws of uncertain linguistic variables and a formula for comparing two uncertain linguistic variables, and propose a new aggregation operator called interval aggregation (IA) operator. Based on the IA operator and the formula for the comparison between two uncertain linguistic variables, we develop a method for MAGDM with uncertain linguistic information. Finally, an illustrative example is given to verify the developed method.

1 Introduction

A multiple attribute group decision making (MAGDM) problem is characterized by several experts, who are called to give their preferences on a predefined set of alternatives with respect to each attribute in a predefined set of attributes. Due to the nature of MAGDM problems, in many situations, the experts can't provide their preference values (attribute values) with numerical values, but with linguistic terms [1-22]. Degani and Bortolan [5] have developed a method based on the extension principle, which makes operations on the fuzzy numbers that support the semantics of the linguistic terms. Delgado et al. [6] have proposed a symbolic method, which makes computations on the indexes of the linguistic terms. Herrera and Martínez [13-15] have developed a fuzzy linguistic representation model, which represents the linguistic information with a pair of values called 2-tuple, composed by a linguistic term and a number. Xu [20-22] have presented some methods, which compute with words directly. All of these methods, however, will fail in dealing with the situations in which the attribute weights and the expert weights are interval numbers, and the attribute values take the form of uncertain linguistic information. The aim of this paper is to develop an approach to overcoming this limitation. To do that, the remainder of this paper is structured as follows. Section 2 introduces some operational laws of uncertain linguistic variables and a formula for comparing two uncertain linguistic values. In Section 3 we propose a new aggregation operator called interval aggregation (IA) operator, and then based on the IA operator and the formula for the comparison between two uncertain linguistic variables, we develop a method for the MAGDM problems, in which the information about the attribute weights and the expert weights are interval numbers, and the attribute values take the form of uncertain linguistic information. In Section 4 we give an illustrative numerical example, and finally, we point out some concluding remarks.

2 A Formula for the Comparison between Two Uncertain Linguistic Variables

Suppose that $S = \{s_\alpha \mid \alpha = -t,...,t\}$ is a finite and totally ordered discrete term set, where s_α represents a possible value for a linguistic variable. For example, a set of nine terms S could be

$$S = \{s_{-4} = extremely\ poor,\ s_{-3} = very\ poor,\ s_{-2} = poor,$$
$$s_{-1} = slightly\ poor,\ s_0 = fair,\ s_1 = slightly\ good,$$
$$s_2 = good,\ s_3 = very\ good,\ s_4 = extremely\ good\}$$

In these cases, it is usually required that there exist the following:

1) The set is ordered: $s_\alpha > s_\beta$ if $\alpha > \beta$;
2) There is the negation operator: $neg(s_\alpha) = s_{-\alpha}$.

To preserve all the given information, we extend the discrete term set S to a continuous term set $\overline{S} = \{s_\alpha \mid a \in [-q,q]\}$, where q ($q > t$) is a sufficiently large positive integer. If $s_\alpha \in S$, then we call s_α the original linguistic term, otherwise, we call s_α the virtual linguistic term. In general, the expert uses the original linguistic terms to evaluate alternatives, and the virtual linguistic terms can only appear in operation [21].

Definition 1. Let $\tilde{\alpha} = [\alpha^-, \alpha^+] = \{x \mid \alpha^- \leq x \leq \alpha^+\}$, where $\alpha^-, x, \alpha^+ \in R$, α^- and α^+ are the lower and upper limits, respectively, then $\tilde{\alpha}$ is called an interval number. Especially, if $\alpha^-, x, \alpha^+ \in R^+$, then $\tilde{\alpha}$ is called a positive interval number.

Definition 2. Let $\tilde{s} = [s_a, s_b]$, where $s_a, s_b \in \overline{S}$, s_a and s_b are the lower and upper limits, respectively, we then call $\tilde{s}$ the uncertain linguistic variable.

Let $\tilde{S}$ be the set of all uncertain linguistic variables, and let Ω be the set of all positive interval numbers. Consider any three uncertain linguistic variables $\tilde{s} = [s_a, s_b]$, $\tilde{s}_1 = [s_{a_1}, s_{b_1}]$ and $\tilde{s}_2 = [s_{a_2}, s_{b_2}]$, and any three positive interval numbers $\tilde{\alpha} = [\alpha^-, \alpha^+]$, $\tilde{\alpha}_1 = [\alpha_1^-, \alpha_1^+]$ and $\tilde{\alpha}_2 = [\alpha_2^-, \alpha_2^+]$, we define their operational laws as follows.

1) $\tilde{s}_1 \oplus \tilde{s}_2 = [s_{a_1}, s_{b_1}] \oplus [s_{a_2}, s_{b_2}] = [s_{a_1} \oplus s_{a_2}, s_{b_1} \oplus s_{b_2}] = [s_{a_1+a_2}, s_{b_1+b_2}]$;

2) $\tilde{\alpha} \otimes \tilde{s} = [\alpha^-, \alpha^+] \otimes [s_a, s_b] = [s_{a'}, s_{b'}]$, where $a' = \min\{\alpha^- a, \alpha^- b, \alpha^+ a, \alpha^+ b\}$, $b' = \max\{\alpha^- a, \alpha^- b, \alpha^+ a, \alpha^+ b\}$;

3) $\tilde{s}_1 \oplus \tilde{s}_2 = \tilde{s}_2 \oplus \tilde{s}_1$;

4) $\tilde{\alpha} \otimes \tilde{s} = \tilde{s} \otimes \tilde{\alpha}$;

5) $\tilde{\alpha} \otimes (\tilde{s}_1 \oplus \tilde{s}_2) = \tilde{\alpha} \otimes \tilde{s}_1 \oplus \tilde{\alpha} \otimes \tilde{s}_2$;

6) $(\tilde{\alpha}_1 \oplus \tilde{\alpha}_2) \otimes \tilde{s} = \tilde{\alpha}_1 \otimes \tilde{s} \oplus \tilde{\alpha}_2 \otimes \tilde{s}$.

Definition 3. Let $\tilde{s}_1 = [s_{a_1}, s_{b_1}]$ and $\tilde{s}_2 = [s_{a_2}, s_{b_2}]$ be two uncertain linguistic variables, and let $l_{\tilde{s}_1} = b_1 - a_1$ and $l_{\tilde{s}_2} = b_2 - a_2$, then the degree of possibility of $\tilde{s}_1 \geq \tilde{s}_2$ is defined as

$$p(\tilde{s}_1 \geq \tilde{s}_2) = \frac{\min\{l_{\tilde{s}_1} + l_{\tilde{s}_2}, \max(b_1 - a_2, 0)\}}{l_{\tilde{s}_1} + l_{\tilde{s}_2}} \tag{1}$$

Similarly, the degree of possibility of $\tilde{s}_2 \geq \tilde{s}_1$ is defined as

$$p(\tilde{s}_2 \geq \tilde{s}_1) = \frac{\min\{l_{\tilde{s}_1} + l_{\tilde{s}_2}, \max(b_2 - a_1, 0)\}}{l_{\tilde{s}_1} + l_{\tilde{s}_2}} \tag{2}$$

From Definition 3, we have the following useful result:

Theorem 1. Let $\tilde{s}_1 = [s_{a_1}, s_{b_1}]$ and $\tilde{s}_2 = [s_{a_2}, s_{b_2}]$ be two uncertain linguistic variables, then

1) $0 \leq p(\tilde{s}_1 \geq \tilde{s}_2) \leq 1$, $0 \leq p(\tilde{s}_2 \geq \tilde{s}_1) \leq 1$;

2) $p(\tilde{s}_1 \geq \tilde{s}_2) + p(\tilde{s}_2 \geq \tilde{s}_1) = 1$. Especially, $p(\tilde{s}_1 \geq \tilde{s}_1) = p(\tilde{s}_2 \geq \tilde{s}_2) = 1/2$.

Proof. 1) Since

$$\max(b_1 - a_2, 0) \geq 0, \ l_{\tilde{s}_1} = b_1 - a_1 \geq 0, \ l_{\tilde{s}_2} = b_2 - a_2 \geq 0$$

then

$$\min\{l_{\tilde{s}_1} + l_{\tilde{s}_2}, \max(b_1 - a_2, 0)\} \geq 0$$

By (1), we have $p(\tilde{s}_1 \geq \tilde{s}_2) \geq 0$. Also since

$$\min\{l_{\tilde{s}_1} + l_{\tilde{s}_2}, \max(b_2 - a_1, 0)\} \leq l_{\tilde{s}_1} + l_{\tilde{s}_2}$$

then by (1), we have $p(\tilde{s}_1 \geq \tilde{s}_2) \leq 1$. Thus $0 \leq p(\tilde{s}_1 \geq \tilde{s}_2) \leq 1$. Similarly, by (2), we can prove that $0 \leq p(\tilde{s}_2 \geq \tilde{s}_1) \leq 1$.

3) Since

$$p(\tilde{s}_1 \geq \tilde{s}_2) + p(\tilde{s}_2 \geq \tilde{s}_1) = \frac{\min\{l_{\tilde{s}_1} + l_{\tilde{s}_2}, \max(b_1 - a_2, 0)\}}{l_{\tilde{s}_1} + l_{\tilde{s}_2}} + \frac{\min\{l_{\tilde{s}_1} + l_{\tilde{s}_2}, \max(b_2 - a_1, 0)\}}{l_{\tilde{s}_1} + l_{\tilde{s}_2}}$$

$$= \frac{\min\{l_{\tilde{s}_1} + l_{\tilde{s}_2}, \max(b_1 - a_2, 0)\} + \min\{l_{\tilde{s}_1} + l_{\tilde{s}_2}, \max(b_2 - a_1, 0)\}}{l_{\tilde{s}_1} + l_{\tilde{s}_2}} \quad (3)$$

then

(i) If $b_1 \leq a_2$, then by (1), we know that $p(\tilde{s}_1 \geq \tilde{s}_2) = 0$. By $b_1 \leq a_2$, we have

$$b_2 - a_1 \geq b_2 - a_1 + b_1 - a_2 = (b_2 - a_2) + (b_1 - a_1) = l_{\tilde{s}_2} + l_{\tilde{s}_1}$$

thus, by (2), we have $p(\tilde{s}_2 \geq \tilde{s}_1) = 1$. Hence, $p(\tilde{s}_1 \geq \tilde{s}_2) + p(\tilde{s}_2 \geq \tilde{s}_1) = 1$.

(ii) If $b_2 \leq a_1$, then, by (2), we know that $p(\tilde{s}_2 \geq \tilde{s}_1) = 0$. By $b_2 \leq a_1$, we have

$$b_1 - a_2 \geq b_1 - a_2 + b_2 - a_1 = (b_2 - a_2) + (b_1 - a_1) = l_{\tilde{s}_2} + l_{\tilde{s}_1}$$

thus, by (1), we have $p(\tilde{s}_1 \geq \tilde{s}_2) = 1$. Hence, $p(\tilde{s}_1 \geq \tilde{s}_2) + p(\tilde{s}_2 \geq \tilde{s}_1) = 1$.

(iii) If $a_1 \leq a_2 < b_1 \leq b_2$, then

$$b_1 - a_2 < b_1 - a_2 + b_2 - a_1 = (b_2 - a_2) + (b_1 - a_1) = l_{\tilde{s}_2} + l_{\tilde{s}_1} \quad (4)$$

and

$$b_2 - a_1 < b_2 - a_1 + b_1 - a_2 = (b_2 - a_2) + (b_1 - a_1) = l_{\tilde{s}_2} + l_{\tilde{s}_1} \quad (5)$$

thus, by (3), we have

$$p(\tilde{s}_1 \geq \tilde{s}_2) + p(\tilde{s}_2 \geq \tilde{s}_1) = \frac{b_1 - a_2 + b_2 - a_1}{l_{\tilde{s}_1} + l_{\tilde{s}_2}} = \frac{(b_2 - a_2) + (b_1 - a_1)}{l_{\tilde{s}_1} + l_{\tilde{s}_2}} = \frac{l_{\tilde{s}_2} + l_{\tilde{s}_1}}{l_{\tilde{s}_1} + l_{\tilde{s}_2}} = 1 \quad (6)$$

(iv) If $a_1 < a_2 \leq b_2 < b_1$, it follows that (4) - (6) hold.
(v) If $a_2 < a_1 \leq b_1 < b_2$, it follows that (4) - (6) hold.
(vi) If $a_2 \leq a_1 < b_2 \leq b_1$, it follows that (4) - (6) hold.

By (i)-(vi), we know that $p(\tilde{s}_1 \geq \tilde{s}_2) + p(\tilde{s}_2 \geq \tilde{s}_1) = 1$ always holds. Especially, if $\tilde{s}_1 = \tilde{s}_2$, i.e., $s_{a_1} = s_{a_2}$ and $s_{b_1} = s_{b_2}$, then

$$p(\tilde{s}_1 \geq \tilde{s}_1) + p(\tilde{s}_1 \geq \tilde{s}_1) = 1 \text{ or } p(\tilde{s}_2 \geq \tilde{s}_2) + p(\tilde{s}_2 \geq \tilde{s}_2) = 1$$

and thus

$$p(\tilde{s}_1 \geq \tilde{s}_1) = p(\tilde{s}_2 \geq \tilde{s}_2) = 1/2$$

This completes the proof of Theorem 1.

3 A Method Based on IA Operator for MAGDM with Uncertain Linguistic Information

Definition 4. An interval aggregation (IA) operator is a mapping $IA: \tilde{S}^n \to \tilde{S}$, such that

$$IA_{\tilde{w}}(\tilde{s}_1, \tilde{s}_2, ..., \tilde{s}_n) = \tilde{w}_1 \otimes \tilde{s}_1 \oplus \tilde{w}_2 \otimes \tilde{s}_2 \oplus \cdots \oplus \tilde{w}_n \otimes \tilde{s}_n$$

where $\tilde{w} = (\tilde{w}_1, \tilde{w}_2, ..., \tilde{w}_n)^T$ is the weighting vector of n arguments $\tilde{s}_j = [s_{a_j}, s_{b_j}]$, $\tilde{w}_j = [w_j^-, w_j^+] \in \Omega$, $j = 1, 2, ..., n$, $\sum_{j=1}^{n} w_j^- \leq 1$, $\sum_{j=1}^{n} w_j^+ \geq 1$.

Example 1. Assume $\tilde{s}_1 = [s_{-1}, s_1]$, $\tilde{s}_2 = [s_{-3}, s_{-2}]$, $\tilde{s}_3 = [s_{-1}, s_0]$ and $\tilde{s}_4 = [s_{-4}, s_{-2}]$, and $\tilde{w} = ([0.1, 0.2], [0.1, 0.3], [0.2, 0.4], [0.3, 0.5])^T$, then

$$IA_{\tilde{w}}(\tilde{s}_1, \tilde{s}_2, \tilde{s}_3, \tilde{s}_4) = [0.1, 0.2] \otimes [s_{-1}, s_1] \oplus [0.1, 0.3] \otimes [s_{-3}, s_{-2}]$$
$$\oplus [0.2, 0.4] \otimes [s_{-1}, s_0] \oplus [0.3, 0.5] \otimes [s_{-4}, s_{-2}]$$
$$= [s_{-0.2}, s_{0.2}] \oplus [s_{-0.9}, s_{-0.2}] \oplus [s_{-0.4}, s_0] \oplus [s_{-2}, s_{-0.6}]$$
$$= [s_{-3.5}, s_{-0.6}]$$

Based on the IA operator and the formula for the comparison between two uncertain linguistic variables, in the following we develop a method for the MAGDM problems, in which the information about the attribute weights and the expert weights are interval numbers, and the attribute values take the form of uncertain linguistic information.

Step 1. For a MAGDM problem with uncertain linguistic information, there exist a finite of alternatives $X = \{x_1, x_2, ..., x_n\}$ and a finite set of attributes $U = \{u_1, u_2, ..., u_m\}$, and there exists a finite set of experts $D = \{d_1, d_2, ..., d_l\}$. Let $\tilde{w} = (\tilde{w}_1, \tilde{w}_2, ..., \tilde{w}_m)^T$ be the weight vector of attributes, where

$$\tilde{w}_i = [w_i^-, w_i^+] \in \Omega, \ i = 1, 2, ..., m, \ \sum_{i=1}^{m} w_i^- \leq 1, \ \sum_{i=1}^{m} w_i^+ \geq 1$$

and let $\tilde{\lambda} = (\tilde{\lambda}_1, \tilde{\lambda}_2, ..., \tilde{\lambda}_l)^T$ be the weight vector of experts, where

$$\tilde{\lambda}_k = [\lambda_k^-, \lambda_k^+] \in \Omega, \ k = 1, 2, ..., l, \ \sum_{k=1}^{l} \lambda_k^- \leq 1, \ \sum_{k=1}^{l} \lambda_k^+ \geq 1$$

Suppose that $\tilde{A}_k = (\tilde{a}_{ij}^{(k)})_{m \times n}$ is the decision matrix with uncertain linguistic information, where $\tilde{a}_{ij}^{(k)} \in \tilde{S}$ is a preference value (attribute value), given by the expert $d_k \in D$, for the alternative $x_j \in X$ with respect to the criterion $u_i \in U$.

Step 2. Utilize the IA operator

$$\tilde{a}_j^{(k)} = IA_{\tilde{w}}\left(\tilde{a}_{1j}^{(k)}, \tilde{a}_{2j}^{(k)}, ..., \tilde{a}_{mj}^{(k)}\right) = \tilde{w}_1 \otimes \tilde{a}_{1j}^{(k)} \oplus \tilde{w}_2 \otimes \tilde{a}_{2j}^{(k)} \oplus \cdots \oplus \tilde{w}_m \otimes \tilde{a}_{mj}^{(k)}$$
$$k = 1,2,..., l;\ j = 1,2,..., n$$

to aggregate the preference information $\tilde{a}_{ij}^{(k)}$ $(i = 1,2,..., m)$ in the jth column of the $\tilde{A}_k$, and then get the preference degree $a_j^{(k)}$ $(j = 1,2,..., n)$ of the alternative x_j over all the other alternatives (corresponding to $d_k \in D$)

Step 3. Utilize the IA operator

$$\tilde{a}_j = IA_{\tilde{\lambda}}\left(\tilde{a}_j^{(1)}, \tilde{a}_j^{(2)}, ..., \tilde{a}_j^{(l)}\right) = \tilde{\lambda}_1 \otimes \tilde{a}_j^{(1)} \oplus \tilde{\lambda}_2 \otimes \tilde{a}_j^{(2)} \oplus \cdots \oplus \tilde{\lambda}_l \otimes \tilde{a}_j^{(l)},$$
$$j = 1,2,..., n$$

to aggregate $\tilde{a}_j^{(k)}$ $(k = 1,2,..., l)$ corresponding to the alternative x_j, and then get the collective preference degree $\tilde{a}_j$ of the alternative x_j over all the other alternatives.

Step 4. To rank these collective preference degrees $\tilde{a}_j$ $(j = 1,2,..., n)$, we first compare each $\tilde{a}_i$ with all $\tilde{a}_j$ $(j = 1,2,..., n)$ by using (1). For simplicity, we let $p_{ij} = p(\tilde{a}_i \geq \tilde{a}_j)$, then we develop a complementary matrix [23-26] as $P = (p_{ij})_{n \times n}$, where

$$p_{ij} \geq 0,\ p_{ij} + p_{ji} = 1,\ p_{ii} = 1/2,\ i, j = 1,2,..., n$$

Summing all elements in each line of matrix P, we have

$$p_i = \sum_{j=1}^{n} p_{ij},\ i = 1,2,..., n$$

Then we rank the $\tilde{a}_j$ $(j = 1,2,..., n)$ in descending order in accordance with the value of p_i $(i = 1,2,..., n)$.

Step 5. Rank all the alternatives x_j $(j = 1,2,..., n)$ and select the best one(s) in accordance with the $\tilde{a}_j$ $(j = 1,2,..., n)$.

Step 6. End.

4 Illustrative Example

In this section, a MAGDM problem of evaluating university faculty for tenure and promotion (adapted from Bryson and Mobolurin [27]) is used to illustrate the developed method.

A practical use of the proposed approach involves the evaluation of university faculty for tenure and promotion. The attributes used at some universities are u_1: teaching, u_2: research, and u_3: service (whose weight vector $w = ([0.20,0.30], [0.30,0.35], [0.35,0.40])^T$). Five faculty candidates (alternatives) x_j ($j = 1, 2, 3, 4, 5$) are to be evaluated using the term set

$$S = \{s_{-4} = extremely\ poor,\ s_{-3} = very\ poor,\ s_{-2} = poor,$$
$$s_{-1} = slightly\ poor,\ s_0 = fair,\ s_1 = slightly\ good,$$
$$s_2 = good,\ s_3 = very\ good,\ s_4 = extremely\ good\}$$

by four experts d_k ($k = 1,2,3,4$) (whose weight vector $\lambda = ([0.15,0.20], [0.23,0.25], [0.30,0.40], [0.15,0.25])^T$ under these three attributes, as listed in Tables 1-4, respectively.

Table. 1. Decision matrix $\tilde{A}_1$

u_i	x_1	x_2	x_3	x_4	x_5
u_1	$[s_2, s_3]$	$[s_0, s_2]$	$[s_2, s_4]$	$[s_3, s_4]$	$[s_1, s_2]$
u_2	$[s_3, s_4]$	$[s_1, s_2]$	$[s_2, s_3]$	$[s_0, s_1]$	$[s_0, s_2]$
u_3	$[s_1, s_3]$	$[s_1, s_2]$	$[s_3, s_4]$	$[s_2, s_3]$	$[s_0, s_1]$

Table. 2. Decision matrix $\tilde{A}_2$

u_i	x_1	x_2	x_3	x_4	x_5
u_1	$[s_0, s_1]$	$[s_1, s_2]$	$[s_0, s_2]$	$[s_3, s_4]$	$[s_2, s_4]$
u_2	$[s_2, s_4]$	$[s_1, s_2]$	$[s_0, s_2]$	$[s_0, s_1]$	$[s_3, s_4]$
u_3	$[s_{-1}, s_0]$	$[s_3, s_4]$	$[s_2, s_3]$	$[s_1, s_3]$	$[s_2, s_4]$

Table 3. Decision matrix $\tilde{A}_3$

u_i	x_1	x_2	x_3	x_4	x_5
u_1	$[s_1, s_3]$	$[s_1, s_2]$	$[s_2, s_3]$	$[s_2, s_4]$	$[s_0, s_1]$
u_2	$[s_3, s_4]$	$[s_0, s_1]$	$[s_1, s_3]$	$[s_{-1}, s_1]$	$[s_0, s_1]$
u_3	$[s_2, s_3]$	$[s_0, s_2]$	$[s_2, s_4]$	$[s_1, s_3]$	$[s_1, s_2]$

Table 4. Decision matrix $\tilde{A}_4$

u_i	x_1	x_2	x_3	x_4	x_5
u_1	$[s_0, s_1]$	$[s_2, s_3]$	$[s_3, s_4]$	$[s_0, s_2]$	$[s_{-1}, s_0]$
u_2	$[s_2, s_3]$	$[s_3, s_4]$	$[s_1, s_3]$	$[s_0, s_1]$	$[s_1, s_2]$
u_3	$[s_2, s_3]$	$[s_2, s_3]$	$[s_2, s_4]$	$[s_0, s_2]$	$[s_0, s_1]$

Step 1. Utilize the IA operator

$$\tilde{a}_j^{(k)} = IA_{\tilde{w}}\left(\tilde{a}_{1j}^{(k)}, \tilde{a}_{2j}^{(k)}, \tilde{a}_{3j}^{(k)}\right) = \tilde{w}_1 \otimes \tilde{a}_{1j}^{(k)} \oplus \tilde{w}_2 \otimes \tilde{a}_{3j}^{(k)} \oplus \tilde{w}_3 \otimes \tilde{a}_{3j}^{(k)},$$
$$k = 1,2,3,4;\ j = 1,2,3,4,5$$

to aggregate the preference information $\tilde{a}_{ij}^{(k)}$ ($i=1,2,3$) in the j th column of the $\tilde{A}_k$, and then get the preference degree $a_j^{(k)}$ of the alternative x_j over all the other alternatives (corresponding to d_k):

$\tilde{a}_1^{(1)} = [s_{1.65}, s_{3.50}]$, $\tilde{a}_2^{(1)} = [s_{0.65}, s_{2.10}]$, $\tilde{a}_3^{(1)} = [s_{2.05}, s_{3.85}]$, $\tilde{a}_4^{(1)} = [s_{1.30}, s_{2.75}]$
$\tilde{a}_5^{(1)} = [s_{0.2}, s_{1.7}]$, $\tilde{a}_1^{(2)} = [s_{0.25}, s_{1.70}]$, $\tilde{a}_2^{(2)} = [s_{1.55}, s_{2.90}]$, $\tilde{a}_3^{(2)} = [s_{0.7}, s_{2.5}]$
$\tilde{a}_4^{(2)} = [s_{0.95}, s_{2.75}]$, $\tilde{a}_5^{(2)} = [s_{2.0}, s_{4.2}]$, $\tilde{a}_1^{(3)} = [s_{1.8}, s_{3.5}]$, $\tilde{a}_2^{(3)} = [s_{0.20}, s_{1.75}]$
$\tilde{a}_3^{(3)} = [s_{1.40}, s_{3.55}]$, $\tilde{a}_4^{(3)} = [s_{0.45}, s_{2.75}]$, $\tilde{a}_5^{(3)} = [s_{0.35}, s_{1.45}]$, $\tilde{a}_1^{(4)} = [s_{1.30}, s_{2.55}]$
$\tilde{a}_2^{(4)} = [s_{2.0}, s_{3.5}]$, $\tilde{a}_3^{(4)} = [s_{1.60}, s_{3.85}]$, $\tilde{a}_4^{(4)} = [s_0, s_{1.75}]$, $\tilde{a}_5^{(4)} = [s_{0.1}, s_{1.1}]$

Step 2. Utilize the IA operator

$$\tilde{a}_j = IA_{\tilde{\lambda}}\left(\tilde{a}_j^{(1)}, \tilde{a}_j^{(2)}, \tilde{a}_j^{(3)}, \tilde{a}_j^{(4)}\right) = \tilde{\lambda}_1 \otimes \tilde{a}_j^{(1)} \oplus \tilde{\lambda}_2 \otimes \tilde{a}_j^{(2)} \oplus \tilde{\lambda}_3 \otimes \tilde{a}_j^{(3)} \oplus \tilde{\lambda}_4 \otimes \tilde{a}_j^{(4)},$$
$$j = 1,2,3,4,5$$

to aggregate $\tilde{a}_j^{(k)}$ ($k=1,2,3,4$) corresponding to the alternative x_j, and then get the collective preference degree $\tilde{a}_j$ of the alternative x_j over all the other alternatives:

$$\tilde{a}_1 = [s_{1.04}, s_{3.16}],\ \tilde{a}_2 = [s_{0.81}, s_{2.72}],\ \tilde{a}_3 = [s_{1.13}, s_{3.78}]$$
$$\tilde{a}_4 = [s_{0.55}, s_{2.78}],\ \tilde{a}_5 = [s_{0.61}, s_{2.25}]$$

Step 3. To rank these collective preference degrees $\tilde{a}_j$ ($j=1,2,3,4,5$), we first compare each $\tilde{a}_j$ with all $\tilde{a}_j$ ($j=1,2,3,4,5$) by using (1), and then develop a complementary matrix:

$$P = \begin{bmatrix} 0.5 & 0.5831 & 0.4256 & 0.6000 & 0.6782 \\ 0.4169 & 0.5 & 0.3487 & 0.6014 & 0.6845 \\ 0.5744 & 0.6513 & 0.5 & 0.6619 & 0.7389 \\ 0.4000 & 0.3986 & 0.3381 & 0.5 & 0.5607 \\ 0.3218 & 0.3155 & 0.2611 & 0.4393 & 0.5 \end{bmatrix}$$

Summing all elements in each line of matrix P, we have

$$p_1 = 2.7869,\ p_2 = 2.5515,\ p_3 = 3.1265,\ p_4 = 2.1974,\ p_5 = 1.8377$$

Then we rank the $\tilde{a}_j$ ($j=1,2,3,4,5$) in descending order in accordance with the value of p_i ($i=1,2,3,4,5$):

$$\tilde{a}_3 > \tilde{a}_1 > \tilde{a}_2 > \tilde{a}_4 > \tilde{a}_5$$

Step 4. Rank all the alternatives x_j ($j=1,2,3,4,5$) in accordance with the $\tilde{a}_j$ ($j=1,2,3,4,5$):

$$x_3 \succ x_1 \succ x_2 \succ x_4 \succ x_5$$

thus the best alternative (school) is x_3.

5 Concluding Remarks

In this paper, we have studied the MAGDM problems, in which the information about the attribute weights and the expert weights are interval numbers, and the attribute values take the form of uncertain linguistic information. We have introduced some operational laws of uncertain linguistic variables and a formula for comparing two uncertain linguistic values, and proposed the interval aggregation (IA) operator. Furthermore, we have developed a method, based on the IA operator and the formula for comparing two uncertain linguistic variables, for MAGDM with uncertain linguistic information. In the future, we shall continue working in the application and extension of the developed method.

Acknowledgement

This work was supported by China Postdoctoral Science Foundation under Project (2003034366).

References

1. Zadeh, L.A.: The Concept of a Linguistic Variable and Its Application to Approximate Reasoning. Part 1,2 and 3, Information Sciences 8(1975) 199-249, 301-357; 9(1976) 43-80.
2. Zadeh, L.A.: A Computational Approach to Fuzzy Quantifiers in Natural Languages. Computers & Mathematics with Applications 9(1983) 149-184.
3. Kacprzyk, J.: Group Decision Making with a Fuzzy Linguistic Majority. Fuzzy Sets and Systems 18(1986) 105-118.
4. Bonissone, P. P., Decker, K.S.: Selecting Uncertainty Calculi and Granularity: An Experiment in Trading-off Precision and Complexity. In: Kanal, L.H., Lemmer, J.F.(eds.): Uncertainty in Artificial Intelligence. North-Holland, Amsterdam (1986) 217-247.
5. Degani ,R., Bortolan, G.: The Problem of Linguistic Approximation in Clinical Decision Making. International Journal of Approximate Reasoning 2 (1988) 143-162.
6. Delgado, M., Verdegay, J.L., Vila, M.A.: On Aggregation Operations of Linguistic Labels. International Journal of Intelligent Systems 8(1993) 351-370.

7. Herrera, F.: A Sequential Selection Process in Group Decision Making with Linguistic Assessment. Information Sciences 85(1995) 223-239.
8. Herrera, F., Herrera-Viedma, E., Verdegay, J.L.: A Model of Consensus in Group Decision Making under Linguistic Assessments. Fuzzy Sets and Systems 78(1996) 73-87.
9. Torra, V.: Negation Functions Based Semantics for Ordered Linguistic Labels. International Journal of Intelligent Systems 11(1996) 975-988.
10. Bordogna, G., Fedrizzi, M., Passi G.: A Linguistic Modeling of Consensus in Group decision Making Based on OWA Operator. IEEE Transactions on Systems, Man, and Cybernetics 27(1997) 126-132.
11. Zadeh, L.A., Kacprzyk, J. Computing with Words in Information/Intelligent Systems-Part 1: Foundations: Part 2: Applications. Heidelberg, Germany: Physica-Verlag, 1999, vol. I.
12. Herrera, F., Herrera-Viedma, E.: Choice Functions and Mechanisms for Linguistic Preference Relations. European Journal of Operational Research 120 (2000) 144-161.
13. Herrera, F., Martínez, L.: A 2-Tuple Fuzzy Linguistic Representation Model for Computing with Words. IEEE Transactions on fuzzy systems 8 (2000) 746-752.
14. Herrera, F., Martínez, L.: A Model Based on Linguistic 2-Tuples for Dealing with Multigranular Hierarchical Linguistic Contexts in Multi-Expert Decision-Making. IEEE Transactions on Systems, Man, and Cybernetics 31 (2001) 227-234.
15. Herrera, F., Martínez, L.: An Approach for Combining Linguistic and Numerical Information Based on 2-Tuple Fuzzy Linguistic Representation Model in Decision-Making. International Journal of Uncertainty, Fuzziness, Knowledge-based Systems 8 (2002) 539-562.
16. Delgado, M., Herrera, F., Herrera-Viedma, E., Martin-Bautista, M.J, Martinez, L., Vila, M. A.: A Communication Model Based on the 2-Tuple Fuzzy Linguistic Representation for a Distributed Intelligent Agent System on Internet. Soft Computing, 6 (2002) 320-328.
17. Cordón, O., Herrera, F., Zwir, I.: Linguistic Modeling by Hierarchical Systems of Linguistic Rules. IEEE Transactions on fuzzy systems 10 (2002) 2-20.
18. Xu, Z.S.: Study on Methods for Multiple Attribute Decision Making Under Some Situations. Ph.D thesis, Southeast University, Nanjing, China (2002).
19. Xu, Z.S., Da, Q.L.: An Overview of Operators for Aggregating Information. International Journal of Intelligent Systems 18 (2003) 953-969.
20. Xu, Z.S.: A Method Based on Linguistic Aggregation Operators for Group Decision taking with Linguistic Preference Relations. Information Sciences 166 (2004) 19-30.
21. Xu, Z.S.: EOWA and EOWG Operators for Aggregating Linguistic Labels Based on Linguistic Preference Relations. International Journal of Uncertainty, Fuzziness and Knowledge-Based Systems 112 (2004) 791-810.
22. Xu, Z.S.: An Approach to Group Decision Making Based on Incomplete Linguistic Preference Relations. International Journal of Information Technology and Decision Making, 4 (2005) 153-160.
23. Xu, Z.S., Da, Q.L.: The Uncertain OWA Operator. International Journal of Intelligent Systems 17 (2002) 569-575.
24. Xu, Z.S., Da, Q.L.: An Approach to Improving Consistency of Fuzzy Preference Matrix. Fuzzy Optimization and Decision Making 2 (2003) 3-12.
25. Xu, Z.S.: Goal Programming Models for Obtaining the Priority Vector of Incomplete Fuzzy Preference Relation. International Journal of Approximate Reasoning 36 (2004) 261-270.
26. Xu, Z.S., Da, Q.L.: A Least Deviation Method to Obtain a Priority Vector of a Fuzzy Preference Relation. European Journal of Operational Research 164 (2005) 206-216.
27. Bryson, N., Mobolurin, A.: An Action Learning Evaluation Procedure for Multiple Criteria Decision Making Problems. European Journal of Operational Research 96 (1996) 379-386.

A New Prioritized Information Fusion Method for Handling Fuzzy Information Retrieval Problems

Won-Sin Hong[1], Shi-Jay Chen[2], Li-Hui Wang[3], and Shyi-Ming Chen[1]

[1] Department of Computer Science and Information Engineering, National Taiwan University of Science and Technology, Taipei, Taiwan, R. O. C.
[2] Department of Information Management, National United University, Miao-Li, Taiwan, R. O. C.
[3] Department of Finance, Chihlee Institute of Technology, Panchiao, Taipei County, Taiwan, R. O. C.

Abstract. In this paper, we present a new prioritized information fusion method for handling fuzzy information retrieval problems. We also present a new center-of-gravity method for ranking generalized fuzzy numbers. Furthermore, we also extend the proposed prioritized information fusion method for handling fuzzy information retrieval problems in the generalized fuzzy number environment, where generalized fuzzy numbers are used to represent the degrees of strength with which documents satisfy particular criteria.

1 Introduction

In [7], [8] and [9], Yager presented a conjunction operator of fuzzy subsets, called the non-monotonic/prioritized operator, to deal with multi-criteria decision-making problems. In [1], Bordogna et al. used the prioritized operator for fuzzy information retrieval. In [4], Chen et al. extended Yager's prioritized operator [8] to present a prioritized information fusion algorithm, based on generalized fuzzy numbers, for aggregating fuzzy information.

In this paper, we present a new prioritized information fusion method for handling fuzzy information retrieval problems. We also present a new center-of-gravity method for ranking generalized fuzzy numbers. Furthermore, we also extend the proposed prioritized information fusion method for handling fuzzy information retrieval problems in the generalized fuzzy number environment, where generalized fuzzy numbers are used to represent the degrees of strength with which documents satisfy particular criteria.

2 Generalized Trapezoidal Fuzzy Numbers

In [2] and [3], Chen presented a generalized trapezoidal fuzzy number $\tilde{A} = (a_1, a_2, a_3, a_4; \hat{w}_{\tilde{A}})$ as shown in Fig. 1, where $0 < w_{\tilde{A}} \leq 1$, a_1, a_2, a_3 and a_4 are positive real numbers, $a_1 \leq a_2 \leq a_3 \leq a_4$, the universe of discourse X of the generalized fuzzy number $\tilde{A}$ is characterized by the membership function $\mu_{\tilde{A}}$, and $\mu_{\tilde{A}} : X \to [0,1]$.

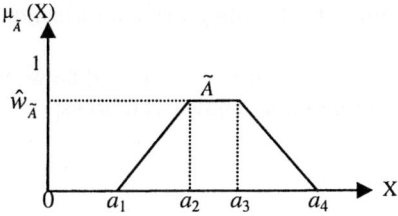

Fig. 1. A generalized trapezoidal fuzzy number $\tilde{A}$.

3 A New Center-of-Gravity Method for Ranking Generalized Fuzzy Numbers

In the following, we present a new method to calculate the center-of-gravity of generalized trapezoidal fuzzy numbers and to rank generalized trapezoidal fuzzy numbers. Assume that there is a generalized fuzzy number $\tilde{A} = (a, b, c, d; w)$, where $0 < w \leq 1$, and a, b, c and d are real numbers. Then we can calculate its center-of-gravity $G(x_{\tilde{A}}, y_{\tilde{A}})$ as follows:

$$G(x_{\tilde{A}}, y_{\tilde{A}}) = (\frac{m}{m+n} \times \frac{a+b+d}{3} + \frac{n}{m+n} \times \frac{b+c+d}{3}, \frac{m}{m+n} \times \frac{1}{3}w + \frac{n}{m+n} \times \frac{2}{3}w), \quad (1)$$

where $m = d - a$ and $n = c - b$. Then, based on the center-of-gravity $G(x_{\tilde{A}}, y_{\tilde{A}})$ of the generalized trapezoidal fuzzy number $\tilde{A}$, we can calculate the ranking value $R(\tilde{A})$ of the generalized trapezoidal fuzzy number $\tilde{A}$, shown as follows:

Case 1: if $a_1 = a_2 = a_3 = a_4$, then the ranking value $R(\tilde{A})$ of the generalized fuzzy number $\tilde{A}$ is calculated as follows:

$$R(\tilde{A}) = \sqrt{a_1^2 + (\frac{w}{2})^2} \times \frac{2}{\sqrt{5}}. \quad (2)$$

Case 2: if $\tilde{A}$ is a generalized trapezoidal fuzzy number, a generalized rectangular fuzzy number, or a generalized triangular fuzzy number, then the ranking value $R(\tilde{A})$ of the generalized trapezoidal fuzzy number $\tilde{A}$ is calculated as follows:

$$R(\tilde{A}) = \sqrt{(x_{\tilde{A}})^2 + (y_{\tilde{A}})^2} \times \frac{2}{\sqrt{5}}. \quad (3)$$

4 A New Prioritized Information Fusion Method for Handling Fuzzy Information Retrieval Problems

Assume that there are n documents $d_1, d_2, ..., d_n$, and assume that there are two different prioritized criteria A_1 and A_2, where A_1 is the first order criterion, A_2 is the second order criterion, $A_1(d_i)$ denotes the degree of strength that document d_i satisfies the first order criterion A_1, $A_1(d_i) \in [0, 1]$, $A_2(d_i)$ denotes the degree of strength that document d_i satisfies the second order criterion A_2, and $A_2(d_i) \in [0, 1]$. The proposed prioritized information fusion method for handling fuzzy information retrieval problems is now presented as follows:

Step 1: Calculate the degree of correlation $r(A_1, A_2)$ [6] between the first order criterion A_1 and the second order criterion A_2, shown as follows:

$$r(A_1, A_2) = \frac{\sum_{i=1}^{n}\left(A_1(d_i) - \overline{A_1(d_i)}\right)\left(A_2(d_i) - \overline{A_2(d_i)}\right)}{\sqrt{\sum_{i=1}^{n}\left(A_1(d_i) - \overline{A_1(d_i)}\right)^2}\sqrt{\sum_{i=1}^{n}\left(A_2(d_i) - \overline{A_2(d_i)}\right)^2}}, \qquad (4)$$

where $A_1(d_i)$ denotes the degree of strength that document d_i satisfies the first order criterion A_1, $A_2(d_i)$ denotes the degree of strength that document d_i satisfies the second order criterion A_2, $\overline{A_1(d_i)}$ denotes the mean value of A_1, $\overline{A_2(d_i)}$ denotes the mean value of A_2, and $r(A_1, A_2) \in [-1, 1]$.

Step 2: Calculate the degree of compatibility $C(A_1, A_2)$ between the first order criterion A_1 and the second order criterion A_2, shown as follows:

$$C(A_1, A_2) = \frac{1+r}{2}, \qquad (5)$$

where r denotes the degree of correlation between the first order criterion A_1 and the second order criterion A_2, and $C(A_1, A_2) \in [0, 1]$.

Step 3: Calculate the fusion result $F(d_i)$ of document d_i, where $1 \leq i \leq n$, shown as follows:

$$F(d_i) = \left(A_1(d_i) \times \frac{1}{1+C(A_1, A_2)}\right) + \left(A_2(d_i) \times \frac{C(A_1, A_2)}{1+C(A_1, A_2)}\right), \qquad (6)$$

where $A_1(d_i)$ denotes the degree of strength that document d_i satisfies the first order criterion A_1, $A_2(d_i)$ denotes the degree of strength that document d_i satisfies the second order criterion A_2, and $C(A_1, A_2)$ denotes the degree of compatibility between the first order criterion A_1 and the second order criterion A_2. The larger the value of $F(d_i)$, the more the degree of satisfaction that document d_i satisfies the user's query, where $1 \leq i \leq n$.

Assume that there are n documents $d_1, d_2, \ldots,$ and d_n, and assume that there are two different prioritized criteria A_1 and A_2, where A_1 is the first order criterion, A_2 is the second order criterion, $\tilde{A}_1(d_i)$ is a generalized trapezoidal fuzzy number denoting the degree of strength that document d_i satisfies the first order criterion A_1, and $\tilde{A}_2(d_i)$ is a generalized trapezoidal fuzzy number denoting the degree of strength that document d_i satisfies the second order criterion A_2. The proposed prioritized information fusion method for handling fuzzy information retrieval problems in the generalized trapezoidal fuzzy number environment is now presented as follows:

Step 1: Based on formula (1), defuzzify generalized trapezoidal fuzzy numbers $\tilde{A}_1(d_i)$ and $\tilde{A}_2(d_i)$, where $1 \leq i \leq n$. Based on formulas (2) and (3), calculate the ranking values $R(\tilde{A}_1(d_i))$ and $R(\tilde{A}_2(d_i))$ of $\tilde{A}_1(d_i)$ and $\tilde{A}_2(d_i)$, respectively, where $1 \leq i \leq n$. Let $A_1(d_i) = R(\tilde{A}_1(d_i))$ and let $A_2(d_i) = R(\tilde{A}_2(d_i))$, where $1 \leq i \leq n$.

Step 2: Based on formula (4), calculate the degree of correlation $r(A_1, A_2)$ between the first order criterion A_1 and the second order criterion A_2.
Step 3: Based on formula (5), calculate the degree of compatibility $C(A_1, A_2)$ between the first order criterion A_1 and the second order criterion A_2.
Step 4: Based on formula (6), calculate the fusion result $F(d_i)$ of document d_i, where $1 \leq i \leq n$. The larger the value of $F(d_i)$, the more the degree of strength that document d_i satisfies the user's query, where $1 \leq i \leq n$.

5 Conclusions

In this paper, we have presented a new prioritized information fusion method for handling fuzzy information retrieval problems. We also have presented a new center-of-gravity method for ranking generalized fuzzy numbers. Furthermore, we also have extended the proposed prioritized information fusion method for handling fuzzy information retrieval problems in the generalized fuzzy number environment, where generalized fuzzy numbers are used to represent the degrees of strength with which documents satisfy particular criteria.

Acknowledgements

This work was supported in part by the National Science Council, Republic of China, under Grant NSC 94-2213-E-011-003.

References

1. Bordogna, G., Pasi, G.: Linguistic Aggregation Operators of Selection Criteria in Fuzzy Information Retrieval. International Journal of Intelligent Systems 10 (1995) 233-248
2. Chen, S.H.: Ranking Fuzzy Numbers with Maximizing Set and Minimizing Set. Fuzzy Sets and Systems 17 (1985) 113-129
3. Chen, S.H.: Ranking Generalized Fuzzy Number with Graded Mean Integration. Proceedings of the Eighth International Fuzzy Systems Association World Congress, Taipei, Taiwan, Republic of China (1999) 899-902
4. Chen, S.J., Chen, S.M.: A Prioritized Information Fusion Algorithm for Handling Multi-criteria Fuzzy Decision-Making Problems. Proceedings of the 2002 International Conference on Fuzzy Systems and Knowledge Discovery, Singapore (2002)
5. Fuller, G., Tarwater, J.D.: Analytic Geometry. Addison-Wesley, Massachusetts (1992)
6. Witte, R.S.: Statistics. Holt, Rinehart and Winston, New York (1989)
7. Yager, R.R.: Non-Monotonic Set Theoretic Operations. Fuzzy Sets and Systems 42 (1991) 173-190
8. Yager, R.R.: Second Order Structures in Multi-Criteria Decision Making. International Journal of Man-Machine Studies 36 (1992) 553-570
9. Yager, R.R.: Structures for Prioritized Fusion of Fuzzy Information. Information Sciences 108 (1998) 71-90

Multi-context Fusion Based Robust Face Detection in Dynamic Environments

Mi Young Nam and Phill Kyu Rhee

Dept. of Computer Science & Engineering Inha University,
253, Yong-Hyun Dong, Incheon, Nam-Gu, South Korea
rera@im.inha.ac.kr, pkrhee@inha.ac.kr

Abstract. We propose a method of multiple context fusion based robust face detection scheme. It takes advantage of multiple contexts by combining color, illumination (brightness and light direction), spectral composition(texture) for environment awareness. It allows the object detection scheme can react in a robust way against dynamically changing environment. Multiple context based face detection is attractive since it could accumulate face model by autonomous learning process for each environment context category. This approach can be easily used in searching for multiple scale faces by scaling up/down the input image with some factor. The proposed face detection using the multiple context fusion shows more stability under changing environments than other detection methods. We employ Fuzzy ART for the multiple context- awareness. The proposed face detection achieves the capacity of the high level attentive process by taking advantage of the context-awareness using the information from illumination, color, and texture. We achieve very encouraging experimental results, especially when operation environment varies dynamically.

1 Introduction

Context, in this paper, is modeled as the effect of the change of application working environment. The context information used here is illumination, color, texture contexts. As environment context changes, it is identified by the multiple context fusion, and the detection scheme is restructured. The detection scheme is restructured to adapt to changing environment and to perform an action that is appropriate over a range of the change. The goal of this paper is to explore the possibility of environment-insensitive face detection by adopting the concept of multiple context fusion. Context, in this paper, is modeled as the effect of the change of application working environment. The context information used here is illumination, color, texture contexts. As environment context changes, it is identified by the multiple context fusion, and the detection scheme is restructured. The detection scheme is restructured to adapt to changing environment and to perform an action that is appropriate over a range of the change.

Even though human beings can detect face easily, it is not easy to make computer to do so. The difficulties in face detection are caused by the variations in illumination, pose, complex background, etc. Robust face detection needs high invariance with regard to those variations. Much research has been done to solve this problem [3,5,6].

Jones et. al. [1] have provided the comparison of nine chrominance spaces and two parametric techniques. Difficulties in detecting face skin color comes from the fact that ambient light, image capturing devices, etc. can affect skin color too much. Recently, the concept of context-awareness has been applied for robust face detection under dynamically changing environments [7]. .

In this paper, we propose multiple context fusion for the adaptive detection scheme using illumination, color, texture information. We employ Fuzzy ART for the context modeling and awareness. It provides the capacity of the high level attentive process by the environment context-awareness in face detection. We achieve very encouraging results from extensive experiments. The outline of this paper is as follows. In section 2, we will present the architecture of the proposed face detection scheme. Adaptive color modeling using illumination context awareness will be discussed in section 3. In section 4, we will present multiple Bayesian classifies based on multiple context-awareness. We will give experimental results in section 5. Finally, we will give concluding remarks.

2 Face Detection Architecture Using Multi-context Fusion

The specific task we are discussing in this paper is a robust face detector which is adaptive against changing illumination environments. The system consists of the multi-resolution pyramid, the multi-context fusion module, face/non-face determination using multiple Bayesian classifier as shown in Fig. 1. The details will be discussed in the following.

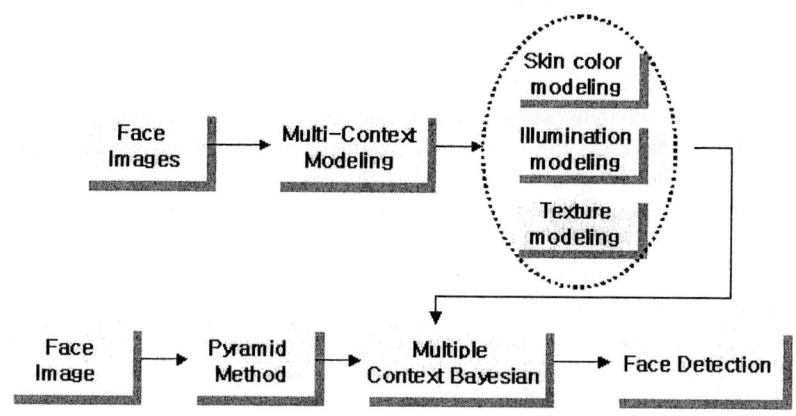

Fig. 1. Face detection system architecture

In the proposed method, the feature space for face detection with multiple scales and varying illumination condition has been partitioned into subspaces properly so that face detection error can be minimized. However, the partitioning of multiple scales is very subjective and ambiguous in many cases. The multi-resolution pyramid generates face candidate region. Pyramid method based face detection is attractive since it could accumulate the face models by autonomous learning process. In our approach, the face object scale is approximated with nine steps. In this module, 20×20 windows are generated. A tradeoff must be considered between the size of multiple scale, view representation of the object and its accuracy.

Finally, multi-context based Bayesian is in a multi classifier structure by combining several Bayesian classifiers. Each Bayesian classifier is trained using face images in each environment context category. Face candidate windows are selected using the multi-context based Bayesian classifiers, and finally face window is filtered by the merging process.

In the learning stage, the candidate face regions are clustered into 4 face models, 6 face-like non-face models, non-face models using combined learning algorithm[10], the threshold parameters of multiple Bayesian classifiers are decided. Initially, seed appearance models of face is manually gathered and classified for training the detector module. The detector with prior classes is trained by the initial seed data set.

3 Environment Modeling and Identification Using Multiple Context-Awareness

In this session, we present environment modeling and identification using multiple contexts such as color, illumination, and texture information

3.1 Illumination Context-Awareness

Environment context is analyzed and identified using the Fuzzy ART. It is hybrid method of supervised and unsupervised learning method. The idea is to train the network in two separate stages. First, we perform an unsupervised training (FuzzyART) to determine the Gaussians' parameters (j, j). In the second stage, the multiplicative weights w_{kj} are trained using the regular supervised approach. Input pattern is vectorized for image size of 20x20 pixels, input node had mosaic of size of 10x10 pixels. 10x10 image is converted 1x100 vector as show in Fig. 2.

Fig. 3 shows images of three clusters various illuminant face dataset, we define 3 step illuminant environment.

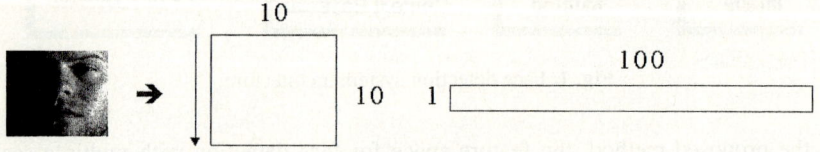

Fig. 2. Face data vectorization is 1x100 dimension

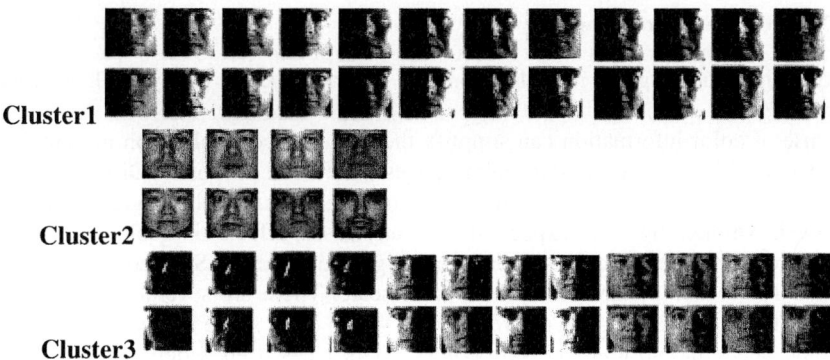

Fig. 3. The three group Discriminant result for illumination conditions

Fig. 4. The five group Discriminant result for illumination conditions

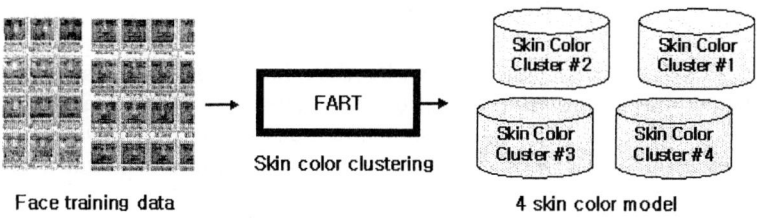

Fig. 5. The details of skin color modeling

3.2 Color Context Modeling

The skin color modeling and identification also is carried out by Fuzzy ART network as shown Fig.5.

The use of color information can simplify the task of face localization in complex environments [9, 10]. Several skin color spaces have been proposed. Difficulties of detecting face skin color comes from no universal skin color model can be determined. Ambient light, un-expectable in many cases, affects skin color. Different cameras produce different colors under same light conditions. Skin color's area is computed 1,200 face images of 20 x 20 pixel scale. There are several color spaces such YCrCb, HSV, YCbCr, CbCr. Even though which color space is most effective in separating face skin and non-face skin colors, YCbCr has been widely used [6]. We choose CrCb color space which almost approaches the performance of YCbCr color space and provides faster execution time than other color spaces. Table 1 is shown context-based skin color histogram.

Table 1. The skin color histogram distribution for each cluster

skin color area	Cluster 1	Cluster 2	Custer 3	Cluster 4
Cr value	125-160	100-130	80-100	60-100
Cb value	90-120	110-140	125-160	100-125

3.3 Texture Context Modeling

We also take advantage of the information of frequency domain using Haar wavelet transform. It is organize the image into sub-bands that are localized in orientation and frequency. In each sub-bands, each coefficient is spatially localized. We use a wavelet transform based on 3 level decomposition producing 10 sub-bands. Generally, low frequency component has more discriminate power than higher. Also too high component has some noisy information. Then we use 3 level decomposition of Haar wavelet transform. The Fuzzy ART network is used for the texture context-awareness using Haar wavelet transform.

4 Multiple Context Based Bayesian Classifiers

We show that illumination context based Bayesian face modeling can be improved by strengthening both contextual modeling and statistical characterization. Multiple Bayesian classifiers are adopted for deciding face and non-face. A Bayesian classifier is produced for each illumination context. We have modeling four Multi class based Bayesian classifiers are made in 4 group for face.

Initially, seed appearance models of face is manually gathered and classified for training the appearance detector module. Each detector, Bayesian classifier here, with prior classes is trained by the training data set. Training set of faces is gathered and divided into several group in accordance with illumination contexts. Multi-resolution

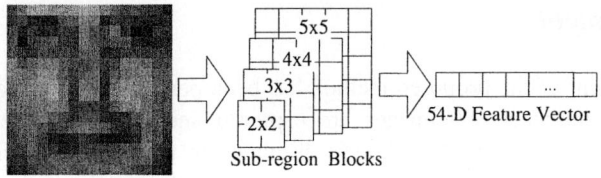

Fig. 6. Face feature extraction[8]

pyramid consists of nine levels by an experiment, and the offset between adjacent levels is established by four pixels. Each training face image is scaled to 20x20 and normalized using the max-min value normalization method, and vectorized. Since the vectorization method affects much to face detection performance, we have tested several vectorization methods and decided 54 dimensional hierarchical vectorization method [11, 12] (see Fig. 6). We use sub-regions for the feature of first classifier. Illumination the training images to a number of predefined clustering group by FART. The detector trained each face illuminant cluster.

Squared Mahalanobis distance[5] a useful method that measures free model and free pattern relationship using similarity degree, is adopted for the Bayesian classifier. The center of cluster is determined by the mean vector, and the shape of the cluster is determined by the covariance matrix Mahalanobis distance as shown in Eq. 1.

$$r^2 = (x - \mu_x)' \Sigma^{-1} (x - \mu_x) \tag{1}$$

where μ_x is the average vector of face's vector, Σ is the covariance matrix of independent elements. We use 4 face group that is shown Fig. 4. This is generated by proposed learning method. The y axis is cluster number and x axis is sample under each cluster. The advantages of the two-stage training are that it is fast and very easy to interpret. It is especially useful when labeled data is in short supply, since the first stage can be performed on all the data (not only the labeled part).

Fig.7, shows the face candidate of Fig.7: blue box is face candidate area and white box is exact face area.

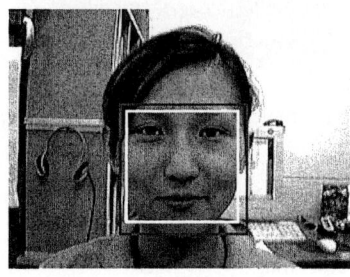

Fig. 7. The face detection using proposed method (skin color area and context based Bayesian)

4 Experiment

The experiment of the proposed method has been performed with images captured in various environments 800 images are captured and used in the experiment. The superiority of the proposed method is discussed in the following. Fig.8 shows face detection in real-time image by using the proposed cascade of context-based skin color and context-based Bayesian methods.

Face detection result shows Table 2-6. Images have size of 320 x 240 pixels and encoded in 256 color levels. We resized to various the size using multi-resolution of 9 steps. Rescaled images are transferred. Each face is normalized into a re-scaled size of 20x20 pixels and each data – training image and test images – is preprocessed by histogram equalization and max-min value normalization. Tables compare the detection rates (AR) for the context-based color system and the numbers of context-based Bayesian classifiers for the three boosting-based systems, given the number of false alarms (FA) and false rejection (FR).

Table 2 shows that the result of face detection between multi context-based Bayesian and single context based Bayesian in our lab gray images.

Table 3 shows that the result of face detection between multi context based Bayesian and single context based Bayesian in our lab color images.

Table 4 shows that the result of face detection between multiple context based Bayesian and single context based Bayesian in FERET fafb dataset.

Table 5 shows that the result of face detection between multiple context based Bayesian and single context based Bayesian in FERET fafc dataset include dark illuminant image.

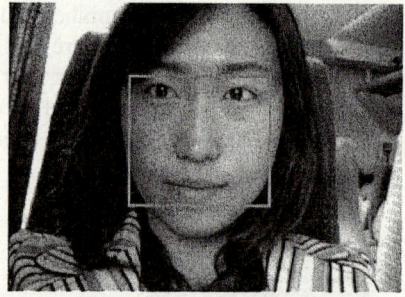

Fig. 8. Examples face detection in color image

Table 2. The face detection our lab gray images

Method	Total Image	AR	FR	FA
Multiple Context Model	800	790	5	5
Single Context Model	800	755	30	16

Table 3. Face detection on our lab in color images

Method	Total Image	AR	FR	FA
Context-based Bayesian	900	895	5	10
Single Bayesian	900	780	45	50

Table 4. Face detection result in FERET dataset normal images

Method	Total Image	AR	FR	FA
Multiple Context Model	1196	1192	5	10
Single Context Model	1196	1130	21	40

Table 5. Face detection result in FERET dataset dark images

Method	Total Image	Accept	FR	FA
Multiple Context Model	194	192	0	3
Normal Context Model	194	130	25	45

From Tables, we know that propose method is good face detection performance other method. By we were going to do studying repeatedly, clustering's performance improved. As seen in tables and figures, error in front side face appears in tilted image. We must consider about tilted image. We could improve pose classification's performance for face through recursive studying like experiment result. The combined effect of eye lasses is also investigated. In this experiment, the factor of glasses and illumination was considered and experimental images were classified by the factor of glasses and illumination. We classified bad illumination images into the image including a partially lighted face, good images into that including a nearly uniformly lighted face. Next figure is shown the face detection in CMU dataset.

5 Concluding Remarks

This paper discusses a cascade detection scheme by combining the color, texture, and feature-based method with the capability of multi context-awareness. Even though much research has been done for object detection, it still remains a difficult issue. The detection scheme aims at robustness as well as fast execution way under dynamically changing context. Difficulties in detecting face coming from the variations in ambient light, image capturing devices, etc. has been resolved by employing Fuzzy ART for multi context-awareness. The proposed face detection achieves the capacity of the high level attentive process by taking advantage of the illumination, color and texture context-awareness. We achieve very encouraging experimental results, especially when illumination condition varies dynamically. Experimental result has shown that the proposed system has detected the face successfully 99% of the offline image under varying face images. The context information used here is illumination, color, texture contexts. As environment context changes, it is identified by the multiple context fusion, and the detection scheme is restructured.

Reference

1. Jones, M.J.; Rehg and J.M.;"Statistical color models with application to skin detection," Computer Vision and Pattern Recognition, 1999. IEEE Computer Society Conference, Vol.1, (1999), pp.23-25
2. B.K.L. Erik Hjelmas, "Face Detection: A Survey," Computer Vision and Image Understanding, Vol. 3, no.3, (2001) pp. 236-274
3. T. V. Pham, et. Al., „Face detection by aggregated Bayesian network classifiers, „ Pattern Recognition Letters 23 (2002) pp. 451-461
4. Li, S.Z.; Zhenqiu Zhang; "FloatBoost learning and statistical face detection," Pattern Analysis and Machine Intelligence, IEEE Transactions on , Vol.26, (2004) pp.1112 – 1123
5. C. Liu, "A Bayesian Discriminating Features Method for Face Detection," IEEE Trans. Pattern Analysis and Machine Intelligence, vol. 25, no. 6, (2003), pp. 725-740
6. H. Schneiderman and T. Kanade, "Object Detection Using the Statistics of Parts," Int'l J. Computer Vision, Vol. 56, no. 3, (2004), pp.151- 177
7. Context Driven Observation of Human Activity
8. M.Y Nam and P.K Rhee, "A Scale and Viewing Point Invariant Pose Estimation", KES2005,(2004), pp. 833-842
9. D. Maio and D. Maltoni, "Real-time face location on gray-scale static images," Pattern Recognition, vol.33, no. 9,(2000), pp. 1525-1539
10. M. Abdel-Mottaleb and A. Elgammal, "Face Detection in complex environments from color images," IEEE ICIP, (1999), pp. 622-626
11. P. Viola and M. Jones, "Robust Real Time Object Detection," IEEE ICCV Workshop Statistical and Computational Theories of Vision, July (2001).
12. S.Z. Li, L. Zhu, Z.Q. Zhang, A. Blake, H. Zhang, and H. Shum, "Statistical Learning of Multi-View Face Detection," Proc. European Conf. Computer Vision, Vol. 4, (2002), pp. 67-81

Unscented Fuzzy Tracking Algorithm for Maneuvering Target

Shi-qiang Hu[1], Li-wei Guo[1], and Zhong-liang Jing[2]

[1] College of Informatics and Electronics, Hebei University of Science and Technology,
Shijiazhuang, 050054, China
sqhu@mail.sjtu.edu.cn
[2] Institute of Informatics and Electronics, Shanghai Jiaotong University,
Shanghai, 200030, China

Abstract. A novel adaptive algorithm for tracking maneuvering targets is proposed in this paper. The algorithm is implemented with fuzzy filtering and unscented transformation. A fuzzy system allows the filter to tune the magnitude of maximum accelerations to adapt to different target maneuvers. Unscented transformation act as a method for calculating the statistics of a random vector. A bearing-only tracking scenario simulation results show the proposed algorithm has a robust advantage over a wide range of maneuvers.

1 Introduction

In the past years, the problem of tracking maneuvering targets has been attracted attention in many fields [1]. The interacting multiple model (IMM) algorithm and the current statistical model and adaptive filtering (CSMAF) algorithm are two effective algorithms for tracking maneuvering targets[2]. In the IMM algorithm, at time k the state estimate is computed under each possible current model using r filters, with each filter using a different combination of the previous model-conditioned estimates. When the range of target maneuvers is wider, the IMM algorithm needs more filters and its computational burden is more considerable. The CSMAF algorithm takes advantage of current model concept to realize adaptive tracking for maneuvering targets using only one filter. It is obvious that the computational amount of this algorithm is much less than that of the IMM algorithm. However, its performance depends on pre-defined parameter maximum acceleration α_{max} indicating the intensity of maneuver. During tracking process, the value of α_{max} is constant so that it is difficult to suit to different movement of target. For solving this problem, we propose a fuzzy system incorporated with CSMAF to online adjust the value of α_{max}.

In tracking applications, the measurements are directly available in the original sensor coordinates, which maybe give rise to nonlinear problem. The classic algorithm to solve the non-linear filtering is extended Kalman filter (EKF). But this filter has two important potential drawbacks. Firstly, the derivation of the Jacobian matrices can be complex causing implementation difficulties. Secondly, these linearization's can introduce large errors and even cause divergence of the filter. To address these limitations, Julier and Uhlmann developed an unscented transformation (UT) instead of linearization method [3], [4]. Unscented filter based on UT has been shown to be a superior alternative to the EKF in a variety of applications.

An unscented fuzzy filtering (UFCSMAF) algorithm for tracking maneuvering targets is proposed in this paper. Fuzzy system is used in CSMAF model to adjust magnitude of maximum accelerations to adapt to different target maneuvers. A UT is combined with fuzzy CSMAF model for tracking maneuvering targets. The structure of our proposed algorithm is shown in Figure 1. Compared with the CSMAF algorithm, this new adaptive algorithm has a robust advantage over a wide range of maneuvers and overcomes the shortcoming of the CSMAF algorithm.

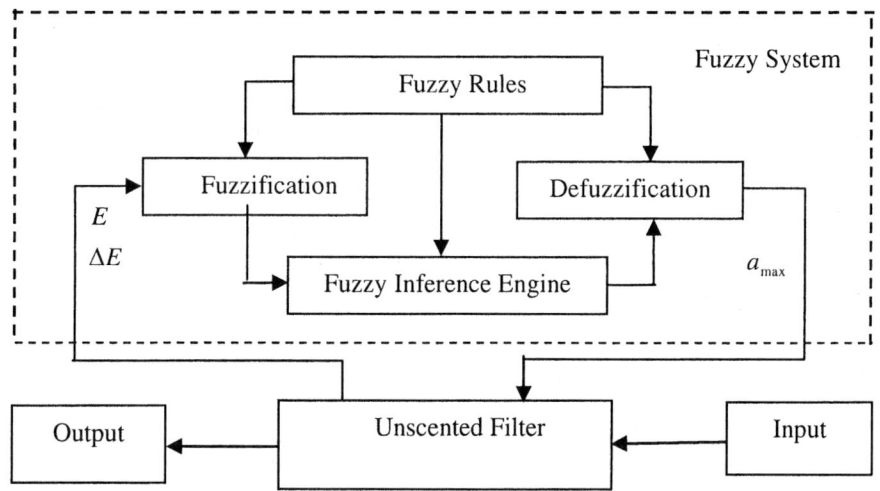

Fig. 1. The structure of Unscented Fuzzy Filter

The remainder of this paper is organized as follows. In section 2, the CSMAF algorithm is analyzed. In section 3, the fuzzy CSMAF is given. An unscented fuzzy CSMAF is proposed in section 4. In section 5 the simulation experiments have carried out to prove the filter performance. Finally, conclusions are given in section 6.

2 Adaptive Filtering and Fuzzy Filtering

2.1 Current Statistical Models and Adaptive Filtering

The CSMAF algorithm assumes that when a target is maneuvering with certain acceleration, its acceleration during the next period is limited within a range around current acceleration. Hence it is not necessary to take all possible values of maneuvering acceleration into consideration when modeling the target acceleration probability. A modified Rayleigh density function whose mean is the current acceleration is utilized, and the relationship between the mean and variance of Rayleigh density is used to set up an adaptive algorithm for the variance of maneuvering acceleration.

The state equation of the current statistical model is

$$\mathbf{x}(k+1) = \mathbf{F}\mathbf{x}(k) + \mathbf{G}\bar{a}(k) + \mathbf{B}_w w(k) \tag{2.1}$$

Considering the one-step-ahead prediction of $\ddot{x}(k)$ as the current acceleration and also as the mean of randomly maneuvering acceleration at the instant kT, we have

$$\bar{a}(k) = \hat{\ddot{x}}(k/k-1) \tag{2.2}$$

$$\sigma_w^2 = 2\alpha\sigma_a^2 = \begin{cases} \dfrac{2\alpha(4-\pi)}{\pi}\left[a_{max} - \bar{a}(k)\right]^2 & \bar{a}(k) > 0 \\ \dfrac{2\alpha(4-\pi)}{\pi}\left[a_{-max} - \bar{a}(k)\right]^2 & \bar{a}(k) < 0 \end{cases} \tag{2.3}$$

From Eqs.(2.3) we see the following facts: (1) when a_{max} is large, the algorithm can keep fast response to target maneuvers because the system variance or system frequency band takes a large value; (2) when a_{max} is a constant, if the target is maneuvering with a smaller acceleration, the system variance will be larger and the tracking precision will be lower; if the target is maneuvering with a larger acceleration, the system variance will be smaller and the tracking precision will be higher. Therefore a_{max} is an important parameter in filter design. But in real application, it is difficult to give an appropriate value to suit an unknown trajectory.

2.2 Fuzzy Filtering Based on Current Statistical Model

Since α_{max} is predefined and affects the system variance, we use a fuzzy system to modify the value of α_{max} in order to get the most appropriate level in every case. Based on the error and change of error in the last prediction, these rules determine the magnitude of α_{max} in CSMAF. The fuzzy decision system consists of two input variables and one output variable. If the dimension of measurement $n_y = 2$, the input variables $E(k)$ and $\Delta E(k)$ at the kth scan are defined by

$$E(k) = \frac{\sqrt{E_1^2(k) + E_2^2(k)}}{\sqrt{2}} \quad , \quad \Delta E(k) = \frac{\sqrt{\Delta E_1^2(k) + \Delta E_2^2(k)}}{\sqrt{2}}$$

Where $E_1(k)$, $E_2(k)$ and $\Delta E_1(k)$, $\Delta E_2(k)$ are normalized error and change of error of each component of measurement respectively. $E_1(k)$ and $\Delta E_1(k)$ are defined by

$$E_1(k) = \begin{cases} \dfrac{y_1(k) - \hat{y}_1(k)}{y_1(k) - y_1(k-1)} & \text{if } |y_1(k) - \hat{y}_1(k)| < |y_1(k) - y_1(k-1)| \\ \dfrac{y_1(k) - \hat{y}_1(k)}{|y_1(k) - \hat{y}_1(k)|} & \text{if } |y_1(k) - \hat{y}_1(k)| > |y_1(k) - y_1(k-1)| \\ 0.0 & \text{if } y_1(k) - \hat{y}_1(k) = y_1(k) - y_1(k-1) \end{cases}$$

$$\Delta E_1(k) = \begin{cases} \dfrac{E_1(k) - E_1(k-1)}{E_1(k-1)} & \text{if } |E_1(k) - E_1(k-1)| < |E_1(k-1)| \\ \dfrac{E_1(k) - E_1(k-1)}{|E_1(k) - E_1(k-1)|} & \text{if } |E_1(k) - E_1(k-1)| > |E_1(k-1)| \\ 0.0 & \text{if } E_1(k) - E_1(k-1) = E_1(k-1) \end{cases}$$

Where $y_1(k)$ and $\hat{y}_1(k)$ are the measurement and prediction respectively of the first component of measurement vector.

The fuzzy sets for the input variables $E(k)$ and $\Delta E(k)$ are labeled as the linguistic terms of LP (large positive), MP (medium positive), SP (small positive), and ZE (zero). The output variable is a_{max}. The fuzzy sets for a_{max} are labeled in the linguistic terms of EP (extremely large positive), VP (very large positive), LP (large positive), MP (medium positive), SP (small positive), and ZE (zero).

With the input and output variables defined above, a fuzzy rule can be expressed as follows in Table 1. The rules were obtained by interviewing some defense experts.

Table 1. Fuzzy rules for α_{max}

		E			
		ZE	SP	MP	LP
ΔE	ZE	VP	SP	EP	EP
	SP	LP	LP	VP	VP
	MP	EP	VP	MP	MP
	LP	VP	ZE	MP	EP

3 Unscented Fuzzy Filtering Based on Current Statistical Model

The discrete state and measurement equations of the target are described by Esq. (2.1). The UFCSMAF equations are as follows:

Initialization:

$$\hat{\mathbf{x}}(k\text{-}1) = E[\mathbf{x}(k-1)], \quad \mathbf{P}(k\text{-}1) = E\left[(\mathbf{x}(k\text{-}1) - \hat{\mathbf{x}}(k-1))(\mathbf{x}(k\text{-}1) - \hat{\mathbf{x}}(k-1))^T\right] \quad (3.1)$$

Calculate sigma points:

$$\chi(k-1) = \left[\hat{\mathbf{x}}(k-1) \quad \hat{\mathbf{x}}(k-1) \pm \sqrt{(n_x + \lambda)\mathbf{P}(k-1)}\right] \quad (3.2)$$

Time update:

$$\chi_i(k|k-1) = \mathbf{F}\chi_i(k-1) + \mathbf{G}[\chi_i(k|k-1)]_3 \Rightarrow \chi_i(k|k-1) = \mathbf{\Phi}\chi_i(k-1) \quad (3.3)$$

$$\mathbf{x}(k|k-1) = \sum_{i=0}^{2n_x} w_i \boldsymbol{\chi}_i(k|k-1) \tag{3.4}$$

$$\mathbf{P}(k|k-1) = \sum_{i=0}^{2n_x} w_i [\boldsymbol{\chi}_i(k|k-1) - \mathbf{x}(k|k-1)][\boldsymbol{\chi}_i(k|k-1) - \mathbf{x}(k|k-1)]^T + 2\alpha \sigma_a^2 \mathbf{B}_w \mathbf{B}_w^T \tag{3.5}$$

Measurement update:

$$\mathbf{P}_{\tilde{y}\tilde{y}^T} = \sum_{i=0}^{2n_x} w_i \left(\boldsymbol{\xi}_i(k|k-1) - \hat{\mathbf{y}}(k|k-1) \right) \left(\boldsymbol{\xi}_i(k|k-1) - \hat{\mathbf{y}}(k|k-1) \right)^T \tag{3.6}$$

$$\mathbf{P}_{x_k y_k} = \sum_{i=0}^{2n_x} w_i \left(\boldsymbol{\chi}_i(k|k-1) - \mathbf{x}(k|k-1) \right) \left(\boldsymbol{\xi}_i(k|k-1) - \hat{\mathbf{y}}(k|k-1) \right)^T \tag{3.7}$$

$$\mathbf{K}(k) = \mathbf{P}_{xy} \mathbf{P}_{\tilde{y}\tilde{y}^T}^{-1} \tag{3.8}$$

$$\begin{cases} E(k) \\ \Delta E(k) \end{cases} \xrightarrow{\text{Fuzzy system}} a_{\max}.$$

4 Simulations and Performance Comparison

In this section, proposed algorithm (UFCSMAF) is tested and compares its performance to UCSMAF. The mean and the standard deviation of the estimation error are chosen as the measure of performance.

The scenario shown in figure 2 is as follows. The target initial position is (10, 15) km and initial velocity is (-0.3, 0) km/s. In first segment from 0 to 30s, the target moves with a constant velocity; in the second segment from 31 to 45s, the target makes a left turn with a centripetal acceleration 70m/s^2; in the third segment from 46 to 60s, the target moves with a constant velocity; in the fourth segment from 61 to 80s, the target makes a right turn with a centripetal acceleration 20m/s^2; in the fifth segment from 81 to 100s, the target moves with a constant velocity. Two distributed observers measure the target line-of-sight (LOS) angles and are located at (0, 0) km and (10, 0) km respectively. So the measured angles are given by

$$\mathbf{y}(k) = \mathbf{h}(\mathbf{x}(k)) + \mathbf{v}(k) = \begin{bmatrix} \arctan\left(\dfrac{y(k) - y_o^1}{x(k) - x_o^1} \right) \\ \arctan\left(\dfrac{y(k) - y_o^2}{x(k) - x_o^2} \right) \end{bmatrix} + \mathbf{v}(k) \tag{4.1}$$

Where $\mathbf{x}(k) = [x(k), \dot{x}(k), \ddot{x}(k), y(k), \dot{y}(k), \ddot{y}(k)]^T$, (x_o^i, y_o^i) is the observer location and $\mathbf{v}(k)$ is Gaussian measurement noise. Table 2 lists the simulation parameters in detail.

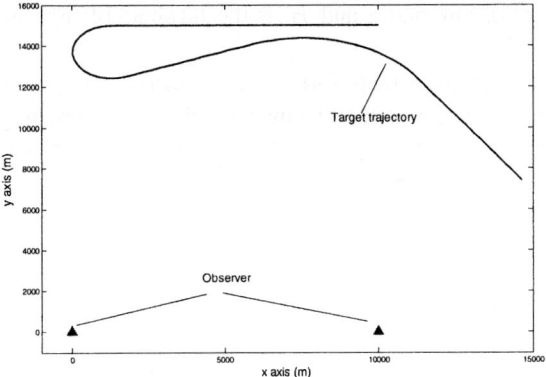

Fig. 2. Tracking scenario

Table 2. Simulation parameters

Item	Description	UCSMAF-I	UCSMAF-II	UFCSMAF
α	Maneuvering frequency	1/20	1/20	1/20
a_{max}	Maximum acceleration	100 m/s^2	500 m/s^2	Adaptation
λ	UT parameter	0	0	0
P_d	Detection probability	1	1	1
T	Sample rates	1 s	1 s	1 s
σ_θ	Sensor noise	0.003 rad	0.003 rad	0.003 rad
L	Simulation length	100	100	100

The theoretical lower-bound (CRLB) plays an important role in algorithm evaluation and assessment of the level of approximation introduced in the filtering algorithms. A derivation of such a bound is derived under the scenario. The CRLB of the state vector components are calculated as the diagonal elements of the inverse information matrix **J**

$$\text{CRLB}\{[\hat{\mathbf{x}}(k)]_j\} = [\mathbf{J}^{-1}(k)]_{jj} \tag{4.2}$$

If the target motion is deterministic (in the absence of process noise and with a non-random initial state vector), for the case of additive Gaussian noise, the information matrix can be calculated using the following recursion

$$\mathbf{J}(k+1) = [\mathbf{F}^{-1}(k)]^T \mathbf{J}(k)\mathbf{F}^{-1}(k) + \mathbf{H}^T(k+1)\mathbf{R}^{-1}\mathbf{H}(k+1) \tag{4.3}$$

Where F is transition matrix and H is the Jacobian of $h(*)$ evaluated at the true state.

It is difficulty to get a tractable form of J in maneuvering scenario. Here we give a conservative bound. Suppose that the maneuvering characteristic is well known, and then we can piecewise calculate J. During the derivation, target state is given by $x(k) = [x(k), \dot{x}(k), y(k), \dot{y}(k)]^T$. The Jacobian is given by

$$\mathbf{H}(k) = \begin{bmatrix} \dfrac{-(y(k)-y_o^1(k))}{(x(k)-x_o^1(k))^2 + (y(k)-y_o^1(k))^2} & 0 & \dfrac{(x(k)-x_o^1(k))}{(x(k)-x_o^1(k))^2 + (y(k)-y_o^1(k))^2} & 0 \\ \dfrac{-(y(k)-y_o^2(k))}{(x(k)-x_o^2(k))^2 + (y(k)-y_o^2(k))^2} & 0 & \dfrac{(x(k)-x_o^2(k))}{(x(k)-x_o^2(k))^2 + (y(k)-y_o^2(k))^2} & 0 \end{bmatrix}$$

When target moves with a constant velocity, $\mathbf{F}$ is give by

$$\mathbf{F} = \begin{bmatrix} 1 & T & 0 & 0 \\ 0 & 1 & 0 & 0 \\ 0 & 0 & 1 & T \\ 0 & 0 & 0 & 1 \end{bmatrix}$$

When target moves with a circle motion

$$\mathbf{F} = \begin{bmatrix} 1 & \dfrac{\sin \Omega T}{\Omega} & 0 & \dfrac{\cos \Omega T - 1}{\Omega} \\ 0 & \cos \Omega T & 0 & -\sin \Omega T \\ 0 & \dfrac{1-\cos \Omega T}{\Omega} & 1 & \dfrac{\sin \Omega T}{\Omega} \\ 0 & \sin \Omega T & 0 & \cos \Omega T \end{bmatrix}$$

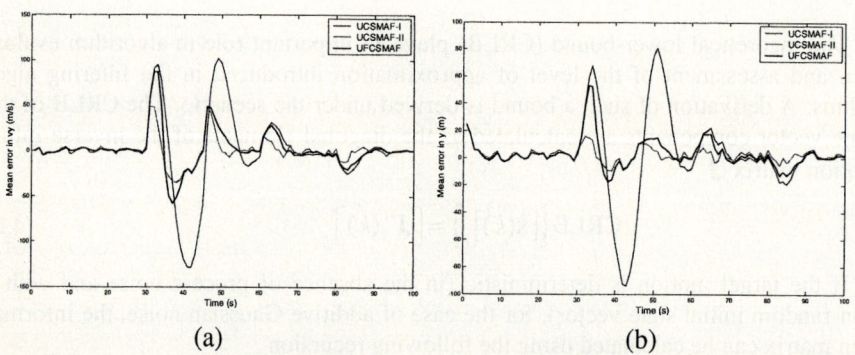

Fig. 4. Mean estimation error (a) y and (b) v_y

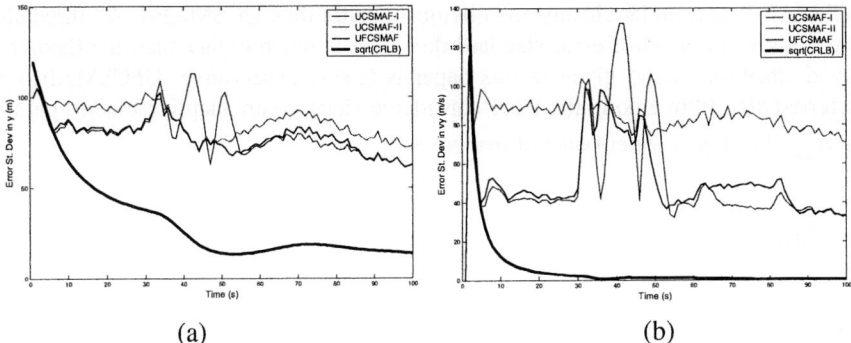

Fig. 5. The Error standard deviation versus square root of CRLB (a) y and (b) v_y

Where Ω is angular rate, the initial $\mathbf{J}(0)$ is given by

$$\mathbf{J}(0) = [\mathbf{P}(0)]^{-1} = \begin{bmatrix} \sigma_x^2 & 0 & \sigma_{xy}^2 & 0 \\ 0 & \sigma_v^2 & 0 & 0 \\ \sigma_{xy}^2 & 0 & \sigma_y^2 & 0 \\ 0 & 0 & 0 & \sigma_v^2 \end{bmatrix}^{-1} \quad (4.4)$$

In (4.4) σ_v is determined by the prior knowledge of target maximum speed. $\sigma_v = 300$ m/s is used in this scenario.

$$\begin{cases} \sigma_x^2 = \dfrac{1}{|C|^2}\left(\dfrac{\cos^2 \varphi_2}{r_2^2}\sigma_{\varphi_1}^2 + \dfrac{\cos^2 \varphi_1}{r_1^2}\sigma_{\varphi_2}^2\right) \\ \sigma_y^2 = \dfrac{1}{|C|^2}\left(\dfrac{\sin^2 \varphi_2}{r_2^2}\sigma_{\varphi_1}^2 + \dfrac{\sin^2 \varphi_1}{r_1^2}\sigma_{\varphi_2}^2\right) \\ \sigma_{xy} = \dfrac{1}{|C|^2}\left(\dfrac{\cos 2\varphi_2}{2r_2^2}\sigma_{\varphi_1}^2 + \dfrac{\cos 2\varphi_1}{2r_1^2}\sigma_{\varphi_2}^2\right) \end{cases} \quad (4.5)$$

In Esq. (4.5) φ_i is the measurement of observer i, $|C| = \dfrac{\sin(\varphi_2 - \varphi_1)}{r_2 r_1}$ and $r_i = \sqrt{(x-x_o^i)^2 + (y-y_o^i)^2}$ $(i=1,2)$.

The mean of estimation error in target position y and velocity v_y is show in figure 4. The standard deviation of error, against the square root of CRLB, is shown in figure 5. Observe from figure 4 three methods all approximately unbiased in non-maneuvering and small maneuvering stages. While in large maneuvering stage, UCSMAF-II is the most accurate, followed by the UFCSMAF and UCSMAF-I. Observe from figure 5 UFCSMAF combines the advantages both UCSMAF-I and UCSMAF-II. In the non-maneuvering and small maneuvering stages the performance of UFCSMAF and UCSMAF-I are similar closely and obviously better than UCSMAF-II. In the large maneuvering stage the performance of UFCSMAF and

UCSMAF-II are similar closely and obviously better than UCSMAF-I. All three algorithms show the level of error standard deviation is much higher than the theoretical bound since the bound given in this paper is fairly conservative. UFCSMAF is the preferred algorithm among the three considered since it can adaptive tune to parameter a_{max} to adapt a wider range of maneuvers.

5 Conclusion

A UFCSMAF algorithm has been presented in this paper for tracking a maneuvering target. The Cramer-Rao lower bound has also been derived. The theoretic analysis and computer simulations have confirmed that the adaptive algorithm has a robust advantage over a wide range of maneuvers and overcomes the shortcoming of the traditional current statistic model and adaptive filtering algorithm.

Future work will need to address the better model characterizing target maneuvers. In addition, more accurate performance bound incorporating uncertainty will be developed.

Acknowledgements

This research was jointly supported by National Natural Science Foundation of China (60375008), Hebei PH.D Discipline Special Foundation (B2004510), Aerospace Supporting Technology Foundation (JD04) and PH.D Foundation of Hebei University of Science and Technology.

References

1. Blackman. S, Popoli. R. Design and Analysis of Modern Tracking System, Artech House, Norwood, MA, 1999.
2. Chan. K, Lee, V. and Leung, H. Radar Tracking for Air Surveillance in a Stressful Environment Using a Fuzzy-gain Filter. IEEE Transactions on Aerospace and Electronic Systems, (1997) 80-89
3. Julier, S. J., Uhlmann, J. K., Unscented Filtering and Nonlinear Estimation. Proceedings of IEEE, (2004) 401-422
4. Julier, S.J, Uhlmann, J.K. A New Method for the Nonlinear Transformation of Means and Covariance in Filters and Estimators. IEEE Transactions on Automatic Control. (2000), 477-482
5. Wan E A, Vander M R. The Unscented Kalman Filter for Nonlinear Estimation. Proceedings of Signal Processing, (2000) 153-158.

A Pixel-Level Multisensor Image Fusion Algorithm Based on Fuzzy Logic[1]

Long Zhao, Baochang Xu, Weilong Tang, and Zhe Chen

School of Automation Science and Electrical Engineering,
Beihang University, Beijing 100083, China
flylong518@tom.com
{xbcyl, welleo}@163.com
chenzhe301@sohu.com

Abstract. A new multisensor image fusion algorithm based on fuzzy logic is proposed. The membership function and fuzzy rules of the new algorithm is defined using the Fuzzy Inference System (FIS) editor of fuzzy logic toolbox in Matlab 6.1. The new algorithm is applied to fuse Charge-Coupled Device (CCD) and Synthetic Aperture Rader (SAR) images. The fusion result is compared with some other fusion algorithms through some performance evaluation measures for the fusion effect, and the comparison results show that the new algorithm is effective.

1 Introduction

Multisensor Image Fusion (MIF) is the technique through which the images for the same target obtained by two or more sensors are processed and combined to create a single composite image which cannot be achieved with a single image source[1]. MIF makes good use of the complementary and redundant information of multisensor images and is widely used in such applications as computer vision, remote sensing, target recognition, medical image processing, and battle filed reconnaissance, etc[2].

Since fuzzy set theory was proposed by Zadeh in 1965, fuzzy logic has been widely used in target recognition, image analysis, and automatic control, etc[3]. And image fusion algorithm based on fuzzy logic has been researched more than ever. Based on the fuzzy logic of both Mamdani model and ANFIS, the medical image fusion algorithm for Infrared and CCD has been discussed by Harpreet, et al[4]. However, there have not been any fuzzy logic based CCD/SAR image fusion references in navigation/guidance application.

On the basis of fuzzy logic, a new algorithm is proposed for CCD/SAR image fusion. The membership function and fuzzy rules of the new algorithm are defined by using FIS editor of fuzzy logic toolbox. The fusion result is compared with some other fusion algorithms through some performance evaluation measures for fusion effect.

---

[1] The work is supported by aeronautical science fund of P. R. China (03D51007).

2 Image Fusion Algorithm Based on Fuzzy Logic

2.1 FIS Based Mamdani Model

FIS is a computing process by using fuzzy logic from input space to output space. FIS includes membership function, fuzzy logic computing, and fuzzy rules. The fuzzy rules of FIS based on Mamdani model are given by:[5]

If input 1 is x, and input 2 is y, then output is z.

2.2 Image Fusion Based on FIS

The membership function and fuzzy rules of CCD/SAR image fusion, based on fuzzy logic, are defined by using FIS editor of fuzzy logic toolbox in Matlab 6.1[5].

The algorithm process of CCD/SAR image fusion using fuzzy logic for pixel-level is given as follows.

- Read CCD image in variable S1 and SAR image in variable S2, respectively.
- Locate the size of CCD image, which has r1 rows and c1 columns.
- Locate the size of SAR image, which has r2 rows and c2 columns.
- Select r and c, which stands for min (r1, r2) and min (c1, c2) respectively, then calculate min (r, c).
- Obtain a new CCD image in variable M1 and a new SAR image in variable M2 from S1 and S2, respectively, and both M1 and M2 are the same in size.
- Make an image fusion.fis file by FIS editor, which contains two input images.
- Decide the number and type of membership function for both CCD and SAR images, and make rules from two input source images to the output fused image.
- Obtain the fused image and the performance evaluation measures of fusion effect.

3 Fusion Experiment and Fusion Effect Analysis

3.1 Experimental Results for CCD/SAR

The image fusion algorithm based on fuzzy logic is applied to CCD/SAR, and a simulating program is designed. In order to prove the effectiveness of the new algorithm, it is compared with other three fusion algorithms[6].

Algorithm 1: weight averaging based on principal component analysis fusion rule.
Algorithm 2: LPD + MAX operator fusion rule.
Algorithm 3: LPD + mean operator fusion rule.

The source images and fused image of CCD/SAR are shown in figure1.

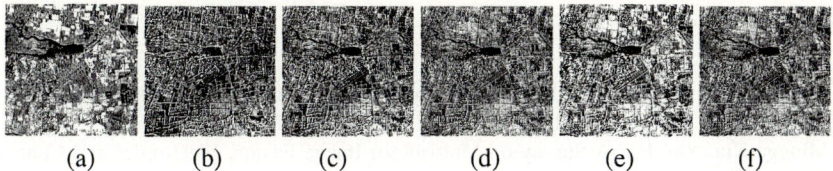

Fig. 1. (a) CCD source image, (b) SAR source image, (c) fused image by fuzzy logic, (d) fused image by algorithm 1, (e) fused image by algorithm 2, and (f) fused image by algorithm 3

Table 1. Performance evaluation measures of fusion effect

Algorithm	Entropy	Mean of Cross Entropy	Mean Square Root of Cross Entropy	Mean	Standard Deviation
New Algorithm	7.645	0.473	0.626	124.02	58.62
Algorithm 1	7.510	0.503	0.589	126.85	45.67
Algorithm 2	7.571	0.312	0.360	149.49	73.90
Algorithm 3	7.504	0.511	0.595	128.14	45.45

3.2 Performance Evaluation of Fusion Effect

In order to analyze the fusion effect and quality of the fused image, image entropy, mean of cross entropy, mean square root of cross entropy, mean, and standard deviation are used as the objective performance evaluation measures [6,7].

The performance evaluation measures of fusion effect for CCD/SAR are shown in table 1.

The experimental results in figure 1 and table 1 show that, because of the great difference in gray level and contrast between CCD image and SAR image, the fused image using algorithm 1 and algorithm 3 cannot obtain more information from source images. And the fusion effect is not good enough. Algorithm 2 and the new algorithm have a better performance than the other two algorithms in fusion effect. However, since the MAX rule is chosen in algorithm 2, the fused image has a stronger aberration, and although it has the smallest cross entropy, it's smaller in entropy than the new algorithm. The new algorithm adopts the algorithm of fuzzy logic, and the fused image is much more tied to human's vision characteristic, and is therefore more effective.

4 Conclusion

A new image fusion algorithm based on fuzzy logic is studied. The membership function and fuzzy rules of the new algorithm is defined using FIS editor of fuzzy logic toolbox in Matlab 6.1, and the new algorithm is easy to apply. The new algorithm is applied to CCD/SAR image fusion, and the fusion result is compared with other three fusion algorithms through some performance evaluation measures of the fusion effect. The comparison results demonstrate that the new fusion algorithm is effective.

References

1. Guoqiang Ni: Study on Multi-Band Image Fusion Algorithms and Its Progressing (I). Photoelectron Technique and Information, 14 (2001) 11-17
2. Mingge Xia, You He: A Survey on Multisensor Image Fusion. Electronics and Control, 9 (2002) 1-7
3. Zadeh L. A.: Fuzzy sets. Information and Control, 8 (1965) 338-353
4. Singh H., Raj J., Kaur G.: Image Fusion Using Fuzzy Logic and Application. Proceedings of IEEE International Conference on Fuzzy Systems, vol. 1. (2004) 337-340
5. Zhixing Zhang, Chunzai Sun: Neuro-Fuzzy and Soft Computing. Xi'an Jiaotong University Press, China Xi'an (2000)
6. Yanli Wang, Zhe Chen: Performance Evaluation of Several Fusion Approaches for CCD/SAR Images. Chinese Geographical Science, 13 (2003) 91-96
7. Yanmei Cui, Guoqiang Ni: Analysis and Evaluation of The Effect of Image Fusion Using Statistics Paraments. Journal of Beijing Institute of Technology, 20 (2000) 102-106

Approximation Bound for Fuzzy-Neural Networks with Bell Membership Function

Weimin Ma[1,2] and Guoqing Chen[1]

[1] School of Economics and Management,
Tsinghua University, Beijing, 100084, P.R. China
{mawm, chengq}@em.tsinghua.edu.cn
[2] School of Economics and Management, Xi'an Institute of Technology,
Xi'an, Shaanxi Province, 710032, P.R. China

Abstract. A great deal of research has been devoted in recent years to the designing Fuzzy-Neural Networks (FNN) from input-output data. And some works were also done to analyze the performance of some methods from a rigorous mathematical point of view. In this paper, the approximation bound for the clustering method, which is employed to design the FNN with the Bell Membership Function, is established. The detailed formulas of the error bound between the nonlinear function to be approximated and the FNN system designed based on the input-output data are derived.

1 Introduction

FNN systems are hybrid systems that combine the theories of fuzzy logic and neural networks. Designing the FNN system based on the input-output data is a very important problem [1,2,3]. Some Universal approximation capabilities for a broad range of neural network topologies have been established by researchers like Ito [4], and T.P.Chen [5]. Their work concentrated on the question of denseness. Some Approximation Accuracies of FNN have also been established in [3]. More results concerning the approximation of Neural Network can be found in [6,7,8].

In paper [2], an approach so called Nearest Neighborhood Clustering was introduced for training of Fuzzy Logic System. The paper [3] studied the relevant approximation accuracy of the clustering method with Triangular Membership Function and Gaussian Membership Function. In this paper, taking advantage of some similar techniques of paper [3], we obtain the upper bound for the approximation by using the clustering method with Bell Membership Function.

2 Clustering Method with Bell Membership Function

Before introducing the main results, we firstly introduce some basic knowledge on the designing FNN systems with clustering method, which was proposed in [2]. Given the input-out data pairs

$$(x_0^q, y_0^q), \quad q = 1, 2, \ldots \tag{1}$$

where $x_0^q \in U = [\alpha_1, \beta_1] \times \ldots \times [\alpha_n, \beta_n] \subset R^n$ and $y_0^q \in U_y = [\alpha_y, \beta_y] \subset R$. If the data are assumed to be generated by an unknown nonlinear function $y = f(x)$, the clustering method in [2] can help us to design a FNN system to approximate the function $f(x)$. For convenience to illustrate the main result of this paper, we describe this method in a brief way as follows.

Step 1. To begin with the first input-output pair (x_0^1, y_0^1), select a radius parameter r, establish a cluster center with letting $x_c^1 = x_0^1$, and set $y_c^1(1) = y_0^1$, $B^1(1) = 1$.

Step 2. For the kth input-out pair (x_0^k, y_0^k), $k = 2, \ldots$, suppose there are M clusters with centers at $x_c^1, x_c^2, \ldots, x_c^M$. Find the nearest cluster center $x_c^{l_k}$ to x_0^k to satisfy

$$|x_0^k - x_c^{l_k}| = \min_l |x_0^k - x_c^l|, \quad l = 1, 2, \ldots, M. \tag{2}$$

Then there are two cases

Case 1. If $|x_0^k - x_c^{l_k}| \geq r$, establish x_0^k as a new cluster center with $x_c^{M+1} = x_0^k$, $y_c^{M+1}(k) = y_0^k$ and $B^{M+1}(k) = 1$, and keep $y_c^l(k) = y_c^l(k-1)$ and $B^l(k) = B^l(k-1)$ for any l.

Case 2. If $|x_0^k - x_c^{l_k}| < r$, do the following:

$$y_c^{l_k}(k) = \frac{y_c^{l_k}(k-1)B^{l_k}(k-1) + y_0^k}{B^{l_k}(k-1) + 1} \tag{3}$$

$$B^{l_k}(k) = B^{l_k}(k-1) + 1 \tag{4}$$

and meanwhile set

$$y_c^l(k) = y_c^l(k-1), \quad B^l(k) = B^l(k-1), \text{ for } l \neq l_k. \tag{5}$$

Step 3. Then the FNN system can be constructed as:

$$\hat{f}_k(x) = \frac{\sum_{l=1}^{\overline{M}} \left[y_c^l(k) \cdot \left(\frac{1}{1 + \frac{|x - x_c^l|^2}{\sigma^2}} \right) \right]}{\sum_{l=1}^{\overline{M}} \left(\frac{1}{1 + \frac{|x - x_c^l|^2}{\sigma^2}} \right)}$$

$$= \frac{\sum_{l=1}^{\overline{M}} \left[y_c^l(k) \cdot \prod_{j=1}^{n} \left(\frac{1}{1 + \frac{|x_j - x_{c,j}^l|^2}{\sigma^2}} \right) \right]}{\sum_{l=1}^{\overline{M}} \prod_{j=1}^{n} \left(\frac{1}{1 + \frac{|x_j - x_{c,j}^l|^2}{\sigma^2}} \right)} \tag{6}$$

where $\overline{M} = M + 1$ for case 1 and $\overline{M} = M$ for case 2.

Step 4. Repeat by going to Step 2 with $k = k + 1$.

The above FNN system is constructed using singleton fuzzier, product inference engine and center-average defuzzifier, as detailed in [2]: the basic idea is to group the data pairs into clusters in terms of their distribution and then use the fuzzy IF-THEN rules for one cluster to construct the fuzzy system; the radius parameter r is selected to determine the size of the clusters, namely the smaller of r, the more number of clusters and vice versa. Some intensive simulation results concerning this method for various problems can also be found in that paper. This paper concentrates on the approximation bound for this method with bell membership function.

3 Approximation Bound with Bell Membership Function

The following theorem gives the approximation bound of FNN system $\hat{f}_k(x)$ of (6) which is constructed by using the clustering method with bell membership function.

Theorem 1. *Let $f(x)$ be a continuous function on U that generates the input-output pairs in (1). Then the following approximation property holds for the FNN system $\hat{f}_k(x)$ of (6):*

$$|f(x) - \hat{f}_k(x)| \leq \left(r + d_x + 2^{n-1} \cdot \overline{M} \cdot (\sigma + \frac{d_x^2}{\sigma})\right) \cdot \sum_{i=1}^{n} \left\|\frac{\partial f}{\partial x_i}\right\|_{\infty} \quad (7)$$

where the infinite norm $\|\cdot\|_{\infty}$ is defined as $\|d(x)\|_{\infty} = \sup_{x \in U} |d(x)|$ and d_x is the distance from x to the nearest cluster, i.e.,

$$d_x = \min_{l} |x - x_c^l| = |x - x_c^{l_x}| \quad (8)$$

Proof. From (6) we have

$$|f(x) - \hat{f}_k(x)| \leq \frac{\sum_{l=1}^{M}\left[|f(x) - y_c^l(k)| \cdot \prod_{j=1}^{n}\left(\frac{1}{1 + \frac{|x_j - x_{c,j}^l|^2}{\sigma^2}}\right)\right]}{\sum_{l=1}^{M} \prod_{j=1}^{n}\left(\frac{1}{1 + \frac{|x_j - x_{c,j}^l|^2}{\sigma^2}}\right)} \quad (9)$$

Paper [3] obtain the following result when the relevant approximation bound for the clustering method with triangular membership function was discussed:

$$|f(x) - y_c^l(k)| \leq \sum_{i=1}^{n}\left(\left\|\frac{\partial f}{\partial x_i}\right\|_{\infty} \cdot (|x_i - x_{c,i}^l| + r)\right) \quad (10)$$

Combining the (9) and (10), we have

$$|f(x) - \hat{f}_k(x)| \leq \frac{\sum_{l=1}^{M}\sum_{i=1}^{n}\left[\left(\left\|\frac{\partial f}{\partial x_i}\right\|_{\infty} \cdot (|x_i - x_{c,i}^l| + r)\right) \cdot \prod_{j=1}^{n}\left(\frac{1}{1 + \frac{|x_j - x_{c,j}^l|^2}{\sigma^2}}\right)\right]}{\sum_{l=1}^{M} \prod_{j=1}^{n}\left(\frac{1}{1 + \frac{|x_j - x_{c,j}^l|^2}{\sigma^2}}\right)}$$

$$\leq \sum_{i=1}^{n}\left\{\left\|\frac{\partial f}{\partial x_i}\right\|_\infty \cdot \left[r + \frac{\sum_{l=1}^{M}\left(|x_i - x_{c,i}^l| \cdot \prod_{j=1}^{n}\left(\frac{1}{1+\frac{|x_j - x_{c,j}^l|^2}{\sigma^2}}\right)\right)}{\sum_{l=1}^{M}\prod_{j=1}^{n}\left(\frac{1}{1+\frac{|x_j - x_{c,j}^l|^2}{\sigma^2}}\right)}\right]\right\} \quad (11)$$

Now, we just focus on analyzing the term

$$\frac{\sum_{l=1}^{M}\left(|x_i - x_{c,i}^l| \cdot \prod_{j=1}^{n}\left(\frac{1}{1+\frac{|x_j - x_{c,j}^l|^2}{\sigma^2}}\right)\right)}{\sum_{l=1}^{M}\prod_{j=1}^{n}\left(\frac{1}{1+\frac{|x_j - x_{c,j}^l|^2}{\sigma^2}}\right)}$$

on the right-hand side of (11). We only consider the case $i = 1$ and the proof remains the same for $i = 1, 2, \ldots, n$.

Employing the similar method in [3], given any point $x = (x_1, x_2, \ldots, x_n) \in U$, we divide space U in to 2^n areas and define some sets concerning the cluster center as follows:

$$\begin{aligned}
U_1^x &= \{\overline{x} \in U : \overline{x}_1 - x_1 \geq 0, \ldots, \overline{x}_n - x_n \geq 0\} \\
U_2^x &= \{\overline{x} \in U : \overline{x}_1 - x_1 \geq 0, \ldots, \overline{x}_n - x_n < 0\} \\
&\cdots \\
U_{2^n-1}^x &= \{\overline{x} \in U : \overline{x}_1 - x_1 < 0, \ldots, \overline{x}_n - x_n \geq 0\} \\
U_{2^n}^x &= \{\overline{x} \in U : \overline{x}_1 - x_1 < 0, \ldots, \overline{x}_n - x_n < 0\}
\end{aligned} \quad (12)$$

And define some sets concerning the cluster centers

$$\begin{aligned}
V^x &= \{\overline{x} \in U : |\overline{x}_1 - x_1| < d_x\}, \\
\overline{V}^x &= \{\overline{x} \in U : |\overline{x}_1 - x_1| \geq d_x\}, \\
V_m^x &= \overline{V}^x \cap U_m^x, \quad (m = 1, \ldots, 2^n).
\end{aligned} \quad (13)$$

Apparently, there are two cases that we need to consider.

Case 1: $x_c^l \in V^x$, which indicates $|x_{c,1}^l - x_1| < d_x$, we have

$$\frac{\sum_{x_c^l \in V^x}\left[|x_1 - x_{c,1}^l| \cdot \prod_{j=1}^{n}\frac{1}{1+\frac{|x_j-x_{c,j}^l|^2}{\sigma^2}}\right]}{\sum_{l=1}^{M}\prod_{j=1}^{n}\left(\frac{1}{1+\frac{|x_j-x_{c,j}^l|^2}{\sigma^2}}\right)}$$

$$< \frac{d_x \cdot \sum_{x_c^l \in V^x}\prod_{j=1}^{n}\left(\frac{1}{1+\frac{|x_j-x_{c,j}^l|^2}{\sigma^2}}\right)}{\sum_{l=1}^{M}\prod_{j=1}^{n}\left(\frac{1}{1+\frac{|x_j-x_{c,j}^l|^2}{\sigma^2}}\right)}$$

$$\leq d_x \quad (14)$$

Case 2: $x_c^l \in \overline{V}^x = \bigcup_{m=1}^{2^n} V^x$. We only consider the case $x_c^l \in V_1^x$. For the cases $x_c^l \in V_2^x, \ldots, V_{2^n}^x$, the same result can be obtained. For any l that satisfied to $x_c^l \in V_1^x$, according to the definition of V_1^x, we have

$$x_{c,1}^l - x_1 \geq d_x \quad \text{and} \quad x_{c,j}^l - x_j \geq 0, \quad j = 1, \ldots, n \tag{15}$$

From (12), (13) and (15), we have

$$\sum_{x_c^l \in V_1^x} \left[|x_1 - x_{c,1}^l| \cdot \prod_{j=1}^n \left(\frac{1}{1 + \frac{|x_j - x_{c,j}^l|^2}{\sigma^2}} \right) \right]$$

$$= \sum_{x_c^l \in V_1^x} \left[|x_1 - x_{c,1}^l| \cdot \left(\frac{1}{1 + \frac{|x_1 - x_{c,1}^l|^2}{\sigma^2}} \right) \cdot \prod_{j=2}^n \left(\frac{1}{1 + \frac{|x_j - x_{c,j}^l|^2}{\sigma^2}} \right) \right]$$

$$\leq \sum_{x_c^l \in V_1^x} \left[\frac{\sigma}{2} \cdot \prod_{j=2}^n \left(\frac{1}{1 + \frac{|0|^2}{\sigma^2}} \right) \right]$$

$$= \frac{\sigma}{2} \cdot \sum_{x_c^l \in V_1^x} 1$$

$$\leq \frac{\overline{M} \cdot \sigma}{2} \tag{16}$$

The second inequality of (16) holds for (15) and the following reason

$$|x_1 - x_{c,1}^l| \cdot \left(\frac{1}{1 + \frac{|x_1 - x_{c,1}^l|^2}{\sigma^2}} \right)$$

$$= \frac{|x_1 - x_{c,1}^l|}{1 + \frac{|x_1 - x_{c,1}^l|^2}{\sigma^2}}$$

$$\leq \frac{1}{\frac{1}{|x_1 - x_{c,1}^l|} + \frac{|x_1 - x_{c,1}^l|}{\sigma^2}}$$

$$\leq \frac{1}{2\sqrt{\frac{1}{|x_1 - x_{c,1}^l|} \cdot \sqrt{\frac{|x_1 - x_{c,1}^l|}{\sigma^2}}}}$$

$$= \frac{\sigma}{2} \tag{17}$$

Considering cases $x_c^l \in V_2^x, \ldots, V_{2^n}^x$, we have

$$\sum_{x_c^l \in \overline{V}^x} \left[|x_1 - x_{c,1}^l| \cdot \prod_{j=1}^n \left(\frac{1}{1 + \frac{|x_j - x_{c,j}^l|^2}{\sigma^2}} \right) \right]$$

$$\leq \frac{2^n \cdot \overline{M} \cdot \sigma}{2}$$

$$= 2^{n-1} \cdot \overline{M} \cdot \sigma \tag{18}$$

On the other hand, it follows from (8) that

$$\sum_{l=1}^{\overline{M}} \prod_{j=1}^{n} \left(\frac{1}{1 + \frac{|x_j - x_{c,j}^l|^2}{\sigma^2}} \right) = \sum_{l=1}^{\overline{M}} \left(\frac{1}{1 + \frac{|x - x_c^l|^2}{\sigma^2}} \right)$$

$$\geq \frac{1}{1 + \frac{d_x^2}{\sigma^2}} \qquad (19)$$

From (18) and (19), we have

$$\frac{\sum_{x_c^l \in \overline{V}^x} \left[|x_1 - x_{c,1}^l| \cdot \prod_{j=1}^{n} \left(\frac{1}{1 + \frac{|x_j - x_{c,j}^l|^2}{\sigma^2}} \right) \right]}{\sum_{l=1}^{\overline{M}} \prod_{j=1}^{n} \left(\frac{1}{1 + \frac{|x_j - x_{c,j}^l|^2}{\sigma^2}} \right)} \leq 2^{n-1} \cdot \overline{M} \cdot \left(\sigma + \frac{d_x^2}{\sigma} \right) \qquad (20)$$

Combining (13), (14) and (20), it can be shown that

$$\frac{\sum_{l=1}^{\overline{M}} \left[|x_i - x_{c,i}^l| \cdot \prod_{j=1}^{n} \left(\frac{1}{1 + \frac{|x_j - x_{c,j}^l|^2}{\sigma^2}} \right) \right]}{\sum_{l=1}^{\overline{M}} \prod_{j=1}^{n} \left(\frac{1}{1 + \frac{|x_j - x_{c,j}^l|^2}{\sigma^2}} \right)} \leq d_x + 2^{n-1} \cdot \overline{M} \cdot \left(\sigma + \frac{d_x^2}{\sigma} \right) \qquad (21)$$

From (11) and (21), we get the desired result. □

In paper [3], by using the same clustering method, an approximation bound, $\left[\overline{d}_x + \left(1 + \frac{\sqrt{n}}{n}\right) r + \frac{2\sqrt{n}\sigma^2}{r^n} \left(2r + \sqrt{n\pi}\sigma\right)^{n-1} \right] \cdot \sum_{i=1}^{n} \left\| \frac{\partial f}{\partial x_i} \right\|_{\infty}$, with Gaussian Membership Function was obtained. However, in fact, by employing the similar techniques of this paper, it can be improved to $\left(r + (2^n \overline{M} + 1)\overline{d}_x \right) \sum_{i=1}^{n} \left\| \frac{\partial f}{\partial x_i} \right\|_{\infty}$. Namely, we have the following theorem [9].

Theorem 2. *Let $f(x)$ be a continuous function on U that generates the input-output pairs in (1). If the Gaussian Membership Function is chosen, the following approximation property holds for the FNN system $\hat{f}_k(x)$ of (6):*

$$|f(x) - \hat{f}_k(x)| \leq \left(r + (2^n \cdot \overline{M} + 1) \cdot \overline{d}_x \right) \cdot \sum_{i=1}^{n} \left\| \frac{\partial f}{\partial x_i} \right\|_{\infty} \qquad (22)$$

where the infinite norm $\|\cdot\|_{\infty}$ is defined as $\|d(x)\|_{\infty} = \sup_{x \in U} |d(x)|$, $\overline{d}_x = \max\left\{ d_x, \frac{\sigma}{\sqrt{2}} \right\}$ and d_x is the distance from x to the nearest cluster.

The proof of the theorem 2 can be found in [9].

4 Concluding Remarks

In this paper, an upper bound, $\left(r + d_x + 2^{n-1} \cdot \overline{M} \cdot (\sigma + \frac{d_x^2}{\sigma})\right) \cdot \sum_{i=1}^{n} \left\|\frac{\partial f}{\partial x_i}\right\|_\infty$, concerning the approximation bound of clustering method with Bell Membership Function is proved in a rigorous mathematical way. The techniques employed in the proof of the theorem are expected to be used to obtain or improve other approximation bound of other methods of FNN.

Acknowledgements. The work was partly supported by the National Natural Science Foundation of China (70401006, 70231010) and China Postdoctoral Science Foundation (2003034014).

References

1. Wang, L.X. and Mendel, J.M., W.: Fuzzy Basis Functions, Universal Approximation, and orthogonal least squares learning. IEEE Trans. Neural Network. **3** (1992) 807–814
2. Wang, L.X.: Training of Fuzzy Logic System Using Nearest Neighborhood Clustering. Proc. 1993 IEEE Int. Conf. Fuzzy Syst. San Francisco, CA, (1993) 13–17
3. Wang, L.X. and Chen, W.: Approximation Accuracy of Some Neuro-Fuzzy Approches. IEEE Trans. Fuzzy Systems. **8** (2000) 470–478
4. Ito, Y.: Approximation of Continuous Functions on R^d by Linear Combination of Shifted Rotations of Sigmoid Function with and without Scaling Neural Networks. Neural Networks. **5** (1992) 105–115
5. Chen, T.P., Chen, H.: Approximation Capability to Functions of Several Variables, Nonlinear Functions, and Operators by Radial Function Neural Networks. IEEE Trans. Neural Networks. **6** (1995) 904–910
6. Maiorov, V., Meir, R.S.: Approximation Bounds for Smooth Functions in $C(R^d)$ by Neural and Mixture Networks. IEEE Trans. Neural Networks. **9** (1998) 969–978
7. Burger, M., Neubauer, A.: Error Bounds for Approximation with Neurnal Networks. J. Approx. Theory. **112** (2001) 235–250
8. Wang, J.J., Xu, Z.B., Xu, W.J.: Approximation Bounds by Neural Networks in L^p_ω. Proc. of 1st International Symposium on Neural Networks, F.Yin, J.Wang, and C.Guo (Eds.): ISSN 2004, LNCS **3173** (2004) 1–6
9. Ma, W.M., Chen, G.Q.: Improvement on the Approximation Bound for Fuzzy-Neural Networks Clustering Method with Gaussian Membership Function. Accepted by the First International Conference on Advanced Data Mining and Applications (ADMA2005).

A Neuro-Fuzzy Method of Forecasting the Network Traffic of Accessing Web Server

Ai-Min Yang[1,2], Xing-Min Sun[1], Chang-Yun Li[1], and Ping Liu[1]

[1] Department of Computer Science, Zhuzhou Institute of Technology,
Hunan 412008, P.R. China
[2] Department of Computing and Information Technology, Fudan University,
Shanghai 200433, P.R. China
amyang18@163.com

Abstract. It is a new idea and approach to forecast Web traffic basing on Neuro-Fuzzy Method. The log files on a Web Server include many useful information about users. In this paper, by analyzing log files the forecasting model is proposed and the basic idea, structure and algorithm of this model are introduced. The Dynamic Clustering Method and Neural Network learning method are introduced also. Experimental results show that the proposed method is very helpful for improving the administration of Web Server and quality of service and forecasting the action of users.

1 Introduction

Web Server has collected lots of log files which are valuable property. Analyzing these historical data with corresponding methods can help us acquire access traffic of Web sites, analyze system performance and understand users' access pattern. This paper analyzes access requests and page requests of every day and every hour on a Web Server by Neuro-Fuzzy[1],[2],[3] approach and constructs corresponding forecasting model.

2 Basic Idea of Model Design

The constructing process of forecasting model is as follows. The number of Web access requests and page requests of every hour and every day is getted from everyday Web log files, then the corresponding training samples and testing samples are produced. The Dynamic Clustering Method is used to create clusters for training samples.For each cluster,a fuzzy rule is created and the corresponding parameters of Neural Network are configured. By learning, the model of traffic forecasting is formed.

2.1 Data Preprocessing

The task of data preprocessing is to get the number of Web access requests and pages requests, then produces corresponding training samples and testing samples. The format which is used to extract data from Web log files is defined as follow:

Time(day/a week or hour) : Number of requests : Number of page requests

Time represents a week or every hour. Samples data should be processed according to forecasting condition and target, otherwise it can't serve the training model. Forecasting condition and target are defined in definition 1.

Definition 1. Provided X(t) represents network traffic, if t is day or hour, the forecasting condition can be represented as X(t), X(t-1), X(t-2) etc. Forecasting target can be represented as X(t+1),X(t+3) etc. For instance, in order to forecast the network traffic of every day, forecasting condition is X(t-1),X(t), forecasting target is X(t+1).

2.2 Dynamic Clustering Method

In this paper, we present a Dynamic Clustering method which is used for data clustering. For each cluster, a fuzzy rule is created.

Notational Conventions:

x_L(L=1,2,…,n) is training samples,C_i denotes the ith cluster (i=1,2,…), C_i^k denotes C_i which has been updated for k times (k=0,1,…). cc_i is the center of C_i,cc_i^k is the center of C_i which has been updated for k times(k=0,1,…).Ru_i is the radius of C_i cluster,Ru_i^k is Ru_i which has been updated for k times (k=0,1,…).d_{iL} is the distance between sample x_L and the center of C_i.dt is the radius threshold for each cluster.

Dynamic Clustering Algorithm:

Step 1: Get first training sample from the training data and C_1 cluster is created. the center of C_1 is self and the following symbols are gived for C_1:C_1^0,cc_1^0,Ru_1^0, and Ru_1^0=0.

Step 2: Get a training sample x_L which is not processed from the training data and calculate the distances d_{iL}=||x_L-cc_i|| (between x_L and created cluster centers of C_i), i=1,…,cn, cn is the number of created clusters.

Step 3: Compare the d_{iL} which is calculated in Step 2 with the corresponding cluster radius Ru_i.If there is the cluster C_g(1≤g≤cn) which satisfies d_{gL}=||x_L-cc_g||=min(||x_L-cc_i||), and d_{iL}≤Ru_i (i=1,2,..,cn),x_L is assigned into C_g,then go to Step 2,else, go to Step 4.

Step 4: Find the cluster C_a which is the nearest to x_L among the created clusters.First,calculate S_{iL},S_{iL}=d_{iL}+Ru_i,i=1,2,…,cn,then choose C_a according to S_{iL} and Eq.(1).

$$S_{aL} = d_{aL} + Ru_a = \min\{S_{iL}\}(i=1,2,...,cn) \tag{1}$$

Step 5: If S_{aL}>2×dt,it indicates that x_L doesn't belong to any created clusters, then a new cluster should be created. the creating method and process are similar to those of Step 1,then go on Step 2,else, go to Step 6.

Step 6: If S_{aL}≤2×dt,the center(cc_a) and radius(Ru_a) of C_a should be updated; the process is as follows,The modification for cc_a:

D_0=S_{aL}-Ru_a,R_0=S_{aL}/2,dist=x_L-cc_a.

for m, from 1 to q, calculate $dx(m)=|dist(m)|\times(R_0/D_0)$,(dist(m) is the value of the mth feature). if dist(m)>0, $cc_a(m)=x_L(m)-dx(m)$,else, $cc_a(m)= x_L(m)+dx(m)$. The modification for Ru_a:$Ru_a=S_{aL}/2$.

If all traing sanples are processed ,stop algorithm,else return to Step 2.

3 Network Structure Based on Neuro-fuzzy Method

A simple Neural Network structure with Neuro-fuzzy method is represented In the Fig.1(a).This structure is composed of three layers: the first represents input variable, for instance X(t), X(t-1),the second is fuzzy rule node layer, the third is output layer.

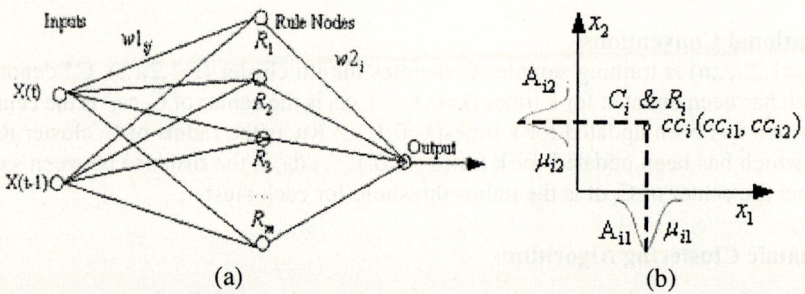

Fig. 1. (a)A simple Neural Network structure diagram, $w1_{ij}$ is weight between input layer and rule nodes layer, i=1,...,q (q is input variable dimensions), j=1,...,m (m is the number of rule nodes, also m is the number of the created clusters). $w2_i$ is weight between rule nodes layer and output layer, i=1,...,m. (b) The relations among cluster, fuzzy partition and fuzzy rule.

Creating Rule Nodes:
Provided cc_i(i=1,2,...,m,m is the number of created clusters) is the ith cluster which was obtained by Dynamic Clustering Algorithm,for each cluster, a fuzzy rule is created and fuzzy rules indicated as follows:

$$R_i: \text{ if } x_1 \text{ is } A_{i1} \text{ and } x_2 \text{ is } A_{i2} \text{ and } \ldots \text{ and } x_q \text{ is } A_{iq}, \text{then } y \text{ is } f_i(x_1, x_2, \ldots, x_q) \quad (2)$$

Here, i = 1, ... m; j = 1, ... q(q is input variable dimensions), are m×q fuzzy propositions as m antecedents for m fuzzy rules respectively; A_{ij}, i = 1,... m; j =1,..., q, are fuzzy sets defined by their fuzzy membership functions u_{ij}:$x_j\rightarrow$[0, 1], i=1,...,m; j=1,..., q.In the consequent parts, y is a consequent variable, and crisp linear functions f_i, i =1,...,m, are employed.The relation between cluster and fuzzy rule was showed in the Fig.1(b).

In the Fig.1(b), fuzzy membership function u_{ij} is defined as follows:

$$A_{ij}, \mu_{ij}(x_j) = \exp\left(-(x_j-cc_{ij})^2 / \delta_{ij}\right)(i=1,2,\ldots,m, j=1,2,\ldots,q) \quad (3)$$

Rule nodes are created as following steps:

Step 1. Find training samples of each cluster. Compute the distance from training samples to the center of cluster, select samples which satisfy Dsc≤k*dt, and regard them as training samples of this cluster. Here, Dsc represents the distance from sample to the center of the cluster, dt is the radius threshold for each cluster. k is a adjustable parameter, its range is between 1.10 and 1.20. If the number of training samples is less than q+h (h>6,q is dimensions of training sample), Inequation Dsc≤k*dt is replaced by Inequation Dsc≤k*k*dt.If the request can't be satisfied still, Inequation Dsc≤k*k*dt is replaced by Inequation Dsc≤k*k*k*dt and so on.

Step 2. Confirm Parameters of rule nodes.Firstly, the initial value of δ_{ij} In the Eq.(3) is confirmed according the Eq.(4). In the Eq.(3) cc_{ij} is confirmed by Dynamic Clustering Algorithm.

$$\delta_{ij} = \max_{x \in C_i} \|x_j - cc_{ij}\| (i=1,2,...,m, j=1,2,...,q) \qquad (4)$$

Secondly, confirm the linear functions f_i(i=1,2,...,m) in the consequence parts. The linear functions f_i is created and updated by linear least-square estimator (LSE) on the training samples of corresponding cluster.We can express function fi as the Eq.(5).

$$y = f_i(x_1, x_2, ... x_q) = \beta_0 + \beta_1 x_1 + \beta_2 x_2 + ... + \beta_q x_q \qquad (5)$$

The least-square estimator (LSE)[4] of $\beta = [\beta_0, \beta_1, ..., \beta_q]^T$ is calculated with on a training data set that is composed of p data pairs {([x_{l1}, x_{l2}, ..., x_{lq}], y_l), l = 1, ..., p} which belong the ith cluster(i=1,...,m).

Setp 3. Calculate link weights(In Fig.1(a)).For an input vector $x^L = [x^L_1, ..., x^L_q]$.The link weight $w1_{ij}$(i=1,...,m, j=1,...,q) is calculated as the Eq.(6).

$$w1_{ij} = \mu_{ij}(x^L_j) = \exp\left(-\left(x^L_j - cc_{ij}\right)^2 / \delta_{ij}\right)(i=1,2,...,m, j=1,2,...,q) \qquad (6)$$

The link weight w2i(i=1,2,...,m) is calculated as the Eq.(7).

$$w2_i = \prod_{j=1}^{q} \mu_{ij}(x^L_j) \quad (i=1,2,...,m) \qquad (7)$$

Step 4. Confirm the expression of the output y.For an input vector $x^L = [x^L_1, x^L_2, ..., x^L_q]$, the result of inference, y^L is calculated as the Eq.(8).

$$y^L = \sum_{i=1}^{m} w2_i f_i(x^L_1, x^L_2, ..., x^L_q) \Big/ \sum_{i=1}^{m} w_i \qquad (8)$$

4 Network Learning

In above section, Forming rule nodes and obtaining its corresponding parameter have been introduced.In this model, there are two kinds learning. One is the learning of cluster center, the other is learning of $W1_{ij}$ and $W2_i$; The learning of $W1_{ij}$ and $W2_i$ is by tuning δ_{ij} in the Eq.(3). Thus, the shape of the membership functions can be changed, as will result in the change of $W1_{ij}$ and $W2_i$.This learning method is similar to parameter learning in BP network. We won't illustrate more here.

If the request of cluster can't satisfy anticipative effect, we can continue to learn in order to ensure learning target.The learning target is illustrated in definition 2.

Definition 2. Provided n is the number of samples which is represented with $Ex(i)(i=1,2,\ldots,n)$. Thus, the Dynamic Clustering Algorithm gets m clusters which cc_j ($j=1,\ldots,m$) is center of C_j.After the hth learning, the sum of the distance from each sample to its cluster center is Dextoc(h) (h=1,2,…,Q, Q is the maximum number of learning).Dextoc(h) is calculated as the Eq.9.

$$Dextoc(h) = \sum_{i=1}^{n} \sum_{j=1}^{rn} \|Ex(i) - cc_j\| \quad (9)$$

The learning target is as the Inequation (10).

$$|Dextoc(h) - Dextoc(h-1)| \leq \varepsilon \quad (in\ general,\ \varepsilon\ is\ 0.0001) \quad (10)$$

The terminating condition of learning: h=Q or satisfying formula (10). The learning algorithm is illustrated as follow:

Step 1. For every training sample, calculate the distances from it to the center of all created clusters,and find out the cluster whose distance to this training sample is minimum. The corresponding information is writed down with $sc_i(i=1,\ldots,m, initially, sc_i=0)$.For example,if this cluster is ith cluster,$sc_i=sc_i+1$.

Step 2. Modify the cluster center.After step (1),for a cluster, if $sc_i \leq 1$ (i= 1,…,m), this cluster center needn't be modified. Otherwise we do the following modification: $N=sc_i$,the cluster center is represented with cc_{old} $(x^{old}_1,\ldots,x^{old}_q)$.the corresponding samples is represented with $Ex(x^l_1,\ldots,x^l_q)$ (l=1,…,N).Now, we should find a new cluster center $cc_{new}(x^{new}_1,\ldots,x^{new}_q)$ which satisfies that the sum of distances from N samples to new cluster center cc_{new} is minimum, and the sum of distances is represented with Dexton as Eq.(11).

$$Dexton(x^{new}_1,\ldots,x^{new}_q) = \sum_{l=1}^{N} \|Ex(x^l_1,\ldots,x^l_q) - cc_{new}(x^{new}_1,\ldots,x^{new}_q)\|$$

$$= \sum_{l=1}^{N} \left(\sqrt{\sum_{j=1}^{q}(x^l_j - x^{new}_j)^2} \Big/ q \right) \quad (11)$$

We use queue quadratic programming(QSP) to find out the cc_{new}.

Step 3. Calculate the sum of the distances from every sample to its cluster center as Eq.(12).

$$Dextoc(h) = \sum_{i=1}^{n} \sum_{j=1}^{m} \left\| Ex(i) - cc_j^{(new)or(old)} \right\| \quad (12)$$

Step 4. Judge terminating condition of this algorithm. If, |Dextoc(h)-Detoc(h-1)|≤ε, the algorithm terminates. or h=Q(Q is the maximum number of learning), the algorithm also terminates,else, go to step 1.

In this paper, the forecasting target of Ex(X(t),X(t-1),...,X(t-q)) is X(t+n) . The result of inference, X(t+n) is calculated as the Eq.(13).

$$X(t+n) = \left(\sum_{i}^{m} W2_i (\beta_0^i + \beta_1^i X(t) + ... + \beta_q^i X(t-q)) \right) \bigg/ \sum_{i=1}^{m} W2_i \quad (13)$$

Here, $W2_i$ is calculated as the Eq.(7). $W1_{ij}$ is calculated as the Eq.(6).

5 Experiment

In this paper, the experiment data is from a Web Server of Monash University. We regard the data from December 30,2002 to June 21,2004 as training samples, and regard the data from June 22,2004 to August 16,2004 as testing samples. We have constructed the model for the number of page access requests in every day. At the same, we forecasted the traffic of the following five state: Provided we have known the number of Web page access requests that day, the model can forecast the number of access requests in next day, next 2 day, next 3 day, next 4 day, next 5 day; The compare between forecasting results and actual results is illustrated in Fig.2.

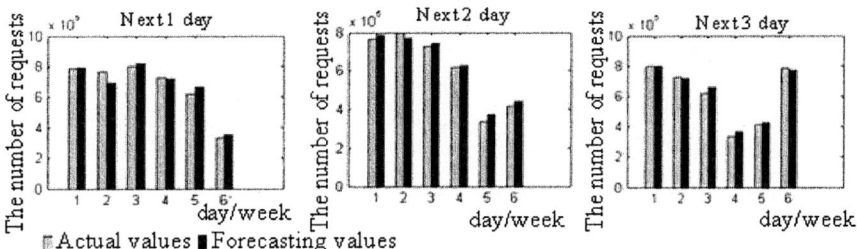

Fig. 2. The comparing diagram of forecasting results and actual results

Acknowledgements

This paper is supported by National Natural Science Fund of China (60373062) and educational committee fund of Hunan province (03C597).

References

1. Abraham A.: Neuro-Fuzzy Systems: State-of-the-Art Modeling Techniques, Connectionist Models of Neurons, Learning Processes, and Artificial Intelligence.Springer-Verlag Germany,Jose Mira and Alberto Prieto (Eds.), Granada, Spain(2001)269-276.
2. Abraham A. and Nath B.:Evolutionary Design of Fuzzy Control Systems - An Hybrid Approach.In Proceedings of The Sixth International Conference on Control, Automation, Robotics andVision, (CD ROM Proceeding), Wang J.L. (Ed.), ISBN 981 0434456, Singapore(2000).
3. Ng A. and Smith K. A.:Web usage mining by a self-organizing map. C. Dagli et al. (Eds.), Smart Engineering System Design: Neural Networks, Fuzzy Logic, Evolutionary Programming, Data Mining and Complex Systems, ASME Press, vol. 10, (2000)495-500.
4. Chiu, S.:Fuzzy Model Identification Based on Cluster Estimation.Journal of Intelligent & Fuzzy System, Vol. 2, No. 3, (1994).

A Fuzzy Neural Network System Based on Generalized Class Cover Problem

Yanxin Huang, Yan Wang, Wengang Zhou, and Chunguang Zhou

College of Computer Science and Technology, Jilin University,
Key Laboratory for Symbol Computation and Knowledge Engineering
of the National Education Ministry, Changchun 130012, China
huangyx@jlu.edu.cn, cgzhou@jlu.edu.cn

Abstract. A voting-mechanism-based fuzzy neural network system based on generalized class cover problem and particle swarm optimization is proposed in this paper. When constructing the network structure, a generalized class cover problem and an improved greedy algorithm are adopted to get the class covers with relatively even radii, which are used to partition fuzzy input space and extract fewer robust fuzzy IF-THEN rules. Meanwhile, a weighted Mamdani inference mechanism is proposed to improve the efficiency of the system output and a particle swarm optimization-based algorithm is used to refine the system parameters. Experimental results show that the system is feasible and effective.

1 Introduction

The first step in designing a fuzzy inference system is partition of input space. A good partition method can implement a small rule base with robust rules[1]. Adam Cannon and Lenore Cowen[2] presented the class cover problem(CCP) firstly, and proved the CCP is the NP-hard. Based on the CCP, a generalized class cover problem(GCCP) and its solution algorithm, which are used to partition input space and extract fewer robust fuzzy IF-THEN rules, are proposed in this paper. Then a voting-mechanism-based fuzzy neural network based on the obtained fuzzy IF-THEN rules and the particle swarm optimization(PSO) is constructed. Experimental results identifying 11 kinds of mineral waters by its taste signals show that the system is the feasible and effective.

2 Generalized Class Cover Problem and Its Solution Algorithm

Adam Cannon and Lenore Cowen defined the CCP as follows[2]: Let B be the set of points in class one, and R be the set of points in class two, with |R|+|B|=n. Then the CCP is:

Minimize K

$$s.t. \max_{v \in B}\{d(v,S)\} < \min_{w \in R}\{d(w,S)\}$$

where $S \subseteq B, |S| = K$, $d(.,.)$ denotes the distance between two points or a point to a set. Carey E. Priebe et al[3] presented a greedy algorithm (we refer to it as the greedy algorithm I below) for the CCP.

Aiming at designing the fuzzy neural network system architecture, we propose the generalized class cover problem (GCCP) as follows:

Minimize K, Var

$$s.t. \begin{cases} r_\beta(v,R) \leq \gamma MaxD_\beta, \forall v \in S \\ |\{w \mid w \in B, and \ \exists v \in S, d(v,w) \leq r_\beta(v,R)\}| \geq \alpha|B| \end{cases}$$

where $S \subseteq B$, $|S| = K$, $Var = \sqrt{\dfrac{1}{|S|-1}\sum_{v \in S}(r_\beta(v,R) - \bar{r})^2}$, $\bar{r} = \dfrac{1}{|S|}\sum_{v \in S}r_\beta(v,R)$, $\gamma, \alpha, \beta \in [0,1]$, $r_\beta(v,R)$ denotes the cover radius centered on the point $v \in S$, and $MaxD_\beta$ denotes the most cover radius in the points in B. The parameter γ is used to control producing the class covers with relatively even radii, which is beneficial to extracting the robust fuzzy IF-THEN rules, while the parameters α and β are used to make the system more noise-resistant, in which, α indicates that at least $\alpha|B|$ points in B must be covered by the class covers produced from S, and β indicates that a class cover centered on a point in B is permitted to cover at most $\beta|R|$ points in R. So, the GCCP is to find a minimum cardinality set of covering balls, with center points in S and relatively radii, whose union contains at least $\alpha|B|$ points in B and each covering ball contains at most $\beta|R|$ points in R. The CCP can be regarded as a special case of the GCCP with $\gamma = 1, \alpha = 1, \beta = 0$.

Toward the GCCP, we propose an improved greedy algorithm (we refer to it as the greedy algorithm II below) as follows.

Let $S = \phi$, and $C = B$:

(1) $\forall x \in B$, computing $d_\beta(x,R)$, which equals to the $\beta|R|+1$th smallest distance from x to the points in R, Let $MaxD_\beta = \max_{x \in B}\{d_\beta(x,R)\}$, then $\forall x \in B$ computing $r_\beta(x,R)$ as: $\begin{cases} r_\beta(x,R) = \gamma MaxD_\beta, if \ d_\beta(x,R) > \gamma MaxD_\beta \\ r_\beta(x,R) = d_\beta(x,R), if \ d_\beta(x,R) \leq \gamma MaxD_\beta \end{cases}$;

(2) producing digraph $G = (B,E)$ as follows: $\forall x \in B$, for all $y \in B$, if $d(x,y) \leq r_\beta(x,R)$, then producing a edge from x to y, namely $(x,y) \in E$;

(3) $\forall x \in C$, computing $cover(x) = \{y \mid y \in C, and \ (x,y) \in E\}$;

(4) taking $z \in C$, and $cover(z) = \max_{x \in C}\{cover(x)\}$, if $|C| < (1-\alpha)|B|$, then output S, and end the algorithm, else Let $S = S \cup \{z\}$, $C = C - \{x \mid x \in C, (z,x) \in E\}$, and go to (3).

Let $|B| = N$, $|R| = M$, and $|S| = \rho(S)$, then the algorithm has the time complexity of $O(NM + N(N-1) + N\rho(S))$. Suppose $N \approx M$, and $\rho(S) \ll N$, then the algorithm has the time complexity of $O(N^2)$ approximately.

3 Extracting the Fuzzy IF-THEN Rules

By projecting the class covers onto each input coordinate axis, the fuzzy subsets (linguistic terms), which are used to produce the initial fuzzy IF-THEN rules, can be obtained. In this paper, the Gauss function defined as Eq. (1)

$$\mu = \exp(-(x-c)^2/\sigma^2) \tag{1}$$

is chosen as the membership functions, and a fuzzy IF-THEN rule is uniquely produced by a corresponding class cover. The detailed procedure is shown as Fig.1(a). Let p_1 and p_2 be the centers of the two class covers with radius r_1 and r_2 on the Shimanto signals, respectively. Projecting p_1 and p_2 onto c_1- axis and c_2- axis respectively, the coordinates: p_{11} and p_{12} on c_1- axis as well as c_{21} and c_{22} on c_2- axis are obtained accordingly, so that the four membership functions can be derived, e.g., the membership function of the fuzzy set A_{11} corresponding to p_{11} is $\mu_{11}(x) = \exp(-(x-p_{11})^2/r_1^2)$. According to the above, two fuzzy IF-THEN rules for the Shimanto can be produced as follows:

$$\text{if } c_1 \text{ is } A_{11} \text{ and } c_2 \text{ is } A_{12} \text{ then } y \text{ is Shimanto}$$

$$\text{if } c_1 \text{ is } A_{21} \text{ and } c_2 \text{ is } A_{22} \text{ then } y \text{ is Shimanto}$$

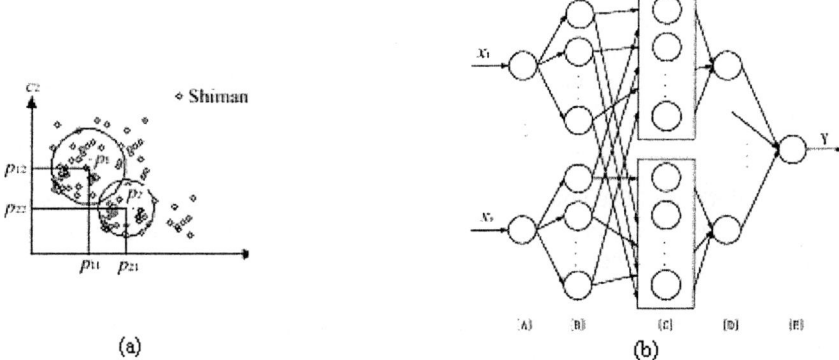

Fig. 1. (a)Projection of two class covers; (b)The VMFNN Architecture

4 Voting-Mechanism-Based Fuzzy Neural Network Model

4.1 Fuzzy Neural Network Architecture

Based on the fact that an object with unknown class tag is generally close to those samples whose class tags are the same with the one while far from the samples that have different class tags, a voting-mechanism-based fuzzy neural network system (VMFNN) is proposed as follows.

Different fuzzy inference mechanism can be distinguished by the consequents of the fuzzy IF-THEN rules[4], such as Mamdani and Takagi-Sugeno inference system. The Mamdani inference mechanism is adopted in this paper. Let $v^{(1)}, v^{(2)}, ..., v^{(r)}$ be the class tags of the training samples. Assume the number of the fuzzy IF-THEN rules with the consequent y is $v^{(j)}$ is a_j, $j = 1, 2, ..., r$, then those fuzzy rules can be represented as:

$$R_{i,j}: \text{if } x_1 \text{ is } A_{1,i} \text{ and } x_2 \text{ is } A_{2,i} \text{ and...and } x_s \text{ is } A_{s,i} \text{ then } y \text{ is } v^{(j)}$$

where $(\bar{x}, y)$ is a training sample, and $\bar{x} = (x_1, x_2, ...x_s)$ is the input feature vector and y is its class tag, $i = 1, 2, ..., a_j$, $j = 1, 2, ..., r$. The matching degree of the fuzzy rule antecedent for the training sample $\bar{x}$ is computed by the multiplication T-norm[5] as:

$$A_T(R_{i,j}) = \prod_{k=1}^{s} A_{k,i}(x_k) \tag{2}$$

Then a subsystem S_j can be constructed using all $R_{i,j}$, $i = 1, 2, ..., a_j$. Its output is defined as:

$$O_j = 1 - \exp(-\sum_{i=1}^{a_j} A_T(R_{i,j})) \tag{3}$$

Finally, the output of the fuzzy neural network system is defined as:

$$Y = v^{(j)}, \text{ Subject to } O_j = \max(O_1, O_2, ..., O_r) \tag{4}$$

According to the above, the VMFNN model is derived as Fig.1(b). Note that the links between nodes in different layers indicate the direction of signal flow, and they have no weights. The nodes in [B] layer have the node parameters, while the ones in other layers have none. The node functions in the same layers are of the same form as described follows:

[A] Input layer. $o_j = I_j = x_j$, $j = 1, 2, ..., s$.
[B] Fuzzification layer. The node parameters include the centers and the radii of the membership functions in the fuzzy rule antecedent. The output of the nodes in the layer are the membership degree calculated by Eq.(1).
[C] Fuzzy rule layer. By Eq.(2), every circle node in the rectangles multiplies the incoming signals and sends the product out. The outputs of the layer make up the input parts of the corresponding subsystems.
[D] Subsystem output layer. The outputs of the layer are calculated by Eq.(3).
[E] Voting output layer. The output of the layer is calculated by Eq.(4).

4.2 Optimization for the Fuzzy Neural Network System

Assume the system has r subsystems, subsystem j contains a_j fuzzy IF-THEN rules, $j = 1, 2, ..., r$, and each fuzzy IF-THEN rule contains $2s$ parameters (i.e., the center and the radius in the Gaussian function), where s is the dimensionality of input feature vectors, then the number of the system parameters is totally $M = 2s \sum_{j=1}^{r} a_j$.

The PSO is used to refine the system parameters, and the PSO algorithm flow can refer to [6][7]. Let Q be the size of the particle swarm (generally set $Q=20$). The initial velocity of the particles are initialized by random numbers uniformly distributed on [-0.3, +0.3], and the initial positions of the particles are initialized by the initial system parameters with 15% noise. Taking use of information included by the particle i, a fuzzy system as Fig.1(b) can be constructed. The misclassification rates of the system constructed by particle i are defined as:

$$E_i = \frac{err_i}{n} \quad (5)$$

where $i = 1, 2, ..., Q$, n is the number of the training samples, and err_i is the misclassification number of the system constructed by particle i. Then the fitness of the particle i can be defined as:

$$f_i = 1 - E_i \quad (6)$$

Set *acceleration coefficients* $W=0.7298$, $C_1=1.42$ and $C_2=1.57$, which satisfy the convergence condition of the particles: $W > (C_1 + C_2)/2 - 1$ [8]. Since $C_2 > C_1$, the particles will faster converge to the global optimal position of the swarm than the local optimal position of each particle. To avoid the premature convergence of the particles, an inactive particle, whose position unchanged in consecutive S epochs (set S=10 in this paper), will be reinitialized.

5 Experimental Results

The taste signals of 11 kinds of mineral waters[1][7] are used as experimental data in this paper. The experimental data consist of 1100 points with 100 points for each taste signal as shown in Fig. 2.

When computing the class covers of a taste signal, let the taste signal be the set of B and all the others the set of R. The class covers of the taste signals obtained by the greedy algorithm I and the greedy algorithm II ($\gamma=0.5, \alpha=0.96, \beta=0.01$) are shown in Fig 3(a) and Fig 3 (b), respectively. From Fig. 3(a), we can see that the obtained class covers are not even in size, and in which many with the radii of about zero (we call them outliers below) distribute over boundary areas between the taste signals, and that are unsuited for extracting robust fuzzy IF-THEN rules. From Fig. 3(b), we can obviously see that the obtained class covers are relatively even in size and have no outliers, that are suited for extracting fewer robust fuzzy IF-THEN rules. Comparison of the greedy algorithm I and the greedy algorithm II with different parameters in terms of K and Var is listed in Tab. 1.

Then the fuzzy neural network system I and II (below which are abridged as Sys I and Sys II) can be constructed using the class covers obtained by the greedy algorithm I and the greedy algorithm II ($\gamma=0.5, \alpha=0.96, \beta=0.01$), respectively. The original taste signals are used for training the systems, and the taste signals polluted by the noise are used for identification experiment. Note that, assume original signal is A, then the signal polluted by the noise is set $A' = A + A \times \eta \times rand$, where $0 \leq \eta \leq 1$, is the noise level, and *rand* is a random number uniformly distributed on [-1, +1].

The PSO is used to refine the system parameters of Sys I and Sys II. The curves of the fitness of the optimal particle of Sys I and Sys II with respect to training epochs and the curves of the misclassification percentages of Sys I and Sys II with respect to noise

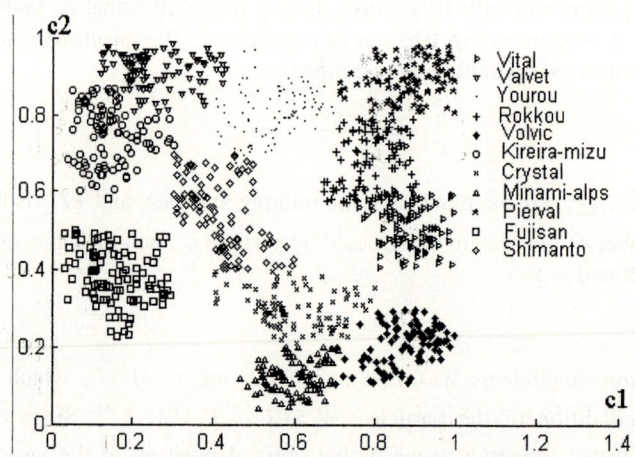

Fig. 2. 11 kinds of taste signals of mineral waters

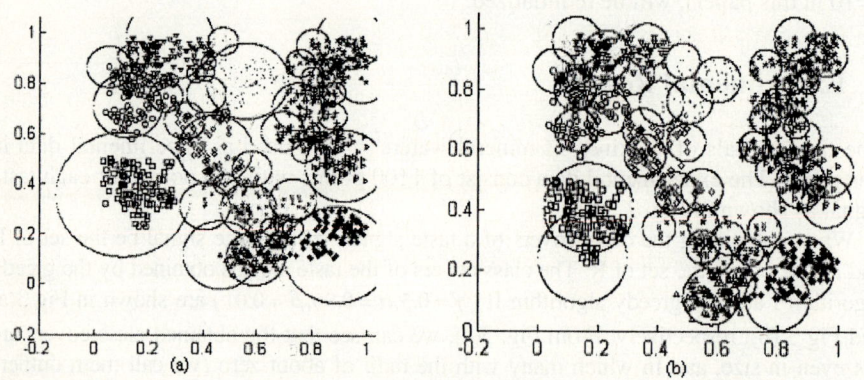

Fig. 3. (a)The class covers of the taste signals obtained by the greedy algorithm I; (b)The class covers of the taste signals obtained by the greedy algorithm II ($\gamma = 0.5, \alpha = 0.96, \beta = 0.01$)

Table 1. Comparison of the greedy algorithm I and the greedy algorithm II in terms of K and Var

	α	β	γ	K	Var
Greedy Algorithm I				70	0.002617
GreedyAlgorithm II	0.97	0	0.4	72	0.000393
	0.96	0.01	0.5	47	0.000647
	0.95	0.02	0.6	36	0.000875

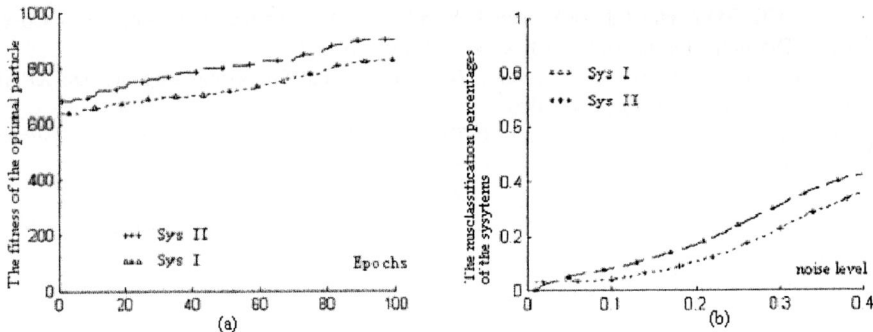

Fig.4. (a) The curves of fitness of optimal particle in Sys I and Sys II with respect to epochs.; (b) The curves of the misclassification percentages in Sys I and Sys II with respect to noise level

level are plotted in Fig.4 (a) and Fig.4 (b), respectively. Obviously, from Fig. 4(a), we can see Sys II shows better learning capability than Sys I, and from Fig. 4 (b), Sys II shows stronger noise-resistance capability than Sys I. On the other hand, because Sys II is constructed by fewer fuzzy IF-THEN rules than Sys I, it needs 71.363 sec. per 10 training epochs, while Sys I needs 106.146 sec. accordingly. Therefore Sys II is better than Sys I in terms of learning capability, error tolerance and running speed.

6 Conclusions and Discussions

By choosing proper parameters of γ, α and β in the greedy algorithm II, fewer robust fuzzy IF-THEN rules can be obtained. Then those rules are used to construct the fuzzy neural network system, which has many perfect characteristics, such as better robustness, learning capability, simplicity and running speed. In our experiences, we set $\gamma = 0.3 \sim 0.7, \alpha = 0.90 \sim 0.97, \beta = 0 \sim 0.05$ as general choices. To applying the system to high-dimensional complex data and to find other more effective optimization methods are our future research works.

Acknowledgement

This paper is supported by the key science-technology project of the National Education Ministry of China under Grant No. 02090 and the National Natural Science Foundation of China under Grant No. 60433020.

References

1. Y. X. Huang, C. G. Zhou: Recognizing Taste Signals Using A Clustering-based Fuzzy Neural Network. Chinese Journal of Electronics, 14(1) (2005) 21-25
2. Cannon A, Cowen L.: Approximation Algorithms for the Class Cover Problem. Annals of Mathematics and Artificial Intelligence, .40(3) (2004) 215-223

3. Priebe CE, Marchette DJ, DeVinney J, Socolinsky D.: Classification using Class Cover Catch Digraphs. Journal of Classification, 20(1) (2003) 3-23
4. J. S. R. Jang, C. T. Sun, E. Mizutani.: Neuro-Fuzzy and Soft computing. Xi An: Xi An Jiaotong University Press, Feb. 2000.
5. Ludmila I. K.: How Good are Fuzzy If-Then Classifiers? IEEE Transactions on Systems, Man, and Cybernetics, Part B: Cybernetics, 30(4) (2000) 501~509
6. Kennedy J., Eberhart R. C.: Particle Swarm Optimization. In Proceeding of IEEE International Conference on Neural Networks, Volume IV, Perth, Australia:IEEE Press. (1995) 1942-1948
7. Y. X. Huang, C. G. Zhou, S. X. Zou, Y. Wang, *et al*.: A Rough-Set-Based Fuzzy-Neural-Network System For Taste Signal Identification. 2004 International Symposium on Neural Networks Proceedings (Part II), Da-Lian, China, (2004) 337-343
8. van den Bergh F.: An Analysis of Particle Swarm Optimizers [PH.D thesis]. Pretoria: Natural and Agricultural Science Department, University of Pretoria, 2001.

A Self-constructing Compensatory Fuzzy Wavelet Network and Its Applications*

Haibin Yu[1], Qianjin Guo[1,2], and Aidong Xu[1]

[1] Shenyang Inst. of Automation, Chinese Academy of Sciences, Liaoning 110016, China
yhb@sia.cn   guoqianjin@sia.cn
[2] Graduate School of the Chinese Academy of Sciences, Beijing 100039, China
xad@sia.cn

Abstract. By utilizing some of the important properties of wavelets like denoising, compression, multiresolution along with the concepts of fuzzy logic and neural network, a new self-constructing fuzzy wavelet neural networks (SCFWNN) using compensatory fuzzy operators are proposed for intelligent fault diagnosis. An on-line learning algorithm is applied to automatically construct the SCFWNN. There are no rules initially in the SCFWNN. They are created and adapted as on-line learning proceeds via simultaneous structure and parameter learning. The advantages of this learning algorithm are that it converges quickly and the obtained fuzzy rules are more precise. The proposed SCFWNN is much more powerful than either the neural network or the fuzzy system since it can incorporate the advantages of both. The results of simulation show that this SCFWNN method has the advantage of faster learning rate and higher diagnosing precision.

1 Introduction

With rapid development of techniques for neural networks and fuzzy logic systems, the neural fuzzy systems are attracting more and more interest since they are more efficient and more powerful than either neural works or fuzzy logic systems [1]. The great important progress in neural fuzzy systems has been made in recent years. From the theoretical point of view, many effective learning algorithms for neurofuzzy systems were proposed [2-5]. In the application aspect, neurofuzzy systems have been widely used in control systems, pattern recognition, consumer products, medicine, expert systems, fuzzy mathematics, game theory, etc [6-10].

Wavelet neural networks (WNN) inspired by both the feedforward neural networks and wavelet decompositions have received considerable attention and become a popular tool for function approximation [12-14]. Incorporating the time-frequency localization properties of wavelets and the learning abilities of general neural network (NN), WNN has shown its advantages over the regular methods such as NN for complex nonlinear system modeling [15-16].

Hybrid intelligent systems, fuzzy wavelet neural networks possess the advantages of both wavelet networks and fuzzy rule-based systems and are particularly powerful

---

* Foundation item: Project supported by the National High-Tech. R&D Program for CIMS, China (Grant No. 2003AA414210).

in handling complex, non-linear and imprecise problems such as fault diagnosis. Their use avoids some of the pitfalls inherent in other more conventional approaches.

However, there exist some difficult but important problems for optimizing neural fuzzy systems [11], Many researchers have been dealing with these problems by using learning algorithms to adjust the parameter of fuzzy membership functions and defuzzification functions [9][11][17]. Unfortunately, for optimizing fuzzy logic reasoning and selecting optimal fuzzy operators, only static fuzzy operators are often used to make fuzzy reasoning. For example, the conventional neural fuzzy system can only adjust fuzzy membership functions by using fixed fuzzy operations such as Min and Max [11]. The compensatory neural fuzzy system with adaptive fuzzy reasoning is more effective and adaptive than the conventional neural fuzzy system with non-adaptive fuzzy reasoning [1][10].

A self-constructing compensatory fuzzy wavelet neural system (SCFWNN) is proposed in this study. The structure and the parameter learning phases are done concurrently and on-line in the SCFWNN. The advantages of this learning algorithm are that it converges quickly and the obtained fuzzy rules are more precise.

2 The Structure of SCFWNN

The basic configuration of the SCFWNN system includes a fuzzy rule base, which consists of a collection of fuzzy IF-THEN rules in the following form

$$FR^l : \text{IF } x_I \text{ is } F_I^l \text{ and ... and } x_m \text{ is } F_m^l \text{ THEN } y_I \text{ is } G_I^l \text{ and ... and } y_n \text{ is } G_n^l \qquad (1)$$

where FR^l is the lth rule $(1 \le l \le R)$, $\{x_i\}_{i=1,...,m}$ are input variables, $\{y_j\}_{j=1,...,n}$ are the output variables of the SCFWNN system, respectively, $F_i^l \in U$ are the labels of the fuzzy sets characterized by the membership functions(MF) $\mu_{F_i^l}(x_i)$, $G_j^l \in V$ are the labels of the fuzzy sets in the output space.

A schematic diagram of the five-layered SCFWNN is shown in Fig. 1. Layer 1 is the input layer, Layer 2 is the fuzzification layer, Layer 3 is the pessimistic-optimistic operation layer, Layer 4 is the compensatory operation layer, and Layer 5 is the defuzzification layer. A compensatory fuzzy neural network is initially constructed layer by layer according to linguistic variables, fuzzy IF-THEN rules, the pessimistic and optimistic operations, the fuzzy reasoning method, and the defuzzification scheme of a fuzzy logic control system. The functions of the nodes in each layer of the SCFWNN model are described as follower.

Layer 1 (input layer): For every node i in this layer, the net input and the net output are related by

$$I_i^{(1)} = x_i$$
$$O_i^{(1)} = I_i^{(1)} \qquad i = 1, 2, ..., m, \qquad (2)$$

where $I_i^{(1)}$ and $O_i^{(1)}$ denote, respectively, the input and output of ith neuron in layer 1

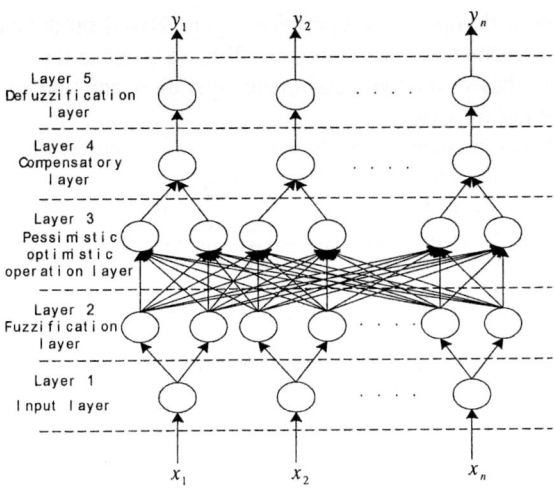

Fig. 1. The structure of SCFWNN

Layer 2 (fuzzification layer): In this layer, each neuron represents the membership function of a linguistic variable. The most commonly used membership functions are in shape of triangle, trapezoid and bell, etc. In this paper, the morlet wavelet basis function is adopted as the membership function [18][19], and five fuzzy sets (very small, small, medium, large, very large) are used for the above-mentioned fuzzy diagnosis rules. The fuzzy membership functions of F_i^j and G^j are defined, respectively, as follows.

$$\mu_{F_i^j}(x_i) = \psi_{\sigma_{ij}, a_{ij}}(x_i) = \psi\left(\frac{x_i - a_i^j}{\sigma_i^j}\right) = \cos(0.25 \cdot \frac{x_i - a_i^j}{\sigma_i^j}) \cdot \exp\left[-\frac{1}{2}\left(\frac{x_i - a_i^j}{\sigma_i^j}\right)^2\right] \quad (3)$$

$$\mu_{G^j}(y) = \psi_{\delta^j, b^j}(y) = \psi\left(\frac{y - b^j}{\delta^j}\right) = \cos(0.25 \cdot \frac{y - b^j}{\delta^j}) \cdot \exp\left[-\frac{1}{2}\left(\frac{y - b^j}{\delta^j}\right)^2\right]. \quad (4)$$

where $i=1,2,\ldots,m$; $j=1,2,\ldots,r$, $r = \prod_{i=1}^{m} p^m = \prod_{i=1}^{m} 5^m$, p ($p=5$) represents the number of linguistic values for each input, a_i^j, b^j and σ_i^j, δ_j are translation parameters and dilation parameters accordingly.

In this layer the relation between the input and output is represented as

Net input: $I_{ij}^{(2)} = O_i^{(1)}$

Net output: $O_{ij}^{(2)} = \mu_{F_i^j}(I_{ij}^{(2)}) = \cos(0.25 \cdot \frac{O_i^{(1)} - a_i^j}{\sigma_i^j}) \cdot \exp\left[-\frac{1}{2}\left(\frac{O_i^{(1)} - a_i^j}{\sigma_i^j}\right)^2\right] \quad (5)$

where $i=1,2,\ldots,m$; $j=1,2,\ldots,r$

Layer 3 (pessimistic-optimistic operation layer): Based on the essence of compensatory operations defined by Zimmermann [20], we propose more extensive compensatory operations based on the pessimistic operation and the optimistic operation which are defined as follows.

Pessimistic fuzzy neurons: The pessimistic fuzzy neuron can map the inputs x_i ($i=1,2,\ldots,m$) to the pessimistic output by making a conservative decision for the pessimistic situation or even the worst case. Actually, the t-norm fuzzy neurons are pessimistic neurons. For example, here, x_i ($i=1,2,\ldots,m$) are in [0,1]

$$P(x_1, x_2, \ldots, x_m) = Min(x_1, x_2, \ldots, x_m) = \prod_{i=1}^{m} x_i \quad . \tag{6}$$

here, the pessimistic operation are defined as follows:

$$u^j = \prod_{i=1}^{m} \mu_{F_i^j}(x_i) \quad \text{where } i=1,2,\ldots,m; j=1,2,\ldots,r \quad . \tag{7}$$

Optimistic fuzzy neurons: On the other hand, the optimistic fuzzy neuron can map the inputs x_i ($i=1,2,\ldots,m$) to the optimistic output by making an optimistic decision for the optimistic situation or even best case. Actually, the t-conorm fuzzy neurons are optimistic neurons. For example, here, x_i ($i=1,2,\ldots,m$) are in [0,1]

$$O(x_1, x_2, \ldots, x_m) = Max(x_1, x_2, \ldots, x_m) = \left[\prod_{i=1}^{m} x_i\right]^{1/m} \quad . \tag{8}$$

here, the optimistic operation are defined as follows:

$$v^j = \left[\prod_{i=1}^{m} \mu_{F_i^j}(x_i)\right]^{1/m} \quad \text{where } i=1,2,\ldots,m; j=1,2,\ldots,r \quad . \tag{9}$$

Layer 4 (compensatory operation layer): The nodes in this layer are compensatory fuzzy nodes. For an input fuzzy set F' in U, the jth fuzzy rule (1) can generate an output fuzzy set G^j in V by the sup-dot composition (10)

$$\mu_{G^j}(y) = \sup_{\underline{x} \in U}\left[\mu_{F_1^j \times \ldots \times F_m^j \to G^j}(\underline{x}, y) \cdot \mu_{F'}(\underline{x})\right] \tag{10}$$

$\mu_{F_1^j \times \ldots \times F_m^j}(\underline{x})$ is defined in a compensatory form (11) using the pessimistic operation (7) and the optimistic operation (9)

$$\mu_{F_1^j \times \ldots \times F_m^j}(\underline{x}) = (u^j)^{1-\gamma}(v^j)^{\gamma} \tag{11}$$

with a compensatory degree $\gamma \in [0,1]$, we have

$$\mu_{F_1^j \times \ldots \times F_m^j}(\underline{x}) = \left[\prod_{i=1}^{m} \mu_{F_i^j}(x_i)\right]^{1-\gamma+\gamma/m} \tag{12}$$

Based on (10-12), we have

$$\mu_{G^j}(y) = \sup_{\underline{x} \in U}\left\{\mu_{G^j}(y)\mu_{F'}(\underline{x})\left[\prod_{i=1}^{m}\mu_{F_i^j}(x_i)\right]^{1-\gamma+\gamma/m}\right\} \tag{13}$$

Since $\mu_{F_i^j}(x) = 1$ for the singleton fuzzifier and $\mu_{G^j}(b^j) = 1$, according to (13) we have

$$\mu_{G^j}(b^j) = \left[\prod_{i=1}^{m} \mu_{F_i^j}(x_i)\right]^{1-\gamma+\gamma/m} \tag{14}$$

where n is dimension number.

Layer 5(defuzzifier layer) is the output layer of the consequent network. Nodes in this layer represent the output variables of the system. Each node acts as a defuzzifier and computes the output value.

Net input:

$$I_k^{(5)} = f(\underline{x}) = \frac{\sum_{j=1}^{r} b^j \delta^j \mu_{G^j}(x_i)}{\sum_{j=1}^{r} \delta^j \mu_{G^j}(x_i)} = \frac{\sum_{j=1}^{r} b^j \delta^j \left[\prod_{i=1}^{m} \mu_{F_i^j}(x_i)\right]^{1-\gamma+\gamma/m}}{\sum_{j=1}^{r} \delta^j \left[\prod_{i=1}^{m} \mu_{F_i^j}(x_i)\right]^{1-\gamma+\gamma/m}} \tag{15}$$

$$\tag{16}$$

Net output: $Y_k = O_k^{(5)} = I_k^{(5)}$ $k=1,2,\ldots,n$

where Y_k represents the kth output to the node of layer 5.

3   The Learning Strategy

Two phases of learning, structure and parameter learning, are used for constructing the neurofuzzy network. In the first phase the structure learning algorithm is used to find proper fuzzy partitions in the input space and create fuzzy logic rules. In the second phase all parameters are tuned using a supervised learning scheme. The back-propagation algorithm to minimize a given cost function adjusts the parameters of membership functions and compensatory degree. There are no rules (i.e., no nodes in the network except the input–output nodes) in the SCFWNN initially. They are created dynamically and automatically as learning proceeds upon receiving on-line incoming training data by performing the structure and parameter learning processes. The procedure of the structure/parameter learning algorithm is through inputting the training pattern to learn successively. Then, we can gain proper rules.

3.1   The Structure Learning Algorithm

The proposition of the structure learning algorithm is to decide proper fuzzy partitions by the input patterns. The procedure of our structure learning algorithm is to find the proper fuzzy logic rules. However, the structure learning algorithm determines whether or not to add a new node in layer 2 via the input pattern data, and decides whether or not to add the associated fuzzy logic rule in layer 3.

After the input pattern is entered in layer 2, the firing strength of the wavelet based membership function will be obtained from Eq.(3), that is used as the degree measure $\mu_{F_i^j}$. For computational efficiency, we can use the firing strength obtained from $\prod \mu_{F_i^j}$ directly as the precondition part's degree measure.

$$P = \prod_{j=1}^{R(t)} \mu_{F_i^j} \qquad (17)$$

where i is input dimension, $i=1,\ldots,m$; j is rule number, $j=1,\ldots,R(t)$, $R(t)$ is the number of existing rules at time t.

To avoid the newly generated membership function being too similarity to the existing one, the similarities between the new membership function and existing ones must be checked. If the new fuzzy rule is different from the existing fuzzy rule, we confirmed the new fuzzy rule would be added in the SCFWNN. It can make neural fuzzy inference system to gain more performance. Therefore, we use similarity measure of membership functions to estimate the rule's similarity degree. Suppose the fuzzy sets to be measured are fuzzy sets A and B with membership function $\mu_A(x) = \psi((x-m_1)/\sigma_1)$ and $\mu_B(x) = \psi((x-m_2)/\sigma_2)$, respectively. Assume $m2 \geq m1$ as in [11], we can compute $|A \cap B|$ by

$$|A \cap B| = \frac{1}{2} \cdot \frac{h^2(m_2 - m_1 + (\sigma_2 + \sigma_1)\sqrt{\pi})}{(\sigma_2 + \sigma_1)\sqrt{\pi}} + \frac{1}{2} \cdot \frac{h^2(m_2 - m_1 + (\sigma_1 - \sigma_2)\sqrt{\pi})}{(\sigma_2 - \sigma_1)\sqrt{\pi}} + \frac{1}{2} \cdot \frac{h^2(m_2 - m_1 + (\sigma_2 - \sigma_1)\sqrt{\pi})}{(\sigma_1 - \sigma_2)\sqrt{\pi}} \qquad (18)$$

where $h(x) = \max\{0,x\}$. So the approximate similarity measure of fuzzy sets is

$$E(A,B) = \frac{|A \cap B|}{|A \cup B|} = \frac{|A \cap B|}{\frac{1}{2}(\sigma_1 + \sigma_2)\sqrt{\pi} - |A \cap B|} \qquad (19)$$

where we use the fact that $|A| + |B| = |A \cap B| + |A \cup B|$.

Using this firing strength and the similarity measure, we can obtain the following criterion for the generation of a new fuzzy rule of new incoming data. The method can be described as follows.

(a) Preset a positive threshold $P_{min} \in (0,1)$ that decays during the learning process.
(b) Find the maximum degree P_{max}

$$P_{max} = \max_{1 \leq j \leq R(t)} P_j \qquad (20)$$

where $R_{(t)}$ is the number of existing rules at time t.

(c) If $P_{max} \leq P_{min}$, the structure learning needs to add a new node in the SCFWNN, and a new rule is generated.
(d) Once a new rule is generated, the next step is to assign initial mean and variance of the new membership function. Since our goal is to minimize an objective function, the mean and variance are all adjustable later in the parameter learning phase. Hence the mean and variance deviation of the new membership function are set as follow:

$$a_{ij}^{(R_{(t+1)})} = x_i \tag{21}$$

$$\sigma_{ij}^{(R(t+1))} = \sigma_{init} \tag{22}$$

where x_i is the new input patte+++rn; σ_{init} is preset constant; i is input dimension; j is rule number.

(e) The similarity measure E between the new membership function and all existing ones are calculated and maximum one E_{max}, is found as follows:

$$E_{max} = \max_{1 \leq j \leq M(t)} E(\mu(a_{new}, \sigma_{new}), \mu(a_j, \sigma_j)) \tag{23}$$

(f) If $E_{max} \leq E_{min}$, where $E_{min} \in (0,1)$ is a prespecified value, then the new fuzzy logic rule is adopted and the rule number is incremented

$$R = R + 1 \tag{24}$$

(g) Therefore, the new fuzzy rule and corresponding parameters are generated.

3.2 The Parameter Learning Algorithm

Once the network structure is established according to the current training data, the network enters the second learning phase to optimally adjust the parameters based on the same training data. A gradient method performing the steepest descent on a surface in the network parameter space is used. The goal of this second learning phase is to adjust both the premise and consequent parameters so as to minimize the error function. Given the y^p as the desired target output of pth input pattern, the cost function can be written as:

$$E^p = \frac{1}{2}(f(x^p) - y^p)^2 \tag{25}$$

Based on (15), we have

$$f(x^p) = \frac{\sum_{j=1}^{R} b^j \delta^j z^j}{\sum_{j=1}^{R} \delta^j z^j} \tag{26}$$

where

$$z^j = \left[\prod_{i=1}^{m} \mu_{F_i^j}(x_i^p) \right]^{1-\gamma+\gamma/m} \tag{27}$$

According to the gradient descending method, we get

1) Training the centers of output membership functions

$$b^j(t+1) = b^j(t) - \eta \left.\frac{\partial E^p}{\partial b^j}\right|_t = b^j(t) - \eta \left.\frac{[f(x^p) - y^p]\delta^j z^j}{\sum_{j=1}^{R} \delta^j z^j}\right|_t \tag{28}$$

2) Training the widths of output membership functions

$$\delta^j(t+1) = \delta^j(t) - \eta \left. \frac{\partial E^p}{\partial \delta^j} \right|_t = \delta^j(t) - \eta \left. \frac{[f(x^p) - y^p][b^j - f(x^p)]\delta^j z^j}{\sum_j^R \delta^j z^j} \right|_t \quad (29)$$

3) Training the centers of input membership functions

$$\frac{\partial E^p}{\partial a_i^j} = \frac{2[f(x^p) - y^p][b^j - f(x^p)][x_i^p - a_i^j][1 - \gamma + \gamma/m]\delta^j z^j}{\sigma_i^{j2} \sum_{j=1}^{R} \delta^j z^j} \cdot \left(0.25 \cdot \frac{\sigma_i^j}{x_i^p - a_i^j} \cdot tg(\frac{x_i^p - a_i^j}{\sigma_i^j}) + 1 \right) \bigg|_t \quad (30)$$

then we have

$$a_i^j(t+1) = a_i^j(t) - \eta \left. \frac{\partial E^p}{\partial a_i^j} \right|_t \quad (31)$$

4) Training the widths of input membership functions

$$\frac{\partial E^p}{\partial \sigma_i^j} = \frac{2[f(x^p) - y^p][b^j - f(x^p)][x_i^p - a_i^j]^2[1 - \gamma + \gamma/m]\delta^j z^j}{\sigma_i^{j3} \sum_{k=1}^{R} \delta^j z^j} \cdot \left(0.25 \cdot \frac{\sigma_i^j}{x_i^p - a_i^j} \cdot tg(\frac{x_i^p - a_i^j}{\sigma_i^j}) + 1 \right) \bigg|_t \quad (32)$$

then we have

$$\sigma_i^j(t+1) = \sigma_i^j(t) - \eta \left. \frac{\partial E^p}{\partial \sigma_i^j} \right|_t \quad (33)$$

5) Training the compensatory degrees.

To eliminate the constraint $\gamma \in [0,1]$, we redefine γ as follows:

$$\gamma = \frac{c^2}{c^2 + d^2} \quad (34)$$

$$\frac{\partial E^p}{\partial \gamma} = \left. \frac{2[f(x^p) - y^p][b^j - f(x^p)][\frac{1}{m} - 1]\ln[\prod_{i=1}^{m} u_{F_i^j}(x_i^p)]\delta^j z^j}{\sum_{j=1}^{R} \delta^j z^j} \right|_t \quad (35)$$

then we have

$$c(t+1) = c(t) - \eta \left\{ \frac{2c(t)d^2(t)}{[c^2(t) + d^2(t)]^2} \right\} \left. \frac{\partial E^p}{\partial \gamma} \right|_t \quad (36)$$

$$d(t+1) = d(t) + \eta \left\{ \frac{2d(t)c^2(t)}{[c^2(t) + d^2(t)]^2} \right\} \left. \frac{\partial E^p}{\partial \gamma} \right|_t \quad (37)$$

$$\gamma(t+1) = \frac{c^2(t+1)}{c^2(t+1) + d^2(t+1)} \quad (38)$$

In all above formulas, η is the learning rate and $t = 0, 1, 2, \ldots$.

4 Experiments for Fault Diagnosis

This work uses a CBM (condition based maintenance) system that we have developed to test SCFWNN method for fault diagnosis of rotating machinery [21]. This system is designed in a modular fashion, and it contains the data acquisition module, signal processing module, feature extraction module, intelligent diagnosis module, training module, and prognostics module, user module etc.

4.1 Data Acquisition

Experiments were performed on a machinery fault simulator, which can simulate the most common faults, such as misalignment, unbalance, resonance, radial rubbing, oil whirling and so on. In this system, the rotor is driven by an electromotor, and the bearing is the journal bearing. The fault samples are obtained by simulating corresponding fault on experiment rotating system. For example, adjusting the simulator plane highness and degree simulates the misalignment faults; adding an unbalance weight on the disc at the normal condition creates the unbalance. A radial acceleration was picked up from an accelerometer located at the top of the right bearing housing. The shaft speed was obtained by one laser speedometer.

The measurements with acceleration, velocity, or displacement data from rotating equipment are acquired by the NI digital signal acquisition module, and then are collected into an embedded controller. A total of six conditions were tested: radial rubbing, twin looseness, misalignment, oil whirling, and unbalance etc. Each condition was measured with a given times continuously. The frequency of used signal is 5000Hz and the number of sampled data is 1024.

4.2 Feature Extraction

The features of vibration signals are extracted with wavelet packet analysis and FFT, the time-frequency spectrum of data is computed and fed into the training stage, in which 6 faults and 7 frequency bounds are selected to form a feature vector. These feature vectors are used as input and output of wavelet based fuzzy neural network.

4.3 Fault Diagnosis and Analysis

4.3.1 Experimental Data Sets and Fuzzy Diagnosis Rules

The network architecture used for fault diagnosis consists of 7 inputs corresponding to the 7 different ranges of the frequency spectrum of a fault signal (listed in table 1) and 6 outputs corresponding to 6 respective faults, such as unbalance, misalignment, oil whirling, oil oscillating, radial rubbing and twin looseness and so on.

In the experiment, 200 groups of feature data are acquired on a machinery fault simulator; 50 groups of feature data are used as the training set, the remaining 150 groups are used as the diagnosis set samples. Feature vectors are used as input and output of neural networks. After all possible normal operating modes of the fault are

Table 1. The sets of training sample

Fault class	<0.40 ω_0	0.4~0.5 ω_0	0.51~0.99 ω_0	1ω_0	2ω_0	3~5 ω_0	>5 ω_0
Unbalance	0.0000	0.0000	0.0000	1.0000	0.0042	0.0035	0.0000
Misalignment	0.0000	0.0000	0.0000	0.8000	1.0000	0.0010	0.0000
Oil whirling	0.0000	0.7100	0.0000	1.0000	0.0200	0.0066	0.0000
Oil oscillating	0.0000	0.9652	0.0000	1.0000	0.0100	0.0050	0.0000
Radial rubbing	0.2100	0.1800	0.1652	1.0000	0.1765	0.1900	0.2000
Twin looseness	0.0000	0.0000	0.0000	0.2000	0.1500	0.4362	0.2533

Table 2. The learned fuzzy logic rules for fault diagnosis

Rules	<0.40 ω_0	0.4~0.5 ω_0	0.51~0.99 ω_0	1ω_0	2ω_0	3~5 ω_0	>5 ω_0	Conclusions
If	VS	VS	VS	VL	VS	VS	VS	Unbalance
If	VS	VS	VS	L	VL	VS	VS	Misalignment
If	VS	L	VS	VL	VS	VS	VS	Oil whirling
If	VS	VL	VS	VL	VS	VS	VS	Oil oscillating
If	S	S	S	VL	S	S	S	Radial rubbing
If	VS	VS	VS	S	S	M	S	Twin looseness

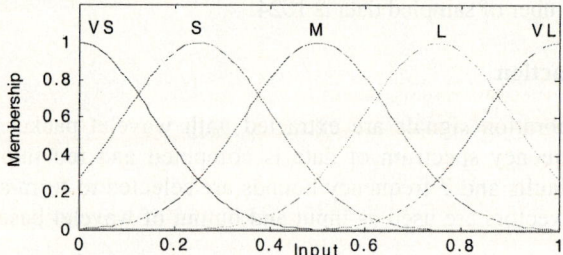

Fig. 2. The membership functions of the input linguistic value

learned, the system enters the fault diagnosis stage, in which the machinery vibration data are obtained and are subjected to the pre-processing and feature extraction methods described in the signal processing and feature extraction stage.

Described as Section 2, five fuzzy sets are used for fuzzy diagnosis rules, corresponding to very small, small, medium, large, very large (labeled as VS, S, M, L and VL respectively). Morlet wavelet function is chosen as the membership functions for these five fuzzy sets. Here, the five fuzzy sets for the input and output linguistic variables of the SCFWNN diagnosis system have both been designed in the same range of [0,1] (shown in Fig. 2).

Once the shape of the fuzzy sets is given, the relationship between input and output variables of the SCFWNN system is defined by a set of linguistic statements that are called fuzzy rules. The corresponding fault diagnosis rules are list in table 2.

4.3.2 Performance of the Fuzzy Network-Based Diagnosis System

To demonstrate the performance of the SCFWNN-based approach on fault classification, comparisons are made with three other types of artificial neural networks, namely compensatory fuzzy neural networks (CFNN), wavelet neural networks (WNN) and BP networks. In CFNN, the parameters and the initial individuals are the same as those used in SCFWNN. Both the WNN networks and BP networks have three layers with 7 inputs corresponding to the 7 sources of information, and 6 outputs corresponding to the 6 faults considered, the number of hidden neurons was fixed at ten, the MSE function is same as Eq.(25), Expecting output error threshold is 0.001, and training processes terminate in given fitness evaluation times. In all experiments, each experiment was run 50 times for given iterations, and the results were averaged to account for stochastic difference.

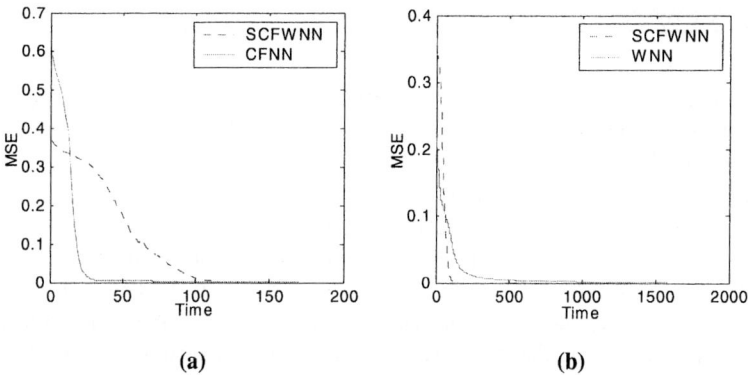

Fig. 3. Convergence curves for SCFWNN, CFNN and WNN

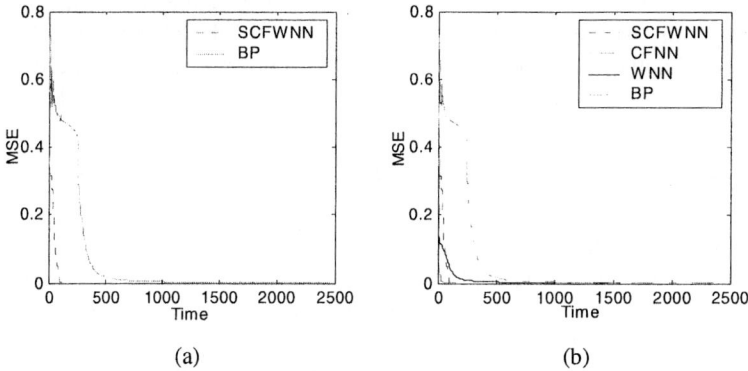

Fig. 4. Convergence curves for SCFWNN, BP, CFNN and WNN

Fig.3 and Fig.4 demonstrate the training history and the performance of the SCFWNN, CFNN, WNN and BP networks respectively. By looking at the shapes of the curves in Fig. 3(a), it is easy to see the SCFWNN trained with structure and parameter learning algorithm (with averaged 135 epochs) converges more quickly than the CFNN trained with GD algorithm (with averaged 175 epochs). As seen in Fig. 3(b) and Fig. 4, it is clear that the simulation time obtained by the SCFWNN trained with structure and parameter learning algorithm is comparatively less compared to the WNN networks trained with GD algorithm (with averaged 1550 epochs) and the BP networks trained with GD algorithm (with averaged 2490 epochs).

Table 3. Comparisons of SCFWNN, CFNN, WNN and BP method

Method	Diagnosis accuracy (%)	Sum error	Epochs
BP	90.67	0.001	2490
WNN	95.33	0.001	1550
CFNN	96.67	0.001	175
SCFWNN	98.67	0.001	135

Table 3 is the comparison results of SCFWNN, CFNN, WNN and BP methods on the fault diagnosis. The second column in this table lists the diagnosis accuracy on the 150 actual sample data. The third column in this table lists the average number of error function evaluations used during training, until running terminated with the network converged or the epochs exceeding the maximum epochs. Compared with the BP method, the SCFWNN method has 8.0% improvement on the diagnosis accuracy, compared with the WNN method, the SCFWNN method has 3.34% improvement on the diagnosis accuracy, and compared with the CFNN method, the SCFWNN method has 2.0% improvement on the diagnosis accuracy. The test results confirm that, in all compared cases, the proposed SCFWNN method has a better capability for generalization than the other methods.

From the comparison of various methods, it can be seen that, the SCFWNN method outperformed all the other architectures. The SCFWNN architecture is decidedly superior, yielding errors that are smaller than the CFNN, WNN and BP architecture. Another benefit of the SCFWNN approach is the reduced training time.

5 Conclusions

A self-constructing compensatory fuzzy wavelet network system (SCFWNN) is proposed in this study. In general, the SCFWNN system is much more powerful than either the wavelet neural networks or the fuzzy system since it can incorporate the advantages of both. A new adaptive fuzzy reasoning method using compensatory fuzzy operators can make the fuzzy logic system more adaptive and effective. The structure and the parameter learning phases are done concurrently and on-line in the SCFWNN. The SCFWNN system with adaptive fuzzy reasoning is more effective and adaptive than the conventional neural fuzzy system with non-adaptive fuzzy reasoning. The system has been validated using the fault diagnosis data sets. The results have shown that SCFWNN based diagnosis system has a better training performance, fast convergence rate, as well as a better diagnosis ability than the other modules to be selected cases.

References

1. Zhang,Y.Q., Kandel,A. : Compensatory Genetic Fuzzy Neural Networks And Their Appli-cations. Singapore: World Scientific.1998.
2. Kosko, B.: Neural Networks and Fuzzy Systems. Englewood Cliffs, NJ: Prentice-Hall,1992.
3. Jang, J.S.R. : ANFIS: Adaptive-network-based fuzzy inference system. IEEE Trans. Systems, Man, and Cybernetics. 23(1993) 665–685.
4. Carpenter,G.A., Groosberg,S., Markuzon,N., Renold,J.H., Rosen, D.B. : Fuzzy ARTMAP: A neural network architecture for incremental supervised learning of analog multidimensional maps. IEEE Trans. On Neural network. 3(1992) 698-713.
5. Wang, L.X. : Adaptive Fuzzy systems and Control. Englewood Cliffs, NJ:Prentice-Hall. 1994.
6. Lin,C., Lee,C.S.G. : Reinforcement structure/parameter learning for neural-network based fuzzy logic control systems. IEEE Transactions on Fuzzy Systems. 2(1994)46-63.
7. Ho,D.W.C., Zhang ,P.A., Xu ,J.: Fuzzy wavelet networks for function learning. IEEE Trans. on Fuzzy Syst. 9 (2001) 200–211.
8. Javadpour,R., Knapp,G.M.: A fuzzy neural network approach to machine condition monitoring. Computers & Industrial Engineering. 45(2003) 323-330.
9. Lin, F. J., Lin,C. H., Shen,P. H.: Self-Constructing Fuzzy Neural Network Speed Controller for Permanent-Magnet Synchronous Motor Drive. IEEE Transactions on Fuzzy Systems. 9(2001)751-759.
10. Zhang,Y.Q.,Kandel,A.: Compensatory Neurofuzzy Systems with Fast Learning Algorithms. IEEE Trans. on Neural Networks. 9(1998) 83-105.
11. Juang ,C. F., Lin, C.T.: An On-Line Self-Constructing Neural Fuzzy Inference Network and Its Applications. IEEE Trans. on Fuzzy Systems. 6(1998)12-31.
12. Zhang,Q. : Using wavelet networks in nonparametric estimation. IEEE Trans. Neural Networks, 8(1997) 227–236.
13. Zhang,Q., Benveniste,A. : Wavelet networks: IEEE Trans. Neural Networks. 3(1992) 889–898.
14. Zhang,J., Walter,G.G., Lee, W.N.W.: Wavelet neural networks for function learning. IEEE Trans. Signal Processing. 43(1995)1485–1497.
15. Pillay,P., Bhattachariee,A. : Application of wavelets to model short-term power system disturbances. IEEE Trans. Power Syst.. 11(1996) 2031–2037.
16. Guo,Q.J, Yu,H.B, Xu,A.D. Wavelet Neural Networks for Intelligent Fault Diagnosis. The International Symposium on Intelligence Computation & Applications.Wuhan,China (2005) 477-485.

17. Lin,C.J., Lin,C.T.: An ART-Based Fuzzy Adaptive Learning Control Network. IEEE Trans. on Fuzzy Systems. 5(1997) 477-496.
18. Daubechies, I.: The wavelet transform, time–frequency localization, and signal analysis. IEEE Trans. Inform. Theory. 36(1990) 961–1005.
19. Mallat ,S.: A theory for multi-resolution signal decomposition: The wavelet representation. IEEE Trans. Pattern Anal. Machine Intell.. 11(1989) 674–693.
20. Zimmermann, H. J., Zysno, P.: Latent connective in human decision. Fuzzy Sets and Systems. 4(1980) 31-51.
21. Guo,Q.J, Yu,H.B, Xu,A.D.: Research and Development on Distributed Condition-Based Maintenance Open System. Computer Integrated Manufacturing Systems.3(2005)416-421.

A New Balancing Method for Flexible Rotors Based on Neuro-fuzzy System and Information Fusion

Shi Liu

Xi'an Jiaotong University, State Key Lab for Manufacturing Systems Engineering,
710049, Xi'an, P.R. China

Abstract. This paper presents a new field balancing method for flexible rotors, which is based on adaptive neuro-fuzzy inference system (ANFIS) and information fusion. Firstly, new method fully utilizes the information supplied from all proximity sensors by holospectral technique for enhancing the balancing efficiency and accuracy. Secondly, a fuzzy model is established to simulate the mapping relationship between vibration responses and balancing weights using the ANFIS. The inputs of ANFIS are the amplitudes and phases of integrated vibration responses, while the outputs are the mass and azimuth of balancing weights. The experimental results show that the fuzzy balancing model based on ANFIS can obtain satisfactory balancing result after a single trial run, and possesses prospects of application in field balancing.

1 Introduction

The field balancing is an ordinary way to reduce rotor vibrations to a given level. The existing balancing methods are still potential for improvements in accuracy and efficiency. On the one hand, most balancing methods need large numbers of trial runs to obtain the vibration responses of trial weights in different planes. On the other hand, the vibration response in each measured section is always taken from a single sensor, and thus is lack of comprehensive vibration information of rotor. In order to overcome above shortcomings of traditional balancing methods, this paper presents a new field balancing method for flexible rotors, which is based on adaptive neuro-fuzzy inference system (ANFIS) [1] and holospectral technique [2]. The simulation study shows that the ANFIS can obtain satisfactory balancing result after a single trial run. The new method does not only decrease test number, but also increases precision and efficiency of balancing. The effectiveness of the new method was validated by the experiments on balancing rig.

2 Information Fusion

The vibration responses can be measured by two eddy current probes mutually perpendicularly mounted across each bearing section. By using holospectral technique, the vibration components with rotating frequency in the i th measuring section can be expressed by initial phase point (IPP) (Fig. 1a):

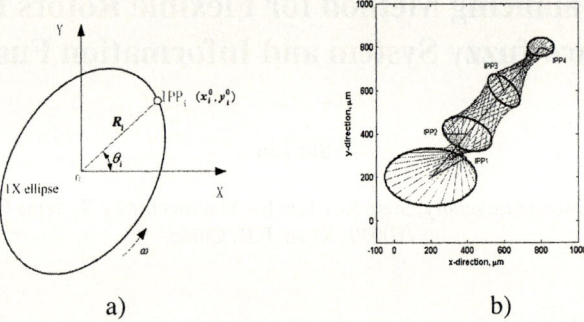

Fig. 1. Information fusion based on holospectrum: a) Initial phase point (IPP) on 1X ellipse in the i th measuring section; b) three-dimensional holospectrum with four measuring section

$$\begin{cases} R_i = \sqrt{[A_i \sin(\alpha_i)]^2 + [B_i \sin(\beta_i)]^2} \\ \theta_i = \arctan\{[B_i \sin(\beta_i)]/[A_i \sin(\alpha_i)]\} \end{cases} \quad (1)$$

The three-dimensional holospectrum integrates all first frequency ellipses, and therefore provide full vibration information of a rotor system (Fig. 1b).

3 Balancing Method Based on ANFIS

3.1 Fuzzy Balancing Model

The balancing rig is shown in Fig. 2. The vibration responses of unbalance are integrated to 4 parameters R_i, θ_i ($i=1,2$) by equation (1) as the inputs of following ANFIS. The balancing weights are represented by the parameters $M_i \angle \Phi_i$ ($i=1,2$) as the outputs of ANFIS. A fuzzy balancing model is established to the mapping relationship between vibration responses and balancing weights. The input R_i is represented with three membership functions, which describing the semantic meaning of the terms 'high', 'medium' and 'low'. The input θ_i is represented with four membership functions, which describing the semantic meaning of the terms 'first quadrant', 'second quadrant', 'third quadrant' and 'fourth quadrant'. The fuzzy IF/THEN rules consist of all 144 (3×4×3×4) possible combinations of inputs. The complete system contains 4 ANFIS structures for estimating the four parameters of balancing weights (Fig. 3).

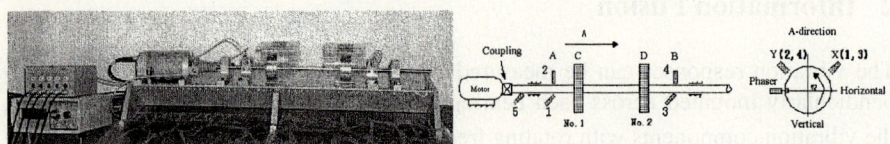

Fig. 2. Configuration of balancing rig ×15

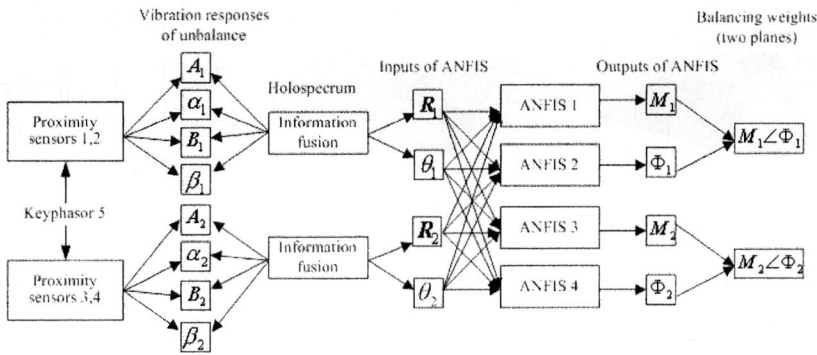

Fig. 3. Fuzzy balancing model using ANFIS

3.2 Collection of Training Data

The trial weights are added on two discs C and D to simulate the unbalance fault, for example, $T_C = 0.4g\angle 45°$, $T_D = 0.8g\angle 315°$. Then, the rig was run up to 4000rpm and the vibration responses were measured as A_i, α_i, B_i and β_i in both X and Y directions. From equation (1), the integrated vibration responses can be obtained as R_1, θ_1, R_2, θ_2. The balancing weights in two balancing planes should be as the mirror images of trial weights T_C and T_D respectively. Thus, the outputs are $M_1 = 0.4g$, $\Phi_1 = 315°$, $M_2 = 0.8g$, $\Phi_2 = 45°$. Then, we get one set of training data. As above method, M_1, Φ_1, M_2 and Φ_2 can be changed to get other sets of training data.

3.3 Experimental Results

Based on experimental data, a combination of least-squares and back-propagation gradient descent methods is then used for adjusting ANFIS membership function and node-parameters [1]. To visualize the effect produced by the input parameters, the surface graphs can be obtained from the fuzzy model, shown in Fig. 4.

To validate the effectiveness of fuzzy balancing model, a set of trial weights are added on two balancing discs to simulate the original unbalance. Rotor vibrations are measured at speed 4000 rpm. Inputting the vibration responses of original unbalance to fuzzy balancing model, the balancing weights can be calculated as $M_1\angle\Phi_1 = 1.04g\angle 121.1°$ and $M_2\angle\Phi_2 = 0.96g\angle 17.8°$. Adding the correcting weights, then running up to 4000rpm and measuring the rotor residual vibrations. The vibration responses expressed by the 3D-holospectra, before and after correction by the fuzzy balancing method, are shown in Fig. 5. The thin lines represent the original vibrations, and the bold lines represent the residual vibrations. Using new method, vibration levels of an experimental rotor were successfully reduced to as much as 20% of their original levels after one test run. It is evidential that the correction weights are reasonable and the new balancing method is effective.

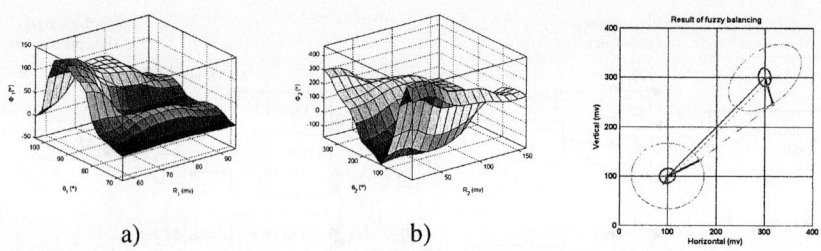

Fig. 4. Surface graphs to visualize the relationship between inputs and outputs: a) inputs: R_1 and θ_1, output: Φ_1; b) inputs: R_2, θ_2, output: Φ_2

Fig. 5. Results of fuzzy balancing

4 Conclusions

By employing ANFIS and holospectrum, we established a fuzzy balancing model to simulate the mapping relationship between vibration responses and balancing weights. Experiment results show that new method can effectively reduce the vibrations of flexible rotors due to unbalance, and decrease the number of test runs.

Acknowledgements

The authors acknowledge supports from NSFC 50475084 and EMDF 20040698017.

References

1. J.-S.R Jang, ANFIS: adaptive-network-based fuzzy inference system, IEEE Trans. Syst. Man Cybern. 23 (3) (1993) 665–685
2. Liangsheng Qu, and Guanghua Xu. One Decade of Holospectral Technique: Review and Prospect [C]. Proceedings of the 1999 ASME Design Engineering Technical Conferences (1999)

Recognition of Identifiers from Shipping Container Images Using Fuzzy Binarization and Enhanced Fuzzy Neural Network

Kwang-Baek Kim

Dept. of Computer Engineering, Silla University, Korea
gbkim@silla.ac.kr

Abstract. In this paper, we propose and evaluate a novel recognition algorithm for container identifiers that effectively overcomes these difficulties and recognizes identifiers from container images captured in various environments. The proposed algorithm, first, extracts the area containing only the identifiers from container images by using CANNY masking and bi-directional histogram method. The extracted identifier area is binarized by the fuzzy binarization method newly proposed in this paper. Then a contour tracking method is applied to the binarized area in order to extract the container identifiers, which are the target for recognition. This paper also proposes an enhanced fuzzy RBF network that adapts the enhanced fuzzy ART network for the middle layer. This network is applied to the recognition of individual codes. The results of experiment for performance evaluation on the real container images showed that the proposed algorithm performs better for extraction and recognition of container identifiers compared to conventional algorithms.

1 Introduction

Recently, the quantity of goods transported by sea has increased steadily since the cost of transportation by sea is lower than other transportation methods. Various automation methods are used for the speedy and accurate processing of transport containers in the harbor. The automation systems for transport container flow processing are classified into two types: the barcode processing system and the automatic recognition system of container identifiers based on image processing. However, these days the identifier recognition system based on images is more widely used in the harbors. The identifiers of transport containers are given in accordance with the terms of ISO standard, which consist of 4 code groups such as shipping company codes, container serial codes, check digit codes and container type codes [1]. The ISO standard prescribes only code types of container identifiers, while it doesn't define other features such as size, position and interval of identifier characters etc. Other features such as the foreground and background colors of containers, the font type, and the size of identifiers, vary from one container to another. These variations in features for container identifiers, makes the process of extraction and recognition of identifiers quite difficult [2].

Since the identifiers are printed on the surface of the containers, shapes of identifiers are often impaired by the environmental factors during the transportation by sea. The damage to the surface of the container may change shapes of identifier characters in container images. So after preprocessing the container images, an additive procedure must be applied, in order to decide whether the results are truly the edges of identifiers or just the noise from the background.

2 Container Identifier Extraction

In this paper, considering the specific attributes of the features of container identifiers, we applied Canny masking to container images for generating edge maps of input images. By applying the bi-directional histogram method to edge maps, identifier areas, which are the minimum rectangles including only all identifiers, are extracted from input images. We used image binarization method to extract individual identifiers from the identifier areas. The container images include diverse colors, globally varying intensity and various types of noise, so that the selection of threshold value for image binarization is not easy. Therefore, we propose a fuzzy binarization method to binarize the identifier areas and apply 4-directional contour tracking to the results for extracting individual identifiers. We also propose an enhanced fuzzy RBF network architecture and apply it for recognizing individual identifier codes.

2.1 Extraction of Container Identifier Areas

For extracting identifier areas from container images, first, we used Canny masking to generate edge maps of input images. The edges extracted by Canny masking are disconnected in several directions and isolated individually. These edge maps are efficient for the separation of identifiers and the background in container images. Canny masking is similar to noise removal carried out using Gaussian masking and edge extraction performed by Sobel masking sequentially.

Since the container images include noise caused by the distortion of the outer surface and shape of containers on the upper and lower areas, the calculation of vertical coordinates of identifier areas ahead of horizontal coordinates can generate more accurate results. Hence, we calculated the vertical coordinates of identifier areas by applying the vertical histogram to edge maps, and applied the horizontal histogram to the block corresponding to the vertical coordinate calculating the horizontal coordinate. Fig. 1 shows an example of extraction results by the proposed algorithm.

Fig. 1. Extraction results of Identifier areas

2.2 Extraction of Individual Identifiers

We extracted container identifiers from identifier areas by binarizing the areas and applying contour tracking algorithm to the binarized areas. Container identifiers are arranged in a single row by calculating Euclidean distances between identifier codes and in turn classified to the three code groups such as shipping company codes, container serial codes and check digit code. Generally, image binarization is used for extraction of recognition targets from input images since those results in data compression in a fashion that is usually loss less as far as relevant information is concerned. However, various features of container identifiers, such as size, position, color etc., are not normalized, and the shapes of identifiers are impaired by the environmental factors during transportation and the container breakdown. Moreover, container images include diverse colors, globally changed intensity and various types of noises, so that the selection of threshold value for image binarization is difficult using traditional methods which use distance measures [3]. Therefore, we propose a novel fuzzy binarization algorithm to separate the background and identifiers for extraction of container identifiers.

The proposed fuzzy binarization algorithm defines I_{Mid} as the mean intensity value of the identifier area for the selection of interval of membership function. I_{Mid} is calculated like Eq.(1).

$$I_{Mid} = \frac{\sum_{i=1}^{W}\sum_{j=0}^{H} I_{ij}}{H \times W} \qquad (1)$$

where I_{ij} is the intensity of pixel (i, j) of identifier area, and H and W are the pixel lengths of height and width of identifier area respectively. I_{Min} and I_{Max} define the minimum intensity value and the maximum value in the identifier area respectively. The algorithm determining the interval of membership function $[I_{Min}^{New}, I_{Max}^{New}]$ in the proposed fuzzy binarization is as follows:

Step 1:
$I_{Min}^{F} = I_{Mid} - I_{Min}$
$I_{Max}^{F} = I_{Max} - I_{Mid}$
Step 2:
If $I_{Mid} \rangle 128$ Then $I_{Mid}^{F} = 255 - I_{Mid}$
Else $I_{Mid}^{F} = I_{Mid}$
Step 3:
If $I_{Mid}^{F} \rangle I_{Max}^{F}$ Then
   If $I_{Min}^{F} \rangle I_{Mid}^{F}$ Then $\sigma = I_{Mid}^{F}$
   Else $\sigma = I_{Min}^{F}$
Else If $I_{Max}^{F} \rangle I_{Mid}^{F}$ Then $\sigma = I_{Mid}^{F}$
Else $\sigma = I_{Max}^{F}$

Step 4: Calculate the normalized I_{Min}^{New} & I_{Max}^{New}.

$$I_{Min}^{New} = I_{Mid} - \sigma$$
$$I_{Max}^{New} = I_{Mid} + \sigma$$

In most cases, individual identifiers are embossed in the identifier area and the noise between identifier codes and the background is caused by shadows. We used the fuzzy binarization algorithm to remove the noise from the shadows.

The degree of membership $u(I)$ in terms of the membership interval $[I_{Min}^{New}, I_{Max}^{New}]$ is calculated using Eq.(2).

$$\begin{aligned} &\text{if } \left(I_{Min}^{New} \leq I < I_{Mid}^{New}\right) \text{ then } u(I) \\ &\text{if } \left(I_{Mid}^{New} \leq I < I_{Max}^{New}\right) \text{ then } u(I) = -\frac{1}{I_{Max}^{New} - I_{Mid}^{New}}\left(1 - I_{Mid}^{New}\right) + 1 \end{aligned} \qquad (2)$$

The identifier area is binarized by applying $\alpha - cut(\alpha = 0.9)$ to the degree of membership $u(I)$. Next, we extracted the container identifiers from the binarized identifier area by using the contour tracking method. In this paper, the 4-directional contour tracking method using 2x2 mask was applied considering the whole preprocessing time of container images. The contour tracking, using 2x2 mask given in Fig.2, scans the binarized identifier area from left to right and from top to bottom to find boundary pixels for identifier codes [4]. If a boundary pixel is found, that pixel is selected as the start position for tracking and placed at the x_k position (see Fig. 2) of the 2x2 mask. By examining the two pixels below the *a* and *b* positions of the mask and comparing them with the conditions in Table 1, the next scanning direction of the mask is determined and the next boundary pixel is selected for tracking. The selected pixels below the x_k position are connected into the contour of an identifier. By generating the outer rectangles including connected contours and comparing the ratio of width to height, the rectangles with the maximum ratio are extracted as individual identifiers.

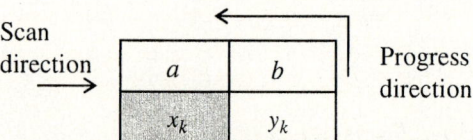

Fig. 2. 2x2 mask for 4-direction contour tracking

The extracted identifiers must be classified into three code groups, shipping company codes, container serial codes and check digit code for the information processing following the identifier recognition. However, extracted identifiers are not normalized in size and position and the vertical coordinates of identifiers placed on the same row are different from each other because of the application of contour tracking to images with distortion caused by the bent surface of containers. As a result, the grouping of related identifiers by using only coordinates of individual

Table 1. Progress direction of a and b by 2x2 mask

	a	b	x_k	y_k
Forward	1	0	a	b
Right	0	1	b	y_k
Right	1	1	a	x_k
Left	0	0	x_k	a

identifiers generates inconsistent results. In this paper, the extracted identifiers are arranged in a single row by using Euclidean distances between identifiers and classified into three code groups. If the row containing the identifiers was distorted, resulting in multiple rows within identifier area, initially, the first identifiers from each row were selected. Then in each row identifiers are arranged according to the Euclidean distance. The Euclidean distance is calculated by measuring the distance between the start pixel of the first identifier and the start pixel of the other identifier having a vertical offset from the first identifier. The vertical offset must be less than one half of the vertical size of the first identifier. Then, by combining identifier sequences in every row, one row of identifiers is created. Finally, identifiers in the row are classified sequentially to code groups according to the ISO standard [1].

3 Identifier Recognition Using an Enhanced Fuzzy RBF Network

We propose an enhanced fuzzy RBF network which constructs the middle layer using the enhanced fuzzy ART network for the recognition of extracted codes. In the traditional fuzzy ART network, the vigilance parameter determines the allowable degree of mismatch between any input pattern and stored patterns[4]. Vigilance parameter is the inverse of degree of tolerance. A large value of vigilance parameter classifies an input pattern to a new category in spite of a little mismatch between the pattern and the stored patterns. On the other hand a small value may allow the classification of the input pattern into an existing cluster in spite of a considerable mismatch. Moreover, because many applications of image recognition based on the fuzzy ART network assign an empirical value to the vigilance parameter, the success rate of recognition may deteriorate[5][6]. To correct this defect, we propose an enhanced fuzzy ART network and apply it to the middle layer in a fuzzy RBF network.

The enhanced fuzzy ART network adjusts the vigilance parameter dynamically according to the homogeneity between the patterns using Yager's intersection operator[7], which is a fuzzy connection operator. The vigilance parameter is dynamically adjusted only in the case that the homogeneity between the stored pattern and the learning pattern is greater than or equal to the vigilance parameter. Also, the proposed fuzzy ART network adjusts the weight of connection for the learning patterns with the authorized homogeneity: Let T^p and T^{p*} be the target value of the learning pattern and the stored pattern respectively. If T^p is equal to T^{p*}, the network decreases the vigilance parameter and adjusts the weight of connection between the input layer and the middle layer. Otherwise, the network increases the vigilance parameter and selects the next winner node.

The algorithm dynamically adjusts the vigilance parameter as follows:

$$\begin{aligned} & if\ (T^p \neq T^{p*})\ then \\ & \quad \rho(t+1) = 1 - \wedge\left(1,\ \left((1-\rho(t))^{-2} + (1-\rho(t-1))^{-2}\right)^{-1/2}\right) \\ & else\ \ \rho(t+1) = 1 - \wedge\left(1,\ \left((1-\rho(t))^{2} + (1-\rho(t-1))^{2}\right)^{1/2}\right) \end{aligned} \quad (3)$$

where ρ is the vigilance parameter.

The authorization of homogeneity for the selected winner node is executed according to Eq.(4).

$$\frac{\|w_{j^*i} \wedge x_i^p\|}{\|x_i^p\|} < \rho \quad (4)$$

If output vector of the winner node is greater than or equal to the vigilance parameter, the homogeneity is authorized and the input pattern is classified to one of the existing clusters. Moreover, in this case, the weight of connection is adjusted according to Eq.(5) to reflect the homogeneity of the input pattern to the weight.

$$w_{j^*i}(t+1) = \beta \times \left(x_i^p \wedge w_{j^*i}(t)\right) + (1-\beta) \times w_{j^*i}(t) \quad (5)$$

where β is the learning rate between 0 and 1.

When the weight is adjusted in the traditional fuzzy ART network, β is set to an empirical value. If a large value of β is chosen, the success rate of recognition goes down since an information loss is caused by the increase in the number of cluster center updates. On the other hand, if the learning is performed with a small value of β, the information of the current learning pattern is unlikely to be reflected in the stored patterns and the number of clusters increases[8]. So, in the enhanced fuzzy ART network, the value of β is dynamically adjusted based on the difference between the homogeneity of the learning pattern to the stored pattern and the vigilance parameter. The adjustment of β is as follows:

$$\beta = \frac{1}{1-\rho} \times \left(\frac{\|w_{j^*i} \wedge x_i^p\|}{\|x_i^p\|} - \rho\right) \quad (6)$$

This paper enhances the fuzzy RBF network by applying the enhanced fuzzy ART algorithm to the middle layer, as shown in Fig. 3.

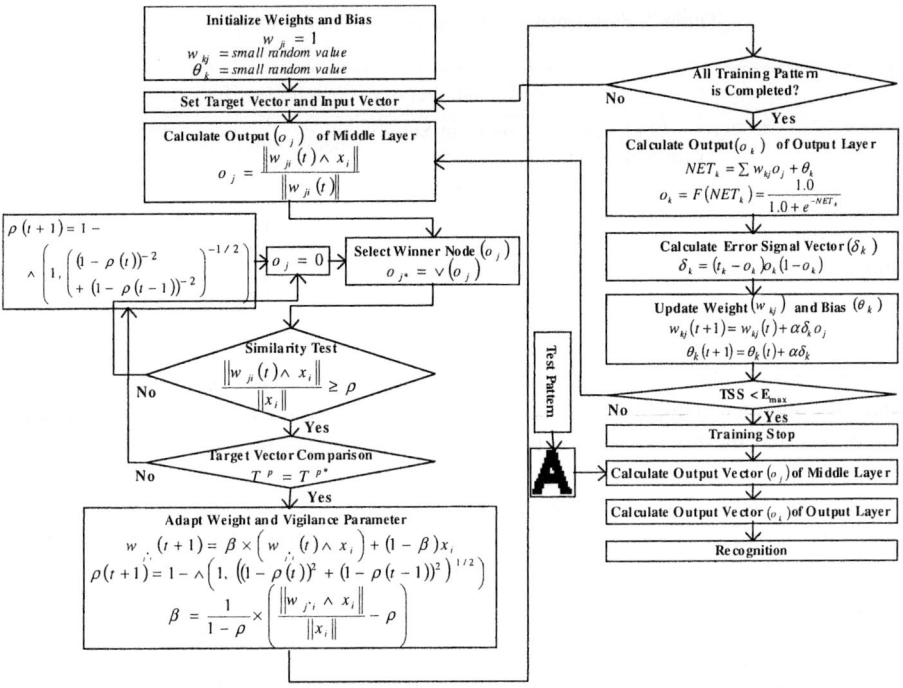

Fig. 3. Learning and recognition algorithm of the enhanced fuzzy RBF network

4 Performance Evaluation

For performance evaluation, we implemented the proposed algorithm and experimented using an IBM-compatible PC with Intel Pentium-IV 2GHz CPU and 256MB RAM. Totally 150 container images of 754x504 pixel size and 256 colors were used in the experiment. In the experiment for identifier extraction, we compared the extraction algorithms proposed in this paper and those obtained by previous researchers [2]. In order to evaluate the recognition performance of enhanced fuzzy RBF network, we compared the results with those obtained using the conventional fuzzy RBF network[9].

4.1 Performance of Individual Identifier Extraction

By using the proposed extraction algorithm for all 150 images the identifier areas were successfully extracted from the images. Applying identifier extraction algorithms proposed in this paper and the histogram based algorithm [2] to the extracted identifier areas, experimental results were summarized and compared in Table 2. As shown in Table 2, the algorithm proposed in the reference [2] is inferior

Table 2. Performance comparison of identifier extraction

	The number of extracted identifiers			
	Shipping Company Codes (600)	Container Serial Codes (900)	Check Digit Code (150)	Total number of identifiers (1650)
Algorithm proposed in Ref.[2]	495	800	90	1385
Proposed extraction algorithm	579	878	135	1592

to our algorithm because it failed to extract identifiers in cases where the background and the container identifiers may not be distinguished from each other or the shape of identifiers and the interval between identifiers are changed by the bent surface of the containers.

Our algorithm, first, distinguished the background and container identifiers by using the proposed fuzzy binarization, and then, extracted identifiers by using the contour tracking. As a result, our algorithm could extract successfully container identifiers in the images where the algorithm of reference [2] failed to extract identifiers. Fig.4 shows an example of the case mentioned above. Note that due to the bent surface of the container the characters in Fig. 4 (a) are not in a straight line. Fig. 4 (b) shows that the histogram based method [2] fails to identify character 3 and it tends to lump the characters in groups of three. At the same time Fig.4(c) shows that the proposed fuzzy binarization and tracking algorithm succeeds in extracting all 15 identifiers.

(a) Extracted identifier area (b) Histogram method in Ref.[3] (c) Proposed method in this paper

Fig. 4. Comparison of identifier extraction results

Fig.5 shows the comparison of experimental results when the mean-intensity based binarization proposed in the reference [2] and the proposed fuzzy binarization were applied to identifier area of Fig.5(a). In Fig.5, the thresholds for the mean-intensity based binarization and for the fuzzy binarization are 117 and 145 respectively. As shown in Fig.5 (c), the fuzzy binarization distinguished clearly the background and container identifiers. At the same time the mean intensity based binarization method (see Figure 5 (b)) for digits 8 and 3. It also failed to remove the background noise.

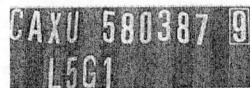

(a) Extracted identifier area

(b) Mean-intensity based image binarization (c) Proposed fuzzy binarization

Fig. 5. Comparison of mean-intensity based binarization and proposed fuzzy binarization

4.2 Performance of Container Identifier Recognition

To evaluate the learning performance of the enhanced fuzzy ART network, this paper compared the number of clusters generated by the conventional fuzzy ART network and the enhanced fuzzy ART network in the learning experiments on individual codes. Table 3 compares learning performances in the experiment that applied the conventional fuzzy ART network and the enhanced Fuzzy ART network to container identifiers extracted by the proposed algorithm mentioned above. In the learning experiment, 1054 container identifiers were used, which consisted of 383 shipping company codes, 582 container serial codes and 89 check digit codes.

Table 3. Comparison of the number of clusters between the fuzzy ART and the proposed fuzzy ART network

		Number of clusters / Number of patterns
Shipping Company Codes	Proposed Fuzzy ART	35 / 383
	Fuzzy ART	78 / 383
Container Serial Codes	Proposed Fuzzy ART	43 / 582
	Fuzzy ART	91 / 582
Check Digit Code	Proposed Fuzzy ART	16 / 89
	Fuzzy ART	27 / 89

As shown in Table 3, the number of clusters in the enhanced fuzzy ART network was much lower than in the traditional fuzzy ART network, so we may know that the enhanced fuzzy ART network refines the classification of the homogenous patterns properly. Table 4 shows the results of the experiment involving enhanced fuzzy RBF network for the 150 container images for recognition. In the experiment, the initial values of the vigilance parameter used for the creation and update of the nodes in the

Table 4. Result of learning and recognition by the proposed fuzzy RBF network

	The number of nodes in middle layer	The number of Epoch	The number of recognition
Shipping Company Codes	35	1204	578 /579
Container Serial Codes	43	1605	877 / 878
Check Digit Codes	16	523	132 / 135

middle layer were set to 0.9, 0.9, 0.85 for the serial code, the shipping company codes and check digit respectively. And as shown in Table 4, the proposed fuzzy RBF network was able to successfully recognize all of the extracted individual codes.

5 Conclusions

In this paper, we have proposed and evaluated a novel recognition algorithm of container identifiers for the automatic recognition of transport containers. Based on the structural attributes of the container image that the identifier areas have more edge information than other areas, the proposed algorithm used Canny masking to generate edge maps of container images. Applying the vertical histogram method and the horizontal one sequentially to the edge map, the identifier area that is the minimum rectangle containing only the container identifiers was extracted. The container images demonstrate certain characteristics, such as irregular size and position of identifiers, diverse colors of background and identifiers, and the impaired shape of identifiers caused by container damages and the bent surface of containers making the identifier recognition by image processing difficult. Hence, we proposed a fuzzy binarization algorithm to separate clearly the background and identifiers and applied it along with the 4-directional contour tracking to the identifier area, extracting individual identifiers. Finally, the extracted identifiers were arranged in a single row by using the Euclidean distances between identifiers and then grouped into code groups such as shipping company code, container serial code and check digit code. For identifier recognition, an enhanced fuzzy RBF network was proposed and applied in the code recognition phase. This algorithm dynamically changes the vigilance parameter in order to improve the clustering performance.

For performance evaluation, experiments applying the proposed identifier extraction and recognition algorithm to totally 150 real container images were performed. All 150 identifier areas were successfully extracted from container images and 1592 identifiers were extracted successfully out of a total of 1650 identifiers. This means that the proposed algorithm performed considerably better than the preprocessing algorithm used by the previous researchers [2]. Moreover, an enhanced fuzzy RBF network recognized effectively the individual container code so that it showed the high success rate of recognition. And the number of clusters created at the learning process of the enhanced fuzzy ART network was much lower than the conventional fuzzy ART network, which means that it is efficient to use the enhanced fuzzy ART network in the construction of middle layer in the fuzzy RBF network. Results of the recognition experiment by applying the conventional fuzzy RBF network and the enhanced fuzzy RBF network to the 1592 extracted identifiers show that the enhanced fuzzy RBF network has a higher rate of recognition compared to the conventional fuzzy RBF network.

References

1. ISO-6346.: Freight Containers-Coding-Identification and Marking. (1995)
2. Kim, K. B.: The Identifier Recognition from Shipping Container Image by using Contour Tracking and Self-Generation Supervised Learning Algorithm Based on Enhanced ART1. Journal of Intelligent Information Systems. Vol.9. No.3. (2003) 65-80
3. Liane C. Ramac and Pramod K. Varshney.: Image Thresholding Based on Ali-Silvey distance Measures. Pattern Recognition. Vol.30. No.7. (1997) 1161-1173
4. Kim, K. B., Jang, S. W. and Kim, C. K.: Recognition of Car License Plate by Using Dynamical Thresholding and Enhanced Neural Networks. Lecture Notes in Computer Science. LNCS 2756. Springer-Verlag. (2003) 309-319
5. Carpenter, G. A., Grossberg, S.: Neural Networks for Vision and Image Processing. Massachusetts Institute of Technology. (1992)
6. Kim, K. B. and Kim, C. K.: Performance Improvement of RBF Network using ART2 Algorithm and Fuzzy Logic System. Lecture Notes in Artificial Intelligence. LNAI 3339. (2004) 853-860
7. Zimmermann, H. J.: Fuzzy set Theory and it's Applications, Kluwer Academic Publishers. (1991)
8. Kim, K. B. , Joo, Y. h. and Cho, J. H.: An Enhanced Fuzzy Neural Network. Lecture Notes in Computer Science. LNCS 3320. Springer-Verlag. (2004) 176-179
9. Kim, K. B. and Yun, H. W.: A Study on Recognition of Bronchogenic Cancer Cell Image using A New Physiological Fuzzy Neural Networks. Japanese Journal of Medical Electronics and Biological Engineering. Vol.13. No.5. (1999) 39-43

Directed Knowledge Discovery Methodology for the Prediction of Ozone Concentration

Seong-Pyo Cheon and Sungshin Kim

School of Electrical and Computer Engineering,
Pusan National University, Changjeon-dong, Busan 609-735, Korea
{buzz74, sskim}@pusan.ac.kr

Abstract. Data mining is the exploration and analysis, by automatic or semiautomatic means, of large quantities of data in order to discover meaningful patterns and rules. Data mining consists of several tasks and each task uses a variety of methodologies. Some of these tasks are suited for a top-down method called hypothesis testing and others are suited for a bottom-up method called knowledge discovery. In this paper, we report our research procedures and results that concern and relate ozone concentration data in various factors and attributes. We use the general steps of directed knowledge discovery methodologies and intelligent modeling techniques. Next, we construct ozone concentration prediction system in order to reduce various adverse effects on human beings and life on the earth.

1 Introduction

Data mining tasks are usually defined as follows [1],[2],[3]: *classification, estimation, prediction, affinity grouping, clustering,* and *description*. Followings are brief definitions for each concept. *Classification*, that is the most common data mining task, relies on human imperative. Classification consists of examining the features of a newly presented object and assigning it to one of a predefined set of classes. *Estimation*, in contrast to classification, deals with continuously valued outcomes such as income, height, or credit card balance. *Prediction* is the same as classification or estimation except that the records are classified according to some predicted future behavior or estimated future value. *Affinity grouping* is one simple approach to generate rules from data, and classify objects by determining which ones go together. *Clustering* is the task of segmenting a heterogeneous population into a number of more homogeneous subgroups or clusters. Clustering is distinguished from classification which it does not rely on predefined classes. *Description*, that is sometimes the purpose of data mining, is to characterize what is occurring in the data in order to clarify of related phenomena. Some of these tasks are best approached in a top-down manner called *hypothesis testing*. Other tasks are best approached in bottom-up manner called *knowledge discovery*. Knowledge discovery can be either directed or undirected. In directed knowledge discovery, the target field is selected and the computer is directed to estimate, classify, or predict. In undirected knowledge discovery, there is no target field. The computer identifies patterns in the data which may be significant.

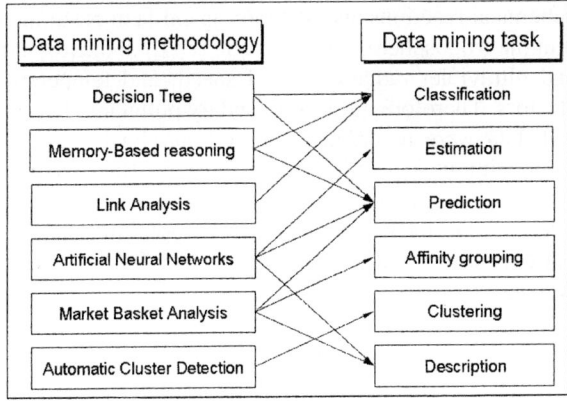

Fig. 1. The relationship of data mining methodologies to tasks

In this paper, we concentrate on how to apply directed knowledge discovery methodology in complex and nonlinear systems. Directed knowledge discovery is aimed at developing methodologies and tools to automate the data analysis process and create useful information and knowledge from data to help in decision making. The goal of directed knowledge discovery is very broad and can describe a multitude of fields of study related to data analysis. Fields related to data analysis include statistics [4], data warehousing [5], pattern recognition [6], artificial intelligence [7] and computer visualization. Directed knowledge discovery draws upon methods, algorithms and technologies from these diverse fields, and the unifying goal is to extract knowledge from data.

As shown Figure 1, there are representative relationship between data mining methodologies and tasks. Many researchers have attempted to find effective methodologies that can easily and compactly extract features from data. This paper will not describe every data mining methodology. We introduced our research procedure and the results of our predictions concerning high-level ozone concentration in summer. As shown in Figure 2, we suggest an entire procedural system for forecasting ozone concentrations. First, we created a small database that included pollution materials and some meteorological data. Second, we prepared the data for analysis in a step called *preprocessing*, using a fuzzy clustering method. Third, we built two prediction models using Dynamic Polynomial Neural Networks (DPNN). One is a low concentration model and the other is a high concentration model. Then, we trained the models with test data until the performance criterion (PC) become satisfied within a predefined tolerance. Next, we added our unique step, called *postprocessing*, which combines low and high concentration models using a fuzzy inference method. Finally, we evaluated our models using unseened data from several areas in Seoul, Korea.

2 Data Identification

One emerging major issue in air pollution is abnormal ozone concentrations in the troposphere in summer. High concentration ozone is the strong oxidizing material which is responsible for various adverse effects on both animals and plants [8][9]. In general, it revealed that the features of ozone distribution and creation are closely related to photochemical reactions and meteorological factors. In the ozone (O_3) reaction mechanism in the troposphere, nitric dioxide (NO_2) and carbon dioxide (CO_2) act as precursors and ultraviolet radiation, wind speed, and temperature are necessary meteorological factors. Therefore, O_3 is a secondary pollutant [10].

Meteorological data were recorded at every hour. Each data stands for weather condition of Seoul without separating small clusters. But, air pollution data were recorded one value per hour at each separated small cluster. The data were collected from 21 areas. We got the meteorological and air pollution data between 1996 and 1997 from Korea Meteorological Administration and National Institute of Environmental Research, respectively.

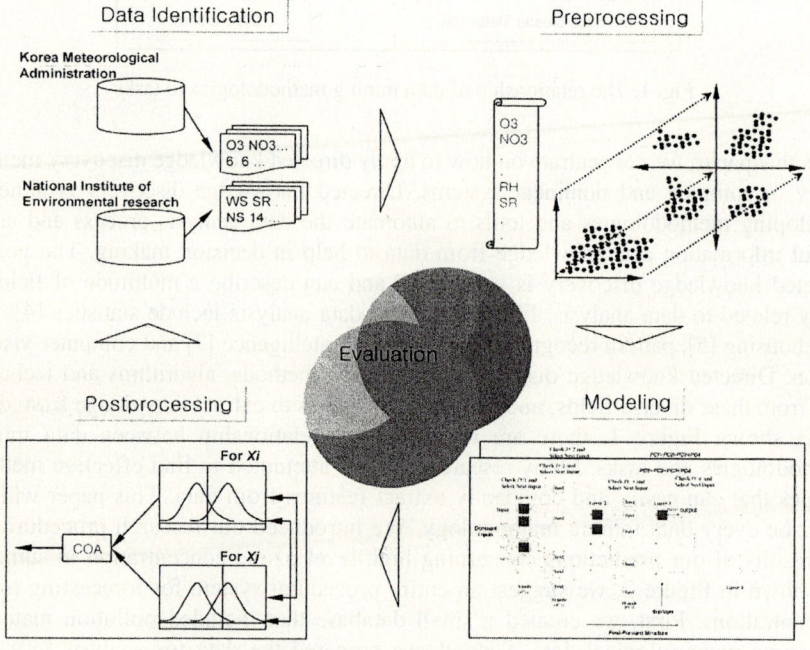

Fig. 2. The structure of the ozone prediction system

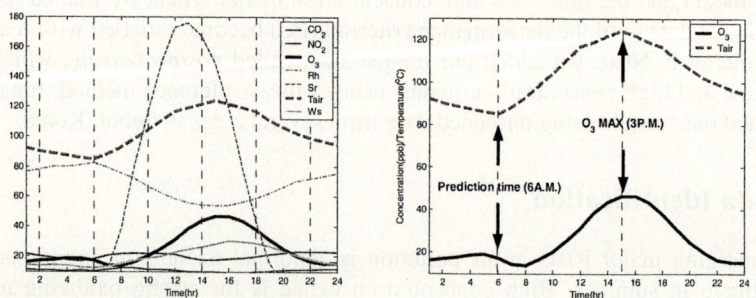

Fig. 3. Hourly distributions of all components in source data

3 Preprocessing: Fuzzy Clustering with Inverse Mapping

The successful knowledge discovery depends on good data. Source data that was originally composed of O_3, CO_2, NO_2, sulfur dioxide (SO_2), total suspended particle (TSP), wind speed (Ws), wind direction (Wd), atmosphere temperature (Tair), solar radiation (Sr), relative humidity (Rh), and rainfall (Rf) are used as the parameters of air pollutants and meteorological factors.

We sifted through the data using a few heuristic approaches. First, we only considered data from June to August because it frequently appears high concentration ozone. Second, we filtered out rainfall because high-level O_3 never appears rainy. Third, Wd is also excluded due to difficulties quantifying it presents. Therefore, O_3, CO_2, NO_2, Ws, Tair, Rh, and Sr were chosen as possible input variables.

Before we decided data structure, we should consider some restrictions. Most of all, we will predict daily O_3-Max in each sector at 6:00 a.m. Therefore, we include cluster representative data in the morning. Next, we analyzed hourly trends between O_3 and others in figure 3. The most similar trend was discovered between Tair and O_3. Finally, the data structure used in the proposed system is as follows: O_3, NO_2, and CO_2 from the morning data, Rh, Sr, Ws and O_3 from 2:00 to 5:00 p.m., and O_3-Max and Tair-Max, from the maximum values in previous day. An example of a possible data structure is shown in Table 1.

Table 1. Data structure of prospective input variables

O_3	NO_2	CO	Rh	Sr	Ws	O_3	O_3 Max	Tair Max
From 6 A.M to 9 A.M. Data			From 2 P.M to 5 P.M. Data				The Maximum values of previous day	

* Tair : Atmosphere temperature, Rh : Relative humidity, Sr : Solar radiation, Ws : Wind Speed, O_3 Max : The Maximum O_3 of previous day, Tair Max : The Maximum atmosphere temperature of previous day.

When O_3 concentration is over 120ppb in some clusters, we should inform O_3 warning sign public. Until now, O_3 warning note has sounded total 110 times and 51 days in Seoul for ten years (1995~2004). So, high concentration O_3 forecasting is a rare data prediction problem. Therefore, if we suggested single prediction model, it would be overfitted by medium- and low-level data. Consequently, we decide to cluster the data using a fuzzy c-mean algorithm (FCM) with inverse mapping and to produce separate models of high- and low-level predictions. This is the reason that take special steps must be taken before and after modeling. Pre- and postprocessing are expected to overcome and compensate for rare data prediction problems. FCM generalizes the hard c-mean algorithm to allow a point to partially belong to multiple clusters. Therefore, it produces a soft partition for a given dataset [11]. FCM of Bezdek [12] is normally applied to fuzzy clustering. In this paper, fuzzy space classification using FCM based on similar features of ozone concentration output data is used to compute classified degrees for use as input variables. Figure 4 illustrates the mapping of the output space as, classified by fuzzy clustering onto input spaces, and then fuzzy clustering was used to relate which features were input variables and output data.

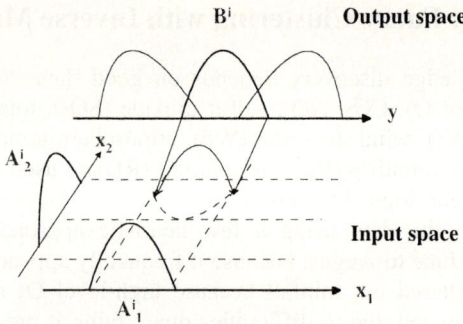

Fig. 4. A map of output variables onto input space

4 Modeling: Dynamic Polynomial Neural Network

4.1 The Basic Structure of DPNN

DPNN uses the GMDH (Group Method Data Handling) method [13] to compose an input/output model based on observed data and variables. This method is widely used for system modeling, prediction, and artificial intelligent control. The simple DPNN structure has four inputs and one output at each node. The following polynomial equations between input and output are used at each node in DPNN. Output y_1 and y_2 at each node are expressed as follows:

$$y_1 = w_{01} + w_{11}x_1 + w_{21}x_2 + w_{31}x_1x_2 + w_{41}x_1^2 + w_{51}x_2^2$$
$$y_2 = w_{02} + w_{12}x_3 + w_{22}x_4 + w_{32}x_3x_4 + w_{42}x_3^2 + w_{52}x_4^2 \quad (1)$$

The final output is represented by a polynomial equation.

$$\hat{y} = w_{03} + w_{13}y_1 + w_{23}y_2 + w_{33}y_1y_2 + w_{43}y_1^2 + w_{53}y_2^2 \quad (2)$$

where, w_{ij} ($i=0,1,2,...,n$, $j=0,1,2,...,k$) is the coefficient. If there are more than three input variables for each node, then other combinations of input variables are added to the above equation.

The least square method is employed to estimate the parameters of each node in DPNN. It searches the solution parameters to minimize the objective function formed by error functions between node outputs and actual target values. Equations (3) and (4) show the objective function and the coefficient, respectively. Parameters are solved by the least square method and polynomial functions of the current node are structured at each of layer. This process is repeated until the criterion is satisfied. The best function for the best performance was determined.

$$J = \sum_{k=1}^{\# \text{ of data}} (y(k) - \hat{y}(k))^2 = \| y - wA \|^2 \quad (3)$$

$$w = (A^T A)^{-1} A^T y \quad (4)$$

4.2 Self-Organization

Another specific characteristic of DPNN is self-organization [14]. The DPNN based on the GMDH method separates data into training data and testing data for modeling [15]. The purposes of this stage are to identify the behavior of the dynamic system and to prevent the overfitting problem.

DPNN estimates the parameters of each node and composes the network structure of the dynamic system using two-separate data sets. The training data set is used to solve the parameters of the functions of each node. The testing data set is used to evaluate the performance of DPNN. The final network structure is constructed by the relationship between errors in the training data and the testing data. Therefore, DPNN selects the input of the next node under a PC based on the relationship between the training and testing errors at each node.

The final network structure is determined as shown in Figure 5. The PC could be determined by following Equation (5), which exists in the $0 \sim 1$ range. The performance of each model is also evaluated by Equation (5). This performance criterion can be applied to both the testing and training data. And also it can be used for the unprepared new data.

$$e_1^2 = \sum_{i=1}^{n_A}(y_i^A - f_A(x_i^A))^2 / n_A,$$

$$e_2^2 = \sum_{i=1}^{n_B}(y_i^B - f_B(x_i^B))^2 / n_B, \qquad (5)$$

$$PC = e_1^2 + e_2^2 + \eta(e_1^2 - e_2^2)^2$$

where, e_1, e_2, n_A, n_B, and y_i indicate training errors, testing errors, the number of training data, the number of testing data and measured outputs, respectively. And where $f_A(x_i^A)$ and $f_B(x_i^B)$ are the separate outputs of training data and testing data separately. The total number of pieces of data is given by $n = n_A + n_B$. From the results, the optimized model structure is constructed at the point of the minimized PC.

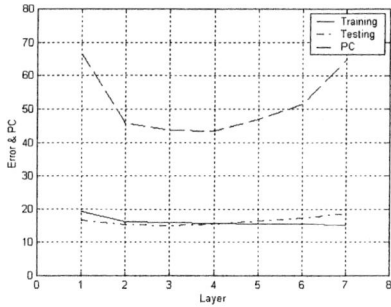

Fig. 5. The variation of model performance corresponding to layer increment

5 Postprocessing: Fuzzy Inference

High- and low-levels were clustered based on ozone concentration data in the postprocessing step and then ozone concentrations are predicted by DPNN. Figure 6 shows results, which are represented as Y_L and Y_H.

In Equation (6), $A_k^i(\cdot)$ and $A_k^j(\cdot)$ ($k = 1,2$) are fuzzy membership functions. α_1, α_2, β_1 and β_2 are fuzzy membership values calculated by two input variables X_i and X_j ($i \neq j$), respectively.

$$\begin{cases} A_1^i(X_i) = \alpha_1 \\ A_2^i(X_i) = \alpha_2 \end{cases}, \quad \begin{cases} A_1^j(X_j) = \beta_1 \\ A_2^j(X_j) = \beta_2 \end{cases} \quad (6)$$

In this case, the two input variables have the greatest difference between high and low level concentrations. It means that the X_i and X_j strongly affect ozone concentrations. The outputs of Y_i and Y_j are influenced by X_i and X_j and are computed by fuzzy inference in Equation (7).

$$Y_i = \frac{(\alpha_1 \times Y_L) + (\alpha_2 \times Y_H)}{\alpha_1 + \alpha_2}$$
$$Y_j = \frac{(\beta_1 \times Y_L) + (\beta_2 \times Y_H)}{\beta_1 + \beta_2} \quad (7)$$

The final prediction, results based upon the decision support system, is given by Equation (8). In this equation, the weights W_{Xi} and W_{Xj} are the relative distances of the input membership functions.

$$\hat{Y} = \frac{(W_{X_i} \times Y_i) + (W_{X_j} \times Y_j)}{W_{X_i} + W_{X_j}} \quad (8)$$

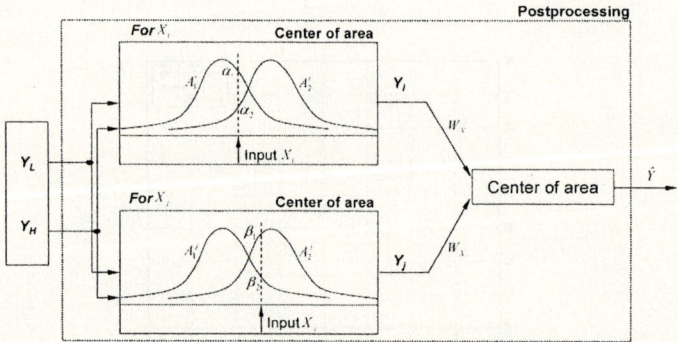

Fig. 6. Fuzzy inference in postprocessing

6 Evaluation: RMSE

The training and testing data are selected by hold-out method. The input variables are classified by the predicted and measured time. For the decision support system, a low concentration model and a high concentration model are constructed by the fuzzy clustering method based on the basic training data.

In this system, mean values and standard deviations are firstly found with respect to the input variables of each model and then required membership functions are selected by the correlative distance of each membership function. A model is chosen based on the RMSE that results from the selected model. This model is determined after clustering the training data.

In the simulations, two to four clusters are used. Basically, high-level ozone used the highest value and low-level ozone consists of the other set. Figure 7 shows the ozone predictions for BangHak-Dong in Seoul, Korea from August 1 to 10, 1997. When four clusters were used, the lowest training RMSE is 27.918 and the prediction RMSE is 20.183. The slope and intercept values for R-square are displayed in the scatter graphs of ozone observation (x-axis) against the predicted values (y-axis) for each model. The two diagonal lines in the plots represent the best-fit regression and the perfect correspondence between observations and predictions [16].

In the second simulation, the predicted area is Ssang-Mun-Dong in Seoul, which is a high-level ozone area in the summer. The predicted period is from May to July in 1999. The training data and testing data are constituted by the data from May to September in 1996, 1997 and 1998. Figure 8 shows the result of the ozone prediction the period from May 20 to July 20 in 1999. When four clusters are used, the lowest training RMSE is 15.634, and the prediction RMSE is 18.034.

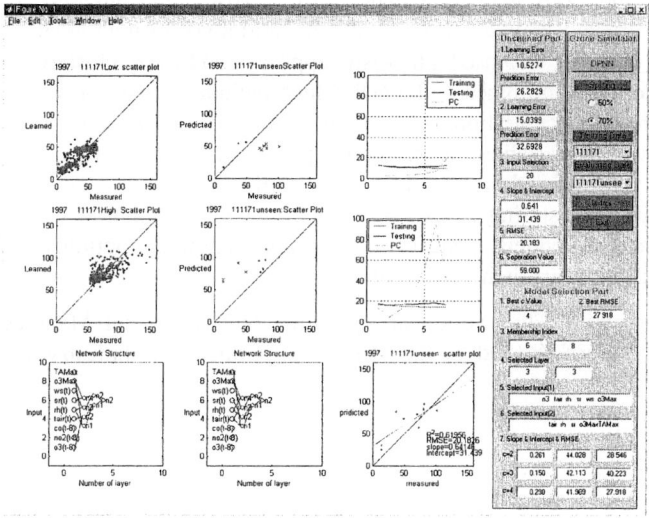

Fig. 7. Predictions for BangHak-Dong

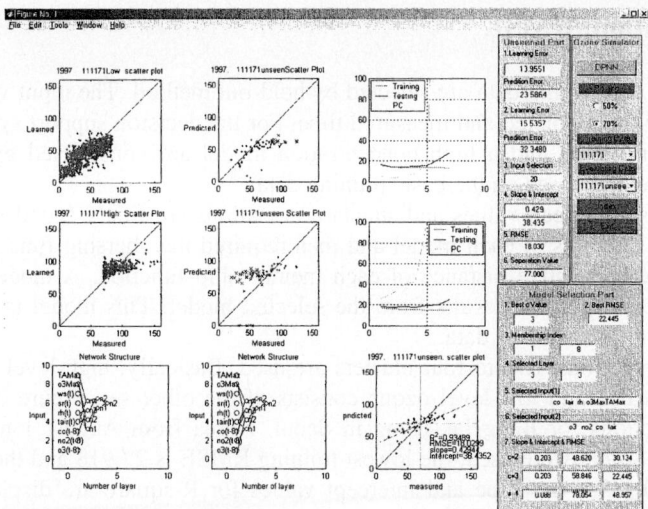

Fig. 8. Predictions for Ssang-Mun-Dong

7 Conclusion

In this paper, we propose a rare data prediction system using directed knowledge discovery methods. Rare data prediction is a particularly challenging application for data mining, such as fraud detection in finance, diagnosis in industry, and affect analysis in chemistry, not only because building cases of training sets is difficult, but also because the cases may have many forms, causes, and unknown relations. We applied the directed knowledge discovery methodology to high-level ozone prediction. A preprocessing step used a heuristic approach and fuzzy clustering with inverse mapping. The model designed by DPNN, which includes a decision support system, is suitable for high concentration ozone prediction. When the models are updated daily based on new input variables, the prediction performances improve and are better than the results given by fixed models. This study confirms that the selection of the cluster number in fuzzy clustering is also very important for the high-level ozone. Finally, combining the two models, using various input selection and optimizing the structure of models will improve their performance.

Acknowledgement. This work was supported by "Intelligent Manufacturing Systems(IMS) Program" hosted by the Ministry of Commerce, Industry and Energy, in Korea.

References

1. Michael, J. A. and Linoff, Gordon.: *Data Mining Techniques*; John Wiley & Sons, Inc. (1997)
2. Wang, Xue Z.: *Data Mining and Knowledge Discovery for Process Monitoring and Control*; Springer-Verlag, Berlin Heidelberg London, (1999)

3. Fayyad, U.M., G. Piatetsky-Shapiro, P. Smyth: From Data Mining to Knowledge Discovery: An Overview. In : Fayyad, U.M. et al: *Advances in Knowledge Discovery and Data Mining*. AAAI Press. Menlo Park, California (1996) 1-34
4. Elder IV, J.F., D. Pregibon.: A Statistical Perspective on Knowledge Discovery in Databases. In Fayyad, U.M. et al: Advances in Knowledge Discovery and Data Mining. AAAI Press/The MIT Press. Menlo Park, California (1996) 1-36
5. Devlin. B.: Data Warehouse from Architecture to Implementation, Addison-Wesley (1997)
6. Earl Gose, Richard Johnsonbaugh, and Steve Jost, Pattern Recognition and Image Analysis, Prentice Hall PTR, 1996.
7. Duc Truong Pham and Liu Xing, Neural Networks for Identification, Prediction and Control, Springer-Verlag Inc., 1995.
8. Burnett, R. T., Smith-Doiron, M., Stieb, D., Raizenne, M. E., Brook, J. R., Dales, R. E., Leech, J. A., Cakmak, S., and Krewski, D., Association between ozone and hospitalization for acute respiratory diseases in children less than 2 years of age. Am. J. Epidemiol. 153, pp. 444-452, 2001.
9. Matyssek, R., Harvranek, W.M., Wieser, G., Innes, J.L., Forest Decline and Ozone, A Comparison of Controlled Chamber and Field Experiments, Springer, Berlin, 1997.
10. Millan, M., Salvador, R., Mautilla, E., Meteorology and photochemical air pollution in southern Europe: experimental results from EC research projects, Atmospheric Environment 30, pp. 1909-1924, 1996.
11. Sungshin Kim, "A Neuro-Fuzzy Approach to Integration and Control of Industrial Processes: Part I," KFIS, pp 58-69, 1998.
12. James C Bezdek, Pattern Recognition with Fuzzy Objective Function Algorithms, Plenum, 1981.
13. A G Ivakhnenko, "The Group Method of Data Handling in Prediction Problem," Soviet Automatic Control, vol 9, no 6, pp 21-30, 1976.
14. S Farlow, ed., Self-Organizing Method in Modeling: GMDH-Type Algorithms, Marcel Deckker Inc., New York, 1984.
15. A G Ivahnenko, "Polynomial theory of complex system," IEEE trans. System. Man and Cybernetic, pp 364-378, 1971.
16. Greg Spellman, "An application of artificial neural networks to the prediction of surface ozone concentrations in the United Kingdom," Applied Geography, pp 123-136, 1999.

Application of Fuzzy Systems in the Car-Following Behaviour Analysis

Pengjun Zheng and Mike McDonald

Transportation Research Group, University of Southampton,
University Road, Highfield, Southampton, SO17 1BJ, UK
{p.zheng, mm7}@soton.ac.uk

Abstract. Realistic understanding and description of car following behaviour is fundamental in many applications of Intelligent Transportation Systems. Historical car following studies had been focused on car following behaviour measured under experiment settings, either at test track or on open road, mainly using statistical analysis. This might introduce errors when they were used to represent everyday driving behaviour because differences might exist between everyday and experiment behaviours, and intelligent data analysis might be necessary in order to identify subtle differences. This paper presents the results of an observation and analysis of driver's car following behaviour on motorway. Car following behaviours were measured under normal driving conditions where drivers were free to follow any vehicles. A time-series database was then established. The data was analysed using neuro-fuzzy systems and driver car following behaviour was quantified using several dynamic behavioural indices, which were combinations of parameters of trained neuro-fuzzy systems. The results indicated that in normal driving conditions, car following was conducted in a 'loose' way in terms of close-loop coupling, and car following performance was slightly 'worse' in terms of tracking error, than in experiment settings.

1 Introduction

Since the first car following experiment [1], many observations have been carried out in an effort to understand car-following behaviour for the single-lane follow-the-leader situation (e.g. [2], [3], [4]). Typical observation involved in a detailed measurement of the variables such as speed, acceleration, headway etc, for a pair or a platoon of vehicles. The data collected was then analysed to estimate parameters of car-following models, (e.g. [1], [3]) or used as a guide for developing/setting Adaptive Cruise Control (ACC) system. (e.g. [4]).

Most of the early observations were conducted on test track or tunnels under controlled condition. Typical settings of experiment were as follows: the subject of following vehicle was asked to follow a leader, which was driven by another experimenter. The leader executed pre-defined manoeuvres during experiment, e.g. relatively severe acceleration/deceleration at some speed, the response of follower relative to leader was recorded and formed a time-series database. Some car following experiments were carried out using driving simulator under controlled condition, (e.g.

Ohio State Experiment, [5]) where more scenarios could be tested in a systematic way, such as the cut-in of leader etc.

The car-following behaviour was also studied in a loosely controlled driving situation on open road so that the pertinent data could be collected in a closer-to-reality situation. McDonald et al. ([6]) conducted a car-following experiment using instrumented vehicle (IV) where only IV was driven by subject. The observation was carried out in two conditions: (1) subject on IV was asked to follow a normal traffic (leader) and observe the leader-IV car following behaviour. (2) subject on IV was asked to drive in front of a normal traffic (follower) and observe IV-follower car following behaviour. In later case, IV could manoeuvre according to experiment setting to observe response of follower.

Very few observations had been conducted under totally uncontrolled situation. Ozaki ([7]) observed car following behaviour of normal traffics on a radial route of the Tokyo Metropolitan Expressway using video camera covering a section of 160 meters, in which neither the car following scenarios nor the population of subjects were controlled.

The advantage of the controlled car following experiment is that car following scenarios of interest can be tested efficiently according to design and data can be collected for interested population of subjects. The loosely controlled experiments, on the other hand, are operated in real-driving conditions; the data collected is therefore more realistic. The car-following scenarios and populations of subjects can also be controlled, although the efficiency of the experiment may be affected by the existence of the uncontrolled traffics. The uncontrolled car following observations are most realistic as both car following scenarios and subjects are natural-formed in real traffic condition. The disadvantage of this approach is that information of subjects can be hardly obtained. The coverage and accuracy of the observation can also be greatly limited in case of the video camera observation.

A basic question concerning car-following behaviour observed under different settings was whether they could represent each other. In a car following experiment involved in both test track and open road situations, Allen et al ([4]) found that open road data appeared to be less constrained and showed larger standard deviation of identified behavioural indexes, such as time headway, gain and time delay etc. By comparing results from several similar experiments of steady-state car following on Ohio State simulator and real driving situation, Rothery ([8]) reported that the range of relative speeds at a given spacing that were recorded under driving condition are much lower than those measured on the simulator. His explanation was that the perceptual worlds of simulator and real traffics were considerably different. A natural extension to their findings was that there might be difference between car following behaviour observed under normal task-demand and experiment task-demand.

Little has been reported on the observation of car following behaviour under normal car following task-demand in real traffic, which should be one of integral parts of the overall knowledge of car following behaviour. In this paper, we reported results of a car following behaviour observation under normal task-demand. The observation was designed to keep subjects from being informed that they are doing car following experiment, thus ensure that all car following situations were naturally formed and car following was performed under normal task demand. The results are compared with that of other experiments under test track and open road.

2 Description of Car Following Behaviour Using Fuzzy Systems

Car following was one type of general manual control tasks, where the driver of the following vehicle had to adjust his or her relative position to the leading vehicle in a reasonable range, neither too far nor too close, in order to maintain a safe separation and keep up with the leader. The only direct control available to driver was the throttle and brake although much information might be obtained through human perception systems to guide his/her control decisions. More precisely, car following was a close-loop tracking task of manual control. There were many ways to achieve the follow-the-leader objective as long as such control could maintain the distance separation between a leader and a follower within a reasonable bound. As to how a driver performed this simple task, a car following model can be used to describe such behaviour. If the model was fitted to the empirical data, the model under fitted parameters should be able to reproduce the car following behaviour with reasonable accuracy. Therefore, a model with parameters could be used to describe car following behaviour and the behavioural differences should be reflected in the parameters of the model.

From every nature of close-loop tracking task of manual control, a conceptual description of car following can be made like this: when driving a vehicle in car-following situation, the driver can perceive the discrepancy or error between the desired state and its actual state. The driver wishes to reduce this discrepancy by adjusting the accelerator or brake position, this in turn produce a change in driving or braking force, and which in turn causes changes in vehicle's actual state. The explicitly desired state in car following was to keep a distance separation of desired headway, then car following could be described by a mapping between perceived error (difference between headway and desired headway) and direct output from driver, that was throttle-move or brake-move. Therefore, car following behaviour can be described by a characteristic headway which drivers choose to keep (desired headway) and a dynamic process describing how drivers manage to keep this headway.

For manual control behaviour, McRuer et al ([9]) successful developed cross-over model under the assumption that the human responded in such a way as to make the total open-loop transfer function (human + system dynamics) behave as a first-order system with gain and effective time delay. In car following situation, that was equal to a mapping between perceived speed error and driver-vehicle system output. If headway tracking error was also considered, then the car following behaviour could be described by a conceptual model as shown in Figure 1, where both perceived speed error (relative speed) and headway error (discrepancy between perceived and desired headway) were system inputs.

A linear form of such open-loop model will be:

$$a(t)=[C_1*DV+C_2*(DX-DX_{dsr})]|t-T_d \qquad (1)$$

where a is acceleration rate of the following vehicle, C_1, C_2 are linear gain, and T_d is time delay, DV and DX denote relative speed and headway respectively. DX_{dsr} is a driver's desired headway.

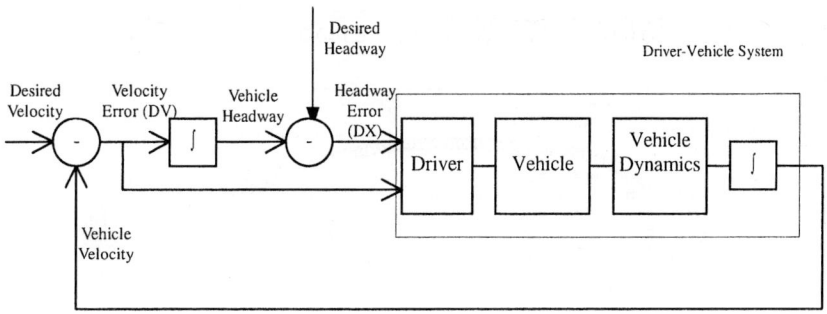

Fig. 1. Conceptual Model of Car Following Behaviour

Many car following research supported the fact that driver did not directly control his/her vehicle to match a desired headway (e.g. [1], [10]). Instead, he/she did so mostly by speed matching (i.e., $C_1>>C_2$). In this case, (1) can be reduced to simple car following model:

$$a(t)=C_1*DV|t-T_d \qquad (2)$$

The physical meaning of T_d was straightforward, which was the sum of perception reaction time (PRT), engine delay, brake delay, etc. and always existed in manual control system. Two linear coefficients, C_1 and C_2 were speed tracking gain and headway tracking gain. Fig 2 illustrated two simulated car following processes using different value of parameters. Headway tracking errors (DX-Desired Headway) and speed tracking errors (DV) were plotted against each other so that small circles were associated with small tracking errors. It can be observed that large C_1 resulted in small speed tracking error and accordingly small headway tracking error. Large C_2 resulted in reduced headway tracking error. Time delay was universally harmful to the performance of any tracking task. However reducing time delay demanded perceptual resources and was constrained by human perception ability. In addition, under large C_1 and C_2, large correction was made to a perceived error, which resulted in stronger acceleration/deceleration manifested as more aggressive car following behaviour. Under large time delay and C_1, C_2 combinations, the system was going to become unstable and tracking error could become very large.

A more accurate representation of car following behaviour should take into account of non-linearity of human response and limitations of human perception system, i.e. drivers may not be able to perceive relative speed and headway accurately, and the decision process (acceleration control) could be highly non-linear. Neuro-fuzzy systems are suitable tool to describe such behaviour. A generic fuzzy inference system (FIS) can map between variables driver can perceive and variables driver can directly control with arbitrary accuracy based on 'fuzzy reasoning'. This makes the system natural and suitable to model a human-in-the-loop system (e.g. [11]).

Fuzzy systems are established on fuzzy logic which is based on the idea that sets are not crisp but some are fuzzy, and these can be described in linguistic terms such as

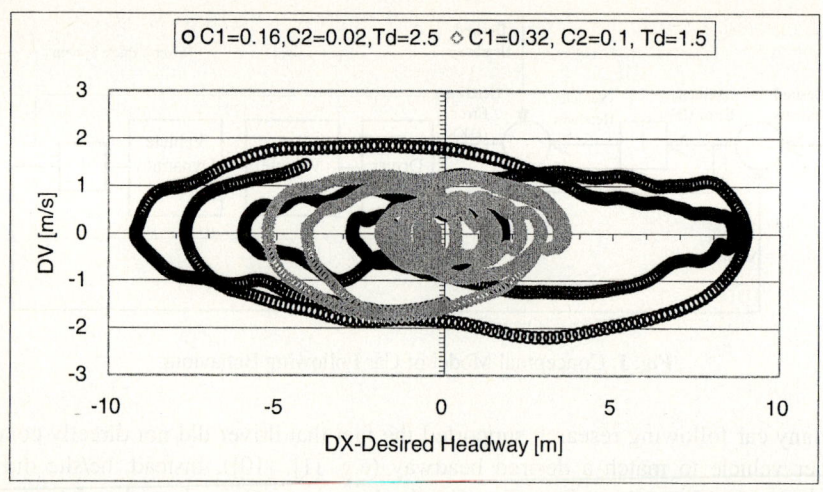

Fig. 2. Car following performances under different combination of parameters

fast, slow and medium [12]. Fuzzy sets have boundaries that are not precise and are described by membership functions that are curves defining how each point in the input space is mapped to a membership grade. The transition from non-membership to membership is gradual rather than abrupt. The value of the membership is between 0 and 1. A membership degree of 1 represents certainty with respect to an entity belonging to a particular set. Fuzzy inference systems use IF THEN ELSE rules to relate linguistic terms defined in output space and input space. Every linguistic term corresponds to a fuzzy set. Neuro-fuzzy systems are a combination of artificial neural networks (ANN) and fuzzy systems. ANN mimics the manner in which the brain processes information, which consists of a number of independent processors that communicate with each other via weighted connections. The most prominent feature of ANN is to learn and adapt to a large quantity of data. The combination of ANN and fuzzy sets offers a powerful method to model human behaviour.

A generic fuzzy inference system (FIS) could then be used to represent the open-loop mapping in car following processes, which was able to approximate any non-linear function with arbitrary accuracy [13] and therefore was able to identify car following behaviour more accurately. The model can be described notational as:

$$a(t) = FUZZY[DV, (DX-DX_{dsr})]|t-T_d \qquad (3)$$

where FUZZY denoting a fuzzy logic mapping.

3 Data Collection and Reduction

A car following experiment was conducted to observe car following behaviour under normal driving conditions. An instrumented vehicle (IV, a Vauxell 2.0 sedan) was used to measure speed, relative speed and headway between a leader and the IV. The instrumentation of the IV included a laser speedometer, front and rear Radar, three video cameras typically facing front, rear and driver, as well as other sensors such as pedal/brake movement, indicators use etc [14]. All outputs from sensors were logged onto a PC at a sampling rate of 10 Hz. The measurements for car following study purpose and their accuracy are summarised in Table1. The video camera recordings were used to help tracking vehicles in later data reduction stage.

Table 1. IV Measurements and the Accuracy

Instrument	Measurement	Accuracy
Laser Speedometer	IV Speed	0.278 m/s
Front Radar	Front Headway	20cm @100m
Rear Radar	Rear Headway	

The survey was carried out at morning peak hours (7:00 –9:00 am) when stable car following condition was frequent at high flow rate. Subjects were asked to drive between two junctions on motorway (M27) repeatedly. The survey involved in a motorway driving of about 7 km when going to and coming back from downstream junction where car following situation was frequent. No instruction was given for the driving so that drivers could follow any leader like in everyday driving. Over the two-month of survey, 590 trials of driving were carried out in 30 experiment days.

Useful car following processes were then extracted from the raw data. The car following situation was assumed to exist if and only if:

(1) both vehicles (IV as the follower) was on the same lane;
(2) distance separation between two vehicles was less than 100 meters;
(3) the above relative position was kept more than 30 second (about 1km);
(4) both vehicles was on motorway (not in round-about or on slip road)

Thirty days of raw data collected on weekday was processed. Totally 1236 segments of car following process were derived.

For each identified car following process, directly measured headway and speed was further processed to reduce measurement errors. Time derivatives such as acceleration, range rate (DV) were calculated from speed and headway measurements. The final reduced data was a time-series of speed, acceleration rate, headway (DX) and relative speed (DV). A detailed data processing procedure was documented elsewhere [15]. A database was formed based on 30 days of observations; typically, about 40 car following processes were identified each day.

4 Results

Neural fuzzy systems were used to extract the necessary information directly from measurements of car following behaviour as it can directly learn and adapt to the pattern from a large quantity of data, i.e., a fuzzy model was generated from the data with parameters trained to the data. The membership functions were obtained by clustering every input variable separately. Generation of fuzzy rules included the generation of the premise and the consequence. Premise was directly derived from the clusters and the obtained membership functions while the consequent was generated by associating clusters to classes. All input variables were linked in a conjunctive rule. The functionality is available through Adaptive Neural Fuzzy Inference System in Matlab for Suegeno-type fuzzy inference system [16]. The description of the adaptive learning algorithm for the neuro-fuzzy system can be found in [17]

Adaptive neural fuzzy training was applied to estimate the parameters of fuzzy inference system, which could be analogue to a least square estimation. All data collected on the same day was grouped as they represented the behaviour of the same driver. 60% of data was used as training data and the remaining 40% as checking data to prevent overfitting. The desired headway of driver (DX_{dsr}) in a car following process was a steady-state where relative speed and acceleration were zero. However, in real traffic, strict defined steady-state can be hardly achieved. A typical car following process lasting for about 2 minutes is illustrated in Figure 3. It showed repeated 'goal seeking' cycles on the relative speed-headway plane. Desired headway can be identified, which was near the centre of each circle on headway axis. In this research, the desired headway was derived from each car following process and was calculated by averaging crossing points (i.e., relative speed from + to – or – to +) at which acceleration rate were below a threshold of 0.1 m/s^2. By varying time delay t, training data was generated to train the fuzzy inference system. The best fit was obtained when minimum root mean square error of acceleration rate was achieved. The trained model was a nonlinear mapping between the inputs (relative speed, headway discrepancy) and the output (acceleration rate), expressed as a fuzzy system. A section plane of two trained model representing two different drivers is shown in Figure 4. The nonlinearity of the response can be clearly noticed.

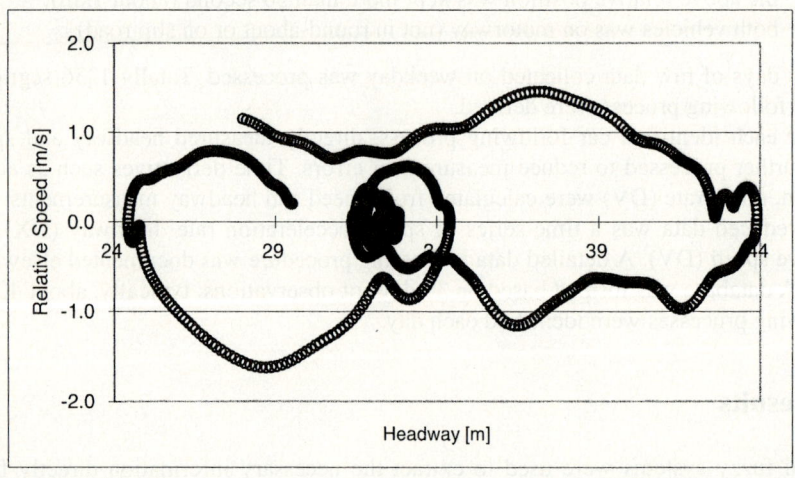

Fig. 3. A car following process, note the desired headway at about 30m headway

For identified non-linear mapping, linear analysis was further performed to determine qualitative/quantitative effects of the equivalent linear parameters as outlined in Equation (1). This made it possible to compare with its linear counterparts. The linearised speed-matching gain (C_1) and headway-matching gain (C_2) were calculated according to Equation near steady-state (DV=0, DX=DX_{dsr}).

$$c_1 = \frac{\partial F(DV, DX - DXdsr)}{\partial DV}, \text{ and } \quad c_2 = \frac{\partial F(DV, DX - DXdsr)}{\partial (DX - DXdsr)}$$

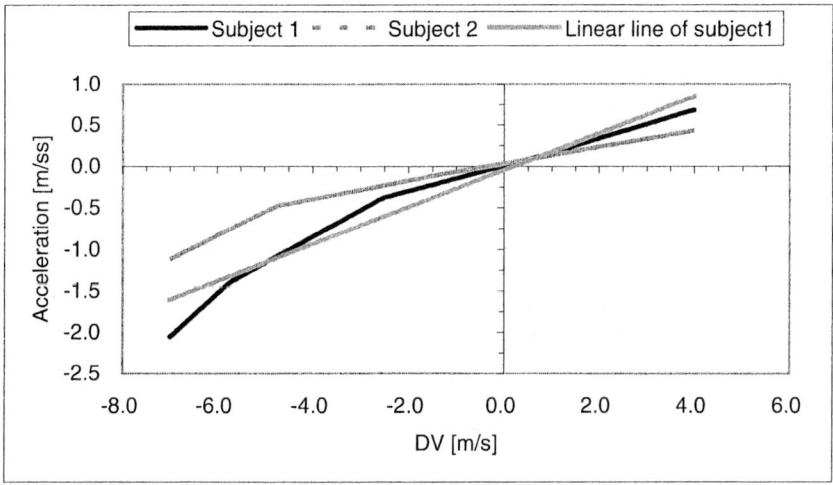

Fig. 4. Section plane of a trained model at DX=DX_{dsr}

The mean and standard deviation of calculated linear gains are shown in Table 2.

Table 2. Mean and Std of Behavioural Indices

Indexes	Time Delay [s]	C_1 [1/s]	C_2 [1/s^2]
Mean	2.5	0.16	0.015
Std	0.81	0.056	0.008

A comparison of behavioural indices between observations was summarised in Table 3. Because of the insignificance of C_2, only C_1 was compared. The fitted value for T_d and C_1 varied considerable between researches, thus made a quantitative comparison difficulty. However, the C_1 obtained in this research was much smaller than all of the others. This implied that car following was performed in a 'loose' manner under normal task-demand, which was intuitively reasonable.

Table 3. A comparison of T_d and C_1 with other observations

Observations	T_d (s)	C_1 (1/s)
Chandler, Test Track	1.55	0.368
Forbes, Helly, Tunnel and Open Road	0.983	0.633
Allen Test Track	3.28	0.21
Allen Open road	4.96 (4.15*)	0.21
McDonald, Zheng, Open Road	2.125	0.296

* If an outlier of 11.47 second was excluded

5 Conclusion

A car following observation had been carried out which differed in its control over driver's task demand of car following compared with other experiments. The task-demand of driver was kept in normal level by a proper manipulation of experiment design. A detailed analysis revealed a property of car following behaviours under normal task-demand, i.e. the dynamic control, measured by open-loop gain, was significantly lower.

Knowledge had accumulated regarding manual car following behaviour through observations under experiment settings. Models of car following behaviour were able to reproduce it because car following behaviour was tightly constrained especially under experiment situations. The results from this research, however, indicated that car following might have been performed in a more 'tight' manner in experiment. While in normal driving, drivers might not always be able to, or unwilling to do a car following in this manner. Fortunately, driver assistance device, such as ACC, was able to do so in a very 'tight' manner because of very short time delay, therefore could be very helpful in improving car-following performance. It was the intent of this research to expand the understanding of manual car following behaviour, which had affected and would keep affecting both traffic modelling and driving-task automation.

References

1. Chandler, F.E., Herman, R., Montroll, E.W.: Traffic Dynamics: Studies in Car Following. Operations Research, 6(1958) 165-184
2. Forbes, T. W., Zagorski, M. J., Holshouser, E. L., Deterline, W. A.: Measurement of Driver Reaction to Tunnel Conditions, Proceedings of the Highway Research Board 37 (1959), 345-357
3. McDonald, M., Wu, J., Brackstone, M.: Development of a Fuzzy Logic Based Microscopic Motorway Simulation Model. In Proceedings of the IEEE Conference on Intelligent Transport Systems. Boston, USA. (1997)
4. Allen, R. W., Magdaleno, R.E., Serafin, C., Eckert, S., Sieja, T.: Driver Car Following Behaviour Under Test Track and Open Road Driving Condition. SAE Paper 970170, SAE. 17 pages. (1997)
5. Todosiev, E. P.: The Action Point Model of the Driver Vehicle System. Report No. 202A-3. Ohio State University. (1963)

6. McDonald, M., Brackstone, M., Sultan, B., Roach, C.: Close Following on the Motorway: Initial Findings of an Instrumented Vehicle Study. Vision in Vehicles VII (Ed, Gale, A. G.), Elsevier, North Holland, Amsterdam. (2000)
7. Ozaki, H.: Reaction and Anticipation in the Car-Following Behaviour, Transportation and Traffic Theory (Ed. Daganzo, C. F.), (1993) 349-366
8. Rothery, R. W.: Car Following Models. In Traffic Flow Theory, TRB Special Report 165, (1999)
9. McRuer, D. T., Krendel, E. S.: The Human Operator as a Servo System Element. Journal of the Franklin Institute, Vol. 267, (1959) 381-403
10. Herman, R., Montroll, E. W., Potts, R. B., Rothery, R. W.: Traffic Dynamics: Analysis of Stability in Car Following, Operational Reseach, Vol. 7, (1959) 86-106.
11. Rouse, W. B.: 'Fuzzy Models of Human Problem Solving.' Advances in Fuzzy Set, Possibility Theory and Application (Ed: Wang, P. P.), Plenum Press, New York (1983)
12. Zadeh, L. A.: Fuzzy Sets, Information and Control, Vol. 8 (1965) 65-70
13. Buckley, J. J.: Sugeno Type Controllers are Universal Controllers, Fuzzy Sets and Systems, No. 53 (1993) 299-304
14. McDonald, M., BrackStone, M, Sultan, B.: Instumented Vehicle Studies of Traffic Flow Models. Proc. of the 3rd International Symposium on Highway Capacity. Copenhagen, Denmark. (1998)
15. Zheng, P., McDonald, M., et al., Identifying Best Predictors for Car Following Behaviour from Empirical Data. Proceedings of 13th ESS, (2001) 158-165
16. MathWorks Inc.: Fuzzy Logic Toolbox User's Guide (version 2). MathWorks Inc (1999)
17. Jang, J. F.: ANFIS: Adaptive-network-based Fuzzy Inference System. IEEE Trans. Syst., Man, Cybern., 23 (1999) 665-685

GA-Based Composite Sliding Mode Fuzzy Control for Double-Pendulum-Type Overhead Crane

Diantong Liu[1], Weiping Guo[1], and Jianqiang Yi[2]

[1] Institute of Computer Science and technology, Yantai University,
Shandong Province, 264005, China
{diantong.liu, weiping.guo}@163.com
[2] Institute of Automation, Chinese Academy of Sciences, Beijing, 100080, China
jianqiang.yi@mail.ia.ac.cn

Abstract. A genetic algorithm (GA) based composite sliding mode fuzzy control (CSMFC) approach is proposed for the double-pendulum-type overhead crane (DPTOC). The overhead crane exhibits double-pendulum dynamics because of the large-mass hook and the payload volume. Its nonlinear dynamic model is built using Lagrangian method. Through defining a composite sliding mode function, the proposed control approach greatly reduces the complexity to design a controller for complex underactuated systems. The control system stability is analyzed for DPTOC. Real-valued GA is used to optimize the parameters of CSMFC to improve the performance of control system. Simulation results illustrate the complexity of DPTOC and the validity of proposed control algorithm.

1 Introduction

Overhead crane works as a robot in many places such as workshops and harbors to transport all kinds of massive goods. It is desired for the overhead crane to transport the payloads to the required position as fast and as accurately as possible without collision with other equipments. Many works have been done in controlling the overhead crane. William et al [1] adopted input shaping control method. Lee [2] proposed feedback control methods. Moreno et al [3] used neural network to tune the parameters of state feedback control. Benhidjeb and Gissinger [4] compared fuzzy control with Linear Quadratic Gaussian Control of an overhead crane. Nalley and Trabia [5] adopted fuzzy logic to both positioning control and swing damping. Mansour and Mohamed [6] used the variable structure control to control the overhead crane. In the above works, the mass of hook was usually ignored and the system was a single-pendulum-type overhead crane (SPTOC). Therefore, the model is easily built and the control algorithm is easily designed. However, when the hook mass is not less so much than the payload mass, the overhead crane performs as a double pendulum system [7]. The dynamics and control of DPTOC are more complex than SPTOC.

In this paper, the overhead crane that exhibits double-pendulum-type dynamics is investigated and a GA-based composite sliding mode fuzzy control algorithm is proposed. The remainder of this paper is organized as follows. In section 2, the system dynamics are given. In section 3, a composite fuzzy sliding mode control algorithm is proposed and its stability is analyzed. In section 4, real-valued GA is designed to optimize the parameters of CSMFC for improving system performance. Section 5 is system simulations and section 6 is conclusions.

2 System Dynamics of DPTOC

Fig.1 shows the schematic representation of DPTOC. The crane is driven by applying a force F to the trolley in the X-direction. A cable of length l_1 hangs below the trolley and supports a hook of mass m_1, to which the payload is attached using some form of rigging. Here the rigging and the payload are modeled as the second cable of length l_2, and mass point m_2. The system dynamic equations can be derived using the Lagrangian method as follows:

$$M(q)\ddot{q} + C(q,\dot{q})\dot{q} + G(q) = \tau . \tag{1}$$

where $q = [x, \theta_1, \theta_2]^T$ is the vector of generalized coordinates, $\tau = [F, 0, 0]^T$ is the vector of generalized forces acting on the system. $M(q)$ is the inertia matrix:

$$M(q) = \begin{bmatrix} m + m_1 + m_2 & (m_1 + m_2)l_1 \cos\theta_1 & m_2 l_2 \cos\theta_2 \\ (m_1 + m_2)l_1 \cos\theta_1 & (m_1 + m_2)l_1^2 & m_2 l_1 l_2 \cos(\theta_1 - \theta_2) \\ m_2 l_2 \cos\theta_2 & m_2 l_1 l_2 \cos(\theta_1 - \theta_2) & m_2 l_2^2 \end{bmatrix} . \tag{2}$$

$G(q)$ is the potential (gravitational/elastic) term:

$$G(q) = [0 \quad (m_1 + m_2)gl_1 \sin\theta_1 \quad m_2 gl_2 \sin\theta_2]^T . \tag{3}$$

$C(q,\dot{q})$ is the vector of velocity (Coriolis/centrifugal) term:

$$C(q,\dot{q}) = \begin{bmatrix} 0 & -(m_1 + m_2)l_1\dot{\theta}_1 \sin\theta_1 & -m_2 l_2 \dot{\theta}_2 \sin\theta_2 \\ 0 & 0 & m_2 l_1 l_2 \dot{\theta}_1 \sin(\theta_1 - \theta_2) \\ 0 & -m_2 l_1 l_2 \dot{\theta}_1 \sin(\theta_1 - \theta_2) & 0 \end{bmatrix} . \tag{4}$$

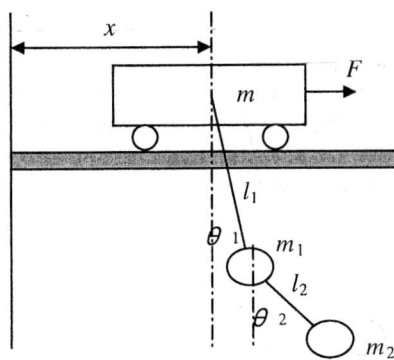

Fig. 1. Scheme of the double-pendulum-type overhead crane

3 Composite Sliding Mode Fuzzy Control

3.1 Sliding Mode Fuzzy Control

Consider a second order nonlinear system:

$$\begin{aligned} \dot{x}_1 &= x_2 \\ \dot{x}_2 &= f(X) + b(X)u \end{aligned} \quad (5)$$

where, $X = [x_1, x_2]^T$ is state vector, $f(X)$ and $b(X)$ are continuous linear or nonlinear functions, u is the control input. The sliding mode function is defined as

$$s = x_2 + \lambda x_1 . \quad (6)$$

which is a measure of the algebraic distance of the current state to the sliding surface $s = 0$; λ is real and positive. The resulting sliding mode fuzzy controller (SMFC) is actually a single-input-single-output fuzzy logic controller [8]. The input to the controller is the sliding mode function s. The output from the controller is the control command u. A typical rule of a SMFC has the following format:

R_i: IF s IS F_i THEN u IS U_i

where F_i is the linguistic value of s in the ith-fuzzy rule, and U_i is the linguistic value of u in the ith-fuzzy rule.

3.2 Composite Sliding Mode Fuzzy Control

The underactuated DPTOC systems described by equation (1) can be represented as:

$$\begin{aligned} \dot{x}_1 &= x_2 \\ \dot{x}_2 &= f_1(X) + b_1(X)u \\ \dot{x}_3 &= x_4 \\ \dot{x}_4 &= f_2(X) + b_2(X)u \\ \dot{x}_5 &= x_6 \\ \dot{x}_6 &= f_3(X) + b_3(X)u \end{aligned} \quad (7)$$

where, $X=(x_1, x_2, x_3, x_4, x_5, x_6)$ is the state variable vector that represents the crane position x, velocity $\dot{x}$, the hook swing angle θ_1 and angle velocity $\dot{\theta}_1$, the payload swing angle θ_2 and angle velocity $\dot{\theta}_2$, $f_1(X)$, $f_2(X)$, $f_3(X)$, $b_1(X)$, $b_2(X)$ and $b_3(X)$ are continuous nonlinear or linear functions, u is the control input.

From equation (7), the system has three coupled subsystems: position subsystem and two (payload and hook) anti-swing subsystems. In order to decouple the system, three sliding mode functions are defined for the three subsystems:

$$\begin{aligned} s_1 &= x_2 + \lambda_1 x_1 \\ s_2 &= x_4 + \lambda_2 x_3 \\ s_3 &= x_6 + \lambda_3 x_5 \end{aligned} \quad (8)$$

where λ_1, λ_2 and λ_3 are positive real numbers.

Based on the three sliding mode functions, a composite sliding mode function can be further defined as

$$s = k_1 s_1 + k_2 s_2 + k_3 s_3 . \tag{9}$$

where k_1, k_2 and k_3 are real numbers.

The control objective is: to determine a feedback control such that the closed-loop system is globally stable in the sense that all the state variables are uniformly bounded and converge to its equilibrium asymptotically. By doing so, the main process is to make the system states move toward the surface $s_1 = 0$, $s_2 = 0$ and $s_3 = 0$, and then converge to the equilibrium point asymptotically. The signs and values of coefficients k_1, k_2 and k_3 can decide the stability and accessibility of the composite sliding mode surface, and their values can adjust the roles of three subsystems in the composite sliding mode function so as to affect their respective proportion in system control.

For DPTOC systems, the composite sliding mode function s works as the input to the sliding mode fuzzy control. To determine the final control action, we design following fuzzy rules:

R_k: IF s IS F_k THEN u IS U^k

where, R_k is the kth rule among p rules, F^k is a fuzzy set of the input variable s, and U^k is a fuzzy set of the output variable u. The output singleton fuzzy sets and the center-of-gravity defuzzification are used:

$$u = (\sum_{k=1}^{p} \mu_{F^k}(s) \times U^k) / \sum_{k=1}^{p} \mu_{F^k}(s) . \tag{10}$$

where $\mu_{F^k}(s)$ is the firing degree of the jth rule and u is the output of the composite sliding mode fuzzy controller.

3.3 Stability Analysis for DPTOC

Now we consider the asymptotical stability of the DPTOC system with the composite sliding mode fuzzy control.

Theorem 1. For the DPTOC system (1), consider the proposed composite sliding mode fuzzy control (10). If in the sliding mode functions, the parameters λ_1, λ_2 and λ_3 are positive real number, in the composite sliding mode functions, parameter k_1 and k_3 are positive real number and k_2 is a negative real number, and make $k_1 b_1(X) + k_2 b_2(X) + k_3 b_3(X) > 0$ hold, then the system is asymptotically stable.

Proof:
The stability of the closed loop system consists of the stability in the sliding mode surface and the accessibility of sliding mode surface.

a. *The stability in the sliding mode surface*
All the sliding mode surfaces $s_1 = 0$, $s_2 = 0$ and $s_3 = 0$ are stable because λ_1, λ_2 and λ_3 are positive real numbers. The stability in the composite sliding mode

surface is decided by the coupling factors k_1, k_2 and k_3 in equation (9) and the definitions of the swing angles and the control input u. Assume positive angles and a positive u are along the positive X-direction as in Fig.1. When the position sliding mode function $s_1 > 0$, a negative control input is required such that the sliding function s_1 approaches $s_1 = 0$. When the payload anti-swing sliding mode function $s_3 > 0$, a negative control input is required such that the sliding function s_3 approaches $s_3 = 0$. However, when the hook anti-swing sliding mode function $s_2 > 0$, a positive driving force is required such that the sliding function s_2 approaches $s_2 = 0$. Therefore, the role of the hook anti-swing sliding mode function in control is contradictive with the others, and the coupling factor k_1 and k_3 should be positive and k_2 should be negative for the stability in sliding mode surface.

b. The accessibility of sliding mode surface

Choose the Lyapunov function candidate

$$V = s^2/2 . \qquad (11)$$

The time derivative of the composite sliding mode function (9) is

$$\dot{s} = k_1(\dot{x}_2 + \lambda_1 \dot{x}_1) + k_2(\dot{x}_4 + \lambda_2 \dot{x}_3) + k_3(\dot{x}_6 + \lambda_3 \dot{x}_5) . \qquad (12)$$

Substituting equation (7) into equation (12)

$$\dot{s} = k_1(f_1(X) + \lambda_1 x_2) + k_2 f_2(X) + k_2 \lambda_2 x_4 + k_3 f_3(X) + k_3 \lambda_3 x_6 \\ + (k_1 b_1(X) + k_2 b_2(X) + k_3 b_3(X))u \qquad (13)$$

It is easy to obtain the time derivative of the Lyapunov function candidate

$$\dot{V} = s\dot{s} = M(X)s + N(X)su . \qquad (14)$$

where,

$$M(X) = k_1(f_1(X) + \lambda_1 x_2) + k_2 f_2(X) + k_2 \lambda_2 x_4 + k_3 f_3(X) + k_3 \lambda_3 x_6 \\ N(X) = k_1 b_1(X) + k_2 b_2(X) + k_3 b_3(X) \qquad (15)$$

If $k_1 b_1(X) + k_2 b_2(X) + k_3 b_3(X) > 0$ holds, it can be realized through the design of the composite sliding mode fuzzy inference that increasing the control input u will result in decreasing $s\dot{s}$ as the sliding mode function s is negative, and decreasing the control input u will result in decreasing $s\dot{s}$ as the sliding mode function s is positive. Therefore, the sliding mode surface $s = 0$ can be accessible.

4 CSMFC Parameters Adjusting by Real-Valued GA

System performance is very sensitive to the slope λ_1 (λ_2 or λ_3) of the sliding mode functions: when the value of λ_1 (λ_2 or λ_3) becomes larger, the rise-time will become smaller, but at the same time, both overshoot and settling-time will become larger, and vice versa. The values of coefficients k_1, k_2 and k_3 not only affect the stability of the closed-loop DPTOC system, but also affect the system control performance and the roles of the three subsystems in the system control.

The speed of motor is limited in practice, and the trolley rated speed v_e that transformed from the motor rated speed is the optimum transporting speed. In order to make the trolley transport with v_e after the acceleration process, the position can be limited as follows when calculating the control variable:

$$\bar{x}_1 = \eta_1 \text{sat}(x_1/\eta_1) . \tag{16}$$

where $\bar{x}_1$ is in the stead of x_1 to obtain s in equation (8), η_1 is the maximum of $\bar{x}_1$ which decides the decelerating distance from current position to desired position. The position sliding mode parameter λ_1 can be obtained by:

$$\lambda_1 = v_e/\eta_1 . \tag{17}$$

It is very difficult to determine the other parameters because these parameters haven't obvious physics meaning and there are complex couplings among them. Therefore, real-valued GA [9] is used to optimize the parameters in the composite sliding mode fuzzy controller. In the application of real-valued GA to the optimizing problem, there are two considerations to be made: fitness function and encoding.

a. Encoding

The chromosome comprising five genes represents a set of parameters:
$P = [\lambda_2 \ \lambda_3 \ k_1 \ k_2 \ k_3]$

b. Fitness Function

The fitness function is the key to use the GA. A simple fitness function that reflects small steady-state errors, short rise-time, low oscillations and overshoots with a good stability is given

$$J = (1 + penalty) \int (\mu_1|x| + \mu_2|\theta_1| + \mu_3|\theta_2| + \mu_4|u|) dt . \tag{18}$$

where μ_1, μ_2, μ_3 and μ_4 are coefficients, which respectively represent the importance degree of position, hook angle, payload angle and control input in the cost function. The *penalty* is a penalty coefficient when position overshoot arises.

c. Steps of Real-Valued GA

The real-valued GA that we implemented can be described as follows:

  Step 1: Generate initial population, P(t) randomly.
  Step 2: While (number of generations <= maximum value) do.
  Step 3: Evaluate each of the strings in P(t) using equation (18) according to the system simulation results.
  Step 4: Select and reproduction operation.
  Step 5: Crossover operation.
  Step 6: Mutation operation.

5 Simulation Studies

In the DPTOC system, the parameters that always change in different transport tasks are m_2, l_1, m_1 and l_2. The effects of these changes and initial condition are considered. In the simulations, the basic parameters are: m=5Kg, m_1=2 Kg, m_2=5Kg, l_1=2m, l_2=1m, and basic initial state is: $\theta_1 = 11.465$ deg, $\theta_2 = 11.465$ deg, $\dot{\theta}_1 = 0$ deg/s, $\dot{\theta}_2 = 0$ deg/s. The system dynamics under the basic parameters and basic initial state

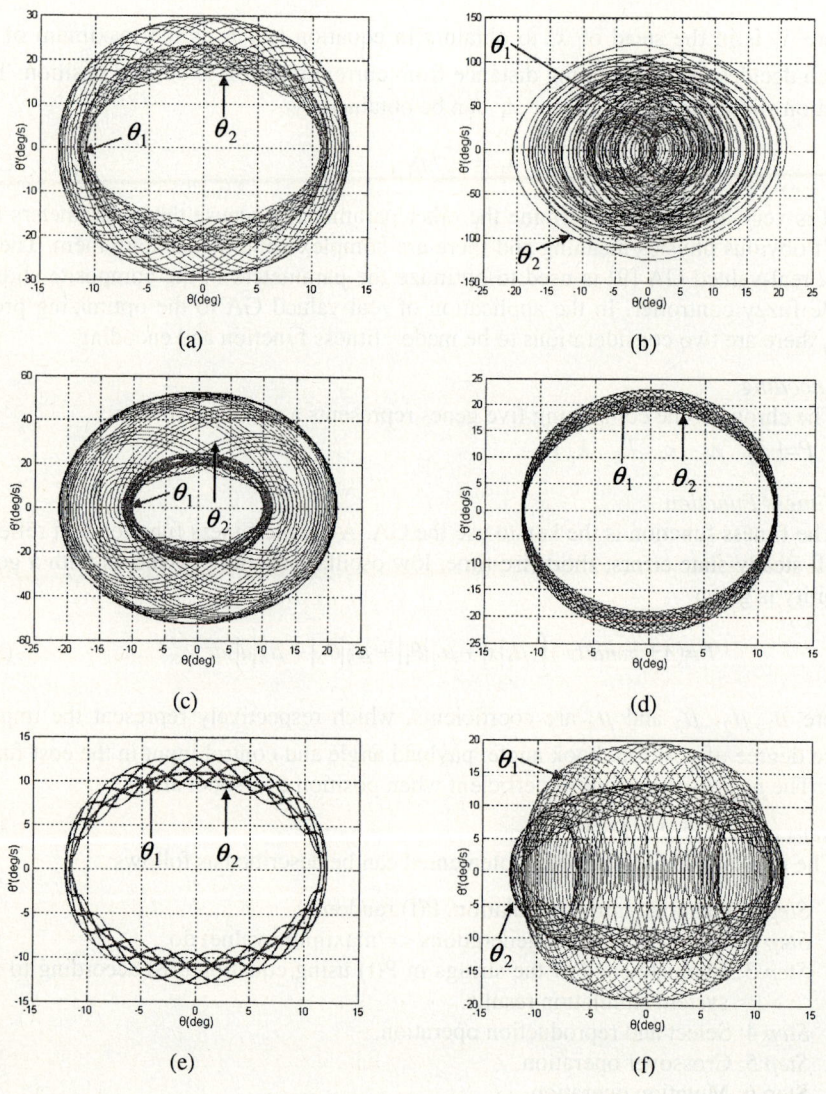

Fig. 2. System dynamic of the DPTOC for different initial condition or system parameters

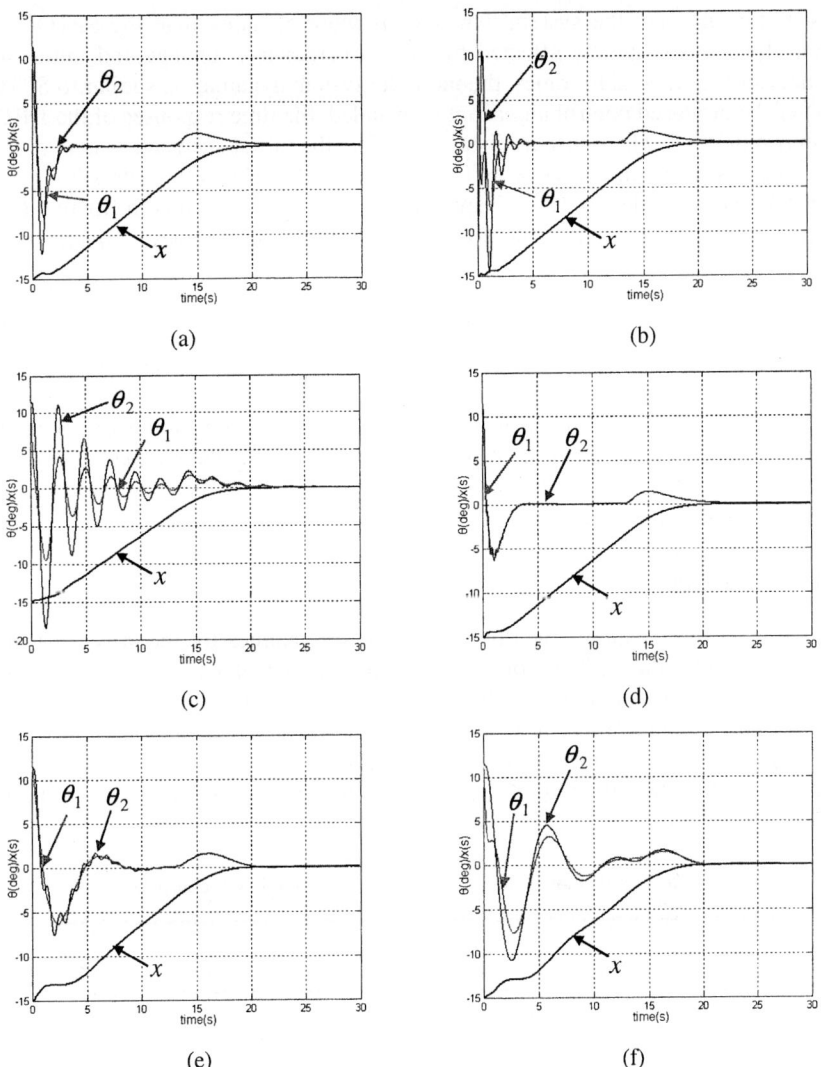

Fig. 3. System dynamics of the DPTOC with the proposed control algorithm for different initial condition and system parameters.

are shown in Fig.2 (a). For the system under the basic parameters and basic initial state, only one parameter of system or initial state is changed in the following simulations. When initial state θ_2 is changed to −11.465degree, the system dynamics are shown in Fig.2 (b). When system parameter m_1 is changed to 20Kg, the system dynamics are shown in Fig.2 (c). When system parameter m_2 is changed to 50Kg, the system dynamics are shown in Fig.2 (d). When system parameter l_1 is changed to 10m, the system dynamics are shown in Fig.2 (e). When system parameter l_2 is changed to 10m, the system dynamics are shown in Fig.2 (f). From the above simulations, it can be seen: the system dynamics is more complex than the SPTOC. The system dynamics is greatly affected by system parameters' change and initial state. The heavier the payload becomes, the more the system dynamics is similar to SPTOC.

After the proposed control algorithm is included, the time responses of the DPTOC are shown in Fig.3. The transport distance is 15m; the initial state and system parameters of each simulation are same as Fig.2 respectively. The composite sliding mode fuzzy rules are provided in Table 1 and the membership function is in figure 4 (a). In the fitness function of GA, the following coefficients are used: $penalty = 0.2$, $\mu_1 = 0.3$, $\mu_2 = 50$, $\mu_3 = 60$ and $\mu_4 = 0.01$. Assume the trolley rated speed $v_e = 1$ and $\eta_1 = 3.5$, then $\lambda_1 = 1/3.5$. In real-valued GA, the evolution of the best solution is illustrated in Fig.4 (b). After the evolution process, the best controller parameters are: $\lambda_2 = 14.9596$, $\lambda_3 = 22.0395$, $k_1 = 73.5339$, $k_2 = 81.1883$, $k_3 = 22.6428$.

From the simulation results, it is clear that the proposed control law can accurately transport the payload to the desired position while damping the payload swing angle especially at the goal. The control law has some robustness to the changes of system's parameters and initial conditions.

6 Conclusions

The overhead crane exhibits the double-pendulum dynamics for the large-mass hook and the payload volume. The nonlinear dynamic model of DPTOC is derived with Lagrangian method. A composite sliding mode fuzzy control method is proposed for DPTOC. The system stability is analyzed via the SMC concept and real-valued GA is used to optimize the parameters of composite sliding mode function. Simulations are performed to illustrate the complexity of DPTOC and the effectiveness of the proposed control algorithm.

Table 1. Rules table of composite sliding mode fuzzy controller

s	NB	NS	ZO	PS	PB
u_f	10	5	0	-5	-10

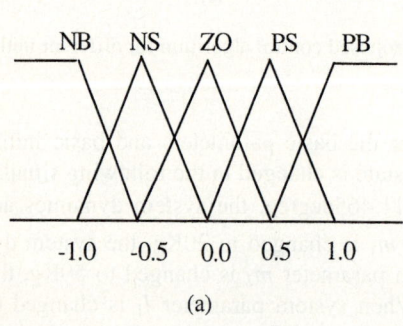
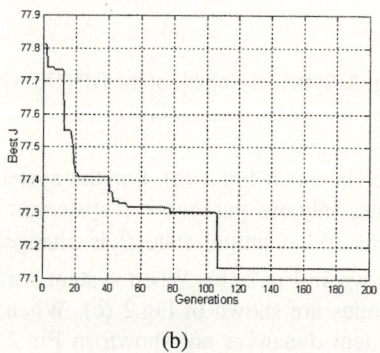

Fig. 4. (a) Membership function of antecedent variable for composite sliding mode fuzzy controller, (b) Fitness function for the best solution as a function of the number of generations

References

1. William S., Lisa P., Michael K., et al: Effects of Hoisting on the Input Shaping Control of Gantry Cranes. Control Engineering Practice, No.8 (2000) 1159-1165.
2. Lee H.H: Modeling and Control of a Three- Dimensional Overhead Crane. Journal of Dynamic System, Measurement and Control, Vol.120 (1998) 471-476.
3. Moreno L., Acosta L., Mendez J.A., Torres S, Hamilton A, Marichal G. N.: A Self-Tuning Neuromorphic Controller: Application to the Crane Problem. Control Engineering Practice, No.6 (1998) 1475-1483.
4. Benhidjeb A. and Gissinger G.L.: Fuzzy Control of an Overhead Crane Performance Comparison with Classic Control. Control Engineering Practice, Vol.12 (1995) 1687-1696.
5. Nalley M. J. and Trabia M. B.: Control of Overhead Cranes Using a Fuzzy Logic Controller. Journal of Intelligent and Fuzzy System, No.8 (2000) 1-18.
6. Mansour, A.K. and Mohamed Z.: Robust Control Schemes for an Overhead Crane. Journal of Vibration and Control, No.7 (2001) 395-416.
7. William E.S. and Samuel T.T.: Double-Pendulum Gantry Crane Dynamics and Control. In: Proceedings of IEEE International Conference on Control Applications, Trieste, (1998) 1205-1209.
8. Lo J.C. and Kuo Y.H.: Decoupled Fuzzy Sliding-Mode Control. IEEE Transactions on Fuzzy Systems, vol.6 (1998) 426-435.
9. Fogel D.B. and Mitchel M.: An introduction to genetic algorithms. Bulletin of Mathematical Biophysics, vol. 59 (1997) 199–205.

A Balanced Model Reduction for T-S Fuzzy Systems with Integral Quadratic Constraints

Seog-Hwan Yoo[1] and Byung-Jae Choi

School of Electronic Engineering, Daegu University, Korea
{shryu, bjchoi}@daegu.ac.kr

Abstract. This paper deals with a balanced model reduction for a class of nonlinear systems with integral quadratic constraints(IQC's) using a T-S(Takagi-Sugeno) fuzzy approach. We define a generalized controllability Gramian and a generalized observability Gramian for a stable T-S fuzzy systems with IQC's. We obtain a balanced state space realization using the generalized controllability and observability Gramians and obtain a reduced model by truncating not only states but also IQC's from the balanced state space realization. We also present an upper bound of the approximation error. The generalized controllability Gramian and observability Gramian can be computed from solutions of linear matrix inequalities.

1 Introduction

For linear finite dimensional systems with high orders, optimal control techniques such as linear quadratic gaussian and H_∞ control theory, usually produce controllers with the same state dimension as the model. Lower dimensional linear controllers are normally preferred over higher dimensional controllers in control system designs for some obvious reasons : they are easier to understand, implement and have higher reliability. Accordingly the problem of model reduction is of significant practical importance in control system design and has been a focus of a wide variety of studies for recent decades(see [1]-[4] and the references therein).

Model reduction techniques have been developed for linear uncertain systems as well. Using LMI machinery, Beck suggested a balanced truncation method for linear uncertain discrete systems with related error bounds[5]. Model reduction techniques for linear parameter varying systems were also reported by several researchers [6]-[8]. However, comparatively little work has been reported for the model reduction of the fuzzy systems.

In recent years, a controller design method for nonlinear dynamic systems modeled as a T-S(Takagi-Sugeno) fuzzy model has been intensively addressed[9]-[11]. Unlike a single conventional model, this T-S fuzzy model usually consists of several linear models to describe the global behavior of the nonlinear system. Typically the T-S fuzzy model is described by fuzzy IF-THEN rules. Based on this fuzzy model, many

---

[1] This work was supported by the Korea Research Foundation Grant.(KRF-2004-041-D00259).

researchers use one of control design methods developed for linear parameter varying system. In order to alleviate computational burden in design phase and simplify the designed fuzzy controller, the state dimension of the T-S fuzzy model should be low.

In this paper, using LMI approaches we develop a balanced model reduction scheme for T-S fuzzy systems with IQC's. The IQC's can be used conveniently to describe uncertain parameters, noises, time delays, unmodeled dynamics, etc.[12] In section 2, we define the T-S fuzzy system with IQC's. A generalized controllability Gramian and a generalized observability Gramian are defined and a balanced realization of the T-S fuzzy system using the generalized controllability and observability Gramian is also presented in section 3. A model approximation bound is derived and a suboptimal procedure is described to get a less conservative error bound in section 4. Finally some concluding remarks are given in section 5.

The notation in this paper is fairly standard. R^n denotes n dimensional real vector space and $R^{n \times m}$ is the set of real $n \times m$ matrices. A^T denotes the transpose of a real matrix A. 0 and I_n denote zero matrix and $n \times n$ identity matrix respectively. $M > 0$ means that M is a positive definite matrix. In a block symmetric matrix, $*$ in (i,j) block means the transpose of the matrix in (j,i) block. Finally $\|\cdot\|_\infty$ denotes the H_∞ norm of the system.

2 T-S Fuzzy System with IQC's

We consider the following fuzzy dynamic system with IQC's.

Plant Rule i ($i = 1, \cdots, r$):
IF $\rho_1(t)$ is M_{i1} and $\cdots$ and $\rho_g(t)$ is M_{ig},
THEN

$$\dot{x}(t) = A_i x(t) + B_i u(t) + \sum_{j=1}^{q} F_{ij} p_j(t),$$

$$y(t) = C_i x(t) + \sum_{j=1}^{q} G_{ij} p_j(t), \qquad (1)$$

$$q_j(t) = H_{ij} x(t) + J_{ij} u(t) + K_{ij} p(t),$$

where r is the number of fuzzy rules. $\rho_j(t)$ and M_{ik} ($k = 1, \cdots, g$) are the premise variables and the fuzzy set respectively. $x(t) \in R^n$ is the state vector, $u(t) \in R^m$ is the input, $y(t) \in R^p$ is the output variable, $p_j(t) \in R^{k_j}$ and $q_j(t) \in R^{k_j}$ are the uncertain variables related to the following IQC's:

$$\int_0^\infty p_j(t)^T p_j(t) dt \leq \int_0^\infty q_j(t)^T q_j(t) dt, \quad j = 1, \cdots, q, \qquad (2)$$

with $p(t) = \begin{bmatrix} p_1(t)^T & \cdots & p_q(t)^T \end{bmatrix}^T$ and $q(t) = \begin{bmatrix} q_1(t)^T & \cdots & q_q(t)^T \end{bmatrix}^T$. In addition, A_i, B_i, $\cdots$, K_{ij} are constant matrices with compatible dimensions. To simplify notation, we define

$$F_i = \begin{bmatrix} F_{i1} & \cdots & F_{iq} \end{bmatrix}, \; G_i = \begin{bmatrix} G_{i1} & \cdots & G_{iq} \end{bmatrix}, \; H_i = \begin{bmatrix} H_{i1}^T & \cdots & H_{iq}^T \end{bmatrix}^T,$$
$$J_i = \begin{bmatrix} J_{i1}^T & \cdots & J_{iq}^T \end{bmatrix}^T, \; K_i = \begin{bmatrix} K_{i1}^T & \cdots & K_{iq}^T \end{bmatrix}^T.$$

Let $\mu_i(\rho(t))$, $i = 1, \cdots, r$, be the normalized membership function of the inferred fuzzy set $h_i(\rho(t))$,

$$\mu_i(\rho(t)) = \frac{h_i(\rho(t))}{\sum_{i=1}^{r} h_i(\rho(t))}, \tag{3}$$

where

$$h_i(\rho(t)) = \prod_{k=1}^{g} M_{ik}(\rho_k(t)), \; \rho(t) = \begin{bmatrix} \rho_1(t) & \rho_2(t) & \cdots & \rho_g(t) \end{bmatrix}^T. \tag{4}$$

Assuming $h_i(\rho(t)) \geq 0$ for all i and $\sum_{i=1}^{r} h_i(\rho(t)) > 0$, we obtain

$$\mu_i(\rho(t)) \geq 0, \; \sum_{i=1}^{r} \mu_i(\rho(t)) = 1. \tag{5}$$

For simplicity, by defining $\mu_i = \mu_i(\rho(t))$ and $\mu^T = \begin{bmatrix} \mu_1 & \cdots & \mu_r \end{bmatrix}$ the uncertain fuzzy system (1) can be written as follows :

$$\begin{aligned}
\dot{x}(t) &= \sum_{i=1}^{r} \mu_i (A_i x(t) + B_i u(t) + F_i p(t)) \\
&= A(\mu) x(t) + B(\mu) u(t) + F(\mu) p(t), \\
y(t) &= \sum_{i=1}^{r} \mu_i (C_i x(t) + G_i p(t)) \\
&= C(\mu) x(t) + G(\mu) p(t), \\
q(t) &= \sum_{i=1}^{r} \mu_i (H_i x(t) + J_i u(t) + K_i p(t)) \\
&= H(\mu) x(t) + J(\mu) u(t) + K(\mu) p(t).
\end{aligned} \tag{6}$$

In a packed matrix notation, we express the fuzzy system (6) as in (7).

$$G = \left[\begin{array}{cc|c} A(\mu) & F(\mu) & B(\mu) \\ H(\mu) & K(\mu) & J(\mu) \\ \hline C(\mu) & G(\mu) & 0 \end{array} \right]. \tag{7}$$

3 A Balanced Realization

In this section we present a balanced realization for the uncertain fuzzy system (7) using generalized controllability and observability Gramians. We define generalized controllability and observability Gramians in lemma 1.

Lemma 1: (Generalized controllability and observability Gramians)
(i) Suppose that there exist $Q = Q^T > 0$ and $R = diag(\tau_1 I_{k_1}, \cdots, \tau_q I_{k_q}) > 0$ satisfying LMI (9), then the output energy is bounded above as follows :

$$\int_0^\infty y(t)^T y(t)dt < x(0)^T Q x(0) \text{ for } u(t) \equiv 0. \tag{8}$$

$$L_{oi} = \begin{bmatrix} A_i^T Q + Q A_i & * & * & * \\ R H_i & -R & * & * \\ F_i^T Q & K_i^T R & -R & * \\ C_i & 0 & G_i & -I \end{bmatrix} < 0, \; i = 1, \cdots, r. \tag{9}$$

(ii) Suppose that there exist $P = P^T > 0$ and $S = diag(\bar{\tau}_1 I_{k_1}, \cdots, \bar{\tau}_q I_{k_q}) > 0$ satisfying LMI (11), the input energy transferring from $x(-\infty) = 0$ to $x(0) = x_0$ is bounded below as follows:

$$\int_{-\infty}^0 u(t)^T u(t)dt > x_0^T P^{-1} x_0, \tag{10}$$

$$L_{ci} = \begin{bmatrix} P A_i^T + A_i P & * & * & * \\ H_i P & -S & * & * \\ S F_i^T & S K_i^T & -S & * \\ B_i^T & J_i^T & 0 & -I \end{bmatrix} < 0, \; i = 1, \cdots, r. \tag{11}$$

Proof: Using the Lyapunov function candidate $V = x(t)^T Q x(t)$ (in the proof of (i)), $V = x(t)^T P^{-1} x(t)$ (in the proof of (ii)) and the S-procedure, the theorem can be easily proved so that we omit the detailed proof.

As in [6], we say Q and P, solutions of LMI's (9) and (11), are generalized observability Gramian and controllability Gramian respectively. While the observability and controllability Gramian in linear time invariant systems are unique, the generalized Gramians of the fuzzy system (7) are not unique. But the generalized Gramians are related to the input and output energy as can be seen in lemma 1.

Using the generalized Gramians, we suggest a balanced realization of the uncertain fuzzy system (7). We obtain a transformation matrix T and W satisfying

$$\begin{aligned} \Sigma &= diag(\sigma_1, \sigma_2, \cdots, \sigma_n) = T^T Q T = T^{-1} P T^{-T}, \\ &\sigma_1 \geq \sigma_2 \geq \cdots \geq \sigma_n, \\ \Pi &= diag(\pi_1 I_{i_1}, \pi_2 I_{i_2}, \cdots, \pi_q I_{i_q}) = W^T R W = W^{-1} S W^{-T}, \\ &\pi_1 \geq \pi_2 \geq \cdots \geq \pi_q, \end{aligned} \tag{12}$$

where i_j ($j = 1, \cdots, q$) is a member of the index set $\{k_1, k_2, \cdots, k_q\}$.

With T and W defined in (12), the change of coordinates in the fuzzy system (7) gives

$$G_b = \begin{bmatrix} A_b(\mu) & F_b(\mu) & B_b(\mu) \\ H_b(\mu) & K_b(\mu) & J_b(\mu) \\ \hline C_b(\mu) & G_b(\mu) & 0 \end{bmatrix} = \begin{bmatrix} T^{-1}A(\mu)T & T^{-1}F(\mu)W & T^{-1}B(\mu) \\ W^{-1}H(\mu)T & W^{-1}K(\mu)W & W^{-1}J(\mu) \\ \hline C(\mu)T & G(\mu)W & 0 \end{bmatrix}. \quad (13)$$

One can easily observe that the state space realization of (13) satisfy following LMI's (14) and (15).

$$L_o(\mu) = \begin{bmatrix} A_b(\mu)^T \Sigma + \Sigma A_b(\mu) + C_b(\mu)^T C_b(\mu) & * & * \\ \Pi H_b(\mu) & -\Pi & * \\ G_b(\mu)^T C_b(\mu) + F_b(\mu)^T \Sigma & K_b^T \Pi & G_b(\mu)^T G_b(\mu) - \Pi \end{bmatrix} < 0, \quad (14)$$

$$L_c(\mu) = \begin{bmatrix} \Sigma A_b(\mu)^T + A_b(\mu)\Sigma + B_b(\mu)B_b(\mu)^T & * & * \\ H_b(\mu)\Sigma + J_b(\mu)B_b(\mu)^T & J_b(\mu)J_b(\mu)^T - \Pi & * \\ \Pi F_b(\mu)^T & \Pi K_b^T & -\Pi \end{bmatrix} < 0. \quad (15)$$

From this reason, we say that the realization (13) is a balanced realization of the fuzzy system (7) and Σ is a balanced Gramian.

4 Balanced Model Reduction

In this section, we develop a balanced model reduction scheme using the balanced Gramian defined in section 3. We also derive an upper bound of model approximation error. We assume that the fuzzy system (7) is already balanced and partitioned as follows :

$$G = \begin{bmatrix} A_{11}(\mu) & A_{12}(\mu) & F_{11}(\mu) & F_{12}(\mu) & B_1(\mu) \\ A_{21}(\mu) & A_{22}(\mu) & F_{21}(\mu) & F_{22}(\mu) & B_2(\mu) \\ \hline H_{11}(\mu) & H_{12}(\mu) & K_{11}(\mu) & K_{12}(\mu) & J_1(\mu) \\ H_{21}(\mu) & H_{22}(\mu) & K_{21}(\mu) & K_{22}(\mu) & J_2(\mu) \\ \hline C_1(\mu) & C_2(\mu) & G_1(\mu) & G_2(\mu) & 0 \end{bmatrix}$$

$$= \sum_{i=1}^{r} \mu_i \begin{bmatrix} A_{i,11} & A_{i,12} & F_{i,11} & F_{i,12} & B_{i,1} \\ A_{i,21} & A_{i,22} & F_{i,21} & F_{i,22} & B_{i,2} \\ \hline H_{i,11} & H_{i,12} & K_{i,11} & K_{i,12} & J_{i,1} \\ H_{i,21} & H_{i,22} & K_{i,21} & K_{i,22} & J_{i,2} \\ \hline C_{i,1} & C_{i,2} & G_{i,1} & G_{i,2} & 0 \end{bmatrix}, \quad (16)$$

where $A_{11}(\mu) \in R^{k \times k}$, $F_{11}(\mu) \in R^{k \times (i_1 + \cdots + i_v)}$ and the other matrices are compatibly partitioned.

From (16) we obtain a reduced order model by truncating $n-k$ states and $q-v$ IQC's as follows:

$$\bar{G} = \begin{bmatrix} A_{11}(\mu) & F_{11}(\mu) & B_1(\mu) \\ H_{11}(\mu) & K_{11}(\mu) & J_1(\mu) \\ \hline C_1(\mu) & G_1(\mu) & 0 \end{bmatrix} = \sum_{i=1}^{r} \mu_i \begin{bmatrix} A_{i,11} & F_{i,11} & B_{i,1} \\ H_{i,11} & K_{i,11} & J_{i,1} \\ \hline C_{i,1} & G_{i,1} & 0 \end{bmatrix}. \quad (17)$$

Theorem 2: The reduced order system (17) is quadratically stable and balanced. Moreover the model approximation error is given by

$$\|G - \bar{G}\|_{\infty} \leq 2\left(\sum_{j=k+1}^{n} \sigma_j + \sum_{j=v+1}^{q} \pi_j \right). \quad (18)$$

Proof: We partition $\Sigma = diag(\Sigma_1, \Sigma_2)$ and $\Pi = diag(\Pi_1, \Pi_2)$ where $\Sigma_1 \in R^{k \times k}$, $\Pi_1 \in R^{(i_1 + \cdots + i_v) \times (i_1 + \cdots + i_v)}$. Then the reduced order system (17) satisfies LMI's (19) and (20).

$$\begin{bmatrix} A_{11}(\mu)^T \Sigma_1 + \Sigma_1 A_{11}(\mu) + C_1(\mu)^T C_1(\mu) & * & * \\ \Pi_1 H_{11}(\mu) & -\Pi_1 & * \\ G_1(\mu)^T C_1(\mu) + F_{11}(\mu)^T \Sigma_1 & K_1(\mu)^T \Pi_1 & G_1(\mu)^T G_1(\mu) - \Pi_1 \end{bmatrix} < 0, \quad (19)$$

$$\begin{bmatrix} \Sigma_1 A_{11}(\mu)^T + A_{11}(\mu) \Sigma_1 + B_1(\mu) B_1(\mu)^T & * & * \\ H_{11}(\mu) \Sigma_1 + J_1(\mu) B_1(\mu)^T & J_1(\mu) J_1(\mu)^T - \Pi_1 & * \\ \Pi_1 F_{11}(\mu)^T & \Pi_1 K_1(\mu)^T & -\Pi_1 \end{bmatrix} < 0. \quad (20)$$

Hence the reduced order system is quadratically stable and balanced. Without loss of generality we consider two cases.
Case1: ($k = n-1$, $v = q$)
Note that in this case $F_{12}(\mu)$, $F_{22}(\mu)$, $H_{21}(\mu)$, $H_{22}(\mu)$, $J_2(\mu)$ and $G_2(\mu)$ are empty matrices. Hence a state space realization of the error system $G^e = G - \bar{G}$ can be written by

$$G^e = \begin{bmatrix} \bar{A}_e(\mu) & \bar{F}_e(\mu) & \bar{B}_e(\mu) \\ \bar{H}_e(\mu) & \bar{K}_e(\mu) & \bar{J}_e(\mu) \\ \hline \bar{C}_e(\mu) & \bar{G}_e(\mu) & 0 \end{bmatrix}$$

$$:= \begin{bmatrix} A_{11}(\mu) & 0 & 0 & F_{11}(\mu) & 0 & B_1(\mu) \\ 0 & A_{11}(\mu) & A_{12}(\mu) & 0 & F_{11}(\mu) & B_1(\mu) \\ 0 & A_{21}(\mu) & A_{22}(\mu) & 0 & F_{21}(\mu) & B_2(\mu) \\ H_{11}(\mu) & 0 & 0 & K(\mu) & 0 & J(\mu) \\ 0 & H_{11}(\mu) & H_{12}(\mu) & 0 & K(\mu) & J(\mu) \\ \hline -C_1(\mu) & C_1(\mu) & C_2(\mu) & -G(\mu) & G(\mu) & 0 \end{bmatrix}. \quad (21)$$

The change of coordinate with M in the error system gives

$$G^e = \left[\begin{array}{cc|c} A_e(\mu) & F_e(\mu) & B_e(\mu) \\ H_e(\mu) & K_e(\mu) & J_e(\mu) \\ \hline C_e(\mu) & G_e(\mu) & 0 \end{array}\right] := \left[\begin{array}{cc|c} M^{-1}\bar{A}_e(\mu)M & M^{-1}\bar{F}_e(\mu) & M^{-1}\bar{B}_e(\mu) \\ \bar{H}_e(\mu)M & \bar{K}_e(\mu) & \bar{J}_e(\mu) \\ \hline \bar{C}_e(\mu)M & \bar{G}_e(\mu) & 0 \end{array}\right], \quad (22)$$

where

$$M = \begin{bmatrix} I & I & 0 \\ I & -I & 0 \\ 0 & 0 & I \end{bmatrix}.$$

It is well known that the existence of $\Sigma_e = \Sigma_e^T > 0$ and $\Pi_e = \Pi_e^T > 0$ satisfying following LMI (23) guarantees $\|G^e\|_\infty \leq \gamma$.

$$L = \begin{bmatrix} \Gamma_{11} & * & * \\ H_e(\mu)\Sigma_e + J_e(\mu)B_e(\mu)^T & J_e(\mu)J_e(\mu)^T - \Pi_e & * \\ \Pi_e F_e(\mu)^T + \gamma^{-2}\Pi_e G_e(\mu)^T C_e(\mu)\Sigma_e & \Pi_e K_e(\mu)^T & \Gamma_{33} \end{bmatrix} < 0, \quad (23)$$

where

$$\Gamma_{11} = \Sigma_e A_e(\mu)^T + A_e(\mu)\Sigma_e + B_e(\mu)B_e(\mu)^T + \gamma^{-2}\Sigma_e C_e(\mu)^T C_e(\mu)\Sigma_e,$$

$$\Gamma_{33} = \gamma^{-2}\Pi_e G_e(\mu)^T G_e(\mu)\Pi_e - \Pi_e.$$

Let $\gamma = 2\sigma_n$, $\Sigma_e = diag(\Sigma_1, \sigma_n^2 \Sigma_1^{-1}, 2\sigma_n)$ and $\Pi_e = \begin{bmatrix} \Pi + \sigma_n^2 \Pi^{-1} & \Pi - \sigma_n^2 \Pi^{-1} \\ \Pi - \sigma_n^2 \Pi^{-1} & \Pi + \sigma_n^2 \Pi^{-1} \end{bmatrix}$. Then LMI (23) can be written as follows :

$$L = \begin{bmatrix} U_1^T & 0 & 0 \\ 0 & U_2^T & 0 \\ 0 & 0 & U_2^T \end{bmatrix} L_c(\mu) \begin{bmatrix} U_1 & 0 & 0 \\ 0 & U_2 & 0 \\ 0 & 0 & U_2 \end{bmatrix}$$

$$+ \begin{bmatrix} V_1^T & 0 & 0 \\ 0 & V_2^T & 0 \\ 0 & 0 & V_2^T \end{bmatrix} L_o(\mu) \begin{bmatrix} V_1 & 0 & 0 \\ 0 & V_2 & 0 \\ 0 & 0 & V_2 \end{bmatrix} < 0, \quad (24)$$

where

$$U_1 = \begin{bmatrix} I & 0 & 0 \\ 0 & 0 & I \end{bmatrix}, \; U_2^T = \begin{bmatrix} I \\ I \end{bmatrix}, \; V_1 = \begin{bmatrix} 0 & \sigma_n \Sigma_1^{-1} & 0 \\ 0 & 0 & -I \end{bmatrix}, \; V_2^T = \begin{bmatrix} \sigma_n \Pi^{-1} \\ -\sigma_n \Pi^{-1} \end{bmatrix}.$$

Case 2: ($k = n$, $v = q-1$)
In this case, $A_{12}(\mu)$, $A_{21}(\mu)$, $A_{22}(\mu)$, $F_{21}(\mu)$, $F_{22}(\mu)$, $H_{12}(\mu)$ and $H_{22}(\mu)$ are empty matrices so that the error system becomes

$$G^e = \begin{bmatrix} \overline{A}_e(\mu) & \overline{F}_e(\mu) & \overline{B}_e(\mu) \\ \overline{H}_e(\mu) & \overline{K}_e(\mu) & \overline{J}_e(\mu) \\ \hline \overline{C}_e(\mu) & \overline{G}_e(\mu) & 0 \end{bmatrix}$$

$$:= \begin{bmatrix} A(\mu) & 0 & F_{11}(\mu) & 0 & 0 & B(\mu) \\ 0 & A(\mu) & 0 & F_{11}(\mu) & F_{12}(\mu) & B(\mu) \\ H_{11}(\mu) & 0 & K_{11}(\mu) & 0 & 0 & J_1(\mu) \\ 0 & H_{11}(\mu) & 0 & K_{11}(\mu) & K_{12}(\mu) & J_1(\mu) \\ 0 & H_{21}(\mu) & 0 & K_{21}(\mu) & K_{22}(\mu) & J_2(\mu) \\ \hline -C(\mu) & C(\mu) & -G_1(\mu) & G_1(\mu) & G_2(\mu) & 0 \end{bmatrix}. \quad (25)$$

The change of coordinate with M in the error system gives

$$G^e = \begin{bmatrix} A_e(\mu) & F_e(\mu) & B_e(\mu) \\ H_e(\mu) & K_e(\mu) & J_e(\mu) \\ \hline C_e(\mu) & G_e(\mu) & 0 \end{bmatrix} := \begin{bmatrix} M^{-1}\overline{A}_e(\mu)M & M^{-1}\overline{F}_e(\mu) & M^{-1}\overline{B}_e(\mu) \\ \overline{H}_e(\mu)M & \overline{K}_e(\mu) & \overline{J}_e(\mu) \\ \hline \overline{C}_e(\mu)M & \overline{G}_e(\mu) & 0 \end{bmatrix}, \quad (26)$$

where $M = \begin{bmatrix} I & I \\ I & -I \end{bmatrix}$.

We define $\gamma = 2\pi_q$ and

$$\Pi = \begin{bmatrix} \Pi_1 & 0 \\ 0 & \pi_q I_{i_q} \end{bmatrix}, \quad \Sigma_e = \begin{bmatrix} \Sigma & 0 \\ 0 & \pi_q^2 \Sigma^{-1} \end{bmatrix}, \quad \Pi_e = \begin{bmatrix} \Pi_1 + \pi_q^2 \Pi_1^{-1} & \Pi_1 - \pi_q^2 \Pi_1^{-1} & 0 \\ \Pi_1 - \pi_q^2 \Pi_1^{-1} & \Pi_1 + \pi_q^2 \Pi_1^{-1} & 0 \\ 0 & 0 & 2\pi_q I_{i_q} \end{bmatrix}.$$

Then LMI (23) can be written as

$$L = \begin{bmatrix} U_1^T & 0 & 0 \\ 0 & U_2^T & 0 \\ 0 & 0 & U_2^T \end{bmatrix} L_c(\mu) \begin{bmatrix} U_1 & 0 & 0 \\ 0 & U_2 & 0 \\ 0 & 0 & U_2 \end{bmatrix}$$

$$+ \begin{bmatrix} V_1^T & 0 & 0 \\ 0 & V_2^T & 0 \\ 0 & 0 & V_2^T \end{bmatrix} L_o(\mu) \begin{bmatrix} V_1 & 0 & 0 \\ 0 & V_2 & 0 \\ 0 & 0 & V_2 \end{bmatrix} < 0, \quad (27)$$

where

$$U_1 = \begin{bmatrix} I & 0 \end{bmatrix}, \quad U_2 = \begin{bmatrix} I & I & 0 \\ 0 & 0 & I \end{bmatrix}, \quad V_1 = \begin{bmatrix} 0 & \pi_q \Sigma^{-1} \end{bmatrix}, \quad V_2 = \begin{bmatrix} \pi_q \Pi_1^{-1} & -\pi_q \Pi_1^{-1} & 0 \\ 0 & 0 & -I \end{bmatrix}.$$

This completes the proof.

In theorem 2, we have derived an upper bound of the model reduction error. In order to get a less conservative model reduction error bound, it is necessary for $n-k$

smallest σ_i's of Σ and $q-v$ smallest π_i's of Π to be small. Hence we choose a cost function as $J = tr(PQ) + \alpha tr(RS)$ for a positive constant α. Thus, we minimize the non-convex cost function subject to the convex constraints (9) and (11). Since this optimization problem is non-convex, the optimization problem is very difficult to solve it. So we suggest an alternative suboptimal procedure using an iterative method. We summarize an iterative method to solve a suboptimal problem.

step 1: Set $i = 0$. Initialize P_i, Q_i, R_i and S_i such that $tr(P_i + Q_i) + \alpha tr(R_i + S_i)$ is minimized subject to LMI's (9) and (11).

step 2: Set $i = i + 1$.
1) Minimize $J_i = tr(P_{i-1} Q_i) + \alpha tr(R_i S_{i-1})$ subject to LMI (9).
2) Minimize $J_i = tr(P_i Q_i) + \alpha tr(R_i S_i)$ subject to LMI (11).

step 3: If $|J_i - J_{i-1}|$ is less than a small tolerance level, stop iteration. Otherwise, go to step 2.

5 Concluding Remark

In this paper, we have studied a balanced model reduction problem for T-S fuzzy systems with IQC's. For this purpose, we have defined generalized controllability and observability Gramians for the uncertain fuzzy system. This generalized Gramians can be obtained from solutions of LMI problem. Using the generalized Gramians, we have derived a balanced state space realization. We have obtained the reduced model of the fuzzy system by truncating not only some state variables but also some IQC's.

References

1. Moore, B.C. : Principal component analysis in linear systems: Controllability, observability and model reduction. IEEE Trans. Automatic Contr., Vol. 26 (1982) 17-32
2. Pernebo, L., Silverman, L.M. : Model reduction via balanced state space representations. IEEE Trans. Automatic Contr., Vol. 27 (1982) 382-387
3. Glover, K. : All optimal Hankel-norm approximations of linear multivariable systems and their error bounds. Int. J. Control, Vol. 39 (1984) 1115-1193
4. Liu, Y., Anderson, B.D.O. : Singular perturbation approximation of balanced systems. Int. J. Control, Vol. 50 (1989) 1379-1405
5. Beck, C.L., Doyle, J., Glover, K. : Model reduction of multidimensional and uncertain systems. IEEE Trans. Automatic Contr., Vol. 41 (1996) 1466-1477
6. Wood, G.D., Goddard, P.J., Glover, K. : Approximation of linear parameter varying systems. Proceedings of the 35th CDC, Kobe, Japan, Dec. (1996) 406-411
7. Wu, F. : Induced L_2 norm model reduction of polytopic uncertain linear systems. Automatica, Vol. 32. No. 10 (1996) 1417-1426
8. Haddad, W.M., Kapila, V. : Robust, reduced order modeling for state space systems via parameter dependent bounding functions. Proceedings of American control conference, Seattle, Washington, June (1996) 4010-4014

9. Tanaka, K., Ikeda, T., Wang, H.O. : Robust stabilization of a class of uncertain nonlinear systems via fuzzy control : Quadratic stabilizability, control theory, and linear matrix inequalities. IEEE Trans. Fuzzy Systems, Vol. 4. No. 1. Feb. (1996) 1-13
10. Nguang, S.K., Shi, P. : Fuzzy output feedback control design for nonlinear systems : an LMI approach. IEEE Trans. Fuzzy Systems, Vol. 11. No. 3. June (2003) 331-340
11. Tuan, H.D., Apkarian, P., Narikiyo, T., Yamamoto, Y. : Parameterized linear matrix inequality techniques in fuzzy control system design. IEEE Trans. Fuzzy Systems, Vol. 9. No. 2. April (2001) 324-332
12. Yakubovich, V.A. : Frequency conditions of absolute stability of control systems with many nonlinearities. Automatica Telemekhanica, Vol. 28 (1967) 5-30

An Integrated Navigation System of NGIMU/ GPS Using a Fuzzy Logic Adaptive Kalman Filter

Mingli Ding and Qi Wang

Dept. of Automatic Test and Control, Harbin Institute of Technology,
150001 Harbin, China
dingml@hit.edu.cn

Abstract. The Non-gyro inertial measurement unit (NGIMU) uses only accelerometers replacing gyroscopes to compute the motion of a moving body. In a NGIMU system, an inevitable accumulation error of navigation parameters is produced due to the existence of the dynamic noise of the accelerometer output. When designing an integrated navigation system, which is based on a proposed nine-configuration NGIMU and a single antenna Global Positioning System (GPS) by using the conventional Kalman filter (CKF), the filtering results are divergent because of the complicity of the system measurement noise. So a fuzzy logic adaptive Kalman filter (FLAKF) is applied in the design of NGIMU/GPS. The FLAKF optimizes the CKF by detecting the bias in the measurement and prevents the divergence of the CKF. A simulation case for estimating the position and the velocity is investigated by this approach. Results verify the feasibility of the FLAKF.

1 Introduction

Most current inertial measurement units (IMU) use linear accelerometers and gyroscopes to sense the linear acceleration and angular rate of a moving body respectively. In a non-gyro inertial measurement unit (NGIMU) [1-6], accelerometers are not only used to acquire the linear acceleration, but also replace gyroscopes to compute the angular rate according to their positions in three-dimension space. NGIMU has the advantages of anti-high g value shock, low power consumption, small volume and low cost. It can be applied to some specific occasions such as tactic missiles, intelligent bombs and so on.

But due to the existence of the dynamic noise of the accelerometer output, it is inevitable that the system error increases quickly with time by integrating the accelerometer output. The best method to solve this problem above is the application of the integrated navigation system. NGIMU/GPS integrated navigation system can fully exert its superiority and overcome its shortcomings to realize the real-time high precision positioning in a high kinematic and strong electrically disturbed circumstance. But when using the conventional Kalman filter (CKF) in the NGIMU/GPS, the filtering results are often divergent due to the uncertainty of the statistical characteristics of dynamic noise of the accelerometer output and the system measurement noise. So, in order to ascertain the statistical characteristics of the noises mentioned above and alleviate the consumption error, a new fuzzy logic adaptive Kalman filter (FLAKF) [7] is proposed in designing a NGIMU/GPS integrated navigation system.

2 Accelerometer Output Equation

As all know, the precession of gyroscopes can be used to measure the angular rate. Based on this principle, IMU measures the angular rate of a moving body. The angle value can be obtained by integrating the angular rate with given initial conditions. With this angle value and the linear acceleration values in three directions, the current posture of the moving body can be estimated.

The angular rate in a certain direction can be calculated by using the linear acceleration between two points. To obtain the linear and angular motion parameters of a moving body in three-dimension space, the accelerometers need to be appropriately distributed on the moving body and the analysis of the accelerometer outputs is needed.

An inertial frame and a rotating moving body frame are exhibited in Fig. 1, where b represents the moving body frame and I the inertial frame.

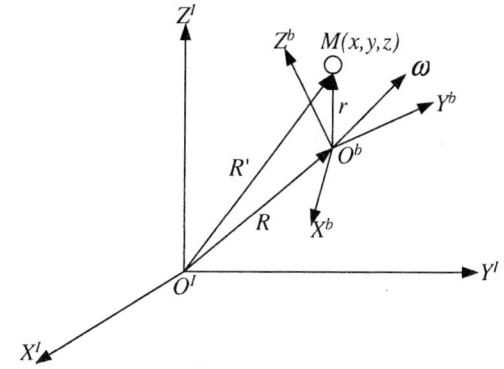

Fig. 1. Geometry of body frame (b) and inertial frame (I)

The acceleration of point M is given by

$$a = \ddot{R}_I + \ddot{r}_b + \dot{\omega} \times r + 2\omega \times \dot{r}_b + \omega \times (\omega \times r), \tag{1}$$

where $\ddot{r}_b$ is the acceleration of point M relative to body frame. $\ddot{R}_I$ is the inertial acceleration of O^b relative to O^I. $2\omega \times \dot{r}_b$ is known as the Coriolis acceleration, $\omega \times (\omega \times r)$ represents a centripetal acceleration, and $\dot{\omega} \times r$ is the tangential acceleration owing to angular acceleration of the rotating frame

If M is fixed in the b frame, the terms $\dot{r}_b$ and $\ddot{r}_b$ vanish. And Eq.(1) can be rewritten as

$$a = \ddot{R}_I + \dot{\omega} \times r + \omega \times (\omega \times r). \tag{2}$$

Thus the accelerometers rigidly mounted at location r_i on the body with sensing direction θ_i produce A_i as outputs.

$$A_i = [\ddot{R}_I + \dot{\Omega} r_i + \Omega\Omega r_i] \cdot \theta_i \qquad (i = 1,2,...,N), \qquad (3)$$

where

$$\Omega = \begin{bmatrix} 0 & -\omega_z & \omega_y \\ \omega_z & 0 & -\omega_x \\ -\omega_y & \omega_x & 0 \end{bmatrix}, \quad \ddot{R}_I = \begin{bmatrix} \ddot{R}_{Ix} \\ \ddot{R}_{Iy} \\ \ddot{R}_{Iz} \end{bmatrix}. \qquad (4)$$

3 Nine-Accelerometer Configuration

In this research, a new nine-accelerometer configuration of NGIMU is proposed. The locations and the sensing directions of the nine accelerometers in the body frame are shown in Fig.2. Each arrow in Fig.2 points to the sensing direction of each accelerometer.

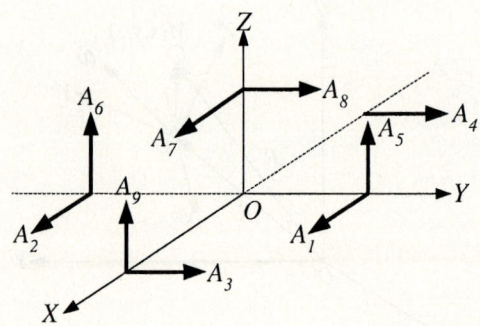

Fig. 2. Nine-accelerometer configuration of NGIMU

The locations and sensing directions of the nine accelerometers are

$$[r_1,\cdots,r_9] = l \begin{bmatrix} 0 & 0 & 1 & -1 & 0 & 0 & 0 & 0 & 1 \\ 1 & -1 & 0 & 0 & 1 & -1 & 0 & 0 & 0 \\ 0 & 0 & 0 & 0 & 0 & 0 & 1 & 1 & 0 \end{bmatrix}, \qquad (5)$$

where l is the distance between the accelerometer and the origin of the body frame.

$$[\theta_1,\cdots,\theta_9] = \begin{bmatrix} 1 & 1 & 0 & 0 & 0 & 0 & 1 & 0 & 0 \\ 0 & 0 & 1 & 1 & 0 & 0 & 0 & 1 & 0 \\ 0 & 0 & 0 & 0 & 1 & 1 & 0 & 0 & 1 \end{bmatrix}. \qquad (6)$$

It is easy to obtain

$$[r_1 \times \theta_1, \cdots, r_9 \times \theta_9] = l \begin{bmatrix} 0 & 0 & 0 & 0 & 1 & -1 & 0 & -1 & 0 \\ 0 & 0 & 0 & 0 & 0 & 0 & 1 & 0 & -1 \\ -1 & 1 & 1 & -1 & 0 & 0 & 0 & 0 & 0 \end{bmatrix}. \quad (7)$$

With Eq.(3), we get the accelerometer output equation

$$A_I = \begin{bmatrix} 0 & 0 & -l & 1 & 0 & 0 \\ 0 & 0 & l & 1 & 0 & 0 \\ 0 & 0 & l & 0 & 1 & 0 \\ 0 & 0 & -l & 0 & 1 & 0 \\ l & 0 & 0 & 0 & 0 & 1 \\ -l & 0 & 0 & 0 & 0 & 1 \\ 0 & l & 0 & 1 & 0 & 0 \\ -l & 0 & 0 & 0 & 1 & 0 \\ 0 & -l & 0 & 0 & 0 & 1 \end{bmatrix} \begin{bmatrix} \dot{\omega}_x \\ \dot{\omega}_y \\ \dot{\omega}_z \\ \ddot{R}_{Ix} \\ \ddot{R}_{Iy} \\ \ddot{R}_{Iz} \end{bmatrix} + \begin{bmatrix} 0 & 0 & 0 & 0 & 0 & l \\ 0 & 0 & 0 & 0 & 0 & -l \\ 0 & 0 & 0 & 0 & 0 & l \\ 0 & 0 & 0 & 0 & 0 & -l \\ 0 & 0 & 0 & l & 0 & 0 \\ 0 & 0 & 0 & -l & 0 & 0 \\ 0 & 0 & 0 & 0 & l & 0 \\ 0 & 0 & 0 & l & 0 & 0 \\ 0 & 0 & 0 & 0 & l & 0 \end{bmatrix} \begin{bmatrix} \omega_x^2 \\ \omega_y^2 \\ \omega_z^2 \\ \omega_y \omega_z \\ \omega_x \omega_z \\ \omega_x \omega_y \end{bmatrix}. \quad (8)$$

With Eq.(8), the linear expressions are

$$\dot{\omega}_x = \frac{1}{4l}(A_3 + A_4 + A_5 - A_6 - 2A_8), \quad (9a)$$

$$\dot{\omega}_y = \frac{1}{4l}(-A_1 - A_2 + A_5 + A_6 + 2A_7 - 2A_9), \quad (9b)$$

$$\dot{\omega}_z = \frac{1}{4l}(-A_1 + A_2 + A_3 - A_4), \quad (9c)$$

$$\ddot{R}_{Ix} = \frac{1}{2}(A_1 + A_2), \quad \ddot{R}_{Iy} = \frac{1}{2}(A_3 + A_4), \quad \ddot{R}_{Iz} = \frac{1}{2}(A_5 + A_6). \quad (9d)$$

4 Conventional Kalman Filter (CKF)

In Eq.(9), the linear acceleration and the angular acceleration are all expressed as the linear combinations of the accelerometer outputs. The conventional algorithm computes the navigation parameters as the time integration or double integrations of the equations in Eq.(9). But a numerical solution for the navigation parameters depends on the value calculated from previous time steps. And if the accelerometer output has a dynamic error, the error of the navigation parameters will inevitably increase with t and t^2 rapidly. So the design of a NGMIMU/GPS integrated navigation system is expected. In this section, the CKF is used in the system.

In order to analyze the problem in focus, we ignore the disturbance error contributed to the accelerometers due to the difference of the accelerometers' sensing directions in three-dimension space. Define the states vector $X(t)$ for the motion as

$$X(t) = [S_e(t) \ \ S_N(t) \ \ V_e(t) \ \ V_N(t) \ \ \omega(t)]^T, \tag{10}$$

where $S_e(t)$ is the estimation eastern position of the moving body at time t with respect to the earth frame (as inertial frame), $S_N(t)$ is the estimation northern position, $V_e(t)$ is the estimation eastern velocity, $V_N(t)$ is the estimation northern velocity and $\omega(t)$ is the estimation angular rate along x axis. Considering the relationship between the parameters, the states equations are then

$$\dot{S}_e = V_e, \ \dot{S}_N = V_N, \ \dot{V}_e = a_e, \ \dot{V}_N = a_N, \ \dot{\omega} = \dot{\omega}_x, \tag{11}$$

where

$$a_e = T_{11}\ddot{R}_{lx} + T_{21}\ddot{R}_{ly} + T_{31}\ddot{R}_{lz}, \ a_N = T_{12}\ddot{R}_{lx} + T_{22}\ddot{R}_{ly} + T_{32}\ddot{R}_{lz}. \tag{12}$$

In Eq.(12), T_{11}, T_{21}, T_{31}, T_{12}, T_{22} and T_{32} are the components of the coordinate transform matrix $T = \begin{bmatrix} T_{11} & T_{12} & T_{13} \\ T_{21} & T_{22} & T_{23} \\ T_{31} & T_{32} & T_{33} \end{bmatrix}$.

We also obtain

$$\dot{\omega}_x = \frac{1}{4l}(A_3 + A_4 + A_5 - A_6 - 2A_8). \tag{13}$$

The system state equation and system measurement equation in matrix form become

$$\dot{X} = \Psi X + Gu + \Gamma W, \tag{14}$$

and

$$Z = HX + \varepsilon. \tag{15}$$

In the system measurement equation Eq.(15), the input vector X is the output of the GPS receiver (position and velocity). In Eq.(14) and Eq.(15), ε and W denote the measurement noise matrix and the dynamic noise matrix respectively.

The preceding results are expressed in continuous form. Equation of state and measurement for discrete time may be deduced by assigning $t = kT$, where $k = 1, 2, \ldots$, and T denotes the sampling period. Straightforward application of the discrete time Kalman filter to (14) and (15) yields the CKF algorithm as outlined below. $\hat{X}_{0/0}$ is the initial estimate of the CKF state vector. $P_{0/0}$ is the initial estimate of the CKF state vector error covariance matrix.

5 Fuzzy Logic Adaptive Kalman Filter (FLAKF)

The CKF algorithm mentioned in the above section requires that the dynamic noise and the system measurement noise process are exactly known, and the noises processes are zero mean white noise. In practice, the statistical characteristics of the noises are uncertain. In the GPS measurement equation Eq.(15), among the measurement noise ε, the remnant ionosphere delay modified by the ionosphere model is just not zero mean white noise. Furthermore, the value of the dynamic error of the accelerometer output also cannot be exactly obtained in a kinematic NGIMU/GPS positioning. These problems result in the calculating error of K_k and make the filtering process divergent. In order to solve the divergence due to modeling error, In this paper, a fuzzy logic adaptive Kalman filter (FLAKF) is proposed to adjust the exponential weighting of a weighted CKF and prevent the Kalman filter from divergence. The fuzzy logic adaptive Kalman filter will continually adjust the noise strengths in the filter's internal model, and tune the filter as well as possible. The structure of NGIMU/GPS using FLAKF is shown in Fig. 3.

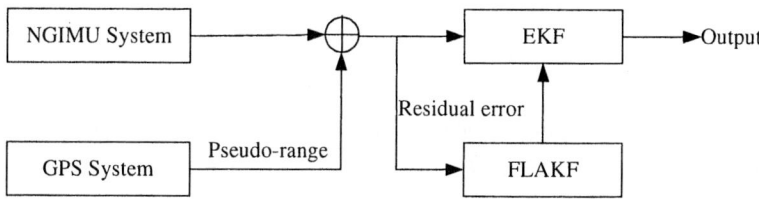

Fig. 3. Structure of NGIMU/GPS using FLAKF

The fuzzy logic is a knowledge-based system operating on linguistic variables. The advantages of fuzzy logic with respect to more traditional adaptation schemes are the simplicity of the approach and the application of knowledge about the controlled system. In this paper, FLAKF is to detect the bias in the measurement and prevent divergence of the CKF. Let us assume the model covariance matrices as

$$\begin{cases} R_k = \alpha R_v \\ Q_k = \beta Q_v \end{cases}, \tag{16}$$

where α and β are the adjustment ratios which are time-varying. The value of α and β can be acquired from the outputs of FLAKF. Let us define $\delta = Z_i - H\hat{X}_{k/k-1}$ as residual error, which reflects the degree to which the NGIMU/GPS model fits the data. According to the characteristic of the residual error of CKF, the variance matrixes of the dynamic noise of the accelerometer output and the measurement noise can be adjust self-adaptively using α and β. If $\alpha = \beta = 1$, we obtain a regular CKF.

The good way to verify whether the Kalman filter is performing as designed is to monitor the residual error. It can be used to adapt the filter. In fact, the residual error δ is the difference between the actual observing results and the measurement predictions based on the filter model. If a filter is performing optimally, the residual error is a zero-mean white noise process. The covariance of residual error P_r relates to Q_v and R_v. The covariance of the residual error is given by:

$$P_r = H(\Psi P_{k-1}\Psi^T + Q_v)H^T + R_v. \tag{17}$$

Using some traditional fuzzy logic system for reference, the Takagi-Sugeno fuzzy logic system is used to detect the divergence of CKF and adapt the filter. According to the variance and the mean of the residual error, two fuzzy rule groups are built up. To improve the performance of the filter, the two groups calculate the proper α and β respectively, and readjust the covariance matrix P_r of the filter.

As an input to FLAKF, the covariance of the residual error and the mean value of the residual error are used in order to detect the degree of the divergence. By choosing n to provide statistical smoothing, the mean and the covariance of the residual error are

$$\bar{\delta} = \frac{1}{n}\sum_{j=t-n}^{t}\delta_j, \tag{18}$$

$$\bar{P}_r = \frac{1}{n}\sum_{j=t-n+1}^{t}\delta_j\delta_j^T. \tag{19}$$

The estimated value $\bar{P}_r$ can be compared with its theoretical value P_r calculated from CKF. Generally, when covariance $\bar{P}_r$ is becoming larger than theoretical value P_r, and mean value $\bar{\delta}$ is moving from away zero, the Kalman filter is becoming unstable. In this case, a large value of β is applied. A large β means that process noises are added and we are giving more credibility to the recent data by decreasing the noise covariance. This ensures that all states in the model are sufficiently excited by the process noise. Generally, R_v has more impact on the covariance of the residual error. When the covariance is extremely large and the mean takes the values largely different from zero, there are presumably problems with GPS measurements. Therefore, the filter cannot depend on these measurements anymore, and a smaller α will be used. By selecting the appropriate α and β, the fuzzy logic controller optimally adapt the Kalman filter and tries to keep the innovation sequence act as a zero-mean white noise. The membership functions of the covariance and the mean value of the residual error are also built up.

6 Simulations and Results

The simulations of NGIMU/GPS using the CKF and the FLAKF are performed respectively in this section. Fig. 4, Fig. 5, Fig. 6 and Fig .7 illustrate the eastern position estimating error, the northern position estimating error, the eastern velocity estimating error and the northern velocity estimating error respectively.

In this simulation, the GPS receiver used is the Jupiter of Rockwell Co.. The initial conditions in position, velocity, posture angle and angular rate are $x(0)=0$ m, $y(0)=0$ m, $z(0)=0$ m, $v_x(0)=0$ m/s, $v_y(0)=0$ m/s, $v_z(0)=0$ m/s, $\alpha_x=0$ rad, $\alpha_y=0$ rad,

$\alpha_z = \pi/3$ rad, $\omega_x(0) = 0$ rad/s, $\omega_y(0) = 0$ rad/s, $\omega_z(0) = 0$ rad/s respectively. The accelerometer static bias is 10^{-5}g and the swing of posture angle is 0.2 rad. Moreover, when using the CKF, assume that W and ε are all Gaussian distribution, the covariance are $Q_v = (0.01)I_{9\times 9}$ and $R_v = (0.01)I_{5\times 5}$ respectively, and $P_{0/0} = (0.01)I_{5\times 5}$. The time required for simulation is 100s, and that for sampling is 10ms.

Comparing the curves in Fig 4 and Fig. 5, it is obvious that the eastern position estimating error and the northern position estimating error of NGIMU/GPS using the two filtering approaches are all leveled off after estimating for some time. And the errors acquired with the FLAKF are less than those with the CKF. In Fig.4, the error drops from 160m to 50m after using the FLAKF at 100s. In Fig.5, that is 220m to 100m. The similar results are also acquired in Fig.6 and Fig. 7 in the velocity estimation. The curves indicate that the NGIMU/GPS with the FLAKF can effectively alleviate the error accumulation of the estimation of the navigation parameters. When a designer lacks sufficient information to develop complete models or the parameters will slowly change with time, the FLAKF can be used to adjust the performance of CKF on-line.

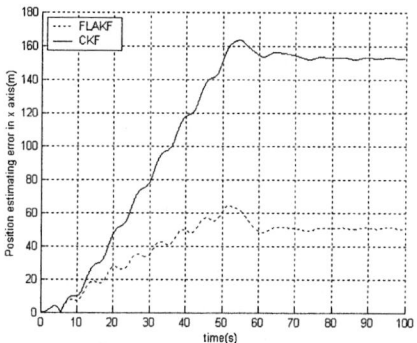

Fig. 4. The estimating error of the eastern position

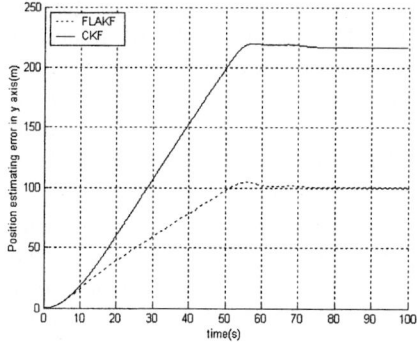

Fig. 5. The estimating error of the northern position

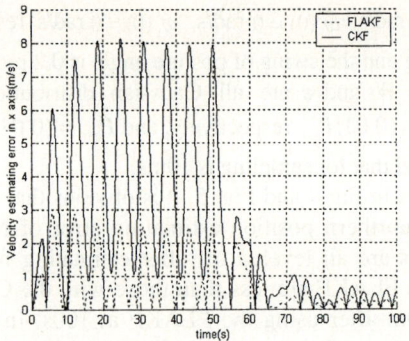

Fig. 6. The estimating error of the eastern velocity

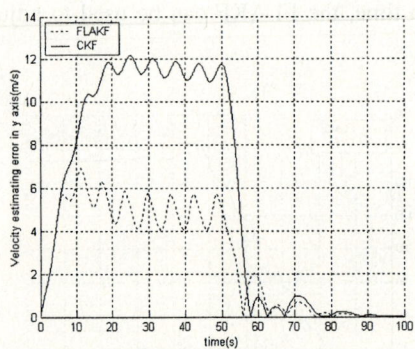

Fig. 7. The estimating error of the northern velocity

6 Conclusions

Due to the existence of the dynamic noise of the accelerometer output, it is inevitable that the navigation parameter estimation error increases quickly with time by integrating the accelerometer output. The use of the FLAKF to design a NGIMU/GPS based on a NGIMU of nine-accelerometer configuration can overcome the uncertainty of the statistical characteristics of the noises and alleviate the errors accumulation speed. By monitoring the innovations sequence, the FLAKF can evaluate the performance of a CKF. If the filter does not perform well, it would apply two appropriate weighting factors α and β to improve the accuracy of a CKF.

In FLAKF, there are 9 rules and therefore, little computational time is needed. It can be used to navigate and guide autonomous vehicles or robots and achieved a relatively accurate performance. Also, the FLAKF can use lower state-model without compromising accuracy significantly. Another words, for any given accuracy, the FLAKF may be also to use a lower order state model.

References

1. L.D. DiNapoli: The Measurement of Angular Velocities without the Use of Gyros. The Moore School of Electrical Engineering, University of Pennsylvania, Philadelphia (1965) 34-41
2. Alfred R. Schuler: Measuring Rotational Motion with Linear Accelerometers. IEEE Trans. on AES. Vol. 3, No. 3 (1967) 465-472
3. Shmuel J. Merhav: A Nongyroscopic Inertial Measurement Unit. J. Guidance. Vol. 5, No. 3 (1982) 227-235
4. Chin-Woo Tan, Sungsu Park: Design of gyroscope-free navigation systems. Intelligent Transportation Systems, 2001 Proceedings. Oakland (2001) 286-291
5. Sou-Chen Lee, Yu-Chao Huang: Innovative estimation method with measurement likelihood for all-accelerometer type inertial navigation system. IEEE Trans. on AES. Vol. 38, No. 1 (2002) 339-346
6. Wang Qi, Ding Mingli and Zhao Peng: A New Scheme of Non-gyro Inertial Measurement Unit for Estimating Angular Velocity. The 29th Annual Conference of the IEEE Industry Electronics Society (IECON'2003). Virginia (2003) 1564-1567
7. J. Z. Sasiadek, Q. Wang, M. B. Zeremba: Fuzzy Adaptive Kalman Filtering For INS/GPS Data Fusion. Proceedings of the 15th IEEE International Symposium on Intelligent Control, Rio, Patras, GREECE (2000) 181-186

Method of Fuzzy-PID Control on Vehicle Longitudinal Dynamics System

Yinong Li[1], Zheng Ling[1], Yang Liu[1], and Yanjuan Qiao[2]

[1] State Key Laboratory of Mechanical Transmission, Chongqing University,
Chongqing 400044, China
ynli@cqu.edu.cn
[2] Changan Automobile CO.LTD Technology Center, Chongqing 400044, China

Abstract. Based on the analysis of the vehicle dynamics control system, the longitudinal control of tracking for the platoon of two vehicles is discussed. A second-order-model for the longitudinal relative distance control between vehicles is presented, and the logic switch rule between acceleration and deceleration of the controlled vehicle is designed. Then the parameters auto-adjusting fuzzy-PID control method is used by adjusting the three parameters of PID to control the varying range of error for the longitudinal relative distance and relative velocity between the controlled vehicle and the navigation vehicle, in order to realized the longitudinal control of a vehicle. The simulation results shown that compared with fuzzy control, the parameters auto-adjusting fuzzy-PID control method decreases the overshoot, enhances the capacity of anti-disturbance, and has certain robustness. The contradictory between rapidity and small overshoot has been solved.

1 Introduction

With the rapid development and increased preservation of the vehicle in recent years, traffic jam and accident has become a global problem. Under the background of increasingly deteriorated traffic conditions, many countries has embarked on the research of the Intelligent Transport System (ITS) in order to utilize the roads and vehicles most efficiently, increase the vehicle security and reduce the pollution and congestion. In the Intelligent Transport System, automatic driving, which is an important component of transport automatization and intelligence, has become popular and important in vehicle research [1-2].

As a main content of the vehicle active safety, the vehicle longitudinal dynamic control system is composed of the super layer system and the under layer system[3]. The super layer of the vehicle longitudinal dynamic control system is to realize the automatic tracking of the longitudinal space of the vehicle platoon. The under layer of the vehicle longitudinal dynamic control system is to transfer the output of the super layer system to the controlled vehicle system, and computed the expected acceleration/deceleration.

The longitudinal dynamic control system is the combination of the super layer system and the under layer system. It is a complicated nonlinear system, thus the actual characteristics of this system are hard to be described exactly by linear system

[4]. For the vehicle dynamic longitudinal control system, intelligent control theory is used to fit the actual system more exactly, to realize real time control, and to respond more quickly. In this paper, parameter auto-adjusting fuzzy-PID control method is adopted to control the established vehicle longitudinal dynamic system, the real-time performance and the validity are studies in order to realize and improve the control performance of the vehicle longitudinal dynamic system.

2 Longitudinal Relative Distance Control Model

2.1 Longitudinal Relative Distance Control Between Two Vehicles

The control quality of the longitudinal relative distance between two vehicles is an important index to evaluate the active security of the vehicle. The control parameter of this system is the longitudinal relative distance between the leader and the follower. Considering the dynamic response characteristic of the system, we take the chang rate of the longitudinal relative distance between the two vehicles as another parameter of this control system to improve the accuracy of the model. The function of the throttle/brake switch logic control system is to determine the switch between the engine throttle and brake main cylinder, and transmit the expected acceleration or deceleration to the under layer of the longitudinal dynamic control system [5]. The logic control for throttle or brake switch is shown in Fig1.

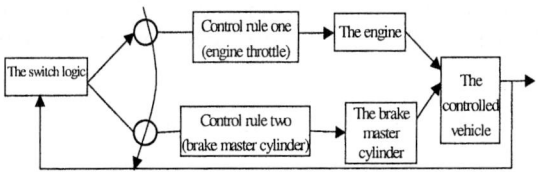

Fig. 1. The logic switch between throttle / brake

The one dimensional control model of the longitudinal relative distance error is described as follow

$$\delta_d^{'} = (x_h - x) - L - H \tag{1}$$

where H is the expected longitudinal relative distance between the two vehicles; L is the length of the vehicle body; x_h, x are the longitudinal dimensions of the back bumper of the leading vehicle and the controlled vehicle, respectively; $\delta d'$ is the error of the longitudinal relative distance.

For this one dimension control model, the structure is simple, the physical meaning is clear, and few information are required, so it is easy to realize the control purpose. However, when in the actual longitudinal running condition, the error of the longitudinal relative distance between the two vehicles is related to the change rate of the longitudinal displacement (i.e. running speed) of the controlled vehicle.

Considering the influence of the controlled vehicle's running speed to the accuracy of the model, some researches proposed the two dimensional model for the longitudinal relative distance control [6]:

$$\delta_d = (x_h - x) - L - H - \lambda \cdot v = \delta_d' - \lambda \cdot v \qquad (2)$$

where δ_d is the error of the longitudinal relative distance between the two vehicles; λ is the compensate time for the controlled vehicle to converge to $\delta_{d'}$; v is the running speed of the controlled vehicle.

2.2 Longitudinal Relative Speed Control

The error of the longitudinal relative speed is expressed by:

$$\delta_v = v_h - v \qquad (3)$$

where δ_v is the error of the longitudinal relative speed; v_h is the speed of the leading vehicle.

2.3 Second-Order Model for the Longitude Relative Distance Control

Combined the two dimensions control model of the longitudinal relative distance and the control model of the relative speed between the two vehicles, a second-order model for the longitudinal relative distance control can be established (rewrite equation (2) and equation (3)), and the state space equations of this model can be written as follows

$$\begin{cases} \delta_d = (x_h - x) - L - H - \lambda \cdot v \\ \delta_v = v_h - v \end{cases} \qquad (4)$$

$$\dot{X} = A \cdot X + B \cdot u + \Gamma \cdot w = \begin{bmatrix} 0 & 1 \\ 0 & 0 \end{bmatrix} \cdot X + \begin{bmatrix} -\lambda \\ -1 \end{bmatrix} \cdot u + \begin{bmatrix} 0 \\ 1 \end{bmatrix} \cdot w \qquad (5)$$

where X is the state variables vector of the control system, $X^T = [x_1 \quad x_2] = [\delta_d \quad \delta_v]$; u is the control variable of the control system (acceleration and deceleration of the controlled vehicle); w is the disturbance variable of the control system (acceleration or deceleration of the leading vehicle).

This second-order control model includes information about the longitudinal displacement, speed, acceleration and deceleration of the leading vehicle and the controlled vehicle respectively. The information can reflect the real time performance, dynamic response and dynamic characteristics of the control system based on the automatic tracking from a vehicle to a vehicle. This control system is a two-input, single-output system, the inputs variables are the relative distance error and the relative speed error, the outputs are the expected acceleration or deceleration of the controlled vehicle.

3 Throttle /Brake Switch Logic Control Model

Throttle/brake switch logic control model is an important component to realize the super layer control of the vehicle longitudinal dynamics. The expected acceleration / deceleration of the controlled vehicle, which are outputs of the second-order longitudinal relative distance controll, are transmitted to the switch logic control system. According to the switch logic rule, we can determine whether the throttle control or the brake control should be executed, and then the acceleration or deceleration will be transformed into engine throttle angle and brake pressure of the master cylinder.

The longitudinal dynamic model of the controlled vehicle can be expressed as:

$$\tau_d - \tau_b - rF_{rr} - rc(v - v_b)^2 = \varepsilon a \qquad (6)$$

where τ_d is the driving torque transmitted to the rear wheel by the driving system; τ_b is the sum of the torques imposed on the rear wheel and the front wheel; r is the wheel radius; c is the coefficient of wind resistance; F_{rr} is rolling resistance force; f is the coefficient of rolling resistance; m is the vehicle mass; v is the running speed of the vehicle; v_b is the speed of wind; a is the acceleration of the vehicle;

$$\varepsilon = \frac{1}{r}(\frac{J_e}{i^2} + mr^2 + J_{wr} + J_{wf}), \quad i = 1/i_{gear} i_i \eta_T \ ;$$

i_{gear} is the speed ratio of the auto transmission; i_i is the main transmission ratio of the driving system; η_T is the mechanical efficiency of the transmission system; J_{wf}、J_{wr} are moments of inertia of the rear wheel and the front wheel, respectively.

When the throttle angle is zero, the minimum acceleration of the controlled vehicle is:

$$a_{min} = \frac{1}{\varepsilon}\left[\frac{T_{em}}{i} - rF_{rr} - rc(v - v_b)^2\right] \qquad (7)$$

where T_{em} is the minimum output torque as the engine throttle is totally closed.

The switch logic rule between the engine throttle and the brake master cylinder can be deduced by the equation of the minimum acceleration.

Control rule for the engine throttle:

$$a_{synth} \geq a_{min} \qquad (8)$$

Control rule for the brake master cylinder:

$$a_{synth} < a_{min} \qquad (9)$$

where a_{synth} is the expected acceleration / deceleration of the controlled vehicle.

The existence of the switch rule between the engine throttle and the brake master cylinder may probably lead to vibration of the control system, thus, it is necessary to introduce a cushion layer near the switch surface to avoid violent vibration of the system and achieve good control. The optimized switch logic rules are:

Control rule for the engine throttle:

$$a_{synth} - a_{min} \geq s \qquad (10)$$

Control rule for the brake master cylinder:

$$a_{synth} - a_{min} < s \qquad (11)$$

where s is the thickness of the cushion layer, let $s=0.05 \text{m/s}^2$.

According to the switch rule between the engine throttle and the brake master cylinder, we can determine whether the engine throttle control or the brake master cylinder control will be executed, and then calculate the expected throttle angle or brake master cylinder pushrod force. For the engine control, the expected longitudinal driving torque can be written as:

$$\tau_d = \varepsilon a_{synth} + rF_{rr} + rc(v - v_b)^2 \qquad (12)$$

For the brake master cylinder control, the engine throttle is totally closed, and then the expected longitudinal brake torque can be written as:

$$\tau_b = -\varepsilon a_{synth} + \frac{T_{em}}{i} - rF_{rr} - rc(v - v_b)^2 \qquad (13)$$

From equation (7), it can be seen the minimum acceleration is a function of the speed of the controlled vehicle. Substituting the simulation parameters of the vehicle in this equation, we can obtain the switch logic rule between the engine throttle and the brake master cylinder by simulation. The simulation result is shown in Fig. 2.

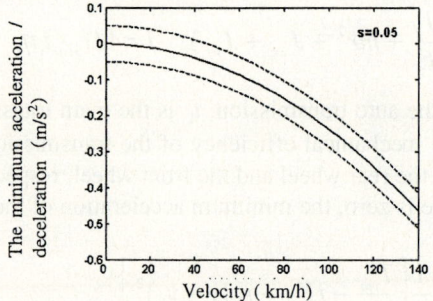

Fig. 2. Throttle/brake master cylinder switch logic law

4 Fuzzy PID Control of the Vehicle Longitudinal Dynamic System

4.1 Parameter Self-adjusting Fuzzy PID Control

Parameter self-adjusting fuzzy PID control is a compound fuzzy control. In order to satisfy the self-adjusting requirement of the PID parameters under different error e and different change rate of the error ec, the PID parameters are modified on line by fuzzy control rules [9]. The accuracy of the traditional PID control combined with the intelligence of the fuzzy control will make the dynamic characteristic and static characteristic of the controlled object well. The basic thought of the parameter self-adjusting fuzzy PID control is: at first, we find the fuzzy relations between the three parameters of the PID and the error e, the change rate of the error ec, respectively, and adjust e and ec through the progress continuously; and then revising the three parameters on line according to the fuzzy control rules, in order to satisfy the different requirement of the control parameters under different e and ec, and achieve the ideal control purpose. The computation is simple, so it is easy to be realized on the single chip processor. The structure of the parameter self-adjusting fuzzy PID controller is shown in Fig.3.

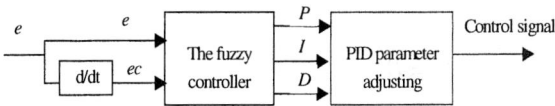

Fig. 3. Parameter self-adjusting fuzzy PID controller

In the parameter self-adjusting fuzzy PID control, self-adjusting rules for the parameters K_P, K_I and K_D under different e and ec can be simply defined as follows:

① If $|e|$ is larger, larger K_P and smaller K_D should be chosen (to accelerate the system response), and let $K_I = 0$ in order to avoid large overshoot, thus the integral effect can be eliminate).
② If $|e|$ is medium, K_P should be set smaller (in order to make the overshoot of the system response relatively small), K_I and K_D should be chosen proper value (the value of K_D has great effect to the system).
③ If $|e|$ is smaller, K_P and K_I should be set larger in the cause of making the system have good static characteristics, K_D should be set properly in order to avoid oscillation near the equilibrium point.

Here it takes the absolute value of the error $|e|_{fuzzy}$ and the change rate of the error $|ec|_{fuzzy}$ as the input language variables, where the variables with subscript "fuzzy" represent the fuzzy quantities of the parameters. Here $|e|_{fuzzy}$ can also be represented as $|E|$. Each of the language variables has three language values, i.e. big (B), medium (M), and small (S). The membership functions are shown respectively in Fig.4.

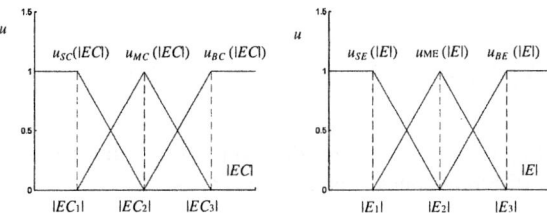

Fig. 4. The membership functions of the fuzzy PID

The membership functions can be adjusted by choosing the different values of $|e_1|$, $|e_2|$, $|e_3|$ and $|ec_1|$, $|ec_2|$, $|ec_3|$ in the different turning points.

4.2 The Process of the Parameter Self-adjusting Fuzzy PID Control

For the fuzzy input variables, there are five combination forms of $|e|$ and $|ec|$ based on the rules designed above.

① $|e|_{fuzzy} = B$

② $|e|_{fuzzy} = M$ 且 $|ec|_{fuzzy} = B$

③ $|e|_{fuzzy} = M$ 且 $|ec|_{fuzzy} = M$

④ $|e|_{fuzzy} = M$ 且 $|ec|_{fuzzy} = S$

⑤ $|e|_{fuzzy} = S$

The membership degree of each form can be computed by expressions as follows:

① $\mu_1(|e|_{fuzzy}, |ec|_{fuzzy}) = \mu_{BE}(|e|_{fuzzy})$

② $\mu_2(|e|_{fuzzy}, |ec|_{fuzzy}) = \mu_{ME}(|e|_{fuzzy}) \wedge \mu_{BC}(|ec|_{fuzzy})$

③ $\mu_3(|e|_{fuzzy}, |ec|_{fuzzy}) = \mu_{ME}(|e|_{fuzzy}) \wedge \mu_{MC}(|ec|_{fuzzy})$

④ $\mu_4(|e|_{fuzzy}, |ec|_{fuzzy}) = \mu_{ME}(|e|_{fuzzy}) \wedge \mu_{SC}(|ec|_{fuzzy})$

⑤ $\mu_5(|e|_{fuzzy}, |ec|_{fuzzy}) = \mu_{SE}(|e|_{fuzzy})$

The three PID parameters can be computed by the following equations based on the measurement of $|e|$ and $|ec|$.

$$K_P = \left[\sum_{j=1}^{5} \mu_j(|e|_{fuzzy}, |ec|_{fuzzy}) \times K_{Pj}\right] / \left[\sum_{j=1}^{5} \mu_j(|e|_{fuzzy}, |ec|_{fuzzy})\right] \quad (14)$$

$$K_I = \left[\sum_{j=1}^{5} \mu_j(|e|_{fuzzy}, |ec|_{fuzzy}) \times K_{Ij}\right] / \left[\sum_{j=1}^{5} \mu_j(|e|_{fuzzy}, |ec|_{fuzzy})\right] \quad (15)$$

$$K_D = \left[\sum_{j=1}^{5} \mu_j(|e|_{fuzzy}, |ec|_{fuzzy}) \times K_{Dj}\right] / \left[\sum_{j=1}^{5} \mu_j(|e|_{fuzzy}, |ec|_{fuzzy})\right] \quad (16)$$

where K_{Pj}, K_{Ij}, K_{Dj} (j=1,2,3...5) are the weights of the parameters K_P, K_I, K_D respectively under different states. They can be set as:

① $K_{P1} = K'_{P1}, K_{I1} = 0, K_{D1} = 0$

② $K_{P2} = K'_{P2}, K_{I2} = 0, K_{D2} = K'_{D2}$

③ $K_{P3} = K'_{P3}, K_{I3} = 0, K_{D3} = K'_{D3}$

④ $K_{P4} = K'_{P4}, K_{I4} = 0, K_{D4} = K'_{D4}$

⑤ $K_{P5} = K'_{P5}, K_{I5} = K'_{I5}, K_{D5} = K'_{D5}$

where $K'_{P1} \sim K'_{P5}$, $K'_{I1} \sim K'_{I5}$ and $K'_{D1} \sim K'_{D5}$ are the adjusting values of the parameters K_P, K_I, K_D respectively under different states by using normal PID parameter adjusting method.

Using the online self-adjusting PID parameters K_P, K_I, K_D, the output control variable u can be computed by the following discrete differential equations of the PID control algorithm.

$$u_n = K_P E_n + K_I \sum E_n + K_D (E_n - E_{n-1}) \qquad (17)$$

4.3 The Selection of Control Parameters and the Research of the Simulation

1) The Selection of Control Parameters

It takes the fuzzy quantitative factors of the inputs e and ec, and the defuzzificational scale factors of the output variables P, I and D respectively as follows:

$K_e = 7.2$, $K_{ec} = 8$, $R_P = 0.6$, $R_I = 1.3$, $R_D = 1.0$

The adjusting values adjusted by normal PID parameter-adjusting method under different states are $K'_{P1} \sim K'_{P5}$, $K'_{I1} \sim K'_{I5}$, $K'_{D1} \sim K'_{D5}$.

1) $K'_{P1} = 1, K'_{I1} = 0, K'_{D1} = 0$
2) $K'_{P2} = 2, K'_{I2} = 0, K'_{D2} = 1.25$
3) $K'_{P3} = 3, K'_{I3} = 0, K'_{D3} = 2.5$
4) $K'_{P4} = 4, K'_{I4} = 0, K'_{D4} = 3.75$
5) $K'_{P5} = 5, K'_{I5} = 2, K'_{D5} = 5$

2) Analysis of the Simulation

Combined the desired parameter-adjusting fuzzy PID controller and the vehicle longitudinal dynamic system, the MATLAB/Simulink platform is used to do the simulation research. The frame of the control is shown in Fig.5. Fig.6~Fig.10 are the simulation results of the parameters self-adjusting fuzzy PID control system of the vehicle longitudinal dynamic system. The simulation parameters are K_e=7.2, K_{ec}=8, R_P=0.6, R_I=1.3, R_D=1.0, λ=2, $L+H$=7.5.

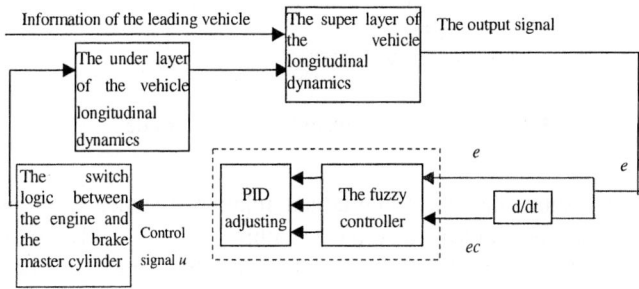

Fig. 5. Fuzzy PID control system of the vehicle longitudinal dynamics

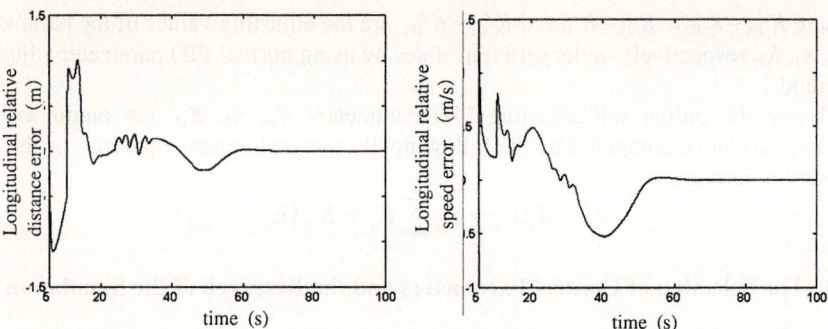

Fig. 6. Longitudinal relative distance error   **Fig. 7.** Longitudinal relative speed error

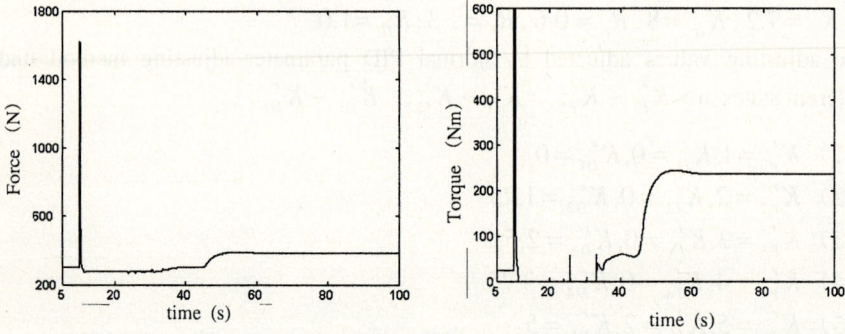

Fig. 8. The push force of the brake master cylinder   **Fig. 9.** The brake torque

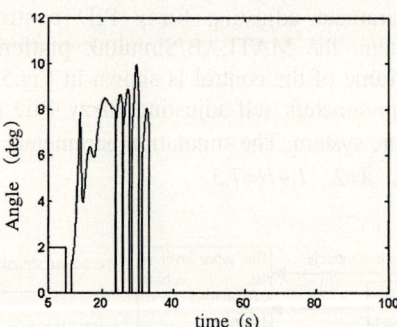

Fig. 10. The open angle of the engine throttle

Fig.6 and Fig.7 show that the longitudinal relative distance error between the leading vehicle and the controlled vehicle within ±1m is achieved through the parameter self-adjusting fuzzy PID control of the controlled vehicle, while the leading vehicle is accelerating or decelerating. The result is better than ±1.4m while only using the fuzzy control method. It also can be seen the longitudinal relative speed error is between 1m/s and –0.6m/s, it is better than ±1m/s while only using the fuzzy control method. Because the change process is smooth, and at the end the error convergent to zero, the auto-tracking of the controlled vehicle to the leading vehicle can be realized. So the parameters self-adjusting fuzzy PID control is more robust than the fuzzy control method. The response of the brake master cylinder pushrod force is shown in Fig.8, it can be seen that there is a break at the moment of 10s when the controller coming to work, and at the rest time the curve is smooth and is accordance with the brake torque curve in Fig.11. Fig.9 and Fig10 show the response of the controlled vehicle's brake torque and the throttle angle. When the controlled vehicle's throttle angle is above zero, the vehicle will accelerate and the brake torque is zero; otherwise, when the throttle open angel is below zero, the vehicle will decelerate and the brake torque is above zero. It can be seen the simulation curve using parameter self-adjusting fuzzy PID control fits the logic, and for most time the transition is smooth. However, it also can be seen the frequent alternation between the acceleration and the deceleration from 20 s to 35 s leads to the frequent switch between the brake torque and the throttle angel.

5 Conclusion

Through above analysis, it can be seen that, for vehicle longitudinal dynamic system, the parameter self-adjusting fuzzy PID control can achieve the control purpose of maintaining the safe distance between two vehicles. It absorbs both the fuzzy control's advantage of expressing irregular events and the adjusting function of the PID control, thus, the overshoot is reduced, the dynamic anti-disturbance capability is strengthened and the robustness is improved. Meanwhile, the parameter self-adjusting fuzzy PID control method also solves the conflict between the quick response and small overshoot. However, the frequent alternation between the acceleration and the deceleration leads to the phenomenon of the frequent switch between the brake torque and the throttle open angel, this should be improved in the future.

Acknowledgments

This research is supported by National Natural Science Foundation of China (No. 50475064) and Natural Science Foundation of Chongqing (No: 8366).

References

1. Hesse, Markus.: City Logistics: Network Modeling and Intelligent Transport Systems. Journal of Transport Geography, Vol.10, No.2 . (2002) 158-159
2. Marell. A, Westin. K.: Intelligent Transportation System and Traffic Safety – Drivers Perception and Acceptance of Electronic Speed Checkers. Transportation Research Part C: Emerging Technologies, Vol. 7, No.2-3. (1999) 131-147
3. Rajamani. R., Shladover.S. E.: An Experimental Comparative Study of Autonomous and Co -Operative Vehicle-Follower Control Systems. Transportation Research Part C: Emerging Technologies, Vol. 9, No. 1. (2001) 15-31

4. Kyongsu,Yi, Young Do Kwon: Vehicle-to-Vehicle Distance and Speed Control Using an Electronic-Vacuum Booster. JSAE Review, Vol. 22. (2001) 403–412
5. Sunan Huang, Wei.Ren: Vehicle Longitudinal Control Using Throttles and Brakes. Robotics and Autonomous Systems, Vol. 26. (1999) 241-253
6. Swaroop, D., Hedrick, J. K., Chien, C.C., Ioannou, P.: A Comparision of Spacing and Headway Control Laws for Automatically Controlled Vehicles. Vehicle System Dynamics., Vol. 23. (1994) 597-625
7. Visioli, A.: Tuning of PID Controllers with Fuzzy Logic. IEE Proceedings-Control Theory Appl, Vol. 148, No. 1. (2001)1-6

Design of Fuzzy Controller and Parameter Optimizer for Non-linear System Based on Operator's Knowledge

Hyeon Bae[1], Sungshin Kim[1], and Yejin Kim[2]

[1] School of Electrical and Computer Engineering, Pusan National University,
30 Jangjeon-dong, Geumjeong-gu, 609-735 Busan, Korea
{baehyeon, sskim, yjkim}@pusan.ac.kr
http://icsl.ee.pusan.ac.kr
[2] Dept. of Environmental Engineering, Pusan National University,
30 Jangjeon-dong, Geumjeong-gu, 609-735 Busan, Korea

Abstract. This article describes a modeling approach based on an operator's knowledge without a mathematical model of the system, and the optimization of the controller. The system used in this experiment could not easily be modeled by mathematical methods and could not easily be controlled by conventional systems. The controller was designed based on input-output data, and optimized under a predefined performance criterion.

1 Introduction

Fuzzy logic can express linguistic information in rules to design controllers and models. The fuzzy controller is useful in several industrial fields, because the fuzzy logic can be easily designed by expert's knowledge for rules. Therefore, the fuzzy logic can be used in un-modeled system control based on the expert's knowledge.

During the last several years, fuzzy controllers have been investigated in order to improve manufacturing processes [1]. Conventional controllers need mathematical models and cannot easily handle nonlinear models because of incompleteness or uncertainty [2], [3]. The controller for the experimental system presented in this study was designed based on the empirical knowledge of the features of a tested system [4].

2 Ball Positioning System and Control Algorithms

As shown in Fig. 1, the experimental system consists of two independent fans operated by two DC motors. The purpose of this experiment was to move a ball to a final goal position using the two fans. This system contains non-linearity and uncertainty caused by the aerodynamics inside the path. Therefore, the goal of this experiment was to initially design the fuzzy controller based on the operator's knowledge and then optimize using the system performance.

In this study, the position of the ball is found by image processing with real-time images. The difference of each image was measured to determine the position of the moving ball [5], [6]. Figure 2 shows the sequence of the image processing to find the ball in the path.

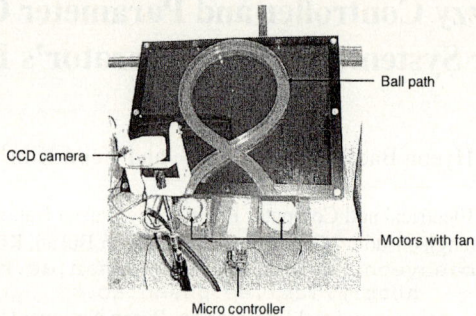

Fig. 1. The experimental system and components

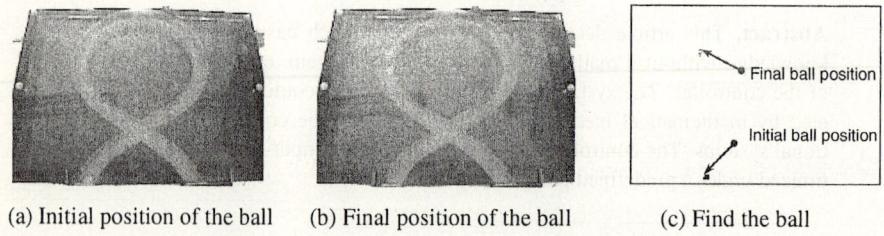

(a) Initial position of the ball   (b) Final position of the ball   (c) Find the ball

Fig. 2. Image processing to determine the ball position

3 Experimental Results

3.1 Fuzzy Controller

In an operator's experiment, the convergence rates and error values were measured to compare the performance. This was achieved by the performance criterion that is, the objective function, and fuzzy rules were fixed based on empirical knowledge. In an operator's test, random parameters were first selected and these parameters were subsequently adjusted according to the evaluated performance. The results of the test are shown in Table 1.

3.2 Optimization

Hybrid Genetic Algorithm: GA and Simplex Method
Thirty percent of crossover and 1.5% mutation rates were implemented for the genetic algorithm in this test. The first graph of Fig. 3 represents the performance with respect to each membership function. The performance values jumped to worse values once in the middle of a graph, but they could be improved gradually with iteration. After a GA process, the performance showed worse values than the previous values for the simplex method, as shown in the second graph of Fig. 3, because performance differences exists in real experiments even though the same membership functions are used. The reason is that all parameters transferred from the GA are used to operate the sys-

tem, and then the performance is evaluated again in the simplex method, so small bits gaps can exist.

Simulated Annealing

As shown in Fig. 4, two graphs represent the results for the SA algorithm. In the initial part of the second graph of Fig. 4, the performance values drop significantly and then lower to better values as the iterations are repeated. SA can search for good solutions quickly even though the repeated iteration times are less than those of the GA. When SA is used for optimization, to select the initial values is very important. In this experiment, the SA starts with the search parameters from the operator's trials. Initially a high temperature coefficient is selected for the global search and then it is improved gradually during the processing for the local search.

Table 1. Parameters of membership functions for experiments

	Error M.F										Derivative Error M.F					
	NB		NS		ZE		PS		PB		N		ZE		P	
	CE	VA	CE	VA	CE	VA	CE	VA	CE	VA	CE	VA	CE	VA	CE	VA
A	-40	20	-20	10	0	10	20	10	70	40	-10	5	0	3	10	5
B	-40	30	-15	15	0	5	15	20	80	50	-4	5	0	5	4	5
C	-50	35	-10	20	0	8	10	20	50	25	-4	5	0	3	4	5
D	-50	25	-20	10	0	10	20	10	50	25	-4	5	0	4	4	5
E	-30	20	-15	10	0	10	15	10	30	20	-4	4	0	3	4	4
F	-35	25	-18	10	0	8	35	25	50	20	-2	2	0	2	2	2

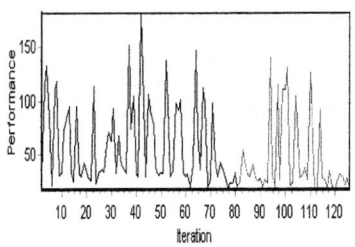
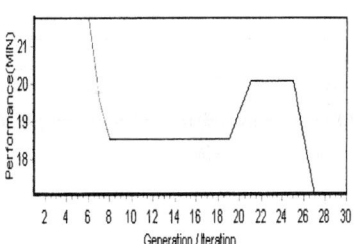

Fig. 3. Performance of the experiment in the case of the Hybrid GA

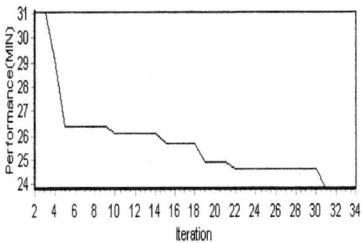
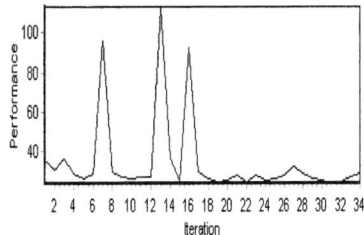

Fig. 4. Performance of the experiment in the case of the SA

4 Conclusions

The primary goal of this study was to design and optimize a fuzzy controller based on the operator's knowledge and running process. The method used characteristics of the fuzzy controller to control systems using fixed rules without mathematical models. Gaussian functions were employed as fuzzy membership functions and 8 functions were used for error and derivative error. Thus, a total of 16 parameters were optimized. SA is better than the hybrid GA considering convergence rate. But it is difficult to determine which one is the better optimization method. It depends on the conditions under which the systems operate.

Acknowledgement

This work was supported by "Research Center for Logistics Information Technology (LIT)" hosted by the Ministry of Education & Human Resources Development.

References

1. Lee, C. C.: Fuzzy Logic in Control Systems: Fuzzy Logic Controller, Part I, II. IEEE Transaction on Systems, Man, and Cybernetics **20**, (1990) 404-435
2. Tsoukalas, L. H. and Uhrig, R. E.: Fuzzy and Neural Approaches in Engineering. John Wiley & Sons, New York (1997)
3. Yen, J. and Langari, R.: Fuzzy Logic: Intelligence, Control, and Information. Prentice Hall, NJ, (1999)
4. Mamdani, E. H. and Assilian, S.: An Experiment in Linguistic Synthesis with a Fuzzy Logic Controller. International Journal of Man-Machine Studies **7**, (1975) 1-13
5. Haralick, R. M. and Shapiro, L. G.: Computer and Robot Vision. Addison Wesley, MA (1992)
6. Jain, R., Kasturi, R. and Schunck, B. G.: Machine Vision. McGraw-Hill, New York, (1995) 30-33

A New Pre-processing Method for Multi-channel Echo Cancellation Based on Fuzzy Control*

Xiaolu Li[1], Wang Jie[2], and Shengli Xie[1]

[1] Electronic and Communication Engineering,
South China University of Technology, 510641, China
[2] Control Science and Engineering Post-doc workstation,
South China University of Technology, 510641, China
pupwang@sohu.com

Abstract. The essential problem of multi-channel echo cancellation is caused by the strong correlation of two-channel input signals and the methods of pre-processing are always used to decorrelate it and the decorrelation degree depends on nonlinear coefficient α. But in most research, α is constant. In real application, the cross correlation is varying and α should be adjusted with correlation. But there is not precise mathematical formula between them. In this paper, the proposed method applies fuzzy logic to choose α so that the communication quality and convergence performance can be assured on the premise of small addition of computation. Simulations also show the effect of validity method.

1 Description of Problem

In free-hand mobile radiotelephone or teleconference system, how to cancel the echo is very important to assure communication. At present, adaptive cancellation technology is mainly adopted in multi-channel echo cancellation. The essential problem of such adaptive multi-channel echo cancellation is caused by strong correlation between input signals[1,2] and the convergence performance is bad. So researchers propose many pre-processing methods[1-4], How to choose the nonlinear transforming coefficient α is the main problem in the method.

Figure 1 is the block of stereophonic echo canceller with nonlinear pre-processing units with two added pre-processing units

Let $x_i(n) = [x_i(n), x_i(n-1), \cdots, x_i(n-L+1)]^T$, $i=1,2$ denote the output of the microphone at time n in the remote room, and let $h_i = [h_{1,0}, h_{1,1}, \cdots, h_{1,L-1}]^T$ denote the true echo path in the local room and

---

* The work is supported by the Guang Dong Province Science Foundation for Program of Research Team (grant04205783), the National Natural Science Foundation of China (Grant 60274006), the Natural Science Key Fund of Guang Dong Province, China (Grant 020826), the National Natural Science Foundation of China for Excellent Youth (Grant 60325310) and the Trans-Century Training Program, the Foundation for the Talents by the State Education Commission of China.

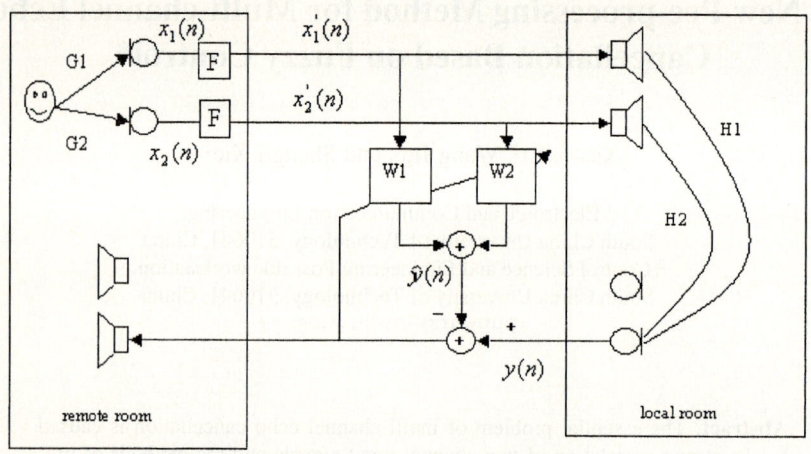

Fig. 1. Block of stereophonic echo canceller with nonlinear pre-processing units

$\hat{w}_i(n) = [\hat{w}_{1,0}(n), \hat{w}_{1,1}(n), \cdots, \hat{w}_{1,L-1}(n)]^T$ denote the adaptive FIR filters at time n respectively, where $i = 1,2$, and L is the length of impulse response and adaptive filters. The echo signal at time n is $y(n)$.

In order to improve convergence, one of pre-processing methods is nonlinearly transforming x_1、 x_2 and using the results as the input of filters, i.e.

$$x_i'(n) = x_i(n) + \alpha f[x_i(n)] \tag{1}$$

where f is a nonlinear function, e.g. half-wave commutate function. Through adjusting the nonlinear coefficient α, we can change the introduced degree of distortion.

2 Adjustment of Nonlinear Component Based on Fuzzy Control

The larger α is, the more added nonlinear component is and the quicker the filters converge, but large α perhaps will influence speech quality. Contrarily, small α brings good speech quality, but perhaps the convergence performance won't be improved much. So adjustment of α should be on the premise of speech quality. So, large α should be adopted when correlation between input signals is strong, and small α should be adopted when the correlation is faint. The relationship is gotten from experience and there is not precise mathematical formula.

Following we propose a sort of LMS algorithm which applies fuzzy logic to nonlinear pre-processing for stereophonic echo cancellation. Fuzzy logic is used to adjust α so that the communication quality and convergence performance can be assured on the premise of small addition of computation. The input of fuzzy inference system(FIR) is the correlation coefficient $\rho_{x_1 x_2}$ of input signals, and the output is the nonlinear component coefficient a. Three linguistic variables: small(S), minimum(M) and large(L) are defined to deal with the magnitude of either of $\rho_{x_1 x_2}$ and a. Complete fuzzy rules are shown in Table 1.

Table 1. Fuzzy Control Rules

Rules	Input: $\rho_{x_1 x_2}$	Output: a
R_0	L	L
R_1	M	M
R_2	S	S

At last, the proposed algorithm can reduce to following equations as follows

$$\rho_{x_1 x_2} = \left| \frac{\text{cov}(x_1(n), x_2(n))}{\sqrt{D(|x_1(n)|)}\sqrt{D(|x_2(n)|)}} \right| \quad ; a(n) = FIS(\rho_{x_1 x_2}(n)) \quad (2)$$

$$x_i' = x_i(n) + \alpha(n) f[x_i(n)] \quad (3)$$

$$x_i'(n) = [x_i'(n), x_i'(n-1), ..., x_i'(n-L+1)] \quad (4)$$

$$d(n) = x^T(n) h(n) \quad , \quad e(n) = d(n) - x^T(n) w(n) \quad (5)$$

$$w_i(n+1) = w_i(n) + 2\mu \frac{e(n) x_i'(n)}{x_1'(n)^T x_1'(n) + x_2'(n)^T x_2'(n)}, i = 1, 2 \quad (6)$$

3 Simulations

Simulations show that when the correlation is strong, $a(n)$ is also large; when weak correlation is caused by background noise or position changing of speakers, the designed fuzzy system will give small $a(n)$ to reduce speech distortion. Fig.2 shows the good convergence performance of mean square error of the adaptive filters.

We adopt Mean Square Error(MSE) as the criterion:

$$MSE = 10 \log_{10} E[e^2(n)].$$

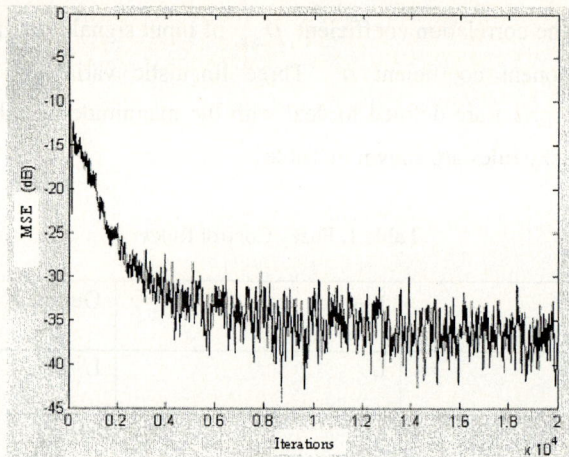

Fig. 2. Mean square error of the adaptive filters

4 Conclusion

The essential problem of multi-channel echo cancellation is caused by the strong correlation, and the methods of pre-processing the input signals are always used to decorrelate it. In real application, the cross correlation of two-channel input signals is varying with time and it also will be influenced by background noise and not be constant. But in most research, α is constant but in fact α should be adjusted with correlation. Their relationship is gotten from experience and there is not precise mathematical formula between them. In this paper the proposed method applies fuzzy logic to nonlinear pre-processing for stereophonic echo cancellation so that the communication quality and convergence performance can be assured on the premise of small addition of computation. Simulations also show the effect of validity method.

References

1. M. Mohan Sondhi, Dennis R. Morgan, Joseph L. Hall, Stereophonic acoustic echo cancellation- an overview of the fundamental problem, IEEE Signal Processing Letters. 2(1995)148-151
2. Jacob Benesty, Dennis R. Morgan and M. Mohan Sondhi, A better understanding and improved solution to the probems of stereophonic acoustic echo cancellation, ICASSP. (1995)303-305
3. Benesty J, Morgan D and Sondhi M.A better understanding and an improved solution solution to the problem of stereophonic acoustic echo cancellation. IEEE Trans. Speech Audio Procesing. 6(1998) 156-165
4. Suehiro Shimauchi and Shoji Makino, Stereo projection echo canceller with true echo path estimation, ICASSP. (1995)3059-3063

Robust Adaptive Fuzzy Control for Uncertain Nonlinear Systems

Chen Gang, Shuqing Wang, and Jianming Zhang

National Key Laboratory of Industrial Control Technology,
Institute of Advanced Process Control, Zhejiang University,
Hangzhou, 310027, P. R. China
gchen@iipc.zju.edu.cn

Abstract. Two different fuzzy control approaches are proposed for a class of nonlinear systems with mismatched uncertainties, transformable to the strict-feedback form. A fuzzy logic system (FLS) is used as a universal approximator to approximate unstructured uncertain functions and the bounds of the reconstruction errors are estimated online. By employing special design techniques, the controller singularity problem is completely avoided for the two approaches. Furthermore, all the signals in the closed-loop systems are guaranteed to be semi-globally uniformly ultimately bounded and the outputs of the system are proved to converge to a small neighborhood of the desired trajectory. The control performance can be guaranteed by an appropriate choice of the design parameters. In addition, the proposed fuzzy controllers are highly structural and particularly suitable for parallel processing in the practical applications.

1 Introduction

Based on the fact that FLS can approximate uniformly a nonlinear function over a compact set to any degree of accuracy, FLS provides an alternative way to modeling and design of nonlinear control system. It provides a way to combine both the available mathematical description of the system and the linguistic information into the controller design in a uniform fashion. In order to improve the performance and stability of FLS, the synthesis approach to constructing robust fuzzy controllers has received much attention.

Recently, some design schemes that combine backstepping methodology with adaptive FLS have been reported. In order to avoid the controller singularity problem, the control gain is often assumed to be known functions [1], [2]. However, this assumption cannot be satisfied in many cases. In [3], a FLS is used to approximate the unknown control gain function, but the controller singularity problem is not solved. The possible controller singularity problem can be avoided when the sign of the control gain is known [4]. The problem becomes more difficult if the sign of the control gain is unknown. Another problem is that some tedious analysis is needed to determine regression matrices [1], [3]. Therefore, the approaches are very complicated and difficult to use in practice. In this note, we will present two approaches to solve the aforementioned problems. One is the robust adaptive fuzzy tracking control (RAFTC) for the low order systems with known sign of the control gain. The other is the robust fuzzy tracking control (RFTC) proposed for the high order systems with unknown sign of the control gain. The control schemes presented in this note have several advantages. First, they can incorporate in an easy way the linguistic information about the system through if-then rules into the controller design. The controllers are highly structural and particularly suitable for parallel processing in the practical applications. Second, the controller singularity problem is completely avoided.

The outline of this note is as follows. In Section 2, formulation of our robust tracking problem of nonlinear system with mismatched uncertainties is presented. RAFTC and RFTC are developed in Section 3 and Section 4, respectively. Finally, the note is concluded in Section 5.

2 System Description and Problem Statement

Consider the n-order nonlinear system of the form

$$\dot{\xi} = f(\xi) + \Delta q(\xi) + bg(\xi)u,$$
$$y = h(\xi), \tag{1}$$

where $\xi \in R^n$ is the state, $u \in R$ is the input, $y \in R$ is the output, and h is a smooth function on R^n. f and g are known functions with $g(\xi) \neq 0, \forall \xi \in R^n$. b is an unknown parameter with $b \neq 0$. $\Delta q(\xi)$ represents uncertainties due to many factors, such as modeling errors, parameter uncertainties, disturbances, and so on. The control objective is to make the output y tracks the desired trajectory $y_r(t)$ in the presence of bounded uncertainties.

Assumption 1. The desired output trajectory $y_r(t)$ and its derivatives up to nth order are known and bounded.

Assumption 2. The uncertainties $\Delta q(\xi)$ satisfy the structural coordinate-free condition $ad_z \Delta q \in \varphi^i$, $\forall z \in \varphi^i$, $0 \leq i \leq n-2$. $\varphi^i = \text{span}\{g, ad_f g, \cdots, ad_f^i g\}$, $0 \leq i \leq n-1$, are involutive and of constant rank $i+1$.

The nominal form of (1) is described by the form

$$\dot{\xi} = f(\xi) + bg(\xi)u,$$
$$y = h(\xi). \tag{2}$$

Lemma 1. If the system (2) has relative degree n and Assumption 2 is satisfied, there exists a global diffeomorphism $x = \phi(\xi)$, transforming the system (1) into the form

$$\dot{x}_i = x_{i+1} + \Delta_i(x_1, \cdots, x_i), \ 1 \leq i \leq n-1,$$
$$\dot{x}_n = \gamma(x) + b\beta(x)u + \Delta_n(x), \tag{3}$$
$$y = x_1.$$

3 Design of RAFTC

Assumption 3. The sign of the unknown parameter b is known. Furthermore, b satisfies $b_0 < |b| < b_1$ for some unknown positive constant b_0, b_1.

In this note, we consider a FLS consisting of the product-inference rule, singleton fuzzifier, center average defuzzifier, and Gaussian membership function. Based on universal approximation theorem, given a compact set $\Omega_{\bar{x}_i} \subset R^i$, the unknown function Δ_i can be expressed as

$$\Delta_i(x_1,\cdots,x_i) = w_i^{*T}h_i(x_1,\cdots,x_i) + v_i(x_1,\cdots,x_i), \tag{4}$$

where w_i^* is an optimal weight vector; h_i is the fuzzy function vector; the reconstruction error v_i is bounded, i.e., there exists unknown constant $\rho_i > 0$ such that $|v_i| < \rho_i$.

The design procedure consists of n steps. At each step i $(1 \leq i < n)$, the direct adaptive fuzzy control techniques are employed to design a fictitious controller α_i. For the residual uncertainties, a robustness term is introduced to compensate them. At the nth step, the actual control u appears and the design is completed. For sparing the space, we omit these steps and directly give the resulting adaptive controller

$$\begin{aligned}
& e_1 = x_1 - y_r, \; e_i = x_i - \alpha_{i-1}, \; i = 2,\cdots,n, \\
& \alpha_1 = -k_1 e_1 - \hat{w}_1^T h_1 - \varphi_1 + \dot{y}_r, \\
& \alpha_i = -k_i e_i - \hat{w}_i^T h_i - \varphi_i - e_{i-1} + \dot{\alpha}_{i-1}, \; i = 2,\cdots,n-1, \\
& u = -\hat{q}\beta^{-1}\left(\gamma + k_n e_n + \hat{w}_n^T h_n + \varphi_n + e_{n-1} - \dot{\alpha}_{n-1}\right), \\
& \varphi_i = \hat{\rho}_i \tanh(e_i/\varepsilon_i), \; i = 1,\cdots,n,
\end{aligned} \tag{5}$$

where $k_i > 0$ and $\varepsilon_i > 0$ for $i = 1,\cdots,n$ are design parameters; $q = b^{-1}$; $\hat{w}_i$, $\hat{\rho}_i$, and $\hat{q}$ denote the estimates of w_i, ρ_i, and q, respectively.

It is well known that for conventional adaptive laws, the unmodeled dynamics, disturbances, and the reconstruction errors may lead to parameter drift and even instability problems. Consider the constraint region $\Omega_{w_i} = \{w_i \mid \|w_i\| \leq c_i\}$ for approximation parameter vector w_i with constant $c_i > 0$. $\hat{w}_i$, $\hat{\rho}_i$, and $\hat{q}$ are updated according to

$$\begin{aligned}
& \dot{\hat{w}}_i = Q_i(h_i e_i - \sigma_i \hat{w}_i), \\
& \dot{\hat{\rho}}_i = -k_c \ell_i(\hat{\rho}_i - \rho_i^0) + \ell_i e_i \tanh(e_i/\varepsilon_i), \\
& \dot{\hat{q}} = g_1(\text{sgn}(b)e_n \phi_1 - \eta\hat{q}),
\end{aligned} \tag{6}$$

where $\phi_1 = \gamma + k_n e_n + \hat{w}_n^T h_n + \varphi_n + e_{n-1} - \dot{\alpha}_{n-1}$; $Q_i = Q_i^T$ is a positive definite matrix; $\ell_i > 0$, $k_c > 0$, $g_1 > 0$, and $\eta > 0$ are constants; ρ_i^0 is initial estimate of ρ_i; the switching parameter σ_i is chosen as

$$\sigma_i = 0, \text{ if } \|\hat{w}_i\| < c_i \text{ or } (\|\hat{w}_i\| \geq c_i \text{ and } \hat{w}_i^T h_i e_i \leq 0)$$
$$\sigma_i = \frac{(\|\hat{w}_i\|^2 - c_i^2)\hat{w}_i^T h_i e_i}{\|\hat{w}_i\|^2 (\delta_i^2 + 2\delta_i c_i)}, \text{ if } \|\hat{w}_i\| \geq c_i \text{ and } \hat{w}_i^T h_i e_i > 0,$$
(7)

with small constant $\delta_i > 0$.

According to (5), we obtain the error dynamics of the closed-loop system

$$\dot{E} = -KE - \tilde{W}^T H + M - B + SE + D,$$
(8)

where $E = (e_1, \cdots, e_n)^T$, $K = \text{diag}(k_1, \cdots, k_n)$, $H = (h_1^T, \cdots, h_n^T)^T$, $\tilde{W} = \text{diag}(\tilde{w}_1, \cdots, \tilde{w}_n)$, $\tilde{w}_i = \hat{w}_i - w_i^*$, $M = (v_1, \cdots, v_n)^T$, $B = (\varphi_1, \cdots, \varphi_n)^T$, $D = (0, \cdots, b\tilde{q}\phi_1)^T$, $\tilde{q} = q - \hat{q}$, and $S \in R^{n \times n}$ has only nonzero elements $s_{i,i+1} = 1$, $s_{i+1,i} = -1$, $i = 1, \cdots, n-1$.

Theorem 1. Consider the closed-loop system consisting of (3) satisfying Assumption 3, controller (5), and the adaptive laws (6). For bounded initial conditions, 1) all signals in the closed-loop system are bounded; 2) the tracking error e_1 can be kept as small as possible by adjusting the controller parameters in a known form.

Proof. Define $Q = \text{diag}(Q_1, \cdots, Q_n)$, $L = \text{diag}(\ell_1, \cdots, \ell_n)$, $\rho = (\rho_1, \cdots, \rho_n)^T$, $\tilde{\rho} = \rho - \hat{\rho}$, $\hat{\rho} = (\hat{\rho}_1, \cdots, \hat{\rho}_n)^T$, $\hat{W} = \text{diag}(\hat{w}_1, \cdots, \hat{w}_n)$. Considering the following Lyapunov candidate

$$V = \frac{1}{2}E^T E + \frac{1}{2}\text{tr}(\tilde{W}^T Q^{-1} \tilde{W}) + \frac{1}{2}\tilde{\rho}^T L^{-1} \tilde{\rho} + \frac{|b|}{2g_1}\tilde{q}^2.$$
(9)

The time derivative of V along the trajectory of (8) satisfies

$$\dot{V} = E^T(-KE - \tilde{W}^T H + M - B + SE + D) + \text{tr}(\tilde{W}^T Q^{-1} \dot{\hat{W}}) - \tilde{\rho}^T L^{-1} \dot{\hat{\rho}} - \frac{|b|}{g_1}\tilde{q}\dot{\hat{q}}$$
$$\leq -E^T KE - \sum_{i=1}^n \sigma_i(\hat{w}_i - w_i^*)^T \hat{w}_i + k_c \sum_{i=1}^n \tilde{\rho}_i(\hat{\rho}_i - \rho_i^0) + |b|\eta\tilde{q}\hat{q} + \sum_{i=1}^n \rho_i \varepsilon_i.$$
(10)

Referring to (7) for $i = 1$, $\sigma_1 = 0$ if the first condition is true. If $\|\hat{w}_1\| \geq c_1$ and $\hat{w}_1^T h_1 e_1 > 0$, then $\sigma_1(\hat{w}_1 - w_1^*)^T \hat{w}_1 \geq 0$, because $\|\hat{w}_1\| \geq \|w_1^*\|$ and $\sigma_1 > 0$. By employing the same procedure, we can achieve the same results $\sigma_i(\hat{w}_i - w_i^*)^T \hat{w}_i \geq 0$,

$i = 2, \cdots, n$. Therefore, we can obtain $\sum_{i=1}^{n} \sigma_i (\hat{w}_i - w_i^*)^T \hat{w}_i \geq 0$. Consequently, there exists

$$\dot{V} \leq -\lambda_{\min}(K)\|E\|^2 - \frac{1}{2}k_c\|\tilde{\rho}\|^2 - \frac{1}{2}|b|\eta\tilde{q}^2 + \frac{1}{2}k_c\sum_{i=1}^{n}(\rho_i - \rho_i^0)^2 + \frac{\eta}{2b_0} + \sum_{i=1}^{n}\rho_i\varepsilon_i, \quad (11)$$

where $\lambda_{\min}(K)$ denotes the minimum eigenvalue of K. We see that $\dot{V}$ is negative whenever $E \notin \Omega_E = \{E \mid \|E\| \leq \sqrt{c_0/\lambda_{\min}(K)}\}$, or $\tilde{\rho} \notin \Omega_{\tilde{\rho}} = \{\tilde{\rho} \mid \|\tilde{\rho}\| \leq \sqrt{2c_0/k_c}\}$, or $\tilde{q} \notin \Omega_{\tilde{q}} = \{\tilde{q} \mid |\tilde{q}| \leq \sqrt{2c_0/(|b|\eta)}\}$, where $c_0 = \frac{k_c}{2}\sum_{i=1}^{n}(\rho_i - \rho_i^0)^2 + \frac{\eta}{2b_0} + \sum_{i=1}^{n}\rho_i\varepsilon_i$.

According to standard Lyapunov theorem extension, these demonstrate the uniform ultimate boundedness of $\|E\|$, $\|\tilde{\rho}\|$, and $|\tilde{q}|$. Since $\hat{w}_i$ $(i=1,\cdots,n)$ and $y_r^{(i)}$ $(i=0,\cdots,n)$ are bounded, all fictitious functions α_i $(i=1,\cdots,n-1)$ and control input u are bounded. Consequently, we can conclude that all signals in the closed-loop system are bounded. According to (11), we know that the tracking error satisfies $\lim_{t \to \infty} |e_1(t)| \leq \sqrt{c_0/\lambda_{\min}(K)}$. The small tracking error can be achieved by increasing control gain k_i and decreasing ε_i, η. The parameter k_c offers a tradeoff between the magnitudes of $\|\tilde{\rho}\|$ and $|e_1|$. It is also shown the closer ρ_i^0 $(i=1,\cdots,n)$ are to ρ_i $(i=1,\cdots,n)$, the smaller $\|\tilde{\rho}\|$ and $|e_1|$ become.

The RAFTC is suitable for the low order system with known sign of control gain. For the high order system, the online computation burden will be heavy. In the next section, we will present an approach for the high order system. The control laws have the adaptive mechanism with minimal learning parameterizations. Furthermore, a priori knowledge of the control gain sign is not required.

4   Design of RFTC

The design procedure is briefly given as follows.

Step i ($1 \leq i < n$): According to (4), the fictitious controllers are chosen as

$$\alpha_1 = -k_1 e_1 - \overline{w}_1^T h_1 - \varphi_1 + \dot{y}_r,$$
$$\alpha_i = -k_i e_i - \overline{w}_i^T h_i - \varphi_i - e_{i-1} + \dot{\alpha}_{i-1}, \quad i = 2,\cdots,n-1, \quad (12)$$

where the nominal vector $\overline{w}_i$ is designed and fixed by a priori knowledge. There exist constants ρ_{wi} and ρ_{vi} such that $\|\overline{w}_i - w_i^*\| \leq \rho_{wi}$, $\|v_i\| \leq \rho_{vi}$. The robustness term φ_i is given by $\varphi_i = \hat{\rho}_{wi}\|h_i\|\tanh(\|h_i\|e_i/\varepsilon_{wi}) + \hat{\rho}_{vi}\tanh(e_i/\varepsilon_{vi})$. $\hat{\rho}_{wi}$ and $\hat{\rho}_{vi}$ denote the estimates of ρ_{wi} and ρ_{vi}, respectively. $k_i > 0$, $\varepsilon_{wi} > 0$, and $\varepsilon_{vi} > 0$ are design parameters. $\hat{\rho}_{wi}$ and $\hat{\rho}_{vi}$ are updated as follows:

$$\dot{\hat{\rho}}_{wi} = -\sigma_{wi}(\hat{\rho}_{wi} - \rho_{wi}^0) + r_{wi} e_i \|h_i\| \tanh(\|h_i\| e_i / \varepsilon_{wi}),$$
$$\dot{\hat{\rho}}_{vi} = -\sigma_{vi}(\hat{\rho}_{vi} - \rho_{vi}^0) + r_{vi} e_i \tanh(e_i / \varepsilon_{vi}), \tag{13}$$

where σ_{wi}, r_{wi}, σ_{vi}, r_{vi} are positive constants.

Step n: In the final step, we will design the actual controller. Since the sign of the control gain is unknown, we will introduce Nussbaum-type gain in the controller design. Choosing the following actual controller

$$u = N(\omega)\overline{\alpha}_n / \beta(x),$$
$$\dot{\omega} = \overline{\alpha}_n e_n, \tag{14}$$
$$\overline{\alpha}_n = k_n e_n + \gamma + \overline{w}_n^T h_n + \varphi_n + e_{n-1} - \dot{\alpha}_{n-1}.$$

$N(\omega)$ is a Nussbaum-type function which has the following properties [6]

$$\lim_{s \to \infty} \sup \int_0^s N(\omega) d\omega = +\infty; \quad \lim_{s \to \infty} \inf \int_0^s N(\omega) d\omega = -\infty. \tag{15}$$

In this note, the Nussbaum function $N(\omega) = \exp(\omega^2)\cos(\pi\omega/2)$ is considered. It should be pointed out that the Nussbaum-type gain technique was firstly proposed in [6] for a class of first-order linear system. Subsequently, the method was generalized to higher order linear systems [7] and nonlinear systems [8], [9].

According to (12) and (14), we obtain the following error dynamics of the closed-loop system

$$\dot{E} = -KE - \widetilde{W}^T H + M - B + SE + D, \tag{16}$$

where $E = (e_1, \cdots, e_n)^T$, $K = \text{diag}(k_1, \cdots, k_n)$, $H = (h_1^T, \cdots, h_n^T)^T$, $\widetilde{W} = \text{diag}(\widetilde{w}_1, \cdots, \widetilde{w}_n)$, $\widetilde{w}_i = \overline{w}_i - w_i^*$, $M = (v_1, \cdots, v_n)^T$, $B = (\varphi_1, \cdots, \varphi_n)^T$, $D = (0, \cdots, (bN(\omega)+1)\overline{\alpha}_n)^T$, and $S \in R^{n \times n}$ has only nonzero elements $s_{i,i+1} = 1$, $s_{i+1,i} = -1$, $i = 1, \cdots, n-1$.

Theorem 2. Suppose Assumption 1 is satisfied. Consider the closed-loop system consisting of (3), controller (12), (14), and the parameter updating laws (13). Given a compact set $\Omega_n \subset R^n$, for any $x(0) \in \Omega_n$, the errors e_i, $i = 1, \cdots, n$, and parameter estimates $\hat{\rho}_{wi}$, $\hat{\rho}_{vi}$, $i = 1, \cdots, n$, are uniformly ultimately bounded (UUB).

Proof. Define $R_w = \text{diag}(r_{w1}, \cdots, r_{wn})$, $R_v = \text{diag}(r_{v1}, \cdots, r_{vn})$, $\rho_w = (\rho_{w1}, \cdots, \rho_{wn})^T$, $\hat{\rho}_w = (\hat{\rho}_{w1}, \cdots, \hat{\rho}_{wn})^T$, $\widetilde{\rho}_w = \hat{\rho}_w - \rho_w$, $\rho_v = (\rho_{v1}, \cdots, \rho_{vn})^T$, $\hat{\rho}_v = (\hat{\rho}_{v1}, \cdots, \hat{\rho}_{vn})^T$, $\widetilde{\rho}_v = \hat{\rho}_v - \rho_v$. Considering the following Lyapunov function candidate

$$V = \frac{1}{2} E^T E + \frac{1}{2} \widetilde{\rho}_w^T R_w^{-1} \widetilde{\rho}_w + \frac{1}{2} \widetilde{\rho}_v^T R_v^{-1} \widetilde{\rho}_v. \tag{17}$$

The time derivative of V along the trajectory of (16) is given by

$$\dot{V} = E^T\left(-KE - \tilde{W}^T H + M - B\right) + (bN+1)\dot{\omega} + \tilde{\rho}_w^T R_w^{-1}\dot{\hat{\rho}}_w + \tilde{\rho}_v^T R_v^{-1}\dot{\hat{\rho}}_v.$$

By completing the squares

$$\tilde{\rho}_{wi}(\hat{\rho}_{wi} - \rho_{wi}^0) = \frac{1}{2}(\hat{\rho}_{wi} - \rho_{wi})^2 + \frac{1}{2}(\hat{\rho}_{wi} - \rho_{wi}^0)^2 - \frac{1}{2}(\rho_{wi} - \rho_{wi}^0)^2,$$

$$\tilde{\rho}_{vi}(\hat{\rho}_{vi} - \rho_{vi}^0) = \frac{1}{2}(\hat{\rho}_{vi} - \rho_{vi})^2 + \frac{1}{2}(\hat{\rho}_{vi} - \rho_{vi}^0)^2 - \frac{1}{2}(\rho_{vi} - \rho_{vi}^0)^2,$$

we obtain

$$\dot{V} \leq -E^T KE - \sum_{i=1}^{n}\left(\frac{\sigma_{wi}}{2r_{wi}}\tilde{\rho}_{wi}^2 + \frac{\sigma_{vi}}{2r_{vi}}\tilde{\rho}_{vi}^2\right) + (bN(\omega)+1)\dot{\omega} + \varsigma,$$

where $\varsigma = \sum_{i=1}^{n}\left(\rho_{wi}\varepsilon_{wi} + \rho_{vi}\varepsilon_{vi} + \frac{\sigma_{wi}}{2r_{wi}}(\rho_{wi} - \rho_{wi}^0)^2 + \frac{\sigma_{vi}}{2r_{vi}}(\rho_{vi} - \rho_{vi}^0)^2\right)$. Thus

$$\dot{V} \leq -\lambda V + \varsigma + (bN(\omega)+1)\dot{\omega}, \tag{18}$$

where $\lambda = \min\{2k_1,\cdots,2k_n,\sigma_{w1},\cdots,\sigma_{wn},\sigma_{v1},\cdots,\sigma_{vn}\}$.

Solving the inequality (18) yields

$$0 \leq V(t) \leq \frac{\varsigma}{\lambda} + e^{-\lambda t}V(0) + \int_0^t (bN(\omega)+1)\dot{\omega}e^{-\lambda(t-\tau)}d\tau, \quad \forall \ t \geq 0. \tag{19}$$

We first show that $\omega(t)$ is bounded on $[0, t_f)$ by seeking a contradiction. Define $P(\omega(0),\omega(t)) = \int_0^t (bN(\omega)+1)e^{-\lambda(t-\tau)}d\omega(\tau)$, $\forall t \in [0, t_f)$. Two cases need to be considered: 1) $\omega(t)$ has no upper bound and 2) $\omega(t)$ has no lower bound.

Case 1: Suppose that $\omega(t)$ has no upper bound on $[0, t_f)$, i.e., there exists a monotone increasing sequence $\{\omega(t_i)\}$ such that $\lim_{i\to\infty}\omega(t_i) = +\infty$ as $\lim_{i\to\infty} t_i = t_f$.

First, we consider the case $b > 0$. Suppose that $4M + 1 > |\omega(0)|$, where M is an integer. We have

$$P(\omega(0), 4M+1) = \int_0^{t_1}(bN(\omega)+1)e^{-\lambda(t_1-\tau)}d\omega(\tau) \tag{20}$$
$$\leq (be^{(4M+1)^2} + 1)(4M+1-\omega(0)).$$

Noting that $N(\omega)$ is negative on intervals $[4M+1, 4M+3]$, thus

$$P(4M+1, 4M+3) = \int_{t_1}^{t_2}(bN(\omega)+1)e^{-\lambda(t_2-\tau)}d\omega(\tau)$$

$$\le be^{-\lambda(t_2-t_1)}\int_{4M+1.5}^{4M+2.5}N(\sigma)d\sigma+2 \le -\sqrt{2}/2 be^{-\lambda(t_2-t_1)+(4M+1.5)^2}+2. \tag{21}$$

According to (20), (21), we have

$$P(\omega(0),4M+3)\le$$
$$e^{(4M+1)^2}\left(-\sqrt{2}/2 be^{-\lambda(t_2-t_1)+4M+1.25}+b(4M+1-\omega(0))+(4M+3-\omega(0))e^{-(4M+1)^2}\right).$$

Hence, $P(\omega(0),4M+3)\to -\infty$ as $M\to\infty$, which yields a contradiction in (19).

Then, we consider the case $b<0$. Suppose that $4M-1>|\omega(0)|$, then

$$P(\omega(0),4M-1)=\int_0^{t_1}(bN(\omega)+1)e^{-\lambda(t_1-\tau)}d\omega(\tau)$$
$$\le \left(|b|e^{(4M-1)^2}+1\right)(4M-1-\omega(0)). \tag{22}$$

Noting that $N(\omega)$ is positive on intervals $[4M-1, 4M+1]$, thus

$$P(4M-1,4M+1)=\int_{t_1}^{t_2}(bN(\omega)+1)e^{-\lambda(t_2-\tau)}d\omega(\tau)$$
$$\le be^{-\lambda(t_2-t_1)}\int_{4M-0.5}^{4M+0.5}N(\sigma)d\sigma+2 \le \sqrt{2}/2 be^{-\lambda(t_2-t_1)+(4M-0.5)^2}+2. \tag{23}$$

According to (22), (23), we have

$$P(\omega(0),4M+1)\le$$
$$e^{(4M-1)^2}\left(\sqrt{2}/2 be^{-\lambda(t_2-t_1)+4M-0.75}+|b|(4M-1-\omega(0))+(4M+1-\omega(0))e^{-(4M-1)^2}\right).$$

Hence, $P(\omega(0),4M+1)\to -\infty$ as $M\to\infty$, which yields a contradiction in (19).

Thus, we conclude that $\omega(t)$ is upper bounded on $[0,t_f)$.

Case 2: Employing the similar method as in case 1, we can prove that $\omega(t)$ is lower bounded on $[0,t_f)$.

Therefore, $\omega(t)$ must be bounded on $[0,t_f)$. As an immediate result, $V(t)$ is bounded on $[0,t_f)$. All the signals in the closed-loop system are bounded, then $t_f\to\infty$. This proves that all the signals of the closed-loop system are UUB, i.e., for any compact set $\Omega_n\subset R^n$, there exist a controller such that as long as $x(0)\in\Omega_n$, e_i, $\hat{\rho}_{wi}$, and $\hat{\rho}_{vi}$ are UUB, respectively. Correspondingly, the tracking error $e_1(t)$ satisfies

$$|e_1(t)|\le\sqrt{2\left(\varsigma/\lambda+\int_0^t(bN(\omega)+1)\dot{\omega}e^{-\lambda(t-\tau)}d\tau+e^{-\lambda t}V(0)\right)}.$$

Thus, by appropriately adjusting the design parameters, the tracking error $e_1(t)$ can be kept as small as possible.

Remark 1. For the low order system with mismatched uncertainties and known sign of control gain, we present a new RAFTC algorithm. In order to avoid the possible divergence of the on-line tuning of FLS, a new adaptive law is proposed to make sure that the FLS parameter vectors are tuned within a prescribed range. By a special design technique, the controller singularity problem is avoided.

Remark 2. For the high order system with mismatched uncertainties and unknown sign of control gain, we present a new RFTC algorithm. The main feature of the algorithm is the adaptive mechanism with minimal learning parameters. The online computation burden is kept to minimum. By employing Nussbaum gain design technique, the controller singularity problem is avoided perfectly.

Remark 3. Both the algorithms are very suitable for practical implementation. The controllers and the parameter adaptive laws are highly structural. Such a property is particularly suitable for parallel processing and hardware implementation in practical applications.

5 Conclusions

In this note, the tracking control problem has been considered for a class of nonlinear systems with mismatched uncertainties, transformable to the strict-feedback form. By combining backstepping design technique with fuzzy set theory, RAFTC for the low order system with known sign of the control gain and RFTC for the high order system with unknown sign of the control gain are proposed. Both algorithms completely avoid the controller singularity problem. The proposed controllers are highly structural and particularly suitable for parallel processing in the practical applications. Furthermore, the RFTC algorithm has the adaptive mechanism with minimal learning parameterizations. It is shown the proposed algorithms can guarantee the semi-globally uniform ultimate boundedness of all the signals in the closed-loop system. In addition, the tracking error can be reduced to arbitrarily small values by suitably choosing the design parameters.

References

1. Lee, H., Tomizuka, M.: Robust Adaptive Control Using a Universal Approximator for SISO Nonlinear Systems. IEEE Trans. Fuzzy Syst. 8 (2000) 95-106
2. Jagannathan, S., Lewis, F.L.: Robust Backstepping Control of a Class of Nonlinear Systems Using Fuzzy Logic. Inform. Sci. 123 (2000) 223-240
3. Wang, W.Y., Chan, M.L., Lee, T.T., Liu, C.H.: Adaptive Fuzzy Control for Strict-Feedback Canonical Nonlinear Systems with $H\infty$ Tracking Performance. IEEE Trans. Syst. Man, Cybern. 30 (2000) 878-885
4. Yang, Y.S., Feng, G., Ren, J.: A Combined Backstepping and Small-Gain Approach to Robust Adaptive Fuzzy Control for Strict-Feedback Nonlinear Systems. IEEE Trans. Syst. Man, Cybern. 34 (2004) 406-420
5. Polycarpou, M.M., Ioannou, P.A.: A Robust Adaptive Nonlinear Control Design. Automatica. 32 (1996) 423-427

6. Nussbaum, D.R.: Some Remarks on a Conjecture in Parameter Adaptive Control. Syst. Contr. Lett. 3 (1983) 243-246
7. Mudgett, D.R., Morse, A.S.: Adaptive Stabilization of Linear Systems with Unknown High Frequency Gains. IEEE Trans. Automat. Contr. 30 (1985) 549-554
8. Ge, S.S., Wang, J.: Robust Adaptive Neural Control for a Class of Perturbed Strict Feedback Nonlinear Systems. IEEE Trans. Neural Networks. 13 (2002) 1409-1419
9. Ding, Z.T.: Adaptive Control of Nonlinear System with Unknown Virtual Control Coefficients. Int. J. Adapt. Control Signal Process. 14 (2000) 505-517

Intelligent Fuzzy Systems for Aircraft Landing Control

Jih-Gau Juang[1], Bo-Shian Lin[1], and Kuo-Chih Chin[2]

[1] Department of Communication and Guidance Engineering
National Taiwan Ocean University, Keelung 20224, Taiwan
jgjuang@mail.ntou.edu.tw
m92670004@mail.ntou.edu.tw
[2] ASUSTeK Computer Inc,
Taipei 101, Taiwan
a8226451@ms33.hinet.net

Abstract. The purpose of this paper is to investigate the use of evolutionary fuzzy neural systems to aircraft automatic landing control and to make the automatic landing system more intelligent. Three intelligent aircraft automatic landing controllers are presented that use fuzzy-neural controller with BPTT algorithm, hybrid fuzzy-neural controller with adaptive control gains, and fuzzy-neural controller with particle swarm optimization, to improve the performance of conventional automatic landing system. Current flight control law is adopted in the intelligent controller design. Tracking performance and adaptive capability are demonstrated through software simulations.

1 Introduction

In a flight, take-off and landing are the most difficult operations in regard to safety issues. The automatic landing system of an airplane is enabled only under limited conditions. If severe wind disturbances are encountered, the pilot must handle the aircraft due to the limits of the automatic landing system. The first Automatic Landing System (ALS) was made in 1965. Since then, most aircraft have been installed with this system. The ALS relies on the Instrument Landing System (ILS) to guide the aircraft into the proper altitude, position, and approach angle during the landing phase. Conventional automatic landing systems can provide a smooth landing which is essential to the comfort of passengers. However, these systems work only within a specified operational safety envelope. When the conditions are beyond the envelope, such as turbulence or wind shear, they often cannot be used. Most conventional control laws generated by the ALS are based on the gain scheduling method [1]. Control parameters are preset for different flight conditions within a specified safety envelope which is relatively defined by Federal Aviation Administration (FAA) regulations. According to FAA regulations, environmental conditions considered in the determination of dispersion limits are: headwinds up to 25 knots; tailwinds up to 10 knots; crosswinds up to 15 knots; moderate turbulence, wind shear of 8 knots per 100 feet from 200 feet to touchdown [2]. If the flight conditions are beyond the preset envelope, the ALS is disabled and the pilot takes over. An inexperienced pilot may not be able to guide the aircraft to a safe landing at airport.

China Airlines Flight 642 had a hard landing at Hong Kong International Airport on 22 August 1999. The lifting wing was broken during the impact that killed 3 passengers and injured 211 people. After 15 months investigation, a crash report was released on November 30, 2000. It showed that the crosswind-correction software 907 on the Boeing MD-11 had a defect. Boeing also confirmed this software problem later and replaced nearly 190 MD-11's crosswind-correction software with the 908 version. According to Boeing's report [3], 67% of the accidents by primary cause are due to human factors and 5% are attributed to weather factors. By phase of flight, 47% accidents are during final approach or landing. It is therefore desirable to develop an intelligent ALS that expands the operational envelope to include more safe responses under a wider range of conditions. The goal of this study is that the proposed intelligent automatic landing controllers can relieve human operators and guide the aircraft to a safe landing in wind disturbance environment.

In this study, robustness of the proposed controller is obtained by choosing optimal control gain parameters that allows wide range of disturbances to the controller. In 1995, Kennedy and Eberhart presented a new evolutionary computation algorithm, the real-coded Particle Swarm Optimization (PSO) [4]. PSO is one of the latest population-based optimization methods, which dose not use the filtering operation (such as crossover and mutation) and the members of the entire population are maintained through the search procedure. This method was developed through the simulation of a social system, and has been found to be robust in solving continuous nonlinear optimization problems [5-7]; they are suitable for determination of the control parameters which give aircraft better adaptive capability in severe environment.

Recently, some researchers have applied intelligent concepts such as neural networks and fuzzy systems to intelligent landing control to increase the flight controller's adaptively to different environments [8-12]. Most of them do not consider robustness of controller due to wind disturbances [8-10]. In [11], a PD-type fuzzy control system is developed for automatic landing control of both a linear and a nonlinear aircraft model. Adaptive control for a wide range of initial conditions has been demonstrated successfully. The drawback is that the authors only set up the wind disturbance at the initial condition. Persistent wind disturbance is not considered. In [12] wind disturbances are included but the neural controller is trained for a specific wind speed. Robustness for a wide range of wind speeds has not considered. In previous works [13-14], we have utilized neural networks to automatic landing control. Environment adaptive capability has been improved but the rate of convergence is very slow. Here, we present three learning schemes, fuzzy-neural controller with BPTT algorithm, fuzzy-neural controller with adaptive control gains, and fuzzy-neural controller with PSO algorithm, to guide the aircraft to a safe landing and make the controller more robust and adaptive to the ever-changing environment.

2   Aircraft Landing System

The pilot descends from the cruise altitude to an altitude of approximately 1200ft above the ground. The pilot then positions the airplane so that the airplane is on a heading towards the runway centerline. When the aircraft approaches the outer airport marker, which is about 4 nautical miles from the runway, the glide path signal is intercepted (as shown in Fig. 1). As the airplane descends along the glide path, its pitch, attitude and speed must be controlled. The aircraft maintains a constant speed along the flight path. The descent rate is about 10ft/sec and the pitch angle is between -5 to +5 degrees. Finally, as the airplane descends 20 to 70 feet above the ground, the glide path control system is disengaged and a flare maneuver is executed. The vertical descent rate is decreased to 2ft/sec so that the landing gear may be able to dissipate the energy of the impact at landing. The pitch angle of the airplane is then adjusted, between 0 to 5 degrees for most aircraft, which allows a soft touchdown on the runway surface.

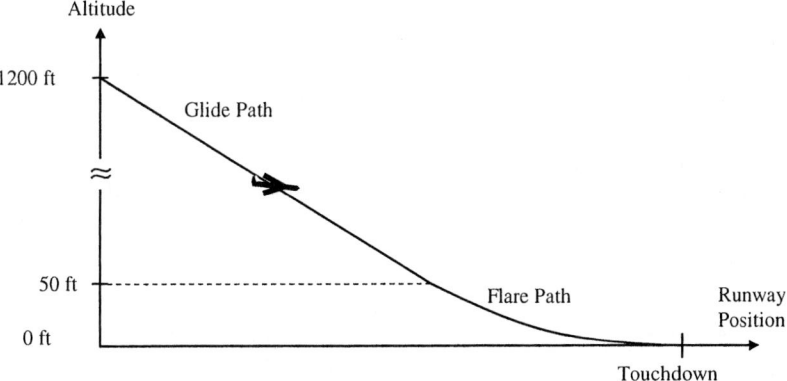

Fig. 1. Glide path and flare path

A simplified model of a commercial aircraft that moves only in the longitudinal and vertical plane is used in the simulations for implementation ease [12]. To make the ALS more intelligent, reliable wind profiles are necessary. Two spectral turbulence forms modeled by von Karman and Dryden are mostly used for aircraft response studies. In this study the Dryden form [12] was used for its demonstration ease. Figure 2 shows a turbulence profile with a wind speed of 30 ft/sec at 510 ft altitude.

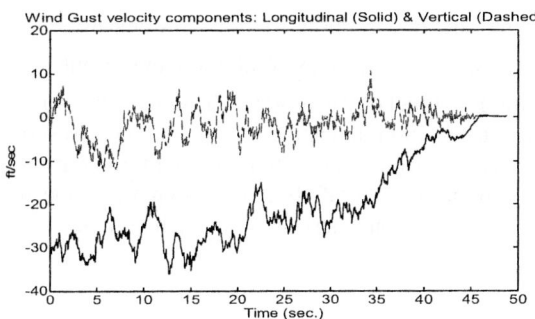

Fig. 2. Turbulence profile

3 Landing Controller Design

In this study, the aircraft maintains a constant speed alone the flight path, we assumed that the change in throttle command is zero. The aircraft is thus controlled solely by the pitch command. Detailed descriptions can be found in [12].

3.1 Fuzzy-Neural Controller with BPTT Algorithm

In this section, we first design and analyze the performance of the fuzzy-neural controller for auto-landing in severe wind condition. The learning process is shown in Fig. 3, where AM is the aircraft model, FNNC is the fuzzy modeling neural network controller, and LIAM is the linearized inverse aircraft model [13]. Every learning cycle consists of all stages from S_0 to S_k. Weight changes in the FNNC are updated using a batch model. The controller is trained by Backpropagation through time (BPTT) algorithm. The inputs for the fuzzy-neural controller are: altitude, altitude command, altitude rate, and altitude rate command. The output of the controller is the pitch command. Detail structure of the fuzzy modeling network can be found in [14-15]. The fuzzy neural controller starts learning without any control rule. The LIAM calculates the error signals that will be used to back propagate through the controller in each stage [13].

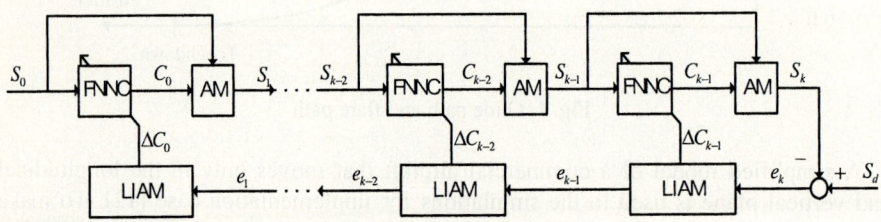

Fig. 3. Learning Process

In the simulations, successful touchdown landing conditions are defined as follows

$-3 \leq \dot{h}(T)$ ft/sec ≤ 0, $200 \leq \dot{x}(T)$ ft/sec ≤ 270

$-300 \leq x(T)$ ft ≤ 1000, $-1 \leq \theta(T)$ degree ≤ 5

where T is the time at touchdown. Initial flight conditions are: $h(0)=500$ ft, $\dot{x}(0) =235$ ft/sec, $x(0) =9240$ ft, and $\gamma_o =-3$ degrees. With the wind turbulence speed at 30 ft/sec, the horizontal position at touchdown is 418.5 ft, horizontal velocity is 234.7 ft/sec, vertical speed is -2.7 ft/sec, and pitch angle is -0.08 degrees, as shown in Fig. 4 to Fig. 6. Table 1 shows the results from using different wind turbulence speeds. The controller can successfully guide the aircraft flying through wind speeds of 0 ft/sec to 45 ft/sec while the conventional controller can only reach 30 ft/sec [12].

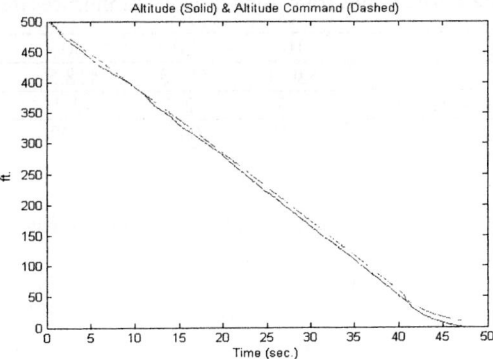

Fig. 4. Aircraft altitude and command

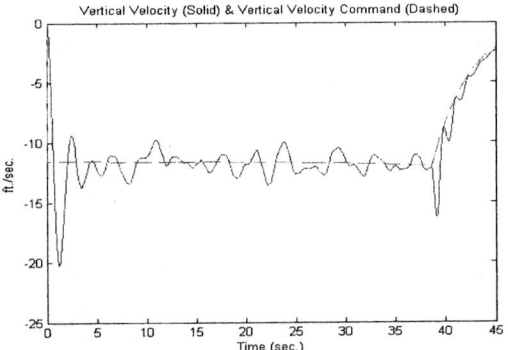

Fig. 5. Aircraft vertical velocity and command

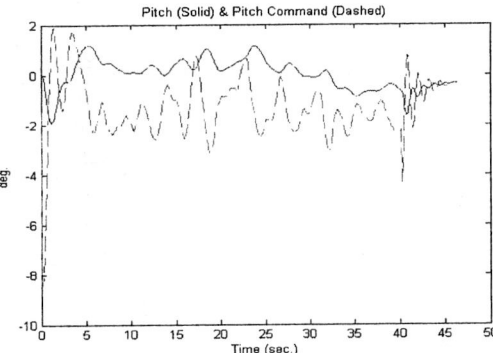

Fig. 6. Aircraft pitch angle and command

Table 1. The results from using different turbulence strength

Wind speed (ft/sec)	10	20	30	40	45
Landing point (ft)	580.3	541.8	418.5	457.3	247.9
Aircraft vertical Speed (ft/sec)	-2.2	-2.3	-2.7	-2.3	-2.9
Pitch angle (degree)	-0.81	-0.63	-0.08	-0.06	0.35

3.2 Fuzzy-Neural Controller with Adaptive Control Gains

In previous section the control gains of the pitch autopilot in glide-slope phase and flare phase are fixed. Robustness of the fuzzy-neural controller is achieved by the BPTT training scheme. In this section, a neural network generates adaptive control gains for the pitch autopilot. Different wind disturbances are the inputs of the neural network, as in Fig. 7. The fuzzy neural controller is trained by BP instead of BPTT. With the wind turbulence speed at 50 ft/sec, the horizontal position at touchdown is 547 ft, horizontal velocity is 235 ft/sec, vertical speed is –2.5 ft/sec, and pitch angle is 0.34 degrees, as shown in Fig. 8 to Fig. 10. Table 2 shows the results from using different wind turbulence speeds. The controller can successfully guide the aircraft flying through wind speeds of 34 ft/sec to 58 ft/sec.

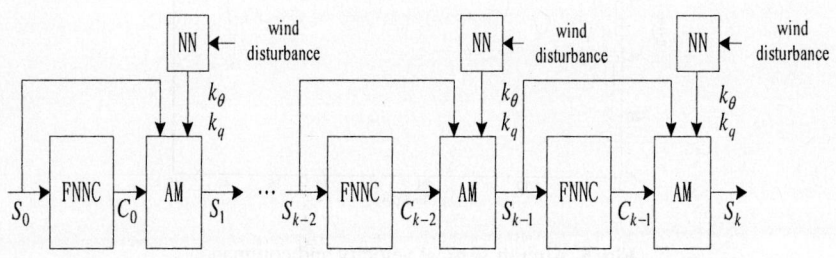

Fig. 7. Learning process – with wind disturbance NN

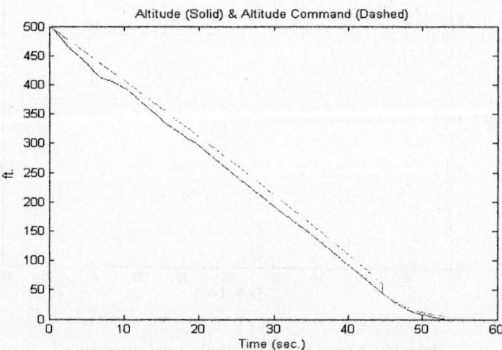

Fig. 8. Aircraft altitude and command

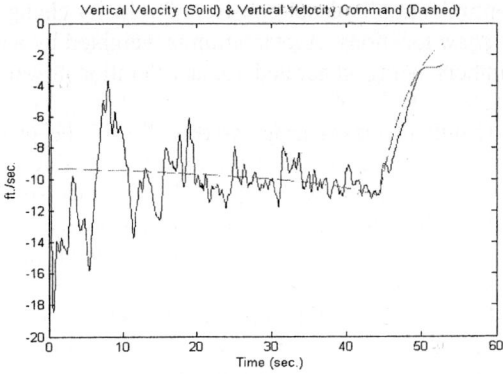

Fig. 9. Aircraft vertical velocity and command

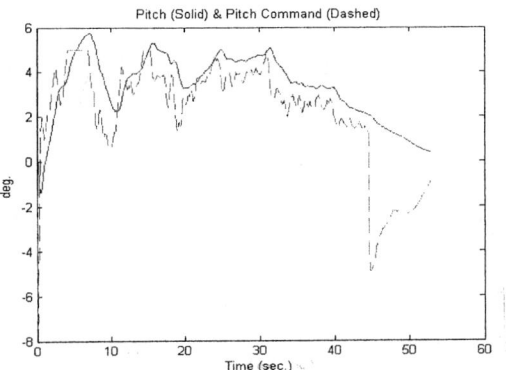

Fig. 10. Aircraft pitch angle and command

Table 2. The results from using different turbulence strength

Wind speed (ft/sec)	34	40	45	50	58
Landing point (ft)	427.9	346.8	853.5	547.5	943.7
Aircraft vertical Speed (ft/sec)	-2.9	-2.9	-2.8	-2.5	-2.8
Pitch angle (degree)	0.15	0.10	0.20	0.34	0.25

3.3 Fuzzy-Neural Controller with Particle Swarm Optimization

In the PSO algorithm, each member is called "particle", and each particle flies around in the multi-dimensional search space with a velocity, which is constantly updated by the particle's own experience and the experience of the particle's neighbors or the experience of the whole swarm. Each particle keeps track of its coordinates in the problem space, which are associated with the best solution (fitness) it has achieved so far. This value is called *pbest*. Another best value that is tracked by the global version of the particle swarm optimizer is the overall best value, and its location, obtained so far by any particle in the population. This location is called *gbest*. At each time step, the particle swarm optimization concept consists of velocity changes of each particle toward its *pbest* and *gbest* locations. Acceleration is weighted by a random term, with separate random numbers being generated for acceleration toward *pbest* and *gbest* locations.

The turbulence strength increases progressively during the process of parameter search. The purpose of this procedure is to search more suitable control gains for $k\theta$ and kq in glide path and flare path. With the wind turbulence speed at 50 ft/sec, the horizontal position at touchdown is 505 ft, horizontal velocity is 235 ft/sec, vertical speed is -2.7 ft/sec, and pitch angle is 0.21 degree, as shown in Fig. 11 to Fig. 13. Table 3 shows the results from using different wind turbulence speeds. The controller can successfully guide the aircraft flying through wind speeds of 30 ft/sec to 90ft/sec. With the same wind turbulence speed at 50 ft/sec, Fig. 14 shows the absolute error for height using fuzzy-neural controller with adaptive control gains (dash line) and hybrid fuzzy-neural controller with PSO (solid line), respectively. It indicates that the performance of hybrid fuzzy-neural controller with PSO is much better.

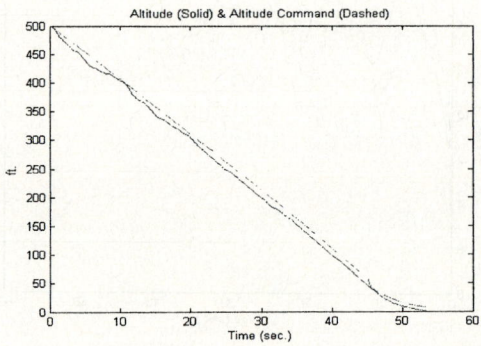

Fig. 11. Aircraft altitude and command

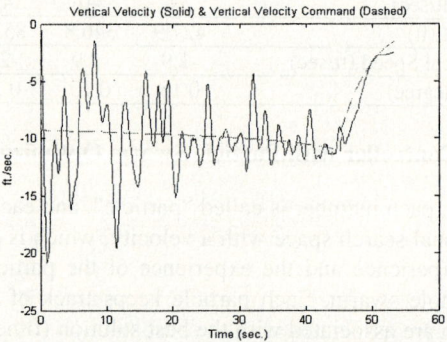

Fig. 12. Aircraft vertical velocity and command

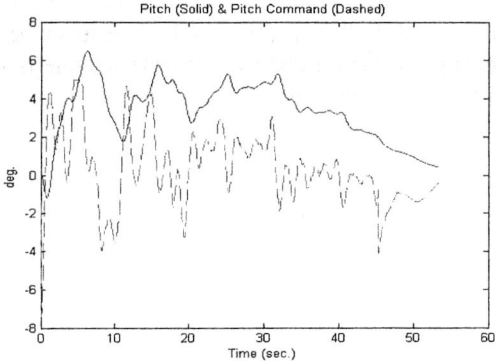

Fig. 13. Aircraft pitch angle and command

Table 3. The results from using different turbulence strength

Wind speed (ft/sec)	30	40	50	70	90
Landing point (ft)	467.2	403.7	505.3	543.0	771.4
Aircraft vertical Speed (ft/sec)	-2.4	-2.5	-2.7	-2.5	-2.8
Pitch angle (degree)	-0.18	0.06	0.21	0.91	1.31

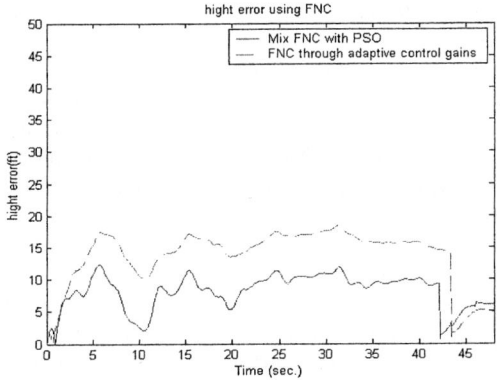

Fig. 14. Performance of Fuzzy-Neural Controller

4 Conclusions

For the safe landing of an aircraft with a conventional controller, the wind speed limit of turbulence is 30 ft/sec. In this study, Control gains are selected by a combination method of a nonlinear control design, a neural network, and particle swarm optimization. Comparisons on different control schemes are given. The hybrid fuzzy-neural controller with adaptive control gains can overcome turbulence to 58 ft/sec. The fuzzy-neural controller with BPTT and the fuzzy-neural controller with PSO algorithm can reach 45 ft/sec and 90 ft/sec, respectively. The fuzzy-neural controller with PSO algorithm has best performance and the convergence rate is also improved. The purpose of this paper has been achieved.

Acknowledgement

This work was supported by the National Science Council, Taiwan, ROC, under Grant NSC 93-2213-E-019 -007.

References

1. Buschek, H., Calise, A.J.: Uncertainty Modeling and Fixed-Order Controller Design for a Hypersonic Vehicle Model. Journal of Guidance, Control, and Dynamics. 20 (1997) 42-48
2. Federal Aviation Administration: Automatic Landing Systems. AC 20-57A (1971)
3. Boeing Publication: Statistical Summary of commercial Jet Airplane Accidents. Worldwide Operations 1959-1999. (2000)
4. Kennedy, J., Eberhart, R.C.: Particle Swarm Optimization. Proceedings of IEEE International Conference on Neural Networks. 4 (1995) 1942-1948
5. Shi, Y., Eberhart, R. C.: Empirical Study of Particle Swarm Optimization. Proceedings of the 1999 Congress on Evolutionary Computation. (1999) 1945-1950
6. Angeline, P. J.: Using Selection to Improve Particle Swarm Optimization. Proceedings of IEEE International Conference on Evolutionary Computation. (1998) 84-89
7. Zheng, Y.L., Ma, L., Zhang, L., Qian, J.: On the Convergence Analysis and Parameter Selection in Particle Swarm Optimization. Proceedings of the Second IEEE International Conference on Machine Learning and Cybernetics. (2003) 1802-1807
8. Izadi, H., Pakmehr, M., Sadati, N.: Optimal Neuro-Controller in Longitudinal Autolanding of a Commercial Jet Transport. Proceedings of IEEE International Conference on Control Applications. (2003) 1-6
9. Chaturvedi, D.K., Chauhan, R., Kalra, P.K.: Application of Generalized Neural Network for Aircraft Landing Control System. Soft Computing. 6 (2002) 441-118
10. Ionita, S., Sofron, E.: The Fuzzy Model for Aircraft Landing Control. Proceedings of AFSS International Conference on Fuzzy Systems. (2002) 47-54
11. Nho, K., Agarwal, R.K.: Automatic Landing System Design Using Fuzzy Logic. Journal of Guidance, Control, and Dynamics. 23 (2000) 298-304
12. Jorgensen, C.C., Schley, C.: A Neural Network Baseline Problem for Control of Aircraft Flare and Touchdown. Neural Networks for Control. (1991) 403-425
13. Juang, J.G., Chang, H.H., Chang, W.B.: Intelligent Automatic Landing System Using Time Delay Neural Network Controller. Applied Artificial Intelligence. 17 (2003) 563-581
14. Juang, J.G.: Fuzzy Neural Networks Approaches for Robotic Gait Synthesis. IEEE Transactions on Systems Man and Cybernetics—Part B: Cybernetics. 30 (2000) 594-601
15. Horikawa, S., Furuhashi, T., Uchikawa, Y.: On Fuzzy Modeling Using Fuzzy Neural Networks with the Back-Propagation Algorithm. IEEE Transactions on Neural Networks. 3 (1992) 801-806

Scheduling Design of Controllers with Fuzzy Deadline[*]

Hong Jin, Hongan Wang, Hui Wang, and Danli Wang

Institute of Software, Chinese Academy of Sciences. Beijing 100080
{hjin, wha, hui.wang, dlwang}@iel.iscas.ac.cn

Abstract. Because some timing-constraints of a controller task may be not determined as a real-time system engineer thinks of, its scheduling with uncertain attributes can not be usually and simply dealt with according to classic manners used in real-time systems. The model of a controller task with fuzzy deadline and its scheduling are studied. The dedication concept and the scheduling policy of largest dedication first are proposed first. Simulation shows that the scheduling of controller tasks with fuzzy deadline can be implemented by using the proposed method, whilst the control performance cost gets guaranteed.

1 Introduction

Existing academic researches on co-designing of control and scheduling assume to know all timing-constraints (*e.g.*, sampling period, deadline, etc.) [1][2][3][4]. However, some timing-constraints of a control loop are usually dependent on requirements for controlled process dynamics and loop performance, *e.g.*, the imprecise clock, overload or computer fault can lead to a relative change for the sampling period all. Moreover, computer hardware, control algorithm, real-time operating system, scheduling algorithm and network delay can cause delay and jitter in control system all [3].

The controller tasks with fuzzy deadline usually appear in control problems although a control engineer does not care of the implementing process of controller task in computer-controlled systems. Lingual terms can be used to describe timing-constraints, *e.g.*, the response time of less than 0.1 seconds and the sampling period of 2 seconds. The later means that 2±0.1% seconds can be considered to be correct and acceptable. Another natural comprehend is that the deadline is admitted to have a small variant around the sampling period which is precise [5], and the small variant can be considered as random, uncertainty or fuzzy. So, it is important and meaningful to study the scheduling of control tasks with uncertain/fuzzy attributes in computer-controlled systems. However, existing researches on the scheduling of tasks with uncertain/fuzzy attributes have not considered control tasks as their scheduling objects besides of the performance cost of a control system [5][6].

Under the precondition of assuring control performance, how to use limited computing resources to achieve the scheduling of controller tasks with fuzzy deadline is the problem cared in this paper. The time-driven sampling is considered. The proposed scheduling algorithm of largest dedication first is introduced in Section 3.

---

[*] This work is supported by China NSF under Grant No. 60374058, 60373055 and 60373056.

2 Controller Model

2.1 Controller Task

In computer-controlled systems [1][3], the dispersion of the reference input $r(t)$ with the system output $y(t)$ is used as the input of a controller which recalculates the manipulated variable used as the input of the controlled system $G(s)$ for every P seconds. The control aim is to make $y(t)$ approximate to $r(t)$ as quickly as possible. The state update of the manipulated variable (Update State) and the calculation of system output (Calculate Output) will make up of the close loop of the whole system.

In the co-design of control and scheduling, the simplest model assumption about a controller task, T_i, is that T_i is periodic, and has a fixed sampling period P_i, a known worst cast execute time and a hard deadline d_i. And its release/arrive time is assumed to be zero in this paper. Moreover, the execute time of T_i is denoted as C_i. In following discussion, the deadline d_i is assumed to be fuzzy uncertain.

2.2 Fuzzy Deadline

For fuzzy description of deadline, which can take any value in some interval with a probability, there are many continuous fuzzy numbers to be referred, *e.g.*, trapezoid/triangle or truncated normal fuzzy number. The continuous trapezoid deadline is used here [6]. Let d_i=*Trapezoid*(a_i,e_i,f_i,b_i), where $a_i \le e_i \le f_i \le b_i \le P_i$. Let $\mu_i(t)$ be $[a_i,b_i]$-cut trapezoid membership function of d_i, then, $\mu_i(t)$ is equal to $(t-a_i)h_i/(e_i-a_i)$ for $a_i<t\le e_i$; h_i for $e_i<t\le f_i$; $(b_i-t)h_i/(b_i-f_i)$ for $f_i<t\le b_i$; 0 for others.

And d_i is also called as $[a_i,b_i]$-cut fuzzy deadline of T_i, where, a_i and b_i are called as the earliest deadline and latest deadline of T_i respectively in this paper. When $h_i=2/[(b_i-a_i)+(f_i-e_i)]$, $\int\mu_i(t)dt=1$. Especially, it becomes the triangle fuzzy number as $e_i=f_i$ or the fuzzy number with the uniform distribution as $a_i=e_i$ and $f_i=b_i$.

3 Largest Dedication First

The proposed scheduling policy, called as Largest Dedication First (*LDF*) in this paper, uses the largest dedication principle to assign the priority for a controller task.

(1) Priority-driven scheduling policy is used, and a task with high dedication can preempt a low dedication task. (2) If two tasks have the same dedication, their latest deadlines are compared and the task with smaller latest deadline will get scheduled first. (3) As a task has not missed its latest deadline, it will have dedication space still and waits for its next scheduling; otherwise, it will be aborted if it has not finished.

3.1 Dedication Function

A fuzzy deadline, varying between its earliest deadline and latest deadline randomly, may influence the dedication index (*DIN*: Dedication INdex). Let F_i be the finish time of T_i, $\forall x \in R^+$, the dedication function of T_i can be defined as follows:

$$Ded_i(x) \equiv Ded_i\{F_i \le x\} \equiv \int_{-\infty}^{x}\mu_i(F)dF \Big/ \int_{-\infty}^{+\infty}\mu_i(F)F$$

Let $C_i(t)$ be the remained execution time of T_i at current time t, for the ideal case that no other tasks preempt resource, T_i will finish at $t+C_i(t)$, which is used instead of x in above integral to calculate the dedication index of T_i: $DIN_i(t)=Ded_i(t+C_i(t))$, which is relative to both the membership function and the remained executing time.

3.2 Priority Assignment

The principle of the largest dedication is that the task with larger dedication value is assigned higher priority level. For any two time t_1 and t_2 ($t_2>t_1$), then $C_i(t_1)-C_i(t_2)\leq t_2-t_1$ or $t_2+C_i(t_2)\geq t_1+C_i(t_1)$. Because $Ded_i(x)$ is an increasing function, then $Ded_i(t_2+C_i(t_2))\geq Ded_i(t_1+C_i(t_1))$ or $DIN_i(t_2)\geq DIN_i(t_1)$, i.e., $DIN_i(t)$ is an increasing function also. Thus, the dedication index can be used as the priority, $p_i(t)$, of a task, i.e., $p_i(t)=DIN_i(t)$, which means that the larger the dedication is, the higher the priority will be.

4 Simulation

Consider a computer-controlled system composed of two servo systems. Each one, described by using the transfer function of $G(s)=1000/[s(s+1)]$, is controlled by a *PD* controller which discrete control algorithm form is [1]: $u(t)=Prob(t)+Der(t)$, where $Prob(t)=K(r(t)-y(t))$, and $Der(t)=\alpha Der(t-P)+\beta[y(t-P)-y(t)]$, $\alpha=M/(NP+M)$, $\beta=NKM/(NP+M)$, K and M are control parameters, N is constant, P is the sampling period.

For Controller 1 (T_1): $P_1=12$ms, $C_1=6$ms, $d_1=Trapezoid(7,8,9,10)$(ms), $K=1$, $M=0.042$, $N=100$; for Controller 2 (T_2): $P_2=10$ms, $C_2=5$ms, $d_2=Trapezoid(6,7,7,8)$ (ms), $K=1.1$, $M=0.035$, $N=50$. Moreover, the simulation time is 2s. Consider a step-function reference input, e.g., $r(t)=1$ as $t\leq 0.5$s or $1s<t\leq 1.5$s, otherwise, $r(t)=-1$.

4.1 Performance Indexes

An integrated control performance cost (*CPC*: Control Performance Cost) can be described as follows: $CPC=\Sigma w_i CPC_i$, where $CPC_i=\int[y_{ideal,i}(t)-y_{actual,i}(t)]^2 dt$ and w_i are the control performance cost and weighed coefficient respectively of the ith control system; $y_{ideal,i}/y_{actual,i}$ is the ideal/actual output of the ith control system.

The performance index (efficient utilization) is defined as the ratio of occupied CPU time by instances met their latest deadlines with the total simulation time *100%.

4.2 Control Performance comparison

Because of the fuzzy uncertainty of d_i, the latest deadline b_i can be used instead of d_i, however the yielded lower utilization bound under the most optimistical condition is larger than 1 still so that the system is overload and no scheduling algorithm can guarantee all instances to meet their latest deadlines.

Due to the limit of space, simulation figures are omitted here. By using *LDF*, System 2 receives the ideal design result ($CPC_2=0$), System 1 has large control performance cost ($CPC_1=55.2715$), and the efficient utilization of CPU reaches to 79.70%.

5 Conclusion

For uncertainty controller tasks with fuzzy deadline, usual dynamic scheduling policies have become inapplicable. By introducing the dedication concept first, the scheduling policy of the largest dedication first is then presented also, which assures to schedule tasks with large dedication first. Simulation shows that the largest dedication first can solve the scheduling problem of controller tasks with fuzzy deadline very well, meanwhile, the control performance cost of every subsystem can get controlled or balanced and the performance stability of the control system can get guaranteed.

References

1. Arzen, K.E., Cervin, A., Eker, J., Sha, L.: An introduction to control and scheduling co-design. *Proc. of the 39th IEEE Conference on Decision and Control.* Sydney, NSW Australia. Piscataway: IEEE Computer Society. Vol.5 (2000) 4865-4870
2. Lin, Q., Chen, P.C.Y., Neow, P.A.: Dynamical scheduling of digital control systems. *Proc. of IEEE International Conference on Systems, Man and Cybernetics.* Washington, D.C., USA. Piscataway: IEEE Computer Society. Vol.5 (2003) 4098-4103
3. Cervin, A., Henriksson, D., Lincoln, B., Eker, J., Arzen, K.E.: How does control timing affect performance? Analysis and simulation of timing using Jitterbug and TrueTime. IEEE Control Systems Magazine, Vol.23, No.3 (2003) 16-30
4. Ryu, M., Hong, S.: Toward automatic synthesis of schedulable real-time controllers. Integrated Computer-Aided Engineering, Vol.5, No.3 (1998) 261-277
5. Terrier, F., Rioux, L., Chen, Z.: Real time scheduling under uncertainty. In Nakanishi, S. ed. *Proc. of the 4th IEEE International Conference on Fuzzy Systems,* Yokohama, Japan. Piscataway: IEEE Computer Society. Vol.3 (1995) 1177-1184
6. Litoiu, M., Tadei, R.: Real-time task scheduling with fuzzy deadlines and processing times. Fuzzy Set and Systems, Vol.117, Iss.1 (2001) 35-45

A Preference Method with Fuzzy Logic in Service Scheduling of Grid Computing

Yanxiang He, Haowen Liu, Weidong Wen, and Hui Jin

School of Computer, State Key Lab of Software engineering,
Wuhan University, Wuhan, Hubei 430072
yxhe@whu.edu.cn, jimlhw@126.com, atu_wen@hotmail.com,
hjin@whu.edu.cn

Abstract. Resource Management and scheduling is the important components of the Grid. It efficiently maps jobs submitted by the user to available resources in grid environment. Most mechanism about the resource scheduling focus on the performance of communication through the network and the load of the services, or the cost the users pay for the service. The performance and the cost are not the total factors the users consider. Much more constraint contained in the requests may make the jobs uncompleted by the services. Our work will consider the preference that the users make to the cost and deadline of his job. Even none of services can fit all the conditions in the request; there is always a service with the most satisfaction for the user.

1 Introduction

Grid infrastructure is a large VO (Virtual Organization) which integrates a large mount of distributed heterogeneous resources and high performance computing capabilities into a super service plat, which can provide huge computing services, storage capability and so on [1, 2]. As the development and maturation of the grid technology, more services registered in the VO will be used for charge, and more users will consider using the service according to his real needs. Service management and scheduling is one of the most important components to implement the aim of Grid. It works efficiently to map applications submitted by the user to available resources in grid environment.

In order to make good use of the resource and satisfy the real need of the users, many mechanisms have been brought forward, such as the Performance Evaluation Method [3], Performance Prediction Method [4], Considering Load Balancing Method [5], Considering deadline and cost-based scheduling mechanism in Nimrod/G [6,7] and GRACE infrastructure for the Nimrod/G resource broker [8].

The methods by Cao in [3,4,5] mainly use the PACE (Performance Analysis and Characterization Environment) as a foundational tool, which provide quantitative data concerning the performance of sophisticated applications running on local high performance resource [9], and use the hierarchy agents system to implement the service advertisement and service discover, any agent in the system will be a service provider or a consumer. But he didn't consider that an agent can represent more then one service, different agents may provide the same service in the agent system. He focuses on the performance of finding or updating a service in the system, including the Discovery Speed, System Efficiency, Load Balancing and so on.

As mentioned in [8], the Nimrod/G resource broker, a global resource management and scheduling system for computational grid, built using Globus services, only support static costing mechanism, which then support deadline based scheduling and dynamic resource trading using Globus and GRACE services. But they also ignore that the less-cost services will be the bottleneck of scheduling. The agents providing these services will get most application requests from the users, so that the application of the users can't be completed because of the constrain even though there are services whose capability can satisfy the users' goals.

Actually, Users express their preferences in their requests, which help choosing available services on different sceneries where they have an urgent job or wish to pay less money. We use a preference method with fuzzy logic to choose the most satisfying service for the users, which may be neither the cheapest nor the quickest, but the most proper one.

This article firstly describes why to consider the preference of the user along with the service economization in the grid world in Section 2. Section 3 discusses administrative level agent architecture and a more detail method with fuzzy logic to choose proper service according to the users' fuzzy regarding on the cost or time. The simulator and results following are in section 4. Section 5 summaries our work and gives a future plan.

2 Why to Consider the User's Preference

In order to satisfy the requests of the users and make good use of the resources, the consumers (using resources) and the providers (providing resources) have their own goals, objectives, strategies and supply-and-demand patterns.

2.1 Service Characters

As we know, a service is provided for a special application, such as the huge information retrieve service that is developed for the application of information retrieve which needs many processes and large storage space, the biology information service (such as genome atlas assembly), the image disposal service, and so on. These services all have its goals and aim to satisfy certain application.

Therefore, a service consists of different kinds of resources. These resources could be computing systems (such as traditional supercomputers, clusters, SMPs, or even powerful desktop machines), visualization platforms, storage devices and software in grid environment.

To a service, the price is not the unique parameter, but one of the most important parameters. We concern mainly the parameters including capability, access price and working time. These parameters are changing parameters due to the instability of the service in dynamic grid.

We abstract a service as an agent, which has the same function, sending requests and providing services. The agent encapsulates the service as a steady entity, which marks different parameters such as capability, cost and working time according the inter-state of its resources. Therefore, two agents providing the same service may have different parameters.

2.2 Users' Preference

As to users, the price and time is the two most important factors concerned. When a user requests a service, he actually hopes to pay less and spend less time as can as possibly. So he will give his condition or constraint about the budget and expected time when he submits the requests on certain application, which means the application must be completed by a certain service, meanwhile, the service must satisfy the requests.

Maybe many services can meet all the conditions of the request and have different price and estimative completive time. How to choose one from the optional services depends on which factor the user play more importance on. For example, the user U_A wants to request a certain service to complete his application in 20 minutes and cost no more than $20. Both service S_A and S_B have the capability to satisfy U_A's conditions in request. But service S_A may consume the user only $5 and 15 minutes; the other service S_B may consume $15, with 5 minutes. The user will choose his preference on price or time cost.

Maybe many services can meet the partial conditions of the request. Taking the same user U_A for example, if U_A wants to find a certain service to complete his application in 10 minutes but cost no more than $10, both service S_A and S_B have the capability to complete U_A's application. But service S_A can't complete it in 10 minutes; SB can't complete it with $10. In order to assume the application to be completed by a service, he must sacrifice his price or time cost.

The above two examples indicate that the user must give an additional instruction of his recognition degree on the price and time cost in order to complete the job. A user paying more attention on price, means that he prefers paying less money to spending less time when choosing the service. If the user pays more attention on time, he may choose to sacrifice his money instead of time.

Actually, the users' preference to the two factors is not clear, we can describe this priority using different probability or express with fuzzy values, expressing a certain extent preference to any one of the costs.

3 The Method Considering the Users' Preference

In this section the hierarchical agent system is based on Cao[3] and we have two different changes. The first is the data structure of ACTs (Agent Capability Table) in leaf nodes, which has the content about his service information including the service name, service capability, access price, possible execution time, and status (Busy, Active and Invalid). Maybe different agents provide the same service, but have different parameters, which can have a different competition capability in the whole system. If the leaf-node in the agent system has the status of Busy, it rejects the request of the users, which can make space for other agents, so as to cost the user more money or time. If the leaf-node is in Active, it can accept more jobs, but return fail when it is invalid. The second difference focuses on the function of the

Coordinator Agent, which not only receives the requests from the user in the local domain, but also evaluates the synthesis values of the satisfaction with the user's request for the available services. The synthesis values depend on the user's request, preference and the service's characters.

We express a user's job request R with a scheme of vector <job description, deadline, deadline preference, budget, budget preference>. We define the deadline and the deadline preference of the job with R_d, P_d respectively, the budget and budget preference with R_b and P_b. Both P_d and P_b vary from 0 to 1. They indicate the fuzzy degrees what the user prefers to time and price. Seeing from the fuzzy degrees, we can judge whether the user has an urgent job or wish to pay less money.

We define the available service collection with S $\{S_1, S_2, ...S_m\}$, and S may be null. Any one service in the collection can complete the execution of the job J, and all has the status of Active. We mainly describe the relative parameter of a service S_i, including the access price with S_{id}, and possible execution time with S_{it}.

There is difficulty in confirming a function or equation with the price and the time cost. We introduce two formulas to respectively express the degree that an attribute of the service S_i can satisfy a condition of the job J.

$$V_{id} = (R_d - S_{it})/R_d \tag{1}$$

$$V_{ib} = (R_b - S_{ib})/R_b \tag{2}$$

Formula one means the departure degree of the possible executed time provided by the Service S_i falling away from the expected deadline condition of the job J. Formula two means the departure degree of the price the Service S_i calls for falling away from the budget condition of the job J.

If V_{id} or V_{ib} is not bigger than zero, it shows that the time or price of the service goes beyond the user's expectation, vice versa. Then the evaluation of service S_i for the request can be denoted as below.

$$V_i = V_{id}*P_d + V_{ib}*P_b \tag{3}$$

Formula three constitutes the evaluation collection V=$\{V_1, V_2, ...V_m\}$. Now we can choose the service with the max synthesis value in set V to complete the job J.

A description of the core of the algorithm follows:

a. The user submits his request with the condition of deadline, budget, and the preference to the deadline, budget to the agent A.
b. The agent A discovers all the available services that are capable and Active.
c. All the information about the services returns to the agent A.
d. Agent A evaluates all the V_i with formula three.
e. If P_d is bigger than P_b, choose service S_k which has the value V_k =max $\{Vi\}$ under the condition that some V_{kd} is equal or bigger than 0; or choose the service S_r which has the value V_r= max $\{Vi\}$, if any V_{id} is smaller than 0.
f. If P_d is smaller than P_b, choose service S_k which has the value V_k= max $\{Vi\}$ under the condition that some V_{kb} is equal or bigger than 0; or choose the service S_r which has the value V_r= max $\{Vi\}$, if any V_{ib} is smaller than 0.
g. The job will be transferred directly to the agent S_k or S_r for execution.

Note that the implementations of the above algorithm ensure the job's competition, as long as there exist the corresponding service which is in Active, irrespective of the service can absolutely satisfy the conditions of the users, but importantly considering the more prior factor the user prefer.

4 Experimentations and Evaluation

For the simulation, we create experiments containing 100 jobs, each with 100 seconds running time, giving a total computation time of 10000 seconds, which is similar with Rajkumar Buyya [10]. For each experiment, we create 10 test services, which have the same service content, but have different capability, access price and execution time. The capability reflects through the amount of the jobs it can accept in one time and complete in a period time. The access price vary with 12,15,18,21,24,27,30,33,36, 39 units/CPU-second, but the corresponding time vary with 170s, 160s, 150s, 140s, 130s, 120s, 110s, 100s, 90s, 80s. The time request of convection in the agent system, the time of discovery all the services and the time of evaluation are uniformly viewed as a delay with 10% of the running time. The experiment is achieved when each service runs 10 jobs in a sequence, giving a max time of 1700s for all the jobs.

We select three deadlines: 1870 seconds (the optimal deadline plus 10%), 3740 seconds (1870×2), and 5610 seconds (1870×3). Also we select three values of budget, the lowest of which is 264000 Units, which are spent to execute 20 jobs on each of the 5 cheapest services. Then a budget of 294000 Units is required to execute 10 jobs on every service. The highest is 330000 Units which allows the scheduler full freedom to schedule over the services without consideration the costs. So the budget of 330000 Units ensures that all jobs can run on the most expensive service.

Each job has a random value in the request with the preference to deadline and budget between 0.4 and 1, in order to look the changes with the relative preference, we set the preference of deadline 0.5 in following figures, and the preference of budget from 0.4 to 1.0. The figure 1 and figure 2 show the total cost and total time under the condition of deadline, budget, and the preferences. The figure 3 shows the jobs each service has completed under different user preference on deadline and budget.

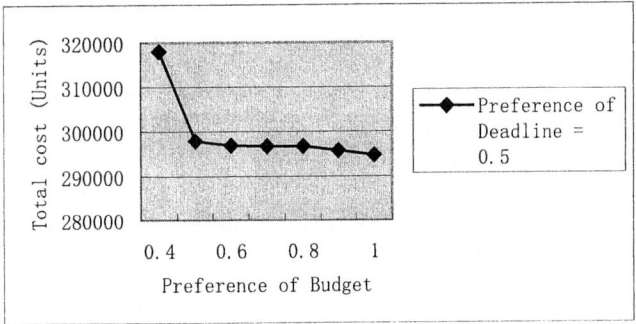

Fig. 1. Relation of total cost with the Budget Preference under the fixed Deadline preference

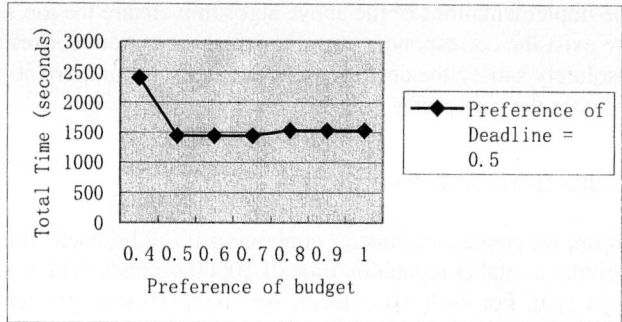

Fig. 2. Relation of total time with the Budget Preference under the fixed Deadline preference

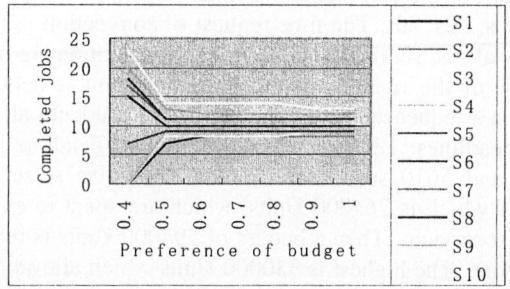

Fig. 3. Completed jobs distributed on different services with the Budget Preference

We can see from the above figures that the highest cost is 318000 Units, and the time-spent is also the most at 2400s, when the preference of budget is 0.4. It is smaller than the preference of deadline with 0.5, which means the user pays more attention to the time; the job will be assigned to the faster services (S_6, S_7, S_8, S_9, S_{10}), which make higher cost, and 23 jobs executed on S_{10} cost the most time with 2400s. With the increase of the preference of budget, the focused factor gradually shifts to budget, which brings the total cost down to 294700 Units, and the total time nearby 1530s. Although the total cost is still beyond the users' requires on budget, we can think that it can fuzzily satisfies the requests in the whole.

While the capability keeps fixed, the more relaxation of the constraint of deadline and budget, the more service have the opportunity to accept the job according to the user's prior concern factor, even though the service calls for more cost or time. If the user pay more attention on the time, the service executes quickly will do the job, whereas, the service has the less access price will do the job, even it will take much more time. But they are still normal within the tolerance scope of the users in the whole.

5 Conclusion

In this paper we have discussed an agent-based method, which mainly considers the real preference of the user with cost and time in the economic service in grid environment. This method will complete the users' request even it will sacrifice some interest about the total cost or execution time, and still not beyond the users' condition. Future work will focuses on the competition among the services, abstaining from the monopolization or the scene that demand exceeds supply, which can protect the user's interest more reliably.

Reference

1. I.Foster, and C. Kesselman, The Grid:Blueprint for a new Computing Infrastructure, Morgan-kaufmann, 1998.
2. II.Foster, C. Kesselman, and S.Tuecke, "The Anatomy of the Grid:Enabling Scalable Virtual Organizations", to appear in Int.J.Supercomputer Applications, 2001
3. Junwei Cao, Darren J.Kerbyson and Graham R.Nudd, "Performance Evaluation of an Agent-based Resource Management Infrastructure for Grid Computing", in Proceeding of 1st IEEE/ACM International Symposium on Cluster Computing and the Grid (CCGrid 2001), Brisbane, Australia, May 2001.
4. Junwei Cao, Stephen A.Jarvis, Daniel P.Spooner, etc, "Performance Prediction Technology for Agent-based Resource Management in Grid Environment", in Proceeding of 11th IEEE Heterogeneous Computing Workshop(HCW 2002) , Fort Lauderdale, Florida, USA, April 2002
5. Junwei Cao, Daniel P.Spooner, Stephen A,etc, "Agent-based Grid Load Balancing Using Performance-Driven Task Scheduling", Proc.17th IEEE Int. Parallel & Distributed Processing Symp., Nice, France 2003
6. Abramson, D., Giddy, J., AND Kotler, L., "High Performance Parametric Modeling with Nimrod/G: Killer Application for the Global Gird?" ,IPDPS'2000, Mexico, IEEE CS Press, USA, 2000.
7. Buyya, R., Abramson, D., and Giddy, J., Nimrod/G: An Architecture for a Resource Management and scheduling System in a Global Computational Grid, HPC ASIA'2000, China, IEEE CS Press, USA, 2000
8. Rajkumar Buyya, David Abramson, and Jonathan Giddy, "An Economy Driven Resource Management Architecture for Globus Computational Power Grids",In PTPTA'2000, Las Vegas, 2000
9. G.R.Nudd, D.J.Kerbyson, E.Papaefstathiou, etc, "PACE – A Toolset for the performance Prediction of Paralled and Distributed systems", International Journal of High Performance Computing Applications, Special Issues on Performance Modelling-Part I, Sage Science Press, Vol.14, No.3, pp.228-251, Fall 2000.
10. Rajkumar Buyya ,Jonathan Giddy, and David Abramson. "An evaluation of Economy-based Resource Trading and scheduling on Computing Power Grids for Parameter Sweep Applications." in The second Workshop on Active Middleware Services AMS2000, conjunction with the HPDC2000.

H_∞ Robust Fuzzy Control of Ultra-High Rise / High Speed Elevators with Uncertainty

Hu Qing[1], Qingding Guo[2], Dongmei Yu[1], and Xiying Ding[2]

[1] School of Electrical Engineering, Shenyang University of Technology,
Postalcode 110023, No.58 Xinghua South Street, Tiexi District,
Shenyang, Liaoning, P.R.China
{aqinghu, yu_dm163}@163.com

[2] School of Electrical Engineering, Shenyang University of Technology,
Postalcode 110023, No.58 Xinghua South Street, Tiexi District,
Shenyang, Liaoning, P.R.China
{guoqd, dingxy}@sut.edu.cn

Abstract. A LMIs (linear matrix inequalities) based H_∞ robust fuzzy control approach to an ultra-high rise/high speed elevator in the presence of uncertainties is presented in this paper. The uncertain nonlinear systems are represented using Takage-Sugeno (T-S) fuzzy models. The proposed controllers, which are in the form of the so-called parallel distributed compensation (PDC), stabilize nonlinear systems and guarantee an induced L_2 norm bound constraint on disturbance attenuation for all admissible uncertainties. Finally, simulation results show the realization of the H_∞ robust fuzzy control.

1 Introduction

Ultra-high rise buildings are becoming increasingly common with buildings such as the Shanghai World Financial Center in China (1510 ft). Ultra-high rise/high speed elevators are uncertain, time varying and nonlinear systems.

In recent years, there has been increasing interest in the study of controlling nonlinear systems based on Takagi-Sugeno (T-S) fuzzy system [1], [2]. There are a few literatures on H_∞ control design for a fuzzy control system. In this paper, H_∞ control design is expanded to include nonlinear systems with parameter uncertainty using fuzzy control. We consider both the stability problem of satisfying the decay rate and the disturbance attenuation. The proposed robust fuzzy control system is represented by T-S fuzzy model with uncertainties. PDC is employed to design the fuzzy controllers from T-S fuzzy model. The stability analysis and control design problems are reduced to LMI problems.

2 Takagi-Sugeno Fuzzy Model

Consider an uncertain nonlinear system described by the following fuzzy model with parameter uncertainties.

Plant rule i:

IF $p_1(t)$ is M_{i1} and $\cdots$ and $p_n(t)$ is M_{in},

THEN $\dot{x}(t) = (A_i + \Delta A_i(t))x(t) + (B_{1i} + \Delta B_{1i}(t))w(t) + (B_{2i} + \Delta B_{2i}(t))u(t)$, (1)

$z(t) = C_i x(t) + D_{1i} w(t) + D_{2i} u(t)$, $i = 1, 2, \ldots, r$.

where $p_i(t)$, $u(t)$, $x(t)$, $w(t)$ and $z(t)$ denote parameter, control input, state, disturbance input and controlled output, respectively. M_{ij} is the fuzzy set, r is the number of IF-THEN rules, and A_i, B_{1i}, B_{2i}, C_i, D_{1i}, and D_{2i} are constant real matrices that describe the nominal system, and $\Delta A_i(t)$, $\Delta B_{1i}(t)$, $\Delta B_{2i}(t)$ are unknown, norm-bounded and possible time-varying parameter uncertainty matrices and have the following structures:

$$[\Delta A_i(t) \quad \Delta B_{1i}(t) \quad \Delta B_{2i}(t)] = HF(t)[E_{1i} \quad E_{2i} \quad E_{3i}], \quad i = 1, 2, \ldots, r. \tag{2}$$

where H, E_{1i}, E_{2i}, E_{3i} are predetermined constant real matrices, and $F(t)$ is an unknown matrix function with Lebesgue-measurable elements and is bounded by

$$F(t) \in \Omega := \{F(t) \mid F^T(t)F(t) \leq I\}. \tag{3}$$

Control rule i:

IF $p_1(t)$ is M_{i1} and $\cdots$ and $p_n(t)$ is M_{in}

THEN $u(t) = K_i x(t)$, $i = 1, 2, \ldots, r$. (4)

The final output of this fuzzy controller is

$$u(t) = \sum_{i=1}^{r} \mu_i(p(t)) K_i x(t). \tag{5}$$

So, the closed-loop system becomes

$$\dot{x}(t) = [\overline{A} + B_p F(t) C_q] x(t) + [B_w + B_p F(t) D_{qw}] w(t)$$
$$z(t) = C_z x(t) + D_{zw} w(t) \tag{6}$$

where

$$\overline{A} = \sum_{i=1}^{r} \sum_{j=1}^{r} \mu_i(p(t)) \mu_j(p(t)) [A_i + B_{2i} K_j],$$

$$B_p = \sum_{i=1}^{r} \mu_i(p(t)) H, \quad B_w = \sum_{i=1}^{r} \mu_i(p(t)) B_{1i},$$

$$D_{qw} = \sum_{j=1}^{r} \mu_i(p(t)) E_{2i}, \quad D_{zw} = \sum_{i=1}^{r} \mu_i(p(t)) D_{1i}, \tag{7}$$

$$C_z = \sum_{i=1}^{r} \sum_{j=1}^{r} \mu_i(p(t)) \mu_j(p(t)) [C_i + D_{2i} K_j],$$

$$C_q = \sum_{i=1}^{r} \sum_{j=1}^{r} \mu_i(p(t)) \mu_j(p(t)) [E_{1i} + E_{3i} K_j].$$

3 H_∞ Robust Controller for T-S Fuzzy Model with Uncertainties

Consider the system (1). Then there exists a robust fuzzy controller (5) such that the closed-loop system (6) is quadratically stable with decay rate α and $\|T_{zw}\|_\infty < \gamma$ (the L_2 gain of system (6)) for all admissible uncertainties, if there exist a common matrix $Q>0$, matrices Y_i, $i=1,2,\ldots,r$, and positive scalar λ satisfying the following LMIs:

$$\begin{bmatrix} G_{ii} & \lambda^{-1}H & B_{1i} & T_{ii}^T & W_{ii}^T \\ \lambda^{-1}H^T & -\lambda^{-1}I & 0 & 0 & 0 \\ B_{1i}^T & 0 & -\gamma^2 I & E_{2i}^T & D_{1i}^T \\ T_{ii} & 0 & E_{2i} & -\lambda^{-1}I & 0 \\ W_{ii} & 0 & D_{1i} & 0 & -I \end{bmatrix} < 0, \quad i=1,2,\ldots,r. \tag{8}$$

$$\begin{bmatrix} G_{ij}+G_{ji} & 2\lambda^{-1}H & B_{1i}+B_{1j} & T_{ij}^T+T_{ji}^T & W_{ij}^T+W_{ji}^T \\ 2\lambda^{-1}H^T & -2\lambda^{-1}I & 0 & 0 & 0 \\ B_{1i}^T+B_{1j}^T & 0 & -2\gamma^2 I & E_{2i}^T+E_{2j}^T & D_{1i}^T+D_{1j}^T \\ T_{ij}+T_{ji} & 0 & E_{2i}+E_{2j} & -2\lambda^{-1}I & 0 \\ W_{ij}+W_{ji} & 0 & D_{1i}+D_{1j} & 0 & -2I \end{bmatrix} < 0, \quad i<j\leq r \tag{9}$$

for all $i,j = 1,2,\ldots,r$. In here

$$\begin{aligned} G_{ij} &= A_i Q + QA_i^T + B_{2i}Y_j + Y_j^T B_{2i}^T + 2\alpha Q, \\ T_{ij} &= E_{1i}Q + E_{3i}Y_j, \quad W_{ij} = C_i Q + D_{2i}Y_j. \end{aligned} \tag{10}$$

Furthermore, state feedback gains are given by

$$K_i = Y_i Q^{-1}, \quad i=1,2,\ldots,r. \tag{11}$$

4 Simulation Results

Elevator dynamic model and representative model parameter values are the same as [3]. The control purpose is to realize the tracking control and the releveling control with desired control performance (6.5 second flight time for the short (one floor) run and 6 mm releveling error). The decay rate and disturbance rejection are utilized in the fuzzy control system design. They can improve the flight time and the releveling error, respectively

In the simulation, 27 cases of parameter changing (rope damping, rope stiffness and gains) of the model are investigated. Fig. 1 and Fig. 2 show the simulation results of the 135th floor (top position) and the 67th floor (middle position) for the short run (one floor run), respectively, where all the 27 cases are shown. The simulation results of the 1st floor (bottom position), which are the similar to Fig. 2, are not shown due to the limited space. The upper and lower parts show the tracking control results and the releveling control results, respectively. The designed fuzzy controller satisfies the required control performance for all the 27 cases. In particular, 5 mm releveling control is realized. These facts mean that the fuzzy control system is robust in practice.

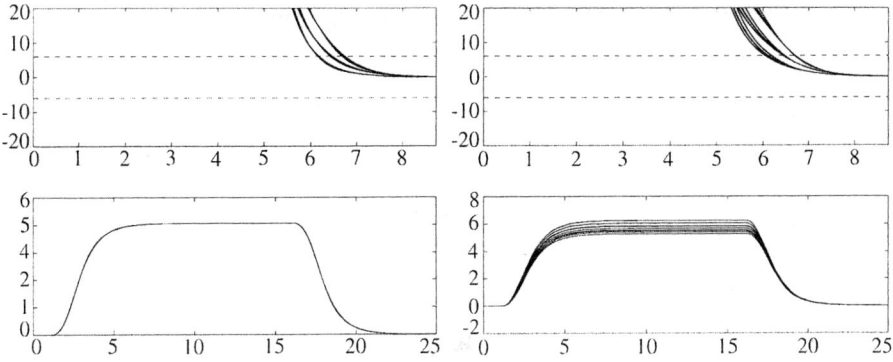

Fig. 1. Simulation results (135th floor) **Fig. 2.** Simulation results (67th floor)

5 Conclusions

This paper has presented a LMIs (linear matrix inequality) based fuzzy control approach to an ultra-high rise/high speed elevator. An H_∞ robust fuzzy controller design method for uncertain nonlinear systems is described by T-S model with uncertainties. Based on the notion of quadratic stabilization with decay rate and an L_2 norm bound, sufficient conditions to solve the H_∞ robust fuzzy control problem have been obtained and solutions have been produced in terms of LMIs. The control design has utilized the concept of parallel distributed compensation. Finally, simulation results have shown the utility of the proposed control strategy.

References

1. Joo, Y. H., Chen, G., Shieh, L. S.: Hybrid State-space Fuzzy Model-based Controller with Dual-rate Sampling for Digital Control of Chaotic Systems. IEEE Trans. Fuzzy Syst. 7 (1999) 394–408.
2. Lee, H. J., Park, J. B., Chen, G.: Robust Fuzzy Control of Nonlinear Systems with Parametric Uncertainties. IEEE Trans. Fuzzy Syst. 9 (2001) 369–379.
3. Roberts, R.: Control of High Rise/High Speed Elevators. Proc. ACC'98. (1998) 3440–3444.
4. Tanaka, K., Sugeno, M.: Stability Analysis and Design of Fuzzy Control Systems. Fuzzy Sets and Systems. 45 (1992) 135–156.

A Dual-Mode Fuzzy Model Predictive Control Scheme for Unknown Continuous Nonlinear System[1]

Chonghui Song[1], Shucheng Yang[2], Hui yang[3], Huaguang Zhang[1], and Tianyou Chai[4]

[1] Department of Information Science and engineering, Northeastern University,
Shenyang 110004 China
chui_song@hotmail.com
[2] Liaoning Seismological Research Institute, Shenyang
[3] Department of electronic engineering, East China Jiaotong University
[4] Research center of automation, Northeastern University

Abstract. In this paper, a method to construct of a stable dual-mode predictive controller of unknown nonlinear system using the fuzzy system as a predictive model is proposed. The dual-mode controller is designed to ensure the stability in this region. In the neighborhood of the origin, a linear feedback controller designed for the linearized system generates the control action. Outside this neighborhood, predictive controller based on the fuzzy model is applied to the real nonlinear system. This method yields a stable closed-loop system when is applied to nonlinear systems under some conditions.

1 Introduction

Predictive controller based on the receding horizon methodology offers a powerful approach to the design of state feedback controllers for nonlinear systems (Garcia *et al.*, 1989; Rawlings *et al.*, 1994). By now there are several predictive control schemes with guaranteed stability for nonlinear systems. See Mayne and Michalska (1990), Mayne and Michalska (1993), Chen and Allegower (1998) and L. Magni *et al.*, (2001). Recently, some papers extended the result of above paper such as Wan and Kothare (2005), Mhaskar *et al.*, (2005) and Gyurkovics and Elaiw (2004).

On the other hand, although the researches in fuzzy modeling and fuzzy control have made a great progress (Cao *et al.* (1997); Ma *et al.* (1998)), it seems that the field of optimal fuzzy control is nearly open (Wu and Lin (2002)).

In the present paper, we develop a new dual-mode predictive control scheme. We show that the open-loop optimal solution for the fuzzy system, is a suboptimal control for the nonlinear system subject to terminal inequality constraint. This method yields a stable closed-loop system when is applied to nonlinear systems under some conditions. An example is also given in section 5.

---

[1] This work is supported by national nature fund. Grant ID: 450474020.

2 Problem Statement

Consider the real nonlinear system

$$\dot{X}_R(t) = f_R(X_R(t), U_R(t)) \quad (1)$$

where $X_R(t) \in \mathbb{R}^m$, $u(t) \in \mathbb{R}^m$, $f_R = (f_R^1, \cdots, f_R^n)^T$ are unknown functions from $\mathfrak{R}^{n+m}$ to $\mathfrak{R}^n$, $f_R(0,0)=0$ and $f_R(\cdot,\cdot)$ satisfies the *Lipschiz* condition. $f_R(\cdot,\cdot)$ is known in the small local area Φ of the equilibrium point and the *Jacobian* linearization of the system (1) exists at the origin and can be stabilized by a local linear state feedback control $U=KX$.

The open-loop finite horizon optimal control problem for the real system (1) is

$$P_R(X_R,t): \min\{J_R^{U_R}(X_R,t) | X_R^{U_R}(t+T_P; X_R, t) \in \bar{B}(0,\varepsilon) \subset \Omega\} \quad (2)$$

where Ω is the neighborhood of the origin. The cost index is defined by

$$J_R^{U_R}(X_R,t) = \int_t^{t+T_P}[\|X_R^{U_R}(\tau;X,t)\|_Q^2 + \|U_R(\tau)\|_L^2]d\tau + \|X_R^{U_R}(t+T_P;X,t)\|_P^2 \quad (3)$$

where P is a positive definite symmetric matrix and let $J_R^U(X,t)$ denote the corresponding optimal value functional due to U for system (1).

Since $f_R(\cdot,\cdot)$ are unknown, its predictive model is a fuzzy system described by

$$R^i: \text{IF } x_1 \text{ is } F_{1i} \text{ and } x_2 \text{ is } F_{2i} \text{ and } \cdots x_n \text{ is } F_{ni}, \text{ THEN}$$
$$\dot{X}_F(t) = A_i(t)X_F(t) + B_i(t)U_F(t) \quad i=1,\cdots,M \quad (4)$$

where $R^i, i=1,2,\cdots M$ is the *i*th rule, $F_{ji}(x_j), j=1,2,\cdots n$ is the fuzzy set in the ith rule, $X(k) \in R^n$ and $U(k) \in R^m$. Formulate equation (4) into a compact form

$$\dot{X}_F(t) = f_F(X_F(t), U_F(t)) \quad (5)$$

The open-loop finite horizon optimal control problem equation (5) is defined as

$$P_F(X_F,t): \min\{J_F^{U_F}(X_F,t) | X_F^{U_F}(t+T_P; X_F(t), t) = 0\} \quad (6)$$

and the cost index is defined by

$$J_F^{U_F}(X_F,t) = \int_t^{t+T_P}[X_F(\tau)^T Q X_F(\tau) + U_F(\tau)^T L U_F(\tau)]d\tau \quad (7)$$

where Q and L are strictly positive definite, symmetric matrices, $X_F(t)$ is the initial state at time t. The minimizer of $P_F(X_F,t)$ is denoted by $U_F^*(\cdot; X_F(t), t)$ and the corresponding optimal value functional is $J_F^{U_F^*}(X_F,t)$.

Let $X_R^U(\cdot; X, t)$ denote the trajectory of the real system (1), due to control U, passing through state X at time t and $X_F^U(\cdot; X, t)$ denotes the trajectory of the fuzzy system (5) due to the control U. T_P denotes the predictive horizon.

3 Preliminary Results for Fuzzy System

Assumption 1: the real system (1) satisfies the *Lipschitz* condition.

If assumption 1 is satisfied, the nonlinear system (1) can be approximated to any specified degree by the fuzzy system (5). For a well-defined fuzzy system (5), suppose θ_g is the adjustable parameter of the fuzzy system (5). For any specified $\varepsilon > 0$ and finite time interval $T > 0$, there must exist a parameter θ_g^* satisfying the following relationship (Cao (1997); Wang (1995)).

Assumption 2: Suppose for a specified $\varepsilon > 0$ and finite time interval $T > 0$, the parameter θ_g^* has been suitably selected to satisfy the relationship (8).

$$\sup_{\tau \in [0,T]} |X_R(\tau) - X_F(\tau)| < \varepsilon \tag{8}$$

4 Fuzzy Dual-Mode Predictive Control Scheme

The fuzzy dual-mode predictive control scheme (FDMPC) involves the determination of a terminal region Ω and a terminal weighted positive symmetry matrix P off-line such that following inequality (38) holds true in the region Ω for system (1)

$$X(t)^T P X(t) - X(t+\delta)^T P X(t+\delta) \geq \int_t^{t+\delta} X(\tau)^T Q^* X(\tau) d\tau \tag{9}$$

where $Q = Q^T > 0$, $Q^* = Q + K^T L K$ and K is the local linear stable feedback gain i.e. $U = KX$ stabilize the linearized system in Ω. The procedure to determine P and Ω can refer to Chen and Allegower (1998) and Magni *et al.*, (2001).

Now we state the FDMPC scheme in the content below.

Off-line:

Step1: Linearize the real system (1) by the *Jacobian* linearization method.

Step2: Determine a local stable linear state feedback gain K, P and $\Omega \subset \Phi$ such that the state $X_R(t) \in \Omega$ satisfies the inequality (12).

Step3: Specify a constant $\varepsilon > 0$ such that the closed ball $\overline{B}(0,\varepsilon) \subset \Omega$

Step4: Train the fuzzy system (5) to satisfy the inequality (8).

On-line:

Step5: Initialization

1) At time $t=0$, let $\overline{X}_0 = X_R(0), i = 0, t_i = 0, n = 1$. IF $X_R(0) \in \Omega$, switch to the local linear control, i.e., employ the linear feedback control law U=KX for all t. **ELSE**:

2) Calculate an optimal control $U_F^*(\cdot, \overline{X}_0, 0)$ for the fuzzy system (5) and the optimal value $J_F^{U_F^*}(\overline{X}_0, 0)$. Apply the control $U_F^*(\cdot, \overline{X}_0, 0)$ to the real system (1) at time interval $[0, n\delta]$.

Controller

1) At any time $n\delta$, IF $X_R(n\delta) \in \Omega$, switch to the local linear feedback control, i.e., employ the linear feedback control law $U=KX$ for all $t \geq n\delta$. **ELSE:**

2) At any time $n\delta$, IF $|X_R(n\delta) - X_F(n\delta)| > \varepsilon$ (It implies the fuzzy system is not trained well enough to satisfy the inequality (8)), then go to **off-line step4**. **ELSEIF** the following switch criteria (13) is satisfied (It implies that a better control sequence is found),

$$J_F^{U_F^*}(X_R(n\delta), n\delta) \leq J_F^{U_F^*}(\overline{X}_0, t_i) - (2T_p + 1)\lambda\varepsilon^2 - (n\delta - t_i)\lambda\varepsilon^2 \\ - \int_{t_i}^{n\delta} \left(\left\| X_F^{U_F^*}(\tau; \overline{X}_0, t_i) \right\|_Q^2 + \left\| U_F^*(\tau; \overline{X}_0, t_i) \right\|_L^2 \right) d\tau \tag{10}$$

Then let $i = i+1, t_i = n\delta, \overline{X}_0 = X_R(n\delta), n = n+1$ (the index $i = 0, 1, 2, \cdots$ is the switch times) and apply the control $U_F^*(\cdot, \overline{X}_0, n\delta)$ to the real system (1) at time interval $[n\delta, (n+1)\delta]$, and then go to **step5 Controller 1)**. **ELSE:**

3) Apply the optimal control $U_F^*(\cdot; \overline{X}_0, t_i)$ at time interval $[n\delta, (n+1)\delta]$ to the real system and let $n = n+1$. Then go to **step5 Controller 1)**.

Lemma 1: Suppose assumption 1 and assumption 2 satisfied. Then for any $t \in [t_{i-1}, t_i]$ and $t + m\delta \in [t_{i-1}, t_i]$ where $m = 1, 2, 3 \cdots$, the objective functional $J_R^{\hat{U}}(X_R(t), t)$ satisfies

$$J_R^{\hat{U}}(X_R(t+m\delta), t+m\delta) \leq J_R^{\hat{U}}(X_R(t), t) \\ - \int_t^{m\delta} \left(\left\| \hat{X}(\tau; X_R(t); t) \right\|_Q^2 + \left\| \hat{U}(\tau; X_R(t); t) \right\|_L^2 \right) d\tau \tag{11}$$

$$J_R^{\hat{U}}(X_R(t), t) - \lambda\varepsilon^2 - \int_t^{m\delta} \left(\left\| \hat{X}(\tau; X_R(t); t) \right\|_Q^2 + \left\| \hat{U}(\tau; X_R(t); t) \right\|_L^2 \right) d\tau \leq J_R^{\hat{U}}(X_R(t+m\delta), t+m\delta) \tag{12}$$

Theorem 3: Suppose assumption 1 and assumption 2 satisfied. Then by using the FDMPC scheme, the closed-loop system is asymptotically stable (i.e. $X_R(t)$ of the real system tends to zero as $t \to \infty$).

5 Illustrative Example

In this section, the fuzzy dual-mode predictive algorithm is applied to the highly nonlinear model of a continuous stirred tank reactor (CSTR), the CSTR is described by the following dynamic model

$$\begin{cases} \dot{C}_A = \frac{q}{V}(C_{Af} - C_A) - k_0 \exp(-\frac{E}{RT})C_A \\ \dot{T} = \frac{q}{V}(T_f - T) + \frac{(-\Delta H)}{\rho C_p} k_0 \exp(-\frac{E}{RT})C_A + \frac{UA}{V\rho C_p}(T_c - T) \end{cases} \quad (13)$$

where C_A is the concentration of A in the reactor, T is the reactor temperature, and T_c is the temperature of the coolant stream. The constraint is $280K \leq T_c \leq 370K$. The objective is to regulate C_A and T by manipulating T_c. The nominal operating conditions, which correspond to an unstable equilibrium $C_A^{eq} = 0.5 \text{mol}/l$, $T^{eq} = 350K$, $T_c^{eq} = 300K$ are: $q = 100 l/\min$, $T_f = 350K$, $V = 100l$, $\rho = 1000 g/l$, $C_p = 0.239 \text{J/g K}$, $\Delta H = -5 \times 10^4$ J/mol, $E/R = 8750 K$, $k_0 = 7.2 \times 10^{10} \min^{-1}$, $UA = 5 \times 10^4$ J/min K. The nonlinear state-space model (1) of system (18) can be obtained by defining the state vector $X = [C_A - C_A^{eq}, T - T^{eq}]^T$ and the input $u = T_c - T_c^{eq}$.

We use following T-S type dynamic fuzzy system to identify the nonlinear part of equation (29)

$$\text{Rule } l : \text{IF } x_1(t) \text{ is } \mu_{1l}^i(x_1(t)) \text{ and } x_2(t) \text{ is } \mu_{2l}^i(x_2(t)) \text{ and } u(t) \text{ is } \mu_{3l}^i(u(t)) \quad (14)$$
$$\text{THEN } \dot{X}(t) = A_l X(t) + B_l u(t)$$

where l is the Rule number and $l = 1, 2, \cdots, M \quad M = 27$.

$J = \int_t^{t+6}(x_F(\tau)^2 + u_F(\tau)^2)d\tau$ and the endpoint constraint is $x(t+6) = 0$. One of the closed loop trajectory is plotted in figure 1.

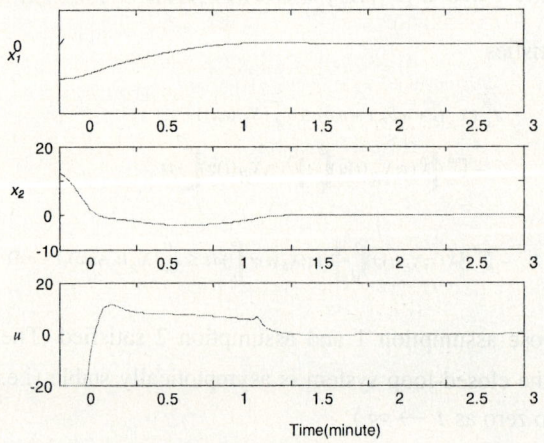

Fig. 1. Closed-loop State Trajectory with Initial State (0.3 363)

6 Conclusion

For the dynamic fuzzy system, a predictive controller with guaranteed stability under some condition is obtained. Then a new dual-mode predictive control scheme for nonlinear system using fuzzy system as a predictive model is proposed in this paper.

References

1. Mayne, D.Q. and H. Michalska: Receding Horizon Control of Nonlinear Systems. IEEE Trans. on Automatic Control 35 (1990) 814-824
2. H. Michalska and Mayne, D.Q.: Robust Receding Horizon Control of Constrained Nonlinear Systems. IEEE Trans. on Automatic Control 38 (1993) 1623-1633
3. Chen H and Allegower F.: A Quasi-Infinite Horizon Nonlinear Model Predictive Control Scheme With Guaranteed Stability. Automatica 34 (1998) 1205-1217
4. Rawling, J.B., E.S. Meadows and K.R. Muske: Nonlinear Model Predictive Control: A Tutorial And Survey. In Proc. IFAC Int. Symp. Adv. Control of Chemical Processes, ADCHEM Kyoto Japan (1994)
5. Mayne, D.Q.: Optimization In Model Based Control. In Proc. IFAC Symposium Dynamics and Control of Chemical Reactor, Distillation Columns and Batch Processes, Heisingor, (1995) pages 229-242
6. Shinq-Jen Wu and Chin-Teng Lin: Discrete-Time Optimal Fuzzy Controller Design: Global Concept Approach. IEEE Trans. on Fuzzy systems 10 (2002) 21-38
7. S.G. Cao, N.W. Rees, and G. Feng: Analysis and Design For a Class af Complex Control Systems, Part 1: Fuzzy Modeling and Identification. Automatica 33 (1997) 1017-1028
8. S.G. Cao, N.W. Rees, and G. Feng: Analysis and Design For a Class of Complex Control Systems Part 1: Fuzzy Controller Design. Automatica 33 (1997) 1029-1039
9. X.J. Ma, Z.Q. Sun, and Y.Y. He: Analysis and Design of Fuzzy Controller and Fuzzy Observer. IEEE Trans. Syst. Man Cybern. 2 (1998) 345-349
10. Jeong, S.C., Park, P.: Constrained MPC Algorithm for Uncertain Time-Varying Systems With State-Delay. IEEE Transactions on Automatic Control 50 (2005) 257-263
11. Zhaoyang Wan andMayuresh V.Kothare: Efficient Scheduled Stabilizing Output Feedback Model Predictive Control for Constrained Nonlinear Systems. IEEE Transactions on Automatic Control 49 (2004) 1172-1177
12. Éva Gyurkovics, AhmedM. Elaiw: Stabilization of Sampled-Data Nonlinear Systems by Receding Horizon Control Via Discrete-Time Approximations. Automatica 40 (2004) 2017-2028

Fuzzy Modeling Strategy for Control of Nonlinear Dynamical Systems

Bin Ye, Chengzhi Zhu, Chuangxin Guo, and Yijia Cao

College of Electrical Engineering, National Laboratory of Industrial Control Technology,
Zhejiang University, Hangzhou 310027, Zhejiang, China
{yebin, yijiacao}@zju.edu.cn

Abstract. This paper presents a novel fuzzy modeling strategy using the hybrid algorithm EPPSO based on the combination of Evolutionary Programming (EP) and Particle Swarm Optimization (PSO) for control of nonlinear dynamical systems. The EPPSO is used to automatically design fuzzy controllers for nonlinear dynamical systems. In the simulation part, one multi-input multi-output (MIMO) plant control problem is performed. The performance of the suggested method is compared to that of EP, PSO and HGAPSO in the fuzzy controllers design. Simulation results demonstrate the superiority of the proposed method.

1 Introduction

The soft computing techniques for the identification and control of nonlinear dynamical systems have attracted considerable interest in the last decade. The most commonly used models are neural networks and fuzzy inference systems. This paper is devoted to a novel fuzzy modeling strategy to the fuzzy inference system for control of nonlinear dynamical systems. In this paper, we will combine EP and PSO to design the fuzzy models, and the new algorithm is called EPPSO. To demonstrate the performance of the proposed algorithm, EPPSO is applied into the control of a Multi-Input Multi-Output control problem.

2 Fuzzy Modeling Strategy Based on the Hybrid of EP and PSO

2.1 T-S Fuzzy Model

The 1st-order T-S fuzzy models consist of linguistic IF-THEN rules which can be represented by the following general form:

$$R^i : if\ x_1\ is\ A_1^i(x_1)\ and\ ...\ x_k\ is\ A_k^i(x_k),\ then\ y^i\ is\ \xi_0^i + \xi_1^i x_1 + ... + \xi_k^i x_k, \quad (1)$$

where $A^i_j(x_j)$ are the fuzzy variables defined as follows:

$$A_j^i(x_j) = \exp[-0.5 * [(x_j - m_j^i)/\sigma_j^i]^2], \quad (2)$$

where m^i_j and σ^j_i are the mean value and the standard deviation of the Gaussian type membership function, respectively. The details of fuzzy reasoning procedure could be referred to Ref. [1].

2.2 Fuzzy Modeling Strategy Based on EPPSO

The evolutionary design of the fuzzy rule base using the hybrid of EP [2] and PSO [3], i.e., EPPSO is described. In this section, the parameters that associated with the fuzzy rule base are firstly described, and then the details of the proposed algorithm EPPSO will be expatiated.

(1) **Parameter Representation.** The parameter matrix, which contains the parameters for defining the membership functions in the premise part and the coefficient parameters in the consequent part of the fuzzy rule base, will be of a two dimensional matrix. The parameters m^i_j, σ^i_j and ξ^a_j described in the above section constitute this matrix named Q. The Gaussian type membership function is used in this paper, thus the size of Q is $\mathbb{C} = r \times (3*k+1)$.

(2) **Initialization.** In this step, M individuals forming the population are initialized, and each consists of $\mathbb{C}$ parameters. These individuals could be regarded as population members in terms of EP and particles in terms of PSO, respectively. In PSO, the velocities of individuals are initialized as random number in [0, 1].

(3) **PSO Operator.** In each generation, after the fitness values of each individual are evaluated, the top 50% individuals are selected as the elites and the others are discarded. Similar to the maturing phenomenon in nature, the individuals are firstly enhanced by PSO and become more suitable to the environment after acquiring the knowledge from the society. The whole elites could be regarded as a swarm, and each elite corresponds to a particle in it. In PSO, individuals (particles) of the same generation enhance themselves based on their own private cognition and social interactions with each other. And this procedure is regarded as the maturing phenomenon in EPPSO. The selected $M/2$ elites are regarded as particles in PSO, and each elite corresponds to a particle. The enhanced elites will be copied to the next generation and also designated as the parents of the generation for EP, copied and mutated to produce the other $M/2$ individuals.

(4) **EP Operator.** To produce better-performing offspring, the mutation parents are selected merely from the enhanced elites by PSO. In the EP operation, the $M/2$ particles in PSO will be $M/2$ population members with parameters $z^\ell_n, 1 \leq \ell \leq M/2; 1 \leq n \leq \mathbb{C}$. The parameter mutation changes the parameters of membership functions by adding Gaussian random numbers generated with the probability of p to them:

$$z^\ell_n = z^\ell_n + \alpha * e^{(F_{max}-F_m)/F_{max}} * N(0,1), \qquad (3)$$

where F_{max} is the largest fitness value of the individuals in the current generation, F_m is the fitness of the mth individual, α is a real value between 0 and 1, and $N(0, 1)$ is a Gaussian random number with mean 0 and variance 1. The combination of the offspring of EP and the elites enhanced by PSO comes to be the new generation of EPPSO, and after the fitness evaluation, the evolution will go ahead again until termination condition is satisfied.

3 Simulation Results

In this section, simulation results of a Multi-Input Multi-Output plant control problem using fuzzy inference systems based on EPPSO are presented.

The MIMO plant to be controlled is the same as described as in [4]:

$$\begin{bmatrix} y_{p1}(k+1) \\ y_{p2}(k+1) \end{bmatrix} = 0.5 * \begin{bmatrix} y_{p1}(k)/[1+y_{p2}^2(k)] \\ y_{p1}(k)*y_{p1}(k)/[1+y_{p2}^2(k)] \end{bmatrix} + 0.5 * \begin{bmatrix} u_1(k-1) \\ u_2(k-1) \end{bmatrix} \quad (4)$$

The controlled outputs should follow desired outputs y_{r1} and y_{r2} as specified by the following 250 pieces of data:

$$y_{r1}(k) = \sin(k\pi/45), \quad y_{r2}(k) = \cos(k\pi/45), \quad 1 \le k \le 250.$$

The inputs to EPPSO controller are $y_{r1}(k)$, $y_{r2}(k)$, $y_{p1}(k)$, $y_{p2}(k)$, and the outputs are $u_1(k)$ and $u_2(k)$.

There are four fuzzy rules in EPPSO fuzzy controller for MIMO control, resulting in 72 free parameters totally. To show the superiority of EPPSO, the fuzzy controllers designed by EP and PSO are also applied to the MIMO control problem.

The RMSE defined for MIMO control performance index is:

$$RMSE = (\sum_{k=1}^{K} [(y_{r1}(k+1) - y_{p1}(k+1))^2 + (y_{r2}(k+1) - y_{p2}(k+1))^2]/K)^{0.5}, \quad (5)$$

where K is the total time steps, $y_{r1}(k+1)$ and $y_{r2}(k+1)$ are the desired output and $y_{p1}(k+1)$, $y_{p2}(k+1)$ are the inferred output.

The evolutions are processed for 100 generations and repeated for 25 runs. The averaged best-so-far RMSE value over 25 runs for each generation is shown in Fig. 1. From the figure, we can see that EP converges with a slower speed compared to PSO and EPPSO, while PSO converges fastest with an unsatisfying RMSE ultimately.

Table 1 gives the control performance for the MIMO control problem based on different methods. From the table, we can see that the fuzzy controller based on EPPSO outperforms all other methods including TRFN controller using HGAPSO [4], which is evolved for 100 generations and repeated for 100 runs. The shortcoming of using HGAPSO in designing TRFN is that there are too many free parameters (156) to be evolved, and the GA is not truly ergodic in a practical sense because of the multiple steps required.

Table 1. MIMO Control Performance Comparisons Using Different Methods

Method	EP	PSO	HGAPSO [4]	EPPSO
RMSE(train_A)	0.1610	0.1110	0.1145	0.0755
RMSE(train_B)	0.0788	0.0617	0.0686	0.0391

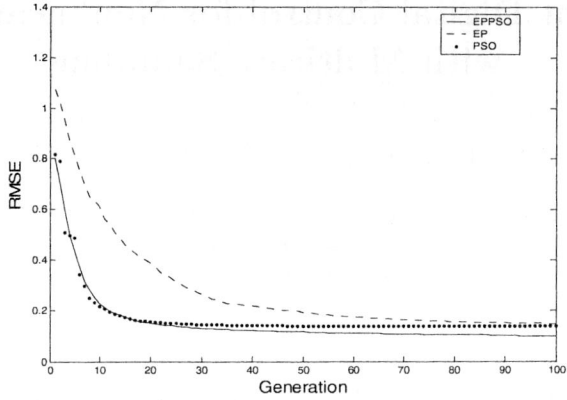

Fig. 1. Average best-so-far RMSE in each generation (iteration) for EPPSO, EP, and PSO

4 Conclusions

A T-S fuzzy model constructing approach based on the hybrid of EP and PSO is applied to the control of nonlinear dynamical systems. EPPSO combines the merits of fast convergence speed in PSO and the ergodic feature of EP, and its control performance greatly outperforms that of either EP or PSO. Comparisons with other methods also show the great effectiveness of the proposed method.

Acknowledgements

This work is supported by the Outstanding Young Scholars Fund (No. 60225006) and Innovative Research Group Fund (No. 60421002) of Natural Science Foundation of China.

References

1. Takagi, T., Sugeno, M.: Fuzzy Identification of Systems and Its Application. IEEE Trans. Systems, Man, and Cybernetics. 15 (1985) 116-132
2. Fogel, D. B.: Evolutionary Computations: Toward a New Philosophy of Machine Intelligence. New York: IEEE (1995)
3. Eberchart, R., Kennedy, J.: A New Optimizer Using Particle Swarm Theory. Proc. Int. Sym. Micro Machine and Human Science, Nagoya, Japan (1995) 39-43
4. Juang, C. F.: A Hybrid of Genetic Algorithm and Particle Swarm Optimization for Recurrent Network Design. IEEE Trans. Syst. Man Cy. B. 34 (2004) 997-1006

Intelligent Digital Control for Nonlinear Systems with Multirate Sampling

Do Wan Kim[1], Jin Bae Park[1], and Young Hoon Joo[2]

[1] Yonsei University, Seodaemun-gu, Seoul, 120-749, Korea
{dwkim, jbpark}@yonsei.ac.kr
[2] Kunsan National University, Kunsan, Chunbuk, 573-701, Korea
yhjoo@kunsan.ac.kr

Abstract. This paper studies an intelligent digital control for nonlinear systems with multirate sampling. It is worth noting that the multirate control design is addressed for a given nonlinear system represented by Takagi–Sugeno (T–S) fuzzy models. The main features of the proposed method are that it is provided that the sufficient conditions for stabilization of the discrete-time T–S fuzzy system derived by the fast discretization method in the sense of Lyapunov stability criterion, which is can be formulated in the linear matrix inequalities (LMIs).

1 Introduction

Drawing upon recent progress in the Takagi–Sugeno (T–S) fuzzy-model-based digital control, it is observed that a number of important works have used a singlerate controller [1, 2, 4, 3, 9] to meet the stability requirements. The digital control problem was conducted as a stabilizing the discretized model of continuous-time T–S fuzzy plant in [1, 2, 3] and a stabilizing the jumped fuzzy system in [9]. However, their discretized model has the approximation error, which is directly proportional to the sampling time. One gets better exact discretized model if one can A/D and D/A conversions faster. But faster A/D and D/A conversions mean higher cost in implementation. In addition, the digital control system is hybrid system involving continuous-time and discrete-time, but their discussion in [1, 2, 3] only contained the stability of the digital control system in the discrete-time domain. A multirate control approach [10, 13, 11, 12] can be an alternative. Interestingly, advantages of applying faster A/D and D/A conversions are obtained by using A/D and D/A at different rates. Furthermore, in [14], the stability of the closed-loop digital system is well guaranteed for sufficiently fast sampling rate if the closed-loop discrete-time nonlinear system.

Motivated by the above observations, we develop an intelligent multirate control for a class of nonlinear systems under the high speed D/A converter. The main contribution of this paper is that we derive some sufficient conditions in terms of the linear matrix inequalities (LMIs), such that the equilibrium point is a globally asymptotically stable equilibrium point of the discrete-time fuzzy model derived by the fast discretization in the sense of Lyapunov stability criterion.

2 Problem Statement

In the following, let T and T' be the sampling period and the control update period, respectively. For convenience, we take $T' = \frac{T}{N}$ for a positive integer N, where N is an input multiplicity. Then, $t = kT + lT'$ for $k \in \mathbb{Z}_{\geq 0}$ and $l \in \mathbb{Z}_{[0,N-1]}$, where the indexes k and l indicate sampling and control update instants, respectively.

Consider a nonlinear digital control system described by

$$\dot{x}(t) = f(x(t), u_d(t)) \tag{1}$$

for $t \in [kT + lT', kT + lT' + T')$, $(k, l) \in \mathbb{Z}_{\geq 0} \times \mathbb{Z}_{[0,N-1]}$, where $x(t) \in \mathbb{R}^n$ is the state vector, and $u_d(t) = u_d(kT, lT') \in \mathbb{R}^m$ is the multirate digital control input. The control actions are switched with T' and N. Moreover, the digital control signals are fed into the plant with the ideal zero-order hold.

To facilitate the control design, we will develop a simplified model, which can represent the local linear input–output relations of the nonlinear system. This type of models is referred as T–S fuzzy models. The fuzzy dynamical model corresponding to (1) is described by the following IF–THEN rules [1, 2, 4, 3, 5, 6, 7, 8]:

$$R_i : \text{IF } z_1(t) \text{ is about } \Gamma_{i1} \text{ and } \cdots \text{ and } z_p(t) \text{ is about } \Gamma_{ip},$$
$$\text{THEN } \dot{x}(t) = A_i x(t) + B_i u_d(t) \tag{2}$$

where $R_i, i \in \mathcal{I}_q = \{1, 2, \ldots, q\}$, is the ith fuzzy rule, $z_h(t), h \in \mathcal{I}_p = \{1, 2, \ldots, p\}$, is the hth premise variable, and $\Gamma_{ih}, (i, h) \in \mathcal{I}_q \times \mathcal{I}_p$, is the fuzzy set. Then, given a pair $(x(t), u_d(t))$, using the center-average defuzzification, product inference, and singleton fuzzifier, the overall dynamics of (2) has the form

$$\dot{x}(t) = A(\theta(t))x(t) + B(\theta(t))u_d(t) \tag{3}$$

where $A(\theta(t)) = \sum_{i=1}^{q} \theta_i(z(t))A_i$, $B(\theta(t)) = \sum_{i=1}^{q} \theta_i(z(t))B_i$, $\theta_i(z(t)) = w_i(z)/\sum_{i=1}^{q} w_i(z)$, $w_i(z) = \prod_{h=1}^{p} \Gamma_{ih}(z_h(t))$, and $\Gamma_{ih}(z_h(t))$ is the grade of membership of $z_h(t)$ in Γ_{ih}.

3 Main Results

To develop the discretized version of (3), we apply the fast discretization technique [11] to (3). In specific, we first derive a multirate discretized version of (3), and then we apply a discrete-time lifting technique to the multirate discrete-time model. Connecting the fast-sampling operator and the fast-hold operator with $[kT + lT', kT + lT' + T')$, $(k, l) \in \mathbb{Z}_{\geq 0} \times \mathbb{Z}_{[0,N-1]}$, to (3) leads the multirate discrete-time plant model.

Assumption 1. *Suppose that $\theta_i(z(t))$ for $t \in [kT + lT', kT + lT' + T')$ is $\theta_i(z(k+l))$. Then, the nonlinear matrices $\sum_{i=1}^{q} \theta_i(z(t))A_i$ and $\sum_{i=1}^{q} \theta_i(z(t))B_i$ of (3) can be approximated as the piecewise constant matrices $A(\theta(k+l))$ and $B(\theta(k+l))$, respectively.*

Proposition 1. *The multirate discrete-time model of (3) can be given by*

$$x(k+l+1) \approx G(\theta(k+l))x(k+l) + H(\theta(k+l))u_d(k+l) \tag{4}$$

for $t \in [kT + lT', kT + lT' + T')$, $(k,l) \in \mathbb{Z}_{\geq 0} \times \mathbb{Z}_{[0,N-1]}$, *where* $G(\theta(k+l)) = \sum_{i=1}^{q} \theta_i(z(k+l))G_i$, $H(\theta(k+l)) = \sum_{i=1}^{q} \theta_i(z(k+l))H_i$, $G_i = \exp(A_i T')$, *and* $H_i = (G_i - I)A_i^{-1}B_i$.

Proof. The proof is omitted due to lack of space.

Assumption 2. *Suppose that the firing strength* $\theta_i(z(t))$, *for* $t \in [kT, kT+T)$ *is* $\theta_i(z(k))$.

Proposition 2. *Given the system (4) for* $l \in \mathbb{Z}_{[0,N-1]}$, *a lifted sampled input*

$$\widetilde{u}_d(k) = \begin{bmatrix} u_d(kT) \\ u_d(kT+T') \\ \vdots \\ u_d(kT + NT' - T') \end{bmatrix} \in \mathbb{R}^{mN} \tag{5}$$

leads a lifted system

$$x(k+1) \approx \widetilde{G}(\theta(k))x(k) + \widetilde{H}(\theta(k))\widetilde{u}(k) \tag{6}$$

for $t \in [kT, kT+T), k \in \mathbb{Z}_{\geq 0}$, *where* $\widetilde{G}(\theta(k)) = G^N(\theta(k))$ *and* $\widetilde{H}(\theta(k)) = [G^{N-1}(\theta(k))H(\theta(k)) \ G^{N-2}(\theta(k))H(\theta(k)) \cdots H(\theta(k))]$.

Proof. The proof is omitted due to lack of space.

We convert the multirate digital control problem to the solvability of LMIs. For the system (6), we consider the following multirate feedback controller $u(k) = K_l(\theta(k))x(k)$ and have the lifted control input represented as

$$\widetilde{u}(k) = \widetilde{K}(\theta(k))x(k) \tag{7}$$

where $\widetilde{K}(\theta(k)) = \left[K_0^T(\theta(k)) \ K_1^T(\theta(k)) \cdots K_{N-1}^T(\theta(k)) \right]^T$, $K_l(\theta(k)) = K_0(\theta(k)) \times (G(\theta(k)) + H(\theta(k))K_0(\theta(k)))^l$, and $K_0(\theta(k)) = \sum_{i=1}^{q} \theta_i(z(k))K_{0i}$.

The next theorem provides the sufficient conditions for the stabilization in the sense of the Lyapunov asymptotic stability for (6).

Theorem 1. *The given system (6) under (7) is globally asymptotically stable in the sense of Lyapunov stability criterion if there exist* $Q = Q^T \succ 0$ *and constant matrices* F_i *such that*

$$\begin{bmatrix} -Q & * \\ G_i Q + H_i F_i & -Q \end{bmatrix} \prec 0 \quad i \in [1,q] \tag{8}$$

$$\begin{bmatrix} -Q & * \\ \frac{G_i Q + H_i F_j + G_j Q + H_j F_i}{2} & -Q \end{bmatrix} \prec 0 \quad i < j \in [1,q] \tag{9}$$

where $*$ *denotes the transposed element in symmetric position.*

Proof. The proof is omitted due to lack of space.

4 Closing Remarks

This paper proposed the multirate control design using the LMI approach for the fuzzy system. Some sufficient conditions were derived for stabilization of the discretized model via the fast discretization. Future work will be devoted to the extension to the nonautomonous system.

References

1. Joo Y. H., Shieh L. S., and Chen G.: Hybrid state-space fuzzy model-based controller with dual-rate sampling for digital control of chaotic systems. IEEE Trans. Fuzzy Syst. **7** (1999) 394-408
2. Chang W., Park J. B., Joo Y. H., and Chen G.: Design of sampled-data fuzzy-model-based control systems by using intelligent digital redesign. IEEE Trans. Circ. Syst. I. **49** (2002) 509-517
3. Lee H. J., Kim H., Joo Y. H., Chang W., and Park J. B.: A new intelligent digital redesign for T-S fuzzy systems: global approach. IEEE Trans. Fuzzy Syst. **12** (2004) 274-284
4. Chang W., Park J. B., and Joo Y. H.: GA-based intelligent digital redesign of fuzzy-model-based controllers. IEEE Trans. Fuzzy Syst. **11** (2003) 35-44
5. Wang H. O., Tanaka K., and Griffin M. F.: An approach to fuzzy control of nonlinear systems: Stability and design issues. IEEE Trans. Fuzzy Syst. **4** (1996) 14-23
6. Tananka K., Kosaki T., and Wang H. O.: Backing control problem of a mobile robot with multiple trailers: fuzzy modeling and LMI-based design. IEEE Trans. Syst. Man, Cybern. C. **28** (1998) 329-337
7. Cao Y. Y. and Frank P. M.: Robust $\mathcal{H}_\infty$ disturbance attenuation for a class of uncertain discrete-time fuzzy systems. IEEE Trans. Fuzzy Syst. **8** (2000) 406-415
8. Tananka K. and Wang H. O.: Fuzzy control systems design and analysis: a linear matrix inequality approach. John Wiley & Sons, Inc. (2001)
9. Katayama H. and Ichikawa A.: H_∞ control for sampled-data fuzzy systems. in Proc. American Contr. Conf. **5** (2003) 4237-4242
10. Hu L. S., Lam J., Cao Y. Y., and Shao H. H.: A linear matrix inequality (LMI) approach to robust H_2 sampled-data control for Linear Uncertain Systems. IEEE Trans. Syst. Man, Cybern. B. **33** (2003) 149-155
11. Chen T. and Francis B.: Optimal Sampled-Data Control Systems. Springer-Verlag (1995)
12. Francis B. A. and Georgiou T. T.: Stability theory for linear time-invariant plants with periodic digital controllers. IEEE Trans. Automat. Contr. **33** (1988) 820-832
13. Shieh L. S., Wang W. M., Bain J., and Sunkel J. W.: Design of lifted dual-rate digital controllers for X-38 vehicle. Jounal of Guidance, Contr. Dynamics **23** (2000) 629-339
14. Nesic D. and Teel A. R.: A framework for stabilization of nonlinear sampled-Data systems based on their approximate discrete-time models. IEEE Trans. Automat. Contr. **49** (2004) 1103-1122

Feedback Control of Humanoid Robot Locomotion

Xusheng Lei and Jianbo Su

Department of Automation & Research Center of Intelligent Robotics,
Shanghai Jiaotong University, Shanghai, 200030, China
{xushenglei, jbsu}@sjtu.edu.cn

Abstract. As an assistant tool for human beings, humanoid robot is expected to cooperate with people to do certain jobs. Therefore, it must have high intelligence to adapt to common working condition. The objective of this paper is to propose an adaptive fuzzy logic control (FLC) method to improve system adaptability and stability, which can adjust hip and ankle joint based on sensor information. Furthermore, it can real time adjust controller parameters to improve FLC performance. Based on sensor information, humanoid robot can get environment and inherent situation and use the adaptive-FLC to realize stable locomotion. The effectiveness of the proposed method is shown with simulations based on the parameters of the "IHR-1" humanoid robot.

1 Introduction

As servants and maintenance machines for people, humanoid robots are expected to assist human activities in our daily life and to replace humans in hazardous places [6]. Thus stable walking is one of the fundamental functions for humanoid robot. However, it is difficult to realize stable locomotion because the humanoid robot is a complex nonlinear system and the common working conditions are full of sudden obstacles. If humanoid robot can not make an intelligent adjustment to error information, it will lose balance with the accumulation of errors. So error compensation is also one of the fundamental problems in walking locomotion. To solve this question, some researchers have introduced reflex control method to compensate errors. According to hip link acceleration and ZMP, John proposed the method to recover the robot posture form sudden slipping and to continue to walk with desired trajectory [5]. Based on sensor information, Boone also proposed a reflex control method to overcome slipping and tripping on a hopping robot [4]. However, these methods need high requirement for system hardware performances, which will increase humanoid robot cost. To improve humanoid robot popularization, low cost is necessary for humanoid robot design and control [10]. However, how to realize a stable walking locomotion based on low cost sensors is an important problem for humanoid robot control.

To solve this problem, we propose an adaptive-FLC method to compensate system errors. FLC have widely used in reality due to its potential in handing with large structural and parametric uncertainties problems [3]. Even if the inputs of an FLC may not be very precisely, it can generate reasonable solution by using proper rules and membership functions [2]. This makes FLC very attractive because the data regarding the presence of obstacles and others got by sensors are not always precise for hardware constraints. However, it is not an easy thing to provide optimal parameters for FLC member function. So GA is chosen as a searching tool to find optimal parameters for FLC. After thorough training, FLC can compensate system error to keep stability in common environment. Furthermore, to improve FLC adaptability, adjustment network (AN) is trained to real time adjust FLC parameters to deal with environment diversity. When there exists big difference between predefined environment and real situation, AN can generate corresponding adjustment parameters to FLC, thus FLC can make a good adaptability to environment changes.

The rest of the paper is organized as follows. In section 2, whole feedback control structure is described. Section 3 gives a detail description of feedback control method, Simulation results are provided in Section 4, followed by Conclusions in Section 5.

2 Feedback Control Structure

As servants and maintenance machines for human beings, the main task for humanoid robot is to assist human activities in daily life. So it should have ability to involve into common working environments, where are full of sudden obstacles. Therefore, humanoid robot need have high intelligent to extract information from environments and make corresponding adjustment to eliminate system errors. The common errors for humanoid robot locomotion can be divided into two schemes. The one is error caused by environment uncertainties. As an assistant robot to human beings, humanoid robot needs work on common living conditions where are full of sudden obstacles. So it is hard to predefine a precise model for environments. The other is from the mechanism constraints. As a complex mechanical system, humanoid robot has so many components. However, not all components play roles in the walking locomotion. So it is common thing to consider main components in humanoid robot model to reduce computation complexity. Therefore, there exists difference between model and real robot. The predefined joint trajectory can not satisfies reality requirement, humanoid robot will lose balance with the accumulation of errors, thus it need error compensation method to real time adjust joint trajectory to keep system stability. The whole feedback compensation structure is shown in Fig. 1. It is a two layer closed loop feedback control, consisting of arm compensation and adaptive- FLC. With the swing of legs, it cause disturbance to system. By doing inverse motion with legs, Arms are used as compensator to reduce disturbance in inner loop. Moreover, to have a good compensation performance, FLC are trained to deal with hip and ankle joint error in outer loop. Furthermore, real time coefficient adjustment method is proposed and stored in an adjustment network (AN) to deal with environment uncertainty. After thorough training, the proposed feedback method can make suitable adjustment to realize stable walking motion.

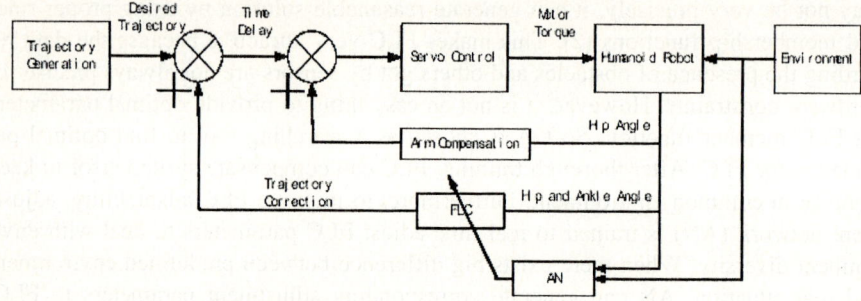

Fig. 1. Block diagram of feedback control system

When humanoid robot walks in reality, it gets environment and inherent information based on sensors. By using adaptive-FLC, humanoid robot eliminates errors to keep system stability. If all situations are the same with the predefined situations, humanoid robot can realize stable locomotion without any modifications. When there exists little difference between predefined environment and reality, Arm compensation and FLC can adjust humanoid robot joint trajectory to eliminate error. Furthermore, if common FLC can not make a good compensation to system error, AN is activated to generate new scale coefficients based on error information to real time adjust FLC member function and weight values to improve FLC performances.

3 Feedback Control

As mentioned above, the whole feedback control structure is divided into two closed-loops. So we will first introduce arm compensation, and then gives the detail description of adaptive-FLC.

3.1 Arm Compensation

Compared with other parts of humanoid robot, the weight of arm shares little proportion in the whole system. Many papers take arms as fixed mechanism to reduce computation complexity. Since arms make little effect to system energy consumption, arms are also separated from gait trajectory generation in our joint trajectory generation process, but used as compensator mechanisms rather than fixed mechanism. With the swing of leg, it will generate rotation moment to humanoid robot system. By doing inverse motion with legs, arms can generate inverse moment to reduce system disturbance. So arms are used as compensator to improve system robustness. The corresponding joint trajectories are generated based on legs motion. When the right leg swings from back to forth, the left shoulder and elbow do the same direction motion and right shoulder and elbow do an inverse direction motion. So the motions of shoulder and elbow are defined as follows:

$$\theta_s = l \cdot (\pi - \theta_h), \qquad (1)$$

$$\theta_e = \pi/2 - g \cdot (\pi - \theta_h), \qquad (2)$$

where θ_s is the angle for the shoulder, θ_e is the angle of the elbow, l and g are scaling factors defined by experiments, θ_h is the hip angle for the leg that is the opposite side of the shoulder and elbow referring to θ_e and θ_s.

3.2 Adaptive-FLC

FLC has widely used in reality due to its potential in handing with large structural and parametric uncertainties problems. It can provide efficient control action for a given purpose quickly by representing expert knowledge in the form of linguistic if-then rules. The soundness of knowledge acquisition and availability of expert knowledge decide the performance of FLC. However, it is a hard thing to define optimal member function and provide systematic procedure for the transformation of expert knowledge into the rule bases. On the other hand, GA represents an optimization searching which is based on the mechanics of natural selections. It can converge to the global optimum results by involving random elements in searching process. However, it need long time to find an optimal solution, which is not fit for real time application. Therefore, a hybrid technique, a fuzzy-GA algorithm is proposed for humanoid robot error compensation problem to take advantage of these two methods. Using GA as searching tool, it can find optimal member function and weight values for FLC.

For humanoid robot, hip and ankle joint play important roles in system stability. A positive acceleration of hip joint in the vertical direction would increase the contact force between the ground and the foot. When the ground friction force becomes larger, the friction force parallel to the ground also increases, which can prevent humanoid robot from the deviation of planed trajectory. Therefore, hip joint plays an important role in system stability. As the direct mechanism to control foot, ankle joint decides the feet landing time and landing position. For common locomotion, if the foot lands the ground too fast or late, the moment of tipping occurs, then the humanoid robot will tip over. Thus ankle is also one key mechanism to decide locomotion stability. By adjusting hip and ankle acceleration, humanoid robot can compensate system errors and realize stable walking locomotion.

In order to reduce the complexity of the humanoid robot error compensation, hip and ankle joint are compensated separately. Thus, there exists two independent FLC, namely, hip FLC and ankle FLC. Each FLC only searches its relevant solution to improve search efficiency. Since hip FLC and ankle FLC has the same structure, so only hip FLC is discussed here to show the structure of FLC. It consists of four layers expressing the fuzzy IF-THEN control process, including input layer, Fuzzier layer, Rule layer and Defuzzier layer. Two inputs, namely Δ_{hip} and Δ_{hip} are fed to the FLC. The Δ_{hip} is the difference between desired hip joint angle and real hip joint angle in sagittal plane. The Δ_{hip} is the change speed of Δ_{hip}. Each input is associated with five fuzzy linguistic variables respectively, i.e. Positive Big (PB), Positive Small (PS), Zero (ZO), Negative Small (NS), and Negative Big (NB). Furthermore, corresponding 25 fuzzy IF-THEN rules constructed the bases of FLC. The corresponding fuzzy rule can be expressed as:

$$R_{ij} : if \ x_1 \ is \ A_i \ and \ x_2 \ is \ B_j \ then \ y \ is \ y_{ij}$$

Every membership function in this controller is defined as a Gaussian function as follows:

$$\mu_{ij} = e^{\frac{-(x_i - c_{ij})^2}{\sigma_{ij}^2}}. \tag{3}$$

The fuzzy layer performs a fuzzification process for input variable. Outputs from this layer are membership values and are fed into the rule layer which perform multiplication operator. Using center of gravity method, the corresponding output y is the adjustment command for humanoid robot.

$$y = \frac{\sum_{i=1}^{5}\sum_{j=1}^{5} \mu_{ij} \cdot w_{ij}}{\sum_{i=1}^{5}\sum_{j=1}^{5} w_{ij}}, \tag{4}$$

where $\mu_{ij} = \mu_{i1} \cdot \mu_{j2}$ $i = 1...5, j = 1...5$

3.2.1 Optimization of FLC

To improve the performance of FLC, there are three types of parameters to optimize: center values c_{ij}, width values σ_{ij} and weight values w_{ij}. A real-value GA was employed in conjunction with the selection, mutation and crossover operator. Every individual represents ten center values, ten width values and twenty-five weight values. The population size is chosen as thirty and the maximum number of generations is used as termination function. Furthermore, to keep system searching performance, the best individuals in the population are selected as offspring directly in the next generation without any genetic operation. The other individuals with high fitness are selected as parents to generate offspring using crossover and mutation operators. Roulette wheel selection method is applied to the selection process. GA moves from generation to generation to select suitable solution until the termination constraints are met.

The ZMP is an important performance index for system stability. For humanoid robot, if actual ZMP keeps close to planed ZMP, it can realize stable locomotion as planed. So the fitness function involves ZMP to evaluate compensation result, which is defined as follows:

$$f = \frac{z}{r + (P_{zmp} - P_{dzmp})^2 + C}, \qquad (5)$$

where f is the fitness value for solution. P_{zmp} and P_{dzmp} are real and desired ZMP respectively, gotten from force sensor in foot mechanisms and angle sensor in body. r is a small positive constant to avoid special situation that fitness value converges to infinity. The constant z is used to enlarge the fitness difference between two individuals. C is punishment factor given as follows:

$$C = \begin{cases} A_1 & \text{if ZMP exceeds stable region} \\ A_2 & \text{if Torque exceeds the limitation} \\ A_3 & \text{if Tempture exceeds the limitation} \\ 0 & \text{otherwise} \end{cases}, \qquad (6)$$

where A_1, A_2 and A_3 are user-defined value. By using punishment factor, system can do a good protection to humanoid robot itself.

If solution can eliminate error quickly, the real ZMP is close to desired ZMP trajectory. Thus it gets higher fitness value and has more chance to generate offspring. In the last, we can get a series of optimal parameter for FLC to deal with system error.

3.2.2 Adjustment of FLC

With the changes of environment, humanoid robot will lose balance without suitable adaptation. However, it is hard thing to build exactly model for environment uncertainty. Taking all environments into consideration will increase the number of rules base and decrease computation efficiency, which is not fit for real time error compensation. In order to cope with this problem, we proposed the real time controller adjustment method stored in an adjustment neural network (AN). It adjusts the FLC member function and weight values by using certain scale coefficients in accordance with the dynamics of the environment. Based on the output of AN, the member function of FLC and weight values make corresponding adjustment to adapt to environment changes.

AN is a three layer RBFNN, which has strong approximate ability to system inputs. It can provide suitable coefficient to adjust FLC to adapt to environment diversity. The difference between desired and actual ZMP, the prediction of ground situation gotten by accelerator sensor, and current velocity are chosen as the inputs of AN, and the corresponding output of AN are the k_1, k_2 and k_3, which are scale coefficients for member function and weights value of FLC. To get a good training result, orthogonal least squares learning algorithm [9] is used to select centers and weights for the AN. In the training process, ten kinds of environment and five kinds of speed and their corresponding adjustment scale coefficients are chosen as train examples. After thorough training, humanoid robot can realize stable locomotion in unexpected environment. Even if new environment has high difference with trained environment, AN can real time provide new approximate scale coefficient to adjust FLC based on

can real time provide new approximate scale coefficient to adjust FLC based on sensor information. The block diagram of adaptive-FLC is shown as follows:

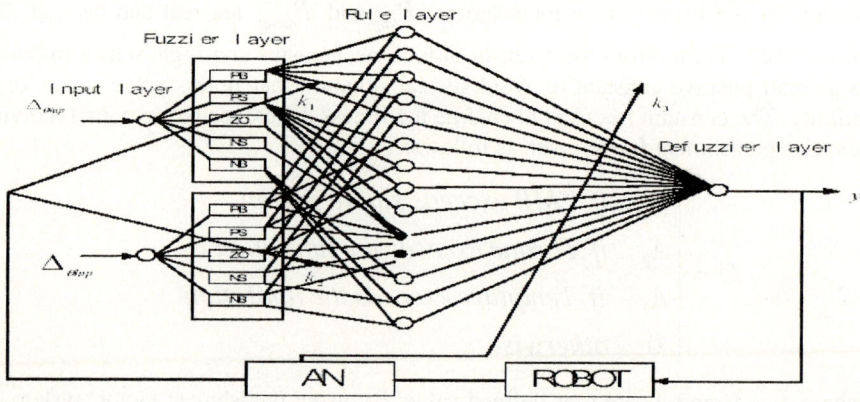

Fig. 2. Block diagram of adaptive-FLC

4 Simulations

A humanoid robot simulation platform is constructed in Automatic Dynamic Analysis of Mechanical System (ADAMS), and the parameters of robot are chosen from "IHR-1" humanoid robot, which is under development in our laboratory. The parameter values are presented in table 1 and the robot is shown in Fig. 3. The "IHR-1" is 0.8 m height and 26 KG weight. It has total of 25 degrees of freedom (DOF), i.e. six DOF for each leg (three for hip joint, one for the knee joint and two for the ankle joint), and five DOF for each arm (three for shoulder joint, one for the elbow joint and one for the wrist joint). There are two DOF in the neck joint to allow the head to turn to different directions and one in trunk joint to realize complex turn motion. Each joint is actuated by maxon brushless motor. Furthermore, CCD, angle sensor, acceleration sensor and FSR are mounted to get environment and inherent information. To have a good estimation for the ground reaction force between the feet and the ground, Young's modulus-coefficient of restitution element is used to build reaction force model in ADAMS [8]. The humanoid robot model is realized in ADAMS and controlled in MATLAB by using interface parameters between MATLAB and ADAMS.

A real-value GA was employed to get optimal parameters in conjunction with the selection, mutation and crossover operators. Thirty individuals are prepared in one generation and error compensation for five kinds of walking speed and step length is performed in this study. The corresponding parameters used in genetic algorithm are shown in Table 2.

Fig. 3. "IHR-1" humanoid robot

Table 1. Parameters of the Humanoid Robot

Link	Weight (kg)	Length (cm)
head	2	10
body	16	30
thigh	1	15
shank	1	15
ankle	0.9	10
foot	0.1	16
arm	0.5	15

Table 2. Funcitons and Parameters of GA

Function name	Parameters
Generation (GN_{max})	60
Heuristic crossover	[2,3]
Arithmetic crossover	2
Uniform mutation	4
Non-uniform mutation	[4, GN_{max} 3]
Normalized geometric selection	0.1

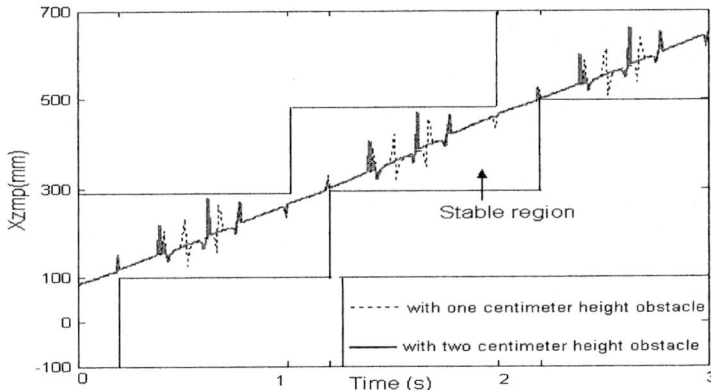

Fig. 4. ZMP trajectory in walking process with one and two centimeter height obstacles

To test the effectiveness of the proposed error compensation method, three boards with one, two and four centimeter height are chosen as obstacles in the simulation and spread on the walking road for humanoid robot. We can see that common FLC can compensate system error quickly for environment with one and two centimeter height board, as shown in Fig. 4. When humanoid robot steps over four-centimeter board, it loses balance quickly without AN real time adjustment. However, if the AN is activated and real time generate adjustment coefficients according to error input information, the humanoid robot can eliminate error quickly and keep system stability. Furthermore, to have a good estimation for the inherent error caused by mechanical constraints, white noises are added to hip joint to simulate real situation. A board with four-centimeter height is chosen as obstacle. The ZMP trajectory in X-axis is shown in Fig. 5. It is easy to see that the ZMP is always in the dotted region, which represents stable region. So the proposed adaptive-FLC method can make a good compensation to errors to keep system stability.

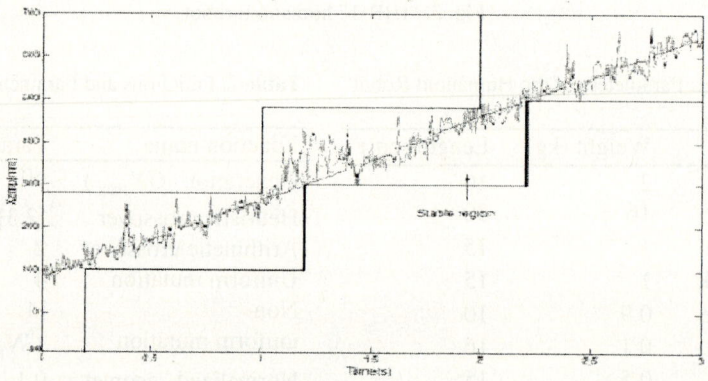

Fig. 5. ZMP trajectory in walking process with four-centimeter height obstacle

5 Conclusions

In this paper, an adaptive-FLC method is proposed to improve system stability. Using GA as searching tool, a FLC is trained to adjust hip and ankle joint to compensate system error based on sensor information. Moreover, the proposed method can real time adjust FLC parameter to adapt to environment diversity. A series of computer simulations show that the proposed algorithm is effective and can make the humanoid robot walk stably.

As a future work, we will test the proposed method on our real humanoid robot to prove their effectiveness.

References

1. C. W. Park, Y. W. Cho. : T-S model based indirect adaptive fuzzy control using online parameter estimation, IEEE Transactions On Systems, Man, and Cybernetics-Part B: Cybernetics Vol. 34 (2004) 2293-2302
2. D. K. Pratihar, K. Deb, A. Ghosh. : Optimal path and gait generations simultaneously of a six-legged robot using a GA-fuzzy approach, Robotics and Autonomous Systems, Vol. 41. (2001) 1-20

3. F. Sun, Z. Sun, L. Li, H. X .Li. : Neuro-fuzzy adaptive control based on dynamic inversion for robotic manipulators, Fuzzy Sets and Systems, Vol. 124 (2003) 117-133
4. G. N. Bonne, J. K. Hodgins. : Slipping and tripping reflexes for bipedal robots. Autonomous Robots, Vol. 4. (1997) 259-271.
5. J.H.Park, O.Kwon. : Reflex control of biped robot locomotion on a slippery surface, IEEE Int. Conf. On Robotics and Automation, Korea. (2001) 4134-4139
6. K.Hirai, M.Hirose, Y.Haikawa, T.Takenaka. : The Development of Honda Humanoid Robot. IEEE Int. Conf. Robotics and Automation, Belgium. (1998) 1321-1326
7. M. Vukobratovic, V. Potkonjak, S. Tzafestas. : Human and humanoid dynamics, Journal of Intelligent And Robotics Systems Vol. 41 (2004) 65-84
8. Q. Huang, K. Yokoi, S. Kajita, K. Kaneko, H. Arai et al. : Planning walking patterns for a biped robot, IEEE Transactions On Robotics And Automation, Vol. 17 (2001) 280-289
9. S.Chen, C.F.N.Cowan, P.M.Grant. : Orthogonal Least Squares Learning Algorithm for Radial Basis Function Networks. IEEE Transactions on Neural Networks, Vol.2. (1991) 302-309
10. T. Ishida, Y. Kuroki. : Sensory system of a small biped entertainment robot, Advanced Robotics, Vol. 18. (2004) 1039-1052

Application of Computational Intelligence (Fuzzy Logic, Neural Networks and Evolutionary Programming) to Active Networking Technology

Mehdi Galily[1], Farzad Habibipour Roudsari[1,2], and Mohammadreza Sadri[2]

[1] Young Researchers Club, Islamic Azad University, Tehran, Iran
[2] Iran Telecom Research Center, Tehran, Iran
mahdijalili@ece.ut.ac.ir

Abstract. Computational intelligent techniques, e.g., neural networks, fuzzy systems, neuro-fuzzy systems, and evolutionary algorithms have been successfully applied for many engineering problems. These methods have been used for solving control problems in packet switching network architectures. The introduction of active networking adds a high degree of flexibility in customizing the network infrastructure and introduces new functionality. Therefore, there is a clear need for investigating both the applicability of computational intelligence techniques in this new networking environment, as well as the provisions of active networking technology that computational intelligence techniques can exploit for improved operation. We report on the characteristics of these technologies, their synergy and on outline recent efforts in the design of a computational intelligence toolkit and its application to routing on a novel active networking environment.

1 Introduction

The events in the area of computer networks during the last few years reveal a significant trend toward open architecture nodes, the behavior of which can easily be controlled. This trend has been identified by several developments [1,2] such as:

- Emerging technologies and applications that demand advanced computations and perform complex operations
- Sophisticated protocols that demand access to network Resources
- Research toward open architecture nodes

Active Networks (AN), a technology that allows flexible and programmable open nodes, has proven to be a promising candidate to satisfy these needs. AN is a relatively new concept, emerged from the broad DARPA community in 1994–95 [1,3,4]. In AN, programs can be "injected" into devices, making them active in the sense that their behavior and the way they handle data can be dynamically controlled and customized. Active devices no longer simply forward packets from point to point; instead, data is manipulated by the programs installed in the active nodes (devices). Packets may be classified and served on a per-application or per-user basis. Complex tasks and computations may be performed on the packets according to the content of the packets. The packets may even be altered as they flow inside the network. Hence, AN can be considered active in two ways [2]. First, the active devices perform customized operations on the data flowing through them. Second, authorized users/applications can "inject" their own programs into the nodes, customizing the way their data is manipulated. Due to these features of AN, an open node architecture is achieved. Custom protocols and services can easily be deployed in active nodes, making the network flexible and adaptive to users' and the network/service administrator's needs.

Computational Intelligence (CI) techniques have been used for many engineering applications [5-10]. CI is the study of the design of intelligent agents. An agent is something that acts in an environment—it does something. Agents include worms, dogs, thermostats, airplanes, humans, organizations, and society. An intelligent agent is a system that acts intelligently: What it does is appropriate for its circumstances and its goal, it is flexible to changing environments and changing goals, it learns from experience, and it makes appropriate choices given perceptual limitations and finite computation. The central scientific goal of computational intelligence is to understand the principles that make intelligent behavior possible, in natural or artificial systems. The main hypothesis is that reasoning is computation. The central engineering goal is to specify methods for the design of useful, intelligent artifacts. There are some concepts of CI like: fuzzy sets and systems, neural networks, neurofuzzy networks and systems, genetic algorithms, evolutionary algorithms, and etc.

Due to highly nonlinear behavior of telecommunication systems and uncertainty in the parameters, using the CI and Artificial Intelligence (AI) techniques in these systems has been widely increased in recent years [11-17]. CI and AI techniques can be used for the design and management of communication networks. In the network-engineering context, such techniques have been utilized to attack different problem. A thorough study of application of CI in traditional networks, such as, IP, ATM, Mobile networks can be found in the literature [16-19]. However, these techniques never really made it into production systems for two basic reasons: the one, that we already mentioned above, is that up to now the primary concern was to address infrastructural issues and algorithmic simplicity, and secondly, researchers hardly had the opportunity to implement their work on real networking equipment, thus giving little feedback on practical issues and little chance of having their algorithms gain in maturity, inside the networking community. We can observe that this is rapidly changing: the trend towards integrating all services over the same network infrastructure introduces new, more complex problems, whose solutions are well within the application domain of CI techniques. In this paper, the application of CI and AI techniques for active networks technology will be studied. CI can be employed to control prices within the market or be involved in the decision process. The environment we provide can act as a testbed for attacking several other control problems in a similar fashion.

2 Active Networks

Active networking technology signals the departure form the traditional store-and-forward model of network operation to a store-compute-and-forward mode. In traditional packet switched networks, such as the Internet, packets consist of a header and data. The header contains information such as source and destination address that is used to forward the packet to the next element that is closer to the destination. The packet format is standardized and processing is limited to looking up the destination address in the routing tables and copying the packet to the appropriate network port. In active networks, packets consist not only of header and data but also of code. This code is executed on the active network element upon packet arrival. Code can be as simple as an instruction to re-send the packet to the next network element toward its destination, or perform some computation and return the result to the origination node. Additionally, it is possible for these packets to install code whose lifetime exceeds the time that is needed for the active packet to be processed. Software modules that are installed in this fashion are called active extensions. Active extensions facilitate for software upgrades, new protocol implementations, system and network monitoring agents. Other potential applications that need control functionality to be installed on demand are also made possible. This is a major breakthrough compared to the current situation where network elements come with a set of configurable, yet pre-installed options at the time the element is shipped. The install new functions, one has to bring the infrastructure off-line to manipulate its functionality.

The high cost of deploying a new function in the infrastructure, required extreme care and experimentation before the whole community would to agree that a standardized protocol or algorithm is good enough. The key component enabling active networking is the *active node*, which is a router or switch containing the capabilities to perform active network processing. The architecture of an active node is shown in Fig. 1, based on the DARPA active node reference architecture [20].

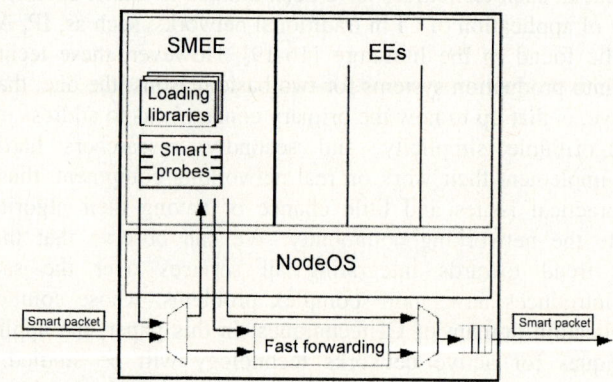

Fig. 1. Architecture of an Active Node

3 Computational Intelligence

Computational Intelligence (CI) [7,10,19] is an area of fundamental and applied research involving numerical information processing (in contrast to the symbolic information processing techniques of Artificial Intelligence (AI)). Nowadays, CI research is very active and consequently its applications are appearing in some end user products. The definition of CI can be given indirectly by observing the exhibited properties of a system that employs CI components [10]:

A system is *computationally intelligent* when it deals only with numerical (low-level) data, has a pattern recognition component, and does not use knowledge in the AI sense; and additionally, when it (begins to) exhibit

- computational adaptivity;
- computational fault tolerance;
- speed approaching human-like turnaround;
- error rates that approximate human performance.

3.1 Artificial Neural Networks

Artificial neural network (ANN) computing is an approach that simulates the human brain's neurons. This machine learning technology has ability to process massive parallel data and recognize patterns [21]. The ability of neural network model to approximate the noisy and incomplete data sample helps explain its current popularity [22]. Over the last decade, the application of ANN has solved many problems like production/operation, finance, human resource, accounting, marketing, and engineering. Moreover, Backpropagation Neural Network (BPN) is one of the most popular learning paradigms of the network models [21]. There are many input and output nodes called processing elements in ANN. The outputs are connected to the inputs through connection weights [23,24]. With these interactive nodes and weights, ANN learns and adjusts until they achieve optimal situation.

3.2 Fuzzy Sets and Systems

Fuzzy Logic, invented in 1965 by Prof. Lotfi A. Zadeh [25], is a branch of mathematics that allows a computer to model the real world the same way that people do. Unlike computers, people are not always precise. People think and reason using linguistic terms such as "hot" and "fast", rather than in precise numerical terms such as "100 degrees" and "70 miles per hour." In addition, people can make "shades of gray" decisions, rather than absolute "black and white" or "yes/no" decisions. Fuzzy Logic provides a computer with the capability to make the same kinds of classifications and decisions that people do. The most common use of Fuzzy Logic is in fuzzy expert systems.

The way Fuzzy Logic works is through the use of *fuzzy sets*, which are different from traditional sets [26,27]. A set is simply a collection of objects. Traditional sets impose rigid membership requirements upon the objects within the set. An object is either completely in the set, or it is not in the set at all. Another way of saying this is an object is a member of a set to degree 1 (completely in the set) or 0 (not in the set at all). For example, the set of "TALL" men could be defined to be all men 6 feet tall or taller. We would say a man 6 feet tall is TALL, but a man 5 feet 11 inches is not TALL. But isn't this unrealistic? Most people would classify a man 5 feet 11 inches tall as somewhat TALL, however the traditional set classifies a man as either completely TALL, or not TALL at all—there is no middle ground. In contrast, fuzzy sets have more flexible membership requirements that allow for partial membership in a set. The degree to which an object is a member of a fuzzy set can be any value between 0 and 1, rather than strictly 0 or 1 as in a traditional set. With a fuzzy set, there is a gradual transition from membership to non-membership. For example, a man 6 feet tall is a member of the fuzzy set TALL to degree .5. A man 5 feet 6 inches tall is TALL to degree .25, and a man 6 feet 6 inches tall is TALL to degree .75.

3.3 Neurofuzzy Systems

Neurofuzzy systems are currently one of the "flavors of the month" in the neural network and fuzzy logic communities. They attempt to combine the structural and learning abilities of a neural network with the linguistic initialization and validation aspects of a fuzzy system [28,29]. Neurofuzzy networks are a particular type of fuzzy system that uses algebraic operators and continuous fuzzy membership functions, as it has been shown that these are generally the best for surface fitting problems. There are several types of neurofuzzy system, and these are categorized by the type of membership functions; the two most common are B-splines and Gaussians. By regarding these fuzzy systems as types of neural networks, the role of the membership function in determining the form of the decision surface is highlighted, rather than its accuracy in modeling the vagueness or uncertainty associated with a particular linguistic term. This interpretation also provides several possibilities for utilizing inductive learning-type techniques to produce parsimonious rule bases. Thus, the mathematical rigor associated with adaptive neural networks can be used to analyze the behavior of learning neurofuzzy systems and improve their performance as data-driven and human-centered approaches are being combined to improve the network's overall performance.

3.4 Evolutionary Programming Algorithms

Stochastic optimization algorithms [30-33], like evolutionary programming, genetic algorithms and simulated annealing, have proved useful in solving difficult optimization problems. In this context, a difficult optimization problem might mean: (1) a non-differentiable objective function, (2) many local optima, (3) a large number of parameters, or (4) a large number of configurations of parameters. Some important topics that can be discussed in this section are: genetic algorithms, simulated annealing, ant colonies, and etc.

Whereas genetic algorithms include a variety of operators (for example, mutation, cross-over and reproduction), evolutionary programs use *only* mutation. As such, an evolutionary program can be viewed as a special case of a genetic algorithm. The basics of evolutionary programming can be described as follows: Given some initial population, proceed as follows:

- Sort the population from best to worst according to the cost function.
- For the worst half of the population, replace each member with a corresponding member in the top half of the population, adding in some 'random noise'.
- Re-evaluate each member according to the cost function.
- Repeat until some convergence criterion is satisfied.

3.5 Combination of Computational Intelligence Techniques

Overall, CI can be regarded as a research endeavor being a home to a number of technologies including genetic, fuzzy, and neural computing [34-36]. In this synergistic combination, each of them plays an important, well-defined, and unique role. As we have already summarized the main technologies, one can easily envision their strong points as well as some deficiencies that require further elimination. The role of synergy between them is to alleviate the existing shortcomings. The bottom line is that the synergy between these key information technologies becomes a necessity. They reflect a way of problem formulation and problem solving. All of these methodologies stem form essential cognitive aspects of fuzzy sets, underlying evolutionary mechanisms of genetic algorithms and biologically sound foundations of neural networks, which provide essential foundations when dealing with engineering problems. With their increasing complexity, it becomes apparent that all of the technologies discussed above should be uses concurrently rather than separately. There are numerous links that give rise to the CI architectures. In the next section, we elaborate on one of them dealing with the synergy between neural networks and fuzzy sets.

4 Application of Computational Intelligence in Active Networks

In the previous sections, we described in summary the foundations of computational intelligence techniques as a set of tools for solving difficult optimization problems, and active networking as a novel networking infrastructure technology. In this section, we will first elaborate on the features of active networking that make it attractive for applying computational intelligence techniques, since the problems that we want to solve are optimization problems, the implementation domain needs to be mapped to the optimization domain. We developed a market-based approach to controlling active network resources that naturally bridged decision and optimization. Based on this, we then provide a number of simple problems that can be effectively dealt within this framework. We also discuss implications and a large set of problems that can be dealt with in a similar fashion. The features of active networks that are appealing to us in our attempt to utilize computational intelligence techniques are mainly:

- *Programmability:* We are able to enhance the network infrastructure on the fly. It is up to us to decide what functions we want to add, where to add them and when to do this. For example, we can easily replace algorithm A with algorithm B when cost or performance indicates such a need.
- *Mobility:* Our function does not need to be located on a single element throughout its lifetime. As is the agent paradigm, the programs we inject into the network might migrate from element to element to perform different tasks.
- *Distributivity:* Our function can be implemented as a distributed algorithm with agents performing tasks and exchanging information throughout the infrastructure.

4.1 Application of Computational Intelligence to Providing an Economic Market Approach to Resource Management

In this section, market-based resource management architecture for active networks will be adopted. Markets provide a simple and natural abstraction that is general enough to efficiently and effectively capture a wide range of different issues in distributed system control. The control problem is cast as a resource allocation problem, where resources are traded within the computational market by exchange of resource access rights. Resource access rights are simply credential that enables a particular entity to access certain resources in a particular way. These credentials result from market transactions, thus transforming an active network into an open service market. Dynamic pricing is used as a congestion feedback mechanism to enable applications to make policy controlled adaptation decisions. This system focuses on the establishment of contracts between user and application as well as between autonomous entities implementing the role of resource traders is emphasized. While this architecture is more oriented to end systems, the enhancements and the mechanisms described are well suited for use in active networks; Market-based control has also been applied to other problems such as bandwidth allocation, operating system memory allocation and CPU scheduling. The control problem can be redefined as the problem of profit or utility maximization for each individual player. The concept of a market is general enough to embrace all aspects of resource control in different contexts and different time scales. From a system engineering perspective, market-based control augments system-level, design-level and administrative-level control.

4.2 Application of Computational Intelligence for Routing

Routing is one of the most fundamental and at the same time most complex control problems in networking. Its function is to specify a path between two elements for the transmission of a packet or the set-up of a connection for communicating packets. There are usually more than one possible paths and the routing function has to identify the one that is most suitable, taking into consideration factors such as cost, load, connection and path characteristics etc. Most routing implementations in modern networks are based on the shortest-path algorithms of Dijkstra and Bellman-Ford, with additional provisions for information exchange developed as standard protocols. In best effort networks, such as the Internet, routing is based on finding the shortest path, where the shortest path is defined as the path with the least number of "hops" between source and destination. For more advanced network services, such as the transmission of multimedia streams that require qualitative guarantees, routing considers factors such as connection requirements (end-to-end delay, delay variation, mean rate) and current (or future) network conditions. Furthermore, the information available to the decision process might be inaccurate or incomplete. Given the above, routing becomes a complex problem, with many aspects, including the perspective of a multi-objective optimization problem. Maximization of resource utilization or overall throughput, minimization of rejected calls, delivery of quality of service guarantees, fault-tolerance, stability, security consideration for administrative policy are just a few of the properties that are requirements for an acceptable solution. Issues of active organization and an approach for quality of service routing restricted to the case of routing connections with specific bandwidth requirements. Our goal here is to address the routing problem using CI. The solution we propose involves the following components:

- Roaming agents are moving from element to element to collect and distribute information on network state. Another difference is that in our work the agents operate within the metaphor of the active network economy rather than as an ant colony where ants are born, feed, reproduce and die. At each element, they communicate with the Resource Brokers or other Roaming Agents to trade topology information and connectivity costs. For efficiency, roaming agents are small-sized programs that take the form of active packets to travel throughout the infrastructure.
- Routing agents, at each network element, are responsible for spawning roaming agents and are also the recipients of the information collected by them. Routing agents also act as resource brokers for connectivity. Routing agents rely on the CI engine for setting their operational parameters and processing the information collected from the roaming agents.
- The CI engine is a set of active extensions that include several subcomponents. These subcomponents from a generic library-like algorithmic infrastructure. For each problem, the Configuration and Information Base (CIB) is used to store information that is specific to the problem domain. This information evolves as the engine learns the problem space. The CI engine can be further populated on demand with active extensions; however the above structure ensures scalability and ease of use by providing a set of commonly used tools.

The components we have currently implemented for the CI engine are:

- An Evolutionary Fuzzy Controller (EFC), which clusters paths according to their current state characteristics. This effectively hides or exposes details on routing information thus effectively controlling information granularity. For example, if information is inaccurate it is encapsulated as fuzzy set membership, or if information irrelevant it is eventually dropped.
- A Stochastic Reinforcement Learning Automation (SELA) which, given a set of input parameters, internal state and a set of possible actions computed the beast action out of the set. Applied to the routing problem, given a set of paths, it computes the "best" route for a given set of connection constraints.
- An Evolutionary Fuzzy Time Series Predictor (EFTSP) that can be used for predicting traffic loads on network links based on past link utilization information.

The features of the model itself are highly appealing. First, the system is highly adaptive. For example, information collection is demand-based. If no guaranteed-service connections are requested, no qualitative parameters such as bandwidth availability are collected by the roaming agents. If, in contrast, very specific guarantees are required, this kind of information is collected and maintained. In a more interesting case, if connection requirements are expressed in fuzzy terms, such terms are used in abstracting routing information as well. This is in sharp contrast with current routing solutions, which communicated and consider a "worst-case" maximum amount of routing information thus must always be present at every node, by design. Second, the system is extensible. New components and improved algorithms can easily be added to the architecture.

5 Conclusions

In this paper, some applications of computational intelligence techniques in active networking technology were presented. One of these possible applications is the novel approach to routing in active networks, which promises to address different aspects of the problem, using the strengths of computational intelligence and the infrastructural provisions of active networks. Another interesting application area is in problems that have a lower feedback cycle, such as traffic shaping and policing. Since such a lower cycle may provide faster convergence the cost of the increased computational intensity becomes apparent, revealing a clear tradeoff, which should be considered. This is our belief that this area of research is open area and this work is a stand stone to tackle the more complex problems.

References

1. Tennenhouse, D.L.: A Survey of Active Network Research, IEEE Communication Magazine, 35(1) (1997) 80–86
2. Boutaba, R., Polyrakis, A.: Projecting Advanced Enterprise Network and Service Management to Active Networks, IEEE Network (2002) 28-33
3. Psounis, K.: Active Networks: Applications, Security, Safety, and Architectures, IEEE Communication Surveys, 2(1) (1999) 65-71
4. Smith, J.M.: Activating Networks: A Progress Report, Comp., 32(4) (1999) 32–41
5. Pedrycz, W.: Fuzzy Sets Engineering, CRC Press, Boca Raton (1995)
6. Jalili-Kharaajoo, M.: Predictive Control of a Continuous Stirred Tank Reactor Based on Neuro-Fuzzy Model of the Process, SICE Annual Conference, Fukui, Japan (2003) 770-775
7. Pedrycz, W.: Computational Intelligence: An Introduction, CRC Press, Boca Raton (1997)
8. Jalili-Kharaajoo, M., Ebrahimirad, H.: Improvement of second order sliding mode control applied to position control of induction motors using fuzzy logic, LNAI, Proc. IFSA2003, Istanbul, Turkey (2003)
9. Moshiri, B., Jalili-Kharaajoo, M., Besharati, F.: Application of fuzzy sliding mode control based on genetic algorithms to Building Structures, in Proc. 9th IEEE Conference on Emerging Technology and Factory Automation, 16-19 September, Lisbon, Portugal (2003)
10. Bezdek, J.C.: What is computational intelligence? In M. Zurada, J., II, R. J. M., and Robinson, C. J., editors, Computational Intelligence Imitating Life, (1994) 1–12
11. Ascia, G., Catania, V., Ficili, G., Palazzo, S., Panno, D.: A VLSI fuzzy expert system for real-time traffic control in ATM networks. IEEE Transactions on Fuzzy Systems, 5(1) (1997) 20–31
12. Catania, V., Ficili, G., Palazzo, S., Panno, D.: A fuzzy expert system for usage parameter control in ATM networks, in Proc. GlobeCom95, (1995) 1338–1342
13. Cheng, R.-G., Chang, C.-J.: Design of a fuzzy traffic controller for ATM networks. IEEE/ACM Transactions on Networking, 4(3) (1996) 460–469
14. Park, Y.-K., Lee, G.: Applications of neural networks in high-speed communication networks. IEEE Communications Magazine, 33(10) (1995) pp.68–74

15. Pitsillides, A., Pattichis, C., Stekercioglu, Y. A., Vasilakos, A.: Bandwidth allocation for virtual paths using evolutionary programming (EP-BAVP), in Proceedings of the International Conference on Telecommunications ICT'97, Melbourne, Australia (1997) 1163–1168
16. Bigham, J., Rossides, L., Pitsillides A., Sekercioglu, A.: Overview of Fuzzy-RED in Diff.-Ser Networks, LNCS 2151 (2001) 1-13
17. Zelezn, F., Zidgek, J., Stepankov, O.: A Learning System for Decision Support in Telecommunications, LNCS 2151 (2001) 83-101
18. Sekercioglu A., Pitsillides A., Vasilakos A.: Computation Intelligence in Management of ATM Networks: a Survey of Current State of Research, Soft Computing, 5 (2001) 279-285
19. Pedrycz, W., Vasilakos, A.: Computational Intelligence in Telecommunication Networks, CRC Press, Boca Raton (2000)
20. Calvert, K., Architectural Framework for Active Networks. Architecture Working Group (1998)
21. Wong, B.K., Lai, V.S., Lam J.: A bibliography of neural network business applications research: 1994-1998. Computer and Operation Research, 27 (2000) 1045-1076
22. Glorfeld, L.W., Hardgrave, B.C.: An improved method for developing neural networks: the case of evaluating commercial loan credit worthiness. Computer and Operations Research, 23(10) (1996) 933-944
23. Mukesh, D.: Hate statistics? Try neural networks. Chemical Engineering: – AST (1997) 96-104
24. Chen, C.L., Kaber, D.B., Dempsey, P.G.: A new approach to applying feedforward neural networks to the prediction of musculoskeletal disorder risk. Applied Ergonomics, 31 (2000) 269-282
25. Zadeh, L.A.: Fuzzy Sets, Information and Control, 8(3) (1965) 338-353
26. Yager, R.: On a general class of fuzzy connectives, Fuzzy Sets and Systems, 4 (1980) 235-242
27. Jalili-Kharaajoo, M., Besharati, F.: Fuzzy variance analysis model, LNCS, Proc. ISCIC2003, Antalya, Turkey (2003)
28. Kosko, K.: Neural Networks and Fuzzy Systems. Prentice-Hall, Englewood Cliffs, NJ (1992)
29. Jang J., Sun, C.: Neuro-fuzzy modeling and control. Proc. IEEE, 83(3) (1995) 378-406
30. Fogel, D.B.: A comparison of evolutionary programming and genetic algorithms on selected constrained optimization problems, Simulation, 64 (6) (1995) 1051-1055
31. Goldberg, D.E.: Genetic Algorithm in Search, Optimization, and Machine Learning. Reading, MA: Addison-Wesley (1989)
32. Jalili-Kharaajoo, M., Besharati, F.: Intelligent Predictive Control of a Solar Power Plant with Neuro-Fuzzy Identifier and Evolutionary Programming Optimizer, in Proc. 9th IEEE Conference on Emerging Technology and Factory Automation, 16-19 September, Lisbon, Portugal (2003)
33. Fogel, D.: Evolving Artificial Intelligence, Ph.D. thesis, University of California, San Diego, 1992.
34. Cho, S.B.: Fusion of neural networks with fuzzy logic and genetic algorithm, Integrated Computer-Aided Engineering, 9 (2002) 363–372
35. Fukuda, T.: Fuzzy-neuro-GA based intelligent robotics, in: Computational Intelligence: Imitating Life, J.M. Zurada, R.J. Marks II, C.J. Robinson, eds, IEEE Press (1994) 252–263
36. Kasabov, N., Watts, M.: Genetic algorithms for structural optimization, dynamic adaptation and automated design of fuzzy neural networks, in Proc. International Conference on Neural Networks ICNN'97, Houston (1997) 1232-1237

Fuel-Efficient Maneuvers for Constellation Initialization Using Fuzzy Logic Control

Mengfei Yang[1,2], Honghua Zhang[2], Rucai Che[2], and Zengqi Sun[1]

[1] Tsinghua University, Beijing 100084, China
yangmfbice@163.com
[2] Beijing Institute of Control Engineering, Zhongguancun, Beijing 100080, China
hhzhang@sina.com

Abstract. This paper deals with fuel-efficient maneuvers for constellation initialization. A new orbital controller based on fuzzy logic is proposed. It is composed of two level controllers: the high level controller is a fuzzy logic based planner, which resolves the conflicting constraints induced from the control efficiency and the control accuracy, while the base level controller is the well-known Folta-Quinn algorithm. The analysis and simulation studies indicate that the algorithm is very effective in reducing fuel cost for constellation formation capturing control.

1 Introduction

The constellation initialization requires that each satellite captures its nominal orbit after its deployment from the same launch vehicle. The fuel-efficient orbital maneuver is necessary because of the limit fuel carried by each satellite and the mission life requirement. The aim of this paper is to design the fuel-efficient orbit maneuver controller for the constellation initialization.

The main points of this paper are summarized as follows: 1) The orbital maneuver of each satellite can be translated as correcting the 6 orbital elements, while the fuel cost of the each orbit element correction depends on the phase (or the position) of each satellite in its orbit. Therefore, the fuel-optimal control rules about the control phase are derived for each orbital element based on the orbital dynamics. 2) The constellation initialization requires that the error between the nominal orbit and the actual orbit should not be too large in the process of orbital maneuver, which may result conflicting constraints with the fuel-optimal orbital element correction. Therefore, the fuzzy logic is utilized to form a new fuzzy logic based orbital controller, which is able to resolve the conflicting constraints between the control error and the control fuel efficiency, and guarantee the fuel cost is almost optimal. 3) The control law proposed in this paper is composed of two level controllers: the high level controller is a fuzzy logic based planner described as above, which resolving the conflicting constraints induced from the control efficiency and control error, while the base level controller is the well-known Folta-Quinn algorithm [1],[2]. 4) The analysis and simulation studies are made for a constellation of 3 satellites to demonstrate the validity of the proposed controller.

It is noted that there are many theoretical results on the optimal orbital maneuver control [3], most of which apply the linearized relative orbital dynamics and the maximum value principle. However, the solution searching related to those results needs repetitious test and optimization, which does not adapt to be realized in the satellite on-board computer. In contrast, the feature of the proposed fuel-efficient controller based on fuzzy logic in this paper is that it is very easy to be realized in the satellite on-board computer, and is able to achieve the almost fuel-optimal cost orbital maneuver by resolving the conflicts. It is also noted that the proposed controller can be further developed through using more recent work on self-tuning fuzzy control as [4] and [5] to get more general results.

2 Mathematics Model

Without loss of generality, consider the constellation of 3 satellites. Let $\sigma(a,e,i,\Omega,\omega,M)$ be the orbital parameters of one satellite, $\sigma_0(a_0,e_0,i_0,\Omega_0,\omega_0,M_0)$ be the nominal orbital parameters, where a,e,i,Ω,ω,M are the semimajor axis, the eccentricity, the inclination, the right ascension longitude of the ascending node, the argument of the perigee, and the mean anomaly, respectively. So the constellation orbital dynamics using Gauss perturbation equation is described as follows[3]:

$$\frac{da}{dt} = \frac{2}{n\sqrt{(1-e^2)}}[F_r e \sin f + F_t(1+e\cos f)] \quad (1)$$

$$\frac{de}{dt} = \frac{\sqrt{(1-e^2)}}{na}[F_r \sin f + F_t(\cos E + \cos f)] \quad (2)$$

$$\frac{di}{dt} = \frac{r\cos(\omega+f)}{na^2\sqrt{(1-e^2)}}F_n \quad (3)$$

$$\frac{d\Omega}{dt} = \frac{r\sin(\omega+f)}{na^2\sqrt{1-e^2}\sin i}F_n \quad (4)$$

$$\frac{d\omega}{dt} = \frac{\sqrt{(1-e^2)}}{nae}[-F_r\cos f + F_t\frac{2+e\cos f}{1+e\cos f}\sin f] - \cos i\frac{d\Omega}{dt} \quad (5)$$

$$\frac{dM}{dt} = n - \frac{1-e^2}{nae}[F_r(\frac{2er}{p}-\cos f) + F_t(1+\frac{r}{p})\sin f] \quad (6)$$

where F_r, F_t, F_n are the control acceleration components generated by thrusters along radial, transverse, and normal directions, respectively; $n = \sqrt{\mu/a^3}$ is the orbital rate; $p = a(1-e^2)$ is the semilatus; f and E are the true anomaly and the eccentric anomaly, respectively; r is the distance from the earth to the satellite.

In case of the thrust duration is far less than the orbital period, the control acceleration generated by the thruster can be simplified as impulsive thrust. In another word, if let the control acceleration be denoted by $\vec{u}(t) = [F_r \quad F_t \quad F_n]^T$, then

$$\vec{u}(t) = \sum_{k=1}^{m} \Delta \vec{v}_k \delta(t - t_k) \tag{7}$$

where $\delta(t - t_k)$ is the Dirac δ function (impulsive function) at the moment $t = t_k$, $\Delta \vec{v}_k$ is the velocity increment of $t = t_k$, m is the number of impulses.

The constellation initialization can be referred to as designing a series of two elements set including control time and the velocity increment $(t_k, \Delta v_k)$ ($k = 1, \cdots, m$) for each satellite, completing the transfer from the actual orbit to the nominal orbit so that

$$|\sigma(a,e,i,\Omega,\omega,M) - \sigma_0(a_0,e_0,i_0,\Omega_0,\omega_0,M_0)| < \delta, \text{ for all time t} \tag{8}$$

$$|\sigma(a,e,i,\Omega,\omega,M) - \sigma_0(a_0,e_0,i_0,\Omega_0,\omega_0,M_0)| < \varepsilon, \text{ for some time } t > t_m \tag{9}$$

where δ, ε are the permission transfer orbit error and the expected terminal orbit error.

3 Control Law

The constellation initialization control comes to the orbital maneuver control of each satellite from the initial orbit to the terminal orbit There are infinite kinds of control law $(t_k, \Delta v_k)$ ($k = 1, \cdots, m$) for the orbital maneuver. However, the different control law results the different transfer orbits (the transfer trajectories), which lead to the different fuel cost. Thereby, it is necessary to design a control law to achieve the fuel-efficient orbital maneuvers.

3.1 Control Efficiency

The orbital maneuver control comes to the 6 orbital parameters correction control. Each parameter correction has an appropriate control phase. The same fuel cost in different phases results in the different effects. The relation between the control phases and the control effects determines control efficiency. Observing the influence coefficients of the control acceleration in Gauss perturbation equations (1)-(6), one can find the following relations between the control efficiency and the satellites phases:

a. Correcting i is most efficient, as the latitude argument satisfies:

$$\cos(\omega + f) = 1, \text{ or } (\omega + f) = k\pi, \quad k = 0,1,\cdots \cdots \tag{10}$$

where the satellite pass the equator.

b. Correcting Ω is most efficient, as the latitude argument satisfies:

$$\sin(\omega+f)=1, \text{ or } (\omega+f)=k\pi+\frac{\pi}{2}, \ k=0,1,\cdots \qquad (111)$$

where the satellite pass the south or north apex.

c. The semimajor axis a can be corrected at any phase. Correcting a by the transverse control force is more efficient than by the radial control force. With the most efficient correction on a by the transverse control force, the true anomaly satisfies:

$$\cos(f)=1, \text{ or } f=2k\pi, \ k=0,1,\cdots \qquad (12)$$

where the satellite is at perigee.

d. The eccentricity e can be corrected at any phase. Correcting e by the transverse control force is more efficient than by the radial control force. With the most efficient correction on e by the transverse control force, the true anomaly satisfies:

$$\cos(f)=\pm 1, \text{ or } f=k\pi, \ k=0,1,\cdots \qquad (13)$$

where the satellite is at apogee or perigee.

e. The argument of perigee ω can be corrected at any phase. With the most efficient correction on ω by the transverse control force, the true anomaly satisfies:

$$\sin(f)=\pm 1, \text{ or } f=k\pi+\pi/2, \ k=0,1,\cdots \qquad (14)$$

where the satellite is at the midpoint between the apogee and the perigee.

f. The mean anomaly M can be corrected at any phase. With the most efficient correction on M by the transverse control force, the true anomaly satisfies:

$$\sin(f)=\pm 1, \text{ or } f=k\pi+\pi/2, \ k=0,1,\cdots \qquad (15)$$

where the satellite is at the midpoint between the apogee and the perigee.

Thereby, the efficiency of orbital parameter correction depends strongly on satellite phases. In the simple case that only one orbital parameter needs to be corrected, the control phase for the orbital maneuver can be determined as above. In the general case that more than one orbital parameter need to be corrected, the control phase for the orbital maneuver can be determined as the plan described in Fig. 1 as follows:

1) The correction of the semimajor axis and eccentricity takes place between the perigee (the initial maneuver position) and the apogee (the terminal maneuver position);
2) The correction of the argument of the perigee and the mean anomaly takes place between two midpoints of perigee and the apogee;
3) The correction of the inclination takes place between the ascending equator and the descending equator;
4) The correction of the right ascension of ascending node takes place between the south apex and the north apex.

Let T denote the orbit period, then the maneuver duration last about 2 times T, thus the plan is called the "2T"-control strategy.

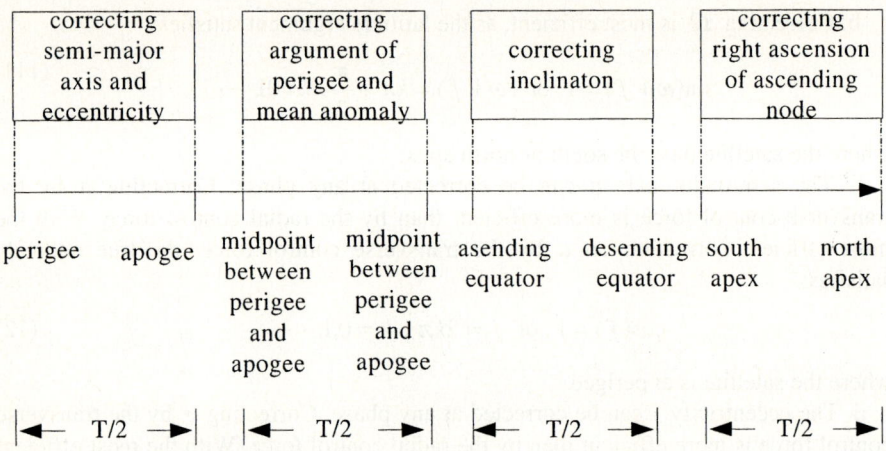

Fig. 1. "2T"-plan of orbital maneuvers control

3.2 Control Rules Based on Fuzzy Logic Rule

The requirements of the control efficiency and the transfer orbit error given by the equation (8) may conflict during the process of the constellation initialization: in some cases, the satellite is at the position requiring orbit maneuver, but the satellite is not at the control efficient phase; while in some other cases, the satellite is at the right control efficient phase, but the satellite is not at the position needed for the orbital maneuver. Obviously, if the bivalent logic is used for the constellation initialization, there may be no solution for the control efficiency. Therefore, the fuzzy logic is proposed to be used to resolve the conflicting constraints of control accuracy and control efficiency.

According to the "2T"-control strategy stated as above, the control rules based on fuzzy logic for each satellite of the constellation are described as follows.

"If 'the semi-major axis is not proper' and 'the satellite is around the perigee', then perform the satellite maneuver, correct the semi-major axis";

"If 'the eccentricity is not proper' and 'the satellite is around the perigee or the apogee', then perform the satellite maneuver, correct the eccentricity";

"If 'the argument of the perigee is not proper' and 'the satellite is around the midpoint between the perigee and the apogee', then perform the satellite maneuver, correct the argument of perigee";

"If 'the mean anomaly is not proper' and 'the satellite is around the midpoint between the perigee and the apogee', then perform the satellite maneuver, correct the mean anomaly";

"If 'the inclination is not proper' and 'the satellite ascend pass the equator or descend pass the equator', then perform the satellite maneuver, correct the inclination";

"If 'the right ascension of ascending node is not proper' and 'the satellite pass by the south or the north apex', then perform the satellite maneuver, correct the right ascension of ascending node";

"If any above condition clause does not hold, then do not perform the corresponding satellite maneuvers".

By the strategies above, if a few conditions cause maneuvers operation meantime, combine all the corresponding strategies to achieve the maneuvers aim: e.g. if both semi-major axis and inclination exceed errors, the satellite's location is at perigee, as well as the perigee is near the equator, then, the maneuver operation will correct both semi-major axis and inclination. If the maneuvers triggered by one or several conditions, all strategies are not available until the maneuvers end. It should be noted that it spends about half orbital period to complete any maneuver, and it spends about "2T" to complete all maneuvers resulted from above rules.

3.3 Control Planner Based on Fuzzy Logic Rule: Determine the Control Moment

The fuzzy logic based planner includes: fuzzifier, fuzzy inference and defuzzifier to every above control rule based on a fuzzy set.

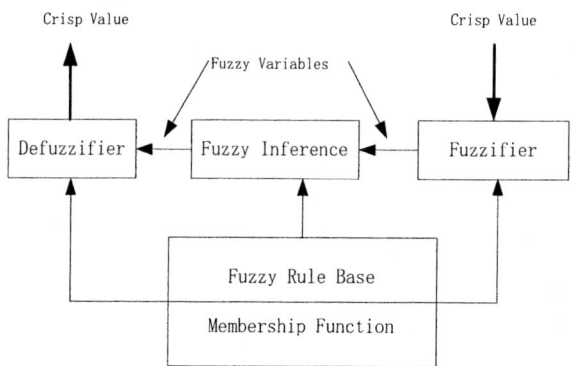

Fig. 2. Planner based on fuzzy logic

3.3.1 Fuzzification

For the fuzzification of the input variables including the orbital parameters and the satellite phases, there are two methods of fuzzification, generally, i.e. single value fuzzification and non-single value fuzzification. In the set formed by single value fuzzification, the 'degree of membership' of given input value is 1, while the 'degrees of membership' of other input values are all 0. In the set formed by non-single fuzzification, the 'degree of membership' of given input value is 1, while the 'degrees of membership' of other input values are decreasing to 0 so that the membership function is of triangle shape. The following Case 1 and Case 2 illustrate the fuzzification for the input variable of semimajor axis and true anomaly.

Case 1, The input of the semimajor axis is a_{input}. The fuzzy set formed by the single fuzzification $AS' =$ {semimajor axis is a_{input} }={ $(a, \mu_{as'}(a))$ },

$$\mu_{as'}(a) = \begin{cases} 1 & a = a_{input} \\ 0 & otherwise \end{cases}$$

Case 2, The input of the true anomaly is $\bar{f}_{input}$. The fuzzy set formed by the single fuzzification $PS' =$ {true anomaly is $\bar{f}_{input}$ }={ $(f, \mu_{ps'}(f))$ },

$$\mu_{ps'}(f) = \begin{cases} 1 & f = \bar{f}_{input} \\ 0 & otherwise \end{cases}.$$

For the fuzzification of each rule stated above, the fuzzy set is formed by the implication relation induced from the rule. The membership function is select as the minimum value among the relative variables. The following Case 3 illustrates the fuzzificatin of rules R1, R2, and R3.

Case 3, The 3 rules are

R_1: If $\bar{a}$ is NAS and $\bar{f}$ is APS, jp is ON;

R_2: if $\bar{a}$ is AS, jp is OFF;

R_3: if $\bar{f}$ is $NAPS$, jp is OFF,

where, NAS is the fuzzy set defined as 'the semimajor axis is not proper', APS is the fuzzy set defined as the 'the satellite is near the perigee', $NAPS$ is the fuzzy set defined as 'the satellite is not near the perigee', ON is the fuzzy set defined as 'the orbital maneuver is on', OFF is the fuzzy set defined as 'the orbital maneuver is off'. The fuzzy sets are

$R_1 = NAS \text{and} APS \to ON = \{(\bar{a}, \bar{f}, jp)\}$

$\mu_{R_1} = \mu_{NAS\text{and}APS} \wedge \mu_{ON} = \min\{\mu_{NAS\text{and}APS}, \mu_{ON}\}$

$R_2 = AS \to OFF = \{(\bar{a}, jp)\}$

$\mu_{R_2} = \mu_{AS} \wedge \mu_{OFF} = \min\{\mu_{AS}, \mu_{OFF}\}$

$R_3 = NAPS \to OFF = \{(\bar{a}, jp)\}$

$\mu_{R_3} = \mu_{NAPS} \wedge \mu_{OFF} = \min\{\mu_{NAPS}, \mu_{OFF}\}$

3.3.2 Fuzzy Inference

By evaluating each input variable in the condition clauses, a lot of new fuzzy sets are constructed. An object set is a union joined by the new fuzzy sets and the fuzzy sets determined by each rule. The "degree of membership" of the object set applies the method "maximum and minimum" in union operation. The following Case 4 illustrates this process.

Case 4, Let $\bar{a}_{input}$ be semi-major axis, $\bar{f}_{input}$ be argument of perigee, the following fuzzy inference is considered:

Input: $\bar{a}$ is AS', $\bar{f}$ is APS'
R_1: If $\bar{a}$ is NAS and $\bar{f}$ is APS, jp is ON ;
R_2: if $\bar{a}$ is AS, jp is OFF ;
R_3: if $\bar{f}$ is $NAPS$, jp is OFF ;
Output: jp is ON'

Now, it is expect to get ON' from AS' and APS'. By the fuzzy set theory, we have the output set resulted from the union operation by input sets and implicit relation sets.

$$ON' = (AS' \text{and} APS') \circ \bigcup_{i=1}^{3} R_i = \bigcup_{i=1}^{3} \left[(AS' \text{and} APS') \circ R_i \right] \tag{16}$$

where the union operation applies "maximum and minimum" method.

$$(AS' \text{and} APS') \circ R_i = \{jp\} \tag{17}$$

$$\mu_{(AS' \text{and} APS') \circ R_i}(jp) = \max \min \{\mu_{(AS' \text{and} APS')}(\bar{a}, \bar{f}), \mu_{R_i}(\bar{a}, \bar{f}, jp)\} \tag{18}$$

$$\mu_{ON'}(jp) = \max \{\mu_{(AS' \text{and} APS') \circ R_1}(jp), \mu_{(AS' \text{and} APS') \circ R_2}(jp), \mu_{(AS' \text{and} APS') \circ R_3}(jp)\} \tag{19}$$

By the union operation on fuzzy set, relation (17)-(19) equals to:

$$\mu_{(AS' \text{and} APS') \circ R_1}(jp) = \alpha_1 \wedge \mu_{ON}(jp) \tag{20}$$

$$\mu_{(AS' \text{and} APS') \circ R_2}(jp) = \alpha_2 \wedge \mu_{OFF}(jp) \tag{21}$$

$$\mu_{(AS' \text{and} APS') \circ R_3}(jp) = \alpha_3 \wedge \mu_{OFF}(jp) \tag{22}$$

where

$$\alpha_1 = \left[\max_{\bar{a}} (\mu_{AS'}(\bar{a}) \wedge \mu_{NAS}(\bar{a})) \right] \wedge \left[\max_{\bar{f}} (\mu_{APS'}(\bar{f}) \wedge \mu_{APS}(\bar{f})) \right] \tag{23}$$

$$\alpha_2 = \left[\max_{\bar{a}} (\mu_{AS'}(\bar{a}) \wedge \mu_{AS}(\bar{a})) \right] \tag{24}$$

$$\alpha_3 = \left[\max_{\bar{f}} (\mu_{APS'}(\bar{f}) \wedge \mu_{NAPS}(\bar{f})) \right] \tag{25}$$

By deriving reasoning above, it leads to: $\alpha_1 = \frac{1}{4} \wedge \frac{1.5}{4} = \frac{1}{4}$, $\alpha_2 = \frac{3}{4}$, $\alpha_3 = \frac{3}{4}$. Then, obtain the output set ON' on Fig.3.

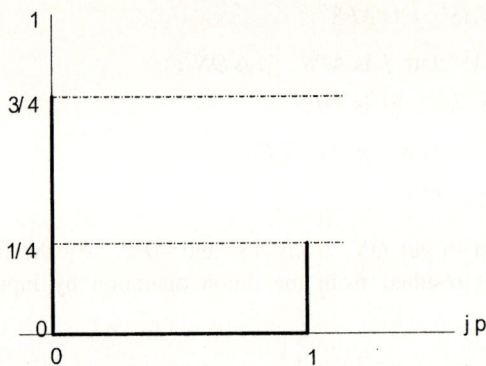

Fig. 3. Output set ON'

3.3.3 Defuzzifier

Defuzzifier is to determine an element (clarity value) and its "degree of membership" according to the object fuzzy set. Then it's possible to determine whether the result clauses (orbital maneuvers operation) are to implement. A simplest method to clarify is called "maximum degree of membership", i.e. by choosing the element of the output set whose value is maximum, define the maximum value as clarity value. This is illustrated in following Case 5.

Case 5, In the output set in case 4, the clarity value is $jp = 0$, relative "degree of membership" is $3/4$. So, the maneuver is not operated by the plan mentioned above.

The above fuzzy logic based planner, which is actually a high level controller of the autonomous control, determines not only the control moment but also the initial state and terminal state in the orbital maneuver control. Then, the control value will be given in the base level controller.

3.4 Base-Level Controller: Determine the Control Value

Fuzzy planner generates the control phases, the initial state, and the terminal state relative to the "2T"-control strategies. Let $(\vec{r}(t_0), \vec{v}(t_0))$ be the initial state, $(\vec{r}(t_f), \vec{v}(t_f))$ be the terminal state, where $\vec{r}$ and $\vec{v}$ represent positon and velocity on the earth centered inertial frame. According to Folta-Quinn algorithm [1],[2], the multi-impulsive control strategy can be given as:

$\Delta \vec{v}(t_0) = V^*(t_0) R^{*-1}(t_0) \delta \vec{r}(t_0) - \delta \vec{v}(t_0)$

$\delta \vec{r}(t_0) = \vec{r}(t_0) - \vec{r}_0(t_0)$, $\delta \vec{v}(t_0) = \vec{v}(t_0) - \vec{v}_0(t_0)$;

$((\vec{r}(t_0), \vec{v}(t_0))$ is measurement value)

$\Delta \vec{v}(t_1) = V^*(t_1) R^{*-1}(t_1) \delta \vec{r}(t_1) - \delta \vec{v}(t_1)$

$\delta \vec{r}(t_1) = \vec{r}(t_1) - \vec{r}_0(t_1)$, $\delta \vec{v}(t_1) = \vec{v}(t_1) - \vec{v}_0(t_1)$;

$((\vec{r}(t_1), \vec{v}(t_1))$ is measurement value)

$$\Delta \vec{v}(t_2) = V^*(t_2)R^{*-1}(t_2)\delta\vec{r}(t_2) - \delta\vec{v}(t_2)$$
$$\delta\vec{r}(t_2) = \vec{r}(t_2) - \vec{r}_0(t_2), \quad \delta\vec{v}(t_2) = \vec{v}(t_2) - \vec{v}_0(t_2);$$

$((\vec{r}(t_2), \vec{v}(t_2)))$ is the measurement value)

....

$$\Delta \vec{v}(t_{n-1}) = V^*(t_{n-1})R^{*-1}(t_{n-1})\delta\vec{r}(t_{n-1}) - \delta\vec{v}(t_{n-1})$$
$$\delta\vec{r}(t_{n-1}) = \vec{r}(t_{n-1}) - \vec{r}_0(t_{n-1}), \quad \delta\vec{v}(t_{n-1}) = \vec{v}(t_{n-1}) - \vec{v}_0(t_{n-1});$$

$((\vec{r}(t_{n-1}), \vec{v}(t_{n-1})))$ is the measurement value)

$$\Delta \vec{v}(t_f) = \vec{v}(t_f) - \vec{v}(t_f^-);$$

($\vec{v}(t_f^-)$ is the measurement value).

where, $(\vec{r}_0(t_i), \vec{v}_0(t_i))$ is the reference orbital state, $\Delta \vec{v}(t_i)$ is the relative velocity increment given by the control force.

The combination of the fuzzy logic based planner and the base-level controller forms the complete control law for the orbital maneuvers of constellation initialization.

4 Simulation Results

The simulation studies are made for the constellation initialization control, which is composed of a leading satellite and two following satellites. The nominal mean orbital elements are given in Table 1. The initial actual mean orbital elements are given in Table 2. The satellites' parameters are given in Table 3.

The control objective is to complete the constellation initialization by making each satellite capture its nominal orbit through the orbital maneuvers. The control law based on fuzzy logic proposed in this paper is utilized. The simulation results are shown by Tables 4-5 and Fig.4. The Fig.4 shows the orbital elements' history of the leading satellite during the orbital maneuvers. The Table 4 indicates the errors between the actual orbit and the nominal orbit after the orbital maneuvers. The Table 5 gives the control velocity increments (related to fuel cost) for each satellite. We compare the fuel cost with that of theoretical fuel-optimal control, which indicates that the proposed controller present almost fuel-optimal for the orbital maneuvers.

Table 1. The nominal mean orbital elements of the satellites

	Leading satellite	Following satellite 1	Following satellite 2
a (km)	7077.732	7077.732	7077.782
e	0.0010444	0.0010444	0.0010444
i (deg)	98.2102	98.2102	98.2102
Ω (deg)	188.297	188.547	189.297
ω (deg)	90.0	90.0	90.0
M (deg)	0.0	-3.645	-3.645

Table 2. The initial mean orbital elements of the satellites

	Leading satellite	Following satellite 1	Following satellite 2
a (km)	7072.732	7072.732	7072.782
e	0.0010444	0.0010444	0.0010444
i (deg)	98.2602	98.2602	98.2602
Ω (deg)	188.287	188.537	189.287
ω (deg)	90.0	90.0	90.0
M (deg)	0.0	-3.645	-3.645

Table 3. The satellites' parameters

	Leading satellite	Following satellite 1	Following satellite 2
Drag area (m^2)	19.0	7.7	7.7
Mass (kg)	2041	529	529
Ballistic coefficient (m^2/kg)	0.0093	0.0146	0.0146

Table 4. The satellites' orbital error after maneuvers

	Leading satellite	Following satellite 1	Following satellite 2
a (km)	3.997173641808331e+000	-3.825943507254124e+000	-3.733077733777463e+000
e	-2.761827728973033e-007	-2.657781140937184e-007	-2.600482751169506e-007
i (deg)	-1.861868721647546e-008	-4.478373658125342e-008	-5.976489974185141e-008
Ω (deg)	-4.378249796752870e-009	-5.423838064955913e-009	-5.604544459737944e-009
ω (deg)	-4.831685062731913e-003	-4.656147215549164e-003	-4.586139368894873e-003
M (deg)	4.827844629503300e-003	4.652108639755664e-003	4.581993999006565e-003

Table 5. The velocity increments of control used for orbital maneuver

	Leading satellite	Following satellite 1	Following satellite 2
Δv correcting a (m/s)	4.757970010946486	4.755761081871730	4.764105520997420
Δv correcting i (m/s)	6.659891731044932	6.660358212506530	6.660588676447985
Δv correcting Ω (m/s)	1.366379351351783	1.365854079243388	1.365907194139736
Total (m/s)	12.78424109334320	12.78197337362165	12.79060139158514

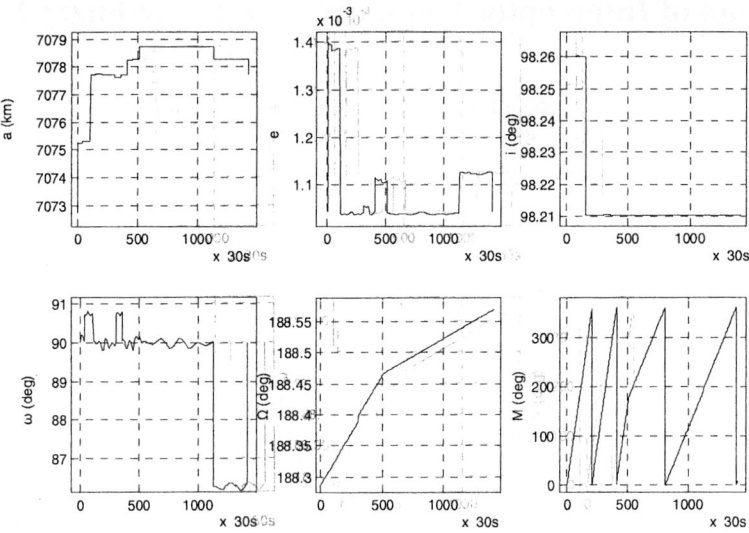

Fig. 4. History of the mean orbital parameters of leading satellite

5 Conclusions

A new orbital controller based on fuzzy logic is proposed. It is composed of two level controllers: the high level controller is a fuzzy logic based planner, which resolves the conflicting constraints induced from the control efficiency and the control accuracy, while the base level controller is the well-known Folta-Quinn algorithm. The analysis and simulation studies indicate that the algorithm is very effective in reducing fuel cost for constellation formation capturing control.

References

1. Joseph, R. Q.: EO-1 Technology Validation Report Enhanced Flying Formation Algorithm (JPL). NASA/GSFC, August 8, (2001)
2. David, F., Hawkins, A.: Results of NASA's First Autonomous Formation Flying Experiment : Earth Observing-1 (EO-1). AIAA 2002-4743, (2002)
3. Battin, R.: An Introduction to the Mathematics and Methods of Astrodynamics, AIAA Education Series, Chapter 9 and 11. (1987)
4. Kiguchi, K., Tanaka, T., Fukuda, T.: Neuro-fuzzy Control of a Robotic Exoskeleton with EMG Signal. IEEE Trans. Fuzzy Systems, 12 (2004) 481-490
5. Wang, L., Frayman, Y.: A Dynamically-generated Fuzzy Neural Eetwork and its Application to Torsional Vibration Control of Tandem Cold Rolling Mill Sprindles. Engineering Applications of Artificial Intelligence, 15 (2003) 541-550

Design of Interceptor Guidance Law Using Fuzzy Logic

Ya-dong Lu, Ming Yang, and Zi-cai Wang

Control & Simulation Center, Harbin Institute of Technology,
150080 Harbin, P.R. China
luyd@hit.edu.cn

Abstract. In order to intercept the high-speed maneuverable targets in three-dimensional space, a terminal guidance law based on fuzzy logic systems is presented. After constructing the model of the relative movement between target and interceptor, guidance knowledge base which including fuzzy data and rules is obtained according to the trajectory performance. On the other hand, considering the time-variant and nonlinear factors in the thrust vector control system, the interceptor's mass is identified in real time. By using the learning algorithms, the logic rules are also revised correspondingly to improve the fuzzy performance index. Simulation results show that this method can implement efficiently the precise guidance of the interceptor as well as preferable robust stability.

Nomenclature

g	Gravity acceleration
m	Interceptor mass
P_x, P_y, P_z	Thrust in body frame
r	Range between interceptor and target
x_M, y_M, z_M	Coordinates of the interceptor
x_T, y_T, z_T	Coordinates of the target
v_M	Interceptor speed
v_T	Target speed
α	Angle-of-attack
β	Sideslip angle
θ_M	Flight path angle of the interceptor
ψ_M	Flight path azimuth angle of the interceptor
θ_T	Flight path angle of the target
ψ_T	Flight path azimuth angle of the target
q_ε	Line-of-sight elevation angle
q_β	Line-of-sight azimuth angle

1 Introduction

As a precise guidance weapon, the role of space interceptor is to engage incoming adversarial target and to destroy it by collision. The interceptor trajectory consists of three relatively guided stages: boost to accelerate, midcourse to steer along a collision trajectory with the target, and terminal to correct for any remaining position and velocity errors. In these stages, guidance law during the terminal stage is a key to a successful intercept.

Some classical guidance laws, such as proportional navigation guidance (PNG)[1], are deduced on the supposition that target flights without maneuvering. But when considering interceptors' time-variant nonlinear factors, it is difficult for interceptors using these laws which are based on precise mathematical model to destroy targets performing uncertain maneuvers.

On the other hand, it is well known that fuzzy logic control have the ability to make use of knowledge expressed in the form of linguistic rules without completely resorting to the precise plant models. In recent years, researchers have also attempted to apply it on missile guidance designs[2]. However, different from common missiles, all of the interceptor actuators are made up of thrust engines and the mass of interceptor varies along with vanishing energy. Therefore interceptor actual overloads are influenced by time-variant mass. Considering these varying factors, a self-tuning fuzzy logic guidance law is designed. In this design scheme, when current mass is estimated on-line, interceptor guidance rules are also revised to improve impact accuracy and trajectory performance. Simulation results show that the proposed guidance scheme has potential for use in the development of high performance interceptors.

2 Interceptor and Target Dynamic Models

The 3-D space can be divided into two planes: vertical plane and horizontal plane. With this arrangement, we can obtain the geometry relations between interceptor and target in the Cartesian inertial frame, as shown in Fig.1. M and T denote mass points of interceptor and target respectively.

Supposing that interceptor flights out of aerosphere during the terminal stage, we consider there are no aerodynamic forces exerting on the interceptor. The interceptor is modeled as a point mass and the equations of motion are described by

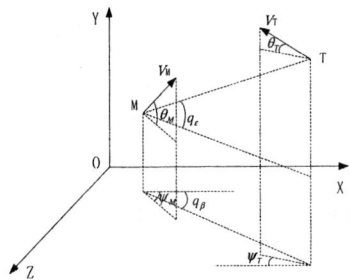

Fig. 1. Intercept Geometry

$$\begin{cases} \dot{x}_M = v_M \cos\theta_M \cos\psi_M \\ \dot{y}_M = v_M \sin\theta_M \\ \dot{z}_M = -v_M \cos\theta_M \sin\psi_M \\ \dot{v}_M = -g\sin\theta_M - \dfrac{P_y \sin\alpha\cos\beta}{m} + \dfrac{P_z \sin\beta}{m} \\ \dot{\theta}_M = \dfrac{-mg\cos\theta_M + P_y \cos\alpha}{mv_M} \\ \dot{\psi}_M = \dfrac{-P_y \sin\alpha\sin\beta - P_z \cos\beta}{mv_M \cos\theta_M} \end{cases} \quad (1)$$

Based on the same supposing, when without considering target's maneuverability, its equations of motion are

$$\begin{cases} \dot{x}_T = v_T \cos\theta_T \cos\psi_T \\ \dot{y}_T = v_T \sin\theta_T \\ \dot{z}_T = -v_T \cos\theta_T \sin\psi_T \\ \dot{v}_T = -g\sin\theta_T \\ \dot{\theta}_T = \dfrac{-g\cos\theta_T}{v_T} \\ \dot{\psi}_T = 0 \end{cases} \quad (2)$$

According to their geometry relations, the line-of-sight elevation angle q_ε and the azimuth angle q_β and their rates can be derived as

$$\begin{cases} q_\varepsilon = \arctan\left[\dfrac{y_r}{\sqrt{x_r^2 + z_r^2}}\right] \\ q_\beta = \arctan\left[\dfrac{-z_r}{x_r}\right] \\ \dot{q}_\varepsilon = \dfrac{(x_r^2 + z_r^2)\dot{y}_r - y_r(x_r \dot{x}_r + z_r \dot{z}_r)}{(x_r^2 + y_r^2 + z_r^2)\sqrt{x_r^2 + z_r^2}} \\ \dot{q}_\beta = \dfrac{z_r \dot{x}_r - x_r \dot{z}_r}{x_r^2 + z_r^2} \end{cases} \quad (3)$$

where the relative range and velocity are defined by $x_r = x_T - x_M$, $y_r = y_T - y_M$, $z_r = z_T - z_M$, $\dot{x}_r = \dot{x}_T - \dot{x}_M$, $\dot{y}_r = \dot{y}_T - \dot{y}_M$, $\dot{z}_r = \dot{z}_T - \dot{z}_M$.

From (3), the line-of-sight angle and rate have non-linear relations with target range. It has been testified that if without exerting orbit control, the interceptor line-of-sight rate will increase in the shape of parabola when the range decreases[3].

3 Fuzzy Logic Guidance Law

In proportional navigation guidance law, the rate of flight path angle is proportional to the rate of line-of-sight angle, which can be described as

$$\frac{d\theta_M}{dt} = K \frac{dq_\varepsilon}{dt} \qquad (4)$$

where the coefficient K should satisfy the requirements as below

1. the minimum value of K should insure that $\dot{q}$ is convergent;
2. the value of K should be limited by missile's usable lateral overload;
3. the selection of K must ensure that the guidance system can run stably.

When proportional navigation guidance is adopted, the shape of trajectory changes according to coefficient K. The more K increases, the flatter trajectory shape is, and the less lateral overload requires. So after satisfying all the requirements above, in order to have perfect maneuverability, the value of K should be increased[4 5].

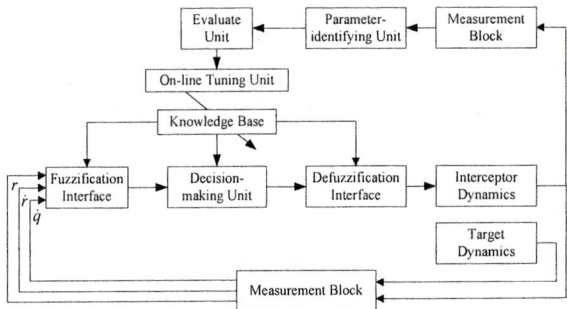

Fig. 2. The Configuration of Self-tuning Fuzzy Logic Guidance System

The fuzzy inference system is next set up to implement above heuristic reasoning. Figure 2 shows the self-tuning fuzzy logic guidance system. In practice, target position and velocity information in the Cartesian inertial frame are obtained from a ground-based radar or satellite. Interceptor position and velocity information in the Cartesian inertial frame are obtained from an inertial reference unit. In this guidance scheme, after these data are acquired, they are converted into linguistic variables through fuzzification interface first. Then these linguistic variables form the inputs to a fuzzy inference engine that uses a knowledge base to generate a set of linguistic outputs. The knowledge base includes a collection of rules that associate each combination of the input linguistic variables into a set of desirable actions expressed in the form of output linguistic variables. After converting into crisp actuator commands through defuzzification interface, the onboard inertial reference unit can measure actual acceleration information. Through these information, interceptor current mass is estimated by parameter-identifying unit. The guidance effects are evaluated correspondingly in evaluation unit. When receiving these effects, the rules in knowledge base are revised by on-line tuning unit. All these proceedings hold on until interceptor destroys target.

The fuzzy inference system is set up with four inputs and two outputs. The inputs to this inference system are the interceptor-target relative range, relative velocity and line-of-sight elevation angle rate and azimuth angle rate. The outputs are the desired thrust forces in the horizon and vertical planes, respectively. Due to non-negative characteristic of relative range and velocity, their linguistic variables are assumed to take three linguistic sets defined as B(Big), M(Middle) and S(Small). Meanwhile, in order to compute conveniently, we choose five linguistic sets defined as PB(Positive Big), PS(Positive Small), ZE(Zero), NS(Negative Small), NB(Negative Big) to express angle rate. To outputs, seven linguistic sets are defined as PB, PM(Positive Middle), PS, ZE, NS, NM(Negative Middle), NB. For these variables, the triangular membership functions are adopted in Fig.3 where the physical domains are set to cover the operating ranges of all variables.

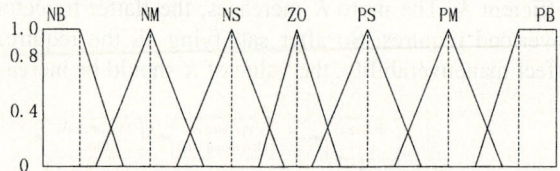

Fig. 3. Membership Functions Used for Inputs and Outputs

The format of fuzzy rules is given by

if $r = A_i$ and $\dot{r} = B_i$ and $\dot{q} = C_i$, then $F_n = D_i$

where A_i, B_i, C_i, D_i represent input and output fuzzy sets. Fuzzy rule table is shown in Tab.1. In this table, there are forty five rules in all.

Table 1. Rule Table for the Fuzzy Logic Guidance Law

	F		$\dot{q}$				
r	$\dot{r}$		NB	NS	ZE	PS	PB
	B		NM	NS	ZE	PS	PM
B	M		NS	NS	ZE	PS	PS
	S		NS	NS	ZE	PS	PS
	B		NB	NM	ZE	PM	PB
M	M		NM	NS	ZE	PS	PM
	S		NS	NS	ZE	PS	PS
	B		NB	NM	ZE	PM	PB
S	M		NB	NM	ZE	PM	PB
	S		NM	NS	ZE	PS	PM

The max-min inference is used to generate the best possible conclusion. For this type of inference is computationally easy and effective, it is appropriate for real-time control applications. The crisp guidance command is calculated by using the center-of-gravity defuzzification which is defined as

$$F = \sum_i W_i F_i / \sum_i W_i \tag{5}$$

When interceptor introduces thrust engines to adjust its orbit and attitudes, the mass of interceptor decreases along with vanishing energy. In order that the trajectory performance can be improved, interceptor mass should be identified on-line during the whole terminal stage[6].

In every guidance period, interceptor current mass can be estimated by using the recursive least square (RLS) method. Then the guidance rules are revised according to the learning algorithms correspondingly. In this proceeding, fuzzy performance index is induced to evaluate guidance performance. Fuzzy performance index is defined as

$$FP \stackrel{def}{=} \mu(e_a) \tag{6}$$

where e_a is the error between desired overload and actual overload. $\mu(\cdot)$ denotes choiceness degree and can be expressed with the triangular membership functions as shown in Fig.3.

Table 2. Revised Rules Used for Guidance Rule Base

	ΔF	$\dot{q}$		
		N	ZE	P
$\ddot{q}$	N	PB	PS	NS
	ZE	PM	ZE	NM
	P	PS	NS	NB

Revised guidance rules are listed in Tab.2. According to Tab.1 and Tab.2, the new value F_i can be expressed as

$$\hat{F}_i = F_i - (1 - FP)\Delta F \cdot w_i^{(k-m)} \tag{7}$$

where $w_i^{(k-m)}$ is the fitness degree of the i rule in the $(k-m)$ guidance period. $\hat{F}_i$ is the revised value of F_i. So we get the guidance output increment ΔF_k^* in the k guidance period.

$$\Delta F_k^* = \frac{\sum_{i=1}^{45} w_i F_i}{\sum_{i=1}^{45} w_i} \tag{8}$$

In the beginning of every period, fuzzy performance FP index is computed. If FP satisfies the ending condition $FP > \theta$, the learning proceeding will be over, otherwise we go on above computing.

4 Simulation Results

The initial conditions of the interceptor are: $m_0=35$kg, $r_0=100$km, $v_{M0}=3400$m/s, $v_{T0}=7400$m/s. In order to test interceptor's performance under different circumstance, the target is assumed to flight in two modes: motion without maneuvering and motion with maneuvering which sine-wave velocity's amplitude is 20m/s^2. In fuzzy logic guidance law, fuzzy set ranges of three inputs(r, $\dot{r}$, $\dot{q}$) are set to be [0, 120000], [9000, 11000], [-2.0, 2.0], respectively. Output ranges are set to be [-120, 120].

Table 3. Simulation Results in Two Target Flight Modes

	Non-maneuvering Target Mode		Maneuvering Target Mode	
	Miss distance(m)	Energy consumption(kg)	Miss distance(m)	Energy consumption (kg)
PNG	0.021	0.481	16.647	1.569
FLG	0.018	0.434	0.392	0.976

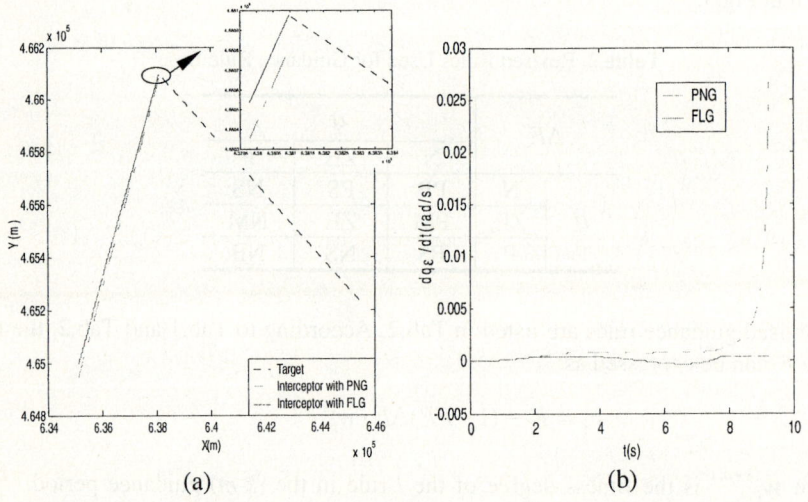

Fig. 4. (a) Interceptor and Target Trajectories (b) Line-of-sight Angle Rates

Tab.3 lists the detailed data of the simulation. When target flights in non-maneuvering mode, both two guidance laws bring on perfect results. But when target is maneuverable, it is obvious that fuzzy logic guidance law causes less miss distance than proportional navigation guidance and the interceptor consumes less engine energy than the latter. Fig.4 shows the interceptor trajectories and the line-of-sight angle rate curves when using different guidance laws. When interceptor approaches to target, the line-of-sight angle rate of interceptor using proportional navigation guidance gets emanative. In the end, interceptor desired overload increases and causes the more miss distance. While in fuzzy logic guidance law, the line-of-sight angle rate maintains convergence in Fig.4. Interceptor's maneuverability is applied sufficiently in flight fore-stage. So interceptor trajectory becomes smooth in the end stage and interceptor has much more maneuverability to destroy target.

5 Conclusions

A self-tuning fuzzy logic guidance law which used in interceptor's terminal stage has been studied. In this novel guidance design scheme, fuzzy inference algorithms synthesize the qualitative aspects of guidance knowledge and reasoning processes. On the other hand, designed guidance rules are revised on-line to suit the changes of onboard status. Interceptor using this fuzzy logic guidance law has much more maneuverability and less desired lateral overloads. Simulation results show the proposed guidance law provides intelligence and robust stability as well as the perfect miss distance performance.

References

1. Wu Wenhai, Qu Jianling, Wang Cunren: An Overview of the Proportional Navigation. Flight Dynamics, Vol. 22. China Test Pilot Institute, Xi'an, PRC (2004) 1-5.
2. Li Shaoyuan, Xi Yugeng, Chen Zengqiang, et al: The New Progresses in Intelligent Control. Control and Decision, Vol. 15. Northeast University, Shenyang, PRC (2000) 1-5.
3. Shih-Ming Yang: Analysis of Optimal Midcourse Guidance Law. IEEE Transactions on Aerospace and Electronic Systems, Vol. 32. IEEE Aerospace and Electronic Systems Society, Sudbury, MA,USA (1996) 419-425.
4. Eun-Jung Song, Min-Jea Tahk: Three-Dimensional Midcourse Guidance Using Neural Networks for Interception of Ballistic Targets. IEEE Transactions on Aerospace and Electronic Systems, Vol. 38. IEEE Aerospace and Electronic Systems Society, Sudbury, MA,USA (2002) 404-414.
5. Chih-Min Lin, Yi-Jen Mon: Fuzzy-Logic-Based Guidance Law Design for Missile Systems. Proceeding of the 1999 IEEE International Conference on Control Applications. IEEE Control Systems Society, Kohala Coast, HI, USA (1999) 421-426
6. Chun-Liang Lin, Hao-Zhen Hung, Yung-Yue Chen, et al: Development of an Integrated Fuzzy-Logic-Based Missile Guidance Law Against High Speed Target. IEEE Transactions on Fuzzy Systems, Vol. 12. IEEE Computational Intelligence Society, Waco, TX, USA (2004): 157-169.

Relaxed LMIs Observer-Based Controller Design via Improved T-S Fuzzy Model Structure

Wei Xie[1], Huaiyu Wu[2], and Xin Zhao[2]

[1] Satellite Venture Business Laboratory, Kitami Institute of Technology,
165 Koencho, Kitami, Hokkaido, 090-8507, Japan
xiewei@mail.kitami-it.ac.jp
[2] College of Information Science and Technology,
Wuhan University of Science and Technology,
Wuhan 430081, Hubei Province, P. R. China
wuhy@mail.wust.edu.cn

Abstract. Relaxed linear matrix inequalities (LMIs) conditions for fuzzy observer-based controller design are proposed based on a kind of improved T-S fuzzy model structure. The improved structure included the original T-S fuzzy model and enough large bandwidth pre- and post-filters. By this structure fuzzy observer-based controller design can be transformed into LMIs optimization problem. Compared with earlier results, it includes the less number of LMIs that equals the number of fuzzy rules plus one positive definition constraint of Lyapunov function. Therefore, it provides us with less conservative results for fuzzy observer-based controller design. Finally, a numerical example is demonstrated to show the efficiency of proposed method.

1 Introduction

As is well known, Takagi-Sugeno (T-S) fuzzy system can be formalized from a large class of nonlinear system. The approach using the T-S fuzzy model [11], considered like a universal approximated fuzzy controller [2], has been investigated extensively. The T-S fuzzy models are described by a set of fuzzy "IF-THEN" rules with fuzzy sets in the antecedents and dynamics LTI systems in the consequent. These submodels are considered as local linear models, the aggregation of which representing the nonlinear system behavior. Despite the fact that the global T-S model is nonlinear due to the dependence of the membership functions on the fuzzy variables, it has a very special formulation, known as Polytopic Linear Differential Inclusions (PLDI) [1], in which the coefficients are normalized membership functions.

A great deal of attention has been focused on the stability analysis and synthesis of these systems. Several researchers have addressed the issue of stability and a substantial amount of progress has been made in stability analysis and design of T-S fuzzy control systems [6, 7, 10, 12-16, 18]. For example, Tanaka and Sugeno presented sufficient conditions for the stability of T-S models [8] using a quadratic Lyapunov approach. The stability depends on the existence of a common positive definite matrix guarantying the stability of all local subsystems. Most of the above mentioned references utilize the interior-point convex optimization methods by solving linear matrix inequalities [1,5]. However, the present results are only sufficient and still include some conservatism, although T-S fuzzy control system is globally asymptotically stable, we maybe fail to find such a common positive definite matrix. Thus it is important to decrease the conservatism of LMI based stability analysis and synthesis conditions of these systems. Some works have been developed in order to establish new stability conditions by relaxing some of the previous constraints. So one way for obtaining relaxed stability conditions consists to use a piecewise quadratic Lyapunov function formulated as a set of LMIs [7]. Using the PI fuzzy controller and the Lyapunov technique, Chai et al. [4] show that asymptotic stability of the Takagi-Sugeno fuzzy systems can be ensured under certain restriction on the control signal and the rate of change of the output. Jadbabaie [8] introduces T-S fuzzy rules that describe state dependent Lyapunov function where each T-S rule has fuzzy sets in the antecedents and quadratic Lyapunov function in the consequent. Recently, a relaxed stability condition for Takagi-Sugeno fuzzy systems is derived by Chadli et al. [3] via non-quadratic Lyapunov function technique and LMIs. A nonlinear control law is investigated by Thierry et al. [17] for Takagi-Sugeno fuzzy models. Some new stability conditions are proposed to increase more design freedom by introducing some non symmetrical matrices in [9].

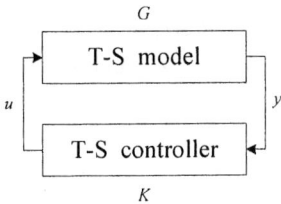

Fig. 1. Relevant fuzzy control structure

Obviously, almost all of above methods are considered to decrease the conservatism of quadratic stability based results by increasing some design freedom. In this paper, some relaxed linear matrix inequalities (LMI) conditions for fuzzy observer-based controllers design are proposed according to a kind of improved T-S fuzzy model structure. In order to increase the conservatism of earlier LMI based stability analysis and synthesis conditions, it is very significant to reduce the number of constrained LMI conditions greatly. From the earlier results, state feedback and observer gain matrices are always dependent on the membership functions on the fuzzy variables, it results in increasing the numbers of LMIs much. Here, we consider constructing a kind of improved T-S fuzzy model structure which is composed of the original T-S fuzzy model and enough large bandwidth stable pre- and post-filters. It will transfer state feedback and observer gain matrices of original plant into state matrix of a new augmented system, and make state feedback and observer gain matrices of new augmented system independent of membership functions on the fuzzy variables. By this trick, it results in less numbers of constrained LMIs conditions, which almost equals the number of fuzzy rules, and less conservative results will be obtained.

2 Preliminary

In this section, firstly the notation regarding T-S fuzzy system is introduced. Useful conception and lemma are recapped.

Definition 1: Takagi-Sugeno Fuzzy System

A dynamic T-S fuzzy model G is described by a set of fuzzy "IF-THEN" rules with fuzzy sets in the antecedents and dynamic LTI systems in the consequent. A general T-S plant rule can be written as follows (for i th plant rule)

IF $x_1(t)$ is M_1^i and $\cdots x_p(t)$ is M_p^i, then $\dot{x}(t) = A_i x(t) + B_i u(t)$ and $y(t) = C_i x(t)$

where $x^T(t) = [x_1(t), x_2(t), \cdots, x_p(t)]$, $u^T(t) = [u_1(t), u_2(t), \cdots, u_q(t)]$, $x_i(t)$ is the state vector, M_p^i is the fuzzy set, $y(t)$ is the output vector, p is the number of the state vector, and q is the number of the input vector.

Using singleton fuzzifier, max-product inference and center average defuzzifier, we can write the aggregated fuzzy model as

$$\dot{x}(t) = \frac{\sum_{i=1}^{p} w_i(x(t))(A_i x(t) + B_i u(t))}{\sum_{i=1}^{p} w_i(x(t))} \text{ and } y(t) = \frac{\sum_{i=1}^{p} w_i(x(t)) C_i x(t)}{\sum_{i=1}^{p} w_i(x(t))} \qquad (1)$$

Where

$$w_i(x(t)) = \prod_{j=1}^{p} g_j^i(x_j(t)) \qquad (2)$$

where g_j^i is the membership grade of the j th state variable $x_j(t)$ to the i th fuzzy set M_j^i. Defining

$$\mu_i(x(t)) = \frac{w_i(x(t))}{\sum_{i=1}^{p} w_i(x(t))} \qquad (3)$$

where $\mu_i(x(t))$ is the normalized membership function in relation with the i th subsystem described by

$$\sum_{i=1}^{p} \mu_i(x(t)) = 1, \qquad (0 \leq \mu_i(x(t)) \leq 1, \ 1 \leq i \leq p) \qquad (4)$$

Therefore, the equation (1) can be represented by

$$\begin{cases} \dot{x}(t) = \sum_{i=1}^{p} \mu_i(x(t))(A_i x(t) + B_i u(t)) \\ y(t) = \sum_{i=1}^{p} \mu_i(x(t)) C_i x(t) \end{cases} \qquad (5)$$

Considering the stability of T-S fuzzy model G, here sufficient conditions are given based on Lyapunov stability theory as follows:

Lemma 1[14]: The continuous time T-S fuzzy system (1) is globally asymptotically stable if there exists a common positive definite matrix P which satisfies the following inequalities:

$$A_i^T P + PA_i < 0, \quad (1 \leq i \leq p) \tag{6}$$

As to T-S fuzzy model (1), it was found in [14] that a stabilizing observed-based controller can be formulated as

$$\dot{\hat{x}} = \sum_{i=1}^{p} \mu_i(x(t))(A_i \hat{x}(t) + B_i u(t) + L_j(C_i \hat{x}(t) - y(t)))$$

$$u = \sum_{j=1}^{p} \mu_j(x(t)) F_j \hat{x}(t) \tag{7}$$

where F_j and L_j are state feedback and observer gain matrices for this plant, respectively. Moreover, $F_j = V_j P_f^{-1}$ and $L_j = P_l^{-1} W_j$ should satisfy the following LMIs, respectively, as

$$\begin{cases} P_f > 0 \\ P_f A_i^T + A_i P_f - B_i V_i - V_i^T B_i^T < 0, \quad (1 \leq i \leq p) \\ P_f (A_i + A_j)^T + (A_i + A_j) P_f - (B_i V_j + B_j V_i) - (B_i V_j + V_j B_i)^T < 0, \quad (j < i \leq p) \end{cases} \tag{8}$$

and

$$\begin{cases} P_l > 0 \\ A_i^T P_l + P_l A_i - W_i C_i - C_i^T W_i^T < 0, \quad (1 \leq i \leq p) \\ (A_i + A_j)^T P_l + P_l (A_i + A_j) - (W_i C_i + W_j C_i) - (W_i C_j + W_j C_i)^T < 0, \quad (j < i \leq p) \end{cases} \tag{9}$$

where matrices V_j, W_j are LMI variables.

Since state feedback and observer gain matrices are dependent on the membership functions on the fuzzy variables, the numbers of LMIs in both (8) and (9) with $r(r+1)+2$ linear matrix inequalities will increase. Therefore, it is important to decouple state feedback and observer gain matrices with the membership functions on the fuzzy variables or make these matrices independent of membership functions.

In the next section, we present an improved T-S fuzzy model structure. Based on this formulation, state feedback and observer gain matrices will be independent of membership functions on the fuzzy variables. By this way the earlier quadratic stability based results become less conservative.

3 Improved T-S Fuzzy Model Structure

A newly-proposed T-S fuzzy model structure is composed of original fuzzy T-S model and enough large bandwidth LTI pre- and post-filters, i.e., G_u and G_y, as shown in Fig.2. These filters are primarily used to make pre-filtering of the control inputs and post-filtering of the measured outputs, respectively.

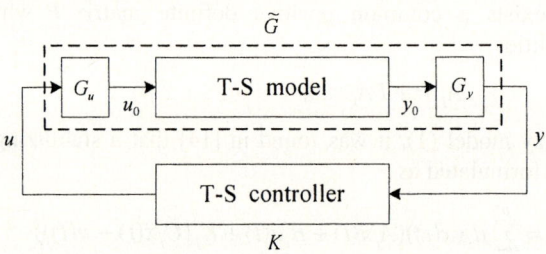

Fig. 2. Augmented fuzzy T-S plant and controller

The state matrix expressions are given by

$$G_u : \begin{cases} \dot{x}_u = A_u x_u + B_u u \\ u_0 = C_u x_u \end{cases} \text{ and } G_y : \begin{cases} \dot{x}_y = A_y x_y + B_u y_0 \\ y = C_y x_y \end{cases} \quad (10)$$

where A_u, A_y are stable coefficient matrices. Thus, the augmented fuzzy T-S model $\tilde{G}$ is described by

$$\begin{cases} \begin{bmatrix} \dot{x} \\ \dot{x}_u \\ \dot{x}_y \end{bmatrix} = \sum_{i=1}^{r} \mu_i(x(t)) \begin{pmatrix} A_i & B_i C_u & 0 \\ 0 & A_u & 0 \\ B_y C_i & 0 & A_y \end{pmatrix} \begin{bmatrix} x \\ x_u \\ x_y \end{bmatrix} + \begin{bmatrix} 0 \\ B_u \\ 0 \end{bmatrix} u \\ y = \begin{bmatrix} 0 & 0 & C_y \end{bmatrix} \begin{bmatrix} x \\ x_u \\ x_y \end{bmatrix} \end{cases} \quad (11)$$

Defining following notations:

$$\tilde{A}_i = \begin{pmatrix} A_i & B_i C_u & 0 \\ 0 & A_u & 0 \\ B_y C_i & 0 & A_y \end{pmatrix}, \tilde{B} = \begin{bmatrix} 0 \\ B_u \\ 0 \end{bmatrix} \text{ and } \tilde{C} = \begin{bmatrix} 0 & 0 & C_y \end{bmatrix} \quad (12)$$

Therefore, the state feedback and observer gain matrices of original plant can be shifted into the new state matrix $\tilde{A}_i$. It can be seen that the new state feedback and observer gain matrices will be independent of membership functions on the fuzzy variables. It should be noted that the filter bandwidth must be chosen larger than the desired system bandwidth. Furthermore, whenever the plant model includes actuator and sensor dynamics, the control and measurement matrices are free from membership functions on the fuzzy variables. Hence the proposed filtering operations are not restrictive in practice. Consequently, to design an observer-based controller for (1) is transferred to an equivalent problem for this augmented fuzzy T-S model (11). As to T-S fuzzy model (11), a fuzzy observer-based controller can be defined as a set of T-S IF-THEN rules, which stabilize this system. A general observer-based controller rule can be written as i th observer-based controller rule, i.e.,

IF $x_1(t)$ is M_1^i and $\cdots x_p(t)$ is M_p^i, then

$\dot{\hat{x}}(t) = \tilde{A}_i\hat{x}(t) + \tilde{B}u(t) + L_i(\hat{y}(t) - y(t))$ and $u(t) = F_i x(t)$

Therefore, we have the following lemma for stabilizing observed-based controller design:

Lemma 2: A stabilizing observed-based controller for fuzzy T-S model (11) can be formulated as

$$\begin{cases} \dot{\hat{x}} = \sum_{i=1}^{p} \mu_i(x(t))(\tilde{A}_i\hat{x}(t) + \tilde{B}u(t) + L_i(\tilde{C}\hat{x}(t) - y(t))) \\ u = \sum_{j=1}^{p} \mu_j(x(t))F_i\hat{x}(t) \end{cases} \quad (13)$$

where matrices F_i and L_i are state feedback and observer gain matrices for each local plant, respectively. Moreover, $F_i = V_i P_f^{-1}$ and $L_i = P_l^{-1} W_i$ should satisfy the following LMIs, respectively, i.e.,

$$P_f > 0, \quad P_f \tilde{A}_i^T + \tilde{A}_i P_f + \tilde{B}V_i + V_i^T \tilde{B}^T < 0, \quad (1 \leq i \leq p) \quad (14)$$

$$P_l > 0, \quad \tilde{A}_i^T P_l + P_l \tilde{A}_i + W_i \tilde{C} + \tilde{C}^T W_i^T < 0, \quad (1 \leq i \leq p) \quad (15)$$

Proof: If the controller (13) is substituted into plant (21), the closed-loop state matrix can be expressed as

$$A_{cl} = \sum_{i=1}^{p} \mu_i(x(t)) \begin{bmatrix} \tilde{A}_i + \tilde{B}F_i & \tilde{B}F_i \\ 0 & \tilde{A}_i + L_i\tilde{C} \end{bmatrix} \quad (16)$$

According to lemma 1, the closed system is also said to be globally asymptotically stable since (14) and (15) make $\tilde{A}_i + \tilde{B}F$ and $\tilde{A}_i + L_i\tilde{C}$ globally asymptotically stable, respectively. Thus, equation (13) can be rewritten as the following equivalent formulation

$$\begin{cases} \dot{x}_k = \sum_{i=1}^{p} \mu_i(x(t))(A_{ki} x_k + B_{ki} y) = \sum_{i=1}^{p} \mu_i(x(t))((\tilde{A}_i + \tilde{B}F_i + L_i\tilde{C})x_k - L_i y) \\ u = \sum_{i=1}^{p} \mu_i(x(t))C_{ki} x_k = \sum_{j=1}^{p} \mu_j(x(t))F_i x_k \end{cases} \quad (17)$$

The original plant and augmented fuzzy T-S controller is shown in Fig.3.

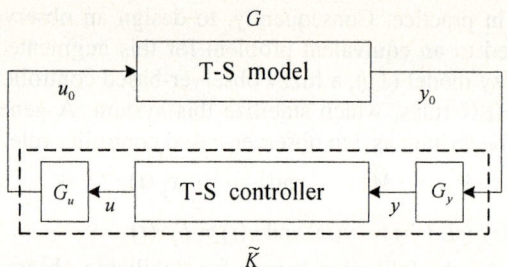

Fig. 3. The original plant and augmented fuzzy T-S controller

It can be seen from Fig.3 that the augmented fuzzy T-S controller is composed of original controller and two LTI stable filters. The state matrix expression of this augmented fuzzy T-S controller is described as:

$$\begin{cases} \begin{bmatrix} \dot{x}_k \\ \dot{x}_y \\ \dot{x}_u \end{bmatrix} = \sum_{i=1}^{p} \mu_i(x(t)) \begin{pmatrix} A_{ki} & B_{ki}C_y & 0 \\ 0 & A_y & 0 \\ B_u C_{ki} & 0 & A_u \end{pmatrix} \begin{bmatrix} x_k \\ x_y \\ x_u \end{bmatrix} + \begin{bmatrix} 0 \\ B_y \\ 0 \end{bmatrix} y_0 \\ u_0 = \begin{bmatrix} 0 & 0 & C_u \end{bmatrix} \begin{bmatrix} x_k \\ x_y \\ x_u \end{bmatrix} \end{cases} \quad (18)$$

In a similar manner to the T-S fuzzy model (11), a fuzzy observer-based controller related to model (18) can be defined as a set of T-S IF-THEN rules. A general observer-based controller rule can be written as i th observer-based controller rule, i.e.,

IF $x_1(t)$ is M_1^i and $\cdots x_p(t)$ is M_p^i, then $\dot{\tilde{x}}_k(t) = \tilde{A}_{ki}\tilde{x}_k(t) + \tilde{B}_k y(t)$ and $u(t) = \tilde{C}_k \tilde{x}_k(t)$

Therefore, we have the following theorem of stabilizing observed-based controller design for model (1) as

Theorem: A stabilizing observed-based controller for fuzzy T-S model (1) can be formulated as

$$\begin{cases} \dot{\tilde{x}}_k(t) = \sum_{i=1}^{p} \mu_i(x(t))(\tilde{A}_{ki}\tilde{x}_k(t) + \tilde{B}_k y(y)) \\ u(t) = \sum_{i=1}^{p} \mu_i(x(t))\tilde{C}_k \tilde{x}_k(t) \end{cases} \quad (19)$$

Where

$$\tilde{A}_{ki} = \begin{pmatrix} A_{ki} & B_{ki}C_y & 0 \\ 0 & A_y & 0 \\ B_u C_{ki} & 0 & A_u \end{pmatrix}, \tilde{B}_k = \begin{bmatrix} 0 \\ B_y \\ 0 \end{bmatrix}, \tilde{C}_k = \begin{bmatrix} 0 & 0 & C_u \end{bmatrix} \quad (20)$$

where $A_{ki} = \tilde{A}_i + \tilde{B}F_i + L_i\tilde{C}$, $B_{ki} = -L_i$, $C_{ki} = F_i$, both $F_i = V_i P_f^{-1}$ and $L_i = P_l^{-1} W_i$ should satisfy the LMIs (14) and (15).

Based on this improved fuzzy T-S model structure, the design of the stabilized fuzzy observer-based controllers can be transformed to solve LMIs optimization problems (14) and (15) only with the number of control rules plus one positive definition constraint of Lyapunov function. Even though the order of controller increases much, it will provide us with less conservative results than that in the situation of inequalities (8) and (9).

4 Numerical Example

Some numerical examples based on the translational oscillation with a rotational actuator (TORA) system by Bupp et al (1995)[19] are provided to verify the effectiveness of the proposed stabilizing fuzzy observer-based controllers design scheme. The TORA system considers a translational oscillator with an attached eccentric rotational proof mass actuator, where the nonlinear coupling between the rotational motion of the actuator and the translational motion of the oscillator provides the control mechanism. The system dynamics can be expressed as the following equation:

$$\dot{x}(t) = f(x) + g(x)u(t)$$

where $u(t)$ is the torque applied to the eccentric mass, and

$$f(x) = \begin{bmatrix} x_2 \\ \dfrac{-x_3 + \varepsilon x_4^2 \sin x_3}{1 - \varepsilon^2 \cos^2 x_3} \\ x_4 \\ \dfrac{\varepsilon \cos x_3 (x_1 - \varepsilon x_4^2 \sin x_3)}{1 - \varepsilon^2 \cos^2 x_3} \end{bmatrix} \text{ and } g(x) = \begin{bmatrix} 0 \\ \dfrac{-\varepsilon \cos x_3}{1 - \varepsilon^2 \cos^2 x_3} \\ 0 \\ \dfrac{1}{1 - \varepsilon^2 \cos^2 x_3} \end{bmatrix} \quad (21)$$

which x_1 is the normalized displacement of the platform from the equilibrium position, $x_2 = \dot{x}_1$, $x_3 = \theta$ is the angle of the rotor, and $x_4 = \dot{x}_3$. The equilibrium point of this system could be any point $[0, 0, x_3, 0]$ among which only the point $[0,0,0,0]$ is the desired equilibrium point. The linearization around the point $[0,0,0,0]$ has two eigenvalues $+i$ and $-i$, which means that this system is a critical nonlinear system.

The resulting T-S model consists of four fuzzy rules as following:

Rule 1: IF $|x_3(t)|$ is near 0, then

$$A_1 = \begin{bmatrix} 0 & 1 & 0 & 0 \\ -\dfrac{1}{1-\varepsilon^2} & 0 & \varepsilon & 0 \\ 0 & 0 & 0 & 1 \\ \dfrac{\varepsilon}{1-\varepsilon^2} & 0 & 0 & 0 \end{bmatrix}, B_1 = \begin{bmatrix} 0 \\ -\dfrac{\varepsilon}{1-\varepsilon^2} \\ 0 \\ \dfrac{1}{1-\varepsilon^2} \end{bmatrix}, C_y = \begin{bmatrix} 1 \\ 0 \\ 1 \\ 0 \end{bmatrix}^T$$

Rule 2: IF $|x_3(t)|$ is near $\dfrac{\pi}{2}$ and $|x_4(t)|$ is small, then

$$A_2 = \begin{bmatrix} 0 & 1 & 0 & 0 \\ -1 & 0 & 0.01\dfrac{2\varepsilon}{\pi} & 0 \\ 0 & 0 & 0 & 1 \\ 0 & 0 & 0 & 0 \end{bmatrix}, B_2 = \begin{bmatrix} 0 \\ 0 \\ 0 \\ 1 \end{bmatrix}, C_y = \begin{bmatrix} 1 \\ 0 \\ 1 \\ 0 \end{bmatrix}^T$$

Rule 3: IF $|x_3(t)|$ is near $\dfrac{\pi}{2}$ and $|x_4(t)|$ is big, 1-3then

$$A_3 = \begin{bmatrix} 0 & 1 & 0 & 0 \\ -1 & 0 & \dfrac{2\varepsilon}{\pi} & 0 \\ 0 & 0 & 0 & 1 \\ 0 & 0 & 0 & 0 \end{bmatrix}, B_3 = \begin{bmatrix} 0 \\ 0 \\ 0 \\ 1 \end{bmatrix}, C_y = \begin{bmatrix} 1 \\ 0 \\ 1 \\ 0 \end{bmatrix}^T$$

Rule 4: IF $|x_3(t)|$ is near π, then

$$A_4 = \begin{bmatrix} 0 & 1 & 0 & 0 \\ -\dfrac{1}{1-\varepsilon^2} & 0 & 0 & 0 \\ 0 & 0 & 0 & 1 \\ \dfrac{\varepsilon}{1-\varepsilon^2} & 0 & 0 & 0 \end{bmatrix}, B_4 = \begin{bmatrix} 0 \\ -\dfrac{\varepsilon}{1-\varepsilon^2} \\ 0 \\ \dfrac{1}{1-\varepsilon^2} \end{bmatrix}, C_y = \begin{bmatrix} 1 \\ 0 \\ 1 \\ 0 \end{bmatrix}^T \tag{22}$$

Since only state feedback matrix includes some parameters dependent of membership functions on the fuzzy variables, a low-pass filter G_u can be added to the input, i.e.,

$$G_u : \begin{cases} \dot{x}_u = -100x_u + u \\ u_0 = 100x_u \end{cases}$$

Then system matrices of the augmented fuzzy T-S model $\tilde{G}$ are described by

$$\tilde{A}_i = \begin{bmatrix} A_i & B_i C_u \\ 0 & A_u \end{bmatrix}, \quad \tilde{B} = \begin{bmatrix} 0 \\ B_u \end{bmatrix} \text{ and } \tilde{C} = \begin{bmatrix} C_y & 0 \end{bmatrix} \tag{23}$$

Using Matlab LMI toolbox, the design of the proposed stabilizing observed-based controller was finished. The observer gain matrices are given by

$$L_1 = \begin{bmatrix} 47.62 \\ -65.15 \\ -66.64 \\ -31.40 \\ 17.71 \end{bmatrix}, L_2 = \begin{bmatrix} 71.74 \\ -94.31 \\ -101.32 \\ -149.34 \\ 128.53 \end{bmatrix}, L_3 = \begin{bmatrix} 54.49 \\ -71.21 \\ -76.53 \\ -96.15 \\ 80.50 \end{bmatrix}, L_4 = \begin{bmatrix} 49.40 \\ -67.08 \\ -68.95 \\ -32.20 \\ 18.01 \end{bmatrix}$$

with the Lyapunov function matrix

$$P_i = \begin{bmatrix} 13.57 & -3.32 & 12.05 & 5.58 & 5.88 \\ -3.32 & 3.24 & -6.17 & 3.82 & 3.71 \\ 12.05 & -6.17 & 16.94 & -10.14 & -9.75 \\ 5.58 & 3.82 & -10.14 & 51.68 & 51.64 \\ 5.88 & 3.71 & -9.75 & 51.64 & 52.04 \end{bmatrix}$$

And the state feedback matrices are given by

$$F_1 = \begin{bmatrix} 261.34 & -105.63 & -94.45 & -192.13 & 97.27 \end{bmatrix}$$
$$F_2 = \begin{bmatrix} 280.45 & -158.11 & -118.68 & -206.02 & 97.14 \end{bmatrix}$$
$$F_3 = \begin{bmatrix} 280.44 & -158.10 & -118.67 & -206.02 & 97.135 \end{bmatrix}$$
$$F_4 = \begin{bmatrix} 261.34 & -105.64 & -94.46 & -192.13 & 97.27 \end{bmatrix}$$

with the Lyapunov function matrix

$$P_f = \begin{bmatrix} 6.66 & -0.0043 & 0.091 & 9.0197 & -0.0045 \\ -0.0043 & 6.66 & -9.023 & 0.0783 & 0.0902 \\ 0.0914 & -9.0225 & 23.26 & -6.347 & -0.2128 \\ 9.0197 & 0.0783 & -6.347 & 22.23 & -0.1030 \\ -0.0045 & 0.0902 & -0.213 & -0.103 & 13.0841 \end{bmatrix}$$

According to Fig.3, a stabilizing T-S fuzzy controller just as described in (19) and (20) can be constructed. From the results of above numerical computation, it is obvious that the proposed structure for T-S fuzzy model provides us less conservative results than earlier work based on LMI conditions (8) and (9).

5 Conclusions

This paper proposed relaxed linear matrix inequality (LMI) conditions for fuzzy observer-based controllers design based on a kind of improved T-S fuzzy model structure. The improved structure is composed of the original T-S fuzzy model and large bandwidth pre- and post-filters. Since fuzzy observer-based controllers design can be transformed into solving LMIs optimization problem, it only deals with the less number of LMIs which almost equals the number of fuzzy rules compared with the earlier approaches. Therefore, some less conservative results for fuzzy observer design can be obtained.

References

1. Boyd, S., Ghaoui, L.E., Feron, E., Balakrishnan, V.: Linear matrix inequalities in systems and control theory. PA: SIAM, Philadelphia (1994)
2. Castro, J.: Fuzzy logic controllers are universal approximator. IEEE Trans. on Systems, Man, Cybernetics, Vol. 25, April (1995) 629-635
3. Chadli, M., Maquin, D., Ragot, J.: Relaxed stability conditions for Takagi-Sugeno fuzzy systems. IEEE International Conference on Systems, Man, and Cybernetics, Nashville, Tennessee, USA, October 8-11, (2000) 3514-3519
4. Chai, J.-S., Tan, S., Chan, Q., Hang, C.-C.: A general fuzzy control scheme for nonlinear processes with stability analysis. Fuzzy Sets and Systems 100, (1998) 179-195
5. Gahinet, P., Nemirovskii, A., Laub, A.J., Chilali, M.: LMI toolbox. Natick, MA: Mathworks, (1995)
6. Jadbabaie, A.: A reduction in conservatism in stability and L2 Gain analysis of T-S fuzzy systems via Linear matrix inequalities. IFAC 1999, 14 ~ triennial World congress, Beijing, P.R. China (1999) 285-289
7. Johansson, M., Rantzer, A., Arz6n, K.: Piecewise quadratic stability for affine Sugeno systems. FUZZ. IEEE'98, Anchorage, Alaska (1998)
8. Jadbabaie, A.: A reduction in conservatism in stability and L2 Gain analysis of T-S fuzzy systems via Linear matrix inequalities. IFAC 1999, 14 ~ triennial World congress, Beijing, P.R. China (1999) 285-289
9. Liu, X.D., Zhang, Q.L.: New approaches to H∞ controller designs based on fuzzy observers for T-S fuzzy systems via LMI. Automatica 39 (2003) 1571 – 1582
10. Narendra, K.S., Balakrishnan, J.: A common Lyapunov function for stable LTI systems with commuting A-matrices. IEEE Trans. on Automatic Control, Vol. 39, No. 12, (1994) 2469-2471
11. Takagi, T., Sugeno, M.: Fuzzy identification of systems and its application to modeling and control. IEEE Trans. on Systems, Man, Cybernetics, Vol. 15, No. 1, (1985) 116-132
12. Takagi, T., Ikeda, T., Wang, H.O.: Fuzzy regulators and fuzzy observers: relaxed stability conditions and LMI-based design. IEEE Trans. on Fuzzy Systems, Vol. 6, No. 2, (1998) 250-256
13. Tanaka, K., Ikeda, T., Wang, H.O.: Robust stabilization of uncertain non-linear systems via fuzzy control: quadratic stability, H control theory, and LMIs. IEEE Trans. on Fuzzy Systems, Vol. 4, No. 1, (1996) 1-12
14. Tanaka, K., Sugeno, M.: Stability and design of fuzzy control systems. Fuzzy Set and Systems, Vol. 45, No. 2, (1992) 135-156
15. Tanaka, K., Nishimuna, M., Wang, H.O.: Multi-objective fuzzy control of high rise/high speed elevators using LMIs. American Control Conference, Philadelphia, Pennsylvanie, June (1998) 3450-3454
16. Teixeira, M.C.M., Zak, S.H.: Stabilizing controller design for uncertain nonlinear systems using fuzzy models. IEEE. Trans. on Fuzzy Systems, Vol.7, No.2, (1999) 133-140
17. Thierry, M.G., Laurent, V.: Control laws for Takagi-Sugeno fuzzy models. Fuzzy Sets and Systems 120 (2001) 95-108
18. Wang, H.O., Tanaka, K., Griffin, M.F.: An approach to fuzzy control of nonlinear systems: stability and design: issues. IEEE. Trans. on Fuzzy Systems, Vol. 4, No. 1, (1996)

19. Bupp, R.T., Bernstein, D.S., Coppola, V.T.: A benchmark problem for nonlinear control design: problem statement, experimental testbed, and passive nonlinear compensation. Proceedings of American Control Conference, 21-23 June 1995, Vol. 6. Seattle, WA, USA, (1995) 4363-4367
20. Wang, L.-X.: Adaptive Fuzzy Systems and Control: Design and Stability Analysis. Prentice Hall, Engelwood Cliffs, N. J., (1994)

Fuzzy Virtual Coupling Design for High Performance Haptic Display

D. Bi[1], J. Zhang[2,*], and G.L. Wang[2]

[1] Tianjin University of Science and Technology, Tianjin, 300222, P. R. China
50001729@alumni.cityu.edu.hk
[2] Sun Yat-Sen University, GuangZhou, 510275, P. R. China
Junzhang@zsu.edu.cn

Abstract. Conventional virtual coupling is designed mainly for stabilizing the virtual environment (VE) and it thus may have poor performances. This paper proposes a novel adaptive virtual coupling design approach for haptic display in passive or time-delayed non-passive virtual environment. According to the performance errors, the virtual coupling can be adaptively tuned through some fuzzy logic based law. The designed haptic controller can improve the "operating feel" in virtual environments, while the system's stability condition can still be satisfied. Experimental results demonstrate the effectiveness of this novel virtual coupling design approach.

1 Introduction

Haptic feedback is a way of conveying information between human and computer [1], [2], [3]. As with most control problems, there are two conflicting goals for haptic display designs, performance and stability. Earlier research focused more on stability than fidelity issues. The stability of haptic system was first addressed by Minsky et al. [4]. In their paper, a continuous time, time-delayed model approximated the effects of sample-and-hold. Colgate et al. [5] used a simple benchmark problem to derive conditions under which a haptic display would exhibit passive behavior. A more general haptic display system design method to guarantee stable operation-- "virtual coupling" structure was introduced by Colgate et al. [6] and Zilles et al. [7] by connecting the virtual environment with the haptic device. The proposed virtual coupling was a virtual mechanical system interposed between the haptic interface and the virtual environment to limit the maximum or minimum impedance presented by the virtual environment in such a way as to guarantee stability.

Correct selection of virtual coupling parameters can guarantee stable haptic display in virtual environments. The virtual coupling parameters can be set empirically or by some theoretical design procedure. One fruitful approach is to use the idea of passivity to design the virtual coupling, as passivity is a sufficient condition for system stability. The major problem with using passivity theory for designing virtual coupling parameters is that it is too conservative. To improve the performance, Adams et al [8] derived

---

* Corresponding author.

a new virtual coupling design approach, by applying impedance or admittance based haptic display, which is less conservative than the passivity based design method. Miller et al. [9] extended the analysis to nonlinear non-passive virtual environments and designed the virtual coupling by considering the whole passivity condition of both the virtual environment and the haptic interface, so that the excessive passivity can be reduced by extracting some constant damping from the haptic interface. Hannaford et al. [10] moved a step further by designing a virtual coupling with adaptively changing parameter value, which was calculated in real time through the virtual environment "passivity observer". A disadvantage of these methods is that they cannot improve the haptic interface performance if the virtual environment is passive.

In this paper, we propose a new virtual coupling design method: two-port network model based adaptive virtual coupling for stable and accurate haptic display. Different from constant-parameter virtual coupling, the parameter values of this two-port network model based virtual coupling can be adaptively tuned according to fuzzy logic algorithm [11], [12]. Due to its simple structure, fuzzy logic algorithm is relatively easy to use and is well understood by a great majority of industrial practitioners and automatic control designers [12], [13]. Comparing with traditional virtual coupling, this two-port network based virtual coupling can increase the performance of haptic interface in addition to stabilizing the whole haptic display system. The adaptive tuning of the virtual coupling can improve the system's response time, increase the haptic display accuracy.

This paper is organized as follows: in Section 2, we briefly review the network based haptic display system and traditional virtual coupling. In section 3, the adaptive nonlinear virtual coupling for haptic display is presented. In section 4, we present some case studies using this fuzzy logic based haptic controller in experiments. Finally the conclusions are drawn in section 5.

2 Network Based Haptic Display

Network models, common in circuit theory, where they are used to characterize the effects of different loading conditions on two terminal electrical networks, are a natural way of describing stability and performance in bilateral teleoperation [2], [14], and in the field of control for haptic display [8], [10]. Haptic display methods can be divided into two categories: Admittance based and Impedance based haptic display. For admittance based haptic display, the operator applies force to the haptic interface, the haptic interface produce a kinematic movement to the operator, whereas the impedance based haptic display works the other way around.

Fig. 1 shows a typical structure of the network based haptic controller for admittance based display. In Fig. 1, the human operator, virtual coupling and virtual environment are modeled as a one-port network model, whereas the haptic interface is modeled as a two-port network model. The human operator contacts the haptic interface with the velocity v_h and force f_h. The virtual environment modulates the angular velocity v_r and force f_r according to the physical law in the virtual world. For impedance based haptic display, the directions of the arrows are inverted.

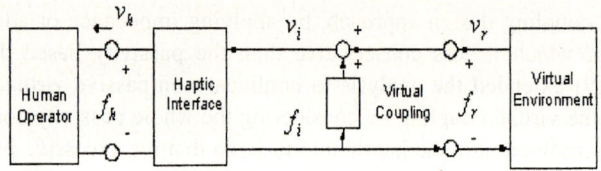

Fig. 1. Typical network based admittance haptic display system

3 Adaptive Virtual Coupling Design for Haptic Display

In most prior research work, the virtual coupling is designed as a damping element connecting the haptic interface and the virtual environment to stabilize the nonlinear VR. It is difficult to achieve at the same time a high performance for the haptic display with such an approach. To overcome such a limitation and to improve the performance of the haptic display, we develop an adaptive nonlinear virtual coupling based on two-port network model and fuzzy logic. By adaptively tuning the parameters of the nonlinear virtual coupling, the fuzzy logic based virtual coupling can result in a stable and high performance haptic display. In what follows, we first introduce the two-port network based virtual coupling. Then we present the design method using fuzzy logic theory. The following derivation is based on the admittance based haptic display.

The proposed two-port network based virtual coupling is different from the traditional virtual coupling structure. As shown in Fig. 2, the two-port network model based virtual coupling is designed as follows: f_r equals the measured interaction force f_h, v_i is the measured velocity from the haptic device, v_r is VE output. Here f_i is the virtual coupling output to the haptic device and its parameter k as shown in equation (1) is adaptively tuned by the fuzzy logic based law. Here only this parameter is tuned, because general fuzzy based PID parameter tuning method has proved its validity and easy to use [11].

From Fig. 2, f_i can be represented in a new form:

$$f_i(t) = K\Delta x(t) + K_i \int \Delta x(t)dt + B\Delta \dot{x}(t) \tag{1}$$

where $\Delta x(t) = x_r(t) - x_i(t)$, is the displacement error between the controller haptic interface output $x_i = \int v_i dt$ and the reference virtual environment displacement output $x_r = \int v_r dt$, $f_i(t)$ is the controller input to the haptic interface. K, B, K_i are constant gains that can be determined by Ziegler-Nichols formula respectively.

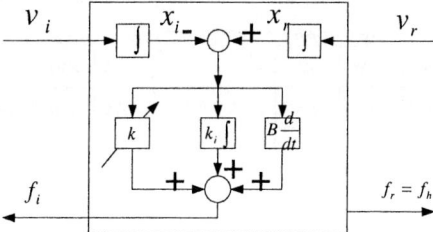

Fig. 2. Adaptive virtual coupling design for haptic display

Converting equation (1) into frequency domain, we have

$$F_i(s) = K\Delta X(s) + K_i \frac{\Delta X(s)}{s} + Bs\Delta X(s) \qquad (2)$$

Using backward difference and the trapezoidal approximation for the derivative and the integral respectively, the discrete-time realization of (2) is

$$F_i(z) = K\Delta X(z) + K_i \frac{T}{2}\frac{1+z^{-1}}{1-z^{-1}}\Delta X(z) + B\frac{1}{T}(1-z^{-1})\Delta X(z) \qquad (3)$$

where $T > 0$ is the sampling period. Here we extended the fuzzy set-point weighting based technique [11] and derive our algorithm called the extended fuzzy set-point weighting technique which is described in the following formulation. Equation (3) can be rewritten as:

$$F_i(nT) = K(\delta(nT)x_r(nT) - x_i(nT)) + K_i \frac{T}{2}\frac{1+z^{-1}}{1-z^{-1}}\Delta x(nT) + B\frac{1}{T}(1-z^{-1})\Delta x(nT) \qquad (4)$$
$$\delta(nT) = 1 + f(nT)$$

Where $f(nT)$ is the output of a fuzzy inference system consisting of triangular and trapezoidal membership functions for the two inputs $\Delta x, \Delta \dot{x}$, and nine triangular functions for the output. The fuzzy rules are listed in Fig. 3.

In Table 1, the definitions of the linguistic variables in the fuzzy inference system are given.

Through the designed controller, we can see, by tuning $f(nT)$, the controller parameter K can be tuned accordingly. While $f(nT)$ is the fuzzy inference of two inputs $\Delta x, \Delta \dot{x}$ according to the fuzzy rule.

The fuzzy based virtual coupling must keep system stable in nonlinear virtual environments. This can be realized by tuning the dead zone width $d > 0$ of the membership function as seen in Fig. 4. In general, the dead zone width can be different. Here we let them be same to simplify the design and notation in the following discussions.

		Δẋ			
	NB	NS	Z	PS	PB
NB	NVB	NB	NM	NS	Z
NS	NB	NM	NS	Z	PS
Δx Z	NM	NS	Z	PS	PM
PS	NS	Z	PS	PM	PB
PB	Z	PS	PM	PB	PVB

		Δẋ			
	NB	NS	Z	PS	PB
NB	PVB	PB	PM	PS	Z
NS	PB	PM	PS	Z	NS
Δx Z	PM	PS	Z	NS	NM
PS	PS	Z	NS	NM	NB
PB	Z	NS	NM	NB	NVB

Fig. 3. Basic rules for the fuzzy inference (upper: when $x_r > 0$, lower: when $x_r < 0$)

Table 1. Definition of the linguistic variables

NVB	Negative very big
NB	Negative big
NM	Negative medium
NS	Negative small
Z	Zero
PS	Positive small
PM	Positive medium
PB	Positive big
PVB	Positive very big

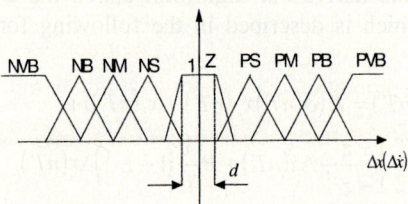

Fig. 4. Stabilizing the haptic interface by tuning the fuzzy dead zone width

4 Implementations

4.1 Experimental Setup

Fig. 5 illustrates the schematic diagram of our experimental test bed. The haptic device is a planar 1-DOF rotating link with a vertical joint connected to a DC motor through a gearbox with a ratio 1:80. The mass of the moment inertial of the link is $I = 2.438 \times 10^{-2} kgm^2$. An optical encoder, with a resolution of 500 pulses per revolution, measures the joint angular displacement. A finger type three-dimensional force sensor is installed on the end of the link. This force sensor has a resolution of $0.005N$ for the force measurement in each direction. A DSP (PS1103 PPC) controller board is used on a host PC for the haptic display control. The PC based virtual environment is connected to the controller for real-time information change.

The first experiment conducted is the performance comparison between the traditional virtual coupling and our adaptive virtual coupling for the case of a virtual wall. The second experiment repeats the first but with the computational time delay and feedback signal time delay incorporated respectively. To obtaining objective results, a point-mass of 0.1 kg is installed on top of the three-dimensional force sensor to replace the human operator.

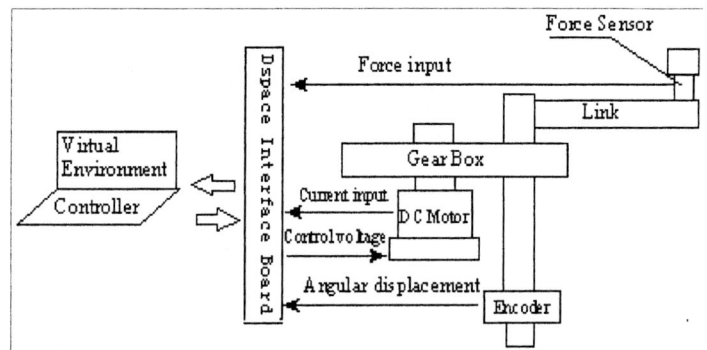

Fig. 5. Schematic diagram of the experimental test bed

4.2 Experimental Results

The first experiment displays a virtual wall without considering any time delay. The virtual finger driven by the haptic device met the virtual wall at $\theta = 1.744$ from its initial position. For admittance based haptic display, we model this ideal "virtual wall contact" behavior as a displacement step function output. Fig. 6 shows the experimental result by using the traditional virtual coupling and our adaptive virtual coupling scheme. The dashed line shows the interaction displacement response of the haptic display system using the adaptive virtual coupling. The dash-dotted line shows the experimental results using the traditional virtual coupling. Comparing the results of the two controllers, we can see that with the adaptive virtual coupling scheme, very little overshoot or fluctuation is observable in the displacement when meeting the virtual wall. Here the initial fuzzy logic based virtual coupling parameters were selected using IAE principle as $K = 3.5, B = 0.12$.

In the next experiment, we considered in the display of a virtual wall with time delay of 0.1s as the disturbance from the feedback signal. The initial virtual coupling parameter values were set at $K = 0.5, B = 0.12$. Fig. 7 shows the corresponding results. The dash-dotted line shows the experimental results using the traditional virtual coupling. The dashed line shows the display result using our adaptive virtual coupling in which case a fast response in the displacement is clearly observable.

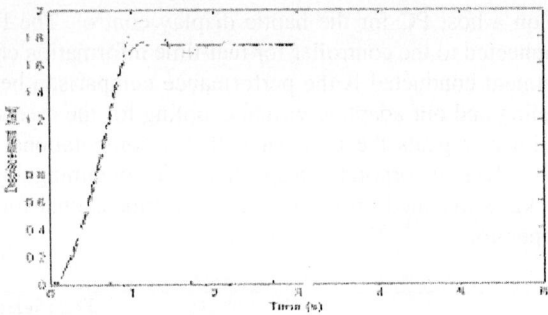

Fig. 6. Interaction with no time delay

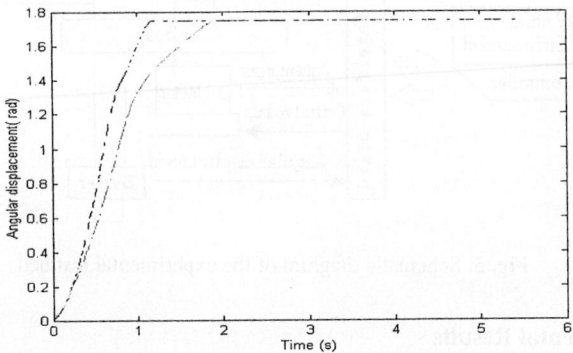

Fig. 7. Interaction with 0.1s delay

Table 2. The IAE values using adaptive virtual coupling and constant virtual coupling

	Adaptive virtual coupling	Constant virtual coupling
Haptic display with no time delay	0.918	1.213
Haptic display with 0.1s delay	1.003	1.308

Table 2 shows the quantitative comparison of the IAE values of the experimental results using the adaptive virtual coupling and traditional virtual coupling approach. From the comparison, we can see that the performance is improved in all the cases when using the adaptive virtual coupling as compared against using the traditional virtual coupling.

5 Conclusions

The conventional virtual coupling design for haptic display is concerned mainly with stability, which usually makes performance conservative. This paper presents an adaptive virtual coupling design approach for haptic display which takes into account both the stability and performance in its design. The studies show that with the adaptive haptic coupling design, improved performance in the haptic display can be achieved, while at the same time the stability can be guaranteed by tuning the parameters when VE is bounded output.

The implementation proved the validity of the developed adaptive virtual coupling design method. The Fuzzy logic based adaptive virtual coupling can increase the system's speed of response and the haptic display accuracy in addition to stabilizing the human-haptic interface interaction. Further explorations in the work include the how to select initial parameters for stable and accurate haptic display.

References

[1] Mandayam A. S., Basdogan C.: Haptics in VE: Taxonomy, Research Status, and Challenges. Computers and Graphics 21 (1997) 393-404
[2] Hannaford B.: A Design Framework for Teleoperators with Kinesthetic Feedback. IEEE J. Robot. Automat. 5 (1989) 426-434
[3] Li Y. F., Bi D.: A Method for Dynamics Identification for Haptic Display of the Operating Feel in Virtual Environments. IEEE Transaction on Mechatronics 8 (2003) 1-7
[4] Minsky M., Ouh Y. M., Steele O., Brooks F. P., Behensky M.: Feeling and Seeing Issues in Force Display. Computer Graphics 24 (1990) 235-243
[5] Colgate J. E., Grafing P. E., Stanely M. C., Schenkel G.: Implementation of Stiff Virtual Walls in Force Reflecting Interface. Proceeding of IEEE Virtual Reality Annual International Symptom, Seattle, WA (1993) 202-208
[6] Colgate J. E., Brown J. M.: Factors Affecting the Z-Width of a Haptic Display. Proceeding of IEEE International Conference on Robot and Automation, Los Alamitos, CA (1994) 3205-3210
[7] Zilles C. B., Salisbury J. K.: A Constraint-Based God-Object Method for Haptic Display. Proceeding of IEEE International Conference on Intelligent Robot system, Pittsburgh, PA (1995) 146-151
[8] Adams R. J., Hannaford B.: Control Law Design for Haptic Interfaces to Virtual Reality. IEEE Transaction on Control System Technology 10 (2002) 3-13
[9] Miller B.E., Colgate J. E., Freeman R. A.: Guaranteed Stability of Haptic Systems with Nonlinear Virtual Environments. IEEE Transaction on Robotics and Automation 16 (2000) 712-719
[10] Hannaford B., Ryu J. H.: Time-Domain Passivity Control of Haptic Interfaces. IEEE Transaction on Robotics and Automation 18 (2002) 1-10
[11] Visioli A.: Fuzzy Logic Based Set-Point Weighting for PID Controllers. IEEE Transactions on System, Man, Cybernatics, Part a 29 (1999) 587-592
[12] Visioli A.: Tuning of PID Controllers with Fuzzy Logic. IEE Proceedings-Control Theory Application 148 (1998) 1-8
[13] Misir D., Maliki H. A., Chen G.: Design and Analysis of a Fuzzy PID Controller. International Journal of Fuzzy Set System 79 (1998) 73-93
[14] Anderson R. J., Spong M. W.: Asymptotic Stability for Force Reflecting Teleoperators with Time Delay. Int. J. Robot. Res. 11 (1992) 135-149
[15] Schaft A. V. D.: L2-Gain and Passivity Techniques in Nonlinear Control. Springer-Verlag London Limited (2000)
[16] Cavusoglu M. C., Tendick F.: Multirate Simulation for High Fidelity Haptic Interaction with Deformable Objects in Virtual Environments. Proceedings of the IEEE International Conference on Robotics and Automation (ICRA 2000), San Francisco, CA (2000) 2458-2465
[17] Margaret L. M., Hespanha J. P., Sukhatme G. S.: Touch in Virtual Environments: Haptics and the Design of Interactive Systems. Prentice Hall (2002)

Linguistic Model for the Controlled Object

Zhinong Miao[1,2], Xiangyu Zhao[2], and Yang Xu[1]

[1] Center of Intelligent Control and Development Southwest Jiaotong University,
Chengdu 610031, Sichuan. P.R. China
miao215@yahoo.com.cn
[2] Department of engineering of electronic information, Panzhihua University,
Panzhihua 617000, Sichuan. P.R. China

Abstract. A fuzzy model representation for describing the linguistic model of the object to be controlled in a control system is prompted. With the linguistic model of controlled object or process to be controlled, we can construct a close loop system representation. Consequently, we can discuss the system appearance with the assistance of the linguistic model as we do using a mathematic model in a conventional control system. In this paper, we discuss the describing ability of a fuzzy model and give a formal representation method for describing a fuzzy model. The combine method for a fuzzy system constructed by multiple fuzzy models is also discussed based on the controller model and the linguistic model of controlled object.

1 Introduction

As a solution to control complex systems in their whole operation range, fuzzy controllers constitute a good offer. Since the first application of fuzzy theory to automatic operating area has to be constrained to small perturbation control in Mamdani's paper in 1975, fuzzy control has gradually been constituted as a powerful technique of control [2] [3]. Fuzzy controllers are non-linear controllers that provide a formal methodology for representing, manipulating and implementing a human's heuristic knowledge about how to control a system. They could be viewed as artificial decision-makers that operate in a closed-loop system in real time [4]. There are many issues discussing the methodology for design and analyzing of fuzzy control system, including discussing the rule base construction, fuzzy modeling, and adaptive fuzzy control [5].

Fuzzy system theory enables us to utilize qualitative, linguistic information about a system to construct a mathematical model for it. For many real-life system, which are highly complex and inherently non-linear, conventional approached to modeling often cannot be applied whereas the fuzzy approach might be the only viable alternative.

For design and analyses of the fuzzy system, conventional fuzzy system theory is lack of the model of controlled object or plant. So the method of conventional fuzzy system design cannot guzarantee the stability and robustness of closed-loop system. To settle these problems, much research has been done and most existing methods for the design and analysis of fuzzy controller can be roughly divided into two groups, one is that use fuzzy information to construct a model of the plant and utilize non-linear theory to synthesize a controller and the other is to treat fuzzy model locally.

Till now few issue address on the linguistic modeling of controlled object. This paper would present a linguistic model describing the characteristics of the controlled object. Actually, as we constructed the rule base of a fuzzy controller, we give an assumption of the characteristics of the controlled object. The potential model is indeed exists in every fuzzy control system but without any syllabify elaborating.

As conventional control system, we can construct a close loop description of a fuzzy control system using the model of object to be controlled. Up to now, fuzzy control system is lack of systemic design and analyses methodology for reason that there is not a complete illustrate of the system because there is not an appropriate model of controlled object. Although it is a great advantage of fuzzy control that we don't need a mathematic model of the controlled plant, it is also a cumber of systemic design and analysis of the system. This paper attempt to prompt an approach to model the controlled plant to provide an appropriate mode illustrating the plant and constructs a close loop system while inheriting the advantage of using human's heuristic knowledge about of the plant instead of using a mathematic model of controlled plant. The new method is based on the state space method of modern control system to construct a fuzzy model for the controlled object or the plant.

The rest of this paper is organized as follows: section II present the linguistic model of controlled object based on human empirical knowledge and the representation of linguistic model using fuzzy state space method. In section III, some combine method of two linguistic model (one is the controller model and another is the linguistic model for controlled object) is prompted section IV includes some conclusion of the future usage for the linguistic model for controlled object.

2 Fuzzy Model of Controlled Object

Fuzzy control use human's intuitional knowledge, we can use these knowledge to construct a fuzzy controller and tune the controller while apply the controller to the practice case or simulation case. Fuzzy control is proved to be an appropriate alternative for traditional control method in many nonlinear and complicated cases which can not be represented by a precise mathematic model. As we construct a fuzzy controller in the form of rule base, it is possible to construct a model of the controlled object using operator's control experience. Just as the linguistic model of controller, the model of controlled object should be linguistic model.

Begin with a typical fuzzy controller with two inputs and one output showed in figure 1. Usually an engineer would construct a fuzzy controller with rules as follows:

IF e is A and $\dot{e}$ is B THEN u is C

Here $e(t) = r(t) - y(t)$ is the error, $\dot{e}(t) = \dfrac{d}{dt}e(t)$ is the change of error and u is the output of the controller. This rule means that if the error e is with value A, and the change of error $\dot{e}$ is with value B, to drive the output of system to the ideal state we should apply a drive single u on the controlled object.

Although theses rules are based on the intuition knowledge of the engineer about control process according to his experience of controlling the controlled process or object, it is actually based on the knowledge of the controlled object. Nobody can take

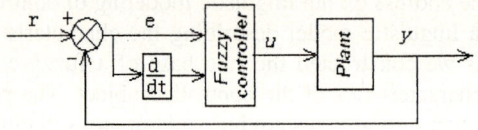

Fig. 1

any control action if he completely has no knowledge about the controlled process. The operator must know that if the controls signal u is applied on the controlled object the response of the controlled object would change to the ideal state. So the rule implies the model of controlled object. That is that the single u would drive the system output change towards the ideal state. That means we can use a rule as follows to represent the dynamic of controlled object.

if u is C then e is A and ė is B

Where C is a fuzzy set defined in the universe of discourse of controls signal u, A is the corresponding fuzzy value of error e and B is the fuzzy value of change of error $\dot{e}$.

So we can construct a linguistic model of controlled object just as we construct a fuzzy controller.

This is the basic idea of the linguistic model of controlled object.

2.1 Describe Ability of Fuzzy Model

Many works have been done on discussing the approximating ability of fuzzy system to linear or nonlinear property.

Theorem: for every continue function g and $\varepsilon > 0$ defined in the set $U \in R^n$, there must exist a fuzzy logic system f with a form as

$$f(x) = \frac{\sum_{j=1}^{m} \overline{y}^l [\prod_{i=1}^{n} a_i^l \exp(-(\frac{x_i - \overline{x}_i^l}{\sigma_i^l})^2)]}{\sum_{j=1}^{m} [\prod_{i=1}^{n} a_i^l \exp(-(\frac{x_i - \overline{x}_i^l}{\sigma_i^l})^2)]} \quad (1)$$

That satisfies

$\sup_{x \in U} |f(x) - g(x)| < \varepsilon$.

The theorem can be extended to discrete domain

Then it can be deduced that for every $g \in L_2(U)$ and $\varepsilon > 0$, there must exist a fuzzy logic system in the form as (1) that satisfy

$$\left(\int_U |f(x) - g(x)|^2 dx \right)^{1/2} < \varepsilon$$

Where $U \in R^n$ and

$$L_2(U) = \left[g : U \to R \mid \int_U |g(x)|^2 dx < \infty \right]$$

According to the description showed above, we can construct a fuzzy model for description of controlled object and the model is able to approximate any linear or nonlinear property of controlled object. The model is also in the form of linguistic representation just as the fuzzy controller. So there is also a rule base in the same form of controller.

With the help of fuzzy model of controlled object, we can construct a close loop of fuzzy control system as showed in figure 2.

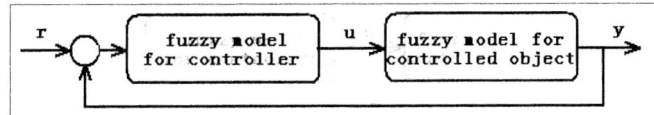

Fig. 2

In the close loop constructed by the controller, controlled object and feedback, the input of fuzzy model of controlled object is the output of the controller u and the output of the fuzzy model of controlled object is the system response y, actually the response is not only related to the input u but also related to the current state of the controlled object. So many issues represent the fuzzy model of the controlled object as

$$x_{k+1} = (u, x_k) \circ R_o \qquad (2)$$

Where x_k is the current state of controlled object and x_{k+1} is the output of the fuzzy model which is also the system response, R_o is the rule relation representing the linguistic model of controlled object.

Also the fuzzy controller can be represented as

$$u = x_k \circ R_c \qquad (3)$$

Where u is the output of fuzzy controller and R_c is the fuzzy relation representing the fuzzy model of fuzzy controller.

In this representation, the rule base should include rules as

R^l : if u is A^l and x_k is B^l
then x_{k+1} is C^l

So the equation (2) and (3) form the base of analysis of fuzzy system for its stability and system performance.

There is some drawback in the rule that in human's intuited knowledge, we always do not know what would the next state (x_{k+1}) be but we know how the state would change. That means by human's intuited knowledge the rule should be in the form as

R^l: if u is A^l and x_k is B^l

then Δx_k is C^l

It prompt a new issue that how we can deduce the next state of controlled object x_{k+1} from the current state x_k and the change of state Δx_k deduced from the rule.

A simple method is to calculate the crisp value of the change of state Δx_k and use regular math's "+" calculator to get the crisp value of x_{k+1}. The crisp value of x_{k+1} is feedback to the controller and the model of controlled object as next input of system. By this way the next state of controlled object is

$$x_{k+1} = x_k + \Delta x$$

$$= x_k + \frac{\sum_{l=1}^{n} \mu(c^l) C^L}{\sum_{l=1}^{n} \mu(c^l)}$$

2.2 Representation of Fuzzy Model for Controlled Object

As with all modeling problems, the first step is to identify the relevant quantities whose interaction the model will specify. These quantities can be classified into input, output, and state variables. Let U, Y, Σ and denote the input, output, and state spaces, respectively. To simplify the discussion, we assume that all these spaces are subsets of Euclidean space of (possibly) different dimensions. Because we are interested in dynamical systems we also need to specify a set $T \subset R$ of times of interest; typically, $T = R$ or R^+ for continuous time systems and $T = \{k\tau \mid \tau > 0$ and $k \in Z$ or $N\}$ for discrete time systems. Given these sets, a general dynamical system is defined in [11] as a quintuple $D = (u, \Sigma, y, s, r)$ where Σ is the state space, u is a set of input functions $u():T \to U$, and y is a set of output functions $y():T \to Y$. The dynamics are encoded by the state transition function s:

$$s: T \times T \times \Sigma \times U \to \Sigma$$

$$(t_1, t_0, x_0, u()) \to x_1 = s(t_1, t_0, x_0, u())$$

that produces the value of the state x_1 at time t_1 given the value of the state x_0 at time t_0 and the input for all times. The map is only defined for $t_1 > t_0$. Finally, r is the read-out function.

$$r: \Sigma \times U \times T \to Y$$

$$(x_t, u(t), t) \to y(t) = r(x_t, u(t), t)$$

that produces the output function at time t given the value of the state and input at time t. To keep the definition consistent two axioms are imposed on the state transition function.

For fuzzy system, the dynamic system can be described as the same form.

In this paper, we restrict our attention to discrete time models and, in particular, models whose time stamps take values in the set $T = \{k\tau \mid k \in N\}$ for some $\tau > 0$. Without loss of generality, we assume $\tau = 1$.

So a fuzzy dynamical system is a quintuple $D = (U^F, \Sigma^F, Y^F, IR, RO)$ where Σ^F is the fuzzy state space, $\Sigma^F \subset I^{\alpha 1} \times \cdots \times I^{\alpha N}$, that is, for every $x^F \in \Sigma^F$,

$$x^F = \begin{bmatrix} x_1^F \\ \vdots \\ x_N^F \end{bmatrix}, \quad x_i^F = \begin{bmatrix} p_1^i \\ \vdots \\ p_{\alpha i}^i \end{bmatrix} \in I^{\alpha i}, \quad \text{with} \quad 0 \leq p_j^i \leq 1,$$

$i \in 1, \cdots, N$ and $j \in 1, \cdots, \alpha i$;

U^F is a set of fuzzy input functions $u^F(\,):T \to U^F \subset I^{b1} \times \cdots \times I^{bm}$

Y^F is a set of fuzzy output functions $y^F(\,):T \to Y^F \subset I^{c1} \times \cdots \times I^{cl}$

$IR = \{IR_1, \cdots, IR_n\}$ is a set of inference rules

$IR_i : \Sigma^F \times U^F \times T \to I^{ai}$

$IR : \Sigma^F \times U^F \times T \to \Sigma^F$

IR produces the value of the state at the next time instant, given the value of the state and the input at the current time instant;

$RO = \{RO_1, \cdots RO_l\}$ is a set of read-out maps

$RO_i : \Sigma^F \times U^F \times T \to I^{ci}$

$RO : \Sigma^F \times U^F \times T \to Y^F$

RO produces the value of the output at the current time given the value of the state and input.

Fuzzy state space is a new concept for fuzzy system introduced from modern control theorem. The basic problem for fuzzy state space is the choice of state variable.

Conventional fuzzy controller always uses the error e and the change of error $\dot{e}$ as it's input variables as showed in figure 1. It is based on human's intuited knowledge about the controlled object and the controller uses the input to deduce an output signal to drive the controlled object. But it is not sufficient to describe the dynamic property of controlled object.

In such kind of system which is common in practice, a given error e, the change of error $\dot{e}$ and the input signal u can not determine the next state of system. So we have to use some concept of state variables in modern control theory for reference. Some important concept is defined as following:

State: state of a dynamic system is a minimized set of variables of the system that it can determine the future performance at the time $t \geq t_0$ if the variables at time $t = t_0$ and the input at the period $t \geq t_0$ are determined.

State variables: state variables for a dynamic system is a minimized set of variables that determining the system state. State variables must efficient to express the system state which means that the state variables can determine the only system behavior at any time. And it also must be necessary which means that the state variables are the minimized set of variables that can be used to represent the system state.

State vector: if representing dynamic system behavior needs n variables; the vector X which takes the n variables as its sub variables, the vector X is the state vector of the dynamic system. As the state at time $t = t_0$ and the input $u(t)$ at the period $t \geq t_0$ are determined, system state $X(t)$ at any time in $t \geq t_0$ is determined.

State space: if x_1 x_2 $\cdots$ x_n is the state variables for a dynamic system, the n dimensions space is the state space. Any state of the system can be represented as a point in the state space.

All these variables form the state vector of input vector of gas burning boiler and the state vector can determine the system state at any time $t \geq t_0$.

The state variables form the state vector for describe the object.

$$X = (x_1, x_2, \cdots, x_n)$$

3 Combination of Fuzzy Model

For conventional fuzzy controller showed in figure 1, the fuzzy controller model uses rules with the form of expression (1) where R^l is the lth rule, x_j is the chosen variables expressing the system's state and u is the considered output variable. The symbols A_j^l are membership functions and C^l is the rule consequent deduced from the R^l.

$$R^l : \text{if} \quad x_j^l \quad \text{is} \quad A_j^l \quad \text{then} \quad u \quad \text{is} \quad C^l \tag{1}$$

$j = 1 \ldots\ldots m$,

For MISO fuzzy controller, suppose that there are n rules in the rule base and the output u can be formulated as

$$u = \frac{\sum_{i=1}^{n} C_i \prod_{j=1}^{m} \mu_j}{\sum_{i=1}^{n} \prod_{j=1}^{m} \mu_j} \tag{2}$$

Where μ_j the membership that variable x_j belongs to the fuzzy subset defined in the universe of discourse and C_i is the center of the fuzzy subset of the consequent deduced from R^i.

In the reference process described above, the model of controlled object is not taken into account. Main reason is that there is not explicit fuzzy model of controlled object. The linguistic model of controlled object makes it possible to analyze the output of plant y based on the input error and change of error.

Based on the model constructed before, it is possible to discuss the combination of the models of controller and controlled object. Basically there are two methods to combine the two fuzzy models; the difference of the two methods lays on that the signal which is transported between the two fuzzy models. One is that the output of the controller u is defuzzificated and the crisp value of u is the input of the fuzzy model of controlled object while the other method takes u as a fuzzy value and translate it to the fuzzy model of controlled object. The combination using these two methods is discussed as following.

3.1 Crisp Value Transfer

The deduced output of the controller is defuzzificated and the crisp value of u is treated as the input of the fuzzy model of controlled object. So the output of model of controlled object which is also the system response is

$$x_{k+1} = x_k + \Delta x_k$$

And Δx_k is deduced from the fuzzy model of controlled object with a input as u and current state of controlled object x_k.

$$\Delta \tilde{x}_k = (\tilde{u}, \tilde{x}) \circ R_o$$

Using the regular combine process and center of gravity defuzzification method, there comes

$$\Delta x_k = \frac{\sum_{i=1}^{R} \Delta x_i \mu_i(u) \prod_{k=1}^{n} \mu_i(x_k)}{\sum_{i=1}^{R} \mu_i(u) \prod_{k=1}^{n} \mu_i(x_k)]}$$

R is the number of the rules in the rule base, and n is the dimension of status variable we choose to represent system state.

For the controller inference process, conventional method use the error and change of error as it's input. Basically fuzzy controller is a discrete controller and the error e can be represented as

$$e_k = x_k - r$$

r is the set value of the system response.

And the change of error $\dot{e}$ can be represented as

$$\dot{e} = e_k - e_{k-1}$$
$$= (x_k - r) - (x_{k-1} - r)$$
$$= \Delta x_k$$

so the output of controller can be represented as

$$u = (e, \dot{e}) \circ R_c$$
$$= (x_k - r, \Delta x_k) \circ R_c$$

As the set value r is a constant value, it can be omitted using a coordinate change on the linguistic value of x_k. Using the same inference mechanism and defuzzification method, we have

$$u = \frac{\sum_{i=1}^{R} u_i \mu_i(x_k) \prod_{k=1}^{n} \mu_i(\Delta x_{k-1})]}{\sum_{i=1}^{R} \mu_i(x_k) \prod_{k=1}^{n} \mu_i(\Delta x_{k-1})]}$$

Based on the description (1) and (2), it is obviously that the next state of controlled object (it is also the system response) is related not only to the current controller output and the current state, but also related to the passed state of controlled object. This is tally with human's knowledge about the process control.

3.2 Fuzzy Value Transfer

As the signal u is transferred between the two fuzzy models, it suggests a method to combine the two models by a kind of rule chain.

Many papers have addressed their goal to combine the two fuzzy models using the signal as a fuzzy variable. But it can only be used as a theoretical analyses method for the reason that in real world the two fuzzy models are physically separated. The signal can not be transferred in fuzzy value.

A typical combine method is:

$$X_{K+1} = (X_K, U_K) \circ R_P$$
$$U_K = X_K \circ R_C$$

R_P is the fuzzy relation of controlled process and R_C is the fuzzy relation of fuzzy controller.

The combination of the two equations

$$X_{K+1} = X_K \circ R_1 \circ R_P$$

4 Conclusions

The model for controlled object and the representation discussed above presents a method to describe the fuzzy control system as a close loop. It constructs a base to study fuzzy control system on systematic design method and analysis for system performance.

Comparing the two combination methods of crisp and fuzzy value transferred, the crisp value method is much complicated but the fuzzy value transferred method can only be used as a theoretic analyses method for the reason that in actual applications the value transferred is crisp instead fuzzy.

Based on the linguistic model of controlled object, the performance of system can be discussed before the great deal of simulation and tuning work. The performance of the system can be determined in the design process.

Acknowledgments

This work is partly supported by the Department of Electronic Engineering and Information Panzhihua University and Department of Mathematics Southwest Jiaotong University. It is also a part of work supported by national nature science fund. Grant number (60474022).

References

1. Y. Lin and G. A. Cunningham : A new approach to fuzzy-neural modeling. IEEE Trans. Fuzzy Syst., vol. 3, pp. 190–197, May1995
2. D. Landau, R. Lozano, and M. M'Saad: Adaptive Control. NewYork: Springer-Verlag, 1998.
3. Lixing Wang : Adaptive fuzzy system and control design and reliability analyses. Prentice Hall 1994
4. John Lygeros,A: Formal Approach to Fuzzy Modeling IEEE TRANSACTIONS ON FUZZY SYSTEMS, VOL. 5, NO. 3, AUGUST 1997
5. F. M. Callier and C. A. Desoer: Linear System Theory. New York:Springer-Verlag, 1991.

Fuzzy Sliding Mode Control for Uncertain Nonlinear Systems

Shao-Cheng Qu and Yong-Ji Wang

Department of Control Science and Engineering,
Huazhong University of Science and technology, Wuhan, China, 430074,
qushaosol1971@163.com

Abstract. The novel fuzzy sliding mode control problem is presented for a class of uncertain nonlinear systems. The Takagi-Sugeno (T-S) fuzzy model is employed to represent a class of complex uncertain nonlinear system. A virtual state feedback technology is proposed to design the sliding mode plane. Based on Lyapunov stability theory, sufficient conditions for design of the fuzzy sliding model control are given. Design of the sliding mode controller based on reaching law concept is developed, which to ensure system trajectories from any initial states asymptotically convergent to sliding mode plane. The global asymptotic stability is guaranteed. A numerical example with simulation results is given to illustrate the effectiveness of the proposed method.

1 Introduction

Since uncertainties are always the sources of instability for nonlinear system, the stabilization problems of uncertain nonlinear systems are extremely important. In the past decade, many researchers have paid a great deal of attention to various control methods in uncertain nonlinear systems [1]. With the development of fuzzy systems, fuzzy control represented by IF-THEN rules has become one of the useful control approaches for complex systems [2]. Takagi and Sugeno et al. proposed a kind of fuzzy inference system so-called Takagi-Sugeno (T-S) fuzzy model [3-6]. It combines the flexibility of fuzzy logic theory and rigorous mathematical theories of linear or nonlinear system into a unified framework. It has been shown that PDC control design and conditions for the stability within the framework of T-S fuzzy model can be formulated into LMIs form [7-9].

Currently, stability analysis and synthetically design for nonlinear system represented in T-S fuzzy models have been well addressed in [8-12]. We would like to point out when the uncertainties are bounded and the bounds are known, sliding mode control (SMC) is a reasonable method to stabilize uncertain fuzzy system. SMC approach, based on the use of discontinuous control laws, is known to be an efficient alternative way to tackle many challenging problems of system synthesis [13]. A sliding mode control system has various attractive features such as fast response, good transient performance. Especially, ideal sliding mode system is insensitive to matched parameter uncertainties and external disturbances. Actually, SMC method has been used to design for linear systems and nonlinear systems long before [14-17]. Although the SMC for uncertain nonlinear systems received increasing attention, there has been very little work to discuss uncertain T-S fuzzy system using SMC approach [18].

In this paper, by using a novel virtual state feedback technique, sufficient conditions for design of robust sliding mode plane are given based on Lyapunov stability approach. The purpose of virtual state feedback control is only helpful to design the robust sliding mode plane. Design of the sliding mode controller based on reaching law concept is developed to guarantee system trajectories from any initial conditions asymptotically convergent to sliding mode plane. Therefore, the asymptotic stability of the global fuzzy sliding mode system is guaranteed.

2 Problem Formulation

The continuous fuzzy dynamics model, proposed by Takagi and Sugeno, is described by fuzzy IF-THEN rules, which represents local linear input-output relations of nonlinear system. Consider an uncertain nonlinear system that can be described by the following uncertain T-S fuzzy model

Plant Rule i: IF $z_1(t)$ is M_{i1} and ... and $z_g(t)$ is M_{ig}

THEN $\quad \dot{x}(t) = (A_i + \Delta A_i(t))x(t) + B_i u(t) + f_i(x,t)$, $i = 1, 2 \cdots n$ (1)

where n is the number of IF-THEN rules, M_{ij} is the fuzzy set; $x(t) \in R^n$ is the state vector, $u(t) \in R^m$ is the input control; A_i and B_i are known constant matrices with appropriate dimensions, $\Delta A_i(t)$ represent parameter uncertainties, and $f_i(x,t)$ is a bounded external disturbance. $z_1(t), z_2(t) \cdots z_g(t)$ are the premise variables, and $\Psi(t)$ is the continuous initial state function.

By using standard fuzzy inference method, i.e., a singleton fuzzifier, product fuzzy inference, and central-average defuzzifier, system (1) can be inferred as

$$\dot{x}(t) = \sum_{i=1}^{n} h_i(z(t))[(A_i + \Delta A_i(t))x(t) + B_i u(t) + f_i(x,t)] \quad (2)$$

where $z(t) = [z_1(t), z_2(t) \cdots z_g(t)]^T$, and $h_i(z(t))$ denotes the normalized membership function which satisfies

$$h_i(z(t)) = w_i(z(t)) / \sum_{i=1}^{n} w_i(z(t)), \quad w_i(z(t)) = \prod_{j=1}^{g} M_{ij}(z_j(t)) \quad (3)$$

and $M_{ij}(z_j(t))$ is the grade of membership of $z_j(t)$ in M_{ij}. It is assumed that $w_i(z(t)) \geq 0$ for all t. Then we can obtain the following conditions

$$h_i(z(t)) \geq 0, \sum_{i=1}^{n} h_i(z(t)) = 1, \; i = 1, 2, \cdots, n \quad (4)$$

In this paper, we make the following assumptions.

Assumption 1: All the matrices B_i are identical, i.e., $B_1 = B_2 = \cdots = B_n := B$. Furthermore, suppose that the matrix B is of full column rank.

Assumption 2: The time-varying parameter uncertain matrices $\Delta A_i(t)$ are defined as follows $\Delta A_i = H_i F_i(t) E_i$, where H_i and E_i are known matrices with appropriate dimensions; and $F_i(t)$ are unknown matrices satisfying $F_i^T(t) F_i(t) \leq I$ for $\forall t$, where the elements of $F_i(t)$ are Lebesgue measurable.

Assumption 3: The external disturbances satisfy the matching conditions and are bounded as $f_i(x,t) = B \bar{f}_i(x,t)$ and $\| \bar{f}_i(x,t) \| \leq \delta_f(t)$.

Let us choose the sliding mode plane

$$S = B^T P x(t) = 0 \qquad (5)$$

where $P \in R^{n \times n}$ is a symmetric positive definite matrix to be chosen later.

The objective of this work is to investigate the stabilization for uncertain T-S fuzzy system in two steps. The first step is to construct suitable sliding mode controller, which to guarantee that system trajectories can be driven to the sliding mode plane in a finite time. Another step is to derive the sufficient conditions for asymptotic stability of the sliding mode system. Before proceeding, we recall some lemmas that will be frequently used throughout the proofs of our main results.

Lemma 1 [19]: Given constant matrices R_1, R_2 and R_3 with appropriate dimensions, where matrices $R_1 = R_1^T$, $R_2 = R_2^T > 0$, then $R_1 + R_3^T R_2^{-1} R_3 < 0$ if and only if

$$\begin{bmatrix} R_1 & R_3^T \\ R_3 & -R_2 \end{bmatrix} < 0$$

Lemma 2 [19]: Given matrices H and E with appropriate dimensions and matrix $F(t)$ satisfies $F^T(t)F(t) \leq I$, then, for any scalar $\varepsilon > 0$, we have

$$HFE + (HFE)^T \leq \varepsilon HH^T + \varepsilon^{-1} E^T E$$

3 Main Conclusions

Theorem 1: Under assumptions 1-4, the trajectories of uncertain T-S fuzzy system from any initial states can be driven to the sliding mode plane described by (5) in a finite time with the control

$$u(t) = u_{eq} + u_n \qquad (6)$$

where the equivalent control is

$$u_{eq} = -\sum_{i=1}^{n} h_i (B^T PB)^{-1} B^T PA_i x(t) \qquad (7)$$

in which, for convenience, we use the briefness notation h_i to denote $h_i(z(t))$. The switching control is

$$u_n = -\sum_{i=1}^{n} h_i \{ (B^T PB)^{-1} [\| B^T PH_i \| \cdot \| E_i x(t) \| + \| B^T PB \| \delta_f(t) + \varepsilon_0] \mathrm{sgn}(S) \} \qquad (8)$$

where ε_0 is a positive constant.

Proof: Consider the Lyapunov function candidate
$$V = 0.5 S^T S \tag{9}$$
which is positive-definite for all $S \neq 0$. The derivative of (9) with respect to time along (2) is

$$\dot{V} = S^T \dot{S} = S^T B^T P \dot{x}(t)$$
$$= \sum_{i=1}^{n} h_i S^T B^T P[(A_i + \Delta A_i(t))x(t) + B_i u(t) + f_i(x,t)]$$

Substituting (6)-(8) into the above equation and noting that assumption 1, we get

$$\dot{V} = \sum_{i=1}^{n} h_i S^T B^T P[(A_i + \Delta A_i(t))x(t) + Bu(t) + f_i(x,t)]$$
$$= \sum_{i=1}^{n} h_i [S^T B^T P \Delta A_i x(t) + S^T B^T P B \bar{f}_i(x,t) + S^T B^T P B u_n]$$

Considering assumption 2-4 and relation (4), $\dot{V}$ in above equation can be expressed as

$$\dot{V} \leq \sum_{i=1}^{n} h_i \| S^T \| [\| B^T P H_i \| \cdot \| E_i x(t) \| + \| B^T P B \| \delta_f(t)] + S^T B^T P B u_n$$
$$= -\sum_{i=1}^{n} h_i \varepsilon_0 S^T \operatorname{sgn}(S)$$
$$= -\varepsilon_0 \| S \|$$
$$\leq 0$$

The last inequality is known to show that the system trajectories will be arrived at the sliding mode plane within a finite time. The reaching mode of the uncertain T-S fuzzy system is guaranteed. The proof is completed.

The next is to design the robust sliding mode plane such that the system trajectories restricted to the sliding mode plane are robust stable in the presence of both parameter uncertainties and external disturbances. The following results can be obtained.

Theorem 2: Under assumptions 1-4, the SMC system (1) will asymptotically stable to sliding mode plane described by (5) with $P = Q^{-1}$ under the control (6) if there exist symmetric positive define matrices $Q > 0$, general matrix L, and scalars $\varepsilon > 0$, satisfying the following LMIs for $i = 1, 2 \cdots n$

$$\begin{bmatrix} W_i & Q E_i^T \\ * & -\varepsilon I \end{bmatrix} < 0 \tag{10}$$

where $W_i = A_i Q - BL + Q A_i^T - L^T B^T + \varepsilon H_i H_i^T$.

Proof: Consider the following Lyapunov function candidate

$$V(x,t) = x^T P x \qquad (11)$$

where symmetric positive define matrices P is value from LMIs (10).

We consider the controller (6) in form

$$u(t) = -Kx + v(t) \qquad (12)$$

where $v(t) = Kx + u_{eq} + u_n$. Substituting (12) into the uncertain fuzzy system (2) and noting that assumption 1, the closed-loop system can be obtained

$$\dot{x}(t) = \sum_{i=1}^{n} h_i(z(t))[(\overline{A}_i + \Delta A_i(t))x(t) + Bv(t) + f_i(x,t)] \qquad (13)$$

where $\overline{A}_i = A_i - BK$.

So the derivative of the Lyapunov function (11) to time along (13) is

$$\dot{V}(x,t) = 2x^T(t)P\dot{x}(t)$$

$$= \sum_{i=1}^{n} h_i \{2x^T P[(\overline{A}_i + \Delta A_i)x(t) + 2x^T PB(v + \tilde{f}_i(x,t))\}$$

Once system is on the sliding mode plane (5), $\dot{V}(x,t)$ can be reduced to the following form

$$\dot{V}(x,t) = \sum_{i=1}^{n} h_i x^T(t) M x(t) \qquad (14)$$

where $M = [(\overline{A}_i + \Delta A_i)^T P + P(\overline{A}_i + \Delta A_i)]$.

By Lemma 2 and Assumption 2, for a scalar $\varepsilon > 0$, we obtain

$$M \leq P\overline{A}_i + \overline{A}_i^T P + \varepsilon P H_i H_i^T P + \varepsilon^{-1} E_i^T E_i \qquad (15)$$

By Lemma 1, inequality (15) is equivalent to

$$M \leq \begin{bmatrix} P\overline{A}_i + \overline{A}_i^T P + \varepsilon P H_i H_i^T P & E_i^T \\ E_i & -\varepsilon I \end{bmatrix} \qquad (16)$$

Pre-multiplying and post-multiplying (16) by N^T and $N = diag\{P^{-1}, I\}$, respectively, then considering relation $\overline{A}_i = A_i - BK$, and defining $P^{-1} = Q, L = KQ$, we obtain $N^T M N \leq \Omega_i$, where

$$\Omega_i = \begin{bmatrix} A_i Q - BL + QA_i^T - L^T B^T + \varepsilon H_i H_i^T & QE_i^T \\ * & -\varepsilon I \end{bmatrix} \qquad (17)$$

If $\Omega_i < 0$, then $N^T M N < 0$. So we can obtain $M < 0$. Furthermore, $\dot{V}(x,t) < 0$. The proof is completed.

In term of the principle of sliding mode control strategy, it can be concluded that the closed-loop system given by (2), (5) and (6) is global asymptotically stable.

4 Numerical Simulation

Consider the uncertain nonlinear system

$$\begin{cases} \dot{x}_1(t) = x_2(t) \\ \dot{x}_2(t) = -0.1x_2^3(t) - 0.02x_1(t) - 0.67x_1^3(t) + f(x,t) + u(t) \end{cases}$$

The purpose of control is to achieve closed-loop stability and to attenuate the influence of the exogenous external disturbance. It is also assumed that states is measurable and $x_1(t), x_2(t) \in [-1.5, 1.5]$. Let us assume that

$$-0.1x_2^3(t) = c(t)x_2(t), \quad c(t) \in [-0.225, 0].$$

Using the same procedure in [5], the nonlinear term can be represented as

$$-0.67x_1^3(t) = M_{11} \cdot 0 \cdot x_1(t) - (1 - M_{11}) \times 1.5075 x_1(t)$$

By solving the equation, the membership functions M_{11} of fuzzy set can be interpreted as

$$M_{11}(x_2(t)) = 1 - \frac{x_2^2(t)}{2.25}, \quad M_{12}(x_2(t)) := 1 - M_{11}(x_2(t)) = \frac{x_2^2(t)}{2.25}.$$

By using these fuzzy sets, the nonlinear system can be presented by the following uncertain T-S fuzzy model.

Plant Rule 1: IF $x_2(t)$ is M_{11} THEN

$$\dot{x}(t) = (A_1 + \Delta A_1(t))x(t) + B_1[u(t) + f_1(x,t)]$$

Plant Rule 2: IF $x_2(t)$ is M_{12} THEN

$$\dot{x}(t) = (A_2 + \Delta A_2(t))x(t) + B_2[u(t) + f_2(x,t)]$$

The model parameters are given as follows

$$A_1 = \begin{bmatrix} 0 & 1 \\ -0.02 & -0.1125 \end{bmatrix}, A_2 = \begin{bmatrix} 0 & 1 \\ -1.5275 & -0.1125 \end{bmatrix}, \Delta A_1 = H_1 F_1(t) E_1,$$

$$\Delta A_2 = H_2 F_2(t) E_2, E_i = [0, 1], H_i = [0, 0.1125]^T, F_i^T(t) F_i(t) \leq I, B_i = [0, 1]^T,$$

$$f_i(x,t) = 0.2\sin(x_1(t) + x_2(t)) \text{ for } i = 1, 2.$$

Next, solving the LMIs (10) for $i = 1, 2$ gives

$$Q = \begin{bmatrix} 0.3285 & -0.2319 \\ -0.2319 & 0.2150 \end{bmatrix}, L = [-0.0131 \quad 0.6995], \varepsilon = 1.0750.$$

It then follows from Theorem 1 and Theorem 2 that the closed-loop system is robust stable. The simulation results for the nonlinear systems are shown in Figs. 1-3. For these simulations, the initial values of the system states are $x_1(0) = -1.2$, $x_2(0) = 1.4$, and control parameter of (12) is $\varepsilon_0 = 0.1$. Obviously, the state trajectories are attracted toward the sliding mode planes and the global closed-loop nonlinear system is asymptotically stable.

5 Conclusion

In this paper, a fuzzy sliding mode control approach has been studied for a class of uncertain nonlinear systems in the presence of both parameter uncertainties and external disturbances. T-S fuzzy model is employed to represent the uncertain nonlinear systems. By using virtual state feedback technique, sufficient conditions for design of robust sliding mode plane are given based on the Lyapunov theory. Design of the sliding mode stabilizing controller based on reaching law concept is proposed, which to guarantee system trajectories convergent to sliding mode plane. The global asymptotically stability of the closed-loop system was proven. It is pointed out that assumption 1 is too strict for general nonlinear system, which is major limitation of the proposed method in this paper. A numerical example with simulation results is given to illustrate the effectiveness of the proposed method.

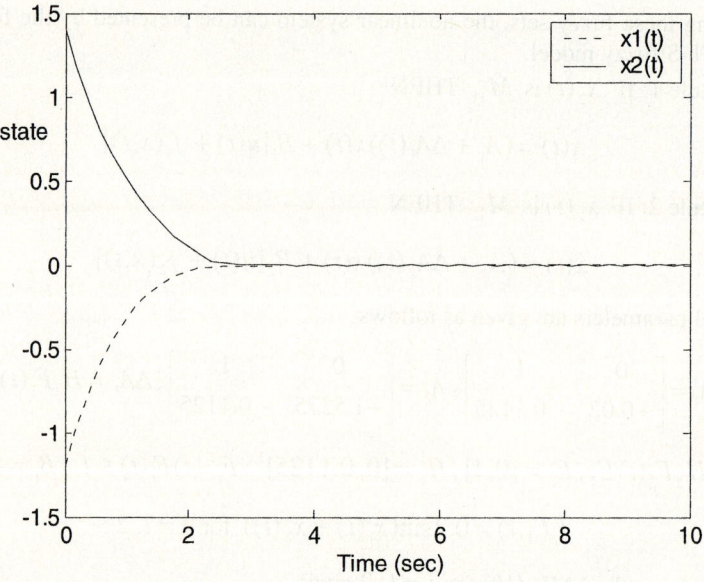

Fig. 1. The state of system under the proposed controller

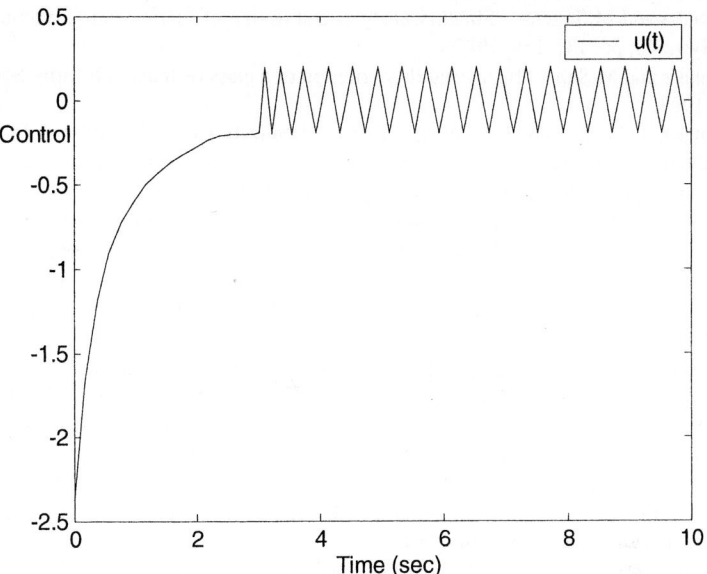

Fig. 2. The proposed controller

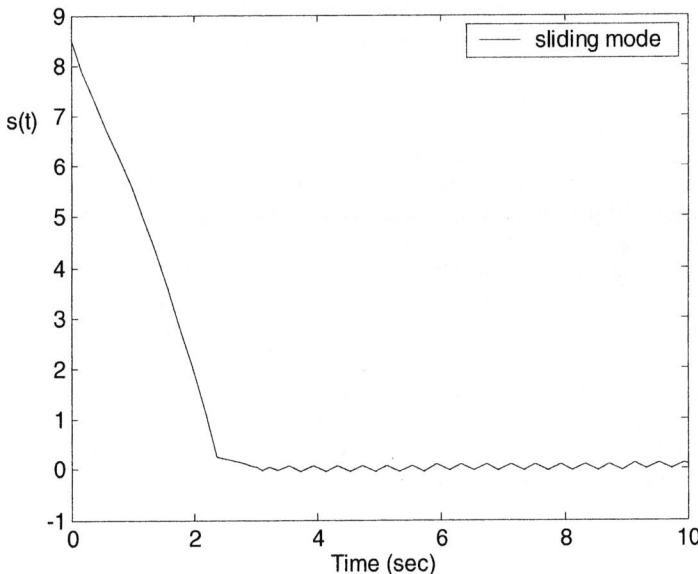

Fig. 3. The sliding mode of system

References

1. M. Vidysagar. Nonlinear Systems Analysis. Englewood Cliffs, NJ: Prentice-Hall, 1993.
2. L. X. Wang, "Adaptive fuzzy systems and control: design and stability analysia", Upper Saddle River, NJ: Prentice-Hall, 1994.

3. K. Tanaka and M.Sugeno. "Stability analysis and design of fuzzy control system", Fuzzy Sets Syst., 45, pp. 135-156, 1992.
4. K. Tanaka and M.Sano. "Fuzzy stability criterion of a class of linear", Inform. Sci, 71, pp. 3-26, 1993.
5. K. Tanaka, T. Ikeda, and H. O. Wang. "Robust stabilization of a class of uncertain nonlinear systems via fuzzy control: Quadratic stability, H∞ control theory and linear matrix inequalities", IEEE Trans. Fuzzy Syst., 4, pp. 1-13, 1996.
6. K. Tanaka, T. Ikeda, and H. O. Wang. "Fuzzy regulators and fuzzy observers: Relaxed stability conditions and LMI-based designs", IEEE Trans. Fuzzy Syst., 6, pp. 250-265, 1998.
7. G. Reng and P. M. Frank, "Approaches to quadratic stabilization of uncertain fuzzy dynamic system", Cir. Syst. 48, pp. 760-769, 2001.
8. K. R. Lee and J. H. Kim, "Output feedback robust H∞control of uncertain fuzzy dynamic systems with time-varying delay", IEEE Trans. Fuzzy Syst, 8, pp. 657-664, 2000.
9. Xing-Ping Guan and Cai-lian Chen, "Delay-dependent guaranteed const control for T-S fuzzy systems with time delays", IEEE Trans. Fuzzy Syst, 12, pp. 236-249, 2004.
10. Zidong Wang, Daniel W. C. and Xiaohui Liu, "A note on the robust stability of uncertain stochastic fuzzy systems with time-delays", IEEE Trans on Syst, man, and cyber-part A: Syst and Huma, 34, pp. 570-576, 2004.
11. Rong-Jyun Wang, Wei-Wei Lin and Wen-June Wang, "Stabilizability of linear quadratic state feedback for uncertain fuzzy time-delay systems ", IEEE Trans on Syst, man, and cyber-part B: Cybernetics, 1, pp. 1-4, 2004.
12. Liu X. and Zhang Q., "New approaches to H∞ controller designs based on fuzzy observers for T-S fuzzy systems via LMI", Automatica, 39, pp. 1571-1582. 2003.
13. Drakunov, S. V. and Utkin, V. I., "Sliding mode control in dynamic system", International Journal of Control, 55, pp.1029-1037, 1992.
14. F. Gouaisbaut, M. Darnbrine and J. P. Richard, "Robust control of delay systems: a sliding mode control design via LMI", Systems & Control Letters, 46, pp.219-230, 2002.
15. Said Oucheriah, "Exponential stabilization of linear delayed systems using sliding-mode controllers". IEEE transaction on circuit and systems, 1: fundamental theory and application, 50 (6), pp.826-830, 2003.
16. Xiaoqiu Li and R. A. Decarlo. "Robust sliding mode control of uncertain time delay systems". International Journal of Control, 76(13), pp.1296-1305, 2003.
17. Y. Niu, J.Lam and and X.Wang. "Sliding-mode control for uncertain neutral delay systems". IEE Proceedings of control theory and applications, 151(1), pp.38-44, 2004.
18. Xinghuo Yu, Ahihong Man and Baolin Wu, "Design of fuzzy sliding mode control systems", Fuzzy Sets and Systems, 95, pp. 295-306, 1998.
19. S. Boyd, L. E. Ghaoui, E. Feron, and V. Balakrishnan, Linear matrix inequalies in system and control theory. Philadelphia, PA: SLAM, 1994.

Fuzzy Control of Nonlinear Pipeline Systems with Bounds on Output Peak[1]

Fei Liu and Jun Chen

Institute of Automation, Southern Yangtze University,
Wuxi, 214122, P.R. China
fliu@sytu.edu.cn

Abstract. A new fuzzy control method for nonlinear pipeline system is discussed in this paper. The nonlinear dynamics of pipeline system is composed by two gravity-flow tanks, and are described by Takagi-Sugeno (T-S) fuzzy model. The controller design is based on overall stability, and is carried out via the parallel distributed compensation (PDC) scheme. To obtain better output dynamic performance, a given bounds is introduced to the output of nonlinear systems. Moreover, by means of linear matrix inequality (LMI) technique, it is shown that the existence of such constrained control system can be transformed into the feasibility of a convex optimization problem. Finally, by applying the designed controller, the simulation results demonstrate the efficiency.

1 Introduction

Pipeline system composed by two gravity-flow tanks is a familiar device in the chemical processes. The control of pipeline system is always attracted lots of scientists' interest because of the nonlinearity of the process dynamics.

Among the existing control approaches, there are two typical methods. The first one is based on strict nonlinear mathematical models or on the given point to linearization, but this method has bad robust performance; while the second one is based on expert experience and intelligent control method, however, it doesn't guarantee the stability of system and lacks universal applicability. Recently, based on passivity theory, paper [4] proposed a control method for pipeline system. Under the precondition of system stability, it provides a class of widely applicable controller design methods.

Also based on stability, this paper presents a new fuzzy control method. Pipeline system [4] is first described by a T-S fuzzy model[1]. It is well known that Takagi-Sugeno (T-S) fuzzy models can provide an effective representation of complex nonlinear systems. Different from the approach depended on fuzzy rule tables, in this type of fuzzy models, the premise is lingual variables, while the conclusion is linear combination with certain input variables. It combines linear system theory with fuzzy theory to deal with the control problem for nonlinear system. The controller, which can guarantee the fuzzy system asymptotical stability in the large, is designed based on the

---

[1] Supported by the Key Project of Chinese Ministry of Education (NO105088).

parallel distributed compensation (PDC) scheme [2]. The stability criterion is come down to find a common symmetric positive definite matrix P to satisfy a set of Lyapunov matrix inequalities. And under the condition of constraints on the system output, design conditions for the stability and output constraints can be reduced to convex problems involving linear matrix inequalities [3] (LMIs). Moreover, due to the interior-point algorithm, the feasibility problem of LMIs can be solved very efficiently by means of MATLAB LMI toolbox. At last, the designed controller is employed in a simulation experiment, and the results demonstrate the efficiency of this method.

2 T-S Fuzzy Model and Stability

Consider the following affine nonlinear system:

$$\dot{x} = f(x) + g(x)u \tag{1}$$

where $x \in R^n$ is the state vector, $u \in R^m$ is the input vector, $f(x)$ and $g(x)$ are both nonlinear functions.

The T-S fuzzy models are described by fuzzy IF-THEN rules, each of which represents the local linear subsystem in a different state-space region. The i th rule of the fuzzy system (1) is given in the following form:

Rule i :

$$\text{IF } z_1(t) \text{ is } M_{i1} \text{ and } \ldots \text{ and } z_p(t) \text{ is } M_{ip}$$
$$\text{THEN } \dot{x}(t) = A_i x(t) + B_i u(t), \ i = 1,2,\ldots,r. \tag{2}$$

where $z_1(t), z_2(t),\ldots, z_p(t)$ are the premise variables. In general, these parameters may be functions of the state variables, external disturbances, and/or time. $M_{ij} (i = 1,2,\ldots,r, j = 1,2,\ldots, p)$ are fuzzy sets, $x(t) \in R^n$ is the state vector, $u(t) \in R^m$ is the input vector, r is the number of IF-THEN rules, and A_i, B_i are some constant matrices of appropriate dimensions.

Using weighted average method for defuzzification, the final output of the fuzzy systems are inferred as follows:

$$\dot{x}(t) = \frac{\sum_{i=1}^{r} w_i(z(t))[A_i x(t) + B_i u(t)]}{\sum_{i=1}^{r} w_i(z(t))}$$

$$= \sum_{i=1}^{r} h_i(z(t))[A_i x(t) + B_i u(t)] \tag{3}$$

where

$$w_i(z(t)) = \prod_{j=1}^{p} M_{ij}(z_j(t)),$$

$$h_i(z(t)) = \frac{w_i(z(t))}{\sum_{i=1}^{r} w_i(z(t))}, \ i = 1,2,\ldots,r.$$

in which $M_{ij}(z_j(t))$ is the grade of membership of $z_j(t)$ in the fuzzy set M_{ij}, $h_i(z(t))$ is the possibility for the i th rule to fire. It is easy to find that $w_i(z(t)) \geq 0$, $i = 1,2,...,r$, $\sum_{i=1}^{r} w_i(z(t)) > 0$ for all t. Therefore $h_i(z(t)) \geq 0$, $i = 1,2,...,r$, and $\sum_{i=1}^{r} h_i(z(t)) = 1$ for all t.

Then the method of the PDC scheme is utilized to design the fuzzy controller to stabilize fuzzy system (2). The idea of the PDC scheme is to design a compensator for each rule of the T-S fuzzy model. The resulting overall fuzzy controller, which is nonlinear in general, is a fuzzy blending of each individual linear controller. The fuzzy controller shares the same fuzzy sets with the fuzzy system (2), so the i th control rule for the fuzzy system (2) is described as follows:

Rule i:

$$\text{IF } z_1(t) \text{ is } M_{i1} \text{ and } ... \text{ and } z_p(t) \text{ is } M_{ip} \tag{4}$$

$$\text{THEN } u(t) = -F_i x(t), \quad i = 1,2,...,r.$$

where F_i is the local state feedback gain matrix. The overall state feedback fuzzy control law is represented by

$$u(t) = -\frac{\sum_{i=1}^{r} w_i(z(t)) F_i x(t)}{\sum_{i=1}^{r} w_i(z(t))} = -\sum_{i=1}^{r} h_i(z(t)) F_i x(t) \tag{5}$$

Now, substituting (5) into (3), we can obtain the following closed-loop system:

$$\dot{x}(t) = \sum_{i=1}^{r} h_i^2(z(t)) G_{ii} x(t) + \sum_{i,j=1, i<j}^{r} h_i(z(t)) h_j(z(t)) (G_{ij} + G_{ji}) x(t) \tag{6}$$

where

$$G_{ij} = A_i - B_i F_j, \quad i,j = 1,2,...,r \text{ and } i < j \leq r.$$

The design of the state feedback fuzzy controller is to determine the local feedback gains F_i such that the above closed-loop fuzzy system is asymptotically stable. Based on the stability theory, it can be seen that the equilibrium of the fuzzy system described by (6) is asymptotically stable in the large if there exist a common symmetric positive definite matrix P such that the following matrix inequalities are satisfied:

$$P > 0 \tag{7}$$

$$G_{ii}^T P + P G_{ii} < 0 \tag{8}$$

$$(G_{ij} + G_{ji})^T P + P(G_{ij} + G_{ji}) < 0 \tag{9}$$

where

$$G_{ij} = A_i - B_i F_j, \quad i,j = 1,2,...,r \text{ and } i < j \leq r.$$

According to Schur complement, it is easy to find that the matrix inequalities (7), (8) and (9) are equivalent to the following LMIs [3], respectively:

$$X > 0 \tag{10}$$

$$A_i X + X A_i^T - B_i Y_i - Y_i^T B_i^T < 0 \tag{11}$$

$$A_i X + X A_i^T + A_j X + X A_j^T - B_i Y_j - Y_j^T B_i^T - B_j Y_i - Y_i^T B_j^T < 0 \tag{12}$$

where

$$X = P^{-1}, \; Y_i = F_i X, \; i, j = 1, 2, \ldots, r \text{ and } i < j \le r.$$

The above conditions are LMIs with respect to variables X, Y_i. This feasibility problem can be solved very efficiently by means of the recently developed interior-point methods. Furthermore, if matrices exist which satisfies these inequalities, then the feedback gains are given by $F_i = Y_i X^{-1}$.

3 Pipeline System Control

Consider the following series arrangement of two identical gravity-flow tanks equipped with outlet pipes (see Fig.1). The dynamic equations of this system are described as follows:

$$\begin{aligned}
\dot{x}_1 &= \frac{A_p g}{L} x_2 - \frac{K_f}{\rho A_p^2} x_1^2 \\
\dot{x}_2 &= \frac{1}{A_t} \left(F_{C\max} \alpha^{-(1-u)} - x_1 \right) \\
\dot{x}_3 &= \frac{A_p g}{L} x_4 - \frac{K_f}{\rho A_p^2} x_3^2 \\
\dot{x}_4 &= \frac{1}{A_t} (x_1 - x_3)
\end{aligned} \tag{13}$$

where x_1 and x_3 are the volumetric flow rates of liquid leaving the tanks via the pipes, x_2 and x_4 are the heights of the liquid in the tank, respectively. $F_{C\max} = 2 m^3/s$ is the maximum value of the volumetric rate of fluid entering the first tank, $g = 9.81 m/s^2$ is the gravitational acceleration constant, $L = 914m$ is the length of the pipes, $K_f = 4.41 Ns^2/m^3$ is the friction factor, $\rho = 998 kg/m^2$ is the density of the liquid, $A_p = 0.653 m^2$ is the cross-sectional area of the pipes, and $A_t = 10.5 m^2$ is the cross-sectional area of each of the tanks. The parameter $\alpha = 5$ is the range ability parameter of the valve and the control input u is the valve position.

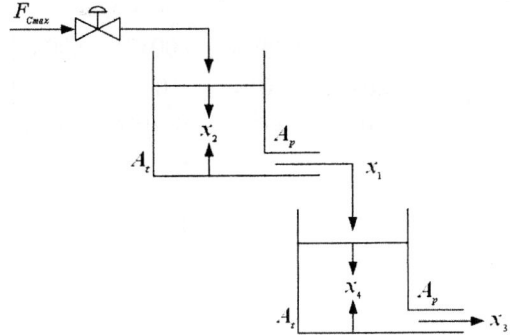

Fig. 1. A series of two gravity flow tanks

In order to represent the system (13) in the format given by (1), we regard the control input term via the following auxiliary variable v

$$v = F_{C\max} \alpha^{-(1-u)} \tag{14}$$

Then the system (13) can be transformed into the following form:

$$\dot{x} = f(x) + g(x)v \tag{15}$$

where

$$f(x) = \begin{bmatrix} \dfrac{A_p g}{L} x_2 - \dfrac{K_f}{\rho A_p^2} x_1^2 \\ -\dfrac{1}{A_t} x_1 \\ \dfrac{A_p g}{L} x_4 - \dfrac{K_f}{\rho A_p^2} x_3^2 \\ \dfrac{1}{A_t}(x_1 - x_3) \end{bmatrix}, \; g(x) = \begin{bmatrix} 0 \\ \dfrac{1}{A_t} \\ 0 \\ 0 \end{bmatrix}.$$

Using the linearization method introduced in paper [5], the system (15) can be illustrated by the following T-S fuzzy models. This linearization method overcomes the disadvantage of the Tailor series approach, and adopts local linearization idea to provide a new linearization method for nonlinear systems.

Rule 1: IF $x_4(t)$ is about 1
THEN $\dot{x}(t) = A_1 x(t) + B_1 v(t)$
Rule 2: IF $x_4(t)$ is about 5
THEN $\dot{x}(t) = A_2 x(t) + B_2 v(t)$

where

$$A_1 = \begin{bmatrix} -0.0153 & 0.0091 & 0.0017 & 0.0021 \\ -0.0952 & 0 & 0 & 0 \\ 0.0017 & 0.0021 & -0.0153 & 0.0091 \\ 0.0952 & 0 & -0.0952 & 0 \end{bmatrix},$$

$$A_2 = \begin{bmatrix} -0.0370 & 0.0101 & 0.0011 & 0.0031 \\ -0.0952 & 0 & 0 & 0 \\ 0.0011 & 0.0031 & -0.0370 & 0.0101 \\ 0.0952 & 0 & -0.0952 & 0 \end{bmatrix},$$

$$B_1 = B_2 = \begin{bmatrix} 0 \\ 0.0952 \\ 0 \\ 0 \end{bmatrix}.$$

Selecting membership functions of the form

$$h_1(x_4) = 1 - (x_4 - 1)/4, \quad h_2(x_4) = 1 + (x_4 - 5)/4,$$

solving LMIs (10), (11), (12), a feasible solution is ontained

$$X = \begin{bmatrix} 1551.5 & -394.8 & 1284.8 & -383.5 \\ -394.8 & 1822.4 & 69.9 & -1028.4 \\ 1284.8 & 69.9 & 1762.9 & 403.4 \\ -383.5 & -1028.4 & 403.4 & 4982.0 \end{bmatrix},$$

$$F_1 = \begin{bmatrix} 3.9787 & 4.0388 & -4.1664 & 1.4611 \end{bmatrix},$$

$$F_2 = \begin{bmatrix} 4.3154 & 4.1636 & -4.4364 & 1.5346 \end{bmatrix}.$$

Then the fuzzy state feedback controller is constructed as follows:

$$u = -(h_1 F_1 + h_2 F_2)(x - x_d) + v_d \tag{16}$$

where (x_d, v_d) is the desired operating point of the system (15), choosing

$$x_d = \begin{bmatrix} 1.8389 & 5.0 & 1.8389 & 5.0 \end{bmatrix}^T, \quad v_d = 1.8389.$$

To test the performance of the controller design, applying the controller (16) to system (15), the simulation results are described in Fig.2 with the initial condition $x_0 = \begin{bmatrix} 1.3 & 2.5 & 1.3 & 2.5 \end{bmatrix}^T$. From Fig.2, it can be seen that the designed fuzzy controller globally stabilizes the system (15).

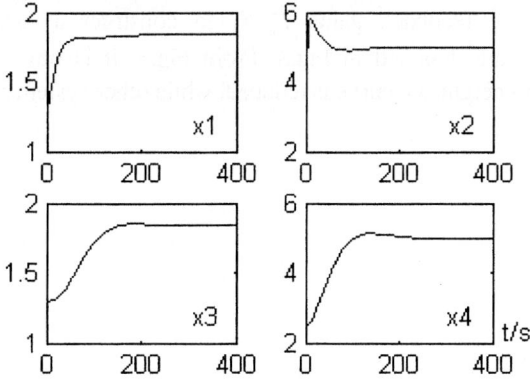

Fig. 2. Simulation results of pipeline system without output constraints

However, the dynamical response of liquid height x_2 is not perfect. To obtain better dynamical performance, output constraints will be introduced to control system. Assume that $z = Cx$ is the controlled output vector, where $x = [x_1 \ x_2 \ x_3 \ x_4]^T$ is system vector, matrix $C \in R^{1 \times 4}$ is selected by requirement. Here, the aim is to limit the liquid height x_2, so we choose $C = [0 \ 1 \ 0 \ 0]$, that is $z = x_2$. Under the known initial condition x_0 and the upper limit $\xi > 0$, the system (15) is asymptotically stable in the large and satisfies the condition $\|z\| \le \xi$ if there exist matrix X such that LMIs (10) - (12) and the following matrix inequalities are satisfied:

$$\begin{bmatrix} X & x_0 \\ x_0^T & \xi \end{bmatrix} > 0 \qquad (17)$$

$$\begin{bmatrix} X & XC^T \\ CX & \xi \end{bmatrix} > 0 \qquad (18)$$

For the sake of comparing with Fig.2, we choose the initial condition and the desired operating point just the same as in Fig.2. Consider that this method has a stated conservativeness, so the value of variable ξ can be tested for several times, such as $\xi = 24$. Then, by solving LMIs (10) - (12) and (17), (18), the feasible solutions are as follows:

$$X = \begin{bmatrix} 322.7 & -5.1 & 290.4 & 8.4 \\ -5.1 & 17.4 & -0.6 & -13.9 \\ 290.4 & -0.6 & 370.7 & 119.7 \\ 8.4 & -13.9 & 119.7 & 1008.4 \end{bmatrix},$$

$$F_1 = [4.0886 \ \ 75.1448 \ \ -4.3597 \ \ 1.5083],$$

$$F_2 = [4.1009 \ \ 75.1490 \ \ -4.3691 \ \ 1.5094].$$

Using the above feedback gains F_1, F_2 to construct a fuzzy controller, the simulation results are depicted in Fig.3. From Fig.3, it is obvious to see that the overshoot of liquid height x_2 curve is reduced, while other system response curves are not influenced.

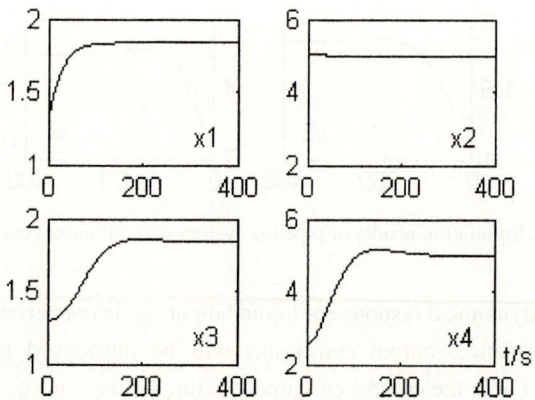

Fig. 3. Simulation results of pipeline system with output constraints

4 Conclusions

In this paper, the control problem for nonlinear pipeline system described by the T-S fuzzy model is carried out. It is shown that the nonlinear system can be represented by a set of local models, and the controllers for each local model are designed via PDC scheme. It is also shown that the stable condition can be transformed into convex problems in terms of LMIs. To improve the response performance of system, control constraints is introduced to the system output variable. The simulation results demonstrate that the method is effective.

References

1. Tanaka, K., Sugeno, M.: Stability Analysis and Design of Fuzzy Control Systems. Fuzzy Sets and Systems, Vol.45, (1992) 135-156
2. Wang, H.O., Tanaka, K., Griffin, M.F.: Parallel Distributed Compensation of Nonlinear Systems by Takagi-Sugeno Fuzzy Model. Proc. FUZZ-IEEE/IFES'95, (1995) 531-538
3. Boyd, S., El Ghaoui, L., Feron, E., Balakrishnan, V.: Linear Matrix Inequalities in System and Control Theory. Philadelphia, PA: SIAM, (1994)
4. Sira-Ramirez, H., Angulo-Nunez, M.I.: Passivity-Based Control of Nonlinear Chemical Processes. Int. J. Control, Vol.68, (1997) 971-996
5. Marcelo C. M. Teixeira, Stanislaw H. Zak: Stabilizing Controller Design for Uncertain Nonlinear Systems Using Fuzzy Models. IEEE Trans. Fuzzy Syst., Vol.7, (1999) 133-142

Grading Fuzzy Sliding Mode Control in AC Servo System

Hu Qing[1], Qingding Guo[2], Dongmei Yu[1], and Xiying Ding[2]

[1] School of Electrical Engineering, Shenyang University of Technology, Postalcode 110023,
No.58 Xinghua South Street, Tiexi District, Shenyang, Liaoning, P.R. China
{aqinghu, yu_dm163}@163.com

[2] School of Electrical Engineering, Shenyang University of Technology, Postalcode 110023,
No.58 Xinghua South Street, Tiexi District, Shenyang, Liaoning, P.R. China
{guoqd, dingxy}@sut.edu.cn

Abstract. In this paper a strategy of grading fuzzy sliding mode control (FSMC) applied in the AC servo system is presented. It combines the fuzzy logic and the method of sliding mode control, which can reduce the chattering without decreasing the system robustness. At the same time, the exponent approaching control is added by grading. The control strategy makes the response of the system quick and no overshoot. It is simulated to demonstrate the feasible of the proposed method by MATLAB6.5 and the good control effect is received.

1 Introduction

Since the 1990's, the combination of the fuzzy logic control and sliding mode control has been widely researched and applied. The most significant aspect is FSMC characterized by illumination [1], [2]. The combination of the fuzzy control and sliding mode control can solve the existed problems and keep the system stability effectively.

To solve the problem of chattering this paper presents the method using the fuzzy reasoning to decide the magnitude of the control. At the same time, the exponent approaching method is adopted in the start-up process of the system, which can speed up the response of the system and make the time of response is secondary priority.

2 System Analysis and Controller Design

2.1 Design of Sliding Mode Controller

The control methods are chosen by the judgment of s to adopt either fuzzy sliding mode control or exponent approaching control.

The speed differential equation of the permanent magnet synchronous motor (PMSM) follows as[3]:

$$\frac{d\omega_r}{dt} = -\frac{B}{J}\omega_r + \frac{P_n}{J}K_t i_q - \frac{P_n}{J}T_L \tag{1}$$

where J, ω_r, B, T_L, P_n and i_q denote inertia, angular velocity, friction factor, load torque, pairs of poles and quadrate-axis current, respectively. K_t is constant.

Suppose $x_1 = \theta_r - \theta$ and $x_2 = \dot{x}_1 = -\omega_r$. Where, θ_r is the reference position of the rotor and θ is the practical position of the rotor. The state equation is

$$\begin{cases} \dot{x}_1 = x_2 \\ \dot{x}_2 = -ax_2 - bu + F(t) \end{cases} \quad (2)$$

where $a = B_0/J_0$, $b = K_t P_n/J_0$, $F(t) = d_0 T_L + g \cdot dw_r/dt + hw_r$, and $u = i_q$ is the input. J_0, B_0 is ideal value, $d_0 = P_n/J_0$, $g = \Delta J/P_n$, $h = \Delta B/P_n$, and ΔJ, ΔB is the practical changing variable of them. The sliding mode line is defined as $s = cx_1 + x_2$.

Using the following control rules:

$$\begin{cases} u = u_1 & \text{if } |s| \geq s_0 \\ u = u_2 & \text{if } |s| < s_0 \end{cases} \quad (3)$$

where u_1 is exponent approaching control and u_2 is fuzzy sliding mode control.

Calculation of the exponent control rules: Supposing $u = u_1$, so u_1 can be derived:

$$u_1 = \frac{1}{b}[(c-a+k)x_2 + c \cdot k \cdot x_1 + \varepsilon \cdot sign(s)] \quad (4)$$

where k, ε are the constants of control.

Calculation of FSMC rules: The reachable condition $s \cdot \dot{s} < 0$ must be satisfied. Suppose $u = u_2$, substituting (2) to $s \cdot \dot{s} < 0$, we have

$$\begin{cases} u_2 > \frac{1}{b}(c-a)x_2 + \frac{1}{b}F & \text{when } s > 0 \\ u_2 < \frac{1}{b}(c-a)x_2 + \frac{1}{b}F & \text{when } s < 0 \end{cases} \quad (5)$$

Define $u_2 = u_{eq} + u_k$. The control law superposes equivalent control u_{eq} and switching control u_k. u_{eq} is the necessary control effort that makes the state trajectory remains on the sliding mode line. The action of u_k is that any initial state point on phase plane can come to the sliding mode line in finite interval. So the achievable property can be satisfied.

2.2 Design of Fuzzy Sliding Mode Controller

The physical meaning of fuzzy controller is that the distance between state point and sliding mode line and the derivative of it reasons out the output.

Chosen of membership: The domain of s, $\dot{s}$, u_k are chosen as $[-3, 3]$. On this domain, the 7 fuzzy subsets are set, [NB, NM, NS, ZO, PS, PM, PB]. The membership of them is shown in Fig. 1.

As shown in Fig. 1, the membership near ZO is lumping and the degree of reasoning is stronger, which makes the control effort soft.

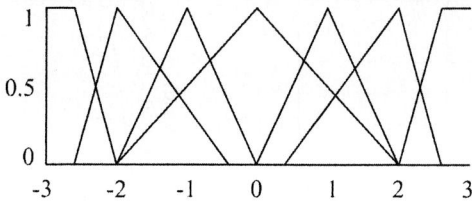

Fig. 1. Fuzzy sets on input and output variables domain

Table 1. Fuzzy reasoning rules

u_k \ $\dot{s}$	NB	NM	NS	ZO	PS	PM	PB
s							
NB	NB	NB	NB	NB	NB	NB	NB
NM	NB	NB	NM	NM	NM	NB	NB
NS	NB	NM	NS	NB	NS	PM	PS
ZO	NB	NM	NS	ZO	PS	PM	PB
PS	NS	NM	PS	PB	PS	PM	PB
PM	PB	PB	PM	PM	PM	PB	PB
PB	PB	PB	PB	PB	PB	PB	PB

Chosen of control rules: The following rules in table 1 are adopted. Where s is the distance between the state point and the sliding mode line and $\dot{s}$ is the velocity of it.

3 Simulations of the System

Mathematical model and parameters of an AC servo motor: $G(s) = P_n/(J+B)$, where $P_n = 2$, $J = 0.0013$ Kgm², $B = 0.0026$ Nm/(rad · s). $K_t = 0.0124$.

Chosen of control parameters: $c = 5$, $K_1 = 3/4$, $K_2 = 3/100$, $K_3 = 5/3$, where K_1, K_2 are the quality factors, K_3 is the anti-quality factor. $s_0 = 10$, $k = 10$, $\varepsilon = 0.2$.

Simulation parameters: the reference position $\theta_r = 5$, the simulation step is 0.01s. At 3s, the step overload is added. PID parameters: $K_P = 10$, $K_I = 2$, $K_D = 0.5$.

Output responses of system with step interruption: Fig. 2 and Fig. 3 are the PID control and FSMC with a step overload, respectively. Compared with the traditional PID methods, as shown in Figures, FSMC has the property of no overshoot, quick response and strong robustness.

Simulations of reducing chattering: Fig. 4 is the phase trajectory of traditional sliding mode control. Fig. 5 is the phase trajectory of FSMC. It is obvious that FSMC has the good control effect in reducing chattering.

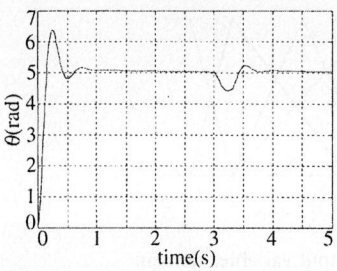
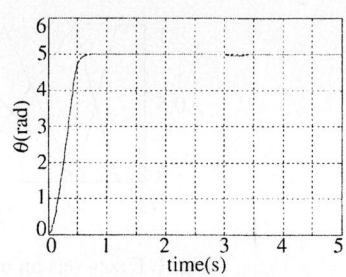

Fig. 2. The PID control with step overload   **Fig. 3.** The FSMC with step overload

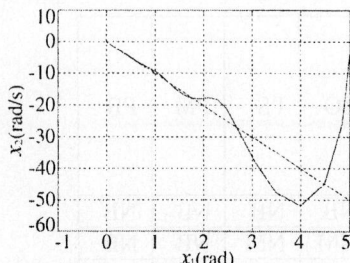
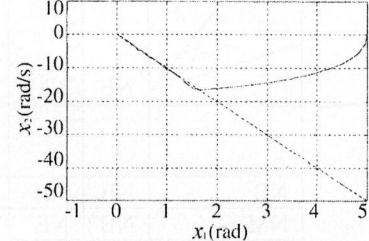

Fig. 4. Error state vector of sliding mode   **Fig. 5.** Error state vector of fuzzy sliding mode

4 Conclusions

The simulations show that the grading FSMC presented by this paper reduces the chattering by controlling the output with the sliding mode variables s and $\dot{s}$. It keeps the robustness to the parameters vary and interruption. Moreover, the exponent approaching control is added by grading, so the response of the system is sped up and the very good effect is received.

References

1. Wang, F. Y.: Sliding Mode Variable Structure Control. Mechanics Industry Press, Beijing (1995).
2. Ha, Q. P., Rye, D.C., Durrant -Whyte, H.F.: Fuzzy Moving Sliding Mode Control with Application to Robotic Manipulators. Automatics, 35 (1999) 607-616.
3. Guo, Q. D., Wang, C. Y.: AC Servo System. Mechanics Industry Press, Beijing (1994).

A Robust Single Input Adaptive Sliding Mode Fuzzy Logic Controller for Automotive Active Suspension System

Ibrahim B. Kucukdemiral[1], Seref N. Engin[1],
Vasfi E. Omurlu[2], and Galip Cansever[1]

[1] Yildiz Technical University, Department of Electrical Engineering, Turkey
{beklan, nengin, cansever}@yildiz.edu.tr,
[2] Yildiz Technical University, Department of Mechanical Engineering, Turkey
omurlu@yildiz.edu.tr

Abstract. The proposed controller in this paper, which combines the capability of fuzzy logic with the robustness of sliding mode controller, presents prevailing results with its adaptive architecture and proves to overcome the global stability problem of the control of nonlinear systems. Effectiveness of the controller and the performance comparison are demonstrated with chosen control techniques including PID and PD type self-tuning fuzzy controller on a quarter car model which consists of component-wise nonlinearities.

1 Introduction

There are two major objectives in the studies of Automotive Active Suspension Systems (AASSs) that are to improve ride comfort by reducing the vertical acceleration of the sprung mass and to increase holding ability of the vehicle by providing adequate suspension deflections. To overcome the problems that originate from the complexity and nonlinearity of vehicle systems, various kinds of Fuzzy Logic Controllers (FLCs) are suggested such as [1,2,3,4,5].In this work, we propose a novel, robust, simple and industrially applicable FLC with a single state feedback for AASSs, where the stability of the controller is secured in the sense of Lyapunov. The other benefit is that, the rule-base for the proposed controller does not need to be tuned by an expert since an adaptation mechanism takes the responsibility for tuning. Online simulations of the proposed system verify that the proposed suspension controller exhibits much better performances compared to other methods namely, the passive suspension, classical Proportional-Integral-Derivative (PID) controller and PD-type self-tuning fuzzy controller (STFPD), when testing with standard bumps and random road profiles.

2 Modelling the Active Suspension System

A typical active suspension system for a quarter car model is illustrated in Fig. 1.

It is assumed that the tire never leaves the ground. The actuator force is denoted with f_u. Nonlinear spring and damping forces are provided as

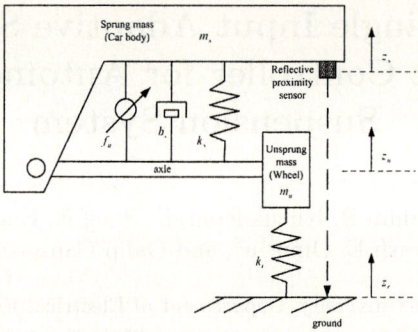

Fig. 1. Quarter car model used

$$f_s = k_s(z_u - z_s)(k_s/4)(z_u^3 - z_s^3)$$
$$f_b = b_s|\dot{z}_u - \dot{z}_s|(\dot{z}_u - \dot{z}_s). \tag{1}$$

Here damping and spring coefficients are denoted by, $b_s = 1000$ Nsec/m and, $k_s = 45000$N/m respectively. Car body displacement, z_s, wheel displacement, z_u, and road displacement, z_r, are all measured from the static equilibrium position. The dynamic equations of the quarter car active suspension system, neglecting nonlinear effects for now, are

$$m_s\ddot{z}_s = k_s(z_s - z_u) + b_s(\dot{z}_s - \dot{z}_u) + f_u$$
$$m_u\ddot{z}_u = -k_s(z_s - z_u) - b_s(\dot{z}_s - \dot{z}_u) - f_u + k_t(z_r - z_u) \tag{2}$$

where $m_s = 250.3$ kg and $m_u = 30.41$ kg are the masses of the car body and the wheel respectively, $\ddot{z}_s$ denotes the acceleration of the car body and $k_t = 150000$ N/m is the tire spring constant.

3 Design of a Robust Single-Input Adaptive Fuzzy Sliding Mode Controller for AASS

In general, a compact dynamic equation for AASS can be regarded as a second order differential equation, such as

$$\ddot{z}_s(t) = -\frac{B}{M}\dot{z}_s(t) - \frac{K}{M}z_s(t) - \frac{T_d}{M} + \frac{F_u}{M} \equiv A_p\dot{z}_s(t) + K_p z_s(t) + D_p T_d + M_p u(t) \tag{3}$$

where M is the total mass of the body, B is the damping coefficient, T_d is the term for total unknown load disturbances and finally F_u is the applied control force to the system. The error of sprung mass displacement is $e(t) = z_s(t) - z_d(t)$ where $z_d(t)$ is the desired displacement of the car body.

The first step in sliding mode controller design process is choosing the sliding surface. Therefore the following sliding surface is employed

$$s(t) = \dot{z}_s(t) - \int_0^t [\ddot{z}_d(\tau) - k_1\dot{e}(\tau) - k_2 e(\tau)] d\tau. \tag{4}$$

Assuming that the system dynamics are well known, a feedback linearization $u^*(t)$ can be applied. However this control law cannot be directly applicable. Thus, one method to overcome this difficulty is to imitate the feedback linearization method given by an adaptive fuzzy logic controller such as

$$u_{fz}(s) = \frac{\sum_{r=1}^{N} w_r \theta_r}{\sum_{r=1}^{N} w_r} \tag{5}$$

where θ_r, $r = 1, 2, \ldots, N$ are the discrete singleton control signals labelled as adjustable parameters and w_r is the firing of r. rule. In the present work, we have chosen Gaussian membership functions for the fuzzification process of s. And N is chosen to be 7. If θ_r is chosen as an adjustable parameter, (5) can be regarded as $u_{fz} = \mathbf{\Theta}^T \mathbf{\Delta}$ where $\mathbf{\Theta} = [\theta_1, \theta_2, \ldots, \theta_N]^T$ and $\mathbf{\Delta} = [\delta_1, \delta_2, \ldots, \delta_N]^T$ are the corresponding vectors. According to the universal approximation property of the fuzzy systems (5) can be approximated by a fuzzy system with a bounded approximation error ε with a bound ρ^* such as $u^*(t) = u_{fz}^*(t) + \varepsilon = \mathbf{\Theta}^{*T} \mathbf{\Delta} + \varepsilon$. Let $\hat{u}_{fz} = \hat{\mathbf{\Theta}} \mathbf{\Delta}$ be the approximation of $u^*(t)$ and $\hat{\mathbf{\Theta}}$ as the approximation of $\mathbf{\Theta}^*$. Moreover, in order to compensate the approximation error $\hat{u}_{fz} - u^*$ we add a variable structure controller term to the control signal which results $u(t) = \hat{u}_{fz} + u_{vs}$. Then, after some algebraic manipulations, one can achieve the error dynamics as $\ddot{e} + k_1 \dot{e} + k_2 e = M_p [\hat{u}_{fz} + u_{vs} - u^*] = \dot{s}$. Denoting $\hat{u}_{fz} - u^* = \hat{u}_{fz} - u_{fz}^* - \varepsilon$ as $\tilde{u}_{fz}$ and $\hat{\mathbf{\Theta}} - \mathbf{\Theta}^*$ as $\tilde{\mathbf{\Theta}}$ then $\tilde{u}_{fz} = \tilde{\mathbf{\Theta}}^T \mathbf{\Delta} - \varepsilon$ is obtained. the variable structure controller is defined as $u_{vs} = -\rho \cdot \text{sat}(s/\Phi)$ where ρ is the variable switching gain and Φ is the thickness of the boundary layer and chosen as 0.0008 for the present application and sat is the saturation function. If ρ^* denotes the equivalent gain and $\hat{\rho}$ denotes the estimated gain then the error for switching gain will be $\tilde{\rho} = \hat{\rho} - \rho^*$. In order to achieve minimum approximation error and to guarantee the existence of sliding mode, we have chosen a Lyapunov function candidate as,

$$V(s, \tilde{\mathbf{\Theta}}, \tilde{\rho}) = \frac{1}{2} + \frac{M_p}{2\gamma_1} \tilde{\mathbf{\Theta}}^T \tilde{\mathbf{\Theta}} + \frac{M_p}{2\gamma_2} \tilde{\rho}^2 \tag{6}$$

where γ_1 and γ_2 are positive constants. then the time derivative of V along the trajectory will be

$$\dot{V}(s, \tilde{\mathbf{\Theta}}, \tilde{\rho}) = s.M_p [-\hat{\rho} \cdot \text{sat}(s/\Phi) - \varepsilon] + \frac{M_p}{\gamma_2} (\hat{\rho} - \rho^*) \dot{\hat{\rho}} \tag{7}$$

Then choosing $\dot{\hat{\rho}} = \gamma_2 \text{sat}(s/\Phi)$, and taking into consideration that $\dot{\tilde{\rho}} = \dot{\hat{\rho}}$ and $\dot{\tilde{\mathbf{\Theta}}} = \dot{\hat{\mathbf{\Theta}}} = \gamma_1 \cdot s \cdot \mathbf{\Delta}$, one can easily bring 7 to result, $\dot{V} = -M_p |s| (\rho^* - |\varepsilon|) \leq 0$ Finally by using Barbalat's lemma, one can easily show that the stability of the proposed controller and adaptation laws are achieved in the sense of Lyapunov. The final structure of the proposed controller is shown in Fig. 2.

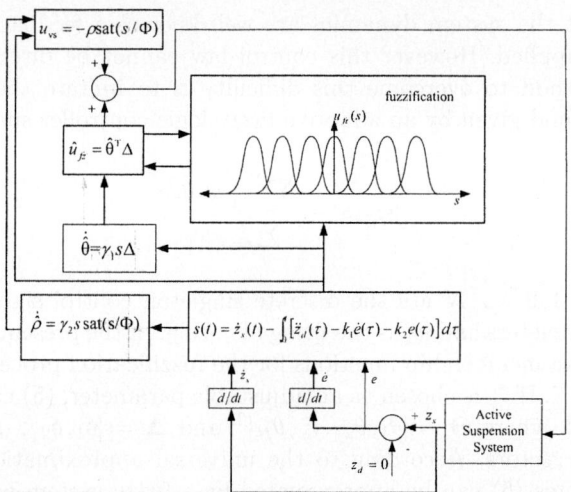

Fig. 2. Block diagram of the proposed controller

4 Simulation Results

To evaluate the proposed controller presented above, a simulation environment of AASS is created for the quarter car model which has component-wise nonlinearities within. Damper and spring element nonlinearities in the system dynamics equations are chosen as in actual components. Vehicle speed of 72 km/h is chosen and two types of road profiles are prepared for controller performance evaluation: standard bump-type surface profile with 10cm length × 10cm height and a random road profile generated to simulate stabilized road with 1cm × 1cm pebbles. Open loop, PID and STFPD are employed along with the proposed controller. The rule-bases for the main FLC and self-tuning mechanism are chosen, as described in [6]. On the other hand, PID controller is, first, tuned with well-known Ziegler-Nichols tuning method and then, with numerous repetitive simulations around previously obtained acceptable PID controller values [7]. Although the initial rule table of the proposed controller is not important because of the self organizing structure of the proposed controller, initial value of the adjustable vector Θ is chosen as $\Theta = [5000\ 3000\ 1000\ 0\ -1000\ -3000\ -5000]^T$. The values of k_1 and k_2 are chosen to be 10.1 and 0.16, respectively.

First, bump-type road profile is applied to the system for four types of controllers. Car body displacements are plotted as shown in Fig. 3. In Fig. 3, the proposed controller clearly produces the shortest response time of 0.85 sec. and the lowest peak value of 0.4 cm. Open loop response has continuing oscillations of 25 sec. and also it has a high peak value. PID controller decreases the peak value while shortening the response time. On the other hand, STFPD still has a higher peak value and slower response time compared to the proposed controller. However, for chosen PID parameters, classical PID seems to perform better than PD type self-tuning fuzzy controller.

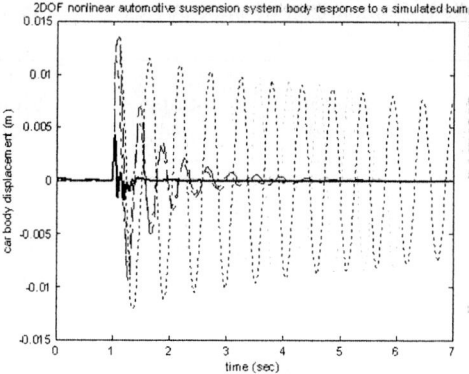

Fig. 3. Car Body Displacement for a simulated bump (Proposed controller: solid bold; passive suspension: dotted; PID controller: dash; STFPD controller: dot-dash)

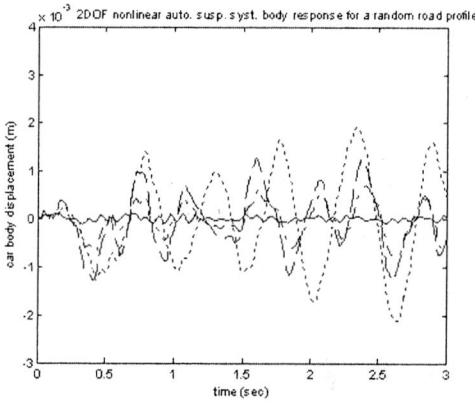

Fig. 4. Body displacement for random road profile (Proposed controller: solid bold; passive suspension: dotted; PID controller: dot-dash; STFPD controller: dash)

As a second experiment, proposed controller and evaluatory ones have been tested using the random road profile. Fig. 4 shows the response of all four controllers for this road profile and it is obvious that the proposed sliding mode controller has overwhelming success over other controllers.

Summary of all responses to two kinds of road conditions can be seen through Table 1.

5 Conclusion

A novel single-input adaptive fuzzy sliding mode controller is proposed and successfully employed to control AASS with component-wise nonlinearities. The strategy is robust and industry applicable since it has a single input FLC as a

Table 1. Comparison of the controller performances

controller	$\ddot{z}_s$ random road (rms)	$\ddot{z}_s$ bumpy road (rms)	Sett. time First peak (meter)	bumpy road (sec.)
Proposed	0.1288	1.2538	0.0040	0.85
Passive (Open loop)	0.4994	1.6369	0.0133	25
STFPD	0.3209	1.4876	0.0135	6
PID	0.2936	1.4097	0.0112	4.15

main controller. Thus, the rule base of the FLC drastically decreases when it is compared with the traditional FLCs. On the other hand, the efficiency of the controller is improved by combining a sliding mode compensator which also has an adaptive structure. The stability of the proposed scheme is achieved in the sense of Lyapunov. In order to demonstrate the effectiveness of the proposed method, the controller is applied to the suspension system in comparison with the passive suspension, PID controller and STFPD. Road profiles that are tested are a simulated random road surface and a bump. The simulation results show that the proposed control scheme improves the ride comfort considerably when compared to the aforementioned controller structures.

References

1. Ting, C.S., Li, T.H.S., Kung, F.C.: Design of fuzzy controller for active suspension system. Mechatronics. **5** (1993) 457–470
2. Yeh, E.C., Tsao, Y.J.: A fuzzy preview control scheme of active suspension for rough road. Int. J. Vehicle Des. **15** (1994) 166–180
3. Rao, M.V.C., Prahald, V.: A tunable fuzzy logic controller for vehicle-active suspension system. Fuzzy Sets and Syst. **85** (1997) 11–21
4. D'Amato, F.J., Viassolo, D.E.: Fuzzy control for active suspensions. Mechatronics. **10** (2000) 897–920
5. Huang, S.J., Lin, W.C.: Adaptive fuzzy controller with sliding surface for vehicle suspension control. IEEE Trans. on Fuzzy Syst. **11** (2003) 550–559
6. Mudi, R.K., Pal, N.R.: A robust self-tuning scheme for PI- and PD-type fuzzy controllers. IEEE Trans. on Fuzzy Syst. **7** (1999) 2–16
7. Omurlu, V.E., Engin, S.N., Kucukdemiral, I.B.: A Robust Single Input Adaptive Sliding Mode Fuzzy Logic Controller for a Nonlinear Automotive Suspension System. MED'04 Conference, June 6-9, (2004), Kusadasi, Aydin, Turkey.

Construction of Fuzzy Models for Dynamic Systems Using Multi-population Cooperative Particle Swarm Optimizer

Ben Niu[1,2], Yunlong Zhu[1], and Xiaoxian He[1,2]

[1] Shenyang Institute of Automation, Chinese Academy of Sciences,
110016, Shenyang, China,
[2] School of Graduate, Chinese Academy of Sciences,
100039, Beijing, China
{niuben, ylzhu}sia.cn

Abstract. A new fuzzy modeling method using Multi-population Cooperative Particle Swarm Optimizer (MCPSO) for identification and control of nonlinear dynamic systems is presented in this paper. In MCPSO, the population consists of one master swarm and several slave swarms. The slave swarms execute Particle Swarm Optimization (PSO) or its variants independently to maintain the diversity of particles, while the particles in the master swarm enhance themselves based on their own knowledge and also the knowledge of the particles in the slave swarms. The MCPSO is used to automatic design of fuzzy identifier and fuzzy controller for nonlinear dynamic systems. The proposed algorithm (MCPSO) is shown to outperform PSO and some other methods in identifying and controlling dynamic systems.

1 Introduction

The identification and control of nonlinear dynamical systems has been a challenging problem in the control area for a long time. Since for a dynamic system, the output is a nonlinear function of past output or past input or both, and the exact order of the dynamical systems is often unavailable, the identification and control of this system is much more difficult than that has been done in a static system. Therefore, the soft computing methods such as neural networks [1-3], fuzzy neural networks [4-6] and fuzzy inference systems [7] have been developed to cope with this problem.

Recently, interest in using recurrent networks has become a popular approach for the identification and control of temporal problems. Many types of recurrent networks have been proposed, among which two widely used categories are recurrent neural networks (RNN) [3, 8, 9] and recurrent fuzzy networks (RFNN) [4, 10].

On the other hand, fuzzy inference systems have been developed to provide successful results in identifying and controlling nonlinear dynamical systems [7, 11]. Among the different fuzzy modeling techniques, the Takagi and Sugeno's (T-S) type fuzzy controllers have gained much attention due to its simplicity and generality [7, 12]. T-S fuzzy model describes a system by a set of local linear input-output relations and it is seen that this fuzzy model can approach highly nonlinear dynamical systems.

The bottleneck of the construction of a T-S model is the identification of the antecedent membership functions, which is a nonlinear optimization problem. Typically, both the premise parameters and the consequent parameters of T-S fuzzy model are adjusted by using gradient descent optimization techniques [12-13]. Those methods are sensitive to the choice of the initial parameters, easily got stuck in local minima, and have poor generalization properties. This hampers the aposteriori interpretation of the optimized T-S model.

The advent of evolutionary algorithm (EA) has attracted considerable interest in the construction of fuzzy systems [14-16]. In [15], [16], EAs have been applied to learn both the antecedent and consequent part of fuzzy rules, and models with both fixed and varying number of rules have been considered. As compared to traditional gradient-based computation system, evolutionary algorithm provides a more robust and efficient approach for the construction of fuzzy systems.

Recently, a new evolutionary computation technique, the particle swarm optimization (PSO) algorithm, is introduced by Kennedy and Eberhart [17, 18], and has already come to be widely used in many areas [19-21]. As already has been mentioned by Angeline [22], the original PSO, while successful in the optimization of several difficult benchmark problems, presented problems in controlling the balance between exploration and exploitation, namely when fine tuning around the optimum is attempted.

In this paper we try to deal with this issue by introducing a multi-population scheme, which consists of one master swarm and several slave swarms. The slave swarms evolve independently to supply new promising particles (the position giving the best fitness value) to the master swarm as evolution goes on. The master swarm updates the particle states based on the best position discovered so far by all the particles both in the slave swarms and its own. The interactions between the master swarm and the slave swarms control the balance between exploration and exploitation and maintain the population diversity, even when it is approaching convergence, thus reducing the risk of convergence to local sub-optima.

The paper is devoted to a novel fuzzy modeling strategy to the fuzzy inference system for identification and control of nonlinear dynamical systems. In this paper, we will use the MCPSO algorithm to design the T-S type fuzzy identifier and fuzzy controller for nonlinear dynamic systems, and the performance is also compared to other methods to demonstrate its effectiveness.

The paper is organized as follows. Section 2 gives a review of PSO and a description of the proposed algorithm MCPSO. Section 3 describes the T-S model and a detailed design algorithm of fuzzy model by MCPSO. In Section 4, simulation results of one nonlinear plant identification problem and one nonlinear dynamical system control problems using fuzzy inference systems based on MCPSO are presented. Finally, conclusions are drawn in Section 5.

2 PSO and MCPSO

2.1 Particle Swarm Optimization

Particle Swarm Optimization (PSO) is inspired by natural concepts such as fish schooling, bird flocking and human social relations. The basic PSO is a population based optimization tool, where the system is initialized with a population of random solutions and searches for optima by updating generations. In PSO, the potential solutions, called particles, fly in a D-dimension search space with a velocity which is dynamically adjusted according to its own experience and that of its neighbors.

The location and velocity for the ith particle is represented as $x_i = (x_{i1}, x_{i2}, ... x_{iD})$ and $v_i = (v_{i1}, v_{i2}, ... v_{iD})$, respectively. The best previous position of the ith particle is recorded and represented as $P_i = (P_{i1}, P_{i2}, ..., P_{iD})$, which is also called *pbest*. The index of the best particle among all the particles in the population is represented by the symbol g, and p_g is called *gbest*. At each time step t, the particles are manipulated according to the following equations:

$$v_i(t+1) = v_i(t) + R_1 c_1 (P_i - x_i(t)) + R_2 c_2 (P_g - x_i(t)) \tag{1}$$

$$x_i(t+1) = x_i(t) + v_i(t) \tag{2}$$

where R_1 and R_2 are random values within the interval [0, 1], c_1 and c_2 are acceleration constants. For Eqn. (1), the portion of the adjustment to the velocity influenced by the individual's own *pbest* position is considered as the cognition component, and the portion influenced by *gbest* is the social component.

A drawback of the aforementioned version of PSO is associated with the lack of a mechanism responsible for the control of the magnitude of the velocities, which fosters the danger of swarm explosion and divergence. To solve this problem, Shi and Eberhart [23] later introduced an inertia term w by modifying (1) to become:

$$v_i(t+1) = w \times v_i(t) + R_1 c_1 (P_i - x_i(t)) + R_2 c_2 (P_g - x_i(t)) \tag{3}$$

They proposed that suitable selection of w will provides a balance between global and local explorations, thus requiring less iteration on average to find a sufficiently optimal solution. As originally developed, w often decreases linearly from about 0.9 to 0.4 during a run. In general, the inertia weight w is set according to the following equation:

$$w = w_{max} - \frac{w_{max} - w_{min}}{iter_{max}} \times iter \tag{4}$$

where $iter_{max}$ is the maximum number of iterations, and $iter$ is the current number of iterations.

2.2 Multi-population Cooperative Particle Swarm Optimization

The foundation of PSO is based on the hypothesis that social sharing of information among conspecifics. It reflects the cooperative relationship among the individuals (fish, bird, insect) within a group (school, flock, swarm). Obviously it is not the case of the nature. In natural ecosystem, many species have developed cooperative interactions with other species to improve their survival. Such cooperative—also called symbiosis—co-evolution can be found in organisms going from cells (e.g., eukaryotic organisms resulted probably from the mutualistic interaction between prokaryotes and some cells they infected) to superior animals (e.g., African tick birds obtain a steady food supply by cleaning parasites from the skin of giraffes, zebras, and other animals) [24, 25].

Inspired by the phenomenon of symbiosis in the natural ecosystem, a master-slave mode is incorporated into the PSO, and a Multi-population (species) Cooperative Optimization (MCPSO) is thus presented. In our approach, the population consists of one master swarm and several slave swarms. The symbiotic relationship between the master swarm and slave swarms can keep a right balance of exploration and exploitation, which is essential for the success of a given optimization task.

The master-slave communication model is shown in Fig.1, which is used to assign fitness evaluations and maintain algorithm synchronization. Independent populations (species) are associated with nodes, called slave swarms. Each node executes a single PSO or its variants, including the update of location and velocity, and the creation of a new local population. When all nodes are ready with the new generations, each node then sends the best local individual to the master node. The master node selects the best of all received individuals and evolves according to the following equations:

$$v_i^M(t+1) = wv_i^M(t) + R_1c_1(p_i^M - x_i^M(t)) + R_2c_2(p_g^M - x_i^M(t)) + R_3c_3(p_g^S - x_i^M(t)) \quad (5)$$

$$x_i^M(t+1) = x_i^M(t) + v_i^M(t) \quad (6)$$

where M represents the master swarm, c_3 is the migration coefficient, and R_3 is a uniform random sequence in the range [0, 1]. Note that the particle's velocity update in the master swarm is associated with three factors:

i. p_i^M : Previous best position of the master swarm.

ii. p_g^M : Best global position of the master swarm.

iii. p_g^S : Previous best position of the slave swarms.

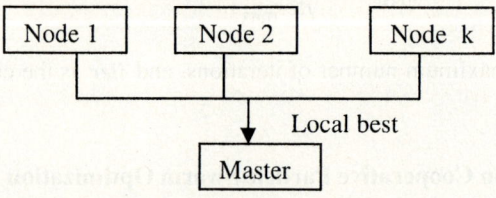

Fig. 1. The master-slave model

As Shown in Eqn. (5), the first term of the summation represents the inertia (the particle keeps moving in the direction it had previously moved), the second term represents memory (the particle is attracted to the best point in its trajectory), the third term represents cooperation (the particle is attracted to the best point found by all particles of master swarm) and the last represents information exchange (the particle is attracted to the best point found by the slave swarms). The pseudocode for the MCPSO algorithm is listed in Fig 2.

Algorithm MCPSO
Begin
    Initialize all the populations
    Evaluate the fitness value of each particle
    Repeat
        Do in parallel
            Node i, $1 \le i \le K$ //K is the number of slaver swarms
        End Do in parallel
        Barrier synchronization //wait for all processes to finish
        Select the fittest local individual p_g^S from the slave swarms
        Evolve the mast swarm
        // Update the velocity and position using (5) and (6), respectively
        Evaluate the fitness value of each particle
    Until a terminate-condition is met
End

Fig. 2. Pseudocode for the MCPSO algorithm

3 Fuzzy Model Based on MCPSO

3.1 T-S Fuzzy Model Systems

In this paper, the fuzzy model suggested by Takagi and Sugeno is employed to represent a nonlinear system. A T-S fuzzy system is described by a set of fuzzy IF-THEN rules that represent local linear input-output relations of nonlinear systems. The overall system is then an aggregation of all such local linear models. More precisely, a T-S fuzzy system is formulated in the following form:

$$R^l : if\ x_1\ is\ A_1^l\ and\ ...\ x_n\ is\ A_n^i,\ then\ \hat{y}^l = \alpha_0^l + \alpha_1^l x_1 + ... + \alpha_n^l x_n, \qquad (7)$$

where $\hat{y}^l (1 \le l \le r)$ is the output due to rule R^l and $\alpha_i^l (1 \le i \le n)$, called the consequent parameters, are the coefficients of the linear relation in the lth rule and will be identified. $A_i^l(x_i)$ are the fuzzy variables defined as the following Gaussian membership function:

$$A_i^l(x_i) = \exp[-\frac{1}{2} * (\frac{x_i - m_i^l}{\sigma_i^l})^2],\qquad(8)$$

where $1 \leq l \leq r, \ldots, 1 \leq i \leq n, x_i \in R$, m_i^l and σ_i^l represent the center (or mean) and the width (or standard deviation) of the Gaussian membership function, respectively. m_i^l and σ_i^l are adjustable parameters called the premise parameters, which will be identified.

Given an input $(x_1^0(k), \cdots, x_n^0(k))$, the final output $\hat{y}(k)$ of the fuzzy system is inferred as follows:

$$\hat{y}(k) = \frac{\sum_{l=1}^{r} \hat{y}^l(k)(\prod_{i=1}^{n} A_i^l(x_i^0(k)))}{\sum_{l=1}^{r}(\prod_{i=1}^{n} A_i^l(x_i^0(k)))} = \frac{\sum_{l=1}^{r} \hat{y}^l(k) w^l(k)}{\sum_{l=1}^{r} w^l(k)},\qquad(9)$$

where the weight strength $w^l(k)$ of the lth rule, is calculated by:

$$w^l(k) = \prod_{i=1}^{n} A_i^l(x_i^0(k)).\qquad(10)$$

3.2 Fuzzy Model Strategy Based on MCPSO

The detailed design algorithm of fuzzy model by MCPSO is introduced in this section. The overall learning process can be described as follows:

(1) Parameter representation

In our work, the parameter matrix, which consists of the premise parameters and the consequent parameters described in section 3.1, is defined as a two dimensional matrix, i.e.,

$$\begin{bmatrix} m_1^1 & \sigma_1^1 & \cdots & m_n^1 & \sigma_n^1 & \alpha_0^1 & \alpha_1^1 & \cdots & \alpha_n^1 \\ m_1^2 & \sigma_1^2 & \cdots & m_n^2 & \sigma_n^2 & \alpha_0^2 & \alpha_1^2 & \cdots & \alpha_n^2 \\ \cdots & \cdots & \cdots & \cdots & \cdots & \cdots & \cdots & \cdots & \cdots \\ m_1^r & \sigma_1^r & & m_n^r & \sigma_n^r & \alpha_0^r & \alpha_1^r & \cdots & \alpha_n^r \end{bmatrix}$$

The size of the matrix can be represented by $D = r \times (3n+1)$.

(2) Parameter learning

a) In MCPSO, the master swarm and the slave swarm both work with the same parameter settings except for the velocity update equation. Initially, $N \times n (N \geq 2, n \geq 2)$ individuals forming the population should be randomly generated and the individuals can be divided into N swarms (one master swarm and $N-1$ slave swarms). Each swarm contains n individuals with random positions and velocities on D dimensions. These individuals may be regarded as particles in terms of PSO. In T-S fuzzy model system, the number of rules, r, should be assigned in advance. In addition, the maximum iterations w_{max}, minimum inertia weight w_{min} and the learning parameters c_1, c_2, the migration coefficient c_3 should be assigned in advance. After initialization, new individuals on the next generation are created by the following step.

b) For each particle, evaluate the desired optimization fitness function in D variables. The fitness function is defined as the reciprocal of RMSE (root mean quadratic error), which is used to evaluate various individuals within a population of potential solutions. Considering the single output case for clarity, our goal is to minimize the error function:

$$RMSE = \sqrt{\frac{1}{K}\sum_{k=1}^{K}(y_p(k+1) - y_r(k+1))^2} \tag{11}$$

where K is the total time steps, $y_p(k+1)$ is the inferred output and $y_r(k+1)$ is the desired reference output..

c) Evaluate the fitness for each particle.

d) Compare the evaluated fitness value of each particle with it's *pbest*. If current value is better than *pbest*, then set the current location as the *pbest* location in D-dimension space. Furthermore, if current value is better than *gbest*, then reset *gbest* to the current index in particle array. This step will be executed in parallel for both the master swarm and the slave swarms.

e) In each generation, after step d) is executed, the best-performing particle p_g^S among the slave swarms should be marked.

f) Update the velocity and position of all the particles in $N-1$ slave swarms according to Eqn. (3) and Eqn. (2), respectively (Suppose that $N-1$ populations of SPSO with the same parameter setting are involved in MCPSO as the slave swarms).

g) Update the velocity and position of all the particles in the master swarm according to Eqn. (5) and Eqn. (6), respectively.

(3) *Termination condition*

The computations are repeated until the premise parameters and consequent parameters are converged. It should be noted that after the operation in master swarm and slaver swarm the values of the individual may exceed its reasonable range. Assume that the domain of the ith input variable has been found to be $[\min(x_i), \max(x_i)]$ from training data, then the domains of m_i^l and σ_i^l are defined as $[\min(x_i) - \delta_i, \max(x_i) + \delta_i]$ and $[d_i - \delta_i, d_i + \delta_i]$, respectively, where δ_i is a small positive value defined as $\delta_i = (\max(x_i) - \min(x_i))/10$, and d_i is the predefined width of the Gaussian membership function, the value is set as $(\max(x_i) - \min(x_i))/r$, the variable r is the number of fuzzy rules.

4 Illustrative Examples

In this section, two nonlinear dynamic applications, including an example of identification of a dynamic system and an example of control of a dynamic system are conducted to validate the capability of the fuzzy inference systems based on MCPSO to handle the temporal relationship. The main reason for using these dynamic systems is that they are known to be stable in the bounded input bounded output (BIBO) sense.

A. Dynamic System Identification

The systems to be identified are dynamic systems whose outputs are functions of past inputs and past outputs as well. For this dynamic system identification, a serial-parallel model is adopted as identification configuration shown in Fig.3.

Example 1: The plant to be identified in this example is guided by the difference equation [2, 4]:

$$y_p(k+1) = f[y_p(k), y_p(k-1), y_p(k-2), u(k), u(k-1)], \tag{12}$$

where

$$f[x_1, x_2, x_3, x_4, x_5] = \frac{x_1 x_2 x_3 x_5 (x_3 - 1) + x_4}{1 + x_3^2 + x_2^2}. \tag{13}$$

Here, the current output of the plant depends on three previous outputs and two previous inputs. Unlike the authors in [1] who applied a feedforward neural network with five input nodes for feeding appropriate past values of $y_p(k)$ and $u(k)$, we only use the current input $u(k)$ and the output $y_p(k)$ as the inputs to identify the output of the plant $y_p(k+1)$. In training the fuzzy model using MCPSO for the nonlinear plant, we use only ten epochs and there are 900 time steps in each epoch. Similar to the inputs used in [2, 3]. The input is an independent and identically distributed (*iid*) uniform sequence over [-2, 2] for about half of the 900 time steps and a single sinusoid given by $1.05 * \sin(\pi k / 45)$ for the remaining time steps. In applying MCPSO to this plant, the number of swarms $N = 4$, the population size of each swarm $n = 20$, are chosen, i.e., 80 individuals are initially randomly generated in a population. The number r of the fuzzy rules is set to be 4. For master swarm, inertial weights $w_{\max}$, $w_{\min}$, the acceleration constants c_1, c_2, and the migration coefficient c_3, are set to 0.35, 0.1, 1.5, 1.5 and 0.8, respectively. In slave swarms the inertial weights and the acceleration constants are the same as those used in master swarm. To show the superiority of MCPSO, the fuzzy identifiers designed by PSO are also applied to the same identification problem. In PSO, the population size is set as 80 and initial individuals are the same as those used in MCPSO. For fair comparison, the other parameters, w_{max}, w_{min}, c_1, c_2 are the same as those defined in MCPSO. To see the identified result, the following input as used in [3, 4] is adopted for test:

$$\begin{aligned}u(k) &= \sin(\pi k/25), \quad k < 250 \\ &= 1.0, \quad 250 \leq k < 500 \\ &= -1.0, \quad 500 \leq k < 750 \\ &= 0.3\sin(\pi k/25) + 0.1\sin(\pi k/32) + 0.6\sin(\pi k/10), \quad 750 \leq k < 1000.\end{aligned} \tag{14}$$

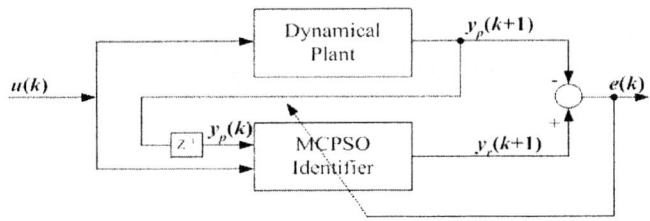

Fig. 3. Identification of nonlinear plant using MCPSO

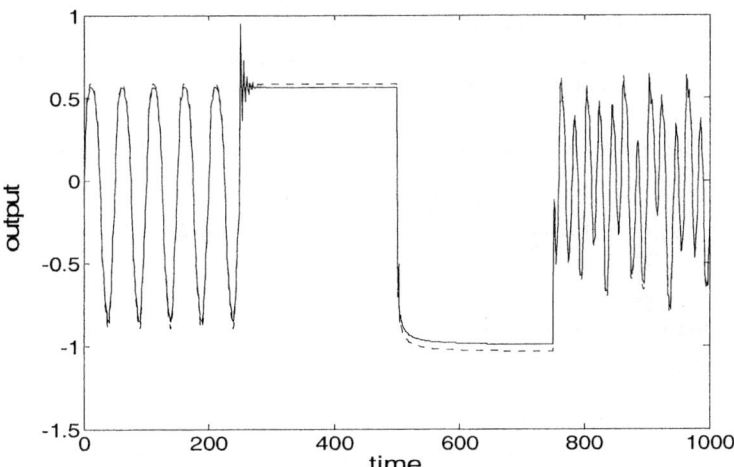

Fig. 4. Identification results using MCPSO in Example 1, where the solid curve denotes desired output and the dotted curve denotes the actual output

Table 1. Performance comparisons with different methods for Example 1

Method	RSONFIN	RFNN	TRFN-S	PSO	MCPSO
RMSE(train)	0.0248	0.0114	0.0084	0.0386	0.0146
RMSE (test)	0.0780	0.0575	0.0346	0.0372	0.0070

Fig.4 shows the desired output (denoted as a solid curve) and the inferred output obtained by using MCPSO (denoted as a dotted curve) for the testing input signal. Table 1 gives the detailed identification results using different methods, where the results of the methods RSONFIN, RFNN and TRFN-S come from literature [5]. From the comparisons, we see that the fuzzy controller designed by MCPSO is superior to the method using RSONFIN, and is slight inferior to the methods using RFNN and TRFN-S. However, among the three types of methods, it achieves the highest identification accuracy in the test part, which demonstrates its better generalized ability. The results of MCPSO identifier also demonstrate the improved performance compared to the results of the identifiers obtained by PSO. The abnormal phenomenon that the test RMSE is smaller than the train RMSE using fuzzy identifier based on PSO and MCPSO may attributes to the well-regulated input data in test part (during time steps [250, 500] and [500, 750], the input data is equal to a constant).

B. Dynamic System Control

As compare to linear systems, for which there now exists considerable theory regarding adaptive control, very litter is known concerning adaptive control of plants governed by nonlinear equations. It is in the control of such systems that we are primarily interested in this section. Based on MCPSO, the fuzzy controller is designed for the control of dynamical systems. The control configuration and input-output variables of MCPSO fuzzy controller are shown in Fig.5, and are applied to one MISO (multi-input-single-output) plant control problem in the following example. The comparisons with other control methods are also presented.

Example 2: The controlled plant is the same as that used in [2] and [5] and is given by

$$y_p(k+1) = \frac{y_p(k)y_p(k-1)(y_p(k)+2.5)}{1+y_p^2(k)+y_p^2(k-1)} + u(k). \tag{15}$$

In designing the fuzzy controller using MCPSO, the desired output y_r is specified by the following 250 pieces of data:

$$y_r(k+1) = 0.6 y_r(k) + 0.2 y_r(k-1) + r(k), 1 \le k \le 250,$$
$$r(k) = 0.5\sin(2\pi k / 45) + 0.2\sin(2\pi k / 15) + 0.2\sin(2\pi k / 90).$$

The inputs to MCPSO fuzzy controller are $y_p(k)$ and $y_r(k)$ and the output is $u(k)$.

There are five fuzzy rules in MCPSO fuzzy controller, i.e., $r = 5$, resulting in total of 35 free parameters. Other parameters in applying MCPSO are the same as those used in Example 1. The fuzzy controller designed by PSO is also applied to the MISO control problem and the parameters in using PSO are also the same as those defined in Example 1.The evolution is processed for 100 generations and is repeated for 50 runs. The averaged best-so-far RMSE value over 50 runs for each generation is shown in Fig.6.

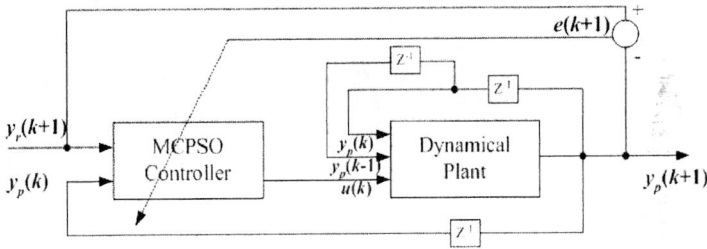

Fig. 5. Dynamical system control configuration with MCPSO fuzzy controller

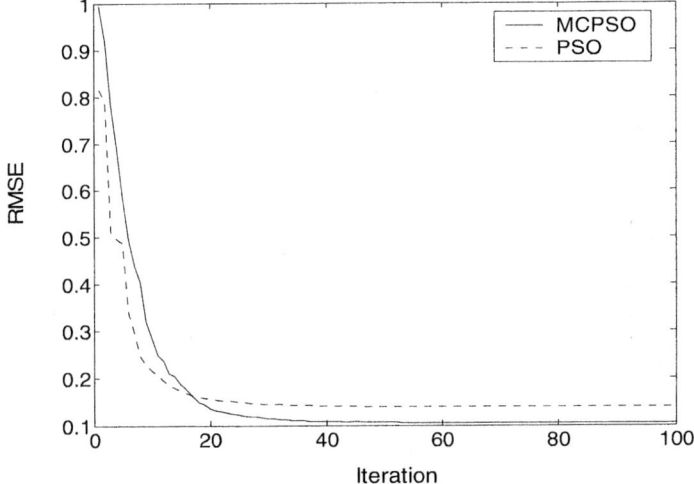

Fig. 6. Average best-so-far RMSE in each generation for PSO and MCPSO in Example 2

From the figure, we can see that MCPSO converges with a higher speed compared to PSO and obtains a better result. In fact, since the competition relationships of the slave swarms, the master swarm will not be influenced much when a certain slave swarms gets stuck at a local optima. Avoiding premature convergence allows MCPSO continue search for global optima in optimization problems

The best and averaged RMSE error for the 50 runs after 100 generations of training for each run are listed in Table 2, where the results of the methods GA and HGAPSO are from [6]. It should be noted that the TRFN controller designed by HGPSO (or GA) is evolved for 100 generations and repeated for 100 runs in literature [6]. To test the performance of the designed fuzzy controller, another reference input $r(k)$ is given by:

$$r(k) = 0.3\sin(2\pi k/50) + 0.2\sin(2\pi k/25) + 0.4\sin(2\pi k/60), 251 \leq k \leq 500.$$

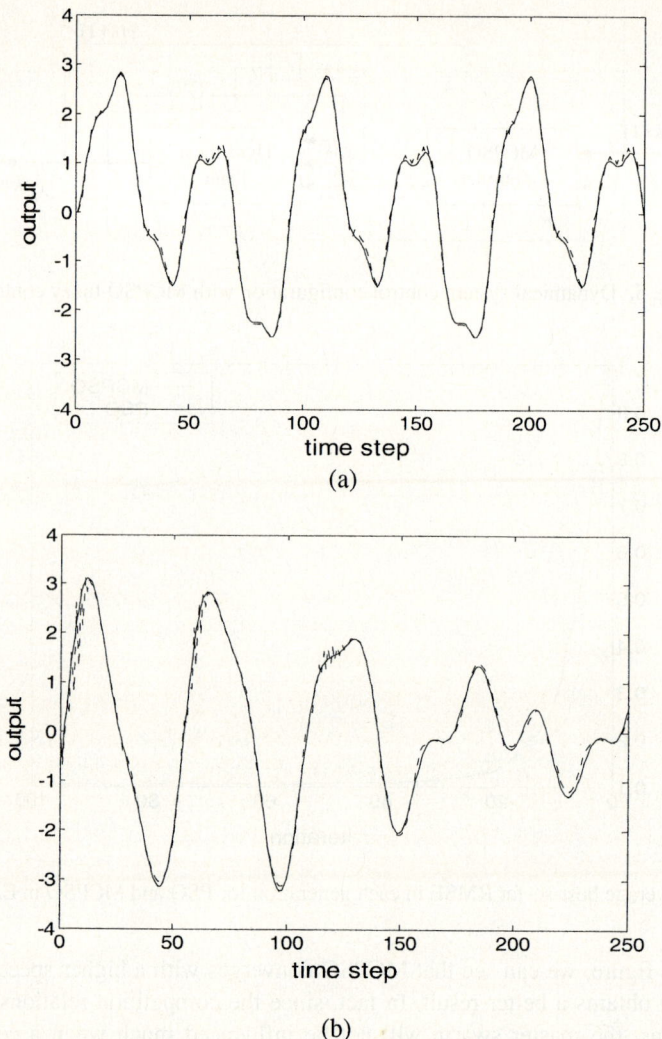

Fig. 7. The tracking performance by MCPSO controller in Example 2 for (a) training and (b) test reference output, where the desired output is dotted as solid curve and the actual output by a dotted curve

The best and averaged control performance for the test signal over 50 runs is also listed in Table 2. From the comparison results, we can see that the fuzzy controller based on MCPSO outperforms those based on GA and PSO greatly especially in the test results, and reaches the same control level with the TRFN controller based on HGPSO.

To demonstrate control performance using the MCPSO fuzzy controller for the MISO control problem, one control performance of MCPSO is shown in Fig.7 for both training and test control reference output.

Table 2. Performance comparisons with different methods for Example 2

Method	GA	PSO	HGAPSO	MCPSO
RMSE (train mean)	0.2150	0.1364	0.0890	0.1024
RMSE (train best)	0.1040	0.0526	0.0415	0.0518
RMSE (test mean)	—	0.1526	—	0.1304
RMSE (test best)	—	0.1024	—	0.0704

5 Conclusions

The paper proposed a multi-population cooperative particle swarm optimizer to identify the T-S fuzzy model for processing nonlinear dynamic systems. In the simulation part, we apply the suggested method to respectively design a fuzzy identifier for a nonlinear dynamic plant identification problem and a fuzzy controller for a nonlinear dynamic plant control problem. To demonstrate the effectiveness of the proposed algorithm MCPSO, its performance is compared to several typical methods in dynamical systems.

Acknowledgements

This work is supported by the National Natural Science Foundation of China (No.70431003) and the National Basic Research Program of China (No. 2002CB312200). The first author would like to thank Prof. Q.H Wu of Liverpool University for many valuable comments. Helpful discussions with Dr. B. Ye, Dr. L.Y. Yuan and Dr. S. Liu are also gratefully acknowledged.

References

1. Narenda, K. S., Parthasarathy, K: Adaptive identification and control of dynamical systems using neural networks. In: Proc. of the 28th IEEE Conf. on Decision and Control, Vol. 2. Tampa, Florida, USA (1989) 1737-1738
2. Narenda, K. S., Parthasarathy, K.: Identification and control of dynamical systems using neural networks. IEEE Trans. Neural Networks 1 (1990) 4-27
3. Sastry, P. S., Santharam, G., Unnikrishnan, K. P.: Memory neural networks for identification and control of dynamical systems. IEEE Trans. Neural Networks 5 (1994) 306-319
4. Lee, C. H., Teng, C. C.: Identification and control of dynamic systems using recurrent fuzzy neural networks. IEEE Trans. Fuzzy Syst. 8 (2000) 349-366
5. Juang, C. F.: A TSK-type recurrent fuzzy network for dynamic systems processing by neural network and genetic algorithms. IEEE Trans. Fuzzy Syst. 10 (2002) 155-170
6. Juang, C. F.: A hybrid of genetic algorithm and particle swarm optimization for recurrent network design. IEEE Trans. Syst. Man Cyber. B 34 (2004) 997-1006
7. Tseng, C. S., Chen, B. S., Uang, H. J.: Fuzzy tracking control design for nonlinear dynamical systems via T-S fuzzy model. IEEE Trans. Fuzzy Syst. 9 (2001) 381-392

8. Chow, T. W. S., Yang, F.: A recurrent neural-network-based real-time learning control strategy applying to nonlinear systems with unknown dynamics, IEEE Trans. Industrial Electronics 45 (1998) 151-161
9. Gan, C., Danai, K.: Model-based recurrent neural network for modeling nonlinear dynamic systems. IEEE Trans. Syst., Man, Cyber. 30 (2000) 344-351
10. Juang, C. F., Lin, C. T.: A Recurrent self-constructing neural fuzzy inference network. IEEE Trans. neural networks. 10 (1999) 828-845
11. Wang, L. X., Mendel, J.M.: Back-propagation fuzzy systems as nonlinear dynamic systems identifiers. In: Proc. IEEE Int. Conf. Fuzzy Syst., San Diego, USA (1992) 1409-1418
12. Takagi, T., Sugeno, M.: Fuzzy identification of systems and its application. IEEE Trans. Syst., Man, Cyber. 15 (1985) 116-132
13. Tanaka K., Ikeda, T., Wang, H. O.: A unified approach to controlling chaos via an LMI-based fuzzy control system design. IEEE Trans. Circuits and Systems 45 (1998) 1021-1040
14. Karr, C. L.: Design of an adaptive fuzzy logic controller using a genetic algorithm. In: Proc. of 4th Int. Conf. Genetic Algorithms, San Diego, USA (1991) 450-457
15. Wang, C. H., Hong, T. P., Tseng, S.S.: Integrating fuzzy knowledge by genetic algorithms. IEEE Trans. Evol. Comput. 2 (1998) 138-149
16. Ishibuchi, H., Nakashima, T. and Murata, T.: Performance evaluation of fuzzy classifier systems for multi dimensional pattern classification problems. IEEE Trans. Syst., Man, Cyber. B 29 (1999) 601-618
17. Eberhart, R. C., Kennedy, J.: A new optimizer using particle swarm theory. In: Proc. of Int. Sym. Micro Mach. Hum. Sci., Nagoya, Japan (1995) 39-43
18. Kennedy, J., Eberhart, .R. C.: Particle swarm optimization. In: Proc. of IEEE Int. Conf. on Neural Networks, Piscataway, NJ (1995) 1942-1948
19. Zhang, C., Shao, H., Li, Y.: Particle swarm optimization for evolving artificial network. In: Proc. of IEEE Int. Conf. Syst., Man, Cyber., Vol.4. Nashville, Tennessee, USA (2000) 2487-2490
20. Engelbrecht, A. P., Ismail, A.: Training product unit neural networks. Stability Control: Theory Appl. 2 (1999) 59-74
21. Mendes, R., Cortez, P. Rocha, M. and Neves, J.: Particle swarms for feedforward neural network training. In: Proc. of Int. Joint Conf. on Neural Networks, Honolulu, USA (2002) 1895-1899
22. Angeline, P. J.: Evolutionary optimization versus particle swarm optimization: philosophy and performance difference. In: Proc. of the 7th Annual Conf. on Evolutionary Programming, San Diego, USA (1998) 601-610
23. Shi, Y., Eberhart, R. C.: A modified particle swarm optimizer. Proc. of IEEE Int. Conf. on Evolutionary Computation, Anchorage, USA (1998) 69-73
24. Moriarty, D., Miikkulainen: Reinforcement learning through symbiotic evolution Machine learning. 22 (1996) 11-32
25. Wiegand, R. P.: An analysis of cooperative coevolutionary Algorithms. PhD thesis, George Mason University, Fairfax, Virginia, USA (2004)

Human Clustering for a Partner Robot Based on Computational Intelligence

Indra Adji Sulistijono[1,2] and Naoyuki Kubota[1,3]

[1] Department of Mechanical Engineering,
Tokyo Metropolitan University,
1-1 Minami-Osawa, Hachioji,
Tokyo 192-0397, Japan
[2] Electronics Engineering Polytechnic Institute of Surabaya – ITS (EEPIS-ITS),
Kampus ITS Sukolilo,
Surabaya 60111, Indonesia
indra-adji@ed.tmu.ac.jp
[3] "Interaction and Intelligence", PRESTO,
Japan Science and Technology Corporation (JST)
kubota@comp.metro-u.ac.jp

Abstract. This paper proposes computational intelligence for a perceptual system of a partner robot. The robot requires the capability of visual perception to interact with a human. Basically, a robot should perform moving object extraction, clustering, and classification for visual perception used in the interaction with human. In this paper, we propose a total system for human clustering for a partner robot by using long-term memory, k-means, self-organizing map and fuzzy controller is used for the motion output. The experimental results show that the partner robot can perform the human clustering.

1 Introduction

Recently, a personal robot is designed to entertain or to assist a human owner, and such a robot should have capabilities to recognize a human, to interact with a human in natural communication, and to learn the interaction style with the human. Especially, visual perception is very important, because the vision might include much information for the interaction with the human.

Image-based human tracking might play a prominent role in the next generation of surveillance systems and human computer interfaces. Estimating the pose of the human body in a video stream is difficult problem because of the significant variations in appearance of the object throughout the sequence [16]. The robot might specify the intention of the human by using a built map, because a task is dependent on its environmental conditions. Furthermore, the robot should learn not only an environmental map, but also human gestures or postures to communicate with the human. In the previous studies, various image processing methods for robotic visual perception such as differential filter, moving object detection, and pattern recognition [14,15]. In order to perform the pattern recognition, the robot needs patterns or templates, but a human-friendly robot cannot know pattern nor templates for the persons beforehand. Therefore, the robot must learn patterns or templates through the interaction with the human. For this, we propose a method for visual tracking of a human. First of all, the robot extracts a human from an image taken by a built-in CCD camera. Since we assume a human can move, a moving object is a candidate for the human. A long-term memory is used in order to extract a human from the taken image. The differential filter is used for detecting a moving object. Next, the color combination pattern is extracted by k-means. A robot classifies the detected human by self-organizing map based on the color pattern. The position of the human is obtained from the SOM node. Finally, the robot moves toward the detected human.

This paper is organized as follows. Section 2 explains a vision-based partner robot, and proposes a human recognition method based on the visual perception. Section 3 shows several experimental results and section 4 summarizes the paper.

2 Visual Perception

2.1 A Partner Robot

We developed a partner robot; MOBiMac as shown in Figure 1. This robot is designed for the usage as a personal computer as well as a partner. Two CPUs are used for PC and robotic behaviors and the robot has two servo motors, four ultrasonic sensors, and a CCD camera. The ultrasonic sensor can measure 2000 mm. Therefore, the robot can take various behaviors such as collision avoiding, human approaching, and visual tracking.

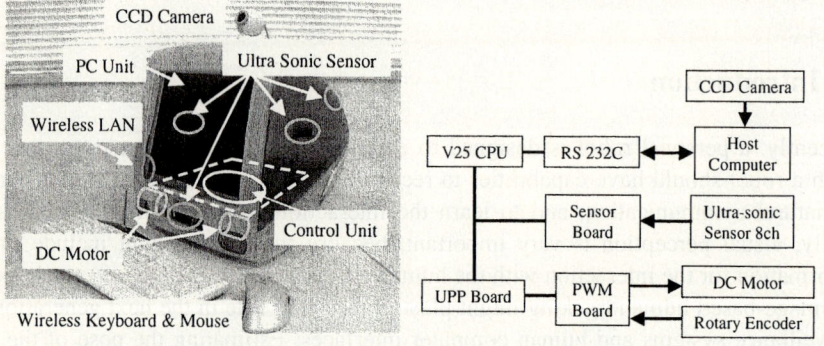

Fig. 1. A partner robot; MOBiMac and its control architecture

The robot takes an image from the CCD camera, and extracts a human. If the robot detects the human, the robot extracts the color patterns of the human. And the robot takes a visual tracking behavior (see also in figure 1). Figure 2 shows the total architecture of visual tracking process and the detailed procedure is explained in the following.

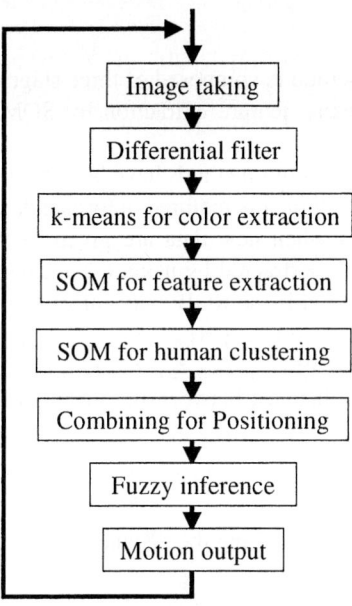

Fig. 2. Total architecture of visual tracking

2.2 Differential Filter

In order to detect a human, we use the recent topic of psychology [3-7]. In this paper, a long-term memory based on expectation is used to detect a human considered as a moving object. The robot has a long-term memory used as a background image $GI(t)$ composed of $\mathbf{g}_i$ ($i=1, 2, ..., l$). A temporal image $TI(t)$ is generated using differences between an image at t and the $GI(t-1)$. Each pixel has a belief value b_i satisfying $0.01 < b_i < 0.99$ ($i=1, 2, ..., l$), and this value is updated as follows,

$$\begin{cases} b_i \leftarrow \alpha \cdot b_i & \text{if } d_i > \gamma \\ b_i \leftarrow \alpha^{-1} \cdot b_i & \text{otherwise} \end{cases} \quad (1)$$

where d_i is the maximal value in the differences of the ith pixel between the current image and the background image; α is a discount rate; l is the number of pixels and γ is expectation parameter. If the change of color of the pixel is small, the belief value becomes large. The values of each pixel of the $GI(t)$ is updated as follows,

$$\mathbf{g}_i \leftarrow b_i \cdot \mathbf{g}_i + (1 - b_i) \cdot \mathbf{p}_i \quad (2)$$

where $\mathbf{p}_i$ is the color values of RGB of the ith pixel at the image t. If there are no moving objects in front of the robot, a background image with high belief values is obtained. Therefore, the robot can be aware that a moving object or a human is in a front of the robot by using $TI(t)$.

2.3 Color Extraction

The proposed learning method is composed of three stages of clustering methods, *i.e.*, color clustering by k-means, feature extraction by SOM, and human clustering by SOM.

Various unsupervised learning methods have been proposed so far [9-11]. In a case of batch learning, a set of all data is required beforehand, but incremental learning can update design parameters when new data are given to the learning system. Here a color pattern of a human is extracted by using a k-means algorithm (first layer). The inputs to the k-means method is the RGB color information (R, G, B) and location (x, y) of the ith pixel ($i=1,2, ..., l$), i.e., $\mathbf{q}_i=(q_{R,i}, q_{G,i}, q_{B,i}, q_{x,i}, q_{y,i})$. The jth reference vector ($j=1, 2, ..., k$) is defined as $\mathbf{r}_j=(r_{R,i}, r_{G,i}, r_{B,i}, r_{x,i}, r_{y,i})$, and the number of input dimensions is 5. The Euclidian distance between the ith input vector and the jth reference vector is defined as

$$x_{j,i} = \|\mathbf{q}_i - \mathbf{r}_j\| \tag{3}$$

Next, the reference vector minimizing the distance is selected by,

$$h_i = \arg\min_j \{\|\mathbf{q}_i - \mathbf{r}_j\|\} \tag{4}$$

After selecting the nearest reference vector to each input, the jth reference vector is updated by the average of the inputs belonging to the jth cluster. If the update is not performed at the clustering process, this updating process is finished. Therefore, the k-means method can extract several segments according to the spatial color distribution of a moving object by using k reference vectors.

2.4 Human Clustering

In order to classify each segment pattern extracted by k-means method, SOM is used for feature extraction based on colors (second layer). SOM is known as an unsupervised learning method based on the incremental learning for extracting a relationship and the hidden topological structure from the learning data [11]. Figure 3 shows a conceptual figure of the SOM. The Euclidian distance is defined as follows,

$$y_{j,i} = \|\mathbf{r}_i - \mathbf{s}_j\| \tag{5}$$

where $\mathbf{s}_j=(s_{R,j}, s_{G,j}, s_{B,j}, s_{x,j}, s_{y,j})$, is the jth reference vector ($j=1,2, ... , n$). We can obtain the mth output unit that minimizes the distance $y_{j,i}$ by

$$m = \arg\min_i \{\|\mathbf{r}_i - \mathbf{s}_j\|\} \tag{6}$$

Furthermore, the jth reference vector is trained by

$$\mathbf{s}_j \leftarrow \mathbf{s}_j + \xi \cdot \varsigma_{m,j} \cdot (\mathbf{r}_i - \mathbf{s}_j) \tag{7}$$

where $\zeta_{m,j}$ and ξ is a neighborhood function and learning rate. Here two-dimensional structure is used to represent the neighboring relationship among output units in SOM. Therefore, SOM plays the role of clustering each color pattern of an image as a feature extraction mechanism.

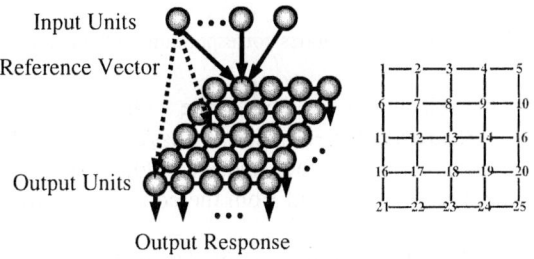

Fig. 3. A self-organizing map and nodes numbering

Furthermore, the combinations of selected nodes corresponding to the color patterns of a human image are used as the input patterns to SOM (third layer). Therefore, the spatially topological patterns can be clustered by SOM, and the selected node corresponds to a specific situation of the combined color patterns. Then, human position can be obtained. From this point of view, we can control the motion of the partner robot by human motion.

The spatially topological patterns can be clustered by SOM, and the selected node corresponds to a specific situation of the combined color patterns. The combination of the selected nodes (s_c) is used to input pattern for SOM as the third layer of clustering. The fth new reference vector is trained by

$$s_c \leftarrow s_c + \xi \cdot \zeta_{f,c} \cdot (s_j - s_c) \qquad (8)$$

where $\zeta_{f,c}$ and ξ is a neighborhood function and learning rate. The fth selected node is obtained by minimizing the distance,

$$f = \arg\min_j \{\|s_j - s_c\|\} \qquad (9)$$

The spatially topological patterns can be clustered by SOM, and the selected node corresponds to a specific situation of the combined color patterns. In order to classify the human from the others, the combinations of selected nodes corresponding to the color patterns of a human image from the second of SOM (the third layer) are used. The combination of selected nodes from the third layer (s_{cn}) is

$$s_{cn} = \sum_{n=1}^{sl} \frac{s_n}{sl} \qquad (10)$$

where s_n is selected node and sl is number of the selected nodes.

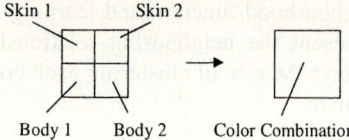

Fig. 4. The selected nodes corresponding to the color of human

Figure 4 shows the selected nodes incorporate with the color of human. For example in this figure, there are 4 selected nodes, we assume that each selected nodes will be represented the color of skin of face and the color of body (wears or body skin), then the color combination is obtained from the combination of color of the selected nodes

Finally, the position of the combination of the selected nodes will be used as the input for the action of the robot. Image processing is performed for detecting a human interacting with the partner robot.

2.5 Fuzzy Control

In this paper, the fuzzy controller is used for target tracing of the motion output. A behavior for a robot can be represented using fuzzy rules based on simplified fuzzy inference. In general, a fuzzy if-then rule is described as follows,

if x_1 is $A_{e,1}$ and ... and x_v is $A_{e,v}$ **then** y_1 is $w_{e,1}$ and ... and y_w is $w_{e,w}$

where x_j and y_u are variables for the jth input and the uth output ($j=1,\cdots, n$; $k=1,\cdots, o$); $A_{e,f}$ and $w_{e,w}$ are a symmetric triangular membership function for the fth input and a singleton for the uth output of the eth rule; v and w are the numbers of inputs and outputs, respectively. Fuzzy inference is generally described by,

$$\mu_{A_{e,f}}(x_f) = \begin{cases} 1 - \frac{|x_f - a_{e,f}|}{b_{e,f}} & |x_f - a_{e,f}| \leq b_{e,f} \\ 0 & \text{otherwise} \end{cases} \quad (11)$$

$$\mu_e = \prod_{f=1}^{v} \mu_{A_{e,f}}(x_f) \quad (12)$$

$$y_t = \frac{\sum_{e=1}^{L} \mu_e \cdot w_{e,u}}{\sum_{e=1}^{L} \mu_e} \quad (13)$$

where $a_{e,f}$ and $b_{e,f}$ are the central value and the width of the membership function $A_{e,f}$; L is the number of rules.

Figure 5 shows the fuzzy controller for the motion output. Inputs for target tracing are the human position extracted by the visual perception. Position of the combination of the selected nodes will be used as the input for the action of the robot. We assume that the human size is obtained from the number of pixels of human moving from the LTM process. Each output is left and right motor speed output. Because the motion output is given to the robot as a motion command, the inference result is transformed into a motion command number.

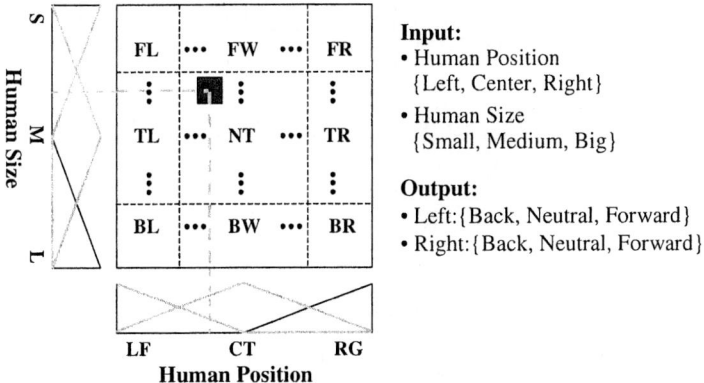

Fig. 5. Fuzzy controller for the motion output

2.6 The Action of Robot

We intend to develop the communication between the robot and human (Figure 6). By using the human approaching, the robot should also moving to keep the attention to the human when the human moving to the left or right direction of the robot. Robot face should direct to the human face and also to keep distance between the robot and human. This is happened when the human are moving forward to close the robot or the opposite direction.

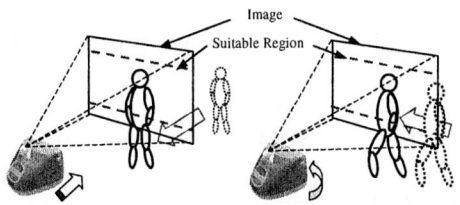

Fig. 6. The action of robot

The combination of the action of robot will be approached according to the human moving. As the partner of human, robot should communicate each other. Communication in here is shown by the attention position each other.

3 Experimental Results

This section shows several experimental results of the proposed method for the partner robot. The number of pixels is 160×90 (l=14,400). The number of output layers is 5×5 (n=25). The number of subjects is 5. An image is taken every a half second.

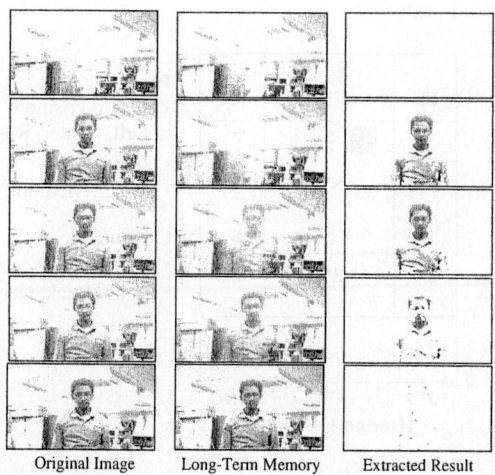

Original Image   Long-Term Memory   Extracted Result

Fig. 7. A series of snapshot pictures on extracted result process

Figure 7 shows a series of snapshot pictures on extracted result process. After the extracted result is obtained, the next process is human clustering process as shown in figure 8. The SOM selected nodes is incorporated with the human. The color shown in the human clustering is the combination color of human face or skin and human body or wears and the position of node shows the position of human from the point view of camera of robot.

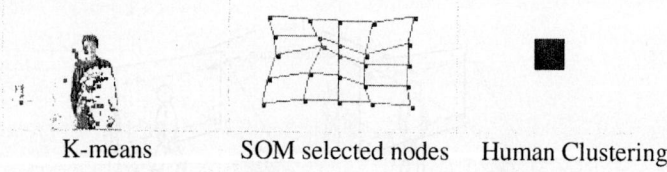

K-means   SOM selected nodes   Human Clustering

Fig. 8. The experimental result the proposed methods

On the top of figure 9 shows the output of motor speed. We used 4 different speeds of DC motor. Negative means the reverse direction. First is initial condition for background taking, the robot is not moving. On the center and the bottom one shows the history of selected nodes in SOM 1 (the second layer) and SOM 2 (the third layer) respectively. The selected nodes number on the SOM1 are more various than SOM 2.

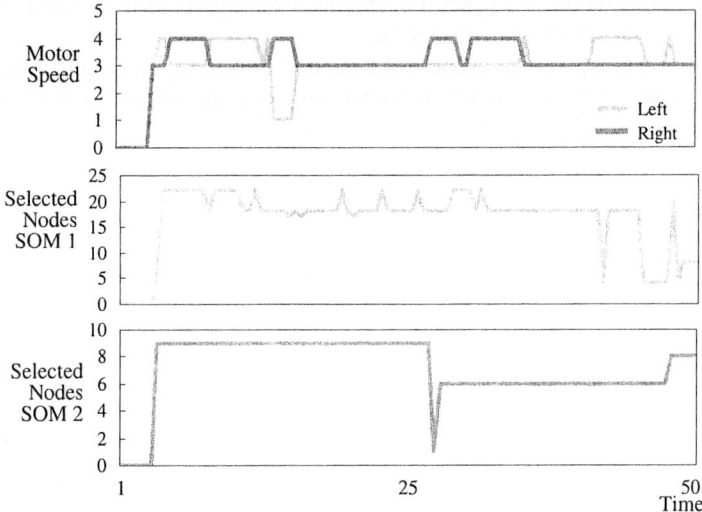

Fig. 9. The output of motor speed (top), the history of selected node in SOM 1 or second layer (center) and the history of selected node in SOM 2 or third layer (bottom)

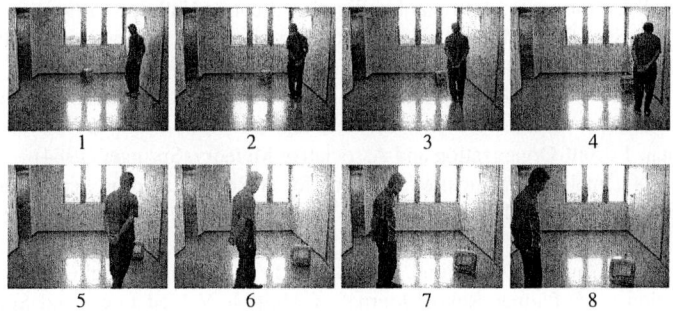

Fig. 10. A snapshot of experimental result of the visual tracking of a human

Figure 10 shows a series of snapshot pictures in the experimental result of visual tracking by a partner robot. At first, the robot capture the human moving and then the robot go forward to the human. When the human move to the right side of the robot then the robot turn right and go forward to the human. The partner robot can perform the human clustering well.

4 Summary

This paper proposed human clustering for a partner robot. We applied long-term memory, a k-means clustering method, and self-organizing map for image processing and fuzzy controller is used for the motion output. The experimental results show that the proposed method can extract human and several parts of the human. The partner robot can perform the human clustering.

As future works, we will propose the method is used for extracting a human intention dependent on the facing environment. Furthermore, we intend to develop a lifetime learning algorithm through interaction with a human.

References

1. Brooks, R.A.: Cambrian Intelligence, The MIT Press (1999)
2. Fukuda, T., Kubota, N.: An Intelligent Robotic System Based on a Fuzzy Approach, Procee-dings of IEEE, 87 (9) (1999) 1448-1470
3. Turvey, M.T., Shaw, R.E.: "Ecological Foundations of Cognition I. Symmetry and Specificity of Animal-Environment Systems", Journal of Consciousness Studies 6, No.11-12 (1999) 95-110
4. Turvey, M.T., Shaw, R.E.: "Ecological Foundations of Cognition II. Degree of Freedom and Conserved Quantities in Animal-Environment Systems", Journal of Consciousness Studies 6, No.11-12 (1999) 111-123
5. Gibson, J.: The Ecological Approach to Visual Perception, LEA (1986)
6. Eysenck, M.: Perception and Attention, Psychology (edited by Michael Eysenck), Prentice Hall (1998) 139-166
7. Marr, D.: Vision, W.H.Freeman, San Francisco, (1982)
8. Russell, S.J., Norvig, P.: Artificial Intelligence, Prentice-Hall, Inc. (1995)
9. Jang, J.S., Sun, C.T., Mizutani, E.: Neuro-Fuzzy and Soft Computing, Prentice-Hall, Inc. (1997)
10. Hastie, T., Tibshirani, R., Friedman, J.: The Elements of Statistical Learning, Springer-Verlag (2001)
11. Kohonen, T.: Self-Organization and Associative Memory, Springer (1984)
12. Kubota, N.: Intelligent Structured Learning for A Robot Based on Perceiving-Acting Cycle, Proc. of the Twelfth Yale Workshop on Adaptive and Learning Systems (2003) 199-206
13. Kubota, N., Hisajima, D., Kojima, F., Fukuda, T.: Fuzzy and Neural Computing for Communication of A Partner Robot, Journal of Multiple-Valued Logic and Soft-Computing 9(2) (2003) 221-239
14. Birchfield, S.: Elliptical Head Tracking Using Intensity Gradients and Color Histograms, Proceedings of the IEEE Conference on Computer Vision and Pattern Recognition, Santa Barbara, California (1998) 232-237
15. Shanahan, M., Randell, D.: A Logic-Based Formulation of Active Visual Perception, Journal American Association for Artificial Intelligence (2004)
16. Bissacco, A., Soatto, S.: "Visual Tracking of Human Body with Deforming Motion and Shape Average", UCLA CSD Technical Report #020046, November (2002)
17. Kubota, N., Sulistijono, I.A., Ito, Y.: "Visual Perception for A Partner Robot Based on Computational Intelligence", Proc. of the International Symposium on Computational Intelligence and Industrial Applications (ISCIIA2004), Haikou-Hainan China, (2004)

Fuzzy Switching Controller for Multiple Model*

Baozhu Jia[1], Guang Ren[1], and Zhihong Xiu[2]

[1] Marine Engineering College, Dalian Maritime University,
116026 Liaoning Province, China
{skysky,reng}@dlmu.edu.cn
[2] Department of Command Control and Information,
Dalian Naval Academy,
116018 Liaoning Province, China
xzhdy@126.com

Abstract. This paper proposes a so-called fuzzy switching multiple (FSM) model which can achieve smooth switching when the control input at switching boundaries. Parallel distributed compensation scheme is employed to design the controller for the FSM. By utilizing fuzzy Lyapunov function, we derive the stabilization condition for closed-loop FSM systems. A design example illustrates the utility of the proposed approach.

1 Introduction

Switching has assumed importance in the fields of multiple model control. A typical problem in switching-based control is nonsmooth switching of control input at switching boundaries. It seriously influences important control performance such as the ride quality etc. The research on smooth switching can improve the performance of controller and make systems consisted of multiple models stabilize. In recent years, many methods were proposed to avoid nonsmooth switching, e.g., sliding-mode control, neural-network control and (adaptive) fuzzy control [1-3].

This paper proposes a so-called fuzzy switching multiple model and constructs the controller by extending the idea of the parallel distributed compensation (PDC). Lyapunov stability condition is derived. FSM system can keep stable and guarantee smooth switching with this condition.

2 Fuzzy Switching Multiple (FSM) Model and Controller

In FSM, the multiple models are switched according to the premise variables. FSM is consisted of regional rule and local plant model. The universe of discourse is divided into P sub-regions by fuzzy sets and switch according to the premise variable $z_1(t) \sim z_p(t)$. The premise variables are permitted to be states, measurable external variables, and/or time. This paper supposes that the rule consequence take the forms of state space. Then, jth rule of the FSM is of the following form:

---

* This work is supported by Ministry of Communication of P.R. China (200332922505) and Doctoral Foundation of Education Committee (20030151005).

Regional Rule j: If $z_1(t)$ is N_{1j} and $\cdots$ and $z_p(t)$ is N_{pj} then

Local Plant model: $\begin{cases} \dot{x}(t)=A_j x(t)+B_j u(t) \\ y(t)=C_j x(t) \end{cases}$ $j=1,2,\cdots,s.$ (1)

where $z(t)=[z_1(t)\ z_2(t)\ldots z_p(t)]$ are the premise variables. s is the number of regions partitioned on the premise variables space. $N_{kj}(z(t))$ is fuzzy set, and has the character as:

$$\sum_{j=1}^{s} N_{kj}(z_k(t)) = 1 \qquad (2)$$

Given a pair of $(x(t), u(t))$ and $z(t)$, the final output of the switching fuzzy model (1) is inferred as follows:

$$\begin{cases} \dot{x}(t)=\sum_{j=1}^{s} h_j(z(t))\{A_j x(t)+B_j u(t)\} \\ y(t)=\sum_{j=1}^{s} h_j(z(t)) C_j x(t) \end{cases} \qquad (3)$$

where,

$$h_j(z(t)) = \prod_{k=1}^{p} N_{kj}(z_k(t)) \Big/ \sum_{j=1}^{s} \prod_{j=1}^{p} N_{kj}(z_k(t)) \qquad (4)$$

The regions j satisfies:

$$\begin{array}{c} \text{Region 1} \cup \text{Region 2} \cup \ldots \text{Region } S = X \\ \text{Region } j_1 \cap \text{Region } j_2 \neq \emptyset \quad (j_1 \neq j_2,\ j_1=1,2,\ldots,s,\ j_2=1,2,\ldots,s) \end{array} \qquad (5)$$

where X denotes the universe of discourse.

The controller for FSM is constructed by extending the idea of the parallel distributed compensation (PDC) [4]. The FSM controller for the systems (1) is the following forms:

Regional Rule j: If $z_1(t)$ is N_{1j} and $\cdots$ and $z_p(t)$ is N_{pj} then

Local Plant Controller: $u(t)=-F_j x(t)$ (6)

Overall fuzzy switching controller is represented by:

$$u(t) = -\sum_{j=1}^{s} h_j(z(t)) F_j x(t) \qquad (7)$$

Close-loop system is:

$$\dot{x}(t) = \sum_{i=1}^{s}\sum_{j=1}^{s} h_i(z(t)) h_j(z(t))\{A_i - B_i F_j\} x(t) \qquad (8)$$

3 Stability Condition

Piecewise Lyapunov function approaches are believed as the effective methods for the stability analysis of T-S fuzzy model [4-6]. Tanaka et al. proposed a fuzzy Lyapunov function (FLF) which is defined by fuzzy blending of quadratic Lyapunov functions. The FLF shares the same membership functions with the Takagi-Sugeno model of the system [3]. This paper utilizes FLF as the candidate Lyapunov function for the stabilization problem of fuzzy switching multiple systems.

Consider the close-loop system (8), the condition for ensuring the stability is given in theorem 1.

Theorem 1: Assume that:

$$\left| \dot{h}_\rho(z(t)) \right| \leq \Phi_\rho \tag{9}$$

The fuzzy system (13) is stable if there exist $\Phi_1 \Phi_2 \cdots \Phi_s$ such that

$$P_i > 0 \text{ and } \sum_{\rho=1}^{s} \Phi_\rho P_\rho + G_{jk}^T P_i + P_i G_{jk} < 0 \tag{10}$$

where: $G_{jk} = A_j - B_j F_k$, and $i, j, k, \rho = 1, 2, \cdots, s$. P is common positive definite matrix and satisfies $P^T = P$.

(Proof) The proof is omitted due to lack of space.

With the given Φ_ρ, the regional feedback gains F_j can be determined with the PDC scheme and LMI approach proposed in ref. [4]. With this condition, the close-loop FSM system can guarantee both stability and smoothness.

Remark: It is difficult to determine $\Phi_\rho s$ so as to satisfy (9), Reference [3] proposes a new PDC design in the case of the time derivatives of membership function is computable from the states. This paper employs the algorithm proposed in ref. [3].

4 Example

Consider a fuzzy switching system which FSM model rules and controller rules are shown as follows:

Model rules: **Regional Rule i:** If $x_2(t)$ is N_i Then

Local Plant Model j: If $x_2(t)$ is M_{ij} Then $\dot{x}(t) = A_{ij} x(t) + B_{ij} u(t)$

Controller rules: **Regional Rule i:** If $x_2(t)$ is N_i Then

Local Controller Model j: If $x_2(t)$ is M_{ij} Then $u = -F_{ij} x(t)$

Where $i = 1, 2; j = 1, 2$. $x(t) = [x_1(t); x_2(t)]$ is state variable. N_i is fuzzy sets. Each regional membership functions are assigned as follows:

$$N_1(x_2(t)) = \begin{cases} 1, & x_2(t) < -3 \\ \frac{-x_2(t)+5}{8}, & -3 \leq x_2(t) < 5 \\ 0, & x_2(t) > 5 \end{cases} \quad N_2(x_2(t)) = \begin{cases} 0, & x_2(t) < -3 \\ \frac{x_2(t)+3}{8}, & -3 \leq x_2(t) < 5 \\ 1, & x_2(t) > 5 \end{cases}$$

By solving (9) and (10) with LMI approach, we can determine the local controller feedback parameters F_{ij}. The control result of close-loop system is shown in Fig.1. It can be seen from Fig.1, without the smooth condition derived in ref.[1], the fuzzy switching systems can keep stable and guarantee smooth switching under theorem 1 too. If the local plant model employs not fuzzy rules, the proposed model shows more applicable compare to ref. [1].

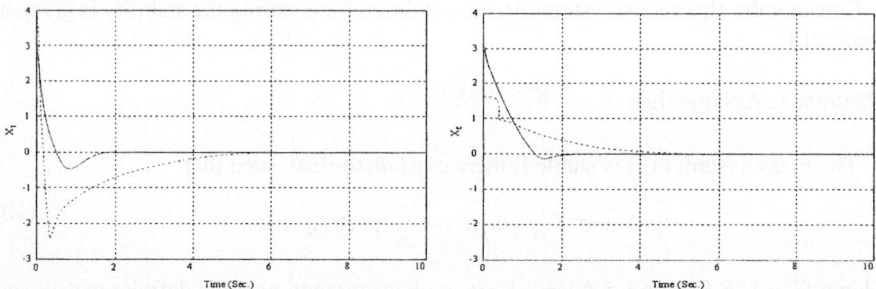

Fig. 1. Comparison of control result (*Dotted line indicate the switching fuzzy control under stability condition in [1], solid line indicate fuzzy switching control under stability condition without smooth condition*)

5 Conclusion

This paper proposes a fuzzy switching multiple model and designs the fuzzy switching controller via PDC scheme. The Lyapunov stabilization condition is derived through FLF. FSM can achieve stability and smooth switching when control input at switching "boundaries" under derived stability condition.

References

1. K. Tanaka, M. Iwasaki and Hua O. Wang: Switching control of an R/C hovercraft: stabilization and smooth switching, IEEE Trans. On SMC-Part B: Cybernetics, Vol. 31, No. 6, (2001) 853-863
2. Chih-Lyang Hwang: A Novel Takagi–Sugeno-Based Robust Adaptive Fuzzy Sliding-Mode Controller, IEEE Trans. On Fuzzy Systems, Vol. 12, No. 5,(2004) 676-687
3. K. Tanaka, T. Ikeda and Hua O. Wang: A multiple Lyapunov function approach to stabilization of fuzzy control systems, IEEE Trans.On Fuzzy Systems, Vol. 11, No. 4, (2003) 582-589
4. H. O. Wang, K. Tanaka, and M. Griffin: An approach to fuzzy control of nonlinear systems: stability and design issues, IEEE Trans. On Fuzzy System, Vol. 4. (1996) 14-23
5. Zhi-Hong Xiu, Guang Ren: Stability analysis and systematic design of Takagi-Sugeno fuzzy control systems, Fuzzy Sets and Systems, Vol. 151, No. 1, (2005) 119-138
6. M. Johansson, A. Rantzer, K.E.Arzen: Piecewise quadratic stability of fuzzy systems, IEEE Trans. On Fuzzy Systems Vol.7, (1999) 713-722

Generation of Fuzzy Rules and Learning Algorithms for Cooperative Behavior of Autonomouse Mobile Robots(AMRs)

Jang-Hyun Kim[1], Jin-Bae Park[2], Hyun-Seok Yang[3], and Young-Pil Park[3]

[1] Department of Electrical Engineering, Yonsei University,
134, Shinchon-Dong, Seodaemun-ku, Seoul, Korea
jhkim@control.yonsei.ac.kr
[2] Department of Electrical Engineering, Yonsei University,
134, Shinchon-Dong, Seodaemun-ku, Seoul, Korea
jbpark@yonsei.ac.kr
[3] Department of Mechanical Engineering, Yonsei University,
134, Shinchon-Dong, Seodaemun-ku, Seoul, Korea
{hsyang,park2814}@yonsei.ac.kr

Abstract. Complex "lifelike" behaviors are composed of local interactions of individuals under fundamental rules of artificial life. In this paper, fundamental rules for cooperative group behaviors, "flocking" and "arrangement" of multiple autonomouse mobile robots are represented by a small number of fuzzy rules. Fuzzy rules in Sugeno type and their related parameters are automatically generated from clustering input-output data obtained from the algorithms for the group behaviors. Simulations demonstrate the fuzzy rules successfully realize group intelligence of mobile robots.

1 Introduction

Research concerned in the artificial life is devoted to understanding life by attempting to abstract the fundamental dynamical principles underlying biological phenomena, and recreating these dynamics in other physical media including computers. In addition to providing new ways to study the biological phenomena associated with biological life(B-life), artificial life(A-life) allows us to extend our concepts to the larger domain of "life-as-it-could-be"[1][2]. Group intelligence in complex lifelike behaviors is considered to be composed of local interactions of individuals with simplefunctions under fundamental rules of artificial life.

In this paper, fundamental rules in cooperative behaviors of multiple autonomous mobile robots(AMRs) are represented by small number of fuzzy system rules. Group intelligence can be observed in such cases as in flocking of birds and in forming a swarm of insects[3][4]. Tow types of group behavior, flocking and arrangement, are considered in this paper as simulations. Flocking refers to the behavior that robots scattered in a space gather to make a squad. Arrangement is regarded as generating a circular shape of robots.

Our goal is to generate a small number of fuzzy rules as fundamental rules that govern complex lifelike group behaviors of mobile robots. Generating the fuzzy rules consists of two steps. First, We realize mathematical algorithms responsible for the group behaviors and collect numerical input-output data from the process. Second, We cluster the input-output data to identify fuzzy system structure and parameters to construct fuzzy rules from the clustering result. The performance of group behaviors governed by the fuzzy rules are compared with those by the algorithm[5].

Section 2 describes the algorithms for group behaviors of flocking and arrangement of autonomouse mobile robots. Section 3 shows the method of generating fuzzy rules from clustering the input-output data. Section 4 simulation results show the fuzzy system with a small number of fuzzy rules successfully realizes cooperative behaviors of autonomouse mobile robots. Final section describes conclusion and future plan for artificial life.

2 Algorithms for the Flocking and Arrangement Behaviors

2.1 Flocking and Arrangement Behaviors

The flocking and the arrangement algorithms are based on the relative angle and distance between a robot and other robots. Each time a robot calculates the farthest robot and the nearest robot according to the relative angle and the distance. When robots are scattered in an open space, they move to the center to implement the flocking behavior. When robots assemble close enough to each other, they make an arrangement as a circular shape to complete the arrangement behavior.

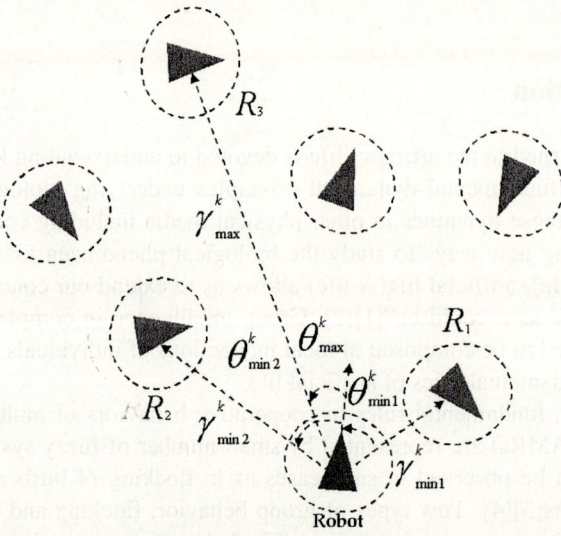

Fig. 1. Movement of a robot in flocking and arrangement group behavior algorithms

Fig. 1 shows definitions of relative distances and angles for the flocking and arrangement group behaviors. F-AMR indicates the farthest robot from a robot, N-AMR1 and N-AMR2 are the nearest and the second nearest robots, respectively. Each step, a robot calculates the relative rotation angle $\hat{\theta}^{k+1}$ and the distance $\hat{\gamma}^{k+1}$. Movement of robot in the flocking behavior requires recognition of the farthest and the nearest robots, while the arrangement behavior needs the nearest and the second nearest robots. When a robot moves to the center, the rotation of robots is divided into three grade to avoid radical rotation of the robot. Such robot moves repeatedly to make arrangement as a circle. Both in the flocking and the arrangement procedures, a robot moves in three different distance steps and in the angle range between 0 and 30 degrees to avoid abrupt movement which is not practically realizable.

2.2 Flocking and Arrangement According to Fuzzy Rules

Complex group behavior are considered as simple local interactions between individuals under fundamental rules of artificial life. Fuzzy system can handle uncertainty involved in natural phenomena in linguistic terms. Fundamental rules involved in the group behavior are represented by a set of fuzzy system rules. If the fuzzy system is modeled in Sugeno form[7], adaptive training algorithms such as clustering can identify the structure and its parameters of the fuzzy system.

Flocking refers to the procedure that gather robots scattered in an open space to form a group of robots. The goal of arrangement is to form a circular shape of robots. Each time a robot calculates the relative distances and angles with other robots. Fig. 2 shows the input and the output variables for the flocking and the arrangement blocks. If robots are widely spread, the flocking behavior gather robots according to the fuzzy system 1, while robots are gathered closely enough, the fuzzy system 2 form a circular shape of robots.

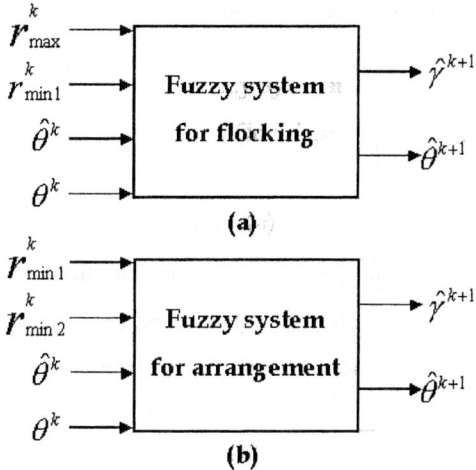

Fig. 2. Fuzzy system for cooperative behaviors and input-output data parameters : (a)Flocking system (b)Arrangement system

Each fuzzy system consists of 4 inputs and 2 outputs. The input variables of the flocking block are the distance γ_{max}^k, γ_{min1}^k and the angles θ_{max}^k, θ_{min1}^k fo the farthest and the nearest robots. The input variables of the arrangement block are the distances γ_{min1}^k, γ_{min2}^k and the θ_{min1}^k, θ_{min2}^k of the nearest and the second nearest robots. The outputs of the fuzzy system are the distance $\hat{\gamma}^{k+1}$ and the rotation angle $\hat{\theta}^{k+1}$ of a individual at the next move. θ^k denotes the on-going direction of the individual. In order to avoid abrupt movement which cannot be implemented practically, $\hat{\theta}^{k+1}$ is set to have values from 0 to 30 degrees and $\hat{\gamma}^{k+1}$ from 0.005 to 0.2 at 0.025. Positive values of θ indicate the rotation of counterclockwise direction. If $\gamma_{max}^k - \gamma_{min1}^k > 1$ then $\hat{\theta}^k - \theta_{max}^k$ both in case of flocking and arrangement. If $\gamma_{max}^k - \gamma_{min1}^k < 1$, $\hat{\theta}^k = \dfrac{\theta_{max}^k + \theta_{min}^k}{2}$ for flocking and $\hat{\theta}^k = \dfrac{\theta_{min1}^k + \theta_{min2}^k}{2}$ for arrangement.

3 Fuzzy Rules Generated from Clustering

3.1 Clustering Algorithm

Subtractive clustering[5] can identify fuzzy system models according to determining cluster centers from numerical input-output data. The number of cluster centers corresponds to the number of fuzzy rules. If we consider the Sugeno-type fuzzy model, the parameters are also determined from the clustering algorithm. The clustering algorithm calculates the potential values P_i from N normalized data obtained from the input-output product-space.

$$P_i = \sum_{k=1}^{N} \exp(-\alpha \|x_i - x_k\|^2), \ \alpha = 4/\gamma_a^2 \qquad (1)$$

Yet, $i = 1, \cdots, N$ and γ_a is a positive constant to set data far apart from a cluster center not to have influence on the potential value. The first cluster center x_1^* corresponds to the largest potential value P_i^*. The second cluster center is calculated after removing the effect of the first cluster center. Eq. 2 shows how to remove the effect of the first cluster center. The second cluster center x_2^* corresponds to the largest potential value of $P_i^{'}$.

$$P_i' = P_i - P_i^* \exp(-\beta\|x_i - x_k\|^2), \quad \beta = 4/\gamma_a^2 \qquad (2)$$

Positive constant γ_b prevents cluster centers to assemble to close. This process repeats until potential values reach a fixed limit $(\varepsilon, \bar{\varepsilon})$.

Cluster centers $\{x_1^*, x_2^*, \cdots, x_M^*\}$ determine M fuzzy rules. They also determine the center position of input membership functions. Widths of membership functions are fixed according to experience. The parameters $\alpha_{i0}, \alpha_{i1}, \cdots, \alpha_{in}$ can be optimized by linear least squares estimation[5] or adaptive training algorithms.

3.2 Generating Fuzzy Rules for Group Intelligence

Sugeno fuzzy system model[5] is used to represent fundamental rules of group intelligence. The MISO type fuzzy rules are of the form given in Eq. 3.

$$\begin{array}{l} IF \ x_1 \ is \ A_{i1} \ and \ x_2 \ is \ A_{i2} \ and \ \cdots \ x_n \ is \ A_{in} \\ THEN \ y_1 = a_{0i} + a_{1i}x_1 + \cdots + a_{ni}x_i \end{array} \qquad (3)$$

A_{ij} is Gaussian membership functions for input fuzzy variables, coefficients $a_{0i}, a_{1i}, \cdots, a_{ni}$ determine the output of the fuzzy system. Fuzzy modeling process based on the clustering of input-output data determines the centers of the membership functions for antecedent fuzzy variables.

In order to develop the fuzzy model, input-output data are obtained from the group behavior algorithms. Fuzzy rules for flocking and arrangement behaviors are generated from clustering the input-output data. We obtained input-output data for 90 times with 5 mobile robots, and for 10 times with 20 mobile robots. The parameters are set to $\gamma_a = 0.5$, $\gamma_b = 2.5$, $\bar{\varepsilon} = 0.3$, and $\varepsilon = 0.1$.

After clustering the data, 10 cluster centers and therefore 10 fuzzy rules are obtained for the flocking block and the 7 fuzzy rules for the arrangement block. The 10 fuzzy rules for flocking are of the form:

$$\begin{array}{l} IF \ \gamma_{max}^k \ is \ A_{i1}, \ \gamma_{min1}^k \ is \ A_{i2}, \ \hat{\theta}^k \ is \ A_{i3}, \ \theta^k \ is \ A_{i4} \\ THEN \ \hat{\gamma}^{k+1} = a_0 + a_1\gamma_{max}^k + a_2 + \gamma_{min1}^k, \quad (i = 1,2,3,\cdots,10) \\ \hat{\theta}^{k+1} = b_0 + b_1\gamma_{max}^k + b_2\gamma_{min1}^k + b_3\hat{\theta}^k + b_4\theta^k \end{array} \qquad (4)$$

θ^k indicates the direction of the robot and $\hat{\theta}^k$ denotes the angle of the robot that must rotate. Table 1 shows the 10 cluster centers and therefore the center locations of the 10 fuzzy rules. Input and output membership functions are Gaussian membership functions

Table 1. Cetner locations of the input membership functions for flocking

Variables / Rules	A_{i1}	A_{i2}	A_{i3}	A_{i4}
1	1.510	0.421	-132.95	-29.90
2	1.603	0.850	68.02	69.30
3	1.506	0.532	-122.84	148.23
4	1.535	0.820	-100.13	-103.36
5	1.544	0.531	114.20	-70.62
6	1.561	0.558	85.64	-40.18
7	1.625	0.537	34.87	81.70
8	1.625	0.609	-129.10	-85.93
9	1.634	1.156	124.34	124.64
10	1.603	1.018	-23.90	-27.12

For arrangement, the subtractive clustering produced 7 fuzzy rules of the form:

$$\text{IF } \gamma^k_{\min 1} \text{ is } A_{i1}, \gamma^k_{\min 2} \text{ is } A_{i2}, \hat{\theta}^k \text{ is } A_{i3}, \theta^k \text{ is } A_{i4}$$
$$\text{THEN } \hat{\gamma}^{k+1} = a_0 + a_1 \gamma^k_{\min 1} + a_2 + \gamma^k_{\min 2}, \qquad (i=1,2,3,\cdots,7) \qquad (5)$$
$$\hat{\theta}^{k+1} = b_0 + b_1 \gamma^k_{\min 1} + b_2 \gamma^k_{\min 2} + b_3 \hat{\theta}^k + b_4 \theta^k$$

Initial position of robots are randomly selected. Table 2 shows the center locations of the 7 fuzzy rules for arrangement. Center locations of the membership functions for $\gamma^k_{\max}$ and $\gamma^k_{\min 1}$ are close, but are not redundant since in the 4 dimensional input space, the centers are apart enough.

Table 2. Center locations of the input membership functions for arrangement

Variables / Rules	A_{i1}	A_{i2}	A_{i3}	A_{i4}
1	0.512	0.684	31.00	29.14
2	0.488	0.731	170.01	169.30
3	0.433	0.679	-78.05	-77.60
4	0.510	0.651	193.99	-167.01
5	0.488	0.567	200.34	-159.96
6	0.524	0.559	129.10	127.20
7	0.465	0.562	31.21	30.43

4 Simulations

According to the flocking algorithm, a group of scattered mobile robots gather and form a flock. Next the arrangement algorithm makes robots form a circle. Fuzzy rules are generated from the input-output data by the algorithms. Group behaviors governed by the fuzzy rules are compared with those by the algorithms. From random initial positions, in case of flocking, robots move toward the center of the flock. A robot calculates the distance to the farthest robot and the distance to the nearest one. If this difference is bigger than a predetermined level, the robot moves toward the farthest robot. If this is less than the level, the robot moves toward the center of the two robots. In case of arrangement, a robot moves toward the center of the nearest and the second nearest robots.

The group behavior algorithms produced ideal movements of flocking and arrangement of 4 mobile robots without collision from a random initial position. Fig. 3 shows an ideal simulation of flocking of 3 robots and 4 robots. Fig. 3(a),(c) show robot trace and Fig. 3(b) and (d) demonstrates the final step of circle-shaping arrangement.

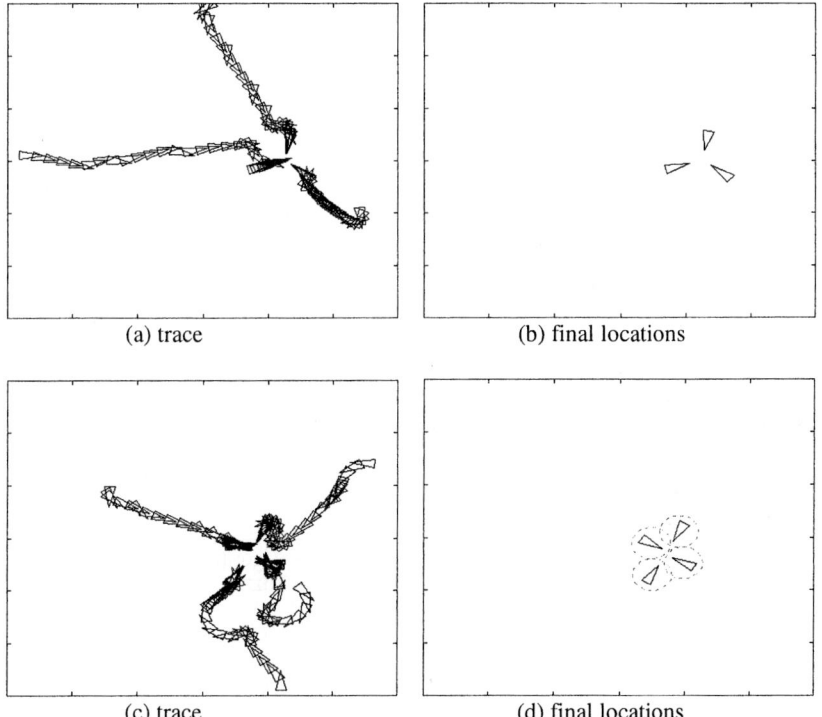

(a) trace (b) final locations

(c) trace (d) final locations

Fig. 3. Flocking and arrangement by the mathematical algorithms. Robot numbers : 3, 4, initial location : random.

Fig. 4 shows the flocking and arrangement result by the fuzzy rules from the same initial positions as in the case of algorithm.

The 10 fuzzy rules for flocking and the 7 fuzzy rules for arrangement, generated from clustering numerical input-output data, successfully realized complex lifelike cooperative behaviors of mobile robots represented by the group behavior algorithms. According to the comparison of the two simulation results, the fuzzy system demonstrated group intelligence of multiple autonomous mobile robots with small number of fuzzy rules.

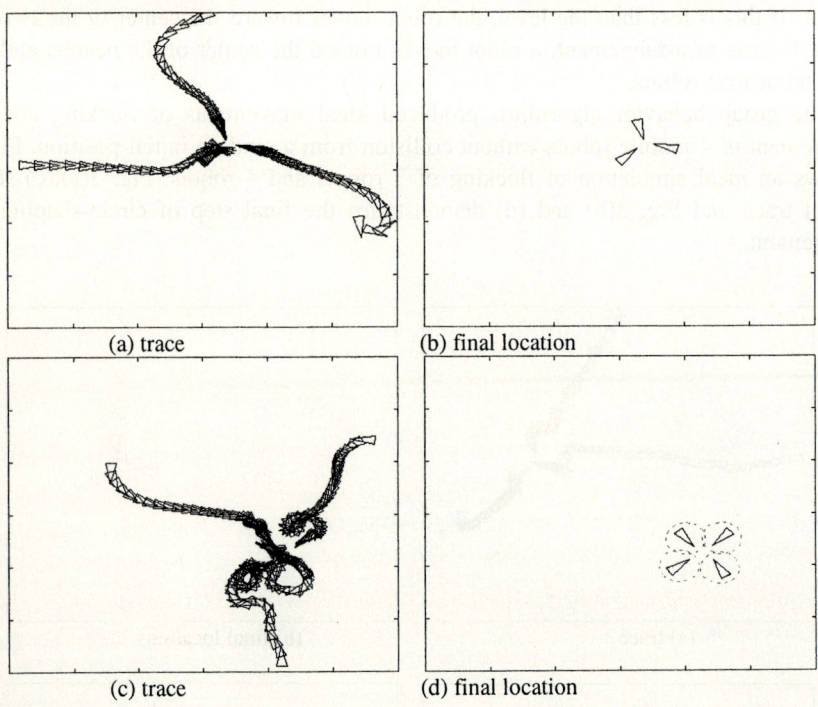

(a) trace (b) final location

(c) trace (d) final location

Fig. 4. Flocking and arrangement by the fuzzy rules. Robot numbers : 4, initial location : random.

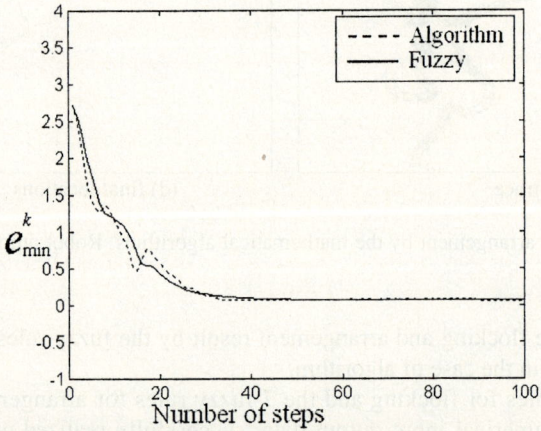

Fig. 5. Performance comparison difference of distance of mathematical algorithms and the fuzzy system

Fig. 5. shows Average of difference of distance of the nearest and the second nearest robots and form given in Eq. 6. Fig 6. and Eq.7 show average of angel variation of all robots

$$e_{min}^{k} = \frac{1}{N}\sum_{i=1}^{N}(\gamma_{min\ 2i}^{k} - \gamma_{min\ 1i}^{k}) \qquad (6)$$

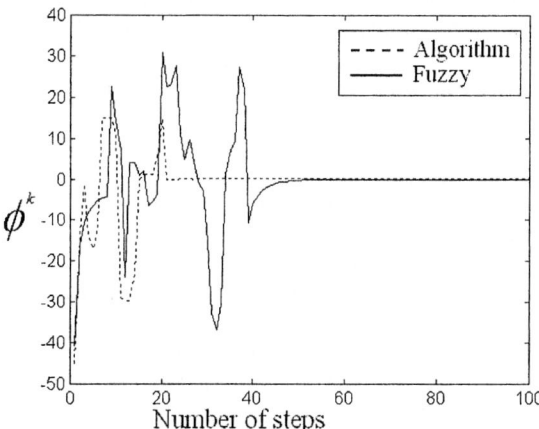

Fig. 6. Performance comparison angle variation average of mathematical algorithms and the fuzzy system

$$\phi^{k} = \frac{1}{N}\sum_{i=1}^{N}\hat{\theta}_{i}^{k} \qquad (7)$$

5 Conclusions

To demonstrate artificial life phenomena, cooperative behaviors of autonomous mobile robots are investigated. Group intelligence of flocking and arrangement behavior are implemented by the group behavior algorithms.

The algorithms produce ideal trace and movement of multiple robots in the flocking and the arrangement. A complex lifelike behavior is believed to consist of a set of simple fundamental rules. The fundamental rules are represented by the fuzzy rules, generated from clustering input-output data obtained from the movement by the algorithms. The fuzzy systems with a small number of fuzzy rules successfully realize cooperative behavior of mobile robots.

Future plan of study includes minimization of the number of rules for group intelligence, obstacle avoidance, and other complex cooperative behaviors.

References

1. C. G. Langton, ed., *Artificial Life II*, Addison Wesley, 1992
2. M. Sipper, "An Introduction to Artificial Life ", *Explorations in Artificial Life*, pp.4-8, 1995.
3. C. W. Reynolds, "Flocks, Herds, and Schools: A Distributed Behaviorsal Model," Computer Graphics, 21(4) pp.25-34, July, 1987
4. D.Teraopoulos, X, Tu, and R. Grzeszczuk, "Artificial Fishes: Autonomous Locomotion, Perception, Behavior, and Learning in a Simulated Physical World," *Artificial Life 1*, pp.337-351, 1994.
5. R.Yager and D.Filev, Essentials of Fuzzy Modeling and Control, John Wiley and Sons, 1994.
6. C. C. Lee, "Fuzzy Logic in Control System : Fuzzy Logic in Controller Part I," IEEE Trans. On Systems, Man and Cyvernetics, Vol. 20, pp. 404-418, 1994.
7. S. Chiu., "Fuzzy Model Identification Based on Cluster Estimation, " Journal of Intelligent & Fuzzy Systems, Vol. 2, No. 3, Sept. 1994.
8. M. Sugeno and G. T. Kang, "Fuzzy Identification of Systems and its Application to Modeling and Control," IEEE Trans. on Systems, Man and Cybernetics, Vol. 15, pp. 116-132, 1985.
9. Han, Jia-Yuan and McMurray, Vincent, "Two-Layer Multiple-Variable Fuqqy Logic Controller," IEEE Trans. on System, man and Cyvernetics, Vol, No. 1, pp. 277-285, 1993.

UML-Based Design and Fuzzy Control of Automated Vehicles

Abdelkader El Kamel[1] and Jean-Pierre Bourey[2]

[1] LAGIS, [2] ERGI,
Ecole Centrale de Lille, BP 48,
F59651 Villeneuve d'Ascq Cedex, France
{abdelkader.elkamel, jean-pierre.bourey}@ec-lille.fr

Abstract. The paper addresses a study case in the frame of ground transportation aiming at improving service quality inside road tunnels. As part of a global project on vehicle automation which aim is to realize a reduced scale multi-sensor platoon of vehicles, a formal specification analysis is carried out and a UML-based design introduced taking into account different riding scenarios inside road tunnels besides a fuzzy control for longitudinal and lateral guidance of a caravan of vehicles. The proposed multimodel fuzzy controllers deal with the grappling and/or the unhooking of the automated train of vehicles for safe tunnels.

1 Introduction

Vehicle automation was introduced since about half a century as one of the prospective solutions to cope with highway congestion and safety. Indeed, by the end of the fifties, RCA (Radio Corporation of America) in collaboration with the GMC (General Motors Corporation) proposed the concept of automated highways, AHS: Automated Highway System [6]. The interest on this topic has largely increased during the last decade in the US, Germany, Japan and France especially with a possible transfer to specific applications such as the safe tunnel issue.

A vehicle automation needs to cope with several issues, among which the following-up of the road or a given trajectory, the detection and the avoidance of obstacles, the insertion and desertion in/out of a traffic, the overtaking, and the maintenance of a secure inter-vehicular distance. It requires a perfect localization and detection of the environment. Based on the information derived from sensors of the surrounding environment, the vehicle automation consists then in generating the accurate control variables [3] [5].

A four-vehicle platoon is realized [8] as a reduced scale (1:10) multi-sensor vehicle prototype and a wireless communication system implemented besides a monitoring system allowing considering different scenarios for the platoon such as the insertion (suppression) of an automated vehicle into (out of) the platoon.

Supported by the French government and the French Nord-Pas-De-Calais Region as part of a global European issue, the main object of this prototype is to test in real conditions and real time different control strategies based on the automotive intelligent control, to introduce on purpose a multi-sensor system and redundancy for safety reasons and sensor failure analysis, to detect and localize any defection and hence propose an automatic reconfiguration process taking into account the failure. Note that a geometrical/timescale similitude was determined in order to transpose the obtained results, valid for reduced-scale 1:10 vehicles, to full-scale regular vehicles or trucks.

The paper addresses a study case in the frame of ground transportation aiming at improving service quality inside road tunnels. A formal specification analysis is carried out considering the global environment before and after the road tunnel and a UML-based design introduced taking into account different riding scenarios inside road tunnels and a fuzzy control for longitudinal and lateral guidance of caravan of vehicles. The proposed multimodel fuzzy controllers deal with the grappling and/or the unhooking of the automated train of vehicles.

The paper is organized as follows: first we focus on the design stage and present an UML-based model [2], then we present different fuzzy controllers introduced to cope with the longitudinal and the lateral control of vehicles.

2 UML-Based Design

2.1 Specifications

The first step in the real development of automated vehicles appears to be their use in a limited zone for instance in the frame of tunnels. Indeed, following the catastrophe of the Mont Blanc tunnel between France and Italy, tough new traffic regulations were imposed to cope with safety inside the tunnel. Regulation authorities imposed alternated one-way traffic, a much larger inter-vehicular distance, a limited number of vehicles insides the tunnel ... These rules enhance safety to the detriment of both congestion and duration.

In this application, we propose to switch to an automated vehicle mode inside the tunnel while allowing the driver to control his vehicle anytime upstream and downstream the tunnel. We have formalized the problem as follows:

Road infrastructure:

- One-way one-line structure is considered which can be extended easily.
- Inside tunnel, vehicles are automated and in a platoon.
- Two buffer zones are considered: one upstream the tunnel allowing vehicles to accelerate till the speed set value and one downstream the tunnel for deceleration.
- The buffer zones are separated from the highway by two dead zones allowing caravan formation and the switching between the automatic and manual control.

Vehicles:

- Vehicles arrive randomly in the dead zone.
- 3-vehicle manual caravan is imposed before entering the upstream zone switching in the tunnel to the automatic mode. The inter-vehicular distance is set a priori (depending on congestion, traffic conditions ...).

- While entering the upstream zone, vehicles accelerate till the caravan speed is set to V1: rapid mode, V2: normal mode or V3: slow mode depending on the desired scenario.
- Inside tunnel, caravan maintains the imposed speed.
- Once in the downstream zone, the caravan decelerates following a deceleration profile.
- Finally, once in the dead zone, manual control is allowed and no caravan is required any more.

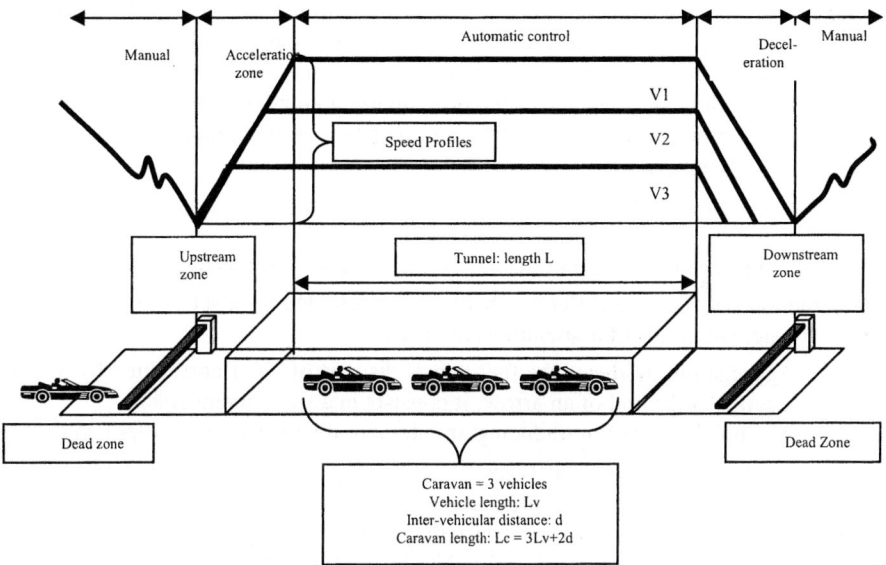

Fig. 1. Safe tunnel control specifications

Operating modes:

- The downstream zone is evacuated only if the dead zone is free.
- Speed inside the tunnel depends on the traffic at the download zone. If it is light, V1 is imposed by default. Otherwise V2 or V3 are required.
- V1, V2 and V3 are determined such us a safe emergency stop can be operated in the downstream zone.
- All caravans evolving in the system could be evacuated to the dead zone. Restrictions are hence imposed at the upstream zone before allowing vehicles to enter the tunnel.
- The number of vehicles in the caravan is set to 3 arbitrarily. It has to be adapted with respect to the characteristics of the tunnel, the density of traffic…

Figure 1 illustrates the different scenarios and specifications.

2.2 UML-Design

As it can be figured out, the project is intrinsically multidisciplinary involving different issues, scenarios and levels. The design stage is therefore highly important to assure the coherence of the different sub-parts of the project, the problem formulation and the human/machine interaction.

Following problem specifications, an UML-based model is introduced and implemented in order to cope with the different aspects related to the global project.

Why UML? Introduced in the nineties, the Unified Modeling Language (UML) [9] became, since then, a standard for system specification and analysis. Based on the object formalism, UML defines a common language and a methodology for user-friendly analysis and communication thanks to the graphical representation of the object solution, the independence of the programming language, the application...

In version 1.5, it introduces nine different diagrams. Among them, we have chosen to use the following ones:

- Class-diagram for the description of the structure (concepts and associations) of the static aspects of the system.
- State transition-diagram for the dynamical part of the system.

Interaction between the different classes is represented by a set of associations. In the following, two kinds of associations are used:

- Aggregation: traduces the fact that X "is part of" Y. Represented by a white lozenge at the end of an arrow, it consists in a weak membership.
- A simple association traducing an interaction between two components.

Once combined, the different types of UML diagrams offer a global view of the static and dynamic aspects of the process.

In figure 2 and 3, we propose the Class-diagram and the State-diagram respectively. We have used the software OBJECTEERING from SOFTEAM to model the system.

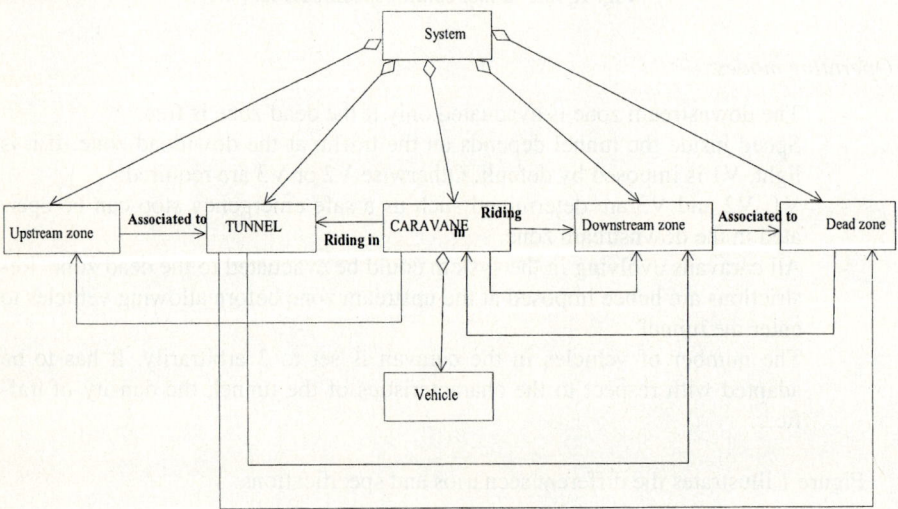

Fig. 2. UML Class-diagram

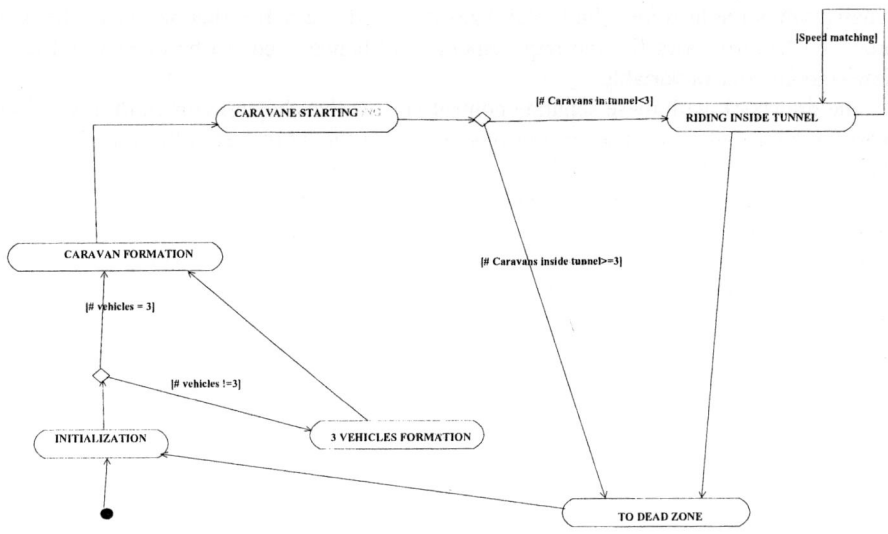

Fig. 3. UML State-diagram

3 Control Strategies

3.1 Longitudinal Control

The purpose of the longitudinal control is to allow a platoon of vehicles follow a leading vehicle while maintaining a constant inter-vehicular distance. Each vehicle will pick up the necessary information, such as velocity and acceleration, derived from the precedent one. A control strategy, based on fuzzy control and a non linear longitudinal model, is carried out [1] [7] [10] [11].

Three separate control laws are developed for the longitudinal guidance: one for acceleration, one for deceleration and finally one for leading vehicle, evolving at roughly constant speed, tracking. The vehicle, following a straight line, is modeled by:

$$\ddot{x} = u - K \dot{x}^2 - h. \tag{1}$$

where u represents the system input, K a coefficient associated with vehicle geometry and aerodynamics, h an effort associated with ground/tire contact.

The European legislation imposes the respect of a security distance between vehicles of at least two seconds, the first one for braking the second due to the driver's reaction time. For automated vehicles inside tunnels, the objective is to reduce this inter-vehicular distance to only one second and hence limit the congestion.

The acceleration control consists in tracking the leading vehicle which accelerates while keeping the secure braking inter-vehicular distance. The deceleration control intervenes when the leading vehicle reduces its speed hence the following vehicle has to adjust its own speed while maintaining the secure braking inter-vehicular distance.

Finally, when the leading vehicle stabilizes its speed, we notice that no one of the two precedent control laws fits the requirements and hence needs to be determined as a new smooth control variable.

The drawback with these separate control laws is that they remain valid in specific working points while we need a control strategy for the whole set of the vehicle working domain. In order to fulfill this requirement we can either commute between these control laws based on a given criteria or apply a multimodel/multicontrol strategy based on a fusion between these control laws. It is obvious to notice that any commutation strategy induces discontinuities in the global control law besides undesirable peaks which may appear very detrimental to the vehicle in terms of quality of service and passenger comfort. We propose then a three level global control law based on the fusion of the single laws. The principle of the control fusion is presented in figure 4.

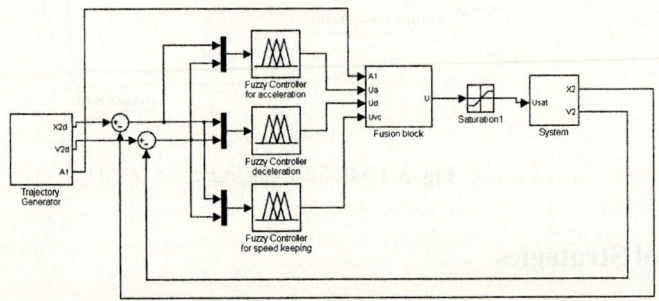

Fig. 4. Principle of the multicontrol fuzzy controller

where X_{2d} and V_{2d} represent respectively the desired position and velocity of vehicle 2 (following vehicle), X_2 et V_2 the real ones, A_1 the acceleration of vehicle 1 (leading vehicle), U_a, U_d and U_{vc} the control variables generated by the fuzzy controller for the acceleration, the deceleration and the stabilization respectively, U et U_{sat} the control variable obtained by fusion and its saturation as an input to the system.

Note that in this scheme we have introduced on purpose a saturation of the control variable in order to take into account the limitations due to physical constraints in a vehicle. The fusion bloc, determining the weighing functions, is based on a fuzzy type strategy to assure a smooth and accurate transition between the working points.

The simulation results for an extreme study case are presented in figure 5: vehicle 1 accelerates at 2.5 $m.s^2$ from 0 to 130 $km.h^{-1}$ (36.1 $m.s^{-1}$) then stabilizes. Its velocity (V1) oscillates with 3.8% overshoot amplitude before the rough deceleration phase at - 8.4 $m.s^{-2}$ till 50 $km.h^{-1}$ (13.9 $m.s^{-1}$) and an almost stabilization phase around 50 km/h. ErreurP and ErreurV represent respectively the position and velocity errors between vehicle 2 and its reference trajectory. Note that at t=0, we consider that the secure inter-vehicular distance is respected.

The multicontrol fuzzy controller, working with 27 rules, allows regulating the vehicle trajectory while maintaining a secure inter-vehicular distance whatever is the velocity profile of the leading vehicle.

The above mentioned multicontrol fuzzy controller uses only two inputs: the position error between the desired and the real positions of vehicle 2, besides a fuzzy fusion between the separate control variables due to acceleration, deceleration and stabilization. We propose to compare these results with those obtained by a classical fuzzy controller using, this time, three inputs: the previous inputs plus the acceleration error between the desired and the real accelerations. The same requirements are imposed for a tracking during acceleration, deceleration and stabilization phases. We notice then that the classical fuzzy controller, using also 27 rules, is less performing than the multicontrol fuzzy controller: the overshoots are more important and time response is longer than previously.

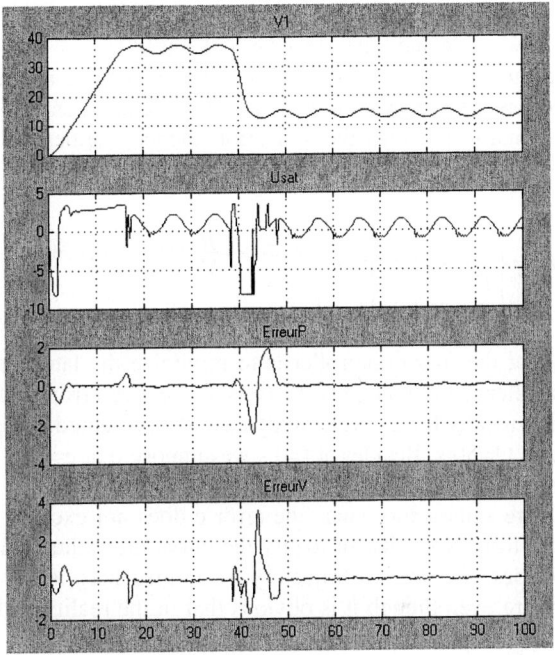

Fig. 5. Simulation results for extreme study case

3.2 Lateral Control

The main object of the lateral control is to allow the vehicle follow a given trajectory and maintain a good ride quality while assuming that the longitudinal guidance is controlled.

We have developed a complete model of a vehicle based on physical analysis of different phenomena which may have an impact on vehicle riding such as tire adhesion, aerodynamic effect ... and by taking into account the three possible movements for the vehicle along the three axes: pumping, pitch and roll. We assumed, as is generally proposed in the literature, that the lateral and longitudinal control are disconnected. A multimodel controller can be introduced at the control level in order to approach the real behavior of vehicles. The global model is non linear and rather complicated to use in real-time control strategies and/or in a virtual reality animation.

Hence, under different classical simplification hypotheses, a partial model taking into consideration just three out of the six degrees of freedom can be expressed in the state space using the state vector $X^T = \begin{bmatrix} \alpha & \psi & \dot{\psi} \end{bmatrix}$ by:

$$\dot{X} = \begin{bmatrix} \dot{\alpha} \\ \dot{\psi} \\ \ddot{\psi} \end{bmatrix} = \begin{bmatrix} -\dot{\psi} \frac{48*\cos(\beta)}{V \cos(\alpha)} \left\{ \sin(\alpha-\beta) + \sin(\alpha) + \frac{2.1}{V} \dot{\psi}(\cos(\beta)-1) \right\} \\ \dot{\psi} \\ 37.5 \left\{ 2.1(\sin(\alpha) - \cos(\beta)\sin(\alpha-\beta)) - \frac{\dot{\psi}}{V}(4.4 + 0.6\sin^2(\beta) + 4.4\cos^2(\beta)) \right\} \end{bmatrix} \quad (2)$$

where ψ is the yaw angle and β the steering-lock angle.

The reference trajectory is a straight line followed by a curvature then another straight line. Note that the curvature is an arc of circle and hence the curvature radius is constant and equal to R with γ the deflection. Hence, the desired yaw angle ψ_d can be expressed by:

$$\psi_d = \begin{cases} 0 & si\ t \in [t_0,\ t_1] \\ \frac{V}{R} t & si\ t \in [t_1,\ t_2] \\ \gamma & si\ t \in [t_2,\ t_3] \end{cases} \text{ with } t_2 = \frac{R}{V}\gamma + t_1,\ t_3 \geq t_2, t_1 \geq 0. \quad (3)$$

The objective of the fuzzy controller is to minimize the lateral error e= ψ_d-ψ. We use a fuzzy controller with two inputs 0-e (0 is the desired error) and ψ_d-ψ.

The simulation results for a severe study case are presented in figure 6: we start at t_1=10 s, γ=π/2 and t3=50 s. Besides at t=0 s we suppose that e=0 m, α=0 rad, and ψ=0 rad, i.e., the vehicle follows a straight line till t_1.

These results are satisfactory since the error e does not exceed, in the worst case, 20 cm during the turn. Note that in reality the curvature radius R is closer to 500 m than 100 m. Besides, the desired trajectory as defined here is the one we generally find in the literature even though it is obvious that in the reality highways do not present precisely such a kind of curves.

4 Conclusion

In this study, a design was carried out at the first stage using the UML modeling language in order to have a precise overview on the different sub-parts of the project. The control of a platoon of vehicles in an automated tunnel is then considered. At first, the longitudinal control issue is divided in three separate stages: acceleration, deceleration and stabilization. Then, a multimodel/multicontrol fuzzy approach is considered in order to deal with the global control. It has the main advantage to take into account the three aspects by introducing a fuzzy fusion of the control variables. The lateral control is also considered by means of a fuzzy controller and gives satisfactory results. Hence, Different scenarios can be considered in the frame of the tunnel operating system. Real time implementation on the reduced-scale vehicles is under way and will be presented in a future work.

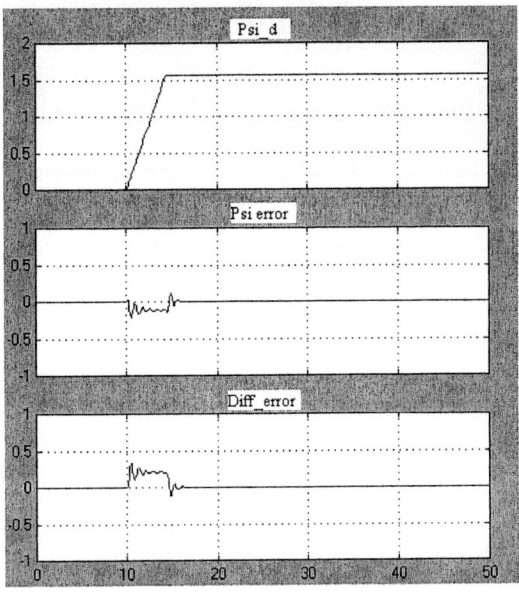

Fig. 6. Simulation results, R=100m, V=130 km.h^{-1}

References

1. Babuska R., Verbruggen H.B.: Identification of composite linear models via fuzzy clustering, ECC Rome (1995), pp. 1207-1212.
2. Booch G., James R., Jacobson I.: Le guide le l'utilisateur UML, Eyrollès (2001), 534p.
3. El Kamel A., Bouslama F, Borne P.: Multimodel Multicontrol of Vehicles in Automated Highways, WAC'2000, ISIAC 057 (2000), Hawaii, USA.
4. El Kamel.A., Dieulot J.Y, Borne P.: Fuzzy Controller for Lateral Guidance of buses, *17th IEEE International Symposium on Intelligence Control*, ROM Proc. (2002), Vancouver, Canada.
5. Frank P.M.: Fault diagnosis in dynamic systems using analytical and knowledge-based redundancy: A survey and some new results, Automatica (1990), Vol. 26, N° 3.
6. Fenton R. E., Mayhon R.J.: Automated highway studies at the Ohio State University: an overview, IEEE Trans. on Vehicular Technology (1991), Vol. 40, N° 1, pp. 100-113.
7. Johansen T.A.: Fuzzy model based control: stability, robustness, and performance issues, IEEE Trans. on Fuzzy Systems (1994), Vol. 2, N° 3, pp. 221-234.
8. Lorimier L., El Kamel A.: Intelligent Instrumentation and Control of An Automated Highway Prototype., IEEE Conference on industrial Electronics IETA (2001), Cairo.
9. OMG Unified Modeling Language Specification: version 1.5 formal/03-03-01, available at http://www.omg.org
10. Sugeno M., Yasukawa T.: A fuzzy-logic-based approach to qualitative modeling, IEEE Trans. on Fuzzy Systems (1993), n° 1, pp. 7-31.
11. ZADEH L.: Outline of a new approach to the analysis of complex systems and decision processes, IEEE SMC (1973), pp. 28-44.

Design of an Analog Adaptive Fuzzy Logic Controller

Zhihao Xu, Dongming Jin, and Zhijian Li

Institute of Microelectronics, Tsinghua University,
Beijing, China, 100084
xzh99@mails.tsinghua.edu.cn
jdm-ime@tsinghua.edu.cn
lizhj@mail.tsinghua.edu.cn

Abstract. An analog Adaptive fuzzy logic controller is proposed. This controller is based on back-propagation algorithm, and designed for the use of on chip learning. This adaptive fuzzy logic controller was composed of an analog fuzzy logic controller and a learning circuit that realizes online learning mechanism. It can tune the consequent parameters automatically with the help of a direction signal. Hspice simulation of functional approximation experiment was held, showing that this controller has the on-chip learning capability.

1 Introduction

Fuzzy logic theory was first introduced by L.A.Zadeh in 1965 [1]. As Fuzzy logic systems have the capability to take decisions based on the experiences of human intelligence, and needs not accurate mathematical models of the physical systems, it has played important roles in improving the dynamical performance of non-linear systems. This is in clear contrast to the design of conventional control systems.

With the developments of VLSI technologies, many hardware implementations of fuzzy logic controller were proposed [2]. As the physical systems to be controlled need a long term solvent, the consequent parameters used by the controller will need to change as time runs. Therefore, it is important to develop techniques of updating fuzzy logic controllers' consequent parameters. Today, many achievements about learning algorithms of fuzzy logic controllers were presented [3], and there have been a special fuzzy logic system called the adaptive fuzzy logic controller.

In general, an adaptive fuzzy logic controller contains two parts, one is a fuzzy logic controller circuit to realize forward fuzzy inference operation for the applications, and the other is a circuit of learning mechanism to update the consequent parameters used by the fuzzy logic controller.

In online applications, the adaptive fuzzy logic controller can manipulate various consequent parameters quickly and dynamically. A design of an analog adaptive fuzzy logic controller was presented in this paper.

The proposed adaptive fuzzy logic controller is composed of a single input single output fuzzy logic controller and a learning circuit cell, and all of the internal subunit circuits are analog circuits. Finally, Hspice simulation of online functional approximation experiments was held, and the results showed a successful online learning capability of the designed adaptive fuzzy logic controller.

2 Back-Propagation Method

The back-propagation algorithm is a powerful training technique that can be applied to networks with feed-forward structure, to turn them into adaptive systems. Since fuzzy logic systems can be represented as a layered feed-forward networks [4], fuzzy logic systems can also be referred as neuro-fuzzy systems, back-propagation algorithm can also be used on them.

The functional map of a fuzzy logic controller is shown in equation (1).

$$f(x) = \frac{\sum_{i=1}^{M} \omega_i(x) \cdot d_i}{\sum_{i=1}^{M} \omega_i(x)} \tag{1}$$

Equation (1) can be easily transformed into an equivalent form shown in equation (2).

$$f(x) = \frac{\sum_{i=1}^{M} \omega_i \cdot d_i}{\sum_{i=1}^{M} \omega_i} = \sum_{i=1}^{M} \left(\frac{\omega_i}{\sum_{i=1}^{M} \omega_i} \right) \cdot d_i = \sum_{i=1}^{M} \varphi_i \cdot d_i \tag{2}$$

$$\varphi_i(x) = \frac{\omega_i(x)}{\sum_{i=1}^{M} \omega_i(x)}$$

$\omega_i(x)$ is the activation degree of the ith rule with the input variable x, and $\varphi_i(x)$ is the corresponding normalized activation degree. d_i is the consequent parameter of the ith rule. Use $E(x)$ to be the error function which should be minimized in the learning stage.

$$E(x) = (1/2) \cdot (Y(x) - f(x, d_i))^2 \tag{3}$$

As $(x^*, Y(x^*))$ is an input-output training pair, the learning rule can be deduced as shown in equation (4)(5).

$$\Delta d_i = -\alpha \cdot \frac{\partial E}{\partial d_i} \tag{4}$$

$$\Delta d_i = \alpha \cdot (Y(x^*) - f(x^*, d_i)) \cdot \frac{\partial f}{\partial d_i}$$
$$= \alpha \cdot (Y(x^*) - f(x^*, d_i)) \cdot \varphi_i \tag{5}$$

As shown in equation (5), the normalized activation degrees should be calculated by hardware first, in order to realize the back-propagation learning mechanism.

3 Circuit Design of the Adaptive Fuzzy Logic Controller

The adaptive fuzzy logic controller is a single input single output one, and the input universe of discourse was divided into five fuzzy sets that represent different input language variables.

The hardware architecture of the designed analog adaptive fuzzy logic controller was shown in Fig.1. The input variable is x and the direction signal is $Y(x)$ which is the expected ideal output of the controller. The normalized activation degrees φ_i ($i=1,2,\ldots,5$) are the key internal signals, which will be used in both the forward fuzzy inference operation and the learning stage.

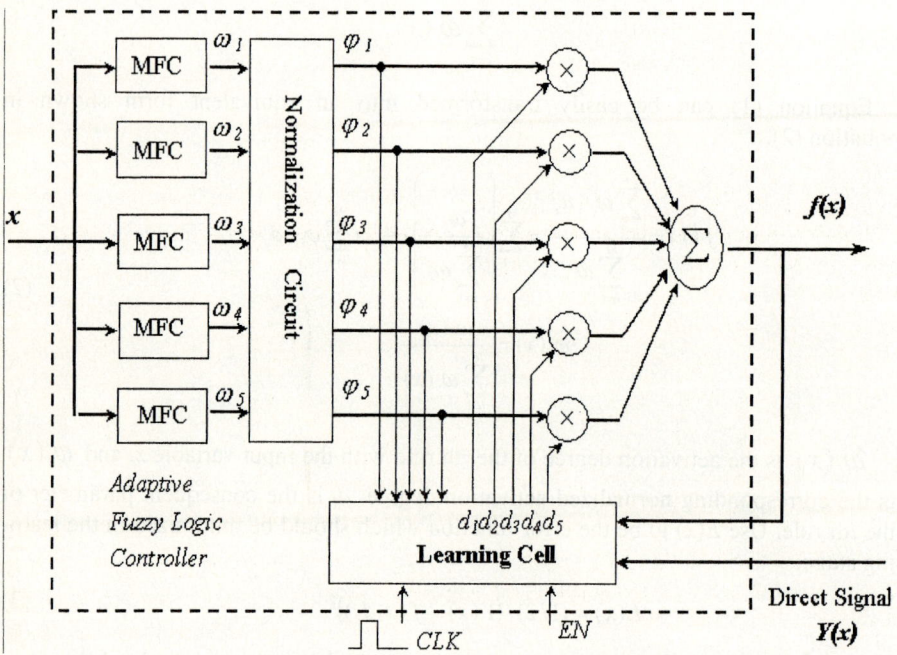

Fig. 1. Hardware architecture of the adaptive fuzzy logic controller

The learning cell is the circuit that realizes the back-propagation learning rule shown in equation (5). The normalized activation degrees φ_i ($i=1,2,\ldots,5$), direction signal $Y(x)$, and the output signal of the controller $f(x)$ are the input signals of the learning cell. The consequent parameters of each rule are the output signals of the learning cell. The adaptive fuzzy logic controller can tune the consequent parameters d_i by itself to realize the expected functional map of the direction signal.

As shown in Fig.1, this controller was composed of four subunits: membership function circuit (MFC), multiplier, normalization circuit, and learning cell. They will be introduced below:

Membership function circuit

The membership function circuit used is shown in Fig.2. This circuit is a transformation of the circuit mentioned in paper [5][6], and here uses two PMOS differential pairs to realize a Gassian shape transfer function. V_{r1}, V_{r2} ($V_{r1}<V_{r2}$) are bias voltages which will decide the location and the shape of the membership function.

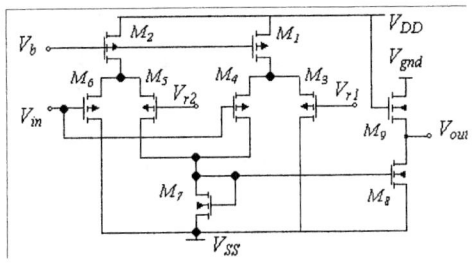

Fig. 2. Membership function circuit

The Hspice simulation result of the MFC was shown in Fig.3. The input signal x is a voltage signal limited in the range of (-0.6v, 0.6v), and the center locations of the membership function of each fuzzy set were set at -0.6v, -0.3v, 0v, 0.3v, and 0.6v.

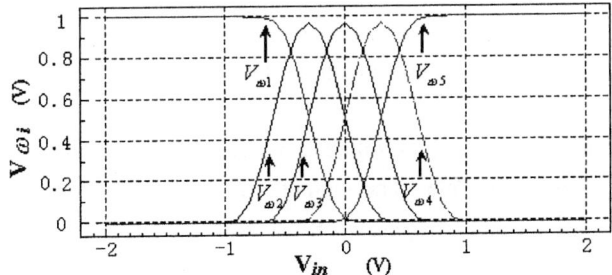

Fig. 3. Hspice simulation result of the MFC

b) Multiplier

The multiplier used in this controller is a folded Gilbert multiplier [7], it is a voltage input current output circuit, which fits the transfer function of equation (6), and this circuit was shown in Fig.4

$$I_{out} = k \cdot V_{in1} \cdot V_{in2} \tag{6}$$

c) Normalization circuit

Use a feedback loop to realize the normalization operation. The circuit was shown in Fig.5

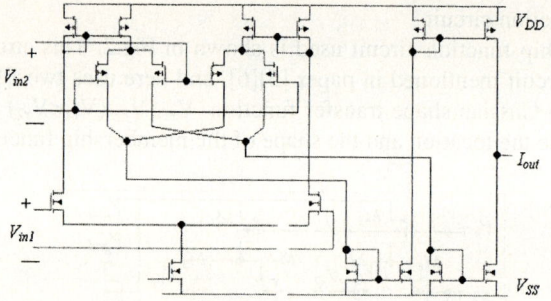

Fig. 4. Folded Gilbert multiplier circuit

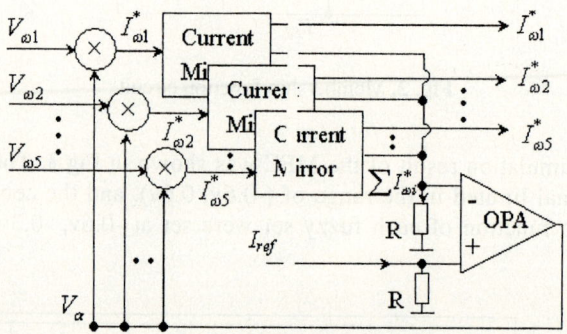

Fig. 5. Normalization circuit

The input signals are a group of voltages: $V_{\omega 1}, V_{\omega 2}, \ldots, V_{\omega 5}$. They represent the activation degrees of each rule. The output signals are a group of current signals: $I^*_{\omega 1}, I^*_{\omega 2}, \ldots, I^*_{\omega 5}$, which represent the normalized activation degree, and it can be seen that the output currents of each multiplier was copied by a current mirror. V_α is the output voltage of the operational amplifier that feedback to one of the input terminal of the folded Gilbert multiplier.

Because of that the two input terminals of the operational amplifier are virtual connected, the relationship shown in equation (7) comes into existence.

$$I_{ref} \cdot R = R \cdot \sum_{i=1}^{5} I^*_{\omega i} \Leftrightarrow I_{ref} = \sum_{i=1}^{5} I^*_{\omega i} \qquad (7)$$

Following the transfer function of the multiplier shown in equation (6), the output currents of each multiplier can be calculated as shown in equation (8)

$$I^*_{\omega i} = k \cdot V_\alpha \cdot V_{\omega i} \qquad (8)$$

Combining equation (7) and equation (8) together, equation (9) can be deduced.

$$V_\alpha = \frac{I_{ref}}{k \cdot \sum_{i=1}^{5} V_{\omega i}} \tag{9}$$

Combining equation (8) and equation (9) together, the expression shown in equation (10) can be deduced. They are the transfer function of the output currents of the normalization circuit.

$$I_{\omega i}^* = I_{ref} \cdot \frac{V_{\omega i}}{\sum_{i=1}^{5} V_{\omega i}} \tag{10}$$

d) Learning cell

This circuit cell was designed to realize the calculation operation shown in equation (5), and equation (5) can be transformed into an equivalent form shown in equation (11)

$$\Delta d_i = \alpha \cdot \|Y(x) - f(x, d_i)\| \cdot \text{sgn}(Y(x) - f(x, d_i)) \cdot \varphi_i \tag{11}$$

It can be seen that φ_i and $\text{sgn}(Y-f)$ are the key variables in the calculation of the increment vector $[\Delta d_1, \Delta d_2, \ldots, \Delta d_5]^T$, because they can determine the direction of the increment vector $[\Delta d_1, \Delta d_2, \ldots, \Delta d_5]^T$ and this direction is a critical value for the learning algorithm.

$\|Y(x) - f(x, d_i)\|$ and the learning step α can only influence the magnitude of vector $[\Delta d_1, \Delta d_2, \ldots, \Delta d_5]^T$, and in order to decrease the hardware design cost, $\alpha \cdot \|Y(x) - f(x, d_i)\|$ can be taken place by a fixed value β in the presented adaptive fuzzy logic controller circuit, and β should be small enough.

Finally, there has been a predigested expression of back-propagation algorithm shown in equation (12), and it is more fitted for the hardware implementation of back-propagation learning mechanism.

$$\Delta d_i = \beta \cdot \text{sgn}(Y(x) - f(x, d_i)) \cdot \varphi_i \tag{12}$$

Because sgn(Y-f) operation can be realized by analog comparator and the normalization circuit has been designed before, the expected learning cell can be designed as shown in Fig.6.

The input signals of the learning cell were the copies of the output currents of the normalization circuit. The output signals of this circuit cell are V_{d1}, V_{d2}, V_{d3}, V_{d4}, V_{d5}, which represents the consequent parameter corresponding to each fuzzy rule, and they are temporal stored on the capacitors $C_1=C_2=\ldots=C_5=C$ when the controller working,.

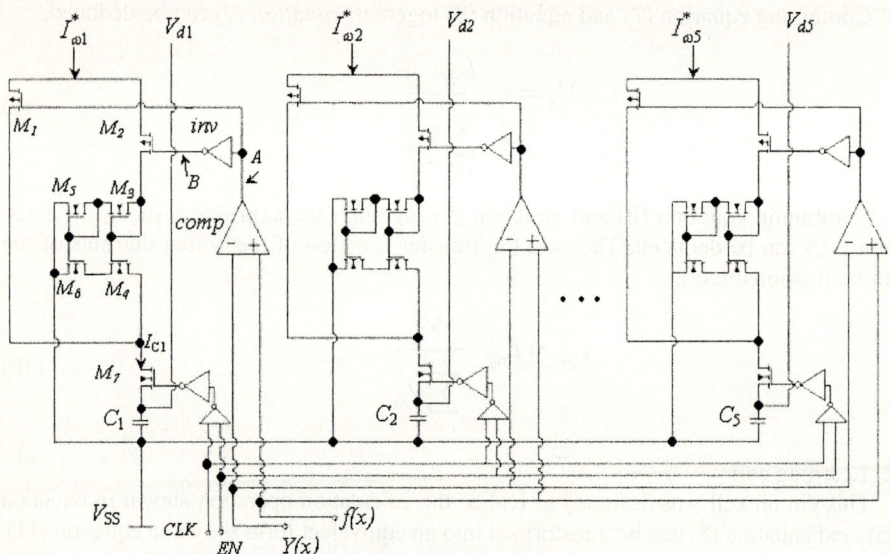

Fig. 6. Circuit of the learning cell

As shown in Fig.6, *comp* was an analog comparator, M_1, M_2, M_7 were PMOS switch, and $M_3 \sim M_6$ were a cascode current mirror that was used to change the direction of the input current.

Use one signal channel of this circuit shown in Fig.6 to describe the principle of the hardware implementation of learning mechanism:

When the signal *EN* is enabled, the learning cell will work.

If $f(x) > Y(x)$, then $\text{sgn}(Y(x)-f(x)) = -1$. The output voltage of the comparator V_A will be V_{DD}, and voltage V_B will be V_{SS}, so that the PMOS switch M_1 will be off, and M_2 will be on. The input current will be copied by the cascode current mirror, and it is can be seen that $I_{C1} = -I_{\omega 1}^*$ at this time.

If $f(x) < Y(x)$, then $\text{sgn}(Y(x)-f(x)) = +1$. Voltage V_A will be V_{SS}, and voltage V_B will be V_{DD}, so that the PMOS switch M_1 will be on, and M_2 will be off, and it is can be seen that $I_{C1} = +I_{\omega 1}^*$ at this time.

Then, the expression formula of current I_{C1} shown in equation (13) comes into existence.

$$I_{C1} = \text{sgn}(Y(x) - f(x)) \cdot I_{ref} \cdot \frac{V_{\omega 1}}{\sum_{i=1}^{5} V_{\omega i}} \tag{13}$$

Assume that there was a time span of ΔT s that the PMOS switch M_7 was on within every CLK signal period, and the capacitor C_1 can be charged or discharged by the current I_{C1} in this time span. The increment of the voltage of the capacitor C_1 was shown in equation (14)

$$\Delta V_{d1} = \frac{\Delta Q}{C} = \frac{I_{C1} \cdot \Delta T}{C}$$
$$= \left(\frac{\Delta T}{C} \cdot I_{ref}\right) \cdot \text{sgn}(Y(x) - f(x)) \cdot \frac{V_{\omega 1}}{\sum_{i=1}^{5} V_{\omega i}} \quad (14)$$

It can be seen that the transfer function shown in equation (14) was fitted for the predigested back-propagation learning rule shown in equation (12), so that the circuit shown in Fig.6 can realize the expected back-propagation learning mechanism.

4 Functional Approximation Experiment

In order to study the online learning capability of this adaptive fuzzy logic controller, the functional approximation experiment was held. Use a triangle wave sweep from −0.6v to 0.6v to be the input signal $x(t)$ of the adaptive fuzzy logic controller, and use a sin wave or a triangle wave with the same frequency to be the direction signal $Y(t)$, so that there has been a sinusoidal function map or a linear function map from $x(t)$ to $Y(t)$. The functional approximation experiment systems used is shown in Fig.7.

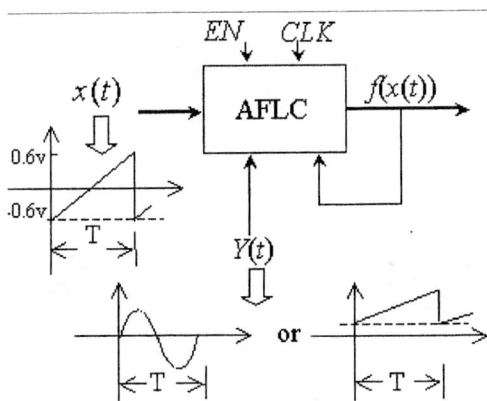

Fig. 7. Sketch map of online learning experiment

As the two signals $x(t)$, $Y(t)$ are periodical signals, the adaptive fuzzy logic controller was trained recursive. The Hspice transient simulation result of the online learning operation was shown in Fig.8.

The time domain variation of the consequent parameter voltage V_{di} was shown in Fig.8, and it can be seen that the learning stage comes into convergence within 5ms.

When the learning stage ended, the consequent parameter voltage V_{di} can be taken out. Using them to be the initial consequent parameters of each rule, a simulation of the feed forward inference operation of the fuzzy logic controller was held, and the results were shown in Fig.9. As shown in Fig.9, $f(x)$ is the Hspice transient simulation result of the output voltage of the forward fuzzy inference computation, and the corresponding ideal output signal $Y(x)$ was shown together. These Hspice simulation results were calculated by CSMC 0.6μm mixed signal process transistor parameters.

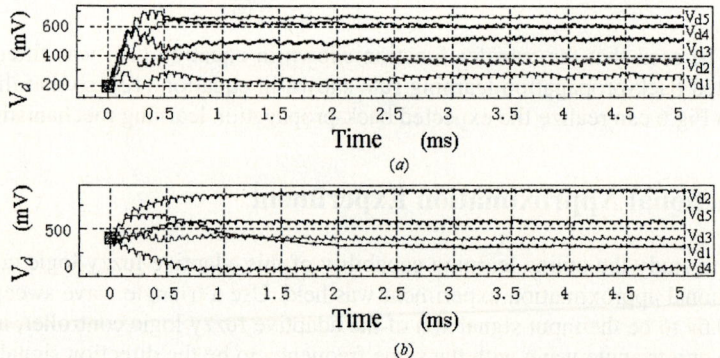

Fig. 8. (*a*) linear function learning (*b*) sinusoidal function learning

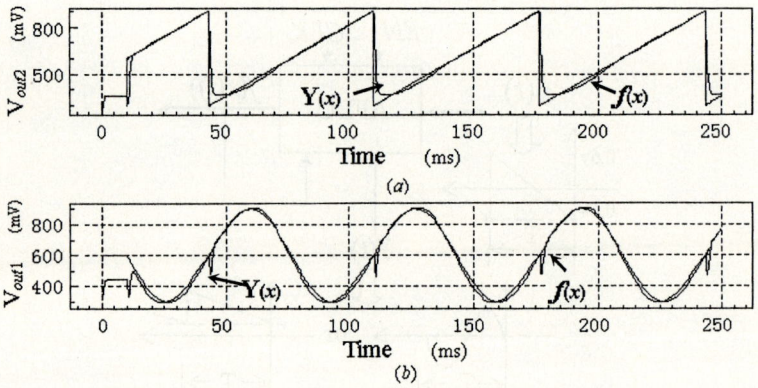

Fig. 9. (*a*) Forward inference result of linear function Learning (*b*) Forward inference result of sinusoidal function Learning

5 Conclusion

In this paper, an analog adaptive fuzzy logic controller was proposed. A linear and a sinusoidal functional approximation experiments were held, and the Hspice simulation results showed that the designed adaptive fuzzy logic controller could achieve the online learning goals, and this characteristic can be use in the construction of robot, machine learning system etc, and as it is an analog chip design, it can be fit for the requirements of high speed, system miniaturization and lower hardware cost.

References

1. L.A.Zadeh, Fuzzy Sets, Information and Control, New York: academic, Vol.8, (1965) 338-353
2. Lee C C., Fuzzy Logic control systems: fuzzy logic controller. Part II, IEEE Trans. on Systems. Man. and Cybernetics. , Vol. 20, No. 2, (1990) 419-435
3. Jerry M. Mendel, George C. Mouzouris, Designing Fuzzy Logic Systems, IEEE Transactions on circuits and systems-II: Analog And Digital Signal Processing, Vol.44, No.11, (1997) 885-895
4. L.X. Wang, Adaptive fuzzy systems and control, Englewood-Cliffs, NJ: Prentice-Hall, (1994)
5. Florin-Adrian Popescu, Nicolae Varachiu, Using an analog fuzzification circuit for real world applications, Proceedings Of International Conference On Semiconductor.Vol.1, (2000) 281-284
6. A.K. Taha, M.M. El-Khatib, Design of a Fuzzy Logic Programmable Membership Function Circuit, 17th National Radio Science Conference (2000) C20/1 - C20/6
7. Chun Lu, Analog VLSI Implementation of BP On-Chip Learning Artifical Neural Networks, Dissertation of Tsinghua University for the Degree of Doctor of Engineering, (2002) 42-45

VLSI Implementation of a Self-tuning Fuzzy Controller Based on Variable Universe of Discourse

Weiwei Shan, Dongming Jin, Weiwei Jin, and Zhihao Xu

Institute of Microelectronics, Tsinghua University,
Beijing 100084, China
{chanww03, jww02, xzh99}@mails.tsinghua.edu.cn
jdm-ime@tsinghua.edu.cn

Abstract. A novel self-tuning fuzzy controller and its VLSI implementation are developed based on variable universe of discourse. This fuzzy controller is constructed by applying a contraction factor before each input of a conventional fuzzy controller, and a self-tuning gain factor after its output, while all the factors are adjusted with the input variables according to a simplified adaptive law. This fuzzy controller has some features: the contraction factors and output gain factor that improve the performance of the controller are based on variable universe of discourse; these factors are simplified for VLSI implementation; only the active rules are processed and the division in the operation of COA defuzzification is omitted by setting the denominator equal to 1. Results of Matlab - ActiveHDL co-simulation indicate that this self-tuning fuzzy controller works successfully in controlling a nonlinear system to track a reference trajectory.

1 Introduction

Fuzzy control has been effectively used in many applications and has attracted more and more attention in control field. Fuzzy logic, on which fuzzy control is based, bears more resemblances in spirit to human thinking and natural language than the traditional logical systems. Experiences show that the fuzzy control yields results superior to many obtained by conventional control algorithms [5]. Hardware implementation of fuzzy controller (FC) becomes more and more widely used too.

However, when the partition of the universe of discourse (UOD) is not compact enough, or when there are not enough control rules, the required control precision cannot be guaranteed. Recent researches on variable UOD presented a type of solution on this problem [1,2]. It did well in controlling nonlinear systems and it successfully controlled a quadruple inverted pendulum through software implementation for the first time in the world [7]. However, this methodology is too complicated for hardware implementation.

In this paper, a novel fuzzy controller based on variable UOD and its VLSI implementation are proposed. Its superiority to those conventional fuzzy controllers based on Mamdani model or Zero-order T-S model lies on its adaptive trait, that is, some parameters change with the input variables to improve the performance. Therefore it does not require many precise control rules.

2 Basic Ideas of the Novel Self-tuning Fuzzy Controller

2.1 Basic Structure of Self-tuning Fuzzy Controller Based on Variable UOD

A so-called variable UOD means that UOD of input and output variables can respectively change along with the changing of input variables. Thus, a smaller error leads to a smaller UOD, acting as if there are more control rules or more refined partition of the UOD so as to increase the control precision.

Therefore, a self-tuning fuzzy controller is constructed by applying a contraction factor before each input of a conventional FC, and a self-tuning gain factor after its output, while all the factors are adjusted with the input variables according to an adaptive law, which will be expatiated in the later section. Take a two-input and one-output fuzzy controller as example, the entire structure is shown in Fig. 1.

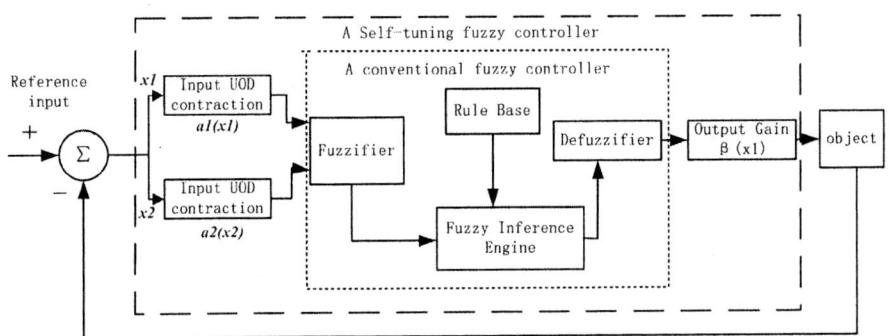

Fig. 1. Structure of self-tuning Fuzzy controller

For the conventional FC, a zero-order T-S model is adopted to complete fuzzy inference. Let $X_i=[-E_i, E_i]$ (i=1,2) be the UOD of input variable x_i (i=1,2) and $Y=[-U,U]$ be the UOD of output variable y. $\mathcal{A}_i=\{A_{ij}\}(1\leq j \leq m)$ is defined as a fuzzy partition on X_i, where A_{ij} is the membership functions. $\{y_j\}(1\leq j \leq m)$ are constants of the consequents. So that a group of fuzzy inference rules are formed as follows:

$$\text{If } x_1 \text{ is } A_{1j} \text{ and } x_2 \text{ is } A_{2j} \text{ then } y \text{ is } y_j, j=1,2\ldots m. \tag{1}$$

The variable UOD can be denoted by equation (2)

$$X_i(x_i) = [-\alpha_i(x_i)E_i, \alpha_i(x_i)E_i], \quad Y(y) = [-\beta(\underline{x})U, \beta(\underline{x})U]. \tag{2}$$

Where $\underline{x}=(x_1, x_2)$, $\alpha_i(x_i)$ is called contraction factor of the universe X_i, and $\beta(\underline{x})$ is the output gain factor that changes with the inputs.

For the conventional FC, input variables are fuzzified as fuzzy singletons, fuzzy AND that combines two antecedents in control rules is defined by production and fuzzy OR which combines all the rules is defined by sum. Use COA (Center of Area) method for defuzzification as equation (3) [6]:

$$y = \frac{\sum_{i=1}^{n} \mu(x_{1i}) \cdot \mu(x_{2i}) \cdot y_i}{\sum_{i=1}^{n} \mu(x_{1i}) \cdot \mu(x_{2i})} . \tag{3}$$

Take symmetrically triangular as the membership function of antecedents with overlapping degrees no more than two. So the denominator of equation (3) always equals to 1 [8]. This characteristic is utilized to omit the divider in hardware architecture. Then the variable UOD fuzzy controller based on the control rules (1) can be represented as formula (4)[1,2]:

$$y = \beta(\underline{x}) \sum_{j=1}^{m} A_{1j}\left(\frac{x_1}{\alpha_2(x_1)}\right) A_{2j}\left(\frac{x_2}{\alpha_2(x_2)}\right) y_j . \tag{4}$$

According to (4), it is easy to find that the effect of contraction of input UOD is to change the membership of input variables in the form of transferring x_i to $x_i/\alpha(x_i)$.

2.2 The Contraction Factors of Input UOD and the Output Gain Factor

A formula for the input contraction-expansion factor is suggested in [2] as equation (5), where λ is a constant nearing to 1.:

$$\alpha(x) = 1 - \lambda \exp(-kx^2), \quad \lambda > 0, \quad k > 0 \tag{5}$$

This complex expression consumes a lot of resource even in software implementation. Another simplified but effective form of factor is given as follows. Given x any increment △x, α should increase △α correspondingly, and △α should be proportional to △x; on the other hand, for the same amount of △x, when x is larger, △α should be larger too. Therefore, △α=k • △x • x, where k is the proportional constant. Thus dα/dx=k • x, so $\alpha(x) = (1/2)k \cdot x^2 + b$, where b is the initial value of α(x), viewed as a parameter. Set k'=(1/2)k, and denote it as k too, then we get: $\alpha(x) = kx^2 + b$, $k>0$, $b>0$.

There are some constraints about choosing the parameters of k and b as follows:

(1) $|x/\alpha(x)| \leq \delta E$, δ is around 1. Input variables after contraction should still in UOD.
(2) $0 < \alpha(x) \leq \alpha_0$, α_0 is around 1. α(x) should be less or not too larger than 1. α(x) is not restricted to no more than 1 because the original UOD is not always the best.
(3) α(x)=1 when x is relatively large because there is no need to contract the UOD where fuzzy rules work well.

Accordingly, α(x) is set as equation(6), where x_0 is the threshold of error that determines whether input UOD should be contracted.:

$$\alpha(x) = \begin{cases} kx^2 + b, & |x| < x_0, \\ 1, & |x| \geq x_0. \end{cases} \tag{6}$$

In equation (6), k determines the steepness of α(x), so k is relative to the width of the UOD. A narrow universe follows a steep α(x) and thus a large k, vise verse. b is the minimum of α(x). Both b and k determines the shape of α(x).

For the output gain factor, a formula is suggested in [1,2] as equation (7):

$$\beta(t) = \int_0^t k_I \underline{e}^T P_n d\tau + \beta(0). \qquad (7)$$

However, this complicated formula with integrator and multiplies consumes too much resource even in software implementation. Therefore, the law of PID control is used for reference to find a simplified form of β: when the system is stable, increasing the proportional coefficient is good for decreasing the stable error [4]. Thus set the output gain factor as equation (8):

$$\beta(x) = \begin{cases} 1/(kx^2 + b), & x < x_0, \\ 1, & x \geq x_0. \end{cases} \qquad (8)$$

Where x_0 is the threshold of input error, k and b are chosen in the way similar to parameters in $\alpha(x)$. Therefore, the output of the simplified self-tuning FC is:

$$y = \frac{1}{k_0 x_1^2 + b_0} U \sum_{j=1}^m A_{1j}\left(\frac{x_1}{k_1 x_1^2 + b_1}\right) A_{2j}\left(\frac{x_2}{k_2 x_2^2 + b_2}\right) y_j. \qquad (9)$$

3 Design and Matlab Simulation of Nonlinear System

The papers [3,2] discuss the control problem of the second order chaotic system-Duffing forced oscillation system:

$$\begin{aligned} \dot{x}_1 &= x_2, \\ \dot{x}_2 &= -0.1 x_2 - x_1^3 + 12 \cos t + u(t). \end{aligned} \qquad (10)$$

Let $\underline{x} = (x1, x2)^T = (x1, x1')^T$ be the state variable. The control objective is to force x1 to follow a given reference signal r(t), that is, the tracking error $\|e(t)\| = \|r(t) - x1(t)\|$ must be as small as possible. Without control, the state of the system will be chaotic. The trajectory of the system with $u(t) \equiv 0$ is shown in Fig. 2.

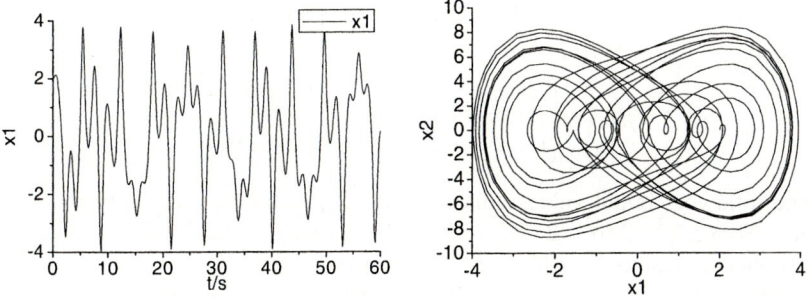

Fig. 2. Output state x1 curve (left) and phase plane (x1,x2) curve(right) with $u(t) \equiv 0$ while x(0,0)=(2,0), T=60s

Now use the simplified self-tuning fuzzy controller expatiated in section 2 to control this system. The UOD of e(t) is taken as [-2,2], ec(t) as [-1,1] and u(t) as

[-3,3]. Partition the UOD of e as Fig. 3 with the fuzzy subsets of NB, NS, ZO, PS, and PB. The fuzzy partition of UOD of ec(t) is the same as that of UOD of e, but in [-1,1]. The control rules are given in Table 1.

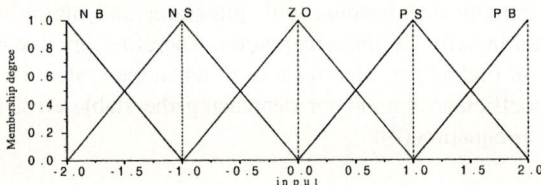

Fig. 3. The fuzzy partition of input variable and its memberships

Table 1. Control rules

	Control quantity	NB	NS	ec ZO	PS	PB
	NB	-3	-3	-3	-2	0
	NS	-2	-2	-2	0	1
e	ZO	-2	-1	0	1	2
	PS	-1	0	2	2	2
	PB	0	2	3	3	3

The fuzzy controller has two inputs e and ec, each is applied a contraction factor as equations (11)(12). The output gain factor is set as equation (13):

$$\alpha_1(e) = \begin{cases} 0.2e^2 + 0.15, & |e| < 0.25 \\ 1, & |e| \geq 0.25 \end{cases} \tag{11}$$

$$\alpha_2(ec) = \begin{cases} 0.01ec^2 + 0.5, & |ec| < 0.125 \\ 1, & |ec| \geq 0.125 \end{cases} \tag{12}$$

$$\beta(e) = \begin{cases} 1/(3.8e^2 + 0.05), & |e| < 0.25 \\ 1, & |e| \geq 0.25 \end{cases} \tag{13}$$

The following initial values almost cover all possible cases in a bounded phase plane: $\underline{x}(0)=(2,2)^T$, $(2,0)^T$, $(2,-2)^T$, $(0,2)^T$, $(0,0)^T$, $(0,-2)^T$, $(-2,2)^T$, $(-2,0)^T$ and $(-2,-2)^T$. The simulation results for $\underline{x}(0)=(2,0)^T$ are shown in Fig.4.

A lot of simulations are done for these initial values and all get quite good results especially compared to the corresponding results in [3] with the setting time about 0.5s much shorter than 60s in [3]. Compared to [2], the same results are achieved but only 25 control rules are used in this paper instead of 49 rules used in [2]. Moreover, this simplified fuzzy controller uses much less resource than the one in [2].

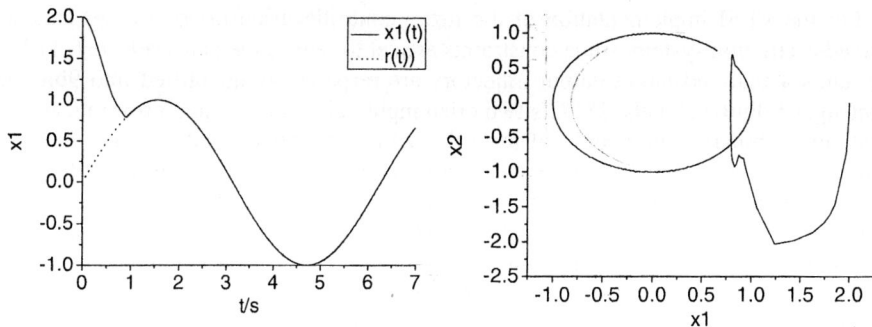

Fig. 4. Output x1 curve (*left*) and (x1,x2) phase plane curve (*right*) for r(t)=sin(t), x(0)=(2,0), T=7s

This type of self-tuning FC is also applied to some other nonlinear systems and gives quite good results too.

4 VLSI Implementation of the Novel Fuzzy Controller

4.1 The Total Chip Architecture

The chip architecture is shown as Fig.5.

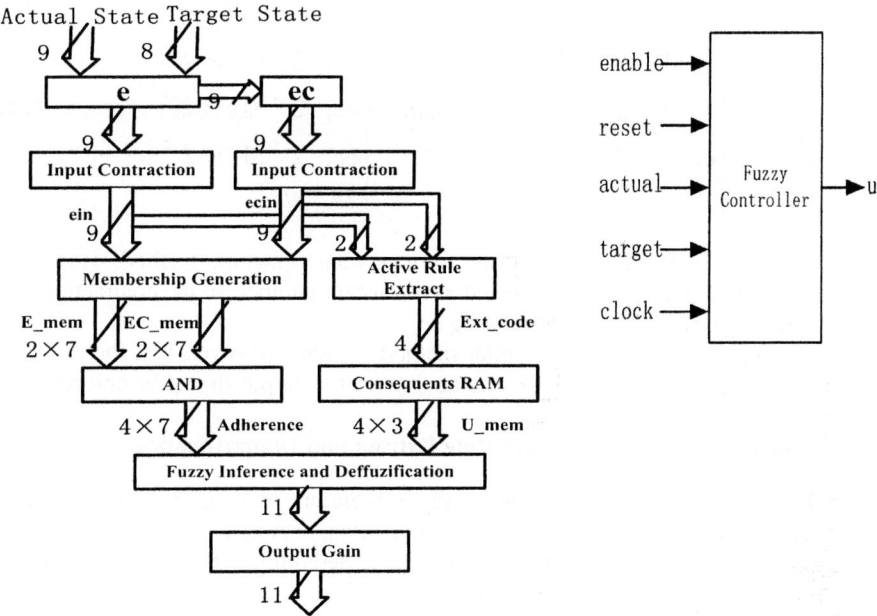

Fig. 5. VLSI architecture of self-tuning fuzzy controller (*left*) and input/output ports (*right*)

For the VLSI implementation of the fuzzy controller used in controlling Duffing forced oscillation system, some constraints are set to reduce the hardware cost. 1) The system's actual and target output trajectory are respectively quantified into 9bit and 8bit signed digital signals. 2) FC's two crisp input values error e and error rate ec, are 9bit signed binary data calculated from the actual and desired output state. 3) Error ein and error ecin derived from e and ec after contraction are still 9bit signed signal. 4) Memberships of ein and ecin are quantified to 7bits unsigned data. 5) All constant values of consequents are of 3bit width. 6) The normal FC's crisp output variable and that after multiplying the output gain factor are 11bit signed binary number.

The "enable" signal is used to sample input actual and target signals from the controlled system. Major features of this architecture lie in units of input contraction, membership generation, active rule extract, defuzzification and output gain.

4.2 The Implementation of Input Contraction Factor and Output Gain Factor

From equation(6), the input variable after contraction is $x/a(x)=x/(ax^2+b)$. It seems that multipliers and divider are required which will consume a lot of resource. In order to simplify these operations for VLSI implementation, we set $1/a(x)$ as some constant numbers that depends on the range of x. Take input error e as example. According to the characteristic curve of $1/a(e)$, set some constants shown as the first part of equation(14) to substitute equation(11). And the input variable after contraction is substituted by ein, shown as the second part of equation (14). In this way, multipliers and dividers are omitted. Only some shift and sum operations are left which greatly reduce the resource for hardware implementation.

$$\frac{1}{\alpha(e)} = \begin{cases} 1, & |e| \geq 0.25 \\ 2, & 2^{-3} \leq |e| < 2^{-2} \\ 4, & 2^{-4} \leq |e| < 2^{-3} \\ 6, & 2^{-5} \leq |e| < 2^{-4} \\ 7, & else \end{cases} \Rightarrow ein = \begin{cases} e, & |e| \geq 0.25 \\ 2e, & 2^{-3} \leq |e| < 2^{-2} \\ 4e, & 2^{-4} \leq |e| < 2^{-3} \\ 6e, & 2^{-5} \leq |e| < 2^{-4} \\ 7e, & else \end{cases} \quad (14)$$

For the input error rate ec, use the similar transformation method. However, $1/a(ec)$ is so smooth that it can be viewed as a constant number, so the contraction can be integrated into the process of evaluating ec.

For the output gain factor, the similar method is used. However, it is the range of input error e that determines whether and how much the output should be enforced.

4.3 Membership Functions, Active Rule Extract and Defuzzification

For triangular memberships as shown in Fig.3, there are two memberships for each input variable. Accordingly, there are four active rules. In Active rule extract unit, two most significant digits of both e and ec are used to determine four active rules and to select relative consequents. The rests of e and ec are used to calculate the memberships of them in Membership functions unit. To expatiate these operations, see Table2. For the variable e, its whole universe [-2,2] is mapped into 9bit signed binary, -255~255, where negative numbers are of complement representation [8].

Table 2. Coding method of e

e	-2<e<-1	-1≤e<0	0≤e<1	1≤e<2
binary	*1*0000000*1*~ *1*0*1*111111	*11*0000000~ *111*111111	*00*0000000~ *00*1111111	*01*0000000~ *01*1111111

And for ec use the same transformation method as e. Thus two most significant digits in each of the four sets are respectively 10, 11, 00 and 01. Accordingly four active rules can be extracted. The rest of e and ec are always from 00 to 7F. Their memberships are denoted as e_mem0, e_mem1, ec_mem0 and ec_mem1, which can be generated through equations (15)(16) for triangular memberships:

$$e\_mem0=e(6:0)\ ,\ e\_mem1=7'1111111-e(6:0)\ . \qquad (15)$$

$$ec\_mem0=ec(6:0),\ ec\_mem1=7'1111111-ec(6:0)\ . \qquad (16)$$

Finally, defuzzification unit actually includes two main operations, implication and aggregation, which are in fact multiplication and sum operations.

4.4 Simulink and Active-HDL Co-simulation Results

The co-simulation results of the VLSI implementation of the self-tuning fuzzy controller for initial state $\underline{x}(0)=(2,0)^T$ are shown in Fig.6.

Compared with the results in Fig.4, Fig.6 shows that the VLSI implementation successfully fulfilled the functions of the self-tuning fuzzy controller designed in section 3.

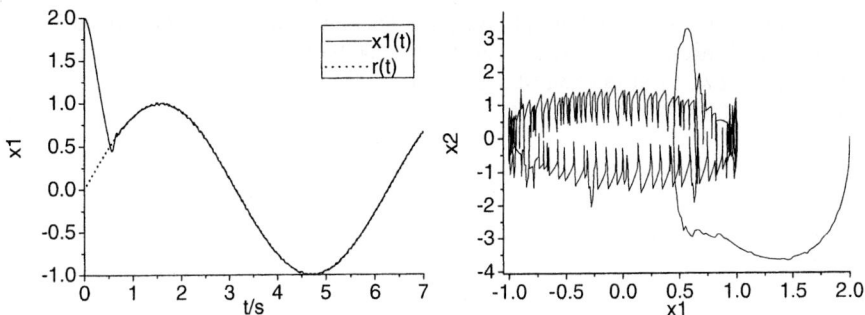

Fig. 6. Output state curve (*left*) and phase plane curve (*right*) for r(t)=sin(t) , x(0)=(2,0)

The VLSI implementation is a discrete system dealing with digital signals, whose attainable precision is determined by the width of signals. Therefore, the tracking precision can be increased by increasing the width of signals.

5 Conclusions

The simplified self-tuning fuzzy controller proposed in this paper performs well in controlling a nonlinear system with less control rules and much less resource compared with [2]. The algorithm of this self-tuning fuzzy controller is suitable for hardware implementation. And its VLSI implementation works successfully in controlling a nonlinear system to track a reference trajectory. The main features of this VLSI FC are: 1) The adaptive factors are simplified so that there are only shift and sum operations instead of multiplication and division. 2) Membership functions of antecedents are calculated instead of being stored in RAM or ROM. 3) Knowledge Base RAM only stores consequents, and the corresponding rules are selected through Active Rule Extract module. The validity of the developed VLSI FLC is verified by co-simulation of Matlab-ActiveHDL environment.

References

1. H.-X. Li.: Adaptive fuzzy controller based on variable universe. Science in China, Ser. E, vol. 42 (1). (1999) 10–20
2. H.-X. Li.: Variable universe stable adaptive fuzzy control of a nonlinear system. Science in China, Ser. E, vol.32 (2) (2002) 211–223
3. L.-X. Wang.: Stable adaptive fuzzy control of nonlinear systems. IEEE Transactions on Fuzzy Systems, vol. 1(2) (1993) 146–155
4. Yonghua Tao et al..: New PID Control and Applications. Mechanical Industry Publication press (1998) 1-135
5. Chuen Chien Lee.: Fuzzy logic in control systems: Fuzzy controller. IEEE Transactions on Systems, Man, and cybernetics, vol.20 (1990) 404-416
6. L.-X. Wang.: A Course in Fuzzy Systems and Control. Tsinghua University Press. (2003) 146-155
7. H.-X. Li et al..: Variable Universe Adaptive Fuzzy Control on the Quadruple Inverted Pendulum. Science in China, Ser. E, vol. 32 (1). (2002) 65–75
8. Jin Weiwei et al. VLSI Design and Implementation of a Fuzzy Logic Controller for Engine Idle Speed. ICSICT2004. 2067-2070

Method to Balance the Communication Among Multi-agents in Real Time Traffic Synchronization

Li Weigang[1], Marcos Vinícius Pinheiro Dib[2], and Alba Cristina Magalhães de Melo[2]

[1] Department of Computer Science University of Brasilia, Brasilia-DF, CEP: 70910-900, Brazil
{Weigang, Albamm}@cic.unb.br
[2] Politec Informatic Ldta., Brasilia-DF, Brazil,
dib@bsb.politec.com.br

Abstract. A method to balance the communication among Multi-Agents in real time traffic synchronization is proposed in this research. The paper presents Air Traffic Flow Management (ATFM) problem and its synchronization property. For such a complex problem, combing grid computing with multi-agent coordination techniques to improve ATFM computational efficiency is the main objective of actual research. To demonstrate the developed model - ATFM in Grid Computing (ATFMGC), the grid architecture, the basic components and the relationship among them are described. At the same time, the function of agents (tactical planning agent etc.), their knowledge representation and inference processes are also discussed. As criteria to measure the effective to reduce quantity of the communication among agents and the delay of the flights, Standard of Balancing among Agents (SBA) is used in the analysis. The simulation shows the efficiency of the developed model and successful application in the case study.

1 Introduction

Air Traffic Flow Management (ATFM) is a kind of Real Time Traffic Synchronization (RTTS) problem [1,2]. Due to large-scale, safety and synchronization characteristics in most of the models and systems for ATFM, computing efficiency is a common critical problem. Since 1970s, scientists from Artificial Intelligence, Operation Research and Air Transportation have worked together to develop more efficient Air Traffic Control (ATC) and Air Traffic Management (ATM) systems, but the computing-based solution still needs to be further enhanced to reduce aircraft delay.

Some Knowledge-Based Systems (KBS) have been developed in ATC/ATM, such as 4D-Planner that is a ground based planning system using a rule-based system for arrival sequencing and scheduling in ATC [3,4]. Gosling [5] pointed out the potential of Artificial Intelligence (AI) application in ATC. A real time knowledge based support for ATM has been developed by IBM Switzerland [6]. In Brazil, an expert system for ATFM has also been investigated to make timetable schedule and centralized flow control [7].

A distributed ATM system has been studied in Australia [9]. The advantages of that approach are inherent distribution, autonomy, communication and reliability. Prevôt from NASA Ames Research Center has studied a distributed approach for operator interfaces and intelligent flight guidance, management and decision support [10]. An application of multi-agent coordination techniques in ATM, which sets up a methodological framework using multi-agent coordination techniques that supports the collaborative work in ATM has also been presented recently by Eurocontrol [11]. It should be mentioned that, the multi-agent coordination technique is a useful methodological framework, however the research in [11] is limited to a software shell. Due to the great quantity of traffic as ATFM, this implementation may be difficult when brought into practical fields.

Recently, grid computing represents a perspective to get the solution for the large-scale computation task such as ATFM. Computational grid has been defined as "coordinated resource sharing and problem solving in dynamic, multi-institutional virtual organizations" [13-17,21]. One of the reasons to choice the Grid computation may be that the huge planning and schedules are in real time. The platform of Grid, however, is efficient half of integration of data and systems legacies [17], allowing the necessary synchronism in real time. They consist of hardware and software infrastructure, which provides dependable, consistent, pervasive, and inexpensive access to high-end computational capabilities. An agent-based resource management system for grid computing shows the importance of the research in both grid computing and AI [12].

Based on the mentioned researches, especially on the papers of [7-11], in this study a multi-agent system in grid computing for real time traffic synchronization is proposed. The main contribution of the paper is to develop an approach of cooperation and negotiation manner to balance the communication among agents using grid computing in RTTS [8]. The paper describes the synchronization concepts in ATFM, the main structure of the system and the relationship among the components of the model of ATFM Grid Computing - ATFMGC. At the same time, the agents functions, their knowledge representation and inference processes are also discussed. ATFMGC was implemented in a network with four computers, which represent four main Brazilian airports. As criteria to measure the effective to reduce quantity of the communication among agents and the delay of the flights, Standard of Balancing among Agents (SBA) is defined in the analyses. As a preliminary study, the paper shows that the investigation of using multi-agent system and grid computing to RTTS is not only a simple application, but it is an important study topic in Artificial Intelligence, grid computing and Air Traffic Control.

The following sections are organized in five parts: soon after this introduction, section 2 presents the basic concept and characteristics about ATFM. In the third section, the definition of the concept of SBA is described. Fourth section presents the implementation of ATFMGC, agents and inference processing. The case study is illustrated in fifth section. And finally, sixth section shows the conclusions.

2 ATFM: A Real Time Traffic Synchronization Problem

ATFM is developed to ensure an optimum flow of air traffic to or through areas within which traffic demands exceeds the available capacity of the ATC system [8] at certain times. ATFM includes four main functions: Strategic Planning (involving long term: days to years), Pre-Tactical and Tactical Planning (involving middle term: from

1 hour to days), Short Term Planning (involving short term: from 20 minutes to 1 hour), and Monitoring and Control (On-Line operation). As mentioned by [1], the ATFM system shows the following two special properties:

- Real time traffic synchronization is an activity, which consists in implementing corrective actions on traffic applicable until the traffic is actually received by controllers to protect.
- Dynamic collaborative decision-making: aimed at achieving prompt dynamic "agreements" between Traffic Managers co-involved in the implementation of corrective actions on traffic transiting from one sector to the other.

In this study, the scope is limited to the ATFM tactical planning, which concerns the planning on the day, i.e. from 1 to 24 hours. This operation analyses the available capacity of the departure, en-route, and arrival airspace, and takes tactical measures in case the predicted demand exceeds the available capacity. The main functions are considered as following [7,9]:

- Creating a schedule for all departing and arriving flights to the airport, while maximizing the utilization of the runway and the terminal;
- Identifying congestion areas with regards of the airspace, aircraft and terminal constraints;
- Negotiating with other agents on the expected traffic flow to and from the airport;
- Re-scheduling of the flights according to the outcome negotiation and communicating the new schedule to other affected agents.

The distributed ATFM task involving multiple airports calls for independent agents at these airports. At each airport, there is an *ATC* agent who is responsible for managing the flow over in and out of the local airport. *Tactical Planning* (TP) agent manages the traffic flow over the entire airspace. *Pre-ATC* agent is used for the simulation of *ATC* agent to serve to the *Tactical Planning* purpose.

3 Definition of the Concept of SBA

The proposed ATFMGC architecture consists of five components: Interface, Open Grid Services Architecture (OGSA), Web Services, Knowledge-Based System (KBS) and Database distributed in a set of airports. The basic components in the ATFMGC architecture and the relationship among the components are shown in [8].

As criteria to measure the effective to reduce quantity of the communication among agents and the delay of the flights, Standard of Balancing among Agents (SBA) is defined as a basic index [9]. It uses a different delay cost function for the landing and taking off of flights. Considering that one of the objectives of ATFM is to minimize the unnecessary time of flight in air, in this research, the function of weight of the delay was simplified for the multiplication of each minute of delay for a weight. For landing flight, the weight is chosen as 5, and take-off the weight is as 1. When an airport is congested, some flights are delayed and the extension of the delay time is proportional to the severity of the congestion. But the flights do not have the same importance and the delay of a flight can have a bigger cost than the delay of others. Thus, according to the time of delay with the weight, factor of importance, and the behind flight has it indication of the load of the airport. They calculate SBAs distinct for the landings and takes-off of the flights, such as:

SBA $(t_1, t_2) = \sum weight\_of\_delay(f_1) \times minutes\_of\_delay/10$

Where: SBA (t_1, t_2): Standard of Balancing among Agents between the instants t_1 and t_2; n: number of flights between the instants t_1 and t_2; f_i: the flight in analysis.
The arrangement of the schedule and the calculation of SBA are carried through in each airport. The schedule is created for one determined interval of time TS (Time Slice), taking in consideration the exits previously set appointments and the esteem time of the entrance flights. The TS is changeable and can be reduced when the load in the airport is high, or be increased when the load if it finds low.

4 Inference Process and Implementation

In this study, *Pre-ATC* agent and *Tactical Planning* (*TP*) agent are developed in every airport of ATFMGC. The former is an expert system for local air traffic control, which is designed just to simulate the situation for ATFM purpose. The later is a distributed KBS for Tactical Planning of ATFM. *ATC* agent is also mentioned in this research but is not described in detail. Figure 1 shows the manner of communication and negotiation among agents in ATFMGC system.

4.1 Inference Process

Knowledge Based System of agents of ATFMGC is represented according to ATC/ATM rules, which have been defined by the Brazilian Department of Civil Aviation - DAC [7, 20]. Six processes have been developed for the knowledge representation of *Pre-ATC* agent: flight data, scheduling and control condition, separation standards, holding assignment, departure and arrival conflict prevention and take-off delay assignment [7,8].

The main inference process Pre-ATC agent is the same as [7]. The following steps are inference processes of *TP* agent:

1. **Web Services** component of *TP* agent at airport A is used to communicate with *Pre-ATC* agent at airport B (and others) to get the departure flight delay requirements at airport A.
2. **Schedule** process of *TP* agent at airport A reschedules the departure flights according to the accepted WCTD value at A.
3. **Diagnosis** process checks the new schedule and stores it in the database at airport A if there is no conflict. When any conflict takes place, the **Schedule** process repeats the work again.
4. **Broadcasts and Negotiation** generates a message to Web Services to send to the *TP* agents at related airports.
5. At the same time, ATFMGC at related airport is also working with the changed schedule.
6. If the system detects the conflicts at a related airport, a new schedule is generated at other airports and the information is sent back to airport A. **Broadcasts and Negotiation** process negotiates with the related airport.

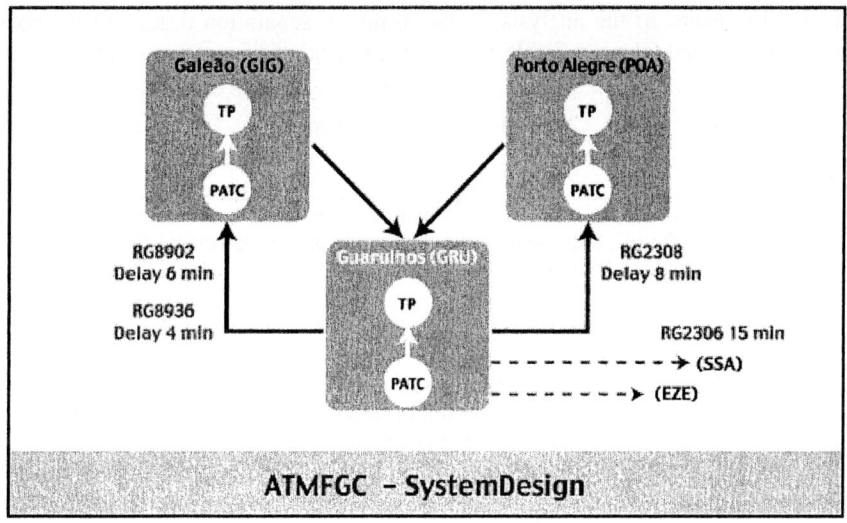

Fig. 1. Communication and Negotiation among agents in ATFMGC system

7. **Evaluation and Validation** processes verify the actual schedule within a certain time period.
8. The *PT* agent at airport A is invoked recursively until there are no more conflicts.

4.2 Implementation

At this moment, a prototype of ATFMGC has been developed using Globus ToolKit (version Alpha 3) on a grid computing. The prototype consists of three personal computers (Pentium 4). Each computer with ATFMGC interface represents the ATFM operation system of an airport. The whole system was codified with JAVA language. The KBS, which consists of the rules and facts was developed using XML language. XML documents use parser JDOM library to for the interpretation of the stored data.

Web Services and client interface also have been integrated with ATFMGC in JAVA. Apache TomCat 4.0 is used as Servlet Container. For the installation and manipulation of the Globus ToolKit, Cygwin is chosen as emulator.

5 ATFM Case Study

Based on the original data from the flight schedule of 19 of July of 2004 [19], we analyze airport capacity and the air traffic flow in the main Brazilian airports (Guarulhos-GRU, Congonhas-CGH, Brasília-BSB, Galeão-GIG).

5.1 Air Traffic Flow in the Four Main Airports

For the main four Brazilian airports, traffic flow and airport capacity are different. Through this analysis, we show the real traffic situation at GRU airport. And we demonstrate the capacity of the ATFMGC to rationalize and creation of the flight schedule, by means of the analysis of the minimum separation times and the operations of landing and take-off. Table 1 shows the legend of of some concepts.

Table 1. Legend of some concepts

Letter	Meaning
SAD	Request of alteration of take-off flight - message sent for the responsible airport for the take-off, requesting to the destination airport agreement for alteration of the schedule of landing of one determined flight.
PAP	Order of landing alteration - asked for of alteration of landing received for the destination airport.
SAP	Request of Alteration of Landing - message sent for the responsible airport for the landing, requesting to the origin airport agreement for alteration of the schedule of take-off of one determined flight.
PAD	Order of Alteration of Take-off flight - asked for of alteration of take-off received for the origin airport.
RPA	Reply the Order of Alteration - message sent for the requested airport agreeing accepting a PAP or PAD
HE	Hour of the Event
HS	Requested hour for the modification
ST	Status
A/N	Accepted / Not accepted

The capacity of GRU airport is 54 flights per hour (landing or take-off). If using 5 minutes as a Time Slice (TS), it means that a capacity is 5 flights in every Time Slice. The real schedule at this airport is 27 flights per hour but the flights are not well distributed per TS. At some TS, more than 5 flights were scheduled. This is the reason of the delay. From the analysis of the flight schedule of GRU airport, we observed:

1). The actual schedule are with a total of 6 hours of delays, in which 55.9% of them are the takes-off flights with the average delays of 2.1 minutes per flight and 31.52% of them are the landing flights with 1.5 minutes per flight.
2). There is problem of actual schedule at 20h55, with 4 landing flights and peak of delay 21h15 for in the takes-off flights. For any meteorological problem, from 20h50 to 21h10, heavy congestions may take place.

5.2 Simulation

For every TP agent, its first task is to get a solution for its conflicts and pass the problems to other airports within the Grid. Using criterion of SBA to measure delay situation in every airport and generate a standard balancing, TP agents can communicate each other to reduce the delay.

For each airport, the SBA is calculated locally, at each TS. It will indicate traffic and the congestion. With the large value of SBA, the congestion situation is going to be more complicated. Based on its traffic capacity of each airport, there is a local acceptable SBA. If the real SBA is more than the local acceptable SBA, the system needs to communicate among the airport to negotiate a suitable solution. It is necessary to establish an acceptable SBA (standard), to unify the interpretation of the ATFMGC. Five accepted SBA cases are considered in the simulation: 0, 0.75, 1, 1.25 and 2.5. Table 2 is the result of the simulation.

Table 2. Messages to airport accord with the accepted SBA

SBA	Airport	SAD	SAP	PAP	PAD	RPA	Sub Total	Total
0	GRU	28	1	17	2	19	67	421
	CGH	133	53	9	7	16	218	
	BSB	64	23	6	4	10	107	
	GIG	9	2	8	1	9	29	
0.75	GRU	11	1	1	1	2	16	98
	CGH	21	5	6	1	7	40	
	BSB	18	9	1	1	2	31	
	GIG	7	2	0	1	1	11	
1	GRU	0	1	1	0	1	3	33
	CGH	0	0	6	1	7	14	
	BSB	5	1	1	1	2	10	
	GIG	3	1	0	1	1	6	
1.25	GRU	0	1	0	0	0	1	21
	CGH	0	0	4	1	5	10	
	BSB	3	1	1	0	1	6	
	GIG	2	0	0	1	1	4	
2.5	GRU	0	0	0	0	0	0	17
	CGH	4	1	2	0	2	9	
	BSB	3	2	0	0	0	5	
	GIG	2	1	0	0	0	3	

1) Accepted SBA is 0. In this case, the system does not allow any delay (even it is impossible in real case). For any delay, the system needs to communicate and negotiate among four airports. 421 communications take place totally within system. Just at CGH, the number of negotiation arrived 218 and with 8 minutes to eliminate any delay in this airport (tables, 2 and 3). It requested 133 messages to eliminate the delays of take-off flights, and 53 messages for landing flights. System processed 9 messages for asking to re-arrangement landing flights and 7 for take-off flights. 16 messages ware the answers to these requirements.
2) Accepted SBA is 1. In this case, the system allows a reasonable delay. For any delay, the system needs to communicate and negotiate among four airports. 33 communications take place totally within system. At CGH, the negotiation times was 14 and with one and half minutes to eliminate any delay in this airport (table 3).
3) Accepted SBA is 2.5. The system allows delay. Just 17 communications take place within the system. At CGH, there were 9 negotiations and with 20 seconds to eliminate any delay in this airport (see table 3).

Table 3. Run time of ATFMGC

Accepted SBA	GRU	CGH	BSB	GIG
0	6:53.408	8:07.354	8:24.220	7:27.628
0.75	0:47.715	1:37.587	0:53.843	0:13.329
1	0:11.896	1:21.377	0:45.431	0:09.794
1.25	0	0:48.478	0:27.106	0:06.800
2.5	0	0:19.453	0:04.366	0:00.360

5.3 Proposal for the Management of Traffic

The ATFMGC analyzes the conflict resolution with 2 hours of antecedence. When there is a conflict between 2 landing flights, one or two of them are still not take off yet. The system will arrange a delay to one of the takes-off flight, which is the same time to maintain this flight at original airport. When this aircraft takes off, its arrival will be synchronized with its window of landing. As this is not in real time, it is assumed that the travel time is always constant.

As the landing flights have absolute priority over the takes-off, these tend to be delayed. As the take-off is at the original airport, with a destination airport, the ATFMGC will negotiate a new schedule of landing at the destination airport. The agents of the ATFMGC will synchronize all the flights of all airports within the Grid. The proposed method considered the synchronization of the flights, as well as the analysis of the behavior of the agents of Tactical Planning and Pre-ATC. The benefit with the process of synchronization of the flights is to reduce flights in air, consequently, the reduction of the possibility of congestions.

Table 4 compares the flight delay from two methods. For the conventional method, the conflicts of the flights are with a delay in air for a total of 20 minutes. This means of an average of 1.7 minutes per flight within one hour. Using proposed method, with the synchronization of the schedules for landing and takes-off flights, there is no more delay in space. Considering the factor of security, the economy of 20 minutes air delay with a hour is significantly.

Table 4. Flight delay from two methods (minute)

Flight	Original	Destination	Delay in air	
			Actual method	Proposed method
JJ3540	CGH	BSB	1	0
				0
G31834	GIG	BSB	0	
RG8920	GIG	GRU	2	0
VP4261	GIG	GRU	2	0
RG2329	BSB	GRU	1	0
G31712	GIG	CGH	5	0
RG2635	BSB	CGH	7	0
VP4281	BSB	GIG	2	0
Total			20	0

6 Conclusions

Multi-agent system using grid computing to the real time air traffic synchronization problem was proposed in this research. In the negotiation and synchronization of the flight schedule among the airports within the Grid, all the schedules converge for a situation of inexistence of conflicts. In the phase of strategically planning, the ATFMGC can assist the controlling of flow of air traffic in the planning and the scheduling of flights, allowing the identification and smoothing the overload and the conflicts. Its structural components allow to the verification and synchronization, in real time, of the schedules of takes-off and landing flights, in all the airports throughout its routes, observing the capacity of airports. Standard of Balancing among Agents – SBA is defined to balance the communication among the airport, but it may use well for same kinds of the problem in multi-agent system.

ATFM is an interesting domain for the application of multi-agent system. Any study about multi-agent system in real time air traffic synchronization problem using grid computing hasn't ever been reported until now. Bearing in mind the advantages of grid computing, this proposal also presents a solution for air transportation. The result obtained in this work may represent a significant contribution to the research of artificial intelligence, grid computing and air traffic transportation [12,18]. For further study, the following aspects may be taken into deep consideration:

- Using multi-agent coordination techniques to orient the formation and implantation of the grid computing platform;
- Constructing an special grid computing for ATFM, to study the main components (both hardware and software), and relationship among computing and air traffic control;

References

1. EUROCONTROL: Future ATFM Measures (FAM) operational Concept, EEC Note No. 13/02, (2002).
2. Stoltz , S., Ky, P.: Reducing Traffic Bunching More Flexible Air Traffic Flow Managemen, 4th USA/Europe ATM R&D Seminar, New_Mexico, (2001).
3. Dippe, D.: 4D-Planner – A Ground Based Planning System for Time Accurate Approach Guidence, DLR- Mitt, (1989), 89-23.
4. Vôlckers, U.: Approach Towards a Future Integrated Airport Surface Traffic Management, DLR- Mitt, (1989), 89-23.
5. Gosling, G. D.: Application of Artificial Intelligence Application in Air Traffic Control, Trans. Res., 21A(1), (1987).
6. Schlatter, U. R.: Real Time Knowledge Based Support for Air Traffic Flow Management, IEEE Expert, (1994), 21-24.
7. Weigang, L., Alves, C. J. P., and Omar, N.: An expert system for Air Traffic Flow Managemen. J. of Advanced Transportation", Vol. 31, No. 3, (1997), 343-361.
8. Weigang L., Dib, M. V. P., Cardoso, D. A.: Grig service agents for real time traffic synchronization, in the proc. of IEEE International Conference on Web Intelligence, pp. 619-623, Beijing, (2004).

9. Tidhar, G., Rao, A., Ljunberg, M.: Distributed Air Traffic Management System, Technical note, No. 2, (1992).
10. Prevôt, T.: Exploring the Many Perspectives of Distributed Air Traffic Management: The Multi Aircraft Control System MACS, S. Chatty, J. Hansman, G. Boy (Eds.), in the Proc. of the International Conference on Human-Computer Interaction in Aeronautics (HCI-Aero), AAAI Press, Menlo Park, CA, (2002), 149-154.
11. Nguyen-Duc, M., Briot, J.-P., Drogoul, A., Duong, V.: An application of Multi-Agent Coordination Techniques in Air Traffic Management, in the Proceedings of the IEEE/WIC International Conference on Intelligent Agent Technology, (2003).
12. Cao, J., Jarvis, S. A., Saini, S., Kerbyson, D. J. and Nudd G. R.: ARMS: an Agent-based Resource Management System for Grid Computing, Scientific Programming, vol. 10, (2002), 135-148.
13. Berman, F., Gox, G., Hey, T.: Grid Computing: Making The Global Infrastructure a Reality, John Wiley & Sons, (2003).
14. Ferreira, L., Berstis, V., Armstrong, J., Kendzierski, M., Neukoetter, A., Takagi, M., Bing-Wo, R., Amir, A., Murakawa, R., Hernandez, O., Magowan, J., Bieberstein, N.: Introduction to Grid Computing with Globus, IBM, 2003.
15. Portella, G. J., Melo, A. C. M. A.: A Load Balancing Strategy to Schedule Independent Tasks in a Grid Environment, in the Proc. of Euro-Par04, Italy, (2004).
16. Sotomayor, B.: The Globus Toolkit 3 Programmer's Tutorial, (2003).
17. Foster, I.: What is the Grid? A Three Point Checklist, (2002).
18. Weiss, G. (Ed.): Multiagent systems, The MIT Press, (2000).
19. Panrotas Editora Ltda: Guia de Horário de Nacionais e Internacionais, No. 370, (2004).
20. DEPV, IMA 100-12, portaria DEPV No. 46 de 30/06/99, Departamento de Previsão de Vôo – DEPV, Ministério de Aeronáutica, (1999).
21. Nabrzyski, J., Schopf, J. M., Weglarz, J. (Eds.): Grid resource management - State of the Art and Future Trends, Kluwer Academic Publishers, (2003).

A Celerity Association Rules Method Based on Data Sort Search

Zhiwei Huang and Qin Liao

Guangzhou South China University of Technology
jerray@126.com

Abstract. Discovering frequent item sets is a key problem in data mining association rules. In this paper, there is a celerity association rules method based on data sort search. Using the plenitude and call terms of frequent item sets, the method efficiency can be improved greatly for the searching time won't increase as the number of item set of the data does, moreover the data can be found by searching the database within 3 times. Using the change between the frequent item sets and standby item sets, the data celerity renew and the min-sup renew can be true.

1 Introduction

Association Rule was mainly applied to the database of transactions to describe how the appearance of object A affects the appearance of object B through quantified number, which was put forward by R.Agrawal in 1993 for the first time [2, 3]. There are so many kinds of arithmetic in working out the Association Rule so far, of which the Apriori arithmetic was the most representative. Other arithmetic such as sample arithmetic, DIC arithmetic [1], FAHA arithmetic [4] and UA arithmetic [5] are improved on the basis of Apriori arithmetic. All the arithmetic have something in common that they adopt the breadth-first search which tends to cause the increase of search times and have to recalculate the newly added data as well as changed min-sup so as not to renew at its own base. Therefore proceeding with the decrease of search times, this paper offers renewable celerity association arithmetic based on the data-sets sort.

2 The Features and Disadvantages of Classic Arithmetic

Digging out the Association Rule is to find out all the frequent item sets and meet the min conf. Apriori arithmetic do utilize set boundary of frequent item sets, adopt breadth-first search, and find out all the frequent item sets[1]. Although Apriori arithmetic could offer the approaches to search for all the frequent item sets, its arithmetic efficiency was restricted by itself and caused such problems as: Firstly, it will bring a segment of inexistent item sets, namely, fake item sets. Secondly, it will search the database many times. Thirdly, couldn't update for the new data or the new min conf. Aiming to these questions, the Apriori arithmetic has been greatly improved, but it still has the deficiency of bringing invalid standup items and frequently searches the database together with failing to provide more effective approaches to renew rules.

3 Features About KFH Arithmetic

The feature the Association Rules based on data sort search referred in this paper are as follows: Firstly, as for any item sets X, it belongs to the Kth item sets, when the Length(X) is K. and we name it as Trade (k) for the database D. Secondly, as for the item sets X, if the transaction T is existing and $X \subset T$, when T is frequent sets, X is also too. It is called the plenitude of frequent sets. Thirdly, because of its set boundary, when the item of the frequent sets is one subset, it is also a frequent set, named the prerequisite of the frequent sets.

Supposing R is the set of all the one-item frequent sets. As the elementary solution, with the plenitude, all the frequent sets are the subset of R, Define frequent sets as KF and standby item sets as KH. According to items number, the sort search name the KF (k) as the Kth frequent sets and KH (k) the Kth standby item sets.

The Association Rules based on the celerity search is called KFH arithmetic.

4 The Flow and Realize of the Association Rules

The process is: Firstly, as for database D, input the min-sup and calculate one-item frequent sets and obtain the element solution R and KF(1),KH(1);Secondly, in the trade(K),as for the transaction T, if count(T)>=min sup, then subset of T can be put into the KF(i), and renew the support in the KF, otherwise put T into the KH(K),and the same for its subset. The cycle will end until the whole database has been searched.

According to the plenitude resulting from the frequent sets, KFH arithmetic can obtain all the KH and KF through two visits to the database. But since it needs to break down every transaction T, it increases the calculating amounts of the arithmetic to some extent. We can further improve the KFH arithmetic with the adoption of searching from up to down. So in Trade (K), if T or its subset belongs to KF, then we needn't to break it down. The bettered KFH arithmetic increase the times of visit to database by one time, but it decrease the process of breaking it down greatly.

5 Renew and Realize the Data of the Association Rules

The KFH arithmetic not only reduces the search times, but also solves the problem of renewing min sup and the added data.

Suppose sets $FH(R) = KF \cup KH = \{X \mid \exists T, X \subset T, X \subset R\}$, FH(R) is element solution R's set of all the subset in the database, set frequent item sets and standby item sets as $KF = \{X \mid \sup port(X) >= \min \sup \quad X \in FH\}$

$KH = \{X \mid \sup port(X) < \min \sup \quad X \in FH\}$

As for the change of min sup or the added data, the renewing process of KFH arithmetic is: on one way when min sup has changed, suppose element solution R1 set $PH(R1 \times R2) = \{X \mid \exists T, X \subset T, X \cap R2 \neq \phi, X \subset (R1 \cup R2)\}$

$$FH(R1) = \begin{cases} FH(R1) = \{X \mid X \in FH(R), X \subset R1\} & \text{if } R1 \subseteq R \\ FH(R1) = \{X \mid X \in (FH(R) + PH(R \times R2)), X \subset R1\} & \text{if } R1 = R \cup R2, R2 \cap R = \phi \\ FH(R1) = \{X \mid X \in (FH(R2) + PH(R2 \times R3)), \text{if } R1 = R2 \cup R3, R2 \subset R, R3 \cap R = \phi \end{cases}$$

On the other way when the data has been added, if the set of the newly added item sets is $D' = \{x_1', x_2'..x_n'\}$, and the new element solution is R1, then new $FH_{D+D'} = FH_D(R1) + FH_{D'}(R1)$.

6 Comparison Between the Apriori and KFH Arithmetic

In general there are two ways to improve the efficiency of the arithmetic: one is reducing the times of searching the database, which could greatly improve the efficiency [1]. The KFH arithmetic is better than the tradition arithmetic in the search and update rules.

Set the database D as Table 1 and the 1, 2, 3, 4 mean the different items. The appearing counts of the transaction is C (T), the min sup is 2.

Table 1. The Database of Transaction

Transaction T	{1,2},{1,3},{2,3},{2,4},{2,3,4}
Count C(T)	1,2,1,1,2,

The process and result of the Apriori arithmetic to search the frequent items sets is in the table 2

Table 2. The process and result of Apriori arithmetic

Set name	Transaction T	Count C(T)
C1	{1},{2},{3},{4}	3,5,3,3
L1	{1},{2},{3},{4}	3,5,3,3
C2	{1,2},{1,3},{1,4},{2,3},{2,4},{3,4}	2,2,0,3,3,2
L2	{1,2},{1,3},{2,3},{2,4},{3,4}	2,2,3,3,2
C3	{1,2,3},{2,3,4},{1,3,4}	0,2,0
L3	{2,3,4}	2

It searches the database 3 times. In the Table 2, {1, 4}, {1, 3, 4}, {1, 2, 3} are the fake item sets.

For the same database, the process and results by the KFH arithmetic is as Table 3. In the table W means Sets of Transaction, KF means frequent item sets; KH means standby item sets. When the minsup is 3, KF' and KH' are both the new ones.

As we can see, since sorting out the data and obtaining the frequent item sets only need to search the database, in any case KFH can obtain the frequent item sets through searching the database 3 times, but the times of the Apriori arithmetic searching the database will increase as the max k increases, so when the max k is bigger, there is a great difference between the KFH arithmetic and Apriori arithmetic which is inferior to the former. Moreover, in the process of updating min sup and data, when the R is not increased, the KFH arithmetic can directly get the new KF, KH from the FH without repeat the process of getting the frequent item sets.

Table 3. The process and result of KFH arithmetic({X :Count(X)})

Name	Kinds		
	3rd kind	2nd kind	1st kind
W	{2,3,4:2}	{1,2:1},{1,3:2},{2,3:1},{2,4:1}	Null
KF	{2,3,4:2}	{1,2:2},{1,3:2},{2,3:3}, {2,4:3},{3,4:2}	{1:3},{2:5},{3:3}, {4:3}
KH	Null	Null	Null
KF'	Null	{2,3:3},{2,4:3}	{1:3},{2:5},{3:3},{4:3}
KH'	Null	{1,2:2},{1,3:2},{3,4:2}	Null

7 Conclusion

In this paper, we can directly get the standby item sets and frequent item sets which can meet the min-sup and avoid the emergence of the "fake items" through classify and sort out the data of transactions logically, which can make the search times of database won't increase as the count of the data item sets increase .It can control the search time of the database for 3 times and consequently enhance the velocity and efficiency of the arithmetic. From the example, the celerity association rules method based on data sort search is superior to the Apriori arithmetic; in particular, the KFH arithmetic can provide a better way for mining the association rule and updating the data of the dynamic huge database.

References

1. Shi, Z, Z.: The discovery of knowledge (2002)
2. Agrawal, R., Imielinski, T., Swami, A.: Database mining: A performance perspective. In: IEEE Trans Knowledge and Data Eng (1993) 914-925
3. Agrawal, R., Imielinski T, Swami A.: Mining association rules between sets of items in large database. In: Proc of ACM SIGMOD Intl Conf on Management of Data (1993) 207-216
4. Qin, J, S., Song, H, T.: Study and Improvement of AprioriHybrid Algorithm in Mining Association Rules. In: Computer Engineering, Vol. 30 (2004)
5. Song, H, S.: An efficient incremental updating algorithm for mining association rules. In: Journal of Lanzhou University, Vol. 40 (2004)

Using Web Services to Create the Collaborative Model for Enterprise Digital Content Portal

Ruey-Ming Chao[1,*] and Chin-Wen Yang[2]

[1] Department of Information Management, Da-Yeh University,
112 Shan-Jiau Rd., Da-Tusen, Chang-hua, 51505, Taiwan,
Tel. no.: +886-48-511-888 ext:3146, fax no.: +886-48-511-500
rueyming@mail.dyu.edu.tw
[2] Department of Information Management, Da-Yeh University,
112 Shan-Jiau Rd., Da-Tusen, Chang-hua, 51505, Taiwan
lucyyang@seed.net.tw

Abstract. In the Knowledge Economy era, trying to promote the whole competition advantage, the electronic businesses utilize information technology and internet to integrate the various kinds of application systems, database, and platform. It becomes common practice to construct an Enterprise Digital Content Portal (EDCP). To vary from minute to minute, coming to the problem of business environment is so difficult to integrate the complicated and huge amount information. By the way of collaboration of EDCP and the information of trade partners, it can provide the customers the real-time information that appearing with dynamic various timing. Through the combine of the information and procedure that between the enterprise and it's business partner, it can use the assisting of information technology, improve the enterprise's internal and external operational procedures, raise the transparency of information in the value chain, achieve the purpose that sharing information with different platform and language. In order to combine the different kinds of platform's information between the different enterprises, we use Java technologies for web services in the construction and development of EDCP, use the Extended Markup Language, and the Web-based communication protocol, it communicates with other software system [1], accomplish the framework of the knowledge service platform that the enterprise deliver and communicate the internal and external information. In this paper, we propose the construction structure of the EDCP that using Web Services, we implement it into one case and utilize WebBench to be the analyzed tool that testing the efficiency when many people connect to the line at the same time, and prove the feasibility of this framework of the module. And we probe into the relevant literatures of collaboration, and make up the deficiency of relevant literatures in the past. Conduct to be the reference of the research that the enterprise and organization establish the relation of collaboration using EDCP in the future.

Keywords: Enterprise Digital Content Portal, Collaboration, Web Services.

---

* Corresponding author.

1 Introduction

1.1 Background and Motivation

With the development of Internet and computer science, trying to promote the efficiency of the operation and the competition advantage, all enterprises march toward the ranks of the electronic business one after another. And utilize information technology to use the various kinds of application systems, databases, and platform properly, the ones that simplified the communication procedure between the enterprise and its supplier and customer, have already become the important thinking direction of increase the competitiveness. [2].

So, we propose the framework that has the collaborative model of combing the business procedures, and the dispersing workflows can be obtained the greatest interests through coordinate. And we use the technology that constructed on the relevant standards of Web Services mainly, and utilize the EDCP that constructed and developed by using Java Technologies for Web Services relevant technology. On the basis of the collaborated model, the system architecture offer to the enterprises the follow-up reference basis that improve the detail of it's EDCP or amend it's strategy. Conduct to be the reference of the research that the enterprise and organization establish the relation of collaboration using the EDCP in the future.

1.2 Background and Motivation

On the basis of above-mentioned research motives, we achieve the three following purposes:

(1) Propose the constructional architecture of an EDCP of using Web Services technology, in order to accord with the enterprise's internal and external user's demands, and obtain the benefit of the electronic business, and implement the EDCP into one case, and explain the feasibility of the system architecture.
(2) Probe into the collaboration relationship between the enterprise and it's supplier while we implement the EDCP, and make up the deficiency of relevant literatures in the past. Conduct to be the reference of the research that the enterprise and organization establish the relation of collaboration using EDCP in the future.

2 Literatures Review

2.1 Enterprise Digital Content Portal

The activities of the Digital Content Industry has officially taken off, the Ministry of Economic Affairs plans to invest around 1.07 billion dollars worth of funding over the next six years to promote Digital Content Development. There are many business issues involved in building an intelligent, efficient, cost-effective solution to store and manage fixed content.

The Enterprise Digital Content Portal [3] is a single-entry portal system operated on a common platform for most of the business services. It is an integrated, secure platform for content storage and archival that can be rapidly deployed. An enterprise digital content portal (EDCP) can provide an environment that can be personalized to delivers only contextual digital content and services to all identified users (employees, customers, suppliers and partners) from any web-enabled device. Successful implementation of a portal ecosystem extending across and beyond the enterprise is a complex task.

2.2 Web Services

Web Services is a service-oriented architecture, and it has the advantage of cross-platform and cross- language. It is a set of common and distributed application program standards applicable to all defined by WS-I organization (ws-i.org), Web Services 's relevant technology used includes XML, SOAP, WSDL, UDDI[4-6].

(1) XML (Extensible Markup Language) [7]:

XML is a standard that opening and extensible markup language label. So, XML is suited and applied to the electronic commerce the file between the different kinds of platform.

(2) SOAP (Simple Object Access Protocol) [8]:

SOAP is the protocol that calling the far-end procedure. It utilize the standard Hypertext Transfer Protocol (HTTP) to transmit information, the XML file of its ground floor can pass through the firewall of enterprises.

(3) WSDL (Web Service Description Language) [9]:

WSDL is a grammar of describing Web Services with XML, enable users to learn service provider's service via this rule, the data type of document and the basic software and hardware demands needed using offered via this specification, it can let the client user know how to use Web Services.

(4) UDDI (Universal Discovery Description and Integration) [10]:

UDDI offers the registration, search, and finding of Web Services, and provides the taking and sharing of Web Services. The service provider can register their service offered via this specification, and the service requester can obtain the service information needed and suitable through the query of service.

The mechanism of the whole Web Service can show with Fig. 1.

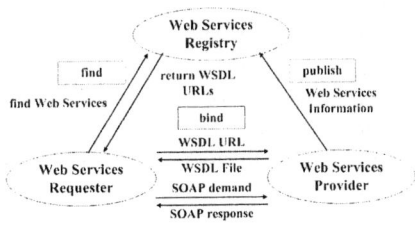

Fig. 1. The operation concept of Web Service

2.3 Collaboration

According to the explanation of The American Heritage Dictionary, the collaboration has two definitions: "1. Work together, especially participate in an achievement of intelligence.", "2. Cooperate rationally, strong just like military strength that an army occupies." Chen and Zhuang [12] propose in "knowledge value chain" that so-called "collaboration" is "Through the assistance of far-end interaction tools, proceed the long distance communication operations likes the exchange of the electronic documents, voice transmission, image transportation, in order to finish face-to-face project execution, multi-user meeting, on-line learning etc." Whether aforesaid far-end interaction tools, include: BBS(Bulletin Board System), zone of discussion, message board, chat-room, video conference system, electronic white board, on-line learning system, project management, etc.

Table 1. The intension of Collaboration

Scholar	Intension
Chen [13]	Under the collaboration, the enterprise combine the procedure and information with the transaction partner (the supplier and customer), and according to the present situation (include the inside and outside) of enterprise resources, do the optimization arrangement, with fast reaction customer, market and competitor.
Huang [14]	No matter the interdepartmental collaboration inside enterprises, among the enterprise and the enterprise (supplier, cooperative partner, seller, service provider, customer, etc.), any kind of collaboration in the commercial exchange (design and research and develop, predict and plan, order and produce, on sale throughout business, etc.), can be considered as the commercial affair.

3 System Design

3.1 The System Architecture of the Enterprise Digital Content Portal

EDCP is a individualized (personalized) platform that have single entries (single gateway), once login (one login), friendly user interface (browser). The knowledge worker can obtain any information needed while working easily, operating the application program that distributed in every place originally. In this case, can promotion of working efficiency, set up the good communicative channel, and take the advantage of which improves the transparency of information while organizing.

The architecture of system is the knowledge service platform that can let enterprise knowledge worker get instant and convenient information make knowledge. By providing information on giving enterprises a single entry form, match the Search engine, file management, etc., it can promote working efficiency, worker of knowledge, let knowledge worker can find needed by it various types of message and file rapidly here through the single entry.

3.2 Operation Architecture of Enterprise Digital Content Portal Based on Web Services

Because of among textile relevant industry, with make specification Cheng often for appear dynamic message of change original price of supplies, for example: Yarn price, decreasing, weaving workers, dyeing the offering that decreased and waited a moment the price and information finely, can change to some extent because of difference clicked in time often, offer instant information these for make planning Cheng important. In order to pick and fetch and deal with these a large number of original prices of supplies and relevant information immediately, in order to adopt Web Services technology as this Enterprise Digital Content entry of key technology of websites, name and act as Web Services-based Enterprise Digital Content Portal, is abbreviated as WS-EDCP. We agree supply of WS-EDCP justice become in textile making all relevant original provider of supplies in the Cheng, for example: Relevant manufacturers of textile industry, such as cotton mill, weaving cotton cloth in the factory or the printing and dyeing mile, etc. WS-EDCP, can let relevant industry of a lot of textile join, and utilize UDDI mechanism, supplier can make their because developed Web Services is it get Enterprise Digital Content entry on the website to announce in conformity with the request of textile industry person, WS-EDCP can utilize the relevant manufacturers of textile industry , such as cotton mill, weaving cotton cloth in the factory or the printing and dyeing mile, etc. use Web Services offered by it to pick the useful information fetched most immediately. And the concept structure of WS-EDCP operation is as Fig. 2 shows.

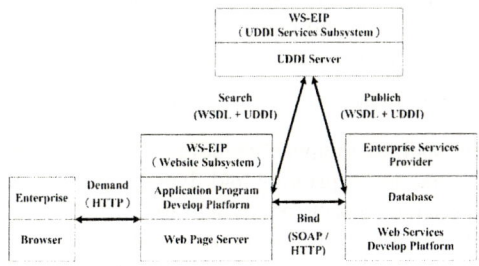

Fig. 2. The Conceptual Relationship of WS-EDCP operation [15]

There are three main participants in Fig. 2, is WS-EDCP, the enterprise and the enterprise service supplier respectively, among them WS-EDCP includes two subsystems, divide and state it as follows:

(1) WS-EDCP website's subsystem:

Enterprises got instant information that the server of the website presents by the webpage browser, it crosses over different platforms and characteristic of the language, enables all inquiry service to go on through the browser. Is it fruit to appear the picture to focus on to need only on website server oneself, the demand functionally, can do the link to the application program server of the back end through the procedure language (JSP, ASP, etc.) The main purpose of developing platform of the application program is running for the bridges of the website server and UDDI service subsystem, after it finding supplier's information from UDDI, and then link the enterprise service supplier's server according to WSDL offered in the catalogue, obtain, combine and carry out its Web Services needed definitely by this.

(2) WS-EDCP UDDI service subsystem:

Have announce, search and function that register through UDDI server, can let enterprises service make them developed to convenient to combine Web Services in conformity with trader have, offer WS-EDCP website's subsystem, will find the source of Web Services and put forward the demand to it the information after also developing WSDL through URL. Do not need to be worried each other Web Services developed to combine in some ways either between service suppliers of enterprise, only need to develop the function information to UDDI server WS-EDCP announced.

(3) Enterprises:

In spinning making a part of the way of planning traditionally, the textile industry person usually gets the original price of supplies and relevant information through the telephone or the fax, and then regard this as the reference basis of price comparison, and then plan textile to make to Cheng, do a deal by way of the telephone or the fax too after finishing, it is thus apt to delay the business opportunity, more ineffecnt too. Through the mechanism of WS-EDCP, all homework can be finished in the internet network, can make information at fastest speed with information entry websites of enterprises, and can offer instant rate of exchange, is it finish order, planning to make Cheng rapidly to help, cooperate with trade mechanism of e-commerce at the same time, finish making an appointment, ordering and waiting a moment for the relevant activity on being on-line directly, efficiency not only improves by a wide margin but also reduce the mistake of the artificial homework.

(4) Service supplier of enterprises:

Relevant industries, such as member including cotton mill among them, weaving cotton cloth in the factory or the printing and dyeing mile, etc., it accords with standard Web Services to cooperate with WS-EDCP to develop, utilize UDDI function in information entry websites of enterprises, announce it that makes use of SOAP/HTTP protocol to convey XML file, the work of communicating with the relevant manufacturer of each textile in order to let WS-EDCP be combined, so as to according to the demands of enterprises, WS-EDCP can offer a lot of different services, for example: Inquire yarn price, fine to decrease, knit worker, is it is it is it make relevant information Cheng to wait a moment to decrease to dye, the service supplier of enterprises can develop the service component that enterprises need for using too.

3.3 The Building and Constructing and Using of Enterprise Digital Content Portal Based on Web Services

On the commercial procedure among enterprises, two enterprise when doing a deal again, one serve as role of Service Requester, another one acts the part of Service Provider, and a just the other person serves as the role of Service Registry and inquires about the necessary service by offering to enterprises. Say with a commercial operation procedure for supplies to cooperative partner, include at least one service routine (service process) of the above, the following is single (order service process), the order is replied (response order service process), the supplies stock is inquired about (material checking service process), and one service process may call out another service process again.

Will explain this festival how to inform the function stipulated of to the service supplier of enterprises, it develop and build step to construct WS-EDCP, if following five point is it state to divide with groundwork included [16]:

(1) Stipulate WS-EDCP demand specification:

WS-EDCP hopes on UDDI that the function that the relevant supplier offers Web Services is announced and got to WS-EDCP, each supplier can learn via the information that is announced which functions Web Services needs to include.

(2) The supplier announces Web Services after developing:

After the supplier makes the information, after developing the function of finishing Web Services, on UDDI that and then announce Web Services and get to WS-EDCP, let WS-EDCP search each supplier Web Services relevant information developed. For example: Web Services after developing them of cotton mill is announced on UDDI, in order to enable WS-EDCP to obtain detailed information of Web Services function, work that can just be linked thus.

(3) WS-EDCP sets up Portal Site:

After linking a supplier with UDDI information, WS-EDCP sets up Portal Site and does a merger of supplier's service offered, enterprises can use the inquiry sample of Portal Site to offer the functions, such as information, etc., and present the information of inquiring about the income through the webpage.

(4) WS-EDCP searches the supplier's Web Services and associated documents:

WS-EDCP, according to the detailed information that the supplier offers on UDDI, works for linking the materials. Announce that must do the detailed and clear expression of their Web Services function released to need on UDDI, help WS-EDCP and supplier pass the good communication.

(5) Link the supplier's Web Services:

WS-EDCP utilizes SOAP/HTTP protocol to communicate with the supplier, can offer abundant instant information on-linly, let enterprises grasp firsthand information through the appearing of the webpage.

4 Case Discussion

4.1 Brief Introduction of Case Company

The case company is the traditional textile industry firm, usually make original supplies relevant information through the telephone and fax, and then regard this as the reference basis of price comparison, and then plan textile to make to Cheng, doing a deal with the telephone, fax too after finishing, the ones that passed this research construction are in coordination with entry websites, the customer can obtain quotation information immediately, can improve the efficiency of enterprise's business procedure, promote the customer in the knowledge value chain transplant to with is it help they getting relevant information they needing fast. By the mechanism of WS-EDCP, Web Services offers one to support environment stepped and served the component and shared in the platform. Web Services as Enterprise Digital Content foundation, entry of website, can spin and weave supplier of industry stand up with system combination of consumer, make to textile industry person planning Cheng necessary fair smooth, have prescroption in textile. The information inside enterprises is combined also for the essential condition of promoting enterprise's competitiveness, in coordination with the cooperative infrastructure, pass the digital information platform applicable to all, is it exchange to circulate fast in enterprise to enable information, in order to shorten products Cheng when beginning to merchandize and reduce products and support cost.

4.2 WS-EDCP Operation Structure and System Are Done in Fact

(1) The architecture of System Operation

The WWW interface which the front browser linked in Web Server through Internet/Intranet, later, used the system to transmit XML information and call out supplier Web Service, and former supplies supplier Web Services in the back end would follow front information, deal with enterprise's demand and change into a form of XML, go back to reach the application system in Web Server, and then present the information of the webpage for the front browser through Internet by Web Server [15].

(2) The quotation among enterprises, in coordination with the cooperative example

The information system of two enterprises, and take enterprise's quotation procedure as an example, the course after information is flowed, dealt with and put in order in enterprises that prove, information is conveyed to the whole operation situation among the cooperative partners again finally, and plan to appear in the future with the possible structure that other systems (such as ERP, CRM, KM) combine inside enterprises [16].

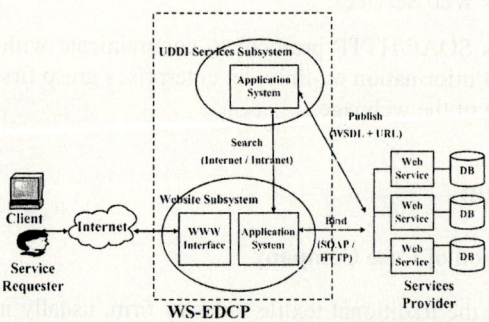

Fig. 3. The architecture of WS-EDCP

(3) System interface

This research expresses WS-EDCP concept through the system, simulation enterprises inquire about the example of the relevant original price of supplies, list in textile making important three supplier Web Services that developing needed in the Cheng as follows. Supplier register get UDDI Web Services that develop, weave cotton cloth inquiry function Web Services which the factory develop as the example with A cotton mill and B, URL that its information registered includes Web Services and Enterprise Digital Content of the cotton mill. WS-EDCP searches the registered information that the cotton mill offers through Internet or Intranet, linked via SOAP/HTTP protocol Web Services which the supplier developed by WS-EDCP application system, present information through WWW interface, in this way, can search original supplies relevant information through Web Services of the cotton mill.

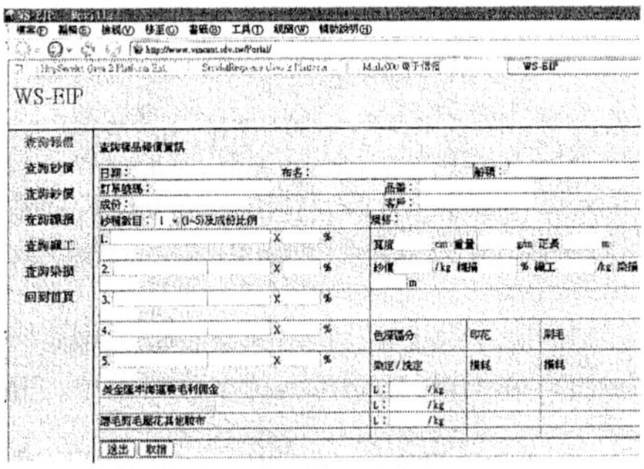

Fig. 5. The example of WS-EDCP's function: quote

5 Conclusions

Under the trend of knowledge economy, knowledge will play an important role in the economic system in the future; enterprises are all devoted to changing knowledge as the management activity of the competitiveness. The EDCP system can help the knowledge worker of enterprises can find various kinds of information and file needed by it rapidly in the single entrance here; Use the cooperative partner whose technology can be correlated with enterprise inside and outside of Web Services too at the same time, combine it within EDCP. Can not merely thus meet various kinds of demands of the knowledge worker's inside enterprises, cooperative partner making enterprises external too, including between the upper reaches supplier and downstream customer, make a profit through the cooperation of tactic, offer to the knowledge worker the more perfect one in coordination with the working environment.

From the gradual progress course of e-commerce and change of the type of operation of enterprise, modern electronic enterprises must use internet network resources to create and promote with the cooperation among the trade partners properly, can fully know the market and on sale throughout the change and answer fast. Now there is must the fast enterprise close exactly, in, enterprise resource of partner not downstream and going on not cooperating in coordination, to achieve the purpose of information transmitting immediately and fast reaction among enterprises. And Web Services is in coordination with the key technology of enterprises of the homework way, it has changed the commercial trade procedure operation way. This research is to systematic information gathering of heterogeneity of the traditional enterprise's system, difficulty summed up, combined and spreading, use the open and mould group advantage of Web Services, and deposit and withdraw the protocol with the simple things of SOAP, the dynamic things needed call out the mechanism to come to electronic enterprises of structure while cooperating in coordination. This research expresses among electronic enterprises to utilize Web Services to go on in coordination with the procedure of the commercial trade and operation way with a simple example of quotation for the goods. Via the case company, we can find out that the commercial processing procedure based on Web Services has advantage with smooth open structure and information transmission.

References

1. You, J.L.: Web service. Information and Computer. 257 (2001) 31-35
2. Chen, Z.M. (ed.): The Relationship Between The Adoption Factor, Implementation Strategy, Application Functions and Implementation Efficiency of The Enterprise Information Portal. The Master's Thesis of National Yunlin University of Science and Technology, Yunlin, Taiwan (2001)
3. Status Internet Co., Ltd.: Power EDCP. http://www.status.com.tw/EDCP_powerEDCP.asp.
4. Zhang, S.Y.: The newest computer operating model – Web Services. Network and Communication. 128 (2002) 4-11
5. Zhang, S.Y.: The impossible mission of Web Services. Network and Communication. 132 (2002) 98-102
6. Liu, W.Y., Huang, W.R.: The integrated revolution of Web Services. Information and Computer. 261 (2002) 19-25
7. W3C: Extensible Markup Language (XML): http://www.w3.org/XML/. (2001)
8. Simple Object Access Protocol (SOAP): http://www.w3c.org/TR/SOAP. (2001)
9. Web Service Definition Language (WSDL): http://www/w3c/org/TR/WSDL. (2001)
10. Universal Description, Discovery, and Integration (UDDI): http://www.uddi.org. (2001)
11. He, Z.F., Hong, M.C.: The study of the web serviced SCORM teaching materials. Proceeding of The 15th International Conference of Information Management, Taipei (2004)
12. Chen, Y.L., Zhuang, Y.C. (ed.): The Knowledge Value Chain,: China Productivity Center, Taipei (2003)
13. Chen, X.B. (ed.): A Study of e-Business Collaboration, The Master's Thesis of National Chengchi University, Taipei, Taiwan (2001)
14. Huang, B.L.: The Strategy and challenge of the business outsourcing services. E-business: businessman report. 18 (2001) 12-21
15. Li. Q.Z., Hong, X.Q.: Utilizing Web Services to establish tourist industry's e-market. Proceeding of The 15th International Conference of Information Management, Taipei (2004)
16. Jiang, X.K., Chen, M.T. (ed.): The Study of The Collaborated E-business to Utilize The Core of Web Service. The Master's Thesis of Da-Yeh University, Changhua, Taiwan (2003)

Emotion-Based Textile Indexing Using Colors and Texture

Eun Yi Kim[1], Soo-jeong Kim[2], Hyun-jin Koo[3], Karpjoo Jeong[1], and Jee-in Kim[1]

[1] CAESIT Department of Internet & Multimedia Engineering,
Konkuk University, Seoul, Korea
{eykim, jeongk, jnkm}@konkuk.ac.kr
[2] Dept. of Computer Eng., Konkuk University, South Korea
cryolite@konkuk.ac.kr
[3] FITI Testing and Research Institute, Seoul, South Korea
koohh@fiti.re.kr

Abstract. For a given product or object, predicting human emotions is very important in many business, scientific and engineering applications. There has been a significant amount of research work on the image-based analysis of human emotions in a number of research areas because human emotions are usually dependent on human vision. However, there has been little research on the computer image processing-based prediction, although such approach is naturally very appealing. In this paper, we discuss challenging issues in how to index images based on human emotions and present a heuristic approach to emotion-based image indexing. The effectiveness of image features such as colors, textures, and objects (or shapes) varies significantly depending on the types of emotion or image data. Therefore, we propose adaptive and selective techniques. With respect to six adverse pairs of emotions such as weak-strong, we evaluated the effectiveness of those techniques by applying them to the set of about 160 images in a commercial curtain pattern book obtained from the Dongdaemoon textile shopping mall in Seoul. Our preliminary experimental results showed that the proposed adaptive and selective strategies are effective and improve the accuracy of indexing significantly depending on the type of emotion.

1 Introduction

There has been a significant amount of research work on the image-based analysis of human emotions in a number of research areas because human emotions are usually dependent on human vision. So far various research efforts has been investigated to analyze relationships between human emotions and images in a number of disciplines[1-4]. Such research work has been focused on what features of image (e.g., colors, textures, or shapes) and how they affect human emotions. However, they usually depend on the manual extraction of image features and therefore, their results can not be directly applied for building emotion-based information systems where data are automatically indexed and retrieved based on emotions.

In computer image processing, there are many effective techniques for indexing images by extracting features from them automatically. Unfortunately, there has been little research work on image indexing based on human feeling.

In this research, we first discuss challenging issues in emotion-based image indexing by using features extracted by computer image processing techniques. In this work, we consider two features: colors and textures. We study relationships between human emotions and these features: first, each feature is independently investigated against human emotions and then their combinations against human emotions.

Based on this work, we propose adaptive and selective strategies for emotion-based indexing images. Our experimental results indicate that the degree of dependency between each feature and human emotions varies significantly according to the type of emotion or image data. Therefore, A straightforward application of traditional analysis techniques such as fuzzy-evaluation methods are likely to produce inaccurate analysis results.

2 Fuzzy-Based Evaluation System

Through the Web evaluation system, we investigate the relationships between emotional features and physical features (color and texture), thereafter we construct the fuzzy rule-based evaluation system to predict the human emotions from the given textile images. Fig. 2 shows the outline of the proposed system, where it is composed of physical feature extraction and fuzzy-based evaluation.

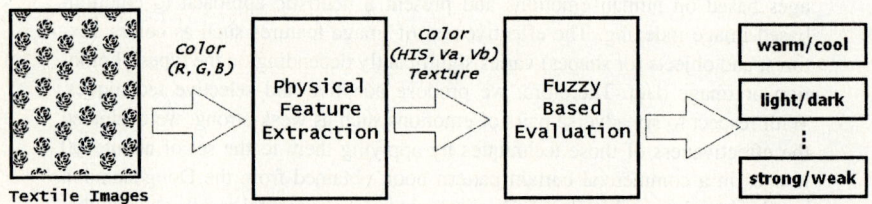

Fig. 1. Outline of the proposed system

Our evaluation system uses six pairs of adverse emotional features expressed as adjective words: {weak/strong, sober/gay, dark/light, dismal/cheerful, warm/cool, soft/hard}. These features are proposed by previous sensory and psychological experiments [5,6]. To predict the emotional features from the physical features of images, we develop a fuzzy-based evaluation system.

Based on these investigations, we construct the relationship between the physical features and human emotional features.

We use both colors and texture for emotion-based image indexing. Then, for the conversion of the physical features such as color and texture to the emotional features, we develop the fuzzy rule-based system.

Then, for the conversion of the physical features to the emotional features, 14 control rules are used; 11 rules out of them are applied from the color properties and the other are applied from texture properties.

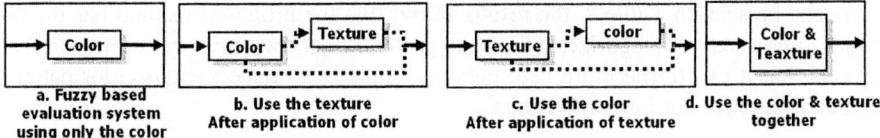

Fig. 2. Fuzzy systems for predicting the emotions

To investigate the influence of the respective features in predicting the emotions, the experiments are performed on four fuzzy systems. One is the system using only color information to predict the emotions, and the others are the systems using both of colors and textures. The difference of the rest of two systems is in the sequence to use the features, which feature is firstly applied in the classification. The last is the system using both color and textures at the same time.

3 Experimental Results

The proposed indexing system has been tested with 160[1] captured textile images. For the respective textile images, seventy peoples are selected to manually annotate according to the emotions that they feels from the images.

Then the half of 160 collected images is used for generating the fuzzy rules and the other is used for test. To generate the fuzzy rules, we use colors and textures. As illustrated in Section 2, color properties gives more hint than the texture to predict the emotions included in a textile image; the colors can be used to distinguish all of the emotion groups, whereas the textures can be used for some of emotion groups.

The performances for these four systems are summarized in Table2. Although the significant difference is not detected, the systems using all of two information show the superior classification rate to the system using only the colors. Moreover, the system to use the color information after application of textures shows the highest performance.

Table 1. Performance of the color properties (%)

cold-warm	sober-gay	dismal-cheerful	dark-light	weak-strong	soft-hard	total
82.98	86.56	100.00	79.88	76.65	93.50	86.59

Table 2. Performance of the proposed method(%)

	Color only	Color & Texture	Color + Texture	Texture + Color
cold-warm	82.98	82.29	85.69	85.64
dark-light	79.88	79.53	82.96	82.94
weak-strong	76.65	83.85	77.33	85.82
Total	79.83	81.89	81.99	84.80

---

[1] This data set is obtained from a commercial curtain pattern book obtained from the Dongdaemon textile shopping mall in Seoul.

As can be seen in Table 2, the results shows that the proposed method has the potential to be used to categorize, index, and search images with respect to human feelings, so that it can furthermore help pattern designers to make various color patterns with consideration for human feelings.

4 Conclusions

This paper presented an approach for labeling images using human emotion or feeling and showed its affirmative results. The proposed system is composed of physical feature extraction and fuzzy-rule-based evaluation. The physical feature extraction stage extracts the color and texture properties from an input image, and the evaluation stage analyzes the extracted features and predicts the included emotions in the input image using fuzzy rules. To assess the effectiveness of the proposed system, it was tested with 160 textile images, and the results shows that it has the potential to be applied for textile industry and e-business.

Acknowledgements

This work was supported by the Ministry of Information & Communications, Korea, under the Information Technology Research Center(ITRC) Support Program

References

1. B. Furht and o. Marques, Content-based Image and Video Retrieval, Kluwer Academic Publishers, 2002.
2. V. N. Gudivada and V. Raghvan, Content-baed Retrieval Systems, IEEE computer, pp. 18-22, 1995.
3. N. Roussopoulos and D. Leifker, Direct Spatial Search on Pictorial Database using Packed R-Trees, Proc. ACM-SIGMOD, p. 17-31, 1985.
4. S. K. Chang, Q. Y. Shi and D. Dimitrof and T. Arndt, An Intelligent Image Database System, IEEE Trans. Software Engng. SE-14, p. 681-688. 1998.
5. N. Kawamoto and T. Soen, Objective Evaluation of Color Design II, Color Res. Appl., vol. 18, p. 260-266, 1993.
6. T. Soen, T. Shimada, and M. Akita, Objective Evaluation of Color Design, Color Res. Appl., vol. 12, pp. 187-194, 1987.
7. Gonzalez et al. Digital Image Processing, Addison-Wesley, 2002.

Optimal Space Launcher Design Using a Refined Response Surface Method

Jae-Woo Lee[1], Kwon-Su Jeon [1], Yung-Hwan Byun [1], and Sang-Jin Kim[2]

[1] Cneter for Advanced e-System Integral Technology, Konkuk University, Seoul, Korea
{jwlee, yhbyun}@konkuk.ac.kr, mizar92@dreamwiz.com
[2] Agency for Defense Development, Daejeon, Korea
bipoo@bcline.com

Abstract. To effectively reduce the computational loads during the optimization process, while maintaining the solution accuracy, a refined response surface method with design space transformation and refined RSM using sub-optimization for the regression model is proposed and implemented for the nose fairing design of a space launcher. Total drag is selected as the objective function, and the surface heat transfer, the fineness ratio, and the internal volume of the nose fairing are considered as design constraints. Sub-optimization for the design space transformation parameters and the iterative regression model construction technique are proposed in order to build response surface with high confidence level using minimum number of experiment points. The derived strategies are implemented to the nose fairing design optimization using the full Navier-Stokes equations. The result shows that an optimum nose fairing shape is obtained with four times less analysis calculations compared with the gradient-based optimization method, and demonstrates the efficiency of the refined response surface method and optimization strategies proposed in this study. The techniques can be directly applied to the multidisciplinary design and optimization problems with many design variables.

1 Introduction

With growing demands on launching earth orbit satellites, recent research directions in the launch vehicle design area are focused on reducing the launching cost.

Important factors to be considered in the aerothermodynamic design of satellite-launching vehicles are the aerodynamic drag, the surface heat transfer and the internal volume for the payload (satellites) accommodation. In early researches, most of the studies implemented analytical or experimental approaches. With the advancement of computational fluid dynamics and the rapid growth of the computing power, optimization techniques utilizing the numerical analyses become the major aerodynamic design tools, and dominate the design problems.

Though many efficient optimization schemes have been developed till now, the basic weakness of the gradient-based optimization techniques (Gradient based method, GBM)[1] is the possibility of deriving a locally optimized design. Global searches, including the genetic algorithm (GA) based on the evolution theory, can be good alternatives to find the global optimum with given design constraints, but these methods require many analysis calculations[2].

Recently, to increase the design efficiency and to curtail huge computational efforts required for the analysis calculations while maintaining global optimum, system approximation techniques like the response surface method (RSM)[3] are utilized to resolve the design optimization problems. As the RSM can enhance design efficiency by improving the approximation errors, it is regarded as an alternative for the complex Multidisciplinary Design and Optimization (MDO) problems[4,5,6].

In this study, an optimum nose fairing configuration of the space launch vehicle satisfying the given design constraints, will be derived by implementing a refined RSM and numerical optimization approaches. To effectively reduce the computational loads while maintaining the solution accuracy, the RSM with design space transformation technique combined with either a GBM or a GA, shall be proposed and applied with sub-optimization strategy.

The methodology to reduce the number of design experimental points and to determine the location and the direction of the design space transformation, an iterative regression model refinement method will be suggested. Using the practically applicable design optimization strategies with the refined RSM, the optimum nose fairing shape will be designed in the flight conditions of Korean three-stage liquid propellant rocket, KSR III.

2 Aerodynamic Analysis Method

2.1 Applied Numerical Method

Axisymmetric Navier-Stokes computer code is developed and applied for this study. The governing equation is given at eqn. (1) in the computational domain, ξ, η. The spatial flux vectors F, G, H and viscous terms F_v, G_v and H_v are given at eqns. (1)~(3) using contravariant velocities U and V. For the two dimensional flow α is zero, and is equal to 1 for the axisymmetric flow.

$$\frac{\partial Q}{\partial t} + \frac{\partial F}{\partial \xi} + \frac{\partial G}{\partial \eta} + \alpha H = \frac{\partial F_v}{\partial \xi} + \frac{\partial G_v}{\partial \eta} + \alpha H_v \qquad (1)$$

$$Q = \frac{1}{J}\begin{bmatrix} \rho \\ \rho u \\ \rho v \\ \rho e_0 \end{bmatrix}, F = \frac{1}{J}\begin{bmatrix} \rho U \\ \rho u U + \xi_x p \\ \rho v U + \xi_y p \\ (\rho e_0 + p)U \end{bmatrix}, G = \frac{1}{J}\begin{bmatrix} \rho V \\ \rho u V + \eta_x P \\ \rho v V + \eta_y p \\ (\rho e_0 + p)V \end{bmatrix}$$

$$F_v = \frac{1}{JR_a}\begin{bmatrix} 0 \\ \xi_x \tau_{xx} + \xi_y \tau_{xy} \\ \xi_x \tau_{xy} + \xi_y \tau_{yy} \\ \xi_x \beta_x + \xi_y \beta_y \end{bmatrix}, G_v = \frac{1}{JR_a}\begin{bmatrix} 0 \\ \eta_x \tau_{xx} + \eta_y \tau_{xy} \\ \eta_x \tau_{xy} + \eta_y \tau_{yy} \\ \eta_x \beta_x + \eta_y \beta_y \end{bmatrix} \qquad (2)$$

$$H = \frac{1}{yJ}\begin{bmatrix} \rho v \\ \rho uv \\ \rho v^2 \\ (\rho e_0 + p)v \end{bmatrix}, H_v = \frac{1}{yJR_a}\begin{bmatrix} 0 \\ \tau_{xy} \\ \tau_{uu} - \tau_{\theta\theta} \\ \beta_y \end{bmatrix}$$

$$\beta_x = u\tau_{xx} + v\tau_{xy} - q_x, \quad \beta_y = u\tau_{xy} + v\tau_{yy} - q_y \qquad (3)$$

Here ρ, e, v, q and τ represent the air density, internal energy per unit mass, kinematic viscosity, heat flux and surface stress, respectively. u and v are the velocity components in x-, y-direction, J is the Jacobian from the coordinate transformation and R_a is the Reynolds number based on the sonic velocity

Roe's FDS(flux difference splitting) scheme is implemented for the spatial discretization with the MUSCL for higher order extension. When implementing Roe's scheme to blunt body such as launcher nose shape, entropy fixing is needed in order to avoid numerical oscillations. Harten-Yee type[7,8] entropy fixing is employed in this study. The minmod limiter is used to remove solution oscillations. Fully implicit LU-SGS scheme is employed for time integration.

To increase the computational efficiency, a grid system is selected by investigating the convergence rate and the converged solutions of several grid systems with baseline configuration. Through extensive grid density study[9], the 81x41 grid system is selected with moderate clustering near the faring surface. The freestream Mach number was 4.6 and the Reynolds number was 1.42×10^7 at the atmospheric condition of 31km altitude. For the consistency of the Stanton number calculation at the stagnation point, the grid space right above the surface at the stagnation point is kept same for all cases.

2.2 Validation of the Numerical Approach: *Flow Around the Compression Corner*

To check the accuracy of the analysis code for the boundary layer flow including separation, flow phenomena of the compression corner are numerically investigated and compared with the experimental results of Holden and Moselle[10].

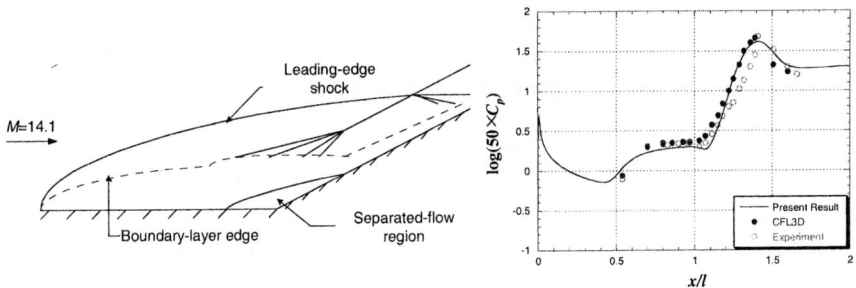

Fig. 1. Compression corner (*left*) and Comparison of the surface pressure(C_P) around the compression corner [13] (*right*, Wedge angle 24°, and M_∞=14.1)

The experiment was performed at Calspan 48 inch shock tunnel, and the model consists of a flat plate and 24° wedge.

Rudy et. al.[11] validated four Navier-Stokes computer codes using the experimental results of Holden and Moselle. Lee et. al.[12] investigated the accuracy of the test time estimation.

For the numerical simulation for this problem, 141×61 grid system is used. At Fig. 2, the surface pressure coefficients are shown and compared with the experiments and CFL3D result[13]. Good agreement is noticed around the separation region, and estimates slightly higher value at the wedge surface than the experimental result. But general tendency is very similar to the CFL3D result.

3 Design Optimization Procedure Using a Refined RSM

3.1 Numerical Optimization Techniques

Both the GBM and the GA are implemented for this study: The Sequential Linear Programming(SLP) and Sequential Quadratic Programming(SQP) in DOT[14] and the GA employed at GENOCOP III[15], developed by Michalewicz et al.

GA which simulates the Darwinian concepts of evolution is an optimization technique based on a stochastic search and natural selection. Because GA do not require gradient information, hence are less dependent upon the initial design, and only carry out an evolution of random tries by 'individuals', a global search is possible.

2.3 Applied RSM

The GBM require the gradient information for each design variable when the objective function is linearly approximated; hence the number of design variables plus one analysis runs is necessary in each design optimization cycle. Therefore, the required analysis runs are increased very rapidly with the increase of the number of design variables. Due to the heavy computational loads, the number of analysis runs is particularly important when the Navier-Stokes equations are utilized for the flow field analysis This prevents from the optimization methods to become a practical aerodynamic design tool. To reduce the number of analysis runs, more efficient method to build a reasonable regression model is necessary and is investigated in this study.

The reliability of the response surface built using the experimental points can be estimated by the adjusted R-square(R^2_{adj})^2, which is defined by eqn. (4)

$$R^2_{adj} = 1 - \frac{SS_E/(n-m)}{SS_y/(n-1)} \tag{4}$$

Where, SS_E, and SS_y are the error sum of squares and the total sum of squares, respectively. n and m are the number of experimental points, and the number of response function coefficients, respectively. R^2_{adj} has a value in-between 0 and 1. When the response surface built approximates the analysis results very closely, it has value near 1.

2.4 Improvement of the RS by Stretching Function (Refined RSM)

Usually a quadratic function is utilized to build a regression model. As the number of design variables increases, the number of coefficients of the regression model, hence the number of required experimental points increases exponentially. Therefore, to increase the order of the regression function above order 2 is not practically useable. In many design problems, the objective function and the constraint functions show highly nonlinear response according to the design variables, so a quadratic polynomial function can not approximate these functions properly even with a large number of experimental points. To improve the accuracy of the optimum solution, either narrowing down of the design space or increasing the order of the regression model can be considered. The former is sometimes very dangerous because the real optimum may be excluded when a detailed information is not available for the given design problem, and the latter requires too many experimental points when increasing the order of the regression model and this causes the increase of the computational expense.

Approaches to effectively reduce the design space without losing the regions containing the optimum point, need to be investigated. To efficiently approximate the general nonlinear response using 2nd order polynomial, design space transformation strategy using the stretching function is suggested.
In this study, we selected following stretching function.

$$x_i = A_i + \frac{1}{\beta_i} \sinh^{-1}\left[\left(\frac{\xi_i}{D_i} - 1\right)\sinh(\beta_i A_i)\right] \tag{5}$$

$$\text{where}, A_i = \frac{1}{2\beta_i} \ln\left[\frac{1 + (e^{\beta_i} - 1)(D_i/H)}{1 + (e^{-\beta_i} - 1)(D_i/H)}\right] \tag{6}$$

Where, β is the stretching parameter and D is the stretching position. H_i is range of the i-th design variable, and subscript i denotes the design variable index.

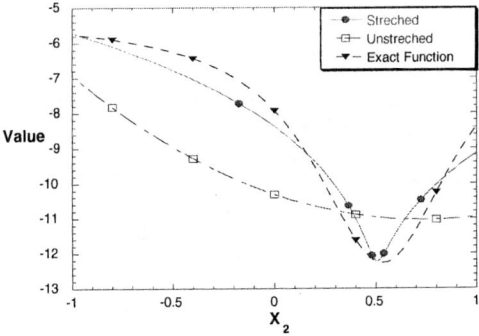

Fig. 2. Effect of design space stretching in the RSM

In order to visualize the design results, a single variable optimization problem is tested. The numerical function known the optimum value is selected for test. Figure 4 demonstrated that the design space transformation technique accurately represents the highly nonlinear behavior near the optimum point using only 2nd order regression model.

2.5 Sub-optimization Procedure for Proper RSM

2.5.1 Determination of Optimal Stretching Parameters

During the process of refined RSM, proper selection of the stretching location and the clustering amount are important for the optimal construction of the response surface. To optimize the RSM, a sub-optimization for the selection of the stretching parameters is proposed to be included during the overall optimization procedure.

For the response surface sub-optimization process, β_i and D_i of each design variable are design variables and R^2_{adj} is selected as the objective function to be maximized. GA is employed for the unconstrained optimization problem.

$$\text{Maximize} \quad R^2_{adj} = f(\beta, D) \quad (7)$$

$$\text{Where,} \quad \beta = \begin{pmatrix} \beta_1 \\ \vdots \\ \beta_k \end{pmatrix}, \quad D = \begin{pmatrix} D_1 \\ \vdots \\ D_k \end{pmatrix} \quad (8)$$

Where, subscript k denotes the number of design variables related to optimize response surface. The upper and lower boundary of stretching parameter β depend on the specific stretching function and the upper and lower boundary of stretching position D is in-between -1 and 1 in the coded variable space.

2.5.2 Overall System Optimization Procedure

To minimize the number of analysis runs while maintaining the reliable regression model, the initial response surface is built with relatively small number of experiment points. Then, the tentative optimum point obtained using the regression model, is

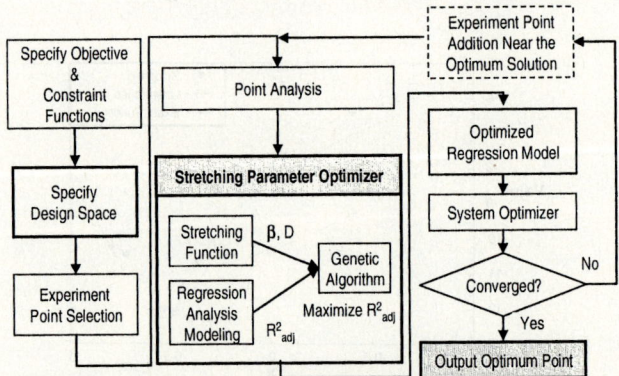

Fig. 3. System optimization procedure

included to the next set of experimental points to build a new regression model. Tentative optimum points (additional experiment points at each design cycle) are kept adding until design convergence. Usually two or three design iterations are enough to obtain the converged optimum solution. With only one additional analysis run at each design iteration, quality regression model can be constructed. Overall system optimization procedure including the sub-optimization is shown at fig. 5.

3 Optimization Approaches for Launcher Space Design

3.1 Shape Representation and Selection of Design Variables

To represent shape of nose faring, simple body shapes are selected as shape functions: minimum drag shapes like the von Karman ogive, the power-law body with power index n of 0.69, and blunt shapes which have small surface heat transfer rate, like the paraboloid, and the sphere-cone[9]. To alleviate the heat flux at the nose tip, the sharp nose tips of the von Karman ogive, and power-law body are replaced with a sphere. The fineness ratio (length/base diameter) was kept the same.

The configuration is expressed as a combination of the shape functions. To maintain the base diameter, a cone shape (y_5) is added as shown at eqn. (9).

$$y = \sum_{i=1}^{4} x_i y_i + (1 - x_1 - x_2 - x_3 - x_4) y_5 \qquad (9)$$

Where, x_i is the design variable and y_i is the shape function.

3.2 Design Space Selection

To implement the RSM refinement strategy explained in the previous section, the design space should be determined appropriately. The accuracy of the response surface is very sensitive to the range of the design space, which implies the importance of the design space selection. The design space is selected in order to exclude the physically unacceptable region with negative radius, hence the accuracy of the constructed response surface model has been improved remarkably. The selected design space in this study is as follows,

$$-1.0 < x_1 < 3.0, \quad -2.0 < x_2 < 3.0,$$
$$-1.0 < x_3 < 3.0, \quad -3.0 < x_4 < 3.0$$

3.3 Design Formulation and Generation of the Response Surface

The optimization problem is formulated as :
Minimize $f(X)$: aerodynamic drag
Subject to
  g_1: surface heat transfer rate (Stanton number, C_h)
  g_2: total heat transfer over the front 40% of the body
  g_3: internal volume of nose fairing
  design constraints, $x_i^l \leq x_i \leq x_i^u : i = 1, 4$
Where, superscript *l* and *u* denote the lower and upper bound.

During the fairing shape design, the design constraints are considered individually or together in order to observe the effect of each constraint. Total surface heat transfer from the stagnation point of the nose to the 40% of the body, where the satellite locates, is calculated, then the corresponding RSM is constructed[16].

Separate response surfaces have been constructed for the drag coefficient, C_D and the Stanton number C_h. . The Stanton number, C_h is defined by the equation below,

$$C_h = \frac{q_w}{\rho_e u_e (h_{aw} - h_w)} \tag{10}$$

C_D is obtained by integrating the surface pressure and the skin friction over the entire launcher body surface. Surface heat transfer rate can be calculated using simple heat conduction equation through the launcher surface, by assuming a constant wall temperature.

4 Design Optimization Results

Drag minimization is performed with given values of fineness ratio, surface heat transfer rate and internal volume constraints of the baseline launcher; 1.578, 0.0124 and 0.1989, respectively. Both a GA and a GBM are applied at free stream Mach number of 4.6 and altitude 43km.

Table 1. The comparison of optimization results with different optimization techniques

Optimization Method	C_D	C_h	Num. of Analysis Runs
GBM	0.2440	0.0107	5×20
RSM(21Points)	0.2475	0.0099	21+1

The optimization results and the number of analysis runs are compared at Table 1 with different optimization methods, and demonstrates the usefulness of the RSM.

From the maximum number of 81 for the 3k model, design points with negative radius, or radius over 125% of the nose faring base, which are physically unacceptable, are excluded. It would be difficult to find a fairing geometry with good aerodynamic characteristics from these rather random or peculiar shapes. Among the design points left, 16 points has been initially selected by utilizing the D-optimality condition.

Table 2. Comparison of drag and heat transfer according to the number of experimental points

Num. of Exp. Points	Num. of Iterations	C_D	C_h	Q	Drag Reduction
16	Iteration 1	0.2692	0.8504	0.4242	7.97 %
	Iteration 2	0.2679	0.8512	0.4317	8.39 %
23	Iteration 1	0.2688	0.8504	0.4306	8.10 %
	Iteration 2	0.2681	0.8510	0.4261	8.34 %
	Iteration 3	0.2674	0.8518	0.4294	8.57 %

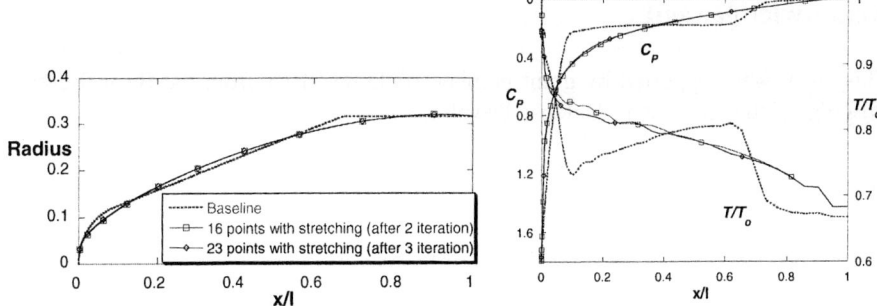

Fig. 7. Results of nose faring configurations (*left*), surface pressure and temperature distributions (*right*)

The refined RSM, the stretching parameter sub-optimization and the iterative regression model revision technique, are utilized to obtain the optimum configuration within minimum number of analysis calculations.

The results are summarized in table 2 and fig. 7. To demonstrate the validity of the results using 16 experimental points, the response surface with 23 experimental points, appropriate number of points for the 15 regression coefficient, is built and the results are compared. As can be seen in table 1, the drag coefficient is decreasing with the addition of the experimental points while satisfying the heat transfer constraints. But the change of the drag coefficient itself is relatively minor, about 0.6% difference. Hence, resulting body shapes, and pressure and temperature distributions over the body surface of 16 point and 23 point cases are almost the same. Compared with the baseline configuration, pressure near the stagnation points are alleviated greatly, which results in the reduction of total drag (fig. 7).

5 Conclusions

A refined RSM is proposed and implemented for the efficient nose fairing shape design of a space launcher using typical GBM and GA. From this study following conclusions can be made.

In order to reduce the computational cost, the RSM with various optimization methods is successfully implemented for the total drag minimization problem with given design constraints. The optimization strategies, the design space transformation technique, the stretching parameter sub-optimization and the iterative regression model revision techniques, have been defined and implemented to obtain the optimum configuration within minimum number of analysis calculations. Same optimization result as that of gradient-based method without any system approximation has been obtained within only a quarter of analysis runs compared with GBM. This demonstrates the effectiveness of the current optimization strategies with the refined RSM. The results of this study can be directly applied to the multidisciplinary system design optimization problems with many design variables.

Acknowledgement

This work was supported by grant number ADD-03-01-01 from the Basic Research Program of the Agency for Defense Development.

References

1. Vanderplaats, G. N., and Hicks, R. M., "Numerical Airfoil Optimization Using a Reduced Number of Design Coordinates", NASA TM X-73, 151, 1976.
2. Y. S. Ong, P. B. Nair, and A. J. Keane. "Evolutionary optimization of computation-ally expensive problems via surrogate modeling", American Institute of Aeronauticsand Astronautics Journal, Vol. 41, No. 4 pp.687-696, 2003.
3. Myers, R. H. and Montgomery, D. C., *Response Surface Methodology*, John Wiley & Sons Inc., 1995.
4. Venter, G., Haftka, Raphael T., Starnes, James H. Jr., "Construction of Response Surface Approximations for Design Optimization," AIAA Journal, Vol. 36, No. 12, 1998, pp.2242-2249.
5. Susan Burgee, Anthony A. Giunta, Vladimir Balabanov, Bernard Grossman, William H. Mason, Robert Narducci, Raphael T. Haftka, Layne T. Watson, " A Coarse-Grained Parallel Variable- Complexity Multidisciplinary Optimization Paradigm," The International Journal of Supercomputer Applications and High Performance Computing, Vol. 10, No. 4, 1996, pp.269-299.
6. Resit Unal, Roger A. Lepsch, and Mark L. McMillin, "Response Surface Model Building and Multidisciplinary Optimization Using D-Optimal Designs," AIAA-98-4759, September 1998.
7. Harten, A., "High Resolution Shemes for Hyperbolic Conservation Laws," Journal of Computational Physics, Vol. 49, No. 3, 1983, pp 357-393.
8. Yee, H. C., Kolpfer, G. H. and Montague, J. L., "High Resolution Shock Capturing Schemes for Inviscid and Viscous Hypersonic Flows," NASA TM 100097, April 1988.
9. J-W Lee, B-Y Min, Y-H. Byun, and S-J Kim, "Multi-Point Nose Shape Optimization of Space Launcher Using Response Surface Method," Journal of Spacecraft and Rockets, accepted for the publication, 2004.
10. Holden, M. S. and Moselle, J. R., "Theoretical and Experimental Studies of the Shock Wave-Boundary Interaction on Compression Surfaces in Hypersonic Flow," AIAA 89-0458, Jan. 1989.
11. Rudy, D. H., Thomas, J. L., Kumar, A., Gnoffo, P., and Chakravarthy, S. R., "A Validation Study of Four Navier-Stokes Codes for High-Speed Flows," AIAA 89-1838, June 1989.
12. J. Y. Lee and M. J. Lewis, "Numerical Study of the Flow Establishment Time in Hypersonic Shock Tunnels," Journal of Spacecraft and Rockets, Vol. 30, No. 2, Mar. 1993, pp. 152-163.
13. Jeong-Soo Tak, Yung-Hwan Byun, Jae-Woo Lee, Jang-Yeon Lee, Chul-Jun Huh, and Byung-Chul Choi, "A Numerical Study on the Performance Analysis of Shock Tunnel," Proceedings of the Korea Society of Computational Fluids Engineering Spring Annual Meeting, May 2000.

14. OT Users Manual, Vanderplaats Research & Development, Inc., 1995.
15. Michalewicz, Z., Genetic Algorithms + Data Structures = Evolution Programs, Springer-Verlag, 1996
16. Chae, Y.-S. et al., "Research and Development of KSR-III(I) Final Report," Korea Aerospace Research Institute, 1998.

MEDIC: A MDO-Enabling Distributed Computing Framework

Shenyi Jin[1,2], Kwangsik Kim[1,2], Karpjoo Jeong[1,2,*], Jaewoo Lee[1,3],
Jonghwa Kim[1,4], Hoyon Hwang[6], and Hae-Gook Suh[1,5]

[1] Center for Advanced E-System Integration Technology, Konkuk Univ., Seoul, Korea
[2] Department of Internet and Multimedia Engineering, Konkuk Univ., Seoul, Korea
[3] Department of Aerospace Engineering, Konkuk Univ., Seoul, Korea
[4] Department of Industrial Engineering, Konkuk Univ., Seoul, Korea
[5] Department of Advanced Technology Fusion, Konkuk Univ., Seoul, Korea
[6] Department of Aerospace Engineering, Sejong Univ., Seoul, Korea
jeongk@konkuk.ac.kr

Abstract. A MDO framework is a collaborative distributed computing environment that facilitates the integration of multi-disciplinary design efforts to achieve the global optimum result among local mutually-conflicting optimum results on heterogeneous platforms throughout the entire design process. The challenge for the MDO framework is to support the integration of legacy software and data, workflow management, heterogeneous computing, parallel computing and fault tolerance at the same time. In this paper, we present a Linda tuple space-based distributed computing framework optimized for MDO which is called MEDIC. In the design of MEDIC, we classify required technologies and propose an architecture in which those technologies can be independently implememnted at different layers. The Linda tuple space allows us to make the MEDIC architecture simple because it provides a flexible computing platform where various distributed and parallel computing models are easily implemented in the same way, multi-agents are easily supported, and effective fault tolerance techniques are available. A prototype system of MEDIC has been developed and applied for building an integrated design environment for super-high temperature vacuum furnaces called iFUD.

1 Introduction

For the last few years, engineering design processes have been changing from sequential to concurrent and parallel methodology. *Multi-disciplinary Design Optimization* (MDO) is one of these methodologies for the design of complex engineering systems that coherently exploits the coordination of mutually interacting phenomena [2,3,4]. The MDO methodology shortens the design processes, saves design costs and improves performance of the final products, and therefore

---

* Corresponding author.

is widely used in complex engineering design projects, such as aircrafts, mobiles, and missiles.

In order to use the MDO methodology, we need an *integrated design environment* which combines various legacy analysis software/data, optimization code, CAD, DBMS, and GUI. The development of such integrated design environment requires a great deal of time and effort because a number of challenging distributed computing issues such as workflow management with legacy software/data support, parallel computing, fault tolerance, and heterogeneous computing must be addressed [4]. A *MDO framework* is a collaborative distributed computing system intended to facilitate the development of MDO-based integrated design environments.

Major challenges for the MDO framework are to support: (1) integration of legacy software and data on heterogeneous platforms, (2) workflow management of various engineering design activities including both computational tasks and human interactions, (3) domain-specific design functionality, (4) distributed and parallel computing, (5) fault tolerance. Until now, there has been little research on how to address those issues together in the IT research community. Furthermore, some of issues such as the support for domain-specific functionality have been neglected.

In this paper, we present a distributed computing framework optimized for MDO which is called MEDIC (MDO-Enabling DIstributed Computing Framework). Novel features of MEDIC are:

- *Linda tuple space-based Design and Implementation.* The Linda model allows us to make the design and implementation of MEDIC simple and flexible due to nice features of the model: time/space decouling, language/platform independence, the availability of effective fault tolerant mechanisms, and easy support for agent technology and workflow [5–7].
- *Technology Classification.* We classify technologies required for MDO frameworks into five classes and propose a layered architecture where independence among them is maintained and changes to one layer does not affect other layers siginificantly.

Because of these features, the design and implementation of MEDIC is simpler and more flexible than other MDO frameworks such as ModelCenter [10], iSIGHT [11], IMAGE [12].

The paper is structured as follows. Section 2 explains requirements for MDO frameworks. In Section 3 and 4, we present the design and implementation of MEDIC, respectively. We discuss related work in Section 5 and conclude this paper in Section 6.

2 Requirements for MDO Frameworks

We summarize a list of requirements for the MDO framework proposed in [4]. In [4], MDO framework requirements are listed in terms of the architectural design, problem formulation construction, problem execution, and information access.

From the aspect of the architectural design, the architecture must be flexible enough to handle needs for system changes. For the problem formulation construction, flexible workflow management mechanisms must be provided to support complicated optimization methods. Effective integration of heterogeneous legacy software and data is also required. In terms of the problem execution, distributed and parallel computing must be required. During the problem execution, failures must be handled and user interaction is also needed. In addition, tasks may be needed to run in a batch mode. From the information access point of view, database and CAD systems must be supported. The user must be able to visualize intermediate and final optimization and analysis results.

These requirements are summarized as following:

- *Complex job execution support such as workflow management.* It is required to support multi-level optimization techniques involving sub-optimizations, or stop, pause and restart the process at runtime.
- *Integration of multi-disciplinary heterogeneous software and data.* Legacy analysis and optimization codes, CAD system, database system and visualization tools.
- *High performance and fault tolerant computing technology support.* Distributed and parallel computing technologies are required to handle large-scaled MDO problems. Fault management and recovery systems are also needed.

3 System Design

3.1 Linda Model

The MEDIC architecture is based on the Linda tuple space model [5,6,7]. Here we give a brief introduction to the model. Linda is based on a shared memory model. In Linda, processes in an application cooperate by communicating through the shared memory called tuple space.

Each tuple in tuple space contains a sequence of data elements of basic types such as integers, floats, characters, and arrays of these types. In Linda, processes access tuple space by using the following four basic operations: **out**, **in**, **rd**, and **eval**. **Out** takes a sequence of typed expressions as arguments. It evaluates them, constructs a tuple from them, and inserts the tuple into tuple space. Like **out**, **eval** creates a tuple from its arguments, but a new process is created to evaluate the arguments. In Linda, this is the only way to create a new process.

In and **rd** take a typed pattern for a tuple as their argument and retrieve a tuple to match the pattern in an associative manner. A pattern is a series of typed fields; some are values and others are typed place-holders. A place-holder is prefixed with a question mark. For example, consider a tuple space operation: ("string", ?f, ?i, y). The first and last fields are values; the middle two fields are place-holders. If there are multiple matching tuples, then one is randomly

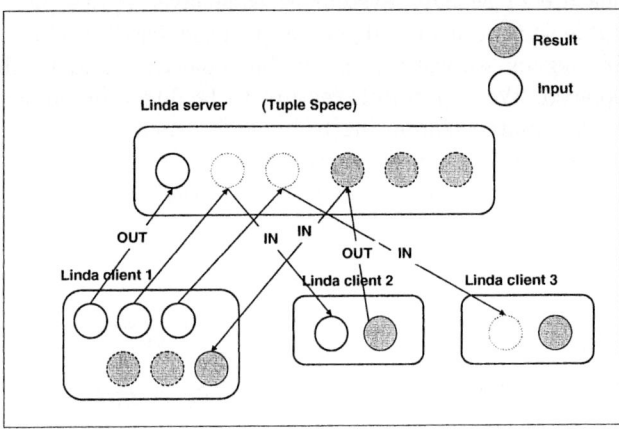

Fig. 1. Linda Model

selected and retrieved. If no matching tuple is found, **in** and **rd** lock until a matching tuple is inserted. The difference is that **in** is destructive (i.e., removes the tuple) while **rd** is not.

The advantages of the Linda model can be summarized as: time decoupling, space decoupling, platform independence and implementation language independence. These advantages facilitate the implementation of distributed applications, like MDO-based design systems, whose components are written in various languages (e.g. Fortran, C, Matlab, etc.) on heterogeneous platforms. In addition, the Linda tuple space model allows us to handle communication, synchronization, shared data management, and process invocation by a single mechanism (i.e., tuple space). This feature facilitates designing and implementating distributed computing systems significantly.

3.2 Layered Architecture

Regarding required technologies for MDO frameworks, we classify required technologies for the MDO framework into five categories: MDO technology, problem-specific technology, distributed computing technology, system integration technology and distributed data management technology.

The problem-specific technology corresponds to diverse legacy or commercial software, such as legacy analysis programs, CAD data and experiment data. the MDO technology is programs and methodologies for concurrent design activities. The distributed computing technology is domain-independent high performance computing systems on heterogeneous platforms. The system integration technology is systems that support seamless integration. Examples are wrapper, agents and data conversion technologies. The distributed data management technology enables the management of legacy data and the global access to data on distributed platforms.

A novel feature of MEDIC system is: each layer is designed as independently as possible so that advanced general purpose distributed middleware and domain-specific legacy software/data can be separately and flexibly managed and easily replaced. We expect this feature to facilitate extending MEDIC to support other distributed computing technologies such as Globus Toolkit [14], Web Services [15] and JINI [16] frameworks.

The focus of this paper is on three layers: distributed computing, system integration, and distributed data management.

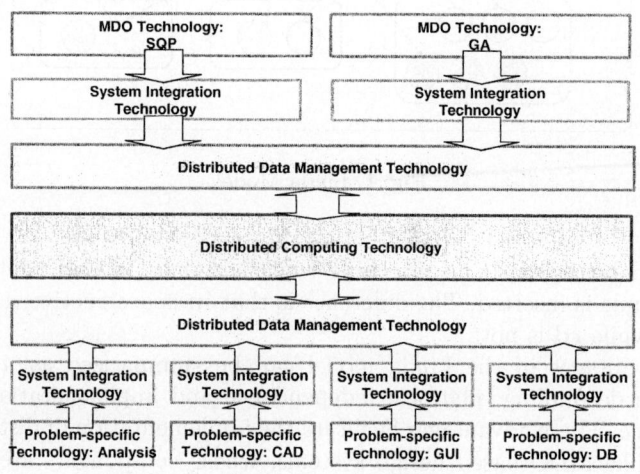

Fig. 2. MEDIC Layered System Structure

3.3 Distributed Computing Layer: Linda Tuple Space-Based Middleware

The scheduler of MEDIC is designed based on the Linda model. Regarding scheduling for a MDO-based design system, the model has a few advantages. First, it provides a tuple space shared storage mechanism where we can easily represent various schedule information as tuples and such schedule information is globally accessible. Second, schedule information represented as tuples can be directly executed by the Linda runtime system and parallel computing is automatically supported.

In MEDIC, we manage schedule information as a set of tuples in tuple space, and the Linda runtime system allows agents to be automatically invoked by these tuples. A simplied version of the data structure for schedule information is as follows. A single tuple is used to represent the status of each task. The first field is the name of the task. the second field shows the number of other tasks that must be finished in order to start the task. Therefore, if the number is zero, then the task is ready to run and the agent responsible for the task can start it at any time. The information in the third field is the list of other tasks to wait for the task to finish. Once the task is done, the agent responsible for it reads the list of waiting tasks in the second field and decrements the number of the second field in the tuple for each task in the list.

Fig. 3 gives a simple example of schedule where there are five jobs (i.e., **A**, **B**, **C**, **D**, and **E**) and dependencies among them (i.e., arrows). Each job can be wrapped application programs, database systems, or optimization programs. This schedule is represented as a set of tuples shown in Fig. 3. In the example, job **A** has no other job to wait and therefore, it can be started immediately. On the other hand, jobs **B**, and **C** must wait for one job to be finished in order to start. Tuples in the box of the right hand side show what jobs are waiting for a given job to finish. In the example, jobs **B** and **C** are waiting for job **A** to finish.

For example, the agent for **A** attempts to retrieve a tuple by the **in** operation in("A", 0, ?WAITING_LIST). If a matching tuple is available (i.e., there is a waiting job), then the agent is able to retrieve the corresponding tuple and to execute the request. Otherwise, the Linda runtime system automatically blocks the agent until a new matching tuple (i.e., a new reay task for the agent) is inserted. In Fig. 3, as soon as the agent for **A** finishes the task **A**, it decrements the numbers in the second fields of the tuples for task **B** and **C**. Then, tasks **B** and **C** become ready to run.

3.4 System Integration Layer: Wrappers and Agents

In addition, we use a wrapper technology that turns legacy software systems into integratible system components. It provides mechanisms for invocation, input preparation, output post-processing, and data structure transformation. The functionality of the wrappers are: (1) make conversion between standard data formats and component-specific data formats, (2) prepare input files and invoke legacy programs, (3) detect the termination of the execution [1]. In MEDIC system, all the programs must be wrapped as MEDIC components by the wrapper system. In the current system, we support five wrapper systems: analysis, optimize, GUI, database and CAD wrappers.

In MEDIC, we use the agent technology to integrate wrappers into distributed computing environments. In the Linda model, we can implemente an agent by tuple space operations easily. First, the tuple space allows agents to communicate with each other in a simple way. Second, a workflow can be represented as a set of tuples and then these tuples can be used to invoke agents. Finally, synchronization among agents can be easily implemented in Linda.

3.5 Distributed Data Management Layer

The distributed data management consists of:

- *Distributed File System*: Real design data and temporary data are stored as files. A distributd file system which is a collection of FTP servers is used to manage those files.

- *Linda tuple space-based Metadata Management*: Metadata about those files are stored as tuples in the tuple space. Therefore, metadata can be accessed by all MEDIC components.

MEDIC components such as agents look for metadata in the tuple space, first and then retrieve real data from the distributed file system.

For access to DB data, MEDIC uses existing commodity DB access methods such as ODBC or JDBC.

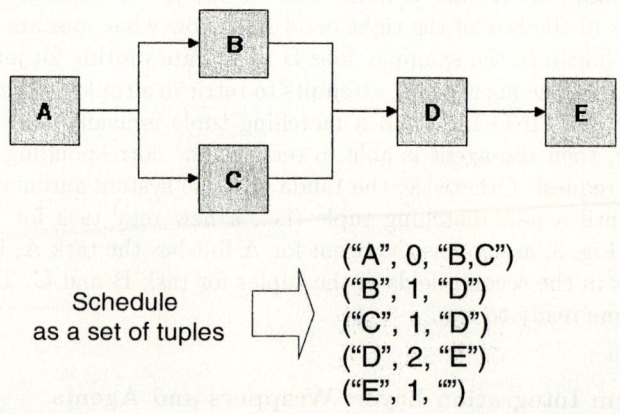

Fig. 3. Schedule Example

4 Implementation and Application

In the current implementation of MEDIC, we chose a fault tolerant version of the shared memory-based Linda model, called Persistent Linda (shortly PLinda [13]) which is implemented by C++. The agent and scheduler, which can be seen as PLinda clients are also implemented by C++. Linux and MS Windows are both supported.

By using MEDIC, we developed a prototype of an integrated design environment for a super high temperature vacuum furnace which is called iFuD (intelligent **F**urnace **D**esign). In this prototype development, one GUI component, three analysis components, two optimum components, one DBMS component, one CAD component and visualization tool are integrated. The MDO components integrated into iFUD are shown in the Table 1.

Fig. 4 shows some snapshots of the iFuD system.

5 Related Work

ModelCenter [10], iSIGHT [11], IMAGE [12] are computational frameworks or problem solving environments that offers the capability to solve MDO problems

Table 1. iFUD Software Components

COMPONENT	DESCRIPTION	IMPLEMENTATION LANGUAGE	SOFTWARE TYPE
Analysis	Conduction and radiation analysis program	Fortran and C mixed	Legacy software
	Conduction, radiation and convection analysis program	Fortran and C mixed	Legacy software
	DB-based analysis program	C	Legacy software
Optimization	Optimization program (GA)	Fortran and C mixed	Legacy software
	Optimization program (DOT)	Fortran and C mixed	Legacy software
GUI	Graphic user interface	MFC/ Visual Basic	Legacy software
CAD	AutoCAD 2002	Visual Basic	Commercial software
Database	MS Access	Visual Basic	Commercial software
Visualization Tools	Formular One	Visual Basic	Commercial software

through networks. Common drawbacks with these frameworks are monolithic-based design in which distributed computing facilities and MDO supports are tightly-coupled. State of art distributed computing systems such as EAI [13], Grid computing [14], Web Services [15] and JINI [16] provide powerful computing features but lack domain-specific support. MEDIC intends to support domain-specific functions on top of general purpose distributed computing system.

6 Conclusions and Future Work

In order to develop a MDO-based design environment, we are required to address challenging distributed computing issues such as legacy software/data support, workflow management with both computational tasks and human interventions, distributed/parallel computing, fault tolerance, and heterogeneous computing. For this reason, the development of an MDO-based integrated design environment for an application domain takes a large amount of time and efforts. MDO frameworks are intended to facilitate the development of such MDO environment.

In this paper, we presented a Linda tuple space-based collaborative distributed computing framework for MDO-based design processes which is called MEDIC. We summarized requirements for MDO frameworks and discussed how the MEDIC system addresses them. We explained how the Linda model simplifies the design and implementation of a MDO framework. A prototype system of MEDIC is implemented and used to develop a super-high temperature vacuum furnace design system called iFUD. In iFUD, one GUI component, three analysis components, two optimization components, one DBMS component, one CAD

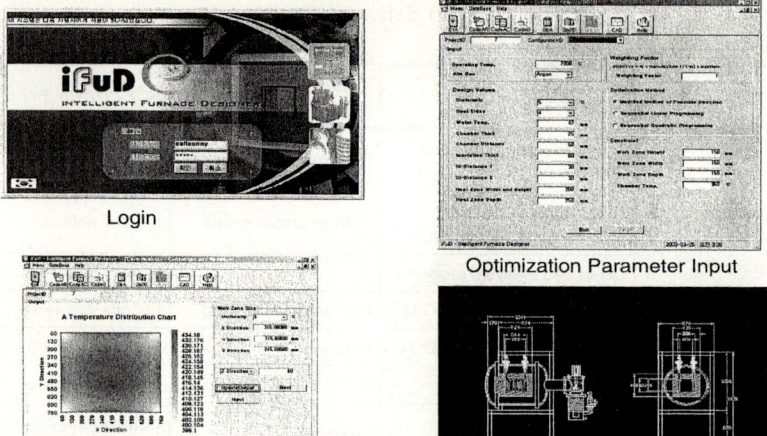

Fig. 4. iFUD Screen Snapshots

component and visualization tool are integrated on heterogeneous platforms including Linux and MS Windows.

The current implementation of MEDIC is based on the PLinda system. We plan to extend the MEDIC architecture to support other distributed computing systems such as JINI/JavaSpaces, Web Services, and Globus. Also, we plan to extend the MEDIC distributed data management system to be based on distributed file systems and DBMSs in an uniform way. Currently, the security support is limited and will be enhanced in the near future.

Acknowledgement

This paper was supported by Konkuk University in 2004.

References

1. Lander, S.: Issues in multiagent design systems. IEEE Expert, pp.18-26 (1997)
2. AIAA MDO Technical Committee: http://www.aiaa.org
3. Sobieszczanski-Sobieski, J., Haftka, R.: Multidisciplinary aerospace design optimization - Survey of recent developments, AIAA-1996-711 (1996)
4. Salas, A., Townsend, J.: Framework Requirements for MDO application development, AIAA-98-4740, (1998)
5. Gelernter, D.: Generative Communication in Linda, ACM Transactions on Programming Languages and Systems, Vol. 7, No. 1, pp.80-112, Jan. 1985.
6. Gelernter, D.: Coordination Languages and Their Significance, Communications of the ACM, Vol. 35, No. 2, pp.96-107, Feb. 1992,

7. N Carriero: Implementing Tuple Space Machines PhD thesis, Yale University, Department of Computer Science (1987)
8. Jeong, K., Shasha, K., Talla, T., Wyckoff, P.: An Approach to Fault-tolerant Parallel Processing on Intermittently Idle, Heterogeneous Workstations, International Symposium on Fault-Tolerant Computing. (1997)
9. Jin, S., Jeong, K., Lee, J., Kim, J., Jin, Y: (2004) MEDICOS: An MDO-Enabling DIstributed COmputing System, KISS, ISSN 1598-5164 (2004)
10. http://www.phoenix-int.com/about/MC6release.html
11. http://www.engineous.com/resources.htm
12. Hale, M. Craig, J.: IMAGE: a design integration framework applied to the high speed civil transport
13. Ruh, W., Maginnis, F., Brown, W.: Enterprise Application Integration A Wiley Tech brief, Wiley computer Publishing, (2001)
14. http://www.globus.org/
15. http://www.w3.org/2002/ws/
16. http://www.sun.com/software/jini/
17. http://www.corba.org

Time and Space Efficient Search for Small Alphabets with Suffix Arrays

Jeong Seop Sim

School of Computer Science and Engineering, Inha University, Incheon, Korea
jssim@inha.ac.kr

Abstract. To search a pattern P in a text, index data structures such as suffix trees and suffix arrays are widely used. It is known that searching with suffix trees is faster than with suffix arrays in the aspect of time complexity. But recently, a few linear-time search algorithms for constant-size alphabet in suffix arrays have been suggested. One of such algorithms proposed by Sim et al. uses Burrows-Wheeler transform and takes $O(|P|\log|\Sigma|)$ time. But this algorithm needs too much space compared to Abouelhoda et al.'s algorithm to search a pattern.

In this paper we present an improved version for Sim et al.'s algorithm. It needs only $2n$ bytes at most if a given alphabet is sufficiently small.

1 Introduction

Suffix trees and suffix arrays are well-known important index data structures in diverse applications of string processing and computational biology. The suffix tree due to McCreight [16] is a compacted trie of all the suffixes of a string T. It was designed as a simplified version of Weiner's position tree [19]. The suffix array due to Manber and Myers [15] and independently due to Gonnet et al. [7] is basically a sorted list of all the suffixes of a string T.

There have been vigorous works on searching patterns with suffix arrays. Ferragina and Manzini [6] developed an $O(|P| + z\log^{1+\epsilon} n)$-time algorithm in a file compressed by the Burrows-Wheeler compression, where $0 < \epsilon < 1$ is a constant and z is the number of occurrences of P in a string of length n. Recently, Abouelhoda et al. [1] and Sim et al. [18] developed linear-time search algorithms in suffix arrays when the alphabet size is a constant. The former simulates suffix trees in suffix arrays and the latter uses backward search developed by Ferragina and Manzini in the context of compressed pattern matching [6]. In general, however, these algorithms take time $O(|P| \cdot |\Sigma|)$ and $O(|P| \cdot \log|\Sigma|)$, respectively. More recently, Choi et al. [3] proposed an algorithm which runs in $O(|P| \cdot \log|\Sigma|)$ time using $O(n\log|\Sigma| + |\Sigma| \cdot n\log\log n/\log n)$-bit space.

In this paper, we propose an improved version of Sim et al.'s algorithm [18] (in the aspect of space usage) that uses at most $3n$ bytes when a given alphabet is not large, that is, $|\Sigma| \leq 256$. Moreover, if the alphabet size is very small, as the case of DNA alphabet, our algorithm needs just $2n$ bytes at most, which is competitive compared with Abouelhoda et al.'s algorithm [1].

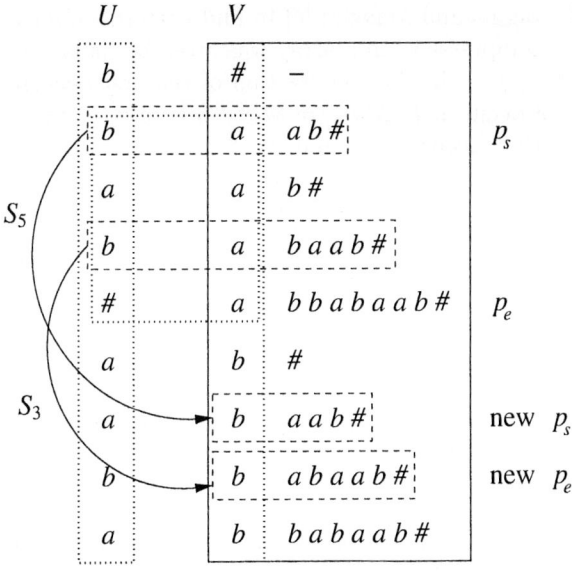

Fig. 1. Arrays U and V when $T = abbabaababbb\#$

This paper is organized as follows. In Section 2, we define some notations. In Section 3, we explain our search algorithm and the data structures. In Section 4, we conclude.

2 Preliminaries

We first give some definitions and notations that will be used in our algorithm. We will use the same notations used in Sim et al.'s algorithm [18]. Consider a string T of length n over an alphabet Σ. $T[i]$ denotes the ith symbol of string T and $T[i,j]$ the substring starting at position i and ending at position j in T. We assume that $T[n]$ is a special symbol $\#$ which is lexicographically smaller than any other symbol in Σ and appears only once in T. Let S_i, $1 \leq i \leq n$, denote the suffix of T that starts at position i. The suffix array A_T is the lexicographically ordered list of all suffixes of T. That is, $A_T[i] = j$ if S_j is lexicographically the ith suffix among all suffixes $S_1, S_2, \ldots, S_n$ of T.

Consider the problem of searching T for a pattern P over alphabet Σ. Let $p = |P|$ and $n = |T|$. Let σ_j be the jth smallest symbol in Σ and assume σ_0 is the special symbol $\#$.

We use two arrays: $V[i] = T[A_T[i]]$ and $U[i] = T[A_T[i] - 1]$ for $1 \leq i \leq n$ (assume $T[0] = \#$ for convenience), i.e., V is the array of the first symbols in the sorted list of all suffixes of T and U is the array of previous symbols of V. See Figure. 1. We use U and V only conceptually; we do not make them but access them in constant time with T and A_T. The idea of searching with U and V was developed by Ferragina and Manzini [6] to find patterns in a compressed file. A similar idea in a compressed suffix array was given by Sadakane [17].

Let $M[\sigma_j]$, $0 \leq j \leq |\Sigma|$, be the position of the first occurrence of σ_j in V. When σ_j does not occur in T, $M[\sigma_j] = M[\sigma_{j'}]$ where $\sigma_{j'}$ is the lexicographically smallest symbol that occurs in T and $\sigma_j < \sigma_{j'}$. Array M logically partitions V by each symbol σ_j occurring in T. We define a function $N : \{0, 1, \ldots, n\} \times \{\sigma_0, \ldots, \sigma_{|\Sigma|}\} \to \{0, 1, \ldots, n-1\}$. $N(i, \sigma_j)$ is the number of occurrences of σ_j in $U[1, i]$. For convenience, we assume $N(0, \sigma_j) = 0$ for $0 \leq j \leq |\Sigma|$. For example, if $T = abbabaab\#$, then $A_T = (9, 6, 7, 4, 1, 8, 5, 3, 2)$, $V = (\#, a, a, a, a, b, b, b, b)$, and $U = (b, b, a, b, \#, a, a, b, a)$. See Fig. 1. Also, $M = (1, 2, 6)$ and $N(1, \#) = 0$, $N(1, a) = 0, \ldots, N(3, b) = 2, \ldots, N(9, b) = 4$.

3 Search Algorithm and Data Structures

3.1 Search Algorithm

We search P in T by exactly the same way as Sim et al.'s algorithm [18]. That is, we search for P from the last symbol to the first symbol of P. Note that P occurs at position i of T if and only if the suffix S_i of T has P as its prefix. Assume we know all the positions where $P[h+1, p]$ occurs in T. Then we can find the positions where $P[h, p]$ appears in T as follows. If there exists any suffix of T that has $P[h+1, p]$ as its prefix and its previous symbol is $P[h]$, then $P[h, p]$ appears in T. Since U is the array of previous symbols of the sorted suffixes and the positions where $P[h+1, p]$ occurs in T are contiguous in A_T, we check if $P[h]$ exists in the corresponding positions by using U. If it does, we find the positions where $P[h, p]$ occurs in T.

Our algorithm is divided into p phases. Assume that M and N are available. At the beginning of the hth phase from $h = p$ to 1, we know all the positions where $P[h+1, p]$ occurs in T. In fact, we maintain the start position p_s and end position p_e of the contiguous block where $P[h+1, p]$ occurs. Initially, $p_s = 1$ and $p_e = n$. (At the beginning when $h = p$.) In the hth phase, we find all the positions where $P[h, p]$ occurs in T. That is, we update p_s and p_e so that all occurring positions of $P[h, p]$ in T are $A_T[p_s], \ldots, A_T[p_e]$. The new start and end positions are set to $M[P[h]] + N(p_s - 1, P[h])$ and $M[P[h]] + N(p_e, P[h]) - 1$, respectively. See [18] for more details.

3.2 Preprocessing and Data Structures of Previous Algorithm

We now explain how to preprocess A_T to construct array M and make data structures for function N. We can construct array M in $O(n)$ time by scanning V once and it needs $O(|\Sigma|)$ space. We first explain the $O(\log |\Sigma|)$-time solution for query $N(i, \sigma_j)$ shown in [18].

For an $O(\log |\Sigma|)$-time solution for query $N(i, \sigma_j)$, we divide U into blocks of size $|\Sigma|$. Let U^i for $1 \leq i \leq n/|\Sigma|$ be the ith block of U and s_i and e_i be the start position and end position of U^i, respectively. First, we make a two dimensional

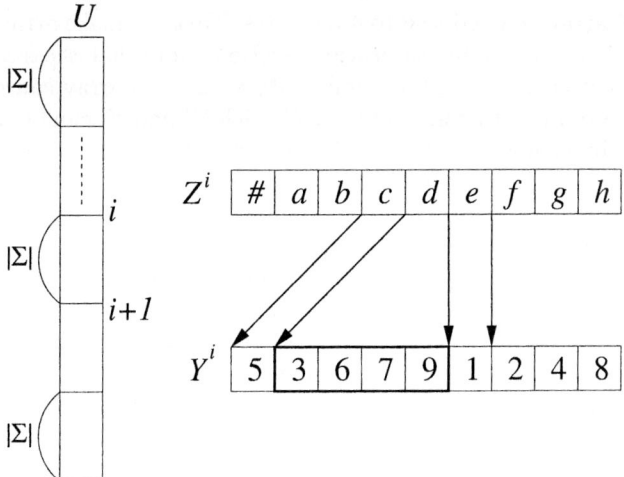

Fig. 2. Arrays Y^i and Z^i

array X of size $O(n)$. $X[i,j]$, $1 \leq i \leq n/|\Sigma|$ and $0 \leq j \leq |\Sigma|$, stores the number of occurrences of σ_j in $U[1, e_i]$. Array X can be made in $O(n)$ time by scanning U. Second, we make two arrays Y^i and Z^i for each U^i. Let n_j^i be the number of occurrences of σ_j in U^i and $a_j^i = \sum_{0 \leq k < j} n_k^i$. For $0 \leq j \leq |\Sigma|$, $Z^i[j]$ stores a_j^i and $Y^i[a_j^i + m]$ stores the place of mth occurrence of σ_j in U^i. Both Y^i and Z^i can be made in $O(|\Sigma|)$ time by scanning U^i since U^i has $|\Sigma|$ elements. See Figure 2.

Now we can answer query $N(i, \sigma_j)$ in $O(\log |\Sigma|)$ time. First, we find the block U^w such that $w = \lceil i/|\Sigma| \rceil$. Then, we access Z^w to find a_j^w and do a binary search on $Y^w[a_j^w + 1], \ldots, Y^w[a_{j+1}^w]$ to find the maximum k such that $Y^w[k] \leq i$. Then, $N(i, \sigma_j) = X[w-1, j] + k - a_j^w$.

3.3 Improved Data Structures

In most real computer systems, an integer ($< 2^{32}$) is represented with four bytes. Thus, if we implement Sim et al.'s algorithm [18] on a general computer system, we need $12n$ bytes in total because X array needs $4n$ bytes and each Y^i and Z^i, $1 \leq i \leq n/|\Sigma|$, needs $4|\Sigma|$ bytes, respectively.

But, when $|\Sigma| \leq 256$ (in mosts cases, $|\Sigma|$ is not so large), we can reduce the size of each Y^i and Z^i to $|\Sigma|$ bytes, and thus the total space required to search patterns is $6n$ bytes [3]. Moreover, we can do much better by the following way. The basic idea is that we divide U into blocks of size $c|\Sigma|$ ($c \geq 1$) to reduce the size of X, which is our main contribution.

First, we logically divide U into blocks of size $c|\Sigma|$ and define X array as follows. The two dimentional array $X[i,j]$, $1 \leq i \leq n/c|\Sigma|$ for $c \geq 1$, stores the number of occurrences of σ_j in $U[1, e_i]$. That is, we store the number of occurrences of σ_j at every $c|\Sigma|$-th character in U. Note that the space needed to implement X array now reduced to $4n/c$ bytes. Next, we make arrays Y^i and Z^i for each U^i ($1 \leq i \leq n/c|\Sigma|$) as previous section. Note that the size of each Y^i and Z^i does not change, but the number stored in each array can be increased by c times. If we take c to make $c|\Sigma| \leq 256$, each Y^i and Z^i can be implemented with still $|\Sigma|$ bytes and thus the total space required to implement all Y^is and Z^is for $1 \leq i \leq n/c|\Sigma|$ is $2n$ bytes.

If $c \geq 4$ and $c|\Sigma| \leq 256$, then we can implement X in n bytes, and every Y^i and Z^i in n bytes respectively. Thus, the total space needed is at most $3n$ bytes. Moreover, if the $c|\Sigma| \leq 16$, we can implement all the arrays in just $2n$ bytes. For example, when $|\Sigma| = 4$ (DNA alphabet) and we choose $c = 4$, X can be implemented in n bytes and each Y^i and Z^i ($1 \leq i \leq n/16$) can be implemented in 16 bytes. Therefore, when DNA alphabet is given, we can implement all the arrays in just $2n$ bytes.

We can answer query $N(i, \sigma_j)$ in $O(\log |\Sigma|)$ time just the way described in Sim et al. [18]. First, we find the block U^w such that $w = \lceil i/c|\Sigma| \rceil$. Then, we access Z^w to find a_j^w and do a binary search on $Y^w[a_j^w + 1], \ldots, Y^w[a_{j+1}^w]$ to find the maximum k such that $Y^w[k] \leq i$. Then, $N(i, \sigma_j) = X[w-1, j] + k - a_j^w$. It is easy to see that the time complexity does not increase because we do binary search on Y^i of size $c|\Sigma|$ which is still $O(|\Sigma|)$.

4 Conclusion

In this paper, we proposed an improved version of Sim et al.'s linear-time search algorithm. This algorithm uses as much space as Abouelhoda et al.'s algorithm [1] when the given alphabet is not large, but runs faster in the aspect of time complexity, that is, our algorithm runs in $O(|P| \cdot \log |\Sigma|)$ time while Abouelhoda et al.'s algorithm [1] runs in $O(|P| \cdot |\Sigma|)$ time.

References

1. M.I. Abouelhoda, E. Ohlebusch, and S. Kurtz: Optimal exact string matching based on suffix arrays, *International Symposium on String Processing and Information Retrieval* (2002), LNCS 2476, 31–43.
2. S. Burkhardt and J. Kärkkäinen: Fast lightweight suffix array construction and checking, *Symp. Combinatorial Pattern Matching* (2003), LNCS 2676, 55–69.
3. Y.W. Choi, J.S. Sim, and K. Park: Time and space efficinet search with suffix arrays, *Journal of Korea Information Science Society*, accepted.
4. M. Farach: Optimal suffix tree construction with large alphabets, *IEEE Symp. Found. Computer Science* (1997), 137–143.
5. M. Farach-Colton, P. Ferragina and S. Muthukrishnan: On the sorting-complexity of suffix tree construction, *J. Assoc. Comput. Mach.* 47 (2000), 987-1011.
6. P. Ferragina and G. Manzini: Opportunistic data structures with applications, *IEEE Symp. Found. Computer Science* (2001), 390–398.
7. G. Gonnet, R. Baeza-Yates, and T. Snider: New indices for text: Pat trees and pat arrays. In W. B. Frakes and R. A. Baeza-Yates, editors, Information Retrieval: Data Structures & Algorithms, pages 66-82. *Prentice Hall* 1992.

8. D. Gusfield: Algorithms on Strings, Trees, and Sequences, *Cambridge Univ. Press* 1997.
9. D. Gusfield: An "Increment-by-one" approach to suffix arrays and trees, *manuscript* 1990.
10. R. Hariharan: Optimal parallel suffix tree construction, *J. Comput. Syst. Sci.* 55 (1997), 44–69.
11. W. Hon, K. Sadakane, and W. Sung: Breaking a time-and-space barrier in constructing full-text indices, *IEEE Symp. Found. Computer Science* (2003), accepted.
12. J. Kärkkäinen and P. Sanders: Simple linear work suffix array construction, *Int. Colloq. Automata Languages and Programming* (2003), LNCS 2719, 943–955.
13. D. Kim, J.S. Sim, H. Park, and K. Park: Linear-time construction of suffix arrays, *Journal of Discrete Algorithms* 3/2-4 (2005), 126-142.
14. P. Ko and S. Aluru: Space efficient linear time construction of suffix arrays, *Symp. Combinatorial Pattern Matching* (2003), LNCS 2676, 200–210.
15. U. Manber and G. Myers: Suffix arrays: A new method for on-line string searches, *SIAM J. Comput.* 22 (1993), 935–938.
16. E.M. McCreight: A space-economical suffix tree construction algorithm, *J. Assoc. Comput. Mach.* 23 (1976), 262–272.
17. K. Sadakane: Succinct representation of lcp information and improvement in the compressed suffix arrays, *ACM-SIAM Symp. on Discrete Algorithms* (2002), 225–232.
18. J.S. Sim, D.K. Kim, H. Park, and K. Park: Linear-time search in suffix arrays, *Journal of Korea Information Science Society*, accepted.
19. P. Weiner: Linear pattern matching algorithms, *Proc. 14th IEEE Symp. Switching and Automata Theory* (1973), 1–11.

Optimal Supersonic Air-Launching Rocket Design Using Multidisciplinary System Optimization Approach

Jae-Woo Lee, Young Chang Choi, and Yung-Hwan Byun

Center for Advanced e-System Integration Technology, Konkuk University,
Seoul, 143-701, Korea
{jwlee, wincyc, yhbyun}@konkuk.ac.kra

Abstract. Compared with the conventional ground rocket launching, air-launching has many advantages. However, comprehensive and integrated system design approach is required because the physical geometry of air launch vehicle is quite dependent on the installation limitation of the mother plane. Given mission objective is to launch 7.5kg nano-satellite to the target orbit of 700km x 700km using the mother plane, F-4E Phantom. The launching altitude and velocity are 12km, Mach number 1.5, respectively. As the propulsion system, a hybrid rocket engine is used for the first stage, and the solid rocket motors are used for the second and third stages. The total mass, length and diameter constraints of the rocket are imposed by the mother plane. The system design has been performed using the sequential optimization method. Gradient based SQP(Sequential Quadratic Programming) algorithm is employed. Analysis modules include mission analysis, staging, propulsion analysis, configuration, weight analysis, aerodynamics analysis and trajectory analysis. As a result of system optimization, a supersonic air launching rocket with total mass of 1272.61kg, total length of 6.43m, outer diameter of 0.60 m and the payload mass of 7.5kg has been successfully designed.

1 Introduction

Because the launch cost per unit mass will grow as the weight of the satellite become smaller, either several nanosats must be launched together or, the nanosat launched with the large satellite. Hence the launch schedule and the operation of the satellite are limited. Therefore, new launching method which can launch the nanosat individually with low launching cost, 'air-launching' can be a solution[1][2].

The air-launching rocket needs multidisciplinary design which considers propulsion, aerodynamics, trajectory and weight analysis at the same time because payload weight, total length, diameter are constrained by mother plane.

In this study, conceptual design process will be defined for the air-launching rocket by including analysis modules like mission analysis, staging, propulsion analysis, configuration, weight analysis, aerodynamics analysis and trajectory analysis. Also, multidisciplinary system design approaches will be studied and implemented for the optimal three-stage air-launching rocket design which satisfies given mission requirement.

2 Optimal Design of the Air-Launching Rocket

2.1 Mission Requirements

For the nanosat air launch vehicle, F-4E Phantom is selected as the mother plane. Dimensions and configuration of air-launching rocket are determined based on the centerline auxiliary fuel tank : 7.11m in length, 0.87m in diameter, and 600gal in volume. Thus, air-launching rocket has the design constraints of 6.5m in length, 0.6m in diameter to avoid any collisions during take-off. The air-launching mission requirements are defined by considering the feasibility of the installation of the payload to F-4E Phantom.

- Payload Mass : 7.5kg
- Target Orbit : 700 km(perigee)×700 (apogee) km, Circular Orbit
- Mother Plane : F-4E Phantom
- Launch Altitude : 12km
- Launch Velocity : Mach 1.5
- Propulsion System
    - First Stage : HTPB + LOX hybrid propulsion system
    - Second and Third Stage : HTPB/AP/Al solid propulsion system
- Launch vehicle constraints imposed by the mother plane[3]
    - Launcher Length : Less than 6.5m
    - Launcher Diameter : Less than 0.6m
    - Launcher Mass : Less than 1800kg

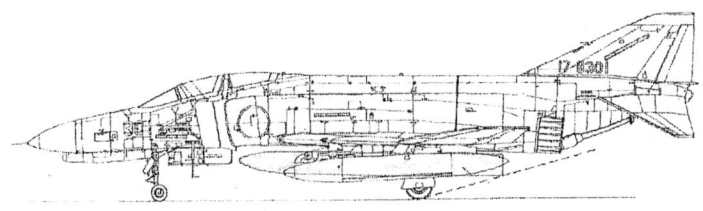

Fig. 1. F-4E Phantom with air-launching rocket

2.2 Multidisciplinary System Optimization

Figure 3 shows the optimization process which includes Step 1(TOGW optimization and Step 2(Trajectory optimization). Step 1 is performed to minimize the take off gross weight and Step 2 is to maximize payload mass, while satisfying the given mission constraints.

Step 1 optimization includes several analyses from mission analysis to aerodynamic analysis and Step 2 optimization includes only trajectory analysis. Optimization formulation of the each step is given below. The optimization algorithm for this study is SQP(Sequential Quadratic Programming)[4].

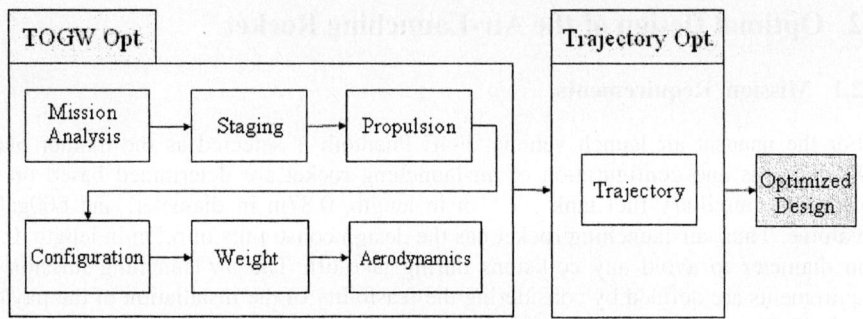

Fig. 2. Multidisciplinary System Optimization Process

• *Step 1 : TOGW Optimization*

Objective Function

   Minimize TOGW, TOGW : Take Off Gross Weight(kg)

Constraints

   $V_1 + V_2 + V_3 = 100.0$ (%), Sum of the Velocity Fractions
   $0.1 \leq Re_i \leq 0.3$, Re_i : Nozzle Exit Radius of Each Stage(m), i=1-3
   $0.55 \leq D \leq 0.6$, D : Max. Diameter of the Rocket
   $6.0 \leq L \leq 6.5$, L : Total Length of the Rocket
   $1000 \leq TOGW \leq 1800$
   $0.10 \leq \lambda_i \leq 0.16$, λ_i : Structure Coefficient of Each Stage, i=1-3

Design Variables

   $30 \leq V_i \leq 40$, V_i : Velocity Fraction of each stage(%), i=1-3
   $100 \leq Gox_i \leq 350$, Gox_i : 1st Stage Initial Oxidizer Mass Flux(kg/m^2-sec)
   $3.0 \leq P_{c1} \leq 6.0$, P_{c1} : 1st Stage Chamber Pressure(MPa)
   $30 \leq e_1 \leq 70$, e_1 : 1st Stage Nozzle Expansion Ratio
   $3.0 \leq P_{c2} \leq 6.0$, P_{c2} : 2nd Stage Chamber Pressure(MPa)
   $30 \leq e_2 \leq 70$, e_2 : 2nd Stage Nozzle Expansion Ratio
   $3.0 \leq P_{c3} \leq 6.0$, P_{c3} : 3rd Stage Chamber Pressure(MPa)
   $40 \leq e_3 \leq 120$, e_3 : 3rd Stage Nozzle Expansion Ratio

• *Step 2 : Trajectory Optimization*

Objective Function
   Maximize M_{pay}, M_{pay} : Payload Weight(kg)

Constraints
$h_p = 700.0$, h_p : Perigee Altitude(km)
$V_p = 7504.0$, V_p : Orbit Insertion Velocity(m/sec)
$\gamma_p = 0.0$, γ_p : Flight Path Angle of Orbit Insertion (°)

Design Variables
$-20 \leq \alpha_i \leq 20$, α_i : Angle of Attack(°), i=1-17
$7.5 \leq M_{pay} \leq 20$

Table 1. Mass Distribution

Stage	Propellant Mass(kg)	Structure Mass(kg)	Payload Mass(kg)	Total Mass(kg)
1	830.87	158.23	283.51	1272.61
2	187.44	42.57	53.50	283.51
3	33.60	11.40	8.50	53.5

Table 2. Propulsion Analysis Results

Stage	Thrust(kgf)	Specific Impulse(sec)	Burn Time(sec)
1	4809.10	282.71	49.0
2	1419.15	265.0	35.0
3	356.20	265.0	25.0

Table 1 shows that the Step 1 and 2 converged within 4 and 18 MDA (Multidisciplinary Analysis) runs and all constraints are satisfied. The designed air-launching rocket has total mass 1272.61kg, payload mass 8.50kg, diameter 0.60m and length 6.43m, which satisfies given design constraints and mission requirement. Table 2 and 3 show mass distribution and propulsion analysis results of each stage. The required propellant mass and thrust of each stage, in order to insert given payload to target orbit are shown in the Table 2 and 3. Figure 6 shows the designed rocket configuration.

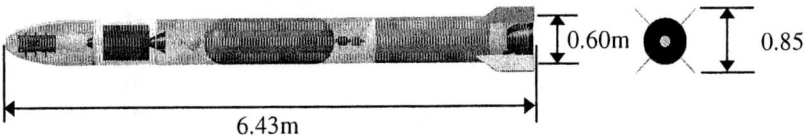

Fig. 3. Configuration of Air-Launching Rocket

3 Conclusions

In this study, the optimal design of the air-launching rocket is performed by multidisciplinary optimization and module which includes mission analysis, staging, propulsion analysis, configuration, aerodynamics, weight analysis and trajectory analysis. From multidisciplinary system optimization, As a result of system optimization, a supersonic air launching rocket with total mass of 1272.61kg, total length of 6.43m, outer diameter of 0.60m and the payload mass of 8.50kg has been successfully designed while satisfying given mission requirements. Moreover, it is demonstrated that the multidisciplinary system optimization approach could be effectively applicable to the air launching rocket design.

Acknowledgement

This work was supported by grant number ADD-03-01-01 from the Basic Research Program of the Agency for Defense Development.

References

1. B. Donahue, "Supersonic air-launch with advanced chemical propulsion", AIAA 2003-4888, 39th AIAA/ASME/SAE/ASEE Joint Propulsion Conference, 2003.
2. J. W. Lee, B. K. Park, Y. H. Byun and C. J. Lee, "Analysis of necessity and Possibility of Air Launching Rocket Development in Korea", *Journal of The Korean Society Aeronautical and Space Sciences,* Vol. 32, No. 4, May 2004.
3. J-W. Lee, J. Y. Hwang, "The Conceptual Design of Air-Launching Micro Space Launcher, Mirinae-1," *Journal of Korean Society for Aeronautical and Space Sciences*, Vol. 29, No. 2, March, 2001.
4. Vanderplaats, Garret N., "Numerical Optimization Techniques for Engineering Design", 2nd Edition, Vanderplaats Research and Development, 1999.

Numerical Visualization of Flow Instability in Microchannel Considering Surface Wettability

Doyoung Byun[1], Budiono[1], Ji Hye Yang[1], Changjin Lee[1], and Ki Won Lim[2]

[1] Konkuk University, School of Mechanical and Aerospace Engineering, Center for Advanced e-System Integration Technology,
1 Hwayang-dong, Gwangjin-gu, Seoul 143-701, Korea
dybyun@konkuk.ac.kr
[2] Korea Research Institute of Standard and Science,
1 Doryong-dong, Yusung-Gu, Daejon 305-600, Korea

Abstract. Recently, researches of microfluidics have been widely studied, because microfluidic systems have been used in several fields such as lab-on-a-chip in medicine, colloid thruster in aerospace, microhydrodynamic in engineering, etc. To handle and control the liquid in the microsystems, the hydrophobic and hydrophilic characteristics on a surface are very important properties. In this study, we performed numerical visualization to investigate the effect of surface wettability in microchannel on the flow characteristics. For the hydrophilic and hydrophobic surfaces arrangement, when the flow reaches the interface region, the meniscus shape changes from concave to convex and the velocity near the centre increases. We can present more efficient method to control the microflow in several micofluidic systems.

1 Introduction

Recently, simulation of microscale thermo-fluidic transport has attracted considerable attention in microelectronic fabrication technologies and the promise of emerging nano-technologies. Fluid-surface interactions such like hydrophobic and hydrophilic characteristics can play a key role in microfluidics.

Due to lower cost, smaller risk in launch, and the ability of formation flying, there exists an increasing demand for micro size and high efficient propulsion systems for micro spacecrafts [1]. Micro-thruster is one example of the application of micro-electro-mechanical systems (MEMS) technology to the space environment that provides a solution to meet the needs of micro spacecrafts. The approach to the direct electrostatic jetting presented to date has two major drawbacks to overcome: 1) the need of high operating voltage, over 1 KV, to be applied and 2) the need of monolithic nozzle with electrodes.

By controlling the surface affinity, the liquid jetting structure is possible to improve the linearity of advance of the liquid droplets and finally generate a very fine and small droplet [2]. It is important to know the fluid flow characteristics in the microchannel for better design of various microsystems. Muller and Gubbins [3] studied the characteristics of hydrophilic and hydrophobic behavior of activated carbon surfaces using molecular dynamic simulation. Until now, no investigation is performed experimentally and theoretically to understand the flow instability and change of meniscus shape depending on the surface affinity in the microchannel.

Therefore, in this paper, we offer the nozzle structure including the interface between the low-affinity region (hydrophobic surface) and the high-affinity region (hydrophilic surface) and investigate the flow instability when the meniscus moves across the interface. Unsteady motion of the meniscus is visualized using the numerical simulation.

2 Method

The micro channel as shown in Fig. 1 has the size of 100μm x 1mm and there exists the interface at the middle of the channel. The liquid is assumed as pure water. The surface tension condition is also included. Navier-Stokes equations are solved using CFD-ACE commercial program. The flow is considered to be a laminar incompressible Newtonian and isothermal flow with velocity field V = (u, v) and governed by the Navier–Stokes equations and continuity equation.

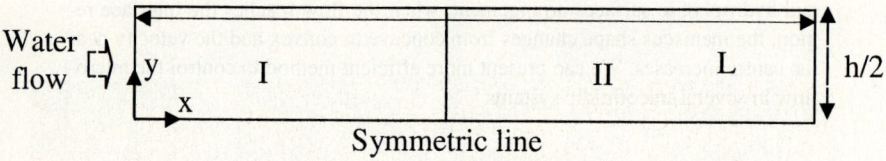

Fig. 1. Schematic of the nozzle with L = 1mm and h/2 = 50μm

$$\frac{\partial u}{\partial x} + \frac{\partial v}{\partial x} = 0 \qquad (1)$$

$$\frac{\partial(\rho u)}{\partial t} + div(\rho uV) = -\frac{\partial p}{\partial x} + div(\mu \ grad \ u) + S_{Mx}$$

$$\frac{\partial(\rho v)}{\partial t} + div(\rho vV) = -\frac{\partial p}{\partial y} + div(\mu \ grad \ v) + S_{My}$$

The boundary conditions on the wall are the no-slip condition and the boundary conditions for the surface affinity can be described by the contact angle. Free surfaces of meniscus are modeled with the Volume of Fluid (VOF) technique, which was reported in Hirt and Nichols [4]. The distribution of the liquid in the grid systems is accounted for using a single scalar field variable, F, which defines the fraction of the volume. The volume fraction distribution can be determined by solving the passive transport equation.

$$\frac{\partial F}{\partial t} + \nabla \cdot \vec{v} F = 0 \tag{2}$$

The numerical simulation of free-surface flows composed of two immiscible fluids involves two coupled tasks: (1) resolving the flow field and (2) updating the position of the interface.

3 Results and Discussions

Fig. 2 shows the contour of volume fraction of liquid when the meniscus moves for the hydrophilic-hydrophobic surfaces arrangement as shown in Fig. 1. The transition of meniscus shape from concave to convex is shown when the liquid is moving across the interface between hydrophilic and hydrophobic surfaces. The contact angles for hydrophilic and hydrophobic surfaces are assumed to be 30 and 120 degrees, respectively. In Fig. 2 (a), the meniscus shape of the fully developed flow is maintained before arriving at the interface. In Fig. 2 (b)-(j), the process of the transition of the meniscus shape at the interface can be described. When the meniscus arrives at the interface, the fluid near the wall stops for a while until the concave shape is disappeared. During the change of the meniscus shape from concave to convex, velocities near the centre increases nearly up to twice of the inlet velocity, as shown in Fig. 3. Therefore using the change of the surface characteristics, we can more easily control and handle the fluid flow in microchannel.

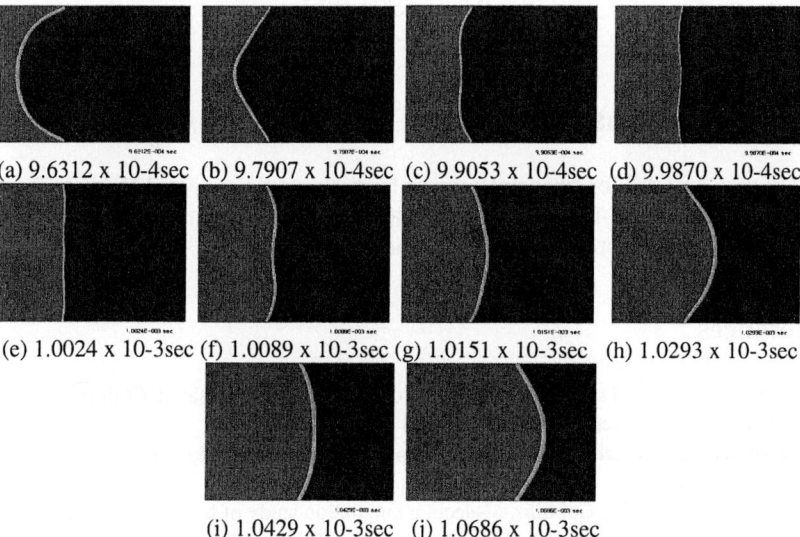

(a) 9.6312×10^{-4}sec (b) 9.7907×10^{-4}sec (c) 9.9053×10^{-4}sec (d) 9.9870×10^{-4}sec

(e) 1.0024×10^{-3}sec (f) 1.0089×10^{-3}sec (g) 1.0151×10^{-3}sec (h) 1.0293×10^{-3}sec

(i) 1.0429×10^{-3}sec (j) 1.0686×10^{-3}sec

Fig. 2. Sequential movement of meniscus across the interface between hydrophilic and hydrophobic surfaces

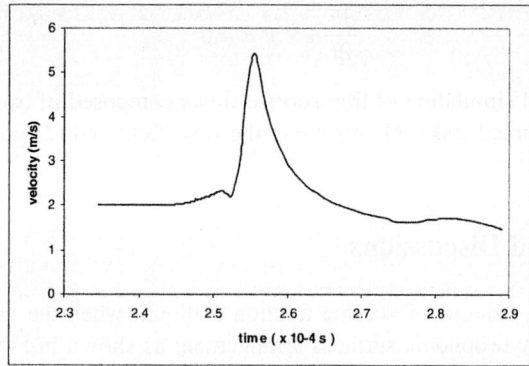

Fig. 3. Variations of the meniscus velocity at center point for the hydrophilic and hydrophobic surfaces arrangement

4 Conclusions

We offer the microchannel structure with the interface between the low-affinity region (hydrophobic surface) and the high-affinity region (hydrophilic surface) to control and handle the fluid flow. Due to this interface, flow instability increases in typical regions. For the hydrophilic and hydrophobic surfaces arrangement, when the flow reaches the interface region, the meniscus shape changes from concave to convex and the velocity near the centre increases. However, for the hydrophobic and hydrophilic surfaces arrangement, when the flow reaches the interface region, the velocity near the wall increases.

Acknowledgement

We are very grateful for financial support from Korea Science and Engineering Foundation Grant (KOSEF-R08-2003-000- 10801-0). And this work is partially supported by Korea Research Institute of Standard and Science.

References

1. J. Xiong, Z. Zhou, D. Sun, X. Ye, Development of MEMS Based Colloid thruster with sandwich structure, Sensors and Actuators 117 (2005) 168-172.
2. Fukushima, Liquid jet structure, inkjet type recording head and printer, U.S. Patent No. 6336697 (2002).
3. E. A. Muller and K. E. Gubbins, Molecular simulation study of hydrophilic and hydrophobic behavior of activated carbon surfaces, Carbon 36 (10) (1998) 1433-1438.
4. C. W. Hirt, and B. D. Nichols, Volume of Fluid (VOF) Method for the Dynamics of Free Boundaries, Journal of Computational Physics 39 (1981) 201-225.

A Interactive Molecular Modeling System Based on Web Service

Sungjun Park[1], Bosoon Kim[1], and Jee-In Kim[2, *]

[1] CAESIT/Department of Computer Science & Engineering,
Konkuk University, Seoul, Korea
{hcipsj, jnkm}@konkuk.ac.kr, foxgap@dreamwiz.com
[2] CAESIT/Department of Internet & Multimedia, Konkuk University, Seoul, Korea

Abstract. We propose a molecular modeling system based on web services. It visualizes three dimensional models of molecules and allows scientists observe and manipulate the molecular models "interactively" through the web. Scientists can examine, magnify, translate, rotate, combine and split the three dimensional molecular models. The real-time simulations are executed in order to validate the operations. We developed a distributed processing system for the real-time simulation. The proposed communication scheme reduces data traffics in the distributed processing system. The new job scheduling algorithm enhances the performance of the system. Thus, the scientists can interactively exercise molecular modeling procedures through the web. For the experiments, HIV-1 (Human Immunodeficiency Virus) was selected as a receptor and fifteen candidate materials were used as ligands. An experiment of measuring performance of the system showed that the proposed system was good enough to be used in molecular modeling on the web.

1 Introduction

Molecular modeling can be used in developing new materials, new drugs, and environmental catalyzers. Scientists can use character strings in order to represent molecules in molecular modeling. The character strings are translated into three dimensional structures of molecules. Then, the scientists examine such three dimensional structures in terms of their shapes, features, stability, and so on. They can also perform docking procedures through which a *receptor* can be combined with a *ligand* at a special position called an *active site*[1]. A real-time simulation is required, since it can computationally prove or disprove if such a chemical operation is valid. The simulation is basically calculating energy minimization equations [1,2,3].

There are some problems when molecular modeling is performed through the Internet. First, the computations of the energy minimization equations are not easily and conven-

---

* Corresponding Author: Jee-In Kim
[1] The term, *receptor,* is used to describe any molecule which interacts with and subsequently holds onto some other molecule. The receptor is the "hand" and the object held by the "hand" is commonly named, *ligand.*

iently provided through the Internet. Since the computations require huge computing power, an ordinary server may not be sufficient to provide a simulation feature through the web. A super computer must be useful. Unfortunately, many scientists can not afford such a high performance computer through the Internet. Second, input data for the computation such as parameters for three dimensional manipulations of molecular models must be transferred to the server in time. Otherwise, the scientists cannot execute real time simulations for molecular modeling. Third, the simulation results must be transferred to the scientists and visualized as three dimensional molecular models in real time. If the energy computations are not properly completed in time or the network latency disturbs the simulation, the scientists would receive incorrect results.

In this paper, we present a molecular modeling system which provides a high computing power of a distributed system. The system can be accessed through web services via the Internet. We develop a new algorithm which enables scientists exercise an interactive and real-time simulation for molecular modeling. Also, the proposed system visualizes three dimensional molecular models through the web and provides a direct manipulation method of the molecular models.

2 Related Works

RASMOL is a molecular modeling tool and mainly visualizes three dimensional structures of molecular models. It is widely used, because it provides fast and simple ways of examining three dimensional structures of molecules through visualization. But it does not provide direct manipulation methods of molecular structures. No real-time simulation function is provided by RASMOL, either.

MDL enhanced RASMOL and developed *Chime* which is a plug-in for a web browser and enabled three dimensional structures of molecules to be visualized through the web. Chime has its own script language and its users issue a command to view three dimensional structures and manipulate the visualized models. This feature makes the Chime plug-in lighter and more portable.

Protein Explorer, *Noncovalent Bond Finder*, *DNA Structure*, *MHC*, *Identifying Spectral*, and *Modes* are web based tools for molecular modeling and they use the Chime plug-in. Protein Explorer is the most popular and widely used tool among them. Protein Explorer [6] has various functions such as presents chemical information of a molecule, displays three dimensional structures of a molecule. Since it can display only one molecular structure at a given time, we cannot visually compare several molecules through the window. We probably need *Insight II* which was commercialized by *Accelys*. Unfortunately, Insight II is not accessible through the web.

3 System

3.1 Overview

We propose a system of molecular modeling using the web service as shown in Figure 1. The molecular modeling system based on the web (called *MMWeb*) consists of clients, a server and a database. The client is a web browser for scientists. The server performs real-time simulations of molecular modeling. The client accesses the server through web services. The database stores and manages files and information about molecules.

The client (*Web Browser*) has four components. *File Manager* reads data about molecules from PDB (*Protein Data Base*) files/Database and exercises parsing the data. *Operation Manager* arranges the parsed data in order to compute energy equations. The results are arranged to be properly displayed by *Rendering Engine* which visualizes three dimensional models of molecules using graphical libraries like OpenGL. Various rendering algorithms are implemented. *Computing Engine* computes energy equations which are essential in the simulation.

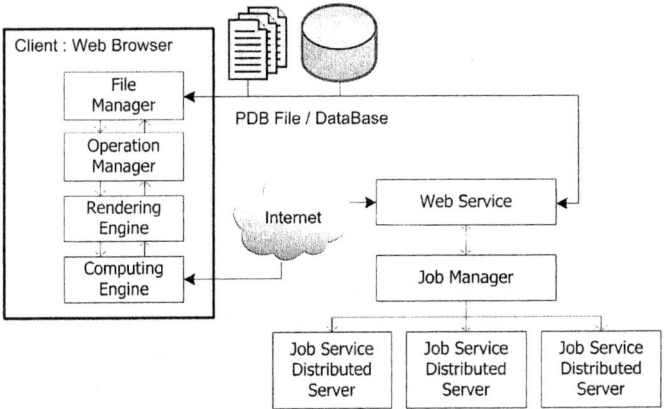

Fig. 1. System Overview

The *Job Service Distributed Servers* communicate with the clients (*Web Browser*) through *Computing Engines* in a form of *Web Services*. They perform real-time simulations of molecular modeling. The computations are basically minimization of energy for molecules. *Job Manager* creates jobs for energy minimization and delivers them to the distributed servers. PDB files and information about molecules are loaded from the database to the servers. The simulation results are sent to the client through web services.

3.2 Web Browser – The Client

We developed Web Browser using OpenGL. The Chime plug-in was not used because we needed to support loading multiple molecular models, docking procedures, bonding and non-bonding procedures, etc.

As shown in Figure 2, the client (Web Browser) displays three dimensional structures of molecular models and allows scientists manipulate the models in interactive and intuitive ways using a mouse. This feature makes the browser more usable than the Chime based browsers which use script language commands to manipulate the models. When the positions and the orientations of the models change, a matrix of transformation is sent to *Web Service*. The matrix becomes an input for real-time simulations. The web service supplies the matrix to the servers each of which has loaded data of molecules from the database and executes computations of energy minimization using the matrix.

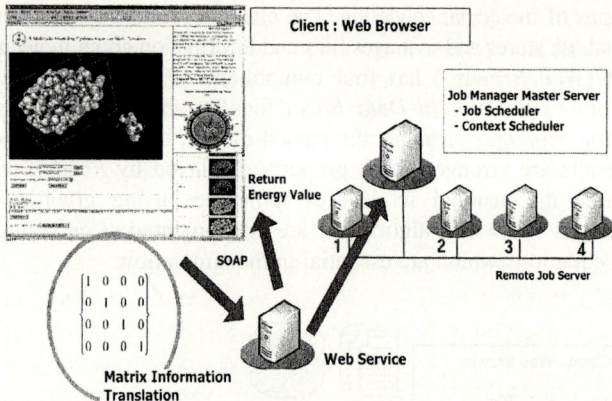

Fig. 2. Communication through web service between a web browser and the servers

3.3 Distributed Processing System

MMWeb requires a high performance computing system in order to exercise real-time simulations. We developed a distributed processing system [5].

Web Service (① in Figure 3) is a gateway which transfers information from the client (*Web Browser*) to *Job Servers* (⑤ in Figure 3) and returns the simulation results from the servers to the client. The information from the clients includes a transformation matrix of molecular models, and information about a receptor and a ligand. SOAP (Simple Object Access Protocol) is used for the communication.

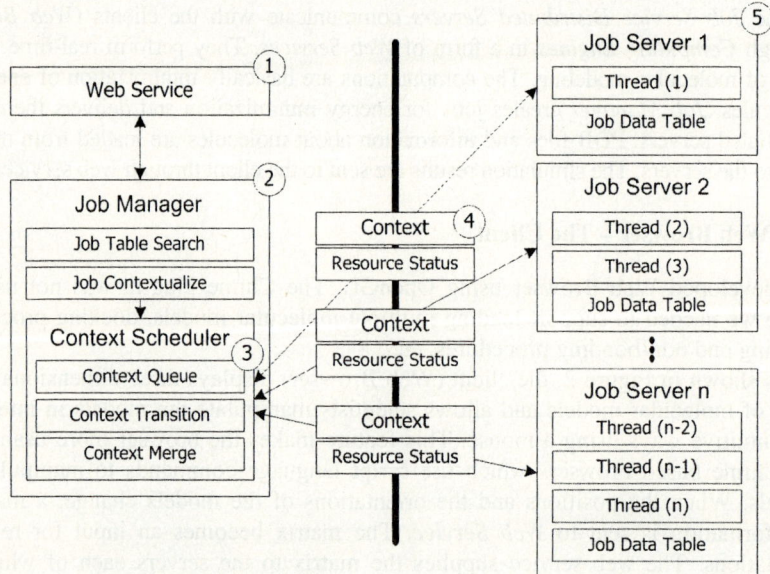

Fig. 3. A block diagram of the distributed processing system

Job Manager (② in Figure 3) is a master server and manages the distributed computing resources. It receives information from the Web Service component and searches a proper job from lookup tables (*Job Table Search* in Figure 3). Then, it partitions the job found in the table into jobs which can be executed by the servers. *Job Manager* also contextualizes the jobs. The procedure aims to store information of processing resources into each job. That is, *Job Contextualizer* assigns status of processing resources for each job.

Context Scheduler (③ in Figure 3) performs job scheduling tasks. There are many jobs from multiple clients. A job of a user is partitioned and contextualized. Then, the contextualized jobs are stored in *Context Queue*. *Context Scheduler* allocates a thread for each job depending on its scheduling policy. The contextualized jobs (④ in Figure 3) are sent to *Job Servers* (⑤ in Figure 3). After the computations, the results from the servers are sent back to the client through *Web Service Context Scheduler* (③ in Figure 3) which performs job scheduling tasks. There are many jobs from multiple clients. A job of a user is partitioned and contextualized. Then, the contextualized jobs are stored in *Context Queue*. *Context Scheduler* allocates a thread for each job depending on its scheduling policy. The contextualized jobs (④ in Figure 3) are sent to *Job Servers* (⑤ in Figure 3). After the computations, the results from the servers are sent back to the client through *Web Service*.

Context Transition means state transition of contexts while they are processed as shown in Figure 4-(a). The context, *C1*, *in Ready Hash Table* is allocated to a resource and stored into *Running Hash Table*. The context, *C2*, in *Running Hash Table* completes its tasks and goes into *Done Hash Table*. The context, *C3*, in *Done Hash Table* is finished. The Contexts, *C4,..., Cn*, in *Ready Hash Table* are waiting for their corresponding resources. When all contexts are processed as shown in Figure 4-(b), they are merged and sent back to the client.

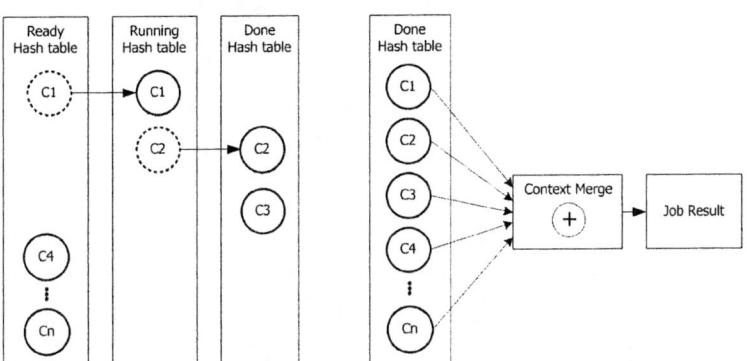

Fig. 4. (a) State Transition of Contexts. (b) Merge the processed Contexts.

3.4 Docking Simulation

The docking simulation checks if a receptor and a ligand can be chemically combined. The main operations of the docking simulation are locating active sites in the receptor and combining the receptor and the ligand at the active site. The energy values of the molecules are minimal when they are stably combined. Figure 5-(a) visualizes the procedure. A series of points for energy computation is shown in Figure 5-(b).

During the interactive simulation, the scientists keep on moving the ligand towards the active site. The positions of the ligand (the matrix information) are transferred to the server (the web service) for calculating energy values. The positions data are combined with other input data and become a job by the server (the web service).

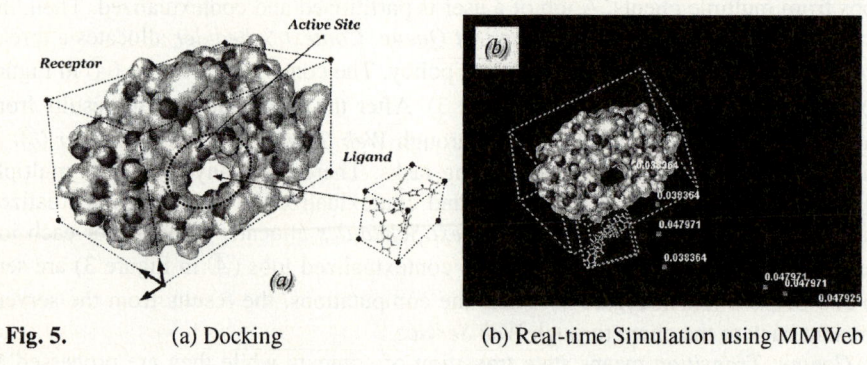

Fig. 5. (a) Docking (b) Real-time Simulation using MMWeb

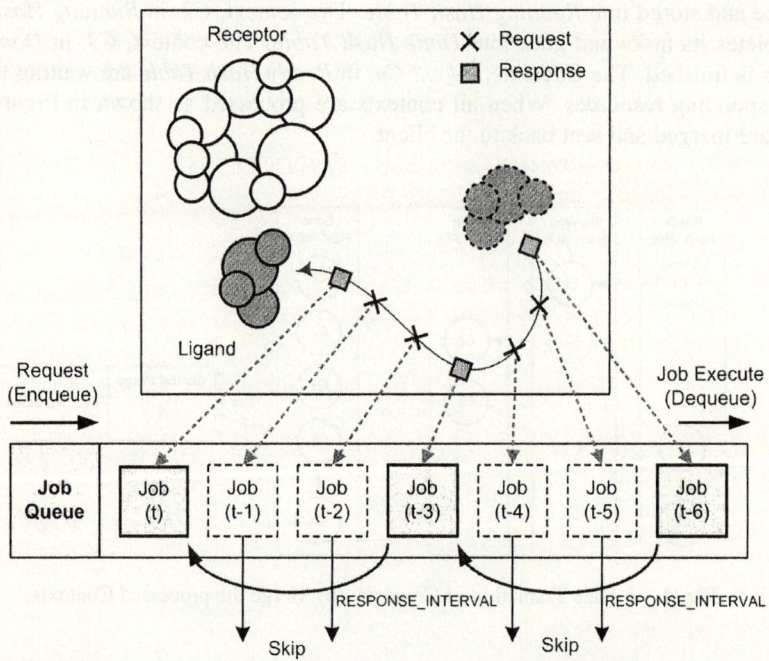

Fig. 6. The job skip operation in the job queue of the web service

The jobs are inserted into the job queue as shown in Figure 6. If the computations are performed for the all jobs, they may not be completed in time. Such a computational latency could deteriorate the quality of the simulation. Notice that the real-time interaction is essential during the simulation. We propose to skip some of the requested jobs in order to satisfy the real-time requirement. As shown in Figure 6, some jobs can be ignored during the simulation as long as the scientists can eventually find the right active site. The scientists should be able to determine the number of jobs to be skipped by setting the value of RESPONSE_INTERVAL. The symbol denotes a number of jobs to be skipped during the energy minimization computation.

4 Experiments

In these experiments, we check 1) if MMWeb computes reliable values after simulation, 2) if MMWeb performs fast enough to exercise real-time simulations. We exercised a docking procedure. HIV-1 (Human Immunodeficiency Virus) [7, 8] was selected as a receptor. Fifteen materials related to reproduction of HIV-1 were chosen as ligands.

First, we checked the reliability of the simulation values. That is, the values of computing energy equations for binding the fifteen ligands with the receptor were calculated. We compared MMWeb with Insight II which is the most popular tool of molecular modeling. Figure 7 presents energy values of fifteen ligands using Insight II and MMWeb. The results show that there are no significant differences between the values of calculating energy equations using Insight II and those using MMWeb. RMSD(Root Mean Squared Deviation) is used as a measure for the reliability. As presented in Table 1, the results of docking procedures using the two tools would not be significantly different.

Secondly, we measured processing times of simulations with a distributed processing system. We compared processing times with respect to different sized receptors, different number of computers for the distributed system, and different number of threads per computer. We used receptors of 3000 atoms, 6000 atoms and 12000 atoms.

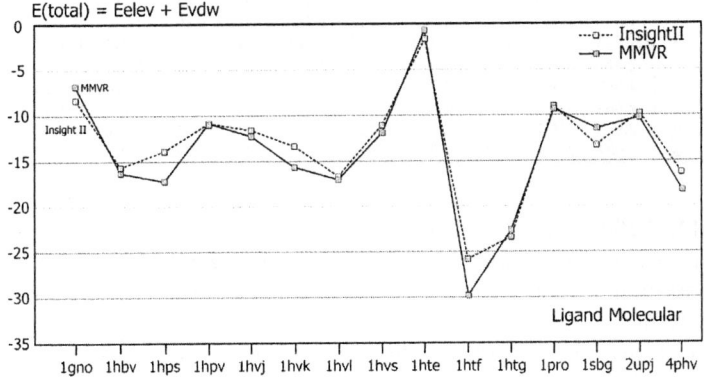

Fig. 7. Energy values of fifteen ligands

Table 1. Computed values of energies using Insight II and MMWeb

$T = Elec + Vdw$

PDB code	Insight II Energy Value				MMWeb Energy Value			
	ΔE^{Vdw}	ΔE^{elec}	ΔE^T	RMSD (A)	ΔE^{Vdw}	ΔE^{elec}	ΔE^T	RMSD (A)
1gno	-7.63	-0.32	-7.95	1.02	-9.08	-0.46	-9.54	0.98
1hbv	-14.73	-1.24	-15.97	0.92	-14.21	-1.14	-15.35	0.86
1hps	-16.87	0.74	-16.13	2.41	-14.64	1.10	-13.54	3.15
1hpv	-10.15	-0.93	-11.08	0.36	-10.28	-0.74	-11.02	0.42
1hvj	-11.85	-0.11	-11.96	1.25	-10.85	-0.21	-11.06	1.28
1hvk	-16.25	0.55	-15.70	0.37	-14.21	0.65	-13.56	0.89
1hvl	-15.43	-1.20	-16.63	0.35	-15.35	-0.98	-16.33	0.39
1hvs	-12.31	-0.24	-12.55	1.66	-11.28	-0.34	-11.62	1.93
1hte	-1.24	-0.23	-1.47	0.39	-1.89	-0.65	-2.54	0.98
1htf	-22.61	-2.30	-24.91	0.32	-18.87	-2.15	-21.02	0.94
1htg	-17.46	-1.23	-18.69	0.49	-18.31	-1.24	-19.55	0.44
1pro	-9.95	0.67	-9.28	1.04	-9.70	0.62	-9.08	1.26
1sbg	-11.29	0.08	-11.21	2.01	-12.99	0.13	-12.86	1.36
2upj	-10.80	0.49	-10.31	1.59	-10.87	0.98	-9.89	1.89
4phv	-17.43	-0.98	-18.41	0.67	-15.64	-1.12	-16.76	0.92

Table 2. Processing times with different receptor sizes vs. Numbers of computers

Receptor Size	Threads Per Computer No.	Localhost			1 Computer			2 Computers			4 Computers		
		1	2	4	1	2	4	1	2	4	1	2	4
3000	1	0.922	0.969	0.937	1.011	0.811	1.002	0.711	0.621	0.711	0.501	0.410	0.511
	2	0.968	0.953	0.938	0.811	1.000	0.911	0.501	0.511	0.611	0.410	0.511	0.510
	3	0.891	0.875	0.969	0.912	0.912	1.012	0.510	0.511	0.711	0.411	0.511	0.410
	4	0.891	0.969	0.921	0.811	0.911	1.012	0.611	0.611	0.711	0.410	0.511	0.511
	5	0.907	0.969	0.938	0.911	0.911	0.901	0.511	0.511	0.711	0.411	0.410	0.510
6000	1	1.671	1.625	1.594	1.612	1.713	1.712	0.902	0.901	1.011	0.611	0.621	0.611
	2	1.578	1.640	1.594	1.612	1.613	1.612	0.912	0.911	1.031	0.501	0.721	0.611
	3	1.562	1.640	1.641	1.522	1.613	1.612	0.811	0.911	1.111	0.611	0.611	0.610
	4	1.531	1.688	1.625	1.612	1.713	1.722	0.911	0.901	1.011	0.611	0.621	0.731
	5	1.563	1.625	1.609	1.713	1.712	1.712	0.821	0.911	1.011	0.511	0.611	0.721
12000	1	3.031	3.047	3.203	2.914	3.315	3.024	1.612	1.622	1.812	1.012	1.022	1.211
	2	2.985	3.000	3.25	2.924	3.224	3.014	1.512	1.713	1.723	1.012	1.122	1.111
	3	2.984	3.016	3.281	3.015	3.115	3.215	1.613	1.622	1.913	0.932	1.282	1.122
	4	2.953	3.281	3.266	2.914	3.115	3.215	1.613	1.612	1.813	0.911	1.282	1.112
	5	2.984	3.204	3.219	2.914	2.925	3.225	1.712	1.622	1.923	1.011	1.122	1.311

The ligand had 100 atoms. The numbers of computers were one, two, and four in the distributed system. The numbers of threads were one, two, and four. For each receptor, a computation of energy equations was executed five times. Table 2 and Figure 8 present the results.

Notice that there is a 0.5 second difference between the case with one computer and the case with four computers in their processing times, when the number of atoms was 3000 in the receptor as shown Table 2. The difference becomes 2 seconds when

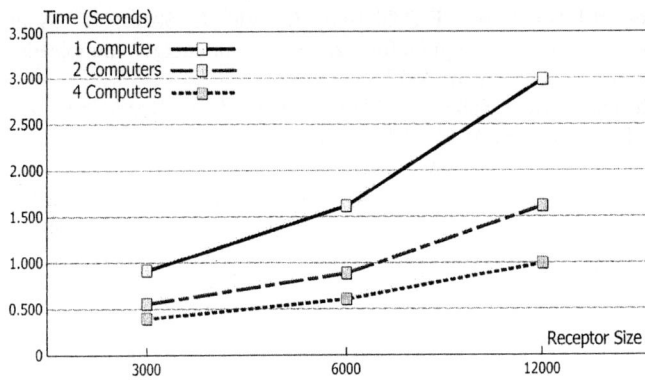

Fig. 8. Average processing times with different numbers of computers and atoms

the number of atoms was 12000. It means that the processing time reduces faster as the number of computer increases when the number of atoms in a molecule is large.

5 Concluding Remarks

In this paper, we proposed a web service based molecular modeling system. Our motivations are 1) to provide an interactive molecular modeling system accessible through the web, 2) to offer direct manipulation methods of molecular models rather than command-based manipulation methods, 3) to develop an affordable high-performance computing system for real-time simulation of molecular modeling. The key features of the system are three dimensional visualization and direct manipulation of molecular models, and interactive real-time simulations.

As a future work, we would like to enhance the proposed system in terms of its interaction methods with scientists. More natural and convenient ways of interacting between the scientists and the system should be researched in order to improve the productivity of the scientists. Secondly, the computing power of the system should be improved. Since the simulation requires high performance computing, we plan to develop faster and more powerful computing system such as grid computing.

Acknowledgements

This work was supported by the Ministry of Information & Communications, Korea, under the Information Technology Research Center (ITRC) Support Program.

References

1. Ekachai Jenwitheesuk and Ram Samudrala, Improved prediction of HIV-1 protease-inhibitor binding energies by molecular dynamics simulations, BMC Structured Biology, 3:2, (2003)
2. James C.Philips, Gengbin Zheng, Sameer Kumar and Laxmikant V.Kale, NAMD:Biomolecular Simulation on Thousands of Processors, IEEE, pp36, (2002)

3. B.R.Brooks, R.E.Bruccoleri, B.D.Olafson, D. Vid J. States, S. Swaminathan, and M.Karplus. CHARMM: A program for macromolecular energy, minimization, and dynamics calculations. J.Comp. Chem., 4:187-217, (1983)
4. Szymin Rusinkiewicz, QSplat : A Multiresolution Point Rendering System for Large Meshes, in Proceedings of ACM 2000, pp 343-352, 2000.
5. Gans J, Shalloway, Qmol: A program for molecular visualization on Windows based PCs Journal of Molecular Graphics and Modelling, Vol.19, pp. 557-559, 2001.
6. Alois Ferscha, Michael Richter, Java based conservative distributed simulation, Proceedings of the Winter Simulation Conference, p381-388, 1997
7. Earl Rutenber, Eric B.Fauman, Robert J.Keenan, Susan Fong, Paul S.Furth, Paul R.Ortiz de Montellano, Elaine Meng, Irwin D.Kuntz, Dianne L.DeCamp, Rafael Salto, Jason R.Rose, Charles S.Craik, and Robert M.Stroud, Structure of a Non-peptide Inhibitor Complexed with HIV-1 Protease, The Journal of Biological Chemistry, Vol. 268, No. 21, pp.15343-15346, 1993.
8. Junmei Wang, Paul Morin, Wei Wang, and Peter A. Kollman, Use of MM-PBSA in Reproducing the Binding Free Energies to HIV-1 RT of TIBO Derivatives and Predicting the Binding Mode to HIV-1 RT of Efavirenz by Docking and MM-PBSA, Journal of American Chemical Society, Vol. 123, pp. 5221-5320, 2001.

On the Filter Size of DMM for Passive Scalar in Complex Flow

Yang Na[1], Dongshin Shin[2], and Seungbae Lee[3]

[1] CAESIT, Dept. of Mechanical Engineering, Konkuk University,
Hwayang-dong 1, Gwangjin-gu, Seoul 143-701, Korea
yangna@konkuk.ac.kr
[2] Dept. of Mechanical Engineering, Hong-Ik University,
Seoul 121-791, Korea
dsshin@wow.hongik.ac.kr
[3] Dept. of Mechanical Engineering, Inha University,
Inchon 402-751, Korea
sbaelee@inha.ac.kr

Abstract. Effect of filter size of dynamic mixed model combined with a box filter on the prediction of passive scalar field has been investigated in complex flow. Unlike in the simple channel flow, the result shows that the model performance depends on the ratio of test to grid filter widths.

1 Introduction

The role of large eddy simulation (LES) in most of engineering applications involving turbulent flows increases everyday. Since direct numerical simulation (DNS) is restricted to a relatively low Reynolds number due to the resolution requirement for the length-scale in dissipation range, it is currently less attractive as an engineering tool and LES becomes more popular as a reasonably accurate and at the same time less expensive methodology.

Even though the significant development has been made to the LES modeling for the prediction of velocity field, relatively much less effort has been done in the calculation of passive scalar transport in spite of its obvious practical importance and this fact is reflected in the difficulty of predicting the passive scalar field with satisfactory accuracy using current LES models. The difficulty of investigating passive scalar transport is possibly due to the fact that errors associated with LES models embedded in a velocity field reduces the accuracy in the prediction of passive scalar in a way not clearly understood. Consequently, more effort should be devoted to the development of LES methodology for more accurate and reliable approach for the design of thermal system.

The present work mainly intended to examine the performance of dynamic mixed model (DMM, Zang et. al. [1]) for passive scalar transport in complex flow (Na [2]). An exhaustive number of LES models has been reported in the literature but DMM was chosen here for the following two reasons: (1) From the perspective of large eddy simulation of engineering flows, computations based on finite difference formulations are certainly of great interest. Thus, DMM with finite difference formulations, which most conveniently use filters in physical space, were considered and tested for turbulent channel flows; (2) DMM has been known to produce good results in a wide range of turbulent flows.

The dynamic mixed model extended to passive scalar transport will be briefly explained and its characteristics are discussed in the case of turbulent channel flow with wall injection.

2 Mathematical Formulation

2.1 Dynamic Mixed Model for Passive Scalar

The filtered governing equations for the LES of a passive scalar T for incompressible flows are given as follow:

$$\frac{\partial \overline{u_i}}{\partial x_i} = 0, \qquad (1)$$

$$\frac{\partial \overline{u_i}}{\partial t} + \frac{\partial}{\partial x_j}(\overline{u_i u_j}) = -\frac{\partial \overline{p}}{\partial x_i} + \frac{\partial}{\partial x_j}(2\nu \overline{S_{ij}} - \tau_{ij}), \qquad (2)$$

$$\frac{\partial \overline{T}}{\partial t} + \frac{\partial}{\partial x_j}(\overline{u_j T}) = \frac{\partial}{\partial x_j}(\alpha \frac{\partial \overline{T}}{\partial x_j} - q_j). \qquad (3)$$

where the grid-filtering operation is denoted by an overbar. The effect of unresolved subgrid scales is represented by the following residual stress tensor τ_{ij} and residual scalar flux vector q_j.

$$\tau_{ij} = \overline{u_i u_j} - \overline{u_i}\,\overline{u_j}, \qquad (4)$$

$$q_j = \overline{T u_j} - \overline{T}\,\overline{u_j}. \qquad (5)$$

Only two terms τ_{ij} and q_j in equations (1)-(3) should be obtained through the appropriate LES models. Details of how to calculate τ_{ij} and q_j using DMM approach are explained in Lee & Na [3] and will not be repeated here.

In order to discretize the grid-scale and the test-scale filters, a box filter in physical space was employed. After the model coefficients C_S and C_T are computed through the least-squares approach, they are averaged locally in space within the test-filtering volume as suggested by Zang et. al [1].

For the actual computation using DMM, the only adjustable parameter is the ratio of test to grid filter width, $\alpha = \widetilde{\Delta}/\overline{\Delta}$. Two commonly used definitions of the effective filter width are (1) $\widetilde{\overline{\Delta}}/\overline{\Delta} = (\widetilde{\overline{\Delta}}_x \widetilde{\overline{\Delta}}_z / \overline{\Delta}_x \overline{\Delta}_z)^{1/3} = 2^{2/3}$ and (2) $\widetilde{\overline{\Delta}}/\overline{\Delta} = (\widetilde{\overline{\Delta}}_x \widetilde{\overline{\Delta}}_z / \overline{\Delta}_x \overline{\Delta}_z)^{1/2} = 2$. The value of $\alpha = 2$ is known to be the optimal choice in the simulation of a turbulent channel flow using a sharp cutoff filter (Germano et al. [4]), but optimal value is likely to depend on the types of grid and test filters used. Thus, the investigation of effect of α on the prediction of passive scalar is the main objective of the present study. The sensitivity of the numerical results to the choice of α was examined for two different values of α (2 and $2^{2/3}$) in a channel with wall injection for Pr=1.

2.2 Computational Domain

In order to test DMM for a passive scalar field, an incompressible flow between two parallel walls driven by the wall injection was considered. Figure 1 shows a schematic diagram of three-dimensional computational domain. The streamwise extent of the domain is $L_x=26h$ and the spanwise extent is $L_z=6.5h$, where h is the half channel height. In terms of wall units (based on the friction velocity at inlet of the computational domain), the domain size is approximately equivalent to 3850 in the streamwise, 296 in the wall-normal, and 963 in the spanwise directions. The Reynolds number based on inlet bulk velocity and half-channel height was set to 2250.

The turbulent structures originated from the flat channel region are lifted by the action of wall injection applied from the location of $x=13.4h$ and this induces a strong mixing layer away from wall. In turn, this formation of strong shear layer causes more turbulent structures to grow in space (Figure 2) and thus, the flow experiences very rapid changes in the mean flow direction.

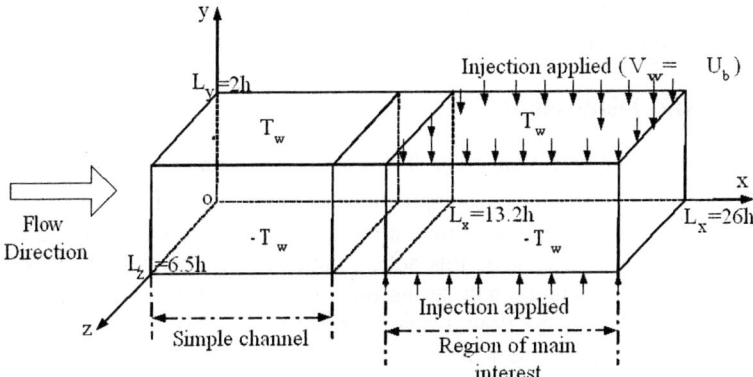

Fig. 1. Flow geometry and computational domain for the test of passive scalar using DMM

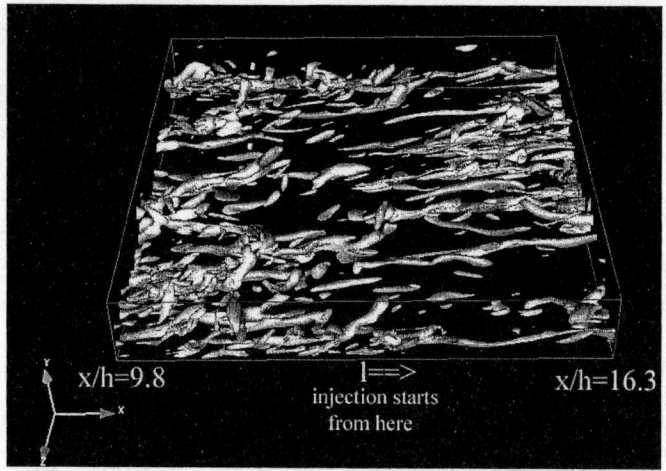

Fig. 2. Turbulent structures generated in a region with a strong injection applied at the wall

2.3 Boundary Condition

No-slip boundary condition was used along the walls except in the region where constant blowing was applied ($x/h > 13.4$). The strength of the wall injection, ε, defined by the ratio of injected velocity to the inlet bulk velocity, was set to 0.05, representing a quite strong injection. It remained constant along both upper and lower walls and the spatial variation of ε was not considered. The bottom wall was cooled ($-T_w$) and the top wall was heated (T_w) at the same rate so that both walls were maintained at constant temperature.

The flow was assumed to be homogeneous in the spanwise direction which allows transform method. The adequacy of the computational domain size and the periodic boundary condition in the spanwise direction was assessed Na [2].

3 Results and Discussion

The effect of α on the performance of DMM for passive scalar was investigated in a channel with wall injection. The computation was done with $129 \times 65 \times 65$ grids. The impact of the injected vertical flow on the turbulent boundary layer is accompanied by the lifted shear layer and this adds complexity to flow. Mass conservation leads to a streamwise acceleration or strong inhomogeneity in the middle of the channel. As shown in Figure 2, the flow is characterized by the formation of increasingly stronger streamwise vortices as it moves downstream. This feature of getting more and more turbulent structures in the streamwise direction is thought to be associated with the growing lifted shear layer.

It would be useful to investigate the effect of α in this type of complex flow to investigate the range of model's utility. For the purpose of comparison, DNS with $513 \times 257 \times 257$ grids were also performed and the data were filtered in physical space to get the filtered statistics.

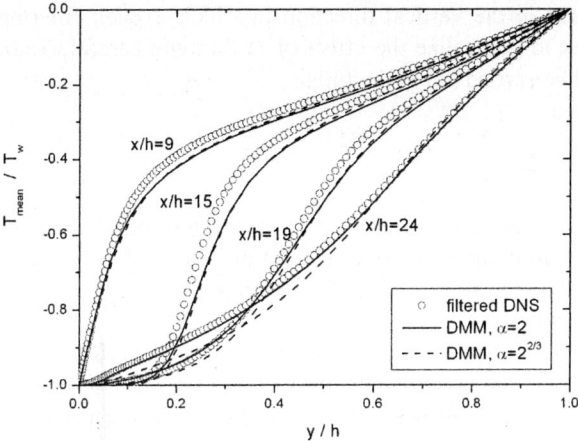

Fig. 3. Time-averaged temperature (passive scalar) profiles at several streamwise locations

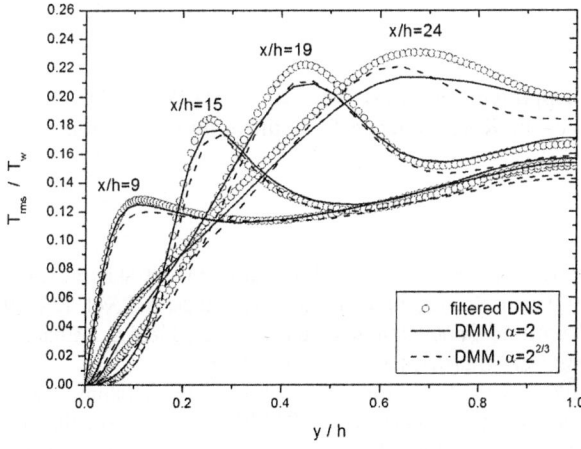

Fig. 4. Root-mean square temperature (passive scalar) profiles at several streamwise locations

Figure 3 shows the comparison of mean temperature profiles at several streamwise locations. Note that the location of x/h=9 is in the simple channel flow. It is clear that the mean temperature is not sensitive to the value of α upstream of wall injection. Even though *a priori* test results suggest the sensitivity of α, this sensitivity does not clearly appear in an actual LES. As the flow moves downstream, however, the choice of $\alpha = 2$ produces much better prediction. Thus, it would be interesting to investigate what feature of DMM generates the difference in the presence of wall injection. A similar behavior can also be found in the rms profiles shown in Figure 4.

As mentioned earlier, the resolution may play an important role in the present flow due to the strong shear layer formed away from wall. As a first step of the research,

65 grids were used in the vertical direction in which explicit filtering is not done. However, in order to generalize the effect of α, a more careful examination on the resolution should be carried out in the future.

4 Summary

The dynamic mixed model extended to the prediction of passive scalar transport was tested in a channel with injection. Since the optimal value of α and its range of utility is likely to depend on the flow, DMM was tested in a strong shear layer generated by the strong wall injection.

A close investigation of the results suggests that the performance of the model showed sensitivity to the size of the effective filter width ratio unlike in a simple channel flow. Overall, the value of $\alpha = 2$ produced a better mean and rms passive scalar statistics for the flow under investigation. However, more work for a variety of complex flows will be required in order to determine the model's range of utility of DMM in its current form.

Acknowledgement

This work was supported by grant No. R01-2004-000-10041-0 from the Basic Research Program of the Korea Science & Engineering Foundation.

References

1. Zang, Y., Street R. L. and Koseff, J. R.: A Dynamic Mixed Subgrid-scale Model and its Application to Turbulent Recirculating Flows, Phys. Fluids, A 5, vol. 12 (1993) 3186-3196.
2. Na, Y.: Direct Numerical Simulation of Turbulent Scalar Field in a Channel with Wall Injection. Numerical Heat Transfer, Part A (2005) 165-181.
3. Lee, G. and Na, Y.: On the Large Eddy Simulation of Temperature Field Using Dynamic Mixed Model in a Turbulent Channel, Trans. KSME B, Vol. 28, No. 10, (2004) 1255-1263.
4. Germano, M. , Piomelli, U., Moin P. and Cabot W. H.: A Dynamic Subgrid-scale Eddy Viscosity Model, Phys. Fluids, A 3, vol. 7 (1991) 1760-1765.

Visualization Process for Design and Manufacturing of End Mills

Sung-Lim Ko, Trung-Thanh Pham, and Yong-Hyun Kim

Department of Mechanical Design and Production Engineering,
Konkuk University, CAESIT(Center for Advanced E-system
Integration Technology), 1 Hwayang-dong, Kwangjin-gu,
Seoul 143-701, South Korea
slko@konkuk.ac.kr

Abstract. The development of CAM system for design and manufacturing of end mills becomes a key approach to save the time and reduce cost for end mills manufacturing. This paper presents the calculation and simulation of CNC machining end mill tools using on 5-axes CNC grinding machine tool. In this study the process of generation and simulation of grinding point data between the tool and the grinding wheels through the machined time are describes. Using input data of end mill geometry, wheels geometry, wheel setting, machine setting the end mill configuration and NC code for machining will be generated and visualized in 3 dimension before machining. The 3D visualizations of end mill manufacturing was generated by using OpenGL in C++.

1 Introduction

The flat end mills are commonly used in industry for high speed machining. They are characterized by a complex geometry with many geometric parameters, which deals with some complicated processes in machining with grinding CNC machine. In the previous studies by Ko [1, 2], the development of CAM system was limited in 2D cross sections of the tools and the results of simulation were restricted to the main fluting operation. Ko already provided a machining simulation for specific 5-axes machines in 2D. Dani Tost presented an approach for the computation of the external shape of the drills through a sequence of coordinated movements of the tool and the wheels on machines of up to 6-axes. The proposed method reduces the 3D problem to 2D dynamic boolean operations followed by a surface tiling [3, 4]. However they only have shown the fluting and gashing simulations. This paper presents the process of generation and simulation of grinding point data between tool and grinding wheels for all grinding processes in 3D. The program was developed for prediction of configuration of end mills. OpenGL was used for developing CAM system. The NC code of all processes for end mill manufacturing will be generated automatically in the program and saved in NC code files. These NC code files will be used for machining in CNC grinding machines.

2 Development of Software for Design and Manufacturing of End Mills

There are several types of CNC grinding machine tools for end mill manufacturing. In this study, the end mill is machined on CNC grinding machine tool, which has 5 degrees of freedom: three axes in translation (X, Y, Z) and two in rotation (B, C) as shown in Fig.1. Tool is fixed at tool holder, which has two degrees of freedom; rotation (C) and translation (Z). The grinding wheels are arranged at the wheel spindle, which has 3 degrees of freedom; two translations (X, Y) and rotation (B).

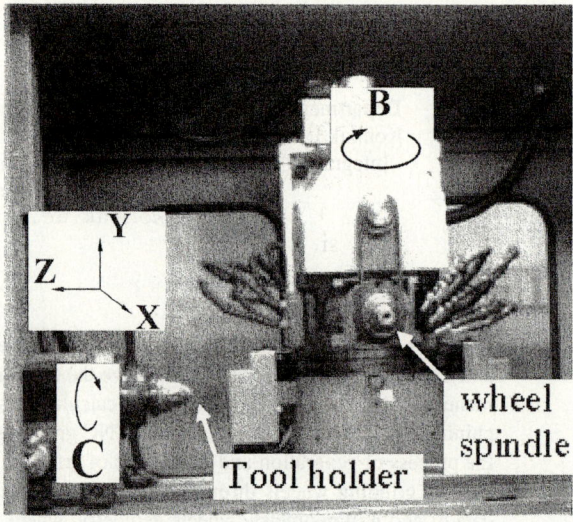

Fig. 1. 5-axis CNC grinding machine

Fig.2 shows the arrangement of grinding wheels for end mill manufacturing. Wheel number 2 and 3 are used for fluting and gashing operations respectively. Wheel number 4 is used for 1st and 2nd clearance operations. The 1st and 2nd end teeth are formed by using wheel number 1.

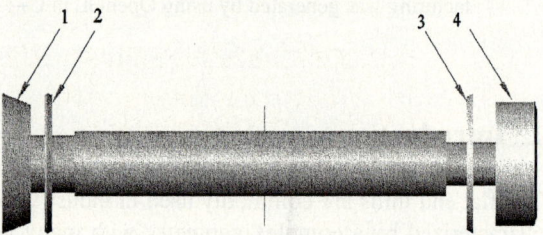

Fig. 2. Grinding wheels for end mill manufacturing

2.1 Basic Algorithm for the Helical Flute Grinding

The simulation algorithm for the helical groove machining in this study was based on the assumption that a wheel with finite thickness consists of a finite number of thin disks. Using the given wheel geometry and the required design parameters of end mill as input data the computer program for predicting and simulating of the helical flute configuration was completed. The program will determine the contact point between circle and ellipse, which is the projection of wheel on the plane normal to axial direction of end mill. And the cross point between wheel and outer end mill diameter are calculated. This point will be reference point to configure the flute shape and another parameters. In this case, the diameter of end mill is fixed and Z-axes of wheel location is also fixed and only Y-axes of wheel is changed to find contact point. The flute configuration was generated from the cross points between the given grinding wheel and end mill. The determination of first contact point is very important. Most of end mill geometry are determined using this point.

2.2 Simulation Results Generation of NC Codes

The end mill shapes are constructed by surface modeling using triangles and quads. The visualization implied for construction of end mill shapes depends on the boundary of the model. Fig.3 and Fig.4 show cross section of helical flute and the simulation results of end mill shapes.

Using the input data for end mill geometry, wheel geometry and machine setting, the computer program will generate the grinding points for clearance faces, endteeth faces, gash face. The visualization of these faces implies the construction of the boundary of the surfaces. The OpenGL graphics system allows creating interactive programs that produce color images of moving three dimension objects.

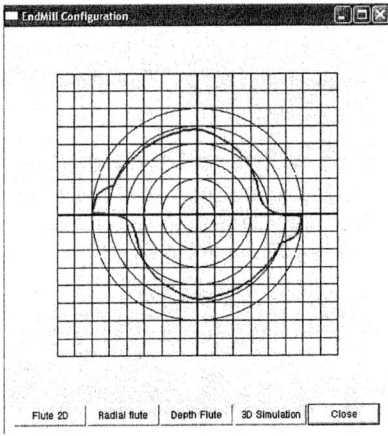

Fig. 3. Cross section of helical flute

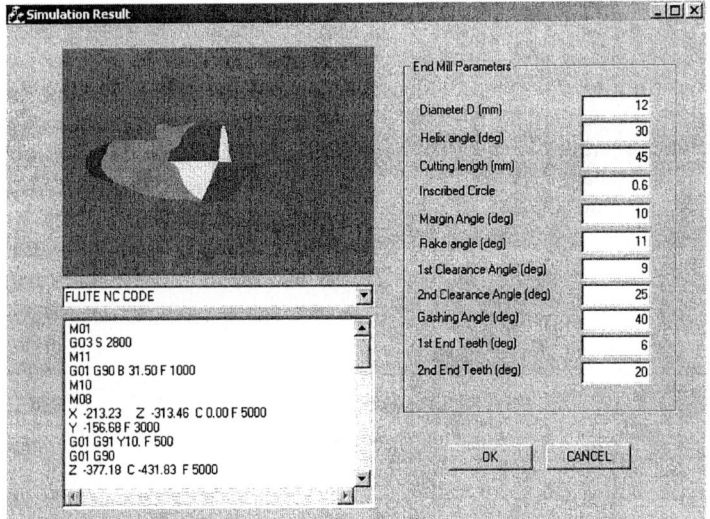

Fig. 4. Simulation results

The visualization of end mill relates to lighting conditions surrounding an object, blending, antialiasing, texture mapping, shading. In addition, OpenGL was used to exploit the possibilities of the projective Z-Buffer and shading techniques, which offer and optimize the visualization delay for whole process.

The NC codes for all processes of manufacturing in 5-axis CNC grinding machine were generated from the computer program based on machine coordinate and stored in the same window with simulation results as shown in Fig.4. Each machining process is carried out by rotation and translation of the axis of wheel and tool. The initial and final positions of all processes were calculated.

3 Conclusion

The CAM system for design and manufacturing of end mills was developed to predict geometric end mill configuration before machining. The NC codes were generated from the CAM system for manufacturing of end mills in 5-axis CNC grinding machine.

Acknowledgement

This work was supported by Konkuk University in 2004.

References

1. Sung-Lim Ko, "Geometrical analysis of helical flute grinding and application to end mill", Trans. NAMRI/SME XXII (1994).
2. Yong-Hyun Kim, Sung-Lim Ko, "Development of design and manufacturing technology for end mill in machining hardened steel", Journal of Materials Processing Technology 130-131 (2002) 653-661.
3. Dani Tost, Anna Puig, Lluis Perez-Vidal, " Boolean operations for 3D simulation of CNC machining of drilling tools", Computer-Aided Design 36 (2004) 315–323
4. Dani Tost, Anna Puig, Lluis Perez-Vidal, "3D simulation of tool machining", Computer & Graphics 27 (2003) 99-106.

IP Address Lookup with the Visualizable Biased Segment Tree

Inbok Lee[1,*], Jeong-Shik Mun[2], and Sung-Ryul Kim[2,**]

[1] Department of Computer Science,
King's College London, London, United Kingdom
iblee@theory.snu.ac.kr
[2] Division of Internet & Media and CAESIT,
Konkuk University
redcroce@hanmail.net
kimsr@konkuk.ac.kr

Abstract. The IP address lookup problem is to find the longest matching IP prefix from a routing table for a given IP address. In this paper we implemented and extended the results of [3] by incorporating the access frequencies of the target IP addresses. Experimental results showed that the number of memory access is reduced significantly.

1 Introduction

An Internet router reads the target address of an incoming packet and chooses a path of the packet using its routing table.

Definition 1. *For m-bit IP addresses, a* routing table T *is a set of pairs (p, hop) such that p is a string in $\{0,1\}^*$ of length at most m (in fact, p is a prefix of an IP address) and hop is an integer in $[1, H]$, where H is the number of the next hops. Let n be the number of IP prefixes stored in the routing table.*

Given an IP address x and a routing table T, the lookup problem is to find $(p_x, hop_x) \in T$ satisfying these two conditions:

1. p_x is a prefix of x.
2. $|p_x| > |p|$ for any other pair $(p, hop) \in T$ such that p is a prefix of x.

Traditionally, patricia tries [5] were used for the IP address lookup problem. Several approaches were proposed to replace the patricia tire, which can be devided into two categories: hardware-based approaches and software-based approaches. We focus on the latter.

---

[*] This work was supported by the Post-doctoral Fellowship Program of Korea Science and Engineering Foundation (KOSEF).
[**] Corresponding Author.

2   Algorithm

Our algorithm is based on Lee et al.'s [3]. It represents the IP prefixes as intervals, which was motivated by [2]. Given a k-bit IP prefix $P = p_1p_2\cdots p_k$, we represent it as an interval $[P_L, P_H]$. where $P_L = p_1p_2\cdots p_k 0\cdots 0$ and $P_H = p_1p_2\cdots p_k 1\cdots 1$ be two integers obtained by appending $m-k$ 0's and 1's at the end of P, where m is the length of an IP address. In this representation, a longer IP prefix has a narrower interval. Since the IP address lookup problem is to find the longest matching IP prefix P of x, our aim is to find the narrowest interval $[P_L, P_H]$ that contains x. These intervals are stored in the segment tree [4]. For each interval $[P_L, P_H]$, we compute four values: $P_L - 1$, P_L, P_H and $P_H + 1$. Let U be the set of these values, where values smaller than 0 or greater than $2^m - 1$ are excluded. Then we update $U \leftarrow U \cup \{0, 2^m - 1\}$ to make sure that U contains both endpoints 0 and $2^m - 1$. The segment tree for the set $U = \{u_1, u_2, \cdots, u_N\}$ ($0 = u_1 < u_2 < \ldots < u_N = 2^m - 1$) is a balanced binary tree. The leaves represent basic intervals $[u_1, u_2], [u_3, u_4], \ldots, [u_{N-1}, u_N]$. A parent node represents the union of the intervals of its children. The height of the segment tree is $O(\log n)$. To perform the IP address lookup, we visit the nodes whose intervals contain the target IP address from the root. We report the shortest interval.

Our new algorithm is based on the frequencies of the intervals. Note that we do not aim to reduce the time for each lookup operation. Rather, we would like to enhance overall performance. In the original segment tree, each leaf has the same depth. We modify the segment tree with a simple rule: *move the frequently accessed leaves close to the root*. For those leaves, the number of memory access is reduced. We will call the resulting tree as *the biased segment tree*. This approach resembles Huffman coding [1] in data compression.

The steps for building the biased segment tree are as follows.

Step 1. Compute the frequencies of the basic intervals. To do so, we first build the normal segment tree and count the number of access for every leaf from the training set which reflects the access pattern.

Step 2. Divide the basic intervals into classes based on the frequency information obtained in Step 1.

Step 3. Put each class into a desired level. Here we use the rule: move the frequently accessed leaves close to the root.

Step 4. From the deepest level, create an internal node between two neighboring node. We keep on doing so until we create a root node.

Example 1. Suppose that $m = 3$ and there are four IP prefixes ϵ, 0, 01 and 11. These can be represented by $[000_{(2)}, 111_{(2)}] = [0, 7]$, $[000_{(2)}, 011_{(2)}] = [0, 3]$, $[010_{(2)}, 011_{(2)}] = [2, 3]$, and $[110_{(2)}, 111_{(2)}] = [6, 7]$. Figure 1 (a) represents them as intervals. Then we represent these intervals as a segment tree. Then basic intervals are $[0, 1]$, $[2, 3]$, $[4, 5]$ and $[6, 7]$. The segment tree for U is given in Figure 1 (b). Now, given an IP address $x = 010_{(2)} = 2$, $x \in [0, 7]$, $x \in [0, 3]$, and $x \in [2,3]$, but $x \notin [6,7]$. It means that ϵ, 0 and 01 are prefixes of x, but 11 is not. After visiting nodes n_1, n_2, and n_5, we can find the intervals which contain x. We report 01 as the longest prefix that matches x. However, if n_7 is heavily visited and that node n_5 and n_6 is rarely visited. We move n_7 close to the root node as in Figure 1 (c). The depth of n_7 is reduced to 1, while those of n_5 and n_6 are increased to 3. In this case, if the number of accessing n_7 is greater than the sum of the numbers of accessing n_5 and n_6, we can get a performance improvement.

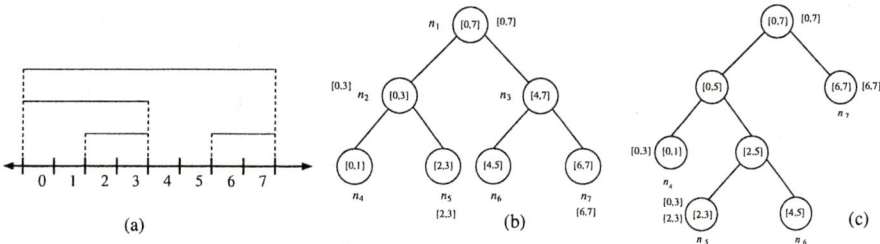

Fig. 1. Examples of (a) intervals, (b) segment tree, and (c) biased segment tree

Now we analyze the effect of modifying the segment tree. We want to know which leaf should be moved closer to the root and how close it should be moved to the root. At first each leave has the same depth at first. Then the number of memory access is $O(\log n)$ where n is the number of leaves. When we move a leaf to k upper level, it has the same effect of adding 2^k leaves. Hence, if the probability that a leaf is accessed is ρ, the following equation should be satisfied to improve the performance:

$$\rho(\log(n + 2^k) - k) + (1 - \rho)\log(n + 2^k) > \log n.$$

It follows

$$\rho > \log(1 + 2^k/n)/k.$$

3 Experimental Results

We implemented the IP address lookup algorithm in [3] and our extension. We compared the number of memory access rather than the running time, since the latter can be changed in a different situation. We used the snapshot of the routing table of ATT Net which contains 77,774 IP prefixes, obtained from http://archive.pch.net/archive. We created two random traffic data set whose sizes are 11,116,474 and 3,131,010, respectively and performed two experiments. In the first experiment, we divided the leaves into three categories. The access ratio among the levels is 1:50:500, that is, a node which belongs to Level 1 is accessed 500 times more frequently than one in Level 3. The depths of the nodes in Level 1 and 2 are reduced by 2 and 1, respectively. In the second experiment, we used five categories and the access ratio is 1:10:50:100:500. The depths of the nodes in Level 1, 2, 3, and 4 are reduced by 4, 3, 2, and 1, respectively. We counted the number of memory access for each level and compute the average value. The results are shown in Figure 2.

	Experiment 1				Experiment 2			
	Data 1		Data 2		Data 1		Data 2	
	[3]	Ours	[3]	Ours	[3]	Ours	[3]	Ours
Level 1	24.60	23.32	24.65	23.14	24.60	22.46	24.84	22.22
Level 2	24.58	24.64	24.57	24.80	24.60	23.93	24.60	23.40
Level 3	24.60	25.32	24.57	26.11	24.57	25.04	24.58	25.02
Level 4	N/A				24.60	25.73	24.61	26.54
Level 5	N/A				24.59	25.93	24.42	27.80
Expected value	24.60	23.44	24.64	23.29	24.60	22.93	24.61	22.72
Average	24.60	23.50	24.64	24.34	24.60	23.22	24.56	22.84

Fig. 2. Experimental results

In the above experiments, we were able to obtain up to 8% speed up in average. Although our modified algorithm is slower than the original algorithm in [3] in the lower levels, the gain obtained from the higher levels can justify our modification.

4 Conclusion

In this paper we showed how to incorporate the frequency information into the routing table and presented the experimental results. We want to know whether other scheme can use the frequency information too. Combining the frequency information and the B-tree data structure will increase the performance.

References

1. D. A. Huffman. A method for the construction of minimum-redundancy codes. *Proceedings of Institute of Electrical and Radio Engineers*, 40(7):1098–1101, 1958.
2. B. Lampson, V. Srinivasan, and G. Varghese. IP lookups using multiway and multicolumn search. *IEEE/ACM Transactions on Networking*, 7(3):324–334, 1999.
3. I. Lee, K. Park, Y. Choi, and S. K. Chung. A simple and scalable algorithm for the IP address lookup problem. *Fundamenta Informaticae*, 56(1-2):181–190, 2003.
4. F. P. Preparata and M. I. Shamos. *Computational Geometry - An Introduction*. Springer-Verlag, 1995.
5. R. Sedgewick. *Algorithms in C++*. Addison-Wesley, 1992.

A Surface Reconstruction Algorithm Using Weighted Alpha Shapes

Si Hyung Park[1], Seoung Soo Lee[2], and Jong Hwa Kim[2]

[1] Voronoi Diagram Research Center, Hanyang University, 17 Haenggang-dong,
Seoungdong-gu, Seoul, 133-791, Korea
shpark@voronoi.hanyang.ac.kr

[2] CAESIT, Konkuk University, 1 Hwayang dong, Gwangjin-gu,
Seoul, 143-701, Korea
{sslee, jhkim}@konkuk.ac.kr

Abstract. This paper discusses a surface reconstruction method using the Delaunay triangulation algorithm. Surface reconstruction is used in various engineering applications to generate CAD model in reverse engineering, STL files for rapid prototyping and NC codes for CAM system from physical objects. The suggested method has two other components in addition to the triangulation: the weighted alpha shapes algorithm and the peel-off algorithm. The weighted alpha shapes algorithm is applied to restrict the growth of tetrahedra, where the weight is calculated based on the density of points. The peel-off algorithm is employed to enhance the reconstruction in detail. The results show that the increase in execution time due to the two additional processes is very small compared to the ordinary triangulation, which demonstrates that the proposed surface reconstruction method has great advantage in execution time for a large set of points.

1 Introduction

Recent advances in measurement systems have enabled more precise measurement of 3-D shapes. Normally a set of points is obtained as the output of measurement systems, but in most cases further processing is required to be useful for engineering purposes. That is, noise should be removed, only useful points should be extracted, and triangular surfaces should be constructed to model the original shape as closely as possible. The triangular surfaces can be used in CAD modeling or computer graphics after surface modeling like NURBS(Non-Uniform Rational B-Spline). The construction of a triangular net can make rendering process easier since it has a simple data structure, and it is easy to access the necessary information to form a surface model. In "surface reconstruction", the closeness to the original shape and the speed of reconstruction is the most important performance measures. This paper presents a construction method of triangular net using 3-D positional data of points in an object without using relational information between the points or the normal vector of surface. We suggest a novel algorithm for surface reconstruction using an enhanced Delaunay triangulation method, which transforms a set of points into a triangular net.

2 Related Works

2.1 Alpha Shapes

The alpha shapes, first described by Edelsbrunner et al.[1-8] are a type of triangulations parameterized by a single global variable, called alpha. The shape of triangular patch is determined by adjusting the value of alpha, which determines the size of a sphere used in a triangulation. If the points are not evenly distributed the value of alpha can be either too small or too large in some regions. To resolve this problem, Edelsbrunner proposed weighted alpha shapes, which assigns different weight for each point in the set. The weight determines the level of details around that point. However, Edelsbrunner described the method only theoretically.

2.2 Crust Algorithm

The crust algorithm proposed by Amenta et al.[9-15] constructs manifold triangulations by interpolating collections of three-dimensional points. In the crust algorithm, Voronoi diagram of the sample points are obtained and the poles in the Voronoi vertices are selected to estimate the medial axis. Then the Delaunay triangulation is performed with the combined point set of the samples and poles and the triangles of which all the vertices are consisted of sample points are chosen. This algorithm is proven to converge to the surface being sampled as the density of point increases. However, it takes long execution time for many practical applications.

2.3 Zero-Set Algorithm

This approach was proposed by Hoppe et al.[16-17]. They estimate tangent planes at each sample point based on the shape of k nearest points, and then the signed distances between the points and the nearest point's tangent plane are calculated. Then the distance function is interpolated and polygonalized by their marching cubes algorithm. The zero-set algorithm constructs an approximate surface rather than interpolated surface. Thus, this method needs low-pass filtering of the data, which is desirable in the presence of noise, but causes some loss of information.

3 The Proposed Surface Reconstruction Algorithm

The Delaunay triangulation method employed in the proposed algorithm is an incremental construction algorithm using the oct-subdivision[18] and it constructs the Delaunay triangles prior to the Voronoi diagram. Thus, the alpha shapes algorithm is more efficient than the crust algorithm to apply the proposed Delaunay triangulation. Although the alpha shapes algorithm works well for the evenly distributed points, it needs to change the value of alpha if the points are distributed irregularly. To resolve the problem, Edelsbrunner suggested the weighted alpha shapes method, in which every point in the set is assigned weight describing the level of details around that point. This method has an advantage in constructing shapes that have different densities, but it is difficult to apply since the weights should be predetermined when the point data is entered. Thus we suggest an algorithm based on the weighted alpha shapes method in which the weights are computed while triangulation, instead of input process of the point data.

3.1 Calculation of Weights

To compute the weights, we first define the following global variable.

$$size = \sqrt[3]{\frac{(x_{max} - x_{min})(y_{max} - y_{min})(z_{max} - z_{min})}{N}}, \qquad (1)$$

where N is the number of points

The above equation makes the average density of the points in the entire grids to be one by adjusting the size of uniform grid. Next, the number of points inside the grid elements is counted. While constructing the Delaunay tetrahedra, the number of grid elements including the smallest box of the empty sphere is computed. For example, if there are 24 uniform grids and 3 oct-subdivided grids, the number of grid elements is 24 + 3/8. Thus, the localized density of points is computed as follows:

$$\rho = \frac{\sum p}{\sum cubic}, \qquad (2)$$

where $\sum p$ is the number of points inside the smallest box
$\sum cubic$ is the number of grid elements in the smallest box

Fig. 1 shows the relationship between the number of points in a cube (500^3 volume) and the average length of edges comprising the Delaunay triangles. The results are obtained with 12 different numbers of points varying the number of points from 50 up to 4,000. For a given number of points, 100 tests are performed and the average length is plotted for each number of points. In each case the points are uniformly distributed. In Fig.1 the average length is in inverse proportion to the cubic root of the number of points (or the density) in the cube. Thus, the relationship between the weighted alpha shapes value and the density of points can be expressed by the following equation.

$$\alpha_w \propto \sqrt[3]{\frac{1}{\rho}} \times size \qquad (3)$$

In the tests, it is observed that the variance of the edge length tends to increase as the irregularity of data points in a grid increases. To adjust that effect, the weight constant, W_{co}, is multiplied in the right hand side of Equation (3) as in Equation (4).

$$\alpha_w = W_{co} \sqrt[3]{\frac{1}{\rho}} \times size \qquad (4)$$

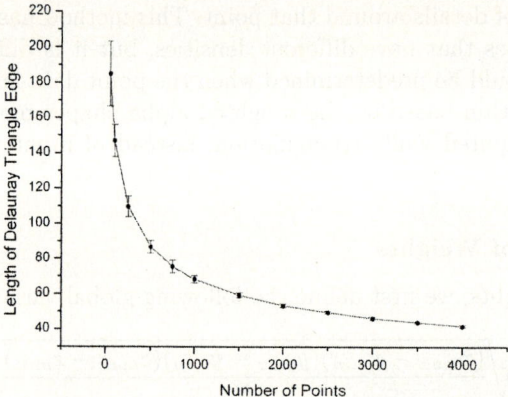

Fig. 1. Relationship between the number of points and length of Delaunay triangle edge

The value of W_{co} is set to be larger than 1 to patch the points that are apart more than the average. The empirical results suggest that the value of W_{co} to be $1.0 \sim 1.2$. Finally, the sensitivity function $f_{st}(\rho)$ is used to correct the error incurred depending on the density of points. Hence Equation (4) is rewritten as Equation (5).

$$\alpha_w = W_{co}(\sqrt[3]{\frac{1}{\rho}} + f_{st}(\rho)) \times size \qquad (5)$$

In Fig. 2 it is observed that the range of edge length is in inverse proportion to the cubic root of the number of points in the cube. That is, if the density of points is very low, the range of edge length becomes larger. Hence the sensitivity function can be expressed as

$$f_{st}(\rho) = S_{co} \times \sqrt[3]{\frac{1}{\rho}}, \qquad (6)$$

where S_{co} is the sensitivity constant

According to the results in Fig. 1, the range of edge length is about 1/10 of the edge length. Thus, the sensitivity constant S_{co} is set to be some value in $(0, 0.1)$. Based on the observations made in Fig.2, the density function $f_{st}(\rho)$ is used in the following two cases:

Case 1: the density of points is very low,

Case 2: the density of points in a grid is larger than but the number of points in the grid is very small due to the small grid size.

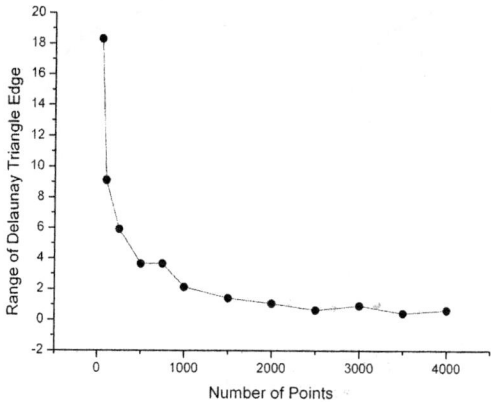

Fig. 2. Relationship between the number of points and range of Delaunay triangle edge

Fig. 3. Relatively uniform point data set

The weight constant and the sensitivity constants are automatically evaluated and determined in the program.

To compare the performance of the constant alpha shapes method and the weighted alpha shapes method, the data points in Fig. 3 are used.

The results by the constant alpha shapes method and the weighted alpha shapes method are presented in Fig. 4 and Fig. 5, respectively. In the constant alpha shapes method, if the value of alpha shapes is small, the details are expressed well but sparse regions are not reconstructed as shown in Fig. 4. On the contrary, if the value of alpha shapes is large, no hole or gap is appearing but details are not described well. The result by the weighted alpha shapes method

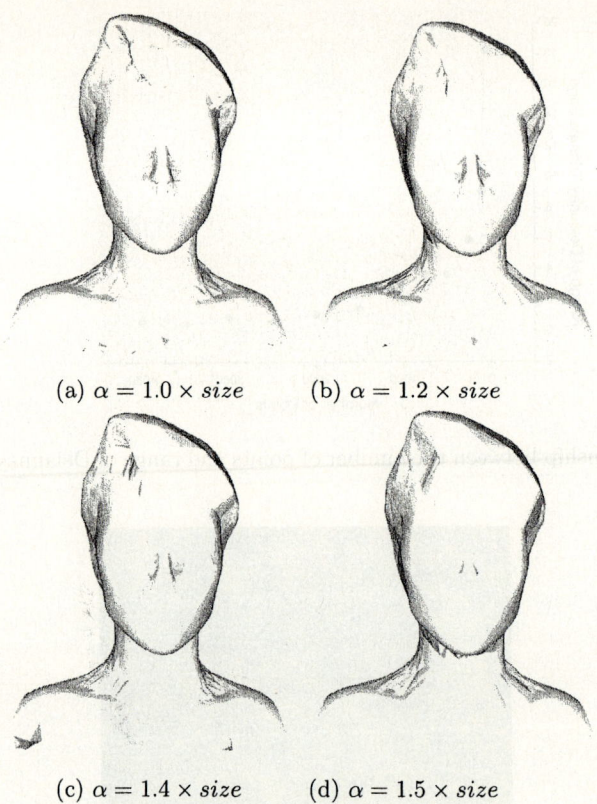

(a) $\alpha = 1.0 \times size$ (b) $\alpha = 1.2 \times size$

(c) $\alpha = 1.4 \times size$ (d) $\alpha = 1.5 \times size$

Fig. 4. Application of constant alpha shapes

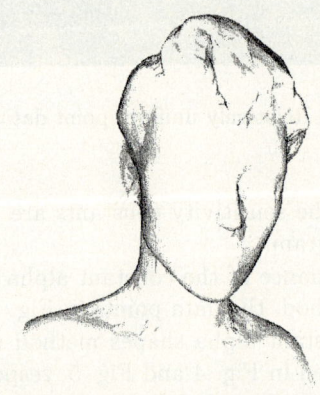

Fig. 5. Application of weighted alpha shapes

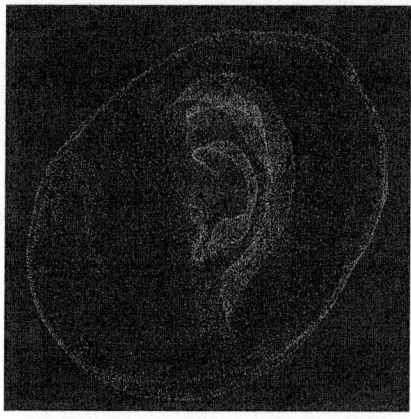

Fig. 6. Non-uniform point data set

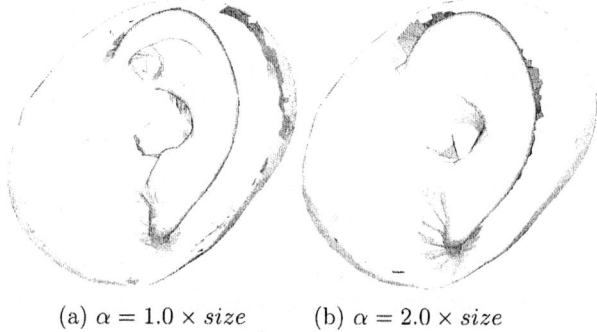

(a) $\alpha = 1.0 \times size$ (b) $\alpha = 2.0 \times size$

Fig. 7. Application of constant alpha shapes

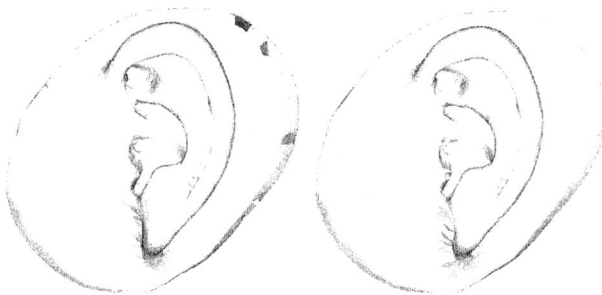

Fig. 8. Application of weighted alpha shapes

shows more detail and less sparse regions compared to the results of the constant alpha shapes method. It also shows a successful triangulation in the region where density changes.

Fig. 6 shows an example of non-uniformly distributed data points compared to Fig. 3. The ratio of the number of high density lattices to that of low density lattices is several hundreds times more in this case. Fig. 7 shows that the constant alpha shapes method results in unsatisfactory shapes. Fig. 8 presents the results by the weighted alpha shapes method with and without sensitivity function. The sensitivity function helps to reduce errors in low density area but describes less detail in the region where density changes.

3.2 Peel-Off Algorithm

Although the weighted alpha shapes method shows better quality than the uniform alpha shapes methods, it still lacks an ability to describe small details. Thus, we develop an additional process for the delicate job, which we call "the peel-off algorithm". The peel-off algorithm can produce fast and accurate results if the following two conditions are satisfied:

1. The peel-off algorithm is applied after the weighted alpha shapes method is applied.
2. There should be no interior point in the data set or if there exists an interior point it should be sufficiently away (bigger than weighted alpha radius of that region) from the points that compose the surface.

The structure of the algorithm is conceptually simple and the pseudo code is presented below.

```
Deletion of discontinuous surfaces;
while(Not all point set(except interior point)exists in the
surface)
{
  Renew current surface;
  while(Check current surface triangle)
  {
    if(The alpha radius of current Delaunay triangle
      is larger than those of three triangle in the
      back)
    {
       Delete the current triangle;
       Set the three triangles in the back to be new
       surfaces;
    }
  }
}
```

The performance of peel-off algorithm is tested for the surface reconstruction of objects that are composed of about 15,000 ∼ 391,000 points. The results

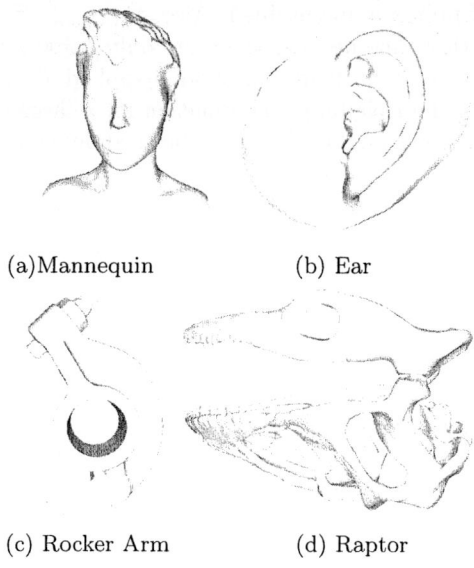

(a) Mannequin (b) Ear

(c) Rocker Arm (d) Raptor

Fig. 9. Results with peel-off algorithm

Table 1. Running time(Test System : P4 2.0 GHz 512 Mbytes)

Model	Number of Points	Time (min)
Mannequin	15K	1 min 2 sec
Ear	21K	1 min 28 sec
Rocker Arm	34K	2 min 21 sec
Raptor	391K	34 min 1 sec

are shown in Fig. 9 and the execution times are summarized in Table 1. The execution times in Table 1 include the execution time of the weighted alpha shapes method and the peel-off algorithm. The execution times increase almost linearly in proportion to the number of points.

4 Conclusion

This paper proposed an improved weighted alpha shapes method, which can be applied if only positional data is given. Also, the weights are not fixed as a single constant but determined according to the density of points. In addition, the peel-off algorithm constructs details successfully by suppressing the formation of wrong triangular surfaces. The alpha shapes method does not require additional computation time because it is a reconstruction method that limits the surface of 3-D Delaunay triangulation. Hence the linearity between the running time and the number of points is maintained. Also, the peel-off algorithm does not add much computation time because there are only a few wrong triangular surfaces after the weighted alpha shapes method is applied. Considering the surface reconstruction using the Delaunay triangulation for a large set of points (larger than 500,000), our reconstruction algorithm has a significant advantage in terms of computation time.

References

1. H.Edelsbrunner. : Surface Reconstruction by Wrapping Finite Sets in Space. Technical Report 96-001. Raindrop Geomagic. Inc. (1996)
2. H.Edelsbrunner, N. Shah. : Triangulating Topological Spaces. Proceedings 10th ACM Symp., Computational Geometry (1994) 285-292
3. H.Edelsbrunner, Ernst P. Mucke : Three-Dimensional Alpha-Shapes. ACM Transactions on Graphics, 13(1) (1994) 43-72
4. H.Edelsbrunner : Weight Alpha Shapes. Technical Report UIUCDCS-R-92-1760, Department of Computer Science. University of Illinois at Urbana-Champaign (1992)
5. H.Edelsbrunner, David G.Kirkpatrick, Raimund Seidel : On the Shape of a Set of Points in the Plane. IEEE Transactions on Information Theory, IT-29(4) (1983) 551-559
6. H.Edelsbrunner, N. Shah : Triangluating Topological Spaces. Proceedings 10th ACM Symposium Computational Geometry. (1994) 43-72
7. H.-L. Cheng, T. K. Dey, H. Edelsbrunner, J. Sullivan : Dynamic Skin Triangulation. Discrete Comput. Geom. 25 (2001) 525-568
8. N. Akkiraju, H. Edelsbrunner : Triangulating the Surface of a Molecule. Discrete Appl. Math. 71 (1995) 5-22
9. Nina Amenta, Sunghee Choi, Guenter Rote : Incremental Constructions con BRIO. ACM Symposium on Computational Geometry (2003) 211-219
10. Sunghee Choi and Nina Amenta : Delaunay Triangulation Programs on Surface Data. The 13th ACM-SIAM Symposium on Discrete Algorithms (2002) 135-136
11. Nina Amenta, Sunghee Choi, Ravi Kolluri : The Power Crust. Proceedings of 6th ACM Symposium on Solid Modeling (2001) 249-260
12. Nina Amenta, Sunghee Choi, Ravi Kolluri : The Power Crust, Unions of Balls, and the Medial Axis Transform. Computational Geometry: Theory and Applications, 19:(2-3) (2001) 127-153
13. Nina Amenta, Marshall Bern : Surface Reconstruction by Voronoi filtering. Discrete and Computational Geometry, 22 (1999) 481-504
14. Nina Amenta, Marshall Bern, David Eppstein : The Crust and the Beta-Skeleton: Combinatorial Curve Reconstruction. Graphical Models and Image Processing, 60/2:2 (1998) 125-135
15. Nina Amenta, Marshall Bern, David Eppstein : Optimal Point Placement for Mesh Smoothing. Journal of Algorithms 30 (1999) 302-322
16. H. Hoppe : Surface Reconstruction from Unorganized Points. Ph.D. Thesis, Computer Science and Engineering, University of Washington (1994)
17. H. Hoppe, T. DeRose, T. Duchamp, J. McDonald, W. Stuetzle : Surface Reconstruction from Unorganized Points. In SIGGRAPH 92 Proceedings (1992) 71-78
18. Si Hyung Park, Seoung Soo Lee, Jong Hwa Kim : The Delaunay Triangulation by Grid Subdivision . ICCSA 2005 Proceedings, LNCS 3482 (2005) 1033-1042

HYBRID: From Atom-Clusters to Molecule-Clusters

Zhou Bing[1], Jun-yi Shen[2], and Qin-ke Peng[2]

[1] Dept. of Computer Science and Engineering,
Northeastern University at Qin Huang-dao 066004,
He Bei, China
zhoub@mail.neuq.edu.cn
[2] School of Electronics and Information Engineering, Xi'an Jiaotong University,
710049, Shaan Xi, China
{jyshen, qkpeng}@mail.xjtu.edu.cn

Abstract. This paper presents a clustering algorithm named HYBRID. HYBRID has two phases: in the first phase, a set of spherical *atom-clusters* with same size is generated, and in the second phase these atom-clusters are merged into a set of *molecule-clusters*. In the first phase, an incremental clustering method is applied to generate atom-clusters according to memory resources. In the second phase, using an edge expanding process, HYBRID can discover molecule-clusters with arbitrary size and shape. During the edge expanding process, HYBRID considers not only the distance between two atom-clusters, but also the closeness of their densities. Therefore HYBRID can eliminate the impact of outliers while discovering more isomorphic molecule-clusters. HYBRID has the following advantages: low time and space complexity, no requirement of users' involvement to guide the clustering procedure, handling clusters with arbitrary size and shape, and the powerful ability to eliminate outliers.

1 Introduction

Since clustering techniques can discover the distributions and patterns of data without any special knowledge about application fields, it is an important task of Data Mining and has been studied widely in the fields of Statistics, Machine Learning and Database communities. There are many different clustering methods, for example, CLARANS[1] and DBSCAN[3] for spatial data clustering, BIRCH[2] and CURE[4] for very large database, etc.

We believe that a good clustering algorithm should be able to solve four problems. First, it should have low time and space complexity. For a clustering algorithm with the time complexity of $O(n^2)$ and the space complexity of $O(n)$, it is very hard to handle a database with millions of data objects due to the high demand on computing resources. Second, the algorithm should require no or very little involvement of users. Because it is hard for users to understand the distribution pattern and interrelationship between data (in fact, this is why we study data mining), it is impractical to require users to provide useful parameters. Normally these subjective parameters cannot represent the real distribution of data objects. Using these parameters to guide the clustering procedure may cause incorrect results. The third problem is how to analyze arbitrary clusters. Many distance-based clustering algorithms use center and radius to represent a cluster. These methods are suitable for discovering the clusters with spherical shape and similar size, but they cannot work well for arbitrary clusters. The last problem is how to eliminate the impact of "noise". It is inevitable that some error data exist in real databases, so detecting and removing these noises are important for obtaining high quality clusters.

This paper attempts to address those problems by proposing a new clustering algorithm named HYBRID. This algorithm has low time and space complexity. It can generate the best clustering results with limited memory by scanning the database only once. Clustering is completed automatically and intelligently without reading in cluster-related parameters from users. HYBRID can discover arbitrary clusters from a dataset containing noises. Our experiments show that this algorithm is a good clustering algorithm.

In the rest of the paper, we first briefly introduce some well-known clustering algorithms. In section 3, our algorithm is described in details. Section 4 shows the experimental results. The last section is our concluding remarks.

2 Related Works

Data clustering has been studied widely in the fields of Statistics, Machine Learn-ing, and Database. This section summarizes some well-known clustering algorithms.

CLARANS[1] attempts to deal with very large-scale databases. In CLARANS, a cluster is represented by its *medoid*, or the most centrally located data point in the cluster. CLARANS requires two parameters: *Numlocal* and *Maxneighbor*. Numlocal represents the expected cluster number K. The clustering problem is how to find Numlocal medoids. The clustering process is formalized as searching a graph in which each node is a K-partition represented by a set of K medoids, and two nodes are neighbors if they only differ by one medoid. Each step of clustering is to find a node with the minimum cost from Maxneighbor neighbors via random searching. This node is called the local minimum node. CLARANS stops when the local minimum node does not change. CLARANS requires the cluster number, as an input parameter, and it cannot discover arbitrary clusters. In addition, it has a high time complexity of $O(n^2)$.

BIRCH[2] is a kind of hierarchical clustering method. In BIRCH, a cluster is represented by its center and radius. BIRCH defines the concept of *Cluster Feature* (CF), and then performs clustering by dynamically building a CF-tree. A CF tree is built when new data objects are inserted into the closest leaf node, making the new radius less than the limitation of radius. While the CF-tree runs out of the memory, the radius limitation can be increased automatically so that a leaf node can "absorb" more data objects, and a new smaller CF-tree can created. The drawback of BIRCH is that it cannot discover arbitrary clusters. However, it requires no input parameters, and it has the ability to handle "noise". Also it is an incremental clustering method that does not require the whole dataset to be read into the main memory in advance. It scans the dataset only once. So the space complexity is very low, and its time complexity is only O(n).

DBSCAN[3] is a kind of clustering algorithm based on density. It needs two parameters: *Eps* and *Minpts*. DBSCAN defines a cluster to be a maximum set of density-connected points, and every core point in a cluster must have at least a minimum number of points (MinPts) within a given radius (Eps). To find a cluster, DBSCAN checks the Eps-neighborhood of each point in the database. If the Eps-neighborhood of a point p contains at least MinPts number of points, a cluster using p as core point will be generated. And then DBSCAN retrieves all density-reachable points from these core points. In this process, some density-reachable clusters will be merged together. While no new point is added into any cluster, the clustering procedure stops. DBSCAN can discover arbitrary clusters and find noise very well. However, its disadvantages are obvious. Firstly, DBSCAN is very sensitive to the parameters. Secondly, its time complexity reaches to $O(n^2)$, if space index is not used. Although the time complexity can reduce to $O(nlogn)$ by using space index, the space complexity will increase.

CURE[4] is also a hierarchical clustering algorithm. In CURE, a cluster is not represented by center and radius, but by a constant number of well-scattered points in a cluster. CURE requires the cluster number K to be inputted as a parameter. At the beginning, each data point is treated as a cluster, and then two clusters with the closest pair of representative points are merged. This process is repeated until obtaining the expected cluster number. Because the representative points can capture the shape and extent of a cluster, CURE can discover arbitrary clusters. At meat time, by shrinking the chosen points towards the center of a cluster, noise can be detected and removed. CURE has the disadvantage of requiring the cluster number as an input parameter. Furthermore, it has a space complexity of $O(n)$, so CURE has to use random sampling to reduce the data amount. However for a very large database, the sample size is still too large to be read into memory in advance. If we restrict the sample size to fit into the memory, clustering error may occur due to too few samples. The time complexity of CURE is $O(n^2 logn)$ at the worst case. For low dimensional data space, the time complexity is $O(n^2)$.

On the contrary, HYBRID has the following advantages. 1. HYBRID has low time and space complexities. 2. HYBRID does not require any input parameters that related to cluster pattern. 3. HYBRID can discover arbitrary clusters. 4.HYBRID has a good ability to handle noises.

3 The HYBRID Clustering Algorithm

3.1 The Basic Idea

HYBRID is based on an observation that any clusters can be considered as an aggregation of some spherical sub-clusters of same size. We call those spherical sub-clusters as *atom-clusters*, and the clusters formed by atom-clusters are called *molecule-cluster*. Atom-clusters and molecule-clusters have the following relationship. The precision of a molecule-cluster is a function of the size of atom-clusters. The smaller the atom-cluster is, the more atom-clusters the molecule can contain and the more precise the molecule-cluster is (as shown in Figure 1). The size of an atom-cluster depends on the size of memory. If all data objects can be hold in the memory, an atom-cluster is a real data point. Otherwise, an atom-cluster is represented by the center and radius of some points. When memory size increases, more atom-clusters can be included in the memory. Then each atom-cluster can stand for a smaller number of real points, and the size of an atom-cluster becomes smaller. This means that more precise molecules can be obtained. Therefore, the precision of molecule-clusters is a function of memory size. In practice, we can obtain the expected quality by toning the size of memory.

Fig. 1. The relationship of atom-cluster and molecule-cluster

Exploiting the above idea, we propose a two-phase clustering algorithm named HYBRID. The first phase is to generate atom-clusters. The second phase is to aggregate atom-clusters to form molecule-clusters. At the first phase, we create atom-clusters according to the size of memory, using the definition of Cluster Feature[2] and an incremental clustering method named CLAP[5]. At the second phase, when merging two atom-clusters, we consider not only their distance but also the similarity of their densities. By merging the atom-clusters with the closest distance and the most similar density, our algorithm provides a powerful ability to eliminate noises and obtain isomorphic molecules clusters.

3.2 Implementation

First phase: Generate spherical atom-clusters with the same size

We use a method named CLAP[5], which is an improvement of BIRCH algorithm[2], to generate atom-clusters. Like BIRCH, CLAP is based on Clustering Feature and CF-tree[2] but introduce a *Separate Factor S* to define the upper bound of the diameter of a leaf node.

A Clustering Feature is defined as a triple: CF={N, $\vec{LS}$, SS}. N is the number of data points in a cluster. $\vec{LS}$ is the linear sum of N data points. SS is the square sum of N data points. CF summarizes the information of data points in a cluster, so that the data in a cluster can be represented by a CF instead of a collection of data points.

A CF-tree is a high balanced tree with two parameters: *branching factor B* and *threshold T*. Branching factor is the maximum number of children of a non-leaf node. Threshold defines the limitation of the diameter of a cluster. The tree size is a function of T. The larger T is, the smaller the tree is.

With the introducing of Separate Factor S, CLAP algorithm modifies the procedure of generating CF-tree. When adding a new child into a leaf node, if the diameter of the new leaf is greater than S, a new leaf node will be generated to hold this new entry, no matter whether the old leaf has a spare place or not. This scheme increases the compactness of leaf nodes and improves the precision of clustering. Like T, S is adjusted dynamically using the following heuristic approach, which is similar to that used in [2]:

1. $N_{i+1}=Min(2N_i, N)$, where N is the total amount of input data. We expect that the new T_{i+1} and S_{i+1} can cause N_{i+1} data to be read into memory. So N_{i+1} is an important parameter to estimate T_{i+1} and S_{i+1}.
2. Generally the average radius of the root cluster in a CF-tree, r, grows with the number of data points N_i, so we can maintain a record of r and the number of points N_i, and estimate r_{i+1} via least squares linear regression. We define an expansion factor $f=Max(1.0, r_{i+1}/r_i)$ to predict the growth rate of T_{i+1} and S_{i+1}.
3. We increase T in order to merge child entries in leaf nodes and increase S in order to merge leaf nodes. We therefore choose the closest two child entries and two leaf nodes to calculate the minimal distance among child entries (entryD_{min}) and the minimal distance among leaf nodes (leafD_{min}).
4. Finally, let $T_{i+1}=Max(entryD_{min}, f* T_i)$, $S_{i+1}=Max(leafD_{min}, f* S_i)$. If $T_{i+1}= T_i$ and $S_{i+1}= S_i$, then we let $T_{i+1}=T_i*(N_{i+1}/N_i)^{1/d}$, $S_{i+1}=S_i*(N_{i+1}/N_i)^{1/d}$ (d is the dimension of the data space) to make T_{i+1} and S_{i+1} greater than T_i and S_i.

The other two improvements of CLAP are using random sampling technique[6] to decrease the running time greatly and combining the global searching and local searching to eliminate the impact of input order, which means that at the beginning CLAP searches all the leaf nodes sequentially to find a proper cluster; and after the first CF-tree rebuilt, CLAP uses CF-tree structure to speedup the searching procedure.

After the first phase, we can obtain lots of atom-clusters represented by CF. We call the set of atom-clusters as ATOM. We define the *centroid C* of an atom-cluster *a* as $C_a= \vec{LS}/N$. According to the property of CF-tree, The radius of all atom-clusters is no lager than T. We define the *radius R* of all atom-clusters as T and the *density D* of an atom-cluster a as $D_a=N/R$.

Next we will merge the atom-clusters into molecule-clusters with arbitrary shape and size.

Second phase: Aggregate the atom-clusters to form molecule-clusters

The process of merging atom-clusters could be considered as the edge expanding process of molecule-clusters, which is an iterative process of forming a new edge from an old edge and then adding the new edge to molecule-cluster. We define E_m, the *edge* of a molecule-cluster m, as a set of atom-clusters, which are neighbors of m and have similar density with m. In order to define the neighbors of molecule-cluster, we first define the neighbors of atom-cluster. We define the neighbors of an atom-cluster a as:

$$G_a=\{x| Distance(C_a,C_x) \leq 2R\}, x \in ATOM \qquad (1)$$

Then the neighbors of a molecule-cluster m are defined as:

$$G_m= \cup G_a, \text{ where } a \in E_m. \qquad (2)$$

Because of the uneven distribution of the density in one molecule-cluster, which may vary gradually, we define the density of a molecule m as the average density of its edge but not the average density of the atom-clusters contained in m, that is:

$$D_m= \Sigma D_a/\|E_m\|, \text{ where } \|E_m\| \text{ is the amount of atom-clusters of } E_m \text{ and } a \in E_m \qquad (3)$$

We define that an atom-cluster *a* has a similar density to a molecule-cluster m, if and only if the relationship between a and m satisfies the following formula:

$$|D_m - D_a| \leq (D_{max} - D_{min})/2 \qquad (4)$$

where D_{max} and D_{min} are the maximal density and the minimal density in ATOM set.

According to above definitions, the process of forming a new edge of molecule-cluster m is to pick out those atom-clusters, which have similar density to m, from G_m. As summary, the procedure of generating molecule-clusters is shown as follows.

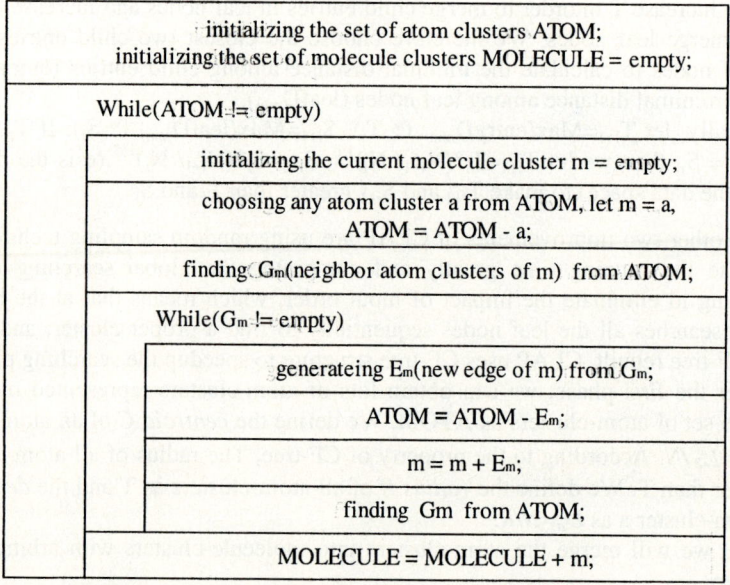

Fig. 2. The procedure of generating molecule-clusters

In the edge expanding procedure, we consider the distance among atom-clusters as well as their densities. Therefore we can obtain more isomorphic molecule-clusters and treat those sparse molecule-clusters as noises.

3.3 Algorithm Analysis

The time complexity of the first phase is $O(n)$[5]. In second phase, suppose we have M atom-clusters at beginning and obtain K molecule-clusters at last. Then each molecule-cluster has M/K atom-clusters on average. When we begin to generate the i-th molecule-cluster, only $(1-(i-1)/K)M$ atom-clusters are left. During the process of edge expanding, the worst case is that only one atom-cluster is found each time, and then we need to search the left atom-clusters M/K times to generate a molecule-cluster. Hence the cost of building the i-th molecule-cluster is $O((1-(i-1)/K)M^2/K)$. The whole cost of phase 2 is about $O(M^2/2)$. The overall time cost of HYBRID is: $O(n+M^2/2)$. HYBRID does not require all the data to be read into memory in advance. The space complexity is $O(M)$, which is independent of the amount of data points.

4 Experimental Results

HYBRID is developed by C language and supported by a database named MYSQL. HYBRID is executed on a PC, which has a CPU of Intel P4 1.7G series and two 256M memory chips. The operation system is LINUX.

4.1 Memory Scalability

In this section, we use one 2-dimensional datasets to evaluate the memory scala-bility. The dataset has four clusters and many noises. We make the experiments three times. At first time, we use all the memory (proximately 512M); at the second time, we limit the memory as half as the first time; and at the third time, the memory was decreased to quarter as much as that of the first time. The results are depicted by figure 3, from which we can see that the precision of molecule-clusters decreases with the decrease of memory. When the memory decreased, an atom-cluster will absorb more points and its diameter becomes larger, which may cause two atom-clusters been merged together wrongly. Then noises are clustered together and cannot be cleared, and even all the atom-clusters are merged to form one molecule-cluster.

(a)　　　　　　(b)　　　　　　(c)　　　　　　(d)

Fig. 3. Memory scalability((a) The original datasets, (b) The result of full memory, (c) The result of half memory, (d) The result of quarter memory)

4.2 Time Scalability

In this section, the time scalability of HYBRID is evaluated by a real dataset that extracted from the 1990 US Census Data, which can be found at the Census Bureau website of the U.S. Department of Commerce[1]. This dataset has 1,106,228 records, and each record has 16 attributes. Two distinct ways of increasing the data size are used. One is to increase the number of 2-dimensional points (The results are depicted by table1 and figure4). The other is to increase the number of dimension with a fixed number of points (The results are depicted by table2 and figure5). With the increase of dimension number, the number of atom-clusters increases very fast. So the time scalability of high dimensional points is not very good.

Table 1. Result of point number increase

2 dimension	1	2	3	4	5
Point Number	614571	737485	860399	983314	1106228
Time (s)	77	87	96	106	117

---

[1] http://www.census.gov/DES/www/des.html

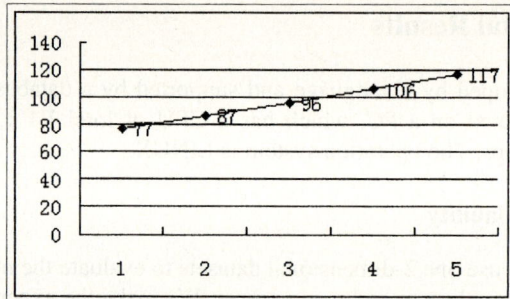

Fig. 4. Time scalability of point number increase

Table 2. Result of dimension number increase

245828 points	1	2	3	4	5
Dimension	3	6	9	12	15
Time(s)	39	145	592	2193	3810
Atom-cluster Numbers	736	7214	19149	51072	89814

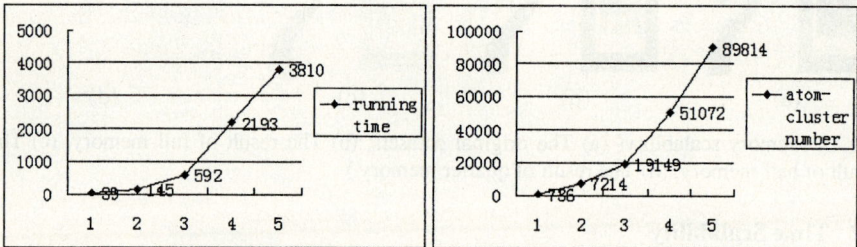

Fig. 5. Time scalability of dimension number increase

4.3 Comparisons with Other Clustering Algorithm

In this section, we compare the performance of HYBRID with that of DBSCAN and CURE. Although HYBRID is applicable to any dataset for which a similarity matrix is available, we perform HYBRID on four two-dimensional datasets. Because clusters in 2D datasets are easy to visualize, and similar datasets have been used to evaluate the performance of DBSCAN and CURE (http://www.cs.umn.edu/~han/-chameleon.html[7]). The geometric shapes of our datasets are shown below.

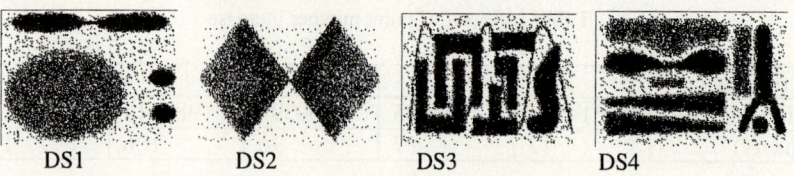

Fig. 6. The four two-dimensional datasets

The first dataset, DS1, has four clusters with different size, shape, and density, and contains noise points. The second dataset, DS2, contains two clusters that are close to each other and different regions of the clusters have different densities. The third dataset, DS3, has six clusters of different size, shape, and orientation, as well as random noise points and special artifacts such as streaks running across clusters. Finally, the forth dataset, DS4, has eight clusters of different shape, size, density, and orientation, as well as random noises. A challenge of this dataset is that clusters are very close to each other and they have different densities.

The results of HYBRID are shown below:

Fig. 7. The clustering results of HYBRID

From above pictures, we can see that HYBRID discovers the cluster patterns in DS1, 2, 4 correctly. For DS3, because of the streaks running across clusters, two clusters are merged improperly, but other clusters are correct. Additionally, at the edges of the clusters, due to the impact of noises and uneven densities, HYBRID generates some small clusters by mistakes. However, the overall results show that HYBRID can discover arbitrary clusters accurately and eliminate noises powerfully.

The following results[7] are obtained by DBSCAN:

Eps=0.5, MinPts=4   Eps=3.5, MinPts=4   Eps=0.4, MinPts=4   Eps=3.0, MinPts=4

Fig. 8. The clustering results of DS1, DS2 obtained by DBSCAN

DBSCAN can find arbitrary shape of clusters, if the right density of the clusters can be determined in a priori and the density of clusters is uniform. DBSCAN finds the right clusters for the datasets DS1, as long as it is supplied the right combination of Eps and Minpts. However, it does not perform well on DS2, as this dataset contains clusters of different densities. The next two pictures show the sensitivity of DBSCAN with respect to the Eps parameter. As we decrease Eps, the natural clusters in the dataset are fragmented into a large number of smaller clusters.

The results[7] of CURE are shown as follows:

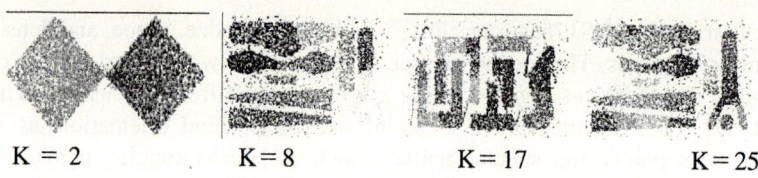

K = 2　　　　　　K = 8　　　　　　K = 17　　　　　　　K = 25

Fig. 9. The clustering results of DS2, DS4 obtained by CURE

CURE is able to find the right clusters for DS2, but it fails to find the right clusters for DS4. Because CURE does not consider the densities of clusters, it may mistakenly merge two clusters with complex shapes under the impact of noises. CURE may select a wrong pair of clusters, which are close to each other but have very different densities, and merge them together. In addition, CURE requires the cluster number as an input parameter. The incorrect parameter will result in wrong clusters.

5 Conclusion

In this paper, we have proposed a new clustering algorithm named HYBRID, which has the following advantages: low time and space complexity, no requirement of cluster-related parameters to guide the clustering procedure, handling clusters with arbitrary size and shape, and the powerful ability to eliminate outliers. The experimental results show that HYBRID is a good clustering algorithm.

Since clustering deals with very large databases and high-dimensional data types, it puts high demands on space and time. As a solution, parallel algorithms can be used to provide powerful computing ability for clustering. Our further research is to study parallel clustering algorithms.

Reference

1. Raymond T., Hau N. J.: Efficient and effective clustering methods for spatial data mining. Proceedings of the 20th VLDB Conference. Santiago Chile (1994) 144-155.
2. Tian Zhang, Raghu Ramakrishnan, Miron Livny: BIRCH: An Efficient Data Clustering Method for Very Large Databases. Proceedings of the ACM SIGMOD Conference on Management of Data. Montreal Canada (1996) 103-114.
3. Ester M., Kriegel H.P., Sander J.: A density-based algorithm for discovering clusters in large spatial database with noise. Proceedings of 2nd Intl. Conf. on Knowledge Discovering in Databases and Data Mining. Portland USA (1996) 226-231.
4. Guha U., Rastogi R., Shim K.: CURE: an efficient clustering algorithm for large databases. Pergamon Information Systems. 26 (2001) 35-58.
5. Zhou Bing, Shen JunYi, Peng QinKe: Clustering Algorithm Based on Random-Sampling and Cluster-Feature. Journal Of Xi'an Jiaotong University. 37 (2003) 1234-1237.
6. Motwani R., Raghavan P.: Randomized Algorithms. Cambridge University Press, London (1995)
7. George Karypis, Eui-Hong (Sam) Han, Vipin Kumar: CHAMELEON: A Hierarchical Clustering Algorithm Using Dynamic Modeling. COMPUTER. 32 (1999) 68-75.

A Fuzzy Adaptive Filter for State Estimation of Unknown Structural System and Evaluation for Sound Environment

Akira Ikuta[1], Hisako Masuike[2], Yegui Xiao[1], and Mitsuo Ohta[3]

[1] Prefectural University of Hiroshima, Hiroshima, Japan 734-8558
[2] NTT Data Chugoku, Hiroshima, Japan 732-086
[3] Hiroshima University, Emeritus, Japan

Abstract. The actual sound environment system exhibits various types of linear and non-linear characteristics, and it often contains an unknown structure. Furthermore, the observations in the sound environment are often contain fuzziness due to several causes. In this paper, a method for estimating the specific signal for acoustic environment systems with unknown structure and fuzzy observation is proposed by introducing a fuzzy probability theory and a system model of conditional probability type. The effectiveness of the proposed theoretical method is confirmed by applying it to the actual problem of psychological evaluation for the sound environment.

1 Introduction

The internal physical mechanism of actual sound environment system is often difficult to recognize analytically, and it contains unknown structural characteristics. Furthermore, the stochastic process observed in the actual phenomenon exhibits complex fluctuation pattern and there are potentially various nonlinear correlations in addition to the linear correlation between input and output time series.

In our previous study, for complex sound environment systems difficult to analyze by using usual structural methods based on the physical mechanism, a nonlinear system model was derived in the expansion series form reflecting various type correlation information from the lower order to the higher order between input and output variables[1]. The conditional probability density function contains the linear and nonlinear correlations in the expansion coefficients, and these correlations play an important role as the statistical information for the input and output relationship.

In this paper, a complex sound environment system including the human consciousness and response for physical sound phenomena is considered. It is necessary to pay our attention on the fact that the observation data in the sound environment system often contain the fuzziness due to several causes. For example, the human psychological evaluation for noise annoyance can be judged by use of 7 scores: 1.very calm, 2.calm, 3.mostly calm, 4.little noisy, 5.noisy,

6.fairly noisy, 7.very noisy[2]. However, each score is affected by the human subjectivity and the borders between two scores are vague. In this situation, in order to evaluate the objective sound environment system, it is desirable to estimate the waveform fluctuation of the specific signal based on the observed data with fuzziness.

As a typical method in the state estimation problem, the Kalman filtering theory and its extended filter are well known[3,4,5]. These theories are originally based on the Gaussian property of the state fluctuation form. On the other hand, the actual signal in sound environment exhibits various types of probability distribution forms apart from a standard Gaussian distribution due to the diversified causes of fluctuation. In our previous studies[6,7], several state estimation methods for sound environment systems with non-Gaussian fluctuations have been proposed on the basis of Bayes' theorem. However, these state estimation algorithms have been realized by introducing the additive model of the specific signal and an external noise. The actual sound environment systems exhibit complex fluctuation properties and often contain unknown characteristics in the relationship between the state variable and the observation. Furthermore, the observation data often contain fuzziness. Thus, it is necessary to improve the previous state estimation methods by taking account of the complexity and uncertainty in the actual systems.

From the above viewpoint, based on the fuzzy observations, a method for estimating adaptively the specific signal for sound environment systems with unknown structural characteristic is theoretically proposed in this study. More specifically, after regarding the human noise annoyance scores as observation data with fuzziness, by adopting an expansion expression of the conditional probability distribution reflecting the information on linear and non-linear correlation between the input and output signals as the system characteristics, an estimation method of the specific signal with non-Gaussian properties is propsed by introducing a fuzzy theory. The proposed estimation method can be applied to an actual complex sound environment system with unknown structure by considering the coefficients of conditional probability distribution as unknown parameters and estimating simultaneously these parameters and the specific signal. Furthermore, the effectiveness of the proposed theory is confirmed experimentally by applying it to the estimation of sound level based on the successive observation of the psychological evaluation for noise annoyance.

2 Sound Environment System with Unknown Structure and Fuzzy Observation

Consider a complex sound environmnet system with an unknown structure that cannot be obtained on the basis of the internal physical mechanism of the system. In the observations of actual sound environment system, the sound level data often contain the fuzziness due to human subjectivity in noise annoyance evaluation, confidence limitations in sensing devices, and quantizing errors in digital observations, etc. Therefore, in order to evaluate adaptively the objective sound environment, it is desirable to estimate the fluctuation waveform of the specific signal based on the observations with fuzziness.

Let x_k and y_k be the input and output signals at a discrete time k for a sound environment system. For example, for the psychological evaluation in sound environment, x_k and y_k denote respectively the physical sound level and human response quantity for it. It is assumed that there are complex nonlinear relationships between them. Since the system characteristics are unknown, a system model in the form of a conditional probability is adopted. More precisely, attention is focused on the conditional probability distribution function $P(y_k|x_k)$ reflecting all linear and nonlinear correlation information between x_k and y_k. Furthermore, let z_k be the fuzzy observation obtained from y_k. For example, z_k expresses the noise annoyance scores (1.very calm, 2.calm, 3.mostly calm, 4.little noisy, 5.noisy, 6.faily noisy, 7.very noisy) taking the individual and psychological situation into consideration for y_k. The fuzziness of z_k is characterized by the membership function $\mu_{z_k}(y_k)$. As the membership function, the following function is adopted.

$$\mu_{z_k}(y_k) = exp\{-\alpha(y_k - z_k)^2\}, \tag{1}$$

where $\alpha(>0)$ is a parameter.

Expanding the joint probability distribution function $P(x_k, y_k)$ in an orthogonal form based on the product $P(x_k)$ and $P(y_k)$, the following expression on the conditional probability distribution function $P(y_k|x_k)$ can be derived.

$$P(y_k|x_k) = \frac{P(x_k, y_k)}{P(x_k)} = P(y_k) \sum_{r=0}^{R} \sum_{s=0}^{S} A_{rs} \theta_r^{(1)}(x_k) \theta_s^{(2)}(y_k), \tag{2}$$

$$A_{rs} = <\theta_r^{(1)}(x_k)\theta_s^{(2)}(y_k)>, \tag{3}$$
$$(A_{00} = 1, \ A_{r0} = A_{0s} = 0 \ (r,s = 1,2,\cdots)),$$

where $<>$ denotes the averaging operation on the variables. The linear and non-linear correlation information between x_k and y_k is reflected hierarchically in each expansion coefficient A_{rs}. The functions $\theta_r^{(1)}(x_k)$ and $\theta_s^{(2)}(y_k)$ are orthonormal polynomials with the weighting functions $P(x_k)$ and $P(y_k)$ respectively, and can be decomposed by using Schmidt's orthogonalization[6]. Though (2) is originally an infinite series expansion, a finite expansion series with $r \leq R$ and $s \leq S$ is adopted because only finite expansion coefficients are available and the consideration of the expansion coefficients from the first few terms is usually sufficient in practice. Since the objective system contains an unknown specific signal and unknown structure, the expansion coefficients A_{rs} expressing hierarchically the statistical relationship between x_k and y_k must be estimated on the basis of the fuzzy observation z_k. Considering the expansion coefficients A_{rs} as unknown parameter vector **a**:

$$\mathbf{a} = (a_1, a_2, \cdots, a_I) = (\mathbf{a}_{(1)}, \mathbf{a}_{(2)}, \cdots, \mathbf{a}_{(S)}),$$
$$\mathbf{a}_{(s)} = (A_{1s}, A_{2s}, \cdots, A_{Rs}), \ (s = 1, 2, \cdots, S), \tag{4}$$

the following simple dynamical model is introduced for the simultaneous estimation of the parameters with the specific signal x_k:

$$\mathbf{a}_{k+1} = \mathbf{a}_k, \tag{5}$$
$$(\mathbf{a}_k = (a_{1,k}, a_{2,k}, \cdots, a_{I,k}) = (\mathbf{a}_{(1),k}, \mathbf{a}_{(2),k}, \cdots, \mathbf{a}_{(S),k})),$$

where $I(=RS)$ is the number of unknown expansion coefficients to be estimated.

On the other hand, based on the correlative property in time domain for the specific signal fluctuating with non-Gaussian property, the following time transition model for the specific signal is generally established.

$$x_{k+1} = Fx_k + Gu_k, \tag{6}$$

where u_k is the random input with mean 0 and variance σ_u^2. Two parameters F and G are estimated by using a system identification method[8]. A method to estimate x_k based on the fuzzy observation z_k is derived in this study by introducing a fuzzy probability theory and an expansion series expression of the conditional probability distribution function.

3 State Estimation Based on Fuzzy Observation

In order to derive an estimation algorithm for a specific signal x_k, based on the successive observations of fuzzy data z_k, we focus our attention on Bayes' theorem[9] for the conditional probability distribution. Since the parameter $\mathbf{a}_k$ is also unknown, the conditional probability distribution of x_k and $\mathbf{a}_k$ is considered.

$$P(x_k, \mathbf{a}_k | Z_k) = \frac{P(x_k, \mathbf{a}_k, z_k | Z_{k-1})}{P(z_k | Z_{k-1})}, \tag{7}$$

where $Z_k(=\{z_1, z_2, \cdots, z_k\})$ is a set of observation data up to a time k. After applying fuzzy probability[10] to the right side of (7), expanding it in a general form of the statistical orthogonal expansion series[11], the conditional probability density function $P(x_k, \mathbf{a}_k | Z_k)$ can be expressed as:

$$P(x_k, \mathbf{a}_k | Z_k) = \frac{\int \mu_{z_k}(y_k) P(x_k, \mathbf{a}_k, y_k | Z_{k-1}) dy_k}{\int \mu_{z_k}(y_k) P(y_k | Z_{k-1}) dy_k}$$

$$= \frac{\sum_{l=0}^{\infty}\sum_{m=0}^{\infty}\sum_{n=0}^{\infty} B_{lmn} P_0(x_k|Z_{k-1}) P_0(\mathbf{a}_k|Z_{k-1}) \psi_l^{(1)}(x_k) \psi_{\mathbf{m}}^{(2)}(\mathbf{a}_k) I_n(z_k)}{\sum_{n=0}^{\infty} B_{00n} I_n(z_k)} \tag{8}$$

with

$$I_n(z_k) = \int \mu_{z_k}(y_k) P_0(y_k | Z_{k-1}) \psi_n^{(3)}(y_k) dy_k, \tag{9}$$

$$B_{lmn} = <\psi_l^{(1)}(x_k)\psi_{\mathbf{m}}^{(2)}(\mathbf{a}_k)\psi_n^{(3)}(y_k)|Z_{k-1}>, \; (\mathbf{m}=(m_1, m_2, \cdots, m_I)). \tag{10}$$

The functions $\psi_l^{(1)}(x_k)$, $\psi_m^{(2)}(\mathbf{a}_k)$ and $\psi_n^{(3)}(y_k)$ are the orthogonal polynomials of degrees l, $\mathbf{m}$ and n with weighting functions $P_0(z_k|Z_{k-1})$, $P_0(\mathbf{a}_k|Z_{k-1})$ and $P_0(y_k|Z_{k-1})$, which cam be artificially chosen as the probability density functions describing the dominant parts of $P(x_k|Z_{k-1})$, $P(\mathbf{a}_k|Z_{k-1})$ and $P(y_k|Z_{k-1})$.

Based on (8), and using the orthonormal relationships of $\psi_l^{(1)}(x_k)$ and $\psi_m^{(2)}(\mathbf{a}_k)$, the recurrence algorithm for estimating an arbitrary $(L, \mathbf{M})$th order polynomial type function $f_{L,\mathbf{M}}(x_k, \mathbf{a}_k)$ of x_k and $\mathbf{a}_k$ can be derived as follows:

$$\hat{f}_{L,\mathbf{M}}(x_k, \mathbf{a}_k) = <f_{L,\mathbf{M}}(x_k, \mathbf{a}_k)|Z_k> = \frac{\sum_{l=0}^{L}\sum_{m=0}^{\mathbf{M}}\sum_{n=0}^{\infty} C_{lm}^{LM} B_{lmn} I_n(z_k)}{\sum_{n=0}^{\infty} B_{00n} I_n(z_k)}, \quad (11)$$

where C_{lm}^{LM} is the expansion coefficient determined by the equality:

$$f_{L,\mathbf{M}}(x_k, \mathbf{a}_k) = \sum_{l=0}^{L}\sum_{m=0}^{\mathbf{M}} C_{lm}^{LM} \psi_l^{(1)}(x_k)\psi_m^{(2)}(\mathbf{a}_k). \quad (12)$$

In order to make the general theory for estimation algorithm more concrete, the well-known Gaussian distribution is adopted as $P_0(x_k|Z_{k-1})$, $P_0(\mathbf{a}_k|Z_{k-1})$ and $P_0(y_k|Z_{k-1})$, because this probability density function is the most standard one.

$$P_0(x_k|Z_{k-1}) = N(x_k; x_k^*, \Gamma_{x_k}), \quad P_0(\mathbf{a}_k|Z_{k-1}) = \prod_{i=1}^{I} N(a_{i,k}; a_{i,k}^*, \Gamma_{a_{i,k}}),$$

$$P_0(y_k|Z_{k-1}) = N(y_k; y_k^*, \Omega_k) \quad (13)$$

with

$$N(x; \mu, \sigma^2) = \frac{1}{\sqrt{2\pi\sigma^2}} exp\{-\frac{(x-\mu)^2}{2\sigma^2}\},$$

$$x_k^* = <x_k|Z_{k-1}>, \quad \Gamma_{x_k} = <(x_k - x_k^*)^2|Z_{k-1}>,$$

$$a_{i,k}^* = <a_{i,k}|Z_{k-1}>, \quad \Gamma_{a_{i,k}} = <(a_{i,k} - a_{i,k}^*)^2|Z_{k-1}>,$$

$$y_k^* = <y_k|Z_{k-1}>, \quad \Omega_k = <(y_k - y_k^*)^2|Z_{k-1}>. \quad (14)$$

Then, the orthonormal functions with three weighting probability density functions in (13) can be given in the Hermite polynomial[12]:

$$\psi_l^{(1)}(x_k) = \frac{1}{\sqrt{l!}} H_l(\frac{x_k - x_k^*}{\sqrt{\Gamma_{x_k}}}), \quad \psi_m^{(2)}(\mathbf{a}_k) = \prod_{i=1}^{I} \frac{1}{\sqrt{m_i!}} H_{m_i}(\frac{a_{i,k} - a_{i,k}^*}{\sqrt{\Gamma_{a_{i,k}}}}),$$

$$\psi_n^{(3)}(y_k) = \frac{1}{\sqrt{n!}} H_n(\frac{y_k - y_k^*}{\sqrt{\Omega_k}}). \quad (15)$$

Accordingly, by considering (1)(13) and (15), (9) can be given by

$$I_n(z_k) = \frac{e^{K_3(z_k)}}{\sqrt{2K_1\Omega_k}} \int \frac{1}{\sqrt{\pi/K_1}} exp\{-\frac{(y_k - K_2(z_k))^2}{1/K_1}\}$$

$$\cdot \frac{1}{\sqrt{n!}} \sum_{r=0}^{n} d_{nr} H_r(\frac{y_k - K_2(z_k)}{\sqrt{1/2K_1}}) dy_k \quad (16)$$

with

$$K_1 = \frac{2\alpha\Omega_k + 1}{2\Omega_k}, \quad K_2(z_k) = \frac{2\alpha\Omega_k z_k + y_k^*}{2\alpha\Omega_k + 1},$$

$$K_3(z_k) = K_1\{K_2^2(z_k) - \frac{2\alpha\Omega_k z_k^2 + y_k^{*2}}{2\alpha\Omega_k + 1}\}, \quad (17)$$

where the fuzzy data z_k are reflected in $K_2(z_k)$ and $K_3(z_k)$. Furthermore, d_{nr} ($r = 0, 1, 2, \cdots, n$) are the expansion coefficients in the equality:

$$H_n(\frac{y_k - y_k^*}{\sqrt{\Omega_k}}) = \sum_{r=0}^{n} d_{nr} H_r(\frac{y_k - K_2(z_k)}{\sqrt{1/2K_1}}). \quad (18)$$

By considering the orthonormal condition of Hermite polynomial[12], (16) can be expressed as follows:

$$I_n(z_k) = \frac{e^{K_3(z_k)}}{\sqrt{2K_1\Omega_k n!}} d_{n0}, \quad (19)$$

where a few concrete expressions of d_{n0} in (19) can be expressed as follows:

$$d_{00} = 1, \quad d_{10} = \frac{1}{\sqrt{\Omega_k}}(K_2(z_k) - y_k^*), \quad d_{20} = \frac{(K_2(z_k) - y_k^*)^2}{\Omega_k} + \frac{1}{2K_1\Omega_k} - 1,$$

$$d_{30} = \frac{3(K_2(z_k) - y_k^*)}{2K_1\Omega_k^{3/2}} - \frac{3(K_2(z_k) - y_k^*)}{\sqrt{\Omega_k}} + \frac{(K_3(z_k) - y_k^*)^3}{\Omega_k^{3/2}},$$

$$d_{40} = \frac{3}{(2\Omega_k K_1)^2} + \frac{3\{(K_2(z_k) - y_k^*)^2 - \Omega_k\}}{\Omega_k^2 K_1}$$

$$+ \frac{(K_2(z_k) - y_k^*)^4 - 6\Omega_k(K_2(z_k) - y_k^*)^2 + 3\Omega_k^2}{\Omega_k^2}. \quad (20)$$

Using the property of conditional expectation and (2), the two variables y_k^* and Ω_k in (20) can be expressed in functional forms on predictions of x_k and $\mathbf{a}_k$ at a discrete time k (i.e., the expectation value of arbitrary functions of x_k and $\mathbf{a}_k$ conditioned by Z_{k-1}), as follows:

$$y_k^* = <<y_k|x_k, Z_{k-1}>|Z_{k-1}> = < \int y_k P(y_k|x_k)dy_k|Z_{k-1}>$$

$$=< \sum_{r=0}^{\infty}\sum_{s=0}^{1} e_{1s} A_{rs}\theta_r^{(1)}(x_k)|Z_{k-1}>$$

$$= \sum_{s=0}^{1} e_{1s} <\mathbf{A}_{(s),k}\Theta(x_k)|Z_{k-1}>, \qquad (21)$$

$$\Omega_k = <\int (y_k - y_k^*)^2 P(y_k|x_k)dy_k|Z_{k-1}> = <\sum_{r=0}^{\infty}\sum_{s=0}^{2} e_{2s} A_{rs}\theta_r^{(1)}(x_k)|Z_{k-1}>$$

$$= \sum_{s=0}^{2} e_{2s} <\mathbf{A}_{(s),k}\Theta(x_k)|Z_{k-1}> \qquad (22)$$

with

$$\mathbf{A}_{(s),k} = (0, \mathbf{a}_{(s),k}), (s = 1,2,\cdots), \quad \mathbf{A}_{(0),k} = (1,0,0,\cdots,0),$$
$$\Theta(x_k) = (\theta_0^{(1)}(x_k), \theta_1^{(1)}(x_k), \cdots, \theta_R^{(1)}(x_k))^T, \qquad (23)$$

where T denotes the transpose of a matrix. The coefficients e_{1s} and e_{2s} in (21) and (22) are determined in advance by expanding y_k and $(y_k - y_k^*)^2$ in the following orthogonal series forms:

$$y_k = \sum_{i=0}^{1} e_{1i}\theta_i^{(2)}(y_k), \quad (y_k - y_k^*)^2 = \sum_{i=0}^{2} e_{2i}\theta_i^{(2)}(y_k). \qquad (24)$$

Furthermore, using (2) and the orthonormal condition of $\theta_i^{(2)}(y_k)$, each expansion coefficient B_{lmn} defined by (10) can be obtained through the similar calculation process to (21) and (22), as follows:

$$B_{lmn} = <\psi_l^{(1)}(x_k)\psi_m^{(2)}(\mathbf{a}_k)\int \psi_n^{(3)}(y_k)P(y_k|x_k)dy_k|Z_{k-1}>$$

$$= \sum_{s=0}^{n} e_{ns} <\psi_l^{(1)}(x_k)\psi_m^{(2)}(\mathbf{a}_k)\mathbf{A}_{(s),k}\Theta(x_k)|Z_{k-1}>, \qquad (25)$$

$$(\psi_n^{(3)}(y_k) = \sum_{i=0}^{n} e_{ni}\theta_i^{(2)}(y_k), \ e_{ni}; appropriate\ coefficients).$$

In the above, the expansion coefficient B_{lmn} can be given by the predictions of x_k and $\mathbf{a}_k$.

Finally, by considering (5) and (6), the prediction step to perform the recurrence estimation can be given for an arbitrary function $g_{L,M}(x_{k+1}, \mathbf{a}_{k+1})$ with (L, M)th order, as follows:

$$g_{L,M}^*(x_{k+1}, \mathbf{a}_{k+1}) = <g_{L,M}(x_{k+1}, \mathbf{a}_{k+1})|Z_k>$$
$$= <g_{L,M}(Fx_k + Gu_k, \mathbf{a}_k)|Z_k>. \qquad (26)$$

The above equation means that the predictions of x_{k+1} and $\mathbf{a}_{k+1}$ at a discrete time k are given in the form of estimates for the polynomial functions of x_k and $\mathbf{a}_k$. Therefore, by combining the estimation algorithm of (11) with the prediction algorithm of (26), the recurrence estimation of the specific signal can be achieved.

4 Application to Psychological Evaluation for Noise Annoyance

To find the quantitative relationship between the human noise annoyance and the physical sound level for environmental noises is important from the viewpoint of noise assessment. Especially, in the evaluation for a regional sound environment, the investigation based on questionnaires to the regional inhabitants is often given when the experimental measurement at every instantaneous time and at every point in the whole area of the region is difficult. Therefore, it is very important to estimate the sound level based on the human noise annoyance data. It has been reported that the noise annoyance based on the human sensitivity can be distinguished each other from 7 annoyance scores, for instance, 1.very calm, 2.calm, 3.mostly calm, 4.little noisy, 5.noisy, 6.fairly noisy, 7.very noisy, in the psychological acoustics[2].

After recording the road traffic noise by use of a sound level meter and a data recorder, by replaying the recorded tape through amplifier and loudspeaker in a laboratory room, 6 female subjects $(A, B, \cdots, F)$ aged of 22-24 with normal hearing ability judged one score among 7 noise annoyance scores (i.e.,$1, 2, \cdots, 7$) at every 5 [sec.], according to their impressions for the loudness at each moment using 7 categories from very calm to very noisy. Two kinds of data (Data 1 and Data 2) were used, namely, the sound level data of road traffic noise with mean values 71.4 [dB] and 80.2 [dB]. The proposed method was applied to an estimation of the time series x_k for sound level of a road traffic noise based on the successive judgments z_k on human annoyance scores after regarding z_k as fuzzy observation data.

Figure 1 shows one of the estimated results of the waveform fluctuation of the sound level. In this figure, the horizontal axis shows the discrete time k of the estimation process, and the vertical axis represents the sound level. For comparison, the estimated result obtained by a method without considering fuzzy theory is also shown in this figure. More specifically, a similar algorithm to our previously reported methods[6,7] based on expansion expressions of Bayes' theorem without consideing the membership function in (8) is applied to estimate the specific signal x_k. There are great discrepancies between the estimates based on the method without considering the membership function and the true values, while the proposed method estimates precisely the waveform of the sound level with rapidly changing fluctuation. The root mean squared errors of the estimation are shown in Table 1 (for Data 1) and Table 2 (for Data 2). It is obvious that the proposed method shows more accurate estimation than the results based on the method without considering fuzzy theory.

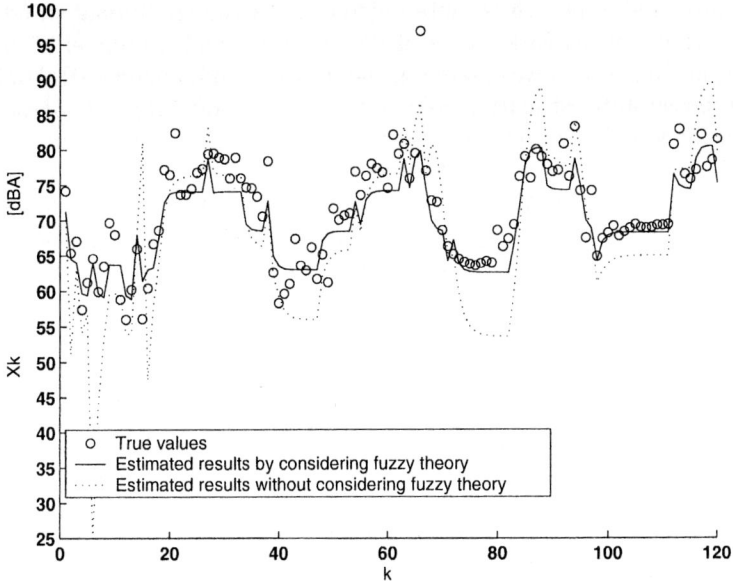

Fig. 1. Estimation results of the fluctuation waveform of the sound level based on the succesive judgement on human annoyance scores by the subject A (for Data 1)

Table 1. Root mean squared error of the estimation in [dB] (for Data 1)

Subject	A	B	C	D	E	F
Proposed Method	3.65	3.63	4.51	4.62	4.89	4.56
Compared Method	7.55	4.10	15.8	5.06	5.13	5.75

Table 2. Root mean squared error of the estimation in [dB] (for Data 2)

Subject	A	B	C	D	E	F
Proposed Method	4.59	4.26	4.82	6.80	7.49	4.65
Compared Method	10.7	7.79	4.96	14.6	11.6	4.64

5 Conclusion

In this paper, based on the observed data with fuzziness, a new method for estimating the specific signal for sound environment systems with unknown structure has been propoesd. The proposed estimation method has been realized by introducing a system model of conditional probability type and a fuzzy probability theory. The proposed method has been applied to the estimation of an actual sound environment, and it has been experimentally verified that better results are obtained as compared with the method without considering fuzzy theory.

The proposed approach is quite different from the traditional standard approaches. It is still at early stage of development, and a number of practical problems are yet to be investigated in the future. These include: (i) Application of the proposed state estimation method to a diverse range of practical estimation problems for sound environment systems with unknown structure and fuzzy observation. (ii) Extension of the proposed method to cases with multi-dimensional state variable and multi-source configuration. (iii) Finding an optimal number of expansion terms in the proposed estimation algorithm of expansion expression type. (iv) Extension of the proposed theory to the actual situation under existence of the external noise (e.g., background noise).

References

1. Ohta, M., Ikuta, A.: An acoustic signal processing for generalized regression analysis with reduced information loss based on data observed with amplitude limitation. Acustica **81** (1995) 129–135
2. Namba, S., Kuwano, S., Nakamura, T.: Rating of road traffic noise using the method of continuous judgement by category. J. Acousti. Soc. Japan **34** (1978) 29–34
3. Kalman, R. E.: A new approach to linear filtering and prediction problem. Trans. ASME, Series D, J. Basic Engineering **82** (1960) 35–45
4. Kalman, R. E., Buch, R.: New results in linear filtering and prediction theory. Trans. ASME, Series D, J. Basic Engineering **83** (1961) 95–108
5. Kushner, H. J.: Approximations to optimal nonlinear filter. IEEE Trans. Autom. Control **AC-12** (1967) 546–556
6. Ohta, M., Yamada, H.: New methodological trials of dynamical state estimation for the noise and vibration environmental system— Establishment of general theory and its application to urban noise problems. Acustica **55** (1984) 199–212
7. Ikuta, A., Ohta, M.: A state estimation method of impulsive signal using digital filter under the existence of external noise and its application to room acoustics. IEICE Trans. Fundamentals **E75-A** (1992) 988–995
8. Eyhhoff, P.: System identification: parameter and state estimation. John Wiley & Sons (1974)
9. Suzuki, Y., Kunitomo, N.: Bayes statistics and its application. Tokyo University Press (1989) 50–52
10. Zadeh, L. A.: Probability measures of fuzzy events. J. Math. Anal. Appl. **23** (1968) 421–427
11. Ohta, M., Koizumi, T.: General treatment of the response of a nonlinear rectifying device to a stationary random input. IEEE Trans. Inf. Theory **IT-14** (1968) 595–598
12. Cramer, H.: Mathematical methods of statistics. Princeton University Press (1951) 133, 221–227

Preventing Meaningless Stock Time Series Pattern Discovery by Changing Perceptually Important Point Detection

Tak-chung Fu[1,2,†], Fu-lai Chung[1], Robert Luk[1], and Chak-man Ng[2]

[1] Department of Computing, The Hong Kong Polytechnic University, Hong Kong.
{cstcfu, cskchung, csrluk}@comp.polyu.edu.hk
[2] Department of Computing and Information Management
Hong Kong Institute of Vocational Education (Chai Wan), Hong Kong.
cmng@vtc.edu.hk

Abstract. Discovery of interesting or frequently appearing time series patterns is one of the important tasks in various time series data mining applications. However, recent research criticized that discovering subsequence patterns in time series using clustering approaches is meaningless. It is due to the presence of trivial matched subsequences in the formation of the time series subsequences using sliding window method. The objective of this paper is to propose a threshold-free approach to improve the method for segmenting long stock time series into subsequences using sliding window. The proposed approach filters the trivial matched subsequences by changing Perceptually Important Point (PIP) detection and reduced the dimension by PIP identification.

1 Introduction

When time series data are divided into subsequences, interesting patterns can be discovered and it is easier to query, understand and mine them. Therefore, the discovery of frequently appearing time series patterns, or called surprising patterns in paper [1], has become one of the important tasks in various time series data mining applications.

For the problem of time series pattern discovery, a common technique being employed is clustering. However, applying clustering approaches to discover frequently appearing patterns is criticized as meaningless recently when focusing on time series subsequence [2]. It is because when sliding window is used to discretize the long time series into subsequences given with a fixed window size, trivial match subsequences always exist. The existing of such subsequences will lead to the discovery of patterns derivations from sine curve. A subsequence is said to be a trivial match when it is similar to its adjacent subsequence formed by sliding window, the best matches to a subsequence, apart from itself, tends to be the subsequence that begin just one or two points to the left or the right of the original subsequence [3]. Therefore, it is necessary to prevent the over-counting of these trivial matches. For example, in Fig.1, the shapes of S_1, S_2 and S_3 are similar to a head and shoulders (H&S) patterns while the

---

[†] Corresponding Author.

shape of S_4 is completely different from them. Therefore, S_2 and S_3 should be considered as trivial matches to S_1 and we should only consider S_1 and S_4 in this case.

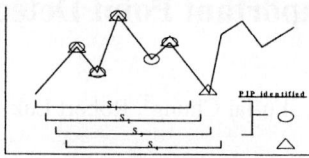

Fig. 1. Trivial match in time series subsequences (with PIPs identified in each subsequence)

References [3,4] defined the problem of enumerating the most frequently appearing pattern (which are called the most significant motifs, 1-motif, in reference [3]) in a time series P is the subsequence S_1 that has the highest count of non-trivial matches. Therefore, the Kth most frequently appearing pattern (significant motif, K-Motif) in P is the subsequence S_K that has the K highest count of non-trivial matches and satisfies $D(S_K, S_i) > 2R$, for all $1 \leq i < K$. However, it is difficult to define a threshold R, to distinguish trivial and non-trivial matches. It is case dependent and there is no general rule for defining this value. Furthermore, reference [2] suggested that applying a classic clustering algorithm in place of subsequence time series clustering to cluster only the motifs discovered from K-motif detection algorithm.

The objective of this paper is to develop a threshold-free approach to improve the segmentation method for segmenting long stock time series into subsequences using sliding window. The goal is redefined as to filter all the trivial matched subsequences formed by sliding window. The remaining subsequences should be considered as non-trivial matches for further frequently appearing pattern discovery process.

2 The Proposed Frequently Appearing Pattern Discovery Process

Given a time series $P = \{p_1, ..., p_m\}$ and fixing the width of the sliding window at w, a set of time series segments $W(P) = \{S_i = [p_i, ..., p_{i+w-1}] \mid i = 1, ..., m - w + 1\}$ can be formed. To identify trivial matches from the matching process of subsequences formed by sliding window, a method based on detecting the changes of the identified Perceptually Important Points (PIPs) is introduced. PIP identification is first proposed in reference [5] for dimensionality reduction and pattern matching of stock time series. It is based on identifying the critical points of the time series as the general shape of a stock time series is typically characterized by a few points. By comparing the differences between the PIP identified between two consequent subsequences, a trivial match occurred if the same set of PIP is identified and the second subsequence can be ignored. Otherwise, both subsequences are non-trivial matched and should be considered as the subsequence candidates of the pattern discovery process. This process carries along from the starting subsequence of the time series obtained by using sliding window till the end of the series. In Fig.1, the same set of PIP is identified in the subsequence S_1, S_2 and S_3. Therefore, the matching of subsequence S_2 and S_3 with subsequence S_1 should be considered as trivial match and subsequence S_2 and S_3

should be filtered. On the other hand, the set of PIP obtained from subsequence S_4 is different from that of subsequence S_3. This means that they are non-trivial match. Therefore, S_1 and S_4 are identified as the subsequence candidates.

After all these trivial matched subsequences are filtered, the remaining subsequences should be considered as non-trivial matches and served as the candidates for further discovery process on frequently appearing patterns. They will be the input patterns for the training of the clustering process. k-means clustering technique can be applied to these candidates. The trained algorithm is expected to group a set of patterns $M_1, ..., M_k$ which represent different structures or time series patterns of the data, where k is the number of the output patterns.

Although the clustering procedure can be applied directly to the subsequence candidates, it will quite time consuming when a large number of data points (high dimension) are considered. By compressing the input patterns with the PIP identification algorithm, the dimensionality reduction can be achieved (Fig.2).

Fig. 2. Examples of dimensionality reduction based on PIP identification process

3 Experimental Result

The ability of the proposed pattern discovery algorithm was evaluated in this section. Synthetic time series is used which is generated by combining 90 patterns with length of 89 data points from three common technical patterns 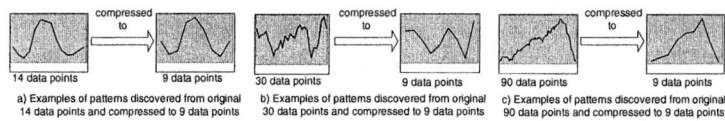 in stock time series as shown in Fig.8 (i.e.30x3). The total length of the time series is 7912 data points. Three sets of subsequence candidate were prepared for the clustering process. They include the (i) original one, the subsequences formed by sliding window, where $w=89$. This is the set which is claimed to be meaningless in reference [2]; (ii) motifs, K-motifs [2,3] formed from the time series, where K is set to 500 based on the suggested method in paper [3] and (iii) proposed PIP method, the subsequence candidates filtered by detecting the change of PIPs and 9 PIPs are used.

The number of pattern candidates and the time needed for pattern discovery are reported in Fig.4a. Only 281 motifs were formed while half of the subsequences were filtered by detecting the change of PIPs. The proposed PIP method is much faster than the other two approaches because the subsequences are compressed from 89 data points to 9 data points. The dimension for the clustering process is greatly reduced. Fig.4b shows the final patterns discovered. Six groups were formed by each of the approaches and it shows that the set of the pattern candidates deducted from the proposed approach is the most similar set to the pattern templates used to form the time series. On the other hand, the patterns discovered from the original subsequences seem not too related to the patterns which are used to construct the time series. Although the motifs approach can also discover the patterns which used to construct

	No. of candidate	Time
(i) original	7833	1:02:51
(ii) motifs	281	0:02:33
(iii) PIP	3550	0:00:03

(a)

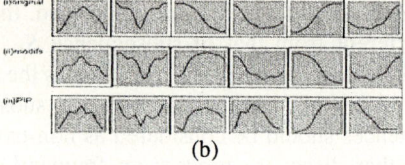

(b)

Fig. 3. (a) Number of patterns and time needed for pattern discovery by using different pattern candidates and (b) Pattern discovered (i) original, (ii) motifs and (iii) PIP

the time series, it smoothed out the critical points of those patterns. Also, uptrends, downtrends and a group of miscellaneous patterns are discovered in all the approaches.

To sum up, meaningless patterns are discovered by applying the clustering process on the time series subsequences (i) whereas both motifs and the proposed approach can partially solve this problem by filtering the trivial matched subsequences. However, it is still difficult to determine the starting point of the patterns and leads to the discovery of the shifting patterns. Additionally, the proposed approach can preserve the critical points of the patterns discovered and speed up the discovery process.

4 Conclusion

In this paper, a frequently appearing pattern discovery process for stock time series by changing Perceptually Important Point (PIP) detection is proposed. The proposed method tackles the main problem of discovering meaningless subsequence patterns with the clustering approach. A threshold-free approach is introduced to filter the trivial matched subsequences, which these subsequences will cause the discovery of meaningless patterns. As demonstrated in the experimental results, the proposed method can discover the patterns hidden in the stock time series which can speed up the discovery process by reducing the dimension and capturing the critical points of the frequently appearing patterns at the same time. We are now working on the problem of determining the optimal number of PIPs for representing the time series subsequences and the results will be reported in the coming paper.

References

1. Keogh, E., Lonardi, S., Chiu, Y.C.: Finding Surprising Patterns in a Time Series Database in Linear Time and Space. Proc. of ACM SIGKDD (2002) 550-556
2. Keogh, E., Lin, J., Truppel, W.: Clustering of Time Series Subsequences is Meaningless: Implications for Previous and Future Research. Proc. of ICDM, (2003) 115-122
3. Lin, J., Keogh, E., Lonardi, S., Patel, P.: Finding Motifs in Time Series. In: Workshop on Temporal Data Mining, at the ACM SIGKDD (2002) 53-68
4. Patel, P., Keogh, E., Lin, J., Lonardi, S, Mining Motifs in Massive Time Series Databases. Proc. of the ICDM (2002) 370-377
5. Chung, F.L., Fu, T.C., Luk, R., Ng, V., Flexible Time Series Pattern Matching Based on Perceptually Important Points. In: Workshop on Learning from Temporal and Spatial Data at IJCAI (2001) 1-7

Discovering Frequent Itemsets Using Transaction Identifiers*

Duckjin Chai, Heeyoung Choi, and Buhyun Hwang

Department of Computer Science, Chonnam National University,
300 Yongbong-dong, Kwangju, Korea
{djchai, hychoi}@sunny.chonnam.ac.kr
bhhwang@chonnam.ac.kr

Abstract. In this paper, we propose an efficient algorithm which generates frequent itemsets by only one database scan. A frequent itemset is a set of common items that are included in at least as many transactions as a given minimum support. While scanning the database of transactions, our algorithm generates a table having 1-frequent items and a list of transactions per each 1-frequent item, and generates 2-frequent itemsets by using a hash technique. $k(k \geq 3)$-frequent itemsets can be simply found by checking whether for all $(k-1)$-frequent itemsets used to generate a k-candidate itemset, the number of common transactions in their lists is greater than or equal to the minimum support. The experimental analysis of our algorithm has shown that it can generate frequent itemsets more efficiently than FP-growth algorithm.

1 Introduction

As an information extraction method, data mining is a technology to analyze a large amount of accumulated data to obtain information and knowledge valuable for decision-making. Because data mining is used to produce information and knowledge helpful in generating profit, it is widely used in various industrial domains such as telecommunication, banking, retailing and distribution for shopping bag analysis, fraud detection, customer classification, and so on [1,4,11]. Data mining technologies include association rule discovery, classification, clustering, summarization, and sequential pattern discovery, etc.[1,4]. Association rule discovery, an area being studied most actively, is a technique to investigate the possibility of simultaneous occurrence of the data.

In this paper, we study a technique to discover association rules that describe the associations among data. Most of previous studies, such as [2,3,5,6,7,8,9,10], have adopted an Apriori-like heuristic approach: if any k-itemset, where k is the number of items in the itemset, is not frequent in the database, a $(k+1)$-itemset containing the k-itemset is never frequent. The essential idea is to iteratively generate the set of $(k+1)$-candidate itemsets from k-frequent itemsets (for $k \geq 1$), and check whether each $(k+1)$-candidate itemset is frequent in the database.

---

* This work was supported by Institute of Information Assessment(ITRC).

The Apriori heuristic method can achieve good performance by significantly reducing the size of candidate sets. However, in situations with prolific frequent itemsets, long itemsets, or quite low minimum support thresholds, where the support is the number of transactions containing the itemsets, an Apriori-like algorithm may still suffer from nontrivial costs[6].

[6] proposed FP-growth algorithm using a novel tree structure(FP-tree). It discovers frequent itemsets by scanning a database twice without generating the candidate itemsets. FP-growth algorithm generates FP-tree and discovers frequent itemsets by checking all nodes of FP-tree. For the construction of FP-tree, FP-growth algorithm needs nontrivial cost for pruning non-frequent items and sorting the rest frequent items per each transaction. Moreover, there can be a space problem because new nodes have to be frequently inserted into the FP-tree if items in each transaction differ from items on the path of FP-tree.

In this paper, we propose an algorithm that can compute a collection of frequent itemsets by only one database scan. We call this algorithm FTL(Frequent itemsets extraction algorithm using a TID(Transaction IDentifier) List table) from now on. By only one database scan, FTL algorithm discovers 1-frequent items and constructs a TID List table of which each row consists of a frequent itemset and a list of transactions including it. Simultaneously, FTL algorithm computes the frequency of 2-itemsets by using a hash function with two 1-frequent items as parameters. This algorithm can reduce much computing cost that is needed for the generation of 2-frequent itemsets. $k(k \geq 3)$-frequent itemsets are discovered by using TID List table.

Most of association rule discovery algorithms spend their time on the scanning of massive database. Therefore, the less the number of database scanning is, the better the efficiency of an algorithm is. Consequently, FTL algorithm can considerably reduce the computing cost since it scans a database only once.

The remaining of this paper is organized as follows. Section 2 presents FTL algorithm. In Section 3, we show the performance of FTL algorithm through the simulation experiment. Section 4 summarizes our study.

2 FTL Algorithm

In association rule discovery, the most efficient method is to discover frequent itemsets by only one database scan. However, previous algorithms must have found frequent itemsets sequentially from 1-frequent items to the final k-frequent itemsets because the frequent itemsets cannot be found easily in the massive database. Therefore, previous algorithms tried to reduce the number of database scans and the number of candidate itemsets.

In this section, we propose an algorithm that can find frequent itemsets by only one database scan. Discovery of frequent itemsets is to find out a set of common items included in transactions of which the number satisfies a given minimum support. Consequently, if there is a data structure having information

Table 1. A transaction database as running example

TID	Items
100	f, a, c, d, g, i, m, p
200	a, b, c, f, l, m, o
300	b, f, h, j, o
400	b, c, k, s, p
500	a, f, c, e, l, p, m, n

about transactions in which each 1-frequent item is included, we can find frequent itemsets by searching the data structure without a database scan.

We generate a data structure which is called TID List table. The TID List table consists of a 1-frequent item and a list of transactions containing it. The TID List table is used to generate $k(k \geq 3)$-frequent itemsets.

The 1-frequent items are discovered when TID List table is constructed and 2-frequent itemsets are extracted using a hash table which is generated when a database is scanned. Therefore, TID List table and hash table are simultaneously constructed when a database is scanned. When a count value of each 2-itemset stored in bucket of hash table is greater than the minimum support, the 2-itemset becomes a 2-frequent itemset.

Many algorithms for generating candidate itemsets focused on the generation of the smaller candidate itemsets [3,10]. Especially, since the number of 2-candidate itemsets is generally big, it fairly improves the performance of mining to reduce the number of 2-candidate itemsets[10]. Our algorithm generates k-candidate itemsets for $k \geq 3$. The performance of our algorithm can be improved since 2-candidate itemsets are not generated.

The generation of candidate itemsets in our algorithm uses a method used in the Apriori[3]. Through probing TID List table, it can be checked whether the generated candidate itemsets are frequent. For all items in a candidate itemset, if the number of common TIDs is greater than the given minimum support, the candidate itemset becomes a frequent itemset.

2.1 The Generation of 1-Frequent Items and the TID List Table

All items of frequent itemset are included together in a transaction represented as one record in a database. That is, k items contained in the k-frequent itemset are included together in some identical transaction. For example, if an itemset {A, B, C} is frequent with the support 3, at least three transactions include all of items A, B, and C.

For each 1-frequent item, FTL algorithm computes a list of transactions that include it, and stores it into the TID List table. Table 2 shows items and their TID lists for an example used at Table 1. 1-frequent items can be discovered by comparing the length of TID list of each item with a minimum support. At Table 2, 1-frequent items which satisfy the minimum support(50%) are shown at Table 3.

Table 2. TID List table

Item	TID List	Length	Item	TID List	Length
a	100, 200, 500	3	j	300	1
b	200, 300, 400	3	k	400	1
c	100, 200, 400, 500	4	l	200, 500	2
d	100	1	m	100, 200, 500	3
e	500	1	n	500	1
f	100, 200, 300, 500	4	o	200, 300	2
g	100	1	p	100, 400, 500	3
h	300	1	s	400	1
I	100	1			

Table 3. 1-TID List table for 1-frequent items

1-frequent item	TID List	Support
a	100, 200, 500	3
b	200, 300, 400	3
c	100, 200, 400, 500	4
f	100, 200, 300, 500	4
m	100, 200, 500	3
p	100, 400, 500	3

2.2 The Generation of 2-Frequent Itemsets Using a Hash Technique

Using a hash function with two parameters, FTL algorithm computes the frequency of 2-itemsets and selects 2-frequent itemsets from them. Two parameters represent two items. When TID List table is constructed through the database scan, at the same time we generate a bucket of each 2-itemset using a hash function having two items as its parameters. Whenever each bucket is referenced by a hash function, the bucket count is increased by 1. Table 4 shows the frequency of 2-itemsets which are calculated by a hash function $f(x,y) = array[x][y]$, where x and y are an item.

Since a pattern in association rule discovery is not a sequential one, items in an itemset are commutative. That is, an itemset {a, b} is the same as an itemset {b, a}. We will maintain items in a itemset in alphabetic order from this moment. A collection of buckets is represented by a two-dimensional array. For example, the frequency count of a itemset {a, b} is stored to the two-dimensional $array[a][b]$. If the frequency count of a bucket for an itemset is greater than the minimum support, it is extracted as 2-frequent itemset. Therefore, we can get 2-frequent itemsets such as Table 4. We assume that the minimum support is 50 percent.

If we use a hash technique to generate all frequent itemsets, the computing cost will be very low. But its space overhead can be high since $\binom{n}{k}$ operations are required, where n is the number of total items in database and k is the number of items included in itemsets. The number of buckets can be increased exponentially

Table 4. Hash table having the frequency of 2-itemsets hash function: $f(x,y) = array[x][y]$

index->	{a,f}	{a,c}	{a,d}	{a,g}	{a,i}	{a,m}	{a,p}	{a,b}	{a,l}	{a,o}	{a,e}	{a,n}
count->	3	3	1	1	1	3	2	1	2	1	1	1
	{b,c}	{b,f}	{b,l}	{b,m}	{b,o}	{b,k}	{b,s}	{b,p}	{c,f}	{c,d}	{c,g}	{c,i}
	2	1	1	1	1	1	1	1	3	1	1	1
	{c,m}	{c,p}	{c,l}	{c,o}	{c,k}	{c,s}	{c,e}	{c,n}	{d,f}	{d,g}	{d,i}	{d,m}
	3	3	2	1	1	1	1	1	1	1	1	1
	{d,p}	{f,g}	{f,i}	{f,m}	{f,p}	{f,l}	{f,o}	{f,h}	{f,j}	{f,o}	{f,e}	{f,n}
	1	1	1	3	2	2	1	1	1	1	1	1
	{g,i}	{g,m}	{g,p}	{i,m}	{i,p}	{l,m}	{l,o}	{l,n}	{l,p}	{m,p}	{m,o}	{m,n}
	1	1	1	1	1	2	1	1	1	2	1	1
	{h,j}	{h,o}	{j,o}	{k,s}	{p,s}	{e,p}	{e,m}	{e,n}	{e,l}	{p,n}		
	1	1	1	1	1	1	1	1	1	1		
	2-frequent itemsets											
			{a,f}	{a,c}	{a,m}	{c,f}	{c,m}	{c,p}	{f,m}			
			3	3	3	3	3	3	3			

according to the increase of k. There is a tradeoff between a computing overhead and space overhead. However, if this hash technique is used partially, frequent itemsets can be computed effectively. Especially, if we use this hash technique in the generation of 2-frequent itemsets or 3-frequent itemsets, the performance of an algorithm may be improved since most computing cost is required for the generation of 2-frequent itemsets or 3-frequent itemsets.

2.3 The Generation of $k(k > 2)$-Frequent Itemsets Using TID List Table

In FTL algorithm, candidate itemsets are generated by using Apriori-gen used at Apriori algorithm. FTL algorithm uses $(k-1)$-TID List table to find k-frequent itemsets($k > 3$). $(k-1)$-TID List table has information about transactions including $(k-1)$-frequent itemsets. Therefore, we can calculate the support of each candidate itemset by computing the number of common transactions that include $(k-1)$-frequent itemsets used to generate the k-candidate itemset.

k-TID List table consists of TID lists. Each of them is a list of transactions containing a k-frequent itemset. k-TID List table is constructed by using $(k-1)$-TID List table since k-frequent itemsets is composed only by the combination of $(k-1)$-frequent itemsets in $(k-1)$-TID List table. Generally, k-TID List table becomes smaller than $(k-1)$-TID List table when k is greater than 3. Thus, it can reduce the number of comparisons since the TID List table size becomes smaller and smaller. We can find the fact by investigating Table 6 and 7.

Table 5 is an example of generating 3-frequent itemsets by using 1-TID Lists of Table 3 and 2-frequent itemsets of Table 4. 2-frequent itemsets are {a,f}, {a,c}, {a,m}, {c,f}, {c,m}, {c,p}, and {f,m} as shown in Table 4.

Table 5. Generation of 3-frequent itemsets using 1-TID List table and 2-frequent itemsets

3-candidate itemset	Item	TID List	Common TIDs
{a, c, f}	a	100, 200, 500	100, 200, 500
	c	100, 200, 400, 500	
	f	100, 200, 300, 500	
{a, c, m}	a	100, 200, 500	100, 200, 500
	c	100, 200, 400, 500	
	m	100, 200, 500	
{a, f, m}	a	100, 200, 500	100, 200, 500
	f	100, 200, 300, 500	
	m	100, 200, 500	
{c, f, m}	c	100, 200, 400, 500	100, 200, 500
	f	100, 200, 300, 500	
	m	100, 200, 500	

Table 6. 3-TID List table for 3-frequent itemsets

3-frequent itemset	TID List	Support
{a, c, f}	100, 200, 500	3
{a, c, m}	100, 200, 500	3
{a, f, m}	100, 200, 500	3
{c, f, m}	100, 200, 500	3

By using Apriori-gen algorithm, FTL algorithm generates 3-candidate itemsets, {a, c, f}, {a, c, m}, {a, f, m}, and {c, f, m}. As shown in Table 5, the candidate itemsets include transactions 100, 200, and 500. These candidate itemsets become 3-frequent itemset since they satisfy the minimum support 50%. Thus, we get Table 6 as 3-TID List table.

Next, we can generate 4-candidate itemset {a, c, f, m} from 3-frequent itemsets. At this time, we use 3-TID List table of Table 6. 4-candidate itemset {a, c, f, m} is generated from the first two 3-frequent itemsets, {a, c, f} and {a, c, m}. Since the remaining two 3-frequent itemsets, {a, f, m} and {c, f, m}, are included in 4-candidate itemset {a, c, f, m}, these itemsets need not be considered.

We have to compute the number of common transactions included in both of TID lists for two 3-frequent itemsets, {a, c, f} and {a, c, m}, and check whether the number satisfies the support. Therefore, for calculating the number of common transactions including all items of k-candidate itemset, the use of $(k-1)$-TID List table is more effective than that of 1-TID List table.

As shown in Table 7, since both TID lists for 3-frequent itemset {a, c, f} and {a, c, m} are transactions 100, 200, and 500, 4-candidate itemset {a, c, f, m} includes transactions 100, 200, and 500, and satisfies the minimum support 50%.

Table 7. Generation of 4-frequent itemsets using 3-TID List table

4-candidate itemset	3-frequent itemset	TID List	Common TIDs
{a, c, f, m}	{a, c, f}	100, 200, 500	100, 200, 500
	{a, c, m}	100, 200, 500	

Algorithm 1. FTL: TID List construction and Frequent Itemset extraction

```
Input :A transaction database DB and a minimum support threshold ε
Output :The complete set of frequent itemsets
Procedure TIDList&2-frequent(DB)
  for each transaction do
    generate 1-TID List table
    generate bucket using hash function
  for all buckets do
    if count value of bucket > ε then
      insert to 2-frequent itemsets
Procedure k-frequent(TID List, k-frequent itemsets)

  for all generate (k + 1)-candidate itemsets do
    for all (k + 1)-candidate itemsets do
      compare TIDs of each item
      if the number of the common TID > ε then
        insert to (k + 1)-frequent itemsets
        update TID List table
    if candidate itemsets can generate then
      call k-frequent(TID List, (k + 1)-frequent itemsets)
```

Thus, the 4-candidate itemset becomes 4-frequent itemset. Since k-candidate itemsets($k > 4$) cannot be generated, FTL algorithm is terminated.

Let the number of items of an itemset be m and the length of TID list be n. Since each TID list in TID List table are sorted by TIDs of transactions, the computing time for calculating the support of an itemset is O(mn). If not being sorted, the computing time is O(m·n^2). For all items in an itemset, TID List table is searched from the first position of each TID list. If a common TID in all their TID lists is found, a count is increased by one. The search is continued until the count is equal to the minimum support or one of the TID lists is ended. If the count is equal to the minimum support, the itemset becomes a frequent itemset. This procedure for computing frequent itemsets can considerably reduce the computing time.

The FTL algorithm generates TID List table having information that 1-frequent items and their list of TIDs including them by only one database scanning. The performance of FTL algorithm can be improved since k-frequent itemsets($k > 2$) can be extracted by searching only $(k - 1)$-TID List table. The FTL algorithm is shown in Algorithm 1.

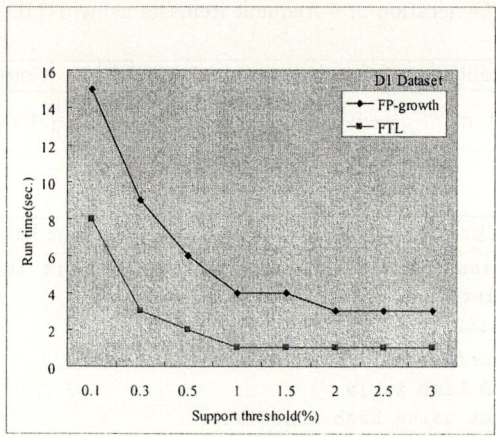

Fig. 1. Scalability with support threshold(Data set : D_1)

3   Experimental Evaluation and Performance Study

In this section, we present a performance comparison of FTL with a recently proposed efficient method, FP-growth algorithm. The experiments are performed on a 2-GHz Pentium PC machine with 1 GB main memory, running on Microsoft Windows/XP. All the programs are written in Microsoft/Visual C++6.0. Notice that we implement their algorithms to the best of our knowledge based on the published reports on the same machine and compare in the same running environment. Please also note that run time used here means the total execution time, i.e., the period between input and output, instead of CPU time measured in the experiments in some literature. As the generator of data set, we have used the generator used for the performance experiment in the previous papers. The data generator can be downloaded from the following URL.
http://www.almaden.ibm.com/software/quest/Resources/index.shtml

We report experimental results on two data sets. The first one is T20.I10. D10K with 1K items, which is denoted as D_1. In this data set, the average size of transactions and the maximal potential size of frequent itemsets are 20 and 10, respectively, while the number of transactions in the data set is set to 10K. The second data set, denoted as D_2, is T20.I10.D100K with 1K items. And the size of transactions and the maximal size of frequent itemsets are 20 and 10, respectively. There are exponentially numerous frequent itemsets in both data sets, as the support threshold goes down. There are long frequent itemsets as well as a large number of short frequent itemsets in them.

The scalability of FTL and FP-growth as the support threshold decreases from 3% to 0.1% is shown in Fig.1 and Fig.2. Each graph shows the runtime of FTL algorithm and FP-growth algorithm while increasing support threshold. Fig.1 and 2 are the result of data set D_1 and D_2, respectively. As shown in Fig.1 and Fig.2, both FP-growth and FTL have good performance when the support

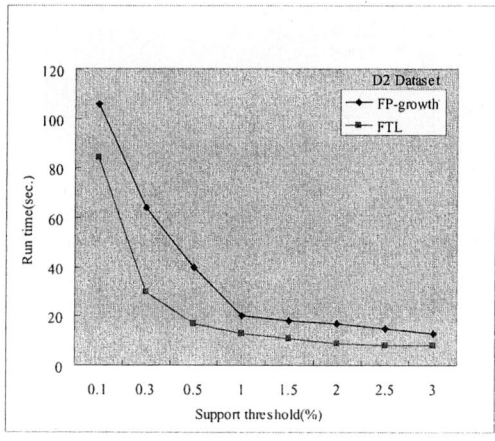

Fig. 2. Scalability with support threshold(Data set : D_2)

threshold is pretty low, but FTL is better. This phenomenon results from the fact that the cost to generate frequent itemsets in FTL algorithm is less than that in FP-growth algorithm and a database is scanned only once. According to the increase of support threshold, the runtime difference of two algorithms becomes much bigger.

To test the scalability with the number of transactions, experiments on data set D_2 are used. The support threshold is set to 0.5%. The results are presented in Fig.3. Both FTL and FP-growth algorithms show linear scalability with the number of transactions from 20K to 80K. As the number of transactions grows up, the difference between the two methods becomes larger and larger. In a database with a large number of frequent items, candidate itemsets can become quite large. However FTL compares only two records of TID List table as shown in Table 7. Therefore, FTL can find frequent itemsets fastly. This explains why FTL has advantages when the number of transactions is large.

4 Conclusions

We proposed the FTL algorithm that can reduce the number of database scans by one and thus discovers efficiently frequent itemsets. FTL algorithm generates TID List table by one database scanning and simultaneously calculates the frequency of 2-itemsets using a hash technique. Since k-TID List table has the information about transactions including k-frequent items, TID List table is the compact database from which k-frequent itemsets can be calculated.

One of advantages of FTL algorithm is that k-frequent itemsets can be computed by only one scan of database. In FTL algorithm, it is easy and efficient to test whether each k-candidate itemset is frequent or not. Since $(k+1)$-candidate itemsets are generated from k-frequent itemsets, a method is necessary to effi-

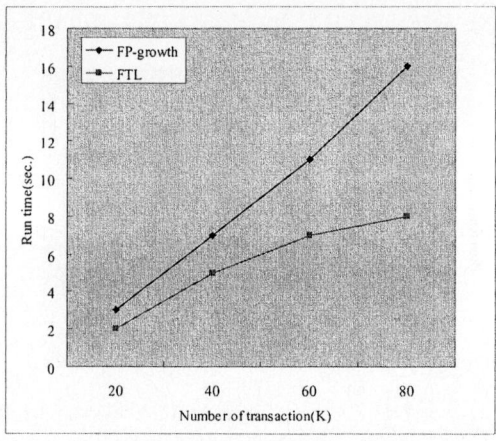

Fig. 3. Scalability with number of transactions

ciently prune candidate itemsets for much more efficiency. Our simulation experiments have shown that FTL algorithm is more efficient than FP-growth algorithm for given two data set.

References

1. Adrians, P., Zantige, D.: Data Mining. Addison-Wesley. (1996)
2. Agrawal, R., Aggarwal, C., and Prasad, V. V. V.: A tree projection algorithm for generation of frequent itemsets. In J. Parallel and Distributed Computing. (2000)
3. Agrawal, R., Srikant, R.: Fast algorithms for mining association rules. In VLDB. (1994) 487–499
4. Berry, M.J.A., Linoff, G.: Data Mining Techniques-For marketing, Sales, and Customer Support. Wiley Computer Publishing. (1997)
5. Grahne, G., Lakshmanan, L., and Wang, X.: Efficient mining of constrained correlated sets. In ICDE. (2000)
6. Han, J., Pei, J., and Yin, Y.: Mining frequent patterns without candidate generation. In ACM SIGMOD. (2000) 1–12
7. Lent, B., Swami, A., and Widom, J.: Clustering association rules. In ICDE. (1997) 220–231
8. Liu, B., Hsu, W., and Ma, Y.: Mining association rules with multiple minimum supports. In ACM SIGKDD. (1999) 337–341
9. Ng, R., Lakshmanan, L. V. S., Han, J., and Pang, A.: Exploratory mining and pruning optimizations of constrained associations rules. In SIGMOD. (1998) 13–24
10. Park, J.S., Chen, M.S., and Yu, P.S.: An effective hash-based algorithm for mining association rules. In ACM SIGMOD. (1995) 175–186
11. Simoudis, E.: Reality Check for Data Mining. IEEE Expert: Intelligent Systems and Their Applications 11 (5), October, (1996)

Incremental DFT Based Search Algorithm for Similar Sequence

Quan Zheng, Zhikai Feng, and Ming Zhu

Department of Automation, University of Science and Technology of China,
Hefei, 230027, P.R. China
qzheng@ustc.edu.cn
fnzhikai@mail.ustc.edu.cn

Abstract. This paper begins with a new algorithm for computing time sequence data expansion distance on the time domain that, with a time complexity of $O(n \times m)$, solves the problem of retained similarity after the shifting and scaling of time sequence on the Y axis. After this, another algorithm is proposed for computing time sequence data expansion distance on frequency domain and searching similar subsequence in long time sequence, with a time complexity of merely $O(n \times fc)$, suitable for online implementation for its high efficiency, and adaptable to the extended definition of time sequence data expansion distance. An incremental DFT algorithm is also provided for time sequence data and linear weighted time sequence data, which allows dimension reduction on each window of a long sequence, simplifying the traditional $O(n \times m \times fc)$ to $O(n \times fc)$.

1 Introduction

In time sequence data mining, the extensively applicable technique, the fundamental issue of time sequence data similarity comparison has a promising prospect. Current methods of data similarity comparison and fast similar subsequence searching include, apart from Euclid technique, frequency domain method [1],[2],[3],[4], segmentation method [5],[6], waveform descriptive language method [7].

Previous studies produced the concept of time sequence expansion distance to preserve the similarity of time sequence data after linear shifting. Major studies on expanded similar sequence searching have been conducted by Chu et al [8] and Agrawal et al [9]. However, the distance proposed in [8] is asymmetrical, which may lead to results against common sense, while the algorithm is basically a costly distance computation on the time domain. The similar sequence searching algorithm in [9] suffers from: 1. simple normalization technique is used to solve the shifting and scaling problems in subsequence similarity comparison, which not universally applicable; 2. complicated and costly.

In this paper, the basic frequency domain method is extended to apply to the search in expanded similar sequence. Main points include: providing an analytical result for computing time sequence data expansion distance on time domain; An innovative computing method for time sequence data expansion distance based on frequency domain analytical solution, and the fast searching technique for relevant similar sequence. Computing on a dimension reduced frequency domain, the technique is highly efficient, with a time complexity of O(n×f_c), and adaptable to expansion distance of time sequence data; In the similar subsequence searching, DFT dimension reduction is necessary for each window of a long sequence. The incremental DFT for each window of a long sequence and the incremental DFT for linear weighted time sequence data proposed below reduce the time complexity from the traditional O(n×m×f_c) to O(n×f_c).

2 Expanded Time Sequence Data Distance and Its Analytical Solution

Definition 1: The expanded asymmetric distance of one dimension time series data of the same length (m) $x=[x_0,x_1,\ldots,x_{m-1}]^T$, $y=[y_0,y_1,\ldots,y_{m-1}]^T$ is defined as:

$$d(x,y) = \min_{a,b}(\sum_{i=0}^{m-1}(x_i - ay_i - b)^2)^{1/2} . \tag{1}$$

The advantage of this definition is that it maintains the similarity of time sequence data after scaling and shifting, applicable to different scaling and shifting amount of transducers. This distance is asymmetrical and against common sense, therefore another definition for time sequence data distance is given by [1], as the minimum value of 2 asymmetric distance d(x,y) and d(y,x).

Although Chu et al [8] proposed an algorithm that maps the time sequence data to a shifting-eliminated plane where the distance is computed, the method is over-complicated. This paper proposes a direct analytical method for computing time sequence data expansion distance through computing the optimum parameter of a, b.

Theorem 1: The optimum analytical solution of the asymmetric distance of one dimension time sequence data x, y of the length m is:

$$d(x,y) = (\sum_{i=0}^{m-1}(x_i - a_t y_i - b_t)^2)^{1/2} . \tag{2}$$

Where: $a_t = \dfrac{\sum_{i=0}^{m-1} x_i y_i - m\overline{x}\,\overline{y}}{\sum_{i=0}^{m-1} y_i^2 - m\overline{y}^2}$, $b_t = \overline{x} - a_t \overline{y}$

The symmetrical is the minimum value of d(x,y) and d(y,x). The time complexity of the search algorithm for relevant similar sequence is, at a perfect matching, O(m), while at subsequence searching, it is O(n×m), where n is the length of the long time sequence and m is the length of subsequence. This algorithm, i.e. computing time sequence data distance on the time domain, avoids searching on the (a,b) plane, and may quickly obtain the analytic solution of the expansion distance according to the values of time sequence data.

3 Computing Time Sequence Data Distance on the Frequency Domain

The search algorithm for similar sequence based on time domain analytic solution, as described in theorem 1, is conducted on the time domain, therefore costly and unsuitable for online application, whereas frequency domain methods are generally not applicable to the expansion distance of time sequence data. Our concern is to somehow extend the frequency domain method for computation of time sequence, and adapt it to the definition of time sequence data expansion distance.

Lemma 1: let the corresponding Fourier coefficient of time sequence data x be X_f, and the time sequence data x after linear transformation be y=a×x+b, then the Fourier parameter Y_f for the number f item in time sequence data y is:

$$Y_f = aX_f + \frac{b}{\sqrt{m}} \frac{1-e^{cfm}}{1-e^{cf}}. \qquad (3)$$

Where $c = -\dfrac{j2\pi}{m}$, X_f, Y_f Are the number f component of time sequence data x and y, respectively.

Theorem 2: The expanded asymmetric distance of time sequence data x and y are approximately:

$$d(x,y) \approx \left(\sum_{f=0}^{f_c-1} \left| X_f - a_t Y_f - \frac{b_t}{\sqrt{m}} \frac{1-e^{cfm}}{1-e^{cf}} \right|^2 \right)^{1/2}. \qquad (4)$$

where:

$$a_t = \frac{\sum_{f=0}^{f_c-1}(X_f \oplus Z_f)\sum_{f=0}^{f_c-1}(Y_f \oplus Z_f) - \sum_{f=0}^{f_c-1}(X_f \oplus Y_f)\sum_{f=0}^{f_c-1}(Z_f \oplus Z_f)}{\left(\sum_{f=0}^{f_c-1}(Y_f \oplus Z_f)\right)^2 - \sum_{f=0}^{f_c-1}(Y_f \oplus Y_f)\sum_{f=0}^{f_c-1}(Z_f \oplus Z_f)}$$

$$b_t = \frac{\sum_{f=0}^{fc-1}(X_f \oplus Y_f) - a_t \sum_{f=0}^{fc-1}(Y_f \oplus Y_f)}{\sum_{f=0}^{fc-1}(Y_f \oplus Z_f)}$$

f_c is the limiting frequency; $Z_f = \dfrac{1-e^{cfm}}{\sqrt{m}(1-e^{cf})}$ is a complex number sequence introduced for convenience's sake; X_f is the Fourier parameter of the number f item in the time sequence data x, and Y_f is the Fourier parameter of the number f item in the time sequence data y; function $\oplus$ is a mapping from complex number to real number, i.e. the product of the real parts of two complex numbers plus the product of their imaginary parts.

The relevant subsequence searching algorithm allows, at the same time, incremental DFT and expansion distance computation with frequency domain analytic solution, thereby to perform similarity comparison. The time complexity of it is $O(n \times f_c)$. Compared to the similar subsequence searching on time domain, the time complexity of which being $O(n \times m)$, this algorithm is more efficient and suitable for online application, because the f_c ranges from 2-5, at 2-3 magnitudes lower than m. Furthermore, this algorithm maintains the similarity of time sequence data after linear shifting, and is therefore adaptable to the expanded definition of distance. To simply the matter, the similar subsequence searching algorithm on the frequency domain that utilizes incremental DFT, and solves the issue of shifting and scaling is henceforth called: Extended frequency domain method.

4 Incremental Fourier Shifting of Time Sequence Data and Linear Weighted Time Sequence Data

Regarding the searching of similar subsequence, the algorithm described in section 3 requires discrete Fourier shifting for each subsequence window. According to traditional DFT formulae, time complexity for obtaining low order f_c Fourier parameters is $O(n \times m \times f_c)$, which is costly. We now present an incremental Fourier shifting algorithm that greatly enhances the efficiency, and is suitable for online application.

The long time sequence x is divided into n-m+1 interlapping time windows at the length m. xw_i represents the number i window, capitalized $XW_{i,f}$ represents the number f frequency component of the time window.

Theorem 3: The relation between $XW_{i,f}$, the number f Fourier parameter of the data time window xw_i, and $XW_{i-1,f}$, the number f Fourier parameter of the previous time window, is:

$$XW_{i,f} = XW_{i-1,f} / e^{cf} + \Delta_{i,f} . \tag{5}$$

Where: $\Delta_{i,f} = \dfrac{1}{\sqrt{m}}(xw_{i,m}e^{cfm} - \dfrac{xw_{i-1,0}}{e^{cf}}) = \dfrac{1}{\sqrt{m}}(x_{i+m}e^{cfm} - \dfrac{x_{i-1}}{e^{cf}})$

On some occasions, if the time sequence data x is closer to the current time (m-1), it is regarded as more important than the more distant points. For convenience sake, we introduce a forgetting function f (t) to contribute to the weight of distance. see (6).

$$f(t) = z + kt = (1 - km + k) + kt \tag{6}$$

Definition 2: The linear forgetting distance $d_w(x,y)$ for 2 one dimensional time sequence data x, y of the length of m is:

$$d_w(x,y) = (\sum_{t=0}^{m-1}(x_t - y_t)^2 f(t)^2)^{1/2} . \tag{7}$$

In the time sequence, the number t datum in the number i window is represented as $xw_{i,t}$. the datum after weighing is $xw'_{i,t}$. Their relations are:

$$xw'_{i,t} = xw_{i,t} f(t) = xw_{i,t}(1 - km + k + kt) . \qquad (8)$$

From definition 2,

$$d_w(x,y) = (\sum_{t=0}^{m-1}(x_t - y_t)^2 f(t)^2)^{1/2} = (\sum_{t=0}^{m-1}(x'_t - y'_t)^2)^{1/2} = d(x', y')$$

Therefore computing the weighted distance between 2 subsequences is equivalent to computing the Euclidean distance between two weighted time sequences. According to Parseval Rules, we may take the first few frequency components from the frequency domain of the weighted time sequence data to perform an approximate distance computation, allowing fast similar sequence search.

The issue now is how to obtain the Fourier parameters of each window after linear weighing, and in an incremental manner. Time window xw'is time window xw after weighing. When the DFT parameters of the previous window $XW_{i-1,f}$, linear weighted Fourier parameter $XW'_{i-1,f}$ and auxiliary parameter $XWT_{i-1,f}$ are given, how to obtain the DFT parameters $XW'_{i-1,f}$ of this linear weighted data window in an incremental manner. Where, $XWT_{i,f} = \frac{1}{\sqrt{m}}\sum_{t=0}^{m-1} tx_{w,t} e^{cft}$, The following lemmas can be obtained:

Lemma 2: $XWT_{i,f} = \frac{1}{e^{cf}}(XWT_{i-1,f} - XW_{i,f}) + \frac{(m-1)x_{w,m}e^{cfm} + x_{w,0}}{e^{cf}\sqrt{m}}$

Lemma 3: The number f Fourier parameter XW i,f of the time sequence window XW'i after linear forgetting is:

$$XW'_{i,f} = (1 - km + k)XW_{i,f} + kXWT_{i,f}$$

Therefore, incremental algorithm for Fourier parameter for the time sequence window $XW'_{i,f}$ after linear forgetting can be obtained, namely, theorem 4. It's easy to prove by combining theorem 3, lemma 2 and lemma 3.

Theorem 4: Recursion formulae for incremental computation of the Fourier parameter of linear forgetting time sequence window are shown in (9)-(13):

$$XW_{0,f} = \frac{1}{\sqrt{m}}\sum_{t=0}^{m-1} x_{0,t} e^{cft} . \qquad (9)$$

$$XWT_{0,f} = \frac{1}{\sqrt{m}}\sum_{t=0}^{m-1} tx_{0,t} e^{cft} . \qquad (10)$$

$$XW_{i,f} = XW_{i-1,f}/e^{cf} + \frac{1}{\sqrt{m}}(x_{w,m} e^{cfm} - \frac{x_{w-1,0}}{e^{cf}}) . \qquad (11)$$

$$XWT_{i,f} = \frac{1}{e^{cf}}(XWT_{i-1,f} - XW_{i-1,f}) + \frac{(m-1)x_{w,m-1}e^{cfm} + x_{w-1,0}}{e^{cf}\sqrt{m}} . \qquad (12)$$

$$XW'_{i,f} = (1-km+k)XW_{i,f} + kXWT_{i,f} .\tag{13}$$

When the weighted Fourier parameters of each window are obtained, the approximate weighted distance of time sequence data can be computed on dimension reduced frequency domain, achieving high efficiency in similar sequence searching. Obviously, the time complexity of incremental DFT algorithmic is O(n×fc), much lower thant the time complexity of traditional DFT algorithm O(n×m×fc).

5 Experiment

A comparison of the running time between extended frequency domain method and time domain method is shown in Table 1 and Table 2. In Table 1, the length of time sequence n = 200000, limiting frequency f_c = 3; in Table 2, length of subsequence m= 2000; limiting frequency in both tables f_c = 3.

The results indicate that with the extended frequency domain method, the time is about 1/10 − 1/50 of the time domain method, greatly improving the efficiency of search algorithm. The former is also adaptable to the definition of expansion distance of time sequence.

Table 1. The running time of time domain method and extended frequency method along with subsequence length

subsequence Length m	Time domain method (Second)	Extended frequency domain method (Second)
500	32.32	3.391
1000	64.44	3.375
1500	98.98	3.359
2000	132.36	3.406
2500	164.97	3.390

Table 2. the running time of time domain method and extended frequency method along with sequence length

Sequence length n	Time domain method (Second)	Extended frequency domain method (Second)
50000	32.65	0.844
100000	69.53	1.687
150000	100.79	2.547
200000	132.36	3.406
250000	169.06	4.297

6 Conclusion

In this paper, an analytical algorithm is proposed for computing time sequence data expansion distance on the frequency domain, offering new techniques for similar subsequence searching. It is proven, through experiment, this algorithm is more efficient than the time domain based algorithm, and suitable for online application, adaptable to the definition of time sequence data expansion distance. An incremental DFT algorithm is also provided for time sequence data and linear weighted time sequence data, which greatly improves the efficiency of DFT dimension reduction on each window of a long sequence, simplifying the traditional $O(n \times m \times f_c)$ time complexity to $O(n \times f_c)$.

Reference

1. R. Agrawal, C.Faloutsos and A.swami: Efficient similarity search in sequence database. In FODO, Evanston, Illinois, October (1993) 69-84
2. Faloutsos Christos. Ranganathan M. and Manolopulos Yannis: Fast subsequence matching in time series databases. Proc ACM SIGMOD, Minneapolis MN, May 25-27 (1994) 419-429
3. D.Rafiei and A.O.Mendelzon: Efficient retrieval of similar time sequences using DFT. In FODO, Kobe,Japan (1998) 203-212
4. K.P.Chan and A.W.C.Fu: Efficient time series matching by wavelets. In ICDE, Sydney, Australia (1999) 126-133
5. Keogh Eamonn, Padhraic smyth: A probabilistic approach to fast pattern matching in time series databases. Proceedings of the Third Conference on Knowledge Discovery in Databases and Data Mining, AAAI Press, Menlo Park, CA (1997) 24-30
6. Keogh Eamonn, Michael J.Pazzani: An Enhanced representation of time series which allow fast and accurate classification, clustering and relevance feedback. Proceeding of the 4th International Conference of Knowledge discovery and Data Mining, AAAI Press, Menlo Park, CA (1998) 239-241
7. Rakesh Agrawal, Giuseppe Psaila, Edward L.Wimmers, Mohamed Zait: Querying shapes of histories. Proceedings of the 21st VLDB Conference, Zurich, Switzerland (1995) 502-514
8. K.K.W. Chu, M.H Wong: Fast time-series searching with scaling and shifting. In proceedings of the 1 g ACM Symposium on Principles of Database Systems, Philadelphia, PA (1999) 237-248
9. Agrawal R, Lin K I, Sawhney H S, Shim K: Fast similarity search in the presence of noise, scaling and translation in time-series databases. In Proc. 1995 Int. Conf. Very Large Data Bases(VLDB'95), Zurich, Switzerland (1995) 490-501

Computing High Dimensional MOLAP with Parallel Shell Mini-cubes

Kong-fa Hu, Chen Ling, Shen Jie, Gu Qi, and Xiao-li Tang

Department of Computer Science and Engineering, Yangzhou University
kfhu@seu.edu.cn

Abstract. MOLAP is a important application on multidimensional data warehouse. We often execute range queries on aggregate cube computed by pre-aggregate technique in MOLAP. For the cube with d dimensions, it can generate 2^d cuboids. But in a high-dimensional cube, it might not be practical to build all these cuboids. In this paper, we propose a multi-dimensional hierarchical fragmentation of the fact table based on multiple dimension attributes and their dimension hierarchical encoding. This method partition the high dimensional data cube into shell mini-cubes. The proposed data allocation and processing model also supports parallel I/O and parallel processing as well as load balancing for disks and processors. We have compared the methods of shell mini-cubes with the other existed ones such as partial cube and full cube by experiment. The results show that the algorithms of mini-cubes proposed in this paper are more efficient than the other existed ones.

1 Introduction

Data warehouses integrate massive amounts of data from multiple sources and are primarily used for decision support purposes. Since the advent of data warehousing and online analytical processing (OLAP), data cube has been playing an essential role in the implementation of fast OLAP operations [1]. Materialization of a data cube is a way to pre-compute and store multi-dimensional aggregates so that multi-dimensional analysis can be performed on the fly. For this task, there have been many efficient cube computation algorithms proposed, such as BUC [2], H-cubing [3], and Star-cubing [4].Those methods have taken effect for the low-dimensional cube in the traditional data warehouse. But in the high-dimensional data cube, it is too costly in both computation time and storage space to materialize a full cube. For example, a data cube of 100 dimensions, each with 10 distinct values, may contain as many as 11^{100} aggregate cells. Although the adoption of Iceberg cube[4], Condensed cube[5],Dwarf[6] or approximate cube[7] delays the explosion, it does not solve the fundamental problem. No feasible data cube can be constructed with such data sets. In this paper we will address the problem of developing an efficient algorithm to perform OLAP on such data sets.

In this paper, we propose a multi-dimensional hierarchical fragmentation of the fact table based on multiple dimension attributes their dimension hierarchical encoding. Such an approach permits a significant reduction of processing and I/O overhead for many queries by restricting the number of fragments to be processed for both the fact table and bitmap data. This method also supports parallel I/O and parallel processing as well as load balancing for disks and processors.

2 Shell Mini-cubes

OLAP Queries tend to be complex and ad hoc, often requiring computationally expensive operations such as joins and aggregation. The OLAP query that accesses a large number of fact table tuples that are stored in no particular order might result to much more many I/Os, causing a prohibitive long response time. To illustrate the method, a tiny data cube PRT, Table 1,is used as a running example.

Table 1. A sample data cube with two measure values

TID	DimProduct			dimRegion			dimTime			Measure	
	Category	Class	Product	Country	Province	City	Year	Month	Day	Count	SaleNum
1	Office	OA	Computer	China	Jiangsu	Nanjing	1998	1	1	1	20
2	Office	OA	Computer	China	Jiangsu	Nanjing	1998	1	2	1	60
3	Office	OA	Computer	China	Jiangsu	Yangzhou	1998	1	2	1	40
4	Office	OA	Computer	China	Jiangsu	Yangzhou	1998	1	3	1	20
...	...	...	...	...	...	...	...	...	...	...	...

The cube PRT have three dimensions ,such as (P,R,T). From the RPT Cube, we would compute eight cuboids:{(P,R,T),(P,R,All), (P,All,T), (All,R,T), (P,All,All), (All,R,All), (All,All,T), (All,All,All)}.To the cube of d dimensions, it would create 2^d cuboids.For the cube with d dimensions $(D_1,D_2,...,D_d)$ and $|D_i|$ distinct values for each dimension D_i, it can generate 2^d cuboids and $\prod_{i=1}^{d}(|D_i|+1)$ cells. But in a high-dimensional cube with many cuboids, it might not be practical to build all these indices. If we consider the dimension hierarchies, the cuboids is vary much. So we can partition all the dimensions of the high-dimensional cube into independent groups, called Cube segments.

For example, for a database of 30 dimensions, D_1, D_2, ..., D_{30}, we first partition the 30 dimensions into 10 fragments(mini-Cubes) of size 3: (D_1,D_2,D_3), (D_4,D_5,D_6), ... , (D_{28},D_{29},D_{30}). For each mini-Cube, we compute its full data cube while recording the inverted indices. For example, in fragment mini-Cube (D_1,D_2,D_3), we would compute eight cuboids:{(D_1,D_2,D_3),...,(All,All,All)}. An inverted encoding index is retained for each cell in the cuboids. The benefit of this model can be seen by a simple calculation. For a base cuboid of 30 dimensions, there are only 8×10 = 80 cuboids to be computed according to the above shell fragment partition. Comparing this to $C_{30}^3 + C_{30}^2 + C_{30}^1$ = 4525cuboids for the partial cube shell of size 3 and 2^{30}= 10^9cuboids for full cube, the saving is enormous.

We propose a novel hierarchical encoding on each dimension table, called dimension hierarchical encoding. It is constructed as a path of values corresponding to dimension attributes that belong to a common hierarchy.

Definition 1. (Dimension hierarchical encoding)

The dimension hierarchical member encoding of the hierarchy L_j^i on the dimension D_i is $BL_j^i : dom(L_j^i) \rightarrow \{<b_{k-1}... b_i...b_0>| b_i \in \{0,1\}, i=0,...,k-1\}$. The dimension hierarchical encoding B^{D_i} of each member on the dimension D_i is defined as formula 1.

$$B^{D_i} = (...((B^{L_1^i} << Bit\, L_2^i | B^{L_2^i}) << Bit\, L_3^i | B^{L_3^i})...) << Bit\, L_h^i | B^{L_h^i} \qquad \text{formula 1}$$

$Bit\, L_j^i$ is the bit number of the $B^{L_j^i}$. $Bit\, D_i$ is the bit number of the member of the dimension D_i.

By using dimension hierarchical encoding, we can register a list of tuples IDs (tids) associated with the dimension members for each dimension. For example, the TID list associated with the *dimProduct*, *dimRegion* and *dimTime* dimension are shown in Table 2, Table 3 and Table 4 in turn. The ID-measure array is shown in Table 5.

Table 2. *dimProduct* dimension TID

$B^{ProductID}$	TID List
0001000010000001	1-2-3-4
...	...

Table 3. *dimRegion* dimension TID

$B^{RegionID}$	TID List
0000001000010001	1-2
0000001000010010	3-4
...	...

Table 4. *dimTime* dimension TID

B^{TimeID}	TID List
001000100001	1
001000100010	2-3
001000100011	4
...	...

Table 5. TID- measure array of Table 1

tid	Count	SaleNum
1	1	20
2	1	60
3	1	40
4	1	20
...	...	...

In our study, the method can rapidly retrieve the matching dimension member hierarchical encoding and evaluate the set of query ranges for each dimension and improve the efficiency of OLAP queries by using encoding prefix.

Definition 2. (Encoding prefix)

The encoding prefix of the member d_k^i is defined as $Bprefix(B^{d_k^i}, L_{m-1}^i) = B^{d_k^i} >> \sum_{l=m}^{j} (Bit\, L_l^i)$, where $m=\{1,...,j\}$.

By using encoding prefix, we can register the dimension hierarchy encoding and its TID list for every dimension hierarchy for each dimension. For each fragment, we compute the complete data cube by intersecting the TID-lists in the dimension and its hierarchies in a bottom-up depths-first order in the lattice (as seen in [3]). For example, to compute the cell{0001000010000001,0000001000010001,0010001}, we intersect the TID lists of $B^{ProductID}$=0001000010000001, $B^{RegionID}$=0000001000010001, and $Bprefix(B^{TimeID}, Month)$= 0010001 to get a new list of {1,2}.

3 Parallel Hierarchical Aggregation Algorithm

Based on the above discussion, the algorithm for shell fragment computation can be summarized as follows.

Algorithm 1 (Shell mini-Cube Parallel Computation)
Input: A base cuboid BC of n dimensions:$(D_1; \ldots ;D_n)$.
Output: A set of mini-Cube fragment partitions $\{P_1;\ldots, P_k\}$ and their corresponding (local) fragment mini-cubes $\{MC_1; \ldots ; MC_k\}$, an ID measure array
{partition the set of dimensions:$(D_1;\ldots;D_n)$ into a set of k fragment mini-Cubes $\{P_1;\ldots;P_k\}$;
 scan base cuboid BC once and do the following with parallel processing
  { insert each <tid, measure> into ID-measure array;
   for each attribute value a_i of each dimension D_i;
    build an dimension hierarchy encoding index entry: <B; TID list>;}
   parallel processing all fragment partition P_i as follows
    build a local fragment mini-cubes MC_i by intersecting their corresponding tid-lists and computing their measures; }

We partition the cube by the queries which may achieve good performance. It is possible to use the above algorithm to compute the full data cube. If we let a single fragment include all the dimensions, the computed fragment cube is exactly the full data cube. If the measure is tuple-count, there is no need to build ID-measure array since the length of the TID-list is tuple-count; for other measures, such as avg(), the needed components should be saved in the array, such as sum() and count().

4 Performance Analysis

Formally, suppose we have a database of |T| tuples and d dimensions. To store it as shown in Table 1 would need $d \times |T|$ bitmap encoding indices. The entire shell Cube segment of size F will create $\sum_{i=1}^{f} C_f^i * d/f = (2^f * d/f)$ cuboids and need $O(|T| * \sum_{i=1}^{f} C_f^i * d/f) = O(|T| * (2^f * d/f))$ storage space, while the partial cube will create $\sum_{i=1}^{f} C_d^i$ cuboids and the full cube will create 2^d cuboids. Their storage space show in the Table 6 with the mini-cube and the partial cube, etc. The Table 6 show the shell fragment mini-cube method has more efficient than other existed ones.

Table 6. The storage space of mini-cube

	Mini-cube	Partial cube	Full Cube
d=6,f=3,T=10^6	60 MB	160MB	250MB
d=9,f=3,T=10^6	75 MB	510MB	2GB
d=18,f=3,T=10^6	150 MB	7GB	1TB
d=30,f=3,T=10^6	320MB	18GB	4PB

5 Conclusion

We have proposed a novel approach for OLAP in high-dimensional datasets with a moderate number of tuples. It partitions the high dimensional space into a set of disjoint low dimensional shell fragment mini-cubes. Using dimension hierarchical encoding and pre-aggregated results, OLAP queries are computed online by dynamically constructing cuboids from the fragment data cubes. With this design, for high-dimensional OLAP, the total space that needs to store such shell fragment mini-cubes is negligible in comparison with a high-dimensional cube, so is the online computation overhead. Moreover, the query I/O costs for large data sets are reasonable and are comparable with reading answers from a materialized data cube, when such a cube is available. We have compared the methods of shell mini-cubes with the other existed ones such as partial cube and full cube by experiment. The analytical and experimental results show that the algorithms of mini-cubes proposed in this paper are more efficient than the other existed ones.

References

1. Gray, J., Chaudhuri, S., Bosworth, A., Layman, A., Reichart, D., Venkatrao, M., Pellow F., Pirahesh, H.: Data Cube: A Relational Aggregation Operator Generalizing Group-by, Crosstab and Subtotals. Data Mining and Knowledge Discovery 1:29-54, 1997
2. Beyer, K., Ramakrishnan, R.: Bottom-up Computation of Sparse and Iceberg Cubes. ACM SIDMOD (1999)359-370
3. Han, J., Pei, J., Dong, G., Wang, K.: Efficient Computation of Iceberg Cubes with Complex Measures. ACM SIGMOD (2001)1-12
4. Xin, D., Han, J., Li, X., Wah, B. W.: Star-cubing:Computing Iceberg Cubes by Top-down and Bottom-up Integration. VLDB(2003)476-487
5. Wang, W., Lu, H., Feng, J., Yu, J. X.: Condensed Cube: An Effective Approach to Reducing Data Cube Size. ICDE(2002)155-165
6. Sismanis, Y. , deligiannakis, A., Kotidis, Y., Roussopoulos, N.: Hierarchical Dwarfs for the Rollup Cube.VLDB(2004)540-551
7. Shanmugasundaram, J., Fayyad, U. M., Bradley, P. S.: Compressed Data Cubes for OLAP Aggregate Query Approximation on Continuous Dimensions. ACM SIGKDD(1999)223-232

Sampling Ensembles for Frequent Patterns[*]

Caiyan Jia[1,2] and Ruqian Lu[1,2,3]

[1] Lab of Intelligent Information Processing, Institute of Computing Technology, Academia Sinica, Beijing, 100080, China
[2] Lab of Intelligent Information Processing, Fudan University, Shanghai, 200433, China,
{cyjia, rqlu}@fudan.edu.cn
[3] Institute of Mathematics, AMSS, Academia Sinica, Beijing, 100080, China

Abstract. A popular solution to improving the speed and scalability of association rule mining is to do the algorithm on a random sample instead of the entire database. But it is at the expense of the accuracy of answers. In this paper, we present a sampling ensemble approach to improve the accuracy for a given sample size. Then, using Monte Carlo theory, we give an explanation for a sampling ensemble and obtain the theoretically low bound of sample size to ensure the feasibility and validity of an ensemble. And for learning the origination of the sample error and therefore giving theoretical guidance for obtaining more accurate answers, bias-variance decomposition is used in analyzing the sample error of an ensemble. According to theoretical analysis and real experiments, we conclude that sampling ensemble method can not only significantly improve the accuracy of answers, but also be a new means to solve the difficulty of determining appropriate sample size needed.

1 Introduction

Discovery of association rules is a prototypical problem in data mining [1-3]. The current algorithms proposed for data mining of association rules need to repeatedly pass over a database to obtain the frequent patterns. As for ever-growing large scale databases, the I/O overhead in scanning the database can be extremely high, and will be beyond the reach of the current computer systems. One possible approach for dealing with huge amounts of data is to take a random sample and do data mining on it, instead of the whole database, to obtain the approximate answers. However, the gain of the efficiency of the algorithm is at the expense of the accuracy of the answers in turn. In general, the larger the sample size, the more the accurate answers will be, but the lower the efficiency will be when sampling strategy is used to find frequent patterns. In fact, it is difficult to determine the appropriate sample size for a given accuracy. In statistics, various formulas have been developed to solve this problem [8-10].

---

[*] This work is supported in part by the Major Research Program of National Natural Science Foundation (No. 60496325) and the National Natural Science Foundation (No. 60303009 and 60375021).

And adaptive sampling method in data mining field is another means [4-7,11-12]. But they seldom consider the problem of whether and how we can improve the accuracy of answers given the ratio of sampling.

In this paper, we present sampling ensemble approach inspired by the ensemble methods in machine learning [13-16] to obtain the high accurate answers at a lower sample size. But we must consider the following basis problems before using sampling ensembles in practice. 1) Why does a sampling ensemble work? 2) When will a sampling ensemble work well? Namely, how much is the sample size needed for ensuring a sampling ensemble works? 3) What is the relationship between the sample error of an ensemble and the sample errors of individual samples of the ensemble? Since Monte Carlo theory can be as an explanation of Bagging [17], and a sampling ensemble is indeed a transformation of Bagging method, we take advantage of the theory to explain sampling ensembles and obtain the theoretical low bound of sample size. It tells us that a sampling ensemble might not work well anymore theoretically when the sample size is smaller than the bound we giving. Moreover, we analyze the error of a sampling ensemble, using the bias-variance decomposition, to get the relationship between the error caused by a sampling ensemble and the errors induced by individual samples in the ensemble. Although we can't conclude that the sample error of an ensemble will be smaller than that of an individual sample in the ensemble according to the theoretical result, fortunately, it implies that the more accurate and the more diverse frequent patterns on the individual samples, the better a sampling ensemble might work.

The paper is organized as follows. In section 2, we give some preliminary knowledge. In section 3, we present the sampling ensemble method, and give some theoretically relevant results by using Monte Carlo theory and bias-variance decomposition. In section 4, we do some experiments to verify the method. Finally, section 5 contains concluding remarks.

2 Preliminaries

Frequent pattern mining is a major step in association rule mining[1-3]. However, finding all frequent patterns contained in a database needs to repeatedly pass over the database. So the speed and scalability of the algorithm is a major concern with ever-growing large scale databases. Sampling is a good way to solve this problem [4-12] since for many data mining applications approximate answers are acceptable. But it is at the cost of the accuracy of answers.

In general, three kinds of error will occur when frequent patterns are mined on a sample set. 1) The false positive frequent pattern: it can be found on the sample set, but is not the frequent pattern of the original database. 2) The false negative frequent pattern: it can't be found on the sample set, but is certain the frequent pattern of the original database. 3) Support parameter error: the support of a frequent pattern mined on the sample set is not equal to the support of the frequent pattern also found on the original database.

In [7], cardinal error, $Cerror(S, D)$, is used to describe the first two kinds of errors. The mean, $\bar{\xi}_{S,D}$, and variance, $V_{S,D}$, of the support parameter error, are used to measure the support parameter error. Where,

$$Cerror(S, D) = 1 - \frac{\|L(S) \cap L(D)\|}{\|L(S) \cup L(D)\|} \tag{1}$$

$$\bar{\xi}_{S,D} = \frac{\sum_{x \in L(S) \cap L(D)} |f(x, S) - f(x, D)|}{\|L(S) \cap L(D)\|} \tag{2}$$

$$V_{S,D} = \sqrt{\frac{\sum_{x \in L(S) \cap L(D)} (|f(x, S) - f(x, D)| - \bar{\xi}_{S,D})^2}{\|L(S) \cap L(D)\|}} \tag{3}$$

$L(S)$ and $L(D)$ are the frequent pattern sets mined on a sample set S and the data set D, respectively. And $f(x, S)$ and $f(x, D)$ are the supports of frequent pattern x in $L(S)$ and in $L(D)$, respectively. Because the measure of support parameter error, (2) or (3), is no more than 1% of cardinal error (1), the sampling error is dominate by $Cerror(S, D)$. For the reasons of simplicity and computational efficiency, the sample error can be represented by cardinal error.

Given an accuracy restriction, it's difficult to determine the appropriate sample size. But it's very easy to find rough patterns on a small sample. Can we improve the accuracy of answers on small samples at a fixed sample size? If this problem is solved, we might obtain answers satisfying a given accuracy at small sample size. Thus, the difficult problem of sample size selection might be avoided.

3 Sampling Ensembles for Frequent Patterns

Enlightened by ensemble methods in machine learning, in this section, we give the sampling ensemble method to improve the accuracy of answers on sample sets, and then, to obtain approximate answers satisfying the requirement of accuracy at a fixed sample size.

Just as its name implies, sampling ensemble is collection of a (finite) number of different sample sets (with the same sampling ratio) generated from the original database randomly. The sets of frequent patterns which are obtained on these samples are gathered to form the pattern ensemble. The final set of frequent patterns is obtained by doing majority voting on the pattern ensemble. After the majority voting is done, the support of the frequent pattern in the final answer set of an ensemble is obtained by averaging the aggregation of the support among those individual pattern sets

But it remains problems that why and when does a sampling ensemble works? What's the relationship between the error caused by a sampling ensemble and the errors induced by individual samples.

3.1 Monte Carlo Theory Explanation and the Theoretically Low Bound

As suggested by [17], Monte Carlo theory can be as an explanation of Bagging. Similarly, sampling ensemble can be viewed as a kind of Monte Carlo algorithm because the sampling ensemble has stochastic property and allows for occasionally incorrect answers (see [8-12] as examples). Monte Carlo theory can also be a theoretical framework of sampling ensembles.

The great interest in Monte Carlo algorithms resides in their amplification ability: given a problem instance π and a Monte Carlo algorithm, MC for short, if MC is run multiple times on the problem instance π and majority answer is taken as the result, then, under mild assumptions, the probability of a correct answer exponentially increases. The conditions are that the algorithm must be consistent, never outputting two distinct correct answers when run two times on the same problem instance, and the probability that it gives a correct answer must be greater than random guess. Thus, the correctness of a frequent pattern x on sampling ensembles can be amplified if: 1. The individual samples in a sampling ensemble are independent; 2. The algorithm for finding the frequent patterns on a sample set is consistent; 3. The probability, that x is a true frequent pattern, $Pr(x) > \frac{1}{2}$.

The first two conditions are naturally satisfied. When it's time to the third condition, in the literature, people usually use the PAC (probably approximately correct) learning model [18] derived from field of machine learning to measure mining quality. That is, for given $\delta > 0$ and $0 < \epsilon < 1$, each frequent pattern x needs to satisfy (4) or (5)

$$Pr[|fr(x,S) - fr(x,D)| < \epsilon] > 1 - \delta. \tag{4}$$

$$Pr[|fr(x,S) - fr(x,D)| < \epsilon fr(x,D)] > 1 - \delta. \tag{5}$$

It means that a sampling algorithm must yield a good approximation of $fr(x,D)$ with reasonable probability. The difference between (4) and (5) is just that (4) is an absolute error measure, while (5) is a relative error measure. In this paper, $Pr(x) = 1 - \delta$. Therefore, if $\delta < \frac{1}{2}$, the sampling ensemble works.

According to the Hoeffding bound, $\forall \delta > 0, 0 < \epsilon < 1$, if the sample size n satisfies the following inequalities (6)/(7), then it satisfies (4)/(5), respectively [10].

$$n > \frac{1}{2\epsilon^2} ln\left(\frac{2}{\delta}\right) \tag{6}$$

$$n > \frac{1}{2fr(x,D)^2\epsilon^2} ln\left(\frac{2}{\delta}\right) \tag{7}$$

Since we require $\delta < \frac{1}{2}$, the sample size n must satisfy:

$$n > \frac{1}{2\epsilon^2} ln 4 \tag{8}$$

$$n > \frac{1}{2fr(x,D)^2\epsilon^2} ln 4 \tag{9}$$

So, we have:

Theorem 1. $\forall\ 0 < \epsilon < 1$, when the sample size n satisfies the inequality (8)/(9), the sampling ensemble works well in terms of the measure of probable error (4)/(5).

As a rule, since we don't know the real support of a frequent pattern, fr(x,D), we always minimize it to the support threshold such that the sample size satisfies all conditions we might encounter. Similar with the result in [9], there exists heavily overestimated phenomenon. However, it has a wonderful function of theoretical direction, especially for large scale databases.

Similarly, according to Chernoff Bound, we have:

Corollary 1. $\forall\ 0 < \epsilon < 1$, when the sample size n satisfies the inequality (10)/(11), the sampling ensemble works well in terms of the measure of probable error (4)/(5).

$$n > \frac{3 fr(x, D)}{\epsilon^2} ln4 \tag{10}$$

$$n > \frac{3}{fr(x, D)\epsilon^2} ln4 \tag{11}$$

According to the Center Limited Theorem in statistic, we have:

Corollary 2. $\forall\ 0 < \epsilon < 1$, when the sample size n satisfies the inequality (12)/(13), the sampling ensemble works well in terms of the measure of probable error (4)/(5).

$$n > \frac{fr(x, D)(1 - fr(x, D))}{\epsilon^2} \left(\phi^{-1}\left(\frac{3}{4}\right) \right)^2 \tag{12}$$

$$n > \frac{(1 - fr(x, D))}{fr(x, D)\epsilon^2} \left(\phi^{-1}\left(\frac{3}{4}\right) \right)^2 \tag{13}$$

where $\Phi(y) = (1/\sqrt{2\pi}) \int_{-\infty}^{y} e^{-x^2/2} dx$ is the normal distribution function. (12)/(13) is the best result under worst-case analysis.

Since sampling ensemble is indeed a kind of Monte Carlo algorithm, we have:

Corollary 3. when the sample size of sampling ensemble is larger the theoretical low bound, with the increase of quantity of individual samples in an ensemble, the probability of correct answers exponentially increases.

3.2 Bias-Variance Decomposition

In order to learn where the performance of a sampling ensemble originates and explain why it works, in this section, we use the bias-variance decomposition [13-16] to analyze the sample error of an ensemble. The sample error is measure by $Cerror(S, D) + \bar{\xi}_{S,D}$ in this section.

Let the error induced by individual samples in an ensemble be:

$$\bar{E} = \frac{1}{N} \sum_{i=1}^{N} \sum_{x \in L(D) \cup L(S_i)} |fr(x, D) - fr(x, S_i)| \qquad (14)$$

where N is the total number of individual samples in an ensemble. For sure the simpleness of the inference, we ignore the factor of normalization in this section, for example, the factor $\frac{1}{\|L(D) \cap L(S_i)\|}$ is ignored in (14).

Similarly, the error of a sampling ensemble is as follows:

$$\bar{E}_b = \sum_{x \in L(D) \cup L(S_{bag})} |fr(x, D) - fr(x, S_{bag})| \qquad (15)$$

where S_{bag} is the final answer set of the ensemble. It's just the bias of the ensemble. And the variance of the ensemble is denoted as follows:

$$\bar{E}_v = \frac{1}{N} \sum_{i=1}^{N} \sum_{x \in L(S_i) \cup L(S_{bag})} |fr(x, S_i) - fr(x, S_{bag})| \qquad (16)$$

It measures the diversity of answers on sample sets in an ensemble.

Seeing that

$$\bar{E} = \frac{1}{N} \sum_{i=1}^{N} \sum_{x \in L(D) \cup L(S_i)} |fr(x, D) - fr(x, S_i)|$$

$$\leq \frac{1}{N} \sum_{i=1}^{N} \sum_{x \in L(D) \cup L(S_i) \cup L(S_{bag})} |fr(x, D) - fr(x, S_i)|$$

$$\leq \frac{1}{N} \sum_{i=1}^{N} \sum_{x \in L(D) \cup L(S_i) \cup L(S_{bag})} |fr(x, D) - fr(x, S_{bag})|$$

$$+ \frac{1}{N} \sum_{i=1}^{N} \sum_{x \in L(D) \cup L(S_i) \cup L(S_{bag})} |fr(x, S_{bag}) - fr(x, S_i)|$$

$$= \bar{E}_b + \bar{E}_v + E_\epsilon$$

where we set

$$E_\epsilon = \frac{1}{N} \sum_{i=1}^{N} \sum_{x \in L(S_i) - (L(D) \cup L(S_{bag}))} |fr(x, D) - fr(x, S_{bag})|$$

$$+ \frac{1}{N} \sum_{i=1}^{N} \sum_{x \in L(D) - (L(S_i) \cup L(S_{bag}))} |fr(x, S_{bag}) - fr(x, S_i)|$$

We have:

Theorem 2. The sample error of sampling ensemble, $\bar{E}_b$, and the average sample error of individual samples, $\bar{E}$, satisfy the following inequality.

$$\bar{E}_b \geq \bar{E} - \bar{E}_v - E_\epsilon \qquad (17)$$

Although we can't draw a conclusion that $E_b \leq E$, the larger the right hand of (17), the larger E_b will be.

For the sample error defined by $Cerror(S,D) + \bar{\xi}_{S,D}$, if we let $\bar{fr}(\cdot,\cdot)$ substitute for $fr(\cdot,\cdot)$, the similar result with (17) will be obtained.

For the two databases, S and D,

$$\bar{fr}(x,D) = \begin{cases} 1 & x \in L(D) \text{ but } x \notin L(S) \\ 0 & x \notin L(D) \text{ but } x \in L(S) \\ fr(x,D) & x \in L(S) \cap L(D) \end{cases}$$

It means that we give a penalty to a false positive or a false negative frequent pattern.

For the three databases, S, S_{bag} and D,

$$\bar{fr}(x,D) = \begin{cases} 1 & \begin{array}{l} x \in (L(D) \cap L(S)) \text{ or} \\ x \in (L(D) \cap L(S_{bag})) \text{ or} \\ x \in (L(S) \cap L(S_{bag})) \text{ but} \\ x \notin L(S) \cap L(S_{bag}) \cap L(D) \end{array} \\ fr(x,D) & x \in L(S) \cap L(S_{bag}) \cap L(D) \\ 0 & others \end{cases}$$

It predicates that the penalty is given to the minority frequent patterns in $L(D)$, $L(S)$ and $L(S_{bag})$.

Thus, the following equation holds.

$$\sum_{x \in L(D) \cup L(S)} |\bar{fr}(x,D) - \bar{fr}(x,S)| = Cerror(S,D) + \alpha \bar{\xi}_{S,D}$$

where $\alpha = \frac{\|L(S) \cup L(D)\|}{\|L(S) \cap L(D)\|}$.

Using similar techniques, we have:

Corollary 4. The sample error of sampling ensemble, $\bar{e}_b$, and the average sample error of individual samples, $\bar{e}$, satisfy the following inequality.

$$\bar{e}_b \geq \bar{e} - \bar{e}_v - e_\epsilon \tag{18}$$

Where

$$\bar{e} = \frac{1}{N} \sum_{i=1}^{N} (Cerror(D, S_i) + \bar{\xi}_{D,S_i}) \tag{19}$$

$$\bar{e}_b = Cerror(D, S_{bag}) + \bar{\xi}_{D,S_{bag}} \tag{20}$$

$$\bar{e}_v = \frac{1}{N} \sum_{i=1}^{N} (Cerror(S_i, S_{bag}) + \bar{\xi}_{S_i,S_{bag}}) \tag{21}$$

and

$$e_\epsilon = \frac{1}{N} \sum_{i=1}^{N} \sum_{x \in L(S_i) - (L(D) \cup L(S_{bag}))} |\bar{fr}(x,D) - \bar{fr}(x,S_{bag})|$$

$$+ \frac{1}{N} \sum_{i=1}^{N} \sum_{x \in L(D) - (L(S_i) \cup L(S_{bag}))} |\bar{fr}(x,S_{bag}) - \bar{fr}(x,S_i)|$$

According to (18), the performance of an ensemble is influenced by the individual samples in an ensemble. It's better to diminish $\bar{e}$ and enlarge $\bar{e}_v$. It means that the more accuracy and the more diversity of the answers on samples in an ensemble, the more accuracy of the ensemble might be (the result will be supported by the experiments in the next section).

4 Some Experiments

In this section, we use both a synthetic and a real-world benchmark database to test the performance of sampling ensembles. The synthetic database is generated using code from the IBM QUEST project [1]. And the parameter settings are as same as those in [1]. In the paper, we choose the T10I4D100k denoted as D_1 as a test database. The real-world database is large sales database derived from Blue Martini Software Inc, named BMS-POS and denoted as D_2. And in order to make a fair comparison we use Apriori written by Christian Borgelt[1] and the fast speed sequential random sampling algorithm, Method D in [19], in all cases to compute the frequent itemsets and perform sampling. For both of the databases, the support threshold is 0.77% for sure there are neither too many nor too few frequent itemsets.

Because of space limitation, we just show the results of sampling ensembles with 3 individual samples at 1% sampling ratios on D_1 and D_2 respectively (table 1). These results indicate that $Cerror(S,D)$, $\bar{\xi}_{S,D}$ and $V_{S,D}$ are all reduced (the similar results can also been obtained at other sampling ratios).

Table 1. The three individual samples with sampling ratio 1%

Name	False	Missed	Cerror	$\bar{\xi}_{S,D}$	$V_{S,D}$
S_1	126	45	0.289831	0.00220286	0.00186835
S_2	173	47	0.345369	0.00243885	0.00202185
S_3	186	48	0.36	0.00246875	0.00179169
$3-Ensemble$	79	40	0.201507	0.00140448	0.00160673

(a) The result on D_1

Name	False	Missed	Cerror	$\bar{\xi}_{S,D}$	$V_{S,D}$
S_1	301	164	0.225745	0.001261	0.0016519
S_2	257	180	0.212239	0.00138286	0.00128904
S_3	307	128	0.206259	0.00117324	0.00113885
$3-Ensemble$	251	108	0.146169	0.000758604	0.00066939

(b) The result on D_2

Figure 1 shows the cardinal errors of sampling ensembles with different number of individual samples at 5% sampling ratio for D_1, 1% sampling ratio for D_2. According to figure 1, with the increase of number of individual samples, the cardinal error is gradually reduced. It means that we can obtain very accurate answers at a small (fixed) sample size when the number of individual samples is proper. It makes the difficult problem of sample size selection, as argued by several researches, can be avoid or covert into the problem of selecting the number of individual samples at small sample size.

---
[1] http://fuzzy.cs.uni-magdeberg.de/~borgelt/

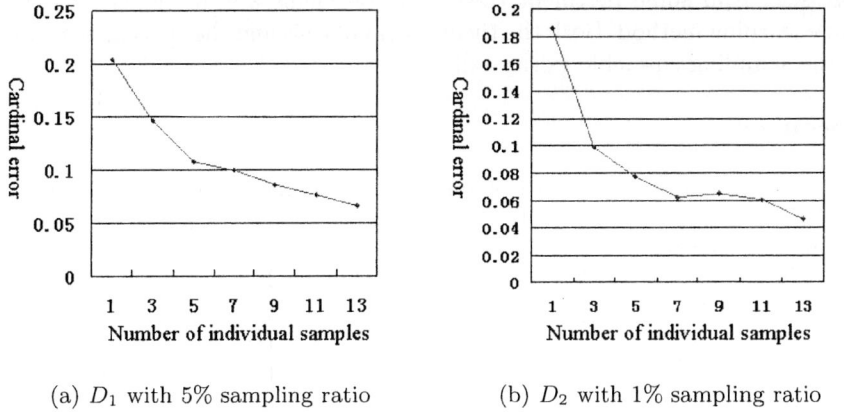

(a) D_1 with 5% sampling ratio   (b) D_2 with 1% sampling ratio

Fig. 1. The cardinal errors of ensembles with different number of individual samples

Since the more diversity of individual samples, the more accuracy of an ensemble might be according to theorem 2. We also test the performance of sampling ensembles with no overlap individual samples (for space limitation, we just show the results on D_1 with the 5% sampling ratio in table 2). According to the experiments, in general, the sample error of a sampling ensemble is sure less than that of a sampling ensemble using sampling Method D directly (it will cause overlap phenomena) at the same sample size.

Table 2. The sample error of a sampling ensemble on D_1

Name	Cerror	$\xi_{S,D}$	$V_{S,D}$
$3 - Ensemble$	0.0987654	0.000938128	0.000818714
$5 - Ensemble$	0.0770833	0.000781941	0.00061186
$7 - Ensemble$	0.0623701	0.000674501	0.000539261

(a) overlap sampling Method D

Name	Cerror	$\xi_{S,D}$	$V_{S,D}$
$3 - Ensemble$	0.0792683	0.000847682	0.000667883
$5 - Ensemble$	0.0695297	0.000655385	0.000504559
$7 - Ensemble$	0.053719	0.00060131	0.000456445

(b) no overlap sampling Method D

5 Conclusion

One of the approaches to improving the scalability of data mining algorithm is to take random sample and do data mining on it. However, it is difficult to determine the appropriate sample size for a given accuracy. Many researchers are dedicate themselves to solve this problem. But they seldom consider how to improve the accuracy of answers at small sample size. In this paper, we give a sampling ensemble method to improve the accuracy of answers at a fixed small sample size. And some pertinent theoretical problems are discussed by using machine learning method. Both the theoretical analysis and the real experiments show the sampling ensemble works well.

References

1. Agrawal, R., Srikant, R.: Fast Algorithms for Mining Association Rules. In VLDB'94, Santiago, Chile (1994) 487-499
2. Zaki, M. J., Hsiao, C.-J.: CHARM: An Efficient Algorithm for Closed Association Rule Mining. Technical Report 99-10, Computer Science Dept., Rensselaer Polytechnic Institute (1999)
3. Han, J., Pei, J., Yin, Y.: Mining Frequent Patterns Without Candidate Generation. In SIGMOD'00, Dallas, TX (2000) 1-12
4. Chen, B., Haas, P., Scheuermann, P.: A New Two-phase Sampling Based Algorithms for Discovery Association Rules. In ACM SIGKDD'02, EDmonton, Alberta, Canada (2002) 462-468
5. Bronnimann, H., Chen, B., et al: Efficient Data Reduction with EASE, In ACM SIGKDD'03, Washington, D.C., USA (2003) 59-68
6. Parthasarathy, S.: Efficient Progressive Sampling for Association Rules. In ICDE'02, Maebashi City, Japan (2002) 354-361
7. Jia, C-Y., Gao, X-P.: Multi-scaling Sampling: An Adaptive Sampling Method for Discovering Approximate Association Rules. to be appeared, Journal of Computer Science and Technology.
8. Toivonen, H.: Sampling Large Databases for Association Rules. In VLDB'96, Mumbai (Bombay), India (1996) 134-145
9. Zaki, M. J., Parthasarathy, S., Li, W., Ogihara, M.: Evaluation of Sampling for Data Mining of Association Rules. In Proceeding of the 7th Workshop on Research Issues in Data Engineer, Birmingham, UK (1997) 42-50
10. Watanabe, O.: Simple Sampling Techniques for Discovery Science. IEICE Trans. Information and Systems, E83-D(1) (2000) 19-26
11. John, G. H., Langley, P.: Static Versus Dynamic Sampling for Data Mining. In ICDM'96, AAAI Press (1996) 367-370
12. Domingo, C., Gavalda, R., Watanabe, O.: Adaptive Sampling Methods for Scaling Up Knowledge Discovery Algorithms. Data Mining and Knowledge Discovery, An International Journal, Vol. 6(2). (2002) 131-152
13. Breiman, L.: Bagging Predictors. Machine Learning, Vol. 24. (1996) 123-140
14. Breiman, L.: Bias, Variance, and Arcing Classifiers. The Annals of Statistics, Vol. 26(3). (1998) 801-849
15. Dietterich, T. G.: Ensemble Methods in Machine Learning. Lecture Notes in Computer Science, Vol. 1857. Springer-Verlag, Berlin Heidelberg New York (2000) 1-15
16. Krogh, A., Vedelsby, J.: Neural Network Ensembles, Cross Validation, and Active Learning. Advances in Neural Infromation Processing Systems 7 (1995) 231-238 (2000) 267-279
17. Esposito, R., Saitta, L.: Monte Carlo Theory As An Explanation of Bagging and Boosting. In IJCAI'03, Acapulco, Mexico (2003) 499-504
18. Valiant, L. G., A Theory of the Learnable. Communications of the ACM 27. (1984) 1134-1142
19. Vitter, J. S.: An Efficient Algorithm for Sequential Random Sampling. In ACM Transactions Mathematical Software, Vol. 13(1). (1987) 58-67.

Distributed Data Mining on Clusters with Bayesian Mixture Modeling[1]

M. Viswanathan, Y.K. Yang, and T.K. Whangbo

College of Software, Kyungwon University,
San-65, Bokjeong-Dong, Sujung-Gu, Seongnam-Si
Kyunggi-Do South Korea
{murli, ykyang, tkwhangbo}@kyungwon.ac.kr

Abstract. Distributed Data Mining (DDM) generally deals with the mining of data within a distributed framework such as local area and wide area networks. One strong case for DDM systems is the need to mine for patterns in very large databases. This requires mandatory partitioning or splitting of databases into smaller sets which can be mined locally over distributed hosts. Data Distribution implies communication costs associated with the need to combine the results from processing local databases. This paper considers the development of a DDM system on a cluster. In specific we approach the problem of data partitioning for data mining. We present a prototype system for DDM using a data partitioning mechanism based on Bayesian mixture modeling. Results from comparison with standard techniques show plausible support for our system and its applicability.

1 Introduction

Data mining research is continually coming up with improved tools and methods to deal with distributed data. There are mainly two scenarios in distributed data mining (DDM): A database is naturally distributed geographically and data from all sites must be used to optimize results of data mining. A non-distributed database is too large to process on one machine due to processing and memory limits and must be broken up into smaller chunks that are sent to individual machines to be processed. In this paper we consider the latter scenario [3].

This paper considers the use of a Bayesian clustering system (SNOB) [1] for dividing large databases into meaningful partitions. These variable-sized partitions are then distributed using a standard MPI based system to a number of local hosts on a cluster. The local hosts run data mining algorithms on the data partition received thus producing local models. These models are combined into a global model by using a model aggregation scheme. The accuracy of the global model is verified through a cross-validation technique.

---

[1] This research was supported by the MIC (Ministry of Information and Communication), Korea, under the ITRC (Information Technology Research Center) support program supervised by the IITA (Institute of Information Technology Assessment).

In light of this paper we typically consider DDM as a four stage process.

- **Data Partitioning** – The large dataset must be partitioned into smaller subsets of data with the same domain according to some function f. The choice of algorithm is dependent on the user, on the DDM system requirements and resources available. Communication needs, message exchange and processing costs are all considerations of this task.
- **Data Distribution** – Once we have partitioned the database the data must be distributed usually over a LAN. An appropriate distribution scheme must be devised depending on number of clusters, size of clusters, content of clusters, size of LAN and LAN communication speeds. For example the user may decide to distribute similar (this measure of similarity is non-trivial) clusters on the same local host machine to save on communication costs or may decide to distribute one cluster per host (assuming the required number of hosts are available on the LAN) to save on processing costs by utilizing absolute parallel processing.
- **Data Modelling** – Once the data is distributed on the LAN, local data models have to be constructed. Many different classification tools and algorithms are widely available. The choice of classification algorithm depends on the nature of the data (numerical, nominal or a combination of both), the size of the data and the destined nature and purpose of the models.
- **Model Aggregation** - All the local models must be combined and aggregated in some optimal way such as to produce one global model to describe the whole database. *Model aggregation* attempts to achieve one of the fundamental aims of DDM that is to come up with a final global data model from the distributed database to match the efficiency and effectiveness of a data model developed from undistributed data. Some existing techniques rely on such methods as voting, combining and meta-classifiers. In order to develop a good and efficient model aggregation framework one must analyse communication needs and costs as well as datasets sizes and distributions. For example if local datasets are reasonably large it would be undesirable to transfer them in their entirety across hosts since the incurring communication costs will effectively halt the whole system, in this case it might be a good idea to transfer sample subsets of the data or maybe just transfer descriptive data [7,8].

2 Data Partitioning and Clustering

As suggested earlier we consider the issue of DDM in a case where the existing database is too big to process on a single machine due to memory and storage constraints. There is a great need in this case to use some intelligent mechanism to break the database into smaller groups of data

2.1 SNOB

SNOB is a system developed for cluster analysis using mixture modeling by Minimum Message Length (MML) [1]. SNOB aims to discover the natural classes in the data by categorising data sets based on their underlying numerical distributions. It does this using the assumption that if it can correctly categorize the data, then the data can be described most efficiently (i.e. using the minimum message length). SNOB uses MML induction, a scale-invariant Bayesian technique based on information theory [2].

To summarize SNOB considers that a database is usually made up of many objects or instances. Each instance has a number of different attributes with each attribute having a particular value. We can think of this as a population of instances in a space with each attribute being a different dimension or variable. SNOB assumes to know the nature, number and range of the attributes. The attributes are also assumed to be uncorrelated and independent of each other. SNOB attempts to divide the population into groups or classes such that each class is tight and simple while ensuring the classes' attribute distributions significantly differ from one another. Snob takes an inductive inference approach to achieve data clustering [2].

2.2 Data Partitioning Sub-system

Our system employs SNOB in the initial process of data partitioning. One of our objectives in this project is to investigate whether using SNOB-based clustering is appropriate and efficient for partitioning in the framework of DDM. In figure 1 we depict the partitioning aspect of out DDM system with Snob being the core of the partitioning operation. Once SNOB is used to cluster the data and generate a report file containing details of the clusters, the complete database is divided into the appropriate clusters using a standard program.

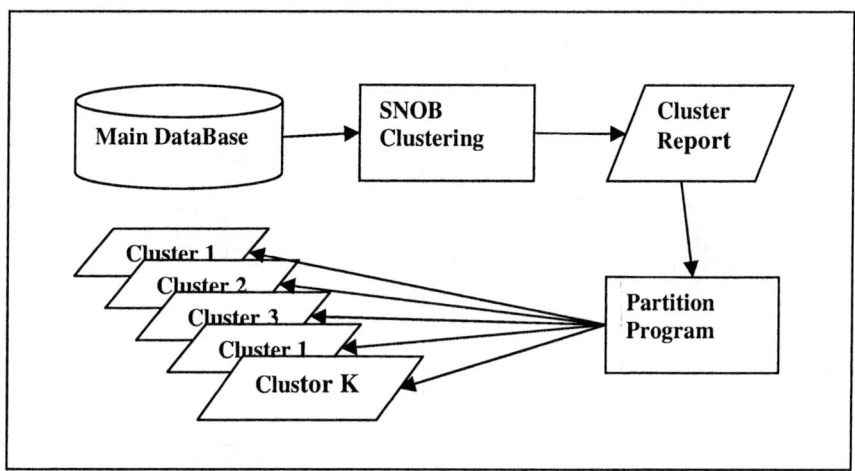

Fig. 1. The partitioning component of the DDM system using SNOB

In order to test the predictive accuracy of the models generated at the end of the DDM process we retain 10% of data records randomly selected from the database for cross-validation purposes. The test dataset is removed from the original dataset so we are finely left with the un-clustered dataset and each cluster containing only training data (for classification). In addition to these there is a separate dataset for testing purposes only.

3 Distributing the Data

Distributing Data over a network such as a LAN is no trivial task. Many considerations must be taken into account before deciding on the distribution strategy. We need to take into consideration the number, size, and nature of clusters and purpose of distributing the data in order to come up with an optimal distribution scheme. Not only that, to further complicate matters every dataset is different and will be clustered differently and so we will have a different distribution scheme that will optimize each of these datasets. This problem is out of the scope of this project.

As a simple solution we used the Message Passing Interface (MPI) to distribute the data in our DDM system. MPI is a library specification for message-passing, proposed and developed as a standard by a broadly based committee of vendors, implementers, and users. MPI is used by programs for inter-process communication and was designed for high performance on both massively parallel machines and on workstation clusters [5].

As can be seen in figure 2 the complete (un-clustered) training dataset resides on the master machine (ID 0) and each cluster is exported to a different host machine (ID>0). The first cluster is sent to host 1, the second cluster to host 2 and so on, although this is not the optimal solution for distribution this solution will still classify the clusters in parallel and cause an overall speedup in the data mining task. It should be noted that the distribution automatically occurs in conjunction with *Data Modelling* (classification) phase in our DDM system. In concept *Data Distribution* and *Data Modelling* are two distinctly different steps however in this project these two steps are combined into one using MPI and the Classification Algorithm.

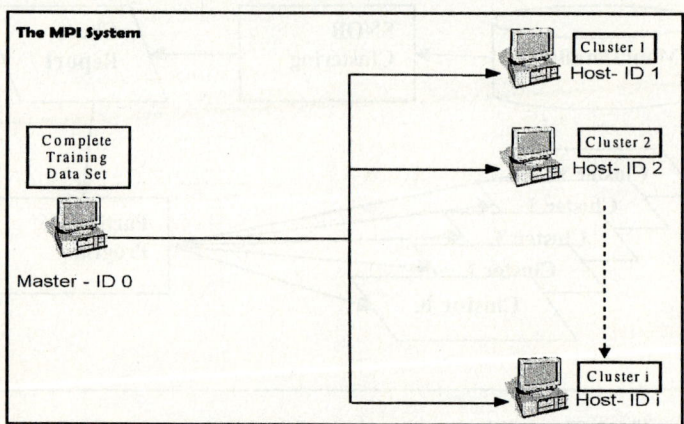

Fig. 2. The MPI Architecture

4 Local Data Mining with C4.5

Creating data models in order to predict new future unseen data is the main task of data mining. There are various types of models that can be constructed however we will use a classification model based on the C4.5 decision tree learning system. C4.5rules is a classification-rule learning algorithm developed by Quinlan that uses decision trees to represent its model of the data. It aims to find a function that can map data to a predefined class correctly thus accurately predicting and classifying previously unseen examples and data. The function is a decision tree and is built from a training data set and is evaluated using a test data set, these sets are usually subsets of the entire data. The algorithm attempts to find the simplest decision tree that can describe the structure of the data [6].

The implementation of the C4.5 that is used for this project is by Ross Quinlan [6]. The original source code was changed to incorporate distribution via MPI and to output human readable rule files. The C4.5 algorithm was embedded with MPI for use in our system. The result is the full dataset is classified by C4.5 on the master host and each cluster is classified by C4.5 on a different host in the network. In case of insufficient number of hosts in the network some hosts will classify more then one cluster. The working directory is copied to all hosts involved in the classification process and the algorithm is run from each host's local copy of this directory. Once the C4.5 program finishes and terminates all the files from all the hosts are copied back to the original directory on the master machine. So as we have mentioned in the previous section the *Data modeling* is combined with *Data Distribution* in our DDM system. It should be noted that our C4.5 classification algorithm is a DDM classification algorithm. Figure 3 shows how C4.5 and MPI work together in our DDM system.

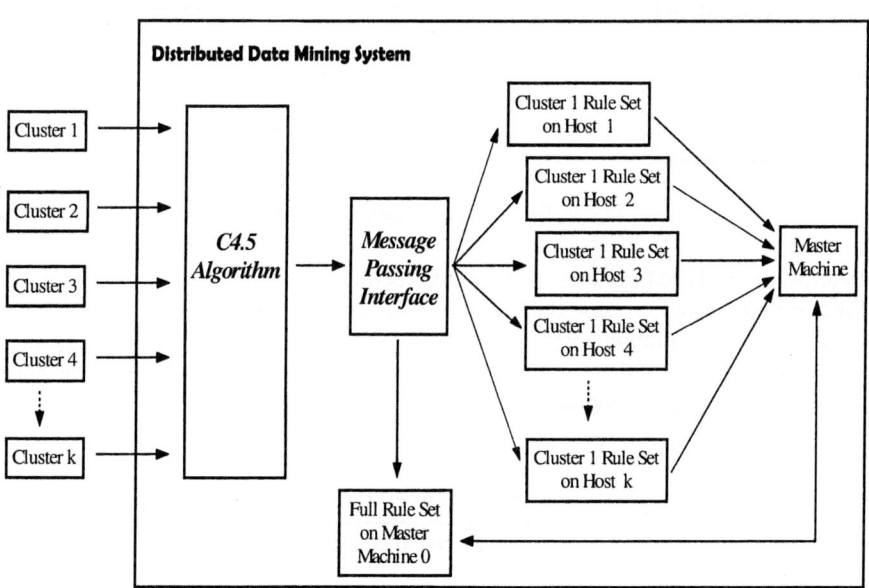

Fig. 3. Distributed Data Modelling using MPI and the C4.5 classification algorithm

5 Model Aggregation Using a Voting Scheme

Model Aggregation is the key step in the DDM process since it uses all distributed data and its associated learned knowledge to generate the all-inclusive and comprehensive data model for the original and complete database. With such a model we can predict the outcome of unseen new data introduced to our system. The *Data Aggregation* task itself is an area of extensive research and many methods and techniques have been developed [7, 8].

We have chosen to use a simple but effective voting technique whereby the classification models (rules) are validated against the unseen test data by means of positive and negative votes. A positive vote is an unseen data record that is classified correctly by a classification rule and the negative vote is an incorrectly classified record. Once the votes are summed up and the rules are thereby ranked, the system can choose what rules to discard and what rules to keep, these rules will form the global data model.

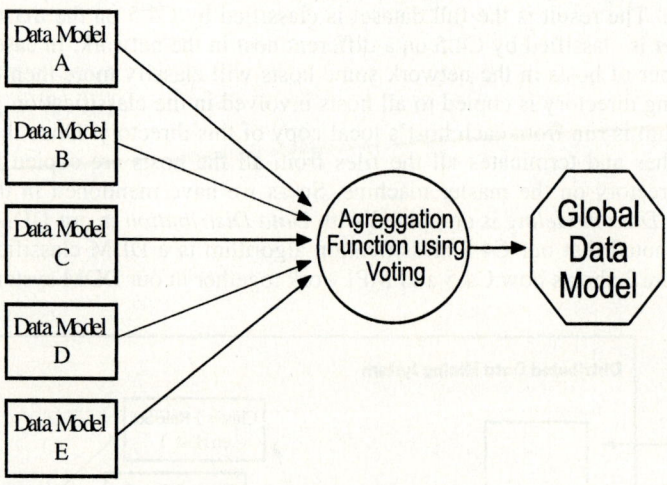

Fig. 4. The Data Aggregation Process

In terms of our DDM system, voting works as follows:

Each Rule is tested on each data record in the test dataset and is given a positive vote for a correct prediction a negative vote for a false prediction and no vote at all if the rule conditions don't hold. After all the records in the test dataset have been scanned the error rate is computed as:

$$\frac{\text{Negative Votes}}{\text{Total Votes}} \quad (1)$$

However if there are no votes at all then the error rate is considered unknown and is set to the value of -1.

6 Distributed Data Mining – System Components

The whole DDM system can finally be discussed in terms of its individual components which have been discussed in the above sections. Our system takes a dataset and using SNOB **partitions** it to clusters. The clusters get **distributed** over the LAN using *MPI*. **Data models** are developed for each cluster dataset using the classification algorithm *C4.5*. Finally the system uses a voting scheme to **aggregate all the data models**. The final global classification data model comprises of the top three rules for each class (where available). We can describe our DDM system in terms of the following algorithm in figure 5.

Report	Snob Report File
N	Number of Clusters produced by Snob
Data	Complete Dataset
MaxHosts	Maximum Number of Hosts on LAN
C_i	Cluster I
R_i	Rule set of I
GModel	Global Classification Model

```
1:      (Report, N) = Snob(Data);
2:      Partition(Data, Report);
3:      For i = 1 to N
4:          R_i = C4.5Classify(MPI(C_i, MaxHosts));
5:      For i = 1 to N
6:          Validate(R_i);
7:      GModel = Aggregate(R_1•R_2•...•R_i);
```

Fig. 5. Functional algorithm

Note that MPI is used in conjunction with the known maximum number of hosts (line 4) to classify the clusters in parallel using the C4.5 classification algorithm. If the number of clusters exceeds the available number of hosts then some hosts will classify multiple clusters (using MPI). Also the aggregation model scans all Rule files (line 7) from all clusters and picks the best rules out of the union of all cluster rule sets.

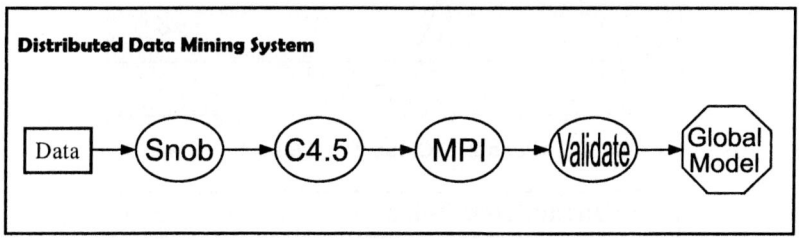

Fig. 6. DDM System Components

During the Classification phase we have also classified the original un-clustered dataset and produced rules modelling this data. To finely ascertain if our DDM system is efficient we compare our Global Model to this un-clustered Data Model. We compare the top three rules for each class from this model with our rules from the Global Model. If our global model is over 90% accurate in comparison to the un-clustered data model we can consider this a good result and conclude that our DDM system is efficient.

7 DDM System Evaluation and Conclusions

Our DDM system was applied to relatively smaller databases in order to test the effectiveness of the concept. The system was used to mine (classify) the original database as well the clustered partitions. The DDM system was designed to output classification rules for the original database and the clustered datasets. In the case of the partitioned clusters the rule were validated and aggregated using our voting scheme. Five real-world databases were compared taking into account the top classification rules for each class from the original database. Thus we could compare the difference in predictive accuracy of the model derived from the original datasets with the aggregated global model from the partitioned database.

As an example of the empirical evaluation we present experiments on the 'Demo' database which is taken from Data for Evaluating Learning in Valid Experiments (Delve) from [13]. The top three rules for each class from the cluster-derived rules are compared against the rules mined from the un-partitioned dataset. The aim of this evaluation is to determine the effect of the clustering process on the efficiency of our classification model and its predictive accuracy. The total average error for the rules from the original dataset is 42.8% while for the clustered partitions it is 41.83%.

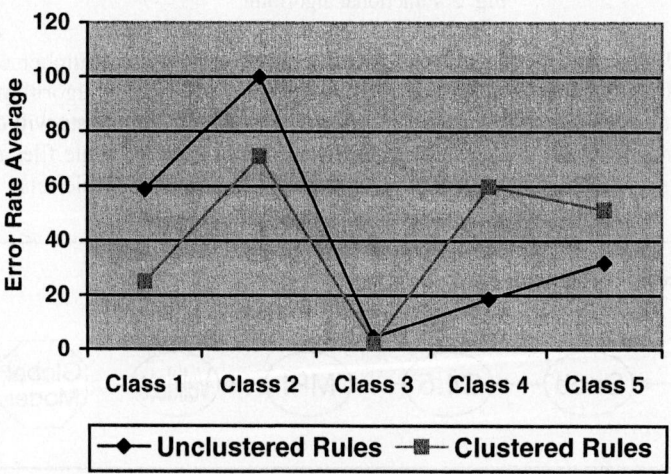

Fig. 7. Accuracy of Un-clustered Rules versus Clustered Rules

From the chart in figure 7 it can be seen that the first two classes of the clustered rules have a lower average error rate then the un-clustered ones. Rules for class three have approximately equal average error rates for both sets of rules and for the last two classes the clustered rules have greater average error rates then the un-clustered ones. If we average out these results both set of rules will result in approximately total equal average error rates. If we had to graph these they would be two straight lines around the 42% average error rate mark.

From the experimental evaluation the total average error rate for the un-partitioned datasets taking all datasets into account, was 30.86%. And the total average error rate for the clustered data taking all datasets into account is 35.46%. Overall in our experiments the clustering caused 4.6% loss of accuracy in our data modeling. Considering the other costs involved in standard schemes we think that this drop in accuracy is not very significant. In general our evaluation gives us plausible evidence supporting SNOB in the framework of Distributed Data Mining.

There are many issues in this project that need further research and enhancement to improve the overall DDM system performance. The use of SNOB to cluster the entire database may be questioned on the premise that it is probably impossible to apply clustering on the entire database due to memory and processor limitations. This being true our primary objective is to demonstrate the effectiveness of SNOB in producing meaningful partitions that enable us to reduce the costs of model aggregation. In other words SNOB could be used to create partitions from arbitrary subsets of the very large database. A more efficient aggregation model may produce more predictive accuracy and hence a better DDM system.

References

1. Wallace C. S., Dowe D. L.: MML clustering of multi-state, Poisson, von Mises circular and Gaussian distributions. Statistics and Computing, 10(1), January (2000) 73-83
2. Wallace C. S., Freeman P. R.: Estimation and Inference by Compact Coding. Journal of the Royal Statistical Society series B., 49(3), (1987) 240-265
3. Park B. H., Kargupta H.: Distributed data mining: Algorithms, systems, and applications. In In Data Mining Handbook, To be published, (2002)
4. Kargupta H., Hamzaoglu I., Stafford B.: Scalable, Distributed Data Mining Using an Agent Based Architecture. Int. Conf. on Knowledge Discovery and Data Mining, August (1997) 211-214
5. Message Passing Interface Forum. MPI: A message-passing interface standard. International Journal of Supercomputer Applications, Vol. 8(3/4), (1994) 165-414
6. Quinlan, J. R.. C4.5: Programs for machine learning. San Mateo, CA: Morgan Kaufmann (1993)
7. Prodromidis A., Chan P.: Meta-learning in Distributed Data Mining Systems: Issues and Approaches. In Hillol Kargupta and Philip Chan, editors, Advances of Distributed Data Mining. MIT/AAAI Press, (2000)
8. Chan P., Stolfo S. J.:.. A Comparative Evaluation of Voting and Meta-learning on Partitioned Data. In Proceedings of Twelfth International Conference on Machine Learning, pages 90–98, (1995)
9. Subramonian R., Parthasarathy S.: An Architecture for Distributed Data Mining. In Fourth International Conference of Knowledge Discovery and Data Mining, New York, NY, (1998) 44–59

10. Fayyad U.M., Piatetsky-Shapiro G., Smyth P.: From data mining to knowledge discovery: an overview. In: U.M. Fayyad, G. Piatetsky-Shapiro, P. Smyth and R. Uthurusamy. Advances in Knowledge Discovery & Data Mining, AAAI/MIT, (1996) 1-34
11. Grigorios T., Lefteris A., Ioannis V.: Clustering Classifiers for Knowledge Discovery from Physically Distributed Databases. Data and Knowledge Engineering, 49(3), June (2004) 223–242
12. Kargupta H., Park B., Hershberger D., Johnson E.: Collective Data Mining: A New Perspective towards Distributed Data Mining. In Advances in Distributed and Parallel Knowledge Discovery, MIT/AAAI Press, (2000) 133–184
13. Neal, R. M.: Assessing relevance determination methods using DELVE', in C. M. Bishop (editor), Neural Networks and Machine Learning, Springer-Verlag (1998) 97-129

A Method of Data Classification Based on Parallel Genetic Algorithm

Yuexiang Shi [1,2,3], Zuqiang Meng [2], Zixing Cai [2], and B.Benhabib [3]

[1] School of information engineer, XiangTan University,
XiangTan 411105, China
[2] School of information engineer, Central South University,
ChangSha 410082, China
[3] Department of Mechanical and Industrial Engineering, Toronto University,
Ontario, Canada M5S 3G8

Abstract. An effectual genetic coding is designed by constructing full-classification rule set. This coding results in full use of all kinds of excellent traits of genetic algorithm in data classification. The genetic algorithm is paralleled in process. So this leads to the improvement of classification and its ability to deal with great data. Some defects of current classifier algorithm are tided over by this algorithm. The analysis of experimental results is given to illustrate the effectiveness of this algorithm.

1 Introduction

With the development of information technology and management, data mining becomes a new fiddle in Computer Science from 1990[1]. Today, data mining has been used in large enterprise, telecom, commercial, bank and so on. And the fields have made the best benefits from it.

Algorithm ID3[2] and C4.5[3] are the first algorithms for data classification. The algorithms and the ramification were based on the decision tree. But there are some defects no way to avoid. Such as it is not powerful for retractility[4] and adjust. There is difficult to get the best decision tree just using the heuristic from information plus. The defect is lack of integer searching strategy.

In fact, constructing the best decision tree is a NP[5]. But parallel genetic algorithm has the specific function to solve those problems. This paper presents a technology based on parallel genetic algorithm to class the data. It can improve and increase the ability of classification algorithm for magnanimity data.

2 Related Definitions

Definition 1 (Date set). For a specific data mining system, the process data is called data set. It was written as $U = \{S_1, S_2, \cdots, S_{size}\}$. The object in U is called as data element. It is also called sample, example or object. The size presents the number of data set.

Definition 2 (Length of attribute). The number was used to describe the attributes of data element is called attribute length.

Definition 3 (Training set). The set of data element being used for the forecast model is called training set. It is written as T and $T \subseteq U$.

Theorem 1. For the attribute length h of data set U, its attributes variables are $A_1, A_2, \cdots, A_h$, the construction depth of decision tree is no more than h. (Proof: omit)

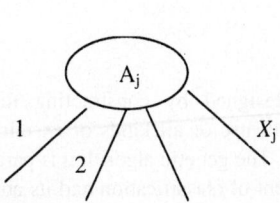

Fig. 1. The decision tree only with one node   **Fig. 2.** The decision tree with multi-nodes

Definition 4 (Decision tree). Tree was called decision tree about B(U), when every interior node can present a test was taken on the attribute of B, every branch presents one test output and every leaf presents a distribution of one class.

Definition 5 (Classification regulation). IF-THEN regulation r was called one classification about $B(\subseteq U)$, when the pre-part of regulation r was consisted of B's attribute which forms the SAT. The post-part is a judge formulation that fits the data elements of SAT. Apparently, one classification regulation corresponds to one class C. Written as $C = r(B)$.

Definition 6 (Completed classification regulation). The set R was called completed classification regulation about $B(\subseteq U)$ that was consisted of some classification regulations and $B = \bigcup_{r \in R} r(B)$.

Theorem 2. For the given data set U, there must be a completed classification regulation about U. That is to say exist R, let $U = \bigcup_{r \in R} r(U)$.(Proof: omit)

Data classification is a prediction model by a limited training set T and directed studying. It was used to describe the reserved data set U. That is to pre-class for U. This model is a completed classification regulation R about U. Assumed T was extracted from U and enough represents U, and U is very large to GB, TB, so T is large too. How to get the best and completed classification regulation R from the magnanimity data about U, avoiding the exist problems, this is the research goal. In order to solve the problem, the genetic algorithm will be used in seeking the objects.

3 The Design of Genetic Algorithm for Data Classification

*Coding.** From theorem 1, for the attribute length h of data set U, the depth of any decision tree is no more than h. Like that, the number of SAT's pre-part is also no more than h from classification regulation export. So for any regulation r, it can be described as IF-THEN:

IF P_1 AND P_2 AND $\cdots$ AND P_h THE P_{h+1}, Among that, $P_i(i=1,2,\cdots h+1)$ all are judgment proposition and corresponds to the test attribute A_i. If there is no test about this attribute, the test was presented with TRUE. Assume A_i have different values: $a_1, a_2, \cdots, a_{X_i}$. T was divided into n class: $C_1, C_2, \cdots, C_n$. So the proposition P_i can be expressed as: the value of A_i is. $j_i\{1,2,\cdots,X_i\}$. This item was coded as j_i. If the value of P_i is TRUE, the item's code is star '*'. Proposition P_{h+1} was presented as: this data element that fits the pre-part SAT belongs to class C_y. This item was coded as j' and $j \in \{1,2,\cdots,n\}$. So the IF-THEN regulation r was coded as: $j_1 j_2 \cdots j_h j'$ and called as chromosome θ_r of regulation r. For the θ_r, the bit's value not equal '*' was called valid genetic bit. The number of valid genetic was called the valid length about this chromosome and was written as genetic(θ_r). It was also called as valid genetic bit and valid genetic length based on regulation r. genetic(θ_r) was written as genetic(r). $j_1, j_2, \cdots, j_h j' \in \{1,2,\cdots, \max\{\max\{X_i \mid i=1,2,\cdots,h\}, n\}, *\}^{h+1}$.

So one completed class regulation set can be coded as:
$\{j_1^{(1)} j_2^{(1)} \cdots j_h^{(1)} j^{(1)'}, j_1^{(2)} j_2^{(2)} \cdots j_h^{(2)} j^{(2)'}, \cdots, j_1^{(q)} j_2^{(q)} \cdots j_h^{(q)} j^{(q)'}\}$ Among these, q represents the size of R and has the relation with specific R. The valid genetic length of classification regulation set was defined as:

$$genetic(R) = \max\{genetic(j_1^{(1)} j_2^{(1)} \cdots j_h^{(1)} j^{(1)'}), genetic(j_1^{(q)} j_2^{(q)} \cdots j_h^{(q)} j^{(q)'})\}$$

*Group set.** Group is an individual set consisted with classification regulation. It was present as:

$$P = \{\{j_1^{(1)} j_2^{(1)} \cdots j_h^{(1)} j^{(1)'}, j_1^{(2)} j_2^{(2)} \cdots j_h^{(2)} j^{(2)'}, \cdots, j_1^{(q)} j_2^{(q)} \cdots j_h^{(q)} j^{(q)'}\} \mid i = 1,2,\cdots, P_{size}\}$$

P_{size} represents the group size.

*Adapt function.** Adapt function is a judge standard given by the algorithm for the individual and shown like +. One excellent completed classification regulation can exactly class for U. There is a contact between classification regulation set and valid genetic bits. According above, for individual $S \in P$, the adapt function was defined as:

$$f(S) = w_1 x_1 + w_2 x_2 + w_3 x_3 + w_4 (1 - x_4)$$

among the formula, $x_1 = genetic(S)$ shows the valid length of genetics.

$x_2 = \sum_{r \in S} genetic(r)$ shows the total of valid genetic bits.

$x_3 = |S|$ shows the size of S. That is to say S has the number of classification regulations.

x_4 = represents the wrong probability based on classification S.

w_x is a weight coefficient corresponds to adapt function.

*Genetic Algorithm

I Select Monte Carlo was used.

II Crossover For any $S_1, S_2 \in P$; random select $\theta_1 \in S_1$, $\theta_2 \in S_2$, and random get a positive integer pos. Then the single crossover was taken at the location pos of θ_1, θ_2. In the same time, the dynamic adjustment was adapted to crossover probability P_c. When the individual tends to single model and heighten the crossover probability P_c optimising ability for the algorithm. Set $P_c = -k(f_{min} - f_{avg})/(f_{max} - f_{avg} + \varepsilon)$. f_{max} is the biggest adapt value at present. f_{min} is the minimum and the f_{avg} is the middle. $k \in (0,1]$, ε is a little positive real number and avoid the denominator equals zero.

III Variation It is mainly to ensure the reach for search space and keeps the multiplicity of individual. The process is random to select $S \in P$, $\theta \in S$, and random select a valid genetic bit(assume j), added a random positive number ram, then models this bit X_j(number of attribute A_j). The variation probability $P_m \in [0, 1]$ and general is 0~0.05[6].

4 Parallel Genetic Algorithms

There are 4 kinds of parallel Genetic Algorithm. Take the characteristics of the database system into consideration, the article selects granularity parallel model(CPGA) which can also be called as acnode. The method is to extract training-set T from dataset U and at the same time, generating P_{size} decision trees through T using C4.5. And above all these, educing P_{size} complete-sorted-rules-sets, then divided them into several subgroup and send them to each acnode as initial colony. The colony evolves themselves independently in succession until they fulfill certain qualifications (for example, time interval or genetic algebra etc.). The acnode exchange each other and continue to evolve until the Algorithm converge. The exchange manner between the acnodes is one of the key method of the Genetic Algorithm. We present a stochastic exchange method in article [8], shown as Fig.3.(annotate: as an example, P_i in the figure represents acnode, i=1,...5; the dotted lines represent possible data exchange fashion). The article selects this method though it maybe inconvenient when implementing program, yet it can avoid the problem of prematurity[8].

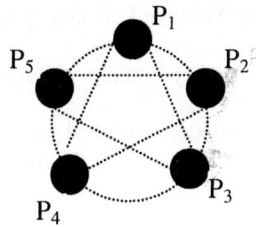

Fig. 3. Improved CPGA

5 Experiment Results

The experiment based upon some filiale of the China Petro Co.. The sample attributes mainly include the following control parameter as temperature, compressive stress, metal etc. The result are divided into 5 kinds and respectively indicate 5 product class of dissimilar quality: particularly excellent, excellent, commonly, not good and waster.

In the experiment, there is a contrast between the CMGA of the parallel Genetic Algorithm and the one based on common Genetic Algorithm. The size of the training-set T is 500, and the testing-set size is 600. The initial value of the Genetic Algorithm Cross-probability is 0.7, and it's dynamic adjust coefficient K is 0.6, the initial value of the aberrance probability Pm is 0.03, the weight coefficient W1, W2, W3 and W4 are respectively denoted as 0.01,0.0001,0.01,0.95, the constringency of the Algorithm is controlled by threshold. The Algorithm will be considered as constringency when the sort mistake rate is under or equal to 0.30%. As to CMPGA, other than denoted as the same parameter as above, the acnodes start an interchange among outstanding individual each 20 generation. The parallel environment is composed of 3 PC connected by internet and the message transfer between courses is based on traditional Socket Programming method. There is a 40%-65% higher of CMPGA than that of CMGA.

6 Conclusions

The article constructs a complete-sorted-rules-set, realizes genetic coding and thereafter reinforces the maneuverability of the Algorithm. The system utterly data management and global optimize ability is advanced at one side because of the effective import of the parallel Genetic Algorithm, and at the other side, enlarging the manage mode from n to O(n3) using the concealed parallel of the Genetic Algorithm. Thus it greatly solves the retractility of the sort algorithm and fit to the mining to the large database.

References

1. Li li, Xu zhanweng. An algorithm of classification regulation and trend regulation. Mini-Micro system, 2000, 21(3): 319-321
2. Quinlan J R. Induction of decsion trees. Machine Learning, 1986, (1):81-106

3. Quinlan J R. C4.5 Programs for Machine Learning. San Mateo, CA: Morgan Kaufmann Publishers, Inc. 1993
4. Han Jianwei, Kamber M. Data Mining: Concept and technology. Peking: Mechanical industry publishers. 2001.8
5. Hong J R. AE1: An extension approximate method for general covering problem. International Journal of Computer and Information Science, 1985,14(6):421-437
6. Xi Yugong, Zi Tianyou. Summary of genetic algorithm. Control theory and application. 1996,13(6):697-708
7. Wang Hongyan. Parallel genetic algorithm research development. Computer Science, 1999,26(6):48-53
8. Meng Zuqiang, Zheng Jinghua. Parallel genetic algorithm based on shared memory and communication. Computer engineering and application. 2000, 36(5):72-74

Rough Computation Based on Similarity Matrix

Huang Bing[1], Guo Ling[2], He Xin[2], and Xian-zhong Zhou[3]

[1] Department of Computer Science & Technology, Nanjing Audit University,
Nanjing 210029, China
[2] Department of Automation, Nanjing University of Science & Technology,
Nanjing 210094, China
[3] School of Engineering Management, Nanjing University, Nanjing 210093, China

Abstract. Knowledge reduction is one of the most important tasks in rough set theory, and most types of reductions in this area are based on complete information systems. However, many information systems are not complete in real world. Though several extended relations have been presented under incomplete information systems, not all reduction approaches to these extended models have been examined. Based on similarity relation, the similarity matrix and the upper/lower approximation reduction are defined under incomplete information systems. To present similarity relation with similarity matrix, the rough computational methods based on similarity relation are studied. The heuristic algorithms for non-decision and decision incomplete information systems based on similarity matrix are proposed, and the time complexity of algorithms is analyzed. Finally, an example is given to illustrate the validity of these algorithms presented.

1 Introduction

The rough set theory proposed by Pawlak[1] is a relatively new mathematic tool to deal with imprecise and uncertain information. Because of the successful application of rough set theory to knowledge acquisition, intelligent information processing, pattern recognition and data mining, many researchers in these fields are very interested in this new research topic since it offers opportunities to discover useful knowledge in information systems.

In rough set theory, a table called information system or information table is used as a special kind of formal language to represent knowledge. In an information system, if all the attribute values of each object are known, this information system is called complete. Otherwise, it is called incomplete. Obviously, it is very difficult to ensure completion of an information system due to the uncertainty of information.

As we know, not all of attributes are indispensable in an information system. The approach to removing all redundant attributes and preserving the classification or decision ability of the system is called attribute reduction or knowledge reduction. Though knowledge reduction has been an important issue in rough set theory, most of knowledge reductions are based on complete information systems [2-4]. To deal with incomplete information, several extended models to the classical rough set are proposed [5-6]. In these extended theories, equivalence relation is relaxed to tolerance relation, similarity relation or general binary relation, respectively. In [7], several Definitions of knowledge reductions based on tolerance relation were presented. However, corresponding definitions and approaches to knowledge reduction based on similarity relation were not studied.

In this paper, we aim at the approach to attribute reduction based on similarity relation in incomplete information systems.

The remainder of this paper is organized as follows. Section 2 reviews rough set model based on similarity relation in incomplete systems and Section 3 gives the definitions of lower approximation reduction and lower approximation matrix in incomplete data tables, and proposes an algorithm for lower approximation reduction. Similar to Section 3, the corresponding contents are studied in incomplete decision tables in Section 4. An illustrative example is given in Section 5 and Section 6 concludes this paper.

2 Incomplete Information Systems and Similarity Relation

Definition 1. $S = (U, A = C \cup D, V, f)$ is called an information system, where U is a non-empty and finite set of objects called the universe, C is a non-empty and finite set of conditional attributes, D is a non-empty and finite set of decision attributes, and $C \cap D = \phi$. In general, D is the set including a single element. The information function, f, is a map from $U \times A$ onto V, a set of attribute values. If every object in the universe has a value on any attribute, this information system is complete. Otherwise, it is incomplete. Considering further, if $D = \phi$, it's a non-decision information system, otherwise a decision table.

For simplicity, we assume that all values of every object on decision attributes are known.

Definition 2. In an incomplete information system $S = (U, A = C \cup D, V, f)$, where $x, y \in U$, the object x is called similar to y with respect to $B \subseteq A$, if $\forall b \in B$, such that $f(x,b) = * \vee (f(x,b) = f(y,b))$, and the relation between x and y is denoted as $S_B(x, y)$.

Obviously, the similarity relation is reflexive and transitive, but not symmetric.

Definition 3. $R_B(x) = \{y : y \in S_B(y, x)\}$; $R_B^{-1}(x) = \{y : y \in S_B(x, y)\}$.

$R_B(x)$ denotes the set of all objects in U which are similar to x with respect to B, and $R_B^{-1}(x)$ denotes the set of all objects in U to which x are similar with respect to B. Obviously, $\forall b \in B \subseteq A$, $R_B(x) \subseteq R_{B \setminus \{b\}}(x)$ and $R_B^{-1}(x) \subseteq R_{B \setminus \{b\}}^{-1}(x)$. In general, $R_B(x) \neq R_B^{-1}(x)$.

Definition 4. $S = (U, A = C \cup D, V, f)$ is an incomplete information system. The upper approximation $\overline{B}(X)$ and lower approximation $\underline{B}(X)$ of $X \subseteq U$ with respect to $B \subseteq A$ are defined as follows:

$$\overline{B}(X) = \bigcup_{x \in X} R_B(x), \underline{B}(X) = \{x : R_B^{-1}(x) \subseteq X\}$$

Theorem 1. $\forall b \in B \subseteq A, X \subseteq U, \overline{B}(X) \subseteq \overline{B \setminus \{b\}}(X), \underline{B \setminus \{b\}}(X) \subseteq \underline{B}(X)$.

Based on the similarity relation, we can define the upper/lower approximation reduction.

3 Similarity Matrix and Approximation Reduction for Non-decision Incomplete Information Systems

Definition 5. $S = (U, C, V, f)$ is an non-decision incomplete information system, where $U = \{x_1, x_2, \cdots, x_n\}$. The lower approximation matrix of $B \subseteq C$ can be defined as $\underline{M}_B = (\underline{m}_{ij}^B)_{n \times n} = \begin{cases} 1 & x_j \in R_B^{-1}(x_i) \\ 0 & \text{else} \end{cases}$.

Definition 6. $S = (U, C, V, f)$ is an non-decision incomplete information system, where $U = \{x_1, x_2, \cdots, x_n\}$. The upper approximation matrix of $B \subseteq C$ can be defined as $\overline{M}_B = (\overline{m}_{ij}^B)_{n \times n} = \begin{cases} 1 & x_j \in R_B(x_i) \\ 0 & \text{else} \end{cases}$.

Theorem 2. The upper/lower approximation matrix has the following attributes

(1) $\underline{m}_{ii} = \overline{m}_{ii} = 1 \quad 1 \leq i \leq n$

(2) $\underline{m}_{ij}^B = 1, \underline{m}_{jl}^B = 1 \Rightarrow \underline{m}_{il}^B = 1$; $\overline{m}_{ij}^B = 1, \overline{m}_{jl}^B = 1 \Rightarrow \overline{m}_{il}^B = 1$

(3) $\underline{m}_{ij}^B = 1 \Leftrightarrow \overline{m}_{ji}^B = 1$

Proof. (1) and (2) are direct results from the self-reflect and attributes of similarity. And (3) holds since $x_j \in R_B^{-1}(x_i) \Leftrightarrow x_i \in R_B(x_j)$.

Definition 7. $S = (U, C, V, f)$ is an non-decision incomplete information system, where $U = \{x_1, x_2, \cdots, x_n\}$ and $B \subseteq C$. If $\forall x_i (1 \leq i \leq n) \in U$ and $R_B^{-1}(x_i) = R_C^{-1}(x_i)$ ($R_B(x_i) = R_C(x_i)$) , B is called lower / upper approximation consistent set. If B is the upper/lower approximation consistent set and any proper subset of B is not, B is called lower/upper approximation reduct.

According to (3) in theorem 2, we'll only discuss lower approximation reduct.

Definition 8. If two matrix $M_1 = (m_{ij}^1)_{m \times n}$, $M_2 = (m_{ij}^2)_{m \times n}$, then $M_1 \leq M_2 \Leftrightarrow m_{ij}^1 \leq m_{ij}^2$ $1 \leq i \leq m, 1 \leq j \leq n$; $M_1 < M_2 \Leftrightarrow M_1 \leq M_2$, and $\exists\ m_{ij}^1 < m_{ij}^2$.

Theorem 3. If $E \subset F \subseteq C$, $\underline{M}_F \leq \underline{M}_E$.

Theorem 4. $B \subseteq C$ is lower similarity reduct $\Leftrightarrow \underline{M}_B = \underline{M}_C$, and $\forall b \in B$, $\underline{M}_B < \underline{M}_{B \setminus \{b\}}$ holds.

Definition 9. $S = (U, C, V, f)$ is a non-decision incomplete system, where $B \subseteq C$. If $\underline{M}_B < \underline{M}_{B \setminus \{b\}}$, we say b is lower-approximation dispensable in B. The set of all lower-approximation dispensable attributes in C is called the core of $S = (U, C, V, f)$, denoted as $\underline{Core}(C) = \{c \in C : \underline{M}_{C \setminus \{c\}} < \underline{M}_C\}$.

Theorem 5. If $\underline{Red}(C)$ is denoted as all the lower-approximation reduct of $S = (U, C, V, f)$, $\underline{Core}(C) = \bigcap \underline{Red}(C)$.

Proof. $\forall c \in \underline{Core}(C)$, then $\underline{M}_{C \setminus \{c\}} > \underline{M}_C$. If $\exists$ a lower approximation reduct, B, such that $c \notin B$, then $B \subseteq C \setminus \{c\}$, hence $\underline{M}_B > \underline{M}_C$. It is contradictive. On the other hand, $\forall c \in \bigcap \underline{Red}(C)$, if $c \notin \underline{Core}(C)$, then $\underline{M}_{C \setminus \{c\}} = \underline{M}_C$, thus there is subset of $C \setminus \{c\}$ which is lower approximation reduct. It is contradictive.

Definition 10. $S = (U, C, V, f)$ is a non-decision incomplete information system. Let $\underline{M}_B = (\underline{m}_{ij}^B)_{n \times n}$ be the lower approximation matrix of $B \subseteq C$. The lower approximation significance of $b \in C \setminus B$ with respect to B and $b \in B$ in B are defined respectively as

$$\underline{Sig}_B(b) = |\underline{M}_B - \underline{M}_{B \cup \{b\}}| \quad \text{and} \quad \underline{Sig}_{B \setminus \{b\}}(b) = |\underline{M}_{B \setminus \{b\}} - \underline{M}_B|,$$

where $|\underline{M}_B| = |\underline{m}_{ij}^B : \underline{m}_{ij}^B = 1, 1 \leq i, j \leq n|$.

According to Definition 5 and 6, the lower and upper approximation reduct are both minimal conditional attribute subsets, which preserve the lower and upper approximation of every decision class respectively.

Definition 11. If $\underline{Sig}_B(b) > 0$, b is called significant with respect to B. If $\underline{Sig}_{B \setminus \{b\}}(b) > 0$, b is called significant in B.

Theorem 6. $\underline{Core}(C) = \{c \in C : \underline{Sig}_{C\setminus\{c\}}(c) > 0\}$.

An algorithm to compute lower approximation reduct of incomplete data table is given as follows.

Algorithm 1.

Input: an incomplete data table $S = (U,C,V,f)$

Output: a lower approximation reduct of $S = (U,C,V,f)$

1. Compute $\underline{M}_C$;
2. Compute $\underline{Core}(C) = \{c \in C : \underline{Sig}_{C\setminus\{c\}}(c) > 0\}$. Let $B = \underline{Core}(C)$;
   (1) Judge $\underline{A}_B = \underline{A}_C$, if it holds, then go to 3;
   (2) Select $c_0 \in \{c : \underline{Sig}_B(c) = \max_{b \in C\setminus B} \underline{Sig}_B(c)\}$, and do $B = B \cup \{c_0\}$. Go to (1).
3. Let k the number of attributes added into $B = \underline{Core}(C)$ in step 2, which is denoted as $\{c_{i_j} : j = 1,2,\cdots,k\}$, where j is the added order. Judge $\underline{Sig}_{B\setminus\{c_{i_j}\}}(c_{i_j}) = 0$ in reverse order. If it holds, then $B = B \setminus \{c_{i_k}\}$.
4. Output B.

Notes: Adding attributes step by step from the core until the attribute set satisfies the condition of lower approximation consistent set. Take the significance of attributes as a measure to be added. The greater the significance is, the more important the attribute is. The attempt aims to get a reduct with the least elements. The last step of algorithm 1 checks the dispensable of every non-core attribute to ensure that the final result is a reduct.

We analyze the time complexity of Algorithm 1 as follows.

In step 1, the time complexity is $O(|C||U|^2)$. In step 2, we need to compute all $\underline{M}_{C\setminus\{c\}}(c \in C)$ and compare them with $\underline{M}_C$, so its time complexity is $O(|C|^2|U|^2)$. The time complexity of (1) is $O(|C|^2|U|^2)$; and that of (2) is $O(|C|^2|U|^2)$ since we need to find all matrices of every $\{c\} \cup B$ and compute the number of nonzero elements. In the worst case, the algorithm loops $|C|$ times and the time complexity of step 2 is $O(|C|^3|U|^2)$. Therefore, the time complexity of Algorithm 1 is $O(|C|^3|U|^2)$.

4 Similarity Matrix and Upper/lower Approximation Reduction for Incomplete Decision Table

Definition 12. $S = (U, A = C \cup D, V, f)$ is an incomplete decision table, where $U = \{x_1, x_2, \cdots, x_n\}$. Define the lower and upper generalized decision function as $\underline{d}_B(x_i) = \{f(x_j, d) : x_j \in R_B^{-1}(x_i)\}$ and $\overline{d}_B(x_i) = \{f(x_j, d) : x_j \in R_B(x_i)\}$ respectively.

Definition 13. $S = (U, A = C \cup D, V, f)$ is an incomplete decision table, where $U = \{x_1, x_2, \cdots, x_n\}$ and $B \subseteq C$. B is a relative lower approximation consistent set if $\forall x_i \in U$, $\underline{d}_C(x_i) = 1 \Rightarrow \underline{d}_B(x_i) = 1$ and B is a relative upper approximation consistent set if $\forall x_i \in U$, $\overline{d}_B(x_i) = \overline{d}_C(x_i)$.

If B is a relative lower/upper approximation consistent set and any proper subsets of B are not, B is relative lower/upper reduct.

According to the above definitions, relative lower/upper approximation reduct is the minimal condition attribution set to keep the lower/upper approximation of each decision class constant. In other words, they keep all certain and possible rules constant, respectively.

Definition 14. $S = (U, C \cup D, V, f)$ is an incomplete decision table, where $U = \{x_1, x_2, \cdots, x_n\}$. The relative lower approximation matrix of $B \subseteq C$ is defined as $\underline{D}_B = (\underline{d}_{ij}^B)_{n \times n}$, where

$$\underline{d}_{ij}^B = \begin{cases} 1 & |\underline{d}_B(x_i)| = 1, f(x_j, d) = f(x_i, d) \\ 0 & else \end{cases}$$

Definition 15. $S = (U, C \cup D, V, f)$ is an incomplete decision table, where $U = \{x_1, x_2, \cdots, x_n\}$. The relative upper approximation matrix of $B \subseteq C$ is defined as $\overline{D}_B = (\overline{d}_{ij}^B)_{n \times n}$, where

$$\overline{d}_{ij}^B = \begin{cases} 1 & f(x_j, d) \in \overline{d}_B(x_i) \\ 0 & else \end{cases}.$$

Theorem 7. (1) $B \subseteq C$ is the relative lower approximation reduct $\Leftrightarrow \underline{D}_B = \underline{D}_C$ and $\forall b \in B, \underline{D}_{B \setminus \{b\}} < \underline{D}_B$;

(2) $B \subseteq C$ is the relative upper approximation reduct $\Leftrightarrow \overline{D}_B = \overline{D}_C$ and $\forall b \in B, \overline{D}_{B \setminus \{b\}} > \overline{D}_B$.

Similarly, we give the matrix representation of relative upper and lower approximation core as follows:

Definition 16. (1) Relative lower approximation core $\underline{Core}^D(C) = \{c \in C : \underline{D}_{C\setminus\{c\}} < \underline{D}_C\}$; (2) Relative upper approximation core $\overline{Core}^D(C) = \{c \in C : \overline{D}_{C\setminus\{c\}} > \overline{D}_C\}$.

Theorem 8. $\underline{Core}^D(C) = \cap \underline{Red}^D(C)$, where $\underline{Red}^D(C)$ is all relative lower approximation reducts. $\overline{Core}^D(C) = \cap \overline{Core}^D(C)$, where $\overline{Red}^D(C)$ is all relative upper approximation reducts.

Definition 17. The relative lower and upper approximation significance of $b \in C \setminus B$ with respect to B are defined as $\underline{Sig}_B^D(b) = |\underline{D}_{B\cup\{b\}} - \underline{D}_B|$ and $\overline{Sig}_B^D(b) = |\overline{D}_B - \overline{D}_{B\cup\{b\}}|$ respectively.

An algorithm to compute relative lower approximation reduct for incomplete decision table is given as follows.

Algorithm 2

Input: an incomplete decision table $S = (U,C,V,f)$

Output: a relative lower approximation reduct of $S = (U,C,V,f)$.

1. Compute $\underline{D}_C$;
2. Compute $\underline{Core}^D(C) = \{c \in C : |\underline{D}_C - \underline{D}_{C\setminus\{c\}}| > 0\}$. Let $B = \underline{Core}^D(C)$;
   (1) Judge $\underline{D}_B = \underline{D}_C$. If it holds, go to 3;
   (2) Select $c_0 \in \{c \in C \setminus B : \underline{Sig}_B^D(b) = \underset{c \in C\setminus B}{Max}\underline{Sig}_B^D(b)\}$ and do $B = B \cup \{c_0\}$. Then go to (1).
3. Let k the number of attributes added into $B = \underline{Core}(C)$ in step 2, which is denoted as $\{c_{i_j} : j = 1,2,\cdots,k\}$, where j is the added order. Judge $\underline{Sig}_{B\setminus\{c_{i_j}\}}^D(c_{i_j}) = 0$ in reverse order. If it holds, then $B = B \setminus \{c_{i_j}\}$.
4. Output B.

The time complexity of algorithm 2 is $O(|C|^3|U|^2)$ through similar analysis with algorithm 1.

The algorithm of relative upper approximation reduct can be given in a similar way.

5 Example Analysis

An incomplete information system is given is table 1, where
$U = \{x_1, x_2, \cdots, x_{12}\}, C = \{c_1, c_2, c_3, c_4, c_5\}, D = \{d\}$.

Table. 1

U\A	c_1	c_2	c_3	c_4	c_5	d
x_1	1	0	1	1	0	1
x_2	1	0	1	*	*	1
x_3	1	1	*	*	*	2
x_4	1	*	*	*	*	2
x_5	1	*	0	*	1	1
x_6	0	1	0	0	1	2
x_7	0	1	0	*	*	2
x_8	0	0	*	*	*	1
x_9	0	*	*	*	*	1
x_{10}	0	*	1	*	0	2

Without considering decision attribute $D = \{d\}$, we use algorithm 1 to obtain the lower approximation reduct for the incomplete data table.

Compute $B = \underline{Core}(C) = \{c_1, c_2\}$, $\underline{M}_{B \cup \{c_3\}} = \underline{M}_C$, $\underline{Sig}_B\{c_3\} = \left|\underline{M}_B - \underline{M}_{B \cup \{c_3\}}\right| = 8$;

$\underline{Sig}_B\{c_4\} = \left|\underline{M}_B - \underline{M}_{B \cup \{c_4\}}\right| = 2$; $\underline{Sig}_B\{c_5\} = \left|\underline{M}_B - \underline{M}_{B \cup \{c_5\}}\right| = 10$ 。 So $B = \{c_1, c_2, c_5\}$ is the minimal reduct.

Similarly, with algorithm 2 we obtain the corresponding lower approximation relative reduct, which is $\{c_1, c_2, c_3\}$ or $\{c_1, c_2, c_5\}$.

6 Conclusions

Knowledge reduction is always a hot topic in rough set theory. However, most of knowledge reductions are based on complete information systems. Because of the existence of incomplete information in real world, the classical theory model of rough sets is extended to that based on similarity relation. In this paper, based on similarity relation, the upper/lower approximation reduction is defined and their approaches to corresponding reduction are examined. Finally, the example analysis shows the validity of this approach.

References

1. Pawlak Z. Rough sets. International Journal of Computer and Information Sciences 1982;11:341-356.
2. Ju-Sheng Mi, Wei-Zhi Wu, Wen-Xiu Zhang. Approaches to approximation reducts in inconsistent decision tables. In G. Wang et al. (Eds.): Rough Set, Fuzzy Sets, Data Mining, and Granular Computing, 2003; 283-286.
3. Wen-Xiu Zhang, Ju-Sheng Mi, Wei-Zhi Wu. Approaches to knowledge reductions in inconsistent systems. International Journal of Intelligent Systems 2003; 18: 989-1000.
4. Zhang Wen-Xiu, Leung Yee, Wu Wei-Zhi. Information systems and knowledge discovery. Beijing: Science Press; 2003.
5. Kryszkiewicz, M. Rough set approach to incomplete information systems. Information Sciences 1998; 112, 39-49.
6. R. Slowinski, D. Vanderpooten. A generalized definition of rough approximations based on similarity. IEEE Transaction on Knowledge and Data Engineering 2000; 12: 331-336.
7. Zhou Xian-zhong Huang Bing. Rough set-based attribute reduction under incomplete information systems. Journal of Nanjing University of Science & Technology, 2003; 27: 630-635.

The Relationship Among Several Knowledge Reduction Approaches

Keyun Qin[1], Zheng Pei[2], and Weifeng Du[1]

[1] Department of Applied Mathematics, Southwest Jiaotong University,
Chengdu, Sichuan 610031, China
keyunqin@263.net
[2] College of Computers & Mathematical-Physical Science,
Xihua University, Chengdu, Sichuan, 610039, China
pqyz@263.net

Abstract. This paper is devoted to the discussion of the relationship among some reduction approaches of information systems. It is proved that the distribution reduction and the entropy reduction are equivalent, and each distribute reduction is a d reduction. Furthermore, for consistent information systems, the distribution reduction, entropy reduction, maximum distribution reduction, distribute reduction, approximate reduction and d reduction are all equivalent.

1 Introduction

Rough set theory(RST), proposed by Pawlak [1], [2], is an extension of set theory for the study of intelligent systems characterized by insufficient and incomplete information. The successful application of RST in a variety of problems have amply demonstrated its usefulness. One important application of RST is the knowledge discovery in information system (decision table). RST operates on an information system which is made up of objects for which certain characteristics (i.e., condition attributes) are known. Objects with the same condition attribute values are grouped into equivalence classes or condition classes. The objects are each classified to a particular category with respect to the decision attribute value, those classified to the same category are in the same decision class. Using the concepts of lower and upper approximations in RST, the knowledge hidden in the information system may be discovered.

One fundamental aspect of RST involves the searching for some particular subsets of condition attributes. By such one subset the information for classification purpose provides is equivalent to (according to a particular standard) the condition attribute set done. Such subsets are called reducts. To acquire brief decision rules from information systems, knowledge reduction is needed.

Knowledge reduction is performed in information systems by means of the notion of a reduct based on a specialization of the general notion of independence due to Marczewski [3]. In recent years, more attention has been paid to knowledge reduction in information systems in rough set research. Many types of knowledge reduction and their applications have been proposed for inconsistent information systems in the area of rough sets [4], [5], [6], [7], [8], [9], [10], [11], [12]. The first knowledge reduction approach due to[6] which is carry out through discernibility matrixes and discernibility functions. This kind of reduction is based on the positive region of the universe and we call it d reduction. For inconsistent information systems, Kryszkiewicz [7] proposed the concepts of distribution reduction and distribute reduction.Zhang [5] proposed the concepts of maximum distribution reduction and approximate reduction and provide new approaches to knowledge reduction in inconsistent information systems. Furthermore, some approaches to knowledge reduction based on variable precision rough set model were proposed [13]. Information entropy is a measure of information involved in a system. Based on conditional information entropy, some knowledge reduction approaches in information systems were proposed in [4].

This paper is devoted to the discussion of the relationship among some reduction approaches of information systems. It is proved that the distribution reduction and the entropy reduction are equivalent, and each distribute reduction is a d reduction. Furthermore, for consistent information systems, the distribution reduction, entropy reduction, maximum distribution reduction, distribute reduction, approximate reduction and d reduction are all equivalent.

2 Preliminaries and Notations

An information system is a quadruple $S = (U, AT \cup \{d\}, V, f)$, where

(1) U is a non-empty finite set and its elements are called objects of S.
(2) AT is the set of condition attributes and d is the decision attribute of S.
(3) $V = \cup_{q \in AT \cup \{d\}} V_q$, where V_q is a non-empty set of values of attribute $q \in AT \cup \{d\}$, called domain of the attribute q.
(4) $f : U \to AT \cup \{d\}$ is a mapping, called description function of S, such that $f(x, q) \in V_q$ for each $(x, q) \in U \times (AT \cup \{d\})$.

Let $S = (U, AT \cup \{d\}, V, f)$ be an information system and $A \subseteq AT$. The discernibility relation $ind(A)$ on U derived from A, defined by $(x, y) \in ind(A)$ if and only if $\forall a \in A, f(x, a) = f(y, a)$, is an equivalent relation and hence $(U, ind(A))$ is a Pawlak approximation space. We denote by $[x]_A$ the equivalent class with respect to $ind(A)$ that containing x and U/A the set of these equivalent classes. For each $X \subseteq U$, according to Pawlak [1], the upper approximation $\overline{A}(X)$ and lower approximation $\underline{A}(X)$ of X with respect to A are defined as

$$\overline{A}(X) = \{x \in U | [x]_A \cap X \neq \emptyset\}, \quad \underline{A}(X) = \{x \in U | [x]_A \subseteq X\}. \tag{1}$$

Based on the approximation operators, Skowron proposed the concept of d reduction of an information system.

Definition 1. *Let $S = (U, AT \cup \{d\}, V, f)$ be an information system and $A \subseteq AT$. The positive region $pos_A(d)$ of d with respect to A is defined as*

$$pos_A(d) = \cup_{X \in U/d} \underline{A}(X).$$

Definition 2. Let $S = (U, AT \cup \{d\}, V, f)$ be an information system and $A \subseteq AT$. A is called a d consistent subset of S if $pos_A(d) = pos_{AT}(d)$. A is called a d reduction of S if A is a d consistent subset of S and each proper subset of A is not a d consistent subset of S.

All the d reductions can be carry out through discernibility matrixes and discernibility functions [6]. Let $S = (U, AT \cup \{d\}, V, f)$ be an information system and $B \subseteq AT$, $x \in U$. We introduce the following notations:

$U/d = \{D_1, \cdots, D_r\}; \quad \mu_B(x) = (D(D_1/[x]_B), \cdots, D(D_r/[x]_B));$

$\gamma_B(x) = \{D_j; D(D_j/[x]_B) = max_{q \leq r} D(D_q/[x]_B)\}; \delta_B(x) = \{D_j; D_j \cap [x]_B \neq \emptyset\};$

$\eta_B = \frac{1}{|U|} \Sigma_{j=1}^r |\overline{B}(D_j)|;$

where $D(D_j/[x]_B) = \frac{|D_j \cap [x]_B|}{|[x]_B|}$ is the include degree of $[x]_B$ in D_j.

For inconsistent information systems, Kryszkiewicz [7] proposed the concepts of distribution reduction and distribute reduction. Based on this work, Zhang [5] proposed the concepts of maximum distribution reduction and approximate reduction. Farther more, the judgement theorems and discernibility matrixes with respect to those reductions are obtained. These reductions are based on the concept of include degree.

Definition 3. Let $S = (U, AT \cup \{d\}, V, f)$ be an information system, $A \subseteq AT$.

(1) A is called a distribution consistent set of S if $\mu_A(x) = \mu_{AT}(x)$ for each $x \in U$. A is called a distribution reduction of S if A is a distribution consistent set of S and no proper subset of A is distribution consistent set of S.

(2) A is called a maximum distribution consistent set of S if $\gamma_A(x) = \gamma_{AT}(x)$ for each $x \in U$. A is called a maximum distribution reduction of S if A is a maximum distribution consistent set of S and no proper subset of A is maximum distribution consistent set of S.

(3) A is called a distribute consistent set of S if $\delta_A(x) = \delta_{AT}(x)$ for each $x \in U$. A is called a distribute reduction of S if A is a distribute consistent set of S and no proper subset of A is distribute consistent set of S.

(4) A is called a approximate consistent set of S if $\eta_A = \eta_{AT}$. A is called a approximate reduction of S if A is a approximate consistent set of S and no proper subset of A is approximate consistent set of S.

[5] proved that the concepts of distribute consistent set and approximate consistent set are equivalent, a distribution consistent set must be a distribute consistent set and a maximum distribution consistent set.

Let $S = (U, AT \cup \{d\}, V, f)$ be an information system, $A \subseteq AT$ and

$U/AT = \{X_i; 1 \leq i \leq n\}, \quad U/A = \{Y_j; 1 \leq j \leq m\}, \quad U/d = \{Z_l; 1 \leq l \leq k\}.$

The conditional information entropy $H(d|A)$ of d with respect to A is defined

$$H(d|A) = -\sum_{j=1}^m (p(Y_j) \cdot \sum_{l=1}^k p(Z_l|Y_j) log(p(Z_l|Y_j))),$$

where $p(Y_j) = \frac{|Y_j|}{|U|}$, $p(Z_l|Y_j) = \frac{|Z_l \cap Y_j|}{|Y_j|}$ and $0 \log 0 = 0$.

Based on conditional information entropy, Wang [4] proposed the concept of entropy reduction for information systems.

Definition 4. Let $S = (U, AT \cup \{d\}, V, f)$ be an information system and $A \subseteq AT$. A is called an entropy consistent set of S if $H(d|A) = H(d|AT)$. A is called an entropy reduction of S if A is an entropy consistent set of S and non proper subset of A is entropy consistent set of S.

3 The Relationship Among Knowledge Reduction Approaches

In this section, we discuss the relationship among knowledge reduction approaches. In what follows we assume that $S = (U, AT \cup \{d\}, V, f)$ is an information system, $A \subseteq AT$ and

$$U/AT = \{X_i; 1 \le i \le n\}, \quad U/A = \{Y_j; 1 \le j \le m\}, \quad U/d = \{Z_l; 1 \le l \le k\}.$$

Theorem 5. Let $Y_j = \cup_{t \in T_j} X_t$, $1 \le j \le m$, where T_j is an index set. For each $1 \le l \le k$,

$$|Z_l \cap Y_j| \log(\frac{|Z_l \cap Y_j|}{|Y_j|}) \le \sum_{t \in T_j} |Z_l \cap X_t| \log(\frac{|Z_l \cap X_t|}{|X_t|}).$$

Proof. Let $T_{j1} = \{X_t; t \in T_j, Z_l \cap X_t \ne \emptyset\}$. If $T_{j1} = \emptyset$, then $Z_l \cap X_t = \emptyset$ for each $t \in T_j$ and hence $Z_l \cap Y_j = \emptyset$, the conclusion holds. If $T_{j1} \ne \emptyset$, by $\ln x \le x - 1$, it follows that

$$|Z_l \cap Y_j| \log(\frac{|Z_l \cap Y_j|}{|Y_j|}) - \sum_{t \in T_j} |Z_l \cap X_t| \log(\frac{|Z_l \cap X_t|}{|X_t|})$$

$$= \sum_{t \in T_j} |Z_l \cap X_t| \log(\frac{|Z_l \cap Y_j|}{|Y_j|}) - \sum_{t \in T_j} |Z_l \cap X_t| \log(\frac{|Z_l \cap X_t|}{|X_t|})$$

$$= \sum_{t \in T_{j1}} |Z_l \cap X_t| \log(\frac{|Z_l \cap Y_j||X_t|}{|Y_j||Z_l \cap X_t|})$$

$$\le \sum_{t \in T_{j1}} |Z_l \cap X_t| (\frac{|Z_l \cap Y_j||X_t|}{|Y_j||Z_l \cap X_t|} - 1) \log e$$

$$= \frac{\log e}{|Y_j|} \sum_{t \in T_{j1}} (|Z_l \cap Y_j||X_t| - |Y_j||Z_l \cap X_t|)$$

$$= \frac{\log e}{|Y_j|} (\sum_{t \in T_{j1}} (|Z_l \cap Y_j||X_t| - \sum_{t \in T_{j1}} |Y_j||Z_l \cap X_t|)$$

$$= \frac{\log e}{|Y_j|} (|Z_l \cap Y_j|(|Y_j| - \sum_{t \in T_j - T_{j1}} |X_t|) - |Y_j||Z_l \cap Y_j|) \le 0.$$

Theorem 6. $H(d|AT) = H(d|A)$ if and only if

$$|Z_l \cap Y_j|log(\frac{|Z_l \cap Y_j|}{|Y_j|}) = \sum_{t \in T_j}|Z_l \cap X_t|log(\frac{|Z_l \cap X_t|}{|X_t|}),$$

for each $1 \leq l \leq k$ and $1 \leq j \leq m$, where $Y_j = \cup_{t \in T_j} X_t$, $1 \leq j \leq m$, and T_j is an index set.

Proof.

$$H(d|AT) = -\sum_{i=1}^{n}(p(X_i) \cdot \sum_{l=1}^{k} p(Z_l|X_i)log(p(Z_l|X_i)))$$

$$= -\frac{1}{|U|}\sum_{l=1}^{k}\sum_{i=1}^{n}(p(X_i)|Z_l \cap X_i|log(\frac{|Z_l \cap X_i|}{|X_i|}),$$

$$H(d|A) = -\sum_{j=1}^{m}(p(Y_j) \cdot \sum_{l=1}^{k} p(Z_l|Y_j)log(p(Z_l|Y_j)))$$

$$= -\frac{1}{|U|}\sum_{l=1}^{k}\sum_{j=1}^{m}(p(X_i)|Z_l \cap Y_j|log(\frac{|Z_l \cap Y_j|}{|Y_j|}).$$

The sufficiency is trivial because each A equivalent class is just a union of some AT equivalent classes.

Necessity: For each $1 \leq l \leq k$ and $1 \leq j \leq m$, by Theorem 5,

$$|Z_l \cap Y_j|log(\frac{|Z_l \cap Y_j|}{|Y_j|}) \leq \sum_{t \in T_j}|Z_l \cap X_t|log(\frac{|Z_l \cap X_t|}{|X_t|}),$$

and hence

$$\sum_{j=1}^{m}|Z_l \cap Y_j|log(\frac{|Z_l \cap Y_j|}{|Y_j|}) \leq \sum_{i=1}^{n}|Z_l \cap X_t|log(\frac{|Z_l \cap X_t|}{|X_t|}).$$

If there exists $1 \leq l \leq k$ and $1 \leq j \leq m$ such that

$$|Z_l \cap Y_j|log(\frac{|Z_l \cap Y_j|}{|Y_j|}) < \sum_{t \in T_j}|Z_l \cap X_t|log(\frac{|Z_l \cap X_t|}{|X_t|}).$$

consequently,

$$\sum_{j=1}^{m}|Z_l \cap Y_j|log(\frac{|Z_l \cap Y_j|}{|Y_j|}) < \sum_{i=1}^{n}|Z_l \cap X_t|log(\frac{|Z_l \cap X_t|}{|X_t|}).$$

and hence $H(d|AT) < H(d|A)$, a contradiction.

Theorem 7. $A \subseteq AT$ is a distribution reduction of S if and only if A is an entropy reduction of S.

Proof. It needs only to prove that A is a distribution consistent set if and only if A is an entropy consistent set.

Sufficiency: Assume that A is a distribution consistent set. For each $1 \leq l \leq k$ and $1 \leq j \leq m$, let $Y_j = [x]_A$. We notice that $J_{x,A} = \{[y]_{AT}; [y]_{AT} \subseteq [x]_A\}$ is a partition of $[x]_A$. By $|Z_l \cap [x]_A| = \sum_{[y]_{AT} \in J_{x,A}} |Z_l \cap [y]_{AT}|$, it follows that

$$\frac{|Z_l \cap [y]_{AT}|}{|[y]_{AT}|} = \frac{|Z_l \cap [y]_A|}{|[y]_A|} = \frac{|Z_l \cap [x]_A|}{|[x]_A|},$$

for each $[y]_{AT} \in J_{x,A}$ and hence

$$|Z_l \cap [x]_A| log(\frac{|Z_l \cap [x]_A|}{|[x]_A|}) = \sum_{[y]_{AT} \in J_{x,A}} |Z_l \cap [y]_{AT}| log(\frac{|Z_l \cap [y]_{AT}|}{|[y]_{AT}|}),$$

it follows by Theorem 6 that $H(d|AT) = H(d|A)$ and A is an entropy consistent set.

Necessity: Assume that A is an entropy consistent set. For each $x \in U$, $J_{x,A} = \{[y]_{AT}; [y]_{AT} \subseteq [x]_A\}$ forms a partition of $[x]_A$. Let $J = J_{x,A} - \{[x]_{AT}\}$. For each $1 \leq l \leq k$, by Theorem 6, it follows that

$$|Z_l \cap [x]_A| log(\frac{|Z_l \cap [x]_A|}{|[x]_A|}) = \sum_{[y]_{AT} \in J_{x,A}} |Z_l \cap [y]_{AT}| log(\frac{|Z_l \cap [y]_{AT}|}{|[y]_{AT}|}).$$

and hence

$$\sum_{[y]_{AT} \in J_{x,A}} |Z_l \cap [y]_{AT}| log(\frac{|Z_l \cap [x]_A|}{|[x]_A|}) = \sum_{[y]_{AT} \in J_{x,A}} |Z_l \cap [y]_{AT}| log(\frac{|Z_l \cap [y]_{AT}|}{|[y]_{AT}|}),$$

$$\sum_{[y]_{AT} \in J_{x,A}} |Z_l \cap [y]_{AT}| log(\frac{|Z_l \cap [x]_A||[y]_{AT}|}{|[x]_A||Z_l \cap [y]_{AT}|}) = 0,$$

that is,

$$-|Z_l \cap [x]_{AT}| log(\frac{|Z_l \cap [x]_A||[x]_{AT}|}{|[x]_A||Z_l \cap [x]_{AT}|})$$

$$= \sum_{[y]_{AT} \in J} |Z_l \cap [y]_{AT}| log(\frac{|Z_l \cap [x]_A||[y]_{AT}|}{|[x]_A||Z_l \cap [y]_{AT}|})$$

$$\leq \sum_{[y]_{AT} \in J} |Z_l \cap [y]_{AT}|(\frac{|Z_l \cap [x]_A||[y]_{AT}|}{|[x]_A||Z_l \cap [y]_{AT}|} - 1)log e$$

$$= \frac{log e}{|[x]_A|} \sum_{[y]_{AT} \in J} (|Z_l \cap [x]_A||[y]_{AT}| - |[x]_A||Z_l \cap [y]_{AT}|)$$

$$= \frac{log e}{|[x]_A|}(|Z_l \cap [x]_A|(|[x]_A| - |[x]_{AT}|) - |[x]_A|(|Z_l \cap [x]_A| - |Z_l \cap [x]_{AT}|))$$

$$= \frac{log e}{|[x]_A|}(|[x]_A||Z_l \cap [x]_{AT}| - |[x]_{AT}||Z_l \cap [x]_A|).$$

It follows that
$$\ln\left(\frac{|Z_l \cap [x]_A||[x]_{AT}|}{|[x]_A||Z_l \cap [x]_{AT}|}\right) \geq \frac{|Z_l \cap [x]_A||[x]_{AT}|}{|[x]_A||Z_l \cap [x]_{AT}|} - 1.$$

By $\ln a \leq a - 1$ for each $a > 0$,
$$\ln\left(\frac{|Z_l \cap [x]_A||[x]_{AT}|}{|[x]_A||Z_l \cap [x]_{AT}|}\right) = \frac{|Z_l \cap [x]_A||[x]_{AT}|}{|[x]_A||Z_l \cap [x]_{AT}|} - 1,$$

and hence
$$\frac{|Z_l \cap [x]_A||[x]_{AT}|}{|[x]_A||Z_l \cap [x]_{AT}|} = 1,$$

because $a = 1$ is the unique root of $\ln a = a - 1$, that is
$$\frac{|[x]_{AT} \cap Z_l|}{|[x]_{AT}|} = \frac{|[x]_A \cap Z_l|}{|[x]_A|}$$

and A is a distribution consistent set.

Theorem 8. *A is an entropy consistent set if and only if*
$$\frac{|[x]_{AT} \cap [x]_d|}{|[x]_{AT}|} = \frac{|[x]_A \cap [x]_d|}{|[x]_A|}$$

for each $x \in U$.

Proof. By Theorem 7, the necessity is trivial.

Sufficiency: We prove that
$$|Z_l \cap Y_j| \log\left(\frac{|Z_l \cap Y_j|}{|Y_j|}\right) = \sum_{t \in T_j} |Z_l \cap X_t| \log\left(\frac{|Z_l \cap X_t|}{|X_t|}\right),$$

for any $1 \leq l \leq k$ and $1 \leq j \leq m$ and finish the proof by Theorem 6, where $Y_j = \cup_{t \in T_j} X_t$.

If $Z_l \cap Y_j = \emptyset$, then $Z_l \cap X_t = \emptyset$ for each $t \in T_j$ and the conclusion holds.

If $Z_l \cap Y_j \neq \emptyset$, suppose that $x \in Z_l \cap Y_j$, it follows that $Z_l = [x]_d$ and $Y_j = [x]_A$. Let $J_{x,A} = \{[z]_{AT}; [z]_{AT} \subseteq [x]_A\}$. Consequently,

$$|Z_l \cap Y_j| \log\left(\frac{|Z_l \cap Y_j|}{|Y_j|}\right) - \sum_{t \in T_j} |Z_l \cap X_t| \log\left(\frac{|Z_l \cap X_t|}{|X_t|}\right)$$

$$= |[x]_d \cap [x]_A| \log\left(\frac{|[x]_d \cap [x]_A|}{|[x]_A|}\right) - \sum_{[z]_{AT} \in J_{x,A}} |[x]_d \cap [z]_{AT}| \log\left(\frac{|[x]_d \cap [z]_{AT}|}{|[z]_{AT}|}\right)$$

$$= \sum_{[z]_{AT} \in J_{x,A}} |[x]_d \cap [z]_{AT}| \log\left(\frac{|[x]_d \cap [x]_A|}{|[x]_A|}\right) - \sum_{[z]_{AT} \in J_{x,A}} |[x]_d \cap [z]_{AT}| \log\left(\frac{|[x]_d \cap [z]_{AT}|}{|[z]_{AT}|}\right)$$

$$= \sum_{[z]_{AT} \in J_{x,A}} |[x]_d \cap [z]_{AT}| \log\left(\frac{|[x]_d \cap [x]_A||[z]_{AT}|}{|[x]_A||[x]_d \cap [z]_{AT}|}\right).$$

Assume that $u \in [x]_d \cap [z]_{AT}$, it follows that $u \in [x]_A$ and hence $[x]_d = [u]_d$, $[z]_{AT} = [u]_{AT}$ and $[x]_A = [u]_A$, consequently,

$$\frac{|[x]_d \cap [x]_A||[z]_{AT}|}{|[x]_A||[x]_d \cap [z]_{AT}|} = \frac{|[u]_d \cap [u]_A||[u]_{AT}|}{|[u]_A||[u]_d \cap [u]_{AT}|} = 1,$$

and hence

$$|Z_l \cap Y_j| \log(\frac{|Z_l \cap Y_j|}{|Y_j|}) = \sum_{t \in T_j} |Z_l \cap X_t| \log(\frac{|Z_l \cap X_t|}{|X_t|}).$$

Theorem 9. Let $S = (U, AT \cup \{d\}, V, f)$ be an information system and $A \subseteq AT$. If A is a distribute consistent set of S, then A is a d consistent set of S.

Proof. Let $A \subseteq AT$ be a distribute consistent set of S and $U/d = \{D_1, D_2, \cdots, D_r\}$. For each $x \in pos_{AT}(d)$, it follows that $[x]_{AT} \subseteq [x]_d$ and hence $\delta_{AT}(x) = \{D_j; D_j \cap [x]_{AT} \neq \emptyset\} = \{[x]_d\} = \delta_A(x)$. Assume that $[x]_d = D_j$, it follows that $D_l \cap [x]_A = \emptyset$ for each $l \leq r, l \neq j$, that is $[x]_A \subseteq D_j = [x]_d$ and $x \in \underline{A}([x]_d) \subseteq pos_A(d)$, it follows that $pos_{AT}(d) \subseteq pos_A(d)$.

$pos_A(d) \subseteq pos_{AT}(d)$ is trivial.

4 Knowledge Reduction for Consistent Information Systems

An information system $S = (U, AT \cup \{d\}, V, f)$ is called to be consistent, if $[x]_{AT} \subseteq [x]_d$ for each $x \in U$. In this section, we discuss knowledge reductions for consistent information systems. We will prove that the concepts of distribution reduction, approximate reduction and d reduction are equivalent for consistent information systems.

Theorem 10. Let $S = (U, AT \cup \{d\}, V, f)$ be an information system. S is consistent if and only if $pos_{AT}(d) = U$.

Proof. If S is consistent, then $[x]_{AT} \subseteq [x]_d$ for each $x \in U$ and hence $x \in \underline{AT}([x]_d) \subseteq \cup_{X \in U/d} \underline{AT}(X) = pos_{AT}(d)$, that is $pos_{AT}(d) = U$.

If $pos_{AT}(d) = U$, then $x \in pos_{AT}(d)$ for each $x \in U$ and hence $x \in \underline{AT}([x]_d)$, that is $[x]_{AT} \subseteq [x]_d$.

Theorem 11. Let $S = (U, AT \cup \{d\}, V, f)$ be an information system. S is consistent if and only if $\delta_{AT}(x) = \{[x]_d\}$ for each $x \in U$.

Proof. If S is consistent, then $[x]_{AT} \subseteq [x]_d$ for each $x \in U$ and hence $\delta_{AT}(x) = \{[x]_d\}$.

If $\delta_{AT}(x) = \{[x]_d\}$ for each $x \in U$, then $[x]_{AT} \cap [y]_d = \emptyset$ for each $[y]_d \neq [x]_d$, that is $[x]_{AT} \subseteq [x]_d$ and S is consistent.

Theorem 12. Let $S = (U, AT \cup \{d\}, V, f)$ be an information system. S is consistent if and only if $H(d|AT) = 0$.

Proof. Assume that

$$U/AT = \{X_1, X_2, \cdots, X_n\}, \quad U/d = \{Y_1, Y_2, \cdots, Y_m\}.$$

It follows that

$$H(d|AT) = -\sum_{i=1}^{n}(p(X_i) \cdot \sum_{j=1}^{m} p(Y_j|X_i)log(p(Y_j|X_i)))$$

$$= -\sum_{i=1}^{n}\sum_{j=1}^{m} \frac{|Y_j \cap X_i|}{|U|} log(\frac{|Y_j \cap X_i|}{|X_i|}).$$

If S is consistent, then for each $i(1 \leq i \leq n)$, there exists unique $j(1 \leq j \leq m)$ such that $X_i \subseteq Y_j$, and hence $\frac{|Y_j \cap X_i|}{|X_i|} = 1$ or $\frac{|Y_j \cap X_i|}{|X_i|} = 0$, consequently, $H(d|AT) = 0$.

If $H(d|AT) = 0$, then

$$-\sum_{i=1}^{n}\sum_{j=1}^{m} \frac{|Y_j \cap X_i|}{|U|} log(\frac{|Y_j \cap X_i|}{|X_i|}) = 0,$$

it follows that

$$\sum_{j=1}^{m} \frac{|Y_j \cap X_i|}{|U|} log(\frac{|Y_j \cap X_i|}{|X_i|}) = 0,$$

for each $i(1 \leq i \leq n)$, that is there exists $j(1 \leq j \leq m)$ such that $X_i \subseteq Y_j$, and S is consistent.

Theorem 13. *Let* $S = (U, AT \cup \{d\}, V, f)$ *be a consistent information system and* $A \subseteq AT$.

(1) *A is an entropy consistent set if and only if $S' = (U, A \cup \{d\}, V, f)$ is consistent.*

(2) *A is a approximate consistent set if and only if $S' = (U, A \cup \{d\}, V, f)$ is consistent.*

(3) *A is a positive domain consistent set if and only if $S' = (U, A \cup \{d\}, V, f)$ is consistent.*

By this Theorem, for consistent information systems, the concepts of distribution reduction, entropy reduction, maximum distribution reduction, distribute reduction, approximate reduction and d reduction are all equivalent.

Acknowledgements

The authors are grateful to the referees for their valuable comments and suggestions. This work has been supported by the National Natural Science Foundation of China (Grant No. 60474022).

References

1. Pawlak, Z.: Rough Sets. International Journal of Computer and Information Science. **11** (1982) 341–356
2. Pawlak, Z.(ed.): Rough Sets: Theoretical Aspects of Reasoning About Data. Kluwer Academic Publishers, Boston (1991)
3. Marczewski.: A General Scheme of Independence in Mathematics. Bulletin de L Academie Polonaise des Sciences–Serie des Sciences Mathematiques Astronomiques et Physiques. **6** (1958) 731–736
4. Wang, G, Y., Yu, H., Yang, D, C.: Decision Table Reduction Based on Conditional Information Entropy. Chinese Journal of Computers (in Chinese). **25** (2002) 759–766
5. Zhang, W, X., Mi, J, S., Wu, W, Z.: Knowledge Reductions in Inconsistent Information Systems. Chinese Journal of Computers (in Chinese). **26** (2003) 12–18
6. Skowron, A., Rauszer, C.: The Discernibility Matrices And Functions in Information System. Intelligent Decision Support Handbook of Applications and Advances of the Rough Sets Theory, Kluwer Academic Publishers, Dordrecht (1992)
7. Kryszkiewicz.: Comparative Study of Alternative Type of Knowledge Reduction in Inconsistent Systems. International Journal of General Systems. **16** (2001) 105–120
8. Beynon, M.: Reducts within The Variable Precision Rough Set Model: A Further Investigation. European Journal of Operational Research. **134** (2001) 592–605
9. Quafatou, M.: RST: A Generalization of Rough Set Theory. Information Sciences. **124** (2000) 301–316
10. Zheng, P., Keyun, Q.: Obtaining Decision Rules And Combining Evidence Based on Modal Logic. Progress in Natural Science (in chinese). **14** (2004) 501–508
11. Zheng, P., Keyun, Q.: Intuitionistic Special Set Expression of Rough Set And Its Application in Reduction of Attributes. Pattern Recognition and Artificial Intelligence(in chinese). **17** (2004) 262–266
12. Slowinski, R., Zopounidis, Dimitras, A. I.: Prediction of Company Acquisition in Greece by Means of The Rough Set Approach. European Journal of Operational Research. **100** (1997) 1–15
13. Jusheng, M., Weizhi, W., Wenxiu, Z.: Approaches to Knowledge Reduction Based on Variable Precision Rough Set Model. Information Sciences. **159** (2004) 255–272

Rough Approximation of a Preference Relation for Stochastic Multi-attribute Decision Problems

Chaoyuan Yue, Shengbao Yao, Peng Zhang, and Wanan Cui

Department of Control Science and Engineering,
Huazhong University of Science and Technology,Wuhan, Hubei, 430074, China

Abstract. Multi-attribute decision problems where the performances of the alternatives are random variables are considered in this paper. The suggested approach grades the probabilities of preference of one alternative over another with respect to the same attribute. Based on the graded probabilistic dominance relation, the pairwise comparison information table is defined. The global preferences of the decision maker can be seen as a rough binary relation. The present paper proposes to approximate this preference relation by means of the graded probabilistic dominance relation with respect to the subsets of attributes.

1 Introduction

Multi-attribute decision making (MADM) is widely applied in many fields such as military affairs, economy and management. When dealing with MADM, it is usual to be confronted to a context of uncertainty. We suppose in this paper that uncertainty is due to the fact that performance evaluations of alternatives on each of the attributes lead to random variables with probability distribution. This kind of problems are called stochastic multi-attribute decision making.

Based on the results in [1]-[3], this short paper presents a method to solve above-mentioned problems. In our approach, we define a dominance relation by grading the probabilities of preference of one alternative over another with respect to the same attribute. The global preference of the DM is approximated by means of the graded probabilistic dominance relation with respect to the subsets of attributes.

This paper is organized as follows. In the next section graded probabilistic dominance relation about attribute is introduced. The global preference is approximated by means of the graded probabilistic dominance relation. Section 3 is devoted to generation of decision rules and section 4 groups conclusion.

2 Rough Approximation of a Preference Relation

The multi-attribute problem that is considered in this paper can be represented by an $<A, Q, E>$ model, where A is a finite set of potential alternatives $a_i (i = 1, 2, \cdots, n)$, $Q = \{q_1, q_2, \cdots, q_n\}$ is a finite set of attributes and E is the set of evaluations X_{ik} expressed by the probability density function $f_{ik}(x_{ik})$ which associates the performance of alternative a_i with respect to the attribute q_k.

For any alternatives a_i and a_j in A, the possibility of preference of a_i over a_j with respect to q_k can be quantified by the probability $p_{ij}^k = p(X_{ik} \geq X_{jk})$. Suppose that random variables X_{ik} and X_{jk} are independent, we have

$$p_{ij}^k = \int\int_{x_{ik} \geq x_{jk}} f_{ij}^k(x_{ik}, x_{jk}) dx_{ik} dx_{jk} = \int\int_{x_{ik} \geq x_{jk}} f_{ik}(x_{ik}) f_{jk}(x_{jk}) dx_{ik} dx_{jk}$$

where $f_{ij}^k(x_{ik}, x_{jk})$ is the joint probability density function of random variables X_{ik} and X_{jk}. Notice that $0 \leq p_{ij}^k \leq 1$ and $p_{ij}^k + p_{ji}^k = 1$ hold.

p_{ij}^k measures the strength of preference of a_i over a_j with respect to q_k. In order to distinguish the strength, we propose to grade the preference according to the value of p_{ij}^k. The following set I of interval is defined:

$$I = \{[0, 0.15], (0.15, 0.3], (0.3, 0.45], (0.45, 0.55), [0.55, 0.7), [0.7, 0.85], [0.85, 1.0]\}$$

In term of above partition, we define the set T_{q_k} of binary graded probabilistic dominance relations on A:

$$T_{q_k} = \{D_{q_k}^h, h \in H\}$$

where $q_k \in Q, H = \{-3, -2, -1, 0, 1, 2, 3\}$. The degree h in the relation $a_i D_{q_k}^h a_j$ corresponds to the interval that p_{ij}^k belongs to one by one. Due to $p_{ij}^k + p_{ji}^k = 1$, $\forall a_i, a_j \in A$, we have

$$a_i D_{q_k}^h a_j \iff a_j D_{q_k}^{-h} a_i.$$

In order to represent preferential information provided by the DM, we shall use the pairwise comparison table (PCT) introduced by Greco et al [2].

The preferential information concerns a set $B \subset A$ of, so called, reference actions, with respect to which the DM is willing to express his/her attitude through pairwise comparisons. Let Q be the set of attributes (condition attributes) describing the alternatives, and D, the decision attribute. The decision table is defined as 4-tuple: $T = <U, Q \cup D, V_Q \cup V_D, g>$, where $U \subseteq B \times B$ is a finite set of pairs of alternatives, $V_Q = \{V_{q_k}, q_k \in Q\}$ is the domains of the condition attributes and V_D is the domain of the decision attribute, and $g: U \times (Q \cup D) \longrightarrow V_Q \cup V_D$ is a total function. This function is such that:

(1) $g[(a_i, a_j), q] = h$, if $a_i D_q^h a_j, q \in Q, (a_i, a_j) \in U$;
(2) $g[(a_i, a_j), D] = P$, if a_i is preferred to $a_j, (a_i, a_j) \in U$;
(3) $g[(a_i, a_j), D] = N$, if a_i is not preferred to $a_j, (a_i, a_j) \in U$,

where "a_i is preferred to a_j" means "a_i is at least as good as a_j". Generally, the decision table can be presented as in Table 1.

Table 1. Pairwise comparison table.

	q_1	q_2	$\cdots$	q_m	D
(a_i, a_j)	$g[(a_i, a_j), q_1]$	$g[(a_i, a_j), q_2]$	$\cdots$	$g[(a_i, a_j), q_m]$	$g[(a_i, a_j), D] = P$
$\cdots$	$\cdots$	$\cdots$	$\cdots$	$\cdots$	$\cdots$
(a_s, a_t)	$g[(a_s, a_t), q_1]$	$g[(a_s, a_t), q_2]$	$\cdots$	$g[(a_s, a_t), q_m]$	$g[(a_s, a_t), D] = N$

The binary relation P defined on A is the comprehensive preference relation of the DM. In this paper, it is supposed that P is a complete and antisymmetric binary relation on B, i.e., 1)$\forall a_i, a_j \in B$, $a_i P a_j$ and/or $a_j P a_i$; 2) both $a_i P a_j$ and $a_j P a_i$ implies $a_i = a_j$.

In order to approximate the global preference of the DM, the following dominance relation with respect to the subset of the condition attributes is defined:

Definition 3.1. Given $a_i, a_j \in A$, a_i positively dominates a_j by degree h with respect to the set of attributes R, denoted by $a_i D^h_{+R} a_j$, if and only if $a_i D^f_{q_k} a_j$ with $f \geq h$, $\forall q_k \in R$; a_i negatively dominates a_j by degree h with respect to the set of attributes R, denoted by $a_i D^h_{-R} a_j$, if and only if $a_i D^f_{q_k} a_j$ with $f \leq h$, $\forall q_k \in Q$.

The above defined dominance relations satisfy the following property:

Property 3.1. If $(a_i, a_j) \in D^h_{+R}$, then $(a_i, a_j) \in D^k_{+S}$ for each $S \subseteq R, k \leq h$; If $(a_i, a_j) \in D^h_{-R}$, then $(a_i, a_j) \in D^k_{-S}$ for each $S \subseteq R, k \geq h$.

In the suggested approach, we propose to approximate the global preference relation by D^h_{+R} and D^h_{-R} dominance relations. Therefore, P is seen as a rough binary relation. The lower approximation of preferences, denoted by $R_*(P)$ and $R_*(N)$, and the upper approximation of preferences, denoted by $R^*(P)$ and $R^*(N)$, are respectively defined as:

$$R_*(P) = \bigcup_{h \in H} \{(D^h_{+R}) \cap U \subseteq P\}, R^*(P) = \bigcap_{h \in H} \{(D^h_{+R}) \cap U \supseteq P\};$$

$$R_*(N) = \bigcup_{h \in H} \{(D^h_{-R}) \cap U \subseteq N\}, R^*(N) = \bigcap_{h \in H} \{(D^h_{-R}) \cap U \supseteq N\}.$$

Let

$h^+_{min}(R) = min\{h \in H : (D^h_{+R}) \cap U \subseteq P\}$, $h^+_{max}(R) = max\{h \in H : (D^h_{+R}) \cap U \supseteq P\}$

$h^-_{min}(R) = min\{h \in H : (D^h_{-R}) \cap U \supseteq N\}$, $h^-_{max}(R) = max\{h \in H : (D^h_{-R}) \cap U \subseteq N\}$

According to the property 3.1 and the definitions of the approximation of the preferences, the following conclusion can be obtained.

Theorem 3.1.

$$R_*(P) = D^{h^+_{min}(R)}_{+R}) \cap U, R^*(P) = D^{h^+_{max}(R)}_{+R}) \cap U;$$

$$R_*(N) = D^{h^-_{max}(R)}_{-R}) \cap U, R^*(N) = D^{h^-_{min}(R)}_{-R}) \cap U.$$

3 Decision Rules

The rough approximations of preferences can serve to induce a generalized description of alternative contained in the information table in term of "if $\cdots$, then $\cdots$" decision rules. We will consider the following two kinds of decision rules:

1. If $a_i D^h_{+R} a_j$, then $a_i P a_j$, denoted by $a_i D^h_{+R} a_j \Longrightarrow a_i P a_j$;
2. If $a_i D^h_{-R} a_j$, then $a_i N a_j$, denoted by $a_i D^h_{-R} a_j \Longrightarrow a_i N a_j$.

Definition 4.1. If there is at least one pair $(a_u, a_v) \in D^h_{+R} \cap U[D^h_{-R} \cap U]$ such that $a_u P a_v [a_u N a_v]$, and $a_s P a_t [a_s N a_t]$ holds for each pair $(a_s, a_t) \in D^h_{+R} \cap U[D^h_{-R} \cap U]$, then $a_i D^h_{+R} a_j \Longrightarrow a_i P a_j [a_i D^h_{-R} a_j \Longrightarrow a_i N a_j]$ is accept as a D_{++}-decision rule [D_{---}-decision rule].

Definition 4.2. A D_{++}-decision rule $a_i D^h_{+R} a_j \Longrightarrow a_i P a_j$ will be called minimal if there is not any other rule $a_i D^k_{+S} a_j \Longrightarrow a_i P a_j$ such that $S \subseteq R$, $k \leq h$; A D_{---}-decision rule $a_i D^h_{-R} a_j \Longrightarrow a_i N a_j$ will be called minimal if there is not any other rule $a_i D^k_{-S} a_j \Longrightarrow a_i N a_j$ such that $S \subseteq R$, $k \geq h$.

The following theorem 4.1 expresses the relationships between decision rules and the approximations of preferences P and N. Both theorem 4.1 and theorem 4.2 can be useful for the induction of the decision rules.

Theorem 4.1.

(1) If $a_i D^h_{+R} a_j \Longrightarrow a_i P a_j$ is a minimal D_{++}-decision rule, then $R_*(P) = D^h_{+R} \cap U$,

(2) If $a_i D^h_{-R} a_j \Longrightarrow a_i N a_j$ is a minimal D_{---}-decision rule, then $R_*(N) = D^h_{-R} \cap U$.

Theorem 4.2. Assuming $U = B \times B$ with $B \subseteq A$, the following conclusions hold:

(1) If $a_i D^h_{+R} a_j \Longrightarrow a_i P a_j$ is a minimal D_{++}-decision rule and $h > 0$, then $a_i D^{-h}_{-R} a_j \Longrightarrow a_i N a_j$ is a minimal D_{---}-decision rule;

(2) If $a_i D^h_{+R} a_j \Longrightarrow a_i P a_j$ is a minimal D_{++}-decision rule and $h \leq 1$, then $a_i D^{-1}_{-R} a_j \Longrightarrow a_i N a_j$ is a minimal D_{---}-decision rule.

4 Conclusion

In this paper a new rough set method for stochastic multi-attribute decision problems was presented. It is based on the idea of approximating a preference relation represented in a PCT by graded probabilistic dominance relations. This methodology supplies some very meaningful "if..., then..." decision rules, which synthesize the preferential information given by the DM and can be suitably applied to obtain a recommendation for the choice or ranking problem. Further research will tend to refine this approach for application.

References

1. Pawlak, Z.: Rough Sets. Theoretical Aspects of Reasoning about Data. Kluwer Academic Publishers, Dordrecht(1991)
2. Greco, S., Matarazzo, B., Slowinski R.: Rough approximation of a preference relation by dominance relations. European Journal of Operational Research.117(1999)63-83
3. Zaras, K.: Rough approximation of a preference relation by a multi-attribute dominance for deterministic, stochastic and fuzzy decision problems. European Journal of Operational Research.159(2004)196-206

Incremental Target Recognition Algorithm Based on Improved Discernibility Matrix

Liu Yong, Xu Congfu, Yan Zhiyong, and Pan Yunhe

College of Computer Science,
Zhejiang University,
Hangzhou 310027, China

Abstract. An incremental target recognition algorithm based on improved discernibility matrix in rough set theory is presented. Some comparable experiments have been completed in our "Information Fusion System for Communication Interception Information (IFS/CI^2)". The results of experimentation illuminate that the new algorithm is more efficient than the previous algorithm.

1 Introduction

It is difficult to recognize kinematic targets, such as communication broadcast station and its embarked platform, accurately and efficiently, in communication intercept. Traditional targets recognition approach is implement by problalities[1]. As it is a mathematical method, it achieves results by complex calculate. In this paper, we provide a new approach for the targets recognition implemented by the data mining method. In our Information Fusion System for Communication Interception Information (IFS/CI^2)[2], the system obtains the identity, attributes, location and so on, analyzes communication parameter, characteristic of communication, neighbor information, manual information etc. together with others, and then evaluates situation of battlefield and minatory degree. So the targets recognition can be regard as a data mining process, mining the targets decision rules from the previous raw data attributes, such as the location of targets, frequency of the broadcast in targets etc., then deduce the new coming data items' type of targets by the previous mining results, normally the results are described as decision rules.

IFS/CI^2 adopts a hierarchical fusion model, which is divided into two parts: the junior associate module and the senior fusion module. The former gets the type of corresponding station, number, platform, and network station and so on, by associating operation of time and location on communication interception data; the latter gets network station attributes, identity of station, types and deploy of arms and so on, and then evaluates situation of battlefield and minatory degree. Since Rough Set theory [3] is effective in data classification, it works as an important tool in the junior associate module of IFS/CI^2. The literature [4] discusses the kinematic target recognition and tracking algorithm based on Rough Set Theory briefly, and obtains satisfied recognition ratio of kinematic targets. However, the result of simulation shows that the efficiency of targets recognition algorithm in [4] is not so ideal, and the essential reason is that the non-incremental algorithm of literature [4] has the following shortages:

(1)Complexity of the non-incremental algorithm is large. Because the communication interception information is incremental, and this means that we cannot obtain all the data at once, the data are produced in batches. When the new batch of data coming, the non-incremental algorithm in [4] has to re-calculate all the data, which includes the historical data and the new one. So the temporal wastage in the computation is huge.

(2) Non-incremental algorithm can't make full use of the existed classification results and the rules. There is no need to compute from scratch when the new batch of data comes. In fact, the new data only affect few classification results and rules, which need to be modified.

Therefore, this paper utilizes the improved discernibility matrix [5] to perfect the incremental targets recognition algorithm, which lays a strong emphasis on improving target recognition efficiency, as well as keeping recognition ratios. Furthermore, this paper introduces confidence factor to solve the problem of quantitative computation of rules, that IFS/CI^2 adopts many subjectively experiential rules, and it is lack of objective criterion to verify the validity and veracity. The algorithm presented in this paper can process the inconsistent data[6] efficiently. It can obtain both certain rules, also named consistent rules, and uncertain rules, also named inconsistent rules.

2 Incremental Algorithm for Target Recognition

This section will focus on the incremental platform recognition algorithm in IFS/CI^2. Sine the principle and process of incremental station recognition algorithm is similar to the platform recognition algorithm, the former will not be discussed. Other algorithm in IFS/CI^2, such as data condense algorithm, target network recognition algorithm, numerical scatter algorithm have been discussed in [2,8]. Before presenting the incremental algorithm, the confidence factor of rule and category of incremental data will be first introduced.

2.1 Confidence Factor and Support Factor of Rule

Since the algorithm presented in this paper will extract both consistent and inconsistent rules, we must first introduce the confidence factor of rules, especially to deal with inconsistent rules. Usually, consistent rules can be denoted by:

$$< Condition\ 1|Condition\ 2|\ ...\ |Condition\ n \rightarrow Conclusion >$$

For inconsistent rules, there may be different conclusions, although the condition attributes and the condition attribute combination are the same, it is necessary to introduce a confidence factor $\alpha(0 < \alpha \leq 1)$ to identify all the inconsistent rules. The α means the probability of the rule on the same condition attributes. And the denotation is:

$$< Condition\ 1 | Condition\ 2 | \ldots | Condition\ n \to Conclusion,\ \alpha >$$

Evidently, consistent rules' confidence factor $\alpha \equiv 1$, while the confidence factor of inconsistent rule will be adjusted during the executing of algorithm. For example, the inconsistent rules, which may appear in communication antagonizing, are:

< there are 4 or more stations in the platform on the ground→the platform is headquarter, 0.95>

<if the platform on the ground makes network with the target in the air → platform is aerodrome, 0.93>

<if there are more than 3 stations in the platform in the air → platform is predictor air, 0.7>

<if the power of the station in the platform in the air is larger than 400W and station number is smaller than 3 → platform is EW aircraft, 0.98>

The computation basis of confidence factor is Rough Operator [8,9] in rough set theory. Rough Operator is the conditional probability of decision ψ, given transcendent probability of condition φ. And it is can be defined as:

$$p(\psi) = \Sigma(p(\varphi) * u(\varphi, \psi)) = \Sigma p(\varphi \wedge \psi) \tag{1}$$

If the data to be recognized in the database matches multiple platform type recognition rules, it is necessary to improve the former formula of Rough Operator. And the improved one is:

$$p'(\psi) = \Sigma(\alpha * (p(\varphi) * u(\varphi, \psi))) = \Sigma(\alpha * p(\varphi \wedge \psi)) \tag{2}$$

Here, α is the confidence factor corresponding to the matched rule. We compute the confidence factors of all matched rules respectively. The larger the Rough Operator is, the higher the matching ratio of the rule. And we choose the rule with largest Rough Operator to judge the platform type.

Normally, there are some noise and distortion in those data from sensors. And this will cause wrong rules based on those distorted data. In our algorithm, we introduce a measure, which is similar to the association rule algorithm[10], named support factor of rules, μ, to filter the noise rules.

2.2 Definition of Improved Discernibility Matrix and Classes of Incremental Data

In the incremental station platform recognition algorithm, we adopt the improved discernibility matrix[5], and its definition is as follows:

Definition 1 Improved Discernibility Matrix: in the information system $S = (U, A)$, where U denotes domain, and A denotes attribute set composed of condition and decision attributes, suppose $B \subseteq A$, $E_i, E_j \in U/IND(B)$, $i, j = 1, 2, ..., n = |U/IND(B)|$; $X_k \in U/IND(D)$, where D is the decision attribute, $j = 1, 2, ..., m$, $m = |U/IND(D)|$. The improved discernibility matrix is an $n \times n$ phalanx $M_S(B) = M_S(i,j)_{n \times n}$, $1 \leq i, j \leq n = |U/IND(B)|$. The unit of the phalanx $M_S(i, j)$ is defined as follows:

If $E_i \subseteq X_k$, $E_j \subseteq X_k$, and $i \neq j$, then $M_S(i, j) = NULL$;

else $M_S(i, j) = \{a \in B : a(E_i) \neq a(E_j)\}, i, j = 1, 2, ..., n$

Where, the $NULL$ means that the difference of the two corresponding items is neglectable. $U/IND(B)$ and $U/IND(D)$ denote the classification of domain on condition attribute B and decision attribute D. From the above definition, it is obvious that the improved discernibility matrix must be a symmetry matrix, and the $NULL$ values in this matrix can be ignored when calculating the discernibility functions and the comparative discernibility functions, so the computational complexity will be decreased greatly.

Furthermore, incremental platform recognition depends on classification of incremental data. It classifies the new data according to the consistency of corresponding condition and decision attributes of new data and existed rules.

Consider information system $S = (U, A)$, and suppose M is the rule set, there is a rule $\phi_i \to \varphi_i$, where i is an element in U, ϕ_i is the antecedent, and φ_i is the consequent. In this new category system, there exist four possible conditions when a new item of data is added to the information system S. They are defined respectively as follows:

Definition 2. **CS category** the new added datum x belongs to CS category, if and only if $\exists (\phi \to \varphi) \in M, \phi_x \to \phi$ and $\varphi_x = \varphi$.

Definition 3. **CN category** the new added datum x belongs to CN category, if and only if $\forall (\phi \to \varphi) \in M, \varphi_x \neq \varphi$.

Definition 4. **CC category** the new added datum x belongs to CC category, if and only if x does not belong to CN category, and $y \in U$ satisfies $\phi_x \equiv \phi_y$ and $\varphi_x \neq \varphi_y$.

Definition 5. **PC category** the new added datum x belongs to PC category, if and only if x does not belong to CN category, and $y \in U$ satisfies $\phi_x \neq \phi_y$.

2.3 Incremental Platform Recognition Algorithm

The raw data is incremental, and the previous algorithm [4] must compute all the data from scratch, it is difficult to deal with huge data in real-time. So the platform recognition algorithm that can process incremental data in real-time is required. The incremental platform recognition algorithm gives each platform recognition rule a confidence factor, if there are several rules matching the new item data, we compute the improved rough operator mentioned previously by formula (2), then select the one with largest rough operator to recognize, and at the same time, the improved discernibility matrix is employed to extract recognition rules contained in the data. When there is new data imported, the forenamed classification definition of incremental data is used to judge and process.

The incremental platform recognition algorithm is constructed by three sub-algorithm, main recognition algorithm, original data rule-extracting algorithm and incremental data rule-extracting algorithm. The original data rule-extracting algorithm is used to deal with the static data in the beginning of the recognition which is same to our previous approach[4]. And the incremental data rule-extracting algorithm processes the incremental data by their categories that is defined in section 2.2.

The recognition algorithm executes after the data condense algorithm, which combines multiple items of data into one item. In data condense algorithm, all the data items describing the same targets at one times-tamp are united into one item, and the unused attributes of raw data are removed from the database. After attributes reduction, there will usual be some conflicts, which is also called inconsistent condition. In this condition, there are more than one items whose condition attributes are all uniform, while they decision attributes are different in the database after data condensing. This may cause by the attributes removed in the condense algorithm. The recognition rules generated by those inconsistent data will be incompatible.

In our recognition approach, the condition equivalence class are calculated by $E_i \in U/IND(C \cup D)$, including the decision attributes into the equivalence categories generating. The conflict items will be classified into different equivalence class, they will produce different rules, and we can choose the rule who has the largest confidence factor.

3 Comparison of Practice Experiment

In IFS/CI2, the processed data arrive batch by batch, it can be regard as a typical incremental data sequence. In the information process system, it is quite important to ensure the system to proceed in real-time and efficiently. The most important criterion to describe the efficiency of this sort of systems is response time, which is the consumed time from the initial data to completing processing of latest data. We design and implement an algorithm comparison experiment based on respond time.

In this section, we discuss the practice comparison experiment between incremental algorithm and non-incremental algorithm emphatically on respond time. The hardware of the experiment is **PC with Pentium IV 866MHZ CPU, 256M ROM and 40G hard disk, whose operating system is Windows 2000 Server.** The comparison experiment results are shown in Figure 1 and Figure 2. In those figures, the time axises are all the response time of fusion center.

From the above comparison experiments results, it can be inferred that there is no notable difference between two algorithms when the size of data is small, however, when the size becomes large, incremental platform recognition algorithm is more effective than no-incremental algorithm for it need not process platform data in the database from the beginning.

Algorithm 1. main recognition algorithm

Data: α_0, μ_0, Preliminary parameter database $P(C,D)$, C is the attributes set of parameters, D is the possible decision.

Result: The recognition rules set whose support factor larger than μ_0 and confidence factor larger than α_0 and the recognition result for each data-item.

begin

    Step 1. Data pretreatment, dividing preliminary database $P(C,D)$ into a number of object equivalent class by condition attribute set C: $E_i \in U/IND(C \cup D), i = 1, 2, ..., |U/IND(C \cup D)|$.

    divide $P(C,D)$ into a number of decision equivalent classes on decision attribute set D:
$X_j \in U/IND(D), j = 1, 2, ..., |U/IND(D)|$

    while *The Preliminary DB not NULL* **do**

        Step 2. Data recognition,

        if *preliminary data item is non-decision* **then**

            As to data in station database without decision, match the existed rules in rule database to judge the type of the platform, which it belongs to. If there are several rules matching the data, compute the Rough Operator of every rules by formula (2), choose the rule with the largest rough operator to match, and then get the type of platform.

        Step 3. Rule extracting,

        if *preliminary data is with decision* **then**

            if *data item is current original one(opposite to incremental data)* **then**

                Call **Algorithm2**(Original Data Rule-extracting Algorithm)

            if *data item is incremental one* **then**

                Call **Algorithm3**(Incremental Data Rule-extracting Algorithm)

end

Algorithm 2. Original data Rule-extracting Algorithm

Data: α_0, μ_0, Data item in Preliminary parameter database $P(C, D)$.
Result: The recognition rules.
begin

    Step 1. Compute improved discernibility matrix $M_S(i,j)$,

        If $E_i \subseteq X_k$, $E_j \subseteq X_k$, and $i \neq j$, then $M_S(i,j) = NULL$;
        else $M_S(i,j) = \{a \in C : a(E_i) \neq a(E_j)\}$, $i, j = 1, 2, ..., n$

    Step 2. Compute relative discernibility function $f(E_i)$ of each equivalent class E_i,
    Step 3. Extract rules from $f(E_i)$,
    if $E_i \subseteq X_j$ then
        Get the consistent rule, $\alpha = 1$
    if $E_i \not\subseteq X_j$ and $E_i \cap X_j \neq 0$ then
        Get the inconsistent rule, and its confidence is:

$$\alpha = \frac{|E_i \cap X_j|}{|E_i|}$$

    And the support factor can be calculated as:

$$\mu = \frac{|E_i \cap X_j|}{|U|}$$

    Step 4. Put the rule whose support factor larger than μ_0 and confidence factor larger than α_0 into the rule database of the recognition algorithm.

end

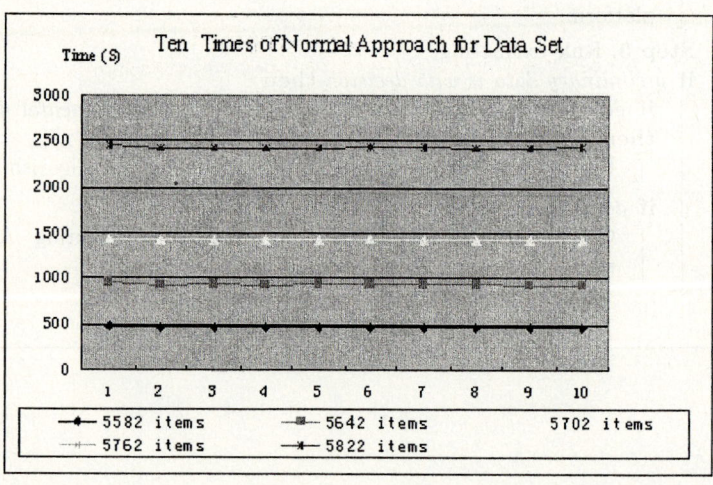

Fig. 1. Response Time of Non-incremental Approach in Platform Data set of IFS/CI[2]

Algorithm 3. Incremental Data Rule-extracting Algorithm

Data: α_0, μ_0, Incremental data item R in Preliminary parameter database $P(C, D)$.
Result: The recognition rules.
begin

> **if** *incremental data R belongs to CS category or CC category* **then**
>> Suppose $R \in E_i$ (object equivalent class), consider E_i, and there are the following two situations:
>> (1) When $E_i \not\subset X_j$ (decision equivalent class) and $E_i \cap X_j \neq 0$, the confidence of $Des(E_i, C) \rightarrow Des(X_j, D)$ is changed to:
>> $$\alpha = \frac{|E_i \cap X_j| + 1}{|E_i| + 1}$$
>> the support factor is changed to:
>> $$\mu = \frac{|E_i \cap X_j| + 1}{|U| + 1}$$
>> as to other X_k, which has the property $k \neq j$ and $E_i \cap X_k \neq 0$, change the confidence of the rule $Des(E_i, C) \rightarrow Des(X_j, D)$ to:
>> $$\alpha = \frac{|E_i \cap X_j|}{|E_i| + 1}$$
>> the support factor is changed to:
>> $$\mu = \frac{|E_i \cap X_j|}{|U| + 1}$$
>> (2) When $E_i \subseteq X_k$, the rule database is not changed.
>
> **if** *incremental data R belongs to CN category or PC category* **then**
>> Add a new line below the last line of improved discernibility matrix $M_S(n, n)$, and add a new column behind the last column, then get a new improved discernibility matrix $M_S(n+1, n+1)$. Compute the new discernibility and new relative discernibility functions according to the new matrix, which are used to extract rules.
>
> Put the rule whose support factor larger than μ_0 and confidence factor larger than α_0 into the rule database of the recognition algorithm.

end

4 Conclusion

The incremental recognition algorithm has more advantages than the non-incremental one[4].

- It processes data incremental, so the recognition time will decrease effectively when dealing with the incremental data.
- It can filter the noise and distorted data by introducing the support factor of rules.

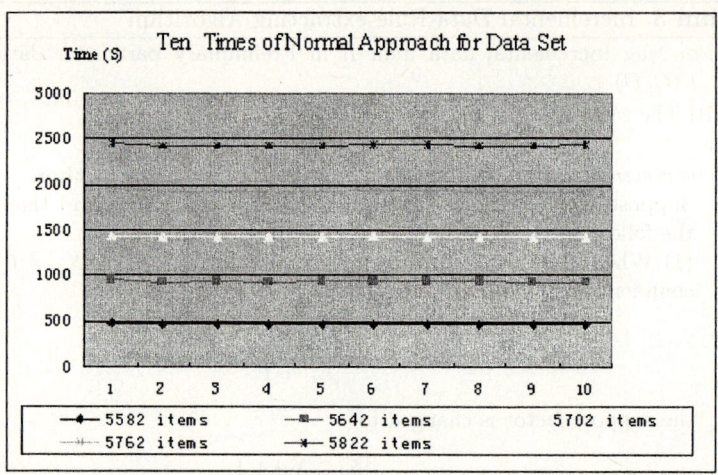

Fig. 2. Response Time of Incremental Approach in Platform Data set of IFS/CI^2

- It can deal with the inconsistent condition,which often occurs after data condensing, using the confidence factor of rules

The experiments results show that the incremental recognition algorithm can is more effective than the previous one, while keeping the same recognition ratio.

Acknowledgements. This paper is sponsored by National Science Foundation of China (No.60402010)and Zhejiang Province Science Foundation(No.M603169), Advanced Research Project of China Defense Ministry (No.413150804), and partially supported by the Aerospace Research Foundation (No. 2003-HT-ZJDX-13).

References

1. Yaakov Bar-Shalom: Tracking Methods in a Multitarget Environment. IEEE Transactions on Automatic Control. 4 (1978) 618–626
2. Xu congfu, Pan Yunhe: IFS/CI2: An intelligent fusion system of communication interception information. Chinese Journal of Electronics and Information Technology. 10 (2002) 1358–1365
3. Pawlak Z.: Rough set. International Journal of Computer and Information Science. 5 (1982) 341–356
4. Liu Yong, Xu congfu, Pan Yunhe: A New Approach for Data Fusion: Implement Rough Set Theory in Dynamic Objects Distinguishing and Tracing. IEEE International Conference on Systems, Man & Cybernetics, Hague (2004) 3318–3323
5. A. Skowron and C. Rauszer: The discernibility matrices and functions in information systems. Fundamenta Informaticae. 2 (1991) 331–362

6. Shan N., Ziarko W.: An incremental learning algorithm for constructing decision rules. In: Kluwer.R.S.(eds.): Rough Sets, Fuzzy Sets and Knowledge Discovery, Springer-Verlag (1994) 326–334
7. Liu Yong, Xu Congfu, Li Xuelan, Pan Yunhe: A dynamic incremental rule extracting algorithm based on the improved discernibility matrix. The 2003 IEEE International Conference on Information Reuse and Integration, USA (2003) 93–97
8. Liu Qing, Huang Zhaohua, Yao Liwen: Rough Set Theory: Present State and Prospects. Chinese Journal of Computer Science. 4 (1997) 1–5
9. Liu Qing, Huang Zhaohua, Liu Shaohui, Yao Liwen: Decision Rules with Rough Operator and Soft Computing of Data Mining. Chinese Jounal of Computer Research and Development. 7 (1999) 800–804
10. Agrawal, R., Srikant, S.: Fast Algorithms for Mining Association Rules in Large Databases. In: VLDB'94, Morgan Kaufmann (1994) 487–499

Problems Relating to the Phonetic Encoding of Words in the Creation of a Phonetic Spelling Recognition Program

Michael Higgins and Wang Shudong

Department of Kansei Design Engineering,
Faculty of Engineering and Technology, Yamaguchi University
2-16-1 Tokiwadai, Ube City, Yamaguchi Prefecture, Japan
{higginsm, peterwsd}@yamaguchi-u.ac.jp

Abstract. A relatively new area of research in centering on the phonetic encoding of information. This paper deals with the possible computer applications of the Sound Approach© English phonetic alphabet. The authors review some preliminary research into a few of the more promising approaches to the application of the processes of machine learning to this phonetic alphabet for computer spell-checking, computer speech recognition etc. Applying mathematical approaches to the development of a data-based phonetic spelling recognizer, and speech recognition technology used for language pronunciation training in which the speech recognizer allows a large margin of pronunciation accuracy, the authors delineate the parameters of the current research, and point the direction of both the continuation of the current project and future studies.

1 Introduction

In 1993-1994, Dr. Michael Higgins of Yamaguchi University, Japan developed and did initial testing on a new system of phonetic spelling of the sounds in English as an aid to learning better English pronunciation and improving listening and spelling skills in English for Japanese students of English. The method, subsequently entitled "A Sound Approach", has been proven to be a very effective English phonetic system [1]. The Sound Approach (SA) alphabet represents without ambiguity all sounds appearing in the pronunciation of English language words, and does so without using any special or unusual symbols or diacritical marks; SA only uses normal English letters that can be found on any keyboard but arranges them so that consistent combinations of letters always represent the same sound, for example, for the word, "this", instead of using IPA (International Phonetic Alphabet) symbols of /ðis/, SA uses /dhis/ to express the pronunciation. Consequently, any spoken word can be uniquely expressed as a sequence of SA alphabet symbols, and pronounced properly when being read by a reader knowing the SA alphabet. For instance, the sentence "One of the biggest problems with English is the lack of consistency in spelling," is written in SA (showing word stress) as: "Wun uv dhu BI-gust PRAA-blumz widh EN-glish iz dhu lak uv kun-SIS-tun-see in SPEL-ing."

2 Project Development

Due to representational ambiguity and the insufficiency of English language characters to adequately and efficiently portray their sounds phonetically, the relationship between a word expressed in SA alphabet and its possible spellings is one to many. That is, each SA sequence of characters can be associated with a number of possible, homophonic sequences of English language characters (e.g. "tuu" is equivalent to "to", "too", and "two"). However, within a sentence usually only one spelling for a spoken word is possible. The major challenge in this context is the recognition of the proper spelling of a homophone/homonym given in SA language.

In addition to the obvious speech recognition OS that would eventually follow, automated recognition of the spelling has the potential for development of SA-based phonetic text editors which would not require the user to know the spelling rules for the language but only being able to pronounce a word within a relatively generous margin of error and to express it in the simple phonetic SA-based form. Speech recognition parameters could also be adjusted to accommodate the wider margin of error inherent in SA which is based on International Broadcast Standard English. As the SA is phonetic (i.e., a clear one-to-one correlation of sound to spelling), the words themselves could be 'regionally tuned' so that one could effectively select the regional accent that they are accustomed to much in the same way that we currently select the keyboard by language groupings. In other words, someone from Australia could select Australia as their text editor and phonetically spell the word 'day' as 'dai' without contextual ambiguity. (See Fig. 1)

I'd like to go to the hospital again today.
1) Ai'd laik tuu go tuu dhu HAAS-pi-tul u-GIN tuu-DEI. (International Standard Broadcast English) 2)Aa'd lak tu go tu dhu HAAS-pi-tul u-GEE-un tu-DEI. (Generic Southern US English) 3) Oi'd laik tuu go tuu dhu HOS-pi-tul u-GEIN tuu-DAI. (Australian-English)

Fig. 1. Regionally tuned sample sentences

The approach adapted in this project involves the application of rough sets [2] in the development of a data-based word spelling recognizer. In this part, the techniques of rough sets, supported by rough-set based analytical software such as KDD-R [3], would be used in the analysis of the classificatory adequacy of the decision tables, and their minimization and extraction of classification (decision) rules to be used in the spelling recognition. While the initial identification and minimization of the required number of information inputs in such decision tables would be one of the more labor intensive aspects of the project, it should be emphasized at this point that the latter stages of the process of minimization and rule extraction would be automated to a large degree and adaptive in the sense that inclusion of new spoken word-context combinations would result in regeneration of the classification rules without human intervention. In this sense the system would have some automated learning ability allowing for continuous expansion as more and more experience is accumulated while being used [4]. The adaptive pattern classification part of the system development is absolutely key to the successful deployment of the system. This aspect of the system, however, only becomes important when some of the more essential problems are solved. Given that, let us briefly outline how we plan to use rough sets in our approach.

In the word spelling recognition problem, in addition to the problem of homophones, one of the difficulties is the fact that many spoken words given in SA form correspond to a number of English language words given in a standard alphabet. To resolve, or to reduce this ambiguity, the context information must be taken into account. That is, the recognition procedure should involve words possibly appearing before, and almost certainly after the word to be translated into Standard English orthography. In the rough-set approach this will require the construction of a decision table for each word. In the decision table, the possible information inputs would include context words surrounding the given word and other information such as the position of the word in the sentence, and so on. (See Appendix)

Several extensions of the original rough sets theory have been proposed recently to better handle probabilistic information occurring in empirical data, and in particular the Variable Precision Rough Sets (VPRS) model [4] which serves as a basis of the software system KDD-R [3] to be used in this project.

In the preliminary testing of SA completed in 1997, a selection of homonyms was put into representative sentences. The words in the sentences were assigned numbers (features) according to a simple, and relatively unrefined, grammatical protocol. These numbers were then inserted into decision tables and using KDD-R it was found that the computer could accurately choose the correct spelling of non-dependent homonyms (i.e., those homonyms for which the simple grammatical protocol was unable to determine the correct spelling from the context) 83.3% of the time, as in the sentence, "The ayes/eyes have it." With dependent homonyms, as in the sentence, "We ate eight meals," the computer could accurately choose the correct spelling more than 98% of the time [4].

Besides the above usages of SA, we are also testing to build HMM/GMM modules of speech recognizers for English pronunciation training based on SA for an online English pronunciation training system, currently, just for Japanese learners of English. As SA encoded speech recognizer allows larger scope of parameter baseline limitation than any current existing speech recognizer, it would more accurately catch the real error of English pronunciation. For example, many acceptable /f/ pronunciations in Japanese English words are judged as wrong or unable to recognize in the current English pronunciation training system using ASR technology [5].

The "regionally tuned" feature of SA can be effectively used for foreign language pronunciation training, too. Usually a non-native speaker's English pronunciation is a mixture of accents of American English, British English, Australian English, etc, and of course colored by his/her own mother tongue. In this case, if ASR modules are built based on SA, then users' input sounds will be encoded to SA with a large margin of speech signal parameters. From SA codes, spoken words are re-coded to regular text on the basis of one-to-one correspondence of SA to regular word, the speech recognition rates will be greatly improved accordingly. Additionally the ASR software training time will be much shorter. Therefore, the speech of foreign language learners who read continuous sentences into a computer will be more easily recognized and translated into correct texts (STT). Also the English learner does not need to worry about what kind of English he/she has to speak.

3 Current Challenges

The adaptive pattern classification part of the system development presents one of the largest difficulties. As alluded to above, a major difficult with the approach suggested is that the practicality of re-coding every word in the English language into a Sound Approach coding is problematic, simply due to the size of the problem, and the fact that it is difficult, at this time, to accommodate any practical public machine learning, whereby users could add new words to the Sound Approach© coded dictionary, which is the way normal spell check systems overcome the problem of an initially small dictionary size. To ease this problem, an already phonetically coded dictionary using the IPA symbols as pronunciation guides has been temporarily adopted, and an interface to link the dictionary (IPA) coding with the Sound Approach coding is being developed. This obviously will save a lot of time, as so many words have already been encoded. However, the problem is that such dictionaries do not contain the phonetic coding of all the inflections of a word, e.g. go = goes = went, or play = playing = played = plays. Therefore, we are still faced with the problem of having to encode many words by hand, before the system can be used as a practical phonetic spell checker. This problem must be solved before we can seriously consider other issues such as how to process words 'in context' and so on.

4 Prospects and Conclusion

If these particular problems can be adequately addressed in the coming year, the authors see no major difficulty for being able to complete the classification of the homonyms into decision tables and, using that as a base, develop a protocol for converting words written in Standard English or IPA symbols into SA characters for ease of encoding. In this way, the interactive database of SA to Standard or Standard to SA could be completed in a relatively short amount of time. This will then make completing the spelling recognition project possible with a more complete regional tunability and, in turn, pave the way to complete voice recognition capability.

References

1. Higgins, M.L, with Higgins M.L and Shima, Y.: Basic Training in Pronunciation and Phonics: A Sound Approach, vol. 19, number 4, The Language Teacher (1995) 4-8.
2. Ziarko,W.: Rough Sets, Fuzzy Sets and Knowledge Discovery. Springer Verlag (1994)
3. Ziarko,W and Shan, N.: KDD-R: A Comprehensive System for Knowledge Discovery Using Rough Sets. Proceedings of the International Workshop on Rough Sets and Soft Computing, San Jose (1994) 164-173.
4. Higgins.M.L, with Ziarko, W.: Computerized Spelling Recognition of Words Expressed in the Sound Approach. New Directions in Rough Sets, Data Mining, and Granular-Soft Computing: Proceedings, 7th International Workshop, RSFDGrC '99. Lecture Notes in Artificial Intelligence 1711, Springer Tokyo (1999). 543-550
5. Goh Kawai and Keikichi Hirose: A Call system for teaching the duration and phone quality of Japanese Tokushuhaku Proceedings of the Joint Conference of the ICA (International Conference on Acoustics) and ASA (Acoustical Society of America) (1998) 2981-2984

Appendix

Values of the observations
(Grammatical Protocol):
0: none   1. verb
2: noun/pronoun 3: adjective
4: adverb   5: article
6: connective   7: number
8: possessive   a: let, please, etc.
b: will, shall, can (modals), etc.
c: prepositions

Head	-2	-1	Spelling
1	2	c	aye
2	2	1	aye
3	0	0	aye
4	0	5	ayes
5	1	8	eye
6	2	5	eye
7	0	8	eyes
8	0	5	eyes
9	1	1	eyes
10	1	5	eye
11	0	2	eyed
12	5	2	i
13	0	0	I

Non-dependent

Fig. 2. Values

Fig. 3. Sample Table 1b — Reduct: "ai"

Table 1. Sample Table 1a: "ai" (IPA symbol: aI); (Sound Spelling: ai)

Head Word	Sentence Number	-5	-4	-3	-2	-1	Spelling
1	15	2	b	1	2	c	aye
2	16	7	2	c	2	1	aye
3	17	0	0	0	0	0	aye
4	18	0	0	0	0	5	ayes
5	19	0	0	2	1	8	eye
6	20	0	2	1	2	5	eye
7	21	0	0	0	0	8	eyes
8	22	0	0	0	0	5	eyes
9	23	0	0	2	1	1	eyes
10	24	1	1	c	1	5	eye
11	25	0	0	0	0	2	eyed
12	26	0	0	0	5	2	i
13	27	0	0	0	0	0	I

Sample Sentences:
15. "I'll love you for aye."
16. "All those in favor say, 'aye'."
17. "Aye, Captain."
18. "The 'ayes' have it."
19. "He injured his eye at work."
20. "He gave me the eye."
21. "Her eyes are blue."
22. "The eyes have it."
23. "She's making eyes at me."
24. "I'm going to keep an eye on you."
25. "He eyed the situation carefully before he went in."
26. "The letter i comes after the letter h and before j."
27. "I want to go out tonight."

Diversity Measure for Multiple Classifier Systems

Qinghua Hu and Daren Yu

Harbin Institute of Technology, Harbin, China
Huqinghua@hcms.hit.edu.cn

Abstract. Multiple classifier systems have become a popular classification paradigm for strong generalization performance. Diversity measures play an important role in constructing and explaining multiple classifier systems. A diversity measure based on relation entropy is proposed in this paper. The entropy will increase with diversity in ensembles. We introduce a technique to build rough decision forests, which selectively combine some decision trees trained with multiple reducts of the original data based on the simple genetic algorithm. Experiments show that selective multiple classifier systems with genetic algorithms get greater entropy than those of the top-classifier systems. Accordingly, good performance is consistently derived from the GA based multiple classifier systems although accuracies of individuals are weak relative to top-classifier systems, which shows the proposed relation entropy is a consistent diversity measure for multiple classifier systems.

1 Introduction

In the last decade, multiple classifier systems (MCS) become a popular technique for building a pattern recognition machine [4, 5]. This system is to construct several distinct classifiers, and then combines their predictions. It has been observed the objects misclassified by one classifier would not necessarily misclassified by another, which suggests that different classifiers potentially offered complementary information. This paradigm is with several names in different views, such as neural network ensemble, committee machine, and decision forest. In order to construct a multiple classifier system, some techniques were exploited. The most widely used one is resampling, which selects a subset of training data with different algorithms. Resampling can be roughly grouped into two classes; one is to generate a series of training sets from the original training set and then trains a classifier with each subset. The second method is to use different feature sets in training classifiers. Random subspace method, feature selection were reported in the documents [2, 4].

The performance of multiple classifier systems not only depends on the power of the individual classifiers in the system, but also is influenced by the independence between individuals [5, 6]. Diversity plays an important role in combining multiple classifiers, which guilds MCS users to design a good ensemble and explain the success of a ensemble systems. Diversity may be interpreted differently from some angles, such as independence, orthogonality or complementarity [7, 8]. Kuncheva pointed that diversity is generally beneficial but it is not a substitute for accuracy [6].

As there are some pair-wise measures, which cannot reflex the whole diversity in MCS, A novel diversity measure for the whole system is presented in the paper, called relation entropy, which is based on the pair-wise measures.

2 Relation Entropy

Here we firstly introduce two classical pairwise diversity measures, Q-statistic and correlation coefficient. Given a multiple classifier system with n individual classifiers $\{C_1, C_2, C_i, \cdots, C_n\}$, the joint output of two classifiers, C_i and C_j, $1 \le i, j \le n$, can be represented in a 2×2 table as shown in table 1.

Table 1. The relation table with classifiers C_i and C_j

	C_j correct (1)	C_j wrong (0)
C_i correct (1)	N^{11}	N^{10}
C_i wrong (0)	N^{01}	N^{00}

Yule introduced Q-statistic for two classifiers defined as

$$Q_{ij} = \frac{N^{11} N^{00} - N^{10} N^{01}}{N^{11} N^{00} + N^{10} N^{01}}$$

The correlation coefficient ρ_{ij} is defined as

$$\rho_{ij} = \frac{N^{11} N^{00} - N^{10} N^{01}}{\sqrt{(N^{11} + N^{10})(N^{01} + N^{00})(N^{11} + N^{01})(N^{10} + N^{00})}}$$

Compute the Q-statistic or correlation coefficient of each pair of n classifiers, a matrix will produce: $M = (r_{ij})_{n \times n}$. Here $r_{ii} = 1$, $r_{ij} = r_{ji}$ and $|r_{ij}| \le 1$. Therefore matrix $|M|$ is a fuzzy similarity relation matrix. the greater the value $|r_{ij}|$, $i \ne j$, is, the stronger the relation between C_i and C_j is and then the weaker of independence between classifiers is. The matrix M surveys the total relation of classifiers in the MCS.

Given a set of classifiers $C = \{C_1, C_2, \cdots, C_n\}$, R is a fuzzy relation on C. It can be denoted as a relation matrix $(R_{ij})_{n \times n}$, where R_{ij} is the relation degree between C_i and C_j with respect to relation R. As we know that the larger R_{ij} is, the stronger the relation of C_i and C_j is. As to correlation coefficient, R_{ij} denotes the degree of correlation between C_i and C_j. If $R_{ij} > R_{ik}$, we say C_i and C_j are more indiscernible than C_i and C_k.

Definition 1. Let R be a fuzzy relation over a set C, w_i the weight of C_i in the ensemble system, $0 \le w_i \le 1$ and $\sum_i w_i = 1$. $\forall C_i \in C$. We define expected relation degree of C_i to all $C_j \in C$ with respect to R as follows:

$$\pi(C_i) = \sum_{j=1}^{n} w_j \bullet r_{ij}$$

Definition 2. The information quantity of relation degree of C_i is defined as

$$I(C_i) = -\log_2 \pi(C_i)$$

It's easy to show that the larger $\pi(C_i)$ is, the stronger C_i is with other classifiers in the ensemble system, and the less $I(C_i)$ is, which shows that the measure $I(C_i)$ describes the relation degree of C_i to all classifiers in system C with respect to relation R.

Definition 3. Given any relation R between individuals in multiple classifier system, and a weight factor series of C, the relation entropy of the pair $<R, w>$ is defined as

$$H_w(R) = \sum_{C_i \in C} w_i \bullet I(C_i) = -\sum_{C_i \in C} w_i \log_2 \pi(C_i)$$

Information entropy gives the total diversity of a multiple classifier system if relations used represent the similarity of outputs of individual classifiers. This measure not only takes the relations between classifiers into account, but also computes the weight factors of individual classifiers in ensemble. The proposed information entropy can applied to a number of pairwise similarity measures for multiple classifier systems, such as Q-statistic, correlation coefficient and so on.

3 Experiments

Searching the optimal ensemble of multiple classifier systems involves combinational optimization. Genetic algorithms make a good performance in this kind of problems. Some experiments were conducted with UCI data. The numbers of reducts range between 5 and 229. All the trees are trained with CART algorithm and two-thirds samples in each class are selected as training set, others are test set. Here, for simplicity, 20 reducts are randomly extracted from the reduct sets of all data sets if there are more than 20 reducts. Subsequent experiments are conducted on the 20 reducts.

The accuracies with different decision forests are shown in table 2. GAS means the forests based on genetic algorithm. TOP denotes the forests with the best trees. We find that GAS ensembles get consistent improvement for all data sets relative to systems combining the best classifiers.

All entropies of Q-statistic and correlation coefficient in two kinds of ensembles as to the data sets are shown in table 3. As the entropies represent the total diversity in systems, we can find GAS based ensembles consistently catch more diversity than top-classifier ensembles.

Table 2. Comparison of decision forests

Data	GAS		TOP	
	size	accuracy	size	accuracy
BCW	10	0.9766	10	0.92642
Heart	6	0.8857	6	0.85714
Ionos	8	0.9901	8	0.94059
WDBC	7	0.9704	7	0.94675
Wine	7	1.00	7	0.97917
WPBC	9	0.75	9	0.70588

Table 3. Relation entropy of multiple classifier systems

	Q-statistic		Correlation coefficient	
	TOP	GAS	TOP	GAS
BCW	0.1385	0.2252	0.6348	0.7719
Heart	0.0031	0.4011	0.0698	0.9319
IONOS	0.0978	0.2313	0.7689	1.2639
WDBC	0.1780	0.2340	1.0296	1.2231
Wine	0	0.0593	0.8740	1.3399
WPBC	1.0541	1.1466	1.7425	1.8319

4 Conclusion

Diversity in multiple classifier systems plays an important role in improve classification accuracy and robustness as the performance of ensembles not only depends on the power of individuals in systems, but also is influenced by the independence between individuals. Diversity measures can guild users to select classifiers and explain the success of the multiple classifier system. Here a total diversity measure for multiple classifier systems is proposed in the paper. The measure computes the information entropy represented with a relation matrix. If Q-statistic or correlation coefficient is employed, the information quantity reflexes the diversity of the individuals. We compare two kinds of rough decision forest based multiple classifier systems with 9 UCI data sets. GA based selective ensembles achieve consistent improvement for all tasks compared with the ensembles with best classifiers. Correspondingly, we find the diversity of GAS with the proposed entropy based on Q-statistic and correlation coefficient is consistently greater than that of top-classifier ensembles, which shows that the proposed entropy can be used to explain the advantage of GA based ensembles.

References

1. Ghosh J.: Multiclassifier systems: Back to the future. Multiple classifier systems. Lecture notes in computer science, Vol.2364. Springer-Verlag, Berlin Heidelberg (2002) 1-15
2. Zhou Z., Wu J., Tang W.: Ensembling neural networks: Many could be better than all. Artificial intelligence 137 (2002) 239-263

3. Ho Ti Kam: Random subspace method for constructing decision forests. IEEE Transactions on Pattern Analysis and Machine Intelligence. 20, (1998) 8, 832-844
4. Czyz J., Kittler J., Vandendorpe L.: Multiple classifier combination for face-based identity verification. Pattern recognition. 37 (2004) 7: 1459-1469
5. Ludmila I. Kuncheval: Diversity in multiple classifier systems. Information fusion. 6, (2005) 3-4
6. Kuncheva L. I. et al.: An experimental study on diversity for bagging and boosting with linear classifiers. Information fusion. 3 (2002) 245-258
7. Hu Qinghua, Yu Daren: Entropies of fuzzy indiscernibility relation and its operations. International journal of uncertainty, fuzziness and knowledge based systems. 12 (2004) 575-589
8. Hu Qinghua, Yu Daren, Wang Mingyang: Constructing rough decision forests. The tenth conference on rough sets, fuzzy sets, data mining and granular computing. 2005

A Successive Design Method of Rough Controller Using Extra Excitation

Geng Wang, Jun Zhao, and Jixin Qian

Institute of Systems Engineering,
Zhejiang University, Hangzhou,
310027, China
jzhao@iipc.zju.edu.cn

Abstract. An efficient design method to improve the control performance of rough controller is presented in this paper. As the input-output data of the history process operation may not be enough informative, extra testing signals are used to excite the process to acquire sufficient data reflecting the control laws of the operator or the existing controller. Using data from the successive exciting tests or excellent operation by operators, the rules can be updated and enriched, which is helpful to improve the performance of the rough controller. The effectiveness of the proposed method is demonstrated through two simulation examples emulating PID control and Bang-Bang control, respectively.

1 Introduction

In industrial process, it is usually very difficult to obtain quantitative models of the complex process. In such cases it is necessary to observe the operation of experts or experienced operators and discover rules governing their actions for control.

Rough set theory [1],[2],[3] provides a methodology for generating rules from process data, which enable to set up a decision-making utility that approximates operator's knowledge about how to control the system. Such a processor of decision rules is referred to as a "rough controller" [4]. The rough controller has been applied to industrial control successfully [5-11]. However, the result of rough control is coarse. One reason is that the rules extracted from history operating data usually are not sufficient, which means the rules may not provide full knowledge for controlling the outputs, therefore the performance of rough controller may be poor.

An improved approach to design rough controller is proposed in this paper, First extra testing signals is used to excite the system to acquire sufficient data reflecting strategy of operators or experts, which is similar to model identification; next, rough set is used to extract rules from the collected data; then the rules are updated and enriched by using data from the successive exciting tests or excellent operations by operators; finally, rough controller is designed based on the rules. This method is applied to emulate two control schemes, a PID control, and a Bang-Bang control, both with good control performance.

2 Design Method of Rough Controller

2.1 Data Acquirement and Excitation

The proposed method to acquire data is shown in Fig.1. Extra testing signals are impose to excite the process. The signals type can be pseudo-random signal or step signals. Both the control signals and output signals are recorded.

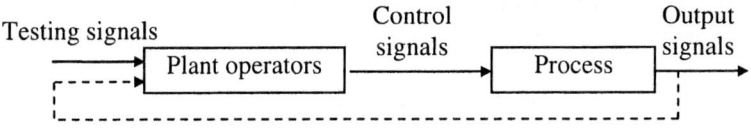

Fig. 1. Data collection system

The testing signals should ensure that the system is excited sufficiently, so that the acquired data can cover the whole input-output space and completely reflect the actions of the operator. For example, in the example of emulating PID controller, the error of set point to output signal and change in error are chosen as condition attributes. Pseudo-random signal is used as the testing signal to ensure enough magnitude and enough frequency. In the example of emulating Bang-Bang controller, to ensure the whole input-output space can be overlapped and the control laws of Bang-Bang control can be implemented exactly, multi-step signal with different magnitude is used.

2.2 Rough Set Analysis

The goal of the analysis stage is to generate rules. The procedure is as follows:

1. Select condition attributes and decision attributes, and discretize these attributes by manual discretization.
2. Check and remove redundant data.
3. Check contradictory data and process them with weighted average method.
4. Generate minimal decision rules from data.

2.3 Rough Rules Update and Complement

As the rules derived from single test may not be sufficient. One or more additional tests need to be performed, and operators' excellent operations can also be referenced. These tests can use the same type of testing signals, but with different magnitude or frequency. Data acquired from these tests or operators' excellent operations are used to generate new rules, and these rules are used to update or enrich original knowledge base successively. The method can be processed through the following operations.

Suppose U representing the rules set in the original knowledge base, x' is a new rule extracted from a new test.

1. if $\forall x \in U$, $\wedge f(x', c_i) \neq \wedge f(x, c_i)$, then x' is a new rule and is appended to set U.
2. if $\exists x \in U$, $\wedge f(x', c_i) = \wedge f(x, c_i)$, and $\wedge f(x', d_j) = \wedge f(x, d_j)$, then x' is ignored.

3. if $\exists x \in U$, $\wedge f(x', c_i) = \wedge f(x, c_i)$, and $\wedge f(x', d_j) \neq \wedge f(x, d_j)$, then x' is processed by weighted average method with these rules x, as follows:

$$\bar{V}_d = \sum_{i=1}^{k} V_{d_k} n_k \bigg/ \sum_{i=1}^{k} n_k \qquad (1)$$

where k is the number of the decisions, V_d the value of the decisions, n the number of their corresponding objects, $\bar{V}_d$ the final decision.

Take that emulating PID controller as example; total 16 tests are performed successively. The numbers of the rules after being updated and enriched are shown in Table 1. It can be seen that the rules are not sufficient only for the first several tests, and the rule number increases with successive tests. The test can be terminated when the rule numbers do not change.

Table 1. The number of the rules after each test

Test	1	2	3	4	5	6	7	8
Rule Number	73	97	106	110	110	112	112	112
Test	9	10	11	12	13	14	15	16
Rule Number	113	113	114	115	115	115	115	115

2.4 Rough Controller Design Based on Rules

The basic structure of the rough controller [4] is shown in Fig. 2. It consists of the following parts: rough A/D converter, inference engine, and knowledge base. Knowledge base is the central part that contains rules derived from test data by rough set.

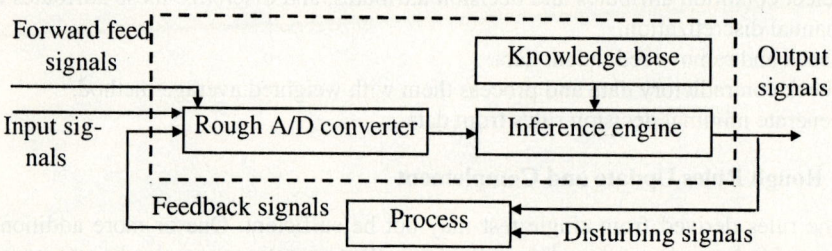

Fig. 2. Rough control system

3 Simulation

Two examples emulating a PID controller and a Bang-Bang controller, respectively, are used to validate the proposed method. In the examples, the PID controller and Bang-Bang controller are assumed as the plant operator.

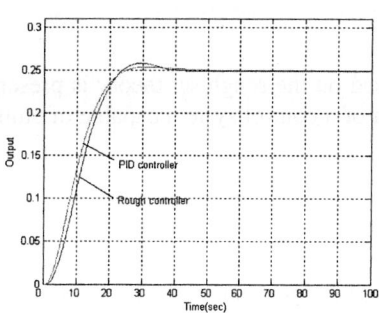 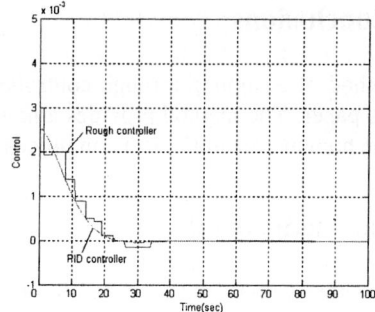

Fig. 3. Comparison between rough controller and PID controller with step input form 0 to 0.25 (Left: output, Right: control move)

In the example of emulating PID controller, the rough controller is expected to response to any input or disturbance distributed in [-0.3, 0.3] like the PID controller. Fig.3 shows the responses of the rough controller and PID controller when tracking step input signal from 0 to 0.25. It is observed that the control move of rough controller is different with the PID controller, but the control performances of the two controllers are very similar.

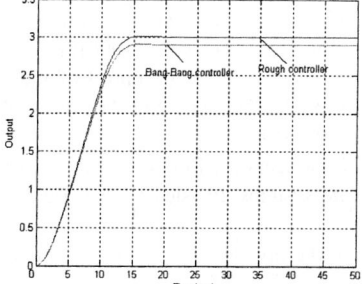 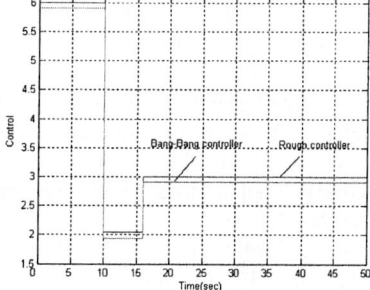

Fig. 4. Comparison between Bang-Bang controller and rough controller with step input form 0 to 2.9 (Left: output, Right: control move)

Bang-Bang control is a special type of control where control move is only allowed to be discrete value, which is very similar to operators' control actions. In the example of emulating Bang-Bang controller, the rough controller is expected to response to any input or disturbance distributed in [0, 5] like the Bang-Bang controller. The response to a step input from 0 to 2.9 of the rough controller and Bang-Bang controller is showed in Fig.4, it is observed that the control scheme of rough controller (e.g. two time switches) is the same as the Bang-Bang controller. The small steady error compared with Bang-Bang controller is just owing to the signal magnitude discretization from 2.9 to 3. When the magnitude of input signal is the same as the exciting signal used in testing procedure, the identical response of the rough controller and the Bang-Bang controller is achieved.

4 Conclusions

A method of designing a rough controller based on the rough set theory is presented in this paper. The method provides a new idea of a control system capable of emulating the human operator's decision process.

Acknowledgement

The authors would like to gratefully acknowledge financial support from the China National Key Basic Research and Development Program under Grant. 2002CB312200.

References

1. Pawlak, Z.: Rough sets. International Journal of Information and Computer Science, **11**(1982)
2. Pawlak, Z.: Rough sets.: Theoretical Aspects of Reasoning about Data. Kluwer Academic Publishers, Dordrecht, The Neatherlands(1991)
3. Mrózek, A.: Rough Sets and Dependency Analysis among Attributes in Computer Implementations of Expert's Inference Models. International Journal of Man-Machine Studies, **30**(1989)
4. Mrózek, A., Planka, L., Kedziera, J.: The methodology of rough controller synthesis. Proceeding of the Fifth IEEE International Conference on Fuzzy Systems, (1996)1135-1139
5. Mrózek, A.: Rough sets in computer implementation of rule-based control of industrial process, Intelligent Decision Support: Handbook of Applications and Advances of the rough sets Theory, Kluwer Academic Publishers, (1992)19-31
6. Mrózek, A., Planka, L.: Rough sets in industrial applications, Rough sets in knowledge discovery 2: applications, case studies and software systems, Physica-Verlag, Heidelberg, (1998)214-237
7. Planka, L., Mrózek, A.: Rule-based stabilization of the inverted pendulum. Computational Intelligence, **11**(1995) 348-356
8. Lamber-Torres, G.: Application of rough sets in power system control center data mining, Power Engineering Society Winter Meeting, **1**(2002) 627-631
9. Munakata, T.: Rough control: A perspective. Rough sets and Data Mining: Analysis of Imprecise Data. Kluwer Academic Publisher, (1997)77-88
10. Huang, J.J., Li, S.Y., Man, C.T.: A T-S type of rough fuzzy controller based on process input-output data. Proceedings of IEEE Conference on Decision and Control, **5**(2003)4729-4734
11. Peters, J.F., Skowron, A., Suraj, Z.: An application of rough set methods in control design, Fundamental Information, **43** (2000)269-290

A Soft Sensor Model Based on Rough Set Theory and Its Application in Estimation of Oxygen Concentration

Xingsheng Gu and Dazhong Sun

Research Institute of Automation, East China University of Science & Technology,
Shanghai, 200237, P. R. China
xsgu@ecust.edu.cn

Abstract. At present, much more research in the field of soft sensor modeling is concerned. In the process of establishing soft sensor models, how to select the secondary variables is still an unresolved question. In this paper, rough set theory is used to select the secondary variables from the initial sample data. This method is used to build the soft sensor model to estimate the oxygen concentration in a regeneration tower and the good result is obtained.

1 Introduction

In many chemical processes, due to the limitations of measurement device, it is often difficult to estimate some important process variables [1] (e.g. product composition variables and product concentration variables), or the variables of interest can only be obtained with long measurement delays. But these variables are often used as feedback signals for quality control [2]. If they are not measured fast and right, the quality of the product can be badly affected. In this case, soft sensor can be used to build a model between easily and frequently obtained variables (secondary variables) and those important process variables (primary variables) are stand for the values of primary variables. Soft sensor is a newly developing technique which is on the basis of inferential control.

An important step of soft sensor modeling is the selection of secondary variables which have functional relationship with the primary variable. But the optimal selection method of secondary variables is still unresolved.

Rough set theory (RST) was proposed by Pawlak as a new method of dealing with fuzzy and uncertain knowledge. In this paper, we apply the rough set theory to the selection of secondary variables in order to propose a new approach for soft sensor technique.

2 Rough Set Theory [3]

An information system is: $S = (U, A, V, f)$, where $U = \{x_1, x_2, \cdots, x_n\}$ is a nonempty and finite set of objects. $A = \{a_1, a_2, \cdots, a_n\}$ is a nonempty and finite set of attributes. $V = \bigcup_{a \in A} V_a$, V_a is the domain set of a. $f: U \times A \to V$ is an information function and

$\forall a \in A$, $x \in U$, $f(x,a) \in V_a$ [3□4]. Let $P \subseteq A$, an indiscernibility relation $ind(P)$ is defined as:

$$ind(P) = \{(x, y) \in U \times U | \forall a \in P, f(x,a) = f(y,a)\} \quad (1)$$

The indiscernibility relation $ind(P)$ is an equivalence relation on U. All the equivalence classes of the relation $ind(P)$ is expressed as

$$U/ind(P) = \{X_1, X_2, \cdots, X_m\} \quad (2)$$

where X_i ($i = 1, 2, \cdots, m$) is called the i-th equivalence class. $X_i \subseteq U$, $X_i \neq \Phi$, $X_i \cap X_j = \Phi$ ($i \neq j; i, j = 1, 2, \cdots, m$), $\bigcup_{i=1}^{m} X_i = U$ [7]. The equivalence class is also called the classification of U. $K = (U, ind(P))$ is called a knowledge base

Knowledge (attributes) reduction is one of the kernel parts of the rough set theory. Some attributes are perhaps redundant. If an attribute is redundant, it can be removed from the information system. Reduction and core are two basic concepts of knowledge reduction.

Definition 1: suppose $R \subseteq A$ is an equivalence relation, $r \in R$, the attribute r is redundant if $ind(R) = ind(R - \{r\})$, otherwise r is indispensable in R.

If every r ($r \in R$) is indispensable in R, R is independent, otherwise R is dependent.

Definition 2: Suppose $P \subseteq R$, if P is independent and $ind(P) = ind(R)$, then P is called a reduction of R, expressed as $red(R)$.

R has many reductions, all the indispensable attributes in R included by all the reductions is called the core of R, expressed as: $core(R) = \bigcap red(R)$.

Definition 3: Suppose P and Q are two equivalence relations of U. $pos_P(Q)$ is called the positive region of Q with respect to P and can be expressed as: $pos_P(Q) = \bigcup_{X \in U/Q} PX$.

A decision table is a special information system $S = (U, A, V, f)$ where $A = C \cup D$, $C \cap D = \Phi$, C is the condition attributes set, D is the decision attributes set and Φ is an empty set. It's a two-dimension table. Each row is an object and each column is an attribute.

Suppose $c \in C$ is a condition attribute, the significance of c with respect to D is defined as:

$$\sigma_{CD}(c) = (|pos_C(D)| - |pos_{C-c}(D)|)/|U| \quad (3)$$

Where $|U|$ represent the cardinality of U.

3 Soft Sensor Modeling Method

In this paper, we consider using the input and output sample data to build a soft sensor model for a MISO system based on the rough set theory. It includes three steps.

3.1 Discretization of Continuous Attribute Values

The input and output data of the real process is continuous. Before using rough set theory to judge the significance of one input attribute with respect to the output attribute, the continuous attribute values should be transferred into discretization values expressed as 1, 2, …, n. There exist a number of discretization methods such as fuzzy c-means clustering method [5] and equal-width-interval method. In this paper, we apply equal-width-interval method to these inputs and output attributes.

The equal-width-interval method is as follows: for one attribute $x \in [x_{min}, x_{max}]$, divide x into n_i part, each partition point is Y_j, $j \in (1,2,\cdots,n_i)$. If $|x - Y_i| = min\{|x - Y_j|\}$, $i = 1, 2, \cdots, n_i$, the discrete value of x is i. Substitute the continuous attribute values with the discrete attribute values can get the decision table [4].

3.2 Removing the Redundant Condition Attributes [3, 4]

Using the above method of judging the significance of the input attributes with respect to the output attribute to remove the redundant condition attributes. $C = \{C_1, C_2, \cdots, C_r\}$ (r is the number of the input attributes) are the condition attributes and $D = \{D_1, D_2, \cdots, D_m\}$ (m is the number of the output attributes) is the output attributes. For each input attribute, compute $\sigma_{CD}(C_i)$, $i = 1, 2, \cdots, r$. If $\sigma_{CD}(C_i)$ is big, the condition attribute C_i is significant to D, otherwise, C_i is less significant to D. If $\sigma_{CD}(C_i) = 0$, C_i is insignificant to D.

3.3 Building Soft Sensor Model

There are several ways to build soft sensor models such as mechanism analysis, regression analysis, artificial neural network and fuzzy techniques and so on. In this paper, we will use the Elman network to build the soft sensor model. Elman network is a dynamic recurrent neural network as shown in figure 1 [6]. It can be seen that in a Elman network, in addition to the input, hidden and the output units, there are also context units. The feedforward weights such as W_{ih}, W_{ch} and W_{ho} are modifiable. The recurrent weights W_{hc} is fixed as a constant one. The output vector is $Y(k) \in R^m$ and the input vector is $u(k-1) \in R$

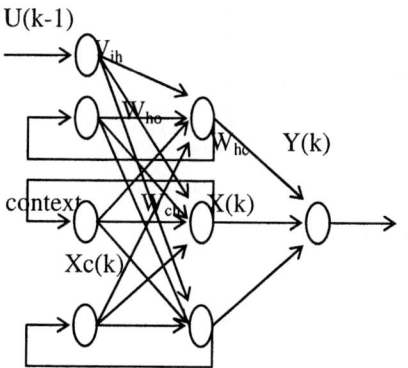

Fig. 1. Elman network

$X(k) \in R^n$ is the hidden layer output vector. $X_c(k)$ is a n×n matrix. The relationship between the input and output can be represented as.

$$X(k) = F(W_{ch}X_c(k) + W_{ih}u(k-1))$$

$$X_c(k) = X(k-1)$$

$$Y(k) = G(W_{ho}X(k))$$

F And *G* are activation functions of hidden units and output units respectively and the dynamic back-propagation (DBP) learning rule is used to train the Elman network.

4 Rough Set Based Soft Sensor Model for Oxygen Concentration

In this part, we will use the rough set theory to select the secondary variables and then build a soft sensor for the oxygen concentration in a regeneration tower. Oxygen concentration is an important variable of the regeneration tower. According to the analysis of technological mechanisms, oxygen concentration is relating to 15 variables as shown in figure 2. Variable 16 is the oxygen concentration. Variables 1 to 4 are gas flux. Variables 5 and 6 are gas pressure. Variables 7 to 13 are gas temperature. Variable 14 is the flux of cycling catalyst. Variable 15 is the coking of the catalyst. But these 15 variables are not equally significant to the oxygen concentration. Perhaps some of them can be removed from the initial data set. Rough set theory is a strong tool to do this work. In this application, $C = \{C_1, C_2, \cdots, C_{15}\}$ and $D = \{D_1\}$. 200 samples are used to judge the significance of each input variable to the oxygen concentration. Using the above method, we can get the following results shown in Table 2.

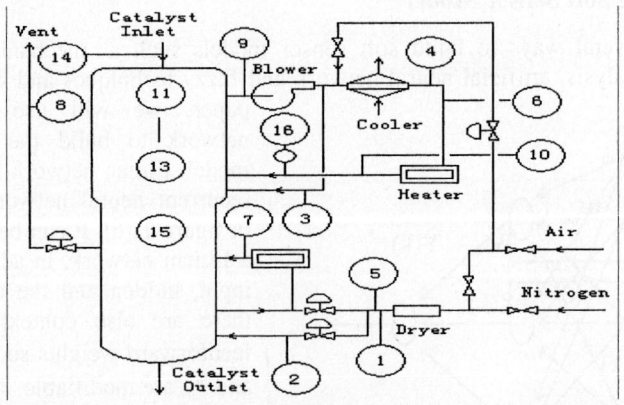

Fig. 2. Regeneration Tower

Table 2. Significance of condition attributes to decision attribute

$\sigma_{CD}(C_1)$	$\sigma_{CD}(C_2)$	$\sigma_{CD}(C_3)$	$\sigma_{CD}(C_4)$	$\sigma_{CD}(C_5)$
0	0	0.003	0	0.01
$\sigma_{CD}(C_6)$	$\sigma_{CD}(C_7)$	$\sigma_{CD}(C_8)$	$\sigma_{CD}(C_9)$	$\sigma_{CD}(C_{10})$
0	0	0	0	0
$\sigma_{CD}(C_{11})$	$\sigma_{CD}(C_{12})$	$\sigma_{CD}(C_{13})$	$\sigma_{CD}(C_{14})$	$\sigma_{CD}(C_{15})$
0.01	0	0.005	0	0.005

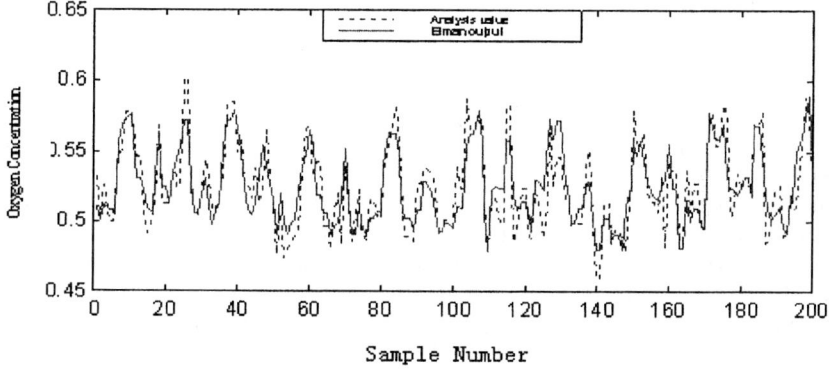

Fig. 3. The training result of the Elman network

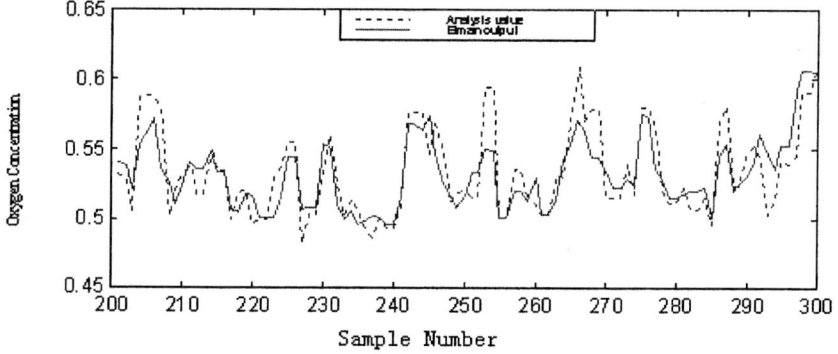

Fig. 4. The generalization result of the Elman network

Table 2 shows that C_3、C_5、C_{11}、C_{13} and C_{15} are significant to D. Using these five variables as the input variables to train the Elman network. The structure of the network is 6-13-1. The train and generalization results are shown in Figure 3 and Figure 4. The root mean square errors for the train and test samples are 0.0184 and 0.0197 respectively which show that the modeling accuracy is high.

5 Conclusions

In this paper, we used rough set theory to judge the significance of the initial selected secondary variables to the primary variables in soft sensor technique. On the basis of rough set theory, some less significance or insignificance variables can be removed. This method is applied to the estimation of oxygen concentration in regeneration tower and the satisfactory result is obtained. The simulation results show the effectiveness of the proposed method.

References

1. Zhong, W., Yu, J.S.: MIMO Soft Sensors for Estimating Product Quality with On-line Correction. Chemical Engineering Research & Design,Transaction of the Insitute of Chemical Engineers. 2000, 78(A): 612-620
2. Tham, M.T., Morris, A.J. et al.: Soft-sensing: a Solution to the Problem of Measurement Delays. Chem Eng Res Des, 1989, 67: 547-554.
3. Zhang, W.X., Wu, W.Z., et al.: Rough Set Theory and Methods. Beijing: Science Press, 2001.
4. Luo, J.X., and Shao, H.H.: Selecting Secondary Measurements for Soft Sensor Modeling Using Rough Set Theory. Proceedings of the 4th World Congress on Intelligent Control and Automation, Press of East China University of Science and Technology, 2002: 415-419.
5. Li, M., and Zhang, H.G.: Research on the Method of Neural Network Modeling Based on Rough Sets Theory. ACTA AUTOMATICA SINICA, 2002, 28(1): 27-33.
6. Tham, D.T., and Liu, X.: Training of Elman Networks and Dynamic System Modeling. International Journal of Systems Science, 1996, 27(2):221-226.

A Divide-and-Conquer Discretization Algorithm

Fan Min, Lijun Xie, Qihe Liu, and Hongbin Cai

College of Computer Science and Engineering, University of Electronic Science and
Technology of China, Chengdu 610051, China
{minfan, xielj, qiheliu, caihb}@uestc.edu.cn

Abstract. The problem of real value attribute discretization can be converted into the reduct problem in the Rough Set Theory, which is NP-hard and can be solved by some heuristic algorithms. In this paper we show that the straightforward conversion is not scalable and propose a divide-and-conquer algorithm. This algorithm is fully scalable and can reduce the time complexity dramatically especially while integrated with the tournament discretization algorithm. Parallel versions of this algorithm can be easily written, and their complexity depends on the number of objects in each subtable rather than the number of objects in the initial decision table. There is a tradeoff between the time complexity and the quality of the discretization scheme obtained, and this tradeoff can be made through adjusting the number of subtables, or equivalently, the number of objects in each subtable. Experimental results confirm our analysis and indicate appropriate parameter setting.

1 Introduction

The majority of machine learning algorithms can be applied only to data described by discrete numerical or nominal attributes (features). In the case of continuous attributes, there is a need for a discretization algorithm that transforms continuous attributes into discrete ones [1]. And the discretization step determines how coarsely we want to view the world [2].

The problem of real value attribute discretization can be converted into the reduct problem in the Rough Set Theory [3][5][6]. But some existing algorithms is not scalable in practice [2][6] when the decision table has many continuous attributes and/or many possible attribute values. Recently we have proposed an algorithm called the *tournament discretization algorithm* for situations where the number of attributes is large. In this paper we propose a divide-and-conquer algorithm that can dramatically reduce the time complexity and is applicable especially for situations where the number of objects in the decision table is large. By integrating these two algorithm together, we can essentially cope with decision tables with any size. A parallel version of this algorithm can be easily written and run to further reduce the time complexity. There is a tradeoff between the time complexity and the quality of the discretization scheme obtained, and this tradeoff can be made through adjusting the number of subtables, or equivalently, the number of objects in each subtable. Experimental results confirm our analysis and indicate appropriate parameter setting.

The rest of this paper is organized as follows: in Section 2 we enumerate relative concepts about decision tables, discretization schemes and discernibility. In Section 3 we analyze existing rough set approaches for discretization and point out their scalability problem. In Section 4 we present and analyze our divide-and-conquer discretization algorithm. Some experimental results are given in Section 5. Finally, we conclude and point out further research works in Section 6.

2 Preliminaries

In this section we emulate relative concepts after Nguyen [3] and Komorowski [2], and propose the definition of discernibility for attributes and cuts.

2.1 Decision Tables

Formally, a *decision table* is a triple $S = (U, A, \{d\})$ where $d \notin A$ is called the decision attribute and elements of A are called conditional attributes. $a : U \to V_a$ for any $a \in A \cup \{d\}$, where V_a is the set of all *values* of a called the *domain* of a.

For the sake of simplicity, throughout this paper we assume that $U = \{x_1, \ldots, x_{|U|}\}$, $A = \{a_1, \ldots, a_{|A|}\}$ and $d : U \to \{1, \ldots, r(d)\}$.

Table 1 lists a decision table.

Table 1. A decision table S

U	a_1	a_2	a_3	d
x_1	1.1	0.2	0.4	1
x_2	1.3	0.4	0.2	1
x_3	1.5	0.4	0.5	1
x_4	1.5	0.2	0.2	2
x_5	1.7	0.4	0.3	2

2.2 Discretization Schemes

We assume $V_a = [l_a, r_a) \subset \Re$ to be a real interval for any $a \in A$. Any pair (a, c) where $a \in A$ and $c \in \Re$ is called a *cut* on V_a. Let P_a be a partition on V_a (for $a \in A$) onto subintervals $P_a = \{[c_0^a, c_1^a), [c_1^a, c_2^a), \ldots, [c_{k_a}^a, c_{k_a+1}^a)\}$ where $l_a = c_0^a < c_1^a < \ldots < c_{k_a}^a < c_{k_a+1}^a = r_a$ and $V_a = [c_0^a, c_1^a) \cup [c_1^a, c_2^a) \cup \ldots \cup [c_{k_a}^a, c_{k_a+1}^a)$. Hence any partition P_a is uniquely defined and often identified as the set of cuts: $\{(a, c_1^a), (a, c_2^a), \ldots, (a, c_{k_a}^a)\} \subset A \times \Re$.

Any set of cuts $P = \cup_{a \in A} P_a$ defines from $S = (U, A, \{d\})$ a new decision table $S^P = (U, A^P, \{d\})$ called *P-discretization* of S, where $A^P = \{a^P : a^P(x) = i \Leftrightarrow a(x) \in [c_i^a, c_{i+1}^a) \text{ for } x \in U \text{ and } i \in \{0, \ldots, k_a\}\}$. P is called a *discretization scheme* of S.

While selecting cut points, we only consider midpoints of adjacent attribute values, e.g., in the decision table listed in Table 1, $(a_1, 1.2)$ is a possible cut while $(a_1, 1.25)$ is not.

2.3 Discernibility

The ability to discern between perceived objects is important for constructing many entities like reducts, decision rules or decision algorithms [8]. The discernibility is often addressed through the discernibility matrix and the discernibility function. In this subsection we propose definitions of discernibility for both attributes and cuts from another point of view to facilitate further analysis.

Given a decision table $S = (U, A, \{d\})$, the discernibility of any attribute $a \in A$ is defined as the set of object pairs discerned by a, this is formally given by

$$DP(a) = \{(x_i, x_j) \in U \times U | d(x_i) \neq d(x_j), i < j, a(x_i) \neq a(x_j)\}, \quad (1)$$

where $i < j$ is required to ensure that the same pair does not appear in the same set twice.

The discernibility of an attribute set $B \subseteq A$ is the union of the discernibility of each attribute, namely,

$$DP(B) = \bigcup_{a \in B} DP(a). \quad (2)$$

Similarly, the discernibility of a cut (a, c) is defined by the set of object pairs it can discern,

$$DP((a,c)) = \{(x_i, x_j) \in U \times U | \; d(x_i) \neq d(x_j), i < j, a(x_i) < c \leq a(x_j) \\ \text{or } a(x_j) < c \leq a(x_i)\}. \quad (3)$$

The discernibility of a cut set P is the union of the discernibility of each cut, namely,

$$DP(P) = \bigcup_{(a,c) \in P} DP((a,c)). \quad (4)$$

It is worth noting that maintaining discernibility is the most strict requirement used in the rough set theory because it implies no loss of information [11].

3 Related Works and Their Limitations

3.1 The Problem Conversion

In this subsection we explain the main idea of the rough set approach for discretization [3] briefly, and we point out that this conversion is independent of the definition of reduction and the reduction algorithm employed.

Given a decision table $S = (U, A, \{d\})$, denote the cut set containing all possible cuts by $C(A)$. Construct a new decision table $S(C(A)) = (U, C(A), \{d\})$ called $C(A)$-cut attribute table of S, where $\forall ct = (a, c) \in C(A)$, $ct : U \to \{0, 1\}$, and

$$ct(x) = \begin{cases} 0, & \text{if } a(x) < c; \\ 1, & \text{otherwise}. \end{cases} \quad (5)$$

Table 2 lists the $C(A)$-cut attribute table, denoted by $S(C(A))$, of the decision table S listed in Table 1.

We have proven the following two theorems [6]:

Table 2. $S(C(A))$

U	$(a_1, 1.2)$	$(a_1, 1.4)$	$(a_1, 1.6)$	$(a_2, 0.3)$	$(a_3, 0.25)$	$(a_3, 0.35)$	$(a_3, 0.45)$	d
x_1	0	0	0	0	1	1	0	1
x_2	1	0	0	1	0	0	0	1
x_3	1	1	0	1	1	1	1	1
x_4	1	1	0	0	0	0	0	2
x_5	1	1	1	1	1	0	0	2

Theorem 1.
$$DP(C(A)) = DP(A). \tag{6}$$

Theorem 2. *For any cut set P,*
$$DP(A^P) = DP(P). \tag{7}$$

Therefore the discretization problem (constructing A^P from A) is converted into the reduction problem (selecting P from $C(A)$). Nguyen [3][5] integrated the conversion process with reduction process and employed the Boolean approach. According to above proved theorems, this conversion maintains the discernibility of decision tables, hence it is independent from the definition of reduction or reduction algorithm employed. For example, if the definition of reduction requires that the positive region is maintained, i.e., the decision tables $(U, C(A), \{d\})$ and $(U, P, \{d\})$ have the same positive region, then S and S^P would have the same positive region; if the definition of reduction requires that the generalized decision is maintained, i.e., $\partial_{C(A)} = \partial_P$, then $\partial_A = \partial_{A^P}$.

3.2 The Scalability Problem of Existing Approaches

Although the above-mentioned approach seems perfect because the discretization problem has been converted to another problem which is solved by many efficient algorithms, using it directly is not scalable in practice when the decision table has many continuous attributes and/or many possible attribute values.

The decision table $S(C(A)) = (U, C(A), \{d\})$ has $|U|$ rows and $|C(A)| = O(|A||U|)$ columns, which may be very large. For example, in the data table WDBC of the UCI library [10] (stored in the file uci/breast-cancer-wisconsin/wdbc.data.txt), there are 569 objects and 31 continuous attributes, each attribute having 300 - 569 different attribute values, and $S(C(A))$ should contain 15,671 columns, which is simply not supported by the ORACLE system run in our computer.

Nguyen [4] also proposed a very efficient algorithm based on Boolean approach. But from our point of view, this approach is not flexible because only object pairs discerned by given cuts are used as heuristic information. Moreover, this approach may not be applicable for mixed-mode data [7].

In previous works [6] we have developed the tournament discretization algorithm. This algorithm has some rounds, during round i the discretization schemes of decision tables containing 2^i conditional attributes (may be less for the last one) are computed on the basis of cut sets constructed in round $i-1$, resulting in $\lceil \frac{|A|}{2^i} \rceil$ cut sets. In this process the number of candidate cuts of current decision table could be kept under a relative low degree.

By using this algorithm we have computed discretized scheme of WDBC, but it took my compute 10,570 seconds for the plain version and 1,886 seconds for the parallel version. Both are rather long. Moreover, this algorithm may be invalid when the number of possible cuts of any attribute exceeds 1000, namely, the upper bound of columns supported by the ORACLE systems.

4 A Divide-and-Conquer Discretization Algorithm

We can use the divide-and-conquer approach to the other dimension of the decision table, namely, the number of objects. In this section we firstly propose the algorithm structure, then analyze parameter setting.

4.1 The Algorithm Structure

Firstly we list our discretization algorithm in Fig. 1.

> $DivideAndConquerDiscretization\ (S=(U,A,\{d\}))$
> {**input:** A decision table S.}
> {**output:** A discretization scheme P.}
> **Step 1.** divide S into K subtables;
> **Step 2.** compute discretization schemes of subtables;
> **Step 3.** compute the discretization scheme P of S based on discretization
> schemes of subtables;

Fig. 1. A Divide-and-Conquer Discretization Algorithm

Generally, in Step 1 we require the family of subtables to be a partition of S. Namely, let the set of subtables be $\{S_1, S_2, \ldots, S_K\}$ where $S_i = (U_i, A, \{d\})$ for all $i \in \{1, 2, \ldots, K\}$, $\bigcup_{i=1}^{K} U_i = U$ and $\forall i,j \in \{1,2,\ldots,K\}, i \neq j, U_i \cap U_j = \emptyset$. Moreover, we require all subtables except the last one to be the same size, namely, $|U_1| = |U_2| = \ldots = |U_{K-1}| = \lceil \frac{|U|}{K} \rceil$. We have these requirements because:

1. Our algorithm is intended for data tables with large amount of rows, subtables containing the same row may not be preferred;
2. Loss of rows, i.e., some rows are not included in any subtable may incur too much loss of information. However, for very huge data tables we may lose this requirement;
3. Subtables with almost the same size is preferred from both statistical and implementation points of view.

Moreover, it is not encouraged to construct subtables using adjacent rows, e.g., $U_1 = \{u_1, u_2, \ldots, u_{\lceil \frac{|U|}{K} \rceil}\}$, because many data tables such as IRIS are well organized according to decision attribute values.

We propose the following scheme to meet these requirement. Firstly, select a prime number p such that $|U|\%p \neq 0$. Then generate a set of numbers $N = \{n_1, n_2, \ldots, n_{|U|}\}$ where $n_j = (j*p)\%|U| + 1$ for all $j \in \{1, 2, \ldots, |U|\}$. Because $|U|\%p \neq 0$, it is easy to prove that $N = \{1, 2, \ldots, |U|\}$. At last we let

$$U_i = \{u_{n_{(i-1)*\lceil \frac{|U|}{K} \rceil + 1}}, u_{n_{(i-1)*\lceil \frac{|U|}{K} \rceil + 2}}, \ldots, u_{n_{i*\lceil \frac{|U|}{K} \rceil}}\}$$

for all $i \in \{1, 2, \ldots K-1\}$ and

$$U_K = \{u_{n_{(K-1)*\lceil \frac{|U|}{K} \rceil + 1}}, u_{n_{(K-1)*\lceil \frac{|U|}{K} \rceil + 2}}, \ldots, u_{n_{|U|}}\}.$$

For example, if $U = 8$, $K = 2$ and $p = 3$, then $U_1 = \{u_1, u_4, u_7, u_2\}$, $U_2 = \{u_5, u_8, u_3, u_6\}$.

It is easy to see that objects of any subtable are distributed in S evenly with no more tendency, and this scheme of constructing subtables can easily break any bias of S by choosing appropriate p (relatively larger ones such as 73, 97 are preferred).

In Step 2 any discretization algorithm could be employed, while in this paper only Nguyen's algorithm [3] (employed while $|A| \leq 8$) and our tournament discretization algorithm [6] (employed while $|A| > 8$) are concerned to keep a unified form.

In Step 3 we use the same idea mentioned in Subsection 3.1. Instead of using $C(A)$, we use cuts selected from all subtables, i.e., $\bigcup_{i=1}^{K} P_i$ where P_i is the discretization scheme of S_i.

4.2 Time Complexity Analysis

Time required for Step 1 is simply ignorable compared with that of Step 2, and time required for Step 3 is also ignorable if K is not very large. It is worth noting that Step 2 can be easily distributed into K computers/processors and run in parallel.

In order to specify its relationship with respective decision table, we use $P(S)$ instead of P to denote the discretization scheme of S, and $P(S_1)$ to denote the discretization scheme of S_1. Obviously, the time complexity of computing discretization scheme of any subtable is equal to that of S_1. The time complexity of the most efficient reduction algorithm is $O(M^2 N \log N)$ where N is the number of objects and M is the number of attribute [9]. We have developed an entropy-based algorithm with the same complexity. In the following we assume that this algorithm is employed and give time complexity of different algorithms or combinations of algorithms:

If we apply Nguey's algorithm [3] directly to S, because $S(C(A))$ has $O(|A||U|)$ cut attributes, the time complexity of computing $P(S)$ is

$$O(|A|^2 |U|^3 \log |U|). \tag{8}$$

If we apply the tournament discretization algorithm [6] directly to S, the time complexity of computing $P(S)$ is

$$O(|A||U|^3 \log |U|), \tag{9}$$

which is reduced to

$$O(\log |A||U|^3 \log |U|) \tag{10}$$

if the parallel mechanism is employed and there are $\lceil \frac{|A|}{2} \rceil$ computers/processors to use [6].

If we apply Nguey's algorithm [3] to subtables, the time complexity of computing $P(S_1)$ is

$$O(|A|^2 (\frac{|U|}{K})^3 \log(\frac{|U|}{K})), \tag{11}$$

which is also the time complexity of the parallel version of computing $P(S)$ if there are K computers/processors to use.

And the time complexity for the plain version of computing $P(S)$ is

$$O(K|A|^2 (\frac{|U|}{K})^3 \log(\frac{|U|}{K})). \tag{12}$$

If we apply the tournament discretization algorithm [6] to subtables, the time complexity of computing $P(S_1)$ is

$$O(|A|(\frac{|U|}{K})^3 \log(\frac{|U|}{K})), \tag{13}$$

which is reduced to

$$O(\log |A|(\frac{|U|}{K})^3 \log(\frac{|U|}{K})). \tag{14}$$

if the parallel mechanism is employed and there are $\lceil \frac{|A|}{2} \rceil$ computers/processors to use. And this is also the time complexity of computing $P(S)$ if there are $\lceil \frac{|A|}{2} \rceil * K$ computers/processors to use.

If the parallel mechanism is not employed, the time complexity for computing $P(S)$ is

$$O(K|A|(\frac{|U|}{K})^3 \log(\frac{|U|}{K})). \tag{15}$$

By comparing equations (8) and (14) or (15) it is easy to see that our algorithms have made great progress on deducing time complexity of the discretization algorithm.

4.3 Tradeoff for Deciding Suitable Parameter

According to above analysis, larger K will incur lower time complexity. But larger K, or equivalently, smaller $\frac{|U|}{K}$ has some drawbacks. When we divide S into subtables, we are essentially losing some candidate cuts. For example, if we divide S listed in Table 1 into 2 subtables $S_1 = (\{x_1, x_2\}, \{a_1, a_2, a_3\}, d)$ and $S_2 = (\{x_3, x_4, x_5\}, \{a_1, a_2, a_3\}, d)$, cuts $(a_1, 1.4)$, $(a_3, 0.35)$ and $(a_3, 0.45)$ will be lost.

When subtables are large enough, the loss of cuts may be trivial and ignorable. But when subtables are relatively small, this kind of loss may be unendurable. For one extreme, if $K = |U|$, any subtable will contain exactly one object, and there will be no candidate cut at all.

Obviously, appropriate K may varies directly as $|U|$ for different applications, and it may be suitable to investigate on appropriate setting of $\frac{|U|}{K}$. In fact, if we keep $\frac{|U|}{K}$ in a certain range for different applications, according to equations (11), (13) and (14), the time complexities of parallel versions are not influenced by $|U|$. We will analyze this issue through examples in the next section.

5 Experimental Results

We are developing a tool called *Rough set Developer's kit* (RDK) using the Java platform and the ORACLE system to test our algorithms and also as a basis of application development. For convenience we run our algorithm in my notebook PC with an Intel Centrino 1.4G CPU and 256M memory. The reducing algorithm employed throughout this paper is entropy based. And the discretization algorithm for subtables is the tournament discretization algorithm. Table 3 lists some results of the WDBC dataset. Specifically, $K = 1$ indicates that we do not divide S into subtables. $POS(S^P)$ denotes the number of objects in the positive region of the discretized decision table S^P. *Total time* indicates time used for the plain version of our algorithm, while *Parallel time* indicates time used for the parallel version of our algorithm. *Processors* indicates the number of computers/processors required for running the parallel version. *Step 3 time* indicates time required for executing Step 3 of our algorithm. Time units are all seconds.

Experiment analysis:

1. While $K \leq 5$, the positive region of S^P is the same as that of S, i.e., all 569 objects are in the positive region. But this no longer hold true for $K > 5$. This difference is essential from the Rough Set point of view.
2. If the discretization quality is estimated by the number of selected cuts $|P|$, suboptimal results could be obtained especially when $K \leq 6$.
3. The *total time* and *parallel time* decreases as K increases, but this trend does not continue significantly when $K > 4$. This is partly because that more subtables will incur more overheads.
4. The run time of Step 3 is no longer ignorable while $K \geq 7$.
5. In this example, $K = 4$ is the best setting.
6. For generalized situations, we recommend that $\frac{|U|}{K}$ to be between 100 and 150 (corresponding to $K = 4$ or 5 in this example) because subtables of such size tend to maintain most cuts, while at the same time easy to discretize.

Table 3. Some results of the WDBC data set

K	$POS(S^P)$	$|P|$	Total time	Parallel time	Processors	Step 3 time
1	569	11	10,570	1,886	16	0
2	569	13	2,858	278	32	11
3	569	13	2,354	207	48	51
4	569	12	1,395	107	64	25
5	569	11	1,189	80	80	24
6	550	13	1,225	108	96	34
7	550	18	920	46	112	16
8	567	13	865	52	128	26
9	566	15	942	58	144	34
10	554	12	994	46	160	21
11	565	12	1093	76	176	49
12	565	13	912	36	192	21

6 Conclusions and Further Works

In this paper we proposed a divide-and-conquer discretization scheme that divides the given decision table into subtables and combine discretization schemes of subtables. While integrated with the tournament discretization algorithm, this algorithm can discretize decision tables with any size. Moreover, the time complexity of parallel versions of our algorithm is influenced by $|U_1|$ rather than $|U|$ if there are $K = \lceil \frac{|U|}{|U_1|} \rceil$ processors/computers to use. We have also given suggestion of $|U_1|$ to be between 100 and 150 through an example.

Further research works include applying our algorithms along with parameter settings on applications to test their validity.

References

1. L. Kurgan and K.J. Cios.: CAIM Discretization Algorithm. IEEE Transactions on Knowledge and Data Engeering 16(2) (2004) 145–153.
2. J. Komorowski, Z. Pawlak, L. Polkowski, A. Skowron.: Rough sets: A Tutorial. S. Pal, A. Skowron (Eds.) Rough Fuzzy Hybridization (1999) 3–98.
3. H.S. Nguyen and A. Skowron.: Discretization of Real Value Attributes. Proc. of the 2nd Joint Annual Conference on Information Science, Wrightsvlle Beach, North Carolina (1995) 34–37.
4. H.S. Nguyen.: Discretization of Real Value Attributes, Boolean Reasoning Approach. PhD thesis, Warsaw University, Warsaw, Poland (1997).
5. H.S. Nguyen.: Discretization Problem for Rough Sets Methods. RSCTC'98, LNAI 1424 (1998) 545–552.
6. F. Min, H.B. Cai, Q.H. Liu, F. Li.: The Tournament Discretization Algrithm, already submitted to ISPA 2005.
7. F. Min, S.M. Lin, X.B. Wang, H. B. Cai, "Attribute Extraction from Mixed-Mode Data," already submitted to ICMLC 2005.

8. R.W. Swiniarski, A. Skowron: Rough Set Methods in Feature Selection and Recognition. Pattern Recognition Letters 24 (2003) 833–849.
9. Liu Shao-Hui, Sheng Qiu-Jian, Wu Bin, Shi Zhong-Zhi, Hu Fei. Research on Efficient Algorithms for Rough Set Methods. Chinese Journal of Computer 26(5) (2003) 524–529 (in Chinese).
10. C.L. Blake, C.J. Merz. UCI Repository of Machine Learning Databases. http://www.ics.uci.edu/ mlearn/MLRepository. html, UC Irvine, Dept. Information and Computer Science, 1998.
11. M. Cryszkiewicz.: Comparative Studies of Alternative Type of Knowledge Reduction in Inconsistent Systems. International Journal of Intelligent Systems 16(1) (2001) 105 – 120.

A Hybrid Classifier Based on Rough Set Theory and Support Vector Machines*

Gexiang Zhang[1], Zhexin Cao[2], and Yajun Gu[3]

[1] School of Electrical Engineering, Southwest Jiaotong University, Chengdu 610031,
Sichuan, China
gxzhang@ieee.org
[2] College of Profession and Technology, Jinhua 321000 Zhejiang, China
[3] School of Computer Science, Southwest University of Science and Technology, Mianyang 621002 Sichuan, China

Abstract. Rough set theory (RST) can mine useful information from a large number of data and generate decision rules without prior knowledge. Support vector machines (SVMs) have good classification performances and good capabilities of fault-tolerance and generalization. To inherit the merits of both RST and SVMs, a hybrid classifier called rough set support vector machines (RS-SVMs) is proposed to recognize radar emitter signals in this paper. RST is used as preprocessing step to improve the performances of SVMs. A large number of experimental results show that RS-SVMs achieve lower recognition error rates than SVMs and RS-SVMs have stronger capabilities of classification and generalization than SVMs, especially when the number of training samples is small. RS-SVMs are superior to SVMs greatly.

1 Introduction

For many practical problems, including pattern matching and classification [1,2] function approximation [3], data clustering [4] and forecasting [5], Support Vector Machines (SVMs) have drawn much attention and been applied successfully in recent years. The subject of SVM covers emerging techniques that have been proven successful in many traditionally neural network-dominated applications [6]. An interesting property of SVM is that it is an approximate implementation of the structure risk minimization induction principle that aims at minimizing a bound on the generation error of a model, rather than minimizing the mean square error over the data set [6]. SVM is considered as a good learning method that can overcome the internal drawbacks of neural networks [7,8,9]. Although SVMs have strong capability of recognizing patterns and good capabilities of fault-tolerance and generalization, SVMs cannot reduce the input data and select the most important information.

Rough Set Theory (RST) can supplement the deficiency of SVMs effectively. RST, introduced by Zdaislaw Pawlak [10] in his seminal paper of 1982, is a new mathematical approach to uncertain and vague data analysis and is also a new

---

* This work was supported by the National EW Laboratory Foundation (No.NEWL51435QT22 0401).

fundamental theory of soft computing [11]. In recent years, RST becomes an attractive and promising research issue. Because RST can mine useful information from a large number of data and generate decision rules without prior knowledge [12,13], it is used generally in many fields [10-16], such as knowledge discover, machine learning, pattern recognition and data mining. RST has strong capabilities of qualitative analysis and generating rules, so it is introduced to preprocess the input data of SVMs so as to extract the key elements to be the inputs of SVMs.

In our prior work, an Interval-Valued Attribute Discretization approach (IVAD) was presented to process the continuously interval-valued features of radar emitter signals [17]. RST was combined with neural networks to design rough neural networks and experimental results verify that rough neural networks are superior to neural networks [18]. Unfortunately, neural networks have some unsolved problems, such as over-learning, local minimums and network structure decision, especially many difficulties for determining the neural nodes of hidden layers [7,8,9]. So this paper incorporates SVMs with RST to design a hybrid classifier called Rough Set Support Vector Machines (RS-SVMs). The new classifier inherits the merits of both RST and SVMs. Experimental results show that the introduction of RST not only enhances recognition rates and recognition efficiencies of SVMs, but also strengthens classification and generalization capabilities of SVMs.

This paper is organized as follows. Section 2 gives feature selection method using RST. Section 3 presents a hybrid classifier based on RST and SVMs. Simulation experimental results are analyzed in section 4. Conclusions are drawn in Section 5.

2 Feature Selection Method

RST can only deal with discrete attributes. In engineering applications, especially in pattern recognition and machine learning, the features obtained using some feature extraction approaches usually vary in a certain range (interval values) instead of fixed values because of some reasons such as plenty of noise. The existing discretization methods based on cut-splitting cannot deal with the information system that contains some interval attribute values effectively, while IVAD can discretize well the interval-valued continuous features. So the IVAD is firstly used to process the features.

The key problem of IVAD is to choose a good class-sepability criterion function. When an attribute value varies in a certain range, in general, the attribute value always orders a certain law. Without loss of generality, suppose the law is approximate Gaussian distribution. This paper uses the below class-separability criterion function in feature discretization.

$$J = 1 - \frac{\int f(x)g(x)dx}{\sqrt{\int f^2(x)dx} \cdot \sqrt{\int g^2(x)dx}} \tag{1}$$

In (1), $f(x)$ and $g(x)$ represent the probability distribution functions of attribute values of two objects in universe U in a decision system, respectively. Using the criterion function and discretization algorithm in [17], the interval-valued features can be discretized effectively. After discretizing continuous features, some methods in

RST can be used to select the most discriminatory feature subset from the original feature set composed of a large number of features. This paper introduces attribute reduction method based on discernibility matrix and logic operation [12,13] to reduce discretized decision table. The detailed reduction algorithm is as follows.

Step 1 Computing discernibility matrix C_D of decision table.

Step 2 For the elements c_{ij} ($c_{ij} \neq 0, c_{ij} \neq \phi$) of all nonempty set in discernibility matrix C_D, construct corresponding disjunction logic normal form.

$$L_{ij} = \bigvee_{a_i \in c_{ij}} a_i \qquad (2)$$

Step 3 Conjunction operation is performed using all disjunction logic normal form L_{ij} and a conjunction normal form L is obtained finally, i.e.

$$L = \bigwedge_{c_{ij} \neq 0, c_{ij} \neq \phi} L_{ij} \qquad (3)$$

Step 4 Transforming the conjunction normal form L into disjunction normal form L' and achieve $L' = \vee L$.

Step 5 The results of attribute reduction is achieved. In disjunction normal form L', each conjunction item corresponds a result of attribute reduction of decision table and the attributes contained in the conjunction item constitute a set of condition attribute after reduction.

Although RST finds all the reducts of the information system, the multi-solution problem brings many difficulties to decide the input features of classifiers of automatic recognition. So this paper introduces the complexity of feature extraction to solve the problem. The complexity is measured using consuming time of feature extraction. In all reducts of the information system, the feature subset with the lowest complexity is considered as the final feature subset.

3 Rough Set Support Vector Machines

SVMs have good classification, fault-tolerance and generalization capabilities. Though, SVMs cannot select and reduce the input data. If the dimensionality of input vector is very high, the training time and testing time of SVMs will be very long. Moreover, high-dimensional input feature vector usually has some redundant data, which may lower the classification and generalization performances of SVMs. So it is very necessary to use some methods to preprocess the input data of SVMs. Fortunately, RST can mine useful information from a large number of features and eliminate the redundant features, without any prior knowledge. To introduce strong capabilities of qualitative analysis and generating rules into SVMs, this section uses RST as preprocessing step of SVMs to design a hybrid classifier called RS-SVMs. The structure of RS-SVMs is shown in Fig. 1. The steps of designing RS-SVM are as follows.

Step 1 Training samples are used to construct a decision table in which all attributes are represented with interval-valued continuous features.

Step 2 IVAD is employed to discretize the decision table and discretized decision table is obtained.

Step 3 Attribute reduction methods are applied to deal with the discrete attribute table. Using attribute reduction method, multiple solutions are usually obtained simultaneously. So the complexity of feature extraction is introduced to select the final feature subset with the lowest cost from the multiple reduction results. After selection again, the final decision rule can be achieved.

Step 4 According to the final decision rule obtained, Naïve Scaler algorithm [12] is used to discretize the attribute table discretized by using IVAD and decide the number and position of cutting points. Thus, all cutting-point values are computed in terms of the attribute table before discretization using IVAD and the discretization rule, i.e. the preprocessing rule of SVMs, is generated.

Step 5 The training samples are processed using the preprocessing rule and then are used to be the inputs to train SVMs.

Step 6 When SVM classifiers are tested using testing samples or SVM classifiers are used in practical applications, the input data are firstly dealt with using preprocessing rule and then are applied to be inputs of trained SVMs.

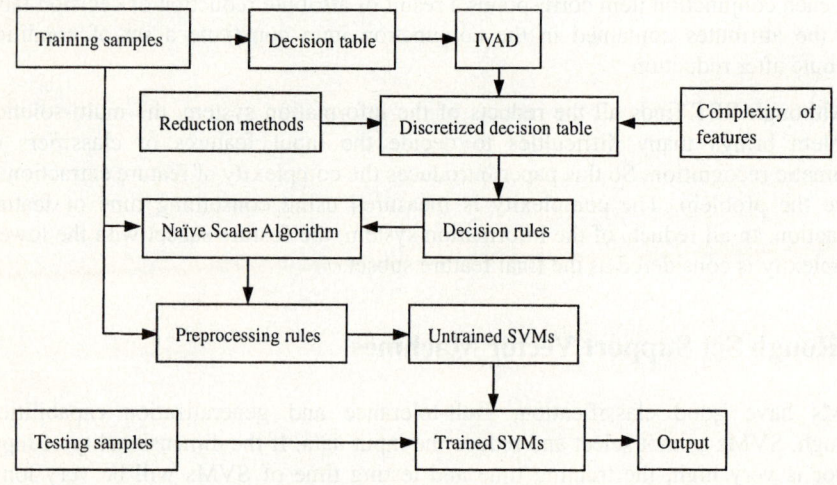

Fig. 1. Structure of RS-SVMs

SVMs were originally designed for binary classification. How to effectively extend it for multiclass classification is still an ongoing research issue [19]. Currently there are two types of approaches for multiclass SVM. One is by constructing and combining several binary classifiers while the other is by directly considering all data in one optimization formulation [19]. The former approach including mainly three methods: one-versus-rest (OVR) [19,20], one-versus-one (OVO) [19,21] and support vector machines with binary tree architecture (BTA) [22]. Some experimental results [1-9,19-22] show that the combinatorial classifier of several binary classifiers is a valid and practical way for solving muticlass classification problem.

4 Simulations

We choose 10 radar emitter signals to make simulation experiments. The 10 signals are represented with $x_1, x_2, \cdots, x_{10}$, respectively. In our prior work, 16 features have been extracted from the 10 radar emitter signals [23,24,25]. The 16 features are represented with $a_1, a_2, \cdots, a_{16}$, respectively. After discretization and reduction, the final result a_5, a_{10} is obtained. To bring into comparison, several feature selection approaches including Resemblance Coefficient method (RC) [25], Class-Separability method (CS) [26], Satisfactory Criterion method (SC) [27], Sequential Forward Search using distance criterion (SFS) [28], Sequential Floating Forward Search using distance criterion (SFFS) [28] and New Method of Feature Selection (NMFS) [29]. The results obtained using RC, CS, SC, SFS, SFFS and NMFS are $a_2 a_7$, $a_4 a_{15}$, $a_5 a_{12}$, $a_1 a_4$, $a_4 a_5$ and $a_6 a_7$, respectively. To test the classification performance of the results obtained by the 7 feature selection methods, BTA is used to construct SVM classifiers to recognize 10 radar emitter signals. Average accurate recognition rates obtained by using the 7 feature selection methods are shown in Table 1.

Table 1. Comparison of recognition rates (RR) obtained by 7 methods

Methods	RC	CS	SC	SFS	SFFS	NMFS	Proposed
Features	$a_2 a_7$	$a_4 a_{15}$	$a_5 a_{12}$	$a_1 a_4$	$a_4 a_5$	$a_6 a_7$	$a_5 a_{10}$
RR (%)	93.89	87.69	95.12	63.27	84.73	77.68	95.32

Table 1 shows that the average recognition rate of the proposed method is higher than other 6 methods, which indicates that the feature selection method using RST is superior to other 6 methods. Simultaneously, the experimental results show that the introduced discretization method is feasible and valid.

The classification and generalization capabilities of RS-SVMs are compared with those of SVMs using the following experiments. 6 classifiers including OVR-SVM, OVO-SVM, BTA-SVM, OVR-RS-SVM, OVO-RS-SVM and BTA-RS-SVM are employed to recognize the 10 radar emitter signals. The inputs of the 6 classifiers uses the selected feature subset obtained by the proposed method, i.e. two features a_5 and a_{10}. Performance criteria including recognition error rate and recognition efficiency are used to evaluate the several classifiers. Recognition efficiency includes training time (Trt) and testing time (Tet). The samples in training group are used to train 6 classifiers and then the samples in testing group are applied to test the trained classifiers. Statistical results of 100 experiments using the 6 classifiers are shown in Table 2, respectively.

To compare the training time and the capabilities of classification and generalization of SVMs with those of RS-SVMs, different samples including 10, 20, 30, 40 and 50 are respectively applied to train OVR-SVM, OVO-SVM, BTA-SVM and OVR-RS-SVM, OVO-RS-SVM, BTA-RS-SVM. Also, testing samples of 5 dB, 10 dB, 15 dB and 20 dB are respectively used to test trained SVM and RS-SVM. After 100 experiments, the changing curves of average recognition rates (ARR) obtained using OVR-SVM and OVR-RS-SVM, OVO-SVM and OVO-RS-SVM, BTA-SVM and BTA-RS-SVM, are shown in Fig.2, Fig.3, Fig.4, respectively. The average training time (ATT) spent by OVR-SVM and OVR-RS-SVM, OVO-SVM and OVO-RS-SVM, BTA-SVM and BTA-RS-SVM, are shown in Fig.5, Fig.6, Fig.7, respectively. All experiments are made using a personal computer (P-IV, CPU: 2.0GHz, EMS memory: 256Mb).

Table 2. Experimental result comparison of 6 classifiers (%)

Signals	SVMs			RS-SVMs		
	OVR	OVO	BTA	OVR	OVO	BTA
x_1	0	0	20.36	0.40	0	0
x_2	13.86	0	26.00	0	0	0
x_3	0	0	0	0	0	0
x_4	0	0	0	0	0	0
x_5	0	0	0	0	0	0
x_6	0	0	0	0	0	0
x_7	0	0	0	0	0	0
x_8	73.20	0	0	0	0	0
x_9	0.33	0	0	0	0	0
x_{10}	0	0	0.39	0	0	0
Error rate	8.74	0	4.68	0.04	0	0
Trt (sec.)	754.74	51.96	101.14	878.11	52.58	108.53
Tet (sec.)	233.58	6.56	4.22	232.38	27.64	25.02

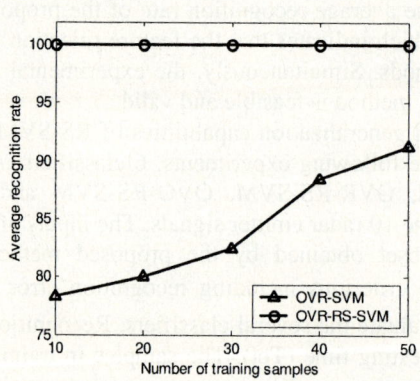

Fig. 2. ARR of OVR-SVM and OVR-RS-SVM

From Table 2 and Fig. 2 to Fig.7, several conclusions can be drawn as follows.

(1) Table 2 shows that recognition rates of 3 RS-SVM classifiers including OVR-RS-SVM, OVO-RS-SVM and BTA-RS-SVM are higher than or not less than those of 3 SVM classifiers including OVR-SVM, OVO-SVM and BTA-SVM. When the number of training samples is 50, three RS-SVM classifiers including OVR-RS-SVM,

OVO-RS-SVM, BTA-RS-SVM and OVO-SVM classifier are good classifiers to recognize the 10 radar emitter signals using the selected features.

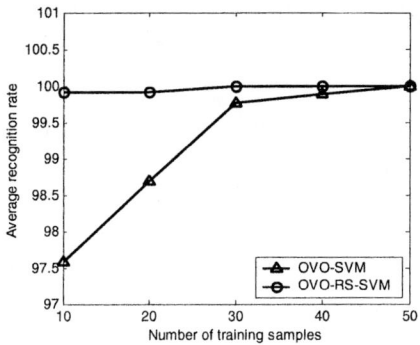

Fig. 3. ARR of OVO-SVM and OVO-RS-SVM

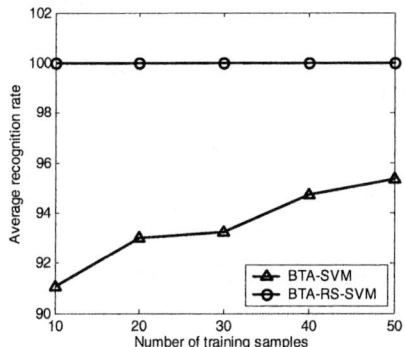

Fig. 4. ARR of BTA-SVM and BTA-RS-SVM

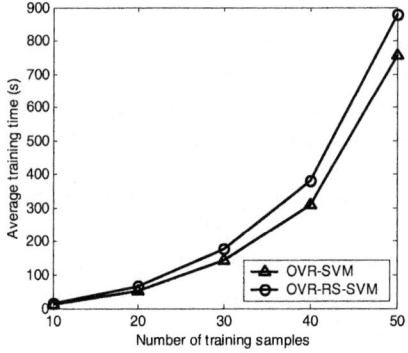

Fig. 5. ATT of OVR-SVM and OVR-RS-SVM

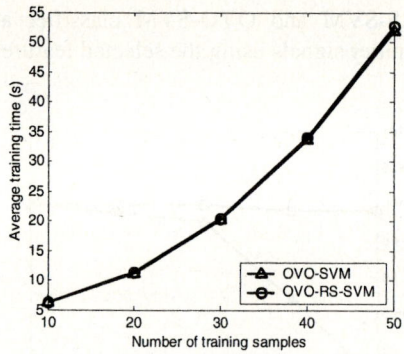

Fig. 6. ATT of OVO-SVM and OVO-RS-SVM

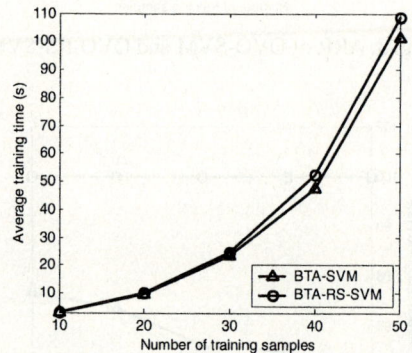

Fig. 7. ATT of BTA-SVM and BTA-RS-SVM

(2) Table 2 and Fig. 5 to Fig.7 show that RS-SVM classifiers need more time than that of SVM classifiers because the discretization procedure in RS-SVM consumes a little time.

(3) From Fig.2 to Fig.4, recognition error rates of 3 RS-SVM classifiers are lower than those of 3 SVM classifiers when the number of training samples varies from 10 to 50. The experimental results indicate that classification and generalization capabilities of RS-SVM classifiers are much stronger than those of SVM classifiers, especially when the number of training samples is small. Fig.2 to Fig.4 also show that classification and generalization capabilities of RS-SVM classifiers when the number of training samples is 10 correspond with those of SVM classifiers when the number of training samples is 50. That is to say, RS-SVM classifiers with 10 training samples are superior to SVM classifiers with 50 training samples because the former have much lower recognition error rates and much shorter training and testing time than the latter. In 3 RS-SVM classifiers, the OVO-RS-SVM classifier is the best from recognition rate and recognition efficiency.

(4) When the number of training samples is 50, OVO-SVM seems to be the best classifier from the evaluation criterions of 6 classifiers. However, Fig.3 and Fig.6 indicate that OVO-RS-SVM is superior to SVM when the number of training samples decreases.

(5) If the same values of evaluation criterions of classifiers are obtained, RS-SVM classifiers need much shorter training time and testing time than that of SVM classifiers because RS-SVM classifiers need smaller training samples.

Therefore, the above analysis indicates that the introduction of rough set theory decreases recognition error rates and enhances recognition efficiencies of SVM classifiers, and strengthens classification and generalization capabilities of SVM classifiers.

5 Conclusions

This paper combines RST with SVMs to design a hybrid classifier. RST is used to preprocess the input data of SVMs both in training procedure and in testing procedure. Because RST selects the most discriminatory features from a large number of features and eliminates the redundant features, the preprocessing step enhances the efficiency of SVMs in training and testing phases and strengthens classification and generation capabilities of SVMs. Experimental results verify that RS-SVMs are much superior to SVMs in recognition capability and in recognition efficiency. The proposed hybrid classifier is promising in other applications, such as image recognition, speech recognition and machine learning.

References

1. Osareh, A., Mirmehdi1, M., Thomas, B., and Markham, R.: Comparative Exudate Classification Using Support Vector Machines and Neural Networks. Lecture Notes in Computer Science, Vol.2489. (2002) 413-420
2. Foody, G.M., Mathur, A.: A Relative Evaluation of Multiclass Image Classification by Support Vector Machines. IEEE Transactions on Geoscience and Remote Sensing, Vol.42, No. 6. (2004) 1335-1343
3. Ma, J.S., Theiler, J., and Perkins, S.: Accurate On-line Support Vector Regression. Neural Computation, Vol.15, No.11. (2003) 2683-2703
4. Ben-Hur, A., Horn, D., Siegelmann, H.T., and Vapnik, V.: Support Vector Clustering. Journal of Machine Learning Research, Vol.2, No.2. (2001) 125-137
5. Kim, K.J.: Financial Time Series Forecasting Using Support Vector Machines. Neurocomputing, Vol.55, No.1. (2003) 307-319
6. Dibike, Y.B., Velickov, S., and Solomatine, D.: Support Vector Machines: Review and Applications in Civil Engineering. Proc. of the 2nd Joint Workshop on Application of AI in Civil Engineering, (2000) 215-218
7. Kecman, V.: Learning and Soft Computing, Support Vector Machines, Neural Networks and Fuzzy Logic Models. The MIT Press, Cambridge, MA (2001)
8. Wang, L.P. (Ed.): Support Vector Machines: Theory and Application. Springer-Verlag, Berlin Heidelberg New York (2005)

9. Samanta, B.: Gear Fault Detection Using Artificial Neural Networks and Support Vector Machines with Genetic Algorithms. Mechanical Systems and Signal Processing, Vol.18, No.3. (2004) 625-644
10. Pawlak, Z.: Rough Sets. Informational Journal of Information and Computer Science, Vol.11, No.5. (1982) 341-356
11. Lin, T.Y.: Introduction to the Special Issue on Rough Sets. International Journal of Approximate Reasoning, Vol.15, No.4. (1996) 287-289
12. Wang, G.Y.: Rough Set Theory and Knowledge Acquisition. Xi'an Jiaotong University Press, Xi'an (2001)
13. Walczak, B., Massart, D.L.: Rough Sets Theory. Chemometrics and Intelligent Laboratory Systems, Vol.47, No.1. (1999) 1-16
14. Dai, J.H., Li, Y.X.: Study on Discretization Based on Rough Set Theory. Proc. of the first Int. Conf. on Machine Learning and Cybernetics, (2002) 1371-1373
15. Roy, A., Pal, S.K.: Fuzzy Discretization of Feature Space for a Rough Set Classifier. Pattern Recognition Letter, Vol.24, No.6. (2003) 895-902
16. Mitatha, S., Dejhan, K., Cheevasuvit, F., and Kasemsiri, W.: Some Experimental Results of Using Rough Sets for Printed Thai Characters Recognition. International Journal of Computational Cognition. Vol.1, No.4. (2003) 109–121
17. Zhang, G.X., Hu, L.Z., and Jin, W.D.: Discretization of Continuous Attributes in Rough Set Theory and Its Application. Lecture Notes in Computer Science, Vol.3314. (2004) 1020-1026
18. Zhang, G.X., Hu, L.Z., and Jin, W.D.: Radar Emitter Signal Recognition Based on Feature Selection Algorithm. Lecture Notes in Artificial Intelligence, Vol.3339. (2004) 1108-1114
19. Hsu, C.W., Lin, C.J.: A Comparison of Methods for Multiclass Support Vector Machines. IEEE Transaction on Neural Networks. Vol.13, No.2. (2002) 415-425
20. Rifkin, R., Klautau, A.: In Defence of One-Vs-All Classification. Journal of Machine Learning Research, Vol.5, No.1. (2004) 101-141
21. Platt, J.C., Cristianini, N., and Shawe-Taylor, J.: Large Margin DAG's for Multiclass Classification. Advances in Neural Information Processing Systems, Vol.12. (2000) 547-553
22. Cheong, S.M., Oh, S.H., and Lee, S.Y.: Support Vector Machines with Binary Tree Architecture for Multi-class Classification. Neural Information Processing-Letters and Reviews, Vol.2, No.3. (2004) 47-51
23. Zhang, G.X., Hu, L.Z., and Jin, W.D.: Intra-pulse Feature Analysis of Radar Emitter Signals. Journal of Infrared and Millimeter Waves, Vol.23, No.6. (2004) 477-480
24. Zhang, G.X., Rong, H.N., Jin, W.D., and Hu, L.Z.: Radar Emitter Signal Recognition Based on Resemblance Coefficient Features. Lecture Notes in Artificial Intelligence, Vol.3066. (2004) 665-670
25. Zhang, G.X., Jin, W.D., and Hu, L.Z.: Resemblance Coefficient and a Quantum Genetic Algorithm for Feature Selection. Lecture Notes in Artificial Intelligence. Vol.3245. (2004) 155-168
26. Zhang, G.X., Hu, L.Z., and Jin, W.D.: Quantum Computing Based Machine Learning Method and Its Application in Radar Emitter Signal Recognition. Lecture Notes in Artificial Intelligence, Vol.3131. (2004) 92-103
27. Zhang, G.X., Jin, W.D., and Hu, L.Z.: A Novel Feature Selection Approach and Its Application. Lecture Notes in Computer Science, Vol.3314. (2004) 665-671
28. Mitra, P., Murthy, C.A., and Pal, S.K.: Unsupervised Feature Selection Using Feature Similarity. IEEE Transactions on Pattern Analysis and Machine Intelligence, Vol.24, No.3. (2002) 301-312
29. Lü, T.J., Wang, H., and Xiao, X.C.: Recognition of Modulation Signal Based on a New Method of Feature Selection. Journal of Electronics and Information Technology, Vol.24, No.5. (2002) 661-666

A Heuristic Algorithm for Maximum Distribution Reduction

Xiaobing Pei and YuanZhen Wang

Department of Computer Science,
HuaZhong University of Science & Technology,
Wuhan, Hubei 430074, China
xiaobingp@tom.com

Abstract. Attribute reduction is one of the basic contents in decision table. And it has been proved that computing the optimal attribute reduction is NP-complete. A lot of algorithms for the optimal attribute reduction were proposed in consistent decision table. But most decision tables are inconsistent in fact. In this paper, the judgment theorem with respect to maximum distribution reduction is obtained and the significance of attributes is defined in decision table, from which a polynomial heuristic algorithm for the optimal maximum distribution reduction is proposed. Finally, the experimental results show that this algorithm is effective and efficient.

1 Introduction

The rough set theory [1] was first proposed by professor Pawlak in 1982, it is an excellent mathematics tool to handle imprecise and uncertain knowledge, it has been successfully applied in many fields such as data mining, decision support [3][4] etc.

Attribute reduction is one of the basic contents in rough set theory. It is well known that an information system or a decision table may usually have irrelevant and superfluous knowledge, which is inconvenient for us to get concise and meaningful decision. When reducing attributes, we should eliminate the irrelevant and superfluous knowledge without losing essential information about the original data in decision table. It is NP-complete problem to computing the optimal attribute reductions [2].

Nowadays, the main algorithms for the optimal reduction have been proposed based on positive region [5], mutual information [6], attribute frequency [7] and attribute ordering [8] etc in consistent decision table. But most decision tables are inconsistent in fact. Zhang et al [9] introduced the concept of maximum distribution reduction in decision table and proposed an algorithm for the set of all maximum distribution reduction based on discernible matrix, but the algorithm is not efficient, its time complexity is exponent. In this paper, the judgment theorem with respect to maximum distribution reduction will be given and the significance of attributes will be defined, from which a polynomial heuristic algorithm for the optimal maximum distribution reduction will be proposed. Finally, experimental results for this algorithm will be given.

2 Basic Concepts of Rough Set Theory

In this section, we introduce only the basic notations from rough set approach used in the paper.

A decision table S is defined by S=<U, A, V, f>, A=C$\bigcup$D is a set of attributes, where C and D are condition attributes and decision attributes respectively, U={$x_1, x_2, ..., x_n$} is a non-empty finite set of objects. V=$\bigcup_{a \in A} V_a$, and V_a is value set of attribute a. f: U×A → V is a total function such that f(x_i, a)∈ V_a for each a∈ A, x_i ∈ U.

For every B ⊆ A defines an equivalence relation denoted by ind(B) called indiscernibility relation as defined below: ind(B)={(x,y): f(x, a_k)=f(y, a_k), $\forall a_k \in B$}.

Thus U/ind(B) is a set of equivalence classes, as defined below, U/ind(B) ={$[x]_B$: x∈ U}, where $[x]_B$ ={y: (x,y)∈ ind(B)} is a equivalence class for an example x with respect to concept B.

If ind(C) ⊆ ind(D) then it can be said that the decision table S is consistent, else inconsistent. Certain rules can be generated in consistent decision table and possible rules or uncertain rules can also be generated from inconsistent decision table.

Let U/ind(D)={$D_1, D_2, ..., D_r$}. The maximum distribution information vector of x∈ U with respect to attribute set C is defined by:
Dinf$_C$(x)={$d_1, d_2, ..., d_r$},

$$d_j = \begin{cases} 1, & P(D_j / [x]_C) = Max\{P(D_k / [x]_C) \mid k = 1, 2, ..., r\} \\ 0, & P(D_j / [x]_C) \neq Max\{P(D_k / [x]_C) \mid k = 1, 2, ..., r\} \end{cases},$$

Where $P(D_j / [x]_C) = \frac{|D_j \cap [x]_C|}{|[x]_C|}$.

For every subset X of U, if Dinf$_C$(x)=Dinf$_C$(y) for every x, y∈ X then the maximum distribution information vector of X with respect to attribute set C is defined by: Dinf$_C$(X)=Dinf$_C$(y), where y∈ X.

Let B is an attribute subset of set C, if Dinf$_B$(x)=Dinf$_C$(x) for every x∈ U, then B can be called maximum distribution consistent set. If B is a maximum distribution consistent set and no proper subset of B is maximum distribution consistent set, then B can be called maximum distribution reduction. There may be more than one maximum distribution reductions in decision table [9].

The maximum distribution reductions are particular subset of attributes with the same preserving maximum decision rules as the full set of attributes.

3 Judgement Theorem of Maximum Distribution Reduction

In this section, the judgment theory with respect to maximum distribution reduction will be obtained.

Theorem 1: Let S=<U, A, V, f> be a decision table, A=C$\cup$D, C and D are condition attributes and decision attributes respectively, B $\subseteq$ C is a attribute set. U/ind(D)={$D_1, D_2, ..., D_r$}, U/ind(C)={$X_1, X_2, ..., X_n$}, U/ind(B) ={$Y_1, Y_2, ..., Y_m$}, Y_j (j$\leq$m)={$X'_{j1}, X'_{j2}, ..., X'_{jt_j}$}, where $X'_{ji} \in$ U/ind(C), $1 \leq i \leq t_j$. Then attributes set B is maximum distribution consistent set with respect to attribute set C if and only if for each j ($1 \leq j \leq m$) satisfy $\text{Dinf}_C(X'_{j1})=...=\text{Dinf}_C(X'_{jt_j})$.

Proof: If B is a maximum distribution consistent set, then $\text{Dinf}_B(x)=\text{Dinf}_B(x)$ for each x$\in$ U. Therefore, we have $\text{Dinf}_C(X'_{ji})=\text{Dinf}_B(Y_j)$ for each X'_{ji} (i=1,2,...,t_j), that is $\text{Dinf}_C(X'_{j1})=...=\text{Dinf}_C(X'_{jt_j})$.

On the other hand, for each x$\in$ U, there exist j and i such that x$\in X'_{ji} \in$ U/ind(C), x$\in Y_j \in$ U/ind(B).

Suppose that there exist $l \leq r$ such that $P(D_l / X'_{ji})=\text{Max}\{P(D_k / X'_{ji}) | k=1,2,...,r\}$, where i=1,2,..., t_j. Since $\text{Dinf}_C(X'_{j1})=...=\text{Dinf}_C(X'_{jt_j})$, it is easy to conclude that
$$\frac{|D_l \cap X'_{ji}|}{|X'_{ji}|} > \frac{|D_m \cap X'_{ji}|}{|X'_{ji}|} \text{ for all } m \leq r, m \neq l.$$

Thus we conclude that $\dfrac{\sum_{k=1}^{t_j}|D_l \cap X'_{jk}|}{\sum_{k=1}^{t_j}|X'_{jk}|} > \dfrac{\sum_{k=1}^{t_j}|D_j \cap X'_{jk}|}{\sum_{k=1}^{t_j}|X'_{jk}|}$, $m \leq r, m \neq l$.

Therefore we have $\text{Dinf}_C(X'_{ji})=\text{Dinf}_B(Y_j)$. That is $\text{Dinf}_B(x)=\text{Dinf}_C(x)$ for each x$\in$ U. Thus we conclude that B is a maximum distribution consistent set with respect to attribute C.

We conclude the judgment theorem of maximum distribution reduction as below according to theorem 1.

Theorem 2: Let S=<U, A, V, f> be a decision table, A=C$\cup$D, C and D are condition attributes and decision attributes respectively, B$\subseteq$C is a attribute set. Then B is maximum distribution reduction if and only if 1) B is maximum distribution consistent set; 2) for each $a \in B$, there is j $(1 \leq j \leq m)$ such that $\text{Dinf}_C(X'_{j1})=...=\text{Dinf}_C(X'_{jt_j})$ is not true, where $U/\text{ind}(B-\{a\})=\{Y_1, Y_2,...,Y_m\}$, Y_j $(j \leq m)=\{X'_{j1}, X'_{j2},...,X'_{jt_j}\}$, $X'_{ji} \in U/\text{ind}(C)$, $1 \leq i \leq t_j$.

Proof: We can directly complete the proof according to theorem 1.

4 Heuristic Algorithm for Maximum Distribution Reduction

4.1 The Significance of Attributes

Definition 1: Let $S=<U, A, V, f>$ be a decision table, $A=C \cup D$, C and D are condition attributes and decision attributes respectively. The significance of attribute $a \in C$ is defined by:

SGF (a, C, D)= $Card$ ($D_{C-\{a\}}$), where $D_{C-\{a\}} =\{[x]_C \mid \text{Dinf}_C([x]_C) \neq \text{Dinf}_{C-\{a\}}(x), x \in U \}$, $Card$ ($D_{C-\{a\}}$) denotes cardinality of set $D_{C-\{a\}}$.

The more big SGF (a, C, D) is, the more important attribute a is, with respect to attributes C. In this paper, we regard SGF (a, C, D) as heuristic information to search for the optimal reduction.

4.2 Heuristic Algorithm for the Optimal Maximum Distribution Reduction

A heuristic algorithm for the optimal maximum distribution reduction will be proposed based on theorem 2 and definition 1.

Algorithm 1: A heuristic algorithm for optimal maximum distribution reduction
Input: Let $S=<U, A, V, f>$, $A=C \cup D$, C is condition attributes and D is decision attributes;

Output: the optimal maximum distribution reduction of decision table S.

Step 1. Calculate $U/\text{ind}(C)=\{X_1, X_2,...,X_n\}$; Calculate $\text{Dinf}_C(X_j)$, where $X_j \in U/\text{ind}(C), j \leq n$;

Step 2. Calculate SGF (a, C, D) for each $a \in C$ and arrange the attribute set C into sort ascending by using SGF (a, C, D).
Step 3. Let RED=C;
Step 4. Judge whether for each attribute $a \in$ RED is required from back to front in attribute set RED:
Step 5.1 if SGF (a, RED, D)=0, then RED=RED-{a};
Step 6. Return RED;

In most case, |D|=1, where |D| denotes cardinality of decision attributes set D. The time complexity of step 1 is $O(|C|^2 |U|^2)$; the time complexity of step 2 is $O(|C|^2 |U|^2)$, the time complexity of step 4 is $O(|C|^2 |U|^2)$. Therefore the time complexity of algorithm 1 is $O(|A|^2 |U|^2)$ according to $|C|<|A|$.

An algorithm [9] has been proposed for the set of all maximum distribution reduction in inconsistent decision table, but it is not efficient and its time complexity is exponent $O(2^{|A|} |U|^2 |A|)$. Therefore, the algorithm 1 is much more efficient in comparison with algorithm [9] for the optimal maximum distribution reduction.

4.3 Experimentation

The heuristic algorithm for the optimal maximum distribution reduction was implemented on Microsoft Access for Windows2000, running on a PC Intel-Pentium, 2GHz, 256MB RAM.

We used algorithm 1 in this paper and the algorithm [9](called algorithm A) respectively on six databases in UCI. Experimental results were obtained in table 1.

Table 1. Experimental results

Decision table	Number of Instances	Number of Condition Attributes	Algorithm 1		Algorithm A	
			Number of attribute	running time (s)	Number of attribute	running time (s)
Post.	90	8	8	0.032	8	2.016
BUPA	345	6	3	0.782	3	95.625
Monk's (1)	432	6	1	0.781	1	339.23
Teaching	152	5	4	0.094	4	7.953
Hayes	132	5	1	0.141	1	4.954
Balance	625	4	4	0.422	4	554.01

According to results above, the heuristic algorithm 1 is effective and it is much more efficient in comparison with algorithm A for the optimal maximum distribution reduction.

5 Conclusions

A lot of algorithms have been proposed for optimal attribute reduction in consistent decision table, but most decision tables are inconsistent in fact. In this paper, the judgment theorem with respect to maximum distribution reduction was obtained and the significance of attributes was defined in decision table, from which a heuristic algorithm for optimal maximum distribution reduction was proposed. The experimental results show that this algorithm is effective and efficient.

References

1. Pawlak Z. Rough sets. International Journal of Computer and Information Science, 1982(5): 341-356
2. Wong S K M, Ziarko W. On optimal decision rules in decision tables. Bulletin of Polish Academy of Sciences, 1985(33): 693-696
3. Liu Qin. Rough Set and Rough Reasoning. Beijing: Science Press(2001)
4. Zhang Wen-Xiu, Wu Wei-Zhi, Liang Ji-Ye, Li De-Yu. Rough set Theory and Methods. Beijing: Science Press(2001)
5. Guan J W, Bell D A. Rough computational methods for information systems. Artificial Intelligences, 1998(105): 77~103
6. Miao Duo-Qian, Hu gui-Rong. A heuristic algorithm for reduction of knowledge. Journal Computer Research and Development, 1999(6): 681-684
7. Wang Jue, Wang Ren, Miao Duo-Qian et al. Data Enriching based on rough set theory. Chinese Journal of Computer, 1998(5): 393-395
8. Wang Jue, Wang Ju. Reduction algorithm based on discernibility matrix: The ordered attributes method. Journal of Computer Science & Technology, 2001(6): 489~504
9. Zhang Wenxiu, Mi jusheng, wu weizhi. Knowledge reduction in inconsistent information systems. Chinese Journal of computers, 2003(1): 12-18

The Minimization of Axiom Sets Characterizing Generalized Fuzzy Rough Approximation Operators

Xiao-Ping Yang

Information College, Zhejiang Ocean University,
Zhoushan, Zhejiang, 316004, P.R. China,
yxpzyp@zjou.net.cn

Abstract. In the axiomatic approach of fuzzy rough set theory, fuzzy rough approximation operators are characterized by a set of axioms that guarantees the existence of certain types of fuzzy binary relations reproducing the operators. Thus axiomatic characterization of fuzzy rough approximation operators is an important aspect in the study of rough set theory. In this paper, the independence of axioms of generalized fuzzy rough approximation operators is investigated, and their minimal sets of axioms are presented.

1 Introduction

Yang et al.[8] examined the minimization of the axiom set for approximation operators. Independence of axioms for rough approximation operators has been investigated. The research of axiomatic approach has also been extended to approximation operators in fuzzy environment [1–7]. The important axiomatic studies for fuzzy rough set are done by Wu [6, 7] who studied generalized rough set approximation operators in fuzzy environment, he examined many axioms on various classes of fuzzy rough approximation operators. In the axiomatic approach, various classes of fuzzy approximation operators are characterized by different sets of axioms, the axiom set of fuzzy approximation operators guarantee the existence of certain types of fuzzy relation producing the same operators.

The above mentioned studies have not solved the problem of the independence and minimization of the axiom set for fuzzy approximation operators. This paper attempts to solve this problem. Independence of axioms for fuzzy rough approximation operators is investigated, and minimal axiom sets corresponding to various generalized fuzzy approximation operators are presented.

2 Fuzzy Rough Approximation Operators

Let U be a finite and nonempty set called the universe of discourse. The class of all fuzzy subsets of U will be denoted by $\mathcal{F}(U)$. A fuzzy subset $R \in \mathcal{F}(U \times U)$ is referred to as a fuzzy binary relation on U, $R(x,y)$ is the degree of relation

between x and y, where $(x,y) \in U \times U$, R is a fuzzy binary relation on U. For any set $A \in \mathcal{F}(U)$, a pair of lower and upper approximations, $\underline{R}(A)$ and $\overline{R}(A)$, are defined respectively as follows: For all $x \in U$,
$$\overline{R}(A)(x) = \bigvee_{y \in U} [R(x,y) \wedge A(y)], \quad \underline{R}(A)(x) = \bigwedge_{y \in U} [(1 - R(x,y)) \vee A(y)].$$
$\underline{R}$, $\overline{R} : \mathcal{F}(U) \to \mathcal{F}(U)$ are referred to as lower and upper generalized fuzzy rough approximation operators respectively.

3 Axiom Sets of the Fuzzy Approximation Operators

In an axiomatic approach, the primitive notion is a system $(\mathcal{F}(U), \wedge, \vee, \sim, L, H)$, where $L, H : \mathcal{F}(U) \to \mathcal{F}(U)$ are fuzzy operators. We call L and H fuzzy approximation operators.

Definition 1. *Let $L, H : \mathcal{F}(U) \to \mathcal{F}(U)$ be two fuzzy operators. They are dual operators if for all $A \in \mathcal{F}(U)$:*
  (L_1) $L(A) = \sim H(\sim A)$,   or   (H_1) $H(A) = \sim L(\sim A)$.

By the duality of L and H, it is sufficient to introduce one operator and to define the other by using L_1 or H_1. For example, one may define the operator H and regard L as an abbreviation of $\sim H \sim$. We are interested in the following conditions (axioms) on L and H for all $A, B \in \mathcal{F}(U), x, y \in U, \alpha \in I$:

(L_2) $L(A \vee \widehat{\alpha}) = L(A) \vee \widehat{\alpha}$,   ($H_2$) $H(A \wedge \widehat{\alpha}) = H(A) \wedge \widehat{\alpha}$;
(L_3) $L(A \wedge B) = L(A) \wedge L(B)$,   ($H_3$) $H(A \vee B) = H(A) \vee H(B)$;
(L_0) $L(\widehat{\alpha}) = \widehat{\alpha}$,   ($H_0$) $H(\widehat{\alpha}) = \widehat{\alpha}$;
(L_6) $L(A) \subseteq A$,   (H_6) $A \subseteq H(A)$;
(L_7) $L(1_{U-\{x\}})(y) = L(1_{U-\{y\}})(x)$,   ($H_7$) $H(1_x)(y) = H(1_y)(x)$;
(L_8) $L(A) \subseteq L(L(A))$,   (H_8) $H(H(A)) \subseteq H(A)$;

where 1_y denotes the fuzzy singleton with value 1 at y and 0 elsewhere. Similarly, $1_{U-\{y\}}$ denotes the fuzzy singleton with value 0 at y and 1 elsewhere.

Definition 2. *Let L and H be dual fuzzy rough approximation operators, if L satisfies axioms L_2 and L_3, or equivalently, H satisfies axioms H_2 and H_3, then the system $(\mathcal{F}(U), \wedge, \vee, \sim, L, H)$ is referred to as a fuzzy rough set algebra, L and H are referred to as lower and upper fuzzy rough approximation operators respectively.*

For completeness, we summarize the main results of axiomatic characterizations of the fuzzy rough approximation operators corresponding to the different fuzzy binary relations in Theorem 3–Theorem 7 [6, 7].

Theorem 1. *Let $L, H : \mathcal{F}(U) \to \mathcal{F}(U)$ be dual fuzzy rough approximation operators. Then there exists a fuzzy binary relation R on U such that $\forall A \in \mathcal{F}(U)$, $L(A) = \underline{R}(A), H(A) = \overline{R}(A)$ iff L satisfies axioms L_2 and L_3 or equivalently, H satisfies axioms H_2 and H_3.*

Theorem 2. Let $L, H : \mathcal{F}(U) \to \mathcal{F}(U)$ be dual fuzzy rough approximation operators. Then there exists a serial fuzzy relation R on U such that $\forall A \in \mathcal{F}(U)$, $L(A) = \underline{R}(A), H(A) = \overline{R}(A)$ if and only if L satisfies axioms L_0, L_2 and L_3, or equivalently, H satisfies axioms H_0, H_2 and H_3.

Theorem 3. Let $L, H : \mathcal{F}(U) \to \mathcal{F}(U)$ be dual fuzzy rough approximation operators. Then there exists a reflexive fuzzy relation R on U such that $\forall A \in \mathcal{F}(U)$, $L(A) = \underline{R}(A), H(A) = \overline{R}(A)$ if and only if L satisfies axioms L_2, L_3 and L_6, or equivalently, H satisfies axioms H_2, H_3 and H_6.

Theorem 4. Let $L, H : \mathcal{F}(U) \to \mathcal{F}(U)$ be dual fuzzy rough approximation operators. Then there exists a symmetric fuzzy relation R on U such that $\forall A \in \mathcal{F}(U)$, $L(A) = \underline{R}(A), H(A) = \overline{R}(A)$ if and only if L satisfies axioms L_2, L_3 and L_7, or equivalently, H satisfies axioms H_2, H_3 and H_7.

Theorem 5. Let $L, H : \mathcal{F}(U) \to \mathcal{F}(U)$ be dual fuzzy rough approximation operators. Then there exists a transitive fuzzy relation R on U such that $\forall A \in \mathcal{F}(U)$, $L(A) = \underline{R}(A), H(A) = \overline{R}(A)$ if and only if L satisfies axioms L_2, L_3 and L_8, or equivalently, H satisfies axioms H_2, H_3 and H_8.

4 Minimal Axiom Sets of Operators

We now discuss independence of axiom sets characterizing serial, reflexive, symmetric, and transitive fuzzy approximation operators. (We only consider H in the theorems by the duality of L and H.)

Theorem 6. *Axioms H_2 and H_3 in Theorem 3 are independent.*

Proof. Firstly, we show that $H_2 \not\Rightarrow H_3$.

Example 1. Let $U = \{x_1, x_2\}$. For all $A \in \mathcal{F}(U)$, define $H(A) = \widehat{\min A(x_i)}$. (The $\min A(x_i)$ is the minimum of all $A(x_i)$ for $x_i \in U$, and the $\widehat{\min A(x_i)}$ denotes the constant fuzzy set with the value of $\min A(x_i)$ for all $x \in U$. Similarly, the $\widehat{\max A(x_i)}$ denotes the constant fuzzy set with the value of $\max A(x_i)$ for all $x \in U$.)

Thus, it is easy to verify that $H(A \wedge \widehat{\alpha}) = H(A) \wedge \widehat{\alpha}$, i.e., H_2 holds.

If $A = 1/x_1 + 0/x_2$, $B = 0/x_1 + 1/x_2$, then $A \vee B = 1/x_1 + 1/x_2$. $H(A \vee B) = \widehat{1}$, $H(A) = \widehat{0}, H(B) = \widehat{0}$. Thus, $H(A \vee B) \neq H(A) \vee H(B)$. This means $H_2 \not\Rightarrow H_3$.

Secondly, we show that $H_3 \not\Rightarrow H_2$.

Example 2. Define $H(A) = \widehat{\beta}, \forall A \in \mathcal{F}(U).$ (β is a constant, $0 < \beta \leq 1$, $\widehat{\beta}$ denotes the constant fuzzy set with the constant value β for all $x \in U$.)

Then, $H(A \vee B) = \widehat{\beta}$, $H(A) \vee H(B) = \widehat{\beta}$, thus, $H(A \vee B) = H(A) \vee H(B)$ for all $A, B \in \mathcal{F}(U)$. But when $\alpha < \beta$, $H(A \wedge \widehat{\alpha}) \neq H(A) \wedge \widehat{\alpha}$. This means that H_2 does not hold. This implies $H_3 \not\Rightarrow H_2$.

Therefore, we have proved that H_2 and H_3 are independent.

Theorem 7. *Axioms* H_2, H_3 *and* H_0 *in Theorem 4 are independent.*

Proof. Firstly, we show that $H_2, H_3 \not\Rightarrow H_0$.

Example 3. Let $U = \{x_1, x_2\}$ and a fuzzy relation R on U be defined by the fuzzy matrix:

$$R = \begin{pmatrix} 0.1 & 0.1 \\ 0.1 & 0.1 \end{pmatrix}$$

Define $H(A)(x) = \overline{R}(A)(x), \forall x \in U$.

It is easy to verify that H_2 and H_3 hold.

When $\alpha = 0.3$, $H(\widehat{\alpha})(x) = \bigvee_{y \in U}(R(x,y) \wedge \alpha) = (\bigvee_{y \in U} R(x,y)) \wedge \alpha = 0.1 \wedge 0.3 = 0.1 \neq \alpha$. This means H_0 does not hold. This implies $H_2, H_3 \not\Rightarrow H_0$.

Secondly, we show that $H_2, H_0 \not\Rightarrow L_3$. By referring to Example 1, we can easily see that H_0, H_2 hold and H_3 does not. This implies that $H_2, H_0 \not\Rightarrow H_3$.

Finally, we show that $H_3, H_0 \not\Rightarrow H_2$.

Example 4. Let $U = \{x_1, x_2\}$. For all $A \in \mathcal{F}(U)$, define

$$H(A) = \begin{cases} \widehat{1} & \max A(x_i) = 1; \\ A & \max A(x_i) < 1. \end{cases}$$

It is easy to verify that H_3 and H_0 are satisfied. If $A = 1/x_1 + 0/x_2$, $\alpha = 0.5$, then, $H(A \wedge \widehat{\alpha}) = A \wedge \widehat{0.5} = 0.5/x_1 + 0/x_2$, $H(A) \wedge \widehat{\alpha} = 0.5/x_1 + 0.5/x_2$. So, $H(A \wedge \widehat{\alpha}) \neq H(A) \wedge \widehat{\alpha}$. i.e., H_2 is not satisfied. This implies $H_3, H_0 \not\Rightarrow H_2$.

Therefore, we have proved that H_2, H_3 and H_0 are independent.

Theorem 8. *Axioms* H_2, H_3 *and* H_6 *in Theorem 5 are independent.*

Proof. Firstly, we show that $H_2, H_3 \not\Rightarrow H_6$. Let U and R be defined as Example 3. By referring to Example 3, it is easy to see that H_2 and H_3 hold.

When $A = \widehat{1}$, $H(A)(x) = \overline{R}(A)(x) = 0.1$,
thus, $A \subseteq H(A)$ does not hold. This implies $H_2, H_3 \not\Rightarrow H_6$.

Secondly, we show that $H_3, H_6 \not\Rightarrow H_2$. Let $H(A)$ be defined as Example 4. It is easy to see that H_3 and H_6 hold. By the definition of $H(A)$, when $A = 1/x_1 + 0/x_2$, and $\alpha = 0.5$, we can verify that H_2 does not hold. This implies $H_3, H_6 \not\Rightarrow H_2$.

Finally, we show that $H_2, H_6 \not\Rightarrow H_3$. Let $U = \{x_1, x_2\}$. Define

$$H(A) = \begin{cases} \widehat{\max A(x_i)} & \min A(x_i) \neq 0; \\ A & \min A(x_i) = 0. \end{cases}$$

H_6 holds by the definition of $H(A)$. Now, we verify the axiom H_2. If $\alpha = 0$, then $H(A \wedge \widehat{\alpha}) = H(\widehat{0}) = \widehat{0}$, $H(A) \wedge \widehat{\alpha} = \widehat{0}$, thus, $H(A \wedge \widehat{\alpha}) = H(A) \wedge \widehat{\alpha}$; If $\min A(x_i) = 0$, by the definition, $H(A) = A$, $H(A \wedge \widehat{\alpha}) = A \wedge \widehat{\alpha}$, thus, $H(A \wedge \widehat{\alpha}) = A \wedge \widehat{\alpha} = H(A) \wedge \widehat{\alpha}$; If $\alpha \neq 0$ and $\min(x_i) \neq 0$, $H(A \wedge \widehat{\alpha}) = \widehat{\max(A \wedge \widehat{\alpha})}(x_i) = \widehat{\max A(x_i)} \wedge \widehat{\alpha} = H(A) \wedge \widehat{\alpha}$. Thus, H_2 holds.

When $A = 0.5/x_1 + 0/x_2$, $B = 0/x_1 + 0.6/x_2$,
$H(A \vee B) = H(0.5/x_1 + 0.6/x_2) = \widehat{0.6}$, $H(A) \vee H(B) = 0.5/x_1 + 0.6/x_2$.
Thus, $H(A \vee B) \neq H(A) \vee H(B)$. This implies H_2, $H_6 \not\Rightarrow H_3$.
Therefore, we have proved that H_2, H_3 and H_6 are independent.

Theorem 9. *Axioms H_2, H_3 and H_7 in Theorem 6 are independent.*

Proof. Firstly, we show that H_2, $H_3 \not\Rightarrow H_7$. Let $U = \{x_1, x_2\}$ and a fuzzy relation R on U be defined by the fuzzy matrix:

$$R = \begin{pmatrix} 0.2 & 0.3 \\ 0.2 & 0.3 \end{pmatrix}$$

Define $H(A)(x) = \overline{R}(A)(x)$, $x \in U, A \in \mathcal{F}(U)$. Obviously, H_2 and H_3 hold. By the definition of $H(A)$, $H(1_x)(y) = R(y, x)$, $H(1_y)(x) = R(x, y)$. Obviously, $R(x_1, x_2) \neq R(x_2, x_1)$, i.e., $H(1_{x_1})(x_2) \neq H(1_{x_2})(x_1)$. Thus, H_7 does not hold. This implies H_2, $H_3 \not\Rightarrow H_7$.

Secondly, we show that H_3, $H_7 \not\Rightarrow H_2$.
Let $U = \{x_1, x_2\}$. Define

$$H(A) = \begin{cases} \widehat{1} & \max A(x_i) = 1; \\ A & \max A(x_i) < 1. \end{cases}$$

By the definition of $H(A)$, $H(1_x) = \widehat{1}$, $H(1_y) = \widehat{1}$, thus, H_7 holds. It is easy to verify that H_3 holds, too. But H_2 does not hold. (see Example 4.) This implies that H_3, $H_7 \not\Rightarrow H_2$.

Finally, we show that H_2, $H_7 \not\Rightarrow H_3$. Define $H(A) = \widehat{\min A(x_i)}$, then $H(1_x) = \widehat{0}$, $H(1_y) = \widehat{0}$. Thus, $H(1_x)(y) = H(1_y)(x)$, i.e., H_7 holds. It is easy to verify that H_2 holds, too.
When $A = 1/x_1 + 0/x_2$, $B = 0/x_1 + 1/x_2$, we see $H(A \vee B) \neq H(A) \vee H(B)$. This implies that H_2, $H_7 \not\Rightarrow H_3$.
Therefore, we have proved that axioms H_2, H_3 and H_7 are independent.

Theorem 10. *Axioms H_2, H_3 and H_8 in Theorem 7 are independent.*

Proof. Firstly, we show that H_2, $H_3 \not\Rightarrow H_8$. Let $U = \{x_1, x_2\}$ and a fuzzy relation R on U be defined by the fuzzy matrix:

$$R = \begin{pmatrix} 0.1 & 0.2 \\ 0.2 & 0.1 \end{pmatrix}$$

For all $A \in \mathcal{F}(U)$, Define $H(A)(x) = \overline{R}(A)(x)$, $x \in U$, then H_2, H_3 hold. When $A = 1/x_1 + 0/x_2$, It is easily to verify that $H(H(A)) \not\subseteq H(A)$. This implies that H_2, $H_3 \not\Rightarrow H_8$.

Secondly, we show that H_2, $H_8 \not\Rightarrow H_3$. Let $U = \{x_1, x_2\}$, $H(A) = \widehat{\min A(x_i)}$. Obviously, H_8 holds, H_2 holds, too. But H_3 does not hold. (See Example 1.) This implies that H_2, $H_8 \not\Rightarrow H_3$.

Finally, we show that H$_3$, H$_8 \not\Rightarrow$ H$_2$. For all $A \in \mathcal{F}(U)$, Define

$$H(A) = \begin{cases} \widehat{1} & \max A(x_i) = 1; \\ A & \max A(x_i) < 1. \end{cases}$$

Obviously, $H(H(A)) = H(A)$, i.e., H$_8$ holds. H$_3$ holds, too. But H$_2$ does not hold. (See Example 4.) This implies that H$_3$, H$_8 \not\Rightarrow$ H$_2$.

Therefore, we have proved that axioms H$_2$, H$_3$ and H$_8$ are independent.

5 Conclusion

We have investigated the independence of axioms characterizing generalized fuzzy approximation operators. The minimal sets of axioms characterizing generalized fuzzy rough approximation operators have been obtained, which complements the results of Wu [7]. Meanwhile, the problem proposed by Yang [8] has been solved.

Acknowledgement

This work was supported by a grant from the National Natural Science Foundation of China (No. 60373078) and the Scientific Research Fund of Zhejiang Provincial Education Department in China (No. 20040538).

References

1. Boixader, D., Jacas, J., Recasens, J.: Upper and lower approximations of fuzzy sets. International Journal of General Systems **29** (2000) 555–568
2. Mi, J.-S., Zhang, W.-X.: An axiomatic characterization of a fuzzy generalization of rough sets. Information Sciences **160** (2004) 235–249
3. Morsi, N. N., Yakout, M. M.: Axiomatics for fuzzy rough sets. Fuzzy Sets and Systems **100** (1998) 327–342
4. Radzikowska, A. M., Kerre, E. E., A comparative study of fuzzy rough sets. Fuzzy Sets and Systems **126** (2002) 137–155
5. Thiele, H.: On axiomatic characterisation of fuzzy approximation operators I, the fuzzy rough set based case. RSCTC 2000, Banff Park Lodge, Bariff, Canada, Oct. 19–19, 2000. Conference Proceedings, 239–247
6. Wu, W.-Z., Mi, J.-S., Zhang, W.-X.: Generalized fuzzy rough sets. Information Sciences **151** (2003) 263–282
7. Wu, W.-Z., Zhang, W.-X.: Constructive and axiomatic approaches of fuzzy approximation operators. Information Sciences **159** (2004), 233–254
8. Yang, X.-P., Li, T.-J.: The minimization of axiom sets characterizing generalized approximation operators. Information Sciences (to appear)

The Representation and Resolution of Rough Sets Based on the Extended Concept Lattice

Xuegang Hu, Yuhong Zhang, and Xinya Wang

Department of Computer Science and Technology, Hefei University of Technology,
Hefei, 230009, China
jsjxhuxg@hfut.edu.cn, zhangyuhong99@163.com,
crystal0085@163.com

Abstract. Rough set (RS) theory is a mathematics tool for handling uncertain problem. It is helpful for KDD, but expensive consumption of time and unclear expression of result are the main problem in practical application. The extended concept lattice (ECL) is a new form of concept lattice which is gotten by introducing equivalence intension into Galois concept lattice (GCL). The ECL is an efficient tool for data analysis and knowledge discovery in database (KDD). Both ECL and RS are based on equivalence class, so the relative between them exists. This paper describes the ECL first, then discusses the relation between the ECL and RS, and describes the implementation of rough set based on ECL.

Keywords: KDD, Concept Lattice, Rule, Rough set

1 Introduction

Rough set (RS) theory[1] is a mathematics tool for handling uncertain problem, and is helpful for KDD. Researchers pay attention to rough set Theory since being proposed, a lot of results in rough set was generated. However, the expensive consumption of time is always the main problem when it is used to deal with the reduction and classification. In addition, the result is not very intuitionistic for users.

R. Wille[2] proposed *concept lattice* (also called *Galois lattice*) according to the binary relation, it is an effective tool for data analysis and rules extraction. Hu X.G improves concept lattice to *extended concept lattice (ECL)* by introducing equivalence relation to the intension of concepts. It can express data and knowledge more clearly, and can extract the rules more easily. Both ECL and RS are based on equivalence class, so there is a hard relation between ECL and RS. Using the frame of ECL to resolve problem of RS, we can gain a high time performance and clearly express the result in form of Hasse map.

Many researchers pay attention to the relation between concept lattice and rough set [10]. This paper discusses the property and relation between ECL and RS, then implements some basic resolution in RS bases on ECL.

2 Extended Concept Lattice

A context is defined as a triple C=(O,D,R), where O is a set of objects, D is a set of attributes, and R is a binary relation between O and D.

Definition 1: Each couple such as (A,B) derived from the context is called a concept, where $A \in 2^O, B \in 2^D$ (2^O and 2^D represent the power set of set O and D respectively). A and B create connection in terms of following properties: A'={$m \in D | \forall g \in A$, gRm}, B'={$g \in O | \forall m \in B$, gRm}, Where A'=B, B'=A. A is called the *extension* of concept, and B is the *intension*.

The concept (A1,B1) and (A_2,B_2) have the *sub-concept* and *sup-concept* relation, if $A_1 \subset A_2$, denoted as $(A_1,B_1)<(A_2,B_2)$.

We use the this concept and relation between of sup-concept and sub-conception to define *Galois concept lattice*(GCL for short).

Definition 2: For a given concept (A,B), if $\exists B_1 \subset B$ satisfies B_1'=A, then B_1 is an *equivalence group* of B, and we define Equ(A)={$B_1 | B_1 \subset A'$ and B_1'=A}.

Definition 3: A couple (A,B) derived from context is called a *elementary concept*, where $A \in 2^O$, and $B \in$ Equ(A). A and B are called extension and intension of concept respectively. Elementary concepts $C_1=(A,B_1)$ and $C_2=(A,B_2)$ are two equivalent concepts, so B_1 and B_2 are equivalent intensions. (A, Equ(A)) is called a *extended concept*, or simply called concept. We use this definition in following content.

Definition 4: In context, immediate sub-concept-sup-concept defined as (A_1,B_1) $<(A_2,B_2)$, if $A_1 \subset A_2$, and there is no concept (A,B) satisfy $A_1 \subset A \subset A_2$.

Extended concepts and the immediate sub-concept- sup-concept relations build a complete concept lattice, called *extended concept lattice* (ECL for short).

3 The Representation and Resolution of Rough Set in the ECL

3.1 The Representation and Resolution of the Terms About RS

The approximation space in rough set model is based on a group of equivalence relations, and every equivalence class deduced by these equivalence relations is a atom of the knowledge representation.

In ECL, C=(X,X') describes that all the objects in extension X satisfy every group of attributes in intension X', and the objects satisfy every group of attributes in intension X' is contained in extension X.

Indiscernibility relation is an important term in rough set. In approximation space K=(U,R), $P \subseteq R$, $P \neq \emptyset$, the intersection of all equivalence relations in **P** form the indiscernibility relation on **P**, denoted as IND(P), every equivalence class defined as $[x]_{IND(P)} = \bigcap_{R \in P} [X]_R$.

From the completeness of the concept lattices and the isomorphism between GCL and ECL, we know that every equivalence class in IND(**P**) correspond to a concept in ECL. Following theorems will describe the relation between them.

Theorem 1: To the discernibility relation IND(**P**) in approximation space **K**=(**U,R**) and nonempty attribute set **P**, the extension of concept C is the equivalence class of IND(**P**), if and only if $\exists B_i \in B(attrib(B_i)=P)$.

Definition 5: The concept C is basic concept of IND(**P**), if the extension of the concept C in ECL is the equivalence class on IND(**P**). And we denoted the set of all the basic concept on IND(**P**) as INDC(**P**).

Definition 6: In ECL, to the attribute set R in approximation space **R**, $r \in R$, if it satisfy INDC(**R**)=INDC(**R-r**), we say r is the dispensable knowledge in **R**.

In approximation space **K**=(**U,R**), $P \subseteq R$, and $P \neq \emptyset$, let $X \subseteq U$, then the upper approximation, lower approximation and boundary of **X** on IND(**P**) are defined as:

$$P_*(X) = \{x : [x]_P \subseteq X\}$$
$$P^*(X) = \{x : [x]_P \cap X \neq \emptyset\}$$
$$BN_P(X) = P^*(X) - P_*(X)$$

If $BN_P(X) \neq \emptyset$, then the set **X** is a rough concept.

By theorem 1, theorems about the basic operation of rough set in ECL as following:

Theorem 2: In ECL, the upper approximation and lower approximation of rough set X can be expressed as following:

$P^*(X) = \cup \{Y|(Y,Y') \in ECL, Y \cap X \neq \emptyset, \exists B_i \in Y'(attrib(B_i)=P)\}$
$P_*(X) = \cup \{Y|(Y,Y') \in ECL, Y \subseteq X, \exists B_i \in Y'(attrib(B_i)=P)\}$

3.2 The Representation and Resolution of the Relation Between Attribute Sets

Then we discuss how to resolve the dependency relationship between attributes in ECL. For convenience, we introduce some items.

Definition 7: The relation '<*' and '<=' between concept C_1 and C_2 are defined as:

$C_1 <^* C_2 \Leftrightarrow (C_1 < C_2)$ or $\exists C(C_1 < C, \ C <^* C_2)$.
$C_1 <= C_2 \Leftrightarrow (C_1 <^* C_2)$ or $(C_1 = C_2)$.

To two attribute sets, we can easily prove the following theorem:

Theorem 3: In approximation space **K**=(**U,R**), $P,Q \subseteq R$, $\exists P,Q \neq \emptyset$, there are:

$P \Leftrightarrow Q$ if and only if INDC(**P**)=INDC(**Q**).
$P \Rightarrow Q$ if and only if $\forall Pc \in INDC(P) \exists Qc \in INDC(Q) (Pc <= Qc)$.

In approximation space **K**=(**U,R**), let $P,Q \subseteq R$, P-positive region of **Q** is defined as
$$POS_P(Q) = \bigcup_{X \in U/Q} P_*(X),$$
P-positive region of **Q** describes that **Q** in terms of **P**.

Because each equivalence class on P,Q corresponding with a concept, the relation is equal to the following relation between two basic concepts in ECL:

Theorem 4: In ECL, the set of **P**-positive region of **Q** in approximation space can be obtained as following: $POSC_P(Q) = \bigcup_{C \in INDC(Q)} \{Y|Y \in INDC(P), (Y<=C)\}$.

Definition 8: In ECL, if the equivalence relation **P** and **Q** satisfy $POSC_P(Q) = POSC_{P-r}(Q)$, $(r \in P)$, we say $r \in P$ is Q-dispensable on **P**, else we say r is Q-indispensable on **P**.

The degree of **Q** depend on **P** in approximation space is $k=\gamma_P(\mathbf{Q})= \dfrac{card(POS_P(\mathbf{Q}))}{card(U)}$, corresponding to $k=\gamma_P(\mathbf{Q})= \dfrac{card(POSC_P(\mathbf{Q}))}{card(U)}$ in ECL.

We know following from above:

① $k=1 \Leftrightarrow \forall C_1 \in INDC(\mathbf{P}) \exists C_2 \in INDC(\mathbf{Q}) (C_1 <= C_2)$
② $k \neq 1 \Leftrightarrow \exists C_1 \in INDC(\mathbf{P}) \forall C_2 \in INDC(\mathbf{Q}) \neg (C_1 <= C_2)$
③ $k=0 \Leftrightarrow \forall C_1 \in INDC(\mathbf{P}) \forall C_2 \in INDC(\mathbf{Q}) \neg (C_1 <= C_2)$

Compare to the degree of dependence in RS, the dependence described by ECL has more advantages. Besides the total degree of dependence, it also give the dependence between INDC(**P**) and INDC(**Q**), it is helpful to describe the dependence furthermore.

4 Conclusion

This paper discussed the corresponding relation between ECL and RS, as well as the basic operation of RS based on ECL. From the discussion above, we know using ECL can visually describe RS, and implement the basic operations of RS conveniently.

References

1. Z.Pawlak, Rough Sets, KLUWER ACADEMIC PUBLISHERS, Dordrecht/Boston/London (1991)
2. Wille R. Restructuring Lattice theory: an approach based on hierarchies of concepts. In: Rival I ed. Ordered Sets. Dordrecht: Reidel, (1982) 445—470
3. Wille R. Knowledge acquisition by methods of formal concept analysis. In: Diday E ed. Data Analysis, Learning Symbolic and Numeric Knowledge. New York: Nova Science Publisher (1989) 365—380
4. Oosthuizen G D. Rough sets and concept lattices. In: Ziarko W P ed. Rough Sets, and Fuzzy Sets and Knowledge Discovery. London: Springer –Verlag (1994) 24—31
5. Godin R, Missaoui R, Alcui H. Incremental concept formation algorithms based on Galois (concept) lattices, Computational Intelligence (1995) 11(2): 246—267
6. Rokia Missaoui, and Robert Gobin . An Incremental Concept Formation. Approach for Learning from Databases. Department de Mathematiques et d' Informatique, UQAM, Montreal Canada.40—52.
7. Gennari, J.H., Langley, P., and Fisher ,D., "Models of Incremental Concept Formation", in Machine Learning: Paradigmas and Methods, J.Carbonell, (Ed.),1990 MIT Press, Amsterdam, The Netherlands,11-62
8. Godin,R, Missaoui, R. and Alaoui, H., "Incremental Algorithms for Updating the Galois Lattices of a Binary Relation ,"Tech .Rep.#155,Dept .of Comp. Science, UQAM, Sep. (1991).
9. Godin, R., Missaoui, R.,and Alaoui ,H., "Learning Algorithms Using a Galois Lattices Structure," Proc. Third Int.Conf .on Tools for Artificial Intelligence ,1991,San Jose, Calif., IEEE Computer Society Press,22-29.
10. Hu xuegang, etc. The Research on Design Knowledge Representation and Acquisition in inTelligence CAD. In Proceedings of Second International Conference on Computer Aided Industrial Design and Conceptual Design. Nov-26-28 (1999) Bangkok, Thailand.

Study of Integrate Models of Rough Sets and Grey Systems

Wu Shunxiang[1,2], Liu Sifeng[1], and Li Maoqing[2]

[1] School of Economics and Management, Nanjing University of Aeronautics and Astronautics, Nanjing 210016,China
`wsx1009@163.com`
[2] Department of Automation, Xiamen University, Xiamen, 361005, China

Abstract. This paper firstly compares rough sets theory with grey system theory, then the concept of grey sets is proposed and the grey degree of grey sets and the basic relationships and operations of grey sets are defined. Next, we set up the models of grey rough sets and wide grey rough sets as well as study their properties. Moreover, the definition of rough grey sets is given and their basic characters are investigated. Furthermore, the relations between rough grey sets and grey rough sets are researched. Finally, we come to the conclusion that it is possibly more effective to deal with some uncertain problems if the theories and methods of rough sets and grey systems are combined.

1 Introduction

Grey system theory initiated by Chinese professor DENG, Ju-Long in the 1980's is a mathematic tool dealing with few data, uncertain information [1]. Grey system refers to systems partly certain and partly uncertain, partly complete and partly incomplete, partly known and partly unknown in information. It is a concept between the "White" and "Black". "White" refers to complete understanding of the information, "Black" refers to complete unknown about the information, while "Grey" means information partly known and partly unknown", "partly clear and partly unclear".[3] The main idea of grey system is to mine the knowledge and rules hidden in few data by grey generation the existent few data. Since its birth, the grey system theory has been widely used in practical areas, it has special uses in the analysis, modeling, forecasting, decision-making, controlling and evaluating of few data, uncertain systems, and has been proved valuable in application.

Rough sets theory is a data analysis theory put forward by Poland mathematician Pawlak Z in the early 1980's [4]. It is a new mathematic method dealing with fuzzy and uncertain information. The main idea is by way of knowledge reduction, educing the decision-making or classifying rule of the problem in the precondition of maintaining the classifying ability [5]. Since its birth, rough sets theory has been widely used in the areas of machine learning, knowledge acquiring, decision-making analysis, knowledge discovery in database, pattern identifying, fuzzy controlling etc.

Grey system and rough sets are two newly developed powerful tools dealing with uncertain information, they have similarities and dissimilarities. The similarity is that they are both different from other theories such as probability and statistics theory, fuzzy theory, unclassical logic and evidence theory in dealing with uncertain problems. They don't need any experience information beyond the given data, and the data they deal with are objective. Their final object are the same, that is mining the valuable information from the existent data. The dissimilarities are grey system is mainly dealing with imprecise or uncertain original data, while rough sets is dealing with precise or certain processed data; in aspect of data disposing, grey system and rough sets are in contrary directions: by grey building, grey system deduct data from large to small, while rough sets increase data form small to large by reduction of decision table. butThe two theories have mutual benefit in a certain way: rough sets does not include the mechanism of dealing with imprecise or uncertain data, while grey system provides abundant and effective methods and ways of dealing with imprecise or uncertain data; grey system mainly studies the actual data and rules, it emphasizes on the information optimization which is a supplement to knowledge dynamic classifying and dynamic reduction of rough sets; on the other hand, grey system can intensify the ability of congregating the incomplete information and covering the grey information by referring to the ways of knowledge classifying of rough sets, in order to discover the hidden rules from the uncertain data or knowledge.

At present, a lot of research work has been done on the combination of rough sets theory and other theory dealing with uncertain information in the world[5,6,7], but the research of integrate of rough sets theory and grey systems is also very important, we will discuss this aspect primarily in the paper.

2 Grey System and Rough Sets

Grey system deals with grey information. Grey information is uncertain information due to the lack of data. Generally speaking, we can only acquire few original data about the research objects as a result of the internal and external restricts. Because the data are so few, our understanding of the original objects are very imprecise. But there may be some valuable information hidden in these few original data, so we must try to mine the approximate precise information about the original data from these few data (mainly by grey generation). In grey system, we usually define the few incomplete data by grey data. Grey data are data that we only known their boundaries but don't know the exact value[3]. In practical application, the mostly used grey data are region grey data. Region grey data refer to grey data that we know their upper and lower boundaries but don't know their real value. We use $[a, b]^{\oplus}$ to denote the region grey data which has a as lower boundary and b as upper boundary.

But in real life, there are a lot of information that can not be described by grey data.

Example 1. A soccer team will attend a match, if we define: A={main force},B={all members},C={the members play in the match}. Apparently, A and B are white information, whereas C seems like a grey information. As we known, main force are sure to be play in the match, while the players will be selected from all the members, so the players playing in this match are uncertain, thus we get $A \subseteq C \subseteq B$.

Example 2. The Congress of a country will vote for a proposal, we define: A={the councilors who brought the proposal}, B={all the councilors}, C={ the councilors who agree to vote}. Apparently, A and B are white information, while C is currently a grey information, and $A \subseteq C \subseteq B$.

In both examples C represent grey information. As the extension of the concept of grey data, we define the aggregate constitute by some concrete elements which we know their approximate ranges but don't know the exact elements as grey sets. We will give the specific definitions of grey sets, range sets and range grey sets below.

Definition 1. Define X as a set on the field U, but the specific elements are unknown, then we call X as a grey set in the field U, for shot grey set.

Definition 2. Suppose U as the discussed field, normal sets $A \subseteq B \subseteq U$, mark $[A,B]=\{C \in 2^U | A \subseteq C \subseteq B\}$, we call [A,B] as the range set which use A as lower boundary and B as upper boundary.

Apparently, the elements included in [A,B] are $2^{|B-A|}$, in which $|B-A|$ represents the radix of set B-A. Specially, we have: $[\phi,\phi]=\{\phi\}, [U,U]=\{U\}, [\phi,U]=2^U$.

We can define the relevant region grey set as the extension of region grey data.

Definition 3. Define X as a grey sets, if there exists a set [A,B] in field U which satisfies $X \in [A,B]$, then we call X as the region grey set that has A as lower boundary and B as upper boundary.

We use $[A,B] \oplus$ to stand for the region grey sets. Obviously, $A \subseteq [A,B] \oplus \subseteq D, [\phi,\phi] \oplus -\phi$, $[U,U] \oplus = U$. If not pointed out what we called grey sets below "≤" refers to region grey sets.

As we known, for grey data [a, b]⊕ exist character a≤b. Here, "$\subseteq$" is the partial sequence among the sets. Thus, the grey sets we defined is a natural extension of grey data.

Generally, the measurement of grey degree (uncertain degree) of grey information is by means of grey degree, the definition of grey degree of grey data can be found in paper [2]. Here we provide the definition of grey degree for grey sets.

Definition 4. Presume[A,B]⊕ is a grey sets, then $G([A,B] \oplus)=1-1/2^{|B-A|}$ is the grey degree of grey sets [A,B]⊕.

Obviously, $0 \leq G([A,B] \oplus) < 1; G([A,A] \oplus)=0$; G([A,B]⊕) increases as |B-A| increases, that is the discrepancy of B and A is becoming more and more large, the grey sets are more large; when $|B-A| \to \infty$, $G([A,B]g) \to 1$. Here, we call the grey sets values zero as white grey sets, and simultaneously take white grey sets as the unusual grey sets that has a grey degree of zero.

We know that there are basic relations such as including, equality etc. and basic operations as union, intersection, complementation etc. in classical sets, also, there are corresponding basic relations as including, equality etc. and basic operations as combine, join, supply etc. in grey sets.

Definition 5. Presume U as the discussed field, $A,B,C,D \in 2^U$, [A,B]⊕ and [C,D]⊕are two grey sets, $C \subseteq A$ and $D \subseteq B$, then we say [C,D]⊕ weakly included in [A,B]⊕or [A,B]⊕ weakly includes [C,D]⊕, mark as $[C,D] \oplus \subseteq w[A,B] \oplus$; if $D \subseteq A$, then we say [C,D]⊕ strongly included in [A,B]⊕or [A,B]⊕ strongly includes [C,D]⊕, mark as $[C,D] \oplus \subseteq s[A,B] \oplus$; when A=C and B=D, we say [A,B]⊕ and [C,D]⊕ grey equal, mark as [A,B]⊕=[C,D]⊕.

Obviously, the including relation of grey sets satisfies reflexivity, asymmetry and transitivity, is a partial relation; while the grey relation of the grey relation satisfies reflexivity, symmetry and transmission, is a equivalent relation. If not pointed out, we use "$\subseteq$" to represent the weakly include and strongly include among grey sets.

Definition 6. Presume U as the discussed field, $A,B,C,D \in 2^U$, $[A,B]^\oplus$ and $[C,D]^\oplus$ are two grey sets, we call $[A,B]^\oplus \cup [C,D]^\oplus, [A,B]^\oplus \cap [C,D]^\oplus, \sim[A,B]^\oplus$ separately the convergence set, joined set and supplemental set of $[A,B]^\oplus$ and $[C,D]^\oplus$. And,

$[A,B]^\oplus \cup [C,D]^\oplus = [A \cup C, B \cup D]^\oplus$,
$[A,B]^\oplus \cap [C,D]^\oplus = [A \cap C, B \cap D]^\oplus$,
$\sim[A,B]^\oplus = [\sim B, \sim A]^\oplus$.

We can conclude from the definition that the empty set and full set are $[\phi, \phi]^\oplus$ and $[U,U]^\oplus$.

Obviously, the operations of union, intersection, complementation satisfy the Idempotent Law, Commutative Law, Distributive Law, Absorptive Law, Associative Law, Involution Law, Dualization Law, Identity Law, but don't satisfy Complementarity Law.

For grey sets, since some elements are uncertain, we can use a certain sets between the upper and lower boundaries of the grey sets to present it for calculation for convenience. We call such sets as White sets, mark as $[A,B]^\oplus$ for grey sets $[A,B]^\oplus$. Apparently $A \subseteq [A,B]^\oplus \subseteq B$.

Theorem 1. If $[C,D]^\oplus \subseteq s[A,B]^\oplus$, then $[C,D]^\oplus \subseteq [A,B]^\oplus$.

Proof: From the definition, it comes into existence obviously.

In grey sets theory, we call a partition (or equivalence relation) of the investigated object U as a knowledge of U, a cluster of partitions called a repository of U. Each equivalence relation generated by the partition of object U is called Information Grain which is the basic unit of knowledge. Meanwhile, we call each subset of the object U as a conception or category.

In real life, we often confront with a lot of unordered phenomena or data, and we usually try to express them by the acquired knowledge. That is, express exactly the concept which we work on using the information grain, in order to find the knowledge and rules hidden in the phenomena and data. But, not all the phenomena and data can be exactly expressed by the acquired knowledge, in most time, we can only define roughly. In rough sets theory, people usually use two exact region (upper approximation and lower approximation) to express (define) the imprecise information. Classic rough sets definitions are given below [5]:

Definition 7. Presume R is an equivalence relation on the discussed field U, for classic (ordinary) set $X \subseteq U$, the lower approximation of X for (U,R) R_X, the upper approximation denoted R☐X, boundary region bnR(X), positive region posR(X), negative region negR(X) is defined below:

$R\_X = \cup \{Y \in U/R \mid Y \subseteq X\}$,
$\overline{R}X = \cup \{Y \in U/R \mid Y \cap X \neq \phi\}$,
$bnR(X) = \overline{R}X - R\_X$,
$posR(X) = R\_X$,
$negR(X) = \sim \overline{R}X$.

R_X(or posR(X)) is constituted by the elements judged definitely belonging to X by knowledge R; $\overline{R}$ X is constituted by the elements judged probably belonging to X by knowledge R; bnR(X) is constituted by the elements that probably belong to X and probably belong to ~X; negR(X) is constituted by the elements that definitely not belong to X;

When R_X=$\overline{R}$ X, we call X and R can be defined (precise set of R); otherwise, we call X is the rough set of R. We usually use r(X,R)=1-(|R_X|/|$\overline{R}$ X|) to express the R rough set of X. The larger its value, the more uncertain (rough)of the set X. Generally, boundary region resembles the uncertainty of information, the lager the boundary region, the more uncertain (rough) of the information. When the boundary region is null, the information is certain.

3 Grey Rough Sets

In the classic rough sets, the concepts being researched are all clear, that is, the concepts being approximated is the classic set. When the concepts being research are grey sets, we will get the models of grey rough sets.

Definition 8. Presume R is an equivalence relation on the discussed field U, [A,B]⊕ is a grey set, then its lower approximation R_([A,B]⊕), upper approximation $\overline{R}$ ([A,B])⊕, boundary region bnR([A,B]⊕), positive region posR([A,B]⊕) and negative region negR([A,B]⊕) about (U,R) is defined below,

R_([A,B]⊕)=R_A,
$\overline{R}$ ([A,B]⊕)=$\overline{R}$ B,
bnR([A,B]⊕)=$\overline{R}$ B-R_A,
posR([A,B]⊕)=R_A,
negR([A,B]⊕)=~$\overline{R}$ B,

where R_A and $\overline{R}$ B are separately the lower approximation and upper approximation of A for (U,R) according to definition 7.

If R_([A,B]⊕)=$\overline{R}$ ([A,B]⊕),then, we say [A,B]⊕ is grey definable, otherwise, we say ([A,B]⊕) is R grey sets.

R_([A,B]⊕) is a set constituted by the elements which are definitely belonging to grey sets [A,B]⊕ according to knowledge R; R-([A,B]⊕) is a set constituted by the elements which are probably belonging to grey sets [A,B]⊕ according to knowledge R; bnR([A,B]⊕) is a set constituted by the elements which are probably belonging to grey sets [A,B]⊕ or probably not belonging to grey sets [A,B]⊕ according to knowledge R; negR([A,B]⊕) is a set constituted by the elements which are definitely not belonging to grey sets [A,B]⊕ according to knowledge R;

When A=B, [A,B]⊕ turns to the white sets (classic sets), and definition 8 turns the definition of the rough sets in classical theory. Thus, definition is the extension of definition 7.

We use r([A,B]⊕,R)=1-(|R_[A,B]⊕|/|R-[A,B]⊕|) to represent the R grey rough degree of [A,B]⊕. The large its value, the large the degree of uncertainty of the incompletion of the knowledge of grey sets [A,B]⊕. Thus, the information uncertainty of grey rough sets is actually determined by two factors: the first factor is that the grey uncertainty if caused by the scarcity of data; the second factor is that the rough uncertainty is caused by the imprecision of knowledge expression. We use $UC([A,B]\oplus,R)=\min\{1,G([A,B]\oplus)+r([A,B]\oplus,R)\}=\min\{1,(1-1/2^{|B-A|})+(1-|R\_A|/|R-B|)\}$ to represent the information uncertainty of grey sets $[A,B]\oplus$ given knowledge R. Obviously, when the discrepancy of sets A and B becomes small, the information uncertainty becomes small. When $A=B, UC([A,B]\oplus,R)=1-|R\_A|/|R-B|=r([A,B]\oplus,R)$, the information uncertainty of grey sets is equal to its rough degree; when $R\_A=R-B$, according to $R\_A \subseteq A \subseteq B \subseteq R\_B$, we know that A=B, here $UC([A,B]\oplus,R)=0$.

Theorem 2. Grey sets $[A,B]\oplus$ can be and only can be defined when A=B and A is definable.

Proof: if A=B and A is definable, then $R\_A=R^-A=R^-B$, according to definition 8, $R\_([A,B]\oplus)=R^-([A,B]\oplus)$, thus $[A,B]\oplus$ is definable; on the contrary, if $[A,B]\oplus$ is definable, then $R\_([A,B]\oplus)=R^-([A,B]\oplus)$, that is $R\_A=R^-B$, if A is indefinable, then $R\_A * A$, while $A \subseteq B \subseteq R^-B$, thus $R\_A \neq R^-B$, contradiction appears. So A is definable. Since A is definable, we can conclude that $R\_A=A$, if $A \neq B$, then $A * B$, thus $R\_A \subseteq B \subseteq R^-B$, so $R\_A \neq R^-B$, contradiction appears. So A=B and A is indefinable.

Theorem 3. Suppose R is a equivalence relation on discussed field U, $[A,B]\oplus$ and $[C,D]\oplus$ are two grey sets on U, then according to the grey rough sets definition 8, exist following properties.

$R\_([A,B]\oplus) \subseteq [A,B]\oplus \subseteq R^-([A,B]\oplus)$;
$R\_([\phi,\phi]\oplus)=R^-([\phi,\phi]\oplus)=[\phi,\phi]\oplus=\phi$,
$R\_([U,U]\oplus)=R^-([U,U]\oplus)=[U,U]\oplus=U$;
$R^-([A,B]\oplus \cup [C,D]\oplus)=R^-([A,B]\oplus) \cup R^-([C,D]\oplus)$;
$R\_([A,B]\oplus \cap [C,D]\oplus)=R\_([A,B]\oplus) \cap R\_([C,D]\oplus)$;
If $[A,B]\oplus \subseteq [C,D]\oplus$,
then $R\_([A,B]\oplus) \subseteq R\_([C,D]\oplus), R^-([A,B]\oplus) \subseteq R^-([C,D]\oplus)$;
$R\_([A,B]\oplus) \cup R\_([C,D]\oplus) \subseteq R\_([A,B]\oplus \cup [C,D]\oplus)$;
$R^-([A,B]\oplus \cap [C,D]\oplus) \subseteq R^-([A,B]\oplus) \cap R^-([C,D]\oplus)$;
$R\_(\sim[A,B]\oplus)=\sim R^-([A,B]\oplus)$;
$R^-(\sim[A,B]\oplus)=\sim R\_([A,B]\oplus)$;
$R\_(R\_([A,B]\oplus))=R^-(R\_([A,B]\oplus))=R\_([A,B]\oplus)$;
$R^-(R^-([A,B]\oplus))=R\_(R^-([A,B]\oplus))=R^-([A,B]\oplus)$.

Proof: It can be easily proofed by definition.

Definition 10. Presume R is an equivalence relation on the discussed field U, $[A,B]\oplus$ and $[C,D]\oplus$ are two grey sets on U, if $R\_[A,B]\oplus=R\_[C,D]\oplus$, then we say $[A,B]\oplus$ and $[C,D]\oplus$ are **R** grey rough lower equal, remark as $[A,B]\oplus=_l[C,D]\oplus$; if $R^-([A,B]\oplus)=R^-([C,D]\oplus)$, then, we say $[A,B]\oplus$ and $[C,D]\oplus$ are R grey rough upper equal, remark as $[A,B]\oplus=_s[C,D]\oplus$; if $R\_[A,B]\oplus=R\_[C,D]\oplus$ and $R^-([A,B]\oplus)=R^-([C,D]\oplus)$, then, we say $[A,B]\oplus$ and $[C,D]\oplus$ are R grey rough equal, remark as $[A,B]\oplus=g[C,D]\oplus$.

Obviously, $=_l$, $=_s$ and $=g$ are equivalent relation.

Theorem 4. $[A,B]\oplus 与 [C,D]\oplus$ are R grey rough lower equal when and only when $R\_A=R\_C$; $[A,B]\oplus$ and $[C,D]\oplus$ are R grey rough upper equal when and only when

$R^- B=R^- D$; $[A,B]\oplus$ 与 $[C,D]\oplus$ are R grey rough equal when and only when $R\_A=R\_C$ and $R^- B=R^- D$.

Proof: It can be easily proofed by definition.

Theorem 5. Presume R is a equivalent relation on field U, $[A,B]\oplus$、$[A',B']\oplus$、$[C,D]\oplus$ 与 $[C',D']\oplus$ are grey sets in U, then for approximate space, exist properties:

$[A,B]\oplus=_l [C,D]\oplus \Leftrightarrow [A,B]\oplus \cap [C,D]\oplus =_l [A,B]\oplus$ and
$[A,B]\oplus \cap [C,D]\oplus =_l [C,D]\oplus$;

$[A,B]\oplus=_s [C,D]\oplus \Leftrightarrow [A,B]\oplus \cup [C,D]\oplus =_s [A,B]\oplus$ and
$[A,B]\oplus \cup [C,D]\oplus =_s [C,D]\oplus$;

$[A,B]\oplus =_l [A',B']\oplus$ and
$[C,D]\oplus =_l [C',D']\oplus => [A,B]\oplus \cap [C,D]\oplus =_l [A',B']\oplus \cap [C',D']\oplus$;

$[A,B]\oplus =_s [A',B']\oplus$ and $[C,D]\oplus =_s [C',D']\oplus =>$
$[A,B]\oplus \cup [C,D]\oplus =_s [A',B']\oplus \cup [C',D']\oplus$;

$[A,B]\oplus =_l [C,D]\oplus => [A,B]\oplus \cap \sim[C,D]\oplus =_l [\phi,\phi]\oplus$;
$[A,B]\oplus =_s [C,D]\oplus => [A,B]\oplus \cup \sim[C,D]\oplus =_s [U,U]\oplus$;
$[A,B]\oplus \subseteq [C,D]\oplus$ and $[A,B]\oplus =_l [U,U]\oplus => [C,D]\oplus =_l [U,U]\oplus$;
$[A,B]\oplus \subseteq [C,D]\oplus$ and $[C,D]\oplus =_s [\phi,\phi]\oplus => [A,B]\oplus =_s [\phi,\phi]\oplus$;
$[A,B]\oplus =_s [C,D]\oplus \Leftrightarrow \sim[A,B]\oplus =_l \sim[C,D]\oplus$;
$[A,B]\oplus =_l [\phi,\phi]\oplus$ or $[C,D]\oplus =_l [\phi,\phi]\oplus => [A,B]\oplus \cap [C,D]\oplus =_l [\phi,\phi]\oplus$;
$[A,B]\oplus =_s [U,U]\oplus$ or $[C,D]\oplus =_s [U,U]\oplus => [A,B]\oplus \cup [C,D]\oplus =_s [U,U]\oplus$;

Proof: It can be easily proofed by definition.

Similar to grey rough equality, we can define grey rough including relation.

Definition 11. Presume R is an equivalence relation on the discussed field U, $[A,B]\oplus$ and $[C,D]\oplus$ are two grey sets on U, if $R\_[A,B]\oplus \subseteq R\_[C,D]\oplus$, then we say $[A,B]\oplus$ is grey rough lower included by $[C,D]\oplus$, mark as $[A,B]\oplus \subseteq l[C,D]\oplus$; if $R^- [A,B]\oplus \subseteq R^- [C,D]\oplus$, then we say $[A,B]\oplus$ is grey rough upper included by $[C,D]\oplus$, mark as $[A,B]\oplus \subseteq s[C,D]\oplus$; if $R\_[A,B]\oplus \subseteq R\_[C,D]\oplus$ and $R^- ([A,B]\oplus) \subseteq R^- ([C,D]\oplus$, then we say $[A,B]\oplus$ is grey rough included by $[C,D]\oplus$, mark as $[A,B]\oplus \subseteq g[C,D]\oplus$.

Obviously, grey rough inclusion comes into existence when and only when grey lower inclusion and grey upper inclusion both exist. It is easily known that grey rough lower inclusion, grey upper inclusion and grey rough inclusion all are lean relation. For grey rough inclusion relation, exists following properties,

Theorem 6. Presume R is an equivalence relation on the discussed field U, $[A,B]\oplus,[A',B']\oplus,[C,D]\oplus, [C',D']\oplus$ and $[E,F]\oplus$ are grey sets on U separately, then for proximity space (U,R), exists following properties,

$[A,B]\oplus \subseteq [C,D]\oplus => [A,B]\oplus \subseteq l[C,D]\oplus, [A,B]\oplus \subseteq s[C,D]\oplus$ and $[A,B]\oplus \subseteq g[C,D]\oplus$;

$[A,B]\oplus \subseteq l[C,D]\oplus$ and $[C,D]\oplus \subseteq l[A,B]\oplus => [A,B]\oplus =_l [C,D]\oplus$;
$[A,B]\oplus \subseteq s[C,D]\oplus$ and $[C,D]\oplus \subseteq s[A,B]\oplus => [A,B]\oplus =_s [C,D]\oplus$;
$[A,B]\oplus \subseteq g[C,D]\oplus$ and $[C,D]\oplus \subseteq g[A,B]\oplus => [A,B]\oplus =g[C,D]\oplus$;
$[A,B]\oplus \subseteq l[C,D]\oplus \Leftrightarrow [A,B]\oplus \cap [C,D]\oplus =_l [A,B]\oplus$;
$[A,B]\oplus \subseteq s[C,D]\oplus \Leftrightarrow [A,B]\oplus \cup [C,D]\oplus =_s [C,D]\oplus$;
$[A,B]\oplus \subseteq [C,D]\oplus, [A,B]\oplus =_l [A',B']\oplus$ and
$[C,D]\oplus =_l [C',D']\oplus =>[A',B']\oplus \subseteq l[C',D']\oplus$;

$[A,B]\oplus \subseteq [C,D]\oplus,[A,B]\oplus=_s [A',B']\oplus$ and
$[C,D]\oplus=g [C',D']\oplus => [A',B']\oplus \subseteq s[C',D']\oplus$;
$[A,B]\oplus \subseteq [C,D]\oplus,[A,B]\oplus=g[A',B']\oplus$ and
$[C,D]\oplus=g[C',D']\oplus => [A',B']\oplus \subseteq g[C',D']\oplus$;
$[A,B]\oplus \subseteq l[A',B']\oplus$ and $[C,D]\oplus \subseteq l[C',D']\oplus =>$
$[A,B]\oplus \cap [C,D]\oplus \subseteq l[A',B']\oplus \cap [C',D']\oplus$;
$[A,B]\oplus \subseteq s[A',B']\oplus$ and $[C,D]\oplus \subseteq s[C',D']\oplus =>$
$[A,B]\oplus \cup [C,D]\oplus \subseteq s[A',B']\oplus \cup [C',D']\oplus$;
$[A,B]\oplus \subseteq l[C,D]\oplus$ and $[A,B]\oplus=_l [E,F]\oplus => [E,F]\oplus \subseteq l[C,D]\oplus$;
$[A,B]\oplus \subseteq s[C,D]\oplus$ and $[A,B]\oplus=_s [E,F]\oplus => [E,F]\oplus \subseteq s[C,D]\oplus$;
$[A,B]\oplus \subseteq g[C,D]\oplus$ and $[A,B]\oplus=g[E,F]\oplus => [E,F]\oplus \subseteq g[C,D]\oplus$.
Proof: It can be easily proofed by definition.

Theorem 7. Presume R is an equivalence relation on the discussed field U, $[A,B]\oplus$ is a grey set on U, the lower approximation and upper approximation of its white set $[A,B]\underline{\oplus}$ for (U,R) are $R_{\_}([A,B]\underline{\oplus})$ and $\overline{R}([A,B])\underline{\oplus})$ separately, then $R_{\_}([A,B]\oplus) \subseteq R_{\_}([A,B]\underline{\oplus}) \subseteq \overline{R}([A,B])\underline{\oplus}) \subseteq \overline{R}([A,B])\underline{\oplus})$ and $r([A,B]\underline{\oplus},R) \leq r([A,B]\oplus,R)$.
Proof: It can be easily proofed by definition.

When the A and B in the grey set $[A,B]\oplus$ are both definable, then we can deduce another type of grey rough set by grey set $[A,B]\oplus$.

Definition 12. Presume R is an equivalence relation on the discussed field U, $[A,B]\oplus$ is a grey set on U, A and B are both R definable, we call set $\{X \in U/R | R\_X=A, R-X=B\}$ is grey rough set deduced by $[A,B]\oplus$, mark as $R([A,B]\oplus)$.
If $U=\{a,b,c,d,e,f,g,h\}$, $R=\{\{a,b\},\{c,d\},\{e,f,g\},\{h\}\}$, $A=\{a,b\}$, $B=\{a,b,e,f,g\}$,
then $R([A,B]\oplus)=\{\{a,b,e\},\{a,b,f\},\{a,b,g\}\}$.

Obviously, grey rough sets are constituted by those sets having the same upper and lower approximations. For grey rough sets, exist following properties,

Theorem 8. Presume R is an equivalence relation on the discussed field U, $[A,B]\oplus$ is a grey set on U, A and B are both R definable, $R([A,B]\oplus)$ is the grey rough sets deduced by $[A,B]\oplus$, then exist following definitions:
If $A=B$, then $R([A,B]\oplus)=\{A\}=\{B\}$;
If $A \neq B$, then $A \notin R([A,B]\oplus)$ and $B \notin R([A,B]\oplus)$;
If $A \neq B, X \notin R([A,B]\oplus)$, then X is an R indefinable (R rough sets);
If $X \in R([A,B]\oplus)$, then $A \subseteq X \subseteq B$;
$\cap X=A, \cup X=B$ ($X \in R([A,B]\oplus)$);
$R([A,B]\oplus) \subseteq [A,B]$;
B-A is the equivalence class made up by none single elements.

We only proof first definition as fellow, and other definitions proof are similar.
Proof: If $A=B$, then for $\forall X \in R([A,B]\oplus)$, as $R\_X=A, R-X=B$, and $R\_X \subseteq X \subseteq R-X$, thus $A \subseteq X \subseteq B$, $X=A=B$, and $R([A,B]\oplus)=\{A\}=\{B\}$.

From the theorems above, we know that grey rough sets are made up with rough sets which having the same upper approximation and lower approximation in a certain definable upper and lower boundary. (except the situation when upper boundary equals with lower boundary)

Sometimes, we call (U,R) a Pawlak approximate space, equivalence class and empty set in R are call the basic sets or atom sets of approximate space (U,R), one or then union of more than one atom sets are called compound sets. Obviously, a set is R definable when and only when it is a compound set (except ϕ). We mark all the sets of compound sets (including empty sets) as Comp((U,R)). Obviously, the number of elements in Comp((U,R)) is $|Comp((U,R))|=2^{|U/R|}$. Similarly, if A,B∈Comp((U,R)) and $A \subseteq B$, we mark all the sets belonging to region sets [A,B] as Comp(([A,B],R)).

If A,B∈Comp((U,R)) and $A \subset B$, we use (B-A)/R to represent all the R equivalence classes included in the set B-A. Obviously, $|Comp(([A,B],R))|= 2^{|(B-A)/R|}$. If we mark (B-A)/R={$C_1,C_2,...,C_k$}, where C_i(i=1,2,...,k) represents the R equivalence classes included in set B-A, k=|(B-A)/R|, then for grey rough sets R([A,B]⊕), we have,

$$|R([A,B]\oplus)|=\prod_{1\leq i\leq k}(2^{|C_i|}-2)=(2^{|C_1|}-2)*(2^{|C_2|}-2)*...*(2^{|C_k|}-2) \qquad (l).$$

We use $r(R([A,B]\oplus))=|R([A,B]\oplus)|/|[A,B]|= (\prod_{1\leq i\leq k}(2^{|C_i|}-2))/2^{|B-A|}$ to represent the rough degree of grey rough sets R([A,B]⊕), we called it grey rough degree.

According to (6) in Theorem 8, we know that, when A≠B, in formula (l) $|C_i|>1$, thus $|R([A,B]\oplus)|>0$, then $r(R([A,B]\oplus))>0$. We define when A=B, $r(R([A,B]\oplus))=0$. Obviously, the size of grey rough sets is related with B-A and R. Specially, when B-A is certain, grey rough sets is tightly related with R (partition). We know that, for the two partitions (equivalence relations) R and R' on the discussed fields U, if for any x∈U, exist $[x]_{R'} \subseteq [x]_R$, then we say R' is finer of R [14]. From the point of knowledge, we called knowledge R' is finer then R. For the relations between grey rough sets and knowledge R, we have theorems below,

Theorem 9. Presume R_1,R_2 are two equivalence relations on the discussed field U, and R_2 is finer than R_1, [A,B]⊕ is a grey set on U. $A \subset B$ and A, B are R definable, $R_1([A,B]\oplus)$ and $R_2([A,B]\oplus)$ are grey rough sets for R_1,R_2 separately, then we have $|R_1([A,B]\oplus)|>|R_2([A,B]\oplus)|$, and thus $r(R_1([A,B]\oplus))>r(R_2([A,B]\oplus))$.

Proof: suppose X=(B-A)/R_1={$X_1,X_2,...,X_p$}, Y=(B-A)/R_2={$Y_1,Y_2,...,Y_q$}, obviously, X,Y are two equivalence relation on B-A, it is clear (B-A)/$R_1 \subseteq U/R_1$, (B-A)/$R_2 \subseteq U/R_2$ according to the definition. For $\forall$ x∈B-A, $[x]_Y \in$(B-A)/R_2, thus $[x]_Y \in U/R_2$, that is $[x]_Y=[x]_{R2}$, since R_2 is finer than R_1, then $[x]_{R2} \subseteq [x]_{R1}=[x]_X$, thus, $[x]_Y \subseteq [x]_X$, that is Y is finer than X. Because for any two equivalence relations R,R' on U, if (B-A)/R=(B-A)/R', according to formula (l), we will have $|R([A,B]\oplus)|=|R'([A,B]\oplus)|$, so we only need to consider the equivalence relations included in (B-A). When we divide a equivalence class in a equivalence relation into two equivalence classes, but keep other equivalence class unchanged, we will get a new equivalence relation and call it a finer of the former equivalence relation. Apparently, Y is gained by many times of fining of X, we might suppose the finer sequence is $X \to X^1 \to X^2 \to ... \to X^i \to X^{i+1} \to ... \to X^n \to Y$, where the equivalence relation after "→" is gained by a fining of the equivalence relation right before it, mark as P((B-A)/R)=R. Obviously, P(X)=R_1, P(Y)=R_2. Without losing generality, suppose X^1={$X_1,X_2,...,X_{p-1},X_{p1},X_{p2}$}, here the p-1 elements before the X^1 is similar with X. While X_{p1},X_{p2} are gained by dividing X_p. Apparently, X^1 is one step of fining of X, according to formula (l), we have,

$|R_1([A,B]\oplus)|=\prod_{1\leq i\leq p}(2^{|X_i|}-2)=(2^{|X_1|}-2)*(2^{|X_2|}-2)*...*(2^{|X_{p-1}|}-2)*(2^{|X_p|}-2)$,

$|P(X^1)([A,B]\oplus)|=(2^{|X_1|}-2)*(2^{|X_2|}-2)*...*(2^{|X_{p-1}|}-2)*(2^{|X_{p1}|}-2)*(2^{|X_{p2}|}-2)$,

the fist p-1 factors in the two formula above are the same, so we only need to compare the size of $(2^{|X_p|}-2)$ and $(2^{|X_{p1}|}-2)*(2^{|X_{p2}|}-2)$. We notice that X_{p1}, X_{p2} are gained by dividing X_p. So, $X_{p1} \cup X_{p2}=X_p$ and $X_{p1} \cap X_{p2}= \phi$, thus, $|X_{p1}|+|X_{p2}|=|X_p|$. Because
$(2^{|X_{p1}|}-2)*(2^{|X_{p2}|}-2)=2^{|X_{p1}|+|X_{p2}|}-2*(2^{|X_{p1}|}+2^{|X_{p2}|})+4=(2^{|X_p|}-2)-2*(2^{|X_{p1}|}+2^{|X_{p2}|}-3)<2^{|X_p|}-2$,
so $|P(X^1)([A,B]\oplus)|<|R_1([A,B]\oplus)|$. Similarly, since X^2 is gained by one step of fining of X^1, thus $|P(X^2)([A,B]\oplus)|<|P(X^1)([A,B]\oplus)|$, deducing by the same way, $|P(Y)-([A,B]\oplus)|< |P(X^n)([A,B]\oplus)|<|P(X^{n-1})([A,B]\oplus)| <...<|P(X^1)([A,B]\oplus)|<|R_1([A,B]\oplus)|$.

Thus $|R_1([A,B]\oplus)|>|R_2([A,B]\oplus)|$ come into existence. Furthery, $r(R_1([A,B]\oplus))> r(R_2([A,B]\oplus))$ come into existence.

From the Theorem above, we know that the rough degree of grey rough sets is related with the fining degree of knowledge. The finer the knowledge (the finer the partition), the smaller the grey rough degree; contrarily, the rougher the knowledge (the rougher the partition), the larger the grey rough sets. When $|(B-A)/R|=1$ 或 $|U/R|=1$, then B-A or U constitute a equivalence class. Knowledge is the roughest, $r(R([A,B]\oplus))=1-2^{1-|B-A|}, r(R([\phi,U]\oplus))=1-2^{1-|U|}$, the rough degree becomes largest.

When the upper and lower boundaries of the grey sets are not probably definable, we can get a more extensive definition of grey rough sets.

Definition 13. Presume (U,R) is a Pawlakan approximate space, $[A,B]\oplus$ is a grey set on U, then we call set $RW([A,B]\oplus)=\{X\in 2^U | A \subseteq X \subseteq B$ and X is R indefinable$\}$ is **extensive grey rough sets** deduced by grey sets $[A,B]\oplus$.

Obviously, $RW([A,B]\oplus)$ is a set made up with all the rough sets using A as upper boundary, B as upper boundary. Specially, when A and B are both R definable, and when $A\neq B$, $R([A,B]\oplus) \subseteq RW([A,B]\oplus)$.

According to definitions, when A and B are both R definable and $A\neq B$, $[A,B]=RW([A,B]\oplus)\cup Comp(([A,B],R))$. The former is the R rough set of A,B, the latter is the exact set of A,B, thus, $R([A,B])\cap Comp(([A,B],R))= \phi$, then we get :
$|[A,B]|=|RW([A,B]\oplus)|+|Comp(([A,B],R))|$, =>
$|RW([A,B])\oplus|=|[A,B]|-|Comp(([A,B],R))|=2^{|B-A|}-2^{|(B-A)/R|}$ (II).

If we mark $r(RW([A,B]\oplus))=|RW([A,B]\oplus)|/|[A,B]|$ as the extensive rough degree of grey sets $[A,B]\oplus$, then
$r(RW([A,B]\oplus))=(2^{|B-A|}-2^{|(B-A)/R|})/2^{|B-A|}=1-2^{|(B-A)/R|-|B-A|}$ (III).

Specially, the extensive grey rough set on the discussed field is $r(RW([\phi,U]\oplus))=1-2^{|U/R|-|U|}$.

Extensive grey rough degree denotes the ratio of the number of indefinable sets (rough sets) in a region to the number of all the sets in the region. When the region extended to the whole discussed fields, the extensive grey rough degree denotes the number of indefinable sets in the whole field to the number of all sets on the discussed field. It resembles the knowledge rough degree of the whole knowledge base to a certain degree,

According to the calculation formula of extensive grey rough sets, we know that, when $|(B-A)/R|=|B-A|$, that is when every element in set B-A constitutes one equivalence class, $r(RW([A,B]\oplus))=0$, it denotes [A,B] has no rough sets; extending to the whole field, if $|U/R|=|U|$, that is when every element on the discussed field constitutes one equivalence class, $r(RW([\phi,U]\oplus))=0$. It denotes that there exist no rough sets on the whole discuss field. When $|(B-A)/R|=1$ or $|U/R|=1$, then the whole B-A or U constitute a equivalence class, $r(R([A,B]\oplus))=1-2^{1-|B-A|}$, $r(R([\emptyset,U]\oplus))=1-2^{1-|U|}$, the rough degree of the extensive grey rough sets become the largest. It denotes that the rough degree of extensive grey rough sets is related with the knowledge finesse, the finer the knowledge, the smaller the rough degree; contrarily, the rougher the knowledge, the larger the rough degree.

Actually, if A and B are both R indefinable, if $A \subseteq X \subseteq B$ and X is R definable, because $R-A \subseteq R-X, R\_X \subseteq R\_B, X=R\_X=R-X => R-A \subseteq X \subseteq R\_B$, then we can use $Comp(([R-A,R\_B],R))$ to denote all the exact sets in A,B, thus can get the radix of the extensive grey rough sets and extensive grey rough degree separately,

$$|RW([A,B])\oplus|=|[A,B]|-|Comp(([R-A,R\_B],R))|=2^{|B-A|}-2^{|(R\_B-R-A)/R|} \quad (IV),$$

$$r(RW([A,B]\oplus))=(2^{|B-A|}-2^{|(R\_B-R-A)/R|})/2^{|B-A|}=1-2^{|(R\_B-R-A)/R|-|B-A|} \quad (V).$$

When A and B are R definable, because $R-A=A, R\_B=B$, then (IV),(V) are equivalent to (II),(III), thus, A and B are both a R definable special situation.

4 Conclusion

Rough sets and grey system are two newly developed powerful mathematical tools dealing with uncertain problems. They deal with uncertain problems from different point of view. Whereas, the two do have some communications and relatively strong mutual benefit whether from theoretic or practical point of view. If we can combine the advantages of the two theories and take the benefits of both, it will be more effective to deal with some uncertain problems. Our paper discussed this combination primarily from theoretic viewpoint, and further research needs to be done.

References

1. Deng JL.Control problems of grey systems.Systems□Control Letters,1982(5):288~294.
2. Liu SF,Guo TB.Grey system theory and applications. Zhengzhou:Henan University Press.
3. Pawlak Z. Rough Sets.International Journal of Computer and Information Sciences, 1982,11(1):341~356.
4. Morsi NN and Yakout MM. Axiomatics for fuzzy rough sets, Fuzzy Sets and Systems,1998,100(1-3):327-342.
5. Radzikowska AM, Kerre EE. A comparative study of fuzzy rough sets. Fuzzy Sets and Systems, 2002, 126(2):137-155.
6. Sarkar M., Rough fuzzy functions in classification. Fuzzy Sets and Systems,2002, 132(3):353-369.
7. Yao, Y.Y., Two views of the theory of rough sets in finite universes, International Journal of Approximation Reasoning, Vol. 15, No. 4, pp. 291-317.

Author Index

Abraham, Ajith II-1067
Adam, Susanne I-662
Afzulpurkar, Nitin V. I-484
Ahn, Heejune II-1170
Amin, M. Ashraful I-484

Bae, Hyeon I-833
Baeg, Seung-Beom II-998
Baek, Jae-Yeon II-186
Bandara, G.E.M.D.C. II-215
Banerjee, Amit I-444
Barsoum, N.N. II-1294
Basir, Otman I-426
Batanov, Dentcho N. I-484
Batuwita, K.B.M.R. II-215
Benhabib, B. I-1217
Bhattacharya, A. II-1294
Bi, D. I-942, II-677
Bie, Rongfang II-1037
Bing, Huang I-1223
Bing, Zhou I-1151
Bloyet, Daniel I-189
Bourey, Jean-Pierre I-1025
Budiono I-1113
Byun, Doyoung I-1113
Byun, Yung-Hwan I-1081, I-1108

Cai, Hongbin I-1277
Cai, Lianhong II-600
Cai, Long-Zheng II-320
Cai, Zixing I-1217, II-921
Cansever, Galip I-981
Cao, Bing-yuan I-156, II-546
Cao, Chunling II-1022
Cao, Fei II-289
Cao, Wenliang II-339
Cao, Yijia I-79, I-882
Cao, Zhe I-285
Cao, Zhen-Fu II-596
Cao, Zhexin I-1287
Chai, Duckjin I-1175
Chai, Tianyou I-876, II-891
Chang, Chin-Chen II-551
Chang, Kuiyu II-1236

Chang, Wen-Kui II-911
Chao, Ruey-Ming I-1067
Chau, Rowena II-768
Che, Rucai I-910
Chen, Dewang II-1008
Chen, Fuzan II-420
Chen, Gang II-452
Chen, Gencai II-961
Chen, Guangsheng II-610
Chen, Guoqing I-721, II-614
Chen, Hanxiong I-584
Chen, Hongqi II-1136
Chen, Jian I-59
Chen, Jun I-969
Chen, Ling II-961
Chen, Shi-Jay I-694
Chen, Shuwei I-276
Chen, Shyi-Ming I-694
Chen, Wei II-742
Chen, Wenbin II-49
Chen, Xia I-130
Chen, Xiaoming II-778
Chen, Xiaoyun II-624
Chen, Xuerong I-672
Chen, Yan Qiu II-81, II-100
Chen, Yanmei II-275
Chen, Yanping I-189
Chen, Yen-Liang II-536
Chen, Yi II-624
Chen, Yi-Fei II-430
Chen, Yiqun II-494, II-710
Chen, Yuehui II-1067
Chen, Yun II-240
Chen, Zhe I-717
Cheng, Lishui I-505
Cheng, Wei II-408
Cheon, Seong-Pyo I-203, I-772
Cheong, Il-Ahn II-160
Chi, Chihong I-594
Chin, Kuo-Chih I-851
Cho, Dong-Sub II-561
Cho, Jinsung II-1166
Cho, Siu-Yeung II-1245
Cho, Tae-Ho II-998

Cho, Yookun II-1154
Choi, Byung-Jae I-802
Choi, Doo-Hyun II-989
Choi, Heeyoung I-1175
Choi, Hyung-Il II-1061
Choi, Su-Il II-329
Choi, Young Chang I-1108
Chon, Tae-Soo II-186
Chu, Yayun II-230
Chun, Myung-Geun II-514, II-1132
Chun, Seok-Ju II-762
Chung, Chan-Soo II-731
Chung, Fu-lai I-1171
Chung, Henry II-677
Congfu, Xu I-1246
Constans, Jean-Marc I-189
Cornez, Laurence II-1281
Cronin, Mark T.D. II-31
Cui, Chaoyuan I-584
Cui, Wanan I-1242

Dai, Honghua II-39, II-368
Dai, Weiwei II-677
Dailey, Matthew N. I-484
Davé, Rajesh N. I-444
Deng, Hepu I-653
Deng, Ke II-362
Deng, Tingquan II-275
Deng, Yingna II-285
Deng, Zhi-Hong I-374
Dib, Marcos Vinícius Pinheiro I-1053
Ding, Jianhua II-1051
Ding, Mingli I-812
Ding, Xiying I-872, I-977
Ding, Zhan II-120
Ding, Zuquan II-869
Dong, Jinxiang II-255
Dong, Xiaoju II-1128
Dong, Yihong I-470
Dong, Zhupeng II-1128
Dou, Weibei I-189
Dou, Wen-Hua I-360
Du, Hao II-81
Du, Lei II-845, II-1184
Du, Weifeng I-1232
Du, Weiwei I-454, II-1
Duan, Hai-Xin II-774
Duan, Zhuohua II-921

Engin, Seref N. I-981
Eric, Castelli II-352
Esichaikul, Vatcharaporn I-484

Fan, Jing II-1103
Fan, Muhui II-677
Fan, Xian I-505
Fan, Xianli II-494
Fan, Xiaozhong I-571
Fan, Yushun I-26
Fan, Zhang II-865
Fan, Zhi-Ping I-130
Fang, Bin II-130
Fang, Yong II-931
Fei, Yu-lian I-609
Feng, Boqin I-580
Feng, Du II-398
Feng, Ming I-59
Feng, Zhikai I-1185
Feng, Zhilin II-255
Feng, Zhiquan II-412
Fu, Jia I-213
Fu, Tak-chung I-1171
Fu, Yuxi II-1128
Fung, Chun Che II-1226
Furuse, Kazutaka I-584

Galily, Mehdi I-900, II-976
Gang, Chen I-841
Gao, Jinwu I-304, I-321
Gao, Kai I-199, II-658
Gao, Shan II-362
Gao, Xin II-524
Gao, Yang II-698
Geng, Zhi II-362
Ghosh, Joydeep II-1236
Glass, David II-797
Goh, Ong Sing II-1226
Gong, Binsheng II-830, II-845
Gu, Wenxiang II-1118
Gu, Xingsheng I-1271
Gu, Yajun I-1287
Gu, Ying-Kui II-897
Gu, Zhimin II-110
Guo, Chonghui II-196, II-801
Guo, Chuangxin I-79, I-882
Guo, Gongde II-31, II-797
Guo, Jiankui II-1051
Guo, Jianyi I-571
Guo, Li-wei I-708

Guo, Maozu II-861
Guo, Ping II-723
Guo, Qianjin I-743
Guo, Qingding I-872, I-977
Guo, Weiping I-792
Guo, Yaohuang I-312
Guo, Yecai I-122
Guo, Yi I-122
Guo, Zheng II-830
Gupta, Sudhir II-811

Han, Byung-Gil II-989
Han, Man-Wi II-186
Han, Qiang II-945
Hang, Xiaoshu II-39
Hao, Jingbo I-629
He, Bo II-723
He, Huacan I-31
He, Huiguang I-436
He, Lin II-1128
Ho, Liping II-503
He, Mingyi II-58
He, Pilian II-67
He, Ruichun I-312
He, Xiaoxian I-987
He, Xing-Jian II-727
He, Yanxiang I-865
Heo, Jin-Seok II-344
Heo, Junyoung II-1154
Higgins, Michael I-1256
Ho, Chin-Yuan II-536
Hong, Choong Seon II-1166
Hong, Dug Hun I-100
Hong, Gye Hang II-1071
Hong, Jiman II-1154, II-1158, II-1170
Hong, Kwang-Seok II-170
Hong, Won-Sin I-694
Hou, Beiping II-703
Hou, Yuexian II-67
Hu, Bo II-442
Hu, Dan II-378
Hu, Desheng II-475
Hu, Hong II-73
Hu, Huaqiang II-120
Hu, Kong-fa I-1192
Hu, Maolin I-148
Hu, Min II-742
Hu, Qinghua I-494, I-1261
Hu, Shi-qiang I-708
Hu, Wei-li I-69

Hu, Xuegang I-1309
Hu, Y.A. II-1174
Hu, Yi II-778
Hu, Yu-Shu II-1012
Hu, Yunfa II-624
Huang, Biao II-265
Huang, Chen I-26
Huang, Hailiang II-577
Huang, Hong-Zhong II-897
Huang, Jin I-59, II-945
Huang, Qian II-483
Huang, Rui II-58
Huang, Xiaochun II-21
Huang, Yanxin I-735
Huang, Yuan-sheng I-635
Huang, Zhiwei I-1063
Huawei, Guo II-398
Hui, Hong II-324
Huo, Hua I-580
Hwang, Buhyun I-1175
Hwang, Changha I-100
Hwang, Hoyon I-1092

Ikuta, Akira I-1161
Im, Younghee I-355
Inaoka, Hiroyuki I-263
Inoue, Kohei I-454, II-1

Jang, MinSeok II-249
Jang, Seok-Woo II-1061
Jeon, Kwon-Su I-1081
Jeong, Karpjoo I-1077, I-1092
Jeong, Ok-Ran II-561
Ji, Ruirui II-293
Ji, Xiaoyu I-304
Jia, Baozhu I-1011
Jia, Caiyan I-1197
Jia, Huibo I-514
Jia, Li-min I-69
Jia, Lifeng II-592
Jia, Limin I-89
Jia, Yuanhua I-118
Jian, Jiqi I-514
Jiang, Jianmin II-483
Jiang, Ping II-483
Jiang, Wei II-852, II-1184
Jiang, Y.C. II-1174
Jiang, Yunliang I-195
Jiao, Zhiping II-302
Jie, Shen I-1192

Jie, Wang I-837
Jin, Dongming I-1034, I-1044
Jin, Hanjun I-213
Jin, Hong I-861
Jin, Hui I-865
Jin, Shenyi I-1092
Jin, Weiwei I-1044
Jing, Zhong-liang I-708
Jing, Zhongliang I-672
Joo, Young Hoon I-406, I-416, I-886
Juang, Jih-Gau I-851
Jun, Byong-Hee II-1132
Jung, Chang-Gi II-989
Jung, In-Sung II-1079

Kajinami, Tomoki II-1208
Kamel, Abdelkader El I-1025
Kan, Li I-531
Kang, Bo-Yeong I-462, II-752
Kang, Dazhou I-232
Kang, Seonggoo II-1150
Kang, Yaohong I-388
Karaman, Özhan II-925
Kasabov, Nikola II-528
Keinduangjun, Jitimon II-1041
Kim, Bosoon I-1117
Kim, Byeong-Man II-752
Kim, Byung-Joo II-581
Kim, Dae-Won I-462
Kim, Deok-Eun II-994
Kim, Dong-Gyu II-170
Kim, Dong-kyoo II-205
Kim, Eun Yi I-1077
Kim, Eunkyo II-1162
Kim, Euntai I-179
Kim, Gye-Young II-1061
Kim, Ho J. II-811
Kim, Il Kon II-581
Kim, Jang-Hyun I-1015
Kim, Jee-In I-1117
Kim, Jee-in I-1077
Kim, Jeehoon II-186
Kim, Jin Y. II-329
Kim, Jong Hwa I-1141
Kim, Jonghwa I-1092
Kim, Joongheon II-1162
Kim, Jung Y. II-1158
Kim, Jung-Hyun II-170
Kim, Jungtae II-205
Kim, Kwang-Baek I-761

Kim, Kwangsik I-1092
Kim, Kyoungjung I-179
Kim, Min-Seok II-344
Kim, Min-Soo II-731
Kim, Minsoo II-160
Kim, Moon Hwan I-406
Kim, Myung Sook II-1093
Kim, Myung Won I-392
Kim, Sang-Jin I-1081
Kim, Soo-jeong I-1077
Kim, Soo-Young II-994
Kim, Sung-Ryul I-1137
Kim, Sungshin I-203, I-772, I-833
Kim, Weon-Goo II-249
Kim, Yejin I-833
Kim, Yong-Hyun I-1133
Kim, Youn-Tae I-203
Ko, Sung-Lim I-1133
Kong, Yong Hae II-1093
Koo, Hyun-jin I-1077
Kubota, Naoyuki I-1001
Kucukdemiral, Ibrahim B. I-981
Kumar, Kuldeep II-316
Kwak, Keun-Chang II-514
Kwun, Young Chel I-1

Lai, K.K. II-931
Lan, Jibin II-503
Le, Jia-Jin II-462
Lee, Changjin I-1113
Lee, Chin-Hui II-249
Lee, Choonhwa II-1162
Lee, Gunhee II-205
Lee, Ho Jae I-406
Lee, Inbok I-1137
Lee, Jae-Woo I-1081, I-1108
Lee, Jaewoo I-1092
Lee, Jang Hee II-1071
Lee, Ju-Hong II-762
Lee, Jung-Ju II-344
Lee, KwangHo II-752
Lee, Sang-Hyuk I-203
Lee, Sang-Won II-170
Lee, Sangjun II-1150
Lee, Sengtai II-186
Lee, Seok-Lyong II-762
Lee, Seoung Soo I-1141
Lee, Seungbae I-1127
Lee, Sukho II-1150
Lee, Vincent C.S. II-150

Author Index

Lee, Wonjun   II-1162
Lei, Jingsheng   I-388
Lei, Xusheng   I-890
Leu, Fang-Yie   II-911
Li, Baiheng   II-1008
Li, Chang-Yun   I-728
Li, Changyun   II-1089
Li, Chao   I-547
Li, Chuanxing   II-845, II-852
Li, Gang   II-368
Li, Gui   I-267
Li, Guofei   I-340
Li, HongXing   II-378
Li, Hongyu   II-49
Li, Jie   II-1190
Li, Jin-tao   II-689
Li, Jing   II-830
Li, Luoqing   II-130
Li, Ming   I-619
Li, Minglu   II-1027
Li, Minqiang   II-420
Li, Ning   II-698
Li, Peng   II-1200
Li, Qing   I-462, II-752
Li, Sheng-hong   II-324
Li, Shu   II-285
Li, Shutao   II-610
Li, Stan Z.   I-223
Li, Sujian   II-648
Li, Xia   II-830, II-836, II-845,
    II-852, II-869, II-1184, II-1190
Li, Xiaoli   I-645
Li, Xiaolu   I-837
Li, Xing   II-774
Li, Xuening   II-778
Li, Yanhong   II-35
Li, Yanhui   I-232
Li, Yinong   I-822
Li, Yinzhen   I-312
Li, Yu-Chiang   II-551
Li, Zhijian   I-1034
Li, Zhijun   I-676
Lian, Yiqun   II-718
Liang, Xiaobei   I-140
Liao, Beishui   II-1089
Liao, Lijun   II-1089
Liao, Qin   I-1063
Liao, Zhining   II-797
Lim, Joon S.   II-811
Lim, Ki Won   I-1113

Lin, Bo-Shian   I-851
Lin, Jie Tian Yao   I-436
Lin, Zhonghua   II-306
Ling, Chen   I-1192
Ling, Guo   I-1223
Ling, Jian   II-718
Ling, Zheng   I-822
Liu, Chun-Sheng   II-897
Liu, Delin   II-362
Liu, Diantong   I-792
Liu, Fei   I-969
Liu, Guoliang   II-567
Liu, Haowen   I-865
Liu, Jian-Wei   II-462
Liu, Jing   II-1031
Liu, Jun   II-907
Liu, Junqiang   I-580
Liu, Lanjuan   II-35
Liu, Linzhong   I-312
Liu, Peide   I-523
Liu, Peng   II-35
Liu, Ping   I-728
Liu, Qihe   I-1277
Liu, Qizhen   II-475
Liu, Shi   I-757
Liu, Wu   II-774
Liu, Xiang-guan   II-667
Liu, Xiao-dong   I-42
Liu, Xiaodong   I-53
Liu, Xiaoguang   II-1031
Liu, Xin   I-662
Liu, Xiyu   I-523
Liu, Yan-Kui   I-321
Liu, Yang   I-822, II-388, II-941
Liu, Yong   I-195
Liu, Yong-lin   I-160
Liu, Yushu   I-531
Liu, Yutian   I-11
Lok, Tat-Ming   II-727
Lu, Chunyan   I-388
Lu, Jianjiang   I-232
Lu, Ke   I-436
Lu, Mingyu   II-196, II-801
Lu, Naijiang   II-638
Lu, Ruqian   I-1197
Lu, Ruzhan   II-778
Lu, Ya-dong   I-922
Lu, Yinghua   II-1118
Lu, Yuchang   II-196, II-801
Luk, Robert   I-1171

Luo, Bin II-140
Luo, Minxia I-31
Luo, Shi-hua II-667
Luo, Shuqian II-524
Luo, Ya II-723
Luo, Zongwei II-698
Lv, Sali II-830, II-836
Lyu, Michael R. II-727

Ma, and Wei I-122
Ma, Cheng I-514
Ma, Cun-bao II-466
Ma, Jing II-1027
Ma, Jixin II-140
Ma, Jun I-276
Ma, Liangyu II-339
Ma, Tianmin II-528
Ma, Weimin I-721
Ma, Yongkang II-289
Ma, Z.M. I-267
Mamun-Or-Rashid, Md. II-1166
Maoqing, Li I-1313
Mariño, Perfecto II-950
Martínez, Emilio II-950
Masuike, Hisako I-1161
Matsumura, Akio II-1208
McDonald, Mike I-782
Melo, Alba Cristina Magalhães de I-1053
Meng, Dan I-175
Meng, Xiangxu II-412
Meng, Zuqiang I-1217
Miao, Dong II-289
Miao, Zhinong I-950
Min, Fan I-1277
Modarres, Mohammad II-1012
Mok, Henry M.K. I-295
Mosavi, Morteza II-976
Muller, Jean-Denis II-1281
Mun, Jeong-Shik I-1137
Murata, Hiroshi II-1216
Murata, Tsuyoshi II-1204

Na, Eunyoung I-100
Na, Seung Y. II-329
Na, Yang I-1127
Nagar, Atulya II-821
Nam, Mi Young I-698
Neagu, Daniel II-31
Nemiroff, Robert J. II-634

Ng, Chak-man I-1171
Ng, Yiu-Kai I-557
Nguyen, Cong Phuong II-352
Ning, Yufu I-332
Niu, Ben I-987
Noh, Bong-Nam II-160
Noh, Jin Soo II-91

Oh, Hyukjun II-1170
Ohbo, Nobuo I-584
Ohta, Mitsuo I-1161
Omurlu, Vasfi E. I-981
Onoda, Takashi II-1216
Ouyang, Jian-quan II-689

Pan, De II-1051
Pan, Donghua I-537
Pan, Yingjie II-293
Park, Chang-Woo I-179
Park, Choon-sik II-205
Park, Daihee I-355
Park, Dong-Chul I-475
Park, Eung-ki II-205
Park, Geunyoung II-1154
Park, Hyejung I-100
Park, Jang-Hwan II-1132
Park, Jin Bae I-406, I-416, I-886
Park, Jin Han I-1
Park, Jin-Bae I-1015
Park, Jong Seo I-1
Park, Mignon I-179
Park, Minkyu II-1154
Park, Si Hyung I-1141
Park, Sungjun I-1117
Park, Young-Pil I-1015
Pastoriza, Vicente II-950
Pedrycz, Witold II-514
Pei, Xiaobing I-1297
Pei, Yunxia II-110
Pei, Zheng I-1232
Peng, Jin I-295
Peng, Ningsong I-370
Peng, Qin-ke I-1151
Peng, Qunsheng II-742
Peng, Wei II-120
Peng, Yonghong II-483
Pham, Thi Ngoc Yen II-352
Pham, Trung-Thanh I-1133
Piamsa-nga, Punpiti II-1041
Piao, Xuefeng II-1154

Ping, Zhao I-383
Poovorawan, Yong II-1041
Purushothaman, Sujita II-821

Qi, Gu I-1192
Qi, Jian-xun I-635
Qian, Jixin I-1266
Qian, Yuntao II-1107
Qiang, Wenyi II-567
Qiao, Yanjuan I-822
Qin, Jie I-360
Qin, Keyun I-1232
Qin, X.L. II-1174
Qin, Zhenxing I-402
Qing, Hu I-872, I-977
Qiu, Yuhui II-972
Qu, Shao-Cheng I-960
Qu, Wen-tao II-324

Rao, Shaoqi II-830, II-845, II-852, II-869, II 1184
Rao, Wei I-122
Ren, Bo II-1103
Ren, Guang I-1011
Ren, Jiangtao II-494
Ren, Yuan I-189
Rhee, Kang Hyeon II-91
Rhee, Phill Kyu I-698
Roudsari, Farzad Habibipour I-900
Ruan, Su I-189
Ryu, Joung Woo I-392
Ryu, Tae W. II-811

Sadri, Mohammadreza I-900
Samuelides, Manuel II-1281
Santamaría, Miguel II-950
Seo, Jae-Hyun II-160
Seo, Jung-taek II-205
Shamir, Lior II-634
Shan, Weiwei I-1044
Shaoqi, Rao II-865
Shen, I-Fan II-49
Shen, Jun-yi I-1151
Shi, Haoshan I-383
Shi, Jiachuan I-11
Shi, Lei II-110
Shi, Qin II-600
Shi, Wenzhong II-614
Shi, Yong-yu II-324
Shi, Yu I-20

Shi, Yuexiang I-1217
Shim, Charlie Y. II-1158
Shim, Jooyong I-100
Shin, Daejung II-329
Shin, Dongshin I-1127
Shin, Jeong-Hoon II-170
Shin, Sung Y. II-1158
Shin, Sung-Chul II-994
Shu, Tingting I-350
Shudong, Wang I-1256
Shunxiang, Wu I-1313
Sifeng, Liu I-1313
Sim, Alex T.H. II-150
Sim, Jeong Seop I-1102
Song, Chonghui I-876, II-891
Song, Jiyoung I-355
Song, Lirong I-571
Song, Qun II-528
Song, Yexin I-676
Song, Young-Chul II-989
Su, Jianbo I-890
Su, Jie II-945
Suh, Hae-Gook I-1092
Sulistijono, Indra Adji I-1001
Sun, Da-Zhi II-596
Sun, Dazhong I-1271
Sun, Jiaguang I-594
Sun, Jiantao II-196, II-801
Sun, Shiliang II-638
Sun, Xing-Min I-728
Sun, Yufang II-408
Sun, Zengqi I-910, II-567
Sun, Zhaocai II-230, II-240

Tai, Xiaoying I-470
Takama, Yasufumi II-1208
Tan, Yun-feng I-156
Tang, Bingyong I-140
Tang, Jianguo I-547
Tang, Shi-Wei I-374
Tang, Wansheng I-332, I-340
Tang, Weilong I-717
Tang, Xiang Long II-1190
Tang, Xiao-li I-1192
Tang, Yuan Yan II-130
Tanioka, Hiroki I-537
Tao, HuangFu II-723
Tawfik, Hissam II-821
Temelta, Hakan II-925
Teng, Xiaolong I-370

Thapa, Devinder II-1079
Theera-Umpon, Nipon II-787
Tse, Wai-Man I-295

Um, Chang-Gun II-989
Urahama, Kiichi I-454, II-1

Vasant, Pandian II-1294
Verma, Brijesh II-316
Viswanathan, M. I-1207

Wan Kim, Do I-416, I-886
Wan, Wunan II-941
Wang, Bingshu II-339
Wang, Danli I-861
Wang, Fei II-1051
Wang, G.L. I-942
Wang, Gang II-1031
Wang, Geng I-1266
Wang, Gi-Nam II-1079
Wang, Haiyun II-869
Wang, Hongan I-861
Wang, Houfeng II-11, II-648
Wang, Hui I-332, I-861, II-797
Wang, J. II-1174
Wang, Juan I-89
Wang, Jue I-223
Wang, Laisheng II-980
Wang, Li-Hui I-694
Wang, Lin II-1142
Wang, Peng I-232
Wang, Qi I-812
Wang, Qian I-676
Wang, Qianghu II-836
Wang, Qiuju II-852
Wang, Shi-lin II-324
Wang, Shizhu I-537
Wang, Shou-Yang II-931
Wang, Shuliang II-614
Wang, Shuqing I-841, II-452
Wang, Tao II-285
Wang, Tong I-619
Wang, Xiangyang I-370
Wang, Xiao-Feng II-320
Wang, Xiaojing II-941
Wang, Xiaorong I-213
Wang, Xinya I-1309
Wang, Xun I-148
Wang, Yadong II-869
Wang, Yan I-735

Wang, Yanqiu II-852
Wang, Yong-Ji I-960
Wang, Yong-quan I-160
Wang, Yongcheng I-199, II-658
Wang, YuanZhen I-1297
Wang, Zhe II-592
Wang, Zhenzhen II-836
Wang, Zhiqi I-199, II-658
Wang, Zhongtuo I-537
Wang, Zhongxing II-503
Wang, Zi-cai I-922
Wang, Ziqiang II-388
Wei, Changhua I-213
Wei, Lin II-110
Wei, Zhi II-677
Weigang, Li I-1053
Wen, Fengxia II-836
Wen, Weidong I-865
Wenkang, Shi II-398
Whang, Eun Ju I-179
Whangbo, T.K. I-1207
Wirtz, Kai W. I-662
Wong, Jia-Jun II-1245
Wu, Aimin II-1255, II-1265, II-1276
Wu, Dianliang II-577
Wu, Huaiyu I-930
Wu, Jian-Ping II-774
Wu, Jiangning I-537, II-176
Wu, Jianping I-118, II-1008
Wu, Ming II-21
Wu, Wei-Dong II-897
Wu, Wei-Zhi I-167

Xia, Delin II-21
Xia, Li II-865
Xia, Yinglong I-243
Xia, Z.Y. II-1174
Xian-zhong, Zhou I-1223
Xiang, Chen I-603
Xiao, Gang I-672
Xiao, Yegui I-1161
Xie, Jiancang II-1136
Xie, Jianying II-907
Xie, Lijun I-1277
Xie, Shengli I-837
Xie, Wei I-930
Xie, Xuehui I-402
Xie, Yuan II-907
Xin, He I-1223
Xin, Jin II-1037

Xin, Zhiyun I-594
Xing, James Z. II-265
Xing, Zong-yi I-69
Xiong, Feng-lan I-42
Xiong, Fenglan I-53
Xiong, Li-Rong II-1103
Xiu, Zhihong I-1011
Xu, Aidong I-743
Xu, Baochang I-717
Xu, Baowen I-232
Xu, De II-1255, II-1265, II-1276
Xu, Jia-dong II-466
Xu, Lin I-350
Xu, Lu II-306
Xu, Song I-336
Xu, Weijun I-148
Xu, Xian II-1128
Xu, Xiujuan II-592
Xu, Yang I-175, I-276, I-950
Xu, Yitian II-980
Xu, Zeshui I-110, I-684
Xu, Zhihao I-1034, I-1044
Xun, Wang I-609

Yamada, Seiji II-1200, II-1216
Yamamoto, Kenichi I-537
Yan, Li I-267
Yan, Puliu II-21
Yan, Shi-Jie II-911
Yan, Weizhen I-285
Yan, Zhang I-388
Yang, Ai-Min I-728
Yang, Aimin II-1089
Yang, Bo II-1067
Yang, Chenglei II-412
Yang, Chin-Wen I-1067
Yang, Chunyan II-891
Yang, Guangfei II-176
Yang, Guifang I-653
Yang, Hui II-891
yang, Hui I-876
Yang, Hyun-Seok I-1015
Yang, Ji Hye I-1113
Yang, Jie I-370, I-505
Yang, Ju II-1067
Yang, Mengfei I-910
Yang, Ming I-922
Yang, Shu-Qiang I-360
Yang, Shucheng I-876
Yang, Tao II-442

Yang, Wu II-723
Yang, Xiao-Ping I-1303
Yang, Xiaogang II-289
Yang, Xu II-1255, II-1265, II-1276
Yang, Y.K. I-1207
Yang, Yongjian II-1022
Yao, Liyue II-67
Yao, Min II-1107
Yao, Shengbao I-1242
Yao, Xin I-253, I-645
Yatabe, Shunsuke I-263
Ye, Bi-Cheng I-619
Ye, Bin I-79, I-882
Ye, Xiao-ling II-430
Ye, Xiuzi II-120
Ye, Yangdong I-89
Yeh, Chung-Hsing II-768
Yeh, Jieh-Shan II-551
Yerra, Rajiv I-557
Yi, Jianqiang I-792
Yi, Sangho II-1154
Yi, Zhang I-603
Yin, Jian I-59, II-494, II-710
Yin, Jianping I-629
Yin, Jianwei II-255
Yin, Minghao II-1118
Yin, Yilong II-230, II-240
Yong, Liu I-1246
Yoo, Seog-Hwan I-802
You, Xinge II-130
Yu, Daren I-494, I-1261
Yu, Dongmei I-872, I-977
Yu, Haibin I-743
Yu, Jinxia II-921
Yu, Sheng-Sheng II-320
Yu, Shiwen II-648
Yu, Shou-Jian II-462
Yu, Zhengtao I-571
Yuan, Hanning II-614
Yuan, Junpeng II-945
Yuan, Weiqi II-306
Yuan, Yueming II-1008
Yue, Chaoyuan I-1242
Yue, Wu I-603
Yun, Ling I-609
Yunhe, Pan I-1246

Zeng, Wenyi I-20
Zeng, Yurong II-1142
Zhan, Xiaosi II-230, II-240

Zhan, Yongqiang II-861
Zhang, Bao-wei I-619
Zhang, Boyun I-629
Zhang, Changshui I-243, I-253, II-638
Zhang, Chao II-466
Zhang, Chengqi I-402
Zhang, Chunkai K. II-73
Zhang, Dan II-130
Zhang, Dexian II-388
Zhang, Gexiang I-1287
Zhang, Guangmei II-836
Zhang, Hao II-324
Zhang, Hong II-972
Zhang, Honghua I-910
Zhang, Huaguang I-876
Zhang, Huaxiang I-523
Zhang, Hui II-255
Zhang, J. I-942
Zhang, Ji II-339
Zhang, Jianming I-841, II-452
Zhang, Jiashun I-336
Zhang, Jie II-845, II-852
Zhang, Jin II-624
Zhang, Jun II-677
Zhang, Junping I-223
Zhang, Ming I-374
Zhang, Peng I-1242
Zhang, Ping II-316
Zhang, Qingpu II-869
Zhang, Runtong II-880
Zhang, Shichao I-402
Zhang, Tao II-742
Zhang, Wei II-600, II-1184
Zhang, Weiguo I-148
Zhang, Wenyin I-547
Zhang, Xiao-hong I-160
Zhang, Xuefeng II-35
Zhang, Xueying II-302
Zhang, Yin II-120
Zhang, Ying-Chao II-430
Zhang, Yingchun II-567
Zhang, Yong I-69
Zhang, Yong-dong II-689
Zhang, Yongjin II-1136
Zhang, Yue II-727
Zhang, Yuhong I-1309
Zhang, Zaiqiang I-175
Zhang, Zhegen II-703
Zhang, Zhizhou II-1128

Zhao, Jianhua I-285, I-304
Zhao, Jieyu I-470
Zhao, Jizhong I-594
Zhao, Jun I-1266
Zhao, Keqin I-195
Zhao, Li II-95
Zhao, Long I-717
Zhao, Min II-667
Zhao, Ruiqing I-336, I-340, I-350
Zhao, Shu-Mei I-360
Zhao, Xiangyu I-950
Zhao, Xin I-930
Zhao, Yongqiang II-1027
Zhao, Zhefeng II-302
Zheng, Guo II-1184
Zheng, Jianhui II-1255, II-1265, II-1276
Zheng, Jie-Liang II-430
Zheng, Min II-600
Zheng, Pengjun I-782
Zheng, Qiuhua II-1107
Zheng, Quan I-1185
Zheng, Su-hua I-42
Zheng, Suhua I-53
Zheng, Wenming II-95
Zheng, Yalin I-243, I-253
Zhicheng, Liu II-865
Zhiyong, Yan I-1246
Zhou, Chang Yin II-100
Zhou, Chunguang I-735, II-592
Zhou, Jing-Li II-320
Zhou, Jun-hua I-635
Zhou, Kening II-703
Zhou, Wengang I-735
Zhou, Yuanfeng I-118
Zhu, Chengzhi I-79, I-882
Zhu, Daoli I-140
Zhu, Guohua II-830
Zhu, Hong II-285, II-293
Zhu, Hongwei I-426
Zhu, Jiaxian II-35
Zhu, Ming I-1185
Zhu, Wen II-703
Zhu, Xiaomin II-880
Zhu, Yunlong I-987
Zhuang, Ling II-39
Zhuang, Yueting I-195, II-718
Zou, Cairong II-95
Zou, Danping II-475
Zou, Xiaobing II-921

Lecture Notes in Artificial Intelligence (LNAI)

Vol. 3632: R. Nieuwenhuis (Ed.), Automated Deduction – CADE-20. XIII, 459 pages. 2005.

Vol. 3626: B. Ganter, G. Stumme, R. Wille (Eds.), Formal Concept Analysis. X, 349 pages. 2005.

Vol. 3625: S. Kramer, B. Pfahringer (Eds.), Inductive Logic Programming. XIII, 427 pages. 2005.

Vol. 3620: H. Muñoz-Avila, F. Ricci (Eds.), Case-Based Reasoning Research and Development. XV, 654 pages. 2005.

Vol. 3614: L. Wang, Y. Jin (Eds.), Fuzzy Systems and Knowledge Discovery, Part II. XLI, 1314 pages. 2005.

Vol. 3613: L. Wang, Y. Jin (Eds.), Fuzzy Systems and Knowledge Discovery, Part I. XLI, 1334 pages. 2005.

Vol. 3607: J.-D. Zucker, L. Saitta (Eds.), Abstraction, Reformulation and Approximation. XII, 376 pages. 2005.

Vol. 3596: F. Dau, M.-L. Mugnier, G. Stumme (Eds.), Conceptual Structures: Common Semantics for Sharing Knowledge. XI, 467 pages. 2005.

Vol. 3587: P. Perner, A. Imiya (Eds.), Machine Learning and Data Mining in Pattern Recognition. XVII, 695 pages. 2005.

Vol. 3584: X. Li, S. Wang, Z.Y. Dong (Eds.), Advanced Data Mining and Applications. XIX, 835 pages. 2005.

Vol. 3581: S. Miksch, J. Hunter, E. Keravnou (Eds.), Artificial Intelligence in Medicine. XVII, 547 pages. 2005.

Vol. 3577: R. Falcone, S. Barber, J. Sabater-Mir, M.P. Singh (Eds.), Trusting Agents for Trusting Electronic Societies. VIII, 235 pages. 2005.

Vol. 3575: S. Wermter, G. Palm, M. Elshaw (Eds.), Biomimetic Neural Learning for Intelligent Robots. IX, 383 pages. 2005.

Vol. 3571: L. Godo (Ed.), Symbolic and Quantitative Approaches to Reasoning with Uncertainty. XVI, 1028 pages. 2005.

Vol. 3559: P. Auer, R. Meir (Eds.), Learning Theory. XI, 692 pages. 2005.

Vol. 3558: V. Torra, Y. Narukawa, S. Miyamoto (Eds.), Modeling Decisions for Artificial Intelligence. XII, 470 pages. 2005.

Vol. 3554: A. Dey, B. Kokinov, D. Leake, R. Turner (Eds.), Modeling and Using Context. XIV, 572 pages. 2005.

Vol. 3539: K. Morik, J.-F. Boulicaut, A. Siebes (Eds.), Local Pattern Detection. XI, 233 pages. 2005.

Vol. 3538: L. Ardissono, P. Brna, A. Mitrovic (Eds.), User Modeling 2005. XVI, 533 pages. 2005.

Vol. 3533: M. Ali, F. Esposito (Eds.), Innovations in Applied Artificial Intelligence. XX, 858 pages. 2005.

Vol. 3518: T.B. Ho, D. Cheung, H. Liu (Eds.), Advances in Knowledge Discovery and Data Mining. XXI, 864 pages. 2005.

Vol. 3508: P. Bresciani, P. Giorgini, B. Henderson-Sellers, G. Low, M. Winikoff (Eds.), Agent-Oriented Information Systems II. X, 227 pages. 2005.

Vol. 3505: V. Gorodetsky, J. Liu, V. A. Skormin (Eds.), Autonomous Intelligent Systems: Agents and Data Mining. XIII, 303 pages. 2005.

Vol. 3501: B. Kégl, G. Lapalme (Eds.), Advances in Artificial Intelligence. XV, 458 pages. 2005.

Vol. 3492: P. Blache, E. Stabler, J. Busquets, R. Moot (Eds.), Logical Aspects of Computational Linguistics. X, 363 pages. 2005.

Vol. 3488: M.-S. Hacid, N.V. Murray, Z.W. Raś, S. Tsumoto (Eds.), Foundations of Intelligent Systems. XIII, 700 pages. 2005.

Vol. 3487: J. Leite, P. Torroni (Eds.), Computational Logic in Multi-Agent Systems. XII, 281 pages. 2005.

Vol. 3476: J. Leite, A. Omicini, P. Torroni, P. Yolum (Eds.), Declarative Agent Languages and Technologies II. XII, 289 pages. 2005.

Vol. 3464: S.A. Brueckner, G.D.M. Serugendo, A. Karageorgos, R. Nagpal (Eds.), Engineering Self-Organising Systems. XIII, 299 pages. 2005.

Vol. 3452: F. Baader, A. Voronkov (Eds.), Logic for Programming, Artificial Intelligence, and Reasoning. XI, 562 pages. 2005.

Vol. 3451: M.-P. Gleizes, A. Omicini, F. Zambonelli (Eds.), Engineering Societies in the Agents World V. XIII, 349 pages. 2005.

Vol. 3446: T. Ishida, L. Gasser, H. Nakashima (Eds.), Massively Multi-Agent Systems I. XI, 349 pages. 2005.

Vol. 3445: G. Chollet, A. Esposito, M. Faundez-Zanuy, M. Marinaro (Eds.), Nonlinear Speech Modeling and Applications. XIII, 433 pages. 2005.

Vol. 3438: H. Christiansen, P.R. Skadhauge, J. Villadsen (Eds.), Constraint Solving and Language Processing. VIII, 205 pages. 2005.

Vol. 3430: S. Tsumoto, T. Yamaguchi, M. Numao, H. Motoda (Eds.), Active Mining. XII, 349 pages. 2005.

Vol. 3419: B. Faltings, A. Petcu, F. Fages, F. Rossi (Eds.), Constraint Satisfaction and Constraint Logic Programming. X, 217 pages. 2005.

Vol. 3416: M. Böhlen, J. Gamper, W. Polasek, M.A. Wimmer (Eds.), E-Government: Towards Electronic Democracy. XIII, 311 pages. 2005.

Vol. 3415: P. Davidsson, B. Logan, K. Takadama (Eds.), Multi-Agent and Multi-Agent-Based Simulation. X, 265 pages. 2005.

Vol. 3403: B. Ganter, R. Godin (Eds.), Formal Concept Analysis. XI, 419 pages. 2005.

Vol. 3398: D.-K. Baik (Ed.), Systems Modeling and Simulation: Theory and Applications. XIV, 733 pages. 2005.

Vol. 3397: T.G. Kim (Ed.), Artificial Intelligence and Simulation. XV, 711 pages. 2005.

Vol. 3396: R.M. van Eijk, M.-P. Huget, F. Dignum (Eds.), Agent Communication. X, 261 pages. 2005.

Vol. 3394: D. Kudenko, D. Kazakov, E. Alonso (Eds.), Adaptive Agents and Multi-Agent Systems II. VIII, 313 pages. 2005.

Vol. 3392: D. Seipel, M. Hanus, U. Geske, O. Bartenstein (Eds.), Applications of Declarative Programming and Knowledge Management. X, 309 pages. 2005.

Vol. 3374: D. Weyns, H. V.D. Parunak, F. Michel (Eds.), Environments for Multi-Agent Systems. X, 279 pages. 2005.

Vol. 3371: M.W. Barley, N. Kasabov (Eds.), Intelligent Agents and Multi-Agent Systems. X, 329 pages. 2005.

Vol. 3369: V. R. Benjamins, P. Casanovas, J. Breuker, A. Gangemi (Eds.), Law and the Semantic Web. XII, 249 pages. 2005.

Vol. 3366: I. Rahwan, P. Moraitis, C. Reed (Eds.), Argumentation in Multi-Agent Systems. XII, 263 pages. 2005.

Vol. 3359: G. Grieser, Y. Tanaka (Eds.), Intuitive Human Interfaces for Organizing and Accessing Intellectual Assets. XIV, 257 pages. 2005.

Vol. 3346: R.H. Bordini, M. Dastani, J. Dix, A.E.F. Seghrouchni (Eds.), Programming Multi-Agent Systems. XIV, 249 pages. 2005.

Vol. 3345: Y. Cai (Ed.), Ambient Intelligence for Scientific Discovery. XII, 311 pages. 2005.

Vol. 3343: C. Freksa, M. Knauff, B. Krieg-Brückner, B. Nebel, T. Barkowsky (Eds.), Spatial Cognition IV. XIII, 519 pages. 2005.

Vol. 3339: G.I. Webb, X. Yu (Eds.), AI 2004: Advances in Artificial Intelligence. XXII, 1272 pages. 2004.

Vol. 3336: D. Karagiannis, U. Reimer (Eds.), Practical Aspects of Knowledge Management. X, 523 pages. 2004.

Vol. 3327: Y. Shi, W. Xu, Z. Chen (Eds.), Data Mining and Knowledge Management. XIII, 263 pages. 2005.

Vol. 3315: C. Lemaître, C.A. Reyes, J.A. González (Eds.), Advances in Artificial Intelligence – IBERAMIA 2004. XX, 987 pages. 2004.

Vol. 3303: J.A. López, E. Benfenati, W. Dubitzky (Eds.), Knowledge Exploration in Life Science Informatics. X, 249 pages. 2004.

Vol. 3301: G. Kern-Isberner, W. Rödder, F. Kulmann (Eds.), Conditionals, Information, and Inference. XII, 219 pages. 2005.

Vol. 3276: D. Nardi, M. Riedmiller, C. Sammut, J. Santos-Victor (Eds.), RoboCup 2004: Robot Soccer World Cup VIII. XVIII, 678 pages. 2005.

Vol. 3275: P. Perner (Ed.), Advances in Data Mining. VIII, 173 pages. 2004.

Vol. 3265: R.E. Frederking, K.B. Taylor (Eds.), Machine Translation: From Real Users to Research. XI, 392 pages. 2004.

Vol. 3264: G. Paliouras, Y. Sakakibara (Eds.), Grammatical Inference: Algorithms and Applications. XI, 291 pages. 2004.

Vol. 3259: J. Dix, J. Leite (Eds.), Computational Logic in Multi-Agent Systems. XII, 251 pages. 2004.

Vol. 3257: E. Motta, N.R. Shadbolt, A. Stutt, N. Gibbins (Eds.), Engineering Knowledge in the Age of the Semantic Web. XVII, 517 pages. 2004.

Vol. 3249: B. Buchberger, J.A. Campbell (Eds.), Artificial Intelligence and Symbolic Computation. X, 285 pages. 2004.

Vol. 3248: K.-Y. Su, J. Tsujii, J.-H. Lee, O.Y. Kwong (Eds.), Natural Language Processing – IJCNLP 2004. XVIII, 817 pages. 2005.

Vol. 3245: E. Suzuki, S. Arikawa (Eds.), Discovery Science. XIV, 430 pages. 2004.

Vol. 3244: S. Ben-David, J. Case, A. Maruoka (Eds.), Algorithmic Learning Theory. XIV, 505 pages. 2004.

Vol. 3238: S. Biundo, T. Frühwirth, G. Palm (Eds.), KI 2004: Advances in Artificial Intelligence. XI, 467 pages. 2004.

Vol. 3230: J.L. Vicedo, P. Martínez-Barco, R. Muñoz, M. Saiz Noeda (Eds.), Advances in Natural Language Processing. XII, 488 pages. 2004.

Vol. 3229: J.J. Alferes, J. Leite (Eds.), Logics in Artificial Intelligence. XIV, 744 pages. 2004.

Vol. 3228: M.G. Hinchey, J.L. Rash, W.F. Truszkowski, C.A. Rouff (Eds.), Formal Approaches to Agent-Based Systems. VIII, 290 pages. 2004.

Vol. 3215: M.G.. Negoita, R.J. Howlett, L.C. Jain (Eds.), Knowledge-Based Intelligent Information and Engineering Systems, Part III. LVII, 906 pages. 2004.

Vol. 3214: M.G.. Negoita, R.J. Howlett, L.C. Jain (Eds.), Knowledge-Based Intelligent Information and Engineering Systems, Part II. LVIII, 1302 pages. 2004.

Vol. 3213: M.G.. Negoita, R.J. Howlett, L.C. Jain (Eds.), Knowledge-Based Intelligent Information and Engineering Systems, Part I. LVIII, 1280 pages. 2004.

Vol. 3209: B. Berendt, A. Hotho, D. Mladenic, M. van Someren, M. Spiliopoulou, G. Stumme (Eds.), Web Mining: From Web to Semantic Web. IX, 201 pages. 2004.

Vol. 3206: P. Sojka, I. Kopecek, K. Pala (Eds.), Text, Speech and Dialogue. XIII, 667 pages. 2004.

Vol. 3202: J.-F. Boulicaut, F. Esposito, F. Giannotti, D. Pedreschi (Eds.), Knowledge Discovery in Databases: PKDD 2004. XIX, 560 pages. 2004.

Vol. 3201: J.-F. Boulicaut, F. Esposito, F. Giannotti, D. Pedreschi (Eds.), Machine Learning: ECML 2004. XVIII, 580 pages. 2004.

Vol. 3194: R. Camacho, R. King, A. Srinivasan (Eds.), Inductive Logic Programming. XI, 361 pages. 2004.

Vol. 3192: C. Bussler, D. Fensel (Eds.), Artificial Intelligence: Methodology, Systems, and Applications. XIII, 522 pages. 2004.

Vol. 3191: M. Klusch, S. Ossowski, V. Kashyap, R. Unland (Eds.), Cooperative Information Agents VIII. XI, 303 pages. 2004.